FLORA OF GREAT BRITAIN AND IRELAND

Planned in five volumes, this critical flora provides a definitive account of the native species, naturalised species, frequent garden escapes and casuals found in the British Isles. Full keys and descriptions will enable the user to name all plants occurring in the wild, plus some ornamental trees and shrubs. For the first time detailed accounts of all the large apomictic genera are given and many infraspecific variants included. Each species entry begins with the accepted Latin name, synonyms and the common English name. A detailed description follows, including information on flowering period, pollination and chromosome number. Separate descriptions are given for infraspecific taxa. Information on the status, ecology and distribution (including worldwide distribution) of the species and infraspecific taxa is also given. Clear black-and-white line drawings illustrate an extensive glossary and also illuminate the diagnostic features in a number of groups of plants.

PETER SELL joined the Herbarium in the University of Cambridge's Department of Plant Sciences in 1944, holding the post of Assistant Curator from 1972 until his retirement in 1997. His work there on this flora continues unabated, together with almost daily visits to the University's Botanic Garden throughout the flowering and fruiting seasons. He is co-author of *Flora of Cambridgeshire* (1964) and *Flora of the Maltese Islands* (1977), and was involved in the whole of the *Flora Europaea* project, also published in five volumes (1964–80) by Cambridge University Press.

GINA MURRELL is Assistant Curator of the Herbarium in the University of Cambridge's Department of Plant Sciences, having previously held the post of Herbarium Technician there. She has worked with Peter Sell over a period of 40 years, and together they have collected a quarter of the British Herbarium's 200,000 specimens.

FLORA
OF
GREAT BRITAIN
AND
IRELAND

VOLUME 3
MIMOSACEAE – LENTIBULARIACEAE

PETER SELL and GINA MURRELL
Herbarium, Department of Plant Sciences
University of Cambridge

CAMBRIDGE
UNIVERSITY PRESS

CAMBRIDGE UNIVERSITY PRESS
Cambridge, New York, Melbourne, Madrid, Cape Town, Singapore, São Paulo, Delhi

Cambridge University Press
The Edinburgh Building, Cambridge CB2 8RU, UK

Published in the United States of America by Cambridge University Press, New York

www.cambridge.org
Information on this title: www.cambridge.org/9780521553377

First published 2009

Printed in the United Kingdom at the University Press, Cambridge

A catalogue record for this publication is available from the British Library

ISBN 978-0-521-55337-7 hardback

To our Mentors

Edred John Henry Corner
James Edgar Dandy
Humphrey Gilbert Carter
Harry Godwin
William Thomas Stearn
Stuart Max Walters
Alexander Stuart Watt
Cyril West

Contents

Foreword to volume 3

by S. M. Walters ScD V.M.H.

It has been one of the continuing satisfactions of my academic career in Cambridge that the University Herbarium, of which I was Curator from 1948 to 1973, has provided an academic base for all my specialist interest in angiosperm taxonomy to develop. Indeed, I count myself doubly fortunate that, 12 years after my retirement from academic life, the Herbarium, with its staff and visitors, still provides such a base where scholarship can be pursued for its own sake. With great pleasure I welcome this volume, the first of a set of five promised to us by Peter Sell and Gina Murrell. My association with Peter goes back more than half a century: though I was 'senior partner' in our happy collaboration in the post-war Herbarium, ours was a symbiotic relationship from which we both greatly benefited, and I was delighted when Gina, who had been part of the team in the 1960s and 1970s, returned to the fold as Herbarium Technician in 1991.

As explained in the Preface, this project to write an entirely new critical flora of the British Isles comes to fruition some 20 years after an earlier scheme, in which the late Professor David Valentine took a leading part, had failed to find any financial support. Both Clive Stace to whose *New Flora of the British Isles* (1997) Peter pays tribute in the Preface, and Peter himself, were enthusiastic supporters of the Valentine project, and were prepared to play major parts in writing the Flora. It is fitting that both these eminent British taxonomists should separately carry on the tradition that David Valentine so enthusiastically advocated.

Two aspects of this new critical flora seem to be especially important. One concerns the acceptance, long overdue, of the 'alien element' in our flora as being equally worthy of taxonomic study: in this respect Stace's *Flora* represents a real change in attitude, which is to my mind unreservedly to be welcomed. The other, interestingly linked to the first by many examples, concerns the taxonomic recognition and treatment of hybrids and infraspecific variants. British botany lacks any single reference work from which the basic information about the variation of British vascular plants can be found, yet this information is increasingly needed by ecologists, conservationists, molecular biologists and biochemists, who will, as the century closes, determine the shape of much botanical study in universities and specialised institutions.

The authors of this impressive work have set themselves a colossal task. They have made an excellent start, and we can only wish them a successful conclusion.

1996

It is for me a very real pleasure to add a further word to welcome this, the second volume of 'Sell & Murrell', as this remarkable Flora is now widely known among British botanists. Of course, this new volume, containing in particular the genus *Hieracium*, must rank as Peter's very own 'labour of love'. One of the very special links that has grown up between Peter and me over our long-standing acquaintance in the pursuit of taxonomic botany must be our steady, persistent enthusiasm for critical apomictic genera. We do not have to explain or justify to each other our passions for, in my case, *Alchemilla*, and his for *Hieracium*. I have to admit, however, that his task, with 412 named and described species of *Hieracium* in this volume, casts my puny efforts with British *Alchemilla* into the shade!

Talking to Peter and Gina about the progress of this remarkable Flora, I am encouraged by what I hear. I really believe that both Peter and I will live to see its completion, in spite of the fact that we both 'creak a little at the joints' – to use one of the common euphemisms we find ourselves using from time to time to describe our state of health!

One final observation. How fortunate Peter is to have such a remarkable fellow-author in Gina! Writing and publishing books involves much more than producing a draft text. Some of the skill is straightforward, if laborious; but some requires real understanding at the level of human relations, and both these skills are possessed in abundance by Gina. So I conclude by saying to both Peter and Gina: keep up the good work to a successful conclusion.

13 February 2002

Sadly, Max died on 11 December 2005. He strongly approved of our whole attitude towards variation and introductions and fully understood the treatment of apomictic genera. In Max, Cambridge had a leader in taxonomic botany and conservation of our flora for over 50 years. See my obituary in *Watsonia* **26**: 215–277 (2007).

Peter Sell
2007

Preface

For over 60 years I have worked in the Herbarium at Cambridge University on the British and European floras. I have collected about 30,000 numbers consisting of some 50,000 specimens from most parts of the British Isles and made many visits to Continental Europe. Particular attention has been given to most critical genera: *Betula*, *Cerastium*, *Chenopodium*, *Conyza*, *Crepis*, *Dactylorhiza*, *Euphrasia*, *Fumaria*, *Hieracium*, *Limonium*, *Pilosella*, *Prunus*, *Rhinanthus*, *Salicornia*, *Salix*, *Scleranthus*, *Sorbus* and *Ulmus*; and in helping friends in various ways I have considered the taxonomy of *Alchemilla*, Batrachian Ranunculi, *Potamogeton*, *Rubus* and *Taraxacum*. I have also spent much time studying ecotypic and geographical variation, in particular a comparison of those variants which occur on the coasts in dunes, shingle and salt-marsh with those growing as arable weeds, and those in mountains. Special attention has also been given to trees and shrubs.

It has long been my wish to publish this information in a critical flora of Great Britain and Ireland. In the 1970s a group of us tried to get a grant to carry this out, but we were unsuccessful. Clive Stace then started work on his *New Flora of the British Isles*, which was first published in 1991, with a second edition in 1997. In it he gives only abbreviated descriptions and omits most of the species in the large apomictic genera and many of the infraspecific variants. Numerous introduced species are included by Stace in a British and Irish flora for the first time, detailed descriptions and specimens of many of which are difficult to find. Stace's flora is to my mind an excellent field guide, which it would be difficult to better, but it does not give the detailed descriptions that are needed to confirm the identification of a plant which is new to you. A good description in my opinion is one in which *a picture of the plant unfolds before you as you read it.*

I considered it was possible for me to write a flora in five volumes which gave a full description of all the species in Stace's flora and to add all the apomicts and many of the infraspecific variants, but it was too large a task to attempt to include all the biological information envisaged by the group in the 1970s. It was necessary, however, to have the help of another author, who lived in Cambridge, to deal with the large amount of work involved. My eye fell upon Gina Murrell who had worked with me in the 1960s and 1970s, when writing accounts for *Flora Europaea*, *Flora of Turkey* and *Flora of the Maltese Islands*. The work of one had complemented the work of the other and we were able to criticise one another without antagonism. We started field work on this flora on 13 May 1987, by describing *Ceratocapnos claviculata,* which was flowering on Dunwich Heath in Suffolk, in a snowstorm. Since then we have as far as possible spent one day a week working in the field or at the Botanic Garden, Cambridge. We started writing Volume 5 in 1992, and completed it by Easter 1994. It was published on 1 April 1997, not 1996 as stated in the volume itself. Volume 4 was published on 1 April 2006.

I have done most of the writing and made the taxonomic and nomenclatural decisions, while Gina has done most of the measuring, sometimes sitting at the microscope dictating the description while I, surrounded by a pile of books, wrote it down, and she has set out and put the whole onto a computer. Gina has also done all the illustrations and organised our field work. Volume 4, probably the largest of the five, took longer than we thought due to my moving house and retiring and having to vacate my room in the Department of Plant Sciences in which I had accumulated much over 50 years. Major alterations to the whole Department have added to the difficulties. Volume 5 contained 28 families, 233 genera, 769 species, 93 subspecies, 148 varieties, 22 formae and 182 hybrids. Volume 4 contained an introduction and full accounts of 7 families, 146 genera, 1098 species, 130 subspecies, 162 varieties, 27 formae and 51 hybrids. It dealt with a whole range of taxa from very variable species which we felt could not be further divided, to species with geographical races, ecotypes, forms and cultivars. The taxa were outbreeding, inbreeding, apomictic or spreading vegetatively.

This volume does not contain any large apomict genera but does include the difficult genera of *Euphrasia* and *Mentha* in its 59 families, 299 genera, 996 species, 187 subspecies, 308 varieties and 102 formae and 235 hybrids.

Peter Sell

Acknowledgements

Philip Oswald has not only translated into Latin the new taxa in this volume, but has answered many general questions concerning Greek and Latin, and of the layout of the flora. The late Max Walters translated much Swedish, German and French as well as discussing many problems of taxonomy. Chris Preston has read the text of the Fabaceae and as well as improving the accounts of the distribution of the taxa, has added their geographical classification which is set out and explained in Preston, C. D. and Hill, M. O., The geographical relationships of British and Irish vascular plants in *Botanical Journal of the Linnean Society* **124**: 1–120 (1997). Geoffrey Kitchener has greatly improved the account of *Epilobium* and brought it in line with his account for the second edition of C.A. Stace (Edit.) *Hybridization and the flora of the British Isles.*

Arthur Chater has acted as a very special sort of adjudicator. Whenever we worked out infraspecific variants which were likely to occur in Cardiganshire, we telephoned him to give him the information, and he would either comment immediately or search for them at the first opportunity. His paper in *Watsonia* **24**: 281–286 (2003) sets out the difficulties in finding information on infraspecific taxa. Charlie Jarvis has helped us while working on the *Hortus Cliffortianus* at the British Museum and Gina Douglas while working on the Linnean collection at Burlington House. Barry Goddard has advised us on computer techniques. To Mrs J. E. Dandy we owe a special debt for giving us the second copy of her husband's manuscript of his detailed work on the nomenclature of the British flora. To Bill and Joan Robinson we are grateful for letting us frequently raid their garden and for allowing some of their vegetables to go to flower and seed so that we were able to collect complete plant specimens. P. D. S. owes a very special debt to Brian and Rosemary Chapman. Every Friday afternoon they have led him over unfamiliar ground at Histon in Cambridgeshire and he has tried to name every plant he has come across down to forma. This has enabled him to keep his eye in with plants in the field as well as in the herbarium.

In the library of the Department of Plant Sciences Richard Savage and Christine Alexander have gone out of their way to get hold of rare and obscure publications for us and to somehow find the books in the library going during major alterations. At the Botanic Garden, Cambridge, Professor John Parker has allowed P.D.S. to continue after his retirement to have a free run of the Garden, its library and herbarium and has endeavoured to answer many of the difficult genetical and biological questions we have put to him. Also at the Garden, James Cullen, Alexander Goodall, Clive King, Peter Kurley, Pete Michna, Sally Petit, Ann Schindler, David Stone, Tim Upson, Norman Villis and Peter Yeo have given us much help. We are grateful to Gwynn Ellis for checking the manuscript and for help with preparing the index.

Professor John Gray and Professor Roger Leigh have allowed P. D. S. to continue to have full use of the herbarium and library after his retirement.

To Clive Stace we owe a very special debt. Had he not written his *New Flora of the British Isles* our task would have been insurmountable.

Introduction

HISTORICAL BACKGROUND

The first real flora of these islands was John Ray's *Catalogus Plantarum Angliae et Insularum Adjacentium* in 1670. The first flora to use the Linnaean binomial system of nomenclature was William Hudson's *Flora Anglica* nearly a hundred years later in 1762. This was followed by William Withering's *Botanical Arrangement of all the Vegetables naturally growing in Great Britain* in 1776–92, the first of many floras written primarily for the amateur.

James Sowerby's *English Botany*, whose text was written by J.E. Smith, was first published between 1790 and 1820. It presented for the first time a complete set of coloured illustrations of our plants, illustrations which are still unsurpassed for line and colour. The third edition, published between 1863 and 1872, has inferior illustrations, but its text, rewritten by James Boswell Syme, is still important for its nomenclature and infraspecific taxa.

Three especially famous floras were produced in the nineteeth century. George Bentham's *Handbook of the British Flora* in 1858 was written as a before-breakfast relaxation. In it keys appeared for the first time in a British flora. It was revised by J.D. Hooker in 1886.

J.D. Hooker's *Student's Flora of the British Islands,* first published in 1870 and finally revised in 1884, had very clear and concise descriptions and was the main flora used by many generations of botanists up until the 1950s. It is also important in that Hooker was one of the first authors to make frequent use of the category of subspecies.

Charles Cardale Babington's *Manual of British Botany* first appeared in 1843 and the tenth edition, revised by A.J. Wilmott, was published in 1922. It contains many critical species and varieties not in other floras, but the descriptions are not clear and without keys it is difficult to use.

C.E. Moss's *Cambridge British Flora* (1914–20) was very detailed and would have supplied a much needed critical flora, but alas only two volumes were published.

The arrival of 'C. T. & W.', A.R. Clapham, T.G. Tutin and E.F. Warburg's *Flora of the British Isles,* in 1952, heralded the beginning of a new era in the study of British plants. It was the first up-to-date treatment this century. A much revised second edition appeared in 1962 and a third in 1987 when D.M. Moore replaced E.F. Warburg. This last edition included the information in Tutin et al., *Flora Europaea* **1–5** (1964–1980). The nomenclature had been brought up to date by J.E. Dandy in his *List of British vascular plants* in 1958, and the work he did on this for *Flora Europaea*. Thus for the first time taxonomy and nomenclature had been brought in line with that of Continental Europe.

The Botanical Society of the British Isles' publication of the *Atlas of the British Flora* in 1962, edited by F.H. Perring and S.M. Walters, and the *Critical supplement to the Atlas of the British Flora* in 1968, edited by F.H. Perring, gave us a much better idea of the distribution of our plants. The second edition of the *Atlas* arrived when most of Volume 4 had been prepared for press, but the fact that Chris Preston had checked most of our distributions meant they were not much out of date. For this volume the *New Atlas of the British and Irish Flora* (2002) was available to us and we were able to bring all distributions up to date. The publication in 2003 of *The vice-county census catalogue of the vascular plants of Great Britain, the Isle of Man and the Channel Islands* edited by C. A. Stace et al. has much helped us to get the distributions up to date.

The arrival of Clive Stace's *New Flora of the British Isles* in 1991 with a second edition in 1997, and D.H. Kent's *List of vascular plants of the British Isles* in 1992 has brought about the end of the C. T. & W. era and given us a completely up-to-date account of our flora. Major changes included the moving over of the main classification to A. Cronquist's *An integrated system of classification of flowering plants* (1981) and the inclusion of almost as many alien species as native ones.

The aim of our Flora is to supply full descriptions of all the species in Stace's flora, to include all the large apomictic genera and as many infraspecific variants as practicable, and to add more information about hybrids for which extensive use has been made of Stace's *Hybridization and the flora of the British Isles* (1975).

THE CONTENTS OF THE FLORA

The flora includes all the vascular plants, Lycopodiophyta (Clubmosses), Equisetophyta (Horsetails), Pteridophyta (Ferns), Pinophyta (Conifers) and Magnoliophyta (Flowering plants). The list of plants is made up of all our native species, including apomicts, and all the introduced plants given in Stace (1991), with some more added, particularly planted trees. E.J. Clement's and M.C. Foster's *Alien plants of the British Isles* arrived in 1994 after we had completed Volume 5, but we went through it and added as much information as possible. It has been used continually while preparing Volumes 4 and 3. These alien taxa may be found to be more widespread when full attention is given to them. In his coverage of alien taxa Stace considers inclusion is merited when an alien is either naturalised (i.e. permanent and competing with other vegetation, or self-perpetuating) or, if a casual, frequently recurrent so that it can be found in most years. These criteria were applied as much to garden escapes or throw-outs as to the unintentionally introduced plants, and rarity was not taken into consideration for any of them. Cultivated species were included if they are field crops or forestry crops, or in the case of trees only, ornamentals grown on a large scale. Stace's aim has been to include **all taxa that the botanist might reasonably be able to find in the wild in any one year**. To these we have added some ornamental trees and shrubs that are planted along streets and roadsides and in parks and estates and which we consider to be part of the landscape. A major attempt is being made in this and the remaining volumes to include the vast number of trees and shrubs planted in new 'woods', by new roads and by streets in the last 40 years. Usually plants in gardens are not mentioned at all, but some species, which seed freely and spread over areas of garden and lawn where they are not planted, are included. Most of the species which Stace has mentioned, but not numbered or included in the keys, are here included, while a few have been left out altogether. We started with Volume 5 because the *The European Garden Flora* had already covered the Monocotyledons, which has made it easier for us to deal with the garden escapes. We followed it with Volume 4, because it contained the large genera *Hieracium* and *Taraxacum*. This volume has given us a brief 'rest' after the difficulties of Volume 4.

GEOGRAPHICAL AREA

The flora deals with the British Isles and includes England, Scotland and Wales, collectively known as Great Britain, Northern Ireland and Eire together forming Ireland, the Isle of Man, and the Channel Islands which include Jersey, Guernsey, Alderney, Sark, Herm and various small islands. In these respects it follows Stace (1991 and 1997).

The smallest geographical area usually referred to is the county. This sometimes includes more than one botanical vice-county for which we use the terminology in J.E. Dandy, *Watsonian vice-counties of Great Britain* (1969). For Great Britain we have used the boundaries adopted by H.C. Watson in 1873 in *Topographical botany* and in Ireland by R.L. Praeger in 1901 in *Irish topographical botany*. These are the vice-counties used by botanists, which have the benefit of not changing at regular intervals as do the political counties. With rare or local species the actual place or area may be given. The extra-limital distributions are those given in Clapham, Tutin and Moore (1987) with as much correcting as we and Chris Preston can give them. Russia and Yugoslavia have been used in the sense of the old U.S.S.R. and Yugoslavia before political disruptions.

CLASSIFICATION AND NOMENCLATURE

The classification follows that of Stace (1991 and 1997) and Kent (1992) which is taken from A. Cronquist, *An integrated system of classification of flowering plants* (1981), with the exception that the main groups are called Divisions and the second groups Classes following H.C. Bold, C. Alexopoulos and T. Deleveryas, *Morphology of plants and fungi*, ed. 4 in 1980 and A. Cronquist, A. Takhtajan and W. Zimmermann, On the higher taxa of embryobionta in *Taxon* **15**: 129–134 in 1966, and set out by one of us, P.D.S., in *The Cambridge cyclopedia of life sciences* in 1985.

One of us (P.D.S.) has specialised in nomenclature for many years and it is here made as accurate as possible according to the latest *International code of botanical nomenclature*. The names of genera and species differ little from those in Stace (1991 and 1997) and Kent (1992). Recent changes in the code of nomenclature have been used to get rid of some names that have been a long-standing source of confusion. New taxa and such changes in nomenclature and taxonomy which do occur are published at the end of the volume.

No rules have been made about the number of synonyms given, as many as possible being included, but an attempt has been made to include all names used in British and Irish floras. The abbreviation *auct.* following a name means only that the name has not been accepted for the plant, it does not mean the type has been checked and the name rejected. Only in the case of a later homonym, which has been checked, does the word *non* and an author follow the name and author. This including of numerous synonyms often shows how a species has moved from one genus to another over the years.

The English name for the species follows Stace (1991 and 1997) as far as possible, and where they are missing from new species they have been created.

In whatever way you arrive at your identified species your plant should fit exactly the detailed description in the text. If a difficult plant has any chance of being identified very detailed notes of every character should be made in the field. All too often a miserable bit of a plant rotting in a plastic bag is all that is available, as after a long day in the field the botanist puts his meal and bed first.

Sitting on the horizon is the so-called **Code of Bionomenclature** which could have a profound effect on the names of our plants in the future. Its aim is to provide a nomenclature for both botany and zoology in which there are no duplication of names. The most up to date account of it is by D.L. Hawksworth and J. McNeill in *Taxon* **47**: 123–150 (1998). I have always though there is much to be said for the zoological idea that all replacing populations have trinomials. It would mean all the subspecies and varieties in this flora would have trinomials and one would no longer have the difficulty of deciding whether a population is ecological or geographical or both, but how many name changes that would require is not known. Only *forma* would then be used in its present context, which is what the zoologists call *morphs*. Perhaps the biggest problem is priority of genera where the same epithet occurs in both botany and zoology. The genus *Oenanthe* for plants in 1753 and for birds in 1816 would mean over 20 familiar bird names would have to be altered. *Prunella* also has these same dates for plants and birds. This results in some 15 name changes. More worrying seems to be the rules for conserving names, which could result in a caucus sitting on a committee telling everyone else what to do.

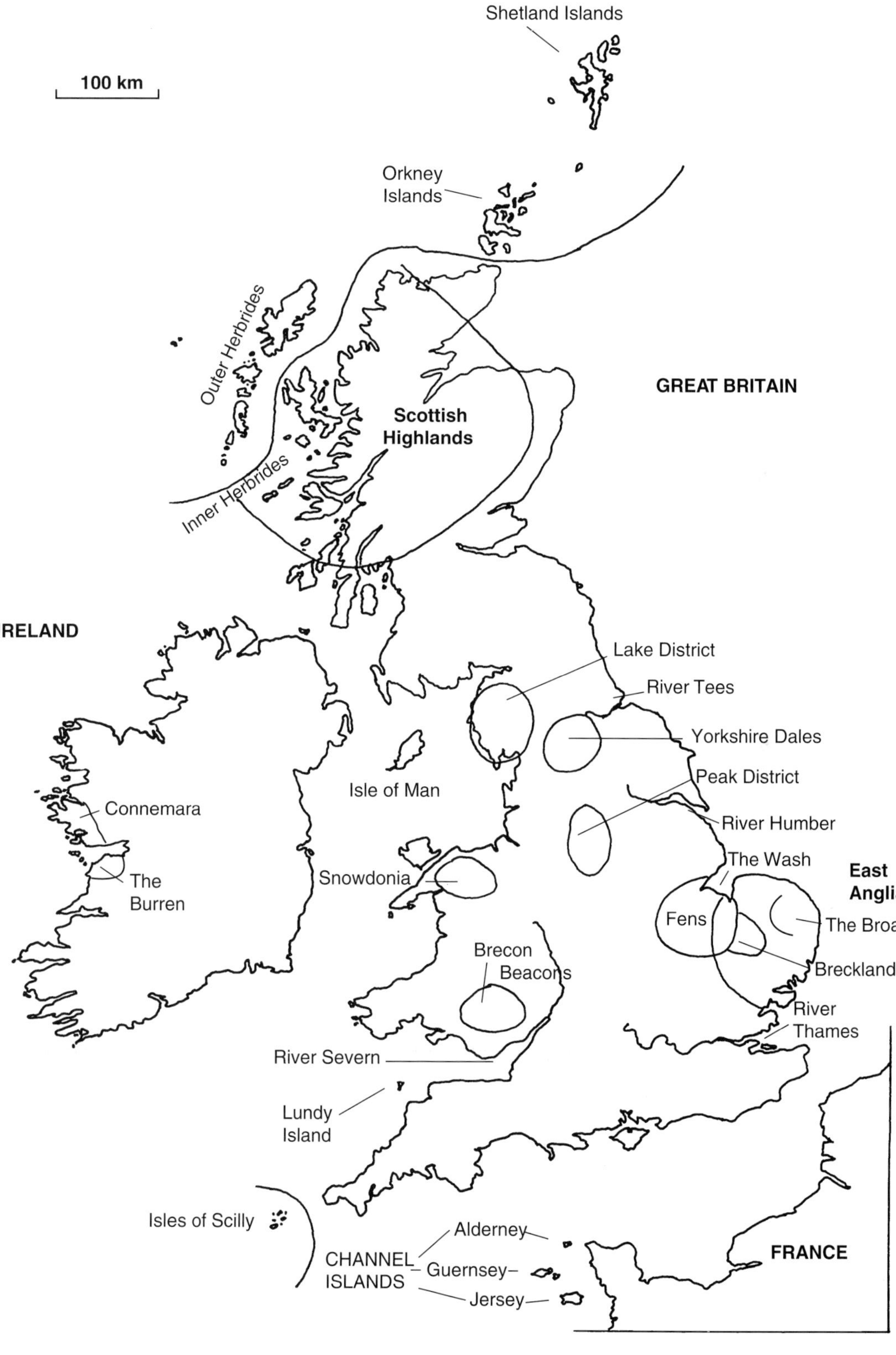

Shetland Islands
100 km
Orkney Islands
Outer Hebrides
GREAT BRITAIN
Scottish Highlands
Inner Hebrides
IRELAND
Lake District
River Tees
Yorkshire Dales
Peak District
Isle of Man
Connemara
River Humber
The Wash
East Anglia
The Burren
Snowdonia
Fens
The Broads
Brecon Beacons
Breckland
River Thames
River Severn
Lundy Island
Isles of Scilly
Alderney
CHANNEL ISLANDS
Guernsey
Jersey
FRANCE

British Isles

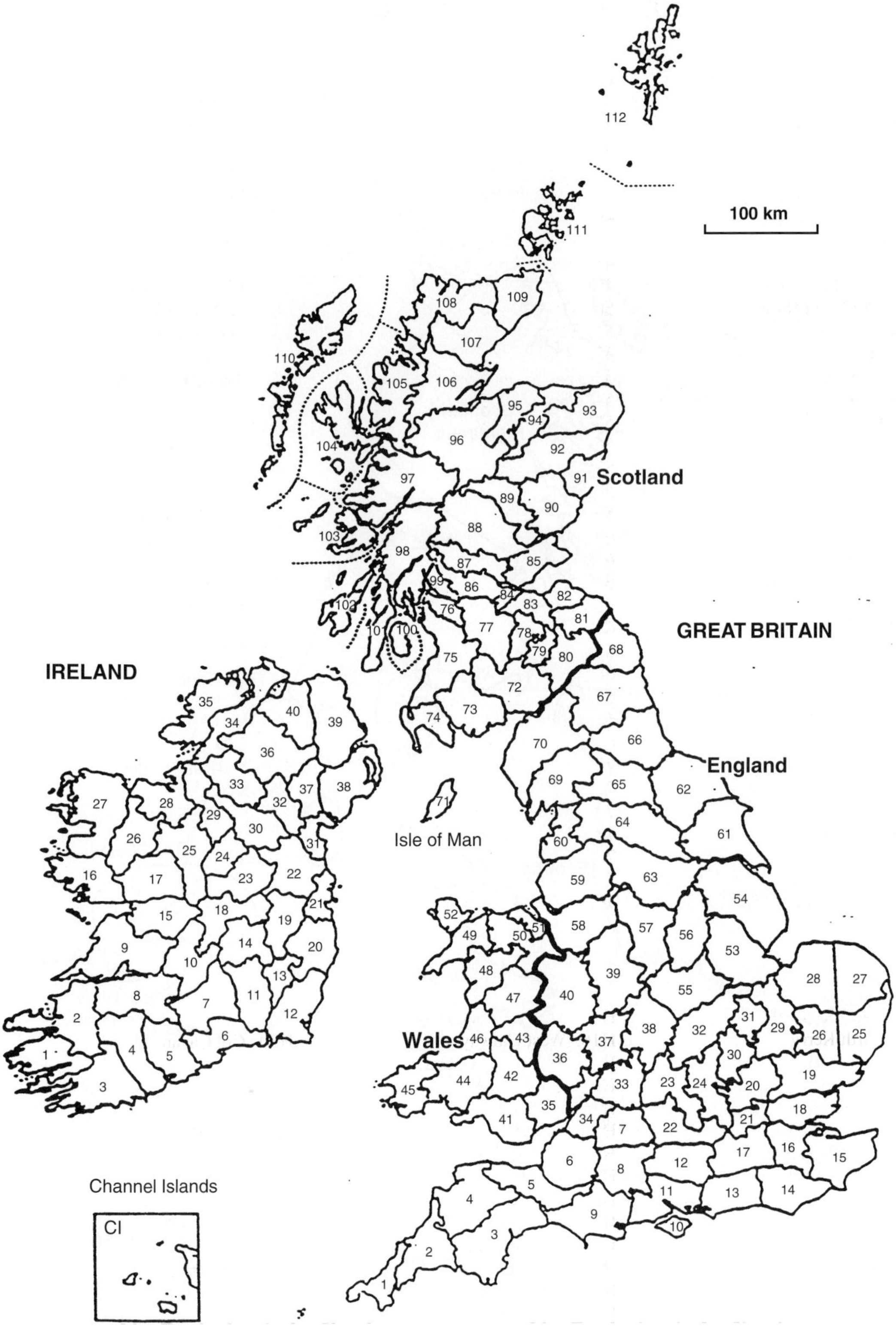

Vice-counties of the British Isles

ENGLAND, WALES, SCOTLAND, ISLE OF MAN

1. West Cornwall
2. East Cornwall
3. South Devon
4. North Devon
5. South Somerset
6. North Somerset
7. North Wiltshire
8. South Wiltshire
9. Dorset
10. Isle of Wight
11. South Hampshire
12. North Hampshire
13. West Sussex
14. East Sussex
15. East Kent
16. West Kent
17. Surrey
18. South Essex
19. North Essex
20. Hertfordshire
21. Middlesex
22. Berkshire
23. Oxfordshire
24. Buckinghamshire
25. East Suffolk
26. West Suffolk
27. East Norfolk
28. West Norfolk
29. Cambridgeshire
30. Bedfordshire
31. Huntingdonshire
32. Northamptonshire
33. East Gloucestershire
34. West Gloucestershire
35. Monmouthshire
36. Herefordshire
37. Worcestershire
38. Warwickshire
39. Staffordshire
40. Shropshire
41. Glamorganshire
42. Breconshire
43. Radnorshire
44. Carmarthenshire
45. Pembrokeshire
46. Cardiganshire
47. Montgomeryshire
48. Merionethshire
49. Caernarvonshire
50. Denbighshire
51. Flintshire
52. Anglesey
53. South Lincolnshire
54. North Lincolnshire
55. Leicestershire
56. Nottinghamshire
57. Derbyshire
58. Cheshire
59. South Lancashire
60. West Lancashire
61. South-east Yorkshire
62. North-east Yorkshire
63. South-west Yorkshire
64. Middle-west Yorkshire
65. North-west Yorkshire
66. Co. Durham
67. South Northumberland
68. Cheviotland
69. Westmorland
70. Cumberland
71. Isle of Man
72. Dumfries-shire
73. Kirkcudbrightshire
74. Wigtownshire
75. Ayrshire
76. Renfrewshire
77. Lanarkshire
78. Peebles-shire
79. Selkirkshire
80. Roxburghshire
81. Berwickshire
82. East Lothian
83. Midlothian
84. West Lothian
85. Fifeshire
86. Stirlingshire
87. West Perthshire
88. Mid Perthshire
89. East Perthshire
90. Forfarshire
91. Kincardineshire
92. South Aberdeenshire
93. North Aberdeenshire
94. Banffshire
95. Morayshire
96. East Inverness-shire
97. West Inverness-shire
98. Main Argyllshire
99. Dunbartonshire
100. Clyde Islands
101. Kintyre
102. South Ebudes
103. Middle Ebudes
104. North Ebudes
105. West Ross-shire
106. East Ross-shire
107. East Sutherland
108. West Sutherland
109. Caithness
110. Outer Hebrides
111. Orkney Islands
112. Shetland Islands

IRELAND

H1. South Kerry
H2. North Kerry
H3. West Cork
H4. Mid Cork
H5. East Cork
H6. Co. Waterford
H7. South Tipperary
H8. Co. Limerick
H9. Co. Clare
H10. North Tipperary
H11. Co. Kilkenny
H12. Co. Wexford
H13. Co. Carlow
H14. Laois
H15. South-east Galway
H16. West Galway
H17. North-east Galway
H18. Offaly
H19. Co. Kildare
H20. Co. Wicklow
H21. Co. Dublin
H22. Meath
H23. West Meath
H24. Co. Longford
H25. Co. Roscommon
H26. East Mayo
H27. West Mayo
H28. Co. Sligo
H29. Co. Leitrim
H30. Co. Cavan
H31. Co. Louth
H32. Co. Monaghan
H33. Fermanagh
H34. East Donegal
H35. West Donegal
H36. Tyrone
H37. Co. Armagh
H38. Co. Down
H39. Co. Antrim
H40. Co. Londonderry

VARIATION

The recording of variation is most important for both ecology and conservation, and even more important for gardeners who go out of their way to both create and conserve prominent variants.

Infraspecific variation is usually recorded by the recognition of subspecies, varieties, formas and cultivars. These taxa differ chiefly in ecology and distribution. A **forma** is a plant with a one- or two-gene difference which occurs with one or more other forms in a mixed population for most or all of its range. A **variety** is when one of these **formas** becomes more or less dominant in a particular ecological area, that is an *ecotype*. A **subspecies** is when one of these formas becomes dominant in a geographical area, that is a *race*. A **cultivar** is a forma which is selected by horticulturalists and perpetuated, usually vegetatively. Because ecotypes and races have become adapted morphologically to different conditions over a long period, it is likely that their physiology and biochemistry, and indeed their whole biology, is different. Also, as they often flower at different periods their pollinators may be different and, if climatic conditions alter, one ecotype may be better able to survive than another. Variation thus becomes very important in conservation. Because the *Biological Flora of the British Isles* has lumped all its information under species it can be highly misleading when applied to individual populations. It is unfortunate that many botanists tend to ignore variation completely, and they will certainly ignore it if it has no name at all; subspecies are usually more often recognised than varieties. Sometimes it is more important to conserve one variety rather than another, e.g. the Chilterns *Orchis militaris* var. *tenuifrons* is endemic, while the Suffolk var. *militaris* occurs in Continental Europe; *Liparis loeselii* var. *ovata* is rare in distribution, but frequent where it occurs, whereas var. *loeselii* is rare in Britain but occurs on the continent. Sometimes the variant will tell us whether the plant is native or not; e.g. *Leucojum aestivum* subsp. *aestivum* is native, subsp. *pulchellum* is a naturalised garden escape. Escaped cultivars are named wherever they can be easily recognised and are considered important. All apomicts, where possible, are treated as species, long experience showing that any sort of lumping deprives them of having an interesting ecology or distribution. Hybrids are dealt with as fully as possible, especially those that spread vegetatively. No serious attempt has been made to decide on the correct infraspecific rank as taxa are often both ecological and geographical. Uniformity of infraspecific rank is often produced in a species or genus, but usually the only important thing considered is that a morphological recognisable infraspecific taxon has a name.

Where species grade gradually into one another over large distances, as the species of *Larix* do round the northern hemisphere, and at given points the whole population is uniformly intermediate it is regarded as a **cline**. Where two populations of ecotypes grow adjacent as in *Geum rivale* and *urbanum* there is often an area in which variable intermediates occur. This is also often called a cline, but it is really only so statistically, and we prefer to call it a **variable hybrid zone**. If you look for these things you will be surprised how often they exist, and clear-cut species, even apomicts, are not so clear-cut as we are made to believe.

Dick Brummitt and Arthur Chater writing in *Watsonia* **23**: 161 (2000) about the genus *Calystegia* say:

> The whole genus, in which some 25 species world-wide may conveniently be recognised, is taxonomically difficult, and few if any of the species are morphologically clear-cut. They mostly vary considerably over their ranges and merge geographically one into another, and division into species and subspecies is of necessity somewhat arbitrary.

We find this true of many groups when their whole range is considered.

If the origin of taxa is considered there are even more difficult problems. During the Quaternary cold stages massive glacierisation from the north caused forests to retreat southwards and come up against other different floras and in some cases to hybridise with them. On climatic amelioration the forests advanced north again, often perhaps by a different path than the one by which they went south, bringing a fresh variety of plants to join those species that survived in rufugia in the cold areas. Richard West in *Plant life of the Quaternary cold stages: evidence from the British Isles* (2001, p 263) writes:

> This overall view shows the cold stage stadial flora to have a long and complex history, originating in the latest Tertiary, occupying a major part of Quaternary time in our area, and surviving short periods of forest dominance at times of climatic amelioration. It is not surprising that the taxonomy of the species concerned is very complex.

Following the forest clearances of the Neolithic about 5,000 years ago and the agricultural revolution which followed, some of the species of open habitats particularly coastal areas, developed ecotypes which became agricultural weeds, or were actually brought in by early Man himself. Indeed, the weeds of Cornwall, East Anglia and Scotland may have been brought in by different races of Man, at different times, from entirely different areas.

To add to this state of affairs Man has brought plants from all over the world into gardens. Sometimes two species which never occur together in nature are grown together in gardens and hybridise, and may even backcross to one or both parents. These plants often escape into the countryside and sometimes our garden plants will hybridise with our native species. To even further confuse the issue Continental races of our native species are introduced to our countryside in packets of wildflower seed or are planted as trees and shrubs in our woodlands.

It can thus be seen that many of our species are far from uniform genetically. Apomictic species and many ecotypic varieties are probably more uniform.

DNA may help us to understand these problems, but will we ever have time and money to look in detail at all our flora?

HERBARIA AND LITERATURE CONSULTED

During the writing of the flora the following books were consulted for every species:

Boreau, A. *Flore du centre de la France*. **1** and **2**. 1857. Paris.
(This contains a large number of segregate species of A. Jordan, which he grew in cultivation for many years, and which are now recognised as infraspecific taxa or apomictic species.)

Clapham, A. R., Tutin, T. G. & Moore, D. M. *Flora of the British Isles*. Ed. 3. 1987. Cambridge.

Dandy, J. E. *List of British vascular plants*. 1958. London.

Hegi, G. *Illustrierte Flora von Mitteleuropa*. Ed. 1. 1906–1931. München. Ed. 2. 1936– München. Ed. 3. 1966– München.

Kent, D. H. *List of vascular plants of the British Isles*. 1992. London.

Perring, F. H. & Walters, S. M. *Atlas of the British flora*. 1962. London & Edinburgh.

Perring, F. H. *Critical supplement to the atlas of the British flora*. 1968. London.

Preston, C. D. & Croft, J. M. *Aquatic plants in Britain and Ireland*. 1997. Colchester.

Preston, C. D. et al. *New atlas of the British and Irish flora*. 2002. Oxford.

Reynolds, S. C. P. *A catalogue of alien plants in Ireland*. 2002. Glasnevin.

Stace, C. A. *Hybridization and the flora of the British Isles*. 1975. London, New York & San Francisco.

Stace, C. *New flora of the British Isles*. Ed. 1. 1991. Cambridge. Ed. 2. 1997. Cambridge.

Stace, C. et al. *Vice-county catalogue of the vascular plants of Great Britain*. 2003. London.

Stearn, W. T. *Botanical Latin*, Ed. 4. 1992. Newton Abbot.
(This book carries an enormous amount of information and may well have been used more often than any other tome.)

Stewart, A., Pearman, D. A. & Preston, C. D. *Scarce plants in Britain*. 1994. Peterborough.

Tutin, T. G., Heywood, V. H., Burges, N. A., Moore, D. M., Valentine, D. H., Walters, S. M. & Webb, D. A. *Flora Europaea*. **1–5**. 1964–1980. Cambridge.

Walters, S. M., Brady, A., Brickell, C. C., Cullen, J., Green, P. S., Lewis, J., Matthews, V. A., Webb, D. A., Yeo, P. F. & Alexander, J. C. M. *The European garden flora* . **1–6.** 1986–2000. Cambridge.

Many other books and journals were consulted, mainly in the Cambridge Department of Plant Sciences, including the N. D. Simpson collection of local floras, and the Cory Library at the Botanic Garden. Where these references were considered to be important for particular plants, we have cited them under the family or genus concerned.

The University herbaria at Cambridge, on which the Flora is mainly based, are ideal for the study of the British flora for the following reasons:

1 The large British collection contains specimens from most of the main collectors of British plants from 1800 onwards, including sets of published exsiccatae and specimens sent through the Botanical Exchange Clubs. Most of the critical species have been named by experts.
2 The British herbarium contains some 50,000 specimens collected by us in the last 50 years. The specimens are accompanied by detailed field notes and are often of critical species or infraspecific taxa. Often a gathering may consist of more than one sheet, particularly of trees which may have been visited three or four times.
3 There is a good herbarium of Continental European plants with which to compare the British plants.
4 The world collection contains over 50,000 sheets of John Lindley's herbarium made when he was secretary of the Royal Horticultural Society, when plants were coming into the country from all parts of the world; and the C. M. Leman collection, named by George Bentham, and put together at the same time. These collections are very important as regards the alien species when considered in conjunction with the Botanic Garden herbarium and recent gatherings of alien specimens.
5 The Botanic Garden herbarium contains a large collection of cultivated plants.

Thus, the libraries, herbaria, our own field notes and plants grown in the Botanic Garden have enabled us to do most of the work in Cambridge. Over many years books and specimens elsewhere have been consulted.

Kingdom **PLANTAE**

Volume 1.

Division **1. LYCOPODIOPHYTA**

Order **1. LYCOPODIALES**

1. LYCOPODIACEAE

Order **2. SELAGINELLALES**

2. SELAGINELLACEAE

Order **3. ISOETALES**

3. ISOETACEAE

Division **2. EQUISETOPHYTA**

Order 1. EQUISETALES

4. EQUISETACEAE

Division **3. PTERIDIOPHYTA**

Order 1. OPHIOGLOSSALES

5. OPHIOGLOSSACEAE

Order 2. OSMUNDALES

6. OSMUNDACEAE

Order 3. PTERIDALES

7. ADIANTACEAE

8. PTERIDACEAE

Order 4. MARSILEALES

9. MARSILEACEAE

Order 5. HYMENOPHYLLALES

10. HYMENOPHYLLACEAE

Order 6. POLYPODIALES

11. POLYPODIACEAE

Order 7. DICKSONIALES

12. CYATHEACEAE

13. DICKSONIACEAE

Order 8. DENNSTAEDTIALES

14. DENNSTAEDTIACEAE

15. THELYPTERIDACEAE

16. ASPLENIACEAE

17. WOODSIACEAE (*ATHYRIACEAE*)

18. DAVALLIACEAE

19. DRYOPTERIDACEAE

20. BLECHNACEAE

Order 9. SALVINIALES

21. AZOLLACEAE

Division 4. PINOPHYTA

Class 1. PINOPSIDA

Order 1. PINALES

22. PINACEAE

23. TAXODIACEAE

24. CUPRESSACEAE

25. ARAUCARIACEAE

Class 2. TAXOPSIDA

Order 1. TAXALES

26. TAXACEAE

Division 5. MAGNOLIOPHYTA

Class 1. MAGNOLIOPSIDA

Subclass 1. MAGNOLIIDAE (*DICOTYLEDONES*)

Order 1. MAGNOLIALES

27. MAGNOLIACEAE

Order 2. LAURALES

28. LAURACEAE

Order 3. ARISTOLOCHIALES

29. ARISTOLOCHIACEAE

Order 4. NYMPHAEALES

30. NYMPHACEAE

31. CERATOPHYLLACEAE

Order 5. RANUNCULALES

32. RANUNCULACEAE

33. BERBERIDACEAE

Order 6. PAPAVERALES

34. PAPAVERACEAE

35. FUMARIACEAE

Subclass 2. HAMAMELIDAE

Order 1. HAMAMELIDALES

36. PLATANACEAE

Order 2. URTICALES

37. ULMACEAE

38. CANNABACEAE

39. MORACEAE

40. URTICACEAE

Order 3. JUGLANDALES

41. JUGLANDACEAE

Order 4. MYRICALES

42. MYRICACEAE

Order 5. FAGALES

43. FAGACEAE

44. BETULACEAE

45. CORYLACEAE

Subclass 3. CARYOPHYLLIDAE

Order 1. CARYOPHYLLALES

46. PHYTOLACCACEAE

46A. NYCTAGINACEAE

47. AIZOACEAE

48. CHENOPODIACEAE

49. AMARANTACEAE

50. PORTULACACEAE

51. BASELLACEAE

52. CARYOPHYLLACEAE

(***ILLECEBRACEAE***)

Order 2. POLYGONALES

53. POLYGONACEAE

Order 3. PLUMBAGINALES

54. PLUMBAGINACEAE

Subclass 4. DILLENIIDAE

Order 1. DILLENIALES

55. PAEONIACEAE

Order 2. THEALES

56. ELATINACEAE

57. CLUSIACEAE (*GUTTIFERAE; HYPERICACEAE*)

Order 3. MALVALES

58. TILIACEAE

59. MALVACEAE

Order 4. NEPENTHALES

60. SARRACENIACEAE

61. DROSERACEAE

Order 5. VIOLALES

62. CISTACEAE

63. VIOLACEAE

64. TAMARICACEAE

65. FRANKENIACEAE

66. CUCURBITACEAE

Order 6. SALICALES

67. SALICACEAE

Volume 2.

Order 7. CAPPARALES

68. CAPPARACEAE

69. BRASSICACEAE (*CRUCIFERAE*)

70. RESEDACEAE

Order 8. ERICALES

71. CLETHRACEAE

72. EMPETRACEAE

73. ERICACEAE

74. PYROLACEAE

75. MONOTROPACEAE

Order 9. DIAPENSIALES

76. DIAPENSIACEAE

Order 10. PRIMULALES

77. MYRSINACEAE

78. PRIMULACEAE

Subclass 5. ROSIDAE

Order 1. ROSALES

79. PITTOSPORACEAE

80. HYDRANGEACEAE

81. ESCALLONIACEAE

82. GROSSULARIACEAE

83. CRASSULACEAE

84. SAXIFRAGACEAE

85. ROSACEAE

Volume 3.

Order 2. FABALES

86. MIMOSACEAE

87. CAESALPINIACEAE

88. FABACEAE (LEGUMINOSAE)

Order 3. PROTEALES

89. ELAEAGNACEAE

Order 4. HALORAGALES

90. HALORAGACEAE

91. GUNNERACEAE

Order 5. MYRTALES

92. LYTHRACEAE

93. THYMELAEACEAE

94. MYRTACEAE

95. ONAGRACEAE

Order 6. CORNALES

96. CORNACEAE

96A. GARRYACEAE

Order 7. SANTALALES

97. SANTALACEAE

98. VISCACEAE

Order 8. CELASTRALES

99. CELASTRACEAE

100. AQUIFOLIACEAE

Order 9. EUPHORBIALES

101. BUXACEAE

102. EUPHORBIACEAE

Order 10. RHAMNALES

103. RHAMNACEAE

104. VITACEAE

Order 11. LINALES

105. LINACEAE

Order 12. POLYGALALES

106. POLYGALACEAE

Order 13. SAPINDALES

107. STAPHYLEACEAE

108. SAPINDACEAE

109. HIPPOCASTANACEAE

110. ACERACEAE

111. ANACARDIACEAE

112. SIMAROUBACEAE

113. RUTACEAE

Order 14. GERANIALES

114. OXALIDACEAE

115. GERANIACEAE

116. LIMNANTHACEAE

117. TROPAEOLACEAE

118. BALSAMINACEAE

Order 15. APIALES

119. ARALIACEAE

120. APIACEAE (*UMBELLIFERAE*)

Subclass 6. ASTERIDAE

Order 1. GENTIANALES

121. GENTIANACEAE

122. APOCYNACEAE

Order 2. SOLANALES

122A. NOLANACEAE

123. SOLANACEAE

124. CONVOLVULACEAE

125. CUSCUTACEAE

126. MENYANTHACEAE

127. POLEMONIACEAE

128. HYDROPHYLLACEAE

Order 3. LAMIALES

129. BORAGINACEAE

130. VERBENACEAE

131. LAMIACEAE (*LABIATAE*)

Order 4. CALLITRICALES

132. HIPPURIDACEAE

133. CALLITRICHACEAE

Order 5. PLANTAGINALES

134. PLANTAGINACEAE

Order 6. SCROPHULARIALES

135. BUDDLEJACEAE

136. OLEACEAE

137. SCROPHULARIACEAE

138. OROBANCHACEAE

139. GESNERACEAE

140. ACANTHACEAE

140A. BIGNONIACEAE

141. LENTIBULARIACEAE

Volume 4.

Order 7. CAMPANULALES

142. CAMPANULACEAE

Order 8. RUBIALES

143. RUBIACEAE

Order 9. DIPSACALES

144. CAPRIFOLIACEAE

145. ADOXACEAE

146. VALERIANACEAE

147. DIPSACEAE

Order 10. ASTERALES

148. ASTERACEAE (*COMPOSITAE*)

Volume 5.

Class 2. LILIOPSIDA (*MONOCOTYLEDONS*)

Subclass 1. ALISMATIDAE

Order 1. ALISMATALES

149. BUTOMACEAE

150. ALISMATACEAE

Order 2. HYDROCHARITALES

151. HYDROCHARITACEAE

Order 3. NAJADALES

152. APONOGETONACEAE

153. SCHEUCHZERIACEAE

156. RUPPIACEAE

157. NAJADACEAE

154. JUNCAGINACEAE

155. POTAMOGETONACEAE

158. ZANNICHELLIACEAE

159. ZOSTERACEAE

Subclass 2. ARECICAE

Order 1. ARECALES

160. ARECACEAE (*PALMAE*)

Order 2. ARALES

161. ARACEAE

162. LEMNACEAE

Subclass 3. COMMELINIDAE

Order 1. COMMELINALES

163. COMMELINACEAE

Order 2. ERIOCAULONALES

164. ERIOCAULONACEAE

Order 3. JUNCALES

165. JUNCACEAE

Order 4. CYPERALES

166. CYPERACEAE

167. POACEAE (*GRAMINEAE*)

Order 5. TYPHALES

168. SPARGANIACEAE

169. TYPHACEAE

Subclass 4. ZINGIBERIDAE

Order 1. BROMELIALES

170. BROMELIACEAE

Subclass 5. LILIIDAE

Order 1. LILIALES

171. PONTEDERIACEAE

172. LILIACEAE (*ALLIACEAE; AMARYLLIDACEAE; TRILLIACEAE*)

173. IRIDACEAE

174. AGAVACEAE

175. DIOSCOREACEAE

Order 2. ORCHIDALES

176. ORCHIDACEAE

Divison **5. MAGNOLIOPHYTA** Cronquist, Takht. & W. Zimm.

Class **1. MAGNOLIOPSIDA** Cronquist, Takht. & W. Zimm.

Subclass 5. ROSIDAE Takht.

Order **2. FABALES** Bromhead

Trees, shrubs or *herbs*. *Leaves* often pinnate or bipinnate, sometimes trifoliolate or simple; stipules present or absent. *Flowers* bisexual, hypogynous to perigynous, actinomorphic to zygomorphic, often large and showy. *Sepals* often 5, often more or less united below into a calyx tube. *Petals* usually 5, rarely absent, occasionally united. *Stamens* often 10, sometimes numerous, often monadelphous or diadelphous. *Ovary* of 1 carpel. *Fruit* a legume, often dehiscent; seeds usually with little or no endosperm, rarely with abundant endosperm; embryo large.

Contains 3 families and some 4,000 species widely distributed throughout the world.

The following 3 families are now accepted as one family in Lewis, G. et al. (2005) *Legumes of the world*. The 3 families are retained here to keep our numbering system and the Fabaceae is brought up to date to follow this fine book. All genera should be consulted in it as it contains a vast amount of information.

86. MIMOSACEAE R. Br. nom. conserv.

Leguminosae subfamily *Mimosoideae* (R. Br.) Taub.

Suckering, deciduous or evergreen, monoecious *trees*. *Leaves* alternate, of 2 kinds, the juvenile bipinnate with numerous leaflets, the adult simple; stipules more or less absent. *Inflorescence* a dense, racemose, spherical head or spiciform raceme; flowers bisexual, hypogynous and actinomorphic. *Calyx* mostly with 5 lobes. *Corolla* 5-lobed. *Stamens* numerous, longer than petals. *Carpel* 1, with several ovules in a row. *Style* 1; stigma capitate. *Fruit* a legume; seeds with a distinct areole, the radicle straight.

Contains 66 genera and over 2,000 species, widespread in tropical and subtropical regions with a few species extending into temperate areas.

Lewis, G. et al. (2005). *Legumes of the world*. Kew.

An easily recognisable family by its balls of flowers of which the stamens are the most conspicuous part.

1. Stamens united into a tube	**1. Albizia**
1. Stamens free or nearly so	**2. Acacia**

1. Albizia Durazz.

Shrubs or small evergreen *trees*. *Leaves* alternate, bipinnate. *Inflorescence* an axillary, spiciform raceme; stipules more or less absent. *Calyx* infundibuliform, with 5 short lobes. *Corolla* infundibuliform, with 5 lobes. *Stamens* numerous; filaments connate in the lower part forming a slender tube. *Style* 1; stigma capitate. *Fruit* a legume, straight and flattened.

Contains about 100–150 species widely distributed in the tropics and subtropics.

Rushforth, K. (1999). *Trees of Britain and Europe*. London.

1. A. lophantha (Willd.) Benth. Plume Albizia
Acacia lophantha Willd.

Shrub or small spreading evergreen *tree* to 7.5 m with fern-like appearance. *Bark* brown. *Branches* spreading, strongly striate or angled, shortly hairy. *Buds* small. *Leaves* alternate, 12–25 × 10–20 cm, oblong in outline, bipinnate; leaflets 20–40 pairs along each pinna, dark green on upper surface, paler beneath, 5–15 × 2.0–4.5 mm, oblong or the terminal obovate, obtuse and shortly mucronate at apex, glabrous or very sparingly hairy; rachis velvety hairy; petiole 4–6 cm, sulcate, usually with a prominent sessile gland near its base. *Inflorescence* an axillary spiciform raceme, solitary or in pairs near the tips of the young shoots; peduncle 1.5–4.0 cm, hairy; pedicels 1–2 mm, hairy; bracteoles deltoid, hairy, caducous; flowers fragrant. *Calyx* infundibuliform, with 5 short, deltoid, obtuse lobes, hairy. *Corolla* about 6 mm, about twice as long as the calyx, greenish, with 5 tongue-shaped, obtuse, hairy lobes. *Stamens* numerous; filaments 13–18 mm, creamy-white, connate for a short distance above the base; anthers whitish, minute. *Style* 1. *Legume* 7–10 × 1.5–2.5 cm, reddish-brown, oblong, flattened, rounded or acute at apex, sutures distinctly thickened; valves glabrous, conspicuously swollen about the seeds; seeds 5–12, about 8.0 × 4.5 mm, oblong, not much compressed, dark blackish-brown, minutely punctate. *Flowers* 6–8. $2n=26$.

Introduced. Planted for ornament in the Isles of Scilly where it produces seedlings. Native of southern and western Australia.

2. Acacia Mill.

Evergreen *shrubs* or small *trees*. *Leaves* alternate, bipinnate, sometimes mixed with phyllodes or only phyllodes. *Inflorescence* of many-flowered, globose heads arranged in axillary racemes. *Calyx* with 5 lobes. *Corolla* white to yellow, 5-lobed. *Stamens* numerous, free. *Style* 1; stigma terminal. *Fruit* a legume.

Contains more than 1,500 species, cosmopolitan but mainly in the southern hemisphere.

Maslin, B. R. et al. (2001). *Acacia* L. in *Flora of Australia*: 11A. Canberra.
More, D. & White, J. (2003). *Trees of Britain and Northern Europe*. Jersey.
Pedley, L. (1986). Derivation and dispersal of *Acacia* (Leguminosae) with particular reference to Australia, and the

recognition of *Senegalia* and *Racosperma*. *Bot. Jour. Linn. Soc.* 92: 219–254.
Rushforth, K. (1999). *Trees of Britain and Europe*. London.

1. Mature plants with bipinnate leaves **1. dealbata**
2. Mature plants with only phyllodes or phyllodes and some bipinnate leaves **2. melanoxylon**

1. A. dealbata Link Silver Wattle

Evergreen *shrub* or *tree* up to 20 m, with a conical crown when young becoming columnar with age, suckering freely. *Trunk* up to 30 cm in diameter, extending into crown. *Bark* bluish-green and smooth in young trees, becoming chocolate brown, then grey or black, often corrugated or fluted. *Branches* spreading or drooping; twigs ribbed; young shoots green, angled, with short silvery hairs. *Buds* 1.5 mm, ovoid; scales blackish-brown, ovate, acute at apex. *Leaves* alternate, broadly oblong in outline, bipinnate; pinnae (12–)20–42, broadly oblong-elliptical in outline, each pinna with 50–80(–100) pinnules, the pinnules 2–4 (–7) × about 0.7 mm, when young golden brown, becoming greyish-green, narrowly oblong, obtuse to subacute at apex, entire; with a few silvery hairs; petioles (5–)10–20 mm; stipules inconspicuous; solitary glands present between each pair of pinnae, usually excepting the basal pair. *Inflorescence* a many-flowered, globose head arranged in axillary, compound racemes. *Calyx* with 5 lobes. *Corolla* bright yellow, fragrant, 5-lobed. *Stamens* numerous, free, exserted; filaments pale; anthers yellow. *Style* 1, filiform; stigma terminal. *Legume* 4–10 × 0.8–1.2 cm, oblong, more or less straight, glabrous, compressed between the seeds. *Flowers* 2–4. $2n = 26$.

Introduced. Frequently planted in estates and parks. Devonshire, Cornwall and Ireland, rare in south-east England. Native of New South Wales, Victoria and Tasmania. Probably the most tolerant of the species cultivated and that most commonly sold as 'mimosa'.

2. A. melanoxylon R. Br. Australian Blackwood
Racosperma melanoxylon (R. Br.) C. Mart.

Tall evergreen *shrub* or *tree* up to 25 m, with a columnar to domed crown. *Trunk* up to 80 cm in diameter, extending through much of tree. *Bark* dark greyish-brown, becoming rough and furrowed. *Branches* spreading; twigs angular, glabrous or hairy; young shoot angular and hairy. *Buds* minute. *Leaves* mostly reduced to phyllodes, or occasionally some of them bipinnate in young trees and mixed with phyllodes, rarely present in old trees; phyllodes 8–10 × 0.6–3.5 cm, dark green on upper surface, slightly paler beneath, oblanceolate or slightly obovate, obtuse to acute at apex, entire, glabrous, with 3–5(–7) longitudinal veins, the venation reticulate and conspicuous; true leaves with a narrowly winged rachis, the pinnae 4–9 cm, bearing about 40 narrowly oblong, obtuse leaflets about 8.0 × 2.5 mm. *Inflorescence* of 1–6 heads, forming short, rather congested, axillary racemes, the heads globose and 5–8 mm in diameter; peduncles usually less than 4 mm. *Calyx* with 5 lobes. *Corolla* white to yellow, 5-lobed. *Stamens* numerous; filaments pale; anthers yellow. *Style* 1; stigma capitate. *Legumes* 7–12 × 0.4–0.7 cm, valves dark brown with paler sutures, linear, compressed, much twisted and curled, obtuse at apex, thickened at sutures; seeds about 4.0 × 2.5 mm, oblong, dark shining brown, the funicle more than 10 mm, bent back on itself and almost completely circling the seed. *Flowers* 5–6. $2n = 26$.

Introduced. Planted for ornament in gardens and parks and locally more or less naturalised in Devonshire, the Isles of Scilly and Cos. Kerry and Cork. Native of Australia and Tasmania and long in cultivation in other parts of the world.

87. CAESALPINIACEAE R. Br. nom. conserv.

Leguminosae Subfamily *Caesalpinioideae* (R. Br.)

Deciduous monoecious *trees* or *shrubs* or perennial herbs. *Leaves* alternate, paripinnate or simple and subrotund. *Inflorescence* of dense lateral clusters of flowers or racemes; flowers bisexual, slightly zygomorphic, but rarely as strongly as in Fabaceae. *Calyx* with 5 lobes as free as the disc. *Corolla* 5-lobed, the upper inside the laterals. *Stamens* usually 10; anthers without an apical gland. *Style* 1. *Fruit* a legume; seeds generally without an areole; radicle normally straight.

Contains 156 genera and over 2,200 species and is widespread in tropical and subtropical regions, with a few species in temperate areas.

Lewis, G. et al. (2005). *Legumes of the world*. Kew.
Petigrew, C. J. & Watson, L. (1977). On the classification of Caesalpinioideae. *Taxon* 26: 57–64.

1. Tree or shrub; leaves subrotund **1. Cercis**
1. Perennial herbs; leaves paripinnate **2. Senna**

1. Cercis L.

Deciduous *trees*. *Leaves* alternate; lamina subrotund. *Inflorescence* of dense, lateral clusters of flowers along mature twigs and branches and sometimes from the trunk. *Calyx* cupular, with 5 lobes. *Corolla* deep rose-purple, rarely paler or white, with 5 free petals, the 2 abaxial ones not connate and forming a keel. *Stamens* 10, free. *Style* 1. *Fruit* a legume, oblong and strongly compressed.

Contains 7 species in the northern hemisphere.

More, D. & White, J. (1993). *Trees of Britain & Northern Europe*. Jersey.
Bean, W. J. (1970). *Trees and shrubs hardy in the British Isles*. Ed. 8. 1. London.

1. C. siliquastrum L. Judas Tree

Deciduous often sprawling *tree* up to 10(–15) m with a spreading, rounded, domed crown. *Trunk* up to 60 cm in diameter, extending well into the crown. *Bark* greyish-brown, smooth or folded, becoming cracked into small, rough squares. *Branches* spreading, twisted and turning at various angles; twigs brown or reddish-brown, angled, glabrous; young shoots green, often tinted red, glabrous. *Buds* 2–3 × 1.0–1.5 mm, ovoid, pointed; scales brownish, ovate, obtuse at apex. *Leaves* alternate; lamina 6–13 × 6–13 cm, tinted copper-colour when opening, when mature medium yellowish-green on upper surface, paler and slightly glaucous beneath, subrotund, slightly emarginate to slightly pointed at apex, entire but slightly sinuate, cordate at base, glabrous; veins prominent beneath; petiole up to 5 cm, green or tinted pink, glabrous; stipules small, deltoid, caducous. *Inflorescence* of dense lateral clusters of flowers along mature twigs and branches, sometimes even protruding from the trunk and appearing slightly in advance of the leaves; pedicels 15–20 mm, deep reddish-purple, glabrous, slender. *Calyx* 5–6 mm, usually deep reddish-purple, rarely pale greenish, cupular, with 5, short, rounded, imbricate lobes. *Corolla* 12–14 mm, deep rose-purple, rarely pale pink or white, all 5 petals free, the 2 abaxial ones not connate and forming a keel, elliptical or ovate and rounded at apex. *Stamens* 10, free; filaments pale purple, curved; anthers yellowish. *Style* 1, pale purple; stigma yellowish. *Legume* 6–12 × 1.5–2.0 cm, narrowly oblong, flattened, acute at apex, the dorsal suture narrowly but distinctly winged; seeds about 6 × 4 mm, dark brown, oblong, strongly compressed, smooth. *Flowers* 5. $2n = 14$.

Introduced. Planted in parks, gardens and amenity areas and seedlings have been recorded in Kent. Southern England, Wales and Ireland.

2. **Senna** Mill.

Perennial herbs or small shrubs. *Leaves* alternate, paripinnate; stipulate. *Inflorescence* of 1–4 axillary flowers. *Sepals* 5, subequal. *Petals* 5, slightly zygomorphic. *Stamens* 10, 6 or 7 fertile, the remainder sterile. *Style* 1, short; stigma terminal, minute. *Fruit* a legume.

The two species described here are fairly frequent casuals from soya-bean waste, but rarely reach flowering. They can, however, be distinguished by their leaves.

1. Leaflets 6–14, broadly elliptical to ovate, acute or acuminate at apex **1. occidentalis**
2. Leaflets 4–6, obovate, rounded-obtuse and mucronulate at base **2. obtusifolia**

1. S. occidentalis (L.) Link Coffee Senna
Cassia occidentalis L.

Shrub up to 2 m. *Stems* erect, branched. *Leaves* alternate, 15–17 cm, broadly oblong in outline, paripinnate; leaflets 6–14, 5–7 × 3–4 cm, medium green on upper surface, paler beneath, broadly elliptical to ovate, acute or acuminate at apex, entire, rounded at base, hairy on both surfaces; petiole 2–4 cm, ribbed, with 1 gland very near the base; stipules acicular and caducous. *Inflorescence* of 2–4 flowers, terminal and in upper leaf axils, paniculate; peduncle 2–5 mm; pedicels 10–15 mm; bracts caducous. *Sepals* 5, green or yellowish, subequal, obovate. *Petals* 5, up to 10 mm, yellow. *Fertile stamens* 6, with 3 adaxial and 1 abaxial staminodes; fertile filaments 2–8 mm, very unequal; fertile anthers 4–6 mm, very unequal, with short beaks. *Style* 1, short; stigma terminal, minute. *Legume* 12–18 × about 0.3 cm, cylindrical, slightly curved, entire; seeds all reniform. *Flowers* 8–9. $2n = 26$.

Introduced. Casual from soya-bean waste. Probably native of the New World but now an aggressive pantropical weed.

2. S. obtusifolia (L.) H. S. Irwin & Barneby
American Sicklepod
Cassia obtusifolia L.

Perennial herb or *subshrub*. *Stems* up to 2 m, erect or spreading, branched. *Leaves* 4–6 cm, alternate, broadly oblong or obovate in outline, paripinnate; leaflets 4–6, 4–5 × 2–3 cm, medium green on upper surface, paler beneath, obovate, rounded-obtuse and mucronulate at apex, entire, narrowed at base, hairy on both surfaces; petiole 1.5–2.0 cm; glands 1–2 between the lowest pairs of leaflets; stipules lanceolate, somewhat persistent. *Inflorescence* of 1–2 axillary flowers; peduncle 2–4 mm; pedicels 15–20 mm; bracts caducous. *Sepals* 5, green or yellowish, subequal, obovate. *Petals* 5, 8–10 mm, yellow. *Fertile* stamens 7, with 3 adaxial staminodes; fertile filaments unequal, 1–4 mm; fertile anthers 3–5 mm, unequal, shortly beaked. *Style* 1. *Legume* 12–18 × about 0.3 cm, entire, curved; seeds dull or lustrous, with a narrow, oblique areole. $2n = 26$.

Introduced. Casual from soya-bean waste. Probably native to the Americas, but now of almost pantropical distribution.

88. FABACEAE Lindl. nom. conserv.

LEGUMINOSAE Juss. nom. altern.

Papilionaceae; *Leguminosae* subfam. *Papilionoideae*; *Leguminosae* subfam. *Lotoideae*

Annual to *perennial* monoecious *herbs*, *shrubs* or trees, sometimes spiny. *Leaves* alternate, simple to palmate, trifoliolate or pinnate, often with tendrils, usually stipulate. *Inflorescence* terminal or axillary, in spikes, racemes or panicles; flowers bisexual, hypogynous, zygomorphic. *Calyx* often 2-lipped, the upper lip of 2 lobes, the lower lip with 3 lobes. *Petals* 5, the upper usually large (*standard*), the 2 laterals free (*wings*) and the 2 lower fused (*keel*), the keel more or less concealing the stamens and carpel. *Stamens* 10, usually fused into a tube below (*monadelphous*), or the uppermost free and the 9 lower fused (*diadelphous*), rarely all 10 free. *Style* 1; stigma capitate. *Carpel* 1, with 1-many ovules in row. *Fruit* a legume, usually dehiscent along 2 sides, but often breaking transversely into 1-seeded units like a schizocarp.

Contains 455 genera and over 10,000 species. Widespread in cold, temperate and tropical regions.

Gunn, C. R. (1969). Genera, types and lectotypes in the tribe *Vicieae* (Leguminosae). *Taxon* **18**: 725–733.
Lewis, G. et al. (2005). *Legumes of the world*. Kew.
Polhill, R. M. & Raven, P. H. (Edits.) (1981). *Advances in legume systematics*. **1** and **2**; Stirton, C. H. (Edit.) (1987). **3**. Kew.
Stirton, C. H. & Zarucchi, J. L. (1989) (Edits.). *Advances in legume biology*. Missouri Botanical Garden.

1. Leaves consisting of a tendril only (but with large leaf-like stipules) **42. Lathyrus**
1. Leaves not consisting of a tendril only 2.
2. At least some leaves paripinnate, imparipinnate or digitate; leaflets 2, 4 or more 3.
2. Leaves simple, 1-foliolate or 3-foliolate, sometimes very small 44.

3. Leaves paripinnate; rhachis often ending in a spine or tendril 4.
3. Leaves imparipinnate or digitate 14.
4. Shrubs or small trees; rhachis often ending in a spine 5.
4. Herbs; rhachis not ending in a spine 6.
5. Pedicels usually less than 5 mm, not articulated **28. Astragalus**
5. Pedicels 5 mm or more, articulated **31. Caragana**
6. Stipules adnate to the petiole; calyx bilabiate, the upper lip with 4 teeth, the lower with 1 tooth **8. Arachis**
6. Stipules not adnate to the petiole; calyx actinomorphic or if bilabiate, the upper lip with 2 teeth, the lower with 3 teeth 7.
7. Stem and leaves glandular-hairy **34. Cicer**
7. Stem and leaves not glandular-hairy 8.
8. Stem winged **42. Lathyrus**
8. Stem not winged 9.
9. Leaflets parallel veined **42. Lathyrus**
9. Leaflets pinnately veined 10.
10. Calyx teeth all equal and at least twice as long as the tube 11.
10. At least 2 calyx teeth less than twice as long as the tube 12.
11. Legume linear **15. Sesbania**
11. Legume strongly compressed **41. Lens**
12. Calyx teeth more or less leaf-like; stipules up to 10 cm **43. Pisum**
12. Calyx teeth not leaf-like; stipules not more than 2 cm 13.
13. Style hairy all round or on the lower side, or glabrous **40. Vicia**
13. Style hairy on upper side only **42. Lathyrus**
14. Principal lateral veins of the leaflets terminating at the margin, often in a tooth 15.
14. Lateral veins of the leaflets anastomosing and not reaching the margin 17.
15. Without or soon losing simple eglandular hairs, without glandular hairs (except to replace bracts) **35. Trifolium**
15. With various amounts of simple eglandular and glandular hairs, sometimes sparse 16.
16. Calyx gibbous at base; stipules free from petiole **34. Cicer**
16. Calyx not gibbous at base; stipules adnate to petiole **36. Ononis**
17. At least some flowers in a terminal or apparently terminal inflorescence 18.
17. All flowers axillary or in axillary inflorescences 20.
18. Leaves digitate **2. Lupinus**
18. Leaves imparipinnate 19.
19. Climbing shrubs; flowers in pendulous racemes; legume velutinous **9. Wisteria**
19. Non-climbing; flowers in erect inflorescences; legume not velutinous **20. Anthyllis**
20. Plant with glandular hairs, at least in part **26. Robinia**
20. Plant without glandular hairs 21.
21. Flowers in umbels or clusters, the pedicels arising more or less from the same point 22.
21. Flowers in racemes or condensed panicles or solitary 30.
22. Legume lomentaceous 23.
22. Legume dehiscent or indehiscent, not lomentaceous 25.
23. Keel obtuse; legume strongly reticulate-veined **22. Ornithopus**
23. Keel acute; legume not or only faintly reticulate-veined 24.
24. Segments of the legume linear or oblong, straight or slightly curved **19. Coronilla**
24. Segments of the legume lunate or horseshoe-shaped to rectangular with a semicircular to orbicular sinus which has a curved protuberance at its base **16. Hippocrepis**
25. Keel beaked 26.
25. Keel not beaked 28.
26. Leaves with 4–7 pairs of leaflets **18. Securigera**
26. Leaves with 2–3 pairs of leaflets or simple 27.
27. All leaves with 2 pairs of leaflets; legume linear or oblong, straight or curved **23. Lotus**
27. Lower leaves simple, upper leaves with 2–3 pairs of leaflets; legume spirally twisted and flattened so that it is circular in outline **21. Hymenocarpus**
28. Keel very dark red or black **24. Dorycnium**
28. Keel not dark red or black 29.
29. Leaves digitate or apparently so **14. Caragana**
29. Leaves imparipinnate **12. Astragalus**
30. Leaflets distinctly parallel veined **42. Lathyrus**
30. Leaflets pinnately veined or the lateral veins obscure 31.
31. Leaflets with a spinescent apex **31. Caragana**
31. Leaflets without a spinescent apex 32.
32. Legume lomentaceous 33.
32. Legume not lomentaceous 34.
33. Racemes 2–4- to many-flowered **32. Hedysarum**
33. Flowers solitary, axillary **16. Hippocrepis**
34. Legume indehiscent, usually toothed or spiny **33. Onobrychis**
34. Legume usually dehiscent, not toothed or spiny 35.
35. Racemes 10 cm or more, pendulous; stipules usually forming spines; leaflets stipulate **2. Robinia**
35. Racemes usually less than 10 cm, erect; stipules not forming spines 26.
36. Legume strongly inflated, membranous 37.
36. Legume not or only slightly inflated, not membranous 38.
37. Acaulescent herbs **28. Astragalus**
37. Shrubs up to 2 m or more **29. Colutea**
38. Keel beaked **23. Lotus**
38. Keel not beaked but sometimes mucronate 39.
39. Keel mucronate at apex 40.
39. Keel not mucronate at apex 41.
40. Mucro on abaxial side of the keel **27. Oxytropis**
40. Mucro on adaxial side of the keel **28. Astragalus**
41. Stamens monadelphous 42.

41. Stamens diadelphous 43.

42. Annual; corolla about 3 mm **28. Astragalus**

42. Perennial; corolla 10–15 mm **30. Galega**

43. Style glabrous **28. Astragalus**

43. Style pubescent on the lower side **40. Vicia**

44. Principal lateral veins on the leaflets terminating at the margin; leaflets often toothed 45.

44. Principal lateral veins of the leaflets anastomosing and not reaching the margin, sometimes obscure; leaflets not toothed (leaves sometimes caducous or reduced to a spine-tipped phyllode) 56.

45. Plant with glandular hairs 46.

45. Plant without glandular hairs (except to replace bracts) 47.

46. Stamens monadelphous; legume straight or very slightly curved **36. Ononis**

46. Stamens diadelphous; legume falcate to spirally coiled, rarely almost straight **39. Medicago**

47. At least some petal-claws adnate to the staminal tube; corolla usually persistent in fruit **35. Trifolium**

47. Petal-claws free from the staminal tube; corolla deciduous 48.

48. Filaments of at least 5 stamens dilated at the apex **35. Trifolium**

48. Filaments all filiform 49.

49. Legumes coiled in 1 or more turns of a spiral **39. Medicago**

49. Legume straight or curved 50.

50. Perennial 51.

50 Annual or biennial 52.

51. Legume obovate or ovate to subglobose **37. Melilotus**

51. Legume oblong, oblong-falcate, oblong-reniform, reniform or variously curved **39. Medicago**

52. Corolla blue **38. Trigonella**

52. Corolla white or yellow 53.

53. Legume linear or oblong, at least 3 times as long as wide **38. Trigonella**

53. Legume ovate or obovate to subglobose or reniform, less than 3 times as long as wide. 54.

54. Legume reniform **39. Medicago**

54. Legume ovate or obovate to subglobose 55.

55. Legume without or with a very short beak and without a membranous wing **37. Melilotus**

55. Legume with a long, curved beak or with a broad membranous wing on the margin **38. Trigonella**

56. Plant spiny 57.

56. Plant not spiny 59.

57. Leaves of adult plants reduced to persistent spine-tipped phyllodes **7. Ulex**

57. Leaves not spine-tipped, often caducous 58.

58. Calyx more or less distinctly bilabiate **5. Genista**

58. Calyx with 5 more or less equal teeth, not or only slightly bilabiate **20. Anthyllis**

59. Leaflets stipellate; leaves 3-foliolate 60.

59. Leaflets not stipellate; leaves simple or 3-foliolate 62.

60. Corolla not more than 7 mm; plant with reddish-brown hairs **10. Glycine**

60. Corolla 10 mm or more; plant glabrous or with whitish hairs 61.

61. Beak of the keel recurved **11. Vigna**

61. Beak of the keel forming $1\frac{1}{3}$–2 turns of a spiral **12. Phaseolus**

62. Leaves simple or 1-foliolate, sometimes very small 63.

62. At least some leaves 3-foliolate 68.

63. Annual herbs 64.

63. Shrubs or perennial herbs, woody at base 65.

64. Leaves obovate or elliptical, not grass-like; legume indehiscent **17. Scorpiurus**

64. Leaves linear, grass-like; legume dehiscent **42. Lathyrus**

65. Calyx split to the base adaxially **6. Spartium**

65. Calyx not split to the base 66.

66. Calyx more or less tubular; legume more or less included in the persistent calyx **20. Anthyllis**

66. Calyx campanulate; legume exserted, or the calyx not persistent 67.

67. Upper lip of calyx with short teeth **4. Cytisus**

67. Upper lip of calyx deeply 2-fid or deeply toothed **5. Genista**

68. Leaflets conspicuously glandular-punctate 69.

68. Leaflets not or very minutely glandular-punctate 70.

69. Fruit never black glandular warty **13. Bitumaria**

69. Fruit conspicuously black glandular warty when mature **14. Cullen**

70. Legume lomentaceous **19. Coronilla**

70. Legume not lomentaceous 71.

71. Annual or perennial herbs, sometimes with a woody stock 72.

71. Shrubs or trees 76.

72. Stamens free; flowers in clusters of 3 arranged in a terminal leafy raceme **1. Thermopsis**

72. Stamens connate; flowers not in a cluster of 3 arranged in a terminal leafy raceme 73.

73. Calyx inflated to 4.5–6.0 mm wide in flower, up to 12 mm wide in fruit and enclosing the legume **20. Anthyllis**

73. Calyx less than 4.5 mm wide, not inflated 74.

74. Keel very dark red or black **24. Dorycnium**

74. Keel not very dark red or black 75.

75. Stipules inserted at the base of the petiole; legume not longitudinally winged **23. Lotus**

75. Stipules inserted on the stem, and adnate to the base of the petiole; legume with 2 or 4 longitudinal wings **25. Tetragonolobus**

76. Legume more or less included in the persistent calyx; calyx with 5 more or less equal teeth **20. Anthyllis**

76. Legume exserted or the calyx deciduous; calyx bilabiate 77.

77. Flowers in pendulous racemes **3. Laburnum**

77. Flowers in erect inflorescences 78.

78. Upper lip of calyx with 2 short teeth **4. Cytisus**

78. Upper lip of calyx deeply 2-fid 79.

79. Pedicel 5–10 mm; legume glabrous **4. Cytisus**

79. Pedicel 1–3 mm; legume hairy **5. Genista**

Tribe **1. Thermopsideae** Yakovlev

Rhizomatous *perennial herbs*. *Leaves* trifoliolate, with entire leaflets. *Inflorescence* a raceme, with small flowers. *Stamens* 10. *Fruits* a longitudinally dehiscent legume, with 2–7 seeds.

1. Thermopsis R. Br.

Perennial herbs with spreading rhizomes. *Leaves* alternate, pinnately 3-foliolate; leaflets entire. *Inflorescence* a lax, terminal raceme with small linear bracts; stipules large, ovate. *Calyx* with 5 lobes. *Corolla* yellow; standard with limb subrotund, deeply emarginate at apex, cordate at base; wings with limb ovate, rounded at apex, with a wrinkled lobe at base and a short claw; keel with limb broadly oblong, with lobe at base and a slender claw. *Stamens* 10, all free. *Style* 1. *Legume* linear, erect, hairy, longitudinally dehiscent; seeds 3–7.

Contains about 23 species in east Asia and North America.

Larisay, M. M. (1940). A revision of the North American species of Thermopsis. *Ann. Miss. Bot. Gard.* **27**: 245–258.

1. T. montana Nutt. ex Torr. & A. Gray False Lupin

Perennial herb with spreading rhizomes. *Stems* 30–70 (–100) cm, pale green, erect, rigid, triangular, hollow, with pale hairs, with long ascending branches, leafy. *Leaves* alternate, pinnately 3-foliolate, broadly ovate in outline; leaflets 30–100 × 14–36 mm, dull medium green on upper surface, paler beneath, elliptical or oblanceolate, acute at apex, entire, attenuate at the sessile or shortly petiolulate base, glabrous on upper surface and with impressed midrib, softly hairy beneath with prominent veins, especially the midrib; petiole up to 35 mm, pale green, sometimes tinted purplish, channelled above, rounded below, hairy; stipules 30–50 × 15–30 mm, ovate, acute at apex. *Inflorescence* a lax terminal raceme of 5–60 flowers; bracts up to 12 mm, linear; pedicels 7–8 mm, densely hairy. *Calyx* 11–12 mm, medium green; lobes 5, unequal, triangular, acute at apex, sericeous-hairy. *Corolla* 18–25 mm, golden yellow; standard with limb subrotund, deeply emarginate at apex, cordate at base; wings with limb ovate, rounded at apex, with a wrinkled lobe at base and a short claw; keel with limb broadly oblong, rounded at apex, with lobe at base and a slender claw. *Stamens* 10, all free; filaments pale yellow; anthers orange. *Style* 1, green, curved at apex; stigma minute. *Legume* 40–60 × 4.5–5.0 mm, linear, erect, hairy, longitudinally dehiscent; seeds 2–7. *Flowers* 5–6. $2n = 18$.

Introduced. Grown in gardens and naturalised on a few old sites or rough grassy places. Fetlar in Sheltand since about 1978; formerly in Northamptonshire by gravel pits at Oundle for at least 25 years. Native of western North America.

Tribe **2. Genisteae** (Adans.) Benth.

Perennial or *annual herbs* or woody *shrubs* or *trees*. *Leaves* simple, entire, ternate or palmate. *Inflorescence* an axillary or terminal raceme, sometimes reduced to 1 or 2 flowers. *Stamens* 10, monadelphous. *Fruits* a longitudinally dehiscent legume, with 2 to many seeds.

2. Lupinus L.

Annual and *perennial herbs* and sometimes a small shrub. *Leaves* alternate, palmately foliolate, petiolate; stipules adnate to the base of the petiole. *Inflorescence* a terminal raceme. *Calyx* 2-lipped, often divided nearly to base. *Corolla* white, yellow or blue to purple; standard erect, subrotund or broadly ovate; wings falcate-oblong or obovate; keel beaked. *Stamens* 10, monadelphous. *Style* 1. *Fruit* a dehiscent, compressed legume, usually constricted between the seeds; seeds 3–12, with a sunken hilum.

Contains about 200 species in America and the Mediterranean region.

Bean, W. J. (1973). *Trees and shrubs hardy in the British Isles*. Ed. 8. **2**. London.
Gorer, R. (1970). *The development of garden flowers*. London.

1. Small shrub with stems woody towards the base and not dying down to the ground in winter 2.
1. Stems herbaceous, dying down to ground in winter 4.
2. Flowers white **1(1). arboreus** forma **albus**
2. Flowers yellow 3.
3. Flowers pale yellow **1(2). arboreus** forma **arboreus**
3. Flowers golden yellow **1(3). arboreus** forma **aureus**
4. Annuals easily uprooted 5.
4. Perennials tuft-forming 7.
5. Leaflets linear **6. angustifolius**
5. Leaflets oblanceolate or obovate-oblong 6.
6. Corolla white with a pale blue keel; seeds 8–14 mm **5. albus**
6. Corolla yellow; seeds 6–8 mm **7. luteus**
7. Basal leaves absent at flowering time; stem with long, shaggy hairs **4. nootkatensis**
7. Basal leaves present at flowering time; stem with sparse, short hairs 8.
8. Stem mostly branched with more than 1 inflorescence; leaflets broad and obtuse-mucronate to acute at apex; corolla various shades of pink, purple, yellow or white **2. × regalis**
8. Stem unbranched with 1 inflorescence; narrower and longer, long acute at apex; corolla blue **3. polyphyllus**

1. L. arboreus Sims Tree Lupin

Short-lived, semi-evergreen *shrub* 1–3 m. *Stems* brownish, much-branched, erect, with rather numerous hairs, leafy. *Leaves* alternate, palmately foliolate; leaflets 5–12, 20–60 × 5–10 mm, medium green on upper surface, paler beneath, obovate-oblong, mucronate at apex, entire, strigose or glabrous on upper surface, strigose beneath; stipules subulate. *Inflorescence* lax, terminal, 10–30 cm, the flowers alternate or subverticellate and scented; peduncles 4–10 cm. *Calyx* 9–10 mm, 2-lipped, the upper lip emarginate, the lower lip entire, with 2 small lobes often present between it and upper lip. *Corolla* 14–17 mm, white to deep yellow, sometimes blue tinged; standard subrotund, rounded at apex, reflexed at sides; wings large, covering the keel and joined together at the point; keel acute and black pointed. *Stamens* 10, monadelphous. *Style* 1, pale. *Legume* 40–80 mm, brown, linear, strigose; seeds 8–12, 4–5 mm, dark brown, ellipsoid, more or less mottled with a pair of spots near the micropyle. *Flowers* 6–8. $2n=40$, 48.

(1) Forma **albus** P. D. Sell
Corolla white.

(2) Forma **arboreus**
Corolla pale yellow

(3) Forma **aureus** P. D. Sell
Corolla golden yellow.

Introduced in 1793. Widely planted on sand dunes and shingle where it covers large areas by virtue of its copiously produced seed. It also occurs on roadsides, railway banks and waste places inland. All round the coast of Great Britain and eastern Ireland and scattered localities inland. Native of California. All 3 colour forms grow together on the Suffolk coast.

2. L. × **regalis** Bergmans — Russell Lupin
L. arboreus × **polyphyllus**

Perennial herb with a tap-root and fibrous side-roots. *Stem* up to 1.5 m, pale green, erect, striate, sparingly branched, with sparse appressed hairs, leafy. *Leaves* alternate, palmate, subrotund in outline; leaflets 8–15, 9–12 × 2–3 cm, medium green on upper surface, paler, slightly glaucous green beneath, elliptical or lanceolate, rounded-mucronate to shortly acute, entire, narrowed at base, glabrous on upper surface, appressed-hairy beneath, sessile or nearly so; petiole up to 30 cm, pale green, appressed-hairy; stipules up to 30 mm, lanceolate, acute at apex. *Inflorescence* a dense, terminal raceme up to 40 cm; pedicels up to 10 mm, densely hairy. *Calyx* 6–8 mm, deeply 2-lipped, upper lip 2-lobed, lower lip 5.8 mm, 3-lobed, very hairy. *Corolla* 12–16 mm, various shades of whitish through yellow to blue; standard subrotund, rounded at apex; wings elliptical, rounded at base; keel upcurved, with a subacute apex. *Stamens* 10, monadelphous; filaments pale; anthers pale yellow. *Style* 1, upcurved. *Legume* 25–50 mm, oblong, silky hairy; seeds almost 4 mm, ellipsoid. *Flowers* 7–9.

Introduced. Commonly grown in gardens and frequently well naturalised on rough ground and rail and roadside banks. Scattered throughout Great Britain with a solitary record in Ireland. Of garden origin, and most records are of garden escapes, but occurs spontaneously in Scotland where both parents occur naturalised. It probably crosses with *L. nootkatensis* but is difficult to distinguish.

3. L. polyphyllus Lindl. — Garden Lupin

Perennial herb. Stem 50–150 cm, pale green, stout, minutely hairy, usually unbranched, leafy. *Leaves* alternate, palmately foliolate, basal present at flowering time; leaflets 9–17, 70–150 × 15–30 mm, medium green on upper surface, paler beneath, obovate-lanceolate, obtuse to acute at apex, entire, usually glabrous on upper surface, sparingly sericeous beneath; stipules subulate. *Inflorescence* a rather dense terminal raceme 15–60 cm, the flowers verticillate; peduncle 3–8 cm. *Calyx* 9–10 mm, 2-lipped, both lips entire, or upper shallowly 2-toothed. *Corolla* 12–14 mm, blue, purple, yellow, pink or white; standard short, subrotund, apiculate, revolute; wings half-oblong, obtuse at apex, convex; keel falcate, with a long acuminate deep purple beak. *Stamens* 10, monadelphous. *Style* 1, subulate. *Legume* 25–40 mm, brown, sparsely hairy; seeds 5–9, about 4 mm, variously spotted. *Flowers* 5–9. $2n=48$.

Introduced. Formerly grown in gardens and naturalised by rivers and railways, sometimes in waste places. Scattered records throughout Great Britain and in north-east Ireland. Native of western North America. Cultivated for ornament and fodder in a large part of Europe.

4. L. nootkatensis Donn ex Sims — Nootka Lupin

Perennial herb with a tap-root and fibrous side-roots. *Stems* up to 1 m, pale green, erect, with long shaggy hairs, leafy. *Leaves* alternate, palmate, subrotund in outline, basal absent at flowering time; leaflets 6–9(–12), 25–50 × 5–10 mm, medium green on upper surface, paler beneath, narrowly elliptical, acute-mucronate at apex, entire, attenuate at base, upper surface glabrous or sparsely hairy, lower surface with numerous hairs; petiole rather longer than leaflet; stipules subulate to linear, acuminate at apex. *Inflorescence* a dense, terminal raceme up to 10 cm; bracts caducous; pedicels up to 10 mm, hairy. *Calyx* up to 13 mm, deeply 2-lipped, upper lip 2-lobed, lower lip 3-lobed, very hairy. *Corolla* 12–16 mm, bluish-purple, sometimes whitish tinged; standard with limb subrotund, rounded at apex; wings with standard broadly elliptical, broadly rounded at apex, rounded at base with a short, narrow claw; keel with limb ovate, upturned at the subacute apex, with a short slender claw. *Stamens* 10, monadelphous; filaments upcurved; anthers yellowish. *Style* 1, upcurved, glabrous; stigma brownish. Legume 40–50 × 6–8 mm, brown, oblong, silky-hairy; seeds 4.5–5.0 mm, ellipsoid. *Flowers* 5–7. $2n=48$.

Introduced. Naturalised on riverside shingle and moorland since at least 1862. Central and north Scotland, from Perthshire to Orkney Islands, especially by the River Tay and River Dee; north-west Ireland. Native of North America and north-east Asia.

× **regalis**
This triple hybrid probably occurs with *L. nootkatensis* and *L.* × *regalis* in Morayshire and Perthshire, but is difficult to identify. It is intermediate and fertile.

5. L. albus L. — White Lupin

Annual herb with fibrous roots. *Stems* up to 120 cm, pale green, shortly hairy, branched, leafy. *Leaves* alternate, palmately 5-foliolate; leaflets of the lower leaves 25–35 × 14–18 mm and obovate, those of the upper leaves 40–50 × 10–15 mm, obovate and cuneate at base, all medium green on upper surface, paler beneath, mucronulate at apex, entire, nearly glabrous on upper surface and sparsely long-hairy beneath; stipules setaceous. *Inflorescence* a terminal raceme 5–10 cm, sessile, with the flowers alternate. *Calyx* 8–9 mm, 2-lipped, the upper lip deeply 2-lobed, the lower lip 2–3-lobed, both lips shallowly dentate. *Corolla* 15–16 mm, white with deep pale blue at apex; standard 18–20 × about 13 mm; wings 17–20 × 8–10 mm; keel 13–15 × about 6.0 mm. *Stamens* 10, monadelphous. *Styles* 1. *Legume* 80–100 × 17–20 mm, yellow, shortly hairy becoming glabrous, and longitudinally rugulose; seeds 4–6, 8–14 mm, dull pale yellow, orbicular-quadrangular, compressed or depressed, smooth. *Flowers* 6–9. $2n = 50$.

Introduced. Casual on tips, at docks and in waste places. Sporadic in a few localities in Great Britain. Subsp. *graecus* (Boiss. & Spruner) Franco & P. Silva (*L. graecus* Boiss. & Spruner) is native of the Balkan Peninsula and the Aegean region. Our plant is subsp. **albus** which is the cultivated variant which is widely grown in central and southern Europe for its edible seeds and for fodder.

6. L. angustifolius L. Narrow-leaved Lupin

Annual or *biennial herb* with fibrous roots. *Stem* 10–50 cm, pale green, appressed hairy, simple or with numerous suberect or ascending branches, leafy. *Leaves* alternate, 2–7 × 2–7 cm, palmately 5-foliolate; leaflets 10–50 × 1–6 mm, dark bluish-green on upper surface, paler beneath, narrowly oblong, rounded at apex, entire, tapered at base, more or less glabrous on upper surface, thinly appressed-hairy beneath; petiole up to 5 cm, channelled above, slender, appressed-hairy; stipules 5–10 mm, linear, caudate at apex. *Inflorescence* lax, terminal raceme 5–10 cm, overtopping the foliage; peduncle very short; pedicels up to 2.5 mm, hairy; bracts 5–8 mm, elliptical, caducous, thinly appressed-hairy; bracteoles 1–2 mm, narrowly obovate or oblanceolate, adnate to upper part of calyx. *Calyx* markedly 2-lipped; upper lip 3–4 mm, divided almost to base into 2 acute lobes; lower lip up to 7 mm, shortly 3-lobed at apex. *Corolla* 12–14 mm, bright purplish-blue with a paler keel; standard with limb ovate, subacute at apex, with 2 marked longitudinal pleats and reflexed margins; wings oblong, distinctly rugose towards the base; keel oblong with an upturned, dark-tipped acuminate apex. *Stamens* 10, monadelphous; filaments not swollen below anther; anthers narrowly oblong. *Style* 1, up-curved, glabrous; stigma barbellate. *Legume* 40–50 × 10–13 mm, oblong, acute and often beaked, strongly compressed laterally, densely appressed-hairy; seeds 3–5, about 6 mm, marbled cream, brown and grey, smooth. *Flowers* 6–9. $2n = 40, 48$.

Introduced. Imported for trial as a seed-crop and scarce casual in docks and waste places. Sporadic records in Great Britain, mainly the south, and Cape Clear in Ireland. Native throughout the Mediterranean area.

7. L. luteus L. Annual Yellow Lupin

Annual herb with fibrous roots. *Stem* 25–80 cm, pale green, erect, hairy, branched, leafy. *Leaves* alternate, palmately 5-foliolate; leaflets 40–60 × 8–12 mm; dark green on upper surface, paler beneath, obovate-oblong, mucronate at apex, entire, sparsely long-hairy; stipules dimorphic, those of the lower leaves about 8 mm and subulate, those of the upper leaves 22–30 × 2–4 mm and linear-obovate. *Inflorescence* a terminal raceme 5–16 cm, the flowers scented and regularly verticillate; peduncle 4–12 cm. *Calyx* 2-lipped, the upper lip 6–7 mm and 2-lobed, the lower lip about 10 mm and shallowly 3-lobed. *Corolla* 13–16 mm, bright yellow; standard 15–18 × 10–11 mm, ovate, subacute; wings 14–17 × 8–9 mm, half-oblong; keel 14–15 × 4–6 mm, oblong. *Stamens* 10, monadelphous. *Style* 1, pale; stigma brownish. *Legume* 40–50 × 10–12 mm, black, linear, usually constricted between the seeds, densely long-hairy; seeds 4–6, 6–8 × 4.5–6.5 mm, black marbled with white, with a white curved line on each side, orbicular-quadrangular, compressed, smooth and dull. *Flowers* 6–9. $2n = 52$.

Introduced. An escape from cultivation, particularly on sandy soil, and a wool casual. Native of the west Mediterranean region. Widely cultivated for fodder and green manure.

3. Laburnum Fabr.

Deciduous *trees*. *Leaves* alternate, 3-foliolate; long petiolate. *Inflorescence* a simple, pendulous raceme on short shoots. *Calyx* 2-lipped, the upper lip 2-lobed, the lower lip 3-lobed. *Corolla* yellow; standard oblong-elliptical or ovate, emarginate at apex, more or less cordate at base with a short claw; wings narrowly oblong, rounded at apex, with a short lobe at base and a short claw; keel oblong, upturned at apex, with a short claw. *Stamens* 10, monadelphous. *Style* 1. *Fruit* a legume; seeds several, poisonous (to humans), dispersed by explosive dehiscence.

Contains 2 species in central and south Europe.

Chater, A. O. (1992). *Laburnum anagyroides* and *L. alpinum* as hedge plants in Cardiganshire v.c. 46. *BSBI Welsh Bulletin* **52**: 4–5.

Bean, W. J. (1973). *Trees and shrubs hardy in the British Isles*. Ed. 8. **2**. London.

Wettstein, R. (1890–1891). Untersuchungen über die Section *Laburnum* der Gattung *Cytisus*. *Östr. Bot. Zeitschr.* **40**: 395–399, 435–439 (1890); **41**: 127–130; 169–173; 261–265 (1891).

1. Racemes lax; corolla 17–23 mm **1. anagyroides**
1. Racemes dense; corolla 13–21 mm 2.
2. Plant hairy to almost glabrous; corolla 15–21 mm; good seeds rarely formed **2. × watereri**
2. Plant glabrous or sparsely hairy; corolla 13–20 mm; good seeds formed **3. alpinum**

1. L. anagyroides Medik. Laburnum

Cytisus laburnum L.; *L. vulgare* Bercht. & J. Presl; *L. laburnum* (L.) Dörfl.

Deciduous *tree* up to 12 m, with an open crown. *Trunk* up to 1.4 m in diameter, only part of way into the crown. *Bark* dark green, becoming brown in old trees, smooth. *Branches* ascending and arching wide; twigs pale brown, slender;

young shoots greyish-green, with dense, appressed, silky hairs. *Buds* 1.0–1.5×0.5–1.0 mm, ovoid, pointed at apex; scales greyish-brown, ovate, pointed at apex, hairy. *Leaves* alternate, 3-foliolate; leaflets 30–80×15–30 mm, medium greyish-green on upper surface, paler and bluish-green beneath, elliptical to elliptical-obovate or oblanceolate, pointed but obtuse at apex, entire, cuneate to rounded at base, shortly petiolulate, glabrous on upper surface, silky appressed hairs beneath when young; petiole 2–8, with short appressed hairs. *Inflorescence* a simple, lax, pendulous raceme, 10–30 cm, on short shoots; pedicels 10–16 mm, appressed-hairy. *Calyx* 1.0–6.5 mm, pale green, 2-lipped, the upper lip 2-lobed with a V-shaped sinus, the lower lip 3-lobed, appressed-hairy. *Corolla* 17–23 mm, medium yellow; standard with limb 18–20 mm, oblong-elliptical, emarginate at apex, subcordate at base with a short claw; wings with limb 18–20 mm, narrowly oblong, rounded at apex, with a short lobe at base and a short claw; keel with limb oblong, upturned to a rounded apex and a narrow claw. *Stamens* 10, monadelphous; filaments pale green; anthers golden yellow. *Style* 1, pale green; stigma shortly bifid, yellow. *Legume* 30–60×7–8 mm, green, turning brown, oblong, pendulous, subterete but bulging over the seeds, the upper suture thickened, appressed-hairy when young; seeds several, explosive dehiscent, poisonous, remaining on the trees all winter. *Flowers* 5–6. Pollination mainly by bumble-bees. $2n=48$.

Introduced. Formerly much planted in gardens and parks and frequently self-sown in rough ground and on banks of roadsides and railways, formerly planted for hedging and still persistent in some areas. Scattered records throughout Great Britain and Ireland. Native of the mountains of south-central Europe.

2. L.×watereri (Wettst.) Dippel Hybrid Laburnum
L. alpinum×anagyroides
L.×vossii auct.; *Cytisus×watereri* Wettst.

Deciduous *tree* up to 12 m, with an open crown. *Trunk* up to 1.4 m in diameter, only part of the way into the crown. *Bark* dark green, becoming brown in old trees, smooth. *Branches* ascending and arching wide; twigs pale brown or greenish-brown, hairy becoming glabrous with age; young shoots pale green, densely appressed-hairy. *Buds* 1.0–1.5×0.5–1.0 mm, ovoid, pointed at apex; scales greyish-brown, ovate, pointed at apex, hairy. *Leaves* alternate, 3-foliolate; leaflets 30–60×15–30 mm, medium green on upper surface, paler beneath, elliptical, ovate or lanceolate, more or less acute at apex, entire, narrowed at base, very shortly petiolulate, glabrous on upper surface, shortly appressed hairy beneath; petiole up to 5 cm, shortly appressed-hairy. *Inflorescence* a simple, dense, pendulous raceme up to 20 cm on short shoots; pedicels 9–11 mm, slender, with short appressed hairs. *Calyx* 4–5 mm, pale green, 2-lipped, the upper lip 2-lobed with a U-shaped sinus, the lower lip 3-lobed, with short appressed hairs. *Corolla* 15–21 mm, medium yellow, with brownish veins at base of standard; standard with limb 18–19 mm, ovate, emarginate at apex, cordate at base to a short, narrow claw; wings 17–18 mm, with limb asymmetrically oblong, rounded at apex, with a small lobe at base and a short, narrow claw; keel with limb oblong, curved up at apex, with a short lobe and a short claw at base. *Stamens* 10, monadelphous; filaments pale green; anthers yellow. *Style* 1, pale green; stigma shortly bifid, yellow. *Legume* 40–60×6–8 mm, green, turning brown, oblong, pendulous, appressed-hairy when young; seeds rarely formed and usually sterile. *Flowers* 5–6. Pollination mainly by bumble-bees.

Introduced. Probably now more widely planted in gardens, by roads and railways and in parks and amenity areas than either of its parents; has been used for hedging in Wales.

3. L. alpinum (Mill.) J. Presl Scottish Laburnum
Cytisus alpinus Mill.

Deciduous *tree* up to 13 m with an open crown. *Trunk* up to 1 m in diameter, only part of the way into the crown. *Bark* dark green, becoming brown in old trees, smooth. *Branches* ascending and arching wide; twigs grey, smooth or wrinkled; young shoots green, glabrous. *Buds* 1.0–1.5×0.1–1.0 mm, ovoid, pointed at apex; scales greyish-brown, ovate, pointed at apex, hairy. *Leaves* alternate, 3-foliolate; leaflets 45–70×15–30 mm, medium yellowish-green on upper surface, slightly paler beneath, elliptical or lanceolate, acute at apex, entire, cuneate at base, sessile or nearly so, hairy only on midrib and margin; petiole up to 6 cm, pale green, hairy. *Inflorescence* a simple, dense, pendulous raceme 15–35 cm; pedicels 8–12 mm, hairy. *Calyx* 5–6 mm, 2-lipped, the upper lip 2-lobed, the lower lip 3-lobed, the lobes triangular and acute at apex. *Corolla* 13–20 mm, medium yellow grading to orange at base; standard subrotund, emarginate at apex, cordate at base with a short claw; wings oblong, obtuse at apex, with a short claw; keel half ovate, turned up at apex. *Stamens* 10, monadelphous; filaments pale green; anthers orange. *Style* 1, green; stigma yellow. *Legume* 30–40×4–5 mm, green, turning brown, linear, pendulous, appressed-hairy. *Flowers* 5–6.

Introduced. Planted on roadsides and in woodland and hedges, occasionally in quantity, rarely reproducing from seed but occasionally forming thickets. Scattered records throughout Great Britain but mainly in the west and north; and locally abundant in one area of north-east Ireland.

4. Cytisus Desf.

Sarothamnus Wimm.; *Lembotropis* Griseb.

Unarmed *shrubs*. *Leaves* alternate, 1- or 3-foliolate, sometimes crowded on older branches; sessile or petiolate. *Inflorescence* of terminal or axillary racemes or 1–few flowers in leaf axils. *Calyx* 2-lipped, the upper lip with 2 lobes, the lower lip with 3 lobes. *Corolla* yellow or white; standard subrotund or ovate; wings broadly oblong to elliptical; keel oblong or ovate. *Stamens* 10, monadelphous. *Style* 1. *Fruit* a linear or oblong, dehiscent legume; seeds usually numerous, usually strophiolate.

Contains about 30 species mostly in Spain and Portugal.

Bean, W. J. (1970). *Trees and shrubs hardy in the British Isles*. Ed. 8. **1**. London.
Böcher, T. W. & Larson, K. (1958). Secondary polyploid and ecological differentication in *Sarothamnus scoparius*. *New Phytol.* **57**: 311–317.

Chater, A. O. (1978). *Cytisus striatus* (Hill) Roth. for 50 years in v.c. 46. *Watsonia* **12**: 191.
Duncan, U. K. (1978). *Cytisus striatus* (Hill) Rothm. in Britain. *Watsonia* **12**: 49.
Gill, J. J. B. & Walker, S. (1971). Studies in *Cytisus scoparius* (L.) Link with particular emphasis on the prostrate forms. *Watsonia* **8**: 345–356.
Heywood, V. H. (1959). *Cytisus* L. *Proc. B. S. B. I.* **3**: 175–176.
Morton, J. K. (1955). Chromosome studies on *Sarothamnus scoparius* (L.) Wimmer and its subspecies *prostratus* (Bailey) Tutin. *New Phytol.* **54**: 68–70.
Wigginton, M. J. (Edit.) (1999). *British red data books*. Vol. 1 *Vascular plants*. Peterborough.

1. All leaves 3-foliolate; most with petioles more than 10 mm; flowers in terminal, leafless racemes **1. nigricans**
1. Lower leaves 3-foliolate, upper leaves usually 1-foliolate; petioles less than 10 mm; flowers 1-few in lateral groups 2.
2. Calyx less than 3 mm; corolla 7–12 mm, white **2. multiflorus**
2. Calyx more than 3 mm; corolla 10–25 mm, yellow to red 3.
3. Twigs 8–10-angled, fragile; legumes densely, long, white hairy **3. striatus**
3. Twigs 5-angled, pliable; legumes with long white hairs on the sutures only 4.
4. Plant erect or spreading; young shoots and leaves glabrous or sparingly hairy **4(a). scoparius** subsp. **scoparius**
4. Plant procumbent, looking as though someone has sat on it; young shoots and leaves densely silky hairy **4(b). scoparius** subsp. **maritimus**

1. C. nigricans L. Black Broom
Lembotropis nigricans (L.) Griseb.

Unarmed *shrub* up to 1.0(–1.5) m. *Stems* erect; twigs up to 45 cm or more and 1–2 mm in diameter, terete, appressed-hairy, flowering in the first year. *Leaves* alternate, all 3-foliolate; leaflets (6–)10–30 × (2–)5–10(–16) mm, medium green on upper surface, paler beneath, obovate to elliptical or linear, obtuse at apex, entire, narrowed at base, appressed-hairy on both surfaces when young, becoming glabrous on upper surface; stipules minute; petioles 5–15 mm, appressed-hairy. *Inflorescence* a pyramidal, terminal, leafless raceme; pedicels 4–8 mm, appressed-hairy, with 1, linear, long-persistent bract. *Calyx* 3–5 mm, campanulate, 2-lipped, the upper lip with 2 lobes, the lower lip with 3 lobes, appressed-hairy. *Corolla* 8–10 mm, yellow, becoming black when dry; standard 7–10 × about 6 mm, subrotund, emarginate; wings 5–7 mm, broadly oblong; keel 7–8 mm, ovate, beaked. *Stamens* 10, monadelphous. Style 1; stigma capitate. *Legume* 20–35 × 5–6(–7) mm, linear-oblong, appressed-hairy, dehiscent; seeds about 6 mm, with a rudimentary strophiole. *Flowers* 5–6. $2n = 48$.

Introduced. Naturalised on waste ground by railways and in gravel pits since 1970. Middlesex, Kent and Ross and Cromarty. Native of central and south-east Europe. Our plant is subsp. **nigricans** which occurs throughout the range of the species.

2. C. multiflorus (L'Hér. ex Aiton) Sweet White Broom
Spartium multiflorum L'Hér. ex Aiton; *C. albus* (Lam.) Link, non Hacq.; *C. lusitanicus* Willk.; *Genista alba* Lam.

Unarmed *shrub* 1–3 m. *Stems* erect and much branched; branches flexible and 5-angled; twigs striate, sericeous when young, becoming glabrous at maturity. *Leaves* alternate, 3-foliolate on lower branches, 1-foliolate on upper branches; leaflets 7–10 × 0.5–2.0 mm, linear-lanceolate or oblong, acute at apex, entire, narrowed at base, silvery-sericeous; stipules minute; petioles very short or absent. *Inflorescence* of profuse fascicles of 1–3 flowers; pedicels up to 10 mm. *Calyx* 2.7–3.5 mm; 2-lipped, the upper lip with 2 lobes, the lower lip with 3 lobes, sericeous. *Corolla* 7–12 mm, white; standard 9–12 mm, ovate, obtuse at apex; wings 10–12 mm, oblong; keel 11–13.5 mm, oblong. *Stamens* 10, monadelphous. *Style* 1. *Legume* 15–25 × 5–9 mm, oblong, strongly compressed, appressed-hairy. *Flowers* 5–6. $2n = 48$, 96.

Introduced. Naturalised on banks of roads and railways. Scattered localities in south England, north Wales and south Scotland. Native of north-west and central Spain and north and central Portugal.

3. C. striatus (Hill) Rothm. Hairy-fruited Broom
Genista striata Hill; *C. pendulinus* L. fil.; *Sarothamnus patens* Webb

Shrub 1–3 m. *Stems* erect and much branched; branches and twigs cylindrical, striate, usually 8- to 10-angled when young, often drying black, fragile, the young branches sericeous or villous, later becoming glabrous and leafless. *Leaves* solitary or sometimes fasciculate, those on the lower branches 3-foliolate and petiolate, those on the middle and upper branches 3- or 1-foliolate and sessile; leaflets 4–16 × 1–6 mm, glaucous on upper surface, paler beneath, ovate or elliptical to linear-lanceolate, acute at apex, entire, glabrous on upper surface, sericeous or villous beneath; stipules minute. *Inflorescence* of solitary or pairs of flowers, rarely in clusters of 3; pedicels as long as or up to twice as long as calyx. *Calyx* 5–7 mm, sericeous; 2-lipped, the upper lip with 2 lobes, the lower lip with 3 lobes. *Corolla* 25–27 mm, yellow; standard 10–25 mm, subrotund, emarginate; wings 18–27 mm, elliptical; keel 19–27 mm, oblong, upcurved at apex. *Stamens* 10, monadelphous. *Style* 1. *Legume* 18–35(–40) × 8–12 mm, oblong-ovate to oblong-elliptical, straight or slightly curved, more or less inflated, densely long white hairy, erect or semi-patent. *Flowers* 5–6. $2n = 46$, 48.

Introduced in 1816. Naturalised on roadside banks where it was planted and now reproduces. Scattered localities throughout Great Britain and one record in the west of Ireland. Native of Portugal and west and central Spain.

4. C. scoparius (L.) Link Broom

Sarothamnus scoparius (l.) W. D. J. Koch; *Genista scoparia* (L.) W. D. J. Koch; *Spartium scoparium* L.; *Genista scoparia* (L.) Lam.; *Genista vulgaris* Gray nom. illegit.; *Sarothamnus vulgaris* Wimm. nom. illegit.

Unarmed *shrub* 60–250 cm. *Branches* numerous, erect, spreading or prostrate; twigs green, 5-angled, pliable, glabrous; young shoots glabrous to silky hairy. *Leaves* alternate, 3-foliolate; leaflets 6–20 × 3–8 mm, green on upper surface, paler beneath, narrowly elliptical to obovate, more or less acute at apex, entire, narrowed at base, sessile, glabrous or with silky appressed hairs on both surfaces; petioles up to 7 mm, glabrous or hairy; stipules absent. *Inflorescence* of 1–2 axillary flowers in lateral groups; pedicels up to 10 mm, slender, glabrous. *Calyx* 6–7 mm, green, 2-lipped, the lower lip with 3 minute lobes, the upper lip with 2 minute lobes, glabrous. *Corolla* 15–20 mm, deep yellow, or with dark red to mauve areas; standard 16–18 mm, the limb ovate, emarginate at apex, rounded to a subcordate base, with a short, narrow claw; wings with limb oblong, rounded at apex, with short, very broad lobe at base and a narrow claw; keel with limb broadly oblong, slightly upturned at the rounded apex, with a short, broad lobe at base and a short, narrow claw. *Stamens* 10, monadelphous; filaments pale; anthers alternately long and basifixed and short and versatile. *Style* 1, long, spirally coiled; stigma capitate. *Legume* 25–50 × 10–12 mm, black, with long brown or white hairs on the sutures, oblong, strongly compressed, 2-valved, valves coiled after dehiscence; seeds several, 3.0–3.5 mm, ellipsoid, dispersed by explosive dehiscence. *Flowers* 5–6. Pollination by large bees.

(a) Subsp. **scoparius**
Plant erect or spreading. *Young shoots* and leaves glabrous or sparingly hairy. $2n = 46$, 48.

(b) Subsp. **maritimus** (Rouy) Heywood
Genista scoparia var. *maritima* Rouy; *Sarothamnus scoparius* subsp. *maritimus* (Rouy) Ulbr.; *Sarothamnus scoparius* subsp. *prostratus* (C. Bailey) Tutin

Plant procumbent, looking as though someone has sat on it. *Young shoots* and leaves densely silky hairy. $2n = 24$, 46, 48.

Native. Throughout most of Great Britain and Ireland. Europe, northwards to southern Sweden, eastwards to the central Ukraine. European Temperate element. Subsp. *scoparius* is a calcifuge of heathland, sandy banks, open woodland and rough ground throughout the range of the species. Subsp. *maritimus* occurs on maritime cliffs and perhaps shingle, in west Wales, south-west England, south-west Ireland and the Channel Islands.

5. Genista L.

Teline Medik.

Spiny or unarmed *shrubs*. *Leaves* alternate, simple or 3-foliolate, sometimes caducous; stipules absent or small. *Inflorescence* of heads, racemes or axillary clusters, rarely solitary flowers. *Calyx* 2-lipped, the upper lip deeply divided, the lower 3-toothed. *Corolla* yellow; standard broadly ovate or elliptical; wings elliptical or oblong; keel oblong. *Stamens* 10, monadelphous. *Styles* 1. *Legume* dehiscent or indehiscent, ovoid to linear-oblong; seeds 1–many.

Contains about 75 species mainly in the Mediterranean region.

Bean, W. J. (1973). *Trees and shrubs hardy in the British Isles.* Ed. 8. **2**. London.
Byfield, A. & Pearman, D. (1996). *Dorset's disappearing heathland flora.* London.
Chater, A. O. (1978). *Genista hispanica* L. naturalised for 50 years in V.C. 46. *Watsonia* **12**: 191.
Hultén, E. & Fries, M (1986). *Atlas of north European vascular plants, north of the Tropic of Cancer.* 3 vols. Königstein.
Perring, F. H. & Sell, P. D. (1967). *Genista tinctoria. Watsonia* **6**: 295.
Wigginton, M. J. (Edit.) (1999). *British red data books.* Vol. 1. *Vascular plants*. Peterborough. [*G. pilosa.*]

1. Leaves trifoliolate 2.
1. Leaves simple 3.
2. Flowers in axillary clusters **1. monspessularna**
2. Flowers in terminal clusters **7. radiata**
3. Plant with branched spines 4.
3. Plant without spines or spines unbranched 5.
4. Branches and leaves with spreading hairs; standard of corolla 6–8 mm **6(a). hispanica** subsp. **hispanica**
4. Branches and leaves with appressed hairs; standard of corolla 8–11 mm **6(b). hispanica** subsp. **occidentalis**
5. Seeds 1–2; leaves most or all fallen at flowering time **8. aetnensis**
5. Seeds 3–12; leaves present at flowering time 6.
6. Calyx hairy on the surface 7.
6. Calyx hairy only on the margin or glabrous 8.
7. Corolla 11–15 mm **3. florida**
7. Corolla 7–11 mm **4. pilosa**
8. Corolla 10–15 mm; legume 15–30 mm 9.
8. Corolla 7–10 mm; legume 12–15(–20) mm 10.
9. Stems ascending or erect; leaves narrowly elliptical **2(a). tinctoria** subsp. **tinctoria**
9. Stems prostrate; leaves elliptic-oblong **2(b). tinctoria** subsp. **littoralis**
10. Plant spiny **5(i). anglica** var. **anglica**
10. Plant without spines **5(ii). anglica** var. **subinermis**

1. G. monspessulana (L.) L. A. S. Johnson
Montpellier Broom

Cytisus monspessulanus L.; *Teline monspessulana* (L.) K. Koch; *Cytisus candicans* (L.) Lam.; *Cytisus kunzeanus* Willk.

Unarmed *shrub*. *Stems* 100–300 cm, brown, erect, branched, striate. *Leaves* alternate, 8–20 mm, 3-foliolate, petiolate; leaflets deep green on upper surface, paler beneath, obovate, emarginate to shortly mucronate at apex, entire, attenuate at base, with dense, patent hairs on both surfaces especially beneath. *Flowers* in axillary clusters; pedicels 2.0–3.5 mm, with 3 linear bracteoles.

Calyx 5–6 mm, 2-lipped, the lower longer than the upper, densely sericeous or with patent hairs. *Corolla* yellow; standard 13.0–16.5 × 10–12 mm, broadly ovate, glabrous; wings 12–15 × 4.0–4.5 mm, glabrous; keel 12–15 × 3.0 mm, sericeous. *Stamens* 10; monadelphous. *Style* 1. *Legume* 15–27 × 5.0–6.4 mm, narrowly oblong, densely sericeous or with patent hairs; seeds 3–6 , ovoid or discoid. *Flowers* 4–5. $2n = 46, 48$.

Introduced. Naturalised on banks, roadsides and rough ground. Scattered records in the Channel Islands and southern England. Native of the Mediterranean region and the Azores.

2. G. tinctoria L. Dyer's Greenweed
G. ovata Waldst. & Kit.

Small unarmed *shrub*. *Stems* 10–200 cm, brown, prostrate, ascending or erect, slender, branched; young twigs green, striate, sparsely hairy. *Leaves* alternate, simple; lamina 10–30 × 3–10 mm, deep green on upper surface, paler beneath, narrowly elliptical to elliptic-oblong or lanceolate-oblong, more or less acute at apex, entire, narrowed to a sessile base, ciliate on margins; stipules 1–2 mm, thin and subulate. *Inflorescence* of solitary axillary flowers, clustered towards the ends of main branches; pedicels 2–3 mm. *Calyx* 5–6 mm, shortly 2-lipped, the upper lip deeply bifid, the lower lip shortly 3-lobed, glabrous. *Corolla* 10–15 mm, yellow; standard with limb broadly ovate, retuse at apex, rounded at base to a short claw; wings with limb oblong, rounded at apex, with a short, broad lobe at base and a long, slender claw, deflexed after flowering; keel with limb oblong, slightly upturned at the rounded apex, with a short broad lobe at base and a long, slender claw, separating and not resilient after deflexion. *Stamens* 10, monadelphous; filaments pale; anthers alternating long and basifixed and short and versatile. *Style* 1, curved; stigma oblique. *Legume* 15–30 × 4–5 mm, oblong, flat, tapering and obtuse at both ends; seeds (3–)4–10, 2.0–2.5 mm, subglobose. *Flowers* 7–9. Pollination by diverse pollen-collecting insects. $2n = 48$.

(a) Subsp. **tinctoria**
Stems ascending or erect. *Leaves* narrowly elliptical. *Legume* glabrous.

(b) Subsp. **littoralis** (Corb.) Rothm.
G. tinctoria var. *littoralis* Corb.; *G. tinctoria* var. *prostrata* Bab.; *G. tinctoria* var. *humifusa* Bab. nom. illegit.

Stems prostrate. *Leaves* elliptic-oblong. *Legumes* hairy or glabrous.

Native. Subsp. *tinctoria* occurs in grassy places, banks and rough ground, and is locally common in Great Britain north to southern Scotland and in Jersey. It occurs through most of Europe from Estonia southwards and eastwards to the Urals, Caucasus and Turkey. Subsp. *littoralis* occurs on cliff-tops in Devonshire, Cornwall and Pembrokeshire. It retains its characters in cultivation. It also occurs in France. The species is a European Temperate element. *G. tinctoria* was formerly widely grown as a dye-plant and naturalised relics have probably extended its native distribution. It gives a yellow colour to wool, which, mordanted with alum, is then dipped is a vat of woad or indigo to give the famous 'Kendal Green'.

3. G. florida L. Floriferous Greenweed
G. polygalaephylla Brot.; *G. polygalaefolia* DC. nom. illegit.; *G. leptoclada* Spach

Shrub up to 2.5 m, erect, much-branched and without spines. *Leaves* simple, 0.5–2.5 × 0.2–0.5 cm, medium green on upper surface, paler beneath, linear to oblanceolate, obtuse to acute at apex, entire, cuneate at base, appressed-hairy; petiole very short, hairy. *Inflorescence* of single flowers borne in the axils of each bract in long, loose racemes; the lowermost bracts leaf-like, the upper reduced. *Calyx* 4–6 mm, 5-lobed, appressed-hairy. *Corolla* yellow; standard 11–15 × 8.0–10.5 mm, ovate, emarginate; wings 11–14 × 3.5–4.0 mm, elliptical; keel 10–14 × 2.8–3.0 mm, oblong, sericeous. *Stamens* 10, monadelphous; filaments pale; anthers yellowish. *Style* 1; stigma oblique. *Legume* 10–25 mm, oblong to oblong-lanceolate, flattened, silky hairy; seeds 2–4, black and shining, ovoid. *Flowers* 4–6. $2n = 46$.

Introduced on the banks of the M1 motorway in Bedfordshire. Native of Spain, Portugal and Morocco.

4. G. pilosa L. Hairy Greenweed

Unarmed *shrub*. *Stems* 10–40(–150) cm, greyish when young, brown when old, prostrate, rather stout, much branched and tortuous; young shoot grooved, hairy. *Leaves* alternate, simple, 3–5 × 2.0–2.5 mm, medium green on upper surface, paler beneath, obovate, obtuse at apex, entire, narrowed at base, glabrous on upper surface, with numerous appressed hairs beneath, subsessile; stipules 0.5 mm, ovate, obtuse at apex, thick. *Inflorescence* of solitary flowers in the axils of leaves and aggregated at the ends of stems and branches; pedicels about 5 mm, densely hairy. *Calyx* 4.5–5.0 mm, campanulate, divided halfway to base into 5 lobes, the lobes triangular-ovate, acute at apex, shortly hairy. *Corolla* 7–11 mm, yellow; standard with limb broadly ovate, emarginate at apex, rounded at base, shortly hairy; wings with limb oblong, rounded at apex, with a short lobe at base and a slender claw; keel with limb oblong, rounded at apex, shortly lobed at base, with a narrow spur. *Stamens* 10, monadelphous; filaments pale; anthers yellowish. *Style* 1, curved upwards; stigma oblique. *Legume* 14–18(–28) × 4–5 mm, oblong, not inflated but bulging over the seeds, acute at apex, rounded at base, hairy; seeds 3–8, 2.0–2.5 mm in diameter, subglobose. *Flowers* 5–6. Pollinated by honey-bees. $2n = 24$.

Native. Cliff-tops and heathland, extremely local and decreasing. Cornwall, Pembrokeshire, Merionethshire, Breconshire and Sussex; formerly in Suffolk and Kent. West and central Europe from southern Sweden to central Italy and Macedonia. European Temperate element.

5. G. anglica L. Petty Whin
Spartium pusillum Salisb. nom. illegit.

Spiny, rarely unarmed *shrub*. *Stems* 10–50(–100) cm, brown, erect or ascending, slender; young shoots terete, glabrous or hairy, spines 10–20 mm, axillary, spreading or recurved, rarely branched, leafy when young. *Leaves* alternate, simple, 2–8 × 2.0–2.5 mm, glaucous-green on upper surface, paler beneath, ovate or elliptical, acute or apiculate at apex, rounded at base; those on the spines linear-lanceolate; all glabrous. *Inflorescence* of solitary flowers in the leaf axils towards the end of the main branches; pedicels about 2 mm, sparsely hairy. *Calyx* 2.5–3.0 mm, campanulate, divided half-way to base into 5 lobes, the lobes triangular-ovate, acute at apex, ciliate. *Corolla* 7–10 mm, yellow; standard with limb elliptical, rounded-apiculate at apex, with rounded base and a short claw; wings with limb oblong, rounded at apex, rounded to a narrow claw at base; with a short keel with limb oblong, rounded at apex, with a short lobe at base and a narrow claw. *Stamens* 10, monadelphous; filaments pale; anthers yellow. *Style* 1, curved upwards; stigma oblique. *Legume* 12–15(–20) × 4–5 mm, broadly oblong, obliquely narrowed and acute at both ends, glabrous; seeds (3–)4–12, 1.5–1.8 mm, subglobose. *Flowers* 5–6. Pollinated by bees. $2n = 42$.

(i) Var. **anglica**
Plant spiny.

(ii) Var. **subinermis** (D. Legrand) Rouy
Plant without spines.

Native. Sandy and peaty heaths and drier areas round bogs on moors. In suitable places throughout most of Great Britain, but declining. Western Europe eastwards to southern Sweden, northern Germany and south-west Italy; north-west Africa. Oceanic Temperate element.

6. G. hispanica L. Spanish Gorse

Shrub 10–50 cm. *Stems* pale brown, decumbent to erect, with branched axillary spines; young shoots green, erect, with long patent or appressed hairs. *Leaves* alternate, 6–10 × 3–5 mm, simple, lanceolate to oblanceolate, more or less acute at apex, entire, rounded at the sessile base, with long, pale, patent or appressed hairs on both surfaces. *Inflorescence* a dense, terminal, subcapitate raceme; pedicels 1.0–2.5 mm, densely long appressed hairy; bracts c. 1.0 mm; bracteoles absent. *Calyx* 4.5–5.0 mm, pale green, divided halfway into 5 unequal lobes, the lobes linear, acute at apex, densely long-hairy. *Corolla* 6–11 mm, golden yellow; standard broadly elliptical, obtuse at apex, truncate to a short claw at base; wings narrowly elliptical, obtuse at apex; keel with limb oblong, obtuse at apex. *Stamens* 10, monadelphous; filaments yellow; anthers yellow. *Style* 1, yellow; stigma oblique. *Legume* 6–9 mm, ovate to deltoid, hairy when young; seeds 1 or 2, brown and somewhat shining. *Flowers* 4–5. $2n = 36$.

(a) Subsp. **hispanica**
Branches and *leaves* with spreading hairs. *Standard* of corolla 6–8 mm.

(b) Subsp. **occidentalis** Rouy
G. occidentalis (Rouy) Coste

Branches and *leaves* with appressed hairs. *Standard* of corolla 8–11 mm.

Introduced. Naturalised on sandy and rocky hills and roadsides, in Cardiganshire since 1927 (Chater, 1978). Scattered records in England, Wales, Scotland and Isle of Man. Most of our plants are subsp. *occidentalis*, but subsp. *hispanica* is recorded for Kent.

7. G. radiata (L.) Scop. Southern Greenweed
Cytisanthus radiatus (L.) O. Lang; *Spartium radiatum* L.; *Cytisus radiatus* (L.) W. D. J. Koch; *Enantiospartium radiatum* (L.) K. Koch

Erect, unarmed *shrub* to 100 cm, with opposite branches at almost every node. *Stems* pale brown. *Leaves* opposite, 3-foliolate; leaflets 5–20 × 2–4 mm, medium green on upper surface, paler beneath, oblanceolate, obtuse at apex, entire, narrowed below, nearly glabrous on upper surface, sericeous-hairy beneath; stipules triangular. *Inflorescence* of subopposite and subsessile flowers of 4–12 in terminal clusters, the lowermost flowers with simple, usually shortly 3-fid, scarious bracts much shorter than the flowers; bracteoles 1–3 mm. *Calyx* 4–6 mm, 2-lipped, the upper lip deeply 2-fid, the lower lip 3-lobed, the lips about as long as the tube. *Corolla* 8–14 mm, yellow; standard 8–14 mm, broadly ovate, as long as or slightly shorter than the keel, glabrous or with a median ridge of sericeous hairs; wings oblong, obtuse at apex; keel oblong, obtuse at apex. *Stamens* 10, monadelphous. *Style* 1. *Legume* 10–12 mm, ovoid, acuminate at apex, densely hairy; seeds 1–2 mm. *Flowers* 4–5.

Introduced. Garden escape recorded in Dorsetshire and Co. Durham. Native of the southern Alps extending to central Italy, south-west Roumania and central Greece.

8. G. aetnensis (Raf. ex Biv.) DC. Mount Etna Broom
Spartium aetnensis Raf. ex Biv.

Unarmed *shrub* up to 5 m, with numerous, erect, alternate, rounded, streaked, pensile branches. *Leaves* simple, linear, sericeous, all caducous before flowering. *Inflorescence* a terminal raceme, the flowers alternate, sweet-scented, sessile or on very short pedicels; without bracts. *Calyx* 2.5–3.0 mm; lobes 5, obtuse, spreading, sparsely hairy. *Corolla* 8–14 mm, yellow; standard subrotund, emarginate at apex, reflexed, glabrous; wings half as long as the standard, bowed; keel equal to the wings, hairy. *Stamens* 10, monadelphous; filaments yellow; anthers yellow. *Style* 1, yellow; stigma oblique. *Legume* 10–13 × 4–5 mm, oblong, with a short beak at apex, hairy at first; seeds 2–3. *Flowers* 7–8. $2n = 54$.

Introduced. Naturalised on waste ground in Kent, Surrey, Berkshire and Warwickshire. Native of Sicily and Sardinia.

6. **Spartium** L.

Non-spiny *shrubs*. *Buds* small. *Leaves* alternate, 1-foliolate, shortly petiolate or sessile. *Inflorescence* a leafless, terminal raceme; pedicels in the axils of caducous bracts, with 2 bracteoles at apex. *Calyx* spathe-like, split down the upper side with 5 short lobes. *Corolla* golden yellow; standard with limb large, ovate and with a very short claw; wings with limb oblong and a tubercle at base with a short claw; keel with limb oblong, curved up at apex. *Stamens* 10, monadelphous. *Style* 1. *Legume* linear-oblong, strongly compressed; seeds many.

Contains 1 species in the Mediterranean region.

Bean W. J. (1973). *Trees and shrubs hardy in the British Isles*. Ed. 8. **4**. London.
Burton, R. M. (1983). *Flora of the London area*. London.

1. S. junceum L. Spanish Broom

Unarmed *shrubs*. *Stems* up to 3 m, brownish below, upper stems and branches greenish-grey, terete with soft pith, erect, striate, flexible and rush-like, smooth, glabrous, branched, leafy. *Buds* about 1 mm, greyish-brown. *Leaves* alternate, 1-foliolate, soon deciduous; lamina 10–30 × 4–10 mm, dull medium green on upper surface, paler beneath, linear to oblanceolate or oblong-linear, rounded at apex, entire, narrowed below, glabrous on upper surface, appressed-sericeous hairy beneath; petiole short or absent. *Inflorescence* a lax, leafless, terminal raceme with up to 30 flowers; pedicels in the axils of small, caducous bracts with 2 bracteoles at apex; flowers fragrant. *Calyx* about 5 mm, brownish, spathe-like, split down the upper side, with 5 short lobes. *Corolla* golden yellow, 20–28 mm; standard with limb 27–23 mm, ovate, mucronate at apex, rounded at base to a very short claw; wings with limb oblong, rounded at apex, with a tubercle at base and a very short claw; keel with limb oblong, curved up at the obtuse apex, with a short claw. *Stamens* 10, monadelphous; filaments pale; anthers brownish-pink. *Style* 1, white to pale green; stigma green. *Legumes* 4–10 cm, linear-oblong, strongly compressed, sericeous-hairy, becoming glabrous and black, erect-patent; seeds many. *Flowers* 6–7. $2n=48$, 52, 54.

Introduced in 1548. Widely planted in gardens and naturalised on sandy roadside banks and rough ground. Scattered localities in central and southern Great Britain and a few records further north. Native of the Mediterranean region and south-west Europe. The yellow flowers have a constituent used in perfume-making and are the source of a yellow dye.

7. Ulex L.

Densely spiny *shrubs*, the spines green and branched. *Leaves* 3-foliolate in young plants, reduced to spines or scales in mature plants; stipules absent. *Inflorescence* of 1–few flowers in lateral clusters. *Calyx* yellow, membranous, bipartite, the lower lip minutely 3-lobed, the upper minutely 2-lobed. *Corolla* yellow; standard with limb ovate or elliptical; wings and keel with limb oblong. *Stamens* 10, monadelphous; anthers alternately long and basifixed and short and versatile. *Style* 1, curved; stigma capitate. *Fruit* a 2-valved, dehiscent legume; seeds 1–6, dispersed by an explosive mechanism.

Contains about 15 species in western Europe and north-west Africa.

Alvarez Martinez, M. J., Fernandez Casado, M. A., Fernadez Prieto, J. A., Nava Fernandez, H. S. & Dela Puente, M. L. (1988). El género *Ulex* en la Cornisa Cantabrica I. *Ulex* gr. *gallii-minor*. *Candollea* **43**: 483–497.
Babington, C. C. (1840). On *Ulex*. *Ann. Nat. Hist*. **5**: 300–304.
Bullock, J. M., Connor, J., Carrington, S. & Edwards, R. J. (1998). Chromosome numbers and flower sizes in *Ulex minor* Roth and *Ulex gallii* Planch. in Dorset. *Watsonia* **22**: 143–152.
Cartroviej, S. F. L. S. & Valdés-Bermejo, E. (1990). On the identity of *Ulex gallii* Planchon. *Bot. Jour. Linn. Soc*. **104**: 303–308.
Grime, J. P. et al. *Comparative plant ecology*. London. [*U. europaeus*.]
Hy, F. (1914). Observations sur les *Ulex* de l'Ouest de la France. *Rev. Gen. Bot*. **25 bis**: 345–358.
Kirschner, F. & Bullock, J. M. (1999). Taxonomic separation of *Ulex minor* Roth and *U. gallii* Planch.: morphometrics and chromosome counts. *Watsonia* **22**: 365–376.
Proctor, M. C. F. (1965). The distinguishing characteristics and geographical distributions of *Ulex minor* and *Ulex gallii*. *Watsonia* **6**: 177–187.
Proctor, M. C. F. (1967). The British species of *Ulex*. *Proc. B. S. B. I*. **6**: 379–380.
Rothmaler, W. (1941). Revision der Genisteen. I. Monografien der Gattungen um *Ulex*. *Bot. Jahrb*. **72**: 69–116.
Stewart, A., Pearman D. A. & Preston, C. D. (1994). *Scarce plants in Britain*. Peterborough. [*U. minor*.]

1. Spines deeply furrowed; bracteoles 1.8–4.5 × 1.5–4.0 mm, more than twice as wide as pedicels — 2.
1. Spines faintly furrowed or striate; bracteoles less than 1.5 × 1.0 mm, less than twice as wide as pedicels — 3.
2. Shrub rather open and erect; stem sparsely hairy — **1(i). europaeus** var. **europaeus**
2. Shrub dense, sometimes almost prostrate; stems very hairy — **1(ii). europaeus** var. **maritimus**
3. Calyx 9–13(–15) mm, lobes of lower lip parallel to convergent; standard (10–)13–18(–22) mm — **2. gallii**
3. Calyx 5.0–9.5(–10) mm, lobes of lower lip divergent; standard 7–12(–13) mm — **3. minor**

1. U. europaeus L. Gorse

U. grandiflorus Pourr.; *U. compositus* Moench; *U. floridus* Salisb. nom. illegit.; *U. strictus* Mackay; *U. hibernicus* G. Don

Densely spiny *shrub*, 60–200(–250) cm. Main *branches* erect or ascending, rarely prostrate, with rather sparse to dense, blackish hairs; spines 15–25 mm, green, rigid, strong, branched and deeply grooved. *Leaves* ternate in young plants, linear, spine-like and hairy in mature plants; stipules absent. *Inflorescence* of 1–few flowers in lateral clusters; pedicels 3–5 mm, densely velvety hairy; bracteoles 1.8–4.5 × 1.5–4.0 mm, more than twice as wide as the pedicels. *Calyx* 10–16(–20) mm, two-thirds the length of the corolla, yellow, membranous, with spreading hairs, bipartite, the lower lip minutely 3-lobed, the lobes convergent. *Corolla* 11–20 mm, bright golden yellow; standard 12–18 mm, with limb broadly elliptical, emarginate at apex, rounded at base to a short, broad claw; wings with limb oblong, rounded at apex, rounded at base to a short, narrow claw; keel with limb broadly oblong, rounded at apex, with a short lobe at base and a narrow claw. *Stamens* 10, monadelphous; filaments pale; anthers alternately long and basifixed and short and versatile. *Style* 1, curved upwards; stigma capitate. *Legumes* (12–)14–17(–19) mm, dehiscing in summer; seeds 2.2–2.5 mm, subglobose. *Flowers* 3–6 and sporadically during a mild winter. Pollinated mainly by bumble-bees. $2n=96$.

(i) Var. **europaeus**
Shrub rather open and erect although leaves and spines on branches dense. *Stem* sparsely hairy.

(ii) Var. **maritimus** Hy
Shrub dense, low and sometimes almost prostrate. *Stem* very hairy.

Native. Grassy places, heathland and open woods, mostly on sandy or peaty soil. throughout Great Britain and Ireland. Western Europe from Holland southwards and eastwards to Italy; introduced in many temperate countries and frequently naturalised. Oceanic Temperate element. Our plant is subsp. **europaeus**. Var. *maritima* occurs by the sea and is also in west France. Var. *europaeus* occurs in all areas even by the sea.

× **gallii**
This hybrid is intermediate between the parents in calyx, corolla and bracteole size and has a seed number of 7–10. It flowers in autumn and winter and is highly fertile.

Occurs in western Great Britain, East Anglia and northern Ireland where the distributions of the parents overlap. It has also been reported from Brittany in France.

2. U. gallii Planch. Western Gorse
U. nanus var. *major* Bab.; *U. bonnieri* Hy;
U. bastardianus Hy; *U. nanus* subsp. *gallii* (Planch.) Syme

Densely spiny *shrub* 10–150(–200) cm. Main *branches* usually ascending, with abundant brown hairs; spines 8–34 mm, green, branched, weakly grooved. *Leaves* ternate on young plants, linear and spine-like and hairy on mature plants; stipules absent. *Inflorescence* of 1–few flowers in lateral clusters; pedicels 3.5 mm, with appressed hairs; bracteoles 0.5–0.8 × 0.6–0.8 mm, usually wider than the pedicels. *Calyx* 9–13(–15) mm, slightly shorter than the corolla, with appressed hairs, yellow, membranous, bipartite, the 3 lobes of the lower lip parallel to convergent. *Corolla* 10–13 mm, yellow, hairy; standard (10–)13–18(–22) mm, the limb ovate, emarginate at apex, rounded to a short claw at base; wings with limb oblong, rounded at apex, with a very short claw; keel with limb oblong, rounded at apex, with a short lobe at base and a short, narrow claw. *Stamens* 10, monadelphous; filaments pale; anthers alternately long and basifixed and short and versatile. *Style* 1, upturned, glabrous; stigma capitate. *Ovules* (3–)4–6(–7). *Legume* (8–)9–13(–14) mm, dehiscing in spring; seed 2.0–2.5 mm, ellipsoid. *Flowers* mainly 4–5, autumn flowers are much withered. $2n=64$, 96.

Native. Grassy places, heathland and open woods, mostly on sandy or peaty soil. In suitable habitats throughout Great Britain and Ireland. Western Europe from Scotland to north-west Spain. Oceanic Temperate element.

3. U. minor Roth Dwarf Gorse
U. nanus T. F. Forster; *U. autumnalis* Thore

Densely spiny *shrub* 5–100(–150) cm. Main *branches* usually procumbent or spreading, with abundant brown hairs; spines up to 10 mm, green, branched, weakly grooved. *Leaves* ternate on young plants, spine-like and hairy in mature plants; stipules absent. *Inflorescence* of 1–few flowers in lateral clusters; pedicels 3–5 mm, with appressed hairs; bracteoles 0.6–0.8 × 0.4–0.6 mm, usually narrower than the pedicels. *Calyx* 6.0–9.5 mm, slightly shorter than corolla, yellow, membranous, bipartite, lower lip minutely 3-lobed, the lobes divergent, with sparse appressed hairs. *Corolla* 8–10 mm, yellow; standard 7–12(–13) mm, with limb ovate, deeply emarginate at apex, rounded at base to a short claw; wings about as long as keel when straight, the limb oblong, rounded at apex, with a very short lobe at base and a short, narrow spur; keel with limb broadly oblong, with a shallow lump, rounded at apex, with a short, broad lobe at base and a narrow claw, hairy. *Stamens* 10, monadelphous; filaments pale; anthers alternately long and basifixed and short and versatile. *Style* 1, curved upwards; stigma capitate. ovules (3–)4–6(–7). *Legume* 6.0–8.5 mm, dehiscing in spring; seed 2.0–2.5 mm, ellipsoid. *Flowers* 7–9. $2n=32$.

Native. Heaths. Mainly in southern England from Kent to Dorsetshire and Wiltshire, and very scattered localities north to Flintshire and Nottinghamshire, Cumberland and Argyllshire; formerly in the Channel Islands; sometimes grown for ornament and naturalised outside that area. Western France, north-west Spain, west Portugal and south-west Spain; introduced in the Azores and Brazil. Oceanic Southern-temperate element.

Tribe **3. Dalbergieae** Bronn. ex DC.
Aeschynomeneae (Benth.) Hutch.; *Hedysareae* subtribe *Aeschynomeninae* Benth.; *Coronilleae* subtribe *Aeschynomeninae* (Benth.) Schulze-Menz

Herbs. *Leaves* paripinnate, with 2 pairs of leaflets. *Inflorescence* an axillary raceme; flowers with a long tubular hypanthium. *Stamens* 10, monadelphous. *Fruit* a cylindrical, torulose, indehiscent legume.

8. Arachis L.

Annual herbs. *Leaves* alternate, usually paripinnate with 2 pairs of leaflets; stipules linear, adnate to the petiole. *Inflorescence* a sessile, axillary raceme, the flowers with a long, tubular hypanthium. *Calyx* 2-lipped, the upper lip (3–)4-lobed, the lower lip entire. *Corolla* yellow, adnate to the base of the stamen tube; standard about 15 mm, ovate; wings oblong, obtuse; keel oblong; after pollination the peduncle elongates and forces the young pod into the soil where it completes its development. *Stamens* 10, monadelphous. *Style* 1. *Fruit* a cylindrical torulose, indehiscent legume.

Contains a few species, native of South America.

In Lewis, G. et al. (2005), *Legumes of the world*, this tribe is included in the Tribe *Dalbergieae* Bronn. ex DC. sens. lat.

Burkhill, I. H. (1966). *A dictionary of the economic products of the Malay Peninsula*. **1**. Kuala Lumpur.

Purseglove, J. W. (1968). *Tropical crops*. London.
Rudd, V. I. (1981). *Aeschynomeneae*. In Polhill, R. M. & Raven, P. H. (Edits.) *Advances in legume systematics*, Vol. 1, pp. 347–354. Kew.
Smith, B. W. (1950). *Arachis hypogaea* aerial flowers and subterranean fruit. *Amer. Jour. Bot.* **37**: 802–815.
Vaughan, J. G. & Geissler, C. A. (1997). *The new Oxford book of food plants*. Oxford.

1. A. hypogaea L. Peanut

Annual herb with fibrous roots. *Stems* up to 60 cm, pale green, prostrate to erect, hairy, much-branched, leafy. *Leaves* alternate, usually paripinnate, with 2 pairs of leaflets; leaflets 25–60 × 15–30 mm, medium green on upper surface, paler beneath, elliptical to obovate, rounded-mucronulate at apex, entire, rounded at base, hairy; stipules linear, adnate to the petiole. *Inflorescence* a sessile, axillary raceme, the flowers with a long, tubular hypanthium. *Calyx* 2-lipped, the upper lip (3–)4-lobed, the lower lip entire. *Corolla* 15–20 mm, yellow, adnate to the base of the stamen tube; standard about 15 × 15 mm, ovate; wings oblong, obtuse at apex; keel oblong; after pollination the peduncle elongates and forces the young pod into the soil where it completes its development. *Stamens* 10, monadelphous, the 2 adaxial ones sterile. *Styles* 1. *Legume* 20–40 × 10–15 mm, cylindrical, markedly reticulate and torulose, indehiscent; seeds (nuts) 1–4, 10–12 × 5–10 mm, ellipsoid, brown. Rarely flowers with us. Self-pollination usual. $2n = 40$.

Introduced. Occasional on rubbish-tips as a bird-seed and tan-bark casual. Native of South America, but widespread as a tropical crop. Also known as Groundnuts and Monkeynuts, the nuts are used in cookery or eaten raw; peanut butter is made from them and also cooking and salad oil.

Tribe **4. Millettieae** Miq.
Galegeae subtribe *Tephrosiinae* Benth.

Climbers. Leaves imparipinnate; leaflets entire. *Inflorescence* a pendulous raceme. *Stamens* 10, diadelphous. *Fruit* a legume; seeds up to 12.

9. Wisteria Nutt.

Deciduous climbers. Buds ovoid, with scales. *Leaves* alternate, imparipinnate; leaflets entire; stipules present. *Inflorescence* of long, pendulous racemes on short, leafy shoots. *Calyx* 2-lipped, the upper lip with 2 lobes, the lower with 3 lobes. *Corolla* bluish or pinkish or white; standard subrotund, with a short claw; wings oblong, curved and with a slender claw; keel half ovate, curved at apex, with a slender claw. *Stamens* 10, diadelphous. *Styles* 1; stigma capitate. *Fruit* a legume. Named after Caspar Wistar (1760–1818).

Contains about 10 species in eastern Asia and North America.

Bean, W. J. (1980). *Trees and shrubs*: hardy in the British Isles. Ed. 8. **4**. London.
Rehder, A. & Wilson, E. H. (1916). *Wisteria*. In Sargent, C. S. (Edit.) *Plantae Wilsonianae*, vol. 2, pp. 509–515. Cambridge, Mass.
Valder, P. (1995). *Wisterias: a comprehensive guide*. Balmain.

1. Leaflets persistently silky hairy on both surfaces **3. venusta**
1. Leaflets hairy at first, becoming nearly glabrous 2.
2. Stems climbing clockwise; leaflets 13–19; flowers opening progressively from the base of the inflorescence upwards when leaves fully expanded **1. floribunda**
2. Stems climbing anticlockwise; leaflets (7–)9–11(–13); flowers all opening together before the leaves are fully expanded **2. sinensis**

1. W. floribunda (Willd.) DC. Garden Wisteria
Glycine floribunda Willd.; *Dolichos japonicus* Spreng.; *Kraunhia floribunda* (Willd.) Taub.; *Milletia floribunda* (Willd.) Matsum.

Deciduous *climber* up to 15 m, spreading. *Trunk* very sinuous. *Bark* grey and very rough, wrinkled when old. *Branches* spreading, twining clockwise; twigs grey and smooth, angled; young shoots green and hairy. *Buds* 4.5–5.0 mm, ovoid, pointed at apex; scales dark reddish-brown to black, ovate, pointed at apex. *Leaves* alternate, oblong-ovate in outline, imparipinnate; leaflets 13–19, 40–60 × 12–28 mm, dull yellowish-green on upper surface, slightly paler beneath, turning yellow in autumn, lanceolate to ovate, bluntly pointed at apex, entire, rounded at base, shortly petiolulate, appressed-hairy on both surfaces becoming nearly glabrous; petioles and rachis yellowish-green, angular and shortly hairy; stipules linear, acute at apex. *Inflorescence* of long, pendulous racemes on short, leafy shoots, flowers strongly scented and opening progressively from the base when leaves fully expanded; pedicels 15–22 mm, shortly hairy. *Calyx* 6–7 mm, reddish-purple, 2-lipped, the upper lip with 2 short lobes, the lower lip with 3 triangular lobes, minutely hairy. *Corolla* 16–20 mm, standard with limb whiter, usually with a bluish or pinkish tinge and a large yellow spot at the base, subrotund, with a short point at apex and a fold down the back, rounded at base, with a pale greenish claw; wings deep violet or rose, oblong, rounded at apex, curved, with a slender claw; keel deep violet or rose, half ovate, curved at apex, with a slender claw. *Stamens* 10, diadelphous; filaments pale green, curved; anthers yellow. *Style* 1, pale green; stigma colourless, capitate. *Legume* 10–12 × 1.5–2.0 cm, dark brown, oblong, pointed at both ends, felted-hairy; seeds 7–8 mm, subrotund, brown, flattened. *Flowers* 4–5. $2n = 16$.

Introduced. A plant grown in gardens and amenity areas, sometimes occurring as an escape or relic persistent on waste ground as at Hextable in Kent. Native of Japan.

2. W. sinensis (Sims) Sweet Chinese Wisteria
Glycine chinensis Sims; *W. consequana* Loudon; *W. polystachya* K. Koch; *Kraunhia sinensis* (Sims) Makino

Deciduous *climber* up to 15 m, spreading. *Trunk* very sinuous. *Bark* grey and very rough. *Branches* spreading, turning anticlockwise; twigs grey and smooth; young shoots green and hairy. *Buds* 4.5–5.0 mm, ovoid, pointed at apex; scales dark reddish-brown, ovate and pointed at apex. *Leaves* alternate, numerous, oblong-ovate in outline, imparipinnate; leaflets (7–)9–11(–13),

6.0–9.5 × 3.5–5.8 cm, dull yellowish-green on upper surface, paler beneath, turning yellow in autumn, ovate or oblong-ovate, acuminate or caudate at apex, entire, rounded at base, shortly petiolulate, hairy on both surfaces at first becoming glabrous or nearly so; petiole up to 8 cm, hairy; rachis hairy; stipules lanceolate, acute at apex. *Inflorescence* a pendulous raceme on short shoots; all flowers opening together before the leaves are fully expanded; pedicels 18–20 mm, hairy. *Calyx* 3–4 mm, violet, 2-lipped, the upper lip with 2 lobes, the lower lip with 3 lobes, glabrous. *Corolla* 20–22 mm, pale violet; standard with limb wider than long, rounded-cordate at base, with a short claw; wings with limb oblong, curved, with a slender claw; keel with limb narrowly oblong. *Stamens* 10, diadelphous; filaments white; anthers yellowish-orange. *Style* 1, pale green; stigma pale. *Legume* 12–15 × 1.6–1.8 cm, dark brown, oblong, pointed at both ends, hairy; seeds 7–8 mm, subrotund, brown. *Flowers* 4–5. $2n=16$.

Introduced in 1816. A garden escape, seedlings being recorded from south London and from Guernsey and a plant in a hedge at Merton, Norfolk. It is occasionally planted in amenity areas and parkland. Native of China.

3. W. venusta Rehder & E. H. Wilson Japanese Wisteria

W. brachybotrys var. *alba* Mill.; *Kraunhia sinensis* var. *brachybotrys* forma *albiflora* Makino

Deciduous *climber* up to 10 m. *Stems* brownish below; young shoots appressed hairy, becoming glabrous, branched, leafy. *Buds* 4.5–5.0 mm, subglobose or ovoid. *Leaves* opposite, 18–35 × 10–15 cm, broadly oblong in outline, imparipinnate; leaflets 9–13, 6–10 × 2.5–5.0 cm, medium green on upper surface, paler beneath, ovate, elliptic-oblong or oblong-lanceolate, shortly acuminate at apex, rounded, truncate or subcordate at base, persistently silky hairy on both sides, densely so beneath, with petiolules 3–4 mm and densely hairy; petiole 4–10 cm, subterete, densely appressed hairy; stipules lanceolate, acute at apex. *Inflorescence* a pendulous raceme 10–15 cm; pedicels 15–35 mm, patent, villous. *Calyx* about 4.5 mm, cupuliform, slightly 2-lipped, upper lip 2-lobed, the lower 3-lobed, the lobes subulate. *Corolla* white or blue, with limb of standard subrotund and truncate at apex, with a short lobe at base; wings with limb oblong, rounded at apex, with a short lobe at base; keel with limb oblong and truncate at apex. *Stamens* 10, diadelphous; filaments 20 mm; anthers yellowish. *Style* 1, glabrous. *Legume* 15–20 × 1.6–2.0 cm, oblong, compressed, densely hairy. *Flowers* 5–6. $2n=16$.

Introduced. Sometimes grown in gardens and a relic on waste ground at Hextable in Kent. Native of Japan.

Tribe **5. Phaseoleae** (Bronn.) DC.
Subtribe *Phaseolinae* Bronn.; Tribe *Erythrineae* (Benth.) Hassk.; Tribe *Glycineae* Burnett

Herbs, often climbing. *Leaves* 3-foliolate, the leaflets entire, with stipels. *Inflorescence* an axillary raceme or cluster. *Stamens* 10, monadelphous or diadelphous. *Fruit* a longitudinally dehiscent legume, more than 2-seeded.

10. Glycine Willd.

Annual herbs with reddish-brown hairs. *Leaves* alternate, 3-foliolate; stipils and small stipules present. *Inflorescence* an axillary raceme with a very short or absent peduncle. *Calyx* campanulate or tubular-campanulate, somewhat 2-lipped. *Corolla* violet, pink or white; standard ovate and emarginate; wings obovate, longer than keel; keel not fused along upper surface. *Stamens* 10, diadelphous or monadelphous. *Style* 1. *Fruit* a pendulous, 2- to 4-seeded legume.

Contains several species in the tropics of the Old World.

Burkill, I. H. (1966). *A dictionary of the economic products of the Malay Peninsula.* **1**. Kuala Lumpur.
Phillips, R. & Rix, M. (1993). *Vegetables.* London.
Purseglove, J. W. (1968). *Tropical crops.* **1**. London.

1. G. max (L.) Merr. Soya Bean

Phaseolus max L.; *Soja max* (L.) Piper

Annual herb with fibrous roots. *Stem* 30–200 cm, pale green, erect, with stiff, reddish-brown hairs, branched, leafy. *Leaves* alternate, broadly ovate in outline, 3-foliolate; leaflets 3–15 × 2–7 cm, dark green on upper surface, paler beneath, ovate-elliptical, pointed at apex, entire, rounded at base, hairy; with stipils and small stipules. *Inflorescence* an axillary raceme, 5- to 8-flowered. *Calyx* (4–)5–7 mm, campanulate or tubular-campanulate, somewhat hairy. *Corolla* (4.5–)6–7 mm, violet, pink or white; standard about 5 mm, longer than the wings, ovate, emarginate; wings longer than the keel, obovate; keel not fused along upper surface. *Stamens* 10, diadelphous or monadelphous. *Style* 1, glabrous; stigma capitate. *Legume* 25–80 × 8–15 mm, linear or oblong, constricted between the seeds and septate, pendulous, dehiscent; seeds 2–4, 6–11 mm. *Flowers* 6–8. Normally self-pollinated. $2n=40$.

Introduced. Found where seed is spilled on tips, waste land and near docks and factories. Scattered records in southern Great Britain. Origin unknown, but possibly derived from *G. soya* Sieb & Zucc. Soya beans are an ancient Chinese crop first grown in the eleventh century BC. Soya bean has become an important source of vegetable oil and protein on a world basis. Sprouts from seedlings grown in the dark, curd made into soya cheese and soya flour mixed with cereal flour for baking are all used. Mature fermented beans boiled with salt, flour and other ingredients and fermented for several days are the basis of Worcester and other English table sauces.

11. Vigna Savi

Annual herbs, erect or climbing. *Leaves* alternate, 3-foliolate; leaflets entire or 2- to 3-lobed. *Inflorescence* axillary, pendulous, 4- to 24-flowered; peduncle up to 9.5 cm; pedicels 2–3 mm. *Calyx* 5-lobed. *Corolla* greenish, with yellow wings and keel tinged reddish. *Stamens* 10, diadelphous. *Style* 1. *Legume* linear-cylindrical, longitudinally dehiscent; seeds (3–)8–14.

Contains about 150 species in tropical regions, particularly of the Old World.

Burkhill, I. H. (1966). *A dictionary of the economic products of the Malay Peninsula.* **2.** Kuala Lumpur. [As *Phaseolus aurea.*]
Vaughan, J. G. & Geissler, C. (1997). *The new Oxford book of food plants.* Oxford.

1. V. radiata (L.) Wilczek — Mung Bean

Phaseolus radiatus L.; *Phaseolus aureus* Roxb.; *Azukia radiata* (L.) Ohwi; *V. mungo* auct.

Annual herb. Stems 20–60 cm, pale green, erect or climbing, with long, spreading, yellowish-brown or brown, rather bristly hairs. *Leaves* alternate, broadly ovate in outline, 3-foliolate; leaflets 5–16 × 3–12 cm, dark green on upper surface, paler beneath, elliptical, ovate or rhomboid, acuminate at apex, entire or 2- to 3-lobed, broadly cuneate or rounded at base, glabrous to bristly hairy on both surfaces; petioles 5–21 cm; rhachis 1.5–4.0 cm; petiolules 3–6 mm; stipules 10–18 × 3–10 mm, ovate peltate, rhomboid or obovate-oblong, with ciliate margins; stipils conspicuous, lanceolate. *Inflorescence* axillary, pendulous, 4- to 25-flowered; peduncle 2.5–9.5 cm; pedicels 2–3 mm; bracts 4–5 mm, ovate-lanceolate; bracteoles 4–7 mm, linear-lanceolate or oblong. *Calyx* with tube 3–4 mm and glabrous, divided into 5 lobes 1.5–4.0 mm, narrowly triangular and ciliate, the upper pair joined to make a bifid lobe. *Corolla* with standard greenish-yellow outside, sometimes pinkish inside, oblate, emarginate; wings yellow; keel green, tinged reddish, incurved. *Stamens* 10, diadelphous. *Style* 1, thickened above and hairy on the internal face. *Legume* 40–90 × 5–6 mm, linear-cylindrical, somewhat constricted between the seeds, covered with dark brown, short, spreading hairs; seeds (3–)8–14, 2.5–4.2 mm, greenish, brown or black, oblong-cylindrical to globose, with raised ridges. *Flowers* 7–8. $2n = 22$.

Introduced. Waste land near docks and factories where seed is spilled on tips. Scattered records in southern Great Britain. An Indian cultivated crop of ancient origin now grown throughout the tropics and subtropics of both hemispheres. This is the pulse most widely used for bean sprouts in North America, Asia and Europe. Bean sprouts may be eaten raw or cooked and are popular in salads.

12. Phaseolus L.

Annual or *perennial*, often climbing *herbs*. *Leaves* alternate, 3-foliolate, the leaflets entire, stipels and stipules usually present. *Inflorescence* an axillary raceme of 1–20 flowers on a long peduncle; bracts mostly small and caducous. *Calyx* with 5 lobes. *Corolla* white, cream or orange red; standard large and subrotund; wings ovate or obovate, rounded at apex; keel small and spirally twisted. *Stamens* 10, diadelphous. *Style* 1. *Legumes* up to 30 cm, linear, compressed, 2-valved, not jointed transversely, longitudinally dehiscent.

MacCarthy, D. (1986). *British food facts and figures.* Cambridge.
Philips, R. & Rix, M. (1993). *Vegetables.* London.
Vaughan, J. G. & Geissler, C. (1997). *The new Oxford book of food plants.* Oxford.

1. Flowers 6–20 per raceme, 30–35 mm, deep orange-red, rarely white — **2. coccinea**
1. Flowers 1–5 per raceme, 20–25 mm, white, cream or rarely mauve or purple — 2.
2. Plant robust, climbing up to 4 m — **1(i). vulgaris** var. **vulgaris**
2. Plant less robust, erect up to 50 cm — **1(ii). vulgaris** var. **nanus**

1. P. vulgaris L. — French Bean

Annual, sometimes climbing *herb* with fibrous roots. *Stem* up to 4 m if spirally climbing, but usually erect and about 50 cm, medium green, roughly hairy, branched, leafy. *Leaves* alternate, ovate in outline, 3-foliolate; leaflets with lamina 6–12(–15) × 3–10 cm, medium yellowish-green on upper surface, paler beneath with prominent veins, subrotund to ovate or rhomboid-ovate, more or less acute to acuminate at apex, entire, rounded or cordate at base, hairy on both surfaces; terminal leaflet with a petiolule up to 4 cm, pale yellowish-green and hairy, lateral leaflets with very short petiolules; petioles long, hairy. *Inflorescence* of 1– to 5-flowered axillary racemes; bracts small and caducous. *Calyx* 5–7 mm, pale green, campanulate, with 5 ovate, emarginate lobes. *Corolla* 20–25 mm, white or cream, sometimes with pale mauve or purplish wings; standard subrotund; wings ovate, rounded at apex; keel small and spirally twisted. *Stamens* 10, diadelphous; filaments white; anthers yellow. *Style* 1, greenish, hairy on inside. *Legume* 5–15 × 1.0–2.5 cm, usually green, sometimes yellow or purple or streaked red, linear, often slightly falcate, compressed, pendulous; seeds (beans) 13–15 × 7–8 mm, milky white, brown, blackish or variously spotted and blotched, reniform or oblong, slightly compressed, smooth. *Flowers* 6–7. Self-pollinating. $2n = 22$.

(i) Var. **vulgaris**
Plants robust, climbing up to 4 m.

(ii) Var. **nanus** (L.) Asch.
P. nanus L.

Plant less robust, erect up to 50 cm.

Introduced. Much grown in gardens as a vegetable and as a crop in southern England, East Anglia and Lincolnshire. It is frequent as an escape on tips and waste land throughout the lowlands. French, Haricot or Kidney Beans are cooked whole in the pod or sliced if they are large, from June to September. Native of central and western Mexico and Guatemala, and in the Andes of Peru, Bolivia and Argentina. All are climbers and the Andean plants have larger seeds. The original country of domestication is uncertain but deposits in Peru date from 6000 BC and central Mexico 4000 BC. Cultivation later spread into North America. *P. vulgaris* was brought to Europe after the Spanish conquest in the early sixteenth century, and was grown in England before 1597. Dwarf plants var. *nanus* were not grown commonly until the early eighteenth century. The seeds have been given a variety of names; berlotto, haricot, canellino, pinto, pea-bean, navy, red kidney, marrow and black bean. They are marketed fresh, canned or frozen and are associated with such products as canned baked beans and in Mexican *chilli con carne*.

2. P. coccineus L. Runner Bean
P. multiflorus Willd.

Perennial climbing *herb* with tuberous roots. *Stems* climbing clockwise to 5 m, pale to medium green, touched dirty purple or brownish-purple, angled, with numerous, short, reflexed hairs. *Leaves* alternate, broadly ovate in outline, 3-foliolate; leaflets with lamina 7–15 × 5–9 cm, medium green on upper surface, with faintly impressed veins, paler beneath with prominent veins, ovate, narrowed to a pointed but obtuse apex, entire, rounded at base, reticulate-veined and shortly hairy on both surfaces, the terminal leaflet with petiolule 3–4 cm, brownish-purple and hairy, the lateral leaflets very shortly petiolulate; petiole up to 11 cm, pale green, touched brownish-purple, channelled above, shortly hairy; stipules small, linear-lanceolate. *Inflorescence* an axillary raceme of up to 20 flowers; peduncles up to 15 cm, pale green, hairy; pedicels up to 10 mm, hairy. *Calyx* 8–10 mm, dull dark red, campanulate, with 5 small, triangular lobes at apex, hairy. *Corolla* 30–35 mm, deep orange-red, rarely white; standard large, rotund or reniform, broadly rounded at apex, cordate at base; wings obovate, rounded at apex; keel pinkish and spirally twisted, small. *Stamens* 10, diadelphous; filaments white; anthers yellow. *Style* 1, green, hairy at top. *Legume* 20–60 × 1.5–2.5 cm, green, broadly linear, compressed, pendulous; seeds (beans) 10–25 × 10–15 mm, usually purplish-brown with black mottling, oblong or reniform, slightly compressed, smooth. *Flowers* 6–7. $2n = 22$.

Introduced. Much grown in gardens as a vegetable, and as a crop, particularly in the Home Counties, Vale of Evesham, East Anglia and Kent. It occurs as a casual on tips and in waste ground. The young legumes can be cooked whole, but are usually stringed and sliced. Native of Mexico, from the mountains west of Durango southwards, and recorded from Guatemala. Deposits have been found from 2200 years ago. It was probably introduced to Europe in the sixteenth century, but one of the first published records in England is 1633, when it was regarded as an ornamental plant.

Tribe **6. Psoraleeae** Lowe
Tribe *Galegeae* subtribe *Psoraleinae* A. Gray

Herbs. *Leaves* 3-foliolate, with dentate leaflets. *Inflorescence* an erect, axillary raceme. *Stamens* 10, monadelphous. *Fruit* 1-seeded, indehiscent.

13. Bituminaria Heist. ex Fabr.

Aspalthium Medik.

Perennial herbs, sometimes with a woody base. *Leaves* alternate, 3-foliolate, glandular-punctate; stipules small, free. *Inflorescence* an axillary head; bracts 3-fid. *Calyx* campanulate, with 5 unequal lobes. *Corolla* blue, violet or white; standard oblong and obtuse at apex; wings oblong and rounded at apex; keel ovate and obtuse at apex. *Stamens* 10, monadelphous. *Style* 1. *Fruit* an indehiscent legume with 1 seed, not glandular-warty.

Contains 2 species endemic to southern Europe and the Mediterranean region, North Africa and Euxine.

Stirton, C. H. (1981) in Polhill, R. M. & Raven, P. H. (Edits.) *Advances in legume systematics*, vol. 1, pp. 337–343. Kew.

1. B. bituminosa (L.) C. H. Stirt. Pitch Trefoil
Aspalthium bituminosum (L.) Fourr.; *Psoralea bituminosa* L.

Perennial herb with a woody base, smelling of bitumen. *Stems* up to 1 m, pale green, conspicuously striate, appressed-strigose, branched mainly at the base, leafy. *Leaves* alternate, 3-foliolate; leaflets of lower leaves 30–50 × 15–40 mm, oblong or subrotund, obtuse, rounded or slightly retuse at apex, entire, rounded at base, the terminal not much larger than the lateral, those of upper 30–60 × 5–20 mm, usually much narrower, oblong, lanceolate or even linear, acute or cuspidate at apex, entire, more or less rounded at base, all thinly strigose to nearly glabrous on both surfaces; petioles of lower leaves up to 12 cm, striate, channelled above and appressed-strigose, those of the upper leaves proportionally shorter; stipules 8–15 mm, subulate, thinly appressed-strigose. *Inflorescence* a dense, many-flowered, axillary head; peduncles 8–30 cm, stout, erect, striate, densely appressed-strigose; bracts 5–15 mm, ovate, long acuminate at apex, appressed-strigose, often connate by their margins; flowers sessile. *Calyx* 12–18 mm, campanulate, divided into 5 unequal, long-acuminate, spine-tipped lobes, appressed-hairy. *Corolla* 15–20 mm, blue or violet; standard with limb oblong, rounded or slightly emarginate at apex, with a distinct lobe at base, narrowing into a rather wide claw; wings with limb narrowly oblong, rounded or subacute at apex, with a long, slender lobe at base and a slender claw; keel more or less adherent to wings, with limb oblong, very blunt at apex, with a short lobe at base, blotched purple and with a slender claw. *Stamens* 10, monadelphous; filaments subequal; anthers oblong. *Style* 1, strigose and dilated below, glabrous and filiform above, sharply upturned; stigma capitate. *Legume* 18–23 × 3–4 mm, falciform, flattened, acute or acuminate at apex, villose below, with an appressed-strigose beak; seed 1, about 6 mm, oblong, compressed, with a permanent pericarp. *Flowers* 6–7. $2n = 20$.

Introduced. A wool casual. Native of southern Europe, the Mediterranean region and Atlantic Islands.

14. Cullen Medik.

Meladenia Turcz.; *Bipontinia* Alef.

Perennial herbs, sometimes with a woody base. *Leaves* alternate, 3-foliolate; stipules linear-subulate, free. *Inflorescence* an axillary raceme. *Calyx* campanulate, 5-lobed. *Corolla* white. *Stamens* 10, monadelphous. *Style* 1; stigma capitate. *Fruit* an indehiscent legume without a beak, conspicuously glandular-warty.

Contains about 35 species, a few in Africa, extending through India and Sri Lanka to Burma, Philippines, Papua New Guinea and Australia.

Stirton, C. H. (1981) in Polhill, R. M. & Raven, P. H.(Edits.) *Advances in legume systematics*, vol. 1, pp. 337–343. Kew.

1. C. americanum (L.) Rydb. Scurfy Pea

Psoralea americana L.; *Psoralea dentata* DC.

Perennial herb sometimes with a woody base. *Stem* 20–100 cm, pale green, erect, sparsely to densely hairy, branched, leafy. *Leaves* alternate, 3-foliolate, leaflets 0.8–5.0 × 0.6–3.5 cm, dark green on upper surface, paler beneath, subrotund-rhombic to ovate, rounded at apex, dentate, hairy. *Inflorescence* an axillary raceme; peduncles about equalling the leaves. *Calyx* 6.0–7.5 mm, campanulate, with 5 unequal lobes, the lobes lanceolate and acute at apex. *Corolla* about 7–8 mm, white, the keel often with a violet tip; standard obovate, emarginate at apex; wings oblong, obtuse at apex; keel ovate, obtuse. *Stamens* 10, monadelphous. *Style* 1; stigma capitate. *Legume* 4.0–5.1 × 3.0–3.6 mm, conspicuous, glandular, warty, without a beak; seed 1, 3.5–4.2 × 2.5–2.7 mm. *Flowers* 6–7. $2n = 22$.

Introduced. Bird-seed alien found on tips. Sporadic in southern Great Britain. Native of the west Mediterranean region.

Tribe **7. Sesbanieae** (Rydb.) Hutch.
Tribe *Galegeae* subtribe *Sesbaniinae* Rydb.

As genus.

15. Sesbania Adans.

Annual herbs. Leaves alternate, paripinnate; leaflets numerous, entire. *Inflorescence* an axillary raceme. *Calyx* campanulate, with 5 nearly equal lobes. *Corolla* yellow; standard subrotund, reflexed, with a short claw; wings with a long claw; keel strongly arching with a elongate claw as long as the limb. *Stamens* 10, diadelphous. *Style* 1. *Fruit* a linear, beaked legume.

Contains about 15 species in warm and tropical regions. *Sesban* is the Arabic name for one of the species.

1. S. exaltata (Raf.) Cory Colorado River-hemp

Darwinia exaltata Raf.; *Sesban exaltata* (Raf.) Rydb.; *S. sonorae* Rydb.; *S. macrocarpa* Muhl. ex Raf.

Annual herb with fibrous roots. *Stems* 30–100 cm, pale green, ascending, striate, glabrous, branches few or absent, leafy. *Leaves* alternate, oblong in outline, paripinnate; leaflets 20–70, 10–30 × 2–6 mm, medium green on upper surface, paler and somewhat glaucous beneath, linear-oblong, rounded at apex, entire, rounded at base, glabrous; petiole short; stipules caducous. *Inflorescence* an axillary raceme, shorter than the leaves; peduncle 2–4 cm, each flower subtended by a caducous bract. *Calyx* about 5 mm, campanulate, divided one-third of way to base into 5 nearly equal lobes, the lobes deltoid and acute at apex. *Corolla* 11–15 mm, yellow; standard subrotund and reflexed, streaked and spotted with purple, with a short claw; wings oblanceolate to oblong, with a long claw; keel strongly arching with an elongate claw as long as the limb, obtuse at apex. *Stamens* 10, diadelphous. *Style* 1. *Legume* 10–20 × 3–4 mm, linear, with much thickened sutures and a beak 5–10 mm, glabrous; seeds 30–40, about 4 mm, narrowly oblong. *Flowers* 7–9. $2n = 12$.

Introduced. An oil-seed casual. Native of North America.

Tribe **8. Loteae** DC.

Usually *herbs. Leaves* mostly distichous, trifoliolate to imparipinnate, with entire leaflets. *Inflorescence* of solitary flowers or an umbel. *Stamens* 10, diadelphous, the free stamen sometimes loosely fused to the others. *Fruit* a dehiscent or indehiscent legume; 1– to many-seeded.

16. Hippocrepis L.

Perennial herbs or *shrubs. Stems* furrowed or ridged. *Leaves* alternate, imparipinnate; leaflets entire. *Inflorescence* an axillary head. *Calyx* 5-lobed. *Corolla* yellow; standard with limb subrotund; wings with limb oblong or obovate; keel with apex upturned. *Stamens* 10, diadelphous. *Style* 1. *Fruit* a legume more or less constricted between the seeds.

Contains about 15 species in Europe, the Mediterranean region, Atlantic islands and western Asia.

Fearn, G. M. (1973). Biological flora of the British Isles. No. 134. *Hippocrepis comosa* L. *Jour. Ecol.* **61**: 915–926.

Lassen, P. (1989). A new delimitation of the genera *Coronilla, Hippocrepis* and *Securigera* (Fabaceae). *Willdenowia* **19**: 49–62.

Lassen, P. (1989). On kroniller, släktet *Coronilla*, i Norden, *Svensk. Bot. Tidskr.* **83**: 83–86.

1. Shrub; leaflets 5–9; corolla (12–)14–20 mm **1. emerus**
1. Perennial herb; leaflets 7–25; corolla 5–10(–14) mm **2. comosa**

1. H. emerus (L.) Lassen Scorpian Senna

Coronilla emerus L.

Deciduous *shrub* up to 1.5(–2.0) m. *Stems* erect, much branched, dark or pale brown, roughly fissured; young shoots greenish, prominently angled, glabrous. *Leaves* alternate, 3–7 × 1.5–5.0 cm, oblong in outline, imparipinnate; leaflets (3–)5–9, 5–30 × 4–18 mm, medium green on upper surface, paler and glaucous beneath, oblong-obovate, rounded, truncate or slightly emarginate at apex, sometimes apiculate, entire, rounded at base, nearly glabrous; petiole up to 25 mm, slender, slightly channelled above, nearly glabrous; stipules 1.0–2.5 mm, ovate, obtuse or subacute at apex, membranous, hairy and usually dark crimson-glandular. *Inflorescence* an axillary cluster of 2–5(–7) flowers, on short, condensed or elongate lateral shoots; bracts small, ovate, subacute at apex, membranous, hairy on margins and often crimson-glandular; peduncles 3–9 cm, elongating in fruit; pedicels 1–5 mm, thinly hairy or glabrous. *Calyx* 4–5 mm, cup-shaped, gibbous, the 5 lobes obscure and broadly deltoid, shortly hairy. *Corolla* (12–)14–20 mm, yellow, the petals together with the stamens attached to a thickened concave disc adnate to the base of the calyx-tube; standard with limb subrotund, truncate and sharply emarginated at apex and with a very slender rigid claw, channelled above and with a small marginal flap or tooth near the middle or towards the base; wings with limb obliquely oblong with a long fold, rounded at apex, shortly and bluntly lobed at base and with a slender claw; keel upcurved almost at a right angle to a slender, acute beak and with a slender claw below. *Stamens* 10, diadelphous; apex of filaments dilated;

anthers oblong. *Style* 1, bent upwards at a right angle; stigma capitate. *Legume* 50–100×2.0–2.5 mm, linear, more or less arcuate, more or less compressed, somewhat constricted between the seeds with 3–12 segments, glabrous; seeds 3–6×1.0–1.5 mm, sausage-shaped, dark brown, smooth. *Flowers* 5–7. $2n=14$.

Introduced. Naturalised on roadsides and banks in several places in England north to Lincolnshire and in Isle of Man; also a hedgerow shrub in Co. Wicklow. Native of southern Europe and the Mediterranean region.

2. H. comosa L. Horseshoe Vetch

Perennial herb with a tap-root and fibrous side-roots. *Stem* 10–30(–50) cm, pale green, procumbent to suberect, glabrous, branched, leafy. *Leaves* alternate, 3–5×1.0–1.5 cm, oblong in outline, imparipinnate; leaflets 7–25, (2–)4–8(–16)×3–5(–8) mm, medium green on upper surface, paler beneath, oblong to obovate, emarginate at apex, entire, narrowed at base, sessile, glabrous; petioles 15–20 mm, glabrous; stipules lanceolate, spreading. *Inflorescence* an axillary head of (2–)4–8(–12) flowers; peduncle up to 8 cm, pale green, exceeding the leaves, glabrous, slender; pedicels short. *Calyx* (3.5–)4–5 mm, divided halfway into 5 subequal lobes, the lobes lanceolate, acute at apex, glabrous. *Corolla* 5–10(–14) mm; standard deep bright yellow, with brownish-orange veins on back which show through on the face, the limb subrotund with short lobes at base and a narrow claw; wings deep yellow, with limb obovate, rounded at apex and with a slender claw; keel deep yellow, ovate with the apex upturned and a narrow claw. *Stamens* 10, diadelphous; filaments with alternate ones dilated towards the top; anthers uniform. *Style* 1, curved; stigma capitate. *Legume* 10–30 mm, with 3–6 segments, the segments horseshoe-shaped and breaking up, minutely papillose; seeds several, 3.0–3.5 mm, semicircular. *Flowers* 5–7. Pollinated mainly by bees. $2n=14$, 28.

Native. Dry calcareous grassland and rock ledges on cliff-tops. Local in Great Britain north to Westmorland, frequent in south and east England. West, central and south Europe, northwards to north-central Germany. European Temperate element.

17. Scorpiurus L.

Annual herbs. Stems furrowed or ridged. *Leaves* alternate, simple, entire. *Inflorescence* an axillary umbel. *Calyx* 5-lobed. *Corolla* yellow, sometimes stained red; standard with limb subrotund or reniform, with a short claw; wings with limb oblong, rounded at apex with a short lobe at base and a short, narrow claw; keel with the limb upcurved into a blunt beak and with a short slender claw. *Stamens* 10, diadelphous. *Style* 1, upcurved. *Fruit* a legume, circinately curled and irregularly twisted and contorted, with 9 prominent, longitudinal ribs; seeds crescent-shaped, rugulose.

Contains 2 species in the Mediterranean region, western Asia; Atlantic Islands and north-east tropical Africa.

1. Legumes with spines and tubercles, rarely smooth **1(i). muricatus** var. **muricatus**
1. Legumes echinate or hispid, at least some of the longitudinal ribs bearing straight or hooked, glabrous or minutely hispidulous bristles **1(ii). muricatus** var. **subvillosus**

1. S. muricatus L. Caterpillar Plant

Annual herb with fibrous roots. *Stems* up to 80 cm, pale green, procumbent to suberect, angular, subglabrous to more or less densely hairy with long spreading or subappressed hairs, branched, leafy. *Leaves* alternate, 2–12×0.5–3.0 cm, medium green on upper surface, paler beneath, lanceolate, narrowly elliptical, oblanceolate, more or less acute at apex, rarely rounded or emarginate at apex, entire, tapered gradually at base, and adnate to the petiole, subglabrous or thinly hairy on both surfaces, often with 3 rather conspicuous longitudinal veins and otherwise obscure venation; petiole long, flattened or somewhat channelled; stipules 5–20 mm, subulate, acuminate at apex, membranous with a greenish midrib, more or less hairy, basal margin adnate to petiole. *Inflorescence* an axillary umbel of (1–)2–5 flowers; peduncle 3–15(–25) cm, rather stout, sulcate; bracts minute, membranous; pedicels 2–3 mm, densely hairy. *Calyx* 3–5 mm, campanulate, divided half of the way to the base into 5 lobes, the lobes triangular and sharply acuminate at apex, glabrous or thinly hairy. *Corolla* 5–12 mm, yellow, sometimes stained red externally; standard with limb subrotund or reniform, abruptly narrowed to a short claw; wings with limbs oblong, rounded at apex, with a short blunt lobe at base and a short narrow claw; keel with limb upcurved into a blunt beak and a short, slender claw. *Stamens* 10, diadelphous; filaments strongly dilated at apex; anthers oblong. *Style* 1, abruptly upcurved; stigma capitate. *Legume* 30–50 × 2–4 mm, circinately curled or irregularly twisted and contorted, rarely rather straight, with 9 prominent longitudinal ribs variously clothed with spines or tubercles and hooked bristles, rarely smooth, the valves distinctly constricted between the seeds and obscurely articulated; seeds 3.0–4.5 × 1.5–2.5 mm, crescent-shaped, rugulose. *Flowers* 6–9. $2n=28$.

(**i**) Var. **muricatus**
Legumes with spines and tubercles, rarely smooth.

(**ii**) Var. **subvillosus** (L.) Lam.
S. subvillosus L.; *S. sulcatus* L.; *S. muricatus* var. *sulcatus* (L.) Lam.; *S. subvillosus* var. *breviaculeatus* Batt. & Trab.

Legumes echinate or hispid, at least some of the longitudinal ribs bearing straight or hooked, glabrous or minutely hispidulous bristles.

Introduced. Bird-seed or wool aliens on tips, rough ground, gardens and parks. Scattered localities in central and southern Great Britain. Native of the Mediterranean region eastwards to Iran; Ethiopia; Atlantic Islands. Both varieties occur in Great Britain.

18. Securigera DC. nom. conserv.

Perennial herbs. Stems furrowed. *Leaves* imparipinnate; leaflets entire; stipules present. *Inflorescence* an axillary umbel on a long peduncle with (5–)10–20 flowers. *Calyx* tubular, divided into 5 short, more or less equal lobes. *Corolla* usually bluish-mauve, sometimes white or pink in general colour; standard with limb subrotund; wings obovate; keel with an ovate limb turned upwards at tip, which is dark purple. *Stamens* 10, diadelphous. *Style* 1. *Fruit* a legume, 4-angled, with 3–8(–12) segments breaking into 1-seeded segments.

Contains 12 species mainly in the Mediterranean area and the Middle East.

Lassen, P. (1989). Om kroniller, släktet *Coronilla*, I Norden. *Svensk. Bot. Tidskr.* **83**: 83–86.

Lassen, P. (1989). A new delimitation of the genera *Coronilla, Hippocrepis* and *Securigera* (Fabaceae). *Willdenowia* **19**: 49–62.

1. S. varia (L.) Lassen — Crown Vetch

Coronilla varia L.

Perennial herb with a tap-root and fibrous side-roots. *Stems* 20–120 cm, pale green, ascending or sprawling, furrowed, glabrous, branched, leafy. *Leaves* oblong in outline, imparipinnate; leaflets 11–25, 6–20 × 4–8 mm, elliptical to oblong, mucronate at apex, entire, rounded at base, sessile or shortly petiolulate, glabrous; petiole short, glabrous; stipules 1.9–2.6 mm, elliptical, acute at apex. *Inflorescence* an axillary umbel of (5–)10–20 flowers; peduncle up to 8 cm, glabrous. *Calyx* 2.8–3.2 mm, tubular, divided halfway into 5 lobes, the lobes triangular, acute at apex. *Corolla* 8–15 mm; standard lilac-mauve with dark purple veins, the limb subrotund, with a short, narrow claw; wings whitish, turning pale bluish-mauve, the limb obovate, rounded at apex with a short narrow claw; keel whitish, turning pale bluish-mauve, tipped dark purple, the limb ovate and curved upwards, the claw short and narrow; sometimes general colour white or pink. *Stamens* 10, diadelphous; filaments white; anthers pale yellow. *Style* 1 white; stigma pale yellow. *Legume* 20–60(–80) × 2–3 mm, linear, 4-angled, with a long tail-like beak at apex and 3–8(–12) segments, breaking into 1-seeded segments; seed 2.8–3.0 mm, oblong. *Flowers* 6–9. Pollinated by bees and flies. $2n = 24$.

Introduced. A garden escape naturalised in quarries, by railways and roadsides and in grassy and waste places. Scattered throughout Great Britain north to central Scotland and in eastern Ireland and Guernsey. Native of central and south Europe and western Asia; introduced in North America.

19. Coronilla L.

Annual herbs or low *shrubs*. *Leaves* alternate, imparipinnate or 3-foliolate; stipules free or connate. *Inflorescence* an axillary head. *Calyx* campanulate, 5-lobed. *Corolla* yellow; standard ovate or elliptical, emarginate at apex; wings round at apex; keel curved and beaked. *Stamens* 10, diadelphous. *Style* 1. *Fruit* a legume, curved to almost straight, beaked, not or scarcely contracted between the (1–)2–11 cylindrical segments.

Contains about 20 species in the Mediterranean region, western Asia and Madeira.

Bean, W. J. (1970). *Trees and shrubs hardy in the British Isles*. Ed. 8. **1**. London.

Jahn, A. (1974). Beitrage zur kentniss der Sippenstruktur einiger Arten der Gattung *Coronilla* L. *Feddes Repert.* **85**: 455–532.

Lassen, P. (1989). A new delimitation of the genera *Coronilla, Hippocrepis* and *Securigera* (Fabaceae). *Willdenowia* **19**: 49–62.

Uhrova, A. (1935). Revision de Gattung *Coronilla* L. *Beih. Bot. Centralb.* **53** (**13**): 1–174.

1. Low shrub; leaves with 5–7 leaflets; corolla 7–12(–14) mm — **1. valentina** subsp. **glauca**
1. Annual herb; leaves with 3 leaflets; corolla 3–8 mm — **2. scorpioides**

1. C. valentina L. — Shrubby Scorpion Vetch

Low deciduous *shrub* up to 1 m. *Stems* erect or spreading, terete, pale brown; young shoots green. *Leaves* alternate, imparipinnate, oblong in outline; leaflets 5–7, 10–20 × 7–10 mm, dark glaucous green on upper surface, paler beneath, obovate, emarginate at apex, entire, glabrous; stipules 5–10 mm, ovate or lanceolate, membranous. *Inflorescence* an axillary head with 4–8(–12) flowers. *Calyx* 2.5–3.0 mm, pale green campanulate, divided into 5 almost equal teeth, glabrous. *Corolla* 7–12(–14) mm, yellow; standard broadly elliptical, emarginate at apex, rounded at base; wings obovate, rounded at apex; keel greenish-yellow, curved. *Stamens* 10, diadelphous. *Style* 1. *Legume* 10–50 mm, with (1–)2–4(–10) segments which are 5–7 mm, cylindrical subcompressed and with 2 obscure angles. *Flowers* 3–5. $2n = 12, 24$.

Introduced. Naturalised on cliffs near Torquay in Devonshire and on the promenade at Eastbourne in Sussex, with a few records elsewhere. Our plant is subsp. **glauca** (L.) Batt. (*C. glauca* L.). Native of the Mediterranean region.

2. C. scorpioides (L.) W. D. J. Koch — Annual Scorpion Vetch

Ornithopus scorpioides L.; *Arthrolobium scorpioides* (L.) DC.; *Astrolobium scorpioides* (L.) DC.

Annual herb with fibrous roots. *Stems* 5–20(–30) cm, glaucous, prostrate, ascending or suberect, subterete or obscurely angled, glabrous, usually much branched, leafy. *Leaves* alternate, (1–)3-foliolate, glaucous; terminal leaflet 0.8–4.0 × 0.3–2.8 cm, oblong, rounded or subtruncate at apex, entire, cordate or broadly cuneate at base; lateral leaflets much smaller than terminal, 2–15 mm in diameter, subrotund or ear-shaped, glabrous; stipules 2–4 mm, leaf-opposed, connate, membranous, with a shortly bifid apex. *Inflorescence* an axillary 2- to 4(–5)-flowered head; peduncle 5–30 mm, usually exceeding the subtending leaf and lengthening in fruit; bracts less than 0.5 mm, deltoid or subulate, membranous, connate; pedicels 1–2 mm, relatively stout. *Calyx* about 2 mm, campanulate, with 5 widely deltoid, obscure lobes. *Corolla* 3–8 mm, yellow, the standard sometimes with brown lines; standard with limb ovate, rounded or somewhat emarginate at apex, narrowed abruptly at base to a short claw; wings with limb strongly concave, rounded at apex, with a short lobe at base and a short claw; keel with limb crescent-shaped, bluntly beaked, with a distinct flap and a short claw. *Stamens* 10, diadelphous; filaments slightly dilated at apex; anthers alternatively basifixed and versatile. *Style* 1, tapering from a slightly swollen base; stigma small and capitate. *Legume* 20–50×2–3 mm, cylindrical, usually much curved, prominently articulated, acute to slightly beaked at apex, glabrous; seeds about 4 mm, sausage-shaped, dark brown, minutely verruculose. *Flowers* 6–8. $2n=12$.

Introduced. Fairly frequent casual, mostly from bird-seed. Scattered records in southern Great Britain and north to Fifeshire and a garden weed in Co. Dublin. Native throughout the Mediterranean region and eastward to Iran.

20. Anthyllis L.

Perennial herbs. Leaves alternate, imparipinnate; leaflets 1–15, entire; stipules small, falling early. *Inflorescence* a cymose, axillary head, surrounded by an involucre, in pairs on long peduncles. *Calyx* with 5 short, unequal lobes, inflated, contracted at the oblique mouth. *Corolla* yellow to orange, rarely red; standard with limb broadly elliptical, retuse at apex, with a long broad claw; wings with limb obovate, rounded at apex, with a narrow lobe at base and a long, thin claw; keel with limb lanceolate and slightly upturned at apex, a short lobe at base and a long, thin claw. *Stamens* 10, diadelphous. *Style* 1. *Legume* subglobose, enclosed in the calyx; seeds 1–2.

Contains about 25 species in Europe, North Africa and western Asia.

Akeroyd, J. R. (1991). *Anthyllis vulneraria* subsp. *polyphylla* (DC.) Nyman, an alien Kidney-vetch in Britain. *Watsonia* **18**: 401–403.

Cullen, J. (1976). The *Anthyllis vulneraria* complex; a résumé. *Notes Roy. Bot. Garden Edinb.* **35**: 1–38.

Cullen, J. (1986). *Anthyllis* in the British Isles. *Notes Roy. Bot. Garden Edinb.* **43**: 277–281.

Hultén, E. & Fries, M. (1986). *Atlas of north European vascular plants north of the Tropic of Cancer.* 3 vols. Königstein.

Rich, T. C. G. (2001). What is *Anthyllis vulneraria* L. subsp. *corbierei* (Salmon & Travis) Cullen (Fabaceae)? *Watsonia* **23**: 469–480.

1. Upper leaves with a large terminal lobe and (0–)1–4 pairs of smaller lateral ones; calyx with lateral teeth not appressed to upper ones 2.
1. Upper leaves with 4–7 pairs of lateral leaflets scarcely smaller than terminal one 4.
2. Calyx hairs more or less patent **1(e). vulneraria** subsp. **lapponica**
2. Calyx hairs sparse and more or less appressed 3.
3. Leaves mainly borne close to the base of the stem; corolla pale yellow **1(d,iv). vulneraria** subsp. **carpatica** var. **vulgaris**
3. Leaves distributed along the stem at least to the middle; corolla deep yellow, or some petals pink **1(d,v). vulneraria** subsp. **carpatica** var. **pseudovulneraria**
4. Stem hairs all patent **1(c). vulneraria** subsp. **corbierei**
4. At least upper part of stem with appressed hairs 5.
5. Stems usually woody at base; leaflets of lower cauline leaves unequal; calyx not red at apex, strongly spreading hairy **1(b). vulneraria** subsp. **polyphylla**
5. Stems usually thin and not woody; leaflets of lower cauline leaves more or less equal in size; calyx red at apex, with weakly spreading hairs 6.
6. Corolla orange, pink or red **1(a,iii). vulneraria** subsp. **vulneraria** var. **coccinea**
6. Corolla yellow 7.
7. Axillary branches absent or few, shorter than the leaves that subtend them **1(a,i). vulneraria** subsp. **vulneraria** var. **vulneraria**
7. Axillary branches many, longer than the leaves which subtend them **1(a,ii). vulneraria** subsp. **vulneraria** var. **langei**

1. A. vulneraria L. — Kidney Vetch

A. pubescens Stokes nom. illegit.; *A. leguminosa* Gray nom. illegit.

Perennial herb with a stout stock. *Stems* up to 60 cm, pale green, prostrate, decumbent or erect, hairy, branched, leafy. *Leaves* alternate, 5–15×2–4 cm, broadly oblong or ovate, imparipinnate; leaflets 1–15, the lower sometimes reduced to the terminal leaflet, those of the later all equal or the terminal much larger, lanceolate, ovate or elliptical, obtuse to more or less acute at apex, entire, rounded at base, hairy on both surfaces; petiole very short or absent; stipules small, falling early. *Inflorescence* a cymose, axillary head up to 4 cm in diameter, surrounded by an involucre of 2 palmate bracts, in pairs on a long peduncle. *Calyx* 8–10 mm, sometimes with a red apex, inflated, contracted at the oblique mouth, with 5 short lobes, the lobes unequal, the 2 lateral lobes may be narrowly triangular, rather large and held below the lower lobes, or they may be more or less linear, narrow and appressed to the upper lobes. *Corolla* 12–15 mm, yellow to orange, rarely red; standard with limb broadly elliptical, retuse at apex, narrowed to a long, broad claw; wings with limb obovate, rounded at apex, with a narrow lobe and a long, thin claw; keel with limb lanceolate and slightly upturned at the pointed apex, with a short lobe at base and a long, thin claw. *Stamens* 10, diadelphous; filaments pale; anthers uniform. *Style* 1, curved upwards, glabrous; stigma discoloured. *Legume* about 3 mm, subglobose, compressed, reticulate, enclosed in the calyx; seeds 1–2, 1.8–2.0 mm, ellipsoid, smooth. *Flowers* 6–9. $2n=12$.

(a) Subsp. vulneraria

Stems appressed-hairy throughout or with subappressed or rarely patent hairs at base. *Leaves* evenly distributed along stems, the upper with 4–7 pairs of leaflets, scarcely or not smaller than the terminal one. *Bract lobes* usually tapering and acute at apex. *Calyx* red-tipped, lateral lobes obscure, pressed to the upper, hairs more or less spreading. *Corolla* yellow, orange, red or pink. $2n=12$.

(i) Var. **vulneraria**
A. vulneraria var. *kerneri* Sagorski; *A. linnaei* Sagorski; *A. vulneraria* subsp. *linnaei* (Sagorski) Jalas

Axillary branches absent or few, shorter than the leaves which subtend them. *Cauline* leaves with equal-sized leaflets. *Bract lobes* clearly tapering and acute at apex. *Corolla* yellow.

(ii) Var. **langei** Jalas
A. vulneraria var. *stenophylla* Lange, non Boiss.; *A. maritima* auct.

Axillary branches many, longer than the leaves which subtend them. *Cauline leaves* with equal-sized leaflets. *Bract lobes* tapering and acute at apex. *Corolla* yellow.

(iii) Var. **coccinea** L.
Axillary branches absent or few, shorter than the leaves which subtend them. *Some cauline leaves* with unequal-sized leaflets. *Bract lobes* tending to be parallel-sided and obtuse at apex. *Corolla* pink or red.

(b) Subsp. **polyphylla** (Ser.) Nyman
A. vulneraria var. *polyphylla* Ser.; *A. polyphylla* (Ser.) Kit. ex G. Don fil.; *A. vulneraria* var. *schiwereckii* Ser.; *A. polyphylla* var. *schiwereckii* (Ser.) Sagorski; *A. schiwereckii* (Ser.) Blocki; *A. macrocephala* Wender.

Stem woody at base, with appressed hairs above and some patent hairs near the base. *Leaves* evenly distributed along the stem, the upper with 4–7 pairs of leaflets scarcely or not smaller than the terminal one. *Bracts* divided for more than half their length, the lobes tapering and acute at apex. *Calyx* usually not red-tipped, lateral lobes obscure, pressed to the upper. *Corolla* yellow. $2n=12$.

(c) Subsp. **corbierei** (C. E. Salmon & Travis) Cullen
A. maritima var. *corbierei* C. E. Salmon & Travis; *A. vulneraria* var. *sericea* Bréb.

Stem with all hairs patent. *Upper leav*es with 4–5 pairs of leaflets scarcely or not smaller than the terminal one. *Bracts* with lobes tapering and acute at apex. *Calyx* usually not red-tipped, lateral lobes obscure, pressed to the upper hairs more or less spreading. *Corolla* yellow.

(d) Subsp. **carpatica** (Pant.) Nyman
A. carpatica Pant.

Stem sparsely appressed-hairy. *Upper leaves* with a large terminal leaflet and (0–)1–4 pairs of small lateral leaflets. *Bracts* with lobes parallel-sided and obtuse at apex. *Calyx* red-tipped or not, lateral lobes distant, not appressed to upper. *Corolla* pale yellow.

(iv) Var. **vulgaris** W. D. J. Koch
A. vulneraria subsp. *communis* (Rouy) Jalas; *A. vulgaris* (W. D. J. Koch) A. Kern.; *A. vulneraria* subsp. *vulgaris* (W. D. J. Koch) Corb.; *A. affinis* Brittinger; *A. vulneraria* var. *affinis* (Brittinger) Wohlf.; *A. vulneraria* prol. *A. communis* Rouy

Leaves mainly borne close to the base of the stem. *Corolla* pale yellow.

(v) Var. **pseudovulneraria** (Sagorski) Cullen
A. pseudovulneraria Sagorski

Leaves distributed along the stem at least to the middle. *Corolla* deep yellow, or some petals pink.

(e) Subsp. **lapponica** (Hyl.) Jalas
A. vulneraria var. *lapponica* Hyl.; *A. kuzenuvae* Juz.

Stem with appressed hairs throughout, or with appressed hairs above and patent hairs below. *Upper leaves* with a large terminal lobe and (0–)1–4 pairs of smaller lateral leaflets. *Bracts* with lobes parallel-sided, obtuse at apex. *Calyx* usually extensively red-tipped, lateral lobes distant, not appressed to upper, with spreading hairs. *Corolla* yellow. $2n=12$.

Native. In dry places on shallow soils. Scattered throughout Great Britain and Ireland, particularly on calcareous soils and near the sea. Throughout Europe, east to the Caucasus; North Africa. The species is a European Boreo-temperate element. Subsp. *vulneraria* var. *vulneraria* occurs in grassland, dunes, cliff-tops and waste ground, usually calcareous, throughout Great Britain and Ireland. It is found in north-west Europe from France to Finland. Subsp. *vulneraria* var. *langei* is intermediate between subsp. *vulneraria* var. *vulneraria* and subsp. *iberica*, but forms stabilised and homogeneous populations. It occurs on the coasts of England, Wales, Ireland and the Channel Islands, and in Denmark, Germany, Belgium and Holland. Subsp. *vulneraria* var. *coccinea* occurs on sea-cliffs in Cornwall, Pembrokeshire and Caernarvonshire and is also found in Gottland and Oeland in Sweden and scattered localities in Denmark. Subsp. *polyphylla* has been introduced in east Scotland and southern England. It is native of central and east Europe, Turkey and Caucasus. It was once cultivated as a forage plant, which may explain its presence in a few localities in western Europe. Subsp. *corbierei* occurs on sea-cliffs in Anglesey, Cornwall and the Channel Islands to which it is possibly endemic. Rich (2001) thinks it should be lumped with subsp. *vulneraria* but Cullen (verbatim) says if it is to be lumped with anything it should be with the south-east European and Turkish subsp. *hispidissima* (Sagorski) Cullen to which it is connected geographically by subsp. *vulnerarioides* (All.) Arcang. from north-east Spain, south-west Alps and central Apennines. All three taxa key out together in *Flora Europaea*. In any case the plant which subsp. *corbierei* grows with is var. *langei*. Subsp. *carpatica* is a polymorphic race of east central Europe. The var. *vulgaris* is really a series of intermediates between subsp. *vulneraria* and subsp. *polyphylla* and probably does not occur with us. Var. *pseudovulneraria* forms stabilised populations, and is the plant occurring in scattered localities over Great Britain, Ireland and the Isle of Man. It is probable that these plants like subsp. *polyphylla*, may have originally been introduced as forage plants. Subsp. *lapponica* is native on banks, cliffs and rock-ledges, mostly in the mountainous area of northern England, west and north-west Ireland and Scotland. It also occurs in Norway, Sweden and Finland. Our plant is var. *lapponica*. Intermediate plants may be found where any of the taxa grow together.

21. Hymenocarpus Savi nom. conserv.

Circinnus Medik. nom. rejic.

Annual herbs. Leaves imparipinnate, exstipulate (the basal leaflets resembling stipules); leaflets entire. *Inflorescence* a small, bracteate cluster, sessile at the top of a distinct peduncle. *Calyx* with 5 subequal deep lobes. *Corolla* yellow or orange; standard with limb subrotund or reniform, abruptly narrowed to a short claw; wings with limb oblong, the base adhering to keel; keel with limb oblong-falcate, upcurved at apex, adaxial margins connate. *Stamens* 10, diadelphous; filaments dilated just below apex. *Style* sharply upcurved at apex; stigma terminal. *Fruit* a legume, flattened, circinate, indehiscent, 2-locular.

Contains 1 species widely distributed in the Mediterranean region and east to Iran.

1. H. circinnatus (L.) Savi Disc Trefoil

Medicago circinnata L.; *Circinnus circinnatus* (L.) Kuntze

Annual herb with fibrous roots. *Stems* 5–30(–40) cm, pale green, prostrate or sprawling, somewhat angular, with rather dense, spreading hairs, much branched, leafy. *Leaves* alternate, irregularly 3- to 7-foliolate, the lower often simple, spathulate-obovate, with a flattened petiole 1–2 cm, the upper usually pinnate and sessile or nearly so, the 2 basal leaflets simulating stipules; terminal leaflet 1.5–4.0 × 0.3–2.0(–3.0) cm, spathulate or narrowly obovate-oblong, rounded or acute at apex, often shortly apiculate; lateral leaflets usually much smaller than terminal, elliptical or oblong, acute at apex, the lowermost occasionally minute and subulate; all thinly to densely appressed-hairy on both surfaces. *Inflorescence* a small axillary cluster of 2–5 more or less sessile flowers; peduncle 1–2 cm, lengthening to 4 cm in fruit, slender, more or less densely clothed with spreading hairs; bracts 3–8 mm, oblong or obovate-elliptical. *Calyx* 4–5 mm, campanulate, divided some two-thirds of the way to the base into 5 lobes, the lobes linear-subulate, subacute at apex, densely hairy. *Corolla* 5–6 mm, yellow or orange; standard with limb subrotund or reniform, narrowed abruptly at base to a short claw; wings with limb oblong, rounded at apex and base adhering to keel, with a short lobe and a slender claw; keel with limb oblong-falcate, upcurved at apex, bluntly beaked, the adaxial margins connate for the greater part of their length. *Stamens* 10, diadelphous; filaments strongly dilated just below the apex; anthers oblong, connective reddish. *Style* 1, upcurved at almost a right angle just below the middle, upper part slightly thickened; stigma small, terminal. *Legume* 10–15 × 8–10 mm, subglobose or reniform, circinnate, flattened, sparsely to densely hairy, obscurely reticulate-veined, dorsal margin entire or variously toothed or erose; seeds 1–2, 3.0–3.5 mm, oblong or oblong-reniform, rich dark brown, smooth. *Flowers* 5–7.

Introduced. A tan-bark casual. Widespread and common in the Mediterranean region and eastwards to Iran.

22. Ornithopus L.

Ornithopodium Mill. nom. illegit.; *Artrolobium* Desv.

Annual herbs. Leaves alternate, imparipinnate. *Inflorescence* an axillary head. *Calyx* divided into 5 lobes. *Corolla* yellow and sometimes veined red, or white or pink; standard obovate or subrotund, obtuse; wings oblong, oblong-obovate or subrotund; keel ovate, obtuse at apex. *Stamens* 10, diadelphous. *Style* 1. *Fruit* a legume, curved to straight, beaked, not or slightly constricted between the 3–12 cylindrical to oblong-ellipsoid segments.

Contains 6 species in Europe, north and tropical Africa, western Asia and south America.

Hultén, E. & Fries, M. (1986). *Atlas of north European vascular plants north of the Tropic of Cancer*. 3 vols. Königstein.
Pinto da Silva, A. R. (1989). *Ornithopus sativus* Brotero (Leguminosae) Proposte de nova neotipo. *Taxon* **38**: 293–295.
Wigginton, M. J. (Edit.) (1999). *British red data books*. Vol. 1. *Vascular plants*. Peterborough.

1. Flower-heads without a bract at the base or with minute bracts **4. pinnatus**
1. Flower-heads with a leaf-like bract at the base 2.
2. Corolla yellow; legume not or scarcely contracted between the segments **1. compressus**
2. Corolla white to pink; legume distinctly contracted between the segments 3.
3. Corolla 6–9 mm; bract about half as long as flowers **2. sativus**
3. Corolla 3–5 mm; bract usually longer than flowers **3. perpusillus**

1. O. compressus L. Yellow Serradella

Annual herb with fibrous roots. *Stems* 2.5–30(–50) cm, pale green, prostrate to decumbent, rarely suberect, somewhat angular, shortly hairy, branched, leafy. *Leaves* alternate, 2.5–7.0 × 0.5–1.5 cm, narrowly oblong in outline, imparipinnate; leaflets 13–37, 2–8 × 1.0–3.5 mm, pale green on upper surface, even paler beneath, oblong to subrotund, obtuse to acute at apex, entire, rounded at base, sessile, shortly hairy on both surfaces; petioles up to 2.5 cm, the upper sessile; stipules up to 7.0 × 1.5 mm, narrowly oblong, acute at apex, adnate along one margin to the petiole. *Inflorescence* a compact axillary cluster of 2–5 flowers with a leaf-like bract at base; peduncle 5–15 mm; bracts minute, dark crimson; bracteoles absent. *Calyx* 4–5 mm, cylindrical, divided one-third of the way to the base into 5 subequal lobes, the lobes lanceolate, acuminate at apex. *Corolla* 5–6 mm, yellow; standard with limb obovate, truncate or slightly emarginate at apex and narrowed to a short, broad claw; wings with limb oblong-obovate, rounded at apex, bluntly lobed at base and with a short claw; keel upcurved to a blunt apex with a slender claw. *Stamens* 10, diadelphous; filaments flattened, alternate ones rather more dilated towards the apex; anthers ovoid. *Style* 1, upcurved, glabrous, tapering from base to apex; stigma capitate. *Legume* 20–50 × 2.5–3.0 mm, cylindrical, not or hardly contracted between the segments, compressed, downcurved, glabrous or slightly hairy; seeds about 3.0 × 1.6 mm, oblong, compressed, smooth. *Flowers* 5–7. $2n = 14$.

Introduced. Bare sandy banks. A rare casual in Great Britain and the Channel Islands, but has occurred sporadically on a roadside bank at Bexley in Kent since 1957. Native of the Mediterranean region and the Atlantic Islands.

2. O. sativus Brot. Serradella
O. roseus Dufour; *O. sativus* subsp. *roseus* (Dufour) Dostál

Annual herb with fibrous roots. *Stems* 20–70 cm, pale green, hairy, branched, leafy. *Leaves* alternate, imparipinnate, oblong in outline; leaflets 19–37, 3.5–4.0 × 1.5–6.0 mm, medium green on upper surface, paler beneath, lanceolate or elliptical to ovate, obtuse at apex, entire, hairy; stipules small, linear, free. *Inflorescence* an axillary head of 2–5 flowers; bracts large with 7–9 leaflets, about half as long as flowers. *Calyx* 2.3–2.8 mm, divided up to half way into 5 equal lobes. *Corolla* 6–9 mm, white or pink; standard 3.5–4.5 × 1.2–1.7 mm, obovate, emarginate at apex; wings 5.5–7.5 × 2.5–3.2 mm, obovate; keel 3–4 × 1.2–1.8 mm, with a claw. *Stamens* 10, diadelphous. *Style* 1. *Legume* 12–25 × 2.0–2.5 mm, distinctly contracted between the segments, the segments 3–7 and ellipsoid-oblong, the beak usually not more than 5 mm, straight and sometimes hooked at the tip. *Flowers* 5–8. $2n=14$.

Introduced. Planted and naturalised on china-clay waste in Cornwall since 1978, a rare casual elsewhere in southern Great Britain. Our plant is subsp. **sativus**, which is native of south-west France, the northern half of the Iberian peninsular and the Azores. It is cultivated as a fodder plant in most of Europe.

3. O. perpusillus L. Bird's-foot
O. nodosus Mill.; *O. pusillus* Salisb. nom. illegit.; *Ornithopodium perpusillum* (L.) Medik.

Annual herb with fibrous roots. *Stems* 2–45 cm, pale green, prostrate to decumbent, very shortly hairy, branched, leafy. *Leaves* alternate, 1.5–5.0 × 0.5–1.0 cm, oblong in outline, imparipinnate; leaflets 9–27, lowest pair often at the base of the rhachis, distant from the others and recurved, 2–4 × 1.5–2.5 mm, pale green on upper surface, even paler beneath, elliptical to linear-oblong, rounded at apex, entire, cuneate or rounded at base and sessile, shortly hairy; petiole short or absent; stipules minute. *Inflorescence* a small, axillary head of 3–8 flowers; peduncle longer or shorter than leaves, filiform, shortly hairy; bract subtending flower head pinnate, sessile and at least as long as the flowers; pedicels very short, stout and hairy. *Calyx* 2.0–2.5 mm, divided almost one-quarter of the way to the base into 5 lobes, the lobes subequal, triangular, acute at apex, hairy. *Corolla* 3–5 mm; standard white with magenta veins, the limb subrotund, retuse at apex, narrowed to a long wide claw with the margins inturned; wings creamy yellow, the limb subrotund with a short lobe at base and a narrow claw; keel deep yellow, ovate, apiculate at apex, with a stubby lobe at base and a long, slender claw. *Stamens* 10, diadelphous; alternate filaments dilated at top; anthers all similar. *Style* 1, upcurved, glabrous; stigma capitate. *Legume* 10–20 × 1.5–1.8 mm, oblong, curved, strongly constricted between the seeds; seeds 1.4–1.5 mm, ellipsoid, smooth. *Flowers* 5–8. Self-pollinated. $2n=14$.

Native. Dry, rather bare, sandy and gravelly ground on rock outcrops, sand-dunes and heathland. Locally common in much of Great Britain especially the south-east, but absent from much of central and north Scotland and Ireland. West and west-central Europe eastwards to Poland and southern Sweden; Macaronesia. Suboceanic Temperate element.

4. O. pinnatus (Mill.) Druce Orange Bird's-foot
Scorpiurus pinnatus Mill.; *O. ebracteatus* Brot.; *Artrolobium ebracteatum* (Brot.) Desv.; *Artrolobium pinnatum* (Mill.) Britton & Rendle.

Annual herb with fibrous roots. *Stem* 10–50 cm, pale green, decumbent to ascending, slender, glabrous to sparsely hairy, leafy. *Leaves* alternate, imparipinnate, oblong or oblong-oblanceolate in outline; leaflets 3–19, 5–7 × 1.5–4.0 mm, medium green on upper surface, paler beneath, linear to oblanceolate, obtuse at apex, entire, narrowed at base, glabrous or sparsely hairy, the lowest pair always distant from the base of the petiole; stipules small, linear and free. *Inflorescence* an axillary head with 1–5 flowers; peduncles filiform, as long as the leaves; ebracteate or with minute, scarious bracts. *Calyx* 3.8–5.0 mm, divided one-third of the way to the base into 5 equal lobes. *Corolla* 6–8 mm, yellow, veined with red; standard 4.5–5.5 × 1.8–2.0 mm, obovate, emarginate at apex; wings 4.5–5.5 × 1.6–2.0 mm; keel 4–5 × 0.9–1.1 mm. *Stamens* 10, diadelphous. *Style* 1. *Legume* 20–35 mm, cylindrical, curved, scarcely contracted between the segments, the segments (6–)8–12 and cylindrical, the beak not more than 5 mm; seeds 0.7–0.8 × 1.5–1.8 mm. *Flowers* 4–8. $2n=14$.

Native. Short turf or open ground on sandy soil. Channel Islands and Isles of Scilly; reports from mainland Cornwall unconfirmed. Atlantic coast of Europe from France to Spain; Mediterranean region; north-west Africa; Macaronesia. Suboceanic Southern-temperate element.

23. Lotus L.

Annual or *perennial herbs*, sometimes woody at base. *Leaves* 5-foliolate, the 2 at the base of the rhachis resembling stipules; leaflets entire; stipules brown, minute. *Flowers* in axillary, pedunculate, cymose heads on long peduncles; bracts 3-foliolate. *Calyx* tubular, 5-lobed. *Corolla* yellow; standard with limb subrotund and upturned; wings with limb broadly oblong; keel incurved with a beak-like apex. *Stamens* 10, diadelphous, alternative filaments dilated at the top. *Style* 1, attenuate at the top. *Fruit* an elongate legume, 2-valved, many-seeded, septate between the seeds, longitudinally dehiscent, not ridged or angled.

Contains about 100 species in temperate Europe, Asia, North and South Africa, North America and Australia.

Borsos, O. (1969). Quantitative anatomical investigations in *Lotus corniculatus* L. aggr. *Acta Bot. Akad. Sci. Hung.* **15**: 227–252.
Brummitt, R. K. (2005). To reject *Lotus glaber* Mill. *Taxon* **54**: 1094.
Chrtková-Zertová, A. (1973). A monographic study of *Lotus corniculatus* L. *Rozpr. Ceskoslov. Akad. Rada Matemat. Prirod.* **83**.
Grime, J.P. et al. (1988). *Comparative plant ecology*. London.
Hultén, E. & Fries, M. (1986). *Atlas of north European vascular plants north of the Tropic of Cancer*. 3 vols. Königstein.

Jones, D. A. & Turkington, R. (1986). Biological flora of the British Isles. No. 163. *Lotus corniculatus* L. *Jour. Ecol.* **74**: 1185–1212.

Larsen, K. (1954). Cytotaxonomical studies in *Lotus* L. I. *Lotus corniculatus* L. sens. lat. *Bot. Tidsskr.* **51**: 205–211.

Miniaev, N. A. (1957). De speciebus generis *Lotus* L. in regionibus occidentali-septentrionalibus partis Europae USSR cresentibus. *Not. Syst. (Leningrad)* **18**: 119–141.

Reynolds, S. C. P. (2002). *A catalogue of alien plants in Ireland.* Glasnevin.

Stewart, A., Pearman, D. A. & Preston, C. D. (1994). *Scarce plants in Britain.* Peterborough. [*L. subbiflorus.*]

Turkington, R. & Franco, G. D. (1980). The biology of Canadian weeds. 41. *Lotus corniculatus* L. *Canad. Jour. Pl. Sci.* **60**: 965–979.

Wigginton, M. J. (Edit.) (1999). *British red data book.* Vol. 1. *Vascular plants.* Peterborough. [*L. angustissimus.*]

1. Annuals with dense hairs; corolla not more than 10 mm — 2.
1. Perennials, glabrous or with various hair clothing — 3.
2. Keel of corolla with an obtusely angled bend on the limb; legume 6–16 mm, with 8–12 seeds — **4. subbiflorus**
2. Keel of corolla with a right-angled bend on the limb; legume 12–40 mm, with more than 12 seeds — **5. angustissimus**
3. Stem hollow — 4.
3. Stem solid — 6.
4. Calyx-teeth erect in bud, the upper 2 with an obtuse sinus between them — **2(iv). corniculatus** var. **sativus**
4. Calyx-teeth recurved in bud, the upper 2 with an acute sinus between them — 5.
5. Whole plant glabrous or with some scattered hairs — **3(a). uliginosus** subsp. **uliginosus**
5. Whole plant with numerous long simple eglandular hairs or with numerous hairs only on the calyx and scattered hairs elsewhere — **3(b). uliginosus** subsp. **vestitus**
6. Leaflets of upper leaves mostly more than 4 times as long as wide — **1. tenuis**
6. Leaflets of upper leaves mostly less than 3 times as long as wide — 7.
7. Longest calyx lobes shorter than tube — **2(vii). corniculatus** var. **carnosus**
7. Longest calyx lobes equalling or longer than tube — 8.
8. Leaflets markedly ciliate or hairy all over — 9.
8. Leaflets glabrous or sparingly hairy — 10.
9. Leaflets markedly long ciliate and sometimes sparingly hairy on the surface — **2(iii). corniculatus** var. **kochii**
9. Leaflets densely short hairy all over — **2(v). corniculatus** var. **hirsutus**
10. Leaflets 2–3 times as long as wide; corolla pale yellow — **2(ii). corniculatus** var. **norvegicus**
10. Leaflets not more than twice as long as wide; corolla usually deep yellow — 11.
11. Leaflets thin and not fleshy, 4–15(–31) × 2–6(–14) mm — **2(i). corniculatus** var. **corniculatus**
11. Leaflets thick and fleshy; 1–12(–14) × 1–6(–9) mm — 12.
12. Calyx lobes broadly triangular at base — **2(vi). corniculatus** var. **crassifolius**
12. Calyx lobes narrowly triangular at base — **2(viii). corniculatus** var. **maritimus**

1. L. tenuis Waldst. & Kit. ex Willd.
Narrow-leaved Bird's-foot Trefoil

L. glaber Mill. nom. rejic.; *L. corniculatus* subsp. *tenuis* (Waldst. & Kit. ex Willd.) Syme; *L. corniculatus* subsp. *tenuifolius* (L.) Hartm.; *L. decumbens* auct.; *L. corniculatus* var. *tenuifolius* L.

Perennial herb with a tap-root and numerous fibrous side-roots. *Stems* 5–90 cm, pale yellowish-green, ascending or erect, slender, glabrous or sparsely hairy, branched, leafy. *Leaves* alternate, 1.0–2.5 × 1–3 cm, 5-foliolate, broadly ovate in outline; leaflets 7–15 × 2–5 mm, medium yellowish-green on upper surface, paler beneath, linear to linear-lanceolate, acute at apex, entire, narrow at base, sessile, glabrous or sparsely hairy; stipules brown, minute. *Inflorescence* an axillary, pedunculate, cymose, head of (1–)2–4(–6) flowers; peduncles up to 5 cm, slender, glabrous; bracts 3-foliolate. *Calyx* 5–6 mm, pale green, tubular, divided halfway to base into 5 lobes, the lobes linear-lanceolate, acute at apex, with numerous, short simple eglandular hairs. *Corolla* 6–10 mm, lemon yellow; standard with limb upturned, subrotund and with a broad claw; wings with limb broadly oblong, rounded at apex and with a short narrow claw; keel with limb ovate, incurved, narrowed at the beak-like apex and obtuse with a short narrow claw. *Stamens* 10, diadelphous; filaments pale; anthers yellow. *Style* 1, pale; stigma yellow. *Legume* 25–30 × 2.5–3.0 mm, linear, straight; seeds 1.0–1.3 mm, globose. *Flowers* 6–8. Pollinated by bees. $2n = 12$.

Native. Dry grassy places by the coast, sand and chalk pits, railway banks and road verges. Scattered records in Great Britain north to central Scotland, north and east Ireland and the Channel Islands but frequent only in south and east England and probably introduced in Ireland, Scotland and much of western England. Most of Europe except the extreme north and north-east; western Asia; North Africa. European Southern-temperate element.

2. L. corniculatus L. Common Bird's-foot Trefoil

Perennial herb with a stout stock, tap-root and numerous side-roots. *Stems* 5–35 cm, pale yellowish to pale glaucous green, decumbent, ascending or erect, glabrous or more or less hairy, branched, leafy. *Leaves* alternate, 1–10 × 1–4 cm, 5-foliolate, broadly ovate in outline; leaflets 1–21(–31) × 1–10(–14) mm, pale to dark, dull to bright green or glaucous-green on upper surface, sometimes with a brownish suffusion, lanceolate or oblanceolate, or ovate to obovate, obtuse or roundish to more or less acute or sometimes emarginate at apex, entire, rounded at the sessile base, glabrous or from sparsely to densely hairy; stipules brown, minute. *Inflorescence* an axillary, pedunculate, cymose, head of 1–6(–8) flowers; peduncles 1–5 cm, glabrous or hairy; bracts 3-foliolate. *Calyx* (4.0–)5.0–8.5 mm, tubular, divided to about halfway to base into 5 lobes, the lobes lanceolate to triangular or awl-shaped, glabrous to hairy. *Corolla* (8–)10–16(–18) mm, pale to deep yellow, often with red in the standard and wings and sometimes with brownish or reddish in the keel; standard with limb upturned and subrotund, slightly emarginate at apex and with a wide claw, curved at apex; wings with limb broadly oblong, rounded-obtuse at apex; wings with limb broadly oblong, rounded-obtuse at apex, with a short, broad lobe at base and a narrow claw; keel with limb ovate, incurved to an obtuse, beak-like apex and with a narrow claw. *Stamens* 10, diadelphous; filaments pale; anthers yellow. *Style* 1, pale; stigma yellow. *Legume* 10–39(–45) × 1.5–3.0 mm, linear, obtuse at apex, straight, glabrous; seeds 1.5–1.8 mm, reniform. *Flowers* 6–9.

The account of this exceedingly variable species is based on the specimens in CGE, all of which have been named by A. Chrtková-Zertová, and we have seen most variants in the field. Although the coastal variants differ greatly, their ecology is difficult to understand.

(i) Var. **corniculatus**
L. corniculatus var. *vulgaris* W. D. J. Koch nom. illegit.; *L. corniculatus* var. *glaber* Opiz; *L. arvensis* Pers.; *L. corniculatus* var. *arvensis* (Pers.) Ser.

Stems decumbent, ascending or erect, solid, without underground shoots. *Leaves* green to glaucous-green; leaflets 4–15(–31) × 2–6(–14) mm, lanceolate to broadly obovate, roundish to acute at apex, thin, glabrous or with sparse, appressed to obliquely patent hairs. *Inflorescence* with (1–)4–6(–8) flowers; peduncles 3–10 cm, glabrous or slightly hairy. *Calyx* 5.0–6.5(–7.0) mm, glabrous or sparsely hairy; lobes awl-shaped to lanceolate, triangular at base, more or less as long as the tube. *Corolla* 10–14(–16) mm, bright to deep yellow, often with red in the standard and wings. *Legume* 10–30(–38) × 1.5–2.5(–3.0) mm; seeds 1.4–2.2 mm, globose or globose-reniform. $2n = 24$.

(ii) Var. **norvegicus** Zertová
Stems ascending to erect, solid, underground shoots present. *Leaves* glaucous to pale green; leaflets (4–)6–14(–17) × (1.5–)3–6(–9) mm, oblanceolate to obovate, round to subacute at apex, thin or slightly fleshy, glabrous or sparsely ciliate. *Inflorescence* with 1–4(–6) flowers; peduncles 3–7 cm, glabrous or with sparse hairs. *Calyx* (4–)5–7(–8) mm, glabrous; lobes narrowly triangular to lanceolate, acute at apex, more or less as long as the tube. *Corolla* (12–)14–15 mm, pale yellow. *Legume* 15–37(–45) × 1.5–2.0 mm; seeds (1.4–)1.5–2.0(–2.5) mm, globose to reniform. $2n = 24$.

(iii) Var. **kochii** Zertová
L. ciliatus Schur; *L. corniculatus* subvar. *ciliatus* Gams

Stems ascending to erect, solid, without underground shoots. *Leaves* pale to bright green; leaflets 5–17(–23) × (2–)3–8 mm, lanceolate to obovate, mostly acuminate at apex, rarely obtuse, thinly, ciliate or sometimes sparsely hairy. *Inflorescence* with (1–)3–5(–6) flowers; peduncles 3–7(–9) cm, glabrous or with sparse, appressed hairs. *Calyx* 6–7 mm, ciliate with long patent hairs; lobes awl-shaped to lanceolate, triangular at base, more or less as long as the tube. *Corolla* 11–13(–14) mm, pale to bright yellow. *Legume* 16–35(–38) × 2–3 mm. $2n = 24$.

(iv) Var. **sativus** Hyl.
L. colocensis Menyh.; *L. corniculatus* subsp. *major* var. *colocensis* (Menyh.) Borsos

Stems erect to ascending, hollow, with no underground shoots. *Leaves* bright green; leaflets (8–)11–21 × (3–)5–10 mm, oblanceolate to obovate, acute at apex, glabrous or sparsely hairy, thin, the network of veins slightly prominent beneath. *Inflorescence* with (1–)3–5(–6) flowers; peduncles (6–)10–15 cm, glabrous or with sparse hairs. *Calyx* (5.0–)7.0–8.5 mm, glabrous or sparsely hairy; lobes awl-shaped, triangular or lanceolate, acute at apex, as long as the tube or a little longer. *Corolla* 14–16 mm, bright or pale yellow. *Legumes* (15–)21–26(–40) × 2.5–3.0 mm. $2n = 24$.

(v) Var. **hirsutus** W. D. J. Koch
L. corniculatus subsp. *hirsutus* (W.D.J. Koch) Rothm.; *L. silvaticus* Wierzb.; *L. ornithopodioides* Schur, non L.

Stems decumbent or ascending, solid, without underground shoots. *Leaves* dark green, glaucous or bright green; leaflets (6–)9–15(–17) × (1.5–)3–6(–9) mm, oblanceolate or obovate, roundish or acute at apex, hairy, thin. *Inflorescence* with (1–)3–5(–7) flowers; peduncles (2.5–)3–7(–9) cm, appressed-hairy. *Calyx* 5–7(–8) mm, with appressed or obliquely patent hairs; lobes awl-shaped, lanceolate or narrowly triangular, acute at apex, more or less as long as the tube. *Corolla* 10–13(–16) mm, bright to dark yellow, the buds sometimes tipped with red. *Legume* 15–25(–35) × 1.5–2.5(–3.0) mm. $2n = 24$.

(vi) Var. **crassifolius** Pers.
L. corniculatus var. *microphyllus* Lange

Stems decumbent to ascending, solid, underground shoots present. *Leaves* dark green to dark glaucous-green or brownish-green; leaflets 1–8(–10) × (1–)2–5 mm, oblanceolate, obovate or broadly ovate, acute, roundish or emarginate at apex, glabrous, ciliate or hairy, very fleshy. *Inflorescence* with (1–)2–4(–6) flowers; peduncles 1–4 cm, glabrous or appressed-hairy. *Calyx* (4–)5–6(–8) mm, glabrous to hairy; lobes broadly or narrowly triangular, obtuse to acute at apex, more or less as long as tube. *Corolla* (8–)11–14(–16) mm, bright yellow. *Legume* 14–39 × 1.8–2.0 mm; seeds (1.2–)1.4–1.7(–1.8) mm, globose to reniform. $2n = 24$.

(vii) Var. **carnosus** Hartm.
L. corniculatus forma *grandiflorus* Druce

Stems ascending, solid, without underground shoots. *Leaves* dark green or glaucous; leaflets (3–)4–7(–10) × (1.5–)2–3(–5) mm, oblanceolate or obovate, roundish or subacute at apex, glabrous to hairy, more or less fleshy. *Inflorescence* with 1–5 flowers; peduncles (3–)4–5(–10) cm, glabrous or appressed-hairy. *Calyx* (4–)5–6(–7) mm, glabrous or appressed-hairy; lobes broadly triangular, shorter than tube. *Corolla* (10–)11–16(–17) mm, bright yellow. *Legume* (14–)18–36 × 2–3 mm. $2n = 24$.

(viii). Var. **maritimus** Rupr.
L. ruprechtii Miniaev

Stems ascending to erect, solid, without underground shoots. *Leaves* green to glaucous; leaflets (5–)8–12 × (3–)5–6(–9) mm, broadly ovate to obovate, roundish to emarginate or obtuse at apex, glabrous or slightly hairy on the margin and beneath, slightly fleshy. *Inflorescence* with (1–)3–5(–7) flowers; peduncle 3–8 cm, glabrous or sparsely appressed-hairy. *Calyx* (5–)6–7(–9) mm, glabrous or with obliquely patent hairs; lobes lanceolate, more or less as long as the tube. *Corolla* (10–)13–16(–18) mm, deep yellow, standard sometimes reddish, the keel tipped with ochre yellow to brown. *Legume* (11–)17–31 × 1.5–2.0 mm. $2n = 24$.

Native. Grassy and waste places, shingle, dunes, and cliff ledges, mostly where well drained. Common throughout Great Britain and Ireland. Almost throughout Europe; Asia; North and East Africa; in the topics only on mountains. Eurasian Southern-temperate element. Var. *corniculatus* occurs throughout Great Britain and Ireland in a great variety of habitats. It also occurs almost throughout the range of the species. Var. *norvegicus* is known with us only from specimens in CGE from Felstead in Surrey and Herne Bay in Kent named by A. Chrtková-Zertová and a specimen collected by Arthur Chater in a conifer forest north of Afon Ystwyth in Cardiganshire. It is widespread in Norway, Sweden, Austria, Poland and Czechoslovakia. It normally grows in marginal areas of coniferous forest so our plants may be introductions. Var. *kochii* is known only from a single specimen from the Skelding Hills in Yorkshire. It is widespread in central Europe and the southern part of northern Europe. Var. *sativa* is introduced in cultivated areas, waysides, meadows and railway banks and has been sown with wild flower seed, recently in grass margins of fields in Cambridgeshire. Also in Ireland in wild flower and amenity seed mixtures. It is probably Mediterranean in origin, but was cultivated because of its large size. Now widespread in western, central and the southern part of northern Europe. Var. *hirsutus* is a plant of dry sandy or limestone grassland, probably widespread. It occurs throughout much of Europe. Var. *crassifolius* occurs on coastal dunes, cliffs and shingle beaches, probably wherever the habitat occurs round Great Britain and Ireland. Widespread in the coastal regions of the Atlantic Ocean. Var. *carnosus* occurs on coastal dunes and shingle where it is not known how its ecology differs from that of var. *crassifolius*, but they can be picked out by their low habit and large flowers. It is most distinct on the north coast of the Baltic Sea while on the south coast it occurs in mixed populations as it does with us. Var. *maritimus* occurs on sandy and stony areas on the coast. Our scattered records are said not to be typical, but they do seem to be distinct plants. Typical var. *maritimus* occurs throughout Finland and the Baltic region of Russia. Intermediates may occur between any variants growing adjacent.

3. L. uliginosus Schkuhr Greater Bird's-foot Trefoil
L. pedunculatus auct.; *L. major* auct.

Perennial herb with a slender stock producing numerous stolons. *Stem* 15–60(–100) cm, pale yellowish-green, erect or ascending, hollow, glabrous or with medium, pale simple eglandular hairs, branched, leafy. *Leaves* alternate, 2–4 × 1.5–3.0 cm, 5-foliolate, broadly ovate in outline; leaflets 15–20 × 5–12 mm, medium green on upper surface, paler beneath, ovate, obovate or elliptical, often obliquely so, rounded-obtuse or mucronate at apex, entire, narrowed or rounded at base, sessile or nearly so, with numerous, short, pale simple eglandular hairs on both surfaces and the margins; petioles up to 10 mm, pale green, with short simple eglandular hairs, stipules brown, minute. *Inflorescence* an axillary, pedunculate, cyme with (1–)5–12(–15) flowers; peduncles up to 15 cm, slender, hairy; bracts 3-foliolate. *Calyx* 6–8 mm, tubular, divided halfway to base into 5 lobes, the lobes lanceolate, narrowly acute at apex, the 2 upper with an acute sinus, spreading in bud, with numerous simple eglandular hairs. *Corolla* 10–18 mm; standard deep rich yellow with red veins, the limb subrotund and upturned with an emarginate apex and a short, broad claw; wings deep rich yellow, the limb oblong, obtuse at apex and with a short, broad lobe at base and a short slender claw; keel pale yellow, ovate, incurved with a narrow, obtuse beak-like apex and a narrow claw, the whole slightly curved. *Stamens* 10, diadelphous; filaments pale; anthers yellow. *Style* 1, pale; stigma yellow. *Legume* 15–35 mm, globose. *Flowers* 6–8. Pollinated by bees. $2n = 12$.

(a) Subsp. **uliginosus**
L. major var. *glabriuscula* Bab.; *L. major* var. *pilosus* Gray

Whole plant glabrous or with some scattered hairs.

(b) Subsp. **vestitus** (Lange) Hansen
L. villosus Thuill., non Forssk.; *L. uliginosus* var. *villosus* Lamotte; *L. uliginosus* var. *vestitus* Lange; *L. major* var. *vulgaris* Bab.

Whole plant with numerous long simple eglandular hairs, or with numerous hairs only on the calyx and scattered hairs elsewhere.

Native. Damp, grassy places, marshes, ditches and pondsides. Throughout most of Great Britain and Ireland. West, central and south Europe, north to 60° N in Fennoscandia and east to about 25° E in Ukraine; North Africa; Canary Islands. The distribution and ecology of the two subspecies is not understood, but they do not seem to grow together. Subsp. *vestitus* seems to be a plant of western Europe, subsp. *uliginosus* more widespread. *L. pedunculatus* Cav. is considered to be a distinct species.

4. L. subbiflorus Lag. Hairy Bird's-foot Trefoil
L. suaveolens Pers.; *L. parviflorus* auct.; *L. hispidus* Desf. ex DC. 1815, non 1805; *L. angustissimus* subsp. *hispidus* Syme

Annual herb with fibrous roots. *Stems* 3–30(–90) cm, pale green, ascending, decumbent or procumbent, with dense, short, spreading simple eglandular hairs, much branched, leafy. *Leaves* alternate, 1.0–2.5 × 1.0–2.5 cm, 5-foliolate, broadly ovate in outline; leaflets 5–20 × 3–7 mm, medium to dark green on upper surface, paler beneath, lanceolate, ovate or oblanceolate, often obliquely so, obtuse at apex, entire, rounded at base, sessile, with numerous to dense, pale, medium simple eglandular hairs on both surfaces and the margins; stipules brown, minute. *Inflorescence* an axillary, pedunculate, cymose, head with (1–)2–4 flowers; peduncles exceeding the leaves; bracts 3-foliolate. *Calyx* 5–6 mm, tubular, divided halfway to the base into 5 lobes, the lobes narrowly linear-lanceolate with a slender, acute apex, with dense, medium, pale simple eglandular hairs. *Corolla* 5–10 mm; standard deep orange-yellow, the limb oblong with a short, pointed lobe at base and a thin claw;

keel pale yellow, turning orange, narrowly ovate, incurved, obtusely angled, pointed at the beak-like apex and with a short, narrow claw. *Stamens* 10, diadelphous; filaments pale; anthers yellow. *Style* 1, pale; stigma yellow. *Legume* 6–15×1.5–2.0 mm, 1.5–3 times as long as calyx, linear, acute at apex, glabrous; seeds 8–12, 0.8–1.0 mm, globose. *Flowers* 7–8. Pollinated by bees. $2n=24$.

Native. Round rock outcrops and in dry grassy places near the sea. Cornwall to Hampshire, Pembrokeshire, south-west and south-east Ireland and the Channel Islands. West Europe from 52° N south and east to Sicily; North Africa. Suboceanic Southern temperate element. Our plant is subsp. **subbiflorus**, which occurs throughout most of the range of the species.

5. L. angustissimus L. Slender Bird's-foot Trefoil
L. angustissimus subsp. *diffusus* (Sm.) Syme;
L. angustissimus var. *linnaeanus* Bab. nom. illegit.;
L. angustissimus var. *seringianus* Bab.; *L. diffusus* Sm.

Annual herb with fibrous roots. *Stem* 3–30(–80) cm, pale green, slender, with numerous short, spreading simple eglandular hairs, much branched, leafy. *Leaves* alternate, 1.5–2.5×1.5–2.5 cm, 5-foliolate, broadly ovate in outline; leaflets 10–20×3–8 mm, medium to dark green on upper surface, paler beneath, ovate, obovate or oblanceolate, more or less acute at apex, entire, narrowed at base, sessile, with numerous, short, pale simple eglandular hairs on both surfaces and the margin; stipules brown, minute. *Inflorescence* an axillary, pedunculate, cymose head with 1–3 flowers; peduncles usually shorter than leaves; bracts 3-foliolate. *Calyx* 5–6 mm, tubular, divided halfway to base into 5 lobes, the lobes narrowly linear-lanceolate, with slender, acute apex, with numerous pale simple eglandular hairs. *Corolla* 5–12 mm, a deep rich yellow with orange-brown veins on the face of the standard and flushed with red outside; standard upturned, the limb subrotund or slightly broader than long, emarginate at apex, narrowed to a medium claw; wings with limb oblong or oblong-obovate, with a short lobe at the base and a slender claw; keel with limb ovate, incurved, obtuse at the beak-like apex, with a right-angled turn halfway along the limb, and a slender claw. *Stamens* 10, diadelphous; filaments pale; anthers yellow. *Style* 1, pale; stigma yellow. *Legume* 12–40×1.0–1.5 mm, 4–7 times as long as calyx, linear, slightly curved, acute at apex, glabrous; seeds more than 12, 0.8–1.0 mm, globose. *Flowers* 7–8. Pollinated by bees. $2n=12$.

Native. Dry grassy places near the sea. Local in Cornwall, Devonshire, Kent and the Channel Islands. Southern Europe, west Asia, North Africa and Macaronesia. European Southern-temperate element.

24. Dorycnium Mill.

Perennial herbs or *small shrubs*. *Leaves* 5-foliolate, the lowest pair simulating stipules; stipules minute, free. *Inflorescence* an axillary head. *Calyx* campanulate, with 5 equal or unequal lobes. *Corolla* white or pink; standard more or less panduriform; wings oblong, obtuse; keel obovate, rounded. *Stamens* 10, diadelphous. *Style* 1. *Fruit* a dehiscent legume.

Contains about 8 species mainly in the Mediterranean region and often included in *Lotus*.

Rikli, M. (1901). Die Gattung *Dorycnium* Vill. *Bot. Jahrb.* **31**: 314–404.

1. Corolla 10–20 mm	**1. hirsutum**
1. Corolla 3–6 mm	2.
2. Leaflets 10–20×2–4 mm	**2. pentaphyllum** subsp. **gracile**
2. Leaflets 6–25×6–18 mm	**3. rectum**

1. D. hirsutum (L.) Ser. Canary Clover
Lotus hirsutus L.; *Bonjeanea hirsuta* (L.) Rchb.

Perennial herb or *small shrub*. *Stems* 20–50 cm, pale green or brownish, long-hairy, branched, leafy. *Leaves* alternate; leaflets without or with a very short rhachis, 7–25×3–8 mm, medium green on upper surface, paler beneath, oblong-obovate, obtuse at apex, entire, hairy; stipules minute, free. *Inflorescence* an axillary head with 4–10 flowers. *Calyx* 8.0–10.5(–12.0) mm, campanulate, with 5 unequal lobes, hairy. *Corolla* 10–20 mm, white or pink; standard 11–18×3–5 mm, panduriform, obtuse at apex, with a short claw; wings (8.5–)9.0–14.0×3.0–4.5(–6.0) mm; keel 6.5–9.5×3.0–4.5 mm. *Stamens* 10, diadelphous. *Style* 1. *Legume* 6–12×3.2–4.5 mm, oblong-obovoid, the valves not contorted at maturity; seeds 1.8–2.5×1.5–2.0 mm. *Flowers* 6–8. $2n=14$.

Introduced. A garden escape on a trackside in Kent and formerly a ballast alien in the Cardiff area. Native of the Mediterranean region.

2. D. pentaphyllum Scop. Badassi
D. suffruticosum Vill. nom. illegit.; *Lotus dorycnium* L.

Perennial herb or *small shrub*. *Stems* 30–80 cm, pale green, appressed-hairy, branched, leafy. *Leaves* alternate; leaflets 5, without a rhachis, 10–20×2–4 mm, medium green on upper surface, paler beneath, linear to obovate-oblong, obtuse at apex, entire, appressed-hairy; stipules minute, free. *Inflorescence* an axillary head of 12–25 flowers on pedicels usually longer than the calyx-tube. *Calyx* 2–3 mm, divided about half way to base into 5 unequal lobes. *Corolla* 3–6 mm, white; standard 3.0–3.5(–4.0)×1.2–2.0 mm, panduriform, obtuse at apex, with a short claw; wings 2.5–3.7×1–2 mm; keel 2.2–3.2×0.8–1.1 mm. *Stamens* 10, diadelphous. Style 1. *Legume* 3–5 mm, ovoid-globose; seeds 1.5–2.0×1.0–1.5 mm, many. *Flowers* 6–8. $2n=14$.

Introduced. A persistent garden escape in Kent and Middlesex. Our plant is subsp. **gracile** (Jord.) Rouy (*D. gracile* Jord.; *D. jordanianum* Willk.; *D. jordanii* Loret & Barrandon) which is native of the Mediterranean coast of France and Spain.

3. D. rectum (L.) Ser. Greater Badassi
Lotus rectus L.; *Bonjeania recta* (L.) Rchb.

Perennial herb with a stout stock. *Stems* up to 1.5 m, pale green, erect, striate or more or less angular, glabrous or thinly hairy, much branched, leafy. *Leaves* alternate, usually 5-foliolate, the rachis 5–10 mm, flattened or narrowly winged; 3 apical leaflets 15–30×7–15 mm, narrowly obovate, rounded and apiculate at apex, entire, narrowly cuneate at base, glabrous or nearly so on upper surface, often thinly hairy beneath; 2 basal leaflets 6–25 × 6–18 mm, broadly ovate, cordate, acute at apex, entire, very shortly petiolulate and stipulate. *Inflorescence* a dense, globose, axillary head of many flowers; peduncles 5–15 mm, elongating to 60 mm or more in fruit, usually densely villose, generally bearing at or near its apex one or more 1- to 3-foliolate leaves; bracts minute, reddish; pedicels 2.0–2.5 mm, villose. *Calyx* 4–5 mm, the tube urceolate, constricted towards its apex, 5-lobed, the lobes subequal, subulate, more or less villous. *Corolla* 5–6 mm, milky–white, stained purple; standard with limb oblong, gradually narrowed at base to a very short claw; wings with limb oblong, rounded at apex, plicate longitudinally, shortly lobed at base and with a slender claw; keel with limb oblong, slightly upcurved at apex which is more or less obtuse, scarcely lobed at base and with a short claw. *Stamens* 10, diadelphous; filaments not dilated at apex; anthers oblong. *Style* 1, slightly upcurved, glabrous; stigma terminal, capitate. *Legume* 10–12×about 2 mm, dark brown, linear-oblong, subterete, twisting spirally on dehiscence; seeds about 1.2 mm in diameter, dark brown, subglobose, smooth. *Flowers* 6–8. $2n=14$.

Introduced. A garden weed, probably from an impurity in flower seed. Persisted for 20 years in a bomb crater on Brockham Hill in Surrey and was formerly persistent at Barry Docks in Glamorganshire. Widely distributed in the northern and eastern Mediterranean region and in western North Africa.

25. **Tetragonolobus** Scop.

Perennial herbs. Leaves 3-foliolate, with 2 large stipules. *Inflorescence* 1- to 2-flowered, on a long peduncle, flowers on short pedicels. *Calyx* 5-lobed, not enclosing fruit. *Corolla* with limb of standard yellow with reddish-brown veins; wings yellow tipped brighter yellow; keel upturned with a greenish-purple tip. *Stamens* 10, diadelphous. *Style* 1, thickened at apex. *Fruit* a 4-winged legume with numerous seeds.

Contains about 6 species mainly in the Mediterranean region.

T. purpureus Moench (*Lotus tetragonolobus* L.) has been recorded as a bird-seed casual.

Daveau, M. J. (1896). Note sur quelques *Lotus* de la Section *Tetragonolobus* L. *Bull. Soc. Bot. Fr.* **43**: 358–369.

1. **T. maritimus** (L.) Roth — Dragon's-teeth

Lotus maritimus L.; *Lotus siliquosus* L. nom. illegit.;
T. siliquosus Roth nom. illegit.; *Lotus erectus* L.;
T. scandalida Scop. nom. illegit.

Perennial herb with a tap-root and fibrous side-roots. *Stems* 10–30(–40) cm, pale green, sometimes suffused brownish-purple, decumbent, shortly hairy, branched, leafy. *Leaves* alternate, 1.5–2.0×2.0–4.0 cm, broadly ovate in outline, 3-foliolate; leaflets 10–30×4–15 mm, medium green on upper surface, paler beneath, asymmetrical, oblanceolate to obovate, rounded or apiculate at apex, entire, cuneate to rounded at base, sessile, with numerous, short hairs on both surfaces and the margins; petioles short, hairy; stipules 8–12 mm, green, ovate. *Inflorescence* 1- to 2-flowered; peduncle longer than subtending leaf, pale green, hairy; bract 3-foliolate; pedicel about 1 mm. *Calyx* 13–15 mm, divided halfway into 5 unequal lobes, the lobes lanceolate and sharply acute at apex, hairy. *Corolla* 25–30 mm; standard with limb yellow with reddish-brown veins, ovate, emarginate at apex and with a medium wide claw; wings yellow tipped with brighter yellow, the limb obovate, rounded at apex and with a narrow claw; keel yellow with a greenish-purple tip, ovate with an upturned apex and a fairly narrow claw. *Stamens* 10, diadelphous. *Style* 1, curved upwards. *Legume* 25–60×4.5–5.0 mm, linear, acute at apex, 4-winged; seeds many, 1.5–2.0 mm, globose. *Flowers* 5–8. Visited by bumble-bees. $2n=14$.

Introduced. Well naturalised in rough calcareous grassland. In a few scattered localities across southern England, a rare casual elsewhere north to Midlothian. Native of Europe from southern Sweden southwards but rare in the Mediterranean region; Caucasus; North Africa.

Tribe **9. Robiniae** (Benth.) Hutch.
Galegeae subtribe *Robiniinae* Benth.; *Galegeae* subtribe *Corynellinae* Benth.

Trees. Leaves imparipinnate, with entire leaflets. *Inflorescence* a pendulous raceme. *Stamens* diadelphous. *Fruit* longitudinally dehiscent; seeds more than 3.

26. **Robinia** L.

Deciduous trees. Buds scaleless. *Leaves* alternate, imparipinnate; stipules spiny; leaflets entire. *Inflorescence* a terminal, pendulous raceme. *Calyx* 5-lobed. *Corolla* white, fragrant; standard subrotund, emarginate, with a short claw; wings oblong, rounded at apex; with a short claw; keel curved. *Stamens* 10, diadelphous. *Style* 1. *Fruit* a legume, linear and compressed, longitudinally dehiscent; seeds 3–14.

Contains about 20 species in North America. Named after John and Vesparian Robin who first grew the tree in Europe.

Bean, W. J. (1980). *Trees and shrubs hardy in the British Isles.* Ed. 8. **4**. London.
Gibbs, V. (1929). *Robinias* at Kew and Aldenham. *Jour. Roy. Hort. Soc.* **54**: 145–158.
Mitchell, A. (1996). *Trees of Britain.* London.
More, D. & White, J. (1993). *Trees of Britain and Northern Europe.* Jersey.

1. **R. pseudoacacia** L. — False Acacia

Deciduous tree 15–30 m, with an open, domed crown, becoming rather flat-topped in old trees, with numerous root-suckers. *Trunk* up to 2 m in diameter, extending through much of the crown. *Bark* brown and fairly smooth in young trees, becoming dull grey, rough and thick, with deep furrows and coarse, forking ridges. *Wood* yellowish, hard and durable. *Branches* spreading or arching; twigs brown, ridged, glabrous; young shoots dark brown or reddish-brown, ridged, glabrous, on strong shoots occur a pair of stipulate spines which are 3–25 mm, stout and recurved. *Buds* small and naked, hidden in the base of the leaf petiole until leaf fall. *Leaves* alternate, 15–30 × 3–5 cm, imparipinnate, oblong in outline; leaflets 7–19, 2.5–6.0 × 1.2–2.0 cm, dark bluish-green to yellowish-green on upper surface, paler greyish-green beneath, whole leaf more or less yellow in some cultivars, turning yellow in autumn, elliptical to oblong, rounded and slightly emarginate at apex with a short bristle in the notch, entire, rounded at base, very shortly petiolulate, initially silky-hairy on both surfaces, soon becoming glabrous; petiole 5–20 mm, slender, glabrous. *Inflorescence* 10–20 cm, a pendulous raceme, terminating the current season's growth. *Calyx* 6–8 mm, green, often spotted red, divided one-third to halfway to base into 5 lobes, the lobes unequal, triangular, acute at apex, with dense, short, unequal simple eglandular hairs. *Corolla* 15–20 mm, white or cream, fragrant; standard 15–20 mm, subrotund, emarginate at apex, with a short claw, and a greenish-yellow spot at base; wings 10–13 mm, oblong, rounded at apex, with a short claw; keel about 10 mm, with a touch of green, curved. *Stamens* 10, diadelphous; filaments white, curved; anthers pale yellow. *Style* 1, green, upturned, hairy all round near the apex. *Legume* 50–100 × 25–30 mm, brown, often turning reddish, linear, compressed, with the upper suture winged and the lower thickened; seeds 3–14, about 4.0–4.5 mm, dark brown, flattened, reniform, red-blotched, smooth. *Flowers* 6. Pollinated by bees. $2n = 20$.

Introduced in the 1630s. Planted in gardens, estates, parks, along streets and in scrub and woodland; naturalised by suckering and less often seeding at least since 1888. Scattered over much of Great Britain and Ireland as a planted tree but more or less naturalised only in the south of Great Britain and rare in Ireland. Native of central and eastern North America, naturalised throughout much of North America, Europe, North Africa, Asia and New Zealand; invasive alien in southern Europe.

Tribe **10. Galegeae** (Bronn.) Dumort.
Subtribe *Galeginae* Bronn.; Tribe *Astragaleae* (DC.) Dumort.

Herbs or *shrubs*. *Leaves* imparipinnate with entire leaflets. *Inflorescence* an acillary raceme. *Stamens* 10, monadelphous. *Fruit* a longitudinally dehiscent legume or more or less indehiscent; few- to many-seeded.

27. Oxytropis DC. nom. cons.

Perennial herbs. *Leaves* imparipinnate; stipules present. *Inflorescence* a leafless raceme arising from the stock. *Calyx* campanulate, 5-lobed. *Corolla* purple or white marked blue or purple. *Stamens* 10, diadelphous. *Style* 1. *Fruit* a many-seeded legume.

Contains about 300 species in north temperate regions.

Jalas, J. (1950). Sur Kausalanalyse der Verbreitung einiger nordischen Os- und Sandpflanzen. *Ann. Bot. Soc. Zool. Bot. Fenn. 'Vanamo'* **24**: 57–64.
Leins, P. & Merxmüller, H. (1966). Zur Gliederung de *Oxytropis campestris* gruppe. *Mitt. Bot. Staatssaml. München* **6**: 19–31.
Wigginton, M. J. (Edit.) (1999). *British red data books*. Vol. 1. *Vascular plants*. Peterborough.

1. Corolla pale purple, the keel tipped with dark purple; legume divided internally by septa from both adaxial and abaxial sutures and thus more or less bilocular 2.
1. Corolla white and cream with faint blue veins and a blue and purple blotch at the end of the keel; legume divided internally by septum from the abaxial suture only and thus semi-bilocular 3.
2. Plant up to 15 cm; leaflets 18–25, 4–8 × 2–4 mm; calyx 7–10 mm, the lobes 1.8–2.5 mm **1(i). halleri** var. **halleri**
2. Plant up to 30 cm; leaflets 18–31, 8–15 × 4–7 mm; calyx 9–12 mm, the lobes 2.5–3.0 mm **1(ii). halleri** var. **grata**
3. Plant with numerous to dense hairs up to 3.5 mm **2(i). campestris** var. **kintyrica**
3. Plant with numerous hairs up to 2 mm 4.
4. Leaflets 10–20 × 1.5–3.0 mm, linear or oblong **2(ii). campestris** var. **perthensis**
4. Leaflets 7–22 × 3–5 mm, oblong or narrowly elliptic-oblong **2(iii). campestris** var. **scotica**

1. O. halleri Bunge ex W. D. J. Koch Purple Oxytropis
O. sericea (DC.) Simonk., non Nutt.; *O. uralensis* auct.; *Astragalus nitens* Host, non *O. nitens* Turcz.

Perennial herb with a stout rootstock and short branches. *Leaves* in a dense tuft round the rootstock, up to 10 cm, imparipinnate, oblong in outline; leaflets 21–31, 5–15 × 3–6 mm, medium green on upper surface, paler beneath, elliptical, ovate or lanceolate, more or less acute at apex, entire, rounded at the sessile base, with numerous hairs on both surfaces, but fewer on the upper; petioles short, very hairy; stipules lanceolate, acute at apex, hairy, persistent and clothing the stock. *Inflorescence* arising on a leafless peduncle from the stock, a raceme of 5–15 flowers; peduncle up to 30 cm, stout, erect, with spreading hairs. *Calyx* 8–12 mm, campanulate, divided into 5 unequal, short lobes, hairy. *Corolla* 15–20 mm, pale purple, the keel tipped with dark purple; standard with limb elliptical or ovate, emarginate at apex and the sides of the upper part slightly inturned, narrowed at base to a long, wide claw; wings with limb narrowly obovate, emarginate at apex, with a fairly broad lobe at base and a long, narrow claw; keel with limb obovate, rounded and mucronate at apex, with a short lobe at base and a long narrow claw. *Stamens* 10, diadelphous; filaments pale green; anthers pale yellow. *Style* 1, upturned at apex; stigma yellow. *Legume* 15–20(–25) × 5–7 mm, oblong, hairy, divided internally by septa from both adaxial and abaxial sutures and thus more or less bilocular; seeds numerous, 2.0–2.5 mm, subglobose. *Flowers* 6–7. $2n = 32$.

(i) Var. halleri
Plant up to 15 cm, with numerous hairs up to 3 mm. *Leaflets* 18–25, 4–8 × 2–4 mm, lanceolate, ovate or elliptical, mostly acute at apex. *Calyx* 7–10 mm; lobes 1.8–2.5 mm.

(ii) Var. **grata** P. D. Sell
Plant up to 30 cm with numerous to dense hairs up to 3 mm. *Leaflets* 18–31, 8–15 × 4–7 mm, ovate, lanceolate or elliptical, acute at apex. *Calyx* 9–12 mm; lobes 2.5–3.0 mm.

Native. Var. *halleri* is very variable; the nearest specimens to continental material are those which originally grew at Queensferry in Fifeshire and those growing on the coast in Ross and Cromarty. The Perthshire and Argyllshire plants are slightly larger and more hairy and some on the north coast of Scotland are similar. Var. *grata* is based on the handsome plants which grow at Melvich and Bettyhill in Sutherland, nothing similar having been seen from continental Europe. Our plants of this species occur on mountain rock ledges and grassy slopes of Dalradian limestone and schists and coastal sites of base-rich sandstone cliffs and calcareous sand-dunes. The species occurs in the Pyrenees and Alps from France to Austria and Carpathians. European Boreal-montane element. All our plants belong to subsp. **halleri**.

2. O. campestris (L.) DC. Yellow Oxytropis
Astragalus campestris L.

Perennial herb with a stout rootstock and short branches. *Leaves* in a tuft round the rootstock, imparipinnate, oblong in outline, up to 15 cm; leaflets 21–31, 7–22 × 1.5–5.0 mm, medium green on upper surface, paler beneath, lanceolate, oblong-lanceolate to oblong or elliptic-oblong, more or less acute at apex, entire, rounded at the sessile base, with numerous long eglandular hairs on both surfaces; petioles up to 30 mm, hairy; stipules ovate, cuspidate, hairy. *Inflorescence* arising on a leafless peduncle from the stock with a raceme of 5–20 flowers; peduncle up to 30 cm, stout and erect, with spreading hairs. *Calyx* 8–10 mm, campanulate, divided into 5 short, unequal lobes, very hairy. *Corolla* 15–20 mm; standard white and creamy towards the centre with faint blue veins, the limb broadly elliptical, emarginate at apex, narrowed at base into a long, broad claw; wings white, the limb narrowly obovate, emarginate at apex, with a rather long lobe at base and long, very narrow claw; keel white with a blue and purple blotch at the apex, obovate with an apiculus at apex, with a long, gradually narrowed claw. *Stamens* 10, diadelphous; filaments pale; anthers yellow. *Style* 1, hairy at base, upcurved at apex; stigma pale green. *Legume* 14–18 × 6–7 mm, ovoid-oblong, hairy, divided internally by septum from abaxial suture only and therefore semi-bilocular; seeds 1.8–2.0 mm, subglobose. *Flowers* 5–7. $2n = 32, 48$.

(i) Var. **kintyrica** P. D. Sell
Plant 8–15 cm, with numerous to dense hairs up to 3.5 mm. *Leaflets* 23–33, 7–13 × 2.5–3.5 mm, lanceolate. *Calyx* 7–8 mm; lobes 1.5–2.0 mm, triangular to linear.

(ii) Var. **perthensis** P. D. Sell
Plant 10–15 cm, with numerous hairs up to 1.5 mm. *Leaflets* 21–25, 10–20 × 1.5–3.0 mm, linear or oblong. *Calyx* 7–10 mm; lobes 2–3 mm, linear.

(iii) Var. **scotica** (Jalas) P. D. Sell
O. campestris subsp. *scotica* Jalas

Plant 8–30 cm, with numerous hairs up to 2 mm. *Leaflets* 17–25, 7–22 × 3–5 mm, oblong or narrowly elliptic-oblong. *Calyx* 7–11 mm; lobes 2–3 mm, linear.

Native. Confined to four localities in Scotland. North Europe and mountains of south and central Europe; North America. All our plants are referable to subsp. **campestris** which occurs through much of Europe except the north. Var. *kintyrica* grows on coastal limestone cliffs from 25–180 m above sea level near Largybaan in Kintyre. It is near to var. *campestris*, which occurs on Oeland but is much more hairy and the hairs are much longer. Var. *perthensis* grows on limestone at Loch Loch and on limestone and calcareous schist at Dun Ban in Perthshire. Var. *scotica* is found on calcareous, hornblende schist in Coire Fee in Glen Clova, Forfarshire. The inland localities are all on the Dalradian series of rocks and are between 500 and 650 m altitude. The species is a member of the Eurasian Arctic montane element.

28. Astragalus L.

Perennial, rarely *annual herbs*. *Leaves* alternate, imparipinnate, stipulate; leaflets entire. *Inflorescence* an axillary raceme; bracteoles small. *Calyx* tubular, with 5 subequal lobes. *Corolla* zygomorphic, papilionate; standard with limb broad and emarginate, shortly clawed; wings with limb ovate or oblong, obtuse at apex; keel with an oblong limb, not beaked and with a narrow claw. *Stamens* 10, diadelphous. *Style* 1. *Fruit* a legume, longitudinally dehiscent, 2- to many-seeded.

Contains about 2,000 species, worldwide except for Australia.

Bunge, A. A. (1868–1869). *Generis Astragali*. St Petersburg.
Druce, G. E. (1916). *Astragalus danicus* Retz. subvar. *parvifolius* Druce in *Rep. Bot. Soc. Exch. Club Brit. Isles* **4**: 194.
Hultén, E. & Fries, M. (1986). *Atlas of north European vascular plants north of the Tropic of Cancer*. 3 vols. Königstein.
Wigginton, M. J. (Edit.) (1999). *British red data books*. Vol. 1. *Vascular plants*. Peterborough. [*A. alpinus*.]

1. Corolla blue to purple or tinged blue to purple — 2.
1. Corolla whitish-cream to yellow — 3.
2. Stipules connate below; flowers erect to erecto-patent; legume with whitish, spreading hairs — **4. danicus**
2. Stipules free; flowers patent to reflexed; legume with dark, appressed hairs — **5. alpinus**
3. Leaflets mostly less than 17 — **6. glycyphyllos**
3. Leaflets more than 17 — 4.
4. Annual herbs; legume more than 20 mm — 5.
4. Perennial herbs; legume less than 20 mm — 6.
5. Most hairs on stem and leaves simple eglandular and basifixed — **1. boeticus**
5. Most hairs on stem and leaves medifixed, with one arm shorter than the other, usually straight and appressed — **2. hamosus**
6. Hairs of stem and leaves simple eglandular and basifixed — **3. cicer**
6. Hairs of stem and leaves medifixed — **7. odoratus**

Subgenus **1. Trimeniacus** Bunge

Annual herbs. *Leaves* imparipinnate. *Hairs* usually simple and eglandular, rarely medifixed.

1. A. boeticus L. Mediterranean Milk Vetch

Annual herb. Stems up to 60 cm, pale green, erect, with sparse simple eglandular hairs, branched, leafy. *Leaves* alternate, imparipinnate, oblong in outline; leaflets 21–31, 8–30 × 7–12 mm, dark green on upper surface, paler beneath, narrowly oblong to oblong-obovate, truncate or emarginate at apex, entire, rounded at base, glabrous on upper surface, with sparse simple eglandular hairs beneath, rarely some hairs medifixed. *Inflorescence* a dense axillary raceme of 5–15 flowers; peduncles half as long as leaves. *Calyx* 5–7 mm, divided halfway to base into 5 lobes. *Corolla* yellow; standard 8–12(–14) mm, oblong, rounded at apex; wings 6.0–7.5 mm, longer than keel; keel 5.0–5.6 mm, oblong, obtuse at apex. *Stamens* 10, diadelphous. *Style* 1. *Legume* 20–40 × (5–)7–8 mm, oblong, triangular in transverse section, grooved beneath, with a hooked beak, the valves keeled, with short, appressed simple eglandular hairs. *Flowers* 7–8. $2n = 16, 30$.

Introduced. A grain casual, formerly naturalised in Gloucestershire. Native of the Mediterranean region.

Subgenus **2. Epiglottis** (Bunge) Willk.

Annual herbs. Leaves imparipinnate. *Hairs* medifixed, straight, appressed, with one arm much shorter than the other.

2. A. hamosus L. Southern Milk Vetch
A. paui Pau

Annual herb with fibrous roots. *Stems* up to 60 cm, pale green, erect, with few medifixed hairs, branched, leafy. *Leaves* alternate, imparipinnate, oblong in outline; leaflets 19–23, 2–25 × 2–10 mm, dark green on upper surface, paler beneath, oblong-obovate, emarginate or truncate at apex, entire, rounded at base, glabrous or nearly so on upper surface, with numerous medifixed hairs beneath. *Inflorescence* a fairly dense, axillary raceme with 5–14 flowers; peduncles about half as long as leaves. *Calyx* 5–6 mm, divided into 5 lobes. *Corolla* yellow; standard (6–)7–11 mm, oblong, obtuse at apex; wings 6–10 mm, oblong, obtuse at apex; keel 5–8 mm, oblong, obtuse at apex. *Stamens* 10, diadelphous. *Style* 1. *Legume* 20–50 × 2–3 mm, linear, acuminate at apex, curved for about a semicircle, laterally compressed, the beak short, the valves not keeled, almost smooth, with short, appressed, medifixed hairs; seeds 1.5 × 2.0–2.5 mm. *Flowers* 7–8. $2n$ = (32, 34, 40, 42, 44) 48.

Introduced. A grain and tan-bark casual. Native of the Mediterranean region and south-west Asia.

Subgenus **3. Hypoglottis** Bunge

Perennial herbs. Leaves imparipinnate; stipules cuneate at base forming a sheath round the stem opposite the petiole. *Flowers* subsessile in dense heads. *Calyx* not inflated in fruit, the mouth not oblique. *Hairs* simple and eglandular, rarely medifixed.

3. A. cicer L. Chick-pea Milk Vetch

Perennial herb. Stems (5–)25–60(–100) cm, pale green, ascending or suberect, with sparse simple eglandular hairs, branched, leafy. *Leaves* alternate, imparipinnate, oblong or oblong-lanceolate in outline; leaflets (17–)21–31, 15–35 × 5–12 mm, medium green on upper surface, paler beneath, lanceolate to ovate-lanceolate, obtuse to subacute at apex, entire, narrowed at base, with appressed, short, often sparse simple eglandular hairs on both surfaces, rarely subglabrous; stipules connate for part of their length, forming a sheath round the stem opposite the petiole. *Inflorescence* a dense, subsessile head; peduncles half to two-thirds as long as leaves. *Calyx* 7–10 mm, not inflated in fruit, divided for one-quarter of the way to the base into 5 lobes, the mouth not oblique. *Corolla* yellow; standard 14–16 mm, ovate, rounded at apex; wings oblong, rounded at apex; keel oblong, obtuse at apex. *Stamens* 10, diadelphous. *Style* 1. *Legume* 10–15 mm, ovoid-globose, inflated, membranous, with short black and white hairs. *Flowers* 6–8. $2n = 64$.

Introduced. An established grain alien known since the 1920s along a hedgebank, near a mill at Feshiebridge, near Edinburgh. Native of Europe and western Asia.

4. A. danicus Retz. Purple Milk Vetch
A. hypoglottis auct.; *A. danicus* subvar. *parvifolius* Druce; *A. danicus* var. *parvifolius* (Druce) Druce; *A. pulchellus* Salisb. nom. illegit.; *A. arenarium* auct.

Perennial herb with a slender, branched stock. *Stems* 5–35 cm, pale green, ascending, with sparse, white, soft simple eglandular hairs, shortly branched, leafy. *Leaves* alternate, 3–7 cm, imparipinnate, oblong in outline; leaflets 13–27, 10–12 × 5–7 mm, medium green on upper surface, paler beneath, lanceolate, narrowed to an obtuse apex, entire, rounded at base, sessile, with sparse, soft, white simple eglandular hairs; petiole 1–2 cm, pale green, with short, pale simple eglandular hairs; stipules 3.5–4.0 × 1.5–2.0 mm, ovate, acute at apex, connate below. *Inflorescence* a more or less erect, axillary, globose or ovoid-oblong head; peduncles 1–9 cm, usually much longer than the subtending leaf, with soft, white simple eglandular hairs below and mixed with black ones above; bracteoles 2.5–3.0 mm, oblong to triangular, obtuse at apex. *Calyx* 6–10 mm, green, tubular, divided one-third of the way into 5 lobes, the lobes subequal, lanceolate, acute at apex, with dense, appressed, mixed black and white simple eglandular hairs. *Corolla* 15–18 mm, bluish-purple, rarely white, erect; standard 15–16 mm, limb elliptical, emarginated at apex, twice as long as the linear claw; wings with limb oblong and obtuse at apex, with a slender claw; keel with limb obovate in outline, obtuse at apex, shortly lobed at base, with a slender claw. *Stamens* 10, diadelphous; filaments pale green; anthers all similar, pale yellow. *Style* 1, pale green; stigma pale green, capitate. *Legume* 7–10 × 2.5–3.0 mm, oblong, covered with white spreading crisped hairs, more or less inflated; seeds 1.0–1.2 mm, black, subglobose, emarginate at apex. *Flowers* 5–7. Pollinated by bees. $2n=16$.

Very variable in size of plant, size of leaflets, degree of hairiness and depth of blue in the flower.

Native. Short grassland on calcareous, well-drained soils, predominantly on chalk and limestone. Local in the east of Great Britain from Bedfordshire to Sutherland, extremely local elsewhere in England and Scotland, and in Ireland only in the Aran Islands of Co. Clare. Denmark and south Sweden eastwards to the Baikal region; south-west Alps; east Asia and North America. Circumpolar Temperate element.

Subgenus **4. Phaca** (L.) Bunge

Phaca L.

Perennial herbs. Leaves imparipinnate; stipules free from each other, but sometimes adnate to the petiole. *Flowers* pedicellate and usually more or less pendulous at anthesis. *Calyx* not inflated at fruiting, the mouth oblique. *Hairs* simple eglandular, usually appressed.

5. A. alpinus L. Alpine Milk Vetch

Phaca alpina L.; *Phaca astragalina* L.

Perennial herb with a slender, branched stock. *Stems* 5–35 cm, pale green, ascending, with sparse, white, soft simple eglandular hairs, shortly branched, leafy. *Leaves* alternate, 3–7 × 1.0–2.5 cm, imparipinnate, oblong in outline; leaflets 13–27, 5–12 × 6–7 mm, medium green on upper surface, paler beneath, elliptical, emarginate at apex, entire, rounded at base, sessile, with sparse, soft, white simple eglandular hairs; petiole 1–3 cm, pale green, with short, pale simple eglandular hairs; stipules 3–7 × 1.5–2.5 mm, ovate, acute at apex, free at base. *Inflorescence* a more or less erect, axillary raceme; peduncle longer than the subtending leaf, with sparse, more or less appressed, pale simple eglandular hairs; flowers pedicellate, patent to reflexed; bracteoles 1.0–1.5 × 0.5–0.7 mm, ovate, obtuse at apex, hairy. *Calyx* 4.5–5.0 mm, green, tubular, divided about halfway into 5 lobes, the lobes subequal, narrowly lanceolate, obtuse at apex, covered with short, dense, appressed simple eglandular hairs, not inflated in fruit, the mouth oblique. *Corolla* 10–14 mm, whitish, tinged with pale mauve to purplish at apex, spreading or reflexed; standard with limb broadly ovate, emarginate at apex and narrowed into a very short claw; wings with limb ovate, obtuse at apex, with a short lobe at base and with a short claw; keel with limb broadly oblong, curved upwards, with a short lobe at base and a narrow claw. *Stamens* 10, diadelphous; filaments pale green; anthers pale yellow. *Style* 1, pale green; stigma pale green. *Legume* 8–12 × 3–4 mm, oblong, covered with dark, appressed hairs; seeds few, 2.0–2.5 mm, broadly oblong. *Flowers* 6–7. $2n=16, 32$.

Native. Grassy calcicolous places and base-rich ledges and rocky outcrops in mountains from 700 to 800 m. In only four localities in Perthshire, Forfarshire and Aberdeenshire. Arctic Europe and on mountains south to the Pyrenees, Alps and Carpathians; Caucasus; temperate Asia; Greenland; North America. Circumpolar Arctic-montane element. Our plant is subsp. **alpinus** which occurs throughout the range of the species except in arctic Europe.

6. A. glycyphyllos L. Wild Liquorice

Perennial herb with a short, stout stock. *Stem* 30–100 (–150) cm, pale green, stout, prostrate, sprawling or ascending, glabrous. *Leaves* alternate, 10–20 × 5–9 cm, imparipinnate, oblong or ovate-oblong in outline; leaflets 7–15, 15–40 × 10–25 mm, medium green on upper surface, paler beneath, elliptical, ovate, ovate-oblong or elliptical-oblong, rounded at apex, entire, rounded or subcordate at base, sessile or very shortly petiolulate, glabrous or with minute hairs beneath; petiole rather short, glabrous or with minute hairs; stipules 15–20 × 7–10 mm, ovate, obtuse at apex, dentate, free at base. *Inflorescence* a more or less erect axillary raceme; peduncles 2–6 cm, much shorter than the subtending leaves; flowers pedicillate; bracteoles 5–6 mm, subulate. *Calyx* 6–7 mm, pale green, tubular, divided halfway to base into 5 lobes, the lobes unequal, triangular, acute at apex, glabrous not inflated in fruit, the mouth oblique. *Corolla* 11–15 mm, whitish-cream with a greenish tinge, spreading; standard with limb almost subrotund, curved, emarginate at apex, with a short broad claw; wing with limb oblong, rounded at apex, with a short lobe at base and a narrow claw; keel with limb very broadly oblong, rounded at apex and a short, broad lobe at base and a narrow claw. *Stamens* 10, diadelphous; filaments 8–10 mm, pale green; anthers pale yellow. *Style* 1, curved round at apex, pale green; stigma pale green. *Legume* 30–40 × 3.5–4.0 mm, linear, more or less laterally compressed, curved, acuminate at apex, clothed with appressed minute hairs; seeds many, 2.0–2.5 mm, suborbicular. *Flowers* 7–8. $2n=16$.

Native. Grassy places and scrub, cliffs, chalk pits, railway banks and road verges mostly on calcareous soils. Scattered records in Great Britain north to Ross and Cromarty. Europe to about 65° N, rare in the Mediterranean region, eastwards to the Altai, Caucasus and Turkey. European Temperate element.

Subgenus **5. Cercidothrix** Bunge

Perennial herbs. Leaves imparipinnate; stipules connate at base. *Flowers* pendulous, subsessile. *Calyx* not inflated in fruit. *Hairs* medifixed.

7. A. odoratus Lam. Lesser Milk Vetch

Perennial herb. Stems up to 30 cm, pale green, ascending to erect, with a few medifixed hairs, branched, leafy. *Leaves* alternate, imparipinnate, oblong in outline; leaflets 19–29, 10–17 × 6–8 mm, dark green on upper surface, paler beneath, lanceolate to elliptic-oblong, obtuse to acute at apex, entire, rounded at base, glabrous on upper surface, with a few medifixed hairs beneath; stipules connate round the stem. *Inflorescence* a dense, axillary raceme; peduncles about equalling the leaves; flowers pendulous. *Calyx* 4–5 mm, divided about one-quarter of the way to the base into 5 lobes, not inflated in fruit. *Corolla* whitish or yellow; standard 10–14 mm, ovate, rounded at apex; wings shorter than standard, oblong, rounded at apex; keel oblong, rounded at apex. *Stamens* 10, diadelphous. *Style* 1. *Legume* 8–10 × 2–3 mm, oblong-lanceolate, laterally compressed, smooth, with sparse, appressed medifixed hairs, becoming glabrous. *Flowers* 7–8. $2n = 16$.

Introduced. Naturalised in grassy places. A few scattered records in central and southern England. Native of the east Mediterranean region.

29. Colutea L.

Deciduous *shrubs*. *Leaves* opposite, imparipinnate, with entire leaflets; stipules small. *Inflorescence* a more or less erect axillary raceme of 2–8 flowers. *Calyx* campanulate, slightly 2-lipped, with 5 small lobes. *Corolla* pale to deep yellow or orange; standard broadly ovate or elliptical; wings oblong, rounded at apex; keel shorter than wings, beakless. *Stamens* 10, diadelphous. *Style* 1; stigma large. *Fruit* a greatly inflated, more or less pendulous legume. Poisonous.

Contains about 28 species in Europe and Asia.

Bean, W. J. (1970). *Trees and shrubs: hardy in the British Isles*. Ed. 8. **1**. London.
Browicz, K. (1963). The genus *Colutea* L. *Monogr. Bot. (Warszawa)* **14**: 1–136.

1. Petioles and leaf-rhachis glabrous; corolla 11–13 mm, orange-red — **3. orientalis**
1. Petioles and leaf-rhachis hairy; corolla 15–25 mm, yellow or orange — 2.
2. Leaflets (7–)9–11(–13), 15–30 × 10–20 mm; corolla 16–25 mm, yellow — **1. arborescens**
2. Leaflets 7–9(–11), 15–25 × 8–14 mm; corolla 15–20 mm, yellow to orange — **2. × media**

1. C. arborescens L. — Bladder Senna

Deciduous *shrub* up to 6 m. *Stem* pale brown, much branched, rough; young shoots pale green, slender, hairy. *Buds* small, globose-ovoid, with 2–4 scales. *Leaves* opposite, oblong in outline, imparipinnate; leaflets (7–)9–11(–13), 15–30 × 10–20 mm, dark bluish-green on upper surface, paler beneath, broadly elliptical, obovate or ovate, emarginate at apex with a mucro in the notch, entire, narrowed at base, glabrous on upper surface, appressed hairs beneath and short ones on the margin, with petiolules 1–2 mm; petiole up to 2 cm, petiole and rhachis shortly hairy; stipules small. *Inflorescence* an axillary raceme of 3–8 flowers, branches and pedicels appressed hairy. *Calyx* 4.0–4.5 mm, campanulate, appressed-hairy, slightly 2-lipped, with 5 small lobes. *Corolla* 16–25 mm, golden yellow with some greenish coloration in the keel; standard broadly elliptical, emarginate at apex, rounded to a short claw; wings narrow, rounded at apex; keel upturned, shorter than wings, without a beak. *Stamens* 10, diadelphous; filaments very pale greenish; anthers orange. *Style* 1, yellowish-green; stigma large, inserted obliquely on the inner edge of the style and surrounded by hairs. *Ovary* glabrous. *Legume* 5–7 × 2.5–3.0 cm, pale green turning brown, broadly oblong, inflated; seeds 7.0–8.5 × 5–6 mm, brown, reniform. *Flowers* 5–9. $2n = 16$.

Introduced in to cultivation by 1568. Formerly much grown in gardens and naturalised in waste and grassy places, on roadsides and railway banks. Frequent in southern Great Britain, especially south-east England and a few records in Ireland, but many of the records may be yellow-flowered *C.* × *media*. South and south-central Europe, extending to north-central France.

2. C. × media Willd. — Intermediate Bladder Senna
C. arborescens × orientalis

Deciduous *shrub* up to 6 m. *Stem* pale brown, much branched, rough; young shoots pale green, slender, glabrous to sparsely hairy. *Buds* small, globose-ovoid. *Leaves* opposite, oblong in outline, imparipinnate; leaflets (7–)9–11, 13–25 × 8–14 mm, dark bluish-green above, paler beneath, elliptical or ovate, rounded or slightly emarginate at apex, entire, rounded at base, glabrous on upper surface, appressed-hairy beneath; petiolules 1–2 mm; petiole up to 15 mm, hairy; stipules small. *Inflorescence* an axillary raceme of 3–8 flowers, branches and pedicels appressed hairy. *Calyx* 4–5 mm, campanulate, with few to numerous appressed hairs, slightly 2-lipped, with 5 small lobes. *Corolla* 15–20 mm, yellow to orange; standard broadly elliptical, slightly emarginate at apex, rounded to a short claw; wings narrow, rounded at apex; keel upturned, sometimes shortly beaked. *Stamens* 10, diadelphous; filaments pale greenish; anthers orange. *Style* 1, yellowish-green. *Ovary* glabrous. *Flowers* 5–9. $2n = 16$.

Introduced. Waste places, rubbish tips and old railway lines. Possibly the commonest *Colutea* in Great Britain and Ireland as defined here, as many yellow-flowered plants have the smaller leaflets and beaked keels of *C.* × *media*. All variants seem to be fully fertile and could be of intermediate rather than hybrid origin.

3. C. orientalis Mill. — Caucasian Bladder Senna

Deciduous *shrub* up to 3 m. *Stems* pale brown, much-branched, rough; young shoots pale green, glabrous. *Buds* small, globose-ovoid, with 2–4 scales. *Leaves* opposite, oblong in outline, imparipinnate; leaflets 7 or 9, 10–20 × 7–17 mm, bluish-green on upper surface, paler beneath, broadly obovate or subrotund, rounded-mucronulate or retuse at apex, entire, narrowed at base, glabrous or nearly so, with petiolules 1–2 mm; petioles very short; rhachis glabrous; stipules small. *Inflorescence* an axillary raceme of 2–5 flowers, branches and pedicels with a few appressed hairs to nearly glabrous. *Calyx* 4.0–4.5 mm, campanulate, with rather few appressed hairs, slightly 2-lipped, with 5 small lobes. *Corolla* 11–13 mm, orange-red; standard very broadly elliptical, retuse at apex, truncate at base to a very short claw; wings oblong or falcate, rounded at apex and slightly upcurved, longer than keel and with a distinct spur on the lower edge; keel oblong, distinctly upturned and beaked at apex. *Stamens* 10, diadelphous; filaments pale greenish; anthers orange. *Style* 1, yellowish-green; stigma large, inserted obliquely on the inner edge of the style and surrounded by hairs. *Legume* 30–40(–50) × 15–20 mm, pale green turning brown, oblong, narrowing and curved upwards towards the apex at which it dehisces; seeds up to 3.0×2.5 mm. *Flowers* 6–9. $2n=16$.

Introduced into cultivation in England in 1710, but we have never seen it. Included here so that the hybrid can be better understood. Native of the Caucasian region.

30. Galega L.

Perennial herbs. Leaves alternate, imparipinnate; stipules half-sagittate to ovate-subrotund. *Inflorescence* an axillary raceme. *Calyx* campanulate, divided to half way or more into 5 subequal lobes. *Corolla* white to pale purplish-blue; standard obovate, obtuse at apex; wings oblong; keel obovate, obtuse at apex. *Stamens* 10, monadelphous. *Style* 1. *Fruit* a cylindrical, dehiscent legume with numerous seeds.

Contains 6 species in Europe, west Asia and North Africa.

This genus is probably continuous from Spain and North Africa to western Asia with intermediates rather than hybrids in intermediate geographical areas. *G. patula* is usually regarded as a hybrid between the other two species given here, but seems to reproduce itself and to occupy an intermediate area.

1. Leaflets oblong-ovate or ovate, acute at apex, 10–30 mm wide; stipules ovate to ovate-subrotund; calyx glandular-hairy; legumes becoming reflexed **3. orientalis**
1. Leaflets oblong or ovate-oblong, obtuse-mucronate or subacute at apex, 5–20 mm wide; stipules half-sagittate to ovate-lanceolate, calyx without glandular hairs; legumes patent or erecto-patent 2.
2. Leaflets linear to linear-lanceolate, up to 15 mm wide; stipules half-sagittate **1. officinalis**
2. Leaflets oblong or elliptic-oblong, up to 20 mm wide; stipules ovate-lanceolate **2. patula**

1. G. officinalis L. Goat's Rue

Perennial herb. Stem 40–150 cm, pale green, erect, stout, glabrous or sparsely hairy, branched, leafy. *Leaves* alternate, broadly oblong in outline, imparipinnate; leaflets 9–17, 15–50×10–15 mm, bluish-green, oblong, narrowly elliptical or lanceolate, subacute or obtuse and mucronate at apex, entire, narrowed or rounded at base, glabrous or hairy beneath; stipules half-sagittate. *Inflorescence* an axillary raceme; bracts setaceous. *Calyx* 4.0–5.0 (–5.5) mm, campanulate, gibbous at base on upper side, glabrous or sparsely hairy, divided about half way into 5 subequal lobes, the lobes setaceous. *Corolla* 10–15 mm, white to pale purplish-blue; standard 8–12.5×7–8 mm, obovate, obtuse at apex; wings 7–11×2.5–3.5 mm; keel 8.0–11×2.8–3.5 mm, obovate. *Stamens* 10, monadelphous. *Style* 1. *Legume* 20–50×2–2.5 mm, cylindrical, torulose, dehiscent, patent or erecto-patent; seeds 4.0–4.2 (–4.5) mm, numerous. *Flowers* 6–7. $2n=16$.

Introduced. Much grown for ornament and frequent on tips and in waste and grassy places. Frequent in south and central Great Britain with scattered records elsewhere. Widespread in Europe; possibly western Asia and north-west Africa but these could be different taxa.

2. G. patula Steven Intermediate Goat's Rue
G. hartlundii Clarke

Perennial herb. Stem 40–150 cm, pale green, erect, stout, glabrous to sparsely hairy, branched, leafy. *Leaves* alternate, oblong in outline, imparipinnate; leaflets 9–17, 30–60×5–20 mm, oblong or elliptic-oblong, rounded-obtuse and mucronate at apex, entire, rounded at base, glabrous or nearly so; stipules ovate-lanceolate. *Inflorescence* an axillary raceme. *Calyx* 4–5 mm, with 5 subequal teeth about equalling the tube, glabrous or sparsely glandular-hairy. *Corolla* 10–15 mm, bluish-violet; standard about 10 mm, obovate; wings about 10 mm, oblong; keel about 10 mm, obovate. *Stamens* 10, monadelphous. *Style* 1. *Legume* 20–50×20–30 mm, cylindrical, torulose, patent; seeds numerous. *Flower* 6–7. $2n=16$.

Introduced. Much grown for ornament and probably frequent on tips and in waste and grassy places. Native of south-east Europe.

3. G. orientalis Lam. Oriental Goat's Rue

Perennial herb. Stem 40–150 cm, pale green, erect, stout, hairy, branched, leafy. *Leaves* alternate, oblong in outline, imparipinnate; leaflets 9–17, 30–60×10–30 mm, bluish-green, oblong-ovate or ovate, acute at apex, entire, rounded at base, glabrous or hairy beneath; stipules ovate to ovate-subrotund, mucronate. *Inflorescence* an axillary raceme; bracts lanceolate. *Calyx* 4–5 mm, with 5 subequal teeth shorter than the tube, glandular-hairy. *Corolla* 10–15 mm, bluish-violet; standard about 10 mm, obovate; wings about 10 mm, oblong; keel about 10 mm obovate. *Stamens* 10, monadelphous. *Style* 1. *Legume* 20–50×2–3 mm, cylindrical, torulose, deflexed; seeds numerous. *Flowers* 6–7. $2n=16$.

Introduced to gardens. Not recorded as an escape, but included to show how little it differs from *G. patula*, and because it could well be an escape from gardens. Native of Caucasus.

Tribe **11. Hedysareae** DC.

Perennial herbs. Leaves imparipinnate, with entire leaflets. *Inflorescence* an axillary raceme. *Stamens* 10, diadelphous. *Fruit* an indehiscent legume; 1-seeded.

31. Caragana Fabr.

Deciduous *shrubs. Leaves* alternate, paripinnate; leaflets entire, without stipils; stipules spine-tipped. *Inflorescence* of fascicles of 2–5 flowers. *Calyx* campanulate, gibbous at base. *Corolla* pale yellow; standard erect, with recurved sides and a long claw; wings with long claws; keel straight. *Stamens* 10, diadelphous. *Style* 1; stigma terminal. *Fruit* a legume, 2-valved; seeds without a strophiole.

Contains about 80 species in eastern Europe to central Asia and China. The generic name is derived from *caragan*, the Mongolian for *C. arborescens*.

Komarov, V. L. (1908). Monografiya roda *Caragana*. *Acta Horti. Petrop.* **29**: 178–362.

1. C. arborescens Lam. Siberian Pea-tree
Robinia caragana L.

Deciduous *shrub* up to 6 m, sometimes with an almost fastigiate habit. *Branches* long with sparse side shoots; young shoots winged and hairy. *Buds* ovoid. *Leaves* alternate, often clustered, 3–7 cm, broadly oblong in outline, paripinnate; leaflets 6–12, 0.8–3.5 × 0.5–1.3 cm, pale green on both surfaces, elliptic-oblong or obovate, rounded and mucronate at apex, entire, hairy when young, becoming glabrous; stipules linear, spine-tipped. *Inflorescence* a fascicle of 2–5 flowers; pedicels 15–60 mm, articulate, hairy. *Calyx* about 10 mm, campanulate, slightly gibbous at base, 5-lobed, the lobes 1.0–1.5 mm, broadly triangular, with hairy margins. *Corolla* 15–22 mm, pale yellow; standard erect with recurved sides and a long claw; wings with long claws; keel straight. *Stamens* 10, diadelphous. *Style* 1, straight or slightly curved; stigma terminal. *Legume* 30–60 × 3–5 mm, linear, acute at apex; seeds usually numerous. *Flowers* 5–7. $2n = 16$.

Introduced. Persistent garden escape at side of old chalkpit at Kemsing in Kent, and seedlings on a roadside at Wrotham Heath in Kent. Native of Mongolia and Siberia.

32. Hedysarum L.

Annual or *perennial herbs*. *Leaves* imparipinnate; stipules free. *Inflorescence* an axillary raceme. *Calyx* campanulate, 5-lobed. *Corolla* pink, purple or violet, rarely white. *Stamens* 10, diadelphous. *Style* 1. *Fruit* a lomentaceous legume, more or less compressed, with 2–4 segments each with 1 seed.

About 100 species in the north temperate region.

Fedtschenko, B. A. (1902). The genus *Hedysarum*. *Acta Horti Petrop.* **19**: 185–375.

1. Corolla 8–11	**2. spinosissimum**
1. Corolla 12–20 mm	2.
2. Perennial; leaflets 15–35 × 12–18 mm	**1. coronarium**
2. Annual; leaflets 5–12 × 2–5 mm	**3. glomeratum**

1. H. coronarium L. Italian Sainfoin

Perennial herb with a stout stock. *Stem* 30–100 cm, pale green, often tinged reddish, with few to numerous, ascending-appressed simple eglandular hairs, branched, leafy. *Leaves* alternate, imparipinnate; leaflets 7–11, 15–35 × 12–18 mm, medium green on upper surface, paler beneath, elliptical to obovate-subrotund, rounded with a mucron at apex, entire, narrowed or rounded below, glabrous or nearly so on upper surface, with numerous, pale simple eglandular hairs beneath; rhachis pale green, hairy; petiole fairly short, pale green, hairy; stipules ovate-cuspidate, free, hairy. *Inflorescence* a dense, axillary raceme of 10–35 flowers. *Calyx* 5–6 mm; lobes about as long as the tube, densely long-hairy. *Corolla* 12–15 mm, bright reddish-purple; standard 15–18(–20) mm, narrowly elliptical, emarginate at apex, narrowed to a fairly broad base; wings oblong, with a lobe and short claw at base; keel 12–14(–15) mm, obtuse. *Stamens* 10, diadelphous. *Style* 1. *Legume* lomentaceous with 2–4 segments, the segments 4–6 × 4.0–5.5 mm, elliptical or subrotund; seeds 2.5–3.0 mm. *Flowers* 5–7. $2n = 16$.

Introduced. Wool casual and garden weed. Native of the central and west part of the Mediterranean region. Cultivated for fodder in south Europe.

2. H. spinosissimum L. Spiny-fruited Sainfoin
H. pallens (Moris) Hal.

Annual herb with a tap-root. *Stems* 15–35 cm, slender, pale green, appressed-hairy, branched, leafy. *Leaves* alternate, imparipinnate; leaflets (5–)9–17, 5–12 × 2–5 mm, medium green on upper surface, paler beneath, elliptical or oblong, obtuse or mucronate at apex, entire, rounded at the shortly petiolulate base, with simple eglandular hairs to nearly glabrous; rhachis pale green, appressed-hairy; petiole up to 6 cm, pale green, appressed-hairy. *Inflorescence* an axillary raceme with 2–10 flowers. *Calyx* 4–6 mm, the lobes as long as or longer than the tube, sparsely hairy. *Corolla* 8–11 mm, white to pale pinkish-purple; standard 7.5–11.5 mm, elliptical, retuse at apex, narrowed to a broad base; wings oblong, lobed at base, obtuse at apex; keel 12–16 mm, oblong-obovate, rounded at apex. *Stamens* 10, diadelphous. *Style* 1. *Legume* lomentaceous, with 1–4 segments, the segments 5.5–11.0 × 5.0–8.5 mm, densely spiny, elliptical or subrotund; seeds 2.5–3.0 mm. *Flowers* 5–7. $2n = 16$.

Introduced. Wool casual. Native of the Mediterranean region.

3. H. glomeratum F. Dietr. Small-leaved Sainfoin
H. capitatum Desf., non. Burm. fil.; *H. spinosissimum* subsp. *capitatum* Asch. & Graebn.

Annual herb with a tap-root and fibrous side-roots. *Stem* 10–40(–50) cm, pale green, often tinted red, appressed-hairy, branched, leafy. *Leaves* alternate, imparipinnate; leaflets (11–)13–19, (4–)5–12 × 2–5 mm, medium green on upper surface, paler beneath, oblong or obovate-oblong, retuse at apex, entire, rounded below and shortly petiolulate, appressed-hairy; rhachis pale green, appressed hairy; petiole up to 4 cm, pale green, appressed-hairy. *Inflorescence* an axillary raceme with (4–)6–11 flowers. *Calyx* 5–6 mm, the lobes about as long as the tube, sparsely hairy. *Corolla* 14–20 mm, 2.5–5.0 times as long as the calyx, pinkish purple; standard 14–18 × 6–7 mm, elliptical, retuse at apex, narrowed to a broad base; wings oblong, lobed at base, obtuse at apex; keel 12–16 mm, obovate, rounded at apex. *Stamens* 10, diadelphous. *Style* 1. *Legume* lomentaceous with (1–)2–4 segments, the segments 5.5–10.0 × 5.5–8.0 mm, elliptical, densely spiny; seeds 3.5–4.0 mm. *Flowers* 5–7. $2n = 16$.

Introduced. Wool casual. Native of the Mediterranean region.

33. Onobrychis Mill.

Perennial herbs. *Stems* prostrate to erect. *Leaves* imparipinnate; leaflets numerous, entire. *Inflorescence* an ellipsoid or ovoid axillary raceme with numerous flowers on a long peduncle. *Calyx* 5-lobed. *Corolla* deep pink; standard with limb obovate with a very short claw; wings half as long as calyx, with narrow claw; keel tinged mauve. *Stamens* 10, diadelphous. *Legume* irregularly subrotund, compressed, strongly reticulately ridged, toothed.

About 130 species in temperate Europe and Asia.

Fearn, G. M. (1987). Exploited plants. Sainfoin. *Biologist* **34**: 93–97.

Grose, D. (1957). *The flora of Wiltshire*. Devizes.

Jordan, A. (1852). *Pugillus Plantarum Novarum praesertion Gallicarum*. Paris.

1. Stems slender, prostrate, leaflets 5–15 × 2–5 mm **1(a). viciifolia** subsp. **collina**
1. Stems decumbent to more or less erect, slender to robust; leaflets 10–35 × 4–8 mm 2.
2. Stems decumbent or ascending, slender; leaflets 10–30 × 2–7 mm, corolla 10–12 mm **1(b). viciifolia** subsp. **decumbens**
2. Stems more or less erect, robust; leaflets 15–35 × 4–8 mm; corolla 12–16 mm **1(c). viciifolia** subsp. **viciifolia**

1. O. viciifolia Scop. Sainfoin

Hedysarum onobrychis L.; *O. onobrychis* (L.) Karst. nom. illegit.; *O. sativa* Lam. nom. illegit.

Perennial herb with a tap-root and fibrous side-roots. *Stems* up to 80 cm, pale green, prostrate to more or less erect, slender to robust, hairy, much branched, leafy. *Leaves* alternate, 8–15 × 3–5 cm, oblong in outline, imparipinnate; leaflets 13–29, 5–35 × 2–8 mm, medium green on upper surface, paler beneath, ovate to oblong, mucronate at apex, entire, narrowed at base, shortly petiolulate, shortly hairy; petiole short, pale green, hairy; stipules ovate, acuminate at apex, scarious. *Inflorescence* an ellipsoid or ovoid raceme of up to 60 dense flowers; peduncle long and stout, exceeding the leaves. *Calyx* 3.5–4.0 mm, tubular, divided up to three-quarters of the way to the base into 5 unequal lobes, the lobes subulate, acute at apex, often woolly. *Corolla* 10–16 mm; standard deep rose-pink, the limb obovate, emarginate, with a very short, broad claw; wings only half as long as calyx, rose-pink, ovate, with a narrow claw; keel rose-pink tinged with mauve, oblanceolate, rounded at apex. *Stamens* 10, diadelphous; filaments pale; anthers yellow. *Legume* 5–8 mm, irregularly subrotund, compressed, strongly reticulately ridged, toothed; seeds 3.5–4.0 mm, broadly oblong-ellipsoid. *Flowers* 6–8. Visited by bees, flies and butterflies. $2n = 28$.

(**a**) Subsp. **collina** (Jord.) P. D. Sell

O. collina Jord.; *O. viciifolia* var. *collina* (Jord.) St Lag.

Stem 15–60 cm, slender, prostrate. *Leaflets* 5–15 × 2–5 mm. *Corolla* 8–12 mm.

(**b**) Subsp. **decumbens** (Jord.) P. D. Sell

O. decumbens Jord.; *O. viciifolia* var. *decumbens* (Jord.) Rouy

Stems up to 60 cm, decumbent to ascending, slender. *Leaflets* 10–30 × 2–7 mm. *Corolla* 10–12 mm.

(**c**) Subsp. **viciifolia**

O. sativa Lam.; *O. sativa* var. *culta* Gren. & Godr.; *O. viciifolia* subsp. *sativa* (Lam.) Thell.

Stems up to 80 cm, more or less erect, robust. *Leaflets* 12–35 × 5–8 mm. *Corolla* 12–16 mm.

Subsp. *collina* is the native plant on remnants of chalk grassland, old tracks by former downs and by chalkpits in south and east England. It occurs in central France and probably elsewhere in Europe. Its relationship to *O. arenaria* (Kit.) DC. needs to be considered. Grose (1957) says that subsp. *collina* retains its character in cultivation. Subsp. *decumbens* is in wild flower seed and seems to have been particularly planted in the sown grass margins of fields and roadsides from which it is likely to be spread on grass cutters. Subsp. *decumbens* is beautifully illustrated in R. Mabey (1996), *Flora Britannica,* p. 220. Subsp. *viciifolia* was formerly commonly grown as a hay crop, at least in East Anglia, but is now rarely seen. It occurs on old tracks, roadsides and waste places where it is a remnant of such crops and is locally frequent in Great Britain north to Yorkshire, and as a scattered casual elsewhere in Great Britain. It is widespread in Continental Europe. The species is a Eurosiberian Temperate element.

Tribe **12. Cicereae** Alef.

Annual herbs. Leaves imparipinnate, with sharply serrate leaflets, without tendrils. *Inflorescence* of solitary, axillary flowers. *Stamens* 10, diadelphous. *Fruit* a legume dehiscing longitudinally, 1- to 2-seeded.

34. Cicer L.

Annual herbs. Stem glandular-hairy. *Leaves* alternate, imparipinnate and glandular-hairy, the leaflets sharply serrate, tendrils absent. *Inflorescence* a solitary axillary flower. Calyx with 5 subequal lobes. *Corolla* white, pink or mauve; standard obovate; wings oblong; keel upcurved. *Stamens* 10, diadelphous. *Style* 1, upcurved, glabrous. *Fruit* an oblong-ellipsoid legume, dehiscing longitudinally; seeds 1–2.

Contains about 20 species in south-east Europe and western and central Asia.

Clarke, G. C. S. & Kupicha, F. K. (1976). The relationship of the genus *Cicer* L. (Leguminosae):

The evidence from pollen morphology. *Bot. Jour. Linn. Soc.* **72**: 35–44.

Phillips R. & Rix, M. (1993). *Vegetables*. London.

Vaughan, J. G. & Geissler, C. (1997). *The new Oxford book of food plants*. Oxford.

1. C. arietinum L. Chick Pea

Annual herb with fibrous roots. *Stem* 15–40(–60) cm, pale green, erect or sprawling, densely glandular-hairy, branched, leafy. *Leaves* alternate, 4–8 × 1–4 cm, oblong in outline, imparipinnate; leaflets 13–17, 5–20 × 2.5–10.0 mm, medium green on upper surface, paler beneath, oblong-obovate, rounded to acute at apex, sharply serrate, cuneate at base, glandular-hairy; petiole usually less than 10 mm, channelled above; stipules 2–12 mm, ovate, acute at apex, coarsely serrate-dentate. *Inflorescence* a solitary axillary flower; peduncle 8–15 mm; bracts 2–3, unequal, linear, glandular-hairy; pedicels 10–15 mm, deflexed after anthesis. *Calyx* 6–10 mm, campanulate with a dorsal hump, divided half-way to base into 5 lobes, the lobes subequal, linear-subulate. *Corolla* 10–12 mm, white, pink or mauve; standard obovate, rounded at apex, tapering to a wide claw; wings with limb oblong, rounded at apex, with a prominent curved lobe at base and a short claw; keel with limb upcurved to a subacute apex, with a distinct short lobe at base and a slender claw. *Stamens* 10, diadelphous; filaments dilated towards the apex; anthers oblong. *Style* 1, sharply upcurved at base, glabrous; stigma small, terminal. *Legume* 17–30 × 10–15 mm, oblong-ellipsoid, more or less acute at apex, the valves pale brown, shortly glandular-hairy; seeds 7.5–8.0 mm in diameter, pale brown, subglobose, with a prominent radical and a shallow median furrow, minutely and sparsely rugulose. *Flowers* 7–8. $2n = 14$, 16.

Introduced. Tips and waste places from seeds imported as food. Scattered records in Great Britain, mainly England. Grown since antiquity for its edible seeds. Probably originated in the west of Asia, but widely cultivated in the Mediterranean region eastwards to India where it is considered to be the most important pulse crop. The small-seeded varieties are usually sold as split peas and made into dhal or flour for poppadoms, and large-seeded varieties are often roasted or eaten whole.

Tribe **13. Trifolieae** (Bronn.) Endl.
Subtribe *Trifoliinae* Bronn.; Subtribe *Trigonelleae* O. E. Schulz; Tribe *Ononideae* Hutch.

Annual or *perennial herbs* or rarely *shrubs*. *Leaves* trifoliolate, main lateral veins of leaflets running whole way to margin, which is often toothed. *Inflorescence* of solitary flowers or a raceme, often greatly condensed into a head. *Stamens* 10, diadelphous or monadelphous. *Fruits* an indehiscent legume or longitudinally dehiscent; with many seeds.

35. Trifolium L.

Annual, *biennial* or *perennial herbs*, rarely somewhat woody. *Leaves* alternate or opposite, trifoliolate, rarely 5-or 7-foliolate; stipules mostly entire and adnate to the petiole. *Inflorescence* axillary or pseudoterminal, generally spicate or capitate, flowers rarely solitary; bracts sometimes small or wanting or sometimes well-developed and conspicuous. *Calyx* persistent; teeth 5, subequal or the lower longer than the upper. *Corolla* white, pink, purple or yellow, marcescent, persistent or deciduous, the claws more or united. *Stamens* 10, diadelphous, the filaments adnate to the petals, often dilated at the apex; anthers uniform. *Style* 1, filiform, upcurved or sometimes sharply inflexed; stigma terminal, usually capitate, sometimes oblique. ovary sessile or stipitate. *Legume* generally small, often concealed by the persistent calyx and petals; walls mostly thin and membranous; seeds estropholiolate, usually smooth.

Contains about 300 species in Europe, Africa, Asia and North America, with the greatest concentration of species in the Mediterranean region.

Afzelius, A. (1791). The botanical history of *Trifolium alpestre, medium* and *pratense*. *Trans. Linn. Soc. London* **1**: 202–248.
Blackstock, T. H. & Roberts, R. H. (1998). *Trifolium occidentale* D. E. Coombe (Fabaceae) in Anglesey (v.c. 52). *Watsonia* **22**: 182–184.
Burdon, J. J. (1988). *Biological flora of the British Isles*. No. 154. *Trifolium repens* L. *Jour. Ecol.* **71**: 307–330.
Coombe, D. E. (1961). *Trifolium occidentale*, a new species related to *T. repens* L. *Watsonia* **5**: 68–87.
Davis, A. (1992). Exploited plants. White Clover. *Biologist* **39**: 129–133.
Druce, G. C. (1897). *Trifolium hybridum* var. *elegans* in *The flora of Berkshire*. p. 141.
Grime, J. P. et al. (1988). *Comparative plant ecology*. London. [*T. dubium, medium, pratense* and *repens*.]
Hultén, E. & Fries, M. (1986). *Atlas of north European vascular plants north of the Tropic of Cancer*. 3 vols. Königstein.
Martin, M. H. & Frost, L. C. (1980). Autecological studies of *Trifolium molinerii* at the Lizard Peninsula, Cornwall. *New Phytol.* **86**: 239–344.
Preston, C. D. (1980). *Trifolium occidentale* D. E. Coombe, new to Ireland. *Irish Nat. Jour.* **20**: 37–40.
Reynolds, S. C. P. (2002). *A catalogue of alien plants in Ireland*. Glasnevin.
Stewart, A., Pearman, D. A. & Preston, C. D. (1994). *Scarce plants in Britain*. Peterborough. [*T. glomeratum, occidentale, ochroleucon, ornithopodioides, squamosum* and *suffocatum*.]
Turkington, R. & Burdon, J. J. (1983). The biology of Canadian weeds. 54. *Trifolium repens* L. *Canad. Jour. Pl. Sci.* **63**: 243–266.
Wigginton, M. J. (Edit.) (1999). *British red data books*. Vol. 1. *Vascular plants*. Peterborough. [*T. bocconei, incarnatum* subsp. *molineri* and *strictum*.]
Zohary, M. & Haller, D. (1984). *The genus* Trifolium. Jerusalem.

1. Leaves often pinnately 3-foliolate; calyx with 5(–6) veins; corolla always persistent and scarious in fruit 2.
1. Leaves digitately 3-foliolate; calyx with more than 5(–6) veins, usually 10, 20 or more; petals deciduous or marcescent, sometimes scarious 14.
2. Flowers more or less umbellate, subtended by membranous bracts; calyx lobes subequal or the 2 upper longer than the rest; corolla before anthesis white, pink or purple; legume 2- to 4-seeded 3.
2. Flowers more or less spicate, the bracts represented by a few, short, red, evanescent glandular hairs; calyx lobes unequal, the 2 upper shorter than the rest; corolla before anthesis yellow, orange, lilac or violet; legume 1(–2)-seeded 8.
3. Calyx lobes unequal, the 2 upper longer than the rest, narrowly lanceolate, separated by narrow, acute sinuses 4.
3. Calyx lobes subequal, subulate, separated by broad, obtuse sinuses 7.
4. Corolla pink or purple 5.
4. Corolla white or cream 6.
5. Corolla pink **3(i). repens** var. **roseum**
5. Corolla purple **3(ii). repens** var. **townsendii**
6. Leaflets up to 20 mm; inflorescence up to 25 mm **3(iii). repens** var. **repens**
6. Leaflets up to 40 mm; inflorescence 30–35 mm **3(iv). repens** var. **grandiflorum**

7. Stems erect, hollow, sparingly branched; inflorescence more than 20 mm in diameter **5(a). hybridum** subsp. **hybridum**
7. Stems solid, usually decumbent, much-branched; inflorescence less than 20 mm in diameter **5(b). hybridum** subsp. **elegans**
8. Corolla violet or reddish-violet before anthesis **16. speciosum**
8. Corolla yellow, rarely orange or lilac before anthesis 9.
9. All leaflets of the upper leaves subsessile, their petiolules subequal 10.
9. Terminal leaflet of the upper leaves distinctly petiolulate, its petiolule longer than that of the lateral leaflets 12.
10. Fruiting pedicels one to one and a half times as long as the upper limb of the calyx-tube; corolla 2–3(–4) mm **21. micranthum**
10. Fruiting pedicels distinctly shorter than the upper limb of the calyx-tube; corolla 5–8 mm 11.
11. Stipules of the upper leaves semicordate-ovate, often auriculate **17. patens**
11. Stipules of the upper leaves lanceolate-ovate, not dilated at the base **18. aureum**
12. Corolla 3.0–3.5 mm, scarcely sulcate **20. dubium**
12. Corolla 4.0–10.0 mm, if less than markedly sulcate 13.
13. Corolla (5–)6–10 mm; legume scarcely exceeding the style **17. patens**
13. Corolla (3–)4–5(–6) mm; legume 3–6 times as long as the style 14.
14. Stems up to 50 cm; inflorescence up to 17 mm in diameter **19(i). campestre** var. **majus**
14. Stems up to 30 cm; inflorescence up to 10 mm in diameter 15.
15. Stem spreading or prostrate **19(ii). campestre** var. **minus**
15. Stem ascending or erect **19(iii). campestre** var. **campestre**
16. Flowers subtended by small, sometimes connate bracts; throat of calyx not closed by a ring of hairs or by an annular or bilabiate callosity; legume usually 2- to 9-seeded, included or exserted 17.
16. Flowers ebracteate but heads sometimes involucrate; throat of the calyx usually more or less closed by a ring of hairs or an annular or bilabiate callosity at maturity; legume 1(–2)-seeded, almost always included in the calyx-tube 40.
17. Calyx-tube not inflated in fruit 18.
17. Calyx-tube slightly to conspicuously inflated or gibbous in fruit 36.
18. Heads 1- to 6-flowered **1. ornithopodioides**
18. Heads 7- to many-flowered 19.
19. All heads sessile or nearly so 20.
19. At least some heads with peduncles 5 mm or more 21.
20. Internodes 10–80 mm; heads more or less remote; corolla longer than the calyx **10. glomeratum**
20. Internodes usually less than 5 mm; heads congested or confluent; corolla shorter than the calyx **11. suffocatum**
21. Stipules denticulate or fimbriate **2. strictum**
21. Stipules not denticulate or fimbriate 22.
22. Stems creeping and rooting at the nodes 23.
22. Stems not creeping or rooting at the nodes 27.
23. Leaflets with opaque lateral veins and usually unmarked; upper calyx lobes ovate-lanceolate or triangular **4. occidentale**
23. Leaflets with translucent lateral veins and usually light or dark markings; upper calyx lobes narrowly lanceolate 24.
24. Corolla pink or purple 25.
24. Corolla white or cream 26.
25. Corolla pink **3(i). repens** var. **roseum**
25. Corolla purple **3(ii). repens** var. **townsendii**
26. Leaflets up to 20 mm; inflorescence up to 25 mm **3(iii). repens** var. **repens**
26. Leaflets up to 40 mm; inflorescence 30–35 mm **3(iv). repens** var. **grandiflorum**
27. Fruiting pedicels shorter than the calyx-tube **8. retusum**
27. Fruiting pedicels as long as or longer than the calyx-tube 28.
28. Plant perennial 29.
28. Plant annual 34.
29. Calyx lobes unequal, the 2 upper longer than the rest, narrowly lanceolate, separated by narrow acute sinuses 30.
29. Calyx lobes subequal, subulate, separated by broad, obtuse sinuses 33.
30. Corolla pink or purple 31.
30. Corolla cream or white 32.
31. Corolla pink **3(i). repens** var. **roseum**
31. Corolla purple **3(ii). repens** var. **townsendii**
32. Leaflets up to 20 mm; inflorescence up to 25 mm **3(iii). repens** var. **repens**
32. Leaflets up to 40 mm; inflorescence 30–35 mm **3(iv). repens** var. **grandiflorum**
33. Stems erect, hollow, sparingly branched; inflorescence more than 20 mm in diameter **5(a). hybridum** subsp. **hybridum**
33. Stems solid, usually decumbent, much-branched; inflorescence less than 20 mm in diameter **5(b). hybridum** subsp. **elegans**
34. All calyx lobes 2–4 times as long as the tube **7. michelianum**
34. Upper 2 calyx lobes equalling or only slightly exceeding the tube 35.
35. Heads 10–20 mm wide; corolla 6–9 mm **6. nigrescens**
35. Heads 8–12 mm wide; corolla 4–5 mm **9. cernuum**
36. Perennial herb 37.
36. Annual herb 38.
37. Inflorescence 10–22 mm in fruit; globose; calyx 4.0–4.5 mm in flower, with teeth longer than tube, 8–10 mm in fruit, concealing all or most of the persistent corolla **13(a). fragiferum** subsp. **fragiferum**
37. Inflorescence (10–)15–25(–35) mm in fruit, subglobose to irregularly cylindrical; calyx 3.5–4.0 mm in flower, the lobes not longer than the tube, 4–6 mm in fruit, the corolla exserted by 2.0–2.5 mm **13(b). fragiferum** subsp. **bonannii**
38. Fruiting calyx glabrous, inflated more or less equally on all sides; bracts prominent, glumaceous, striate; heads pseudoterminal **12. spumosum**

38. Fruiting calyx hairy, tomentose or lanate, adaxially gibbous; bracts inconspicuous and more or less concealed; heads lateral 39.
39. Fruiting heads more or less pedunculate; calyx pyriform, hairy to tomentose, becoming glabrous, its 2 upper lobes evident, divergent **14. resupinatum**
39. Fruiting heads subsessile; more or less globose, lanate, its 2 upper lobes more or less concealed **15. tomentosum**
40. Fertile flowers 2–12; inner flowers consisting only of sterile calyces developing either at or after anthesis 41.
40. Fertile flowers usually numerous; sterile flowers absent 42.
41. Stems up to 15 cm; leaflets 2–10 mm wide; calyx covering most of ripe legume **45(i). subterraneum** var. **subterraneum**
41. Stems 15–45 cm; leaflets 10–22 mm wide; calyx covering only half of ripe legume **45(ii). subterraneum** var. **oxaloides**
42. All or at least some calyces evidently 20-veined, or the 20 veins completely obscured by dense hairs 43.
42. Calyx 10-veined, or some calyces with up to 14 veins 46.
43. Plant perennial; corolla 12–20 mm **35. medium**
43. Plant annual; corolla 4–10 mm 44.
44. Fruiting heads shortly pedunculate, not involucrate; calyx-tube glabrous or becoming so **32. lappaceum**
44. Heads sessile with an involucre formed by the upper stipule; calyx-tube hairy 45.
45. Free part of the stipules, except of the uppermost leaves, long linear-lanceolate, straight; corolla exceeding the calyx **33. hirtum**
45. Free part of the stipules, except of the uppermost leaves, shortly ovate-lanceolate, often incurved; corolla not exceeding calyx **34. cherleri**
46. Plant perennial 47.
46. Plant annual 54.
47. Lowest calyx lobe about 2–3 times as long as the other four, linear; corolla yellowish-white 48.
47. Calyx teeth subequal or the lowest not more than 1.5 times as long as the other four, setaceous, filiform or linear; corolla usually red or purple, sometimes white 49.
48. Peduncles not more than 25 mm; corolla 15–20 mm **38. ochroleucon**
48. Peduncles 40–80 mm; corolla 40–80 mm **39. pannonicum**
49. Free part of stipules of cauline leaves linear to lanceolate and green more or less to the apex **35. medium** subsp. **medium**
49. Free part of stipules of cauline leaves triangular-ovate and abruptly narrowed to a brown bristle-like point 50.
50. Flowers on long pedicels (monstrosity) **30(i). pratense** var. **parviflorum**
50. Flowers not on long pedicels 51.
51. Inflorescence 15–25 mm in diameter 52.
51. Inflorescence 25–40 mm in diameter 53.
52. Upper part of stem and calyx with dense long appressed hairs **30(ii). pratense** var. **villosum**
52. Upper part of stem and calyx hairy but not densely so **30(iii). pratense** var. **pratense**
53. Stem sparsely appressed-hairy to glabrous; calyx with long hairs; corolla pale to deep red or pink, sometimes white **30(iv). pratense** var. **sativum**
53. Stem with numerous stiff, spreading hairs; calyx with dense, spreading hairs; corolla usually deep red, rarely white **30(v). pratense** var. **americanum**
54. All leaves alternate 55.
54. At least the two uppermost leaves opposite 59.
55. Heads of flowers sessile, axillary or terminal, involucrate 56.
55. Heads of flowers pedunculate, axillary or terminal 61.
56. Lateral veins of the leaflets recurved, often more or less thickened towards the margins **27. scabrum**
56. Lateral veins of the leaflets more or less straight 57.
57. Calyx lobes subequal, all longer than the tube, divergent in fruit **25. gemellum**
57. Calyx lobes unequal, only the lowest equalling or slightly exceeding the tube, connivent or somewhat divergent in fruit 58.
58. Fruiting calyx readily abscissing, with more or less inflated tube and erecto-patent lobes **22. striatum**
58. Fruiting calyx not readily abscissing, tube not inflated, lobes straight or connivent **24. bocconei**
59. Calyx lobes subulate, setaceous or filiform, more or less straight and erect in fruit **31. diffusum**
59. Calyx lobes triangular-lanceolate(-acuminate) or subulate from an expanded triangular base, spreading or recurved in fruit 60.
60. Throat of the calyx closed with a ring of hairs **25. gemellum**
60. Throat of the calyx closed by a bilabiate callosity, leaving only a narrow vertical slit **44. squamosum**
61. Upper leaves alternate 62.
61. At least the two uppermost leaves opposite 68.
62. Leaflets of upper leaves lanceolate, linear or linear-oblong 63.
62. Leaflets of upper leaves obovate-cuneate or obcordate 65.
63. Heads 10–25 mm, usually numerous; calyx-throat not closed by a bilabiate callosity **23. arvense**
63. Heads 20–110 mm, one or few, usually terminal; calyx-throat at maturity narrowed to a vertical slit by a bilabiate callosity 64.
64. Corolla 10–12 mm, not or scarcely exceeding the calyx **36. angustifolium**
64. Corolla 16–25 mm, much exceeding the calyx **37. purpureum**
65. Heads capitate, more or less globose in fruit; stipules denticulate **28. stellatum**
65. Heads spicate, oblong, cylindrical or conical in fruit; stipules entire or obscurely dentate 66.
66. Stipules lanceolate, entire; calyx-tube obconical or campanulate; corolla much shorter than the calyx **26. ligusticum**
66. Stipules ovate, at least at apex entire or obscurely dentate or angled; calyx-tube ovoid or globose; corolla equalling or exceeding the calyx 67.

67. Stem with sparse, spreading hairs; corolla blood red, rarely pure white, equalling or slightly exceeding the calyx **29(a). incarnatum** subsp. **incarnatum**
67. Stem with dense appressed hairs; corolla usually yellowish-white, rarely pink, much exceeding the calyx **29(a). incarnatum** subsp. **molineri**
68. Legume exserted slightly from the mouth of the calyx-tube **40. alexandrinum**
68. Legume not exserted, but concealed by the closed bilabiate callosity at the mouth of the calyx-tube 69.
69. Calyx lobes subequal **43. leucanthum**
69. Calyx lobes unequal 70.
70. Calyx lobes 3-veined to the middle or above **44. squamosum**
70. Calyx lobes 1-veined or 3-veined only at the base 71.
71. Lowest calyx lobe scarcely longer than tube; corolla pale yellow **41. constantinopolitanum**
71. Lowest calyx lobe twice as long as the tube; corolla pink or cream **42. echinatum**

Subgenus **1. Falcatula** (Brot.) Coombe
Falcatula Brot.

Flowers bracteate. *Calyx-throat* open, without a ring of hairs or a callosity. *Legume* oblong, slightly curved, exceeding the calyx, dehiscent. *Seeds* 5–9.

1. T. ornithopodioides L. Bird's-foot Clover
Trigonella purpurascens Lam. nom. illegit.; *Trigonella ornithopodioides* (L.) DC.; *Falcatula ornithopodioides* (L.) Brot. ex Bab.

Annual or short-lived *perennial herb* with a slender tap-root and fibrous side-roots. *Stems* 5–10(–20) cm, pale green, procumbent, glabrous. *Leaves* alternate, digitated, 3-foliolate; leaflets 0.4–1.0(–1.4) × 0.3–0.6 cm, medium green on upper surface, paler beneath, obovate or triangular-obovate, deeply emarginate at apex, serrate, cuneate at base, glabrous, very shortly petiolulate; petioles 20–40 (–50) mm, up to 100 mm in winter in shallow water when the leaves are floating and pale green, glabrous; stipules 7–10 mm, lanceolate, acuminate at apex, entire, sheathing at base. *Inflorescence* a short raceme, (1–) 2- to 4 (–6)-flowered, axillary, bracteate; peduncles up to 8 mm, glabrous. *Calyx* 5–6 mm, divided to more than halfway into 5 subequal, narrowly triangular, acuminate lobes, the tube narrowly conical, glabrous, open at the throat, without hairs or a callosity. *Corolla* 6–8 mm, white often tinged flesh-pink; standard with limb oblong in outline, narrowed in the middle, obtuse at apex and narrowed to a rounded base; wings with an oblong, obtuse limb and a slender claw, longer than the limb; keel with limb lanceolate-oblong and obtuse at apex, abruptly narrowed to a slender claw longer than the limb. *Stamens* 10, diadelphous. *Style* 1. *Legume* 6–8 mm, oblong, slightly curved, exceeding the calyx, slightly hairy, exserted, dehiscent; seeds 5–9, about 1.2 mm. *Flowers* 5–9. $2n=18$.

Native. Acid sandy, semi-open ground, mainly near the sea and frequent in car- parks, and on trackways and picnic sites. Scattered round the coasts of Great Britain and Ireland north to Norfolk and Lancashire, and in the Isle of Man and on the coast of south-east Ireland. South and west Europe from Holland southwards and eastwards to Italy; north-west Africa. Suboceanic Southern-temperate element.

Subgenus **2. Lotoidea** Pers.

Flowers subtended by free or united bracts. *Calyx-throat* open, without a ring of hairs or callosity. *Legume* included in the calyx or exserted. *Seeds* (1–)2–4(–10).

Section 1. **Paramesus** (C. Presl) Endl.
Paramesus C. Presl; *Trifolium* section *Involucrarium* auct.

Annual herbs. Inflorescence of 1–4 closely superimposed whorls, each subtended by a minute involucre of connate bracts. *Calyx* slightly inflated in fruit. *Legume* with a swollen indurated wall, indehiscent, exceeding the calyx tube, (1–)2-seeded.

2. T. strictum L. Upright Clover
Paramesus strictus (L.) C. Presl; *T. laevigatum* Poir.

Annual herb with fibrous roots. *Stem* 3–15(–25) cm, pale green, erect or ascending and rather stiff, glabrous, shortly branched, leafy. *Leaves* alternate, digitately 3-foliolate; leaflets 5–20(–25) mm, medium green on upper surface, paler beneath, linear to oblong-elliptical, acute at apex, denticulate, narrowed at base to a short petiolule, the veins ending in glandular hairs; petioles 5–10 (–40) mm, pale green, slender; stipules conspicuous, whitish, broadly ovate or rhombic, acuminate at apex, glandular-denticulate. *Inflorescence* an axillary or pseudoterminal ovoid head 7–10 mm, with the connate involucral bracts exceeding the pedicels; peduncles 1–2 times as long as the leaves. *Calyx* 3–5 mm, not swollen in fruit, without a ring of hair; tube 1.4–1.9 mm, strongly 10-ribbed and angled, about half as long as the 5 spinescent lobes. *Corolla* 5–6 mm, pinkish-purple; standard oblong with a constriction in the middle, rounded at apex, cuneate at base; wings elliptical, rounded at apex, long clawed at base; keel broadly elliptical, rounded at apex, clawed at base. *Stamens* 10, diadelphous; filaments whitish; anthers cream. *Style* 1, pale; stigma brownish. *Legume* more or less enclosed in calyx, with swollen indurated wall, 3–4 mm, nearly orbicular, dorsally gibbous, beaked, indehiscent; seeds 1–2, about 1.0 mm, ovoid to lentiform. *Flowers* 5–7. $2n=16$.

Native. Known only from about nine sites in the Lizard peninsula in Cornwall, on the Stanner Rocks in Radnorshire and in Jersey. At the Lizard it occurs in species-rich, short-grazed grassland on the upper and middle slopes of cove valleys, with some populations on large rock outcrops in intensively managed grasslands. In Radnorshire it occurs in grasslands overlying basalt and the population is very small. Grazing by livestock appears essential for the survival of most of the Lizard populations. At St Brelades in Jersey it is known from a single locality although it occurred in others in the past. Formerly in Guernsey and recorded as a casual in the Isle of Man in 1994. Western and southern Europe northwards to Great Britain and from Spain and Portugal eastwards to Bulgaria, Greece and Turkey. Submediterranean-Subatlantic element.

Section **2. Lotoidea** Pers.

Annual or *perennial herbs. Inflorescence* an umbel or rarely a spike, the flowers numerous, pedicellate and subtended by lanceolate, membranous bracts. *Calyx* hardly inflated in fruit, the teeth subequal or unequal. *Legume* (1–)2- to 4(–5)-seeded.

3. T. repens L. White Clover

Amoria repens (L.) C. Presl

Perennial herb with wiry roots. *Stems* up to 30 cm, pale green, prostrate, rooting at the nodes, glabrous or nearly so, much branched, leafy. *Leaves* alternate, pinnately 3-foliolate; leaflets 4–40 × 3.5–20.0 mm, medium to dark on upper surface usually with a whitish or dark angled band towards the base, paler beneath, obovate or oblong-elliptical, rounded or sometimes subacute or emarginate at apex, distinctly veined, the veins excurrent into short teeth, glabrous on both surfaces; petioles up to 20 cm, slightly channelled above, glabrous; stipules about 10 × 1.5–2.0 mm, connate below, narrowly oblong, the basal part membranous, distinctly veined, subulate to almost filiform at apex. *Inflorescence* an umbellate, globose head, 14–35 mm in diameter, flowers scented, reflexed after fertilisation; peduncles 4–40 cm, glabrous; bracts 15–20 × about 5 mm, membranous, lanceolate, midrib excurrent as a short mucro; pedicels 1–4 mm. *Calyx* 4–5 mm, narrowly campanulate, hardly inflated in fruit; tube 1.5–2.0 mm, membranous, 10-veined; lobes 5, lanceolate, the 2 adaxial about 2 mm, the others a little shorter, separated by narrow, acute sinuses. *Corolla* 7–10 mm, white or tinged pink, rarely purple; standard with limb oblong, obtuse at apex, often denticulate, narrowed at base to a wide claw up to 3 mm, occasionally slightly auriculate, claw free almost to base; wings with limb oblong, rounded at apex, auriculate at base, with a claw about 3 mm; keel with limb oblong, rounded at apex, abruptly narrowed at base to a claw about 3 mm. *Stamens* 10, diadelphous; filaments about 5 mm, whitish; anthers cream to orange, oblong. *Style* 1, 2.0–2.5 mm, green, glabrous; stigma greenish, capitate. *Legume* 4–5 × about 1.8 mm, protruding from calyx tube, membranous, oblong, usually 2-seeded; seed about 1.4 mm, yellowish-brown, subrotund-reniform, compressed, smooth. *Flowers* 6–9. $2n = 32$.

(i) Var. **carneum** Gray

Trifolium repens var. *roseum* Peterm.

Leaflets up to 10 mm. *Inflorescence* up to 20 mm. *Corolla* slightly to deeply tinged pink.

(ii) Var. **townsendii** Beeby

Leaflets up to 20 mm. *Inflorescence* up to 30 mm. *Corolla* purple.

(iii) Var. **repens**

Leaflets up to 20 mm. *Inflorescence* up to 25 mm. *Corolla* white.

(iv) Var. **grandiflorum** Peterm.

Leaflets up to 40 mm. *Inflorescence* 30–35 mm. *Corolla* white.

Native. Grassy and rough ground. Throughout Great Britain and Ireland. Europe to 71° N, ascending to 2750 m; north and west Asia; North Africa; introduced to Macronesia, South Africa, east Asia and North and South America. Eurosiberian Boreo-temperate element. Although long cultivated in Continental Europe it may not have been grown much with us until the eighteenth century when its chief value was as a pasture plant for grazing. Var. *carneum*, seen most often on heaths and by the sea, is probably untainted by agricultural selection as also may be similar small white-flowered plants. Var. *repens*, the most widespread plant, is a mixture of wild plants and those produced by selection for grazing. Var. *grandiflorum* is the plant formerly grown as a crop for hay from fields of which it still occurs as an escape. It is also much planted on disturbed roadsides and is included in wild flower seed. Var. *townsendii* is found on Tresco, Bryher, St Martin's, Samson, St Mary's, St Helen's and Tean in the Isles of Scilly. It is not known if the occasional records of purple-flowered plants recorded elsewhere are the same thing. Sometimes the pedicels of the flowers elongate and the flowers are replaced by small leaves. It is a monstrosity which has been called var. *proliferum* Gray. *Trifolium repens* has a strong claim to be the Shamrock of the Irish, but others contend that the shamrock is *Oxalis acetosella*. The great value to agriculture of *T. repens* is that it has the capacity to fix nitrogen in habitats where other major nutrients are not limiting. It regenerates almost exclusively by rooted stolons in closed communities to form large diffuse patches, but colonises new sites mainly by seed.

4. T. occidentale D. E. Coombe Western Clover

Perennial herb with wiry roots. *Stems* up to 26 cm, mostly pale green, but sometimes reddish, prostrate, rooting at the nodes, usually glabrous, sometimes sparsely eglandular hairy especially near the nodes, branched, leafy. *Leaves* alternate, digitately 3-foliolate; leaflets 6–10 × 6–10 mm, matt dark bluish-green on upper surface without any markings and papillose under the microscope on upper surface with opaque lateral veins, paler and glossy beneath, thicker than most plants of *T. repens*, subrotund or obovate, rounded or slightly emarginate at apex, with rather few, not prominent teeth or subentire, rounded or cuneate at base, glabrous; petioles and petiolules with sparse but persistent, flexuous, erecto-patent or patent hairs especially at anthesis; stipules deep vinous red, the margins hyaline at maturity. *Inflorescence* an umbellate globose head with 20–40 flowers; peduncles green, sparsely to densely clothed with erecto-patent eglandular hairs; pedicels conspicuously reddish on the upper side, sparsely eglandular hairy, about twice as long as the translucent, pinkish bracteoles at anthesis; flower scentless. *Calyx* green and slightly eglandular hairy at the base, hardly inflated in fruit; tube reddish on the upper side, the main veins red or green; lobes 5, green except at their red apices and at their conspicuously hyaline basal margins, the 2 upper 1.5–2.0 mm, ovate-lanceolate or triangular, the base of each 1.5 mm wide, often with 1 or 2 prominent teeth on the upper margins, the lower margin broadly hyaline at base, the upper margins touching or overlapping at base, so leaving no sinus and more or less parallel or convergent at anthesis, the 3 lower 1.0–1.5 mm, more or less divergent, not much more than half the length of the others, ovate-lanceolate, the hyaline margins meeting or overlapping at their widest part and slightly constricted at their bases. *Corolla* 8–9 mm, creamy-white, becoming pale to dark brown; standard 8–9 mm, with limb broadly elliptical, emarginate at apex, narrowed at base; wings with limb oblong, notched or slightly toothed at apex, rounded at base; keel with limb elliptical, rounded at apex, rounded at base. *Stamens* 10, diadelphous; filaments whitish; anthers cream. *Style* 1, greenish; stigma greenish, capitate. *Legume* 3.0–3.2 mm; seeds 1.0–1.3 mm. *Flowers* 3–6. $2n=16$.

Native. Dry, species-rich coastal grasslands, cliff slopes by rock outcrops or stabilised sand. Coasts of the Channel Islands, Cornwall, Devonshire, Glamorganshire, Pembrokeshire, Caernarvonshire, Anglesey and east Ireland. It is frequent in north-west France and rare in northern Spain and Portugal. Oceanic Temperate element.

5. T. hybridum L. Alsike Clover
Amoria hybrida (L.) C. Presl; *T. fistulosum* Gilib.

Perennial herb with wiry roots. *Stems* (5–)20–40(–90) cm, pale green, erect or ascending and lax, rarely densely caespitose and procumbent, but not rooting at the nodes, glabrous or nearly so, shortly branched above, leafy. *Leaves* alternate, 3-foliolate; leaflets 10–20(–35) × 10–15(–20) mm, medium green on upper surface, paler beneath, obovate, emarginate at apex, denticulate, cuneate or rounded at base, sessile; petioles up to 10 cm, pale green, glabrous; stipules 10–25 mm, greenish, ovate to ovate-lanceolate, gradually narrowed into a subulate apex, not denticulate. *Inflorescence* globose, with numerous flowers, pseudoterminal and axillary; pedicels 4–5 mm, up to twice as long as the calyx-tube; peduncles up to 15 cm, longer than leaves, green, glabrous. *Calyx* 3–4 mm; tube 1.0–1.5 mm, campanulate, not inflated, whitish, 5 veins distinct, the other 5 often obscure; lobes 5, 2–3 mm, the 2 upper slightly longer than the other 3, subulate, separated by broad, obtuse sinuses. *Corolla* (5–)7–10 mm, purple or white at first, pink later, becoming brown; standard folded over the legume, broadly elliptical, rounded at apex, with a claw at base, wings oblong, bluntly pointed at apex, with a long, slender claw; keel elliptically rounded at apex, with a slender claw. *Stamens* 10, diadelphous; filaments pale; anthers cream. *Style* 1, pale, curved; stigma brownish. *Legume* 3–4 mm, included within or slightly exserted from the calyx, oblong; seeds 2–4, ovoid, reddish, tuberculate. *Flowers* 5–9. $2n=16$.

(a) Subsp. **hybridum**
Stems erect, hollow, sparingly branched. *Inflorescence* more than 20 mm in diameter.

(b) Subsp. **elegans** (Savi) Asch. & Graebn.
T. elegans Savi
Stems solid, usually decumbent, much-branched. *Inflorescence* less than 20 mm in diameter.

Introduced. Formerly much grown as a forage crop. Commonly naturalised in grassy and rough ground. Throughout most of Great Britain and Ireland. Most of our plants are subsp. hybridum. Native subspecies occur in southern Europe and south-west Asia, but the most widespread plant is the cultivated subsp. *hybridum* which is naturalised further north and elsewhere. Both subspecies seem to occur in wild flower seed. Subsp. *elegans* has scattered records throughout Great Britain.

6. T. nigrescens Viv. Mediterranean Clover

Annual herb with fibrous roots. *Stems* 6–35 cm, pale green, frequently tinged purple, erect or sprawling, glabrous, usually much branched, angular, leafy. *Leaves* alternate, digitately 3-foliolate; leaflets 3–20 × 3–17 mm, medium green on upper surface, paler beneath, broadly obovate, distinctly veined, the veins exserted from short teeth, usually rounded at apex, sometimes truncate or slightly emarginate, cuneate at base, glabrous; petioles of lower leaves up to 15 cm, shallowly channelled on upper side, those of the upper leaves becoming progressively shorter; stipules 10–13 × 2–4 mm, connate at base, narrowly oblong, membranous at base, subulate-filiform at apex, distinctly veined, not toothed. *Inflorescence* axillary 10–20 mm in diameter, globose; flowers numerous, reflexed after fertilisation, peduncles 15–50 mm; bracts narrowly lanceolate-acuminate at apex; pedicels 1–3 mm. *Calyx* 4.0–4.5 mm, narrowly campanulate, not inflated; tube about 2.0 mm, pale, membranous, 10-veined; lobes 5, green, erect, subulate, the 2 adaxial 2.0–2.5 mm, the rest rather shorter. *Corolla* 6–9 mm, white or pink-tinged; standard with limb oblong, folded longitudinally, obtuse or emarginate at apex, tapering at base to a wide claw about 2 mm; wings with limb oblong, rounded at apex, conspicuously auriculate at base, the claw about 2.5 mm; keel with limb oblong, rounded at apex, abruptly narrowed at base to a claw about 2.5 mm. *Stamens* 10, diadelphous; filaments about 4 mm; anthers suborbicular. *Style* 1, about 2.5 mm, upcurved towards apex; stigma capitate. *Legume* 2–4 × 1.0–1.2 mm, membranous, oblong, strongly or slightly constricted between the seeds; seeds 0.8–1.3 × about 0.8 mm, brown or olive green, oblong or subglobose, smooth. *Flowers* 6–9. $2n=16$.

Introduced. A grain and tan-bark casual. Widespread in the Mediterranean region.

7. T. michelianum Savi Micheli's Clover
Amoria micheliana (Savi) C. Presl; *T. macropodum* Guss.

Annual herb with fibrous roots. *Stems* up to 65 cm, 2–6 mm thick, pale green, erect, hollow, striate, often constricted at the nodes, branching, leafy. *Leaves* alternate, 3-foliolate; leaflets 10–30 × 15–20 mm, medium green on upper surface, paler beneath, oblong or obovate, dentate, narrowed at base; petioles up to 70 mm, pale green; stipules 5–15 mm, ovate, acuminate at apex, not toothed. *Inflorescence* a many-flowered, globose head 20–25 mm in diameter; peduncles equalling or exceeding the leaves, bracteate; pedicels 3–6 mm; flowers deflexed after anthesis. *Calyx* 5–6 mm, not inflated in fruit; tube 1.0–1.5 mm; throat not closed by a ring of hairs; lobes 5, subequal, linear-subulate, 2–4 times as long as the tube. *Corolla* 8–11 mm, pink; standard ovate, rounded at apex, shortly and broadly clawed; wings oblong, rounded at apex, clawed; keel oblong, rounded at apex, clawed at base. *Stamens* 10, diadelphous; filaments pale; anthers brownish. *Style* 1, pale; stigma cream. *Legume* about 4 mm, obovate or orbicular, stipitate, hairy, exserted from tube; seeds 2, about 2 mm, ovoid-oblong, brown, smooth. *Flowers* 6–9. $2n = 16$.

Introduced. A wool casual and formerly a grain alien. Native of the Mediterranean region.

8. T. retusum L. Small-flowered Clover
T. parviflorum Ehrh.; *Amoria parviflora* (Ehrh.) C. Presl

Annual herb with fibrous roots. *Stems* numerous 10–20(–40) cm, pale green, procumbent or ascending, glabrous or nearly so, branched, leafy. *Leaves* alternate, digitately 3-foliolate; leaflets 8–18 × 4–9 mm, medium green on upper surface, paler beneath, the lower obovate-lanceolate, the upper oblong, mucronate at apex, denticulate, the veins curved and prominent, glabrous or nearly so; petioles 10–70 mm; stipules triangular, acuminate at apex, membranous, not toothed. *Inflorescence* a dense globose, many-flowered head 8–11 mm in diameter; the upper flowers nearly sessile, the lower with peduncles up to 30 mm; pedicels about 1 mm or less, much shorter than the bracts or the calyx-tube, not or slightly deflexed in fruit. *Calyx* 4–5 mm, not inflated in fruit; tube about 2 mm; throat not closed by a ring of hairs; lobes 5, very unequal, finally recurved, the upper longer than the tube. *Corolla* 4–5 mm, white or pink; standard with limb ovate, rounded at apex, narrowed to a short claw; wings ovate, pointed at apex, with a long claw; keel lanceolate, narrowed at base to a slender claw. *Stamens* 10, diadelphous; filaments pale; anthers cream. *Legume* about 2 mm, ovate-oblong, membranous; seeds 2, 1.0 mm, brown, ovoid-reniform, finely granulate. *Flowers* 5–6. $2n = 16$.

Introduced. A tan-bark and grass seed casual. Native of central and south Europe and south-west Asia.

9. T. cernuum Brot. Nodding Clover
T. serrulatum Lag.

Annual herb with fibrous roots. *Stems* numerous, 5–40 cm, pale green, procumbent or ascending, much-branched, leafy. *Leaves* alternate, digitately 3-foliolate; leaflets 4–15 × 4–10 mm, medium green on upper surface, paler beneath, obovate, rounded or emarginate at apex, coarsely dentate, cuneate at base, prominently veined beneath, glabrous; petioles long, glabrous; stipules about 10 mm, triangular-lanceolate, long-acuminate at apex, white-membranous. *Inflorescence* about 10 mm in diameter, axillary, globose, 8- to 20-flowered; peduncles thin, shorter than the subtending leaves, the upper ones almost sessile; bracts about 1.0 mm, lanceolate; pedicels about 2 mm, and long as the calyx tube, strongly deflexed and thickened after anthesis. *Calyx* about 4 mm; tube about 2 mm; lobes 5, subequal, triangular at base, subulate above, the 2 upper ones longer than the 3 lower ones, glabrous. *Corolla* 4–5 mm, pink; standard oblanceolate, emarginate at apex, narrowed at base; wings shorter than standard; keel shorter than standard. *Stamens* 10, diadelphous; filaments pale; anthers cream. *Style* 1. *Legume* about 4 mm, exserted from calyx, long-ovoid, membranous, opening by prominent sutures; seeds 1–4, about 1 mm, yellow, ovoid. *Flowers* 5–7. $2n = 16$.

Introduced. A persistent wool and grain alien. Scattered records throughout Great Britain. Native of south-west Europe.

10. T. glomeratum L. Clustered Clover
Amoria glomerata (L.) Soják; *Micrantheum glomeratum* (L.) C. Presl

Annual herb with fibrous roots. *Stems* numerous, (2–)10–20(–35) cm, internodes 10–80 mm, pale green, often flushed purple, prostrate or ascending, glabrous, much-branched, leafy. *Leaves* alternate, digitately 3-foliolate; leaflets 3–12(–20) × 2.5–7.0 mm, medium green on upper surface, paler beneath, obovate, rounded or rarely subacute at apex, venation distinct, veins downcurved, excurrent into short sharp marginal teeth, glabrous or almost so; petioles 1–2(–7) cm, distinctly channelled above; stipules 4–10 × 1–2 mm, ovate or oblong, membranous, obscurely veined, filiform or acuminate at apex. *Inflorescence* 6–12 mm in diameter, axillary, globose, sessile; flowers densely crowded, more or less sessile; bracts about 0.6 mm, narrowly ovate or subulate, membranous. *Calyx* about 4 mm; tube about 2.5 mm, not inflated in fruit; throat not closed by a ring of hairs, whitish, glabrous, 10-veined; lobes 5, about 1.5 mm, subequal, ovate, acuminate at apex, green and strongly recurved, with 3 veins and a narrow, membranous margin, glabrous or thinly eglandular hairy. *Corolla* 4–5 mm, longer than calyx, creamy white or pink-tinged; standard with limb oblong-ovate, obtuse at apex, infolded longitudinally, sometimes obscurely dentate, expanding into a very wide gibbous claw about 2.0–2.5 mm, and rather wider than the limb; wings with limb narrowly oblong, rounded at apex, shortly auriculate at base, with a claw 2.0–2.5 mm; keel with limb oblong, obtuse at apex, narrowed at base rather abruptly to a claw about 2 mm. *Stamens* 10, diadelphous; filaments about 3 mm; anthers shortly oblong. *Style* 1, about 2 mm, tapering from base to apex; stigma capitate. *Legume* about 1.8 × 1.2 mm, membranous, 1–2-seeded; seeds about 1 mm, subglobose or very shortly oblong, bright yellowish-brown, minutely verruculose. *Flowers* 6–8. $2n = 16$.

Native. In short, open communities on sandy drought-prone soils near the sea, on the tops of stone walls, beside sandy tracks and rides, newly sown lawns, sandy arable fields, golf course fairways and railway sidings, and in the Isles of Scilly it is frequent as a bulb-field weed. South and east coast of England north to Norfolk, south-east Ireland and the Channel Islands; a rare casual elsewhere in England; formerly commoner. More or less restricted to southern and western Europe from the Mediterranean region and the Canary Islands northwards to western France, southern Great Britain and south-east Ireland; North Africa; south-west Asia. Mediterranean-Atlantic element.

11. T. suffocatum L. Suffocated Clover
Micrantheum suffocatum (L.) C. Presl; *Amoria suffocata* (L.) Soják

Annual herb with fibrous roots. *Stems* 1–3(–5) cm, often not developed, mostly subterranean, glabrous or thinly hairy, procumbent; internodes rarely reaching 5 mm. *Leaves* alternate, digitately 3-foliolate; leaflets 3–8 × 2–5 mm, medium green on upper surface, paler beneath, obovate, usually emarginate at apex, veins rather obscure, produced into short marginal teeth, cuneate at base, glabrous on both surfaces; petioles 1–6 cm, shallowly channelled or rather flat above, glabrous; stipules 3–4 × about 1.0 mm, ovate or oblong, acuminate at apex. *Inflorescence* 4–5 mm in diameter, more or less globose, densely clustered close to the ground at the base of the petioles; flowers numerous, sessile or nearly so; bracts minute, narrowly ovate, membranous. *Calyx* about 4 mm, cylindrical, not inflated in fruit, thinly eglandular hairy to nearly glabrous; tube about 2.5 mm, membranous, 10-veined, throat not closed by a ring of hairs; lobes about 1.5 mm, lanceolate or subulate, 3-veined, with a narrow membranous margin. *Corolla* 3–4 mm, white; standard with limb oblong, acute or subacute at apex, longitudinally infolded, tapering at base to a rather slender claw 1.5–1.7 mm; wings with limb oblong, rounded at apex, scarcely auriculate at base, narrowed abruptly to a slender claw about 2 mm; keel with limb oblong, rounded at apex, tapering at base to a claw about 2 mm. *Stamens* 10, diadelphous; filaments about 2.5 mm; anthers oblong. *Style* 1, about 1.5 mm; stigma capitate. *Legume* about 1.0 mm, bright yellowish-brown, minutely verruculose. *Flowers* 3–5 with a second flowering in 8. $2n = 16$.

Native. Occurs on sunny, very thin, dry soils on rocky coasts, or on acidic compacted sand and shingle, and is common in car parks, picnic sites and trampled lawns. South and east coasts of England north to Yorkshire and Isle of Man, and in the Channel Islands also a rare alien in wool shoddy. Widespread in the Mediterranean region, extending north along the coasts of Europe to Great Britain and south to the Atlantic islands. Mediterranean-Atlantic element.

Section **3. Mistyllus** (C. Presl) Godr.

Annual herbs. Inflorescence a pseudoterminal head with prominent, glumaceous, striate, free bracts. *Calyx* inflated more or less equally on all sides in fruit, 20- to 34-veined, lobes setaceous and recurved. *Legumes* included in calyx 1- to 4-seeded.

12. T. spumosum L. Pink Clover
Mistyllus spumosus (L.) C. Presl

Annual herb with fibrous roots. *Stems* 8–25(–40) cm, pale green, prostrate or ascending, glabrous or nearly so, usually much branched, leafy. *Leaves* alternate below, uppermost opposite, digitately 3-foliolate; leaflets 10–30 × 5–25 mm, medium green on upper surface, paler beneath, broadly obovate, rounded or occasionally somewhat emarginate at apex, minutely serrulate, the lateral veins running into the serrulations; petioles of lower leaves 6–14 cm, slightly channelled above, those of the upper leaves becoming progressively shorter; stipules 8–20 × 3–7 mm, membranous, ovate or oblong, caudate-acuminate at apex. *Inflorescence* pseudoterminal, 2–3 cm in diameter, globose; peduncle 1.5–4.0 cm, stout; bracts 5–8 × 2.5–4.0 mm, whitish, scarious, ovate, aristate at apex; flowers crowded. *Calyx* 10–13 mm, ovoid, inflated, glabrous, throat not closed by a ring of hairs; tube 8–9 mm, 20- to 35-veined, pale, membranous, with obscure transverse striations; lobes 5, 2–4 mm, subequal, setaceous, greenish, recurved. *Corolla* 7–8 mm, pink; standard with limb narrowly oblong, acute or acuminate at apex, tapering at base to a wide claw about 10 mm, wings with limb narrowly oblong, acute at apex, strongly auriculate at base with a claw about 10 mm; keel with limb oblong, acute or shortly beaked at apex, narrowed abruptly to a claw about 10 mm. *Stamens* 10, diadelphous; filaments about 12 mm; anthers oblong. *Style* 1, 7–8 mm, glabrous; stigma oblique. *Fruiting calyx* inflated, whitish or pale brown, distinctly reticulate with well-marked transverse striations; legume about 2.0 × 2.5–3.0 mm, membranous, compressed; seed 3–4 in each pod, 1.5–2.0 × 1.0–1.5 mm, bright brown, oblong or subglobose, verruculose. *Flowers* 6–8. $2n = 16$.

Introduced. A tan-bark, grain and grass seed casual. Widespread in the Mediterranean region and eastward to Iraq; Atlantic islands.

Section **4. Vesicaria** Crantz

Annual or *perennial herbs. Inflorescence* an axillary head with subsessile flowers, with bracts free or united into a small involucre. *Calyx* inflated in fruit, the upper lip externally densely hairy or rarely glabrous, scarious and reticulately veined, its 2 teeth often setaceous. *Legume* 1- to 2-seeded.

13. T. fragiferum L. Strawberry Clover
Galearia fragifera (L.) C. Presl

Perennial herb with a tough rootstock. *Stems* (2–)10–30(–40) cm, pale green, often rooting at the nodes, prostrate, glabrous or very sparsely eglandular hairy, usually much branched, leafy. *Leaves* alternate, digitately 3-foliolate; leaflets (3–)6–25 × 4–18 mm, medium green on upper surface, paler beneath, oblong-obovate, rounded or shallowly emarginate at apex, cuneate at base, glabrous on upper surface, thinly eglandular hairy or nearly glabrous beneath, the veins prominent, curved near the margins of the leaflets and excurrent as minute, sharp teeth, shortly petiolulate; petioles up to 15 cm, often slender, glabrous or thinly eglandular hairy; stipules 10–25(–30) × 2–6 mm, narrowly oblong, the basal part pale and scarious, long acuminate at apex, entire, distinctly veined, glabrous. *Inflorescence* axillary, 10–15 mm in diameter at anthesis, subglobose; flowers numerous, crowded, very shortly pedicillate; peduncles 3–20 cm, usually exceeding petioles, glabrous or thinly hairy; involucre distinct, the bracts about 2.0 × 0.5–0.8 mm, narrowly oblong, acute at apex, pale with greenish veins, the floral bracts much smaller, usually less than 1.0 × 0.5 mm, pale, cuspidate-acuminate at apex. *Calyx* at anthesis 3.5–4.5 mm, narrowly campanulate inflated in fruit, throat not closed by a ring of hairs, the tube about 2 mm, obscurely 10-veined and the adaxial surface rather densely eglandular hairy, the lobes about 1.5 mm, erect, subequal, subulate, the 2 adaxial ones shorter and wider than the others. *Corolla* 6–7 mm, whitish or becoming pink with age; standard with limb bent upwards, emarginate at apex, gradually narrowed at base to a broad claw 2.0–2.5 mm; wings with limb narrowly oblong, obtuse at apex, and distinctly auriculate at base, the claw 2.5–3.0 mm; keel with limb oblong, obtuse at apex and scarcely auriculate at base, the claw about 2.5 mm. *Stamens* 10, diadelphous; filaments about 4.5 mm; anthers broadly oblong. *Style* about 2.5–3.0 mm, upcurved at apex, glabrous; stigma capitate. *Fruiting calyx* 4–10 × 4–6 mm, reflexed, inflated, papery, distinctly reticulate, adaxial surface markedly gibbous; legume enclosed by calyx, brownish, membranous; seeds usually 2 in each legume, about 1.5 × 1.2 mm, olive-brown, flecked with small fuscous blotches, shortly oblong. *Flowers* 7–9.

(a) Subsp. **fragiferum**
Inflorescence 10–22 mm in fruit, globose. *Calyx* 4.0–4.5 mm in flower, with teeth longer than the tube, 8–10 mm in fruit, concealing all or most of the persistent corolla. $2n = 16$.

(b) Subsp. **bonannii** (C. Presl) Soják
T. bonannii C. Presl; *T. neglectum* C. A. Mey.

Inflorescence (10–)15–25(–35) mm in fruit, subglobose to irregularly cylindrical. *Calyx* 3.5–4.0 mm in flower, the teeth not longer than the tube, 4–6 mm in fruit, the corolla exserted by 2.0–2.5 mm.

Native. Grassy places, often on heavy or brackish soils. Records rather scattered but locally common in Great Britain and Ireland north to south Scotland; in Ireland mostly coastal. Widespread in Europe and the Mediterranean region eastwards to Afghanistan and Pakistan, also in north Ethiopia. Eurosiberian Southern-temperate element. The common plant is subsp. *fragiferum*. Subsp. *bonannii* occurs in southern England and southern Europe north to Poland.

14. T. resupinatum L. Reversed Clover
Galearia resupinata (L.) C. Presl; *T. bicorne* Forssk.

Annual herb with fibrous roots. *Stems* 20–30(–80) cm, pale green, ascending, sprawling or prostrate, glabrous or nearly so, usually much-branched, leafy. *Leaves* alternate, digitately trifoliate; leaflets 8–25 × 4–15 mm, bright medium green on upper surface, paler beneath, obovate, oblong or elliptical, acute at apex, rarely rounded or emarginate, finely serrulate, the lateral veins protruding from the serrulations as short cusps, glabrous on both surfaces; petioles of lower leaves often more than 10 cm, slightly channelled above, those of the upper leaves progressively shorter, glabrous; stipules 10–17 × 2–4 mm, basal part pale and distinctly veined, oblong, caudate at apex. *Inflorescence* axillary, 8–15 mm in diameter, more or less globose; peduncles generally exceeding petioles; flowers crowded; bracts minute, membranous, truncate; pedicels less than 0.5 mm, thinly hairy or glabrous. *Calyx* at anthesis narrowly pyriform, 2.5–3.5 mm, distinctly bilabiate, conspicuously inflated throat not closed by a ring of hairs, tube about 1 mm, membranous, about 13-veined, adaxial surface villose, abaxial surface becoming glabrous or nearly so; 2 adaxial lobes spreading, 0.8–1.2 mm, filiform-subulate; the 3 abaxial lobes erect, 1.5–2.5 mm, subulate. *Corolla* 5–8 mm, rose-pink or purple, resupinate, petals with claws united into a twisted tube; standard with limb oblong, truncate or shallowly emarginate at apex, narrowed at base to a broad claw; wings with limb oblong, subacute at apex, prominently auriculate at base with a claw about 2.5 mm; keel with limb oblong, acute at apex, abruptly narrowed to a claw 2.0–2.5 mm. *Stamens* 10, diadelphous; filaments 2.5–4.0 mm; anthers oblong. Style 1, 2.0–2.5 mm, glabrous, with sharply upturned apex; stigma capitate. *Fruiting calyx* 8–10(–12) × 4–5 mm, reflexed, thinly hairy or subglabrous, the adaxial part becoming inflated, gibbous, papery and conspicuously reticulate, the 2 terminal teeth lengthening and becoming upcurved and antenna-like; legume enclosed within calyx, 2.0–2.5 × 1.2–1.8 mm, oblong, hard and brown, usually 1-seeded; seed 1.3–1.5 × 1.2–1.3 mm, dark brown, oblong, smooth. *Flowers* 6–8. $2n = 14, 16, 32$.

Introduced. Rather frequent casual from wool and other sources. Scattered records throughout Great Britain and formerly in Ireland. Throughout the Mediterranean region and eastwards to Afghanistan and central Asia; now widely naturalised in southern Europe, Africa, South America and elsewhere.

15. T. tomentosum L. Woolly Clover
Galearia tomentosa (L.) C. Presl

Annual herb with fibrous roots. *Stems* 5–25 cm, pale green, prostrate or sprawling, glabrous or nearly so, much-branched, leafy. *Leaves* alternate, digitately trifoliate; leaflets 5–17 × 3–9 mm, medium green on upper surface, paler beneath, obovate to almost elliptical, rounded or subacute at apex, finely and sharply serrulate, the lateral veins protruding from the serrulations as short cusps, glabrous on upper surface, occasionally thinly eglandular hairy beneath; petioles of lower leaves 2–5 cm, slender, slightly channelled above, glabrous or nearly so, those of the upper leaves progressively shorter; stipules 10–15 × 2–3 mm, narrowly oblong, caudate-acuminate at apex, basal part pale and distinctly veined. *Inflorescence* axillary, 6–9 mm in diameter, more or less globose, lanate; peduncles shorter than petioles; bracts minute, membranous, truncate or emarginate; pedicels very short, thinly eglandular hairy; flowers crowded. *Calyx* 2.0–2.5 mm, narrowly campanulate, distinctly 2-lipped, inflated, throat not closed by a ring of hairs; tube 1.0–1.5 mm, membranous, 12- to 13-veined, adaxial surface densely eglandular hairy, abaxial surface glabrous or nearly so; 2 adaxial teeth 0.7–1.0 mm, erect or spreading and subulate, 3 abaxial teeth 0.5–1.0 mm, erect and subulate. *Corolla* 3–6 mm, pale or bright pink; the petals with claws united into a twisted tube; standard with limb oblong or narrowly obovate, emarginate at apex, the base narrowed into a broad claw; wings with limb oblong, rounded at apex, distinctly auriculate at base, with a slender claw about 2 mm; keel with limb oblong, acute or obtuse at apex, abruptly narrowed at base to a claw 1.8–2.0 mm. *Stamens* 10, diadelphous; filaments 2.5–3.0 mm; anthers oblong. *Style* 1, 1.5–1.7 mm, glabrous; stigma capitate. *Fruiting calyx* 3–7 × 3–5 mm, reflexed, hairy or glabrous, the adaxial part strongly inflated, gibbous and papery, distinctly reticulate veined in glabrous forms, the 2 terminal teeth hidden or rather short and upcurved; legume enclosed within calyx, 1.5–2.0 × about 1 mm, membranous, oblong, 1- to 2-seeded; seeds 1.2–1.5 × 1.0–1.3 mm, broadly oblong, brown or yellowish, smooth. *Flowers* 5–7. $2n = 16$.

Introduced. Casual from wool and other sources. Sporadic, records throughout Great Britain. Throughout the Mediterranean region and eastwards to Iran; Atlantic Islands.

Section **5. Chronosemium** Ser.
Chrysaspis Desv.

Annual herbs. Inflorescence of terminal and axillary heads, the bracts represented by a few, short, red glandular hairs. *Calyx* 5-veined, upper teeth shorter than the lower. *Legume* stalked, slightly exceeding calyx, 1(–2)-seeded.

16. T. speciosum Willd. Large-flowered Clover
T. grandiflorum Schreb. nomen; *Amarenus speciosus* (Willd.) C. Presl

Annual herb with fibrous roots. *Stems* 10–30 cm, pale green, erect, solid, with appressed or patent eglandular hairs, poorly branched, leafy. *Leaves* alternate, pinnately 3-foliolate; leaflets 10–18(–24) × 4–8 mm, medium green on upper surface, paler beneath, oblong-elliptical, emarginate at apex, the terminal petiolulate, glabrous or eglandular hairy; petioles up to 20 mm, pale green, glabrous; stipules 10–20 mm, semi-ovate or oblong. *Inflorescence* a lax, ovoid spike up to 30 mm in diameter in fruit; peduncles 2–3 times as long as the leaves; pedicels about 1 mm, about as long as the upper limb of the calyx tube. *Calyx* 4–5 mm; tube 1.5–2.0 mm; lobes 5, the 2 upper equalling or shorter than the upper limb of the tube, the 3 lower 2–3 times longer than the upper. *Corolla* 8–10 mm, violet or reddish-violet; standard broadly ovate, pointed at apex, narrowed to a broad claw, wings oblong, rounded at apex, with a long claw; keel oblong, rounded at apex, shorter than wings. *Stamens* 10, diadelphous; filaments pale; anthers cream. *Style* 1, pale; stigma brownish. *Legume* 1.5–2.0 mm, ovoid-ellipsoid, membranous, 1(–2)-seeded, slightly stalked; seed about 1.2 mm, yellowish-brown. *Flowers* 4–5. $2n = 16$.

Introduced. A tan-bark casual. Native of the Mediterranean region.

17. T. patens Schreb. Small-fruited Clover
Amarenus patens (Schreb.) C. Presl

Annual herb with fibrous roots. *Stems* 20–50 cm, pale green, erect or ascending, flexuous, sparsely eglandular hairy, branched, leafy. *Leaves* alternate, 3-foliolate; leaflets 5–18 × 3–4 mm, medium green on upper surface, paler beneath, narrowly elliptic-obovate, rounded to acute or truncate at apex, cuneate at base, the terminal one subsessile or with a petiolule up to 2 mm; petiole up to 2 cm, pale green, glabrous; stipules 5–10 mm, ovate, acute at apex, dilated and rounded at base. *Inflorescence* a head 10–15 mm in diameter in fruit; peduncles 20–50 mm, slender, usually much longer than the leaves; flowers almost sessile. *Calyx* 2.5–3.0 mm, tube 0.6–1.0 mm, glabrous; lobes 5, the upper shorter than the upper limb of the tube, the 3 lower equal to or longer than the lower limb of the tube. *Corolla* 5–7 mm, yellowish; standard with limb oblong-ovate, rounded at apex, sulcate, with a claw at base; wings oblong, rounded at apex, with a claw at base; keel oblong, acute at apex. *Stamens* 10, diadelphous; filaments pale; anthers brownish. *Style* 1, pale; stigma brownish. *Legume* 1.5–2.0 mm, ovoid-ellipsoid, stalked; seeds 1(–2), yellow to brown, ellipsoid, shiny. *Flowers* 6–8. $2n = 16$.

Introduced. A wool casual. Introduced with wild flower seed to Cork City in 1994, but gone by 1997. Native of central and south Europe and south-west Asia.

18. T. aureum Pollich Large Trefoil
Chrysaspis aurea (Pollich) B. D. Greene; *T. agrarium* L. nom. rejic.; *T. strepens* Crantz nom. illegit.; *Chrysaspis candollei* Desv.; *Amarenus agrarius* (L.) C. Presl

Biennial herb with fibrous roots. *Stems* 15–30 (–40) cm, pale green, erect or ascending, usually appressed-hairy, branched, leafy. *Leaves* alternate, 3-foliolate; leaflets 15–25 × 6–8 mm, medium green on upper surface, paler beneath, oblong-ovate or rhombic, widest near the middle, rounded at apex, denticulate, rounded at base, hairy, the terminal one nearly sessile; petioles 5–15 mm; stipules 12–15 mm, lanceolate-ovate, acuminate at apex, not dilate below. *Inflorescence* of dense, many-flowered terminal and axillary heads 12–20 mm in diameter; peduncles up to 50 mm, stout, equalling or exceeding the leaves; flowers nearly sessile. *Calyx* 2.5–3.0 mm; tube 1.0–1.5 mm; lobes 5, the upper shorter than the upper limb of the tube, the lower 1–2 times as long as the lower limb of the tube. *Corolla* 6–8 mm, golden yellow; standard with limb obovate, rounded at apex, sulcate, with a claw; wings oblong, rounded at apex, with a claw; keel oblong, rounded at apex, with a claw. *Stamens* 10, diadelphous; filaments pale; anthers yellow. *Style* 1, pale; stigma brownish. *Legume* 1.5–2.0 mm, oblong ; seeds 2, 1.0–1.2 mm. *Flowers* 7–8. $2n = 14$, 16.

Introduced in 1815. Naturalised in grassy and rough ground, formerly frequent, now very local. Scattered records throughout much of Great Britain, but now well naturalised only in central and south Scotland. Along the banks of the River Blackwater in Ireland. Most of Europe except the extreme north, most of the west and the Mediterranean region; Turkey.

19. T. campestre Schreb. Hop Trefoil
T. glaucescens Hausskn.; *T. procumbens* auct.; *T. agrarium* auct.; *Chrysaspis campestris* (Schreb.) Desv.; *Amarenus procumbens* auct.

Annual herb with fibrous roots. *Stem* 3–30(–50) cm, pale greenish or flushed purple, prostrate, sprawling or ascending, thinly clothed with white, appressed hairs, usually much branched from the base, leafy. *Leaves* alternate,

3-foliolate; leaflets 3–18 × 1.5–10 mm, variable, medium green on upper surface, paler beneath, sometimes bluish or tinged purple, broadly or narrowly obovate or oblong, rounded, truncate or slightly emarginate at apex, cuneate or rounded at base, glabrous or thinly hairy on both surfaces, the terminal leaflet with a petiolule up to 5 mm; petiole usually less than 10 mm, slender, scarcely channelled above, thinly hairy to nearly glabrous; stipules 1.5–7.0 × 0.5–4.0 mm, oblong or ovate-acuminate at apex, entire or sinuate, glabrous or sparsely hairy. *Inflorescence* axillary, 4–17 × 7–12 mm, globose or ovoid; peduncle up to 25 mm, slender, or sometimes almost wanting; flowers 20–30, crowded, reflexed or spreading after anthesis; pedicels usually less than 1 mm, or flowers nearly sessile. *Calyx* 1–2 mm, membranous, glabrous or with the lobes tipped with a few long hairs; tube about 0.5–1.0 mm; lobes unequal, the 2 adaxial 0.3–0.5 mm, deltoid or shortly linear, obtuse or acute at apex, the 3 abaxial 0.5–1.0 mm, linear, obtuse at apex. *Corolla* (3–) 4–5(–6) mm, pale yellow, becoming brown and scarious after anthesis; standard with limb subrotund, flattish or concave, rarely keeled, rounded or shortly emarginate at apex, distinctly erose-dentate, especially towards the base, the claw very short and wide; wings with limb oblong, with an expanded, rounded apex, minutely and obscurely erose, strongly auriculate at base with a slender claw about 1 mm, the sinus between claw and auricle rounded or rather rectangular; keel with limb oblong, slightly auriculate at base, narrowed abruptly to a claw about 0.8 mm. *Stamens* 10, diadelphous; filaments 2.0–2.5 mm, pale green; anthers yellow, minute, subglobose. *Style* 1, 0.5–0.6 mm, green, sharply upcurved at apex; stigma green, capitate. *Legume* about 1.3 × 1.0 mm, oblong-ellipsoid, compressed, membranous; seed 1.0–1.2 × 0.6–0.8 mm, yellowish-brown, oblong, smooth. *Flowers* 6–9. $2n = 14$.

(i) Var. **minus** (W. D. J. Koch) Gremli
T. procumbens var. *minus* W. D. J. Koch; *T. pseudoprocumbens* C. C. Gmel.; *T. agrarium* var. *pseudoprocumbens* (C. C. Gmel.) Lloyd; *T. campestre* subsp. *pseudoprocumbens* (C. C. Gmel.) Asch. & Graebn.

Stems up to 12 cm, spreading or prostrate, with short branches. *Inflorescence* up to 10 mm in diameter.

(ii) Var. **campestre**
Stems up to 30 cm, ascending or erect, with longer, slender branches. *Inflorescence* up to 10 mm in diameter.

(iii) Var. **majus** (W. D. J. Koch) P. D. Sell
T. procumbens var. *majus* W. D. J. Koch

Stems up to 50 cm, usually erect, usually much branched. *Inflorescence* up to 17 mm in diameter.

Native. Grassy and rather bare places on dry, relatively infertile neutral or base-rich soils. Throughout much of Great Britain and Ireland, but absent from the mountains and coastal in north Scotland and north-west Ireland. Europe except the extreme north and east; west Asia; North Africa; Macaronesia; introduced in North America, Eurosiberian Southern- temperate element. Var. *minus* is the plant of bare places, particularly coastal and on sand. the other two varieties are of grassy places. Var. *majus* has recently been included in wild flower seed planted round the grass margins of fields in Cambridgeshire.

20. T. dubium Sibth. Lesser Trefoil
T. procumbens auct.; *T. filiforme* auct.; *Chrysaspis dubia* (Sibth.) Desv.; *T. flavum* C. Presl; *Amarenus flavus* (C. Presl) C. Presl.; *T. minus* Sm.

Annual herb with fibrous roots. *Stems* 3–30 (–45) cm, pale green, prostrate or decumbent, thinly subappressed hairy to nearly glabrous, much-branched, leafy. *Leaves* alternate, 3-foliolate; 3–8(–12) × 2–6(–8) mm, medium green on upper surface, paler beneath, narrowly obovate, rounded or truncate, sometimes emarginate at apex, often distinctly denticulate, cuneate at base, glabrous or thinly hairy, the median leaflet usually very shortly petiolulate; petiole usually less than 8 mm, slender, thinly hairy; stipules 2–7 × 1–3 mm, narrowly ovate, acuminate at apex, entire or obscurely denticulate. *Inflorescence* axillary, 5–10 mm in diameter, subglobose; flowers (4–)5–12(–20), reflexed and scarious after anthesis; peduncle 2–20(–35) mm, slender, more or less appressed-hairy; pedicels very short or almost wanting. *Calyx* 1.7–2.0 mm, narrowly campanulate, glabrous or nearly so; tube 1.0 mm, membranous; lobes 5, unequal, glabrous or tipped with hairs, the 2 adaxial about 0.5 mm, deltoid, blunt at apex, the 3 abaxial 0.7–1.0 mm, linear, blunt at apex. *Corolla* 3–4 mm, yellow, becoming brown with age; standard with limb obovate, strongly concave, the sides infolded over the wings and keel, obtuse or slightly emarginate at apex, tapering at base to an indistinct claw; wings with limb oblong, rounded and distinctly gibbous at apex, strongly auriculate at base, the auricle slender and subulate with a slender claw about 1.2 mm, the sinus between the auricle and claw rounded; keel with limb oblong, acute at apex, shortly auriculate at base, with a claw about 1 mm. *Stamens* 10, diadelphous; filaments about 2.5 mm, pale green; anthers pale yellow, minute. *Style* 1, about 0.5 mm, green, sharply upcurved at apex; stigma green, capitate. *Legume* 1.8 × 1.3 mm, membranous, oblong, 1-seeded; seed 1.0–1.3 × about 0.8 mm, light brown, oblong, smooth. *Flowers* 5–10. $2n = 32$.

(i) Var. **microphyllum** (Ser.) P. D. Sell
T. minus var. *microphyllum* Ser.; *T. filiforme* var. *pygmaeum* Soy.-Will.; *T. filiforme* var. *minimum* Gaudin; *T. filiforme* var. *pauciflorum* Coss. & Germ.

Stems up to 10 cm, spreading or prostrate. *Leaves* small, mostly under 5 mm. *Inflorescence* up to 7 mm in diameter.

(ii) Var. **dubium**
Stems up to 30(–45) cm, ascending or erect. *Leaves* up to 12 mm. *Inflorescence* up to 10 mm in diameter.

Native. Grassy and open ground. Common throughout most of Great Britain and Ireland in meadows, lawns, waste places, rock outcrops, quarry spoil and railway ballast. Widespread in Europe and eastwards to northern Iran; Atlantic Islands. European Temperate element. Often confused with *Medicago lupulina* when not in fruit, but easily distinguished by its leaflets which are without an apiculus which is present in *Medicago*. Var. *microphyllum* is a plant of rather bare ground or short turf especially near the sea. Var. *dubium* occurs throughout the range of the species.

21. T. micranthum Viv. Slender Trefoil

Chrysaspis micrantha (Viv.) Hendrych; *T. filiforme* L. nom. rejic.; *Amarenus filiformis* auct.

Annual herb with fibrous roots. *Stems* 2–10(–20) cm, pale green, procumbent or ascending, sparsely hairy, sparsely branched, leafy. *Leaves* alternate, 3-foliolate; leaflets 5–8 × 3–6 mm, medium green on upper surface, paler beneath, obovate, often emarginate at apex, denticulate, cuneate at base, glabrous or nearly so, with 4–9 pairs of lateral veins, the terminal one subsessile; stipules 2–3 mm, oblong to ovate, dentate. *Inflorescence* of 1–6-flowered heads about 4 mm in diameter; peduncles capillary; pedicels capillary, as long as or longer than the upper limb of the calyx-tube. *Calyx* 1.3–1.5 mm; lobes 5, unequal, the lower longer than the lower limb of the tube. *Corolla* 2–3(–4) mm, yellow, becoming yellowish-brown after anthesis; standard with limb oblong and boat-shaped, emarginate at apex, rounded at base; wings with limb oblong, rounded at apex, auricled and clawed at base, keel ovate, rounded at apex, minutely auricled and long clawed at base. *Stamens* 10, diadelphous; filaments pale; anthers cream. *Style* 1, pale; stigma brownish. *Legume* 2.0–2.5 mm, ovoid; seeds 2, dull brown, ovoid-reniform, smooth. *Flowers* 6–7. $2n = 16$.

Native. Short turf, especially close-cut lawns and sandy areas by the sea. Common in the Channel Islands and much of England, the lowlands of Wales, scattered localities in eastern coastal areas of Ireland and areas of eastern coasts of Scotland. West and central Europe; North Africa; Caucasus. Submediterranean Subatlantic element.

Subgenus **3. Trifolium**

Flowers ebracteate. *Calyx-throat* usually more or less closed with a ring of hairs or an annular or 2-lobed callosity. *Legume* nearly always included in the calyx tube. *Seeds* 1- to 2-seeded.

Section **6. Trifolium**

Annual or *perennial herbs*. *Inflorescence* usually a spicate or rarely capitate head, the flowers usually sessile, all fertile and ebracteate.

22. T. striatum L. Knotted Clover

Annual herb with fibrous roots. *Stem* 4–30 (–50) cm, pale green, prostrate or sprawling, with more or less dense, long, spreading hairs, much branched, rarely simple, leafy. *Leaves* usually all alternate, the uppermost rarely opposite, 3-foliolate; leaflets 5–16 × 3–10 mm, medium green on upper surface, paler beneath, narrowly obovate, rounded or retuse at apex, cuneate at base, appressed hairy on both surfaces, lateral veins running straight into the margin or very little curved at apex; petioles of lower leaves up to about 7 cm, slender, channelled above, thinly hairy, those of the upper leaves becoming progressively shorter; stipules 4–6 × 2–3 mm, membranous, ovate, aristate at apex, villous. *Inflorescence* axillary and terminal, sessile or nearly so, 8–15 × 7–10 mm, shortly oblong or ovoid, usually not paired; flowers ebracteate. *Calyx* 4–5 mm, shortly cylindrical; tube about 2.5 mm, distinctly 10-veined, villous externally, glabrous internally except at the shortly hairy throat; lobes 5, 2–3 mm, subequal, subulate, narrowing above to a very slender, aristate apex, suberect in both flower and fruit. *Corolla* 4–5 mm, pink; standard with limb oblong, obtuse to subacute at apex, scarcely auriculate at base, the claw about 2 mm long and free; wings with limb narrowly oblong, obtuse or subacute at apex, minutely auriculate at base, the claw about 2.5 mm and united with the claw of the keel; keel with limb oblong, subacute at apex, indistinctly auriculate at base, with a claw about 2.5 mm. *Stamens* 10, diadelphous; filaments about 3.5 mm; anthers oblong. *Style* 1, sharply upcurved towards base, slender above; stigma oblique. *Legume* concealed within calyx; seed about 1.8 mm, yellow or pale brown, subglobose, smooth. *Flowers* 5–7. $2n = 14$.

Native. Short grassland and open places on sandy ground, especially near the sea and round rock outcrops and on banks and road verges. Locally frequent through Great Britain north to central-east Scotland, eastern Ireland and the Channel Islands. South, west and central Europe northwards to southern Sweden; Caucasus; north-west Africa; Atlantic Islands. European Southern-temperate element.

23. T. arvense L. Hare's-foot Clover

Annual herb with fibrous roots. *Stems* 4–40 cm, pale green, erect or spreading, usually much-branched, rarely simple, densely hairy, leafy. *Leaves* alternate, 3-foliolate; leaflets greyish-green on upper surface, paler beneath, the lower 5–12 × 3–6 mm, broadly oblong, obovate or elliptical and rounded or emarginate at apex, the upper 7–20 × 1.5–4.0 mm, narrowly linear-oblong, subacute and cuspidate at the sometimes erose apex, all shortly to rather long hairy on both surface; petioles up to 35 mm, very slender, slightly channelled above, rounded beneath, hairy, becoming progressively shorter towards the apex; stipules 5–10 mm, ovate, distinctly veined, hairy, with slender, filiform apex. *Inflorescence* 10–35(–40) × 10–15 mm, an oblong or ovoid spike, flowers densely crowded without an involucre; peduncles 15–40(–60) cm, erect, slender, hairy. *Calyx* 3.5–7.0(–9.0) mm, the tube about 2 mm, the 5 lobes 1–3(–5) times as long as the tube, reddish, subequal, filiform, with long hairs. *Corolla* about 4 mm, whitish, becoming pinkish with age, much shorter than the calyx-teeth; standard with limb oblong, rounded or truncate at apex, scarcely auriculate at base, the claw proportionately very short; wings with limb oblong, rounded at apex and shortly auriculate at base; keel with limb oblong and subacute at apex. *Stamens* 10, diadelphous; filaments about 3.5 mm, the margins of the fused ones joined near their apex to form a short tube round the style; anthers suborbicular. *Style* 1, about 2.3 mm, glabrous, upturned at apex; stigma capitate. *Legume* completely concealed by the persistent calyx; seed about 1.4 × 0.8 mm, pale yellowish-brown, ovoid, smooth. *Flowers* 6–9. $2n = 14$.

Native. Rather bare ground on sandy soils such as acidic heathlands, sea-cliffs and dunes, introduced with railway ballast and on waste land. Locally frequent in Great Britain north to central Scotland, and in coastal Ireland.

24. T. bocconei Savi Twin-headed Clover

Annual herb with fibrous roots. *Stems* several, 1.5–6.0(–15) cm, pale green, spreading or ascending, with dense, ascending or spreading, pale simple eglandular hairs, sparingly branched, leafy. *Leaves* alternate, 3-foliolate; leaflets 3–11 × 2–7 mm, greyish-green on upper surface, paler beneath, mostly obovate, sometimes oblanceolate, truncate or rounded at apex, denticulate, cuneate at base, glabrous on upper surface, hairy at first beneath becoming almost glabrous or with hairs on the midrib and margin; petiolules very short, glabrous or hairy; petioles up to 6 mm, hairy; stipules up to 5 mm, ovate-oblong or lanceolate, abruptly contracted above with subulate, ciliate points. *Inflorescence* terminal and axillary, in dense, globose to cylindrical heads 9–15 mm, the terminal ones often paired but unequal; flowers without bracts. *Calyx* 4–5 mm; tube about 3 mm, cylindrical, hairy, strongly 10-ribbed, the throat thickened and slightly hairy; lobes 5, subulate, erect or connivent, unequal, the lowest equalling the tube. *Corolla* 4–6 mm, pinkish or white; standard oblong, rounded at apex, slightly outcurved toward base and then cuneate; wings shortly oblong, with a short lobe at base and a slender spur; keel oblong, rounded at apex, with a short lobe and a long spur at base. *Stamens* 10, diadelphous; filaments white; anthers cream. *Style* 1, hooked at apex, slender; stigma brownish. *Legume* completely concealed by the persistent calyx, 1–2 mm, yellowish, ovoid. *Flowers* 4–6. $2n = 14$.

This species is very variable over its total range. Our material, however, is very uniform. Two specimens have been seen which were collected by Paolo Savi, son of Gaetano Savi who named the species. They are not identical with one another and neither of them is our plant. One is var. *tenuifolium* (Ten.) Griseb. which is regarded as a distinct species in *Flora Europaea*. The illustration of var. *bocconei* in Zohary and Heller's monograph is nothing like our plant. The name that best fits our plant, until it can be properly sorted out, is var. **semiglabrum** (Brot.) Merino (*T. semiglabrum* Brot.) which occurs in Galicia and the Algarve.

Native. On soils overlying serpentine and rarely on schist in species-rich grasslands on south-facing valley slopes and around rock outcrops with a southerly aspect, the soils moist in winter, severely droughted in summer. Cove valleys that dissect the Lizard Peninsula, Cornwall and to a lesser extend on the upper sections of sea-cliffs, with few small populations on rock outcrops further inland. Grazing by livestock appears to be essential for the survival of most populations. It is still in one locality in turf in Jersey and was formerly in another. The species occurs in southern and western Europe extending northwards to the Lizard and Jersey and eastwards to Bulgaria and Greece; Turkey.

25. T. gemellum Pourr. ex Willd. African Clover

Annual herbs with fibrous roots. *Stems* several, 10–35 cm, pale green, diffuse, erect or ascending, with spreading or deflexed hairs, sparsely branched, leafy. *Leaves* alternate 3-foliolate; leaflets 6–15 mm, medium green, obovate or oblong-obovate, rounded, truncate or retuse at apex, denticulate towards apex, narrowed at base, sessile, hairy; petioles long; stipules up to 10 mm, lanceolate, acuminate at apex, adnate to the petioles for up to two-thirds of their length. *Inflorescence* terminal and axillary, of dense ovoid to oblong heads 8–18 × 6–10 mm, surrounded by the upper leaves, sessile or with short peduncles. *Calyx* 5–6 mm; tube 2.0–2.5 mm; lobes 5, subulate, setaceous, 1-nerved, plumose, the throat open, but provided with annular, hairy or glabrous thickening. *Corolla* 4–5 mm, pink; standard 3.2–4.2 mm. *Stamens* 10, diadelphous. *Style* 1. *Legume* ovoid, compressed, membranous, 1-seeded; seeds about 1 mm, yellow to brown, almost globose. *Flowers* 5–6. $2n = 14$.

Introduced. A tan-bark casual. Native of south-west Europe and north-west Africa.

26. T. ligusticum Balb. ex Loisel. Ligurian Clover

Annual herb with fibrous roots. *Stems* 10–40 (–60) cm, dark green, ascending or diffuse, branched, with sparse, patent simple eglandular hairs, leafy. *Leaves* alternate, 3-foliolate; leaflets 10–20 × 5–13 mm, medium green on upper surface, paler beneath, broadly obovate, rounded at apex, denticulate, cuneate at base, glabrous or hairy; petioles up to 25; stipules ovate or oblong, with a setaceous apex. *Inflorescence* an ovoid or oblong head 6–15 mm, often paired, then one axillary and long-pedunculate, the other terminal but laterally displaced and shortly pedunculate; flowers ebracteate. *Calyx* 4–6 mm, mouth of the tube closed with a callosity; lobes 5, subequal, setaceous, ciliate, ultimately divergent, up to twice as long as the tube. *Corolla* 3–4 mm, whitish, rarely pink, much shorter than the calyx; standard 3.0–4.5 mm, obtuse at apex; wings with limb obtuse at apex; keel with limb obtuse at apex. *Stamens* 10, diadelphous; filaments pale; anthers brown. *Style* 1; stigma brown. *Legume* included in the calyx, indehiscent, ovoid; seed 0.8–1.1 mm, brown, globose. *Flowers* 5–7. $2n = 12$, 14.

Introduced. A wool casual. Native of west Mediterranean region and Atlantic Islands.

27. T. scabrum L. Rough Clover

Annual herb with fibrous roots. *Stems* 4–25 cm, pale green, often tinted purple, prostate or ascending, flexuous, frequently zigzag in fruiting specimens, hairy, usually much-branched, leafy. *Leaves* alternate, 3-foliolate; leaflets 5–10 × 2–5(–8) mm, coriaceous, medium green on upper surface, paler beneath, obovate, rounded at apex, minutely denticulate or subentire, cuneate at base, thinly hairy on both surfaces, lateral veins recurved and prominent at the margins; petioles slender, the lower 2–5 cm canaliculate, hairy, those of upper leaves progressively shorter and stouter, the uppermost subsessile;

stipules 3–5 × 1.0–3.5 mm, narrowly ovate, acuminate at apex, the acumen often recurved, the basal part strongly veined. *Inflorescences* numerous, 5–12 × 5–8 mm, globose or ovoid, mostly axillary, attenuate and scarcely clasped at the base by the stipules; flowers ebracteate. *Calyx* 6–7 mm, densely hairy externally, glabrous internally with a hairy throat; tube 2.5–3.0 mm, cylindrical or narrowly campanulate, distinctly 10-veined; the lobes 5, 3–4 mm, lanceolate, acuminate at apex, the lowest longer than the tube, all rather rigid, with a strong midrib, suberect or recurved in fruit. *Corolla* 4–5 mm, whitish, rarely pink; standard with oblong limb, rounded at apex or slightly emarginate, scarcely auriculate at base, the claw rather wide; wings with limb oblong, rounded at apex, prominently auriculate at base, with a slender claw about 4 mm; keel with limb oblong, obtuse at apex, shortly auriculate at base, with claw about 4 mm. *Stamens* 10, diadelphous; filaments 5–6 mm, united almost to apex; anthers oblong. *Legume* concealed within the persistent calyx; seeds about 1.5 × 0.9 mm, yellowish or pale brown, oblong, smooth. *Flowers* 5–7. $2n = 10$.

Native. Short grassland and open places on sandy ground, especially near the sea. Locally frequent through Great Britain north to central-east Scotland, east Ireland and the Channel Islands. South and west Europe, Mediterranean region; North Africa; west Asia to Iran; Atlantic Islands. Submediterranean-Subatlantic element.

28. T. stellatum L. Starry Clover

Annual herb with fibrous roots. *Stems* (3–)5–20(–35) cm, pale green, erect or sprawling, with rather dense, spreading, silky simple hairs, sometimes simple, but usually much branched at the base, leafy. *Leaves* alternate, 3-foliolate; leaflets 4–15 × 4–15 mm, medium green on upper surface, paler beneath, obovate, emarginate at apex, minutely denticulate, cuneate at base, thinly villous on both surfaces; petioles 2–8 cm, channelled above, rounded beneath, thinly long-hairy; stipules 5–15 × 3–10 mm, conspicuous, obovate, rounded or subacute at apex, entire or denticulate, villous, prominently veined. *Inflorescence* about 3 cm in diameter, terminal, broadly ovoid or subglobose; peduncle usually well developed to 5 cm, more or less densely clothed with spreading or appressed, silky hairs; flowers ebracteate. *Calyx* 12–15 mm, narrowly campanulate, with dense, long hairs externally, glabrous or shortly hairy internally except at the densely hairy throat; tube 3–5 mm, obscurely 10-veined; lobes 6–9 mm, subulate, subequal, suberect at anthesis, spreading star-like in fruit. *Corolla* 8–12 mm, pink, rarely purple or yellow; standard with limb oblong, rounded or slightly emarginate at apex, scarcely auriculate at base, with the claw about 5 mm; wings with limb oblong, rounded at apex, prominently auriculate at base, with a claw 7–8 mm; keel with limb oblong, subacute at apex, bluntly auriculate at base, with claw about 8 mm. *Stamens* 10, diadelphous; filaments about 10 mm, united almost to apex; anthers oblong. *Style* 1, about 10 mm, glabrous. *Legume* hidden within the calyx; seed about 2.0 × 1.5 mm; pale brown, broadly oblong, smooth. *Flowers* 5–7. $2n = 14$.

Introduced. Naturalised since at least 1804 on shingle at Shoreham in Sussex; an infrequent casual elsewhere in southern Great Britain. Native of the Mediterranean region east to Iran; Atlantic Islands.

29. T. incarnatum L.

Annual herb with fibrous roots. *Stems* few to several, (10–)20–50 cm, pale green, erect or ascending, with simple eglandular hairs usually patent below and appressed above, simple or branching only from the base, leafy. *Leaves* alternate, 3-foliolate; leaflets 5–35 × 8–23 mm, medium green on upper surface, paler beneath, obovate or subrotund, emarginate at apex, denticulate towards the apex, cuneate at base, very shortly petiolulate; appressed hairy; petioles up to 8 cm, pale green, patent or appressed hairy; stipules 5–12 mm, often green, ovate, obtuse at apex, angled or obscurely dentate. *Inflorescence* a solitary, dense, oblong-ovoid to cylindrical head up to 6 cm, subtended by a solitary leaf; peduncles long, hairy. *Calyx* 9–10 mm; tube 2.7–5.0 mm nearly cylindrical, ribbed and villous, throat somewhat hairy but scarcely thickened; lobes 5, linear, acute at apex, patent in fruit. *Corolla* 9–16 mm, blood-red, pink, cream or white; standard elliptical, obtuse at apex, narrowed at base; wings oblong, obtuse at apex, with a narrow claw; keel oblong, mucronate at apex, truncate, with a narrow claw. *Stamens* 10, diadelphous; filaments pale; anthers yellowish. *Style* 1, pale; stigma brownish. *Legume* about 2.5 mm, included within the calyx, membranous in the lowest part, thickened above, often rupturing at the junction of the two parts; seed about 2 mm, greenish-yellow, ovoid. *Flowers* 5–9.

The following two subspecies are very distinct in north-west Europe but are connected by intermediates in the south.

(a) Subsp. **incarnatum** Crimson Clover

Stem usually 30 cm or more, often unbranched, with sparse spreading hairs. *Corolla* blood-red, rarely pure white, equalling or slightly exceeding the calyx. $2n = 14$.

(b) Subsp. **molineri** (Balb. ex Hornem.) Ces. Long-headed Clover

T. molineri Balb. ex Hornem.

Stems usually less than 20 cm, with dense appressed hairs. *Corolla* usually yellowish-white, rarely pink, much exceeding the calyx. $2n = 14$.

Subsp. *incarnatum* was introduced by 1596 and widely grown as a crop plant in the nineteenth century and as a winter annual for forage, particularly for sheep. It was sown in the early autumn, was fast growing and produced a good crop early in spring. Now rarely grown and a sporadic casual in the Channel Islands, Isle of Man and Great Britain; formerly in Ireland. Native of the Mediterranean region but extending to the extreme north of Europe as a crop plant and relic. Subsp. *molineri* is native in Great Britain in about five localities on cliff slopes, extending inland for only about 200 m along the south-facing slopes of the cove valleys that dissect the coasts of the Lizard Peninsula in Cornwall. It is confined to freely drained soils on hornblende and mica schists derived from periglacial material and most sites have a southerly aspect. In Jersey it is found on the islet in Portelet Bay and cliffs of the north coast. It is also native of western and southern Europe, extending eastwards from Iberia to Greece northwards to Great Britain and Jersey.

30. T. pratense L. Red Clover

Caespitose *perennial herb. Stems* 5–100 cm, pale green, erect or sprawling, thinly to densely hairy to nearly glabrous, rather sparingly branched, chiefly at the base, leafy. *Leaves* alternate, 3-foliolate; leaflets 15–50 × 10–30 mm, medium to rather dark, often bluish-green on upper surface, often white-blotched, paler beneath, oblong-ovate, obtuse or subacute at apex, sometimes emarginate, entire, cuneate at base, thinly hairy or nearly glabrous on upper surface, more densely hairy beneath; petioles of lower leaves usually very long, often exceeding 15 cm, becoming progressively shorter upwards, the uppermost leaves subsessile or very shortly petiolate; stipules 10–25 × 3–5 mm, pale and conspicuously veined, with anastomosing green veins, oblong-ovate, abruptly narrowed to a slender filiform-subulate apex, thinly hairy to nearly glabrous. *Inflorescence* a dense, pseudoterminal, subglobose, sessile spike 15–40 mm in diameter, subtended by 2 subopposite, reduced, bract-like leaves with very broad stipules; flowers sessile or subsessile. *Calyx* tube about 3 mm, obscurely 10-veined, 5-lobed, the teeth filiform, rigid and thinly hairy, the lowermost about 7 mm, the others about 3.5 mm, separated by broad sinuses, usually appressed-hairy. *Corolla* 12–15 mm, usually reddish-purple or pink, rarely cream or white; standard with limb narrowly ovate-oblong, truncate at apex, shortly auriculate at base, with a claw about 8 mm; wings with limb narrowly oblong, obtuse or subacute at apex, the base conspicuously auriculate; keel with an acute apex; all the petals united by the claws into a tube. *Stamens* 10, diadelphous; filaments about 12 mm, green; anthers orange or yellow, oblong. *Style* 1, green, filiform; stigma green, capitate. *Legume* completely concealed by persistent calyx and corolla; seeds about 1.7 × 1.2 mm, rich brown, ovoid, compressed, smooth. *Flowers* 5–9. $2n=14$.

(i) Var. **parviflorum** Bab.
T. pratense var. *micropetalum* Lange

Stems 15–35 cm, with few appressed hairs. *Leaflets* up to 20 mm. *Inflorescence* 15–25 mm long, shortly pedunculate; flowers pedicellate, sometimes bracteate and corolla not exceeding the calyx. *Calyx* with few to numerous, long hairs.

(ii) Var. **villosum** Wahlb.
T. pratense var. *maritimum* Zabel

Stems up to 20 cm, procumbent to ascending, with dense, long, appressed hairs in upper part. *Leaflets* up to 15 mm. *Inflorescence* 15–25 mm in diameter. *Calyx* with dense, long appressed hairs.

(iii) Var. **pratense**
Long-lived *perennial. Stems* 20–40 cm, solid, procumbent or ascending, appressed-hairy or more rarely with patent hairs above. *Leaflets* up to 25(–40) mm, emarginate at apex. *Inflorescence* 15–25 mm in diameter, often solitary. *Calyx* with numerous, long hairs. *Corolla* often deep red, rarely white.

(iv) Var. **sativum** Afzel.
T. sativum (Afzel.) Crome

Short-lived *perennial. Stems* 40–70(–100) cm, more or less erect, sparsely appressed-hairy to glabrous or becoming so below, hollow. *Leaflets* up to 50 mm or more, not emarginate at apex. *Inflorescence* 25–40 mm in diameter, often paired. *Calyx* with long hairs. *Corolla* pale to deep red or pink, sometimes white.

(v) Var. **americanum** Harz
T. expansum Waldst. & Kit.; *T. diffusum* Baumg., non Ehrh.

Stems up to 80 cm, with numerous, stiff, spreading hairs. *Leaflets* up to 50 mm, never emarginate. *Inflorescence* 25–40 mm in diameter. *Calyx* with dense spreading hairs. *Corolla* usually deep red, rarely white.

Native. Grassy places, waste and rough ground. Common throughout Great Britain and Ireland. Most of Europe; west and central Asia; North Africa; Atlantic Islands; introduced elsewhere. Eurosiberian Temperate element. Var. *pratense* is the commonest native plant, especially in non-agricultural areas. Var. *sativum* was formally grown in agricultural areas where it escaped and was widely naturalised in Great Britain and Ireland. Very little is now grown, but the variety is still naturalised. It has recently been included in wild flowers seed planted round the grass margins of fields in Cambridgeshire. Var. *americanum* is native of parts of south-east Europe. It was introduced into cultivation in the rest of Europe via North America. Specimens have been seen from as far apart as southern England and Shetland, and Co. Donegal and Co. Mayo, but it is not known if it still occurs. It appears also to be in wild flower seed. Specimens of var. *villosum* have been seen from South Uist and Shetland in central areas. It also occurs on the coast of the south Baltic. Var. *parviflorum* is a widely occurring monstrosity which is included because it is often confused with unrelated species. Red Clover was formerly one of our commonest field crops and was usually sown with corn in the spring, and allowed to grow up after the crop was harvested, pastured by sheep and cattle, and either cut for hay the following year or mowed several times and used for cattle feed. Nitrogen-fixing root nodules are formed in conjunction with *Rhizobium trifolii*.

31. T. diffusum Ehrh. Diffuse Clover
T. purpurascens Roth; *T. ciliosum* Thuill.; *T. pratense* subsp. *diffusum* (Ehrh.) Gibelli & Belli

Annual herb with fibrous roots. *Stems* 20–50 cm, pale green, ascending or erect, long-hairy, branched, leafy. *Leaves* alternate, 3-foliolate; leaflets 20–30 × 8–15 mm, medium green on upper surface, paler beneath, obovate, broadly elliptical or oblong, obtuse at apex, denticulate towards the apex, cuneate at base, hairy; petioles long and slender; stipules oblong, membranous, green-veined, connate for over half their length, the free portion lanceolate or subulate, hairy. *Inflorescence* of globose, terminal heads up to 30 mm in diameter; sessile or shortly peduncled; flowers ebracteate. *Calyx* 13–15 mm, tubular-campanulate, the tube 10-veined, long-hairy, throat open with a ciliate ring; lobes subulate, setaceous, with triangular, 3-veined base, obtuse at apex. *Corolla* 8–14 mm, purplish-pink; standard ovate, obtuse at apex; wings oblong, obtuse; keel oblong, obtuse at apex. *Stamens* 10, diadelphous; filaments pale; anthers yellowish. *Style* 1, pale; stigma brownish. *Legume* membranous, with cartilaginous operculum. *Flowers* 4–6. $2n=16$.

Introduced. Wool casual. Native of central and south Europe and the Caucasus.

32. T. lappaceum L. Bur Clover

Annual herb with fibrous roots. *Stems* 6–35(–40) cm, pale green, erect or decumbent, glabrous or thinly hairy especially towards the apex, branched especially towards the base, leafy. *Leaves* alternate, 3-foliolate; leaflets 5–15 × 2–8 mm, medium green on upper surface, paler beneath, obovate, rounded or slightly emarginate at apex, obscurely denticulate, narrowly cuneate at base, thinly hairy on both surfaces; petioles 10–20 mm, pale green, slender channelled above, rounded beneath, glabrous or thinly hairy, the uppermost leaves subsessile; stipules 6–8 × 1–2 mm, pale, linear-subulate at apex, spreading, papery and distinctly veined, often thinly hairy. *Inflorescence* terminal, 10–15(–20) mm in diameter; peduncles short; flowers ebracteate. *Calyx* 5–7 mm, narrowly campanulate, the tube 20-veined, glabrous or hairy, usually with a ring of hairs in the throat; lobes 5, longer than the tube, linear-setaceous and elongating somewhat in fruit. *Corolla* 4–8 mm, whitish or pinkish; standard with limb oblong, rounded or truncate at apex and abruptly narrowed at apex; wings with limb oblong, obtuse at apex and auricled at base; keel with limb oblong, obtuse at apex and not auricled at base. *Stamens* 10, diadelphous; filaments with the connate ones so for almost their complete length, sometimes with the margins of the fused filaments joined near the apex and forming a short tube round the style; anthers subglobose. *Style* 1, 4–5 mm, glabrous; stigma capitate. *Legume* more or less concealed by the persistent calyx; seeds 1.3–1.8 mm, medium brown, subglobose or broadly ovoid, glossy and smooth. *Flowers* 6–7. $2n=16$.

Introduced. On tips and waste places, mainly casual from bird-seed. Sporadic in scattered localities over Great Britain and formerly in Ireland. Native of southern Europe and the Mediterranean region.

33. T. hirtum All. Rose Clover

Annual herb with fibrous roots. *Stems* 10–35 cm, pale green, erect or decumbent, often with patent branches from the base, with dense spreading hairs, leafy. *Leaves* alternate, trifoliate; leaflets 8–20 × 4–15 mm, medium green on upper surface, paler beneath, blotched or unblotched, obovate, rounded at apex, rarely emarginate, minutely denticulate, cuneate at base, densely appressed-hairy on both surfaces; petioles up to 30 mm, pale green, slender, channelled above, rounded beneath, hairy; stipules 10–14 × 2–3 mm, pale, lanceolate, with a slender, elongate almost filiform apex, with spreading hairs. *Inflorescence* 17–20(–25) mm in diameter, globose, solitary, sessile, with an involucre formed of dilated stipules and 1 or 2, 3-foliolate leaves. *Calyx* 10–11 mm, campanulate, 20-veined, with 5 lobes twice as long as the tube, setaceous and densely long-hairy. *Corolla* 12–17 mm, purple; standard with a very narrowly oblong, longitudinally folded limb with an acute apex; wings with an oblong limb, an ovate apex and a distinctly auriculate base; keel with an acute or acuminate apex. *Stamens* 10, diadelphous; filaments 7.0–7.5 mm; anthers oblong. *Style* 1, 6–7 mm, glabrous; stigma capitate. *Legume* concealed by the persistent calyx and corolla; seeds 2.0–2.5 × 1.5–1.8 mm, pale brown, broadly ovoid, smooth. *Flowers* 6–7. $2n=10$.

Introduced. A wool casual on tips and waste ground. Sporadic in scattered localities over Great Britain. Widespread native of southern Europe and the Mediterranean region, eastwards to Iraq.

34. T. cherleri L. Cherler's Clover

Annual herb with fibrous roots. *Stems* 4–15 cm, pale green, decumbent, densely hairy with spreading hairs, much branched at base, leafy. *Leaves* alternate, 3-foliolate; leaflets 2–10 × 1.5–6.0 mm, medium green on upper surface, paler beneath, obovate, emarginate at apex, minutely denticulate, the veins running into the teeth, narrowly cuneate at base, hairy on both surfaces; petioles 5–20(–30) mm, slender, channelled above, rounded beneath, hairy; stipules 4–5 × 1.5–2.5 mm, whitish towards the base and conspicuously veined, ovate, acuminate at apex, glabrous below and hairy towards apex. *Inflorescence* pseudoterminal, globose, 10–17 mm in diameter, sessile, subtended by a conspicuous involucre of whitish, prominently veined, broadly ovate or subrotund, overlapping stipule-bracts 5–7 mm. *Calyx* 6–7 mm, narrowly campanulate, 20-veined, asymmetrical, densely hairy, the 5 lobes about 4 mm, often purplish, setaceous and long-hairy. *Corolla* 8–10 mm, whitish or very pale pink becoming reddish with age; standard with limb narrowly oblong, longitudinally folded, acute at apex, sometimes mucronate, scarcely auriculate at base; wings with limb narrowly oblong, obtuse at apex and prominently auriculate at base; keel with apex acute and base scarcely auriculate. *Stamens* 10, diadelphous; filaments 6–7 mm; anthers oblong. *Style* 1, about 5 mm, glabrous; stigma capitate. *Legume* quite concealed by persistent calyx and corolla; seeds 1.5–2.0 mm, pale brown, broadly ovoid or subglobose, smooth. *Flowers* 5–6. $2n=10$, 16.

Introduced. A wool and tan-bark casual; formerly a ballast alien. Mediterranean region, eastwards to west Iran; Atlantic Islands. Named after Johann Heinrich Cherler (1570–1610).

35. T. medium L. Zigzag Clover

T. alpestre Pollich; *T. bithynicum* Boiss.

Perennial herb with extensive, slender rhizomes which form clonal patches. *Stems* (10–)30–45(–60) cm, pale green, straggling and ascending, more or less flexuous, sparsely appressed-hairy, sometimes becoming glabrous, often branched, leafy. *Leaves* alternate, 3-foliolate; leaflets 20–50×(5–)9–20(–35) mm, medium green on upper surface, often with a faint, whitish spot, paler beneath, ovate, obovate or elliptical, obtuse or acute at apex, scarcely toothed, ciliate; petioles up to 8 cm and usually exceeding the stipules, pale green; stipules 20–25 mm, oblong or linear-oblong, the upper ovate, usually adnate by less than half their length to the petioles, the free part subulate and spreading. *Inflorescence* a terminal, globose head 25–35 mm in diameter, ultimately shortly pedunculate and subtended by a pair of leaves. *Calyx* 8–10 mm; tube more or less globose, 10-veined, hairy but not thickened in the throat, glabrous or becoming so; lobes 5, filiform, the upper 4 not longer than the tube, ciliate, spreading in fruit. *Corolla* 12–20 mm, 2–3 times as long as calyx and tardily deciduous, purplish-red; standard with limb ovate above, rounded at apex, and oblong in lower half; wings with limb oblong, obtuse at apex, with a lobe and short spur at base; keel with limb, oblong, obtuse at apex, with a long narrow spur. *Stamens* 10, diadelphous; filaments pale; stigma brownish. *Legume* 1.8–2.0 mm, membranous, obovoid, asymmetrical, truncate at apex, dehiscing longitudinally; seeds about 1.7 mm. *Flowers* 6–9. $2n=80$.

Native. Grassy places, hedgerows and wood borders and ruderal habitats such as spoil heaps and railway banks. Throughout most of Great Britain and Ireland, but local and decreasing in lowland areas. Europe except the extreme north and south; Caucasus; west Siberia; naturalised in North America. Our plant is subsp. **medium**, which occurs throughout the range of the species in Europe except perhaps Greece.

36. T. angustifolium L. Narrow-leaved Clover

Annual herb with fibrous roots. *Stems* 8–50 cm, pale green, erect, appressed-hairy, simple or sparingly branched at the base, leafy. *Leaves* alternate, 3-foliolate; leaflets (10–)20–80×(1–)2–5 mm, medium green on upper surface, pale beneath, narrowly oblong or linear, the lower obtuse or rounded at apex, the upper acute or acuminate, entire, thinly appressed-hairy on both surfaces; petioles up to 35 mm, pale green, slightly channelled above, rounded beneath, thinly hairy, the uppermost leaves often subsessile; stipules 10–25×3–5 mm, pale, conspicuously veined, with a slender caudate appendage at apex 5–10 mm, narrowly oblong and connate and tubular at base, thinly hairy. *Inflorescence* terminal and solitary, 15–60×15–25 mm, oblong-cylindrical; peduncles usually less than 4 cm, densely appressed-hairy; flowers ebracteate. *Calyx* about 10×2 mm, narrowly campanulate, 10-veined, densely hairy externally, with a ring of hairs in the throat, 5-lobed, the 3 abaxial 5–6 mm, the other a little shorter, subulate, rigid, sharply pointed at apex. *Corolla* 10–13 mm, pink or purplish; standard a little longer than wings, the limb oblong, longitudinally infolded, obtuse or subacute at apex; wings with limb oblong, obtuse at apex, with a prominent auricle at base and a very long, slender claw; keel about as long as wings, the limb oblong and subacute at apex. *Stamens* 10, diadelphous; filaments 9–10 mm; anthers oblong-subglobose. *Style* 1, about 7 mm, glabrous; stigma incurved. *Legume* concealed within the persistent calyx; seed 2.0–2.5×about 1.5 mm, yellowish, oblong, smooth. *Flowers* 5–6. $2n=14$, 16.

Introduced. Tips and waste ground where it is a casual from wool and other sources. Sporadic in scattered localities in Great Britain. Native of south Europe and the Mediterranean region, east to Iran; Atlantic Islands.

37. T. purpureum Loisel. Purple Clover

T. angustifolium subsp. *purpureum* (Loisel.) Gibelli & Belli; *T. loiseleuri* Rouy

Annual herb with fibrous roots. *Stems* 10–30 (–50) cm, erect or ascending, rarely decumbent, appressed-hairy, branched above, leafy. *Leaves* alternate, 3-foliolate; leaflets (10–)20–40 (–60)×2–10 mm, medium green on upper surface, paler beneath, oblong-lanceolate to linear, acute and mucronulate at apex, slightly toothed in upper part, narrowed at base, sessile, mostly hairy only at the margins; petioles getting shorter up the stem; stipules 10–15 mm, oblong-lanceolate, many-veined, membranous, adnate below, the free portion long-subulate. *Inflorescence* of ovoid to conical heads in flower, becoming ovoid-oblong to cylindrical in fruit, sessile or nearly so, flowers ebracteate. *Calyx* 8–10 mm, the tube almost tubular, the lobes unequal, subulate-setaceous, obtuse, the lowermost one slightly longer than to twice as long as the others, subappressed- to antrorse-hairy. *Corolla* 10–25 mm, purple, lilac or whitish; standard with limb oblong, obtuse at apex; wings oblong, obtuse at apex; keel with limb oblong, obtuse at apex. *Stamens* 10, diadelphous; filaments pale; anthers yellowish. *Style* 1, glabrous. *Legume* ovoid, membranous, with a cartilaginous apex; seed 1, about 1 mm, brown, ovoid. *Flowers* 5–6. $2n=14$.

Introduced. A wool casual. Native of the Mediterranean region.

38. T. ochroleucon Huds. Sulphur Clover

Perennial herb with a short rhizome or caespitose. *Stem* 20–50 cm, pale green, erect or ascending, more or less hairy, shortly branched, leafy. *Leaves* alternate, 3-foliolate; leaflets 15–30(–50)×5–8 mm, medium green on upper surface, paler beneath, oblong-elliptical or lanceolate, obtuse or sometimes emarginate at apex, entire, narrowed at base, more or less hairy; petioles up to 10 cm, pale green; stipules 10–30 mm, oblong or ovate-oblong, with a linear-lanceolate, herbaceous apex. *Inflorescence* a terminal globose to ellipsoid head, becoming subtended by a pair of nearly sessile leaves; flowers ebracteate. *Calyx* 12–25 mm, tube campanulate, strongly ribbed, the throat hairy, becoming somewhat thickened in fruit; lobes 5, the 4 upper equalling or shorter than the tube, the lowest usually longer and deflexed in fruit, with one distinct central vein, rarely with 2 lateral veins. *Corolla* 15–20 mm, longer than calyx, eventually deciduous, yellowish-white; standard with limb lanceolate in upper part, oblong below, acute at apex, rounded at base; wings with limb oblong, subobtuse at apex, with a lobe and short claw below;

keel with limb oblong, obtuse at apex, with a long claw. *Stamens* 10, diadelphous; filaments pale; anthers yellowish. *Style* 1, slender, with a stout base; stigma brownish. *Legume* 20–25 mm, obovoid; seed 1.0–1.8 mm, obovoid. *Flowers* 6–7. $2n = 16$.

Native. Grassy places on chalky boulder clay and more rarely chalk in pastures, roadside verges, trackways and wood borders. Very local and decreasing in East Anglia west to Northamptonshire; casual elsewhere in scattered localities in England. West central and east Europe; north-west Africa; Caucasus. European Temperate element. Our plant is var. **ochroleucum** with yellowish-white corolla. Plants in south Europe have a pink corolla and are var. **roseum** (C. Presl) Guss. (*T. roseum* C. Presl).

39. T. pannonicum Jacq. Hungarian Clover

Perennial herb. Stems 20–40 (–60) cm, pale green, erect or ascending, striate, leafy, sparsely branched. *Leaves* alternate, 3-foliolate; leaflets 20–30(–45) × 5–12 mm, dark dull green on upper surface, paler beneath, obovate to elliptical or oblong-lanceolate, obtuse at apex, entire, narrowed below, appressed-hairy; petioles long in lower leaves, shorter in upper; stipules somewhat membranous with green veins, linear-oblong, the free portion linear-subulate, hairy. *Inflorescence* an ovoid to ovoid-oblong, many-flowered head 20–40(–50) × 20–30(–50) mm, short or long peduncled and often subtended by 2 opposite, shortly mucronate leaves. *Calyx* 16–20 mm, cylindrical at first, later campanulate, the tube with 10 veins, long-hairy, the 5 lobes subulate, sharply acute at apex, the upper ones about as long as the lower, the fruiting calyx with throat closed by a bilabiate callosity and stellately spreading or deflexed teeth. *Corolla* 20–25 mm, yellowish-white; standard elliptical or oblong, obtuse at apex; wings lanceolate or oblong-lanceolate, obtuse at apex; keel oblong and obtuse at apex. *Stamens* 10, diadelphous; filaments pale; anthers yellowish. *Style* 1, pale; stigma brownish. *Legume* included in the calyx; seed about 2 mm. *Flowers* 4–6. $2n = 126, 128, 130$.

Introduced. A scarce casual as a grass-seed contaminant, established on a railway bank at Forty Green, near Beaconsfield in Buckinghamshire. Sporadic in a few localities in Great Britain. Native of southern Europe. Our plant is subsp. **pannonicum**.

40. T. alexandrinum L. Egyptian Clover

Annual herb with fibrous roots. *Stems* 40–70 cm, pale green, erect, sparsely appressed-hairy, branching, leafy. *Leaves* alternate, 3-foliolate; leaflets 15–25 × 5–10 mm, medium green on upper surface, paler beneath, oblong or lanceolate, mucronate at apex, denticulate, narrowed at base, appressed-hairy; petiole up to 3 cm, pale green, hairy; stipules 17–22 mm, adnate at base, free part subulate, marginal hairs of the upper ones dilated at base. *Inflorescence* an ovoid or oblong-conical head 15–25 mm in diameter; peduncles up to 30 mm; flowers ebracteate. *Calyx* 7–9 mm, hairy; tube obconical; lobes 5, unequal, triangular-subulate, spiny, the lowest 3-veined, at least at the base, about as long as the tube, others shorter and 1-veined. *Corolla* 8–10 mm, about twice as long as the calyx, cream; standard with limb oblong and slightly waisted, obtuse at apex, narrowed below; wings with limb oblong, obtuse at apex, narrowed at base; keel with limb oblong; obtuse at apex. *Stamens* 10, diadelphous; filaments pale; anthers yellowish. *Style* 1, pale; stigma brownish. *Legume* 2.2–2.5 mm, coriaceous, apex when mature slightly exserted, not concealed by the ring of hairs in the calyx throat, which is devoid of a bilabiate callosity. *Flowers* 4–6. $2n = 16$.

Introduced. Casual mainly from grass-seed mixtures, in newly sown grass by roads and in parks. Guernsey in Channel Islands and rare in southern England. Possibly native in eastern Mediterranean region and widely cultivated in the warmer parts of Europe.

41. T. constantinopolitanum Ser. Constantinople Clover

Annual herb with fibrous roots. *Stems* (10–)15–35(–60) cm, erect or ascending, hairy, leafy, branched. *Leaves* alternate below, opposite above, 3-foliolate; leaflets (8–)15–20(–30) × 4–6(–8) mm, medium green on upper surface, paler beneath, elliptical to obovate-oblong, narrowed to an acute or obtuse apex, denticulate towards the apex, cuneate at the sessile base, appressed-hairy; petioles long and hairy; stipules adnate below, lanceolate or subulate above, with green or blackish veins. *Inflorescence* an ovoid or obconical terminal head 12–22 × 10–15(–22) mm, long-peduncled; flowers ebracteate. *Calyx* tube cylindrical, with 10 prominent nerves; lobes 5, narrowly lanceolate to subulate, purple-tipped, 1-nerved, rarely 3-nerved, lower lobe 1.5 times as long as the others, densely hairy at base. *Corolla* 8–12 mm, cream, the keel often purple-tipped; standard oblong, obtuse at apex; wings oblong, obtuse at apex; keel obtuse at apex. *Stamens* 10, diadelphous; filaments pale; anthers yellowish. *Style* 1. *Legume* oblong, with coriaceous, apiculate apex, membranous; seeds about 1.5 mm, yellowish-brown, ovoid. *Flowers* 4–6. $2n = 16$.

Introduced. A wool, grain and tan-bark casual. Native of Switzerland, France and Syria, introduced in Italy and Algeria; sometimes cultivated in Turkey.

42. T. echinatum M. Bieb. Hedgehog Clover

T. supinum Savi; *T. leucanthum* auct.; *T. procerum* Rochel; *T. trichostomum* Godr.; *T. reclinatum* Waldst. & Kit.

Annual herb with fibrous roots. *Stems* 5–50 cm, purplish, erect or sprawling, thinly hairy, usually much-branched, leafy. *Leaves* alternate below, opposite above, trifoliate; leaflets medium green on upper surface, paler beneath, 5–20 × 2.5–8 mm, obovate or elliptical, rounded or slightly emarginate at apex and often mucronate, entire, cuneate at base, thinly appressed hairy on both surfaces; petiole usually less than 20 mm, narrowly channelled above, rounded beneath, hairy and often purplish; stipules 10–12 × 1–2 mm, pale with conspicuous purple veins, narrowly lanceolate, subentire or shortly serrulate, thinly long-hairy. *Inflorescence* 10–20 mm, subglobose or shortly oblong, axillary though appearing terminal; peduncle 10–40 cm, more or less densely appressed-hairy, flowers 8–15 mm, ebracteate. *Calyx* 4–7 mm, campanulate, thinly hairy outside with a ring of hairs in the throat, distinctly 10-veined; lobes 5, narrowly subulate, the 3 abaxial 2–5 mm, distinctly longer than the 2 adaxial, suberect or spreading in fruit. *Corolla* cream, pink or purple; standard with a narrowly oblong limb, the apex rounded or truncate and sometimes erose; wings with limb narrowly oblong, subacute at apex and strongly swollen and auriculate at base; keel with limb with an obtuse apex, a slightly auriculate base and a long, slender claw. *Stamens* 10, diadelphous; filaments 5–7 mm, connate nearly to the apex; anthers yellow, oblong. *Style* 1, 7–8 mm, pale, glabrous; stigma oblique. *Legume* completely concealed within the calyx; seeds about 1.5 × 1.3 mm, brown, reniform, smooth. *Flowers* 6–7. $2n=16$.

Introduced. Casual on tips and rough ground, mostly from bird-seed. Sporadic in a few scattered localities in Great Britain. Native of south-east Europe and south-west Asia east of Iran and Iraq.

43. T. leucanthum M. Bieb. White-flowered Clover
T. dipsaceum subsp. *leucanthum* (M. Bieb.) Gibelli. & Belli

Annual herb with fibrous roots. *Stems* (5–)15–30 cm, pale green, erect, with dense, patent hairs, branched, leafy. *Leaves* alternate, 3-foliolate; leaflets 10–25 × 7–9 mm, medium green on upper surface, paler beneath, oblong or elliptical, emarginate at apex, cuneate below; petioles up to 40 mm; stipules 18–20 mm, adnate at base, with a long, linear, free apex. *Inflorescence* of nearly globose, often paired heads 10–15 mm in diameter; peduncles 30–120 mm, appressed-hairy; flowers ebracteate. *Calyx* 8–10 mm, urceolate when fruiting, densely hairy, the veins reaching the base of the 5 subequal, lanceolate, 3-veined acute lobes. *Corolla* 6–9 mm, more or less equalling the calyx, white or pink; standard with limb oblong, obtuse at apex, narrowed at base; wings with limb oblong, obtuse at apex, narrowed at base; keel with limb oblong, obtuse at apex, narrowed at base. *Stamens* 10, diadelphous; filaments pale; anthers yellowish. *Style* 1, pale; stigma brownish. *Legume* included, ovoid, apiculate at apex; seeds 1.6–2.0 × 1.0–1.6 mm, globose or obovoid. *Flowers* 4–7. $2n=14, 16$.

Introduced. A wool and tan-bark casual. Native of the Mediterranean region.

44. T. squamosum L. Sea Clover
T. maritimum Huds.

Annual herb with fibrous roots. *Stems* 10–40 cm, pale green, procumbent, ascending or erect, more or less hairy, sometimes becoming glabrous, branched, leafy. *Leaves* alternate, 3-foliolate; leaflets 10–20 × 6–8 mm, medium green on upper surface, paler beneath, narrowly obovate or oblong, often apiculate at apex, entire, cuneate at base, hairy; petioles up to 10 mm, pale green; stipules 15–20 mm, linear or oblong, the free part herbaceous and longer than the rest. *Inflorescence* an ovoid, terminal head 10–20 mm in diameter, shortly pedunculate and subtended by a pair of leaves. *Calyx* 2.6–5.5 mm; tube campanulate, glabrous or thinly hairy above, strongly 10-veined with furrows disappearing below the dilated mouth, and coriaceous in fruit, throat minutely hairy becoming strongly thickened in fruit; lobes 5, green, lanceolate, acuminate at apex, spreading, the lowest distinctly 3-veined, about half as long as the tube. *Corolla* 5.0–7.5 mm, exceeding the calyx, pale pink; standard with limb oblong and slightly waisted, emarginate at apex, cuneate at base, wings with limb oblong, rounded at apex, auriculate at base; keel with limb oblong, rounded at apex, truncate at base. *Stamens* 10, diadelphous; filaments pale; anthers yellowish. *Style* 1, pale; stigma brownish. *Legume* 2.0–2.5 mm, obovoid, membranous in the lower part, thickened above, often rupturing at the junction of the two parts; seeds about 2 mm, obovoid. *Flowers* 4–7. $2n=16$.

Native. Edges of salt-marshes and eroded saltings, on sea walls, in brackish meadows, on drained estuarine marshes and on tidal rivers or creeks; rarely on sand or on low limestone cliffs and as a casual in waste places and by railways; the soil is usually a salty clay. Very local in Great Britain north to south Wales and Essex, formerly to Suffolk and Lincolnshire and in Guernsey in the Channel Islands, its farthest inland site is about 10 km on the tidal Sussex Ouse. Western and southern Europe; western Asia; North Africa. It is at its northern limit in Great Britain. Mediterranean-Atlantic element.

Section 7. **Trichocephalum** W. D. J. Koch
Annual herbs. Inflorescence a capitate head, the outer flowers fertile, the inner consisting of only sterile calyces.

45. T. subterraneum L. Subterranean Clover
Calycomorphum subterraneum (L.) C. Presl

Annual herb with fibrous roots. *Stem* up to 45 cm, pale green, prostrate, sparsely to densely villous, usually much branched, leafy. *Leaves* alternate, 3-foliolate; leaflets 3–17 × 2–22 mm, medium green on upper surface, paler beneath, broadly obovate, usually emarginate at apex, entire or obscurely denticulate, cuneate at base, subappressed hairy on both surfaces; petioles up to 10 cm, slender, slightly channelled above, rounded below, subglabrous or slightly villous; stipules 4–14 × 2–4 mm, membranous, oblong, caudate at apex, subglabrous or thinly clothed with long hairs. *Inflorescence* axillary, with 1–5 fertile flowers; peduncle at first 5–30 mm, lengthening and recurving after fertilisation and burying the infructescence, thinly or rather densely villous. *Calyx* 4–5 mm; tube 1.5–2.0 mm, narrowly cylindrical, often purplish, obscurely veined, subglabrous, the lobes 2–3 mm, subequal, thinly villous. *Corolla* of fertile flowers 8–14 mm, creamy-white, sometimes tinged pink; standard with limb oblong, rounded or slightly emarginate at apex, obscurely auriculate at base, the claw about 3 mm; wings with limb oblong, rounded at apex, strongly auriculate at base, the claw about 4 mm; keel with limb oblong, obtuse at apex, scarcely auriculate, with the claw about 4 mm. *Stamens* 10, diadelphous; filaments 1.0–1.5 mm, glabrous; anthers brownish. *Style* 1, about 5 mm, glabrous; stigma oblique. *Legume* concealed by numerous, reflexed sterile flowers which develop after fertilisation, partly, but wholly, enclosed within enlarged fruiting calyx, minutely and obscurely veined; seed 2.5–3.0×2.0–2.5 mm, dark brown, broadly oblong, smooth. *Flowers* 5–6. 2*n* = 16.

(i) Var. **subterraneum**
Stems up to 15 cm. *Leaflets* 2–10 mm wide. *Calyx* covering most of ripe legume.

(ii) Var. **oxaloides** (Bunge ex Nyman) Rouy
T. subterraneum subsp. *oxaloides* Bunge ex Nyman

Stems 15–45 cm. *Leaflets* 10–22 mm wide. *Calyx* covering only half of ripe pod.

Native. Short turf and rather bare places on sandy soils, especially by the sea. Scattered localities in the Channel Islands and Great Britain north to Lincolnshire and Isle of Man, and Co. Wicklow in Ireland; also a frequent wool alien. Widespread in Europe and the Mediterranean region eastwards to Iran; Atlantic Islands. Submediterranean-Subatlantic element. Var. *subterraneum* is the native plant. Var. *oxaloides* is the plant of wool shoddy. Var. *oxaloides* is native of south-east Europe and the Caucasus, but may have come via Australia where it is introduced.

36. Ononis L.

Annual or *perennial herbs* or *dwarf shrubs*, often glandular-hairy. Leaves alternate, 3-foliolate, sometimes simple or imparipinnate; leaflets usually dentate; stipules adnate to the petiole. *Inflorescence* of solitary flowers or terminal racemes. *Calyx* campanulate or tubular. *Corolla* yellow or pink, rarely white. *Stamens* 10, monadelphous. *Style* 1. *Fruit* a legume, oblong or ovate, straight, dehiscing longitudinally; seeds 1–many.

Contains about 70 species in Europe, western Asia, North Africa and the Atlantic Islands.

Greuter, G. (1986). *Ononis* L. *Willdenowia* 16: 113.
Hultén, E. & Fries, M. (1986). *Atlas of north European vascular plants north of the Tropic of Cancer.* 3 vols. Königstein.
Ivimey-Cook, R. B. (1969). Investigations into the phenetic relationships between species of *Ononis* L. *Watsonia* **7**: 1–23.
Lang, H. J. (1977). *Ononis reclinata* in v.c. 74 (Wigtown). *B. S. B. I. News* **17**: 29.
Morisset, P. (1978). Chromosome numbers in *Ononis* L. series *Vulgares* Sirj. *Watsonia* **12**: 145–153.
Stephens, C. E. (1978). Variation in the terminal leaflet shape of *Ononis repens* L. in the British Isles. *Watsonia* **12**: 165–166. (The leaves of coastal colonies with lower indices are referable to *O. spinosa* subsp. *maritima*.)
Wigginton, M. J. (Edit.) (1999). *British red data books.* Vol. 1. *Vascular plants*. Peterborough. [*O. reclinata*.]

1. Perennials, with stems woody at least below — 2.
1. Annuals, not woody below — 6.
2. Corolla yellow, often streaked red; legume (11–)13–25 mm — **1. natrix**
2. Corolla pink or rose; legume 5–10 mm — 3.
3. Stems erect — 4.
3. Stems decumbent or prostrate — 5.
4. Stem with two lines of simple eglandular hairs — **3(a). spinosa** subsp. **spinosa**
4. Stem sparingly to densely hairy, the hairs a mixture of simple eglandular and glandular — **3(b). spinosa** subsp. **intermedia**
5. Stems decumbent or prostrate with numerous simple eglandular and glandular hairs; leaflets 10–15(–20) mm — **3(c). spinosa** subsp. **procurrens**
5. Stems flat to the ground, with dense simple eglandular and glandular hairs — **3(d). spinosa** subsp. **maritima**
6. Racemes not leafy, borne on bare peduncles — **7. baetica**
6. Racemes leafy, borne on leafy peduncles — 7.
7. Calyx 7–10 mm, with broadly lanceolate or oblong, acuminate lobes; seeds 1.5–2.0 mm — **4. mitissima**
7. Calyx 10–13 mm, with linear-setaceous lobes — **5. alopecuroides**

Section **1. Natrix** (Moench) Griseb.
Natrix Moench
Flowers in panicles, the primary branches 1- to 3-flowered; fruiting pedicel more or less deflexed. *Legume* oblong.

1. O. natrix L. — Yellow Restharrow

Dwarf *shrub. Stems* 20–60 cm, greenish to brownish, erect, densely glandular-hairy, much-branched, leafy. *Leaves* alternate, 3-foliolate, the lower rarely pinnate; leaflets 5–30×3–20 mm, medium or dark green on upper surface, paler beneath, elliptical, oblong-elliptical or oblong-obovate, truncate at apex, more or less dentate, glandular-hairy; petioles fairly short; stipules lanceolate, acute at apex. *Inflorescence* a lax, leafy panicle; peduncle 2–15 mm, glandular-hairy. *Calyx* 7–17 mm, divided into 5 lobes as long as tube, glandular-hairy. *Corolla* 11–25 mm, yellow, with red or violet streaks; standard glabrous; subrotund, retuse at apex, with a short claw; wings oblong, rounded at apex; keel obovate, with a beak. *Stamens* 10, monadelphous; filaments pale; anthers yellowish. *Style* 1, pale. *Legume* (11–)13–25 mm, oblong, glandular-hairy; seeds 1.5–2.1 mm, subreniform, tuberculate. *Flowers* 6–7. 2*n* = 28, 30, 32, 64.

Introduced. Naturalised in rough ground at Greenham Common in Berkshire since 1947 and recorded for a few other counties. Native of south and west Europe. Our plant is the subsp. **natrix** of *Flora Europaea*.

2. O. reclinata L. — Small Restharrow

Annual herb with fibrous roots. *Stem* 2–15 cm, pale green, erect or ascending, viscid-glandular, branched from near the base, leafy. *Leaves* alternate, pinnately 3-foliolate; leaflets 3–10×2–6 mm, dull medium green on upper surface, paler beneath, oblong or narrowly obovate, truncate or rounded at apex, usually sharply denticulate, narrowly cuneate at base, with dense simple eglandular hairs and glandular-hairy; petioles up to 8 mm, slightly channelled above; stipules 1–4 mm, ovate or oblong, obtuse to acute at apex, sometimes obscurely denticulate or serrulate towards the apex, conspicuously veined, clasping the stem, hairy. *Inflorescence* a rather loose terminal panicle; peduncle less than 8 mm, slender, glandular-hairy,

1-flowered. *Calyx* 5–8 mm, tubular, deeply divided into 5 linear or linear-subulate, entire or bluntly and obscurely toothed lobes, densely glandular-hairy. *Corolla* 5–7(–10) mm, deep rose-pink; standard with limb broadly obovate, rounded at apex and more or less apiculate, narrowed at base to a short, broad claw; wings with limb oblong, rounded at apex, with a blunt lobe at base and a short slender claw; keel with limb curved almost at a right angle to the short rather blunt beak. *Stamens* 10, monadelphous; filaments pale; anthers oblong. *Style* 1, glabrous; stigma terminal, capitate. *Legume* 5–14 × 3–4 mm, pale or dark brown, blunt at apex; oblong, deflexed, glandular-hairy; seeds 10–20, 0.8–1.3 mm in diameter, pale brown, covered with blunt glistening tubercles. *Flowers* 4–10. $2n=60$.

Native. Rather bare, stony, sand or eroding limestone, rare and very local. Devonshire, Glamorganshire, Pembrokeshire, Wigtownshire and Alderney; extinct in Cumberland, Midlothian and Guernsey. Throughout the Mediterranean region and eastwards to Iran, Ethiopia and Atlantic Islands. Mediterranean-Atlantic element. Our plant is subsp. **reclinata**.

Section **2. Ononis**
Section *Bugrana* Griseb.

Flowers in racemes or very condensed panicles with the primary branches not more than 1.5 mm. *Legume* usually erect or patent, ovate or rhombic.

3. O. spinosa L. typ. conserv. Common Restharrow

Rhizomatous *perennial herb* or *subshrub*. *Stem* up to 80 cm, pale green, often suffused reddish-purple, prostrate, decumbent spreading or erect, from few to numerous or dense simple eglandular hairs or a mixture of them with glandular hairs, often spiny, much branched, leafy. *Leaves* alternate, broadly ovate in outline, 1- or 3-foliolate; leaflets 5–25 × 2–8 mm, rather dark green on upper surface, paler beneath, narrowly elliptical, oblong-oblanceolate, obovate or subrotund, obtuse or rounded at apex, sharply denticulate, narrowed or rounded at base, hairy on both surfaces, shortly petiolulate; petiole short, hairy; stipules 2.5–3.0 × 4–5 mm. *Inflorescence* of flowers, solitary or in pairs, along the upper parts of the branches, shortly pedicillate or subsessile. *Calyx* 5–10 mm, more or less densely glandular-hairy, with 5 narrow linear-subulate, subfalcate lobes 3–6 mm. *Corolla* 5–24 mm, rose-pink, the wings white with purple veins; standard with limb broadly elliptical or subrotund, rounded or shortly mucronate at apex, with a rounded base and a short claw; wings with limb narrowly obovate, rounded at apex, with a short lobe at base and a narrow claw; keel with limb obovate, narrowed and upturned at apex, with a short lobe at base and a narrow claw. *Stamens* 10, monadelphous; filaments white, curved up at apex; anthers yellowish, narrowly oblong. *Style* 1, white, curved up at apex, glabrous; stigma yellowish, capitate. *Legume* 6–9 × 5–7 mm, ovate, strongly compressed, acute or obtuse at apex; seeds 1-few, 2–3 mm in diameter, dark brown, globose-reniform, verruculose or smooth. *Flowers* 6–9.

We agree with W. Greuter (1986) that *O. repens* and *O. spinosa* are best regarded as one species with several subspecies.

(**a**) Subsp. **spinosa**
O. repens subsp. *spinosa* Greuter nom. illegit.; *O. repens* subsp. *campestris* (W. D. J. Koch & Ziz); *O. campestris* W. D. J. Koch & Ziz; *O. arvensis* auct.

Stem up to 80 cm, erect, usually with 2 lines of simple eglandular hairs, usually spiny, not rooting. *Leaflets* 8–25 × 4–8 mm, oblong-oblanceolate. *Corolla* 10–24 mm.

(**b**) Subsp. **intermedia** (Rouy) P. Fourn.
O. vulgaris forme *O. intermedia* Rouy; *O. repens* subsp. *intermedia* (Rouy) Asch. & Graebn.; *O. pseudohircina* Schur

Stem up to 60 cm, usually erect, sparingly to densely hairy the hairs simple eglandular and glandular, sometimes in lines, often spiny. *Leaflets* 8–17 × 3–7 mm, narrowly elliptical or oblong-lanceolate. *Corolla* 15–24.

(**c**) Subsp. **procurrens** (Wallr.) Briq.
O. procurrens Wallr.; *O. mitis* Mill.; *O. repens* L.; *O. spinosa* subsp. *repens* (L.) Hook. fil.; *O. inermis* Huds. nom. illegit.; *O. arvensis* auct.

Stems up to 60 cm, prostrate or decumbent, with numerous simple eglandular and glandular hairs, sometimes spiny, sometimes rooting. *Leaflets* 10–15(–20) × 4–6 mm, oblong-oblanceolate or obovate. *Corolla* 12–15(–20) mm.

(**d**) Subsp. **maritima** (Dumort.) P. Fourn.
O. maritima Dumort.; *O. procurrens* var. *maritima* (Dumort.) Gren. & Godr.; *O. repens* subsp. *maritima* (Dumort.) Asch. & Graebn.

Stems up to 60 cm, prostrate, with dense simple eglandular and glandular hairs, sometimes spiny, often rooting. *Leaflets* 5–10 × 2–7 mm, obovate or subrotund, very glandular-hairy. *Flowers* 5–12.

Native. Rough grassy places. Scattered throughout Great Britain and Ireland and locally common. Most of Europe; Asia. Subsp. *spinosa* occurs on well-drained soils especially clays and is locally frequent in Great Britain north to south Scotland, but mostly in south and central England. It occurs in most of Europe except the extreme north and the high mountain regions; also in Asia. Subsp. *procurrens* is a plant of dry grassland especially on calcareous soils. It is locally common in Great Britain and Ireland and occurs also in west and central Europe. Subsp. *intermedia* is probably of hybrid origin between subsp. *spinosa* and subsp. *procurrens*, and is fertile and forms stable populations. It occurs in a few places from Cambridgeshire to Co. Durham. In Europe it is known only from north Germany and France. P.D.S. has known the population at Orwell in Cambridgeshire for about 50 years. Subsp. *maritima* is the plant of coastal sands and cliff-tops. It is easily recognisable by being flat to the ground, covered with glandular and simple eglandular hairs and giving off a strong scent when trodden on.

4. O. mitissima L. Mediterranean Restharrow

Annual herb with fibrous roots. *Stems* (15–)20–60 cm, pale green, erect or spreading, robust, thinly glandular-hairy or nearly glabrous, much branched, leafy. *Leaves* alternate, pinnately 3-foliolate; leaflets 10–25×5–15 mm, medium green on upper surface, paler beneath, obovate, obtuse or rounded at apex, finely and regularly serrulate, narrowed at base; petiole under 10 mm, channelled above, thinly glandular-hairy; stipules 5–10 mm, oblong, acute at apex, entire or shortly toothed above, glandular-hairy. *Inflorescence* a terminal spike of crowded flowers; floral leaves much reduced and often bract-like, with conspicuous pale green or whitish stipules; pedicels very short or absent. *Calyx* 7–10 mm, narrowly campanulate, with 5 broadly lanceolate or oblong, acuminate lobes, many-veined, sparingly glandular-hairy or nearly glabrous. *Corolla* 9–12 mm, pink or whitish; standard with limb narrowly obovate, rounded at apex, tapering at base to a long narrow claw; wings with limb narrowly oblong, rounded at apex, and with a distinct lobe at base; keel with limb narrow, the upper margin more or less straight, the lower margin gently curved to a short subacute beak. *Stamens* 10, monadelphous; filaments strongly dilated; anthers shortly oblong. *Style* 1, glabrous; stigma terminal, capitate. *Legume* 7–8×4.0 mm, pale brown, ovate, much compressed, acute at apex, sparingly glandular-hairy; seeds 2–3, 1.5–2.0 mm, dark brown, subglobose-reniform, compressed, densely papillose-veniculose. *Flowers* 6–9. $2n=30$.

Introduced. Occasional bird-seed alien on tips. Sporadic in England, Glamorganshire and Midlothian. Native throughout the Mediterranean region and east to Iraq; Atlantic Islands.

5. O. alopecuroides L. Salzmann's Restharrow
O. salzmanniana Boiss. & Reut.; *O. baetica* auct.

Annual herb with fibrous roots. *Stems* 12–65 cm, pale green, erect to prostrate, fistulose, glabrous or with hairs above, simple or branched, leafy. *Leaves* alternate, always 1-foliolate, the leaflet up to 6×3.5 cm, oblong, with a broad, rounded apex, glabrous on surface with a minutely serrulate margin; petiole up to 35 mm, bordered on either side by the conspicuous, oblong, wing-like stipules. *Inflorescence* a terminal spike of dense flowers; floral leaves much reduced and bract-like. *Calyx* 10–13 mm, the tube narrowly campanulate, with 5 linear-setaceous lobes, densely glandular-hairy, with long spreading hairs. *Corolla* 13–15 mm, pink; standard with limb oblong; wings with limb narrowly oblong; keel narrower and slightly longer than wings, gently upcurved to a short beak. *Stamens* 10, monadelphous; filaments pale; anthers yellowish. *Style* 1, pale; stigma brownish. *Legume* 8–10×4–5 mm, pale brown, ovate, compressed, subacute at apex; seeds 2–3, 2.0–2.5 mm in diameter, dark brown, smooth. *Flowers* 6–9. $2n=30$, 32.

Introduced. Occasional bird-seed alien on tips. Few scattered records in Great Britain. Widely but sparsely distributed in the Mediterranean region.

6. O. baetica Clemente Andalusian Restharrow

Annual herb with fibrous roots. *Stems* up to 50 cm, pale green, erect or ascending, shortly glandular-hairy to glabrous, much branched, leafy. *Leaves* alternate, 3-foliolate; leaflets 3–15×1–6 mm, medium green on upper surface, paler beneath, oblong-lanceolate to elliptical, dentate, glandular-hairy; petiole short, glandular-hairy; stipules ovate or oblong-elliptical, acute at apex, dentate. *Inflorescence* a short, more or less leafless terminal panicle. *Calyx* 4–8 mm, campanulate, densely glandular-hairy; tube 1.5–2.5 mm; lobes 5, 2.5–6.5 mm, linear-lanceolate. *Corolla* (7–)10–17 mm; standard rose, subrotund, rounded at apex, with a short claw; wings white, elliptical; keel white, upturned at end with a short beak. *Stamens* 10, monadelphous; filaments pale; anthers yellowish. *Style* 1, pale. *Legume* 5–7 mm, ovoid; seeds 5, 1.0–1.5 mm, reniform, tuberculate. *Flowers* 6–7.

Introduced. Occasional bird-seed alien on tips. Sporadic in England. Native of the west Mediterranean region.

37. Melilotus Mill.

Annual or *biennial*, rarely *perennial herbs*, often smelling of coumarin when dried. *Stems* erect or sprawling. *Leaves* alternate, pinnately 3-foliolate; stipules adnate to petiole. *Inflorescence* a few- to many-flowered axillary raceme, the flowers pendulous on short pedicels with minute bracts. *Calyx* campanulate, divided into 5 lobes. *Corolla* white or yellow; standard with a broad limb; wings oblong, more or less adherent to keel. *Stamens* 10, usually monadelphous, rarely diadelphous; filaments not dilated at apex; anthers uniform. *Style* 1, upcurved; stigma terminal. *Fruit* an indehisent or tardily dehiscent legume, globose or ovoid.

Contains about 20 species widely distributed in Europe, Asia and North Africa, and found as adventives through much of the world.

M. infestus Guss. has been recorded as a bird-seed and grain casual.

Hultén, E. & Fries, M. (1986). *Atlas of north European vascular plants north of the Tropic of Cancer*. 3 vols. Königstein.
Schultz, O. E. (1901). Monographie der Gattung *Melilotus*. *Bot. Jahrb.* **29**: 660–735.
Turkington, R. A., Cavers, P. B. & Rempel, E. (1978). The biology of Canadian weeds. 29. *Melilotus alba* Desv. and *M. officinalis* (L.) Lam. *Canad. Jour. Pl. Sci.* **58**: 523–537.

1. Corolla white **1. albus**
1. Corolla yellow 2.
2. Legumes with strong, concentric ridges 3.
2. Legumes reticulate or rugose 4.
3. Stipules laciniate; legumes 3.0–3.5 mm, obtuse at apex **5. sulcatus**
3. Stipules dentate; legumes 6–7 mm, acute at apex **6. messanensis**
4. Corolla 2.0–3.5 mm; legume 2.0–2.5 mm **2. indicus**
4. Corolla 4–7 mm; legume 3–7 5.
5. Keel of corolla more or less equalling wings; legumes 5–7 mm, mostly 2-seeded, black when ripe, hairy **3. altissima**
5. Keel of corolla shorter than wings; legumes 3–5 mm, mostly 1-seeded, brown when ripe, glabrous **4. officinalis**

1. M. albus Medik. White Melilot
M. leucanthus W. D. J. Koch nom. illegit.

Annual or *biennial herb* with fibrous roots. *Stems* 30–200 cm, pale green, erect, glabrous or slightly hairy, angular, usually much-branched, leafy. *Leaves* alternate, pinnately 3-foliolate; leaflets 10–30 × 5–15 mm, medium green on upper surface, paler beneath, oblong, obovate or elliptical, rounded, apiculate or subacute at apex, irregularly serrate, especially towards the apex, sometimes almost entire, narrowed at base, shortly petiolulate, glabrous on upper surface, appressed-hairy beneath; petioles 1–2(–3) cm, somewhat flattened, channelled above, glabrous or sparsely hairy; stipules 3–10 mm, filiform or narrowly subulate, entire, more or less glabrous. *Inflorescence* an elongate, dense, many-flowered axillary raceme up to 7 cm; peduncles 1.5–4.0 cm, generally straight and erect; pedicels 0.5–1.5 mm, pendulous. *Calyx* 1.5–2.5 mm, tubular, divided about halfway into 5 lobes, the lobes deltoid, acute at apex and subequal, glabrous. *Corolla* 4–5 mm, white; standard with limb obovate, emarginate at apex, tapering abruptly to a very short claw; wings with limb oblong, a little shorter than standard, rounded at apex, with a distinct lobe at base and a short claw; keel with limb ovate, rounded at apex, with a shallow lobe at base and a rather short claw. *Stamens* 10, monadelphous; filaments united for half their length; anthers ovate. *Style* 1, filiform, upcurved; stigma capitate. *Legume* 3–4 × about 2 mm, ovoid, slightly compressed, irregularly reticulate-rugose; seeds 1–2, 2.0–2.5 mm, ovoid, slightly compressed, brown, smooth. *Flowers* 7–8. $2n = 16$.

Introduced and first recorded in 1822. Naturalised in open grassland and rough ground, casual in waste places. Rarely grown as green fodder and has arrived as a contaminant of lucerne. Frequent in south and central England, scattered records in the west, in east and south Ireland and north to central Scotland. Throughout most of Europe but doubtfully native; North Africa; Asia eastwards to Tibet; introduced in America and Australia.

2. M. indicus (L.) All. Small Melilot
Trifolium indicum L.; *M. parviflorus* Desf.

Annual or *biennial herb* with fibrous roots. *Stems* (10–) 20–50 cm, pale green, erect or sprawling, angled, more or less glabrous, much-branched, leafy. *Leaves* alternate, pinnately 3-foliolate; leaflets 5–25 × 4–15 mm, medium green on upper surface, paler beneath, obovate or oblanceolate, truncate or emarginate at apex, often minutely apiculate, irregularly serrulate especially towards the apex, narrowed below, glabrous on upper surface, thinly hairy beneath; petioles up to 4 cm, channelled above, slender, more or less glabrous; stipules 3–8 mm, linear-subulate, entire or sometimes 1–2-toothed near the expanded base. *Inflorescence* an elongate, dense, many-flowered axillary raceme 1–3 cm, elongating to 5 cm in fruit; peduncles 1–3 cm, erect, more or less glabrous; pedicels 0.5–0.8 mm, recurved; bracts 0.3–0.5 mm, subulate. *Calyx* 1.0–1.3 mm, divided halfway into 5 lobes, the lobes deltoid, more or less glabrous. *Corolla* 2.0–3.5 mm, yellow; standard with limb narrowly obovate, emarginate at apex, with the claw barely developed; wings with a rounded apex, a short lobe at base and a rather long, slender claw; keel about as long as wings, obtuse at apex, with a slender claw. *Stamens* 10, diadelphous; filaments pale; anthers subglobose. *Style* 1, filiform, upcurved; stigma capitate. *Legume* 2.0–2.5 × about 2 mm, pale brown, broadly ovoid to subglobose, irregularly reticulate-rugose; seed 1, about 1.7 mm, dark brown, subglobose, minutely tuberculate. *Flowers* 6–10. $2n = 16$.

Introduced. Rough ground and waste places, usually casual from bird-seed and wool, but sometimes naturalised. Scattered through Great Britain north to central Scotland, locally common in the Channel Islands and a few records in Ireland. Native throughout the Mediterranean region and eastwards to central Asia and India; Atlantic Islands.

3. M. altissimus Thuill. Tall Melilot
M. officinalis auct.

Biennial or short-lived *perennial herb* with a tap-root and fibrous side-roots. *Stems* 60–150 cm, pale green, erect, glabrous, branched, leafy. *Leaves* alternate, pinnately 3-foliolate; leaflets 15–20(–30) × 5–12 mm, medium green on upper surface, paler beneath, oblong, obovate or oblanceolate, rounded or emarginate at apex, serrate, narrowed at base, shortly petiolulate, glabrous or slightly hairy; petioles rather short, glabrous; stipules 5–6 mm, subulate, dentate at base. *Inflorescence* an elongate, rather lax, many-flowered, axillary raceme up to 50 mm, lengthening in fruit; peduncles up to 5 cm, glabrous, erect; pedicels 1–2 mm, hairy. *Calyx* 3.0–3.5 mm, campanulate, divided halfway to base into 5 lobes, the lobes narrowly triangular or subulate, subacute at apex, hairy. *Corolla* 5–7 mm, yellow; standard with limb subrotund, emarginate at apex, abruptly narrowed at base into a short, broad claw; wings equal to standard and keel, the limb oblong, rounded at apex, with a short, broad lobe at base and a very slender claw; keel broadly oblong-obovate, slightly upturned at the rounded apex, with a slight swelling at base and a slender claw. *Stamens* 10, monadelphous; filaments not dilated; anthers yellowish. *Style* 1, swollen towards apex; stigma brownish. *Legume* 5–7 × 3–4 mm, ovoid or obovoid, reticulate, compressed, black when ripe, shortly hairy; seeds mostly 2, 2.5–3.0 mm, subglobose, smooth. *Flowers* 7–9. $2n = 16$.

Introduced. Naturalised in open grassland, rough ground and by waysides, casual in waste places. Frequent in south and central England, in scattered localities in the rest of Great Britain north to central Scotland; scarce in eastern and southern Ireland. Throughout most of Europe.

4. M. officinalis (L.) Pall. Ribbed Melilot
Trifolium officinale L.; *M. arvensis* Wallr. nom. illegit.; *M. petitpierranus* Willd.

Biennial herb with fibrous roots. *Stems* 60–150 cm, pale green, decumbent or erect, glabrous, branched, leafy. *Leaves* alternate, pinnately 3-foliolate; leaflets 10–20(–30) × 4–10 mm, medium green on upper surface, paler beneath, oblong, oblong-obovate, or elliptical, rounded at apex, serrate, narrowed at base, shortly petiolulate, glabrous or nearly so; petiole up to 2 cm, glabrous;

stipules 5–7 mm, subulate, dentate at base. *Inflorescence* an elongate, rather lax, many-flowered axillary raceme up to 50 mm, lengthening in fruit; peduncles up to 5 cm, erect, glabrous; pedicels 1–2 mm, pendulous. *Calyx* 3.0–3.5 mm, campanulate, divided halfway to base into 5 lobes, the lobes subulate, long acute at apex, glabrous. *Corolla* 4–7 mm, yellow; standard with limb broadly elliptical, emarginate at apex, abruptly narrowed to a short claw; wings equal to standard, longer than keel with limb oblong, rounded at apex, with an obvious lobe at base and a very narrow spur; keel shorter than wings with limb broadly oblong, rounded at apex, with a short, broad lobe at base and a narrow spur. *Stamens* 10, monadelphous; filaments not dilated; anthers yellowish. *Style* 1, curved upwards; stigma brownish. *Legume* 3–5 × 2.5–3.0 mm, brown when ripe, ovoid, slightly compressed, transversely rugose, glabrous; seeds mostly 1, 2.0–2.5 mm, brown, smooth. *Flowers* 7–9. $2n = 16$.

Introduced. Naturalised in open grassland, field borders and rough ground and casual on rubbish tips and in waste places. Frequent in south and central England, scattered localities north to central Scotland and in east and south Ireland. Most of Europe, though often only a weed in cultivation, and eastwards to western China.

5. M. sulcatus Desf. Furrowed Melilot

Annual herb with fibrous roots. *Stems* 10–40 cm, pale green, erect or sprawling, angled, sparsely hairy to nearly glabrous, much branched, leafy. *Leaves* alternate, pinnately 3-foliolate; leaflets 10–25 × 5–13 mm, medium green on upper surface, paler beneath, obovate or oblanceolate, rounded at apex, finely and regularly denticulate, glabrous on upper surface, thinly appressed-hairy beneath; petioles 8–25 mm, channelled above, slender, thinly hairy to nearly glabrous; stipules 5–8 mm, ovate, acuminate at apex, laciniate with narrow linear-filiform lobes. *Inflorescence* an axillary, pedunculate raceme 10–20 mm, the flowers spreading or pendulous; peduncles 3–15 mm, erect, thinly hairy or nearly glabrous; bracts about 1 mm, subulate-filiform; pedicel 1.0–1.5 mm, recurved. *Calyx* about 2 mm, narrowly campanulate, divided halfway into 5 lobes, the lobes deltoid, acute at apex, nearly glabrous. *Corolla* 3.0–4.5 mm, yellow; standard with limb ovate, strongly folded longitudinally, subacute at apex, often obscurely denticulate, narrowed abruptly to a short distinct claw; wings with limb oblong, rounded at apex, with a lobe at base and a slender claw; keel more or less equal in length to standard, the wings much shorter, the apex obtuse, the base with a short lobe and distinct claw. *Stamens* 10, diadelphous; filaments 3.4 mm; anthers broadly ovate. *Style* 1, upcurved at apex; stigma minute and capitate. *Legume* 3.0–3.5 mm, oblong or subglobose, with conspicuous close concentric ridges, rounded at apex; seeds about 2.5 mm, pale brown, oblong-subglobose, not much compressed, minutely tuberculate. *Flowers* 6–7. $2n = 16$.

Introduced. Fairly frequent bird-seed alien on tips and waste land. Sporadic in scattered localities over Great Britain and a few records from Ireland. Throughout the Mediterranean region; Atlantic Islands.

6. M. messanensis (L.) All. Sicilian Melilot

Trifolium messanense L.; *Trifolium siculum* Turra ex Vitm.; *M. sicula* (Turra ex Vitm.) B. D. Jackson

Annual herb with fibrous roots. *Stems* 15–30(–45) cm, pale green, decumbent or sprawling, rather succulent, glabrous, much branched, especially near the base, leafy. *Leaves* alternate, pinnately 3-foliolate; leaflets 10–20 × 4–15 mm, medium green on upper surface, paler beneath, rather thick, obovate or oblanceolate, rounded or shortly emarginate at apex, sharply and somewhat irregularly serrulate, narrowly cuneate at base, shortly petiolulate, glabrous on upper surface, thinly appressed-hairy beneath; petioles 8–30 mm, flattened, narrowly winged; stipules 5–8 mm, ovate, acuminate at apex, shortly and irregularly dentate, glabrous or thinly hairy. Inflorescence a short, few-flowered axillary raceme; peduncle usually less than 5 mm; pedicels 0.8–1.0 mm, recurved; bracts less than 0.5 mm, subulate-deltoid. *Calyx* about 2 mm, campanulate, divided halfway into 5 lobes, the lobes deltoid, acute at apex, thinly pilose to nearly glabrous. *Corolla* 4–5 mm, yellow; standard with limb narrowly oblong-obovate, slightly emarginate at apex, tapering to a distinct claw at base; wings with limb oblong, rounded at apex, with a prominent lobe at base and a slender claw; keel broadly oblong, with an obtuse apex, an inconspicuous lobe at base and a slender claw. *Stamens* 10, diadelphous; filaments pale; anthers yellowish, oblong. *Style* 1, pale; stigma capitate. *Legume* 6–7 × 4.0–4.5 mm, pale brown when ripe, ovoid, acute at apex, distinctly compressed, keeled, closely marked with prominent, concentric ridges; seeds 1–2, 2.5–4.0 mm, oblong to subglobose, dark brown, minutely rugulose-tuberculate. *Flowers* 7–9.

Introduced. A bird-seed casual. Widespread native in the Mediterranean region.

38. Trigonella L.

Annual herbs. Leaves alternate, pinnately 3-foliolate; leaflets usually dentate. *Inflorescence* an axillary raceme or solitary flowers in leaf axils. *Calyx* without glandular hairs, lobes equal or unequal. *Corolla* yellow, blue or purplish, free from the staminal tube and deciduous. *Stamens* 10, diadelphous or monadelphous; filaments not dilated. *Style* 1. *Legume* linear or oblong, straight or curved, indehiscent or dehiscing along one suture; seeds 1–many.

Contains about 80 species, mainly Mediterranean but also Macaronesia, South Africa, Western Asia and Australia.

Širjaev, G. (1928–1934). Generis *Trigonella* revisio critica. *Publ. Fac. Sci. Univ. Masaryk Brno* **1(1)**: 1–56 (1928); **1(2)**: 1–37 (1929); **1(3)**: 1–31 (1930); **1(4)**: 1–33 (1931); **1(5)**: 1–43 (1932); **1(6)**: 1–37 (1933); **2(1)**: 1–15 (1934).
Vassilczenko, L. T. (1953). The species of the genus *Trigonella* L. *Acta Inst. Bot. Acad. Sci. U.R.S.S.* ser. 1(10): 124–269.

1. Flowers 1(–2) in leaf axils; corolla 12–18 mm; legume 70–120 mm; seeds 10–20 **3. foenum-graecum**
1. Flowers in axillary racemes; corolla 6–7 mm; legume 4–16 mm; seeds fewer than 8 2.

2. Racemes elongated; corolla yellow; legumes more than 8 mm, 4–8 seeded **1. corniculata**
2. Racemes subcapitate; legumes less than 8 mm, 1–3 seeded **2. caerulea**

Subgenus 1. **Trigonella**

Calyx usually campanulate. *Legume* not inflated.

1. T. corniculata (L.) L. Sickle-fruited Fenugreek
Trifolium (*Melilotus*) *corniculatum* L.

Annual herb with fibrous roots. *Stem* 10–55 cm, pale green, procumbent to erect, glabrous or nearly so, branched, leafy. *Leaves* alternate, 3-foliolate; leaflets 10–40 × 7–35 mm, medium green on upper surface, linear-lanceolate to obovate, obtuse and sometimes emarginate at apex, dentate, narrowed at base, glabrous or nearly so; stipules lanceolate, incise-dentate. *Inflorescence* an axillary raceme with 8–15 flowers; peduncles up to 60 mm; pedicels about 3 mm. *Calyx* 3–4 mm, campanulate, 2-lipped. *Corolla* 6–7 mm, yellow; standard obovate, emarginate; wings oblong, obtuse at apex; keel rounded, with a claw. *Stamens* 10, diadelphous. *Style* 1. *Legume* 10–16 × (1.5–)2–3 mm, linear, acuminate, compressed, not inflated, somewhat curved, glabrous, with thin transverse veins, pendulous; seeds 4–8, 1.0–1.5 mm, oblong, tuberculate. *Flowers* 6–8. $2n = 16$.

Introduced. Frequent bird seed alien on tips and waste ground in southern Great Britain, casual in Ireland. Native of the Mediterranean region.

Subgenus 2. **Trifoliastrum** (Moench) Beck

Calyx campanulate. *Legume* inflated.

2. T. caerulea (L.) Ser. Blue Fenugreek
Trifolium caeruleum L.; *Melilotus procumbens* Besser; *T. besseriana* Ser.; *T. procumbens* (Besser) Rchb.

Annual herb. Stems 20–60(–100) cm, pale green, erect, hollow, sparsely hairy, branched, leafy. *Leaves* alternate, broadly ovate in outline; leaflets 20–50 × 5–20 mm, medium green on upper surface, paler beneath, ovate to oblong, emarginate at apex, denticulate, narrowed at base, sparsely hairy; stipules broad, denticulate. *Inflorescence* a dense, many-flowered, axillary, globose head; peduncles 20–50 mm; pedicels about 1 mm. *Calyx* 3–4 mm, campanulate, divided into 5 lobes about equalling the tube. *Corolla* 6–7 mm, blue or white; standard oblong, emarginate; wings oblong, obtuse; keel rounded, with a claw. *Stamens* 10, diadelphous. *Style* 1. *Legume* 4–5 × 2.5–3.0 mm, rhomboid-obovate, inflated, abruptly contracted to a beak about 2 mm, erect or patent; seeds 1–3, about 2 mm, brown, ovoid, finely tuberculate. *Flowers* 6–8. $2n = 16$.

Introduced. Rather sparse wool, grain and bird-seed alien on tips and waste land. Very scattered records in Great Britain, formerly more common; casual in Ireland. Perhaps native in the east Mediterranean region, but cultivated for fodder throughout Europe and widely naturalised or casual as a weed.

Subgenus 3. **Foenum-graecum** Sirj.

Calyx tubular. *Legume* not inflated.

3. T. foenum-graecum L. Fenugreek

Annual herb with fibrous roots. *Stems* 15–40(–50) cm, pale green, erect or spreading, glabrous or thinly hairy, mostly unbranched above, but branched towards the base, leafy. *Leaves* alternate, pinnately 3-foliolate; segments 10–25(–35) × 4–10(–20) mm, deep bluish-green on upper surface, paler beneath, narrowly to broadly obovate, rounded or occasionally shallowly retuse at apex, subentire to acutely denticulate, narrowed below, glabrous on upper surface, thinly hairy to nearly glabrous beneath; petioles up to 10(–40) mm, flattened, more or less distinctly winged, thinly hairy to nearly glabrous; stipules 5–7 mm, narrowly ovate-acuminate, more or less entire, thinly hairy. *Inflorescence* of solitary, erect, axillary flowers; peduncle and bracts absent; pedicels usually less than 1 mm, hairy. *Calyx* 6–7 mm, tubular, membranous, with 5 subulate lobes, thinly hairy. *Corolla* 12–18 mm, yellowish-white; standard with limb narrowly obovate, apex sharply emarginate, tapered below to a very short claw, distinctly constricted about one-third of the way up; wings with limb oblong, rounded at apex, somewhat gibbous at base and with a conspicuous lobe and slender claw; keel with limb almost subrotund and a long, slender claw. *Stamens* 10, diadelphous; filaments pale; anthers small and oblong. *Style* 1, very short, upcurved; stigma capitate. *Legume* 70–120 × 4–5 mm, linear, not inflated, rather straight or curved downwards, taping gradually to a slender beak, very thinly glandular-hairy; seeds 10–20, 3.0–4.5 mm, rich brown, oblong, more or less compressed, rather sparsely verruculose. *Flowers* 6–8. $2n = 16$.

Introduced. Frequent bird-seed and spice alien on tips and waste land and cultivated on a small scale. Sporadic in central and south England. Throughout southern Europe and the Mediterranean region, eastward to central Asia; cultivated elsewhere.

39. Medicago L.

Annual, biennial or *perennial herbs*, rarely *subshrubs*. *Leaves* alternate, pinnately 3-foliolate; stipules adnate to the petiole. *Inflorescence* of 1–many flowers in axillary racemes; bracts small and membranous; pedicels short and slender. *Calyx* campanulate, with 5 subequal lobes. *Corolla* yellow, rarely purple or violet, petals free from the staminal tube, caducous; standard obovate or oblong, rounded at apex, narrowed to a cuneate base, usually longer than keel; wings oblong, usually longer than keel; keel obtuse at apex. *Stamens* 10, diadelphous; filaments not dilated at apex; anthers 2-thecous. *Ovary* sessile or stipitate, straightish or coiled. *Style* glabrous; stigma terminal or oblique, more or less capitate. *Fruit* a legume, indehiscent or tardily dehiscent, exserted from the persistent calyx, normally falcately curved or coiled spirally, often spinous; the spirally coiled legumes have a variously thickened vein or border on the outer edge of each coil; just inside this vein, on each face of the coil, is a variously thickened submarginal vein; between these 2 veins is the usually channelled submarginal border; the base of each spine originates from across both these veins.

Contains about 50 species, widely distributed in Europe and the Mediterranean region and extending eastwards to China; often introduced and naturalised elsewhere.

There appears to be no characters in isolation that will define the genera *Medicago*, *Trigonella* and *Melilotus*.

Grime, J. P. et al. (1988). *Comparative plant ecology*. London. [*M. lupulina*.]

Heyn, C. C. (1963). The annual species of *Medicago*. *Scripta Hierosolymitana Sci. Rep.* **12**: 1–154.

Hultén, E. & Fries, M. (1986). *Atlas of north European vascular plants north of the Tropic of Cancer*. 3 vols. Königstein.

Lesins, K. A. & Lesins, I. (1979). Genus *Medicago* (Leguminosae). *A taxonomic study*. The Hague.

Norton, J. A. (2005). More Mediterranean 'aliens' in Gosport. *B. S.B.I. News* **100**: 46–48.

Small, E. & Jomphe, M. (1989). A synopsis of the genus *Medicago* (Leguminosae). *Canad. Jour. Bot.* **67**: 3260–3294.

Small, E., Lessen, P. & Brookes, B. S. (1987). An expanded circumscription of *Medicago* (Leguminosae, Trifolieae) based on explosive flower tripping. *Willdenowia* **16**: 415–437.

Smart, J. & Simmonds, N. W. (Edits.) (1995). *Evolution of crop plants*. Ed. 2. London. [*M. sativa* subsp. *sativa*.]

Stewart, A., Pearman, D. A. & Preston, C. D. (1994). *Scarce plants in Britain*. Peterborough. [*M. minima, polymorpha, sativa* subsp. *falcata*].

Trist, P. J. O. (1971). *A survey of the agriculture of Suffolk*. London. [*M. sativa* subsp. *sativa*.]

Turkington, R. & Cavers, P. B. (1979). The biology of Canadian weeds. 33. *Medicago lupulina* L. *Canad. Jour. Pl. Sci.* **59**: 99–110.

Urban, I. (1873). Prodromus einer Monographie der Gattung *Medicago*. *Verh. Bot. Ver. Brandenb.* **15**: 1–85.

1. Margin of legume with one longitudinal vein, with which the transverse veins join, sometimes running into spines, without a strong submarginal vein close to the marginal one 2.
1. Legumes with strong submarginal vein or with a wide veinless border 15.
2. Legume more or less falcate or reniform 3.
2. Legume spirally coiled 9.
3. Legume up to 3 mm, reniform; seed solitary 4.
3. Legume more than 3 mm, more or less falcate; seeds usually several 8.
4. Leaflets 15–30 × 10–22 mm; petioles up to 9 cm; stipules up to 15 mm; flowers 4.0–4.5 mm **2(v). lupulina** var. **cupaniana**
4. Leaflets 3–20 × 7–15 mm; petioles up to 4 cm; stipules up to 5 mm; flowers 2.5–4.0 mm 5.
5. Legume glandular-hairy **2(iv). lupulina** var. **willdenowiana**
5. Legume not glandular-hairy 6.
6. Legume and often rest of plant densely hairy **2(i). lupulina** var. **eriocarpa**
6. Legume and rest of plant glabrous or slightly hairy 7.
7. Plant prostrate, decumbent or spreading; leaflets 5–12 × 5–10 mm; flowers 2–3 mm; legumes 2.0–2.5 mm **2(ii). lupulina** var. **lupulina**
7. Plant more or less erect or spreading; leaflets (5–)10–20 × 5–15 mm; flowers 2.5–4.0 mm; legumes 2–3 mm **2(iii). lupulina** var. **major**
8. Corolla yellow; legume nearly straight to curved; seeds 2–5 **3(a). sativa** subsp. **falcata**
8. Corolla yellow, mauve, purple, green or blackish; legumes curved to spiralled; seeds 3–8 or abortive **3(b). sativa** subsp. **varia**
9. Shrubs or perennials with a stout woody stock 10.
9. Annual herbs 12.
10. Shrub 100–400 cm **4. arborea**
10. Perennial herbs less than 100 cm 11.
11. Corolla yellow, pale mauve to purple, green or blackish; legumes curved or spiralled in 0.5–1.5 complete turns; seeds 3–8 or abortive **3(b). sativa** subsp. **varia**
11. Corolla pale mauve to violet; legume spiralled in 2–3(–4) complete turns; seeds 10–20 **3(c). sativa** subsp. **sativa**
12. Margin of legume with distinct spines 13.
12. Margin of legume without spines, sometimes with small rounded projections 14.
13. Leaflets 6–15 mm; legume 9–12 mm, glabrous **15(i). intertexta** var. **intertexta**
13. Leaflets 15–20 mm; legume 12–15(–17) mm, dorsal surface and spines densely hairy **15(i). intertexta** var. **ciliaris**
14. Legume (5–)7–9(–10) mm in diameter, orbicular-discoid **1. orbicularis**
14. Legume 9–18 mm in diameter, pelviform **5. scutellata**
15. Legume in a lax spiral, the young legume projecting from the calyx as soon as the petals have fallen 16.
15. Legume in a very close spiral, the young legume concealed within the calyx when the petals fall 22.
16. Legume without apparent transverse veins or with a veinless border one-quarter to one-third as wide as the radius of the spiral 17.
16. Legume with prominent transverse veins and no wide veinless border 18.
17. Legumes armed with well-developed spines 3–4 mm **9(i). disciformis** var. **disciformis**
17. Legumes with very short triangular spines less than 1 mm **9(ii). disciformis** var. **apiculata**
18. Leaflets nearly always with a dark spot; marginal vein of legume sulcate and margin therefore with 3 conspicuous grooves **11. arabica**
18. Leaflets never with a dark spot; marginal vein of legume not sulcate 19.
19. Groove between the marginal and submarginal veins of the legumes very wide, visible when the legume is viewed from the edge 20.
19. Groove between the marginal and submarginal veins narrow, not visible when the legume is viewed from the edge 21.
20. Legume subglobose, sparsely villous and often glandular; transverse veins sigmoid, not anastomosing freely **8. minima**
20. Legume discoid to shortly cylindrical, usually glabrous or nearly so; transverse veins curved but not sigmoid, anastomosing freely **10. polymorpha**
21. Leaflets dentate near the apex; transverse veins of legume strongly curved but not sigmoid, anastomosing freely near the submarginal vein **6. praecox**

21. Leaflets usually incise-dentate or almost pinnatifid; transverse veins of legume sigmoid, sparingly branched, not anastomosing **7. laciniata**
22. Legume with a veinless border one-quarter to one-third as wide as the radius of the spiral 23.
22. Legume without a wide veinless border 24.
23. Inflorescence usually 5- to 8-flowered; legume unarmed, almost smooth or at most very shortly rugose-tuberculate **14(i). turbinata** var. **turbinata**
23. Inflorescence 1–2 (–3)-flowered; legume spinose, with short spines usually much thickened at the base **14(ii). turbinata** var. **aculeata**
24. Marginal and submarginal veins of legume separated by a distinct groove at maturity and so forming 3 keels **13. truncatula**
24. Marginal and submarginal veins of legume confluent at maturity and forming a single acute or convex keel 25.
25. Legumes spiny, the spines 1–4 mm, erect or occasionally loosely appressed to legume **12(i). littoralis** var. **littoralis**
25. Legumes unarmed or with very short tubercles, not spinous **12(i). littoralis** var. **inermis**

Section **1. Orbiculares** Urb.

Annual herbs. Legume coiled spirally with more than 2 seeds in each coil; seeds tuberculate; the radicle as long as the cotyledons.

1. M. orbicularis (L.) Bartal. Button Medick
M. polymorpha var. *orbicularis* L.

Annual herb with fibrous roots. *Stems* 10–40(–50) cm, pale green, prostrate or straggling, glabrous, usually much branched, leafy. *Leaves* alternate, pinnately 3-foliolate; leaflets 6–12(–17) × 5–10(–13) mm, medium green on upper surface, paler beneath, obovate, truncate or emarginate at apex, margins serrulate in the upper third, entire or subentire towards the base, cuneate at base, glabrous, or thinly hairy beneath; petioles 10–50 mm, slender, flattened above, glabrous; stipules 2–6 mm, laciniate, with narrow filiform lobes. *Inflorescence* of 1–2(–5) flowers, scattered along the upper parts of the branches; peduncles 8–15 mm, slender, elongating in fruit; bracts 1.0–1.5 mm, filiform, dilated at base, membranous, often tinged purplish; pedicels 2–3 mm, slender, patent, glabrous or nearly so. *Calyx* about 3.5 mm, campanulate, divided over halfway into 5 subulate lobes, thinly appressed-hairy to nearly glabrous. *Corolla* 4–5 mm, yellow; standard with limb broadly obovate, emarginate at apex, cuneate at base; wings with limb oblong, rounded at apex, with a prominent, broad lobe at base and a slender claw; keel oblong, rounded at apex. *Stamens* 10, diadelphous; filaments pale; anthers oblong. *Style* 1, thick and slightly twisted; stigma capitate with papillose margin. *Legume* 7–17(–20) mm in diameter, orbicular-discoid, with 2.5–6.0 anticlockwise coils, marginal vein present, submarginal vein absent, lateral surfaces convex, with a few transverse veins with few, usually weakly anastomosing branches, not spiny, but somewhat glandular-hairy. *Flowers* 6–9. $2n = 16$.

Introduced. A grain and tan-bark casual. Native throughout the Mediterranean region.

Section **2. Lupularia** Ser.

Annual, *biennial* or *perennating herbs. Legumes* 1-seeded, reniform, distal portion coiled, without spines; seeds smooth; radicle about half as long as cotyledons.

2. M. lupulina L. Black Medick

Annual or short-lived *perennial herb* with fibrous roots. *Stems* 15–80 cm, pale green, procumbent, decumbent, ascending or suberect, angular, glabrous to densely hairy, sometimes glandular-hairy, usually much branched, leafy. *Leaves* alternate, 3-foliolate; leaflets 3–30 × 3–22 mm, medium green on upper surface, paler beneath, broadly ovate or obovate-elliptical, shortly retuse and apiculate at apex, minutely serrulate, narrowed at base and shortly petiolulate, appressed-hairy on both surfaces; petioles 5–90 mm, channelled above, more or less hairy; stipules 3–15 mm, ovate, acuminate at apex, subentire or minutely denticulate. *Inflorescence* a dense globose or oblong-cylindrical, axillary cluster of numerous flowers along the greater length of the stems; peduncles 20–30 mm, slender, hairy and sometimes glandular-hairy; pedicels less than 1 mm, hairy; bracts minute, membranous, cuspidate. *Calyx* about 2 mm, campanulate, divided about halfway into 5 lobes, the lobes subulate, usually more or less hairy. *Corolla* 1.5–4.5 mm, yellow; standard with limb broadly obcordate, emarginate at apex, narrowed to a very short claw; wings with limb oblong, obtuse at apex, with a prominent lobe at base and a slender claw; keel a little longer than wings, obtuse at apex, with a rather broad claw. *Stamens* 10, diadelphous, filaments pale; anthers subglobose. *Style* 1, much thickened, upturned; stigma capitate. *Legume* 2.0–3.5 mm, reniform, distinctly reticulate-veined, with one marginal vein, glabrous, simply hairy or glandular-hairy, the base concealed by the persistent calyx, the apex usually twisted clockwise; seeds 1, about 1.8 mm, reniform, pale brown, smooth. *Flowers* 4–8. $2n = 16$.

(i) Var. **eriocarpa** (Rouy) P. D. Sell
M. lupulina subvar. *eriocarpa* Rouy; *M. lupulina* forma *maritima* Corb.

Annual. Stems prostrate, plant often with a dense habit, or dense with long creeping stems spreading outwards, more or less densely hairy. *Leaflets* 3–11 × 3–11 mm, very hairy. *Petioles* 5–15 mm. *Corolla* 1.5–2.5 mm. *Legume* 2.0–2.5 mm, with dense simple eglandular hairs.

(ii) Var. **lupulina**
M. lupulina var. *prostrata* R. Keller

Annual. Stems prostrate, decumbent or spreading, glabrous to fairly hairy. *Leaflets* 5–12 × 5–10 mm, glabrous to sparingly hairy. *Petioles* 5–15 mm. *Corolla* 2–3 mm. *Legume* 2.0–2.5 mm, with few simple eglandular hairs.

(iii) Var. **major** G. Mey.
Annual. Stems more or less erect to spreading. *Leaflets* (5–) 10–20 × 5–15 mm, glabrous to fairly hairy. *Petioles* 5–40 mm. *Corolla* 2.5–4.0 mm. *Legume* 2–3 mm, glabrous or slightly hairy.

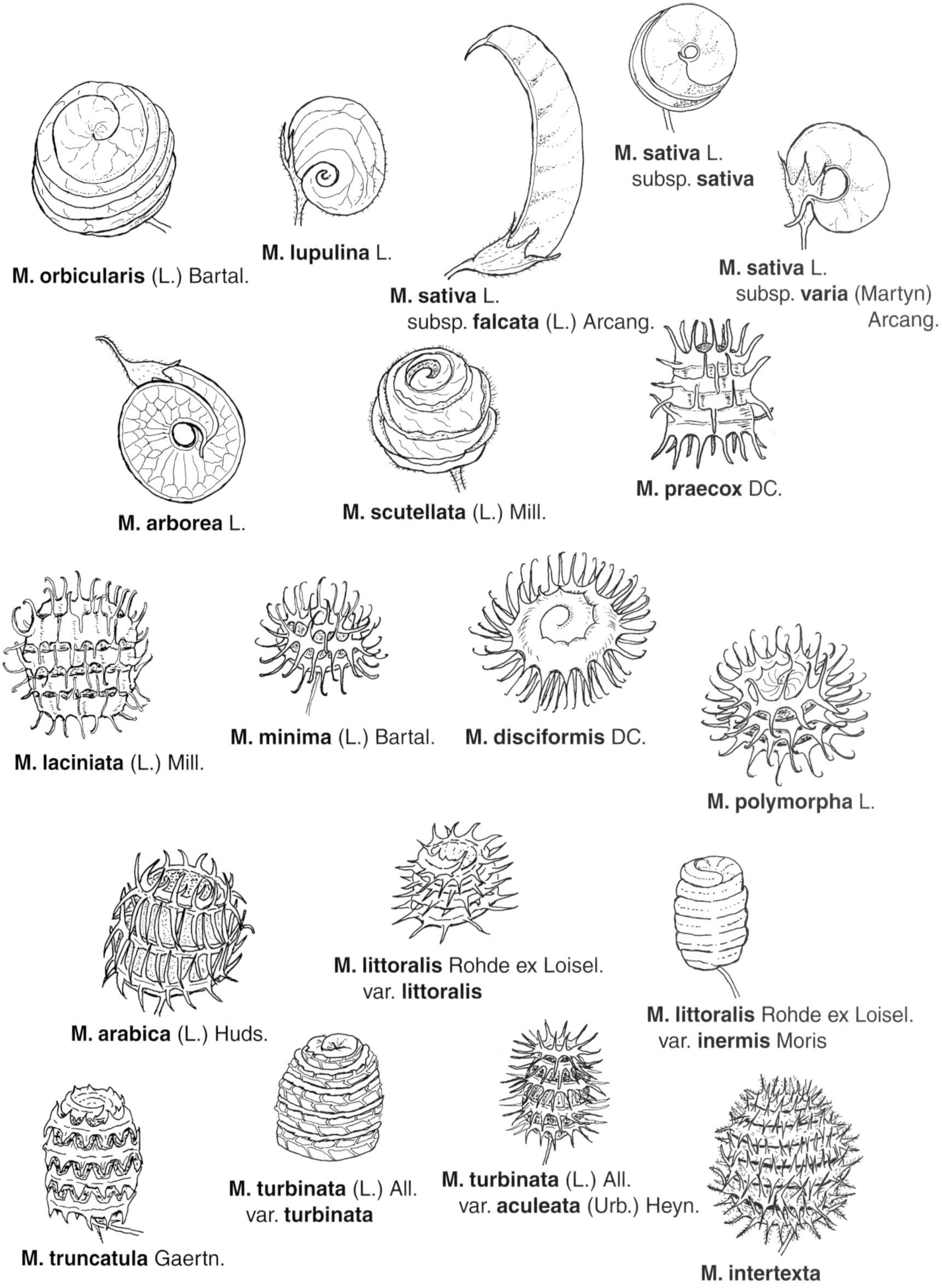
M. orbicularis (L.) Bartal.
M. lupulina L.
M. sativa L. subsp. falcata (L.) Arcang.
M. sativa L. subsp. sativa
M. sativa L. subsp. varia (Martyn) Arcang.
M. arborea L.
M. scutellata (L.) Mill.
M. praecox DC.
M. laciniata (L.) Mill.
M. minima (L.) Bartal.
M. disciformis DC.
M. polymorpha L.
M. arabica (L.) Huds.
M. littoralis Rohde ex Loisel. var. littoralis
M. littoralis Rohde ex Loisel. var. inermis Moris
M. truncatula Gaertn.
M. turbinata (L.) All. var. turbinata
M. turbinata (L.) All. var. aculeata (Urb.) Heyn.
M. intertexta var. ciliaris (L.) Heyn.

Fruits of **Medicago** L.

(iv) Var. **willdenowiana** (Boenn.) W. D. J. Koch
M. willdenowiana Boenn.; *M. lupulina* var. *glandulosa* Mert. & W. D. J. Koch; *M. lupulina* subvar. *glandulosa* (Mert. & W. D. J. Koch) Rouy

Annual. Stems more or less erect to spreading. *Leaflets* 5–15×3–16 mm, more or less hairy, sometimes glandular-hairy. *Petioles* 5–15 mm. *Corolla* 2.5–3.0 mm. *Legume* 2.5–3.5 mm, with numerous, long glandular hairs.

(v) Var. **cupaniana** (Guss.) Boiss.
M. cupaniana Guss.; *M. lupulina* var. *perennans* Halácsy

Perennial. Stems spreading, more or less hairy. *Leaflets* 15–30×10–22 mm, with few to numerous hairs. *Petioles* up to 90 mm. *Corolla* 4.0–4.5 mm. *Legume* 2.5–3.0 mm, with simple or glandular hairs.

Native. Grassy and waste places up to 400 m. Common throughout most of Great Britain and Ireland except in parts of Scotland and the north of Ireland. Widespread in Europe and Asia, introduced and naturalised in many temperate regions of the world. Var. *eriocarpa* occurs in sandy places by the sea and on sandy heaths inland, and is the most obviously native of all the varieties. It occurs in similar localities in Continental Europe. Var. *lupulina* is a plant of grassy roadsides and tracks and rough ground, probably throughout the range of the species. It may not be native in Great Britain and Ireland. Var. *major* is the plant formerly much grown in hay crops and often called 'Trefoil', at least in East Anglia. It occurs in wild flower seed and can still be found as an escape in cultivated areas, but seems no longer to be grown as a crop. It has recently been planted in wild flower seed in the grass margins of Cambridgeshire fields. Var. *willdenowiana* also occurs in wild flower seed and may also be involved in cultivation, though it occurs in some habitats where it might be native. Var. *cupaniana* we have collected in a car park at Ely and on the Devil's Dyke in Cambridgeshire where it is presumably introduced. It is a native of southern Europe.

Section **3**. **Medicago**

Perennial herbs. Legumes falcate or forming an open spiral, often without spines; seeds 2 or more in each legume, smooth, the radicle about half as long as the cotyledons.

3. M. sativa L.

Perennial herb with a tap-root. *Stems* 25–90 cm, pale to medium green, decumbent to more or less erect, angular above, glabrous to thinly hairy. *Leaves* pinnately 3-foliolate; leaflets 8–30×2–12 mm, medium green on upper surface, paler beneath, narrowly oblong-oblanceolate to oblong, rounded or truncate at apex, entire or with a few irregular teeth, entire below and tapering to base, glabrous or nearly so on upper surface, thinly hairy beneath; petiole 4–10 mm, flattened above, slender; stipules 4–10 mm, lanceolate, acuminate at apex, entire or with a few irregular teeth, shortly adnate to the petiole. *Inflorescence* a globose to elongated raceme of few to numerous flowers; peduncle 10–40 mm, appressed-hairy; bracts 1–2 mm, membranous, setaceous; pedicels 2–5 mm, slender, erecto-patent. *Calyx* 3–6 mm, campanulate, with 5 narrowly subulate, subequal lobes, thinly hairy. *Corolla* 6–11 mm, yellow, pale mauve to purple, green, blackish or violet; standard with limb obovate, shortly retuse at apex and cuneate at base; wings with limb oblong, rounded at apex, with a conspicuous lobe at base and slender claw; keel with limb oblong, rounded at apex and with a long claw. *Stamens* 10, diadelphous; filaments pale; anthers broadly oblong. *Style* 1, slightly upcurved; stigma conspicuous, oblique. *Legume* 6–10×2.5–3.0 mm, oblong, straight, falcate or twisted spirally into 1.5–3.5 open coils with one marginal vein, glabrous or thinly hairy, obscurely veined; seeds 2–10, 1.5–2.5 mm, bright brown, oblong, smooth. *Flowers* 6–7.

(a) Subsp. **falcata** (L.) Arcang. Sickle Medick
M. falcata L.

Corolla 6–9 mm, yellow; legume nearly straight to curved in less than half a circle; seeds 2–5. $2n=16$, 32.

(b) Subsp. **varia** (Martyn) Arcang. Sand Lucerne
M. varia Martyn; *M. media* Pers.; *M. subfalcata* Schur; *M. silvestris* Fr.; *M. falcata* subsp. *silvestris* (Fr.) Syme
Corolla 7–10 mm, usually yellow, pale mauve to purple, green or blackish. *Legumes* curved or spiralled in 0.5–1.5 complete turns; seeds 3–8, or abortive.

(c) Subsp. **sativa** Lucerne
Corolla 8–11 mm, pale mauve to violet. *Legume* spiralled in 2–3(–4) complete turns; seeds 10–20. $2n=32$.

Native. Subsp. *falcata* occurs locally in grassy places and rough or waste ground in East Anglia, particularly the area known as Breckland. Formerly in Ireland. It is widely distributed in Europe and is said to occur eastwards through central Asia to China. Up until the Second World War subsp. *sativa* was a common hay crop known as lucerne or alfalfa, originally introduced about 1650. Since then improved strains were developed in both Europe and the United States. It is now rarely seen as a crop, but remains as a relic on roadsides, rough grassland, waste places and particularly old tracks. It is probably indigenous to central Asia and China but is widely naturalised in Europe and western Asia. It has recently been included in wild flower seed planted round the grass margins of Cambridgeshire fields. Subsp. *varia* is of hybrid origin from the other two species. It is established or casual on sandy or rough ground in scattered localities in Great Britain and Ireland north to central Scotland. It can arise *in situ* when the parents grow together or be introduced as hybrid seed, particularly in grass margins of fields, as it is partly fertile, and can backcross. It is recognised as a distinct subspecies as it has an independent existence in many places. It is widely distributed in Continental Europe and the Mediterranean region and is often used as a crop plant.

Section **4**. **Dendrotalis** (Lar. N. Vassiljeva) Lassen
Trigonella section *Dendrotelis* Lar. N. Vassiljeva; *Medicago* section *Arboreae* K. A. Lesins & I. Lesins

Evergreen shrub. Legume a spiral of 1.0–1.5 turns, submarginal vein absent, lateral surfaces reticulate veined, not spiny.

4. M. arborea L. Tree Medick

Evergreen *shrub* 100–400 cm. *Stems* pale brown, young shoots green, with grey silky hairs, branched, leafy. *Leaves* alternate, 3-foliolate; leaflets 10–20×8–18 mm, medium green on upper surface, paler beneath, ovate to obovate, entire or toothed at apex, cuneate at base; sericeous hairy; stipules lanceolate, entire. *Inflorescence* a very short, almost capitate, 4- to 8-flowered raceme. *Calyx* 4.5–6.0 mm, divided into 5 lobes shorter than the tube. *Corolla* 12–15 mm; standard oblong-obovate, obtuse at apex; wings oblong, obtuse at apex; keel oblong, obtuse at apex. *Stamens* 10, diadelphous. *Style* 1. *Legume* 12–15 mm in diameter, subrotund, with 1.0–1.5 anticlockwise coils, with a hole through the centre, marginal vein present, submarginal vein absent, lateral surfaces reticulately veined, not spiny. *Flowers* 6–9. $2n=32$.

Introduced. Has survived on a cliff at Clevedon in Somerset since 1973. Relic or escape from cultivation at Dalkey station in Co. Dublin in 1880. Native of the Mediterranean region.

Section **5. Spirocarpos** Ser.

Annual herbs, rarely perennating. *Legume* forming a congested spiral usually of 2 or more coils, the dorsal suture smooth or spinose; seeds never more than 2 in each coil, smooth; radicle about half as long as cotyledons.

5. M. scutellata (L.) Mill. Snail Medick
M. polymorpha var. *scutellata* L.

Annual herb with fibrous roots. *Stems* 25–50 cm, pale green, suberect or sprawling, angled, or longitudinally rigid, more or less densely glandular-hairy, much-branched, leafy. *Leaves* alternate, pinnately 3-foliolate; leaflets 10–25 × 5–15 mm, medium green on upper surface, paler beneath, oblong-obovate, rounded or subacute at apex, margins sharply serrulate in upper half, entire below, narrowly cuneate at base, glabrous on upper surface, glandular-hairy beneath intermixed with scattered, simple eglandular hairs; petiole 10–25 mm, channelled above, glandular-hairy; stipules 7–12 × 2.5–6.0 mm, ovate-deltoid, acuminate at apex, shortly lobed or 4- to 5-toothed, glandular-hairy. *Inflorescence* 1(–3)-flowered; peduncle less than 10 mm, lengthening in fruit, glandular-hairy; bract 1.0–1.5 mm, subulate-filiform, membranous, abruptly dilated at base. *Calyx* 3.5–4.0 mm, campanulate, divided to about halfway into linear-subulate lobes, glandular-hairy. *Corolla* 5–8 mm, dull yellow; standard with limb obovate, emarginate at apex, broadly cuneate at base; wings with rounded apex, with a blunt lobe at base and a long, slender claw; keel with a rounded apex. *Stamens* 10, diadelphous; filaments pale; anthers subglobose. *Style* 1, thick, upcurved, glabrous; stigma capitate, with narrow, fimbriate wing. *Legume* 9–18 mm in diameter, ovoid, with 4–8, pelviform, anticlockwise, imbricate coils, marginal vein strong, submarginal vein absent, lateral surfaces with numerous, conspicuous, freely anastomosing transverse veins which join the marginal vein, without spines, but more or less glandular-hairy; seeds 5–7 × 2.5–4.0 mm, pale brown, reniform or semi-lunate, smooth. *Flowers* 6–9. $2n=16, 32$.

Introduced. A wool casual. Native throughout the Mediterranean region.

6. M. praecox DC. Early Medick

Annual herb with fibrous roots. *Stem* (5–)10–20 cm, pale green, often flushed purplish, prostrate or decumbent, angular, thinly hairy becoming glabrous, much-branched especially at base, leafy. *Leaves* alternate, 3-foliolate; leaflets 2–6 × 2–4 mm, medium green on upper surface, paler beneath, obovate, truncate or slightly emarginate at apex, apiculate, minutely but sharply toothed towards apex, entire towards base, glabrous and distinctly veined on upper surface, thinly strigose-hairy beneath; petioles 2–10 mm, very slender, channelled above, more or less glabrous; stipules 2–3 mm, ovate, sharply laciniate-dentate, thinly hairy or nearly glabrous. *Inflorescence* of 1–2 flowers scattered along the greater length of the branches; peduncle usually less than 2 mm, glabrous or thinly hairy; pedicels about 1 mm, hairy; bracts subulate-cuspidate, with a dilated sometimes dentate base. *Calyx* 1.5–2.0 mm, divided in 5 short lobes, the lobes narrowly deltoid, thinly hairy. *Corolla* 2–5 mm, yellow; standard subrotund; wings rounded at apex, with an erect rounded lobe at base almost as large as the limb and with a slender claw and a well-developed tooth; keel with an obtuse apex and a scarcely developed lobe at base. *Stamens* 10, diadelphous, the free stamen sterile; anthers suborbicular. *Style* 1, turned up in the middle at a right angle, glabrous; stigma capitate, minutely papillose. *Legume* when young projecting from the calyx when the petals fall, 3–4 mm in diameter, subglobose to cylindrical, pale brown, with 2.5–5.0 anticlockwise lax coils, marginal vein broad and rounded, submarginal vein separated from the marginal vein by a narrow groove not visible when the legume is viewed from the edge, lateral surfaces with strong, curved, transverse veins anastomosing near the submarginal vein and forming a tangentially elongated network, spinose, the spines up to as long as the diameter of the legume, usually somewhat curved and uncinate, sparsely hairy; seeds 1–2 in each coil, about 2.5 mm, ellipsoid-reniform, pale brown, smooth. *Flowers* 3–8. $2n=16$.

Introduced. Casual on tips and rough ground, mostly from wool. Sporadic throughout most of Great Britain. Widely distributed in the Mediterranean region from Spain to Lebanon, but usually rare.

7. M. laciniata (L.) Mill. Tattered Medick
M. aschersoniana Urb.; *M. polymorpha* var. *laciniata* L.

Annual herb with fibrous roots. *Stems* up to 40 cm, pale green, erect to procumbent, shortly hairy, branched, leafy. *Leaves* alternate, 3-foliolate; leaflets 5.0–7.5 × 2.5–4.0 mm, medium green on upper surface, paler beneath, obovate, emarginate at apex, dentate to incise-dentate or almost pinnatifid, narrowed at base, slightly hairy; stipules ovate to ovate-lanceolate in outline, dentate to pectinate; stipules laciniate. *Inflorescence* a 1- to 2-flowered raceme. *Calyx* 2–3 mm, campanulate, with 5 lobes. *Corolla* 3.5–5.0 mm, yellow; standard ovate, obtuse and emarginate at apex; wings oblong, obtuse at apex; keel oblong, obtuse at apex. *Stamens* 10, diadelphous. *Style* 1. *Legume* when young projecting from the calyx as soon as the petals have fallen, 2.5–5.0 mm in diameter, globose to ellipsoid, with

3–7 anticlockwise, lax coils, marginal vein present, submarginal vein broad, separated from the flat or convex marginal vein by a narrow groove, lateral surfaces with sigmoid transverse veins sparingly branched, spinose, the spines up to as long as the diameter of the legume, nearly straight but uncinate and deeply sulcate at the base, glabrous or hairy. *Flowers* 6–9. $2n = 16$.

Introduced. Common wool alien and from other sources on tips and waste ground, rarely persisting. Sporadic throughout most of Great Britain. Native of North Africa and western Asia and naturalised locally in the Mediterranean area.

8. M. minima (L.) Bartal. Bur Medick
M. polymorpha var. *minima* L.

Annual herb with fibrous roots. *Stems* 5–25(–30) cm, pale green, decumbent, prostrate or ascending, angular, with more or less dense eglandular or glandular hairs, much-branched near the base, leafy. *Leaves* alternate, pinnately 3-foliolate; leaflets 3–10 × 2–7 mm, medium green on upper surface, paler beneath, obovate, rounded or obscurely emarginate at apex and often apiculate, denticulate above with the lower half entire, hairy on both surfaces; petioles 5–50 mm, slender, channelled above, with more or less dense eglandular or glandular hairs; stipules 3–6 mm, narrowly ovate, acuminate at apex, entire or obscurely denticulate, hairy on both surfaces. *Inflorescence* with (2–)3–6(–8) flowers, scattered along the greater length of the branches; peduncles up to 12 mm, hairy; bracts about 0.5 mm, membranous, cuspidate at apex, with a broad oblong base; pedicels very short and hairy. *Calyx* 2–3 mm, campanulate with 5 subulate lobes, villous. *Corolla* 2.5–4.0 mm, yellow; standard with limb obovate, emarginate at apex and broadly cuneate at base; wings with limb oblong, rounded at apex, with a conspicuous lobe at base and a slender claw; keel with limb ovate, rounded at apex, with a short broad lobe at base and a slender claw. *Stamens* 10, diadelphous; filaments pale; anthers broadly oblong. *Style* 1, thick, upcurved at apex; stigma oblique, capitate. *Legume* 3–5(–6) mm in diameter, dark brown, subglobose or shortly cylindrical, with 3–5 anticlockwise, lax coils, marginal vein present, submarginal vein wide, separated from the narrow, flat or concave marginal vein by a wide groove, lateral surfaces with slender, curved sigmoid transverse veins not anastomosing, with very short to longer spines which are patent, usually uncinate and deeply sulcate at base, sparsely hairy and glandular. Flowers 5–7. $2n = 16$.

Native. Sandy heaths and dunes and shingle by the sea. Very local in eastern England from Kent to Norfolk and in Jersey; common casual throughout much of Great Britain north to southern Scotland, often as a wool alien. Throughout the Mediterranean region and much of southern and central Europe; adventive in many parts of the world. Eurosiberian Southern temperate element; widely naturalised outside its native range.

9. M. disciformis DC. Disc-fruited Medick

Annual herb with fibrous roots. *Stems* 10–20 cm, pale green, prostrate or ascending, angular, more or less densely eglandular or glandular-hairy, much branched especially at base, leafy. *Leaves* alternate, pinnately 3-foliolate; leaflets 4–12 × 2–9 mm, medium green on upper surface, paler beneath, obovate, rounded, retuse or apiculate at apex, sharply denticulate above, entire below, broadly to narrowly cuneate at base; hairy on both surfaces; petioles 5–20(–40) mm, channelled above, slender, hairy; stipules 4–6 mm, broadly ovate, acuminate at apex, sharply serrate, with simple eglandular or glandular hairs. *Inflorescence* with 1–3 flowers; peduncle 10–25 mm, erect or erecto-patent, with eglandular or glandular hairs; bracts minute, cuspidate, with a broad, oblong base; pedicels 1–3 mm, slender, hairy. *Calyx* 2.0–3.5 mm, campanulate, with 5 subulate lobes, hairy. *Corolla* 5.2–7.8 mm, yellow; standard with limb broadly obovate, emarginate at apex, broadly cuneate at base; wings with limb oblong, obtuse at apex, with a conspicuous, elongate lobe at base and a slender claw; keel with limb oblong, rounded at apex, with a distinct lobe at base and a broad claw. *Stamens* 10, diadelphous; filaments pale; anthers subglobose. *Style* 1, thickened, glabrous, curved upwards at a right angle; stigma capitate. *Legume* when young projecting from the calyx as soon as the petals have fallen, 5–6 mm in diameter, discoid or shortly cylindrical, rich brown when ripe, with 5–7 anticlockwise, lax coils, marginal vein present, submarginal vein parallel with and close to the marginal vein, lateral surfaces obscurely veined towards the ventral margin, with a broad flat area toward the dorsal margin, with short triangular spines or longer well-developed ones, glabrous. *Flowers* 6–9. $2n = 16$.

(**i**) Var. **disciformis**
Legume armed with well-developed spines 3–4 mm.

(**ii**) Var. **apiculata** Urb.
Legume with very short triangular spines less than 1 mm.

Introduced. A wool casual. Native of the northern Mediterranean region from Spain to Turkey. The varieties apparently often grow together, but it is not known whether both occur with us.

10. M. polymorpha L. Toothed Medick
M. denticulata Willd.; *M. hispida* Gaertn.; *M. lappacea* Desr.; *M. polycarpa* Willd.; *M. nigra* (L.) Krock.; *M. muricata* auct.

Annual herb with fibrous roots. *Stems* 10–40 cm, pale green, prostrate or ascending, glabrous or sparsely hairy, branching from the base, leafy. *Leaves* alternate, pinnately 3-foliolate; leaflets 5–20 × 4–16 mm, medium green on upper surface, paler beneath, obovate, retuse or emarginate and shortly apiculate at apex, denticulate in upper part, entire towards the base, narrowed at base, glabrous on upper surface, sometimes thinly hairy beneath; petiole 8–50 mm, channelled above, glabrous; stipules 4–7 mm, ovate, acuminate at apex, laciniate with slender filiform lobes. *Inflorescence* with (1–)2–10 flowers, scattered along the greater length of the stems; peduncles usually less than 10 mm, glabrous; bracts minute, membranous, cuspidate, with a widely dilated base; pedicels seldom exceeding 1 mm, glabrous or thinly hairy. *Calyx* about 2.5 mm, divided in 5 short, deltoid lobes, glabrous or thinly hairy. *Corolla* 3.0–4.5 mm, yellow; standard with limb obovate, emarginate at apex, and broadly cuneatc at base; wings with limb oblong, rounded at apex, with a small lobe at base and a short, wide claw; keel rounded at apex, with a short basal lobe. *Stamens* 10, diadelphous; filaments pale; anthers subglobose. *Style* 1, short and thick, sharply upcurved; stigma capitate. *Legume* when young projecting from the calyx as soon as the petals have fallen, 3–8(–10) mm in diameter, discoid or cylindrical, with 1.5–6.0 anticlockwise, lax coils, marginal vein present, submarginal vein conspicuous, separated from the marginal by a deep groove, the margin consequently with 3 keels separated by 2 grooves, lateral surfaces with strong, curved transverse veins, anastomosing freely, at least near the submarginal vein, spines absent or present and up to more than the diameter of the legume in length and arising from both the marginal and submarginal veins and therefore deeply sulcate, glabrous or hairy. *Flowers* 6–9. $2n=14$, 16.

Native. Very local in sandy ground near the sea. South of England north to Somerset and Norfolk. Common casual throughout Great Britain and Ireland in waste places, often from wool. Throughout the Mediterranean region and eastwards to the western Himalayas; naturalised almost everywhere in Europe and in many other temperate countries. Submediterranean-Subatlantic element.

11. M. arabica (L.) Huds. Spotted Medick
M. polymorpha var. *arabica* L.; *M. maculata* Sibth. nom. illegit.

Annual herb with fibrous roots. *Stems* 20–50(–60), pale green, prostrate or sprawling, rather robust, sharply angular, subglabrous or thinly hairy, usually more frequent at the nodes, usually much branched, leafy. *Leaves* alternate, pinnately 3-foliolate; leaflets 7–20(–30) × 7–20(–30) mm, medium to rather dark green, usually with a conspicuous dark central blotch on upper surface, paler beneath, broadly obovate, rounded, truncate or emarginate at apex, entire or denticulate, cuneate or cordate at base, glabrous on upper surface, thinly hairy beneath; petiole 2–8(–12) cm, flattened or slightly channelled above, hairy; stipules 5–12 mm, ovate, acuminate at apex, dentate or shortly laciniate, with sharp, narrowly deltoid teeth. *Inflorescence* with (1–)2–5(–6) flowers; peduncle 5–15(–20) mm, very slender, hairy; bracts about 1 mm, ovate, acuminate at apex; pedicels up to 1 mm, nearly glabrous to hairy. Calyx about 3 mm, campanulate, with 5 subulate-deltoid lobes, thinly hairy. *Corolla* 4–6 mm, yellow; standard with limb broadly obovate, emarginate at apex, with a short, rather broad claw; wings with limb broadly ovate, blunt at apex, with a conspicuous lobe at base and a short slender claw; keel with limb ovate, with a short basal lobe and a long claw. *Stamens* 10, diadelphous; filaments pale; anthers subglobose. *Style* 1, thick, upturned at a right angle; stigma oblique and conspicuous. *Legume* 4–10 mm in diameter, subglobose to shortly ellipsoid, somewhat flattened at both ends, with 3–7 anticlockwise, lax coils, marginal vein present, submarginal parallel with a vein close to the marginal forming 3 grooves, the lateral deeper and wider than the central, lateral surfaces with curved, anastomosing veins forming a tangentical elongated network near the margin, spiny, the spines usually one-half to three-quarters as long as the diameter of the legume and deeply sulcate, glabrous. *Flowers* 5–9. $2n=16$.

Native. Grassy and rather bare places, especially on light, sandy and gravelly soils particularly near the sea. Throughout Great Britain and the east coast of Ireland but probably native only in the Channel Islands and south and central Great Britain and coastal Wales, elsewhere a common casual especially of wool. Europe and most of the Mediterranean region to eastern Asia. Submediterranean-Subatlantic element.

12. M. littoralis Rohde ex Loisel. Shore Medick
M. striata Bastard

Annual herb with fibrous roots. *Stems* 10–30 cm, pale green, often purplish, prostrate or ascending, angular, hairy, sometimes becoming glabrous, much-branched, leafy. *Leaves* alternate, pinnately 3-foliolate; leaflets medium green on upper surface, paler beneath, 3–7(–15) × 2–5(–12) mm, obovate, truncate, emarginate or rounded at apex, sharply denticulate, cuneate at base, appressed-hairy on both surfaces; petiole 5–15(–20) mm, slender, channelled above, hairy; stipules 3–5 mm, narrowly ovate, acuminate at apex, sharply dentate-laciniate. *Inflorescence* with 1–3 flowers, scattered along the stems; peduncles 5–15 mm, hairy; bracts 1.5 mm, cuspidate, with a broad oblong base; pedicels very short or wanting, more or less densely hairy. *Calyx* 3.0–3.5 mm, campanulate, often purplish, divided into 5 subulate teeth, hairy. *Corolla* 5–7 mm, yellow or orange; standard with limb obovate, emarginate at apex, cuneate at base; wings with limb oblong, rounded at apex, with a prominent lobe at base and a slender claw; keel with limb blunt at apex, with a short lobe at base. *Stamens* 10, diadelphous; filaments pale; anthers subglobose or broadly oblong. *Style* 1, thickened, slightly upcurved; stigma capitate, oblique. *Legume* when young concealed within the calyx when the petals fall, 3–7 mm in diameter, discoid to cylindrical, with 2–6 usually clockwise, closely appressed coils, margin thick and marginal vein present, submarginal vein at first separated from the marginal by a shallow groove, becoming confluent with it and forming a 3-keeled margin when fully ripe, lateral surfaces obscurely veined, the veins nearly straight and scarcely anastomosing except near the submarginal vein, spines or tubercles arising from the submarginal vein, the tubercles varying from short and conical to spines half as long as the diameter of the legume, glabrous. *Flowers* 6–9. $2n=14$, 16.

(i) Var. **littoralis**
Legumes spiny, the spines 1–4 mm, erect or occasionally loosely appressed to legume.

(ii) Var. **inermis** Moris
M. littoralis var. *subinermis* Boiss.; *M. littoralis* var. *tricycla* (DC.) Holmboe

Legume unarmed or with very short tubercles, not spinose.

Introduced. Recorded as a ballast and wool alien and a casual on sand and granite heaps at Gloucester docks. Native throughout the Mediterranean region.

13. M. truncatula Gaertn. Strong-spined Medick
M. tribuloides Desr.

Annual herb with fibrous roots. *Stems* 15–30(–40) cm, pale green, prostrate or ascending, angular, glabrous or hairy, much branched especially at the base, leafy. *Leaves* alternate, pinnately 3-foliolate; leaflets 7–13 × 4–10 mm, medium green on upper surface, paler beneath, obovate, rounded, subacute or rarely emarginate at apex, sharply denticulate, hairy on both surfaces; petioles up to 15 mm, channelled above, slender, hairy; stipules 3–5 mm, narrowly ovate, acuminate at apex, deeply laciniate near the base, Inflorescence with 1–2(–3) flowers; peduncles slender; bracts about 1.5 mm, cuspidate; pedicels short, hairy. *Calyx* 4–5 mm, with 5 subulate lobes, hairy. *Corolla* 6–8 mm, with limb obovate, emarginate at apex, cuneate at base; wings oblong, rounded at apex, with a prominent lobe and a slender claw; keel obtuse at apex, with a distinct lobe at base. *Stamens* 10, diadelphous; filaments pale; anthers subglobose or oblong. *Style* 1, thickened and upcurved; stigma oblique, capitate. *Legume* when young concealed within the calyx when the petals fall, 5–8 mm in diameter, cylindrical, with 3–6 usually clockwise, closely appressed coils, margin usually thick and marginal vein present, submarginal vein thick, parallel with and close to the slender marginal vein, but separated from it by a groove, thus forming 3 keels, lateral surfaces obscurely veined and scarcely anastomosing, spinose, the spines arising partly from the marginal and partly from the submarginal vein and up to half as long as the diameter of the legume and curved or uncinate, nearly always sparsely long-hairy. *Flowers* 6–9. $2n = 16$.

Introduced. A wool, tan-bark and grain casual. Scattered records throughout Great Britain. Native throughout the Mediterranean region; Atlantic islands; introduced elsewhere.

14. M. turbinata (L.) All. Top Medick
M. polymorpha var. *turbinata* L.; *M. tuberculata* Willd.

Annual herb with fibrous roots. *Stems* 15–50 cm, pale green, often purplish or reddish, prostrate or ascending, strongly angular, more or less hairy with long hairs, usually much branched, leafy. *Leaves* alternate, pinnately trifoliolate; leaflets 10–25(–35) × 5–15(–25) mm, medium green on upper surface, paler beneath, obovate or broadly elliptical, occasionally obcuneate or lanceolate, rounded or acute, rarely truncate or emarginate at apex, finely and sharply serrate, hairy on both surfaces; petioles 1–4 cm, distinctly channelled above, thinly hairy; stipules 5–10 mm, ovate, acuminate at apex, sharply laciniate, the basal laciniae long and slender, rarely subentire, thinly hairy. *Inflorescence* with (1–)3–8(–12) flowers, frequently capitate; peduncles 15–35 mm, usually longer than the subtending petioles, thinly hairy; bracts about 15 mm, cuspidate and with an oblong base. *Calyx* about 3.5 mm, campanulate, with 5 subulate lobes, thinly hairy. *Corolla* 6.0–6.5 mm, golden-yellow or orange-yellow; standard with limb broadly obovate, rounded at apex, often minutely apiculate, broadly cuneate at base; wings oblong, rounded at apex, upcurved, with a long, narrow lobe at base and a slender claw; keel oblong, obtuse, often truncate at apex. *Stamens* 10, diadelphous; filaments pale; anthers subglobose. *Styles* upcurved, slightly thickened; stigma oblique, small, capitate. *Legume* when young concealed within the calyx when the petals fall, 5–8 mm in diameter, cylindrical or somewhat conical, or barrelshaped, with flattened ends, with 5–7(–9) clockwise or anticlockwise coils, margin vein present and forming a keel, submarginal vein parallel with and close to the marginal one, lateral surfaces with slender, curved transverse veins, curved, but not or rarely anastomosing and ending in a veinless margin one-quarter to one-third as wide as the radius of the spiral, dorsal margin smooth or with short, broad tubercles or with spines up to half as long as the diameter of the legume; glabrous. *Flowers* 6–9. $2n = 14, 16$.

(**i**) Var. **turbinata**
Inflorescence usually 5- to 8-flowered. Legume unarmed, almost smooth or at most very shortly rugose-tuberculate.

(**ii**) Var. **aculeata** (Moris) Heyn.
M. tuberculata var. *aculeata* (Moris) Urb.; *M. tuberculata* var. *spinosa* H. S. Thomps. nom. nud.

Inflorescence 1–2 (–3) -flowered. *Legume* spinose, with short spines usually much thickened at the base.

Introduced. A grain casual. Native of the northern Mediterranean region. It is not known if only one or both varieties have occurred.

15. M. intertexta (L.) Mill. Hedgehog Medick
M. echinus Lam. ex DC.

Annual herb with fibrous roots. *Stems* 30–50 cm, robust, ascending or decumbent, angular, glabrous or thinly clothed with simple glandular or eglandular hairs, usually much branched, leafy. *Leaves* alternate, pinnately trifoliolate; leaflets 6–20 × 5–15 mm, medium green on upper surface, sometimes blotched red, paler beneath broadly or narrowly obovate, acute, rounded, truncate or rarely retuse at apex, minutely serrulate, glabrous on both surfaces or thinly hairy beneath; petioles 10–30 mm, flattened, sometimes narrowly winged; stipules 8–15 mm, dentate or laciniate, with narrow, sharply pointed teeth. *Inflorescence* with 1–4(–10) flowers along the greater length of the branches; peduncles about as long as petiole; bracts 1–2 mm, narrowly ovate, with a long apical cusp; pedicels about 2 mm, filiform. *Calyx* about 4 mm, campanulate, with 5 subulate lobes, glabrous or thinly hairy. *Corolla* 5–8 mm, mustard yellow; standard with limb subrotund, retuse at apex, with a short claw; wings with limb oblong, obtuse at apex, with an acute lobe at base and a filiform claw; keel with limb oblong, rounded at apex, with a long slender claw. *Stamens* 10, diadelphous; filaments pale; anthers subglobose. *Style* 1, distinctly thickened, glabrous; stigma capitate. *Legume* 9–15(–17) mm in diameter, ovoid, cylindrical or rarely discoid, convex at both ends, with (3–)6–10 anticlockwise coils, marginal vein present, submarginal vein absent, lateral surfaces with strong transverse veins, anastomosing and forming a tangentially elongated network, dorsal margin strongly spinose, the spines (1–)3–4(–8) mm, curved and appressed to the legume, glabrous except often for the spines. *Flowers* 6–9. 2*n* = 16.

(**i**) Var. **intertexta**
M. polymorpha var. *intertexta* L.

Leaflets 6–15 mm. *Legume* 9–12 mm, glabrous.

(**ii**) Var. **ciliaris** (L.) Heyn
M. polymorpha var. *ciliaris* L.; *M. ciliaris* (L.) All.

Leaflets 15–20 mm. *Legume* 12–15(–17) mm in diameter, dorsal surface and spines densely clothed with hairs.

Introduced. A wool and bird-seed casual. Native of the Mediterranean region and Atlantic islands. The form with blotched leaves is occasionally cultivated in gardens as Calvary Clover and is generally named *M. echinus* Lam. ex DC.

Tribe **14. Fabeae**
Tribe *Vicieae* (Bronn.) DC.

Annual or *perennial herbs*. Leaves usually imparipinnate, often with tendrils at apex, rarely simple or reduced to a tendril, with usually entire leaflets.

40. Vicia L.

Annual or *perennial herbs*, often climbing by means of tendrils. *Stems* not winged, but often ridged. *Leaves* paripinnate, usually with a tendril, with (1–)2–many, pinnately veined leaflets; stipules smaller than leaflets. *Inflorescence* of solitary axillary flowers, or axillary fascicles or racemes. *Calyx* with at least 2 of the 5 lobes less than twice as long as tube. *Corolla* whitish, yellow, lilac, red or purple; standard elliptical, obovate or subrotund, often emarginate at apex, usually with a claw; wings oblong or oblanceolate, obtuse at apex, lobed at base and with a narrow claw; keel oblong or ovate, with a lobe at base and a claw. *Stamens* 10, diadelphous. *Style* 1, glabrous, hairy all round or hairy only on lower side. *Fruit* a more or less oblong, compressed legume, dehiscent; seeds usually 2 or more. Often cross-pollinated by bees.

Contains about 140 species in north temperate regions and South America.

Aarssen, I. W. et al. (1986). The biology of Canadian weeds. 76. *Vicia angustifolia* L., *V. cracca* L., *V. tetrasperma* (L.) Schreb. and *V. villosa* Roth. *Canad. Jour. Pl. Sci.* **66**: 711–737.
Akeroyd, J. R. (1996). Coastal ecotypic variants of two vetches, *Vicia sepium* L. and *V. sylvatica* L. (Fabaceae) in Britain and Ireland. *Watsonia* **21**: 71–78.
Ascherson, P. & Graebner, P. (1909). *Synopsis der Mitteleuropäischen Flora*. **6**: 902–995.
Boreau, A. (1857). *Flore du center de la France et cu Bassin de la Loire*. Ed. 3. **1** and **2**. Paris.
Burtt, B. L. & Lewis, P. (1950). *Vicia monantha* Retzius. *Kew Bull.* **4**: 497–515.
Chrtková-Zertová, A. (1973). A cytotaxonomic study of the *Vicia cracca* complex I. *Folia Geobot. Phytotaxon.* **8**: 67–93.
Chrtková-Zertová, A. (1973). A cytotaxonomic study of the *Vicia cracca* complex II. *Folia Geobot. Phytotaxon.* **8**: 249–254.
Druce, G. C. (1907). *Vicia sativa* L. var. *canescens* mihi and *V. angustifolia* L. var. *Garlandii* mihi in Notes on the flora of the Channel Islands. *Jour. Bot. (London)* **45**: 420.
Druce, G. C. (1920). Described botanical varieties and subspecies of *V. sativa* L. *Rep. Bot. Soc. Exch. Club Brit. Isles* **7**: 768–769.
Forster, E. (1830). Observations of the *Vicia angustifolia* of the English Flora of Sir James Edward Smith F. L. S. *Trans. Linn. Soc. London* **16**: 435–444.
Grime, J. P. et al. (1988). *Comparative plant ecology*. London. [*V. cracca, sepium*.]
Hanelt, P. & Mettin, D. (1966). Cytosystematische Untersuchungen in der Artengruppe um *Vicia sativa* L. 2. *Kulturpflanze* **1**.
Hollings, E. & Stace, C. A. (1974). Karyotype variation and evolution in the *Vicia sativa* L. aggregate. *New Phytol.* **73**: 195–208.
Hollings, E. & Stace, C. A. (1978). Morphological variation in the *Vicia sativa* L. aggregate. *Watsonia* **12**: 1–14.
Hultén, E. & Fries, M. (1986). *Atlas of north European vascular plants north of the Tropic of Cancer.* 3 vols. Königstein.
Killick, H. J. (1975). The decline of *Vicia sativa* L. sensu stricto in Britain. *Watsonia* **10**: 288–289.
Kupicha, F. K. (1976). The infrageneric structure of *Vicia*. *Notes Roy. Bot. Garden. Edinb.* **34**: 287–326.
Mettin, D. & Hanelt, P. (1964). Cytosystematische Untersuchungen in der Artengruppe um *Vicia sativa* L. I. *Kulturpflanze* **12**: 163–225.
Phillips, R. & Rix, M. (1993). *Vegetables*. London.
Stewart, A., Pearman, D. A. & Preston, C. D. (1994). *Scarce plants in Britain*. Peterborough. [*V. bithynica, lutea, orobus, parviflora*.]
Vaughan, J. G. & Geissler, C. A. (1997). *The new Oxford book of food plants*. Oxford.
Yamamoto, K. (1966). Studies on the hybrids among *Vicia sativa* L. and its related species. *Mem. Fac. Agric. Kagawa Univ.* **21**: 1–104.
Zertová, A. (1962). *Vicia oreophila*, montane Art aus der Gruppe *Vicia cracca* L. *Novit. Bot. Hort. Univ. Carol. Praha* **1962**: 51–53.
Zohary, D. & Hopf, M. (2000). *Domestication of plants of the Old World*. Ed. 3. Oxford. [*V. faba, sativa*.]

1. Inflorescence sessile or with the peduncle much shorter than flower 2.
1. Inflorescence pedunculate, the peduncle much longer than flower 21.
2. Standard hairy on the back 3.
2. Standard glabrous on the back 4.
3. Calyx lobes subequal; corolla 14–22 mm, dirty brownish-yellow **13. pannonica**
3. Calyx lobes unequal; corolla 18–30 mm, sulphur yellow **16. hybrida**
4. All leaves without a tendril, the rhachis terminated with a small point 5.
4. At least the upper leaves terminated by tendrils 9.
5. Perennials; leaflets more than 10; peduncles more than 3 cm; inflorescence with more than 6 flowers **1. orobus**
5. Annuals; leaflets less than 10; peduncles absent or less than 3 cm; inflorescence with fewer than 6 flowers 6.
6. Leaflets less than 20 mm; corolla less than 10 mm; legumes less than 30 mm **15. lathyroides**
6. Leaflets more than 20 mm; corolla more than 10 mm; legumes more than 30 mm 7.

7. Seeds 20–30 mm, clear, bright, pale brown, ovoid, compressed **20(i). faba** var. **faba**
7. Seeds 10–18 mm, dark brown or white, compressed or subglobose 8.
8. Seeds 15–18 mm, compressed and much thickened in the region of the hilum **20(ii). faba** var. **equina**
8. Seeds 10–14 mm, subglobose or slightly oblong, not compressed **20(iii). faba** var. **minor**
9. Mouth of the calyx-tube oblique or the calyx lobes unequal, the lowest lobe much longer than the upper lobes 10.
9. Mouth of the calyx-tube not oblique, the calyx lobes more or less equal 13.
10. Corolla pale yellow, often purple tinged **17. lutea**
10. Corolla reddish or purple 11.
11. Annual; leaflets 2–6; legume 10–15 mm wide, glabrous or hairy on margin **19. narbonensis**
11. Perennial; leaflets 5–18; legume 5–8 mm wide, glabrous 12.
12. Stems 30–120 cm, ascending, climbing or trailing; leaflets 8–18, 8–30(–40) × 4–12 mm, ovate to ovate-oblong; tendrils 1- to 4-branched **12(i). sepium** var. **sepium**
12. Stems 5–20(–35) cm, procumbent, decumbent or weakly ascending; leaflets 6–14, 5–13 × 3–5 mm, ovate, elliptical or subrotund; tendrils 1- to 2-branched or unbranched **12(ii). sepium** var. **hartii**
13. Seeds tuberculate **15. lathyroides**
13. Seeds smooth 14.
14. Wings and keel of corolla whitish **18. bithynica**
14. Wings and keel of corolla magenta-purple 15.
15. Legumes 6–11 mm wide 16.
15. Legumes 3–6 mm wide 17.
16. Legumes 6–9 mm wide, contracted between the seeds **14(f). sativa** subsp. **sativa**
16. Legumes 9–11 mm wide, not contracted between the seeds **14(g). sativa** subsp. **macrocarpa**
17. Calyx 12–16 mm, the lobes 6–9 mm **14(e). sativa** subsp. **cordata**
17. Calyx 4.5–15 mm, the lobes 2–6 mm 18.
18. Leaflets of upper leaves 1–6 mm wide; calyx 7–15 mm 19.
18. Leaflets of upper leaves 1–3 mm wide; calyx 4.5–7 mm 20.
19. Calyx 7–10 mm, the lobes 3–5 mm **14(c). sativa** subsp. **nigra**
19. Calyx 10–15 mm, the lobes 5–6 mm **14(d). sativa** subsp. **segetalis**
20. Leaflets of upper leaves about 1.0 mm wide, very narrowly linear; calyx 4.5–5.5 mm, the lobes 2.0–2.5 mm **14(a). sativa** subsp. **uncinata**
20. Leaflets of upper leaves 1–3 mm, often slightly wider in the middle than at the ends; calyx 5–7 mm, the lobes 3.0–3.5 mm **14(b). sativa** subsp. **bobartii**
21. Flowers 1–8 per raceme; corolla 2–8(–9) mm, white to purple, not blue 22.
21. Corolla 8–20 mm; if less than 10 mm, then blue and more than 8 flowers per raceme 24.
22. Calyx lobes equal, all at least as long as the tube; legume hairy, usually 2-seeded **9. hirsuta**
22. Calyx lobes unequal, at least the upper shorter than tube; legume glabrous, 3- to 6-seeded 23.
23. Flowers 1–4(–5) per raceme; seeds 4–6(–8) per legume, with hilum little longer than wide and less than one-third the seed circumference **10. parviflora**
23. Flowers 1–2 per raceme; seeds (3–)4(–5) per legume, with hilum more than twice as long as wide and about one-fifth the seed circumference **11. tetrasperma**
24. Leaves with 4–6 leaflets; flowers 1–3 per raceme **18. bithynica**
24. Leaves with more than 6 leaflets; flowers usually more than 4 per raceme 25.
25. Corolla with limb of standard shorter than claw 26.
25. Corolla with limb of standard equalling or longer than claw 34.
26. Corolla reddish-purple with a blackish tip **7. benghalensis**
26. Corolla blue or violet or purple, sometimes with white or yellow wings 27.
27. Stipules bipartite 28.
27. Stipules entire 29.
28. Corolla 10.0–14.5 mm; legume 23–33 mm **8(a). monantha** subsp. **monantha**
28. Corolla 14.5–19.0 mm; legume 32–49 mm **8(b). monantha** subsp. **triflora**
29. Stems villous; lower calyx lobes longer than tube **6(a). villosa** subsp. **villosa**
29. Stems glabrous or appressed-hairy; lower calyx-teeth shorter than tube 30.
30. Legume hairy 31.
30. Legume glabrous or becoming so 32.
31. Racemes 5- to 20-flowered **6(c). villosa** subsp. **eriocarpa**
31. Racemes 2- to 6-flowered **6(d). villosa** subsp. **microphylla**
32. Racemes 10- to 30-flowered **6(b). villosa** subsp. **varia**
32. Racemes 2- to 10-flowered 33.
33. Corolla with wings purple, violet or white; legume usually sparsely hairy when young **6(d). villosa** subsp. **microphylla**
33. Corolla with wings usually yellow; legume glabrous **6(e). villosa** subsp. **ambigua**
34. Lower lobe of stipules strongly toothed; corolla white with blue or purple veins; seeds with hilum more than half total circumference 35.
34. Lower lobe of stipules entire; corolla blue to purple or violet; seeds with hilum less than total circumference ??
35. Stems 50–200 cm, trailing or scrambling; leaflets 12–20(–24), 8–18 × 4–8 mm; raceme with 8–20 flowers; corolla 15–20 mm **5(i). sylvatica** var. **sylvatica**
35. Stem 20–50 cm, procumbent to decumbent; leaflets 8–14, 6–10 (–12) × 3–6 mm, raceme with 4–8 (–12) flowers; corolla 13–18 mm **5(ii). sylvatica** var. **condensata**
36. Corolla (10–)12–18 mm 37.

36. Corolla 8–13 mm 40.
37. Leaflets 3–10 mm wide, oblong-lanceolate **2(iv). cracca** var. **pulchra**
37. Leaflets 0.5–6.0 mm, linear to almost subulate 38.
38. Corolla 8–10 mm, pale bluish-mauve **2(iv). cracca** var. **leptophylla**
38. Corolla (10–) 12–18 mm, lilac, rich violet-blue or purple 39.
39. Leaflets linear; inflorescence 20- to 30-flowered; corolla (10-)12–15 mm **3(i). tenuifolia** var. **tenuifolia**
39. Leaflets narrowly linear-subulate; inflorescence (8–)10- to 20-flowered; corolla 15–18 mm **3(ii). tenuifolia** var. **laxiflora**
40. Lower lobe of calyx distinctly shorter than tube; seeds 1–3 **4. cassubica**
40. Lower lobe of calyx more or less equalling tube; seeds 2–6(–8) 41.
41. Plant with numerous short hairs and dense longer sinuous hairs **2(ii). cracca** var. **sericea**
41. Plant with few to fairly numerous short hairs and few longer hairs **2(iii). cracca** var. **cracca**

Section **1. Cracca** (L.)

Leaflets usually numerous. *Flowers* usually numerous in a long pedunculate raceme. *Calyx* 2-lipped, somewhat gibbous at base. *Corolla* usually large. *Style* equally hairy all round.

1. V. orobus DC. Wood Bitter Vetch
Orobus sylvaticus L., non *Vicia sylvatica* L.

Perennial herb with a tap-root and fibrous side-roots. *Stems* 10–60 cm, pale green, erect, stout, more or less hairy, branched, leafy. *Leaves* 5–10 × 2.5–4.0 cm, oblong or lanceolate-oblong in outline, paripinnate, without tendrils; leaflets 12–30, 10–20 × 6–8 mm, medium green on upper surface, paler beneath, elliptical or lanceolate, mucronate at apex, entire, rounded at base, sessile, more or less hairy; tendril absent; petiole short, hairy; stipules half arrow-shaped, slightly toothed. *Inflorescence* a dense, secund, axillary raceme of 6–20 flowers; peduncle more than 3 cm, about as long as leaves, hairy; pedicels short, the flowers drooping. *Calyx* 5–6 mm, tubular, divided halfway into 5 unequal lobes, the lobes lanceolate, acute at apex, hairy. *Corolla* 12–15(–20) mm; standard white tinged with purple, marked with purple veins at apex, the limb elliptical-oblong, emarginate at apex, and with a short wide claw; wings white tinged purple, the limb oblong with a long, narrow claw; keel white tinged purple, the limb ovate and upturned with a short lobe at base and a long slender claw. *Stamens* 10, more or less diadelphous; filaments pale; anthers yellow. *Style* 1, hairy round the upper half, pale; stigma brownish. *Legume* 20–30 × 5–6 mm, oblong or oblong-lanceolate, acute at both ends, glabrous; seeds 4–5, 4.5–5.0 mm, ellipsoid. *Flowers* 6–9. Pollinated by bees. $2n = 12$.

Native. Grassy and rocky places, banks, edges of fields and scrub usually between 200 and 300 m. Scattered localities through western England, Wales, Scotland and local in central, and west and north Ireland. Western Europe from northern Spain to Norway. Suboceanic Temperate element. We have a significant proportion of the world population of this species.

2. V. cracca L. Tufted Vetch
Cracca major Gren. & Godr.; *Ervum cracca* (L.) Trautv.

Perennial herb with a tough rootstock. *Stem* 60–200 cm, pale green, scrambling or climbing, glabrous to more or less hairy, ribbed, branched, leafy. *Leaves* alternate, 4–8 × 1.5–3.5 cm, oblong in outline, paripinnate; leaflets 10–30, 5–30 × 3–10 mm, medium green on upper surface, paler beneath, linear or oblong-lanceolate to linear-lanceolate, acute or mucronate at apex, entire, more or less rounded at base, sessile, more or less hairy; tendrils branched; petioles very short or absent; stipules half-arrow-shaped, entire. *Inflorescence* 2–10 cm, a dense axillary raceme of 10–30(–40) flowers; peduncle 2–10 cm, stout, hairy; pedicels short and pendulous. *Calyx* 2.5–3.0 mm, divided nearly halfway into 5 lobes, the lobes very unequal, the upper often minute, triangular-ovate, acute at apex, hairy. *Corolla* 8–15 mm; standard pale bluish-mauve flushed with deeper mauve, with a subrotund limb, emarginate at apex, narrowed to a broad claw almost as long as limb, which has the sides folded inwards; wings pale bluish-mauve, flushed deeper mauve, the limb broadly oblong, rounded at apex, with a lobe at base and a narrow claw; keel very pale mauve with a dark mauve blotch at the apex, broadly oblong, curved upwards, with a long claw; all petals turning bluish with age. *Stamens* 10, diadelphous; filaments pale; anthers all similar, yellow. *Style* 1, curved upwards, hairy all round at apex, but with longer hairs on one side than on the other; stigma brownish. *Legume* 10–25 × 5–7 mm, oblong, obliquely truncate, glabrous; seeds 2–6(–8), 3.0–3.2 mm, subglobose, the hilum one-quarter to one-third the total circumference. *Flowers* 6–8. $2n = 28$.

(i) Var. **leptophylla** Fr.
Vicia cracca var. *linearis* Peterm.

Plant with few to fairly numerous short hairs and few longer hairs. *Leaflets* 5–15 × 1–3 mm, linear, sometimes some leaves a little larger. *Corolla* 8–10 mm, usually pale bluish-mauve.

(ii) Var. **sericea** Peterm.
V. incana Thuill., non Vill.; *V. cracca* var. *incana* Coss. & Germ.; *V. cracca* var. *incana* Asch. & Graebn.

Plant with numerous short hairs and dense longer sericeous hairs. *Leaflets* 8–25 × 2–5 mm, oblong-lanceolate. *Corolla* 9–12 mm, often deep bluish-mauve.

(iii) Var. **cracca**
Plant with few to fairly numerous short hairs and few longer hairs. *Leaflets* 10–25 × 2–6 mm, linear-lanceolate to oblong-lanceolate. *Corolla* 9–12 mm, pale to rather deep bluish-mauve.

(iv). Var. **pulchra** (Druce) P. D. Sell
V. cracca forma *pulchra* Druce; *V. oreophila* Zertová

Plant with numerous short hairs mixed with longer hairs. *Leaflets* 15–30×3–10 mm, oblong-lanceolate. *Corolla* 11–15 mm, usually deep indigo blue. $2n=28$.

Native. Grassy and bushy places and hedgerows, river and canal banks. Common throughout Great Britain and Ireland. Almost all Europe; Asia to Sakhalin and Japan; Greenland; introduced in North America. Originally Eurasian Boreo-temperate element now Circumpolar Boreo-temperate. The most widespread variety is var. *cracca* which is certainly the common plant of the clays and fens. In eastern England var. *leptophylla* is the plant of the chalk. It also occurs in Surrey, but its distribution elsewhere is unknown. Var. *sericea* is known from a few scattered localities, but its ecology is not understood. Var. *pulchra* is indeed a beautiful plant and is possibly the only variant of western and northern Scotland and probably also in the east. It also occurs in Norway, Iceland and Faeroes. In the field all the variants are easily recognisable, but it is more difficult to identify poor herbarium specimens.

3. V. tenuifolia Roth — Fine-leaved Vetch
V. cracca subsp. *tenuifolia* (Roth) Gaudin

Perennial herb with a tough rootstock. *Stem* 30–100 cm, pale green, strongly ribbed, flexuous, thinly appressed-hairy, much branched near the base, leafy. *Leaves* alternate, 3–10×1.5–8.0 cm, oblong in outline, paripinnate; leaflets 8–30, 8–35×0.5–6.0(–8.0) mm, medium greyish-green on upper surface, paler beneath, linear to almost subulate, acute, acuminate or often shortly cuspidate at apex, entire, narrowed and shortly hairy, petiolulate at base, glabrous on upper surface, thinly subappressed-hairy beneath; tendrils branched, short and slender; petiole very short or absent; stipules 4–10 mm, linear or semisagittate. *Inflorescence* a lax to rather dense, axillary raceme with (8–)10–30 flowers; peduncle 4–10 cm, often stout, erect, thinly appressed-hairy, the rhachis sometimes with glandular hairs; pedicels 1–2 mm. *Calyx* 3–4 mm, shortly cylindrical, divided to about halfway into 5 lobes, the lobes unequal, narrowly deltoid to deltoid, acuminate at apex, ciliate with simple eglandular and glandular hairs. *Corolla* (10–)12–18 mm, lilac, rich violet-blue or purple; standard with limb obovate, emarginate at apex, with a claw about half as long as limb; wings with limb closely adherent to keel, oblong, rounded at apex, sharply lobed at base, and with a slender claw; keel with limb oblong, curved to a rounded apex, bluntly lobed at base and with a narrow claw. *Stamens* 10, diadelphous; filaments pale; anthers oblong, yellow. *Style* 1, curved upwards at a right angle, hairy all round apex; stigma capitate. *Legume* 20–35×6–9 mm, pale brown, oblong-elliptical, compressed, abruptly acute at apex; seeds 3–5 mm, dark brown, smooth, hilum one-fifth to one-quarter total circumference. *Flowers* 6–8. $2n=24$.

(**i**) Var. **tenuifolia**
Leaflets linear. *Inflorescence* 20- to 30-flowered. *Corolla* (10–)12–15 mm.

(**ii**) Var. **laxiflora** Griseb.
V. elegans Guss.; *V. tenuifolia* var. *stenophylla* Boiss.; *V. cracca* subsp. *stenophylla* (Boiss.) Velen.; *V. cracca* subsp. *elegans* (Guss.) Holmboe; *V. tenuifolia* var. *elegans* (Guss.) Dinsm.

Leaflets narrowly linear-subulate. *Inflorescence* (8–)10- to 20-flowered. *Corolla* 15–18 mm.

Introduced. Naturalised in grassy places and rough ground, especially by railways and in wild flower seed. Scattered records through most of England and rare in Scotland, often only casual, but has been known in some areas for 50 years. The species is native from southern Italy, through the Balkan Peninsula, Turkey, Syria and Palestine eastwards to Afghanistan. Sometimes only one variety is found, at others both with intermediates. Eurosiberian Southern-temperate element.

4. V. cassubica L. — Danzig Vetch
Vacilla cassubica (L.) Schur

Perennial herb. Stems 30–60(–100) cm, pale green, hairy to nearly glabrous, branched, leafy. *Leaves* alternate, paripinnate, oblong in outline, with tendrils; leaflets 10–32, 7–30×3–10 mm, dark green on upper surface, paler beneath, linear-lanceolate to oblong or elliptical, obtuse at apex, entire, glabrous to hairy; stipules entire. *Inflorescence* an axillary raceme of 4–15 flowers. *Calyx* 5–6 mm, 2-lipped, the upper lip slightly longer than lower, the lower lip more or less as long as tube, somewhat gibbous at base. *Corolla* 9–13 mm, purple or pink with whitish wings and keel; standard 9–13 mm, obovate, rounded at apex; wings oblong, obtuse at apex; keel oblanceolate, obtuse at apex. *Stamens* 10, diadelphous. *Style* 1. *Legume* 15–30×6–8 mm, yellow, oblong-rhombic, glabrous; seeds 1–3, the hilum one-third of the circumference. *Flowers* 6–8. $2n=12$.

Introduced. Naturalised in a chalkpit near Greenhithe in Kent between 1931 and 1964. Native of south, central and east Europe.

5. V. sylvatica L. — Wood Vetch

Perennial herb with a rhizome. *Stem* 20–200 cm, weak, procumbent to decumbent, scrambling or climbing, glabrous, branched, leafy. *Leaves* alternate, paripinnate; leaflets 8–20(–24), 6–18×3–8 mm, medium green on upper surface, paler beneath, elliptical, oblong-elliptical or lanceolate-elliptical, ovate or subrotund, rounded-mucronate at apex, entire, rounded at base, glabrous; tendrils much branched; petioles very short; stipules lanceolate, acute at apex, semicircular with many teeth at base. *Inflorescence* an axillary raceme with up to 20 flowers, secund and rather lax, with the flowers pedicellate and drooping; peduncles up to 18 cm, rather stout. *Calyx* 4–5 mm, tubular, with 5 short lobes, the lobes setaceous, shorter than the tubes, the upper about half the length of the lower. *Corolla* 13–20 mm; standard white, tinged with lilac and veined with purple, the limb broadly elliptical, emarginate at apex, rounded at base into a broad, folded claw; wings white, tinged lilac, veined with purple, the limb broadly oblong, rounded at apex, with a broad lobe at base and a slender claw; keel with a purple blotch at the apex, the limb ovate, upturned, with a short lobe at base and a slender claw. *Stamens* 10, diadelphous; filaments upturned at apex; anthers yellow. *Style* 1, curved upwards, hairy all round at apex. *Legume* 25–30 mm, oblong or oblong-lanceolate, acuminate at both ends, glabrous; seeds 4–5, subglobose, black. *Flowers* 6–8. $2n = 14$.

(**i**) Var. **sylvatica**
Stems 50–200 cm, weak, trailing or scrambling. *Leaves* 4–8 cm; leaflets 12–20 (–24), more or less distant, 8–18 × 4–8 mm, ovate to ovate-oblong or elliptical, thin in texture; tendrils 2–7 cm, 2- to 4-branched. *Raceme* usually distinctly exceeding leaves, lax, with 8–20 flowers; peduncles 5–18 cm. *Corolla* 15–20 mm.

(**ii**) Var. **condensata** Druce
V. sylvatica var. *maritima* Lange

Stems 20–50 cm, somewhat rigid, procumbent to decumbent, forming compact patches and low hummocks. *Leaves* 2.5–5.0 cm; leaflets 8–14, rather crowded, 6–10(–12) × 3–6 mm, ovate to subrotund or broadly elliptical, slightly fleshy, often glaucous; tendrils 1–4 cm, unbranched or 1- to 3-branched. *Racemes* about as long as or only slightly exceeding the leaves, compact, with 4–8(–12) flowers; peduncles 2–5 cm. *Corolla* 13–18 mm.

Native. Open woods, scrub, rocky slopes, and sea-cliffs and shingle. Scattered throughout much of Great Britain and Ireland, but local. North, central and east Europe, southwards to Italy, east to the Baikal region of Siberia. The most frequent plant is var. *sylvatica*. Var. *condensata* occurs on coastal shingle and sea-cliffs in scattered localities in western Great Britain, particularly Galloway, and at White Park Bay, Co. Antrim in Ireland. It also occurs in Denmark on the Djurland Peninsula in eastern Jutland. Intermediates between the two varieties are recorded.

6. V. villosa Roth — Fodder Vetch

Annual, *biennial* or *perennial herb* with a stout stock. *Stems* 50–200 cm, pale green, sprawling or climbing, sharply angular, glabrous to more or less densely villous with spreading hairs, branched, leafy. *Leaves* alternate, 1–12 × 0.5–7.0 cm, oblong in outline, paripinnate; leaflets 4–12, 3–30 × 1.5–9.0 mm, medium green on upper surface, paler beneath, linear-lanceolate, oblong, elliptical or subrotund, obtuse, acute or acuminate at apex, entire, shortly petiolulate or subsessile, subglabrous to thinly villous; tendril branched or rarely simple; petiole very short, usually less than 1 cm; stipules 1–12 mm, semi-sagittate. *Inflorescence* a lax to dense, axillary raceme with few to many flowers; peduncle 3–10 cm, spreading; pedicels 1.0–1.5 mm, slender. *Calyx* 10–20 mm, blue to purple or violet, the wings sometimes white or yellow; standard with limb much smaller than claw, oblong, recurved, emarginate at apex, the claw 8–13 mm, often slightly dilated about the middle or tapering to a rather broad base; wings with limb narrowly oblong, sometimes widening towards the blunt or rounded apex, closely adherent to keel with a distinct tooth and a long, blunt lobe at base and a long slender claw; keel with limb oblong, slightly upcurved, obtuse to subacute and blotched dark purple at apex, with a distinct lobe at base and a broad claw. *Stamens* 10, diadelphous; filaments pale; anthers oblong, yellow. *Style* 1, sharply upcurved, bearded for some distance below the apex; stigma obliquely capitate. *Legume* 20–40 × 8–10 mm, oblong, compressed, acute or shortly beaked, pale brown, glabrous or hairy; seeds 2–8, 4.0–4.5 mm, dark blackish-brown, sometimes obscurely marbled, smooth, hilum small. *Flowers* 6–8. $2n = 14$.

(**a**) Subsp. **villosa**
Plant villous. *Leaflets* (8–)10–35 × 2–8 mm. *Racemes* 10- to 30-flowered. *Calyx* lobes plumose, the lower as long as or longer than the tube. *Corolla* 10–20 mm; wings variously coloured. *Legume* glabrous.

(**b**) Subsp. **varia** (Host) Corb.
V. varia Host; *Cracca varia* (Host) Godr. & Gren.; *V. dasycarpa* Ten.; *V. polyphylla* Desf.; *V. glabrescens* (W. D. J. Koch) Heimeri

Plant glabrous or appressed-hairy. *Leaflets* (8–)10–30 × 2–8 mm. *Racemes* 10- to 30-flowered. *Calyx* lobes glabrous or appressed-hairy, all shorter than the tube. *Corolla* 10–16(–18) mm; wings violet, purple, blue or white. *Legume* glabrous.

(**c**) Subsp. **eriocarpa** (Hausskn.) P. W. Ball
V. eriocarpa (Hausskn.) Halácsy; *V. varia* var. *eriocarpa* Hausskn.

Plant glabrous or appressed-hairy. *Leaflets* 5–15(–20) × 1–5 mm. *Racemes* 5- to 20-flowered. *Calyx* lobes glabrous or appressed-hairy, all shorter than the tube. *Corolla* 10–16(–18) mm; wings violet, purple, blue or white. *Legume* hairy, at least when young.

(**d**) Subsp. **microphylla** (D'Urv.) P. W. Ball
V. microphylla D'Urv.; *V. salaminia* Boiss.

Plant glabrous or appressed-hairy. *Leaflets* 3–10 × 1–4 mm. *Racemes* 2- to 6-flowered. *Calyx* lobes glabrous or appressed-hairy, all shorter than the tube. *Corolla* 10–16(–18) mm; wings violet, purple, blue or white. *Legume* glabrous or hairy.

(**e**) Subsp. **ambigua** (Guss.) Kerguélen
V. ambigua Guss.; *V. villosa* subsp. *pseudocracca* (Bertol.) Rouy; *V. pseudocracca* Bertol.; *V. varia* subsp. *ambigua* (Guss.) Nyman; *V. littoralis* Salzm., non Jacq.; *V. elegantissima* Shuttlew.

Plant glabrous or appressed-hairy. *Leaflets* 5–20 × 1–5 mm. *Racemes* 3- to 10-flowered. *Corolla* 10–16(–18) mm; wings usually yellow. *Legume* glabrous. $2n = 14$.

Introduced. Naturalised in grassy places, tips, rough and waste places, and more often casual. Scattered records in Great Britain and a solitary record for Ireland. All five subspecies have been recorded. The species occurs in south-west Europe, Sicily, Greece, Aegean Islands, Turkey, Syria, Lebanon and eastwards to Iraq. Subsp. *villosa* occurs throughout the range of the species and subsp. *varia* almost so. Subsp. *eriocarpa* occurs in Greece, Aegean region and Sicily, subsp. *microphylla* in southern Greece and the Aegean and subsp. *ambigua* in south-west Europe.

7. V. benghalensis L. — Purple Vetch
V. atropurpurea Desf.

Annual or short-lived *perennial herb*. *Stem* 20–80 cm, pale green, hairy, branched, leafy. *Leaves* alternate, paripinnate, oblong in outline; leaflets 10–18, 10–25 × 1.5–6.0 mm, medium green on upper surface, paler beneath, linear, oblong or elliptical, pointed at apex, entire, hairy; stipules entire or dentate. *Inflorescence* an axillary raceme with 2-flowers. *Calyx* 6.0–10.5 mm, strongly gibbous at base, 2-lipped, the upper lip slightly longer than the lower. *Corolla* 10–18 mm, reddish-purple with a blackish tip; standard 15–18 × 6–7 mm, ovate, rounded, upturned; wings 13–16 × 2.5–3.0 mm; keel 12–14 × 2.1–2.7 mm. *Stamens* 10, diadelphous. *Style* 1. *Legume* 25–40 × 8–12 mm, brown, oblong, shortly stipitate, hairy at least on the sutures; seeds 3.5–5.5 mm, hilum round one-fifth of the circumference. *Flowers* 5–8. $2n = 12$, 14.

Introduced. Occasional casual, sometimes briefly persisting on tips and rough and waste ground. Sporadic since about 1910, mostly in central and southern Great Britain. Native of the Mediterranean region.

8. V. monantha Retz. Few-flowered Vetch
V. calcarata Desf.

Annual herb with fibrous roots. *Stems* (10–)15–40(–50) cm, pale green, trailing or climbing, sharply angular, thinly subappressed-hairy. *Leaves* alternate, 2–4(–5) × 1.0–3.5 cm, oblong in outline, paripinnate; leaflets 8–14(–18), 5–20 × 1–6 mm, medium green on upper surface, paler beneath, oblong, rounded or obscurely emarginate at apex, entire, cuneate at the shortly petiolulate base, glabrous on the upper surface, thinly villous beneath; tendrils branched or unbranched; petioles very short, usually less than 5 mm; stipules 2–4 mm, bipartite, semi-sagittate, glabrous or thinly villous. *Inflorescence* a lax, axillary raceme with 1–3(–5) flowers; peduncle 7–30(–60) mm, erect; pedicels usually less than 2 mm, long and thinly hairy. *Calyx* 3–5(–6.5) mm, campanulate, divided half way to the base into 5 lobes, the lobes subequal, deltoid, thinly villous. *Corolla* 12–20 mm, purplish or pinkish; standard with limb oblong, slightly emarginate at apex, scarcely narrowed at base, claw very broad; wings with limb oblong, rounded at apex, with a prominent blunt lobe at base and a slender claw; keel with limb oblong, curved upwards to a blunt apex, with a short, blunt lobe at base and a long claw. *Stamens* 10, diadelphous; filaments pale; anthers broadly oblong, yellow. *Style* 1, upcurved almost at a right angle, villous all round near the apex; stigma capitate. *Legume* 23–50 × 8–12 mm, oblong, strongly compressed, with a short, sharp upcurved or downcurved beak at apex, finely reticulate-veined; seeds 2.5–5.0 mm, oblong or subglobose, slightly compressed, smooth, minutely rugulose or pitted. *Flowers* 6–8.

(**a**) Subsp. **monantha**
V. gracilis Banks & Sol; *V. cinerea* M. Bieb. *V. calcarata* var. *cinerea* (M. Bieb.) Boiss.; *V. griffithii* Baker; *V. cossoniana* Batt.; *V. monantha* subsp. *cinerea* (M. Bieb.) Maire

Corolla with standard 10.0–14.5 mm; wings 9.0–12.5 mm; keel 9.0–11.5 mm. *Legume* 23–33 × 6.0–8.5 mm; seeds not exceeding 3.5 mm, brownish.

(**b**) Subsp. **triflora** (Ten.) B. L. Burtt & P. Lewis
V. calcarata Desf.; *V. biflora* Desf.; *V. triflora* Ten.; *Cracca calcarata* (Desf.) Gren. & Godr.; *Ervum calcaratum* (Desf.) Trautv.; *V. calcarata* subsp. *triflora* (Ten.) Nyman

Corolla with standard 14.5–19.0 mm; wings 9.0–12.5 mm; keel 12–14 mm. *Legume* 32–49 × 8.5–12.0 mm; seeds usually exceeding 3.5 mm, blackish.

Introduced. A grain casual. Widespread in the Mediterranean region and eastwards to Afghanistan and Pakistan. Subsp. *triflora* seems to be the race most recorded.

Section **2. Ervum** (L.) Gray
Ervum L.

Leaflets usually numerous. *Flowers* few in long pedunculate racemes. *Calyx* not gibbous at base. *Corolla* usually less than 10 mm. *Styles* glabrous or equally hairy all round.

9. V. hirsuta (L.) Gray Hairy Tare
Ervum hirsutum L.

Annual herb with fibrous roots. *Stems* 20–80 cm, pale green, slender and trailing or scrambling, nearly glabrous, branched, leafy. *Leaves* alternate, 3–6 × 1.5–2.5 cm, oblong in outline, paripinnate; leaflets 8–20, 5–20 × 1.5–3.0 mm, medium green on upper surface, paler beneath, linear or oblong, truncate to emarginate and often mucronulate at apex, entire, rounded at base, glabrous; tendrils usually branched; petioles short; stipules often 4-lobed. *Inflorescence* an axillary raceme of (1–)2–7(–9) flowers; peduncles 1–3 cm, slender. *Calyx* 2.0–2.5 mm, tubular, divided halfway to base into 5 subequal lobes, the lobes subulate, acute-mucronate at apex, hairy. *Corolla* (2–)3–5 mm; standard bluish-white, the limb broadly elliptical, rounded at both ends without a claw; wings bluish-white, the limb oblong, rounded at apex, with a short lobe at base and a slender claw; keel dull bluish-white with a purple blotch at apex, the limb ovate and upturned, with a slender claw. *Stamens* 10, diadelphous; filaments pale; anthers yellow. *Style* glabrous; stigma brownish. *Legume* 6–11 × 2–4 mm, oblong, hairy; seeds almost always 2, 1.0–1.2 mm, globose. *Flowers* 5–8. $2n = 14$.

Native. Waste and grassy places, particularly roadsides, railways and tracks and by margins of arable fields, sea-cliffs and consolidated shingle beaches. Throughout lowland Great Britain and Ireland, but rare in north and central Scotland. Almost all Europe, west Asia and north Africa. A weed of cultivation throughout much of the world. European Temperate element.

10. V. parviflora Cav. Slender Tare
V. laxiflora Brot. nom. illegit.; *V. tetrasperma* subsp. *gracilis* Hook. fil.; *V. tenuissima* auct.; *V. gracilis* Loisel., non Banks & Sol.; *Ervum varium* Brot.; *V. varia* (Brot.) Lacaita

Annual herb with fibrous roots. *Stems* 30–60 cm, pale green, slender and trailing or scrambling, glabrous or slightly hairy, branched, leafy. *Leaves* alternate, 3–5 × 2–4 cm, broadly oblong in outline, paripinnate; leaflets 4–8(–10), 10–25 × 1.5–2.5 mm, medium green on upper surface, paler beneath, linear, acute or acuminate at apex, entire, narrowed at base, sessile, glabrous or slightly hairy; tendrils usually unbranched; petioles short; stipules half arrow-shaped. *Inflorescence* an axillary raceme of 1–4(–5) flowers; peduncles up to 8 cm, slender, longer than leaves. *Calyx* 2.0–2.5 mm, divided one-third of the way to the base into 5 unequal lobes, the lobes linear-lanceolate, acute at apex, hairy. *Corolla* (5–)6–9 mm; standard blue turning mauvish-blue, obovate, with an emarginate apex and a short very broad claw; wings blue turning mauvish-blue, the limb broadly oblong, rounded at apex, with a narrow lobe at base and a long slender claw; keel pale greenish-white with a blue tip, the limb broadly oblong, slightly upturned at apex, with a short broad lobe at base and a long slender claw. *Stamens* 10, diadelphous; filaments pale; anthers yellow. *Style* 1, pale; stigma brownish. *Legume* 12–17 × 3.0–3.5 mm, oblong, glabrous; seeds 4–6(–8), 2.5–3.0 mm, globose, the hilum little longer than wide and less than one-third the seed circumference. *Flowers* 6–8. $2n = 14$.

Native. Grassy places on calcareous clay soils in hedgerows, grassy banks, tracks, verges and coastal cliffs. Local in southern England north to Huntingdonshire and a single record in Lincolnshire. South and west Europe; Macaronesia; introduced in central Europe. Submediterranean, Subatlantic element.

11. V. tetrasperma (L.) Schreb. Smooth Tare
Ervum tetraspermum L.; *V. gemella* Crantz

Annual herb with fibrous roots. *Stems* (15–)30–60 cm, pale green, slender and scrambling or trailing, branched, glabrous, leafy. *Leaves* alternate, 2.5–4.0 × 2.5–4.0 cm, broadly oblong in outline, paripinnate; leaflets 6–12(–16), 12–25 × 2–3 mm, medium green on upper surface, paler beneath, linear, obtuse-mucronate to acute at apex, entire, narrowed at base, sessile, glabrous or sparsely hairy; tendrils usually simple; pedicels short; stipules half arrow-shaped. *Inflorescence* an axillary raceme of 1–2(–4) flowers; peduncles 2–4 cm, about equalling the leaves, slender. *Calyx* 3.0–3.5 mm, tubular, divided halfway to the base into 5 unequal lobes, the lobes narrowly triangular, acute at apex, hairy. *Corolla* 4–8 mm, standard very pale mauve with darker veins, the limb obovate, emarginate with a short broad claw; wings pale blue, the limb oblong, rounded at apex, with a short, narrow lobe at base and a slender claw; keel very pale mauve with a bluish-mauve blotch at apex, the limb oblong-ovate with a slender claw. *Stamens* 10, diadelphous; filaments pale; anthers yellow. *Style* 1, pale; stigma greenish. *Legume* 9–16 mm, oblong, glabrous; seeds (3–)4(–5), 1.0–1.2 mm, globose, with hilum more than twice as long as wide and about one-fifth the seed circumference. *Flowers* 5–8. $2n = 14$.

Native. Grassy places by roadsides, on railway banks and in margins of woods, scrub and fields. Widespread in England, Wales and the Channel Islands, scattered records and probably introduced in Scotland and Ireland. Europe north to 62° N, west Asia, Japan, North Africa, Madeira, Canary islands. European Temperate element but widely naturalised and now Circumpolar Temperate.

Section **3. Vicia**
Leaflets usually more than 6. *Flowers* solitary, axillary or in few-flowered, sessile or shortly pedunculate racemes. *Corolla* usually large, more than 10 mm. *Style* hairy on the lower side beneath the stigma.

12. V. sepium L. Bush Vetch
V. sordida Salisb. nom. illegit.; *V. dumetorum* auct.

Perennial herb with a tap-root and fibrous side-roots. *Stems* 5–120 cm, pale green, procumbent, decumbent, climbing or trailing, glabrous, branched, leafy. *Leaves* alternate, 2–7 × 3–5 cm, oblong or lanceolate-oblong in outline, paripinnate; leaflets 6–18, 5–30(–40) × 3–12(–18) mm, medium green on upper surface, paler beneath, ovate to elliptical, or subrotund, obtuse, truncate or emarginate at apex and mucronate, entire, rounded at base, very shortly petiolulate, shortly hairy on both surfaces and the margins; at least the tendrils of the lower leaves often branched sometimes absent or vestigial; petioles short; stipules half arrow-shaped, sometimes toothed, marked with a dark spot. *Inflorescence* an axillary raceme of 2–6 flowers; peduncles very short or subsessile. *Calyx* 5.0–5.5 mm, reddish-purple, tubular, divided halfway to base into 5 unequal lobes, the lobes triangular, acute at apex, hairy. *Corolla* 12–15 mm; standard reddish-purple with dark purple veins, the limb subrotund and emarginate at apex, with a very broad claw; wings paler or bluish, the limb obovate, rounded at apex, with a narrow lobe at base and a long thin claw; keel reddish, limb ovate, upturned, with a long slender claw; all petals fading to dull blue or greenish-blue. *Stamens* 10, diadelphous; filaments pale; anthers yellow. *Style* 1, bearded on lower side, stigma brownish. *Legume* 20–35 × 5–8 mm, oblong, beaked, glabrous; seeds 3–10, 3.0–3.5 mm black, globose. *Flowers* 5–8. Pollinated by bees. $2n = 14$.

(i) Var. **sepium**
Stems 30–120 cm, ascending, climbing or trailing. *Leaves* 3–7 cm; leaflets 8–18, 8–30(–40) × 4–12(–18) mm, ovate to ovate-oblong, subacute, obtuse or truncate at apex; tendrils 1–3 cm, 1- to 4-branched.

(ii) Var. **hartii** Akeroyd
V. sepium forma *prostrata* Druce; *V. sepium* forma vel var. *dunensis* Druce nom. nud.

Stems 5–20(–35) cm, procumbent, decumbent or weakly ascending, forming mats and low hummocks, sometimes climbing *Ammophila arenaria* stems. *Leaves* 2–4 cm; leaflets 6–14, 5–13 × 3–5 mm, ovate to elliptical or subrotund, truncate to rounded; tendrils 0.5–2.0 cm, usually unbranched or 1- to 2-branched, often absent or vestigial, sometimes replaced by a terminal leaflet.

Native. Grassy places, hedgerows, waysides, scrub and wood borders, rarely in sand-dunes. Throughout Great Britain and Ireland. Almost all Europe, but rare and local in the Mediterranean region; temperate Asia and Kashmir; Greenland. Eurosiberian Boreo-temperate element. Var. *sepium* is the common variety. Var. *hartii* is the plant of sand-dunes in widely scattered localities on the island of Coll in the Hebrides, north coasts of Sutherland and

Caithness and Co. Mayo and Co. Donegal in the north-west of Ireland and is endemic. Intermediates are recorded between the two varieties. Var. *hartii* is named after Henry Chichester Hart (1847–1908).

13. V. pannonica Crantz Hungarian Vetch

Annual herb with fibrous roots. *Stems* 30–60 cm, pale green, conspicuously striate, thinly hairy to nearly glabrous, usually much branched at base, leafy. *Leaves* alternate, paripinnate, 3–9 × 1.5–4.0 cm, oblong in outline; leaflets 10–20, 6–20 × 2–5 mm, medium green on upper surface, paler beneath, oblong, truncate, emarginate, rounded or subacute at apex, usually apiculate, entire, thinly appressed-villous on both surfaces; petiole short and rather stout; tendril branched; stipules 2–4 mm, oblong or ovate, acute at apex, 2- to 3-veined with a median, dark, hairy, glandular patch. *Inflorescence* with 2–3(–5), pendulous flowers in a congested raceme or cluster; peduncle very short; bracts absent; pedicels 1–2 mm. *Calyx* 8–9 mm, cylindrical, slightly gibbous at base, with 5 subequal, subulate lobes, thinly villous. *Corolla* 14–22 mm, dirty brownish-yellow; standard with limb subrotund, with a long claw, villous on back; wings with limb oblong, rounded at apex, with a blunt lobe at base and a long claw; keel with limb subrotund or very shortly oblong, base lobed and with a long claw. *Stamens* 10, diadelphous; filaments pale; anthers oblong. *Style* 1, with a beard. *Legume* 25–30 × 8–10 mm, oblong, compressed, abruptly narrowed at apex, shortly beaked, thinly villous; seeds 3–5 mm in diameter, dark brown, subglobose, smooth. *Flowers* 6–9. $2n = 12$.

Introduced. Frequent casual in waste places and on roadside banks in England and Wales. Has been naturalised in the Dartford to Northfleet area of Kent since 1971. Native of central and southern Europe and Turkey.

14. V. sativa L. Common Vetch

Annual herb with fibrous roots. *Stem* 15–120 cm, pale green, diffuse, sprawling or climbing, angular, with few to fairly numerous, short, spreading hairs, branched, leafy. *Leaves* alternate, paripinnate, 1.5–8.0 × 1–5 cm, oblong in outline; leaflets 6–16, 3–30 × 2–15 mm, medium green on upper surface, paler beneath, linear to oblong, acute, rounded or emarginate at apex, entire, rounded at base, very shortly petiolulate, thinly hairy on both surfaces to nearly glabrous; tendril simple or branched; petioles usually less than 10 mm; stipules 3.8 mm, semi-sagittate, with a conspicuous dark glandular median blotch, acutely dentate-laciniate. *Inflorescence* of 1 or rarely 2, shortly pedicelled, axillary flowers. *Calyx* 8–15 mm, cylindrical, with 5 subequal lobes, the lobes subulate, thinly hairy or ciliate. *Corolla* 14–26(–30) mm, usually bright magenta-purple or reddish-purple, rarely white or yellow; standard with limb broadly elliptical, emarginate at apex, rounded at base to a broad claw which has inturned margins; wings with limb obovate, rounded at apex, with a narrow lobe at base and a slender claw; keel with limb ovate, upturned, with a short lobe at base and a narrow claw. *Stamens* 10, diadelphous; filaments pale; anthers yellowish. *Style* 1, upturned at right angles, shortly hairy on one side just below the apex; stigma brownish. *Legume* 25–70 × 3–11 mm, linear or oblong, beaked, glabrous or sparsely hairy; seeds 4–12, 1.4–5.0 mm, subglobose. *Flowers* 5–9.

(**a**) Subsp. **uncinata** (Rouy) P. D. Sell
V. uncinata Desv. ex Boreau, non Rchb.; *V. communis* var. *uncinata* Rouy

Plant 10–20 cm, often more or less prostrate. *Leaves* with 4–8 leaflets, the lower shorter than upper, 2–5 × 1–4 mm, obovate or subrotund, obtuse at apex, the upper 8–12 × about 1.0 mm, narrowly linear. *Calyx* 4.5–5.5 mm, the lobes 2.0–2.5 mm, linear-subulate. *Corolla* 10–12 mm. *Legumes* 20–30 × 3–4 mm, not contracted between the seeds.

(**b**) Subsp. **bobartii** (E. Forst.) P. D. Sell
V. bobartii E. Forst.; *V. angustifolia* var. *bobartii* (E. Forst.) W. D. J. Koch; *V. sativa* var. *bobartii* (E. Forst.) Burnat

Plant 10–20 cm, more or less ascending. *Leaves* with 4–8 leaflets, the lower shorter than upper, 2–7 × 1–7 mm, obovate or subrotund, emarginate, the upper 5–12 × 1–3 mm, linear, obtuse at apex. *Calyx* 5–7 mm, the lobes 3.0–3.5 mm, linear-subulate. *Corolla* 10–15(–20) mm. *Legumes* 20–30 × 3–4 mm, not contracted between the seeds.

(**c**) Subsp. **nigra** (L.) Ehrh.
V. sativa var. *nigra* L.; *V. forsteri* Jord.

Plant up to 50 cm. *Leaves* with 8–14 leaflets, the lower 4–12 × 3–5 mm, obovate, obtuse to emarginate, the upper or all 10–30 × 1–6 mm, linear or linear-lanceolate, obtuse at apex. *Calyx* 7–10 mm, the lobes lanceolate-subulate. *Corolla* 15–20 mm. *Legume* 30–40 × 4–5 mm, not contracted between the seeds.

(**d**) Subsp. **segetalis** (Thuill.) Gaudin
V. segetalis Thuill.; *V. angustifolia* var. *segetalis* (Thuill.) W. D. J. Koch; *V. sativa* var. *segetalis* (Thuill.) Ser.

Plant up to 60 cm. *Leaves* with uniform leaflets or lower broader, 7–25 × 2–8 mm, narrowly elliptical, ovate or lanceolate, obtuse or emarginate at apex. *Calyx* 10–15 mm, the lobes 5–6 mm, lanceolate-subulate. *Corolla* 15–20(–22) mm. *Legumes* 30–50 × 3–6 mm, sometimes contracted between the seeds. $2n = 12$.

(**e**) Subsp. **cordata** (Wulfen ex Hoppe) Arcang.
V. cordata Wulfen ex Hoppe

Plant up to 80 cm. *Leaves* with mostly uniform leaflets, 10–28 × 3.5–13.0 mm, oblong or obovate-cuneate, obtuse or emarginate. *Calyx* 12–16 mm, the lobes 6–9 mm, triangular-subulate. *Corolla* 20–24 mm. *Legumes* 39–62 × 5–6 mm, not contracted between the seeds. $2n = 10$.

(**f**) Subsp. **sativa**
Faba sativa (L.) Bernh.

Plant up to 120 cm. *Leaves* with 8–14 more or less uniform leaflets 8–31 × 1.5–4 mm, oblong, oblanceolate or elliptical, obtuse or emarginate. *Calyx* 11.5–18.0 mm, the lobes 6–11 mm, triangular-subulate. *Corolla* 20–22 mm. *Legumes* 35–70 × 6–9 mm, contracted between the seeds. $2n = 12$.

(**g**) Subsp. **macrocarpa** (Moris) Arcang.
V. sativa var. *macrocarpa* Moris

Plant up to 120 cm. *Leaves* with 8–14 more or less uniform leaflets 10–40 × 5–15 mm, obovate or oblanceolate, obtuse or emarginate. *Calyx* 11–17 mm, the lobes 4–10 mm, triangular-subulate. *Corolla* 20–22 mm. *Legumes* 54–57 × 9–11 mm, not contracted between the seeds. $2n = 12$.

Native. Grassy and rough places, sandy banks, heathland, field borders and coastal sand and shingle. Throughout Great Britain and Ireland except in mountainous areas. Throughout Europe to 69° N in Russia, temperate Asia, North Africa, introduced in North America. European southern-temperate element. Subsp. *uncinata* is a plant of open sandy heaths and dunes, often growing with and confused with *V. lathyroides*. It is common in the Breckland of East Anglia and may have a general distribution similar to *V. lathyroides* as given in the second edition of the Atlas. It also occurs in central and south Europe. Subsp. *bobartii* occurs in similar habitats, but is more widespread and is found in more closed habitats. It also is widespread in Continental Europe. Subsp. *nigra* is the common plant of closed grassy places by roadsides, on tracks and footpaths, along hedgerows and in meadows. It is widespread in Continental Europe. Subsp. *segetalis* occurs in similar habitats to subsp. *nigra* and also in waste places. The original description refers to it as 'Habitat inter segetes' and not 'in segetes'. We have never seen it grown as a crop and see no reason why it is not native. The inclusion of subsp. *cordata* is based on plants grown as a crop in the Cambridgeshire area and occuring on adjacent grassland, on specimens from the Lizard, Cornwall, Breconshire in Wales and Galway in Ireland and on being present in wild flower seed. The plant G. C. Druce described as var. *garlandii* from Jersey and the plant recorded by J. E. Lousley in the *Flora of the Isles of Scilly* which is similar, may also belong here. Subsp. *cordata* is a plant of southern Europe. In Spain it is cultivated as a nitrophilous crop. Fifty years ago subsp. *sativa* was commonly grown as a crop plant both for hay and for seed. It was used as a similar crop plant throughout Europe. It is now rarely seen as a crop plant and only rarely found on trackways, roadsides, waste places and field margins. Subsp. *macrocarpa* is an alien of which we have only seen records from Bristol and Knutsford in Cheshire. It is native of the Mediterranean region. The above infraspecific taxa usually occur in uniform populations and seem to retain their characters by inbreeding.

15. V. lathyroides L. Spring Vetch
Lathyrus angulatus auct.

Annual herb with fibrous roots. *Stem* 5–20 cm, pale green, slender, spreading, shortly hairy, branched, often low down, leafy. *Leaves* alternate, 2–3 × 1.5–2.0 cm, oblong in outline, paripinnate, without a tendril; leaflets 4–8, 4–14 × 3–5 mm, medium green on upper surface, paler beneath, linear-oblong or narrowly obovate, obtuse or emarginate and mucronulate, entire, narrowed at base, sessile, shortly hairy on both surfaces; tendrils small and unbranched or absent; petiole short; stipules small, half arrow-shaped, entire. *Inflorescence* a solitary axillary flower, with a short peduncle. *Calyx* 2.5–3.0 mm, tubular, divided about one-third of the way to the base into 5 more or less equal lobes, the lobes narrowly lanceolate, acute at apex, hairy. *Corolla* (5–)6–9 mm, violet or light purple; standard with limb subrotund, shallowly emarginate or mucronulate at apex, with a winged claw; glabrous; wings with limb obovate, rounded at apex, with a narrow lobe at base and a slender claw; keel obovate, with a long, slender claw. *Stamens* 10, diadelphous; filaments pale; anthers yellow. *Style* 1, bearded on lower side; stigma capitate. *Legume* 15–30 × 3–4 mm, oblong, tapering at both ends, glabrous; seeds 6–12, 1.5–1.7 mm, globose, tuberculate. *Flowers* 5–6. $2n = 10, 12$.

Native. Maritime sand and inland sandy heaths and old walls. Scattered localities over much of Great Britain and Ireland except west and south Ireland and north-west Scotland, but mainly coastal. Widely distributed throughout Europe, north to south-west Finland and east to the Caucasus and Turkey; North Africa. European Temperate element.

16. V. hybrida L. Hairy Yellow Vetch

Annual herb with fibrous roots. *Stems* 15–50 cm, pale green, sharply angular, becoming glabrous, usually much branched especially at base, leafy. *Leaves* alternate, paripinnate, 2–5 × 1–2 cm, oblong in outline; leaflets 6–14, 3–12 × 2–8 mm, medium green on upper surface, paler beneath, oblong or obovate, truncate, emarginate or rounded at apex, usually apiculate, thinly hairy on both surfaces; petiole usually less than 10 mm; stipules 1–4 mm, semi-sagittate, becoming glabrous, often with a small, dark medium gland; upper leaves with long branched tendrils, the basal without a tendril or the rhachis ending in a short bristle. *Inflorescence* of solitary pendulous flowers in the axils of the upper leaves; pedicels usually less than 5 mm, villous; bracts absent. *Calyx* about 10 mm, cylindrical, distinctly gibbous at base, with 5 unequal, filiform-dentate lobes which have dark blackish apices, thinly hairy. *Corolla* 18–30 mm, sulphur yellow, pencilled or stained dark brown, or rarely purplish; standard with limb obovate, emarginate, margins reflexed, with a long, wide claw, hairy externally; limb of wings narrowly obovate, rounded at apex, strongly falcate-lobed at base with a long claw; keel with limb shortly oblong, rounded at apex, base lobed and with a slender claw. *Stamens* 10, diadelphous; filaments pale; anthers oblong, alternatively basifixed and versatile. *Style* 1, bearded near the apex; stigma small, capitate. *Legume* 20–30 × 9–14 mm, oblong, strongly compressed, acute and shortly beaked at apex, thinly hairy; seeds 5–6 mm in diameter, dark brown, smooth, sometimes obscurely marbled. *Flowers* 6–8.

Introduced. A grain and tan-bark casual. Formerly established for about 150 years on Glastonbury Tor in Somersetshire. Native throughout the Mediterranean region and eastwards to Iraq and Iran.

17. V. lutea L. Yellow Vetch
V. laevigata Sm.; *V. lutea* subsp. *laevigata* (Sm.) Syme

Annual herb with fibrous roots. *Stem* 10–45(–60) cm, tufted, glabrous or hairy, branched, leafy. *Leaves* alternate, paripinnate; leaflets 6–16, 10–25×3–5 mm, medium green on upper surface, paler beneath, linear-oblong to narrowly elliptical, obtuse and mucronate, rarely acute at apex, entire, rounded at base, with numerous hairs on both surfaces; tendrils simple or branched; petioles short or absent; stipules small, triangular, the lower with a basal lobe. *Inflorescence* of 1–2(–3) axillary flowers; pedicels short. *Calyx* 5–7 mm, tubular, with 5 unequal lobes, the lower longer than the tube, setaceous, hairy. *Corolla* 15–25(–30) mm, pale straw-coloured; standard with limb elliptical, emarginate at apex, attenuate into a long, very broad claw, folded down the middle, glabrous; wings with limb oblong, rounded at apex, with a short lobe at base and a narrow claw; keel with limb ovate, upcurved, with a very short lobe at base and a long, slender claw. *Stamens* 10, diadelphous; filaments upturned at end; anthers yellow. *Style* 1, upturned at right angles, longer hairy on one side than the other near apex; stigma brownish. *Legume* 20–40×7–8 mm, oblong-lanceolate, beaked, very hairy; seeds 4–8, 3.5–4.0 mm, subglobose. *Flowers* 6–8. $2n=14$.

Native. Maritime shingle and cliffs. Scattered round the coasts of the Channel Islands and Great Britain north to Scotland, most frequent in southern England and Channel Islands; frequent casual inland in Great Britain with a few records in Ireland. South and west Europe; Caucasus; North Africa; Macaronesia. Submediterranean-Subatlantic element. Our plant is subsp. **lutea**.

Section **4. Faba** (Mill.) Gray
Faba Mill.

Leaflets 2–6. *Flowers* solitary and axillary or in few-flowered, sessile or pedunculate racemes. *Corolla* large. *Style* hairy on the lower side beneath the stigma.

18. V. bithynica (L.) L. Bithynian Vetch
Lathyrus bithynicus L.

Annual herb with a tap-root and fibrous side-roots. *Stems* 30–60 cm, pale green, climbing or trailing, more or less glabrous; branched, leafy. *Leaves* alternate, 4–6×6–9 cm, broadly ovate in outline, paripinnate; leaflets (2–)4–6, 20–50×10–15 mm, medium green on upper surface, paler beneath, elliptical or ovate, obtuse-mucronate to acute at apex, entire, narrowed below, sessile, glabrous or hairy; tendrils branched; petioles up to 20 mm; stipules large, ovate, acuminate at apex, dentate. *Inflorescence* of 1–2 axillary flowers on peduncles up to 5 cm and short pedicels. *Calyx* 8–10 mm, divided halfway to base or more into 5 lobes, the lobes narrowly triangular, acute at apex, hairy. *Corolla* 16–20 mm; standard purple, the limb subrotund, emarginate at apex, with a very broad claw; wings very pale almost white, the limb elliptical, rounded at apex, with a narrow claw; keel very pale, almost white, blotched with blue, ovate, curved upwards, rounded at apex, with a narrow claw. *Stamens* 10, diadelphous; filaments pale; anthers yellowish. *Style* 1, bearded on lower side, curved; stigma brownish. *Legume* 25–50×7–10 mm, oblong, abruptly beaked, hairy; seeds 4–8, 4.5–5.0 mm, globose. *Flowers* 5–6. $2n=14$.

Probably native. Scrub, rough grassland and hedgerows and on railway banks. By and near the coast in the Channel Islands and Great Britain north to Wigtownshire in very scattered localities; also a frequently introduced casual which becomes naturalised. South and west Europe north to Scotland; Caucasus; Algeria. Mediterranean-Atlantic element.

19. V. narbonensis L. Narbonne Vetch
V. serratifolia Jacq.; *V. narbonensis* subsp. *serratifolia* (Jacq.) Nyman

Annual herb with fibrous roots. *Stems* (10–)20–50 (–60) cm, pale green, stout, more or less erect, climbing, sharply angled, more or less densely ciliate along the angles to nearly glabrous, simple or sparingly branched at the base, leafy. *Leaves* alternate, 2–6, 1–7×2.5–8.0(–10.0) cm, oblong in outline, paripinnate, with a tendril; leaflets 2–6, 10–50×6–35 mm, medium green on upper surface, paler beneath, oblong, ovate, obovate or broadly elliptical, rounded or acute at apex, often shortly apiculate, entire or bluntly toothed, rounded at the subsessile or very shortly petiolulate base, glabrous or very thinly hairy on both surfaces, the upper surface often with small sessile glands, ciliate with bulbous-based hairs; tendrils usually restricted to the uppermost leaves, slender, unbranched or branched at apex; petiole up to 2 cm, stout; stipules 5–18×3–9 mm, oblong or broadly semi-sagittate, obtuse to acute, sharply toothed towards the base, usually ciliate with bulbous-based hairs, often with a small dark glandular spot near the apex. *Inflorescence* of 1–3(–6) flowers in the axils of the upper leaves; peduncle 10–30 mm, or sometimes absent; pedicels 2–5 mm, slender, hairy. *Calyx* 9–14 mm, cylindrical, divided into 5 unequal lobes, the lobes deltoid or narrowly ovate, acuminate at apex, glabrous or very thinly hairy with ciliate margins. *Corolla* 10–30 mm; whitish, veined purple or purplish all over, the wings often purple-blotched at apex; standard with limb obovate, emarginate at apex, tapering to a broad claw, glabrous; wings with limb oblong, broadly rounded apex, a blunt lobe at base and a slender claw; keel with limb oblong, rounded at apex, with a short tooth on the adaxial margin and a blunt lobe at base and a long claw. *Stamens* 10, diadelphous; filaments not dilated; anthers oblong. *Style* 1, bent upright at a right angle, densely bearded at apex; stigma small. *Legume* 30–60(–70)×10–15 mm, oblong, flattened, acutely beaked at apex, glabrous or thinly hairy on margin; seeds 4–7, about 5 mm in diameter, dark brown, subglobose, smooth. *Flowers* 6–7. $2n = 14$.

Introduced. Frequent casual from grain on waste land. Scattered localities in Great Britain north to Midlothian. Widely distributed in southern Europe and the Mediterranean region, eastwards to Iran.

20. V. faba L. Broad Bean

Annual herb with fibrous roots and a characteristic smell. *Stems* 20–100 cm, pale yellowish-green, erect, square, glabrous, simple, leafy. *Leaves* alternate, 14–20×13–16 cm, broadly oblong in outline, paripinnate, without tendrils; leaflets 2–8, 60–100×30–80 mm, dull greyish-green on upper surface, slightly paler and greener beneath, broadly elliptical to subrotund, rounded or shortly acute at apex, entire, rounded below and subsessile, glabrous, with 5 parallel veins; rhachis pale green, glabrous, without tendrils and replaced by a weak point; petiole short to medium, glabrous; stipules 10–20 × 8–15 mm, ovate or semi-sagittate, acute at apex, with few teeth. *Inflorescence* of 1–6 subsessile, fragrant flowers in the axils of the leaves. *Calyx* 12–15 mm, cylindrical, divided into 5 subequal lobes, the lobes green in the centre with pale margins, narrowly triangular, acute at apex, glabrous. *Corolla* 30–40 mm; standard 20–25 mm, creamy-white with pinkish-brown veins toward the purple-tinted base, limb almost subrotund, but folded into 2 with 4 rounded lobes at apex, narrowed to a short claw, glabrous; wings 13–15 mm, with a large purplish-black centre surrounded by a white margin, the limb half-ovate, rounded at apex, narrowed to a long claw; keel about 4 mm, cream and forming a narrow ridge. *Stamens* 10, diadelphous; filaments pale green; anthers yellowish. *Style* 1, pale green, with a bunch of hairs round the apex; stigma yellowish. *Legume* 10–30 × 1.5–4.0 cm, green, turning brownish or blackish when ripe, oblong, with thick smooth valves; seeds 4–8, 10–30 × 10–30 mm, pale brown with a black hilum, compressed or subglobose. *Flowers* 4–7. Pollinated by bees, but can be selfed. $2n = 12$.

(**i**) Var. **faba** Broad Bean
V. faba var. *megalosperma* Alef.

Seeds 20–30 mm, clear, bright pale brown, ovoid, compressed.

(**ii**) Var. **equina** Pers. Horse Bean
Faba equina (Pers.) Rchb.

Seeds 15–18 mm, dark brown or white, compressed and much thickened in the region of the hilum.

(**iii**) Var. **minor** Peterm. Tick Bean
Seeds 10–14 mm, subglobose or slightly oblong, not compressed.

Introduced. The wild ancestor of the Broad Bean is not now known, but it probably originated in the Mediterranean or Middle East in Neolithic times. Primitive varieties had small black seeds and it is most closely related to *V. narbonensis*. The earliest evidence of cultivation by Man comes from pre-pottery Neolithic B levels at Jericho and in Hungary. Beans have been found in Egypt dating from 1800 BC and are mentioned in Greek and Roman literature. They are found in Iron Age deposits in lake dwellings near Glastonbury and in northern Europe. Var. *faba* is commonly grown in gardens and as a crop. The very young pods can be eaten whole, but mostly it is the green, immature seeds which are cooked as a vegetable. Var. *equina* is grown as animal feed and var. *minor* is used as a food for tame pigeons. All three varieties occur as relics from crops and the beans fall from trailers onto tracks and roadsides. Common in southern and middle England, with scattered localities elsewhere.

41. **Lens** Mill. nom. conserv.

Ervum L. nom. rejic.

Annual herbs. Leaves alternate, paripinnate, the rhachis terminating in a short awn or a longer simple or branched tendril. *Inflorescence* an axillary, few-flowered raceme. *Calyx* deeply divided into 5 subequal lobes. *Corolla* whitish, blue or pale violet; standard with limb squarish or subrotund, abruptly narrowed to a rectangular claw; wings with limb narrowly obovate, with a slender claw; keel with limb oblong, rather inflated and longitudinally plicate, with a short claw. *Stamens* 10, diadelphous, with an oblique apex; anthers uniform. *Style* 1, flattened towards apex, bearded on adaxial surface. *Fruit* a 2-seeded, strongly compressed legume.

Five species in southern Europe, the Mediterranean region and western Asia.

L. culinaris is more widely distributed as a cultivated plant or a relict of cultivation.

Ferguson, M. E., Maxted, N., Slageren, M. van & Robertson, L. D. (2000). A re-assessment of the taxonomy of *Lens* Mill. (Leguminosae, Papilionoideae, Vicieae). *Bot. Jour. Linn. Soc.* **173**: 41–59.
Phillips, R. & Rix, M. (1993). *Vegetables*. London.
Vaughan, J. G. & Geissler, C. A. (1997). *The new Oxford book of food plants*. Oxford.
Zohary, D. & Hopf, M. (2000). *Domestication of plants of the Old World*. Ed. 3. Oxford.

1. L. culinaris Medik. Lentil
Ervum lens L.; *Vicia lens* (L.) Coss. & Germ.; *Lens esculenta* Moench

Annual herb with fibrous roots. *Stems* 20–40 cm, more or less erect, sparingly clothed with crispate hairs and small, sessile glands, angular, usually much-branched, leafy. *Leaves* alternate, 2–4 × 1–3 cm, oblong in outline, paripinnate; leaflets (6–)10–16, 5–15 × 2–5 mm, medium green on upper surface, paler beneath, oblong, rounded to subacute at apex, entire, narrowed below, thinly subappressed hairy on both surfaces; tendril simple or occasionally branched; petiole very short or absent; stipules 5–7 mm, lanceolate, acuminate at apex, more or less entire, thinly hairy. *Inflorescence* an axillary raceme of (1–)2–3(–4)-flowered; peduncle 15–35 mm; pedicels 2–3 mm, arcuate. *Calyx* 5–7 mm, campanulate, deeply divided into 5 lobes, the lobes filiform-subulate and subequal. *Corolla* 4–6 mm, milky white or with pale blue veins; standard with limb squarish or subrotund, deeply emarginate at apex, abruptly narrowed to a rectangular claw; wings with limb narrowly obovate, rounded at apex, with a short sharp lobes at base and a distinct gibbosity, and a slender claw; keel with limb oblong, rather inflated and longitudinally plicate, subacute at apex, obscurely lobed at base, with a short claw. *Stamens* 10, diadelphous, filaments with an oblique apex; anthers oblong. *Style* 1, slender, sharply upcurved, flattened towards apex, adaxial surface bearded; stigma capitate. *Legume* 10–14(–17) × 8–10 mm, oblong-rhomboid, flattened, acute or shortly beaked at apex; seeds usually 1, 5–6 mm in diameter, compressed-subrotund, reddish-brown, smooth. *Flowers* 6–8. $2n = 14$.

Introduced. Fairly frequent grain alien on tips and rough ground. Scattered records in southern Great Britain and

Dublin in Ireland. Widely distributed as a cultivated plant in Europe, Asia and North Africa. Possibly a cultivated derivative of *L. orientalis* (Boiss.) Hand.-Mazz. It is one of the most ancient of crops in the eastern Mediterranean, grown since at least 6700 BC.

42. Lathyrus L.

Annual or *perennial herb*s often climbing by tendrils. *Stems* angled or winged. *Leaves* usually paripinnate with 2 to many leaflets which are pinnately or parallel veined, sometimes reduced to a simple lamina or a simple tendril; terminal tendril present or absent; stipules variable. *Inflorescence* an axillary raceme or flowers solitary; bracts very small and caducous. *Calyx* with tube frequently oblique and somewhat gibbous; lobes 5, subequal or the adaxial shorter than the abaxial. *Corolla* yellow or purplish; standard usually broad with a short claw; wings oblong or obovate; keel obtuse or shortly acute. *Stamens* 10, diadelphous. *Style* 1, usually hairy on the upperside. *Legume* dehiscing into 2 valves; seeds globose or angular, rarely compressed. The flowers have nectar and are visited by bees.

Contains about 140 species widely distributed in the northern hemisphere, and in the mountains of tropical Africa and South America.

Several species have extreme variation in the width of the leaves, but a population in a given area seems always to be uniform. The *L. heterophyllus, latifolius, sylvestris* complex is particularly difficult.

Bässler, M. (1966). Die Stellung des Subgen. *Orobus* (L.) Baker in der Gattung *Lathyrus* L. und seine systematische Gliederung. *Feddes Repert.* **72**: 69–97.

Brightmore, D. & White, P. H. F. (1963). Biological flora of the British Isles. No. 94. *Lathyrus japonicus* Willd. *Jour. Ecol.* **51**: 795–801.

Brunsberg, K. (1977). Biosystematics of the *Lathyrus pratensis* compex. *Opera Botanica* **41**: 1–78.

Cannon, J. F. M. (1964). Infraspecific variation in *Lathyrus nissolia* L. *Watsonia* **6**: 28–35.

Duncan, U. K. (1970). *Lathyrus maritimus* Bigel. Subsp. *acutifolius* (Bab.) Pedersen in eastern Scotland. *Watsonia* **8**: 87.

Fernald, M. L. (1932). *Lathyrus japonicus* versus *L. maritimus*. *Rhodora* **34**: 177–187.

Grime, J. P. et al. (1988). *Comparative plant ecology*. London. [*L. linifolius, pratensis*.]

Hultén, E. & Fries, M. (1986). *Atlas of north European vascular plants north of the Tropic of Cancer.* 3 vols. Königstein.

Kupicha, F. K. (1975). Observations on the vascular anatomy of the tribe Vicieae (Leguminosae). *Bot. Jour. Linn. Soc.* **70**: 231–242.

Kupicha F. K. (1981). Tribe Vicieae. In Polhill, R. M. & Raven, P. H. (Edits.) *Advances in legume Systematics,* Part I, pp. 377–438. Kew.

Kupicha, F. K. (1983). The infrageneric structure of *Lathyrus*. *Notes Roy. Bot. Garden Edinb.* **41**: 209–244.

Melderis, A. (1957). *Lathyrus heterophyllus* var. *unijugus* Koch. *Proc. B.S.B.I.* **2**: 238–240.

Munro-Smith, D. (1964). *Lathyrus palustris* var. *pilosus* (Cham.) Ledeb. in north Somerset. *Proc. B.S.B.I.* **4**: 41.

Randall, R. E. (1977). The past and present status and distribution of Sea Pea (*Lathyrus japonicus* Willd.) in the British Isles. *Watsonia* **11**: 247–251.

Shenstone, J. C. (1910). *Lathyrus tuberosus* in Britain. *Jour. Bot. (London)* **48**: 327–331.

Stewart, A., Pearman, D. A. & Preston, C. D. (1994). *Scarce plants in Britain*. Peterborough. [*L. aphaca, japonica* and *palustris*.]

Valero, M. & Hassaert-Mckey (1991). Discriminant alleles and discriminant analysis: effieient characters to separate closely related species: the example of *Lathyrus latifolius* L. and *Lathyrus sylvestris* L. (Leguminosae). *Bot. Jour. Linn. Soc.* **107**: 139–161.

Vaughan, I. M. (1978). *Lathyrus palustris* L. var. *pilosus* (Cham.) Ledeb. in v.c. 44. *B.S.B.I. News* **20**: 14.

The following classification of the epidermal cells of the leaves used in the descriptions of the sections was first proposed by Bässler (1966) and added to by Kupicha (1983).

Hypostomatic epidermal cells are isodiametric with strongly wavy walls.

Hypo-amphistomatic epidermal cells are isodiametric or slightly elongated, with wavy or straight walls.

Epi-amphistomatic epidermal cells are slightly elongated and usually with wavy walls on the upper leaf surface and elongated with straight walls on the lower surface.

1. Leaves without leaflets (but with large leaf-like stipules or phyllodes) 2.
1. At least the upper leaves with 1 or more pairs of leaflets 4.
2. Rhachis forming a tendril; stipules ovate-hastate; corolla yellow **17. aphaca**
2. Rhachis forming a grass-like phyllode; stipule minute; corolla reddish-magenta 3.
3. Legume glabrous or with some minute glands **21(i). nissolia** var. **nissolia**
3. Legume covered with glandular and eglandular hairs **21(ii). nissolia** var. **pubescens**
4. Stem winged, at least in the upper part 5.
4. Stem not winged 24.
5. Lower leaves without leaflets the rhachis broadly winged and resembling a leaf 6.
5. All leaves with one or more pairs of leaflets 7.
6. Upper leaves with 4–8(–10) leaflets; corolla purple with violet, lilac or pink wings; dorsal suture of legume not winged **18. clymenum**
6. Upper leaves with 2–4 leaflets; corolla yellow; dorsal suture of legume with 2 wings **19. ochrus**
7. At least some leaves with 4 or more leaflets; racemes 2- to many-flowered 8.
7. All leaves with only 2 leaflets; rarely some with 4 and then the flowers solitary 14.
8. Leaves without a tendril; rhachis mucronate 9.
8. Leaves with a tendril 11.
9. Leaflets 10–25 mm wide, narrowly elliptical to elliptical, obtuse-mucronate at apex **5(iii). linifolius** var. **montanus**
9. Leaflets 0.5–10 mm wide, long-linear to narrowly elliptical, mostly pointed at apex 10.
10. Leaflets 0.5–3.0 (–5.0) mm wide, long-linear, narrowly acute at apex **5(i). linifolius** var. **linifolius**

10. Leaflets 3–10 mm wide, narrowly elliptical, those of the lower leaves usually broader than those of the upper, mostly pointed at apex **5(ii). linifolius** var. **varifolius**
11. Rhachis of leaf at least 4 mm wide, broadly winged **9(i). heterophyllus** var. **heterophyllus**
11. Rhachis of leaf not more than 3 mm wide 12.
12. Plant hairy in all its parts **4(iii). palustris** var. **pilosus**
12. Plant glabrous or nearly so 13.
13. Leaflets up to 3.5 mm wide **4(i). palustris** var. **linearifolius**
13. Leaflets 4–15 mm wide **4(ii). palustris** var. **palustris**
14. Racemes (3–)5- to many-flowered 15.
14. Racemes 1- to 3(–4)-flowered 19.
15. Stipules narrowly ensiform, their width less than half as wide as stem 16.
15. Stipules semi-sagittate-ovate, their width at least half as wide as stem 17.
16. Leaflets up to 15 mm wide, long-acute **8(i). sylvestris** var. **sylvestris**
16. Leaflets 10–25 (–35) mm wide, narrowed but often obtuse at apex **8(ii). sylvestris** var. **latifolius**
17. Flowers 15–22 mm **9(ii). heterophyllus** var. **unijugus**
17. At least some flowers more than 22 mm and up to 30 mm 18.
18. Leaflets up to 30 mm wide **10(i). latifolius** var. **latifolius**
18. Leaflets up to 60 mm wide **10(ii). latifolius** var. **rotundifolius**
19. Corolla yellow **11. annuus**
19. Corolla not yellow 20.
20. Corolla 20 mm or more 21.
20. Corolla less than 20 mm 22.
21. Peduncles more than 70 mm; legume not winged **12. odoratus**
21. Peduncles not more than 60 mm; legume with 2 wings on the dorsal suture **14. sativus**
22. Calyx lobes not or only slightly longer than tube **13. hirsutus**
22. Calyx lobes 1.5–3.0 times as long as tube 23.
23. Peduncles 30–60 mm; legumes with 2 wings on the dorsal suture **14. sativus**
23. Peduncles 5–30 mm; legume with 2 keels on the dorsal suture **15. cicera**
24. At least the upper leaves with a tendril 25.
24. All leaves without a tendril 32.
25. Corolla more than 25 mm **6. grandiflorus**
25. Corolla less than 25 mm 26.
26. Flowers solitary **20. inconspicuus**
26. Racemes 2- to many-flowered 27.
27. Corolla yellow 28.
27. Corolla purple or bluish 30.
28. Plant glabrous or with a few hairs **16(ii). pratensis** var. **pratensis**
28. Plant with numerous hairs 29.
29. Leaflets 10–25 × 1–5 mm; corolla 10–18 mm **16(i). pratensis** var. **velutinus**
29. Leaflets 20–40 × 5–7 mm; corolla 18–20 mm **16(iii). pratensis** var. **speciosus**
30. Leaflets 2 **7. tuberosus**
30. Leaflets 4–10 (–12) 31.
31. Leaflets 12–45 × 8–22 mm, broadly elliptical, obtuse to acute at apex **3(i). japonicus** subsp. **maritimus** var. **glaber**
31. Leaflets 10–40 × 4–10 mm, narrowly elliptical, acute at apex **3(ii). japonicus** subsp. **maritimus** var. **acutifolius**
32. Leaflets acute or acuminate at apex; stipules 10–25 mm **1. vernus**
32. Leaflets obtuse-mucronate or subacute at apex; stipules 4–10 mm **2. niger**

Section **1. Orobus** (L.) Gren. & Godr.
Orobus L.; *Lathyrus* subgen. *Orobus* (L.) Peterm.; *Lathyrus* section *Lathyrobus* (Tamamsch.) Czefr.

Perennial herbs. Stems usually not winged. *Leaves* hypostomatic to amphistomatic, multijugate, with or without tendrils; leaflets broadly ovate to elliptical, lanceolate or linear; venation pinnate or rarely parallel; stipules semi-sagittate or rarely hastate. *Inflorescence* with few to many flowers; flowers brownish-yellow to bluish or reddish-purple. *Calyx* lobes unequal. *Style* curved upwards. *Legumes* oblong, not stipitate; seeds smooth.

1. L. vernus (L.) Bernh. Spring Pea
Orobus vernus L.

Perennial herb. Stem 20–40(–60) cm, pale green, not winged, glabrous or sparsely hairy, branched, leafy. *Leaves* alternate, paripinnate, broadly elliptical in outline, without a tendril; leaflets (2–)4–8, 30–70(–100) × (1–)10–30 mm, medium green on upper surface, paler beneath, lanceolate or ovate, acute or acuminate at apex, entire, narrowed at base, feebly parallel-veined; stipules 10–25 × 2–8 mm, ovate-lanceolate, rarely linear or semi-sagittate. *Inflorescence* an axillary raceme with 3–10 flowers. *Calyx* (5.5–)7.5–9.0 (–10.5) mm, 2-lipped, the upper lip with lobes triangular, the lower lip with lobes linear-lanceolate. *Corolla* 13–20 mm, reddish-purple; standard 14.0–17.5 mm, emarginate at apex; wings 13–16 × 3.5–5.5 mm; keel 12–13 mm, falcate. *Stamens* 10, diadelphous. *Style* 1. *Legume* 40–60 × 5–8 mm, brown, oblong, glabrous; seeds 8–14, smooth, hilum about one-quarter of the circumference. *Flowers* 6–8. $2n = 14$.

Introduced. A persistent garden escape and relic. Native of most of Europe except the islands, and western Asia.

2. L. niger (L.) Bernh. Black Pea
Orobus niger L.

Perennial herb with a short rhizome. *Stem* 15–90 cm, pale green, erect, angled but not winged, glabrous or sparsely hairy, branched, leafy. *Leaves* alternate, paripinnate, oblong in outline, without a tendril; leaflets 6–12(–22), 10–48 × 4–20 mm, medium green on upper surface, paler beneath, lanceolate to elliptical, obtuse-mucronate or subacute at apex, entire, rounded at base, glabrous or with a few hairs, more or less pinnately-veined; stipules 4–10 × 1–2 mm, linear or semi-sagittate, entire. *Inflorescence* an axillary raceme of 2–10 flowers; peduncles hairy with appressed, crisped hairs. *Calyx* 4–6 mm, 2-lipped, the upper lip with triangular lobes, the lower lip with linear-lanceolate lobes, with appressed, crisped hairs.

Corolla 10–15 mm, reddish-purple; standard 10–15 mm, oblong, emarginate; wings 11.5–13.5 × 2.5–4.5 mm, ciliate; keel 9–13 × 3.0–4.5 mm. *Stamens* 10, diadelphous. *Style* 1. *Legume* 35–60 × 4–6 mm, black, oblong-oblanceolate, rugose, glabrous; seeds 6–10, smooth, hilum round one-quarter of the circumference. *Flowers* 6–7. $2n = 14$.

Introduced. Garden escape, naturalised in grassy, rocky and scrubby places. Scattered records in England and Scotland. Most of Europe but absent from much of the north-east and many islands, east to the Caucasus; very rare in north-west Africa.

3. L. japonicus Willd. Sea Pea

Pisum marinum Huds. nom. illegit.; *Pisum multiflorum* Stokes nom. illegit.

Perennial herb with an extensive root system. *Stem* 30–90 cm, pale glaucous-green, ascending and creeping, angled, not winged, much-branched, glabrous. *Leaves* alternate, paripinnate, leaflets 4–10(–12), 10–45 × 4–22 mm, glaucous on upper surface, paler beneath, narrowly to broadly elliptical, obtuse at apex, entire, rounded at base, glabrous, sessile or nearly so; tendrils simple or branched, sometimes absent; petioles very short; stipules 10–25 mm, semi-hastate. *Inflorescence* an axillary raceme of 2–10(–15) flowers; peduncles usually shorter than leaves, glabrous. *Calyx* 5–7 mm, tubular, divided into 5 unequal lobes, the lower lobes narrowly triangular and acute, the upper subulate. *Corolla* (12–)15–25 mm; standard reddish-purple with dark veins, changing to bluish-purple, the limb very broadly elliptical, emarginate at apex, narrowed to a short very broad claw which has its margins inturned; wings very pale bluish-mauve, with limb ovate, rounded at apex, irregularly lobed at base and with a long slender claw; keel pale, tinged with pale reddish-purple and green, the limb ovate, upcurved, with a small blunt lobe at base and a slender claw. *Stamens* 10, diadelphous; filaments upturned, pale; anthers yellowish. *Style* 1, rigidly upturned, minutely hairy on one side at apex; stigma brownish. *Legume* 30–50 × 7–8 mm, pale green, turning brown, oblong, glabrous; seeds 4–8, 4.0–4.2 mm, subglobose, smooth. *Flowers* 6–8. $2n = 14$.

Our plant is subsp. **maritimus** (L.) P. W. Ball (*L. maritimus* (L.) Fr., non Bigelow.; *Pisum maritimum* L.; *Pisum marinum* Huds. nom. illegit.; *Pisum multiflorum* Stokes nom. illegit.). It can be divided into 2 varieties.

(**i**) Var. **glaber** (Ser.) Fernald

Pisum maritimum var. *glabrum* Ser.; *L. maritimus* var. *glaber* (Ser.) Eames; *Orobus maritimus* (L.) Rchb.

Leaflets 12–45 × 8–22 mm, broadly elliptical, obtuse to acute at apex, narrowed or rounded at base.

(**ii**) Var. **acutifolius** (Bab.) P. D. Sell

L. maritimus var. *acutifolius* Bab.; *L. japonicus* forma *acutifolius* (Bab.) Fernald; *L. maritimus* subsp. *acutifolius* (Bab.) Pedersen

Leaflets 10–40 × 4–10 mm, narrowly elliptical, acute at apex, cuneate at base.

Native. Bare maritime shingle or sand where it forms large and conspicuous patches. Very local on the coasts of Great Britain from the Isles of Scilly and Kent to Shetland, and in Guernsey; frequent in south-east England. The species is circumpolar, China, Japan and North America. European Boreo-arctic Montane element. Our plant is subsp. **maritima** which occurs in western Europe, the Baltic region and subarctic Russia. Our main plant is var. *glaber*. Var. *acutifolius* was described from Burrafirth in Shetland where it is now apparently extinct. It has more recently been found in some quantity on the dunes near Arbroath (cf. Duncan, 1970).

4. L. palustris L. Marsh Pea

Perennial herb with a tap-root. *Stem* 60–120 cm, bluish-green, scrambling, winged, glabrous or hairy, branched, leafy. *Leaves* alternate, paripinnate; leaflets 4–6(–10), 35–70 × 1–15 mm, bluish-green on upper surface, paler beneath, narrowly elliptical-oblong, obtuse to acute and mucronate at apex, entire, rounded at base and sessile, glabrous or hairy; tendrils branched; rhachis less than 3 mm wide; petiole up to 20 mm; stipules 10–20 mm, semi-sagittate. *Inflorescence* an axillary raceme with 2–6 flowers; peduncles usually longer than leaves. *Calyx* 7–9 mm, tubular, very unequally 5-lobed, the lower lobes triangular, the upper subulate, acute at apex, glabrous or hairy. *Corolla* 12–20 mm, a delicate mauve, the wings and eventually the whole flower turning greenish; standard with limb very broadly elliptical, emarginate at apex, narrowed to a short wide claw the margins of which are turned inwards; wings with limb irregularly ovate, rounded at apex, with a short broad lobe at base and a long slender claw; keel very irregular and upturned with a wide lobe at base and a long slender claw. *Stamens* 10, diadelphous; filaments pale; anthers yellowish. *Style* 1, bent upwards at right angles, hairy only on one side. *Legume* 25–60 × 7–10 mm, oblong, compressed, glabrous or hairy; seeds 3–12, 3.5–4.0 mm, subglobose, smooth. *Flowers* 5–7. $2n = 42$.

(**i**) Var. **linearifolius** Ser.

L. palustris var. *leariformis* Wallr.

Plant glabrous. *Leaflets* up to 3.5 mm wide.

(**ii**) Var. **palustris**

L. palustris var. *latifolius* Lamb.

Plant glabrous. *Leaflets* 4–15 mm wide.

(**iii**) Var. **pilosus** (Cham.) Ledeb.

L. pilosus Cham.

Plant hairy in all its parts.

Native. Fens and tall, damp grassland. Of local occurrence in England, Wales, south-west Scotland and Ireland, most frequent in East Anglia, but generally decreasing. Europe, rare in the Mediterranean region; arctic Russia and Siberia east to Sakhalin and Japan; North America. Circumpolar Boreo-temperate element. Our plant is subsp. **palustris**. Var. *palustris* is the normal variety. Var. *linearifolius* occurs in Cambridgeshire, Yorkshire and Kintyre. Var. *pilosus* has been recorded from Berrow in Somersetshire and Pembrey in Carmarthenshire. It is a common plant of North America and northern Asia.

5. L. linifolius (Reichard) Bässler Bitter Vetch
Orobus linifolius Reichard; *Orobus tuberosus* L., non *L. tuberosus* L.; *L. montanus* Bernh.

Perennial herb with creeping, tuberous rhizome. *Stem* 15–50 cm, pale green, erect, winged, glabrous, branched, leafy. *Leaves* alternate, paripinnate; rhachis mucronate; leaflets (2–)4–8, 18–50(–100) × 0.5–25 mm, medium green on upper surface, paler beneath, finely linear to elliptical, obtuse-mucronate to acute at apex, entire, narrowed to rounded at base, shortly petiolulate, glabrous; tendrils absent; petiole short; stipules 5–25 mm, linear to lanceolate, semi-sagittate, variable but usually somewhat toothed towards base. *Inflorescence* an axillary raceme with 2–6 flowers; peduncles longer than leaves. *Calyx* 4–6 mm, divided to halfway into 5 unequal lobes, the lower lobes triangular and acute at apex, the upper subulate. *Corolla* 10–16 mm, pale bluish-mauve, the keel somewhat deeper in colour and greenish at base; standard with limb very broadly elliptical, emarginate at apex, rounded to a broad rather long claw with the margins inturned; wings with limb obovate, rounded at apex, with an irregular lobe at base and a long narrow claw; keel with limb broadly ovate and upturned, with a short broad lobe at base and a long slender claw. *Stamens* 10, diadelphous, filaments upturned, pale; anthers yellowish. *Style* 1, turned up at right angles, hairy on one side at apex. *Legume* 25–45 × 5–7 mm, oblong and subcylindrical, glabrous; seeds 4–10, 3.0–3.5 mm, subglobose, smooth. *Flowers* 4–7. $2n = 14$.

(i) Var. **linifolius**
Orobus tenuifolius Roth, non *L. tenuifolius* Desf.; *Orobus graminifolius* Becker; *L. rothii* Rouy; *Orobus tuberosus* var. *tenuifolius* (Roth) Bab.; *L. montanus* var. *tenuifolius* (Roth) F. Hanb.

Leaflets 20–50(–100) × 0.5–3.0(–5.0) mm, long-linear, narrowly acute at apex.

(ii) Var. **varifolius** (Martrin-Donos) P. D. Sell
L. macrorhizus var. *varifolius* Martrin.-Donos; *L. montanus* var. *varifolius* (Martrin-Donos) Asch. & Graebn.

Leaflets 18–50 × 3–10 mm, narrowly elliptical, those of the lower leaves usually broader than those of the upper, mostly pointed at apex.

(iii) Var. **montanus** (Bernh.) Bässler
L. montanus Bernh.; *Orobus tuberosus* L., non *L. tuberosus* L.; *L. macrorhizus* Wimm.

Leaflets 25–50 × 10–25 mm, narrowly elliptical to elliptical, obtuse-mucronate at apex.

Native. Moist, infertile, neutral and acidic soils in heathy meadows, lightly grazed pastures, grassy banks, open woodlands and rock ledges, generally lowland but reaching 760 m in Perthshire. Widespread in Great Britain and Ireland, but absent from East Anglia and much of central Ireland. South, west and central Europe, extending to the Baltic region and White Russia; North Africa. The distribution and ecology of the three varieties is not understood, though they all appear to be widespread including in Continental Europe. European Temperate element. Var. *varifolius* is intermediate between the other two varieties and could be a hybrid, but none of the varieties seem to grow together and all seem to be fertile.

6. L. grandiflorus Sm. Two-flowered Everlasting Pea
L. tingitanus auct.

Perennial herb with a long, thin tap-root. *Stem* up to 2 m, pale green, climbing, twisting, not winged, glabrous, leafy. *Leaves* paripinnate; leaflets 2, dull medium green on upper surface, paler beneath, ovate, rounded at apex, entire, often sinuate, rounded at base with a very short petiolule, minutely hairy on both surfaces; tendrils branched; petiole up to 5 cm, pale green, channelled, shortly hairy; stipules 10–14 mm, half sagittate, sharply acute, minutely hairy. *Inflorescence* of 1–3 flowers together in a leaf axil; peduncle up to 10.0 mm; pedicels 10–25 mm; flowers 40–50 mm. *Calyx* 11–12 mm, with 5 unequal lobes, the upper divided into 2, the lower into 3, large. *Corolla* with standard 25–50 mm across, deep rose, reniform, emarginate at apex, cordate at base; wings with limb 20–25 mm, deep brownish-red, obovate, rounded at apex, narrowed at base with a short lobe and a short, narrow, green claw; keel about 22 mm, whitish, grading to pink with a green tip, obovate, curved upwards. *Stamens* 10, diadelphous; filaments whitish-green; anthers pale yellow. *Style* 1, curved upwards, twisted, flattened, hairy all round in upper half, yellowish-green; stigma deeper coloured. *Legume* 60–90 × 6–7 mm, brown, oblong, acute at apex, glabrous; seeds 15–20, elliptical, smooth. *Flowers* 6–8. $2n = 14$.

Introduced. Persistent garden escape in hedges, waste ground near old gardens and coastal sands. Scattered records throughout Great Britain and in the Isle of Man and Alderney; two records in Ireland. Native of the central Mediterranean region and the Balkans.

Section **2. Lathyrus**
Lathyrus section *Cicercula* (Medik.) Godr.; *Cicercula* Medik.; *Lastila* Alef.; *Navidura* Alef.; *Lathyrus* section *Lentiformia* Zohary

Annuals or *perennial herbs*, sometimes with a tuberous rootstock. *Stems* usually winged. *Leaves* hypo-amphistomatic, unijugate, with tendrils; leaflets subrotund to narrowly lanceolate, linear or narrowly elliptical; venation pinnate, parallel or intermediate; stipules semi-sagittate. *Inflorescence* with 1–numerous flowers; flowers with a wide range of colours. *Calyx* lobes equal or unequal. *Style* upcurved. *Legume* oblong, sometimes with winged sutures; seeds often with a rough texture.

7. L. tuberosus L. Tuberous Pea

Perennial herb with roots bearing tubers. *Stems* 30–120 cm, pale green, scrambling or climbing, angled, not winged, more or less glabrous, branched, leafy. *Leaves* alternate, paripinnate; leaflets 2, 15–45 × 7–20 mm, dull medium green on upper surface, paler beneath, elliptical to obovate, rounded at apex, entire, narrowed or rounded at base, very shortly petiolulate, glabrous or nearly so; tendrils simple or branched; petioles up to 20 mm, glabrous; stipules up to 20 mm, narrowly semi-sagittate.

Inflorescence an axillary raceme, with 2–7 flowers; peduncle usually exceeding leaves, glabrous. *Calyx* 6–7 mm, tubular, divided halfway to base into 5 lobes, the lobes triangular and acute at apex, glabrous. *Corolla* 12–20 mm, bright reddish-purple; standard with limb broadly reniform, emarginate at apex, cordate at base, with a short, broad claw; wings with limb obovate, with a broad rounded apex, a folded lobe at base and a narrow claw; keel with limb broadly ovate and upturned, rounded at apex, with a short broad lobe at base and a boat-shaped claw. *Stamens* 10, diadelphous; filaments upturned, pale green; anthers greenish. *Style* 1, pale green, upturned, hairy all round at apex; stigma greenish. *Legume* 20–40 × 6–8 mm, oblong, glabrous; seeds 3.5–4.0 mm, subglobose. *Flowers* 7–8. $2n = 14$.

Introduced. Cornfields, hedgerows, waste places and roadsides. Naturalised in Essex since 1859 and casual and shortly persisting in scattered localities in Great Britain north to central Scotland. Most of Europe and western Asia, but absent from the islands. Eurosiberian Temperate element. It was being cultivated in gardens by 1596, but is a common cornfield weed in Continental Europe and may have been introduced before this date as a seed contaminant.

8. L. sylvestris L. Narrow-leaved Everlasting Pea

Perennial herb with a tuberous rootstock. *Stem* 100–200 cm, bluish-green, scrambling or climbing, broadly winged, glabrous, branched, leafy. *Leaves* alternate, paripinnate; leaflets 25–150 × 8–25(–30) mm, bluish-green on upper surface, paler beneath, ensiform, oblong-lanceolate or narrowly elliptical, acute at apex, entire, narrowed or rounded at base, very shortly petiolulate, glabrous; tendrils branched; petioles up to 30 mm, glabrous, with wings narrower than stem; stipules up to 30 mm, narrowly ensiform with a spreading basal lobe, less than half as wide as stem. *Inflorescence* an axillary raceme of 3–12 flowers; peduncles 10–20 cm, glabrous. *Calyx* 5–7 mm, divided less than halfway to base into 5 subequal lobes, the lobes triangular and acute at apex, glabrous. *Corolla* 13–28 mm; standard rose-pink tinged with greenish, with limb subrotund, emarginate at apex, rounded at base, with a short claw with an incurved margin; wings rose, becoming purplish towards the apex, limb obovate, rounded at apex with a short lobe at base and a slender claw; keel greenish-white, ovate, upcurved, with a short lobe at base and a rather thick claw. *Stamens* 10, diadelphous; filaments upcurved; anthers yellowish. *Style* 1, markedly upcurved, with hairs all round just below the apex; stigma brownish. *Legume* 40–70 × 8–10 mm, pale brown, oblong, compressed, narrowly winged along the upper side, glabrous; seeds 8–14, 5.0–5.5 mm, subglobose. *Flowers* 6–8. $2n = 14$.

(i) Var. **sylvestris**
L. angustifolius Medik.; *L. sylvestris* forma *angustifolius* (Medik.) Morris; *L. sylvestris* var. *ensifolius* Buek, non Badaro; *L. sylvestris* var. *liniefolius* Borbás; *L. sylvestris* var. *linearifolius* Saut.

Leaflets up to 15 mm wide, long-acute.

(ii) Var. **latifolius** Peterm.
L. platyphyllus Retz.; *L. sylvestris* var. *platyphyllus* (Retz.) Asch.; *L. intermedius* Wallr.

Leaflets 10–25(–35) mm wide, narrowed but often obtuse at apex.

Native. Scrub, wood borders, hedgerows and rough ground. Scattered through Great Britain north to central Scotland, but said to be introduced in many places. Most of Europe east to the Caucasus; north-west Africa. European Temperate element. The distribution of the two varieties is not known, but var. *sylvestris* seems to be the more common.

9. L. heterophyllus L. Norfolk Everlasting Pea

Perennial herb with a tuberous rootstock. *Stem* 100–200 cm, bluish-green, scrambling or climbing, broadly winged, glabrous, branched, leafy. *Leaves* alternate, paripinnate; leaflets 2 or 4, 5–10 × 1–3 mm, bright bluish-green on upper surface, paler beneath, oblong-elliptical to narrowly elliptical, obtuse to subacute at apex, entire, narrowed at base, glabrous, very shortly petiolulate; petioles 10–40 mm, winged, narrower (with wings) than the winged stem; stipules up to 35 mm, semi-sagittate-ovate, long-acute at apex, with a spreading linear-lanceolate basal lobe, reticulate-veined, from half as long as to as long as the petiole. *Inflorescence* an axillary raceme of up to 10 flowers; peduncles up to 14 cm, glabrous. *Calyx* 9–10 mm; lobes 5, unequal, the lower much longer than upper, linear-lanceolate, long-acute at apex, glabrous. *Corolla* 15–22 mm, pale crimson-purple; standard with limb subrotund, emarginate at apex, with a short claw; wings with obovate limb, rounded at apex, with a short lobe at base and a slender claw; keel greenish, ovate, upcurved with a short lobe at base and a thick claw. *Stamens* 10, diadelphous; filaments upcurved, pale; anthers yellowish. *Style* 1, markedly upcurved, with hairs just below the apex; stigma brownish. *Legume* 40–80 × 8–10 mm, pale brown, oblong, compressed, narrowly winged on the upper side, glabrous; seeds about 5 mm, subglobose. *Flowers* 6–8. $2n = 14$.

(i) Var. **heterophyllus**
Leaflets 4.

(ii) Var. **unijugus** W. D. J. Koch
Leaflets 2.

Introduced. Var. *unijugus* has been long established on dunes in Norfolk and there appears to be a few other records. It sits between *L. sylvestris* var. *latifolius* and *L. latifolius* var. *latifolius* having the smaller flowers of the former and the larger stipules of the latter. Var. *heterophyllus* has not been recorded but is included to make the variation complete. The species occurs in south-west and central Europe.

10. L. latifolius L. Broad-leaved Everlasting Pea

Perennial herb with a large tuberous rootstock. *Stems* up to 3 m, medium green, scrambling, broadly winged, glabrous, leafy. *Leaves* alternate, paripinnate; leaflets 2,

7–10 × 3–6 cm, medium green on upper surface, paler and bluish-green beneath, subrotund, ovate, ovate-lanceolate or elliptical, rounded-mucronulate at apex, entire, rounded at base, sessile, glabrous on both surfaces, with 5 longitudinal veins; tendrils branched, the branches long and slender; petioles up to 7 cm, winged, glabrous; stipules up to 20 mm, semi-sagittate-ovate, acute at apex. *Inflorescence* an axillary raceme of 5–15 flowers; peduncles up to 20 cm, striate, glabrous; pedicels 11–13 mm, pale green, glabrous, bracteoles linear. *Calyx* 12–14 mm, with 2 short, ovate lobes and 3 long-lanceolate ones, glabrous. *Corolla* 15–30 mm; standard bright pinkish-purple, with white at the very base and veins slightly darker, broadly reniform, emarginate at apex, with a short, broad claw; wings bright pinkish-purple, paler towards the base, obovate, rounded at apex, with a pale green lobe at base; keel white with a touch of pink and green, ovate. *Stamens* 10, diadelphous; filaments pale green; anthers yellowish-brown. *Style* 1, green; stigma green with hairs on 2 sides. *Legume* 90–100 × 10–12 mm, oblong, compressed, glabrous. *Flowers* 6–8. Pollinated by bees. $2n = 14$.

(i) Var. **latifolius**
Leaflets up to 3 cm wide, ovate-lanceolate or elliptical.

(ii) Var. **rotundifolius** Rchb.
L. rotundifolius (Rchb.) Janke, non Willd.

Leaflets up to 6 cm wide, subrotund or elliptical.

Introduced. Persistent garden escape in hedges and on roadsides, railway banks and waste places. Scattered records throughout Great Britain north to central Scotland and a few records in east and north Ireland. Native of central and southern Europe. The distribution of the two variants is not known, but both occur frequently.

11. L. annuus L. Fodder Pea

Annual herb with fibrous roots. *Stems* 20–100 cm, pale green, climbing or sprawling, distinctly winged, glabrous, often rather flexuous, much branched, leafy. *Leaves* alternate, narrowly 2-foliolate, the lowermost cirrhose, the upper with leaflets 15–150 × 2–20 mm, linear or narrowly lanceolate, bright green on upper surface, paler beneath, tapering to a slender apex, glabrous, distinctly parallel-veined with a well-developed, branched tendril, the petioles 5–30 cm, the stipules 10–30 mm, semi-sagittate, with subulate or almost filiform lobes. *Inflorescence* a lax, axillary 1- to 3-flowered raceme; peduncle 15–100 mm, rigid erecto-patent, glabrous or very sparsely glandular-hairy; pedicels about 5 mm, straight or arcuate, glabrous. *Calyx* 5–8 mm, campanulate, divided into 5 lobes, the lobes subequal and narrowly deltoid-acuminate, with narrowly scarious margins, distinctly veined and sometimes sparsely clothed with brown, club-shaped glands. *Corolla* 12–18 mm, yellow, orange or salmon-coloured, the keel often greenish; standard with limb flabellate, shallowly emarginate at apex, with 2 indistinct gibbosities at the base, tapering to a wide claw; wings with limb oblong, rounded at apex, shortly lobed at base with a small tooth and a slender claw; keel with limb rather sharply upcurved to a rounded apex, scarcely lobed at base and with a slender claw. *Stamens* 10, diadelphous; filaments filiform; anthers oblong. *Style* 1, up to 3 mm. *Legume* 50–70 × 7–14 mm, oblong, flattened, acute or beaked at apex, the ventral sutures thickened but not winged, thinly glandular-hairy, reticulately veined; seeds 4–5 mm in diameter, subglobose, dark brown very strongly tuberculate or rugose. *Flowers* 6–8. $2n = 14$.

Introduced. Tips and waste places from bird-seed and other sources. Sporadic in southern England. Native of Portugal and the Mediterranean region east to Iraq.

12. L. odoratus L. Sweet Pea

Annual herb. Stem climbing to 2.5 m, bluish green, with broad wings, glabrous, branched, leafy. *Leaves* alternate, broadly oblong in outline, paripinnate; leaflets 2 or 4, 4–5 × 1.5–4.0 cm, bluish-green on upper surface, paler and greener beneath, narrowly to broadly elliptical, narrowed to an obtuse apex, entire, narrowed at base, glabrous; tendrils slender, pinnate with long branches; petiole up to 6 cm, winged, glabrous; stipules sagittate, lobes narrow. *Inflorescence* an axillary raceme of 1–3(–4) flowers; peduncle up to 15 cm, curved over at top, hairy; pedicels short, hairy; flowers strongly scented. *Calyx* 11–14 mm, greyish-green, divided half-way to the base into 5 subequal lobes, the lobes linear-lanceolate, acute and curved down, hairy. *Corolla* 20–35 mm, white, cream, orange, red, blue, purple, rose or several-coloured; standard with limb large, subrotund, rounded with a small mucro at apex, cordate at base; wings with limb half-rotund, rounded at apex; keel with limb small, ovate, curved. *Stamens* 10, diadelphous; filaments whitish; anthers yellow. *Style* 1, pale green; stigma greenish. *Legume* 50–70 × 10–12 mm, brown, oblong, hairy; seeds about 8, smooth, hilum about one-quarter of circumference. *Flowers* 6–8. $2n = 14$.

Introduced. Widely grown in gardens for ornament and a frequent casual on tips and in waste places. Scattered records in Great Britain, mainly the south. Native of southern Italy and Sicily.

13. L. hirsutus L. Hairy Vetchling

Scrambling or climbing *annual herb* with fibrous roots. *Stem* 20–120 cm, pale green, winged, glabrous or slightly hairy, branched, leafy. *Leaves* of 1 pair of leaflets with branched tendrils; leaflets 15–80 × 3–20 mm, medium green on upper surface, paler beneath, linear or oblong, acute-mucronate at apex, entire but wavy, narrowed at base, sparsely hairy or glabrous; stipules 10–18 × 1–2 mm, linear, semi-sagittate. *Inflorescence* an axillary raceme, with 1–3(–4) flowers; peduncles exceeding leaves. *Calyx* 4–8 mm, divided into 5 unequal lobes shorter than the tube. *Corolla* 15–20 mm, red, with pale blue wings; standard 7–15 × 7–14 mm, obovate or subrotund; wings 5.5–11.0 × 2.5–5.5 mm; keel 5.0–9.5 × 2.5–4.0 mm. *Stamens* 10, diadelphous. *Style* 1. *Legume* 20–50 × 5–10 mm, brown, oblong, tuberculate, silky hairy; seeds 5–10, rough. *Flowers* 5–7. $2n = 14$.

Introduced. Grassy, rough and waste places, formerly naturalised but now only casual. Very scattered localities throughout Great Britain. Native of central and northern Europe; North Africa; temperate Asia.

14. L. sativus L. Indian Pea

Annual herb with fibrous roots. *Stems* 8–100 cm, pale green, narrowly winged, sprawling or climbing, glabrous, usually much branched especially near the base, leafy. *Leaves* alternate, 2-foliolate; leaflets (8–)25–70 × 2–10 mm, medium green on upper surface, paler beneath, linear, lanceolate or narrowly elliptical, acuminate-cuspidate, entire, subglabrous or very sparsely hairy, distinctly parallel-veined; the lower leaves aristate, the upper with a well-developed, branched tendril; petiole 4–20 mm, narrowly winged, glabrous or ciliate; stipules (4–)10–29(–30) mm, semi-sagittate or semi-hastate, usually with a minute median tooth. *Inflorescence* of solitary, axillary flowers; peduncles often 30–60 mm; pedicels 5–8 mm, arcuate. *Calyx* 8–10 mm, campanulate, deeply and equally 5-lobed, the lobes lanceolate, acuminate at apex. *Corolla* 12–24 mm, white, pink or bluish; standard with limb subrotund or flabelliform, emarginate at apex, without gibbosities at base and with a broad, very short claw; wings with limb obovate, rounded at apex, with a prominent, acute lobe at base and a short claw; keel with limb deltoid, upturned at a right angle to a subacute apex, the adaxial margin minutely papillose, distinctly lobed at base and with a short claw. *Stamens* 10; diadelphous; filaments filiform; anthers oblong. *Style* 1, upcurved, narrowly spathulate. *Legume* 20–50 × 15–20 mm, oblong, flattened, shortly beaked, with 2 wings on the dorsal suture, when mature distinctly reticulate-veined; seeds 5–8 mm wide, squarish in outline, creamy brown, smooth. *Flowers* 6–8. $2n = 14$.

Introduced. Fairly frequent on tips, mainly from birdseed. Sporadic records in England and Midlothian. Widespread in central and southern Europe and in the Mediterranean region eastwards to India; temperate and tropical Africa and the Atlantic Islands.

15. L. cicera L. Red Vetchling

Annual herb with fibrous roots. *Stems* 8–30(–50) cm, sprawling or climbing, narrowly winged, glabrous, usually much-branched especially near the base, leafy. *Leaves* alternate, 2-foliolate; leaflets (8–)25–70 × 2–10 mm, medium green on upper surface, paler beneath, linear, lanceolate or narrowly elliptical, acuminate-cuspidate at apex, entire, subglabrous or very sparsely hairy, distinctly parallel-veined; the lower leaves cristate, the upper with a well-developed, branched tendril; petiole 4–20 mm, narrowly winged, glabrous or ciliate; stipules (4–)10–20(–30) mm, semi-sagittate or semi-hastate, usually with a minute median tooth. *Inflorescence* a solitary axillary flower; peduncle 5–30 mm, suberect, glabrous or thinly hairy; pedicel 3–10 mm; bracts small, subulate or narrowly deltoid, often hairy. *Calyx* 8–10 mm, campanulate, deeply divided into 5 equal lobes, the lobes narrowly deltoid or lanceolate, acuminate at apex, glabrous, distinctly veined, at least 1.5 times longer than tube. *Corolla* (5–)10–14(–20) mm, orange or brick-red, occasionally salmon-pink, turning purplish on drying; standard with limb subrotund, emarginate at apex, basal gibbosities scarcely developed, with a short claw; wings with limb oblong-obovate, rounded at apex, narrowly and bluntly lobed, with a slender claw; limb of keel oblong, gently upcurved to a blunt apex, adaxial margin distinctly papillose, shortly lobed at base, with a short claw. *Stamens* 10, diadelphous; anthers oblong. *Style* 1, incurved, distinctly flattened, twisted. *Legume* 20–40 × about 10 mm, oblong, slightly compressed, acute at apex, shortly beaked, with 2 keels on the dorsal suture; seeds about 5 mm wide, squarish in outline, more or less wedge-shaped in cross-section, bright brown, blotched pale brown, smooth or minutely rugulose. *Flowers* 6–8. $2n = 14$.

Introduced. A tan-bark casual and possibly a bird-seed alien; formerly a cotton alien. Widespread native in southern Europe and the Mediterranean region eastwards to Iran.

Section **3. Pratensis** Bässler

Perennial herbs with a creeping rootstock. *Stems* not winged. *Leaves* hypo-amphistomatic, unijugate, with tendrils or a mucro; leaflets linear-lanceolate to narrowly elliptical; venation parallel; stipules hastate. *Calyx* lobes unequal. *Style* linear, straight. *Legume* oblong; seeds smooth.

16. L. pratensis L. Meadow Vetchling
Orobus pratensis (L.) Stokes

Perennial herb with a rhizome. *Stem* 30–120 cm, pale green, angled, not winged, glabrous to finely hairy, branched, leafy. *Leaves* alternate, paripinnate; leaflets 2, 10–40 × 1–10 mm, medium green on upper surface, paler beneath, lanceolate, linear-lanceolate or narrowly elliptical, acute at apex, entire, narrowed at base, sessile, shortly hairy to glabrous or nearly so on both surfaces; tendrils simple or branched; petioles up to 20 mm, glabrous or shortly hairy; stipules 10–30 mm, sagittate, acute at apex. *Inflorescence* an axillary raceme with (2–)5–12 flowers; peduncle exceeding the leaves, stout. *Calyx* 5–7 mm, divided into 5 unequal lobes, the lower lobes triangular-subulate, acute at apex, the upper longer and narrower, glabrous to hairy. *Corolla* 10–20 mm, rich yellow, with veins on the standard greenish; standard with limb obovate, retuse at apex, narrowed at base to a short wide spur which has the margins inturned; wings with limb obovate, rounded at apex, with a narrow lobe at base and a slender spur; keel with limb ovate, curved upwards, rounded at apex, with a small lobe at base and a long slender spur. *Stamens* 10, diadelphous; filaments curved upwards, pale; anthers yellowish. *Style* 1, curved upwards at right angles, hairy on 2 sides near the apex; stigma brownish. *Legume* 25–38 × 5–6 mm, oblong, compressed, glabrous or hairy; seeds 5–10, 3.0–3.2 mm, subglobose, smooth. *Flowers* 5–8. $2n = 14$, 28.

(**i**) Var. **velutinus** DC.
L. pratensis var. *gracilis* Druce

Plant hairy. *Leaflets* 10–25 × 1–5 mm. *Corolla* 10–18 mm. *Legumes* 30–35 mm.

(**ii**) Var. **pratensis**
Plant glabrous or nearly so. *Leaflets* 12–35 × 4–10 mm. *Corolla* 10–18 mm. *Legumes* 25–38 mm.

(**iii**) Var. **speciosus** (Druce) Druce
L. pratensis forma *speciosa* Druce

Plant more or less hairy. *Leaflets* 20–40 × 5–7 mm. *Corolla* 18–20 mm. Mature legumes not seen.

Native. Roadside and railway banks, hedges, unimproved pastures, hay meadows and other grassy places and stabilised dunes. Throughout Great Britain and Ireland but generally in the lowlands, though reaching 450 m in Co. Durham. Almost all Europe, Siberia to the Arctic Circle, south to the Himalaya; North Africa; Ethiopia; introduced in North America. Eurosiberian Boreo-temperate element, but widely introduced outside its natural range. The distribution of the variants is not known, but var. *speciosus* seems to be mainly in the north and the other two more widespread.

Section **4. Aphaca** (Mill.) Dumort.
Aphaca Mill.

Annual herbs. Stems not winged. First 2 seeding *leaves* with a pair of elliptical, parallel-veined leaflets; leaves of adult plant without leaflets, but with strong, simple tendrils and large hastate stipules. *Inflorescence* of 1–2 flowers; flowers bright yellow. *Calyx* lobes subequal. *Style* linear and nearly straight. *Legume* narrowly oblong; seeds smooth.

17. L. aphaca L. Yellow Vetchling
L. aphyllus Gray nom. illegit.

Annual herb with fibrous roots. *Stems* 7–40(–100) cm, pale green to glaucous, trailing or climbing, angular but not winged, glabrous, much branched especially at the base. *Leaves* reduced to a simple filiform tendril; stipules 5–30 × 5–20 mm, glaucous on upper surface, paler beneath, ovate-hastate, acute and often apiculate at apex, with spreading, deltoid lobes at base, glabrous. *Inflorescence* of 1(–2) axillary flowers; pedicel 10–50 mm, erecto-patent, filiform, articulated about 5 mm from apex, glabrous. *Calyx* 5–10 mm, campanulate, divided up to halfway into 5 lobes, the lobes subequal, linear-lanceolate, acuminate at apex, with longitudinal veins and a narrow, scarious margin, glabrous. *Corolla* 10–13 mm, bright yellow; standard with limb subrotund, emarginate at apex, often streaked with violet, smooth externally, densely and minutely papillose internally, with 2 gibbous folds just above the long claw; wings with limb broadly spathulate, rounded at apex, with a long narrow lobe at base and a slender claw, papillose internally; keel with limb boat-shaped, apiculate at apex, with a short lobe and short claw at base. *Stamens* 10, diadelphous; filaments pale; anthers oblong. *Style* 1, slightly upcurved, strongly flattened towards the apex and bearded below the apex. *Legume* 25–30 × about 7 mm, brown, narrowly oblong, straight or slightly curved, strongly compressed, acute or shortly beaked at apex, obscurely veined; seeds 3.5–4.0 mm, broadly oblong, rich crimson-brown, distinctly compressed, smooth. *Flowers* 6–8. $2n = 14$.

Possibly native. Dry banks, grassy places, waste ground, cliffs by the sea, and ancient tracks. South and east England and the Channel Islands, casual north to central Scotland and in Ireland. Widespread in Europe and the Mediterranean region and eastwards to India; introduced elsewhere. Most early records suggest it was an arable weed. Submediterranean-Subatlantic element.

Section **5. Clymenum** (Mill.) DC. ex Ser.
Clymenum Mill.; *Lathyrus* section *Gloeolathyrus* Warb. & Eig

Annual. Stems strongly winged. *Leaves* hypo-amphistomatic; juvenile phyllodic and at first without tendrils, the later with several pairs of leaflets, with tendrils; leaflets linear to elliptical to lanceolate or ovate; venation of leaflets and phyllodes pinnate; stipules semi-hastate or minute. *Inflorescence* 1- to few-flowered; flowers purplish, pink or yellow. *Calyx* lobes subequal. *Styles* spathulate at apex. *Legume* oblong, sometimes broadly winged at sutures; seeds smooth.

18. L. clymenum L. Climbing Vetchling
L. articulatus L.; *L. clymenum* subsp. *articulatus* (L.) Ball

Annual herb with fibrous roots. *Stems* 30–100 cm, pale green, scrambling or climbing, strongly winged, glabrous, branched, leafy. *Leaves* with a broad leaf-like petiole and rhachis, the lower linear-lanceolate, without leaflets, the upper paripinnate with 4–8(–10) leaflets; leaflets 20–60(–80) × (3–)6–11(–20) mm, medium green on upper surface, paler beneath, linear to elliptical or lanceolate, obtuse at apex, entire, glabrous; stipules 9–18 × 2–6 mm, linear to ovate, semihastate. *Inflorescence* an axillary raceme of 1–5 flowers. *Calyx* 5.0–10.5 mm, divided into 5 more or less equal lobes shorter than the tube. *Corolla* 12–22 mm; standard 12–22(–25) × 9.5–20 mm, crimson, subrotund; wings 12–21 × 3.5–9.5 mm, violet or lilac; keel 8.5–15 × 3.5–6.5 mm. *Stamens* 10, diadelphous. *Style* 1, spathulate at apex. *Legume* 30–70 × 5–12 mm, brown, channelled not winged on the dorsal suture; glabrous; seeds 5–7, smooth, the hilum round up to one-sixth of the circumference. *Flowers* 6–8. $2n = 14$.

Introduced. Bird-seed and spice casual. Native of the Mediterranean region.

19. L. ochrus (L.) DC. Winged Vetchling
Pisum ochrus L.

Annual herb with fibrous roots. *Stems* 15–40 cm, slightly glaucous, strongly winged, sprawling or subclimbing, glabrous, unbranched or sparingly branched near the base, leafy. *Leaves* alternate, the lower without leaflets, the lowermost up to 10 × 5 mm, narrowly oblong, acute at apex and cirrhose, the middle 5–8 × 1–3 cm, slightly glaucous, oblong, with an emarginate apex, terminating in 3 simple tendrils, and with long-decurrent bases, the upper similar to middle but terminating in a single branched tendril and generally bearing 1–4(–5) leaflets 13–40 × 8–20 mm, ovate or oblong-elliptical and acute at apex, all glabrous. *Inflorescence* of 1(–2) axillary flowers; pedicel 10–30 mm, transversely articulate about the middle, usually curved at apex, elongating considerably in fruit, glabrous. *Calyx* 8–9 mm, tubular, divided to about halfway into 5 lobes, the lobes narrowly deltoid to narrowly subulate. *Corolla* 9–10 mm, whitish or pale yellow; standard with limb subrotund, emarginate at apex, with 2 prominent gibbosities at base and a short claw; wings with limb obovate, free or slightly adherent to the keel, rounded at apex, with a short lobe at base and a slender claw; keel with limb boat-shaped, curved upwards to an apiculate apex and with a short claw. *Stamens* 10, diadelphous; filaments filiform; anthers oblong. *Style* 1, sharply upcurved, spathulate towards the apex and bearded adaxilly. *Legume* 40–50 × 10–15 mm, narrowly oblong, strongly compressed, acute at apex, dorsal sutures narrowly winged; seeds 5–7 mm in diameter, dark or reddish-brown, subglobose; smooth. *Flowers* 6–7. $2n = 14$.

Introduced. A bird-seed casual and formerly a grain and ballast alien. Widespread in southern Europe and the Mediterranean region.

Section **6. Linearicarpus** Kupicha
Graphiosa Alef.

Annual herbs. Stems unwinged. *Leaves* epi-amphistomatic, unijugate, mucronate or with a simple tendril; leaflets linear-lanceolate; veins parallel; stipules semisagittate. *Inflorescence* 1-flowered; flowers violet or brick-red. *Calyx* with subequal lobes. *Style* straight. *Legume* oblong; seeds smooth.

20. L. inconspicuus L. Inconspicuous Vetchling
L. erectus Lag.; *Orobus inconspicuus* (L.) A. Br.

Annual herb with fibrous roots. *Stems* 10–30 cm, pale green, scrambling, not winged, glabrous, branched, leafy. *Leaves* alternate, paripinnate, without or with a usually simple tendril; leaflets 2, 25–40 × 1–4 mm, medium green on upper surface, paler beneath, linear-lanceolate, acute at apex, entire, narrowed at base, glabrous; stipules 7–10 × 0.5–2.0 mm, linear or lanceolate, semi-sagittate. *Inflorescence* a solitary axillary flower; peduncles 2–5 mm, articulated near the base or middle. *Calyx* 3.5–5.0 mm, divided into 5 subequal lobes as long as the tube. *Corolla* 4.0–9.5 mm, violet or brick-red; standard 4.0–9.5 × 2–7 mm, elliptical; wings 4–9 × 1–3 mm, obtuse; keel 4.0–6.5 × 1.3–3.5 mm. *Stamens* 10, diadelphous. *Style* 1, straight. *Legume* 30–60 × 2–5 mm, pale brown, oblong, densely hairy when young, becoming glabrous; seeds 5–14, smooth, hilum going a very small distance round the circumference. *Flowers* 6–8. $2n = 14$.

Introduced. A bird-seed and tan-bark casual in both Great Britain and Ireland. Native of the Mediterranean region.

Section **7. Nissolia** (Mill.) Dumort.
Nissolia Mill.

Annual herbs. Stems unwinged. *Leaves* phyllodic throughout life cycle, linear-lanceolate and grass-like; veins parallel, without tendrils, epi-amphistomatic; stipules minute. *Inflorescence* 1-flowered; flowers reddish-magenta. *Calyx* lobes subequal. *Style* linear, straight. *Legume* oblong; seeds tuberculate.

21. L. nissolia L. Grass Vetchling
L. gramineus Gray nom. illegit.

Annual herb with fibrous roots. *Stems* 30–90 cm, erect or ascending, slender, nearly glabrous, sparsely branched. *Leaves* and tendrils absent; stipules minute. *Phyllodes* 50–150 × 2–6 mm, narrowly linear-lanceolate and grass-like, narrowed to a fine point at apex, entire, rounded at base, sessile, veins parallel, glabrous or nearly so. *Inflorescence* of 1–2 axillary flowers; peduncles shorter than to equalling phyllodes. *Calyx* 5–6 mm, tubular with 5 lobes shorter than the tube, the lobes triangular, subequal, and acute, glabrous. *Corolla* 8–18 mm, erect; standard bright reddish-magenta and pearly white at base, the limb subrotund, emarginate at apex, rounded at base to a broad claw which has its sides inturned; wings bright reddish-magenta and pearly white at base, the limb broadly oblong, rounded at apex, with an incurled lobe at base and a narrow claw; the keel pale greenish-white flushed with pale mauve on upper edge, the limb irregularly elliptical with a beak at apex, a short rounded lobe at base and a very slender claw. *Stamens* 10, diadelphous; filaments pale; anthers yellowish. *Style* 1, sharply upturned, and straight, hairy all round just below the apex. *Legume* 50–60 × 4.0–4.5 mm, oblong, shortly acuminate at apex; seeds 2.0–2.6 mm, subglobose, tuberculate. *Flowers* 5–7. $2n = 14$.

(i) Var. **nissolia**
Legume glabrous but rough or sometimes with some minute glandular hairs.

(ii) Var. **pubescens** Beck
L. nissolia var. *lanceolatus* Rouy

Legume with numeous glandular and eglandular hairs.

Native. Dry grassy places on chalk and clay and road verges, railway banks, woodland rides and shingle. Local in England and south Wales, north to Lincolnshire, casual north to central Scotland. East, central and south Europe, east to the Caucasus and Syria; North Africa; introduced in Belgium, Holland and North America. European Temperate element. Our plant is mostly var. *nissolia*. Var. *pubescens* has only been recorded from the Isle of Wight (see Cannon, 1964.)

43. Pisum L.

Annual herbs. Stems pale glaucous, angled but not winged. *Leaves* alternate, paripinnate; leaflets entire or dentate; tendrils branched; stipules larger than leaflets. *Inflorescence* a lax, axillary raceme of 1–3 flowers. *Calyx* 5-lobed. *Corolla* white, lavender, pink or purple; standard broadly obcordate; wings with subrotund limb and long claw; keel with oblong limb upcurved at a right angle. *Stamens* 10, diadelphous. *Style* 1, hairy only on upper side, proximal part with reflexed margins. *Fruit* a large linear-oblong legume.

Contains 2 or possibly 3 species in southern Europe, the Mediterranean region, western and central Asia and east tropical Africa.

Phillips, R. & Rix, M. (1993). *Vegetables*. London.
Vaughan, J. G. & Geissler, C. (1997). *The new Oxford book of food plants*. Oxford.

1. Racemes exceeding the leaves; corolla lilac or purple; seeds globose and granular **1(b). sativum** subsp. **elatius**

1. Racemes shorter than or only slightly exceeding leaves; flowers white to purple; seeds globose to somewhat angular, smooth or rugose 2.
2. Flowers white **1(a, i). sativum** var. **sativum**
2. Flowers lavender, pink or dark purple, the wings often with a dark purple limb **1(a, ii). sativum** var. **arvense**

1. P. sativum L. Garden Pea

Annual herb with fibrous roots. *Stems* 15–140(–200) cm, usually pale glaucous, climbing or sprawling, striate but not winged, glabrous, simple or branched, leafy. *Leaves* alternate, 2–20 × 2–10 cm, oblong in outline, paripinnate; leaflets 2–6, 10–50 × 5–35 mm, usually glaucous, paler beneath, oblong, elliptical or obovate, usually apiculate at apex, entire or dentate, rounded at base, glabrous; tendril well-developed and usually branched; petiole 1–6 cm, flattened or slightly channelled above, but not winged; stipules 1–6 × 0.5–5.0 cm, glaucous, ovate or oblong, obtuse or subacute at apex, dentate at base, entire or dentate above, glabrous. *Inflorescence* a lax axillary raceme of 1–3 flowers; peduncle 1–15(–20) cm, rather stout, suberect; bracts deltoid or subulate, glabrous or thinly hairy, caduceus; pedicels 5–15 mm. *Calyx* 7–10(–15) mm, divided over halfway to the base into 5 lobes, the lobes narrowly deltoid, acute at apex, more or less glabrous, sometimes sparsely red glandular-hairy within. *Corolla* 15–35 mm, white, lavender, pink or purple; standard with limb broadly obcordate, emarginate and apiculate at apex, with 2 distinct median gibbosities at base and a fairly long claw; wings with limb subrotund, rounded at apex, with a blunt gibbous lobe at base and a long claw; keel with limb oblong, upcurved almost at a right angle to a short acute apex, the lower margin distinctly cristate with a long claw. *Stamens* 10, diadelphous; filaments slightly dilated towards apex; anthers oblong. *Style* 1, upcurved at a right angle, hairy on upper side; stigma subterminal. *Legume* 3–12 × 0.8–1.0 cm, linear-oblong, compressed or subterete, shortly beaked, valves pale brown and distinctly veined at maturity; seeds 5–10 mm in diameter, subglobose or angular, green or pale brown. *Flowers* 6–7. Mostly self-pollinated. $2n = 14$.

(**a**) Subsp. **sativum**
Racemes shorter than or only slightly exceeding the leaves. *Flowers* white to purple. *Seeds* globose to somewhat angular, smooth or rugose.

(**i**) Var. **sativum** Garden Pea
Flower white. *Legumes* 6–12 cm.

(**ii**) Var. **arvense** (L.) Poir. Field Pea
P. arvense L.

Flowers lavender, pink or dark purple, the wings often with a dark purple limb. *Legumes* 3–8 cm.

(**b**) Subsp. **elatius** (M. Bieb.) Asch. & Graebn.
P. elatius M. Bieb.

Racemes exceeding the leaves. *Corolla* lilac or purple. *Seeds* globose and granular.

Introduced. Both varieties of subsp. *sativa* are grown as field crops and var. *sativum* in vegetable gardens. Both occur as casuals by roads and tracks and in waste places and on tips. Throughout much of Great Britain and Ireland. The species is probably native in the Mediterranean regions and south-west Asia. Peas were one of the earliest vegetables grown by Man and have been discovered in Stone Age dwellings. Ancient Greeks and Romans were partial to them and there have been periods in European history when they have been very much a luxury. They have been boiled as a vegetable, added to a stew or casserole, sieved to a pulp or made into a thick soup. There are three main types of var. *sativum*, the normal kind in which the peas are shelled, the *mange tout* varieties where the pod remains flat and is cooked whole, and the sugar peas in which the peas swell but the pod is still cooked whole. Var. *arvense* is grown as a crop for animal fodder. Subsp. *elatius* has been recorded at Galashiels in Selkirk.

Order **3. Proteales** Lindl.

Trees or *shrubs*. *Leaves* alternate, exstipulate. *Flowers* unisexual or bisexual, hypogynous to strongly perigynous, actinomorphic. *Hypanthium* present, segments 2 or 4. *Stamens* 2 or 4. *Style* 1. *Fruit* a drupe-like achene, completely or partially surrounded by the hypanthium; 1-seeded.

Consists of 2 families and more than 2,000 species.

89. ELAEAGNACEAE Juss. nom. conserv.

Deciduous or evergreen, monoecious or dioecious *shrubs* or small *trees* sometimes with thorns. *Leaves* alternate, simple, shortly petiolate, exstipulate, with dense silvery to brown scales or stellate hairs. *Flowers* in small, axillary clusters, unisexual or bisexual, with a prominent hypanthium, actinomorphic. *Perianth segments* 2 or 4, green, round the rim of the hypanthium. *Stamens* 2 or 4, inserted in the throat of the hypanthium. *Style* 1, long; stigma linear to capitate. *Fruit* a drupe-like achene surrounded by the succulent hypanthium; 1-seeded.

Includes 3 genera and about 45 species, mostly in temperate parts of Europe, Asia and North America, with a few in tropical Asia and extending to north-eastern Australia.

1. Dioecious shrub with unisexual flowers; perianth segments 2; hypanthium very short **1. Hippophae**
2. Usually monoecious shrub with bisexual flowers; perianth segments 4; hypanthium long **2. Elaeagnus**

1. Hippophae L.

Deciduous, dioecious *shrubs* or small *trees*; twigs spine-tipped. *Leaves* alternate, simple, entire, shortly petiolate, exstipulate, with dense, silvery to reddish-brown scales. *Flowers* unisexual in small, axillary, raceme-like clusters; male with short hypanthium and 2 large perianth segments at apex, a small disk and 4 stamens enclosed in the hypanthium; female with conspicuous elongated hypanthium with 2 minute perianth segments at the apex, the disc absent and 1 cylindrical style. *Fruit* a drupe-like achene, the dry true fruit surrounded by the lower half of the hypanthium, which becomes fleshy; endosperm absent or scanty.

Contains 3 species in Europe and Asia.

Bean, W. J. (1973). *Trees and shrubs hardy in the British Isles*. Ed. 8. **2**. London.

Hultén, E. & Fries, M. (1986). *Atlas of north European vascular plants north of the Tropic of Cancer*. 3 vols. Königstein.

Pearson, M. C. & Rogers, J. A. (1962). *Biological flora of the British Isles*. No. 85. *Hippophae rhamnoides L. Jour. Ecol.* **50**: 501–513.

Ranwell, D. S. (Edit.) (1972). *The management of sea buckthorn* Hippophae rhamnoides *on selected sites in Great Britain*. Norwich.

Stewart, A., Pearman, D. A. & Preston, C.D. (1994). *Scarce plants in Britain*. Peterborough.

1. H. rhamnoides L. Sea Buckthorn

H. littoralis Salisb. nom. illegit.

Deciduous, dioecious *shrub* or small *tree* up to 3(–13) m, suckering freely, normally a rounded shrub or thicket of shrubs, but in cultivation tending to form a slender, columnar crown. *Stems* up to 30 cm in diameter. *Bark* brown, with scaly ridges. *Branches* numerous, spreading; twigs dark grey or brown; young shoots greyish- to reddish-brown, stiff and usually ending in a sharp spiny point 20–50 mm, densely covered with silvery-grey or brownish-grey scales which persist for 2 to 3 years. *Buds* 2–5 × 2–4 mm, more or less globose, rounded at apex; scales reddish-brown. *Leaves* alternate, crowded on the lateral branches; lamina 2–8 × 0.2–1.3 cm, bluish-green on upper surface, paler beneath, turning slightly yellowish in autumn, linear to linear-lanceolate, tapering to a slender, acute apex, entire, cuneate at base, with a dense clothing of silvery-grey scales on the upper surface with a groove along the midrib silvery-white with brownish-scales beneath and with a raised, rounded midrib; petiole about 2 mm, channelled and covered in brownish scales. *Flowers* 3–4 mm in diameter, unisexual, in small axillary clusters, appearing before the leaves on the wood of the previous year; male with a short hypanthium, bearing 2 large, greenish perianth segments and 4 stamens, the anthers orange; female with conspicuous elongated hypanthium, bearing 2 minute perianth segments and a yellowish style with cylindrical stigma. *Achene* drupe-like, 6–10 mm, orange, subglobose or ovoid. *Flowers* 3–4. Wind pollinated. $2n = 24$.

Native. Dunes and other sandy places by the sea. Round the coast of Great Britain and Ireland, but perhaps native only in the east from Sussex to central Scotland; widely planted by the sea, often for sand-binding, and along roads inland where often self-sown. Coasts of Atlantic and Baltic Seas from Norway to north Spain, central Italy and Bulgaria; river shingles in central Europe in the Rhone Valley and Alps; Black Sea coast; temperate Asia to Kamchatka, Japan and the north-west Himalaya. European Boreo-temperate element. Much eaten by birds and although edible to humans is very astringent. The species varies geographically in stature and scaliness and the taller tree-like plants are probably all introduced.

2. Elaeagnus L.

Deciduous or evergreen, monoecious or rarely dioecious *shrubs* or *trees*; twigs often with scales and sometimes spiny. *Leaves* alternate, simple, entire, shortly petiolate, densely covered at least beneath with peltate scales or stellate hairs. *Inflorescence* of solitary or few flowers in axillary umbels which are bisexual or rarely unisexual. *Hypanthium* tubular, enclosing and more or less constricted above the ovary, then opening out again. *Perianth segments* 4, round the top of the hypanthium. *Stamens* 4, almost sessile, inserted in the throat of the hypanthium, alternating with the perianth segments. *Nectariferous disc* present above the constriction in base of upper hypanthium and inconspicuous in most species. *Style* 1, with an oblique stigma. *Ovary* superior, solitary; ovule 1. *Fruit* a drupe-like achene, stone ribbed and embedded in the fleshy or mealy, swollen base of the hypanthium. The ovary appears to be inferior because of the constriction in the hypanthium.

About 40 species mostly from temperate eastern Asia, one from Europe and one from North America.

Bean, W. J. (1973). *Trees and shrubs hardy in the British Isles*. Ed. 8. **2**. London.

Rao, V. S. (1974). The nature of the perianth in *Elaeagnus* on the basis of floral anatomy, with some comments on the systematic position of Elaeagnaceae. *Jour. Indian Bot. Soc.* **53**: 156–161.

Rehder, A. (1940). *Manual of cultivated trees and shrubs*. Ed. 2. New York.

1. Leaves deciduous; flowering in spring or summer 2.
1. Leaves evergreen; flowering in autumn 4.
2. Upper surface of leaves with persistent, dense, silvery scales or stellate hairs; achene mealy **1. commutata**
2. Upper surface of leaves soon becoming green; achene juicy or mealy 3.
3. Hypanthium abruptly constricted above the ovary; achene 12–16 mm, oblong or ellipsoid, their stalks 10–40 mm **2. multiflora**
3. Hypanthium tapered to base; achene 5–8 mm, subglobose, their stalks 3–7 mm **3. umbellata**
4. Scrambling shrub with long, flexible shoots; leaves without wavy margins, brownish beneath **4. glabra**
4. Stiffly branched shrub; leaves with more or less wavy or crisped margins, silvery or whitish beneath 5.
5. Leaves broadly ovate to almost subrotund, abruptly pointed **5. macrophylla**
5. Leaves elliptical, oblong or ovate, obtuse or gradually tapering to a point 6.
6. Twigs with scattered stout spines; leaves oblong or ovate, with wavy or often crisped margins, underside dull whitish with a brown midrib 7.
6. Twigs without spines; leaves elliptical, with more or less wavy but not crisped margins, underside silvery with only slightly darker midrib 8.
7. Leaves deep green **6(1). pungens** forma **pungens**
7. Leaves variegated yellow **6(2). pungens** forma **variegata**
8. Leaves deep green **7(1). × ebbingei** forma **ebbingei**
8. Leaves variegated yellow **7(2). × ebbingei** forma **flavistriata**

1. E. commutata Bernh. Silverberry
E. argentea Pursh, non Moench

Deciduous, monoecious, suckering *shrub* up to 4 m, upright and with stolons. *Branches* ascending or spreading; twigs almost black; young shoots brown, densely scaly. *Buds* small. *Leaves* alternate; lamina 1–8 × 1–3 cm, silvery on both surfaces, elliptical, mostly acute at apex, entire, rounded or cuneate at base, with dense, silvery, fimbriate, peltate scales on both surfaces with some brown ones mixed with them on lower surface; lateral veins not visible; petiole short. *Inflorescence* of 1–3 flowers in an axillary cluster; pedicels 1–2 mm; flowers bisexual, sweet-scented. *Hypanthium* 6–7 mm, pale yellow, gradually getting wider upwards. *Perianth segments* 4, about 5 mm, pale yellow, triangular-ovate, pointed at apex. *Stamens* 4; filaments white; anthers cream. *Style* 1, white, with stellate hairs towards the base; stigma brownish. *Achene* about 10 mm, broadly ovoid or obovoid, mealy with silver scales, with dry flesh. *Flowers* 5–6. $2n=28$.

Introduced. Planted in gardens, estates, parks and amenity areas. Native of Alaska, southern Canada and northern United States.

2. E. multiflora Thunb. Cherry Oleaster
E. longipes A. Gray; *E. edulis* Carrière; *E. edulis* Sieber; *E. rotundifolia* Gagnebin

Deciduous, monoecious *shrub* up to 3 m, often widely spreading. *Branches* numerous, spreading and rigid; twigs brown, sometimes with spines; young shoots densely covered with fimbriate, peltate, silver and brown scales. *Buds* small, ovoid. *Leaves* alternate; lamina 3–8 × 2–4 cm, medium green on upper surface, silvery-white beneath, broadly elliptical to obovate-oblong, obtuse to acute at apex, entire, rounded or broadly cuneate at base, with deciduous stellate hairs on upper surface and dense brown and silver, fimbriate scales beneath; veins 4–5 pairs, not prominent; petiole 4–8 mm, with dense, brown scales. *Inflorescence* of 1 or 2 flowers in the leaf-axils; pedicels 10–20 mm, with dense scales; flowers bisexual, sweet-scented. *Hypanthium* 12–16 mm, pale yellow, ellipsoid where it covers the ovary, constricted and tubular above, then dilating and campanulate, with scales. *Perianth segments* 4, pale yellow, ovate, pointed at apex, with scales. *Stamens* 4, white; anthers small, cream. *Style* 1, white; stigma brownish. *Achene* 12–16 mm, yellowish-red, dotted, flesh yellowish, oblong or ellipsoid, on stalks 10–40 mm. *Flowers* 4–5.

Very variable in leaf shape, presence or absence of spines, length of pedicels and size of fruit. It is probable that several races occur.

Introduced. Planted in gardens, estates, parks and amenity areas. Native of China, Korea and Japan.

3. E. umbellata Thunb. Spreading Oleaster
E. crispa Thunb.

Deciduous, monoecious *shrub* up to 4 m, spreading broadly. *Branches* spreading; twigs dark brown, with dense scales; young shoots whitish, with very dense scales. *Buds* small, ovoid, with few outer scales. *Leaves* alternate; lamina 2–9 × 1–4 cm, pale green on upper surface, grey beneath, narrowly elliptical or oblanceolate, narrowed to an obtuse apex, entire, cuneate at base, with a few scales at first on upper surface, with dense, silvery, fimbriate scales beneath; veins 4–6 pairs, not prominent; petiole up to 10 mm, slender, with dense scales. *Inflorescence* of 1–7 flowers in axillary clusters; pedicels 1–3 mm, with dense scales; flowers bisexual, scented. *Hypanthium* about 7 mm, tapered gradually to base, with dense, silvery scales. *Perianth segments* 4, 3.0–3.5 mm, whitish-yellow, triangular-ovate, mucronate at tip. *Stamens* 4; filaments white; anthers cream. *Style* 1, white; stigma brownish. *Achene* 5–8 mm, red, with some silver scales, subglobose, with stalk 3–7 mm. *Flowers* 5–7. $2n=28$.

Introduced. Frequently grown in gardens and by roads. Used as a stock plant for *E. macrophylla* in Guernsey and Sark and often surviving when the latter dies. Bird-sown plants are naturalised in southern Great Britain. Native in Asia from the Himalaya to China and Japan.

4. E. glabra Thunb. Scrambling Oleaster
E. buisanensis Hayata; *E. daibuensis* Hayata; *E. paucilepidota* Hayata; *E. oiwakensis* Hayata

Evergreen, monoecious, scrambling *shrub* up to 6(–10) m, with a broad crown. *Stems* brown, much-branched. *Branches* spreading; twigs shining grey or brown; young shoots with scales. *Leaves* alternate; lamina 4–7 × 2.5–3.5 cm, glossy deep green on upper surface, shining brown beneath, elliptic-ovate, tapering to a sharp point at apex, simple and not wavy, rounded or broadly cuneate at base, glabrous on upper surface, with dense, brownish, fimbriate, peltate scales beneath; veins 4–5 pairs, not prominent; petioles 5–10 mm, brown, scaly. *Inflorescence* of up to 5 flowers in an axillary cluster; pedicels 3–4 mm; flowers bisexual, fragrant. *Hypanthium* 10–12 mm, abruptly contracted above the ovary into an elongated tube, with scales. *Perianth segments* 4, 4–5 mm, pale yellow, ovate, pointed at apex. *Stamens* 4; filaments white; anthers cream. *Style* 1, white; stigma brownish. *Achenes* 12–20 mm, silvery or reddish when ripe, ellipsoid. *Flowers* 10–11.

Introduced. Planted in gardens, parks, estates and amenity areas. Native of China and Japan.

5. E. macrophylla Thunb. Broad-leaved Oleaster

Evergreen, monoecious *shrub* up to 4 m, with a broad crown, often wider than high. *Stems* brown, much divaricately branched. *Branches* stiffly spreading; twigs brown, with dense, fimbriate scales; young shoots pale brown, with dense fimbriate scales. *Leaves* alternate; lamina 3–12 × 2–9 cm, semi-glossy deep green on upper surface, silvery-white beneath, subrotund or broadly ovate, abruptly pointed at apex, wavy or crisped on margin, rounded at base, with scattered scales or stellate hairs on upper surface and rather dark midrib; veins 4–5 pairs, not prominent; petiole up to 20 mm, with dense, fimbriate, peltate scales. *Inflorescence* of 4–6 flowers in a cluster in the axils of the leaves; pedicels up to 7 mm, with dense scales; flowers bisexual, very fragrant. *Hypanthium* 6–7 mm, constricted above the ovary, broadly inflated above, covered with fimbriate, peltate hairs. *Perianth segments* 4, 2.5–3.0 mm, silvery-white with brown spots, triangular-ovate, pointed at apex, densely covered with peltate-fimbriate scales. *Stamens* 4; filaments white; anthers cream. *Style* 1, white; stigma brown. *Achene* about 16 mm, silvery-scaly, red when ripe, ellipsoid or ovoid. *Flowers* 10–12.

Introduced. Frequently grown as a hedging plant in Guernsey and Sark and long persistent. Also in parks, gardens and amenity areas, rarely self-sowing in southern England. Native of Japan and Korea.

6. E. pungens Thunb. Spiny Oleaster

Evergreen, monoecious *shrub* up to 5 m, widely spreading. *Stems* brown, much-branched. *Branches* stiffly spreading; twigs silvery-brown, with scattered spines; young shoots brown with dense, fimbriate, peltate scales. *Leaves* alternate; lamina 4–10 × 1.5–4.0 cm, glossy deep green on upper surface, dull whitish beneath with a brown midrib, often variegated, elliptical or oblong, shortly pointed at apex, with wavy or crisped margins, rounded at base, with a few scales on upper surface and dense, silvery, fimbriate, peltate scales beneath; lateral veins not visible; petiole up to 10 mm, dark brown, with dense scales. *Inflorescence* usually of axillary clusters of 3 flowers; pedicels short; flowers bisexual, very fragrant. *Hypanthium* 9–10 mm, constricted above the ovary, widened again above, with dense, silvery scales. *Perianth segments* 4, 3.0–3.5 mm, ovate, pointed at apex, with scales. *Stamens* 4; filaments white; anthers cream. *Style* 1, white; stigma brownish. *Achene* 12–16 mm, becoming red and shedding scales when ripe, ovoid. *Flowers* 10–11. $2n=28$.

(1) Forma **pungens**
Leaves deep green.

(2) Forma **variegata** (Bean) P. D. Sell
E. pungens var. *variegata* Bean

Leaves variegated yellow.

Introduced. Planted in gardens, estates, parks and amenity areas. Native of Japan.

7. E. × ebbingei Doorenbos Hybrid Oleaster
E. macrophylla × pungens
?*E. submacrophylla* Serrettaz

Evergreen, monoecious *shrub* or semi-evergreen in colder areas, up to 3 m, forming a broad crown. *Stems* several, brown, much divaricate. *Branches* spreading; twigs medium brown to grey, densely covered with fimbriate, peltate scales; young shoots slightly paler brown, similarly covered with scales. *Buds* small, ovoid, with few outer scales. *Leaves* alternate; lamina 4–13 × 2.5–6.0 cm, dark shining green on upper surface, silvery-white on lower surface with a pale brown midrib, sometimes variegated yellow, elliptical to oblong-elliptical, gradually narrowed to an acute apex, the entire margin more or less wavy, rounded or cuneate at base, with scattered, fimbriate, peltate scales on upper surface and dense, silvery, fimbriate, peltate scales beneath; veins 5–8 pairs, faint; petiole up to 14 mm, with dense, brown, fimbriate, peltate scales. *Inflorescence* a 3- to 6-flowered umbel in the leaf-axils; pedicels up to 5 mm, with dense, brownish peltate scales; flowers bisexual, strongly scented. *Hypanthium* 4–5 mm, white, spotted reddish-brown, broadly inflated. *Perianth segments* 4, 4–5 mm, white, spotted reddish-brown, triangular-ovate, acute at apex. *Stamens* 4; filaments white; anthers cream. *Style* 1, white; stigma brown. *Achene* rarely produced, red, elliptical to ovoid. *Flowers* 10–11.

(1) Forma **ebbingei**
Leaves deep green on upper surface.

(2) Forma **flavistriata** P. D. Sell
Leaves variegated yellow on upper surface.

Introduced. This name was given to a batch of seedlings raised by S. G. A. Doorenbos at the Hague in 1929. It is found in gardens, parks, estates and amenity areas. *E. submacrophylla* may be the correct name of this group of hybrids. It may also be sold in nurseries as *E. macrophylla*.

Order 4. HALORAGALES Novák

Perennial herbs. Leaves alternate, opposite or whorled; without stipules or with large scales looking like stipules. *Sepals* 2 or 4, minute. *Petals* 2 or 4, minute or absent. *Stamens* (1–)2–8. *Styles* 4 or more or less absent; stigmas 2 or 4. *Fruit* a drupe or nut.

Contains 2 families and fewer than 200 species.

90. HALORAGACEAE R. Br. nom. conserv.

Perennial, dioecious or monoecious, mainly subaquatic *herbs*. *Stems* weak and trailing. *Leaves* opposite and simple or in whorls of 3–6 and finely pinnatisect, more or less sessile, without stipules. *Inflorescence* a terminal spike, the flowers whorled, opposite or alternate, unisexual or bisexual, epigynous with 2 minute bracteoles and one bract at base. *Sepals* 4, minute. *Petals* 4, small in male or bisexual flowers, minute or absent in female flowers. *Stamens* (4–)8. *Styles* 4 or absent; stigmas 4, clavate to plumose. *Ovary* 1- or 4-celled, with 1 apical ovule in each cell. *Fruit* a nut or group of up to 4 drupes.

Contains 8 genera with about 200 species, cosmopolitan but with a concentration of species in the southern hemisphere.

1. Leaves opposite, simple, not or slightly toothed **1. Haloragis**
1. Leaves in whorls of 3–6, finely pinnate with capillary segments **2. Myriophyllum**

1. Haloragis J. R. & G. Forst.

Perennial monoecious *herbs* forming mats, rooting at the lowest nodes. *Leaves* opposite, entire or shallowly incised; exstipulate. *Inflorescence* a terminal raceme, with a bract and 2 bracteoles which fall early and leave the flowers in opposite pairs; flowers unisexual or bisexual. *Sepals* 4, minute. *Petals* 4. *Stamens* 8. *Styles* 4. *Fruit* an achene.

Contains about 27 species in the southern hemisphere extending to south-east Asia.

Allan, H. H. (1961). *Flora of New Zealand*. **1**. Wellington.
Cook, C. D. K. (1990). *Aquatic plant book*. (Plate). Amsterdam.
Green, P. (1989). *Haloragis micrantha* – Creeping Raspwort. *B.S.B.I. News* **51**: 48.
Schindler, A. K. (1905). Halorrhagaceae. In Engler, A. (Edit.) *Das Pflanzenreich*: **V.225**. Leipzig.

1. H. micrantha (Thunb.) R. Br. ex Siebold & Zucc. Creeping Raspwort

Goniocarpus micranthus Thunb.; *H. tenella* Brongn., non DC.; *Goniocarpus citriodorus* A. Cunn.; *H. minima* Colenso

Perennial monoecious *herb* forming mats 1 m across. *Stems* up to 10 cm, many, pale, obscurely tetragonus, slender, glabrous, rooting from the lowest nodes. *Leaves* opposite, 3–10 mm, membranous, subrotund, ovate, acute at apex, margin cartilaginous and entire or with 4–5 shallow incisions on each side, almost sessile. *Inflorescence* a terminal raceme from which the bract and 2 bracteoles fall early leaving the flowers in opposite pairs, the flowers unisexual or bisexual, drooping. *Sepals* 4, minute, broadly triangular, persistent. *Petals* 4, about 1.5 mm in male or bisexual flowers, minute or absent in female flowers, reddish, hooded. *Stamens* 8. *Styles* 4, short, erect, cylindrical; stigmas of long hairs. *Fruit* about 0.5–1.0 mm, dark red, with 8 vertical ribs, faces between the ribs concave and smooth; pericarp not hardened; one loculus with one seed. *Flowers* 9–10. $2n = 12$.

Introduced. Naturalised on bare peat in a bog near Lough Bola in West Galway. Native of south-east Asia and Australia.

2. Myriophyllum L.

Perennial, monoecious, rarely dioecious, aquatic *herbs*, with rhizomes. *Leaves* in whorls, pinnatisect, with capillary segments; aerial leaves and bracts sometimes simple and toothed or entire; exstipulate. *Inflorescence* a leafy or bracteate, terminal spike, or with sessile flowers in axillary whorls; upper flowers commonly male and lower female, sometimes with bisexual flowers in between. *Sepals* 4, round the top of the ovary. *Petals* 4, usually only present in male and bisexual flowers, soon caducous. *Stamens* usually 8, sometimes 4 or 6. *Styles* very short or absent; stigmas 4, persistent. *Ovary* inferior. *Fruit* a schizocarp, separating into 4, 1-seeded mericarps. Anemophilous.

Contains about 45 species, cosmopolitan.

Aiken, S. G., Newroth, P. R. & Wile, I. (1979). The biology of Canadian weeds. 34. *Myriophyllum spicatum* L. *Canad. Jour. Pl. Sci.* **59**: 201–215.
Brennan, J. P. M. & Chapple, J. F. G. (1949). The Australian *Myriophyllum verrucosum* Lindley in Britain. *Watsonia* **1**: 63–70.
Harris, S. A., Maberly, S. C. & Abbott, R. J. (1992). Genetic variation within and between populations of *Myriophyllum alterniflorum* DC. *Aquatic Bot.* **44**: 1–21.
Hultén, E. & Fries, M. (1986). *Atlas of north European vascular plants north of the Tropic of Cancer.* 3 vols. Königstein.
Milner, J. M. (1979). *Myriophyllum aquaticum* (Velloso) Verdc. in East Sussex. *Watsonia* **12**: 259.
Orchard, A. E. (1981). A revision of the South American *Myriophyllum* (Halagoraceae) and its repercussions on some Australian and North American species. *Brunonia* **8**: 173–291.
Pearsall, W. H. (1934). The British species of *Myriophyllum*. *Rep. Bot. Soc. Exch. Club Brit. Isles* **10**: 619–621.
Praeger, R. L. (1938). A note on Mr Pugsley's *Myriophyllum alterniflorum* var. *americanum*. *Jour. Bot. (London)* **76**: 53–54.
Preston, C. D. & Croft, J. M. (1997). *Aquatic plants of Britain and Ireland.* Colchester.
Preston, C. D. (1998). in Rich, T. C. G. & Jermy, A. C. *Plant Crib*, pp. 192–193. London.
Pugsley, H. W. (1938). A new variety of *Myriophyllum alterniflorum* DC. *Jour. Bot. (London)* **76**: 51–53.
Schindler, A. K. (1905). Halorrhagaceae. In Engler, A. (Edit.) *Das Pflanzenreich*: **V.255**. Leipzig.
Stewart, A., Pearman, D. A. & Preston, C. D. (1994). *Scarce plants in Britain*. Peterborough. [*M. verticillatum.*]
Walsh, H. (1944). *Myriophyllum heterophyllum* Michx and M. spicatum L. in the Halifax Canal. *Naturalist (Hull)* **1944**: 143–144.
Weber, J. A. & Noodén, l. D. (1974). Turion formation and germination in *Myriophyllum verticillatum* phenology and its interpretation. *Michigan Bot.* **13**: 151–158.
Weber, J. A. & Noodén, l. D. (1976). Environmental and hormonal control of turion germination in *Myriophyllum verticillatum*. *Amer. Jour. Bot.* **63**: 936–944.

1. Stamens 4 — **6. heterophyllum**
1. Stamens 8 — 2.
2. Uppermost bracts deeply serrate to pinnatisect — 3.
2. Uppermost bracts simple, entire or minutely serrate — 4.
3. Emergent leaves not glaucous, with sparse sessile glands — **1. verticillatum**
3. Emergent leaves glaucous, with dense, translucent, sessile glands — **2. aquaticum**
4. Submerged leaves usually 3 in a whorl; flowers usually solitary — **3. verrucosum**
4. Submerged leaves (3–)4(–5) in a whorl; flowers usually more than one together — 5.
5. Flower spike 5–15 cm — **4. spicatum**
5. Flower spike 1–2(–3) cm — 6.
6. Leaves 10–25 mm, the segments 6–20 mm — **5(i). alterniflorum** var. **alterniflorum**
6. Leaves 3–5 mm, the segments 2–4 mm — **5(ii). alterniflorum** var. **americanum**

1. M. verticillatum L. Whorled Water-milfoil

Perennial monoecious, aquatic *herb*, with an elongated rhizome creeping in the muddy substratum; perennation and vegetative reproduction by clavate turions. *Leafy shoots* 50–300 cm, branched. *Leaves* (4–)5(–6) in a whorl; lamina 25–45 mm, often longer than the internodes, simply pinnatisect, with 25–35, rather distant segments, the emergent ones with sparse, sessile glands.

Inflorescence an emergent spike 7–25 cm; flowers usually in whorls of 5 in the axils of shortly pinnate or pectinate bracts of very variable length, from a little shorter than the leaves to a little longer than the flowers, but never entire and never shorter than the flowers even at the tip of the spike, the flowers unisexual or bisexual. *Male flowers* with 4, triangular, dentate, minute sepals; 4, obovate, greenish-yellow or reddish, caducous petals; and 8 stamens, the filaments pale, the anthers yellow, and a rudimentary ovary. *Female flowers* with 4 sepals, usually without petals and with 4 subsessile, feathery, persistent stigmas. *Bisexual flowers* with all parts including petals present usually occur between the male and female flowers. *Schizocarp* about 2 mm, subglobose, 4-lobed, at length separating into 4, 1-seeded mericarps. *Flowers* 7–8. $2n=28$.

Native. Clear or slightly turbid, still or slowly flowing, calcareous water in lakes, streams, canals and ditches over both peaty and inorganic substrates. Scattered over England, the Welsh border area and mainly central and northern Ireland. Has steadily decreased over the last century. Widespread in the temperate regions of the northern hemisphere. Circumpolar Temperate element.

2. M. aquaticum (Vell.) Verdc. Parrot's-feather

Enydria aquatica Vell.; *M. brasiliense* Cambess.; *M. proserpinacoides* Gillies ex Hook. & Arn.

Perennial dioecious, aquatic *herb* often woody at the base and rhizomatous. *Leafy shoots* up to 200 cm, submerged and emergent, ascending, glaucous, glabrous, rooting at the lower nodes. *Leaves* (4–)5(–6) in a whorl; lamina (17–)35–45 mm, usually exceeding the internodes, simply pinnatisect with 8–30 segments, the segments up to 7 mm, linear-subulate, rather stiff, the emergent ones glaucous and densely covered with minute, hemispherical glands. *Inflorescence* an emergent spike with solitary, axillary, unisexual flowers, only the female found so far in Great Britain and Ireland; bracts similar to leaves and glaucous. *Female flowers* on a pedicel 0.2–0.4 mm; sepals 4; 0.4–0.5 mm, white, deltoid, denticulate; petals absent; styles 4; ovary 4-ribbed. *Male flowers* with 4 sepals; petals 4, about 5 mm, white; stamens 8, and a rudimentary ovary. *Schizocarp* 0.8–1.2 mm, ovoid, finely tuberculate, 4-lobed, at length dividing into 4, 1-seeded mericarps. *Flowers* 7–8.

Introduced. Shallow, still or very slowly moving water in a range of lowland habitats such as ponds, reservoirs, flooded gravel pits, streams, canals and ditches. Scattered records in England and Wales, particularly near London and just getting into Scotland and Ireland. Native of the lowlands of central South America and naturalised elsewhere in South America, Africa, Asia, Australia, New Zealand, North and Central America, Hawaii and France and Austria in Europe.

3. M. verrucosum Lindl. Australian Water-milfoil

Perennial monoecious, aquatic *herb*, with a rhizome. *Leafy shoots* up to 50 cm, submerged except for the inflorescence or almost entirely immersed when growing on mud, in submerged plants often more or less defoliated and blackish below, whitish-green above, often rooting at the lowermost nodes and branched from the base upwards, in immersed plants up to 10 cm, pale green to bright reddish-purple, decumbent, slender, profusely branched, especially below and rooting freely from the lower nodes. *Leaves* in submerged plants in whorls of 3(–5); lamina longer than the internodes on barren shoots, less crowded and shorter than the internodes on flowering shoots, simply pinnatisect with 8–18 segments; on immersed plants usually pale green or purple-tinged, ternate to paired and opposite or even alternate, with a broader rhachis and shorter, broader and often fewer segments. *Inflorescence* of submerged plants terminal and elongating up to about 18 cm; bracts 2–8 × 1.2–3.0 mm, the lower pale glaucous green, the upper pale glaucous-pink, broadly ovate to elliptical, obtuse at apex, the lower incise-serrate, the upper more or less minutely serrate or entire; flowers only recorded as bisexual in England, but said to be male to top of inflorescence, female at bottom and bisexual in between in Australia, solitary in each bract axil, 1–3 per node; bracteoles 2, about 0.7–1.0 mm, whitish, narrowly ovate to oblong, more or less serrate. *Sepals* 4, almost 0.4 mm, whitish, oblong or more or less triangular, more or less serrate. *Petals* 4, except in female flowers, about 2.7 mm, pinkish-green, obovate, entire, hooded at apex. *Stamens* 8; filaments pale; anthers yellow. *Styles* 4, almost smooth a first, later clothed with stigmatic hair-like papillae towards apex. *Ovary* inferior, cup-shaped, 4-sulcate. *Schizocarp* up to 1.5 × 1.7 mm, broadly ovate-truncate, bluntly 4-lobed, with persistent styles, ultimately separated into 4, 1-seeded mericarps; mericarps prominently, bluntly and longitudinally ridged on back, minutely and bluntly tubercled on the outer face, purplish-red on faces and cream to grey on ridges. *Flowers* 6–9.

Introduced. Gravel pits near Eaton Socon in Bedfordshire between 1944 and 1946. Probably introduced with wool shoddy. Native of Australia.

4. M. spicatum L. Spiked Water-milfoil

Perennial monoecious, aquatic *herb* with rhizomes. *Leafy shoots* 50–250 cm, branched, naked below through decay of leaves. *Leaves* (3–)4(–5) in a whorl; lamina up to 35 mm, simple pinnatisect with 13–35 segments, glabrous. *Inflorescence* an emergent spike for 5–15 cm, erect throughout even in bud; flowers usually in whorls of 4 in the axils of bracts, all but the lowest of which are entire and shorter than the flowers, the lowest usually pectinate and somewhat larger than the flowers; about 4 basal whorls are of female flowers, then 1 whorl of bisexual flowers and the upper whorls of male flowers; bracts ovate, obtuse at apex, minutely dentate; bracteoles ovate, acuminate at apex, toothed. *Male flower* with 4 sepals, about 0.3 mm, erect and broadly triangular, acuminate at apex. *Petals* 4, about 2.5 mm, pink or dull red, obovate and rounded at apex. *Stamens* 8, and a rudimentary ovary. *Female flowers* with 4 sepals, petals absent, styles 4, very short and stigmas plumed. *Bisexual flowers* with all parts including petals. *Schizocarp* about 3.0 mm, suborbicular, 4-lobed, at length dividing into 4, 1-seeded mericarps. *Flowers* 6–7. $2n = 28$.

Native. Meso-eutrophic or eutrophic and often calcareous waters of lakes, ponds, rivers, canals and ditches. Throughout Great Britain and Ireland except in the mountains of the north and Wales. In the broad sense the species is widespread in temperate regions of the northern hemisphere extending south to North Africa, the Himalaya and Japan. The native American populations are sometimes treated as a distinct species, *M. exalbescens* Fernald. The European plant has also been introduced in North America. Temperate element or Circumpolar Temperate element.

5. M. alterniflorum DC. Alternate Water-milfoil

Perennial monoecious, aquatic *herb* with a rhizome. *Leafy shoots* 20–120 cm, naked below through decay of older leaves. *Leaves* usually (3–)4 in a whorl; lamina 3–25 mm, about equalling the internodes, simply pinnatisect with 6–18 segments 2–20 mm. *Inflorescence* an emergent spike 1–2(–3) cm, its tip drooping in bud; basal whorl of flowers female in the axils of leaf-like pinnatisect bracts; then other female flowers, solitary or in groups of 2–4 in the axils of short pectinate bracts; next bisexual flowers; and in the upper half of the spike male flowers, usually about 6, solitary or in opposite pairs in the axils of entire bracts which are shorter than the flowers. *Sepals* 4, 0.2 mm. *Petals* 4, about 2.5 mm, yellow with red streaks, obovate, hooded at apex. *Stamens* 8; filaments pale; anthers yellow. *Stigmas* 4, plumed. *Schizocarp* 1.5–2.0 mm, oblong, separating into 4, 1-seeded mericarps. *Flowers* 5–8. $2n = 14$.

(i) Var. **alterniflorum**
Leaves 10–25 mm, the segments 6–20 mm.

(ii) Var. **americanum** Pugsley
Leaves 3–5 mm, the segments 2–4 mm.

Native. Standing and flowing waters, including rapidly flowing, peaty streams and rivers in which few other macrophytes grow. Throughout much of Great Britain and Ireland but more common in the west. In Scotland and Ireland it occurs in a wide range of habitats including calcareous sites, but in south-east England it is confined to acidic, mesotrophic or oligotrophic waters. The most widespread plant is var. *alterniflorum*. Var. *americanum* occurs around the shores of Lough Beg and Lough Neagh in Ireland in shallow water over sandy substrate in apparent absence of the type, and in the Hebrides. The species occurs in boreal and temperate zones of the northern hemisphere and North Africa, eastern and western North America and Greenland. The common American plant is said to be var. *americanum*. Boreo-temperate element.

6. M. heterophyllum Michx American Water-milfoil
Potamogeton verticillatum Walter

Perennial monoecious, aquatic *herb* with a rhizome. *Leafy shoots* up to 100 cm, the leaves rather sparse. *Leaves* usually in whorls of 4 or 6; lamina of submerged leaves 20–50 mm with 5–12 segments; emergent leaves lanceolate to elliptical, entire or toothed; leaves in the transitional region pinnately cut. *Inflorescence* an emergent spike 3–35 cm, with flowers in whorls of 4 or 6, the upper male, the lower female and the intermediate bisexual; bracts lanceolate, entire or dentate. *Sepals* 4, 0.5–0.7 mm, triangular, acuminate at apex, serrulate. *Petals* 4, about 1.5 mm, boat-shaped. *Stamens* 4, filaments pale; anthers yellow. *Stigmas* 4, diverging. *Schizocarp* 1.0–1.5 mm, subglobose, separating into 4, 1-seeded, beaked mericarps with the outer face 2-ridged and minutely papillose. *Flowers* 6–9.

Introduced. Discovered in a lowland canal between Halifax and Salterhebble in Yorkshire in 1941, where it stayed until the canal was drained in 1947 or 1948. Native of North America where it extends from south-west Quebec, Ontario and North Dakota south to Florida and New Mexico.

91. GUNNERACEAE Meisn. nom. conserv.

Huge *perennial herbs*. *Stems* wholly rhizomatous. *Leaves* alternate, but clustered, rhubarb-like, simple, palmately 5- to 9-lobed, with jagged, serrate lobes, long, stout petioles and numerous, large, jagged intravaginal scales looking like stipules. *Inflorescence* a huge, compound, erect, catkin-like panicle, usually with male, female and bisexual flowers mixed, flowers epigynous. *Sepals* 2, minute. *Petals* 2, small, or absent. *Stamens* (1–)2. *Style* more or less absent; stigmas 2, linear. *Fruit* a small drupe.

Contains a single genus with about 50 species.

1. Gunnera L.

As family.

Contains about 50 species in Central and South America, South and south-east Africa, Madagascar, East Indies, New Guinea, Tasmania and New Zealand.

As regards the species of *Gunnera* every book seems to have descriptions different from every other book. The following two species have been seen in gardens. Specimens of *G. manicata* have been seen from Cornwall and Co. Kerry. We do not know what most records refer to, nor do we know if the names applied to the two species here are the correct ones. *Gunnera* species occur in damp rough grassland, woodland, shady places near lakes and rivers and on sheltered sea-cliffs in scattered localities in Great Britain and Ireland, particularly near the coast.

Reynolds, S. C. P. (2002). *A catalogue of alien plants in Ireland*. Glasnevin.
Schindler, A. K. (1905). Halorrhagaceae. In Engler, A. (Edit.) *Das Pflanzenreich* **IV.225**. Leipzig.

1. Leaves palmately lobed, the lobes rounded to an acute apex; petioles with numerous small spines which give a reddish appearance **1. tinctoria**
1. Leaves pedately lobed, the lobes narrowed towards an acute apex; petioles with numerous small spines which although having a small red tip make the whole appear green **2. manicata**

1. G. tinctoria (Molina) Murb. Giant Rhubarb
Panke tinctoria Molina; *G. chilensis* Lam.

Huge *perennial* monoecious *herb* looking like a giant rhubarb, with many large leaves covering the compound spikes and a huge rhizome covered with debris. *Stems* wholly rhizomatous. *Leaves* alternate, but clustered; lamina 90–150 cm in diameter, dull yellowish-green on upper surface and very rough, slightly paler beneath, rotund to reniform in outline, palmately 5- to 13-lobed, the lobes rounded to a point, margin of lobes with larger acuminate teeth and smaller teeth in between, all ending in short yellow spines, base of leaf cordate, upper surface rough with papillae, lower surface with veins covered with prickles which are often brownish; veins strongly impressed on upper surface, very prominent beneath; petiole up to 180 cm, with dense short prickles with a red tip which turns brownish, making the whole look reddish-brown; intravaginal scales which look like large laciniate and filamentous stipules adnate to petiole. *Inflorescence* a large, erect panicle with a central rhachis and dense catkin-like branches of male, female and bisexual flowers. *Sepals* 2. *Petals* absent. *Stigmas* 2. *Drupe* red. *Flowers* 5–6. $2n=34$.

Introduced. Grown in gardens and parks and sometimes escaping. Scattered records in Great Britain and particularly in the west and south-west of Ireland. Native of Chile.

2. G. manicata Linden ex Andre
Brazilian Giant Rhubarb

Huge *perennial* monoecious *herb* up to 4 m, looking like a giant rhubarb, with many large leaves covering great compound spikes and a huge rhizome covered with debris. *Stems* wholly rhizomatous. *Leaves* alternate, but clustered; lamina 90–200 cm in diameter, dull medium rather yellowish-green on upper surface, slightly paler beneath, subrotund to reniform in outline, pedately 5- to 13-lobed, the lobes ovate and gradually narrowed to an acute, spine-tipped apex, margin of lobes has large, acuminate teeth and smaller teeth in between, both kinds of teeth ending in short yellow spines, base of leaf cordate, upper surface covered with small papillae making it rough, lower surface with short hairs on the veins; veins strongly impressed on upper surface, very prominent beneath; petiole up to 2.4 m and 5 cm thick, pale green, covered with short green spines with a minute red tip, but the whole looks green; intravaginal scales which look like large laciniate and filamentous stipules adnate to petiole. *Inflorescence* a large, erect panicle with a central rhachis and dense catkin-like branches of male, female or bisexual flowers. *Sepals* 2, minute. *Petals* 2, small, free. *Stamens* 2; filaments very short, pale green; anthers red, turning brown. *Stigmas* 2, pale or slightly pink, plumed with glandular hairs. *Drupe* small, rounded, slightly fleshy, reddish-green. *Flowers* 5–6. $2n=34$.

Introduced. Grown in gardens and parks and sometimes escaping. Native of Brazil and Columbia.

Order 5. MYRTALES Lindl.

Herbs or *shrubs*. *Leaves* opposite, alternate or in whorls; exstipulate. *Flowers* unisexual or bisexual, actinomorphic or zygomorphic, perigynous or epigynous. *Hypanthium* often prolonged well beyond the ovary. *Sepals* often round top of hypanthium. *Petals* alternating with the sepals. *Stamens* numerous. *Style* 1. *Fruit* variable.

Contains about 12 families and 9,000 species.

92. LYTHRACEAE St-Hil. nom. conserv.

Annual or *perennial* monoecious *herbs*. *Leaves* opposite or in whorls of 3 or alternate, simple, entire, sessile or petiolate, without stipules. *Inflorescence* of solitary flowers in the leaf axils towards the apex of the stem, or in clusters; flowers bisexual, perigynous, actinomorphic, monomorphic or trimorphic. *Hypanthium* tubular to funnel- or cup-shaped. *Epicalyx* segments 4–6, on the apex of the hypanthium. *Sepals* (4–)6, borne on the hypanthium. *Petals* 0–6, borne near the apex of the hypanthium. *Stamens* usually 6 or 12, sometimes fewer. *Style* 1; stigma 1, capitate. *Ovary* 2-celled, each cell with many ovules on axile placentas. *Fruit* a capsule, opening by 2 valves.

Contains 25 genera and about 550 species, throughout the world except for cold regions.

1. Lythrum L.

Peplis L.

As family.

About 35 species, worldwide.

Allen, D. E. (1954). Variation in *Peplis portula* L. *Watsonia* **3**: 85–91.
Callaghan, D. A. (1996). The conservation status of *Lythrum hyssopifolia* L. in the British Isles. *Watsonia* **21**: 179–186.
Darwin, C. (1865). On the sexual relations of the three forms of *Lythrum salicaria*. *Jour. Linn. Soc.London (Bot.)* **8**: 169–196.
Hultén, E. & Fries, M. (1986). *Atlas of north European vascular plants north of the Tropic of Cancer*. 3 vols. Königstein.
Mal, T. et al. (1991). The biology of Canadian weeds. 100. *Lythrum salicaria*. *Canad. Jour. Pl. Sci.* **72**: 1305–1330.
Preston, C. D. (1989). The ephemeral pools of south Cambridgeshire. *Nat. Cambridgeshire* **31**: 2–11. [*L. hyssopifolia*.]
Preston, C. D. & Whitehouse, H. L. K. (1986). The habitat of *Lythrum hyssopifolia* L. in Cambridgeshire, its only surviving English locality. *Biological Conservation* **35**: 41–62.
Preston, C. D. & Croft, J. M. (1997). *Aquatic plants in Britain and Ireland*. Colchester. [*L. portula*.]
Sell, P. D. (1967). *Lythrum portula* (L.) D. A. Webb. *Watsonia* **6**: 296.
Wigginton, M. J. (Edit.) (1999). *British red data books*. Vol. 1. *Vascular plants*. Ed. 3. Peterborough. [*L. hyssopifolia*.]

1. Petals 5–10 mm; stigma or some stamens exceeding sepals 2.
1. Petals up to 3 mm or absent; stigma and stamens not reaching apex of sepals 3.
2. Flowers clustered in whorls; petals 8–10 mm **1. salicaria**
2. Flowers 1–2 in each axil; petals 5–6 mm **2. junceum**
3. Leaves linear to oblong; hypanthium tubular; capsule cylindrical **3. hyssopifolium**
3. Leaves obovate-spathulate; hypanthium funnel or cup-shaped; capsule subglobose 4.

4. Epicalyx segments up to 0.5 mm **4(a). portula** subsp. **portula**
4. Epicalyx segments 1.5–2.0 mm **4(b). portula** subsp. **longidentata**

1. L. salicaria L. Purple Loosestrife
L. spicatum Gray nom. illegit.

Perennial monoecious *herb* with a tufted rootstock. *Stem* up to 150 cm, pale green, often suffused brownish-purple, erect, sharply angled, densely short hairy, little branched, leafy. *Leaves* opposite or whorled; lamina 2.5–8.0 × 1.2–2.2 cm, medium green on upper surface, paler beneath, often suffused dull red, ovate-lanceolate to ovate, more or less acute at apex, entire, cordate at the sessile, semi-amplexicaul base, minutely hairy on both surfaces and the margins, midrib impressed above, prominent beneath. *Inflorescence* of clusters of bisexual flowers in whorls with bracts between the whorls, spike-like and occupying the upper one-third of the plant, sometimes with side branches; bracts lanceolate and hairy. *Hypanthium* 4–5 × 1.5–2.0 mm, pale green, suffused brownish-purple, tubular, ribbed, hairy. *Epicalyx* with 6 lobes, the lobes 2.5–3.0 mm, subulate, acute at apex. *Sepals* 6, about 1.6 mm, triangular-ovate, mucronate at apex. *Petals* 6, 8–10 mm, inserted at the apex of the hypanthium, purple, oblanceolate, with wavy margins. *Stamens* 12; filaments white or purple; anthers yellow or purple. *Style* 1, green; stigma brown. Stamens and style arranged in 3 different ways: style short and stamens long and medium, style medium with stamens long and short, or style long with stamens short and medium (Darwin, 1865). *Capsule* 3–4 mm, oblong-obovoid, enclosed in the hypanthium; seeds about 1.1 mm, oblanceolate, rounded at apex, cuneate at base. *Flowers* 6–8. Visited by bees and hoverflies. $2n = 60$.

(i) Var. **salicaria**
Leaves from nearly glabrous to with many hairs beneath, but never completely covering under surface.

(ii) Var. **tomentosum** DC.
L. salicaria var. *cinereum* Dumort.; *L. salicaria* var. *canescens* W. D. J. Koch; *L. salicaria* var. *pubescens* Coss. & Germ., non DC.

Leaves densely hairy beneath covering the whole surface.

Native. Along the margins of water and in marshes and fens. Common throughout most of Great Britain and Ireland except north Scotland. Almost throughout Europe; Asia; North Africa; naturalised in North America. Eurasian Temperate element, but naturalised in North America so now Circumpolar Temperate. Our common plant is var. *salicaria*. Specimens collected in St Ouens, Jersey in 1842 and Branscombe, Devon in 1839 are var. *tomentosum*. this variety also occurs across southern Europe and may be a distinct race. Very tall bright-flowered plants of unknown origin escape from gardens and are also found in wild flower seed.

2. L. junceum Banks & Sol. False Grass-poly
L. graefferi Ten.; *L. flexuosum* auct.

Annual to *perennial* monoecious *herb* with fibrous roots. *Stems* up to 70 cm, pale green, often tinted reddish-purple, erect to decumbent, often much branched especially near the base, the branches long, lax and straggling, subterete or distinctly angled, glabrous, leafy. *Leaves* opposite towards the base of the plant, alternate above; lamina (0.5–)1–2(–5) × (0.2)0.4–0.6 cm, bluish-green on upper surface, paler beneath, narrowly oblong, rounded, obtuse or subacute at apex, entire, shortly and bluntly auriculate at base, glabrous, bisexual, subsessile or very shortly petiolate. *Inflorescence* of solitary flowers in the axils of upper cauline leaves; flowers subsessile and trimorphically heterostylous; bracteoles 2, 1.0–1.5 mm, narrowly subulate and membranous. *Hypanthium* about 7 × 2 mm, narrowly tubular, wide at apex and tapering to base, often tinged reddish. *Epicalyx* with 6 lobes, the lobes 0.7–0.5 mm, deltoid, acute at apex, as long as or longer than the sepals, thickish, opaque, often reddish. *Sepals* 6, broadly deltoid, membranous. *Petals* 6, inserted at the apex of the hypanthium, 5–6 × about 2 mm, purple, obovate, tapering at the base to a very short claw. *Stamens* 12, included or shortly exserted, attached at different levels towards the middle of the hypanthium; filaments very unequal; anthers yellow. *Style* 1, filiform, varying in length; stigma capitate and papillose. *Capsule* about 5 × 1.5 mm, included in the persistent hypanthium, dark brown, cylindrical; seeds about 0.8 mm, shining brown, broadly ovoid, gibbous or strongly convex on the dorsal surface, flattish or slightly convex on the ventral, minutely rugulose. *Flowers* 6–7. $2n = 10$.

Introduced. Bird-seed alien in parks and waste places, sometimes from other sources. Sporadic but frequent in Great Britain, one record for Ireland. Widespread in the Mediterranean region; Atlantic Islands.

3. L. hyssopifolia L. Grass-poly

Annual monoecious *herb* with fibrous roots. *Stems* 6–25 cm, pale green, often tinged reddish, erect or decumbent, terete or somewhat angled, glabrous, branched, leafy. *Leaves* opposite; lamina 5–15 × 1.0–2.5 mm, bluish-green on upper surface, paler beneath, linear or narrowly oblong, obtuse or subacute at apex, entire, not auriculate at base, sessile or subsessile, glabrous. *Inflorescence* of solitary flowers in the axils of the upper cauline leaves; flowers bisexual, subsessile and homostylous; bracteoles 2, 1.0–2.5 mm, subulate, membranous or herbaceous. *Hypanthium* about 4.5 × 2.0 mm, tubular or narrowly funnel-shaped. *Epicalyx* with 4–6 lobes, the lobes about 0.7 × 0.4 mm, thickish, narrowly deltoid and opaque. *Sepals* 4–6, about 0.4 × 0.7 mm, much shorter than epicalyx lobes, membranous. *Petals* 4–6, 2.5–3.0 × about 0.8 mm, attached to the apex of the hypanthium, magenta-pink, oblanceolate. *Stamens* (1–)2–6(–8), included; filaments about 1.5 mm; anthers shortly oblong. *Style* 1; stigma capitate, papillose. *Capsule* 4.0–4.5 × about 0.7 mm, cylindrical, dark brown; seeds about 0.8 mm, medium brown, ovoid, almost equally convex on both surfaces, shortly winged at the distal end, minutely rugulose. *Flowers* 6–7. $2n = 20$.

Native. Disturbed ground flooded during the winter months such as hollows, ruts and other low-lying areas in arable fields and winter-flooded ground disturbed in summer by waterfowl. A population of several hundred thousand plants occurs at Slimbridge in Gloucestershire and it continues to spread. Substantial populations of up to 10,000 plants occur in hollows in arable fields south of Cambridge and smaller populations are known in Dorsetshire, Oxfordshire and Sussex; it also occurs in Jersey and a casual in Ireland. Central and south Europe; temperate Asia; North Africa; probably introduced in America and Australia. Eurosiberian Southern-temperate element.

4. L. portula (L.) D. A. Webb Water Purslane

Peplis portula L.; *Portula palustris* Gray; *Portula diffusa* Moench

Annual monoecious herb with slender, pale, fibrous roots. *Stems* up to 25 cm, pale green to reddish, floating, prostrate or creeping on half dried mud, sometimes matted, rarely erect, bluntly quadrangular, glabrous, branched, leafy, rooting at the nodes. *Leaves* opposite; lamina 10–12 (–20) × 2–10 mm, bright green or reddish on upper surface, paler beneath, obovate-spathulate, rounded at apex, entire, narrowed at base, rather thick, glabrous; stipules minute and gland-like; petiole short. *Inflorescence* of solitary bisexual flowers in each leaf-axil, nearly sessile at first becoming shortly peduncled, with a pair of subulate bracts nearly as long as the epicalyx. *Hypanthium* funnel or cup-shaped. *Epicalyx* with 6 segments 0.5–2.0 mm, linear-lanceolate, with a long acute apex. *Sepals* 6, triangular-ovate, apiculate at apex. *Petals* 0.7–0.8 × 1.0–1.2 mm, pink, very broadly elliptical, rounded at apex, with a very short claw. *Stamens* 6 or 12, inserted below the petals; filaments usually inflexed in bud. *Style* 1. *Capsule* about 1.5 mm, subglobose; seeds about 0.7 mm, obovoid. *Flowers* 6–10. Usually self-pollinated. $2n=10$.

(a) Subsp. **portula**

Epicalyx segments up to 0.5 mm.

(b) Subsp. **longidentata** (Gay) P. D. Sell

Peplis portula var. *longidentata* Gay; *Peplis portula* subsp. *longidentata* (Gay) Nyman; *Peplis longidentata* (Gay) Batt.; *Peplis fradinii Pomel*

Epicalyx segments 1.5–2.0 mm.

Native. Open or bare ground, by or in water, or in damp trackways, occasionally in water up to 1 m deep. It is a distinct calcifuge which is typically found over base-poor mineral soils, but avoids the most acidic and nutrient-poor soils. Scattered throughout most of Great Britain and Ireland, but has decreased in some areas, particularly south-east England. It is primarily a European species extending from Scandinavia to southern Europe and occurs also in North Africa and west Asia; introduced in California, central and south America and New Zealand. The subspecies are geographically divided, subsp. *portula* being widespread in Europe and extending to south Italy, Sardinia and Corsica, while subsp. *longidentata* is western in its distribution being recorded from France, Portugal, Spain, Algeria, the Azores and western Great Britain and Ireland. Intermediates occur in the area between the two subspecies and they do not seem to have a different ecology. The species is a European Temperate element.

93. THYMELAEACEAE Juss. nom. conserv.

Evergreen or deciduous, monoecious, early flowering, poisonous *shrubs*. *Leaves* usually alternate, simple or entire, sessile or shortly petiolate; exstipulate. *Inflorescence* a terminal or axillary raceme or cluster; *flowers* bisexual, perigynous, actinomorphic or more or less so; basically cup-shaped, the hollowed-out receptacle (*hypanthium*) is green or brightly coloured and forms a deep tube with the floral parts mostly arranged at the rim. *Sepals* 4, green or coloured like the hypanthium. *Petals* absent in our species, but present though inconspicuous in other members of the family. *Stamens* 8, inserted near the top of the inside of the hypanthium. *Stigma* 1, large and capitate, the style absent or short. *Ovary* 1-celled, with 1 apical ovule. *Fruit* a 1-seeded drupe; endosperm sparse.

Contains about 500 species in 58 genera, worldwide but especially well represented in Africa. Both our species belong to the Subfamily *Thymelaeoideae*.

1. Daphne L.

Mezereum C. A. Mey.

As family.

Contains about 70 species in Europe, North Africa and Australia.

Daphne is the Greek name for the Bay Tree or Laurel (*Laurus nobilis*), the name being later transferred to the present genus. According to mythology it was named after a nymph changed by the gods into a Bay Tree to save her from the pursuit by Apollo, but the name itself may be derived from an Indo-European root meaning 'odour' (fide W. T. Stearn).

Brickell, C. D. & Mathew, B. (1976). Daphne. The genus in the wild and in cultivation. Alpine Garden Society Publication. Surrey.

Hultén, E. & Fries, M. (1986). *Atlas of north European vascular plants north of the Tropic of Cancer*. 3 vols. Königstein.

Marshall, E. S. (1903). West Sussex plant notes for 1902. Jour. Bot. (London) **41**: 227–232.

Marshall, E. S. (1910). *Daphne laureola* × *mezereum* in N. Somerset. Jour. Bot. (London) **48**: 79.

Stewart, A., Pearman, D. & Preston, C. D. (1994). Scarce plants in Britain. Peterborough. [D. mezereum.]

1. Evergreen shrub; leaves thick, dark green; flowers (hypanthium) yellowish-green, in the axils of the persistent leaves; drupe black **2. laureola**
1. Deciduous shrub; leaves thin, pale green; flowers (hypanthium) pinkish-purple or rarely white, appearing before the leaves; drupe bright red or yellow 2.
2. Flowers pinkish-purple; drupe bright red **1(1). mezereum** forma **mezereum**
2. Flowers white; drupe yellow **1(2). mezereum** forma **alba**

1. D. mezereum L. Mezereon
Mezereum officinerum C. A. Mey.; D. *florida* Salisb.

Deciduous, monoecious, poisonous *shrub*. *Stems* up to 200 cm, pale brown, smooth, woody, with erect branches from the base. *Twigs* short and erect, pale and heavily brown-spotted, glabrous. *Leaves* mainly in the top half of the plant; lamina 3–10×0.8–2.5 cm, thin, pale green on upper surface, with midrib impressed, pale green beneath with prominent midrib, oblanceolate or oblong-oblanceolate, more or less acute at apex, entire, gradually narrowed at base to a short petiole, glabrous or ciliate. *Inflorescence* with flowers appearing before the leaves in a subsessile cluster in the axils of a fallen leaf of the previous year and forming intercalary spikes; bracteoles reddish, ovate, acute at apex; flowers 8–12 mm in diameter, bisexual, with a heavy, cloying scent. *Hypanthium* 5–10 mm, usually pinkish, rarely white, purple, tubular, appressed-hairy. *Sepals* 4, on top of the hypanthium, 4–5×4–5 mm, unequal, broadly ovate, apiculate at apex, spreading but not flat. *Petals* absent. *Stamens* 8, inserted near the top of the hypanthium; filaments very short; anthers brownish-yellow. *Stigma* 1, pale yellow, capitate, sessile. *Ovary* 1-celled. *Drupe* usually bright red, rarely yellow, ovoid. *Flowers* 2–4. Pollinated by early-flying Lepidoptera and long-tongued bees. $2n=18$.

(**1**) Forma **mezereum**
Flowers pinkish-purple. *Drupe* bright red.

(**2**) Forma **alba** (Aiton) Schelle
D. mezereum var. *alba* Aiton
Flowers white. *Drupe* yellow.

Native. Ancient and recent secondary woodland on calcareous soils, chalk pits and wet, rich fens. It usually fails to flower if a dense canopy develops. Mainly in south and central England where it was once more common and in a few localities north to Lancashire and Yorkshire. Widely grown in gardens in both forms, from which it escapes and becomes naturalised which makes its native distribution difficult to assess. Widespread in Europe and Asia from Spain east to the Altai Mountains and naturalised as a garden escape in North America. It is poisonous to humans and pigs. Eurosiberian Boreo-temperate element.

2. D. laureola L. Spurge Laurel
D. *sempervirens* Salisb. nom. illegit.

Evergreen, monoecious, poisonous *shrub*. *Stems* 40–100 cm, pale brown, erect, woody, smooth, usually branched only in upper half. *Twigs* green, glabrous. *Leaves* usually only in the top half of the plant; lamina 4–12×1.5–3.0 cm, thick, dark glossy green on upper surface, midrib impressed and laterals not visible, pale green beneath with midrib even paler yellowish-green and prominent, oblanceolate or obovate-lanceolate, subacute or acute at apex, entire, gradually narrowed at base to a short petiole, glabrous. *Inflorescence* a short, axillary, 5- to 10-flowered raceme, in clusters among the leaves at the top of the plant; peduncle 10–15 mm, green; pedicels 1–3 mm, green; bracteoles obovate-lanceolate, obtuse at apex, entire, caducous; *flowers* bisexual, faintly honey fragrant. *Hypanthium* 8–9 × 2.0–2.5 mm, pale yellowish-green, tubular, glabrous. *Sepals* on top of the hypanthium, 4, 2 × 2 mm, triangular-ovate, pale yellowish green, obtuse at apex, glabrous. *Petals* absent. *Stamens* 8, inserted near the top of the hypanthium; filaments pale green; anthers yellow, included. *Stigma* 1, yellow, capitate, sessile or nearly so. *Ovary* 1-celled. *Drupe* 10–13 mm, black, ellipsoid. *Flowers* 1–4. Pollinated by Lepidoptera and bumblebees. $2n=18$.

Native. Woods, mostly on calcareous or clayey soils. Locally frequent in England north to Cumberland and Durham; North Wales, Pembrokeshire, Glamorganshire and the Channel Islands; introduced in Scotland, Isle of Man and eastern Ireland. West, south-central and south Europe northwards to Hungary; south-west Asia; rare in North Africa; Azores. Our plant is subsp. **laureola** which occurs throughout the range of the species. The species is a Submediterranean-Subatlantic element.

× **mezereum** = **D.** × **houtteana** Lindl. & Paxton
Intermediate between the parents in most characters; the leaves appear after the flowers, are usually deciduous, but sometimes evergreen and are thicker in texture and more shining than in *D. mezereum*; the flowers are mostly in threes and less markedly terminal than in *D. laureola*; the hypanthium and sepals are whitish-green, often tinged with red outside and glabrous as in *D. laureola*. Sterile.

Native. There are records for Somersetshire (1907), Sussex (1902) and Yorkshire (1954). Widely grown in gardens, but endemic.

94. MYRTACEAE Juss. nom. conserv.

Usually evergreen, rarely deciduous, monoecious or dioecious *trees* or *shrubs* with abundant, scattered secretory cavities containing aromatic oils. *Leaves* opposite, or alternate, simple, exstipulate. *Flowers* solitary or in terminal or axillary clusters. *Calyx* 4- or 5-lobed. *Corolla* 4- or 5-lobed, sometimes united to form a bud-like cap which is shed when the flower opens. *Stamens* numerous, free or in bundles, usually bent inwards in bud. *Style* 1; stigma 1, capitate. *Ovary* inferior or half-inferior; 2- to 5-celled; ovules 2–numerous on axile placentas. *Fruit* a capsule or berry; seeds few to many.

Contains about 155 genera and 3,850 species from tropical and subtopical areas, mainly America and Australia.

1. Leaves opposite; ovary inferior; fruit a fleshy berry 2.
1. Leaves usually alternate when plant is mature, sometimes opposite when young; fruit a woody capsule 3.
2. Calyx 4-lobed; petals 4; ovary 2-celled **1. Luma**
2. Calyx 5-lobed; corolla 5-lobed; ovary 3-celled **2. Ugni**
3. Leaves more than 2 cm; calyx and corolla each or together united to form a bud-cap **3. Eucalyptus**
3. Leaves less than 2 cm; calyx and corolla with free lobes **4. Leptospermum**

1. **Luma** A. Gray

Amomyrtus auct.

Evergreen, monoecious *trees* or *shrubs*. *Leaves* opposite, dense, leathery, simple and entire, exstipulate. *Flowers* 1–3 in leaf axils, bisexual. *Calyx* 4-lobed. *Petals* 4, free, white. *Stamens* numerous. *Style* 1. *Ovary* inferior, 2-celled; each cell with 6–14 ovules. *Fruit* a fleshy berry, with 1–16 seeds.

Contains 2 species in Chile and Argentina.

Landrum, L. R. (1988). The Myrtle family (Myrtaceae) in Chile. *Proc. Calif. Acad. Sci.* **45**: 277–317.

1. L. apiculata (DC.) Burret Chilean Myrtle
Eugenia apiculata DC.; *Murceugenia apiculata* (DC.) Nied.; *Myrtus luma* auct.; *Amomyrtus luma* auct.

Evergreen, monoecious *tree* or *shrub* rarely to 18 m, with a conical or columnar crown, but broader when there are several trunks. *Trunk* up to 40 cm in diameter, through much of the tree. *Bark* orange or cinnamon coloured, flaking, developing white streaks in old trees. *Branches* erect or ascending; twigs ash-grey, stout, glabrous; young shoots reddish-brown, slender, with pale, curved hairs. *Buds* in pairs, opposite, 1–2 mm, conical, pointed at apex; scales green or dark red, ovate, pointed at apex, hairy. *Leaves* opposite, dense and often curved upwards; lamina 1.5–4.5 × 0.5–3.5 cm, dark, leathery, more or less shiny green on upper surface, pale, shiny green beneath, broadly elliptical to subrotund, tapering to a short, acute, apiculate apex, entire, cuneate or rounded at base, with scattered glands and an indented midrib on upper surface, with scattered glands and numerous eglandular hairs on the margin, a prominent, hairy midrib and scattered glands below; petiole up to 2 mm, pink, channelled above and hairy. *Flowers* bisexual, 1–3 in leaf axils, 20–30 mm in diameter, globose in bud, opening cup-shaped. *Calyx* green, 4-lobed, the lobes ovate-triangular. *Petals* 4(–5), 3–5 mm, white or pink-tinged, cup-shaped, hairy on margins. *Stamens* numerous; filaments white; anthers yellow. *Style* 1; stigma purplish-pink. *Berry* 6–10 mm, dark purple, globose; seeds 1–16. *Flowers* 8–9.

Introduced. Thriving only in very mild areas, but self-sown in semi-natural woodland. South-west England, Guernsey, south-west Ireland and Isle of Man. Native of the temperate rain forests in central Chile and the adjoining regions of Argentina.

2. **Ugni** Turcz.

Evergreen, monoecious *shrubs*. *Leaves* opposite, leathery, elliptical to lanceolate, exstipulate. *Flowers* solitary, bisexual, nodding, arising from the axil of a leaf or bract. *Calyx* usually with 5 lobes. *Corolla* usually 5-lobed. *Stamens* numerous. *Style* 1. *Ovary* usually 3-celled. *Fruit* a fleshy berry.

Depending on taxonomy from 5 to 15 species from tropical to warm America. *Ugni* is a native Chilean name.

1. U. molinae Turcz. Molina's Myrtle
Eugenia ugni Hook. & Arn.; *Myrtus ugni* Molina

Evergreen, monoecious *shrub* up to 2 m. *Young stems* erect, with whitish hairs. *Leaves* opposite; lamina 1.4–3.6 × 1–2 cm, dark green on upper surface, paler beneath, lanceolate or elliptical, leathery, acute to acuminate at apex, entire, narrowed below, glabrous or with a few scattered hairs; midrib impressed above, prominent beneath; petiole 2–4 mm, shallowly channelled above. *Inflorescence* of numerous, solitary, nodding flowers, borne in the axils of leaves or bracts; flowers bisexual, fragrant. *Calyx* with 5 lobes, the lobes 2–5 × 1.0–2.5 mm, triangular. *Corolla* campanulate, 5-lobed, the lobes 5–8 mm, pink, subrotund, fleshy. *Stamens* numerous, not exceeding the corolla; filaments 2–4 mm. *Style* 1. *Berry* about 10 mm in diameter, dark reddish-brown, subglobose; seeds pale brown. *Flowers* 6–8.

Introduced. Garden escape, seedlings recorded from woodland in Channel Islands and on Tresco, Isles of Scilly. Native of Chile and Argentina. Named after Juan Ignazio Molina (1740–1829).

3. **Eucalyptus** L'Hér.

Evergreen or deciduous, usually monoecious *trees* and *shrubs*. *Bark* fibrous, stringy, fleshy or tassellated. *Leaves* usually of different shapes in seedling, juvenile, intermediate or adult phases; when juvenile often opposite and often more than 3 cm wide; when adult usually alternate, lanceolate, ovate or falcate. *Flower bud* solitary or in small umbels, usually axillary and bisexual. *Hypanthium* obconical, campanulate, cylindrical or urceolate, adnate to the ovary at the base or rarely to the top, truncate, entire, smooth or ribbed or with 4 minute teeth; the orifice closed by a hemispherical, conical or elongated operculum covering the stamens in bud and falling off entire when the stamens expand; the operculum, usually simple, was formed of the connate petals, thin or more rarely thick, fleshy or woody, the separation from the hypanthium usually, but not always, marked in the bud by a distinct line; there is also frequently in the very young bud a very thin external operculum more continuous with the hypanthium, and very rarely this external one persists nearly as long as the internal one and is as thick or nearly so. *Stamens* numerous, in 2 or several irregular rows, free or very rarely united at the extreme base into 4 clusters; anthers versatile or adnate, the cells parallel and distinct or divergent and confluent at the apex, opening in longitudinal slits, lateral or in terminal pores, the connective obscure or enlarged, as long or nearly as long as the cells, usually bearing a globose or ovoid gland on the back or at the apex. *Style* 1, simple, subulate or rarely subclavate, with a small truncate or capitate stigma. *Ovary* inferior, flat, convex or conical, glabrous, 2- to 7-celled, with numerous ovules in each cell in 2–4 rows, on an adnate, oblong and peltate axile placenta. *Fruit* consisting of the more or less enlarged, truncate hypanthium enclosing the capsule, usually of a hard and woody texture and interspersed with resinous receptacles, the persistent disc usually thin and lining the orifice of the hypanthium when the capsule is deeply sunken; concave, horizontal, convex or conically projecting and more or less contracting the orifice when the capsule is not much shorter than or longer than the hypanthium; the capsule thin and woody, always adnate to the hypanthium, although often readily separable from it when quite ripe and dry, very rarely protruding from the orifice left by the disc before maturity, but opening at the apex in as many valves as there are cells which often protrude, especially when acuminate by the persistent and split bases of the style. *Seeds* often abortive, when perfect ovate or globose and compressed, sometimes variously shaped and angular.

Contains over 500 species, mostly in Australia and Tasmania.

Species in this genus are usually maintained by geographical or ecological isolation and where two taxa meet there are often intermediates. When brought together in cultivation there is a much greater chance of them hybridising. No information as to exactly what is grown was found, especially in Ireland. We have thus done our best to give an account of those species in Brooker & Evans (1983).

Blakely, W. F. (1965). *A key to the Eucalypts*. Ed. 3. Camberra.
Brooker, M. I. H. & Evans, J. (1983). *A key to Eucalypts in Britain and Ireland with notes on growing Eucalypts in Britain.* Forestry Commission Booklet **50**. Edinburgh.
Brooker, M. I. H. & Kleinig, D. A. (1983). *Field guide to Eucalypts* **1**. Camberra.
Chippendale, G. M. (1988). *Eucalyptus* L'Her. *Flora of Australia* **19**: 1–543.
Evans, J. (1980). Prospects for eucalypts as forest trees in Great Britain. *Forestry* **53**: 129–143.
Johnson, L. A. S. (1976). Problems of species and genera in *Eucalyptus* (Myrtaceae). *Pl. Syst. Evol.* **125**: 155–167.
Ladiges, P. Y. el al. (1981). Pattern of geographic variation, based on seedling morphology in *Eucalyptus ovata* Labill. and *E. brookerana* A. M. Gray and comparison with some other *Eucalyptus* species. *Austral. Jour. Bot.* **29**: 593–603.
Ladiges, P. Y. el al. (1983). Cladistic relationships and biogeographic patterns in the peppermint group of *Eucalyptus* (informal subseries *Amygdaliniae*, subgenus *Monocalyptus*) and the description of a new species, *E. willisii*. *Austral. Jour. Bot.* **31**: 565–584.
Ladiges, P. Y. et al. (1984). Seedling characters and phylogenetic relationships in the informal series *Ovatae* of *Eucalyptus* subgenus *Symphyomyrtus*. *Austral. Jour. Bot.* **32**: 1–13.
More, D. & White, J. (2003). *Trees of Britain and northern Europe*. London.
Potts, B. M. & Reid, J. B. (1985). Variation in the *Eucalyptus gunnii–archeri* complex. 1. Variation in the adult phenotype. *Austral. Jour. Bot.* **33**: 337–359.
Pryor, L. D. & Johnson, L. A. S. (1971). *A classification of the Eucalypts*. Camberra.
Shaw, M. J. et al. (1984). Variation within and between *Eucalyptus nitida* Hook. f. and *E. coccifera* Hook. f. *Austral. Jour. Bot.* **32**: 641–654.
Wiltshire, R. J. E. & Reid, J. B. (1987). Genetic variation in the Spinning Gum, *Eucalyptus perriniana* F. Muell, ex Rodway. *Austral. Jour. Bot.* **35**: 35–47.

1. Flower buds single 2.
1. Flower buds 3 or more together 3.
2. Shrub or small tree; juvenile leaves small, elliptical, green; bark smooth; adult leaves opposite, less than 5 cm; fruits less than 10 mm wide **22. vernicosa**
2. Tree; juvenile leaves conspicuous, large, ovate, glaucous; bark mostly smooth; adult leaves alternate, sometimes falcate, up to 28 × 4 cm; fruit more than 10 mm wide **21(a). globulus** subsp. **globulus**
3. Flower buds in 3s 4.
3. Flower buds in 7s or more 18.
4. Shrub or small tree; adult leaves opposite, green, less than 5 cm; flower buds and capsule sessile; bark smooth **22. vernicosa**
4. Shrub to tall tree; adult leaves alternate, green, more than 5 cm, or if leaves opposite glaucous 4.
5. Flower buds and capsule on distinct pedicels; peduncles up to 2.5 cm; fruits up to 1.8 × 1.1 cm, urceolate; bark mostly smooth and often pink **30. urnigera**
5. Fruits sessile or with short pedicels, not urceolate; peduncles less than 10 mm 6.
6. Juvenile leaves connate; leaves, flower buds and capsule glaucous; bark smooth **31. perriniana**
6. Juvenile leaves not connate 7.
7. Stem subject to frost-split; bark hanging in ribbons above, smooth or rough below; juvenile leaves narrowly lanceolate, green; adult leaves undulate **25(a). viminalis** subsp. **viminalis**
7. Juvenile leaves ovate to subrotund, green or glaucous 8.
8. Juvenile leaves green 9.
8. Juvenile leaves glaucous or greyish-green 11.
9. Juvenile leaves pale green, entire; adult leaves undulate; bark mostly smooth, white, grey or greenish; fruits hemispherical or cupular, to 8 × 9 mm, smooth **26(a). dalrympleana** subsp. **dalrympleana**
9. Juvenile leaves bright shiny green, crenulate, glandular, thick; bark mostly smooth, pink or pinkish-grey; fruits obconical, hemispherical or campanulate 10.
10. Fruits hemispherical to campanulate, up to 6 × 9 mm; operculum conical **23. subcrenulata**
10. Fruits hemispherical to obconical, up to 8 × 13 mm, often angled; operculum flattened and beaked **24. johnstonii**
11. Operculum flattened 12.
11. Operculum conical or beaked 13.
12. Adult leaves 5–10 × 1–2 cm **12. coccifera**
12. Adult leaves 12–28 × 1.2–4.0 cm **21(b). globulus** subsp. **bicostata** and **21(c). globulus** subsp. **pseudoglobulus**
13. Shrub or small tree with much juvenile, subrotund or ovate, cordate-based glaucous foliage 14.
13. Small to tall trees with a canopy of green adult leaves 16.
14. Bark rough, fibrous and thick **34. cinerea**
14. Bark smooth, sometimes flaking 15.
15. Stems square in section; juvenile leaves crenulate; fruit up to 13 × 13 **32. cordata**
15. Stems not square in section; juvenile leaves entire; fruit up to 9 × 9 mm **33. pulverulenta**
16. Bark rough at base, smooth above; juvenile leaves entire; fruit cylindrical to barrel-shaped, up to 12 × 10 mm; operculum very short **27. glaucescens**
16. Bark mostly smooth, pink or pinkish-grey; juvenile leaves crenulate; cylindrical to obconical, up to 9 × 7 mm 17.

17. Flower buds glaucous; usually cylindrical **28. gunnii**
17. Flower buds not glaucous; cylindrical to obconical **29. archeri**
18. Bark rough at base, smooth above, crown often with ribbons of imperfectly decorticated bark; juvenile leaves green; adult leaves oblique, green; any peduncles in pairs in the axils; fruit obconical **1. regnans**
18. Juvenile leaves green or glaucous 19.
19. Adult leaves with parallel venation 20.
19. Adult leaves with a distinct midrib and side veins at an angle to the midrib, not parallel to the midrib 24.
20. Bark rough over lower half of trunk, smooth above; adult leaves elliptical to broadly lanceolate, less than 10 cm long, with a midrib and 2 fainter side veins parallel to midrib and 2 more veins parallel to edge; flower buds 6 × 2 mm in dense stellate clusters, many more than 7; operculum sharply conical **7. stellulata**
20. Bark smooth; adult leaves lanceolate to broadly lanceolate with several parallel veins; flower buds up to 15 × 7 mm, often warty, in loose clusters of 7 or more; operculum hemispherical or shortly conical 21.
21. Flower buds with sharp angles, sessile, glaucous **4(c). pauciflora** subsp. **debeuzevillei**
21. Flower buds without angles, sessile or pedicellate, clavate, green or glaucous 22.
22. Shrub; with narrow juvenile leaves to 2.5 cm wide; peduncles to 1 cm long **5. gregsoniana**
22. Small or medium sized tree; juvenile leaves ovate, to 8 cm wide; peduncles to 16 mm long 23.
23. Adult leaves lanceolate to falcate, up to 16 cm long; flower buds glaucous or green **4(a). pauciflora** subsp. **pauciflora**
23. Adult leaves lanceolate or oblong, not or scarcely falcate, to 10 cm long, uncinate; flower buds always glaucous **4(b). pauciflora** subsp. **niphophila**
24. Operculum flattened 25.
24. Operculum hemispherical or conical 26.
25. Adult leaves 5–10 cm **12. coccifera**
25. Adult leaves 12–28 cm **21(d). globulus** subsp. **maidenii**
26. Flower buds in 7s or more, when mature without a ring scar formed by the loss of the outer operculum 27.
26. Mature flower buds with a ring scar formed by the loss of the outer operculum 35.
27. Juvenile leaves alternate, petiolate, pendulous, oblique, ovate, up to 20 × 10 cm 28.
27. Juvenile leaves opposite, sessile and green 31.
28. Bark rough and fibrous over the whole trunk; juvenile leaves green and shiny; capsule barrel-shaped **2. obliqua**
28. Bark rough on lower part of trunk only; juvenile leaves glaucous 29.
29. Fruit urceolate **6. fraxinoides**
29. Fruit hemispherical, ovoid or campanulate 30.
30. Flower buds 5–27 together; peduncle 9–20 mm **3. delegatensis**
30. Flower buds 7 together; peduncle 3–8 mm **26(b). dalrympleana** subsp. **heptantha**
31. Bark smooth or thinly fibrous 32.
31. Bark fibrous over most of trunk 33.
32. Lamina 4–6 cm, linear to linear-lanceolate **9. pulchella**
32. Lamina 2.5–3.5 cm, ovate **11. nitida**
33. Flower buds 7 together **25(b). viminalis** subsp. **cygnetensis**
33. Flower buds 7–23 together 34.
34. Adult leaves 7–15 × 0.7–1.5 cm **8. radiata**
34. Adult leaves 7–12 × 0.5–1.3 cm **10. amygdalina**
35. Leaves on mature tree opposite and sessile or subopposite and shortly petiolate, held somewhat stiffly and not pendulous **17. parvifolia**
35. Leaves on mature tree alternate, pendulous 36.
36. Juvenile leaves up to 17 cm **20. nitens**
36. Juvenile leaves less than 10 cm 37.
37. Adult leaves less than 2 cm wide 38.
37. Adult leaves 2.5–5.0 cm wide 40.
38. All leaves green; juvenile leaves elliptical to broadly lanceolate **16. aggregata**
38. All leaves bluish-green; juvenile leaves linear 39.
39. Bark rough, reddish-brown, fibrous **18. nicholii**
39. Bark smooth, white and powdery **19. mannifera**
40. Juvenile leaves usually emarginate; adult leaves elliptical up to 5 cm wide **14. camphora**
40. Juvenile leaves not emarginate; adult leaves broadly lanceolate, up to 2.5 cm wide 41.
41. Juvenile leaves entire; adult leaves without prominent oil glands **13. ovata**
41. Juvenile leaves crenulate; adult leaves with prominent oil glands **15. brookeriana**

Series **1. Regnantes** Chippend.

Tall *trees*. *Bark* rough, persistent. *Juvenile leaves* alternate, green. *Adult leaves* alternate, green. *Inflorescence* axillary, simple; umbels paired, 9- to 15-flowered. *Flower* buds clavate.

1. E. regnans F. Muell. Australian Mountain Ash
E. amygdalina var. *regnans* (F. Muell.) F. Muell.

Evergreen *tree* up to 75(–100) m. *Trunk* tall and straight extending through much of tree. *Wood* pale, strong and hard, but not durable. *Branches* ascending; twigs and young shoots glabrous. *Bark* up to 15 m on trunk rough, persistent, brown and fibrous, above smooth, white or greyish-green. *Juvenile leaves* alternate; lamina 3–8 × 2–6 cm, green on upper surface, and beneath, ovate to broadly lanceolate, acute or obtuse at apex, entire, rounded or narrowed at base, the petiole 8–12 mm. *Adult leaves* alternate; lamina 9–14 × 1.6–2.7 cm, green on upper surface and beneath, lanceolate or broadly lanceolate, acuminate at apex, entire, narrowed at base, often oblique; lateral veins at 15°–30°, conspicuous, the intramarginal vein up to 4 mm from margin; petiole 12–22 mm, channelled. *Inflorescence* axillary and simple; umbels paired, 9- to 15-flowered; peduncle 5–13 mm, angular; pedicels 2–4 mm. *Flower buds* clavate; hypanthium about 3.0 × 3–4 mm, obconical; *operculum*

2–3 × 3–4 mm, conical; anthers adnate, reniform, opening in divergent slits, gland terminal. *Fruits* 5–9 × 4–7 mm, obconical to pyriform, pedicellate; disc broad, level or ascending; valves 3, at rim level or slightly exserted; seeds brown, more or less pyramidal. *Flowers* summer.

Introduced. Planted in gardens and parks and in Ireland for forestry. Native of the mountains of southern Victoria and in the Huon and Derwent Rivers valley, Tasmania. The tallest tree species in Australia and the tallest hardwood in the world. The wood has been used for building, flooring, plywood and pulp and paper-making.

Series **2. Eucalyptus**
Trees and *shrubs*. *Bark* smooth throughout. *Juvenile leaves* alternate, green or glaucous. *Adult leaves* alternate, green or glaucous. *Inflorescence* axillary and simple; umbels 3- to many-flowered. *Flower* buds clavate, ovoid or fusiform.

2. E. obliqua L. Hér. Messmate Stringybark
E. pallens DC.; *E. procera* Dehnh.; *E. fabrorum* Schltdl. *E. heterophylla* Miq.; *E. falcifolia* Miq.; *E. nervosa* F. Muell. ex Miq., non Hoffmanns.

Evergreen *tree* up to 90 m. *Trunk* up to 3 m in diameter, extending through much of crown. *Bark* fibrous, stringy, furrowed throughout, grey to reddish-brown. *Wood* pale. *Juvenile leaves* alternate; lamina 6–8 × 3–4 cm, shining green on upper surface, ovate, often shortly acuminate at apex, entire or slightly denticulate, oblique, narrowed or rounded at base, the midrib glandular-scabrid, petiolate. *Intermediate leaves* which are obliquely lanceolate occur. *Adult leaves* alternate; lamina 10–15 × 1.5–3.3 cm, dark shining green on upper surface, broadly lanceolate, acuminate at apex, entire, narrowed at base; lateral veins at 20°–30°, just visible, the intramarginal up to 1 mm from margin; petiole 7–17 mm, channelled. *Inflorescence* axillary and simple; umbels 11 or more-flowered; peduncle 4–15 mm, angular or flattened; pedicels 1–6 mm. *Flower buds* clavate; hypanthium 2–4 × 2–3 mm, obconical; operculum 1–2 × 2–3 mm, hemispherical, apiculate; anthers adnate, reniform, opening by divergent slits, the gland terminal. *Fruit* 6–11 × 5–9 mm, ovoid, subglobose, barrel-shaped or ureolate; disc level or steeply descending; valves 3 or 4, level to included; seeds brown, pyramidal. *Flowers* summer.

Grown in Ireland in parks and for forestry. Native of South Australia, Queensland, New South Wales, Victoria and Tasmania. Used for joinery, flooring, furniture and pulp production.

3. E. delegatensis R. T. Baker Alpine Ash

Evergreen *tree* up to 40(–80) m with an open crown. *Trunk* straight and tapering. *Bark* grey to brown and fibrous on lower half of trunk, smooth and white above. *Wood* white or pinkish-white, light and strong. *Twigs* glaucous or dark red. *Juvenile leaves* alternate; lamina dull green or glaucous on upper surface, ovate to subrotund, sometimes with a prominent drip-tip, sometimes rounded and entire at base, petiolate. *Adult leaves* alternate, pleasantly aromatic; lamina 5–22 × 1.1–4.0 cm, shining green on upper surface, narrowly to broadly lanceolate or falcate, oblique, entire, narrowed at base; lateral veins at 15°–35°, conspicuous, the intramarginal up to 2 mm from margin; petiole 8–45 mm, flattened to channelled. *Inflorescence* axillary and simple; umbels 5- to 27-flowered; peduncle 9–20 mm, terete or angular; pedicels 2–7 mm. *Flower buds* clavate; hypanthium 3–4 × about 3 mm, obconical; operculum about 2 × 3 mm, hemispherical and apiculate; anthers adnate, reniform, opening in confluent slits, glands small and terminal. *Fruits* 8–19 × 6–11 mm, ovoid or pyriform, sometimes hemispherical; disc descending, sometimes level; valves 3–5, included; seeds brown, pyramidal.

(a) Subsp. **delegatensis**
E. obliqua var. *alpina* Maiden

Seedling stems glabrous. *Juvenile leaves* with lamina ovate. *Adult leaves* with lamina 9–22 × 1.3–4.0 cm; petiole 8–45 mm.

(b) Subsp. **tasmaniensis** Boland
E. risdonii var. *elata* Benth.; *E. tasmanica* Blakely; *E. gigantea* Hook. fil., non Desf.

Seedling stems warty with conspicuous oil glands. *Juvenile leaves* with lamina subrotund with a prominent drip-tip. *Adult leaves* with lamina 5–17 × 1.1–3.6 cm; petiole 10–40 mm.

Introduced. Planted in gardens and parks and for trial forestry. Native of New South Wales, Australian Capital Territory, Victoria and Tasmania. Both subspecies have been used for flooring, plywood, veneers, furniture, panelling and turnery, as well as pulp.

4. E. pauciflora Sieber ex Spreng. Snow Gum
E. coriacea A. Cunn. ex Schauer; *E. phlebophylla* F. Muell. ex Miq.; *E. submultiplinervis* Miq.

Evergreen *tree* or shrub up to 20 m, often with a round crown. *Trunk* up to 60 cm in diameter, extending well into crown. *Bark* white to pale grey or sometimes brownish-red, smooth, shredding in irregular patches or stripes giving a mottled appearance, sometimes with scribbles. *Wood* pale, full of gum veins. *Branches* erect or spreading; twigs dark red or orange-red; young shoots green or greyish-green, with a glaucous bloom or dark red. *Juvenile leaves* alternate; lamina 2.5–6.0 × 1.2–3.0 cm, dull bluish-green on upper surface, thick and leathery, lanceolate to ovate, acute at apex, entire, narrowed at base, petiolate. *Adult leaves* alternate; lamina 5–16 × 1.2–4.5 cm, green to bluish-green on upper surface, lanceolate to ovate, acute to acuminate or uncinate at apex, entire, cuneate at base; lateral veins parallel to the midrib or at 20°, the intramarginal up to 3 mm from margin; petiole 10–20 mm, channelled, angular or flattened. *Inflorescence* axillary and simple; umbels 7- to 15-flowered; peduncle 3–16 mm, terete, flattened or angular; pedicels up to 3 mm, or absent. *Flower bud* clavate or ovoid, sometimes angular; hypanthium 4–8 × 4–6 mm, obconical or angular; operculum 2–4 × 3–6 mm, conical, triangular or hemispherical; anthers very thin, reniform, opening in front with broad cells. *Fruits* 5–15 × 5–11 mm, glaucous, hemispherical, pyriform, obconical or subglobose; disc level; valves 3 or 4; seeds black, pyramidal. $2n = 22$.

(a) Subsp. **pauciflora**

Tree to 20 m. *Adult leaves* with lamina 7–16 × 1.2–3.2 cm. *Flower buds* green. *Fruits* 6–10 × 5–9 mm.

(b) Subsp. **niphophila** (Maiden & Blakely) L. A. S. Johnson & Blaxell

E. niphophila Maiden & Blakely; *E. pauciflora* var. *alba* Ewart

Small *tree* or *shrub* to 6 m. *Adult leaves* with lamina 5–8 × 1.2–2.0 cm. *Flower buds* glaucous, not angular. *Fruits* 5–10 × 6–9 mm.

(c) Subsp. **debeuzevillei** (Maiden) L. A. S. Johnson & Blaxell

E. debeuzevillei Maiden

Many-stemmed *shrub* or small *tree* up to 9 m. *Adult leaves* 7.5–15.0 × 1.3–4.5 cm. *Flower buds* glaucous, angular. *Fruits* 10–15 × 7–11 mm.

Introduced. Planted in gardens, parks and estates, especially in Ireland. All three subspecies are recorded. Native of Queensland, New South Wales, Victoria, South Australia and Tasmania.

5. E. gregsoniana (L.) L. A. S. Johnson & Blaxell Mallee Snow Gum

E. pauciflora var. *nana* Blakely

Evergreen *shrub* to 5 m. *Stems* many. *Bark* white to pale grey, smooth. *Wood* pale. *Branches* slender. *Juvenile leaves* alternate; lamina to 2.5 cm wide, bluish-green on upper surface, paler beneath, lanceolate, entire, petiolate. *Adult leaves* alternate; lamina 7–11 × 1.5–2.5 cm, greyish-green on upper surface, paler beneath, thick, lanceolate, acuminate at apex, entire, symmetrical or slightly oblique at base; lateral veins almost parallel to midrib, distinct, the intramarginal up to 1 mm from the margin; petiole 8–12 mm, thick, flattened or channelled. *Inflorescence* axillary and simple; umbels 7- to 11-flowered; peduncle up to 1 cm, terete or absent; pedicels up to 2 mm or absent. *Flower buds* green or glaucous, clavate without angles; hypanthium 5–6 mm; obconical; operculum 2–3 × about 4 mm, hemispherical or conical; anthers very thin, reniform, opening in front with broad cells. *Fruits* 6–8 × 6–8 mm, truncate-pyriform, sometimes wrinkled; disc level or convex; valves 3, just included; seeds black, pyramidal.

Introduced. Planted in gardens and parks and for forestry, particularly in Ireland.

Native of New South Wales.

Series **3. Fraxinales** Blakely

Trees. *Bark* fibrous below, smooth above. *Juvenile leaves* alternate, pale grey or bluish-green. *Adult leaves* alternate, green on both surfaces. *Inflorescence* axillary and simple; umbels 7- to 11-flowered. *Flower buds* clavate.

6. E. fraxinoides H. Deane & Maiden White Mountain Ash

E. virgata var. *fraxinoides* (H. Deane & Maiden) Maiden

Slender evergreen *tree* to 40 m. *Bark* fibrous, dark grey and compact up to 10 m, smooth and white above, usually with insect scribbles. *Wood* pale, light and strong. *Young shoot* with internodes very glandular. *Juvenile leaves* alternate; lamina 4–10 × 2–4 cm, pale grey or bluish-green on both surfaces, broadly lanceolate to ovate, acute at apex, entire, cuneate at base, petiolate. *Intermediate leaves* alternate and with a broadly lanceolate lamina. *Adult leaves* alternate; lamina 8–16 × 1–2 cm, green on both surfaces, narrowly lanceolate to lanceolate, falcate, acuminate or uncinate at apex, usually oblique; lateral veins at 15°–30°, faint, the intramarginal 1–2 mm from margin; petiole 10–15 mm, flattened or channelled. *Inflorescence* axillary and simple; umbels 7- to 11-flowered; peduncle 5–18 mm, angular or flattened; pedicels 1–6 mm. *Flower buds* clavate; hypanthium 3–4 × 3–4 mm, hemispherical or obconical; operculum 1–2 × 3–4 mm, conical or hemispherical, apiculate, often warty; anthers adnate and reniform. *Fruits* 7–11 × 6–11 mm, urceolate or subglobose; disc moderately broad, descending; valves 4 or 5, included; seeds black.

Introduced. Grown for forestry in Ireland. Native of New South Wales, Victoria and Lord Howe Island in Australia.

Series **4. Longitudinales** Blakely

Shrubs or small *trees*. *Bark* smooth. *Juvenile leaves* alternate or opposite, green. *Adult leaves* alternate, green. *Inflorescence* axillary and simple; umbels 7- to many-flowered. *Flower buds* fusiform.

7. E. stellulata Sieber ex DC. Black Sallee

Evergreen *tree* up to 15 m with a spreading crown. *Bark* dark grey, greyish-black or olive green, smooth. *Wood* pale, not durable. *Juvenile leaves* opposite; lamina 3–6 × 2.5–6.0 cm, green on upper surface, paler beneath, subrotund to ovate, acute at apex, entire, narrowed at base, sessile, becoming shortly petiolate. *Adult leaves* alternate; lamina 5–9 × 1.3–2.3 cm, shining green on upper surface, paler beneath, thick, elliptical to ovate, acute at apex, entire, narrowed at base; lateral veins at 10°–15°, conspicuous with 3 main veins, extramarginal vein 1–3 mm from margin; petiole 4–9 mm, terete or flattened. *Inflorescence* axillary and simple; umbels 7- to 23-flowered; peduncle 1–5 mm, terete. *Flower buds* sessile, fusiform; hypanthium 2–3 × 2–3 mm, obconical; operculum 3–4 × 2–3 mm, conical, acute; anthers reniform; gland terminal. *Fruit* 3–5 × 3–5 mm, globose, sessile; disc level; valves 3, included; seeds reddish-brown, more or less pyramidal.

Introduced. Planted in gardens, parks and estates in Ireland and for forestry. Native of New South Wales and Victoria.

Series **5. Radiatae** Chippend.

Eucalyptus series *Piperitales* sensu Blakely

Trees or *shrubs*. *Bark* smooth throughout or fibrous throughout. *Juvenile leaves* opposite, green, greyish-green or bluish-green. *Adult leaves* alternate, green, aromatic with a peppermint scent. *Inflorescence* axillary and simple; umbels 7- to many-flowered. *Flower buds* clavate.

P. Y. Ladiges et al. (1983) have discussed the relationships and biogeographic patterns in this series.

8. E. radiata DC. Narrow-leaved Peppermint-gum
E. australiana R. T. Baker & H. G. Sm.; *E. radiata* var. *australiana* (R. T. Baker & H. G. Sm.) Blakely; *E. phellandra* R. T. Baker & H. G. Sm.; *E. radiata* var. *subplatyphylla* Blakely & McKie; *E. radiata* var. *subexerta* Blakely

Evergreen *tree* up to 50 m, with a spreading crown. *Bark* greyish-brown and fibrous throughout. *Wood* pale, soft and not very durable. *Juvenile leaves* opposite; lamina 4–7 × 1–2 cm, green, greyish-green or bluish-green on upper surface, paler beneath, narrowly to broadly lanceolate, entire at base. *Adult leaves* alternate, with peppermint aroma; lamina 7–15 × 0.7–1.5 cm, green to greyish-green or slightly glaucous on upper surface, paler beneath, acuminate at apex, entire; lateral veins at 20°–40°; the intramarginal up to 2 mm from margin; petiole 5–15 mm, terete. *Inflorescence* axillary and simple; umbels 7- to 23-flowered; peduncle 2–8 mm; pedicels 1–5 mm. *Flower buds* clavate; hypanthium 2–3 × 1–3 mm, obconical; operculum 1–2 × 1–3 mm, hemispherical to conical; anthers reniform, opening in more or less confluent slits; gland small and terminal. *Fruits* 3–7 × 4–7 mm, hemispherical, subglobose or pyriform; disc more or less level; valves 3 or 4, more or less level; seeds reddish-brown, more or less pyramidal.

(a) Subsp. **radiata**
Tree to 30 m. *Juvenile leaves* green. *Adult leaves* green with lamina narrowly lanceolate to almost linear.

(b) Subsp. **robertsonii** (Blakely) L. A. S. Johnson & Blaxell
E. robertsonii Blakely

Tree to 50 m. *Juvenile leaves* greyish-green to bluish-green. *Adult leaves* greyish-green or subglabrous, with lamina narrowly to broadly lanceolate.

Introduced. Planted in gardens and parks and for trial forestry in Ireland. Native of New South Wales, Australian Capital Territory and Victoria.

9. E. pulchella Desf. White Peppermint-gum
E. linearis Dehnh.; *E. amygdalina* var. *angustifolia* F. Muell. ex L. H. Bailey

Evergreen *tree* to 21 m. *Bark* yellow to white or grey, smooth throughout. *Wood* white. *Juvenile leaves* opposite; lamina 4–6 × 0.2–0.3 cm, green on upper surface, paler beneath, linear to linear lanceolate, entire, usually sessile. *Adult leaves* alternate, with a peppermint aroma; lamina 5–12 × 0.2–0.7 cm, green on upper surface, and beneath, aromatic with a peppermint-like smell, linear, acuminate or uncinate at apex, entire; veins at 10°–20°, faint, the intramarginal up to 1 mm from margin; petiole 4–7 mm, more or less terete. *Inflorescence* axillary and simple; umbel with 15 or more flowers; peduncle 3–8 mm, angular or flattened; pedicels 1–3 mm. *Flower buds* 4–5 mm, clavate, shortly pedicellate; hypanthium 2–3 × 2–3 mm, obconical; operculum 1–2 × 2–3 mm, hemispherical; anthers reniform. *Fruits* 4–6 × 5–7 mm, subglobose or subpyriform; disc broad, level or descending; valves 4, more or less level; sepals reddish-brown, more or less pyramidal. $2n = 22$.

Introduced. Planted for ornament in the extreme south-west of England and Ireland and self-sown in the Isles of Scilly. Native of central and south-eastern Tasmania. Highly suitable as a street tree.

10. E. amygdalina Labill. Black Peppermint
E. salicifolia auct.; *E. glandulosa* Desf.

Slender evergreen *tree* 15–30 m, occasionally smaller and up to 10 m. *Bark* finely fibrous, greyish-brown on trunk and larger branches, then smooth, salmon-pink or white to grey above. *Wood* pale brown, moderately durable and strong. *Juvenile leaves* opposite; lamina 3–6 × 0.7–1.3 cm, green to slightly glaucous, lanceolate, often crenulate, sessile to shortly petiolate. *Adult leaves* alternate, with a strong peppermint-like aroma; lamina 7–12 × 0.5–1.3 cm, dull green on both surfaces, narrowly lanceolate, falcate, acuminate or uncinate, entire, thin, glabrous; lateral veins at 15°–25°, faint, the intramarginal up to 1 mm from margin; petiole 7–10 mm, flattened or angular. *Inflorescence* axillary and simple; umbels with 7–15 or more flowers; peduncle 4–10 mm, terete or angular; pedicels 1–3 mm. *Flower buds* clavate; hypanthium 2–3 × 2–3 mm, subpyriform; operculum 1–2 mm, hemispherical, sometimes apiculate. *Fruits* 4–7 × 5–7 mm, hemispherical or obconical; disc broad, level, sometimes ascending; valves 4, level.

Introduced. Grown in parks and estates. Native of Tasmania.

11. E. nitida Hook. fil. Smithton Perppermint
E. amygdalina var. *nitida* (Hook. fil.) Benth.; *E. simmondsii* Maiden

Evergreen *tree* up to 40 m with a narrow crown. *Wood* pale. *Branches* smooth and white; young shoots with internodes very glandular-hispid. *Juvenile leaves* opposite; lamina 2.5–3.5 × 1.5–3.0 cm, greyish-green on both surfaces, ovate to broadly lanceolate, rounded at apex, entire, more or less hispid with numerous oil glands. *Adult leaves* alternate, smelling of peppermint; lamina 6.5–13.0 × 0.8–1.7 cm, green on both sides, lanceolate or narrowly lanceolate, acuminate or uncinate at apex, entire, usually thick, narrowed at base; lateral veins at 15°–25°, faint, the intramarginal 1–2 mm from margin; petiole 7–12 mm, terete or angular. *Inflorescence* axillary and simple; umbels with 11 or more flowers; peduncle 2–8 mm, terete or angular; pedicels absent or up to 4 mm. *Flower buds* clavate; hypanthium 2–4 × 2–3 mm, subpyriform; operculum 1–2 × 2–3 mm, hemispherical or conical; anthers adnate, reniform. *Fruits* 4–8 × 5–9 mm, hemispherical or conical; disc broad, level or ascending; valves 3–5, more or less level.

Introduced. Grown in parks and estates especially in Ireland. Native of Tasmania.

Intermediates occur between *E. nitida* and *E. coccifera* and are discussed by M. J. Shaw et al. in *Austral. Jour. Bot.* **32**: 641–654 (1984).

12. E. coccifera Hook. fil. Tasmanian Snow Gum
E. daphnoides Miq.; *E. coccifera* var. *parviflora* Benth.

Evergreen *shrub* or *tree* up to 25 m, when a tree has a rounded crown. *Trunk* up to 1.2 m in diameter, through much of crown. *Bark* whitish-grey, or yellow or pink when fresh, smooth throughout. *Wood* pale. *Young shoots* yellow or reddish-brown, often with a waxy bloom; glandular-hairy. *Juvenile leaves* opposite; lamina 2.5–5.0 × 1.5–3.5 cm, slightly glaucous on upper surface and beneath, thin, broadly elliptical or subrotund, apiculate at apex, entire, rounded at base, sessile or shortly petiolate. *Adult leaves* alternate, when crushed smelling of peppermint; lamina 5–10 × 1–2 cm, greyish-green on upper surface, slightly paler beneath, thick, elliptical or lanceolate, markedly uncinate or shortly pointed at apex, rounded or cuneate at base; lateral veins at 15°–30°, faint, the intramarginal distinct, up to 1 mm from margin; petiole 10–15 mm, whitish, terete or channelled. *Inflorescence* axillary and simple; umbels 3-flowered, sometimes up to 9-flowered; peduncle 5–10 mm, terete to angular; pedicels up to 4 mm, or absent. *Flower buds* glaucous, clavate; ribbed or angled; hypanthium 4–5 × 3–5 mm, glaucous and obconical; operculum 1–2 × 3–5 mm, glaucous, flattened-hemispherical, wrinkled and warty; anthers adnate, reniform. *Fruits* 7–11 × 10–13 mm, glaucous, hemispherical or obconical, often 2-ribbed; disc broad, level; valves 3 or 4, level.

Introduced. Planted in gardens, parks, estates and arboreta especially in Ireland. Produces coppice shoots and has reproduced from seed. Native of the central plateau area of Tasmania. Said to form a clinal change to *E. nitida* (see M. J. Shaw et al. (1984)).

Series **6. Foveolatae** Maiden
Eucalyptus subseries *Semidecorticatae* Blakely

Trees. *Bark* smooth throughout or fibrous below or throughout. *Juvenile leaves* alternate, green. *Adult leaves* alternate, green. *Inflorescence* axillary and simple; umbels 7-flowered. *Flower buds* fusiform to obovoid.

13. E. ovata Labill. Swamp Gum
E. gunnii var. *ovata* (Labill.) H. Deane & Maiden; *E. stuartiana* F. Muell. ex Miq.; *E. muelleri* Naudin; *E. paludosa* R. T. Baker; *E. stuartiana* var. *longifolia* Benth.; *E. acervula* auct.; *E. gunnii* var. *acervula* auct.; *E. ovata* var. *grandiflora* Maiden

Evergreen *tree* up to 30 m. *Trunk* extending through crown. *Bark* white, grey or pinkish-grey, smooth throughout, often with accumulate decorticating bark forming a rough trunk base. *Wood* pale, soft and not very durable. *Juvenile leaves* alternate; lamina 4–8 × 3–7 cm, dull green on upper surface, paler beneath, elliptical to ovate, obtuse to acute at apex, entire, rounded or cuneate at base; petiolate. *Adult leaves* alternate; lamina 8–15 × 1.7–3.0 cm, lanceolate to ovate, acuminate at apex, undulate, rounded to narrowed at base; lateral veins at 25°–40°, distinct, the intramarginal vein up to 2 mm from the margin; petiole 17–25 mm, terete. *Inflorescence* axillary and simple; umbels 7-flowered; peduncle 3–14 mm, terete or angular; pedicels 1–4 mm, sometimes absent. *Flower buds* fusiform; hypanthium 3–4 × 4–5 mm, obconical; operculum 3–5 × 4–5 mm, conical or slightly rostrate. *Anthers* versatile, obcordate, opening in parallel slits, dorsal gland ovate, large. *Fruits* 5–7 × 4–7 mm, obconical; disc broad, more or less level; valves 3 or 4, level or slightly exerted; seeds brownish-grey, irregular, more or less flat, shallowly reticulate. $2n = 22$.

Introduced. Planted in gardens and parks. Native of South Australia, Victoria, New South Wales and Tasmania. Shows considerable geographical variation (see P. Y. Ladiges et al. (1981)).

14. E. camphora R. T. Baker Mountain Swamp Gum
E. ovata var. *camphora* (R. T. Baker) Maiden; *E. ovata* var. *aquatica* Blakely

Evergreen *tree* up to 22 m. *Trunk* extending into crown. *Bark* grey to brownish-grey or almost black, smooth throughout though often with accumulated decorticating bark at base of trunk. *Wood* pale. *Juvenile leaves* alternate; lamina 5–7 × 6–8 cm, green on upper surface, paler beneath, ovate, emarginate at apex, entire; petiolate. *Adult leaves* alternate; lamina 6–13 × 3.8–5.0 cm, shining green on upper surface, paler beneath, lanceolate to ovate, sometimes emarginate, entire; lateral veins at 35°–40°, distinct, the intramarginal up to 2 mm from the margin; petiole 25–40 mm, terete. *Inflorescence* axillary and simple; umbels 7-flowered; peduncle 10–18 mm, terete; pedicels 2–5 mm. *Flower buds* fusiform; hypanthium 2–3 × 3–5 mm, obconical; operculum 4–5 × 3–5 mm, conical or slightly rostrate; anthers versatile, obovate, opening in parallel slits, gland ovate, large, protruding in front. *Fruits* 4–6 × 4–6 mm, obconical; disc narrow, level or slightly ascending; valves 3 or 4, level or slightly exserted; seeds irregular, more or less flat, shallowly reticulate.

Introduced. Planted in gardens and parks. Native of Queensland, New South Wales and Victoria.

15. E. brookeriana A. M. Gray Brooker's Gum

Evergreen *tree* up to 40 m. *Trunk* extending into crown. *Bark* grey, rough and fibrous up to 6 m, sometimes tessellated, then orange-red, olive green or cream and smooth. *Wood* pale. *Juvenile leaves* alternate; lamina slightly shining bright green on upper surface, paler beneath, broadly ovate to subrotund, strongly crenulate, petiolate. *Adult leaves* alternate; lamina 6.5–14.0 × 1.5–3.0 cm, green on upper surface, paler beneath, lanceolate to ovate, acute at apex, usually crenulate, often undulate, narrowed at base; lateral veins at 25°–45°, faint, the intramarginal up to 5 mm from the margin; petiole 10–30 mm, terete. *Inflorescence* axillary and simple; umbels usually 7-flowered, sometimes more; peduncle 5–12 mm, angular; pedicels 1–5 mm. *Flower buds* fusiform to almost obovoid; hypanthium 2–4 × 3–5 mm, obconical; operculum 3–5 × 3–5 mm, conical or almost rostrate. *Fruits* 5–8 × 5–7 mm, obconical or almost hemispherical; disc narrow, level or ascending; valves 3 or 4, level or exserted.

Introduced. Planted in gardens and parks. Native of Victoria and Tasmania.

16. E. aggregata H. Deane & Maiden Black Gum
E. rydalensis R. T. Baker & H. G. Sm.

Evergreen *tree* up to 18 m. *Trunk* extending into crown. *Bark* grey, rough, fibrous and hard throughout. *Wood* white. *Juvenile leaves* alternate; lamina 4–6 × 1.5–3.5 cm, green on upper surface, paler beneath, elliptical, ovate or lanceolate, rounded at apex, entire, rounded at base, petiolate. *Adult leaves* alternate; lamina 5–12 × 1–2 cm, green on upper surface, paler beneath, linear-lanceolate to lanceolate, acuminate at apex, entire, narrowed at base; lateral veins at 15°–40°, faint, the intramarginal up to 3 mm from margin; petiole 4–10 mm, terete. *Inflorescence* axillary and simple; umbels 7-flowered; peduncle 3–4 mm, terete; pedicels up to 2 mm or absent. *Flower buds* ovoid to fusiform; hypanthium about 2.0 × 2–3 mm, obconical; operculum 2–3 × 2–3 mm, conical; anthers versatile, obovate, opening in parallel slits, dorsal gland globose. *Fruits* 2–3 × 3–5 mm, obconical; disc narrow, level; valves 3 or 4, exserted or level; seeds brownish-grey, irregular, more or less flat, shallowly reticulate. $2n = 22$.

Introduced. Planted in gardens and parks especially in Ireland. Native of New South Wales and Victoria.

Series **9. Microcarpae** Blakely

Trees. *Bark* smooth or rough. *Juvenile leaves* opposite or alternate, green to slightly glaucous or greyish-green. *Adult leaves* alternate or subopposite, green to glaucous or greyish-green. *Inflorescence* axillary and simple; umbels 7-flowered. *Flower buds* ovoid or fusiform, rarely clavate.

17. E. parvifolia Cambage Kybean Gum

Evergreen *tree* up to 9m. *Trunk* 30–40 cm in diameter. *Bark* dull grey, grey-green or sometimes pink, smooth throughout. *Wood* pink, soft and brittle. *Juvenile leaves* opposite; lamina 2–3 × 1.0–1.5 cm, green or slightly glaucous on upper surface and paler beneath, elliptical, acute at apex, entire, rounded at base, sessile, often persisting on mature trees. *Adult leaves* subopposite; lamina 5–7 × 0.6–1.0 cm, green on upper surface, paler beneath, lanceolate, acute at apex, entire, narrowed at base; lateral veins at 15°–20°, the intramarginal up to 1 mm from the margin; petiole 2–6 mm, more or less flattened. *Inflorescence* axillary and simple; umbel 7-flowered; peduncle 4–7 mm, thick and terete; pedicels absent. *Flower bud* ovoid; hypanthium 2–3 × 2–3 mm, hemispherical; operculum 1–2 × 2–3 mm, conical, anthers versatile, nearly globose, opening in longitudinal slits. *Fruits* 3–4 × 3–4 mm, hemispherical to conical; disc slightly ascending; valves 3 or 4, level; seeds brownish-grey, irregularly-shaped, more or less flat, shallowly reticulate.

Introduced. Planted in large gardens and parks, especially in Ireland. Native of New South Wales.

18. E. nicholii Maiden & Blakely
Narrow-leaved Peppermint
E. acaciiformis var. *linearis* H. Deane & Maiden

Evergreen *tree* up to 15 m. *Trunk* extending into crown. *Bark* yellowish-brown to greyish-brown, rough and fibrous throughout. *Wood* pale red, soft, not very durable. *Juvenile leaves* alternate, crowded; lamina 2–5 × 2–5 cm, greyish-green on upper surface, paler beneath, linear, acute at apex, entire or crenulate, narrowed at base, shortly petiolate. *Adult leaves* alternate; lamina 6–13 × 0.5–1.0 cm, slightly bluish-green to greyish-green on upper surface, paler beneath, linear-lanceolate, acuminate at apex, entire; lateral veins at an angle of 35°–45°, faint, the intramarginal up to 1 mm from margin; petiole 7–12 mm, terete. *Inflorescence* axillary and simple; umbels 7-flowered; peduncle 5–7 mm, terete; pedicels 2–3 mm. *Flower buds* ovoid or more or less fusiform; hypanthium 2–3 × about 2 mm, hemispherical to subglobose; operculum 1–3 × about 2 mm, conical; anthers versatile, almost ovate, emarginate, opening in broad parallel slits, dorsal gland globose and large. *Fruits* 2–5 × 3–4 mm, hemispherical to obconical; disc level; valves slightly exserted.

Introduced. Planted in gardens and parks. Native of New South Wales.

19. E. mannifera Mudie Mottled Gum

Evergreen *tree* up to 25 m. *Trunk* clean and straight. *Bark* white, cream or grey, sometimes with patches of red, smooth throughout, usually powdery. *Wood* pink, soft and not very durable. *Juvenile leaves* alternate; lamina 3–9 × 3–6 cm, pale green to more or less glaucous on upper surface, paler beneath. *Adult leaves* alternate; lamina 6–25 × 0.8–3.0 cm, bluish-green to greyish-green or pale green on upper surface, paler beneath, more or less lanceolate, acuminate at apex, entire, narrowed at base; lateral veins at 30°–40°, the intramarginal up to 1 mm from margin; petiole 6–27 mm, terete or channelled. *Inflorescence* axillary and simple; umbels 7-flowered; peduncle 5–10 mm, terete; pedicel 1–5 mm. *Flower buds* usually ovoid, rarely clavate; hypanthium 2–3 × 3–5 mm, obconical to hemispherical; operculum 2–3 × 2–5 mm, hemispherical to conical; anthers versatile, broadly obovate, opening in parallel slits. *Fruits* 5–7 × 4–7 mm, subglobose, conical or ovoid; disc level, ascending or convex; valves slightly exserted; seeds brownish-grey, irregular, flat, shallowly reticulate.

(a) Subsp. **mannifera**
Juvenile leaves glaucous, linear to lanceolate. *Adult leaves* 1.5–3.0 cm wide. *Peduncle* 5–6 mm. *Flower buds* green. *Hypanthium* obconical or hemispherical; disc convex.

(b) Subsp. **elliptica** (Blakely & McKie) L. A. S. Johnson
E. mannifera var. *elliptica* Blakely & McKie

Juvenile leaves with lamina subglaucous, subrotund, ovate or lanceolate. *Adult leaves* with lamina 1.5–3.0 cm wide. *Peduncle* 5–10 mm. *Flower buds* glaucous. *Hypanthium* hemispherical.

(c) Subsp. **maculosa** (R. T. Baker) L. A. S. Johnson
E. gunnii var. *maculosa* R. T. Baker

Juvenile leaves with lamina glaucous, elliptical to narrowly lanceolate. *Adult leaves* with lamina 0.8–2.0 cm wide. *Peduncle* 5–10 mm. *Flower buds* not glaucous. *Hypanthium* hemispherical.

(d) Subsp. **praecox** (Maiden) L. A. S. Johnson
E. praecox Maiden; *E. lactae* R. Baker

Juvenile leaves with lamina slightly glaucous, subrotund. *Adult leaves* with lamina 1–2 cm wide. *Peduncle* 5–8 mm. *Flower buds* often glaucous. *Hypanthium* hemispherical.

(e) Subsp. **gullickii** (R. T. Baker & H. G. Sm.) L. A. S. Johnson
E. gullickii R. T. Baker & H. G. Sm.

Juvenile leaves with lamina pale green, elliptical to broadly lanceolate. *Adult leaves* with lamina 0.6–1.0 cm wide. *Peduncle* 5–8 mm. *Flower buds* green. *Hypanthium* obconical.

Introduced. Planted in gardens and parks, which of the subspecies is not known. Native of New South Wales and Victoria.

Series **10. Viminales** Blakely
Trees or *shrubs*. *Bark* smooth or rough. *Juvenile leaves* opposite, green, greyish-green or glaucous. *Adult leaves* alternate, green, less often greyish-green or glaucous. *Inflorescence* axillary and simple; umbels 3- to many-flowered. *Flower buds* very variable.

A variable group difficult to define which provides the greatest number of species introduced to our flora.

20. E. nitens (H. Deane & Maiden) Maiden — Shining Gum
E. goniocalyx var. *nitens* H. Deane & Maiden

Evergreen *tree* up to 70(–90) m. *Trunk* extending into crown. *Bark* yellowish-white or grey, smooth throughout or sometimes rough, flaky and grey to black at base of trunk. *Wood* pale, moderately light and strong. *Juvenile leaves* opposite; lamina 7–10 × 5–9 cm, glaucous on upper surface, paler beneath, lanceolate to ovate, acute at apex, entire, rounded at base, sessile and amplexicaul. *Adult leaves* alternate; lamina 13–24 × 1.5–2.5 cm, shining green on upper surface and beneath, linear-lanceolate to lanceolate, acuminate at apex, entire, narrowed at base; lateral veins at 20°–45°, distinct, the intramarginal up to 3 mm from margin; petiole 15–22 mm, terete or channelled. *Inflorescence* axillary and simple; umbels 7-flowered; peduncle 6–15 mm, slightly flattened; pedicels absent. *Flower buds* ovoid to cylindrical, angular or ribbed; hypanthium 3–4 × 3–4 mm, cylindrical or angular; operculum 2–3 × 3–4 mm, conical; anthers versatile, clavate, opening in parallel slits, with a large, globose dorsal gland. *Fruits* 4–7 × 4–6 mm, shining, cylindrical or ovoid, often slightly ribbed; disc narrow, descending; valves 3 or 4, level or slightly exserted; seeds irregular, reticulate.

Introduced. Planted in gardens and parks especially in Ireland and sometimes producing coppice shoots. Native of New South Wales and Victoria.

21. E. globulus Labill. — Southern Blue Gum

Evergreen *tree* up to 45(–70) cm, with a conical crown when young, becoming tall and domed. *Trunk* up to 2 m in diameter, with branches only in upper part. *Bark* white to cream, yellow or grey, but accumulated, greyish-brown, undecorticated bark at base, otherwise smooth. *Wood* pale, hard and durable. *Branches* large and heavy, ascending; twigs brown; young shoots square in section, often with a waxy bloom. *Juvenile leaves* opposite; lamina 7–16 × 4–9 cm, greyish-green to glaucous on upper surface, paler beneath, ovate, rounded at apex, entire; cordate at base, sessile and amplexicaul. *Adult leaves* alternate; lamina 12–28 × 1.2–4.0 cm, green on upper surface, paler beneath, leathery, linear-lanceolate to lanceolate, sometimes falcate, acute or acuminate at apex, entire, cuneate at base; lateral veins at 20°–45°, more or less conspicuous, the intramarginal up to 2 mm from margin; petiole 15–50 mm, terete, channelled or flattened. *Inflorescence* axillary and simple; umbels 1-, 3- or 7-flowered; peduncles 1–25 mm, sometimes absent, flattened or terete; pedicels 1–8 mm, or absent. *Flower buds* single, glaucous, turbinate to obconical, warty; hypanthium 5–12 × 5–17 mm, obconical, ribbed or more or less smooth; operculum 3–15 × 5–17 mm, flattened-hemispherical, shortly umbonate; anthers versatile, obovate, opening in broad parallel slits with a globose gland. *Fruits* 5–21 × 6–24 mm, green or glaucous, obconical to hemispherical or subglobose, warty; disc broad, level to ascending; valves 3–5, level or exserted; seeds irregular and reticulate. *Flowers* winter to autumn. $2n = 22$.

Very variable. Four subspecies have been recognised all of which have been recorded as introduced.

(a) Subsp. **globulus**
E. gigantea Desf.

Adult leaves with lamina 12–25 × 1.7–3.0 cm; petiole 20–30 mm. *Umbels* 1-flowered; pedicels absent or very short. *Fruits* 10–21 × 14–24 mm.

(b) Subsp. **bicostata** (Maiden, Blakely & J. Sim) Kirkp.
E. bicostata Maiden, Blakely & J. Sim; *E. globulus* var. *bicostata* (Maiden, Blakely & J. Sim) Ewart

Adult leaves with lamina 14–25 × 2–3 cm; petiole 30–50 mm. *Umbel* 3-flowered; pedicels absent. *Fruits* 8–17 × 10–20 mm.

(c) Subsp. **pseudoglobulus** (Naudin ex Maiden) Kirkp.
E. pseudoglobulus Naudin ex Maiden; *E. globulus* var. *stjohnii* R. T. Baker

Adult leaves with lamina 13–25 × 1.5–3.0 cm; petiole 15–35 mm. *Umbels* 3-flowered; pedicels 1–5 mm. *Fruits* 7–11 × 9–16 mm.

(d) Subsp. **maidenii** (F. Muell.) Kirkp.
E. maidenii F. Muell.

Adult leaves with lamina 12–28 × 1.2–2.5 cm; petioles 15–35 mm. *Umbels* 7-flowered; pedicels up to 8 mm or absent. *Fruits* 5–11 × 6–10 mm.

Introduced. Planted for ornament in gardens, parks and estates and rarely forestry. Self-sown in western Ireland and Isles of Scilly. Native of New South Wales, Victoria and Tasmania.

22. E. vernicosa Hook. fil. — Varnished Gum

Shrub or small *tree* up to 3.5 m. *Trunks* numerous. *Bark* grey and smooth throughout. *Juvenile leaves* opposite; lamina 1.5–2.0 × 1–2 cm, shining green on upper surface, paler beneath, thick, coriaceous, elliptical to ovate or oblong-lanceolate, obtuse-mucronate at apex, serrulate, cuneate to rounded at base, sessile or nearly so, glabrous. *Adult leaves* opposite; lamina 1.5–2.5 × 1.0–1.5 cm, shining green on upper surface, paler beneath, thick, elliptical or ovate, mucronate, entire, rounded to cuneate at base; lateral veins at 45°–50°, faint, the intramarginal vein up to 1 mm from margin; petiole 5–8 mm, channelled. *Inflorescence* axillary and simple; umbels 3-flowered, usually only 1 maturing; peduncle up to 2 mm, or absent, terete; pedicels absent. *Flower buds* ovoid, single; hypanthium 3–5 × 3–6 mm, campanulate; operculum 3–5 × 3–6 mm, conical; anthers versatile, obcordate, opening in parallel slits, with a very small dorsal globular gland. *Fruits* 5–8 × 5–9 mm, hemispherical to campanulate, ribbed; disc broad and descending; valves 3 or 4, level or just exserted.

Introduced. Planted in parks or estates. Native of western Tasmania.

23. E. subcrenulata Maiden & Blakely Tasmanian Alpine Yellow Gum

Evergreen *tree* up to 18 m. *Trunk* often crooked. *Wood* pale. *Bark* grey to white or yellowish-green, smooth throughout. *Juvenile leaves* opposite; lamina 6–8 × 2–4 cm, pale glossy green on upper surface, slightly paler beneath, elliptical, ovate or subrotund, rounded at apex, crenulate, rounded at base, sessile or shortly petiolate. *Adult leaves* alternate; lamina 6–10 × 1.5–2.5 cm, green on upper surface and beneath, lanceolate to ovate, acuminate at apex, subcrenulate, narrowed at base; lateral veins at 20°–35°, intramarginal vein up to 2 mm from margin; petiole 15–25 mm, terete. *Inflorescence* axillary and simple; umbels 3-flowered; peduncle 2–6 mm, terete; pedicels up to 2 mm or absent. *Flower buds* ovoid, slightly wrinkled; hypanthium 3–4 × 4–5 mm, campanulate; operculum 3–4 × 4–5 mm, hemispherical to conical. *Anthers* versatile, obovate, opening in parallel slits, with a large dorsal gland. *Fruits* 5–8 × 5–8 mm, hemispherical to campanulate, sometimes 2-ribbed, sessile; disc narrow to moderately broad, descending; valves 3 or 4, level or exserted.

Introduced. Planted in gardens, parks and estates. Native of western and central Tasmania.

24. E. johnstonii Maiden Tasmanian Yellow Gum
E. muelleri T. Moore, non Miq., nec Naudin

Evergreen *tree* up to 40 m, with a narrow, conical crown. *Trunk* up to 1 m in diameter, extending through much of the crown. *Bark* orange-red or yellowish-green to grey or yellowish-bronze, smooth throughout. *Wood* pale red, very hard and strong. *Juvenile leaves* opposite becoming alternate; lamina 4–6 cm, shining dark green on upper surface, paler beneath, elliptical, rounded at apex, entire or crenulate and sessile, becoming alternate, subrotund and petiolate. *Adult leaves* alternate; lamina 5–13 × 2–3 cm, shining dark green on upper surface and beneath, leathery, lanceolate to ovate, acute or acuminate at apex, crenulate, cuneate at base; lateral veins at 25°–30°, intramarginal vein up to 4 mm from the margin; petiole 15–30 mm, terete, pale brown. *Inflorescence* axillary and simple; umbels 3-flowered; peduncle 3–9 mm, flattened or angular; pedicels up to 2 mm or absent. *Flower buds* ovoid, slightly wrinkled; hypanthium 5–6 × 6–9 mm, obconical, angular or 2-ribbed; operculum 4–6 × 6–9 mm, low, hemispherical, shortly umbonate; anthers versatile, obovate, opening in parallel slits, with an ovate gland. *Fruit* 7–8 × 9–13 mm, obconical to hemispherical, wrinkled, 2- or 3-ribbed; disc broad, level or slightly ascending; valves 3 or 4, exserted. $2n = 22$.

Introduced. Planted in gardens, parks and estates especially in Ireland and for small forest trials. Native of south-east Tasmania.

25. E. viminalis Labill. Ribbon Gum

Evergreen *tree* up to 50 m with a broad dome. *Trunk* up to 1.2 m in diameter, extending through much of crown. *Bark* grey, white or yellowish-white and smooth throughout, or rough and fibrous on the lower or whole trunk. *Wood* pale, light and brittle, moderately durable. *Branches* spreading; twigs brown; young shoots dark red, warty. *Buds* not visible. *Juvenile leaves* opposite; lamina 5–10 × 1.5–3.0 cm, pale green on upper surface, paler beneath, lanceolate, acute at apex, entire, sessile and cordate or amplexicaul at base. *Adult leaves* alternate; lamina 12–20 × 0.8–2.5 cm, pale green on upper surface, similar beneath, linear-lanceolate to lanceolate, acuminate at apex, entire, sinuate, cuneate at base; lateral veins at 30°–40°, just visible to distinct, intramarginal vein up to 2 mm from margin; petiole 10–25 mm, terete or slightly flattened. *Inflorescence* axillary and simple; umbels 3- or 7-flowered; peduncle 4–13 mm, terete or angular and flattened; pedicels absent or very short. *Flower buds* ovoid; hypanthium 2–3 × 3–5 mm, hemispherical or campanulate; operculum 3–5 × 3–5 mm, conical or hemispherical, apiculate; anthers versatile, obovate, emarginate, opening in narrow parallel slits, with an ovate gland. *Fruits* 4–8 × 5–9 mm, hemispherical to subglobose; disc broad, ascending; valves 3–4, exserted; seeds irregular, reticulate. $2n = 22$.

(a) Subsp. **viminalis**
E. angustifolia Desf. ex Link; *E. viminalis* var. *rhynchocorys* F. Muell. ex Maiden; *E. pilularis* auct.

Bark white or yellowish-white and smooth throughout, or rough and undecorticated at base or on most of trunk. *Umbels* 3-flowered.

(b) Subsp. **cygnetensis** Boomsma
Bark greyish-brown, rough, thick and fibrous on trunk and larger branches, then smooth and greyish-white above. *Umbels* 7-flowered.

Introduced. Planted in gardens and parks especially in Ireland, and has been planted on a small scale for forestry. Native of South Australia, Queensland, Victoria and Tasmania. It is not known to which subspecies our plants belong. The wood has been used for building, and it secretes a sugary material from the bark.

26. E. dalrympleana Maiden
Broad-leaved Kindlingbark

Evergreen *tree* up to 40 m with a narrow, conical crown. *Trunk* up to 1 m in diameter, extending into crown. *Bark* blotched white and grey to yellowish-white and sometimes pink, green or olive, smooth throughout or with about 1 m of accumulated undecorticated bark at base of trunk. *Wood* pink, light and not very durable. *Branches* ascending to erect; twigs brown; young shoots glabrous. *Juvenile leaves* opposite; lamina 4–6 × 4–6 cm, pale green to subglaucous on upper surface, similar beneath, ovate to subrotund, obtuse at apex, entire, rounded or cordate at base. *Adult leaves* alternate; lamina 10–22 × 1.5–3.0 cm, green, often shining, on upper surface, paler beneath, linear-lanceolate to lanceolate or falcate, acute or acuminate at apex, entire but wavy, sometimes undulate, cuneate at base; lateral veins at 30°–45°, with intramarginal vein up to 1 mm from the margin; petiole 15–27 mm, terete. *Inflorescence* axillary and simple; umbels 3- or 7-flowered; peduncle 3–8 mm, angular or slightly flattened; pedicels 1–4 mm, or absent. *Flower buds* ovoid; hypanthium 2–4 × 2–5 mm, hemispherical to obconical; operculum 2–4 × 3–5 mm, conical or sometimes almost hemispherical; anthers versatile, opening in parallel slits, with a small dorsal gland. *Fruits* 5–8 × 5–9 mm, hemispherical, ovoid or campanulate; disc moderately broad, ascending, convex or sometimes more or less level; valves 3 or 4, exserted. $2n = 22$.

(a) Subsp. **dalrympleana**
Adult leaves with lamina 10–20 × 1.5–2.5 cm. *Umbels* 3-flowered; peduncles 3–8 mm. *Hypanthium* 3–4 × 3–5 mm; operculum 2–4 × 3–5 mm. *Fruits* 5–8 × 5–9 mm.

(b) Subsp. **heptantha** L. A. S. Johnson
Adult leaves with lamina 10–22 × 1.5–3.0 cm. *Umbels* 7-flowered; peduncles 4–7 mm. *Hypanthium* 2–3 × 3–4 mm; operculum 2–4 × 3–4 mm. *Fruits* 5–7 × 5–7 mm.

Introduced. Planted in gardens and parks and for trial forestry. Native of New South Wales, Victoria and Tasmania. Which subspecies is introduced is not known. The species is susceptible to silver leaf disease, *Chondrostereum purpureum*.

27. E. glaucescens Maiden & Blakely Tingiringi Gum
E. gunnii var. *glauca* H. Dean & Maiden

Evergreen *tree* or *shrub* up to 45 m. *Trunk* extending into crown. *Bark* rough, fibrous and dark grey on lower trunk of larger trees, becoming greenish-grey and smooth above, or more or less smooth throughout on small trees or shrubs. *Juvenile leaves* opposite; lamina 1.5–3.0 × 1.5–3.0 cm, slightly glaucous on upper surface, paler beneath, subrotund, emarginate at apex, entire, rounded at base, sessile. *Adult leaves* alternate, pendulous; lamina 6–13 × 1.2–2.0 cm, dull green or greyish-green on upper surface and beneath, lanceolate, acute or acuminate at apex, entire, narrowed at base; lateral veins at 35°–40°, just visible, intramarginal vein up to 1 mm from margin; petiole 15–25 mm, pink, terete. *Inflorescence* axillary and simple; umbels 3-flowered; peduncle 2–6 mm, terete; pedicels absent or obscured. *Flower buds* glaucous and shiny, cylindrical; hypanthium 3–4 × 3–4 mm, cylindrical; operculum 2–3 × 3–4 mm, conical or slightly rostrate. *Anthers* versatile, opening in long parallel slits, with an ovate dorsal gland. *Fruits* 6–12 × 6–10 mm, often glaucous, cylindrical to slightly ovoid; disc narrow, level or convex to descending; valves 3 or 4, just included.

Introduced. Planted in parks and estates especially in Ireland and can produce coppice shoots. Native of the Howe region of Australia.

28. E. gunnii Hook. fil. Cider Gum
E. gunnii var. *montana* Hook. fil.; *E. whittingehamei* Landsb.; *E. divaricata* McAulay & Brett ex Brett

Evergreen *tree* up to 25 m, when young narrowly conical. *Trunk* ascending throughout much of the tree or columnar, growing fast and becoming domed in old trees. *Bark* white, grey or greyish-green, smooth throughout, or sometimes about 1 m of persistent, undecorticated, flaky bark at the base of the trunk. *Wood* pale. *Branches* erect or ascending, pale brown; twigs green and brown; young shoots green or pinkish-purple, with warts in first winter, often with a waxy bloom. *Juvenile leaves* opposite; lamina 2.5–4.5 × 2–4 cm, greyish-green to glaucous on upper surface, cream to whitish beneath, ovate to subrotund, emarginate at apex, crenulate, rounded at base, sessile and amplexicaul. *Adult leaves* alternate and pendulous, leathery, with a curry-like smell when crushed; lamina 4–8 × 1.2–3.0 cm, greyish-green or glaucous on upper surface and beneath, elliptical, ovate or broadly lanceolate, acute, acuminate or apiculate at apex, entire, cuneate or rounded at base; lateral veins at 30°–45°, faint, intramarginal vein up to 1 mm from margin; petiole 11–20 mm, yellow, terete, wrinkled, glandular. *Inflorescence* axillary and simple; umbels 3-flowered; peduncle 5–9 mm, slightly angular; pedicels 1–2 mm, or absent. *Flower buds* 6–8 mm, usually glaucous, shortly pedicellate, cylindrical to urn-shaped; hypanthium 4–5 × 3–5 mm, obconical; operculum 2–3 × 3–5 mm, hemispherical, slightly umbonate; filaments pale yellow; anthers pale yellow, versatile, oblong to globose, opening in parallel slits, with a large, globose, dorsal gland. *Fruits* 7–10 × 4–9 mm, usually glaucous, cylindrical or suburceolate; disc broad, level or slightly descending; valves 3 or 4, included. *Flower* 6–7. $2n = 22$.

Introduced. Planted for full-scale forestry and for ornament in gardens, parks and estates. Widely grown in Ireland and south and west Great Britain; persistent in Essex since 1887. The most hardy species we grow and sometimes self-sowing. Native of the central plateau of Tasmania. For intergradation with *E. archeri* see B. M. Potts & J. B. Reid (1985).

29. E. archeri Maiden & Blakely Alpine Cider Gum

Evergreen *shrub* or small straggly *tree* to 9 m. *Trunk* extending into crown. *Bark* white, grey or greyish-green, usually smooth throughout, but sometimes with accumulated decorticating bark at trunk base. *Juvenile leaves* opposite; lamina 3–5 × 2.5–3.5 cm, greyish-green on upper surface and beneath, ovate, rounded or sometimes apiculate at apex, entire, rounded at base, sessile. *Adult leaves* alternate;

lamina 5–8 × 1–2 cm, greyish-green on upper surface and beneath, thick, lanceolate, acuminate at apex, entire, narrowed at base; lateral veins at 40°–50°, faint, intramarginal vein up to 2 mm from margin; petiole 10–15 mm, terete. *Inflorescence* axillary and simple; umbels 3-flowered; peduncles 1–4 mm, thick, flattened; pedicels absent. *Flower buds* green, obovoid; hypanthium 3–4 × 4–5 mm, hemispherical to obconical; operculum 2–3 × 4–5 mm, hemispherical, slightly apiculate. *Anthers* versatile, oblong, with an ovate gland. *Fruits* 6–8 × 6–8 mm, hemispherical to subcampanulate, slightly wrinkled; disc moderately broad, descending; valves 3 or 4, level or included.

Introduced. This is included because it has apparently been brought into cultivation and is very close to *E. gunnii* from which it is distinguished by its green, not glaucous buds, narrower leaves, shorter, thicker peduncle and slightly wrinkled fruits. Like *E. gunnii*, with which it intergrades, it is native of the central plateau of Tasmania.

30. E. urnigera Hook. fil. Urn-fruited Gum
E. urnigera var. *elongata* Rodway

Evergreen *tree* up to 30 m, with a domed crown. *Trunk* up to 1.2 m in diameter, extending well into crown. *Bark* whitish-grey, greyish-yellow or yellowish-brown or pink, smooth throughout. *Wood* pale. *Branches* drooping or spreading; young shoot warty. *Juvenile leaves* opposite; lamina 2.5–4.5 × 2.5–4.5 cm, shining green on upper surface, paler beneath, subrotund, shortly pointed often emarginate, at apex, crenulate, rounded at base, sessile and amplexicaul. *Adult leaves* alternate; lamina 5–10 × 1.3–2.5 cm, green or slightly glaucous on upper surface and beneath, lanceolate to ovate, acute at apex, entire, narrowed at base; lateral veins at 45°–50°, intramarginal vein up to 2 mm from margin; petiole 13–25 mm, red, terete. *Inflorescence* axillary and simple; umbels 3-flowered; peduncle 8–25 mm, angular, recurved when fruiting; pedicels 2–7 mm. *Flower buds* often glaucous, urceolate; hypanthium 6–8 × 5–6 mm; operculum 2–4 × 5–6 mm, flattened hemispherical, slightly rostrate. *Anthers* versatile, ovate, opening in broad parallel slits, with a large globose gland. *Fruits* 10–16 × 8–10 mm, often glaucous, urceolate; pedicellate; disc broad, descending; valves 3 or 4, included. *Flowers* 2–4.

Introduced. Planted in gardens, parks and estates particularly in Ireland, and for small forestry trials. Native of south-eastern Tasmania.

31. E. perriniana F. Muell. ex Rodway Spinning Gum

Evergreen *shrub* or straggly *tree* up to 9 m. *Trunk* small. *Bark* bronze, whitish-green or grey, smooth throughout or with a short stacking of persistent bark at the trunk base. *Wood* pale. *Branches* straggling. *Juvenile leaves* opposite and connate; lamina 5–6 × 4–10 cm, glaucous on upper surface, paler beneath, subrotund, obtuse or mucronate at apex, entire, rounded at base, sessile, often persisting on mature plants. *Adult leaves* alternate or subopposite; lamina 8–13 × 1.2–2.5 cm, dull greyish-green or bluish-green on upper surface and beneath, lanceolate, acuminate at apex, entire, rounded at base; lateral veins at 25°–45°, faint, intramarginal vein up to 2 mm from the margin; petiole 10–15 mm, terete or flattened. *Inflorescence* axillary and simple; umbels 3-flowered; peduncle 2–5 mm, glaucous, terete; pedicels up to 2 mm or absent. *Flower buds* glaucous and ovoid; hypanthium 3–4 × 3–4 mm, obconical; operculum 2–3 × 3–4 mm, hemispherical to conical. *Anthers* versatile, broadly ovoid to subglobose, opening in parallel slits, with a small, globose dorsal gland. *Fruits* 5–7 × 5–7 mm, glaucous, hemispherical, sessile; disc narrow, level or descending; valves 3–5, more or less level.

Introduced. Planted in gardens, parks and estates, particularly in Ireland. Native of Australian Capital Territory, New South Wales, Victoria and Tasmania. Its common name refers to the juvenile leaf pairs spinning in the wind. For an account of its genetic variation see R. J. E. Wiltshire & J. B. Reid (1987).

32. E. cordata Labill. Silver Gum

Evergreen *shrub* up to 3 m or a *tree* to 21 m. *Bark* white, grey, green or yellowish-green and purple, smooth throughout. *Wood* pale yellow. *Juvenile leaves* opposite; lamina 4–5 × 3–6 cm, glaucous on upper surface, paler beneath, subrotund, rounded at apex, crenulate, sessile and amplexicaul, usually persisting on mature trees. *Adult leaves* usually only in upper parts of large trees, alternate; lamina 7.5–13.0 × 2.0–3.8 cm, dull greyish-green or glaucous on upper surface and beneath, lanceolate, acute at apex, entire; lateral veins at 40°–45°, faint, intramarginal vein very faint, up to 2 mm from margin; petiole 15–20 mm, flattened. *Inflorescence* axillary and simple; umbels 3-flowered; peduncle 5–8 mm, glaucous, flattened; pedicels absent. *Flower buds* obovoid or turbinate, glaucous; hypanthium 6–7 × 8–10 mm, obconical or campanulate; operculum 4–5 × 8–10 mm, low-hemispherical and slightly rostrate; anthers versatile, obovate, opening in parallel slits, with a large, globose dorsal gland. *Fruits* 10–13 × 10–13 mm, glaucous, subglobose or cylindrical, smooth or 2-ribbed; disc narrow, descending; valves 3 or 4, included or level. $2n = 22$.

Introduced. Planted in parks, gardens and estates, especially in Ireland. Native of a small area of south-eastern Tasmania.

33. E. pulverulenta Sims Silver-leaved Mountain Gum
E. cordata Lodd., non Labill.; *E. pulvigera* A. Cunn.

Evergreen *tree* or straggly *shrub* up to 5 m. *Trunk* extending into crown. *Bark* bronze to grey, smooth throughout. *Wood* pale, very hard. *Juvenile leaves* opposite; lamina 4–6 × 3.0–4.5 cm, glaucous on upper surface and beneath, subrotund to ovate, rounded at apex, entire, at base, sessile and amplexicaul, usually persisting on mature trees. *Adult leaves* rarely seen, alternate; lamina 5–10 × 1.2–1.9 cm, glaucous on upper surface, and beneath, lanceolate or oblong, apiculate at apex, entire; lateral veins at 20°–35°, faint, intramarginal vein up to 1 mm from margin; petiole 10–18 mm, terete. *Inflorescence* axillary and simple; umbels 3-flowered; peduncle 5–8 mm, terete; pedicels absent. *Flower buds* ovoid and glaucous; hypanthium 5–6 × 5–6 mm, obconical; operculum 2–3 × 5–6 mm, conical. *Anthers* versatile, obovate, opening in parallel slits, with a rather large, globose dorsal gland. *Fruits* 5–9 × 6–10 mm, glaucous, hemispherical; disc narrow, level; valves 3 or 4, level or slightly exserted. $2n = 22$.

Introduced. Planted in gardens and parks especially in Ireland. Native of Nepean and Howe regions of Australia. Extensively grown in California for the ornamental young leaves.

34. E. cinerea Benth. Argyll Apple
E. pulverulenta var. *lanceolata* Howitt; *E. stuartiana* var. *cordata* R. T. Baker & H. G. Smi. nom. illegit.; *E. pulverulenta* auct.

Evergreen *tree* up to 16 m. *Trunk* extending into crown. *Bark* reddish-brown, rough and fibrous below, reddish-brown or grey above, or sometimes rough throughout. *Wood* reddish. *Juvenile leaves* opposite; lamina 3.5–4.5 × 3.5–5.0 cm, glaucous on upper surface, subrotund, entire, cordate at base, sessile and amplexicaul or shortly petiolate, usually persisting on the adult tree. *Intermediate leaves* opposite; glaucous to almost green on upper surface, paler beneath, broadly ovate, entire, more or less cordate at base, sessile and sometimes amplexicaul or shortly petiolate. *Adult leaves* alternate; lamina 7.5–11.5 × 1.5–2.5 cm, glaucous on upper surface and beneath, thick, broadly lanceolate, acuminate at apex, entire, at base; lateral veins at 30°–45°, faint, intramarginal vein up to 2 mm from margin; petiole 5–11 mm, flattened. *Inflorescence* axillary and simple; umbels 3-flowered; peduncle 2–6 mm, terete; pedicels absent. *Flower buds* glaucous and fusiform; hypanthium 3–4 × 4–5 mm, obconical; operculum 2–3 × 4–5 mm, conical. *Anthers* versatile, oblong, emarginate, opening in long parallel slits, with a small, ovate gland. *Fruits* 5–8 × 5–9 mm, obconical to hemispherical; disc broad, level and ascending; valves 3–5, slightly exserted. $2n = 22$.

Introduced. Planted in gardens and parks especially in Ireland. Native of Nepean and Howe regions of Australia. Cultivated as an ornamental or street tree.

4. Leptospermum J. R. & G. Forst.

Evergreen, monoecious or dioecious *shrubs* or small *trees*. *Bark* smooth and flaking, fibrous or papery. *Leaves* alternate, entire, often aromatic, exstipulate. *Flowers* solitary, unisexual or bisexual, axillary or terminal, with translucent bracts; hypanthium campanulate or obconical, fused to the ovary below. *Sepals* 5. *Petals* 5, spreading, deciduous, white, pink or red. *Stamens* numerous, arranged in bundles opposite a petal. *Style* 1. *Ovary* 5-celled; ovules anatropous. *Fruit* a woody capsule with flaps opening at the top.

Contains about 79 species or more, confined to Australia, New Zealand, New Caledonia and Malaysia.

Hybrids seem to occur widely in the genus and we do not know what will happen when these species are brought together as in the Isles of Scilly. This account is of the species as they occur in their native localities. Plants which do not fit the descriptions exactly should be considered for hybridisation.

Thompson, J. (1989). A revision of the genus *Leptospermum* (Myrtaceae). *Telopea* **3**(3): 301–448.

1. Leaves 5–7 × 1–2 mm, smelling of lemon **1. liversidgei**
1. Leaves 3–20 × 2–10, not smelling of lemon 2.
2. Valves of fruit not woody; seeds 1.5–2.0 mm **4. sericeum**
2. Valves of fruit very woody; seeds 2.0–3.5 mm 3.
3 Leaves abruptly narrowed at apex; sepals persistent on fruit **2. lanigerum**
3. Leaves tapered to a pungent point; sepals deciduous as soon as flowers fade **3. solarium**

1. L. liversidgei R. T. Baker & H. G. Sm. Lemon-scented Tea-tree
L. flavescens var. *citriodorum* Bailey; *L. polygalifolium* var. *citriodorum* (Bailey) Domin

Compact evergreen, monoecious *shrub* up to 4m. *Stems* slender, with short, often crisped hairs. *Bark* close and fibrous. *Branches* more or less erect. *Leaves* alternate, dense, erect to spreading, smelling of lemon, thick; lamina 5–7 × 1–2 mm, narrowly obovate, broadly acute to obtuse at apex, often with an umbo behind the somewhat recurved tip, entire and somewhat incurved on margin, attenuate at base, glabrous; petiole very short. *Inflorescence* of single, bisexual flowers on modified shoots at the ends of short, few-leaved branches and occasionally on long, slender, leafy branches; bracts pale and scarious, short and broad; bracteoles reddish-brown and scarious, concave, enclosing the flat-topped young bud, but soon shed so that only the bracts remain; flowers 10–12 mm in diameter. *Hypanthium* about 2.5 mm, dark coloured, at first evenly tapering, later very broad above, glabrous or minutely hairy with rather conspicuous glands; pedicel 1.0–1.5 mm, broad. *Sepals* 5, persistent or tardily deciduous, about 2 mm, dark with pale scarious margins, almost hemispherical to deltoid, somewhat erose at apex, glabrous. *Petals* 5, about 5 mm. *Stamens* in bundles of 5–7; filaments 1.5–2.5 mm. *Style* 1, rather stout and tapering slightly; stigma very large. *Capsule* persistent, often with the base sunk into the stems, 7–10 mm in diameter, much wider than deep, with a broad, very woody rim, below which it tapers to the base, the surface firm and wrinkled, the valves very thick and woody; seeds 1.5–2.0 mm, irregularly cuneiform, striate. *Flowers* 6–8. $2n = 22$.

Introduced. A garden escape in Abbey Wood on Tresco in the Isles of Scilly. Native of Queensland and New South Wales.

2. L. lanigerum (Aiton) Sm. Woolly Tea-tree
Philadelphus laniger Aiton; *Philadelphus laniger* var. *canescens* Aiton; *Philadelphus laniger* var. *piliger* Aiton; *L. pubescens* Willd., non Lam.; *L. lanigerum* var. *pubescens* DC. nom. illegit.; *L. lanigerum* var. *pubescens* Hook. fil.; *L. australe* Salisb. nom. illegit.; *L. microphyllum* F. Muell. ex Miq., non Hoffmanns; *L. microphyllum* var. *viride* F. Muell. ex Miq.; *L. microphyllum* var. *glaucum* F. Muell. ex Miq.; *L. pubescens* forma *angustifolia* Miq.; *L. pubescens* forma *minor* Miq.; *L. lanigerum* var. *montanum* Rodway

Evergreen, monoecious *shrub* or small *tree* up to 5 m. *Stems* when young stout and densely hairy and when very young with long spreading hairs. *Bark* fibrous, close and firm. *Branches* spreading. *Leaves* alternate, diverging or spreading, rather thick; lamina 3–15×2–4 mm, oblong to narrowly oblanceolate, abruptly narrowed at apex and infolded behind a short pungent or blunt point, the margins often recurved, attenuate at base, sometimes grey-hairy on both surfaces, sometimes glabrous on upper surface, rarely glabrous on both; petiole short and broad-based. *Inflorescence* of single, bisexual flowers on modified shoots at the end of short, densely leafy side-branches, the subtending leaves often longer-petiolate, the new growth very dense and extending from beyond the flowers after flowering; bracts pale brown or reddish-brown, broad and stiff, the inner somewhat larger than the broad bracteoles; flowers about 15 mm in diameter. *Hypanthium* 3–5 mm, the upper part expanded, the lower shallowly rounded to broadly tapered, densely white villous; pedicel minute and narrow. *Sepals* 5, persistent on fruit, 2–4 mm, deltoid to long-deltoid, often folded or banded at apex, densely long silky-hairy. *Petals* 5, about 6 mm. *Stamens* in bundles of about 7; filaments 2–3 mm. *Style* 1, slender but broad-based; stigma small, often absent. *Capsule* persistent, 5–10 mm in diameter, the rim not or scarcely extended, the lower part broadly rounded but often flat-based, the valves very woody; seeds about 2.5 mm, narrowly linear-cuneiform, curved, striate. *Flowers* 6–8. $2n=22$.

Introduced. Naturalised on Tresco in the Isles of Scilly and widely planted in gardens and parks in Ireland. Native of Australia.

3. L. solarium J. R. & G. Forst. Broom Tea-tree
Melaleuca scoparia (J. R. & G. Forst.) L. fil.; *Philadelphus scoparius* (J. R. & G. Forst.) Aiton; *Philadelphus scoparius* var. *linifolius* Aiton; *Melaleuca scoparia* var. *diosmatifolia* Wendl.; *L. scoparium* var. *linifolium* (Aiton) Aiton; *L. linifolium* (Aiton) Dum.-Cours.; *L. scoparium* var. *forsteri* Schauer; *Philadelphus scoparius* var. *myrtifolius* Aiton; *Melaleuca scoparia* var. *myrtifolia* (Aiton) Wendl. fil. & Schrad.; *L. scoparia* var. *myrtifolium* (Aiton) Aiton; *Philadelphus aromaticus* Aiton; *L. floribundum* Salisb. nom. illegit.; *Philadelphus floribundus* Usteri ex Roem. & Usteri; *L. scoparium* var. *parvum* Kirk; *L. scoparium* var. *prostratum* Kirk; *L. nichollsii* Dorr. Sm.; *L. scoparium* var. *nichollsii* Turrill; *L. scoparium* var. *eximium* B. L. Burtt

Evergreen, monoecious *shrub* up to 2(–4) m. *Stems* when young with a long, fine, silky hairiness, but soon becoming glabrous. *Bark* close and firm. *Branches* at 45°. *Leaves* alternate, widely diverging, spreading or even deflexed, variable; lamina 6–20×2–6 mm, with those of new shoots longer, narrowly to broadly elliptical, broadly lanceolate or oblanceolate, tapering at apex to an acuminate, pungent point, margins incurved or infolded especially in upper part, the base attenuate, silvery-hairy when young, soon becoming glabrous, firm and thick and usually minutely tuberculate; petiole short and stout-based or negligible. *Inflorescence* of single, rarely several, bisexual flowers on modified shoots on short, leafless or few-leaved, occasionally many-leaved branches, with new growth dense after flowering; bracts reddish-brown, broad and scarious, the inner and bracteoles larger and enclosing the young bud; flowers 8–12 mm in diameter. *Hypanthium* 2–3(–4) mm, the upper part widely expanded, the lower tapering or somewhat rounded, glabrous or rarely with sparse, short hairs, sometimes with obvious glands. *Sepals* 5, deciduous, about 2 mm, oblong to broadly deltoid and sometimes extended at the base, scarious, glabrous or with minutely ciliate margins. *Petals* 5, 4–7 mm, white, or rarely pink or red. *Stamens* in bundles of 5–7(–9); filaments 2.5–3.5 mm. *Style* 1, stout and straight-sided; stigma large. *Capsule* long-persistent and enlarging, 6–10 mm in diameter, widest at the scarcely extended rim, the lower part rounded or broadly rounded, the valves very woody; seeds 2.0–3.5 mm, irregularly narrowly linear-cuneiform or sigmoid, curved, striate. *Flowers* 6–8. $2n=22$.

This very variable species clearly contains a number of ecotypes and races and a number of cultivars have been selected in horticulture.

Introduced. Confined by frost to the extreme south-west of Great Britain and more frequent in gardens and parks in Ireland. Freely self-sowing and naturalised in Tresco in the Isles of Scilly. Native of Australia and New Zealand.

4. L. sericeum Labill. Swamp Tea-tree
Kunzea sericea (Labill.) Turcz.

Evergreen, monoecious *shrub* up to 3 m. *Stems* when young stout, with short hairs. *Bark* close. *Branches* at about 45°. *Leaves* alternate, erect at first, but soon diverging or reflexed; lamina 10–20×5–10 mm, obovate, broadly rounded at apex often with a short point behind it, entire, attenuate at base, with dense, silvery-grey hairs at first with the upper surface sometimes becoming glabrous; petiole short and broad. *Inflorescence* of single flowers or occasionally 2 together, mostly on modified shoots at the ends of short, leafy side-branches in leaf axils, the terminal buds developing with or immediately after flowering; bracts reddish-brown, few and broad, nearly glabrous; bracteoles similar, all shed as or before the flower opens; flowers bisexual, 15–25 mm in diameter. *Hypanthium* 4–6 mm, with the upper part spreading somewhat and tapering below, with dense, shining, spreading hairs. *Sepals* persistent, 2–3 mm, deltoid, densely silkily hairy and not distinguishable from the hypanthium. *Petals* 6–12 mm, pink. *Stamens* in bundles of about 7; filaments 3.5–4.0 mm. *Style* 1, broad-based and tapering above; stigma small. *Capsule* deciduous, about 7 mm in diameter, with a broad rim above, distinctly lobed in cross-section, the valves not woody; seeds 1.5–2.0 mm, irregularly obovoid-cuneiform, coarsely reticulate. *Flowers* 6–8.

Introduced. Seedlings recorded in Abbey Wood on Tresco in the Isles of Scilly. Native of Western Australia.

95. ONAGRACEAE Juss. nom. conserv.

Annual, biennial or *perennial,* monoecious *herbs*, rarely *shrubs*. *Leaves* opposite or alternate, simple, sessile or petiolate, without or with small, early-falling stipules. *Inflorescence* of axillary or terminal racemes or solitary flowers, the flowers bisexual, epigynous and actinomorphic to usually zygomorphic. *Hypanthium* absent or a short to very long tube, sometimes possibly adnate to the ovary. *Outer perianth segments* (sepals) 2 or 4 at apex of hypanthium, sometimes coloured, when present. *Inner perianth segments* (petals) 2 or 4, or rarely absent. *Stamens* 2, 4, or 8. *Style* 1; stigma capitate to clavate or 4-lobed. *Ovary* 1-, 2- or 4-celled, each cell with 1–many ovules on axile placentas. *Fruit* a 4-celled capsule or berry, or 1- to 2-seeded nut.

About 20 genera with a cosmopolitan distribution.

1. Outer perianth segments 2; inner perianth segments 2; stamens 2; ovary 1- to 2-celled; fruit with hook-tipped bristles	**8. Circaea**
1. Outer perianth segments 4; inner perianth segments 4 or rarely absent; ovary 4-celled; fruit without hooked bristles	2.
2. Shrub; fruit a berry	**7. Fuchsia**
2. Herb; fruit a capsule	3.
3. One row of perianth segments only; stamens 4	**3. Ludwigia**
3. Two rows of perianth segments present; stamens 8 or 10	4.
4. Inner perianth segments 5; stamens 10	**3. Ludwigia**
4. Inner perianth segments 4; stamens 8	5.
5. Inner perianth segments yellow, sometimes streaked or tinged reddish	**5. Oenothera**
5. Inner perianth segments pink, red or purple	6.
6. All leaves alternate	7.
6. At least some leaves opposite	8.
7. Seeds with a hairy plume	**2. Chamerion**
7. Seeds without a hairy plume	**6. Clarkia**
8. Hypanthium not forming a funnel shape and not having scales inside	**1. Epilobium**
8. Hypanthium forming a tube, globose at base and bearing within the narrow part 8 bractlike appendages	**4. Zauschneria**

1. Epilobium L.

Chamaenerion Adans. nom. illegit.

The original draft was prepared by P. D. S. but Geoffrey Kitchener has been through it and greatly corrected and amended it in line with his account for the second edition of *Hybridization and the flora of the British Isles* edited by C. A. Stace.

Perennial monoecious herbs. *Leaves*, at least the lower, usually opposite, or in whorls of 3, the upper leaves and bracts usually alternate. *Inflorescence* of more or less erect flowers, solitary and axillary or in terminal, bracteate racemes; flowers bisexual, actinomorphic, perigynous, usually considered to have a very short hypanthial tube. It is possible, however, the hypanthium starts below the ovary and elongates round it as the capsule grows and extends for a short way about it (see Bunniger & Weberling (1968)). *Sepals* 4. *Petals* 4, pink to purple, rarely white, usually emarginate. *Stamens* 8. *Style* 1; stigma 4-lobed to entire, capitate to clavate. *Fruit* a long, linear capsule, dehiscing loculicidally by 4 valves; seeds with a chalazal plume of long hairs.

Contains about 200 species in temperate and Arctic-Alpine regions of both hemispheres.

Plants vary very greatly in stature, leaf-size and degree of branching and hairiness, but type of hairs, leaf-shape and dentation, seed-coat ornamentation and presence of terminal appendage are more constant.

Hybrids occur commonly where two or more species occur together, especially if in quantity, for several years in disturbed ground. They are recognisable by their larger and branched stature, longer flowering season, unusually large or small flowers markedly more darkly coloured at the petal tips and partially or entirely abortive fruits. Most seeds are abortive, but some are fertile and backcrossing may rarely occur. Meiosis in the hybrids is normal and sterility results from early abortion in developing embryos. A valuable morphological feature for recognising hybrids is the stigma. Hybrids between species with 4-lobed stigmas and species with entire stigmas show obscure or irregular lobing. The presence of long, spreading eglandular stem hairs, often with their distal ends turned up or down, suggests *E. hirsutum* or *E. parviflorum* as one parent, flower size also being informative with these two species. The presence of abundant spreading glandular hairs on the stem suggests *E. hirsutum*, *E. parviflorum*, *E. ciliatum* or *E. roseum* or to a lesser degree *E. montanum* or *E. palustre* as one parent. Toothing of leaves is useful for example in hybrids between *E. hirsutum* and *E. parviflorum.* Features of leaf shape, petiole length, perennating structures and depth of flower colour all need to be used to identify hybrids between *E. obscurum*, *E. palustre* and *E. tetragonum.* Hybrids from the same two parents can vary considerably. Many are intermediate and seem to be the dominant form. However, there may be hybrids nearer one or other parent which are difficult to recognise and are not normally recorded.

When examining a large colony of *Epilobium* the first thing to do is establish which species are present. This can only be done by finding plants which have **all** the characters of the species. Hybrids can then be looked for. In some cases, however, hybrids can be found without one or even both the parents. This regularly happens in more than one year set-aside fields where seeds may have blown a considerable distance across open fields.

That our species do not seem to be in danger of losing their individuality is probably due to ecological requirements and especially the prevalence of self-pollination.

In 2004 a large set-aside field at Bassingbourn (v.c. 29) was so thick with *Epilobium* you could not walk through it. Here, hardly any hybrids were seen. Later in the same year in the same village another set-aside field had a great deal of variation and hybrids of among the six species involved. There one could also see the forms given in Haussknecht's monograph (1884). Some of these plants were in large identical patches but it was difficult to tell if the spread was vegetative or from seed (the set-aside was for more than one year).

Åkerman, Å. (1921). Untersuchungen über Bastarde zwischen *Epilobium hirsutum* und *Epilobium montanum*. *Hereditas* **2**: 99–112.

Ash, G. M. (1934). *Epilobium adenocaulon* Hausskn. *Rep. Watson Bot. Exch. Club.* **4**: 218–219.

Ash, G. M. (1937). *Epilobium hirsutum* L.×*tetragonum* L. *Rep. Bot. Soc. Exch. Club Brit. Isles* **11**: 401.

Ash, G. M. (1937). *Epilobium obscurum* Schreb.×*palustre* L. and *E. obscurum* Schreb.×*parviflorum* Schreb. *Rep. Bot. Soc. Exch. Club Brit. Isles* **11**: 402.

Ash, G. M. (1947). *Epilobium adenocaulon* Hausskn.×*obscurum* Schreb. *Rep. Bot. Soc. Exch. Club Brit. Isles* **13**: 160.

Ash, G. M. (1953). *Epilobium adenocaulon* in Britain. In Lousley, J. E. (Edit.) *The changing flora of Britain*, pp. 168–170. Oxford.

Ash, G. M. & Sandwith, N. Y. (1935). *Epilobium adenocaulon* Hausskn. in Britain. *Jour. Bot. (Lond.)* **73**: 177–184.

Brockie, W. B. (1970). Artificial hybridisation in *Epilobium* involving New Zealand, European, and North American species. *New Zealand Jour. Bot.* **8**: 94–97.

Bunniger, L. & Weberling, F. (1968). Unterschungen über die morphologische Natur des Hypanthiums bei Myrtales-Familien. 1. Onagraceae. *Beitr. Biol. Pflanzen* **44**: 447–477.

Compton, R. H. (1911). Notes on *Epilobium* hybrids. *Jour. Bot. (Lond.)* **49**: 158–163.

Compton, R. H. (1913). Further notes on *Epilobium* hybrids. *Jour. Bot. (Lond.)* **51**: 79–85.

Compton, R. H. (1914). *Epilobium hirsutum* var. *villosissimum* Koch. *Rep. Bot. Soc. Exch. Club Brit. Isles* **3**: 469.

Compton, R. H. & Marshall, E. S. (1913). *Epilobium hirsutum* L. ♀ and *E. parviflorum* Schreb. ♂. *Rep. Bot. Soc. Exch. Club Brit. Isles* **3**: 254.

Compton, R. H. & Marshall, E. S. (1913). *Epilobium montanum* L. ♀ and *E. parviflorum* Schreb. ♂. *Rep. Bot. Soc. Exch. Club Brit. Isles* **3**: 254–255.

Compton, R. H. & Marshall, E. S. (1913). *Epilobium hirsutum* L. ♀ and *E. montanum* L. ♂. *Rep. Bot. Soc. Exch. Club Brit. Isles* **3**: 254.

Compton, R. H. & Moss, C. E. (1911). *Epilobium hirsutum* L. ♀ and *E. tetragonum* ♂. *Rep. Bot. Soc. Exch. Club Brit. Isles* **2**: 562.

Compton, R. H. & Moss, C. E. (1911). *Epilobium montanum* ♂×*E. tetragonum* ♀. *Rep. Bot. Soc. Exch. Club Brit. Isles* **2**: 563.

Davey, A. J. (1961). Biological flora of the British Isles. *Epilobium nerterioides* A. Cunn. *Jour. Ecol.* **49**: 753–759.

Doogue, D., Kelly, D. L. & Wyse Jackson, P. S. (1985). The progress of *Epilobium ciliatum* Rafin. (*E. adenocaulon* Hausskn.) in Ireland, with some notes on its hybrids. *Irish Nat. Jour.* **21**: 444–446.

Fitter, A. H. (1980). Hybridization in *Epilobium* (Onagraceae): the effect of clearance and re-establishment of fen carr. *Biol. Jour. Linn. Soc.* **13**: 331–339.

Geith, K. (1924). Experimentell-systematische Unterschungen an der Gattung *Epilobium* L. *Bot. Arch.* **6**: 123–186.

Grime, J. P. et al. (1988). *Comparative plant ecology*. London. [*E. ciliatum, hirsutum, montanum, obscurum, palustre* and *parviflorum*.]

Håkansson, A. (1924). Beiträge zur Zytologie eines *Epilobium*-Bastardes. *Bot. Not.* **1924**: 269–278.

Hall, P. M. (1936). *Epilobium adenocaulon* Hausskn.×*obscurum* Schreb. *Rep. Bot. Soc. Exch. Club Brit. Isles* **11**: 29.

Hall, P. M. & Sledge, W. A. (1937). *Epilobium adenocaulon* Hausskn.×*hirsutum* L. *Rep. Bot. Soc. Exch. Club Brit. Isles* **11**: 223.

Harrison, S. G. (1968). A New Zealand Willow-herb in Wales. *Nature in Wales* **11**: 74–78.

Haussknecht, C. (1884). *Monographie der Gattung* Epilobium. Jena.

Holyoak, D. T. & Kitchener, G. D. (2001). *Epilobium*×*kitcheneri* McKean (Onagraceae) in East Cornwall. *Watsonia* **23**: 452–453.

Hultén, E. & Fries, M. (1986). *Atlas of north European vascular plants north of the Tropic of Cancer.* 3 vols. Königstein.

Kitchener, G. D. (1991).Willow-herb hybrids and the Great Storm of 1987. *Trans. Kent Field Club* **11**: 69–73.

Kitchener, G. D. (1992).Willow herbs. Part II. *Wild Flower Mag.* **424**: 28–32.

Kitchener, G. D. (1997). A triple hybrid willowherb: *Epilobium ciliatum*×*E. hirsutum*×E. *parviflorum*. *B.S.B.I. News* **75**: 66–67.

Kitchener, G. D. (1998). *Epilobium*×*montaniforme. B.S.B.I. News* **78**: 90.

Kitchener, G. D. (2003). A new *Epilobium* (Onagraceae) hybrid: *Epilobium brunnescens* (Cockayne) Raven & Engelhorn×*Epilobium parviflorum* Schreber (*E.*×*argillaceum*). *Watsonia* **24**: 519–523.

Kitchener, G. D. (2003). The relationship between hybridization in *Epilobium* and Cornish china clay and other mining waste. *Botanical Cornwall* **12**: 20–32.

Kitchener, G. D. & McKean, D. R. (1998). Hybrids of *Epilobium brunnescens* (Cockayne) Raven & Engelhorn (Onagraceae) and their occurrence in the British Isles. *Watsonia* **22**: 49–60.

Krahulec, F. (1999). Two new hybrids of *Epilobium ciliatum* (Onagraceae). *Preslia* **71**: 241–248.

Kytövuori, I. (1969). *Epilobium davuricum* Fisch. (Onagraceae) in Eastern Fennoscandia compared with *E. palustre* L. A morphological, ecological and distributional study. *Ann. Bot. Fenn.* **6**: 35–58.

Kytövuori, I. (1972). The Alpinae group of the genus *Epilobium* in northernmost Fennoscandia. A morphological, taxonomical and ecological study. *Ann. Bot. Fenn.* **9**: 163–203.

Kytövuori, I. (1976). Biosystematics of the genus *Epilobium* groups Alpinae and Palustriformes (Onagraceae). I. Dwarfism in crosses of the Fennoscandian species. *Ann. Bot. Fenn.* **13**: 69–96.

Lehmann, E. (1918). Über reziproke Bastarde zwischen *Epilobium roseum* und *parviflorum. Zeitschr. Bot.* **10**: 497–511.

Lehmann, E. (1924). Über Sterilitätserscheinungen bei reziprok verschiedenen Epilobiumbastarden. *Biol. Zentr.* **44**: 243–254.

Lehmann, E. (1925). Die Gattung *Epilobium. Bibl. Genet.* **1**: 363–416.

Lehmann, E. & Schwemmle, J. (1927). Genetische Untersuchungen in der Gattung *Epilobium. Bibl. Bot.* **95**: 1–156.

Léveille, H. (1910). Iconographie du genre *Epilobium*. Paris.

Little, J. E. & Pugsley, H. W. (1923). *Epilobium palustre*×*parviflorum*? *Rep. Watson Bot. Exch. Club* **3**: 216–217.

Marshall, E. S. (1888). Notes on Highland plants. *Jour. Bot. (Lond.)* **26**: 149–156.

Marshall, E. S. (1889). Notes on *Epilobia*. *Jour. Bot. (Lond.)* **27**: 143–147.

Marshall, E. S. (1890). *Epilobium* notes for 1889. *Jour. Bot. (Lond.)* **28**: 2–10.

Marshall, E. S. (1891). *Epilobium* notes for 1890. *Jour. Bot. (Lond.)* **29**: 6–9.

Marshall, E. S. (1895). Two hybrid *Epilobia* new to Britain. *Jour. Bot. (Lond.)* **33**: 106–108.

Marshall, E. S. (1899). Notes on West Surrey plants. *Jour. Bot. (Lond.)* **37**: 249.

Marshall, E. S. (1913). *Epilobium hirsutum* L.×*E. montanum* L. *Rep. Bot. Soc. Exch. Club. Brit. Isles* **3**: 254.

Marshall, E. S. (1916). A new hybrid willow-herb. *Jour. Bot. (Lond.)* **54**: 75–76.

Marshall, E. S. (1916). *Epilobium hirsutum*×*palustre* and *E. palustre*×*tetragonum* in E. Kent. *Jour. Bot. (Lond.)* **54**: 114–115.

Marshall, E. S. (1918). *Epilobium hirsutum*×*roseum* in Surrey. *Jour. Bot. (Lond.)* **56**: 332–333.

McClintock, D. (1972). New Zealand *Epilobiums* in Britain. *Watsonia* **9**: 140–142.

McClintock, D. (1973). *Epilobium komarovianum* Leveille. *Watsonia* **9**: 274. (det. by McClintock in *Plant Records*.)

McKean, D. R. (1999). A new Epilobium hybrid from Scotland, *E. pedunculare* A. Cunn. × *E. montanum* L. *Watsonia* **22**: 417–419.

Melville, R. (1960). *Epilobium pedunculare* A. Cunn. and its allies. *Kew Bull.* **14**: 296–300.

Michaelis, P. (1944). Untersuchungen an reziprok verschiedenen Artbastarden bei *Epilobium*. 1. Über die Bastarde verschiedener Sippen der Arten *E. hirsutum* mit *E. parviflorum*, resp. *E. montanum*. *Flora* **137**: 1–23.

Michaelis, P. & Ross, H. (1944). Untersuchungen an reziprok verschiedenen Artbastarden bei *Epilobium*. 2. Über Abanderungen an reziprok verschiedenen und reziprok gleichen *Epilobium* Bastardes -Artbästarden. *Flora* **137**: 24–56.

Myerscough, P. J. & Whitehead, F. H. (1966). Comparative biology of *Tussilago farfara* L., *Chamaenerion angustifolium* (L.) Scop., *Epilobium montanum* L. and *Epilobium adenocaulon* Hausskn. I. General biology and germination. *New Phytol.* **65**: 192–210; (1967) II. Growth and ecology. **66**: 785–823.

Preston, C. D. (1989). The spread of *Epilobium ciliatum* Raf. in the British Isles. *Watsonia* **17**: 279–288.

Raven, P. H. & Moore, D. M. (1964). Chromosome numbers of *Epilobium* in Britain. *Watsonia* **6**: 36–38.

Raven, P. H. & T. E. (1976). The genus *Epilobium* (Onagraceae) in Australasia: a systematic and evolutionary study. *Bull. New Zealand Dept. Sci. Ind. Research* **216**.

Salter, T. Bell (1852). On the fertility of certain hybrids. *Phytologist* **4**: 737–742.

Schmitz, U. K. (1988). Dwarfism and male sterility in interspecific hybrids of *Epilobium*. I. Expression of plastid genes and structure of the plastome. *Theoret. and Appl. Genetics* **75**: 350–356.

Schnitzler, O. (1933). Untersuchungen über reziprok verschiedene Bastarde in der Gattung *Epilobium*. *Zeitschr. Induktive Abstammungs- und Verchungslehre* **63**: 305–356.

Seavey, S. R. & Raven, P. H. (1977). Chromosomal evolution in *Epilobium* sect. *Epilobium*. (Onagraceae). *Pl. Syst. and Evol.* **127**: 107–119; **128**: 195–200.

Shamsi, S. R. & Whitehead, F. H. (1974). Comparative eco-physiology of *Epilobium hirsutum* L. and *Lythrum salicaria* L. 1. General biology, distribution and germination. *Jour. Ecol.* **62**: 279–290.

Shaw, H. K. Airy (1951). An interesting *Epilobium* population. *Yearbook B.S.B.I.* **1951**: 78.

Sledge, W. A. (1945). *E. alsinifolium × montanum*. *Naturalist* **812**: 24.

Smejkal, M. (1974). *Epilobium × novae-civitatis* hybr. nova (*E. adenocaulon × hirsutum*), ein neuer Bastard. *Preslia* **46**: 64–66.

Smejkal, M. (1995). Sieben neue Bastarde in der Gattung *Epilobium* L. (Onagraceae) *Acta Musei Moraviae Sci. Nat.* **79**: 81–84.

Smejkal, M. (1997). *Epilobium* L. in Slavik, B. ed. *Květena* +České republiky. **5.** Prague.

Stewart, A., Pearman, D. A. & Preston, C. D. (1994). *Scarce plants in Britain*. Peterborough. [*E. alsinifolium* and *lanceolatum*.]

Townsend, C. C. (1953). *Epilobium*. *Watsonia* **2**: 412.

Walters, S. M. (1979). *Epilobium lanceolatum* Seb. & Mauri – a plant to look for in your garden. *Watsonia* **12**: 399.

Wolley Dod, A. H. (1893). *Epilobium hirsutum × obscurum* in Cheshire. *Jour. Bot. (Lond.)* **31**: 372.

1. Stems quite prostrate or only tips ascending, slender and rooting at nodes; leaves usually 2–10 mm, broadly ovate to subrotund; flowers solitary and axillary — 2.
1. Stems usually more or less erect, sometimes ascending from a decumbent base, never wholly prostrate; leaves more than 12 mm; flowers in terminal and axillary racemes, rarely solitary and then terminal — 4.
2. Leaves distinctly dentate with 3–14 teeth on each side — **13. pedunculare**
2. Leaves entire or with a few obscure teeth — 3.
3. Leaves with obscure veins on upper surface — **12. brunnescens**
3. Leaves with more or less prominent veins on upper surface — **14. komarovianum**
4. Stems with spreading hairs, mostly eglandular but some shorter glandular hairs on upper part of stem; stigma 4-lobed — 5.
4. Stems glabrous or with more or less appressed eglandular hairs; spreading hairs if present, glandular only; stigma either 4-lobed or entire — 6.
5. Leaves semiamplexicaul and slightly decurrent; petals 12–16 mm, deep rose — **1. hirsutum**
5. Leaves neither amplexicaul or decurrent, petals 6–9 mm, pale rose — **2. parviflorum**
6. Stigma 4-lobed — 7.
6. Stigma entire, capitate to clavate — 8.
7. Leaves ovate-lanceolate to ovate, the whole margin toothed including the rounded base; flowers pink in bud and when expanded — **3. montanum**
7. Leaves elliptical to elliptical-lanceolate, basal part entire; flowers white in bud, pink when expanded — **4. lanceolatum**
8. Leaves with petioles 3–20 mm — **7. roseum**
8. Leaves sessile or with petioles not exceeding 3 mm — 9.
9. Plant producing very slender stolons in summer, which finally end in bulbil-like buds; stems with no distinct ridges or raised lines decurrent from the leaves, but sometimes with 2 rows of crisped hairs; leaves narrowly lanceolate to linear-lanceolate — **9. palustre**
9. Plant with or without stolons; stems with 2 or 4 ridges or raised lines decurrent from the leaves; leaves usually lanceolate to ovate, but if narrowly lanceolate not markedly narrowed at base — 10.
10. Stems erect or ascending, 25–100(–150) cm; top of inflorescence and unopened flowers erect or slightly drooping — 11.
10. Small plants of upland flushes and streamsides with decumbent or ascending stems only 4–20(–30) cm; top of inflorescence and unopened flowers markedly drooping — 14.
11. Upper part of stems with numerous, slender, spreading glandular hairs as well as more or less appressed crisped hairs — **8. ciliatum**
11. Spreading glandular hairs absent from upper part of stems — 12.
12. Elongating epigeal stolons formed near the base of the plant in summer; plant with few, spreading glandular hairs on the hypanthial tube; capsule 4–6(–6.5) mm — **6. obscurum**
12. Almost sessile leaf-rosettes formed at the base of the plant in autumn; plant wholly without spreading glandular hairs; capsule 6.5–9.0(–11.0) mm — 13.

13. Leaves sessile, decurrent as lines on the stem; flowers 6–10 mm in diameter **5(a). tetragonum** subsp. **tetragonum**
13. Leaves shortly petiolate and not decurrent on the stem; flowers 10–12 mm in diameter **5(b). tetragonum** subsp. **lamyi**
14. Plant producing in summer numerous more or less prostrate, slender, epigeal stolons with distant pairs of small green leaves; stems 1–2 mm in diameter; leaves 1.0–1.5 cm, yellowish-green, more or less narrowly ovate-oblong, obtuse at apex **10. anagallidifolium**
14. Plant producing in summer slender, yellowish, hypogeal stolons with distant pairs of yellowish scale-leaves; stems 2–3 mm in diameter; leaves 1.5–4.0 cm, bluish-green, ovate to ovate-lanceolate, acute at apex **11. alsinifolium**

1. E. hirsutum L. Great Willowherb
E. ramosum Huds.

Perennial herb in summer producing white, fleshy, underground stolons. *Stems* 80–150 cm, pale yellowish-green to pale brownish, erect, almost terete, with numerous, long, spreading, pale simple eglandular hairs and above with more or less dense, short, pale glandular hairs intermixed, branched above and sometimes throughout with ascending branches, very leafy. *Leaves* gradually becoming smaller upward and on the branches, mostly opposite; lamina 4–12 × 1–3 cm, rather shiny, medium yellowish-green on upper surface, slightly paler beneath, oblong, oblong-lanceolate or lanceolate, more or less acute at apex, serrate, with unequal, incurved teeth, semiamplexicaul at base and slightly decurrent, with very short to long, pale, soft simple eglandular hairs on both surfaces and the margins, the veins impressed above and prominent beneath. *Inflorescence* of more or less corymbose racemes terminating the main stem and branches; pedicels 5–20 mm, erect, densely glandular-hairy; flowers 5–25 mm in diameter, erect in bud. *Hypanthium* see under genus. *Sepals* 4, 7–9 mm, pale yellowish-green, sometimes tinted pink, lanceolate, hooded and shortly apiculate, with dense, short, pale glandular hairs and rather few, longer, pale simple eglandular hairs. *Petals* 4, 12–16 mm, deep rose-purple, broadly obovate, distinctly but not deeply emarginate. *Stamens* 4 long and 4 short; filaments whitish, tinged purple; anthers cream. *Style* white; stigma of 4 revolute lobes, exceeding all the stamens. *Capsules* 5–8 cm, pale green, linear, with some long simple eglandular hairs and dense, short, pale glandular hairs; seeds about 0.8–1.2 mm, brownish-red, oblong-obovoid, truncate at hairy end, minutely and uniformly papillose. *Flowers* 7–8. Protandrous and visited chiefly by bees and hoverflies. $2n = 36$.

(i) Var. **hirsutum**
Plant with numerous to dense, short glandular hairs and few to fairly numerous, longer, white simple eglandular hairs.

(1) Forma **hirsutum**
Flowers deep rose.

(2) Forma **albiflorum** Hausskn.
Flowers white.

(ii) Var. **villosissimum** W. D. J. Koch
E. hirsutum var. *intermedium* auct.; *E. hirsutum* var. *tomentosum* auct.

Plant with numerous, short glandular hairs and dense, long, white simple eglandular hairs especially on the upper stem, pedicels and calyx. The dense hairiness is retained in cultivation.

Native. Characteristically forming dense stands along ditches and streams and often dominant over large areas of marshes and fens and damp hollows in meadows; sometimes forming dwarf plants in drier places, even streets, and often common in set-aside fields. Common throughout Great Britain and Ireland except in north and central Scotland. Europe northwards to south Sweden; temperate Asia; north, east and southern Africa; introduced in North America. Eurasian Southern-temperate element. Var. *villosissimum* is recorded in scattered records over England north to Yorkshire.

× **lanceolatum** Sebast. & Mauri = **E.** × **surreyanum** E. S. Marshall
The specimens from Worplesdon in Surrey on which this hybrid is based are considered by Kitchener to be very doubtful.

× **montanum** = **E.** × **erroneum** Hausskn.
Perennial herb up to 100 cm with close-growing rosettes, sometimes with weak stolons. *Stems* pale green, often suffused brownish-purple, with dense, unequal, short glandular hairs and varying amounts of longer simple eglandular hairs. *Leaves* 3–12 × 0.5–3.5 cm, linear-lanceolate to lanceolate or ovate, usually acute at apex, serrulate, the teeth sometimes incurved, narrowed or rounded at base, sometimes semiamplexicaul, usually with some long simple eglandular hairs especially on the margin. *Flowers* can be minute, poorly formed and dry up on opening, although parts are present in rudimentary condition, with every gradation to plants with large showy petals which are usually deep rose-purple. *Stigma* 4-lobed. Good seeds are not usually produced.

Native. Scattered records in England, Wales and southern Scotland. Also widespread in central Europe.

× **obscurum** = **E.** × **anglicum** E. S. Marshall
Perennial herb with fleshy, greenish-white stolons at or below ground level. *Stems* pale green, sometimes tinted reddish, with few, curled simple eglandular hairs below and numerous above, mixed with some glandular hairs, with wide-spreading side-branches throughout the stem. *Leaves* linear to oblong-lanceolate, acute at apex, the teeth sometimes incurved. *Flowers* usually deep rose-purple, may be paler in centre, large and showy. *Stigmas* obscurely lobed. *Capsules* up to 7 cm; seeds generally sterile.

Native. Rare endemic. First found in 1890 from Surrey by E. S. Marshall and later from Cheshire. In 2004 it was found in Co. Durham.

× **palustre** = **E.** × **waterfallii** E. S. Marshall
Perennial herb with slender stolons. *Stems* pale green, sometimes tinted reddish, slender, with few to numerous branches, with numerous, short glandular hairs and numerous, longer simple eglandular hairs. *Leaves*

3–6 × 0.5–1.0 cm, narrowly lanceolate, acute at apex, denticulate or serrulate, the teeth sometimes incurved, narrowed or rounded below, sometimes semiamplexicaul, with numerous, long, white simple eglandular hairs. *Flowers* often deep rose-purple, rather large and showy but smaller than *E. hirsutum*. *Capsules* up to 40 mm, with numerous, short glandular hairs and dense, longer, white simple eglandular hairs. *Stigma* often obscurely 4-lobed. *Seeds* generally shrunken and sterile, but may include better developed ones up to 1.6 mm, reflecting some of the length and shape of those of *E. palustre*.

Native. Recorded in Devonshire, Kent, Sussex and Cheshire. Czech Republic, Poland and Russia. Named after Charles Waterfall (1851–1938).

× **parviflorum** = **E.** × **subhirsutum** Gennari
E. × *intermedium* Ruhmer, non Mérat

Perennial herb with fibrous roots and sometimes stolons. *Stems* pale green, sometimes tinted reddish, rather slender, with numerous, short glandular hairs in the upper part and varying amounts of simple eglandular hairs more like those of *E. parviflorum*. *Leaves* 1–13 × 0.8–2.2 cm, oblong or oblong-lanceolate, narrowed to an obtuse or acute apex, denticulate, the teeth sometimes incurved, semiamplexicaul and often slightly decurrent at base, with numerous, long, white simple eglandular hairs on both surfaces. *Flowers* often medium in size between the parents, often deep rose-purple. *Stigma* 4-lobed. *Seeds* at least partially fertile. *Flowers* 7–8.

Native. River-banks, waste or disturbed ground, marshes, gravel pits and quarries. Scattered records in England, Wales, Scotland, northern Ireland and Guernsey. Also widespread in mainland Europe.

× **roseum** = **E.** × **goerzii** Rubner
Perennial herb sometimes with stolons. *Stems* pale green to reddish, with numerous glandular hairs and various amounts of long simple eglandular hairs in upper part. *Leaves* oblong or oblong-lanceolate to ovate-lanceolate, pointed at apex, serrulate, the teeth sometimes incurved, sessile or sometimes semiamplexicaul at base. *Flowers* usually twice the size of *E. roseum* and deep pink to rose-purple. *Stigma* obscurely to distinctly 4-lobed.

Native. Records all before 1920 in Surrey, Middlesex, Monmouthshire and Derbyshire. Also scattered records in central and eastern Europe.

× **tetragonum** = **E.** × **brevipilum** Hausskn.
Perennial herb without the long runners of *E. hirsutum* but sometimes with short ones. *Stems* pale green, sometimes tinted reddish, robust, with few, curled simple eglandular hairs below and numerous to dense above, usually mixed with various amounts of glandular hairs. *Leaves* 3–8 × 0.5–1.0 cm, linear to oblong-lanceolate, mostly acute at apex, denticulate, the teeth sometimes incurved, narrowed or rounded below and sometimes semiamplexicaul, with numerous, short to long, more or less appressed, whitish simple eglandular hairs to nearly glabrous. *Flowers* smaller than in *E. hirsutum* and like those of *E. tetragonum*, but with the deep colouring of *E. hirsutum*. *Stigma* often obscurely lobed and may be almost clavate. Usually sterile.

Native. Waste and arable land, quarries and sand works. Scattered records in England mostly in central and southern area. Also in France, Germany and Holland.

2. E. parviflorum Schreb. Hoary Willowherb
Chamaenerion parviflorum Schreb; *E. villosum* Curt.

Perennial herb with overwintering rosettes or short, leafy stolons. *Stems* 20–100 cm, pale green or tinted red, erect, more or less terete, with dense, long, spreading simple eglandular hairs below mixed with short glandular hairs above, sometimes branched at the base, but frequently unbranched except at the apex, leafy. *Leaves* mostly opposite; lamina 1.5–10 × 0.5–2.5 cm, medium green on upper surface, paler beneath, oblong, ovate, lanceolate or oblong-lanceolate, obtuse or acute at apex, remotely and irregularly serrulate, rounded at the sessile base but not amplexicaul, with numerous to dense, curled simple eglandular hairs. *Inflorescence* of solitary, axillary flowers subtended by much-reduced alternate leaves in more or less corymbose terminal racemes; pedicels 3–10 mm, erecto-patent; flowers 6–13 mm in diameter, erect in bud. *Hypanthium* see under genus. *Sepals* 4, 4–6 mm, pale green, sometimes tinted red, oblong, acute at apex, with few to numerous simple eglandular and glandular hairs, connate for a short distance above the base. *Petals* 4, 6–9 mm, pale purplish-rose with darker veins, obcordate, rather deeply and bluntly 2-lobed, narrowing to a very short claw. *Stamens* 4 long and 4 short; filaments white; anthers cream or yellow. *Style* white, tinted mauve; stigma of 4 non-revolute fairly upright lobes about equalling the stamens. *Capsules* 30–80 × about 1.5 mm, tetragonous, with short glandular and some eglandular hairs; seeds 0.8–1.1 mm, dark brown, oblong, minutely and uniformly papillose, truncate at hairy end. *Flowers* 7–8. Visited sparingly by hive bees, but often self-pollinated. Homogamous. $2n = 36$.

Very variable in leaf shape of which forma ***longifolia*** Hausskn. ex E. S. Marshall is the most striking.

Native. Stream-banks, marshes, fens and other damp places up to 380 m in Derbyshire, and as a weed in waste ground and streets. Throughout lowland Great Britain and Ireland. Widespread in mainland Europe and Asia; North Africa and Atlantic Islands; introduced in New Zealand. European Temperate element.

× **roseum** = **E.** × **persicinum** Rchb.
Perennial herb sometimes producing short stolons. *Stems* usually much branched, pale green, with curved, subappressed simple eglandular hairs below and longer simple eglandular hairs above mixed with glandular hairs. *Leaves* opposite below and often alternate above; lamina 2–11 × 0.5–3.0 cm, linear-lanceolate or oblong-lanceolate to broadly lanceolate, mostly pointed at apex, serrulate, narrowed or rounded at base, with long simple eglandular hairs on both surfaces. The flowers are variable in size, mostly intermediate or nearer *E. roseum*, but sometimes as large as *E. parviflorum*. *Stigma* often obscurely lobed. *Sepals* and *capsule* with long simple eglandular and shorter glandular hairs.

Native. Roadsides, gardens, shrubberies and disturbed woodland. Scattered records in England and Wales. Also recorded across mainland Europe.

× **tetragonum** = **E. × palatinum** F. W. Schultz
E. × weissenburgense F. W. Schultz

Perennial herb sometimes producing short stolons. *Stems* usually much-branched, pale green to brownish, with numerous to dense, medium simple eglandular hairs, mixed with occasional shorter glandular hairs above. *Leaves* 2–9 × 0.5–1.5 cm, linear or linear-lanceolate to narrowly lanceolate, mostly pointed at apex, serrulate, usually narrowed at the sessile base, with more or less numerous, long simple eglandular hairs on both surfaces. The flowers vary in size between both parents. *Stigma* entire or often obscurely 4-lobed. *Sepals* and *capsules* with long simple eglandular hairs and shorter glandular ones.

Native. Disturbed or open ground in woodland, shrubberies, fallow fields and quarries. Scattered records in southern and central England and recorded in Guernsey and south Wales. Occurs widely across central Europe. Triple hybrids involving *E. obscurum* may also occur.

3. E. montanum L. Broad-leaved Willowherb
Chamaenerion montanum (L.) Scop.; *E. sylvaticum* Boreau

Perennial herb producing in late autumn short stolons which may be underground with fleshy pink and white scales, or above ground, very short and terminating in subsessile leaf-rosettes. *Stems* (5–)20–70 cm, pale green, often tinted reddish, erect, slender, terete, branched, leafy, with sparse short, curved simple eglandular hairs or nearly glabrous below mixed with a few to fairly numerous glandular hairs above. *Leaves* mostly opposite or occasionally in whorls of 3; lamina 4–7 × 1.5–3.0 cm, medium green on upper surface, paler beneath, ovate-lanceolate to ovate, acute at apex, whole margin sharply and irregularly toothed, rounded at base, subglabrous but usually with hairs on the margins and veins; petioles up to 6 mm, with narrow, slightly connate wings. *Inflorescence* of solitary axillary flowers forming a terminal, leafy raceme; pedicels 2–18 mm, with short, curved eglandular hairs and minute glandular hairs; flowers 6–9 mm in diameter, more or less drooping in young bud. *Hypanthium* see note under genus. *Sepals* 4, 5.0–6.5 mm, more or less reddish, lanceolate, more or less acute at apex, with curved simple eglandular and glandular hairs. *Petals* 4, 8–10 mm, pale at first then pink, obcordate, deeply 2-lobed at apex, narrowed at base to a short claw. *Stamens* 4 long and 4 short; filaments pale; anthers cream or yellow. *Style* 1, white; stigma of 4 short, non-revolute lobes, exceeded by the longer stamens. *Capsules* 40–80 × 1–2 mm, tetragonous, with short, curved simple eglandular hairs mixed with minute glandular hairs; seeds 1.0–1.2 mm, reddish-brown, narrowly obovoid, truncate at hairy end, minutely and uniformly papillose. *Flowers* 6–8. Homogamous. Sparingly visited by insects and commonly self-pollinated. $2n = 36$.

Native. Shady places, walls, rocks and cultivated ground. Common throughout Great Britain and Ireland. Europe to Norway and Finland; west Asia, Siberia and Japan. European Temperate element; also in temperate and east Asia.

× **obscurum** = **E. × aggregatum** Čelak.
Perennial herb sometimes with short stolons. *Stems* pale green or reddish tinted, usually much branched, with short, curled, appressed simple eglandular hairs throughout, densely so above, sometimes with glandular ones. *Leaves* opposite or whorled; lamina 1.5–9.0 × 0.3–2.5 cm, lanceolate to ovate-lanceolate, more or less pointed at apex, serrulate, narrowed or rounded at base, rather sparsely short eglandular hairy. *Flowers* 7–9 mm. *Stigma* often obscurely lobed. *Sepals* and *capsules* with dense, curled simple eglandular and glandular hairs. At least partially sterile.

Native. Disturbed ground, gardens, roadsides, quarries and woodland margins. Scattered records over lowland Great Britain and Ireland and probably one of our commonest hybrids. Also recorded across mainland Europe. It is capable of producing F_2 progeny.

× **obscurum × parviflorum**
This triple hybrid was said by Thakur (1965) to exist in various herbaria among specimens of *E. obscurum × parviflorum*.

× **palustre** = **E. × montaniforme** Knaf. ex Čelak.
Perennial herb sometimes producing filiform stolons. *Stems* pale green or reddish tinted, usually branched, with short, curled simple eglandular hairs. *Leaves* usually opposite; lamina ovate or ovate-lanceolate, more or less pointed at apex, serrulate, narrowed or rounded at base, rather sparsely clothed with simple eglandular hairs. *Flowers* 7–9 mm. *Stigma* sometimes obscurely lobed. *Sepals* and *capsules* with curled simple eglandular and some glandular hairs. At least partially sterile, but when seeds well-developed 1.3–1.8 mm, intermediate between those of the parents and extending into the range of *E. palustre*.

Native. Roadsides and quarries. Recorded for Leicestershire, Cumberland, Perthshire and Ross and Cromarty. Also recorded across central Europe.

× **parviflorum** = **E. × limosum** Schur
Perennial herb sometimes with short stolons. *Stems* pale green, usually robust, with long simple eglandular hairs below, mixed with short glandular hairs above. *Leaves* mainly opposite, sometimes alternate or whorled above; lamina 1.5–8.0 × 0.3–2.5 cm, lanceolate to ovate-lanceolate, obtuse to acute at apex, serrulate, narrowed or rounded at base, sessile, rather shortly hairy. *Flowers* 7–9 mm, pale pink. *Stigma* 4-lobed, the lobes often upright. *Sepals* and *capsules* with long simple eglandular hairs and short glandular hairs. Usually at least partially sterile.

Native. Gardens, roadsides, waste land, quarries and woodland margins. Scattered records in Great Britain and the north and east of Ireland. Recorded across central Europe. It is capable of producing F_2 progeny.

× **parviflorum × roseum**
This triple hybrid was discovered by E. S. Marshall near Worplesdon, Surrey in 1889 and determined by

C. Haussknecht. The parentage of *E. roseum* is apparent from the long petioles and changing flower colour; the upper stem hairs show some of the length of *E. parviflorum*; and though *E. montanum* is less clear may be reflected in the rounding of the leaf base (fide Kitchener).

× **pedunculare** = **E.** × **kitcheneri** D. R. McKean

Perennial herb. Stems about 20 cm, semi-prostrate, rooting at the lower nodes and occasionally branching, terete, generally clothed in dense, short, curled eglandular hairs. *Leaves* mostly opposite, but upper alternate; lamina up to 2.5 × 1.8 cm, slightly bronze-coloured on the underside, ovate, obtuse at apex, with 4–11 prominent teeth on each side, rounded at base, glabrous except for the ciliate margins; petioles 3–5 mm. *Sepals* about 0.3 mm, linear-lanceolate, acute at apex. *Petals* about 7 mm, pale pink. *Stamens* 8, 4 long ones slightly overtopping stigma, 4 short ones just reaching the stigma base and capable of shedding pollen there. *Style* 0.4 mm; stigma clavate or shortly 4-lobed. *Capsules* about 17 mm, with scattered, short, curled eglandular hairs; seeds mainly sterile, but with a few, well-developed ones 1.0–1.3 mm.

E. montanum is native in the British Isles and *E. pedunculare* is native to New Zealand. This endemic hybrid was described from a dampish hillside track in an oak wood, off Duke's Pass Road, near Aberfoyle in Perthshire. It was later found in Cornwall. Named after Geoffrey Kitchener (b. 1949).

× **roseum** = **E.** × **heterocaule** Borbás

E. × *mutabile* Boiss. & Reut.

Perennial herb sometimes with stolons. *Stems* pale green, usually robust, usually much branched and leafy, with short, curved simple eglandular hairs throughout and mixed with short glandular hairs above. *Leaves* opposite; lamina 1.5–10.0 × 0.5–5.0 cm, lanceolate, elliptical or ovate, mostly acute at apex, serrulate, narrowed or rounded at base, shortly hairy, mainly on the veins; petioles 3–6 mm. *Flowers* 5–8 mm. *Stigma* often obscurely lobed. *Sepals* and *capsules* with rather long simple eglandular hairs and short glandular hairs. At least partially sterile.

Native. Waste ground and gardens. Scattered records over Great Britain and in Co. Dublin in Ireland. Also recorded through mainland and northern Europe.

× **tetragonum** = **E.** × **haussknechtianum** Borbás

E. × *beckhausii* Hausskn.

Perennial herb. Stems pale green, much branched, leafy, often rather slender, with curled, subappressed simple eglandular hairs throughout, dense above. *Leaves* opposite below, alternate above; lamina 1.5–5.0 × 0.4–1.7 cm, linear-lanceolate to lanceolate, obtuse to acute at apex, serrulate, rounded or cuneate at base, with minute hairs on the veins and margins. *Flowers* 5–7 mm. *Stigma* often obscurely lobed. *Sepals* and *capsules* with dense, short, appressed simple eglandular hairs. At least partially sterile.

Native. Quarries, gardens and shrubberies. Scattered records in southern and central England. Also recorded from central Europe. Named after Heinrich Carl Haussknecht (1838–1903).

4. E. lanceolatum Sebast. & Mauri

Spear-leaved Willowherb

Perennial herb producing in late autumn, short, above-ground stolons terminating in spreading leaf-rosettes. *Stems* 20–60(–90) cm, pale green, often suffused reddish, simple or with slender, lateral branches, leafy, rather slender, erect, with 4 hardly raised lines, subglabrous below and downy with short, curled simple eglandular hairs below and mixed with small glandular hairs above. *Leaves* with only lower opposite; lamina 1–8 × 0.3–2.5(–3.0) cm, medium green on upper surface, paler beneath, elliptical to elliptical-lanceolate, obtuse at apex, with more or less equal, rather distant marginal teeth except at the entire cuneate base which narrows gradually into a petiole 4–8 mm, with short simple eglandular hairs on the veins and margins. *Inflorescence* of solitary, axillary flowers, each subtended by a small, much reduced leaf or bract and forming a terminal raceme; pedicels 8–35 mm, with simple eglandular hairs intermixed with small, shining glandular ones; flowers 6–7 mm in diameter, drooping in bud. *Hypanthium* see under genus. *Sepals* 4, 2.5–4.0 mm, often reddish, oblong-lanceolate, more or less acute, connate for a short distance above the base, with small glandular hairs and curled eglandular ones. *Petals* 4, 6–8 mm, white turning pale pink and becoming deeper in colour, obcordate, 2-lobed at apex, narrowed at base to a short claw. *Stamens* 4 long and 4 short; filaments pale; anthers yellow. *Style* 1; stigma of 4 short, spreading lobes. *Capsules* 50–70 × about 1.5 mm, tetragonous, straight or slightly arcuate, with short, curled simple eglandular and short glandular hairs; seeds about 1.2 mm, pale to dark brown, narrowly oblong-ellipsoid, truncate at hairy end, minutely and uniformly papillose. *Flowers* 7–9. Probably self-pollinated. $2n = 36$.

Native. Waysides, walls, dunes, quarries, gardens and waste places. Locally frequent in southern Great Britain north to Leicestershire and Merionethshire, and in the Channel Islands. West and south Europe from France and Belgium to the Balkans; North Africa; Caucasus to Iran. Submediterranean-Subatlantic element.

× **montanum** = **E.** × **neogradense** Borbás

Perennial herb. Stems often suffused reddish, rather robust, usually much branched, leafy, with fairly numerous, short, curled simple eglandular hairs throughout, and some small glandular ones. *Leaves* opposite or sometimes alternate above; lamina 1.5–6.0 × 0.5–2.0 cm, lanceolate to ovate, more or less acute at apex, serrulate, cuneate or rounded at base, with very short hairs on the veins and sometimes the surfaces, with a petiole up to 6 mm. *Flowers* 5–10 mm, pale in bud. *Stigma* 4-lobed. *Sepals* and *capsules* with numerous, short, curled simple eglandular intermixed with short glandular hairs. At least partially sterile.

Native. Gardens, quarries, roadsides and banks. Few scattered records in southern England and Wales, particularly Devonshire, Cornwall and Surrey. Also recorded for western and central Europe.

× **obscurum** = **E.** × **lamotteanum** Hausskn.

Perennial herb. Stems often suffused reddish, often branched, slender to robust, with numerous, short, curled simple eglandular hairs throughout mixed with short glandular hairs above which extending on to the floral tube. *Leaves* mostly opposite, sometimes alternate above; lamina 1.5–7.0×0.4–2.0 cm, linear to lanceolate, obtuse to long acute at apex, serrulate, rounded or narrowed at base, with few to fairly numerous, short, curly simple eglandular hairs; shortly petiolate or sessile. *Flowers* 4–7 mm. *Stigma* often obscurely 4-lobed. *Sepals* and *capsules* with dense, short, curled simple eglandular hairs and intermixed with short glandular hairs. At least partially sterile; some well-developed seeds produced.

Native. Gardens, quarries and disturbed ground. Scattered records in southern England and Caernarvonshire. Also recorded from Spain, France, Germany and Holland. Named after Martial Lamotte (1820–1883).

× **palustre** = **E.** × **langeanum** Hausskn.
In Great Britain and Ireland, known only from E. S. Marshall's garden at Milford vicarage, Surrey. There is considerable doubt about the identification.

× **parviflorum** = **E.** × **aschersonianum** Hausskn.
Perennial herb. Stems often suffused reddish, long-branched, rather robust, with short, curled simple eglandular hairs below, becoming longer above and mixed with short glandular hairs. *Leaves* opposite or alternate above; lamina 1.5–5.0×0.3–1.2 cm, linear-lanceolate to lanceolate, more or less acute at apex, prominently toothed, narrowed at base, shortly hairy, particularly on the veins, petiole up to 4 mm. *Flowers* 6–10 mm, pale in bud. *Stigma* 4-lobed. *Sepals* and *capsules* with rather long, curved simple eglandular hairs mixed with short glandular hairs. At least partially sterile.

Native. Recorded in the nineteenth century from south Devonshire near Plymouth. Also recorded from France and Holland. Named after Paul Friedrich August Ascherson (1834–1913).

× **roseum** = **E.** × **abortivum** Hausskn.
Perennial herb. Stems often suffused reddish, branched, faintly 4 lined, with short, curled simple eglandular hairs mixed with short glandular hairs above. *Leaves* mainly opposite with some alternate above; lamina 2–7×0.3–1.7 cm, greyish-green, lanceolate or ovate-lanceolate, obtuse to acute at apex, serrulate, cuneate at base, shortly hairy mainly on the veins; petioles up to 12(–15) mm. Flowers 5–10 mm, white in bud turning pink. *Stigma* shortly and obscurely 4-lobed. *Sepals* and *capsules* with dense, short, curled simple eglandular hairs mixed with short glandular ones. At least partially sterile.

Native. Recorded from Kent and possibly Shropshire in England. Also recorded for Germany. As *E. lanceolatum* prefers dry habitats and *E. roseum* damp ones they do not often come together.

× **tetragonum** = **E.** × **fallacinum** Hausskn.
E. × *ambigens* Hausskn.

Perennial herb. Stems often suffused reddish, often branched, slender to robust, with short, curled simple eglandular hairs throughout and more numerous above. *Leaves* mostly opposite, but some alternate above; lamina 1.5–9.0×0.2–1.5 cm, linear or linear-lanceolate, pointed at apex, irregularly serrulate, narrowed or rounded at base with short petioles, glabrous or with a few hairs on the veins and margins. *Flowers* 5–8 mm, usually poorly developed. *Stigma* often obscurely 4-lobed. *Sepals* and *capsules* with short, appressed simple eglandular hairs and some small glandular hairs. At least partially sterile.

Native. Recorded for Surrey, Oxfordshire and Buckinghamshire in England and Guernsey in the Channel Islands. Also recorded from France, Germany and Holland.

5. E. tetragonum L. Square-stalked Willowherb
E. adnatum Griseb.

Perennial herb producing in late autumn several lax leaf-rosettes on very short stolons. *Stems* 25–90 cm, pale green, sometimes tinted brownish-red, erect, firm, tough, with (2–)4 conspicuously raised lines or often wings decurrent from the leaves, glabrous below, but with silky, more or less appressed, white simple eglandular hairs above, branched, leafy. *Leaves* with the lower and middle usually opposite and the rest alternate; lamina 2.0–7.5×0.3–1.5 cm, dull medium green on upper surface, paler beneath, sometimes tinted reddish, linear, oblong or oblong-lanceolate, obtuse to more or less acute at apex, strongly and irregularly denticulate, narrowed at base, glabrous or slightly eglandular hairy on the veins, the midrib prominent beneath, sessile or with a short petiole. *Inflorescence* a bracteate raceme, the bracts alternate, like the leaves but smaller, the flowers 6–12 mm in diameter, the buds erect. *Hypanthium* see under genus. *Sepals* 4, 4.0–4.5 mm, medium green, lanceolate, acute at apex, with short, appressed, acute-tipped simple eglandular hairs. *Petals* 4, 5–7 mm, mauve, ovate, deeply emarginate. *Stamens* 4 long and 4 short; filaments white; anthers cream. *Style* 1; stigma clavate, entire, equalling the style. *Capsules* (6.5–)7–9(–11) cm, pale green, linear, with short, appressed, curled eglandular hairs; seeds 1.0–1.2 mm, reddish-brown, truncate at hairy end, minutely and uniformly papillose. *Flowers* 7–8. Homogamous, the anthers dehiscing before the flowers open. Self-pollinated. $2n=36$.

(**a**) Subsp. **tetragonum**
Leaves sessile, decurrent as lines on the stem. *Flowers* 6–10 mm in diameter.

(**b**) Subsp. **lamyi** (F. W. Schultz) Nyman
E. lamyi F. W. Schultz

Leaves shortly petiolate and not decurrent on the stem. *Flowers* 10–12 mm in diameter.

More needs to be known about the two subspecies, especially if they occur in populations and if their ecology is different. Intermediates are said to occur. The identification of a single specimen of subsp. *lamyi* here and there is not helpful.

Native. Hedgebanks, woodland clearings, streams and ditch-sides, cultivated and waste land. Locally common in south and central Great Britain and very scattered records further north and in Ireland. Europe northwards to southern Sweden and in western Asia. Eurosiberian Temperate element. Subsp. *tetragonum* occurs throughout the range of the species. Subsp. *lamyi* is recorded mainly from southern Great Britain and is widespread in mainland Europe, Turkey and Madeira.

6. E. obscurum Schreb. Short-fruited Willowherb
E. tetragonum subsp. *obscurum* (Schreb.) Hook. fil.; *E. virgatum* Lam.

Perennial herb producing in late summer slender, more or less elongated stolons above or below ground, bearing fairly distant pairs of small leaves which do not form distinct rosettes. *Stems* 30–60(–80) cm, pale green, erect from a curving base, with 4(2) distinctly raised lines, slender to robust, glabrous below, but with more or less appressed, whitish simple eglandular hairs above. *Leaves* with the lower opposite, the upper mostly alternate; lamina 3–7 × 0.8–1.7 cm, dull medium green on upper surface, paler beneath, lanceolate to ovate-lanceolate, or the lower oblong-lanceolate, tapering to an obtuse apex, with a few, distant, irregular teeth, rounded at the sessile base and suddenly contracted to become decurrent into the raised lines of the stem, glabrous or hairy only on the margins or veins. *Inflorescence* a bracteate raceme, the bracts alternate, like the leaves but smaller, the flowers 7–9 mm in diameter, the buds erect and acute. *Hypanthium* see under genus. *Sepals* 4, 3–4 mm, lanceolate, acute at apex. *Petals* 4, 5–6 mm, deep rose, obcordate, shortly 2-lobed, narrowed at the base. *Stamens* 4 long and 4 short; filaments mauve; anthers cream. *Style* 1, pale; stigma clavate and entire, equalling the style. *Capsules* 4–6 cm, tetragonous, with short, curled hairs; seeds (0.85–)0.9–1.0 (–1.1) mm, reddish, narrowly oblong-ellipsoid, truncate at hairy end, minutely and uniformly papillose. *Flowers* 7–8. Homogamous and self-pollinated. $2n=36$.

Native. Hedgerows, open woods, by water and cultivated and waste land. Throughout Great Britain and Ireland but less common than *E. tetragonum* in the south. Europe northwards to central Norway; Turkey; North Africa; Madeira; Caucasus; New Zealand. European Temperate element.

× **palustre** = **E.** × **schmidtianum** Rostk.
E. × *ligulatum* Baker

Perennial herb often with stolons. *Stems* with raised lines below, much-branched, usually slender, with few, short, curled simple eglandular hairs below, becoming numerous above where glandular ones may also occur. *Leaves* opposite or alternate; lamina 2–10 × 0.3–1.3 cm, linear-lanceolate to lanceolate, gradually drawn out to a pointed apex, entire, wavy or slightly serrulate, narrowed at base, with short eglandular hairs especially on the veins and margins, sessile or nearly so. *Flowers* 4–7 mm, often deeply coloured. *Stigma* entire and clavate. *Sepals* and *capsules* with dense, appressed simple eglandular hairs and shorter glandular hairs. At least partially sterile, but the developed seeds can exceed 1.3 mm and extend through the whole length of the range of *E. palustre*.

Native. Marshes and ditches in acid conditions, particularly where there has been disturbance. Scattered records over much of Great Britain and Ireland. It is widespread in Continental Europe.

× **palustre** × **parviflorum**
This origin was suggested by Fitter (1980) as the identity of a willowherb growing with all three parents in an area of cleared carr near York.

× **parviflorum** = **E.** × **dacicum** Borbás
Perennial herb sometimes with short epigeal stolons. *Stems* often branched, sometimes reddish, with rather long, curved simple eglandular hairs mixed with short glandular hairs above. *Leaves* opposite below and alternate above; lamina 2–8 × 0.3–1.5 cm, lanceolate to oblong-lanceolate, gradually narrowed to an acute apex, serrulate, rounded or narrowed below, sessile or nearly so. *Flowers* 7–10 mm. *Stigma* obscurely 4-lobed or sometimes entire and clavate. *Sepals* and *capsules* with rather long simple eglandular hairs mixed with short glandular hairs. *Seeds* mainly abortive.

Native. Young plantations, quarries, marshes and roadsides. Scattered records over much of Great Britain and Ireland. Occurs widely in mainland Europe.

× **roseum** = **E.** × **brachiatum** Čelak.
Perennial herb. Stems pale green, branched, slender to robust, with curved simple eglandular hairs throughout, mixed with short glandular hairs above. *Leaves* opposite or sometimes alternate above; lamina 2–8 × 0.2–2.2 cm, lanceolate, oblong or elliptical, obtuse to acute at apex, serrulate, narrowed or rounded at base, glabrous or shortly hairy on the veins, sessile or shortly petiolate. *Flowers* 5–7 mm, can be whitish in bud. *Stigma* entire and clavate. *Sepals* and *capsules* with dense, appressed simple eglandular hairs mixed with short glandular hairs.

Native. Disturbed ground, ditches and shrubberies. Scattered records in England and Wales and Lanarkshire in Scotland. Also recorded through central Europe.

× **tetragonum** = **E.** × **semiobscurum** Borbás
E. × *thuringiacum* Hausskn.

Perennial herb. Stems usually branched, pale or suffused brownish-red, with short, dense, appressed, curled simple eglandular hairs. *Leaves* opposite below and alternate above; lamina 1.5–6.0 × 0.2–1.2 cm, linear, oblong or lanceolate, mostly acute, serrulate, rounded or narrowed at base, with short hairs mainly on the veins, sessile or very shortly petiolate. *Flowers* 5–7 mm. *Stigma* entire and clavate. *Sepals* and *capsules* with dense, short simple eglandular hairs mixed, on the floral tube with shorter glandular hairs. At least partially sterile, the seed surfaces papillose.

Native. Quarries, pits and gardens. Scattered records over Great Britain north to Yorkshire. Occurs also in western and central Europe.

7. E. roseum Schreb. Pale Willowherb
E. tetragonum var. *roseum* (Schreb.) Gray

Perennial herb producing in late autumn short stolons and lax, subsessile leaf-rosettes from very short stolons. *Stems* 25–60(–80) cm, pale green, more or less branched, slender to robust, erect, fragile, usually with 2 distinct and 2 indistinct raised lines, glabrous below, with whitish, curled eglandular hairs and numerous, spreading glandular hairs above. *Leaves* with lower opposite or all alternate; lamina 3–8×1.5–3.0 cm, medium green on upper surface, paler beneath, ovate-elliptical to lanceolate-elliptical, narrowed to the acute apex, finely and sharply toothed, cuneate at base, glabrous or with eglandular hairs only on the prominent veins; petiole 3–20 mm. *Inflorescence* a bracteate raceme, the bracts like the leaves only smaller, the flowers 4–6 mm in diameter, their buds cuspidate and drooping. *Hypanthium* see under genus. *Sepals* 4, 3.0–3.5 mm, pale green, lanceolate, acute at apex, with short, curved eglandular and short glandular hairs. *Petals* 4, 4–7 mm, at first white, then streaked with rose-pink, shortly 2-lobed. *Stamens* 4 long and 4 short; filaments mauve; anthers cream. *Style* 1; cream; stigma entire and clavate, about equalling the style. *Capsules* 4–7 cm, pale green, linear, downy with curved eglandular and glandular hairs; seeds 1.0–1.1 mm, oblong-obovoid, more or less truncate at hairy end, minutely and uniformly papillose. *Flowers* 7–8. $2n=36$.

Native. Shady places, damp ground, cultivated and waste land. Scattered throughout most of Great Britain and Ireland and locally frequent but apparently decreasing. Europe northwards to south Scandinavia; Turkey. Eurosiberian Temperate element.

× **tetragonum** = **E.×borbasianum** Hausskn.
E.×dufftii Hausskn.

Perennial herb. Stem pale green sometimes tinted red, with curled, appressed simple eglandular hairs and spreading glandular hairs above. *Leaves* mostly opposite, elliptic-lanceolate, prominently toothed, narrowed at base. *Flowers* 4–7 mm, can be white in bud. *Stigma* clavate. *Sepals* and *capsules* with curved simple eglandular and short glandular hairs. Usually partially sterile.

Native. Recorded for Kent, Surrey, Middlesex and Buckinghamshire. Also occurs from Belgium, France, Holland and Spain to Poland. Named after Vincent von Borbás (1844–1905).

8. E. ciliatum Raf. American Willowherb
E. adenocaulon Hausskn.; *E. watsonii* Barbey

Perennial, non-stoloniferous *herb*, with sessile or subsessile basal rosettes formed in late summer, having at first small, fleshy, rounded leaves, but later developing normal ones. *Stems* (30–)60–100(–150) cm, pale green, often suffused reddish, often much-branched, leafy, erect, with 4 raised lines except at the base where there are only 2, more or less glabrous below, more or less densely clothed above with short, curled simple eglandular hairs and short, spreading glandular hairs. *Leaves* all but the uppermost opposite; lamina 3–10×0.5–3.0 cm, rather shiny, medium yellowish-green above, paler beneath, sometimes flushed red, oblong-lanceolate, lanceolate or linear-lanceolate, gradually tapering to an acute apex, with numerous, irregular, forwardly directed teeth, suddenly narrowed to a rounded or subcordate base, more or less glabrous; midrib very prominent and lateral veins fairly prominent beneath; petiole 1.5–3.0 mm. *Inflorescence* a bracteate raceme, the bracts like the leaves, but smaller, the flowers 6–8 mm in diameter, erect in bud. *Hypanthium* see under genus. *Sepals* 4, 3–4 mm, pale green, sometimes red-tinted especially along the margin, lanceolate, obtuse at apex, with short glandular hairs. *Petals* 4, 3–4 mm, rose with darker veins, ovate, with a rounded, deeply emarginate apex, the 2 lobes more or less parallel. *Stamens* 4 long and 4 short; filaments mauve; anthers cream. *Style* 1, cream; stigma pale yellow, clavate. *Capsules* 4.0–7.5 cm, pale green, linear, with numerous, curled simple eglandular hairs and spreading, short glandular hairs, or becoming glabrous when ripe; seeds 1.0–1.3 mm, pale reddish-brown, when ripe with a usually paler translucent appendage below the plume of hairs, with longitudinal papillose ridges. *Flowers* 6–8. Probably self-pollinated. $2n=36$.

Introduced. Damp woods, copses, streamsides, railway banks, cultivated and waste places. First recorded in 1891 since when it has spread rapidly over most of Great Britain and Ireland and is often the commonest *Epilobium* in areas in southern England. A detailed account of its spread in Great Britain and Ireland is given by Preston (1988). Native of North America. Naturalised over a large part of Europe, far eastern Russia, north-east China, Japan, Korea, Australasia, Chile and Argentina.

× **hirsutum** = **E.×novae-civitatis** Smejkal
Perennial herb. Stems often branched, often suffused reddish, with long, spreading simple eglandular hairs throughout mixed with long and short, slender, spreading glandular hairs in the upper part. *Leaves* opposite; lamina 2–8×0.4–2.0 cm, oblong or lanceolate-oblong, acute at apex, serrulate, sometimes with incurved teeth, rounded and clasping at base, with long and short hairs. *Flowers* 5–8 mm, often deep-coloured but smaller than *E. hirsutum*. *Stigma* often obscurely lobed. *Sepals* and *capsules* with spreading simple eglandular hairs mixed with long, spreading glandular hairs; seeds when developed sufficiently showing the surface influenced by that of *E. ciliatum*, with evidence of papillose ridging.

E. ciliatum is native of North America and *E. hirsutum* is native in Great Britain and Ireland. Disturbed ground in quarries, set-aside, gardens, ditches and felled woodland. The hybrid has occurred scattered over England, Wales, central Scotland and Co. Dublin in Ireland. Also in France and Czech Republic.

× **lanceolatum**
Perennial herb. Stems often suffused reddish, covered with short, curved simple eglandular hairs and spreading glandular hairs, with 4 raised lines. *Leaves* opposite; lamina elliptical to lanceolate, acute at apex, serrulate, glabrous or with hairs on veins and margins. *Flowers* 5–7 mm. *Stigma* often obscurely lobed. *Sepals* and *capsules* with short, curved simple eglandular hairs and spreading glandular hairs. Seeds when developed sufficiently show papillose ridging of *E. ciliatum*.

E. ciliatum is native of North America and *E. lanceolatum* is native in southern Great Britain. Disturbed ground such as gardens and quarries. The hybrid is endemic to scattered localities in southern England.

× **montanum** = **E.×interjectum** Smejkal
Perennial herb. Stems pale green, sometimes branched, covered with numerous, long, curved simple eglandular

hairs mixed with spreading glandular hairs above. *Leaves* opposite below and alternate above; lamina 2–6×0.5–2.0 cm, ovate to lanceolate, acute at apex, serrulate, rounded at base, sessile or shortly petiolate, sparsely short-hairy especially on the veins and margins. *Flowers* 4–7 mm, numerous, pink flushed purple. *Stigma* sometimes obscurely lobed. *Sepals* and *capsules* with curled simple eglandular hairs mixed with long, spreading glandular hairs. At least partially sterile, but when seeds developed sufficiently show papillose ridging of *E. ciliatum*.

E. ciliatum is native of North America and *E. montanum* is native of Great Britain and Ireland. Disturbed ground on roadsides, quarries, felled woodland, shrubberies and amenity plantings. The hybrid is widespread in Great Britain and scattered in Ireland. Also in France, Switzerland and the Czech Republic, but is probably under-recorded.

× **obscurum** = **E.** × **vicinum** Smejkal
Perennial herb sometimes with stolons. *Stems* pale green or reddish, sometimes widely branched, covered with numerous, curved simple eglandular hairs to nearly glabrous below, mixed with spreading glandular hairs above. *Leaves* mostly opposite, or sometimes alternate above; lamina 2–6×0.5–1.5 cm, lanceolate or oblong-lanceolate, more or less acute at apex, serrulate, rounded at base, glabrous or with short hairs on the veins and margins. *Flowers* 5–7 mm, numerous, flushed purple. *Stigma* clavate. *Sepals* and *capsules* with curled simple eglandular hairs, and spreading glandular hairs. At least partially sterile, but when seeds are sufficiently developed they show signs of papillose ridging of *E. ciliatum*.

E. ciliatum is native of North America and *E. obscurum* is native of Great Britain and Ireland. Waste ground, roadsides, quarries, set-aside, gardens and planted areas. Widespread in Great Britain and scattered in Ireland. Also in Finland and Czech Republic.

× **palustre** = **E.** × **fossicola** Smejkal
Perennial herb sometimes with filiform stolons. *Stems* pale green to reddish, with curved simple eglandular hairs and spreading glandular hairs. *Leaves* opposite below, alternate above, linear-lanceolate to lanceolate, obtuse to acute at apex, slightly serrulate and revolute margins, sometimes glabrous or with a few hairs on veins and margins. *Flowers* 4–6 mm, numerous. *Stigma* clavate. *Sepals* and *capsules* with curved eglandular hairs and spreading glandular hairs. At least partially sterile, but seeds can exceed 1.3 mm and show ridging of *E. ciliatum*.

E. ciliatum is native of North America and *E. palustre* is native of Great Britain and Ireland. Ditches, disturbed marshy ground and loch and pond sides. The hybrid has scattered records through Great Britain, and there is one record for Ireland. Also occurs in France, Holland and Slovenia.

× **parviflorum** = **E.** × **floridulum** Smejkal
Perennial herb sometimes with short stolons. *Stems* pale green or reddish, branched above, covered with numerous, long simple eglandular hairs mixed with long glandular hairs above. *Leaves* opposite below, alternate above; lamina 1–6×0.3–1.2 cm, lanceolate to ovate-lanceolate, acute at apex, serrulate, rounded at base, with long simple eglandular hairs. *Flowers* 4–7 mm, deep pink towards apex. *Stigma* obscurely 4-lobed or nearly clavate. *Sepals* and *capsules* with curved simple eglandular hairs and spreading glandular hairs. At least partially sterile, but when seeds sufficiently developed showing evidence of the papillose ridging of *E. ciliatum*.

E. ciliatum is native of North America and *E. parviflorum* is native of Great Britain and Ireland. Disturbed places such as quarries, abandoned fields, orchards, roadside and felled woodland. The hybrid occurs scattered through England and Wales with a few records in Scotland and Ireland. It has also been recorded across Continental Europe.

× **roseum** = **E.** × **nutantiflorum** Smejkal
Perennial herb, sometimes producing stolons. *Stems* often reddish, covered with curled eglandular hairs and spreading glandular hairs. *Leaves* opposite below and alternate above, lanceolate-elliptical, glabrous or nearly so, often distinctly petiolate. *Flowers* 4–6 mm, pale in bud. *Stigma* clavate. *Sepals* and *capsules* with curled eglandular hairs and spreading glandular hairs. Usually partially sterile, but when seeds sufficiently developed showing evidence of the papillose ridging in *E. ciliatum*.

E. ciliatum is native of North America and *E. roseum* native of Great Britain and Ireland. Gardens and shrubberies. The hybrid occurs in scattered localities throughout Great Britain. Also in Slovenia, Finland and Czech Republic.

× **tetragonum** = **E.** × **mentiens** Smejkal
E. iglaviense Smejkal nom. nud.

Perennial herb. Stems often purplish, much branched, covered with numerous, curled eglandular hairs and spreading glandular hairs above. *Leaves* opposite below and alternate above; lamina 4–6 cm, linear-lanceolate to lanceolate, obtuse to acute at apex, serrulate, glabrous or nearly so. *Flowers* 2–4 mm, pink, flushed purple. *Stigma* clavate. *Sepals* and *capsules* with curled eglandular hairs and spreading glandular hairs. Sometimes partially sterile, but when seeds are sufficiently developed they show papillose ridging of *E. ciliatum*.

E. ciliatum is native of North America and *E. tetragonum* is native of Great Britain and Ireland. Waste or disturbed ground, quarries, gardens and cleared woodlands. The hybrid is recorded for scattered localities throughout England and Wales and possibly for Perthshire in Scotland. Also in Europe as far north as Finland.

9. E. palustre L. Marsh Willowherb

Perennial herb producing in summer filiform, hypogeal stolons bearing distant pairs of yellowish scale-leaves and terminating in autumn in a bulbil-like bud with fleshy scales. *Stems* 15–60 cm, pale green, simple or branched, leafy, erect from a curved base, terete and without raised lines but often with 2 rows of curved eglandular hairs, subglabrous or downy with short curved eglandular hairs. *Leaves* mostly opposite; lamina 2–7×0.4–1.0(–1.5) cm, medium yellowish-green on upper surface, paler beneath, narrowly lanceolate to linear-lanceolate, narrowed to an obtuse apex, entire or very obscurely denticulate, cuneate at base, subglabrous or with eglandular hairs on the margins and veins, more or less sessile or the uppermost very shortly petiolate. *Inflorescence* a bracteate raceme, the bracts like the leaves but smaller, the flowers 4–6 mm in diameter and held almost horizontally, the buds blunt, initially erect, but soon drooping so that the raceme hangs over to one side. *Hypanthium* see note under genus. *Sepals* 4, about 4 mm, green, lanceolate, acute at apex, with curled eglandular hairs. *Petals* 4, 5–7 mm, pale rose or lilac, rarely white, emarginate. *Stamens* 4 long and 4 short, clavate, shorter than style. *Capsules* 5–8 cm, pale green, linear, downy with short, curled eglandular hairs; seeds 1.6–1.8(–2.0) mm, pale reddish-brown, more or less fusiform, rounded above and with a short appendage formed by the projecting inner integument, minutely and uniformly papillose. *Flowers* 7–8. $2n=36$.

Native. Marshes, fens and ditches. Frequent throughout Great Britain and Ireland, Europe northwards to Iceland and Lapland; Asia; North America and Greenland. Circumpolar Boreo-temperate element.

× **parviflorum** = **E. × rivulare** Wahlenb.

Perennial herb often with filiform stolons. *Stems* pale green or tinted reddish-brown, with slender branches, covered with some curved and spreading, long simple eglandular hairs and glandular hairs above. *Leaves* opposite or alternate above; lamina 2–6 × 0.3–1.4 cm, lanceolate or linear-lanceolate with more or less revolute margins, serrulate, narrowed or rounded at base, sessile or nearly so, rather long-hairy. *Flowers* 5–7 mm. *Stigmas* obscurely 4-lobed or entire. *Sepals* and *capsules* with rather long, curled and spreading simple eglandular hairs mixed with glandular hairs above. *Seeds* can be up to 2mm.

Native. Marginal or disturbed habitats, especially semi-open areas with marshes, fens or ponds. Recorded in scattered localities throughout Great Britain and the northern half of Ireland.

× **roseum** = **E. × purpureum** Fr.

The E. S. Marshall collection from Surrey on which the only record was based has been determined by Kitchener as straight *E. palustre*.

× **tetragonum** = **E. × laschianum** Hausskn.
E. probstii H. Lév.

Perennial herb; short filiform stolons may develop. *Stems* pale green, branched, covered with short, curved simple eglandular hairs, dense above. *Leaves* opposite or alternate; lamina 1.5–5.0 × 0.2–0.7 cm, linear to linear-lanceolate, acute at apex, sparingly serrulate and weakly revolute, narrowed at the sessile base, with few hairs on veins and margins, shortly petiolate. *Flowers* 5–7 mm, rose. *Stigma* clavate. *Sepals* and *capsules* with numerous, crisped simple eglandular hairs and some minute glandular hairs. *Seeds* may develop up to 1.5 mm.

Native. Has been recorded from Kent and Surrey. Also occurs in Spain, France, Holland and through central Europe. Named after Wilhelm Gottfried Lasch (1787–1863).

10. E. anagallidifolium Lam. Alpine Willowherb
E. alpinum auct.

Perennial alpine *herb* producing in summer numerous, more or less prostrate, slender, epigeal stolons, at first forming small, subsessile rosettes but soon elongating and then with distant pairs of small, green leaves. *Stems* 4–10(–20) cm, slender and 1–2 mm in diameter, ascending from a decumbent base, more or less glabrous except for 2 lines of curled hairs down the 2 faint ridges. *Leaves* mostly opposite; lamina 10–25 mm, often yellowish-green, lanceolate or elliptic-lanceolate, entire or faintly and distantly sinuate-toothed, gradually narrowed below into a short, stalk-like base, more or less glabrous. *Inflorescence* a bracteate raceme, the bracts like the leaves, the flowers 3.0–4.5 mm in diameter, the top of the raceme drooping in flower and young fruit. *Hypanthium* see note under genus. *Sepals* 4, 2.5–4.0 mm, reddish, lanceolate, glabrous or with few, curled hairs. *Petals* 4, 3.5–4.5 mm, rose-red. *Stamens* 4 long and 4 short; filaments pale; anthers cream. *Style* 1; stigma clavate. *Capsules* 2.5–4.0 cm, reddish, linear, on a stalk 2.5–5.0 cm, glabrous or nearly so; seeds (at least in Great Britain) 0.8–1.1 mm, obovoid, with an appendage at the hairy end, obscurely reticulate, not papillose. *Flowers* 7–8. Probably self-pollinated. $2n=36$.

Native. Mountain flushes and streamsides. Locally frequent in northern England and Scotland. North Europe and mountains of central Europe; Asia; Greenland and North America. Circumpolar Arctic-montane element.

× **obscurum** = **E. × marshallianum** Hausskn.

Perennial, often semi-prostrate *herb* with very leafy stolons. *Stems* pale green, sometimes obscurely 4-angled, often rather slender, with very slender branches, with lines of short, curled simple eglandular hairs below, more numerous above. *Leaves* opposite below, alternate above; lamina 0.7–5.0 × 0.2–2.0 cm, linear to lanceolate, acute at apex, sparsely serrulate, rounded at base; petiole up to 5 mm. *Flowers* 4–7 mm. *Stigma* clavate. *Sepals* and *capsules* with dense, curled simple eglandular hairs mixed with fine glandular hairs. Generally sterile, but seeds do occur from 0.9–1.1 mm.

Native. Recorded from Stirlingshire, Banffshire and Sutherland. Also recorded from France and Germany. Named after Edward Shearburn Marshall (1858–1919).

× **palustre** = **E. × dasycarpum** Fr.

Perennial, semi-prostrate *herb* with filiform stolons. *Stems* pale green, slender, not branched, covered with numerous, curled simple eglandular and glandular hairs. *Leaves* opposite below, alternate above; lamina 0.6–2.5 × 0.3–0.6 cm, ovate, lanceolate or narrowly elliptical, more or less obtuse at apex, with more or less revolute margin, narrowed at base, with a few hairs on the veins and margins. *Flowers* 4–6 mm. *Stigma* clavate. *Sepals* and *capsules* with curled simple eglandular hairs and fine glandular hairs. Mainly sterile, but seeds do occur from 1.30–1.45 mm, showing influence of *E. palustre*.

Native. Recorded from Inverness-shire, Ross and Cromarty and Sutherland. There are a few records from Continental Europe in Scandinavia, Austria, Czech Republic, France and Spain.

11. E. alsinifolium Vill. Chickweed Willowherb
E. alpinum auct.

Perennial, alpine *herb* producing in summer slender, yellowish, hypogeal stolons with distant pairs of yellowish scale-leaves. *Stems* 5–20(–30) cm, 2–3 mm in diameter, pale green, rather slender, ascending from a decumbent base, more or less glabrous except for 2 rows of hairs down the 2 faint ridges and a few, curled hairs above. *Leaves* mostly opposite; lamina 1.5–4.0 cm, somewhat bluish-green and shining on upper surface, pale beneath, ovate to ovate-lanceolate, obtuse at apex, distantly sinuate-toothed, rounded at base, more or less glabrous; petiole short. *Inflorescence* a bracteate raceme of 2–5 flowers, the top of the raceme drooping in flower and young fruit; flowers 8–11(–12) mm in diameter. *Hypanthium* see note under genus. *Sepals* 4, 4–6 mm, lanceolate. *Petals* 4, 7–9 mm, bluish-red. *Stamens* 4 long and 4 short; filaments pale; anthers cream. *Style* 1; stigma clavate. *Capsules* 3–5 cm, linear, glabrous or nearly so, on an erect stalk 2–3 cm; seeds (1.3–)1.4–1.8 mm, narrowly obovoid, with an appendage below the plume of hairs, obscurely reticulate, not papillose. *Flowers* 7–8. Probably self-pollinated. $2n = 36$.

Native. Mountain flushes and streamsides. Locally frequent in northern England and Scotland, and in Caernarvonshire. North and central Europe. European Arctic-montane element; also in Greenland.

× **anagallidifolium** = **E.** × **boissieri** Hausskn.
Perennial herb with leafy stolons. *Stems* pale green, unbranched, with 4 lines of short, curled eglandular hairs below. *Leaves* opposite below, alternate above; lamina 0.7–5.0 × 0.5–1.6 cm, lanceolate to ovate, obtuse to acute at apex, entire or nearly so, narrowed at base, shortly eglandular hairy on the veins and margins. *Flowers* 5–10 mm long. *Stigma* clavate. *Sepals* and *capsules* with numerous, curved simple eglandular hairs mixed with glandular hairs.

Native. Recorded from mountains in central and north Scotland, from 500 to 1000 m. Also known from montane habitats across central Europe. Named after Pierre Edmond Boissier (1810–1885).

× **lanceolatum**
E. S. Marshall said a spontaneous hybrid of this came up in his garden in 1895. Kitchener believes the plants on which the record is based are more likely to be *E. lanceolatum* × *obscurum*.

× **montanum** = **E.** × **facchinii** Hausm.
E. grenieri Rouy & E. G. Camus ex Focke

Perennial herb. Stems pale green or suffused brownish-red, with slender branches, covered with dense, curled simple eglandular hairs mixed with glandular hairs above. *Leaves* opposite below, alternate above; lamina 1–3 × 0.4–1.0 cm, ovate to ovate-lanceolate, subobtuse at apex, serrulate, rounded to cuneate at base, very sparsely hairy; petiole short. *Flowers* 5–7 mm. *Stigma* obscurely 4-lobed. *Sepals* and *capsules* with rather sparse simple eglandular and glandular hairs.

Native. Occurs by montane streams. Recorded from northern England and Scotland. It occurs also across central Europe. Named after Francesco Facchini (1788–1852).

× **obscurum** = **E.** × **rivulicola** Hausskn.
Perennial upright *herb* which may have underground stolons. *Stems* pale green, touched reddish, slightly branched, covered with numerous, curled simple eglandular hairs mixed with glandular hairs above. *Leaves* opposite below, alternate above; lamina 1–4 × 0.2–1.5 cm, lanceolate to ovate, acute at apex, entire or slightly serrulate, rounded below, glabrous or slightly hairy round the margin. *Flowers* 5–10 mm long. *Stigma* entire. *Sepals* and *capsules* with curled simple eglandular and glandular hairs. Usually sterile, but seeds do occur 1.2–1.3 mm with the appendage of *E. alsinifolium*.

Native. By montane streams. Recorded for Perthshire and Banffshire. Also recorded from central Europe and Spain.

× **palustre** = **E.** × **haynaldianum** Hausskn.
Perennial herb. Stems pale green, unbranched, covered with numerous, short, curved simple eglandular hairs mixed with small glandular hairs. *Leaves* opposite below, alternate above; lamina 1.5–5.0 × 0.3–1.5 cm, lanceolate, oblong-lanceolate or linear-lanceolate, gradually narrowed at apex, entire or obscurely serrulate, rounded or narrowed at base, with hairs mainly on the veins and margins. *Flowers* 6–8 mm. *Stigma* clavate. *Sepals* and *capsules* with curled simple eglandular hairs and glandular hairs. *Seeds* may develop up to 2.0 mm showing the conspicuous appendage of *E. alsinifolium*.

Native. Scattered records in northern England and Scotland. Recorded for Finland and Spain and through central and eastern Europe. Named after Stephan Franz Ludwig Haynald (1816–1891).

12. E. brunnescens (Cockayne) P. H. Raven & Engelhorn
New Zealand Willowherb
E. pedunculare auct.; *E. nerteroides* auct.; *E. pedunculare* var. *brunnescens* Cockayne

Perennial, prostrate, mat-forming *herb* up to 2 m across. *Stems* up to 20 cm, reddish, creeping and rooting at the nodes, with lines of short, appressed hairs, much-branched, the branches often with ascending tips. *Leaves* mostly opposite; lamina 2.5–10.0 mm, dull green or somewhat bronzed and smooth on the upper surface, often more or less flushed with purple and with an obvious midrib, but faint lateral veins beneath, broadly ovate to subrotund, rounded to acute at apex, entire to obscurely sinuate-toothed, glabrous; petioles 0.5–6.0 mm. *Inflorescence* of solitary, axillary flowers but usually in only 1 leaf axil of a pair; pedicels 0.5–5.5 cm, erect or briefly drooping, lengthening up to 7.5 cm in fruit, glabrous. *Sepals* 4, 1.5–3.5 mm, reddish, lanceolate, glabrous. *Petals* 4, (2.3–)3.2–7.0 mm, white or pale pink, obovate, emarginate. *Stamens* 8; filaments 0.5–2.5 mm; anthers yellow. *Style* 1.2–5.2 mm, white; stigma white, clavate. *Capsules* 1.5–6.1 cm, linear, glabrous, on a pedicel 3.5–8.0 cm; seeds 0.6–0.9 mm, truncate at hairy end, minutely and uniformly papillose, with a tuft of hairs, detaching or persistent. *Flowers* 5–10. Self-pollinated. $2n = 36$.

Introduced. Damp, barish ground, especially gravelly hillsides, railway sidings and waste tips. First collected in

1908 and now over most of Great Britain and Ireland and still spreading, but rare in the south and south-east. Native of New Zealand.

× **ciliatum**

= **E. × brunnatum** Kitchener & D. R. McKean

Perennial herb with prostrate, leafy runners. *Stems* 8–20(–30) cm, curving up to an erect position, much-branched below, little branched above, except in cultivation when growth habit is more erect; with 2 lines of short, crisped eglandular hairs, descending from the nodes and beginning between a pair of petioles and descending to the next node, these hairs becoming generally scattered on the stem upper parts and may be absent near the base. *Leaves* mainly opposite, but largely alternate in the upper parts, green on younger growth, especially the upper surface; lamina (0.5–)0.8–1.2(–2.3) × (0.3–)0.4–0.6(–1.1) cm, ovate-lanceolate to elliptical, obtuse at apex, with a few, obscure teeth, rounded at base, glabrous except on the margin, those of runners tending to be more elliptical; petioles not exceeding 1.5 mm. *Sepals* 2.5–3.0 × 0.7–1.2 mm, lanceolate, with crisped eglandular and glandular hairs. *Petals* 5–6 mm, very pale pink, rarely darker. *Stamens* 8, 4 longest projecting to or just below top of stigma, their anthers 0.5–0.8 mm, 4 shortest extending as far as the stigma base. *Style* 1.3–2.8 mm, longer than stigma, erect, white; stigma 1.2–1.9 mm, entire, clavate. *Capsules* 12–30 mm, covered with dense glandular and crisped eglandular hairs, the glandular hairs extending to but less frequent on the pedicels, the capsules sometimes twisted and shrivelled, mostly sterile, containing shrunken abortive seeds 0.3–0.4 mm, also a few larger seeds, either 0.6–0.7 mm and malformed, or 0.8–0.9 mm and full formed and fertile with rows of tubercles, often with the surface texture tending towards the ridged rows of *E. ciliatum*; the larger seeds sometimes bearing a short neck or appendage at the point of attachment of the coma. *Flowers* 7–8.

Both parents introduced, *E. brunnescens* from New Zealand and *E. ciliatum* from North America. Described from Cornwall and recorded for Carmarthenshire. Also recorded from New Zealand where one parent is native.

× **lanceolatum**

= **E. × cornubiense** Kitchener & D. R. McKean

Perennial herb. Stems 6.5–25 cm, little branched above except in larger plants, with a fairly uniform covering of crisped, short eglandular hairs. *Leaves* opposite, but mainly alternate on inflorescence, reddish; cauline with lamina 0.5–1.1(–1.5) × 0.7(–0.9) cm; basal crowded, ovate to broadly elliptical, obtuse at apex, with a few, obscure teeth, glabrous except on the margin and a few hairs on the midrib below; all sessile or with petioles not exceeding 1 mm. *Sepals* 2.8–3.5 × 1.9–1.4 mm, lanceolate, with short, crisped eglandular hairs and patent glandular hairs. *Petals* about 5 mm, white, pale pink or very pale pink and the colour may vary on the same plant, but not on the same flower. *Stamens* 8, the 4 long ones extending to, or just below the stigma, their anthers about 0.6 × 0.3 mm, the 4 short ones reaching well below the stigma. *Style* 3.5–4.0 mm; stigma variable with confused partial lobing 0.8–1.0 mm. *Capsules* 22–31 mm, mostly sterile, containing shrunken abortive seeds 0.3–0.5 mm, their surface with rows of platelets, sometimes reticulate, seeds occasionally fully formed, cylindrical, 0.6–0.9 mm, with rows of tubercles. *Flowers* 7–9.

E. brunnescens is a native of New Zealand and *E. lanceolatum* is a native of Europe. The hybrid is known in both east and west Cornwall. Endemic.

× **montanum**

= **E. × confusilobum** Kitchener & D. R. McKean

Perennial herb. Stems 5–10(–30) cm, reddish, creeping at first and then curving up to become erect, branched at ground level but scarcely above, with a fairly uniform covering of short, crisped eglandular hairs diminishing in the upper parts and with occasional glandular hairs. *Leaves* opposite, largely alternate on inflorescence; basal leaves crowded; cauline 0.5–1.0 × 0.2–0.5 cm; lamina ovate to broadly elliptical, obtuse at apex, with a few, obscure teeth, with short, crisped eglandular hairs on the veins and midrib beneath and a scattering of hairs elsewhere, the upper surface hairy becoming glabrous. *Sepals* 1.4–2.9 × 0.7–1.0 mm, lanceolate, with a few, scattered glandular hairs. *Petals* about 3.5 mm, pale purplish-pink. *Stamens* 8, with 4 long ones projecting to top of stigma, their anthers being 0.4 × 0.3 mm, the 4 short ones extending to the mid level or base of stigma. *Style* longer than the stigma; stigma with confused partial lobing. *Capsules* 16–26 mm, with dense, short glandular hairs, containing shrunken abortive seeds 0.3–0.4 mm, surface flattish, marked in rows and sometimes bearing low tubercles, occasional larger seeds more fully formed, 0.6–0.9 mm, cylindrical with a tubercular surface. *Flowers* 7–9.

E. brunnescens is native of New Zealand and *E. montanum* is a common native species of Europe. The hybrid was described from East Cornwall where it is known in at least six localities and also in Co. Antrim. Endemic.

× **obscurum**

= **E. × obscurescens** Kitchener & D. R. McKean

Perennial herb. Stems 8–23 cm, conspicuously reddish, creeping at first and then curving up to an erect position, well-branched at ground level, sharply quadrangular and with raised lines running from one node to the next, with fairly sparse, short, crisped eglandular hairs, more numerous in upper parts. *Leaves* mainly opposite, but largely alternate on upper part of stems; lamina ovate-lanceolate to elliptic, obtuse at apex, with a few, obscure teeth, the cauline 0.6–1.5 × 0.3–0.6 cm. *Sepals* 3.0–3.5 × about 1.0 mm, with crisped eglandular hairs. *Petals* 5–6 mm, very pale pink. *Stamens* 8, the 4 long ones projecting to the top of the stigma, their anthers 0.4–0.5 × 0.3 mm, the 4 short ones extending to the mid-level or base of the stigma. *Style* longer than the stigma; stigma with confused partial lobing. *Capsules* 16–26 mm, with dense, short eglandular hairs, occasional glandular hairs present but not extending to the pedicel, shrunken abortive seeds 0.3–0.4 mm, surface flattish, marked in rows and sometimes bearing low tubercles, occasional larger seeds more fully formed 0.6–0.9 mm, cylindrical with tubercular surface. *Flowers* 7–9.

E. brunnescens is native of New Zealand and *E. obscurum* is native of Great Britain and Ireland. This hybrid was described from spoil tips in a quarry at Magheramorne, Co. Antrim and occurs in similar places in Co. Tyrone and Cornwall, Dorsetshire and Glamorganshire. Endemic.

× **palustre** = **E.** × **chateri** Kitchener & D. R. McKean
Perennial herb with prostrate, leafy runners. *Stems* 20–25 cm, reddish, erect, scarcely branched above, with 2 broad lines of short, crisped eglandular hairs, descending from each node, the density of the hairs increasing in the upper part of the stem. *Leaves* opposite, largely alternate on the flowering part of the stem, reddish; cauline 0.6–1.2 × 0.2–0.4 cm, narrow lanceolate, more or less pointed at apex, with a few, obscure teeth, glabrous except sometimes on the margin or upper side of midrib; leaves on runners similar but smaller and more distinctly petiolate. *Sepals* about 2.2 mm, lanceolate, with appressed hairs. *Petals* very pale pink. *Stamens* 8, the 4 long ones projecting to lower part of stigma, their anthers 0.4–0.5 mm, the 4 short ones extending to style below. *Style* 1.4–2.3 mm, erect; stigma 0.9–1.7 mm, entire, clavate. *Capsules* 15–33 mm, with numerous, crisped eglandular hairs and some patent glandular hairs, mostly sterile, containing shrivelled, abortive seeds 0.4 mm, also a few, large seeds 1.0–1.3 mm, with tubercled surface, generally part collapsed longitudinally, but occasionally fully formed and fertile, the larger seeds bearing a neck or appendage at the point of attachment of the coma. Flowers 7–9.

E. brunnescens is native of New Zealand and *E. palustre* is native of Europe. This hybrid is known only from damp, acidic shaley soil by a Forestry Commission road, south of Hadre, Llyn Brianne, Cardiganshire. Named after its discoverer, Arthur O. Chater (b. 1933).

× **parviflorum** = **E.** × **argillaceum** Kitchener
Perennial herb. Stems up to 15(–30) cm, reddish, branched, secondary stems emerging at the lowest nodes and curving upwards to a more or less erect position, flexuous to rigid, with a dense covering of patent eglandular hairs, often some glandular hairs in upper part, the longest hairs 0.2–0.3 mm. *Leaves* opposite below, alternate on the inflorescence, reddish; cauline with lamina 1.1–2.8 × 0.7–1.7 cm, ovate to broadly elliptical, obtuse at apex, with few teeth, with some short, crisped eglandular hairs beneath, especially on the midrib and veins and minutely hairy becoming glabrous on upper surface; generally with 4 lateral veins on each side; petioles up to 4.5 mm, hairy, the lower leaves often sessile. *Sepals* 3–4 × 0.9–1.2 mm, lanceolate, with short, patent and crisped eglandular hairs and some glandular hairs. *Petals* (5.5–)6.0–7.0(–8.0) mm, very pale pink with purple veins. *Stamens* 8, the 4 long ones projecting to the level of the stigma, the 4 short ones below; anthers 0.4–0.6 mm, yellow, indehiscent. *Style* longer than the stigma; stigma with confused partial lobing. *Capsule* 19–31(–38) mm, generally shrivelling up rather than ripening and containing shrunken abortive seeds 0.3–0.4 mm. *Flowers* 7–9.

E. brunnescens is native of New Zealand and *E. parviflorum* is a native of Europe. This hybrid is recorded only from Wheal Remfrey china clay works, south-east of Indian Queens, Cornwall, from where the binomial was described. Endemic.

13. E. pedunculare A. Cunn. Rockery Willowherb
E. linnaeoides Hook. fil.; *E. nummularifolium* var. *pedunculare* (A. Cunn.) Hook. fil.; *E. caespitosum* Hausskn.; *E. nummularifolium* var. *caespitosum* (Hausskn.) T. Kirk

Perennial herb forming loose mats up to 0.5 m in diameter. *Stem* up to 20 cm, creeping and rooting at the nodes, with short, curled hairs in lines decurrent from the petiole, branched, leafy. *Leaves* opposite, 2–14 × 2–15 mm, green to coppery and dull or somewhat shining, the lateral veins inconspicuous, very broadly ovate to subrotund, subacute to rounded at apex, serrate with 3–14 teeth on each side, rounded to truncate at base, glabrous; petioles 0–5 mm. *Inflorescence* of solitary, axillary flowers; pedicels 3–50 mm, erect. *Sepals* 4, 1.7–3.5 mm, pale, lanceolate, glabrous, not keeled. *Petals* 4, 3.5 mm, white, occasionally pink, obovate, emarginate. *Stamens* 8; filaments 0.5–2.0 mm; anthers yellow. *Styles* 0.7–1.8 mm, white; stigma white, clavate. *Capsules* 2–5 cm, glabrous on a pedicel 4.5–10 cm; seeds 0.6–0.8 mm, brown, narrowly obovoid, the tuft of hair detaching readily. *Flowers* 5–10. $2n = 36$.

Introduced. A weed of barish, damp ground. Naturalised in Cornwall, Surrey, Glamorganshire, Pembrokeshire, Yorkshire, Wigtownshire, Perthshire, Co. Galway and Co. Mayo. Rare garden weed in England and Wales. Native of New Zealand.

14. E. komarovianum H. Lév. Bronzy Willowherb
E. inornatum Melville; *E. nummularifolium* var. *nerterioides* auct.; *E. nummularifolium* var. *brevipes* Hook. fil.; *E. pedunculare* auct.; *E. nummularifolium* subsp. *nerterioides* auct.; *E. nummularifolium* var. *minimum* T. Kirk; *E. nerterioides* var. *minimum* (T. Kirk) Cockayne; *E. nerterioides* auct.; *E. inornatum* var. *brevipes* (Hook. fil.) Melville

Creeping *perennial herb* forming mats up to 0.5 m in diameter. *Stems* up to 20 cm, pale green or reddish, prostrate, rooting at the nodes, with sparse, appressed-ascending, eglandular hairs in 2 lines, decurrent from the petioles, branched, leafy. *Leaves* opposite, distant to crowded and imbricate, frequently reflexed; lamina 2–12 × 1–9 mm, often reddish to copper-coloured and rugose with prominent veins on upper surfaces, subrotund, oblong or ovate, subacute to obtuse at apex, entire or with 1–3, remote, weak teeth on each side, attenuate to rounded at base, sometimes sparsely hairy on both surfaces; petiole up to 3 mm, or sessile. *Inflorescence* of solitary axillary flowers; pedicels 1–7(–38) mm, the flowers falling before fully elongated. *Sepals* 4, 1.5–2.5 mm, lanceolate, not keeled. *Petals* 4, 2–4(–5) × 0.9–2.5(–3.0) mm, white; obovate, emarginate. *Stamens* 8; filaments 0.2–1.2 mm; anthers yellow. *Styles* 1.1–1.8(–3.0) mm; stigma white, clavate or capitate. *Capsules* 4–30 mm, linear, on a pedicel up to 9.3(–13.5) mm; seeds 0.5–0.9(–1.1) mm, brown, obovoid, smooth, with a tuft of hair at apex readily detaching. *Flowers* 5–10. $2n = 36$.

Introduced. Naturalised as a garden weed. Recorded in a few places in England, Scotland and the north of Ireland. Native of New Zealand.

2. Chamerion (Raf.) Raf.

Chamaenerion auct.; *Epilobium* subgen. *Chamerion* Raf.

Perennial, monoecious, rhizomatous *herbs. Leaves* alternate. *Inflorescence* of dense flowers in terminal racemes; flowers bisexual, slightly zygomorphic. *Hypanthium* shortly cylindrical, but see *Epilobium. Outer perianth* segments 4. *Inner perianth* segments 4, rose-purple, rarely white. *Stamens* 8. *Style* 1; stigma finally exceeding the anthers, at first spreading then recurved or revolute. *Ovary* 4-celled. *Fruit* a linear capsule; seeds with a chalazel plume of long hairs.

Contains about 8 species in temperate and arctic regions of the northern hemisphere.

Broderick, D. H. (1990). The biology of Canadian weeds. 93. *Epilobium angustifolium* L. (Onagraceae). *Canad. Jour. Pl. Sci.* **70**: 247–259.
Grime, J. P. et al (1988). *Comparative plant ecology*. London.
Harvey, M. J. (1966). An experiment with *Epilobium angustifolium. Proc. B. S. B. I.* **6**: 229–231.
Hultén, E. & Fries, M. (1986). *Atlas of north European vascular plants north of the Tropic of Cancer.* 3 vols. Königstein.
Myerscough, P. J. (1980). Biological flora of the British Isles. No. 148. *Epilobium angustifolium* L. *Jour. Ecol.* **68**: 1047–1074.

1. Leaves oblong-lanceolate or oblong-elliptical **1. angustifolium**
1. Leaves linear **2. dodonaei**

1. C. angustifolium (L.) Holub Rosebay Willowherb
Epilobium angustifolium L.; *Chamaenerion angustifolium* (L.) Scop.; *Epilobium spicatum* Lam.; *Epilobium salicifolium* Stokes nom. illegit.; *Chamaenerion spicatum* (Lam.) Gray; *Epilobium macrocarpum* Stephens; *Epilobium brachycarpum* Leight., non C. Presl

Perennial monoecious *herb* with long, horizontally spreading roots which give rise to leafy stems. *Stems* 30–120 cm, pale green, erect, subterete, glabrous below, more or less hairy above, leafy. *Leaves* all alternate, numerous, more or less ascending; lamina 5–15 × 1.0–3.5 cm, dark green on upper surface, glaucous or greyish beneath with conspicuous veins, the lateral veins numerous, joining into a continuous, wavy, intramarginal vein, narrowly oblong-lanceolate or oblong-elliptical, narrowed to a pointed apex, entire or with small and distinct horny teeth, often more or less waved at the margins, narrowed below, glabrous on upper surface, sometimes thinly hairy beneath. *Inflorescence* a long, rather dense, spike-like, bracteate raceme; flowers bisexual, 20–30 mm in diameter; pedicels 10–15 mm, ascending, hairy; bracts up to 10 mm, narrowly subulate or filiform. *Hypanthium* either absent or very short at the top of the ovary, or as considered by some authors adnate to the ovary. *Outer perianth* segments 4, 8–12 mm, usually reddish-purple, lanceolate, acuminate at apex. *Inner perianth* segments 10–16 × 8–10 mm, rose-purple, rarely white, obovate, rounded or obscurely emarginated at apex, shortly clawed, the upper slightly smaller than the lower. *Stamens* 8; filaments 6–9 mm, dilated towards the base, glabrous, at first spreading, becoming deflexed; anthers 2.5–3.0 mm, reddish-purple. *Style* 1, about 8 mm, hairy at base, at first deflexed with the immature stigma lobes connivent and clavate, at maturity suberect with cruciform stigma-lobes. *Capsule* 2.5–8.0(–12.0) × 0.2–0.3 cm, obscurely tetragonous, sparsely to densely appressed crispate-hairy; seeds 0.3–2.0 mm, pale brown, oblong-ellipsoid, smooth or slightly rugulose, with a white plume at apex. *Flowers* 7–9. Protandrous, visited by various insects for nectar. $2n = 36$.

Efforts to relate differences in capsule length to supposed wild and naturalised forms in Great Britain have failed. Although certain populations, especially in mountainous districts, are native, the status of the common lowland plants is in doubt. There are, however, no constant morphological differences between them.

Native. Waste land, woodland clearings, embankments, rocky places and screes on mountains. Throughout Great Britain and Ireland and often abundant. A century ago the species was a local plant, though scattered through the countryside, especially in rocky places and on scree. Its phenomenal spread in the last 50 years may be related to the increasing areas of cleared woodland and waste places. Widespread in Europe, Asia and North America, but a recent introduction in many areas. Circumpolar Boreo-temperate element.

2. C. dodonaei (Vill.) Holub Dodoens's Willowherb
Epilobium dodonaei Vill.; *Chamaenerion dodonaei* (Vill.) Schur; *C. rosmarinifolium* Moench; *Chamaenerion angustissimum* (Weber) Sosn.

Perennial monoecious *herb* with a rhizome. *Stems* 20–110 cm, pale green suffused dark red, erect, rigid, unbranched, with dense, short, curved, mostly appressed simple eglandular hairs. *Leaves* alternate, numerous; lamina 2–6 × 0.1–0.4 cm, medium green, linear, pointed at apex, entire, narrowed at base, with numerous, appressed simple eglandular hairs on both surfaces, sessile or with very short petiole. *Inflorescence* a terminal raceme, slightly nodding before anthesis; pedicels with dense, appressed simple eglandular hairs; flowers bisexual. *Hypanthium* either very short or absent or starting at the base of the ovary. *Sepals* 4, 13–15 × 2.5–3.0 mm, pale brownish, lanceolate, acute at apex, with dense, appressed simple eglandular hairs. *Petals* 4, unequal, 2 larger about 15 × 12 mm, 2 smaller about 15 × 10 mm, pinkish-purple, elliptical, rounded at apex. *Stamens* 8; filaments white; anthers pale brown or yellow. *Style* 1, 7–15 mm, purplish; stigma 4-lobed, at first deflexed but later erect after the anthers have dehisced. *Capsule* 50–70 mm, linear, densely appressed-hairy; seeds 1.5–2.0 mm, papillose. *Flowers* 7–9. $2n = 36$.

Introduced. Recorded as a rare casual or garden escape. Native of central and east Europe and west Asia.

3. Ludwigia L.

Isnardia L.

Annual to *perennial*, monoecious, more or less aquatic *herbs. Leaves* alternate or opposite. *Inflorescence* of solitary bisexual flowers in leaf axils; flowers actinomorphic.

Hypanthium absent. *Perianth segments* 4 or 10. *Stamens* 4 or 10. *Style* 1; stigma entire. *Ovary* 4-celled. *Fruit* a short, slightly angled, scarcely dehiscent capsule with persistent perianth segments; seeds without plumes.

Contains about 20 species in temperate and warmer regions, especially of the New World.

Clement, E. J. (2000). *Ludwigia × kentiana* E. J. Clement: a new hybrid aquatic. *Watsonia* **23**: 167–172.

Cox, J. (1997). Hampshire Purslane found in Dorset. *Recording Dorset* **7**: 20–21.

Preston, C. D. & Croft, J. M. (1997). *Aquatic plants in Britain and Ireland*. Colchester.

Raven, P. H. (1963). The Old World species of *Ludwigia* (including *Jussiaea*), with a synopsis of the genus (Onagraceae). *Reinwardtia* **6**: 327–427.

Salisbury, E. J. (1972). *Ludwigia palustris* (L.) Ell. in England with special reference to its dispersal and germination. *Watsonia* **9**: 33–37.

Wigginton, M. J. (Edit.) (1999). *British red data books*. Vol. 1. *Vascular plants*. Ed. 3. Peterborough.

1. Inner perianth segments 12–20 mm	**2. grandiflora**
1. Inner perianth segments absent or 0.5 mm	2.
2. Outer perianth segments as long as broad; inner perianth segments absent	**1. palustris**
2. Outer perianth segments longer than broad; inner perianth segments about 0.5 mm	**× kentiana**

1. L. palustris (L.) Elliott — Hampshire Purslane
Isnardia palustris L.; *L. apetala* Walter

Annual or short-lived *perennial,* monoecious, aquatic *herb* often creeping or free-floating. *Stems* 5–30(–60) cm, reddish, slender, prostrate and rooting below, ascending or floating above, glabrous. *Leaves* opposite and decussate; lamina 1–3(–5) × 0.3–2.0 cm, shining medium green on upper surface, paler beneath, ovate to broadly elliptical, widest near the middle, acute or shortly acuminate at apex, entire, narrowed abruptly at base to a short petiole, glabrous, the leaves of wholly submerged shoots elliptical to oblanceolate. *Inflorescence* of solitary, bisexual, sessile flowers in leaf axils, each with 2 bracteoles about 0.5 mm at the base of the ovary. *Perianth segments* 4, about as long as broad, green, often with red margins, broadly ovate, abruptly acuminate or cuspidate at apex, persistent in fruit. *Stamens* 4, opposite the sepals but shorter. *Style* 1, short; stigma entire. *Capsule* 2–5 mm, with 4, blunt, green angles between the yellowish faces, oblong-obovoid to subglobose, truncate above and crowned by the horizontally spreading perianth segments; seeds less than 1 mm, non-plumed. *Flowers* 6. Probably self-pollinated. $2n = 16$.

Native. An emergent species in shallow water in ponds, streams, ditches and old gravel, sand, clay or marl pits and as a terrestrial plant on mud at the edge of ponds, the dried-up beds of ponds and streams and damp ground in marshy hollows. Terrestrial plants flower and fruit freely; aquatic plants flower less freely but perennate. Frequent in the New Forest, Hampshire. It has also been recorded as native in Epping Forest, Essex on mud at the edge of a pond. Elsewhere, it was recorded in a garden pond at Tonbridge in 1989 and a dew pond at Seaford Head in 1991. It formerly occurred as a native in Sussex, Jersey and Dorsetshire.

× **repens** J. R. Forst. = **L. × kentiana** E. J. Clement
L. × muelleri auct.

Differs from *L. palustre* in its leaves being widest in the uppermost third, its bracteoles about 1.0 mm, its outer perianth segments longer than broad, inner perianth segments present, about 0.5 mm and its capsule cylindrical and caducous.

Recorded from Hampshire, Sussex, Kent, Surrey, Essex and Lancashire. The epithet commemorates Douglas Henry Kent (1920–1998).

2. L. grandiflora (Michx) Greuter & Burdet — American Purslane
Jussiaca grandiflora Michx; *Jussiaea repens* subsp. *grandiflora* (Michx) P. Fourn.; *Jussiaea uruguayensis* Cambess.; *L. uruguayensis* (Cambess.) Hara

Perennial, monoecious aquatic *herb* with widespread rhizomes, often forming mats up to 2 m across. *Stems* prostrate and freely rooting or ascending, reddish, slightly hairy to glabrous. *Leaves* alternate; lamina up to 1.2 cm, medium green, elliptical to oblong-spathulate, lanceolate or linear-lanceolate, narrowed to an obtuse apex, entire, narrowed at base; petioles 2–3 cm, slender to winged. *Inflorescence* a solitary, axillary flower; pedicel 1–3 cm, often bracteolate just below the flower. *Outer perianth* segments 5, 6–13 mm. *Inner perianth* segments 5, 12–20 mm, clear deep yellow, oblong-ovate, slightly retuse and very shortly clawed. *Stamens* 10, 5–8 mm. *Style* 1; stigma capitate. *Capsule* up to 25 × 3–4 mm, more or less cylindrical, 10-veined; seeds about 1.5 × 1.5 mm, pendulous. *Flowers* 6–7.

Introduced. Recorded from Hampshire, Surrey and Buckinghamshire. Native of America.

4. Zauscheria C. Presl

Perennial monoecious *herbs* with a woody base. *Leaves* with lower opposite and upper sometimes alternate. *Inflorescence* spicate; flowers bisexual. *Hypanthium* forming a tube, globose at the base and narrowed into a tube bearing within the narrow part 8 lobe-like appendages, 4 erect and 4 deflexed. *Outer perianth segments* 4, red. *Inner perianth segments* 4, red. *Stamens* 8. *Style* 1. *Fruit* a capsule.

Contains 4 species. Named after Johann Jozef Zauschner (1737–1799).

1. Z. californica C. Presl — Californian Fuchsia
Z. cana Greene; *Z. mexicana* C. Presl; *Epilobium canum* (Greene) P. H. Raven

Perennial monoecious *herb* with a woody base. *Stems* 30–90 cm, green, often much-branched, grey-hairy and glandular-hairy. *Leaves* with lower opposite, the upper sometimes alternate; lamina 5–40 × 1.5–6.0 mm, green to greyish-green on upper surface, paler beneath, lanceolate to linear-lanceolate or linear-oblong, obtuse at apex, entire or remotely denticulate, narrowed at base, hairy. *Inflorescence* spicate; flowers bisexual. *Hypanthium* forming a tube, globose at base, narrowed into a long tube bearing within the narrow part 8 lobe-like appendages, 4 erect and 4 deflexed. *Outer perianth segments* 4, 8–10 mm, erect, lanceolate, red. *Inner perianth segments* 4, 2-cleft, 8–15 mm, red. *Stamens* 8, 4 shorter than the others. *Style* 1. *Capsule* 15–20 mm, sessile to shortly stalked, linear, 4-angled, 8-veined, often curved, with a beak; seeds about 1.5 mm, with a tuft of hair at apex. *Flowers* 6–8.

Introduced. A garden relic persistent for many years at Sand Point, Kewstoke, Somersetshire. Native of California and Mexico.

5. Oenothera L.

Annual to *biennial*, rarely *perennial*, monoecious *herbs*. *Leaves* all alternate, without stipules. *Inflorescence* of solitary or sometimes 2 flowers in the axils of the upper leaves to form a large leafy spike; flowers bisexual, more or less actinomorphic. *Hypanthium* a long narrow tube. *Sepals* 4, at first ascending, the bud sometimes open at the tips, later strongly reflexed and often falling early. *Petals* 4, broad and overlapping. *Stamens* 4+4 equal, or the inner shorter than the outer. *Stigma* entire or 4-lobed. *Ovary* 4-celled. *Fruit* a many-seeded, loculicidal, more or less cylindrical or clavate capsule; seeds not plumed. *Flowers* usually opening and becoming fragrant in the evening when they are visited by moths. Some, however, are diurnal and some cleistogamous, and most are self-compatible.

Contains about 100 species, mainly in the New World.

A critical genus where species limits are a matter of opinion, made difficult by the variable breeding mechanisms. Many are genetically and cytologically normal with the chromosomes forming seven pairs at meiosis. Several, however, are permanently heterozygous for segmental interchanges affecting all the chromosomes and form a 14-membered ring at meiosis with alternate members passing to the same pole. As a result each individual gives rise to genetically distinct gametes. Zygotes from the fusion of similar gametes fail to survive. Such species are true-breeding although highly heterozygous but occasional chiasma formation within the ring gives rise to genetically very different combinations. Between these extremes are other species forming many chromosome pairs but with occasional rings of four or more at meiosis.

Bowra, J. C. (1992, 1997). Hybridization of *Oenothera* subgen. *Oenothera* in Britain. *B.S.B.I. News* **61**: 19–33; **76**: 64–71.
Bowra, J. C. (1999). *Oenothera* (evening primroses) – the way forward. *B.S.B.I. News* **81**: 24–26.
Davis, M. B. (1926). The history of *Oenothera biennis* Linnaeus, *Oenothera grandiflora* Solander and *Oenothera lamarckiana* of De Vries in England. *Proc. Amer. Philos. Soc.* **65**: 349–378.
Dietrich, W. (1991). The status of *Oenothera cambrica* Rostański and *O. novae-scotiae* Gates (Onagraceae). *Watsonia* **18**: 407–408.
Hall, I. V. et al. (1988). The biology of Canadian weeds. 84. *Oenothera biennis* L. *Canad. Jour. Pl. Sci.* **68**: 163–173.
Hultén, E. & Fries, M. (1986). *Atlas of north European vascular plants north of the Tropic of Cancer*. 3 vols. Königstein.
McClintock, D. (1978). *Oenothera* in Britain. *Watsonia* **12**: 164–165.
Munz, R. A. (1965). Onagraceae. *Nort. Amer. Flora* **11:** 1–231. New York.
Rostański, K. (1977). Some new taxa in the genus *Oenothera* L. subgenus *Oenothera*. *Fragm. Fl. Geobot.* **23**: 285–293.
Rostański, K. (1982). The species of *Oenothera* L. in Britain. *Watsonia* **14**: 1–34.
Rostański, K. & Ellis, G. (1979). Evening primroses (*Oenothera* L.) in Wales. *Nature Wales* **16**: 238–249.

1. Capsule oblong, fusiform or cylindrical, without wings; petals yellow 2.
1. Capsule clavate, the basal part sterile and narrowed, the upper part thicker, fertile and ribbed or winged **11. rosea**
2. Capsule cylindrical, usually somewhat tapered upwards, about 6–8 mm wide at base; seeds prismatic, sharply angled 3.
2. Capsule oblong-fusiform, usually enlarged towards apex, about 2–4 mm wide at base; seeds not angled 18.
3. Capsule-teeth obtuse or truncate, rarely somewhat emarginated; ovary with stiff eglandular hairs and glandular hairs 4.
3. Capsule-teeth distinctly emarginated; ovary and young capsule whitish, appressed strigose, without glandular hairs at least in lower part of the inflorescence; petals 7–25 mm 17.
4. Tip of stem erect; cauline leaves various; sepal-tips appressed at least below, their apices usually arcuate-divergent; petals 10–50 mm 5.
4. Tip of stem more or less nodding before anthesis then usually erect; cauline leaves lanceolate; sepal tips erect, separated from their bases in bud; petals less than 20 mm 16.
5. Stem without red bulbous-based hairs on green parts; sepals always green; petals 15–30 6.
5. Stem, rhachis and ovaries with red bulbous-based hairs 9.
6. Cauline leaves lanceolate; petals more or less as broad as long; lower capsules without glandular hairs **2(ii). cambrica** var. **impunctata**
6. Cauline leaves elliptic or elliptic-lanceolate; petals broader or narrower than long; rhachis and capsules with numerous glandular hairs 7.
7. Petals narrower than long **1(3). biennis** forma **leptomeres**
7. Petals broader than long 8.
8. Petals yellow **1(1). biennis** forma **biennis**
8. Petals pale yellow **1(2). biennis** forma **sulphurea**
9. Sepals red-striped; petals 30–50 mm; style long with stigma lobes spreading above the anthers; capsules red-spotted, with numerous eglandular and glandular hairs **3. glazioviana**
9. Sepals red-striped or not; petals 10–35 mm; style with stigma lobes spreading between anthers or at their apices; capsule variably hairy 10.
10. Sepals red-striped 11.
10. Sepals green; midrib of leaf red 14.
11. Rhachis reddened at tip; all capsules glandular hairy with some red bulbous-based hairs 12.
11. Rhachis green; lower capsules with only eglandular hairs, upper ones with eglandular and glandular hairs 13.
12. Leaves elliptical or elliptic-lanceolate, often crinkled, with red and white midribs; petals 20–30 mm, broader than long; capsules densely hairy **4. fallax**

12. Leaves lanceolate or narrowly lanceolate, flat, with red midribs; petals 10–20 mm, as broad as long; capsules with glabrous spaces along the valves **6(ii). perangusta** var. **rubricalyx**
13. Petals 15–20 mm, slightly hairy outside at base or glabrous **2. cambrica × 4. fallax**
13. Petals 25–35. distinctly hairy outside at base **2. cambrica × 3. glazioviana**
14. Leaves lanceolate, flat; inflorescence rhachis always green at tip; petals 20–30 mm; lower capsules with eglandular hairs only; capsule-teeth up to 2 mm, obtuse **2(i). cambrica** var. **cambrica**
14. Inflorescence rhachis red or reddened at tip; petals 10–20 mm; all capsules with numerous glandular and eglandular hairs; capsule teeth shorter 15.
15. Leaves elliptical or elliptic-lanceolate, wavy; hypanthium 15–25 mm; capsule hairy **5. muricata**
15. Leaves lanceolate, flat; hypanthium 30–32 mm; capsules with glabrous spaces along the valves **6(i). perangusta** var. **perangusta**
16. Sepal tips 2–3 mm, distinctly separated in bud; sepals green, sometimes turning red at late flowering stage; petals 6–12 mm; capsule with glandular and eglandular hairs **9. parviflora**
16. Sepal tips 2–4, less separated in bud; buds reddened between sepal tips from start of flowering; petals 12–18 mm; capsule with mostly glandular hairs **10. rubricuspis**
17. Stem and rhachis with red papillae; cauline leaves oblong-lanceolate with wavy margins, curved tips and reddish midribs; inflorescence loose, young rhachis red; flowers often cleistogamous **7. salicifolia**
17. Stem usually with green papillae; cauline leaves lanceolate, flat or channelled with white midrib; inflorescence compact, tip of rhachis green; flowers open **8. canovirens**
18. Stems usually branched, decumbent; leaves sinuate-pinnatifid; mature buds nodding, the younger ones erect; petals 5–18 mm **12. laciniata**
18. Stem erect; buds erect; petals 15–40 mm 19.
19. Cauline leaves 1.5–6.0 × 1–3 cm; hypanthium 60–100 mm; petals 20–40 mm **13. longiflora**
19. Cauline leaves 0.6–18.0 × 0.6–2.5 mm; hypanthium 20–45 mm; petals 15–35 mm **14. stricta**

Subgenus **1. Oenothera**

Oenothera subg. *Euoenathera* Munz

Annual or *biennial herbs* usually making a rosette of leaves in the first year and erect flowering shoots in the second. *Leaves* dentate. *Flowers* yellow, vespertine. *Hypanthium* long cylindrical. *Capsule* sessile, more or less tapering upwards; seeds prismatic, angled.

1. O. biennis L. Common Evening Primrose

Onagra biennis (L.) Scop.; *Onagra europaea* Spach

Biennial monoecious *herb* with fibrous roots. *Stems* 100–150 cm, erect at tip, pale green, with appressed, arcuate hairs and longer stiff hairs with bulbous, green or red bases, simple or branched, leafy. *Leaves* alternate; basal in a rosette, the lamina 7–20 × 1–5 cm, medium green with a red midrib (except in shade), elliptical to elliptic-lanceolate or oblanceolate, obtuse at apex, slightly denticulate, narrowed at base and flat, the cauline lanceolate to elliptical more or less acute at apex; all hairy on both surfaces and margins; petioles of the stem leaves short, the upper ones sessile. *Inflorescence* of 1–2, bisexual flowers in the axils of each of the upper leaves forming an erect spike; rhachis green at tip and strongly glandular-hairy. *Hypanthium* 28–35 mm, glandular-hairy. *Sepals* 4, 23–26 mm, green, linear-lanceolate, acute at apex, appressed at least below, their apices usually arcuate-divergent, becoming reflexed, glandular-hairy. *Petals* 4, 15–30 × 18–35 mm, yellow, sometimes pale yellow, obovate, emarginate at apex, narrowed at base. *Stamens* 8; anthers 5–10 mm. *Stigma* 4-lobed, the lobes 5–15 mm, spreading between the anthers. *Ovary* with stiff eglandular hairs and glandular hairs. *Capsule* 20–35 × 6–8 mm, green, more or less cylindrical, without wing but tapering upwards, glandular-hairy with arcuate simple or bulbous-based, stiff eglandular hairs; teeth obtuse; seeds prismatic, sharply angled. *Flowers* 6–9. $2n = 14$.

(1) Forma **biennis**

Petals yellow, obovate.

(2) Forma **sulphurea** de Vries

Petals pale yellow, obovate.

(3) Forma **leptomeres** (Bartlett) P. D. Sell

O. biennis var. *leptomeres* Bartlett

Petals deep yellow, linear.

Introduced. Naturalised on sand-dunes, waste ground, wayside and railway banks. First recorded in about 1650 and now frequent in Great Britain north to central Scotland. Native of temperate Europe, eastern Asia and northern Japan. Apparently now utilised as a crop for oil.

× cambrica

This hybrid has the stem and rhachis green, but red-spotted on green parts, the leaves lanceolate or elliptic-lanceolate with a white or pink midrib, the hypanthium 23–33 mm, the petals 13–30 mm, the stigma lobes 4–10 mm and the capsule green, slightly eglandular and glandular-hairy with glabrous spaces along the valves.

Introduced. Occurs with both parents in central and southern England and south Wales. When *O. cambrica* is the female parent the petals and stigma lobes are much larger than when *O. biennis* is the female parent.

× cambrica × glazioviana

This hybrid has the stem green, not spotted, the leaves elliptic-lanceolate with a white midrib, glandular hairs only in the upper part of the inflorescence, hypanthium about 30 mm and green, petals 25–35 mm, anthers 8–11 mm and stigma lobes 6–10 mm.

It is recorded only from Emscota in Warwickshire.

× fallax

This hybrid differs from *O. biennis* only in its red-spotted stem and rhachis.

It is recorded only from Birkdale, Lancashire.

× glazioviana × albivelutina Renner nom. nud.
This hybrid has the stem and rhachis red-spotted or rarely not, leaves elliptical to elliptic-lanceolate, hypanthium (25–)30–35(40) mm, petals about 35 mm, anthers 6–10(–13) mm, and stigma lobes 4–8(–10) mm.

Introduced. Widely scattered records in England. When *O. biennis* is the male parent the hybrid is constant in its characters and is treated as a distinct species *O. fallax*.

2. O. cambrica Rostański

Welsh Evening Primrose

O. novae-scotiae auct.; *O. parviflora* auct.

Biennial monoecious *herb* with fibrous roots. *Stems* 60–100(–150) cm, green or reddish, strongly red-blotched, erect at tip, with numerous stiff hairs on red bulbous bases, often branched in lower half with branches lying on ground with upturned tips, leafy. *Leaves* alternate; lamina 4–12 × 1–4 cm, medium green with reddish midrib on upper surface, paler beneath, elliptic-lanceolate or lanceolate, obtuse at apex, entire to slightly denticulate, narrowed at base, flat, all hairy on both surfaces and the margins; petioles up to 4.5 cm. *Inflorescence* of 1–2 bisexual flowers in the axils of each of the upper leaves forming an erect spike; rhachis green, usually red-spotted with glandular hairs only in the upper part. *Hypanthium* 25–35 mm, covered with glandular and stiff eglandular hairs. *Sepals* 4, 18–22 mm, green, their tips appressed at least below, their apices usually arcuate-divergent, linear-lanceolate, acute at apex, reflexed, with glandular and stiff eglandular hairs. *Petals* 4, 20–30 × 21–28 mm, yellow, obovate, emarginate at apex, glabrous. *Stamens* 8; anthers 6–12(–14) mm. *Stigma* 4-lobed the lobes 6–16 mm, spreading between the anthers. *Ovary* with stiff eglandular and glandular hairs. *Capsule* 30–40(–45) × 6–8 mm, cylindrical, hairs usually with red-bulbous bases but sometimes green, in the upper part with glandular hairs as well; teeth up to 2 mm, obtuse; seeds prismatic, sharply angled. *Flowers* 6–9. $2n = 14$.

(i) Var. **cambrica**
Stem with red blotches and hairs with red, bulbous bases. *Inflorescence* rhachis with red blotches. *Capsule* hairs with red, bulbous bases.

(ii) Var. **impunctata** Rostanski
Stem without red blotches and hairs without red bases. *Inflorescence* rhachis without red blotches. *Capsule* hairs without red bases.

Introduced, probably from Canada in the eighteenth century, but origin uncertain. Naturalised on sandy places near the sea and waste places inland. Frequent by the coast in Great Britain north to southern Scotland, and in the Channel Islands.

3. O. glazioviana P. Micheli ex Mart.

Large-flowered Evening Primrose

O. erythrosepala Borbás; *O. lamarckiana* auct.; *O. vrieseana* H. Lév.; *O. grandiflora* subsp. *erythrosepala* (Borbás) Å. & D. Löve

Biennial monoecious *herb* with fibrous roots. *Stems* 30–180 cm, green or reddish with red blotches, erect at tip, with spreading short hairs and longer stiff hairs with red bulbous bases, simple or branched, leafy. *Leaves* alternate; lamina 7–15 cm, medium green with a white or reddish midrib on upper surface and pale beneath, elliptical to oblong-lanceolate, obtuse at apex, denticulate, narrowed at base, fairly hairy on both surfaces, often strongly crinkled and rarely flat; petiole up to 2 cm. *Inflorescence* of 1–2, bisexual flowers in the axils of each of the upper leaves forming an erect spike, rhachis green, but red-blotched and reddened at apex, with numerous glandular hairs and stiff hairs with red bulbous bases. *Hypanthium* 30–40 mm. *Sepals* 4, 36–40 mm, red-striped, rarely green, long linear-lanceolate, tips appressed at least below, their apices usually arcuate, divergent, strongly glandular-hairy, finely reflexed. *Petals* 4, 30–50 × 32–58 mm, yellow, broadly obovate, emarginate at apex, slightly hairy outside. *Stamens* 8; anthers 10–13 mm. *Stigma* 4-lobed, the lobes 6–10 mm. *Ovary* with stiff eglandular hairs and glandular hairs. *Capsule* 20–35 × 6–8 mm, green or red-striped when young, cylindrical, densely glandular-hairy and with long, stiff hairs with red bulbous bases; teeth concave to more or less obtuse; seeds prismatic, sharply angled. *Flowers* 6–9. $2n = 14$.

Introduced about 1858. Sand-dunes, roadsides, railway, tracks and waste places. Common in Great Britain and eastern Ireland, but absent from north and central Scotland. Our most common Evening Primrose. Native of North America. Apparently not utilised as a crop for oil.

4. O. fallax Renner Intermediate Evening Primrose
O. cantabrigiana B.M. Davis; *O. velutirubata* Renner

Biennial monoecious *herb* with fibrous roots. *Stem* 80–150 cm, green or somewhat reddened, distinctly red-blotched, tip erect, with numerous, long, stiff hairs with bulbous red bases and shorter, arcuate hairs, often with branches from low down, leafy. *Leaves* alternate; lamina 4–17 × 1–6 cm, medium green with a white or reddish midrib on upper surface, paler beneath, elliptical to ovate-lanceolate, obtuse at apex, denticulate, narrowed at base, hairy on both surfaces and the margins, the lower ones crinkled; petiole up to 4 cm. *Inflorescence* of 1–2 bisexual flowers in the axils of each of the upper leaves forming an erect spike; rhachis green, red-blotched and reddened at apex, with numerous glandular hairs and stiff hairs with red bulbous bases. *Hypanthium* 30–40 mm, with stiff eglandular and glandular hairs. *Sepals* 4, 20–23 mm, red-striped, linear-lanceolate, acute at apex, tips appressed at least below their apices, usually arcuate-divergent, with stiff eglandular and glandular hairs. *Petals* 4, 20–30 × 22–34 mm, yellow, obovate, emarginate at apex, narrowed at base. *Stamens* 8; anthers 3–10 mm. *Stigmas* 4-lobed, the lobes 5–10 mm, spreading between the anthers. ovary with stiff eglandular hairs. *Capsule* 20–30 × 6–8 mm, green, red-striped when young, red-blotched, cylindrical, with numerous glandular hairs; teeth obtuse to emarginate; seeds prismatic, sharply angled. *Flowers* 6–9. $2n = 14$.

This species arose as a hybrid between male *O. biennis* and female *O. glazioviana*, but behaves as a constant species reproducing freely.

Native and escaped from cultivation. Sand-dunes and waste places. Scattered over England from Hampshire to Ross and Cromarty, and two records for Ireland. Also recorded from Germany, Poland and Czech Republic.

× **glazioviana**
This hybrid has stem, rhachis and capsules red-spotted with glandular and eglandular hairs, leaves lanceolate or elliptic-lanceolate, flat or crinkled, with a white or red midrib, sepals more or less red-striped, hypanthium 20–34 mm, petals 30–37 mm, anthers 7–11 mm and stigma lobes 3–6 mm.

Recorded only from Birkdale dunes, Lancashire. When *O. glazioviana* is the female parent it has crinkled leaves with a red midrib, but when *O. fallax* is the female parent it has flat leaves with a white midrib.

5. O. muricata L. Blotch-stemmed Evening Primrose
O. rubricaulis Klebahn; *O. biennis* var. *parviflora* Abrom.; *O. muricata* var. *latifolia* Asch. *O. biennis* subsp. *rubricaulis* (Klebahn) Stomps

Annual or *biennial* monoecious *herb*. *Stem* 100–150 cm, green or reddish and strongly red-blotched, erect at tip, with numerous, long, stiff hairs with red bulbous bases and shorter appressed hairs, often branched in lower half, leafy. *Leaves* alternate; lamina 5–10 × 1.5–3.5 cm, medium green with a red midrib on upper surface, paler beneath, elliptic-lanceolate or lanceolate, obtuse at apex, denticulate, hairy and flat or crinkled; petiole short. *Inflorescence* of 1–2 bisexual flowers in the axils of each of the upper leaves forming an erect spike; rhachis green below, reddish at tip, red-spotted with numerous glandular hairs and stiff hairs with red bulbous bases. *Hypanthium* 15–25 mm. *Sepals* 4, 20–23 mm, tips appressed at least below, their apices usually arcuate-divergent, green, long linear-lanceolate, acute at apex; finally reflexed, glandular-hairy. *Petals* 4, 10–20 × 9–18 mm, yellow, obovate, emarginate at apex, narrowed at base. *Stamens* 8; anthers 5–8. *Stigma* 4-lobed, the lobes 5–8 mm, spreading between the anthers. *Ovary* with stiff eglandular and glandular hairs. *Capsule* 20–30 × 6–8 mm, green with red stripes when young, cylindrical, strongly glandular-hairy and with stiff eglandular hairs; teeth obtuse; seeds prismatic, sharply angled. *Flowers* 6–9.

Introduced. Casual in waste ground. Few scattered records in England and occurring in grassland over a wide area between Swanscombe and Greenhithe in Kent. Native of sandy shores of rivers and lakes in eastern Europe to northern and central Russia and eastern Asia, introduced elsewhere.

6. O. perangusta R. R. Gates
Narrow-leaved Evening Primrose

Annual or *biennial* monoecious *herb*. *Stem* 50–100 cm, green with a red suffusion, red-spotted, erect at tip, with scattered, appressed hairs and long, stiff hairs with red bases, leafy. *Leaves* alternate; lamina 5–10 × 1–3 cm, medium green with a red midrib on upper surface, paler beneath, narrowly lanceolate, obtuse at apex, slightly denticulate, narrowed below, flat, hairy. *Inflorescence* of 1–2 bisexual flowers in the axils of the upper leaves, forming an erect spike; rhachis green or somewhat reddened at tip, red spotted, with numerous glandular hairs and stiff, sometimes recurved hairs with red bulbous bases. *Hypanthium* 30–32 mm. *Sepals* 4, green or sometimes red-striped, their tips appressed at least below, their apices usually arcuate-divergent, linear-lanceolate, acute at apex, finally reflexed, with glandular and stiff eglandular hairs. *Petals* 10–20 × 10–20 mm, yellow, obovate, emarginate at apex, narrowed below. *Stamens* 8; anthers about 6 mm. *Stigma* 4-lobed, the lobes about 6 mm, spreading between the anthers. *Ovary* with stiff eglandular and glandular hairs. *Capsule* 25–35 × 6–8 mm, green, red-spotted, cylindrical but tapering upwards, with stiff hairs with red bulbous bases and glandular hairs, but glabrous on midribs of valves; teeth short, somewhat emarginate; seed prismatic, sharply angled. *Flowers* 6–9.

(i) Var. **perangusta**
Sepals green.

(ii) Var. **rubricalyx** R. R. Gates
Sepals red-striped.

Introduced. A rare casual in waste places in England and Wales. Described from Canada, but also known in northern Europe.

7. O. salicifolia Desf. ex G. Don
Willow-leaved Evening Primrose
Onagra salicifolia (Desf. ex G. Don) Spach; *O. depressa* Greene; *Onagra depressa* (Greene) Small; *O. hungarica* Borbás; *O. bauri* Boedijn; *O. strigosa* var. *depressa* (Greene) R. R. Gates; *O. biennis* subsp. *bauri* (Boedijn) Tischler; *O. strigosa* subsp. *hungarica* (Borbás) Å. & D. Löve; *O. villosa* auct.

Annual or *biennial* monoecious *herb*. *Stem* 100–200 cm, green, usually tinged red below, red-spotted, erect, with soft, crispate, short hairs and long, stiff hairs with small, red, cone-like bases, usually unbranched, leafy. *Leaves* alternate; lamina 7–12 × 1–4 cm, medium green with midrib white at first turning red at base later on upper surface, paler beneath, lanceolate, with a twisted, obtuse apex, appressed-hairy, crinkled; petiole short. *Inflorescence* of 1–2 flowers in the axils of each of the upper leaves forming an erect spike; rhachis green or reddish below, with a reddened tip, slightly red-spotted, with arcuate-ascending hairs with short appressed hairs and glandular hairs developing later; flowers bisexual, cleistogamous or chasmogamous. *Hypanthium* 28–35 mm. *Sepals* 4, red-tipped, linear-lanceolate, reflexed, whitish hairy. *Petals* 4, 15–20 × 15–20 mm, yellow, obovate, emarginate at apex, glabrous. *Stamens* 8; anthers 5–8 mm. *Stigma* 4-lobed, the lobes 6–11 mm, spreading between the anthers. *Capsule* 30–45 × 6–8 mm, greyish-green, red-spotted when young, cylindrical, with whitish, appressed hairs and the upper ones also with glandular hairs; teeth emarginate; seeds prismatic, sharply angled. *Flowers* 6–9.

Introduced. Recorded in a few localities as a rare casual and persisted for some years on a railway tip at Edinburgh. Native of North America but occurring widely in Europe as an introduction.

8. O. canovirens Steele White-ribbed Evening Primrose
O. renneri H. Scholz; *O. strigosa* subsp. *canovirens* (Steele) Munz

Annual or *biennial* monoecious *herb. Stem* 60–100 cm, green or slightly reddened, usually not spotted, erect, with soft, crispate, spreading, short hairs and long, stiff hairs with small green or rarely red, cone-like bases, mostly simple. *Leaves* alternate; lamina 3–10×0.5–2.5 cm, medium green with a white midrib or rarely a somewhat pink base, lanceolate, obtuse at apex, slightly denticulate, flat, usually softly hairy; petiole short. *Inflorescence* of 1–2, bisexual flowers in the axils of each of the upper leaves forming an erect spike; rhachis green or somewhat reddened at tip, not or slightly red-spotted, with soft, arcuate, spreading hairs. *Hypanthium* 20–35 mm. *Sepals* 4, green or red-striped, whitish-hairy, linear-lanceolate, acute at apex, reflexed. *Petals* 7–25×7–25 mm, yellow, obovate, emarginate, narrow at base. *Stamens* 8; anthers 4–9 mm. *Stigma* 4-lobed, the lobes 4–9 mm, spreading between the anthers. *Capsule* 25–40 mm, grey-green, cylindrical, with whitish-appressed hairs; teeth emarginate. *Flowers* 6–9.

Introduced. Recorded as a casual and apparently also utilised as a crop for oil. A native of North America rarely found introduced in northern Europe and the Far East.

9. O. parviflora L. Small-flowered Evening Primrose
O. pachycarpa Renner ex Rudloff; *O. muricata* subsp. *parviflora* (L.) Tischler

Annual or *biennial* monoecious *herb. Stems* 100–150 cm, green to dark red, sometimes slightly red-spotted, erect, bur more or less nodding at tip, becoming erect, with short, arcuate hairs and long, stiff hairs with green or pinkish bulbous bases, simple or branched, leafy. *Leaves* alternate; lamina 3–10×0.5–2.5 cm, medium green on upper surface with a red midrib, paler beneath, lanceolate, obtuse at apex, denticulate, slightly hairy; petiole short. *Inflorescence* of 1–2, bisexual flowers in the axils of each of the upper leaves forming an erect spike; rhachis green, slightly red-spotted, with glandular hairs and long, stiff hairs with green or pink, bulbous bases; lower bracts longer than flowers. *Hypanthium* 30–40 mm. *Sepals* 4, green, but those developing late in the season brownish-red, tips erect, 2–3 mm, separated from their bases in bud with glandular and stiff eglandular hairs, linear-lanceolate, reflexed. *Petals* 6–12×6–12 mm, yellow, obovate, emarginate at apex, narrowed at base, glabrous. *Stamens* 8; anthers 4–6 mm. *Stigma* 4-lobed, the lobes 4–6 mm, spreading but often falling short of anthers. *Capsule* 20–30 × 6–8 mm, green, with a thick wall, cylindrical with glandular and stiff eglandular hairs; teeth somewhat emarginate; seeds prismatic, sharply angled. *Flower* 6–9.

Introduced. Locally plentiful in a chalkpit at Stone in Kent and a few other records as a casual although some are referable to *O. cambrica*. Native of eastern North America. apparently now utilised as a crop for oil.

10. O. rubricuspis Renner ex Rostański
Red-stemmed Evening Primrose
O. muricata subsp. *rubricuspis* (Renner ex Rostański) Weihe

Annual or *biennial* monoecious *herb. Stem* up to 100 cm, turning red, red-spotted, tip nodding at first becoming erect, with short arcuate hairs and long stiff ones with red bulbous bases, leafy. *Leaves* alternate; lamina dark green on upper surface with a red midrib, paler beneath, lanceolate, obtuse at apex, denticulate or nearly entire, narrowed below, slightly hairy to nearly glabrous. *Inflorescence* of 1–2 bisexual flowers in the axils of each of the cauline leaves forming an erect spike; lower bracts shorter than the flowers or nearly so; rhachis green, red-spotted, with many glandular hairs and scattered stiff hairs with red bulbous bases. *Hypanthium* 25–35 mm. *Sepals* 4, turning red in upper part, tips 2–4 mm erect separated from their bases in bud, linear-lanceolate, acute at apex. *Petals* 4, 12–18 × 12–18, yellow, obovate, emarginate at apex, narrowed below, glabrous. *Stamens* 8; anthers 3–6 mm. *Stigma* 4-lobed, the lobes 5–8 mm, spreading between the anthers or below them. ovary with stiff eglandular hairs. *Capsule* 18–25(–30)×6–8 mm, green with a thin wall, cylindrical with hairs almost all glandular teeth obtuse or truncate; seed prismatic, angular. *Flowers* 6–9.

Introduced. Known only from Cardiff in 1922 and 1923, but it is a very rare plant which may turn up. It is a North American species the origin of which is unknown. It has turned up elsewhere in Europe in Belgium and Germany.

Subgenus **2. Hartmannia** (Spach) Munz
Hartmannia Spach

Perennial herbs. Leaves subentire to pinnatifid. *Flowers* yellow, or white to rose or purplish, vespertine or diurnal. *Hypanthium* funnel-shaped. *Capsule* clavate, ribbed or winged, attenuate at base into a sterile pedicel-like portion or truly pedicelled; seeds obovoid, not angled.

11. O. rosea L'Hér. ex Aiton Rosy Evening Primrose

Annual or *biennial* monoecious *herb. Stems* 10–50 cm, green, erect, ascending or decumbent, more or less strigulose-hairy throughout, leafy. *Leaves* alternate; lamina 1.5–3.0 cm, medium green on upper surface, paler beneath, oblanceolate to oblong-ovate, obtuse to acute at apex, subentire to sinuate-denticulate and even pinnatifid towards base, narrowed below. *Inflorescence* of 1–2 bisexual flowers in the axils of each of the upper leaves forming an erect spike; bracts linear-lanceolate; flowers diurnal. *Hypanthium* 4–8 mm, funnel-like, strigulose-hairy. *Sepals* 4, linear-lanceolate, acute at apex. *Petals* 4, 5–10 mm, rose to reddish-violet, broadly obovate, narrowed at base. *Stamens* 8; anthers 2.5–4.0 mm. *Style* 1, equalling stamens; stigma 4-lobed, the lobes about 2 mm. *Capsule* 8–10×3–4 mm, obovoid, somewhat winged, narrowed at base into a hollow, ribbed pedicel 5–20 mm; seeds about 0.6 mm, oblong-obovoid. *Flowers* 6–9.

Introduced. A persistent grain alien. Native of Brazil, Uruguay and Argentina.

Subgenus **3. Raimannia** (Rose) Munz

Annual or *perennial herbs. Leaves* entire to pinnatifid. *Flowers* yellow or white, usually vespertine. *Hypanthium* cylindrical. *Capsule* linear to oblong-fusiform, usually enlarged upwards, sessile; seeds subcylindrical or narrowly obovoid, not sharply angled.

12. O. laciniata Hill Cut-leaved Evening Primrose
O. sinuata L.

Annual or *perennial* monoecious *herb. Stem* 20–30(–50) cm, pale green, decumbent, usually finely strigose, with or without stiff, spreading hairs, usually branched, leafy. *Leaves* alternate; lamina 2–6 × 0.5–1.5 cm, medium green on upper surface, paler beneath, oblanceolate to oblong-lanceolate, obtuse at apex, sinuate-pinnatifid or sinuate-dentate, sometimes entire, narrowed below; lower petiolate, the upper sessile. *Inflorescence* of solitary bisexual flowers in the axils of the upper leaves. *Hypanthium* 15–35 mm, cylindrical. *Sepals* 4, 9–10 mm, linear-lanceolate, acute at apex. *Petals* 4, 5–18 mm, yellow, drying red, broadly obovate, sometimes emarginate at apex. *Stamens* 8. *Style* 1, equalling or exceeding stamens; stigma 4-lobed, the lobes 2–4 mm. *Capsule* 10–35 × 2–4 mm, oblong-fusiform, usually somewhat arcuate and divaricate, sessile or shortly stalked; seeds 1.0 mm, pitted, not angled. *Flowers* 6–9.

Introduced. Scattered records in western England and Wales, but not recorded since 1928. Native of North America.

13. O. longiflora L. Long-flowered Evening Primrose

Annual or *biennial* monoecious *herb. Stem* 40–80 cm, pale green, tinted red, erect or ascending, with long simple eglandular hairs mixed with short glandular hairs above, with arcuate-ascending branches arising from near the rosette, leafy. *Leaves* alternate; lamina 1.5–6.0 × 1–3 cm, medium green on upper surface with a white midrib, paler beneath with a prominent white midrib, oblong to elliptical or more or less ovate, shortly acute at apex, irregularly serrate, with flat or undulate margin, truncate to subcordate at base, sessile or nearly so, shortly hairy on both surfaces and the margins. *Inflorescence* a solitary bisexual flower at the end of each stem or branch; bracts 10–30 mm, subobtuse, sessile. *Hypanthium* 6–10 cm, pale green, streaked with red, cylindrical. *Sepals* 4, 28–30 mm, pale yellow, striped red, narrow linear-lanceolate, acute at apex, reflexed, with eglandular and glandular hairs. *Petals* 4, 20–40 mm, yellow with a red spot at base, obovate, rounded at apex, narrowed at base. *Stamens* 8; filaments yellow; anthers 7–13 mm, yellow. *Style* 1, yellow; stigma 4-lobed, the lobes 6–12 mm, spreading between or above anthers. *Capsule* 30–45 × 3–4 mm, curved, oblong-fusiform, hairy, the hairs with red, bulbous bases; seeds 1.5–2.0 mm, elliptical. *Flowers* 6–9.

Introduced. This plant was recorded by Rostański from Galashiels in Selkirkshire but is said to be in error for *O. affinis*. Native of South America.

14. O. stricta Ledeb. ex Link
Fragrant Evening Primrose
O. agari R. R. Gates

Annual or *biennial* monoecious *herb. Stem* 35–150 cm, green, erect or decumbent, hairy below, with long eglandular and glandular hairs above, unbranched or with side-branches arching upwards, leafy. *Leaves* alternate; lamina 0.6–18 × 0.6–2.5 cm, medium green on upper surface, paler beneath, lanceolate, acute at apex, remotely serrate, flat or slightly undulate, sessile. *Inflorescence* of 1 or few bisexual flowers in the axils of each upper cauline leaf forming an erect spike; rhachis greenish but usually reddened at tip; bracts 20–35 × 7–15 mm, green with reddish margins, lanceolate to ovate, acute at apex, truncate to subcordate at base, sessile. *Hypanthium* 20–45 mm, cylindrical. *Sepals* 4, green or reddened, linear-lanceolate, reflexed. *Petals* 4, 15–35 mm, yellow with a red spot at base, broadly obovate. *Stamens* 8; anthers 5–11 mm. *Style* short; stigma 4-lobed, the lobes 3–6 mm. *Capsule* 30–50 × 3–4 mm, linear, but enlarged in upper half, shortly hairy; seeds 1.3–1.8 mm, brown, elliptical, smooth. *Flowers* 6–9. $2n = 14$.

Introduced. Naturalised in sandy places, mostly on the coast. Locally frequent in the Channel Islands and Great Britain north to Norfolk and south Wales, a rare casual further north. Native of Chile.

6. Clarkia Pursh

Godetia Spach

Annual monoecious *herbs. Leaves* alternate, entire. *Inflorescence* a loose terminal raceme; flower buds erect to pendulous; flowers bisexual. *Hypanthium* funnel-shaped. *Sepals* 4. *Petals* 4, white, pink, crimson, mauve or purple. *Stamens* 8, in 2 whorls; anthers attached at base; stigma 4-lobed. *Ovary* 4-celled. *Fruit* a linear capsule; seeds not plumed.

Contains about 44 species in western North America and one in Chile. Named after Captain William Clark (1770–1838).

Lewis, H. & Lewis, M. E. (1955). The genus *Clarkia*. *Univ. Calif. Publ. Bot.* **20**: 241–392.

1. Flower buds pendulous; hypanthium 2–5 mm **1. unguiculata**
1. Flower buds erect; hypanthium 20– 30 mm **2. amoena**

1. C. unguiculata Lindl. Clarkia

Annual monoecious *herb* with fibrous roots. *Stem* up to 1 m, erect to ascending, glaucous, glabrous, branched, leafy. *Leaves* alternate; lamina 1–6 × 1–3 cm, bluish-green on upper surface, paler beneath, lanceolate to elliptical or ovate, acute at apex, entire at base, glabrous; petioles up to 10 mm. *Inflorescence* a loose, terminal raceme, erect; flower buds pendulous; flowers bisexual. *Hypanthium* 2–5 mm, narrowly tubular. *Sepals* 4, 10–16 mm. *Petals* 4, lavender-pink to salmon or dark reddish-purple, triangular or diamond-shaped to subrotund. *Stamens* 8, the outer 4 with red anthers, the inner with whitish anthers. *Style* 1; stigma projecting beyond the anthers, 4-lobed. *Capsule* 15–30 mm, straight or curved. *Flowers* 5–6. $2n = 18$.

Introduced. Commonly grown in gardens and a frequent casual on tips, in parks and waste places. Scattered records in Great Britain. Native of California.

2. C. amoena (Lehm.) A. Nelson & J. F. Macbr. Godetia
Oenothera amoena Lehm.; *Godetia amoena* (Lehm.) G. Don

Annual monoecious *herb* with fibrous roots. *Stems* up to 50 cm, becoming woody towards the base, pale green, suberect, minutely hairy, branching freely. *Leaves* alternate; lamina 1.5–4.0×0.3–1.2 cm, medium green on upper surface, paler beneath, narrow linear to lanceolate, acute at apex, entire, attenuate at base, minutely hairy; petiole up to 10 mm, often indistinct. *Inflorescence* a short terminal raceme of bisexual flowers; pedicels short at anthesis, erect, shortly hairy; flower buds erect. *Hypanthium* 5–30 mm. *Sepals* 4, 10–23 mm, reflexing at anthesis, mucronate at apex. *Petals* 4, (15–)20–30 mm, pink to crimson, with central area often more deeply coloured, obovate, claw well developed. *Stamens* 8; anthers usually mauve, linear. *Styles* 1; stigmatic lobes 4, linear. *Capsule* up to 30 mm, narrowly fusiform, terete, with dense minute hairs; seeds brown, minutely scaly. *Flowers* 5–6. $2n=14$.

Introduced. Commonly grown in gardens and a frequent casual on tips, in parks and waste places. Scattered records in England. Native of western North America.

7. Fuchsia L.

Deciduous, monoecious *shrubs*. *Leaves* opposite or sometimes in whorls of 3 or 4. *Flowers* bisexual solitary in the leaf axils, actinomorphic, pendulous on long pedicels. *Hypanthium* broadly tubular. *Outer perianth segments* 4. *Inner perianth segments* 4 or absent, pink to purple or violet, rarely white. *Stamens* 8. *Ovary* 4-celled. *Fruit* a cylindrical, black berry.

Contains about 110 species occurring in cool moist habitats in South and Central America, Hispaniola, New Zealand and Tahiti. Named after Leonard Fuchs (1501–1565).

The account of the 4 taxa which follows should allow you to name your plant. There is however some question or doubt as to whether it is the correct name or rank.

Bean, W. J. (1973). Trees and shrubs hardy in the British Isles Ed. 8. **2**. London.
Paxton, J. (1843). Fuchsia×exoniensis. Mag. Bot. **10**: 153.

1. Outer perianth segments 25–50 mm	**2. ×exoniensis**
1. Outer perianth segments 12–30 mm	2.
2. Leaves 0.5–3.0×0.3–1.2 cm	**1(i). magellanica** var. **magellanica**
2. Leaves 2–5×0.8–2.0 cm	3.
3. Pedicels up to 3 cm	**1(ii). magellanica** var. **riccartonii**
3. Pedicels 3–6 cm	**1(iii). magellanica** var. **macrostema**

1. F. magellanica Lam. Fuchsia

Deciduous, monoecious *shrub* about 4 m. *Main stems* with peeling, brown, papery bark. *Twigs* red, smooth, glabrous; young shoots red, minutely hairy. *Leaves* opposite; lamina 2–9×0.5–4.0 cm, medium green on upper surface, paler beneath, often red on the veins, lanceolate to ovate or elliptical, acute to acuminate at apex, dentate or sinuate-dentate to almost entire, cuneate or rounded at base, glabrous or minutely hairy on margin; petioles up to 15 mm, very slender, reddish, minutely hairy. *Flowers* bisexual, solitary in upper leaf axils; pedicels (1–)2–6 cm, filiform, reddish. *Hypanthium* 6–12 mm, narrow to broadly cylindrical, red or rose, constricted at the junction with ovary. *Outer perianth segments* 4, 12–30×(2.5–)3–7 mm, bright red, lanceolate to ovate-lanceolate, acute at apex. *Inner perianth segments* 4 or absent, (7–)9–17 mm, usually purple, sometimes pale mauve or pink, obovate, imbricate. *Stamens* 8, 4 borne on the perianth segments (0.7–)20–35 mm, 4 borne on the petals about 5 mm shorter; filaments pink or red; anthers yellowish. *Style* 1, (20–)30–55 mm, rose, hairy or glabrous; stigma more or less 4-lobed, red. *Berry* 10–20 mm, black, oblong, glabrous. *Flowers* 6–9. $2n=22, 44$.

(i) Var. **magellanica**
F. discolor Lindl.

Leaves 0.5–3.0×0.3–1.2 cm, lanceolate, acute to acuminate at apex, denticulate, rounded or cuneate at base; petiole 3–7 mm. *Pedicels* up to 3 cm.

(ii) Var. **riccartonii** L. H. Bailey
Leaves 2–5×0.8–2.0 cm, lanceolate to ovate, acute at apex, denticulate to dentate, rounded at base; petioles 5–10 mm. *Pedicels* up to 3 cm.

(iii) Var. **macrostema** (Ruiz & Pav.) Munz
F. gracilis Lindl.; *F. magellanica* var. *gracilis* (Lindl.) L. H. Bailey; *F. macrostema* Ruiz & Pav.
Leaves 2–5×0.8–2.0 cm, ovate, acute at apex, dentate, rounded at base; petioles 5–10 mm. *Pedicels* 3–6 cm.

Introduced. Planted in hedges, and by abandoned cottages and naturalised in hedgerows, scrub, by streams, among rocks and on walls. Thought to have been introduced in 1788, but probably did not arrive until the 1820s. Widespread in our whole area but especially in southern and western Ireland. All three varieties occur but var. *riccartonii* is probably the most common. Native of South America.

2. F.×exoniensis Paxton Large-flowered Fuchsia
F. cordifolia Benth.×**globosa** Lindl.
F. Cv. Corallina

Spreading, monoecious *shrub* up to 1.5(–3.0) cm. *Twigs* dark brown, glabrous. *Leaves* opposite; lamina 3–9×0.8–4.5 cm, dark green on upper surface, paler beneath, lanceolate to ovate, acute to acuminate at apex, minutely serrate, rounded at base, glabrous; petiole short. *Flowers* bisexual, solitary in upper leaf axils; pedicels up to 40 mm, reddish. *Hypanthium* 7–10 mm, green. *Outer perianth segments* 25–50 mm, deep red. *Inner perianth segments* 15–30 mm, dark bluish-purple. *Stamens* 8; filaments pink; anthers pink. *Style* 1, 30–40 mm, pink. *Berry* oblong or ovoid-globose, 4-celled, many-seeded.

Garden origin. Naturalised on the Lleyn Peninsula, Caernarvonshire and Lundy Island off Devonshire.

8. Circaea L.

Perennial monoecious *herbs* with rhizomes and/or stolons. *Leaves* all opposite. *Inflorescence* a loose, terminal raceme. *Flowers* bisexual, more or less zygomorphic, more or less held horizontally. *Hypanthium* very short. *Sepals* 2, pale green to white. *Petals* 2, deeply 2-lobed, white or pinkish.

Stamens 2. *Style* 1. *Ovary* 1- to 2-celled, each cell with 1 seed. *Fruit* a 1- to 2-seeded achene.

Contains about 10 species in north temperate regions of Europe, Asia and North America.

Grime, J. P. et al. (1988). *Comparative plant ecology*. London. [*C. lutetiana*.]

Hultén, E. & Fries, M. (1986). *Atlas of north European vascular plants north of the Tropic of Cancer.* 3 vols. Königstein.

Raven, P. H. (1963). *Circaea* in the British Isles. *Watsonia* **5**: 262–272.

Stewart, A., Pearman, D. A. & Preston, C. D. (1994) *Scarce plants in Britain*. Peterborough. [*C. alpina*.]

1. Petioles glabrous; open flowers in a terminal cluster; pedicels and sepals glabrous; inner perianth segments less than 1.4 mm; no nectar-secreting ring round the style; achene 1-celled **3. alpina**
1. Petioles hairy at base above; open flowers spaced on the elongating inflorescence axis; pedicels and outer perianth segments at least sparsely glandular-hairy; inner perianth segments 1.8–4.0(–5.0) mm, nectar-secreting ring visible round the base of the style; achene with 2 equal or unequal cells 2.
2. Plant without stolons; petioles hairy all round; ovary with 2 equal cells; achene ripening **1. lutetiana**
2. Plant with stolons produced from the lower leaf-axils; petioles hairy only on upper side, subglabrous on lower side; ovary with 1 large and 1 small cell; achene not ripening **2.×intermedia**

1. C. lutetiana L. Enchanter's Nightshade
C. ovatifolia Stokes nom. illegit.

Perennial monoecious *herb* with non-tuberous, overwintering rhizomes; stolons from lower axils of aerial stem usually absent. *Stem* 20–60 cm, pale green, erect or ascending, swollen at nodes, more or less densely eglandular hairy with some glandular hairs. *Leaves* opposite; lamina 4–10×1–6 cm, dull medium green on upper surface, thin, paler and shining beneath, ovate, gradually acuminate at apex, sinuate-toothed or distantly denticulate, truncate or slightly cordate at base, glabrous or with margins and veins on lower surface eglandular hairy; petiole furrowed above, hairy all round. *Inflorescence* a terminal raceme, axis elongating before petals drop so that the bisexual flowers are well spaced; pedicels densely glandular-hairy, usually without bracts, reflexed in fruit. *Hypanthium* 1.0–1.2 mm. *Outer perianth segments* 2, pale green, glandular-hairy, caducous. *Inner perianth segments* 2, 2–4(–5) mm, white or pinkish, truncate to rounded at base, deeply notched at apex. *Stamens* 2; filaments 2.5–5.5 mm, white; anthers yellow. *Style* 1; stigma deeply 2-lobed. *Achenes* 3–4×2.5–2.5 mm, obovoid, equally 2-celled, densely covered with stiff, hook-tipped, white bristles, most persisting and ripening seed. *Flowers* 6–8. Nectar secreted from a prominent dark green ring round the style just above the hypanthium and flowers visited by small Diptera. $2n=22$.

Native. Woods, hedgerows and other shady places, and also as a garden weed. Common throughout Great Britain and Ireland, but rare in north Scotland. Most of Europe except the north-east and eastwards to central Asia; North Africa. European Temperate element.

2. C.×intermedia Ehrh. Upland Enchanter's Nightshade
C. alpina×lutetiana

Perennial monoecious *herb* with overwintering rhizome more slender than in *C. lutetiana* and with stolons produced from the lower axils of the aerial stems. *Stem* 10–45 cm, pale green, more or less erect, rather sparsely glandular-hairy, leafy. *Leaves* opposite; lamina 3–8×1–5 cm, medium green on upper surface, paler beneath, thin, ovate, abruptly acuminate at apex, dentate, shallowly cordate at base, nearly glabrous; petioles furrowed and hairy above, nearly glabrous beneath. *Inflorescence* a terminal raceme, its axis elongating before the petals drop so that the open bisexual flowers are well spaced; pedicels sparsely glandular-hairy, somewhat reflexed in fruit; often with very small setaceous bracts. *Hypanthium* 0.5–1.2 mm. *Outer perianth segments* 2, whitish, sparsely glandular-hairy, caducous. *Inner perianth segments* 2, 1.8–4 mm, white or pinkish, deeply notched at apex, rounded to cuneate at base. *Stamens* 2; filaments 2–5 mm. *Stigma* more or less deeply 2-lobed. *Achene* up to 2×1.2 mm, obovoid, unequally 2-celled, falling without ripening seed. *Flowers* 7–8. Some nectar is secreted from an inconspicuous ring round the style, but the pollen is sterile and no good seed is set. $2n=22$.

Native. Woods and shady and rocky places, often on mountains. Locally frequent in north and west Great Britain south to south Wales and Derbyshire, with very scattered records in north Ireland, often in the absence of *C. alpina* or both parents, it continuing to survive vegetatively. North-west and central Europe eastwards to central Russia and in North America.

3. C. alpina L. Alpine Enchanter's Nightshade
C. cordifolia Stokes nom. illegit.

Perennial monoecious *herb* overwintering only as fusiform tubers formed at the tips of short-lived, slender stolons, some from the axils of the aerial stem. *Stem* 5–30 cm, pale green, erect or ascending, glabrous at least below. *Leaves* opposite; lamina 2–6(–8)×0.5–3.0 cm, thin and translucent, somewhat shining medium green on upper surface, paler beneath, glabrous; petioles flat above, glabrous. *Inflorescence* a terminal raceme, its axis elongating only after the petals have dropped so that the open bisexual flowers are in a terminal cluster; pedicels glabrous, little reflexed in fruit; often with small, setaceous bracts. *Hypanthium* 0.1–0.2 mm. *Outer perianth segments* 2, white, glabrous, caducous. *Inner perianth segments* 0.6–1.4 mm, white or pinkish, often only shallowly notched at apex, cuneate at base. *Stamens* 2; filaments 1.0–1.5 mm, anthers yellow. *Stigma*, usually entire or shallowly 2-lobed, rarely deeply 2-lobed. *Achene* about 2×1 mm, narrowly obovoid, 1-celled, covered with soft bristles which are sparser, shorter and less consistently hook-tipped than in *C. lutetiana*, freely ripening seed although it sometimes falls prematurely. *Flowers*

7–8. No nectar-secreting ring visible, and probably largely self-pollinated; late in the season the flowers often cleistogamous, but produce seed. $2n=22$.

Native. Shaded rocky places and upland woods, very local. Very scattered localities in Wales, Lake District and central and western Scotland. North Europe and some mountains in central Europe. Circumpolar Boreal-montane element.

Order 6. CORNALES Lindl.

Shrubs, small *trees* or *herbs* with creeping rhizomes and annual stems. *Leaves* opposite, less commonly alternate, simple, usually without stipules. *Inflorescence* a terminal panicle, often umbel-like. *Flowers* bisexual or unisexual, epigynous, actinomorphic, without a hypanthium. *Sepals* 4 or 5. *Petals* 4 or 5. *Styles* 1 or 3; stigma capitate. *Ovary* inferior. *Disc* above ovary. *Fruit* a drupe.

Contains 4 families and about 150 species.

96. CORNACEAE Dumort. nom. conserv.

Deciduous or evergreen, monoecious or dioeceous *shrubs*, small *trees* or *perennial herbs*. *Leaves* opposite or alternate, simple, exstipulate. *Flowers* in terminal panicles, often more or less umbel-like, bisexual or unisexual, epigynous, actinomorphic, without a hypanthium. *Sepals* 4 or 5, small. *Petals* 4 or 5, free. *Stamens* 4 or 5, alternate with the petals. *Styles* 1 or 3; stigma capitate. *Ovary* inferior, 1- to 2-celled, each cell with 1 apical ovule, or ovary 3-celled with 2 cells empty and 1 with an apical ovule; ovule anatropous, integument 1. *Disc* above ovary and usually cushion-like. *Fruit* a drupe, with one 1- to 2-celled stone; embryo straight; endosperm copious.

Contains 14 genera and about 100 species, mainly in north temperate regions and south-east Asia, a few in Africa, South America and New Zealand.

Benthamidia florida (L.) Spach has been recorded as a garden relic at Thorndon Park in Essex.

1. Dioecious, evergreen shrubs; leaves thick and glossy — 2.
1. Monoecious, deciduous trees, shrubs or herbs; leaves not thick and glossy — 3.
2. Leaves opposite; petals 4, dark purple — **4. Aucuba**
2. Leaves alternate; petals 5, yellowish-green — **5. Griselinia**
3. Herbs with rhizomes; petals (not the large, white petal-like bracts) purple — **3. Chamaepericlymenum**
3. Shrubs or small trees; petals white or yellow — 4.
4. Inflorescence appearing before the leaves, with 4 yellow, petal-like bracts at base; petals yellow — **1. Cornus**
4. Inflorescence appearing after the leaves, without petal-like bracts; petals white — **2. Swida**

1. Cornus L.

Deciduous, monoecious *shrubs* or small *trees*. *Leaves* opposite, entire and slightly wavy. *Flowers* in umbels, appearing before the leaves, yellow, 4-merous, bisexual, subtended by 4, yellowish-green bracts. *Fruit* an ellipsoid or oblong drupe.

Four species in central and south Europe to Japan and Korea, and in California.

Bean, W. J. (1970). Trees and shrubs hardy in the British Isles. Ed. 8. **1**. London.
Sims, J. (1826). Cornus mascula. Bot. Mag. **53**: tab. 2675
Weaver, R. E. (1976). The Cornelian cherries. Arnoldia **36**: 50–56.

1. C. mas L. — **Cornelian Cherry**

C. macula Zorn; C. erythrocarpa St Lag.; *C. flava Steud.; C. homerica Bubani; C. nudiflora* Dumort.; *C. praecox Stokes; C. vernalis Salisb.*

Deciduous, monoecious *shrub* or small *tree* up to 4(–8) m with a broad, open crown. *Trunk* solitary or several, often branched from base. *Bark* reddish-brown when young, becoming rather pale brown and scaling in longitudinal plates showing an orange blaze. *Branches* erect-ascending to spreading, dull brown, rough and scaly; twigs divaricate at a wide angle, dull brown, rough and scaly; young shoots green or purplish, minutely appressed-hairy. *Buds* 5–7 × 3–4 mm, ovate, pointed at apex; scales dull brown, acute at apex, shortly appressed-hairy. *Leaves* opposite; the lamina 3.5–10.0 × 2–5 cm, dull, rather dark green on upper surface, paler beneath, turning reddish-purple in autumn, lanceolate, ovate or elliptical, acute to bluntly acuminate at apex, entire or slightly wavy, cuneate to rounded at base, with pale, short, appressed simple eglandular hairs on both surfaces; veins 3–5 pairs, curved upwards; petiole up to 8 mm, pale green, shortly hairy. *Inflorescence* of small umbels on old wood appearing before the leaves; bracts 4, 6–7 × 6–7 mm, pale yellowish-green tinted pink on outside, broadly ovate, shortly acuminate at apex, minutely hairy; flowers bisexual, fusty-smelling, 4–5 mm in diameter; peduncle 7–10 mm, pale green, with ascending-appressed hairs. *Sepals* 4–5, minute, triangular, green. *Petals* 4–5, 2.0–2.5 × 1.0–1.5 mm, pale yellow, oblanceolate, apiculate at apex. *Stamens* 4–5; filaments 1–2 mm, pale yellow; anthers deep yellow. *Style* 1, about 1 mm, green; stigma green. *Drupe* 12–20 × 10–15 mm, bright shiny red, ellipsoid or oblong. *Flowers* 12–3. $2n=18, 54$.

Introduced by 1596, it is grown in hedges and along road verges as well as in gardens and can be long persistent. Scattered over much of Great Britain and Ireland, rarely north to central Scotland and in Ireland. Native of central and southern Europe and south-west Asia. Its fruit is edible and it is sometimes sold in European markets.

2. Swida Opiz

Cornus section *Thelycrania* Dumort.; *Thelycrania* (Dumort.) Fourr.; *Cornus* subgenus *Kraniopsis* Raf.

Deciduous, monoecious *shrubs*. *Leaves* opposite, entire. *Inflorescence* a corymbose cyme, appearing with the leaves; flowers bisexual. *Sepals* 4, small. *Petals* 4, white or cream. *Stamens* 4. *Style* 1. *Fruit* a globose or oblate-spheroid drupe.

Contains about 45 species in north temperate regions and Peru and Bolivia.

S. alternifolia (L. fil.) Small has been recorded as a garden relic at Warley Place Gardens in Essex.

This interpretation of species is based on a large number of living plants and is rather different from most accounts in British reference books. Intermediates may or may not be hybrids and may simply reproduce themselves. Large numbers of shrubs of this genus appear to have been planted all over our countryside.

Bean, W. J. (1970). Trees and shrubs hardy in the British Isles. **1**. Ed. 8. London.

Fosberg, F. H. (1942). Cornus sericea L. (C. stolonifera Michx.). Bull. Torrey Bot. Club **69**: 583–589.

Hultén, E. & Fries, M. (1986). Atlas of north European vascular plants north of the Tropic of Cancer. 3 vols. Königstein.

Kelly, D. L. 91990). *Cornus sericea* L. in Ireland: an incipient weed of wetlands. Watsonia **18**: 33–36.

Rehder, A. (1940). Manual of cultivated trees and shrubs hardy in North America. Ed. 2. New York.

1.	Hairs on underside of leaf mostly curved upwards	2.
1.	Hairs on underside of leaf flat to the surface and medifixed	3.
2.	Leaves 4–9 × 2–6 cm	**1. sanguinea**
2.	Leaves 5–13 × 2.5–7.0 cm	**2. koenigii**
3.	Leaves with 2–4 pairs of veins; drupes black	**3. australis**
3.	Leaves with 4–7 pairs of veins; drupes white or bluish	4.
4.	Twigs bright yellow or bright red in autumn and winter	5.
4.	Twigs becoming dark brownish-red in autumn	6.
5.	Twigs becoming bright yellow in autumn and winter	**4(i). alba** var. **flaviramea**
5.	Twigs becoming bright red in autumn and winter	**4(iii). alba** var. **sibirica**
6.	Plant not stoloniferous; leaves 5–15 × 3–10 cm	**4(ii). alba** var. **alba**
6.	Plant stoloniferous; leaves 4–9 × 2.5–6.0 cm	**5. sericea**

1. S. sanguinea (L.) Opiz Dogwood
Cornus sanguinea L.; Thelycrania sanguinea (L.) Fourr.; *Cornus citrifolia* Wahlenb.; *Cornus latifolia* Bray

Deciduous, monoecious *shrub* or rarely a small *tree* up to 6 m with a rounded crown, suckering freely. *Stems* usually several, up to 20 cm in circumference. *Bark* brown, slightly rough. *Branches* ascending or spreading; twigs pale brown to purplish-red; young shoots green, or pinkish on the upper side and green beneath, with short, white appressed hairs. *Buds* opposite, 7–10 mm, ovoid, acute at apex; scales dark brown, ovate, acute at apex, with very short, pale hairs. *Leaves* opposite; lamina 4–9 × 2–6 cm, dull medium yellowish-green on upper surface, slightly paler beneath, usually turning dull purplish-red in autumn, ovate, ovate-lanceolate or elliptical, acuminate or cuspidate at apex, entire, but undulate, rounded at base, glabrous on upper surface or with very short hairs, with numerous, short, spreading, upcurved pale simple eglandular hairs beneath; veins 3–5 pairs from below the middle of the leaf, curving round towards the apex, impressed above, prominent beneath; petiole up to 15 mm, pale green, sometimes suffused reddish, with very short simple eglandular hairs. *Inflorescence* a many-flowered, fragrant, more or less flat-topped, corymbose cyme; flowers bisexual; peduncle 2.5–3.5 cm, very shortly appressed-hairy; pedicels 4–6 mm, pale green, very shortly appressed-hairy. *Sepals* 4, about 1 mm, green, triangular, acute at apex. *Petals* 4, 4–7 mm, cream, narrowly lanceolate, obtuse at apex, appressed-hairy outside. *Stamens* 4; filaments 2–3 mm, white; anthers cream. *Style* 1, green; stigma clavate. *Drupe* 6–7 mm in diameter, globose, ripening from green to purplish-black, with a single seed. *Flowers* 6–7(–9). Pollinated by various insects. Spreads by suckers and bird-distributed seeds. $2n=22$.

Native. Woodland edges, scrub and hedgerows on limestone or base-rich clays. Common in most of south and central lowland Great Britain, very local in southern Ireland and an escape elsewhere which may well be *S. australis* or *S. koenigii* or hybrids. Most of Europe, except the north-east and extreme north, very rare in south-west Asia. European temperate element. Dogwood leaves, if pulled slowly from each end, split, leaving a number of elastic tissues joining the two pieces. Dogwood is probably a corruption of Dagwood, a dag being a spike or skewer for which the wood was used The medifixed hairs or much larger leaves should distinguish all the introduced species and hybrids from our native *S. sanguinea.*

2. S. koenigii (C. K. Schneid.) Pojark. Asian Dogwood
Cornus koenigii C. K. Schneid.; *Thelycrania koenigii* (C. K. Schneid.) Sanadze; *Cornus australis* var. *koenigii* (C. K. Schneid.) Wangerin

Deciduous, monoecious *shrub* up to 4 m. *Stems* erect-spreading, brown, glabrous; twigs reddish-brown, glabrous; young shoots green to reddish-brown, shortly medifixed appressed-hairy. *Buds* 5–8 × 1.5–2.5 mm, narrowly ovoid; scales dark brown, with appressed hairs. *Leaves* opposite; lamina 5–13(–18) × 2.5–7.0 cm, medium to dark yellowish-green on upper surface, paler beneath, turning red or pinkish-red in autumn, ovate, elliptical or subrotund, rounded to a short acumen at apex, entire, but wavy, broadly cuneate to rounded at base, with scattered, appressed, medifixed hairs on upper surface and numerous, upcurved and some spreading simple eglandular hairs beneath; veins 4–5 pairs, impressed on upper surface, very prominent beneath; petiole 5–20 mm, pale green, channelled above, rounded beneath, with numerous spreading and appressed hairs. *Inflorescence* a terminal cyme up to 50 mm wide; pedicels pale yellowish-green, sometimes tinted pinkish-red, with appressed hairs; flowers bisexual, 14–15 mm, in diameter. *Sepals* 4, 0.5 mm, pale green, ovate, rounded at apex, glabrous. *Petals* 4, 6–7 mm, white, oblong, rounded at apex. *Stamens* 4; filaments white; anthers cream. *Style* 1, greenish; stigma clavate. *Drupe* 6–7 mm, dark purplish-black, subglobose; stone 4.0–5.5 mm, subglobose. *Flowers* 6–7.

Introduced. Planted in amenity areas, parks and by roadsides, sometimes as infills in existing hedges. Frequent in Cambridgeshire. Sometimes the leaves have some hairs flat to the surface and medifixed. These may be hybrids with *S. australis* which is planted much more frequently in similar places. Native of Transcaucasia.

3. S. australis (C. A. Mey.) Grossh. Southern Dogwood
Cornus australis C. A. Mey.; *Cornus sanguinea* var. *australis* (C. A. Mey.) Koehne; *Thelycrania australis* (C. A. Mey.) Sanadze; *Cornus sanguinea* subsp. *australis* (C. A. Mey.) Jáv.

Deciduous, monoecious *shrub* up to 4m. *Stems* erect-spreading, brown, glabrous; branches spreading, glabrous; twigs yellowish-brown, glabrous; young shoots green to purple, very shortly medifixed appressed-hairy. *Leaves* opposite; lamina 3–13×2–7cm, yellowish- green on upper surface, paler beneath, turning red or pinkish-red in autumn, ovate or lanceolate, acute at apex, entire but wavy, cuneate at base, with appressed medifixed hairs on both surfaces, more numerous beneath; veins 3–4 pairs, impressed on upper surface, prominent beneath; petiole 8–15mm, red, channelled above, rounded beneath, appressed- hairy. *Inflorescence* a terminal cyme up to 50mm wide; pedicel very short, pale green, with appressed white and brown hairs; flowers bisexual, 10–12mm in diameter, scentless. *Sepals* 4.0–4.5mm, pale green, ovate, rounded at apex, glabrous. *Petals* 4, 4–5mm, *white*, oblong, rounded at apex. *Stamens* 4; filaments white; anthers cream. *Style* 1, pale green; stigma cream. *Drupes* 6–7mm in diameter, dark purplish-black, subglobose; stone 5.0–5.5mm, subglobose. *Flowers* 6–7(–9).

Introduced. Planted in new woods, parks, by motorways, around amenity areas and sometimes as infills in existing hedgerows in the countryside. Common in Cambridgeshire; also in Cardiganshire (fide A. O. Chater) and should be looked for elsewhere.

Native around the Black and Caspian Seas.

× **koenigii**
Plants with both types of hairs on under surface of leaves and leaves up to 10cm occur in Cambridgeshire and Cardiganshire.

× **sanguinea**
Plants with both types of hair on under surface of leaves and much smaller leaves than the above also occur in Cambridgeshire and Cardiganshire and may be of this origin, or they may be intermediates reproducing themselves.

4. S. alba (L.) Holub White Dogwood
Cornus alba L.; *Thelycrania alba* (L.) Pojark.; *C. tatarica* Mill.

Deciduous, monoecious *shrub* up to 3 m, sometimes rooting and suckering. *Stems* usually several, reddish-brown, red or green, erect or spreading. *Bark* rough. *Branches* brown to reddish-brown; twigs dark reddish-brown, glabrous; young shoots green to dark red, bright red or bright yellow, glabrous. *Buds* 5–6×2–3mm, narrowly ovoid, pointed at apex; scales dark reddish-brown, lanceolate, acute at apex. *Leaves* opposite; lamina 5–15×3–10cm, dull medium yellowish-green on upper surface, paler and whitish-green beneath, usually turning bright red or dark red in autumn, sometimes variegated, ovate or broadly elliptical, acute or acuminate at apex, more or less entire, rounded or cuneate at base, with short, medifixed, appressed hairs on both surfaces; veins 4–7 pairs, curving towards the apex, slightly impressed on upper surface and slightly prominent beneath; petiole 8–25mm, green to red, glabrous. *Inflorescence* a terminal cyme 3–8cm in diameter, more or less flat-topped; peduncles and pedicels green, shortly hairy; flowers bisexual, 8–9mm in diameter. *Sepals* 4, about 0.5mm, 4 small points, green, glabrous. *Petals* 4, 3.0–4.0mm, white, elliptical or ovate, obtuse at apex, incurved along margins. *Stamens* 4; filaments white; anthers cream. *Style* 1, cream; stigma cream. *Nectary disc* yellow, becoming striped red. *Drupe* 7–8mm in diameter, white or bluish, oblate-spheroid; stone ellipsoid, longer than wide and narrowed above and below. *Flowers* 5–6(–10). $2n=22$.

(i) Var. **flaviramea** (Späth) P. D. Sell
Cornus alba var. flaviramea Späth; *Cornus stolonifera* var. *flaviramea* (Späth) Rehder

Young shoots yellowish-green, becoming bright yellow in winter. *Leaves* 8–14×3–9cm, not turning red, hairs beneath appressed; veins 5 or 6 pairs. *Drupe* white.

(ii) Var. **alba**
Young shoots dull dark red. *Leaves* 8–14×4–8cm, turning dull red in autumn, hairs beneath appressed; veins 5–6 pairs. *Drupe* white.

(iii) Var. **sibirica** (Lodd. ex Loudon) P. D. Sell
Cornus sibirica Lodd. ex Loudon; *Cornus alba* var. *sibirica* (Lodd. ex Loudon) Loudon

Young shoots bright red. *Leaves* 5–10×3–6cm, turning bright red in autumn, hairs beneath appressed; veins 5–7 pairs. *Drupe* bluish.

Introduced. All taxa are much grown in gardens, parks and amenity areas and are naturalised on waste land, old railways and roadsides. The species would appear to be native of Asia and perhaps North America. Var. ***gouchaultii*** (Carrière) Rehder with variegated leaves also occurs in gardens but has not been seen naturalised.

5. S. sericea (L.) Holub Red-osier Dogwood
Cornus sericea L.; *Thelycrania sericea* (L.) Dandy; *Cornus stolonifera* Michx; *Thelycrania stolonifera* (Michx) Pojark.; *Swida stolonifera* (Michx) Pojark.

Deciduous, monoecious *shrub* up to 3m, stems rooting and suckering and forming patches. *Stems* numerous, spreading or prostrate. *Bark* greenish-yellow to dark red in winter. *Branches* spreading; twigs reddish-brown, divaricate; young shoots blood red, sometimes greenish beneath, glabrous or hairy. *Buds* 5–6×2–3mm, narrowly ovoid, pointed at apex; scales dark reddish-brown, lanceolate, acute at apex. *Leaves* opposite; the lamina 4–9×2.0–6.0cm, dull yellowish-green on upper surface, pale greyish-green beneath, turning dull red in autumn, ovate or elliptical, rounded to a shortly acuminate at apex, entire, rounded or shortly cuneate at base, very shortly appressed hairy on both surfaces, becoming glabrous on upper surface; veins 5–6 pairs, deeply impressed on upper surface, very prominent beneath; petioles up to 25mm, pale green, often reddish on upper surface, sparsely short-hairy. *Inflorescence* a terminal cyme, 3–5cm in diameter, more or less, flat-topped; peduncles and pedicels pale green, glabrous or sparsely hairy; flowers bisexual, 7–8mm in diameter, slightly scented. *Sepals* very small, yellowish-green, forming 4 triangular points.

Petals 4, 3.0–4.0 mm, white, lanceolate, obtuse at apex, incurved along margins. *Stamens* 4; filaments white; anthers cream. *Style* 1, white; stigma cream. *Nectary disc* dark red. *Drupe* 4–7 mm in diameter, wider than long, an oblate-spheroid; stone broadly ovoid, wider than long with a flat base. *Flowers* 5–11. Pollinated by various insects. $2n = 22$.

Introduced. Much grown for ornament and game cover, in parks, estates and along roadsides and frequently spreading by suckers and forming dense thickets. Scattered over much of lowland Great Britain and Ireland. Native of North America. It is possible the Irish plants referred to here are *S. alba* var. *sibirica* as their stems are said to be bright blood red.

3. **Chamaepericlymenum** Hill

Cornus subgenus *Arctocrania* (Endl.) Rchb.

Perennial herbs with creeping rhizomes and annual stems. *Leaves* opposite, entire. *Flowers* in umbel-like cymes appearing with the leaves, bisexual, surrounded by 4 large petal-like white bracts. *Sepals* 4, *Petals* 4. *Stamens* 4. *Style* 1. *Fruit* a globose drupe.

Contains 2 species in the Arctic and mountains of the north temperate zone.

C. canadensis (L.) Asch. & Graebn. was recorded as a nursery weed at Seal in Kent.

Hultén, E. & Fries, M. (1986). Atlas of north European vascular plants north of the Tropic of Cancer. Königstein.

Taylor, K. (1999). Biological flora of the British Isles. No. 209. Cornus suecica L. (Chamaepericlymenum suecicum (L.) Ascherson & Graebner. Jour. Ecol. **87**: 1068–1077.

1. C. suecicum (L.) Asch. & Graebn. Dwarf Cornel

Cornus suecica L.; *Cornus biramis* Stokes nom. illegit.; *Cornus borealis* Gorter; *Cornus herbacea* L.

Perennial herb 6–20 cm, with a creeping rhizome and annual stems. *Stems* erect, often a few together, pale green, tinted brownish-red, glabrous or appressed-hairy, simple or with short, axillary branches from the uppermost pair of leaves. *Leaves* opposite, the lamina 1–3 × 0.5–2.0 cm, medium green on upper surface, paler and bluish-green beneath, ovate or ovate-elliptical, acute or very shortly acuminate at apex, entire and wavy, rounded or shortly cuneate at base, sessile or shortly petioled, with short, appressed hairs on the upper surface and glabrous beneath; veins 3–5 pairs from near or at the base and curved upwards, impressed on upper surface, prominent beneath. *Inflorescence* a terminal, umbel-like cyme; flowers 8–25, bisexual, surrounded by 4 large, 5–8 mm, white, ovate bracts which make them look flower-like; pedicels 1–2 mm, appressed-hairy. *Sepals* 4, about 0.5 mm, deltoid, acute at apex, recurved, appressed-hairy. *Petals* 4, 1–2 mm, blackish-purple, triangular-ovate, acute at apex, reflexed. *Stamens* 4; filaments 2 mm, white; anthers yellow. *Style* 1, white; stigma yellow. *Drupe* 5–10 mm in diameter, bright red, oblate-spheroid or ovoid. *Flowers* 7–8. Visited by hoverflies; self-pollination possible. $2n = 22$.

Native. Upland moors usually under *Calluna* or *Vaccinium myrtillus* up to 915 m. Locally frequent in the mainland part of west and central Scotland and extremely local in northern England. Northern Europe, extending southwards to Holland; northern Asia and North America. European Boreo-arctic Montane element.

4. **Aucuba** Thunb. ex Murray

Evergreen, dioecious *shrubs*. *Leaves* opposite, simple, entire to remotely serrate. *Flowers* unisexual, the male in erect, terminal panicles, the female in small, terminal clusters. *Sepals* minute or absent. *Petals* 4. *Stamens* 4. *Style* 1. *Ovary* 1-celled. *Fruit* a 1-seeded drupe.

Contains 3 species in Asia; much cultivated elsewhere.

Bean, W. J. (1970). Trees and shrubs hardy in the British Isles. Ed. 8. **1**. London.

Rehder, A. (1940). Manual of cultivated trees and shrubs hardy in North America. Ed. 2. New York.

Wangerin, W. (1910). Aucuba Thunb. ex Murray. In Engler, A.(Edit.) *Das Pflanzenreich* **IV.229**: 38–41.Leipzig.

1. Leaves with pale yellow spots and blotches **1(3). japonica** forma **variegata**
1. Leaves without spots or blotches 2.
2. Leaves 6–15 × 1.0–2.5 cm, lanceolate or narrowly elliptical **1(1). japonica** forma **longifolia**
2. Leaves 6–20 × 2–8 cm, lanceolate, ovate, elliptical or ovate-oblong **1(2). japonica** forma **japonica**

1. A. japonica Thunb. ex Murray Spotted Laurel

Eubasis dichotoma Salisb. nom. illegit.

Evergreen, dioecious *shrub* up to 5 m, with a dense, spreading habit. *Stems* erect, much- branched. *Bark* pale brown. *Branches* spreading; twigs brown or green, terete, smooth, glabrous; young shoots green, smooth and glabrous. *Buds* 15–30 × 5–10 mm, narrowly ovoid, pointed at apex; scales shining medium green, tinted brownish-purple, lanceolate, acute at apex, glabrous. *Leaves* opposite; lamina 6–20 × 1–8(–12) cm, medium yellowish-green, usually with pale yellow spots and blotches on upper surface with a slightly paler midrib, slightly paler and often similarly blotched and spotted beneath, coriaceous, lanceolate, ovate, narrowly to broadly elliptical or oblong-ovate, gradually narrowed to a subacute or acuminate apex, entire to remotely serrate and slightly undulate, cuneate or rounded at base, glabrous; midrib prominent beneath, lateral veins obscure; petiole up to 20 mm, pale green, glabrous. *Male flowers* in erect terminal panicles on pedicels 3–5 mm, hairy, with small, ovate bracts; sepals minute and triangular or absent; petals 4, 3.5–5.0 × 2.5–3.0 mm, dull brownish-purple on inner surface, pale yellowish-green on outer surface, ovate, acuminate-cuspidate at apex, glabrous; stamens 4, filaments pale yellowish-green, anthers brownish with yellow pollen. *Female flowers* in small terminal clusters on pedicels 2–3 mm and hairy, with large, oblong bracts; sepals minute or absent; petals 2–3 × 1–2 mm, brownish-purple on inner surface, pale yellowish-green on outer surface, ovate, acuminate at apex, glabrous; style 1, 1.0–1.5 mm, green, stigma green, capitate; ovary 1-celled. *Fruit* 12–18 × 11–14 mm, pale green turning deep brownish-orange or red, ellipsoid; seed 12–13 × 8–9 mm, pale brown. *Flowers* 3–4. $2n=32$.

(**1**) Forma **longifolia** (T. Moore) Schelle
Cv. Longifolia

A. japonica var. *angustifolia* Regel; *A. japonica* var. *salicifolia* Bean; *A. japonica longifolia* T. Moore.

Leaves 6–15 × 1.0–2.5 cm, green without spots or blotches, lanceolate or narrowly elliptical.

(**2**) Forma **japonica** Cv. Concolor

A. japonica var. *concolor* Regel; *A. japonica viridis* Bull. ex Hamb.

Leaves 6–20 × 2–8 cm, green without spots or blotches, lanceolate, ovate, elliptical or oblong-ovate.

(**3**) Forma **variegata** (Dombrain) Rehder
Cv. Variegata

A. japonica variegata Dombrain; *A. japonica picturata* T. Moore; *A. japonica* forma *picturata* (T. Moore) Rehder

Leaves 6–20 × 1–8 cm, green with yellowish spots and blotches, lanceolate, ovate, narrow to broad elliptical and oblong-ovate.

Introduced in 1783. Widely planted in gardens, shrubberies, parks, plantations, amenity areas and especially churchyards throughout Great Britain and Ireland north to mid Scotland. Rarely self-sown. Forma *variegata* is the most common variant. Native of Japan.

5. Griselinia G. Forst.

Evergreen, dioecious *shrubs*. *Leaves* alternate, simple and entire. *Flowers* unisexual in axillary racemes or panicles. *Sepals* 5. *Petals* 5. *Styles* 3. *Ovary* 1- to 2-celled. *Fruit* a 1-seeded drupe.

Contains 6 species in New Zealand, Chile and Brazil.

Bean, W. J. (1973). Trees and shrubs hardy in the British Isles. **2**. Ed. 8. London.
Wangerin, W. (1910). Griselinia G. Forst. in Engler, A.(Edit.) *Der* Pflanzenreich **IV.229** pp. 94–99. Leipzig.

1. G. littoralis (Raoul) Raoul New Zealand Broadleaf
Pukateria littoralis Raoul

Evergreen, dioecious *shrub* up to 3 m, or rarely a *tree* up to 15 m. *Stems* erect, gnarled. *Bark* grey, smooth. *Branches* erect; twigs brown, rough; young shoots yellow to brownish-green, shiny, glabrous. *Buds* ovate. *Leaves* alternate; lamina 3–10 × 2.5–6.0 cm, medium to dark shining green on upper surface, pale yellowish-green beneath, coriaceous, broadly ovate, broadly elliptical or almost subrotund, rounded to acute at apex, entire, cuneate or rounded at base, glabrous; only midrib visible; petiole up to 2.5 cm, pale yellowish-green, glabrous. *Inflorescence* axillary, a simple raceme or panicle; pedicels pale green and minutely hairy. *Male flowers* with sepals 5, 1–2 mm, pale greenish, broadly elliptical, rounded at apex; petals 5, 1–2 mm, yellowish-green, elliptic-oblong, rounded at apex; stamens 5, filaments 0.5–0.7 mm, pale green, anthers greenish-yellow. *Female flowers* with sepals 5, 1–2 mm, pale greenish, broadly elliptical, rounded at apex; petals 5, 1.5–2.0 mm, yellowish-green, elliptic-oblong, rounded at apex. *Styles* 3, pale green, stigmas colourless. Ovary 1- to 2-celled. *Drupe* 6–8 × 3–4 mm, green turning dark purple, ellipsoid, obtuse at apex. *Flowers* 4–5. $2n=36$.

Introduced. Commonly planted in western Great Britain north to central Scotland, Isle of Man and Ireland, especially near the sea where it is persistent and sometimes self-sown. Native of New Zealand.

96A. GARRYACEAE Lindl. nom. conserv.

Evergreen, dioecious *shrubs* or small *trees. Leaves* opposite, entire or wavy. *Flowers* in terminal or axillary, catkin-like, pendulous racemes. *Male flowers* with 4 introrse anthers with short filaments, surrounded by 4 valvate perianth segments. *Female flowers* naked or with 2–4, small, decussate bracts united to form a cup and sometimes interpreted as sepals or perianth segments. *Ovary* inferior, of 2–3 united carpels, with 1 locule and 2 pendulous ovules on parietal placentas. *Fruit* a dryish, 2-seeded berry and with abundant endosperm.

Contains 1 genus and about 18 species in south-west North America, northern Central America and the West Indies. Named after Nicholas Garry, Secretary of the Hudson's Bay Company, who assisted David Douglas in his exploration of the Pacific North-west.

1. Garrya Douglas ex Lindl.

As for family.

Bean, W. J. (1973). Trees and shrubs hardy in the British Isles. **2**. Ed. 8. London.
Rehder, A. (1940). Manual of cultivated trees and shrubs hardy in North America. Ed. 2. New York.

1. G. elliptica Douglas ex Lindl. Silk-tassel

Evergreen, dioecious *shrub* or small *tree* up to 5 m, with a bushy habit. *Stems* erect, irregular. *Bark* pale greyish-brown, smooth. *Branches* erect or spreading; twigs thick, greyish, rough, glabrous; young shoots green and hairy. *Buds* small. *Leaves* opposite, lamina 3.5–8.5 × 2–5 cm, dull dark green with a pale midrib on upper surface, paler beneath with an even paler midrib, elliptical to subrotund, rounded at apex, sometimes with a short, abrupt tip, entire but wavy and undulate, cuneate to rounded at base, glabrous on upper surface, densely short-hairy beneath; veins 7–9 pairs, only the midrib prominent; petiole 3–10 mm, eglandular hairy. *Flowers* densely crowded on slender, patent, becoming pendulous catkins 75–150 mm, in a cluster towards the end of the shoot and in the leaf axils; bracts in male plant, cup-shaped, silky-hairy and enclosing the base of the stamens; bracts in female plants longer and narrower. *Stamens* 4; filaments short, surrounded by 4, valvate perianth segments. *Female flowers* naked; styles 2, slender. *Berry* 7–8 × 7–8 mm, green becoming purplish, ellipsoid or ovoid, densely hairy; seeds 2, embedded in a dark red juice. *Flowers* 2–3. $2n = 22$.

Introduced. A garden relic in Essex and Gloucestershire and sometimes planted by roadsides. Native of California and Oregon.

Order 7. SANTALALES Lindl.

Trees, shrubs and *herbs*, often parasitic on other angiosperms. *Leaves* usually opposite, sometimes scale-like; without stipules. *Flowers* bisexual or unisexual, epigynous, actinomorphic. *Calyx* with valvate lobes or often reduced, sometimes to a ring. *Petals* present or absent, sometimes united into a tube. *Stamens* same number as calyx lobes and opposite them or opposite petals when present. *Ovary* 1-celled; ovules few. *Fruit* a drupe or berry, less frequently a nut; seeds with endosperm.

Consists of 10 families and about 2,000 species.

97. SANTALACEAE R. Br. nom. conserv.

Hemiparasitic *perennial herbs* with haustoria. *Leaves* alternate, simple, entire, sessile, exstipulate. *Inflorescence* a small dichotomous cyme with a bract and 2 bracteoles; flowers bisexual, epigynous, actinomorphic, with a short, funnel-shaped hypanthium. *Perianth segments* 5, free, on the hypanthium, yellowish. *Stamens* 5, opposite the perianth segments. *Style* 1; stigma capitate. *Ovary* 1-celled, with 3 ovules. *Fruit* a 1-seeded nut.

Contains about 30 genera and 400 species in tropical and temperate regions.

1. Thesium L.

As family.

Contains about 300 species, widely distributed.

Stewart, A., Pearman, D. A. & Preston, C. D. (1994). *Scarce plants in Britain.* Peterborough.

1. T. humifusum DC. Bastard Toadflax
T. linophyllum auct.

Hemiparasitic *perennial herb* with haustoria on their roots by means of which they attach themselves to the roots of other plants and with a woody stock. *Stems* 5–20 cm, dull green, spreading or prostrate, slender, angled, the angles slightly rough, branched, leafy below. *Leaves* alternate, distant; lamina 5–15(–25) × 0.2–2.0 mm, yellowish-green, linear, obtuse to acute at apex, the lower scale-like, glabrous, 1-veined. *Inflorescence* of small dichotomous cymes, branched and terminal; bract longer than flower, linear; bracteoles 2, linear-lanceolate, serrulate, inserted at the base of the short, stout pedicel; flowers about 3 mm in diameter. *Hypanthium* short and funnel-shaped. *Perianth segments* 5, triangular, acute at apex, yellowish. *Stamens* 5, opposite the perianth segments; filaments white; anthers yellow. *Style* 1; stigma capitate. *Nut* about 3 mm, green, ovoid, ribbed, crowned by the persistent, inrolled perianth segments. *Flowers* 6–8. $2n =$ c. 26.

Native. Species-rich chalk and limestone grassland. Very local in England north to Gloucestershire and Lincolnshire, and in Jersey and Alderney. West Europe. Oceanic Temperate element.

98. VISCACEAE Miers

Hemiparasitic, dioecious, evergreen *shrubs* growing on tree branches. *Leaves* opposite, simple, without stipules. *Inflorescence* a compact cyme of 3–5 flowers. *Flowers* actinomorphic, epigynous. *Sepals* 4, minute. *Petals* 4, yellowish-green, nearly free. *Stamens* 4, sessile, epipetalous. *Style* absent; stigma sessile. Rudimentary ovary is usually present in male flowers and staminodes in female. *Ovary* inferior; ovules 2, usually not differentiated from the massive placenta. *Fruit* a 1-seeded berry.

Contains about 36 genera and 1,300 species in tropical and temperate regions.

1. Viscum L.

Hemiparasitic, usually dioecious, evergreen *shrubs* growing on tree branches. *Stems* apparently dichotomously branched. *Leaves* usually opposite, simple, without stipules. *Inflorescence* a compact cyme of 3–5 flowers, bracteate. *Sepals* 4, minute. *Petals* 4, yellowish-green, male larger than female. *Anthers* 4, sessile and occupying most of the inner surface of the petals. *Stigma* sessile and capitate. *Fruit* a 1-seeded berry.

Contains about 60 species in the Old World.

Bean, W. J. (1980). *Trees and shrubs hardy in the British Isles.* **4**. Ed. 8. London.
Briggs, J. (1999). *Kissing goodbye to Mistletoe.* Plantlife and the Botanical Society of the British Isles.
Hultén, E. & Fries, M. (1986). *Atlas of north European vascular plants north of the Tropic of Cancer.* 3 vols. Königstein.
Mabey, R. (1996). *Flora Britannica.* London.
Perring, F. H. (1973). Mistletoe. In P. S. Green (Edit.) *Plants wild and cultivated*, pp. 139–145. Hampton.

1. V. album L. Mistletoe

Somewhat woody, usually dioecious, evergreen, hemiparasitic *shrub* on the branches of trees. *Stems* up to 1 m, green, much-branched, apparently dichotomous, glabrous, forming a more or less spherical mass up to 2 m across. *Leaves* usually opposite; lamina 2–8 × 0.5–2.0 cm, dull yellowish-green, oblong-oblanceolate to oblong-obovate, rounded at apex, entire, narrowed at base, glabrous, thick and leathery; veins parallel; without stipules. *Inflorescence* a small, compact cyme of 3–5 subsessile flowers; bracts united to the short pedicels. *Male flowers* fragrant, with sepals minute; petals 4, about 3 × 2.5 mm, yellowish-green, leathery, ovate; anthers 4, sessile, opening by pores and occupying most of the inner surface of the petal; rudimentary ovary sometimes present. *Female flowers* less fragrant, with sepals of 4 teeth; petals 4, about 1 × 1 mm, yellowish-green, ovate, obtuse at apex, inclining towards the centre; stigma about 0.5 mm, capitate; nectaries forming a ring between the petals and the stigma; staminodes sometimes present. *Berry* 6–10 mm in diameter, globose, pale green turning pearly-white; seed solitary. *Flowers* 2–4. Pollinated by bees and flies. *Fruits* 11–12. $2n = 20$.

Native. Semi-parasitic on the branches of many trees and shrubs including: *Acer campestre, Acer pseudoplatanus, Aesculus, Amelanchier, Betula, Chaenomeles, Corylus, Cotoneaster, Crataegus, Cupressus, Fraxinus, Juglans, Laburnum, Malus, Populus, Prunus, Pyrus, Quercus, Robinia, Sorbus aucuparia, Tilia* and *Ulmus*. England and Wales north to Yorkshire, and a few records further north and in north-east Ireland. Mostly local, but in a wide circle of land round the Severn estuary, where the moist valleys are sheltered from the wind and there is a long tradition of fruit growing, it is abundant on apple trees. Most of Europe except the extreme north and east; Caucasus. European Temperate element. Our plant is subsp. **album** which occurs throughout the range of the species. In early times the milk-white berries between splayed leaves probably signalled an aphrodisiac and fertility potion, and women who wished to conceive would tie a sprig round their waist or arm, and in modern times a folk ritual has developed of kissing under the mistletoe at Christmas. It was Pliny's account of mistletoe and the Gallic Druids which became commercialised in the eighteenth and early nineteenth centuries, particularly by one William Stukeley (1687–1765), which brought about stories of the oak, mistletoe and the Druids.

Order **8. CELASTRALES** Wettst.

Trees or *shrubs*. *Leaves* simple, often entire; stipules small or absent. *Flowers* usually bisexual, rarely unisexual, hypogynous to perigynous or slightly epigynous, actinomorphic, usually small. *Sepals* 4–5. *Petals* 4–5. *Ovary* 3- to 5-celled. *Fruit* various; seeds usually with abundant endosperm.

Consists of 11 families and about 2,000 species.

99. CELASTRACEAE R. Br. nom. conserv.

Evergreen to deciduous, usually monoecious *shrubs* or woody *climbers*. *Leaves* opposite or alternate, simple, petiolate, exstipulate or with small stipules. *Inflorescence* a small axillary cyme; flowers unisexual to bisexual, hypogynous to slightly epigynous, actinomorphic. *Sepals* 4–5, free or fused at base. *Petals* 4–5, free. *Stamens* 4–5, alternating with petals outside a ring-like nectar-secreting disc. *Style* 1; stigma capitate or 3-lobed. *Ovary* 3- to 5-celled, each cell with (1–)2 ovules on axile placenta. *Fruit* an often more or less succulent, 3- to 5-angled dehiscent capsule with 1(–2) seeds in each cell; seeds covered in a bright orange to red aril.

Contains about 60–70 genera and 1,300 species, worldwide except for the Arctic.

1. More or less erect shrubs or small trees; leaves opposite; capsule 4- to 5-celled **1. Euonymus**
1. Woody climber; leaves alternate; capsule 3-celled **2. Celastrus**

1. Euonymus L.

Evergreen or deciduous, usually monoecious, more or less erect *shrubs* or small *trees*. *Leaves* opposite, simple, stipulate. *Inflorescence* an axillary cyme; flowers bisexual. *Sepals* 4–5. *Petals* 4–5, free, greenish yellow or white. *Stamens* 4–5, round the edge of nectar-secreting disc. *Style* 1; stigma capitate. *Ovary* 4- to 5-celled, each cell with 2 ovules on axile placenta. *Fruit* a capsule with 4–5 distinct rounded to winged lobes; seeds covered with a bright orange to red aril.

Contains about 160 species in Europe, North Africa, Madagascar, Asia, North and Central America and one in Australia.

Euonymus species are some of the hosts of *Aphis fabae* Scop., which attacks sugar beet (Hull, R. (1949). 'Sugar Beet Diseases'; their recognition and control. *Min. Agric. Fish. Pamph.* no. 142, p. 11). Many shrubs were rooted out because of this during the Second World War. The shrubs are poisonous in all their parts, especially the fruits, but birds seem immune and eat the fruits and distribute the seeds. *Euonymus* is feminine in Greek, but the masculine endings of the specific epithets are retained because of botanical tradition.

Bean, W. J. (1973). *Trees and shrubs hardy in the British Isles*. Ed. 8. **2.** London.
Blakelock, R. A. (1951). A synopsis of the genus *Euonymus* L. *Kew Bull.* **6**: 210–290.
Hültén, E. & Fries, M. (1986). *Atlas of north European vascular plants north of the Tropic of Cancer*. 3 vols. Königstein.
Mabey, R. (1996). *Flora Britannica*. London.
Rehder, A. (1940). *Manual of cultivated trees and shrubs hardy in North America*. Ed. 2. New York.
Roper, I. M. (1912). *Euonymus europaeus* with white fruit. *Jour. Bot. (London)* **50**: 377.

1. Petals 5; capsules usually 5-lobed **5. latifolius**
1. Petals usually 4; capsules usually 3- or 4-lobed 2.
2. Plant deciduous 3.
2. Plant evergreen or semi-deciduous 6.
3. Veins 8–9 pairs; anthers brownish-purple; seeds blood-red **2. sieboldianus**
3. Veins 5–8 pairs; anthers yellow; seeds white 4.
4. Capsule pure white **1(2). europaeus** forma **leucocarpus**
4. Capsule pink or purplish-red 5.
5. Leaves 4–6 × 1.5–2.0 cm, lanceolate, elliptical or oblong **1(1). europaeus** forma **europaeus**
5. Leaves 6–13 × 2–4 cm, ovate **1(3). europaeus** forma **intermedius**
6. Leaves dark green in centre with broad cream margins **3(2). fortunei** forma **gracilis**
6. Leaves dark shiny green throughout 7.
7. Shrub with spreading habit; leaves 1.0–2.5 cm wide, sharply crenate-serrate **3(1). fortunei** forma **fortunei**
7. Shrub or small tree, but often planted as a hedge; leaves 2–4 wide, bluntly serrate-crenate 8.
8. Leaves 3.0–4.5 × 2.0–2.5 cm **4(1). japonicus** forma **japonicus**
8. Leaves 4.0–5.5 × 3.0–4.0 cm **4(2). japonicus** forma **macrophyllus**

Subgenus **1. Euonymus**

Winter buds small, ovoid, acute at apex. *Capsule* not winged.

Section **1. Euonymus**
Section *Biloculares* Rouy & Foucaud

Leaves deciduous. *Capsule* with angled lobes.

1. E. europaeus L. Spindle
E. vulgaris Mill.

Deciduous, monoecious *shrub* or small tree up to 8 m, with an ovoid crown when in the open, but usually in a hedge where it is often trimmed. *Trunk* up to 25 cm in diameter when grown into a tree, rather short and not very far into the crown. *Bark* pale brown or grey and rough. *Wood* whitish, firm and smooth. *Branches* ascending and slender; twigs brown or green, stiff, sometimes becoming square with ridged angles; young shoots green and minutely hairy. *Buds* opposite, 2–4 × 1.5–2.5 mm, flattened against the shoot, ovoid and more or less 4-angled, acute at apex; scales greenish, ovate, acute at apex, glabrous. *Leaves* opposite; lamina 3–13 × 1.5–3.5 cm, a characteristic medium green on upper surface, paler beneath, turning reddish or reddish-purple in autumn, lanceolate, elliptical, ovate or oblong, obtuse or acute at apex, minutely crenate-apiculate, cuneate at base, glabrous or with some minute hairs beneath; veins 6–8 pairs, more or less impressed on upper surface, prominent beneath; petioles 6–12 mm, pale green, glabrous. *Inflorescence* an axillary, dichotomous, pedunculate cyme; peduncles and pedicels green, with some minute hairs when young. *Flowers* 3–10 together, usually bisexual but unisexual ones occur with vestiges of the opposite sex organs, 8–10 mm in diameter, scentless. *Sepals* 4, 1.2–1.5 mm, green, broadly ovate, rounded at apex, glabrous. *Petals* 4, 4.0–4.5 mm, pale green, spathulate, rounded at apex. *Disc* wide, green and fleshy. *Stamens* 4, round the edge of the nectar-secreting disc; filaments white; anthers yellow. *Style* 1, short, pale green; stigma paler. *Capsule* 10–18 mm in diameter, pendulous, 4-lobed, green turning deep pink or purplish-red; seeds about 6 mm, white, covered by a bright orange aril. *Flowers* 4–6. Pollinated by small insects. Self-pollination can occur if there are no insect visitors. $2n = 32$.

(1) Forma **europaeus**

Leaves 4–6 × 1.5–2.0 cm, lanceolate, elliptical or oblong. *Capsule* 10–15 mm in diameter, deep dull pink or purplish-red.

(2) Forma **leucocarpus** (DC.) Hegi
E. europaeus var. *leucocarpus* DC.; *E. europaeus* subvar. *leucocarpus* (DC.) Rouy

Leaves with lamina 4–6 × 1.5–2.0 cm, lanceolate or elliptical. *Capsule* 8–15 mm in diameter, pure white.

(3) Forma **intermedius** (Gaudin) Borza
E. europaeus var. *intermedius* Gaudin; *E. europaeus* var. *ovatus* Dippel; *E. vulgaris* var. *macrophyllus* Rouy; E. *vulgaris* var. *intermedius* (Gaudin) C. K. Schneid.

Leaves with lamina 6–13 × 2–4 cm, ovate, often very deeply coloured in autumn. *Capsule* 15–18 mm in diameter, bright pinkish-red.

Native. Hedges, scrub and open woods on calcareous or base-rich soils up to 365 m. Frequent in Great Britain and Ireland, especially southern England, north to central Scotland. Most of Europe except the extreme north and much of the Mediterranean region; western Asia. European Temperate element. The usual native plant is forma *europaeus*. Forma *leucocarpus* is recorded for Somersetshire and Surrey. Forma *intermedius* occurs naturally in Switzerland and Italy with forma *europaea* and is grown in gardens for its large-sized leaves with fine autumn colouring, and is spread into hedgerows by birds cropping the seeds. It also widely occurs in planted woodlands and infills in hedgerows. Once familiar with *Euonymus europaeus* it can be recognised at a glance by its distinctive green colour. In the sixteenth century William Turner did not know an English name for the plant and said the Dutch called it Spilboome because they made spindles from it for hand-spinning raw wool. It does not, however, seem to have been a well-favoured wood for that purpose in England, though the name has stuck. Wandering gypsies used to use it for making the pegs they sold, hence one of its popular names, Pegwood, and they knew where the shrubs were in the villages they passed through regularly. It was also used for making skewers, toothpicks and knitting needles. The fruits are strongly purgative, and were also baked and powdered and rubbed into hair as a remedy for head-lice.

2. E. sieboldianus Blume Siebold's Spindle
E. hamiltonianus subsp. *sieboldianus* (Blume) Kom.;
E. semiexertus Koehne

Deciduous, monoecious *shrub* or small *tree* up to 12 m, with a spreading crown. *Trunk* up to about 57 cm in circumference. *Bark* greyish-brown, rough and ridged. *Branches* spreading, arching and hanging down; twigs pendulous, grey and smooth; young shoots green, glabrous. *Buds* 2–3 mm, narrowly ovoid, pointed at apex; scales green with brown tips, ovate, glabrous. *Leaves* opposite; lamina 4.0–13.5 × 1.5–7.0 cm, dull dark green on upper surface, paler beneath, turning pink or red in autumn, lanceolate or ovate-lanceolate, long-acute or acuminate at apex, minutely serrate-dentate, cuneate or attenuate at base, glabrous; veins 8–9 pairs, impressed on upper surface with the midrib paler, prominent below especially the midrib; petiole up to 22 mm, green, glabrous. *Inflorescence* axillary and near the central branch, with up to 10 bisexual flowers which are 9–11 mm in diameter and scentless; peduncles and pedicels green and glabrous. *Sepals* 4, 2.0–2.5 mm, green, broadly ovate, acute at apex, rolled outwards, glabrous. *Petals* 4, 5.0–5.5 mm, pale yellowish-green, oblanceolate or obovate, rounded at apex. *Stamens* 4, round edge of disc; filaments short, green; anthers brownish-purple. *Style* 1, green; stigma slightly darker. *Capsule* about 10 mm in diameter, 4-lobed, the lobes angled, rose-pink; seeds blood-red, partly covered by orange, orange-red or blood-red aril. *Flowers* 5–6.

Introduced. Grown in gardens, parks and estates and persistent on a Surrey roadside. Native of Russia, Sakhalin, Japan and Korea. Named after Philipp Franz van Siebold (1796–1866).

Section **2. Ilicifolia** Nakai

Evergreen *shrub*. *Capsule* with rounded lobes.

3. E. fortunei (Turcz.) Hand.-Mazz. Fortune's Spindle
Elaeodendron fortunei Turcz.

Low, monoecious evergreen *shrub* with spreading habit. *Stems* numerous, up to 2 m, when young trailing or creeping, brown, rough, erect and spreading when mature. *Bark* brown, rough. *Branches* spreading; twigs brown; young shoots green or pale yellow, much wrinkled, glabrous. *Buds* 6–7 × 2–3 mm, narrowly ovoid, pointed at apex; scales green and brown, ovate, pointed at apex, glabrous. *Leaves* opposite; lamina 3.5–6.0 × 1.0–2.5 cm, dark rather shiny green on upper surface, pale green beneath, or often dark green in centre with irregular cream margins on upper surface and similar but much paler beneath, narrowly elliptical, elliptical, lanceolate or oblanceolate, narrowed to an obtuse or acute apex, shallowly and sharply crenate-serrate, cuneate or attenuate at base, glabrous and rather coriaceous; midrib slightly paler on upper surface, slightly prominent beneath, the lateral veins 5–6 pairs, and obscure; petioles up to 20 mm, pale green or pale yellow, glabrous. *Inflorescence* of axillary cymes with up to 40 bisexual flowers which are 6–7 mm in diameter; peduncle up to 40 mm, pale green, slender, glabrous; pedicels very short, pale green, glabrous. *Petals* 4, 2.5–3.0 mm, ovate, rounded at apex, glabrous. *Petals* 4, 2.5–3.0 mm, pale yellowish-green, ovate or rhombic, rounded at apex, glabrous. *Stamens* 4, round edge of disc; filaments pale green; anthers greenish-cream. *Style* green; stigma green, capitate. *Capsule* 7–8 × 8–15 mm, 4-lobed, the lobes rounded, pale green turning deep pink; seeds 6–7 mm, covered by a bright orange aril. *Flowers* 5–7. Pollinated by small insects. $2n = 32$.

(1) Forma **fortunei**
Leaves dark shiny green.

(2) Forma **gracilis** (Regel) Rehder Cv. Variegatus
E. japonicus var. *gracilis* Regel

Leaves dark green in centre with broad cream margins.

Introduced. Much grown in gardens and planted on roadsides and railway banks, parks, estates and amenity areas. Forma *gracilis* is probably the more common form. Native of Japan. Named after Robert Fortune (1812–1880). Forma *fortunei* is difficult to distinguish from *E. japonicus* and *E. fortunei* should possibly not be treated as a different species.

4. E. japonicus L. fil. Evergreen Spindle

Evergreen, monoecious *shrub* or small *tree* up to 5(–8) m, the crown domed and often wider than it is high. *Stems* 1–several, more or less erect. *Bark* pale greyish-brown and smooth. *Branches* erect or spreading; twigs thick, weakly angled, greenish with pale brown lines; young shoots pale green, thick, glabrous. *Buds* in pairs, opposite, 10–20 × 3–5 mm, narrowly ovoid, acute at apex; scales pale yellowish-green, with a narrow, scarious margin which is sometimes touched with brown, very broadly ovate, rounded to a short point at apex. *Leaves* retained for 1–2 years, opposite; lamina 2–5 × 2–4 cm, dark shining green on upper surface with a pale midrib, often variegated white or yellow, pale dull green beneath, oblanceolate, obovate or elliptical, rounded-obtuse or with an abrupt, short point at apex, finely and bluntly serrate-crenate, rounded, cuneate or attenuate at base, glabrous; midrib prominent near the base on both surfaces, otherwise veins obscure; petioles 10–15 mm, pale yellowish-green, glabrous. *Inflorescence* an axillary cyme on current season's growth; peduncles and pedicels pale yellowish-green, glabrous; flowers bisexual in clusters of 5–12, 6–11 mm in diameter. *Sepals* 4, 2 smaller, 0.9–2.0 mm, wider than long, pale green rounded at apex. *Petals* 4, 1.4–2.5 mm, greenish-white, broadly ovate, rounded at apex. *Stamens* 4; filaments pale green; anthers cream. *Styles* 1, yellow, cone-like; stigma green. *Capsule* 7–8 × 8–10 mm, ovoid with rounded angles; green in November; seeds completely covered by the orange aril. *Flowers* 5–6. $2n = 32$.

(1) Forma **japonicus**
Leaves 3.0–4.5 × 2.0–2.5 cm.

(2) Forma **macrophyllus** (Regel) Beissn. Cv. Macrophylla
E. japonicus var. *macrophyllus* Regel

Leaves 4.0–5.5 × 3.0–3.5 cm.

Introduced in 1804. Planted in gardens, parks, cemeteries, churchyards and for hedging, especially near the sea. South and west Great Britain, Channel Islands and Isle of Man and scattered in south-west Ireland. Self-sown mainly in the south and Isle of Man. Native of Japan. Both forms occur and forma *macrophylla* is particularly used for hedging.

Subgenus **2. Kalonymus** Beck
Kalonymus (Beck) Prokh.

Winter buds large, conical and very acute. *Capsule* winged.

5. E. latifolius (L.) Mill. Large-leaved Spindle
E. europaeus var. *latifolia* L.; *Kalonymus latifolia* (L.) Prokh.; *Pragmotassera latifolius* (L.) Pierre

Deciduous, monoecious *shrub* or small *tree* up to 6 m, with a broad, open crown. *Trunk* nearly to top of tree. *Bark* grey and smooth becoming rough. *Branches* numerous, ascending or spreading and arching; twigs dark brown above and green beneath, smooth and glabrous; young shoots green to dull brownish-red, glabrous. *Buds* 7–22 × 4.0–5.0 mm, narrowly ovoid or fusiform, acute at apex; scales green to brownish-red, narrowly ovate, acute at apex, glabrous. *Leaves* opposite; lamina (4–)8–16 × 2.0–4.8 cm, soft, medium yellowish-green on upper surface, slightly paler beneath, turning pinkish-purple in autumn, elliptical, ovate or obovate, shortly but sharply acute at apex, minutely serrulate or dentate, rounded or subtruncate at base, glabrous; veins 5–8 pairs, slightly impressed above, prominent beneath particularly the midrib; petiole up to 15 mm, pale green to reddish-purple, glabrous. *Inflorescence* a cyme, 4- to 12-flowered, axillary; peduncles up to 7 cm, green to dark red, very slender, glabrous; pedicels 3–10 mm, green to dark red, slender, glabrous; flowers 7–8 mm in diameter, scentless. *Sepals* about 1.2×1.5 mm, green, very broadly ovate, rounded at apex, glabrous. *Petals* 5, about 1.5–2.0 mm, brown with a cream margin, broadly ovate, rounded at apex. Disc green. *Stamens* 4–5, round the edge of nectar-secreting disc; filaments green; anthers cream. *Style* 1, green, short; stigma pale brownish, shorter than anthers. *Capsule* 15–25 mm wide, bright red-purple, 5-lobed, 5-locular, narrowly winged on the angles; seeds 8–10×4–6 mm, wrinkled, covered by a bright orange-red, fleshy aril. *Flowers* 4–5. Pollinated by small insects. $2n = 64$.

Introduced. Planted in gardens and hedges and sometimes naturalised from bird-sown seeds. Scattered records in England and Scotland. Native of south-central and south-east Europe extending to central Italy and southern France; south-west Asia.

2. Celastrus L.

Deciduous, dioecious or monoecious, woody *climber*. Outer scales of buds on vigorous young shoots transformed into spines. *Leaves* alternate, single. *Inflorescence* an axillary cyme; flowers yellowish-green, often unisexual. *Sepals* 4–5. *Petals* 4–5. *Stamens* 4–5. *Style* 1; stigma 3-lobed. *Ovary* 3-celled. *Fruit* a 3-celled capsule; seeds immersed in a bright red aril.

Contains about 30 species in tropical and warm temperate areas mostly in the northern hemisphere.

Airy Shaw, H. K. (1935). *Celastrus orbiculatus. Bot. Mag.* no. 9394. London.
Bean, W. J. (1970). *Trees and shrubs hardy in the British Isles.* Ed. 8. **1**. London.
Hou, D. (1955). A revision of the genus *Celastrus*. *Ann. Missouri Bot. Gard.* **42**: 215–302.

1. C. orbiculatus Thunb. ex Murray Staff Vine
C. articulatus Thunb.

Deciduous, dioecious or monoecious *climber* or scrambler up to 12 m. *Stems* pale to dark brown, terete or striate, smooth to lenticellate. *Branches* short; young shoots glabrous. *Buds* small, ovoid or subglobose, rounded at apex; scales brownish, those of the vigorous young shoots often with outermost ones transformed into sharp, recurved spines 1–2 mm. *Leaves* alternate, extremely variable in size and shape; lamina 4–8×3–6 cm, pale green on upper surface, even paler beneath, obovate, elliptical, oblong or subrotund, rounded to acuminate at apex, usually closely and conspicuously crenate-dentate, cuneate at base, glabrous on upper surface, sometimes with hairs on the veins beneath; petioles up to 3 cm, pale green; stipules 1–2 mm, divided to the base into 2–3 filiform laciniae. *Inflorescence* an irregular, axillary cyme with 1–8, unisexual flowers; arranged often rather densely along the shoots of the current year; pedicels slender; glabrous. *Male flowers* with calyx 2–3 mm, obconical or turbinate, glabrous, the lobes short, erect, ovate-deltoid, more or less obtuse; petals 3–4 mm, pale yellowish-green, narrowly oblong to oblanceolate, rounded to subacute at apex, glabrous; stamens 4–5; filaments 2–3 mm; anthers about 1 mm, ovate-orbicular, yellow; female parts rudimentary. *Female flowers* with calyx about 2 mm, cupular-turbinate, the lobes deltoid and obtuse; petals as in male flower or reduced or absent; non-functional stamens inserted on the margin of a cupular disc. *Style* 1, stigmas 3-lobed. *Capsule* 6–8 mm, bright yellow, subglobose; seeds 6, 3–4 mm, oblong-ovoid, entirely enclosed in the scarlet fleshy aril. *Flowers* 4–6. $2n=46$.

Introduced. Naturalised at Shottenmill in Surrey since 1985. Also recorded for Somersetshire and Hampshire. Native of Sakhalin, Japan, Korea, North Kiang-si, Anhwei and North Kiang-su.

100. AQUIFOLIACEAE Bartlett nom. conserv.

Evergreen, dioecious *trees* or *shrubs*. *Leaves* alternate, simple, usually at least some with spines on the margin, petiolate; stipules minute. *Flowers* in small, axillary cymes, hypogynous, actinomorphic, unisexual but abortive parts of other sex present. *Calyx* 4-lobed. *Petals* 4, fused at base or more or less free. *Male flower* with 4 stamens alternating with the petals and a vestigial ovary. *Female flowers* with 4-celled ovary, a sessile or very shortly stalked, 4-lobed stigma and usually 4 abortive stamens. *Fruit* a (2–)4-seeded drupe.

Contains 2 genera and over 400 species distributed in the tropical, subtropical and temperate regions of the world.

1. Ilex L.

As family.

Contains over 400 species distributed as family.

It is impossible to write a key to the hollies of Great Britain and Ireland unless all the cultivars are recognised. *I. aquifolium* in a wild habitat with lower leaves very spiny and upper leaves less so, with bright red berries is reasonably likely to be correct, but even there the chance of one of the cultivated forms being introduced is great. *I. pernyi* when typical is distinct, but it can apparently hybridise with *I. aquifolium*. In gardens, parks and estates the variation in *I.×altaclerensis* is immense, running into *I. aquifolium* variants on one side and the *I. perado* aggregate on the other. The enormous range of cultivars in *I. aquifolium* does not help, especially as many are male and have no fruit. Plants which have been called *I. balearica* Desf. are probably best in *I. aquifolium* but have the appearance of the *I. perado* group and its hybrid with *I. aquifolium*. We do not know how much cross-pollination takes place, nor how successful it is.

Bean, W. J. (1973). *Trees and shrubs hardy in the British Isles.* Ed. 8. **2**. London.
Dallimore, W. (1908). *Holly, Yew and Box.* London.
Hume, H. H. (1953). *Hollies*. New York.
Loesner, T. (1901 and 1908). *Nova Acta Acad. Leop.-Carol. German. Nat. Cur.* **78**: 1–598 (1901); **89**: 1–313 (1908).
Mabey, R. (1996). *Flora Britannica*. London.
Mitchell, A. (1996). *Trees of Britain*. London.
More, D. & White, J. (2003). *Trees of Britain and northern Europe*. London. (Many illustrations of cultivars.)
Peterken, G. F. & Lloyd, P. S. (1967). Biological flora of the British Isles. No. 108. *Ilex aquifolium* L. *Jour. Ecol.* **55**: 841–858.

Peterken, G. F. (1975). Holly survey. *Watsonia* **10**: 297–299.

Rackham, O. (2003). *Ancient woodland: its history, vegetation and uses in England*. Ed. 2. London.

Railey, J. (1961). Holly as a winter feed. *Agric. Hist. Rev.* **9**: 89–92.

Rehder, A. (1940). *Manual of cultivated trees and shrubs hardy in North America*. Ed. 2. New York.

Roberts, A. N. & Boller, C. A. (1948). Pollination requirements of English Holly, *Ilex aquifolium*. *Proc. Amer. Soc. Hort. Sci.* **52**: 501–509.

1. I. aquifolium L. Holly

I. vulgaris Gray nom. illegit.

Evergreen, dioecious *shrub* or small *tree* up to 24 m, with a conical crown at first, tending to become hemispherical at maturity at the top. *Trunk* up to 1 m in diameter, extending through much of the crown. *Bark* pale grey, smooth for many years, but eventually finely fissured. *Wood* ivory white, heavy and hard. *Branches* spreading at a wide angle and tending to become pendulous in shade; twigs dark green or purplish, stout, glabrous; young shoots yellowish-green or purple, with dense, very short, pale, rigid hairs. *Buds* 1–3 × 0.7–2.0 mm, terminal slightly larger than axillary, ovoid, pointed; scales yellowish-green to purplish-brown, lanceolate, acute at apex, with dense, very short, pale, rigid simple eglandular hairs. *Leaves* alternate; lamina 3–12 × 1–7 cm, dark glossy green on upper surface, paler and duller beneath, sometimes variegated yellow, thick and coriaceous, turning yellow after 2–5(–8) years and mainly falling in midsummer, ovate, elliptical or oblong, spinose-acuminate or spinose-cuspidate at apex, mostly with an undulate, strongly spinose margin, but sometimes (commonly on upper branches of old trees, and rarely over most of the tree) flat and entire, rarely spinose on upper surface, with a narrow, cartilaginous border, cuneate at base, glabrous; venation pinnate-reticulate and looped; petioles up to 15 mm, green, with a wide, shallow channel; stipules minute, black and triangular. *Inflorescence* an axillary cyme, crowded on old wood; flowers 6–8 mm in diameter. *Calyx* yellowish-green, lobes 4, 1.0–1.2 mm, broadly ovate, more or less obtuse at apex, with dense, very short, pale, rigid simple eglandular hairs on surface and margin. *Petals* 4, 2.0–2.5 × 2.0–2.5 mm, white, tinted pink on back, broadly obovate or subrotund, rounded at apex. *Stamens* 4, inserted with the petals; filaments greenish or brownish; anthers yellow; female flowers bear full-sized filaments but small anthers. *Style* very short or absent; stigma 4-lobed. *Ovary* 4-celled, small and vestigial in male flowers; nectar is secreted in tissue at its base. *Drupe* 6–10(–12) × 6–10 mm, usually bright red, rarely yellow or whitish, globose or broadly ellipsoid; seeds 5–6 × 2.5–3.5 mm, pale brown, hard, plano-convex, with strong, irregular, longitudinal ridges and occasional intersecting ones. *Flowers* 9–3. Visited by honey bees. $2n = 40$.

Upwards of 140 variants have been named of which many are of horticultural origin, but are occasionally found in wild populations. They are based on habit, colour of bark, colour of berries, variegation of foliage, size, shape and curvature of leaf and the number of spines and their distribution on the leaf surface. A detailed account of the total variation of the plant is given by Dallimore (1908).

(**1**) Forma **aquifolium**

Drupes bright red.

(**2**) Forma **chrysocarpa** Loes.

Drupes yellow or whitish.

Native. An understorey component of many types of woodland and wood pasture including *Quercus robur, Quercus robur–Carpinus betulus, Quercus petraea, Quercus petraea–Carpinus betulus, Fagus sylvatica, Taxus baccata, Pinus sylvestris, Alnus glutinosa, Fraxinus excelsior and Fraxinus excelsior–Betula*. Pure woods of it are found in the New Forest, Shropshire, the Black Mountains and Scotland. It also occurs in scrub communities as in western Ireland where it grows with *Corylus avellana* and *Arbutus unedo*. At Dungeness in Kent it forms climax scrub on shingle banks and also occurs on cliffs. As a hedgerow tree it has often been planted but may be of ancient origin on boundary hedges. *I. aquifolium* occurs throughout most of Great Britain and Ireland up to 520 m in Argyllshire and 550 m in Co. Kerry on a wide variety of soils from acid heath to chalk and limestone. It occurs throughout north-western, central and southern Europe, North Africa and south-west Asia. The common form is forma *aquifolium*. Forma *chrysocarpa* is only occasional. The timber of Holly was used for making machine cogs and whips, and for carving and turning. Holly coppices and pollards freely and is very difficult to kill. Its leaves and twigs are palatable to many animals, the axillary buds of browsed shoots developing to form a dense mass of foliage. It is one of the best of all firewoods, burning freely even when green. The drupes are purgative and emetic. Together with Ivy it is the best known of all Christmas decorations, a habit associated with religious festivals throughout Europe since the pre-Christian era. Many place-names scattered over England refer to Holly, but there is a confusion with other similar words.

2. I. × altaclerensis (Loudon) Dallim. Highclere Holly

I. aquifolium × perado Aiton

Evergreen, usually dioecious *tree* up to 12 m with rather narrow, irregular crown. *Trunks* sometimes several, throughout the height of the tree. *Bark* pale grey, smooth. *Branches* spreading or erect; twigs thick, green, with dark lenticels, glabrous; young shoots fresh green or purple, stout and flattened, glabrous. *Buds* minute, green, at the end of a shoot. *Leaves* alternate, variable; lamina 6–12 × 2–7 cm, dark glossy green with a pale green midrib, sometimes boldly margined rich yellow on upper surface, pale green beneath, mostly elliptical or ovate, sometimes oblong-elliptical, flat, rounded with an apiculate, sometimes spiny apex, entire to variably spiny, sometimes with just 1 spine, less often spiny along both margins, cuneate to rounded at base, glabrous; midrib prominent beneath, the lateral veins also sometimes showing; petioles up to 15 mm, pale green. *Inflorescence* on axillary cyme; flowers few. *Calyx* yellowish-green; lobes 4,

1.0–1.5 mm, green, triangular-lanceolate, rounded at apex. *Petals* 4, 2–3 mm, greenish-white, ovate, rounded at apex. *Stamens* 4, in female flowers usually abortive; filaments short, white; anthers green, suffused pink. *Stigma* 4-lobed, sessile. *Ovary* 4-celled, with 1(–2) ovules per cell, and nectar is secreted in tissue at its base. *Drupe* 8–10×7–8 mm, slightly longer than wide, orange-red, broadly ellipsoid; seeds about 8×5 mm.

Introduced. Widely planted in gardens, parkland and estates. Bird-sown in hedges and woodland or as relics of cultivation. Scattered records over Great Britain, Isle of Man and Ireland, and is probably under-recorded. Of garden origin, there being two forms, one with green shoots and one with purple shoots (Mitchell, 1996). Named *altaclerensis* because it was raised at Highclere in Hampshire.

3. I. perado Aiton — Madeira Holly

Evergreen, dioecious *tree* up to 15 m, with a columnar crown. *Trunk* extending through much of the crown. *Bark* greyish-brown and smooth. *Branches* spreading at a wide angle or ascending; twigs rough with lenticels which erupt into small pimples; young shoots green and purple, terete, glabrous, with small, buff, linear lenticels. *Buds* that are terminal are up to 6 mm, conical, slender and pointed at apex; while those lateral are up to 1 mm, ovoid and obtuse at apex; scales green or purple and finely warty. *Leaves* alternate; lamina 4–13(–16)×3.5–8.6(–11.5) cm, dark glossy green on upper surface, pale green beneath, ovate, obovate, spathulate, oblong or subrotund, short, slender and pointed at apex, often entire and spineless or variably spiny and often with only spines near the apex, cuneate to subcordate at base; veins showing clearly on upper surface, midrib prominent beneath; petioles up to 15 mm, green, channelled, often winged; stipules minute. *Inflorescence* an axillary cyme. *Calyx* yellowish-green; lobes 4, broadly ovate, more or less obtuse. *Petals* 4, 2.0–2.5 mm, white, tinted pink, obovate, rounded at apex. *Stamens* 4, inserted with the petals; filaments greenish; anthers yellow. *Styles* very short or absent; stigma 4-lobed. *Drupe* 8–10 mm, subglobose or obovoid. *Flowers* 4–6.

(**a**) Subsp. **azorica** (Loes.) Tutin — Azores Holly
I. perado var. *azorica* Loes.; *I. azorica* Gand.

Leaves with lamina 4.0–6.6×3.5–5.0 cm, subrotund or ovate, often spineless.

(**b**) Subsp. **perado**
I. *maderensis* Lam.; *I. perado* var. *maderensis* (Lam.) Loes.

Leaves with lamina 6–8×4.0–5.6 cm, ovate or obovate, spathulate in cultivation, often spineless.

(**c**) Subsp. **platyphylla** (Webb & Berthel.) Tutin
I. platyphylla Webb & Berthel.; *I. perado* var. *platyphylla* (Webb & Berthel.) Loes.

Leaves with lamina 8.5–13.0 (–16)×5.0–8.6 (–11.5) cm, broadly ovate or oblong, somewhat spiny.

Introduced. This species occurs in cultivation in Ireland and may survive in some estates in Great Britain. It is one of the parents of *I.×altaclerensis* the boundary with which is difficult to define. Subsp. *azorica* is native of the Azores, subsp. *perado* native of Madeira and subsp. *platyphylla* native of Tenerife and Gomera in the Canary Islands.

4. I. pernyi Franch. — Perny Holly

Evergreen, dioecious *shrub* or small *tree* up to 15 m with a columnar or conical crown, becoming rounded in old trees. *Trunk* up to 30 cm in diameter, extending through much of the crown. *Bark* greyish-brown and smooth. *Branches* spreading or ascending; twigs brown, with fissures; young shoots green and densely short-hairy. *Buds* up to 2 mm, conical, with a slender, pointed apex; scales green, hairy. *Leaves* alternate, spreading pectinately either side of the shoot; lamina 1.2–3.4×1.3–2.4 cm, medium to dark glossy green on upper surface, pale green beneath, broadly triangular-ovate, with a sharp, long, yellow, brown-tipped spine at each of the 5 corners, occasionally with another pair in the centre, or only 3, hyaline at the margin, cupped and truncate or subcordate at base; midrib impressed above, prominent beneath; petiole up to 3 mm, green and hairy. *Inflorescence* an axillary cyme. *Calyx* yellowish-green; lobes 4, ovate, more or less obtuse at apex. *Petals* 4, yellowish. *Stamens* 4, inserted with the petals; anthers yellow. *Style* very short or absent; stigma 4-lobed. *Drupe* 4–7 mm in diameter, red, subglobose; seeds 4. *Flowers* 5–6.

Introduced. Planted in parks and estates. Native of central and west China. A hybrid occurs with *I. aquifolium*, *I.×aquipernyi* Gable. Named after Paul Hubert Perny (1818–1907).

Order 9. EUPHORBIALES Lindl.

Trees, *shrubs* or *herbs*. *Leaves* usually alternate, simple or compound, sometimes reduced. *Flowers* unisexual, hypogynous, actinomorphic. *Sepals* 4, or absent. *Petals* absent. *Ovary* usually 3-celled; ovules 1–2 in each cell, placentation axile. *Fruit* a capsule or drupe.

Consists of 4 families and about 8,000 species.

101. BUXACEAE Dumort. nom. conserv.

Evergreen, monoecious *shrubs* or small *trees*. *Leaves* opposite or alternate, simple, entire or dentate, petiolate, without stipules. *Inflorescence* of small clusters of flowers, the flowers unisexual but both sexes occurring in the same cluster, the flowers opening in early spring, hypogynous and actinomorphic. *Male flowers* with 4 sepals, without petals, and 4 stamens. *Female flowers* with more than 4 sepal-like bracteoles without stalks, with 2–3 short styles with linear of bilobed stigmas. *Ovary* 2- to 3-celled with 2 apical ovules per cell. *Fruit* a 2- to 3-celled capsule with 2 seeds per cell or a more or less succulent drupe.

Contains 4 genera and about 100 species, scattered over tropical and temperate regions.

1. Erect shrub or tree with opposite, entire leaves; fruit 3-horned — **1. Buxus**
1. Stoloniferous dwarf shrub with alternate, dentate leaves; fruit 2-horned — **2. Pachysandra**

1. **Buxus** L.

Erect *shrub* or *tree*. *Leaves* opposite, entire, glabrous or sparsely hairy. *Inflorescence* an axillary cluster, usually with one female and 5–8 male flowers. *Male* flowers with 4 sepals, without petals and with 4 stamens. *Female flowers* with up to 4 sepal-like bracteoles, without petals, with 3 short styles. *Ovary* 3-celled. *Fruit* a capsule, with persistent styles as 3 horns.

Contains about 70 species in western Europe, Mediterranean region, temperate east Asia, Socatra, Madagascar, West Indies and North and Central America.

Bean, W. J. (1970). *Trees and shrubs hardy in the British Isles*. **1**. Ed. 8. London.
Mabey, R. (1996). *Flora Britannica*. London.
Rehder, A. (1940). *Manual of cultivated trees and shrubs hardy in North America*. Ed. 2. New York.
Wigginton, M. J. (Edit.) (1999). *British red data books*. Vol. 1. *Vascular plants*. Peterborough.

1. Leaves 1.2–2.5 cm, elliptical or oblong
1(1). sempervirens forma **sempervirens**
1. Leaves up to 4 cm, ovate, boat-like and emarginate
1(2). sempervirens forma **hansworthiensis**

1. **B. sempervirens** L. Box

Evergreen, monoecious *shrub* or small *tree*, rather sprawling as a shrub, often as a hedge, with narrowly columnar crown as a tree. *Trunk* up to 30 cm in diameter, reaching into centre of tree, several stems usually present when a shrub. *Wood* yellow, hard, even-textured and heavy. *Bark* pale or whitish-brown, becoming cracked into small squares. *Branches* ascending or spreading in trees, tending to flop in shrubs; twigs brownish, square and slightly winged, with 2 lines of pale hairs between the decurrent leaf bases, forming rather flat sprays; young shoots green for several years. *Buds* opposite, 1–2 mm, cylindrical, pointed at apex; scales 2, brown, hairy. *Leaves* opposite; lamina 1.2–2.5(–4.0) × 0.5–1.1 cm, glossy, medium to dark green on upper surface with a raised midrib, paler beneath, ovate, elliptical or oblong, tapering to a rounded or emarginate apex with a short mucro, entire, cuneate at base; petiole 1–2 mm, flat and finely hairy. *Inflorescence* in clusters of 5–8 male flowers and a single female flower. *Male flowers* with 4, greenish-white sepals; without petals; stamens 4, exserted, the anthers yellow. *Female flowers* with up to 4 sepal-like bracteoles, without petals, with 3 short styles and linear or bilobed stigmas. *Capsule* 7–11 mm, bluish-green, subglobose to oblong, with persistent styles as 3 horns; seeds 5–6 mm, shiny black, violently expelled as the capsule opens. *Flowers* 4–5. Pollinated by bees and flies. $2n = 28$.

(1) Forma **sempervirens**
Leaves with lamina 1.2–2.5 cm, elliptical or oblong.

(2) Forma **handsworthiensis** (A. Henry) Rehder
B. sempervirens var. *handsworthiensis* A. Henry

Leaves with lamina up to 4 cm, ovate, boat-like and emarginate.

Native. Woods and scrub on chalk and limestone. Extremely local in Kent, Surrey, Berkshire, Buckinghamshire and Gloucestershire. Common also in gardens of all kinds, parks, estates, churchyards, amenity areas and as hedges. South-west and west central Europe and in the mountains of North Africa. Submediterranean Subatlantic element. Forma *sempervirens* is the common plant. Forma *handsworthiensis* originated in the Handsworth nursery of Fischer, Son and Sibray before 1872. It is possibly fairly widespread in parks and estates. It should not be confused with *B. balearica* Lam., which also has large leaves and is occasionally planted, and which can be distinguished by its shoots being flattened at the nodes, large flowers and the female having 3 or 4 horn like styles which in fruit are as long as the capsule. When there were large trees the wood of Box was long used for high-class wood carving and in the eighteenth century was particularly in demand by artists to use as blocks for wood engravings. Many slight varieties are available in nurseries including ones with yellow-tipped leaves. Many place-names in south England are formed from the word Box.

2. **Pachysandra** Michx

Dwarf, evergreen, stoloniferous *shrub*. *Leaves* alternate, dentate distally, glabrous. *Inflorescence* terminal on previous year's growth, with male flowers terminal and the lower female. *Male flowers* with 4 sepals and 4 stamens. *Female flowers* with 4 or more sepals and 2 or 3 styles. *Ovary* 3-celled, with 1 or 2 ovules in each cell. *Fruit* a more or less succulent drupe with persistent styles as 2 horns.

Contains 4 species in east Asia and eastern North America.

Bean, W. J. (1976). *Trees and shrubs hardy in the British Isles*. **3**. Ed. 8. London.

1. **P. terminalis** Siebold & Zucc. Carpet Box

Evergreen, procumbent or ascending *shrub* with stolons. *Stem* 20–30 cm, creeping and ascending, pale green, at first shortly hairy, thick, elongate, sparsely branched, leafy. *Leaves* alternate and fasciculate, persisting for 2 to 3 years, each years crop being produced in a whorl-like cluster at the end of its growth, and separated from the previous one by a length of naked stem; lamina 3–7 × 1.5–3.0 cm, dull medium to dark green on upper surface, paler beneath, thick, oblanceolate, obovate or rhombic-ovate, obtuse at apex, crenate-dentate in the upper half, attenuate at base, glabrous; veins 3; petioles up to 2 cm, pale green, glabrous or minutely hairy on veins above. *Inflorescence* an erect, terminal spike 2–4 cm; flowers unisexual, female below the male, white. *Male flowers* with 4 sepals and 4 stamens, the filaments about 8 mm and thick. *Female flowers* shortly stalked, with 4 or more sepals and 2 or 3, spreading or suberect styles with minute tubercles or papillae. *Sepals* 2.5–3.5 mm, broadly ovate, ciliate. *Drupe* 10–15 mm, white, glabrous, with persistent styles as 2 or 3 horns; stones about 5 mm, ovoid. *Flowers* 4–5. $2n = 48$.

Introduced. Much grown in public places as a ground cover, sometimes running wild. A few sites in Kent, Berkshire, Buckinghamshire and Yorkshire. Native of Japan.

102. EUPHORBIACEAE Juss. nom. conserv.

Annual to *perennial,* monoecious or dioecious *herbs*, rarely woody, often with white latex. *Leaves* opposite or alternate, sometimes whorled, simple and entire or serrate, rarely palmately lobed, petiolate or sessile, with or without stipules. *Inflorescence* of catkin-like, axillary spikes or axillary clusters, or flowers in a cup-like cyathium; flowers unisexual or bisexual, hypogynous, actinomorphic. *Perianth* absent or of 3 (–5) sepal-like free lobes or both petals and sepals present. *Male flowers* with 1–many stamens, the filaments simple, jointed or branched. *Female flowers* with 2–3 styles, the stigmas strongly papillose or branched. *Ovary* 2- to 3-celled, with 1 ovule per cell. *Fruit* a 2- to 3-celled capsule; seeds with abundant endosperm and large embryo, usually carunculate.

Contains about 300 genera and 5,000 species, mainly tropical. Our genera are not closely related and give an inadequate idea of a large and varied family.

1. Leaves peltate and palmately lobed **2. Ricinus**
1. Leaves not peltate, rarely slightly lobed 2.
2. Both sepals and petals present **3. Chrozophora**
2. Only one row of perianth segments present 3.
3. Plant with watery sap, usually dioecious; stamens numerous; ovary and fruit 2-celled **1. Mercurialis**
3. Plant with copious white latex; monoecious; stamen 1; ovary and fruit 3-celled **4. Euphorbia**

1. Mercurialis L.

Dioecious, rarely monoecious, *annual* to *perennial herbs* with a watery sap, but no latex. *Leaves* opposite, not lobed, crenate-serrate. *Male inflorescence* of catkin-like, more or less erect, axillary spikes, the female of fewer flowers in axillary clusters. *Perianth segments* 3, green. *Stamens* numerous, with free, simple filaments. *Styles* 2. *Ovaries* 2-celled. *Fruit* a didymous capsule.

Contains 8 species, mainly Mediterranean, but also in north and central Europe; Atlantic Islands; south-west and east Asia.

Curtis, T. G. F. (1981). A further station for *Mercurialis perennis* L. in the Burren with comments on its status there. *Irish Nat. Jour.* **20**: 184–185.

Grime, J. P. et al. (1988). *Comparative plant ecology*. London.

Hültén, E. & Fries, M. (1986). *Atlas of north European vascular plants north of the Tropic of Cancer.* 3 vols. Königstein.

Martin, M. R. (1968). Conditions affecting the distribution of *Mercurialis perennis* in certain Cambridgeshire woodlands. *Jour. Ecol.* **56**: 777–793.

Mukerji, S. K. (1927). New variety of *Mercurialis perennis* L. *Jour. Bot. (London)* **65**: 56.

Pannell, J. R. et al. (2004). Polyploidy and the sexual system: what can we learn from *Mercurialis annua*? *Biol. Jour. Linn. Soc.* **82**: 547–560.

Saunders, J. (1883). Monoecious and hermaphrodite *Mercurialis perennis*. *Jour. Bot. (London)* **21**: 181–182.

Williams, I. A. (1925). Monoecious form of *Mercurialis perennis* L. *Jour. Bot. (London)* **63**: 179–180.

1. M. perennis L. Dog's Mercury

M. cynocrambe Scop.; *M. nemoralis* Salisb.; *M. sylvatica* Hoppe; *M. longifolia* Host; *M. perennis* var. *salisburgana* Mukerji

Perennial, dioecious, very rarely monoecious *herb* with long, creeping rhizomes and a watery sap but without latex, becoming dominant over large areas. *Stems* 15–40 cm, deep green, erect, more or less hairy, simple, leafy. *Leaves* opposite; lamina 3–8 × 1.5–3.0 cm, deep green on upper surface, paler beneath, elliptical, elliptical-ovate or elliptical-lanceolate, more or less acute at apex, crenate-serrate, cuneate or rounded at base, more or less hairy on both surfaces; petioles 3–10 mm, hairy; stipules small. *Inflorescence* of male flowers in long, axillary spikes, of female solitary or in small axillary clusters. *Male flowers* with perianth segments 3, about 2 × 1.8 mm, green, ovate, pointed but obtuse at apex; stamens numerous, the filaments pale, the anthers yellow. *Female flowers* with perianth segments 3, similar to those of male flowers; disc of 2, filiform glands; ovary bilocular, hairy; styles 2, the stigmas prominently papillose and white. *Capsule* didymous; seeds 2, 6–8 mm in diameter, subglobose, hairy. *Flowers* 2–4. $2n = 64$.

Native. Woods on good soils where it often forms large clones by vegetative spread, and is frequently dominant in beechwoods on the chalk; also frequent in hedgerows and among shady mountain rocks up to 1035 m on Ben Lawes in Perthshire. Common over much of Great Britain but absent from the Isle of Man, Orkney and Shetland Islands and very local in Jersey and in Ireland, where, except for the Burren, it is mainly introduced. Most of Europe northwards to about 66° N in Norway; Caucasus; south-west Asia; Algeria. European Temperate element.

2. M. annua L. Annual Mercury

Annual, dioecious, rarely monoecious *herb* with fibrous roots and a watery sap but no latex. *Stem* 5–50 cm, green, erect, somewhat thickened at the nodes, more or less glabrous, much branched, leafy. *Leaves* opposite; lamina 1.0–7.5 × 0.3–3.5 cm, bright medium green on upper surface, paler beneath, ovate to elliptic-lanceolate, obtuse or obtusely-acuminate at apex, crenate-serrate, the teeth usually sparingly ciliate, cuneate, rounded or shallowly cordate at base, basal glands minute; petiole 1–40 mm; stipules 1–2 mm, ovate-deltoid to lanceolate, whitish. *Inflorescence* of male flowers an axillary spike, of female flowers either solitary or in fascicles of 1–4. *Male flowers* with perianth segments 3, 1.5–2.0 mm, pale yellowish-green, translucent, broadly ovate, acute at apex, glabrous; stamens 8–12, the filaments pale, the anthers yellow. *Female flowers* with sepals 3, 1–2 mm, pale green or whitish, ovate, more or less acute at apex, glabrous; disc glands about 1.0 mm; ovary bilocular, sparingly to evenly tuberculate, with each tubercle bristle-tipped; styles 2, less than 1.0 mm. *Capsule* 3–4 mm; seeds about 2.0 mm, shiny greyish-brown, ovoid or subglobose, tuberculate-rugose, with a small, white, keel-shaped caruncle. *Flowers* 7–10. A polyploid complex which varies in its sexual system from dioecy through androdioecy to functional bisexuality. $2n = 16$.

Introduced. Cultivated ground, especially gardens and waste places. Frequent in southern England and the Channel Islands, with scattered records in the rest of Great Britain north to central Scotland and in Ireland. Most of Europe but introduced in much of the north and west, doubtfully native in the Azores. Submediterranean subatlantic element. The monoecious plant has been called var. **ambigua** (L. fil.) Duby (*M. ambigua* L. fil.; *M. ladanum* Hartm.) but needs further study.

2. Ricinus L.

Annual or *perennial herb* with a stout rootstock and without milky latex. *Stems* thick and hollow. *Leaves* alternate, peltate and palmately deeply 5- to 11-lobed. *Inflorescence* an interrupted, leaf-opposed or subterminal panicle with male flowers in the lower part and female in the upper. *Sepals* 5, greenish-yellow. *Petals* absent. *Stamens* numerous. *Styles* 3. *Fruit* a trigonous, echinate or rarely smooth, 3-celled, septicidally dehiscent, 3-seeded regma (capsule); seeds compressed-ovoid.

A monotypic genus probably native in north-east tropical Africa, but widely cultivated throughout the tropics, subtropics and warm temperate regions, often becoming naturalised.

Burkill, I. H. (1966). *A dictionary of economic products of the Malay Peninsula*. **2**. Kuala Lumpur.

1. R. communis L. Castor-oil Plant

Annual or *perennial herb* with a stout rootstock, without milky latex. *Stems* up to 5 m, somewhat glaucous, thick, hollow, glabrous, simple or branched, leafy. *Leaves* alternate; lamina (6–)10–50(–100) × (6–)10–50(–100) cm, glaucous-green, peltate, palmately deeply 5- to 11-lobed, the lobes lanceolate or ovate-lanceolate, acuminate at apex and coarsely serrate or biserrate, the teeth gland-tipped, glabrous; petiole equalling or somewhat exceeding the lamina in length, with one or more stipitate, subsessile or sessile glands on the adaxial surface; stipular sheath up to 20 mm, ovate. *Inflorescence* an interrupted, leaf-opposed or subterminal panicle up to 30 cm with male flowers in the lower part and female in the upper; bracts 5–8 mm, broadly triangular-lanceolate and membranous. *Male flowers* on articulate pedicels up to 20 mm, the flowers 20 mm in diameter, the sepals 5, 7–10 mm, greenish-yellow, ovate-lanceolate to triangular-ovate; petals absent; stamens numerous, the anthers yellow. *Female flowers* with pedicels 5–10(–20) mm; sepals like those of male flowers; styles 3, 5–10 mm, dark red, plumose-papillose. *Capsule* 10–20(–24) × 10–15 mm, trigonous, echinate or rarely smooth; seeds 8–15(–20) mm, greyish, mottled dark brown, compressed-ovoid, smooth, the caruncle bilobate. *Flowers* 6–8. $2n = 20$.

Introduced. Casual on tips and in waste places as a garden throw-out or oil-seed alien, often not reaching either flowering or fruiting. Scattered records over Great Britain and the Channel Islands. Probably native to north-east tropical Africa but widely cultivated in the warmer parts of the world, often becoming naturalised. It is very variable as regards size and ornamentation of the capsule and colour and patterning of the seeds.

3. Chrozophora A. Juss.

Annual herbs without milky latex. *Leaves* alternate; repand-dentate, rarely slightly 3-lobed, stipulate. *Inflorescence* a congested axillary panicle with male flowers above and female below. *Sepals* 5, valvate. *Petals* 5, connate. *Disc* 5-lobed, minute. *Stamens* 3–12, the filaments united into a column. *Styles* 3, more or less free. *Ovary* 3-locular, with 1 ovule per loculus. *Fruit* a trigonous, 3-celled, septicidally dehiscent, 3-seeded regma (capsule).

Contains about 7 species from the Mediterranean region, tropical Africa and south-west Asia.

1. C. tinctoria (L.) Raf. Turn-sole

Croton tinctorius L.; *Croton obliquus* Vahl; *Croton verbascifolius* Willd.; *Croton villosus* Sm.; *C. hierosolymitana* Spreng.; *C. obliqua* (Vahl) A. Juss. ex Spreng.

Annual herb with a woody stock, without milky latex. *Stem* (10–)30–100 cm, pale green, erect, ascending or procumbent, sparingly stellate-hairy to felty-tomentose, divaricately branched, leafy. *Leaves* alternate; lamina 2–9 × 1–7 cm, medium green on upper surface, paler beneath, ovate-rhombic to ovate-lanceolate, obtuse or acute at apex, shallowly repand-dentate, occasionally slightly 3-lobed, cuneate to subcordate at base, sparingly to rather densely hairy; petioles 1–9 cm; stipules 2–5 mm. *Inflorescence* a congested, axillary panicle with male flowers above and female below. *Male flowers* with sepals 5, about 4 mm, linear-lanceolate, stellate-hairy; petals 5, about 4 mm, yellowish-green, elliptic-lanceolate, connate, hairy; stamens 3–12, biseriate, the column about 3.5 mm and minutely papillose, the anthers 2-lobed. *Female flowers* on pedicels 3–5 mm, extending to 30–40 mm in fruit; sepals and petals both resembling those of the male; ovary densely lepidote; styles 3, stellate-hairy. *Capsule* (regma) 5–8 mm in diameter, purple, trigonous; seeds about 4 mm, grey, ovoid, angular, smooth. *Flowers* 6–9. $2n = 48$.

Introduced. A tan-bark casual. Mediterranean region; south-west Asia, east to the Tien Shan and Pakistan, and south to Yemen and Socotra.

4. Euphorbia L.

Tithymalus Ség., non Gaertn.; *Galarhoeus* Haw.; *Esula* Haw.; *Anisophyllum* Haw., non Jacq.; *Characias* Gray; *Chamaesyce* Gray

Annual, *biennial* or *perennial*, monoecious *herbs* with white latex. *Leaves* opposite, alternate or whorled, unlobed, entire or serrate. *Inflorescence* of small groups of flowers, composed of 1 female and a few male flowers together in a cup-like, axillary structure termed a cyathium, which has 4–5 small teeth alternating with conspicuous glands at the top, the cyathia solitary in leaf axils, or several together each on a stalk (raylet), subtended by a raylet bract; the whole on paired or whorled branches (rays) each subtended by a bract which is leafy, but often different in shape from the leaves. *Male flowers* consist of a single stamen with a jointed filament. *Female flowers* have 3 styles, the stigmas often bifid. *Ovary* 3-celled, on a pedicel which elongates in fruit; ovules 1 in each cell. *Fruit* a 3-valved capsule. Probably all pollinated by flies.

Contains about 2,000 species in tropical and temperate regions.

Baiges, J. C. & Blanche, C. (1991). Morphologie des graines des espèces ibers-baleariques du genre *Euphorbia* L. (Euphorbiaceae); 2. Subgen. *Esula* Pers. *Bull. Soc. Bot. Fr.* **138**: *Lettre Bot.* 321–327.

Best, K. F., Bowes, G. G., Thomas, A. G. & Maw, M. G. (1980). The biology of Canadian weeds. 39. *Euphorbia esula* L. *Canad. Jour. Pl. Sci.* **60**: 651–663.

Britton, C. E. (1910). *Euphorbia cyparissias* Linn. *Rep. Bot. Soc. Exch. Club Brit. Isles* **2**: 469–470.

Clement, E. J. (1976). Prostrate spurges in Britain. *B.S.B.I. News* **13**: 21.

Clement, E. J. (1997). Can *Euphorbia robbiae* be revived?. *B.S.B.I. News* **76**: 58–60.

Clement, E. J. (2000). *Euphorbia myrsinites* established in Cambs. (v.c. 29). *B.S.B.I. News* **85**: 39.

Hultén, E. & Fries, M. (1986). *Atlas of north European vascular plants north of the Tropic of Cancer.* 3 vols. Königstein.

Juan, R. & Pastor, J. (1995). Contribucion al estudio cariológico del género *Euphorbia* L. subgen. *Esula* Pers. *Bol. Soc. Brot.* sér. 2, **66**: 41–54.

Lousley, J. E. (1971). *Flora of the Isles of Scilly.* Newton Abbot. [*E. peplis*.]

Radcliffe-Smith, A. (1985). Taxonomy of North American leafy spurges. In Watson, A. K. (Edit.) *Monograph Series of the Weed Science Society of America* **3**: 14–25.

Stewart, A., Pearman, D. A. & Preston, C. D. (1994). *Scarce plants in Britain.* Peterborough. [*E. paralias, platyphyllos* and *portlandica*.]

Wigginton, M. J. (Edit.) (1999). *British red data books.* Vol. 1. *Vascular plants*. Peterborough. [*E. hyberna* and *serrulata*.]

1. Plant a shrub **29. mellifera**
1. Plant a herb 2.
2. Plants usually procumbent; stipules present; bracts and leaves similar 3.
2. Plants usually erect or ascending; stipules absent; bracts and leaves often different 5.
3. Stems and capsules hairy **3. maculata**
3. Stems and capsules glabrous 4.
4. Glands of cyathium subglobose **1. peplis**
4. Glands of cyathium saucer-shaped **2. serpens**
5. Leaves on main stem opposite **14. lathyris**
5. Leaves on main stem alternate 6.
6. Glands on cyathia entire and rounded on outer edge 7.
6. Glands on cyathia concave on outer edge, emarginate or prolonged into 2, rarely 4 horns 15.
7. Ovary and capsule smooth to granulose 8.
7. Ovary and capsule conspicuously warty or papillose 10.
8. Leaves glabrous or nearly so **12. helioscopia**
8. Leaves hairy at least beneath 9.
9. Capsule 4–8 mm, subglobose, glabrous or sparsely hairy **4. villosa**
9. Capsule 3–4 mm, broadly ellipsoid, densely long-hairy **5. corallioides**
10. Annuals with a simple root system 11.
10. Perennials with rhizomes 12.
11. Umbels with 5 main branches (rays); capsules with hemispherical papillae **10. platyphyllos**
11. Umbels with 2–5 main branches (rays); capsules with cylindrical papillae **11. serrulata**
12. Capsule with flattened-conical tubercles 13.
12. Capsule with prominent cylindrical tubercles 14.
13. Leaves oblong to lanceolate-oblong or oblanceolate-oblong, entire **6. ceratocarpa**
13. Leaves narrowly obovate, serrulate **8. oblongata**
14. Stems without scales near base; bracts yellowish; capsule 5–6 mm **7. hyberna**
14. Stems with scales near the base; bracts green; capsule (2–)3–4 mm **9. dulcis**
15. Stems hairy; opposite pairs of bracts fused at base 16.
15. Stem glabrous or nearly so; opposite pairs of bracts not fused at base 20.
16. Primary branches (rays) of topmost whorl of inflorescence 10–20; capsule glabrous or hairy 17.
16. Primary branches (rays) of topmost whorl of inflorescence 4–12; capsule glabrous 19.
17. Leaves 18–25 × 11–15 mm **13. myrsinites**
17. Leaves (14–)30–130 × 4–10(–17) mm 18.
18. Glands of cyathium dark brown, rarely yellow, emarginate or with short horns **28(a). characias** subsp. **characias**
18. Glands of cyathium yellowish with long horns **28(b). characias** subsp. **wulfenii**
19. Rhizomes short or absent; leaves of 1-year's stems dull, pale to medium green, more or less hairy on lower side and margin **27(a). amygdaloides** subsp. **amygdaloides**
19. Rhizomes long; leaves of 1-year's stem more or less coriaceous, dark, more or less shiny green and more or less glabrous **27(b). amygdaloides** subsp. **robbiae**
20. Annuals to perennials, stems sometimes buried in sand 21.
20. Perennials with rhizomes 26.
21. Annuals with thin leaves, bracts and leaves similar, rarely on maritime sands 22.
21. Biennials to perennials with more or less succulent leaves, bracts and leaves markedly different, on maritime sands 25.
22. Leaves ovate, subrotund to obovate **16. peplus**
22. Leaves linear to linear-lanceolate or oblong 23.
23. Capsule granulate-rugose **17. segetalis**
23. Capsule smooth 24.
24. Stem simple or with an occasional slender branch from base; leaves up to 15(–20) mm **15(i). exigua** var. **exigua**
24. Stem with numerous, fairly robust branches from the base making a dense domed clump; leaves up to 30 mm **15(ii). exigua** var. **diffusa**
25. Leaf with midrib beneath prominent; seeds pitted **18. portlandica**
25. Leaf with midrib beneath obscure; seeds smooth **19. paralias**
26. Leaves less than 2.5(–3.0) mm wide 27.
26. Leaves more than 2.5 mm wide 28.

27. Leaves of sterile branches 10–12 mm, dense; bracts of main umbel 10–15 × 1.0–2.5 mm **26(i). cyparissias** var. **cyparissias**
27. Leaves of sterile branches 12–25 × 0.7–0.8 mm; bracts of main umbel 15–20 × 3–4 mm **26(ii). cyparissias** var. **esuloides**
28. Leaves oblanceolate to oblong-oblanceolate, widest above the middle, attenuate at base 29.
28. Leaves linear to oblong-lanceolate, widest at or below the middle, not attenuate at base 30.
29. Leaves more than 4 mm wide **21. esula**
29. Leaves less than 4 mm wide **24. × pseudoesula**
30. Leaves rounded to broadly cuneate-rounded at base 31.
30. Leaves abruptly narrowed to a cuneate base 32.
31. Leaves 4–12 mm wide **20. boisseriana**
31. Leaves 3.5–5.5 mm wide **22. waldsteinii**
32. Leaves mostly 4–5 mm wide **22. × pseudovirgata**
32. Leaves mostly 2–3 mm side **25. × gayeri**

Subgenus **1. Chamaesyce** (Gray) Prokh.

Annual herbs with decumbent or procumbent stems. *Leaves* opposite, distichous, usually asymmetrical at base, petiolate, with stipules. *Cyathia* axillary or clustered, not in umbels. *Glands of cyathium* often with petaloid appendages. *Seeds* without a caruncle.

1. E. peplis L. Purple Spurge

Anisophyllum peplis (L.) Haw.; *Chamaesyce maritima* Gray nom. illegit.

Annual monoecious *herb* with fibrous roots. *Stems* pale glaucous green often suffused purplish-crimson, procumbent, glabrous, often with 4 branches from the base up to 9 cm, leafy. *Leaves* opposite, distichous; lamina 3–15 × 2.5–10.0 mm, glaucous green, more or less oblong,obtuse or emarginate at apex, more or less entire, obliquely truncate at base, glabrous; petioles 2–3 mm; stipules 1.0–1.5 mm, filiform. *Inflorescence* with cyathia in the forks and axils of the dichotomous branches; bracts similar to leaves. *Glands of cyathium* reddish-brown, subglobose, entire, with narrow, white or pinkish appendages. *Stamen* 1; anther yellow. *Styles* 3. *Capsule* 4–5 mm in diameter, often purplish, trigonous, glabrous; seeds 2.5–3.0 mm, pale grey, ovoid-pyriform, smooth, ecarunculate. *Flowers* 7–9.

Native. Formerly on sandy or shingly beeches but now probably extinct. Last record 1976 in Alderney. Formerly in southern Great Britain from Kent to Cardiganshire, in Waterford in Ireland and in the Channel Islands. South-west Europe; north-west Africa; Mediterranean Basin; Black Sea; south Caspian. Mediterranean Atlantic element.

2. E. serpens Kunth Hierba de la Colondrina

Chamaesyce serpens (Kunth) Small

Annual monoecious *herb* with fibrous roots. *Stem* 5–30(–50) cm, pale green, glabrous, prostrate or decumbent and often rooting at the nodes, with slender, filiform branches from the base. *Leaves* opposite, distichous; lamina 2–6 × 2–6 mm, medium green on upper surface or slightly glaucous, paler beneath, subrotund, subrotund-ovate, or elliptical, obtuse or emarginate at apex, entire and revolute, rounded or subcordate at base, glabrous; petiole short; stipules united into a white to pink membranous scale round the stem. *Inflorescence* of solitary cyathia in the axils; bracts similar to leaves. *Glands of cyathium* 4, saucer-shaped, with 2 minute, irregularly crenate appendages. *Stamen* 1. *Styles* 3. *Capsules* 1 mm, ovoid-oblong, with sharp angles, glabrous; seeds about 1 mm, narrowly oblong, quadrangular, pale grey, obtusely 4-angled, smooth, without a caruncle. *Flowers* 6–9.

Introduced. A bird-seed and wool casual. Widespread in tropical and temperate America.

3. E. maculata L. Spotted Spurge

Chamaesyce maculata (L.) Small; *Chamaesyce supina* (Raf.) Raf.

Annual monoecious *herb* with fibrous roots. *Stems* 10–50 cm, pale green, decumbent or procumbent, long hairy, with branches up to 50 cm, leafy. *Leaves* opposite; lamina (0.2–)0.4–0.7(–1.3) × 0.1–0.2(0.4) cm, dark green on upper surface, paler beneath, ovate-oblong to oblong, slightly curved, obtuse or subacute at apex, serrulate near the apex, obliquely truncate at base, sparingly hairy on upper surface, more densely so beneath, usually with a purple blotch on the midrib; petiole 0.5–1.5 mm, hairy; stipules about 1.0 mm, triangular-subulate. *Inflorescence* of 1 to few cyathia in leaf axils and stem forks; bracts similar to leaves. *Glands of cyathium* 4, transversely ovate, with small purplish appendages. *Stamen* 1, anthers yellow. *Styles* 3. *Capsule* 1.0–1.3 × 1.2–1.5 mm, ovoid-triangular, shallowly sulcate, smooth, sparsely covered with closely appressed hairs; seeds 0.8–1.0 mm, brownish, ovoid-quadrangular, with 3–4 transverse furrows on each face, without a caruncle. *Flowers* 5–6. $2n=28$.

Introduced. A more or less naturalised weed of nurseries and quarries. Scattered records in England, south Wales and Jersey. Native of North America.

Subgenus **2. Esula** Pers.

Annual or *perennial herbs* or *shrubs*. *Leaves* usually alternate, symmetrical at base, more or less sessile, without stipules. *Cyathia* usually in umbels. *Glands of cyathium* without petaloid appendages. *Seeds* usually with a caruncle.

4. E. villosa Waldst. & Kit. ex Willd. Hairy Spurge

E. pilosa auct.

Perennial monoecious *herb* with a stout rhizome. *Stems* more or less numerous, 30–120 cm, pale green, stout, erect, glabrous or eglandular-hairy, with axillary, sterile branches above, leafy. *Leaves* alternate, dense; lamina 4–10 × 1–4 cm, medium green on upper surface, paler beneath, oblong or oblong-lanceolate to elliptical, obtuse to subacute at apex, serrulate near the apex, entire near the base, broadly cuneate at the sessile base, softly eglandular-hairy on both sides or nearly glabrous on upper surface, the uppermost leaf less than its own length below the umbel. *Inflorescence* of cyathia arranged in umbels. *Rays* of terminal umbel 4–6, often with axillary inflorescence branches below; ray leaves about 2 cm, ovate, obtuse and mucronate, raylet leaves 0.8–1.2 cm, yellowish, ovate or subrotund, obtuse-mucronate at apex, rounded at base, glabrous or sparsely eglandular-hairy. *Glands of cyathium* entire and rounded. *Stamen* 1; anthers yellow. *Styles* 3. *Capsule* 4–8 mm, subglobose, scarcely grooved, glabrous or sparsely eglandular-hairy, smooth or minutely tuberculate; seeds 2.5–3.2 mm, brown, smooth, with a caruncle. *Flowers* 5–6. $2n=18$.

Introduced. Known in a wood and neighbouring hedgebanks near Bath in Somersetshire from 1576 to about 1924. South-east, south and east-central Europe extending northwards to central Russia and north-west France; Algeria.

5. E. corallioides L. Coral Spurge

Caespitose *perennial* monoecious *herb* with rather stout rhizome. *Stems* 40–60 cm, few, rather slender, pale green, sometimes tinted red, erect, with few eglandular hairs to nearly glabrous, with few, axillary branches above, leafy. *Leaves* alternate; lamina 4–8 × 1–2 cm, bluish-green on upper surface, paler beneath, oblong to oblanceolate, obtuse at apex, entire to serrulate, narrowed to a rounded, semiamplexicaul base, with soft, long eglandular hairs on both surfaces, uppermost leaf well below the umbels. *Inflorescence* of cyathia arranged in umbels. *Rays* of terminal umbel 3–5; glabrous; ray leaves like the cauline but wider; raylet leaves 1.5–2.5 cm, green or red-tinged, ovate, obtuse at apex, rounded at base, with long eglandular hairs on both surfaces. *Glands of cyathium* transversely elliptical, entire. *Stamen* 1; anther yellow. *Styles* 3. *Capsule* 3–4 mm, broadly ellipsoid, finely granulate, densely long-hairy; seeds about 2.5 mm, reddish-brown, obovoid, minutely punctate, with a caruncle. *Flowers* 5–6. $2n=26$.

Introduced. Naturalised in woods and hedgerows at Slinfold in Sussex since about 1808 and along a laneside in Somersetshire. Also recorded in a few other localities in England and Glamorganshire in Wales. Native of central and south Italy and Sicily.

6. E. ceratocarpa Ten. Horned Spurge

Perennial monoecious *herb* with a rhizome. *Stems* 70–150 cm, pale green, sometimes tinted red, erect, striate, glabrous, branched only in the inflorescence, leafy. *Leaves* alternate, numerous; lamina 3–9 × 0.5–1.6 cm, pale to medium green on upper surface, paler beneath, turning pinkish-red, oblong to lanceolate-oblong or oblanceolate-oblong, obtuse or mucronate at apex, entire, narrowed at base, sessile, glabrous or nearly so. *Inflorescence* of cynthia arranged in umbels. *Rays* of terminal umbel 5–6, often with axillary inflorescence branches below; ray leaves 15–25 mm, ovate, mucronate at apex, entire; raylet leaves similar but slightly smaller. *Glands of cyathium* ovate, entire, rounded on outer edge. *Stamen* 1; anther yellow. *Styles* 3. *Capsule* 4–5 mm, subglobose, glabrous, with long, flattened-conical tubercles, sulcate; seeds about 3 mm, dark grey, with a caruncle. *Flowers* 4–7.

Introduced. Persisted for some years at Barry docks in Glamorganshire. Native of south Italy and Sicily.

7. E. hyberna L. Irish Spurge
Tithymalus hybernus (L.) Hill; *Galarhoeus hybernus* (L.) Haw.

Perennial monoecious *herb* with a thick rhizome. *Stems* numerous 30–60 cm, pale green, erect, simple, glabrous or nearly so, leafy, not scaly at base. *Leaves* alternate; lamina 5–10 × 1.0–2.5 cm, medium green on upper surface, paler beneath, eventually turning red, oblong, elliptical-oblong or oblanceolate-oblong, obtuse or retuse at apex, entire, cuneate at the more or less sessile base, glabrous on upper surface, sparsely eglandular-hairy beneath especially near the midrib or more rarely glabrous. *Inflorescence* of cynthia arranged in umbels. *Rays* of terminal umbel 4–6, often with axillary inflorescence branches below; ray leaves 3–6 cm, elliptical-oblong; raylet leaves 0.8–3.0 cm, yellowish, ovate or ovate-elliptical, obtuse to subacute, entire, rounded or subcordate at base. *Glands of cyathium* 5, yellowish, finally brown, reniform, entire rounded on outer edge. *Stamen* 1; anther yellow. *Styles* 3. *Capsule* 5–6 × 6–7 mm, subglobose, grooved, glabrous, with prominent cylindrical tubercles; seeds 3.4–3.8 mm, pale brown, smooth, with a caruncle. *Flowers* 4–7. $2n=36$.

Native. Woods, hedgerows, grassy places and stream-banks. Very local in Cornwall, Devonshire and Somersetshire, frequent in most of south-west Ireland; naturalised in Cardiganshire. West and central Europe eastwards to northern Italy. Suboceanic Southern temperate element. Our plant is subsp. **hyberna** which occurs throughout the range of the species.

8. E. oblongata Griseb. Balkan Spurge

Perennial monoecious *herb* with strong rhizomes. *Stems* 40–90 cm, pale yellowish-green, erect, more or less hairy, branched above, leafy. *Leaves* alternate; lamina 30–50 × 1.5–3.0 cm, narrowly obovate, obtuse at apex, serrulate, cuneate to a sessile base, more or less eglandular-hairy. *Inflorescence* of cyathia in umbels. *Rays* of terminal umbels 5, pale green, hairy, about equalling or somewhat exceeding the ray leaves, often with axillary rays below the terminal umbel; ray leaves ovate, obtuse at apex, serrulate, narrowed or rounded at base. *Glands of cyathium* yellowish-green, often only 2–3, ovate and entire, rounded on outer edge. *Stamen* 1; anther yellow. *Styles* 3. *Capsule* 3.8–4.3 mm, broadly ellipsoid, sparsely covered with short, more or less hemispherical, flattened tubercles, glabrous; seeds about 2.5 mm, brown, smooth, with a caruncle. *Flowers* 4–7.

Introduced. Naturalised on a grass bank in Hampshire since at least 1993, and recorded in Isle of Wight, Essex and Cambridgeshire. Native of the Balkans and Aegean region.

9. E. dulcis L. Sweet Spurge
Galarhoeus dulcis (L.) Haw.

Perennial monoecious *herb* with strong, creeping, swollen and jointed rhizome. *Stems* more or less numerous,

20–50 cm, pale green, erect, slender, glabrous or sparsely hairy, simple, leafy, scaly at base. *Leaves* alternate; lamina 3–5 × 0.8–2.0 cm, medium green on upper surface, paler beneath, oblanceolate-oblong or oblong, obtuse at apex, entire or finely serrulate near the apex, tapered at base, subsessile, glabrous or with a few, scattered eglandular hairs beneath and occasionally above. *Inflorescence* of cyathia in umbels. *Rays* of terminal umbel usually 5, pale green, often with axillary rays below; ray leaves 2–3 cm, oblong-elliptical, the upper ones 0.7–2.0 cm, green, ovate-deltate, subacute at apex, denticulate, truncate at base. *Glands of cyathium* green, some turning purple, obovate-subrotund, entire, rounded on outer edge. *Stamen* 1; anther yellow. *Styles* 3. *Capsule* 2–4 mm, subglobose, grooved, glabrous, with prominent cylindrical tubercles; seeds brown, smooth, with a caruncle. *Flowers* 5–7. $2n$ = 12, 18, 24, 28.

Introduced. Naturalised in shady places among woodland and scrub. Scattered records throughout Great Britain. Known since 1894 by the Conan River in Ross and Cromarty. Native of west and central Europe; extending to central Italy and Macedonia. European Temperate element.

10. E. platyphyllos L. Fat-leaved Spurge

Galarhoeus platyphyllos (L.) Haw.; *Galarhoeus strictus* Haw.; *E. verrucosa* auct.; *E. stricta* auct.

Annual monoecious *herb* with a single vertical root. *Stem* 15–80(–134) cm, pale green, erect, simple, glabrous or hairy, leafy. *Leaves* alternate; lamina 2.0–7.0 × 1.0–2.5 cm, matt green with a pale midrib on upper surface, paler beneath, often turning reddish, obovate or oblong-lanceolate, acute or subobtuse at apex, serrulate except at base, deeply cordate or auricled at base, glabrous or slightly eglandular-hairy. *Inflorescence* of cyathia arranged in umbels. *Rays* of terminal umbel usually 5, with axillary branches below, the secondary forks often 3-rayed; ray leaves 2–3 cm, elliptical-oblong; raylet leaves 0.5–1.5 cm, yellowish-green, deltate, acute or obtuse at apex, mucronate, the lowest not differing from the upper and markedly different from the ray leaves. *Glands of cyathium* yellowish-green to orange, suborbicular, entire. *Stamen* 1; anther yellow. *Styles* 3, yellowish-green. *Capsule* 2–4 mm, subglobose, shallowly grooved, with hemispherical papillae; seeds about 2 mm, olive-brown, smooth. *Flowers* 6–10. Visited by beetles and ants. $2n$=30.

Introduced. Cultivated and rough ground. Formally locally frequent in south and east England with a few more localities further north, now very local in southern England north to Cambridgeshire and Worcestershire. South, west and central Europe; west Caucasus. European southern Temperate element.

11. E. serrulata Thuill. Upright Spurge

E. stricta L. nom. illegit.; *E. verrucosa* L. nom. rejic.; *E. platyphyllos* subsp. *stricta* Hook. fil.; *Galarhoeus platyphyllus* var. *stricta* Gray

Annual or *biennial* monoecious *herb* with a tap-root. *Stems* 1(–3), 30–80(–100) cm, pale green, sometimes suffused reddish, slender to robust, erect, usually simple, glabrous, leafy. *Leaves* alternate, scattered; lamina 1–6 × 0.5–1.5 (–2.0) cm, medium yellowish-green on upper surface, paler beneath, narrowly elliptical, lanceolate, oblanceolate or oblong, obtuse to subacute at apex, serrulate, cordate at base, glabrous. *Inflorescence* of cyathia arranged in umbels. *Rays* of terminal umbel (2–)4–5, glabrous; ray leaves oblong; raylet leaves 7–22 mm, ovate, acute at apex, gradually becoming narrower downwards and grading into the ray leaves. *Glands of cyathium* subglobose, entire, rounded on outer edge. *Stamen* 1; anther yellow. *Styles* 3. *Capsule* 2.0–2.5 mm, deeply sulcate, covered with cylindrical papillae which are longer than wide; seeds 1.2–1.5 mm, reddish-brown, ovoid, smooth and shining. *Flowers* 6–9. $2n$=20, 28.

Native. Limestone woods in about 24 localities in Gloucestershire and Monmouthshire; naturalised rarely in southern England from Kent to Somersetshire, south Wales and northern England. From west and central Europe to Iran and Turkestan. Southern Temperate element.

12. E. helioscopia L. Sun Spurge

Galarhoeus helioscopius (L.) Haw.

Annual herb with a pale brown tap-root and fibrous side-roots. *Stem* usually single, 10–50 cm, pale green, erect, glabrous, simple or with a few branches below, leafy. *Leaves* alternate; lamina 10–50 × 3–25 mm, obovate or obovate-spathulate, rounded-obtuse at apex, serrulate in the upper half, attenuate from near the apex to a narrow base, sessile, glabrous. *Inflorescence* of cyathia in umbels. *Rays* of terminal umbel 5, pale green, glabrous, without axillary rays, each ray trichotomous then dichotomous; ray leaves like the cauline leaves but smaller; raylet leaves also like the cauline but smaller; rays and raylets glabrous. *Glands of cyathium* green, transversely ovate and entire, rounded on outer edge. *Stamen* 1; anther yellow. *Styles* 3. *Capsule* 2.5–3.5 × 2.5–3.5 mm, subglobose and somewhat trigonous, smooth; seeds about 2 mm, brown, ovate, reticulate-rugose. *Flowers* 5–10. $2n$=42.

Introduced. Cultivated ground and waste places. Common throughout lowland Great Britain and Ireland. Almost throughout Europe, but a casual in the extreme north; east to central Asia. Circumpolar Southern-temperate element.

13 E. myrsinites L. Broad-leaved Spurge

Galarhoeus myrsinites (L.) Haw.

Perennial monoecious *herb* with a tap-root. *Stems* several, up to 40 cm, pale bluish-yellowish-green, fleshy, spreading or ascending, stout, glabrous, simple, densely leafy. *Leaves* all cauline, alternate; lamina 18–25 × 11–15 mm, glaucous, slightly paler beneath, obovate to subrotund, cuspidate to mucronate at apex, entire, often narrowed at base, glabrous, fleshy, sessile. *Inflorescence* of cyathia in umbels at top of stem; primary branches of topmost whorl 10–12; rays (1–)4–12, once or twice dichotomous, variable in length, pale yellowish-green, glabrous; ray leaves 1.8–2.5 × 1.1–1.5 cm, obovate to subrotund, mucronate at apex; raylets 10–15 mm, pale yellowish-green, glabrous; raylet leaves 1.0–1.2 cm, subrotund or ovate, mucronate at apex, cordate at base, glabrous. *Glands of cyathium* yellowish-green, broad and concave on outer edge with short, thick pointed horns. *Stamen* 1; filament pale yellowish-green; anther yellow. *Styles* 3, green; stigmas colourless. *Capsule* (4–)5–7 mm, trigonous, glabrous, smooth or minutely tuberculate; seeds (2–)3–4 mm, vermiculate-rugose or rarely smooth, greyish-brown. *Flowers* 4–5. $2n=20$.

Introduced. Commonly grown in gardens and recorded as an escape in Surrey, Kent and Cambridgeshire. Native of the Mediterranean region.

14. E. lathyris L. Caper Spurge

E. decussata Salisb. nom. illegit.; *Galarhoeus lathyris* (L.) Haw.; *Galarhoeus decussatus* Gray nom. illegit.

Biennial herb with a thick, white tap-root and numerous, white, fibrous roots forming a short, erect, leafy stem in the 1 year, elongating and flowering in the 2nd. Stem 30–200 cm, pruinose-glaucous over yellowish-green, often purplish towards the base, terete, erect, smooth, hollow, emitting copious, white latex, very leafy. *Leaves* all cauline, opposite or subopposite, decussate; lamina 4–12 × 1–5 cm, getting gradually larger upwards, dark bluish-green with a pale midrib, the upper sometimes yellowish-green, much paler yellowish-glaucous beneath, linear, oblong or oblong-lanceolate, obtuse and often mucronulate at apex, entire, cordate at base, sessile, glabrous. *Inflorescence* of cyathia in umbels at top of stem with some short branches lower down. Rays 2–6, up to 200 mm, pruinose, glabrous; ray leaves 90–180 × 25–40 mm, dark bluish-green with a pale midrib, much paler beneath, oblong-lanceolate, obtuse at apex, glabrous; raylets 2, 15–45 mm, pruinose, glabrous; raylet leaves 15–70 × 12–30 mm, bluish-or yellowish-green, triangular-ovate, acute at apex, entire, cordate at base, glabrous. Glands of cyathium with a yellowish-green saucer-like base with projections at the two upper corners which have a pale yellowish-green stalk and a yellowish-green head. *Stamen* 1; filaments pale yellowish-green; anthers deep yellow. *Styles* 3; pale green; stigmas yellowish. *Capsule* 8–20 mm, green with purple lines between the segments, trigonous, glabrous, smooth; seeds about 5 mm, ellipsoid, greyish-brown, reticulate. *Flowers* 6–7. Visited by bees. $2n=20$.

Possibly native in shady places and woods in a few places in southern England. Frequent casual and naturalised alien in waste places and a weed in gardens over much of Great Britain and the Channel Islands and scattered records in Ireland. Formerly cultivated for its fruit. South Europe from Spain, northern Italy and Greece to France and Germany, but perhaps only native in the east and central Mediterranean region; Morocco rare; Azores. European southern-temperate distribution.

15. E. exigua L. Dwarf Spurge

Esula exigua (L.) Haw.; *Esula tricuspidata* Fourr.

Annual or rarely *biennial herb* with fibrous roots. *Stems* 5–30 cm, pale green, erect, slender to fairly robust, simple to much-branched from the base, glabrous, leafy. *Leaves* alternate; lamina 5–30 × 0.3–5.0 mm, pale to slightly bluish green on upper surface, paler beneath, linear or linear-lanceolate, more or less acute to truncate, retuse or tricuspidate at apex, entire, narrowed at base, glabrous. *Inflorescence* of cyathia in umbels. *Rays* of terminal umbel 3–5, up to 8 times dichotomous; axillary rays 0–1(–6); ray-leaves 2–20(–25) × 0.3–3.0 mm, linear, linear-oblong or linear-lanceolate, acute or subacute at apex, or occasionally truncate, retuse or tricuspidate; raylet leaves 2–10(–15) × 1–4 mm, linear-lanceolate to triangular-ovate, more or less acute at apex, rounded or shallowly cordate at base; rays and raylets glabrous. *Glands of cyathium* golden-brown, lunate, with long, slender horns. *Stamens* 1; anther yellow. *Styles* 3, branched. *Capsule* 1.5 mm in diameter, trigonous, smooth; seeds 1.2 mm, pale grey, ovoid-quadrangular, tuberculate-rugose, with a bilobate, white caruncle. *Flowers* 6–10. $2n=24$.

(i) Var. **exigua**

E. exigua var. *acuta* L.; *E. exigua* var. *retusa* L.; *E. exigua* var. *truncata* W. D. J. Koch

Plant *annual. Stem* simple or with an occasional slender branch from the base. *Leaves* up to 15(–20) mm, linear.

(ii) Var. **diffusa** (Jacq.) Beck

E. diffusa Jacq.

Probably *biennial. Stem* with numerous, fairly robust branches from the base making a dense domed clump. *Leaves* up to 30 mm, linear to linear-lanceolate.

Introduced. Arable and waste land and gardens. Formerly common in south and east England with scattered records elsewhere, except Scotland, but it seems to be much declining even in eastern England. West Europe; North Africa; Atlantic Islands, east to Palestine and northern Iran. European Southern-temperate element. Var. *exigua* is the common plant. Var. *diffusa* occurs in scattered localities. It grows true from seed, is uniform and seems to be biennial or even shortly perennial. There may be an earlier name for the variety, but without seeing types there is no way of telling.

16. E. peplus L. Petty Spurge

Esula peplus (L.) Haw.; *Esula rotundifolia* Gray nom. illegit.

Annual monoecious *herb* with a pale brown tap-root and fibrous side-roots. *Stems* single, 10–40 cm, pale green, erect, glabrous, branched, leafy. *Leaves* alternate; lamina 2–30 × 3–17 mm, dull medium green on upper surface, paler beneath, ovate, subrotund or obovate, rounded or emarginate at apex, entire, rounded at base, glabrous; veins obscure; petiole up to 8 mm, glabrous. *Inflorescence* of cyathia arranged in umbels. *Rays* of terminal umbel 3, pale green, glabrous, with 0–3 axillary rays, up to 5 times dichotomous; ray leaves like the cauline but with shorter petioles; raylet leaves smaller and slightly obliquely ovate; rays and raylets glabrous. *Glands of cyathium* yellowish- green, concave on outer edge with 2 filiform horns. *Stamen* 1; anther yellow. *Styles* 3. *Capsule* 2–3 × 2–2.5 mm, trigonous, smooth, each valve with 2 dorsal ridges; seeds 1.1–1.4 mm, pale grey, ovoid-hexagonal, sulcate ventrally and pitted dorsally, darker in the depressions. *Flowers* 4–11. $2n=16$.

Introduced. Cultivated and waste ground, common in gardens up against houses. Common throughout most lowlands of Great Britain and Ireland. Most of Europe and across Siberia to Lake Baikal. European Southern-temperate element.

17. E. segetalis L. Corn Spurge
E. tetraceras Lange; *Esula segetalis* (L.) Haw.

Annual monoecious *herb* with fibrous roots. *Stems* up to 35 cm, pale green, erect, glabrous, simple or branched from the base, leafy. *Leaves* alternate; lamina 1–3(–6) × 0.1–0.3 cm, medium green on upper surface, paler beneath, linear to linear-lanceolate, obtuse at apex, entire. *Inflorescence* of cyathia in umbels. *Rays* of terminal umbels 5(–6), up to 5 times dichotomous; ray leaves elliptic-oblong; raylet leaves deltate-rhombic, obtuse at apex, cuneate to subcordate at base. *Glands of cyathium* emarginate or with 2, rarely 4, horns. *Stamen* 1; anther yellow. *Styles* 3. *Capsule* 2.5–3.0 × 3.0–3.5 mm, deeply sulcate, granulate-rugulose on the keels; seeds 1.5–2.0 mm, pale grey, ovoid. *Flowers* 5–9. $2n = 16$.

Introduced. A bird-seed casual. Native of south-west Europe and Mediterranean region.

18. E. portlandica L. Portland Spurge
E. segetalis subsp. *portlandica* (L.) Litard.; *E. segetalis* var. *portlandica* (L.) Cout.; *E. segetalis* var. *littoralis* Lange; *Tithymalus portlandicus* (L.) Hill; *Esula portlandica* (L.) Haw.

Biennial or short-lived *perennial*, monoecious *herb* with a more or less vertical root. *Stems* usually several, all flowering or some sterile, 5–40 cm, pale yellowish-green, sometimes suffused deep reddish-purple, erect, glabrous, much-branched, leafy. *Leaves* alternate, numerous and dense; lamina 5–25 × 2–6 mm, bluish-green and somewhat coriaceous, obovate to narrowly oblanceolate, usually obtuse or apiculate at apex, entire or rarely denticulate near the apex, attenuate to base, sessile, glabrous, midrib prominent beneath. *Inflorescence* of cyathia arranged in umbels. *Rays* of terminal umbel (3–)4–5(–6), pale green, glabrous, pattern and number of axillary rays very variable, each ray up to 4 times dichotomous; ray leaves usually like the upper cauline, rarely ovate-subrotund; raylet leaves 5–8 mm, deltate-rhombic, obtuse-mucronate at apex, entire, glabrous; rays and raylets glabrous. *Glands of cyathium* lunate, yellowish, with 2 long horns. *Stamen* 1; anther yellow. *Styles* 3. *Capsule* 2.8–3.0 × 3.0–3.2 mm, trigonous, deeply sulcate, granulate on the keels; seeds 1.5–1.8 mm, pitted, pale grey, darker in the pits, ovoid. *Flowers* 5–9. $2n = 16$.

Native. Maritime sand-dunes. Rather local on coasts of Ireland, Channel Islands and south and west Great Britain from Hampshire to Kintyre. West coast of Europe from Portugal to France and reaching its northern limit in Kintyre; formerly in the Hebrides. Mediterranean-Atlantic element.

19. E. paralias L. Sea Spurge
Galarhoeus paralias (L.) Haw.; *Esula paralias* (L.) Gray

Perennial monoecious *herb* with a short, woody stock and vertical root. *Stems* several, simple, fertile and sterile, pale yellowish-green to glaucous, fleshy, simple, leafy, ascending, glabrous. *Leaves* alternate, often imbricate, thick and fleshy; lamina 5–25(–30) × 2–10(–15) mm, glaucous, linear-oblong, oblong or oblong-lanceolate, the uppermost ones ovate, subacute or obtuse at apex, entire, adaxially concave, rounded at base and sessile, glabrous, midrib beneath obscure. *Inflorescence* of cyathia arranged in umbels. *Rays* of terminal umbel 3–6(–8), glaucous, glabrous, with 0–10 axillary rays; ray leaves ovate-lanceolate to broadly ovate; raylet leaves 5–15 × 7–17 mm, subrotund-rhombic to reniform, cuneate to truncate at base, strongly adaxially concave; rays and raylets glabrous. *Glands of cyathium* orange, lunate, with 2 short horns. *Stamen* 1; anther yellow. *Styles* 3. *Capsule* 5–6 mm in diameter, strongly trigonous, granulate-rugulose; seeds about 3 mm, pale grey or whitish, broadly ovoid, smooth, with a minute caruncle. *Flowers* 7–10. $2n = 16$.

Native. Maritime dunes and shingle. Coasts of Ireland, Channel Islands and Great Britain north to Norfolk and Wigtownshire. West Europe; Mediterranean and Black Sea coasts; Atlantic Islands. Mediterranean-Atlantic element.

× **portlandica**
This hybrid has a low, branched but open habit with a diffuse habit like *E. portlandica* but is more robust and longer-lived, often producing barren overwintering shoots and is very glaucous.

Recorded from Kenfig Burrows in Glamorganshire, Morfa Dyffryn and Morfa Harlech in Merionethshire, Newborough Warren in Anglesey and Curracloe in Co. Wexford. Endemic.

20. E. boissieriana (Woronow) Prokh. Boissier's Spurge
Tithymalus boissierianus Woronow

Perennial herb with a rhizome. *Stems* up to 60 cm, pale green, sometimes reddish towards the base, erect, shallowly channelled, glabrous or nearly so. *Leaves* alternate; lamina 20–60 × 4–12 mm, widest at or below the middle, medium green on upper surface, paler beneath, linear-lanceolate or lanceolate, rounded at apex, entire, rounded at base, glabrous, midrib prominent beneath, sessile. *Inflorescence* of cyathia is umbels. *Rays* of terminal umbel 4–9, pale green, glabrous, often with axillary rays below; raylet leaves greenish-yellow, reniform, entire, glabrous. *Glands of cyathium* yellowish, lunate, with 2 short horns. *Stamens* 1; anther yellow. *Styles* 3. *Capsule* 2.5–3.0 mm, globose, granulate on the keel, glabrous; seeds smooth. *Flowers* 6–9.

Introduced. Recorded by roadsides and in grassy places in a few scattered localities in England. It is native of eastern Europe. Named after Pierre Edmund Boissier (1810–1885).

× **esula**
Has been recorded from railway sidings at Adderley Park in Birmingham, Warwickshire.

21. E. esula L. Leafy Spurge
Esula angustifolia Haw.; *Esula ararica* Fourr.; *Esula dalechampii* Haw.; *Esula vulgaris* Fourr.

Perennial monoecious *herb* with a rhizome. *Stems* up to 60 cm, pale green, sometimes suffused reddish-purple, erect,

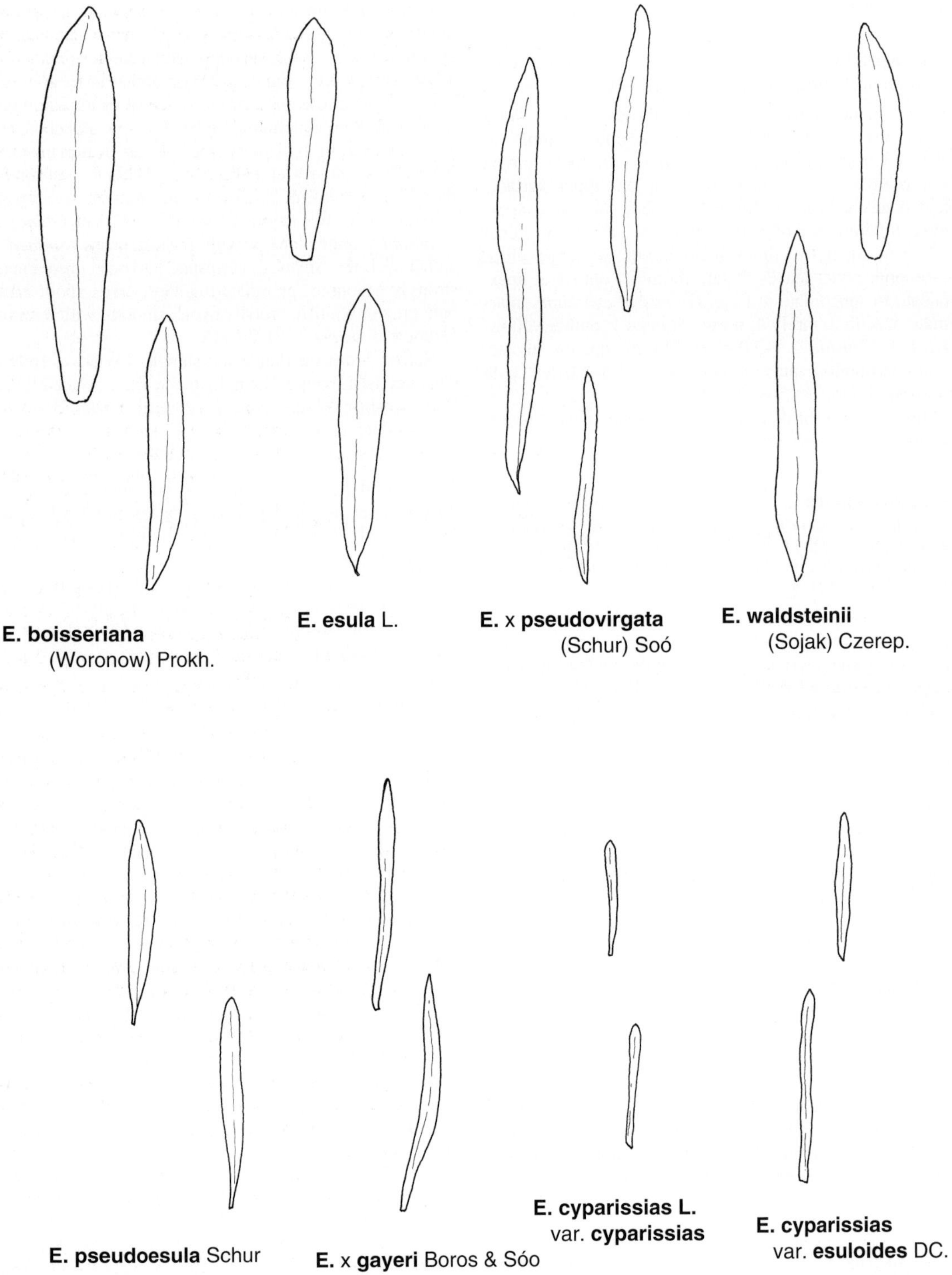

Leaves of **Euphorbia esula/cyparissias** complex

channelled, glabrous or nearly so. *Leaves* alternate; lamina 20–60×(3–)5–12 mm, widest about the middle, medium green on upper surface, paler beneath, oblanceolate, rounded at apex, entire, gradually narrowed to a cuneate base, glabrous, midrib prominent beneath, sessile. *Inflorescence* of cyathia arranged in umbels. *Rays* of terminal umbel 3–12, pale green, glabrous, often with axillary rays below; raylet leaves greyish-yellow, reniform, with entire, undulate margins, glabrous. *Glands of cyathium* yellowish, lunate, with 2 short horns. *Stamens* 1; anther yellow. *Styles* 3. *Capsule* 2.5–3.0 mm, globose or ellipsoid, granulate on the keel, glabrous; seeds smooth. *Flowers* 5–9. 2*n*=60.

Introduced. Naturalised in grassy and open places in scattered localities in Great Britain, but more frequent in parts of southern Scotland. Native of Continental Europe.

22. E.×pseudovirgata (Schur) Soó Twiggy Spurge
E. uralensis auct.; *E. esula* subsp. *tommasiniana* auct.; *E. virgata* auct.; *E. virgata* var. *pseudovirgata* Schur; *E.×podperae* Croizat, non Fisch. ex Link; *E. esula×waldsteinii*?

Perennial monoecious *herb* with rhizomes. *Stems* up to 1 m, pale yellowish-green, often suffused reddish, erect, smooth and glabrous. *Leaves* with lamina 25–70×4–5 mm, dull medium green on upper surface, often turning reddish, linear to lanceolate or sometimes more or less oblanceolate, mostly widest at or below the middle, acute or acuminate at apex, entire, narrowed to a cuneate base, sessile, glabrous, midrib prominent beneath. *Inflorescence* of cyathia arranged in umbels. *Rays* of terminal umbel 5–9, pale green, glabrous, with up to 2–12 axillary rays below, each ray once or twice dichotomous; ray leaves shorter and wider than cauline; raylet leaves rhombic, deltate or reniform, entire, glabrous; rays and raylets glabrous. *Glands of cyathium* yellow, with 2 fairly long horns. *Stamen* 1; anther yellow. *Styles* 3. *Capsule* 2.5–3.0×3.0–3.5 mm, globose, deeply sulcate, granulate on the keels; seeds 2 mm, grey or brownish, ovoid. *Flowers* 4–8. 2*n*=60.

Introduced. Frequently naturalised in grassy and waste places in Great Britain north to central Scotland and in Jersey. Both supposed parents are native of Continental Europe.

23. E. waldsteinii (Soják) Czerep. Waldstein's Spurge
Tithymelus waldsteinii Soják; *E. virgata* Waldst. & Kit., non Desf.

Perennial herb with a rhizome. *Stem* up to 1 m, pale green, erect, shallowly channelled, glabrous or nearly so. *Leaves* alternate; lamina 20–70×3.5–5.5 mm, medium green on upper surface, paler beneath, linear to oblong-lanceolate, widest below the middle, cuspidate to acuminate at apex, entire, rounded or rounded-cuneate at base, glabrous, midrib prominent beneath, sessile. *Inflorescence* of cyathia arranged in umbels. *Rays* of terminal umbel 5–14, pale green, glabrous, often with axillary rays below; raylet leaves greyish-green, ovate, entire, glabrous. *Glands of cyathium* yellowish, lunate, with 2 short horns. *Stamen* 1; anther yellow. *Styles* 3. *Capsules* 2.5–3.0 mm, globose, granulate on the keel, glabrous; seeds smooth. *Flowers* 6–9.

Introduced. Naturalised in scattered localities in Great Britain north to Lanarkshire. Native of central and south-east Europe and south-west Asia. Named after Franz de Paula Adam Waldstein (1759–1823).

24. E.×pseudoesula Schur Figert's Spurge
E. cyparissias×esula

Perennial, monoecious *herb* with a rhizome. *Stem* up to 60 cm, pale green, sometimes tinted purplish, erect, shallowly channelled, glabrous or nearly so. *Leaves* alternate; lamina 30–40×2–4 mm, medium green on upper surface, paler beneath, linear, oblanceolate or oblong-oblanceolate, at least some widest at or above the middle, rounded or subacute-mucronate at apex, entire, narrowed to a cuneate base, glabrous, midrib prominent beneath, sessile. *Inflorescence* of cyathia arranged in umbels. *Rays* of terminal umbel 6–9, pale green, glabrous, often with axillary rays below; raylet leaves greyish-yellow, reniform, with entire, undulate margins, glabrous. *Glands of cyathium* yellowish, lunate, with 2 short horns. *Stamen* 1; anther yellow. *Styles* 3. *Capsule* 2.5–3.0 mm, globose, granulate on the keel, glabrous; seeds smooth. *Flowers* 6–9.

Introduced. Roadsides, trackways and open spaces. Possibly widespread in eastern England and perhaps elsewhere in southern England. Native of Continental Europe.

25. E.×gayeri Boros & Soó Gáyer's Spurge
E. cyparissias×waldsteinii

Perennial monoecious *herb* with a rhizome. *Stem* up to 60 cm, pale green, erect, shallowly channelled, glabrous or nearly so. *Leaves* alternate; lamina 20–45×2–3 mm, linear or oblong to linear-lanceolate, widest at or below the middle, sharply acute or acuminate at apex, entire, abruptly narrowed at base, glabrous, midrib prominent beneath, sessile. *Inflorescence* of cyathia arranged in umbels. *Rays* of terminal umbel 6–10, pale green, glabrous, often with axillary rays below; raylet leaves yellowish, reniform, with entire, undulate margins, glabrous. *Glands of cyathium* yellowish, lunate, with 2 short horns. *Stamen* 1; anther yellow. *Styles* 3. *Capsule* 2.5–3.0 mm, globose, granulate on the keel, glabrous; seeds smooth. *Flowers* 5–7.

Introduced. Roadsides and grassy places. Naturalised in Cumberland and Breconshire; recorded elsewhere. Native of Continental Europe. Named after Gy Gáyer (1883–1932).

26. E. cyparissias L. Cypress Spurge
Esula cyparissias (L.) Haw.; *Esula cupressina* Gray nom. illegit.

Perennial herb with long, creeping rhizomes often forming large patches. *Stems* numerous, 10–60 cm, pale yellowish-green with a slightly glaucous hue, often flushed purple, erect, glabrous, often with sterile branches above. *Leaves* very numerous, alternate, spaced on stem below, numerous on sterile branches above; lamina 15–40×1–3 mm, slightly bluish-green on upper surface, paler beneath, linear, obtuse at apex, entire, sessile, glabrous, midrib prominent beneath. *Inflorescence* of cyathia arranged in umbels. *Rays* of terminal umbel (5–)9–18(–22), pale green, glabrous, often with axillary rays below, each ray once or twice dichotomous; ray leaves linear to oblong; raylet leaves yellow, reniform, rhombic or subrotund, entire, glabrous; rays and raylets glabrous. *Glands of cyathium* lunate, yellowish, with 2 short horns. *Stamens* 1; anthers yellow. *Styles* 3. *Capsule* 3.0–3.5 × 3.0–3.5 mm, globose, deeply sulcate, granulate on the keels; seeds 1.5–1.7 mm, grey and shiny, ovoid, smooth. *Flowers* 4–8. $2n = 20, 40$.

(i) Var. **cyparissias**

Stem up to 40 cm, often suffused purple. *Leaves* 1.5–30 × about 3 mm. *Sterile* branches of upper stem 3.5–7.0 cm, their leaves 10–12 × 0.5–0.6 mm and dense. *Bracts* of main umbel 10–15 × 1.0–2.5 cm, linear.

(ii) Var. **esuloides** DC.

E. cyparissias var. *longibracteata* Schröt.; *E. cyparissias* var. *major* Boiss.; *E. pseudocyparissias* auct.

Stem up to 60 cm, green. *Leaves* 3.0–4.0 × about 2.5 mm. *Sterile branches* of upper stem 5–10 cm, their leaves 12–25 × 0.7–0.8 mm and more open. *Bracts* of main umbel 15–20 × 3–4 mm, lanceolate.

Possibly native in Kent and East Anglia on calcareous grassland as var. *cyparissias*; elsewhere naturalised in rough grassland and waste places usually as var. *esuloides*. Throughout Great Britain and Ireland. Most of Europe except the extreme north and the extreme south, but only as an introduction in most of the north. This species is often infected with the orange-coloured rust *Uromyces scutellatus* (Schrank) Lév.

27. E. amygdaloides L. Wood Spurge

Characias amygdaloides (L.) Gray; *Esula amygdaloides* (L.) Haw.

Perennial monoecious *herb* with a thick stock, tufted or with rhizomes. *Stems* biennial, with inflorescences arising from the top in 2 year, 30–90 cm, pale to dark green, erect, robust, solid, smooth, glabrous or eglandular-hairy, unbranched. *Leaves* of 1 year with lamina 3–8 × 1.5–2.5 cm, dark green on upper surface with a pale midrib or medium green, much paler beneath, oblanceolate to oblong-oblanceolate, obtuse to subacute at apex, entire, attenuate at base to a short petiole, glabrous or hairy beneath, midrib very prominent beneath; leaves of the flowering stem alternate, oblong or obovate-oblong, obtuse or subacute at apex, entire, more or less sessile, glabrous. *Inflorescence* of cyathia arranged in umbels. *Rays* of terminal umbel (3–)4–12, with axillary branches below, each ray up to 4 times dichotomous; ray leaves broadly ovate, rounded at apex; raylet leaves yellowish, reniform, entire, connate in pairs for about half the width of their bases; rays and raylets pale yellowish-green and glabrous. *Glands of cyathium* yellowish, with 2 converging horns. *Stamen* 1; anther yellow. *Styles* 3. *Capsule* 3–4 × 2.5–4.0 mm; globose, deeply sulcate, punctate; seeds 2.0–2.5 mm, blackish, ovoid. *Flowers* 3–5.

(a) Subsp. **amygdaloides**

E. × *turneri* Druce

Rhizomes short or absent. *Leaves* of 1-year stems dull pale to medium green, more or less hairy on lower sides and margins. $2n = 20$.

(b) Subsp. **robbiae** (Turrill) Stace

E. robbiae Turrill; *E. amygdaloides* var. *robbiae* (Turrill) Radel.-Sm.

Rhizomes long. *Leaves* of 1-year stems more or less coriaceous, dark, more or less shiny green and more or less glabrous.

Native. Subsp. *amygdaloides* is native in damp woods and shady hedgerows. It is common in southern Great Britain north to Flintshire and Norfolk, and in the Channel Islands. Further north and in Ireland it is a rare introduction. Coastal plants tend to be more dwarf, fleshier and more densely hairy, but it is difficult to separate them technically, though they look very different in the field, especially those at Portland, Dorset. Central, south and west Europe extending to Ukraine and Caucasus, and in the mountains of Algeria. European Temperate element. Subsp. *robbiae* has been introduced into gardens and is naturalised in woods and other shady places. There are scattered records in south-east and central England and it is native of north-west Turkey.

28. E. characias L. Mediterranean Spurge

Esula characias (L.) Haw.

Tufted, *perennial* monoecious *herb. Stems* numerous, biennial, 50–150 cm, pale yellowish-green, erect, robust, solid, smooth, fleshy, terete, shortly and densely eglandular-hairy, unbranched. *Leaves* of 1 year's growth usually larger than those of the second, dense, and all round the stem; lamina (14–)30–130 × 4–10(–17) mm, medium, slightly bluish-green on upper surface, with a pale midrib, slightly paler bluish-green beneath, linear to linear-oblanceolate, or rarely linear-obovate, obtuse to acute at apex, entire, gradually narrowed to a sessile base, densely and minutely eglandular-hairy on both surfaces, midrib only visible, more prominent beneath than on upper surface. *Inflorescence* of cyathia arranged in umbels. *Rays* of terminal umbel 10–20, with 13–30(–40) axillary rays below, each ray usually twice, but up to 4 times dichotomous; rayleaves similar to upper cauline or oblanceolate to obovate; raylet leaves yellowish-green, subrotund-deltate, entire and usually connate at the base; rays and raylets densely and shortly eglandular-hairy. *Glands of cyathium* dark brown to yellowish, emarginate or with short to long horns. *Stamen* 1, anther yellow. *Styles* 3. *Capsule* 4–7 × 5–6 mm, globose, deeply sulcate, smooth, densely hairy; seeds 2.5–3.8 mm, silver-grey, ovoid. *Flowers* 3–5.

(a) Subsp. **characias**

Stems up to 80 cm. *Glands of cyathium* dark brown, rarely yellow, emarginate or with short horns. $2n = 20$.

(b) Subsp. **wulfenii** (Hoppe ex W. D. J. Koch) Radel.-Sm.

E. wulfenii Hoppe ex W. D. J. Koch

Stems up to 180 cm. *Glands of cyathium* yellowish, with long horns.

Introduced. Both subspecies are grown in gardens and are naturalised on old garden sites and in waste places. There are numerous records for southern England and outlying records in the Isle of Man and southern Scotland. Subsp. *characias* is native of the west Mediterranean region and subsp. *wulfenii* of the east Mediterranean.

29. E. mellifera Aiton — Honey-bearing Spurge
E. longifolia Lam.

Monoecious *shrub* up to 2 m. *Stems* pale, erect, striate, glabrous, branching at the top. *Leaves* alternate; lamina 8–16 × 1.5–3.5 cm, dark green with a pale midrib on upper surface, paler beneath, fleshy, lanceolate or elliptical, rounded-mucronate at apex, entire, narrowed at base, glabrous, sessile or shortly petiolate. *Inflorescence* of cyathia arranged in panicles at the ends of stems and branches. *Rays* of terminal umbel numerous, up to 40 mm, glabrous; ray leaves small; raylets up to 10 mm, glabrous. *Glands of cyathium* subrotund, entire. *Stamen* 1; filament pale green; anther orange. *Styles* 3. *Capsule* 5–6 × 7–8 mm, broadly ovoid, muricate; seeds 3.0–4.5 × 3.0–3.5 mm, pale brown, keeled, smooth. *Flowers* 4–5.

Introduced. A garden escape on cliffs at Mousehole, Cornwall and seedlings on walls near Tresco Abbey gardens in Isles of Scilly. Native of Madeira and Canary Islands.

Order **10. RHAMNALES** Lindl.

Deciduous or evergreen, monoecious or dioecious *shrubs* or *trees* or *woody climbers*. *Leaves* alternate or subopposite, usually with stipules. *Flowers* solitary or in axillary fascicles or racemes, unisexual or bisexual. *Stamens* 4–5. *Fruit* usually a drupe or berry.

Contains 3 families and about 1700 species.

103. RHAMNACEAE Juss. nom. conserv.

Deciduous or evergreen, monoecious or dioecious *shrubs* or small *trees*. *Leaves* alternate or subopposite, simple, petiolate, stipulate. *Inflorescence* of solitary flowers or axillary fascicles or racemes; flowers unisexual or bisexual, perigynous, actinomorphic. *Hypanthium* bell-shaped. *Sepals* 4–5. *Petals* 4 or 5, or absent. *Stamens* 4 or 5, arising from an intrastaminal disc which lines or rims the hypanthium. *Styles* 1, with 2–3 lobed capitate stigma or divided into 2–4 distally , each arm with a capitate stigma. *Ovary* 2- to 4-celled, each cell with one basal ovule. *Fruit* a drupe, eventually black; with 2–4 seeds (pyrenes).

Contains about 58 genera and more than 900 species, cosmopolitan.

1. Leaves serrate; winter buds with scales — **1. Rhamnus**
1. Leaves entire; winter buds without scales — **2. Frangula**

1. Rhamnus L.

Deciduous, dioecious *shrubs* or small *trees*. *Winter buds* with scales. *Leaves* alternate or subopposite, crenate-serrate or serrate. *Inflorescence* of solitary flowers or axillary fascicles or racemes; flowers unisexual or bisexual, the unisexual flowers usually showing vestigial parts of the other sex. *Hypanthium* bell-shaped or obconical. *Sepals* 4 or 5, green, attached to the hypanthium. *Petals* 4 or 5, greenish-yellow, sometimes minute or absent in female flowers. *Stamens* 4 or 5, staminodes in female flowers. *Style* 1, divided into 3 or 4 distally, each arm with a capitate stigma. *Fruit* a drupe, with 3–4 pyrenes. *Cotyledons* epigeal.

Contains about 100 species, cosmopolitan.

Bean, W. J. (1976). *Trees and shrubs hardy in the British Isles*. 3. Ed. 8. London.
Godwin, H. (1943). Biological flora of the British Isles. Rhamnaceae. *Rhamnus cathartica* L., *Frangula alnus* Miller (*Rhamnus frangula* L.). *Jour. Ecol.* **31**: 66–92.
Hultén, E. & Fries, M. (1986). *Atlas of European vascular plants north of the Tropic of Cancer*. 3 vols. Königstein.

1. Evergreen shrub or small tree without spines; sepals and petals mostly 5; drupe 5–6 mm, obovoid — **2. alaternus**
1. Deciduous shrub or small tree with spines; sepals and petals mostly 4; drupe 6–10 mm, more or les globose — 2.
2. Plant prostrate — **1(i). cathartica** var. **prostrata**
2. Plant erect — 3.
3. Leaves and petioles densely eglandular-hairy — **1(v). cathartica** var. **schroeteri**
3. Leaves glabrous or sparsely eglandular-hairy — 4.
4. Leaves narrowly elliptical or lanceolate — **1(iii). cathartica** var. **hydrensis**
4. Leaves elliptical to ovate or subrotund — 5.
5. Leaves (4–)5–8 × 3–6 cm, broadly ovate to subrotund; petioles 1.5–3.0 cm — **1(vi). cathartica** var. **longipetiolata**
5. Leaves 1.5–9.0 × 1.2–3.0 cm, elliptical; petioles 0.5–1.5(–2.0) cm — 6.
6. Leaves 1.5–3.5 × 1.2–1.5(–2.0) cm — **1(ii). cathartica** var. **ambigua**
6. Leaves 3–9 × 2–3 cm — **1(iv). cathartica** var. **carthartica**

Section **1. Rhamnus**

Armed, deciduous *shrubs* or small trees. *Flowers* 4-merous.

1. R. cathartica L. — Buckthorn

Deciduous, dioecious *shrub* or small *tree* up to 8(–10) m with a narrow, but rounded crown. *Trunk* soon dividing into branches. *Wood* hard, the sapwood yellow, the heartwood reddish-brown. *Bark* pale brown, with numerous lenticels, fissured and scaling on old branches; blaze orange. *Branches* ascending or spreading, many short, lateral branches ending in thorns; twigs greyish-brown with lenticels, glabrous; young shoots green touched with brown above, pale yellowish-green beneath, slender, glabrous. *Buds* 3.0–3.5 × 1.2–1.5 mm, ovoid-conical, acute at apex; scales dark brown, ovate, acute at apex. *Leaves* opposite or subopposite; lamina 4–9 × 0.5–3.5 cm, rather pale, dull, yellowish-green on upper surface, not much paler beneath, sometimes turning dull yellow in autumn, ovate, broadly elliptical or subrotund, shortly and bluntly acuminate at apex, minutely crenate-serrate, rounded, sometimes asymmetrically at base, rarely cuneate, minutely hairy becoming glabrous on upper surface, shortly eglandular-hairy, particularly on the veins beneath, becoming glabrous, or more densely hairy; veins 2–4(–5) pairs, ascending, impressed on upper surface, prominent beneath; petiole 6–25 mm, pale green, with few to numerous, short, pale simple eglandular hairs; stipules small and deciduous. *Inflorescence* of solitary flowers or axillary fascicles or 2–5 flowers on the previous year's wood of the short shoots; flowers 5–7 mm in diameter, unisexual, but vestigial stamens on female flowers and variable stigmas on male flowers, erect on pale yellowish-green, slender pedicels up to 12 mm, with few to numerous eglandular hairs. *Hypanthium* pale yellowish-green, more or less bell-shaped, glabrous. *Sepals* 4, 1.0–1.2 mm, yellowish-green, triangular-ovate, shorter and narrower in female flowers, glabrous, attached to the hypanthium. *Petals* 4, 2.3–3.0 mm, pale yellowish-green, triangular-ovate, acute at apex, shorter and narrower in female flowers. *Stamens* 4; filaments pale green; anthers reddish. *Style* very short, divided into a 2- to 3-lobed capitate stigma. *Drupe* 6–10 mm, green turning black, more or less globose; seeds 3–4, about 6 mm, purplish-black. *Flowers* 5–6. Pollinated by various insects. $2n=24$.

(i) Var. **prostrata** Druce
Plant prostrate.

(ii) Var. **ambigua** J. Murr
Plant erect. *Leaves* 1.5–3.5 × 1.2–1.5(–2.0) cm, elliptical, eglandular-hairy, the petioles 0.5–1.5 cm.

(iii) Var. **hydrensis** (Hacq.) DC.
Plant erect. *Leaves* 2–6 × 0.7–2.0 cm, narrowly elliptical or lanceolate, cuneate at base, eglandular-hairy when young becoming nearly glabrous, but the petioles hairy and 0.5–1.5 cm.

(iv) Var. **cathartica**
Plant erect. *Leaves* 3–9 × 2–3 cm, elliptical, sparsely eglandular-hairy; petioles up to 1.5 (–2.0) cm.

(v) Var. **schroeteri** DC.
R. cathartica var. *pubescens* Bean

Plant erect. *Leaves* 2–7 × 1.5–3.5 cm, broadly elliptical or subrotund, eglandular-hairy on both surfaces with densely hairy petioles 0.5–1.5(–2.0) cm.

(vi) Var. **longipetiolata** Grubov
Plant erect. *Leaves* (4–)5–8 × 3–6 cm, broadly ovate to subrotund; petiole 1.5–3.0 cm.

Native. Hedgerows, scrub and open woods on peat and base-rich soils; planted by motorways elsewhere. Locally common in England, scattered records in Wales and Ireland, a planted tree elsewhere. Most of Europe to 61° 45′ N in Sweden and eastwards to the River Ob; North Africa on high mountains in Morocco and Algeria. Eurosiberian Temperate element. Var. *catharticus* and var. *ambiguus* appear to be widespread. Var. *hydrensis* is in scattered localities and var. *schroeteri* in the north of England. The distribution and ecology of these varieties is not understood. Var. *longipetiolata* is planted by motorways and probably elsewhere and seems to come from eastern Europe and Russia.

Section **2. Alaternus** DC.

Unarmed evergreen *shrubs* or small *trees*. *Flowers* 5-merous.

2. R. alaternus L. Mediterranean Buckthorn

Alaternus phylica Mill.; *Alaternus glabra* Mill.; *Alaternus latifolia* Mill.; *Alaternus angustifolia* Mill.

Evergreen, unarmed, dioecious *shrub* or small *tree* up to 5 m with a dense, sprawling to rounded crown. *Trunk* up to 20 cm in diameter, soon dividing into branches. *Bark* pale brown, becoming ridged, with pale brown fissures. *Branches* spreading; twigs greyish-brown, eglandular-hairy; young shoots green, with short, pale eglandular hairs. *Buds* about 2 mm, on a short, peg-like extension, ovoid, pointed; scales brown, ovate. *Leaves* alternate or subopposite; lamina 2–6 × 1.0–2.5 cm, dark shining green on upper surface, pale green and shiny beneath, leathery, linear-elliptical, lanceolate, ovate or elliptical, sharply acute at apex, entire and narrowly revolute or with small, sharp, gland-tipped serrations, cuneate at base, glabrous except for some axillary tufts of eglandular hairs in the axils of the veins beneath; mainly 3- to 6-veined from the base, raised on upper surface, the midrib prominent beneath; petiole up to 8 mm, glabrous; stipules about 2 mm, deltoid, acuminate at apex, thinly eglandular-hairy, caducous. *Inflorescence* unisexual, flowers with characteristic smell; male flowers forming dense, cylindrical racemes about 1.0–1.5 cm, the peduncle very short to almost wanting, the rachis angular and shortly eglandular-hairy, the bracts about 8 mm, ovate-deltoid, acute at apex and shortly eglandular-hairy and often with dark sessile glands on the dorsal surface and margin, the pedicels about 1.0 mm and glabrous; female flowers forming a similar raceme to male, but shorter. *Hypanthium* obconical, glabrous. *Sepals* 5, about 1.5 mm, green, ovate, acute at apex, 3-veined, reflexed at anthesis in male flowers, erect or spreading in female flowers. *Petals* 5, about 1.0 mm, greenish-yellow, narrowly oblong and subacute at apex in male flowers, minute or absent in female flowers. *Stamens* 5; filaments green; anthers yellow, reduced to minute staminodes in female flowers. *Style* 1, very short, divided into 2–3 filiform lobes; stigma terminal and lateral. *Drupe* 5–6 mm, attached basally to the persistent hypanthium, black when ripe, obovoid; seed about 5 mm, pale brown, obovoid, with a distinct median ridge on the ventral surface and a broad median furrow, surrounded by a narrow horseshoe-shaped marginal furrow on the dorsal surface, verruculose. *Flowers* 4–5. Pollinated by various insects. $2n=24$.

Variable in leaf shape and size.

Introduced. Well naturalised in scrub near the sea in Caernarvonshire and Denbighshire and recorded in Isle of Wight, Gloucestershire and Glamorganshire. Widely planted in gardens and parks. Native throughout the Mediterranean area, but more frequent in the western half.

2. Frangula Mill.

Deciduous, monoecious *shrubs* or small *trees*. *Winter buds* without scales. *Leaves* alternate or subopposite, entire but undulate. *Inflorescence* of axillary fascicles of 2–10 flowers on the young wood; flowers bisexual. *Hypanthium* greenish, bell-shaped. *Sepals* 5, white inside, green outside. *Petals* 5, white, hooded. *Stamens* 5. *Style* 1; stigma 2-lobed. *Fruit* a drupe, with 2–3 pyrenes. *Cotyledons* hypogeal.

Contains about 50 species, cosmopolitan.

Godwin, H. (1943). Biological flora of the British Isles. No. 11. *Frangula alnus* Miller (*Rhamnus frangula* L.). *Jour. Ecol.* **31**: 66–92.
Hultén, E. & Fries, M. (1986). *Atlas of north European vascular plants north of the Tropic of Cancer.* 3 vols. Königstein.
Rushforth, K. (1999). *Trees of Britain and Europe*. London.
Webb, D. A. & Scannell, M. J. P. (1983). *Flora of Connemara and the Burren*. Cambridge.

1. Shrub prostrate **1(iii). alnus** var. **prostrata**
1. Shrub erect 2.
2. Leaves 5–9(–12) × 3–6 cm, dark green; drupe 8–10 mm **1(ii). alnus** var. **latifolia**
2. Leaves 2–8 × 0.5–4.0 cm; drupe 6–8 3.
3. Leaves obovate, 1.5–4.0 cm wide **1(i,1). alnus** var. **alnus** forma **alnus**
3. Leaves oblanceolate, 0.5–2.0(–2.5) cm wide **1(i,2). alnus** var. **alnus** forma **angustifolia**

1. F. alnus Mill. Alder Buckthorn
Rhamnus frangula L.; *Rhamnus alnoides* Gray nom. illegit.

Deciduous, monoecious *shrub* or small *tree* up to 6 m, usually with a rounded crown. *Stems* usually several, up to 20 cm in circumference, without spines. *Bark* dark brown or blackish-brown, when cut shows a lemon-yellow blaze beneath. *Branches* subopposite, ascending to an acute angle to the main stem, without a marked distinction into long and short shoots; twigs dark grey, glabrous; young shoots green, minutely appressed-hairy. *Buds* up to 5 mm, conical or ovoid-conical, without scales, densely covered with brownish eglandular hairs. *Leaves* alternate or subopposite; lamina 2–9(–12) × 0.5–6.0 cm, sub-shiny medium to dark green on upper surface, slightly paler beneath, turning yellow in autumn, elliptical to obovate, shortly pointed at apex, entire but undulate, cuneate or rounded at base, glabrous on upper surface, eglandular-hairy when young beneath especially on the veins, becoming glabrous; veins 7–9 pairs, impressed on upper surface, prominent beneath; petiole up to 15 mm, slender, grooved, yellowish-green, eglandular-hairy becoming glabrous. *Inflorescence* of axillary fascicles on the young wood, of 2–10 bisexual flowers 5–6 mm in diameter. *Hypanthium* greenish, bell-shaped. *Sepals* 5, 1.5–1.6 mm, white inside, green outside, triangular-ovate, obtuse at apex. *Petals* 5, 1.0–1.2 mm, white, each folded round a stamen like a monk's cowl, each side of the fold half-ovate, with an emarginate apex between the two. *Stamens* 5; filaments very short, whitish; anthers violet. *Style* 1, green; stigma green, 2-lobed. *Drupe* 6–10 mm in diameter, changing from green to red and then violet-black, obovoid, with the calyx persisting as a rim at the base and the style as a short prickle at the tip, juicy; seeds 2–3, about 5 mm, pale brown, elliptical. *Flowers* 5–6. Pollinated by various insects, especially bees. $2n = 20$.

(i) Var. **alnus**
Erect *shrub*. *Leaves* 2–8 × 0.5–4.0 cm, medium yellowish-green, usually more or less obovate or oblanceolate, narrowed at base. *Drupe* 6–8.

(1) Forma **alnus**
Leaves 2.5–8.0 × 1.5–4.0 cm, obovate.

(2) Forma **angustifolia** (Loudon ex Bean) Schelle
Rhamnus frangula var. *angustifolia* Loudon ex Bean
Leaves 2–5(–8) × 0.5–2.0(–2.5) cm, oblanceolate.

(ii) Var. **latifolia** Dippel
Erect shrub. *Leaves* 5–9(–12) × 3–6 cm, dark green, broadly elliptical to slightly obovate-elliptical, rounded or broadly cuneate at base. *Drupe* 8–10 mm.

(iii) Var. **prostrata** P. D. Sell
Prostrate *shrub*. *Leaves* 1.5–4.0 × 1.3–2.5 cm.

Native. Scrub on fen peat and around the margins of raised and valley bogs, on moist heaths and commons, in limestone scrub and as undergrowth in open woods, usually on damp and more or less peaty soils; occasionally planted by foresters and charcoal burners. Locally common in England and Wales and more common in the west than *Rhamnus cathartica*; very scattered records in Ireland and Scotland. The common plant is var. *alnus* forma *alnus*. Forma *angustifolia* occurs in Wicken Fen, Cambridgeshire and is in cultivation. Var. *latifolia* has been planted in recent years as food for the larva of the Brimstone butterfly. Var. *prostrata* is the dominant plant of populations of the Burren lowlands and Connemara, but var. *alnus* is found with it by Lough Mask. It retains its habit in cultivation. Europe north to about 67° in Sweden and 66° 50′ in Russian Lapland, eastwards to the Urals and Siberia; Morocco and Algeria; rare in the Mediterranean region. Eurosiberian Temperate element. The charcoal made by burning the wood is regarded to be the finest for making gunpowder. The berries and bark when fresh cause vomiting. Natural dyes in shades of yellow or brown come from the bark, while the fruit yields green or bluish-grey dyes. Butchers used to favour its hard, easily sharpened wood for skewers which were called dogs, hence one of its common names, Black Dogwood.

104. VITACEAE Juss. nom. conserv.

Deciduous, woody, monoecious or dioecious *climbers* with leaf-opposed tendrils. *Leaves* alternate, simple and palmately lobed or palmate, petiolate, stipulate. *Flowers* small, reddish to greenish, in leaf-opposed cymes, bisexual or mostly so, hypogynous, actinomorphic. *Sepals* 5, very short, fused into a more or less lobed rim. *Petals* 5, free or fused distally. *Stamens* 5. *Style* 1; stigma capitate. *Ovary* 2-celled, each cell with 2 nearly basal ovules. *Fruit* a berry with up to 4 seeds.

Contains about 13 genera and 800 species, mostly in warm temperate or tropical regions.

1. Leaves simple; tendrils not ending in discs; petals fused distally, falling as flowers open **1. Vitis**
1. Leaves palmate or simple, if simple then tendrils ending in discs; petals free 2.
2. Tendrils with adhesive discs at the end of each branch; nectariferous disc inconspicuous **2. Parthenocissus**
2. Tendrils sometimes swollen at the end of each branch but without adhesive discs; nectariferous disc prominent in flower **3. Cissus**

1. Vitis L.

Deciduous, monoecious or dioecious *woody climbers* with tendrils not ending in discs. *Leaves* simple, palmately lobed. *Sepals* 5, united except at the top where they form 5 lobes. *Petals* fused distally, forming a cap in bud which drops as the flowers open. *Fruit* a berry, 6–22 mm, a bluish-black, green, yellow, red or purplish-black, ellipsoid to globose; seeds 0–4, subglobose to pyriform.

Contains about 65 species in the northern hemisphere.

Barty-King, H. (1997). *A tradition of English wine*. Oxford.
Vaughan J. G. & Geissler, C. (1997). *The new Oxford book of food plants*. Oxford.
Zohary, D. & Hopf, M. (2000). *Domestication of plants of the Old World*. Ed. 3. Oxford.

1. Leaves 9–30 cm, very shallowly 3-lobed; stem and leaves with rusty eglandular hairs **2. coignetiae**
1. Leaves 5–15 cm, palmately 5- to 7-lobed one-third of way to base; stem and leaves glabrous or with pale eglandular hairs. 2.
2. Dioecious, with dimorphic foliage, the leaves of male plants more deeply lobed; fruit acid tasting; seeds subglobose **1(a). vinifera** subsp. **sylvestris**
2. Flowers bisexual; fruit sweet; seeds pyriform **1(b). vinifera** subsp. **vinifera**

1. V. vinifera L. Grape Vine

Deciduous, monoecious or dioecious *woody vine* climbing to 20 m. *Stems* brown; bark flaking in strips; pith brown, interrupted by nodal diaphragms; climbing by leaf-opposed, branched tendrils often absent at every third node; young shoots pale green, glabrous, woolly or downy. *Leaves* alternate; lamina 5–15 × 5–15 cm, medium yellowish-green, paler beneath, dark green, greyish-green or claret-red in some variants, subrotund, broadly ovate or nearly reniform in outline, palmately divided one-third of the way to the base into 5 to 7 lobes, the lobes triangular and irregularly triangular-dentate, cordate at base, the sinus narrow and the lobes often overlapping, glabrous or becoming so on the upper surface, glabrous, eglandular-hairy becoming glabrous or persistently tomentose beneath; veins often paler on upper surface, prominent beneath; petiole up to 10 cm, pale green, glabrous or becoming so. *Inflorescence* a large, dense panicle which replaces the tendrils in the upper part of the stem. *Sepals* 5, united except at the top where they form 5 short lobes, the lobes triangular, rounded at apex. *Petals* 5, about 5 mm, pale green, cohering at the apex and falling at anthesis without separating. *Ovary* with a 5-lobed glandular disc present at base. *Stamens* 5; filaments pale; anthers yellow. *Style* 1, greenish; stigma capitate. *Berry* (grape) 6–22 mm, bluish-black, green, yellow, red or purplish-black, ellipsoid to globose; seeds subglobose to pyriform.

(a) Subsp. **sylvestris** (C. C. Gmel.) Hegi
V. sylvestris C. C. Gmel.

Dioecious, with dimorphic foliage, the leaves of male plants more deeply lobed. *Berry* about 6 mm, bluish-black, ellipsoid, acid-tasting; seeds usually 3, subglobose, with a short truncate beak.

(b) Subsp. **vinifera** Cultivated Vine
V. vinifera subsp. *sativa* Hegi

Flowers bisexual. *Berry* 6–22 mm, green, yellow, red or purplish-black, sweet; seeds 0–2, pyriform, with a rather long beak. $2n = 38$.

The cultivated vine was brought to Great Britain by the Romans and helped to supply wine to the occupying armies. Viticulture continued to prosper after the Romans left and is given specific mention by Bede in 731. Its decline was brought about after Henry II acquired a large slice of France, including the Bordeaux area, bringing in a flood of cheap wine. From Tudor times until after the Second World War vine growing was in the hands of the few. In 1946 Barrington Brock of Oxted in Surrey set out to discover which of the 2,000 varieties of grape would grow and ripen in the English climate, and from which a drinkable wine could be made. Major growing areas now are Suffolk, Kent, Essex, Sussex, Hampshire, Wiltshire and Berkshire. The vines of Europe are certainly derived, at least in part, by selection from subsp. *sylvestris*, which is native to a large part of central and south-east Europe, though the practice of cultivation probably originated in south-west Asia. The situation has been complicated in the last century by the introduction to Europe of many species of *Vitis* from North America. These were more or less resistant to the attacks of *Viteus vitifolii* ('phylloxera'), a parasitic aphid which did immense damage to European vines from 1867 onwards. Several cultivars with dark green, grey-green and claret-red foliage are grown for ornament and probably belong to subsp. *sylvestris*. The species is naturalised in hedges and scrub and by tips in the Channel Islands, southern and central England and south and west-central Wales.

2. V. coignetiae Pulliat ex Planch. Crimson-glory Vine

Deciduous *woody vine* climbing to 10 m or more. *Main stem* brown, rough and scaling; young shoots pale green, mottled and suffused brownish-purple, rather rough, with rusty-brown, cobwebby hairs; tendrils branched, pale green touched with purple, glabrous and smooth; pith interrupted by nodal diaphragms. *Leaves* alternate; lamina 9–30 × 9–30 cm, dull medium yellowish-green on upper surface, paler beneath, turning scarlet to carmine-red in autumn, ovate or nearly reniform in outline, shallowly 3-lobed, shortly acuminate at apex, crenate-dentate, cordate at base, the lobes overlapping, glabrous on upper surface, with a network of brown, cobwebby hairs beneath; veins impressed on upper surface, very prominent beneath; petiole fairly long, yellowish-green, suffused reddish, covered with cobwebby hairs. *Inflorescence* a large panicle up to 20×8 cm; flowers male on some plants, bisexual on others. *Sepals* 5, minute, united. *Petals* 5, cohering at the apex and falling at anthesis without separating. *Stamens* 5; filaments pale green; anthers orange. *Style* 1, short. *Berry* about 8 mm in diameter, black with a purple bloom, globose; seeds 2–4, pyriform. *Flowers* 6–7.

Introduced. Grown in gardens for ornament. A persistent escape or relic on a bombed site at Ludgate Hill, Middlesex, derelict gardens at Easton Lodge in Essex and on roadsides at Hextable in Kent. It also occurs in parks and on estates. Native of Japan and Korea.

2. **Parthenocissus** Planch.

Deciduous, monoecious, *woody climbers* with tendrils ending in adhesive discs. *Bark* not peeling in long shreds. *Leaves* simple, palmately lobed or palmate, lamina falling before the petiole. *Calyx* small, of 5 fused sepals forming a cup. *Petals* 5, greenish, free, the margins inturned and the tip hooded to form a narrow cup. *Stamens* 5, lying in the cup of the petal. *Style* 1; stigma 1. *Ovary* superior and 2-locular. *Disc* adnate to the base of the ovary and not visible as a distinct structure. *Fruit* a berry, black with a blue bloom, globose or obpyriform; seeds 1–4.

Contains 10 species from North America, east Asia and the Himalaya.

Bean, W. J. (1976). *Trees and shrubs hardy in the British Isles*. **3.** Ed. 8. London.

1. Leaves with (3–)5(–7) leaflets **1. quinquefolia**
1. Leaves simple or with 3 lobes or rarely with 3 leaflets **2. tricuspidata**

1. P. quinquefolia (L.) Planch. Virginia Creeper
Hedera quinquefolia L.; *Ampelopsis quinquefolia* (L.) Michx; *Cissus quinquefolia* (L.) Sol. ex Sims; *Vitis hederacea* Ehrh.; *Vitis quinquefolia* (L.) Lam.

Woody, monoecious, deciduous *climber* up to 30 m, trailing or ascending by numerous tendrils which have 5–8 branches, each developing an expanded, adhesive disc at its apex. *Stems* pale brown, rough and knobbly, glabrous; young shoots pale green tinged red, terete, glabrous. *Leaves* alternate, 3–10×8–12 cm, dull green on upper surface, paler and bluish beneath, turning bright red in autumn, broadly ovate in outline, palmate, the leaflets (3–)5(–7), 4–10×2–5 cm, oblong-obovate or elliptical, long acute or acuminate at apex, serrate, with coarse forwardly pointing teeth and cuneate at base with short, red petiolules, glabrous, the petioles pale green, turning red and glabrous or slightly eglandular-hairy. *Inflorescence* a compound thyrsoid panicle opposite the leaves; flowers about 6–8 mm in diameter; peduncles and pedicels pale green, turning red, glabrous. *Calyx* about 0.5 mm, of 5 fused sepals forming a cup, green with a red rim. *Petals* 5, 2.5–4.0 mm, pale green, oblanceolate, incurved on the sides, hooded at the apex, reflexed. *Stamens* 5, included within the cup of the petal; filaments whitish; anthers cream. *Style* 1, pale green; stigma slightly paler. *Berry* about 6 mm, turning red, then black with a bluish bloom, globose. *Flowers* 6–10. Visited by bees and wasps. The stigma does not become receptive until the petals and stamen have dropped off. $2n=40$.

Various varieties with different sized leaves, number of tendrils and hairiness have been described, but not studied in the area of our flora.

Introduced. Much planted and naturalised on houses, old walls and tips and in hedges and scrub. Scattered records in Great Britain, particularly round London, north to south-west Scotland and a few records in Ireland. Native of the eastern United States; formerly widely planted in Europe, but now largely replaced by *P. tricuspidata* and *Cissus verticillata*.

2. P. tricuspidata (Siebold & Zucc.) Planch. Boston Ivy
Ampelopsis tricuspidata Siebold & Zucc.; *Vitis inconstans* Miq.; *Vitis thunbergii* auct.

Woody, monoecious, deciduous *climber* to 20 m, trailing or ascending by tendrils, the tendrils short, well-branched, each branch developing an expanded adhesive disc at its apex. *Stems* pale brown, knobbly, glabrous. *Leaves* alternate, of 2 kinds; mostly small and palmate ones, glossy medium green on upper surface, much paler beneath, turning yellow, through orange to dark red in autumn; simple with lamina 2–9×1.5–7.5 cm, ovate, acute to acuminate at apex, shallowly sinuate-crenate, more or less cordate at base, the petioles up to 5 cm, pale green turning reddish and glabrous or with a few eglandular hairs; lobed ones with lamina 12–19×9.5–16 cm, obovate in outline, 3-lobed to less than halfway, the middle lobe slightly longer than the 2 laterals, the lobes ovate, acuminate at apex and crenate-serrate, the lower part of the leaf crenate-serrate with a truncate or subcordate base; veins impressed above, prominent beneath, petioles up to 22 cm, pale green turning red, with fairly numerous hairs; all glabrous on upper surface, with short eglandular hairs on the veins beneath. *Inflorescence* a compound, thyrsoid panicle opposite the leaves; flowers 8–10 mm in diameter; peduncles and pedicels red. *Calyx* about 3 mm, of 5 fused sepals forming a cup, green turning dark red. *Petals* 5, 3–4 mm, pale green, oblanceolate, incurved on the sides, hooded at the apex, reflexed. *Stamens* 5, included within the cup of the petal; filaments greenish; anthers pale brownish-cream. *Style* 1, greenish; stigma pink. *Berry* 5–6×6–7 mm, black with a bluish bloom, obpyriform with faint 5-lobing. *Flowers* 6–10. Visited by bees and wasps. $2n=40$.

Introduced. Much planted against houses and walls and naturalised on tips and in hedges and scrub. Scattered records in south-east England, Cornwall and Jersey extending north to Lancashire. Native of Japan, China and Korea.

3. **Cissus** L.

Deciduous, monoecious, *woody climbers* with tendrils sometimes swollen at ends but without adhesive discs.

Bark not peeling in long shreds. *Leaves* palmate. *Calyx* small, of 5 fused sepals. *Petals* 5, greenish with a cream margin, the margins inturned and hooded at apex, forming a narrow cup. *Style* 1; stigma 1. *Ovary* superior and 2-locular. *Disc* prominent in flower. *Fruit* a berry, black with a bluish bloom, ovoid to globose; seed 1.

Contains about 350 species in tropical and subtropical forests, particularly in South America, South Africa, south-east Asia and Australia.

Bean, W. J. (1976). *Trees and shrubs hardy in the British Isles*. **3**. Ed. 8. London. (As *Parthenocissus inserta.*)

Nicolson, D. H. & Jarvis, C. (1984). *Cissus verticillata* a new combination for *C. cicyoides* (Vitaceae). *Taxon* **33**: 726–727.

1. C. verticillata (L.) Jarvis — False Virginia Creeper

Viscum verticillatum L.; *C. sicyoides* L.; *Vitis sicyoides* (L.) Baker; *C. albonitens* André; *C. argenteus* Linden; *Parthenocissus inserta* (A. Kern.) Fritsch; *Ampelopsis inserta* A. Kern.; *Parthenocissus vitacea* (Knerr) Hitchc.; *Phoradendron verticillatum* (L.) Druce; *Phoradendron trinervium* (Lam.) Griseb.

Woody, monoecious, deciduous *climber* up to 30 m, trailing or ascending by numerous tendrils which have 3–5 branches not developing an adhesive disc at apex but clinging by coiling or by swelling of the apex inside a crevice. *Stems* pale yellowish-green, often suffused reddish-purple, glabrous. *Leaves* alternate; lamina 3–15 × 8–15 cm, dark green with paler veins on upper surface, paler and slightly bluish or rarely silvery-grey beneath, turning yellow to red in autumn, broadly ovate in outline, palmate, the leaflets 5, 3–15 × 3–5 cm, elliptical or obovate, acuminate at apex, serrate, cuneate or attenuate to a short red petiolule, glabrous or with a few eglandular hairs; petioles up to 15 cm, pale green turning reddish, glabrous. *Inflorescence* a compound, thyrsoid panicle opposite the leaves and conspicuously dichotomously branched; peduncles and pedicels pale green turning reddish, glabrous. *Calyx* 1.0–1.2 mm, of 5 fused sepals forming a cup, pale yellowish-green. *Petals* 5, 3.0–3.5 × 1.5–1.7 mm, pale green, with a narrow, cream margin on the outside, incurved on the margin and hooded at the apex, reflexed. *Stamens* 5, included within the cup of the petal; filaments white, tinted pink; anthers pale yellow. *Style* 1, cream; stigma paler. *Berry* 5–8 mm in diameter, shining bluish-grey, ovoid to globose. *Flowers* 6–10. Visited by bees and wasps. $2n = 40$.

Introduced. Planted and naturalised on houses and old walls, on tips and in hedges and scrub. In scattered localities in southern and central Great Britain and occasional records north to Morayshire. Native of north and west parts of the United States.

Order **11. LINALES** Cronquist

Annual to *perennial*, monoecious *herbs*. Leaves opposite or alternate, simple, entire, without stipules. *Inflorescence* a terminal cyme; flowers bisexual, hypogynous, actinomorphic. *Sepals* 4 or 5. *Petals* 4 or 5. *Stamens* 4, without staminodes, or 5 with staminodes. *Styles* 4 or 5. *Fruit* a loculicidal capsule.

Consists of 5 families and about 550 species.

105. LINACEAE Gray nom. conserv.

Annual to *perennial*, monoecious *herbs*. *Leaves* opposite or alternate, simple, entire, sessile, without stipules. *Inflorescence* a terminal cyme; flowers bisexual, hypogynous, actinomorphic. *Sepals* 4 or 5, free, usually imbricate, persistent. *Petals* 4 or 5, free, usually contorted, often shed early. *Stamens* 4, without staminodes, or 5, usually alternating with filiform staminodes; filaments more or less connate at base; anthers introrse. *Styles* 4 or 5, usually free; stigmas capitate. *Ovary* with 2 ovules per cell on axile placenta, or cells nearly halved by a false septum and thus more or less 8- or 10-celled with 1 ovule per cell. *Fruit* a loculicidal capsule, dehiscing also down the false septum and so with twice as many lines of dehiscence as carpels.

Contains about 12 genera and 290 species, both tropical and temperate.

1. Sepals, petals and stamens 5; sepals entire to minutely serrate at apex; capsule with 10 lines of dehiscence **1. Linum**
1. Sepals petals and stamens 4; sepals deeply 2- to 4-toothed at apex; capsule with 8 lines of dehiscence **2. Radiola**

1. Linum L.

Annual or *perennial*, monoecious *herbs*. Leaves opposite or alternate. *Inflorescence* a terminal cyme; flowers bisexual. *Sepals 5*, imbricate, entire or minutely serrate at apex. *Petals* 5, blue, red or white, clawed. *Stamens* 5, united into a tube at the base, with tooth-like staminodes between. *Styles* 5. *Ovary* 5-celled, ovules 2 in each cell, separated by a false septum. *Fruit* a capsule, with 10 lines of dehiscence; seeds flat.

Contains about 230 species, subtropical and temperate, mainly in the northern hemisphere.

Druce, G. C. (1918). *Linum catharticum L. forma dunense. Rep. Bot. Soc. Exch. Club Brit. Isles* 5: 17.

Grime, J. P. et al. (1988). *Comparative plant ecology*. London. [L. catharticum.]

Harvey, J. (1981). *Mediaeval gardens*. London.

Hultén, E. & Fries, M. (1986). *Atlas of north European vascular plants north of the Tropic of Cancer*. 3 vols. Königstein.

Lancaster, Mrs (1864). In Syme, J. T. B. (Edit.). *English Botany*, Ed. 3, vol. 2, pp. 180–189. London.

Ockendon, D. J. (1967). *Linum perenne* group. *Regnum Veg*. 74: 20–22.

Ockendon, D. J. (1968). Biological flora of the British Isles. No. 113. *Linum perenne* subsp. *anglicum* (Miller) Ockendon. *Jour. Ecol*. 56: 871–882.

Stewart, A., Pearman, D. A. & Preston, C. D. (1994). *Scarce plants in Britain*. Peterborough. [*L. perenne*.]

Thirsk, J. (1997). *Alternative agriculture*. Oxford.

1. Leaves, at least the lower, opposite; petals white, less than 7 mm 2.
1. Leaves alternate; petals more than 7 mm, usually red or blue, very rarely white 3.
2. Stems 5–30 cm, usually branched only in the inflorescence; leaves 5–12 × 2–3 mm, oblong or oblong-elliptical **6(i). catharticum** var. **catharticum**
2. Stems up to 6 cm, densely branched from the base; leaves 2–5 × 2–3 mm, elliptical to broadly elliptical **6(ii). catharticum** var. **dunense**

3. Flowers dark red **5. grandiflorum**
3. Flowers blue, very rarely white 4.
4. Stigma capitate; usually heterostylous 5.
4. Stigmas linear or clavate; homostylous 6.
5. Pedicels erect, more or less straight **3. perenne** subsp. **anglicum**
5. Pedicels deflexed or flexuous **4. austriacum**
6. Usually biennial or perennial; petals 8–12 mm, pale blue; capsule 5–6 mm **1. bienne**
6. Annual; petals 10–15 mm, blue; capsules 6–9 mm 7.
7. Stems solitary, branched only in upper part; leaves linear-lanceolate, seeds dark **2(i). usitatissimum** var. **usitatissimum**
7. Stems often several, often branched to base; leaves linear; seeds paler **2(ii). usitatissimum** var. **crepitans**

Section **1. Linum**

Leaves alternate. *Petals* large, blue, red or rarely white.

1. **L. bienne** Mill. Pale Flax
L. angustifolium Huds.

Perennial, sometimes *annual* or *biennial*, monoecious herb with a tough, woody rootstock. *Stems* several, (5–)15–60 cm, pale green or slightly glaucous, ascending or suberect, rather rigid, often flexuous, usually more or less branched, sometimes from the base, sometimes simple, obscurely longitudinally ridged, glabrous. *Leaves* numerous, alternate; lamina 2–25 × 0.5–4.0 mm, medium green on upper surface, slightly paler beneath, oblong or linear, acute or acuminate at apex, entire, glabrous, with 3 parallel veins, sessile. *Inflorescence* an irregular, lax, branched, leafy cyme; pedicels 10–30 mm, slender, more or less erect, glabrous; bracts like leaves but smaller; flowers bisexual. *Sepals* 5, 5–6 mm, green, the inner with a scarious margin, ovate, acuminate at apex, concave, glabrous and with a keeled midrib. Petals 5, 8–12 mm, pale blue, obovate, with a short claw. Stamens 5, united at base; filaments pale, slender; anthers yellow. Styles 5; stigma clavate. *Capsule* 5–6 mm, pale brown, subglobose; seeds 2.5–3.0 mm, medium brown, ovoid, dehiscing by 10 lines, not beaked. *Flowers* 5–9. Homostylous. $2n = 30$.

Native. Dry grassy places, especially near the sea and on path and field boundaries inland. Local in England, Isle of Man and east and south Ireland and mostly coastal in Wales; introduced in central England and north and central Ireland. South Europe extending further north only in the west to west and central France; Madeira and Canary Islands. Mediterranean-Atlantic element.

2. L. usitatissimum L. Flax

Annual monoecious herb with a whitish, tapering root and a few, lateral, fibrous ones. *Stems* single or several, 30–90 cm, slightly glaucous or greyish-green, fairly robust, erect, glabrous, simple or branched, leafy. *Leaves* alternate, numerous; lamina 1.0–4.0 × 0.1–0.5 cm, medium greyish- or bluish-green on upper surface, paler beneath, linear, acute at apex, entire, 3-veined, glabrous, sessile. *Inflorescence* a lax cyme; bracts like leaves; pedicels 10–20 mm, slender; flowers bisexual, 11–22 mm, without odour. *Sepals* 5, 4–9 mm, imbricate, ovate, acuminate at apex, with a conspicuous pale midrib, the inner with a scarious, serrulate margin and sometimes glandular-ciliate. *Petals* 5, 10–15 × 7–11 mm, blue with darker veins, obovate, with a short claw. *Stamens* 5, united into a tube at the base, with tooth-like staminodes between; filaments white tinted blue; anthers blue. *Styles* 5; stigmas linear. *Capsule* 6–9 mm, subglobose, beak about 1 mm; seeds 4.0–4.5 mm, ovate, with a short, obtuse beak. *Flowers* 5–9. Visited by flies. Homostylous. $2n = 30$.

(i) Var. **usitatissimum** Flax
Stems solitary, up to 90 cm, branched only in the upper part. *Leaves* linear-lanceolate. *Capsule* indehiscent, with more or less glabrous septa; seeds dark.

(ii) Var. **crepitans** Boenn. Linseed
L. crepitans (Boenn.) Dumort.; *L. humile* Mill.; *L. usitatissimum* var. *humile* (Mill.) Pers.

Stems often several, up to 40 cm, often branched to base. *Leaves* linear. *Capsule* subdehiscing with ciliolate septa; seeds paler.

Introduced. The countries of the Mediterranean were the first to use the fibres of flax, var. *usitatissimum*, for linen, and from there its use for this purpose spread throughout Europe. The first use of the plant for its fibre in England does not seem to have occurred until after the Norman Conquest, in 1240. In 1531 a law was passed to compel one rood in every 60 acres of arable land to be sown with flax or hemp. In Ireland also there was much encouragement to grow it. In the nineteenth century it was grown in great quantity in Lincolnshire and the eastern counties, but it almost died out as a crop in the twentieth century. As well as being grown for fibre, a different plant, var. *crepitans*, is grown for its seed, from which is extracted linseed oil. The 'cake' left over when the oil has been extracted is used to feed cattle, and has also been used as a manure. There was a return to growing linseed as an alternative crop towards the end of the twentieth century, but now seems to have almost stopped. When these crops are harvested, seed is dropped on tracks and roadsides, but the resulting plants are mostly casual. It is also a frequent bird-seed casual. Recorded throughout much of Great Britain and Ireland. Not known as a wild plant, but perhaps derived from *L. bienne*.

3. L. perenne L. Perennial Flax

Perennial monoecious *herb*. *Stems* several, up to 60 cm, pale bluish-green, decumbent or ascending, rather rigid, often curved, glabrous, more or less branched, leafy. *Leaves* alternate, numerous; lamina 10–20 × 1.0–3.5 mm, bluish-green above, paler beneath, linear, acute at apex, entire, 1-veined, glabrous, sessile. *Inflorescence* a lax cyme; pedicels 10–20 mm, slender, erect, more or less straight, glabrous; bracts like leaves but smaller; flowers bisexual, 23–25 mm in diameter. *Sepals* 5, 3.5–6.5 mm, green, the inner with a narrow pale margin, elliptical to ovate, the inner rounded, the outer mucronate at apex. *Petals* 5, fugacious, 10–20 mm, blue, obovate, rounded at apex, with a short claw. *Stamens* 5, united into a tube at base; filaments blue; anthers cream; pollen grains with pores. *Styles* 5, blue; stigma capitate, cream. *Capsule* 5.5–7.5 mm, globose or ellipsoid; seeds 4–5 mm, matt, oblong-ovate, obscurely beaked. *Flowers* 6–7. Heterostylous. 2n = 36.

Native. Base-rich grassland over chalk, limestone or calcareous sand in open, sunny, well-drained habitats on road verges, dry banks and lightly grazed grassland. Very local in eastern and northern England, with two large populations, one in Cambridgeshire and one in Co. Durham; and another in Kirkcudbrightshire in Scotland. Our plant is the endemic subsp. **anglicum** (Mill.) Ockendon (*L. anglicum* Mill.). Other subspecies occur in east and central France southwards to central Spain, northwards to north Germany and eastwards to the Urals. European Temperate element. Subsp. **montanum** (DC.) Ockendon (*L. montanum* DC.), which is a native of the Alps and Jura, has been recorded as a casual in Kent and Berkshire. It differs from subsp. *anglicum* in having acute inner sepals.

4. L. austriacum L. Austrian Flax

Perennial monoecious *herb* with a tap-root and fibrous side-roots. *Stems* (6–)10–60 cm, several to one plant, pale green, erect, glabrous or minutely hairy above, simple or branched, leafy. *Leaves* alternate, numerous, or in whorls close together; lamina 10–32 × 2–3 mm, rather greyish medium green on upper surface, paler beneath, linear, rounded at apex, entire, dotted white, sessile, obscurely 1–3 veined, glabrous. *Inflorescence* a many- to 1-flowerd cyme; pedicels 10–20 mm, deflexed or flexuous, glabrous; flowers bisexual, 28–30 mm in diameter. *Sepals* 5, 3.4–5.5 mm, green with a narrow, pale margin, ovate, outer acute at apex, the inner obtuse, glabrous. *Petals* 5, 14–20 × 14–15 mm, pale blue, obovate, broadly rounded at apex. *Stamens* 5; filaments blue; anthers cream, with tooth-like staminodes in between. *Styles* 5; stigmas capitate. *Capsule* 3.5–5.0 mm, subglobose, the beak very short or absent. *Flowers* 6–7. Heterostylous. $2n = 18$.

Introduced. Casual or persistent garden escape; possibly also introduced with grass and clover seed mixtures. Native of central and south Europe.

5. L. grandiflorum Desf. Crimson Flax

Annual monoecious *herb* with fibrous roots. *Stems* up to 45 cm, pale yellowish-green, much branched, the branches ascending, glabrous. *Leaves* alternate; lamina 15–30 × 3–8 mm, slightly bluish medium green on upper surface, only slightly paler beneath, linear to lanceolate, acute at apex, entire, rounded at base, sessile or nearly so, glabrous. *Inflorescence* of 1 or 2 bisexual flowers at the ends of branches, up to 35 mm in diameter and scentless; pedicels up to 30 mm, slender but rigid, glabrous. *Sepals* 5, 10–15 mm, pale green at base, dark green towards apex and with a narrow pale margin, lanceolate, sharply acute at apex, glabrous except for a ciliate margin. *Petals* 5, 18–25 × 13–15 mm, dark red with an even darker area at base, obovate, rounded at apex. *Stamens* 5; filaments red; anthers black. *Styles* 5, red. *Capsule* 6–7 mm, ovoid, dehiscing along 10 lines. *Flowers* 7–9. $2n = 16$.

Introduced. Bird-seed casual and garden escape. Native of North Africa.

Section **2. Cathartolinum** (Rchb.) Griseb.

Leaves opposite at least below. *Petals* small, white.

6. L. catharticum L. Fairy Flax

Annual monoecious *herb* with a slender tap-root and fibrous roots. *Stems* usually solitary, sometimes several, 5–30 cm, pale green, erect, wiry and usually simple except for the inflorescence. *Leaves* distant, opposite; lamina 5–12 mm, medium green on upper surface, paler beneath, oblong or obovate, subobtuse at apex, entire, 1-veined, sessile, glabrous. *Inflorescence* a lax, often dichasial cyme; pedicels 5–10 mm, slender, glabrous; bracts partly alternate, the upper small and linear, the lower passing into the leaves; flowers numerous, bisexual, 5–6 mm in diameter. *Sepals* 5, 2–3 mm, lanceolate or ovate-lanceolate, acuminate at apex, glandular-ciliate. *Petals* 5, 4–6 mm, white, narrowly obovate, entire or shallowly emarginate, with a yellow claw. *Stamens* 5, connate at base, alternating with 5 staminodes; filaments pale; anthers yellow. *Styles* 5; stigmas capitate. *Capsule* 2–3 mm, subglobose; seeds 1.2–1.4 mm, compressed. *Flowers* 6–9. Pollinated by various insects or selfed. Homogamous. $2n = 16$.

(i) Var. **catharticum**

Stems 5–30 cm, usually solitary, usually branched only in the inflorescence. *Leaves* 5–12 × 2–3 mm, oblong or oblong-elliptical.

(ii) Var. **dunense** (Druce) Druce

L. catharticum forma *dunense* Druce

Stems up to 6 cm, densely branched from the base. *Leaves* 2–5 × 2–3 mm, elliptical to broadly elliptical.

Native. Grassland, heaths, moors, rock-ledges and dunes up to 855 m, especially common on calcareous grassland, but not confined to basic soils. Frequent throughout Great Britain and Ireland. Europe northwards to 69° N in Fennoscandia and mainly on mountains in the south; Caucasus; Turkey; Iran. European Temperate element. Our most widespread plant is var. *catharticum*. Var. *dunense* occurs on coastal sands and in the Breckland of East Anglia and is probably all round our coasts. The whole of the plant is cathartic and was formerly much used in medicine and called Purging Flax.

2. Radiola Hill

Millegrana Adans.

Annual monoecious *herbs*. *Leaves* opposite. *Inflorescence* a terminal cyme; flowers bisexual. *Sepals* 4, deeply 2-to 4-toothed at apex. *Petals* 4, white, about as long as sepals. *Stamens* 4. *Styles* 4. *Ovary* 4-celled, ovules 2 in each cell, separated by a false septum. *Fruit* a capsule with 8 lines of dehiscence.

Contains 1 species as below.

Byfield, A. & Pearman, D. (1996). *Dorset's disappearing heathland flora*. London.
Hultén, E. & Fries, M. (1986). *Atlas of north European vascular plants north of the Tropic of Cancer.* 3. vols. Königstein.

1. R. linoides Roth Allseed
R. millegrana Sm.; *Millegrana radiola* (L.) Druce; *Linum radiola* L.

Annual monoecious *herb* with fibrous roots. *Stems* 1.5–10.0 cm, pale green, filiform, glabrous, simple or dichotomously branched so that the habit is bushy, leafy. *Leaves* opposite; lamina 1–3 mm, medium green on upper surface, paler beneath, ovate-elliptical or elliptical, subobtuse to acute at apex, entire, 1-veined, glabrous, sessile. *Inflorescence* a dichasial cyme; bracts similar to leaves; pedicels short; flowers bisexual. *Sepals* 4, about 1 mm, (2–)3(–4)-lobed or toothed at apex. *Petals* 4, about 1 mm, white, obovate. *Stamens* 4. *Styles* 4. *Capsule* 0.7–1.0 mm, globose; seeds ovoid. *Flowers* 7–8. Probably always self-pollinated. $2n=18$.

Native. Seasonally damp, bare, peaty or sandy acid ground in open places or in woodland rides. Scattered over much of Great Britain and Ireland, but mostly near the coast. Most of Europe but not the north-east and extreme north; North Africa; Madeira; Tenerife; temperate Asia; mountains of tropical Africa. European Temperate element.

Order 12. POLYGALALES Benth. & Hook. fil.

Perennial monoecious *herbs,* often woody at base. *Leaves* opposite or alternate; stipules small or absent. *Flowers* bisexual, hypogynous to subperigynous, zygomorphic. *Sepals* 5. *Petals* 5. *Stamens* 8. *Ovary* syncarpous; placentation axile or apical. *Fruit* a capsule.

Consists of 7 families and about 2,300 species.

106. POLYGALACEAE R. Br. nom. conserv.

Perennial monoecious *herbs*, often woody at base. *Leaves* opposite or alternate, simple, entire, sessile or shortly petiolate, without stipules. *Inflorescence* usually a terminal raceme; flowers bisexual, hypogynous, zygomorphic. *Sepals* 5, the 2 inner much larger than the 3 outer. *Petals* 5, the 2 outer free or united with lower, the 2 upper free, or minute or absent. *Stamens* 8, monadelphous for more than half their length, rarely free, the tube split above and often adnate to the petals; anthers usually opening by an apical pore. *Style* 1, simple. *Ovary* superior, 2-celled, with 1 apical ovule per cell. *Fruit* a 2-seeded capsule.

Contains about 12 genera and 800 species throughout the world except New Zealand, Polynesia and the Arctic.

1. Polygala L.

As family.

Some 500–600 species throughout the world except New Zealand, Polynesia and the Arctic.

The varieties under *P. serpyllifolia* and *P. vulgaris* may have some ecological significance although intermediates and difficult plants to place can occur. We have found it impossible to work out the correct nomenclature for these varieties without seeing the types. Self-pollination may help to keep them apart.

Babington, C. C. (1853). Remarks on British plants. *Ann. Mag. Nat. Hist.* ser. 2, **11**: 265–273.
Bennett, A. W. (1877). Review of the British species and subspecies of *Polygala*. *Jour. Bot. (London)* **15**: 168–174.
Davy, F. H. (1906). New variety of *Polygala serpyllacea*. *Jour. Bot. (London)* **44**: 34–35.
Druce, G. C. (1912). *Polygala babingtonii*, mihi. *Rep. Bot. Soc. Exch. Club Brit. Isles* **3**: 12.
Du Mortier, B. (1868). Bouquet de littoral Belge (*Polygala* on pp. 335–345). *Bull. Soc. Roy. Bot. Belg.* **7**: 318–371.
Fearn, G. M. (1974). Variation of *Polygala amarella* Crantz in Britain. *Watsonia* **10**: 371–383.
Glendinning, D. R. (1954). The British *Polygala* species. *Proc. Bot. Soc. Exch. Club Brit. Isles* **1**: 259–260.
Huebl, G. R. (1984). Systematische Untersuchungen an Mitteleuropaischen *Polygala*-arten. *Mitt. Bot. München* **20**: 205–428.
Hultén, E. & Fries, M. (1986). *Atlas of north European vascular plants north of the Tropic of Cancer.* 3 vols. Königstein.
McNeill, J. (1968). Taxonomic and nomenclatural notes on *Polygala* in Europe. *Feddes Repert.* **79**: 23–34.
Salmon, C. E. (1915). *Polygala dunensis*. *Jour. Bot. (London)* **53**: 279.
Salmon, C. E. (1929). A new variety of *Polygala serpyllifolia* J. A. C. Hose (*serpyllacea* Weihe). *Jour. Bot. (London)* **67**: 193–194.
Stewart, A., Pearman, D. A. & Preston, C. D. (1994). *Scarce plants in Britain*. Peterborough. [*P. calcarea*.]
Yeo, P. F. (1952). A possible hybrid between *Polygala vulgaris* L. and *P. calcarea* F. Schultz. *Year Book B. S. B. I.* **1952**: 36.

1. Leaves near the base of stems larger than those above, more or less obtuse, congested into a rosette; inner sepals with veins not anastomosing or sparingly so and not around the edges 2.
1. Leaves near the base of the stems smaller than those above, more or less acute, not congested into a rosette; inner sepals with veins anastomosing around the edges 4.
2. Stems with more or less leafless portion below the leaf-rosette; flowers 6–7 mm **3. calcarea**
2. Stems with leaf-rosettes at or very near the base; flowers 2–5 mm 3.
3. Stems up to 4(–8) cm; leaves 10–20 × 2–7 mm, more or less dense; flowers blue or pink **4(i). amarella** var. **amarella**
3. Stems 6–11(–16) cm; leaves up to 25 × 2–7 mm, more or less spaced; flowers blue or greyish **4(ii). amarella** var. **austriaca**
4. Lower stem leaves (sometimes fall early, but scars are left) more or less opposite; 3 outer sepals acute at apex 5.
4. All leaves alternate; 3 outer sepals obtuse at apex 7.
5. Majority of leaves including stem leaves more or less opposite **2(i). serpyllifolia** var. **vincoides**
5 Only the lower leaves more or less opposite, middle and upper leaves obviously alternate 6.
6. Leaves 3–15 × 2–4 mm, elliptical or lanceolate **2(ii). serpyllifolia** var. **serpyllifolia**

6. Leaves 12–22×4–7 mm, oblong-elliptical **2(iii). serpyllifolia** var. **decora**
7. Stems erect 8.
7. Stems spreading or prostrate 9.
8. Leaves 1–4 mm wide; flowers 5–7 mm **1(iii). vulgaris** var. **vulgaris**
8. Leaves 5–10 mm wide; flowers 7–10 mm **1(v). vulgaris** var. **ballii**
9. Leaves very dense especially towards the base **1(i). vulgaris** var. **caespitosa**
9. Leaves not dense towards the base 10.
10. Leaves 1–4 mm wide **1(ii). vulgaris** var. **dunensis**
10. Leaves 3–7 mm wide **1(iv). vulgaris** var. **intermedia**

1. P. vulgaris L. Common Milkwort

Perennial, monoecious *herb* with a woody stock. *Stems* 7–35 cm, pale green, ascending to erect, glabrous or sparsely hairy, branched, leafy. *Leaves* alternate, not bitter, the lamina medium green on upper surface, paler beneath, the lower 6–25×1–10 mm, obovate to elliptical, and more or less acute, entire, the upper longer, up to linear-lanceolate, acute at apex and entire. *Inflorescence* a dense terminal raceme with 10–40 bisexual flowers, conical at first, elongating in fruit; bracts membranous except for the midrib, scarcely exceeding pedicels in flower and shorter than flower-buds, caducous, the flowers blue, pink or white. *Sepals* 5, unequal, the 2 inner (wings) 4.0–8.5×2–5 mm, with 3 anastomosing veins, ovate and apiculate at apex, the 3 outer about 3 mm, green with coloured margins, obtuse at apex. *Corolla* 4–10 mm; tube usually longer than upper lobes, 3-lobed, the lower (keel) of different form from the 2 upper and bearing a fimbriate crest. *Stamens* 8; filaments partly united into a tube which is partly adnate to the corolla tube. *Style* usually 1.0–1.5 times as long as stigma; stigma 2-lobed, only the posterior lobe receptive. *Capsule* 4.0–8.5 mm, ovate, compressed, with a marginal wing and emarginate at apex; seeds 2.5–3.0 mm, oblong-ellipsoid, lobes of strophiole about one-third as long as seed. *Flowers* 5–9. $2n=56$, 58.

(i) Var. **caespitosa** Pers.

Stems up to 15 cm, spreading. *Leaves* dense especially at base, 6–15×1–4 mm, linear or linear-lanceolate. *Flowers* 4–7 mm.

(ii) Var. **dunensis** (Dumort.) Buchenau
P. dunensis Dumort.; *P. vulgaris* var. *scotica* F. N. Williams

Usually with numerous, spreading stems. *Leaves* 10–15×1–4 mm, linear to narrowly linear-lanceolate, numerous but not dense at base. *Flowers* 5–6 mm.

(iii) Var. **vulgaris**

Stems erect, up to 30 cm. *Leaves* 10–25×1–4(–5) mm, linear to linear-lanceolate, usually widely spaced, not dense. *Flowers* 5–7 mm.

(iv) Var. **intermedia** Chodat

Stems spreading. *Leaves* 15–25×3–7 mm, narrowly elliptical, numerous and rather dense. *Flowers* 5–7 mm.

(v) Var. **ballii** Nyman ex A. Benn.
P. vulgaris var. *grandiflora* sensu Bab., non DC.; *P. buxifolia* Ball nom. nud.; *P. ballii* Nyman nom. in syn.; *P. babingtonii* Druce; *P. grandiflora* Druce

Stems erect, up to 30 cm. *Leaves* 10–35×5–10 mm, lanceolate or narrowly elliptical, usually fairly dense. *Flowers* large, 7–10 mm.

Native. Usually found in short, moderately infertile neutral to basic grassland on banks, hill-slopes and sand dunes. Throughout Great Britain and Ireland. Most of Europe, west Asia and North Africa. European Temperate element. Var. *caespitosa* is a plant of grassland. Var. *intermedia* is also of grassland, as well as coastal. Var. *dunensis* is a plant of the coast particularly of sand-dunes. Var. *vulgaris* is the most widespread variant on grassland, fens and river-banks; often on chalk or limestone. Var. *ballii* is known only from Ben Bulben in Co. Sligo. All our plants are referable to subsp. **vulgaris**.

2. P. serpyllifolia Hosé Heath Milkwort
P. serpyllacea Weihe; *P. depressa* Wender.

Perennial monoecious *herb* not woody at base. *Stems* 6–25 cm, pale green, slender, decumbent to ascending, glabrous. *Leaves* alternate to more or less opposite; lamina 3–22×2–7 mm, the lower elliptical to obovate or rarely subrotund, obtuse to subacute at apex, entire and narrowed below, the upper lanceolate to linear-lanceolate or rarely broader to subrotund, obtuse to acute at apex, entire, alternate to more or less opposite. *Inflorescence* a terminal or pseudolateral raceme, with 3–10 flowers; bracts shorter than pedicels at anthesis. *Sepals* 5, the outer 1.5–2.5 mm, acute at apex, the 2 inner 4.5–5.5 mm, oblanceolate to elliptical; veins anastomosing. *Corolla*, usually blue, the 2 upper lobes usually longer than the lower, the crest of the lower lobe 10–to 25–lobulate. *Stamens* 8. *Styles* 1. *Capsule* 4–5 mm, shorter and wider than the upper petals; seeds about 2–5 mm, ovoid; lateral lobes of stophiole about one-third as long as seed. *Flowers* 5–8. $2n=32$, 34, about 68.

(i) Var. **vincoides** (Chodat ex Davy) P. D. Sell
P. serpyllacea var. *vincoides* Chodat ex Davy

Leaves 8–10×3.5–6.0 mm, those of upper stem more or less opposite, elliptical, shortly pointed.

Specimens of dwarf plants with rounded leaves from Ben Lawes in Perthshire and Ashill, Mayo will probably key out here but appear to be a distinct infraspecific taxon.

(ii) Var. **serpyllifolia**

Leaves 5–15×2–4 mm, those of upper stem obviously alternate, elliptical or lanceolate, rounded to pointed.

(iii) Var. **decora** C. E. Salmon

Leaves 12–22×4–7 mm, those of upper stem obviously alternate, oblong-elliptical, pointed at apex. *Flowers* usually larger than the other two varieties.

Native. Acidic grassland, moors, heaths and mires up to 1035 m in Perthshire. Throughout Great Britain and Ireland, but declining in southern England. West

Var. **caespitosa** Pers.

Var. **dunensis** (Dumort.) Buchenau

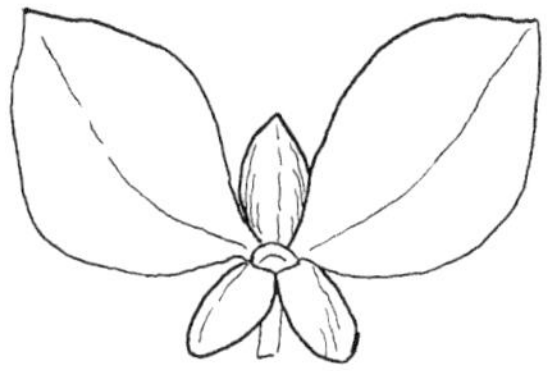

(Inner and outer sepals)

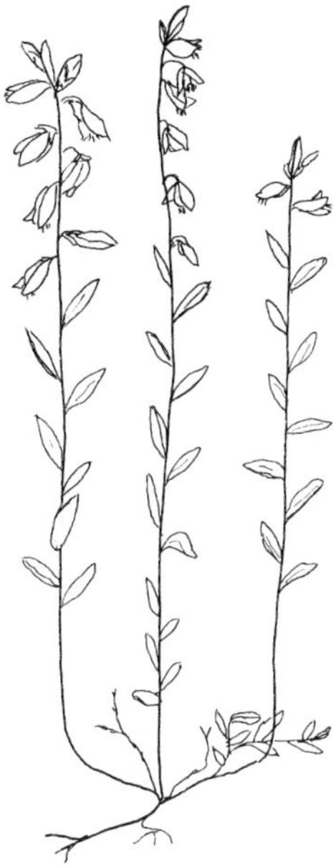

Var. **vulgaris**

Var. **intermedia** Chodat

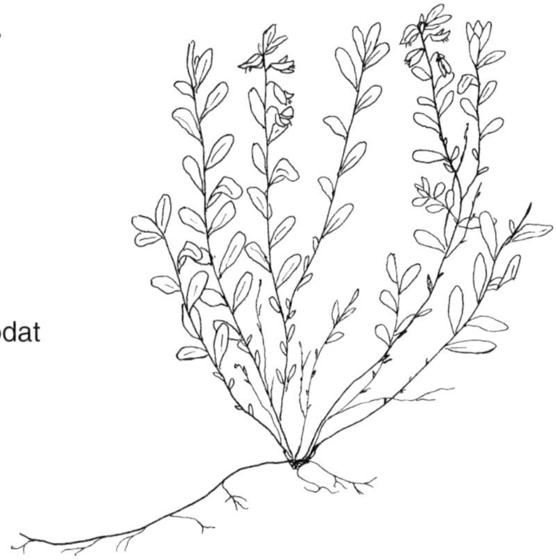

Var. **ballii** Nyman ex A. Benn.

Polygala vulgaris L.

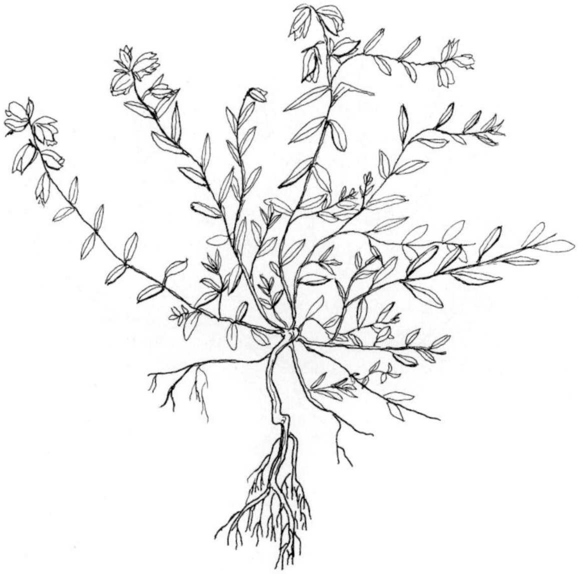

Var. **vincoides** (Chodat ex Davy) P. D. Sell

Achill, Mayo plant

(Inner and outer sepals)

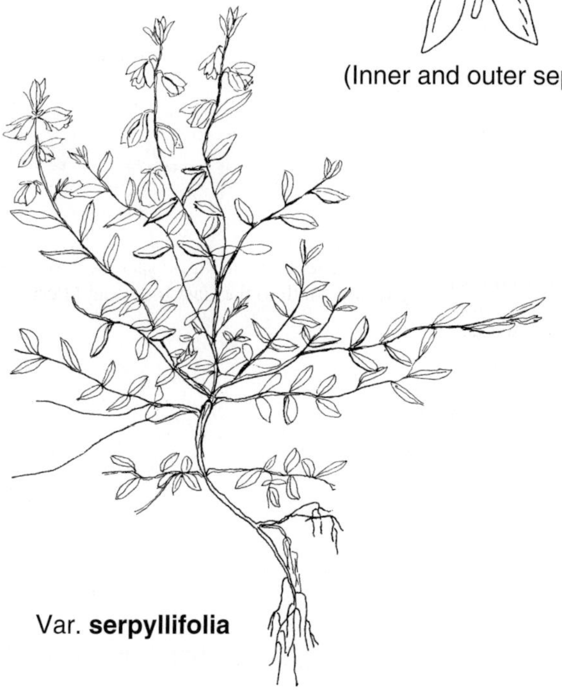

Var. **serpyllifolia**

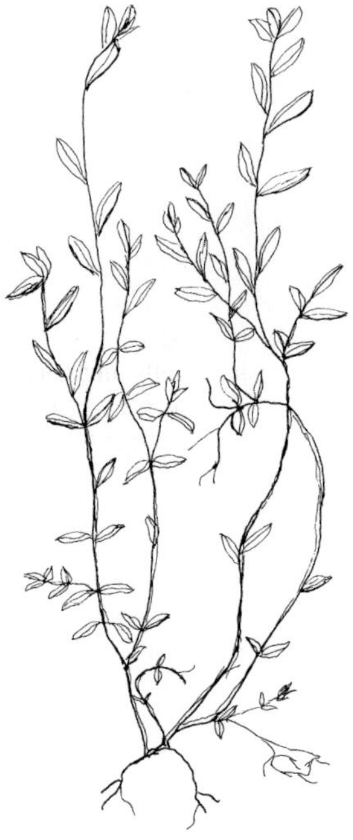

Var. **decora** Salmon

Polygala serpyllifolia Hosé

and central Europe; Greenland. Suboceanic Temperate element. Var. *serpyllifolia* is the widespread plant. Var. *vincoides* occurs in Cornwall and Staffordshire. Var. *decora* is recorded from mountainous areas in all three countries of Great Britain and in Ireland.

3. P. calcarea F. W. Schultz Chalk Milkwort

Perennial monoecious *herb* with a somewhat woody base. *Stems* with decumbent, usually leafless stolons terminating in leaf-rosettes, from which arise a number of almost erect flowering stems 10–20 cm, pale green and glabrous; non-flowering shoots also present, arising from the stock or from the rosette. *Leaves* medium green on upper surface, paler beneath; those of rosette with lamina 1–2 × 0.5–1.0 cm, spathulate to obovate, rounded at apex, entire, narrowed at base to a short petiole; cauline much smaller, linear-lanceolate, obtuse at apex, entire, narrowed at base, sessile or nearly so; all glabrous or nearly so. *Inflorescence* a raceme of 6–20 flowers; bracts linear-lanceolate. *Sepals* 5, the 2 inner about 5 mm, obovate to oblong-elliptical, the veins not anastomosing. *Corolla* 6–7 mm, blue or white, exceeding the inner sepals. *Stamens* 8. *Style* 1. *Capsule* 4–6 mm; seeds ovoid-oblong; lateral lobes of strophiole about half as long as seed. *Flowers* 5–6. $2n=34$.

Native. Grazed chalk and limestone grassland, especially on warm, south-facing slopes, disappearing when the coarser grasses become dominant. Southern England and south Lincolnshire-Leicestershire oolite. Western Europe. Oceanic Southern-temperate element.

× **vulgaris**

This hybrid is intermediate in characters and sterile. Native. It has been recorded for a few localities in southern England. Endemic.

4. P. amarella Crantz Dwarf Milkwort

P. amara auct.

Perennial monoecious *herb* up to 11(–16) cm, woody at base, bitter-tasting. *Stems* pale green, erect or ascending, rather stout, usually unbranched, sometimes very short so that the inflorescence is condensed and subsessile. *Leaves* medium green on upper surface, paler beneath, all alternate; basal with lamina 5–25 × 2–7 mm, obovate, obtuse at apex, entire, narrowed at base, forming a rosette at base; cauline smaller, narrowly obovate to narrowly lanceolate or linear, acute or obtuse at apex, entire, narrowed below; all glabrous or nearly so. *Inflorescence* a raceme of many flowers; bracts linear-lanceolate. *Sepals* 5, the 2 inner 2–5 mm in fruit, oblong to obovate, about half as wide as capsule, their veins sparingly branched, not anastomosing. *Corolla* 2–5 mm, usually pink, greyish-white or blue. *Stamens* 8. *Style* 1. *Capsule* 3–4 mm; seeds 1.5–2.3 mm, oblong. *Flowers* 6–8. $2n=34$.

(i) Var. **amarella**

P. amara auct.; *P. uliginosa* Rchb.

Stems up to 4(–8) cm. *Leaves* 10–20 × 2–7 mm, obovate or oblanceolate, obtuse to acute at apex, more or less dense. *Flowers* blue or pink.

(ii) Var. **austriaca** (Crantz) Beck

P. austriaca Crantz

Stems 6–11(–16) cm. *Leaves* up to 25 × 2–7 mm, linear to oblanceolate, more or less obtuse at apex, more or less spaced. *Flowers* blue or greyish-white.

Native. Var. *amarella* occurs on damp mountain pasture on limestone in Yorkshire and Durham. Var. *austriaca* occurs on chalk grassland in Kent and Surrey. The species occurs in much of Europe (except areas of the south) where there is even more variation. European Boreo-temperate element.

× **vulgaris**

P. × skrivanckii auct.

This hybrid has large lower leaves which have a little taste as in *P. amarella*, but the plant has much larger leaves and an intermediate corolla size and is partially fertile.

Native. It occurs in Kent with both parents. It is also recorded for Czechoslovakia.

Order **13. SAPINDALES** Benth. & Hook. fil.

Deciduous *trees* or *shrubs*. *Leaves* alternate or opposite, usually pinnate and without stipules. *Flowers* unisexual or bisexual, hypogynous to slightly perigynous, sometimes zygomorphic. *Sepals* 4–5. *Petals* usually 4–5, rarely absent. *Stamens* often twice as many as petals. *Disc* present. *Ovary* usually syncarpous with 1–2 ovules in each cell; placentation axile. *Fruit* various; seeds usually without endosperm; embryo curved or bent.

Contains 15 families and about 5,400 species.

107. STAPHYLEACEAE Lindl. nom. conserv.

Deciduous, monoecious *shrubs* or small *trees*. *Leaves* opposite, imparipinnate or ternate, petiolate, with stipels and stipules when young. *Inflorescence* a pendulous terminal panicle; flowers bisexual, hypogynous and actinomorphic. *Sepals* 5, free. *Petals* 5, free. *Stamens* 5. *Styles* 2–3; stigmas capitate. *Ovary* 2- to 3-celled with carpels free distally; ovules numerous on axile placentas. *Fruit* a much inflated, 2- to 3-celled capsule with many seeds.

Five genera and about 60 species in the north temperate zone and south to the East Indies and Peru.

1. Staphylea L.

As family.

Contains about 10 species in the north temperate zone.

Bean, W. J. (1980). *Trees and shrubs hardy in the British Isles*. Ed. 8. **4**. London.

Rehder, A. (1940). *Manual of cultivated trees and shrubs hardy in North America*. Ed. 2. New York.

1. Sepals 5–6 mm; petals 7–8 mm **1. pinnata**
1. Sepals 12–14 mm; petals 14–16 mm **2. colchica**

1. S. pinnata L. Bladdernut

Deciduous, monoecious *shrub* or small *tree* up to 6 m. *Stems* many from one base, erect and arching. *Bark* greyish-brown, longitudinally striate. *Branches* ascending and arching;

twigs pale brown; young shoots pale greenish-brown, glabrous. *Buds* 5–10 mm, narrowly ovoid, pointed; scales green, tinged reddish, ovate, acute at apex. *Leaves* opposite, imparipinnate or ternate, broadly ovate in outline; leaflets 3–7, the lamina 4–10 × 1.5–4.0 cm, medium yellowish-green on upper surface, paler beneath, lanceolate or ovate, acuminate at apex, entire, rounded at the sessile base with tiny stipels, glabrous; veins 6–7 pairs, impressed above, prominent beneath; petiole up to 5 cm, pale green, glabrous as in the rhachis; stipules present when young. *Inflorescence* a terminal pendulous panicle; flowers bisexual, 9–10 mm in diameter, sweet-scented, disc present; peduncle and pedicels glabrous; bracts linear, acute at apex. *Sepals* 5, 5–6 × 2–3 mm, very pale yellowish-green, ovate, rounded at apex, glabrous. *Petals* 5, 7–8 × 4.0–4.5 mm, white, broadly ovate, rounded at apex. *Stamens* 5; filaments white; anthers yellow. *Styles* 2–3, pale yellow; stigmas slightly darker. *Capsule* 25–30 mm, subglobose, much inflated, 2- to 3-lobed. *Flowers* 5–6. Visited by Diptera. $2n = 24, 26$.

Introduced. Planted in gardens, parks and amenity areas. Naturalised in hedges and on banks. Scattered records throughout much of Great Britain. Native of central Europe.

2. S. colchica Steven Caucasian Bladdernut

Deciduous, monoecious *shrub* or small *tree* up to 6 m. *Stems* many from 1 base, erect and straight. *Bark* greyish-brown, fairly smooth but longitudinally striate. *Branches* ascending; twigs greyish-brown, spreading; young shoots green, glabrous. *Buds* 8–10 mm, narrowly ovoid, pointed; scales greenish, ovate, acute at apex. *Leaves* opposite, ternate, broadly ovate in outline; leaflets 3, with lamina 4–11 × 1.5–5.5 cm, medium yellowish-green on upper surface, slightly paler beneath, lanceolate, ovate or elliptical, acuminate at apex, minutely serrate, cuneate or rounded at base, the terminal petiolulate, the lateral sessile, glabrous; veins 6–7 pairs, impressed above, prominent beneath; petiole up to 5 cm, green; stipels and stipules present when young. *Inflorescence* a pendulous terminal panicle; flowers bisexual, 20–25 mm in diameter, very sweet-smelling, disc present; peduncle and pedicels pale green and glabrous; bracts pale green, linear, acute at apex. *Sepals* 5, 12–14 × 4–6 mm, white, linear-lanceolate, obtuse at apex. *Petals* 5, 14–16 × 5–6 mm, white, oblong, rounded at apex. *Stamens* 5; filaments white; anthers yellow. *Styles* 2, pale green to white; stigma white. *Capsule* 4–8 mm, obovoid, 2- to 3-lobed, much inflated; seed about 8 mm. *Flowers* 4–6. $2n = 52$.

Introduced. Grown in gardens, parks, estates and amenity areas. Recorded as garden relic at Headington, Oxfordshire. Native of the Caucasus. This is perhaps the most common of the species replacing *S. pinnata* in gardens.

108. SAPINDACEAE Juss. nom. conserv.

Deciduous, monoecious *trees* or *shrubs*. *Leaves* alternate, imparipinnate to bipinnate, petiolate, without stipules. *Inflorescence* a large terminal panicle, flowers bisexual, hypogynous, zygomorphic. *Sepals* 5, unequal, fused proximally. *Petals* 4, free, all upturned, with basal appendages. *Stamens* 8, with hairy filaments. *Style* 1; stigmas 3, minute. *Ovary* 3-celled, each cell with 1 ovule on axile placenta. *Fruit* a much-inflated, 3-celled capsule with 3 seeds.

Contains about 130 genera and 1,000 species, mostly restricted to tropical and subtropical regions.

1. Koelreuteria Laxm.

As family.

About 8 species in China, Japan and Fiji.

Bean, W. J. (1973). *Trees and shrubs hardy in the British Isles*. Ed. 8. **2**. London.

More, D. & White, J. (2003). *Trees of Britain and northern Europe*. London.

Rehder, A. (1940). *Manual of trees and shrubs hardy in North America*. Ed. 2. New York.

Rushforth, K. (1999). *Trees of Britain and Europe*. London.

1. K. paniculata Laxm. Pride of India

K. pentlinioides L'Her.; *Sapindus chinensis* Murray

Deciduous, monoecious *tree* up to 20 m, with a narrow, rounded crown. *Trunk* up to 50 cm in diameter, extending through much of the crown. *Bark* pale brown or purplish-brown, with rough ridges and narrow fissures. *Wood* soft and pithy. *Branches* ascending and arching; twigs dull pale brown, smooth, with numerous round lenticels and minutely eglandular-hairy; young shoots buff or pale coppery brown, with small, raised lenticels and minutely eglandular-hairy. *Buds* 5–6 × 2–3 mm, axillary, ovoid, pointed; scales green and brown, ovate, pointed at apex. *Leaves* alternate; lamina 15–50 × 9–20 cm, oblong-elliptical to oblong-lanceolate in outline, imparipinnate, rarely bipinnate; leaflets 7–15, 4–9 × 2–6 cm, opening pink, reddish or yellowish, becoming dark rather shiny green on upper surface and pale green beneath, turning yellow in autumn, ovate or oblong-ovate, acute or acuminate at apex, shallowly lobed, the lobes crenate, rounded, cuneate or truncate at base, very shortly petiolulate, glabrous on upper surface except for a few minute eglandular hairs on the midrib, shortly hairy beneath, the veins prominent beneath and with a raised midrib above; rhachis pale green, red at the junction of the leaflets, shortly hairy; petiole up to 9 cm, pale, shortly hairy and swollen at base. *Inflorescence* a broad, loose, terminal, pyramidal panicle 20–40 cm; flowers bisexual, numerous, 10–15 mm in diameter. *Calyx* unequally 5-lobed, the lobes about 2 mm, ovate or ovate-oblong and acute at apex. *Petals* 4, bright yellow, lanceolate, obtuse at apex, clawed, the limb with 2 orange upturned appendages at the cordate base *Stamens* 8; filaments pale yellow, anthers pale brownish-yellow, 3, minute. *Capsule* 40–50 mm, oblong-ovoid, gradually narrowed towards the mucronate apex, much inflated and 3-valved; seeds 3, about 7 mm in diameter, dark brown to black, globose.

Introduced. Widely planted in parks and estates in south and south-east England and Ireland and naturalised as saplings in waste land in Kent, Surrey and Middlesex and recorded in a few other counties north to Yorkshire. Native of China.

109. HIPPOCASTANACEAE DC. nom. conserv.

Deciduous, monoecious *trees* and *shrubs*. *Leaves* opposite, palmate, petiolate, exstipulate. *Flowers* in large, terminal, pyramidal panicles, bisexual and male in each panicle, hypogynous, zygomorphic. *Calyx* 5-lobed. *Petals* 4–5, free. *Stamens* 5–9. *Style* 1; stigma minute. *Ovary* 3-celled, each cell with 2 ovules on axile placentas. *Fruit* a large capsule (conker) with 3 valves and 1–3 large seeds with a large, pale hilum.

Contains 2 genera and about 25 species in North America, the Sino-Himalayan area, a small part of south-east Europe, and Central and South America.

1. Aesculus L.

As family.

Contains about 23 species in North America, the Sino-Himalayan area and a small part of south-east Europe.

To understand the white-flowered European and Asiatic Horse-chestnuts you need to mark a tree and collect a series of specimens of flowers, fruits and leaves and make copious notes. *A. hippocastanum* was first known in western Europe by way of Constantinople in 1576, when seeds were sent to the Dutch botanist Charles de l'Ecluse (Clusius) who was working in Vienna, whence it was sent to England and France early in the seventeenth century. This was presumably the tree with round fruits covered with prickles until mature and with obovate leaflets without drawn-out bases. It is native in the high mountains where Greece, Yugoslavia and Albania meet. A similar tree with obturbinate fruits occurs in Cambridgeshire, but its origin is unknown. In the last 50 years two other species have been widely planted in Cambridgeshire and have spread naturally. They are to be found by field margins as well as along roads. Both have prickles on their fruits when young but lose most or all of them when mature, leaving them rough. *A. turbinata* from Japan has obturbinate fruits and leaflets like *A. hippocastanum* but more elongate. *A. chinensis* has more or less globose fruits a truncate apex and narrow leaflets more or less petiolulate. There are trees with fruits that retain a few prickles, which are thought to be hybrids. The most curious tree seen had its obturbinate fruit in fives, four of which were without prickles and the other covered with them. Its leaves were *A. turbinata*. These problems could only be solved if a collection were made from Europe to Japan.

In 1985 a moth larva was found in Macedonia at Lake Ohrid close to the Albanian border which made blotches on the leaves of *Aesculus*. In 1986 it was described as a new species, *Cameraria ohridella* Deschka & Dimic. The genus is not represented by any other species in Europe, but several are found in Asia and North America. The most likely explanation is that it was introduced by Chinese settlers. It spread into the rest of Yugoslavia and was introduced into Austria. It reached Germany and France in 1992 and reached Poland by 1998. It is now widespread in Europe and has reached Great Britain and Ireland. All the taxa do not seem to have been attacked, so it is possible the moths' normal host is not *A. hippocastanum*. The first generation attacks the lower part of the crown and the second and third generations the higher parts. Heavy infestation causes browning of leaves and their early fall. The long-term impact of defoliation in not known.

Bean, W. J. (1917). *Aesculus turbinata*. *Bot. Mag.* ser. 4, **13**: tab. 8713.
Bean, W. J. (1970). *Trees and shrubs hardy in the British Isles*. London. Ed. 8. **1**. London.
Csoka György (2003). *Leaf mines and leaf minors*. Forest Research Institute, Budapest.
Elwes, H. J. & Henry, A. (1907). *The trees of Great Britain and Ireland*. **2**. Edinburgh.
Mabey, R. (1996). *Flora Britannica*. London.
Mitchell, A. (1996). *Trees of Britain*. London.
More, D. & White, J. (2003). *Trees of Britain and northern Europe*. London.
Rehder, A. (1940). *Manual of trees and shrubs hardy in North America*. Ed. 2. New York.
Rushforth, K. (1999). *Trees of Britain and Europe*. London.
Sargent, C. S. (1913). *Plantae Wilsoniana*. **1**: 499–500. Cambridge, Mass.

1. Flowers pink or red 2.
1. Flowers white or yellow 4.
2. Flowers deep pinkish-red to deep red **5. carnea**
2. Flowers pale pink 3.
3. Leaves dark dull green **7. plantierensis**
3. Leaves medium, slightly shining green with pale veins **2. indica**
4. Flowers yellow **1. flava**
4. Flowers white 5.
5. Capsules with numerous prickels when mature 6.
5. Capsules without or with occasional prickles when mature, sometimes with prickles when young 7.
6. Leaflets up to 20 × 10 cm, irregularly crenate-dentate; inflorescence up to 25 cm **6(i). hippocastanum** var. **hippocastanum**
6. Leaflets up to 35(–46) × 18(–20) cm, irregularly jagged serrate-dentate; inflorescence up to 40 cm **6(ii). hippocastanum** var. **memmingeri**
7. Leaflets 7–10 cm **9. parviflora**
7. Leaflets 20–35 cm 8.
8. Capsules subrotund, truncate at apex, rounded at base **3. chinensis**
8. Capsules turbinate or obturbinate, narrowed at either apex or base 9.
9. Capsules narrowed at apex with a mucro **4. wilsonii**
9. Capsule narrowed at base, rounded at apex **5. turbinata**

1. A. flava Sol. ex Hoppe Yellow Buckeye
A. octandra Marshall; *A. lutea* Wangenh.; *Pavia lutea* (Wangenh.) Poir.

Deciduous, monoecious *tree* up to 35 m with a narrow, domed crown. *Trunk* up to 1 m in diameter extending through much of the crown. *Bark* pinkish-grey or reddish-brown, smooth, but cracking into thin, scaly plates at the base. *Branches* ascending, spreading and arching down; twigs medium to dark brown, thick, smooth and glabrous; young shoots greyish-green or greyish-brown, smooth, glabrous and with small, rounded lenticels. *Buds* opposite, 9–15 × 6–10 mm, ovoid, pointed at apex; scales pale brown, broadly ovate, mucronate at apex. *Leaves* opposite; lamina 10–35 × 15–35 cm, broadly ovate to subrotund in outline, palmate; leaflets 5, 8–25 × 3–8 cm, dull or slightly shiny, dark green with a pale midrib on upper surface, much paler beneath, turning bright orange-red in autumn, lanceolate, oblanceolate or elliptical, acuminate at apex, crenate-serrate, the teeth upcurved but blunt, cuneate at base, glabrous or sparsely eglandular-hairy on upper surface, densely short eglandular-hairy beneath, the veins 14–22 pairs, the petiolules up to 12 mm and pale yellowish-green; petiole up to 18 cm, pale yellowish-green, shallowly channelled above, rounded beneath, minutely eglandular-hairy and spotted. *Inflorescence* an erect, terminal, cylindrical panicle 10–35 cm; rhachis and peduncles with dense, short glandular hairs; pedicels 4–7 mm with dense short glandular hairs; flowers 25–30 mm in diameter, bisexual, glandular-hairy. *Calyx* 5–10 mm; lobes broadly ovate, obtuse at apex, with numerous, small glandular hairs. *Petals* 4, free, the upper 3 greenish-yellow, 13–20 mm, elliptical, with a hairy claw, the lower elliptical or subrotund, with a winged, hairy claw. *Stamens* 7–8; filaments whitish or pale green, more or less hairy; anthers orange. *Style* 1, clavate; stigma minute. *Capsule* 50–80 mm in diameter, golden yellow, ovoid, warty but without prickles, splitting along 2 or 3 sutures; seeds 1–3, 20–35 mm in diameter, chestnut brown, with a pale hilum half the diameter of the seed. *Flowers* 5–6. $2n = 40$.

Introduced. Large gardens and parks. England and Ireland. Native of the eastern United States from Pennsylvania south to Alabama and Georgia and west to Illinois.

2. A. indica (Cambess.) Hook. Indian Horse-chestnut
Pavia indica Cambess.

Deciduous, monoecious *tree* 15–25 m, the crown rather spiky when young, becoming rounded and domed. *Trunk* with a diameter up to 80 cm and extending well into the crown. *Bark* pinkish-brown or buff, rough and warty, scaly near the base in old trees. *Branches* ascending or spreading, covering the tree to the base; twigs shiny pinkish-brown, stout; young shoots green or brown with raised elliptical lenticels, glabrous. *Buds* opposite, 10–15 mm, ovate, obtuse at apex; scales green or pink, somewhat resinous, ovoid, rounded at apex, glabrous. *Leaves* opposite; lamina 15–30 × 17–40 cm, broadly ovate in outline, acute at apex, palmately divided into 5 or 7, rarely 8 leaflets; leaflets 7–30 × 1.5–10.0 cm, slightly shiny medium green on upper surface with pale green veins, pale green to slightly glaucous beneath, the midrib even paler, narrowly to rather broadly elliptical or oblong-elliptical or obovate-oblong, acute or acuminate at apex, regularly serrate, with acute, forward-pointing teeth, cuneate at base, glabrous on both surfaces, the veins 15–20 pairs, fairly prominent beneath, the petiolules up to 15 mm, pale yellowish-green, with brownish-red bases and glabrous; petioles 7–18 cm, pale yellowish-green, suffused brownish-red and glabrous. *Inflorescence* a terminal, cylindrical panicle 25–40 cm; rhachis and peduncles pale yellowish-green, tinted brownish-red, minutely eglandular-hairy; pedicels short, pinkish, minutely hairy; flowers 15–25 mm in diameter, sweet-smelling. *Calyx* 5–8 mm, pink, glabrous; lobes 5, triangular, acute at apex. *Petals* 4, free, the upper 3 white or pale pink, 8–10 mm, obovate and emarginate at apex, the lower 2, 7–8 mm, white with yellow guide spots, oblanceolate and emarginate. *Stamens* 8; filaments 16–18, white, curving in upper half; anthers orange. *Style* 1, whitish; stigma entire. *Capsule* 30–40 mm in diameter, globose, rough but not spiny, with a leathery skin; seeds 25–33 mm, dark, glossy chocolate or blackish-brown, with a small, pale hilum where attached to the outer wall, wrinkled. *Flowers* 6–7, 5 or 6 weeks later than *A. hippocastanum*. Visited by bumble-bees. $2n = 40$.

Introduced in 1851. Widely planted in large gardens, parks, estates and by roads, recorded as self-sown in Sussex, Essex, Berkshire, Worcestershire, Yorkshire and Middlesex. Native of the Himalayas from west Nepal to Afghanistan.

3. A. chinensis Bunge Chinese Horse-chestnut

Deciduous, monoecious *tree* up to 30 m, with a broad, rounded crown. *Trunk* up to 6 m in circumference, not extending far into crown. *Bark* grey, smooth when young, splitting into longitudinal plates and scaling in old trees. *Wood* pale reddish-brown, soft. *Branches* spreading and arching, the lower drooping; twigs greyish-brown, thick; young shoots green to dark brown, minutely eglandular-hairy. *Buds* 15–25 × 10–15 mm, ovoid, pointed at apex, sticky; scales brown, broadly ovate, rounded at apex. *Leaves* opposite; 25–40 × 30–50 cm, broadly obovate in outline, palmately divided into (5–)7 leaflets; leaflets 10–25 × 5–10 cm, dull dark green on upper surface, paler beneath, turning yellowish or brownish in autumn, oblong-oblanceolate to oblong-obovate, rounded to an acuminate apex, unequally serrulate-crenate, the teeth ascending but rounded, gradually attenuate at base, glabrous on upper surface or with an occasional eglandular hair on the veins, with numerous, brown simple eglandular hairs on the main veins beneath with fewer hairs on the smaller veins; veins 13–22 pairs, impressed above and prominent beneath; petiolules up to 10 mm, often hidden with a matt of brown hair; petiole 6–12 cm, pale green, with brown hair early in year. *Inflorescence* subcylindrical, elongated and gradually getting a little narrower near the apex; rhachis with dense brown eglandular hair when young; pedicels and peduncles very eglandular-hairy when young; flowers 20–22 mm in diameter, scented. *Calyx* 5–7 mm, campanulate, 5-lobed, the lobes obtuse at apex, minutely hairy. *Petals* 4, 9–11 × 1.5–2.5 mm, white, the 2 upper oblanceolate, with yellow or pink marks at the base, the 2 lower obovate, minutely hairy. *Stamens* 5–7; filaments white; anthers pinkish-orange. *Style* 1, white. *Capsule* 30–70 × 30–70 mm, yellowish-green, subrotund, truncate at apex, spiny when young, but losing all or most spines when mature; much inflated and 1- to 2-valved; peduncle up to 16 cm, minutely hairy; seed 20–25 mm in diameter, dark brown, the hilum half the size of the seed, white. Flowers 5–6. Visited by bumble-bees.

Introduced in 1912. Planted by roads and recreation grounds in a few places in Cambridgeshire in the last 30 years. Probably elsewhere. Native of western China.

4. A. wilsonii Rehder Wilson's Horse-chestnut

Deciduous, monoecious *tree* up to 25 m, with a rather narrow crown. *Trunk* up to 4 m in circumference, not extending far into crown. *Wood* pale. *Bark* greyish, smooth when young, becoming rough when old. *Branches* ascending; twigs thick; young shoots with dense minute hairs. *Buds* 15–25 × 12–14 mm, ovoid, pointed at apex; scales dark brown, ovate, rounded at apex, resinous. *Leaves* opposite; leaflets 5–7, 10–30 × 8–13 cm, dark green on upper surface with a pale midrib, paler beneath and slightly glaucescent, oblong-obovate, shortly acuminate at apex, serrate-dentate, gradually narrowed at base, glabrous or nearly so on upper surface, densely minutely hairy beneath; veins 23–30 pairs; petiolules obvious; petiole up to 30 cm, pale green, glabrous. *Inflorescence* 20–30 × 8–11 cm, a conical erect panicle of numerous flowers; pedicels and peduncles hairy; flowers fragrant. *Calyx* 6–7 mm, tubular, shortly hairy; lobes 5, unequal, obtuse at apex. *Petals* 4, 12–14 mm, white; upper about 3 mm wide, oblong-spathulate, attenuate, with a yellow blotch; lateral about 4.5 mm, obovate-oblong, cuneate at base. *Stamens* usually 7, unequal, up to 30 mm long; filaments glabrous. *Style* 1, hairy. *Capsule* 30–60 × 30–40 mm, yellowish green, ovoid or subrotund with a mucro at apex, and a broad, rounded base, when mature without or with occasional spines, rough; seed 3.0–3.5 cm, chestnut-brown, subglobose, with a large white hilum. *Flowers* 4–5.

Introduced in 1908. A rare tree of gardens and collections, but a tree by a country road in Bassingbourn, Cambridgeshire suggests that it possibly has been planted but not noticed. This tree fits the original detailed description but we suspect it may not be this species. Native of China. Named after Ernest Henry Wilson (1876–1930).

5. A. turbinata Blume Japanese Horse-chestnut
A. dissimilis Blume; *A. chinensis* auct.

Deciduous, monoecious *tree* up to 35 m with a rounded, but not domed crown. *Trunk* up to 6 m in girth, extending through much of crown. *Bark* grey, smooth when young, splitting into longitudinal plates and scaling in old trees. *Wood* pale reddish-brown, soft. *Branches* ascending and arching, the lower drooping; twigs greyish-brown, thick; young shoots green to dark brown, minutely eglandular-hairy. *Buds* 20–25 × 10–15 mm, ovoid, pointed at apex, sticky; scales brown, ovate, rounded at apex. *Leaves* opposite 25–40 × 35–60 cm and broadly obovate in outline, palmately divided into 5–7 leaflets; leaflets with lamina 15–40 × 5–15 cm, dark yellowish-green on upper surface with a pale midrib, paler beneath, turning yellow in autumn, narrowly oblong-obovate, rounded to a point at apex, crenate-serrate, unequal near the apex, more regular in lower half, gradually narrowed below to a sessile base, glabrous above, with small patches of brown hairs in the axils of the veins beneath and a few eglandular hairs on the veins themselves; veins 14–25 pairs impressed above, prominent beneath; petiole up to 25 cm, pale green, glabrous or slightly eglandular-hairy. *Inflorescence* a pyramidal panicle 15–25 cm; pedicels up to 30 mm, pale green, with dense minute hairs and some longer brown ones; flowers bisexual, 15–25 mm in diameter, scented. *Calyx* 6–7 mm, pale green, unequally divided into 4–5, ovate lobes, rounded at the apex, minutely hairy. *Petals* 4, 14–15 × 14–15 mm, crinkled, white, the 2 upper with yellow or pink marks at the base, ovate, rounded at apex, minutely eglandular-hairy. *Stamens* 7; filaments white; anthers orange. *Style* 1, white; stigma pink. *Capsule* 45–50 × 40–50 mm, broadly obpyriform, narrowed at apex, prickly when young, but loses its prickles and is only warty when mature; seeds 25–38 mm, brown, with a large whitish hilum covering half of the seed. *Flowers* 4–5.

Introduced. Several trees are planted as a street tree around Cambridge and as a hedgerow tree and seedlings in Bassingbourn and it might be widespread in streets, parks and amenity areas. Native of Japan. Trees with capsules warty, but with a few spines, may be hybrids with *A. hippocastanum*, **A. × hemiacantha** Topa.

6. A. hippocastanum L. Horse-chestnut

Deciduous, monoecious *tree* up to 25 (rarely to 40) m, with a broad, dense crown, sometimes layering in vast circles. *Trunk* up to 2.2 m in diameter, stout, sometimes slightly fluted, not extending far into the crown. *Bark* dark greyish-brown or reddish-brown, smooth to rough, becoming scaly and finally cast in squarish plates, often with a spiral pattern indicating spiral grain in the wood below. *Wood* pale brown to whitish, smooth and soft. *Branches* erect above, the lower typically sweeping down in long, graceful curves and bending up again at their tips, but planted trees often trimmed below; twigs dark grey or brown with large horseshoe-shaped traces, breaking easily; young shoots dark brown or pinkish-brown, with numerous, minute simple eglandular hairs and scattered lenticels. *Buds* in pairs; scales dark brown, very viscid especially in spring. *Leaves* opposite; lamina 8–30(–40) × 25–30 cm and obovate in outline, palmate; leaflets (3–)5–7, 8–35 (–46) × 3–18(–20) cm, dark green with pale veins on upper surface, paler beneath with even paler veins, quickly turning yellow, gold or orange, rarely red, then brown in mid-September to early November, obovate-oblanceolate or oblanceolate, rounded to a cuspidate apex, irregularly crenate-dentate to deeply and jaggedly serrate-dentate, long attenuate at base, sessile, with minute glandular hairs on the veins, otherwise glabrous on upper surface, with numerous, very short simple eglandular hairs on the veins and scattered on the surface with tufts of hairs in the vein axils beneath, often glabrescent, the veins 10–30 pairs; petiole 3–25(–36) cm, pale green tinted reddish-brown, glabrous. *Inflorescence* 15–40 × 9–15 cm, a conical, erect panicle with 15–45 flowers; pedicels and peduncle pale green, tinged pink, with dense, minute, pale crisped hairs. *Calyx* 5–7 mm, yellowish-green the lobes with a narrow brown margin, deeply divided, the lobes obtuse, with dense, minute, colourless simple eglandular and glandular hairs. *Petals* 5, with broad, obtuse apex, white outside and covered with dense, minute, unequal, colourless simple eglandular hairs, white inside with blotch (nectar guide) at base which is at first orange and later turns deep pink, with dense, unequal colourless (yellowish on blotch) simple eglandular hairs; upper 3 with limb 9–11 × 9–11 mm, broadly ovate in outline with an undulate margin, truncate, the claw about 3 × 1.5 mm; lower 2 with limb 12–13 × 8–9 mm, quadrangular-oblong with undulate-plicate margins, truncate to the short claw. *Stamens* 5–9; filaments 15–18 mm, whitish, tinged pink, sigmoid, with short, pale, unequal simple eglandular hairs; anthers pinkish-orange. *Style* 1, clavate, eglandular-hairy; stigma red-tipped. *Capsule* 4–6 × 4–6 cm, yellowish-green, turning brown when ripe, subglobose or rarely obturbinate, covered with curved prickles; seeds (conkers) 1 or 2, 20–40 × 15–25 mm, dark shining chestnut brown, with a large white hilum. *Flowers* 4–6. Pollinated by *Bombus* spp. Andromonoecious and protogynous. $2n = 40$.

(i) Var. **hippocastanum**
Leaflets 8–20 × 3–10 cm, irregularly crenate-dentate. *Inflorescence* up to 25 cm.

(ii) Var. memmingeri (K. Koch) P. D. Sell
A. memmingeri K. Koch

Leaflets up to 35(–46) × 18(–20) cm, very characteristically irregular and jagged serrate-dentate. *Inflorescence* up to 40 cm.

Introduced. First planted in London by John Tradescant about 1633. Now commonly planted for ornament along streets, in squares, gardens, parkland, churchyards and particularly village greens in England and Wales and less so in Scotland and Ireland. Often self-sown and sometimes solitary trees are found along hedgerows and field margins where the seeds may have been dropped by Rooks or Grey Squirrels. It is native to the central parts of the Balkan Peninsula from Albania and Yugoslavia to Greece with a single locality in Bulgaria. It is extensively planted for ornament and as a shade tree in most of Europe except the extreme north. The common tree is var. *hippocastanum*. Var. *memmingeri* occurs in scattered localities and is possibly quite frequent. The obturbinate fruited plant should perhaps be recognised as a distinct variety. Var. *memmingeri* is named after Edward Read Memminger (1850s). The wood is soft and easy to work but is of poor strength and durability. In heavy rain when in leaf its boughs easily break from the extra weight of water on the leaves. Deer and cattle feed on the seeds and Rooks and Grey Squirrels carry them about. Aesculin extracted from the fruits is a common constituent of bubble baths, while arscin another constituent is a powerful remedy for sprains and bruising. It is one of the commonest components of street names. Children particularly, and sometimes adults, gather the seeds to play the game of conkers, each seed being threaded on a string, the participants taking it in turn to hit one another's conkers. Starting in 1965 a World Conker championship was held at Ashton in Northamptonshire, site of a great avenue of Horse-chestnuts.

7. A. plantierensis André Pink Horse-chestnut

Deciduous, monoecious *tree* up to 20 m, with a rounded crown. *Trunk* up to 1 m in diameter, stout, not extending far into crown. *Bark* greyish-brown, rather rough. *Wood* pale brown. *Branches* spreading or drooping; twigs brown; young shoots grey, very shortly hairy. *Buds* 7–10 mm, ovoid, pointed; scales brown. *Leaves* opposite, broadly obovate in outline, palmate; leaflets 5, 10–20 × 5–10 cm, dark green on upper surface, paler beneath, turning brown in autumn, broadly elliptical to broadly obovate, rounded to a cuspidate-acuminate apex, irregularly crenate-serrate, the teeth small, rounded to cuneate at base, glabrous on upper surface, sometimes with tufts of hair in the axils of veins beneath; veins 10–15 pairs, very prominent beneath; petiolules up to 5 mm, red; petiole up to 20 cm, red, with scattered hairs. *Inflorescence* 10–25 cm, a conical panicle, with up to 50 flowers; rhachis with dense, crinkly glandular and eglandular hairs; peduncles and pedicels similarly clothed. *Calyx* 10–11 mm, campanulate, 5-lobed, the lobes rounded at apex, covered with dense, minute glandular and eglandular hairs. *Petals* 5, 20–25 mm, salmon-pink or whitish-pink, with a yellow to orange blotch at base, broadly rounded at apex, covered with glandular and eglandular hairs. *Stamens* with filaments white; anthers orange. *Style* 1, pink. *Capsules* variable, 30–70 × 22–50 mm, brown, with prickles when young, losing most or all of them when mature; seeds 1–2, 30–40 mm, brown; hilum covering three-quarters of seed, white. *Flowers* 5–6. This tree is said to be *A. carnea* × *hippocastanum*, *A. carnea* being a hybrid between *A. hippocastanum* and *pavia* which has doubled its chromosomes. *A. plantierensis* is thought not to have produced seed, but we have collected them and they are very variable. This fine tree can be recognised immediately by the colour of its flowers. It is said to be $2n = 40$.

Introduced. Planted in a few scattered localities by streets, on greens and in parks. Originated in the nursery of Messrs Simon-Louis Frères at Plantières near Metz.

8. A. carnea Hayne Red Horse-chestnut

Deciduous, monoecious *tree* up to 15 (rarely 25) m, with a low, broad crown. *Trunk* up to 1 m in diameter, stout, sometimes fluted, not extending far into the crown. *Bark* greyish-brown, later reddish-brown, rather rough, often with large, brown cankers. *Wood* pale brown to whitish, smooth and soft. *Branches* often spreading and twisted above, the lower often drooping; twigs brown, fissured, glabrous; young shoots pale grey, thick, much divaricate, with numerous, minute simple eglandular hairs and a few, orange-buff lenticels. *Buds* 7–30 × 4–14 mm, ovoid, acute at apex; scales greyish-green tinted dull reddish-brown with a darker marginal area, ovate, rounded at apex, covered with minute hairs and scarcely sticky. *Leaves* opposite, broadly obovate in outline, palmate; leaflets 5–7, with lamina 5–25 × 3–18 cm, dark green with paler veins on upper surface, paler with even paler veins beneath, turning yellow, then brown in autumn, obovate-oblanceolate or oblanceolate, acuminate or cuspidate at apex, irregularly crenate-serrate or serrate, attenuate at base, sessile or nearly so, glabrous on upper surface, with tufts of pale, wavy simple eglandular hairs in the axils of the veins beneath; veins 10–25 pairs; petiole 4–23 cm, yellowish-green, often reddish-brown on one side. *Inflorescence* 12–28 × 10–18 cm, a conical panicle with 10–30 flowers; rhachis, peduncles and pedicels with pale crisped hairs. *Calyx* 10–15 mm, yellowish-green, tinged pink, 5-lobed, the lobes obtusely-truncate at apex, with short or very short, pale simple eglandular hairs and glandular hairs on the surface and ciliate with red-stalked, pale-headed glandular hairs at apex. *Petals* with broad, obtuse apex, the outside pinkish to bright red and covered with red-stalked, broad-based, often pale-headed glandular hairs and colourless simple eglandular hairs; the inside pink or red at apex and clothed as outside, at base yellowish-orange with a deep pinkish-orange central vein and clothed with yellowish and colourless glandular and simple eglandular hairs; upper 3 petals with limb 13–15 × 12–15 mm, nearly quadrangular in outline with undulate margin, cuneate into the 7 × 2 mm claw; lower 2 petals 20–24 × 8–10 mm, oblanceolate, with undulate-plicate margins, gradually narrowed from apex to base with an abrupt contraction at the start of the claw. *Stamens* 5–9; filaments 18–22 mm, curved, pale greenish with a tinge of pink and numerous, wavy simple eglandular hairs; anthers pinkish-orange with deep coloured apex and base, with wavy simple eglandular hairs. *Styles* clavate, hairy; stigma red-tipped. *Capsule* 3–4 × 3–4 cm, green, soon turning brownish, subglobose, with numerous stiff, hooked glandular hairs when young, with few or no prickles with age; seeds 1–2, 20–35 × 15–25 mm, brown, with large white hilum. *Flowers* 5–6. Visited by *Bombus* spp. An allotetraploid of garden origin from hybridisation between *A. hippocastanum* L. ($2n=40$) and *A. pavia* L. ($2n=40$), with chromosome doubling so that it behaves as a fertile species ($2n=80$).

Introduced. Commonly planted along streets and avenues and in parkland and gardens as an ornamental and commemorative tree. Seeds often germinate where they fall. It is widely planted in Europe as an ornamental tree. The original sterile diploid is said to still occur in gardens where it is kept going by grafting.

9. A. parviflora Walter — Bottlebrush Buckeye

A. macrostachya Michx; *Pavia macrostachya* (Michx) Loisel.; *Pavia alba* Poir.

Deciduous, monoecious *tree* or *shrub* up to 5 m, with a broad crown. *Trunk* or stems extending into the crown. *Bark* dark brown or grey, smooth. *Branches* spreading or arching; twigs brown or grey, glabrous; young shoots pale green, glabrous. *Buds* 8–11 × 6–7 mm, narrowly ovoid, pointed at apex; scales pale brown, ovate, pointed at apex. *Leaves* opposite, 10–35 × 15–30 cm, decussate, palmate, broadly obovate in outline; leaflets 5–7, with lamina 5–23 × 5–12 cm, dark green on upper surface, paler beneath, elliptic-oblanceolate to obovate, shortly acuminate or cuspidate at apex, shallowly crenate or crenate-serrate, rounded or cuneate at base, glabrous on upper surface, with dense, short simple eglandular hairs beneath; veins 10–15 pairs; petiolules up to 3 cm, pale green, glabrous; petioles up to 15 cm, pale green, glabrous. *Inflorescence* a cylindrical panicle up to 30 cm; rhachis with dense, very short glandular and simple eglandular hairs; pedicels up to 11 mm, with dense, very short glandular and simple eglandular hairs. *Calyx* 5–7 mm, nearly tubular, with dense, short glandular hairs; 5-lobed, the lobes oblong-ovate, obtuse at apex. *Petals* 4, white, the upper pair linear-spathulate, obtuse at apex, narrowed into long-winged, hairy claw, the lower pair longer, fiddle-shaped and truncate or retuse at apex. *Stamens* 6–7; filaments 30–40 mm, pinkish-white, thread-like; anthers red. *Style* 1. *Capsule* 25 30 × 25–30 mm, globose or obovoid, without prickles when mature; seed dark brown. *Flowers* 7–8. $2n=40$.

Introduced in 1785. A garden relic at Stone and Hextable in Kent and is widely planted for ornament. Native of the south-east United States.

110. ACERACEAE Juss. nom. conserv.

Deciduous, monoecious or dioecious *trees* or small *shrubs*. *Leaves* opposite, palmately lobed, ternate or pinnate, rarely simple and unlobed, petiolate; exstipulate. *Inflorescence* a terminal corymb or raceme-like panicle. *Flowers* functionally unisexual, hypogynous or the male slightly perigynous, actinomorphic. *Sepals* 4–5, free. *Petals* (0–)4–5, free, often greenish. Sometimes only 1 row of perianth segments. *Stamens* usually 8. *Styles* 2; stigma long and linear, 1-sided. *Fruit* a schizocarp splitting into 2 samaras (winged mericarps) each with a nutlet containing 1 seed. *Endosperm* absent; embryo with flat, folded or rolled cotyledons and a long radicle.

Contains 2 genera and about 120 species mainly from temperate and subtropical areas of the northern hemisphere.

1. Acer L.

Saccharodendron (Raf.) Nieuwl.; *Rufacer* Small; *Saccharosphendamnus* Nieuwl.; *Argentacer* Small; *Rulac* Adans.

As family.

About 120 species widely distributed in temperate regions of the northern hemisphere and subtropical south-east Asia, with a few species in tropical Asia.

When choosing specimens of leaves from a tree try to get an average size one and a small and large one. Matching leaves with the illustrations is probably a surer way of identification than using the key. *Flowers* can be of at least 5 kinds and appear before or with the leaves. 1. Entirely female flowers. 2. The first developed flowers are female and the later ones male. 3. The earliest flowers at the apex are male, followed by male and female, and the last to open are mostly male. 4. Male flowers are first developed, followed by female. 5. All the flowers are male. Only one of these varieties of inflorescence is found on most trees, but two or even three of them may be associated on the same tree in exceptional cases. All female flowers possess stamens that appear normal, as do the numerous pollen-grains contained in the anthers, but never dehisce. The filaments are considerably shorter in female flowers than they are in male flowers. Some species can have the tree female one year and male the next, or male when young trees and female when they are older. The stamens are arranged on a honey disc. When the disc is entire and the stamens inserted in the centre, or lobed and the stamens in the centre of the lobes, or more or less enclosed between the lobes the stamens are said to be amphistaminal. When the stamens are on the outer margin of the disc they are said to be extrastaminal. *Fruits* are occasionally parthenocarpic, that is the 2 nutlets remain empty. These seedless empty fruits are practically indistinguishable from seed-bearing mericarps, especially in those species where the wall of the seed is very woody.

Bean, W. J. (1970). *Trees and shrubs hardy in the British Isles*. Ed. 8. **1**. London.
Delendick, T. J. (1990). The chemotaxonomy of the Aceraceae. *Int. Dendr. Soc. Yearbook* **1990**: 22–41.
Deschênes, J. M. (1970). The history of the genus Acer, a review. *Nature Canada* **97**: 51–59.
Elwes, H. J. & Henry, A. (1908). *The trees of Great Britain and Ireland*. **3**. Edinburg.
Forest, M. (1985). *Trees and shrubs cultivated in Ireland*. Clifden.
Gelderen, D. M. van, Jong, P. C. de & Oterdoom, H. J. (1994). *Maples of the World*. Portland, Oregon.
Grime, J. P. et al. (1988). *Comparative plant ecology*. London.
Hultén, E. & Fries, M. (1986). *Atlas of north European vascular plants north of the Tropic of Cancer*. 3 vols. Königstein.
Hunt, D. R. (1978). *Acer capillipes*. *Bot. Mag*. **182**: 113–115.
Hunt, D. R. (1978). *Acer griseum*. *Bot. Mag*. **183**: 13–15.
Jones, E. (1832). *The Acer Saccharinum*. London.
Jones, E. W. (1945). Biological flora of the British Isles. No. 13. *Acer* L. genus (pp. 215–219). *Acer pseudoplatanus* (pp. 220–237), *Acer platanoides* L. (p. 238), *Acer campestre* L. (pp. 239–252). *Jour. Ecol*. **32**: 215–252.
Jones, S. G. (1925). Life history and cytology of *Rhytisma acerinum* (Pers.) Fries. *Ann. Bot*. **39**: 41–75.
Jong, P. C. de (1976). Flowering and sex expression in *Acer* L. A biosystematic study. *Meded. Landb. Univ. Wageningen* **76**(2): 1–201.
Krüssmann, G. (1984). *Manual of cultivated broad-leaved trees and shrubs*. Beaverton, Oregon.
Loudon, J. C. (1838). *Arboretum et Fruiticetum Britannicum*. London. Ed. 2 (1844). London.
Mitchell, A. (1996). *Trees of Britain*. London.
More, D. & White, J. (2003). *Trees of Britain and northern Europe*. London.
Murray, A. E. (1970). Key to the sections and series of *Acer* subgenus *Acer*. *Kalmia* **2**: 3–4.
Murray, A. E. (1970). A checklist of species of *Acer*. *Kalmia* **2**: 22–45.
Murray, A. E. (1979). Afrasian and European maples. *Kalmia* **9**: 2–39.
Nelson, E. C. (1985). *Trees and shrubs cultivated in Ireland*. Dublin.
Pax, F. (1885–1886). Monographic der Gattung *Acer. Bot. Jahrb*. **6**: 287–347; **7**: 177–263.
Pax, F. (1890). Nachträge und Ergänzungen zu der Monographie der Gattung *Acer. Bot. Jahrb*. **11**: 72–83.
Rackham, O. (1980). *Ancient woodland, its history, vegetation and uses in England*. New ed. (2003). Colvend. [*A. campestre*.]
Rehder, A. (1940). *Manual of cultivated trees and shrubs*. Ed. 2. New York.
Rushforth, K. (1999). *Trees of Britain and Europe*. London.
Schwerin, E. Graf von (1893). Der Varietäten der Gattung *Acer*. *Gartenflora* 42: 161–714.
Townsend, A. M. (1972). Geographic variation in fruit of *Acer rubrum*. *Bull. Torrey Bot. Club* **99**: 122–126.
Vertrees, J. D. (1978). *Japanese Maples*. Beaverton, Oregon. Ed. 2. (1987). Portland, Oregon.

1. Leaves with leaflets 2.
1. Leaves simple, but usually lobed 5.
2. Leaves pinnate 3.
2. Leaves trifoliolate 4.
3. Leaflets olive green **6(1). negundo** forma **negundo**
3. Leaflets variegated creamy white **6(2). negundo** forma **variegatum**
4. Bark heavily scaling in small plates; leaflets 3–8 cm, shortly toothed or lobed in upper half **14. griseum**
4. Bark smooth, not peeling; leaflets 5–20 cm, sinuate or obscurely dentate **15. maximowiczianum**
5. At least some of the leaves with more than 5 lobes 6.
5. Leaves with up to 5 lobes, or unlobed 10.
6. At least some leaves with more than 7 lobes **1. japonicum**
6. Leaves with not more than 7 lobes 7.
7. Leaves with more or less obtuse lobes **16. macrophyllum**
7. Leaves with acute, acuminate, cuspidate or caudate lobes 8.
8. Leaves serrate, incise-serrate or with a few teeth or lobules **2. palmatum**
8. Leaves entire or nearly so 9.
9. Leaves 7–20 cm wide; wing of samara 15–30 mm **19. cappadocicum**
9. Leaves 8–15 cm wide; wing of samara 14–16 mm **20. pictum**
10. Leaves unlobed 11.
10. Leaves lobed 14.
11. Leaves whitish beneath 12.
11. Leaves green or bluish-green beneath 13.
12. Leaves gradually narrowed to an acute apex **4(a). davidii** subsp. **davidii**
12. Leaves rounded to a short, acute apex **4(b). davidii** subsp. **grosseri**
13. Leaves 5–11 cm wide **3. capillipes**
13. Leaves 3–5 cm wide **21. tataricum**
14. Lobes of leaves more or less obtuse at apex, or rounded to an acute tooth 15.
14. Lobes of leaves sharply long acute or acuminate or even cuspidate 28.
15. Wings of samaras nearly horizontal, slightly upturned or slightly downturned 16.
15. Wing of samara hanging down and almost parallel 20.
16. Some leaves on tree more than 10 cm long **18. miyabei**
16. All leaves on tree less than 10 cm long 17.
17. Samaras glabrous or nearly so **17(b). campestre** subsp. **leiocarpum**
17. Samaras very hairy especially on the seeds 18.
18. Centre lobe of leaves usually much longer than wide **17(a,iii). campestre** subsp. **campestre** var. **oxytomum**
18. Centre lobe of leaves about as long as wide, or wider than long 19.
19. Leaves near the inflorescence with lobes pointed, the centre lobe about as long as wide **17(a,i). campestre** subsp. **campestre** var. **campestre**
19. Leaves near the inflorescence with the centre lobe often wider than long and with a rounded appearance **17(a,ii). campestre** subsp. **campestre** var. **trilobatum**
20. Leaves 6–15 cm wide; wing of samara 15–25 mm **12. opalus**
20. Leaves 11–25 cm wide; wing of samara 12–60 cm 21.

21. Wing of samara 50–60 mm **7(ii). pseudoplatanus** var. **macrocarpum**
21. Wing of samara 20–40(–50) mm 22.
22. Leaves when mature with dense hairs on whole of lower surface 23.
22. Leaves when mature with hairs only in the axils of the veins or along the veins on the lower surface 24.
23. Short twigs forming a network at end of branches; bud scales green with a brown margin; inflorescence pendulous **8. villosum**
23. Twigs rather long, not forming a network at ends of branches; bud scales dark brown; inflorescence erect **9. velutinum**
24. Bud scales dark brown; leaves 7–20×8–25 cm **10. vanvolxamii**
24. Bud scales green with a brown margin; leaves 7–15×7–15 cm 25.
25. Leaves variegated whitish or yellowish on upper surface **7(i,2). pseudoplatanus** var. **pseudoplatanus** forma **variegatum**
25. Leaves uniform green on upper surface 26.
26. Leaves reddish to purplish beneath **7(i,3). pseudoplatanus** var. **pseudoplatanus** forma **purpureum**
26. Leaves paler green or bluish-green beneath 27.
27. Young fruits green **7(i,1). pseudoplatanus** var. **pseudoplatanus** forma **pseudoplatanus**
27. Young fruits red **7(i,4). pseudoplatanus** var. **pseudoplatanus** forma **erythrocarpum**
28. Leaves whitish or silvery beneath 29.
28. Leaves paler green or bluish-green beneath 32.
29. Leaves unlobed to markedly 3- to 5-lobed on the same tree, lobes and teeth small 30.
29. Leaves deeply and sharply 3- to 5-lobed 31.
30. Leaf gradually narrowed to an acute apex **4(a). davidii** subsp. **davidii**
30. Leaves rounded to a short, acute apex **4(b). davidii** subsp. **grosseri**
31. Leaves 7–11 cm wide; flowers with both petals and sepals **23. rubrum**
31. Leaves 9–16 cm wide; flowers with only 5 perianth segments **24. saccharinum**
32. Leaves 3–5 cm wide **22. tataricum**
32. Leaves more than 5 cm wide 33.
33. Wing of samara 12–25 mm 34.
33. Wing of samara 20–50 mm 36.
34. Buds 7–10×5–7 mm **11. trautvetteri**
34. Buds 4–7×3–4 mm 35.
35. Lobes of leaves shallowly crenate-serrate **3. capillipes**
35. Lobes of leaves sharply and unequally serrate **5. rufinerve**
36. Twigs grey, densely covered with lenticels; leaf lobes not lobed again; buds conical, pointed, the scales medium brown; stamens extrastaminal **13. saccharum**
36. Twigs dull, dark brown, with few lenticels; buds dark reddish-brown, ovoid, obtuse; leaf lobes often lobed again; stamens amphistaminal 37.
37. Leaves reddish to purplish **21(3). platanoides** forma **schwedleri**
37. Leaves green or variegated yellow or white 38.
38. Leaves green **21(1). platanoides** forma **platanoides**
38. Leaves variegated yellow or white **21(2). platanoides** forma **drummondii**

Section **1. Palmata** Pax

Deciduous *trees* or *shrubs*. *Leaves* 5- to 11-lobed. *Bud scales* always 4-paired. *Inflorescence* terminal and corymbose. *Sepals* 5, red or greenish-red. *Petals* mostly white and rolled inwards. *Disc* extrastaminal. *Stamens* 8.

1. A. japonicum Thunb. ex Murray Japanese Maple
A. circumlobatum Maxim.; *A. japonicum* var. *macrophyllum* Hort. ex Nicholson; *A. circumlobatum* var. *insulare* Pax; *A. insulare* (Pax) Pax, non Makino; *A. kobakoense* Nakai; *A. japonicum* var. *kobakoense* (Nakai) Hara; *A. japonicum* var. *insulare* (Pax) Ohwi

Deciduous, monoecious *shrub* or small *tree* up to 12 m, with a broadly ovoid crown. *Trunk* up to 40 cm in diameter, branching strongly at about 1 m. *Bark* grey or greyish-green and smooth. *Branches* spreading then upcurved; twigs reddish-brown, glabrous; young shoots green or pinkish to reddish-brown with a bloom above, greenish beneath, glabrous or with a few hairs and not sticky, rather thick. *Buds* 4–5×2–3 mm, ovoid-conical, sharply acute at apex; scales few, dark red or green, fringed grey, ovate, acute at apex. *Leaves* opposite; lamina 8–15(–22)×11–17(–24) cm, bright medium green on upper surface, paler beneath, in autumn turning bright gold to scarlet or dappled green, yellow or pink and finally ruby-red, subrotund or broadly obovate in outline, palmately divided one-quarter of the way to the base into 7–11 lobes, the lobes ovate-triangular, acute at apex, and coarsely and irregularly crenate-serrate, in cultivars more deeply divided, more or less cordate at base, when unfolding covered with silky eglandular hairs, but by midsummer only with hairs on some veins beneath; veins prominent beneath; petioles 3–6 cm, hairy at least at first. *Inflorescence* of long-stalked, nodding corymbs, conspicuous before the leaves; flowers 11–25, appearing before the leaves, about 15 mm in diameter. *Sepals* 5, 5–6 mm, red or greenish-red, lanceolate, acute at apex. *Petals* 5–6 mm, rolled inwards, reddish to rose-purple, pointed at apex. *Disc* extrastaminal. *Stamens* 8; filaments pale yellow; anthers yellow. *Styles* 2, greenish; stigma green. *Samara* with nutlet elliptical-globose, hairy at first; wing 20–25 mm, horizontal or at a wide angle, bright green; peduncle and pedicel red. *Flowers* 4. $2n=26$.

A large number of cultivars have been named, in particular those with deeply divided leaves and those which turn different colours in autumn.

Introduced in 1864. Common in gardens, parkland and estates. Native of mountain forests in northern Japan from 900 to 1800 m.

2. A. palmatum Thunb. ex Murray Smooth-leaved Maple
A. polymorphum var. *palmatum* (Thunb. ex Murray) K. Koch; *A. formosum* Carrière; *A. jacundum* Carrière; *A. japonicum* var. *polymorphum* auct.; *A. palmatum* var. *thunbergii* Pax; *A. polymorphum* var. *thunbergii* (Pax) Dieck; *A. palmatum* var. *spectabile* Koidz.; *A. palmatum* var. *amabile* Koidz.; *A. polymorphum* auct.

Deciduous monoecious *shrub* or *tree* up to 16 m, with a broad, domed crown. *Trunk* rarely more than 2 m and up to 40 cm in diameter, and sinuous. *Bark* a rich brown, sometimes striped pale buff, becoming grey in old trees, smooth but longitudinally striate. *Branches* ascending and curving outwards; twigs greyish-brown, smooth, glabrous; young shoots greenish to reddish-brown, slender, glabrous. *Buds* always in pairs, 2–4 × 2–4 mm, ovoid, acute at apex; scales 4-paired, green to bright red, ovate, obtuse at apex. *Leaves* opposite; lamina 5–10 × 5–10 cm, fresh apple green on upper surface, paler beneath, turning brilliant yellowish-orange in autumn, some cultivars have red to purple or variegated leaves, broadly obovate or reniform in outline, palmately 5- to 7-lobed to more than halfway to base, the lobes lanceolate or narrowly elliptical, long acute or acuminate at apex and sharply and finely serrate, truncate at base, often slightly hairy when young; veins prominent beneath; petioles 2–5 cm, green to pinkish-purple, glabrous. *Inflorescence* a small, terminal, spreading or erect corymb appearing with the leaves; flowers 12–15, 6–8 mm in diameter. *Sepals* 5, 2.0–2.5 mm, red, reddish-purple, or greenish-red, linear to obtuse at apex. *Petals* 2.0–2.5 mm, white or pink, rolled inwards, linear-lanceolate, obtuse at apex. *Disc* extrastaminal. *Stamens* 8; filaments pale green to pink; anthers dark red. *Styles* 2, green; stigma green. *Samara* with nutlet reddish-purple, elliptical-globose; wing about 10–20 mm, often underdeveloped, more or less red, diverging at a wide angle; pedicels dark red. *Flowers* 5–6. $2n=26$.

Extremely variable with a very large number of cultivars, especially in Japan.

Introduced in 1820. Commonly planted in gardens of all kinds, amenity areas, parks and estates. Native of Japan, Korea, Taiwan and eastern China.

Section **2. Macrantha** Pax

Deciduous *trees* or *shrubs*. *Leaves* undivided or 3- to 7-lobed. *Bud scales* 2-paired, red or greenish-red. *Inflorescence* a terminal or axillary raceme, male and female flowers usually on different branches. *Sepals* 5, greenish-yellow. *Petals* 5, pale green or greenish-yellow. *Disc* extrastaminal. *Stamens* 8.

3. A. capillipes Maxim Red Snake-bark Maple
A. pensylvanicum subsp. *capillipes* (Maxim.) Wesm.

Deciduous, monoecious *tree*, or sometimes a large *shrub* with several trunks, up to 12 m, with a broad, dome-shaped crown. *Trunk* straight to base of crown. *Bark* pale greyish-brown, with whitish longitudinal lines. *Branches* erect in centre of tree, spreading at base of crown, side-branches straight and forward-projecting, with white longitudinal lines; twigs reddish-brown with white lines, glabrous; young shoots brownish-red or purplish-red, longitudinally striate and glabrous. *Buds* always paired and valvate, 4–7 × 3–4 mm, narrowly ovoid or cylindrical, acute at apex; scales purple to red, angled, glabrous. *Leaves* opposite; lamina 6–15 × 5–11 cm, often red when unfolding, later dull yellowish-green on upper surface, paler beneath, turning crimson or dark red, rarely orange in autumn, ovate, sometimes broadly so in outline, gradually acute or acuminate at apex, sometimes caudate, 3- to 5-lobed or unlobed, the lateral lobes smaller than the terminal and acute at apex, very shallowly crenate-serrate, cordate at base, glabrous; veins 8–9 pairs, prominent beneath, often red; petioles 3–10 cm, purplish-red, grooved above, glabrous. *Inflorescence* a narrow, hanging, terminal or axillary raceme with 10–25 unisexual flowers, 6–15 cm, male and female flowers usually on different branches. *Sepals* 5, 2.0–4.0 mm, greenish-yellow, ovate to narrowly oblong, acute at apex, glabrous. *Petals* 5; 2.0–2.5 mm, greenish-yellow, ovate, rounded at apex. *Disc* extrastaminal. *Stamens* 8; filaments pale green; anthers green. *Styles* 2, green; stigma green. *Samaras* copious on older trees; nutlet convex or flat; wing 15–25 mm, very pale yellowish-green, turning pink then crimson, spreading almost horizontally, rounded at tips, glabrous. *Flowers* 5–6. $2n=26$.

Introduced in 1894 and now occurring in large areas of garden, estates, parkland and arboreta. Native of mountain forests of the Chichibu Range, Honshu and the Shikoku Islands, Japan.

4. A. davidii Franch. David's Maple

Deciduous, monoecious *shrub* or *tree* up to 15 m, with a broad or narrow crown. *Trunks* short, sometimes several. *Bark* reddish or bright olive green with broad white stripes made up of minute lines of brilliant bluish-white densely massed in the middle and more spread towards the edges, smooth, some older trees cracking into dark grey fissures or with some triangular pits. *Branches* ascending and arching or radiating; twigs brown; young shoots olive green or dark red, sometimes striped whitish, sometimes pale pink, glabrous. *Buds* 2.0–2.5 × 1.5–2.0 mm, narrowly conical, acute at apex, appressed; scales green to dark red, ovate. *Leaves* opposite; lamina 7–15 × 3–10 cm, unfolding shiny or olive-green to rich orange, becoming dark green, sometimes with yellowish-green veins on upper surface, pale whitish-green beneath, turning yellow to red in autumn, leathery, lanceolate to broadly ovate, gradually narrowed to an acute or shortly acuminate apex, very variable in toothing and lobing, unlobed to markedly 3- to 5-lobed leaves can be found on the same tree, the lobes acute and shallowly and unevenly crenate, the leaf rounded, truncate or more or less cordate at base, young leaves sometimes with brown hairs, but soon glabrous, sometimes with peg-like processes in the main vein axils; veins prominent beneath; petiole up to 6 cm, green or bright red, glabrous. *Inflorescence* a pendulous raceme 5–10 cm, female racemes longer than those of male, one branch may bear only male flowers and another branch bear only female flowers, but these branches may alternate or at random change the sex of their flowers from one year to the next; flowers 10–30, unfolding with the leaves. *Sepals* 5,

2.5–3.0 mm, greenish-yellow, ovate, rounded at apex, glabrous. *Petals* 5, 2.5–3.0 mm, greenish-yellow, lanceolate to ovate, narrowed to an obtuse apex. *Disc* extrastaminal, green. *Stamens* 8; filaments greenish-yellow; anthers greenish-yellow. *Styles* 2, greenish-yellow; stigma greenish. *Samara* with nutlet small and flattened; wing 25–30 mm, pale green, becoming tinged pink. *Flowers* 4–6.

(**a**) Subsp. **davidii**
A. cavaleriel Lév.; *A. laxiflorum* var. *ningpoense* Pax; *A. sikkimense* var. *davidii* (Franch.) Westm.; *A. sikkimense* var. *serrulatum* Pax

Young trunks often reddish with white stripes and striped green and white when mature. *Leaves* ovate-lanceolate, gradually narrowed to an acute apex, usually only shallowly crenate. $2n=26$.

(**b**) Subsp. **grosseri** (Pax) De Jong
A. grosseri Pax; *A. pavolinii* Pamp.; *A. hersii* Rehder; *A. grosseri* var. *hersii* (Rehder) Rehder; *A. teogmentosum* subsp. *grosseri* (Pax) A. E. Murray; *A. tomentosum* var. *hersii* (Rehder) A. E. Murray

Young trunks green with white stripes, mature trunks less conspicuously striped. *Leaves* broadly triangular-ovate, rounded to a short, acuminate apex, often shortly 3- to 5-lobed as well as shallowly crenate. $2n=26$.

Introduced in 1879. Both subspecies are widely planted in parkland and gardens and it seems to be good for street planting. Subsp. *davidii* is widely distributed in central and western China between 1200 and 3000 m, usually on acid, moist soils. It was named after Father Armand David (1826–1900). Subsp. *grosseri* occurs in northern and central China in the province of Hunan and Shaanxi. It is named after W. C. H. Grosser (1869–1942).

5. A. rufinerve Siebold & Zucc. Grey-budded Maple
A. pensylvanicum subsp. *rufinerve* (Siebold & Zucc.) Wesm.; *A. cucullobracteatum* Lév. & Vaniot

Deciduous, monoecious *shrub* or small *tree* up to 15 m, with a fairly narrow crown. *Trunk* rather short. *Bark* either green with greyish-white stripes or distinctly grey with pink stripes, becoming dull grey and rough, losing the stripes and becoming pitted. *Branches* of young trees are covered with a bluish bloom, later green to silvery green with distinctive white stripes; twigs green, glabrous; young shoots with a bluish-white bloom, then lilac, finally dull green, striped white. *Buds* 4–7 × 4–6 mm, ovoid, acute at apex; scales bright, waxy bluish-grey, ovate, on a stalk 3–4 mm. *Leaves* opposite; lamina 6–15 × 6–15 cm, unfolding yellowish, then dark dull bluish-green on upper surface, paler beneath, turning orange-yellow to dark red or crimson in autumn, obovate in outline, acute at apex, shallowly 3-lobed, the terminal lobe broadly triangular-ovate and shortly acuminate at apex, the lateral narrower, about halfway up the leaf and acute at apex, sharply and unequally serrate, truncate or shallowly cordate at base, with rusty hairs in the vein axils which disappear in summer; veins impressed on the upper surface, prominent beneath; petiole 3–6 cm, pink, grooved, glabrous. *Inflorescence* a terminal, erect raceme; flowers appearing with the unfolding leaves. *Sepals* 5, 3.5–4.0 mm, greenish-yellow, oblong, obtuse at apex. *Petals* 5, 5–6 mm, pale green, obtuse, at apex. *Disc* extrastaminal. *Stamens* 8; filaments pale; anthers yellow. *Style* 2, greenish; stigma green. *Samara* with thick, rounded nutlet; wing about 20 mm, at an obtuse angle. *Flowers* 4. $2n=26$.

Introduced in 1879. In many gardens and estates. Native of mountain forests in Honshu, Shikoku, Kiyushu and Yakushima Islands, Japan.

Section **3. Negundo** (Boehm.) Maxim.
Negundo Boehm.

Deciduous, dioecious *trees*. *Leaves* imparipinnate, 3- to 7(–9)-foliolate. Bud scales 2- to 3-paired. *Inflorescence* axillary, male and female on different trees. *Perianth segments* 4, greenish. *Disc* absent. *Stamens* 4–6.

6. A. negundo L. Ashleaf Maple
Negundo aceroides Moench; *Negundo fraxinifolium* (Stokes) DC.; *A. trifoliatum* Raf.; *A. negundo* subsp. *typicum* Wesm.; *A. negundo* subsp. *vulgare* (Pax) Schwer.; *Rulac negundo* (L.) Hitchc.; *A. fauriei* Lév.; *Rulac nuttallii* Nieuwl.; *A. orizabense* (Rydb.) Standley

Deciduous, dioecious *tree* up to 20 m, with an irregular, domed crown, often made more dense by sprouts from the branches and trunk. *Trunks* often more than 1 rising to the base of the crown. *Bark* smooth and greyish-brown when young, becoming dark grey, often greened with algae and becoming shallowly cracked with broad, rounded ridges. *Wood* is creamy-white, light, soft and close-grained. *Branches* stout, erect-spreading; twigs brittle, green or pale purple with a violet bloom, glabrous; young shoots pale green, straight, and glabrous or slightly eglandular-hairy. *Buds* 3–5 × 2.0–2.5 mm, ovate, obtuse to acute at apex; scales 2- to 3-paired, silky white, ovate, rounded at apex. *Leaves* opposite, imparipinnate, broadly ovate in outline, 3- to 7(–9)-foliolate; leaflets 5–10 × 3–5 cm, olive green on upper surface, paler beneath, turning yellow in autumn, lanceolate to ovate, acute or acuminate at apex, entire to remotely dentate or one leaflet sometimes 3-lobed, glabrous, basal pair often petiolulate; petiole 6–10 cm, pale yellow or pink, glabrous, the enlarged bases often furnished with minute stipule-like hairs. *Inflorescence* axillary, male and female on separate trees, the males with 12–16 flowers, both forming a pendulous raceme. *Perianth* cut over halfway into 5-lobes, 5–6 mm, yellowish-green, the lobes, ovate, obtuse at apex. *Disc* absent or rudimentary. *Stamens* 4–6; filaments pale; anthers red. *Styles* 2, greenish; stigmatic lobes long and whitish. *Samara* with small elliptic-globose to rather flat, acute, veined nutlet; wing 20–25 mm, yellowish, incurved at an acute angle. *Flowers* 3–4. $2n=26$.

(**1**) Forma **negundo**
Leaflets olive green.

(**2**) Forma **variegatum** (Jacques) Kuntze
Cv. Variegatum
A. negundo var. *variegatum* Jacques

Leaflets variegated creamy-white.

Introduced in 1688. Common in gardens, parks, car parks, and by roads and railways, sometimes self-sown when both sexes grow together. Native of Canada where it sometimes forms immense impenetrable thickets, and in mixed forests in the eastern and middle regions of the United States, westwards to the Rocky Mountains and south to Guatemala. Our tree is subsp. **negundo**. Forma *variegatum* is the most common form.

Section **4. Acer**

Deciduous *trees* or *shrubs*. *Leaves* 3- to 5-lobed. *Bud scales* 5- to 13-paired. *Inflorescence* terminal or axillary and corymbose. *Sepals* 5, yellowish-green. *Petals* 5, yellowish-green. *Disc* extrastaminal. *Stamens* 8.

The first five species of this section may all be recorded as *A. pseudoplatanus*. We discovered at the last minute that they are planted in Cambridgeshire, but do not know their total distribution; it could be widespread. They all look like Sycamore at a glance and can be difficult to work out until you know them. Fortunately we have specimens from M. T. Masters' garden in the Elwes and Henry collection at the Cambridge Herbarium and all five species grow in the University Botanic Garden.

7. A. pseudoplatanus L. Sycamore

A. montanus Garsault; *A. opulifolium* Thuill., non Vill.;
A. procerum Salisb.; *A. ramosum* Schwer.;
A. platanophyllum St Lag. ex Keegan

Deciduous, very hardy, monoecious *tree* up to 30 m, with a broad, dense, irregular domed crown casting heavy shade. *Trunk* straight and passing though much of the crown. *Bark* grey, smooth for a long period, finally scaling in shallow, irregular plates. *Wood* hard, pale and fine-grained, white to yellowish with a silky lustre, clean and compact. *Branches* spreading parallel to the ground and slightly drooping at the ends, several coming off of the trunk at the same point; twigs dull medium brown, with many lenticels; young shoots paler shining brown with lenticels, glabrous, twigs and young shoots forming a network at and of branches. *Buds* 8–10×5–8 mm, ovoid, rounded at apex; scales green with blackish or brownish margins, broadly ovate, apiculate at apex, shortly hairy near the tip. *Leaves* opposite; lamina 10–18×11–25 cm, dull, dark olive green on upper surface, sometimes variegated whitish or yellowish, often covered with the large black blotches of the fungus, *Rhytisma acerina* (Pers.) Fr., paler and almost glaucous beneath, sometimes reddish to purplish, in autumn turning a dingy yellow from the edges inwards, then brown, broadly ovate or almost reniform in outline, palmately lobed up to halfway to base, the lobes (3–)5, broadly ovate, more or less acute at apex and shallowly dentate near the apex, the leaf cordate at base, glabrous or nearly so on upper surface, with greyish or brownish hairs along the veins and in the vein axils when young, usually becoming nearly glabrous or with only hairs in the vein axils when old; petioles up to 25 cm, without milky juice, usually rather dark red above and greener beneath, glabrous and smooth. *Inflorescence* a narrow pendulous raceme 6–20 cm and appearing after the leaves; flowers 60–100 in stalked clusters 4–6 mm in diameter. *Sepals* 5, 1.2–1.5 mm, yellowish-green, narrowly triangular-lanceolate, obtuse at apex, glabrous. *Petals* 5, 1.5–2.5 mm, yellowish-green, oblong, obtuse at apex. *Disc* extrastaminal. *Stamens* 8; filaments greenish-yellow, hairy at base, long exserted in male flowers; anthers greenish-yellow. *Styles* 2, greenish-yellow; stigma greenish. *Samara* with nutlet 6–10 mm, ovoid and convex; wings 12–60 mm, green or more rarely red, pointing down and almost parallel, on pedicels 10–20 mm. *Flowers* 4–6. Pollinated mainly by bees. $2n=52$.

(**i**) Var. **pseudoplatanus**

Leaves 10–15×11–20 cm, with occasional minute hairs on lower surface and longer ones on veins. *Samara* with wing 20–40 mm, pointing down and more or less parallel with the wing of the other samara.

(**1**) Forma **pseudoplatanus**

Leaves uniform green above, paler bluish-green beneath. *Samara* green.

(**2**) Forma **variegatum** (Weston) Rehder
Cv. Variegatum

A. pseudoplatanus var. *variegatum* Weston

Leaves variegated whitish or yellowish. *Samara* green.

(**3**) Forma **purpureum** (Loudon) Rehder
Cv. Atropurpureum

A. pseudoplatanus var. *purpureum* Loudon

Leaves green on upper surface, reddish to purplish beneath. *Samara* green to reddish or purplish.

(**4**) Forma **erythrocarpum** (Carrière) Pax
Cv. Erythrocarpum

A. pseudoplatanus var. *erythrocarpum* Carriére

Leaves green often more yellowish-green on upper surface, paler beneath. *Samara* red.

(**ii**) Var. **macrocarpum** Spach

Leaves 15–18×17–25 cm, with occasional minute hairs on lower surface and longer ones on veins. *Samara* with wing 50–60 mm, and often subhorizontal.

Introduced. Prefers deep, moist, well-drained, rich soils, but will grow on all but the very poor soils, and is tolerant of both exposure and salt spray. Probably introduced sometime during the fifteenth or sixteenth century. Now fully naturalised and in a wide range of habitats throughout Great Britain and Ireland including railway banks and waste land, plantations, woods, parkland, estates, large gardens and roadsides as well as field margins. It produces abundant seed and spreads rapidly unlike the other species in the group. It is tolerant of pollution and is planted in town parks and gardens. In nature reserves it can become a weed and has to be eliminated. Native of south and central Europe and widely planted further north. It is much in demand for its smooth beautifully marked veneers and is used in turnery, carving, furniture making, joinery and violin making, ladles, spoons, bowls, plates and rollers. It makes an excellent firewood. Many large trees have become famous and been given pet names.

8. A. villosum C. Presl Mediterranean Sycamore

A. pseudoplatanus subsp. *villosum* (C. Presl) Parl.;
A. pseudoplatanus var. *tomentosum* Tausch

Deciduous, very hardy, monoecious *tree* up to 30 m, with a broad, dense, irregular, domed crown. *Trunk* straight and passing through much of the crown. *Bark* grey, smooth for a long period, finally scaling in shallow, irregular plates. *Wood* hard and fine-grained. *Branches* spreading and slightly drooping at ends; twigs dull medium brown, striate, covered with round lenticels, glabrous or nearly so; young shoots pale, shining brown, striate with lenticels, glabrous; twigs and young shoots forming a network at the ends of branches. *Buds* 7–10×5–8 mm, ovoid, rounded at apex; scales green, with a brown, shortly hairy margin, broadly ovate, with a short mucro at apex. *Leaves* opposite; lamina 5–18×5–22 cm, dark olive green on upper surface, pale green beneath, sometimes with spots of *Rytisma acerina*, in autumn turning brown from the edges inwards, broadly ovate to reniform in outline, palmately lobed up to halfway to base, the lobes (3–)5, more or less ovate, more or less obtuse at apex, rather coarsely dentate, glabrous or nearly so on upper surface, with brownish to greyish, more or less appressed hairs on the undersurface especially on the veins, throughout the year, densely so early on becoming much less dense by autumn; petiole up to 18 cm, green, glabrous or with an occasional hair. *Inflorescence* a narrow, pendulous raceme 6–20 cm, appearing after the leaves; pedicels with occasional hairs, up to 2.5 mm; flowers numerous in stalked clusters. *Sepals* 5, 1.2–2.5 mm, yellowish-green, oblong, obtuse at apex. *Petals* 5, 1.5–2.5 mm, yellowish-green, oblong, obtuse at apex. *Disc* extrastaminal. *Stamens* 8; filaments pale greenish-yellow, hairy at base, long exserted in male flowers; anthers greenish-yellow. *Styles* 2, greenish-yellow; stigma greenish. *Samara* with nutlets 6–9 mm, ovoid and convex, hairy; wings 15–35 mm, green, pointing down but slightly spreading. *Flowers* 4–6. Pollinated by bees.

Introduced. By roads, parks, estates, hedgerows and round fields. There are some very old trees as well as many planted in the last 40 years. Common in Cambridgeshire, probably widespread. Native of south Italy, Sicily and Dalmatia. If a sycamore was introduced by the Romans it is most likely to have been this species.

9. A. velutinum Boiss. Downy Sycamore

Deciduous monoecious *tree* up to 20 m with a rather narrow crown. *Trunk* up to 80 cm in diameter. *Bark* greyish-brown, rather smooth, becoming scaly below in old trees. *Branches* ascending or spreading; twigs dull dark brown; young shoots paler, twigs and young shoots spreading and not forming a network at apex of branches. *Buds* 8–15×4–7 mm, conical, acute at apex; scales dark brown, broadly ovate, pointed to rounded, very shortly hairy round the margin. *Leaves* opposite, 9–17×11–17 cm, dull dark olive green on upper surface, paler beneath, turning dingy yellow in autumn, broadly ovate or almost reniform in outline, palmately lobed up to halfway to base, the lobes (3–)5, broadly ovate, obtuse to acute at apex, shallowly dentate near the apex, the leaf cordate at base, glabrous or nearly so on upper surface, usually covered with pale, becoming reddish-brown eglandular hairs beneath, denser in the leaf axils; petiole 10–25 cm, green, with white latex. *Inflorescence* an erect, many-flowered corymbose, pyramidal panicle. *Sepals* 5, about 2.0×0.7 mm yellowish-green, ovate-oblong. *Petals* 5, about 1.2 mm, yellowish-green, linear. *Disc* extrastaminal. *Stamens* 8; filaments greenish-yellow; anthers greenish-yellow. *Samaras* with nutlet 6–7 mm, ovoid, hairy; wings 25–35 mm, diverging at 45°. *Flowers* 4–6.

Introduced. Planted in large gardens, estates and by roadsides, possibly more widespread than recorded. Native of Georgia and the eastern and central Caucasus.

10. A. vanvolxemii Mast. Van Volxem's Sycamore
A. velutinum var. *vanvolxemii* (Mast.) Rehder

Deciduous, hardy, monoecious *tree* up to 30 m with a rather narrow crown. *Trunk* straight and extending through much of the crown. *Bark* grey and rather smooth. *Branches* spreading to ascending; twigs dull, dark brown with dense lenticels and striations; young shoot slightly paler brown and with dense lenticels. *Buds* 7.5–12.0×3.0–5.5 mm, conical, acute at apex; scales dark brown, broadly ovate, pointed to rounded, very shortly hairy round the margin. *Leaves* opposite; lamina 7–20×8–25 cm, rather pale green on upper surface, pale and sometimes slightly glaucous beneath, broadly ovate in outline, palmately 3- to 5-lobed up to halfway to base, the lobes mostly rounded-acute or very shortly acuminate or the smaller obtuse at apex and shallowly crenately toothed, the leaf more or less cordate at base, more or less glabrous on upper surface, with pale to brown hairs beneath only in the axils of the main veins with an occasional hair on the veins elsewhere; petiole up to 25 cm, pale green, glabrous. *Inflorescence* an erect pyramidal corymb of numerous flowers; pedicels hairy. *Sepals* 5, 2–3×1.2–1.5 mm, yellowish-green, oblong, obtuse at apex, hairy. *Petals* 5, 3–4×0.4–0.5 mm, yellowish-green, oblong, obtuse at apex. *Disc* amphistaminal. *Stamens* 8; filaments very pale greenish-yellow; anthers greenish-yellow, exserted. *Styles* 2, very pale greenish-yellow. *Samara* with nutlet 5–7 mm, ovoid and convex, hairy; wings 25–32×15–17 mm, half-spreading. *Flowers* 5–6.

Introduced. Large gardens and estates, roadsides and field margins and probably elsewhere. Native of the Caucasus.

11. A. trautvetteri Medw. Red-bud Sycamore
A. heldreichii subsp. *trautvetteri* (Medw.) A. E. Murray

Deciduous, monoecious *tree* 15–25 m with a narrow crown. *Trunk* up to 80 cm in diameter, extending some way into the crown. *Bark* grey-brown and rather smooth. *Branches* ascending or spreading; twigs dark red at maturity; young shoot glabrous. *Buds* 7–10×5–7 mm, ovoid, obtuse at apex; scales red to brownish-red, ovate, rounded at apex. *Leaves* opposite; lamina 9–15×4–19 cm, dark green on upper surface, pale beneath, broadly ovate in outline, usually 5-lobed, the lobes lanceolate or ovate-lanceolate, almost caudate-acuminate at apex, with rather large irregular obtuse teeth; glabrous or nearly so or with brownish hairs in the axils of the veins and sometimes along the veins; leaf cordate at base; petiole 4–19 cm, pale green or reddish on upper surface, with white latex. *Inflorescence* an upright obpyramidal panicle. *Sepals* 5, yellowish-green, ovate-oblong, obtuse at apex. *Stamens* 8; filaments greenish-yellow, glabrous. *Styles* 2, greenish-yellow. *Samaras* with nutlets ovoid and glabrous; wings 30–40×10–18 mm, usually red or reddish,

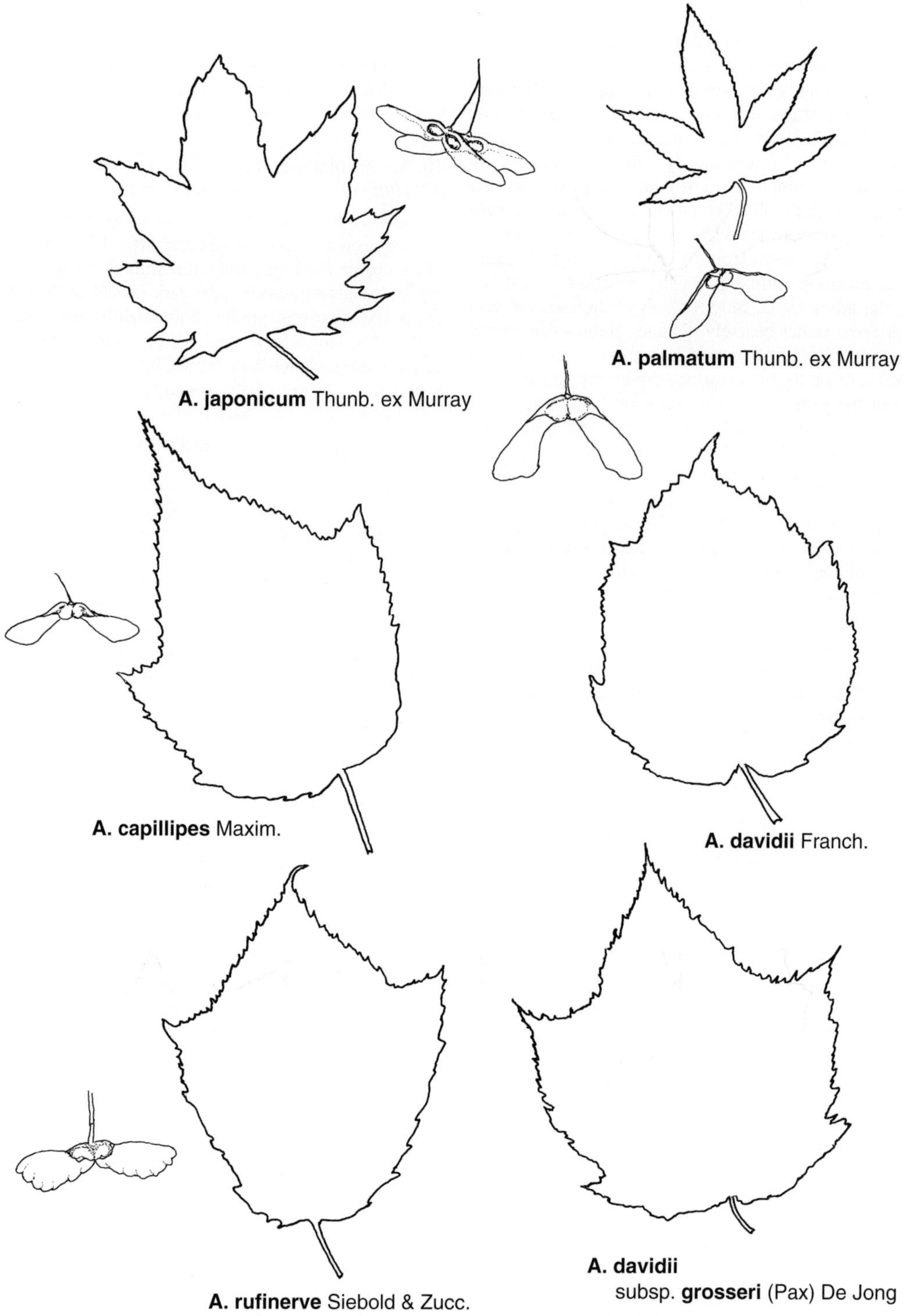
A. japonicum Thunb. ex Murray
A. palmatum Thunb. ex Murray
A. capillipes Maxim.
A. davidii Franch.
A. rufinerve Siebold & Zucc.
A. davidii
subsp. grosseri (Pax) De Jong

Leaves of **Acer** L.

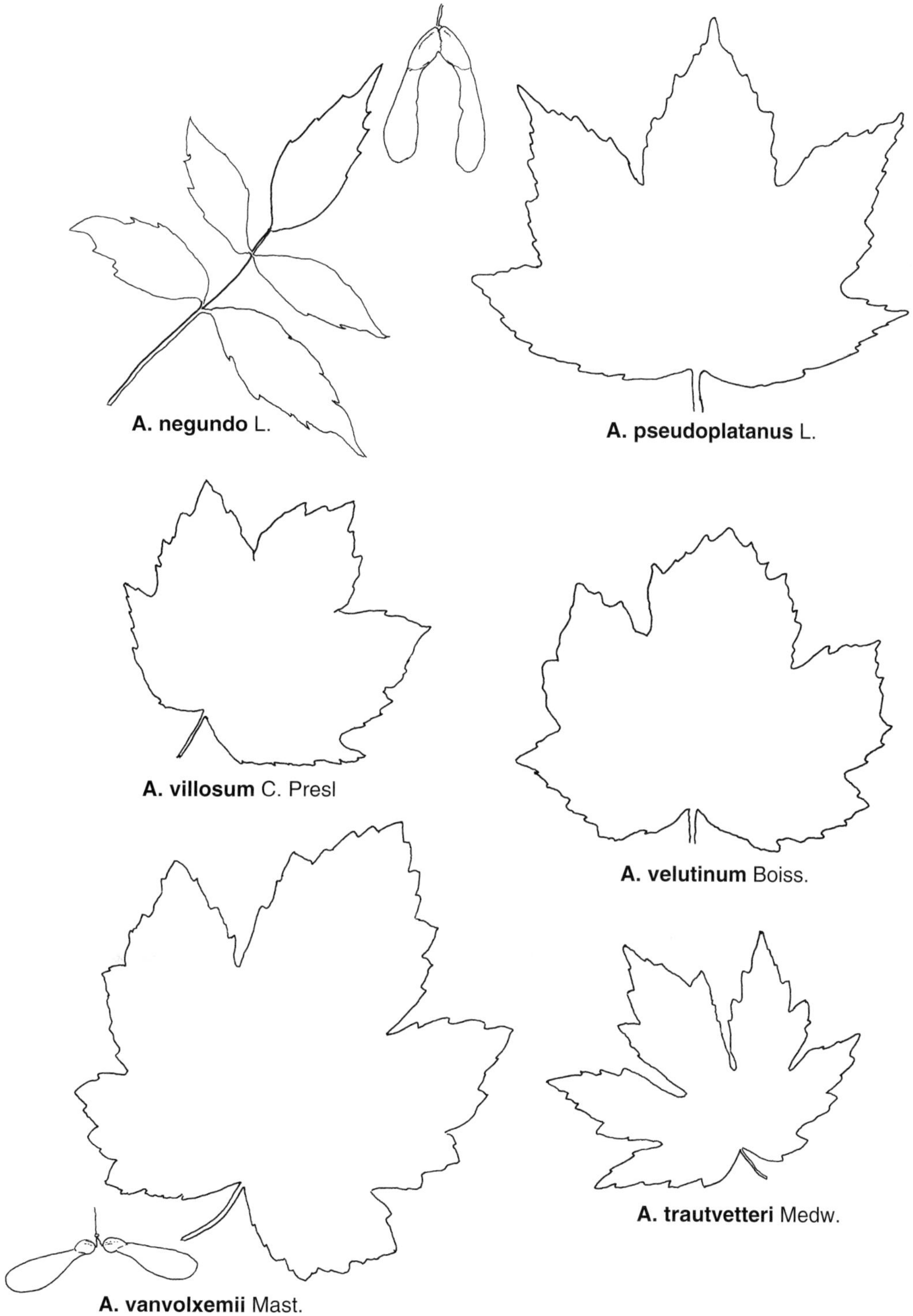

Leaves of **Acer** L.

pointed downwards and parallel or slightly diverging. *Flowers* 4–6.

Introduced. As well as being planted in estates and parks this species is planted by roadsides, in new woods and along field margins in Cambridgeshire. It is possibly widespread. Some trees have typical leaves of *A. trautvetteri* but do not have the upright inflorescence. They should not be lumped with *A. pseudoplatanus* or *A. villosum* neither of which they resemble. Native of the Caucasus to Iran.

12. A. opalus Mill. Italian Maple
A. italum Lauth; A. montanum Daléchamps ex Lam.; A. opulifolium Vill.; A. vernum Carrière ex Lam.; A. rotundifolium Lam.; A. opolifolium Pers.; A. opulifolium var. opalus (Mill.) K. Koch; A. italicum Lauche; A. rupicola Chabert

Deciduous, monoecious *shrub* or *tree* up to 20 m, with a broad, dome-like crown. *Trunk* erect, well into the crown. *Bark* dark pinkish-grey, young trees with small squares scaling away to leave orange patches, old trees with long plates adhering in middle and curving outwards at each end and coarsely shaggy. *Branches* low, spreading and twisting; twigs brown; young shoots dark reddish-brown with pale lenticels, glabrous. *Buds* 5–8 × 3–4 mm, narrowly ovoid-conical, acute at apex; scales 5–13 paired, imbricate, pale and dark brown, ovate, acute at apex. *Leaves* opposite; lamina 6–12 × 6–15 cm, dark green on upper surface, pale bluish-green beneath, broadly obovate in outline, palmately 3- to 5-lobed, the lobes triangular-ovate, rounded at apex and with coarse, obtuse, irregular teeth, cordate at base, hairy beneath when young, remaining so along the veins; veins impressed above, prominent beneath; petiole up to 10 cm, red above, green beneath, glabrous. *Inflorescence* terminal, in short corymbose umbels; flowers appearing before or with the leaves, the pedicels 30–40 mm. *Sepals* 5, 4–5 mm, yellowish-green, linnear-lanceolate, obtuse at apex. *Disc* extrastaminal. *Stamens* 8; filaments pale; anthers yellowish, exserted in male flowers. *Styles* 2, greenish; stigma greenish. *Samara* with nutlet ovoid; wing 15–25 mm, pale green, tinged pink, hanging down and nearly parallel to the other samara. *Flowers* 4. $2n=26$.

Introduced in 1752. Planted in parks, by roads and in shrubberies, sometimes producing seedlings. Native of the Jura, Pyrenees and Apennines eastwards to the Caucasus and in Morocco and Algeria. Our plant is subsp. **opalus**.

13. A. saccharum Marshall Sugar Maple
A. palmifolium Borkh.; *A. saccharophorum* K. Koch; *A. palmifolium* var. *barbatum* (Michx) Schwer.; *Saccharodendron barbatum* (Michx) Nieuwl.; *A. trelaeaseanum* Bush; *Saccharodendron saccharum* (Marshall) Moldenke

Deciduous, monoecious *tree* up to 40 m, with an irregularly domed, open crown. *Trunk* straight and erect through much of the centre of the tree. *Bark* smooth, grey and finely ridged until the tree is fairly large, when long wide fissures develop and broad ridges lift up into large, shaggy plates. *Branches* wide-spreading, rather pale; twigs greyish-brown with numerous lenticels and striations; young shoots bright, pale green with pale lenticels and often a band of purplish-red by each pair of leaves or buds, slender and glabrous. *Buds* 5–6 × 2–6 mm, ovoid-conical, acute at apex; scales brown with pale margins, broadly ovate, rounded at apex, with pale brown hairs. *Leaves* opposite; lamina 8–15 × 8–15 cm, thin, unfolding pale green, later dark green on upper surface, pale green or slightly bluish beneath, turning yellow to orange or scarlet, often all these colours mixed in the autumn, broadly ovate in outline, deeply palmately 3- to 5-lobed, the lobes ovate, acute at apex, with 2–3 large teeth and many smaller teeth, cordate at base, glabrous or with some hairs on the veins beneath; veins prominent beneath; petioles up to 12 cm, pale green, sometimes red, glabrous. *Inflorescence* a corymb; flowers appearing before the leaves, unisexual or bisexual. *Perianth segments* 5, pale yellow, obtuse at apex. *Disc* extrastaminal. *Stamens* 8; filaments white; anthers yellowish. *Styles* 2, greenish; stigma reddish. *Samara* with ovoid, often keeled nutlet; wing 30–50 mm, forming a U, or in Europe there are often no fruits at all, or if seeds are present they are not viable. *Flowers* 5–6. $2n=26$, 52, 78.

A very variable species with a wide range to which several subspecies have been described. Our plant, as described above is subsp. **saccharum.**

Introduced in 1735. Widely planted in gardens, parks and along roads and may be misidentified as *A. platanoides* from which it may only be easily distinguished by the extrastaminal disc. Native of eastern North America to Mexico and Guatemala, where it forms forests with *Tsuga, Betula* and *Prunus*. *A. saccharum* is the true Sugar Maple which is famous for producing sap from which maple syrup is made, and for its spectacular autumn colours.

Section **5. Trifoliata** Pax

Deciduous *trees* or *shrubs*. *Leaves* 3-foliolate. *Bud scales* 11- to 15-paired. *Inflorescence* corymbose. *Sepals* 5, yellowish-green. *Petals* 5, yellowish-green. *Disc* extrastaminal. *Stamens* 10–13.

14. A. griseum (Franch.) Pax Chinese Paperback Maple
A. nikoense var. *griseum* Franch.; *Crula grisea* (Franch.) Nieuwl.; *A. pedunculatum* K. S. Hao

Deciduous, monoecious *tree* up to 15 m, with a slender habit and narrow crown. *Trunk* straight, through much of the tree. *Bark* chestnut to orange-brown, heavily scaling in small plates. *Branches* spreading and arching; twigs dull brown; young shoots pale brown, sparsely hairy becoming glabrous. *Buds* about 1 mm, ovoid-conical, acute at apex; scales dark greyish-brown to nearly black, ovate, acute at apex. *Leaves* opposite, broadly ovate in outline, 3-foliolate; leaflets with lamina 3–8 × 4–8 cm, unfolding pale orange-buff, later dull greyish-green on upper surface, pale grey beneath, turning yellow to orange with scattered red ones, or finally deep crimson or red in autumn, irregularly lanceolate to ovate, acute to acuminate at apex, shortly toothed or lobed in upper half with blunt teeth or lobes, sometimes toothed or lobed below on one side, with few hairs on upper surface and numerous soft hairs beneath; veins prominent beneath; petioles up to 44 mm, often reddish, with dense, spreading, pale

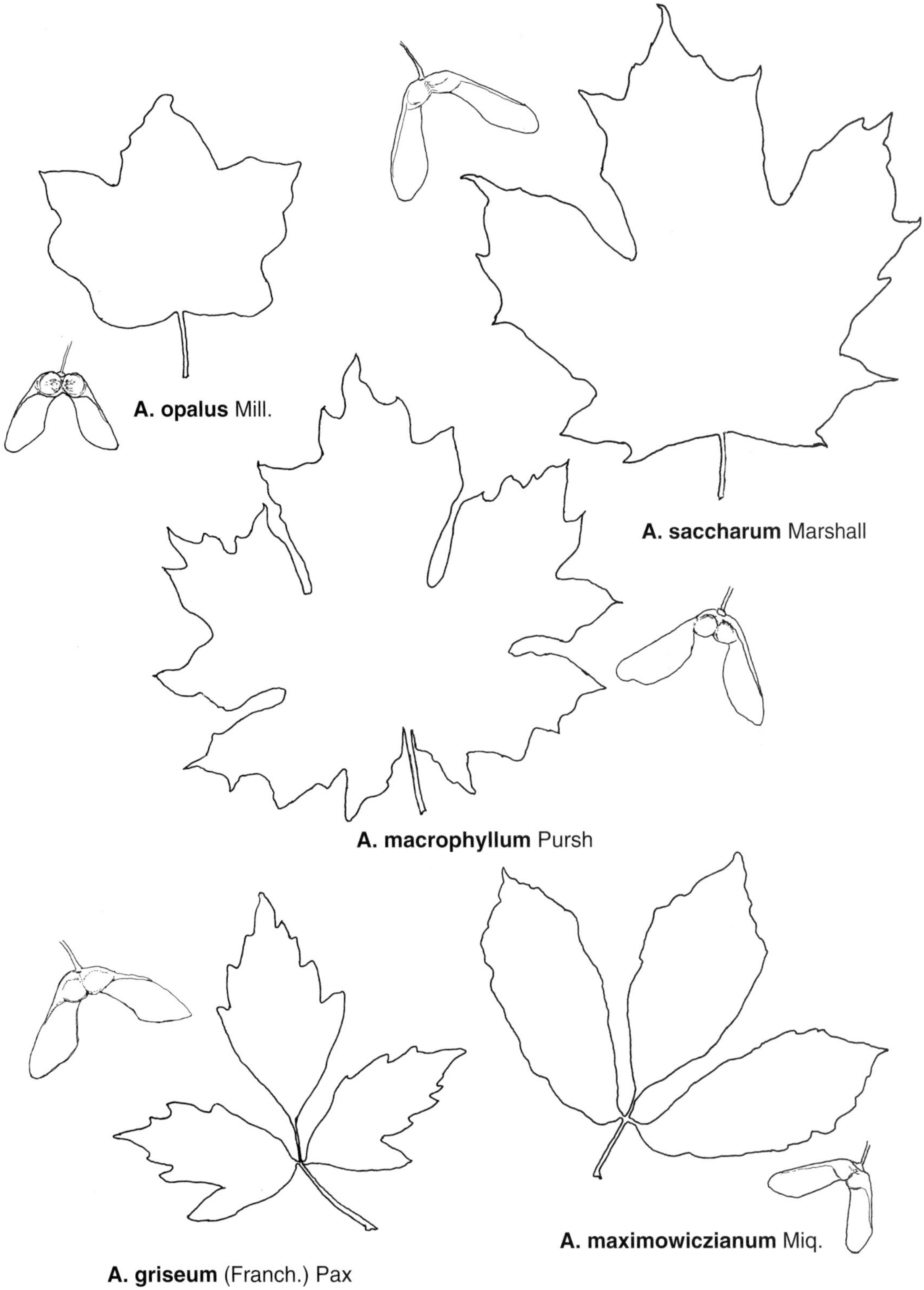

Leaves of **Acer** L.

simple eglandular hairs. *Inflorescence* shortly corymbose, few-flowered, pendulous; flowers opening as the leaves unfold; pedicels up to 30 mm, hairy. *Sepals* 5, 4–5 mm, yellowish-green, lanceolate, obtuse at apex. *Petals* 5, 5–6 mm, pale yellowish-green, oblong, rounded at apex. *Disc* extrastaminal. *Stamens* 10–13; filaments pale; anthers yellowish. *Styles* 2. *Samara* with dark brown, spherically convex and hairy nutlet, mostly thick, lignified and sterile; wing 25–33 mm, pale yellowish-green, pointing downwards and nearly parallel with the other samara. *Flowers* 5–6. $2n=26$.

A very uniform species with no described varieties or cultivars.

Introduced in 1901. Much planted in gardens, parkland and estates, but not as common as its handsome appearance suggests it ought to be. Native of China in the provinces of Shaanxi, Sichuan, Hubei, Henan, Guizhou, Jiangxi, Anhui and Hunan, but becoming scarce, though it is rather widely planted in some of China's major cities.

15. A. maximowiczianum Miq. Nikko Maple
Negundo nikoense auct.; *A. nikoense* auct.; *Crula nikoense* auct.; *A. shensiense* Fang & Hu; *A. maximowiczianum* subsp. *megalocarpum* (Rehder) A. E. Murray

Deciduous, dioecious or occasionally polygamodioecious *tree* up to 15 m, with a broadly conical crown and level branches below, becoming domed with age and flat-topped. *Trunk* often much divided. *Bark* greenish-grey, then pinkish-grey finely speckled with reddish, smooth, not peeling. *Branches* mostly erect or ascending; twigs grey; young shoots rather dark purplish-grey, with short, dense hairs. *Buds* 3–5 × 2–3 mm, ovoid-conical, narrowed to an obtuse apex; scales blackish-purple with pale grey tips where they are hairy, ovate, obtuse at apex. *Leaves* opposite, 3-foliolate; leaflets with lamina 5–20 × 2–5 cm, dark green on upper surface, paler bluish-green beneath, becoming reddish-orange to deep red in autumn, elliptic-oblong, acute at apex, entire, sinuate or obscurely dentate, cuneate at base, the middle leaflet shortly petiolulate, hairy beneath and on the margin at first, later becoming nearly glabrous; midrib nearer the inner margin, veins prominent beneath; petioles 4–6 cm, stout, green tinted pink, long-hairy. *Inflorescence* of 3-flowered fascicles, appearing well after the leaves, terminal on short shoots, the flowers 10–12 mm in diameter. *Sepals* 4–5 mm, yellowish-green, oblong, obtuse at apex. *Petals* 6–7 mm, yellowish-green, oblong, obtuse at apex. *Stamens* 10–13; filaments pale; anthers yellow. *Styles* 2. *Disc* extrastaminal. *Samara* with nutlet large, hairy and spherically convex; wing 40–50 mm, at more than 90° from other samara, glabrous. *Flowers* 4–5. $2n=26$.

Introduced. Widely planted in gardens, parks and large estates. Native of Japan in the mountain forests on Honshu, Kyushu and the Shikoku Islands; widely distributed in the western Hubei and Anhui provinces of China.

Section **6. Lithocarpa** Pax

Large deciduous *trees*. *Leaves* 5- to 7-lobed. *Bud scales* 5- to 8-paired. *Inflorescence* terminal and axillary, corymbose-paniculate. *Sepals* 5–6. *Petals* greenish-yellow. *Disc* amphistaminal. *Stamens* 8–12.

16. A. macrophyllum Pursh Oregon Maple
A. murrayanum Hort.; *A. speciosum* Hort.; *A. auritum* Greene; *A. coptophyllum* Greene; A. *dactylophyllum* Greene; *A. flabellatum* auct.; *A. hemionites* Greene; *A. leptodactylum* Greene; *A. platyterum* Greene; *A. politum* Greene; *A. stellatum* Greene

Deciduous, monoecious *shrub* or small *tree* up to 12(–25) m, when older forming a wide, compact, rounded crown. *Trunk* strong, sometimes with strong basal suckers. *Bark* at first brownish-purple with small pink lenticels, later with smooth, dark grey ridges networked above orange-brown or dark orange, finally thick, rough and deeply fissured into square-plated ridges. *Branches* thick, ascending and arching; twigs brown, glabrous; young shoots stout, deep green with small, white lenticels, sometimes green only beneath and deep purple above. *Buds* 4–5 × 3–4 mm, conical, acute at apex; scales with reddish-brown bases and green margins, ovate, obtuse at apex. *Leaves* opposite; lamina 19–27 × 20–35 cm, dark, shining, yellowish-green on upper surface, paler beneath, often turning dull brown in autumn, but sometimes yellow to orange, thin, broadly obovate in outline, palmately divided into 5–7 lobes three-quarters of the way or more towards the base, the lobes oblanceolate, narrowed to a more or less obtuse apex and with 2–5 obtuse teeth, cordate at base, often hairy when young, becoming glabrous but with tufts of hairs in the axils of each main vein; petioles up to 30 cm, reddish, with a milky sap. *Inflorescence* corymbose-paniculate, of 30–80 flowers, terminal and axillary, the flowers fragrant and about 6–9 mm in diameter, developing with the young leaves. *Sepals* 5–6, 4.5–5.5 mm, ovate, rounded at apex. *Petals* 5–6, 5.5–6.5 mm, pale greenish-yellow, ovate, rounded at apex. *Disc* amphistaminal. *Stamens* 8–12; filaments whitish; anthers yellowish. *Styles* pale, greenish. *Samara* with nutlet suborbicular, keeled, covered with dense, whitish, prickly hairs; wing 5–7 cm, horizontal, the pedicels short, thick and pale bright green. *Flowers* 4–5. $2n=26$.

Introduced. Planted in large gardens, parks and estates. Native in North America from Alaska to southern California, mixed with conifers in the forests of the Pacific coast. It bears the largest leaves of the genus.

Section **7. Platanoidea** Pax

Deciduous *trees* or *shrubs*. *Leaves* 3- to 7-lobed. *Bud scales* 5- to 10-paired. *Sepals* 5, yellowish-green to reddish or purplish. *Petals* 5, yellowish-green to reddish or purplish. *Disc* amphistaminal. *Stamens* 5–8.

17. A. campestre L. Field Maple
A. vulgare Borkh.; *A. heterolobum* Opiz

Deciduous, monoecious, *shrub* or *tree* up to 26 m, with a round crown. *Trunk* solitary or several, usually going well up into the crown, sometimes with bosses. *Bark* pale brown with wide orange fissures or cracked into squares, in older trees greyish-brown or dark grey, with fine cracks and dark corky ridges. *Wood* white and smooth with a close texture and fine grain. *Branches* spreading and turned upwards at the ends; twigs rough and striate, often becoming corky and winged; young shoots dark brown above, pale brown

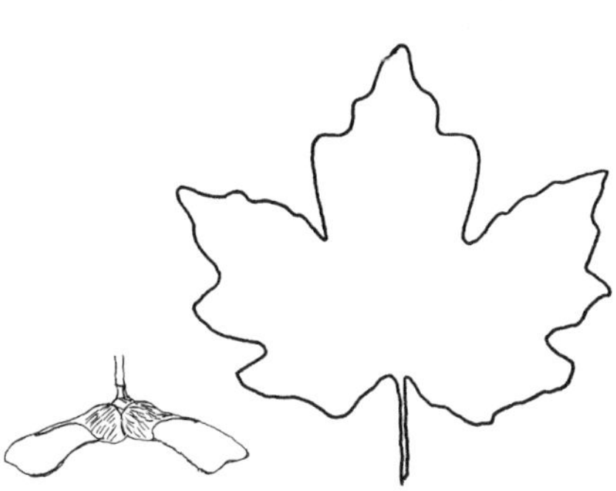

A. campestre L.
subsp. **campestre**
var. **campestre**

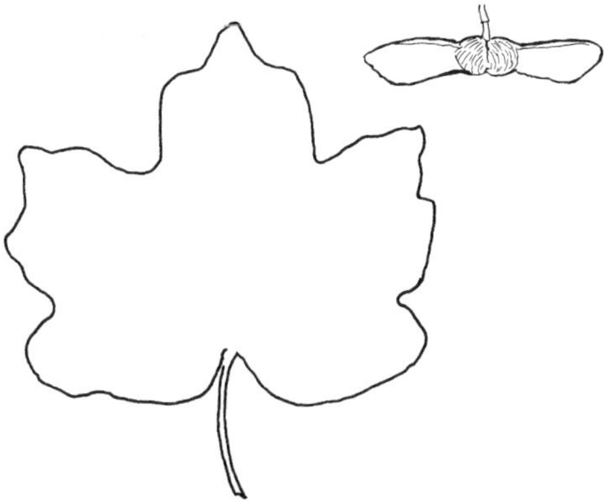

A. campestre L.
subsp. **campestre**
var. **lobatum** Pax

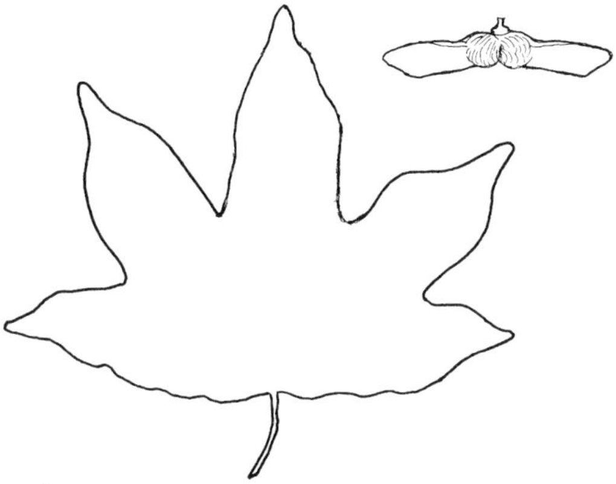

A. campestre L.
subsp. **campestre**
var. **oxytomum** Borbas

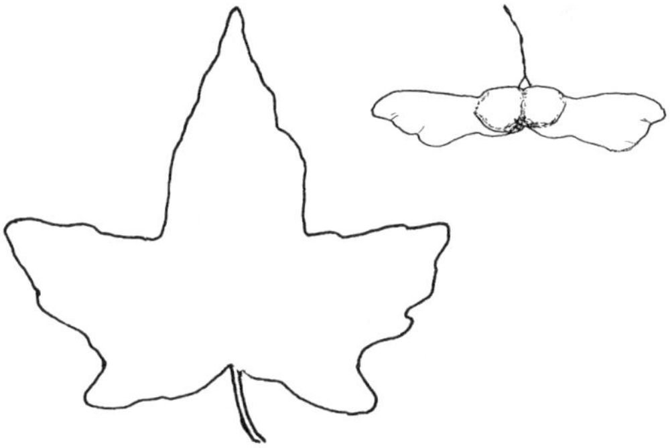

A. campestre L.
subsp. **leiocarpum** (Opiz) Pax

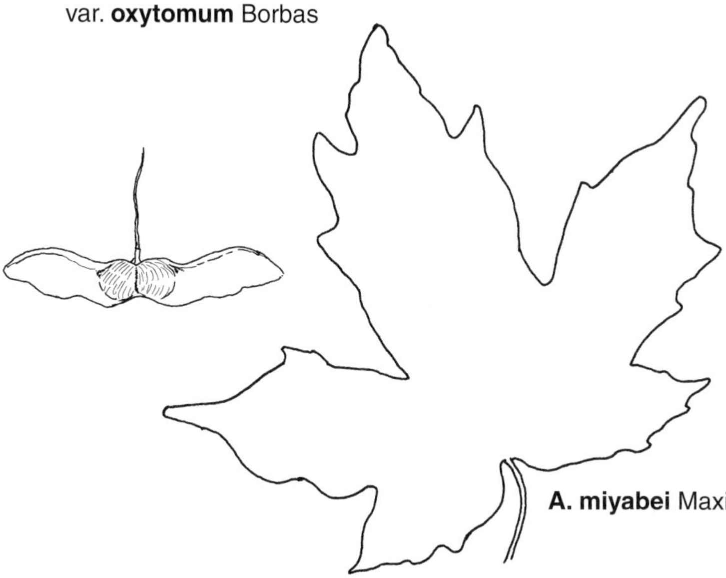

A. miyabei Maxim.

Leaves of **Acer** L.

beneath, slender, shortly eglandular-hairy. *Buds* 3.0–3.5×1.5–2.0mm, ovoid, obtuse at apex; scales brown or reddish-brown, ovate, acute at apex, hairy at tip. *Leaves* opposite; lamina 3–10×6–12cm, unfolds pinkish, turning bright then dark green on upper surface, paler beneath, changing to brownish, bronze, pale to golden yellow or red to purplish in autumn, broadly ovate in outline, 3- to 5-lobed, two-thirds of the way to base, the lobes ovate or lanceolate or oblong, rounded-obtuse or pointed-obtuse at apex, often narrowing to base, bluntly toothed and often asymmetrical, truncate or cordate at base, with tufts of hairs in the vein axils, and hairs along the veins above and below; petioles up to 8cm, exuding a milky sap when broken, slender, green or bright pink, glabrous. *Inflorescence* of more or less erect terminal corymbs; flowers coming out with the unfolding leaves. *Sepals* 5, 2–3mm, yellowish-green, triangular-ovate, obtuse at apex, long-hairy. *Petals* 5, 2–3mm, yellowish-green, oblong, obtuse at apex, long-hairy. *Disc* amphistaminal. *Stamens* 5–8; filaments pale green; anthers greenish-yellow. *Styles* 2, greenish; stigma pale brownish. *Samara* with flat shortly hairy or glabrous nutlet; wing 25–50mm, yellowish-green to crimson, nearly horizontal or slightly turned up, shortly hairy, rounded at ends. *Flowers* 4–5. $2n=26$.

(a) Subsp. **campestre**
A. campestre subsp. *hebecarpum* (DC.) Pax; *A. campestre* var. *hebecarpum* DC.

Samaras hairy.

(i) Var. **campestre**
A. campestre forma *gracile* Beldie

Leaves mostly, at least those near the inflorescences, up to 5(–6)×6(–7)cm, 5-lobed, the lobes pointed and the centre lobe about as long as wide. *Wings of samaras* usually spreading or slightly curved down.

(ii) Var. **lobatum** Pax
Leaves up to 8cm, 5-lobed, the centre lobe often wider than long and with a more rounded appearance than those of the other two varieties. *Wings of samaras* usually spreading.

(iii) Var. **oxytomum** Borbas
Leaves up to 8×10cm, 5-lobed, the centre lobe usually longer than wide and pointed. *Wings of samaras* usually spreading or slightly turned up.

(b) Subsp. **leiocarpum** (Opiz) Pax
A. leiocarpum Opiz; *A. austriacum* Tratt.; *A. campestre* var. *leiocarpum* (Opiz) Wallr.; *A. campestre* var. *austriacum* (Tratt.) DC.

Samaras glabrous or nearly so.

Subsp. *campestre* var. *campestre* is possibly our only native maple. It is a component of woods both on wet clay and dry light calcareous soils, and reaches its best development on the downs of south-east England. In open areas it is a tree of hedgerows, often left with individual trees of other species when the hedgerow is regularly trimmed, or is part of the trimmed hedge itself. Common in England and Wales north to Co. Durham. The remaining variants are all probably introduced, particularly in the last 30 years. Thousands if not millions have been planted in new woods, infills in hedgerows, in amenity areas and by motorways. They have even been given by Councils to farmers who have planted them around field boundaries. They produce copious good seed and rapidly growing young trees while our native tree seems rarely to produce offspring. All the introduced varieties seem to come into flower at least three weeks before our native plants and to be in young fruit before they break bud. The species is widespread in Europe and western Asia and in a few places in North Africa, and is mainly a tree of plains, steppes, lower hills and river-banks and is infrequent in mountains. Its wood was formerly much used for carving, particularly bowls. Two other taxa from the Balkans and Turkey, *A. hyraanum* Fisch. & C. A. Mey. subsp. *hyrcanum* and subsp. *tauricola* (Boiss. & Balansa) Yalt., look very like *A. campestre*, but have the petiole not lactiferous and the wings more downturned. They may well occur among our plantings.

18. A. miyabei Maxim. Miyabe Maple

Deciduous *tree* up to 30m with a round crown. *Trunk* up to 25cm in diameter, extending well into the crown, often fluted. *Bark* grey and wrinkled but not lumpy. *Branches* ascending and arching; twigs medium brown, striate, glabrous; young shoots pale brown, with dense hairs of various lengths. *Buds* 5–6×2.0–2.5mm, ovoid, acute at apex; scales pale brown, ovate, acute at apex, eglandular-hairy. *Leaves* opposite; lamina 7–13×8–14cm, dark green on upper surface, paler beneath, turning golden in autumn, broadly ovate in outline, palmately 5-lobed, the lobes ovate or elliptical, pointed but obtuse at apex, the terminal with 1 or 2 small lobes or teeth on each side, leaf cordate at base, shortly hairy on the veins on both surfaces; petioles up to 10cm, reddish-brown, densely short eglandular-hairy. *Inflorescence* a dense terminal cyme; peduncles and pedicels densely eglandular-hairy. *Sepals* 5, 2.5–3.0mm, yellowish-green, oblong or spathulate, rounded at apex, ciliate on margin. *Petals* 5, greenish, spathulate, rounded at apex. *Stamens* 10; filaments pale; anthers greenish. *Disc* yellowish-green. *Styles* 2. *Samara* 60–70×9–10mm, horizontal or slightly upturned, eglandular-hairy. *Flowers* 4–5.

Introduced. Widely planted in East Anglia and probably elsewhere in new 'woods', along motorways and in hedgerows and amenity areas. It reproduces freely from seed. Native of Japan. There is a possibility it is not this species but we can match it with nothing else.

19. A. cappadocicum Gled. Cappadocian Maple
A. laetum C. A. Mey.; *A. colchicum* Booth ex Gordon; *A. pictum* forma *caucasicum* K. Koch; *A. hederifolium* Rupr.; *A. cappodocicum* var. *indicum* (Pax) Rehder; *A. pictum* var. *colchicum* (Booth ex Gordon) A. Henry; *A. pictum* auct.

Deciduous, monoecious *tree* up to 30 m, with a rather narrow to broad, dense crown. *Trunk* fairly straight, but nobbly at the branches and going through much of the crown, the base surrounded by a dense thicket of suckers. *Bark* pale grey, slightly grooved and longitudinally striate. *Branches* all spreading, irregularly angled; twigs dull pale brown, glabrous; young shoots pale green, often with a greyish bloom, glabrous. *Buds* 5–6×3–4mm, ovoid, obtuse at apex; scales green with a dark margin, very broad and

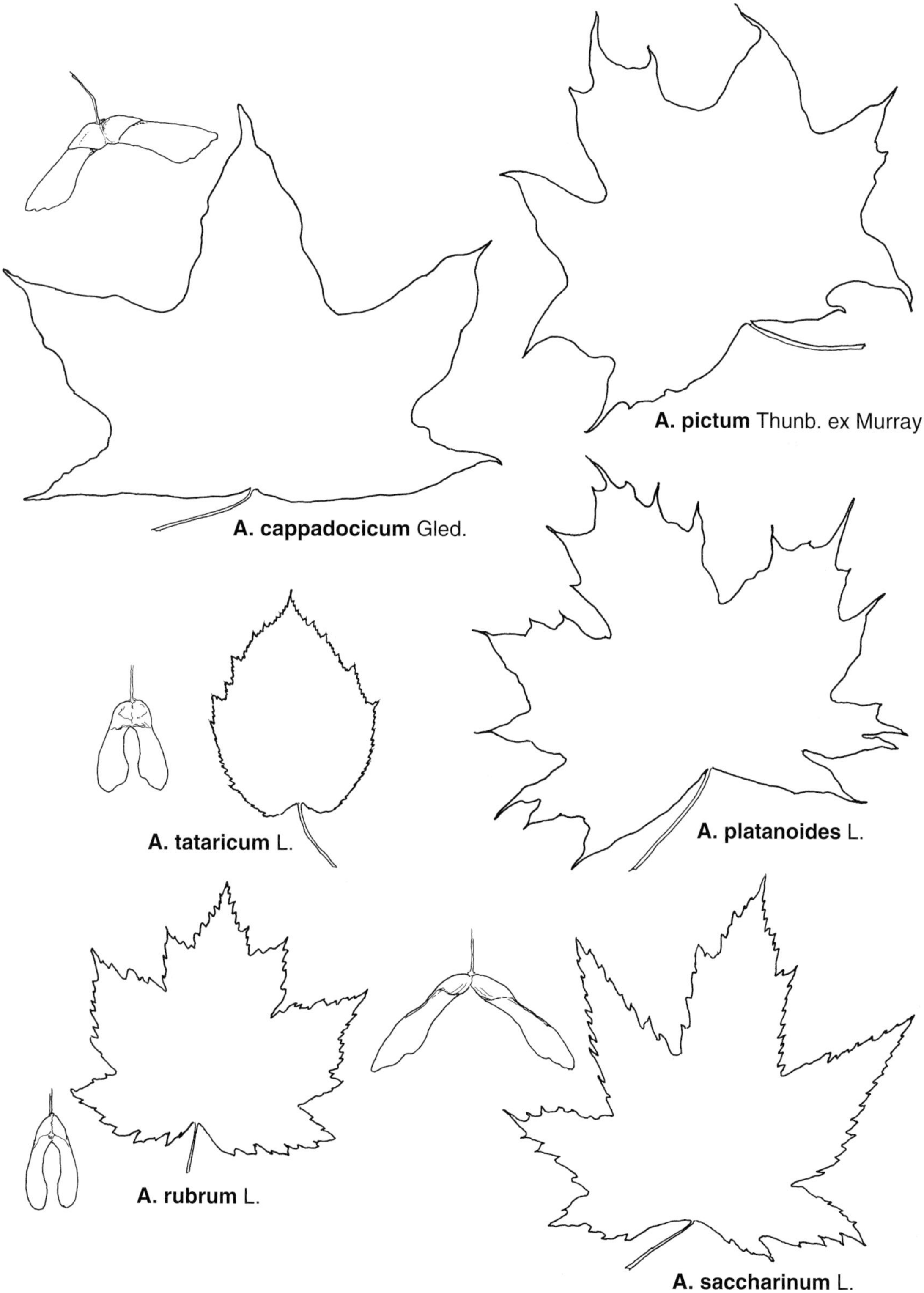

Leaves of **Acer** L.

rounded. *Leaves* opposite; lamina 7–11 × 7–20 cm, green or reddish when unfolding, later rather dark, dull green on upper surface, paler and yellowish-green beneath, turning golden yellow in autumn, ovate to very broadly ovate, 5–7 palmately lobed one-third of the way to the base, the lobes narrowly to broadly triangular-ovate, narrowly acuminate to caudate at apex and entire but undulate, shallowly cordate to truncate at base, with hair tufts in the axils of the veins beneath, otherwise glabrous; primary veins only prominent on lower surface; petioles up to 12 cm, with a milky-white latex when broken, pale yellowish-green becoming reddish tinted, glabrous. *Inflorescence* of 15–20 flowers in a wide, erect corymb, often hidden away behind unfolding leaves. *Sepals* 5, 2.0–2.5 mm, yellowish-green, narrowly triangular-lanceolate, obtuse at apex, glabrous. *Petals* 5, 2.5–3.0 mm, pale greenish-yellow, oblong to narrowly elliptical, rounded at apex. *Disc* yellowish-green, amphistaminal. *Stamens* 5–8; filaments yellowish-green; anthers yellowish-green. *Styles* greenish; stigma greenish. *Samara* with a flat nutlet; wing 15–30 × 8–12 mm, pale yellowish-green, sloping down midway between spreading and down-turned, glabrous. *Flowers* 4–5. $2n = 26$.

Introduced. Frequently planted in parkland and by roads. Self-sown and extensively suckering in south-east England, rarely north to Yorkshire. Native of south-west Asia. Our tree is subsp. **cappadocicum**.

20. A. pictum Thunb. ex Murray Painted Maple

A. mono Maxim.; *A. laetum* var. *parviflorum* Regel; *A. pictum* var. *mono* (Maxim.) Franch.; *A. lobelii* var. *mono* (Maxim.) Wesm.; *A. mono* var. *horizontale* Nakai; *A. mono* var. *savatieri* (Pax) Nakai; *A. mono* forma *septemlobum* Fang & Soong.; *A. mono* var. *tashiroi* Hisauti; *A. mono* var. *trunculatum* Nakai; *A. mono* subsp. *savatieri* (Pax) Kitam.; *A. truncatum* subsp. *mono* (Maxim.) A. E. Murray

Deciduous, monoecious or dioecious *tree* up to 25 m, with a broadly pyramidal crown. *Trunk* up to 60 cm in diameter, straight and going well into the crown. *Bark* yellowish-grey, smooth and unfissured or somewhat fissured. *Branches* spreading; twigs yellowish-grey, with lenticels, glabrous; young shoots yellowish-brown, glabrous or hairy. *Buds* 4–5 × 2–3 mm, ovoid, pointed at apex; scales dark red or purple, ovate, pointed at apex. *Leaves* opposite; lamina 6–12 × 8–15 cm, medium green on upper surface, paler beneath, turning yellow in autumn, broadly obovate in outline, 5- to 7-lobed, the lobes broadly triangular-ovate, acute or acuminate at apex and entire or nearly so, cordate at base; petioles long. *Inflorescence* of terminal corymbs; flowers 10–60, appearing before or with the leaves, trees tending to produce all male or all female flowers. *Sepals* 5, 4–5 mm, oblong, obtuse at apex. *Petals* 5, 5–6 mm, greenish-yellow, oblong, obtuse at apex. *Disc* amphistaminal. *Stamens* 5–8; filaments pale; anthers yellowish. *Styles* 2. *Samara* with flat nutlet up to 10 mm; wing 14–16 mm, sloping down and rounded-obtuse. *Flowers* 3–4. $2n = 26$.

Introduced in 1901. Planted in parks, estates and by roads. Native of China, Manchuria, Mongolia, south-eastern Siberia, Sakhalin and common in Japan. Our plant is subsp. **pictum**.

21. A. platanoides L. Norway Maple

A. platanifolium Stokes; *A. lactescens* Steud.; *A. rotundum* Dulac; *A. laetum* var. *cordifolium* Kanitz; *A. dobrudschae* Pax; *A. fallax* Pax

Deciduous, monoecious *tree* up to 30 m, with a wide, rounded crown. *Trunk* rather short, soon dividing into large branches. *Bark* pale grey, smooth and finely folded, or shallowly ridged in a network, becoming blackish-brown. *Branches* erect or spreading; twigs green to dull, dark brown, glabrous with few lenticels; young shoots pinkish-brown to olive brown, glabrous. *Buds* 5–9 × 3–6 mm, narrowly ovoid, obtuse at apex; scales dark red or reddish-brown, ovate, obtuse at apex. *Leaves* opposite; lamina 8–12 × 10–25 cm, thin, shining green on upper surface, paler beneath, turning yellow to orange-red or rarely scarlet in autumn, some cultivars red to purple throughout and some variegated yellow throughout, broadly ovate in outline, 5-lobed, the lobes broadly ovate, acute-cuspidate at apex and with a few, large, acuminate teeth or lobules, cordate at base, with small, whitish tufts of hairs in the axils of the veins beneath, otherwise glabrous; petiole up to 20 cm, sometimes red, glabrous. *Inflorescence* of erect terminal corymbs; flowers 30–40, 8–10 mm in diameter, out before the leaves, but lasting until they are out. *Sepals* 5, 5–6 mm, creamy-yellow, lanceolate, narrowed to an obtuse apex. *Petals* 5, 5–6 mm, creamy yellow, or purplish in cultivars with purplish leaves, elliptical, rounded at apex. *Disc* amphistaminal, large, green. *Stamens* 5–8; filaments pale yellowish-green; anthers yellowish cream to brownish. *Styles* 2, pale greenish; stigma greenish. *Samara* with a flat, green nutlet; wing 20–50 mm, horizontal, yellowish-green to reddish-purple. *Flowers* 3–4. $2n = 26, 39$.

(1) Forma **platanoides**

Leaves shining green above, pale beneath. *Flowers* and samaras yellowish-green.

(2) Forma **drummondii** Hegi Cv. Dummondii

Leaves variegated yellow or white. *Flowers* and samaras yellowish-green.

(3) Forma **schwedleri** K. Koch Cv. Schwedleri

A. platanoides forma *rubra* (Herder) Pax

Leaves reddish to purplish. *Flowers* and samaras reddish to purplish.

Introduced before 1683. Abundantly planted and often self-sown in rough grassland, scrub, hedges, woodland, gardens, shelter-belts, parks and streets. Throughout the lowlands of Great Britain and Ireland. Native of northern Europe from Norway to Russia, but not crossing the Urals, south to central Europe and the Caucasus and Crimea, but not in western Europe, though widely cultivated there and in North America. All three forms are widely cultivated. It is possible some records are *A. saccharum* as most books do not clearly distinguish it.

Section **8. Ginnala** Nakai

Deciduous *shrubs* or small *trees*. *Leaves* undivided or 3-lobed. *Bud scales* 5- to 10-paired. *Inflorescence* terminal or axillary and corymbose. *Sepals* 5, greenish. *Petals* 5, greenish-white. *Disc* extrastaminal. *Stamens* 8.

22. A. tataricum L. Tartar Maple
A. tartaricum Moench; *A. cordifolium* Moench; *A. tartarinum* Poir.

Deciduous, monoecious, small *tree* or shrub up to 10 m, with a narrow crown. *Trunk* irregular and up to about half-way up the tree. *Bark* pale grey, smooth, striped pale buff. *Branches* crowded, erect-spreading or spreading, angled; twigs dull brown, glabrous; young shoots brownish-green, glabrous. *Buds* small and inconspicuous; scales few, greyish-brown, ovate. *Leaves* opposite; lamina 5–9 × 3–5 cm, slightly shiny, medium yellowish-green on upper surface, slightly paler beneath, turning ochre, then reddish to brownish-red in autumn, or sometimes they just shrivel and fall, lanceolate, ovate or obovate, acuminate at apex, serrate-dentate, or the larger with 2- to 3- pairs of shallow lobes, narrowed or rounded at base; veins 5–7 pairs, impressed above, prominent beneath, sometimes eglandular-hairy; petioles 2–5 cm, green, whitish or reddish, grooved. *Inflorescence* a terminal, suberect panicle; flowers 20–30, appearing with the opening leaves. *Sepals* 5, 3.0–3.5 mm, yellowish-green, ovate, rounded at apex. *Petals* 5, 3.0–3.5 mm, pale green to white, oblong or oblanceolate, rounded at apex. *Disc* pale green, extrastaminal. *Stamens* 8; filaments white or very pale green; anthers green. *Styles* 2, greenish; stigma reddish. *Samarae* in large clusters; nutlet rounded and somewhat wrinkled; wing 20–25 mm, yellowish-green, turning deep red in autumn, but with green outer edge, pointing downwards and nearly parallel. *Flowers* 5–6. Visited by flies. $2n = 26$.

Introduced in 1759. Planted in parks, estates and by roads. Native of south-east Europe and south-west Asia. Our plant is subsp. **tataricum.**

Section **9 Rubra** Pax

Deciduous *trees*. *Leaves* 3- to 5-lobed. *Bud scales* 4- to 7-paired. *Inflorescence* axillary and corymbose. *Sepals* 5, red. *Petals* 5, red, or corolla 5-lobed. *Disc* extrastaminal. *Stamens* 5–8.

23. A. rubrum L. Red Maple
A. carolinianum Walter; *A. barbatum* Michx; *A. sanguineum* Spach; *Saccharodendron barbatum* (Michx) Nieuwl.; *Rufacer rubrum* (L.) Small

Deciduous, monoecious or rarely dioecious *tree* up to 20(–40) m, with a narrow to very broad, open crown. *Trunk* up to 1 m in circumference, straight and rising well into the crown. *Bark* pale grey and smooth when young, becoming darker and cracking into long plates when old. *Wood* pale brown, tinged red, very heavy and close-grained. *Branches* ascending and spreading; twigs brownish-red to grey; young shoots red-brown to coppery-brown with numerous lenticels, slender and glabrous. *Buds* 3–35 × 1–8 mm, narrowly ovate, obtuse at apex; scales dark reddish-brown, ovate, rounded at apex, ciliate. *Leaves* opposite; lamina 6–10 × 7–11 cm, opening reddish-green and shiny on both surfaces, turning yellowish-green then dark green on upper surface, silvery beneath, turning scarlet and gold and finally wine red in autumn, ovate or elliptical in outline, 3- to 5-lobed, the lobes triangular to ovate, acute at apex and crenate-serrate or biserrate, rounded, truncate or slightly cordate at base; petiole 7–9 cm, red above, yellowish-green beneath, glabrous. *Inflorescences* of fascicles lateral along twigs; flowers appearing abundantly before the leaves and lasting some time, male and female often on different trees. *Sepals* 2.0–2.5 mm, red, oblong, obtuse at apex, connate. Petals 2.0–2.5 mm, red, linear to oblong, obtuse at apex. *Disc* extrastaminal, red. *Stamens* 5–8; filaments and anthers red. *Styles* 2, connate at base; stigmatic surface long. *Samara* with nutlet small and convex; wing 1.2–2.5 cm, red or brown pointing down and narrowly parted; ripening early and germinating immediately after falling. Flower 3–4. $2n = 78$, rarely 91 or 108.

Introduced in 1656. Planted in gardens, parks, churchyards and along roadsides. Native of eastern North America, west to Winnipeg, the Dakotas and the Great Plains.

24. A. saccharinum L. Silver Maple
A. sylvestre Young; *A. glaucum* Marshall; *A. floridanum* Lézermes; *A. rubrum* var. *pallidum* Aiton; *A. dasycarpum* Ehrh.; *A. eriocarpum* Michx; *A. tomentosum* Steud.; *A. coccineum* Loudon; *A. floridum* Loudon; *A. macrocarpum* Loudon; *A. album* Nicholson; *A. palmatum* Nicholson, non Thunb.; *A. douglasii* Dieck; *A. lutescens* Pax; *A. macrophyllum* Pax; *A. spicatum* Pax, non Lam.; *Saccharosphendamnus saccharinus* (L.) Nieuwl.; *Argentacer saccharinum* (L.) Small

Deciduous, monoecious or dioecious *tree* up to 25(–40) m, with an open, rather narrow or spreading crown. *Trunk* up to 1 m in circumference, straight and well up into the crown. *Bark* smooth, orange-brown to grey with a glaucous bloom when young, becoming reddish-brown and developing a few shallow fissures and frequent burrs and sprouts, later flaking into large, thin scales and may be slightly shaggy. *Wood* hard, strong and close-grained, pale in colour with a tinge of brown. *Branches* graceful, slender and pendulous; twigs greenish-brown tinged red; young shoots at first pale green with numerous lenticels, becoming bright brownish-red with some grey bloom and then deep purple. *Buds* 3–20 × 2–6 mm, oblong-ovoid, slightly angled; scales 4–7 paired, bright red, ovate, apiculate at apex and minutely ciliate. *Leaves* opposite; lamina 8–15 × 9–16 cm, opening orange or dark red, becoming pale brown then bright greenish-yellow to dark green on upper surface, silvery-white beneath, turning pale yellow in autumn, broadly ovate in outline, deeply 5-lobed, sometimes even to base, the lobes lanceolate, ovate-lanceolate or oblong-lanceolate, sharply acute or acuminate at apex, serrate or incise-serrate, the central lobes sometimes with 3 acute lobules, cordate or truncate at base, thickly hairy beneath when young; petiole 6–13 cm, pink to bright red, slender. *Inflorescence* an axillary corymb at the nodes of shoots of the previous year; flowers appear before the leaves. *Perianth* 2.0–2.5 mm, 5-lobed, greenish to dark red, narrow in male flowers, broad in female flowers, the lobes rounded at apex. *Disc* extrastaminal. *Stamens* 5–8; filaments pale green, long and slender; anthers pale yellow. *Styles* 2, united at the base only;

stigmatic lobes long. *Samara* with small, convex nutlet; wing 30–60 mm, yellowish-green with thick, pink veins, twisted, broad at the ends, subhorizontal; pedicel up to 30–50 mm, dark red. *Flowers* 3–4. $2n=52$.

Introduced in 1725. Frequent in large gardens, by roads and in parks and squares. Native of eastern North America from Quebec to Florida, west to Minnesota, Kansas and Oklahoma.

111. ANACARDIACEAE Lind. nom. conserv.

Deciduous, dioecious or monoecious *shrubs* or small *trees*. *Leaves* alternate, imparipinnate, or simple, exstipulate. Inflorescence a large terminal panicle. Flowers bisexual or unisexual, sometimes mixed in the same panicle, hypogynous, actinomorphic. *Calyx* 5-lobed. *Petals* 5. *Stamens* 5. *Styles* 3; stigma capitate. *Ovary* 1-celled, with 1 basal ovule. *Fruit* a small 1-seeded drupe.

Contains about 73 genera and 600 species, mostly tropical or subtropical, but with some genera from the Mediterranean and temperate North America.

1. Leaves imparipinnate **1. Rhus**
1. Leaves simple **2. Cotinus**

1. Rhus L.

Deciduous, dioecious or monoecious shrubs or small trees. Leaves alternate, imparipinnate, exstipulate. *Inflorescence* a large, terminal, pyramidal panicle with small flowers. *Calyx* 5-lobed. *Petals* 5. *Stamens* 5, reduced to staminodes in female flowers. *Stigmas* 3, at apex of ovary. *Fruit* a drupe.

Contains about 150 species from temperate and subtropical North America, southern Africa, subtropical east Asia and north-east Australia.

Bean, W. J. (1976). *Trees and shrubs hardy in the British Isles*. Ed. 8. **2**. London.
Coombes, A. J. (1994). Cut-leaved sumacs. *The New Plantsman* **1**(2): 107–113.
More, D. & White, J. (2003). *Trees of Britain and northern Europe*. London.
Rushforth, K. (1999). *Trees of Britain and Europe*. London.

1. Leaflets serrate and sessile **1. typhina**
1. Leaflets entire and shortly petiolulate **2. verniciflua**

1. R. typhina L. Stag's-horn Sumach
R. hirta (L.) Sudw.

Deciduous, dioecious or monoecious *shrub* or small *tree* with a broad, rounded, rather open crown. *Stems* 1–several, up to 25 cm in diameter, not usually extending into the crown, often suckering. *Bark* greyish-brown, very rough and scaling. *Wood* pale orange-coloured and soft. *Branches* erect or arching, brown; twigs brown, thick, shortly eglandular-hairy; young shoots thick, velvety, with dense, short, brown simple eglandular hairs. *Buds* about 3 mm, conical, pointed; scales pale brown, ovate, acute, enclosed in the base of the petiole at first. *Leaves* alternate, 15–60 × 10–22 cm, oblong in outline, imparipinnate; leaflets with lamina 11–27, 4–16 × 0.8–4.0 cm, deep, slightly shiny yellowish-green on upper surface, pale greyish-green beneath, turning yellow then deep orange-red in autumn, oblong or elliptic-oblong, long acute or acuminate at apex, serrate with broad teeth, rounded at the sessile base, eglandular-hairy on both surfaces at first but soon becoming glabrous, the veins up to 22 pairs, impressed above and prominent below; rhachis pale green, becoming suffused purple, shortly eglandular-hairy; petiole up to 12 cm, pale green, often suffuse brownish-purple, densely eglandular-hairy. *Inflorescence* 10–20 cm, a terminal, pyramidal panicle; flowers bisexual and unisexual mixed, or unisexual. *Calyx* 2–3 mm, 5-lobed, the lobes ovate, obtuse at apex. *Petals* 5, more or less free, greenish-yellow in male flowers, rusty-red in female, ovate, obtuse at apex. *Stamens* 5; filaments pale; anthers yellow. *Styles* 3; stigmas capitate. *Drupe* about 3 × 2 mm, brown, ovoid, acute at apex, densely covered with long, red simple eglandular hairs; seed about 3 × 2 mm, white to grey, broadly oblong, rounded at both ends. *Flowers* 5–6. $2n=30$.

Introduced in 1629 by John Parkinson, apothecary to James I, who kept medicinal gardens in London's Long Acre. An extract of the roots was used to treat fevers. Now much planted on verges and banks by roads and railways where it suckers extensively but is rarely or never self-sown. Scattered records outside gardens throughout Great Britain but particularly in the south. Native of North America.

2. R. verniciflua Stokes Varnish Tree

Deciduous, monoecious or dioecious *tree* up to 20 m, with a conical crown when young, becoming broader and domed when old. *Trunk* up to 80 cm in diameter, straight, extending well into the centre of the crown. *Bark* dark grey with black diamond shapes which coalesce to form fissures, in old trees becoming somewhat flaky. *Branches* ascending or spreading; twigs brown, thick; young shoots pale grey, rough with orange-brown lenticels. *Buds* about 10 mm, ovoid, pointed; scales chestnut-brown, ovate. *Leaves* alternate, 50–80 cm, broadly oblong in outline, imparipinnate, aromatic; leaflets with lamina 3–19, 10–20 × 5–10 cm, glossy green on upper surface, turning red or crimson in autumn, paler beneath, ovate, with a narrow, acute apex, entire, narrowed or rounded at base, shortly petiolulate, glabrous on upper surface, densely, shortly and softly hairy beneath; rhachis pale green, tinted reddish-brown, leathery; petiole pale green, often tinted reddish-brown. *Inflorescence* a large, lax panicle 25 × 15 cm, in the axils of the current season's shoots in the upper crown. *Calyx* 2–3 mm, with 5 lobes, the lobes ovate, obtuse. *Petals* 5, yellowish-white, *Stamens* 5; filaments pale; anthers yellow. Styles 3; stigmas capitate. *Drupe* up to 6 × 8 mm, glossy creamy-brown to yellow, only sometimes produced. *Flowers* 7–8. $2n=30$.

Introduced. Planted in gardens, parks and estates. Mainly in southern England and Ireland. This is the tree that yields the famous varnish or lacquer of Japan. Native of the Himalaya and China, but cultivated and possibly native in other parts of east Asia and Malaysia.

2. **Cotinus** Adans.

Deciduous, usually dioecious *shrubs* or small *trees*. *Leaves* alternate, simple, entire, exstipulate. *Inflorescence* a large, loose, spreading or drooping panicle, densely feathery and giving the appearance of a smoky haze. *Calyx* 5-lobed, the lobes overlapping. *Petals* 5. *Stamens* 5. *Stigma* on side of ovary. *Ovary* 1-celled. *Fruit* a drupe.

Contains 3 species occurring from southern Europe to northern China and the south-eastern United States.

Bean, W. J. (1970). *Trees and shrubs hardy in the British Isles*. Ed. 8. **1**. London.
Rushforth, K. (1999). *Trees of Britain and Europe*. London.

1. Leaves and inflorescence green
1(1). coggygria forma **coggygria**
1. Leaves and inflorescence deep reddish-purple
1(1). coggygria forma **atropurpurea**

1. C. coggygria Scop. Smoke Tree
Rhus cotinus L.

Deciduous *shrub* or small *tree* up to 6 m, with an ovoid to domed crown. *Trunk* short. *Bark* greyish-brown, fissured. *Branches* ascending or spreading; twigs grey and rough; young shoots yellow and red, glabrous. *Buds* small, ovoid, pointed; scales brown, ovate, acute at apex, glabrous. *Leaves* alternate; lamina 3–8 × 2.5–4.0 cm, dull yellowish-green on upper surface, slightly paler beneath, turning red and yellow, or dark maroon-purple turning red in autumn, broadly elliptical, ovate or obovate, rounded at apex, entire but slightly wavy, cuneate at base, glabrous; veins 8–10 pairs, impressed above with the midrib pale, prominent beneath, especially the midrib; petiole up to 4 cm, green or reddish, glabrous. *Inflorescence* a large, loose, terminal panicle; flowers 3–6 mm in diameter; peduncle and pedicels green tinted red, or purple, the pedicels sometimes with red hairs, when fruiting brown or purplish then grey with a multitude of peduncles and pedicels covered with plume-like hairs, the whole appearing like smoke above the leaves in summer and autumn. *Sepals* about 1.5 mm, green with a pale margin or purple, ovate, acute at apex, glabrous. *Petals* 5, 1.5–3.0 mm, greenish or brownish, triangular, pointed at apex, spreading. *Stamens* 5, filaments green or red; anthers yellow. *Styles* 3, reddish; stigmas capitate. There are numerous intermediate stages between purely male flowers, bisexual flowers and purely female flowers on the same plant. *Female flowers* are smallest (3–4 mm in diameter) and male flowers largest (5–6 mm in diameter). Some female flowers possess anthers shaped like those in male flowers, but have abnormal pollen grains; others have only vestigial stamens. There is a yellow or orange disc at the base of all flowers on which nectar lies exposed. *Drupe* 3–4 mm, brown, obovoid, net-veined. *Flowers* 6–7.

(1) Forma **coggygria**
Leaves and *inflorescence* green.

(2) Forma **atropurpurea** (Dippel) C. K. Schneid
C. coggygria var. *atropurpurea* Dippel; *Rhus cotinus* var. *purpureus* Dup.-Jam.; *C. coggygria* forma *purpureus* (Dup.-Jam.) Rehder

Leaves and *inflorescence* deep reddish-purple.

Introduced. Commonly grown in gardens, parks, estates, amenity areas, and on road and railway banks, occasionally producing seedlings. Native of southern Europe eastwards to the Himalayas and northern China. The young branches have been used for tanning leather and the bark for dyeing. The plant is poisonous. Both forms are equally grown.

112. SIMAROUBACEAE DC. nom. conserv.

Deciduous, monoecious or dioecious *trees*. *Leaves* alternate, imparipinnate, petiolate, exstipulate. *Inflorescence* a large terminal panicle; flowers unisexual and bisexual mixed, hypogynous, actinomorphic. *Calyx* 5-lobed nearly to the base. *Petals* 5, free. *Stamens* 10. *Style* 5(–6); stigmas peltate. *Ovary* of 5(–6) carpels loosely fused, each carpel with 1 axile ovule. *Fruit* a group of 1–5(–6) winged samaras each containing a single seed; endosperm scarce or absent.

Contains 20 genera and about 120 species mostly in the topics and subtropics.

1. **Ailanthus** Desf.

As family.

Contains 10 species from temperate and tropical Asia and Australasia.

Bean, W. J. (1970). *Trees and shrubs hardy in the British Isles*. Ed. 8. **1**. London.
More, D. & White, J. (2003). *Trees of Britain and northern Europe*. London.
Rushforth, K. (1999). *Trees of Britain and Europe*. London.

1. A. altissima (Mill.) Swingle Tree of Heaven
Toxicodendron altissimum Mill.; *A. glandulosa* Desf. nom. illegit.

Deciduous, monoecious or dioecious *tree* 15–30 m, with a narrow crown and often suckering. *Trunk* up to 90 cm in diameter, cylindrical, soon dividing into large branches. *Bark* greyish-brown to blackish-brown with white streaks in young trees, becoming shallowly fissured. *Wood* pale brown. *Branches* large, spreading and then arching down; twigs grey, thick; young shoots reddish-grey or coppery-brown, stout, with lenticels. *Buds* 1.5–2.0 × 3.5–4.0 mm, ovoid, rounded at apex; scales reddish-brown, becoming scarlet as the bud swells, ovate. *Leaves* alternate, strongly foetid when bruised, 30–60(–90) × 20–30 cm, ovate-oblong in outline, imparipinnate; leaflets 7–25, subopposite, with lamina 7–15 × 3.5–9.0 cm, dull medium to rather dark green on upper surface, with very pale green veins, and with bronze and crimson tints when opening, very pale glaucous-green beneath, without autumnal colour, ovate to ovate-oblong, bluntly acuminate at apex, entire but slightly sinuate, truncate or subcordate at base and glabrous; veins 7–8 main pairs which are prominent and with a fine network of lesser ones between; petiolules 10–15 mm, pale green and reddish where they join the rhachis; petioles up to 11 cm, pale green, glabrous; rhachis pale brownish-yellow, glabrous; with 1–3, large glands on the margin of the lamina near the base which act as extrafloral nectaries and dispense nectar to ants who keep the foliage clear of insects and caterpillars. *Inflorescence* a large terminal panicle 10–20 cm, male and female flowers on different trees or with bisexual and unisexual flowers on same tree, the male foul-smelling. *Calyx* 1.5–2.0 mm, green, divided nearly to the base into 5 lobes, the lobes ovate, acute at apex. *Petals* 5, 2.0–3.5 mm, greenish-white, elliptical, pointed at apex. *Stamens* 10, filaments pink; anthers green. *Styles* 5(–6); stigma peltate, greenish. *Samarae* borne in bunches, 25–40 × 10–13 mm, pale yellowish-green turning orange-red, the wing narrowly elliptical, shortly and bluntly acuminate at apex and entire; seed 1, in centre. *Flowers* 6–7. Visited by bees. $2n = 64, 80$.

Introduced in 1751. Frequently planted in large gardens, parks and estates in south-east England, especially in streets, squares and parks of Greater London where it often forms extensive suckers and is frequently self-sown; rare further north. Native of China. The name *Ailanthus* is derived from the Moluccan name *ailanto*, for another species of this genus. Its vernacular name means 'sky tree' hence Tree of Heaven, because of its great height.

113. RUTACEAE Juss. nom. conserv.

Deciduous or evergreen *shrubs*. *Leaves* opposite or alternate, simple, ternate, pinnate or deeply pinnately lobed, petiolate, without stipules. *Inflorescence* of variously arranged groups of flowers, the flowers bisexual, hypogynous, actinomorphic, with a well-developed nectariferous disc. *Sepals* 4 or 5, free or fused above base and only free at base and at apex. *Petals* as many as sepals, free or fused above base and free only at base and apex. *Stamens* twice as many as sepals. *Style* 1; stigma 1 or as many as carpels, capitate. *Ovary* of fused carpels as many as sepals and visible as obvious lobes, each carpel with 2–many ovules. *Fruit* a capsule, 4-, 5- or 8-lobed.

Contains about 140 genera widely distributed in temperate and tropical regions and particularly well represented in South Africa and Australia.

1. Herbaceous perennials or small shrubs; leaves pinnatisect	**1. Ruta**
1. Shrubs or trees; leaves simple	2.
2. Leaves ternate	**2. Choisya**
2. Leaves simple	3.
3. Leaves alternate; without stellate hairs	**3. Skimmia**
3. Leaves opposite; with stellate hairs	**4. Correa**

1. Ruta L.

Aromatic perennial *herbs* becoming woody at base. *Leaves* alternate, pinnatisect, covered with sessile glands. *Inflorescence* cymose. *Sepals* 4. *Petals* 4–5, yellow. *Stamens* 8–10. *Style* 1; stigmas 4. *Fruit* a 4-lobed capsule.

Contains 5 species in Europe and western Asia and 2 in Macaronesia.

Bean, W. J. (1980). *Trees and shrubs hardy in the British Isles*. Ed. 8. **4**. London.
Grieve, M. (1931). *A modern herbal*. London.

1. R. graveolens L. Rue

Aromatic *perennial herb* becoming woody towards the base. *Stems* up to 50 cm, pale glaucous-green, erect, glabrous. *Leaves* alternate; lamina deeply pinnatisect, oblong in outline; the segments 2–9 mm wide, bluish-green, lanceolate to narrowly oblong or obovate, obtuse at apex, covered with sessile translucent glands which give off a characteristic smell; petiole of lower long, the uppermost almost sessile. *Inflorescence* cymose, rather lax; pedicels as long as or longer than the capsule; bracts lanceolate and leaf-like; flowers 15–20 mm in diameter. *Sepals* 4, lanceolate, acute at apex. *Petals* 4–5, yellow, oblong-ovate, denticulate, undulate. *Stamens* 8–10; filaments pale; anthers yellow. *Styles* 1; stigmas 4, capitate. *Disc* lobed. *Capsule* 4-lobed, the lobes somewhat narrowed above to an obtuse apex. *Flowers* 6–7. $2n = 72$.

Introduced. Much grown in gardens as an ornamental and sometimes producing seedlings in shrubberies. East Mediterranean region and naturalised from gardens throughout south and south-central Europe. Grown for many centuries as a medicinal herb, but dangerous if taken in large quantities. The sap can cause a severe rash on skin if exposed to sunlight.

2. Choisya Kunth

Evergreen, monoecious *shrub*. *Leaves* opposite, ternate. *Inflorescence* a cluster of axillary corymbs. *Sepals* 5, yellowish-green. *Petals* 5, white. *Stamens* 10. *Style* 1; stigma 5-lobed. *Capsule* of 5 fused carpels, each 2-lobed and with 2 ovules.

Contains about 7 species in Mexico and the south-western United States. Named after Jaques-Denis Choisy (1797–1859).

Turrill, W. B. (1958). *Choisia ternata*. *Bot. Mag.* **172**: tab. 318.
Bean, W. J. (1970). *Trees and shrubs hardy in the British Isles*. Ed. 8. **1**. London.

1. C. ternata Kunth Mexican Orange
C. grandiflora Regel; *Juliania caryophillata* La Llave

Evergreen *shrub* up to 2(–3) m, often wider than high with an irregular bushy habit. *Stems* erect to widely spreading and much divaricately branched. *Bark* dull, dark brown, rough. *Branches* terete, spreading or erect; twigs grey; young shoots pale yellowish-green, with numerous to dense stellate and very short simple eglandular hairs. *Leaves* opposite, 5–12 × 6–20 cm, broadly ovate in outline, ternate; leaflets with lamina 2.5–8.0 × 1.0–3.5 cm, dark, shining green with a pale midrib on upper surface when mature, young leaves yellowish-green, much paler beneath with an even paler midrib, oblanceolate, obovate or elliptical, rounded at apex, entire, gradually narrowed to a sessile base, pitted on both surfaces with numerous oil glands, which when crushed give off an unpleasant odour, with some short hairs on margin and veins beneath; midrib prominent beneath, lateral veins 5–7 pairs, prominent but not as much as midrib; petiole 5–45 mm, pale green, with numerous stellate and very short simple eglandular hairs. *Inflorescence* a cluster of axillary corymbs at the end of a short branch, each corymb 3- to 6-flowered; peduncles and pedicels pale green, with dense, short, pale simple eglandular hairs and very short, pale glandular hairs; bracts and bracteoles 4–7 × 2–5 mm, lanceolate to ovate, more or less acute, hairy; flowers 20–30 mm in diameter, strongly sweet-scented. *Sepals* 5, 4.5–5.0 × 2.5–3.0 mm, pale yellowish-green, ovate, subacute at apex, with very short hairs. *Petals* 5, free, 12–15 × 7–10 mm, white, elliptical or ovate, rounded at apex. *Stamens* 10; filaments 5–6 mm, white; anthers yellow. *Styles* 5, united; pale yellowish-green; stigmas 5. *Capsule* of 5 carpels, hairy, each 2-lobed and with 2 ovules. Not known to fruit in Great Britain. *Flowers* 4–5, but often up to 9 and sometimes to 12, and in Devon and Cornwall it is at its best in 11–12.

Introduced. Widely planted in gardens, parks, roadsides and public places. Very widespread, even near London. Sometimes self-sowing and naturalised in south-east England and Isle of Man. Native of Mexico introduced in 1825.

3. Skimmia Thunb. ex Murray

Evergreen, usually dioecious, aromatic *shrubs. Leaves* alternate, simple, entire, without stipules. *Inflorescence* a terminal panicle. *Calyx* shortly 4- to 5-lobed. *Petals* 4 or 5, white. *Stamens* 4 or 5. *Style* 1. *Fruit* a drupe.

Contains 4 species from China, Japan and the Himalayas.

Bean, W. J. (1980). *Trees and shrubs hardy in the British Isles*. Ed. 8. **4**. London.
Thiselton-Dyer, W. T. (1905). *Skimmia japonica. Bot. Mag.* tab. 8038.

1. S. japonica Thunb. ex Murray — Skimmia
S. oblata T. Moore; *S. fragrans* Carrière

Evergreen, usually dioecious, aromatic shrub up to 2 m, about as wide as high. Branches dense, spreading, pale greyish-brown; twigs greyish-brown; young shoots red, glabrous. *Leaves* alternate; lamina 5.0–12.5 × 1.2–1.8 cm, pale to dark, shining green on upper surface, paler beneath, coriaceous, obovate-oblong to lanceolate or elliptic, obtuse, acute or acuminate at apex, entire, attenuate at base, covered with transparent sessile glands; veins 4–8(–11) pairs, obscure except for the midrib; petiole short and thick. *Inflorescence* a terminal thyrsoid panicle; pedicels short and thick; bracts small and opposite. *Calyx* shortly 4- or 5-lobed, the lobes triangular and obtuse, often suffused red. *Petals* 4 or 5, 5–6 mm, white, oblong, rounded at apex. *Stamens* 4 or 5; filaments white; anthers yellow. *Style* 1, green, short and thick; stigma capitate. *Drupe* 8–10 × 7–9 mm, dark red when ripe, ellipsoid to nearly obovoid; seeds 3, whitish, ellipsoid. *Flowers* 3–4. $2n = 36$.

Introduced. Garden escape on Brading Down, Isle of Wight. Commonly planted in parks, estates and amenity areas. Native of east Asia.

4. Correa J. Kenn.

Evergreen, monoecious *shrubs* or bushy *trees*. *Leaves* opposite, entire, stellate-hairy. *Inflorescence* of solitary flowers or few-flowered cymes. *Calyx* cup-shaped, not or hardly lobed. *Corolla* infundibuliform, 4-lobed. *Stamens* 8, 4 of them with winged filaments. *Style* 1; stigma of 4 short lobes. *Fruit* a drupe, splitting into 4 carpels except at base, each with 1–2 seeds.

Contains about 12 species confined to Australia.

Bean, W. J. (1970). *Trees and shrubs hardy in the British Isles*. Ed. 8. **1**. London.
Turrill, W. B. (1957). *Correa backhousiana* Hook. *Bot. Mag.* nov. ser. **171**: tab. 289.
Wilson, P. G. (1961). A taxonomic revision of the genus *Correa* (Rutaceae). *Trans. Roy. Soc. South Australia* **85**: 21–53.

1. C. backhouseana Hook. — Tasmanian Fuchsia
C. speciosa var. *backhouseana* (Hook.) Benth.

Evergreen, monoecious *shrub* or bushy tree up to 5 m. *Branches* spreading, greyish-brown, terete and rather slender; twigs more or less glabrous; young shoots densely covered with appressed stellate hairs. *Leaves* opposite; lamina 1.5–3.5 × 1.2–2.2 cm, dark green on upper surface, paler beneath, ovate-elliptical to subrotund, obtuse to rounded at apex, entire, rounded or slightly cordate at base, with scattered, small stellate hairs on upper surface, covered with rusty stellate hairs beneath; midrib prominent beneath, lateral veins obscure; petiole 3–8 mm, densely and shortly stellate-hairy. *Inflorescence* of solitary flowers or in few-flowered cymes; pedicels 5–8 mm, densely and shortly stellate-hairy. *Calyx* 5–6 mm, cup-shaped, not or hardly lobed, densely stellate-hairy. *Corolla* 2–6 cm, infundibuliform, 4-lobed, the lobes triangular-ovate and acute at apex, the tube green passing into tawny or russet above, densely stellate-hairy. *Stamens* 9; filaments 26–29 mm, 4 of them winged; anthers greenish-yellow. *Styles* 1, about 25 mm; stigma of 4 short lobes. *Drupe* splitting into 4 carpels, except at base, each with 2 seeds. *Flowers* 6–8. $2n = 32$.

Introduced. Garden escape abundantly naturalised in woods on Tresco in the Isles of Scilly. Native of Tasmania.

Order 14. GERANIALES Lindl.

Herbs or small *shrubs*. *Leaves* usually alternate or basal, usually with stipules. *Flowers* bisexual or rarely unisexual, hypogynous, actinomorphic to zygomorphic. *Sepals* imbricate. *Petals* often clawed. *Stamens* as many as 3 times the numbers of sepals or petals. *Ovary* syncarpous, 3- to 5-celled; placentation axile. *Fruit* various, but very rarely fleshy; seeds usually with endosperm.

Contains 5 families and about 2,600 species.

114. OXALIDACEAE R. Br. nom. conserv.

Perennial, rarely *annual*, monoecious, often slightly succulent *herbs*, sometimes with bulbs and bulblets. *Stems* often rhizomatous. *Leaves* all basal or alternate, usually 3-foliolate, sometimes palmate, petiolate, with or without stipules. *Inflorescence* of 1–several-flowered, axillary, often umbelliform cymes; flowers bisexual, hypogynous, actinomorphic, often trimorphic. *Sepals* 5. *Petals* 5. *Stamens* 10, sometimes not all with anthers. *Styles* 5; stigmas minute. *Ovary* 5-celled, each cell with many ovules on axile placentas. *Fruit* a 5-celled capsule.

Contains 6 genera and about 950 species, distributed worldwide from tropical to cold regions.

1. Oxalis L.

As family.

Contains 3 genera and nearly 900 species, tropical and temperate.

Grime, J. P. et al. (1988). *Comparative plant ecology*. London.
Hultén, E. & Fries, M. (1986). *Atlas of north European vascular plants north of the Tropic of Cancer.* 3 vols. Königstein.
Knuth, R. (1930) in Engler, A. (Edit.). 130. Oxalidaceae. *Das Pflanzenreich* **IV**. Leipzig.
Lourteig, A. (1975). Oxalidaceae extra-austroamericanae I. *Oxalis* L. Sectio *Thamnoxys* Planchon. *Phytologia* **29**: 449–471.
Lourteig, A. (1982). Oxalidaceae Extra-Austroamericanae IV. *Oxalis* L. Section *Articulatae* Knuth. *Phytologia* **50:** 130–142.
Packham, J. B. (1978). Biological flora of the British Isles. *Oxalis acetosella* L. *Jour. Ecol.* **66**: 669–693.
Reid, J. A. (1982). Differences in the flowering behaviour of *Oxalis corniculata* L. and *O. exilis* A. Cunn. *Watsonia* **14**: 63–65.
T. M. Salter (1944). The Genus *Oxalis* in South Africa. *Jour. South African Bot.* (Suppl.) **1**.
T. M. Salter (1944). The Genus *Oxalis* in South Africa. *Jour. South African Bot.* (Suppl.) **4**.
D. P. Young (1958). *Oxalis* in the British Isles. *Watsonia* **4**: 51–69.

1. Petals yellow 2.
1. Petals red, pink, mauve or white 10.
2. Three outer sepals cordate at base; leaves succulent **8. megalorrhiza**
2. No sepals cordate at base; leaves thin and not succulent 3.
3. Aerial stems absent; bulbils present at or below soil level **16. pes-caprae**
3. Aerial stems present; bulbils absent 4.
4. Stems procumbent, rooting freely at nodes 5.
4. Stems decumbent to erect, not or very sparsely rooting 7.
5. Inflorescences always 1-flowered; capsules 3.0–4.5 mm, with 3–4 seeds per cell; usually 5 stamens with anthers and 5 without **4. exilis**
5. At least most inflorescences 2–8(–12)-flowered; capsules (4–)8–20 mm, with more than 4 seeds per cell; usually all 10 stamens with anthers 6.
6. Stems, stipules and leaves green; capsule densely glandular-hairy **3(i). corniculata** var. **corniculata**
6. Stems, stipules and leaves violet-purple; capsule glabrous to slightly hairy **3(ii). corniculata** var. **atropurpurea**
7. Petals 12–16 mm, with purple veins **1. valdiviensis**
7. Petals 5–11 mm, not purple-veined 8.
8. Petals up to 6 mm **7. perennans**
8. Petals 5–11 mm 9.
9. Inflorescence an umbel; pedicels patent or reflexed in fruit; vegetative parts with only white simple hairs **5. dillenii**
9. Inflorescence cymose; pedicels erect in fruit; vegetative parts with septate as well as white simple hairs **6. stricta**
10. Leaves with 5–11 leaflets **15. decaphylla**
10. Leaves with 3–4 leaflets 11.
11. Plant with a more or less erect stem 12.
11. Plant with stem absent or as a rhizome at or below soil level 13.
12. Bulbs absent; inflorescence more than 1-flowered **2. rosea**
12. Stem arising from a bulb below ground and with axillary aerial bulbs above; inflorescence with a single flower **17. incarnata**
13. Bulbs absent; stem a rhizome 14.
13. Leaves arising from a bulb at or below soil level; bulb often producing thin rhizomes 17.
14. Rhizomes thick, with dense, papery scales; flowers in umbelliform cymes 15.
14. Rhizome slender, with distant, fleshy scales 16.
15. Hairs on plant long; leaves linear-lanceolate **9(a). articulata** subsp. **articulata**
15. Hairs shorter; leaves elliptical **9(b). articulata** subsp. **rubra**
16. Leaflets 2.5–9.0 mm; petals pure white **10. magellanica**
16. Leaflets 10–30 mm; petals white with lilac to pale purple or violet veins **11. acetosella**
17. Leaves with 4 leaflets **14. tetraphylla**
17. Leaves with 3 leaflets 18.
18. Leaflets widest at or near apex, without spots near the margin **13. latifolia**
18. Leaflets widest about the middle, with orange or dark spots near the margin on the lower side 19.
19. Petals salmon-pink to brick red; styles with glandular hairs **12(i). debilis** var. **debilis**
19. Petals pinkish-mauve; styles without glandular hairs **12(ii). debilis** var. **corymbosa**

1. O. valdiviensis Barnéoud Chilean Yellow Sorrel
Acetosella valdiviensis (Barnéoud) Kuntze

Annual monoecious *herb* with fibrous roots and without bulbs. *Stem* up to 30 cm, sometimes almost absent, erect or decumbent, succulent, not or little branched, glabrous. *Leaves* scattered alternately up the stem; leaflets 3, with lamina 8–9 × 8–9 mm, medium green on upper surface, paler beneath, obovate, emarginate at apex, entire, narrowed below, glabrous or nearly so; petioles up to 14 cm, glabrous or sometimes minutely hairy. *Flowers* bisexual, in long, forked cymes; peduncles much longer than the leaves; pedicels 5–10 mm, glabrous; bracts linear-subulate. *Sepals* 5, about 4 mm, lanceolate or ovate, obtuse at apex, glabrous. *Petals* 5, 12–16 mm, yellow, with purple veins, obovate, rounded at apex. *Stamens* 10. *Styles* 5. *Capsule* about 6 mm, subglobose, pedicels reflexed, the capsules pointing downwards. *Flowers* 6–8. $2n = 18$.

Introduced. Garden weed, reproducing from seed. Scattered records in Great Britain. Native of Chile.

2. O. rosea Jacq. Annual Pink Sorrel
O. floribunda Lindl., non Lehm.; *O. rubra* auct.

Annual monoecious *herb*, with fibrous roots and without bulbs. *Stems* up to 25 cm, erect to ascending, glabrous, usually much branched and leafy. *Leaves* alternate; leaflets 3, with lamina 4–11×4–11 mm, pale green, sometimes reddened beneath, ovate, deeply emarginate at apex, entire, cuneate at base, glabrous or nearly so; petioles up to 3 cm. *Inflorescence* composed of 1–3 bisexual flowers in loose cymes; pedicels reddish. *Sepals* 5, red-tipped, without apical warts, pale mauve-pink with darker veins and a white throat, rarely entirely white, obovate, rounded at apex. *Petals* 5, 11–17 mm, pale mauve-pink. *Stamens* 10. *Styles* 5. *Capsule* rounded. *Flowers* 6–8. $2n=12$, 14.

Introduced. Garden weed, reproducing by seed. Jersey, Cornwall, Leicestershire and Northumberland. Native of Chile.

3. O. corniculata L. Procumbent Yellow Sorrel
Xanthoxalis corniculata (L.) Small

Perennial monoecious *herb* with fibrous roots and without bulbs. *Stems* many, up to 50 cm, pale green or flushed violet, prostrate with the tips ascending, internodes variable in length, rooting at nodes, much branched, eglandular-hairy. *Leaves* alternate, sometimes in fascicles; leaflets 3, with lamina 4–20×7.5–20 mm, medium green on upper surface, paler beneath, or suffused violet-purple, obovate, deeply divided at apex with rounded, obtuse lobes, cuneate below, glabrous on upper surface, eglandular-hairy beneath, often only on one half of leaf; petioles up to 7 times longer than lamina; stipules 0.8–3.0×1–3 mm, truncate, ciliate. *Inflorescence* of umbelliform 2–8(–12)-flowered cymes; peduncles up to 21 cm, ascending, loosely eglandular-hairy, bracts 0.3–3.0 mm, elongating in fruit, linear-triangular; pedicels 4–15 mm, lengthening in fruit and deflexed; bracteoles smaller than bracts, linear, ciliate. *Sepals* 2.5–5.5×4–15 mm, greenish or violet, linear to narrowly ovate, acute at apex, sparingly eglandular-hairy. *Corolla* yellow, divided to halfway; lobes 5–11×1.0–2.5 mm, oblong-subspathulate. *Stamens* 10, glabrous, the longer about 6 mm, the shorter about 4 mm, connate for the basal third. *Styles* 5, micro-, meso- and macrostylous, up to 8 mm, long-connate, appressed-retrorse-hairy; stigma 2-lobed, papillose. *Ovary* oblong, acute at apex, retrorse-hairy, the cells with 4–15 ovules. *Capsule* (4–)8–20×1–3 mm, cylindrical, acute at apex, densely retrorse-glandular-hairy, rarely glabrous; seeds more than 4 per cell, about 1.0 mm, reddish-brown, ovoid, apiculate, flattened, 6-ribbed, transversely striate and finely tuberculate at the intersections. *Flowers* 6–9. Reproducing mainly by seed. $2n=24$.

(i) Var. **corniculata**
Stems, stipules and leaves green. Micro-, meso- or macrostylous. Capsule densely glandular-hairy.

(ii) Var. **atropurpurea** Planch.
O. corniculata var. *viscidula* Wiegand.

Stems, stipules and leaves violet-purple. *Micro-* and macrostylous. *Capsule* glabrous to slightly hairy.

Introduced. Pernicious weed of gardens, paths, walls and waste ground. Common in most of England, Wales, Isle of Man and Channel Islands and very scattered localities in Scotland and Ireland. Originated in the European Mediterranean region but now distributed worldwide. It is well adapted to all climates, requiring only some humidity and reaching its largest size in cultivated areas. It can become weedy and hybridise with autochthmous species of the same section. Our plants belong to subsp. **corniculatus**. Var. *corniculatus* is the common plant. Var. *atropurpurea* is of European cultivated origin and is found in gardens and greenhouses and often escapes.

4. O. exilis A. Cunn. Least Yellow Sorrel
O. corniculata var. *microphylla* Hook. fil.; *O. repens* auct.

Perennial monoecious *herb* with a slender to stout primary root, without bulbs. *Stem* to 15 cm, pale green, more or less filiform, creeping to prostrate, rooting at the nodes, much branched and often mat-forming, glabrous to densely hairy. *Leaves* alternate; leaflets 3, lamina 1–9×1.5–12.0 mm, medium green on upper surface, often purplish beneath, obovate or ovate, often emarginate at apex, entire, narrowed at base, glabrous to hairy on both surfaces, the hairs beneath appressed, sometimes ciliate; petiole 5–40 mm, with sparse to dense, reflexed hairs; petiolules very short; stipules adnate, truncate or rounded at apex. *Flowers* bisexual, usually solitary, very rarely paired; peduncles glabrous to densely reflexed-hairy, generally geniculate at apex; pedicels when present (4–)8–25 mm; bracts 1–3 mm, linear-subulate, hairy, arising from the middle to base of the peduncles. *Sepals* 5, 2.5–3.5(–4.0) mm, oblong, oblong-lanceolate or ovate-elliptical, acute at apex, glabrous to reflexed-hairy. *Petals* 5, 4.5–9.0(–13.0) mm, yellow, often pink-flushed, oblong-obovate, glabrous. *Stamens* 10, in 2 series, 1 without anthers; filaments united at base. *Styles* 5, densely eglandular-hairy. *Capsule* 3.0–4.5 mm, broadly cylindrical, conical or cylindrical-ovoid, glabrous to densely eglandular-hairy, sometimes glandular; seed 3–4 per cell, 1.0–1.3(–1.5) mm, broadly ellipsoid, with 7–13 transverse ridges, sulcate at apex. *Flowers* 6–8.

Introduced. Grown in garden rockeries and escaping into disturbed ground, path-sides, by walls and crevices in pavements. Scattered throughout Great Britain and northern Ireland; most populations are casual but persist in some localities for many years. Native of Australia and New Zealand.

5. O. dillenii Jacq. Sussex Yellow Sorrel
O. stricta auct.; *O. corniculata* auct.; *O. navierei* Jord.

Perennial monoecious *herb* with a fibrous root and without bulbs, behaving as an annual. *Stem* up to 20 cm, pale green, erect or decumbent, occasionally rooting but without specialised stolons, often branched from the base, rather fleshy, strigose-hairy. *Leaves* subopposite or in groups or fascicles, trifoliolate; leaflets 1.0–1.8 cm, obovate, emarginate at apex, entire, narrowed below, glabrous on both sides or sometimes with a few simple hairs beneath; petioles strigose-hairy. *Inflorescence* a (1–)2–3(–4)-flowered umbel, not a cyme; peduncles longer than the leaves, strigose hairy; pedicels 8–25 mm, patent or reflexed in fruit, strigose-hairy. *Sepals* 5, up to 5 mm, narrowly lanceolate. *Petals* 5, 7–11 mm, yellow, oblanceolate. *Stamens* 10, in 2 series; filaments glabrous. *Styles* 5, minutely hairy. *Capsule* 15–25 × about 3.0 mm, finely and densely appressed-hairy. *Flowers* 8–10. $2n = 16–24$.

Introduced. Weed in sandy arable fields. Known near Pulborough in Sussex from 1950 to 1984 and recorded at Herm in the Channel Islands, Sussex, Hertfordshire, Nottinghamshire, Lanarkshire, Berwickshire and Clyde Islands. Native of North America.

6. O. stricta L. Upright Yellow Sorrel

O. fontana Bunge; *O. europaea* Jord.

Short-lived *perennial* or *annual herb* with a fibrous root. *Stem* 5–40 cm, pale green, usually erect, with underground stolons from the base and sometimes horizontal but not rooting, branches above ground, glabrous or with septate hairs. *Leaves* subopposite, fascicled or whorled, 3-foliolate; leaflets 12–30 mm wide, medium green on upper surface, paler beneath, sometimes purple, obovate, emarginate at apex, entire, narrowed at base, with septate as well as simple hairs; stipules none or a narrow wing adnate to the enlarged basal joint of petiole; petioles glabrous or strigose-hairy. *Inflorescence* of 2–3 flowers, often cymose; pedicels 8–10 mm, erect in fruit. *Sepals* 5, lanceolate. *Petals* 5, 5–10 mm, yellow, oblanceolate. *Stamens* 10, in 2 series; filaments glabrous. *Styles* hairy. *Capsule* 8–12 × 2.5–3.5 mm, cylindrical, gradually acuminate, glabrous or with only a few scattered hairs. *Flowers* 7–10. $2n = 18–28$.

Introduced. Weed of gardens and arable fields, reproducing mainly by seed. Scattered over most of Great Britain and Ireland and common in parts of southern England. Native of North America.

7. O. perennans Haw. Woody Oxalis

Bushy, *perennial* monoecious *herb* with a very stout primary root, without bulbs. *Stems* numerous, up to 20 cm, pale green, mostly more or less erect, slender, sometimes the outer prostrate, glabrous to densely clothed with reflexed hairs. *Leaves* alternate; leaflets 3, with lamina 3–14 × 1–4 mm, medium green on upper surface, sometimes purplish, shaped like the tail of a fish, glabrous or sometimes with appressed hairs beneath, ciliate; petioles up to 5 cm, glabrous to densely clothed with reflexed hairs; stipules about 1.5 mm, nearly completely adnate to the petiole, ciliate. *Flowers* 1–3; peduncles 2–8 cm, glabrous or hairy; pedicels 5–15 mm, glabrous or hairy; bracteoles 1–2 mm, linear-subulate, hairy, attached at the base of the pedicels. *Sepals* 5, 3–4 mm, ovate-elliptical or elliptic-oblong, with appressed, reflexed hairs. *Petals* 5, up to 6 mm, yellow, more or less obovate, glabrous. *Stamens* 10, in 2 series; filaments united in lower half, glabrous. *Styles* 5, densely hairy. *Capsule* 10–20 mm, narrowly cylindrical, often falcate, more or less densely clothed with short, retrorse hairs; seeds about 1.5 mm, oblong-ellipsoid, with (8–)10–11 transverse ridges. *Flowers* 5–6.

Introduced. A wool casual. Native of Australia and New Zealand.

8. O. megalorrhiza Jacq. Fleshy Yellow Sorrel

O. carnosa auct.; *Acetosella megalorhiza* (Jacq.) Kuntze

Perennial monoecious *herb* with a stout root, without bulbs. *Stems* up to 20 cm, erect to ascending, subterranean and aerial, pale green, thick and succulent, more or less unbranched, glabrous. *Leaves* alternate; leaflets 3, with lamina 1.1–1.8 × 1–2 cm, bluish-green and succulent, obovate, entire, rounded to very shallowly notched at apex, narrowed to base, glabrous, covered with translucent cells on the lowerside; petiole up to 8 cm, glabrous. *Flowers* bisexual, 2–5 in an umbel; peduncles up to 10 cm long. *Sepals* 5, about 5 mm, triangular, inner narrower than the 3 cordate outer. *Petals* 5, about 15 mm, bright yellow. obovate, emarginate. *Stamens* 10, in 2 series; filaments glabrous or slightly hairy; anthers yellow. *Styles* 5, filiform; stigmas green. *Capsule* about 7 × 4 mm, oblong, greenish. *Flowers* 5–6.

Introduced. Naturalised on walls and banks. Isles of Scilly since about 1936 and one locality in Cornwall. Native of Chile.

9. O. articulata Savigny Pink Sorrel

O. floribunda auct.; *O. semiloba* auct.; *O. rosea* auct.

Perennial monoecious *herb* with rhizomes and sometimes tubers, but without arial stems. *Rhizome* up to 15 cm by 2 cm in diameter, tortuous, woody, brown, dark, bearing the scars of the bases of stipules adnate to the petioles of decayed leaves; with dense papery scales. *Roots* tuberous, rather woody, cylindrical, knobbly or articulate, dark and thick, sometimes with a long branch bearing a tuber at its apex. *Leaves* in fascicles of 3–30; leaflets 3, with lamina 5–30 × 7–45 mm, medium green on upper surface, paler beneath, widely obovate, slightly retuse or acute at apex, incised to one-fifth of its length, with straight, appressed, uniform hairs, dense on both surfaces or only below, or shorter and looser, a purple basal callus sometimes present and orange, unequal calluses abundant near the margins, unequal and rarely having them patterned on the surface; petioles 13–30 cm, with ascending, dense, short, appressed hairs to glabrous; petiolules 0.5 mm, thick, densely eglandular-hairy; stipules 4–16 × 2.0–6.5 mm, ovate or triangular, narrowed towards the apex, with ascending-appressed hairs and connate to the petiole. *Inflorescence* of an umbelliform, simple cyme, with 2–5 bisexual flowers, or bifid, asymmetrical cymes with up to 20 flowers, or bifid cymes composed of umbelliform cymes; peduncles 2.5–40 cm, longer than the leaves, eglandular-hairy; bracts 2–3 mm; pedicels 10–55 mm, long-filiform, articulate near the base; bracteoles 0.5–1.0(–2.0) mm, oblong or linear, appressed-hairy, with 1–2 orange calluses at the apex. *Sepals* 5, 2.5–6.5 × 1–2 mm, elliptical to linear-lanceolate, acute except for the internal sepal at apex, often with 3–4, unequal calluses, densely appressed-hairy, the hairs straight. *Corolla* 2–3 times the length of the sepals, divided to halfway, usually deep pink, rarely pale pink or white; lobes 5, obovate-oblong, clawed, the areas exposed in bud covered with white, sericeous-appressed hairs. *Stamens* 10; filaments white, sometimes tinted pink, linear, basally broad and abruptly narrowed about the middle, the longer ones about 3.5 mm and hairy, the shorter ones about 2 mm and glabrous; anthers golden yellow, ovoid. *Styles* about 5.5 mm, greenish, densely hairy; stigma small, green, capitate. *Ovary* hairy at least at the apex, cells with 4–8 ovules. *Capsule* 8–11 mm, cylindrical or oblong, acute at apex, densely ascending-appressed hairy throughout or only at the apex of the carpels; seeds 1.0–1.2 mm, pale brown, ovoid or ellipsoid, acute at both ends, 8- to 9-ribbed in zigzag, transverse striae with 4–8 deep pits. *Flowers* 5–10.

(a) Subsp. **articulata**
Hairs on plant long. *Sepals* linear-lanceolate.

(b) Subsp. **rubra** (A. St Hil.) Lourteig
O. rubra A. St Hil.

Hairs on plant shorter. *Sepals* elliptical.

Introduced. Subsp. *articulata* is much grown in gardens and is an established escape in waste and stony places, sandy ground, roadsides, banks and seashores, often in closed vegetation. Frequent in south-west England and the Channel Islands, with scattered localities in central and south Great Britain and Ireland. Originally from Uruguay, extreme southern Brazil and some localities in Argentina. It has been introduced and cultivated in other American countries and in the Old World where it has escaped and become naturalised. Some of our plants may be subsp. *rubra*. A specimen collected by J. E. Lousley at Bryher in the Isles of Scilly in 1938 was so named by A. Lourteig.

10. O. magellanica G. Forst. White Sorrel
O. lactea Hook. fil.

Perennial monoecious *herb* with a slender, scaly rhizome, sometimes mat-forming, without bulbs. *Leaves* alternate, the basal rosette occurring in tufts at intervals; leaflets 3, with lamina 2.5–9.0 × 2.5–9.0 mm, medium green on upper surface, often dark reddish to deep purple beneath, broadly obovate, emarginate at apex, entire, narrowed below, glabrous to hairy; petiole (5–)15–50(–90) mm, often dark reddish; stipules 3–5 mm, broad and scarious. *Flowers* bisexual, solitary; peduncles 1–6(–11) cm, slender and hairy; bracts 2.5–4.0 mm, linear-lanceolate, glabrous. *Sepals* 5, 3–6 mm, more or less elliptical, with more or less appressed hairs. *Petals* 5, 8–15(–20) mm, white, broadly obovate, usually glabrous, sometimes ciliate. *Stamens* 10, in 2 series; filaments not dilated at base. *Styles* glabrous. *Capsule* 4–5 mm, subglobose; seeds 5–7 mm, ellipsoid-ovoid, with prominent raised reticulations. *Flowers* 6–8.

Introduced. Known only as a relic of cultivation in flower beds in the Isle of Man. Native of South America, Australia and New Zealand.

11. O. acetosella L. Wood Sorrel
Perennial monoecious *herb* with a slender, creeping rhizome clothed with distant, fleshy, scales without a bulb. *Leaves* alternate, scattered; leaflets 3, with lamina 10–30 × 15–30 mm, bright medium yellowish-green on upper surface, paler beneath, obovate, emarginate at apex, narrowed below, ciliate and with scattered, appressed hairs on the surfaces; petiole 5–15 cm. *Flowers* bisexual, solitary; peduncles about equalling petioles; bracts 2, small, in the middle of the peduncle. *Sepals* 5, oblong-oblanceolate, subobtuse at apex. *Petals* 5, 8–16 mm, white with lilac to pale purple or violet veins, obovate, rounded at apex, late flowers without petals and cleistogamous. *Stamens* 10; filaments glabrous; anthers yellow. *Style* glabrous. *Capsule* 3–4 mm, ovoid to subglobose. *Flowers* 4–5. $2n = 22$.

Native. Woodland, hedgerows, banks and other moist shady places, also rough mountain pasture and grikes in limestone from sea level to 1160 m. Throughout most of Great Britain and Ireland. Most of Europe, but rarer in the south; north and central Asia to Sakhalin and Japan. Eurasian Boreo-temperate element.

12. O. debilis Kunth Large-flowered Pink Sorrel
Perennial monoecious *herb* with roots, bulb and bulblets. *Root* vertical, thick with fibrous branches; bulb 10–20 × about 30 mm, globose, ovoid or oblong, formed of nutritious scales, the scales 7–11 × 3–5 mm, brown, often covered with rigid, acuminate, orange calluses, the margin hyaline, ciliate with long, smooth hairs up to 4 mm, rusty, the 3 longitudinal veins connate below the apex; when mature the bulb develops into a large mass of bulblets each 3–6 mm and sessile in the axils of the old scales. *Leaves* arising from bulb; leaflets 3, with lamina 12–55 × 15–47 mm, medium green on upper surface, paler beneath, subrotund or broadly obovate, incised up to one-fifth of its length, cuneate at base, with sparse hairs on both surfaces or only beneath and minute punctiform translucid orange or violaceous oxalate crystals scattered on the surfaces, more visible below and frequently with a larger and more prominent row along the margin; petiolules 0.5–1.0 mm, fleshy, eglandular-hairy; petioles up to 30 cm, with sparse, rarely dense hairs; stipules about 13 × 2 mm, brown, scarious, linear, 3-veined, the veins, convergent just below the apex, sometimes with 2 additional thinner lateral veins, running from the central vein into the petiole, the margins ciliate like the scales, the bulblets covered by similar smaller scales. *Inflorescences* of bifid cymes, the branches unequal, sometimes twice bifid, generally asymmetrical or umbelliform, loosely 3- to 15-flowered, the flowers bisexual; peduncle to 45 cm; bracts opposite, 2–5 × 2.0–2.5 mm, ovate or rhombic, hyaline; pedicels 4–20 mm, articulate 1–3 mm from the insertion; bracteoles 0.2–1.0 × 1–3 mm, inserted near the base of the pedicel on the hairy articulation. *Sepals* 5, 4–7 × 0.7–1.5 mm, subequal, greenish with a hyaline margin, linear to elliptical, obtuse to acute at apex, ciliate at apex, the calluses orange, linear or enlarged towards the apex, parallel or oblique, rarely convergent or divided, giving 3 or 4 calli. *Petals* 5, 8–12 × 14–21 mm, salmon-pink, brick-red or pinkish-mauve, obovate, clawed. *Stamens* 10; filaments enlarged towards the base, the longer about 4.5 mm, hairy towards the apex and with some short glandular hairs mixed, the shorter about 3.0 mm, glabrous, connate for one-third of its length. *Style* macrostylous or mesostylous, rarely microstylous, with scattered ascending eglandular-hairs sometimes mixed with short glandular hairs; stigma bifid, subcapitate. *Capsule* up to 17 mm, cylindrical; seeds 3–10, brownish, ovoid, flattened, 12-ribbed. *Flowers* 7–9. $2n = 14–28$.

(i) Var. **debilis**
Petals salmon-pink to brick-red. *Styles* with glandular hairs.

(ii) Var. **corymbosa** (DC.) Lourteig
O. corymbosa DC.; *O. martiana* Zucc.; *O. urbica* A. St Hil.; *O. bipunctata* Graham ex Hook.; *Acetosella martiana* (Zucc.) Kunze; *Ionoxalis martiana* (Zucc.) Small

Petals pinkish-mauve. *Styles* without glandular hairs.

Introduced. Weed of gardens and other open ground, reproducing by bulblets only. Var. *corymbosa* is frequent in southern England and the Channel Isles, with scattered records in central and southern Great Britain and eastern Ireland. Var. *debilis* occurs in a few gardens in southern England. The species is native of South America, but now occurs throughout the world, usually reproducing by abundant bulblets, but occasionally producing good seed.

13. O. latifolia Kunth — Garden Pink Sorrel
O. vespertilionis A. Gray, non Zucc.; *O. intermedia* Rich.; *O. mauritiana* Lodd.; *Ionoxalis latifolia* (Kunth) Rose

Perennial monoecious *herb* with underground stem reduced to a disc, the leaves and peduncles having their bases modified to form nutritive bulbil scales. *Bulb* 10–15 mm, ovoid or globose; scales brown, the outer few protective only, papyraceous and ciliate; the inner all nutritive, vertically ridged; bulblets numerous, produced on short runners from the base of the bulb. *Leaves* 1–5; leaflets 3, with lamina 10–45 × 20–75 mm, green on upper surface, violet beneath, triangular, shortly mucronate or broken in a broad, obtuse angle and the lobes divergent, rarely incised quarter of the way to the midrib, glabrous or sparsely hairy on the veins beneath and on the margins, small calluses of various shapes also present; petiolules about 1 mm, hairy; petioles 5–15 cm, glabrous or somewhat hairy; stipules rusty-brown, ovate or rectangular, auriculate at apex, ciliate with viscid, rusty hairs, 3 prominent veins arising from the base and with orange calluses converging just below the apex, connate to the petioles. *Inflorescence* of umbelliform cymes often twice as long as the leaves, up to 30 cm, loosely 5- to 20-flowered; pedicels 14–20 mm, thin; bracteoles about 1 mm; bracts about 2 × 2 mm, hyaline, broadly ovate to rhombic or elliptical, acute at apex, 1-veined, softly crenate and clasping the peduncle. *Sepals* 5, 3–5 × 0.5–1.5 mm, green with a hyaline margin, linear-elliptical to oblong, narrowed to the obtuse or subacute apex, glabrous, with orange or violet calli at the apex. *Petals* 5, 2.5–3.5 times the length of the calyx, the pink to purplish limb with a white base, obovate, spathulate, clawed. *Stamens* 10, filaments enlarged towards the base, hairy in the upper half, the longer about 4 mm, the shorter about 3 mm, connate for one-quarter of their length. *Style* microstylous or rarely macrostylous, 4.5–5.5 mm, glabrous; stigma bifid, papillose. *Capsule* 4.0–8.5 mm, oblong, acute at apex, glabrous; seeds 1.0–1.1 mm, brownish, ellipsoid, slightly flattened, 8- to 9-ribbed. *Flowers* 5–9. $2n = 14, 24$.

Introduced. Weed of gardens and open ground. Frequent in southern England and Channel Islands, scattered records elsewhere including one in Ireland. Native of Mexico, Costa Rica, the Antilles and the Andes of South America south to Bolivia; introduced in Brazil, Europe, Africa, Australia and New Zealand.

14. O. tetraphylla Cav. — Four-leaved Pink Sorrel
O. deppei Lodd. ex Sweet; *Acetosella tetraphylla* (Cav.) Kuntze; *Ionoxalis tetraphylla* (Cav.) Rose; *Ionoxalis deppei* (Lodd. ex Sweet) Small

Perennial herb with vertical, fasciculate, turnip-like or cylindrical roots, which sometimes have profuse and very long fibrous branches; bulb 4 × 3 cm, globose or ovoid, formed by the reduced conical or discoid stem, the scales and the stipules connate to the petioles; outer scales scarious, brownish or with a hyaline margin, the outermost row 4–18 × 1–12 mm, linear, 3- to 5-veined, sharp at apex, the other 15–17 × 5–8 mm, ovate or ovate-triangular, abruptly acute at apex, (5–)7- to 14-veined; inner nutritious scales 3–7 × 2–7 mm, yellow, thick, rigid, flat-convex or subconcave, 3- to 5(–7)-veined; bulbils distant on thin stolons. *Leaves* arising from the bulb; leaflets (3–)4, with lamina 20–70 × 25–75 mm, medium green with a purplish to blackish centre on upper surface, paler beneath with reddish veins, obovate or subtriangular, rarely subrotund or with divergent lobes, truncate, roundish or slightly incised at the apex, the midrib sometimes mucronate, hairy, with small, unequal, translucid or dark oxalate spots near the incision on the margin or scattered on the upper part of the lamina; petioles 5–30 cm, glabrous or sparsely eglandular hairy. *Inflorescences* of bifid, umbelliform cymes, with 2–3 bisexual flowers; pedicels up to 17 mm, articulate at the base; bracteoles hyaline, ovate to rhombic, generally keeled, the midrib lengthened to a mucro, sometimes covered with oxalate calli, irregularly long ciliate; peduncles up to 50 cm; bracts 2–4 × 2.0–2.5 mm. *Sepals* 5, 4.0–7.5 × 1–2 mm, green, oblong to elliptical, with a hyaline margin, obtuse, truncate or subacute at apex, glabrous or with a few glandular or eglandular hairs, with 2–4, unequal, linear calluses. *Petals* 5, pink to violet, the lower part, greenish-white, 2–5 times the length of the sepals. *Stamens* 10; filaments in 2 series, longer filaments 2.5–4.0 mm and with short glandular and eglandular hairs, shorter filaments 1.5–2.5 mm, glabrous. *Styles* mostly macrostylous, rarely mesostylous or microstylous, with glandular and eglandular hairs; stigma bifid. *Capsule* about 13 mm, cylindrical, acute at apex, glabrous; seeds about 1.2 mm, tawny, ovoid, flattened, with 9–14 zig-zag ribs and 12–16 transverse striae. *Flowers* 5–9. $2n = 56$.

Introduced. Weed of gardens and arable land, reproducing only by bulblets. Jersey, Isles of Scilly, Somersetshire, Dorsetshire, Hampshire, Surrey and Berkshire. Native of Mexico.

15. O. decaphylla Kunth — Ten-leaved Pink Sorrel
O. lasiandra auct.; *O. grayi* (Rose) Kunth; *Ionoxalis decaphylla* (Kunth) Rose; *Ionoxalis grayi* Rose; *Acetosella decaphylla* (Kunth) Kuntze; *Ionoxalis confusa* Rose; *Inoxalis furcata* Rose; *Ionoxalis jaliscana* Rose; *Ionoxalis occidentalis* Rose; *Ionoxalis painteri* Rose; *O. painteri* (Rose) Knuth; *O. confusa* (Rose) Knuth; *O. furcata* (Rose) Knuth; *O. jaliscana* (Rose) Knuth; *O. zacatesasensis* Knuth

Perennial monoecious *herb* with a bulb and rarely bulblets. *Bulbs* 10–20(–30) × 10–15(–30) mm; scales (3–)5- to 20 (–30)-veined, pale brown to reddish-brown, papyraceous, ovate, with ascending white hairs usually at base adaxially and sometimes with glandular hairs on the margins; bulblets seldom formed. *Leaves* arising from the bulb; leaflets with lamina (3–)5–11, (10–)17–38(–72) × (2–)8–25(–40) mm, medium green on upper surface, paler beneath, obtriangular, obovate or oblanceolate, usually lobed to one-third to three-fifths of way to base, with a more or less acute claw, glabrous on both surfaces or with sparse eglandular hairs; petiolules 0.5–1.0 mm, purple, pinkish-purple, green or brown, glabrous or eglandular-hairy; petioles (4–)9–32(–46) cm, usually glabrous, occasionally hairy. *Inflorescence* of (2–)6–11(–15) flowers; peduncles 6–18(–24) cm; bracts 1.5–4.0 mm, glabrous or with a few hairs; pedicels (7–)11–27(–33) mm, glabrous or hairy. *Sepals* 5, (2–)3–6 × 0.8–2.0(–2.5) mm, oblong, narrowly ovate or ovate, rounded, acute or attenuate at apex, 5- to 7-veined, ciliate, with orange, oxalate apical deposits. *Petals* 5, (2–) 4–9(–13) mm, pinkish-purple, pink or lavender. *Stamens* 10; short-styled flowers with filaments of shorter whorl 2.5–5.0 mm, entire, with or without sparse glandular hairs and filaments of longer whorl 3–6 mm, entire or appendaged and with glandular and eglandular hairs, the styles 0.5–1.0 mm, glabrous and the stigmas 0.2–0.4 mm wide; long-styled flowers with filaments of the shorter whorl 1.0–2.5(–3.0) mm, entire and rarely with glandular hairs, the filaments of the longer whorl 1.5–3.0(–4.0) mm, entire, with sparse glandular hairs and few to numerous eglandular hairs, the styles 1.5–3.0 mm, with numerous glandular hairs and stigmas 0.3–0.5 mm wide. *Capsules* 3–11 mm, ellipsoid, glabrous or rarely with a few hairs; seeds 0.8–1.2, with 7–8 longitudinal ridges and 8–13 transverse ridges. *Flowers* 5–10. $2n = 28$.

Introduced. Weed in flower beds in Isle of Man. Native of south-western United States south to Volcánica Transversal in Mexico.

16. O. pes-caprae L. Bermuda Buttercup

O. cernua Thunb.; *O. libica* Viv.; *O. burmannii* Jacq.; *O. kuibisensis* Knuth; *O. cernua* var. *namaquana* Sond.

Perennial monoecious *herb* with rhizome, with bulblets and bulb deeply buried below ground. *Rhizome* long, thick and fleshy in the upper part; bulb up to 25 × 10 mm, ovate-oblong, attenuate at apex; tunic rather pale brown, indistinctly longitudinally ridged; bulblets 5–10 mm; root tuberous and contractile. *Leaves* 10–40; leaflets 3, with lamina medium green and often purple-zoned or spotted on upper surface, paler beneath, very broadly cuneate-obcordate and sub-bilobed or more rarely triangular in outline and deeply bilobed, usually glabrous on upper surface and sparsely appressed-hairy beneath; petioles 3–12 cm. *Peduncles* up to 30 cm, with 3–12 flowers in umbelliform cymes, twice as thick as the petioles; bracts linear, usually callose-tipped; pedicels 5–20 mm, articulate near the base, often with glandular hairs. *Sepals* 5, 5–7 mm, lanceolate or oblong-lanceolate, often with 2 orange apical calluses. *Corolla* 15–25 mm, yellow, rarely cream, with a broadly funnel-shaped concolorous tube; lobes 5, broadly cuneate, obliquely subtruncate at apex. *Stamens* 10, in 2 series, the outer with filaments 3.0–4.5 mm, the inner 5–7 mm, with glandular hairs and prominent teeth. *Styles* with simple eglandular and glandular hairs mixed; stigma 2-fid. *Ovary* oblong, hairy in the upper half, the cells usually with 9 ovules. *Capsule* exserted; seeds endospermous. *Flowers* 5–6. Reproducing by bulblets and tuberised roots. $2n = 28, 34$.

Introduced. Common weed of arable land, particularly bulb-fields in the Isles of Scilly and rare in Devon and the Channel Islands, Oxfordshire, Cardiganshire and Yorkshire. Native of South Africa, but has become a troublesome cosmopolitan weed.

17. O. incarnata L. Pale Pink Sorrel

Perennial monoecious *herb* with rhizome, bulblets and bulbils. *Rhizome* long and slender; *bulb* up to 2 cm, ovoid or narrowly ovoid, with a short, curved beak; outer tunic pale brown and minutely puberulous; bulblets forming both on the rhizome and bulbils in the leaf axils, the latter up to 1 cm. *Stem* 10–30 cm, erect, glabrous or sparsely hairy, branching, leafy. *Leaves* mostly in pseudowhorls of 4–10, the lower sometimes opposite; leaflets 3, with lamina 5–15 × 8–20 mm, medium green on upper surface, often livid beneath, broadly or very broadly obcordate, with rounded margins and an obtuse to subacute sinus, cuneate at base, folding downwards at night, more or less glabrous; petiole 2–6 cm, slender, dilated below the articulation. *Flowers* solitary, terminal or in the upper leaf axils; peduncles 3–7 cm, with 2 opposite, callose bracts at an articulation above the middle. *Sepals* 5, 4–6 mm, oblong, subacute at apex, with several, converging, pale brown calluses near the apex. *Corolla* 12–20 mm, white or very pale lilac with darker veins, with a rather broad, pale greenish tube; lobes 5, obliquely subcuneate, rounded or somewhat truncate at apex, slightly attenuate into a claw and minutely papillate at the base. *Stamens* 10; in 2 series; the outer with filaments 3–4 mm, the inner 4.0–6.5 mm, hairy and with conspicuous, swollen, obtuse teeth; anthers 0.4 mm. *Styles* hairy; stigmas small, green. *Ovary* hairy on the upper half, each cell with 1 ovule. *Capsule* not produced in Great Britain; seeds endospermous. *Flowers* 5–7. Reproducing by bulblets and bulbils. $2n = 14$.

Introduced. Weed of cultivated ground, walls and banks. Frequent in south-west England and the Channel Islands, with scattered records elsewhere in Great Britain and Ireland. Native of South Africa.

115. GERANIACEAE Juss. nom. conserv.

Annual to *perennial* monoecious *herbs*, sometimes woody below. *Leaves* alternate, simple and variously, often very deeply, palmately or pinnately divided, or palmate to imparipinnate, at least the lower petiolate, stipulate. *Inflorescence* of 1–several-flowered, terminal or axillary, often umbellate cymes; flowers bisexual, hypogynous, actinomorphic to zygomorphic. *Sepals* 5, free but in *Pelargonium* the upper one with a spur fused to the pedicel. *Petals* 5, rarely absent. *Stamens* 5 or 10 but sometimes 3 or 5 without anthers, or rarely 15. *Style* 1; stigmas 5, linear. *Ovary* 5-celled, each cell with 2 ovules, elongated distally into a sterile column. *Fruit* a dry, 5-celled schizocarp, with 1 seed per cell, each mericarp and its apical sterile beak separating from the column and splitting open at maturity in a variety of ways.

Contains 5 genera and about 900 species widely distributed but chiefly in temperate regions of the northern hemisphere and in South Africa.

1. Uppermost sepal with inconspicuous spur tightly fused to pedicel; corolla zygomorphic, the 2 upper petals much wider than others **4. Pelargonium**
1. Sepals not spurred; corolla actinomorphic to weakly zygomorphic, the petals scarcely different in width 2.
2. Stamens 15, all with anthers, fused to about half-way into 5 groups of 3 each **2. Monsonia**
2. Stamens 10, or 5 alternating with anther-less staminodes, not fused 3.
3. Leaves palmate or palmately lobed; beaks of mericarps curved or loosely spirally twisted for 1 or 2 turns at maturity **1. Geranium**
3. Leaves pinnate or pinnately or rarely ternately lobed **3. Erodium**

1. Geranium L.

Annual to *perennial herbs. Leaves* alternate, simple and palmately lobed or palmate, at least the lower petiolate, stipulate. *Flowers* 1–several in terminal or axillary, often umbellate cymes, actinomorphic or more or less so, bisexual, hypogynous. *Sepals* free. *Petals* 5, free. *Stamens* 10, free, sometimes the outer 5 without anthers. *Style* 1; stigmas 5, linear. *Ovary* 5-celled, each cell with 2 ovules, elongated distally into a sterile column. *Fruit* a dry, 5-celled schizocarp, the beaks of the 5 mericarps encircling the column, usually rolling upwards at dehiscence and remaining attached by its apex, releasing the seeds, rarely the beak twisting spirally into 1 or 2 large loops and the mericarp separating whole from the column.

About 400 species, mainly in temperate regions.

G. potentilloides L'Her. ex DC. (*G. microphyllum* auct.) occurred as a wool alien by the railway at Borden Station, Hampshire, but is now gone.

Aedo, C., Navarro, C. & Alarcón, M. L. (2005). Taxonomic revision of *Geranium* sections *Andina* and *Chilensia* (Geraniaceae). *Bot. Jour. Linn. Soc.* **149**: 1–68.

Baker, H. G. (1955). *Geranium purpureum* Vill. and *G. robertianum* L. in the British Flora. I. *Geranium purpureum. Watsonia* **3**: 160–167.

Baker, H. G. (1956). *Geranium purpureum* Vill. and *G. robertianum* L. in the British flora. II. *Geranium robertianum. Watsonia* **3**: 270–279.

Baker, H. G. (1957). Genecological studies in *Geranium* (Section Robertiana). General considerations and the races of *G. purpureum* Vill. *New Phytol.* **56**. 172–192.

Böcher, T. W. (1947). Cytogenetic and biological studies in *Geranium robertianum. Biologiske Meddelelser* **20**(8): 1–29.

Forty, J. (1980). A survey of hardy Geraniums in cultivation. *The Plantsman* **2**: 67–78.

Grime, J. P. et al. (1988). *Comparative plant ecology*. London [*G. molle* and *robertianum*.]

Harz, K. (1921). *Geranium phaeum* L. + *G. reflexum* L. = *G. monacense* Harz. *Mitt. Bayer. Bot. Gesell.* **4**: 7.

Hylander, N. (1961). Kungsnaven, vara trädgardars praktfullaste *Geranium. Lustgarden.* **1961**: 109–114.

Kent, D. H. (1959). *Geranium retrorsum* L'Herit. ex DC. *Proc. B. S. B. I.* **3**: 284.

Knuth, R. (1912). Geraniaceae. In Engler, A. (Edit.) *Das Pflanzenreich* **IV. 129**. Leipzig.

Lousley, J. E. (1962). *Geranium retrorsum* L'Herit. *Proc. B. S. B. I.* **4**: 413–414.

Terracciano, A. (1890). Specie rare critiche di Geranii italiani. *Malphighia* **4**.

Townsend, C. C. (1961). *Geranium macrophyllum* Hook. fil. as an adventive plant in Britain. *Watsonia* **5**: 43–44.

Townsend, C. C. (1964). More on the introduced Bordon and Alderney geraniums. *Proc. B. S. B. I.* **5**: 224–226.

Wigginton, M. J. (1999). (Edit.) *British red data books*. Vol. 1. *Vascular plants*. Peterborough. [*G. purpureum*.]

Wilmott, A. J. (1921). *Geranium purpureum* T. F. Forster. *Jour. Bot. (London)* **59**: 93–101.

Yeo, P. F. (1970). The *Geranium palmatum* group in Madeira and the Canary Isles. *Jour. Roy. Hort. Soc.* **95**: 410–414.

Yeo, P. F. (1984). Fruit-discharge type in *Geranium* (Geraniaceae): its use in classification and its evolutionary implications. *Bot. Jour. Linn. Soc.* **89**: 1–36.

Yeo, P. F. (1985). *Hardy Geraniums*. London.

Yeo, P. F. (2001). A new identity for the Alderney Crane's-bill: *Geranium herrerae* Kunth *B.S.B.I. News* **89**: 24–25.

Yeo, P. F. (2001). *Hardy Geraniums*. New ed. London.

While preparing this account we have had available to us the large collection of *Geranium* in the Cambridge Botanic Garden brought together by Peter Yeo while writing his book *Hardy Geraniums*. He has also personally answered many questions we have asked about the genus, and has read through the final account.

The main divisions of the leaves are called lobes. The second division lobules and the third division teeth. Leaves of *Geranium* species often turn red when dying. Flowers are produced in pairs, each flower on a pedicel, the common stalk to the two pedicels is a peduncle. The 5 sepals have a projecting point, the mucro. The 5 petals may or may not be emarginate at apex. The 10 stamens are in two whorls. The outer whorl is opposite the petals. At the base of each of the inner whorl is a swollen gland, the nectary. The 5 carpels are joined to a central column, but not to each other. They are topped by a common style, which is thickened at the base and divided at the top into 5 stigmatic branches. The 5 inner anthers burst later than the 5 outer ones. After another day the stigmas curl back and are receptive. This adaptation to cross-pollination applies to the larger-flowered species. The smaller-flowered species have the stigmas spread when the flower opens and are mainly self-pollinated, although they can be cross-pollinated. After fertilisation there is a rapid growth in the length of the thickened base of the style to form a beak. An explosive break-up of the beak disperses the seeds. Five strips, the awns, one from each carpel, peel away from the bottom upwards and leave behind a central column. Previous to this happening, each carpel is freed from the lower part of the central column and remains attached only to the lower end of the awn. In view of this the carpels move only when the awns move. There are three different ways in which the fruit breaks up. The first is called *seed ejection*. On discharge the awn curls back and the seed is thrown out of the mericarp. Species with this seed ejection belong to the subgenus *Geranium*. The second method is termed *carpel projection*. The mericarp without the awn is thrown off to a distance of 1–2 m and the awn drops away at the moment of discharge. Species with this seed ejection belong to subgenus *Robertium*. The third method is when the awn with the seed inside comes away from the central column being thrown some distance by the explosion. The awn becomes coiled, but is not plumed. Species with this type of seed ejection belong to the subgenus *Erodioideae*.

1. Petals narrowed at base into a distinct claw some at least half as long as limb to longer than limb 2.
1. Petals without a claw, or with a claw less than half as long as the limb 18.
2. Most petals more than 14 mm 3.
2. Most petals less than 14 mm 5.
3. Perennial herb with a thick, fleshy rhizome; sepal with mucro 1.0–4.5 mm about half as long as sepal **23. macrorrhizum**
3. Perennial or biennial herb with thin, fibrous roots; sepal with mucro about 2 mm 4.
4. Stamens nowhere near twice as long as sepals **27. yeoi**
4. Stamens about twice as long as sepals **28. reuteri**
5. Leaves divided less than three-quarters of the way to the base 6.
5. Leaves divided more or less to the base 12.
6. Leaves shining green to pinkish-red, glabrous or very sparsely hairy; sepals strongly keeled on back **24. lucidum**
6. Leaves greyish-green, obviously hairy; sepals rounded on back 7.
7. Mericarp appressed-hairy **20. pusillum**
7. Mericarp glabrous 8.
8. Mericarp not ribbed **22(iii). molle** var. **aequale**
8. Mericarp with numerous ribs 9.
9. Petals 8.0–12.5 mm **21. brutium**
9. Petals 4–8 mm 10.
10. Plant forming a circle flat on the ground **22(i). molle** var. **arenarium**
10. Plant with erect or ascending stems 11.
11. Flowers 5–8 mm in diameter **22(ii). molle** var. **molle**
11. Flowers 9–13 mm in diameter **22(iv). molle** var. **grandiflorum**
12. Petals 5–9 mm; anthers yellow; mericarps with dense wrinkle-like ribs and 2–3(–4) deep collar-like ridges at apex 13.
12. Petals 8–14 mm; anthers pinkish-orange or purple; mericarps with sparse, fine ribs towards the base and 0–1(–2) deep collar-like ridges at the apex 14.
13. Stems erect or strongly ascending **26(a). purpureum** subsp. **purpureum**
13. Stems procumbent, the tips ascending **26(b). purpureum** subsp. **forsteri**
14. Stems reddish only at the nodes; flowers pale pink or whitish **25(a). robertianum** subsp. **celticum**
14. Stems suffused red; flowers deep pink 15.
15. Stems erect or ascending; fruits more or less hairy **25(c). robertianum** subsp. **robertianum**
15. Stems procumbent or arcuate-ascending 16.
16. Leaves glabrous or nearly so **25(b,i). robertianum** subsp. **maritimum** var. **maritimum**
16. Leaves more or less hairy 17.
17. Plant fairly hairy **25(b,ii). robertianum** subsp. **maritimum** var. **intermedium**
17. Plant densely hairy **25(b,iii). robertianum** subsp. **maritimum** var. **hispidum**
18. Annuals to biennials or sometimes perennials more or less without a rhizome; petals mostly less than 10 mm, rarely to 22 mm 19.
18. Perennials with distinct thick and/or elongated rhizome; petals mostly more than 10 mm 30.
19. Petals more than 15 mm 20.
19. Petals less than 12.5 mm 21.
20. Plant usually less than 60 cm; flowers less than 33 mm in diameter; claw of petal 6–7 mm **27. yeoi**
20. Plant usually more than 60 cm; flowers more than 40 mm in diameter; claw of petal 2.0–2.5 mm **29. maderense**
21. Seeds smooth 22.
21. Seeds pitted or reticulately ridged 27.
22. Mericarps glabrous 23.
22. Mericarps hairy 26.
23. Mericarps without ribs **22(ii). molle** var. **molle**
23. Mericarps ribbed 24.
24. Plant prostrate forming a circle flat on the ground **22(i). molle** var. **arenarium**
24. Plant erect 25.
25. Flowers 5–8 mm in diameter **22(ii). molle** var. **molle**
25. Flowers 9–13 mm in diameter **22(iv). molle** var. **grandiflorum**
26. Petals 7–10 mm; all 10 stamens with anthers **19. pyrenaicum**
26. Petals 2–4 mm; outer 5 stamens lacking anthers **20. pusillum**
27. Leaves divided less than halfway to base; sepals with a mucro up to 0.5 mm at apex; petals rounded or slightly retuse at apex **5. rotundifolium**
27. Leaves divided more than three-quarters of the way to the base; sepals with a mucro more than 0.5 mm; petals distinctly emarginate or rarely rounded 28.
28. Pedicels more than 2.5 cm; mericarps glabrous or sparsely hairy **13. columbinum**
28. Pedicels less than 2.5 cm; mericarps hairy 29.
29. Leaves divided almost to base; glandular hairs frequent on upper parts of plant **14. dissectum**
29. Leaves divided about three-quarters of way to base; glandular hairs absent **15. core-core**
30. Flowers mostly all solitary on pedicel or peduncle 31.
30. Flowers at least mostly in pairs, that is 2 pedicels to 1 peduncle 33.
31. Flowers pale flesh-pink **12(i). sanguineum** var. **striatum**

31. Flowers purplish-red or rose-pink 32.
32. Plant prostrate **12(ii). sanguineum** var. **prostratum**
32. Plant erect **12(iii). sanguineum** var. **sanguineum**
33. Petals not emarginate, usually apiculate or with a small triangular point at apex, sometimes creased or subentire; mericarps pointed at base, with 2–4 collar-like ridges at apex 34.
33. Petals rounded or emarginate at apex, sometimes with a small point in the notch; mericarps rounded at base, with 0–1 collar-like ridges at apex 38.
34. Petals nearly as wide as long or wider, widely spreading 35.
34. Petals about two-thirds as wide as long or less, eventually reflexed just above the base 36.
35. Petals dark pinkish-lilac to nearly black, with silvery-white or nearly white bases, the white area has a jagged edge and the flower by transmitted light thus has a star-shaped white centre, the apex is cuspidate or sometimes lobulate **30(i). phaeum** var. **phaeum**
35. Petals bluish, lilac or pink, rather pale and with a white base smaller than in var. *phaeum*, the base and the lowest part of the coloured portion traversed by short bluish or violet veins which become diffuse as they enter the coloured zone, giving rise to a bluish halo and in the principal colour there may be a noticeable change from pink toward blue during the life of each flower **30(ii). phaeum** var. **lividum**
36. Petals bright rose-pink, about half as wide as long; filaments with only very fine hairs on the edges at the base **32. reflexum**
36. Petals plum-purple to lilac-pink, about two-thirds as wide as long; filaments with long, spreading, glistening hairs on lower half 37.
37. Leaves often with dark brown blotches; petals dark purplish-red, the whitish base relatively large and with a toothed edge, while the bluish-violet zone above it is inconspicuous **31(ii).×monacense** nothovar. **monacense**
37. Leaves without blotches; petals pinkish-lilac, the whitish base small and straight-edged, while the bluish-violet zone above it is wide, conspicuous and strongly veined. **31(ii).×monacense** nothovar. **anglicum**
38. Stalked glands absent or with stalks less than 0.3 mm 39.
38. Sepals and pedicels and usually peduncles and upper parts of stem, with stalked glands more than 0.3 mm long 44.
39. Petals violet-blue; base of mericarps without tufts of bristles on inside **16. ibericum**
39. Petals whitish to bright or purplish-pink; base of mericarps with tuft of apically pointed bristles on inside, directed on the seed or into cavity 40.
40. Main leaf lobes toothed, with teeth up to about 5 mm long; hairs on pedicels, peduncles and upper parts of stems less than 0.2 mm and appressed **4. nodosum**
40. Main leaf lobes with sub-lobes or deep teeth more than (5–)10 mm long; pedicels, peduncles and upper parts of stem with patent hairs more than 0.5 mm 41.
41. Petals with veins darker than ground colour; beaks of mericarps with hairs about 0.1 mm 42.
41. Petals with veins paler than or same colour as ground colour; beaks of mericarps with some hairs more than 0.5 mm 43.
42. Petals spreading, not curved out at apex, making the flowers funnel-shaped **2.×oxonianum**
42. Petals curved outwards at apex making the flowers trumpet-shaped, with white to very pale pink ground-colour **3. versicolor**
43. Petals deep bright pink; fruit with style 2.5–3.0(–4) mm **1. endressii**
43. Petals usually mid- mauve-pink, often variable on one plant; fruit with style (3–)4–6 mm **2.×oxonianum**
44. Stalked glands sparse, confined to floral parts and pedicels 45.
44. Stalked glands abundant on pedicels, peduncles and upper parts of stems 46.
45. Petals deep bright pink; fruit with style 2.5–3.0 (–4.0) mm **1. endressii**
45. Petals usually mid-mauve-pink, often variable on one plant; fruit with style (3–)4–6 mm **2.×oxonianum**
46. Petals more or less emarginate; base of mericarps without tuft of bristles inside 47.
46. Petals rounded at apex; base of mericarps with tuft of apically pointed bristles on inside directed on the seed or into cavity 48.
47. Basal leaves divided from two-thirds to seven-eighths of the way to the base, their main lobes widened well below the apex; flowers pointing upwards; seeds not ripening **17.×magnificum**
47. Basal leaves divided about halfway to base, their main lobes widened near the apex; flowers pointing horizontally; seeds ripening **18. platypetalum**
48. Petals blue to violet-blue, with a white base; flowers and immature schizocarps pointing sideways or more or less downwards; schizocarps with styles more than 4 mm 48.
48. Petals pinkish-purple to magenta, with black or white base; flowers and immature schizocarps pointing obliquely or vertically upwards; schizocarps with styles less than 4 mm 50.
49. Sterile hybrid, schizocarps scarcely developing **9.×johnsonii**
49. Fertile schizocarps developing freely on reflexed pedicels 50.
50. Rootstock not creeping; leaves divided more than five-sixths of the way to the base; sepals with mucro more than one-fifth as long as main part; flowers 35–45 mm in diameter **8. pratense**
50. Rootstock more or less creeping; leaves divided as far as six-sevenths or more to the base; sepals with mucro less than one-fifth as long as the main part; flowers 40–60 mm in diameter **10. himalayense**
51. Petals 15–20 mm, with a black patch at base **7. psilostemon**
51. Petals 10–18 mm, with a white patch at base 52.
52. Leaves divided up to four-fifths of the way to the base; sepals 5–7 mm **6. sylvaticum**
52. Leaves divided half-way to base; sepals 7.0–9.5 mm **11. eriostemon**

Subgenus **1. Geranium**

Petals more or less without a claw. *Seed ejection* is through a hole in the carpel which is carried out by explosive recurvature of the awn; the awn may or may not remain attached to the central column of the beak. *Mericarps* without ribs and rounded at base.

Section **1. Geranium**

Annual or *perennial herbs*. *Retention of seed* in the mericarp during the pre-explosive interval by bristles attached to a horny tubercle at the lower end of the mericarp. *Awn* remaining attached to central column of the beak after the discharge.

1. G. endressii J. Gay French Crane's-bill

Perennial monoecious *herb* with extensive, elongated rhizomes on or just below the surface. *Stems* 20–50 cm, pale yellowish-green, erect, with dense, unequal, very short to medium, pale, soft, spreading simple eglandular hairs, branched, leafy. *Leaves* alternate; lamina 3–10 × 3–10 cm, gradually declining in size and length of petiole upwards but changing little in shape, dull medium green on upper surface, paler beneath, soft, subrotund or broadly ovate in outline, acute at apex, palmately divided four-fifths or five-sixths of the way to the base into 5 lobes, the lobes rhombic, acute at apex and divided about halfway to the midrib into lobules which have 2 or 3 acute teeth on each side, with numerous, short, pale simple eglandular hairs on both surfaces; veins impressed on upper surface, prominent beneath; petioles up to 12 cm, with numerous, unequal, short to medium, pale, spreading simple eglandular hairs. *Inflorescence* rather dense; flowers bisexual, erect, funnel-shaped, scentless, 23–30 mm in diameter; pedicels up to 2.5 times as long as the sepals, yellowish-green, with dense, unequal, short to medium, pale, patent simple eglandular hairs. *Sepals* 5, 7–9 mm, pale green, linear-lanceolate, with a mucro 1–2 mm. *Petals* 5, 16–18 × 8–9 mm, bright deep pink, becoming darker and redder with age, base and lower parts of veins colourless, translucent, apical parts of veins slightly darker than ground colour and slightly netted, oblanceolate, more or less emarginate at apex, more or less without a claw, base with hairs extending across the front surface and on the margins. *Stamens* 10; filaments 4–6 mm, white tinged with pink, curving outward at the tips, hairy in the lower half; anthers yellow or purplish. *Style* 1, 2.5–3.0(–4.0) mm, pale green; stigmas 2.5–3.5 mm, pink or reddish, bristling on the backs. *Immature schizocarps* and their pedicels erect; mericarps 5, 3.0–3.5 mm; beak 18–21 mm, including the stylar portion which is 2.5–3.0 mm, rounded and sometimes with a ridge at base, with dense, minute hairs and a few longer ones; a cluster of stiff hairs at the lower end of the carpel orifice hold back the seed until it is discharged from the mericarp by the awn curling back. *Flowers* 6–9. $2n = 26, 28$.

Introduced. Much grown in gardens and frequently naturalised in grassy places and on waste ground. Scattered records from most of Great Britain, rare in Ireland. Native of the Western half of Basses Pyrénées, mainly in France, but just extending into Spain.

2. G. × oxonianum Yeo Druce's Crane's-bill
G. endressii × versicolor
G. endressii var. *armitageae* Turrill

Perennial herb with extensive rhizomes on or below the surface. *Stems* 50–80 cm, pale yellowish-green, erect to spreading, with dense, spreading eglandular hairs below and very dense glandular ones above, branched, leafy. *Leaves* alternate; lamina 5–20 × 5–20 cm, gradually declining in size upwards, dull medium to dark green on upper surface, much paler beneath, sometimes brown-blotched when young, more or less wrinkled, subrotund or broadly ovate in outline, acute at apex, palmately divided five-sixths of the way to the base into 5 lobes, the lobes rhombic or elliptical, acute at apex, divided almost half-way to the midrib into lobules with 2 or 3 shortly acute teeth on each side, numerous soft eglandular and minute glandular hairs on both surfaces; veins impressed on upper surface, prominent beneath; petioles up to 10 cm, hairy. *Inflorescence* lax; flowers 35–40 mm in diameter, bisexual erect, funnel-shaped; pedicels up to 1.5 times as long as the sepals, yellowish-green with dense, very short glandular and longer eglandular hairs. *Sepals* 5, 8–11 mm, pale green, lanceolate, acute at apex, ribbed, with dense, very short hairs. *Petals* 5, 20–26 × 14–15 mm, mauve-pink, sometimes with darker net-veining, oblanceolate to obovate, emarginate at apex, without a claw. *Stamens* 10; filaments white below, pink above, hairy; anthers pink. *Style* 1, (3–)4–6 mm, pale green; stigmas 5, brownish, hairy beneath. *Mericarps* 5, 3.0–3.5 mm; beak 18– 27 mm, its stylar portion 3–4 mm, hairy, sometimes with a collar-like ridge at apex; a cluster of stiff hairs at the lower end of the carpel orifice hold back the seed until it is discharged from the mericarp by the awn curling back. *Flowers* 6–9.

Introduced. Often grown in gardens and naturalised in grassy places and open woodland independent of parents. Scattered records over much of Great Britain and a few in Ireland. Of garden origin, a fertile hybrid which produces a large range of variants some of which have cultivar names.

3. G. versicolor L. Pencilled Crane's-bill

Low-growing *perennial* monoecious *herb* with a rather compact rootstock. *Stems* up to 60 cm, but spreading so that the plant is usually low and bushy, pale green, with more or less dense, pale, short to long, bristly hairs and dense, minute glandular hairs, branched, leafy. *Leaves* alternate; lamina 5–20 × 5–20 cm, gradually diminishing in size up the stem, medium yellowish-green on upper surface, paler beneath, brown-blotched between the lobes, broadly ovate in outline, subacute at apex, divided palmately two-thirds to four-fifths of the way to the base into 5 lobes, the lobes tapered both ways from about the middle, pinnately or palmately lobulate about halfway to the midrib, the lobules as long as broad or less with 1–3 teeth, the teeth obtuse or acute at apex, with short, pale, bristle-like eglandular hairs on both surfaces; petioles up to 15 cm, pale green, with dense, pale, spreading eglandular hairs. *Inflorescence* diffuse; pedicels up to about 1.5 times as long as the sepals, but 1 of each pair nearly always shorter than the sepals with dense, very short glandular hairs and eglandular hairs; flowers 25–30 mm, in diameter, bisexual, erect, trumpet-shaped. *Sepals* 5, 7–9 mm, pale green, erect but not appressed to petal bases, lanceolate, recurved at tips with a mucro about 1.5 mm, with a few, long hairs. *Petals* 5, 13–18×6–9 mm, twice as long as broad, white or very pale pink, with a close network of fine magenta-coloured veins which fade with age, obovate, emarginate at apex, erect at base and without a claw, spreading above and recurved at tips, base thinly hairy on margins and front surface. *Stamens* 10, slightly longer than sepals; filaments white with pink tips, sparsely hairy to beyond the middle; anthers bluish. *Styles* red; stigmas 3.5 mm, red with whitish receptive surface. *Immature schizocarps* and their pedicels erect; mericarps 5, about 3.5 mm; beak including its stylar portion of about 6 mm is about 18 mm, with dense minute hairs and not bristles, rounded at base, sometimes with collar-like ridge at apex; a cluster of stiff hairs at the lower end of the carpel orifice hold back the seed until it is discharged from the mericarp by the awn curling back. *Flowers* 5–10. $2n = 28$.

Introduced. Grown in gardens and naturalised in grassy places and waste ground. Scattered records in central and south Great Britain especially the south-west peninsula, Ireland and the Channel Islands. Central and south Italy, Sicily, southern part of Balkan Peninsula including Greece, Yugoslavia and Albania.

4. G. nodosum L. Knotted Crane's-bill

Perennial monoecious *herb* with elongated rhizomes on or just below the surface. *Stem* 20–50 cm, pale yellowish-green, sometimes slightly tinted brownish-purple, glabrous or with short, appressed hairs. *Leaves* alternate; lamina 3–8×13–20 cm, leaves gradually decreasing in size upwards, medium yellowish-green on upper surface, paler beneath, reniform in outline, more or less acute at apex, palmately divided two-thirds of the way to the base into 5 lobes, the lobes ovate or elliptical, more or less acute at apex, scarcely lobulate but unevenly toothed, the basal pair of lobes widely splayed; with numerous, short and medium, pale simple eglandular hairs on both surfaces and the margins; veins impressed on upper surface, prominent beneath; petiole pale green with small, appressed hairs. *Inflorescence* unequally forked and diffuse; flowers 25–30 mm in diameter, bisexual, erect and funnel-shaped; pedicels with short, appressed hairs. *Sepals* 5, 8–10 mm, pale green, linear, with a mucro 1.5–2.0 mm, with very short, pale, appressed eglandular hairs. *Petals* 5, 16–22 mm, bright purplish-pink or tending to violet with carmine veins at base, oblanceolate, deeply emarginate, without a claw, with numerous hairs across the front surface on the margins. *Stamens* 10, longer than sepals; filaments white, hairy for two-thirds of their length; anthers blue. *Styles* red; stigmas about 2 mm, red, glabrous. *Immature schizocarps* and their pedicels horizontal or slightly nodding; mericarps 5, about 3.5 mm, rounded at base; beak including stylar portion of about 3 mm is about 22 mm, with minute hairs; a cluster of stiff hairs at the lower end of the carpel orifice hold back the seed until it is discharged from the mericarp by the awn curling back. *Flowers* 6–10. $2n = 28$.

Introduced. Grown in gardens and naturalised in hedgerows and woodlands. Very scattered records from central and southern Great Britain. Native of central France to the Pyrenees and in central Italy and central Yugoslavia.

5. G. rotundifolium L. Round-leaved Crane's-bill

Annual, monoecious *herb* with fibrous roots. *Stems* 10–40 cm, pale green, erect, ascending or decumbent, with dense, short but unequal, red-tipped glandular hairs, branched from the base, leafy. *Leaves* alternate; lamina 3–7 cm in diameter, dull yellowish- or greyish-green on upper surface, paler beneath, subrotund or subrotund-reniform in outline, divided to about halfway into 7–9 lobes, the lobes with rounded teeth, the base of the sinuses between the lobes often stained reddish, with numerous, short glandular hairs and longer eglandular ones on both surfaces; petioles long and slender; stipules 2–5×1–2 mm, purplish, ovate, acuminate at apex. *Inflorescence* of 2 bisexual flowers arising from the axil of the upper leaves; peduncles up to 2 cm, with dense, short, unequal, red-tipped glandular hairs; pedicels up to 15 mm, clothed like peduncles. *Sepals* 5, 3.5–6.0 mm, pale green, oblong-ovate, with a mucro up to 0.5 mm at apex, 3(–5)-veined, with dense, short but unequal, red-tipped glandular hairs and some longer simple eglandular hairs. *Petals* 5, 5–7×2.5–3.0 mm, white at base with the terminal one-third pink, narrowly obovate, entire or slightly retuse at apex, wedge-shaped with a short claw, with a few, short hairs at base, erect with spreading tips. *Stamens* 10; filaments about 5 mm, hairy, gradually tapering upwards; anthers broadly oblong, bluish. *Style* 1, about 1.5 mm, hairy; stigmas purple. *Ovary* densely hairy. *Immature schizocarps* erect on sharply reflexed pedicels; mericarps dark olive-brown, thinly hairy, smooth; beak 13–14 mm, including stylar portion which is about 2.5 mm; seeds 1.5–2.0 mm, dark brown, broadly oblong or almost globose, distinctly reticulately ridged; a cluster of stiff hairs at the lower end of the carpel orifice hold back the seed until it is discharged from the mericarp by the awn curling back. *Flowers* 6–7. $2n = 26$.

Native. Banks, walls and stony and sandy ground. Local in central and south England, southern Wales, southern Ireland and Channel Islands; introduced further north. Europe and Mediterranean region eastwards to central Asia and the Himalayas; Atlantic Islands; introduced elsewhere. Eurosiberian Southern-temperate element.

6. G. sylvaticum L. Wood Crane's-bill

Perennial monoecious *herb* with a compact, stout, oblique rootstock. *Stems* 30–80 cm, pale green, erect or ascending, with short, deflexed glandular hairs below and dense glandular and eglandular hairs above. *Leaves* alternate; lamina 7–20×7–20 cm, gradually becoming smaller upwards, reniform or polygonal in outline, medium green on upper surface, paler beneath, divided palmately up to four-fifths of the way to the base into 7 or 9 lobes, the lobes obovate-rhombic in outline, pinnately lobulate, the lobules triangular in outline and deeply and coarsely dentate, the teeth more or less acute, hairy on both surfaces; petioles up to 25 cm on the lower leaves, becoming shorter and almost absent in the uppermost. *Inflorescence* of pairs of bisexual flowers on axillary peduncles forming a lax cyme; pedicels erect after flowering, with numerous glandular hairs; flowers 22–30 mm in diameter, cup-shaped. *Sepals* 5, 5–7 mm, ovate-lanceolate, with a mucro almost one-fifth as long

G. x **oxonianum** Yeo

G. endressii J. Gay

G. versicolor L.

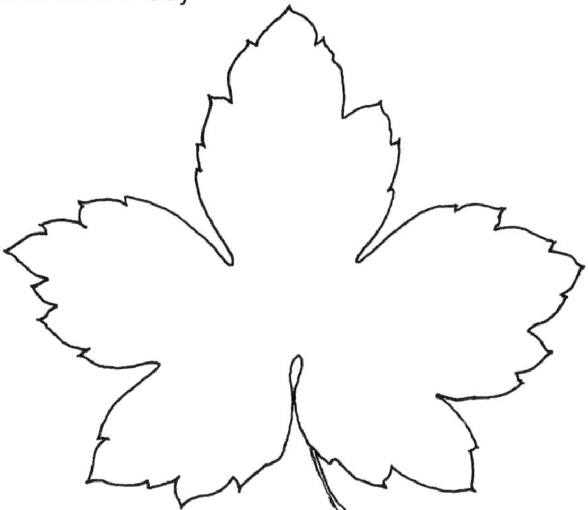

G. nodosum L.

G. rotundifolium L.

Basal leaves of **Geranium** L.

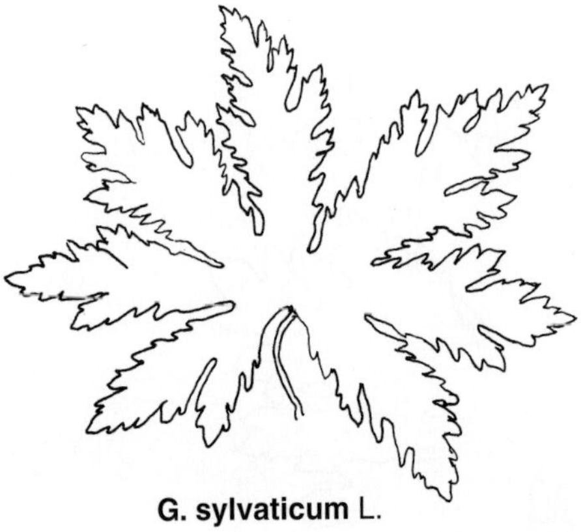

G. sylvaticum L.

G. himalayense Klotzsch

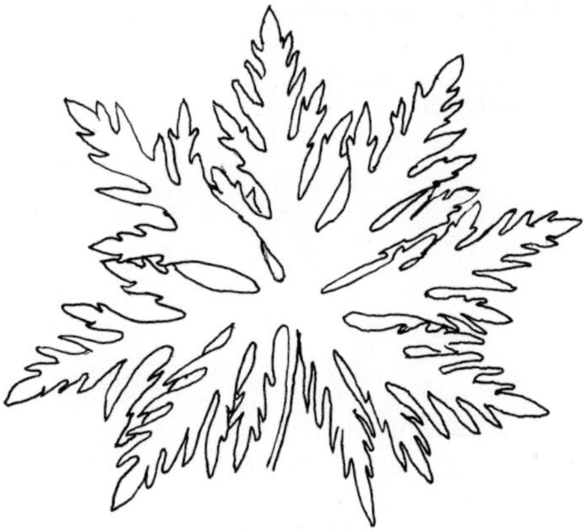

G. pratense L.

G. x **johnsonii** P. D. Sell
(G. himalayense x **pratense)**

G. psilostemon Ledeb.

Basal leaves of **Geranium** L.

as the sepal, with numerous glandular hairs. *Petals* 5, 12–18×8–12 mm, purplish-violet with a white base, obovate, rounded or retuse at apex, hairy, with a short claw. *Stamens* 10; filaments divergent, pinkish, with a lanceolate base; anthers bluish. *Stigmas* 2–3 mm, purplish. *Immature schizocarps* and their pedicels erect; mericarps about 4 mm, smooth, rounded at base, glandular-hairy; beak 17–21 mm including the stylar portion which is 1.5–2.5 mm; seeds minutely reticulate; a cluster of stiff hairs at the lower end of the carpel orifice hold back the seed until it is discharged from the mericarp by the awn curling back. *Flowers* 6–7. Pollinated by various insects; protandrous; homogamous and unisexual flowers also recorded. $2n$ =28.

Native. Woods and hedges in the lowlands, rock ledges, stream gullies and meadows in the uplands. Locally common in northern Great Britain south to Yorkshire, very local in central England, Wales and northern Ireland, rarely naturalised elsewhere. Most of Europe, but only on mountains in the south and absent from many islands; north Turkey. Eurosiberian Boreal-montane element.

7. G. psilostemon Ledeb. Armenian Crane's-bill
G. armenum Boiss.; *G. backhouseanum* Regel

Perennial, monoecious *herb* with a compact rootstock. *Stems* several 80–120 cm, pale green, sometimes tinted brownish-purple, erect, stout, solid, with appressed simple eglandular hairs below and a mixture of medium to long, pale simple eglandular hairs and very unequal, red-tipped glandular hairs above, branched above, leafy. *Leaves* alternate; lamina 7–25×11–25 cm, diminishing gradually in size and length of petiole upwards, dull medium yellow-green on upper surface, pale green beneath, broadly obovate or subrotund in outline, acute at apex, divided as far as four-fifths of the way to the base into 5–7 lobes, the lobes obovate, tapered to both apex and base, each with 3 lobules which are elliptical in outline and with several acute teeth, cordate at base, with soft, short simple eglandular hairs on both surfaces; veins impressed above and very prominent beneath; petiole up to 30 cm, pale green, with appressed simple eglandular hairs. *Inflorescence* rather loose, made up of little clusters of 2–3 flowers; flowers 30–40 mm in diameter, bisexual, shallowly bowl-shaped; pedicels up to 30 mm, erect, with dense, short, appressed simple eglandular hairs and dense, unequal, short to long, very slender, white, red-tipped glandular hairs. *Sepals* 5, 8–12 mm, pale green, lanceolate, with a mucro up to 3 mm at apex, with dense, unequal, short and medium, white, red-tipped glandular hairs. *Petals* 5, 15–20×15–20 mm, deep, bright magenta with black veins and a black patch at base which has a jagged apex, broadly obovate, retuse at apex, not clawed with a few marginal hairs at base. *Stamens* 10; filaments mainly blackish, but with pale dilations on either side of base, glabrous; anthers purplish-black. *Style* greenish-yellow at base, becoming purplish-black at apex, glabrous; stigmas 2.5–3.0 mm, dark reddish-purple. *Immature schizocarps* erect on more or less reflexed pedicels; the whole green tinted brownish-purple and covered with numerous, short and medium, white, red-tipped glandular hairs; mericarps 5, about 5 mm, rounded at base; beak 27–30 mm, including the stylar portion which is about 3 mm; a cluster of stiff hairs at the lower end of the carpel orifice hold back the seed until it is discharged from the mericarp by the awn curling back. *Flowers* 5–8.

Introduced. Commonly grown in gardens and naturalised in grassy places. A few scattered records in England and Scotland. Native of north-east Turkey and the south-west Caucasian region.

8. G. pratense L. Meadow Crane's-bill

Perennial monoecious *herb* with a compact, stout, oblique rootstock. *Stem* 30–80(–130) cm, pale green, erect or ascending, with short, deflexed eglandular hairs below, dense glandular and eglandular hairs above, branched, leafy. *Leaves* alternate; lamina 7–20×7–20 cm, slowly decreasing in size upwards, medium green on upper surface, paler beneath, circular, polygonal or reniform in outline, palmately divided more than five-sixths of the way to the base into 7 or 9 lobes, the lobes ovate-rhombic in outline, acute at apex and pinnately lobulate, the lobules with more or less oblong, acute or apiculate teeth, appressed hairy on both surfaces; petioles up to 30 cm, hairy; stipules up to 30 mm, narrow and tapered to a fine point. *Inflorescence* of pairs of flowers on axillary peduncles forming a dense cluster; peduncles mostly 2–10 cm; pedicels short, one or both shorter than the sepals; flowers 35–45 mm in diameter, bisexual, saucer-shaped. *Sepals* 5, 7–12 mm, ovate, mucro 1.5–3.5 mm, forming a slightly bladdery calyx after flowering. *Petals* 5, 16–24×13–20 mm, deep violet-blue or white, usually white at extreme base, with translucent, sometimes pinkish veins, the base with a dense tuft of hairs on either side, obovate, rounded at apex, claw very short. *Stamens* 10; filaments abruptly widened at the base, more or less deep pink, fringed with hairs in the lower part; anthers dark violet or bluish-black. *Style* 1, more than 4 mm, stigmas 2.0–2.5 mm, greenish, tinged with pink or dull brownish, purple or crimson. *Immature schizocarps* reflexed on reflexed pedicels; mericarps 5, 4.5–5.0 mm, rounded at base, smooth, glandular-hairy; beak 23–29 mm, including the distinct stylar portion which is 7–8 mm; seeds minutely reticulate; a cluster of stiff hairs at the lower end of the carpel orifice hold back the seed until it is discharged from the mericarp by the awn curling back. *Flowers* 6–9. Pollinated mainly by Hymenoptera, protandrous. $2n$=28.

Native. Meadows, roadsides, open woodland and damp places on calcareous soil. Abundant in much of Great Britain but absent from much of northern Scotland and parts of Wales and southern England, very local in Ireland, naturalised elsewhere. Most of Europe and the Altai Mountains of central Asia, possibly west and east Siberia and western China. Plants which ought to be described as subspecies occur elsewhere in Asia. Eurasian Boreo-temperate element.

9. G.×johnsonii P. D. Sell Johnson's Blue Crane's-bill
G. himalayense Klotzsch×**pratense** L.

Perennial monoecious *herb* spreading moderately by underground rhizomes. *Stems* up to 70 cm, pale yellowish-green, erect, slender, branched above, leafy, with numerous, short, pale reflexed-appressed simple eglandular hairs. *Leaves* alternate, gradually getting smaller upwards; lamina 2–7×5–20 cm, dull medium green on upper surface, paler beneath, subrotund or reniform in outline, acute at apex, divided three-quarters or seven-eights of the way to the base into 7 lobes, the lobes obovate or oblanceolate in outline and divided into lobules, the lobules with small teeth and the lobules and teeth more or less acute, the lamina cordate at base, with short, pale simple eglandular hairs on the upper surface, margin and veins beneath; veins impressed above, prominent beneath; petioles up to 12 cm, with short, pale, reflexed-appressed simple eglandular hairs. *Inflorescence* loose or with bisexual flowers mostly in pairs; peduncles 3–10 cm, erect, with numerous, short and medium, pale glandular hairs, mixed with pale simple eglandular hairs; pedicels about twice as long as sepals, spreading, with dense, very short and short, pale, red-tipped glandular hairs; bracts like upper leaves; bracteoles linear-lanceolate, acute at apex, with simple eglandular hairs. *Sepals* 5, 7–11×5–6 mm, green with a purplish stain at base, the mucro 1–2 mm, lanceolate to narrowly ovate, ribbed, with dense, very short to short, pale, red-tipped glandular hairs. *Petals* 5, 15–25×10–17 mm, deep blue, paler and pinker towards the base with translucent and almost colourless veins, obovate, rounded at apex, with a short claw. *Stamens* 10, slightly longer than sepals; filaments more or less deep pink, finely fringed with hairs below; anthers dark violet, bluish-black or dark brown and not opening. *Styles* more than 4 mm, stigmas about 3 mm. *Mericarp* rounded at base; beak up to 28 mm with stylar portion about 9 mm; seeds not developing. *Flowers* 5–8.

Introduced. This hybrid, of garden origin, appeared about 1950 among plants raised by Mr B. Ruys, of Dedemsvaart in Holland, from seed of *G. pratense* sent from Mr A. T. Johnson, of Tyn-y-Groes in North Wales. It has become a popular garden plant known as Cv. Johnson's Blue, and may have escaped and been confused with either of its parents.

10. G. himalayense Klotzsch Himalayan Crane's-bill
G. meeboldii Briq.; *G. grandiflorum* auct.

Perennial monoecious *herb* spreading by rhizomes and mat-forming. *Stems* up to 70 cm, pale yellowish-green, sometimes brownish-purple towards the base, erect, solid, with numerous to dense, short, white simple eglandular hairs and some glandular hairs above. *Leaves* alternate, gradually getting smaller up the stem; lamina 2.5–20.0×2.5–20.0 cm, dull medium to dark green on upper surface, paler beneath, broadly ovate or reniform in outline, palmately divided at least six-sevenths of the way to the base into 5 or 7 lobes, the lobes obovate, obtuse at apex, divided into 3 lobules, the lobules obovate with incise-dentate teeth, with sparse eglandular hairs on the upper surface and numerous, short, pale simple eglandular hairs beneath, especially on the veins and margins; petioles up to 40 cm, pale green, with numerous, very short, appressed hairs. *Inflorescence* forming a diffuse panicle; peduncles 4–18 cm; pedicels short; flowers 40–60 mm in diameter, bisexual, saucer-shaped. *Sepals* 5, 8–12 mm, medium green, often stained with purple at base and veins, lanceolate, with a mucro 0.5–1.5 mm, with numerous, unequal, short or very short, pale glandular hairs. *Petals* 5, 20–31×18–25 mm, deep blue, often flushed pinkish near the base and white at extreme base, obovate, rounded at apex, with a short claw at base, with a tuft of hairs either side at the base. *Stamens* 10; filaments white, tinged with pink, with an enlargement at base, fringed with hairs; anthers dark blue. *Style* greenish; stigmas 2.0–3.5 mm, pink to purplish. *Immature schizocarps* reflexed on reflexed pedicels; mericarps 4.5–5.0 mm, rounded at base; beak 27–30 mm including stylar portion which is 7–10 mm; a cluster of stiff hairs at the lower end of the carpel orifice hold back the seed until it is discharged from the mericarp by the awn curling back. *Flowers* 6–10.

Introduced. Grown in gardens and naturalised in grassy places. A few scattered records in England and Scotland. Native of the Himalayas from north-east Afghanistan to central Nepal and in the Pamir Region of Russia.

11. G. eriostemon DC. Asiatic Crane's-bill
G. platyanthum Duthie

Perennial herb with a thick, compact rootstock. *Stem* 30–50 cm, pale green, with coarse, more or less spreading eglandular hairs mixed with shorter glandular hairs, leafy. *Leaves* alternate or opposite, decreasing slowly upwards; lamina 5–20×5–20 cm, pale green sometimes wrinkled and sometimes with a narrow red edge on upper surface, paler beneath, broadly ovate or reniform in outline, divided halfway to the base into 5 or 7 lobes, the lobes obovate, shortly tapered to the apex, shallowly 3-lobulate or only unevenly toothed with pale eglandular hairs on both surfaces; petioles long, with dense, spreading eglandular hairs. *Inflorescence* very dense, with the flowers in umbel-like clusters; flowers 25–32 mm in diameter, bisexual, flat or nearly so. *Sepals* 5, 7.0–9.5 mm, pale green, flushed with brownish-red, lanceolate, with a mucro 0.5–1.0 mm, with numerous glandular and eglandular hairs. *Petals* 5, 10–16×10–16 mm, pale violet-blue fading to white at base, obovate, rounded or emarginate at apex, rounded at base, with a dense tuft of hairs either side at base. *Stamens* 10; filaments appressed to the style, white at the moderately enlarged base, the remainder blackish-purple, the basal half covered with long, coarse, spreading hairs with recurved tips; anthers dull bluish. *Style* 1, 8.5–10.5 mm, usually glabrous; stigmas 1.5–3.0 mm, greenish to dull red. *Immature schizocarps* erect on erect pedicels, standing above the currently opening flowers, glandular-hairy; mericarps 3.5–4.0 mm, nearly black, rounded at base; beak 22–28 mm including stylar portion which is 5–6 mm; a cluster of stiff hairs at the lower end of the carpel orifice hold back the seeds until they are discharged from the mericarp by the awn curling back. *Flowers* 4–6.

Introduced. Persistent garden escape at St Bernard's Weir, below Upper Helensburgh, Dunbartonshire. Native of north-east Asia from eastern Siberia eastwards, eastern Tibet and western China, Korea and Japan.

12. G. sanguineum L. Bloody Crane's-bill

Low bushy *perennial* monoecious *herb* with shortly spreading, stout rhizomes. *Stems* 10–40 cm, pale green, prostrate to erect, with numerous, long, spreading or

G. **eriostemon** DC.

G. sanguineum L.

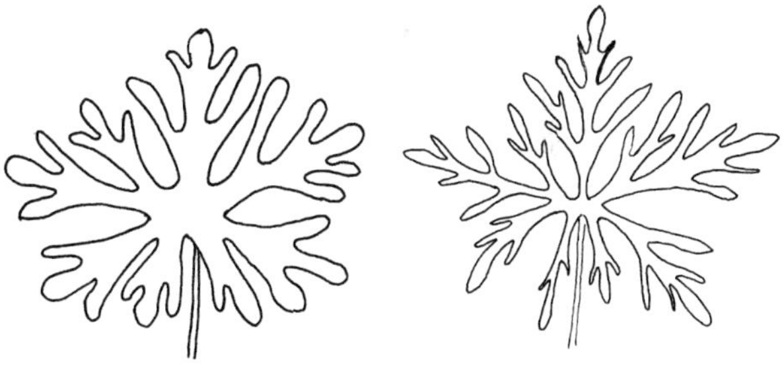

G. columbinum L.

G. dissectum L.

G. core-core Steud.

Basal leaves of **Geranium** L.

G.ibericum Cav.

G. x **magnificum** Hyl.

G. x **magnificum** Hyl.
(upper leaves)

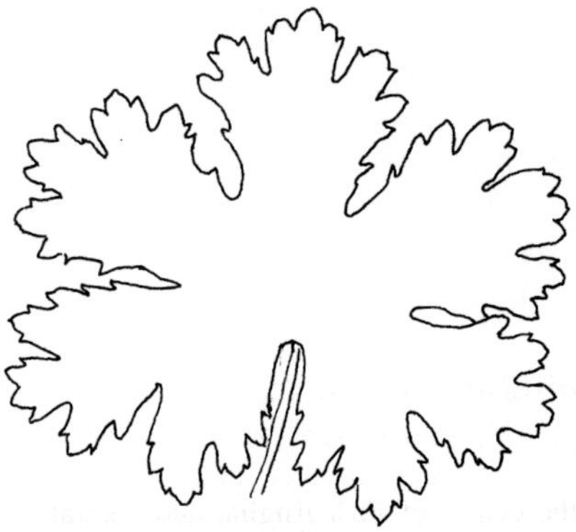

G. platypetalum Fisch. & C. A. Mey.

Basal leaves of **Geranium** L.

slightly deflexed hairs, often branched from the base and geniculate at the nodes, leafy. *Leaves* alternate, mostly cauline; lamina 2–10 cm in diameter, subrotund in outline, divided up to three-quarters of the way to the base into 5 or 7 lobes, the lobes deeply 3-lobulate up to half their length, the lobules entire or with 1–2 teeth, the middle lobule longer than the 2 lateral, with stiff, appressed eglandular hairs on both surfaces; petioles fairly long; stipules sometimes fused in pairs. *Inflorescence* of single axillary bisexual flowers or rarely in pairs; branches forming a very diffuse and leafy inflorescence of long duration; peduncles 4–7 cm; pedicels 2–5 times as long, as sepals; with long, spreading hairs and sessile glands; flowers 25–42 mm in diameter, saucer-shaped, more or less erect. *Sepals* 5, 6.5–10.0 mm, elliptical, with mucro 1.5–3.0 mm, with long eglandular hairs. *Petals* 5, 14–21 × 13–17 mm, commonly intense purplish-red and white at the extreme base, rarely white, rose-pink or pale flesh pink, usually with darker veins, obovate, shallowly emarginate at apex, not clawed, with a dense tuft of hairs on both edges at the base. *Stamens* 10; filaments gradually enlarged towards the base, coloured more or less like the petals, edges fringed with hairs in the lower half; anthers bluish. *Style* 1, about 5 mm; stigmas 3.5–4.5 mm, red or flesh-coloured. *Immature schizocarps* erect on erect pedicels; mericarps 4.0–4.5 mm, rounded at base; beak 22–32 mm, including stylar portion, 1–2 mm; a cluster of stiff hairs at the lower end of the carpel orifice hold back the seed until it is discharged from the mericarp by the awn curling back. *Flowers* 7–8. Pollinated by various insects, protandrous but capable of self-pollination. $2n=84$.

(i) Var. **striatum** Weston
G. sanguineum lancastriense G. Nicholson; *G. lancastriense* Mill.; *G. sanguineum* var. *lancastriense* (Mill.)

Stems prostrate. *Flowers* pale flesh pink.

(ii) Var. **prostratum** (Cav.) Pers.
G. prostratum Cav.

Stems prostrate. *Flowers* purplish-red or rose-pink.

(iii) Var. **sanguineum**
Stems erect. *Flowers* purplish-red or rose-pink.

Native. Grassland, rocky places, sand-dunes, and open woods on calcareous soils. Locally native in north and west Great Britain and the centre of East Anglia, and central Ireland and a widespread garden escape. Var. *sanguineum* is the most frequent plant and includes a number of cultivars that escape from gardens. Var. *prostratum* is a native plant of coastal sands and var. *striatum* is endemic on Walney Island in Lancashire. The species occurs in most of Europe, the Caucasus and north Turkey. European Temperate element.

13. G. columbinum L. Long-stalked Crane's-bill

Annual monoecious *herb* with fibrous roots. *Stems* 7–13 cm, pale green, sometimes tinted red, erect or sprawling, appressed-hairy, usually much branched from base, leafy. *Leaves* medium green on upper surface, paler beneath; lower alternate with lamina 2–4 × 2–4 cm, subrotund-reniform in outline, divided three-quarters of the way to the base into 5 or 7 lobes, the lobes obovate in outline and lobulate, the lobules oblong and obtuse at apex, the petioles up to 10 cm or more, appressed-hairy; upper often more deeply divided, with shorter petioles and usually opposite; all appressed-hairy on both surfaces; stipules 5–7 mm, reddish, linear to almost filiform. *Inflorescence* of pairs of flowers arising from the upper nodes; peduncles 2–12 cm, elongating in fruit, thinly hairy to nearly glabrous; pedicels up to 6 cm, slender. *Sepals* 5, 6–8 mm, enlarging to 8–9 in fruit, ovate, 3-veined, with a mucro up to 2 mm, appressed-hairy, with membranous, ciliate margins. *Petals* 5, 7–12 × 4.0–4.5 mm, pale to deep reddish-pink and white at base, obovate, rounded or irregularly toothed at apex, without a claw, sparsely ciliate at base. *Stamens* 10; filaments about 4 mm, glabrous or slightly hairy, enlarged at base; anthers bluish, oblong. *Stigma* arms about 1 mm, purplish. *Immature schizocarps* erect on erect pedicels; mericarps 3–4 mm, sparsely hairy or glabrous, smooth; beak 16–17 mm with distinct stylar portion about 4 mm; a cluster of stiff hairs at the lower end of the carpel orifice hold back the seed until it is discharged from the mericarp by the awn curling back; seeds about 2.5 mm, bluish-brown, oblong, reticulate. *Flowers* 6–7. $2n=18$.

Native. Grassy places, banks and scrub, mostly on calcareous soils which include cliff slopes, sandy dunes, hedgebanks, field margins, railway banks and old quarries. Locally frequent in much of Great Britain especially the south-west and scattered in Ireland, but rare in the north. Europe, except the extreme north; Algeria, Tunisia; naturalised in North America.

Section **2. Dissecta** Yeo

Annual or *biennial*, monoecious *herbs*. Retention of seed in mericarp during the pre-explosive interval is by a part of the mericarp wall which projects as a prong. *Awn* remaining attached to central column of the beak after discharge.

14. G. dissectum L. Cut-leaved Crane's-bill

Annual monoecious *herb* with fibrous roots, but usually overwintering. *Stems* 10–60 cm, pale green, decumbent, erect or sprawling, sparsely to densely glandular-hairy, often much branched from the base, leafy. *Leaves* dull or medium greyish-green on upper surface, paler beneath; lower alternate with lamina 2–7 × 2–7 mm, subrotund-reniform in outline, divided almost to base into 5 or 7 lobes, the lobes oblanceolate and divided to the midrib into narrowly oblong or linear-lobules which are obtuse or subacute, the petioles long and slender and sparsely hairy; upper opposite or alternate, smaller than the lower and deeply divided into linear, entire or irregularly pinnatisect lobes and more shortly petiolate; all more or less appressed-hairy especially on the veins; stipules 6–9 × 3–4 mm, brownish, membranous, thinly hairy. *Inflorescence* of paired flowers arising from the upper nodes; peduncles 5–10 mm, glandular-hairy; pedicels 10–15 mm, densely glandular-hairy towards the apex. *Sepals* 5, 5–6 mm, narrowly ovate, with a mucro 1.0–1.5 mm, 3-veined, accrescent in fruit, glandular-hairy.

Petals 5, 4.5–6.0 × about 3 mm, pale to deep pink, obovate, shallowly emarginate, without a claw, with hairs either side at the base, but not across the front surface. *Stamens* 10; filaments about 2.5 mm, enlarged at base, the enlarged part ciliate; anthers bluish, subrotund. *Stigma* arms up to 0.5 mm, reddish or purplish. *Immature schizocarps* erect on erect pedicels; mericarps smooth, rounded at base, hairy; beak 12–13 mm, including the distinct stylar portion which is 1–2 mm; a flexible prong on the lower edge of the orifice of the mericarp holds back the seed until it is discharged from the mericarp by the awn curling back and remaining attached to the central column; seeds about 2 mm, dark brown, subglobose, strongly and conspicuously reticulate. *Flowers* 5–8. Monogamous and protogynous, probably usually selfed, insect visitors few. $2n=32$.

Perhaps native in coastal localities, elsewhere an introduced weed. Grassy, waste and cultivated places. Common throughout most of Great Britain and Ireland. Europe and the Mediterranean region eastwards to Iran; Atlantic Islands; introduced elsewhere. European Southern-temperate element.

Section **3. Chilensia** Knuth

Perennial monoecious *herbs* with napiform roots. *Petals* emarginate. *Immature schizocarps* erect on erect pedicels. *Fruits* of seed-ejection type.

15. G. core-core Steud. nom. conserv. Alderney Crane's-bill
G. submolle auct.; *G. retrorsum* auct.; *G. hererae* Knuth; *G. rapulum* A. St Hil. & Naudin nom. rejic.; *G. commutatum* Steud.; *G. ochsenii* Phil.; *G. moorei* Phil.; *G. melanopotamicum* Speg.; *G. argentinum* Knuth; *G. subsericeum* Knuth

Perennial monoecious *herb* with napiform roots. *Stems* 15–75 cm, pale green, flushed red, decumbent or ascending and straggling, with numerous, short, retrorse, appressed simple eglandular hairs, much branched, leafy. *Leaves* alternate below and soon dying, the cauline slowly decreasing in size upwards; lamina 1.5–4.5 × 2–5 cm, reniform in outline, divided up to three-quarters of the way to the base into 5–7 lobes, the lobes oblanceolate or obovate, with 3 or 5 obtuse teeth or short lobules, with minute, appressed eglandular hairs on both surfaces; petioles up to 3 cm, pale green, often tinted red and with numerous, short, appressed eglandular hairs; stipules lanceolate, with a long filamentous tip. *Inflorescence* of pairs of flowers in the axils of upper leaves; peduncles up to 15 mm, with numerous, short, appressed eglandular hairs; pedicels up to 10 mm, with numerous, short, appressed eglandular hairs; flowers 7–8 mm in diameter. *Sepals* 5, 4.5–5.0 mm, pale green, ovate, with a mucro about 0.5 mm, midrib prominent, with numerous, minute, retrorse, appressed eglandular hairs. *Petals* 5, 2.5–6.0 × 2.0–3.5 mm, purplish-rose with white veins, oblanceolate or obovate, deeply emarginate. *Stamens* 10; filaments pale green; anthers whitish to pink. *Styles* 1, whitish; stigmas pale brown. *Immature schizocarps* erect on erect pedicels; mericarps 3.5–4.0 mm, rounded at base; beak 10–11 mm; seeds reticulate, retention during the pre-explosive interval is by a twist in the junction of the awn and mericarp, so that the orifice faces sideways; awn not remaining attached to the central column after discharge. *Flowers* 6–7.

Introduced. Hedgerows and grassy and waste places. Naturalised in the Channel Islands, Guernsey since 1926, Alderney since 1938 and a rare casual in Jersey; also in Cornwall. Native of South America from Ecuador to Chile and Argentina.

Section **4. Tuberosa** Reiche

Perennial herbs. Petals more or less emarginate. *Stamens* with filaments with long hairs. *Immature fruits* erect on erect pedicels. Retention of seed in the mericarp during the pre-explosive interval is by a twist in the junction of the awn and mericarp, so that the orifice faces sideways; awn not remaining attached to the central column after discharge.

16. G. ibericum Cav. Caucasian Crane's-bill

Perennial monoecious *herb* with a short rhizome. *Stamens* up to 50 cm, pale yellowish-green, erect, with dense, short to very long, pale simple eglandular hairs, branched, leafy. *Leaves* alternate; lamina 2–10 × 2–12 cm, dull medium green on upper surface, paler beneath, reniform in outline, divided three-quarters or seven-eights of the way to the base into 9 or 11 lobes, the lobes obovate and pinnately divided into acutely dentate lobules, the lobes more or less overlapping and forming a dense leaf which is soft to the touch and with dense, short to medium, pale simple eglandular hairs on both surfaces; petiole medium, the upper leaves sessile. *Inflorescence* of groups of flowers in the upper leaf axils; peduncles few or none; pedicels hairy; flowers 45–50 mm in diameter, bisexual, slightly sweet-scented. *Sepals* 5, 13–16 mm, pale green, tinted brownish-purple, lanceolate, mucro 2–3 mm, with dense, short to long, pale simple eglandular hairs. *Petals* 5, 23–26 × 16–17 mm, deep violet blue with darker veins and white near the base, obovate, deeply emarginate with a mucro in the centre, without a claw. *Stamens* 10; filaments purple; anthers black. *Style* and stigma red. *Immature schizocarps* erect on erect pedicels; mericarps without a tuft of bristles at base; rounded below; beak 27–35 mm including stylar portion which is 4–7 mm; seed retention is achieved by a twist at the point where the mericarp joins the awn, bringing the open side of the mericarp into a sideways facing position, the mericarp and awn dropping away after the seed is ejected. *Flowers* 6–7. $2n=56$.

Introduced. Naturalised for many years in an old churchyard at Llangwyryfon in Cardiganshire and on a wooded area of sand dunes near Longniddry in East Lothian; also recorded from Norfolk, Denbighshire, Derbyshire, Yorkshire, Isle of Man and Shetland. Native of north-east Turkey, Caucasus and perhaps north Iran.

17. G. × magnificum Hyl. Purple Crane's-bill
G. ibericum subsp. **ibericum × platypetalum**

Perennial monoecious *herb* with a thick, compact rootstock. *Stems* 50–75 cm, pale yellowish-green, erect, robust, solid, with numerous, medium to long, pale, spreading simple eglandular hairs, leafy and branched. *Leaves* alternate; lamina 10–15 × 10–15 cm, dull dark green on upper surface, pale green beneath, subrotund or reniform in outline, acute at apex, divided two-thirds or seven-eights of the way to the base into 5–9 lobes, the lobes oblanceolate, acute at apex, divided into about 3 lobules halfway to the midrib, the lobules lanceolate with more or less acute teeth, cordate at base, the basal lobes overlapping, with dense, short and medium, pale, soft simple eglandular hairs on both surfaces; veins slightly impressed on upper surface, very prominent beneath; petioles up to 30 cm, pale yellowish green, with dense, medium and long, pale, spreading simple eglandular hairs. *Inflorescence* rather dense; flowers bisexual, erect, saucer-shaped, scentless, 40–50 mm in diameter; pedicels 7–15 mm, pale green, with dense, short to medium, pale, red-tipped glandular hairs and fewer medium to long, pale simple eglandular hairs; bracts like leaves; bracteoles 5–9 mm, linear, with long hairs. *Sepals* 5, 12–14 mm, lanceolate, with a mucro 2–3 mm, with dense, short to medium, pale, red-tipped, spreading glandular hairs and numerous, long, pale simple eglandular hairs. *Petals* 5, 22–24 × 16–22 mm, overlapping, deep purple-violet, sometimes with more red, sometimes more blue, with darker veins, obovate, emarginate, with point in the notch, without a claw. *Stamens* 10; filaments 6–7 mm, purple, white or pink at the base; anthers sometimes normal, sometimes with yellowish anthers which do not burst. *Style* red; stigmas deep coloured. *Schizocarp* not fully developed; beak partly developed but usually reaching full size, the stylar portion 4–7 mm; seeds not ripening. *Flowers* 5–6. $2n = 42$.

Introduced. Probably the commonest *Geranium* grown in gardens and frequently naturalised in grassy places, roadsides, railway banks and on waste land. Scattered records for much of Great Britain and in Guernsey. Of garden origin and sterile. Its place of origin is unknown and the earliest specimen found was gathered in the Botanic Garden of Geneva in 1871.

18. G. platypetalum Fisch. & C. A. Mey. Glandular Crane's-bill

Perennial herb with a thick, compact rootstock. *Stem* up to 40 cm, pale yellowish-green, deep brownish-red at base, sometimes tinted so above, with a dense mixture of short to very long, pale glandular and simple eglandular hairs, leafy. *Leaves* alternate or opposite, gradually reduced in size upwards; lamina 2–16 × 2–20 cm, dull medium green on upper surface, paler beneath, subrotund in outline, divided palmately about halfway to the base into 7 or 9 lobes, the lobes obovate, divided in 3 or 4 lobules, the lobules with obtuse teeth; with dense, short to long, pale glandular and simple eglandular hairs on both surfaces; petioles up to 20 cm, pale green, glandular-hairy. *Inflorescence* a dense panicle; peduncles up to 8 cm; pedicels not more than 3 times as long as sepals; bracteoles 7–15 mm; flowers 30–45 mm in diameter, bisexual, flat or saucer-shaped. *Sepals* 5, 8–12 mm, green, tinted brownish-purple, lanceolate, with a mucro 2.0–4.5 mm, with dense, short to very long, pale, red-tipped glandular hairs. *Petals* 5, 16–22 × 13–19 mm, deep violet or bluish-violet, usually slightly paler or pinkish at base, with very dark violet, embossed, glossy and forking veins, broadly obovate, emarginate at apex, narrowed at base but without a claw. *Stamens* 10; filaments violet, paler at base, slightly curved outwards at the tips at first, arching back towards petals later, with many long hairs; anthers bluish-black. *Style* red; stigmas about 2.5 mm, dark red. *Immature schizocarps* erect on erect pedicels; mericarps about 5 mm, rounded at base, without tuft of bristles inside; beak 24–30 mm including stylar portion which is 4–5 mm; seed retention is achieved by a twist at the point where the mericarp joins the awn, bruising the open side of the mericarp into a sideways-facing position, the mericarp and awn dropping away after the seed is ejected. *Flowers* 6–7. $2n = 28, 42$.

Introduced. Rarely grown in gardens. Naturalised in Dunbartonshire. Native of Caucasus including north-east Turkey and north-west Iran.

Subgenus **2. Robertium** Picard

Annual or *perennial* monoecious *herbs*. *Petals* usually with a distinct claw. *Carpels containing seed* are thrown off explosively and the awn drops away at the moment of the explosion; mericarp with rounded base, without a horny point or blunt tubercle.

Section **5. Batrachioides** W. D. J. Koch

Annual or *biennial* monoecious *herbs* usually soft hairy. *Sepal* with a very short mucro. *Petal* emarginate. *Pollen* bluish. *Immature schizocarps* erect on reflexed pedicels; mericarps ribbed, though sometimes only faintly and only near the midrib at the top.

19. G. pyrenaicum Burm. fil. Hedgerow Crane's-bill

Perennial monoecious *herb* with an ill-defined, short rootstock. *Stems* 25–60 cm, pale green, erect or trailing, with numerous, short glandular hairs mixed with few to numerous, long, spreading eglandular hairs, branched, leafy. *Leaves* medium green on upper surface, paler beneath; basal alternate, 5–10 × 5–10 cm, subrotund in outline, divided half to two-thirds of the way to the base into 7 or 9 lobes, the lobes obovate in outline and divided at the apex for about one-third of their length into oblong, obtuse lobules, usually with 2 lateral teeth, with a long petiole and crimson, ovate, acute stipules; cauline leaves opposite, becoming gradually smaller in size and length of petiole upwards, the uppermost with narrow, acute, toothless lobes; all with minute glandular and longer simple eglandular hairs on both surfaces. *Inflorescence* of bisexual flowers in pairs forming a loose cyme; peduncles 15–50 mm, slender, hairy; pedicels 10–20 mm, very slender, hairy. *Sepals* 5, 4–5 mm, green, ovate-oblong, with a very small mucro, hairy. *Petals* 5, 7–10 mm, purple with a white haze, rarely whitish, veins darker, obovate-cuneiform, deeply emarginate, with claw less than half as long as limb. *Stamens* 10; filaments pale pink, all with anthers; anthers and pollen bluish. *Stigmas* about 1.5 mm, pale yellow, sometimes

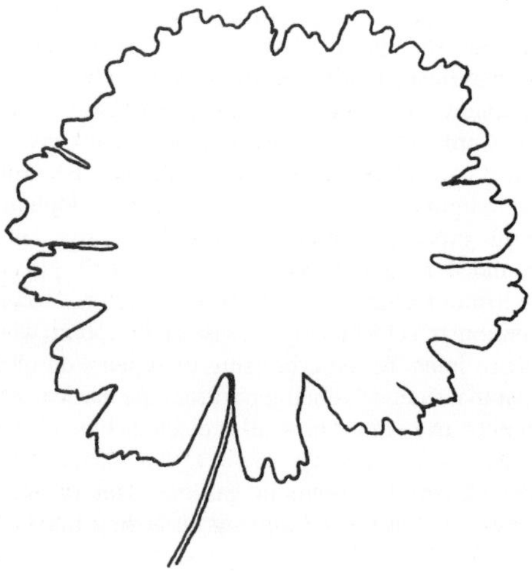

G. **pyrenaicum** Burm. fil.

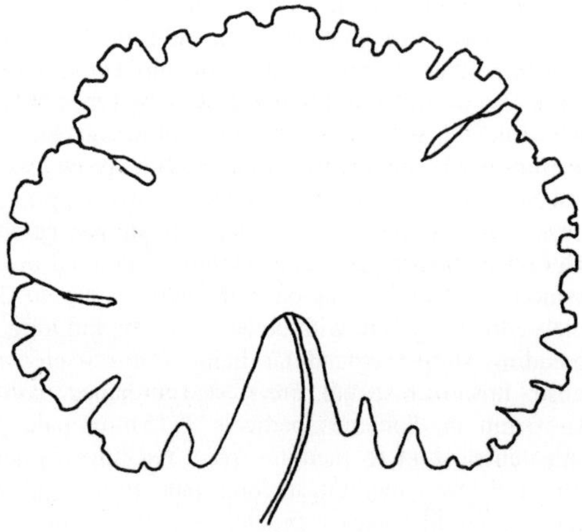

G. **brutium** Gasp.

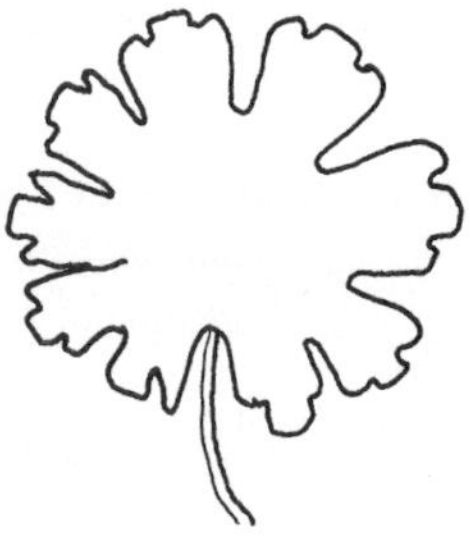

G. **molle** L.

G. **pusillum** L.

G. **macrorrhizum** L.

G. **lucidum** L.

Basal leaves of **Geranium** L.

tinged pink. *Immature schizocarps* erect on sharply reflexed pedicels; mericarps about 2.5 mm, with short appressed hairs and a few raised ribs in the immediate vicinity of the midrib at the top, rounded at base; beak 12–13 mm, the stylar portion not more than 0.5 mm; the mericarp without the awn is thrown off to a distance of a metre or more and the awn drops off at the moment of discharge; seeds smooth. *Flower* 6–8. Pollinated by various insects, protandrous but capable of self-pollination. $2n=28$.

Probably introduced. Hedgerows, roadsides, grassy places and rough ground. Locally frequent in much of Great Britain and Ireland but absent from much of the north and west. Possibly native only in south-west and west Europe, east to the Caucasus and introduced and naturalised further north. European Temperate element.

20. G. pusillum L. Small-flowered Crane's-bill

Annual monoecious *herb* with fibrous roots. *Stems* 10–40 cm, pale green, often suffused brownish-purple, spreading, ascending or suberect, with numerous, very short, more or less equal glandular hairs, branched from the base, leafy. *Leaves* opposite; lamina 1–5 × 1–5 cm, dull medium greyish-green on upper surface, paler beneath, more or less subrotund in outline, divided two-thirds to three-quarters of the way to the base into 7–9 lobes, the lobes narrowly obovate and each with 3 or 4 obtuse lobules or teeth, with dense, very short to medium, soft, pale simple eglandular hairs on both surfaces; veins faint above and fairly prominent beneath; petioles up to 8 cm, with very short glandular and slightly longer simple eglandular hairs. *Inflorescence* of numerous bisexual flowers in pairs and 4–6 mm in diameter; pedicels up to 15 mm, spreading or ascending with dense, short, more or equal glandular hairs. *Sepals* 5, 2–4 mm, green, ovate, acute at apex with a very short mucro, with numerous, very short glandular hairs and longer, pale simple eglandular hairs. *Petals* 5, 2–4 mm, pale dingy lilac, broadly obovate, emarginate at apex, with a short claw. *Stamens* 10, but only 5 have anthers; filaments white; anthers and pollen bluish. *Style* greenish; stigmas dark. *Immature schizocarps* erect on more or less reflexed pedicels; mericarps about 2 mm, rounded at base, with short appressed hairs and a few raised ribs in the immediate vicinity of the midrib at the top; beak 5–7 mm, without a stylar portion; the mericarp without the awn is thrown off to a distance of a metre or so feet and the awn drops off a the moment of discharge; seeds smooth. *Flowers* 6–9. Insect visitors few and probably usually selfed. Protogynous. $2n=26$.

Native. Cultivated and waste ground and open habitats in dry grassland in scattered localities in some of which it is frequent. Throughout much of Great Britain but absent from much of the north and west; in a few mainly coastal localities in Ireland. Most of Europe, except the extreme north; Turkey to the Caucasus, Himalayas and Israel; naturalised in North America. Eurosiberian Temperate element.

21. G. brutium Gasp. Mediterranean Crane's-bill
G. molle subsp. *brutium* (Gasp.) P. H. Davis

Annual monoecious *herb* with fibrous roots. *Stems* up to 50 cm, pale green, sometimes suffused purplish, erect or trailing, the plant becoming bushy, with short glandular and longer eglandular hairs, branched, leafy. *Leaves* medium green on upper surface, paler beneath; basal alternate with lamina 5–10 × 5–10 cm, subrotund or reniform in outline, divided nearly to halfway to the base into 9 lobes, the lobes obovate and divided for less than one-third of the way to the base into obtuse lobules, the lobules with 1 tooth; cauline leaves alternate becoming rapidly reduced in size and length of petiole upwards; all with glandular and simple eglandular hairs on both surfaces. *Inflorescence* diffuse; peduncles and pedicels very slender, hairy; flowers bisexual. *Sepals* 5, 4–6 mm, lanceolate, with hardly a mucro, rounded on back, glandular-hairy. *Petals* 5, 8.0–12.5 mm, pinkish-red, with darker veins, oblanceolate, emarginate at apex, narrowed below to a distinct claw, base with a dense fringe of wavy hairs on both sides. *Stamens* 10; filaments pink; anthers and pollen bluish-black. *Stigmas* 1.5 mm, pink or purple. *Immature schizocarps* erect on sharply reflexed pedicels; mericarps 2.0–2.5 mm, acute at apex, rounded at base, glabrous, densely covered with slanting ribs; beak 7–9 mm; seed smooth; the mericarp without the awn is thrown off to a distance of a metre or so and the awn drops off at the moment of discharge. *Flowers* 6–7.

Introduced. Rare garden escape and also found in wild flower seed. Native of the central and east Mediterranean region.

22. G. molle L. Dove's-foot Crane's-bill

Annual monoecious *herb* with fibrous roots. *Stems* 10–50 cm, pale green, short and prostrate, spreading or tall and erect, with long, soft simple eglandular hairs mixed with numerous, short glandular hairs above, branched from the base, leafy. *Leaves* dull or greyish-green on upper surface, paler beneath; lower alternate, the lamina 2.5–5.0 × 2.5–5.0 cm, subrotund-reniform in outline, divided about halfway to the base into 5 to 9 lobes, the lobes obovate, bluntly toothed and cuneate at base, often stained red in the sinuses between the lobes, on long petioles, the stipules 5–6 mm, pale brown, ovate, acute or shortly acuminate at apex and often laciniate or frayed; upper usually alternate, smaller and more deeply lobed than the lower with shorter petioles; all shortly and softly hairy on both surfaces. *Inflorescence* of numerous flowers in pairs; peduncles 10–30 cm, slender, with dense glandular hairs mixed with longer simple eglandular ones; pedicels 8–15 mm, hairy like the peduncle. *Sepals* 5, 3–5 mm, oblong-ovate, 3-veined, obtuse or subacute at apex, with numerous glandular hairs. *Petals* 5, 4–8 mm, purplish-pink with a white base or pale pink to white, obovate, deeply emarginate, narrowed at base to a short claw, with a fringe of wavy hairs either side at the base. *Stamens* 10; filaments with a wide base, more or less glabrous; anthers and pollen bluish, oblong. *Stigma* arms purple or crimson. *Immature schizocarps* erect on reflexed pedicels; mericarps about 1.5 mm, rounded at base, glabrous, ribbed, or rarely smooth, nearly spherical; beak 5–8 mm, the stylar portion about 0.5 mm; the mericarp without the awn is thrown off to a distance of a metre or so and the awn drops off at the moment of discharge; seeds smooth. *Flowers* 4–9. Visited by various insects, more or less homogamous, probably often selfed. $2n=26$.

(i) Var. **arenarium** N. Terracc.
Plant forming a circle flat on the ground. *Leaves* 0.5–3.0 cm in diameter. *Flowers* 5–6 mm in diameter. *Mericarps* ribbed.

(ii) Var. **molle**
Plant erect or ascending. *Leaves* 1.5–5.0 cm in diameter. *Flowers* 5–8 mm in diameter. *Mericarps* ribbed.

(iii) Var. **aequale** Bab.
Plant erect. *Leaves* 1–3 cm in diameter. *Flowers* 5–6 mm in diameter. *Mericarps* smooth, without ribs.

(iv) Var. **grandiflorum** Vis.
Plant erect. *Leaves* 3–6 cm in diameter. *Flowers* 9–13 mm in diameter. *Mericarp* with ribs.

Native. Cultivated and waste land, rather open grassland, sandy and calcareous heaths and shingle, ascending to 550 m. Common throughout Great Britain and Ireland. Europe, except the extreme north; south-west Asia to the Himalaya; North Africa; Macaronesia; naturalised in North and South America, New Zealand and Japan. European southern-temperate element. The common plant is var. *molle*. Var. *arenarium* is an ecotype of sandy places by the sea, sandy heaths inland and occasionally elsewhere. It seems also to occur on the coasts of western and southern Continental Europe. The distribution of var. *aequale* is unknown. Var. *grandiflorum* occurs in scattered localities across southern Great Britain and in Continental Europe. Its flowers approach those of *G. pyrenaicum* in size, but it does not look like that plant in general appearance.

Section **6. Unguiculata** Reiche
Perennial herbs, with hairs mostly glandular. *Sepals* erect, usually forming a swollen calyx. *Petals* with a distinct limb and claw, scarcely emarginate. *Pollen* yellow. *Immature schizocarps* erect, on erect pedicels. *Mericarps* with a pattern of ribs.

23. G. macrorrhizum L. Rock Crane's-bill

Perennial monoecious *herb* with fleshy, underground rhizomes. *Stems* 30–50 cm, pale yellowish-green, tinted brownish-purple, ascending or erect, rigid, thick, with scattered short and medium, simple eglandular hairs and numerous minute glandular hairs, branched above and with few leaves. *Leaves* alternate; lamina 3–10 × 6–20 cm, decreasing rapidly in size and length of petiole upwards, matt medium yellowish-green on upper surface, paler beneath, subrotund to reniform in outline, obtuse at apex, divided two-thirds to three-quarters of the way to the base into 7 lobes, the lobes oblanceolate, obtuse, shallowly lobulate in upper half, gradually attenuate and entire in the lower half, the lobules with several obtuse teeth, with dense, minute glandular hairs and longer simple eglandular hairs on both surfaces; veins prominent beneath; petioles up to 10 cm, with minute glandular and simple eglandular hairs. *Inflorescence* dense, usually with umbel-like clusters of flowers above the first few nodes; flowers 20–25 mm in diameter, bisexual; pedicels shorter than or a little longer than sepals, pale green often suffused brownish-purple, with dense, short glandular hairs. *Sepals* 5, 6–9 mm, reddish with 3 green ribs, forming a bladdery calyx, elliptical, the mucro at apex 1.0–4.5 mm, quarter to half as long as the sepal, with dense, minute glandular hairs and some longer ones. *Petals* 5, 15–18 × 10–15 mm, usually pinkish-purple, sometimes white, broadly obovate, limb, rounded at apex, with a cuneate claw, slightly asymmetrically spreading, with 2 rounded ridges and a central channel, hairy on the back and the front of the sides, the hairs sometimes extending on to the tops of the ridges and overhanging the channel. *Stamens* 10; filaments 18–24 mm, usually purplish-red, sometimes pink, displaced to the lower side of the flower and slightly turned up at the tips; anthers orange-red to dull red. *Style* about 22 mm, purplish-red, displaced and curved like the filaments; stigmas 1.0–1.5 mm, yellowish. *Immature schizocarps* and their pedicels erect; mericarps 2.5–3.0 mm, with wavy, horizontal ribs; beak about 30–34 mm, the stylar portion about 18 mm; the mericarp without the awn is thrown off to distance of a metre or so and the awn simply drops off at the moment of discharge. *Flowers* 4–6 and usually again later. $2n=46$, c. 92.

Introduced. Commonly grown in gardens and naturalised on walls, banks and in grassy places. Naturalised in Postbridge on Dartmoor in Devonshire since 1890 and recorded in other scattered localities throughout Great Britain.. Native of the south side of the Alps, Apennines, Balkan Peninsula and south and east Carpathians. It has long been used for medicinal purposes and as a source of oil of Geranium, which is used in perfumery. It has also been used in tanning.

× **dalmaticum** (Beck) Rech. fil.
= **G. × cantabrigiense** Yeo

This hybrid has long-lived, trailing stems like those of *G. dalmaticum*. The leaves are 3–9 cm, divided into 7 lobes, each lobe divided again into 3 which are entire or with few teeth. The flowers are similar to *G. macrorrhizum*, with hair-tipped teeth on the filaments like *G. dalmaticum*. The petals are bright pink or white. The plant is sterile, but spreads by runners.

Introduced. A. O. Chater has recorded three large colonies, well naturalised, presumably from garden throw-outs, on a grassy, disused quarry slope, Bryn-y-mor Road, on the outskirts of Aberystwyth, Cardiganshire. This hybrid has arisen several times in gardens and was described from those in the Botanic Garden, Cambridge.

Section **7. Lucida** Knuth
Annual monoecious *herbs* with mostly glandular hairs. *Sepals* erect, with lengthwise keels and transverse flaps between them. *Petals* with a distinct limb and claw, rounded at apex. *Pollen* yellow. *Immature schizocarps* and their pedicels erect; mericarps with lengthwise ribs bearing glandular hairs.

24. G. lucidum L. Shining Crane's-bill

Annual monoecious *herb* with fibrous roots. *Stem* 10–50 cm, pale green, usually suffused with red, rather slender, but succulent, erect, glabrous, often branched from base, leafy. *Leaves* alternate; lamina 1.0–4.5 × 1–5 cm, pale shining green turning pinkish-red on upper surface, paler beneath, subrotund in outline, obtuse at apex, palmately divided about two-thirds of the way to the base into 5 lobes, the lobes obovate with obtuse lobules and teeth, glabrous or nearly so; veins obscure above, prominent beneath; petioles medium in length, often red-tinted, glabrous. *Inflorescence* rather open; flowers 7–9 mm in diameter, bisexual, erect; peduncles and pedicels glabrous. *Sepals* 5, 5–7 mm, pale yellowish-green with a red line round the margin, oblong-ovate, acuminate at apex, keeled on back, glabrous. *Petals* 5, 8–9 mm, deep pink or rarely white, glabrous , the limb oblanceolate and rounded at apex, the claw longer than the limb and ridged so as to form nectar passages. *Stamens* 10, exceeding throat of corolla by about 1 mm; filaments white; anthers yellow; pollen yellow. *Style* whitish, minutely bristly; stigma whitish. *Immature schizocarps* and their pedicels erect; mericarps about 2.2 mm, broadest at the top, with a network of raised ribs which run parallel and vertically at the top, the ridges bearing glandular hairs; beak about 11 mm, its stylar portion about 4 mm; the mericarp without the awn is thrown for a distance of a metre or so and the awn drops away at the moment of discharge. *Flowers* 5–8. $2n = 20$.

Native. Rocks, walls, bare ground and stony banks, mainly on calcareous soils; also widespread in artificial habits. Locally common in most of Great Britain and Ireland except northern Scotland, and also a garden escape in southern England. Europe except the north-east; south-west Asia to the Himalaya; North Africa. Submediterranean-Subatlantic element.

Section **8. Roberta** Dumort.

Annual, biennial or *perennial* monoecious *herbs* with mostly glandular hairs. *Sepals* with a distinct mucro. *Petals* with a distinct limb and claw, scarcely emarginate. *Pollen* yellow. *Immature schizocarps* and pedicels upwardly inclined; mericarps with a pattern of ribs.

25. G. robertianum L. Herb Robert

Rosette-forming, succulent *annual* or *biennial* monoecious *herb*, usually overwintering and with a strong disagreeable smell. *Stems* 10–50 cm, bright or dark green, usually tinged and often suffused with red, especially at the nodes, arising in a cluster from the rosette or in one or more whorls on a central stem emerging from the rosette, decumbent, ascending or prostrate, forking regularly but slightly unequally below, more unequally above, with numerous, very short to long, pale, spreading simple eglandular hairs throughout or only below, mixed with very short and short glandular hairs above and sometimes a few below, much branched, leafy. *Leaves* alternate; lamina 2.5–11.0 × 2.5–11.0 cm, gradually decreasing in size upwards, bright green on upper surface, slightly paler beneath, broadly ovate in outline, acute at apex, palmately divided into 5 petiolulate leaflets, the leaflets ovate in outline, pinnately lobed almost or quite to midrib, the lobes with obtuse teeth and often mucronulate, shortly eglandular-hairy on both surfaces; veins faint above, fairly prominent beneath; petiole long, green or red, with medium to long, pale simple eglandular hairs and short glandular hairs. *Inflorescence* diffuse; pedicel ascending after flowering, mostly straight, with numerous, very short and short glandular hairs; flowers bisexual, in pairs, 12–16 mm in diameter. *Sepals* 5, 4–6 mm, pale green, often suffused red, oblong-lanceolate, the mucro 0.7–1.2 mm, with long simple eglandular hairs and short glandular ones. *Petals* 5, 8–14 mm with claw, deep pink with paler veins towards the base, the limb 3.5–5.5 mm, the claws slightly shorter than the limb, each with a double ridge abutting a stamen so that 5 nectar passages of approximately circular cross-section are formed. *Stamens* 10, projecting about 1.5 mm from throat of flower; filaments white; anthers red to pinkish-orange or purple, pollen yellow. *Stigmas* 0.5–1.5 mm, pale to deep pink. *Immature schizocarps* and their pedicels upwardly inclined; mericarps 2.0–2.8 mm, with a fine network of ribs, sparse towards the base, and 1 or 2 collar-like rings round the apex; beak 12–15 mm, the stylar portion 4–5 mm; with medium and long simple eglandular hairs and short glandular hairs; the mericarp without the awn is thrown for a distance of a metre or so and the awn drops away at the moment of discharge. *Flowers* 5–9. Visited by various insects, but self-pollination possible.

(a) Subsp. **celticum** Ostenf.
G. robertianum var. *celticum* (Ostenf.) Druce

Usually *annual*. *Stems* ascending, reddish only at the nodes and petiole bases. *Flowers* pale pink or whitish. *Stigmas* pink or white. *Schizocarps* hairy. $2n = 64$.

(b) Subsp. **maritimum** (Bab.) H. G. Baker
Usually *biennial*. *Stems* procumbent or arcuate-ascending, usually suffused red. *Flowers* deep pink. *Stigmas* deep pink. *Schizocarps* usually glabrous. $2n = 64$.

(i) Var. **maritimum** Bab.
G. purpureum var. *littorale* Rouy

Plant glabrous or nearly so.

(ii) Var. **intermedium** Wilmott
G. raii Lindl.

Plant fairly hairy.

(iii) Var. **hispidum** Druce
Plant densely hairy.

(c) Subsp. **robertianum**
Usually *biennial*. *Stems* ascending, usually suffused red. *Flowers* deep pink. *Stigmas* deep pink. *Schizocarps* more or less hairy. $2n = 64$.

Native. Woods, hedgebanks, among rocks, on shingle, ditch and streamsides and a weed in gardens. Common throughout Great Britain and Ireland up to 700 m. Europe except the extreme north, north-west Africa in the mountains, Canary Islands, western Asia, Himalayas and south-west China, probably introduced in eastern North America. European Temperate element. Subsp. *robertianum* is the common inland form throughout the range of the species. Subsp. *celticum* occurs on sunny limestone rocks in Glamorganshire, Breconshire, Carmarthenshire, Somersetshire, Co. Clare and Co. Galway; apparently endemic. Subsp. *maritimum* occurs all around the costs of Great Britain and Ireland. All plants are prostrate in habit but show much local variation in other characters, which are maintained by self-pollination. More variants almost certainly need to be described.

26. G. purpureum Vill. Little Robin
G. robertianum subsp. *purpureum* (Vill.) Nyman

Annual or *biennial* monoecious *herb* with fibrous roots, usually overwintering. *Stems* 10–40 cm, pale green, but usually deeply suffused red or becoming so, slender, erect, ascending or prostrate, with few to numerous, medium to long, pale simple eglandular hairs below with fewer or none above, often branched from the base, leafy. *Leaves* alternate; lamina 2–5 × 2–5 cm, pale green on both surfaces, usually becoming suffused with red, broadly ovate in outline, divided to the base into 3 petiolulate segments, the segments ovate in outline and divided to the base into 5 lobes, the lobes ovate and divided nearly to the base into lobules, which are obtusely toothed, glabrous or with a few to numerous short hairs; petiole up to 2 cm, pale green or red, with short hairs. *Inflorescence* rather diffuse; flowers bisexual, erect, 10–12 mm in diameter; pedicels reddish, with dense, short, curled simple eglandular hairs. *Sepals* 5, 4–5 mm, suffused red, lanceolate, gradually narrowed above, mucro 1.5–2.0 mm, with numerous short to medium glandular and eglandular hairs. *Petals* 5, 5–9 mm, purplish-pink, oblanceolate, narrowed at base to a claw half as long as blade. *Stamens* 10; filaments white; anthers yellow; pollen yellow. *Style* whitish; stigma red. *Immature schizocarps* upwardly inclined; mericarps with dense wrinkle-like ribs and 2–3(–4) deep, collar-like ridges at apex; with short to long, glandular and eglandular hairs; beak 8–10 mm, with dense, short hairs; the mericarp without the awn is thrown to a distance of a metre or so and the awn drops away at the moment of discharge. *Flowers* 5–9. $2n = 32$.

(a) Subsp. **purpureum**
G. modestum Jord.; *G. lobelii* Boreau; *G. minutiflorum* Jord.; *G. robertianum* var. *modestum* (Jord.) Syme; *G. intricatum* Gren.; *G. scopulicola* Jord. nom. nud.

Stems erect or strongly ascending.

(b) Subsp. **forsteri** (Wilmott) H. G. Baker
G. purpureum var. *forsteri* Wilmott

Stems procumbent, the tips ascending.

Native. Subsp. *purpureum* is found in sunny, rocky and stony places mainly near the sea. It is very local from Cornwall and Gloucestershire to Sussex in southern England, Carmarthenshire in Wales, Cork and Waterford in Ireland and the Channel Islands. Also in south-west Europe. Mediterranean-Atlantic element. Subsp. *forsteri* is endemic to stabilised shingle beaches in Hampshire, Sussex and Guernsey. Its schizocarps approach those of *G. robertianum* and it possibly arose from a cross with that species. In cultivation it flowers a fortnight later than subsp. *purpureum*.

27. G. yeoi Aedo & Muñoz Garmendia
Greater Herb Robert
G. anemonifolium auct.; *G. lowei* nomen; *G. rubescens* Yeo, non Andrews

Biennial herb with thin, fibrous roots forming rosettes. *Stems* 20–60 cm, beetroot red, especially below, robust, with numerous, short to very long, pale, red-tipped glandular hairs, branched, leafy. *Leaves* alternate; lamina 2–22 × 2–22 cm, gradually decreasing in size upwards, medium yellowish-green on upper surface, slightly paler beneath, subrotund or broadly ovate in outline, divided to the base into 5 leaflets, the leaflets broadly ovate and divided into 5–7 lobes, the lobes lanceolate, mucronate at apex and with numerous, acute or mucronate teeth, with scattered to fairly numerous, medium, pale simple eglandular hairs on both surfaces; veins impressed above, slightly prominent beneath; petiole rather long and thick, beetroot red above, pale green beneath, with numerous eglandular hairs. *Inflorescences* arising on a cluster of stems from the rosette, forking regularly; flowers 22–33 mm in diameter, bisexual; peduncles and pedicels with dense, short to very long, pale, red-tipped glandular hairs. *Sepals* 5, 7–10 mm, medium green with a red base, oblong-lanceolate, with a mucro about 2 mm, with dense, short to long, pale, red-tipped, glandular hairs. *Petals* 5, 12–22 mm, bright purplish-pink with paler veins near the base, throat of flower dark red, obovate, with a claw 6–7 mm. *Stamens* 10; filaments white; anthers purple; pollen yellow. *Style* and stigmas pale red. *Immature schizocarps* and pedicels upwardly inclined; mericarps 3.3–3.7 mm, with a pattern of ribs and 1 or 2 collar-like rings at the top and no hairs or tangle-strands; beak about 20 mm, of which 6–7 mm is the stylar portion; the mericarp without the awn is thrown off to a distance of a metre or so, and the awn simply drops away at the moment of discharge. *Flowers* 5–8.

Introduced. Naturalised on rough ground in Guernsey since 1968 and on the Isle of Man since the 1930s, and recorded for the Isles of Scilly and Nottinghamshire. Native of Madeira. Named after Peter Frederick Yeo (b. 1929).

Section **9. Anemonifolia** Kunth

Short-lived *perennial herbs* with mostly glandular hairs. *Sepals* with a short mucro. *Petals* with a distinct limb and claw. *Immature schizocarps* variously orientated; mericarps with a pattern of ribs and 1–3 collar-like rings at the top.

28. G. reuteri Aedo & Muñoz Garmendia
Canary Herb Robert
G. canariense Reut., non (Willd.) Poir.

Perennial monoecious *herb* with fibrous roots forming rosettes. *Stems* 25–60 cm, more or less succulent, pale green, sometimes tinted red. *Leaves* alternate, broadly ovate in outline; lamina 5–25 × 5–25 cm, medium green on upper surface, paler beneath, divided to the base into 3 lobes, the lobes divided again nearly to the midrib into 5 divisions, sometimes with secondary toothed divisions, with few to numerous simple eglandular hairs on both surfaces; veins impressed above, prominent beneath; petiole dull brownish to purplish, long. *Inflorescences* arising

G. robertianum L.

G. purpureum Vill.

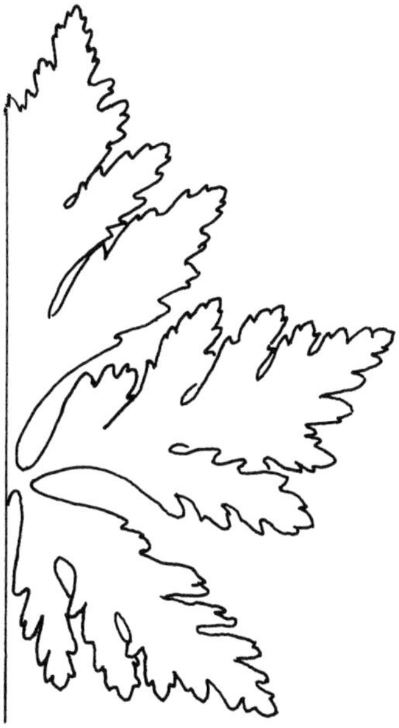

G. reuteri Aedo & Muñoz Garmendia

G. yeoi Aedo & Muñoz Garmendia

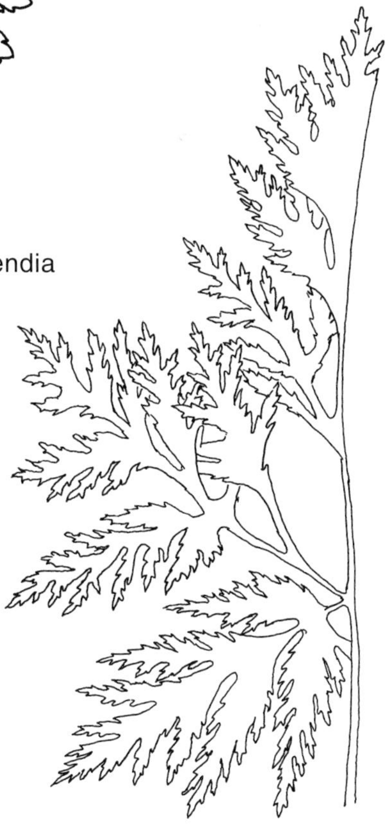

G. maderense Yeo

Leaves of **Geranium** L.

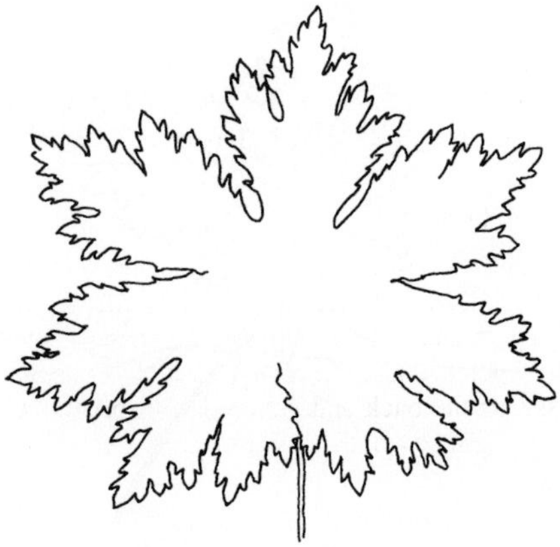

G. phaeum L. var. **phaeum**

G. phaeum L. var. **lividum** (L'Her.) Pers.

G. x **monacense** Harz
nothovar. **monacense**

G. x **monacense** Harz
nothovar. **anglicum** Yeo

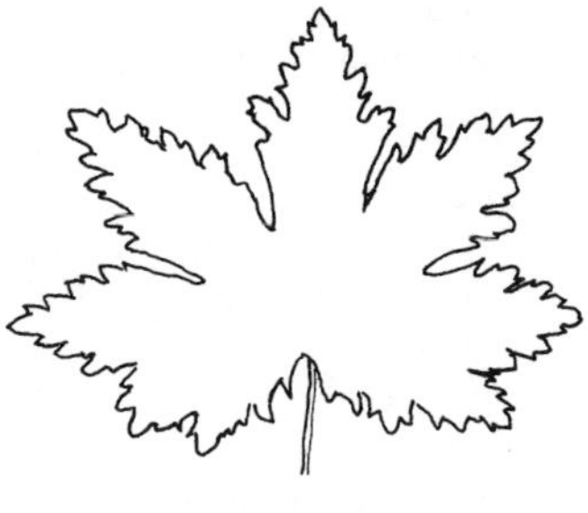

G. reflexum L.

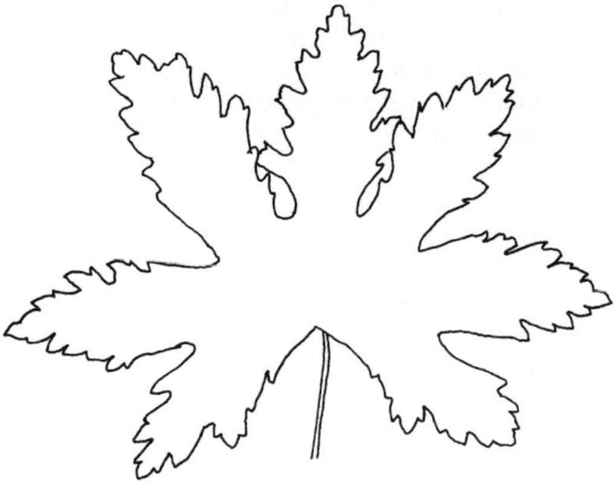

G. reflexum L.

Basal leaves of **Geranium** L.

from axils in the rosette, and in old plants with dense, bisexual flowers; the peduncles and pedicels with dense, purple glandular hairs. *Sepals* 5, 8–10 mm, medium green, oblong-lanceolate, with a mucro at the apex about 1 mm with glandular hairs. *Petals* 5, 17–24 mm, deep pink, the blade 6–9 mm wide and 1.2–2.5 times as long as wide, the claw half as long as blade or less. *Stamens* 10; filaments twice as long as sepals; anthers scarlet or dark red. *Style* white; stigmas pink or white. *Immature schizocarps* slightly nodding or spreading, or with deflexed pedicels; mericarps about 3.5 mm, with a uniform network of ribs and 1 or 2, sometimes 3, small collar-like rings at apex, glabrous or with fine hairs at the top; beak about 30 mm, including stylar portion of 16–18 mm; the mericarp without the awn is thrown off to a distance of a few metres and the awn simply drops away at the moment of discharge. *Flowers* 5–8.

Introduced. Naturalised along roadside hedgebanks, Menacuddle, north of St Austell, Cornwall. Native of the Canary islands. Named after George François Reuter (1805–1872).

29. G. maderense Yeo — Giant Herb Robert

Large, monoecious *herb* usually taking 2 years to flower and either dying after flowering or continuing to grow by means of lateral branching. *Stems* up to 1.5 m, medium green, more or less succulent, erect, with dense, short to very long, pale and purple glandular hairs, much branched, leafy. *Leaves* alternate; lamina 5–60 × 5–60 cm, in a large rosette, the cauline rapidly decreasing in size upwards and less divided, fresh medium green on upper surface, paler beneath, subrotund or reniform in outline, palmately divided into 5 leaflets, the leaflets broadly ovate, acute at apex and divided to the midrib into 9 lobes and petiolulate, the lobes lanceolate, acute at apex and divided into acute lobules and teeth, glabrous on both surfaces; veins slightly impressed above, the midrib very prominent beneath, the lateral veins less so; petioles very long, dull reddish-brown, with dense, minute glandular hairs. *Inflorescence* arising from the centre of the rosette, consisting of a short central stem bearing 1 or 2 whorls of branches which give rise to great mounds of flowers, the lower parts of which are reddish-brown, and the upper parts with dense, purple glandular hairs, the branches regularly forking; flowers 40–45 mm in diameter, bisexual. *Sepals* 5, 7–10 mm, appressed to the petal bases, medium green, lanceolate, with a mucro about 1 mm, with dense, short to long glandular hairs. *Petals* 5, 19–21 mm, glabrous, the limb 13–18 × 13–18 mm, purplish-pink with a network of pale elevated veins and darker towards the throat which is blackish-purple, the claw 2.0–2.5 mm, with a double ridge abutting a stamen so that 5 nectar passages of elongate cross-section are formed. *Stamens* 10; filaments about as long as sepals, purplish or dark red, curved outwards at the tips; anthers dark red. *Styles* dark red; stigmas about 2.5 mm, dark red. *Immature schizocarps* indefinitely orientated, approximately in line with their pedicels; mericarps 4.0–4.5 mm, with a network of ribs and 1–3 collar-like rings at the top, glabrous; rostrum including stylar portion about 20 mm, stylar portion about 7 mm; the mericarp without the awn is thrown off to a distance of a metre or so, and the awn simply drops away at the moment of discharge. *Flowers* 5–6.

Introduced. Well naturalised on cliffs in dense, low vegetation and increasing. Scilly Islands. Native of Madeira.

Subgenus **3. Erodioidea** Yeo

Perennial herbs. Petals without a claw. *Schizocarps* containing the seed discharged together with the awn as one unit, the awn becoming coiled after discharge; mericarps with the base tapered and terminating in a horny point, which is glossy on the back and covered with bristles on the sides with transverse ridges near the apex.

Section **10. Erodioideae** Picard

Tall branching *perennial herbs. Flowers* nodding, with widely spreading or reflexed petals. *Mericarp* with apex conical and 3–5 keels or ridges round it.

30. G. phaeum L. — Dusky Crane's-bill

Perennial monoecious *herb* with a stout rootstock growing on the soil surface. *Stems* 40–80 cm, pale yellowish-green, often purple-dotted, slender, oblique, striate, with long, pale simple eglandular hairs and dense, minute glandular hairs. *Leaves* alternate, yellowish-green on upper surface, paler beneath; basal with lamina 10–20 × 10–20 cm, subrotund or reniform in outline, divided one-fifth to two-thirds, occasionally four-fifths of the way to the base into 7 or 9 lobes, often blotched in the notches, the lobes tapered both ways from about the middle, the lobules about as long as broad, straight and curved outwards and with 2–5 teeth a proportion of which are usually shallow, the teeth and tips of the lobes acute or obtuse, the stipules large, thin and rounded; cauline and sometimes lower inflorescence leaves solitary, the upper inflorescence leaves paired, unequal, often with lobes up to one and a half times as long as broad with very sharp teeth and sessile, the stipules thin, brownish, conspicuous and usually slashed at the tip; all with short to medium, appressed, pale simple eglandular hairs and minute glandular hairs on both surfaces and the margins. *Inflorescence* loose with 1–few, long branches with recurved tips, each with only 1 or 2 flowers open at a time, the flowers directed to 1 side; peduncles and pedicels upwardly inclined with long spidery eglandular hairs and dense short glandular ones; flowers bisexual, nodding or with axis horizontal. *Sepals* 5, 6–11 mm, green and usually purplish at the base, oblong-lanceolate, mucro up to 0.5 mm, spreading, with long, spidery eglandular hairs and dense, short glandular ones. *Petals* 5, 11–14 × 11–14 mm, dull lilac or pinkish to deep violet, dull purple, maroon or nearly black, with a white base, all white occur occasionally, subrotund-obovate, apiculate at apex, with a very short claw, with a few hairs across the front surface of the base and tufts at the sides. *Stamens* 10; filaments curved outwards at first, becoming appressed to the style as the anthers open but still spreading at the tips, bearing very long, spreading and glistening hairs with recurved tips in their lower halves; anthers whitish with a purple edge. *Stigmas* 1.2–2.0 mm, yellowish or greenish. *Immature schizocarps* upwardly inclined, pedicels likewise or sometimes more nearly horizontal; mericarps 5.0–5.5 mm, bristly with 4 or 5 keels or ridges about the top; beak about 15 mm, the stylar portion about 3–4 mm; the mericarp with the seed inside and the awn attached comes away from the central column, being thrown a short distance by the explosion, the awn becoming coiled but not plumed. *Flowers* 5–6. Pollinated by bees, protandrous. $2n=28$.

(i) Var. **phaeum**

Petals dark pinkish-lilac to nearly black, with silvery white or nearly white bases, the white area has a jagged edge and the flower by transmitted light thus has a star-shaped white centre, the apex is cuspidate or sometimes lobulate.

(ii) Var. **lividum** (L'Her.) Pers.
G. lividum L'Her.

Petals bluish, lilac or pink, rather pale and with a white base smaller than in var. *phaeum*, the base and the lowest part of the coloured portion are traversed by short, bluish or violet veins which become diffuse as they enter the coloured zone, giving rise to a bluish halo; in the principal colour there may be a noticeable change from pink towards blue during the life of each flower.

Introduced. Commonly grown in gardens and naturalised in shady places in hedges and wood borders. Scattered records throughout Great Britain and very local in Ireland except the north. Mountains of south and central Europe from the Pyrenees through the Alps to north Yugoslavia, south-east Germany, Czechoslovakia, Poland and western Russia. Var. *lividum* extends from Croatia westwards along the southern side of the Alps reaching the French Alps. Var. *phaeum* occupies the remainder of the natural geographical range. A range of intermediates can be found in some gardens and thus may escape as do both varieties.

31. G. × **monacense** Harz — Munich Crane's-bill
G. phaeum × **reflexum**
G. punctatum auct.

Perennial monoecious *herb* with a stout rootstock and spreading by rhizomes. *Stems* 40–60 cm, pale green, erect or spreading, robust, with dense, minute glandular hairs and scattered, long simple eglandular hairs, branched, leafy. *Leaves* yellowish-green on upper surface, sometimes with brown patches, with impressed veins, paler beneath with prominent veins; basal with lamina 9–20 × 9–20 cm, broadly ovate or reniform in outline, divided three-quarters of the way to the base into 7 to 9 lobes, the lobes oblanceolate, obtuse at apex, with obtuse teeth and short lobules, the petioles up to 50 cm, with dense, minute glandular hairs and longer simple eglandular hairs; cauline becoming slowly smaller upwards, with the petioles shorter; all soft with numerous, unequal simple eglandular hairs and very short glandular hairs on both surfaces. *Inflorescence* loose with 1–few, long branches, each with 1 or 2 bisexual flowers open at a time; peduncles and pedicels with dense, very short glandular hairs and numerous, long simple eglandular hairs. *Sepals* 5, 9–10 mm, lanceolate, with a red mucro up to 2 mm, with dense, very short glandular hairs and numerous, long simple eglandular hairs. *Petals* 5, 11–14 mm, dark, dull purplish-red or pinkish-lilac, with a white basal zone and above it a dull, bluish-violet zone which is traversed by darker violet veins, reflexed, the margins of the base with dense hair tufts. *Stamens* 10; filaments pinkish-purple, with long simple eglandular hairs; anthers cream, turning brownish. Style and stigma pale green. *Immature schizocarps* inclined slightly upwards and their pedicels slightly deflexed; mericarps 5.0–5.5 mm; beak about 15 mm; the mericarp with the seed inside and the awn attached comes away from the central column being thrown a short distance by the explosion, the awn becoming coiled, but not plumed. *Flowers* 5–6.

(i) Nothovar. **monacense**
G. phaeum var. **phaeum** × **reflexum**

Leaves often with dark brown blotches. *Petals* dark purplish-red, the whitish base relatively large and with a toothed edge, while the bluish-violet zone above it is inconspicuous.

(ii) Nothovar. **anglicum** Yeo — Cv. Eric Clement
G. phaeum var. **lividum** × **reflexum**

Leaves without blotches. *Petals* pinkish-lilac, the whitish base is small and straight-edged, while the bluish-violet zone above it is wide, conspicuous and strongly veined.

Introduced. Nothovar. *monacense* was described by E. Harz from his garden in Bavaria. Nothovar. *anglicum* is naturalised on a roadside verge at Hurst Green in Sussex since 1975. However, both varieties have arisen in different places at different times. The nothospecies is recorded elsewhere from Surrey, Berkshire, Bedfordshire and Staffordshire and Edinburgh but it is not known to which nothovar. they belong.

32. G. reflexum L. — Reflexed Crane's-bill

Perennial monoecious *herb* with a stout rootstock. *Stems* 40–80 cm, pale yellowish-green, slender, with long, slender, pale simple eglandular hairs and numerous, minute glandular ones. *Leaves* alternate, dull yellowish-green on upper surface, paler beneath; basal with lamina 10–20 × 10–20 cm, subrotund or reniform in outline, divided up to two-thirds of the way to the base into 5 or 7 lobes, the lobes ovate or elliptical, subacute at apex, and dentate or shortly lobulate, the petioles long and hairy with simple eglandular hairs and dense minute glandular ones; cauline alternate or opposite, gradually smaller and with shorter petioles upwards; all with unequal, appressed simple eglandular hairs on both surfaces and the margins. *Inflorescence* loose with 1–few, long branches, each with 1 or 2 flowers open at a time; peduncles and pedicels with long, spidery eglandular hairs and dense, short glandular ones; flowers bisexual, more or less inverted. *Sepals* 5, 6–11 mm, more or less red-flushed, oblong-lanceolate, with a mucro up to 0.2 mm, spreading, with long, spidery simple eglandular hairs and dense, short glandular ones. *Petals* 5, 11–13 mm, twice as long as broad or more, brighter rose-pink with a white base occupying about one-quarter of the length of the petal and with a bluish band between the white and pink areas, sometimes with a triangular point and often with one or more incisions at the tip, very strongly reflexed a short way above the erect base, the tips more spreading when they are pressed outwards by the sepals, without hairs across the front surface or the base and with a strong tuft of hairs on each side. *Stamens* 10; filaments appressed to the style, with small, fine hairs at the base; anthers bluish. *Style* and stigma greenish. *Immature schizocarps* deflexed or horizontal on deflexed or horizontal peduncles, glandular-hairy; mericarps with 3 or 4 keels or ridges around the top; the mericarp with the seed inside and the awn attached comes away from the central column, being thrown a short distance by the explosion, the awn becoming coiled but not plumed. *Flowers* 5–6.

Introduced. Recorded as a garden escape at Morland in Westmorland and by the Little Ouse in Norfolk. It is native of south-east Europe.

2. Monsonia L.

Annual monoecious *herbs*. *Leaves* alternate, simple, crenate-serrate, stipulate. *Inflorescence* with few, bisexual flowers. *Sepals* 5. *Petals* 5, bluish or white. *Stamens* 15, all with anthers, their filaments fused for half their length into 5 groups of 3. *Seeds* dispersed inside mericarps with beaks attached, the beaks tightly spiralled (twisted) on their own axes.

Contains about 30 species in Africa.

Knuth, R. (1912). Geraniaceae. In Engler, A. (Edit.) *Das Pflanzenreich*. **IV.129**. Leipzig.

1. Sepals 10–12 mm; mericarps 50–70 mm	**1. angustifolia**
1. Sepals 6–7 mm; mericarps 22–30 mm	**2. brevirostrata**

1. M. angustifolia E. Mey. ex A. Rich.

Narrowed-leaved Dysentery-herb

M. biflora auct.

Annual monoecious *herb* with rather woody base and a pale brown tap-root. *Stem* 15–40 cm, pale green, ascending to more or less erect, much branched, the branches ascending, sparsely short to long hairy, leafy. *Leaves* alternate; lamina 2–3 × 0.4–0.5 cm, medium green on upper surface, paler beneath, linear-oblong, obtuse at apex, crenate-serrate, cuneate at base, with sparse short and long often glandular hairs; petiole shorter than lamina; stipules 5–10 mm, setaceous, hairy. *Inflorescence* 2-flowered; peduncles with minute hairs; bracts like the stipules; pedicels about 25 mm, with short and long hairs; flowers bisexual. *Sepals* 5, 10–12 mm, spathulate or lanceolate-oblong, margin broadly membranous, with a mucro 1–3 mm at apex, slightly hairy. *Petals* 5, a little longer than sepals, bluish or white, obovate-cuneate. *Stamens* 15. *Style* 1. *Mericarps* 50–70 mm, valves hairy; beak very short, densely hairy. *Flowers* 6–8.

Introduced. A wool casual. Native of east and South Africa.

2. M. brevirostrata R. Knuth

Short-fruited Dysentery-herb

Annual monoecious *herb* with a brown tap-root. *Stem* 10–40 cm, pale green, ascending or base procumbent, sparsely short to long hairy, branched, the branches ascending, leafy. *Leaves* alternate; lamina 1.6–3.0 × 0.4–0.5 cm, medium green on upper surface, paler beneath, lanceolate or narrowly ovate, obtuse at apex, irregularly serrate, cuneate at base, more or less glabrous on upper surface, hairy on the veins beneath; petiole long, puberulous with some longer hairs; stipules up to 4 mm, subulate, hairy. *Inflorescence* 1–3-flowered; peduncles 10–20 mm, puberulous; bracts 2 or 4, about 2 mm; pedicels 7–8 mm, puberulous with a few longer hairs, recurved; flowers bisexual. *Sepals* 5, 6–7 mm, margin membranous, linear-lanceolate, mucronate at apex, hairy. *Petals* 5, 5–8 mm, bluish, obovate, rounded at apex. *Stamens* 15, all with anthers; filaments fused for about half their length into 5 groups of 3. *Styles* 1. *Mericarps* 22–30 mm, hairy; beak very short, puberulous. *Flowers* 6–8.

Introduced. Fairly frequent wool alien in fields and waste places. Few scattered records in Great Britain. Native of South Africa.

3. Erodium L'Hér.

Annual to *perennial* monoecious *herbs*, usually with bisexual flowers, rarely dioecious. *Leaves* mostly opposite, often pinnatifid to pinnate. *Inflorescence* an umbel, rarely reduced to a single flower, subtended by 2 or more, usually scarious bracts; flowers actinomorphic or slightly zygomorphic. *Sepals* 5. *Petals* 5. *Stamens* 5, with a nectary at the base of the filament, alternating with 5 scale-like staminodes. *Stigmas* 5, filiform. *Fruit* a schizocarp of 5, indehiscent mericarps, separating from the base upwards, retaining during the dispersal the outer part of the style as a long beak, which in most species becomes twisted into a spiral at maturity, the pitch of the spiral varying with the humidity.

Contains about 100 species in temperate areas, especially the Mediterranean region and west Asia.

All our species belong to the Section **Borbata** Boiss.

Alston, A. H. G. (1948). *Erodium cicutarium* (L.) Hérit. *Watsonia* **1**: 170–171.

Andreas, C. M. (1947). De inheemsche Erodia van Nederland. *Ned. Kruidk. Arch.* **54**: 138–231.

Benoit, P. M. (1966). *Erodium cicutarium* agg. *Proc. B. S.. B. I.* 6: 364–366.

Carolin, R. C. (1958). The species of the genus *Erodium* L'Her. endemic to Australia. *Proc. Linn. Soc. New South Wales* **82**: 92–100.

El-Oqlah, A. A. (1989). A revision of the genus *Erodium* L'Heritier in the Middle East. *Feddes Repert.* **100**: 97–118.

Hultén, E. & Fries, M. (1986). *Atlas of north European vascular plants north of the Tropic of Cancer*. 3 vols. Königstein.1

Knuth, R. (1912). Geraniaceae. In A. Engler (Edit.), *Das Pflanzenreich*: **IV 129**. Leipzig.

Larsen, K. (1958). Cytological and experimental studies on the genus *Erodium* with special reference to the collective species *E. cicutarium* (L.) L'Herit. *Biol. Meddr.* **23**(6): 1–25.

Leslie, A. C. (1980). The hybrid of *Erodium corsicum* Lém. with *E. reichardii* (Murray) DC. *The Plantsman* **2**: 117–126.

Stewart, A., Pearman, D. A. & Preston, C. D. (1994). *Scarce plants in Britain*. Peterborough. [*E. maritimum* and *moschatum*.]

1. Leaves mostly pinnate 2.
1. Leaves simple, shallowly to deeply lobed, or if more or less compound then ternate 6.
2. Primary leaflets divided less than one-quarter of way to midrib; apical pits of mericarp with sessile papillose glands **16. moschatum**
2. Primary leaflets divided nearly to base; apical pits of mericarp without glands 3.
3. Bracts green, fused into a cupule; petals 15–20 mm; schizocarp with beak 4–8 cm **18. manescavii**
3. Bracts brown, several, free; petals 4–12 mm; schizocarp with beak 1–4 cm 4.
4. Flowers 2–4(–5) per umbel, mostly less than 10 mm in diameter; apical pits of mericarp not delimited by sharp ridge and groove, overarched by hairs from main part of mericarp **15. lebelii**
4. Flowers 3–7 per umbel, mostly more than 10 mm in diameter; apical pits of mericarp separated from main part of mericarp by a sharp ridge and groove, not overarched by hairs 5.
5. Stems up to 60 cm, more or less erect, glandular hairs few or none in whole plant; umbel 4- to 7-flowered; flowers 12–15 mm in diameter; beak of mericarp 25–40 mm **14(a). cicutarium** subsp. **cicutarium**
5. Stems up to 30 cm, spreading or prostrate; glandular hairs numerous; umbels mostly 3- to 5-flowered; flowers about 12 mm in diameter; beak of mericarp 5–6 mm **14(b). cicutarium** subsp. **dunense**
6. Beak of mericarp not more than 17 mm 7.
6. Beak of fruit more than 17 mm 9.
7. Leaves at least 3 cm wide **3. malacoides**
7. Leaves less than 3 cm wide 8.
8. Petals 3 mm or less **8. maritimum**
8. Petals 7–11 mm **9. × variabile**
9. Perennial **1. chium**
9. Annual or biennial 10.
10. Bracts at base of umbel 2, subrotund to reniform **2. laciniatum**
10. Bracts at base of umbel at least 3, ovate to lanceolate 11.
11. Beak of mericarp less than 45 mm 12.
11. Beak of mericarp more than 45 mm 16.
12. Pedicels and sepals tomentose 13.
12. Pedicels and sepals hairy but not tomentose 14.
13. Leaves variously lobed; apical pits of meriocarp hairy **1. chium**
13. Leaves deeply 2- to 3-pinnatisect; apical pits of meriocarp glabrous **17. stephanianum**
14. Leaves with 3 principal lobes **6. aureum**
14. Leaves with several lobes 15.
15. Apical pits of mericarp glandular-papillose **3. malacoides**
15. Apical pits of mericarp without glands **4. geoides**
16. Mericarp less than 9 mm 17.
16. Mericarp more than 9 mm 18.
17. Pedicels with eglandular hairs **7. crinitum**
17. Pedicels with dense glandular hairs **11. brachycarpum**
18. Petals 20–25 mm **12. gruinum**
18. Petals much less than 20 mm 19.
19. Pedicels with only appressed eglandular hairs **5(a). cygnorum** subsp. **cygnorum**
19. Pedicels with numerous glandular hairs 20.
20. Leaves 3-lobed **5(b). cygnorum** subsp. **glandulosum**
20. Leaves pinnatifid to pinnatisect 21.
21. Sepals 7–8 mm **10. botrys**
21. Sepals 12–14 mm **13. ciconium**

1. E. chium (L.) Willd. Three-lobed Stork's-bill
Geranium chium L.

Annual or *biennial* monoeocious *herb*, rarely *perennial*. *Stems* 5–50 cm, pale green, ascending to suberect, with deflexed simple eglandular hairs near the base, more or less branched, leafy. *Leaves* opposite; lamina up to 4 cm long and wide, reniform or ovate in outline, variously lobed, the lobes crenate-dentate, the basal with long petioles, the upper leaves more finely lobed with short petioles; all with patent, soft simple eglandular hairs often appressed; stipules often 3 mm long and broad, broadly ovate and obtuse at apex. *Inflorescence* of umbels with 2–8 bisexual flowers; peduncles up to 8 cm; pedicels 7–17 mm, with crisp simple hairs and densely tomentose. *Sepals* 5, 5–7 × 2–3 mm, oblong or ovate, mucronate at apex, 5-veined, tomentose with soft, white simple eglandular hairs. *Petals* 5, 5–9 mm, purplish-pink, obovate, rounded at apex. *Stamens* 5; staminodes 5, glabrous or hairy. *Styles* 5; stigmas filiform. *Schizocarp* of 5, 1-seeded mericarps, the mericarp 3.5–4.5 mm, apical pits small, but rather deep, covered with minute glandular hairs, without a furrow at the base; beak 20–40 mm. *Flowers* 6–9. $2n = 20$.

Introduced. Infrequent wool alien. Recorded from Hampshire, Middlesex, Bedfordshire and Glamorganshire. Native of the Mediterranean region.

2. E. laciniatum (Cav.) Willd. Cut-leaved Stork's-bill
Geranium laciniatum Cav.; *E. alexandrinum* Delile; *E. reflexum* Delile; *E. hispidum* C. Presl

Annual or *biennial* monoecious *herb* with a strong taproot. *Stems* 7–50 cm, pale green, ascending, branched, with deflexed hairs at least near the base. *Leaves* opposite; lamina 2–7 cm, medium green, oblong to broadly ovate, very variously dissected, undivided and irregularly serrate, or with 3 pinnatifid lobes, or almost bipinnatisect with linear-lanceolate segments, sparsely appressed-hairy; stipules up to 8 mm, more or less broadly ovate, obtuse at apex, membranous. *Inflorescence* of 4–9 bisexual flowers in umbels; bracts 2, brown, subrotund to reniform; pedicels with simple eglandular hairs. *Sepals* 5, about 7 mm, oblong, with a mucron to 1 mm. *Petals* 5, 7–10 mm, purplish, obovate, rounded at apex. *Stamens* 10. *Styles* 5. *Schizocarps* of 5, 1-seeded mericarps, the mericarps 4.5–6.5 mm, with short, whitish hairs; apical pits shallow, without glands, without a furrow at the base; beak 35–90 mm. *Flowers* 6–9. $2n = 20$.

Introduced. Wool and esparto casual. Native of the Mediterranean region.

3. E. malachoides (L.) L'Hér. Soft Stork's-bill
Geranium malachoides L.

Annual or *biennial* monoecious *herb* with fibrous roots. *Stems* 6–40(–50) cm, pale green, erect or sprawling, sparingly to densely clothed with retrorse bristles interspersed with shorter glandular hairs, usually much-branched, leafy. *Leaves* alternate; lamina of lower 1.5–6.0 × 1.5–5.5 cm, broadly ovate in outline, rounded or obtuse, obscurely pinnatifid, with 5–7, rounded, dentate lobes and a cordate base; lamina of upper leaves similar, but often small and rather acute, with sharply toothed lobes; all thinly appressed-hispidulous on both surfaces and with conspicuous, shining, sessile glands; petioles 1–5 cm, long glandular-hairy, the uppermost leaves sometimes subsessile; stipules 4–10 × 3–6 mm, membranous, ovate, obtuse at apex, stiffly hairy. *Inflorescence* umbellate, (2–)4– to 8-flowered; peduncles 2–4(–7) cm, slender, densely glandular-hairy; bracts 3–6, about 2.0 × 1.5 mm, membranous, ovate, ciliate; pedicels about 10 mm, filiform, erect at anthesis, later spreading with upcurved tips, more or less densely glandular-hairy; flowers bisexual. *Sepals* 5, the outer about 5 × 1.8 mm, oblong, obtuse, more or less cucullate and caudate-cuspidate with a very narrow membranous margin, the inner narrower, 3-veined and with a wide membranous margin. *Petals* 5, about 6 × 4 mm, magenta, pink or purplish, rounded at apex, with a very short, sparsely hairy or subglabrous claw. *Stamens* 10; fertile 5, with filaments about 4 mm, lanceolate-subulate, the anthers oblong; 5 staminodes about 2 mm, oblong-ovate. *Style* 1, stigma arms 0.8–1.0 mm. *Schizocarp* of 5, 1-seeded mericarps, the mericarps about 4 × 1 mm, narrowly spindle-shaped, with a very pointed base and a blunt apex, more or less densely covered with suberect hairs, the apical pits glandular-papillose within, surrounded by a distinct concentric ridge; beak 18–35 mm; seeds 2.5–3.0 mm, bright brown, clavate, minutely pitted. *Flowers* 5–7. $2n = 20$, 40.

Introduced. Infrequent wool alien. Recorded from Devonshire, Bedfordshire, Worcestershire, Glamorganshire, Yorkshire, Edinburgh and West Lothian. Throughout the Mediterranean area and eastwards to Iran; Atlantic Islands.

4. E. geoides A. St Hil. Argentine Stork's-bill

Annual or *biennial* monoecious *herb* with a tap-root. *Stems* 10–30 cm, pale green, ascending, much-branched, with numerous, patent glandular and simple eglandular hairs. *Leaves* opposite; lamina 3–5 × 1.5–2.0 cm, medium green, oblong in outline, deeply lobed, the lobes broad, acutely crenate-dentate, cordate at base, stiffly hairy beneath and on the margin; petioles short to long, glandular-hairy; stipules 3–6 mm, ovate or triquetrous, obtuse at apex. *Inflorescence* 3- to 5-flowered; peduncle up to 4 cm; pedicels 14–20 mm, often recurved, glandular-hairy; bracts 2–3 mm, ovate, obtuse to subacute at apex; flowers bisexual. *Sepals* 5, 5–6 mm, oblong or oblong-lanceolate, mucronate at apex. *Petals* 5, lilac, obovate, rounded at apex. *Stamens* 10. *Styles* 5. *Schizocarp* of 5, 1-seeded mericarps, hairy; apical pits without glands; beak up to 30 mm. *Flowers* 6–9.

Introduced. A wool casual. A few scattered records. Native of Uruguay and Brazil.

5. E. cygnorum Nees Western Stork's-bill
E. peristeroides Turcz.

Annual or *biennial* monoecious *herb* with a long, thick, fleshy tap-root. *Stems* 7–60 cm, pale green, simple or branched, decumbent or ascending, densely reflexed-hairy. *Leaves* opposite, medium green; basal more or less numerous, the lamina 1.5–6.0 × 0.5–4.0 cm, oblong in outline, deeply 3-lobed, the lobes deeply lobulate, the lobules denticulate or dentate, cordate at base, the petiole 1–3 times as long as the lamina; upper leaves similar but smaller; all with appressed, short eglandular hairs; stipules about 5 × 3 mm, ovate, obtuse at apex, hairy only on the margin. *Inflorescence* 1- to 6-flowered; peduncle 1–10 cm; bract 2–4 × 1–2 mm, ovate or ovate-lanceolate, acute or acuminate at apex, hairy only on margin; pedicels 10–25 mm, glabrous or appressed-reflexed glandular-hairy. *Sepals* 5, 5–13 mm, ovate, obtuse-mucronate at apex, with short, rigid, appressed simple eglandular or glandular hairs. *Petals* 5, bluish-purple with red veins, obovate, rounded at apex. *Stamens* 10; anthers red or yellow. *Styles* 5, stigmas red or green. *Schizocarp* of 5, 1-seeded mericarps about 10 mm, apical pits without glands and delimited by 2 grooves, glabrous to hairy; beak 50–100 mm. *Flowers* 6–9. $2n = 60$.

(a) Subsp. **cygnorum**
Pedicels glabrous or with few simple eglandular hairs. *Sepals* with only more or less appressed simple eglandular hairs. *Petals* bluish with yellow or white veins. *Mericarps* with sparse simple eglandular hairs. $2n = 60$.

(b) Subsp. **glandulosum** Carolin
Pedicels with many patent glandular hairs. *Sepals* with many glandular hairs. *Petals* bluish with red veins. *Mericarps* densely hairy. $2n = 60$.

Introduced. Wool alien. Native of west Australia. Subsp. *cygnorum* is recorded from Somersetshire, Hampshire, Bedfordshire, Worcestershire, Lancashire, and Yorkshire. Subsp. *glandulosum* is recorded from Worcestershire.

6. E. aureum Carolin Australian Stork's-bill

Annual or short-lived *perennial* monoecious *herb* with a stout stock. *Stems* up to 20 cm, pale green, erect, with numerous glandular and eglandular hairs. *Leaves* opposite; lamina 1–3 × 0.7–1.5 cm, medium green on upper surface, paler beneath, ovate in outline, with 3 principal lobes, but not divided to the midrib, the lobes dentate, with scattered glandular and eglandular hairs on both surfaces; petioles slender, with scattered hairs; stipules acute at apex. *Inflorescence* in umbels of (2–)3(–5) flowers; pedicels covered with numerous glandular and fewer simple eglandular hairs. *Sepals* 5, 5–8 × 2–3 mm, oblong-lanceolate, shortly mucronate, membranous towards margin, with long glandular and simple eglandular hairs the latter often more numerous towards the apex. *Petals* 5, 8 × 6 mm, bluish purple with reddish veins and bases, ovate. *Stamens* 10, filaments often purple with a long cristate apex, the 5 sterile shorter than the 5 fertile; anthers red. *Styles* 5; stigmas yellow. *Schizocarps* of 5, 1-seeded mericarps 5 mm long, covered with short golden brown simple eglandular hairs lying diagonally across the long axis; pits 2, with often 1–2 folds beneath each pit; beak about 35 mm. *Flowers* 8–9. $2n=20$.

Introduced. A wool casual. Native of Australia.

7. E. crinitum Carolin — Eastern Stork's-bill

Annual or *biennial* monoecious *herb. Stem* 20–50 cm, pale green, often tinted red, erect, branched, leafy, with numerous, spreading or deflexed, pale, rigid eglandular hairs. *Leaves* opposite; lamina 1.2–7.0 × 1.2–3.0 cm, decreasing in size upwards, medium green on upper surface, paler beneath, ovate in outline, deeply palmately 3-lobed, the lobes dentate-mucronate, with short, stiff, more or less appressed, pale simple eglandular hairs; petioles up to 20 cm, with numerous, deflexed, pale simple eglandular hairs. *Inflorescence* of umbels of 2–6 bisexual flowers; peduncle 2–7 cm; pedicels up to 3.5 cm, with numerous, deflexed simple eglandular hairs. *Sepals* 5, 6–10 mm, oblong, with a mucro at apex up to 1.0 mm, with dense, pale, short, rigid simple eglandular hairs. *Petals* 5, up to 10 mm, blue to purple with yellow to white veins and base. *Stamens* 10, 5 fertile and 5 sterile. *Styles* 5. *Schizocarp* of 5, 1-seeded mericarps 6.5 mm long; pits shallow, usually with 2 folds beneath them; beaks 40–70 mm, with minute hairs. *Flowers* 8–9. $2n=40$.

Introduced. A wool casual. Recorded from Hampshire, Kent, Hertfordshire, Huntingdonshire, Worcestershire, Yorkshire, Co. Durham and Roxburghshire. Native of Australia.

8. E. maritimum (L.) L'Hér. — Sea Stork's-bill
Geranium maritimum L.

Annual or *biennial* monoecious *herb* with a tap-root. *Stems* up to 30 cm, but often almost absent, decumbent, pale green, with stiff, white hairs. *Leaves* opposite but mostly basal; lamina 0.5–2.5 cm, medium green, ovate, pinnatifid or obtusely dentate, the lobes lobulate or dentate, with white, appressed simple eglandular hairs on both sides; petioles long. *Inflorescence* of a solitary bisexual flower or rarely 2; peduncles about equalling the leaves; bracts brown, ovate; pedicels with appressed, often glandular hairs. *Sepals* 5, 3.0–4.5 mm, oblong, mucronate at apex, with white hairs. *Petals* 5, 3 mm or less, pink, or often absent. *Schizocarp* of 5, 1-seeded mericarps, each about 3 mm, with short, brownish hairs and a few longer ones near the apex; apical pits deep, usually eglandular, with a pronounced furrow at the base, delimited by a high, narrow ridge; beak about 10 mm. *Flowers* 6–9. $2n=20$.

Native. Fixed maritime dunes and rather bare places in short grassland and on walls and pavement; rarely inland. Coasts of Channel Islands, east and south Ireland and Great Britain from Cornwall and Kent to Wigtownshire; formerly inland in Worcestershire; casual in a few places. West coast of France from the Somme to the Vendée and north Spain along the Mediterranean to Sicily. Suboceanic Southern-temperate element.

9. E. × variabile Leslie — Hybrid Stork's-bill
E. corsicum Léman × **reichardii** (Murray) DC.

Perennial monoecious *herb* with a strong tap-root and fibrous side-roots. *Stems* present or absent, up to 20 cm, leafy, pale green. *Leaves* opposite; lamina 0.5–2.3 × 0.4–2.5 cm, dark or greyish-green on upper surface, paler beneath, ovate, lobed or unlobed, cordate at base, with simple eglandular and glandular hairs; petioles up to 10 cm. *Inflorescence* of 1 or 2 bisexual flowers; peduncles 15–30 mm. *Sepals* 5. *Petals* 5, 7–11 × 4–8 mm, reddish-purple with darker veins or white with reddish-purple veins, unblotched, obovate or broadly obovate. Pits of mericarps glandular, not ringed by a furrow. *Flowers* 6–9.

Introduced. A casual or persistent garden escape. Of garden origin. Hybrids of this parentage are often sold as one or other of the parents.

10. E. botrys (Cav.) Bertol. — Mediterranean Stork's-bill
Geranium botrys Cav.

Annual monoecious *herb* with fibrous roots. *Stems* (7–) 15–30(–50) cm, pale green, erect or spreading, more or less densely covered with retrorse bristles, usually much branched, especially towards the base, leafy. *Leaves* alternate, medium green on upper surface, paler beneath; basal forming a loose rosette, the lamina 1.5–4.0 × 0.8–3.0 cm, oblong in outline, obtuse at apex, shortly pinnatifid with bluntly toothed lobes, or sometimes with the lobes barely distinct, truncate or shallowly cordate at base, thinly strigose on both surfaces, the petioles 1–10 cm, with conspicuous retrorse bristles; cauline oblong or narrowly deltoid, deeply pinnatisect, with 7–9 narrowly oblong, sharply toothed or laciniate lobes, usually stiffly hairy, and often glandular-hairy, the petioles shorter than the basal leaves; stipules 3–5 × about 3 mm, ovate-oblong, acute at apex, membranous with ciliate margins. *Inflorescence* umbellate, 2(–4)-flowered; peduncles 2–8 cm, often densely glandular-hairy, especially towards the apex; bracts normally free, 2–3 × 1.5–2.0 mm, ovate, acute at apex, ciliate; pedicels 10–20 mm, often densely glandular-hairy, at first erect, later spreading with upcurved tips; flowers bisexual. *Sepals* 5, 7–8 × 2–3 mm, green, oblong, obtuse at apex, cuspidate, more or less densely glandular hairy, the 3 outer with 5 the 2 inner with 3 veins and wide membranous, ciliate margins. *Petals* 5, 7–8 × 3.5–4.0 mm, violet or purplish, obovate, rounded at apex, tapering at base to a very short, hairy claw. *Stamens* about 4.5 mm; filaments free, ovate-oblong and about 1.5 mm wide for half their length, subulate above; anthers about 1.0 × 0.8 mm; staminodes lanceolate or narrowly deltoid. *Stigma* arms about 0.5 mm. *Schizocarp* of 5, 1-seeded mericarps; mericarps about 3 mm; beak up to 50–100 mm; with 2 deep apical pits, each pit surrounded by 3 distinct concentric ridges, covered with short, more or less appressed bristles, with scattered, longer bristles up to 20 mm, in the lower half; seeds 4.5 × 1.0–1.5 mm, pale brown, apparently smooth but minutely pitted and striatulate. Flowers 6–9. $2n=40$.

Introduced. Common wool alien. Scattered records in Great Britain. Native in the Mediterranean region and particularly frequent in the western half; Atlantic Islands.

11. E. brachycarpum (Godr.) Thell. Hairy-pitted Stork's-bill
E. obtusiplicatum (Maire, Weiller & Wilcz.) Howell; *E. botrys* var. *brachycarpum* Godr.

Annual monoecious *herb* with a strong tap-root. *Stem* up to 40cm, pale green, often tinted red, erect, branched, leaf, with sparse simple eglandular and glandular hairs below and dense glandular hairs above. *Leaves* opposite; lamina 2–8×2–4cm, medium green on upper surface, paler beneath, ovate or oblong-ovate in outline, pinnatisect nearly to midrib, lobes obovate, incise-dentate, with scattered to rather numerous, stiff, pale, appressed hairs; petioles up to 20cm, pale green, with slightly reflexed, pale hairs. *Inflorescence* 2- to 4-flowered; peduncles up to 7cm, densely glandular-hairy; pedicels up to 1.5cm, densely glandular-hairy; flowers bisexual. *Sepals* 5, 8–9mm, oblong-lanceolate, pointed at apex, with dense glandular hairs and longer simple eglandular hairs. *Petals* 5, bluish, obovate, rounded at apex. *Stamens* 10. *Styles* 5. *Schizocarp* of 5, 1-seeded mericarps 5–6mm, apical pits with some bristles and 1 blunt-rimmed groove; beak to 70(–100)mm. Flowers 8–9. $2n=40$.

Introduced. A frequent wool alien. Scattered records in Great Britain. Native of central and East Australia.

12. E. gruinum (L.) L'Hér. Long-beaked Stork's-bill
Geranium gruinum L.

Annual or *biennial* monoecious *herb* with fibrous roots. *Stems* 10–30(–50)cm, pale green, erect or spreading, strigose with stiff, white simple eglandular hairs, without glandular hairs, but sometimes with scattered sessile glands, unbranched or branched especially at base, leafy. *Leaves* alternate; basal 2–5×1–3cm, medium green on upper surface, paler beneath, oblong, crenate or obscurely lobed, cordate at base; thinly strigose, the petioles 2–8cm; upper cauline up to 10(–12)×5cm, distinctly 3-lobed, the middle lobe much larger than the laterals and often distinctly pinnatifid, the petioles up to 15cm, the uppermost leaves in luxuriant plants sometimes pinnatisect, with broadly oblong, sharply dentate lobes; stipules 5–10 × 3–8 mm, ovate, acute or acuminate at apex, conspicuous, membranous, glabrous or with minutely ciliate margins. *Inflorescence* umbellate, generally 2-flowered, but occasionally 1- or 3- to 5-flowered; peduncles 3–12 cm, erect, stigma with spreading hairs; bracts 4, connate in pairs, occasionally more, about 5×4mm, membranous, conspicuously ribbed, ciliolate; pedicels 20–45mm, strigose, at first more or less erect, later spreading with sharply upturned tips; flowers bisexual, showy. *Sepals* 5, 10–18mm, oblong or oblong-ovate, with a mucro 2–7mm at apex, more or less stiffly hairy. *Petals* 5, 20–25mm, violet, broadly obovate, rounded at apex. *Stamens* 10. *Styles* 5. *Schizocarp* of 5, 1-seeded mericarps about 14mm, with numerous, ascending, whitish hairs; apical pits deep with a wide furrow at base; beak 60–110mm. *Flowers* 6–9. $2n=40$.

Introduced. A wool casual. Few scattered records. Native of the east Mediterranean region.

13. E. ciconium (L.) L'Hér. Southern Stork's-bill
Geranium ciconium L.; *Herodium ciconium* (L.) Rchb.

Annual or *biennial* monoecious *herb* with a thick tap-root. *Stems* 10–70cm, pale green, ascending or erect, robust, with usually deflexed and glandular hairs. *Leaves* opposite; lamina 6–9×4–5cm, medium green on upper surface, paler beneath, ovate, obtusish at apex, pinnatisect, the lobes ovate or oblong-ovate and pinnatisect, the lobules incise, hairy; rhachis between pinnae lobulate and dentate; petiole 1–3 times longer than lamina; stipules 5–6×2–3mm, deltoid-lanceolate or deltoid-ovate, subacute at apex, glabrous or nearly so. *Inflorescence* 3- to 10-flowered, peduncles 4–12cm; bracts 3–4mm, ovate or rotund-ovate, abruptly acuminate, acute or mucronate at apex, hairy; pedicels 10–20mm, glandular-hairy; flower 14–18mm in diameter, bisexual. *Sepals* 5, 12–14mm, ovate, obtuse at apex, mucronate, the mucro 2–4mm, veins 5, with long, patent glandular hairs. *Petals* 5, about 8×4mm, bluish or lilac, obovate, rounded at apex. *Stamens* 10. *Styles* 5; stigma blackish-purple. *Schizocarp* of 5, 1-seeded mericarps 9–11mm, apical pits deep, densely glandular; beak 50–100mm. *Flowers* 6–9. $2n=18$.

Introduced. A wool casual. Native of the Mediterranean region north to east-central Europe.

14. E. cicutarium (L.) L'Hér. Common Stork's-bill
Geranium cicutarium L.; *Geranium viscosum* Mill.; *Herodium circutarium* (L.) Rchb.; *E. salzmannii* Delile

Annual monoecious *herb* with a strong tap-root and fibrous side-roots. *Stems* 10–60cm, erect to decumbent or prostrate, pale yellowish-green, sometimes tinted reddish, terete, with numerous, medium to long, pale simple eglandular hairs and sometimes pale glandular hairs which become more numerous upwards. *Leaves* 2–20×1–6cm, dark greyish-green, slightly paler beneath, tinged red on the margins, with numerous, pale glandular and eglandular hairs, oblong or elliptical in outline, pinnate; leaflets oblong or ovate, mostly divided more than halfway to the midrib, lobes oblong, aristate at apex, petiolate or the upper sessile; petiole and rhachis pale yellowish-green. *Inflorescence* a terminal umbel of 3–7 bisexual flowers; peduncles equalling or exceeding leaves, with numerous, short, pale glandular hairs and sometimes longer, simple eglandular hairs; bracts brown, free; pedicels 8–10mm, suffused purplish-red, with numerous, unequal, pale glandular hairs and sometimes fewer, longer, pale simple eglandular hairs; flowers usually more than 10mm in diameter. *Sepals* 5, 2–7×1–2mm, margin and midrib medium green, in between pale green, lanceolate, mucronate at apex. *Petals* 5, 4–12mm, purplish-pink, lilac or white, upper 2, often with a blackish basal patch, obovate, rounded or retuse at apex. *Stamens* 10. *Styles* 5. *Schizocarp* of 5, 1-seeded mericarps, the mericarps 4–7mm, with ascending hairs, apical pits separated from main part of mericarp by a sharp ridge and groove not over-arched by hairs; beak 10–70mm. *Flowers* 6–9. $2n=40$.

(a) Subsp. **cicutarium**
Stems up to 60cm, usually erect, with sparse simple eglandular hairs and rare or no glandular hairs, usually very robust at maturity. *Peduncles* mostly 4- to 7-flowered,

sparsely hairy, the hairs mostly or all eglandular. *Flowers* 12–15(–17) mm in diameter, the upper 2 petals often with a basal black patch. *Mericarp* about 6 mm, furrow conspicuous; beak 25–40 mm. $2n=40$.

(b) Subsp. **dunense** Andreas
Stems up to 30 cm, spreading or prostrate, with rather dense glandular hairs and few simple eglandular hairs, not becoming very robust. *Peduncles* mostly 3- to 5-flowered, densely glandular-hairy with few simple eglandular hairs. *Flowers* about 12 mm, the upper 2 petals rarely with a basal black patch. *Mericarp* 5–6 mm, furrow inconspicuous or absent; beak 22–28 mm. $2n=40$.

Native. Well-drained rocky and sandy places, sand dunes, summer-parched grasslands and heaths, also on roadsides, stone walls and railway ballast. Throughout Great Britain but absent from most of the mountain areas; all round the perimeter of Ireland. Throughout Europe to about 70° N in Norway, but perhaps introduced in much of the centre, north and east; temperate Asia to north-west Himalaya and Kamchatka; North Africa to Abyssinia; Macaronesia; naturalised in North and South America. Eurosiberian South-temperate element, but widely naturalised and now Circumpolar Southern temperate. Subsp. *dunense* is the plant of coastal dunes and the Breckland of East Anglia. Subsp. *cicutarium* occurs elsewhere and adjacent to but rarely on the coast. Subsp. *dunense* often grows with *E. lebelii* and could be said to be intermediate between that and subsp. *cicutarium*. However, all these taxa seem to reproduce from seed.

15. E. lebelii Jord. Sticky Stork's-bill
E. glutinosum Dumort.; *E. cicutarium* subsp. *bipinnatum* auct.

Annual monoecious *herb* with a tap-root and fibrous side-roots. *Stem* up to 15(–25) cm, pale yellowish-green, sometimes tinted reddish, terete, densely glandular-hairy especially at the top. *Leaves* opposite; lamina 3–10 × 1–3 cm, dark greyish-green, paler beneath, deeply pinnatisect to more or less pinnate, hairy. *Inflorescence* a terminal umbel of 2–4(–5) flowers; pedicels densely glandular-hairy, bracts brown, free; flowers less than 10 mm in diameter. *Sepals* 5, 2–6 mm, lanceolate, mucronate, densely glandular-hairy. *Petals* 5, pale pink to white, without a black patch. *Stamens* 10. *Styles* 5. *Schizocarp* of 5, 1-seeded mericarps, the mericarps 5–7 mm, with hairs arising from black tubercles, apical pits large, not delimited by a sharp ridge and groove, overarched by hairs from main part of mericarp; beak up to 22 mm. *Flowers* 5–9. $2n=20$.

Native. Rather bare places on fixed dunes. Coasts of Great Britain, probably more widespread than recorded; Ireland from Co. Cork to Co. Antrim, and in Jersey and Sark. Suboceanic Southern-temperate element. Named after Jacques Eugène Lebel (1801–1878).

16. E. moschatum (L.) L'Hér. Musk Stork's-bill
Geranium cicutarium var. *moschatum* L.

Annual or *biennial* monoecious *herb. Stems* up to 50 cm, pale green, decumbent or sprawling, thick and rather fleshy, more or less strigose-hairy, with numerous, small, scattered, sessile glands, branched, leafy. *Leaves* alternate or opposite, 10–30 × 3–8 cm, oblong in outline, imparipinnate; leaflets 5–11, alternate or opposite, 1–4 × 0.4–3.0 cm, medium green on upper surface, paler beneath, oblong, rounded to acute at apex, more or less deeply lobed to one-quarter of way to midrib and serrate, broadly cuneate or rounded at base, shortly petiolulate or sessile, thinly strigose-glandular on both surfaces; petioles up to 15 cm; stipules 5–13 × 4–10 mm, ovate-oblong, obtuse at apex, papery or membranous, glabrous or with ciliate-strigose margins. *Inflorescence* umbellate with 6–12 bisexual flowers; peduncles 8–10 cm, elongating greatly in fruit, more or less densely glandular-strigose; bracts about 3.0 × 2.5 mm, broadly ovate, membranous, thinly glandular-strigose to nearly glabrous; pedicels less than 10 mm, lengthening to 20 mm or more in fruit, becoming patent, with sharply upcurved tips, densely glandular-hairy. *Sepals* 5, 5–8 mm, oblong-ovate, cuspidate-caudate at apex, concave, cucullate, 5-vein, glandular-hairy, with membranous margins. *Petals* 5, 5–7 × 2–3 mm, mauve-pink, oblong, rounded at apex, narrowed below to a shortly hairy claw. *Stamens* that are fertile 5, the filaments 3–4 mm, free with a broad, flattened base and a subulate apex; the anthers oblong; staminodes 5, about 2 mm, oblong, with an acute or emarginate apex. *Style* 1; stigma arms about 0.5 mm. *Schizocarp* oblong; mericarps about 6.0 × 1.5 mm, narrowly spindle-shaped, with a slender pointed base covered with suberect bristles and with a blunt, thinly bristly apex with 2 conspicuous pits, glandular-papillose within; beak 40–50 mm, thinly appressed-strigose with a few longer hairs; seeds about 3.5 × 1.0 mm, bright brown, clavate, apparently smooth but minutely pitted under the microscope. *Flowers* 5–7. $2n=20$.

Probably native on the coast but sometimes considered to be introduced everywhere. Rough ground and bare places in short grassland mainly near the sea, but also a frequent wool alien. Coasts of Channel Islands, western Great Britain from Cornwall to Sussex and to the Isle of Man, and the south and east coast of Ireland; widespread in Great Britain as a casual. Mediterranean region north along the Atlantic coastal area to Great Britain; Atlantic Islands; Ethiopia; now widely distributed as a weed. Mediterranean-Atlantic element.

17. E. stephanianum Willd. Stephan's Stork's-bill
E. stevenii M. Bieb.; *Geranium stephanianum* (Willd.) Poir.; *Geranium stevenii* (M. Bieb.) Poir.; *Geranium multifidum* Patrin ex DC.

Annual or *biennial* monoecious *herb* with fibrous roots. *Stems* 15–50 cm, pale green, ascending or erect, with ascending branches from the base, leafy, puberulous. *Leaves* 10– 30 cm; basal soon deciduous; cauline pale green, with lamina ovate or ovate-triquetrous, 2- to 3-pinnatisect, the laciniae broadly linear, rather obtuse to acute at apex; stipules 6–7 mm, linear-lanceolate, more or less acute at apex. *Inflorescence* 2–4 flowered; peduncle 6–7 cm; pedicels 9–20 mm; flowers bisexual. *Sepals* 2–3 mm, ovate or obovate, obtuse at apex and mucronate, softly hairy. *Petals* 5, broadly rounded-ovate, rounded at apex, purplish. *Stamens* 10. *Styles*

5. *Schizocarp* of 5, 1-seeded hairy mericarps, the apical pits glabrous; beak 30–43 mm. *Flowers* 6–9.

Introduced. A wool casual. Few scattered records. Native of central Asia and China. Named after Christian Friedrich Stephan (1757–1814).

18. E. manescavii Coss. Garden Stork's-bill

Perennial monoecious *herb* with branched, woody stock. *Leaves* all basal; lamina 15–25 × 6–10 cm, pinnate, without subsidiary leaflets, the leaflets up to 30, 2–5 × 1.5–2.5 cm, medium yellowish-green on upper surface, slightly paler beneath, ovate or lanceolate in outline, pinnatisect and dentate, sessile and glandular-hairy, the rhachis pale green and glandular-hairy; petiole shorter than lamina, pale green and glandular-hairy mixed with longer simple eglandular hairs. *Flowering stem* 10–50 cm, pale green, sometimes touched brownish-purple, mixed glandular and simple eglandular hairs below and dense, unequal glandular hairs above; bracts green, fused into a cupule; pedicels 2–6 cm, pale green, tinted reddish, with dense, unequal glandular hairs; flowers 5–20, 25–35 mm in diameter, bisexual. *Sepals* 5, 10–15 mm, pale green, lanceolate, subacute at apex, ribbed, with dense, unequal glandular hairs. *Petals* 5, 15–20 × 7–16 mm, purplish-pink with darker veins, the 2 upper blotched, broadly obovate, rounded with a small mucro at apex, narrowed below. *Stamens* 10; filaments purple; anthers deep purple. *Style* simple, 5-lobed; green; stigma dark purple. *Schizocarp* of 5, 1-seeded mericarps; mericarps 10–12 mm, with patent, brown or white hairs, the apical pits eglandular, without or with a slight furrow at the base; beak 4–7 cm. *Flowers* 5–7. $2n = 40$.

Introduced. Garden escape naturalised in hedgerows and open ground. Kent, formerly Surrey and Somersetshire. Native of west and central Pyrenees. The garden plant seems to be var. **luxurians** Rouy.

× **castellanum** (Pau) Guitt. = **E.** × **hybridum** Knuth
This hybrid differs from *E. manescavii* in the narrow segments of its leaflets and smaller less deeply coloured flowers.

Occurs in waste areas in the Cambridge Botanic Garden and may be a garden escape elsewhere. It appears to reproduce itself.

4. Pelargonium L'Hér. ex Aiton

Perennial monoecious *herbs*, more or less woody at base. *Leaves* alternate or opposite, palmately lobed, stipulate. *Inflorescence* an umbel; flowers bisexual, zygomorphic, with the upper 2 petals wider than the others and the upper sepal with a backward-directed spur adnate to the pedicel. *Sepals* 5. *Petals* 5. *Stamens* 10, but only 7 with anthers. *Style* 1. *Fruit* a schizocarp; seeds dispersed inside mericarps.

Contains over 200 species in Africa.

Knuth, R. (1912). Geraniaceae. In Engler, A. (Edit.) *Das Pflanzenreich* **IV.129**. Leipzig.

1. Leaves velvety white tomentose; petals 7.0–8.5 mm **1. tomentosum**
1. Leaves glandular-hairy; petals 18–22 mm **2. × hybridum**

1. P. tomentosum Jacq. Peppermint-scented Geranium
P. macranthum Eckl. & Zeyh.; *P. corymbarum* Turcz.

Woody, monoecious *perennial* smelling strongly of peppermint. *Stems* up to 75 cm, with a velvety white tomentum and glandular scales beneath the hairs. *Leaves* opposite; lamina up to 12 × 15 cm, medium green, broadly obovate in outline, rather shallowly 3-lobed, with each lobe usually lobulate, serrate-dentate, usually shallowly cordate at base, with more or less velvety white tomentum; petiole up to 30 cm, with velvety white tomentum; stipules brown and membranous, ovate, acuminate at apex. *Inflorescence* of several umbels, each with 5–8 bisexual flowers; peduncles densely hairy, most hairs long and simple eglandular and with some short glandular ones; pedicels 2.0–3.5 cm, generally spreading widely in umbels. *Sepals* 5, 5.5–6.5 mm, green, more or less triangular-ovate, with dense, long, white hairs. *Petals* 5, white or pinkish; upper 2, 7.0–8.5 mm, more or less oblong, crimson-purple marked in lower part, usually with an asymmetric base; lower 3, 8.0–10.5 mm, linear or ligulate. *Stamens* 10. *Style* 1, crimson or purplish; stigmas crimson or purplish. *Mericarp* 20–22 mm; beak with white, patent hairs. *Flowers* 6–8. $2n = 44$.

Introduced. Naturalised in a woodland clearing on Tresco in the Isles of Scilly. Native of South Africa.

2. P. × hybridum Aiton Scarlet Geranium
P. inquinans (L.) Aiton × *P. zonale* (L.) Aiton

Perennial monoecious *herb*. *Stem* up to 50 cm, pale green, stout, erect, hairy, branched, leafy. *Leaves* alternate; lamina 5–10 × 5–10 cm, pale yellowish-green marked with a brownish-red ring on upper surface, paler beneath, subrotund or reniform, broadly palmately 5-lobed, each lobe rounded and crenate, cordate at base, shortly and softly glandular-hairy on both surfaces; petioles up to 11 cm, hairy. *Inflorescence* an umbel; peduncle up to 25 cm, hairy; pedicels short, pale green, tinted red, with dense, short glandular hairs; flowers bisexual. *Sepals* 5, 7–8 mm, medium green, ovate, acute at apex, with dense, short glandular hairs. *Petals* 5, 18–20 × 20–22 mm, white, pink or red, broadly obovate, narrowed at base. *Stamens* 10, but only 7 with anthers; filaments pale; anthers dark. *Style* 1, glabrous; stigmas 5. *Mericarps* hairy. *Flowers* 6–8.

Garden origin. Widely planted in gardens, pots and window boxes. Sometimes persists on rubbish tips until the first frosts.

116. LIMNANTHACEAE R. Br. nom. conserv.

Annual monoecious, slightly succulent, glabrous *herbs*. *Leaves* alternate, pinnate to deeply pinnately lobed, petiolate, exstipulate. *Flowers* solitary in leaf axils, bisexual, hypogynous, actinomorphic. *Sepals* 5. *Petals* 5. *Stamens* 10. *Style* 1; stigmas 5. *Fruits* separating into indehiscent, 1-seeded mericarps.

Contains 2 genera in North America.

1. Limnanthes R. Br.

As family.

Contains 7 species in North America.

1. L. douglasii R. Br. Meadow-foam
Floerkea douglasii (R. Br.) Baill.

Annual monoecious *herb* with fibrous roots. *Stems* branched from the base, 8–35(–40) cm, erect to ascending, pale yellowish-green, terete, slightly fleshy, glabrous. *Leaves* all cauline, alternate; lamina 5–12 × 2–5 cm, oblong-lanceolate in outline, pinnate or deeply pinnately lobed, glabrous, the pinnae medium yellowish-green, ovate in outline, deeply divided into lanceolate or oblong, acute-mucronate lobes; petioles long, pale yellowish-green, glabrous. *Flowers* 15–30 mm in diameter, bisexual, strong-smelling, solitary in leaf axils; peduncles 50–100 mm, pale yellowish-green, glabrous. *Sepals* 5, 6–10 × 2.5–3.0 mm, yellowish-green, lanceolate, slightly hooded at apex, glabrous. *Petals* 5, 8–16 × 7–12 mm, white in upper half, yellow in basal half, obovate, emarginate and shallowly sinuate-dentate at apex, gradually narrowed to base, with prominent veins, long-hairy at base within. *Stamens* 10; filaments 5–8 mm, yellow; anthers yellow. *Style* pale yellow; stigmas 5, whitish, capitate. *Mericarp* 2.5–4.0 mm, dark brown, ovoid, smooth to somewhat triangular-tuberculate. *Flowers* 4–6. Visited by bees. $2n=10$.

Introduced in 1833. Commonly grown for ornament and occasionally found on sea-cliffs, beaches, tips and by roads and lakes. In gardens it can spread rapidly by seed and become a weed. Scattered records throughout Great Britain and in north and south Ireland. Native of California. Our plant is var. **douglasii**. In its native distribution the species shows geographical variation in flower colour. Var. ***sulphurea*** (Loudon) C. T. Mason (*L. sulphurea* Loudon) has all yellow petals, in var. *nivea* C. T. Mason the petals are white often with purple veins and var. ***rosea*** (Hartw. ex Benth.) C. T. Mason (*L. rosea* Hartw. ex Benth.) has the petals white with pink veins or white with a cream base and ageing rose pink. These variants do not seem to have been recorded in Great Britain and Ireland.

117. TROPAEOLACEAE DC. nom. conserv.

Annual or *perennial* monoecious *herbs*, climbing. *Leaves* alternate, palmately lobed or divided into leaflets; with or without stipules. *Flowers* usually solitary in leaf axils, bisexual, slightly perigynous, zygomorphic, borne on long peduncles. *Sepals* 5, more or less free, the upper 1 or 3 with a spur. *Stamens* 8. *Styles* 1; stigmas 3. *Ovary* 3-celled. *Fruit* usually a schizocarp with 3 mericarps.

Contains 3 genera from Central and South America.

1. Tropaeolum L.

As family.

Contains 86 species endemic to the New World, from Mexico to temperate South America.

Sparre, B. & Andersson, A.A. (1991). Taxonomic revision of the Tropaeolaceae. *Opera Bot.* **108**: 1–139.

1. Stipules large and persistent **1. speciosum**
1. Stipules very small and falling early, or absent 2.
2. Spur of sepals 8–12 mm; flower up to 25 mm in diameter **2. peregrinum**
2. Spur of sepals up to 40 mm; flowers 25–60 mm in diameter **3. major**

1. T. speciosum Poepp. & Endl. Flame Nasturtium

Perennial monoecious *herb* with a thick, fleshy rhizome. *Stems* slender and sometimes high climbing (up to several m) by means of coiling petioles. *Leaves* digitate, leaflets 5, sessile or nearly so; lamina of terminal leaflet (10–)15–35 × 7–16 mm, medium green above, paler beneath, obovate or oblong-obovate, rounded to emarginate at apex, entire, narrowly cuneate at base, other leaflets smaller, especially the basal pair, glabrous or sparsely hairy beneath; petiole up to 7 cm; stipules large and persistent. *Flowers* 18–25 mm in diameter, bisexual, solitary; pedicels up to 15 cm, often purplish. *Sepals* 5, 5.5–8.0 mm, reddish-green, ovate-lanceolate, accrescent and deep pink at fruiting; spur 20–30 mm, curved and tapering, the proximal half reddish, the distal green. *Petals* 5, usually scarlet, occasionally rose; upper 2 petals about 17 mm, obovate, broadly emarginate at apex, cuneate at base; lower 3 petals with limb (10–)13–15 × (10–)15–16 mm, squarish, broadly emarginate at apex. *Stamens* 8; filaments pink above; anthers green. *Style* 1; stigmas 3. *Schizocarp* thinly fleshy, deep blue. *Flowers* 7–9.

Introduced. Naturalised among bushes and in hedges. Sussex, Yorkshire, Isle of Man, Ayrshire, Colonsay in the South Ebudes, Sutherland, Caithness and Shetland. Co. Antrim and Co. Londonderry. Native of Chile.

2. T. peregrinum L. Canary Creeper
T. canariense auct.

Annual monoecious *herb*. *Stems* up to 2 m or more, climbing, glabrous. *Leaves* alternate; lamina 3.5–8.0 cm wide, pale green, subrotund in outline, palmately 5-lobed, glabrous; petiole 4–11 cm. *Inflorescence* of single, axillary, bisexual flowers up to 25 mm in diameter; peduncles 6–12 cm. *Sepals* 5, the 3 upper smaller than the 2 lower; spur 8–12 mm, green, stout, hooked at tip. *Petals* 5, bright yellow, the 2 upper 15–20 mm, often spotted with red or purple at the base, obovate, fringed, long-clawed, the claw recurved with a tooth at the bend, the 3 lower 8–10 mm, linear, with long marginal hairs. *Stamens* 8. *Style* 1. *Schizocarp*, 10–14 mm, dark brown or bluish, stalked, divided into 3 mericarps. *Flowers* 7–9. $2n=24$.

Introduced. A persistent garden escape. A few scattered records. Native of Peru and Ecuador.

3. T. majus L. Nasturtium

Annual monoecious *herb* with fibrous roots. *Stems* several, up to 2 m, pale yellowish-green, erect or trailing, fleshy, glabrous, much branched, leafy. *Leaves* alternate; lamina 35–60 mm in diameter, medium yellowish-green on upper surface, paler and bluish beneath, subrotund and peltate, broadly 5-angled with a wavy margin; glabrous on upper surface, densely short-hairy beneath; petiole up to 15 cm, pale yellowish-green, attached nearer one margin than other, the veins radiating from it, shortly hairy. *Inflorescence* a single axillary flower, plant often covered with flowers; peduncles up to 15 cm, pale yellowish-green, sometimes with a few hairs; flowers 25–60 mm in diameter, bisexual. *Sepals* 5, 15–20 mm, pale yellowish-green, lanceolate, obtuse to acute at apex, densely short-hairy, one of the lobes with a spur up to 40 mm, the spur narrow, curved and gradually tapering. *Petals* 5, unequal in size, yellow, pale to deep orange, red or reddish-brown, rarely pink, purple or blue, the limb subrotund to obovate, rounded and wavy at apex, broadly cuneate to a frilled base and a narrow claw, the veins in the throat of the smaller petals deeply coloured. *Stamens* 8; filaments reddish; anthers yellow. *Style* 1, yellow; stigmas 3. *Schizocarp*, 3-celled, breaking into 3 indehiscent, more or less succulent mericarps. *Flowers* 7–9. $2n=28$.

Introduced. Commonly cultivated and a frequent casual on tips and waste ground. Scattered records in Great Britain and the Channel Islands. Native of Peru. *T. majus* probably arose as a spontaneous hybrid between ***T. ferreyae*** Sparre and ***T. minus*** L. Some of the garden plants might also be crossed with ***T. peltophorum*** Benth.

118. BALSAMINACEAE Rich. nom. conserv.

Slightly succulent *annual* monoecious *herbs. Leaves* alternate, opposite or in whorls of 3, simple, petiolate; without stipules or with stipules represented by basal glands. *Inflorescence* an axillary or terminal cyme; flowers bisexual, sometimes some cleistogamous, hypogynous, zygomorphic. *Sepals* 3, more or less petaloid, the lowest sac-like with a conspicuous backward-directed or variously bent spur. *Petals* 5, but apparently 3, the uppermost free, each of the 2 lower ones fused to each of the 2 lateral ones to form 2 compound flanges. *Stamens* 5; filaments fused distally; anthers fused round ovary. *Style* 1; stigma 1 or 5, minute to linear. *Ovary* 5-celled, each cell with numerous ovules; placentation axile. *Fruit* an explosive, 5-celled capsule with many seeds.

Consists of only 2 genera with about 450 species, mainly in tropical Asia and Africa, but with a few species in temperate regions of both the Old and New Worlds.

1. Impatiens L.

Monoecious *herbs. Leaves* opposite, alternate or in whorls of 3; stipules absent or represented by glands. *Inflorescence* of solitary flowers or racemes; flowers strongly zygomorphic, bisexual, hypogynous. *Sepals* usually 3, the lower spurred. *Petals* 5. *Stamens* 5, alternate with the petals; filaments connate above; anthers connate round the ovary. *Style* 1. *Fruit* an explosive 5-celled capsule with many seeds.

Contains over 500 species, mainly in Asia and Africa, very few in the north temperate zone and South Africa.

Beerling, D. J. & Perrins, J. M. (1993). Biological flora of the British Isles. No. 177. *Impatiens glandulifera* Royle. *Jour. Ecol.* **81**: 367–382.

Coombe, D. E. (1956). Biological flora of the British Isles. No. 60. *Impatiens parviflora* DC. *Jour. Ecol.* **44**: 701–713.

Grime, J. P. et al. (1988). *Comparative plant ecology*. London. [*I. glandulifera*.]

Halliday, G. (1997). *A flora of Cumbria*. Lancaster. [*I. noli-tangere*.]

Hultén, E. & Fries, M. (1986). *Atlas of north European vascular plants north of the Tropic of Cancer.* 3 vols. Königstein.

Stewart, A., Pearman, D. A. & Preston, C. D. (1994). *Scarce plants in Britain*. Peterborough. [*I. noli-tangere*.]

Trewick, S. & Wade, P. M. (1986). The distribution and dispersal of two alien species of *Impatiens*, waterway weeds in the British Isles. *Proc. EWRS/AAB 7th symposium on Aquatic Weeds*, pp. 351–356.

1. Flowers pink, pinkish purple or white — 2.
1. Flowers pale yellow to orange — 3.
2. Leaves opposite or in whorls; sepal spur 5–7 mm — **4. glandulifera**
2. Leaves alternate; sepal spur more than 10 mm — **5. balfourii**
3. Larger leaves with more than 20 teeth on each side; flowers including spur less than 15(–20) mm — **3. parviflora**
3. Larger leaves with less than 20 teeth on each side; flowers including spur all or many more than 20 mm — 4.
4. Flowers yellow; sepal spur held at about 90° to the rest of sepal in a fresh state — **1. noli-tangere**
4. Flowers orange; sepal spur held more or less parallel to rest of sepal in fresh state — **2. capensis**

1. I. noli-tangere L. Touch-me-not Balsam

Annual monoecious *herb* with fibrous roots, sometimes forming pure stands. *Stems* 20–180 cm, pale green, erect, simple or branched, swollen at nodes, glabrous. *Leaves* alternate; lamina 1.5–12 × 1.0–1.8 cm, bluish-green on upper surface, paler beneath, oblong, ovate-elliptical or ovate-lanceolate, obtuse to acute at apex, serrate to crenate, with 7–16(–20) teeth on each side, cuneate to subcordate at base, glabrous; petiole short. *Inflorescence* raceme-like with 2–4 flowers, axillary; flowers (15–)20–35 mm in diameter, bisexual, the early ones often cleistogamous; peduncles 8–40 mm. *Sepals* 3, the lower deeply pouched, more or less funnel-shaped and constricted into a short, recurved spur 6–12 mm, yellow with brown dots, rarely bent through 90° or more. *Petals* 5, yellow, with or without brown marks; upper petal hooded, upper lateral petals about half the size of the lower. *Stamens* 5. *Style* 1. *Capsule* about 15 mm, linear, glabrous. *Flowers* 7–9. Pollinated by bees. $2n=20$.

Native. By streams and wet ground in woods. Lake District and north Wales. Occurs in many other places in Great Britain where it is introduced and often only casual. Most of Europe, but absent from the extreme north and parts of the south; temperate Asia to the Pacific. Eurasian Temperate element.

2. I. capensis Meerb. Orange Balsam

I. biflora Walter; *I. fulva* Nutt.

Annual monoecious *herb* with fibrous roots. *Stems* 20–150 cm, pale green, erect, simple or branched, swollen at nodes, glabrous. *Leaves* alternate; lamina 3–8 cm, bluish-green on upper surface, paler beneath, ovate-oblong or elliptical-oblong, obtuse to acute at apex, with

5–12(–14) teeth on each side, the margins somewhat concave at base which is cuneate to attenuate, glabrous or with a few hairs along the midrib; petioles up to 3 cm, glabrous. *Inflorescence* an axillary cyme; pedicels to 25 mm, glabrous; bracts linear-subulate; flowers 20–35 mm, often all cleistogamous. *Sepals* 3, the 2 outer about 10 mm, broadly elliptical, the lower 10–20 mm, plus spur 5–10 mm which is held parallel to the rest of the sepal and has dark red spots inside which show faintly on the outside. *Petals* 5, but appearing to be 3, the 2 lower ones fused to the lateral ones, the lateral ones irregular, pale orange, blotched with scarlet inside, the upper pale orange on outer surface, blotched with deep orange inside. *Stamens* 5; filaments pale and fused distally; anthers fused round the ovary. *Style* 1; stigma minute. *Capsule* oblong, explosive; seeds numerous, about 5 mm, ellipsoid. *Flowers* 6–8. $2n=20$.

Introduced, probably in the very early nineteenth century. Naturalised by rivers, canals and adjacent reservoirs. Widespread in central and southern England with one record in Ireland. Native of North America; also naturalised in France and Germany. The original author wrongly thought it was introduced to Europe from the Cape of Good Hope and called it *I. capensis*.

3. I. parviflora DC. Small Balsam

Annual monoecious *herb* with fibrous roots. *Stems* 30–100 cm, pale green, sometimes tinted brownish-red, erect, angled, zigzag, glabrous, smooth, simple or branched, leafy. *Leaves* alternate; lamina 5–15 × 3.5–6.0 cm, dull medium yellowish-green on upper surface, paler beneath, ovate to oblong-ovate, shortly acuminate at apex, serrate or crenate-serrate, with 20–35 teeth on each side, attenuate at base, glabrous; veins impressed on upper surface, prominent beneath; petiole up to 5 cm, pale green, sometimes tinted reddish, glabrous. *Inflorescence* an axillary raceme which often elongates in fruit, flowers including spur less than 15(–20) mm, bisexual; pedicels to 15 mm, slender, glabrous; bracts 2.5–3.0 mm, green, lanceolate, obtuse at apex. *Sepals* 3, the 2 upper 2.5–3.0 mm, green, lanceolate and obtuse at apex, the lower yellowish, more or less conical, sac-like, gradually tapered into a more or less straight red-tipped spur in total 4–12 mm. *Petals* 5, but appearing to be 3, the 2 lower ones fused to the 2 lateral ones, pale yellow, the 2 lower obovate, rounded at apex and with orange streaks at base, the 2 lateral ovate and rounded and the upper subrotund and keeled. *Stamens* 5, with filaments white and fused distally; anthers fused round the ovary. *Style* 1; stigma minute. *Capsule* up to 20 mm, oblong, explosive; seeds numerous, 4–6 mm, whitish, oblong, striate. *Flowers* 7–11. $2n=24, 26$.

Introduced. Damp shady places in woods and hedgerows and on disturbed or cultivated soils. Scattered localities throughout most of Great Britain and locally common. Native of central Asia; naturalised in north and central Europe.

4. I. glandulifera Royle Indian Balsam
I. roylei Walp.

Annual monoecious *herb* with fibrous roots. *Stems* up to 2.5 m, pale green or reddish, erect, succulent, ribbed, branched above, swollen at the nodes, glabrous. *Leaves* opposite or in whorls of about 3; lamina 10–20 × 3–8 cm, bluish-green on upper surface, paler beneath, lanceolate to ovate, acuminate at apex, sharply serrate, cuneate to attenuate at base, glabrous; veins impressed above, prominent beneath; petioles up to 8 cm, pink above, narrowly winged with scattered, purplish, elongated glands in upper part. *Inflorescence* a diffuse cyme in the upper leaf axils of branches; bracts 5–10 mm; flowers bisexual. *Sepals* 3; the 2 lateral 5–10 mm, pink, ovate, acuminate at apex; the posterior forming a large backwards projecting hood 22–30 mm (excluding spur), pink or rose with dark spots inside; spur 5–7 mm, green, rather stout. *Corolla* 2-lipped; petals 5, white or pale pink, the uppermost petal 15–20 mm, subreniform, truncate or slightly emarginate at apex, the 2 lateral petals about 35 mm, very asymmetric, sharply bent in the middle with the lower half forming a circular orifice enclosing the mass of anthers. *Stamens* 5; filaments pale; anthers yellow. *Style* 1; stigma 5-lobed. *Capsule* 20–30 mm, usually purplish on exposed side, constricted in upper half, with 5 acute ridges; seeds 3.5–5.0 mm, shining black, more or less broadly ovoid, with a single ridge down one side, slightly beaked at apex and truncate at base. *Flowers* 7–10. Pollinated by bumble-bees. $2n=18$.

Introduced. Banks of rivers and canals where it often forms continuous stands and in other damp places and waste ground. Locally common throughout much of Great Britain and Ireland. Native of the Himalayas.

5. I. balfourii Hook. fil. Himalayan Balsam

Annual monoecious *herb* with fibrous roots. *Stems* 40–80 cm, pale green, erect, simple or branched, glabrous. *Leaves* alternate; lamina 2–13 × 1.5–7.0 cm, bright medium green on upper surface, paler beneath, ovate, ovate-lanceolate or ovate-oblong, long-acuminate at apex, serrate, with 20–40 teeth on each side, cuneate at base and shortly decurrent, glabrous. *Inflorescence* of 3–8 flowers in an axillary raceme; flowers 25–40 mm in diameter, bisexual. *Sepals* 3, lowest forming a sac 8–9 × 6–8 mm, longer than wide, gradually contracted to the spur; spur 12–18 mm, straight or slightly curved. *Petals* 5, pinkish-purple; upper petal hooded, upper lateral petals ovate, one-third the size of the lower, lower petals asymmetrically elliptical and 2-lobed. *Stamens* 5; anthers connate. *Style* 1. *Capsule* 20–40 mm, linear to subclavate, glabrous. *Flowers* 6–9. $2n=14$.

Introduced. Grown in gardens and occasionally recorded as an escape. Native of the Himalayas.

Order **15. APIALES** Nakai

Trees, shrubs or *herbs*. *Leaves* usually alternate, often much divided; stipules present or absent. *Inflorescence* usually an umbel or head. *Flowers* unisexual or bisexual, epigynous and actinomorphic to weakly zygomorphic. *Calyx* small, truncate or lobed. *Petals* 4–5, rarely absent. *Stamens* usually the same number as petals and alternate with them. *Ovary* usually 1- to 2-celled, sometimes many-celled; ovules solitary in each cell, pendulous from the apex; seeds usually with copious endosperm.

Consists of 2 families and about 3,700 species of cosmopolitan distribution.

119. ARALIACEAE Juss. nom. conserv.

Evergreen, woody *climbers*, herbaceous *perennials* or evergreen or deciduous *shrubs*. *Leaves* alternate, simple and usually palmately lobed or 1- to 2-pinnate, petiolate and without stipules. *Inflorescence* an umbel, often numerous and grouped into panicles. *Flowers* white to greenish, bisexual or andromonoecious, epigynous, actinomorphic; hypanthium absent. *Calyx* with 5 small teeth near the top of the ovary. *Petals* 5, free. *Disc* at top of the ovary. *Stamens* 6. *Styles* 1 or 5; stigma minute. *Ovary* 2- to 5-celled, with 1 apical ovule per cell. *Fruit* usually a black berry, with 2–5 seeds.

This family is closely related to the Apiaceae and some authorities have advocated their union. It can be distinguished by being mostly woody, by the 1 or 5 styles and the fruit a berry.

Contains 50 or so genera and at least 1,400 species, mainly in the tropics, subtropics of warm temperate areas of both hemispheres with the greatest diversity in Asia, Malaysia, Australia and the Pacific Islands.

1. Deciduous shrub or herbaceous perennial; leaves 1- to 2-pinnate **3. Aralia**
1. Evergreen shrub or climber; leaves simple, usually palmately lobed 2.
2. Climber or scrambler; leaves less than 25 cm wide; style 1 **1. Hedera**
2. Upright shrub; most or all leaves more than 25 cm wide; styles 5, free **2. Fatsia**

1. Hedera L.

Woody, evergreen, climbing *perennials* which become arborescent at maturity; the climbing juvenile stage has stems supported by aerial rootlets which are present in various degrees of abundance, and the scrubby stage which has no aerial rootlets. *Leaves* alternate, in juvenile stage, mostly creeping over the ground or lowest part of climbing plant, usually conspicuously lobed and cordate, when adult, often at the top of the climbing plant, becoming entire with a cordate or broadly cuneate base, intermediate leaves often present between the two stages. *Inflorescence* a solitary or compound umbel. *Calyx* 5-lobed. *Petals* 5, yellowish-cream. *Stamens* 5, alternate with the petals. *Style* 1; stigma obscurely 5-lobed. *Ovary* 5-, rarely 4-celled, each cell with an ovule, half-inferior, terminated by a convex, nectariferous disc having 5 or 10, often obscure, radiating ribs. *Fruit* a berry, 4- to 5-seeded, usually flattened on top, with persistent style. *Flowers* in autumn, fruits in spring.

Contains about 20 species in Europe, Asia, North Africa and Macaronesia.

The hairs on the young shoots are characteristic of the taxa. They may be stellate with the arms elevated from the surface on which they are borne, or scale-like with a flattened centre and spreading branches.

The taxonomy of ivies seems to be fundamentally flawed in that it is based on chromosomes and hairs and the flowers and fruits, considered so important in all other groups, have been completely ignored. This account has taken them into consideration with the result the distribution of taxa is not known. An adequate collection of ivy requires several herbarium sheets and at least two visits to the plant for flowers and fruits. Most material in herbaria is inadequate.

American Ivy Society (1975). *Preliminary check-list of cultivated* Hedera.

Bean, W. J. (1973). *Trees and shrubs hardy in the British Isles*. Ed. 8. **2.** London.

Carriére, E. A. (1890). Une importante collection de Lierres. *Revue Hort.* **1890**: 162–165.

Druce, G. C. (1913). *Hedera helix* L. var. *borealis* Druce. and *Hedera helix* L. var. *sarniensis* Druce *Rep. Bot. Soc. Exch. Club Brit. Isles* **3**: 162–163.

Hibberd, S. (1864). The Ivy. *Floral World* **7**: 7–10.

Hibberd, S. (1870). *Garden oracle*. 123–125.

Hibberd, S. (1872). *The Ivy: a monograph*. London.

Huxley, A. et al. (Edits.) (1992). *Dictionary of gardening*, vol. 2, pp. 510–512. London.

Jacobsen, P. (1954). Chromosome numbers in the genus Hedera L. *Hereditas* **40**: 252–254.

Key, H. (1997). *Ivies*. London.

Koch, K. (1868). *Wochenscrift für Gartnerei und Pflanzenkunde* **13**: 403.

Lawrence, G. H. M. & Schulze, A. E. (1942). The cultivated *Hederas*. *Gentes Herbarium* **6**(3): 107–173.

Lawrence, G. H. M. (1956). The cultivated ivies. *Morris Arb. Bull.* **7**.

Lum, C. & Maze, J. (1989). A multivariate analysis of the trichomes of *Hedera* L. *Watsonia* **17**: 409–418.

McAllister, H. & Rutherford, A. (1983). The species of ivy. *Ivy Jour.*: **9**(4): 45–54.

McAllister, H. & Rutherford, A. (1990). *Hedera helix* L. and H. *hibernica* (Kirchner) Bean (Araliaceae) in the British Isles. *Watsonia* **18**: 7–15.

McAllister, H. A. & Rutherford, A. (1997). Hedera L. In Cullen, J. et al., *The European garden flora*, vol. 5, pp. 375–380. Cambridge.

Petzold, E. & Kirchner, G. (1864). *Arboretun Muscaviense*. Gotha.

Pierot, S. W. (1974). *The Ivy book*. London.

Paul, W. (1867). *Gardeners' chronicle and agricultural gazette* **1867**: 12–15.

Rose, P. Q. (1980). *Ivies*. Poole.

Rutherford, A. (1984). The history of the Canary Islands ivies and its relatives. *Ivy Jour.* **10**(4): 13–18.

Rutherford, A. (1989). The ivies of Andalusia (Southern Spain). *Ivy Jour.* **15**(1): 7–17.

Rutherford, A. (1995). Cases of mistaken identity or *Hedera headaches*. *B.S.B.I. News* **69**: 24–26.

Rutherford, A., & McAllister, H. (1983). *Hedera hibernica* and its cultivar. Ivy Jour. **9**: 23–28.

Rutherford, A., McAllister, H. A. & Mill, R. R. (1993). New ivies from the Mediterranean area and Macaronesia. *The Plantsman* **15**: 115–128.

Seeman, B. (1864). Revision of the natural order Hederaceae. *Jour. Bot. (London)* **2**: 235–250; 289–309.

Seemann, B. (1865). On *Hedera canariensis* as an Irish plant. *Jour. Bot. (London)* **3**: 201–203.

Tobler, F. (1912). Die Gattung *Hedera*. Jena.

Tobler, F. (1927). Die Gartenformen der Gattung *Hedera. Mitt. Deutsch. Dendrol. Ges.* **38**.

1. Hairs on young leaves and stems stellate, white to yellowish brown, the hairs with mostly 4–8(–10) rays fused only at the very base 2.
1. Hairs on young leaves and stems semipeltate, white to orange-brown, the hairs with 8–25 rays fused for quarter to half their length 14.
2. Berries yellowish-orange when ripe **1(c). helix** subsp. **poetarum**
2. Berries black when ripe 3.
3. Hairs on young leaves and shoots whitish, with rays lying parallel to the leaf surface and also away from it 4.
3. Hairs on young leaves and shoots often pale yellowish-brown, their rays all more or less lying parallel to the leaf surface and in one plane 9.
4. Leaves dull greyish-green edged with cream **1(a,i,5). helix** subsp. **helix** var. **helix** forma **cavendishii**
4. Leaves not edged with cream 5.
5. Juvenile leaves with long narrow lobes, the terminal lobe 3–6 times as long as wide **1(a,i,1). helix** subsp. **helix** var. **helix** forma **pedata**
5. Juvenile leaves with lobes not as above 6.
6. At least some juvenile leaves oblong-ovate with lobes arranged hastately **1(a,i,6). helix** subsp. **helix** var. **helix** forma **sagittifolia**
6. Juvenile leaves not so 7.
7. Juvenile leaves 2.5–5.0 cm, spaced, dull dark green with pale veins, mostly 5-lobed **1(a,i,4). helix** subsp. **helix** var. **helix** forma **helix**
7. Juvenile leaves 1–4 cm, dense, dull dark green, sometimes with pale veins, sometimes 2-ranked, sometimes without lobes or 4-lobed 8.
8. Juvenile leaves 3–4 × 2–3 cm, 4(–5)-lobed, the lobes triangular **1(a,i,2). helix** subsp. **helix** var. **helix** forma **minima**
8. Juvenile leaves 1–3 × 2–4 cm, often 2-ranked, often without lobes, sometimes very shortly 2- to 5-lobed **1(a,i,3). helix** subsp. **helix** var. **helix** forma **conglomerata**
9. Umbels of flowers 50–70 mm in diameter, dense **1(b,iii). helix** subsp. **hibernica** var. **sarniensis**
9. Umbels of flowers 20–50 mm in diameter, open 10.
10. Adult leaves lanceolate **1(b,i). helix** subsp. **hibernica** var. **borealis**
10. Adult leaves ovate 11.
11. Leaves variegated yellow **1(b,ii,3). helix** subsp. **hibernica** var. **hibernica** forma **aureovariegata**
11. Leaves entirely green 12.
12. Juvenile leaves with narrow lobes, the terminal lobe 3–6 times as long as wide **1(b,ii,1). helix** subsp. **hibernica** var. **hibernica** forma **lobatomajor**
12. Juvenile leaves with broad lobes 13.
13. Leaves shining medium to rather dark green, not thick and leathery **1(b,ii,2). helix** subsp. **hibernica** var. **hibernica** forma **hibernica**
13. Leaves dull dark green, thick and leathery **1(b,ii,4). helix** subsp. **hibernica** var. **hibernica** forma **irica**
14. Hairs white 15.
14. Hairs orange or reddish 16.
15. Juvenile leaves 5- to 7-lobed **4. azorica** cv. São Miguel
15. Juvenile leaves hardly lobed, just angled **4. azorica** cv. Pico
16. Juvenile leaves wider than long 15.
16. Juvenile leaves longer than wide 18.
17. At least some juvenile leaves more than 10 cm wide **3. algeriensis**
17. Juvenile leaves up to 10 cm wide **5. maderensis**
18. Leaves completely green 19.
18. Leaves with yellow or cream blotches or stripes 20.
19. Leaves entire **2(1). colchica** forma **colchica**
19. Leaves remotely denticulate **2(2). colchica** forma **dentata**
20. Leaves with broad whitish or cream margins **2(3). colchica** forma **dentatovariegata**
20. Leaves with centre and veins streaked and blotched yellow **2(4). colchica** forma **flava**

1. H. helix L. Ivy

Evergreen woody *climber*, with a disagreeable smell, sometimes climbing to 30 m or creeping along the ground and forming carpets, flowering only in the sun at the top of whatever it is climbing on and not flowering when creeping on the ground. *Stems* up to 25 cm in diameter, pale brown, densely clothed with adventitious roots, young shoots densely stellate-hairy. *Leaves* usually dark or medium green, often with pale veins on upper surface, sometimes bluish or yellowish-green, much paler beneath, in cultivated variants blotched or otherwise marked or suffused with cream, white or yellow; lamina of leaves of creeping or climbing stems 2–12(–15) × 2–12(–15) cm, broadly ovate in outline, palmately 3- to 5-lobed, with more or less triangular, obtuse to acute, entire lobes, in cultivated forms sometimes more divided and irregularly lobed; lamina of leaves of flowering stems, 4–12 × 2–12 cm, ovate or rhombic, acute at apex and entire; all glabrous; petiole up to 20 cm, pale green or brownish-purple, glabrous. *Inflorescence* a globose, terminal umbel; peduncles and pedicels stellate-hairy. *Calyx* yellowish-green, with 5, small, deltate lobes. *Petals* 5, 3–4 mm, yellowish-green, triangular-ovate, obtuse and somewhat hooded at apex, entire and with a broad base, recurved, surrounding a nectar-secreting, yellowish-green disc. *Stamens* 5, at the margin of the disc and inwardly inclined; filaments 2–3 mm, pale green; anthers bright yellow, turning to a brownish-yellow and quickly dropping off after dehiscence. *Style* 1, about 1 mm, arising from the middle of the disc, green. *Berry* 6–8 mm, green turning black or rarely yellow when ripe; pyrenes about 5 mm, ovoid-prismatic or trigonous; seed brownish. *Flowers* 9–11. Visited by flies and wasps; homogamous or protandrous.

(a) Subsp. **helix** Common Ivy

Stems slow growing; internodes 3–8 cm; young shoots often wiry with dense stellate hairs which have their arms of unequal lengths and ascending in all directions, the sap with a weak, acrid odour. *Leaves* 3–8 cm, dark, often blackish-green on upper surface, the veins on the creeping or climbing ones pale or silvery white and raised on the surface, hyaline margin not apparent due to them being downturned, the sinuses shallow to deep with surface of the lamina flat; petioles usually bronzy green, but sometimes ruby. *Berry* 6–8 mm in diameter, black when ripe.

(1) Forma **pedata** (Hibberd) Tobler Cv. Pedata
H. helix pedata Hibberd; *H. helix* var. *pedata* (Hibberd) Rehder

Plant climbing. *Juvenile leaves* 4–5 × 5–6 cm, dark green, usually 5-lobed, the lobes narrow, the terminal lobe 3–6 times as long as wide.

(2) Forma **minima** (Hibberd) Tobler Cv. Congesta
H. helix minima Hibberd; *H. helix* var. *minima* (Hibberd) Rehder

Plant climbing, sometimes erect. *Juvenile leaves* 3–4 × 2–3 cm, dense, dark green, usually with pale veins, 4(–5)-lobed, lobes triangular.

(3) forma **conglomerata** (Nichols) Tobler
H. helix conglomerata Nichols

Plant erect. *Juvenile leaves* dense, often 2-ranked, 1–3 × 2–4 cm, dark green, ovate, often without lobes, sometimes shortly 3- to 5-lobed.

(4) Forma **helix**
H. communis Gray nom. illegit.

Plant climbing. *Juvenile leaves* 2.5–5.0 cm, dull dark green with pale veins, mostly 5-lobed.

(5) Forma **cavendishii** (Paul) Tobler Cv. Glaciale
H. helix var. *cavendishii* Paul

Plant climbing. *Juvenile leaves* 5–6 × 6–7 cm, dull greyish-green, edged with cream, with grey markings on upper surface, the petiole brownish-purple, 5-lobed.

(6) Forma **sagittifolia** (Hibberd) Tobler
H. sagittifolia Hibberd; *H. helix* var. *sagittifolia* (Hibberd) Paul

Plant climbing. *Juvenile leaves* 3–5 × 4–6 cm, dull dark green, ovate to ovate-oblong, 3- to 5-lobed, the side lobes obscure and hastately arranged.

(b) Subsp. **hibernica** (Kirschner) D. C. McClint. Atlantic Ivy
H. helix var. *hibernica* Kirschner; *H. hibernica* (Kirschner) Bean

Stem fast and vigorous growing; internodes up to 18 cm or more; young shoots thick and succulent with sparse stellate hairs which have their arms appressed to the shoot and often tinted fawn or sometimes orange-brown, the sap with a strong odour, often pink-like and sweet. *Leaves* up to 10 cm, frequently fleshy, waxy, turning pinkish to light bronze in cold weather, the veins of the creeping or climbing ones with apple green margins and rarely raised above the leaf surface, hyaline margin to edge of leaf prominent due to upturned edges, the sinuses often deep and strongly arched and sometimes funnel shaped; petioles often ruby red. *Umbels* 20–70 mm in diameter. *Berry* 7–12 mm in diameter, black when ripe.

(i) Var. **borealis** Druce
Adult leaves medium yellowish-green, lanceolate. *Umbels* 20–40 mm in diameter, open.

(ii) Var. **hibernica** Kirschner
Adult leaves medium to dark green, broadly ovate. *Umbels* 30–50 mm in diameter, open.

(1) Forma **lobatomajor** (G. H. M. Lawr.) P. D. Sell Cv. Lobata Major
H. helix var. *lobatomajor* G. H. M. Lawr.

Juvenile leaves dark green, 3- to 5-lobed, the lobes narrow, the terminal lobe 3–6 times as long as wide, often the only leaves present.

(2) Forma **hibernica** (Kirschner) P. D. Sell
Juvenile leaves rather shining green and not thick and leathery, 5-lobed, the lobes broad, often not present, but mature leaves same colour.

(3) Forma **aureovariegata** (Weston) P. D. Sell
H. helix var. *aureovariegata* Weston

Juvenile leaves shiny medium to dark green, splashed with yellow, 5-lobed, the lobes broad.

(4) Forma **irica** P. D. Sell Cv. Irish Ivy
Juvenile leaves dark dull green, thick and leathery, 5-lobed, the lobes very broad.

(iii) Var. **sarniensis** Druce
Adult leaves dark green, broadly ovate. *Umbels* 40–70 mm in diameter, dense.

(c) Subsp. **poetarum** Nyman Poet's Ivy
H. chrysocarpa Walsh; *H. poetarum* Bertol. nom. illegit.; *H. helix* var. *chrysocarpa* (Walsh) Ten. ex Caruel; *H. helix* var. *poetica* Halácsy

Stem of climbing habit; internodes 4–5 cm; young shoots pale pink to green, with sparse stellate hairs which have their arms spreading in all directions. *Leaves* up to 8 cm, pale glossy green on upper surface; the veins of the creeping or climbing ones pale or silvery white, with wavy margins, the sinuses rounded, the petioles pinkish-green. *Flower clusters* 25–30 mm in diameter. *Fruit* 7–9 mm in diameter, yellowish-orange when ripe.

Native on trees, banks, rocks and sprawling over the ground. Common throughout Great Britain and Ireland. Subsp. *helix* is said to be the common subspecies over east, central and northern Great Britain, and its distribution further west is uncertain. In East Anglia, however, where a large percentage of the ground is arable, it is confined to ditches, hedgerows, woods and copses and is mostly forma *helix*. In towns and villages, forma *pedata*, *minima*, *conglomerata*, *sagittifolia* and *cavendishii* spill out of gardens onto highways, railway banks, play areas and amenity areas and may even be found in more open areas where their origin is questionable. Subsp. *hibernica* is said to be the commoner taxon in the Channel Islands, west and south-west Great Britain north to south-west Scotland, and Ireland. The original description of var. *hibernica* Kirchner says only that the leaves are larger than var. *helix* and it is common in gardens and Scotland. It is scarcely validly published, but we preferred not to cause even greater confusion by rejecting it. The common plant of western Great Britain and Ireland with shining medium green leaves and a small open inflorescence is regarded as var. *hibernica* forma *hibernica*. The common plant of gardens with large, dark green, thick, leathery leaves is named forma *irica*. It will certainly flower and fruit in the right situation, when its umbels are small and open like those of forma *hibernica*. Forma *lobatomajor* is the var. *hibernica* equivalent of var. *helix* forma *pedata*. Forma *aureovariegata* is a common plant of gardens and spills out onto similar places to subsp. *helix* forms. Var. *sarniensis* is a large dark-leaved plant with a large, dense umbel described from Guernsey and occurring in parks and gardens around Cambridge and also in old areas of fen or former fen where it may be native. Var. *borealis* is a variant of subsp. *hibernica* with small bright green leaves and the characteristic small open inflorescence. It has been re-collected in its original localities of Silverdale in Lancashire and Skye, and may be a coastal or mountain ecotype. Hibberd (1872) mentions observing very small to very large leaved subsp. *hibernica* in the Conway Valley in Caernarvonshire. In Cambridgeshire subsp. *helix* forma *helix*, forma *pedata*, forma *minima*, forma *conglomerata* and forma *sagittifolia* and subsp. *hibernica* var. *hibernica* forma *hibernica*, forma *irica* and forma *aureovariegata* and var. *sarniensis* all grow in quantity and flower and fruit freely. Numerous insects go from one to the other but hybrids have never been recorded although we have seen plants with abortive fruits. Presumably the high chromosome numbers are difficult to count exactly and hybrids may be missed. The species occurs in Europe from Norway and Latvia southwards and in south-west Asia. European Southern-temperate element. Ivy manufactures all its own nourishment and does not do serious damage to the trees it climbs until it gets massive and makes the tree top-heavy, so that it is more easily blown down during a gale. For this reason the scutchers when trimming a hedgerow would cut through the ivy ascending a trunk, but now all trimming is done by machine.

Ivy flowers form the last main source of nectar and pollen for the winter stores of insect visitors. Ivy has always been browsed by domestic animals and is sometimes still used as an emergency winter food. During the eighteenth and nineteenth centuries it was included in festive garlands brought into the house at Christmas. Ivy was sacred to the Greek God Dionysus, more commonly known as Bacchus, who was depicted crowned with ivy and carrying a thyrsus entwined with ivy. Mythology also linked Dionysus with the Egyptian God Osiris, who also carried an ivy-wreathed thyrsus, which was originally an emblem of virtue. Early worship of Dionysus was wild and orgiastic, and female devotees, Bacchantes, wore ivy garlands and were reputed to obtain a religious frenzy by chewing ivy leaves, but modern chemistry does not support this possibility. The term Bacchanalian has passed into our language as a term for wild drinking and orgiastic behaviour. A more sober image was developed in Sparta where the victorious athletes were crowned with the orange-berried variety, subsp. *poetarum*, which is sometimes cultivated and occasionally escapes. Subsp. *poetarum* is native of south-east Europe and south-west Asia. In Chaucer's day a clump of ivy attached to a pole and suspended outside a house indicated a tavern. These alestakes as they were called were restricted to seven feet by an Act of Parliament, and were the forerunners of the public house signs of today. The interest of gardeners in ivy during the nineteenth century died out with the onset of the First World War. Nevertheless, subsp. *hibernica* can still be frequently found in the residences of Victorian gentlemen, especially vicarages. The popularity of ivy-growing, however, continued in the United States where new cultivars were developed. After the Second World War interest again developed in Great Britain. The American Ivy Society was founded in 1973 and the British Ivy Society in 1974.

2. H. colchica (K. Koch) K. Koch Persian Ivy
H. helix var. *colchica* K. Koch; *H. regneriana* Hibberd; *H. cordifolia* Hibberd; *H. colchica* var. *arborescens* Paul; *H. coriacea* Hibberd

Perennial monoecious woody *climber* or *creeper* with a sweet, resinous smell. *Stems* pale brown, densely branched, creeping, scrambling or weakly climbing up to 10 m; young shoots pea-green to pale yellow, densely hairy, the hairs yellow to orange-brown, peltate with 18–30 rays. *Leaves* alternate, emitting a characteristic odour when crushed that is suggestive of celery or other umbelliferous plants; lamina 7–25 × 3.5–15 cm, dark, dull to shiny green on upper surface, sometimes with yellow veins and yellow central area, sometimes with whitish or cream round the margin, paler beneath and sometimes with a yellow hue and deeper yellow veins, thickly coriaceous, ovate, sometimes broadly so, acute to acuminate at apex, entire or very shallowly lobed, sometimes remotely denticulate, rounded or shallowly cordate at base, with orange-brown, peltate hairs with 18–30 rays held flat to the surface, becoming glabrous; petioles up to 13 cm, pale green. *Inflorescence* a globose umbel 35–60 mm in diameter; peduncles and pedicels green to yellow, with numerous peltate hairs as above; flowers bisexual. *Calyx* 5-lobed, the lobes 1.5–1.0 mm, green turning blackish or purple, triangular. *Petals* 5, 6–7 × 2.5–3.0 mm, yellowish-green with a touch of metallic-purple, lanceolate, subacute at apex, fringed with peltate hairs on outside. *Stamens* 5; filaments 5–6 mm, green; anthers pale yellow, tinted pink. *Style* 1; stigma green. *Berries* up to about 60, 9–11 × 10–13 mm, dull black with a metallic greenish hue, oblate-spheroid, with flat top and persistent style. *Flowers* 9–11. Visited by many flies, wasps and Lepidoptera. $2n = 192$.

(1) Forma **colchica**
Leaves dark green, entire.

(2) Forma **dentata** (Hibberd) Hibberd Cv. Dentata
H. coriacea dentata Hibberd; *H. helix dentata* (Hibberd) Nichols

Leaves dark green, remotely denticulate.

(3) Forma **dentatovariegata** (Schulze) P. D. Sell Cv. Dentata-Variegata
H. colchica var. *dentatovariegata* Schulze

Leaves dark green with broad whitish or cream margins, remotely denticulate.

(4) Forma **flavovariegata** P. D. Sell Cv. Sulphur Heart
Leaves dark shining green, the centre and veins blotched and streaked with yellow, entire.

Introduced. Much grown in shrubberies and on walls and spreading onto waste ground, railway banks, footpaths and roadsides. All forms occur but the two variegated ones are the most common both flowering and fruiting even on the ground. Scattered localities over Great Britain and Isle of Man where it is fully naturalised, but in many other localities where it is on the outside of walls and only just out of gardens. Native of the Caucasus and north Turkey.

3. H. algeriensis Hibberd Algerian Ivy
H. canariensis auct.

Perennial monoecious woody *climber* or *creeper*. *Stems* pale brown or reddish-brown, much branched, climbing to about 5m; young shoots pale bright green, with orange-brown semi-peltate hairs with 8–18 rays held parallel to the surface. *Leaves* alternate; lamina 8–19×6–20 cm, from medium to dark glossy green on upper surface, sometimes with a silvery grey margin, marbled or spotted cream or bronzed, paler beneath, lanceolate to ovate, pointed but not acute at apex, entire or 3-lobed, the lobes wide and usually acute, rounded, cordate or truncate at base, with orange-brown, peltate hairs with 8–18 rays held parallel to the surface; petioles up to 6 cm, usually red. *Inflorescence* of compound, very open umbels; peduncles and pedicels with dense hairs with 8–18 rays, flat to the surface; flowers bisexual. *Calyx* yellowish-green, with 5, small, deltate lobes. *Petals* 5, about 2.5×2.0 mm, yellowish-green, triangular, obtuse at apex. *Stamens* 5, at the margin of the disc and inwardly inclined; filaments about 2.5 mm, pale; anthers yellow. *Style* 1, arising from the middle of the disc, green. *Ovary* terminated by a dull, dark brown disc with 10 radiating ribs. *Berry* 7–8×8–10 mm, bright green turning dull black with a grey bloom. *Flowers* 10–11. $2n=96$.

Introduced. Much grown in conservatories and on patios and walls, sometimes persisting and spreading. Scattered records in south and west Great Britain north to west-central Scotland and Isle of Man. Native of North Africa. The commonest indoor variegated ivy usually with distinctive red petioles.

4. H. azorica Carrière Azorean Ivy

Evergreen monoecious *climber* clinging to buildings and trees by root-like growths from the stem. *Stems* far-reaching, pale brown, rough; young shoots downy with whitish, spreading hairs. *Leaves* dull medium green, paler beneath; juvenile 11–15×11–15 cm, broadly ovate in outline, 3- to 7-lobed, the lobes very broad, rounded or bluntly acuminate with an obtuse to acute tip, truncate, rounded or cordate at base, with white, spreading hair flat on the surface when young, becoming glabrous with age; petiole long; adult 3.5–9.0×3–9 cm, ovate, rarely subrotund, entire, pointed at apex, rounded at base. *Inflorescence* of compound, open umbels; peduncles and pedicels with dense 3- to 8-rayed white peltate hairs; flowers bisexual. *Calyx* yellowish-green with 5 short lobes. *Petals* 5, 2.5–3.0 mm, yellowish green, triangular, subacute at apex. *Stamens* 5; filaments pale; anthers pale yellow. *Style* 1, arising from the middle of the disc. *Ovary* terminated by a disc with 10 radiating ribs. *Berry* 7–9×7–9 mm, purplish-black, obovoid. *Flowers* 10–11.

(i) Var. **São Miquel**
Leaves 5- to 7-lobed, the lobes obtuse.

(ii) Var. **Pico**
Leaves hardly lobed, but slightly angled.

Introduced. The São Miquel plants occurs on the walls and trees of Abbey Farm at Histon in Cambridgeshire and the Pico plant under pines by the sea at Holkham in Norfolk. They should be looked for elsewhere. Both are native of the Azores.

5. H. maderensis A. Rutherf. Madeira Ivy

Perennial monoecious woody *climber* or *creeper*. *Stems* pale brown or reddish-brown, much-branched, climbing; young shoots pale brown, with peltate hairs with numerous rays. *Leaves* alternate; juvenile leaves very variable; lamina 2–10×3–12 cm, dark green on upper surface, paler beneath, broadly ovate or reniform in outline, 3- or 5-lobed, the lobes often shallow, rounded or shortly pointed, entire, lamina cordate at base, young leaves with peltate, few-rayed, brownish hairs flat to the leaf beneath; petioles up to 13 cm, pale green. *Inflorescence* of compound, open umbels 30–35 mm in diameter; flowers bisexual; peduncles and pedicels with dense peltate hairs with numerous, unequal rays flat to the surface. *Calyx* yellowish-green, with 5 short lobes. *Petals* 5, 2.5–3.0 mm, yellowish-green, triangular, subacute at apex. *Stamens* 5; filaments 2.5–3.0 mm, pale; anthers pale yellow. *Style* 1, arising from the middle of the disc, green. *Ovary* terminated by a disc with 10 radiating ribs. *Berry* 5–8×7–9 mm, dull black, subglobose. *Flowers* 10–11.

Introduced. On an old wall and farm buildings spreading onto nearby trees at Abbey Farm, Histon in Cambridgeshire. With other ivies it is grown in gardens and should be looked for elsewhere. Native of Madeira.

2. Fatsia Decne & Planch.

Evergreen, monoecious non-spiny *shrubs* without roots along the stems. *Leaves* alternate, mostly more than 25 cm, deeply palmatisect into 7–9(–11) lobes; petiole 8–30 cm. *Inflorescence* up to 46 cm, a large panicle of umbels, those at the end of the central axis and primary branches bisexual, those along the primary branches usually male. *Calyx* with 5 small teeth. *Petals* 5, edge to edge in bud. *Disc* convex, verrucose. *Stamens* 5. *Styles* 5. *Fruits* globose, bluish-black when mature; seeds 5. *Flowers* in autumn. Fruit a *berry*, in spring.

Contains 3 species in east and north-east Asia from Taiwan to Japan.

Bean, W. J. (1973). *Trees and shrubs hardy in the British Isles*. Ed. 8. **2**. London.
Rehder, A. (1940). *Manual of cultivated trees and shrubs*. Ed. 2. New York.

1. F. japonica (Thunb. ex Murray) Decne & Planch. Fatsia
Aralia japonica Thunb. ex Murray; *A. sieboldii* de Vriese

Evergreen, monoecious *shrub* up to 3(–6) m, broader than high, dense. *Stems* several, thick, erect or spreading, covered with old leaf scars. *Bark* pale grey, smooth to slightly rough. *Branches* near the top of the stem and similar to it; young shoots pale green and fleshy. *Leaves* congested round the top of the stem, 10–40 × 10–40 cm, dark glossy green, becoming medium yellowish-green with a pale midrib on upper surface, paler beneath, leathery, ovate or reniform in outline, palmatisect, the segments 7–9(–11), oblong, lanceolate or oblong-elliptical, more or less acute at apex, shallowly crenate-serrate and each with its own midrib, meeting with the rest at the apex of the petiole, glabrous; midribs prominent beneath; petioles 8–30 cm, pale yellowish-green, striate and glabrous. *Inflorescence* up to 46 cm, at the end of branches and stems; rhachis pale creamy-brown; branches spreading and same colour as rhachis, each bearing an umbel, the umbel subtended by large bracts which soon fall. *Umbels* 2.5–3.7 cm in diameter, those at the end of the central axis and primary branches bisexual, those along the primary branches usually male; pedicels jointed at the receptacle. *Calyx* campanulate, about 2.5 mm, 5-ribbed. *Petals* 5, about 3 mm, ovate, apiculate at apex. *Disc* convex, verrucose. *Stamens* 5; filaments and anthers cream. *Styles* 5; cream, slightly thickened at the stigma. *Berry* 7–9 mm in diameter, globose, green, maturing bluish-black, smooth when fresh, topped by the persistent styles; seeds 5. *Flowers* 10–12. Visited by various flies. $2n = 24$, 48.

Introduced. Much used in formal plantings and surviving as relics in a few scattered localities in Great Britain and as relics and self-sown plants in the Isle of Man. Native of Japan and Ryukyu Islands.

3. **Aralia** L.

Herbaceous, monoecious *perennials* or deciduous spiny *shrubs* without roots along the stems. *Leaves* alternate, mostly more than 50 cm, 1- to 2-pinnate. *Inflorescence* 20–60 cm, a large panicle of umbels. *Hypanthium* adnate to ovary. *Sepals* 5, minute, on hypanthium. *Petals* usually 5, overlapping in bud. *Disc* at top of ovary. *Ovary* 2- to 5-celled. *Stamens* 5. *Styles* 5. *Fruit* a globose berry, bluish-black when mature; seeds 5. *Flowers* in summer. *Fruits* in autumn.

Contains 55 species from the Americas, eastern and southern Asia and Malaysia.

Bean, W. J. (1970). *Trees and shrubs hardy in the British Isles*. Ed. 8. **1**. London.

Rehder, A. (1940). *Manual of cultivated trees and shrubs*. Ed. 2. New York.

1. Perennial herb up to 2 m, spineless **3. racemosa**
1. Deciduous shrubs often over 2 m, spiny 2.
2. Panicle more or less conical 3.
2. Panicle umbrella-shaped with short main rhachis and 3–15 branches up to 60 cm, the branches with short side-branches terminated by umbels 4.
3. Leaves hairy beneath **1(i). chinensis** var. **chinensis**
3. Leaves glabrous beneath except for a few hairs on the midrib **1(ii). chinensis** var. **nuda**
4. Leaflets sessile, broadly ovate; primary branches of inflorescence 5–15 and up to 60 cm; flower pedicels 1–6 mm **2(i). elata** var. **elata**
4. Leaflets usually oblong-ovate, with petiolules 2–8 mm; primary branches of inflorescence 3–5 and up to 45 cm; flowers pedicels 5–10 mm **2(ii). elata** var. **mandshurica**

1. A. chinensis L. Chinese Angelica Tree

Deciduous, monoecious *shrub* up to 3(–6) m, with suckers. *Stems* sparsely spiny. *Bark* greyish. *Branches* spiny. *Leaves* alternate, 40–80 cm, broadly ovate in outline, bipinnate; leaflets 5–10 cm, medium green on upper surface, paler beneath, ovate to broadly ovate, acute at apex, finely serrate, rounded at base, hairy or glabrous beneath; veins dividing before reaching the margin and ending in the teeth; petiolules short; petioles rather short, spiny. *Inflorescence* usually an elongated, conical, solitary, axillary panicle 25–40 cm, with terminal umbels of bisexual flowers. *Hypanthium* adnate to the ovary. *Sepals* 5, broadly ovate. *Petals* 5, white, ovate. *Stamens* 5. *Styles* 5. *Berry* about 3 mm in diameter. *Flowers* 8–10.

(i) Var. **chinensis**

Leaves hairy beneath.

(ii) Var. **nuda** Nakai

Leaves glaucous and glabrous except for a few hairs along the midrib.

Introduced. Planted on banks and in shrubberies and spreading by suckers. Naturalised in Kent, Isle of Man and Midlothian and recorded from a few other localities as well as being planted elsewhere. Native of China. Our naturalised plant is said to be var. *nuda*, but it is not known what is planted in estates and amenity areas.

2. A. elata (Miq.) Seem. Japanese Angelica Tree

Dimorphanthus elatus Miq.; A. chinensis var. elata (Miq.) Lavallée; A. japonica auct.

Deciduous, monoecious *shrub* or *tree* up to 14 m with a spreading crown. *Stems* numerous, some actually branching from the base, 12–20 cm in diameter, more or less erect, simple or sparsely branched above. *Bark* grey, rough, usually prickly. *Branches* few, short, spreading, flexuous, grey, glabrous; young shoots hairy. *Leaves* alternate, clustered near the stem and branch apices, 40–80 × 35–80 cm, broadly ovate in outline, 1- to 2-pinnate; leaflets 7–11 on each side of branch from rhachis, 6–18 × 2.5–8.0 cm, dark green on upper surface, paler and slightly bluish-green beneath with even paler veins, lanceolate to broadly ovate or oblong-ovate, acuminate at apex, shallowly serrate or crenate-serrate, rounded or slightly cordate at base, with scattered short hairs mainly on the veins on the upper surface or glabrous, with numerous hairs mainly on veins beneath to glabrous; veins 8–14 pairs, gradually curving and dividing before reaching the margin, fairly prominent beneath; sessile or with petiolules up to 8 mm; rhachis brownish-red, hairy; petiole brownish-purple, triangular and shortly hairy. *Inflorescence* a pendulous, terminal panicle, the panicle umbrella-like, with 3–15 branches, each branch up to 60 cm and consisting of a rhachis with numerous short side-branches at the end of which are umbels, the peduncles 30–35 mm, the pedicels 1–10 mm, the branches, peduncles and pedicels pale greenish-yellow touched with pink, with dense, short, curved hairs; flowers bisexual. *Hypanthium* adnate to the ovary. *Sepals* 5, broadly triangular, acute at apex, glabrous. *Petals* 5, about 0.2–1.2 mm, yellowish-white, ovate. *Stamens* 5, longer than the petals; filaments white; anthers ovoid. *Styles* 5, 1–5 mm, white, spreading; stigma reddish. *Berry* 3–4 × 3–5 mm, pale green, turning purplish-black, oblate-spheroid, with persistent styles; seeds 5, ovoid, smooth. *Flowers* 8–10. $2n = 14$.

(i) Var. **elata**
Leaflets more or less sessile, broadly ovate, with large teeth, usually hairy beneath. *Primary branches of inflorescence* 5–15 and up to 60 cm. *Flowers* with pedicel 1–6 mm.

(ii) Var. **mandshurica** (Rehder) Wen
A. chinensis var. *mandshurica* Rehder

Leaflets with petiolules 2–8 mm, usually oblong-ovate, with small teeth, glabrous beneath. *Primary branches of inflorescence* 3–5 and up to 45 cm. *Flower* with *pedicel* 5–10 mm.

Introduced. Planted on banks and in shrubberies and spreading by suckers. Naturalised in Cheshire, Isle of Man and Jersey and recorded from a few other localities in Great Britain. Native of eastern Asia. Both varieties probably occur.

3. A. racemosa L. American Spikenard
Perennial monoecious spineless *herb* up to 2 m. *Stem* pale green, suffused brownish-purple, erect, angled, branched, leafy. *Leaves* up to 45 × 60 cm, broadly ovate in outline, 2-pinnate; leaflets with lamina 10–17 × 8–11 cm, medium to dark green on upper surface, paler beneath, the midrib often tinted brownish-purple, ovate or elliptical, acute or acuminate at apex, shallowly serrate, cordate, often asymmetrically, at base, glabrous or with few hairs beneath; petiolules up to 4 cm; petiole up to 20 cm, swollen at base; petiolules, rhachis and petiole, brownish-purple, glabrous. *Inflorescence* an elongated, panicle of umbels; rhachis, branches and pedicels pale green turning brownish-purple and glabrous; pedicels 5–8 mm. *Hypanthium* adnate to the ovary. *Sepals* 5, minute, green, triangular. *Petals* 5, about 0.7 mm, yellowish-green, triangular, acute at apex, spreading. *Stamens* 5; filaments white; anthers pale yellow. *Styles* 5. *Berry* 1.5–2.0 mm, pale yellowish-green at first, turning purplish-black when ripe, oblate-spheroid. Flowers 8–9. $2n = 24$.

Introduced. Planted in shrubberies and open woodland. Naturalised in Shropshire and Lanarkshire. Native of eastern North America.

120. APIACEAE Lindl. nom. conserv.

UMBELLIFERAE Juss. nom. altern.

Annual or *perennial*, monoecious, rarely dioecious *herbs* or rarely *shrubs*. *Leaves* alternate, simple to palmate, pinnate or ternate, often several times so, sessile or petiolate, the petiole often much widened and sheathing at base, stipules rarely present. *Inflorescence* a lateral or terminal umbel, the umbels sometimes simple, but usually compound, rarely with whorls of flowers below the main umbel; flowers bisexual or frequently functionally andromonoecious, epigynous, actinomorphic or with zygomorphic petals, those away from the centre of the umbel larger; hypanthium absent. *Sepals* 5, usually represented by small teeth near the top of the ovary, or absent. *Petals* 5, usually white, pink or yellow. *Stamens* 5. *Styles* 2, often with swollen base, the stylopodium; stigma minute to capitate. *Ovary* 2-celled, with 1 apical ovule per cell. *Fruit* a dry, 2-celled schizocarp, the 2 mericarps usually separating from the sterile carpophore (or carpophore lacking) but each remaining indehiscent at maturity.

Contains about 428 genera and about 2,800 species, worldwide, but chiefly in the north temperate region.

Hedge, I. C. (1973). Umbelliferae in 1672 and 1972. *Notes Roy. Bot. Gardess. Edinb.* **32**: 151–158.
Heywood, V. H. (Ed.) (1970). The biology and chemistry of the Umbelliferae. *Bot. Jour. Linn. Soc.* **64**: Suppl. 1.
Ingram, T. (1993). *Umbellifers*. Pershore.
Knees, S. G. (1989). *Umbellifers of the British Isles*. Princes Risborough.
Koso-Poliansky, B. M. (1916) Sciadophytorum systematis lineamenta. *Bull. Soc. Nat. Mosc.* nov. ser. **29**: 93–221.
Tutin, T. G. (1980). *Umbellifers of the British Isles*. B.S.B.I. Handbook 2. London.
Wolff, H. (1910–1927). Umbelliferae. In Engler, A. (Edit.) *Das Pflanzenreich* **IV.228**, **43**: 1–214 (1910); **61**: 1–305 (1913); **90**: 1–398 (1927).

Most species of the Apiaceae can be recognised as belonging to that family at a glance, from their compound leaves and umbellate inflorescences, and their smell is often characteristic. In the majority of the species the leaves are pinnately divided, mostly frequently two or more times. Simply pinnate or palmately lobed leaves are much less frequent and entire ones rare. The ultimate, distinctly stalked divisions of a leaf are referred to as lobes which may be entire, toothed or pinnatifid. The leaves are usually petiolate and spirally arranged. The petioles of the lower leaves are usually long and have a sheathing base. The upper leaves usually have short, often entirely sheathing and sometimes greatly inflated petioles. There are no stipules except in the Subfamily Hydrocotyloideae. The commonest type of inflorescence is a compound umbel, which consists of a number of branches arising at the same level called *rays*, and each bearing several flowers which also arise from the same level, forming a *partial umbel*. In the subfamilies Hydrocotyloideae and Saniculoideae simple umbels occur and in *Hydrocotyle* the flowers are often in whorls. In *Eryngium* the flowers are sessile and crowded into thistle-like heads with each individual flower subtended by a bracteole. The umbels terminating a stem or branch, the oldest one of which being on the main stem and usually having the greatest number of rays, are referred to as *terminal umbels*. The peduncle may be long so that the terminal umbel overtops the lateral umbels, or it may be almost non-existent as in *Torilis nodosa* where each successive umbel appears to be lateral and leaf-opposed. Where the rays arise from the peduncle there are sometimes small leaves called bracts. Similar small leaves subtending the partial umbel are called *bracteoles*. The individual flowers are small, but the umbel often large. The 5 sepals are free and usually smaller than the petals and are often absent altogether. If the sepals are present the 5 petals alternate with them. Usually the petals are concave with an inflexed apex. The outer petals of the flowers at the periphery of the umbel are sometimes much larger than the remainder when they are referred to as *radiating*, otherwise the flowers are actinomorphic. An oil gland is often found along the middle of the petal. The 5 stamens alternate with the petals in a single whorl. The filaments are much shorter than to longer than petals. The 2 styles arise from a nectar-secreting disc at the summit of the ovary and the often thickened base is called the *stylopodium*. The stigmas may be unthickened, slightly and gradually thickened, a more or less globose knob, or capitate. The fruit, a schizocarp with 2 mericarps, contains most of the important characters for recognising both genera and species. The size of the fruit and its ridges and wings show little variation within a species. It may be glabrous, hairy, covered with papillae or with straight or hooked bristles or spines. They may be semicircular in cross-section of dorsally compressed if parallel to the commissure or laterally compressed if at right angles to the commissure. The 2 mericarps are joined by a *commissure* along which they separate at maturity. The usually bifid *carpophore* lies between 2 oil canals called *vittae* on the commissural face of each mericarp. The carpophore and commissural vittae are absent in the subfamilies Hydrocotyloideae and Saniculoideae. On the outer face of the mericarp between each pair of vascular bundles there is usually a single vitta, which may be absent or split into several small vittae. Sometimes there are secondary ridges which alternate with and may be more prominent than the primary ridges and which may hide the vittae. In some genera the mericarps have a distinct beak.

1. Leaves all simple and entire **28. Bupleurum**
1. Leaves simple to compound; if simple at least toothed 2.
2. Stem leaves spiny **4. Eryngium**
2. Leaves not spiny 3.
3. Leaves all more or less rotund and shallowly lobed, with a long petiole and stipulate **1. Hydrocotyle**
3. Leaves without stipules, if all simple then either lobed more than halfway to the base or sessile 4.
4. Leaves all simple with small teeth only **11. Smyrnium**
4. Leaves simple to compound, if all simple then divided more than halfway to base 5.
5. Basal leaves all ternately or palmately lobed almost to base 6.
5. Basal leaves ternate, palmate or pinnate, often compoundly so 7.
6. Flowers subsessile, with inconspicuous bracteoles; fruits with hooked bristles **2. Sanicula**
6. Flowers distinctly pedicellate, with bracteoles at least as long; fruits covered with bifid scales **3. Astrantia**
7. Schizocarp with an apical beak more than twice as long as seed-bearing part **7. Scandix**
7. Schizocarp beakless or with beak shorter than seed-bearing part 8.
8. Stem and often basal leaves with white, flexuous subterranean part arising from a brown tuber 9.
8. Stem and basal leaves if present arising from roots at ground level, or from rhizome, or plant aquatic 10.
9. Stem solid at fruiting; schizocarp with curved styles suddenly contracted from stylopodium **12. Bunium**
9. Stem hollow at fruiting; schizocarp with more or less erect styles gradually narrowed from stylopodium **13. Conopodium**
10. Petals yellow 11.
10. Petals white to pink or purplish, or greenish-white 21.
11. Schizocarp compressed dorsally, i.e. distinctly wider in dorsal view than in lateral view 12.
11. Schizocarp not compressed dorsally, or scarcely so 15.
12. Bracts more than 3; bracteoles connate at base **43. Levisticum**
12. Bracts 0–3; bracteoles absent or free 13.
13. Leaves 1-pinnate, with ovate lobes **46. Pastinaca**
13. Leaves 2–several times pinnate or ternate, with linear lobes 14.
14. Stem not surrounded by fibrous remains of petiole; cauline leaves with wide sheathing petiole; fruits 10–16 mm **44. Ferula**
14. Stem surrounded by fibrous remains of petioles at base; cauline leaves with narrow petiole; fruits 5–8 mm **45. Peucedanum**
15. Ultimate leaf-lobes filiform, less than 0.5 mm wide 16.
15. Ultimate leaf-lobes flat, linear to ovate, more than 0.5 mm wide 18.
16. Fresh plants not smelling of aniseed; schizocarps 1.5–2.5 mm, compressed laterally **34. Ridolphia**
16. Fresh plants smelling of aniseed; schizocarps 3–6 mm, not compressed or compressed dorsally 17.
17. Firmly rooted perennial; stems solid at first, becoming more or less hollow; schizocarps oblong-ellipsoid, scarcely compressed, mericarps with prominent, thick ribs **22. Foeniculum**
17. Easily uprooted annual; stems hollow; schizocarps elliptical, strongly compressed dorsally, mericarps with low, slender dorsal ridges and winged marginal ridges **23. Anethum**
18. Schizocarps laterally compressed, about twice as wide in lateral view as in dorsal view 19.
18. Schizocarps not or scarcely compressed, about as wide in lateral view as in dorsal view 20.
19. Bracts and bracteoles 0–2, very short; mericarps with 3 sharp, dorsal ridges **11. Smyrnium**
19. Bracts 1–3, often lobed; bracteoles more than 3; mericarps with 3 rounded, dorsal ridges **33. Petroselinum**
20. Leaves succulent; bracts more than 4; mericarps not winged **18. Crithmum**
20. Leaves not succulent; bracts 0–3; mericarps with narrow, lateral wings **24. Silaum**

21. Schizocarps with spikes, bristles, hairs or conspicuous tubercles 22.
21. Schizocarps glabrous, more or less smooth 32.
22. Schizocarps strongly compressed dorsally, i.e. distinctly wider in dorsal view than in lateral view, including projections 23.
22. Schizocarps not compressed dorsally or scarcely so 24.
23. Stems 1 m; umbels with more than 30 rays; schizocarps more than 8 mm **47. Heracleum**
23. Stem less than 1.5 m; umbels with fewer than 20 rays; schizocarps less than 7 mm **48. Tordylium**
24. Schizocarps with conspicuous tuberculate papillae **32. Trachyspermum**
24. Schizocarps with spines, bristles or hairs 25.
25. Schizocarps with hairs or weak or minute bristles 26.
25. Schizocarps with usually stout, terminally hooked or barbed spines 28.
26. Fresh plant smelling of aniseed; schizocarps more than 3 times as long as wide **8. Myrrhis**
26. Fresh plant not smelling of aniseed; schizocarps less than 3 times as long as wide 27.
27. Perennial with base of stem sheathed by a mass of fibres; bracts and bracteoles each more than 5; schizocarps less than 3.5 mm **19. Seseli**
27. Slender annual without fibres at base of stem; bracts and bracteoles each less than 5; schizocarps more than 3.5 mm **30. Cuminum**
28. Bracts deeply pinnately or ternately divided **53. Daucus**
28. Bracts absent or simple 29.
29. Schizocarps with spineless but ridged beak below the stylopodium; sepals absent **6. Anthriscus**
29. Schizocarps without a beak, with spines up to the base of the stylopodium 30.
30. Sepals small but persistent; mericarps with slender, ciliate ridges and spines and tubercles in the furrows **50. Torilis**
30. Sepals not persistent; mericarps with broad-based spines arranged in rows on the ridges 31.
31. Mericarps with 3 of the primary ridges with uniseriate cilia, the secondary ridge thickened with aculeate spines **51. Caucalis**
31. Mericarps with the 2 primary, marginal ridges each with a single row of spines or tubercles and the remaining primary and secondary ridges similar to each other, with spines in 2–3 rows **52. Turgenia**
32. Schizocarps strongly compressed dorsally, distinctly wider in dorsal view than in lateral view 33.
32. Schizocarps not compressed dorsally or scarcely so 41.
33. Leaves and stems hairy **47. Heracleum**
33. Leaves and main parts of stems glabrous or nearly so, sometimes coarsely papillose 34.
34. Easily uprooted annuals 35.
34. Firmly rooted perennials 36.
35. Sepals conspicuous and persistent; outer petals more than 2 mm; mericarps with low ridges **9. Coriandrum**
35. Sepals absent; outer petals less than 1.5 mm; mericarps with prominent keeled ridges **21. Aethusa**
36. Leaf-lobes less than 20 × 10 mm 37.
36. Leaf-lobes more than 20 × 10 mm 38.
37. Stems solid; bracts absent, or few and soon falling; bracteoles not or weakly reflexed **40. Selinum**
37. Stems hollow; bracts more than 3, reflexed; bracteoles reflexed **45. Peucedanum**
38. Larger leaves 2- to 3-pinnate, smaller ones sometimes 2- to 3-ternate or pinnate-ternate **42. Angelica**
38. All leaves 1- to 3-ternate 39.
39. Umbels with more than 20 rays **45. Peucedanum**
39. Umbels with fewer than 20 rays 40.
40. Schizocarps with 3 prominent, acute dorsal ridges **41. Ligusticum**
40. Schizocarps with low dorsal ridges and broadly winged lateral ridges **49. Laser**
41. Base of stem sheathed by a mass of fibres which are the remains of petioles 42.
41. Stem without mass of fibres at base 46.
42. Schizocarps didymous, retuse at apex, cordate at base **10. Bifora**
42. Schizocarps not didymous 43.
43. Basal leaves 1-pinnate, the segments deeply palmatisect, the lobes filiform and appearing as though whorled **39. Carum**
43. Basal leaves 1- to 4-pinnate; if 1-pinnate then lobes not divided more or less to base into filiform segments 44.
44. Basal leaves 3- to 4-pinnate, with filiform ultimate lobes; schizocarps 4–10 mm **25. Meum**
44. Basal leaves 1- to 3-pinnate, with ovate to linear ultimate lobes; schizocarps 2–4 mm 45.
45. Plant monoecious; leaf segments variable, some often ovate, very rarely all linear; umbels usually with more than 10 rays **14. Pimpinella**
45. Plant usually dioecious; leaf segments linear; umbels with less than 10 rays **29. Trinia**
46. Schizocarps more than twice as long as widest width 47.
46. Schizocarps less than twice as long as widest width 52.
47. Sepals absent, or minute, or a vestigial rim, not or scarcely visible at top of schizocarp 48.
47. Sepals more than 0.2 mm, distinctly visible at top of schizocarp 50.
48. Stems hollow; mericarps not ridged in mid or basal regions **6. Anthriscus**
48. Stems solid or hollow; mericarps ridged along whole length 49.
49. Fresh plant not smelling of aniseed; stems solid; mericarps with low, rounded ridges **5. Chaerophyllum**
49. Fresh plant smelling of aniseed; stem hollow; mericarps with prominent ridges **8. Myrrhis**
50. Umbels with 1–5 rays; at least some bracts more than half as long as rays **30. Cacuminum**
50. Umbels with more than 6 rays; if fewer then bracts absent or much less than half as long as rays 51.
51. Plant often aquatic; leaf-lobes various, but if linear or lanceolate and more than 50 mm, then entire to distantly and irregularly toothed; schizocarps not compressed, or only slightly so dorsally **20. Oenanthe**

51. Plant not aquatic; lobes of lower leaves more than 50 mm and linear-lanceolate with regularly and sharply serrate margins; schizocarps laterally compressed **38. Falcaria**
52. Sepals more than 0.2 mm, distinctly visible at top of schizocarp 53.
52. Sepals absent or minute, or a vestigial rim, not or scarcely visible at top of schizocarp 58.
53. Schizocarps subglobose, in lateral view about as wide as long 54.
53. Schizocarps in lateral view distinctly longer than wide 58.
54. Easily uprooted annuals; mericarps remaining fused even at maturity **9. Coriandrum**
54. Firmly rooted perennials; mericarps splitting apart at maturity 55.
55. Stems solid; petioles not widened at base, nor sheathing lateral stems; schizocarps more than 2.5 mm **26. Physospermum**
55. Stems hollow; petioles widened at base and sheathing lateral stems; schizocarps less than 2.5 mm **36. Cicuta**
56. Not aquatic; lobes of lower leaves more than 50 mm, linear-lanceolate and regularly and sharply serrate **38. Falcaria**
56. Often aquatic; leaf-lobes various, but if linear or lanceolate and more than 50 mm, then entire to distantly and irregularly toothed 57.
57. Lobes of lower leaves mostly more than 40 × 20 mm; schizocarps laterally compressed **16. Sium**
57. Lobes of lower leaves less than 40 × 20 mm; schizocarps not compressed or slightly dorsally so **20. Oenanthe**
58. Lower leaves simply pinnate, the lobes not divided as far as the midrib 59.
58. Lower leaves 2- to 4-pinnate or 1- to 2-ternate 65.
59. Bracts absent or sometimes up to 2, if constantly present then stem mostly procumbent with only leaves and peduncles erect 60.
59. Bracts 2–numerous, at least apical part of stem erect to ascending 62.
60. Plant often aquatic, at least lower part of stem procumbent and rooting **31. Apium**
60. Plant more or less never aquatic; stem erect, not rooting 61.
61. Schizocarps more than 2 mm, or if less than 2 mm at least the rays hairy **14. Pimpinella**
61. Schizocarps less than 2 mm and plant glabrous **31. Apium**
62. Styles at fruiting at least as long as stylopodium 63.
62. Styles at fruiting much shorter than stylopodium 64.
63. Lowest leaves with 4–12 pairs of leaflets each 5–35 mm; longest bracts more than half as long as shortest rays **33. Petroselinum**
63. Lowest leaves with 2–5 pairs of leaflets each 30–60 mm; all bracts less than half as long as rays **35. Sison**
64. Stems hollow; all leaves 1-pinnate; bracts lobed, but usually not to base, less than half as long as rays **17. Berula**
64. Stems solid; upper leaves more than 1-pinnate; bracts divided to base into linear to filiform lobes, the longest more than half as long as the rays **37. Ammi**
65. Bracts more than 4 66.
65. Bracts absent to 2(–3) 67.
66. Stems hollow; bracts less than quarter as long as the rays, undivided; schizocarps with prominent, more or less undulate-crenulate ridges **27. Conium**
66. Stems solid; bracts more than half as long as rays, deeply divided; schizocarps with low, smooth ridges **37. Ammi**
67. Plant in water or in mud; stems procumbent and rooting at least near the base; styles at fruiting much shorter than stylopodium **31. Apium**
67. Plant not aquatic; stems erect, but rhizomes sometimes produced; styles at fruiting at least as long as stylopodium 68.
68. Plant rhizomatous; leaves 1- to 2-ternate **15. Aegopodium**
68. Plant not rhizomatous; leaves 2- to 3-pinnate 69.
69. Easily uprooted annuals; bracteoles long and strongly reflexed **21. Aethusa**
69. Firmly rooted biennials or perennials; bracteoles absent, or if present scarcely reflexed 70.
70. Bracts absent; usually at least some leaf-lobes ovate, if all linear to linear-lanceolate then styles not appressed to the stylopodium **14. Pimpinella**
70. Usually some umbels with 1–2(–3) bracts; all ultimate leaf-lobes linear to linear-lanceolate; styles more or less appressed to the stylopodium **39. Carum**

Subfamily **1. Alydrocotyloideae** Drude

Leaves simple, more or less subrotund, not or shallowly lobed, stipulate. *Flowers* in simple axillary umbels, often with whorls below. *Schizocarp* with a woody inner wall, without a carpophore and no vittae. *Basic chromosome number 8.*

1. Hydrocotyle L.

Perennial monoecious *herbs. Stems* thin, procumbent, rooting at nodes. *Leaves* simple, more or less subrotund, not or shallowly lobed; petioles thin, usually erect, as long as or much longer than the axillary peduncles; stipulate. *Flowers* very small, in simple, axillary umbels. *Schizocarp* more or less subrotund in side view, strongly laterally compressed; vittae absent.

Contains about 70 species in temperate and tropical regions of the world.

Grime, J. P. et al. (1988). *Comparative plant ecology*. London.

Hultén, E. & Fries, M. (1986). *Atlas of north European vascular plants north of the Tropic of Cancer.* 3 vols. Königstein.

Preston, C. D. & Croft, J. M. (1997). *Aquatic plants in Britain and Ireland*. Colchester. [*H. ranunculoides*.]

Webb, D. A. (1959). *Hydrocotyle moschata* Forst. fil. *Proc. B. S. B. I.* **3:** 288.

Webb, C. J. (1990). New Zealand species of *Hydrocotyle* (Apiaceae) naturalised in Britain and Ireland. *Watsonia* **18**: 93–95.

Westall, C. B. (1988). *Hydrocotyle sibthorpioides* Lam. *B.S.B.I. News* **48**: 36.

1. Leaves glabrous on upper surface 2.
1. Leaves hairy on upper surface 4.
2. Leaves not or scarcely lobed **1. vulgaris**
2. Leaves lobed one-third to half-way to the base 3.
3. Leaves up to 80 mm; schizocarps 2–3 **2. ranunculoides**
3. Leaves 3–12(–20) mm; schizocarps 1.0–1.5 mm **3. sibthorpioides**

4. Leaf-lobes serrate; schizocarps 1.3–1.6 × 1.0–1.5 **4. moschata**
4. Leaf-lobes crenate; schizocarps 2.0–3.5 × 1.8–2.5 mm **5. novae-zeelandiae**

1. H. vulgaris L. Marsh Pennywort

Perennial monoecious *herb. Stems* up to 30 cm, pale green, creeping, slender, rooting at the nodes, glabrous. *Leaves* alternate; lamina 0.8–3.5 cm in diameter, medium yellowish-green on upper surface, paler beneath, subrotund, peltate, shallowly crenate, with 6–9 main veins radiating from the junction of the lamina and petiole, glabrous; petiole 1–25 cm, pale green erect, glabrous or with medium to long, pale simple eglandular hairs especially near the apex; stipules scarious, often laciniate. *Inflorescence* axillary, usually about half as long as the subtending leaf, a head-like umbel about 3 mm in diameter, sometimes with 1–3 whorls of flowers below; bracteoles present. *Sepals* minute or absent. *Petals* 5, greenish-white, tinged with pink, with an incurved point. *Stamens* 5; filaments whitish; anthers pink. *Styles* not thickened at base; spreading horizontally or somewhat recurved in fruit, about twice as long as the disc; stigma truncate. *Schizocarp* about 2 mm wide, broadly ellipsoidal, wider than long, covered with brownish, resinous dots, strongly compressed laterally; commissure narrow; carpophore absent; mericarps with slender, rather prominent ridges; vittae absent, at least when the schizocarp is mature. *Flowers* 6–7. $2n = 96$.

Native. Bogs, fens, marshes and sides of lakes. Throughout Great Britain and Ireland, locally common but absent from large areas of the Midlands and northern Scotland. West, central and south Europe to about 60° N in Scandinavia; Caucasus; north-west Africa, but no recent records. Southern Temperate element.

2. H. ranunculoides L. fil. Floating Pennywort

Perennial monoecious *herb. Stems* up to 30 cm, pale green, free-floating or creeping and rooting on mud, slender to somewhat thickened, rooting at the nodes, glabrous. *Leaves* alternate; lamina up to 8 × 8 cm, medium green on upper surface, paler beneath, subrotund or reniform, not peltate, 5- to 6-lobed to about halfway to base, the lobes crenate or lobulate, cordate, glabrous; petiole up to 35 cm, slender, glabrous; stipules laciniate. *Inflorescence* a simple umbel, with 5–10 flowers, rays 1–3 mm, spreading and ascending; peduncles shorter than the leaves, tinted red, glabrous. *Sepals* minute or absent. *Petals* 5, greenish-white. *Stamens* 5; filaments whitish; anthers pink or pale yellow. *Styles* 2, not thickened at base. *Schizocarp* 1–3 × 2–3 mm, subglobose, the dorsal surface rounded, with ribs; commissure narrow; carpophore absent; mericarps without ridges; vittae absent. *Flowers* 6–7. Plants growing on mud flower and set seed while floating plants do not flower but reproduce vegetatively. $2n = 24$.

Introduced. Naturalised in rivers and canals. First recorded in 1990 and apparently spreading rapidly. Scattered records in Great Britain north to Lancashire. Native of North America.

3. H. sibthorpioides Lam. Sibthorp's Pennywort

Perennial monoecious *herb* forming patches. *Stems* pale green, filiform, creeping, rooting at the nodes. Leaves with lamina 5–12(–20) mm, medium green, reniform, crenately lobed, with a deep basal sinus, glabrous; petiole 5–20 mm, filiform, hairy. *Inflorescence* a simple capitate umbel with 3–8(–10) subsessile flowers. *Sepals* minute or absent. *Petals* 5, white. *Stamens* 5; filaments pale; anthers pink. *Styles* 2, curved; stigma minute, stylopodium absent. *Schizocarp* 1.0–1.5 mm, ovoid, with prominent lateral ridges. *Flowers* 6–7.

Introduced. Persistent as a weed in paving slabs at Highfields School, Wolverhampton in Staffordshire since about 1970. A widespread tropical weed.

4. H. moschata G. Forst. Hairy Pennywort

Perennial monoecious *herb* forming patches up to 30 cm in diameter. *Stems* up to 30 cm, slender, glabrous or hairy. *Leaves* with lamina 5–20(–25) mm in diameter, usually dull green, often with a reddish margin, subrotund in outline, divided one-quarter to one-third (to one half) of the radius into 5–7 segments, the segments broadly obovate, each again usually 3-lobed and serrate, the teeth with blunt or sharp, cartilaginous points, the margins flat, with a deep basal sinus, moderately to densely hairy on upper surface, glabrous or hairy beneath; petiole more or less glabrous or moderately to densely hairy. *Inflorescence* a simple umbel with numerous, subsessile flowers. *Sepals* minute or absent. *Petals* 5, greenish-white. *Stamens* 5; filaments pale; anthers pink. *Styles* 2, curved; stigma minute; stylopodium absent. *Schizocarp* 1.3–1.6 × 1.0–1.5 mm, reddish-brown, crowded, acute on dorsal edge; mericarps with raised, slender ribs; vittae absent. *Flowers* 6–7. $2n = 48$.

Introduced. In lawns and on grassy banks. Well naturalised on Valencia Island, Co. Kerry, less so in Cornwall, Sussex and Ayrshire. Native of New Zealand. Like *Sibthorpia europaea* in general appearance.

5. H. novae-zeelandiae DC. New Zealand Pennywort
H. microphylla auct.

Perennial monoecious *herb* forming large patches. *Stems* stout or slender. *Leaves* 5–35 mm in diameter, dull green, greenish-brown or bright glossy green, reniform in outline, divided one-fifth to one-third of the radius into 5–7 segments, triangular to broadly oblong-obovate, each again sometimes obscurely 3-lobed and crenate, the teeth rounded and sometimes with short hair-tips, the margins flat, cordate at base with an open sinus, sparsely to densely hairy on upper surface and beneath; petiole with a few retrorse hairs near the lamina to densely hairy. *Inflorescence* axillary, a simple umbel with 5–many flowers, the flowers subsessile or shortly pedicelled. *Sepals* minute or absent. *Petals* 5, greenish-white. *Stamens* 5; filaments pale; anthers pink. *Styles* 2, curved; stigma minute; stylopodium absent. *Schizocarp* (1.8–)2.0–3.5 × 1.8–2.5 mm, usually rounded on dorsal edge; mericarps with ribs usually slightly engraved into the surface, rarely slightly raised; vittae absent. *Flower* 6–7. $2n = 48$.

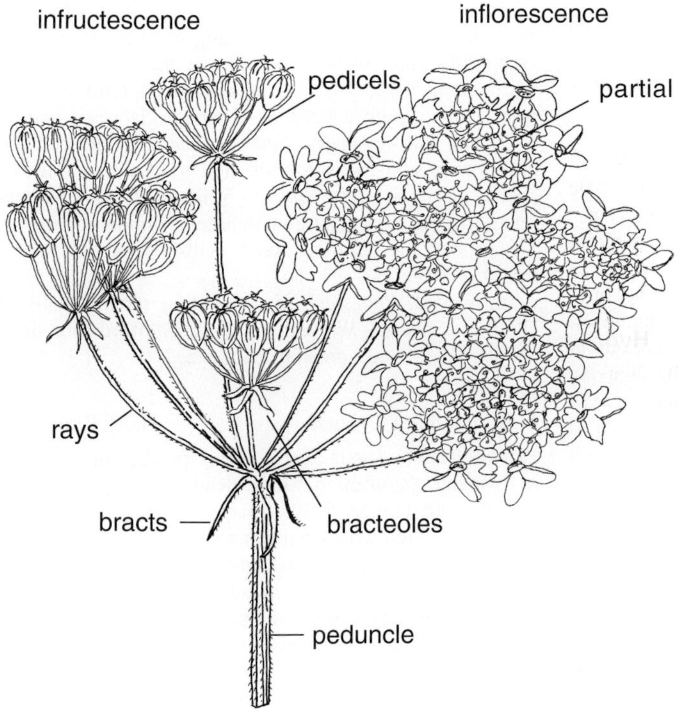

Heracleum sphondylium L.

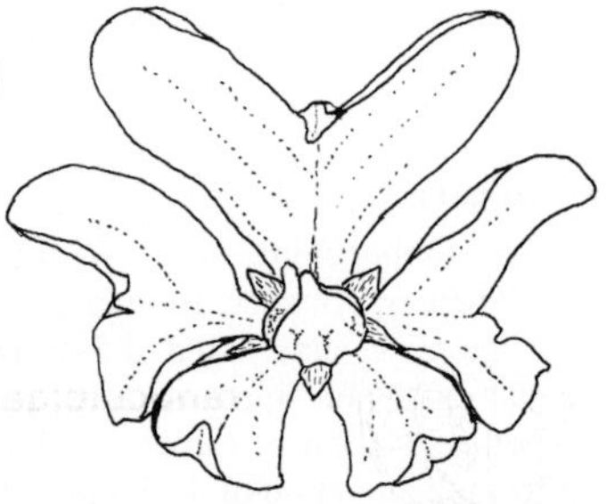

outer flowers

inner flowers

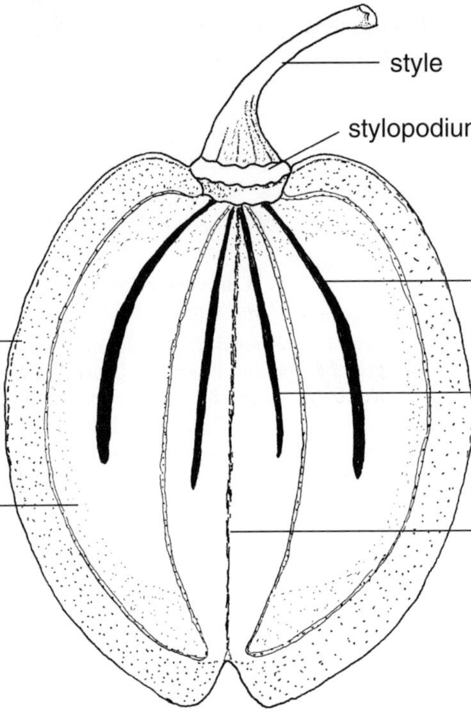

schizocarp

Structure of Apiaceae

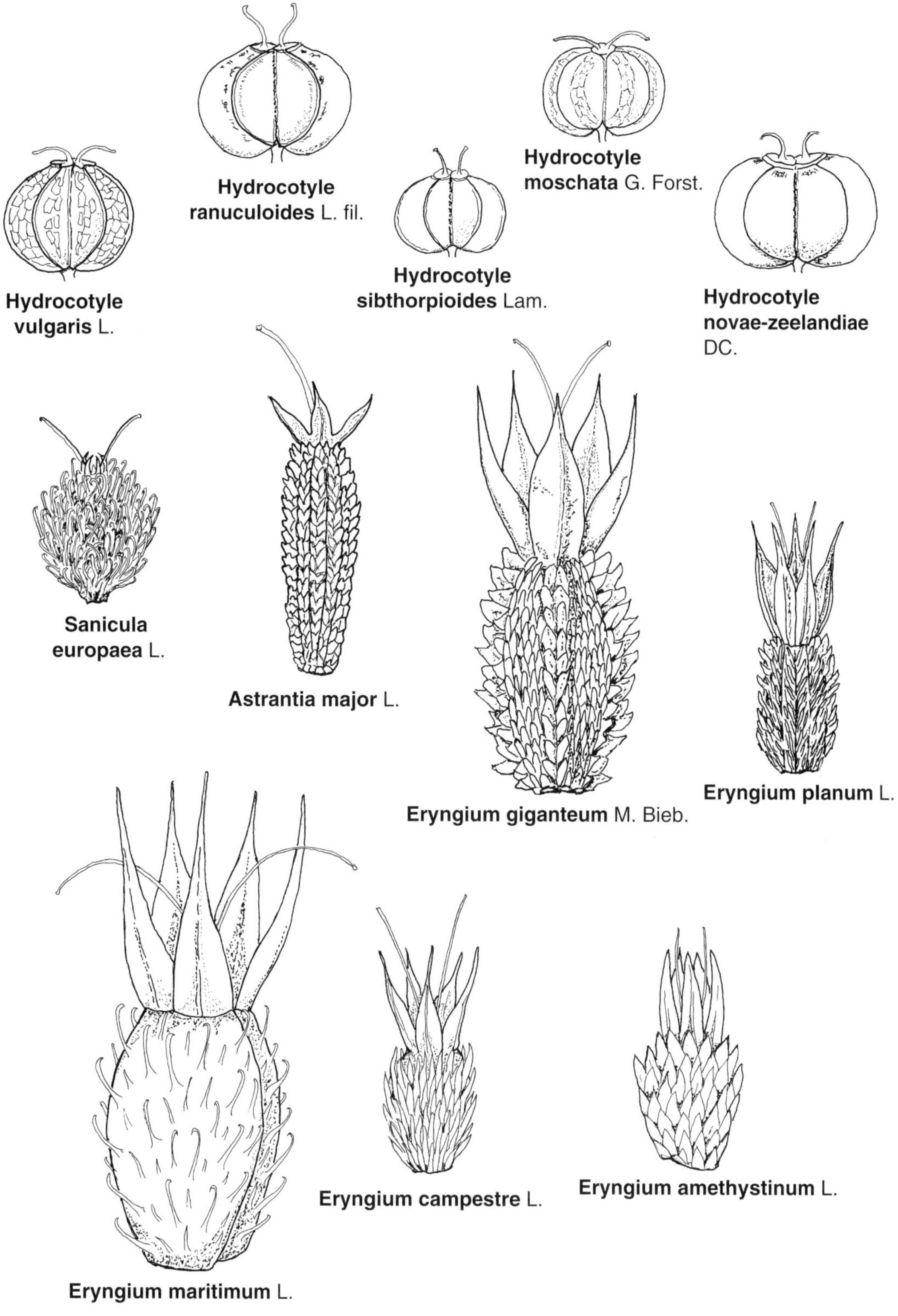
Hydrocotyle vulgaris L.
Hydrocotyle ranuculoides L. fil.
Hydrocotyle sibthorpioides Lam.
Hydrocotyle moschata G. Forst.
Hydrocotyle novae-zeelandiae DC.
Sanicula europaea L.
Astrantia major L.
Eryngium giganteum M. Bieb.
Eryngium planum L.
Eryngium maritimum L.
Eryngium campestre L.
Eryngium amethystinum L.

Schizocarps of **Apiaceae.**

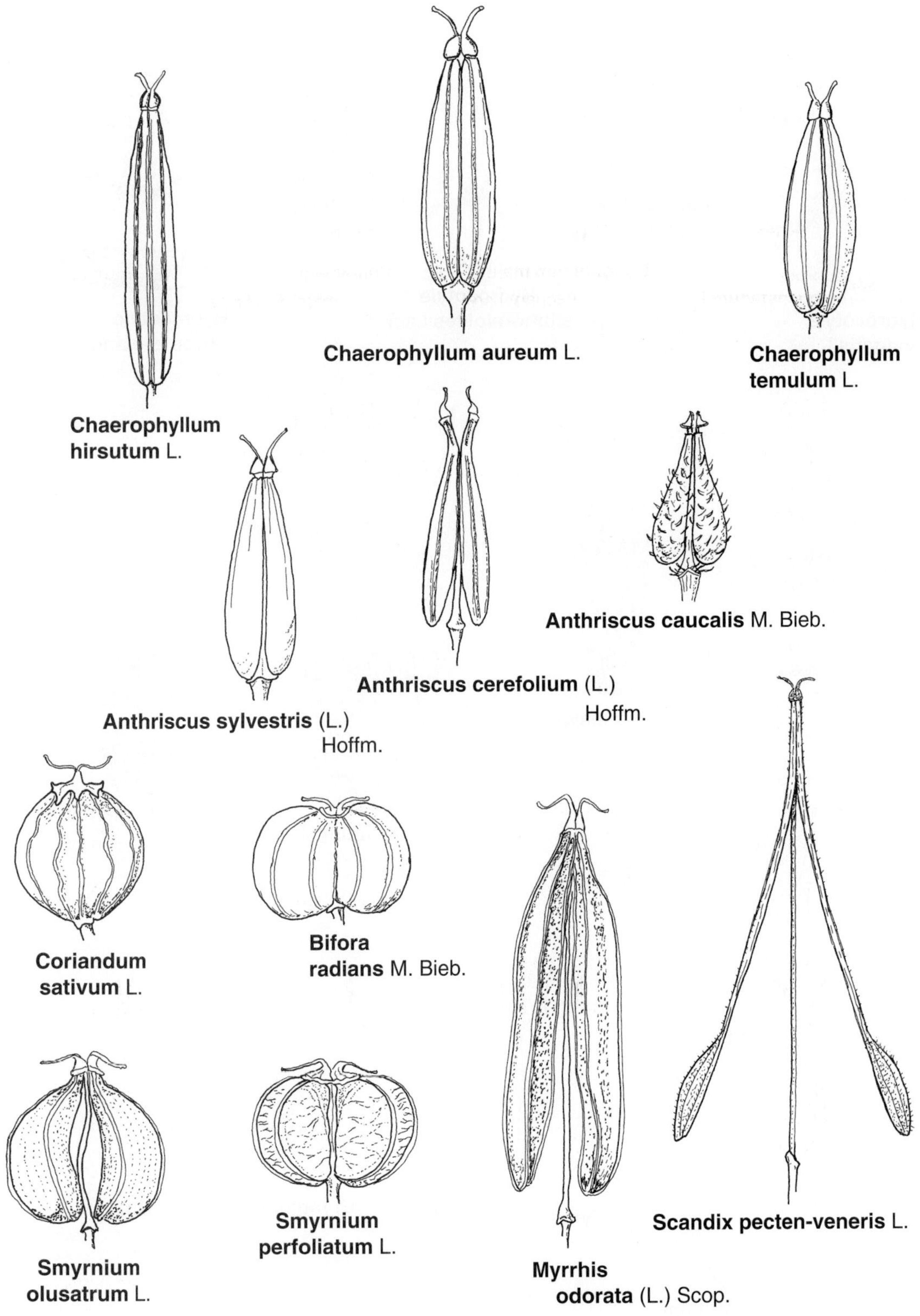
Chaerophyllum aureum L.
Chaerophyllum temulum L.
Chaerophyllum hirsutum L.
Anthriscus caucalis M. Bieb.
Anthriscus cerefolium (L.) Hoffm.
Anthriscus sylvestris (L.) Hoffm.
Coriandum sativum L.
Bifora radians M. Bieb.
Smyrnium olusatrum L.
Smyrnium perfoliatum L.
Myrrhis odorata (L.) Scop.
Scandix pecten-veneris L.

Schizocarps of **Apiaceae**

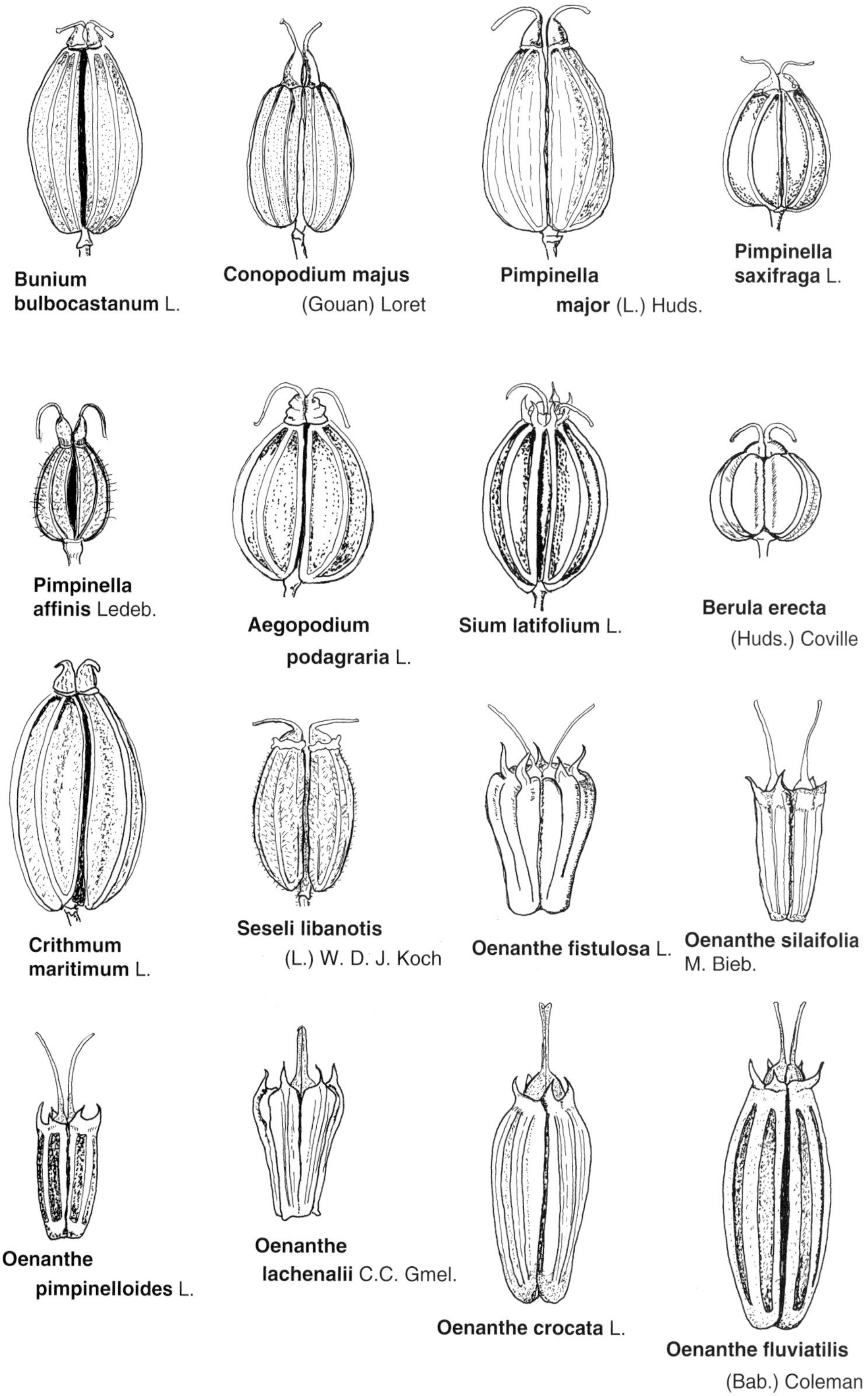

Bunium bulbocastanum L.
Conopodium majus (Gouan) Loret
Pimpinella major (L.) Huds.
Pimpinella saxifraga L.
Pimpinella affinis Ledeb.
Aegopodium podagraria L.
Sium latifolium L.
Berula erecta (Huds.) Coville
Crithmum maritimum L.
Seseli libanotis (L.) W. D. J. Koch
Oenanthe fistulosa L.
Oenanthe silaifolia M. Bieb.
Oenanthe pimpinelloides L.
Oenanthe lachenalii C.C. Gmel.
Oenanthe crocata L.
Oenanthe fluviatilis (Bab.) Coleman

Schizocarps of **Apiaceae**

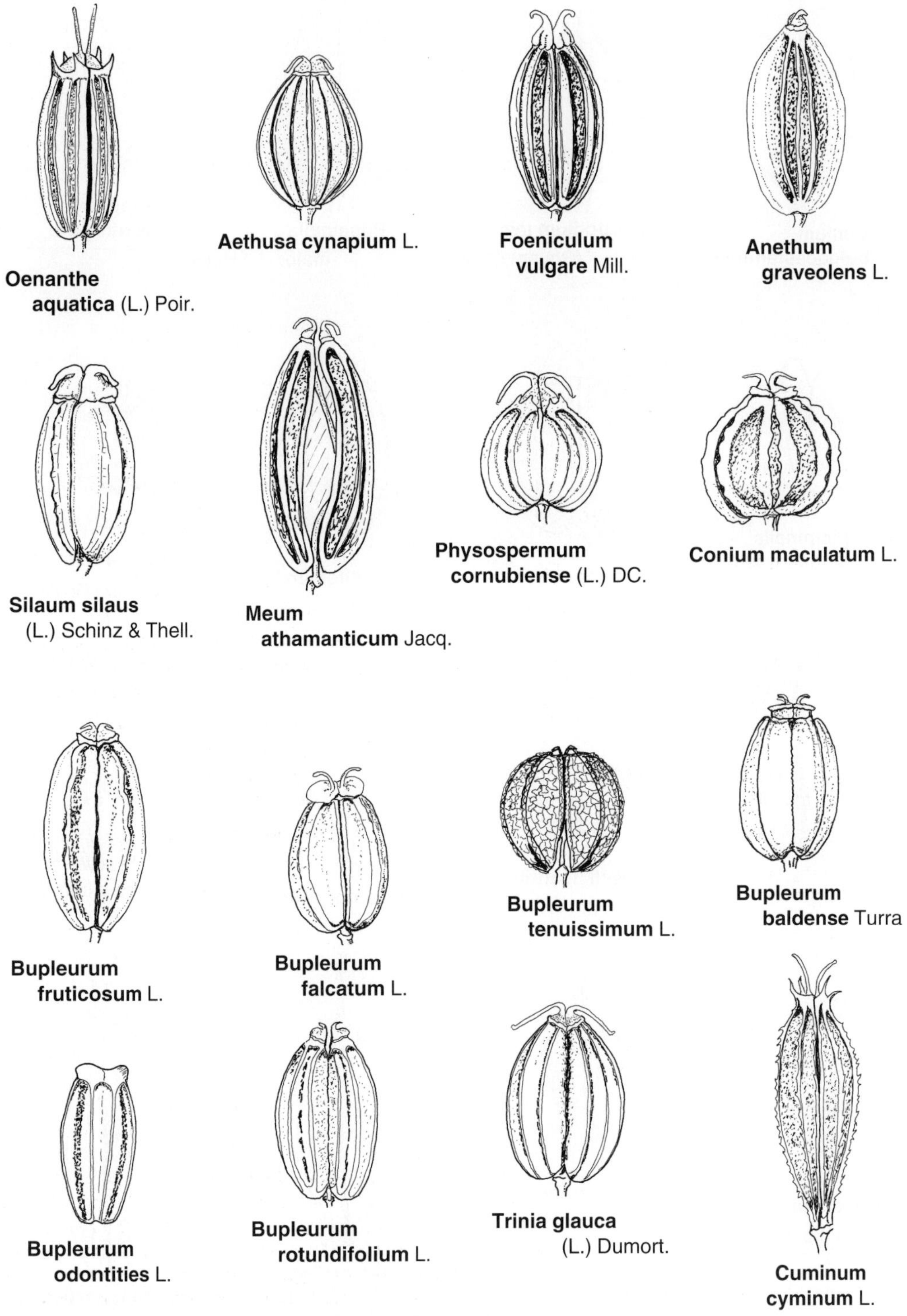

Oenanthe aquatica (L.) Poir.
Aethusa cynapium L.
Foeniculum vulgare Mill.
Anethum graveolens L.
Silaum silaus (L.) Schinz & Thell.
Meum athamanticum Jacq.
Physospermum cornubiense (L.) DC.
Conium maculatum L.
Bupleurum fruticosum L.
Bupleurum falcatum L.
Bupleurum tenuissimum L.
Bupleurum baldense Turra
Bupleurum odontities L.
Bupleurum rotundifolium L.
Trinia glauca (L.) Dumort.
Cuminum cyminum L.

Schizocarps of **Apiaceae**

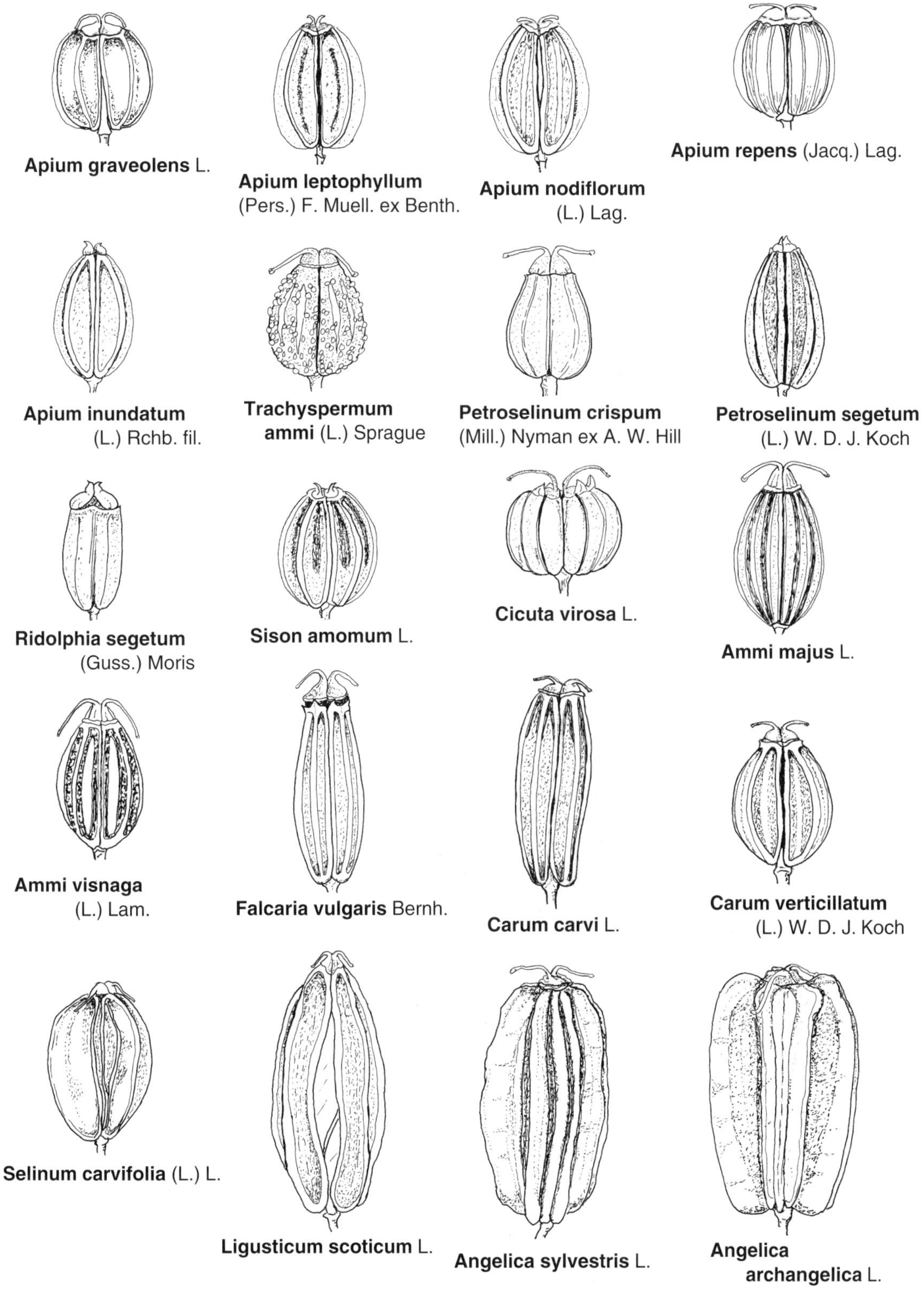
Apium graveolens L.
Apium leptophyllum (Pers.) F. Muell. ex Benth.
Apium nodiflorum (L.) Lag.
Apium repens (Jacq.) Lag.
Apium inundatum (L.) Rchb. fil.
Trachyspermum ammi (L.) Sprague
Petroselinum crispum (Mill.) Nyman ex A. W. Hill
Petroselinum segetum (L.) W. D. J. Koch
Ridolphia segetum (Guss.) Moris
Sison amomum L.
Cicuta virosa L.
Ammi majus L.
Ammi visnaga (L.) Lam.
Falcaria vulgaris Bernh.
Carum carvi L.
Carum verticillatum (L.) W. D. J. Koch
Selinum carvifolia (L.) L.
Ligusticum scoticum L.
Angelica sylvestris L.
Angelica archangelica L.

Schizocarps of **Apiaceae**

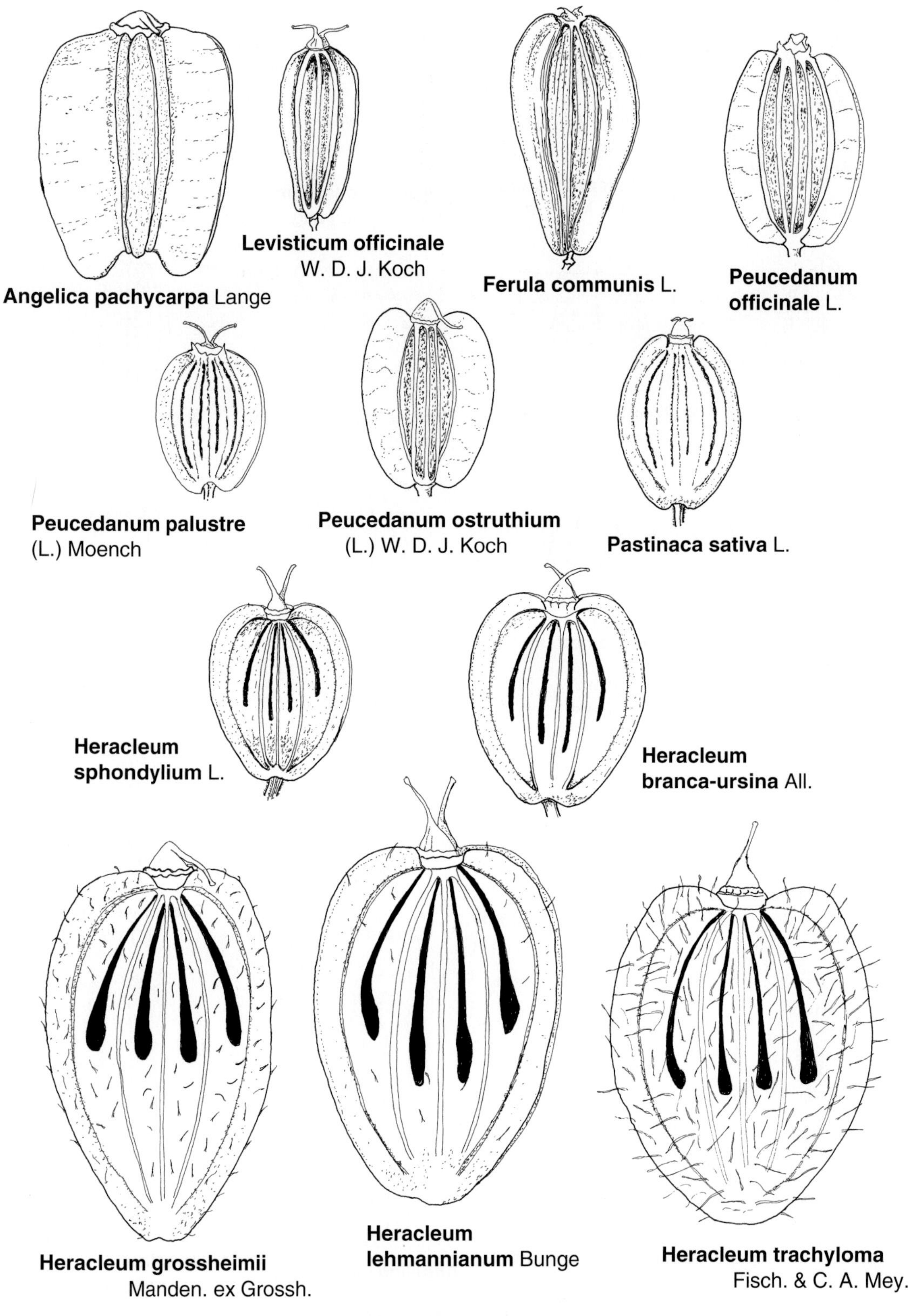
Angelica pachycarpa Lange
Levisticum officinale W. D. J. Koch
Ferula communis L.
Peucedanum officinale L.
Peucedanum palustre (L.) Moench
Peucedanum ostruthium (L.) W. D. J. Koch
Pastinaca sativa L.
Heracleum sphondylium L.
Heracleum branca-ursina All.
Heracleum grossheimii Manden. ex Grossh.
Heracleum lehmannianum Bunge
Heracleum trachyloma Fisch. & C. A. Mey.

Schizocarps of **Apiaceae**

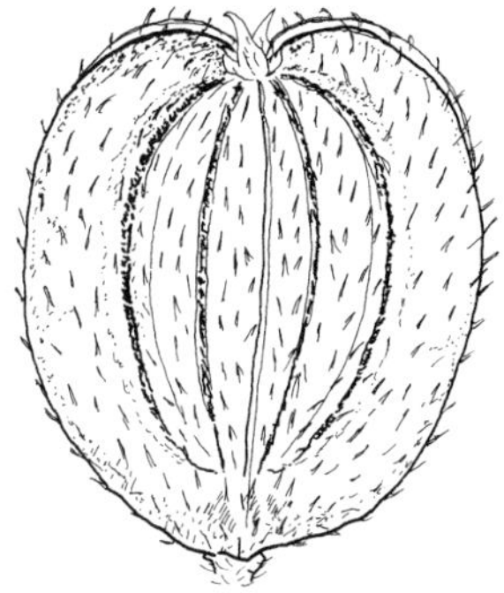

Tordylium maximum L.

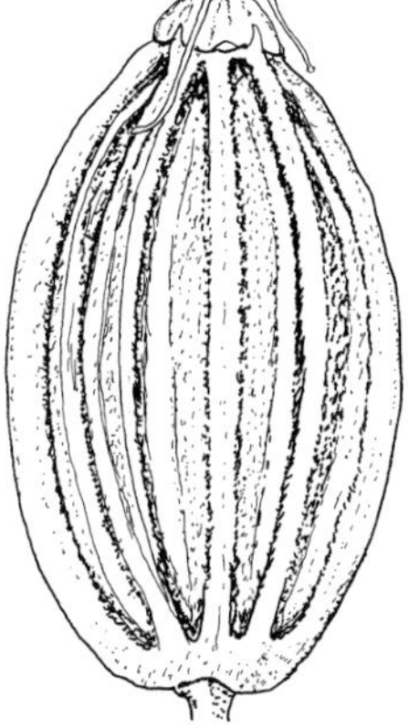

Laser trilobum (L.) Borkh. ex P. Gaertn., B. Mey. & Scherb.

Torilis japonica (Houtt.) DC.

Torilis ucrania Spreng.

Torilis arvensis (Huds.) Link

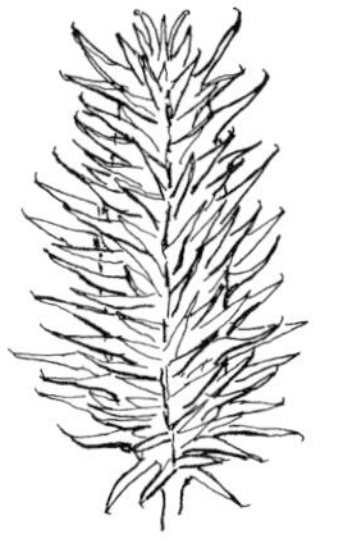

Torilis leptophylla (L.) Rchb. fil.

Torilis nodosa (L.) Gaertn.

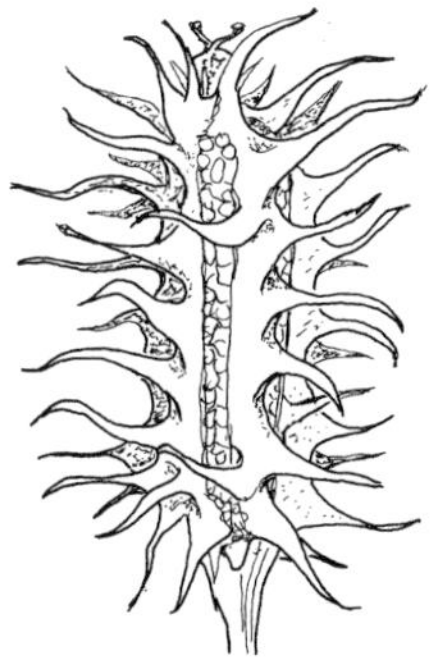

Caucalis platycarpos L.

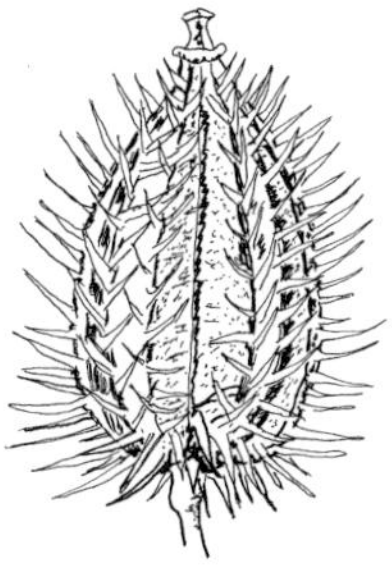

Turgenia latifolia (L.) Hoffm.

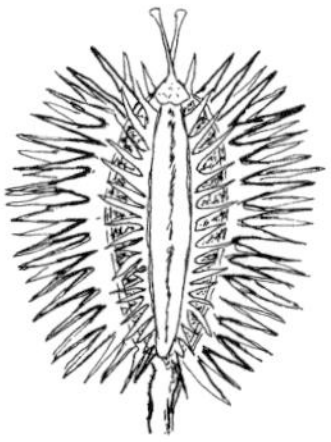

Daucus carota L.

Daucus carota subsp. **sativus**

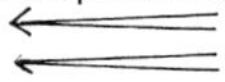

Daucus carota subsp. **carota**

Daucus carota subsp. **major**

Daucus carota subsp. **gummifer**

Daucus glochidiatus

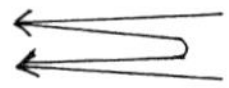

Daucus muricatus

Schizocarps of **Apiaceae**

Introduced. On lawns and golf courses, naturalised in the turf. Cornwall and Forfarshire. Native of New Zealand. All records are referable to var. **montana** Kirk.

Trachemene cyanopetala (F. Muell.) Benth. and **T. pilosa** Sm. have been recorded as wool casuals and **Bowlesia incana** Ruiz & Pav. as a wool and bird-seed casual.

Subfamily **2. Saniculoideae** Drude

Leaves mostly simple, palmately or ternately lobed, rarely pinnately lobed and then spiny, without stipules. *Flowers* in simple umbels or heads. *Schizocarp* with soft inner walls, without a carpophore, with vittae. *Basic chromosome number* 7 or 8.

2. Sanicula L.

Perennial monoecious *herbs*. *Stems* erect. *Leaves* mostly in a basal rosette, palmately lobed almost to base, long petiolate; without stipules. *Flowers* in small, loosely aggregated umbels. *Fruit* a schizocarp, subterete, with rigid, forward-pointing, hooked bristles; commissure broad; carpophore absent; vittae absent.

Contains about 40 species in temperate regions and on tropical mountains; absent from Australia.

Grieve, M. (1931). *A modern herbal*. London.
Grime, J. P. et al. (1988). *Comparative plant ecology*. London.
Hultén, E. & Fries, M. (1986). *Atlas of north European vascular plants north of the Tropic of Cancer*. 3 vols. Königstein.
Ingre, O. & Tamm, C. O. (1985). Survival and flowering of perennial herbs. 4. The behaviour of *Hepatica nobilis* and *Sanicula europaea* on permanent plots during 1943–1981. *Oikos* **45**: 400–420.

1. S. europaea L. Sanicle
Caucalis capitata Stokes nom. illegit.

Perennial monoecious *herb* with a stout stock. *Stems* 20–60 cm, pale yellowish-green, often tinted reddish, erect, glabrous. *Leaves* shiny dark green on upper surface, paler beneath; basal numerous, 2–6 cm in diameter, subrotund or reniform in outline, deeply, palmately 5-lobed, the lobes obovate, shallowly 3-lobulate at apex, sharply serrate, with the teeth ending in a bristle up to 0.5 mm, and long cuneate below, the petioles 5–25 cm, glabrous; cauline few or absent, similar to basal but smaller, shortly petiolate or sessile; all glabrous; stipules absent. *Inflorescence* of a number of simple umbels about 5 mm across arranged in an irregular cyme and often giving the appearance of a compound umbel; bracts simple, 3-fid or sometimes leaf-like; bracteoles shorter than the flowers, simple or 3-fid. *Male flowers* pedicellate, bisexual flowers 3–6, sessile. *Sepals* 5, conspicuous, longer than the petals. *Petals* 5, white or pinkish. *Stamens* 5; filaments white; anthers pink. *Styles* 2, about 3 mm, divergent to somewhat recurved, not thickened at base; stigma capitate. *Ovary* with a more or less flat disc at the apex. *Schizocarp* about 3 mm, subterete, covered with rigid, forward-pointing, hooked bristles; commissure broad; carpophore absent; vittae present or absent. *Flowers* 5–8. Pollinated by small flies and beetles; self-pollination is also possible. $2n=16$.

Native. In woods, forming societies in chalk beechwoods and oakwoods on loams. Throughout most of Great Britain and Ireland. Wooded regions of Europe, south, central and east Asia and North Africa, only in the mountains in the Mediterranean region; mountains of Tropical Africa; South Africa. European Temperate element. Used as a vulnerary by herbalists.

Hacquetia epipactis (Scop.) DC. is a long-persistent garden relic near Landford in Wiltshire.

3. Astrantia L.

Perennial monoecious *herbs*. *Stems* erect. *Leaves* mostly in a basal rosette, ternately or deeply palmately lobed almost to base; petioles long; without stipules. *Flowers* in simple umbels. *Fruit* a schizocarp, oblong, scarcely compressed, covered with bidentate, vascular scales; commissure broad; carpophore absent; vittae solitary.

Contains about 10 species in Europe and western Asia.

1. Bracteoles up to 30 mm, with inconspicuous cross-veins **2. maxima**
1. Bracteoles up to 22 mm, with conspicuous cross-veins 2.
2. Bracteoles 5-veined; schizocarps strongly papillose **1(c). major** subsp. **elatior**
2. Bracteoles 3-veined; schizocarps not papillose 3.
3. Bracteoles 6–15 mm, more or less equalling the umbel **1(a). major** subsp. **major**
3. Bracteoles 15–22 mm, up to twice as long as umbel **1(b). major** subsp. **carinthiaca**

1. A. major L. Astrantia

Densely tufted, *perennial* monoecious *herb*, with a rhizome and fibrous roots. *Stems* 30–100 cm, pale yellowish-green, sometimes tinted reddish, erect, robust, glabrous, usually branched, with the uppermost branches in a whorl of 3 or more. *Leaves* medium yellowish-green on upper surface, paler beneath; basal few, with lamina 6–20 cm in diameter, broadly ovate or reniform in outline, ternately or deeply palmately divided, the segments 3–5(–7), oblong to obovate, obtuse at apex, irregularly toothed and often somewhat lobed, the middle segment free for at least two-thirds of its length, the petioles long and glabrous; cauline usually few, smaller but similar to basal, the lower shortly petiolate, the upper sessile; all glabrous. *Inflorescence* of simple umbels up to 50 mm in diameter, the terminal of bisexual flowers with a few male, the lateral with mostly male flowers; bracteoles 6–30 mm, as long as or much longer than the flowers, erect, lanceolate, acute at apex, pink or white, 3- or 5-veined, strongly reticulately cross-veined or not. *Male flowers* with pedicels 4–10 mm, those of bisexual flowers 2–5 mm. *Sepals* 5, conspicuous, triangular, acuminate at apex, longer than petals. *Petals* 5, whitish or pinkish, inflexed, emarginate. *Stamens* 5; filaments white; anthers pink. *Styles* 2, 4–5 mm, divergent or somewhat recurved, not thickened at base; stigma capitate. *Ovary* with a more or less flat disc at the apex. *Schizocarp* 3–8 mm, about 3 times as long as wide, oblong, scarcely compressed, sometimes papillose, covered with bifid, vesicular scales arranged in 5 longitudinal rows on each mericarp; commissure broad; carpophore absent; vittae solitary. *Flowers* 5–7. Pollinated by various insects especially beetles. $2n=14$, 28.

(a) Subsp. **major**

Bracteoles 6–15 mm, more or less equalling the umbels, 3-veined. *Schizocarp* not papillose.

(b) Subsp. **carinthiaca** (Hoppe ex W. D. J. Koch) Arcang.

A. carinthiaca Hoppe ex W. D. J. Koch; *A. major* var. *involucrata* W. D. J. Koch; *A. pallida* C. Presl

Bracteoles 15–22 mm, twice as long as the umbels, at least in the terminal umbel, 3-veined. *Schizocarp* not papillose.

(c) Subsp. **elatior** (Friv.) K. Maly

A. elatior Friv.; *A. major* var. *elatior* (Friv.) Murb.; *A. carinthiaca* Wettst.

Bracteoles 5-veined. *Schizocarp* strongly papillose.

Introduced. Garden escape naturalised in grassy and shady places. Scattered records throughout Great Britain. Central and east Europe extending to White Russia. Subspp. *major* and *carinthiaca* occur more or less equally. Subsp. *elatior* is grown in gardens but has not been seen as an escape. The plant originally found at Stokesay Castle in Shropshire in 1841 is subsp. *major*. Most, if not all, recent collections from that area are from a different locality and are subsp. *carinthiaca*. These two subspecies are very different and may be better regarded as distinct species.

2. A. maxima Pall. Caucasian Astrantia

Perennial monoecious *herb. Stems* up to 90 cm, pale yellowish-green, terete, ridged, hollow, glabrous. *Leaves* medium yellowish-green on upper surface, paler beneath; basal with lamina 6–20 cm in diameter, broadly ovate or reniform in outline, ternately or palmately divided, the segments 3–5, elliptic-ovate, unequally serrate or crenate with apiculate teeth; petiole much longer than lamina; cauline smaller and sessile. *Inflorescence* of simple umbels, the terminal of bisexual flowers with a few male, the lateral with mostly male flowers; bracteoles 9–12, 10–30 × 3–12 mm, white or pinkish, elliptical to ovate, acute to acuminate at apex, serrulate above, more or less connate below, cross-veins inconspicuous; pedicels 2–16 mm. *Sepals* 5, 2–3 mm, linear-setaceous, persistent. *Petals* 5, 1.5–2.0 mm, white or pinkish. *Stamens* 5; filaments white; anthers pink. *Styles* 2; not thickened at base; stigma capitate. *Schizocarp* 5–6 × 2.0–2.5 mm, oblong-cylindrical, slightly dorsally compressed, covered with vesicular scales; commissure broad; carpophore absent; vittae solitary. *Flowers* 5–7. $2n=14$.

Introduced. This species is grown in gardens, and is said to escape, but we have seen no examples. Native of Caucasia and north-west Iran.

4. Eryngium L.

Perennial monoecious *herbs. Stems* erect. *Leaves* mostly glaucous, at least the upper spiny, the basal petiolate, the upper cauline sessile; without stipules. *Flowers* sessile in globose to ovoid heads with leaf-like spiny bracts at the base; bracteoles entire or tricuspidate. *Fruits* a schizocarp, obovoid, scarcely compressed, densely scaly; commissure broad, carpophore absent; vittae very slender.

Contains about 230 species in temperate and subtropical regions of the world, especially South America; absent from South Africa. The flowers have nectar and are pollinated by various insects.

Hultén, E. & Fries, M. (1986). *Atlas of north European vascular plants north of the Tropic of Cancer*. 3 vols. Königstein.
Wigginton, M. J. (Edit.) (1999). *British red data books*. Vol. 1. *Vascular plants*. Peterborough.
Wolff, H. (1913) *Eryngium*. In Engler, A. (Edit.) *Das Pflanzenreich* **IV.228.**

E. tripartitum Desf. formerly occurred as a garden escape on a roadside at Headley Mill in Hampshire and was recorded from waste ground at Sandwich Bay in Kent.

1. Basal leaves and lower cauline leaves not lobed or lobed less than halfway to midrib 2.
1. Basal leaves and lower cauline leaves pinnately or ternately divided or lobed almost to midrib 4.
2. Upper cauline leaves and bracts palmate or palmately lobed almost to base **2. planum**
2. Upper cauline leaves toothed or lobed less than halfway to base 3.
3. Basal leaves and at least lower cauline leaves shallowly toothed, not spiny **1. giganteum**
3. Basal and cauline leaves with deep, strongly spiny teeth **3. maritimum**
4. Basal leaves with petiole unwinged, about as long as its lamina; plant not blue **4. campestre**
4. Basal leaves with petiole broadly winged, much shorter than its lamina; plant strongly blue-tinged **5. amethystinum**

1. E. giganteum M. Bieb. Tall Eryngo

E. asperifolium Delarbre; *E. haussknechtii* Bornm.

Perennial monoecious *herb* with a thick, napiform root. *Stem* up to 2 m and 2 cm thick, pale green, glabrous, branched, leafy. *Leaves* alternate, coriaceous, glaucous; basal with the lamina up to 15 × 10 cm, ovate, acute at apex, crenate-serrate or crenate-dentate, cordate at base, glabrous; cauline variable, the lower more or less subrotund, spinulose-serrate, palmately 7–9(–13) veined and long-petiolate, the upper more or less deeply lobed, with long, rigid, spinose teeth and sessile, or with a short, sheathing petiole. *Inflorescence* of pedunculate, cylindrical or ovoid-cylindrical heads up to 10 × 4 cm, arranged in a broad open panicle; bracts up to 5 × 5 cm, incise-lobate with long spines; bracteoles up to 10 × 4(8) cm, lanceolate or obovate. *Sepals* 5, 3–4 mm, ovate or ovate-lanceolate, with long spines. *Petals* 5, about 3 × 1 mm, obovate-lanceolate, slightly emarginate. *Stamens* 5; filaments pale; anthers yellow. *Styles* 2, pale blue; stigma cream. *Ovary* with a more or less flat disc at apex. *Schizocarp* up to 10 × 4 mm, ellipsoid, densely scaly; mericarps slightly ridged; carpophore absent; vittae absent. *Flowers* 8–9. $2n=16$.

Introduced. Grown in gardens for ornament and well naturalised on waste ground near Otley in Yorkshire since 1986 and recorded from a number of other scattered localities in Great Britain. Native of the Caucasus.

2. E. planum L. Blue Eryngo

Perennial monoecious *herb,* with a tap-root. *Stem* 25–180 cm, pale green, sometimes suffused with purple, erect, striate, smooth, glabrous, branched, leafy. *Leaves* slightly coriaceous, medium to dark green or glaucous on upper surface, paler beneath; basal persistent, numerous; the lamina 5–22×3–11 cm, broadly oblong or oblong-ovate, rounded at apex, crenate-dentate, the teeth broadly and shallowly mammiform, and cordate at base, the petioles unwinged, pale green, channelled on upper surface, rounded beneath and glabrous; cauline gradually getting smaller up the stem, the lower ovate, rounded at apex, crenate-dentate and sessile, the upper ovate in outline, divided almost to base into 5 linear or lanceolate, acute, spiny-toothed lobes and sessile; all glabrous. *Inflorescence* usually bluish, of pedunculate, globose to ovoid heads 10–20×10–15 mm, arranged in a broad, subcorymbose panicle; bracts 6–7, 15–25 mm, longer than the head, dark greyish-green, linear or linear-lanceolate, with 1–4 pairs of spinose teeth; bracteoles entire or 3-cuspidate. *Sepals* 5, about 2 mm, ovate-lanceolate, aristate. *Petals* 5, blue, emarginate. *Stamens* 5; filaments pale blue; anthers yellow. *Styles* 2, pale blue, exceeding anthers; stigma cream. *Ovary* with a more or less flat disc at apex. *Schizocarps* 5–6 mm, subglobose, more or less densely covered with overlapping scales; mericarps plano-convex, slightly ridged; carpophore absent; vittae usually slender. *Flowers* 8–9. $2n=16$.

Introduced. Grown for ornament and sometimes naturalised in waste places. Scattered localities over Great Britain, notably on sandy ground at Littlestone in Kent, and in Lanarkshire. Native of central and south-east Europe and south-west Asia.

3. E. maritimum L. Sea Holly

Perennial monoecious *herb*, perhaps sometimes monocarpic, with strong tap-root. *Stems* 15–60 cm, pale below but usually bluish above, erect, robust, glabrous, rigid, solid and branched. *Leaves* coriaceous, intensely glaucous; basal several, 4–10×5–15 cm, subrotund in outline, 3(–5)-lobed, the lobes with large spinose teeth, truncate or cordate at base, with thick, prominent veins and a thick, cartilaginous margin, the petiole about as long as the lamina with a broad, swollen base; cauline smaller but similar and sessile; all glabrous. *Inflorescence* of pedunculate, subglobose heads 15–30 mm in diameter; bracts about as long as the head, ovate or ovate-lanceolate, similar to the leaves in texture, spininess and colour; bracteoles tricuspidate, spiny, usually purplish-blue, longer than the flowers. *Sepals* 5, 4–5 mm, longer than the petals, bluish, lanceolate, with a prominent midrib, which is excurrent as a stout spine. *Petals* 5, bluish-white, oblong, emarginate at apex. *Stamens* 5; filaments bluish; anthers yellow. *Styles* 2, about 6 mm, divergent or somewhat recurved, slightly thickened at the base; stigma tapering. *Ovary* with a more or less flat disc at apex. *Schizocarp* 13–15 mm, obovoid, scarcely compressed, covered in papillae which become longer towards the apex; commissure broad; carpophore absent; vittae very slender. *Flowers* 6–9. $2n=16$.

Native. Sand-dunes and less frequently shingle a little above high-tide mark. Formerly around all the coasts of Great Britain and Ireland, but now gone from most of Scotland and north-east England before 1930. Europe north to about 60° N, and in North Africa and south-east Asia. European Southern-temperate element.

4. E. campestre L. Field Eryngo
E. vulgare Lam.

Perennial monoecious *herb* with a tap-root. *Stems* 30–60 cm, pale greyish-green, erect, rigid, solid, glabrous, branched, leafy. *Leaves* coriaceous, greyish-glaucous on upper surface, paler beneath; basal 5–20×5–15 cm, triangular-ovate in outline, acute at apex, pinnately divided almost to midrib, the primary divisions decurrent on the rhachis, the lobes oblong, spinose-serrate, with a thick, prominent midrib ending in a stout spine and with a thick, cartilaginous margin, the petiole unwinged and about as long as the lamina; cauline progressively smaller upwards and less divided, with a broad, spiny-margined, sessile, semi-amplexicaul base. *Inflorescence* of numerous, pedunculate, ovoid heads 10–15 mm in diameter; bracts 5–8, 1.5–3.0 times as long as the heads, linear-lanceolate, spinose at apex, entire or with 1(–2) pair of lateral spines; bracteoles entire, longer than the flowers. *Sepals* 5, about 2.5 mm, longer than the petals, lanceolate, with a prominent midrib which is excurrent as a stout spine. *Petals* 5, white, emarginate at apex. *Stamens* 5; filaments white; anthers pale yellow. *Styles* 2, about 4 mm, divergent to somewhat recurved, slightly thickened at the base; stigma tapering. *Ovary* with a more or less flat disc at the apex. *Schizocarp* about 5 mm, obovate, scarcely compressed, densely covered in very acute, white scales; commissure broad; carpophore absent; vittae very slender. *Flowers* 7–8. $2n=14$, 28.

Long introduced. Open habits, mostly near the sea, especially calcareous. First recorded by John Ray in Devonshire in 1662. Now established in a few places in south and south-west Great Britain, casual elsewhere in Great Britain and the Channel Islands north to Northumberland. Rare in south and central Europe, North Africa and south-west Asia. European Southern-temperate element.

5. E. amethystinum L. Italian Eryngo

Perennial monoecious *herb* with a tap-root. *Stems* 20–45 cm, pale bluish-grey, erect, rigid, solid, striate, glabrous, branched above, leafy. *Leaves* dark green with paler central area of segments on upper surface, hardly paler beneath, coriaceous; basal usually persistent, the lamina 10–15 cm, obovate in outline, palmatisect above and pinnatisect below, the segments 2- or 3-pinnatisect with linear-lanceolate and spinous-serrate segments, the petiole broadly winged; cauline gradually smaller and less divided upwards, the petioles winged and amplexicaul; all glabrous. *Inflorescence* usually bluish, of numerous, ovoid heads 18–22 mm in diameter, arranged in further umbels at the ends of branches; bracts 2–5 cm, much longer than the flowers, linear, with a spiny tip and 1–4 pairs of spines; bracteoles entire or 3-cuspidate. *Sepals* 5, 1.5–2.5 mm, ovate-lanceolate, shortly aristate. *Petals* 5, blue, emarginate at apex. *Stamens* 5; filaments blue; anthers pale blue. *Styles* 2, slightly thickened at base; stigma tapering. *Ovary* with a more or less flat disc at the apex. *Schizocarp* about 5 mm, obovate, scarcely compressed, sparsely covered in white scales; commissure broad; carpophore absent; vittae very slender. *Flowers* 7–8. $2n=14$, 16.

Introduced. Grown for ornament. Naturalised on dunes in Caernarvonshire since 1963; a rare relic of gardens elsewhere. Native of the Balkan Peninsula and Aegean region, Italy and Sicily.

Subfamily 3. Apioideae

Leaves various, often much divided, not spiny, without stipules. *Flowers* usually in compound umbels. *Fruit* a schizocarp with soft inner wall, carpophore and usually oil bodies. *Chromosome base number* various, often 11.

5. Chaerophyllum L.

Biennial or *perennial* monoecious *herbs*. *Stems* erect, solid. *Leaves* 2- to 3-pinnate. *Inflorescence* of compound umbels; bracts absent, or rarely 1–3. *Sepals* absent or minute. *Petals* 5, white or pink, the outer not radiating. *Fruit* a schizocarp, oblong, slightly compressed laterally; mericarps with broad, rounded ridges; constricted at commissure; carpophore present; vittae solitary.

Contains about 40 species in north temperate regions of the world.

C. aromaticum L. has occurred as a garden escape in Cambridge.

Clapham, A. R. (1953). Human factors contributing to a change in our flora: the former ecological status of certain hedgerow species. In Lousley, J. E. (Edit.) *The changing flora of Britain*, pp. 26–39. Arbroath. [*C. temulentum*.]

Hultén, E. & Fries, M. (1986). *Atlas of north European vascular plants north of the Tropic of Cancer.* 3 vols. Königstein.

1. Petals usually pinkish, minutely hairy at margin; styles suberect, forming an angle of less than 45° **1. hirsutum**
1. Petals white, glabrous; styles more or less recurved forming an angle more than 45° 2.
2. Leaf segments acute; schizocarps 8–12 mm **2. aureum**
2. Leaf segments obtuse or abruptly contracted to an acute apex; schizocarps 4.0–6.5 mm **3. temulum**
3. Stem with spreading, long eglandular hairs and sometimes shorter ones intermixed, but never dense **3(i). temulum** var. **temulum**
3. Stems at least in lower half densely short eglandular hairy **3(i). temulum** var. **canescens**

1. C. hirsutum L. Hairy Chervil

Perennial monoecious *herb* with a stout root. *Stem* up to 120 cm, pale yellowish-green, often suffused brownish-purple, erect, robust, solid, with numerous, rather rough, reflexed simple eglandular hairs or dense, short eglandular hairs. *Leaves* pale to medium yellowish-green on upper surface, paler beneath; basal; 4–20 × 5–25 cm, broadly ovate in outline, 2- to 3-pinnate, the segments triangular-ovate and acute at apex, the lobes 10–70, lanceolate, acute at apex and toothed or lobed again; cauline gradually smaller and less divided upwards; all nearly glabrous to more or less hairy on both surfaces. *Inflorescence* of compound umbels with 10–17 rays, the rays 15–60 mm, glabrous or nearly so, the peduncles usually longer than the rays; bracts absent; bracteoles 7–8, lanceolate, with a long, tapering apex, deflexed. *Sepals* minute. *Petals* 5, usually pinkish, hairy round the margin, obovate. *Stamens* 5; filaments white; anthers cream. *Styles* 2, nearly erect, enlarged at base into the stylopodium; stigma capitate. *Schizocarp* 8–12 mm, oblong, but narrowed near the apex; mericarps with broad rounded ridges; constricted at the commissure; carpophore present; vittae solitary; pedicels without a ring of hairs at the apex. *Flowers* 6–7.

Introduced. Naturalised on a grassy roadside verge in Westmorland since 1979, and on a river-bank in Lanarkshire since 1989. Native of central and south Europe.

2. C. aureum L. Golden Chervil

Myrrhis aurea (L.) All.; *C. maculatum* Willd. ex DC.

Perennial monoecious *herb* with stout roots. *Stem* up to 150 cm, pale green, sometimes with purple spots, erect, robust, solid, swollen below the nodes, with numerous, soft to rather rough, pale simple eglandular hairs. *Leaves* pale yellowish-green on upper surface, paler beneath; basal 6–12 × 6.0–15 cm, triangular in outline, 3-pinnate, the segments triangular and acute, the lobes 10–40 mm, lanceolate in outline, deeply toothed or lobed and the teeth gradually narrowed at the apex, long petiolate; cauline gradually smaller and less divided upwards; all appressed-hairy to nearly glabrous on both surfaces. *Inflorescence* of compound umbels, with 12–25 rays, the rays 15–30(–45) mm and glabrous, the peduncle usually longer than the rays and hairy, the terminal umbel with mostly bisexual flowers, overtopped by the lateral umbels which have mostly male flowers; bracts absent or rarely 1–3; bracteoles 5–8, about as long as the pedicels in flower, hairy, ultimately deflexed. *Sepals* minute. *Petals* 5, pure white, the outer not radiating, emarginate, apex inflexed. *Stamens* 5; filaments white; anthers cream. *Styles* 2, recurved, the enlarged base forming the stylopodium which is about half as long as the style; stigma capitate. *Schizocarp* 8–12 mm, oblong, but narrowed rather abruptly near the apex; mericarps with broad, rounded ridges; constricted at the commissure; carpophore present; vittae solitary; pedicels without a ring of hairs at the apex. *Flowers* 6–7. $2n=22$.

Introduced. Naturalised in grassy places. Scattered records in England and Scotland. Occurs in abundance since 1909 in water-meadows near Callander in Perthshire and locally frequent at Calton Hill in Edinburgh. Found in wild flower seed. Native of central and south Europe and south-west Asia.

3. C. temulum L. Rough Chervil

C. temulentum L.; *Myrrhis temula* (L.) All.; *Scandix temula* (L.) Roth; *Scandix natans* Moench; *Polgidon temulum* (L.) Raf.; *Bellia temulenta* (L.) Bubani; *Selinum temulum* (L.) E. H. L. Krause

Biennial monoecious *herb* with stout roots, sometimes very stout. *Stems* up to 100 cm, pale dull green with purple spots to almost entirely purple, erect, solid, swollen below the nodes, with numerous, stiff, more or less appressed simple eglandular hairs and sometimes shorter ones intermixed, sometimes with dense short hairs only. *Leaves* dull, dark green on upper surface, paler beneath; basal few, 4–11 × 4–11 cm, 2- to 3-pinnate, the segments ovate, obtuse or abruptly contracted at apex, often petiolulate; lobes 10–30 mm, ovate, deeply toothed, the teeth abruptly contracted at the apex, long-petiolate; cauline similar and gradually decreasing in size upwards, petiolulate; all with medium to long, pale, appressed simple eglandular hairs on both surfaces. *Inflorescence* of compound umbels, with (4–)6–12(–17) rays 15–50 mm, with short simple eglandular hairs, the peduncle longer than the rays with similar hairs, the terminal umbel with mostly bisexual flowers, overtopped by the lateral umbels which have mostly male flowers; bracts absent, or rarely 1 or 2; bracteoles 5–8, shorter than the pedicels, ciliate, eventually deflexed. *Sepals* absent. *Petals* 5, white, the outer not radiating, emarginate, the apex inflexed. *Stamens* 5; filaments white; anthers cream. *Styles* 2, recurved, the enlarged base forming the stylopodium about as long as the style; stigma capitate. *Schizocarp* 4.0–6.5 mm, often purplish, oblong but narrowed towards the apex, slightly compressed laterally, glabrous; mericarps with broad, rounded ridges; constricted at commissure; carpophore present; vittae solitary, conspicuous; pedicels without a ring of hairs at the apex. *Flowers* 5–7. $2n = 14, 22$.

(i) Var. **temulum**

Plant often short with a slender rootstock. *Lower stem* with spreading long eglandular hairs and short ones in between, but never dense. *Inflorescence rays* usually up to 8, rarely to 12.

(ii) Var. **canescens** (Benitez ex Thell.) P. D. Sell

C. temulum forma *canescens* Benitez ex Thell.

Plant often tall with a stout rootstock. *Lower stem* with dense, short, white eglandular hairs, less so upwards. *Inflorescence rays* up to 17.

Native. Along hedges and the borders of woods and in grassy and waste places. Common in much of Great Britain and Ireland, but sparse in the Channel Islands and Ireland and absent from most of west and north Scotland. This poisonous plant is the commonest roadside umbellifer in flower, just after *Anthriscus sylvestris,* in much of England, and is easily recognised by its hairy, purple-spotted stem and swollen top of the internodes. Most of Europe, but rare in the Mediterranean region; south-west Asia; north-west Africa. European Temperate element. In western and northern Europe the fruit is always glabrous, forma **temulum**. Plants with hairy fruits, forma **eriocarpum** Guss., occur locally in south and south-east Europe. Var. *canescens* occurs in great quantity in a planted wood at Bassingbourn in Cambridgeshire where it may have been planted in wild flower seed or brought in with grasscutters. It occurs also along hedges in Cambridge where it may also have been introduced by grasscutters. This variety occurs across central Europe and needs to be compared with ***C. angelicifolium*** M. Bieb. from Caucasus and Georgia of which no authentic specimens have been seen. See also under *Torilis ucranica* Spreng.

6. Anthriscus Pers. nom. conserv.

Cerefolium Fabr.

Annual to *perennial* monoecious *herbs. Stems* erect to decumbent, hollow. *Leaves* 2- to 3-pinnate. *Inflorescence* of compound umbels; bracts usually absent, rarely 1; bracteoles present. *Sepals* absent or minute. *Petals* 5, white, the outer not radiating. *Fruit* a schizocarp, oblong-ovoid or ovoid, slightly compressed laterally, with a short, ridged beak, glabrous or with hooked bristles; mericarps with ridges only on the beak; commissure narrow or with a constriction; carpophore present; vittae absent or solitary.

Contains about 20 species in Europe, temperate Asia and North Africa.

Clapham, A. R. (1953). Human factors contributing to a change in our flora: the former ecological status of certain hedgerow species. In Lousley, J. E. (Edit.) *The Changing flora of Britain,* pp. 26–39. Arbroath.

Druce, G. C. (1917). *Anthriscus sylvestris* Hoffm. *Rep. Bot. Soc. Exch. Club Brit. Isles* **4**: 412–414.

Druce, G. C. (1918). *Anthriscus sylvestris* (L.) Hoffm. *Rep. Bot. Soc. Exch. Club Brit. Isles* **5**: 8–31.

Grime et al. (1988). *Comparative plant ecology*. London. [*A. sylvestris*.]

Hultén, E. & Fries, M. (1986). *Atlas of north European vascular plants north of the Tropic of Cancer.* 3 vols. Königstein.

1. Schizocarp less than 5 mm **3. caucalis**
1. Schizocarp more than 5 mm 2.
2. Perennial; rays more or less glabrous; schizocarps with scarcely differentiated beak less than 1 mm 3.
2. Annual; rays hairy; schizocarps with well differentiated beaks 1–4 mm 4.
3. Whole plant more open and sparsely hairy; leaves more rigid with the ultimate segments 10–25 mm wide, not overlapping, the lobes linear or linear-lanceolate and gradually narrowed into a conspicuous mucro **1(i). sylvestris** var. **sylvestris**
3. Whole plant denser and hairier; leaves softer, the ultimate segments 20–25 mm wide and overlapping; the lobes lanceolate or ovate and rounded-obtuse-mucronate **1(ii). sylvestris** var. **latisecta**
4. Schizocarp glabrous **2(i). cerefolium** var. **cerefolium**
4. Schizocarp with numerous hooked bristles **2(ii). cerefolium** var. **longirostris**

1. A. sylvestris (L.) Hoffm. Cow Parsley
Chaerophyllum sylvestre L.; *Cerefolium sylvestre* (L.) Besser; *A. torquata* Coste; *Myrrhis sylvestris* (L.) Spreng.

Plant perennating by buds in the axils of the basal leaves, which develop a tap-root and separate from the flowering stem when it dies. *Stems* 60–155 cm, pale green, sometimes purplish towards the base, erect, hollow, furrowed, with more or less numerous, short, pale simple eglandular hairs. *Leaves* pale and matt to darker and more shiny green on upper surface, paler beneath, rarely tinted reddish; basal 2.5–5.0 × 1.5–3.0 cm, triangular in outline, 3-pinnate, the lowest segments much smaller than the rest of the leaf, narrowly triangular and acute at apex, the lobes 10–50 mm, pinnatifid and often serrate, petiolate; cauline similar and gradually smaller up the stem; all with short to medium, pale, appressed simple eglandular hairs on both surfaces and margins to glabrous or nearly so. *Inflorescence* of compound umbels, with (3–)6–12 rays which are 15–30 mm and glabrous, the peduncles about as long as the rays and more or less glabrous, the terminal umbel with male flowers in the middle of each partial umbel and surrounded by bisexual flowers, the lateral umbels with mostly or entirely male flowers and equalling or slightly overtopping the terminal umbel; bracts absent; bracteoles 4–6, ovate, aristate, ciliate and patent or deflexed. *Sepals* minute. *Petals* 5, creamy white, the outer not radiating, emarginate, apex inflexed. *Stamens* 5; filaments white or cream; anthers pale yellow or cream. *Styles* 2, divergent, with enlarged base forming the stylopodium which is not quite as long as the style; stigma capitate. *Schizocarp* 6–10 mm, oblong-ovoid, slightly compressed laterally; mericarps smooth, with a very short, ridged beak; constricted at the commissure; carpophore present; vittae solitary; pedicels with a ring of short, stout hairs at the apex. *Flowers* 4–6. $2n = 16$.

(i) Var. **sylvestris**
A. sylvestris var. *angustisecta* Druce; *Cerefolium sylvestre* var. *angustisectum* (Druce) Druce

Whole *plant* more open and sparsely hairy. *Leaves* more rigid, ultimate segments 10–25 mm wide, not overlapping; lobes linear or linear-lanceolate and gradually narrowed into a conspicuous mucro.

(ii) Var. **latisecta** Druce
Cerefolium sylvestre var. *latisectum* (Druce) Druce

Whole *plant* denser and hairier. *Leaves* softer, ultimate segments 20–25 mm wide, overlapping; lobes lanceolate or ovate, rounded-obtuse-mucronate.

Native. Grassy places, hedgerows and wood margins, the first of the umbels to line the roadsides in April and May. Abundant throughout most of Great Britain and Ireland. Europe, but rare in the south; temperate Asia and North Africa. Eurasian Boreo-temperate element. Var. *sylvestris* seems to be the only variety in Scotland and it comes as far south as Derbyshire and in Wales and is found on ancient tracks and grassland elsewhere. It was probably the main plant of northern Continental Europe. Var. *latisecta* is the common plant of England south of Derbyshire. It is not known if there is an intermediate zone but intermediate plants occur. It is not known what happens in Ireland. Clapham (1953) believed that the northern var. *sylvestris* might be native and the southern var. *latisecta* might have immigrated at a later date and perhaps through human agency. It is interesting that the plant on Gun's Way at Histon in Cambridgeshire is var. *sylvestris* but it is replaced by var. *latisecta* just before you reach the village. Arthur Chater collecting between Wales and London found more or less var. *sylvestris* in Wales and a mixture nearer to London.

2. A. cerefolium (L.) Hoffm. Garden Chervil
Cerefolium cerefolium (L.) Schinz & Thell.; *Cerefolium sativum* Besser; *Chaerophyllum cerefolium* (L.) Crantz; *Scandix cerefolium* L.; *Chaerophyllum sativum* Lam. nom. illegit.; *Scandix tenuifolia* Salisb.; *Myrrhoides cerefolium* (L.) Kuntze; *Selinum cerefolium* (L.) E. L. H. Krause

Wiry *annual* monoecious *herb* with a pale brown tap-root and fibrous side-roots. *Stem* up to 30 cm, medium green, striate, erect, hollow, with numerous simple eglandular hairs. *Leaves* medium green on upper surface, paler beneath; basal 2–5 × 1.5–3.0 cm, ovate in outline, 3-pinnate; the segments ovate, pointed or blunt at apex, lobed, the lobes 6–15 mm, triangular-ovate or lanceolate, pointed, the petioles inflated at base, more or less hairy; cauline similar but decreasing in size upwards; all glabrous or nearly so on the upper surface, with short or medium simple eglandular hairs beneath. *Inflorescence* of compound umbels, with 2–6 rays which are more or less hairy, the peduncle longer than the rays and hairy, the umbels leaf-opposed, the flowers bisexual; bracts absent; bracteoles 4–5, linear, pointed. *Sepals* absent. *Petals* 5, white, not radiating, emarginate, apex inflexed. *Stamens* 5; filaments white; anthers cream. *Styles* 2, longer than the swollen base which forms the stylopodium and nearly erect. *Schizocarp* 7–10 mm, linear with a prominent, slender beak up to 4 mm, glabrous or with numerous hooked bristles; mericarps with a ridge only on the beak; commissure constricted; carpophore present; vittae solitary; pedicels becoming at least as thick as the rays in fruit, with a ring of hairs at apex. *Flowers* 5–6. $2n = 18$.

(i) Var. **cerefolium**
Schizocarp glabrous.

(ii) Var. **longirostris** (Bertol.) Cannon
A. longirostris Bertol.

Schizocarp with numerous hooked bristles.

Introduced. Formerly cultivated as a herb, now rarely so. Naturalised on a rock face in Ross, Herefordshire since at least 1867; casual in waste places but decreasing. South and central Great Britain and Guernsey. Native of south-east Europe and western Asia. The cultivated form is used rather like parsley in salads, soups, omelettes and for garnishing.

3. A. caucalis M. Bieb. Bur Chervil
A. scandicina Mansf. nom. illegit.; *A. vulgaris* Pers., non Bernh.; *Chaerophyllum anthriscus* (L.) Crantz; *Cerefolium anthriscus* (L.) Beck; *A. neglecta* Boiss. & Reut.; *Scandix anthriscus* L.; *Caucalis scandix* Scop.; *Myrrhis chaerophyllea* Lam. nom. illegit.; *Caucalis scandicina* Weber nom illegit.; *Caucalis aequicolorum* All. nom. illegit.; *Scandix laeta* Salisb. nom. illegit.; *A. scandix* (Scop.) Asch., non M. Bieb.; *A. chaerophyllum* Druce nom. illegit.

Wiry, *annual* monoecious *herb* with a sweetish smell and with tapering, brown tap-root and fibrous side-roots. *Stem* 25–70(–100) cm, delicate, pale yellowish-green, sometimes striped purple and purplish towards the base, erect to decumbent, hollow, usually freely branched, glabrous or with a few simple eglandular hairs. *Leaves* a characteristic pale yellowish-green on upper surface, paler beneath; basal 1.5–6.0 × 1.5–6.0 cm, triangular or oblong in outline, 2- to 3-pinnate, the segments oblong or narrowly triangular and acute at apex, the lobes 5–10 mm, ovate, dentate or pinnatifid, the petioles inflated and long-hairy towards the base; cauline similar but gradually decreasing in size upwards; all usually glabrous or nearly so on the upper surface, with short or medium simple eglandular hairs beneath. *Inflorescence* of compound umbels, with 3–6 rays which are usually about 10 mm and glabrous, the peduncle shorter than the rays and glabrous, the umbels are all leaf-opposed and all the flowers bisexual; bracts absent, rarely 1; bracteoles usually 4–5, ovate, acuminate at apex and ciliate. *Sepals* absent. *Petals* 5, white, the outer not radiating, emarginate, apex inflexed. *Stamens* 5; filaments greenish-white; anthers pale greenish-yellow. *Styles* 2, very short and connivent, with enlarged base forming the stylopodium; stigma slightly thickened. *Schizocarp* 2.9–3.2 mm, ovoid, slightly compressed laterally, shortly beaked, covered in hooked bristles; mericarps with ridges only in the beak; commissure narrow; carpophore present; vittae absent; pedicels becoming at least as thick as the rays in fruit, with a ring of stout hairs at the apex. *Flowers* 5–6. $2n = 14$.

Native. Hedgebanks, waste places, open ground on sandy or shingly soils especially near the sea. It often appears where sand has been dumped inland. Common in parts of eastern England and scattered over the rest of Great Britain and Ireland except north and north-west Scotland. West, central and south Europe; temperate Asia; North Africa; introduced in North America and New Zealand. European Temperate element. The plant described above is var. **caucalis**, the only variant recorded in Great Britain and Ireland. Var. **neglecta** (Boiss. & Reut.) P. Silva & Franco (*A. neglecta* Boiss. & Reut.) has a smooth schizocarp.

7. **Scandix** L.

Pectinaria Bernh., non Haw.

Annual herb. Stems erect or spreading, hollow when fruiting. *Leaves* 2- to 4-pinnate. *Inflorescence* of simple umbels; bracts usually absent; bracteoles present. *Sepals* 5, very small, but persistent. *Petals* 5, white, the outer not radiating, but very unequal. *Fruit* a *schizocarp,* more or less cylindrical, slightly compressed laterally, with beak 3–5 times as long as seed-bearing portion; constricted at the commissure; carpophore present; vittae solitary and conspicuous.

Contains about 15 species, mainly in the Mediterranean region.

S. stellata Banks & Sol. has been recorded as a bird-seed casual.

Hultén, E. & Fries, M. (1986). *Atlas of north European vascular plants north of the Tropic of Cancer.* 3 vols. Königstein.

Stewart, A., Pearman, D. A. & Preston, C. D. (1994). *Scarce plants in Britain. Peterborough.*

1. S. pecten-veneris L. Shepherd's-needle

S. pectinifera Stokes nom. illegit.; *S. vulgaris* Gray nom. illegit.; *S. rostrata* Salisb.; *S. pectiniformis* St Lag.; *Chaerophyllum pecten-veneris* (L.) Crantz; *Chaerophyllum rostratum* Lam.; *Pastinaca pecten-veneris* (L.) Lam.; *Myrrhis pecten-veneris* (L.) All.; *Pectinaria vulgaris* Bernh.; *Wylia pecten-veneris* (L.) Bubani

Annual monoecious *herb* with a tapering tap-root and fibrous side-roots. *Stems* 15–50 cm, pale green, erect or spreading, becoming hollow when old, glabrous or nearly so, branched, leafy. *Leaves* yellowish-green on upper surface, paler beneath; basal numerous, 2–6 × 2.0–6.5 cm, oblong-ovate in outline, 2- to 4-pinnate, the segments triangular and acute at apex, the lobes up to 15 mm, narrow and entire to pinnatifid, the petioles long, widened at the base and with a scarious, usually ciliate margin; cauline few and smaller; all glabrous or nearly so. *Inflorescence* of simple umbels, with 1–3 rays which are 5–40 mm, stout and glabrous or sparsely hairy, the peduncle very short or absent, the terminal umbel with bisexual flowers, the lateral umbels with varying proportions of male and bisexual flowers; bracts usually absent; bracteoles usually 5, longer than the pedicels, simple or often deeply and irregularly divided. *Sepals* 5, very small. *Petals* 5, white, the outer not radiating, often very unequal, apex incurved. *Stamens* 5; filaments white; anthers cream. *Styles* 2, erect, with an enlarged base forming the stylopodium which is half to quarter as long as the styles; stigma tapering. *Schizocarp* 30–70 mm, more or less cylindrical, slightly compressed laterally, with a strongly dorsally flattened beak 3–5 times as long as and clearly distinct from the seed-bearing portion; constricted at the commissure; mericarps ribbed, scabrid and with forward-pointing bristles on the margins; carpophore present; vittae solitary, conspicuous; pedicels almost as thick as the rays, glabrous at the apex. *Flowers* 5–8. $2n = 16$.

Possibly native. A lowland species which was once widespread as a weed of arable land and so abundant at times it impeded harvesting. There was a rapid decline after 1955 with the introduction of herbicides. Formerly common in England and scattered over the rest of Great Britain and Ireland. Now occurs as occasional plants, except in East Anglia where it is still sometimes abundant. It sometimes occurs in coastal habitats where it was possibly native. Mediterranean region extending north to Denmark and casual in Scandinavia; North Africa; south-west Asia to the borders of India; introduced in North and South America, southern Africa, Australia and New Zealand. Eurosiberian Southern-temperate element. Our plant is subsp. **pecten-veneris**, which occurs throughout the range of the species.

Osmorhiza chilensis Hook. & Arn. has been recorded as a wool casual.

8. **Myrrhis** Mill.

Perennial monoecious *herb* smelling of aniseed when crushed. *Stems* erect, hollow. *Leaves* 2- to 4-pinnate. *Inflorescence* of compound umbels; bracts usually absent; bracteoles present. *Sepals* 5 and minute or absent. *Petals* 5, white, the outer slightly radiating. *Fruit* a schizocarp, linear-oblong, slightly laterally compressed, with a short beak, mericarps sharply ridged, with short, forward-pointing bristles on the ridges; commissure broad; carpophore present; vittae slender or absent.

One species in Europe, widely naturalised in other parts of the world.

Grieve, M. (1931). *A modern herbal*. London.
Grime, J. P. et al. (1988). *Comparative plant ecology*. London.
Hultén, E. & Fries, M. (1986). *Atlas of north European vascular plants north of the Tropic of Cancer.* 3 vols. Königstein.
Lhotska, H. (1977). Notes on the ecology of germination in *Myrrhis odorata*. *Folia Geobot. Phytotax.* **12**: 209–213.

1. M. odorata (L.) Scop. Sweet Cicely
Scandix odorata L.; *Chaerophyllum odoratum* (L.) Crantz; *Lindera odorata* (L.) Asch.; *Selinum myrrhis* E. H. L. Krause

Perennial monoecious *herb* smelling of aniseed when crushed. *Stems* 60–200 cm, pale green, erect, hollow, somewhat grooved, stout, with sparse hairs. *Leaves* medium greyish-green on upper surface, paler beneath and usually with some whitish markings; basal 6–40 × 6–35 cm, 2- to 4-pinnate, broadly triangular in outline, acute at apex, the segments triangular, the lobes 10–30 mm, oblong-lanceolate, deeply serrate or pinnatifid; cauline gradually smaller up the stem, with narrow, sheathing petioles; all glabrous or with sparse to numerous, short hairs. *Inflorescence* of compound umbels, with 4–10(–21) rays which are 15–30 mm and hairy, those bearing male flowers slender, the others stout, the peduncles usually longer than the rays and hairy, the terminal umbel with male and bisexual flowers, the lateral umbels usually with male flowers only; bracts usually absent; bracteoles about 5, lanceolate, with a long, slender apex and whitish. *Sepals* 5, minute. *Petals* 5, white, outer slightly radiating, cuneate-obovate, apex short and inflexed. *Stamens* 5; filaments white; anthers greenish. *Styles* 2, patent or recurved in fruit, with enlarged base forming the stylopodium which is much shorter than the styles; stigma capitate. *Schizocarp* 15–25 mm, linear-oblong, slightly laterally compressed, with a short beak; mericarp sharply ridged, dark brown and shiny when mature; commissure broad; carpophore present; vittae slender or absent; pedicels about 5 mm and sparsely hairy. *Flowers* 5–6. $2n=22$.

Introduced. Hedgebanks, waste and grassy places and pathsides especially near houses. Commonly naturalised in much of the north of Great Britain from Derbyshire northwards, rare in southern Wales, southern England and north and central Ireland. It is the commonest spring-flowering umbellifer on many northern roadsides. Pyrenees, Alps, Apennines and west part of the Balkan Peninsula. Its roots have been boiled and eaten like parsnips, its leaves have been included as a mild sweetener and used to polish oak. Medicinally it is used as a carminative, an expectorant and for stomach ailments.

9. Coriandrum L.

Annual monoecious *herbs*. *Stems* erect, solid. *Leaves* simple to 3-pinnate. *Inflorescence* of compound umbels; bracts 0–1(–2), entire; bracteoles present. *Sepals* 5, conspicuous and unequal, persistent. *Petals* 5, white to purplish, the outer radiating and deeply 2-lobed. *Fruit* a schizocarp, ovoid or globose; mericarps not separating at maturity, with primary and secondary ridges; commissure broad; carpophore present; vittae under the secondary ridges.

Contains 2 species in south-west Asia and North Africa.

Vaughan, J. G. & Geissler (1997). *The new Oxford book of food plants*. Oxford.
Zohary, D. & Hopf, M. (2000). *Domestication of plants in the Old World*. Ed. 3. Oxford.

1. C. sativum L. Coriander
C. majus Gouan; *C. globosum* Salisb.; *Selinum coriandrum* E. H. L. Krause

A fetid *annual* monoecious *herb* with fibrous roots. *Stems* 15–50(–70) cm, pale green, erect, solid, almost smooth and glabrous. *Leaves* medium green on upper surface, paler beneath; basal few, with lamina 0.7–1.5 × 0.6–1.6 cm, subrotund, ovate or obovate, 3-pinnate or simple with broad, incise-serrate lobes about 10 mm, long petiolate, soon withering; lower cauline 1–5 × 1–6 cm, 2- to 3-pinnate with narrow lobes, long-petiolate; all glabrous. *Inflorescence* of compound umbels, with 2–5(–10) rays which are 5–15 mm and glabrous, the peduncle longer than the rays and sometimes leaf-opposed, all the umbels with a mixture of male and bisexual flowers; bracts 0–1(–2), entire; bracteoles usually 3, linear. *Sepals* 5, conspicuous and unequal. *Petals* 5, purplish or white, the outer radiating and deeply 2-lobed, the apex inflexed. *Stamens* 5; filaments white; anthers pale yellow. *Styles* 2, recurved in fruit, with enlarged base forming the stylopodium which is much shorter than the style; stigma capitate. *Schizocarp* 2–6 × 2.0–5.5 mm, reddish-brown, ovoid or globose, hard; mericarps not separating at maturity, with primary and secondary ridges, both low, the former narrower than the latter; commissure broad; carpophore present; vittae under the secondary ridges; pedicels of bisexual flowers of 4–5 mm, those of the male flowers shorter. *Flowers* 7–10. $2n=22$.

Introduced. A casual at tips and on waste ground. Scattered records throughout Great Britain. Perhaps native in North Africa and western Asia. Grown on a small scale as a green salad and the fruits are used as a condiment in curries, meat dishes, bread and confectionery. It probably originated in the eastern Mediterranean and was known in Neolithic times in Israel. It was used in medicine and was utilised by the ancient Greeks and Romans.

10. Bifora Hoffm.

Annual monoecious *herb*. *Stems* erect, hollow. *Leaves* 2- to 3-pinnate. *Inflorescence* of compound umbels; bracts 0–1, filiform; bracteoles 0–2, filiform. *Sepals* absent. *Petals* 5, white, the outer radiating. *Schizocarp* didymous, retuse at apex, cordate at base; mericarps globose, more or less rugose, without conspicuous ridges; commissure narrow; carpophore present; vittae absent.

Contains about 6 species in Europe, Asia and North America.

1. B. radians M. Bieb. Didymous-fruited Umbel
Doriandrum radians (M. Bieb.) Prantl; *Anidrum radians* (M. Bieb.) Kuntze; *Coriandrum radians* Prantl

Annual monoecious *herb* with a pale, slender tap-root and fibrous side-roots. *Stem* 10–80 cm, pale green, sometimes tinted purplish, erect, hollow, slender, ridged, glabrous, branched, leafy. *Leaves* yellowish-green on upper surface, paler beneath; basal and lower cauline with lamina 8–16 × 3–7 cm, oblong in outline, 2-pinnate, the segments oblong and incised, long-petiolate; upper cauline 2- to 3-pinnate, the segments filiform, sessile; all glabrous. *Inflorescence* of compound umbels, flowers bisexual or male; rays (2–)4–8, 15–25 mm, more or less equal, arcuate, ascending, glabrous; bracts 0–1, 4–7 mm, filiform; bracteoles 0–2, 1.0–2.5 mm, filiform. *Sepals* absent. *Petals* 5, white, the outer radiating to 4–5 mm, the inner 1.5–2.0 mm, bifid and often with only 1 lobe developing. *Stamens* 5; filaments white; anthers green. Styles 2, 1.0–1.5 mm, swollen at base to form the stylopodium which is only half as long as the style. *Schizocarp* didymous, 2.0–3.5 × 4–7 mm, retuse at apex, cordate at base, pale brown; mericarps globose, more or less rugose, without conspicuous ridges; commissure narrow; carpophore present; vittae absent. *Flowers* 6–8. $2n = 22$.

Introduced. A rare grain casual. Native of south and central Europe east to Caucasia and north and west Iran.

11. Smyrnium L.

Biennial (to *perennial*) monoecious *herbs*. *Stems* erect, solid. *Leaves* simple to 3-pinnate or ternate. *Inflorescence* of compound umbels; bracts and bracteoles 0–2 and small or absent, entire. *Sepals* 5, very small. *Petals* 5, yellow, the outer not radiating. *Fruit* a schizocarp broadly ovoid and blackish, laterally compressed; mericarps with 3 prominent, sharp ridges; constricted at the commissure; carpophore present; vittae numeous.

Contains 7 species in Europe and North Africa.

Salisbury, E. J. (1961). *Weeds and aliens*. London.

1. Upper cauline leaves ternate, with a greatly expanded sheathing petiole **1. olusatrum**
1. Upper cauline leaves subentire to minutely crenate-serrate and amplexicaul with rounded lobes **2. perfoliatum**

1. S. olusatrum L. Alexanders
S. vulgare Gray nom. illegit.

Biennial monoecious *herb* with a fleshy, tuberous tap-toot in 1 year. *Stems* 50–200 cm, pale yellowish-green, stout, solid, becoming hollow, distinctly ridged, especially above, branched, leafy, glabrous. *Leaves* a characteristic dark shiny green on upper surface, paler beneath; basal 40–60 × 30–40 cm, broadly oblong in outline, 2- to 3-pinnate or ternate, the segments 25–80 mm, ovate to rhombic, rounded at apex, serrate and sometimes lobed, the teeth obtuse, the petioles long; lower cauline similar, the upper often opposite, ternate, with a greatly expanded, pale green sheathing petiole with purplish-brown stripes; all glabrous. *Inflorescence* of compound umbels, with 4–15 rays which are 30–40 mm and glabrous, the peduncle usually longer than the rays and glabrous, the terminal umbel with male and bisexual flowers, the lateral umbels with mostly male flowers; bracts and bracteoles 0–2, small and sometimes absent. *Sepals* 5, very small, somewhat accrescent in fruit. *Petals* 5, yellow, the outer not radiating, the apex inflexed. *Stamens* 5; filaments green; anthers greenish. *Styles* 2, colourless at first, patent and usually appressed to the yellow stylopodium in fruit, with an enlarged base forming the stylopodium which is a little shorter than the style; stigma capitate. *Schizocarp* 7–8 mm, black, broadly ovoid, laterally compressed; mericarps with 3 prominent, sharp ridges; constricted at the commissure; carpophore present; vittae numerous; pedicels somewhat longer than the fruit and papillae on the inner angles, the fruit smelling strongly of aniseed when crushed. *Flowers* 4–6. $2n = 22$.

Introduced. Hedgebanks, cliffs, grassy roadsides, ditches and waste places mostly by the sea. Common on the coast of Great Britain and Ireland north to central Scotland with scattered records inland. Dwarf plants occur on Portland, Dorsetshire, but are possibly phenotypes. Europe northwards to south-west France; North Africa; Macaronesia; south-west Asia. Probably introduced by the Romans as an all-purpose vegetable and tonic, almost every part of the plant being used, the young flower-buds picked like miniature cauliflowers and the young stems cooked like celery. It was cultivated in monastic herb gardens which accounts for its present at old religious sites such as Steepholm on the Bristol Channel.

2. S. perfoliatum L. Perfoliate Alexanders
S. dioscoridis Spreng.; *Selinum dioscoridis* (Spreng.) E. H. L. Krause

Biennial monoecious *herb* with a tuberous, turnip-like tap-root. *Stems* (20–)40–150 cm, pale yellowish-green, erect, solid, strongly angled and narrowly winged on the angles, particularly near the middle, the wings with sparse, small stellate hairs, branched, leafy. *Leaves* dark, shiny yellowish-green on upper surface, paler beneath; basal 6–8 × 5–6 cm, broadly ovate in outline, 2- to 3-pinnate or ternate, the segments ovate, dentate or somewhat lobed; cauline alternate, 3–12 × 2–9 cm, ovate, rounded to a more or less acute apex, subentire to minutely crenate-serrate, amplexicaul with rounded lobes; all glabrous. *Inflorescence* of compound umbels, with 5–10 rays which are 1–4 cm long, glabrous and papillose near the apex, the peduncle winged on the angles and longer or shorter than the rays, the terminal umbel with mostly bisexual flowers and the lateral umbels with mostly male flowers; bracts and bracteoles absent. *Sepals* 5, very small, scarcely accrescent in fruit. *Petals* 5, yellow, the outer not radiating, the apex inflexed. *Stamens* 5; filaments greenish; anthers yellow. *Styles* 2, recurved and appressed in fruit, with enlarged base forming the stylopodium which is shorter than the style; stigma capitate. *Schizocarp* 3.0–3.5 × 5.0–5.5 mm, brownish-black, much broader than long, laterally compressed; mericarps with 3, slender, dorsal ridges; constricted at the commissure; carpophore present; vittae numerous, slender. *Flowers* 5–6. $2n = 22$.

Introduced. Naturalised in grassy places and flower-borders. Recorded in scattered localities in England and at one locality in Kirkcudbrightshire. Native of southern Europe north to Czechoslovakia; North Africa; south-east Asia.

12. Bunium L.

Perennial herbs with a solitary, globose tuber. *Stems* erect, solid. *Leaves* 2- to 3-pinnate, with linear lobes. *Inflorescence* of compound umbels; bracts 5–10, entire. *Sepals* a minute rim. *Petals* 5, white, the outer not radiating. *Fruit* a schizocarp, oblong-ellipsoid, laterally compressed; mericarps with slender pale ridges; constricted at the commissure; carpophore present; with a solitary, brown vitta on each mericarp.

Contains about 40 species in Europe and temperate Asia.

Dony, J. G. (1953). *Flora of Bedfordshire*. Luton.
Wigginton, M. J. (Edit.) (1999). *British red data books*. Vol. 1. *Vascular plants*. Peterborough.

1. B. bulbocastanum L. Great Pignut
Ligusticum bulbocastanum (L.) Crantz; *B. majus* Vill., non Gouan; *Pimpinella bulbocastanum* (L.) Stokes; *Carum bulbocastanum* (L.) W. D. J. Koch

Perennial monoecious *herb* with a solitary, dark brown, globose tuber 10–30 mm in diameter. *Stems* 30–100 cm, pale green, the subterranean part flexuose and tapering downwards, the aerial part usually erect, solid, glabrous, sparsely branched and sparsely leafy. *Leaves* pale green; basal few, 6–11 × 3–5 cm, broadly triangular in outline, 2- to 3-pinnate, the segments 5–10 mm, linear with a cartilaginous apex, mostly withered at anthesis, the petioles long, slender and abruptly expanded into a sheathing base; cauline leaves few, similar to basal, the lower petiolate like basal, the upper leaves with petiole sheathing throughout; all glabrous. *Inflorescence* of compound umbels, with (5–)10–20 rays which are 15–48 mm, scabrid on the angles and the peduncle longer than the rays, most umbels with bisexual and a few male flowers, the latest developed lateral ones with all male flowers; bracts 5–10, linear-lanceolate, acuminate at apex, entire; bracteoles 5–10, linear to lanceolate, entire. *Sepals* a minute rim. *Petals* 5, white, the outer not radiating, obcordate, inflexed at apex. *Stamens* 5; filaments greyish-yellow; anthers yellowish. *Styles* 2, recurved in fruit and suddenly contracted from the enlarged base forming the stylopodium which is about as long as the styles; stigma capitate. *Schizocarp* 3.5–4.5 mm, oblong-ellipsoid, laterally compressed; mericarps with slender, pale ridges; constricted at the commissure; carpophore present; with a solitary, conspicuous reddish-brown vitta on each mericarp; pedicels 2–6 mm. *Flowers* 6–7. $2n = 22$.

Native. Chalk grassland and on banks. Very local in Hertfordshire, Buckinghamshire (Ivinghoe Hills), Bedfordshire and Cambridgeshire; very occasionally elsewhere as an introduction. Europe, northwards to England and eastwards to central Germany and north-western Yugoslavia; north-west Africa. Southern Temperate element. The seedling has only one cotyledon through abortion of the other during development. The tuber arises from the hypercotyl of the seedling and is edible raw or cooked.

13. Conopodium W. D. J. Koch

Perennial monoecious *herbs. Stems* erect, hollow after flowering. *Leaves* 2- to 3-pinnate, with linear lobes. *Inflorescence* of compound umbels; bracts absent or rarely 1–2; bracteoles 2–5, linear. *Sepals* absent. Petals 5, white, the outer not or scarcely radiating. *Fruits* a schizocarp ovoid-oblong, slightly laterally compressed; mericarps with slender ridges; constricted at the commissure; carpophore present; with 2–3 reddish-brown vittae in each groove between the ridges of the mericarp.

Contains about 20 species in Europe, North Africa and temperate Asia.

Grime, J. P. et al. (1988). *Comparative plant ecology*. London.
Hultén, E. & Fries, M. (1986). *Atlas of north European vascular plants north of the Tropic of Cancer.* 3 vols. Königstein.

1. C. majus (Gouan) Loret Pignut
Bunium majus Gouan; *C. denudatum* W. D. J. Koch nom. illegit.; *Pimpinella flexuosa* Stokes nom. illegit.; *Bunium denudatum* DC.; *Bunium minus* Gray, non Gouan; *Carum majus* (Gouan) Britten & Rendle

Perennial monoecious *herb* with a dark brown, solitary, rather irregular tuber 8–35 mm in diameter. *Stems* 8–60(–90) cm, pale green, erect, hollow after flowering, the subterranean part flexuous and tapering downwards, glabrous, branched above, leafy. *Leaves* medium green on upper surface, paler beneath; basal few 3.5–5.0 × 3.5–5.0 cm, broadly triangular in outline, 3-pinnate, the segments triangular, the lobes 3–5 mm, narrowly linear-lanceolate, acute at apex, the petioles medium; cauline few, 2-pinnate, the segments triangular, the lobes linear, the petiole short and sheathing; all glabrous. *Inflorescence* of compound umbels with 6–12 rays which are 10–60 mm and smooth, the peduncle longer than the rays, the umbels mostly with bisexual and a few male flowers, the latest developed lateral umbels with male flowers; bracts absent, rarely 1 or 2; bracteoles 2–5, linear. *Sepals* absent. *Petals* white, the outer not or scarcely radiating, obcordate, apex inflexed. *Stamens* 5; filaments greenish; anthers yellowish. *Styles* 2, erect or divergent in fruit, gradually narrowed from an enlarged base forming the stylopodium which is about as long as the style; stigma truncate. *Schizocarp* 2.5–4.5 mm, ovoid-oblong, almost terete, slightly laterally compressed; mericarps with slender ridges; constricted at the commissure; carpophore present; with 2–3 reddish-brown vittae in each groove between the ridges of the mericarps; pedicels longer than the fruit. *Flowers* 5–6. $2n = 22$.

Native. Grassland, roadsides, hedgerows and woods. Common throughout much of Great Britain and Ireland, but absent from the Fens and scarce on chalk soils. Confined to western Europe from Norway southwards and extending eastwards to Italy. Oceanic Temperate element. The tuber is edible, raw or cooked, and has a pleasant nutty flavour.

14. Pimpinella L.

Tragoselinum Mill.

Perennial monoecious *herbs. Stems* erect, solid or hollow. *Leaves* 1–2(–3) pinnate. *Inflorescence* of compound umbels; bracts and bracteoles absent. *Sepals* absent.

Petals 5, white or pinkish, the outer not radiating. *Fruit* a schizocarp, ovoid, laterally compressed; mericarps with prominent ridges; commissure broad; carpophore present; vittae reddish-brown.

Contains about 150 species mainly in the north temperate region of the world. The flowers are strongly protandrous and are pollinated by insects.

Druce, G. C. (1918). *Pimpinella saxifraga var. poteriifolia* forma *nana* mihi. *Rep. Bot. Soc. Exch. Club Brit. Isles* **5**: 108.
Grime, J. P. et al. (1988). *Comparative plant ecology*. London.
Hegi, G. (1926). *Illustrierte Flora von Mittel-Europa* **5**(2): 1196–1212.
Hultén, E. & Fries, M. (1986). *Atlas of north European vascular plants north of the Tropic of Cancer*. 3 vols. Königstein.
Wolff, H. (1927). *Pimpinella* L. In Engler, A. (Edit.) *Das Pflanzenreich* **IV.228**: 219–319.

1. Rays hispid; schizocarp about 1.5 mm **3. affinis**
1. Rays glabrous or minutely hairy; schizocarp 2–4 mm 2.
2. Stems hollow, strongly ridged or angled; lower leaves with 3–4 pairs of segments; schizocarp 3–4 mm 3.
2. Stems solid, subterete; lower leaves with (2–)4–7 pairs of segments; schizocarp 2–3 mm 4.
3. Segments of lower leaves ovate or oblong-ovate, coarsely serrate **1(i). major** var. **major**
3. Segments of lower leaves linear or linear-lanceolate **1(ii). major** var. **dissecta**
4. Segments of leaves glabrous 5.
4. Segments of leaves hairy 6.
5. Segments of lower leaves ovate or subrotund and dentate **2(b,v). saxifraga** subsp. **saxifraga** var. **ovata**
5. Segments of lower leaves oblanceolate-cuneate **2(a). saxifraga** subsp. **alpestris**
6. Segments of lower leaves deeply divided with a narrow entire portion 7.
6. Segments of lower leaves dentate or incise-dentate with a broad entire portion 8.
7. Segments of lower leaves with linear lobes and linear teeth **2(b,i). saxifraga** subsp. **saxifraga** var. **dissecta**
7. Segments of lower leaves lanceolate with deeply incised lobes **2(b,iii). saxifraga** subsp. **saxifraga** var. **intercedens**
8. Plants up to 15 cm **2(b,ii). saxifraga** subsp. **saxifraga** var. **nana**
8. Plants taller 9.
9. Segments of basal leaves 5–15, fairly hairy **2(b,iv). saxifraga** subsp. **saxifraga** var. **saxifraga**
9. Segments of basal leaves 12–35 mm, often densely hairy **2(c). saxifraga** subsp. **nigra**

1. P. major (L.) Huds. Greater Burnet-saxifrage
P. saxifraga var. *major* L.; *P. magna* L. nom. illegit.; *Tragoselinum majus* (L.) Lam.; *Tragoselinum magnum* (L.) Moench nom. illegit.; *P. media* Weber; *P. austriaca* Mill.; *P. orientalis* Gouan; *Apium pimpinella* Caruel.

Perennial monoecious *herb* arising from a stout stock with no fibrous remains of petioles. *Stems* up to 120 (–200) cm, pale green, often purplish towards the base, erect, prominently ridged or angled, hollow, glabrous or nearly so, branched above, leafy. *Leaves* dull medium green on upper surface, paler beneath; basal and lower cauline 7–30×5–12 cm, oblong-ovate or ovate in outline, simply pinnate, very rarely 2(–3)-pinnate, with 3–4 pairs of segments up to 80(–100) mm, ovate or oblong-ovate, acute at apex, coarsely serrate or sometimes lobed, the teeth usually with fine cartilaginous points, truncate or subcordate at base and shortly petiolulate, the petioles long; upper cauline with 1–2(–3) pairs of segments, similar to the lower but smaller and often cuneate at the base, the petiole short and sheathing; all usually strigulose on the veins, at least beneath, and sparsely hairy between them. *Inflorescence* in compound umbels, with 10–20 rays which are 10–40 mm and smooth, the peduncle longer than the rays, the terminal umbel with mostly bisexual flowers, the lateral with mostly or entirely male flowers; bracts absent; bracteoles usually absent. *Sepals* absent. *Petals* 5, white or pinkish, the outer not radiating, not or slightly emarginate, the apex inflexed. *Stamens* 5; filaments white; anthers yellow. *Styles* 2, erect in flower, elongating later but often breaking off as the fruit ripens, with enlarged base forming the stylopodium; stigma capitate. *Schizocarp* 3–4 mm, ovoid, laterally compressed; mericarps with prominent, slender ridges; commissure broad; carpophore present; vittae usually 3, reddish-brown, deeper in colour than the ridges of the mericarps, the pedicels about 5 mm and glabrous. *Flowers* 7–8. $2n=20$.

(i) Var. **major**
Segments of lower leaves ovate or oblong-ovate, coarsely serrate.

(ii) Var. **dissecta** (Lej.) Fiori & Paol.
P. bipinnata Boberski ex Georgi; *P. laciniata* Thore; *P. dissecta* Lej.; *P. magna* var. *dissecta* (Lej.) Spreng.; *Carum dissectum* (Lej.) Baill.

Segments of lower leaves linear or linear-lanceolate.

Native. Locally common at the margins of woods, and on grassy roadsides, by hedges and often where woods existed in the past. Central, east, south-east and south-west England, south and west-central Ireland; rare casual in Wales and Scotland; very rare in Guernsey. Most of Europe, except the extreme north and south. European Temperate element. The common variant is var. *major*. Var. *dissecta* is rare and can easily be mistaken for *P. saxifraga*.

2. P. saxifraga L. Burnet-saxifrage
Apium tragoselinum Crantz; *Tragoselinum minus* Lam.; *P. hircina* Mill.; *Tragoselinum saxifragum* (L.) Moench; *Carum saxifragum* (L.) Baill.; *Selinum pimpinella* E. H. L. Krause; *Apium saxifragum* (L.) Calest.

Perennial monoecious *herb* arising from a stock which often has on it fibrous remains of petioles. *Stems* up to 100 cm, pale green, erect, rather slender, solid, subterete, developing a small hollow in fruit, glabrous or minutely hairy, branched above, sparingly leafy. *Leaves* medium dull green on upper surface, paler beneath; basal and lower cauline few to numerous, 3.0–7.5(–11)×3–5(–6) cm broadly ovate to oblong-lanceolate in outline, pinnate, with (2–)4–6(–7) pairs of segments which are up to 25 mm, broadly ovate to lanceolate in outline, coarsely serrate to pinnatifid, the teeth with very slender cartilaginous points, the petiole long; upper cauline with up to 3 pairs of small, very narrow segments, the petiole longer than the lobes, inflated and usually purplish; all glabrous to more or less densely hairy. *Inflorescence* in compound umbels, with 10–22 rays which are (10–)20–30(–45) mm and smooth, the peduncle longer than the rays, the umbels mostly with bisexual flowers; bracts and bracteoles absent. *Sepals* absent. *Petals* 5, white, rarely pinkish, not or slightly emarginate, the apex inflexed. *Stamens* 5; filaments whitish; anthers yellowish. *Styles* 2, shorter than petals and usually diverging when in flower, elongating later, but often breaking off as the fruit ripens, with enlarged base forming the stylopodium; stigma capitate. *Schizocarp* 2–3 mm, ovoid, laterally compressed; commissure broad; carpophore present; mericarps with slender ridges; vittae reddish-brown, darker than the slender ridges of the mericarps; pedicels about 5 mm, glabrous. *Flowers* 7–8. $2n = 40$.

(a) Subsp. **alpestris** (Spreng.) Vollm.
P. saxifraga var. *alpestris* Spreng.; *P. alpestris* (Spreng.) Schult.; *P. saxifraga* forma *alpestris* (Spreng.) Schube; *P. alpina* Wulfen ex Vest; *Carum alpinum* (Wulfen ex Vest) Baill.; *P. saxifraga* subsp. *alpina* (Wulfen ex Vest) K. Malý; *P. lucida* Schur

Plant slender, up to 40 cm. *Basal leaves* with segments oblanceolate-cuneate, glabrous.

(b) Subsp. **saxifraga**
Plant slender to robust, up to 100 cm. *Basal leaves* with segments linear to ovate, usually hairy, sometimes glabrous.

(i) Var. **dissecta** With.
P. hircina Mill.; *Tragoselinum hircinum* (Mill.) Moench; *P. saxifraga* var. *hircina* (Mill.) DC.; *P. saxifraga* var. *laciniata* G. Mey.; *Tragoselinum dissectum* (With.) Moench; *P. saxifraga* forma *dissecta* (With.) Schube; *Apium tragoselinum* var. *dissectum* (With.) Caruel; *P. genevensis* Vill.; *P. pratensis* Thuill.; *P. saxifraga* var. *pratensis* (Thuill.) Corb.; *P. tenuifolia* Schweigg. & Körte; *P. saxifraga* var. *dissectifolia* W. D. J. Koch; *P. saxifraga* var. *roseiflora* E. & H. Drabble

Plant slender, up to 100 cm. *Basal leaves* with segments dissected into linear lobes, hairy.

(ii) Var. **nana** (Druce) P. D. Sell
P. saxifraga forma *nana* Druce; *P. saxifraga* forma *arenaria* Bryhn ex Blytt

Plants small, up to 15 cm. *Basal leaves* with segments up to 12 mm, ovate, incise-dentate, hairy.

(iii) Var. **intercedens** Thell.
P. saxifraga var. *intermedia* Hagenb.

Plant medium. *Basal leaves* with segments lanceolate, sharply dentate to incise-dentate, hairy.

(iv) Var. **saxifraga**
P. saxifraga var. *minor* Spreng.; *P. saxifraga* var. *poteriifolia* W. D. J. Koch

Plant medium. *Basal leaves* with segments 5–15 mm, ovate, dentate, hairy.

(v) Var. **ovata** Spreng.
P. saxifraga var. *major* Mert. & W. D. J. Koch; *P. saxifraga* var. *rotundifolia* Beck

Plant medium. *Basal leaves* with segments 8–20 mm, ovate, dentate or slightly lobed, glabrous.

(c) Subsp. **nigra** (Mill.) Gaudin
P. nigra Mill.; *Carum nigrum* (Mill.) Baill.; *P. saxifraga* var. *nigra* (Mill.) Spreng.; *P. saxifraga* var. *pubescens* auct.; *P. saxifraga* var. *nemoralis* Wimm. & Grab.; *P. saxifraga* var. *cinerea* Lamotte

Plant robust. *Basal leaves* with segments 20–40 mm, obovate, dentate or incise-dentate, hairy to densely hairy.

Native. Grassland and open rocky places. Common throughout most of Great Britain and Ireland. Most of Europe; southwest Asia. Eurosiberian Temperate element. Subsp. *saxifraga* is the common plant. Var. *nana* was described from Clogwyn ddur Arddu, Snowdon, and occurs on Freshwater Down on the Isle of Wight. These plants must have different origins but seem to be morphologically inseparable. A similar Norwegian plant has been called forma *arenaria* Bryhn ex Blytt. Var. *dissecta* occurs in chalk and limestone grassland, sandhills, greensand and streamsides. Var. *intercedens* also occurs on chalk and sand as does var. *saxifraga*. It is possible that var. *intercedens* originates from a cross between var. *dissecta* and var. *saxifraga*, but it is not known if they grow together or what their breeding mechanism is. The only specimens we have seen of var. *ovata* are from a shingly beach at Shoreham in Sussex, and Bettyhill in Sutherland. It was recorded by J. H. Salter on a cliff slope above the beach at Penbryn in Cardiganshire and has recently been re-found there by Arthur Chater. It should be looked for in other coastal localities. Subsp. *alpestris* is known only from a specimen collected at Yspytly Cynfyn in Cardiganshire which fits European material. Arthur Chater has recently re-found it there. Subsp. *nigra* is a large, hairy plant sometimes mistaken for *P. major* which occurs in widely scattered localities, perhaps usually in shady and damp places.

3. P. affinis Ledeb. Caucasian Burnet-saxifrage
P. gracilis Bisch.; *P. reuteriana* Boiss.; *P. peregrina* Kotschy ex Boiss.

Biennial monoecious *herb* arising from a rather stout stock. *Stems* up to 60 cm, pale green, erect, strict, glabrous and shortly hairy. *Leaves* dull medium green on upper surface, paler beneath; basal oblong or oblong-ovate in outline, simply pinnate, the segments 3–4 × 2–3 cm, rounded-ovate, obtuse at apex, deeply and obtusely crennate-serrate, obliquely cuneate at base, glabrous or slightly hairy, sessile or slightly petiolulate, the petioles long; cauline with entire or serrate, lanceolate segments. *Inflorescence* in compound umbels, with 10–12 rays, the rays hispid; bracts and bracteoles absent. *Sepals* absent. *Petals* 5, white, the outer non-radiating. *Stamens* 5. *Styles* 2, with enlarged base forming an elongate-conical stylopodium. *Schizocarp* about 1.5 mm, ovoid, compressed; commissure broad; carpophore present. *Flowers* 7–8.

Introduced. Naturalised at Nelson Hill, Southampton in Hampshire since 1990, but perhaps now gone. Native of north and west Iran, Caucasia and Transcaspia.

15. **Aegopodium** L.

Perennial monoecious *herbs* with long, slender rhizomes. *Stems* erect, hollow. *Leaves* 1- to 2-ternate. *Inflorescence* of compound umbels; bracts and bracteoles usually absent. *Sepals* absent. *Petals* 5, white, the outer somewhat larger than the inner. *Fruits* a schizocarp, ovoid, laterally compressed; mericarps with slender ridges; commissure broad; carpophore present; vittae absent when the fruit is mature.

Contains 5 to 7 species in Europe and Asia.

Clapham, A. R. (1953). Human factors contributing to a change in our flora: the former ecological status of certain hedgerow species. In Lousley, J. E. (Edit.) *The changing flora of Britain*, pp. 26–39. Arbroath.

Hultén, E. & Fries, M. (1986). *Atlas of north European vascular plants north of the Tropic of Cancer.* 3 vols. Königstein.

1. A. podagraria L. Ground-elder

Pimpinella angelicaefolia Lam. nom. illegit.; *A. angelicaefolia* Salisb. nom. illegit.; *Apium biternatum* Stokes nom. illegit.; *Ligusticum podagraria* (L.) Crantz; *Seseli aegopodium* Scop.; *Sium podagraria* (L.) Webber ex Wigg.; *Sium vulgare* Bernh.; *Pimpinella podagraria* (L.) Lestib.; *Tragoselinum angelica* Lam.; *Podagraria aegopodium* (Scop.) Moench; *Podagraria erratica* Bubani; *Apium podagraria* (L.) Caruel; *Sison podagraria* (L.) Spreng.; *Carum podagraria* (L.) Roth; *Selinum podagraria* (L.) E. H. L. Krause

Perennial monoecious *herb* with long, slender rhizomes. *Stems* up to 120 cm, pale green, more or less robust, erect, hollow, grooved, glabrous, branched above, leafy. *Leaves* pale yellowish-green on upper surface, paler beneath; basal and lower cauline numerous, 6–15 × 11–17 cm, ovate in outline, 1- to 2-ternate, the segments up to 10.5 × 5.6 cm, ovate, rounded to an acute apex, serrate, the teeth cartilaginous and often unequal at the base, long-petiolate; upper cauline much smaller and with short, sheathing petioles; all glabrous. *Inflorescence* of compound umbels which are hemispherical with 10–21 rays which are 10–40 mm and smooth, the peduncle longer than the rays, the umbels with most flowers bisexual; bracts and bracteoles usually absent. *Sepals* absent. *Petals* 5, white, the outer somewhat larger than the inner, obcordate, the apex inflexed. *Stamens* 5; filaments white; anthers pink. *Styles* 2, short and erect in flower, elongating and becoming appressed to the nature fruit with enlarged base forming the stylopodium; stigma capitate. *Schizocarp* 3–4 mm, ovoid, laterally compressed; mericarps with slender ridges; commissure broad; carpophore present; vittae absent when the fruit is mature; pedicels 3–8 mm, glabrous. *Flowers* 5–6. $2n = 22, 42, 44$.

Probably always introduced. Grassy places near buildings, waste places and other open ground and an extremely persistent garden weed. Common throughout Great Britain and Ireland. A native woodland plant in most of Europe and temperate Asia. Almost certainly introduced to Great Britain by the Romans as a pot-herb and a medicine against gout. Hence one of its popular names, Goutweed.

16. **Sium** L.

Perennial monoecious *herbs. Stems* erect, hollow. *Leaves* of 2 kinds; submerged leaves in spring 2- to 3-pinnate; aerial leaves simply pinnate, with 3–8(–16) pairs of lanceolate to ovate segments. *Inflorescence* of compound umbels; bracts 2–6, sometimes large and leaf-like; bracteoles 3–6, lanceolate. *Sepals* 5, lanceolate. *Petals* 5, white, the outer not radiating. *Fruit* a schizocarp, ovoid or ellipsoid, laterally compressed; mericarps with thick, prominent ridges; constricted at the commissure; carpophore present; vittae 3, superficial.

Contains 10 to 15 species, widely distributed, but absent from South America and Australia.

Hultén, E. & Fries, M. (1986). *Atlas of north European vascular plants north of the Tropic of Cancer.* 3 vols. Königstein.

Stewart, A., Pearman, D. A. & Preston, C. D. (1994). *Scarce plants in Britain.* Peterborough.

1. S. latifolium L. Greater Water Parsnip

Pimpinella latifolia (L.) Stokes; *Drepanophyllum palustre* Hoffm. nom. illegit.; *S. lancifolium* Schrank; *S. berula* J. F. Gmel.; *S. sulcatum* Pers.; *S. longifolium* J. & C. Presl; *Coriandrum latifolium* (L.) Crantz; *Cicuta latifolia* (L.) Crantz; *Drepanophyllum latifolium* (L.) K.-Pol.; *Sisarum palustre* Bubani; *Selinum sium* E. H. L. Krause

Perennial monoecious *herb* which can live submerged for some years without flowering. *Stems* up to 150 (–200) cm, pale green, erect, robust, hollow, grooved, glabrous, little branched above, leafy. *Leaves* pale green, even paler beneath; submerged leaves present only in spring, 2- to 3-pinnate with linear segments; aerial leaves 15–37 × 6–26 cm, ovate-oblong in outline, simply pinnate, with 3–8(–16) pairs of segments, the segments up to 15 × 6 cm, lanceolate to ovate, more or less acute at apex, serrate with cartilaginous margin and teeth and with a rounded, unequal base, the lower leaves with long petioles the upper gradually decreasing in size with short, sheathing petioles; all glabrous. *Inflorescence* of compound umbels, with 20–30 rays, the rays 15–40 mm and smooth, the peduncle shorter than to longer than the rays, often leaf-opposed and smooth, the terminal umbel with bisexual flowers, the lateral umbel with the flowers almost entirely male; bracts 2–6, sometimes large and leaf-like; bracteoles 3–6, lanceolate. *Sepals* 5, about 1 mm, lanceolate. *Petals* 5, white, obcordate, apex inflexed, papillose beneath, the outer not radiating. *Stamens* 5; filaments and anthers whitish. *Styles* 2, very short and stout, recurved in fruit, with an enlarged base forming the stylopodium; stigma spathulate. *Schizocarp* 3–4 mm, ovoid, or ellipsoid, laterally compressed; mericarps with thick, prominent ridges; constricted at the commissure; carpophore present; vittae 3, superficial; pedicels 5–10 mm. *Flowers* 7–8. Produces abundant seed but new individuals are rare. $2n = 20$.

Native. Its natural habitat is very wet, species-rich, tall herb fen which develops as a semi-floating raft on the margins of lakes and large rivers. After drainage and reclamation of many fens it is now most often found in drainage ditches in more species-poor reed-swamp, and is frequent in old peat cuttings. It prefers shallow, still or slow-moving water which is alkaline, rich in nitrogen and on a peat or alluvial soil. Very local and decreasing in England and central Ireland, formerly north to central-east Scotland and now mostly in central-east England; there is one recent record for Jersey. Widespread in Europe, but very rare near the Mediterranean and absent from Portugal; extends into temperate Siberia and is recorded from south-eastern Australia, but is unknown in Japan and the Americas. Eurosiberian Temperate element.

17. **Berula** Besser ex W. D. J. Koch

Perennial monoecious *herbs* with stolons. *Stems* erect or decumbent, hollow. *Leaves* pinnate, those submerged remaining green all winter, but little different from aerial. *Inflorescence* of compound umbels; bracts 4–7, often 3-fid or pinnatifid; bracteoles 4–7, sometimes 3-fid. *Sepals* 5, triangular, not persistent. *Petals* 5, white, the outer not radiating. *Fruit* a schizocarp, globose, somewhat laterally compressed; mericarps with slender ridges; constricted at the commissure; carpophore present; vittae sunk in the pericarp.

Contains 3 species in the north temperate regions of the world.

Haslam, S. M. (1978). *River plants*. Cambridge.
Haslam, S. M., Sinker, C. A. & Wolseley, P. A. (1975). British Water plants. *Field Studies* **4**: 243–351.
Hultén, E. & Fries, M. (1986). *Atlas of north European vascular plants north of the Tropic of Cancer.* 3 vols. Königstein.
Preston, C. D. & Croft, J. M. (1997). *Aquatic plants in Britain and Ireland*. Colchester.

1. B. erecta (Huds.) Coville Lesser Water Parsnip
Sium erectum Huds.*; Sium angustifolium* L. nom. illegit.; *Sium berula* Gouan; *Apium berula* (Gouan) Caruel; *B. angustifolia* Mert. & W. D. J. Koch nom. illegit.; *Sium nodiflorum* Oed., non L.; *Sium incisum* Pers.; *Apium sium* Crantz; *Berla monspeliensium* Bubani; *Selinum berula* (Gouan) E. H. L. Krause

Perennial monoecious *herb* with stolons. *Stems* up to 100 cm and 10 mm wide, pale green, erect or decumbent, hollow, grooved, glabrous, branched, leafy. *Leaves* medium yellowish-green on upper surface, paler beneath, submerged leaves remaining green all winter; basal and lower cauline, 10–30 × 4–11 cm, broadly oblong in outline, pinnate, with 5–9(–14) pairs of segments, the segments up to 5 × 2 cm, oblong-lanceolate to ovate, acute at apex, 1- to 2-serrate, sometimes lobed, the teeth acute, cartilaginous or sometimes rounded with a cartilaginous mucro and cuneate to rounded and somewhat unequal at base, long-petiolate with a ring-mark below the lowest pair of leaflets; upper cauline similar, but smaller with short, sheathing petioles; all glabrous. *Inflorescence* of compound umbels with 7–18 rays which are 5–25 mm and smooth, the peduncle as long as or somewhat longer than the rays and usually leaf-opposed, the umbels all with bisexual flowers; bracts 4–7, lanceolate, 3-fid or pinnatifid; bracteoles usually 4–7, lanceolate or 3-fid. *Sepals* 5, triangular, sometimes unequal. *Petals* 5, white, smooth beneath, the outer not radiating, obcordate, the apex inflexed. *Stamens* 5, recurved, with enlarged base forming the stylopodium which is a little shorter than the style; stigma capitate. *Schizocarp* 1.3–2.0 mm, globose, somewhat compressed laterally; mericarps with slender ridges; constricted at the commissure; carpophore present; vittae sunk in the pericarp; pedicels 3–5 mm, smooth. *Flowers* 7–9. $2n = 12$, 18.

Native. In and by water in ditches, streams, marshes, lakes and rivers. Frequent over much of Ireland and Great Britain north to central Scotland. Europe from southern Scandinavia southwards although rare in the Mediterranean region; extends eastwards to central Asia and is widespread in temperate North America; also recorded as an introduction in Africa. European Temperate element and elsewhere. Submerged clumps often fail to flower, but terrestrial plants flower more freely. The species can become established from small floating plants and vegetative fragments, but little in known how frequently they reproduce from seed. It can be distinguished from *Apium nodiflorum* with which it is often confused in a vegetative state by the presence of a ring-mark on the petiole some way below the lowest pair of leaflets.

18. **Crithmum** L.

Perennial monoecious *herbs. Stems* woody below, solid, erect to decumbent. *Leaves* 2- to 3-pinnate, fleshy. *Inflorescence* of compound umbels; bracts 5–10, broadly membranous and ultimately deflexed; bracteoles 6–8, deflexed in fruit. *Sepals* 5, very small. *Petals* 5, yellowish-green, the outer not radiating. *Fruits* a schizocarp, ellipsoid, not compressed; mericarps with thick, prominent ridges; commissure broad; carpophore present; vittae several in each group of the spongy mesocarp.

Contains 1 species on the Atlantic coasts of Europe, the Mediterranean, the Black Sea and Macaronesia.

Malloch, A. J. C. & Okusanya, O. T. (1979). An experimental investigation into the ecology of some maritime cliff species. I. Field Observations. *Jour. Ecol.* **67**: 283–292.
Okusanya, O. T. (1979). An experimental investigation into the ecology of some maritime cliff species. II. Germination studies. *Jour. Ecol.* **67**: 293–304; III. Effects of water on growth. *Jour. Ecol.* **67**: 579–590; IV. Cold sensitivity and competition studies. *Jour. Ecol.* **67**: 591–600.

1. C. maritimum L. Rock Samphire
Cachrys maritima (L.) Spreng.

Perennial monoecious *herb* smelling of furniture polish when crushed. *Stems* 15–45 cm, greyish-green, erect to decumbent, robust and woody below, solid, striate, branched, leafy. *Leaves* 9–15 × 7–11 cm, greyish-green, fleshy; lower deltate in outline, 2- to 3-pinnate, the segments subulate to linear-lanceolate, acute at apex, entire, glabrous, with long petioles, shortly sheathing at the base; upper not much smaller than lower, with entirely sheathing petioles and glabrous. *Inflorescence* of compound umbels, with 8–36 rays which are 10–40 mm, smooth and thickening in fruit, the peduncles usually longer than the rays, all umbels with bisexual flowers; bracts 5–10, lanceolate, broadly membranous and ultimately deflexed; bracteoles 6–8, lanceolate, deflexed in fruit. *Sepals* 5, very small. *Petals* 5, yellowish-green, the outer not radiating, obcordate, the apex inflexed. *Stamens* 5; filaments and anthers greenish. *Styles* 2, divergent, with enlarged base forming the stylopodium which is longer than the styles; stigma truncate. *Schizocarp* 3.5–5.0 mm, ellipsoid, not compressed; mericarps with thick, prominent ridges; commissure broad; carpophore present; vittae several in each group of the spongy meriocarp; pedicels 5–10 mm. *Flowers* 6–8. $2n=20$.

Native. Sea-cliffs and rocks, more rarely on sand or shingle near the sea. Coasts of Great Britain and Ireland, north to Suffolk on the east coast and southern Scotland on the west. Coasts of the northern Atlantic, the Mediterranean, the Black Sea and Macaronesia. Mediterranean-Atlantic element. It was formerly preserved in brine and sent by cask from the south coast to London for distribution.

19. **Seseli** L.

Libanotis Hill

Biennial or monocarpic *perennial herbs*. *Stems* erect, solid, surrounded at the base by fibrous remains of petioles. *Leaves* 2- to 3-pinnate. *Inflorescence* of compound umbels; bracts usually 8 or more, entire, erect or more or less deflexed; bracteoles 10–15. *Sepals* 5, small, but conspicuous. *Petals* 5, white, the outer not radiating. *Fruit* a schizocarp ovoid or ellipsoid, not or scarcely compressed, minutely hairy; mericarps with broad ridges; commissure broad; carpophore present; vittae solitary.

About 80 species, mainly in Europe and temperate Asia.

Wigginton, M. J. (Edit.) (1999). *British red data books*. Vol.1. *Vascular plants*. Peterborough.

Hultén, E. & Fries, M. (1986). *Atlas of north European vascular plants north of the Tropic of Cancer.* 3 vols. Königstein.

1. S. libanotis (L.) W. D. J. Koch Moon Carrot

Athamanta libanotis L.; *Libanotis daucifolia* (Scop.) Rchb.; *Libanotis montana* Crantz

Biennial or monocarpic *perennial herb*. *Stems* up to 60 (–100) cm, pale green, erect, robust, solid, strongly ridged, more or less hairy, branched above, leafy and surrounded at the base by numerous, fibrous remains of petioles. *Leaves* dull medium green on upper surface, paler beneath; basal 9–18 × 2–9 cm, triangular-ovate or oblong in outline, usually 2- to 3-pinnate, the pinnae oblong or oblong-lanceolate, the segments 5–15 mm, lanceolate to ovate in outline, deeply serrate to pinnatifid, the lobes narrow, long-petiolate; cauline small, with short, inflated, sheathing petioles; all with short hairs. *Inflorescence* of compound umbels, with (10–)20–60 rays, the rays 20–30 mm and more or less shortly hairy, the peduncle longer than the rays and more or less hairy at least at the apex; all umbels with bisexual flowers; bracts usually 8 or more, linear, entire, erect or more or less deflexed; bracteoles 10–15. *Sepals* 5, conspicuous, deciduous, linear to ovate-lanceolate. *Petals* 5, white, outer not radiating. *Stamens* 5; filaments and anthers whitish. *Styles* 2, divergent to recurved, with an enlarged base forming the stylopodium which is shorter than the style; stigma capitate. *Schizocarp* 2.5–3.5 mm, ovoid or ellipsoid, scarcely compressed, rather sparsely to densely, short-hairy; mericarps with broad ridges; commissure broad; carpophore present; vittae solitary; pedicels 2–7 mm, hairy. *Flowers* 7–8. $2n=18, 22$.

Very variable in leaf-dissection and hairiness.

Native. Rough grassy and bushy places on chalk. Very local and stable in Sussex, Hertfordshire, Cambridgeshire and Bedfordshire. Most of Europe, western Asia and North Africa. Eurasian Temperate element. *S. libanotis* has been split into many species, subspecies and varieties, but it seems to be impossible to distinguish them. Our plant is subsp. **libanotis** which is mainly found in western and central Europe.

20. **Oenanthe** L.

Phellandrium L.

Annual to *perennial* monoecious *herbs*, often with tuberous roots. *Stems* erect to ascending, or floating, hollow or solid. *Inflorescence* of compound umbels; bracts absent, to several; bracteoles usually numerous. *Sepals* 5, conspicuous and persistent. *Petals* 5, white, the outer somewhat or scarcely radiating. *Fruit* a schizocarp, cylindrical to obovoid or ovoid, not or very slightly compressed; mericarps with usually rather prominent ridges; commissure broad; carpophore present; vittae solitary.

Contains about 35 species in the north temperate regions of the Old World.

Coleman, W. H. (1844). Observations on a new species of *Oenanthe*. *Ann. Mag. Nat. Hist.* **13**: 188–191, with plate.

Cook, C. D. K. (1983). Aquatic plants endemic to Europe and the Mediterranean. *Bot. Jahrb.* **103**: 539–582. [*O. fluviatilis*.]

Hroudová, Z., Zákravsky, P., Hrouda, L. & Ostry, I. (1992). *Oenanthe aquatica* (L.) Poiret; seed reproduction, population structure, habitat conditions and distribution in Czechoslovakia. *Folio Geobot. Phytotax.* **27**: 301–355.

Hultén, E. & Fries, M. (1986). *Atlas of north European vascular plants north of the Tropic of Cancer.* 3 vols. Königstein.

Preston, C. D. & Croft, J. M. (1997). *Aquatic plants in Britain and Ireland*. Colchester. [*O. aquatica,crocata, fistulosa* and *fluviatilis*.]

Stewart, A., Pearman, D. A. & Preston, C. D. (1994). *Scarce plants in Britain*. Peterborough. [*O. fluviatilis, pimpinelloides* and *silaifolia*.]

1. Some umbels leaf-opposed; peduncles shorter than rays; styles less than one-quarter as long as the mature schizocarp 2.
1. All umbels terminal; peduncles longer than rays; styles more than quarter as long as the mature schizocarp 3.
2. Stems usually floating at least at base; schizocarp 5–6.5 mm **6. fluviatilis**
2. Stems ascending to erect, often terrestial; schizocarp 3.0–4.5 mm **7. aquatica**
3. All leaves usually with petioles larger than the divided part; ultimate umbels of ripe fruit globose; schizocarps sessile **1. fistulosa**

3. All leaves usually with petioles shorter than the divided part; ultimate umbels of ripe fruit not globose; some schizocarps with pedicels 4.
4. Segments of middle cauline leaves ovate to more or less subrotund, less than twice as long as wide; schizocarps 4.0–5.5 mm **5. crocata**
4. Segments of middle cauline leaves more or less linear, more than 3 times as long as wide 5.
5. Stems at maturity hollow, straw-like, with walls about 0.5 mm thick; bracts absent on most umbels; rays more than 1 mm thick at fruiting **2. silaifolia**
5. Stems at maturity solid to hollow with walls more than 0.5 mm thick; bracts 1–5; rays less than 1 mm thick at fruiting 6.
6. Root-tubers ovoid or ellipsoid, the proximal part of the root not thickened; rays and pedicels thickening at fruiting, the pedicels more than 0.5 mm thick **3. pimpinelloides**
6. Root-tubers cylindrical or fusiform, more or less gradually widening from base of root; rays and pedicels scarcely thickening when fruiting, the pedicels less than 0.5 mm thick **4. lachenalii**

1. O. fistulosa L. Tubular Waterdropwort
O. fistulifolia Stokes nom. illegit.; *O. lanceolata* Poir.; *O. filipendula* Dumort.; *O. meifolia* Schloss. & Vuk.; *O. biloba* Dulac; *Phellandrium fistulosum* (L.) Clairv.; *Phellandrium dodonaei* Bubani; *Selinum fistulosum* (L.) E. H. L. Krause

Perennial monoecious *herb* with stolons and cylindrical to fusiform, tuberous roots. *Stems* 30–80 × about 0.5 cm, pale bluish-green, erect, fistular, thin-walled, often constricted at the nodes, striate, glabrous, sparsely branching, leafy, sometimes rooting at the lower nodes. *Leaves* medium green on upper surface, paler beneath; basal 3–15 × 2–4 cm, oblong or oblong-lanceolate in outline, 3-pinnate, the segments linear to lanceolate, lobed and obtuse at apex, petiolate, sometimes submerged with filiform segments and soon withering; cauline mostly 1-pinnate, the segments 5–20 mm, obtuse at apex and entire, the petioles fistular and longer than the lamina; all glabrous. *Inflorescence* of compound umbels, with 2–4 rays, the rays 10–30 mm, thickening after flowering, the peduncle longer than the rays; terminal umbels with bisexual and some male flowers, the lateral umbels with male flowers; bracts absent; bracteoles 7–16, linear; partial umbels globose in fruit. *Sepals* 5, conspicuous, acute at apex, persistent. *Petals* 5, white or pinkish, the outer somewhat radiating, emarginate, the apex long and inflexed. *Stamens* 5; filaments white; anthers greenish. *Styles* 2, at least as long as the fruit, erect, the enlarged base forming the stylopodium; stigma a small knob. *Schizocarp* 2.5–4 mm, cylindrical to obovoid; mericarps with inflated, corky ridges; commissure broad; carpophore present; vittae solitary; pedicels present only on periphery of partial umbels, not thickened in fruit. *Flowers* 7–9. $2n=22$.

Native. Meadows and pastures on the flood plain of rivers, marshes and fens, with a species-rich sward, and in the fringing vegetation at the edge of lakes, ponds, rivers, streams, canals and ditches in the lowlands, which are flooded for at least part of the winter. Also as an emergent in shallow water, particularly in grazing marsh ditches and in dune slacks. Widespread, but not common in England, Wales and Ireland and rare in southern Scotland. Much decreased since 1960 because of conversion of grassland to arable. Europe from Scandinavia southwards and in adjacent parts of North Africa and south-west Asia. European Temperate element.

2. O. silaifolia M. Bieb. Narrow-leaved Waterdropwort
O. media Griseb.; *O. smithii* H. C. Watson; *O. caucasica* Simon; *O. radiata* Sakalo

Perennial monoecious *herb* with narrowly obovoid or fusiform tubers which taper towards their junction with the stem. *Stems* 30–100 × about 0.7 cm, pale bluish-green, solid at the base, hollow above with walls about 0.5 mm thick, erect, grooved, striate and glabrous, branched above, leafy. *Leaves* medium green on upper surface, paler beneath; lower 5–12 × 2–7 cm, triangular-ovate in outline, 2- to 4-pinnate, the segments 10–30 mm, linear-lanceolate and acute at apex, long petiolate but shorter than the divided part, soon withering; upper 1- to 2-pinnate, the segments linear-lanceolate and acute at apex, the petiole shorter than the lamina; all glabrous. *Inflorescence* of compound umbels, with 4–8(–10) rays, the rays 15–30 mm, smooth and thickening after flowering to more than 1.0 mm, the peduncle longer than the rays, the terminal umbels with long-pedicellate male flowers and shortly pedicellate bisexual flowers, the lateral umbels with male flowers; bracts usually absent; bracteoles 10–17, lanceolate and acute at apex; partial umbels not flat-topped in fruit. *Sepals* 5, conspicuous, acute at apex, persistent. *Petals* 5, white, the outer somewhat radiating, emarginate, the apex long and inflexed. *Stamens* 5; filaments and anthers whitish. *Styles* 2, shorter than the fruit, erect to somewhat divergent, with enlarged bases forming the stylopodium; stigma tapering. *Schizocarp* 2.5–3.5 mm, cylindrical; mericarps with prominent ridges; commissure broad; carpophore present; vittae solitary; pedicels thickened in fruit, sometimes very short and occasionally absent in centre of partial umbels. *Flowers* 6. $2n=22$.

Native. Unimproved damp meadows, primarily those used for hay on riverside alluvium which receives calcareous floodwater in winter. Some colonies are very large, but the plant is only readily detected during its brief flowering period. Most plants are cut for hay before shedding seed and there is no second flowering or further growth until the next winter rosette. The cluster of tubers undergoes replenishment, enabling a plant to exist at the same spot for 20 or more years. Scattered localities in central and southern England north to Yorkshire and just into Wales, decreasing. Quite widespread in western central and southern Europe east to Russia; north-west Africa; Turkey and possibly further east, but confusion with other *Oenanthe* species. European Southern-temperate element.

3. O. pimpinelloides L. Corky-fruited Waterdropwort
Phellandrium mathioli Bubani

Perennial monoecious *herb* with ovoid or ellipsoid root tubers very variable in size and distant from the base of the stem. *Stems* 30–100 × about 0.5 cm, pale bluish-green, often purplish at base, solid, strongly grooved, erect, branched, leafy, glabrous. *Leaves* medium green on upper surface, paler beneath; lower 7–12×5–9 cm, triangular-ovate in outline, 2-pinnate, the segments about 5 mm, lanceolate to ovate, deeply toothed or pinnatifid and cuneate at base, long-petiolate but shorter than the divided part; upper 1- to 2-pinnate, the segments 10–30 mm, linear and entire, at least as long as the petiole; all glabrous. *Inflorescence* of compound umbels, with 6–15 rays, the rays 10–20 mm, smooth, thickening after flowering, the peduncles longer than the rays, the terminal umbels with long-pedicillate male flowers and shortly pedicellate bisexual flowers, the lateral umbels with male flowers; bracts 1–5, linear to linear-lanceolate; bracteoles 12–20, linear to linear-lanceolate; partial umbels flat-topped in fruit. *Sepals* 5, conspicuous, acute at apex, persistent. *Petals* 5, white, the outer somewhat radiating, emarginate, the apex long and inflexed. *Stamens* 5; filaments whitish; anthers cream. *Styles* 2, whitish, about as long as the fruit, erect, with enlarged base forming the stylopodium; stigma a small knob. *Schizocarp* 3.0–3.5 mm, cylindrical; mericarps with prominent ridge; commissure broad; carpophore present; vittae solitary; pedicels thickening after flowering to more than 0.5 mm, especially near their glabrous apex. *Flowering* 6–8. $2n=22$.

Native. Lowland, neutral pastures and hay meadows, and on grassy banks and roadsides, extending on drier ground than other British species of the genus. Locally common from Devonshire and Somersetshire to Hampshire, very local across the remainder of southern England to southern Essex and Suffolk and north to Worcestershire and Monmouthshire; formerly in one locality in Co. Cork. Widespread in southern and central Europe north to western Ireland, southern Great Britain and Belgium, and eastwards to Asia Minor. Mediterranean-Atlantic element.

4. O. lachenalii C. C. Gmel. Parsley Waterdropwort

Perennial monoecious *herb* with cylindrical or fusiform tuberous roots, more or less gradually widening from root base. *Stems* 30–100×0.4 cm, pale green, purplish at base, solid, sometimes developing a small cavity when old, with walls more than 0.5 mm thick, erect, striate, glabrous, branched, leafy. *Leaves* medium green on upper surface, paler beneath; lower 5–10×2–5 cm, oblong in outline, (1–)2(–3) -pinnate, the segments 10–20 mm, linear to spathulate or rarely narrowly obovate and entire to pinnatifid, long-petiolate, but shorter than the divided part, soon withering; upper 1- to 2-pinnate, the segments 15–50 mm and linear to linear-lanceolate, the petiole shorter than the lamina; all glabrous. *Inflorescence* of compound umbels with 5–9(–20) rays, the rays 10–30 mm, smooth and not thickening after flowering, the peduncle longer than the rays, the umbels with male and bisexual flowers, the proportion of the latter decreasing in the latter lateral umbels; bracts up to 5, subulate or linear-lanceolate; bracteoles 5–7, oblong-lanceolate, acute at apex; partial umbels not flat-topped in fruit. *Sepals* 5, conspicuous, acute at apex, persistent. *Petals* 5, white, the outer somewhat radiating, emarginate, the apex long and inflexed. *Stamens* 5; filaments whitish; anthers purplish. *Styles* 2; shorter than the fruit, divergent or recurved, with enlarged base forming the stylopodium; stigma tapering. *Schizocarp* 2.5–3.0 mm, obovoid; mericarps with prominent, slender ridges; commissure broad; carpophore present; vittae solitary; pedicels not thickening after flowering and less than 0.5 mm thick. *Flowers* 6–9. $2n=22$.

Native. Marshes, fens, ditches and dykes, mostly near the sea and often where the water is brackish. Coasts of Great Britain and Ireland except north and east Scotland; scattered localities inland in England. Western Europe, extending eastwards to Poland and Yugoslavia; very rare in Algeria. Southern Temperate element. Named after Werner de Lachenal (1736–1800).

5. O. crocata L. Hemlock Waterdropwort
O. apiifolia Brot.

Perennial monoecious *herb*, the roots with cylindrical-obovoid tubers up to 60×10 mm. *Stems* 50–150×1.0 cm, pale yellowish-green, robust, hollow, grooved, erect, but often falling over after flowering, branched, glabrous, leafy. *Leaves* medium yellowish-green above, paler beneath, remaining green in the winter; lower 14–35×8–30 cm, ovate, oblong or ovate-oblong in outline, 3(–4)-pinnate, the segments 10–40 mm, ovate to subrotund in outline, crenate to pinnatifid, and usually cuneate at base, with mostly sheathing petioles shorter than the divided part; upper 1- to 2-pinnate, usually with narrow lobes and a short, sheathing petiole; all glabrous. *Inflorescence* of compound umbels, with (7–)12–40 rays, the rays (15–)30–80 mm, smooth and not thickening after flowering, the peduncle longer than the rays, the terminal umbels with mostly bisexual flowers, the lateral umbels with largely or wholly male flowers; bracts about 5, linear to 3-fid; bracteoles 6 or more; partial umbels not flat-topped in fruit. *Sepals* 5, conspicuous, ovate to triangular, acute at apex, persistent. *Petals* 5, white, the outer scarcely radiating, emarginate, the apex long and inflexed. *Stamens* 5; filaments white; anthers pinkish. *Styles* 2; about half as long as the fruit, with enlarged base forming the stylopodium; stigma a small knob. *Schizocarp* 4.0–5.5 mm, cylindrical or narrowly obovoid; mericarps with slender ridges; commissure broad; carpophore present; vittae solitary; pedicels not thickening after flowering. *Flowers* 6–7. Reproducing by seed. $2n=22$.

Native. Shallow water in streams, on the banks of rivers, streams, lakes, ponds and canals, in marshes and wet woodland, in crevices and waterside masonry, among flushed stones and boulders at the top of beaches and on dripping or flushed sea-cliffs. Most frequent in or by acidic waters, or where calcareous waters flow over acidic substrates but it also occurs in calcareous habitats. It is a lowland species rarely recorded above 300 m altitude. Common in western and southern Great Britain, frequent in places near the east coast and with few localities in central England and central and north Scotland; in Ireland absent from much of the centre. Western Europe from Belgium and Great Britain south to Spain, Portugal and Morocco, extending into the western Mediterranean region as far as Italy. Suboceanic southern-temperate element. The tubers are sweetish-tasting, but very poisonous due to a series of polyacetylenes. The active principle is oenanthetoxin, a convulsant poison which can cause rapid death with few symptoms. It has been fatal to humans when they have mistaken the leaves for celery or the tubers for parsnips. Cattle poisoning is fairly common.

6. O. fluviatilis (Bab.) Coleman River Waterdropwort
O. phellandrium var. *fluviatilis* Bab.; *O. phellandrium* subsp. *fluviatilis* (Bab.) Hook. fil.

Perennial monoecious *herb* with roots of mature plants not tuberous although young plants may be tuberous, the tubers disappear as the stem elongates and fibrous roots develop at the nodes. *Stems* 30–100×1.0 cm, pale green, floating or erect to ascending, hollow, striate, branched, leafy. *Leaves* medium green on upper surface, paler beneath, remaining green in winter; lower submerged, 5–23×4–23 cm, triangular-ovate in outline, 2-pinnate, the segments oblanceolate-cuneate and deeply cut into linear to filiform lobes, the petiole long, aerial 1- to 3-pinnate, the segments about 10 mm, ovate to subrotund, shallowly lobed and cuneate at base, the petiole with a sheathing base; all glabrous. *Inflorescence* of compound umbels, leaf-opposed and terminal, with 5–10(–15) rays, the rays 10–30 mm, smooth or slightly scabrid and not thickening after flowering, the peduncle usually shorter than the rays, most umbels with male and bisexual flowers; bracts usually absent; bracteoles 5–8, linear-lanceolate; partial umbels not flat-topped in fruit. *Sepals* 5, conspicuous, lanceolate, acute at apex, persistent. *Petals* 5, white, outer scarcely radiating, emarginate, the apex long and inflexed *Stamens* 5; filaments white; anthers greenish. *Styles* 2, less than quarter as long as the fruit, divergent, with enlarged base forming the stylopodium; stigma a small knob. *Schizocarp* 5.0–6.5 mm, cylindrical or narrowly ellipsoid; mericarps with prominent, slender ridges; commissure broad; carpophore present; vittae solitary; pedicels not thickened in fruit. *Flowers* 7–9. $2n=22$.

Native. Most frequent in crystal-clear meso-eutrophic water of calcareous streams and rivers, where it can form large beds. It is also found in canals, fenland lodes and ditches, but only rarely grows in ponds. It occurs in still or sluggish waters or where the flow is more rapid, from shallow water to at least 1.5 metres deep in fenland lodes. In a broad band stretching from the Wash to the Thames across to the Severn and the Dorsetshire coast, with a gap south of the Thames except for east Kent, and with a few scattered localities in the east north to Yorkshire. In Ireland it is in scattered localities in the central area. *O. fluviatilis* is endemic to western Europe and is known on the mainland of Denmark and from a compact area between the Seine and the Rhine in Belgium, France and Germany. It is declining in north-west France and is apparently extinct in Germany. Oceanic Temperate element. In rapidly flowing water the stems root at the nodes and form large vegetative clumps which rarely flower and reproduction is by vegetative fragmentation. Plants in still or sluggish water flower more freely and if not broken off below water produce viable seed, but it is not known how frequently the species reproduces from seed.

7. O. aquatica (L.) Poir. Fine-leaved Waterdropwort
Phellandrium aquaticum L.; *O. phellandrium* Lam. nom. illegit.; *Ligusticum phellandrium* Crantz; *Stephanorossia palustris* Chiov.; *Selinum phellandrium* (Crantz) E. H. L. Krause

Annual or *biennial* monoecious *herb* with fleshy roots, the young plants with tubers which disappear as the plant flowers. *Stems* 30–150×1.0 cm, pale green, robust, erect to ascending, hollow, striate, branching, leafy, glabrous. *Leaves* medium green on upper surface, paler beneath; lower 5–12×2.5–6.0 cm, oblong-lanceolate in outline, 3- to 4-pinnate, finely divided with filiform segments when submerged; aerial leaves lanceolate or oblong in outline, 2- to 3-pinnate, the segments about 5 mm, lanceolate to ovate in outline, pinnatifid, the lobes acute, the petiole with a sheathing base. *Inflorescence* of compound umbels, leaf-opposed and terminal, with (4–)6–16 rays, the rays 10–40 mm, usually scabrid, not thickening after flowering, the peduncle usually shorter than the rays, most umbels with male and bisexual flowers; bracts usually absent; bracteoles usually 4–8, linear-lanceolate; partial umbels not flat-topped in fruit. *Sepals* 5, conspicuous, narrowly triangular, acute at apex, persistent. *Petals* 5, white, outer scarcely radiating, emarginate, the apex long and inflexed. *Stamens* 5; filaments white; anthers pink. *Styles* 2, less than a quarter as long as the fruit, divergent, with an enlarged base forming the stylopodium; stigma a small knob. *Schizocarp* 3.0–5.5 mm, narrowly ovoid; mericarps with prominent, rather thick ridges; commissure broad; carpophore present; vittae solitary; the pedicels not thickening in fruit. *Flowers* 6–9. $2n=22$.

Native. Most frequent in shallow ponds and ditches, where it often grows on deep, silty and often eutrophic substrates where the vegetation is kept open by trampling cattle. It is also found in open vegetation at the edge of sheltered lakes, reservoirs, canals, sluggish streams and rivers, in marshes and seasonally flooded depressions. It is a common species throughout England, inner Wales and the extreme south of Scotland. In Ireland it is missing from many coastal areas. It has clearly decreased in some areas of eastern Great Britain and may be over-recorded in southern England instead of *O. fluviatilis*. *O. aquatica* is widespread in Europe and western Asia, especially in the middle latitudes, and extends north to Scandinavia. It also occurs in a few, scattered sites in central Asia and is introduced in New Zealand. Eurosiberian Temperate element. The seeds of *O. aquatica* germinate soon after they are shed and swards of seedlings can be found on ground from which water has receded. Plants may behave as winter annuals, germinating in autumn and flowering the next summer, or as summer annuals, flowering in the same year as they germinate.

21. Aethusa L.

Annual monoecious *herbs*. *Stems* erect, hollow. *Leaves* 2- to 3-pinnate. *Inflorescence* of compound umbels; bracts usually absent; bracteoles 3–4, linear, deflexed. *Sepals* absent. *Petals* 5, white, the outer scarcely radiating. *Fruit* a schizocarp, broadly ovoid or globose, somewhat compressed dorsally; mericarps with prominent, keeled ridges; commissure broad; carpophore present; vittae solitary.

Contains 1 variable species in Europe, North Africa and south-west Asia; introduced in North America. Usually contains coniine and cynapine and is thus poisonous.

Hultén, E. & Fries, M. (1986). *Atlas of north European vascular plants north of the Tropic of Cancer.* 3 vols. Königstein.

1. Stems up to 80 cm; leaf-lobes ovate 2.
1. Stems over 100 cm; leaf-lobes linear-lanceolate or oblong 3.
2. Stems up to 20 cm; bracteoles shorter than to as long as partial umbels **1(a). cynapium** subsp. **agrestis**
2. Stems up to 80 cm; bracteoles 1–2 times as long as partial umbels **1(b). cynapium** subsp. **cynapium**
3. Leaf-lobes linear-lanceolate; bracteoles 1–2 times as long as partial umbels **1(c). cynapium** subsp. **giganteum**
3. Leaf-lobes linear to oblong; bracteoles 2–3 (–4) times as long as partial umbels **1(d). cynapium** subsp. **cynapioides**

1. A. cynapium L. Fool's Parsley
A. toxicaria Salisb. nom. illegit.; *A. tenuifolia* Gray, non Salisb.; *A. cicuta* Necker; *Coriandrum cynapium* (L.) Crantz; *Cicuta cynapium* (L.) Targ.; *Selinum cynapium* (L.) E. H. L. Krause

Annual monoecious *herb* with fusiform tap-root and fibrous side-roots. *Stems* 5–200 cm, medium, shining green, sometimes pruinose, erect, hollow, finely striate, glabrous, branched, leafy. *Leaves* dark green on upper surface, paler beneath; lower 3–12 × 2–10 cm, broadly triangular-ovate, 2- to 3-pinnate, the segments 5–15 mm, lanceolate to ovate in outline, acute at apex, pinnatifid and with antrorsely scabrid margins, the petioles short, mostly sheathing with red at time of flowering; cauline similar and only slightly smaller. *Inflorescence* of compound umbels, leaf-opposed and terminal, with 4–20 rays, the rays 5–30 mm and antrorsely serrulate on the angles, the peduncles longer than the rays, the umbels with almost entirely bisexual flowers; bracts usually absent; bracteoles 3–4, on the outer side of the partial umbels, linear, long-acute at apex and deflexed. *Sepals* absent. Petals 5, white, the outer scarcely radiating, obcordate, the apex inflexed. *Stamens* 5; filaments white; anthers cream or pinkish. *Styles* 2, white becoming purplish, with enlarged base forming the stylopodium to which the style, that is as long as it, is closely appressed; stigma capitate. *Schizocarp* 2.5–4.0 mm, broadly ovoid or globose, somewhat compressed dorsally; mericarps with prominent, keeled ridges; commissure broad; carpophore present; vittae solitary; pedicels slender. *Flowers* 6–11. $2n=20$.

The four following subspecies appear to be ecotypes which are adapted to various conditions in Man's agricultural practices.

(a) Subsp. **agrestis** (Wallr.) Dostál
A. cynapium forma *agrestis* (Wallr.) Schube; *A. cynapium* var. *agrestis* Wallr.; *A. segetalis* Boenn.; *A. cynapium* forma *segetalis* (Boenn.) Rchb.; *A. cynapium* var. *pumila* Roth; *A. cynapiumn* var. *humilior* Spenn.; *A. cynapium* var. *pygmaea* W. D. J. Koch

Annual. Stems 5–20 cm, angled, shining green. *Leaf-lobes* ovate. *Bracteoles* shorter than to as long as partial umbels. *Outer pedicels* usually shorter than, to as long as their schizocarp. *Schizocarp* 2.5–3.0 × 2.5–3.0 mm, globose, flowering 8–11.

(b) Subsp. **cynapium**
A. cynapium var. *domestica* Wallr.; *A. cynapium* var. *hortensis* Bernh.; *A. cynapium* var. *vulgaris* Rchb.; *A. cynapium* var. *genuina* Ducommen; *A. cynapium* var. *typicum* Beck

Annual. Stems 30–80 cm, angled, shining green. *Leaf-lobes* ovate. *Bracteoles* 1–2 times as long as partial umbels. *Outer pedicels* about twice as long as schizocarp. *Schizocarp* 3.0–3.5 × 2.0–2.5 mm, ovoid. *Flowers* 6–7.

(c) Subsp. **gigantea** (Lej.) P. D. Sell
A. elata Friedl.; *A. cynapium* var. *elata* (Friedl.) Peterm., non Gaudin; *A. cynapium* var. *nemorum* Lamotte; *A. cynapium* var. *gigantea* Lej.

Annual. Stems 100–200 cm, terete, striate, often pruinose, branched from base. *Leaf-lobes* linear-lanceolate. *Bracteoles* 1–2 times as long as partial umbels. *Outer pedicels* up to twice as long as their schizocarp. *Schizocarp* 2.0–2.5 × 2.0–2.5 mm, globose. *Flowers* 6–7.

(d) Subsp. **cynapioides** (M. Bieb.) Arcang.
A. cynapioides M. Bieb.; *A. cynapium* var. *cynapioides* (M. Bieb.) Ficinus & Heynh.; *A. cynapium* var. *elata* Gaudin; *A. cynapium* var. *sylvestris* Godr.; *A. cynapium* var. *elatior* Döll

Biennial. Stem 100–200 cm, terete, striate, pruinose. *Leaf-lobes* linear to oblong, narrowing towards their apex. *Bracteoles* 2–3(–4) times as long as partial umbel. *Outer pedicels* more than twice as long as their schizocarp. *Schizocarps* 2.0–2.5 × 2.0–2.5 mm, globose.

Native. Waste and cultivated places. Throughout Great Britain and Ireland except much of central and northern Scotland. Europe, North Africa and south-west Asia; introduced in North America. European Temperate element. Subsp. *agrestis* is common throughout East Anglia and probably other agricultural areas of Great Britain and Ireland. It is usually very dwarf and comes into flower in the stubble-land after harvest when the cereal crops have been cut, and usually has ripe seed by when the ground is ploughed. It also occurs in root crops, waste places and gardens where it still flowers and fruits late. It was possibly brought to Great Britain by early Man from Continental Europe. Subsp. *cynapium* is the plant of root crops and waste land, where it flowers and fruits earlier than subsp. *agrestis*. This is also the case in cultivation at the Cambridge Botanic Garden where the two subspecies did not overlap. Subsp. *cynapium* rarely grows in cereal crops and then usually along the field margin. Subsp. *cynapioides* was first seen amongst some bulb plants at Bassingbourn in Cambridgeshire and was also found in a field at Meldreth in Cambridgeshire, in which the bulbs originated. In 1998 a single plant was found among chrysanthemums in a garden at Bassingbourn and it has occurred at intervals several times since. In 1998 also, subsp. *giganteum* was found in some quantity in a wheat field at Bassingbourn and then a wheat field at Histon, both in Cambridgeshire. In both localities subsp. *agrestis* also grew, but subsp. *giganteum* was in ripe fruit before subsp. *agrestis* came into flower. A single plant of subsp. *giganteum* occurred in the entrance to the Botanic Garden at Cambridge, but it was not known in the Garden. In 2007 a single plant was also found in another part of Bassingbourn. Tall plants have also been recorded from Jersey and Kent. In Continental Europe both subsp. *cynapioides* and subsp. *giganteum* grow in woods, on banks, in meadows and in shady places. Subsp. *cynapioides* has also been found in wild flower seed.

22. **Foeniculum** Mill.

Perennial herbs smelling strongly of aniseed. *Stems* erect, solid, later developing a small hollow. *Leaves* 3- to 4-pinnate, the segments linear or filiform. *Inflorescence* of compound umbels; bracts and bracteoles absent. *Sepals* absent. *Petals* 5, yellow, the outer not radiating. *Fruit* a schizocarp, oblong-ellipsoid, scarcely compressed; mericarps with prominent thick ribs; commissure broad; carpophore present; vittae solitary.

Contains 1 species in Europe and the Mediterranean region.

Grieve, M. (1931). *A modern herbal*. London.
Phillips, R. & Rix, M. (1988). *Vegetables*, pp. 130–131. London.
Vaughan, J. G. & Geissler, C. (1997). *The new Oxford book of food plants*. Oxford.

1. Leaf segments rarely more than 10 mm, rigid and rather fleshy; rays 4–10 **1(c). vulgare** subsp. **vulgare**
1. Leaf segments more than 10 mm, flaccid; rays 12–25(–40) 2.
2. Stems 100–250 cm; petioles not markedly swollen **1(a). vulgare** subsp. **sativum**
2. Stems up to 30 cm; petioles greatly swollen at base and forming a sort of false bulb as big as a fist **1(b). vulgare** subsp. **dulce**

1. F. vulgare Mill. Fennel

F. officinale All.; *Anethum foeniculum* L.; *Anethum rupestre* Salisb.; *Ligusticum foeniculum* (L.) Crantz; *Meum foeniculum* (L.) Spreng.; *Ozodia foeniculacea* Wight & Arn.; *Selinum foeniculum* (L.) E. H. L. Krause

Biennial to *perennial* monoecious *herb* smelling strongly of aniseed, with a tap-root and fibrous side-roots. *Stems* up to 250 cm, more or less glaucous, erect, solid but developing a small hollow when old, finely striate, branched, leafy. *Leaves* waxy, bluish-green; lower 12–25 × 12–18 cm, more or less triangular in outline, 3- to 4-pinnate, the segments 5–50 mm, filiform, with a cartilaginous apex, the petioles with a sheathing base, sometimes markedly swollen; upper similar and only slightly smaller, usually widely spaced and not all lying in the same plane, the petioles entirely sheathing; all glabrous. *Inflorescence* of compound umbels, leaf-opposed and terminal, with 4–40 rays, the rays 10–60 mm and smooth, the peduncles usually longer than the rays, all the umbels with bisexual flowers; bracts and bracteoles usually absent. *Sepals* absent. *Petals* 5, yellow, the outer not radiating, oblong, scarcely narrowed to the involute apex. *Stamens* 5; filaments white; anthers yellow. *Styles* 2, divergent or recurved, the enlarged base forming the stylopodium which is longer than the styles; stigma capitate. *Schizocarp* 4–6 mm, often not developing, greenish, yellowish-brown or greyish with yellow ridges, oblong-ellipsoid, scarcely compressed; mericarps with prominent, thick ribs; commissure broad; carpophore present; vittae solitary; pedicels 2–5 mm and slender. *Flowers* 7–10. $2n = 22$.

(a) Subsp. **sativum** (C. Presl) Bertol. Wild Fennel

F. officinale subsp. *sativum* (C. Presl) Arcang.; *F. vulgare* subsp. *silvestre* (Brot.) Janch.; *F. vulgare* var. *silvestre* (Brot.) C. Presl; *F. silvestre* Brot.; *F. vulgare* var. *sativum* C. Presl; *F. dulce* Mill.; *F. vulgare* var. *dulce* (Mill.) Thell.

Biennial herb. Stems 100–250 cm. *Leaves* with segments usually more than 10 mm, flaccid; petioles not markedly swollen. *Terminal umbel* not overtopped by lateral ones; rays 12–25(–40). *Schizocarps* sweet-tasting.

(b) Subsp. **dulce** (DC.) Bertol. Florence Fennel

Anethum dulce DC.; *F. officinale* subsp. *dulce* (DC.) Arcang.; *F. dulce* (DC.) DC., non Mill.; *F. officinale* var. *dulce* (DC.) Alef.; *F. azoricum* Mill.; *F. vulgare* var. *azoricum* (Mill.) Thell.

Biennial herb. Stems up to 30 cm, robust. *Leaves* with segments usually more than 10 mm, flaccid; petioles greatly swollen at base and forming a sort of false bulb as big as a fist. *Terminal umbel* not overtopped by lateral ones; rays 12–25. *Schizocarps* sweet-tasting.

(c) Subsp. **vulgare** Carosella

F. piperitum (Ucria) Sweet; *Anethum piperitum* Ucria; *F. vulgare* subsp. *piperitum* (Ucria) Cout.; *Anethum foeniculum* auct.; *F. officinale* All. nom. illegit.; *Ligusticum foeniculum* Roth; *Meum foeniculum* Spreng.

Perennial herb with a deep tap-root. *Stems* up to 100 cm. *Leaves* with segments rarely more than 10 mm, rigid and rather fleshy, sometimes purple or bronze; petioles not swollen at base. *Terminal umbel* often overtopped by lateral ones; rays 4–10. *Schizocarps* sharp-tasting.

Probably introduced by the Romans. On open ground, roadsides and waste places especially near the coast where it also occurs on cliffs. Great Britain north to the Isle of Man and Yorkshire and in Ireland; scarce and mainly casual further north. Most of Europe except the north, but probably native only in the Mediterranean region. The common plant is subsp. *sativum*. The fruits, which contain the essential oil anethol, are widely used in culinary flavouring, particularly for fish dishes, marinades, bouillons, sauces and stuffing. Sometimes offered after meals as a digestive and breath freshener. The stem and leaves, fresh or dried, are also used as a herb for flavouring dishes and in the preparation of various liqueurs. Subsp. *dulce* was introduced to England from Italy where it probably originated. Philip Miller wrongly thought it originated from the Azores. It is usually eaten, both cooked and raw, with cheese. Cv. *Purpurescens* with deep purple-bronze stem and foliage is grown in gardens for ornament. Subsp. *vulgare* is native to rocky places in the Mediterranean. Its flowering stems are usually served as an hors d'oeuvre.

There are specimens in herbaria from the east coast of Great Britain and the Channel Islands.

23. Anethum L.

Annual monoecious *herbs* smelling strongly of aniseed. *Stems* erect, hollow. *Leaves* 3- to 4-pinnate, with filiform segments. *Inflorescence* of compound umbels; bracts and bracteoles absent. *Sepals* absent. *Petals* 5, yellow, the outer not radiating. *Fruit* a *schizocarp,* elliptical, strongly compressed dorsally; mericarps with low, slender dorsal ridges and winged marginal ridges; commissure broad; carpophore present; vittae solitary.

Contains 1 or possibly 2 species in south-west Asia.

Grieve, M. (1931). *A modern herbal.* London.
Vaughan, J. G. & Geissler, C. (1997). *The new Oxford book of food plants.* Oxford.
Zohary, D. & Hopf, M. (2000). *Domestication of plants in the Old World.* Ed. 3. Oxford.

1. A. graveolens L. Dill
Selinum anethum Roth; *Pastinaca anethum* (Roth) Spreng.; *Peucedanum graveolens* (L.) Benth. & Hook. fil.

Annual monoecious *herb* with fibrous roots, smelling strongly when crushed. *Stems* 20–60(–100) cm, pale green, erect, hollow, striate, glabrous, branched, leafy. *Leaves* slightly bluish or greyish dark green on upper surface, paler beneath; lower 5–17×4–10 cm, broadly ovate in outline, 3- to 4-pinnate, the segments about 15 mm, filiform and mucronate at apex, the petiole shorter than the lamina, sheathing; upper similar and only a little smaller with short sheathing petioles; all glabrous. *Inflorescence* of compound umbels, with 7–30 rays, the rays 20–90 mm, subequal and smooth, the peduncle longer than the rays, the umbels mostly with bisexual flowers; bracts and bracteoles absent. *Sepals* absent. *Petals* 5, yellow, the outer not radiating, oblong, apex incurved. *Stamens* 5; filaments white; anthers yellow. *Styles* 2, recurved and appressed, with enlarged base forming the stylopodium which is about twice as long as the styles; stigma capitate. *Schizocarps* 3.0–6.0 mm, elliptical, strongly compressed dorsally, dark brown with a paler wing; mericarps with low, slender dorsal ridges and winged marginal ridges; commissure broad; carpophore present; vittae solitary; pedicels 4–10 mm and slender. *Flowers* 7–8. $2n=22$.

Introduced. Cultivated and sometimes persisting for a few years on waste ground and rubbish-tips where it is a casual from bird-seed or grain; also in wild flower seed. Scattered records in Great Britain and could be widespread in wild flower seed especially round field margins. Probably native of south-west Asia, but widely cultivated and naturalised. Both the leaves and fruits are used for flavouring. In Great Britain it is used in pickles and sauces, also with fish, and the seeds, which are rather like those of Caraway, as a condiment.

24. Silaum Mill.
Silaus Bernh. nom. illegit.

Perennial monoecious *herbs. Stems* erect, solid. *Leaves* 1- to 4-pinnate. *Inflorescence* of compound umbels; bracts 0–3, entire; bracteoles 5–11. *Sepals* absent. *Petals* 5, yellowish, the outer not radiating. *Fruit* a schizocarp, oblong-ovoid, scarcely compressed; mericarps with prominent, slender ridges, the lateral forming narrow wings; commissure broad; carpophore present; vittae numerous and inconspicuous.

Contains about 10 species in Europe and temperate Asia.

Hultén, E. & Fries, M. (1986). *Atlas of north European vascular plants north of the Tropic of Cancer.* 3 vols. Königstein.

1. S. silaus (L.) Schinz & Thell. Pepper Saxifrage
Peucedanum silaus L.; *S. flavescens* (Bernh.) Hayek; *S. pratensis* Besser; *Selinum silaus* (L.) Crantz; *Ligusticum silaus* (L.) Vill.; *Silaus flavescens* Bernh.; *Seseli selinoides* Jacq.; *Ligusticum tripartitum* Dumort.; *Sium silaus* (L.) Roth; *Crithmum silaus* (L.) Wibel; *Cnidium silaus* (L.) Spreng.

Perennial, monoecious, strongly aromatic *herb* with a cylindrical, stout, woody tap-root with a few fibrous remains at the top. *Stem* 30–100 cm, pale green, erect, solid, striate, glabrous, branched, leafy. *Leaves* medium yellowish-green on upper surface, paler beneath; basal 9–25×4–20 cm, triangular to lanceolate in outline, 1- to 4-pinnate, the segments, ovate in outline, pinnately divided and long-petioluled, the lobes 10–15 mm, linear to lanceolate, obtuse, mucronate or acuminate, finely serrulate, with prominent midrib and the apex often reddish, long-petiolate; cauline often only 1-pinnate, simple or reduced to sheaths; all glabrous. *Inflorescence* of compound umbels, with 4–15 rays, the rays sharply angled, the peduncle longer than the rays, the top of the peduncle and the rays papillose, the umbels mostly with bisexual flowers; bracts 0–3, entire; bracteoles 5–11, linear or lanceolate, with scarious margins. *Sepals* absent. *Petals* 5, yellowish, the outer not radiating, ovate, the apex short and involute. *Stamens* 5; filaments white; anthers yellow. *Styles* 2, greenish, with enlarged base forming the greenish-yellow stylopodium, the styles about as long as and appressed to the stylopodium; stigma capitate. *Schizocarp* 4–5 mm, oblong-ovoid, scarcely compressed; mericarps with very prominent, slender ridges, the lateral forming narrow wings; commissure broad; carpophore present; vittae numerous and inconspicuous, the pedicels 2–3 mm and rather stout. *Flowers* 6–8. $2n=22$.

Native. Meadows and grassy banks. Mainly in south and east England with a few localities in Wales and south-east Scotland, not in Ireland. West, central and east Europe, northwards to Netherlands and Sweden, but absent from Portugal. Eurosiberian Temperate element.

25. Meum Mill.

Perennial monoecious *herbs. Stems* erect, hollow. *Leaves* 3- to 4-pinnate, the ultimate lobes filiform. *Inflorescence* of compound umbels; bracts few and entire or absent; bracteoles 3–8, linear, sometimes lobed. *Sepals* absent. *Petals* 5, white, sometimes pink-tinged, the outer not radiating. *Fruit* a schizocarp, ovoid or elliptical, scarcely compressed; mericarps with prominent, thick ridges; commissure broad; carpophore present; vittae 3–5 in each furrow.

Contains 1 species in the mountains of western Europe.

Hultén, E. & Fries, M. (1986). *Atlas of north European vascular plants north of the Tropic of Cancer.* 3 vols. Königstein.
Stewart, A., Pearman, D. A. & Preston, C. D. (1994). *Scarce plants in Britain.* Peterborough.

1. M. athamanticum Jacq. Spignel
Ligusticum meum (L.) Crantz; *Athamanta meum* L.; *Aethusa meum* (L.) L.; *Aethusa tenuifolia* Salisb. nom. illegit.; *Carum meum* (L.) Stokes; *Meum meum* (L.) Karst.

Strongly aromatic, *perennial,* monoecious *herb* with a tough rootstock. *Stems* 7–60 cm, hollow, striate, erect, glabrous, branched, slightly leafy, the base surrounded by abundant fibrous remains of petioles. *Leaves* medium green; basal numerous 5–18 × 2.5–18 cm, lanceolate to ovate in outline, 3- to 4-pinnate, the segments about 5 mm, filiform and crowded, the petioles slender and abruptly expanded into a largely scarious sheath at the base; cauline few, small, oblong-lanceolate, much divided and the petiole entirely sheathing; all glabrous. *Inflorescence* of compound umbels, with 6–15 rays, the rays 10–50 mm, unequal and somewhat scabrid, the peduncle longer than the rays and papillose, the terminal umbel with bisexual and a few inner male flowers, the lateral umbels with mostly male flowers; bracts few or sometimes absent, linear to ovate, sometimes lobed, very variable in length; bracteoles 3–8, linear, sometimes lobed. *Sepals* absent. *Petals* 5, white, sometimes pink-tinged, the outer not radiating, ovate, the apex more or less inflexed. *Stamens* 5; filaments white; anthers pinkish. *Styles* 2, divergent or somewhat recurved, with enlarged base forming the stylopodium which is a little shorter than the style; stigma truncate. *Schizocarp* (4–)5–7(–10) mm, ovoid or elliptical, scarcely compressed; mericarps with very prominent, thick ridges; commissure broad; carpophore present; vittae 3–5 in each furrow; pedicels 1–7 mm, papillose on the angles. *Flowers* 6–7. $2n = 22$.

Native. Dry neutral or acidic grassland in unimproved pastures, hay meadows and roadside banks in upland areas, but is not normally found over 300 m, though it can reach 610 m. It has probably suffered due to changes in agricultural practice but persists on banks too steep to cut. Local in north Wales, northern England and central and south Scotland. Western Europe from southern Norway to northern Spain, east to Poland, Czechoslovakia and the Balkans. European Boreal-montane element.

26. Physospermum Cusson ex Juss.

Danaa All., non Sm.; *Pseudospermum* Spreng. ex Gray

Perennial, monoecious *herbs. Stems* erect, solid. *Leaves* 2-ternate. *Inflorescence* of compound umbels; bracts 4–7, membranous; bracteoles usually fewer than bracts. *Sepals* 5, triangular, conspicuous. *Petals* white, the outer not radiating. *Fruit* a schizocarp broader than long and inflated; mericarps smooth; commissure narrow; carpophore present; vittae solitary.

Contains about 4 species in Europe and western Asia.

Wigginton, M. J. (Edit.) (1999). *British red data books.* Vol. 1. *Vascular plants.* Peterborough.

1. P. cornubiense (L.) DC. Bladderseed
Ligusticum cornubiense L.; *Danaa cornubiensis* (L.) Burnat; *P. aquilegifolium* var. *cornubiense* (L.) Lange; *Danaa nudicaulis* (M. Bieb.) Grossh.; *Ligusticum tenuifolium* Salisb. nom. illegit.; *P. commutatum* Spreng. nom. illegit.; *Pseudospermum commutatum* Gray nom. illegit.

Perennial monoecious *herb* with a large, cylindrical tap-toot. *Stems* 30–120 cm, pale green, solid, striate, branched, leafy. *Leaves* medium green on upper surface, pale beneath; basal 5–15 × 10–23 cm, broadly rhombic in outline, 2-ternate, the primary segments long-stalked, the secondary segments 15–50 mm, ovate in outline with a cuneate base, pinnatifid or incise-serrate, the petiole long and slender; cauline few, small and often simple; all puberulous on the margin and larger veins on both surfaces. *Inflorescence* of compound umbels, with 10–14 rays, the subequal rays 20–50 mm, smooth and slender, the peduncle usually longer than the rays, the terminal umbel with bisexual flowers, the lateral with male and bisexual flowers; bracts 4–7, lanceolate to linear, acute at apex, membranous; bracteoles similar, but usually fewer. *Sepals* 5, triangular, acute at apex. *Petals* 5, white, the outer not radiating, obovate, emarginate, the apex inflexed. *Stamens* 5; filaments white; anthers pale yellow. *Styles* 2, recurved, with enlarged base forming the stylopodium which is about half as long as the styles; stigma truncate. *Schizocarp* 3–4 mm, dark brown, inflated, broader than long, didymous; mericarps smooth; commissure narrow; carpophore present; vittae solitary; pedicels 5–8 mm, smooth. *Flowers* 7–8. $2n = 22$.

Native. Open woods and scrub, arable fields and hedgebanks. Very local in Cornwall and Devonshire; naturalised in one locality in Buckinghamshire. Southern Europe and temperate west Asia. European Temperate element.

27. Conium L.

Biennial monoecious *herbs. Stems* erect, hollow, with brownish-purple spots. *Leaves* 2- to 4-pinnate. *Inflorescence* of compound umbels; bracts 5–6, deflexed, with wide scarious margins; bracteoles 3–6. *Sepals* absent. *Petals* 5, white, the outer not radiating. *Fruit* a schizocarp, broadly ovoid, somewhat compressed; mericarps with prominent, more or less undulate-crenate ridges; commissure narrow; carpophore present; vittae several.

Contains 2 species in north temperate regions of the world and South Africa.

Hultén, E. & Fries, M. (1986). *Atlas of north European vascular plants north of the Tropic of Cancer.* 3 vols. Königstein.

1. C. maculatum L. Hemlock
Coriandrum cicuta Crantz; *Coriandrum maculatum* (L.) Roth; *Cicuta officinalis* Crantz; *Cicuta maculata* (L.) Lam.; *Cicuta major* Lam.; *Sium conium* Vest; *Selinum conium* (Vest) E. H. L. Krause

Gregarious, monoecious, winter *annual* or *biennial herb,* with a tap-root and a mousey smell. *Stems* 50–250 cm, usually pruinose with brownish-purple spots at least below, erect, hollow, striate below, furrowed above, glabrous, branched, leafy. *Leaves* medium green on upper surface, paler beneath; basal 2–35×2–30 cm, triangular in outline, 2- to 4-pinnate, the segments 10–20 mm, oblong in outline, pinnatifid or incise-serrate, the lobes and teeth with a cartilaginous apex, the petioles long, with a shortly sheathing base; cauline similar but gradually smaller upwards, the petioles short and entirely sheathing; all glabrous. *Inflorescence* of compound umbels, with 10–20 rays, the rays subequal, 10–35 mm and scabrid, the peduncle equalling or longer than the rays, the terminal umbel with bisexual flowers, the lateral with male and bisexual flowers; bracts 5–6, narrowly triangular to lanceolate, deflexed and with a wide, scarious margin; bracteoles 3–6, on the outside of the partial umbel and widened and often connate at the base. *Sepals* absent. *Petals* 5, white, the outer not radiating, obcordate, emarginate, the apex inflexed. *Stamens* 5; filaments greenish; anthers yellowish. *Styles* 2, horizontal or recurved, with enlarged base forming the stylopodium which is about as long as the styles; stigma a small knob. *Schizocarp* 2–4 mm, broadly ovoid, somewhat compressed; mericarps with prominent, more or less undulate-crenulate ridges; commissure narrow; carpophore present; vittae several, slender; pedicels 1–8 mm. *Flowers* 6–7. $2n=22$.

Native. Damp places, roadsides, ditches, rubbish-tips and waste ground where it forms large patches or swards. Common over much of Great Britain and Ireland except in west and central Scotland. Europe and temperate Asia; Macaronesia; introduced in eastern North America, California, Mexico, West Indies, temperate South America and New Zealand. Highly poisonous due to the presence of coniine. Hemlock is notorious as the poison given to Socrates at his execution, which causes paralysis, respiratory failure and stupor.

28. Bupleurum L.

Agrostana Hill

Annual to *perennial* monoecious *herbs* or small *shrubs*. *Stems* erect to procumbent, sometimes woody at base, hollow or solid. *Leaves* simple, entire. *Inflorescence* of compound umbels; bracts 2–6, or absent; bracteoles 4–6. *Sepals* absent. *Petals* 5, yellow, the outer not radiating. *Fruit* a schizocarp, oblong-ellipsoid, ellipsoid or subglobose, somewhat compressed laterally; mericarps with slender, prominent ridges which are sometimes winged and sometimes papillose between the ridges; commissure broad; carpophore present; vittae inconspicuous.

Contains about 150 species in Europe, temperate Asia, Africa and North America.

Coombe, D. E. (1994). 'Maritime' plants of roads in Cambridgeshire (v.c. 29). *Nature in Cambridgeshire* **36**: 37–60. [*B. tenuissimum*.]

Field, M. H. (1994). The status of *Bupleurum falcatum* L. (Apiaceae) in the British flora. *Watsonia* **20**: 115–117.

Stewart, A., Pearman, D. A. & Preston, C. D. (1994) *Scarce plants in Britain*. Peterborough. [*B. tenuissimum*.]

Wigginton, M. J. (Edit.) (1999). *British red data books*. Vol. 1. *Vascular plants*. Peterborough. [*B. baldense*, and *B. falcatum*.]

1. Upper leaves rounded-perfoliate 2.
1. Upper leaves not rounded-perfoliate; bracts present 3.
2. Leaves mostly less than twice as long as wide; umbels with 4–8 rays; mericarps smooth between the ridges **6. rotundifolium**
2. Leaves mostly more than twice as long as wide; umbels with 2–3 rays; mericarps tuberculate between the ridges **7. subovatum**
3. Firmly rooted perennial; at least some leaves often more than 10 mm wide 4.
3. Easily uprooted annual; leaves all less than 10 mm wide 5.
4. Evergreen shrub; leaves with a prominent midrib, pinnate veins and a close reticulum of veins between them; schizocarp 4.5–7.0 mm **1. fruticosum**
4. Herb, but often woody at extreme base; leaves with 3–5 main parallel veins; schizocarp 2.5–3.5 mm **2. falcatum**
5. Bracteoles not concealing the flowers or schizocarps **3. tenuissimum**
5. Bracteoles concealing the flowers and schizocarps 6.
6. Bracteoles without cross-veins or with inconspicuous ascending, not recurved ones **4. baldense**
6. Bracteoles with numerous, conspicuous, ascending and then abruptly recurved cross-veins **5. odontites**

Section **1. Coriaceae** Godr.

Shrubs. *Leaves* evergreen, pinnately veined, with a well-marked midrib.

1. B. fruticosum L. Shrubby Hare's-ear

Evergreen, monoecious *shrub* with a woody base, but liable to frost damage. *Stems* up to 250 cm, pale green, solid, smooth, glabrous, leafy. *Leaves* 4–9×1.5–2.5 cm, dark bluish-green on upper surface, paler beneath, oblanceolate to oblong-obovate, rounded-obtuse-mucronate at apex, entire, with a cartilaginous margin, narrowed to a sessile base, with a prominent midrib, pinnate veins and a close reticulum of fine veins between them, glabrous, *Inflorescence* of compound umbels, with 3–25 rays, the rays subequal and smooth, the peduncle as long as or longer than the rays, all flowers bisexual; bracts usually 5–6, ovate to obovate, with 5–7 longitudinal veins and conspicuous cross-veins, deciduous; bracteoles 5–6, obovate, 4–5(–7)-veined, not concealing the flowers, deciduous. *Sepals* absent. *Petals* 5, yellow, the outer not radiating; not emarginate, the apex inflexed. *Stamens* 5; filaments and anthers yellowish. *Styles* 2, about 0.3 mm, divergent, with an enlarged base forming the stylopodium; stigma truncate. *Schizocarp* 4.5–7.0 mm, oblong-ellipsoid, somewhat compressed laterally, smooth; mericarps with slender, prominent, narrowly winged ridges; commissure broad; carpophore present; vittae inconspicuous; pedicels about 5 mm. *Flowers* 7–8. $2n=14$.

Introduced. Grown in gardens for ornament and naturalised on roadside and railway banks since before 1909. Suffolk, Kent, Devonshire and Worcestershire. Native of southern Europe and north-west Africa.

Section **2. Isophyllum** (Hoffm.) Dumort.

Annual or *perennial*. *Lower leaves* sessile and linear or wider and petiolate; veins 3–many, more or less parallel, the lateral usually few, short and inconspicuous, the marginal vein more or less distinct.

Subsection **1. Nervosa** (Godr.) Briq.
Section *Nervosa* Godr.

Usually *perennial. Lower leaves* often elliptical, the upper sessile and amplexicaul.

2. B. falcatum L. Sickle-leaved Hare's-ear

Biennial or short-lived *perennial* monoecious *herb* with a stout, woody stock. *Stems* 30–100 cm, pale green, hollow, striate, glabrous, branched, leafy. *Leaves* bluish-green, paler beneath; basal 2–12 × 0.2–2.0 cm, elliptical, obtuse at apex, entire, with 3–5 main parallel veins, a cartilaginous margin and long petiole; lower cauline linear-lanceolate, often somewhat falcate and narrowed into the petiole; upper linear, sessile and semiamplexicaul; all glabrous. *Inflorescence* of compound umbels, with 5–11 rays, the rays up to 30 mm, smooth and subequal, the peduncle longer than the rays, all flowers bisexual; bracts 2–5, very unequal, lanceolate to linear, 3- to 5-veined, not concealing the flowers and fruits. *Sepals* absent. *Petals* 5, yellow, the outer not radiating, not emarginate, the apex inflexed. *Stamens* 5; filaments and anthers yellowish. *Styles* 2, about 0.5 mm, divergent or appressed, enlarged base forming the stylopodium; stigma truncate. *Schizocarp* 2.5–3.5 mm, ellipsoid, somewhat compressed laterally, smooth; mericarps with slender, prominent, narrowly winged ridges; commissure broad; carpophore present; vittae inconspicuous; pedicels 1–3 mm, slender. *Flowers* 7–10. $2n = 16$.

First recorded in 1831 and possibly not native. It formerly grew abundantly by the roadside and borders of fields for some distance between Ongar and Chelmsford in Essex and was scattered more or less over that neighbourhood for several miles. It became extinct in 1962 when hedgerow clearance and ditch cleaning destroyed the last colony. Stanley Jermyn had grown stock of the plant in his garden and from there it was planted in Sawbridgeworth Marsh Nature Reserve, Hertfordshire. South, central and east Europe; temperate Asia. Eurasian Southern-temperate species. The Essex plant is subsp. **falcatum** which occurs throughout the range of the species. It is however, a very variable species and it is probable that more infraspecific taxa need to be described.

Subsection **2. Trachycarpa** (Lange) Briq.
Section *Trachycarpa* Lange
Annual. Leaves narrow; veins 3–7, more or less parallel. *Bracteoles* herbaceous, flat, 3-veined. *Schizocarp* rugulose-papillose.

3. B. tenuissimum L. Slender Hare's-ear
Agrostana tenuissima (L.) Bute ex Gray; *Odontites tenuissima* (L.) Hoffm.

Annual monoecious *herb* with fibrous roots. *Stems* 30–50(–70) cm, pale green, solid, flexuous, striate, branched, leafy. *Leaves* 0.5–6.0 × 0.1–0.4 cm, bluish-green, linear to narrowly oblanceolate, widest above the middle, acute at apex, scarcely sheathing at the base, glabrous; veins 5–7 parallel, with some slender cross-veins. *Inflorescence* of compound umbels at the ends of long branches, shortly pedunculate or sessile, with (1–)2–3 rays, the rays up to 10 mm and unequal, partial umbels with few (often 1–4) flowers, all flowers bisexual; bracts 3–5, up to 6 mm, linear-lanceolate, 3-veined, herbaceous; bracteoles similar to the bracts and not concealing the flowers and fruit. *Sepals* absent. *Petals* 5, yellow, the outer not radiating, not emarginate, the apex inflexed. *Stamens* 5, subglobose; filaments and anthers yellowish. *Styles* 2, about 0.1 mm, horizontal, with an enlarged base forming the stylopodium; stigma truncate. *Schizocarp* 1.5–2.0 mm, somewhat compressed laterally, rugulose-papilose; mericarps with narrowly winged, undulate-crenulate ridges; commissure broad; carpophore present; vittae inconspicuous; pedicels very short. *Flowers* 7–9. $2n = 16$.

Native. Dry, usually brackish grassland on sea walls, drained grazing marshes and less frequently on the disturbed parts of upper salt-marshes. South coast from Dorsetshire eastwards, coast of southern East Anglia and the Thames estuary, around the Wash and the Humber estuary, the Severn estuary and Flintshire. It has disappeared from its inland localities in south-central England and along the Lancashire coast, but still occurs inland in Worcestershire. Coasts of western Europe northwards to southern Scandinavia and locally inland; in southern Europe it occurs by the Mediterranean eastwards to the Middle East. Our plant is subsp. **tenuissimum** which occurs throughout the range of the species except the southern part of the Balkan peninsula and the south-east part of the Russia.

Subsection **3. Aristata** (Godr.) Briq.
Section *Aristata* Godr.
Annual. Leaves narrow; veins 3–5. *Bracteoles* lanceolate or ovate, with 3–9 veins and acuminate or aristate.

4. B. baldense Turra Small Hare's-ear
B. aristatum auct.; *B. opacum* (Ces.) Lange;
B. divaricatum Lam. nom. illegit.; *Agrostana divaricata* Gray nom. illegit.

Annual monoecious *herb* with fibrous roots. *Stems* up to 25 cm, but usually less than 10 cm, solid, with raised or very narrowly winged angles, glabrous, simple or divaricately branched, leafy. *Leaves* 1–5 × 0.3–0.6 cm, greyish-green, simple, linear to narrowly oblanceolate, widest above the middle, acute at apex, entire, sessile, with a somewhat scarious, sheathing base, glabrous; veins 3–5, parallel, without any apparent cross-veins. *Inflorescence* of compound umbels, with (1–)2–4 rays, the rays unequal and shorter than the bracts, the peduncle longer than the rays, all flowers in terminal and lateral umbels bisexual; bracts about 4, 5–10 mm, lanceolate, acuminate, glaucous, with 3–5 conspicuous main veins and a narrow, scarious margin; bracteoles similar to bracts but rather smaller and usually 3-veined, acuminate or aristate, concealing the flowers and fruits. *Sepals* absent. *Petals* 5, yellow, the outer not radiating, not emarginate, the apex inflexed. *Stamens* 5; filaments and anthers yellow. *Styles* 2, about 1 mm, horizontal, with enlarged base forming the stylopodium. *Schizocarp*

1.5–2.5 mm, ellipsoid-oblong, somewhat compressed laterally, often pruinose; mericarps with slender ridges; commissure broad; carpophore present; vittae solitary; pedicels about 1 mm. *Flowers* 6–7. $2n=16$.

Native. In more or less open habitats, such as shorter turf on rocky ground and grey dunes and always near the sea. Sussex, Devonshire and the Channel Islands; rarely casual elsewhere. West and south Europe from England to Roumania. Our plant is subsp. **baldense** which occurs throughout the range of the species except for Italy and the Balkan Peninsula. Mediterranean Atlantic element.

5. B. odontites L. Mediterranean Hare's-ear
B. fontanesii Guss.

Annual monoecious *herb* with fibrous roots. *Stems* up to 50 cm, solid, with raised angles, glabrous, divaricately branched, leafy. *Leaves* 1–7 × 0.2–0.5 cm, greyish-green, simple, linear and often somewhat falcate, acuminate at apex, entire, the upper sessile, the lower more or less petiolate, with 3–5 more or less parallel veins, glabrous. *Inflorescence* of compound umbels, with 5–7 rays, the rays unequal and shorter than or equalling the bracts, the peduncles longer than the rays, all flowers in terminal and lateral umbels bisexual; bracts half as long to as long as the longer rays, lanceolate, acuminate at apex, whitish and translucent in fruit; bracteoles like the bracts but smaller, acuminate, exceeding the flowers and connate at base, with 3–9, stout veins connected by numerous cross-veins which are ascending and then usually abruptly recurved. *Sepals* absent. *Petals* 5, yellow, the outer not radiating, the apex inflexed. *Stamens* 5; filaments and anthers yellow. *Styles* 2, horizontal, with the enlarged base forming the stylopodium. *Schizocarp* 1.5–1.7 mm, oblong-ellipsoid, somewhat compressed laterally; mericarps with very slender ridges; commissure broad; carpophore present; vittae solitary; pedicels short. *Flowers* 6–7.

Introduced. A rare grain and bird-seed casual. Casual *B. baldense* should always be checked to make sure it is not this species. Native of the Mediterranean region.

Section **3. Bupleurum**
Section *Perfoliata* Godr.
Annual. Lower leaves sessile or petiolate, the upper rounded-perfoliate; veins numerous, slender, radiating, anastomosing near the margin and connected by the cross-veins elsewhere. *Bracts* absent. *Bracteoles* usually 5.

6. B. rotundifolium L. Thorow-wax
B. perfoliatum Lam. nom. illegit.

Annual monoecious *herb* with fibrous roots. *Stems* 15–30 cm, pale greyish-green, often purple-tinged, erect, hollow, smooth, glabrous, branched above, leafy. *Leaves* bluish or greyish-green, paler beneath; lower 1–5(–8) × 1–4(–6) cm, ovate, apiculate at apex, entire, attenuate into a short petiole; upper 2–6 × 1.5–4.0 cm, elliptic-ovate to subrotund, apiculate at apex, entire, rounded-perfoliate at base; all glabrous, with a cartilaginous margin and numerous, slender veins radiating from the base, anastomosing near the margin and connected by slender cross-veins elsewhere. *Inflorescence* of compound umbels, with 4–8 rays, the rays unequal, up to 10 mm and smooth, the peduncle as long as or a little longer than the rays, all the flowers in both terminal and lateral umbels bisexual; bracts absent; bracteoles usually 5, 5–12 mm, unequal, oblanceolate to ovate, acuminate at apex, patent in flower, connivent in fruit, shortly connate at base, the venation similar to that of the leaf. *Sepals* absent. *Petals* 5, yellow, the outer not radiating, not emarginate at apex, the apex inflexed. *Stamens* 5; filaments and anthers yellow. *Styles* 2, 0.2–0.3 mm, with an enlarged base forming the stylopodium which is much longer than the styles; stigma truncate. *Schizocarp* 3.0–3.5 mm, pruinose, ellipsoid-oblong, somewhat compressed laterally; mericarps with slender, prominent ridges, smooth between the ridges; commissure broad; carpophore present; vittae inconspicuous; pedicels about 1 mm. *Flowers* 6–7. $2n=16$.

Introduced. Formerly common in cornfields where it probably needed repeated introduction with cereal seed. Formerly common in much of England especially the centre and south. Extinct since the 1960s except as a very rare casual. Central and south Europe and Russia southwards from about 52° N; western Asia.

7. B. subovatum Link ex Spreng. False Thorow-wax
B. intermedium (Loisel. ex DC.) Steud.;
B. lancifolium auct.

Annual monoecious *herb* with fibrous roots. *Stems* 15–30 cm, pale greyish-green, sometimes purple-tinted, erect, hollow, smooth, glabrous, branched above, leafy. *Leaves* bluish or greyish-green, paler beneath; lower ovate, apiculate at apex, entire and narrowed to short petiole; upper 2.5–6.0 × 1.5–2.8 cm, ovate to oblong-ovate, obtuse-apiculate at apex, entire, rounded-perfoliate at base; all glabrous, with a cartilaginous margin and numerous, slender veins radiating from the base, anastomosing near the margin and connected by slender cross-veins elsewhere. *Inflorescence* of compound umbels, with 2–3 rays, the rays subequal and smooth, the peduncle as long as or a little longer than the rays, all the flowers in both terminal and lateral umbels bisexual; bracts absent; bracteoles usually 5, 5–12 mm, unequal, oblanceolate to ovate, acuminate at apex, patent in flower, connivent in fruit, shortly connate at base, the venation similar to that of the leaf. *Sepals* absent. *Petals* 5, yellow, the outer not radiating, not emarginate at apex, the apex inflexed. *Stamens* 5; filaments and anthers yellow. *Styles* 2, 0.2–0.3 mm, with an enlarged base forming the stylopodium which is much longer than the style; stigma truncate. *Schizocarp* 3.5–5.0 mm, pruinose, ellipsoid-oblong, somewhat compressed laterally; mericarps with slender, prominent ridges, strongly tuberculate between the ridges, the tubercles develop soon after flowering and become more conspicuous as the fruit matures; commissure broad; carpophore present; vittae inconspicuous; pedicels about 1 mm. *Flowers* 6–10. $2n=16$.

Introduced. First recorded about 1870 and showed a marked increase about 1950 onwards due to the greater availability of seeds for feeding wild birds. Scattered records over Great Britain and Ireland. Native of southern

Europe, south-west and central Asia and North Africa; widespread casual elsewhere.

29. Trinia Hoffm. nom. conserv.
Apinella Raf. nom. illegit.

Dioecious *biennial* to monocarpic *perennial herbs. Stems* erect, solid. *Leaves* 1- to 3-pinnate. *Inflorescence* of compound umbels; bracts 0–1, 3-fid; bracteoles 0–several, entire to 2- to 3-fid. *Sepals* absent. *Petals* 5, white, the outer not radiating. *Fruit* a schizocarp, ovoid, somewhat compressed; mericarps with prominent, broad ridges; commissure narrow; carpophore present; vittae 5 within the ridges and 4 smaller ones between them.

Contains about 12 species in southern Europe and temperate Asia.

Wigginton, M. J. (Edit.) (1999). *British red data books*. Vol. 1. *Vascular plants*. Peterborough.

1. T. glauca (L.) Dumort. Honewort
Pimpinella glauca L.; *T. vulgaris* DC. nom. illegit.; *Apinella glauca* (L.) Caruel; *Seseli pumilum* L.; *Peucedanum minus* Huds.; *Pimpinella pumila* (L.) Jacq.; *Pimpinella dioica* L. nom. illegit.; *T. glaberrima* Hoffm. nom. illegit.; *Apinella glaberrima* Druce nom. illegit.

Usually dioecious *biennial* or *perennial herb*, probably always monocarpic, with a stout root. *Stems* 3–20 cm, pale green, erect, solid, grooved, glabrous, much branched, the branches widely spreading, surrounded by abundant, fibrous remains of petioles at the base. *Leaves* glaucous-green on upper surface, paler beneath; lower 2–20 × 1–4 cm, ovate to ovate-oblong in outline, 2- to 3-pinnate, the segments 5–125 mm, linear, acute and with a cartilaginous apex, the petiole slender and with a sheathing base; upper smaller and less divided; all glabrous. *Inflorescence* of compound umbels, the male with 4–8 rays, the rays 5–10 mm, smooth and subequal, the female similar but with very unequal rays up to 30 mm, the peduncle longer than the rays; bracts 0–1, 3-fid; bracteoles 0–several, simple or 2-to 3-fid; partial umbels of male plants dense and with numerous, shortly pedicellate flowers, those of the female plants lax, with few, long-pedicellate flowers. *Sepals* absent. *Petals* 5, white, the outer not radiating, the apex incurved. *Stamens* 5; filaments white; anthers pale yellow. *Styles* 2, recurved and appressed, with enlarged base forming the stylopodium which is half to one-third as long as the style; stigma capitate. *Schizocarp* 2.3–3.0 mm, ovoid, somewhat compressed laterally, smooth; mericarps with prominent, broad ridges; commissure narrow; carpophore present; vittae, 5 within the ridges and 4 smaller ones between them; pedicels up to 5 times as long as the mature fruit. *Flowers* 5–6. $2n = 18$.

Native. Dry, stony limestone grassland. Very local in Devonshire, Somersetshire and Gloucestershire. Central and south Europe northwards to England; south-west Asia. European southern-temperate element. Our plant is subsp. **glauca** which occurs throughout the range of the species.

30. Cuminum L.

Annual monoecious *herbs. Stems* erect, solid. *Leaves* 2-ternate. *Inflorescence* of compound umbels; bracts 2–4, entire and 3-fid with long, filiform lobes; bracteoles usually 3, very unequal. *Sepals* absent. *Petals* 5, white, the outer not radiating. *Fruit* a schizocarp, ovoid-oblong, slightly dorsally compressed; mericarps with prominent narrow ridges; commissure broad; carpophore present; vittae solitary.

Contains 2 species in southern Europe, North Africa and central Asia.

Grieve, M. (1931). *A modern herbal*. London.

1. C. cyminum L. Cumin
C. officinale Garsault; *C. odorum* Salisb.; *Ligusticum cuminum* Crantz; *Cuminia cyminum* (L.) J. F. Gmel.; *Cyminon longeinvolucellatum* St Lag.; *Luerssenia cyminum* (L.) Kuntze; *Selinum cuminum* (Crantz) E. H. L. Krause

Annual monoecious *herb* with fibrous roots. *Stems* 10–50 cm, pale green, erect, slender, solid, ridged, glabrous, branched, leafy. *Leaves* with lamina bluish-green, up to 5 cm, ovate in outline, 2-ternate, the lobes 20–50 mm, filiform, acute at apex, glabrous; petiole widened and slightly sheathing. *Inflorescence* of compound umbels; rays 1–5, 3–5 mm, glabrous; bracts 2–4, filiform or 3-fid, usually longer than the rays; bracteoles usually 3, very unequal. *Sepals* 5, subulate, conspicuous. *Petals* 5, white or pink, emarginate, apex long, not radiating. *Stamens* 5; filaments pale; anthers yellow. *Styles* 2, recurved, with enlarged base forming the stylopodium; stigma capitate. *Schizocarp* 4–6 mm, ovoid-oblong, dorsally compressed, glandular-hairy, with prominent filiform ridges, the secondary more conspicuous; commissure broad; carpophore present; vittae solitary. *Flowers* 6–7. $2n = 14$.

Introduced. Increasingly frequent casual from bird-seed and from use as a spice, found on tips and waste places. Scattered records in southern England. Native of North Africa and south-west Asia and cultivated eastwards to Japan and Indonesia. The seeds have a pungent aroma and flavour and are used in curries, for pickling, and in soups and casseroles.

31. Apium L.

Helosciadium W. D. J. Koch

Biennial or *perennial* monoecious *herbs. Stems* hollow or solid, erect to procumbent. *Leaves* 1- to 3-pinnate. *Inflorescence* of compound umbels; bracts and bracteoles absent or several, entire. *Sepals* absent. *Petals* 5, white, the outer not radiating. *Fruit* a schizocarp, ovoid, ellipsoid, ellipsoid-oblong or suborbicular, laterally compressed; mericarps with prominent, slender to thick ridges; commissure narrow; carpophore present; vittae solitary.

Contains about 6 species in Europe, temperate Asia, North America and the circum-Arctic zone.

Druce, G. C. (1912, 1914). *Apium moorei* (Syme) mihi. *Rep. Bot. Soc. Exch. Club Brit. Isles* **3**: 20,324–325, 470.
Grassly, N. C., Harris, S. A. & Cronk, Q. C. B (1996). British *Apium repens* (Jacq.) Lag. (Apiaceae) status assessed using random amplified polymorhic DNA (RAPD). *Watsonia* **21**: 103–111.
Grime, J. P. et al. (1988). *Comparative plant ecology*. London. [*A. nodiflorum*.]

Hultén, E. & Fries, M. (1986). *Atlas of north European vascular plants north of the Tropic of Cancer.* 3 vols. Königstein.
Phillips, R. & Rix, M. (1988). *Vegetables*, pp. 124–127. London. [*A. graveolens*.]
Preston, C. D. & Croft, J. M. (1997). *Aquatic plants in Britain and Ireland.* Colchester.
Riddelsdell, H. J. (1914). *Helosciadium moorei. Irish Nat.* **23**: 1–11.
Riddelsdell, H. J. (1917). *Helosciadium* in Britain. *Rep. Bot. Soc. Exch. Club Brit. Isles* **4**: 409–412.
Riddelsdell, H. J. & Baker, E. G. (1906). British forms of *Helosciadium nodiflorum* Koch. *Jour. Bot. (London)* **44**: 185–190.
Sprague, T. A. (1923). *Apium leptophyllum. Jour. Bot. (London)* **61**: 129–133.
Thommen, G. H. & Westlake, D. F. (1981). Factors affecting the distribution of populations of *Apium nodiflorum* and *Nasturtium officinale* in small chalk streams. *Aquat. Bot.* **11**: 21–36.
Wigginton, M. J. (Edit.) (1999). *British red data books.* Vol. 1. *Vascular plants*. Peterborough.

1. Bracts and bracteoles absent; fresh plant often smelling of celery 2.
1. Bracts 0–7; bracteoles 3–7; fresh plant not smelling of celery 5.
2. Segments of leaves filiform **2. leptophyllum**
2. Segments of leaves deltate to rhombic 3.
3. Petioles and stock not swollen **1(a). graveolens** subsp. **graveolens**
3. Petioles swollen or base of stem swollen 4.
4. Biennial, in 1 year forming an upright rosette of leaves with closely appressed, swollen, succulent petioles **1(b). graveolens** subsp. **dulce**
4. Base of stem swollen, subglobose and about 10 cm in diameter, the petioles not or little swollen **1(c). graveolens** subsp. **rapaceum**
5. Lower leaves 2- to 3-pinnate, with more or less filiform segments if submerged; stylopodium about twice as long as the style **5. inundatum**
5. All leaves 1-pinnate, even if submerged; styles longer than stylopodium 6.
6. Leaf segments about as long as wide; bracts (1–)3–7; fruits slightly wider than long **4. repens**
6. Leaf segments longer than wide; bracts 0–4; fruits longer than wide or undeveloped 7.
7. Fruits well-developed **3. nodiflorum**
7. Fruits undeveloped 8.
8. Rays 5–30; bracts 1–4 **3.×4. nodiflorum×repens = ×longipedunculatum**
8. Rays 2–3; bracts 0–2 **5.×3. inundatum×nodiflorum = ×moorei**

1. A. graveolens L. Celery
A. maritimum Salisb.; *Seseli graveolens* (L.) Scop.; *Seseli apium* Roth; *Sium graveolens* (L.) Vest; *Smyrnium laterale* Thunb.; *Heliosciadium ruta* DC.; *Heliosciadium rutaceum* St Lag.; *Celeri graveolens* (L.) Britten; *Selinum graveolens* (L.) E. H. L. Krause

Biennial herb with a strong smell of celery. *Stems* 30–100 cm, pale yellowish-green, erect, solid, strongly grooved, glabrous, branched, leafy. *Leaves* medium yellowish-green on upper surface, paler beneath; lower 10–30×5–15 cm, broadly oblong to ovate-oblong in outline, simply pinnate, the 1–3 pairs of segments 10–50 mm, deltate to rhombic in outline, coarsely toothed and sometimes pinnately lobed, the slender petiole with a sheathing base; upper mostly 3-fid or simply pinnate with 1 pair of segments, the petioles sheathing for all or most of their length; all glabrous. *Inflorescence* of compound umbels, with 4–12 rays, the rays 10–30 mm, smooth and subequal, the peduncles shorter than the rays or almost absent and often leaf-opposed, all umbels with bisexual flowers; bracts and bracteoles absent. *Sepals* absent. *Petals* 5, white with a greenish tinge, the outer not radiating, not emarginate and apex sometimes inflexed. *Stamens* 5; filaments white; anthers yellow. *Styles* 2, recurved or appressed, with enlarged base forming the stylopodium which is about as long as the style; stigma a small knob. *Schizocarp* 1.4–1.6 mm, broadly ovoid or ellipsoid, laterally compressed; smooth; mericarps with prominent, slender ridges; commissure narrow; carpophore present; vittae solitary; pedicels 2–4 mm. *Flowers* 6–8. $2n=22$.

(a) Subsp. **graveolens** Wild Celery
Stems, petioles and stock not swollen. *Petioles* green.

(b) Subsp. **dulce** (Mill.) Bertol. Celery
A. dulce Mill.; *A. graveolens* var. *dulce* (Mill.) DC.

Biennial which in 1 year forms an upright rosette of leaves with closely appressed, swollen, succulent petioles, white from blanching or tinged with pink.

(c) Subsp. **rapaceum** (Mill.) P. D. Sell Celeriac
A. rapaceum Mill.; *A. graveolens* var. *rapaceum* (Mill.) DC.

Base of stem swollen, subglobose and about 10 cm in diameter, with a brown skin and white flesh; petioles not or little swollen.

Native. Subsp. *graveolens* occurs in damp, rather bare, usually brackish places, mostly near the sea; rarely inland. Coasts of Great Britain and Ireland, north to southern Scotland, with a few scattered records inland. Europe to about 56° N; south-west Asia; Macaronesia; North Africa; introduced elsewhere. Eurosiberian Southern-temperate element. It has been used as a pot-herb since classical times. Subsp. *dulce* was first developed in Italy by selection and reached France in the seventeenth century, but remained almost unknown in Great Britain until the nineteenth century. As a crop it is largely confined to the Fenlands of East Anglia where the soil is rich and moist. It is planted in trenches and later earthed up to blanch the stems. It can be eaten raw, but is rather indigestible. Its culinary value is as a flavouring of savoury dishes, or as a vegetable eaten raw or cooked. The seeds are often used as a flavouring agent. Subsp. *rapaceum* is thought to have been developed later, but has long been a popular winter vegetable on the continent of Europe, but has only become widely available in Great Britain in recent years. The swollen stem is best eaten as a salad, grated and dressed with mayonnaise, but can also be eaten as a cooked vegetable. In Great Britain it has never equalled subsp. *dulce* in popularity, but is grown commercially on a small scale. Both subsp. *dulce* and subsp. *rapaceum* sometimes occur as crop remnants.

2. A. leptophyllum (Pers.) F. Muell. ex Benth. Slender Celery

Pimpinella leptophylla Pers.; *Ciclospermum leptophyllum* (Pers.) Britton & E. E. Wilson; *Apium ammi* Urb. nom. illegit.; *A. tenuifolium* (Moench) Thell. nom. illegit.; *Sison ammi* auct.; *Aethusa leptophylla* (Pers.) Spreng.; *Heliosciadium leptophyllum* (Pers.) DC.; *Selinum leptophyllum* (Pers.) E. H. L. Krause; *Pimpinella lateriflora* Link; *Heliosciadium lateriflorum* (Link) W. D. J. Koch; *Sison lateriflorum* (Link) Bertol.

Annual monoecious *herb* with a slender tap-root, smelling of celery when fresh. Stem 10–70 cm, pale green, hollow, erect, slightly grooved, glabrous, branching, leafy. *Leaves* medium green on upper surface, paler beneath; basal 2–10 × 1–4 cm, ovate in outline, 1- to 4-pinnate or ternate, the ultimate lobes 2–15 mm, linear to filiform and entire, petiolate; cauline similar to basal, but 2- to 3-ternate and shortly petiolate. *Inflorescence* of compound umbels, with 1–3(–4) rays; bracts and bracteoles absent. *Sepals* absent. *Petals* 5, white, the outer not radiating, with an acute, inflexed apex. *Stamens* 5; filaments white; anthers yellow. *Styles* 2, recurved, with enlarged base forming the stylopodium; stigma a small knob. *Schizocarp* 1.5–3.0 mm, dark brown, ovoid-oblong, slightly compressed laterally; mericarps with 5, broad, thickened, pale brown ridges; commissure narrow; carpophore present; vittae solitary in furrows. *Flowers* 8–9. $2n = 14$.

Introduced. A rare wool and grain alien. North and South America; Europe; Japan; China; Australia and New Zealand, often behaving as a weed.

3. A. nodiflorum (L.) Lag. Fool's Watercress

Pimpinella nodiflora (L.) Stokes; *Sium nodiflorum* L.; *Helosciadium nodiflorum* (L.) W. D. J. Koch; *Cicuta nodiflora* (L.) Crantz; *Seseli nodiflorum* (L.) Scop.; *Sison nodiflorum* (L.) Brot.; *Tordylium cyrenaicum* Spreng.; *Helodium nodiflorum* Dumort.; *Selinum nodiflorum* (L.) E. H. L. Krause

Perennial aquatic *herb* with dense, fibrous roots, normally emergent, but can form submerged patches. *Stems* (4–)30–100 cm, pale yellowish-green, slender or stout, procumbent or ascending, furrowed, rooting at the lower nodes, hollow, glabrous, branching, leafy. *Leaves* (1.5–)10–30 × (1–)3–15 cm, shiny, medium bright green on upper surface, paler beneath, broadly oblong in outline, simply pinnate, with (1–)2–4(–6) pairs of segments, the segments (3–)15–60(–100) mm, lanceolate to ovate, obtuse at apex, serrate or crenate and often somewhat lobed, glabrous and sessile, the petioles long and slightly sheathing. *Inflorescence* of compound umbels, with 3–15 rays, the rays (3–)10–20 mm, usually subequal, scabrid and spreading or recurved, the peduncles shorter than the rays or absent, leaf-opposed; all umbels with bisexual flowers; bracts usually absent, rarely 1–2; bracteoles 4–7, linear-lanceolate or ovate, usually as long as or longer than the flowers. *Sepals* absent. *Petals* 5, greenish-white, the outer not radiating, not emarginate at apex, the acute point sometimes inflexed. *Stamens* 5; filaments white; anthers pink. *Styles* 2, recurved, with enlarged base forming the stylopodium which is shorter than the style; stigma a small knob. *Schizocarp* 1.5–2.5 mm, ovoid, laterally compressed, smooth; mericarps with prominent, thick ribs; commissure narrow; carpophore present; vittae solitary; pedicels 1–2 mm. *Flowers* 7–8. $2n = 22$.

Extremely variable from small plants (var. *pseudorepens* H. C. Watson) growing on mud in winter-wet areas to very large plants (var. *nodiflorum*) growing in deep water, with all intermediates (var. *ochreatum* DC.).

Native. Shallow water in marshes, springs, streams and ditches, where it may form dense, pure stands, and at the edges of rivers, lakes, ponds and canals. It is highly palatable to cattle and can be eliminated by grazing. Common in the lowlands of Great Britain and Ireland north to southern Scotland; very local in central and north Scotland. West, central and southern Europe, reaching its northern limit in Scotland; south-west Asia; North Africa; introduced, naturalised and spreading in North and South America. Eurosiberian South-temperate element. *Berula erecta* is very similar in appearance though more often submerged, and can be distinguished by having a ring-mark on the petiole some way below the lowest pair of leaflets.

× **repens**
= **A. × longipedunculatum** (F. W. Schultz) Rothm.

This hybrid has a more procumbent and extensively rooting stem than *A. nodiflorum*, with the leaf segments longer than wide, the peduncles shorter than to about as long as the rays, 1–4 bracts and sterile. $2n = 20$.

Native. Has been found in some of the Oxfordshire localities of *A. repens* and still exists in Cambridgeshire, Yorkshire, Fifeshire and perhaps elsewhere where *A. repens* has become extinct. Endemic.

4. A. repens (Jacq.) Lag. Creeping Marshwort

Sium repens Jacq.

Perennial monoecious *herb* with creeping stems rooting at the nodes and fibrous roots. *Stems* long, pale green, prostrate, slender, glabrous. *Leaves* 1–5 × 1.2–1.5 cm, medium yellowish-green on upper surface, paler beneath, from the nodes of the creeping stem, oblong in outline, simply pinnate, usually with 2–5 pairs of segments, the segments 2–10 mm, subrotund, rounded at apex, incise-toothed or -lobed, glabrous. *Inflorescence* of compound umbels, with 4–7 rays, the rays 5–30 mm, usually subequal and smooth, the peduncles longer than the rays, leaf-opposed, all umbels with bisexual flowers; bracts 4–7, lanceolate to ovate, deflexed; bracteoles like the bracts, usually shorter than the flowers. *Sepals* absent. *Petals* 5, white, the outer not radiating, entire at apex, which is acute, the point sometimes shortly inflexed. *Stamens* 5; filaments white; anthers pale yellow. *Styles* 2, recurved, with an enlarged base forming the stylopodium which is half as long as the style; stigma a small knob. *Schizocarp* 0.7–1.0 mm, suborbicular or broadly ellipsoid, slightly broader than long, laterally compressed, smooth; mericarps with prominent, slender ribs; commissure narrow; carpophore present; vittae solitary; pedicels 1–2 mm. *Flowers* 7. $2n = 16$.

Native. Old damp meadows, ditches and shallow ponds. Known only from four localities in Oxfordshire, in only one of which is it probably the pure species. It may also occur in Cambridgeshire. The general distribution is uncertain, but it appears to be central Europe. European Temperate element. The species is unsatisfactorily defined.

5. A. inundatum (L.) Rchb. fil. Lesser Marshwort
Sison inundatum L.; *Helosciadium inundatum* (L.) W. D. J. Koch; *Sium inundatum* (L.) Lam.; *Hydrocotyle inundata* (L.) Sm.

Straggling, often submerged or floating *perennial*, monoecious *herb* with fibrous roots. *Stems* 10–75 cm, pale green, slender, glabrous, decumbent and rooting at nodes at the base, then ascending, branched. *Leaves* 1–8 × 0.5–6.0 cm, yellowish-green, oblong or triangular-oblong in outline, pinnate, the submerged leaves and lower leaves of terrestrial plants 2- to 3-pinnate and the segments filiform, the floating and aerial leaves with linear segments, the upper leaves pinnate with lanceolate to ovate, often 3-lobed sessile segments and the lobes about 5 mm, glabrous. *Inflorescence* of compound umbels, with 2(–4) rays, the rays 5–10 mm, smooth, the peduncles short and leaf-opposed, all umbels with bisexual flowers; bracts absent; bracteoles 3–6, lanceolate, not concealing the flowers and fruit. *Sepals* absent. *Petals* 5, white, the outer not radiating, entire at apex, which is acute, the point sometimes shortly inflexed. *Stamens* 5; filaments white; anthers yellow. *Styles* 2, spreading, with enlarged base forming the stylopodium which is twice as long as the style; stigma a small knob. *Schizocarp* 2.5–3.0 mm, ellipsoid-oblong, laterally compressed; mericarps with prominent ridges, commissure narrow; carpophore present; vittae solitary; pedicels very short. *Flowers* 6–8. Probably self-pollinated. $2n = 22$.

Native. Edges of lakes and reservoirs, pools and dune slacks where the water-level fluctuates and periodic desiccation occurs. Also in more permanent shallow water in streams, ditches, canals, the backwaters of rivers and marshes. It is confined to oligotrophic or mesotrophic habitats over sand, silt, gravelly stones and peat. Scattered over Great Britain and Ireland, but primarily a lowland species and appears to be declining. A suboceanic species found in western and central Europe from south Scandinavia and Poland south to the Iberian Peninsula and Sicily. Suboceanic Temperate element.

× nodiflorum **A. × moorei** (Syme) Druce
Helosciadium inundatum var. *moorei* Syme

This hybrid resembles *A. inundatum*, but the lower leaves are 1-pinnate and linear or ligulate instead of filiform, it flowers much less freely than the parents and appears to be completely sterile.

Native. In shallow water or on damp mud. Scattered over most of Ireland and local in eastern central England. Endemic.

32. Trachyspermum Link

Annual monoecious *herbs. Stems* erect, hollow. *Leaves* 2- to 3-pinnatisect. *Inflorescence* of compound umbels; bracts 4–5, entire or lobed; bracteoles 3–6. *Sepals* 5, small, not persistent. *Petals* 5, white, the outer not radiating. *Fruit* a schizocarp, ovoid, laterally compressed, densely covered with grey tuberculate papillae; mericarps with prominent ridges; commissure narrow; carpophore present; vittae solitary.

Contains about 20 species in Africa and Asia.

1. T. ammi (L.) Sprague Ajowan
Sison ammi L.; *T. copticum* (L.) Link; *Ammi copticum* L; *Carum aromaticum* (L.) Druce, non Salisb.; *Carum copticum* (L.) Benth. & Hook. fil.; *Apium ammi* (L.) Crantz; *Bunium aromaticum* L.; *Daucus copticus* (L.) Lam.; *Daucus anisodorus* Blanco; *Ammios muricata* Moench; *Seseli foeniculaceum* Poir.; *Ligusticum ajawain* Roxb. ex Fleming; *Ptychotis coptica* (L.) DC.; *Ptychotis ajawain* (Roxb. ex Fleming) DC.; *Athamanta ajowian* (Roxb. ex Fleming) Wall. ex DC.; *Carum copticum* (L.) C. B. Clarke; *Carum ajowan* (Roxb. ex Fleming) Benth. & Hook. fil.; *Carum korolkowii* Regel & Schmalh.; *Pituranthos korolkowii* (Regel & Schmalh.) Schinz; *Selinum copticum* (L.) E. H. L. Krause

Annual monoecious *herb* with fibrous roots. *Stems* 10–30 cm, pale green, erect, hollow, striate, glabrous, much-branched, leafy. *Leaves* medium green on upper surface, paler beneath, triangular-ovate in outline, the lower withered by anthesis, triangular-ovate in outline, with a long petiole, the lower 2- to 3-pinnatisect, the segments 10–20 mm and linear or filiform and long-petiolate, the upper similar but smaller and sometimes simply pinnatisect and shortly petiolate, the petiole with a short, sheathing base; all glabrous. *Inflorescence* of compound umbels, with 5–10(–20) rays, the rays about 10 mm, smooth or sparsely puberulent, the peduncles longer than the rays; all umbels with bisexual flowers; bracts 4–5, linear, sometimes lobed; bracteoles 3–6, linear-lanceolate, sometimes sparsely puberulent. *Sepals* 5, small, acute at apex. *Petals* 5, white, hairy beneath, the outer not radiating. *Stamens* 5; filaments white; anthers yellow. *Styles* 2, recurved, with enlarged base forming the stylopodium which is about half as long as the style; stigma capitate. *Schizocarp* 1.5–2.0 mm, ovoid, laterally compressed, irregularly covered with grey tuberculate papillae; mericarps with prominent ridges; commissure narrow; carpophore present; vittae solitary; pedicels 1–4 mm, hairy. *Flowers* 9. $2n = 18$.

Introduced. An occasional casual on rubbish-tips. Scattered records in England and Lanarkshire. Grown in eastern Mediterranean countries for its aromatic fruits which are used in flavouring, but it is not known in the wild. Its classification has caused much trouble as can be seen by the number of names if has received.

Spermolepis divaricatus (Walter) Britton and **S. echinatus** (Nutt.) Heller have been recorded as wool casuals.

33. Petroselinum Hill

Annual or *biennial* monoecious *herbs. Stems* erect, solid. *Leaves* 1- to 3-pinnate. *Inflorescence* of compound umbels; bracts 1–5, entire or 2- to 3-fid; bracteoles 2–8. *Sepals* absent. *Petals* 5, white or yellow, the outer not radiating. *Fruit* a schizocarp, ovoid to ellipsoid, laterally compressed; mericarps with prominent thick to narrow ridges; commissure narrow; carpophore present; vittae solitary and conspicuous.

Contains 2 species in Europe, western Asia and North Africa.

Grieve, M. (1931). *A modern herbal*. London.
Philips, R. & Rix, M. (1988). *Vegetables*, pp. 128–129. London. [*P. crispum*.]

1. Leaves 1-pinnate; flowers white **2. segetum**
1. Leaves 3-pinnate; flowers yellowish 2.
2. Root dingy-white, thick, fleshy and rather like a small parsnip with a flavour like that of celeriac; leaves flat **1(c). crispum** subsp. **tuberosum**
2. Plant with a taproot and fibrous side roots 3.
3. Leaves flat **1(a). crispum** subsp. **foliosum**
3. Leaves crisped **1(b). crispum** subsp. **crispum**

1. P. crispum (Mill.) Nyman ex A. W. Hill Garden Parsley
P. sativum Hoffm.; *Carum petroselinum* (L.) Benth.; *Apium crispum* Mill.; *Apium petroselinum* L.; *Apium vulgare* Lam. nom. illegit.; *P. vulgare* Lag. nom. illegit.; *Apium laetum* Salisb.; *P. romanum* Sweet; *Apium romanum* (Sweet) Zucc.; *Selinum petroselinum* (L.) E. H. L. Krause

Biennial monoecious *herb* with a tap-root and fibrous side-roots or rarely with a tuberous root, all its parts with a familiar parsley aroma. *Stems* 30–75 cm, bluish-green, erect, solid, striate, often with some sheathing remains at the base, glabrous. *Leaves* 2–15 × 1–8 cm, bright, dark green on upper surface, paler beneath, the lower triangular in outline, 3-pinnate, the segments 10–30 mm, ovate in outline, more or less 3-fid, toothed, crispate in many plants, cuneate at base, and with a long petiole, sheathing near the base, the upper cauline much smaller, 1- to 2-pinnate, with short, mostly or entirely sheathing and usually broadly scarious petioles, all glabrous. *Inflorescence* of compound, flat-topped umbels 20–50 mm in diameter, with 8–20 rays, the rays 5–30 mm, smooth and subequal, the peduncle longer than the rays, the terminal umbel with bisexual flowers, the lateral umbels with male and bisexual flowers, the latest to develop with almost entirely male flowers; bracts 1–3, entire or 2- to 3-fid and more or less leaf-like; bracteoles 5–8, linear-oblong to ovate, cuspidate at apex, usually with a broad, scarious margin. *Sepals* absent. *Petals* 5, yellowish, the outer not radiating, scarcely emarginate and with a small, inflexed point at apex. *Stamens* 5; filaments white; anthers yellow. *Styles* 2, recurved and appressed, with enlarged base forming the stylopodium which is about as long as the style; stigma a small knob. *Schizocarp* 2.0–2.5 mm, ovoid to ellipsoid, laterally compressed, smooth; mericarps with 3 rather slender, low, rounded, dorsal ridges; commissure narrow; carpophore present; vittae solitary, conspicuous; pedicels 2–4 mm, rather stout. *Flowers* 6–8. $2n=22$.

(a) Subsp. **foliosum** (Alef.) Janch. Wild Parsley
P. sativum var. *foliosum* Alef.; *P. hortense* var. *foliosum* (Alef.) Thell.

Plant with tap-root and fibrous side-roots. *Leaves* flat.

(b) Subsp. **crispum** Garden Parsley
P. segetum var. *crispum* (Mill.) Gaudin

Plant with tap-root and fibrous side-roots. *Leaves* crisped.

(c) Subsp. **tuberosum** (Bernh.) Janch. Turnip-rooted Parsley
Apium tuberosum Bernh.

Root dingy-white, thick, fleshy and rather like a small parsnip with a flavour like that of celeriac. *Leaves* flat and less compound.

Introduced. Commonly grown on a small scale and a frequent escape onto tips and in waste places. Scattered records throughout Great Britain and Ireland except most of central and north Scotland. Its origin is uncertain, but it is probably native of south-east Europe. It was used by the ancient Greeks and Romans and later spread right through Europe and most temperate countries of the world. It has been grown in Great Britain since the sixteenth century. Subsp. *foliosum* was the early variant. Subsp. *crispum* came in at a later date and is now more popular. Both these subspecies are used in cooking, for garnishing a variety of dishes and for window-dressing in the more traditional butcher's and fishmonger's shops. It is used chopped, fresh or dried, to flavour sauces, soups, salads, omelettes and stuffings. Subsp. *tuberosum* is not an ancient vegetable, but the history of its development is not known. It is grown to a small extend, but is not as popular in Great Britain as in European markets.

2. P. segetum (L.) W. D. J. Koch Corn Parsley
Seseli segetum Crantz; *Apium junceum* Stokes nom. illegit.; *Sison segetum* L.; *Carum segetum* (L.) Benth. ex Hook. fil.; *Sium segetum* (L.) Lam.

Annual to *biennial* monoecious *herb* with a slender, fusiform tap-root and fibrous side- roots. *Stems* 30–100 cm, medium bluish-green, erect, solid, striate, glabrous, divaricately branched, leafy. *Leaves* smelling of parsley when crushed, 2–13 × 0.5–4.5 cm, bluish medium green on upper surface, paler beneath, oblong in outline, simply pinnate, mostly near the base of the plant, with 4–12 pairs of segments which are 5–35 mm, lanceolate to subrotund, serrate or sometimes shallowly lobed, the obtuse teeth with cartilaginous margin and a mucro up to 0.5 mm, glabrous, the petiole slender and not much dilated at the base. *Inflorescence* of compound umbels, with 3–10 rays, the rays up to 30 mm, smooth and very unequal, the peduncle longer than the rays, all umbels with bisexual flowers; bracts and bracteoles 2–5, subulate, glabrous. *Sepals* 5, small. *Petals* 5, white, the outer not radiate, scarcely emarginate and with a small inflexed point at apex. *Stamens* 5; filaments white; anthers purple. *Styles* 2, divergent, with enlarged base forming the stylopodium which is much longer than the styles; stigma a small knob. *Schizocarp* 2.5–3.0 mm, ovoid, laterally compressed, smooth; mericarps with prominent, stout ridges; commissure narrow; carpophore present; vittae solitary, conspicuous; some pedicels very short, others up to 15 mm. *Flowers* 8–9. $2n=18$.

Native. Rather bare or grassy places in arable fields, pastures, hedgerows and on banks. Local in southern Great Britain north to Yorkshire and Cardiganshire, and in Jersey. West and south Europe from Holland to Portugal and eastwards to central Italy. Everywhere rare and apparently decreasing. Suboceanic Southern-temperate element.

Easily mistaken for *Sison amomum* but has extremely unequal length rays, more leaflets, thicker ridges on the mericarps and lacks the characteristic smell of that species.

34. **Ridolphia** Moris

Annual monoecious *herbs. Stems* erect, solid. *Leaves* 3- to 4-pinnate. *Inflorescence* of compound umbels; bracts and bracteoles absent. *Sepals* absent. *Petals* 5, yellow, the outer not radiating. *Fruit* a schizocarp, ellipsoid or ovoid-cylindrical, compressed laterally; mericarps with slender, scarcely prominent ridges; commissure broad; carpophore present; vittae solitary and slender.

Contains 1 species in southern Europe, North Africa and Lebanon.

1. R. segetum (Guss.) Moris False Fennel
Carum segetum (Guss.) Benth. ex Arcang.; *Meum segetum* Guss. exclud.; *Anethum segetum* L.

Annual monoecious *herb* with a tap-root and fibrous side-roots. *Stems* 40–100 cm, pale green, solid, erect, glabrous, branched, leafy. *Leaves* 3–6 × 3–6 cm, without smell when crushed, medium green on upper surface, paler beneath, broadly ovate in outline, 3- to 4-pinnate, the segments long and divaricate, filiform, glabrous, with an inflated petiole, the upper often reduced to just the petiole. *Inflorescence* of compound umbels, with 10–60 rays, the rays slender and nearly equal; bracts and bracteoles absent. *Sepals* absent. *Petals* 5, yellow, the outer not radiating, ovate, truncate and inflexed at apex. *Stamens* 5; filaments pale; anthers yellow. *Styles* 2, recurved and appressed, with enlarged base forming the stylopodium which is twice as long as the styles; stigma capitate. *Schizocarp* 1.5–2.5 mm, ellipsoid or ovoid-cylindrical, compressed laterally; mericarps with slender, scarcely prominent ridges; commissure broad; carpophore present; vittae solitary, slender; pedicels glabrous. *Flowers* 7–8. $2n = 22$.

Introduced. Rather infrequent bird-seed alien on tips and waste ground. Scattered localities in southern England. Native of southern Europe; North Africa and Lebanon.

35. **Sison** L.

Biennial monoecious *herbs. Stems* erect, solid. *Leaves* 1- to 2-pinnate. *Inflorescence* of compound umbels; bracts 2–4; bracteoles 2–4. *Sepals* absent. *Petals* 5, white or greenish-white, the outer not radiating. *Fruit* a schizocarp, subglobose, somewhat laterally compressed; mericarps with fairly prominent ridges; commissure narrow; carpophore present; vittae about half as long as the mericarp, solitary and conspicuous.

Contains 2 species in west and south Europe, south-west Asia and North Africa.

1. S. amomum L. Stone Parsley
Cicuta amomum (L.) Crantz.; *Seseli amomum* (L.) Scop.; *Sium aromaticum* Lam. nom. illegit.; *Sium amomum* (L.) Roth; *Sison erectum* Salisb. nom. illegit.; *Apium amomum* (L.) Stokes; *S. heterophyllus* Moench

Biennial monoecious *herb* with a nauseous, petrol-like smell when crushed and a strong tap-root. *Stems* 50–100 cm, pale to medium green, solid, striate, erect, glabrous, branched, leafy, sometimes turning deep wine purple with age. *Leaves* medium green on upper surface, paler beneath; lower 5–15 × 1.5–8.0(–15.0) cm, oblong or elliptical-oblong in outline, simple to 2-pinnate, the lowest segments sometimes pinnatifid or rarely pinnatisect, with 2–5 pairs of segments which are 30–60 mm, lanceolate to ovate-lanceolate, serrate, the obtuse teeth with cartilaginous margin and short, acute mucro, cuneate at base and more or less sessile, and the petiole long with a gradually dilated base; upper mostly with 1–2 pairs of segments which are linear to narrowly spathulate and toothed, with a short, sheathing petiole; all glabrous. *Inflorescence* of compound umbels, with 3–6 smooth rays, the rays up to 30 mm, one of which is usually shorter than the remainder, the peduncle longer than the rays, all the umbels with bisexual flowers; bracts 2–4, linear-lanceolate to subulate; bracteoles 2–4, usually ovate-lanceolate. *Sepals* absent. *Petals* 5, white or greenish-white, the outer not radiating, subrotund-cordate, emarginate, with an inflexed point. *Stamens* 5; filaments white; anthers yellow. Styles 2, divergent to recurved, with enlarged base forming the stylopodium which is twice as long as the styles; stigma a small knob. *Schizocarp* 1.5–3.0 mm, subglobose, but somewhat longer than wide, somewhat compressed laterally and smooth; mericarps with fairly prominent ridges; commissure narrow; carpophore present; vittae about half as long as the mericarp, solitary, conspicuous; pedicels 1–5 mm, somewhat unequal. *Flowers* 7–9. $2n = 14$.

Very like *Petroselinum segetum* but distinguished by its petrol-like smell when crushed, larger leaf segments of the lower leaves, finely divided upper leaves and less unequal rays of the umbels.

Native. Hedgebanks, grassland and roadsides. Locally frequent in Great Britain north to Cheshire and Yorkshire, but seems to be declining in East Anglia; rare casual further north. West and south Europe; south-west Asia; North Africa. Submediterranean-Subaltantic element. In the last few years it has become frequent on some Cambridgeshire roadsides and round fields where wild flower seeds have been planted. It is slightly different from our normal plant in being taller, more branched and with more (5–7) rays. It turns deep wine-purple in autumn and the seeds have the characteristic petrol-like smell. It is not the only other named species in the genus *S. exaltatum* Boiss. No name has been found for it at any rank. It appears to be carried from one area to another on grasscutters, even into nature reserves.

36. **Cicuta** L.

Cicutaria Lam.

Perennial monoecious *herbs. Stems* erect, hollow. *Leaves* 2- to 3-pinnate; petioles widened at base and sheathing. *Inflorescence* of compound umbels; bracts absent; bracteoles numerous. *Sepals* 5, conspicuous and persistent. *Petals* 5, white, the outer not radiating. *Fruit* a schizocarp, subglobose or broader than long, laterally compressed; mericarps with low, very wide ridges.

Contains about 10 species in north Temperate regions of the world.

Hultén, E. & Fries, M. (1986). *Atlas of north European vascular plants north of the Tropic of Cancer*. 3 vols. Königstein.
Mulligan, G. A. & Munro, D. B. (1981). The biology of Canadian weeds. 48. *Cicuta maculata* L., C. *douglasii* (DC.) Coult. & Rose and C. *virosa* L. *Canad. Jour*. Pl. Sci. **61**: 93–105.
Stewart, A., Pearman, D. A. & Preston, C. D. (1994). *Scarce plants in Britain*. Peterborough.

1. C. virosa L. Cowbane
Cicutaria aquatica Lam. nom. illegit.; *Cicutaria virosa* (L.) Clairv.; *Coriandrum cicuta* Roth.; *C. angustifolia* Kit.; *C. tenuifolia* Froel.; *C. aquatica* Dumort.; *C. pumila* Behm.; *C. orientalis* Degen & Bald.; *C. sachalinensis* Koidz.; *Sium cicuta* Webb; *Selinum virosum* (L.) E. H. L. Krause

Perennial herb with an ovoid or shortly cylindrical, septate stock. *Stems* 30–150 cm, pale green, erect, hollow, somewhat ridged or striate, glabrous, branched, leafy. *Leaves* 7–40×7–25 cm, medium green on upper surface, paler beneath, deltoid or triangular in outline, 2- to 3-pinnate, the segments 30–90 mm, linear-lanceolate, acute at apex, remotely and deeply serrate, unequal at the base, midrib scabrid above, smooth or almost so beneath, the petiole long, hollow, sheathing but not conspicuously dilated at the base, the upper cauline leaves are similar, but smaller than the lower. *Inflorescence* of compound umbels, with 10–30 rays, the rays 20–60 mm, subequal and smooth, the peduncle somewhat longer than the rays, all umbels with mostly bisexual flowers; bracts absent; bracteoles numerous, linear-oblong. *Sepals* 5, ovate and conspicuous. *Petals* 5, white, the outer not radiating, with inflexed points. *Stamens* 10; filaments white; anthers pale yellow. *Styles* 2, curved, with enlarged base forming the stylopodium which is about one-third to half as long as the styles; stigma a small knob. *Schizocarp* 1.7–2.0 mm, subglobose or broader than long, laterally compressed, smooth; mericarps with low, very wide ridges; commissure wide; carpophore present; vittae slender, solitary, conspicuous; pedicels up to 6 mm, slender. *Flowers* 7–8. $2n=22$.

Native. Shallow water in ponds and ditches or the sides of rivers, lakes and lodes. Local and mainly in East Anglia, Shropshire, Cheshire and south-east Scotland with a few, scattered localities elsewhere. It has disappeared from many places in eastern England through drainage. Widespread from Great Britain and Ireland east to eastern Siberia and Japan, becoming more scattered in the western part of its range and virtually absent from the Mediterranean region. Closely related species occur in North America. Boreo-temperate element. It contains a convulsant poison, cicutoxin, which is frequently fatal to cattle, and has perhaps been selectively eliminated from some areas by Man.

37. Ammi L.

Visnaga Vill.

Annual or *biennial herbs*. *Stems* erect, solid. *Leaves* 1- to 3(–4)-pinnate. *Inflorescence* of compound umbels; bracts several, pinnately divided; bracteoles numerous. *Sepals* absent. *Petals* 5, white, the outer slightly radiating. *Fruit* a schizocarp, ovoid or ellipsoid, slightly compressed laterally; mericarps with slender, prominent ridges; commissure narrow; carpophore present; vittae solitary and conspicuous.

Contains about 10 species in the Mediterranean region and Atlantic Islands.

Wurzell, B. (1994). Year of the Bullwort. *B.S.B.I. News* **66**: 32.

1. Rays 15–60, patent and slender in both flower and fruit; schizocarp 1.5–2.0 mm **1. majus**
1. Rays up to 150, slender and patent in flower becoming erect, stout and rigid in fruit; schizocarp 2.0–2.5 mm **2. visnaga**

1. A. majus L. Bullwort
A. diversifolium Noulet

Annual monoecious *herb* with a tap-root and fibrous side-roots. *Stems* 15–100 cm, pale green, erect, solid, striate, glabrous, branched, leafy. *Leaves* very variable, 4–20×3–15 cm, medium green of upper surface, paler beneath; lower narrowly to broadly triangular in outline, 1- to 2-pinnate, the segments linear to ovate, obtuse at apex and often serrate; the upper smaller, usually 2-pinnate, the segments linear to lanceolate, acute at apex, often deeply toothed or pinnatifid; petiole gradually widened to the sheathing base; all glabrous. *Inflorescence* of compound umbels with 9–40 rays, the rays 30–70 mm, slender, subequal and somewhat scabrid, the peduncle longer than the rays, all umbels with bisexual flowers; bracts several, one-third to three-quarters as long as the rays, usually leaf-like, 3-fid to pinnatisect, with linear lobes, occasionally some entire; bracteoles numerous, usually lanceolate, with a long filiform apex and broad scarious margin. *Sepals* absent. *Petals* 5, white, the outer slightly radiating, obovate, irregularly and unequally 2-lobed, with an inflexed, emarginate apex. *Stamens* 5; filaments white; anthers yellow. *Styles* 2, recurved, with enlarged base forming the stylopodium which is half as long as the styles; stigma a small knob. *Schizocarp* 1.5–2.0 mm, ellipsoid, slightly compressed laterally, smooth; mericarps with slender, prominent ridges; commissure narrow; carpophore present; vittae solitary, conspicuous; pedicels 1–6 mm, slightly scabrid. *Flowers* 6–10. $2n=22$.

Var. *majus* (var. *serratum* Mutel) which has elliptical or lanceolate leaf segments, var. *glaucifolium* (L.) Noulet which has linear leaf segments and var. *intermedium* Gren. & Godr. which has intermediate leaf segments. All occur in Great Britain, but it is not known if they have any ecological significance.

Introduced. A casual on tips, waste ground and in fields, mainly from bird-seed and wool. Scattered records throughout Great Britain, rarely further north. It is native of the Mediterranean region.

2. A. visnaga (L.) Lam. Toothpick Plant
Daucus visnaga L.; *Visnaga daucoides* P. Gaertn; *Apium visnaga* (L.) Crantz; *A. dilatatum* St Lag.; *Sium visnaga* Stokes; *Visnaga vera* Raf.; *Selinum visnaga* (L.) E. H. L. Krause

Annual or *biennial* monoecious *herb* with a tap-root and fibrous side-roots. *Stems* 30–100 cm, pale green, solid, striate, erect, robust, glabrous, branched, leafy. *Leaves* 4–12×3–6 cm, medium green on upper surface, paler beneath; lower triangular in outline, pinnate, the segments narrowly linear or filiform; upper similar to lower but 2- to (3–)4-pinnate, and slightly smaller; all glabrous. *Inflorescence* of compound umbels, with up to 150 rays, the rays slender and patent in flower, becoming erect, stout and rigid in fruit, all umbels with bisexual flowers; bracts 1- to 2-pinnatisect, equalling or exceeding the rays, becoming strongly reflexed in fruit; bracteoles subulate. *Sepals* absent. *Petals* 5, white, the outer slightly radiating, obcordate, apex inflexed. *Stamens* 5; filaments white; anthers yellow. *Styles* 2, recurved, with enlarged base forming the stylopodium, which is half as long as the styles; stigma a small knob. *Schizocarp* 2.0–2.5 mm, ovoid, slightly compressed laterally, smooth; mericarps with slender, prominent ridge; commissure narrow; carpophore present; vittae solitary; pedicels erect, stout and rigid in fruit. *Flowers* 6–10. $2n=20, 22$.

Introduced. Bird-seed and wool casual, but rarer than *A. major*. Very scattered records throughout Great Britain. Native of the Mediterranean region and Portugal; frequent weed further north in Europe.

38. **Falcaria** Fabr. nom. conserv.

Prionitis Adans., non J. Agardh nom. conserv.; *Drepanophyllum* Wibel nom illegit.; *Critamus* Besser, non Hoffm.

Perennial, rarely *annual* or *biennial*, monoecious *herbs. Stems* erect, solid. *Leaves* 2- to 3-fid. *Inflorescence* of compound umbels; bracts and bracteoles 4–15, entire. *Sepals* 5, conspicuous, persistent. *Petals* 5, white, the outer not radiating. *Fruit* a schizocarp, oblong, laterally compressed; mericarps with low ridges, wider than the grooves; commissure narrow; carpophore present; vittae solitary and slender.

Contains about 3 species in Europe and Asia.

Hultén, E. & Fries, M. (1986). *Atlas of north European vascular plants north of the Tropic of Cancer*. 3 vols. Königstein.

1. F. vulgaris Bernh. Longleaf

F. rivinii Host; *F. sioides* Asch. nom. illegit.; *Sium falcaria* L.; *Seseli falcaria* (L.) Crantz; *Drepanophyllum sioides* Wibel nom. illegit.; *Drepanophyllum falcaria* (L.) Desv.; *Critamus agrestis* Besser; *Bunium falcaria* (L.) M. Bieb.; *Prionitis falcaria* (L.) Dumort.; *Prionitis falcata* Delarbre nom. illegit.; *F. falcaria* (L.) Karst.; *F. glauca* Dulac; *F. persica* Stapf & Wettst.; *F. serrata* St Lag.; *Sium falcatum* Dubois; *Critamus falcaria* (L.) Rchb.; *Helosciadium falcaria* (L.) Hegetsch.; *Carum falcaria* (L.) Lange; *Selinum falcaria* (L.) E. H. L. Krause

Perennial, rarely *annual* or *biennial*, monoecious *herb* with a stout root. *Stems* 30–90 cm, erect, bluish-green, solid, striate, glabrous, branched, leafy. *Leaves* 1.5–30×1.0–10 cm, bluish-green, paler beneath, obovate in outline, 3-fid, one or more of the segments often again 2- or 3-fid, the segments up to 30 cm, linear-lanceolate to linear, often somewhat falcate, acute at apex, regularly and sharply serrate, often puberulous beneath, with a thick, cartilaginous margin, the midrib usually with another vein parallel with and close to it on each side, the petiole slender and gradually widened to the sheathing base. *Inflorescence* of compound umbels, with 9–18 rays, the rays 20–40 mm, slender, smooth and subequal, the peduncle usually longer than the rays, often leaf-opposed, the terminal umbels with bisexual flowers, the lateral with male flowers; bracts and bracteoles 4–15, subulate, entire. *Sepals* 5, conspicuous. Petals 5, whitish, the outer not radiating, obcordate with an inflexed point. *Stamens* 5; filaments white; anthers cream. *Styles* 2, recurved, with enlarged base, forming the stylopodium which is half as long as the style; stigma a small knob. *Schizocarp* 3–4 mm, oblong, laterally compressed, smooth; mericarps with low ridges, wider than the grooves; commissure narrow; carpophore present; vittae solitary, slender; pedicels 4–5 mm, slender. *Flowers* 7–9. $2n=22$.

Introduced. Locally naturalised in grassy and waste places and by hedgerows. Scattered localities in the Channel Islands and south and central Great Britain especially East Anglia. Europe from northern France and central Russia southwards; western Asia.

39. **Carum** L.

Biennial or *perennial* monoecious *herbs. Stems* erect, hollow. *Leaves* 1- to 3-pinnate. *Inflorescence* of compound umbels; bracts and bracteoles absent to numerous. *Sepals* absent or small and not persistent. *Petals* 5, white, sometimes pink or red, the outer sometimes slightly radiating. *Fruit* a schizocarp, ellipsoid, laterally compressed; mericarps with narrow ridges; commissure narrow; carpophore present; vittae solitary and wide.

Contains about 30 species in Europe, temperate Asia and North Africa.

Blackstock, T. H., Howe, E. A. & Rimes, C. A. (1991). Whorled Caraway, *Carum verticillatum* (L.) Koch in Lleyn. *Welsh Bull. B. S. B. I.* **51**: 8–10.
Grieve, M. (1931). *A modern herbal*. London.
Hultén, E. & Fries, M. (1986). *Atlas of north European vascular plants north of the Tropic of Cancer.* 3 vols. Königstein.
Marren, P. (1981). A possible origin of *Carum verticillatum* (L.) Koch in north-eastern Scotland. *Watsonia* **13**: 323.

1. C. carvi L. Caraway

C. officinale Gray nom. illegit.; *Apium carvi* (L.) Crantz.; *Seseli carum* Scop.; *Ligusticum carvi* (L.) Roth; *Bunium carvi* (L.) M. Bieb.; *C. aromaticum* Salisb.; *C. rosellum* Woron.; *Seseli carvi* (L.) Lam.; *Aegopodium carum* Wibel; *Foeniculum carvi* (L.) Link; *Falcaria carvifolia* C. A. Mey.; *Pimpinella carvi* (L.) Jessen; *Carvi careum* Bubani; *Selinum carvi* (L.) E. H. L. Krause

Biennial monoecious *herb* with a fusiform tap-root. *Stems* 30–60 cm, pale green, hollow, erect, striate, glabrous, branched, leafy. *Leaves* 3–17×2–7 cm, medium green on upper surface, paler beneath, all basal in 1 year, narrowly triangular to oblong in outline, 2- to 3-pinnate, the segments often pinnatifid, the lobes 5–10 mm, linear-lanceolate to linear, acute with a cartilaginous apex, the petiole rather short, with a broad, scarious, sheathing base. *Inflorescence* of compound umbels, with 5–16 rays, the rays 10–60 mm, smooth and markedly unequal, the peduncle shorter to longer than the rays and often leaf-opposed, the umbels with mostly bisexual flowers; bracts and bracteoles absent or few, the former occasionally leaf-like. *Sepals* absent. *Petals* 5, usually whitish, sometimes pink or red, the outer slightly radiating, deeply emarginate at apex with an inflexed point. *Stamens* 5; filaments white; anthers pink. *Styles* 2, recurved and appressed, with enlarged base forming the stylopodium which is a little shorter than the styles; stigma capitate. *Schizocarp* 3–5 mm, greyish-brown, ellipsoid, laterally compressed, smooth, with a distinctive smell when crushed; mericarps with narrow, low ridges; commissure narrow; carpophore present; vittae solitary, wide; pedicels 5–15 mm. *Flowers* 6–7. $2n=20$.

Introduced. Naturalised in fields, on roadsides and in waste places. Scattered records throughout Great Britain and Ireland. Most of Europe; temperate Asia; north-west Africa; introduced elsewhere. The plant has been grown for its fruits ('seeds') since early times. These are used for flavouring cakes, bread, cheese and soups. The fruits contain an aromatic oil which is a remedy for flatulence, and which contributes a distinctive flavour to Kummel liqueur.

2. C. verticillatum (L.) W. D. J. Koch Whorled Caraway
Sison verticillatum L.; *Sium verticillatum* (L.) Lam.; *Bunium verticillatum* (L.) Gren. & Godr.; *Apium verticillatum* (L.) Caruel

Perennial monoecious *herb* with fusiform roots which are thickened downwards. *Stems* 30–60 cm, pale green, hollow, striate, glabrous, slightly branched, sparsely leafy and surrounded at the base by fibrous remains of petioles. *Leaves* mostly basal, 4–20 × 0.2–2.0 cm, medium green, narrowly oblong in outline, simply pinnate, with usually more than 20 pairs of deeply palmatisect segments, the lobes filiform, appearing as if whorled, the lowest about 1 mm, the upper to about 10 mm, the petiole short and scarcely enlarged at the base; all glabrous. *Inflorescence* of compound umbels, with 8–14 rays, the rays 15–40 mm, smooth and subequal, the peduncle longer than the rays and leaf-opposed, the umbels mostly with bisexual flowers; bracts up to 10, linear to lanceolate, acute at apex, deflexed; bracteoles like the bracts, but not deflexed. *Sepals* 5, small. *Petals* 5, white, the outer not radiating, entire, inflexed at apex. *Stamens* 5; filaments green; anthers red. *Styles* 2, recurved, with enlarged base forming the stylopodium, slightly shorter than the styles; stigma slightly thickened. *Schizocarp* 2.0–3.0 mm, ellipsoid, laterally compressed, smooth; mericarps with conspicuous narrow ridges; commissure narrow; carpophore present; vittae solitary, wide; pedicels 3–5 mm. *Flowers* 7–8. $2n=?$ 18 (20, 22).

Native. Calcifuge in marshes, streams and damp meadows. Locally frequent in northern and south-west Ireland, western parts of Great Britain from Cornwall to Inverness-shire; Jersey; very rare elsewhere. Western Europe northwards to Scotland and Holland. Southern Temperate element.

40. Selinum L.

Perennial monoecious *herbs. Stems* erect, solid. *Leaves* 2- to 3-pinnate. *Inflorescence* of compound umbels; bracts usually absent, sometimes few; bracteoles numerous, not or weakly deflexed. *Sepals* 5, minute. *Petals* 5, white, the outer not radiating. *Fruit* a schizocarp, ovoid-oblong, dorsally compressed; mericarps with 3 dorsal ridges narrowly winged and the lateral broadly winged; commissure narrow; carpophore present; vitttae solitary.

Contains about 4 species in Europe and Asia.

Hultén, E. & Fries, M. (1986). *Atlas of north European vascular plants north of the Tropic of Cancer.* 3 vols. Königstein.
Meade, M. (1989). Year by year observations of *Selinum carvifolia, Parnassia palustris* and other species on Sawston Hall Moor. *Nat. Cambridgeshire* **31**: 43–53.
O'Leary, M. (1989). The habitat of *Selinum carvifolia* in Cambridgewhire. *Nat. Cambridgeshire* **31**: 36–43.
Walters, S. M. (1956). *Selinum carvifolia* in Britain. *Proc. B. S. B. I.* **2**: 119–122.
Wigginton, M. J. (Edit.) (1999). *British red data books.* Vol. 1. *Vascular plants.* Peterborough.

1. S. carvifolia (L.) L. Cambridge Milk Parsley
Seseli carvifolia L.; *S. palustre* Crantz; *S. pseudocarvifolia* Crantz; *S. angulatum* Lam.; *S. tenuifolium* Salisb.; *Angelica carvifolia* (L.) Vill.; *Athamanta carvifolia* (L.) Weber; *Mylinum carvifolia* (L.) Gaudin; *Carum sulcatum* Steud.; *Ligusticum carvifolia* (L.) Caruel

Perennial monoecious *herb* with a stock and fibrous roots. *Stems* 30–100 cm, pale green, erect, solid, grooved and strongly angled, the angles narrowly winged, slightly branched, glabrous, leafy. *Leaves* 1–20 × 1–10 cm, medium green on upper surface, paler beneath, lanceolate or ovate-lanceolate in outline, 2- to 3-pinnate, the segments 3–10 mm, linear-lanceolate to ovate, minutely serrulate, sometimes lobed and with an acute, cartilaginous apex, the petiole long and somewhat dilated at the base. *Inflorescence* of compound umbels, with 15–25 rays, the rays 15–40 mm, somewhat unequal and papillose on the angles, the peduncles longer than the rays and papillose on the angles near the top, the umbels with mostly bisexual flowers; bracts usually absent, sometimes few and soon falling; bracteoles about 10, subulate to linear-lanceolate, not or weakly deflexed. *Sepals* 5, minute. *Petals* 5, white, the outer not radiating, emarginate, with an inflexed point. *Stamens* 5; filaments white; anthers cream. *Styles* 2, recurved and usually appressed, with enlarged base forming the stylopodium which is half as long as the styles; stigma capitate. *Schizocarp* 3–4 mm, ovoid-oblong, dorsally compressed, smooth; mericarps with the 3 dorsal ridges narrowly winged and the lateral broadly winged; commissure narrow; carpophore present; vittae solitary; pedicels as long as to twice as long as the bracteoles and papillose. *Flowers* 7–10. $2n=22$.

Native. Calcareous fens and damp meadows. Known only in three localities in Cambridgeshire: Chippenham Fen, Snailwell meadows and Sawston Hall meadows; formerly in single localities in Nottinghamshire and Lincolnshire and one other in Cambridgeshire. Much of Europe eastwards to central Asia, but absent from most of the Mediterranean region; introduced in North America. European Temperate element.

41. Ligusticum L.

Haloscias Fr.

Perennial monoecious *herbs. Stems* erect, hollow. *Leaves* 1- to 2-ternate. *Inflorescence* of compound umbels; bracts 1–5, entire; bracteoles several. *Sepals* 5, conspicuous, persistent. *Petals* 5, greenish-white, sometimes tinged with pink, the outer not radiating. *Fruit* a schizocarp, oblong-ovoid, not compressed; mericarps with prominent, narrowly winged ridges; commissure broad; carpophore present; vittae several.

Contains about 30 species in north temperate regions of the world.

Grieve, M. (1931). *A modern herbal.* London
Hultén, E. & Fries, M. (1986). *Atlas of north European vascular plants north of the Tropic of Cancer.* 3. vols. Königstein.
Palin, M. A. (1988). Biological flora of the British Isles. No. 164. *Ligusticum scoticum* L. *Jour. Ecol.* **76**: 889–902.

1. L. scoticum L. Scots Lovage

Angelica scotica (L.) Lam.; *L. boreale* Salisb. nom. illegit.; *L. biternatum* Stokes nom. illegit.; *Haloscias scotica* (L.) Fr.; *Cenolophium scoticum* (L.) Cav.

Perennial monoecious *herb* with a stout tap-root and fibrous side-roots. *Stems* 50–90 cm, green but often reddish towards the base, erect, hollow, striate, glabrous, branched, leafy. *Leaves* 2–15 × 2–13 cm, bright green on upper surface, paler beneath, broadly ovate in outline, 1- to 2-ternate, the segments ovate and long-stalked, the lobes 20–50 mm, rhombic to ovate, coarsely toothed or shallowly lobulate in the upper half, entire in the lower half, the teeth and apex obtuse, the petiole of the lower leaves long with a broadly sheathing base, those of the upper entirely sheathing. *Inflorescence* of compound umbels, with 8–14 rays, the rays 15–40 mm, subequal and papillose near the top, the peduncle longer than the rays and papillose near the top, the umbels with mostly bisexual flowers; bracts 1–5, linear, membranous; bracteoles usually about 7, linear to lanceolate-linear, *Sepals* 5, triangular. *Petals* 5, greenish-white, sometimes tinged with pink, the outer not radiating, entire, inflexed at apex. *Stamens* 5; filaments white; anthers cream. *Styles* 2, recurved, with enlarged base forming the stylopodium which is about as long as the styles; stigma truncate. *Schizocarp* 4–7 mm, oblong-ovoid, not compressed, smooth; mericarps with 3 prominent, narrowly winged ridges; commissure broad; carpophore present; vittae several; pedicels about as long as the bracteoles and papillose. *Flowers* 6–7. $2n = 22$.

Native. Cliffs and rocky places near the sea. Frequent around the whole coast of Scotland, local in north and west Ireland, formerly in Northumberland. Coasts of northern Europe, Greenland and eastern North America. European Boreo-arctic Montane element. Very susceptible to grazing by sheep. Formerly eaten as a pot-herb and the root is aromatic and pungent.

42. Angelica L.

Archangelica Wolf

Perennial monoecious *herbs. Stems* erect, hollow. *Leaves* 2- to 3-pinnate. *Inflorescence* of compound umbels; bracts absent or few and caducous; bracteoles numerous. *Sepals* 5, small. *Petals* 5, white, greenish-white or pinkish-white, the outer not radiating. *Fruit* a schizocarp, ovate or ovate-oblong, dorsally compressed; mericarps with prominent dorsal ridges and broadly winged lateral ridges; commissure narrow; carpophore present; vittae solitary.

Contains about 60 species in north temperate regions of the world and in New Zealand.

Grieve, M. (1931). *A modern herbal.* London.
Grime, J. P. et al. (1988). *Comparative plant ecology.* London. [*A. sylvestris.*]
Hultén, E. & Fries, M. (1986). *Atlas of north European vascular plants north of the Tropic of Cancer.* 3 vols. Königstein.
Vaughan, J. G. & Geissler, C. A. (1997). *The new Oxford book of food plants.* Oxford.

1. Schizocarps with thick, corky wings, 5–10 mm 2.
1. Schizocarps with thin, membranous wings, 4–5 mm 3.
2. Leaflets acute to acuminate; schizocarps 5–6 mm **2. archangelica**
2. Leaflets subacute to acute; schizocarps 5–10 mm **3. pachycarpa**
3. Leaflets 2–7 × 1–4 cm, broadly elliptical, or subrotund, abruptly acute at apex; peduncles and rays densely hairy **1(i). sylvestris** var. **villosa**
3. Leaflets 8–12 × 1–5 cm, lanceolate, elliptical or ovate, narrowed at apex; peduncles and rays less densely hairy 4.
4. Leaves 2–7 × 1–4 cm, dentate or serrate, the teeth mostly small **1(ii). sylvestris** var. **sylvestris**
4. Leaves 4–12 × 3–5 cm, broadly elliptical or ovate, serrate or biserrate, the teeth large **1(iii). sylvestris** var. **grossedentata**

1. A. sylvestris L. Wild Angelica

A. villosa Lag.; *A. macrophylla* Schur; *Selinum sylvestre* (L.) Crantz; *Selinum angelica* Roth; *Selinum pubescens* Moench; *Athamanta sylvestris* Weber; *Peucedanum angelica* Caruel

Perennial monoecious *herb* with a tap-root. *Stems* 10–200 cm, pale green and often purplish and pruinose, hollow, erect, striate, glabrous, branched, leafy. *Leaves* 6–30 × 5–20 cm, yellowish-green on upper surface, paler beneath, broadly ovate in outline, 2- to 3- pinnate, the primary divisions long-stalked, the segments 15–80 mm, obliquely lanceolate to ovate, usually acute or cuspidate at apex, serrate or biserrate, with cartilaginous teeth and usually with short hairs on both surfaces, at least on the veins, the lower leaves with long, laterally compressed petioles deeply channelled on the upper side and inflated at the base, the upper with a strongly inflated petiole with the lamina very small or absent. *Inflorescence* of compound umbels with 15–40 rays, the rays 20–80 mm and subequal, the rays and peduncle densely hairy on the ridges and sparsely so between them, the umbels with bisexual flowers; bracts absent or few and caducous; bracteoles 6–10, linear and puberulent. *Sepals* 5, minute. *Petals* 5, white or pinkish, the outer not radiating, lanceolate, suberect and incurved. *Stamens* 5; filaments white; anthers cream. *Styles* 2, recurved, with enlarged base forming the stylopodium which is about one-third as long as the styles; stigma capitate. *Schizocarp* 4–5 mm, ovate, dorsally compressed, smooth; mericarps with prominent, obtuse, dorsal ridges and broadly winged lateral ridges, the wings being wider than the mericarp, undulate and not closely appressed to one another; commissure narrow; carpophore present; vittae solitary; pedicels 5–10 mm, very shortly hairy. *Flowers* 7–9. $2n=22$.

(i) Var. **villosa** (Lag.) Willk. & Lange
A. villosa Lag.

Stem up to 60 cm. *Leaflets* 2–5 × 2–5 cm, broadly elliptical to subrotund, abruptly acute or apiculate at apex, serrulate to serrate, hairy on both surfaces. *Peduncles* and rays with dense, short, stiff hairs.

(ii) Var. **sylvestris**
A. sylvestris var. *decurrens* auct.

Stem up to 100 cm. *Leaflets* 2–7 × 1–4 cm, lanceolate, ovate or elliptical, narrowed at apex, dentate or serrate-dentate, the teeth mostly regular and small, shortly hairy on both surfaces. *Peduncle* and rays with short, stiff hairs, the rays sometimes sparsely so.

(iii) Var. **grossedentata** Rouy & Camus
A. sylvestris var. *elatior* auct.

Stem up to 2 m. *Leaflets* 4–12 × 3–5 cm, broadly elliptical or ovate, narrowed at apex, serrate or biserrate with large teeth, with hairs mainly on the veins. Peduncles and rays with short, stiff hairs.

Native. Damp grassy places, fens, marshes, by streams, ditches and ponds and in damp open woods. Common throughout Great Britain and Ireland. Almost throughout Europe but rare in the south; temperate Asia; introduced in North America. Eurosiberian Boreo-temperate element. Var. *villosa* occurs by the sea from Cornwall to Shetland. Var. *grossedentata* occurs in a few scattered localities. All the remaining plants are var. *sylvestris*.

2. A. archangelica L. Garden Angelica
A. officinalis Moench nom. illegit.; *Ligusticum angelica* Stokes nom. illegit.; *Archangelica officinalis* Hoffm.; *Archangelica archangelica* (L.) Karst. nom. illegit.; *Archangelica sativa* Mill.; *Archangelica procera* Salisb.; *Selinum archangelica* (L.) Vest

Perennial monoecious *herb* with a large tap-root. *Stems* 50–200 cm, bright medium green, erect, robust, hollow, striate, glabrous, branched, leafy. *Leaves* 6–40 × 5–30 cm, bright medium green on upper surface, paler beneath, broadly ovate in outline, 2- to 3-pinnate, the primary divisions with rather short stalks, the segments up to 150 mm, lanceolate to ovate, the apex acute with the terminal often 3-fid, deeply toothed and often irregularly lobed, the margin and apex of the teeth cartilaginous and usually glabrous on both surfaces, the lower with a long petiole inflated at the base, the upper with strongly inflated petiole and the lamina very small or absent. *Inflorescence* of compound umbels, with about 40 rays, the rays 40–80 mm, subequal and shortly hairy, the peduncle longer than the rays and glabrous, the umbels all with bisexual flowers; bracts absent or few and caducous; bracteoles numerous, linear. *Sepals* 5, small. *Petals* 5, greenish, the outer not radiating, lanceolate, incurved. *Stamens* 5; filaments white; anthers cream. *Styles* 2, recurved, with enlarged base forming the stylopodium which is half as long as the styles. *Schizocarp* 5–6 mm, ovate-oblong, dorsally compressed, smooth; mericarps with prominent, acute dorsal ridges and winged lateral ridges, the wings narrower than the mericarp, rather thick and corky and not closely appressed to one another; commissure narrow; carpophore present; vittae solitary; pedicels 10–15 mm, minutely papillose. *Flowers* 5–6. $2n=22$.

Introduced. Naturalised on river-banks and waste places. Scattered records throughout Great Britain and more or less frequent in the London area. Native of northern and east Europe, eastwards to central Asia; Greenland; and naturalised in other parts of Europe. The plants in Great Britain are subsp. **archangelica** which occurs almost throughout the range of the species. Pieces of the young stems and leaf-stalks crystallised with sugar are used by confectioners for flavouring and decoration, their bright green colour giving the finishing touch to cakes and sweets. The roots and seeds are rich in essential oils, the former being used with juniper berries in gin making and the seeds in the making of vermouth and chartreuse.

3. A. pachycarpa Lange Portuguese Angelica

Perennial monoecious *herb* with a long, thick fusiform root. *Stem* up to 100 cm, robust, pale green, fleshy, erect, hollow, striate, branched, leafy. *Leaves* 5–25 × 5–15 cm, dark green on upper surface, paler beneath, broadly triangular-ovate in outline, 2- to 3-pinnate, the primary divisions long-stalked, the segments lanceolate to broadly elliptical, often 3-lobed, the margins serrate to rather coarsely dentate, fleshy, sometimes shortly hairy; basal long-petiolate and shortly sheathing, the cauline smaller, subsessile and broadly sheathing. *Inflorescence* of compound umbels, with numerous rays, the rays robust and very shortly hairy; bracts 6–8, linear; bracteoles linear. *Sepals* 5, minute. *Petals* 5, white with green striations, the outer not radiating, obovate-elliptical, long-acuminate, reflexed. *Stamens* 5; filaments white; anthers cream. *Styles* 2, recurved, with enlarged base forming the stylopodium; stigma capitate. *Schizocarp* 5–10 mm, broadly oblong, dorsally compressed; mericarps with prominent dorsal ridges, the wings as wide as the mericarps and rather thick and corky, projecting beyond the stylopodium when mature; commissure narrow; carpophore present; vittae solitary. *Flowers* 7–9.

Introduced. Naturalised in a hedgebank at St Peter-in-the-Wood, Guernsey since 1993. Native of north-west Spain and Portugal.

43. **Levisticum** Hill nom. conserv.

Perennial monoecious *herbs* smelling of celery when crushed. *Stems* erect, hollow. *Leaves* 2- to 3-pinnate. *Inflorescence* of compound umbels; bracts and bracteoles numerous, entire, the latter connate at base. *Sepals* 5, minute. *Petals* 5, yellowish, the outer not radiating. *Fruit* a schizocarp, broadly elliptical, dorsally compressed; mericarps with prominent dorsal ridges and winged lateral ridges; commissure broad; carpophore present; vittae solitary.

Contains only 1 species.

Grieve, M. (1931). *A modern herbal*. London.
Hultén, E. & Fries, M. (1986). *Atlas of north European vascular plants north of the Tropic of Cancer.* 3 vols. Königstein.

1. L. officinale W. D. J. Koch Lovage
L. levisticum (L.) Karst.; *L. paludapifolium* Asch.; *Ligusticum levisticum* L.; *Angelica paludapifolia* Lam. nom. illegit.; *Angelica levisticum* (L.) All.; *Selinum levisticum* (L.) E. H. L. Krause; *Hipposelinum levisticum* (L.) Britton & Rose

Perennial monoecious *herb* with a tap-root, smelling strongly of celery. *Stems* 60–250 cm, medium green, erect, hollow, striate, glabrous, branched, leafy, surrounded at the base by numerous, scale-like remains of petioles. *Leaves* 50–70 × 45–65 cm, medium green on upper surface, paler beneath, broadly ovate or rhombic in outline, 2- to 3-pinnate, the segments up to 11 cm, rhombic in outline, the upper half coarsely incise-serrate to shortly lobed, the teeth and lobes with a cartilaginous apex, the cuneate basal half entire, the petiole hollow and slightly inflated near the base. *Inflorescence* of compound umbels, with 12–20 rays, the rays stout, unequal and papillose on the adaxial surface, the peduncles usually longer than the rays, those of the lateral umbels often opposite or in whorls of 3–4, all umbels with bisexual flowers; bracts numerous, linear-lanceolate, long-acute at apex, deflexed, with a scarious margin; bracteoles similar in shape, not deflexed, almost entirely scarious, connate at least near the base. *Sepals* 5, minute. *Petals* 5, yellowish, the outer not radiating, elliptical, obtuse at the inflexed apex. *Stamens* 10; filaments white; anthers yellowish. *Styles* 2, recurved, with enlarged base forming the stylopodium; stigma truncate. *Schizocarp* 4–7 mm, broadly elliptical, dorsally compressed, smooth; mericarps with prominent dorsal ridges and winged lateral ridges, the wings narrower than the mericarp; commissure broad; carpophore present; vittae solitary; pedicels usually shorter than the schizocarp. *Flowers* 7–8. $2n = 22$.

Introduced. Rough ground, by walls and paths usually not permanent and sometimes only a relic of cultivation. Scattered records in Great Britain, mostly in Scotland and northern England but south to Surrey and Kent. Probably native of Iran, but naturalised in much of Europe, mainly in mountain regions. Lovage is an old-fashioned vegetable and herb, formerly blanched and used like celery, and also used as a herbal tea. The young stems were candied like those of Angelica.

44. Ferula L.

Perennial monoecious *herbs*. *Stems* erect, solid. *Leaves* 4- to 6-pinnate. *Inflorescence* of compound umbels; bracts absent; bracteoles few, soon falling. *Sepals* minute. *Petals* 5, yellow, the outer not radiating. *Fruits* a schizocarp, ellipsoid, strongly dorsally compressed; mericarps with low dorsal and winged lateral ridges; commissure broad; carpophore present; vittae numerous.

Contains about 170 species in the Mediterranean region east to central Asia.

Grenfell, A. L. (1988). *Ferula communis* in Suffolk. *B.S.B.I. News* **50**: 30–32.

1. F. communis L. Giant Fennel
F. nodiflora L.

Perennial monoecious *herb* with a thick rootstock with a fibrous collar. *Stem* 2–5 m, pale green, very robust up to 3 cm in diameter, subterete or obscurely channelled, glabrous, usually branched, leafy. *Leaves* alternate; basal 25–45 × 20–30 cm, broadly triangular-ovate in outline, 4- to 6-pinnate, the ultimate segments 15–50 × 0.8–3.0 mm, green on both sides or glaucous beneath, linear or filiform, acute at apex, with a petiole up to 15 cm; cauline leaves increasingly smaller upwards, the uppermost with the lamina reduced to a few filiform segments, more or less sessile with very large, conspicuous, inflated basal sheaths, often up to 15 × 5 cm; all glabrous. *Inflorescence* large, of compound umbels, elongate, the central umbel short or sessile; rays (16–)20–30(–45), subequal 30–40 mm, glabrous; bracts absent; bracteoles few, linear-lanceolate, deciduous. *Sepals* minute or absent. *Petals* 5, about 2 × 1.5 mm, yellow, ovate, with an incurved, acute or acuminate apex. *Stamens* 5; filaments 2.5–3.0 mm, pale; anthers yellow. *Styles* 2, about 1.5 mm, slender, strongly reflexed in fruit; stylopodium conical; stigma capitate. *Schizocarp* 10–16 × 7–16 mm, ellipsoid to suborbicular, dorsally compressed; mericarps with filiform dorsal ridges with winged margins 1.0–1.5 mm wide; commissure broad; carpophore bipartite; dorsal vittae (1–)2–3 per vallecula. *Flowers* 4–6. $2n = 22$.

Introduced. On a roadside near Barton Mills in Suffolk since 1988; formerly in Northamptonshire between 1956 and 1988, and occasionally elsewhere in England. Native of southern Europe; North Africa; south-west Asia and the Canary Islands.

45. Peucedanum L.

Selinum L. (1753), non (1762) nom. conserv.; *Imperatoria* L.; *Thysselinum* L.

Biennial to *perennial* monoecious *herbs*. *Stems* erect, hollow or solid. *Leaves* 1- to 6-ternate or 2- to 4-pinnate. *Inflorescence* of compound umbels; bracts absent, few or numerous; bracteoles few to several. *Sepals* 5, minute to conspicuous. *Petals* 5, white or yellow, the outer not radiating. *Fruit* a schizocarp, elliptical, obovate or suborbicular; dorsally compressed; mericarps with rather low, obtuse dorsal ridges and winged lateral ridges; commissure broad; carpophore present; vittae solitary.

Contains about 100 species, widely distributed in Europe, Asia and Africa.

Grieve, M. (1931). *A modern herbal*. London.
Harvey, H. J. & Meredith, T. C. (1981). Ecological studies of *Peucedanum palustre* and their implications for conservation management at Wicken Fen, Cambridgeshire. In Synge, H. (Edit.) *The biological aspects of rare plant conservation*, pp. 365–378. Chichester.
Hultén, E. & Fries, M. (1986). *Atlas of north European vascular plants north of the Tropic of Cancer.* 3 vols. Königstein.

Meredith, T. C. & Grubb, P. J. (1993). Biological flora of the British Isles. No. 179. *Peucedanum palustre* (L.) Moench. *Jour. Ecol.* **81**: 813–826.
Randall, R. E. & Thornton, G. (1996). Biological flora of the British Isles. No. 191. *Peucedanum officinale* L. *Jour. Ecol.* **84**: 475–485.
Stewart, A., Pearman, D. A. & Preston, C. D. (1994). *Scarce plants in Britain*. Peterborough. [*P. palustre*.]
Wigginton, M. J. (Edit.) (1999). *British red data books*. Vol.1. *Vascular plants*. Peterborough. [*P. officinale*.]

1. Stems solid, with a sheath of fibres at base; leaf segments more than 40 mm and linear; petals yellow **1. officinale**
1. Stems hollow, without basal fibres at base; leaf segments not linear; petals white 2.
2. Leaves 2- to 4-pinnate; bracts more than 2 **2. palustre**
2. Leaves 1- to 2-ternate; bracts 0(–2) **3. ostruthium**

1. P. officinale L. Hog's Fennel

Perennial monoecious *herb* with a stout stock. *Stems* 60–200 cm, medium green, erect, solid, striate, sometimes weakly angled, glabrous, branched, leafy, surrounded by the fibrous remains of petioles at the base. *Leaves* medium green on upper surface, paler beneath, 6–20 × 2–20 cm, broadly ovate in outline, the lower 3–6 times ternately divided, the primary segments long-stalked, the segments 40–100 mm, linear, attenuate at both ends, not all lying in the same plane, with a narrow, cartilaginous margin and usually serrulate near the apex, the petiole long and not much inflated at the base, the upper smaller and less divided with an oblong, sheathing petiole, all glabrous. *Inflorescence* of compound umbels with 15–45 rays, the rays 15–80 mm, smooth and unequal, the peduncle longer than the rays and glabrous, the umbels all with bisexual flowers; bracts absent or few, linear and usually caducous; bracteoles several, linear. *Sepals* 5, conspicuous, acute at apex. *Petals* 5, yellow, the outer not radiating, with a long, inflexed point. *Stamens* 5; filaments white; anthers yellow. *Styles* 2, recurved, with enlarged base forming the stylopodium which is about as long as the styles; stigma spathulate. *Schizocarp* 5–8 mm, elliptical to obovate, dorsally compressed, smooth; mericarps with rather low, obtuse dorsal ridges and winged lateral ridges, the wings flat and closely appressed; commissure broad; carpophore present; vittae solitary; at least some pedicels several times as long as the schizocarp. *Flowers* 7–9. $2n = 66$.

Native. Rough, brackish grassland and banks of creeks and path-sides near the sea. Extremely local in Kent and Essex and at Southwold in Suffolk; formerly in Sussex. Central and south Europe. European Southern-temperate element. Our plant is subsp. **officinale** which occurs throughout the range of the species.

2. P. palustre (L.) Moench Milk Parsley

Selinum palustre L.; *Thysselinum palustre* (L.) Hoffm.; *P. sylvestre* DC.; *Selinum thysselinum* Crantz; *Selinum lactescens* Lam.; *Selinum intermedium* Besser; *Selinum schiwereckii* Besser; *Calestania palustris* (L.) K.-Pol.; *Thysselinum plinii* Spreng.; *Thysselinum schiwereckii* Besser; *Thysselinum angustifolium* Rchb.; *Thysselinum sylvestre* (DC.) Rchb.; *Callisace cantabrigiensis* Hoffm. ex Steud.

Biennial monoecious *herb* with a tap-root and a watery latex in the young parts. *Stems* 50–150 cm, pale green, often tinted purplish, hollow, strongly grooved and angled, glabrous, branched, leafy, without fibres at the base. *Leaves* medium green on upper surface, paler beneath; lower 3–25 × 3–20 cm, lanceolate to ovate in outline, 2- to 4-pinnate, the segments 5–20 mm, lanceolate to ovate in outline, pinnatifid, with a finely serrulate, cartilaginous margin and acute, cartilaginous apex, the petiole long, strongly canaliculate above and often very shortly hairy beneath, with a short, brown or purple, usually auriculate, sheathing base; upper smaller, less divided and with an entirely sheathing, short petiole. *Inflorescence* of compound umbels, with 15–40 rays, the rays 15–50 mm, somewhat unequal and papillose-puberulent on the inner side, the peduncle longer than the rays, usually scabrid at the top, the terminal umbel with bisexual flowers, the lateral umbels with about equal numbers of male and bisexual flowers; bracts 4 or more, linear-lanceolate or sometimes divided, deflexed; bracteoles several, linear-lanceolate, deflexed. *Sepals* 5, obtuse. *Petals* 5, white, the outer not radiating, with a long, inflexed point. *Stamens* 5; filaments white; anthers cream. *Styles* 2, recurved, with enlarged base forming the stylopodium which is slightly shorter than the styles; stigma capitate. *Schizocarp* 3–5 mm, elliptical, dorsally compressed, smooth; mericarps with rather low, obtuse dorsal ridges and narrowly winged lateral ridges, the wings flat, closely appressed to one another and rather thick; commissure broad; carpophore present; vittae solitary; pedicels longer than the fruit and papillose. *Flowers* 7–9. $2n = 22$.

Native. One of the most characteristic plants of species-rich, tall-herb fen which develops where there is a high water-table in summer and is often flooded in winter, and the peat has a pH of 5.0–6.5. *P. palustre* has markedly declined over the last two centuries and is now most frequent in the Fens of East Anglia with outlying localities in Somersetshire, Sussex and Yorkshire. Throughout most of Europe, but absent from the extreme south and most of the islands, and extending into central Asia. Eurosiberian Boreo-temperate element. Although producing much seed it persists by the development of axillary buds from the stem base. It is renowned as the larval foodplant of the Swallowtail Butterfly (*Papilio machaon* L.).

3. P. ostruthium (L.) W. D. J. Koch Masterwort

Imperatoria ostruthium L.; *Selinum imperatoria* Crantz; *Imperatoria major* Gray nom. illegit.; *P. imperatoria* Endl.; *Selinum ostruthium* (L.) Wallr.

Perennial monoecious *herb* with a tap-root and fibrous side-roots. *Stems* 60–100 cm, pale green, often purplish below, hollow, striate, glabrous, branched, leafy, without fibres at the base. *Leaves* medium green on upper surface, paler beneath; lower 3–20 × 4–30 cm, broadly ovate in outline, 1- to 2-ternate, the segments 5–10 cm, lanceolate to ovate, the middle segment often 3-fid, the lateral acute at apex, or sometimes 2-fid, irregularly biserrate, the teeth with a long cartilaginous point and the base often unequal, the petiole long, with a slightly inflated, sheathing base; upper small, often with pinnatifid lobes and the petiole greatly inflated and entirely sheathing; all glabrous. *Inflorescence* of compound umbels, with 30–60 rays, the rays 10–50 mm, somewhat unequal and papillose on the adaxial side, the peduncle longer than the rays, somewhat minutely hairy at the top, the terminal umbel with bisexual flowers, the lateral with male flowers; bracts absent, rarely 1 or 2; bracteoles few and linear. *Sepals* 5, minute. *Petals* 5, white, the outer not radiating, with a long, inflexed point. *Stamens* 10; filaments white; anthers pink. *Styles* 2, slender and recurved, with enlarged base forming the stylopodium which is half as long as the styles; stigma capitate. *Schizocarp* 3–5 mm, suborbicular, dorsally compressed, smooth; mericarps with rather low, obtuse dorsal ridges and winged lateral ridges, the wings about as wide as the mericarp, flat, closely appressed to one another, thin; commissure broad; carpophore present; vittae solitary; pedicels mostly longer than the fruit and smooth. *Flowers* 7–8. $2n=22$.

Introduced. Naturalised in grassy places, marshes and riversides and sometimes established near farm buildings. In numerous localities in northern Ireland and Great Britain north from Lincolnshire and Staffordshire; a few scattered localities farther south. Native of the mountains of central and south Europe. Formerly cultivated as a pot-herb, and also used for veterinary purposes.

46. Pastinaca L.

Biennial monoecious *herbs* with a strong characteristic smell. *Stems* erect, hollow or solid. *Leaves* simply pinnate, with ovate lobes. *Inflorescence* of compound umbels; bracts and bracteoles 0–2, entire. *Sepals* absent. *Petals* 5, yellow, the outer not radiating. *Fruit* a schizocarp, broadly elliptical or subrotund, dorsally compressed; mericarps with slender dorsal ridges and narrowly winged lateral ridges; commissure broad; carpophore present; vittae solitary.

Contains about 15 species in Europe and temperate Asia.

Grieve, M. (1931). *A modern herbal*. London.
Phillips, R. & Rix, M. (1988). *Vegetables*, pp. 114–115. London.
Zohory, D. & Hopf, M. (2000). *Domestication of plants in the Old World*. Ed. 3. Oxford.

1. Stem terete, with short, straight hairs to nearly glabrous; rays of terminal umbel 5–7, nearly equal **1(a). sativa** subsp. **urens**
1. Stem angled, with long, soft, flexuous hairs or sparse, short hairs; rays of terminal umbel 9–20, very unequal 2.
2. Root not swollen; stem with long, soft, flexuous hairs; upper surface of leaves grey hairy **1(b). sativa** subsp. **sylvestris**
2. Root swollen; stem with sparse, short, straight hairs; upper surface of leaves with sparse, short hairs **1(c). sativa** subsp. **sativa**

1. P. sativa L. Parsnip

Peucedanum sativum (L.) Benth. ex Hook. fil.; *P. vulgaris* Bubani; *Selinum pastinaca* Crantz; *Anethum pastinaca* (Crantz) Wibel

Biennial monoecious *herb* with a stout, sometimes fleshy root and a strong smell of parsnip. *Stems* 30–180 cm, pale green, erect, hollow or solid, furrowed, terete or angled, with sparse, short hairs or numerous, long, flexuous hairs, branched, leafy. *Leaves* 3–30 × 2–20 cm, yellowish-green on upper surface, paler beneath, elliptical-oblong or ovate-oblong in outline, simply pinnate, with (2–)5–11 pairs of segments, the segments 20–50(–100) mm, ovate, rounded to acute at apex, coarsely serrate, occasionally pinnatifid, cuneate to rounded at the base, usually hairy on both surfaces, the petiole of the lower leaves inflated near the base, that of the upper leaves inflated throughout its length. *Inflorescence* of compound umbels, with 5–20 rays, the rays 15–80 mm, somewhat to very unequal, papillae hairy, the peduncles variable in length, the terminal umbel with bisexual flowers towards the outside and male flowers towards the middle, the lateral umbels with male flowers only; bracts and bracteoles 0–2, caducous. *Sepals* 5, absent. *Petals* yellow, the outer not radiating, with a truncate involute point. *Stamens* 5; filaments white; anthers pink. *Styles* 2, rather stout and recurved, with enlarged base forming the stylopodium which is shorter than the styles; stigma capitate. *Schizocarp* 5–7 mm, broadly elliptical or subrotund, dorsally compressed and smooth; mericarps with slender, low, dorsal ridges and rather narrowly winged lateral ridges, the wings flat and closely appressed to one another; commissure broad; carpophore present; vittae solitary, usually rather shorter than the mericarp and tapering at the ends; pedicels as long as or longer than the fruit, papillose hairy. *Flowers* 7–8. $2n=22$.

(a) Subsp. **urens** (Req. ex Godr.) Čelak. Eastern Parsnip

P. urens Req. ex Godr.; *P. umbrosa* Steven ex DC.; *P. teretiuscula* Boiss.

Root not swollen. *Stem* terete, with short, straight hairs to nearly glabrous. *Upper surface of leaves* grey-hairy, the segments broad, obtuse, crenate-dentate or shallowly lobed. *Rays of terminal umbel* 5–7, nearly equal.

(b) Subsp. **sylvestris** (Mill.) Rouy & Camus Wild Parsnip

P. sylvestris Mill.; *P. sativa* var. *sylvestris* (Mill.) DC.

Root not swollen. *Stem* angled, with long, soft, flexuous hairs. *Upper surface of leaves* grey-hairy, the segments broad, obtuse, crenate-dentate or shallowly lobed. *Rays of terminal umbel* 9–20, very unequal.

(c) Subsp. **sativa** Parsnip

P. sativa var. *hortensis* Gaudin; *P. fleischmannii* Hladnik

Root swollen. *Stem* strongly angled, with sparse, short, straight hairs. *Upper surface of leaves* with sparse, short hairs, the segments often narrow and more or less acute, often cuneate at base and more or less pinnatisect in lower part. *Rays of terminal umbel* 9–20, very unequal.

Subsp. *sylvestris* is the native plant of grassland, roadsides and rough ground especially on chalk and limestone. It is common in England south-east of a line from the River Humber to the River Severn, and is very local elsewhere in England, Scotland Wales and Ireland where it may always be introduced. Most of Europe and western Asia. Subsp. *sativa* is of cultivated origin and is commonly grown in vegetable gardens and as a crop in fields and market gardens. The root is eaten as a cooked vegetable, and is improved by prolonged exposure to frost. Parsnip wine is sometimes made. Parsnips were probably known to the ancient Greeks and Romans and were certainly grown as a vegetable in Germany in the mid-sixteenth century. It is now widely cultivated throughout Europe and can sometimes be found as an escape in waste and grassy places. Subsp. *urens* has been found on the Suffolk coast where it might be native. Its main distribution is in south, central and east Europe.

47. Heracleum L.

Sphondylium Mill. nom. illegit.

Biennial or *perennial* foetid *herbs. Stems* erect, hollow. *Leaves* simple and pinnately or ternately divided, or 1- to 2-pinnate. *Inflorescence* of compound umbels; bracts 0–several, entire; bracteoles several. *Sepals* 5, minute or conspicuous and persistent. Petals white to purplish or greenish-white, the outer more or less radiating. *Fruit* a schizocarp, obovate, subrotund or elliptical, strongly dorsally compressed; mericarps with slender, low, dorsal ridges and more or less broadly winged lateral ridges; commissure broad; carpophore present; vittae solitary.

Contains about 70 species in north temperate regions of the world and on mountains in the tropics.

It appears that *Heracleum* starts with relatively small plants on the west coast of Europe and makes its way eastwards across the continent steadily getting larger until in Asia there are plants up to 5 m high. In some areas there are obvious intermediates between taxa and in others natural barriers between them. If all the taxa were brought together, however, they would almost certainly hybridise with one another. As far as Europe is concerned the subspecific treatment of R. K. Brummitt (1968) is followed and in Asia the specific account by I. P. Mandenova (1951). *H. branca-ursina* is halfway between the two groups looking like one and having the size of the other. None of this is logical, but is the best that can be done at the present time. It is a difficult genus to study because of the size of the plants and the bits of them needed for identification. They are in fact more difficult to collect than most trees. As many as 12 sheets with photographs may be needed for one gathering of the large species. Once the species are defined a photograph of a spread-out basal leaf before it becomes moribund and ripe fruits before they lose their hairs is just about adequate. We have long known there is more than one species of the large type introduced and have steadily been accumulating material over 30 years. This account is based on some 320 sheets, over 50 colour slides and many black-and-white photographs, plus over 100 sheets of *H. branca-ursina*, all in CGE, plus the fact that nearly all the taxa have been growing in the Cambridge Botanic Garden over the last 30 years. To sort out the distribution of the taxa in Great Britain and Ireland will require much hard work, as it is necessary to visit a colony more than once. No type material of *H. mantagazzianum* Sommier & Levier has been traced, but specimens grown by the original authors have been seen and they are not like any material seen from Great Britain and Ireland.

The sap from the large species contains a toxic chemical which sensitises the skin and may cause blistering or severe dermatitis when exposed to bright sunlight.

Brummitt, R. K. (1968). *Heracleum* L. In *Flora Europaea* **2**: 364–366.

Drever, J. C. & Hunter, J. A. A. (1970). Giant Hogweed dermatitis. *Scot. Med. Jour.* **15**: 315–319.

Grime, J. P. et al. (1988). *Comparative plant ecology*. London. [*H. sphondylium*.]

Mandenova, I. P. (1951). *Heracleum* L. In Shischkin, B. K. (Edit.) *Flora of the U. S. S. R.* **17**: 222–259. (Translated by R. Lavoott (1974) pp. 161–185.)

Sheppard, A. E. (1991). Biological flora of the British Isles. No. 171. *Heracleum sphondylium* L. *Jour. Ecol.* **79**: 235–258.

Stewart, F. & Grace, J. (1984). An experimental study of hybridisation between *Heracleum mantegazzianum* Somm. & Levier and *H. sphondylium* L. subsp. *sphondylium* (Umbelliferae). *Watsonia* **15**: 73–83.

Tiley, G. E. D., Dodd, F. S. & Wade, P. M. (1996). Biological flora of the British Isles. No. 190. *Heracleum mantagazzianum* Sommier & Levier. *Jour. Ecol.* **84**: 297–319.

1. Umbel with 10–20 rays; schizocarp 7–12 mm — 2.
1. Umbel with 40–60 rays; schizocarp 10–15 mm — 8.
2. Schizocarp 9–12 mm; outer flowers up to 20 mm in diameter — **2. branca-ursina**
2. Schizocarp 7–8 mm; outer flowers up to 12 mm in diameter — 4.
3. Flowers greenish-white or yellowish-green, the outer not or scarcely radiating — **1(b). sphondylium** subsp. **sibiricum**
3. Flowers white, pink or purple, the outer radiating — 3.
4. Flowers purple — **1(a,3). sphondylium** subsp. **sphondylium** forma **rubiflorum**
4. Flowers white or pink — 5.
5. Leaves with segments less than 30 mm wide — 6.
5. Leaves with segments up to 80 mm wide — 7.
6. Leaves with segments less than 20 mm wide — **1(a,1). sphondylium** subsp. **sphondylium** forma **angustisecta**
6. Leaves with segments up to 30 mm wide — **1(a,2). sphondylium** subsp. **sphondylium** forma **stenophyllum**
7. Leaves with medium sized segments up to 40 mm wide — **1(a,4). sphondylium** subsp. **sphondylium** forma **sphondylium**
7. Leaves with broad rounded segments up to 80 mm wide — **1(a,5). sphondylium** subsp. **sphondylium** forma **latifolium**
8. Leaves with broad segments which overlap so that you can hardly see through the leaf; fruits with long hairs up to 2 mm — **5. trachyloma**
8. Leaves with narrow segments so that you can easily see through the leaf; fruits with few hairs not more than 1 mm — 9.
9. Schizocarp 10–15 × 5–7 mm, oblong, cuneate at base, vittae only reaching halfway to base — **3. grossheimii**
9. Schizocarp 10–14 × 6–8 mm, oblong-obovoid or obovoid, rounded at base, 2 central vittae reaching nearly to base, outer vittae shorter — **4. lehmannianum**

1. H. sphondylium L. Hogweed
Pastinaca sphondylium (L.) Calest.; *Sphondylium vulgare* Gray nom. illegit.; *H. fastuosum* Salisb. nom. illegit.

Biennial monoecious *herb. Stems* up to 200(–300) cm, pale green, erect, hollow, ridged, hairy, the hairs often deflexed but not closely appressed, particularly below, branched above, leafy. *Leaves* alternate, broadly ovate in outline, imparipinnate; segments 5(–9), the segments very variable in shape, subrotund, ovate, oblong or linear, obtuse or acute at apex, shallowly lobed, crenate or serrate, hairy on both surfaces, especially beneath; petiole often purplish, greatly inflated and sheathing, more or less hairy. *Umbels* compound; rays 10–20, 2–12 cm, somewhat unequal, hairy; peduncle longer than rays, more or less puberulent; terminal umbel with bisexual flowers, the lateral umbels with male and bisexual flowers or only male flowers; outer flowers up to 12 mm in diameter; bracts few or absent; pedicels about as long as or longer than the fruit; bracteoles usually present, linear, stiffly hairy and often deflexed. *Sepals* 5, very small. *Petals* 5, white, greenish-white, pink or rarely purple, outer radiating or not. *Stamens* 5. *Styles* 2, with enlarged base, forming the stylopodium and about twice as long as it, divergent or somewhat recurved; stigma capitate. *Schizocarp* (6–)7–8 mm, obovate to subrotund, dorsally compressed, smooth or sparsely hairy; commissure broad; mericarps with slender, low, dorsal ridges and rather broadly winged lateral ridges, the wings flat and closely appressed to one another; carpophore present; vittae solitary, about half to three-quarters as long as the fruit, up to 0.4 mm wide, widest near the lower end, 2 outer longer than the 2 inner *Flowers* 6–9. $2n=22$.

(a) Subsp. **sphondylium**
Petals white, pink or rarely purple, those of the outer flowers strongly radiating.

(1) Forma **angustisectum** Gremli
Leaves deeply divided with segments up to 20 mm wide. *Petals* white.

(2) Forma **stenophyllum** (Gaudin) P. D. Sell
H. sphondylium var. *stenophyllum* Gaudin

Leaves deeply divided with segments up to 30 mm wide. *Petals* white.

(3) Forma **rubriflorum** (Schröt.) Thell.
H. sphondylium var. *rubriflorum* Schröt.

Leaves with medium sized segments up to 40 mm wide. *Petals* deep red to purple.

(4) Forma **sphondylium**
H. sphondylium forma *roseum* Druce

Leaves with medium sized segments up to 40(–60) mm wide. *Petals* white or pink.

(5) Forma **latifolium** (Mert. & W. D. J. Koch) P. D. Sell
H. sphondylium var. *latifolium* Mert. & W. D. J. Koch

Leaves with broad, rounded segments up to 80 mm wide. *Petals* white.

(b) Subsp. **sibiricum** (L.) Simonk.
H. sibericum L.

Petals greenish-white or yellowish-green, the outer not or scarcely radiating.

Native. Grassy places, hedgerows, ditchsides, open woods, cliff ledges and roadsides. Common throughout Great Britain and Ireland. As a species it occurs throughout the north temperate region in a large number of subspecies. Eurasian Boreo-temperate element. Subsp. *sphondylium* is our common subspecies, which is mainly confined to north-west Europe. It seems to have become more abundant on roadsides in East Anglia since rotary mowers have been used. In agricultural areas where hedgerows have been cut down or removed between fields it is often found in dense lines. The common plant with us is forma *sphondylium*. One or more of the other forms may be growing with it. Forma *rubiflorum* is rare. This name has been included in *H. branca-ursina* to which it has a similar leaf shape, but smaller flowers and fruits. The others forms are widespread. Rather dwarf plants of forma *latifolium* are found in some Cornish coastal areas and peninsulas in populations and may be distinct ecotypes. Dwarf plants in mountains also need more research. Subsp. *sibiricum* has been recorded for Norfolk and occurs in the wild areas of the Cambridge Botanic Garden. It is mainly a plant of north-east and east-central Europe but also occurs in central and south-east France.

× **trachyloma**
What appears to be this hybrid has been seen in the Cambridge area, but it is difficult to know which of the large species form the other parent of *H. sphondylium* hybrids, which are widespread.

2. H. branca-ursina All. Bear's Breech Hogweed
H. asperum M. Bieb.

Biennial foetid, monoecious *herb. Stems* up to 200 cm, pale green, sometimes tinted or suffused brownish-purple, deeply ridged and channelled, hollow, covered with short, rough hairs throughout, branched above, leafy. *Leaves* alternate, 15–25 × 20–35 cm, dark green above, paler beneath, broadly ovate in outline; upper leaves divided into 3 lobes, each lobe narrow or rounded to an acute tip and with rounded teeth; lower leaves with 3 leaflets, the terminal deeply divided with acute lobes and rounded teeth, the lateral less divided with more rounded lobes and rounded teeth; all with hairs on both surfaces, densely so beneath and especially on the veins; petioles densely hairy with a long, broad sheath prominently veined and densely hairy. *Umbels* compound; rays 8–20, 6–15 cm, hairy; pedicels several together, much longer than rays, hairy; outer flowers up to 20 mm in diameter; bracts few, linear, with a long narrow tip; pedicels 15–40, 10–30 mm, hairy; bracteoles linear. *Sepals* 5, small. *Petals* 5, white, the outer radiating. *Stamens* 5; filaments white; anthers green. *Styles* 2, the enlarged base forming the stylopodium, *Schizocarp* 9–12 × 7–9 mm, mostly obovate; vittae about three-quarters as long as the schizocarp, 2 inner longer than the 2 outer. *Flowers* 6–11.

Introduced. This plant occurs commonly in several places in Bassingbourn, Cambridgeshire near where crops of vegetables, flowers and root crops have been grown, and may be widespread in agricultural areas. Recently it spread to roadsides at Shingay, Cambridgeshire, probably carried by grasscutters. It also grew where wild flower seed had been sown in the Cambridge Botanic Garden. It looks like a very large *H. sphondylium* with thick ridged stem, rounded leaf lobes, large fruits and characteristic hairiness. It does not look like any of the giant hogweeds nor their hybrids with *H. sphondylium*. It seems to best fit the plant called H. *sphondylium* var. *branca-ursina* in *Flora Rep. Pop. Romine* **6**: 629, *H. sphondylium* subsp. *australe* subvar. *branca-ursina* in Hegi, *Ill. Fl. Mitteleur.* **5**(2): 1438 and to match the specimen of *H. asperum* in Reichenb. *Fl. Gem. Escicc.* no. 1874 of which CGE has two sheets. Although we have made a large collection of specimens it was too late in the day to sort the very complicated taxonomy and nomenclature. *H. branca-ursina* All. is probably either invalid or illegitimate. Reichenbach's *H. asperum* may not be identical with that of Marshall von Bieberstein. A colony of what seems to be this plant was collected at Abberton reservoir in Essex in 1988. It had deep pink flowers and this is recorded in *Flora Rep. Pop. Romine* under the name forma *rubriflorum* which we have applied to a form of *H. sphondylium*.

× **sphondylium**
In the large roadside colonies at Bassingbourn there are plants which may be this hybrid.

3. H. grossheimii Manden. ex. Grossh. Grossheim's Hogweed
H. mantegazzianum auct.

Biennial or *perennial* monoecious strong-smelling *herb*. *Stem* 100–200 cm, pale yellowish-green, spotted brownish-purple, deeply furrowed and ribbed, erect, densely, shortly and stiffly spreading, hairy, branched, leafy. *Leaves* alternate, 40–88 × 40–88 cm, ovate in outline, ternately or pinnately divided; lamina yellowish-green on upper surface, paler beneath, the primary segments ovate or obovate, pinnately divided into ovate-oblong secondary segments, the secondary segments divided nearly to midrib into lanceolate-oblong, sharply acute, very irregularly dentate lobes, the terminal segment narrowed to a broadly winged petiolule, glabrous on upper surface, with minute, stiff hairs on the veins beneath; petioles up to 60 cm long, hairy, the cauline with a large, hairy sheath; *Umbels* large, with 40–50 rays, the rays glandular-hairy; bracts linear-subulate. *Petals* 5, the terminal one of outer flowers 10 × 3.5 mm, lateral 8 × 3 mm, the lower 2 × 2 mm, all white. *Stamens* 5; filaments about 1.5 mm, whitish; anthers pale yellowish-green. *Styles* with enlarged base forming the stylopodium, 3–4 times as long as stylopodium, greenish, stigma capitate. *Schizocarp* 10–15 × 5–7 mm, pale green, oblong, cuneate at base, dorsally compressed, commissure very broad; vittae conspicuous and swollen at their lower ends, only half the length of fruit, very hairy when young with few long hairs later. *Flowers* 6–9.

Introduced. Waste and grassy places. This species seems to have been first collected by C. C. Babington at Shelford in Cambridgeshire in 1828 and it is still there. It also occurs in the wild areas of the Botanic Garden in Cambridge, behind the tennis courts at Kew Gardens and at Furneux Pelham in Essex. Native of Caucasus and Transcaucasia.

4. H. lehmannianum Bunge Lehmann's Hogweed
H. mantegazzianum auct.

Biennial or *perennial* monoecious strong-smelling herb with a large, branched root. *Stem* 100–250 cm, hollow, pale green, suffused brownish-purple, deeply furrowed and ribbed, erect, with dense, spreading simple eglandular hairs, with spreading branches in the upper half, leafy. *Leaves* alternate, 30–150 × 30–100 cm, obovate in outline, pinnately divided; lamina dull medium to dark green on upper surface, paler beneath, the main segments 5 or 7, not overlapping, obovate or ovate in outline, acute at apex, pinnatisect almost to midrib, the secondary segments oblong, narrowly and sharply acute at apex, with large lobes, the lobes ovate, acute at apex and irregularly serrate-dentate, the lower main segments with short petiolules, the others sessile, glabrous on upper surface, sparingly hairy beneath; petioles up to 70 cm long, flattened, pale green, hollow, roughly hairy, the cauline with broad sheathing bases with purple veins. *Umbel* 20–60 cm in diameter with 40–60 rays, the rays pale green, hairy; bracts and bracteoles linear-lanceolate. *Petals* 5, the terminal one of outer flowers about 11 × 9 mm, its arm asymmetrical, the lateral 11 × 3 mm, the lower 2.5–1.5 mm, all white, sometimes tinged pink. *Stamens* 5; filaments 2.0–2.5 mm, white; anthers cream. *Styles* 2, 3–5 mm, white tinted pink with an enlarged base forming the stylopodium. *Schizocarp* 10–14 × 6–8 mm, pale green, obovoid or oblong-obovoid, dorsally compressed; commissure very broad; vittae conspicuous and swollen at their lower ends, the 2 outer much shorter than the 2 inner, with few hairs about 1 mm. *Flowers* 6–9.

Introduced. Waste and grassy places. Recorded from Yorkshire and Cambridgeshire. Native of central Asia.

× **sphondylium**
In 1958 P. D. S. collected *H. lehmannianum* at Milton in Cambridgeshire and noted that it was hybridising with *H. sphondylium*, but did not collect the hybrid.

5. H. trachyloma Fisch. & C. A. Mey. Giant Hogweed
H. pubescens var. *trachyloma* (Fisch. & C. A. Mey.) Boiss.; *H. mantagazzianum* auct.

Biennial or *perennial* monoecious, strong-smelling *herb*. *Stem* 100–500cm, hollow, pale yellowish-green, furrowed and ribbed, erect, with numerous, prickly-based spreading hairs, branched from near the base, leafy. *Leaves* alternate, 15–75 × 11–65 cm, broadly ovate in outline, pinnately divided; lamina medium yellowish-green on upper surface, paler beneath, the main segments 5 or 7, more or less overlapping, ovate, acute at apex, pinnatisect nearly to midrib, the secondary segments oblong, acute at apex, lobed halfway to midrib, the lobes with large irregular teeth, the lower main segments with short petiolules the others sessile, glabrous or with a few minute hairs on upper surface and more numerous beneath; petiole up to 45 cm, pale green, hollow, hairy. *Umbels* large, with 40–50 rays, the rays with dense spreading simple eglandular hairs; bracts lanceolate; bracteoles ovate-lanceolate. *Petals* 5, white or pinkish, terminal of outer flowers 14 × 13 mm, lateral petals 13 × 3 mm, lower 4.0–5.5 mm. *Stamens* 5; filaments about 2 mm, white; anthers yellowish. *Styles* 2, about 5 mm, as long as or twice as long as stylopodium, white or pink tinged. *Schizocarp* 10–12 × 7.0–8.5 mm, pale green, obovoid, dorsally compressed; commissure very broad; vittae conspicuous and swollen at their lower ends, all 4 more or less the same length, with numerous long hairs up to 2 mm. *Flowers* 6–9.

Introduced. Waste and grassy places. Possibly the commonest of our plants which we have seen from Kent, East Anglia, Breconshire and Edinburgh. Native of Caucasus and Transcaucasia.

48. Tordylium L.

Annual or *biennial* monoecious *herbs*. *Stems* erect, hollow to more or less solid. *Leaves* simply pinnate. *Inflorescence* of compound umbels; bracts and bracteoles 4–7, entire. *Sepals* 5, conspicuous, persistent. *Petals* 5, white, the outer radiating. *Fruit* a schizocarp, elliptical, dorsally compressed. hairy; mericarps with slender, low, dorsal ridges and broadly whitish-winged lateral ridges; commissure broad; carpophore present; vittae solitary, as long as the fruit.

Contains 16 to 20 species in Europe, south-west Asia and North Africa.

Wigginton, M. J. (Edit.) (1999). *British red data books*. Vol. 1. *Vascular plants*. Peterborough.

1. T. maximum L. Hartwort

T. officinale auct.; *Heracleum tordylium* Spreng.; *Pastinaca tordylium* (Spreng.) Calest.

Annual or *biennial* monoecious *herb* with a tap-root and fibrous side-roots. *Stems* 20–130 cm, pale green, hollow to almost solid, erect, ridged, with short, closely appressed, deflexed bristles, branched, leafy. *Leaves* dull medium green on upper surface, paler beneath; outer basal 4–20 × 2–15 cm, lanceolate or ovate, simple; lower ovate in outline, simply pinnate, the segments 3–5, 10–80 mm, ovate to subrotund and crenate, with a cordate base; upper with 5–11 segments which are ovate-lanceolate to linear-lanceolate with coarsely serrate lobes which are cuneate at the base and sometimes reduced to the terminal lobe only; all with appressed, subrigid simple eglandular hairs on both surfaces; petioles of basal and lower cauline leaves scarcely sheathing, those of upper narrow, but sheathing. *Inflorescence* of compound umbels, with (3–)5–15 rays, the rays 5–15 mm and with dense, short, stiff hairs, the peduncle longer than the rays, with short, closely appressed, deflexed bristles, the umbels with all flowers bisexual; bracts and bracteoles 4–7, linear, entire with stiff hairs. *Sepals* 5, conspicuous, about half as long as the petals. *Petals* 5, white, the outer radiating, with an incurved point. *Stamens* 5; filaments white; anthers cream. *Styles* 2, stout and diverging, with enlarged base forming the stylopodium; stigma capitate. *Schizocarp* 5–6 mm, elliptical, dorsally compressed, covered with short, stiff hairs; mericarps with slender, low dorsal ridges and broadly whitish-winged lateral ridges; commissure broad; carpophore present; vittae solitary, as long as the schizocarp; pedicels much shorter than the schizocarp, stiffly hairy. *Flowers* 6–7. $2n = 20$.

Possibly native. Unstable, south-facing, sunny banks in rabbit-grazed scrub and grassland. Possibly only surviving in Hadleigh Country Park near the top of Benfleet Downs in Essex. First found some time before 1670 between St James and Chelsea in Middlesex and between Isleworth and Twickenham. In 1875 it was found near Tilbury Fort in Essex where it persisted until 1984. In 1949 and again in 1966 it was found in two separate localities near Benfleet in Essex and still survives in one of them. All its localities are along the Thames valley where it may be overlooked. Widespread in the Mediterranean region including North Africa, extending northwards to Great Britain and east to the Caucasus and north Iran.

49. Laser Borkh. ex. P. Gaertn., B. Mey. & Scherb.

Perennial monoecious *herbs*. *Stems* erect, solid, with fibrous remains at base. *Leaves* 2- to 3-ternate. *Inflorescence* of compound umbels; bracts absent or few; bracteoles few, soon falling. *Sepals* 5, conspicuous. *Petals* 5, white, the outer radiating. *Fruit* a schizocarp, ovoid or oblong, dorsally compressed; mericarps with long dorsal ridges and broadly winged lateral ridges; commissure broad; carpophore present; vittae more or less conspicuous.

Contains 1 species in west, south and central Europe and south-west Asia.

1. L. trilobum (L.) Borkh. ex P. Gaertn., B. Mey & Scherb. Laser

Laserpitium trilobum L.; *Siler trilobum* (L.) Crantz; *Laserpitium aquilegifolium* Jacq., non DC.; *L. aquilegifolium* Röhl. ex Steud.; *L. carniolicum* Bernh. ex Steud.; *Angelica aquilegifolia* (Röhl. ex Steud.) Lam.

Perennial monoecious *herb* with a sturdy tap-root crowned with a fibrous collar. *Stem* 50–120 cm, pale bluish-green, erect, solid, minutely hairy, branched above, leafy. *Leaves* 25–45 × 16–25 cm, broadly ovate in outline, 1- to 2-pinnate or 1- to 2-ternate; segments 3–7 × 2.5–6.5 cm, bluish-green on upper surface, paler beneath, sometimes tinted purplish, broadly ovate to subrotund, more or less obtuse at apex, deeply or shallowly ternately lobed, the lobes crenate-dentate and mucronate, entire and rounded at base, glabrous and conspicuously reticulately veined, petiolules up to 12 cm; petioles up to 20 cm, pale bluish-green. *Inflorescence* of compound umbels, with 12–25 rays, the rays 4–10 cm, equal or not, glabrous; bracts 0–2, up to 2 cm, ovate-lanceolate; bracteoles about 5, lanceolate, long acuminate, soon deciduous; pedicels 3–12 mm, more or less equal. *Sepals* 5. *Petals* 5, white, apex inflexed, clawed. *Stamens* 5; filaments whitish; anthers cream. *Styles* 2, recurved, with enlarged base forming the stylopodium; stigma capitate. *Schizocarp* 7–8 × 3–4 mm, oblong; mericarps with 5 main ridges and 4 secondary ridges, the lateral ridges somewhat winged; commissure broad; carpophore present; vittae more or less conspicuous, 4 dorsal and 2 commissural. *Flowers* 6–8.

Introduced. Formerly naturalised for many years at Cherry Hinton in Cambridgeshire, and occurred as a casual at King's Lynn in Norfolk. West, south and central Europe; Caucasia, northern Iran and Lebanon.

50. Torilis Adans.

Annual or rarely *biennial herbs. Stems* solid, erect, decumbent or procumbent. *Leaves* 1- to 3-pinnate. *Inflorescence* of compound umbels; bracts absent to numerous, entire; bracteoles several. *Sepals* 5, persistent, hidden by spines at fruiting. *Petals* 5, white to purplish white, the outer sometimes radiating. *Fruit* a schizocarp, oblong-ovoid or ovoid, subterete, not compressed, variously furnished with curved or hooked spines; mericarps with slender ciliate ridges; commissure narrow; carpophore present; vittae solitary.

Contains 8 species found in Europe and temperate Asia.

T. stocksiana (Boiss.) Grossh. has been recorded as a wool casual.

Grime, J. P. et al. (1988). *Comparative plant ecology.* London. [*T. japonica.*]
Stewart, A., Pearman, D. A. & Preston, C. D. (1994). *Scarce plants in Britain.* Peterborough. [*T. arvensis.*]

1. Peduncles less than 1 cm; rays less than 5 mm, more or less hidden by flowers or schizocarps; schizocarps with dimorphic mericarps, 1 with spines, 1 tuberculate **5. nodosa**
1. Peduncles more than 1 cm; rays more than 5 mm; both mericarps with spines 2.
2. Rays 2–3 **4. leptophylla**
2. Rays 4–13 3.
3. Bracts 4–6(–12); schizocarp with curved spines not hooked at the ends 4.
3. Bracts 0–1; schizocarp with more or less straight spines minutely hooked at the end 5.
4. Leaves medium to dark green on upper surface; outer petals scarcely larger than inner; schizocarp 3–4 mm, oblong-ovoid; style twice as long as stylopodium **1. japonica**
4. Leaves rather pale yellowish-green on upper surface; outer petals much larger than inner; schizocarp 2–3 mm, subglobose; style more than twice as long as stylopodium **2. ucranica**
5. Plant often little branched, but sometimes much widely branched; style 2–3 times as long as the stylopodium **3(a). arvensis** subsp. **arvensis**
5. Plant much branched; style 3–6 times as long as the stylopodium **3(b). arvensis** subsp. **neglecta**

1. T. japonica (Houtt.) DC. Upright Hedge Parsley

Caucalis japonica Houtt.; *Torilis anthriscus* (L.) C. C. Gmel., non Gaertn.; *Caucalis anthriscus* (L.) Huds.; *Tordylium anthriscus* L.; *T. rubella* Moench; *T. stricta* Wibel; *T. elata* Spreng.; *T. scabra* DC.; *T. persica* Bois & Buhse; *T. praetermissa* Hance; *T. convexa* Dulac; *T. verecundum* Salisb.; *Caucalis aspera* Lam.; *Chaerophyllum scabrum* (DC.) Thunb.; *Chaerophyllum hispidum* Thunb. ex Miq.; *Daucus anthriscus* Baill.; *Selenum torilis* E. H. L. Krause

Annual or rarely *biennial* monoecious, slightly aromatic *herb* with a slender, tapering tap-root and fibrous side-roots. *Stems* 5–125 cm, pale green, often much tinted purplish, solid, erect, striate, branched, leafy, with closely appressed, reflexed bristles. *Leaves* 3–10 × 2–6 cm, medium to dark green on upper surface, paler beneath, ovate in outline, 1- to 3-pinnate, the segments 10–30 mm, lanceolate to ovate in outline, pinnatifid or coarsely serrate, with appressed, forward-pointing bristles on both surfaces; petioles slender, with mostly deflexed bristles. *Inflorescence* of compound umbels, with (4–)6–10(–13) rays, the rays 5–20 mm, somewhat unequal, with appressed, forward-pointing hairs, the peduncle longer than the rays, with closely appressed, deflexed hairs, the umbels with bisexual flowers only; bracts 4–6(–12), linear, unequal and roughly hairy; bracteoles several, linear to lanceolate, roughly hairy, the longer about equalling the pedicels. *Sepals* 5, conspicuous, acute, persistent. *Petals* 5, pinkish to purplish-white, outer one slightly longer than inner, with an inflexed point, hairy beneath. *Stamens* 5; filaments white; anthers purplish. *Styles* 2, recurved, glabrous, with enlarged base forming the often purplish stylopodium which is about half as long as the styles; stigma capitate. *Schizocarp* 3–4 mm, oblong-ovoid, subterete; both mericarps with slender, almost glabrous ridges and rough, tapering, forward-curved, but not hooked spines of varied length; commissure narrow; carpophore present; vittae solitary; pedicels with a ring of hairs at apex. *Flowers* 7–8. $2n = 16$.

Native. Grassy and waste places, hedgerows, wood borders and clearings. Frequent throughout Great Britain and Ireland (except north and north-west Scotland) where it is the commonest roadside umbellifer flowering in July. Most of Europe; temperate Asia; Japan; North Africa. Eurasian Temperate element.

2. T. ucranica Spreng. Eastern Hedge Parsley

T. microcarpa Besser; *Caucalis grandiflora* var. *aculeata* (Boiss.) Druce

Annual or rarely *biennial* monoecious, slightly aromatic *herb* with a slender tapering tap-root and fibrous side-roots. *Stem* 20–130 cm, pale green, sometimes slightly tinted purplish, solid, erect, striate, branched, leafy, with closely appressed, reflexed bristles. *Leaves* 3–10 × 2–6 cm, rather pale yellowish-green on upper surface, paler beneath, lanceolate to ovate in outline, 1- to 3-pinnate, the segments 10–40 mm, lanceolate or linear in outline, pinnatifid to coarsely serrate, with appressed, forward-pointing bristles on both surfaces; petioles very slender, with mostly deflexed bristles. *Inflorescence* of compound umbels, with 5–12 rays, the rays 5–20 mm, slightly unequal, with appressed, forward-pointing hairs, the peduncle longer than the rays, with closely appressed, deflexed hairs, the umbels with bisexual flowers only; bracts 4–6(–12), linear, unequal and roughly hairy. *Sepals* 5, conspicuous, acute and persistent. *Petals* 5, pinkish to purplish-white, outer much larger than inner, with an inflexed point, hairy beneath. *Stamens* 5, filaments whitish, with enlarged base forming the sometimes purplish stylopodium which is much shorter than the long, curved, often purplish style; stigma capitate. *Schizocarp* 2–3 mm, subglobose, both mericarps with slender almost glabrous ridges and rough, tapering, forward-curved but not hooked spines of varied length; commissure narrow; carpophore present; vittae solitary. *Flowers* 6–7.

Introduced. In quantity in a recently planted wood in Bassingbourn in Cambridgeshire with *T. japonica* which it closely resembles. For at least 70 years before the wood was planted it was an arable field. Most of the shrubs planted in the wood are not native. With *T. ucranica* grows *Torilis nodosa*, *Chaerophyllum temulum* var. *canescens*, *Silene latifolia* subsp. *latifolia* and subsp. *alba* and their hybrid with *S. dioica*, *Tragopogon pratensis* subsp. *pratensis* and subsp. *minor* and *Tragopogon porrifolius* subsp. *australis* and *Heracleum branca-ursina*. All must have been brought in while planting the trees and shrubs, or while cutting the grass paths, or planted with wild flower seed. *T. ucranica* is native of central and east Europe and western Asia. It seems to come into flower and go over earlier than our native plant.

3. T. arvensis (Huds.) Link Spreading Hedge Parsley
Caucalis arvensis Huds.; *Caucalis segetum* Thuill.; *Caucalis purpurea* Ten.; *Caucalis segetalis* Steud.; *Anthriscus arvensis* (Huds.) K.-Pol.

Annual or rarely *biennial* monoecious *herb* with a faint smell and slender, tapering tap-root with fibrous roots. *Stems* 5–40(–100) cm, erect, solid, pale green, often purplish at base, terete and often glabrous below, somewhat angled and with sparse to dense, closely appressed, deflexed bristles above, often freely branched, with the branches spreading at a wide angle, leafy. *Leaves* 1–6 × 1–4 cm, medium green on upper surface, paler beneath, ovate in outline, 1- to 2-pinnate or sometimes 3-lobed, the segments up to 50 mm, lanceolate in outline, coarsely serrate to pinnatifid, with appressed, forward-pointing bristles on both surfaces; petiole slender, with appressed, forward-pointing bristles. *Inflorescence* of compound umbels, with 4–12 rays, the rays 5–15 mm, somewhat unequal, with appressed, forward-pointing hairs, the peduncle longer than the rays, with closely appressed, deflexed hairs, the umbels with bisexual flowers only; lanceolate, with stiff hairs, about equalling the shortly pedicellate flowers. *Sepals* 5, acute, persistent. *Petals* 5, white or pinkish, the outer radiating, with an inflexed point. *Stamens* 5; filaments white; anthers purplish. *Styles* 2, patent or recurved, usually with a few hairs near the base, with enlarged base forming the often purplish stylopodium which is one-sixth as long as to as long as the styles; stigma capitate. *Schizocarp* 4–6 mm, oblong-obovoid, subterete; both mericarps with slender, ciliate ridges and long rough, tapering spines, thickened near the apex and with a slender hook; commissure narrow; carpophore present; vittae solitary; pedicels with a ring of hairs at the apex. *Flowers* 7–8. $2n = 12$.

(a) Subsp. **arvensis**
T. arvensis subsp. *divaricata* Thell.; *T. helvetica* (Jacq.) C. C. Gmel.; *T. divaricata* Moench; *Caucalis helvetica* Jacq.

Stem either ascending and often little-branched, the branches forming a narrow angle with the stem, or low and much-branched, the branches making a wide angle to the stem. *Outer petals* not more than 1.5 mm, very slightly radiating. *Styles* 2–3 times as long as the stylopodium.

(b) Subsp. **neglecta** (Schult.) Thell.
T. neglecta Schult.; *T. infesta* (L.) Clairv.; *T. radiata* Moench; *Scandix infesta* L.; *Caucalis infesta* (L.) Curtis.

Stem erect, much-branched. *Outer petals* up to 2 mm or more, distinctly radiating. *Styles* 3–6 times as long as the stylopodium.

Subsp. *arvensis* is a rare probably introduced weed of heavy, calcareous, clay soils, almost exclusively found in winter-sown cereal crops. It also occurs occasionally in other crops and on waste land in open, well-drained situations. Once widespread on the chalk and limestone soils of much of Great Britain, it has diminished considerably since the 1950s, and is very vulnerable to herbicides. West, south and central Europe; south-west Asia. Subsp. *neglecta* is a wool and bird-seed casual, and is native of central and southern Europe.

4. T. leptophylla (L.) Rchb. fil. Slender Hedge Parsley
Caucalis leptophylla L.; *T. erythrotricha* Rchb.; *Caucalis erythrotricha* (Rchb.) Boiss. & Hausskn.; *Caucalis leptophylla* var. *erythrotricha* (Rchb.) Post; *Caucalis pumila* Lam.; *Caucalis parviflora* Lam.; *Caucalis humilis* Jacq.; *Daucus leptophylla* (L.) Scop.; *Nigera parviflora* (Lam.) Bubani; *Anthriscus leptophylla* (L.) K.-Pol.; *Selinum humile* (Jacq.) E. H. L. Krause

Annual monoecious *herb* with tap-root. *Stems* 5–40 cm, pale green, erect or spreading, shallowly channelled, with rather sparse, reflexed, appressed simple eglandular hairs to nearly glabrous, much branched, leafy. *Leaves* rather distant; basal 2–6 × 0.6–3.0 cm, ovate to narrowly deltoid-oblong in outline, 2- to 3-pinnate, the ultimate segments 1–3 mm, oblong, acute at apex, entire or toothed, thinly appressed-hairy, the petiole 3–5 cm, slender, channelled above, thinly hairy; cauline smaller, the upper narrowly sheathing. *Inflorescence* of leaf-opposed, compound umbels; rays 2–3; peduncle 1–4(–5) cm, divaricate, thinly appressed-hairy to almost glabrous; bracts absent; bracteoles about 3 mm, lanceolate, acute at apex, concave, with membranous margins, hairy; pedicels up to 6 mm, appressed-hairy. *Sepals* 5, small, persistent. *Petals* 5, up to 2 mm, white, pink or purplish, unequal, broadly obovate or subrotund, deeply emarginate. *Stamens* 5; filaments about 0.4 mm, pinkish or whitish; anthers pinkish. *Styles* 2, very short or stigmas sessile on the conical stylopodium. *Schizocarps* 4–6 mm, linear-cylindrical, laterally compressed; mericarps with 4 secondary ridges covered with bristle-like glandular hairs; commissure narrow; carpophore present; vittae solitary. *Flowers* 5–6. $2n=12$.

Introduced. A rare bird-seed, grain and wool alien. Native of west and southern Europe; Caucasia; Khorassan, northern Iraq, Syrian desert, Turkistan and western Pakistan.

5. T. nodosa (L.) Gaertn. Knotted Hedge Parsley
Tordylium nodosum L.; *Caucalis nodosa* (L.) Scop.; *T. nodiflorus* (Lam.) Bubani; *T. stocksiana* (K.-Pol.) K.-Pol.; *Caucalis nodiflora* Lam.; *Lappularia nodosa* (L.) Pomel; *Daucus nodosa* (L.) E. H. L. Krause; *Anthriscus stocksiana* K.-Pol.; *Anthriscus nodiflorus* (Lam.) K.-Pol.

Annual monoecious *herb* with a faintly sweetish smell and a slender tap-root with fibrous roots. *Stems* 20–50 cm, pale green, erect, decumbent or procumbent, solid, striate, with rather sparse, closely appressed, deflexed bristles, branched, leafy. *Leaves* 1–5 × 1–2 cm, medium greyish-green on upper surface, paler beneath, ovate-oblong in outline, 1- to 2-pinnate, the segments up to 30 mm, lanceolate to ovate in outline and pinnatifid, with appressed forward-pointing bristles on both surfaces, the petiole of lower leaves slender, with a short, sheathing base and forward-pointing, appressed bristles, the upper leaves subsessile or with a short, sheathing petiole. *Inflorescence* of compound umbels, leaf-opposed, with 2–3 rays, the rays short and stout and generally hidden by the flowers or schizocarps, the peduncle less than 10 mm, often almost absent, with rather sparse, appressed, deflexed bristles, the umbels with bisexual flowers only; bracts absent; bracteoles linear, longer than the subsessile flowers, with patent or forward-pointing bristles. *Sepals* 5, small and acute at apex. *Petals* 5, pinkish-white, the outer not radiating, bristly beneath, with an inflexed point. *Stamens* 5; filaments white; anthers pink or purplish. *Styles* 2, colourless, erect or somewhat divergent, with enlarged base forming the stylopodium which is twice as long as the style; stigma capitate. *Schizocarp* 2.5–3.5 mm, ovoid; mericarps with slender ciliate ridges, the outer with long, glochidiate spines and the inner densely tuberculate towards the ridges; commissure narrow; carpophore present; vittae solitary; peduncles with a ring of hairs at the apex. *Flowers* 5–7. $2n=24$.

Var. *pedunculata* Rouy & Camus, with a longer peduncle to the umbels, does not seem to be worth recognising, but the dwarf plant called var. *nana* Bréb., which occurs on the coast from the Channel Isles northwards, may be a distinct ecotype.

Native. Arable land and rather bare ground, especially near the sea. Scattered records in Great Britain north to south-east Scotland. South and west Europe; west Asia; North Africa. Mediterranean-Atlantic element, but naturalised north of its native range so that its distribution is now Submediterranean Subatlantic.

51. Caucalis L.

Annual monoecious *herbs. Stems* erect, solid. *Leaves* 2- to 3-pinnate. *Inflorescence* of compound umbels, leaf-opposed; rays 2–5; bracts 1–3; bracteoles similar to bracts. *Sepals* 5. *Petals* 5, white, pink or purple. *Stamens* 5. *Styles* 2; stylopodium conical. *Fruit* a schizocarp, oblong-ellipsoid, somewhat compressed laterally; mericarps with primary ridges thinly hispid, secondary ridges conspicuously spinose, with smooth, usually uncinate spines; commissure narrow; carpophore generally terete and undivided.

Contains 1 species in Europe, Mediterranean region and eastwards to Iran. Tropical African species included in *Caucalis* probably belong to a different genus.

1. C. platycarpos L. Small Bur Parsley
C. daucoides L. (1767), non L. (1763); *C. lappula* Grande nom. illegit.; *Daucus lappula* Weber nom. illegit.; *C. royeni* (L.) Crantz; *Orlaya platycarpos* (L.) W. D. J. Koch; *Daucus platycarpos* (L.) Scop.; *Daucus royenii* (Willd.) Baill.; *Daucus caucalis* E. H. L. Krause; *Conium royenii* Willd.; *Nigera daucoides* Bubani

Annual monoecious *herb* with a tap-root. *Stems* 10–40(–50) cm, green, erect, solid, distinctly ribbed, with few, stiff hairs especially at the nodes, usually much branched, leafy. *Leaves* alternate, 1.5–7.0 × 1.5–7.0 cm, broadly deltoid in outline, 2- to 3-pinnate; ultimate segments 2–4 × 1.0–1.5 mm, narrow linear-oblong, acute at apex, incurved on margin, nearly glabrous; the basal with petioles up to 4 cm, slender and channelled, the upper subsessile with a broad, stiffly hairy, sheathing base. *Inflorescence* of compound umbels, the umbels leaf-opposed; rays 2–5, unequal or subequal; peduncle up to 10 cm, stout and ribbed; bracts 1–3, sometimes absent, deciduous, unequal, 2–4 × 0.5–1.0 mm, linear-subulate, acute at apex, margins stiffly hairy and narrowly membranous; pedicels up to about 3 mm, minutely scabrid; bracteoles similar to bracts but persistent; flowers about half functionally male and sterile. *Sepals* 5, 1.0–2.5 mm, narrowly deltoid. *Petals* 5, about 5 mm, white, pink or purple, broadly obovate, deeply emarginate at apex, slightly radiating. *Stamens* 5; filaments about 1.5 mm; anthers yellow. *Styles* 2, swollen at base to form the conical stylopodium. *Schizocarp* 8–9 mm, crowned with a persistent calyx, oblong-ellipsoid, somewhat compressed laterally; mericarps with primary ridges thinly hispid, secondary ridges conspicuously spinose, with smooth usually uncinate spines; commissure narrow; carpophore generally terete and undivided; vittae present. *Flowers* 6–7. $2n = 20$.

Introduced. A grain, bird-seed and wool casual; formerly a long-persistent weed of arable land now rarely found. Native of the Mediterranean region; Caucasia, Iran and Khorassan.

52. Turgenia Hoffm.

Annual monoecious *herbs. Stems* erect, hollow. *Leaves* alternate; simply imparipinnate. *Inflorescence* of compound umbels; bracts (2–)3–5; bracteoles 5–7, with a wide, scarious margin. *Sepals* 5, small and unequal. *Petals* 5, white, pink or purplish, the outer little radiating. *Stamens* 5. *Styles* 2. *Fruit* a schizocarp, ellipsoid to ovoid, compressed laterally; mericarps with the 2 primary marginal ridges each with a single row of spines or tubercles, and the remaining primary and secondary ridges similar to each other, with spines in 2–3 rows; commissure narrow; carpophore present; vittae present.

Contains 2 species in Europe and the Mediterranean region east to central Asia.

1. T. latifolia (L.) Hoffm. Great Bur Parsley
Caucalis latifolia L.; *Tordylium latifolum* L.; *Daucus latifolius* (L.) Baill.; *Daucus turgenia* E. H. L. Krause

Annual monoecious *herb* with fibrous roots. *Stem* 12–40 (–60) cm, pale green, prominently ribbed, erect, hollow, with short and long stiff hairs, usually much branched, leafy. *Leaves* alternate, 4–15 × 3–10 cm, dull green on upper surface, paler beneath, oblong in outline, simply imparipinnate, the 3–6 pairs of segments 10–50 × 3–15 mm, linear-oblong, acute at apex, and coarsely serrate lobed, with stiff hairs, the petiole up to 10 cm, channelled and with a sheathing base, the upper leaves subsessile with a sheathing membranous-margined base. *Inflorescence* of compound umbels, unequally or subequally 2- to 5-rayed, the rays 10–60(–100) mm, spreading, rigid-hairy; peduncles up to 13 mm, sulcate, rigid-hairy; bracts (2–)3–5, about 6 × 2 mm, narrowly oblong, membranous margined, persistent; bracteoles 5–7, similar to bracts but usually with broader, membranous margins and often with a rigid-hairy midrib. *Sepals* 5, small, unequal, deltoid. *Petals* 5, purplish, pink or whitish, broadly obovate, deeply 2-lobed, the outer a little radiating. *Stamens* 5; filaments about 1 mm; anthers yellow. *Styles* 2, less than 1 mm, thick, swollen into a conical stylopodium; stigmas truncate. *Schizocarp* 8–12 mm, ellipsoid to ovoid, compressed laterally, shortly stalked; mericarps with the 2 primary marginal ridges each with a single row of spines or tubercles, and the remaining primary and secondary ridges similar to each other, with spines in 2–3 rows; commissure narrow; carpophore terete or shortly bifid; vittae present. *Flowers* 6–8.

Introduced. A grain and bird-seed casual, decreasing and now rarely found. Native of Europe and the Mediterranean region, east to central Asia.

53. Daucus L.

Annual to *biennial* monoecious *herbs*. *Stems* erect to procumbent, solid. *Leaves* (1–)2- to 3-pinnate. *Inflorescence* of compound umbels; bracts numerous, usually longer than rays, pinnately or ternately divided into filiform lobes; bracteoles numerous. *Sepals* 5, small, scarcely visible in fruit. *Petals* 5, white, but central flower of umbel often dark purple, the outer slightly radiating. *Fruit* a schizocarp, ovoid, strongly dorsally compressed; mericarps with slender, ciliate primary ridges and broader secondary ridges bearing a single row of spines; commissure wide; carpophore present; vittae solitary, under the secondary ridges.

Contains about 20 species in the temperate parts of the northern hemisphere. The characteristic scent of carrots is caused by asarone, carotol, limonene and pinene.

D. guttatus Sibth. & Sm., **D. montanus** Humb. & Bonpl. ex Schult. & Schult. fil. and **D. pusillus** Michx have been recorded as wool casuals.

D. carota subsp. *gummifer* is our native carrot. *D. carota* subsp. *sativa* was probably introduced by the Romans and our other inland carrots derived from that. *D. carota* subsp. *major* and *D. muricatus* have been introduced with wild flower seed.

Dales, H. M. (1974). The biology of Canadian weeds. 5. *Daucus carota*. *Canad. Jour. Pl. Sci.* **54**: 673–685.
Grieve, M. (1931). *A modern herbal*. London.
Grime, J. P. et al. (1988). *Comparative plant ecology*. London.
Hultén, E. & Fries, M. (1986). *Atlas of north European vascular plants north of the Tropic of Cancer.* 3 vols. Königstein.
Malloch, A. J. & Okusanya, O. T. (1979). An experimental investigation into the ecology of some maritime cliff species. I. Field observations. *Jour. Ecol.* **67**: 283–292.
Okusanya, O. T. (1979). An experimental investigation into the ecology of some maritime cliff species II. Germination studies. *Jour. Ecol.* **67**: 293–304; III. Effects of water on growth. *Jour. Ecol.* **67**: 579–590. IV. Cold sensitivity and competition studies. *Jour. Ecol.* **67**: 591–600.
Phillips, R. & Rix, M. (1988). *Vegetables*, pp. 116–123. London.
Vaughan, J. G. & Geissler, C. A. (1997). *The new Oxford book of food plants*. Oxford.
Zohary, D. & Hopf, M. (2000). *Domestication of plants in the Old World*. Ed. 3. Oxford.

1. Rays up to 8; schizocarps, with dense, slender stellate-tipped spines **3. glochidiatus**
1. Rays more than 10; schizocarps, with stout, straight spines entire to stellate-tipped 2.
2. Spines on secondary ridges of the mericarp dilated and confluent at the base, longer than the width of the mericarp **2. muricatus**
2. Spines on the secondary ridges of the mericarp not confluent at the base, nor longer than the width of the mericarp 3.
3. Umbels convex to slightly concave in fruit; the rays with long patent to recurved hairs 4.
3. Umbels very contracted and concave in fruit; the rays glabrous or shortly hairy 5.
4. Plant 3–30 cm; umbels few, 2–7 cm in diameter **1(c,iii). carota** subsp. **gummifer** var. **acaulis**
4. Plant 30–100 cm; umbels numerous; 5–9 cm in diameter **1(c,iv). carota** subsp. **gummifer** var. **gummifer**
5. Root usually orange, swollen and fleshy; many spines of schizocarp stellate at tip **1(a). carota** subsp. **sativus**
5. Root usually whitish not swollen or fleshy; spines of schizocarp 1 or 2 pointed at tip, but not stellate 6.
6. Terminal umbels up to 15 cm in diameter; schizocarps with many spines 2-pointed **1(b). carota** subsp. **major**
6. Terminal umbels up to 9 cm in diameter; schizocarps rarely with spines 2-pointed 7.
7. Plant 30–100 cm; umbels few to numerous, 40–90 mm in diameter **1(b,i). carota** subsp. **carota** var. **carota**
7. Plant 3–30 cm; umbels few, 1.5–5.0 cm in diameter **1(b,ii). carota** subsp. **carota** var. **nanus**

1. D. carota L. Carrot
D. sylvestris Mill.; *Caucalis carota* (L.) Huds.; *Caucalis daucus* Crantz; *Caucalis carnosa* Roth; *Carota sylvestris* (Mill.) Roth

Biennial monoecious *herb* with a white, yellow, orange or purplish-brown sometimes fleshy tap-root and a characteristic smell. *Stems* 3–100 cm, pale green, solid, striate or ridged, more or less erect, with more or less dense, patent to deflexed bulbous-based, stiff, eglandular hairs, more or less branched, leafy. *Leaves* 2–20 × 1–10 cm, dull medium green on upper surface, paler beneath, lanceolate, ovate or oblong-lanceolate in outline, (1–)2- to 3- pinnate, the segments 5–30 mm, lanceolate to ovate in outline, pinnatifid or coarsely serrate, obtuse and mucronate to acuminate at apex, more or less hairy on both surfaces, the petiole of the lower leaves slender, of the upper leaves sheathing but not greatly expanded. *Inflorescence* of compound umbels, with numerous rays, the rays 10–50 mm, somewhat unequal and more or less hairy; bracts 7–13, pinnately or ternately divided into linear lobes; bracteoles 7–10, linear-lanceolate, almost entirely scarious, ciliate and those of the partial umbels sometimes 3-fid. *Sepals* 5, triangular. *Petals* 5, often pink in bud, white at anthesis, those of the centre flower of the umbel usually dark purple, the outer somewhat radiating, emarginate. *Stamens* 5; filaments white; anthers pink. *Styles* 2; divergent, with enlarged base forming the stylopodium which is one-third or one-quarter as long as the styles; stigma capitate. *Schizocarp* 2–3(–4) mm, ovoid, somewhat compressed dorsally; mericarps with slender, ciliate primary ridges and broader secondary ridges bearing a single row of spines, the spines flattened, broad-based, smooth or hooked to stellate; commissure wide; carpophore present; vittae solitary, under the secondary ridge; pedicels without a ring of hairs at the apex. *Flowers* 6–8. $2n=18$.

(a) Subsp. **sativus** (Hoffm.) Arcang. Cultivated Carrot
D. carota var. *sativus* Hoffm.; *D. sativus* (Hoffm.) Röhl.; *Carota sativa* (Hoffm.) Rupr.

Tap-root swollen, fleshy, orange, yellow or whitish. *Stem* erect. *Leaves* bright green, thin, sparsely hairy. *Umbels* strongly contracted in fruit; rays sparsely hairy. *Schizocarp* spines often with stellate tips. $2n=18$.

(b) Subsp. **major** (Vis.) Arcang.
D. major Vis.

Tap-root tough, but not fleshy or swollen, whitish. *Stem* erect. *Leaves* dull dark to greyish-green, segments of lower ovate-lanceolate to lanceolate, dentate or pinnatifid, sparsely long hairy. *Umbels* strongly contracted in fruit; terminal umbel up to 15 cm in diameter; pedicels glabrous or shortly hairy. *Schizocarps* with many spines 2-pointed.

(c) Subsp. **carota** Wild Carrot
Tap-root tough, but not fleshy or swollen, whitish to purplish-brown. *Stem* erect or with branches spreading from the base. *Leaves* dull dark to greyish-green, deeply pinnatifid or pinnatisect, shortly rough-hairy. *Umbels* strongly contracted in fruit; rays glabrous or shortly hairy. *Schizocarp* spines with a bent tip, occasionally 2-pointed, but without stellate tip. $2n=18$.

(i) Var. **carota**
Plant 30–100 cm, erect. *Umbels* few–numerous, 40–90 mm in diameter.

(ii) Var. **nanus** (Druce) Druce
D. carota forma *nanus* Druce

Plant 3–30 cm, the stem short or absent and branches spreading from the base. *Umbels* few, 15–50 mm in diameter.

(d) Subsp. **gummifer** (Syme) Hook. fil. Sea Carrot
D. gummifer Lam., non All.; *D. carota* var. *gummifer* Syme; *D. maritimus* With., non Lam.

Root tough and whitish, not fleshy. *Stem* erect, or branches spreading from the base. *Leaves* glossy dark green and glabrous or sparsely hairy on upper surface. *Umbels* not contracted in fruit; rays with long patent to recurved hairs. *Schizocarp* spines single pointed.

(iii) Var. **acaulis** (Bréb.) P. D. Sell
D. gummifer var. *acaulis* Bréb.

Plant 3–30 cm, often branching from the very base. *Umbels* few, 20–70 mm in diameter.

(iv) Var. **gummifer** Syme
Plant 30–100 cm, erect. *Umbels* few–numerous, 50–90 mm in diameter.

Native. Subsp. *carota* occurs on grassy and waste places, mostly on chalky soils and also near the sea, throughout most of Great Britain and Ireland but is mainly coastal in north and west Great Britain. Throughout most of Europe, temperate Asia and North Africa. Eurosiberian Temperate element. Subsp. *gummifer* occurs on cliffs, dunes and rocky places near the sea from Kintyre to Kent, the Channel Islands and south and south-east Ireland. It also occurs on the Atlantic coast of France and north Spain. The ecology of the two varieties, which also occur in France, is not understood and more experimental research is required. The most characteristic var. *acaulis* is in the south, plants becoming less typical northwards, Kintyre being the most northerly locality which can be included. Dwarf plants north of that are best placed in var. *nana* which also occurs in Co. Cork. Var. *acaulis* is more common in herbaria than var. *gummifer* but in the field they seem to be equally common. Subsp. *sativus* is the cultivated carrot, which is widely grown in gardens and in some areas, particularly East Anglia, as a crop. It possibly originated from subsp. **maximus** (Desf.) Ball of the Mediterranean and west Asia and has the same stellate-tipped spines on the schizocarp. Carrots may be eaten raw, cooked as a vegetable, with numerous culinary uses from soups to cakes. They frequently occur as casuals in waste places, on tips and as a relic of crops. It is likely carrots were first cultivated in the eastern Mediterranean region and were certainly grown by the Romans. Wild carrots with red and purple roots are found in Afghanistan, and it may be there the coloured cultivated carrot originated. Yellow carrots were first recorded in Turkey in the tenth century, and both yellow and purple ones were grown throughout Europe until the seventeenth century, when the orange carrot was developed in Holland. By the eighteenth century four varieties, Cv. Long Orange, Late Half Long, Early Half Long and Early Scarlet Horn were distinguished and from these all modern Western carrots are derived. The total distribution of the species, *Daucus carota*, depends on your taxonomy but subsp. *sativus* is cultivated everywhere and in North America has escaped from cultivation and become a pernicious weed. There will be intermediates between all these taxa if they grow together, but their origins are almost certainly different.

2. D. muricatus (L.) L. Mediterranean Carrot
Artedia muricata L.

Annual herb with a whitish tap-root and a characteristic smell. *Stems* up to 60 cm, pale green, sometimes tinted brownish-red, solid, more or less erect, ridged, with numerous, patent to deflexed, bulbous-based, stiff eglandular hairs, leafy, more or less branched. *Leaves* 2–20 × 1–10 cm, dull greyish-green on upper surface, paler beneath, lanceolate to ovate in outline, 3-pinnate, the segments up to 30 mm, lanceolate to linear-lanceolate in outline, pinnatifid to coarsely serrate, mucronate to acuminate at apex, stiffly eglandular hairy particularly on the margins and veins; petiole of lower leaves slender, of the cauline sheathing but not greatly expanded. *Inflorescence* of compound umbels, with numerous rays, the rays 10–15 mm, markedly unequal, with rather sparse, short, stiff eglandular hairs; bracts up to 12, pinnatisect, the segments linear-setaceous, later deflexed; bracteoles up to 8, entire or 3-fid. *Petals* 5, white, the outer strongly radiate. *Stamens* 5; filaments white; anthers cream. *Styles* 2, divergent, with an enlarged base forming the stylopodium which is about one-third as long as the style; stigma capitate. *Schizocarp* 5–10 mm, with spines; spines on secondary ridges silvery-white, 1–2 times as long as the width of the mericarp, strongly dilated and confluent at the base and with forked tips. *Flowers* 6–8.

Introduced. Large patches on the roadside by the large roundabout south of Arrington in Cambridgeshire where it was probably sown with wild flower seed. *D. carota* subsp. *carota* var. *carota* occurred with it. Neither is likely to be native there. *D. muricata* is native of the west Mediterranean region. If from wild flower seed it should be looked for elsewhere.

3. D. glochidiatus (Labill.) Fritsch, C. A. Mey. & Avé-Lall. Australian Carrot
Scandix glochidiatus Labill.; *D. brachiatus* Sieber ex DC.

Annual monoecious *herb* with a slender, whitish tap-root. *Stems* 5–60 cm, pale yellowish-green, sometimes tinted brownish-purple, slender, erect or ascending, solid, striate, glabrous or with pale, subrigid simple eglandular hairs, much branched to base, leafy. *Leaves* slightly bluish-green on upper surface, paler beneath; basal and cauline similar, 5–9 × 1.5–3.5 cm, oblong to lanceolate in outline, obtuse at apex, 2- to 3-pinnate, the segments deeply pinnately lobed, the lobes linear, mucronulate at apex and entire, with sparse, short simple eglandular hairs or nearly glabrous; petioles up to 14 cm, slender, glabrous. *Inflorescence* of terminal compound umbels; umbels with 2–5 rays of unequal length; bracts pinnately divided; bracteoles narrow and entire. *Sepals* 5, small and triangular. *Petals* 5, pinkish, rarely red, the outer somewhat radiating, apex inflexed. *Stamens* 5; filaments white; anthers pink. *Styles* 2, divergent with enlarged base forming the stylopodium; stigma capitate. *Schizocarp* 3–5 mm, ovoid to oblong, somewhat compressed dorsally, dark brown; mericarps with slender hairs on the primary ridges and dense, rather slender spines on the secondary ridges which have reflexed, stellate tips; commissure wide; carpophore present; vittae solitary, under the secondary ridges. *Flowers* 8–9. $2n = 44$.

Introduced. Rather frequent wool alien, but often does not flower, but can be recognised if you know the look of the plant. Scattered records in Great Britain. Native of Australia.

Agrocharis gracilis Hook. fil. and **A. melanantha** Hochst. have been recorded as wool casuals.

Subclass 6. ASTERIDAE Takht.

Trees, shrubs or *herbs*. *Leaves* simple or variously compound or dissected. *Flowers* hypogynous to epigynous, but not strongly perigynous. *Corolla* usually sympetalous. *Stamens* usually attached to the corolla tube. *Carpels* mostly 2–5, generally united to form a compound ovary. *Seeds* with or without endosperm.

Consists of 11 orders, 49 families and nearly 60,000 species.

Order 1. GENTIANALES Lindl.

Trees, shrubs or *herbs*. *Leaves* simple, often opposite, usually without stipules. *Flowers* bisexual or rarely unisexual, hypogynous, actinomorphic. *Calyx* tubular or rarely composed of separate sepals or absent. *Corolla* usually 4- to 5-lobed, rarely of free petals or absent, lobes contorted or valvate in bud. *Stamens* 4 or 5. *Style* 1. *Ovary* mostly 1–2-celled; ovules numerous or 1 in each cell; placentation parietal or axile. *Fruit* various; seeds with endosperm; embryo straight, often small.

Consists of 6 families and about 5,500 species.

121. GENTIANACEAE Juss. nom. conserv.

Annual to *perennial* monoecious *herbs*. *Leaves* opposite, simple, entire, without stipules, sessile or nearly so. *Inflorescence* a terminal, dichasial cyme, sometimes monochasial or flowers solitary; flowers actinomorphic, bisexual and hypogynous. *Calyx* 4–5(–8)-lobed. *Corolla* 4–5(–8)-lobed, persistent in fruit. *Stamens* as many as petals, borne on the corolla tube. *Styles* 1–2, if one usually divided into 2 at apex, sometimes more or less absent; stigmas (1–)2, capitate to various expended. *Fruit* a capsule, dehiscing into 2 valves.

Contains about 75 genera and about 1,000 species, cosmopolitan.

1. Petals purple to blue, rarely white, very rarely pinkish; style and stigmas persistent in fruit 2.
1. Petals pink, yellow or white; stigmas and sometimes style falling before fruiting 3.
2. Calyx-tube without a membranous part; corolla without small inner lobes but with long fringes **5. Gentianella**
2. Distal part of calyx-tube membranous between the calyx lobe origins; corolla with small inner lobes alternating with the main ones **6. Gentiana**
3. Pairs of stem leaves fused at base; flowers 6- to 8-merous **4. Blackstonia**
3. Stem leaves not fused in pairs; flowers 4- to 5-merous 4.
4. Calyx lobes shorter than calyx-tube; corolla yellow **1. Cicendia**

4. Calyx lobes longer than calyx-tube, corolla pink or white 5.
5. Calyx lobes 4, not keeled; corolla lobes 4, less than 2 mm; anthers not becoming twisted **2. Exaculum**
5. Calyx lobes (4–)5, keeled; corolla lobes (4–)5, more than 2 mm; anthers twisting after flowering **3. Centaurium**

1. **Cicendia** Adans.

Microcala Hoffmanns. & Link nom. illegit.; *Franquevillia* Salisb. ex Gray nom. illegit.

Small *annual* monoecious *herbs*. *Leaves* opposite, entire. *Calyx* campanulate, with 4 short, deltate lobes. *Corolla* with ovoid tube and 4 short, spreading lobes, yellow. *Stamens* 4; anthers cordate, not twisted. *Style* 1, filiform; stigma peltate, caducous. *Fruit* a capsule.

Contains 2 species, the second in California and temperate South America.

Hultén, E. & Fries, M. (1986). *Atlas of north European vascular plants north of the Tropic of Cancer.* 3 vols. Königstein.
Stewart, A., Pearman, D. A. & Preston, C. D. (1994). *Scarce plants in Britain*. Peterborough.

1. C. filiformis (L.) Delarbre Yellow Centaury
Gentiana filiformis L.; *Microcala filiformis* (L.) Hoffmanns. & Link; *Exaceum filiforme* (L.) Sm.; *Franquevillia minima* Gray nom. illegit.

Annual monoecious *herb. Stems* 3–12 cm, pale green, sometimes tinted red, erect, simple or somewhat branched, slender, glabrous. *Leaves* opposite, few; lamina 2–6×0.2–1.0 mm, bright medium green, linear, pointed at apex, entire, narrowed at base, glabrous. *Flowers* solitary at the ends of stems and branches, bisexual; pedicels 10–50 mm, glabrous. *Calyx* campanulate, 3.0–3.5 mm, 4-lobed, the lobes triangular, acute at apex, glabrous. *Corolla* 3–7 mm, yellow, 4-lobed, the lobes elliptical, rounded at apex, opening only in the sun. *Stamens* 4; filaments pale; anthers yellow. *Style* 1, filiform; stigma peltate. *Capsule* 4–5 mm, ovoid. *Flowers* 6–10. $2n=26$.

Native. Open heathlands on sandy and peaty soils of relatively high base-status which are damp in winter and spring; also in damp pasture, dune slacks and cliffs and woodland rides. Cornwall, Devonshire, Dorsetshire, Hampshire, Sussex, Norfolk, Pembrokeshire, Caernarvonshire, Lincolnshire, Channel Islands, Co. Kerry, Co. Cork and Co. West Mayo. Extinct in some areas and much declined in others but may also be refound. West and south Europe, Turkey, North Africa and Azores. Submediterranean Subatlantic element.

2. **Exaculum** Caruel

Annual monoecious *herbs. Leaves* opposite, entire. *Flowers* bisexual, solitary at the ends of stems and branches. *Calyx* with 4 flat, linear lobes. *Corolla* cylindrical, 4-lobed. *Stamens* 4; anthers not becoming twisted. *Styles* 1, becoming bifid near apex into 2 stigmas. *Fruit* a fusiform capsule.

Contains a single species.

McClintock, D. (1975). *The wildflowers of Guernsey*. London.

1. E. pusillum (Lam.) Caruel Guernsey Centaury
Cicendia pusilla (Lam.) Griseb.; *Gentiana pusilla* Lam.; *Exacum pusillum* (Lam.) DC.

Annual monoecious *herb* with fibrous roots. *Stem* 3–12 cm, pale green, very slender, procumbent to ascending, glabrous. *Leaves* opposite; lamina up to 7 mm, medium green, linear, pointed at apex, entire, narrowed at base, glabrous. *Flowers* bisexual, solitary at the ends of stems or branches; pedicel slender. *Calyx* deeply divided into 4 lobes, the lobes linear and flat. *Corolla* 3–6 mm, pink or white, cylindrical, 4-lobed, the lobes about 1.5 mm, flat and linear. *Stamens* 4; filaments pale; anthers ovate, not becoming twisted. *Styles* 1; bifid near apex into 2 stigmas, caducous. *Capsule* fusiform. *Flowers* 7–9. $2n=20$.

Native. Discovered in 1850. Moist, open, short turf in coastal dune slacks, with winter flooding and rabbit grazing maintaining the open conditions it requires. Known only from a few localities in Guernsey and varying in quantity from year to year. South-west Europe from west France to Spain and Italy and in North Africa. Suboceanic Southern-temperate element.

3. **Centaurium** Hill

Erythraea Borkh. nom. illegit.

Annual, biennial or *perennial* monoecious *herbs. Leaves* opposite, simple, entire, without stipules, sessile or almost so. *Inflorescence* a terminal, dichasial cyme. *Calyx* (4–)5-lobed, the lobes linear and keeled, longer than the tube. *Corolla* pink or white, (4–)5-lobed. *Stamens* (4–)5, borne on the corolla tube; anthers becoming twisted at fruiting. *Style* 1, divided near the apex; stigmas 2. Ovary 1-celled, each cell with many ovules. *Fruit* a capsule dehiscing into 2 valves.

Contains about 50 species worldwide, but chiefly in temperate regions.

Druce, G. C. (1919). *Centaurium scilloides* Druce var. *portense* Brot. *Rep. Bot. Soc. Exch. Club Brit. Isles* **5**: 290–295.
Gilmour, J. S. L. (1937). Notes on the genus *Centaurium*. The nomenclature of the British species. *Kew Bull.* **1937**: 497–502.
Jakobsen, K. (1960). *Centaurium glomeratum. Danmark. Bot. Tidsskr.* **56**: 89–104.
Melderis, A. (1932). Genetical and taxonomic studies in the genus *Erythraea* Rich. *Acta Horti Bot.Univ. Latv.* **6**: 123–156.
Melderis, A. (1972). Taxonomic studies on the European species of the genus *Centaurium* Hill. *Bot. Jour. Linn. Soc.* **65**: 224–250.
O'Connor, W. M. T. (1955). Variation in *Centaurium* in West Lancashire. In Lousley, J. E. (Edit.) *Species studies in the British flora*, pp. 119–125. London.
Rich, T. C. G. et. al. (2005). Distribution of the western European endemic *Centaurium scilloides* (L. f.) Samp. (Gentianaceae), Perennial Centaury. *Watsonia* **25**: 275–281.
Rich, T. C. G. & Jermy, A. C. (1998). *Plant crib*. London.
Salmon, C. E. & Thompson, H.S. (1902). West Lancashire notes. *Jour. Bot. (London)* **40**: 293–295.
Townsend, F. (1881). On a *Erythraea* new to England, from the Isle of Wight and south coast. *Jour. Linn. Soc. London (Bot.)* **18**: 398–405.
Ubsdell, R. (1972). The status of some intermediates between *Centaurium littorale* and *Centaurium erythraea* from the Lancashire coast. *Watsonia* **9**: 204.
Ubsdell, R. (1974). A natural hybrid in *Centaurium. Watsonia* **10**: 231–232.

Ubsdell, R. A. E. (1976). Studies on variation and evolution in *Centaurium erythraea* Rafn and *C. littorale* (D. Turner) Gilmour in the British Isles. 1. Taxonomy and biometrical studies. *Watsonia* **11**: 7–31; 2. Cytology. *Watsonia* **11**: 33–43.

Ubsdell, R. A. E. (1979). Studies on variation and evolution in *Centaurium erythraea* Rafn and *C. littorale* (D. Turner) Gilmour in the British Isles. 3. Breeding systems, floral biology and general discussion. *Watsonia* **12**: 225–232.

Vohra, J. N. (1970). Natural hybridization in *Centaurium erythraea* and *C. littorale* at Freshfield and Ainsdale, South Lancashire. *Bull. Bot. Surv. India* **12**: 144–150.

Wheldon, J. A. & Salmon, C. E. (1925). Notes on the genus *Erythraea*. *Jour. Bot. (London)* **63**: 345–52.

Wigginton, M. J. (Edit.) (1999). *British red data books*. Vol. 1. *Vascular plants*. Ed. 3. Peterborough. [*C. scilloides* and *tenuiflorum*.]

Zeltner, L. (1961). Contribution à l'étude cytologique des genres *Blackstonia* Huds. et *Centaurium* Hill (Gentianacées). *Bull. Soc. Bot. Suisse* **71**: 17–24.

Zeltner, L. (1962). Deuxième contribution à l'étude cytologique des genres *Blackstonia* Huds. et *Centaurium* Hill (Gentianacées). *Bull. Soc. Neuchâtel. Sci. Nat.* **86**: 93–100.

Zeltner, L. (1970). Recherches de biosystématique sur les genres *Blackstonia* Huds. et *Centaurium* Hill (Gentianaceae). *Bull. Soc. Neuchâtel. Sci. Nat.* **93**: 1–164.

1. Perennial herb with procumbent to decumbent non-flowering stems; corolla lobes 7–9 **1. scilloides**
1. Annual to biennial herbs without procumbent or decumbent non-flowering stems; corolla lobes less than 7 mm 2.
2. Usually biennials, normally with a basal leaf-rosette at flowering time; flowers with 1–2 bracts at base of calyx, the pedicels 0–1 mm 3.
2. Usually annuals, normally without basal leaf-rosette at flowering time; flowers with pedicels 1–5(–10) mm between base of calyx and bracts 13.
3. Stem leaves obovate to elliptical, acute to subacute at apex; calyx usually more than three-quarters as long as the corolla tube; stigmas narrowly rounded to nearly conical 4.
3. Stem leaves narrowly linear to linear-ovate, almost parallel-sided, obtuse at apex; calyx more than three-quarters as long as the corolla tube; stigmas broadly rounded to nearly flat at apex 10.
4. Basal leaves broadly elliptical to subrotund; corolla lobes 3–4 mm **2. latifolium**
4. Basal leaves obovate or oblanceolate; corolla lobes 4–6 mm 5.
5. Corolla lobes 5.5–5.6 mm; pollen 27–28 µm; stigma apex a rounded dome or flat **4. intermedia**
5. Corolla lobes 4.5–5.4 mm; pollen 24–26 µm; stigma conical 6.
6. Plant tall, up to 50 cm 7.
6. Plant short, up to 8 cm 9.
7. Cauline leaves 1–4 mm wide, linear **3(iii). erythraea** var. **sublitorale**
7. Cauline leaves broader 8.
8. Flowers in a short-peduncled cluster; stems usually several **3(i). erythraea** var. **fasciculare**
8. Flowers in a long-peduncled cluster; stems usually solitary **3(ii). erythraea** var. **erythraea**
9. Stamens inserted at apex of corolla-tube **3(iv). erythraea** var. **subcapitatum**
9. Stamens inserted at base of corolla-tube **3(v). erythraea** var. **capitatum**
10. Stems 10–25 cm; inflorescence fastigiate **5(iv). littorale** var. **occidentale**
10. Stems 2–10 cm; inflorescence compact 11.
11. Inflorescence equalled by or exceeded by bracts; carpels as long as calyx **5(iii). littorale** var. **baileyi**
11. Inflorescence longer than bracts; capsule longer than calyx 12.
12. Cauline leaves lanceolate to ovate-lanceolate **5(ii). littorale** var. **turneri**
12. Cauline leaves linear **5(i). littorale** var. **littorale**
13. Main stem with 5–9 internodes; upper branches arising at 20–30° and forming a dense inflorescence; corolla usually white **7. tenuiflorum**
13. Main stem with 2–4 internodes; all branches arising at 30–45° and forming a rather open inflorescence; corolla usually pink 14.
14. Stems simple with few flowers **6(i). pulchellum** var. **pulchellum**
14. Stems branched; flowers numerous 15.
15. Stems with few branches and fairly numerous flowers **6(ii). pulchellum** var. **intermedium**
15. Stems much branched from the base with numerous flowers **6(iii). pulchellum** var. **ramosissimum**

1. C. scilloides (L. fil.) Samp. Perennial Centaury
Gentiana scilloides L. fil.; *Gentiana portensis* Brot.; *Erythraea portensis* (Brot.) Hoffmanns. & Link; *Erythraea massonii Sweet; Erythraea diffusa* Woods; *Erythraea scilloides* (L. fil.) Chaub. ex Puel; *C. scilloides var. portense* (Brot.) Druce; *C. portense* (Brot.) Butcher

Perennial monoecious *herb. Stems* numerous, some decumbent and sterile, others ascending and fertile, glabrous, leafy. *Leaves* opposite, medium green on upper surface, paler beneath, those of the sterile stems with lamina 2–10 × 2–10 mm, subrotund to rhombic, rounded at apex and narrowed into a short petiole, those of the flowering stems with lamina oblong to lanceolate, obtuse at apex, and sessile; all glabrous. *Inflorescence* a terminal, dichasial cyme; flowers bisexual, in groups of 1–6, pedicellate. *Calyx* 6.5–7.5 mm, divided in (4–)5, linear, keeled lobes. *Corolla* pink; lobes (4–)5, 7–9 mm. *Stamens* (4–)5, inserted at the top of the corolla-tube. *Style* 1. *Capsule* longer than calyx. *Flowers* 7–8. $2n = 20$.

Native. Grassy cliff-tops and dunes by the sea. Now known only in Pembrokeshire, formerly in Cornwall. Naturalised as a lawn-weed from garden plants in Kent, Hampshire and Sussex, and Co. Kerry. Atlantic coast from Pembrokeshire and north-west France to north-west Portugal; Azores. Oceanic Southern-temperate element. Azores plants usually have white flowers (*C. scilloides*; *Erythraea massonii*), while those elsewhere in Europe mostly have pink flowers (*Erythraea portensis; Erythraea diffusa*). No other character seems to go with the flower colour.

2. C. latifolium (Sm.) Druce Broad-leaved Centaury
Erythraea latifolia Sm.; *Erythraea centaurium* subsp. *latifolia* (Sm.) Hook. fil.

Biennial monoecious *herb* with fibrous roots. *Stems* solitary or several, 5–10 cm, pale green, erect, with few, ascending or spreading branches, glabrous. *Leaves* medium green on upper surface, paler beneath; basal in a rosette, the lamina 1.0–1.5 × 1.0–1.5 cm, broadly ovate to subrotund, rounded to pointed at apex, entire; cauline opposite, the lamina 1.0–1.5 × 1.0–1.5 mm, ovate, pointed at apex, entire; all glabrous. *Inflorescence* a compact corymb-like cyme; flowers bisexual, sessile; bracts 5–6 mm, ovate, pointed at apex. *Calyx* 2.5–3.0 mm, divided to three-quarters of way to base into 5 lobes, the lobes lanceolate and sharply acute at apex. *Corolla* pink; tube about 4 mm; 5-lobed, the lobes about 1–2 mm, ovate, obtuse at apex. *Stamens* 5, inserted on the corolla tube. *Style* 1. *Capsule* oblong-cylindrical. *Flowers* 7–8.

Extinct native. Formerly occurred on sandhills near Ainsdale, 3 miles south of Southport; Seaforth Common, north of Liverpool; and Bootle in Lancashire. Endemic.

3. C. erythraea Rafn Common Centaury
C. centaurium (Borkh.) Druce nom. illegit.; *Erythraea centaurium* Borkh.; C. *minus* auct.; *C. umbellatum* auct.; *Erythraea vulgaris* auct.; *Chironia centaurium* auct.

Annual or *biennial* monoecious herb with fibrous roots. *Stems* 5–50 cm, pale green, erect, with 4–6 nodes, sharply angled, glabrous, unbranched or with ascending branches, leafy. *Leaves* opposite, those at the base usually forming a persistent rosette; lamina 1–6 × 0.6–2.0 cm, medium green on upper surface, paler beneath, obovate, oblanceolate, elliptical or oblong-elliptical, never parallel-sided, more or less acute at apex, entire, narrowed at base, glabrous, sessile, 3- to 7-veined. *Inflorescence* a lax or congested, usually many-flowered, terminal, subcorymbose cyme; pedicels 0–1 mm, stout, glabrous or scabridulous; bracts leafy, resembling the uppermost cauline leaves; flowers bisexual. *Calyx* 5–8 mm, usually more than half as long as the corolla tube, divided almost to the base into 5 lobes, the lobes narrowly linear-subulate, acute at apex, keeled. *Corolla* bright pink, sometimes white; tube 7–12 mm, 5-lobed, the lobes 4.5–5.4 mm, ovate or narrowly oblong. *Stamens* 4 or 5, inserted at the apex or base of the corolla tube; filaments 1.2–7.0 mm, filiform, glabrous; anthers yellow, twisting spirally after dehiscence; pollen 25–26 µm. *Style* 1, 1.5–5.0 mm; stigma conical. *Capsule* about 10 × 2 mm, fusiform; seeds 0.3–0.5 mm, dark brown, irregularly oblong-subglobose, coarsely and irregularly reticulate. *Flowers* 6–10. Pollinated by insects or more usually selfed. $2n=40$.

(i) Var. **fasciculare** (Duby) Ubsdell
C. umbellatum var. *fasciculare* Duby; *Erythraea centaurium* var. *conferta* Wheldon & C. E. Salmon

Plant tall. *Stems* usually several. *Leaves* broadly ovate-oblong. *Inflorescence* a compact clusters of flowers in a short-peduncled cluster. *Stamens* inserted at apex of corolla tube.

(ii) Var. **erythraea**
C. umbellatum var. *centaurium* auct.

Plant 9–30 cm. *Stem* usually single; nodes 4–6, internodes long with well-spaced leaves. *Leaves* ovate-oblong. *Inflorescence* a dense, corybiform, long-peduncled cyme. *Stamens* inserted at apex of corolla tube. $2n=40$.

(iii) Var. **sublitorale** (Wheldon & C. E. Salmon) Ubsdell
Erythraea centaurium var. *sublitoralis* Wheldon & C. E. Salmon; *C. umbellatum* var. *sublitorale* (Wheldon & C. E. Salmon) Druce

Plant 4–20 cm. *Stems* usually solitary; nodes 4–6; internodes long. *Leaves* oblong or obovate-spathulate below, oblong to linear above. *Inflorescence* a few-flowered, corymbiform cyme. *Stamens* inserted at the apex of the corolla tube.

(iv) Var. **subcapitatum** (Corb.) Ubsdell
Erythraea centaurium var. *subcapitata* Corb.; *C. umbellatum* var. *subcapitatum* (Corb.) Gilmour

Plants up to 8 cm. *Inflorescence* capitate, of at least 5 crowded cymes. *Stamens* inserted at apex of corolla tube. $2n=40$.

(v) Var. **capitatum** (Willd.) Melderis
Erythraea capitata Willd.; *Erythraea centaurium* var. *capitata* (Willd.) W. D. J. Koch; *Erythraea centaurium* subsp. *capitata* (Willd.) Hook. fil.; *C. capitatum* (Willd.) Borbás; *C. umbellatum* subsp. *capitatum* (Willd.) P. Fourn.; *C. minus* var. *capitatum* (Willd.) Zeltner

Plant up to 8 cm. *Leaves* narrow. *Inflorescence* capitate. *Stamens* inserted at base of corolla tube. $2n=40$.

Native. Occurring in a wide range of habitats including chalk and limestone grassland, heathland, woodland rides, open scrub, dune grassland, quarries, spoil heaps and road verges up to 330 m in mildly acidic to calcareous, well-drained soils. Throughout Great Britain and Ireland except for mountainous regions in the north. Europe from Sweden southwards; Mediterranean region; Azores; south-west Asia to the Pamir-Alai region; naturalised in North America and New Zealand. European Southern-temperate element. All our plants are referable to subsp. **erythraea**. Var. *fasciculare* occurs on a few coastal dunes. Var. *erythraea* is widespread both here and in Continental Europe. Both var. *subcapitatum* and var. *capitatum* are probably widespread in coastal habitats both here and in Continental Europe. Var. *sublitorale* is widespread in coastal localities in England and Wales. All these varieties retain their characteristics in cultivation.

× **littorale**
This hybrid is intermediate between the parents and can be distinguished from *C. intermedium* by being highly sterile with a chromosome number of $2n=60$.

Native. Coasts of Lancashire and Anglesey. Known also from Germany.

× **pulchellum**
This hybrid is intermediate between the parents in pedicel length and corolla lobe length and is highly fertile.

Native. Occurs with the parents on the coasts of Somersetshire, Essex and Lancashire.

4. C. intermedium (Wheldon) Druce
Intermediate Centaury
Erythraea littoralis var. *intermedia* Wheldon

Biennial monoecious *herb* with fibrous roots. *Stems* up to 25 cm, pale green, erect, minutely scaberulous, often branched from the base, leafy. *Leaves* medium green on upper surface, paler beneath; basal with lamina 1.5–2.5 × 0.3–0.5 cm, linear-elliptical, sides never parallel, acute at apex, entire, indistinctly 3-veined, usually forming a rosette; cauline similar but can be longer, linear-spathulate, acute at apex, entire, in opposite pairs; all glabrous or semi-scabrid. *Inflorescence* clustered in a more or less dense corymb-like cyme; pedicels about 4 mm; bracts up to 20 mm, linear-lanceolate. *Calyx* 8–10 mm, divided nearly to the base into 4 or 5 lobes, the lobes linear-subulate, acute at apex. *Corolla* pink; tube 7–10 mm; lobes 4 or 5, 5.5–6.5 mm, ovate, obtuse at apex. *Stamens* 4 or 5, inserted at the top of the corolla tube; anthers yellow; pollen 24–32 µm. *Style* 1; stigma apex a flat dome. *Capsule* about 10 mm, pale brown; seeds brown. *Flowers* 7–8. $2n = 60$.

Native. Coastal sand. Hightown, Freshfield and Ainsdale on the Lancashire coast. Endemic.

5. C. littorale (Turner ex Sm.) Gilmour
Seaside Centaury
Chironia littoralis Turner ex Sm.; *C. minus* Moench; *C. vulgare* Rafn nom. illegit.; *Gentiana centaurium* L.; *C. lineariifolium* auct.; *Erythraea littoralis* (Turner ex Sm.) Fr.; *Erythraea centaurium* subsp. *littoralis* (Turner ex Sm.) Hook. fil.

Biennial monoecious *herb* with fibrous roots. *Stems* several or solitary, 2–25 cm, pale to medium green, erect, scaberulous to glabrous, simple or branched above, leafy. *Leaves* shining medium green on upper surface, paler beneath; basal with lamina 1–2 × 0.3–0.5 cm, linear or linear-spathulate, obtuse at apex, entire, 1-veined, usually forming a rosette; cauline shorter, linear to linear-lanceolate, obtuse at apex, scabridulous to glabrous. *Inflorescence* clustered in a more or less dense, corymbose-like cyme; flowers bisexual, relatively few; pedicels short; bracts 7–8 × 2–4 mm, lanceolate or linear, acute at apex. *Calyx* 6.0–6.5 mm, more than three-quarters as long as the corolla tube, divided nearly to the base into 4 or 5 lobes, the lobes linear-subulate, acute at apex. *Corolla* bright pink or white; tube 6–7 mm, slightly constricted below the limb; lobes 4 or 5, (4.5–) 5.0–6.6 × 3–4 mm, ovate, concave, obtuse at apex. *Stamens* 4 or 5, inserted at the apex of the corolla tube; filaments 3–4 mm, white, filiform; anthers yellow; pollen 29–32 µm. *Style* 1, 2.5–4.0 mm; stigmas oblong, bifid, broadly rounded to nearly flat at apex. *Capsule* much exceeding calyx. *Flowers* 7–8. $2n = 40$.

(i) Var. **littorale**
Erythraea compressa var. *friesii* forma *minor* (Hartm.) Wheldon & C. E. Salmon; *Erythraea littoralis* var. *minor* Hartm.; *C. littorale* var. *minor* (Hartm.) Gilmour

Stems often numerous, 2–5 cm, subscabridulous or glabrous, fairly robust. *Leaves* of basal rosettes obovate-spathulate, the cauline linear, exceeding the internodes. *Inflorescence* a compact cyme. *Capsule* longer than the calyx.

(ii) Var. **turneri** (Weldon & C. E. Salmon) Ubsdell
Erythraea turneri Wheldon & C. E. Salmon; *C. turneri* (Wheldon & C. E. Salmon) Druce

Stems 1–(2–5), 3–6 cm, subscabridulous or glabrous, often with only 1 node. *Leaves* lanceolate or ovate-lanceolate. *Inflorescence* a lax to dense, few-flowered, compact cyme.

(iii) Var. **baileyi** (Wheldon & C. E. Salmon) Gilmour
Erythraea compressa var. *baileyi* Wheldon & C. E. Salmon

Stems 4–10 cm, several, usually robust, much branched, with short internodes, scabridulous. *Leaves* linear to linear-spathulate, the basal up to 60 × 5 mm, scabridulous. *Inflorescence* of closely crowded compact heads of short peduncled flowers which are subequalled or exceeded by the long leaf-like bracts. *Capsule* as long as calyx.

(iv) Var. **occidentale** (Wheldon & C. E. Salmon) Gilmour
Erythraea compressa var. *occidentalis* Wheldon & C. E. Salmon

Stems 10–25 cm, solitary or many, fairly robust, narrowly winged, the wings often ciliate above, rigid. *Leaves* linear, linear-oblong to linear-lanceolate, scabrous-papillose on the margins, the cauline contiguous or distant. *Inflorescence* a few-flowered, fastigiate cyme. *Capsule* one-third longer than calyx.

Native. Coastal dunes, sandy turf, upper levels of salt-marsh and calcareous, humus-rich turf near the sea where the vegetation is kept open by grazing or trampling. Coastal areas of western Great Britain from South Wales northwards and on the east coast from northern England northwards; rare on the coast of west and north Ireland. Coasts of western Europe from central Scandinavia to north-west France; inland in central Europe from Austria to south-east Russia. European Temperate element. Var. *littorale* occurs on the coasts of Scotland and in northern Continental Europe. Var. *turneri* also occurs on the coasts of Scotland. Var. *baileyi* occurs on the coasts of Anglesey and Lancashire. Var. *occidentale* occurs on the coasts of north Wales, north-western England and south-western Scotland. All these varieties retain their characters in cultivation, presumably by self-pollination. They all belong to subsp. **littorale**.

6. C. pulchellum (Sw.) Druce Lesser Centaury
Gentiana pulchella Sw.; *Erythraea pulchella* (Sw.) Fr.

Annual monoecious *herb* with fibrous roots. *Stems* (3–)5–20(–42) cm, pale green, erect, simple or branched, leafy. *Leaves* medium green on upper surface, paler beneath; basal with lamina usually less than 10 × 8 mm, rarely forming loose rosettes and usually withering before anthesis, oblong-spathulate, obtuse or rounded at apex, entire, attenuate at base; cauline rather remote and much shorter than the internodes in well-developed plants, crowded and overlapping in starved specimens 0.5–3.0 × 0.3–1.5 cm, narrowly to broadly oblong, obtuse to shortly acute at apex, narrowing rather abruptly at base; all glabrous. *Inflorescence* a lax dichasial cyme with (1–)3–70 bisexual flowers, with erecto-patent branches often arising from near the base of the plant; pedicels

1–4 mm at anthesis, occasionally elongating to 10 mm in fruit; bracts 2.5–15 × 0.8–5.0 mm, lanceolate or narrowly elliptical. *Calyx* 8–9 mm, divided nearly to the base into 4 or 5 lobes, the lobes linear-subulate, tapering to a fine acute apex and sharply keeled dorsally. *Corolla* bright or pale pink, rarely white; tube 8–10 mm, slightly constricted below the limb; lobes 4 or 5, 2–4 × 2.0–2.5 mm, concave, oblong-ovate, obtuse or subacute at apex, usually minutely erose, sometimes irregularly emarginate. *Stamens* 4 or 5, inserted at the apex of the corolla tube; filaments 3–4 mm, filiform; anthers yellow, oblong. *Style* about 2.5 mm; stigmas oblong. *Capsule* 10–12 × 2.0–2.5 mm, pale brown; seeds 0.3–0.2 mm, brown, coarsely reticulate. *Flowers* 6–9. Pollinated by insects or selfed. $2n = 34, 36$.

(i) Var. **pulchellum**
Erythraea pulchella var. *palustris* Gaudin; *C. pulchellum* var. *palustris* (Gaudin) Druce; *Erythraea swartziana* Wittr.

Stems simple and flowers few.

(ii) Var. **intermedium** (Mérat) Gilmour
Chironia intermedia Mérat; *Erythraea pulchella* var. *subelongata* Wheldon & C. E. Salmon

Intermediate between var. *pulchellum* and var. *ramosissima* in stems, branching and number of flowers.

(iii) Var. **ramosissima** (Vill.) Gilmour
Erythraea ramosissima (Vill.) Pers.; *Erythraea pulchella* var. *ramosissima* (Vill.) Gaudin; *Gentiana ramosissima* Vill.

Much branched from the base with numerous flowers.

Native. Great Britain north to Cumberland and Yorkshire. Very variable. Europe to China.

7. C. tenuiflorum (Hoffmanns. & Link) Fritsch
Slender Centaury
Erythraea tenuiflora Hoffmanns. & Link

Annual herb with fibrous roots. *Stems* (15–)20–40 cm, pale green, erect, glabrous, branched, leafy. *Leaves* medium green on upper surface, paler beneath; basal with lamina usually less than 10 × 8 mm, not obviously forming a rosette and often withering before anthesis, oblong-spathulate, obtuse or rounded at apex , entire, attenuate at base; cauline 0.2–2.0 × 0.3–0.8 cm, subequal, oblong, obtuse to acute at apex, often in numerous pairs along the straight, unbranched lower part of stem, often remote and shorter than the internodes, but not always so; all glabrous. *Inflorescence* a dense, many (60–180) flowered, repeatedly branched, fastigiate, corymbose cyme, usually branched but not always from near the apex of the stem; pedicles 1.0–2.5 mm; bracts 2–15 × 0.5–3.0 mm, lanceolate or almost linear, smooth or minutely papillose. *Calyx* 5–6 mm, usually more shorter than the corolla tube and more or less adherent to it, divided into 4 or 5 lobes, the lobes linear-subulate, tapering to a fine acumen and sharply keeled dorsally. *Corolla* usually white with us, but often pink elsewhere; tube 7–9 mm, distinctly constructed below the limb; lobes 5 , 3.0–4.5 × 1.0–1.5 mm, concave, oblong, obtuse at apex, often a little erose or minutely emarginate. *Stamens* 5, inserted at apex of the corolla tube; filaments 2–3 mm, filiform; anthers yellow, oblong. Style 1, 1.5–2.5 mm; stigmas oblong-spathulate. *Capsule* 7–10 × 1.5–2.0 mm, pale to dark brown; seeds 0.3 × 0.2–0.3 mm, brown , irregularly oblong or subglobose, coarsely reticulate. *Flowers* 7–9. $2n$ =40.

Native. Damp grassy places near the sea. Very rare in Dorsetshire, formerly in the Isle of Wight until 1953, and in the Channel Islands previous to 1840. Southern Europe and Mediterranean region. Atlantic coasts of western Europe and the coasts of the Mediterranean region.

4. Blackstonia Huds.

Annual monoecious *herbs. Cauline leaves* glaucous, perfoliate. *Inflorescence* a lax or crowded terminal dichasial cyme; bracts leafy. *Calyx* divided almost to the base into linear, flat lobes. *Corolla* yellow, lobes usually 8. *Stamens* usually 8, attached just below the sinuses of the corolla; anthers not becoming twisted. *Style* 1, with 4 arms, the stigmas decurrent along the arms. *Fruit* a capsule.

Contains 4 species in Europe and the Mediterranean region to south-west Asia.

Zeltner, L. (1970). Recherches de biosystématique sur les genres *Blackstonia* Huds. et *Centaurium* Hill (Gentianaceae). *Bull. Soc. Neuchâtel. Sci. Nat.* **93**: 1–164.

1. B. perfoliata (L.) Huds. Yellow-wort
Gentiana perfoliata L.; *Chlora perfoliata* (L.) L.

Annual monoecious *herb* with fibrous roots. *Stems* 15–45 cm, pale green, erect, terete, glabrous, usually unbranched, remotely leafy. *Leaves* often more or less glaucous, slightly paler beneath; basal crowded into an irregular, rather persistent rosette, the lamina 1–5 × 0.5–2.5 cm, ovate, obovate or oblong, obtuse to acute at apex, entire, narrowed at base; cauline 0.5–6.0 × 0.5–5.0 cm, remote, sharply acute or shortly cuspidate at apex, perfoliate; all glabrous with obscure veins. *Inflorescence* a lax or crowded terminal dichasial cyme; pedicels usually less than 20 mm, slender, elongating in fruit; bracts leafy, 2–10 × 2–8 mm, sessile or connate; flowers bisexual. *Calyx* 7–16 mm, divided almost to the base into 8 linear-filiform lobes, the lobes convex or obscurely keeled dorsally, and glabrous. *Corolla* 2.5–8.0 mm long and 10–15 mm in diameter, bright golden yellow; tube 2.5–8.0 mm; lobes usually 8, 4–14 × 2.5–6.0 mm, narrowly obovate-elliptical, acute at apex, spreading at anthesis, afterwards erect and persisting until the fruits are ripe. *Stamens* usually 8; filaments 1.5–3.0 mm, attached just below the sinuses, linear-filiform; anthers yellow, linear. *Style* 1, filiform, twice bifid towards apex into 4 arms 1.0–1.5 mm, the stigmas decurrent along arms. *Capsule* 5–9 × 3.5–5.0 mm, shining brown, compressed-ellipsoid; seeds 0.3–0.5 mm, dark brown, oblong-ovoid or subglobose, conspicuously reticulate-foveolate, the surfaces within the pits minutely and regularly verruculose. *Flowers* 6–10. Self-pollinated. $2n = 40$.

Native. Calcareous grassland, bare chalk and dunes. Locally frequent in Great Britain and Ireland north to Co. Sligo, Westmorland and Northumberland. West, central and south Europe; south-west Asia; Morocco. Submediterranean-Subatlantic element. Our plant is subsp. **perfoliata** of west and central Europe.

5. **Gentianella** Moench

Annual or *biennial* monoecious *herbs*. *Leaves* opposite. *Inflorescence* of terminal and axillary flowers. *Calyx* more or less tubular, with 4 or 5 lobes, the lobes not joined by a membrane. *Corolla* blue or dark to whitish-purple, rarely pink or whitish; lobes 4 or 5 with long fringes at the margin or at base on inner face and 5- to 9-veined. *Stamens* 4 or 5, not twisted, versatile. *Stigmas* 2, persistent on the capsule. *Ovary* gradually tapering into the style or style absent. *Nectaries* on the corolla. *Fruit* a capsule.

About 125 species in the northern hemisphere, South America; south-east Australia, Tasmania and New Zealand.

Druce, G. C. (1893). *Gentian germanica* Willd. *Rep. Bot. Soc. Exch. Club Brit. Isles* **1**: 379.

Druce, G. C. (1896). The occurrence of a hybrid gentian in Britain. *Ann. Bot.* **10**: 621–622.

Gulliver, R. L. (1998). Population sizes of *Gentianella uliginosa* (Willd.) Boerner, Dune Gentian, on Colonsay (v.c. 102), in 1996. *Watsonia* **22**: 111–113.

Halliday, G. (1997). *A flora of Cumbria*. Lancaster.

Holyoak, D. T. (1999). *Gentianella uliginosa* (Willd.) Börner (Gentianaceae) rediscovered in north Devon. *Watsonia* **22**: 428–429.

Hultén, E. & Fries, M. (1986). *Atlas of north European vascular plants north of the Tropic of Cancer.* 3 vols. Königstein.

Knipe, P. R. (1988). *Gentianella ciliata* (L.) Borkh. in Buckinghamshire. *Watsonia* **17**: 94–95.

Lousley, J. E. (1950). The habitats and distribution of *Gentiana uliginosa* Willd. *Watsonia* **1**: 279–282.

Murbeck, S. (1892). Studien Gentianen aus der Gruppe Endotricha Froel. *Acta Horti Berg*. **2**(3).

Pritchard, N. M. (1959). *Gentianella* in Britain. I. *Gentianella amarella, G. anglica* and *G.uliginosa. Watsonia* **4**: 169–193.

Pritchard, N. M. (1960). *Gentianella* in Britain. II. *Gentianella septentrionalis* (Druce) E. F. Warb. Watsonia **4**: 218–237.

Pritchard, N. M. (1961). *Gentianella* in Britain. III. *Gentianella germanica* (Willd.) Börner. *Watsonia* **4**: 290–303.

Pugsley, H. W. (1924). *Gentiana uliginosa* Willd. in Britain. *Jour. Bot. (London)* **62**: 193–196.

Pugsley, H. W. (1936). *Gentiana amarella L.* in Britain. *Jour. Bot. (London)* **74**: 163–170.

Rich, T. C. G. (1996). *Gentianella uliginosa* (Willd.) Boerner (Gentianaceae) present in England. *Watsonia* **21**: 208–209.

Rich, T. C. G. (1997). Early gentian (*Gentianella anglica* (Pugsley) E. F. Warb. present in Wales. *Watsonia* **21**: 289–290.

Rich, T. C. G. & Jermy, A. C. (1998). *Plant crib.* London.

Rich, T. C. G., Holyoak, G. T., Margetts, L. J. & Murphy, R. J. (1997). Hybridisation between *Gentianella amarella* (L.) Boerner and *G. anglica* (Pugsley) E. F. Warb. (Gentianaceae). *Watsonia* **21**: 313–325.

Rose, F. (1998). *Gentianella uliginosa* (Willd.) Boerner (Gentianaceae) found in Colonsay (v.c. 102), new to Scotland. *Watsonia* **22**: 114–116

Stewart, A., Pearman, D. A. & Preston, C. D. (1994). *Scarce plants in Britain*. Peterborough. [*G. germanica; amarella* subsp. *amarella* var. *anglica.*]

Stratton, F. (1878). On an Isle of Wight Gentian. *Jour. Bot. (London)* **16**: 263–265.

Wettstein, R. von (1896). Die europäischen Arten der Gattung *Gentiana* aus der Section *Endotricha* Froel. *Pamphlets Acad. Vienna* **1896**: 309–382.

Wigginton, M. J. (Edit.) (1999). *British red data books*. Vol. 1. *Vascular plants*. Peterborough. [*G. amarella* subsp. *amarella* var. *uliginosa; ciliata.*]

1. Corolla lobe with long narrow fringes along sides, not at base on inner side **1. ciliata**
1. Corolla lobes with long narrow fringes at base on inner side, but not along the sides 2.
2. Calyx with 4 lobes, the 2 outer several times wider than the 2 inner and overlapping and enclosing them **2. campestris**
2. Calyx with 4 or 5 lobes (often on the same plant), the widest less than twice as wide as the rest 3.
3. Plant with 9–15 internodes; corolla (15–)25–35 mm, more than twice as long as calyx **3. germanica**
3. Plant with 2–11 internodes; corolla 12–22 mm, less than twice as long as calyx 4.
4. Corolla creamy-white within, purplish-red on the outside 5.
4. Corolla bluish-purple inside and out, rarely pale blue, pink or whitish 6.
5. Internodes (4–)6–7; middle and upper cauline leaves ovate to ovate-lanceolate, more or less widened at base; calyx lobes markedly unequal **4(c,v). amarella** subsp. **septentrionalis** var. **calycina**
5. Internodes 2–5(–6); middle and upper cauline leaves ovate-lanceolate to linear-lanceolate, not markedly widened at base; calyx lobes more or less equal **4(c,vi). amarella** subsp. **septentrionalis** var. **druceana**
6. Flowers May and June 7.
6. Flowers late July to September 8.
7. Internodes 2–3(–4); corolla 13–16 mm **4(a,i). amarella** subsp. **amarella** var. **praecox**
7. Internodes 3–5; corolla (15–)17–20 mm **4(a,ii). amarella** subsp. **amarella** var. **cornubiensis**
8. Internodes 0–2(–3), the terminal one and the terminal pedicel together usually forming from half to seven-eighths of the plant **4(a,iv). amarella** subsp. **amarella** var. **uliginosa**
8. Internodes 4–11, the terminal one and the terminal pedicel together usually forming less than half the height of the plant 9.
9. Internodes 4–9(–10); corolla (14–)16–18 mm **4(a,iii). amarella** subsp. **amarella** var. **amarella**
9. Internodes 7–11; corolla (17–)19–22 mm **4(b). amarella** subsp. **hibernica**

Section **1. Crossopetalae** (Froel.) N. M. Pritch.
Crossopetalae Froel.

Calyx lobes 4, equal, not overlapping. *Corolla* 25–50 mm, obconical, 4-lobed, the lobes ovate, obtuse at apex, fringed on margin. *Style* distinct. *Capsule* stipitate.

1. G. ciliata (L.) Borkh. Fringed Gentian
Gentiana ciliata L.

Biennial monoecious *herb. Stems* 5–30 cm, pale green, erect, glabrous; internodes usually 4–6, more or less equal and about equalling the terminal pedicel. *Lower leaves* spathulate, obtuse at apex, entire, glabrous; cauline 10–30 mm, lanceolate or linear-lanceolate, acute at apex, entire, glabrous. *Calyx* 12–20 mm; with 4 equal, non-overlapping lobes, about half as long as the corolla. *Corolla* 25–50 mm, blue, obconical, fringed at the throat; lobes 4, the lobes 10–18 mm, ovate to oblanceolate, obtuse-apiculate at apex and with long, narrow fringes along the sides.

Stamens 4; filaments pale; anthers versatile. *Style* 1, distinct; stigmas 2. *Capsule* stipitate. *Flowers* 8–10. $2n=44$.

Native or introduced. First reported as *G. ciliata* from Wendover in Buckinghamshire in 1875. These plants were subsequently dismissed as *Campanula glomerata* until the species was rediscovered, in perhaps the same field in 1982. This, the only known extant population, is variable in size, being largest in number during the late 1980s, but declining to almost none. Specimens in herbaria have been found from Wiltshire in 1892 and Surrey in 1910. Eurosiberian Temperate element. Our plant is subsp. **ciliata**.

Section **2. Gentianella**

Calyx lobes 4 or 5, equal or unequal. *Corolla* usually more than 15 mm, obconical or cylindrical, fringed in the throat, 4- or 5-lobed, the lobes erect or spreading. *Style* absent. *Capsule* stipitate or sessile.

Plants in this section vary in leaf shape, number of internodes, density of branching and ratio of leaf length to internode length. These characters are more or less correlated with flowering time so that plants flowering before the middle of August have few branches, (0–)2–6 internodes which are longer than the leaves and usually obtuse middle cauline leaves, while those flowering after the middle of August have many branches, 6–12(–15) internodes which are shorter than the leaves and usually more or less acute middle cauline leaves. Plants are also variable in duration. Most are overwintering annuals, germinating in late summer and autumn to form rosettes of narrow, acute leaves which mostly wither during winter leaving only a small bud which in spring forms a secondary rosette. Early flowering populations tend to grade into late flowering ones and produce a mixture of characteristics which suggest that there are hybrid swarms which because of the great variation can be exaggerated by statistical analysis. However, breeding appears to be mostly by self-pollination and most of the variation could be explained by inbreeding lines.

2. G. campestris (L.) Börner Field Gentian
Gentiana campestris L.; *G. baltica* auct.; *Gentiana baltica* auct.

Usually *biennial* monoecious *herb*. *Stems* (5–)7–35(–50) cm, pale green, sometimes flushed purple, simple or branched, the branches erect; internodes 2–12(–15). *Basal leaves* 1.0–2.5 cm, medium green, ovate, lanceolate or spathulate, obtuse or subacute at apex; cauline 2–3 cm, lingulate to oblong or lanceolate, obtuse to acute at apex, entire; all glabrous. *Inflorescence* racemose or subcorymbose; flowers bisexual. *Calyx* divided nearly to base into 4 lobes, the 2 outer lobes several times wider than the 2 inner and overlapping, the outer ovate to ovate-lanceolate and acute or acuminate at apex, the inner lanceolate or 1 or both absent, with a flat, papillose to ciliate margin. *Corolla* (12–)15–30 mm, bluish-lilac or white; tube up to twice as long as calyx, more or less cylindrical, fringed in the throat; with 4, oblong-ovate lobes 6–11 mm. *Stamens* 4; anthers not twisted and versatile. *Stigmas* 2, persistent on the capsule. *Capsule* sessile or stipitate. *Flowers* 7–10. $2n=36$.

Native. Pastures and dunes, usually acid or neutral, ascending to 790 m. Locally common in the north, absent from most of southern Ireland and south and central Great Britain. North and central Europe eastwards to north-west Russia and central Austria, and southwards to east Spain and central Italy. European Boreo-temperate element. Our plant is subsp. **campestris**.

3. G. germanica (Willd.) Börner Chiltern Gentian
Gentiana germanica Willd.

Annual or *biennial* monoecious *herb*, producing in the first year a rosette of leaves which die in the autumn. *Stems* (5–)7–35(–50) cm, pale green, sometimes suffused purple, erect, usually branched about the middle sometimes simple; internodes 9–15, all equal or the uppermost shorter. *Basal leaves* in a rosette; lamina spathulate, obtuse at apex, entire, usually dead at anthesis; cauline with lamina 1.0–2.5 cm, ovate or ovate-lanceolate, more or less acute at apex, entire, subcordate at base and tapering from a wide base to the apex; all glabrous. *Inflorescence* consists of terminal and axillary bisexual flowers sometimes in clusters. *Calyx* 10–16 mm, 5-lobed, the lobes more or less equal and more or less spreading. *Corolla* 15–35 mm, bright bluish-purple, at least twice as long as the calyx tube, the tube obconical, fringed in the throat the lobes narrowly triangular-ovate. *Stamens* 5; anthers not twisted and versatile. *Stigmas* 2, persistent on the capsule. *Capsule* about 20 mm, oblong, sessile; seed about 1 mm. *Flowers* (8–)9–10. Visited by bumble-bees. $2n=36$.

Native. Chalk grassland, frequently new tracks and chalkpits and sometimes in open scrub or on woodland margins. Very local in south-central England from Hampshire to Hertfordshire and Bedfordshire; formerly more widespread. Belgium and north-east France to the southern Alps and east Carpathians. European Temperate element.

4. G. amarella (L.) Börner Autumn Gentian
Gentiana amarella L.

Annual or *biennial* monoecious *herbs* with fibrous roots. *Stems* 3–50, pale green or flushed purple, erect, simple or branched from the base, glabrous, leafy. *Leaves* opposite; lamina 10–20(–30) × 7–12 mm, medium yellowish-green on upper surface, often flushed purple, paler beneath, ovate to linear, more or less obtuse to acute at apex, narrowed or rounded below, glabrous; internodes 0–9(–10). *Inflorescence* consisting of solitary, bisexual flowers and branched clusters of flowers both axillary and terminal. *Calyx* 12–15 mm, 4- or 5-lobed, the lobes unequal to subequal and more or less appressed to the corolla, linear-lanceolate, long-acute, glabrous. *Corolla* 10–22 mm, dull reddish-purple, sometimes whitish inside, cylindrical, fringed in the throat, divided into 4 or 5 lobes, the lobes broadly ovate and rounded at apex. *Stamens* 4 or 5; anthers not twisted, versatile. *Stigmas* 2, persistent on the capsule. *Capsule* about 20 mm, oblong, sessile or rarely shortly stipitate; seed about 1 mm, subglobose. *Flowers* (3–)4–9(–10). Pollinated by bumble-bees. $2n=36$.

This very variable species can be divided into a number of infraspecific taxa of which three are geographical and the rest ecological. These taxa appear to be inbreeding lines and it is probable that var. *amarella* could be even further divided. Hybrid swarms have been suggested, but supposed hybrids appear to be fully fertile and reproduce themselves.

(a) Subsp. **amarella**
Annual or *biennial. Stem* up to 50cm. *Internodes* 0–9(–10), variable. *Cauline leaves* lanceolate or ovate-lanceolate, more or less acute at apex. *Corolla* 10–23 mm, dull bluish-purple.

(i) Var. **praecox** (F. Towns.) P. D. Sell
Gentiana amarella var. *praecox* F. Towns.; *Gentiana lingulata* var. *praecox* (Towns.) Wettst.; *Gentiana anglica* Pugsley; *G. anglica* (Pugsley) E. F. Warb.

Plant 4–20cm. *Internodes* 2–3(–4); terminal internode about 1.5 times as long as average. Terminal pedicel forming about half the height of the plant, frequently even more. *Basal leaves* narrowly spathulate and obtuse at apex; middle and upper cauline lanceolate, acute at apex, the lower ones more or less linear, more or less obtuse at apex. *Corolla* 13–16mm, about 1.5 times as long as the calyx lobes which are markedly unequal. *Flowers* (4–)5–6(–7).

(ii) Var. **cornubiensis** (N. M. Pritch.) P. D. Sell
G. anglica subsp. *cornubiensis* N. M. Pritch.

Plant 4–15cm. *Internodes* 3–5; all internodes more or less equal in length. *Terminal pedicels* rather longer than the internodes, but less than one-third of the height of the plant. *Basal leaves* broadly spathulate or rosulate and obtuse at apex; cauline lanceolate, more or less obtuse at apex. *Corolla* (15–)17–20mm, about 1.5 times as long as the subequal calyx lobes. *Flowers* 4–5(–6).

(iii) Var. **amarella**
Gentiana amarella L.; *Gentiana axillaris* (F. W. Schmidt) Rchb.; *G. davidiana* T. C. G. Rich

Plant 3–50cm. *Internodes* 4–9(–10), more or less equal, or else the terminal one reduced to about 1.0mm. *Basal leaves* of 1st year lanceolate to lingulate, of 2nd year obovate or spathulate; cauline lanceolate or ovate-lanceolate, more or less acute at apex. *Corolla* (14–)16–18(–20)mm, dull blue-purple, about 1.2 times as long as the calyx which has subequal lobes. *Flowers* (7–)8–9.

(iv) Var. **uliginosa** (Willd.) P. D. Sell
Gentiana uliginosa Willd.; *G. uliginosa* (Willd.) Börner

Plant up to 15cm. *Internodes* 0–2(–3), the terminal one and the terminal pedicel very markedly elongated and together usually forming from half to seven-eights of the height of the plant; in the larger biennial plants usually with long flowering branches from the base which give the whole plant a characteristically pyramidal habit; the smaller annual plants usually consist of no more than a basal rosette of 4–8 leaves with 1 or 2 flowers arising directly from the rosette. *Basal leaves* lanceolate (annuals) or obovate or spathulate (biennials); cauline ovate or ovate-lanceolate, more or less acute at apex, more or less widened at base. *Corolla* about 10mm in annuals, up to 20(–23)mm , bluish-purple, about as long as or shorter than the calyx which has very unequal lobes. *Flowers* (7–)8–11.

(b) Subsp. **hibernica** N. M. Pritch.
Plant 10–50cm. *Internodes* 7–11. *Cauline leaves* linear-lanceolate, more or less acute at apex. *Calyx lobes* subequal. *Corolla* (17–)19–22mm, dull bluish-purple. *Flowers* (7–)8–9.

(c) Subsp. **septentrionalis** (Druce) N. M. Pritch.
Gentiana septentrionalis (Druce) Druce;
G. septentrionalis (Druce) E. F. Warb.

Plant 4–30cm. *Internodes* 2–7. *Middle and upper cauline leaves* ovate-lanceolate to linear-lanceolate. *Calyx* lobes unequal or subequal. *Corolla* (12–)14–16mm, creamy-white within, purplish-red on outside, the lobes more or less erect. *Flowers* 7–9.

(v) Var. **calycina** (Druce) P. D. Sell
Gentiana amarella var. *calycina* Druce; *Gentiana amarella* forma *multicaulis* Lange ex Beeby

Internodes (4–)6–7. *Middle and upper cauline leaves* ovate to ovate-lanceolate, more or less widened at base. *Calyx* lobes markedly unequal. *Flowers* 7–9.

(vi) Var. **druceana** (N. M. Pritch.) P. D. Sell
G. amarella subsp. *druceana* N. M. Pritch.

Internodes 2–5(–6). *Middle and upper cauline leaves* ovate-lanceolate to linear-lanceolate, not markedly widened at base. *Calyx* lobes usually more or less equal. *Flowers* 7–9.

Native. Well-drained basic soils, typically grazed chalk and limestone grassland, calcareous dunes and machair, spoil tips and quarries. Throughout much of Great Britain and Ireland in the lowlands, but up to 750m in Westmorland. North and central Europe eastwards to the eastern Ukraine, the Caucasus and the Yenisei region. Subsp. *amarella* is mostly in England and Wales, Subsp. *hibernica* is endemic to Ireland and subsp. *septentrionalis* is endemic in northern Great Britain. Var. *praecox* is the early flowering plant of southern England, var. *cornubiensis* is the early-flowering plant of Cornwall and var. *amarella* is the late-flowering plant of England. Var. *uliginosa* occurs on coastal dunes, dune slacks and machair at Braunton Burrows in Devonshire and in South Wales. Molecular research supports the recognition of these varieties, but there must be some reason why they retain their characters and the early-flowering ones remain early-flowering when grown from seed in cultivation. Subsp. *septentrionalis* can be divided into two varieties; var. *calycina* occupies the western and northern part of the range and var. *druceana* the eastern southern part. The *G. amarella* aggregate in Europe is even more difficult.

x **germanicus** = **G.** × **pamplinii** (Druce) E. F. Warb.
Gentiana × *pamplinii* Druce

Even in this hybrid where plants tend to have either the large flowers of *G. germanicus* and the habit of *G. amarella* or vice versa, it may be that they are only variants of one or other of the parents. They appear to produce plants of good seed, but we have found no report of uniform or variable offspring.

6. Gentiana L.

Annual to *perennial* monoecious *herbs*. *Leaves* opposite. *Inflorescence* of terminal and axillary bisexual flowers, sometimes forming a densely flowered plant. *Calyx* more or less tubular with 4 or 5 lobes, the lobes joined by a membrane which forms the upper part of the tube. *Corolla* usually blue, with 5 large, 3-veined inner lobes and 5 small ones between them. *Stamens* 5; anthers neither twisted nor versatile. *Stigmas* 2, persistent on the capsule. *Ovary* gradually tapering into the style or the style absent. *Nectaries* at the base of the ovary. *Fruit* a capsule.

Contains about 400 species, mainly in the mountains of the northern hemisphere, with a few in the Andes.

Chapman, S. B., Rose, R. J. & Clarke, R. T. (1989). The behaviour of populations of the marsh gentian (*Gentiana pneumonanthe*): a modelling approach. *Jour. Appl. Ecol.* **26**: 1059–1072.
Elkington, T. T. (1963). *Gentiana verna* L. *Jour. Ecol.* **51**: 755–767.
Halda, J. J. (1996). *The Genus* Gentiana. Dobré.
Halliday, G. (1997). *A flora of Cumbria*. Lancaster.
Hultén, E. & Fries, M. (1986). *Atlas of north European vascular plants north of the Tropic of Cancer.* 3 vols. Königstein.
Simmonds, N. W. (1946). Biological flora of the British Isles. No. 16. *Gentiana pneumonanthe* L. *Jour. Ecol.* **33**: 295–307.
Stewart, A., Pearman, D. A. & Preston, C. D. (1994). *Scarce plants in Britain*. Peterborough. [*G. pneumonanthe*.]
Webb, D. A. & Scannell, M. J. P. (1983). *Flora of Connemara and the Burren*. Cambridge. [*G. verna*.]

1. Leaves all or most more than 15(–20) mm; corolla tube more than 10 mm wide, widening distally 2.
1. Leaves all less than 15 mm; corolla tube less than 8 mm wide and more or less cylindrical 4.
2. Leaves crowded in a basal rosette, few reduced ones up stem **3. clusii**
2. Leaves spread more or less evenly up the stem 3.
3. Leaves ovate to lanceolate, more than 10 mm wide; with 3–5 veins **1. asclepiadea**
3. Leaves linear to linear-lanceolate, less than 10(–15) mm wide; with 1 vein **2. pneumonanthe**
4. Rhizomatous perennial with several rosettes of leaves; corolla lobes more than 8 mm **4. verna**
4. Annual, with or without 1 basal leaf rosette; corolla lobes less than 6 mm **5. nivalis**

Section **1. Pneumonanthe** (Gled.) Link
Pneumonanthe Gled.

Leaves 1–13 cm, those at the base of the stem scale-like. *Stems* with 1–several flowers. *Calyx* green, 5-lobed. *Corolla* obconical, plicate, 5-lobed.

1. G. asclepiadea L. Willow Gentian
G. schistocalyx K. Koch; *G. asclepiadea* subsp. *schistocalyx* (K. Koch) I. P. Zakh.; *G. asclepiadea* var. *schistocalyx* (K. Koch) Grossh.; *G. asclepiadea* var. *macrocalyx* Sommier & Lév.; *Gentianusa asclepiadea* (L.) Pohl

Perennial monoecious *herb*. *Stem* 30–100 cm, medium green, erect or ascending, glabrous, with dense leaves. *Leaves* opposite, all cauline; lamina 5–13 × 1.5–5.0 cm, medium green, ovate or lanceolate, acuminate at apex, entire, sessile, with 3–5 veins, scabrous. *Inflorescence* of 1–30 bisexual flowers, opposite, sessile and axillary. *Calyx* 8–30 mm, medium green, divided halfway to base into 5 lobes, the lobes 2.2–9.0 mm, unequal, linear, acute at apex. *Corolla* 30–50 × 12–16 mm, usually deep blue, rarely pale blue, pinkish or white, usually with purple spots inside, with 5 short appendages alternating with 5 large lobes, the lobes triangular. *Stamens* 5, alternating with the large corolla lobes; anthers neither twisted nor versatile. *Style* sometimes absent; stigmas 2, persistent on the capsule; nectaries at the base of the ovary. *Capsule* stipitate; seeds 1–2 mm, elliptical, widely winged. *Flowers* 7–9. $2n=32$, 36, 44.

Introduced. Grown in gardens and self-sowing freely. Naturalised by streams and in shady places. Sussex, Yorkshire and central Scotland. Native of the mountains of Europe, Turkey, Caucasus and Talish on mountain slopes, open woods and alpine meadows.

2. G. pneumonanthe L. Marsh Gentian
G. ascendens Schmidt; *G. linearifolia* Lam.; *Gentianusa pneumonanthe* (L.) Soják

Perennial monoecious *herb*. *Stems* numerous, up to 80 cm, simple or rarely branched, erect, glabrous, leaves not dense. *Leaves* opposite, all cauline; lamina 3–10 × 0.2–1.5 cm, dark green on upper surface, paler beneath, the lower scale-like, the upper linear to linear-lanceolate, obtuse at apex, entire, with 1 vein, more or less sheathing at the base and densely covering the stem. *Inflorescence* of 1–10(–28) bisexual flowers, opposite, sessile and axillary and terminal; pedicels up to 20 mm. *Calyx* 15–20 mm, medium green, campanulate, divided into 5 lobes, the lobes 5–10 mm, linear to linear-lanceolate, acute at apex. *Corolla* 35–55 mm, usually blue, rarely white or pink, with 5 green lines outside, inside greenish spotted, tube obconical with 5 lobes 4–6 mm, ovate or widely ovate and acute at apex with 5 small appendages in between. *Stamens* 5, alternating with the large corolla lobes; anthers not twisted nor versatile. *Style* sometimes absent; stigmas 2. *Capsule* oblong-lanceolate, stipitate; seeds about 1.5 mm, winged. *Flowers* 7–10. Pollinated by bumble-bees; protandrous. $2n=26$.

Native. Wet heathland and damp, acid grassland. Very local from Dorsetshire and Sussex to Yorkshire and Westmorland, decreasing. Europe, west Asia and Siberia in meadows and open woods. Eurosiberian-Temperate element.

Section **2. Megalanthe** Gaudin
Gentiana section *Thylacites* Griseb.

Leaves up to 12 cm, crowded at the base of the stem. *Flowers* solitary, terminal. *Calyx* herbaceous, with 5 lobes. *Corolla* obconical, plicate, 5-lobed, the appendage in the sinus small. *Anthers* connate. *Stigma* 2-lobed. *Seeds* not winged.

3. G. clusii E. P. Perrier & Songeon Trumpet Gentian
Ciminalis clusii (E. P. Perrier & Songeon) Holub

Perennial monoecious *herb* forming clumps up to 15 cm across. *Stems* 7–15 cm, medium green, erect, glabrous.

Leaves opposite; basal crowded in a rosette with lamina 6–12×0.6–1.5 cm, lustrous olive or medium green on upper surface, paler beneath, coriaceous, narrowly elliptical, acute at apex, entire, the margin scabrous; cauline small and scale-like. *Inflorescence* a solitary bisexual flower at the end of a stem. *Calyx* up to 25 mm, herbaceous, campanulate, 5-lobed, the lobes up to 18 mm, lanceolate and acute at apex. *Corolla* 40–70 mm, bright azure blue, whitish inside tube with green spots, obconical, plicate, 5-lobed, the lobes up to 15×15 mm, broadly triangular, acute-mucronate at apex, the appendages 2–3×4–6 mm, rounded or truncate and entire. *Stamens* 5, alternating with the corolla lobes; anthers not twisted, connate or versatile. *Style* sometimes absent; stigmas 2. *Capsule* about 30×4 mm, oblong; seeds about 1.0×0.5 mm, pale brownish-yellow, elliptical, minutely reticulate, not winged. *Flowers* 4–7. $2n=36$.

Introduced. Planted and naturalised since 1960 in chalk grassland in three localities in Surrey. Native of the Alps.

Section **3. Calathianae** Froel.

Leaves up to 3 cm, sometimes crowded towards the base of the stem. *Stems* usually with solitary terminal flowers. *Calyx* green, with 5 lobes. *Corolla* tube cylindrical, plicate, with 5 patent lobes, the appendage in the sinus small. *Anthers* free. *Stigma* lobes contiguous, forming a circular, slightly concave disc. *Seeds* not winged.

4. G. verna L. Spring Gentian
Hippion vernum Schmidt; *G. serrata* Lam.; *G. arctica* Grossh.; *G. angulosa* M. Bieb.; *Calathiana verna* (L.) Delarbre; *Hippion aestivum* Schumach.; *Gentianusa verna* (L.) Pohl

Perennial monoecious *herb* with few or many underground stems from a short stock, each ending in a rosette of persistent leaves. *Stem* up to 7 cm in flower, 12 cm in fruit, medium green, glabrous, erect. *Leaves* opposite; basal in a rosette, the lamina 0.7–1.5(–3.0)×0.4–0.8 cm, medium green, ovate, elliptic-lanceolate or elliptical, acute or acuminate at apex and entire; cauline 1–2 pairs, shorter and narrower than basal, shortly sheathing; all glabrous. *Inflorescence* of 1 bisexual flower at the top of each stem. *Calyx* 15–20 mm, tubular, 5-angled, green and more or less inflated, with wings 1(–2) mm wide, 5-lobed, the lobes 3–5 mm, narrowly lanceolate, acute at apex. *Corolla* 30–40 mm, blue, the tube about 20 mm and tubular-hypocrateriform, with 5 lobes 8–12×4–8 mm, ovate or rounded-ovate and obtuse at apex, 1.5–2.0 mm, triangular and bilobed. *Stamens* 5, alternating with inner corolla lobes; anthers free, not twisted nor versatile. *Style* sometimes absent; stigmas 2, white, contiguous. *Capsule* narrowly lanceolate or oblong, subsessile; seeds 0.5–0.7 mm, dark brown, elliptical, not winged. *Flowers* 4–6. Pollinated by Lepidoptera. $2n=28$.

The northern England populations have a shorter calyx and a shorter, wider rosette of leaves than the Irish ones.

Native. Grassland on limestone, calcareous glacial drift and fixed dunes. Extremely local in north England where it is centred on the Upper Tees and frequent in the Burren of western Ireland extending northwards to near Lough Carra and eastwards to Atheary. Europe, western Asia, Russia, north Siberia and central Asia. European Arctic montane element. Our plant is subsp. **verna** which occurs through much of the range of the species.

5. G. nivalis L. Alpine Gentian
G. humilis Rochel, non Stev.; *G. prostrata* Schur; *Calathiana nivalis* (L.) Delarbre; *Ericoila nivalis* (L.) Borkh.; *Gentianusa flava* Pohl; *Hippion nivale* (L.) Schmidt

Annual monoecious *herb. Stem* 3–16 cm, medium green, erect, simple or branched, glabrous. *Leaves* opposite; basal forming a rosette, the lamina 2–5 mm, medium green, elliptic-obovate and obtuse at apex; cauline 3–9×2–4 mm, and ovate; all glabrous. *Inflorescence* of solitary, terminal, bisexual flowers. *Calyx* herbaceous, the tube about 8 mm, more or less cylindrical, 5-angled, 5-lobed, the lobes triangular-lanceolate. *Corolla* 10–15 mm, deep blue; tube narrowly cylindrical, plicate, with 5 lobes, the lobes ovate and acute at apex, the appendage in the sinus small. *Stamens* 5, alternating with the inner corolla lobes; anthers free, not twisted nor versatile. *Styles* sometimes absent; stigmas 2, contiguous. *Capsule* 10–15 mm, cylindrical; seeds about 0.5 mm, dark brown, wingless. *Flowers* 6–8. Probably usually self-pollinated. $2n=14$.

Native. Rock ledges in mountains between 730 and 1005 m. Perthshire and Forfarshire. North Europe, mountains of central Europe and southern Europe; northern Turkey; Caucasus; Arctic North America; Greenland. European Arctic-montane element.

122. APOCYNACEAE Juss. nom. conserv.

Slightly woody, evergreen, monoecious *perennials. Leaves* opposite, simple, entire, shortly petiolate, without stipules. *Flowers* solitary in leaf axils, bisexual, actinomorphic, hypogynous. *Calyx* 5-lobed. *Corolla* 5-lobed, blue, rarely white. *Stamens* 5, inserted on corolla tube, not exserted. *Style* 1; stigma capitate-peltate. *Ovaries* 2, free, each with many ovules, united by the style. *Fruit* rarely produced, of 2 follicles.

Contains 2 well-marked subfamilies, about 200 genera and 2,000 species, mainly tropical and subtropical, few genera and species in temperate areas.

1. Vinca L.

Creeping, woody, evergreen, monoecious *perennials. Leaves* opposite. *Flowers* bisexual, solitary in the axils of leaves. *Calyx* 5-lobed. *Corolla* blue or white, with 5 broad, asymmetric lobes and an obconical tube, fluted and hairy within. *Stamens* 5, with short, sharply kneed filaments and introrse anthers ending in broadly triangular, hairy, connective flaps which meet over the stylar head. *Styles* united in a column, slender below, tapering upwards to an enlarged head with a plume of white hairs; stigmatic surface as a broad band round the stylar head. *Ovary* of 2 free carpels united only by their styles. *Fruit* of 2 follicles each with several long, narrow seeds.

Contains 5 species in Europe, North Africa and west Asia. There are 2 fleshy nectaries at the base of the gynoecium, alternating with the carpels, and the flowers are insect-pollinated.

Harvey J. (1981). *Medieval gardens*. London.
Stearn, W. T. (1932). Notes from the University Herbarium, Cambridge. *Jour. Bot. (London)* **70**: App. 2.
Stearn, W. T. (1972). *Vinca difformis* subsp. *sardoa* W. T. Stearn. *Bot. Jour. Linn. Soc.* **65**: 253–256.

1. Calyx lobes 3–5 mm, narrowly ovate to narrowly triangular, glabrous **1. minor**
1. Calyx lobes 5–17 mm, very narrowly triangular or almost linear 2.
2. Hairs of calyx 0.2 mm or less **2. difformis**
2. Hairs of calyx 0.5–1.0 mm 3.
3. Leaves lanceolate; corolla lobes narrow and acute at apex **3(ii). major** var. **oxyloba**
3. Leaves ovate; corolla lobes obovate and obliquely and asymmetrically truncate from each side 4.
4. Leaves green **3(i,1). major** var. **major** forma **major**
4. Leaves with broad pale yellow margins **3(i,2). major** var. **major** forma **variegata**

1. V. minor L. Lesser Periwinkle

Procumbent, evergreen, slightly woody, monoecious *perennial* spreading widely and rooting at most of the nodes. *Non-flowering stems* up to 1 m, green with a channel down the 2 sides, glabrous. *Flowering stems* much shorter and more or less erect, green, glabrous. *Leaves* opposite; lamina 1.5–4.5 × 0.5–1.5 cm, dark green on upper surface, paler beneath, coriaceous, narrowly elliptical, lanceolate or ovate, obtuse or pointed at apex, entire, cuneate or rounded at base, glabrous; petiole very short. *Flowers* 25–30 mm in diameter, bisexual, solitary in the leaf axils. *Calyx* divided over halfway into 5 lobes, glabrous; the lobes 3–5 mm, narrowly ovate to narrowly triangular, glabrous. *Corolla* purplish-blue or violet-blue, darker at base; 5-lobed, the lobes obovate and obliquely truncate at apex. *Stamens* 5; filaments pale, sharply kneed; anthers yellow. *Style* 1, pale; stigma cream. *Fruit* of 2 follicles. *Flowers* 3–5. Pollinated by long-tongued bees. $2n=46$.

Introduced. Grown in gardens in Great Britain by the year 995, now in woodlands, roadside banks and verges, waste places and rubbish tips. Occurring throughout most of Great Britain and Ireland. Apparently native of south, west and central Europe, central and south Russia and the Caucasus, but the limits of its natural range obscured by its spread in cultivation.

2. V. difformis Pourr. Intermediate Periwinkle

V. media Hoffmans. & Link

Evergreen, slightly woody overwintering, monoecious *perennial. Non-flowering stems* up to 200 cm, green, procumbent or ascending, glabrous. *Flowering stems* up to 30 cm; pedicels shorter than the subtending leaves. *Leaves* opposite; lamina 2.5–7.0 × 1.5–4.5 cm, dark green on upper surface, paler beneath, ovate to lanceolate, mostly rather narrowly lanceolate, pointed at apex, entire, rounded at base then attenuate to the petiole, glabrous or with a minutely ciliate margin; petiole very short. *Flowers* 30–70 mm in diameter, bisexual. *Calyx* 5-lobed, the lobes 5–14 mm, very narrowly triangular, glabrous or the margin with minute hairs up to 0.2 mm. *Corolla* pale blue or almost white, the tube 12–18 mm; 5-lobed, the lobes 5–14 mm, acute or obliquely truncate at apex. *Stamens* 5; filaments pale; anthers yellow. *Style* 1, pale; stigma cream. *Fruit* of 2 follicles. *Flowers* 3–5. $2n=$ about 46.

Introduced. Naturalised on a bank in Chevening Park in Kent and a white-flowered form recorded in Cornwall. The species is also recorded for Hertfordshire. Native of south-west Europe.

3. V. major L. Greater Periwinkle

Semi-procumbent, evergreen, slightly woody, monoecious *perennial* spreading widely and dying down in winter. *Non-flowering stems* up to 1.5 m, trailing or ascending, green, with a channel down 2 sides, glabrous. *Flowering stems* much shorter, pale yellowish-green, sometimes tinted brownish-purple, angled, with numerous, pale, medium simple eglandular hairs towards the tip. *Leaves* opposite; lamina 2–9 × 1–6 cm, dark dull to shining green on upper surface, paler beneath, sometimes with broad, pale yellow margins, ovate or rarely lanceolate, more or less acute at apex, entire, rounded or cordate at base and glabrous on both surfaces with short, pale, ascending hairs round the margin, the petiole up to 20 mm, with numerous, pale, medium simple eglandular hairs. *Flowers* 30–50 mm in diameter, bisexual, solitary in the leaf axils. *Calyx* 7–17 mm, pale green, tinged brownish-purple, divided almost to base; lobes 5, narrowly linear, obtuse at apex, lax but incurved at the top, with minute simple eglandular hairs. *Corolla* purplish-blue or violet-blue, paler towards and in the throat; tube 12–18 mm; lobes 5, 20–25 mm, obovate, obliquely and asymmetrically truncate from each side and spreading, or narrower and acute. *Stamens* 5; filaments white, sharply kneed; anthers cream or yellow. *Style* 1, cream or orange; stigma white or orange, surrounded by a plume of short, white hairs. *Fruit* of 2 follicles 25–50 mm, rarely seen in Great Britain and Ireland, divergent, each follicle with 1–4 seeds; seeds about 8 × 2.5 mm, blackish, oblong. *Flowers* 4–6. Pollinated by long-tongued bees. $2n=92$.

(i) Var. **major**

Leaves ovate. *Corolla* lobes purplish-blue, obovate and obliquely and asymmetrically truncate from each side.

(1) Forma **major**

Leaves all green.

(2) Forma **variegata** (Loudon) P. D. Sell Cv. Variegata

Leaves with broad, pale yellow margins.

(ii) Var. **oxyloba** (Boiss.) Stearn

V. oxyloba Boiss.; V. *major* subsp. *hirsuta* auct.

Leaves lanceolate. *Corolla* lobes violet-blue, narrow and acute at apex.

Introduced. Naturalised in hedgebanks, shrubberies and old railways and in rough ground and woodland. Scattered over Great Britain and Ireland, frequent in the south, rare in the north. Native of central and south Europe and North

Africa. The common plant is forma *major*. Forma *variegata* is common round amenity areas and often escapes. Var. *oxyloba* is of rare occurrence and is native to north Turkey and the Caucasus.

Order 2. **SOLANALES** Lindl.

Herbs, less frequently *trees*, *shrubs* or woody *climbers*, usually monoecious. *Leaves* often opposite, usually without stipules. *Flowers* usually bisexual, hypogynous or rarely epigynous, actinomorphic to zygomorphic. *Calyx* visually 4- to 5-lobed, sometimes 2-lipped. *Corolla* 4- to 5-lobed, often 2-lipped. *Stamens* as many or fewer than the corolla lobes, and alternate with them. *Style* 1. *Ovary* usually 1- to 2-celled; ovules numerous to 1 in each cell; placentation usually axile, parietal or basal. *Fruits* various; seeds with or without endosperm.

Consists of 8 families and about 5,000 species.

122A. NOLANACEAE Dumort. nom. conserv.

Woody-based *shrubs* or *herbs*. *Leaves* usually of 2 kinds, basal and cauline, simple, alternate or whorled; stipules absent. *Flowers* solitary or in clusters in the leaf axils, bisexual. *Calyx* tubular to campanulate to urceolate, 4–5-lobed. *Corolla* campanulate or funnel-shaped, 5-lobed, blue to purplish-blue, pink or white. *Stamens* 5, unequal, alternating with the corolla lobes; anthers opening by longitudinal slits. *Nectary disc* well developed, fleshy, often lobed. *Style* 1, terminal; stigma with 2–5 lobes. *Ovary* superior; carpels more or less united, each carpel with 1–several ovules, placentation axile. *Fruit* a schizocarp or cluster of nutlets; seeds 1–several.

Contains a single genus with 18 species.

1. Nolana L.

As for family.

Contains 18 species from desert, semi-desert and coastal areas of Chile and Peru and 1 species from the Galapagos Islands.

Johnston, I. M. (1936). A study of the Nolanaceae. *Contr. Gray Herb.* **112**.
Mesa, A. (1981). *Flora Neotropica*. **26**.

1. Stems 15–25 cm; calyx up to 20 mm; corolla up to 3.5×5.0 cm, bright blue, rarely white with a yellow or white throat **1(a). paradoxa** subsp. **paradoxa**
1. Stems up to 10 cm; calyx up to 15 mm; corolla up to 3×4 cm, white, violet or blue with a yellow or white throat **1(b). paradoxa** subsp. **atriplicifolia**

1. N. paradoxa Lindl. Chilean Bellflower

Perennial herb. Stems up to 25 cm, usually decumbent, usually downy or glandular-hairy. *Basal leaves* in rosettes, up to 5.5×2.0 cm, succulent, ovate, elliptical or spathulate, obtuse at apex; cauline opposite, elliptical to linear, acuminate at apex, petiolate to nearly sessile. *Calyx* up to 20 mm, campanulate to obconical, 5-lobed, the lobes ovate-acuminate to lanceolate-acuminate. *Corolla* blue with a yellow or white throat, 5-lobed, opening only in full sun. *Fruit* of 12–27, large nutlets in unequal 2 or 3 series.

(a) Subsp. **paradoxa**

Stems 15–25 cm. Basal leaves up to 5.5 x 2.0 cm; cauline elliptic to narrowly linear. Calyx up to 20 mm. *Corolla* up to 3.5 x 5.0 cm, bright blue, rarely white.

(b) Subsp. **atriplicifola** (D. Don) Mesa

N. atriplicifolia D. Don; *N. grandiflora* G. Don; *N. rupicola* Gaudich.; *Alona baccata* Lindl.; *Alona longifolia* Lindl.; *Sorema atriplicifolia* (D. Don) Lindl.; *Sorema acuminata* Miers; *N. baccata* (Lindl.) Dunal; *N. acuminata* (Miers) Dunal; *N. longifolia* (Lindl.) Dunal

Stems up to 10 cm. *Cauline leaves* elliptic to linear. *Calyx* up to 15 mm. *Corolla* up to 3 x 4 cm, white, violet or blue with a yellow or white throat.

Introduced. Both subspecies occur as garden casuals. Native of Peru to Chile.

123. SOLANACEAE Juss. nom. conserv.

Herbs or *shrubs*. *Leaves* alternate, sometimes paired through adnation; without stipules. *Calyx* (3–)5(–6)-lobed, usually persistent. *Corolla* usually 5-lobed. *Stamens* inserted on the corolla tube, alternating with the corolla lobes. *Style* 1. *Ovary* superior, 2-locular, cells sometimes divided by false septa; ovules numerous, placentation axile. *Fruit* a capsule or berry; seeds with endosperm; embryo straight or bent.

Contains about 90 genera and more than 2,000 species, widely distributed but chiefly tropical.

D'Arcy, W. G. (1986). *Solanaceae: biology and systematics*. New York.
Hawkes, J. G., Lester, R. N. & Skelding, A. D. (1979). *The biology and taxonomy of the Solanaceae*. London.

1. Anthers opening by apical pores **12. Solanum**
1. Anthers opening by longitudinal slits 2.
2. Leaves pinnate **11. Lycopersicon**
2. Leaves simple 3.
3. Corolla urceolate **8. Salpichroa**
3. Corolla not urceolate 4.
4. Corolla more than 13 cm **13. Datura**
4. Corolla less than 13 cm 5.
5. Fruit a capsule 6.
5. Fruit a berry 12.
6. Stamens unequal in length 7.
6. Stamens equal in length 8.
7. All leaves alternate; calyx lobes mostly less than two-thirds of the total length of calyx; flowers opposed to or in axils of small bracts **14. Nicotiana**
7. Upper leaves appearing opposite; calyx teeth more than three-quarters of the total length of calyx; flowers solitary at leaf nodes **15. Petunia**
8. Calyx detaching near base of capsule by a line around its circumference often reflexed; capsule usually spiny or tuberculate **13. Datura**

8. Calyx persistent; capsule smooth 9.
9. Capsule opening by flaps 10.
9. Capsule opening by a lid, splitting along a line around its circumference 11.
10. Plants completely glabrous, not glaucous; seeds around 3 mm **3. Vestia**
10. Plants glandular-hairy or glaucous; seeds 0.5–1.0 mm **14. Nicotiana**
11. Flowers solitary or 2 or 3 together **6. Scopolia**
11. Flowers more than 3 in panicles **7. Hyoscyamus**
12. Calyx hardly expanding in fruit or if expanding not enveloping the berry **10. Capsicum**
12. Calyx expanding in fruit, often enveloping or partly enveloping berry 13.
13. Plant a woody shrub or small tree 14.
13. Plant a herb or shrubby herb, but not woody 15.
14. Spiny shrub **2. Lycium**
14. Shrub without spines **4. Iochroma**
15. Calyx not papery and inflated **5. Atropa**
15. Calyx papery and inflated, enclosing fruit 16.
16. Calyx lobes separate; ovary 3- to 5-celled **1. Nicandra**
16. Calyx lobes fused along margins; ovary 2-celled **9. Physalis**

1. Nicandra Adans.

Glabrous, monoecious *annual herbs. Stems* much branched, fleshy. *Leaves* large, simple, toothed or more or less shallowly lobed. *Flowers* solitary, bisexual, axillary. *Calyx* deeply 5-lobed, fused along the margins, papery, later enlarging and enclosing the fruit, lobe keeled on the back. *Corolla* mauve and white, campanulate, shallowly lobed. *Stamens* 5, anthers separated and not forming a cone-shaped group around the style. *Ovary* 3- to 5-celled. *Style* 1. *Fruit* a rather dry berry.

Contains 1 species, native of Peru, but widely naturalised.

Bitter, G. (1903). Die Rassen der *Nicandra physaloides. Beih. Bot. Centralbl.* **14**: 145–176.
Darlington, C. D. & Janaki-Ammal, E. K. (1945). Adaptive isochromosomes in *Nicandra. Ann. Bot.* **9**: 267–281.
Gill, L. S. (1971). Chromosome numbers in certain West-Himalayan bicarpellate species. *Bull. Torrey Bot. Club* **98**: 281.
Horton, P. (1979). Taxonomic account of *Nicandra* in Australia. *Jour. Adelaide Bot. Gard.* **1**: 351–356.
Sharma, A. & Sarkar, A. E. (Edits.) (1967–68). Chromosome number reports of plants in Annual Report Cytogenetics Laboratory, Depart of Botany, University Calcutta. *The Research Bulletin* **2**: 38–48.
Sinha, N. P. (1951). The somatic chromosomes of *Nicandra. Jour. Indian Bot. Soc.* **30**: 92–94.

1. N. physalodes (L.) Gaertn. Apple-of-Peru
Atropa physalodes L.; *Physalis peruviana* Mill.; *N. violacea* André ex Lemoine; *Physalis daturaefolia* Lam.; *Calydermos erosus* Ruiz & Pavon

Annual monoecious shrubby *herb* with fibrous roots. *Stems* up to 150 cm, pale green, sometimes flushed brownish-purple, robust, fleshy, angled and longitudinally ridged, with many long branches from base upwards, glabrous, leafy. *Leaves* subopposite; lamina 5–20(–31) × 5–20 cm, dull medium yellowish-green on upper surface, paler beneath, ovate, obtuse to acute at apex, irregularly dentate to incise-dentate or shallowly lobed, the teeth broadly mammiform, rounded to a cuneate base, decurrent on the petiole, glabrous or with few eglandular hairs; veins pinnate, paler on upper surface and impressed, main ones very prominent beneath; petiole up to 10 cm, pale green, narrowly winged on upper surface, rounded beneath, glabrous, becoming shorter and more decurrent on the upper ones. *Flowers* 20–70 mm in diameter, bisexual, solitary in the axils of leaves; peduncles 10–16 mm in flower, lengthening to 30 mm in fruit. *Calyx* 20–25 mm in flower, lengthening to 35 mm in fruit, yellowish-green, papery, divided two-thirds of the way to the base into 5 lobes, the lobes ovate, acute at apex with cuspidate tip, keeled on the back and forming a ridge where they are joined at the base, glabrous. *Corolla* 20–40 mm, whitish at base with 5 trifid blotches of mauve in the throat, pale mauve in the upper third, campanulate, undulate round the apex and very shallowly 5-lobed. *Stamens* 5; filaments 5–6 mm, white, white-hairy at base; anthers pale yellow. *Style* 1, shorter than anthers, white; stigma pale yellow, capitate. *Capsule* 12–20 mm in diameter, globular with flattened top and base, pale green, becoming brownish; seeds numerous, 1.5–2.0 mm, yellowish-green, reniform, surrounded by a narrow wing. *Flowers* 8–9. Self- or cross-pollinated. Visited by flies and bees. $2n = 20$.

(i) Var. **physaloides**
Plant mostly green. *Corolla* 20–40 mm in diameter, white or with a pale mauve band round the upper part.

(ii) Var. **violacea** (André ex Lemoine) P. D. Sell
N. violacea André ex Lemoine
Plant with much brownish- or blackish-purple in stems, branches, pedicels and calyx. *Corolla* 40–70 mm in diameter, with a deep mauve band round the top and 5 trifid blotches in the throat.

Introduced and grown in Britain by 1930. Frequent wool and bird-seed casual of waste and cultivated ground, and on tips, sometimes persistent. Scattered records in Great Britain and the Channel Islands and two in Ireland. Native of southern Peru, widely naturalised in both temperate regions. Both varieties occur.

Several authors have obtained the somatic chromosome number $2n = 20$. Darlington & Janaki-Ammal (1945) however, sometimes found $2n = 19$ by loss of a pair of isochromosomes. Sinha (1951) obtained counts of $2n = 21$ for a white-flowered variant and var. *violacea*. Gill (1971) found the haploid number to be 10 + 1B, and Sharma & Sarkar (1967–68) reported a count of $n = 11$. The above varieties seem worth recording as they seem to breed true, but detailed experimental work is badly needed.

2. Lycium L.

Spinous or rarely unarmed, monoecious *shrubs. Leaves* alternate or fascicled, entire, without stipules. *Calyx* campanulate, (3–)5-lobed, expanding in fruit, papery and inflated. *Corolla* 5-lobed, funnel-shaped. *Stamens* inserted at mouth of corolla tube, long-exserted; anthers short, opening by longitudinal slots. *Style* 1. *Fruit* a berry.

Contains about 90 species in temperate and subtropical regions.

Bean, W. J. (1973). *Trees and shrubs hardy in the British Isles*. Ed. 8. **2**. London.
Coats, A. M. (1963). *Garden shrubs and their histories*. London.
Feinbrun, N. & Stearn, W. T. (1964). Typification of *Lycium barbarum* L., L. *afrum* L., and L. *europaeum* L. *Israel Jour. Bot.* **12**: 114–123.
Hitchcock, C. L. (1932). A monographic study of the genus *Lycium* of the western hemisphere. *Ann. Missouri Bot. Gard.* **19**: 179–374.

1. Stems ascending; leaves not particularly dense, narrowly elliptic or lanceolate; corolla divided to about halfway **1. barbarum**
1. Stems characteristically arched over; leaves dense, ovate to lanceolate, gradually getting smaller up the stem; corolla divided to more than halfway **2. chinense**

1. L. barbarum L. Duke of Argyll's Teaplant
L. halimifolium Mill.; *L. vulgare* Dunal

Deciduous, scrambling, monoecious *shrub*. *Stems* up to 2.5 m, ascending or erect, pale brown to pale greyish, glabrous with few, slender spines. *Branches* slender and smooth; twigs smooth and shiny; young shoots smooth and glabrous. *Leaves* alternate or fascicled; lamina 2–10×0.6–3.0 cm, medium greyish-green on upper surface, paler beneath, narrowly elliptical or lanceolate, obtuse to acute at apex, entire, cuneate at base, glabrous. *Flowers* bisiexual, solitary or few in axillary clusters; pedicels long and slender. *Calyx* about 4 mm, campanulate, (3–)5-lobed, irregularly 2-lipped. *Corolla* 10–15 mm, purple, becoming brownish, infundibuliform, the tube narrowly cylindrical at base for 2.5–3.0 mm, 5-lobed, the lobes about 4 mm. *Stamens* 5, inserted at mouth of corolla tube, long-exserted; filaments with a dense tuft of hairs at base. *Style* 1; stigma 2-lobed. *Berry* 10–20 mm, red, ellipsoid. *Flowers* 6–9, $2n=24$.

Introduced before 1696 and known since 1848 in the wild. It is grown as a hedge and escapes on waste places, along roads and railways, in hedges and on walls. This we believe is the main inland form which occurs throughout England and scattered records elsewhere. It is native of China. The plants common name results from a muddle. Archibald Campbell, third Duke of Argyll, and a famous plant collector, was sent a true teaplant, *Camellia sinensis*, and a *Lycium* with their labels mixed and he unwittingly or as a joke continued to grow them under their wrong names.

2. L. chinense Mill. Chinese Teaplant
L. rhombifolium (Moench) Dippel

Deciduous, scrambling, monoecious *shrub*. *Stems* up to 2.5 m, ascending in a great arc its dense leaves often bending it over so that its tip nearly touches the ground, pale brown or greyish, glabrous. *Branches* and twigs slender and smooth, with few slender spines, young shoots glabrous. *Leaves* alternate, dense; lamina 1–14×0.5–6.0 cm, large at the bottom, gradually getting smaller on the stem until they are very small at the tip, lanceolate to ovate, obtuse to acute at apex, entire, rounded at base. *Flowers* bisexual, solitary or few in axillary clusters. *Calyx* about 3 mm, 5-lobed, irregularly 2-lipped. *Corolla* 10–15 mm, purple, becoming brownish, infundibuliform, the tube narrowly cylindrical at base for about 1.5 mm, 5-lobed, the lobes 5–8 mm. *Stamens* 5, inserted at mouth of corolla tube, long exserted; filaments bearded at base. *Style* 1; stigma 2-lobed. *Berry* 10–20 mm, red, ellipsoid. *Flowers* 6–9. $2n=24$.

Introduced. Long known in gardens and now widely naturalised, particularly round the coast, throughout Great Britain and Ireland. Native of eastern Asia.

3. Vestia Willd.

Glabrous, evergreen, monoecious *shrubs*. *Leaves* alternate, entire, attenuate at base. *Flowers* bisexual, 1–4 in upper nodes, pendulous. *Calyx* campanulate, 5-lobed, expanding in fruit. *Corolla* greenish-yellow to purplish-yellow, tubular or funnel-shaped, 5-lobed, spreading. *Stamens* 5, equal, inserted in the lower quarter of the corolla; anthers opening by longitudinal slits. *Style* 1, protruding. *Ovary* 2-celled, partly enveloped by the persistent corolla base. *Fruit* a capsule, opening by 4 flaps which become reflexed, the lower half enveloped by the tightly fitting expanded calyx; seeds numerous, about 3 mm.

Contains a single species from southern South America.

1. V. foetida (Ruiz & Pavon) Hoffmanns. Vestia
Periphragmos foetidus Ruiz & Pavon; *V. lycioides* Willd.; *Cantua ligustrifolia* Juss.; *Cantua foetida* (Ruiz & Pavon) Pers.

Glabrous, evergreen, monoecious *shrub* up to 3.5 m. *Stems* and branches erect. *Leaves* alternate; lamina 1.2–5.0×0.6–1.8 cm, bright glossy green, often tinged purple on upper surface and strong smelling, oblong to oblong-elliptical at apex, entire, attenuate at base, glabrous. *Flowers* bisexual, 1–4 in upper nodes, pendulous. *Calyx* campanulate, 5-lobed, the lobes short, expanding in fruit. *Corolla* 25–35 mm, greenish-yellow or purplish-yellow, tubular or funnel-shaped, 5-lobed, the lobes 6–9 mm, ovate to triangular and spreading, with short hairs on margin. *Stamens* 5, equal, inserted in the lower quarter of the corolla, hairy at the base and protruding; anthers about 2 mm, opening by longitudinal slits. *Style* 1, protruding. *Ovary* 2-celled, partly enveloped by the persistent corolla base. *Capsule* ovoid, opening by 4 flaps which become reflexed, the lower half enveloped by the tightly fitting expanded calyx; seeds 3.0×1.5 mm, numerous, wrinkled and wingless. *Flowers* 5–7.

Introduced. Grown in hedges in Cornwall where it may persist. Native of Chile.

4. Iochroma Benth.

Deciduous, woody, monoecious *shrubs*. *Stems* arise from centre of the bush, grow vertically until the bush is overtopped, then spread out horizontally. *Leaves* alternate, entire, with petiole. *Inflorescence* of 1–3 axillary, bisexual flowers. *Calyx* campanulate, 5-lobed, expanding in fruit. *Corolla* deep blue, purple or white with a pink tinge, 5-lobed. *Stamens* 5, inserted near the base of the corolla tube;

filaments flattened towards the base; anthers opening by longitudinal slits. *Style* 1, curved at first becoming straight; stigma bilobed. *Fruit* a berry, with a persistent style.

Contains 16 to 20 species and natural hybrids from Columbia, Ecuador, Galapagos, Peru, Bolivia and north-western Argentina.

Shaw, J. M. H. (1998). *Ichroma*: a review. *New Plantsman* **5**(1): 154–192.

1. I. australe Griseb. Argentine Pear

Acnistus australis (Griseb.) Griseb.; *I. grandiflorum* auct.

Deciduous, monoecious *shrub* up to 5 m. *Stems* arise from the centre of the bush, growing vertically until the bush is overtopped, then spreading out horizontally; in subsequent years branches gradually become weighed down and replaced by newer ones from above. *Bark* silvery-grey with numerous lenticels and fissures, rather corky. *Leaves* alternate; lamina 4–8 × 1.0–2.5 cm, dark green on upper surface, paler beneath, lanceolate or narrowly elliptical, at apex, entire, cuneate at base, sparsely hairy on upper surface and more numerous hairs beneath; veins impressed on upper surface, prominent beneath; petiole up to 20 mm. *Inflorescence* of 1–3 axillary flowers; pedicels up to 30 mm, finely hairy. *Calyx* 5–6 mm, campanulate, 5-lobed, the lobes about 1.0 mm, mucronate at apex, with prominent ribs. *Corolla* about 25 mm, deep blue, purple or white faintly pink-tinged, 5-lobed, with hairy to papillate margin, inner surface only pigmented in the upper half, glabrous except for basal third, outer surface velvety. *Stamens* 5, inserted near the base of the corolla tube; filaments flattened towards the base, elongating during flowering from 10–25 mm; anthers about 1 mm, pale yellow. *Style* 1, up to 30 mm, curved at first, becoming straight, expanded below the bilobed stigma. *Berry* about 12 mm yellowish-green, with persistent style base becoming yellowish when ripe; seeds numerous, 1.5 × 2.0 mm, flattened, surface with minute pits; sclerotic granules spherical or lobed, 1–2 mm across. *Flowers* 5–6. $2n = 24$.

Introduced. Grown for pharmaceutical purposes since at least 1975, now naturalised in waste places. Nottinghamshire. Native of Argentina.

5. Atropa L.

Tall, much-branched, monoecious *herbs*. *Leaves* alternate, entire. *Flowers* bisexual, axillary and solitary. *Calyx* 5-lobed, expanding in fruit, but not inflated. *Corolla* campanulate, slightly zygomorphic. *Stamens* inserted at base of corolla tube, alternating with corolla lobes; anthers open by longitudinal slits. *Style* 1; stigma peltate. *Ovary* 2-celled. *Fruit* a many-seeded, 2-celled berry subtended by the spreading calyx.

Contains 4 species in Europe, western Asia and North Africa.

Butcher, R. W. (1947). Biological flora of the British Isles. No. 18. *Atropa belladonna* L. *Jour. Ecol.* **34**: 345–353.

1. Leaves medium green; corolla purplish-brown **1(i). belladonna** var. **belladonna**
1. Leaves pale yellowish-green; corolla yellowish to greenish-yellow **1(ii). belladonna** var. **lutea**

1. A. belladonna L. Deadly Nightshade

Perennial monoecious *herb*. *Stems* numerous, up to 1.5 m, pale brown, erect, smooth, branched, leafy, the young shoots pale yellowish-green, zigzagged, with numerous, short, fine glandular hairs and a few longer simple eglandular hairs. *Leaves* alternate; lamina 7–20 × 4–12 cm, dull medium green with impressed veins on upper surface, pale green with prominent veins beneath, broadly ovate, abruptly narrowed to an acute or shortly acuminate apex, entire, slightly sinuate, cuneate at base, minutely hairy especially beneath; petiole up to 15 mm, pale green, with very short glandular hairs and an occasional simple eglandular hair; without stipules. *Inflorescence* of 1–2 bisexual flowers per axil spread along the branches; pedicels up to 15 mm, drooping, pale green, with dense, unequal glandular hairs and few, longer simple eglandular hairs; bracts leaf-like but smaller. *Calyx* 13–15 mm, pale green, campanulate, divided almost to base into 5 lobes, the lobes erect in flower, spreading in fruit, ovate or ovate-lanceolate, acute at apex, with dense, short glandular hairs and a few, longer simple eglandular ones. *Corolla* 20–25 mm, purplish-brown, yellow or greenish-yellow, paler below, shortly glandular-hairy, divided quarter of the way to base into 5 lobes, the lobes broadly triangular with a subacute apex. *Stamens* 5, inserted at base of the corolla tube; filaments white; anthers pale brown. *Style* 1, short; stigma peltate. *Berry* 12–20 mm in diameter, shining medium green turning shiny black. *Flowers* 6–8. $2n = 72$.

(i) Var. **belladonna**

Belladonna trichotoma Scop.; *Belladonna baccifera* Lam.; *A. lethalis* Salisb. nom. illegit.

Leaves medium green. *Corolla* purplish-brown.

(ii) Var. **lutea** Döll

A. acuminata Royle; *A. lutescens* Jacq. ex C. B. Clarke; *A. pallida* Bornm.; *A. belladonna* var. *flava* Páter

Leaves pale yellowish-green. *Corolla* yellow to greenish-yellow.

Native. Open woods, scrub and rough and cultivated ground. Locally frequent in central and southern England, scattered records elsewhere in Great Britain and in Ireland, but probably only native in central and southern Great Britain on the chalk and limestone. The usual plant is var. *belladonna* which occurs in west, central and southern Europe; west Asia; North Africa. Var. *lutea* comes from the Himalaya and is grown in gardens and is presumably the reason for plants being recorded with greenish corollas. *A. belladonna* is a powerful narcotic and is very poisonous. It contains the alkaloids atropine and hyoscyamine.

6. Scopolia Jacq.

Perennial, monoecious *herbs* with fleshy, horizontal rhizome. *Leaves* alternate, entire or dentate. *Flowers* bisexual solitary at nodes or branch forks or 2 or 3 together, pendulous; pedicels thread-like. *Calyx* campanulate, 5-lobed, enlarging and enclosing fruit. *Corolla* dark brownish-violet to reddish-violet, campanulate, 5-lobed. *Stamens* 5, equal, inserted at the base of the corolla tube, included, opening by longitudinal slits. *Style* 1; stigma clavate. *Fruit* a capsule, opening by a line round the circumference.

Contains 2 species from central and south-east Europe, Korea and Japan.

1. S. carniolica Jacq. Scopolia

Perennial monoecious *herb* with fleshy, horizontal rhizome. *Stems* 20–60 cm, erect, branching, with scale leaves at base, glabrous. *Leaves* alternate; lamina up to 20 × 8 cm, medium green on upper surface, paler beneath, elliptic to ovate or obovate, long-pointed at apex, entire or dentate, cuneate at base, glabrous, petiolate. *Flowers* bisexual, solitary at nodes or branch forks or 2 or 3 together, pendulous; pedicels thread-like, 20–40 mm. *Calyx* about 10 mm, campanulate, 5-lobed, the lobes unequal, triangular and acute at apex, enlarging and enclosing fruit. *Corolla* 15–25 mm, dark brownish-violet to reddish-violet outside and yellowish- to brownish-green within, radially symmetric, cylindrical to campanulate, the limb 5-lobed, the lobes short and inconspicuous. *Stamens* 5, equal, inserted at the base of the corolla tube, included, opening by longitudinal slits. *Style* 1; stigma clavate. *Capsule* about 10 mm, subglobose, opening by a line round the circumference. *Flowers* 5–6.

Introduced. A persistent garden escape at Wisley in Surrey and in a shrubbery at Brinton in Norfolk. Native of central Europe.

7. Hyoscyamus L.

Annual or biennial monoecious *herbs*. *Leaves* alternate, simple, toothed to more or less lobed. *Inflorescence* a scorpioid cyme. *Calyx* campanulate or urceolate, enlarging later and becoming swollen at base to accommodate the fruit; broadly 5-lobed. *Corolla* infundibuliform, deeply 5-lobed. *Stamens* 5, equal, anthers opening by slits. *Ovary* 2-celled. *Fruit* a capsule, dehiscing by a lid.

Contains about 20 species in Europe, Asia and North Africa.

1. Stem leaves petiolate; corolla without purple veins **1. albus**
1. Upper stem leaves sessile; corolla with purple veins **2. niger**

Hultén, E. & Fries, M. (1986). *Atlas of north European vascular plants north of the Tropic of Cancer*. 3 vols. Königstein.
Stewart, A., Pearman, D. A. & Preston, C. D. (1994). *Scarce plants in Britain*. Peterborough.

1. H. albus L. White Henbane

Annual or *biennial* monoecious *herb* with fibrous roots. *Stems* up to 1 m, pale green, erect, somewhat viscid with a mixed indumentum of long and short, spreading hairs, usually much-branched, leafy. *Leaves* alternate; lamina 5–10 × 3–8 cm, medium green on upper surface, paler beneath, broadly ovate, obtuse or shortly acute at apex, coarsely but rather bluntly dentate-lobulate, sometimes strongly undulate, broadly cuneate, truncate or subcordate at base, thinly hairy or glabrous; petiole 2–8 cm, stout, channelled, hairy. *Inflorescence* of lower bisexual flowers often solitary in the axils of the branches and the upper crowded into a scorpioid cyme; pedicels very short or almost wanting, accrescent to 10 mm in fruit; bracts up to 15 × 5–15 mm, narrowly oblong to oblanceolate and sessile. *Calyx* about 13 mm, campanulate, 8–10 mm wide at anthesis, accrescent to 20 mm or more and becoming urceolate and conspicuously ribbed in fruit; lobes 5, 1–3 mm, sharply deltoid and unequal, sometimes with small secondary lobes in the sinuses. *Corolla* pale greenish-white, broadly infundibuliform, strongly oblique, thinly hairy; tube 15–25 mm; lobes 5, adaxial up to 8 mm, the abaxial distinctly smaller. *Stamens* 5, inserted near the base of the corolla tube; filaments 10–20 mm, glabrous or hairy; anthers shortly exserted, yellow, narrowly oblong. *Ovary* sessile on a small, deeply lobed disc. *Style* 16–23 mm, glabrous or thinly hairy; stigma capitate. *Capsule* about 10 × 8 mm, ovoid; seeds about 1.5 mm, pale grey, deeply and closely foveolate with sinuous ridges. *Flowers* 6–8.

Introduced. A rare casual. Native of the Mediterranean region.

2. H. niger L. Henbane

Annual to *biennial* monoecious *herb* with a strong narcotic odour and a thick, white, fleshy, fusiform root. *Stems* up to 80 cm, pale green, rather woody at base, erect, with viscid glandular hairs, much branched, leafy. *Leaves* alternate; lamina 15–20 × 5–15 cm, dull medium greyish-green on upper surface, paler beneath, the lower oblong-ovate or triangular-ovate, acute at apex, sinuate-dentate or pinnatifid, with sharp, spreading, lobe-like teeth, soft and pliant and strongly veined, glabrous, the lower shortly petiolate, the upper sessile and semiamplexicaul. *Inflorescence* of subsessile, bisexual flowers in 2 rows forming a scorpioid cyme; bracts leaf-like. *Calyx* 10–15 mm, urceolate, strongly ribbed, 5-lobed, the lobes broadly triangular, erect, with a sharp, hard point. *Corolla* 20–30 mm in diameter, pale yellow, usually veined with purple, infundibuliform, 5-lobed, the lobes rounded and unequal. *Stamens* 5, inserted near the base of the corolla tube; filaments pale; anthers violet or purple. *Style* 1, smooth and purplish; stigma roundish, flat and hairy. *Capsule* about 10 mm, broadly ovoid, enclosed in the calyx and constricted in the middle; seeds numerous, greyish, reniform, compressed, covered with angular reticulations. *Flowers* 6–8. $2n = 34$.

Continuously recorded from the Bronze Age onwards. Maritime sand and shingle, rough and waste ground inland, especially that manured by rabbits or cattle. Numerous records in Great Britain mainly central and the south, and around the coast of Ireland. Almost all Europe; west Asia; North Africa. Poisonous and narcotic, containing the alkaloids hyoscyamine and scopolamine.

8. Salpichroa Miers

Hairy, *perennial,* monoecious *herbs,* woody at base. *Leaves* usually opposite, occasionally alternate or in whorls of 3, simple and entire. *Inflorescence* an axillary corymb or flowers solitary. *Calyx* urceolate, 5-lobed nearly to base, not enlarging. *Corolla* white, with 5 short lobes. *Stamens* 5; anthers opening by longitudinal slits. *Style* 1. *Ovary* 2-celled. *Fruit* an ovoid-oblong berry.

Contains 17 species in South America.

1. S. origanifolia (Lam.) Thell. Cock's-eggs
Physalis origanifolia Lam.; *S. rhomboidea* (Gillies & Hook.) Miers

Perennial monoecious *herb,* woody at base. *Stems* up to 1.5 m, pale green, angled, sprawling, hairy. *Leaves* usually opposite, occasionally alternate or in whorls of 3; lamina 0.5–5.0 × 0.5–3.5 cm, medium green on upper surface, paler beneath, ovate-rhombic, obovate or subrotund, rounded to obtuse at apex, entire, attenuate at base, with appressed hairs on both surfaces; petiole 5–20(–35) mm. *Inflorescence* of axillary corymbs or solitary bisexual flowers; pedicels slender and hairy; flowers pendulous. *Calyx* urceolate, deeply 5-lobed, the lobes narrowly triangular, acute at apex, hairy. *Corolla* 6–10 mm, urceolate, white, 5-lobed, the lobes acute at apex, ultimately reflexed. *Stamens* 5, equal; anthers opening by longitudinal slits. *Style* 1. *Berry* ovoid-oblong; seeds numerous, compressed. *Flowers* 6–8. $2n = 24$.

Introduced. Grown for ornament and naturalised in rough ground and open places on the south and south-east coasts of England since 1927 and Guernsey since 1946, a rare casual elsewhere. Native of South America.

9. Physalis L.

Annual to *perennial* monoecious *herbs. Leaves* alternate, opposite or in whorls of 3, simple. *Flowers* bisexual, solitary and axillary. *Calyx* campanulate, 5-lobed, accrescent. *Corolla* 5-lobed, rotate or broadly campanulate. *Stamens* 5, exserted, inserted near the top of the short corolla tube. *Stigma* capitate. *Fruit* a globose berry, surrounded and usually much exceeded by the inflated calyx; seeds suborbicular to reniform.

Contains about 100 species of wide distribution, particularly diverse in Mexico.

Rydberg, P. A. (1896). The North American species of *Physalis* and related genera. *Mem. Torrey Bot. Club* **4**: 279–374.
Waterfall, U. T. (1958). A taxonomic study of the genus *Physalis* in North America north of Mexico. *Rhodora* **60**: 107–114; 128–142; 152–173.
Waterfall, U. T. (1967). *Physalis* in Mexico, Central America and the West Indies. *Rhodora* **69**: 82–120; 202–239; 319–329.

1. Corolla whitish sometimes with a yellow centre 2.
1. Corolla yellow with brownish markings 3.
2. Corolla whitish with a yellow centre; berry purple when ripe **1. acutifolia**
2. Corolla whitish; berry orange to red **2. alkekengi**
3. Berry yellow or orange 4.
3. Berry green or purple 5.
4. Calyx glabrous and smooth **3. angulata**
4. Calyx densely hairy **5. peruviana**
5. Calyx 3.5–4.5 mm, 15–20 mm in fruit, the lobes deltoid **4. ixocarpa**
5. Calyx 4–10 mm, 30–50 mm in fruit, the lobes ovate **6. philadelphica**

1. P. acutifolia (Miers) Sandwith Wright's Ground Cherry
Saracha acutifolia Miers; *P. wrightii* A. Gray

Annual monoecious *herb* with fibrous roots. *Stems* 20–100 cm, pale green, erect or ascending, branched, glabrous to sparsely appressed-hairy above. *Leaves* alternate; lamina 2–7 cm, medium green on upper surface, paler beneath, ovate-lanceolate, lanceolate, pointed at apex, deeply sinuate-dentate, cuneate at base, glabrous or nearly so; petioles 2–7 cm. *Flowers* bisexual, solitary in the axils of leaves; pedicels 30–40 mm. *Calyx* 3–5 mm in flower, 20–30 mm in fruit, campanulate, 5-lobed, the lobes narrowly deltate, acuminate in fruit. *Corolla* 10–16 mm, whitish with a yellow centre, rotate. *Stamens* 5, inserted near the base of the corolla tube; anthers yellow with a blue tinge. *Style* 1, slender; stigma faintly 2-lobed. *Berry* purple. *Flowers* 7–8.

Introduced. A wool casual. Native of the USA and Mexico.

2. P. alkekengi L. Japanese Lantern
P. franchetii Mast.

Perennial monoecious *herb* with a horizontal rhizome. *Stems* 25–60(–100) cm, pale green often tinted brownish-purple, erect, glabrous, simple or branched, leafy. *Leaves* alternate or opposite, 4–15 × 2–8 cm, dull dark green on upper surface, paler beneath with even paler midrib, broadly ovate, acute to acuminate at apex, entire or with a few, coarse teeth, sinuate, truncate to broadly cuneate at base, sparsely hairy; petioles 1.5–6.0 cm, pale green, sometimes tinted brownish-purple. *Inflorescence* of solitary, axillary, bisexual flowers 15–25 mm in diameter; pedicels 5–15 mm. *Calyx* 5–18 mm, divided one-third of the way to the base into 5 lobes, the lobes narrowly lanceolate, acute at apex and recurved, tomentose. *Corolla* dirty white, rotate, distinctly 5-lobed, the lobes broadly ovate and cuspidate at apex. *Stamens* 5, inserted near the top of the corolla tube, exserted; filaments white; anthers about 2 mm, yellow. *Style* 1; stigma capitate. *Berry* 12–17 mm, red to orange, globose, surrounded by the much enlarged fruiting calyx which is 25–50 mm, red to orange, like an inflated lantern and papery when ripe; seeds about 3 mm, very pale brown and reniform. *Flowers* 7–8. $2n = 24$.

Introduced. Grown in gardens for ornament and naturalised in shrubberies and on waste land and roadsides. Scattered records throughout England and Wales and into south Scotland. Native of southern Europe. The red berry is edible, but the calyx should not be eaten.

3. P. angulata L. Cut-leaved Ground Cherry

Annual monoecious *herb* with fibrous roots. *Stems* up to 90 cm, angular, pale green, erect, glabrous or with a few, appressed hairs. *Leaves* alternate; lamina 5–10 cm, medium green on upper surface, paler beneath, lanceolate to ovate or rather oblong, acute at apex, sinuate-dentate, cuneate to truncate at base; petiole up to 5 cm, glabrous. *Flowers* bisexual, solitary in the angle of the leaves; peduncles 10–15 mm, glabrous. *Calyx* smooth; 3–5 mm in flower, 5-lobed, the lobes 2–3 mm, triangular to lanceolate, generally shorter than the tube, about 30 mm in fruit. *Corolla* 4–10 mm, yellowish, 5-lobed, the lobes broadly triangular. *Stamens* 5; filaments slender, anthers purplish-tinged. *Styles* slender; stigma 2-lobed. *Berry* yellow.

Introduced. A wool and oil-seed casual. Native of tropical North America.

4. P. ixocarpa Brot. ex Hornem. Tomatillo
P. aequata J. Jacq. ex Nees

Annual monoecious *herb,* with fibrous roots. *Stems* up to 60 cm, pale green, erect to spreading, glabrous or young parts sparsely hairy, much-branched. *Leaves* alternate; lamina 1.5–6.0 cm, medium green on upper surface, paler beneath, ovate, or ovate-lanceolate, pointed at apex, entire to sinuate-dentate, cuneate or cordate at base, glabrous or becoming so; petiole 1.5–6.0 cm. *Flowers* bisexual, solitary in the axils of leaves; pedicels 4–8 mm. *Calyx* 3.5–4.5 mm in flower, becoming 15–26 mm in fruit, 5-lobed, the lobes deltoid and shorter than the tube. *Corolla* 8–15 mm, yellow, with a dark centre. *Stamens* 5, inserted near the base of the corolla tube, straight; anthers blue. *Style* 1, slender; stigma faintly 2-lobed. *Berry* purple. *Flowers* 6–8. $2n=24$.

Introduced. Wool and bird-seed casual. Scattered records throughout Great Britain. Native of Mexico.

5. P. peruviana L. Cape Gooseberry

Annual or short-lived, monoecious *perennial. Stem* (15–)20–150(–200) cm, sprawling or spreading, pale green, densely hairy. *Leaves* alternate; lamina 5–16(–20) × 3–10 (–17) cm, medium green on upper surface, paler beneath, usually broadly ovate, acuminate at apex, entire or slightly sinuate-dentate, more or less cordate at base, hairy on both surfaces. *Flowers* bisexual, solitary. *Calyx* 5-lobed, the lobes 3–5 mm, narrowly triangular, acuminate at apex, densely hairy. *Corolla* (15–)20–40 mm in diameter, yellow with a purplish-brown base, 5-lobed, the lobes short and obtuse at apex. *Stamens* 5; anthers 3.0–3.5 mm, 10-ribbed, hairy. *Berry* 10–20 mm in diameter, orange, flesh sweet; seed 1.5–2.0 mm in diameter, broadly oblong-ellipsoid. *Flowers* 6–8. $2n=24, 48$.

Introduced. Imported as a fruit, and casual on tips. Scattered records throughout Great Britain and east Ireland. Native of South America.

6. P. philadelphica Lam. Large-flowered Tomatillo

Annual monoecious *herb* with fibrous roots. *Stems* 45–60 cm, pale green, erect, with a few hairs, branched, leafy. *Leaves* opposite; lamina 2–10 × 1–4 cm, dark green on upper surface, paler beneath, ovate to ovate-lanceolate, more or less acuminate at apex, entire or sinuate to somewhat dentate towards the cuneate base, with a few, scattered hairs; petiole (1–)2–5 cm. *Inflorescence* of solitary, axillary, bisexual flowers; pedicels 5–10 mm. *Calyx* 4–10 mm, green and often purple-veined, accrescent and 30–50 mm in fruit, campanulate, divided up to halfway to the base into 5 lobes, the lobes ovate, with a few hairs. *Corolla* 5–30 mm in diameter, yellow, with brownish-purple markings at the throat, subentire. *Stamens* 5, exserted; filaments purple; anthers 1.2–4.0 mm, purple, curved after dehiscence. *Style* 1; stigma capitate. *Berry* 13–40(–60) mm, green to purple, filling and sometimes splitting calyx; seeds suborbicular. *Flowers* 7–8. $2n=24$.

Introduced. A food refuse and grain casual. Native of North and South America.

10. Capsicum L.

Shrubby, *perennial* monoecious *herbs* but usually grown in gardens as annuals. *Leaves* alternate, entire. *Inflorescence* of 1–4 bisexual flowers mostly in the forks of branches or leaf nodes; pedicels thickened in fruit. *Calyx* campanulate, truncate at apex, obscurely 5-lobed. *Corolla* white or rarely bluish-white, with 5–9 lobes. *Stamens* 5, inserted at the base of the corolla; anthers opening by slits. *Style* 1. *Fruit* a berry; seeds flattened.

Contains 10–25 species from tropical America, a complex mixture of wild species and cultivated variants.

Terpo, A. (1966). Kritische Revision de wild wachsenden Arten und der kultivierten Sorten der Gattung *Capsicum*. 1. *Feddes Repert*. **72**: 155–191.

1. C. annuum L. Sweet Pepper

Shrubby, *perennial* monoecious *herb*, but usually grown in gardens as an annual. *Stems* up to 2 m, glabrous. *Leaves* alternate; lamina 3.5–15.0 × 1–5 cm, medium green on upper surface, paler beneath, ovate to lanceolate, acute to acuminate at apex, entire, cuneate at base and often equal; petiole short. *Inflorescence* of 1–4 bisexual flowers mostly in the forks of branches or at leaf nodes; pedicels thickened in fruit. *Calyx* 3–4 mm, campanulate, truncate, lobes absent or 4 very short lobes. *Corolla* 10–25 mm in diameter, white, rarely bluish-white, star-shaped, with 5–9, deep triangular, obtuse lobes. *Stamens* 5, inserted at the base of the corolla; anthers yellow or purple. *Style* 1. *Berry* erect or hanging, 1–30 cm, very variable in colour and shape; seeds 3–4 mm, yellow, flattened. *Flowers* 7–9. $2n=12, 24$.

Introduced. Imported as bird-seed and used as a vegetable in cooking and salads. Casual on tips and sewerage works. Scattered records in England and Isle of Man. Native of Tropical America.

11. Lycopersicon Mill.

Annual or short-lived *perennial* monoecious *herbs*. *Leaves* pinnate with variable-sized leaflets. *Inflorescence* of leaf-opposed cymes. *Calyx* star-shaped, divided nearly to the base into 5 lobes, slightly enlarging and lobes reflexed when fruiting. *Corolla* yellow, star-shaped, deeply 5-lobed. *Stamens* 5; anthers dehiscing by slits. *Style* 1. *Ovary* 2–3(–5)-celled. *Fruit* a succulent, depressed-globose or globose berry.

Differs from the closely allied genus *Solanum* by its anthers opening by slits and not by pores.

Contains about 10 species all of which are native to western South America.

Phillips, R. & Rix, M. (1993). *Vegetables*. London.
Smart, J. & Simmonds, N. W. (Edits.) (1995). *Evolution of crop plants*. Ed. 2. London.
Vaughan, J. G. & Geissler, C. A. (1997). *The new Oxford book of food plants*. Oxford.

1. L. esculentum Mill. Tomato
Solanum lycopersicum L.; *L. lycopersicum* (L.) Karst.

Short-lived, monoecious *perennial,* behaving as an *annual,* with fibrous roots. *Stems* up to 2 m, pale yellowish-green, erect to decumbent or scrambling, angled, with dense, pale simple eglandular hairs of various lengths and short glandular hairs, branched, leafy. *Leaves* alternate, 10–45 × 10–40 cm, imparipinnate, elliptical in outline; leaflets 5–25 × 2–8 cm, large and small mixed, dull dark green on upper surface, paler beneath, ovate in outline, pinnately lobed, the lobes ovate and mostly obtuse at apex, glabrous or with short hairs on upper surface, densely hairy beneath; rhachis and fairly long petiolules pale yellowish-green and densely glandular-hairy; petiole short. *Inflorescence* a leaf-opposed cyme; pedicels pale green and densely glandular-hairy; flowers bisexual, 15–25 mm in diameter. *Calyx* 13–15 mm, divided nearly to the base in 5, linear-lanceolate, pointed lobes, glandular-hairy. *Corolla* yellow, tubular, with 5, triangular, pointed lobes. *Stamens* 5, borne on the corolla tube; filaments and anthers green. *Style* 1, greenish. *Berry* 0.6–12 cm in diameter, green, usually turning red, sometimes yellow or orange, rarely white or pale pink, normally obulate-spheroid, sometimes elongated, plum-shape or square, fleshy. *Flowers* 6–9. $2n = 24$.

The cultivated tomato, var. *esculentum*, in all cultivars, is almost certainly derived from the wild yellow-fruited plant var. *cerasiforme* (Dunal) Alef. This wild plant is very closely allied to the small cherry tomatoes in cultivation and in fact there is a complete series of fruits through to the very large beef tomatoes. The early introductions to Europe were probably from Mexico in the Vera Cruz and Puebla areas. They were probably brought to Europe soon after the completion of the conquest of Mexico by Cortes in 1523, but the earliest actual record is by the Italian botanist Petrus Andreas Matthiolus in 1544 who described a yellow-fruited variety. Tomatoes became known as an aphrodisiac and received the name Love Apple or Pomme d'Amour. Most cultivars of tomatoes can be divided into three groups; the large-fruited beef tomatoes about 10 cm in diameter, those with medium-sized fruits about 5 cm and the small-fruited cherry tomatoes about 1.5 cm across. Cultivars with the fruits elongated, plum-shaped, square, striped, hollow, white and pale pink occur as well as the common red, flattened spheroid fruit. Fruits are consumed raw or cooked in a great variety of dishes or in products such as juice, soup, sauces, ketchups, puree, pastes and powder. Tomatoes are grown in glasshouses and gardens and as a crop particularly in the Lea Valley, Sussex, Lancashire, Surrey, Essex and Yorkshire. Escapes are found on tips, waste land and sewerage works throughout Great Britain and Ireland.

12. Solanum L.

Monoecious *herbs* or *shrubs*. *Leaves* alternate or in pairs. *Flowers* bisexual, in cymes, rarely reduced to 1 flower, white, purple or blue. *Calyx* usually 5-lobed. *Corolla* rotate, usually 5-lobed. *Stamens* inserted on the throat of the corolla-tube. *Style* 1. *Ovary* 2(–4)-celled. *Fruit* a succulent or dry berry.

Contains about 1,500 species, mainly in the tropics.

S. capsicoides All., **S. melongena** L., **S. pygmaeum** Cav., **S. pyracanthum** Jacq. and **S. toruum** Swartz have been recorded as casuals.

Bassett, I. J. & Munro, D. B. (1983). The biology of Canadian weeds. 67. *Solanum ptycanthum* Dun, *S. nigrum* L. and *S. sarrachoides* Sendt. *Canad. Jour. Pl. Sci.* **65**: 401–414.
Baylis, G. T. S. (1958). A cytogenetical study of New Zealand forms of *Solanum nigrum* L., *S. nodiflorum* Jacq. and *S. gracile* Otto. *Trans. Roy. Soc. New Zealand* **85**(3): 379–385.
Bitter, G. (1913). Solana africana I. *Bot. Jahrb.* **49**: 560–569.
Bitter, G. (1917). Solana africana II. *Bot. Jahrb.* **54**: 416–506.
Bitter, G. (1921). Solana africana III. *Bot. Jahrb.* **57**: 248–256.
Cooper, M. R. & Johnson, A. W. (1984). *Poisonous plants in Britain and their effects on animals and man.* London.
Correll, D. S. (1962). *The potato and its wild relatives.* Renner, Texas.
Edmonds, J. M. (1972). A synopsis of the taxonomy of *Solanum* Sect. *Solanum* (*Maurella*) in South America. *Kew Bull.* **27**: 95–114.
Edmonds, J. M. (1977). Taxonomic studies on *Solanum* L. Section *Solanum* (*Maurella*). *Bot. Jour. Linn. Soc.* **75**: 141–178.
Edmonds, J. M. (1978). Numerical taxonomic studies on *Solanum* L. section *Solanum* (*Maurella*). *Bot. Jour. Linn. Soc.* **76**: 27–51.
Edmonds, J. M. (1979). Nomenclatural notes on some species of *Solanum* L. found in Europe. *Bot. Jour. Linn. Soc.* **78**: 213–233.
Edmonds, J. M. (1981). The artificial synthesis of *Solanum* × *procurrens* Leslie (*S. nigrum* L. × *S. sarrachoides* Sendtn. *Watsonia* **13**: 203–207.
Edmonds, J. M. (1982). Epidermal hair morphology in *Solanum* L. section *Solanum*. *Bot. Jour. Linn. Soc.* **85**: 153–167.
Edmonds, J. M. (1983). Seed coat structure and development in *Solanum* L. section *Solanum*. (Solanaceae). *Bot. Jour. Linn. Soc.* **87**: 229–246.
Edmonds, J. M. (1984). Pollen morphology of *Solanum* L. section *Solanum*. *Bot. Jour. Linn. Soc.* **88**: 237–251.
Edmonds, J. M. (1984). *Solanum* L. section *Solanum* – a name change in *S. villosum* Miller. *Bot. Jour. Linn. Soc.* **89**: 165–170.
Edmonds, J. M. (1986). Biosystematics of *Solanum sarrachoides* Sendtn. and *S. physalifolium* Rusby (*S. nitidibaccatum* Bitter). *Bot. Jour. Linn. Soc.* **92**: 1–38.
Edmonds, J. M. & Chweya, J. A. (1997). Black Nightshades. *Solanum nigrum* L. and related species. *Int. Pl. Genet. Inst.* **15**: 1–113.
Grime, J. P. et al. (1988). *Comparative plant ecology*. London. [*S. dulcamara*.]
Grun, P. (1979). Evolution of the cultivated potato: a cytoplasmic analysis. In Hawkes, J. G., Lester, R. N. & Skelding, A. D. (Edits.) *The biology and taxonomy of the Solanaceae*. London.
Hultén, E. & Fries, M. (1986). *Atlas of north European vascular plants north of the Tropic of Cancer.* 3 vols. Königstein.
Hawkes, J. G. (1990). *The Potato*. London.
Hawkes, J. G. & Hjerting, J. P. (1969). *The potatoes of Argentina, Brazil, Paraguay, and Uruguay: a biosystematic study.* Oxford.

Jackson, M. T. (1986). Exploited plants. The potato. *Biologist* **33**: 161–167.
Karschon, R. & Horowitz, M. (1985). Infraspecific variation of *Solanum nigrum* L. and *S. villosum* Miller, important crop weeds in Israel. *Phytoparasitica* **13**: 63–67.
Leslie, A. C. (1978). The occurrence of *Solanum nigrum* L.× *S. sarachoides* Sendtn. in Britain. *Watsonia* **12**: 29–32.
Phillips, R. & Rix, M. (1993). *Vegetables*. London.
Salaman, R. N. (1985). *The history and social significance of the potato*. Cambridge. Reissued with a critical introduction by J. G. Hawkes in 1985.
Venkateswarlu, J. & Rao, M. K. (1972). Breeding systems, crossability relationships and isolating mechanisms in the *Solanum nigrum* complex. *Cytologia* **37**: 317–326.
Wessely, I. (1960). Die Mitteleuropaischen Sippen der Gattung *Solanum* Sektion *Morella*. *Feddes Repert*. **63**: 290–321.

1. Stems and leaves with strong spines 2.
1. Spines absent 4.
2. Corolla yellow; 1 anther longer than 4 others **15. rostratum**
2. Corolla whitish to bluish-purple; 5 anthers of equal length 3.
3. Rhizomatous perennial; leaves lobed less than halfway to midrib, with more or less entire lobes **13. carolinense**
3. Annual herb; leaves mostly divided halfway to midrib, with toothed or lobed lobes; berry red. **14. sisymbriifolium**
4. Perennials with stems more or less woody below or a small shrub 5.
4. Annual to perennial with entirely herbaceous stems 10.
5. Corolla 40–50 mm in diameter; berry 23–30 mm **12. laciniatum**
5. Corolla 10–15 mm in diameter; berry 5–15 mm 6.
6. Stems scrambling to procumbent; corolla usually purple, rarely white 7.
6. Stems erect; corolla white 8.
7. Stem scrambling, glabrous or nearly so; leaves not succulent **9(i). dulcamara** var. **dulcamara**
7. Stem prostrate and hairy; leaves succulent **9(ii). dulcamara** var. **marinum**
8. Berry 5–7 mm, black or blackish-purple **5. chenopodioides**
8. Berry 8–15 mm, red 9.
9. Inflorescence of 2–4 flowers in a cluster or solitary; berry about 14 mm in diameter **16. diflorum**
9. Inflorescence of 20–30 densely congested flowers; berry 9–10 mm **17. abutiloides**
10. Perennial with subterranean stem tubers; leaves pinnate 11.
10. Annual herbs; leaves entire to deeply pinnately lobed 12.
11. Stems green to slightly purple-tinged; leaves sparsely hairy on lower side; corolla white to purple **10. tuberosum**
11. Stems strongly purple-blotched; leaves grey-tomentose on lower side; corolla purplish-violet **11. vernei**
12. Leaves pinnately lobed more than three-quarters of way to base **8. triflorum**
12. Leaves entire or toothed less than halfway to base 13.
13. Stems scrambling with weak spine-like outgrowths; anthers brownish-yellow **4. scabrum**
13. Stems erect to decumbent, without spine-like outgrowths, but sometimes with dentate angles; anthers bright yellow to orange 14.
14. Ripe berries yellow to red 15.
14. Ripe berries black, brownish-purple or green 16.
15. Stem with rounded, entire ridges; glandular hairs mixed with eglandular hairs **3(a). villosum** subsp. **villosum**
15. Stem with angled, slightly dentate ridges; hairs eglandular, mostly appressed and often sparse **3(b). villosum** subsp. **miniatum**
16. Plant without glandular hairs 17.
16. Plant with many glandular hairs 20.
17. Cyme more or less umbellate; corolla 5–8 mm in diameter; seeds 1.0–1.5 mm; fruiting pedicels erect **2. americanum**
17. Flowers usually in short racemes; corolla 5–12 mm in diameter; pedicels deflexed in fruit 18.
18. Plant prostrate **1(a,i). nigrum** subsp. **nigrum** var. **prostratum**
18. Plant erect 19.
19. Berries black **1(a,ii). nigrum** subsp. **nigrum** var. **nigrum**
19. Berries greenish **1(a,iii). nigrum** subsp. **nigrum** var. **chlorocarpum**
20. Calyx not enlarging in fruit, with obtuse teeth; berries usually black, rarely green, without groups of stone cells in the flesh **1(b). nigrum** subsp. **schultesii**
20. Calyx enlarging in fruit with acute teeth, berries green to purplish-brown with more than 2 groups of stone cells in the flesh 21.
21. Calyx lobes less than 2 mm in flower, less than 4 mm in fruit, usually shorter than berry; corolla lobes 2–4 mm wide; berries with less than 34 seeds **6. physalifolium**
21. Calyx lobes more than 3 mm in flower, more than 5 mm in fruit, usually at least as long as berry; corolla lobes 5–7 mm wide; berries with more than 50 seeds **7. sarachoides**

1. S. nigrum L. Black Nightshade

Annual or short-lived monoecious *perennial herb* with fibrous roots. *Stems* up to 1 m, pale green or suffused purplish, erect or straggling, terete or obscurely angled or winged, subglabrous to densely hairy, the hairs eglandular or glandular, branched, leafy, without spines. *Leaves* alternate; lamina 2–10(–13)×0.8–5.0(–7.5) cm, dark green on upper surface, paler beneath, ovate, obtuse, acute or acuminate at apex, shallowly lobed or coarsely dentate, narrowed at base, with sparse to numerous hairs on both surfaces; petioles up to 3 cm, narrowly winged, with few to numerous hairs. *Inflorescence* extra-axillary, umbellate or shortly racemose, 3–8(–12)-flowered; peduncle 1–3 cm, lengthening in fruit, erecto-patent, thinly to densely hairy; pedicels 5–15 mm at anthesis, at first spreading, becoming deflexed in fruit, thinly to densely hairy. *Calyx* with 5 lobes, the lobes about 2 mm, oblong, obtuse at apex, erect, hairy, not enlarging in fruit. *Corolla* white, rotate or shallowly infundibuliform, 5–12 mm in diameter, shortly hairy; tube about 2 mm, lobes 5, about 5×3 mm, ovate, acute at apex. *Stamens* 5; filaments 1.0–1.5 mm, flattened dorsiventrally; anthers about 2 mm, bright yellow, oblong. *Style* 1, about 3 mm, shortly hairy towards the base; stigma capitate or obscurely 2-lobed. *Berry* 7–10 mm in diameter, dull or shining purplish-black, rarely green when ripe, with stone cells in flesh, subglobose; seeds about 1.8 mm, pale brown, subreniform, strongly compressed, foveolate. *Flowers* 7–9.

(a) Subsp. **nigrum**
Hairs eglandular, mostly appressed, often sparse. $2n=72$.

(i) Var. **prostratum** F. Gérard
Plant prostrate. *Berries* black.

(ii) Var. **nigrum**
S. nigrum var. *sinuatum* Druce

Plant erect. *Berries* black.

(iii) Var. **chlorocarpum** Spenn.
Plant erect. *Berries* greenish.

(b) Subsp. **schultesii** (Opiz) Wessely
S. schultesii Opiz

Hairs frequent to abundant, mostly glandular, mostly patent.

Native. Subsp. *nigrum* is common in waste and cultivated ground in the Channel Islands, most of England and south Wales and is local in the rest of England, Wales and the Isle of Man, with scattered records as a casual in Scotland and Ireland. Throughout most of Europe and a cosmopolitan weed of cultivation. Euroasian Southern-temperate element. Var. *prostratum* and var. *chlorocarpum* seem mainly coastal. Subsp. *schultesii* is an introduction which is naturalised in scattered localities in waste and cultivated places in south and east England north to Yorkshire. It is native of southern Europe.

× **physalifolium** = **S.** × **procurrens** A. C. Leslie
This hybrid is intermediate in calyx characters and its black berries have 0–few seeds.

It occurs with the parents in cultivated ground in Kent, Suffolk, Cambridgeshire and Bedfordshire.

2. S. americanum Mill. Small-flowered Nightshade
S. adventitium Polg.; *S. calvum* Bitter; *S. caribaeum* Dunal; *S. curtipes* Bitter; *S. gollmeri* Bitter; *S. inconspicuum* Bitter; *S. microtatanthum* Bitter; *S. minutibaccatum* Bitter; *S. nigrum* var. *dillenii* A. Gray; *S. nigrum* var. *nodiflorum* A. Gray; *S. nodiflorum* Jacq.; *S. oleraceum* Dunal; *S. photeinocarpum* Nakai & Odash.; *S. scrophilum* Bitter; *S. tenellum* Bitter

Annual monoecious *herb* with fibrous roots. *Stems* up to 1 m, pale green, slender, glabrous or with fairly numerous, appressed simple eglandular hairs, with entire or minutely dentate ridges and diverging branches, leafy, without spines. *Leaves* alternate; lamina (2.2–)3.0–6.2(–11.0) × 1.1–4.0(–6.6) cm, pale green and membranous on upper surface, paler beneath, ovate-lanceolate or lanceolate, shortly acuminate at apex, entire to sinuate, rarely sinuate-dentate, rounded to a shortly cuneate base, sometimes with axillary leaflets. *Inflorescence* of few-flowered umbels; peduncles slender, up to 20 mm; pedicels erect in fruit. Calyx about 2 mm, accrescent; 5-lobed, the lobes ovate to almost elliptical, usually strongly reflexed in fruit. *Corolla* 5–8 mm in diameter, white or very pale mauve; 5-lobed, the lobes more or less triangular. *Stamens* 5, filaments pale; anthers bright yellow. *Style* 1. *Berry* 5–8 mm, glossy black, globose; seeds 1.0–2.5 mm, broadly obovoid to almost globose. *Flowers* 6–10.

Introduced. Rare wool and oil-seed alien. Native of the tropics.

3. S. villosum Mill. Red Nightshade
S. nigrum var. *villosum* L.; *S. villosum* (L.) Lam.

Annual or short-lived monoecious *perennial herb* with fibrous roots. *Stems* up to 50 cm, pale green, erect or sprawling, terete or sometimes obscurely angled, densely clothed with coarse, spreading, white eglandular hairs and sometimes mixed with glandular hairs, usually much-branched, leafy, without spines. *Leaves* alternate; lamina 2–6 × 1–5 cm, medium green on upper surface, paler beneath, broadly ovate, usually acute at apex, irregularly sinuate-dentate, abruptly contracted at base, hairy on both surfaces; petiole 1–2(–3) cm, slender, hairy and scarcely winged. *Inflorescence* extra-axillary, umbellate or shortly racemose, with 2–6(–8) bisexual flowers; peduncle 5–15(–20) mm, erecto-patent, lengthening a little in fruit, usually with spreading hairs; pedicels 5–8 mm, slender, hairy, lengthening and becoming strongly deflexed in fruit. *Calyx* with 5 lobes, the lobes about 2 mm, oblong, erect, hairy, usually joined by a membranous web for almost half their length. *Corolla* white, rotate or shallowly infundibuliform, shortly hairy especially towards the apex of the lobes or nearly glabrous; tube about 2.5 mm; lobes 5, about 5 × 3–4 mm, ovate, acute at apex. *Stamens* 5; filaments about 1.5 mm, flattened dorsiventrally; anthers bright yellow, narrowly oblong. *Style* 1, about 3 mm, hairy towards base; stigma capitate or obscurely 2-lobed. *Berry* 8–9 mm in diameter, orange-yellow or red when ripe, subglobose; seeds about 2 × 1.5 mm, reniform, strongly compressed, pale brown, foveolate. *Flowers* 7–9. $2n=48$.

(a) Subsp. **villosum**
S. luteum Mill.

Stems with rounded, entire ridges. *Glandular hairs* mixed with eglandular hairs.

(b) Subsp. **miniatum** (Bernh. ex Willd.) Edmonds
S. villosum subsp. *alatum* (Moench) Edmonds; *S. luteum* subsp. *alatum* (Moench) Dostál; *S. miniatum* Bernh. ex Willd.; *S. alatum* Moench

Stems with angled, slightly dentate ridges. *Hairs* eglandular, mostly appressed and often sparse.

Introduced. Casual from wool, bird-seed and oil-seed, very rarely naturalised. Native of central and southern Europe and the Mediterranean region eastwards to Iran. Both subspecies are infrequent in Great Britain.

4. S. scabrum Mill. Garden Huckleberry
S. melanocerasum All.; *S. nigrum* var. *guineense* L.

Annual or short-lived *perennial* monoecious *herb* with fibrous roots. *Stems* up to 1 m, pale green, erect or straggling, prominently and dentately winged, with spine-like outgrowths, sparsely hairy, branched, leafy. *Leaves* alternate; lamina 4–8 × 4.0–6.5 cm, dark green on upper surface, paler beneath, ovate, obtuse to acute at apex, shallowly lobed or dentate, narrowed at base, with sparse hairs on both surfaces;

petioles up to 5 cm, narrowly winged, with few hairs. *Inflorescence* extra-axillary, umbellate or shortly racemose; peduncles lengthening in fruit; flowers bisexual. *Calyx* with 5 lobes, the lobes oblong, obtuse at apex. *Corolla* 10–13 mm in diameter, white, rotate, 5-lobed, ovate, obtuse at apex. *Stamens* 5; filaments flattened; anthers brownish-yellow. *Style* 1; stigma capitate. *Berry* 12–17 mm, shiny black, subglobose; seeds strongly compressed. *Flowers* 7–9. $2n=48$.

Introduced. A rare casual from fruit or cultivated plants on tips, sewage farms and fields spread with sludge. A few records in southern England, naturalised in Hertfordshire. Possibly native of Africa.

5. S. chenopodioides Lam. Tall Nightshade
S. sublobatum Willd. ex Roem. & Schult.

Perennial monoecious *herb* or *subshrub*. *Stems* up to 3.5 m, greyish, erect, unarmed, with 2 ridges, velvety-tomentose. *Leaves* alternate; lamina (1.5–)2.5–11.0 × (0.5–) 1–6 cm, medium green on upper surface, paler beneath, ovate or lanceolate-ovate, acute to acuminate at apex, entire or nearly so, attenuate at base, tomentose on both surfaces; petiole up to 3 cm. *Inflorescence* an umbellate cyme; flowers bisexual; peduncles very variable in length; pedicels becoming strongly deflexed in fruit. *Calyx* 2–3 mm, 5-lobed, the lobes very small, lanceolate to oblong or elliptical, somewhat accrescent. *Corolla* 8–13 mm in diameter, white, 5-lobed, the lobes lanceolate or narrowly triangular, densely hairy outside, soon becoming reflexed. *Stamens* 5; anthers 1.5–3.0 mm. *Style* 1. *Berry* 5–7 mm in diameter, dull purplish-black or purple, globose; seeds 1.0–1.5 mm, subrotund. *Flowers* 6–8. $2n=24$.

Introduced. Naturalised on rough ground. Occurs in Jersey, Guernsey since 1958 and the London area since 1989, and a rare casual elsewhere. Native of South America.

6. S. physalifolium Rusby Green Nightshade

Annual monoecious *herb* with fibrous roots. *Stems* 10–40(–90) cm, pale green, prostrate, decumbent or sprawling, terete, smooth, with entire of scarcely dentate ridges, with numerous to dense unequal, spreading, glandular hairs, without spines. *Leaves* solitary, alternate or in unequal pairs; lamina 1.9–5.3 × 1.3–3.5 cm, pale to dark green on upper surface, paler beneath, ovate, ovate-lanceolate or trullate, occasionally lanceolate, acute to subobtuse at apex or rarely obtuse, regularly sinuate-dentate with 2–6 obtuse to acute, antrorsely directed teeth on each side, occasionally sinuate or entire, truncate or truncate-cordate or attenuate to the petiole, with few to numerous glandular hairs on both surfaces; petioles 6–17(–22) mm, pale green, with glandular hairs. *Inflorescence* a lax extended cyme of 4–8(–10) bisexual flowers; peduncles leaf-opposed or extra-axillary, when often subtended by ovate leaves, ascending or spreading in flower, usually deflexed or occasionally patent in fruit, glandular-hairy; pedicels 4–10(–15) mm, recurved to erect in bud and flower, spreading or occasionally reflexed in fruit, glandular-hairy. *Calyx* (1.5–)2.5–3.0(–4.0) mm in fruit, often flecked with anthocyanin, glandular-hairy; lobes 5, 2.5–3.5(–4.0) ×3–4 mm, broadly triangular, with subacute apices and enlarged and indistinctly veined when fruiting. *Corolla* rotate, white or white-tinged with purple especially on the abaxial surfaces of the lobes, with a distinct purple or brown and yellow basal star; lobes 5, (2.0–)2.5–3.5(–5.0) × 2–4 mm, broadly triangular, acute at apex, glandular-hairy. *Stamens* 5; filaments 1.5–2.0 mm, inner surfaces hairy; anthers bright yellow. *Style* 1, 3–4 mm, straight. *Berry* 5.5–8.0 × 6–9 mm, dark green, greenish-brown or purple-brown, globose or broadly ovoid, with stone cells, covered with appressed and enlarged calyces whose lobes often reflex away from the berry at full maturity; seeds 5–34, brown, embedded in a purple placenta. *Flowers* 7–9. $2n=16$.

Introduced. Naturalised in cultivated and waste ground. Central and southern England and the Channel Islands; casual from wool and other sources elsewhere. Native of South America, introduced in western and central Europe, North and Central America, Australia and New Zealand. The above description and distribution is of var. **nitidibaccatum** (Bitter) Edmonds (*S. nitidibaccatum* Bitter; *S. sarachoides* auct.; *S. atriplicifolium* auct.; *S. chenopodioides* auct.). Var. **physalifolium** (*S. nitidibaccatum* var. *robusticalyx* Bitter) has the inflorescence with 3–5 flowers, the pedicels 8–14 mm, the sepals with subobtuse apex and prominently veined, and berries about 10 mm in diameter is known only from northern Argentina and Bolivia.

7. S. sarachoides Sendtn. Leafy-fruited Nightshade
S. atiplicifolium var. *minus* Gillies ex Nees; *S. styleanum* Dunal; *S. justi-schmidtii* E. H. L. Krause; ?*S. sarachidium* Bitter; *S. hirtulum* E. H. L. Krause; *S. sarachoides* var. *sarachidium* (Bitter) Morton

Annual monoecious *herb* with fibrous roots. *Stems* 25–60 cm, pale green, terete, smooth, with unequal, spreading, viscid glandular hairs, without spines. *Leaves* solitary, alternate or in unequal pairs; lamina (3.2–)3.9–7.6 (–11.2) × (2.7–)3.1–5.1(–8.0) cm, pale green, ovate-lanceolate to lanceolate, acute at apex, regularly sinuate-dentate with 3–9 obtuse to acute, antrorsely directed teeth on each side, truncate, cordate or attenuate at base, covered with viscid glandular hairs; petioles 16–32(–68) mm, pale green, glandular-hairy. *Inflorescence* of a single, umbellate cyme with 3–4(–5), bisexual flowers; peduncles usually leaf-opposed, occasionally extra-axillary, ascending in flower, ascending to spreading, rarely deflexed in fruit, glandular-hairy; pedicels 7–11 mm, recurved to erect in bud and flower, reflexed in fruit, glandular-hairy. *Calyx* 3–6 mm in flower, glandular-hairy; lobes 5, (2–)3–5 × 1(–2) mm, oblong-lanceolate, acute at apex, enlarging and becoming 5.5–8.0 × 3.5–4.0 mm in fruit. *Corolla* semi-stellate to pentagonal, white or yellow with a translucent basal star; lobes 3.0–4.5 × 5–7 mm, broadly triangular, acute at apex, glandular-hairy. *Stamens* 5, filaments 1.0–1.5 mm, inner surfaces hairy; anthers bright yellow to orange. *Style* 3.0–3.5 mm, straight, lower half hairy. *Berry* 6–9 mm, pale, shiny green becoming dull black, subglobose, with stone cells in flesh, usually completely enveloped by the enlarged calyx, which when fully matured becomes papery exposing the berries; seeds (23–)59–69(–93), pale yellow, embedded in green placenta. *Flowers* 7–9. $2n=24$.

Introduced. Casual on tips and waste ground and a weed in crops. Scattered records in Great Britain. Recorded in Europe and North America. Native of South America.

8. S. triflorum Nutt. Small Nightshade

Annual monoecious, foetid *herb* with fibrous roots. *Stems* up to 1 m, pale green, erect or sprawling, more or less hairy, with simple eglandular hairs, branching diffusely, sometimes from the base, leafy, without spines. *Leaves* alternate; lamina 2–4 × 1–2 cm, oblong, acute at apex, deeply pinnatifid, with linear lobes and rounded sinuses, glabrous; petiole up to 15 mm. *Inflorescence* of 1–3, bisexual flowers; pedicels soon reflexed. *Calyx* 4–5 mm, scarcely accrescent, 5-lobed, the lobes triangular-ovate, acute at apex. *Corolla* less than 10 mm in diameter, slightly exceeding calyx, white, stellate. *Stamens* 5; filaments pale; anthers yellow. *Style* 1. *Berry* 10–15 mm in diameter, marbled green and white, globose. *Flowers* 7–9. $2n=24$.

Introduced. Casual in cultivated and rough ground. Scattered records in Great Britain; naturalised in East Anglia. Native of western North America.

9. S. dulcamara L. Bittersweet

Woody, monoecious *perennial scrambler*. *Stems* 30–200 (–700) cm, pale green or purplish, scrambling or prostrate, glabrous, hairy or tomentose, much branched, leafy, without spines. *Leaves* alternate; lamina (3–)5–8 × (1.5–) 2.5–5.0(–9.0) cm, medium green on upper surface, pale beneath, lanceolate to ovate, acute at apex, entire or with 1–4 deep lobes or stalked pinnae at the rounded, cordate or hastate base, glabrous or hairy; petiole up to 6 cm. *Inflorescence* a pedunculate, leaf-opposed, branched, umbellate cyme with 10–25 bisexual flowers; pedicels erect in flower, recurved in fruit. *Calyx* 2.5–3.0 mm, 5-lobed, the lobes broad, shallow and rounded. *Corolla* 10–15(–20) mm in diameter, purple or very rarely white; 5-lobed, at first spreading, then revolute. *Stamens* 5; filament pale; anthers pale yellow, cohering in a cone. *Style* 1. *Berry* 8–15 × 7.5–10.0 mm, bright red, ovoid or ovoid-ellipsoid. *Flowers* 6–9. $2n=24$.

(i) Var. **dulcamara**

Stem scrambling, glabrous or nearly so. *Leaves* not succulent.

(ii) Var. **marinum** Bab.
S. marinum (Bab.) Pojark.

Stems prostrate and hairy. *Leaves* succulent.

Native. Var. *dulcamara* occurs on walls, hedges, woods and particularly ditch or stream sides and rough ground. Var. *marinum* occurs on coastal sand or shingle. Throughout lowland Great Britain and Ireland except northern Scotland. Most of Europe; Asia; North Africa; naturalised in North America. Eurasian Southern-temperate element.

10. S. tuberosum L. Potato

Perennial monoecious *herb* with tuber-bearing (potatoes) stolons. *Stems* up to 1 m, pale yellowish-green, sometimes slightly purple tinged, erect, fleshy, angled, ridged and narrowly winged with scattered simple eglandular hairs, branched, leafy, without spines. *Leaves* 15–45 × 15–25 cm, alternate, obovate in outline, interruptedly imparipinnate; leaflets 6–8, 6–15 × 4–8 cm, medium yellowish- to dark green on upper surface, paler beneath, subrotund, elliptical or ovate, bluntly acuminate at apex, entire or shallowly crenate, rounded or cordate at base, sparingly, shortly and stiffly hairy on both surfaces, veins inpressed on upper surface, prominent beneath; rhachis and petiole pale yellowish green, hairy; smaller leaflets occur between the longer leaflets. *Inflorescence* a lax terminal cyme, many-flowered; flowers bisexual, heavily scented; pedicels 20–35 mm, usually articulated in the middle. *Calyx* 15–20 mm, unequally divided for three-quarters of the way to the base into 5 lobes, the lobes linear-lanceolate, acuminate at apex and densely hairy. *Corolla* 25–35(–40) mm in diameter, white or mauve, rotate-pentagonal, 5-lobed, the lobes folded and pointed at apex. *Stamens* 5; filaments white; anthers yellow to deep orange. *Style* 1, pale green; stigma dark green. *Berry* 20–40 mm in diameter, greenish to purplish, more or less globose, succulent. *Flowers* 6–9. $2n=48$.

There is no single ancestor of the potato grown in Europe that can be found wild. J. G. Hawkes has suggested the earliest wild potato to be cultivated was *S. leptophes* Bitter which occurs in rocky places in the mountains of northern Bolivia. From this species, when brought into cultivation, evolved the diploid *S. stenotomum* Juz. & Buk, which can still be found cultivated in Peru. Hybrids between *S. stenotomum* and another wild diploid species, *S. sparsipilum*, became by chromosome doubling the fertile tetraploid *S. tuberosum* subsp. *andina* Hawkes, the ancestor of the common potato. It is assumed that potatoes were among the crops cultivated by Man when the Andes were first colonised 7,000–10,000 years ago. Remains of potatoes occur in deposits near Lima in Peru dating from 4000 BC. The potato was widely cultivated in Inca times about AD 1400 and its success was due to its ability to grow in icy, windswept highlands. The first report of potatoes by Europeans was when Cieza de Leon published accounts of his travels in 1537 and in 1553. The first record of potatoes in Europe is usually said to be found in the accounts of the La Sangre hospital in Seville in 1573. Clusius recorded a drawing of potatoes sent him from James Garret who lived in England between 1589 and 1593. A specimen of potato from this early period was preserved by Bauhin and has been identified by Hawkes as his subsp. *andina*. The potatoes of Europe were selected and bred from those forms which happened to arrive in Europe at an early date. Growers repeatedly raised new varieties from seed and competed with one another for early-producing varieties. It was only in the eighteenth century, however, that the potato became an important item of diet in Great Britain and Ireland. So important was it in Ireland that outbreaks of blight (*Phytophora infestans*) in 1845 and 1846 caused severe famine among the people between 1846 and 1851. Potatoes are now produced throughout Great Britain and

Ireland in small gardens as well as main crops in fields. They are discarded on the tips and fall from transport when being carted from the fields. At the present time only potatoes of a certain size are harvested and the remainder are left in the fields. These grow in the cereal crops which follow and are not all destroyed by herbicides. They also grow on fields left uncultivated for set-aside.

11. S. vernei Bitter & Wittm. Purple Potato
S. ballsii Hawkes

Perennial monoecious *herb* with tuber-bearing (potatoes) stolons. *Stems* 50–100(–200) cm, purple mottled, moderately to densely hairy, generally winged, the wings 1–5 mm, straight or crisped, and often deep purple, without spines. *Leaves* alternate; lamina 7–40 × 5.0–18.5 cm, dark to medium green on upper surface, whitish-grey beneath, divided pinnately into leaflets, the terminal longer than laterals; leaflets ovate-lanceolate to lanceolate, obtuse to acuminate at apex; with short, appressed hairs on upper surface, more or less densely whitish-grey tomentose beneath. *Inflorescence* a terminal cyme, 2- to 15(–25)-flowered; peduncle 1.5–13 cm; pedicels 15–30(50) mm, with short, often crisped hairs. *Calyx* 6–10 mm in diameter, 5-lobed, the lobes 1.0–2.5 × 3–5 mm, lanceolate, dark purple and hairy. *Corolla* 25–45(–50) mm in diameter, deep violet purple, 5-lobed, the lobes 8–12 × 8–18 mm, long-acuminate at apex up to 5 mm. *Stamens* 5; filaments pale; anthers yellow. *Style* 10–13 mm, papillose in the lower one-third to one-half; stigma subglobose to conical. *Berry* green or green with white slightly raised spots, more or less globular. *Flowers* 7–9. $2n=24$.

Introduced. Established as a weed in shrubbery beds in Whiteknights Park at Reading in Berkshire from 1985 onwards. Native of Argentina.

12. S. laciniatum Aiton Kangaroo Apple
S. aviculare auct.

Soft-wooded, monoecious *shrub*. *Stems* up to 3 m, often purplish or greenish-purple, unarmed, glabrous. *Leaves* alternate; 10–40 cm, entire to pinnatisect with 1–4 pairs of lobes sometimes nearly reaching the midrib, both types of leaf found on the same plant; lamina of entire leaves up to 5 cm wide, lanceolate or linear-lanceolate, sometimes more or less elliptical, obtuse to acuminate at apex, the base decurrent on the petiole; lobes of pinnatisect leaves up to 2 cm wide; all glabrous and petiolate. *Inflorescence* of 2- to 10-flowered cymes; peduncles to 18 cm, slender; pedicels pendulous at fruiting; flowers bisexual. *Calyx* 5–8 mm, 5-lobed, the lobes very broadly ovate-triangular, mucronate at apex, accrescent. *Corolla* 40–50 mm in diameter, violet or purple, 5-lobed, the lobes very broad and rather shallow, more or less emarginate. *Stamens* 5, anthers 3–4 mm. Style 1. *Berry* 23–30 mm, ovoid or ellipsoid, usually yellow or pale orange, pendulous; seeds 2.2–2.5 mm in diameter, more or less obovoid, but somewhat asymmetric. *Flowers* 6–8. $2n=48, 92$.

Introduced. Rough ground, tips and maritime sand, mostly casual, but sometimes naturalised. Channel Islands and very scattered records in Great Britain. Native of Australia.

13. S. carolinense L. Horse-nettle

Perennial monoecious *herb* with a creeping rhizome. *Stems* up to 1 m, greenish, erect, spiny and loosely hairy throughout with 4- to 8-rayed, sessile stellate hairs, branched, leafy. *Leaves* alternate; lamina 4.5–12 × 2–5 cm, medium green on upper surface, paler beneath, ovate to ovate-elliptical, typically with several large teeth or shallow lobes on each side, rounded at base, more or less spiny along the main veins; petiole up to 3 cm. *Inflorescence* of several, bisexual flowers, elongating at maturity to form a simple, racemiform cluster. *Calyx* 5–7 mm, 5-lobed, the lobes lanceolate, acuminate at apex. *Corolla* 20–30 mm in diameter, pale violet to white, 5-lobed, the lobes ovate, obtuse at apex, covered with stellate hairs. *Stamens* 5, equal in length; anthers 6–8 mm. *Style* 1. *Berry* 10–20 mm in diameter, yellow when mature, globose. *Flowers* 6–10.

Introduced. Oil-seed casual. Recorded for Kent and Suffolk. Native of eastern North America.

14. S. sisymbriifolium Lam. Red Buffalo-bur

Annual monoecious *herb* with fibrous roots. *Stems* up to 70 cm, pale green, suberect, stout, long-hairy with simple eglandular or more or less glandular and viscid hairs, with some few-rayed stellate hairs, and with long-subulate, straight, unequal orange-red prickles. *Leaves* alternate; lamina 12–14 × 7–8 cm, oblong-ovate in outline, deeply pinnatifid, the lobes oblong, sinuate or even somewhat pinnatifid, clothed like the stem; petioles up to 4 cm. *Inflorescence* a terminal raceme of few to numerous, bisexual flowers; pedicels spreading at maturity. *Calyx* 5–8 mm, 5-lobed, the lobes, lanceolate to ovate-lanceolate, acute at apex, long glandular-hairy. *Corolla* up to 35 mm in diameter, white to pale blue, 5-lobed, the lobes lanceolate or ovate, densely stellate hairy outside. *Stamens* 5, equal in length; anthers yellow. *Style* 1. *Berry* 15–20 mm in diameter, red, globose, loosely and completely or incompletely enclosed by the spiny calyx; seeds minutely reticulate-pitted. *Flowers* 6–8.

Introduced. A wool, oil-seed, bird-seed and agricultural seed casual. Scattered records in England and Wales. Native of South America.

15. S. rostratum Dunal Buffalo-bur

Annual monoecious *herb* with fibrous roots. *Stems* up to 70 cm, pale green, erect or sprawling, with numerous to dense, forked and stellate hairs and numerous, unequal, yellowish, straight spines up to 7 mm, branched, leafy. *Leaves* alternate; lamina 6–20 × 3–14 cm, medium green on upper surface, paler beneath, ovate, elliptical or oblong-elliptical in outline, obtuse at apex, 1- to 2-pinnatisect, the lobes ovate, subrotund or oblong, the margins rounded and wavy, densely stellate-hairy on both surfaces and spiny on some of the veins; petiole up to 2 cm, densely stellate-hairy and with numerous spines. *Inflorescence* a terminal raceme; flowers bisexual; pedicels soon ascending, with dense stellate and unequal glandular hairs and a few yellowish spines. *Calyx* 5–7 mm, cup-shaped, 5-lobed, the lobes lanceolate, acute at apex, almost completely hidden by spine-like prickles. *Corolla* about 25 mm in diameter, yellow, 5-lobed, the lobes short, broadly ovate, obtuse at apex. *Stamens* 5; filaments short, declined; lowest anther much larger and longer than the others and with an incurved beak. *Style* 1. *Berry* wholly enclosed by the close-fitting and often adherent calyx; seeds coarsely undulate-rugose. *Flowers* 6–8. $2n=24$.

Introduced. Rather frequent casual from wool, bird-seed and other sources in arable fields, waste places and on tips. Scattered records in Great Britain and the Channel Islands. Native of the United States from Nebraska to Texas and introduced eastwards and northwards.

16. S. diflorum Vell. Winter Cherry
S. capsicastrum Link ex Schauer

Small, bushy, monoecious *shrub*. *Stems* up to 50 cm, pale green, with few-branched hairs, unarmed. *Leaves* alternate; lamina 2.5–14(–20) × 0.7–8.0(–11.0) cm, dull green on upper surface, paler beneath, lanceolate to lanceolate-oblong or elliptical, obtuse at apex, entire but sinuate, attenuate at base, with dense stellate hairs beneath when young, sometimes with simple eglandular hairs as well, becoming less so with age and with few on the midrib above; petiole up to 3.5 cm. *Inflorescence* of clusters of 2–4 bisexual flowers or flowers solitary; pedicels up to 15 mm, stellate-hairy, more or less erect at fruiting. *Calyx* about 5 mm, deeply divided into 5 lobes, the lobes linear or linear-lanceolate, with few stellate hairs. *Corolla* 10–14 mm in diameter, white, 5-lobed, the lobes ovate. *Stamens* 5; anthers 2.0–2.5 mm. *Style* 1. *Berry* about 15 mm in diameter, scarlet or orange-red, usually globose; seeds 2.5–3.0 mm, obovoid, irregular. *Flowers* 6–8.

Introduced. A common pot plant which has been found self-sown in pavements and sewage farms in south-east England. Native of temperate South America.

17. S. abutiloides (Griseb.) Bitter & Lillo
Abutilon solanum
Cyphomandra abutiloides Griseb.

Monoecious *shrub* up to 2 m. *Stems* and branches green, clothed with pale yellow stellate hairs and longer glandular hairs; internodes 1.5–4.0 cm. *Leaves* alternate; lamina 7–17 × 5.0–12.5 cm, broadly subreniform-ovate, attenuate-acuminate at apex, entire, cordate at base, with short, sparse stellate hairs on upper surface, and dense, pale yellow stellate hairs beneath. *Inflorescence* of 20–30 bisexual flowers densely congested; pedicels 7–10 mm, up to 15 mm in fruit. *Calyx* campanulate, 5-lobed, the lobes about 7 × 3.5–4.0 mm, lanceolate, densely stellate-hairy, and with eglandular and glandular hairs. *Corolla* 13–15 mm in diameter, white, 5-lobed, the lobes 5 × 3 mm, becoming 5–7 × 3.5 mm, with stellate hairs. *Stamens* 5, equal; filaments about 1.5 mm, glabrous; anthers yellow. *Style* 1. *Berry* 9–10 × 9–10 mm, globose, red, with stellate hairs. *Flowers* 8–10.

Introduced. Occurs sporadically in waste and cultivated ground in Nottingham. Native of South America.

12. Datura L.

Annual monoecious *herbs*, sometimes foetid. *Leaves* alternate, dentate or lobed. *Inflorescence* of solitary bisexual flowers in branch forks or leaf axils, or in leafy cymes. *Calyx* with cylindrical tube, 5-lobed, detaching near base of capsule by a line round its circumference. *Corolla* with 5 or 10 lobes. *Stamens* 5, equal, inserted at or below middle of the corolla tube, anthers opening by longitudinal slits. *Style* 1; stigma 2-lobed. *Ovary* 2-celled, with small spines, nectary ring present at base. *Fruit* a spiny or tuberculate capsule, opening by 4 valves; seeds 100 or more per capsule, D-shaped.

Contains 8–10 species centred on the south-west United States and Mexico, with several species naturalised elsewhere.

Blakeslee, A. F. (1959). *The genus* Datura. New York.
Dewolf, G. P. (1956). Notes on cultivated Solanaceae. 2. *Datura*. *Baileya* **4**: 12–23.

1. Capsule deflexed **3. inoxia**
1. Capsule erect 2.
2. Capsule with fewer than 60 spines of unequal length, some at least 13 mm **2. ferox**
2. Capsule with more than 100 spines of similar length (6–15 mm. or spines replaced with tubercles 3.
3. Capsule without spines **1(iii). stramonium** var. **inermis**
3. Capsules with numerous spines 4.
4. Corolla white **1(i). stramonium** var. **stramonium**
4. Corolla purple **1(ii). stramonium** var. **chalybaea**

1. D. stramonium L. Thorn-apple

Foetid-smelling *annual* monoecious *herb* with fibrous roots. *Stems* 50–150(–200) cm, pale green, stout, terete, erect, glabrous or slightly hairy, usually much-branched, leafy. *Leaves* alternate; lamina 5–20 × 2.5–8.0(–14.0) cm, medium green on upper surface, paler beneath, ovate, elliptical or ovate-oblong, acute at apex, sharply lobed, cuneate at base, glabrous or thinly hairy; petiole 1–6 cm, glabrous. *Inflorescence* of solitary, bisexual flowers in the bifurcations of the branches; peduncle very short, lengthening to 10 mm in fruit. *Calyx* 20–50 × 6–10 mm, pale green, narrowly cylindrical; lobes 5, 4–5 × 1.0–1.5 mm, rather unequal, subulate, minutely ciliate. *Corolla* 70–100 mm, white, mauve or purplish, infundibuliform, the tube spreading rather abruptly near the apex; lobes 5, each 7 × 7 mm, acuminate-caudate. *Stamens* 5, included; filaments about 25 mm, glabrous; anthers yellow, narrowly oblong. *Style* about 50 mm, glabrous; stigma 2-lobed. *Capsule* 3–7 × 2–4 cm, ovoid, woody, erect, usually spiny, splitting almost to the base into 4 lobes; seeds about 3.5 × 2.5 mm, blackish, reniform, strongly compressed, closely foveolate. *Flowers* 7–10. $2n=24$.

(i) Var. **stramonium**
Corolla white. *Capsule* with slender spines 2–15 mm.

(ii) Var. **chalybaea** W. D. J. Koch
D. stramonium var. *tatula* (L.) Torr.; *D. tatula* L.

Corolla mauve or purple. *Capsule* with slender spines 2–15 mm.

(iii) Var. **inermis** (Juss. ex Jacq.) Schinz & Thell.
Corolla white or pale mauve. *Capsule* without spines.

Introduced. Casual on tips and waste and cultivated ground, especially manured places, from bird-seed, wool and soya beans. Sporadic more or less throughout Great Britain and Ireland. All the varieties have been recorded. A cosmopolitan weed, probably of American origin. Contains the alkaloids hyoseyamine, hyoscine and scopolamine and is very poisonous.

2. D. ferox L. Angel's Trumpets

Somewhat foetid *annual* monoecious *herb* with a tap-root and fibrous side-roots. *Stems* up to 1 m, pale green, erect, more or less hairy especially when young, leafy. *Leaves* alternate; lamina 5–18(–25) × 4–16(–20) cm, medium green on upper surface, paler beneath, broadly ovate, shallowly lobed, the lobes obtuse to subacute at apex, somewhat shortly hairy on the veins especially on the upper surface; petiole up to 8 cm. *Inflorescence* tending to form a leafy cyme, more or less erect; flowers bisexual. *Calyx* 25–35 mm, 5-lobed, the lobes 3–5 mm, subequal, reflexed at fruiting. *Corolla* about 60 mm, white, funnel-like, 5-lobed, the lobes 1–2 mm, aristate. *Stamens* 5, inserted near the base of the corolla tube. *Stigma* 2-lobed. *Capsule* 5(–6) × 3(–4) cm (excluding spines), narrowly elliptic-ovoid, erect, with a moderate number of spines up to 2.5(–3.0) mm, stout and the uppermost longer than lower; seed about 4 mm, black, subreniform, rugose. *Flowers* 7–10. $2n = 24$.

Introduced. Mainly a wool-alien on tips and in fields. Occasional as a casual in England and Wales, mainly the south. Native of east Asia.

3. D. innoxia Mill. Recurved Thorn-apple
D. metel auct.

Annual monoecious *herb* with fibrous roots. *Stems* up to 2 m, erect, densely clothed with soft, clammy hairs, usually repeatedly branched, leafy. *Leaves* alternate; lamina 3–14 × 3–10 cm, medium green on upper surface, paler beneath, broadly ovate, acute at apex, subentire or irregularly dentate, abruptly tapered at base or sometimes truncate or subcordate, densely clothed with soft, clammy hairs; petiole 1–7 cm. *Inflorescence* of solitary, bisexual flowers in the bifurcations of the branches; peduncles about 10 mm, stout. *Calyx* with tube 6–7 cm, narrowly cylindrical, densely hairy; lobes 5, 5–6 mm, subulate. *Corolla* white, infundibuliform, with a tube 15–16 cm, spreading gradually towards the apex and terminating in 10 small, subulate lobes. *Stamens* 5, included; filaments about 6 cm, glabrous; anthers yellow, narrowly oblong. *Style* 10–12 cm, glabrous; stigma 2-lobed. *Capsule* 4–5 × 3–4 cm, ovoid, nodding, the base surrounded by a frill formed from the much accrescent base of the circumscissile calyx-tube; fruiting peduncle strongly arcuate-decurved; seeds numerous, about 5 × 4 mm, pale brown, reniform, strongly compressed, minutely and closely foveolate with a number of larger, deeper pits along the dorsal surface.

Introduced. A wool casual. Native of Central America, widely naturalised in tropical and subtropical regions.

14. Nicotiana L.

Annual to *perennial* monoecious *herbs*, rarely *shrubs*, often viscid-hairy. *Leaves* alternate, the lamina entire. *Inflorescence* a terminal panicle; flowers scentless or fragrant in the evening. *Calyx* 5-lobed, somewhat enlarged in fruit. *Corolla* divided into the tube proper, the widened portion above called the throat, and the 5-lobed limb. *Stamens* 5, equal or unequal in length; anthers dehiscing with a longitudinal suture. *Style* terminal; stigma slightly grooved. *Ovary* 2-celled, the base adnate to a thick, sometimes nectariferous, annular, hypogynous disc. *Fruit* a capsule; seeds minute.

Contains about 60 species in North and South America, Australia and the south Pacific region; naturalised and cultivated throughout the temperate tropical regions of the world. The genus was named after Jean Nicot (1530–1600), French consul in Portugal who introduced *Nicotiana* to France.

Comes, O. (1899). *Monographic du genre Nicotiana comprenant le classement botanique des tabacs industriels*. Naples.
Goodspeed, T. H. (1954). The genus *Nicotiana*. *Chron. Bot.* **16**. Waltham.
Watson, W. (1904). *Nicotiana sanderae*. *Flora and Sylva* **2**: 216.

1. Shrub; limb of corolla entire, not divided into lobes **1. glauca**
1. Annual or perennial herb; limb of corolla divided into 5 lobes 2.
2. Petiole not winged **2. rustica**
2. Petiole winged 3.
3. Stem hairy; corolla limb more than 20 mm across 4.
3. Stem nearly glabrous; corolla limb less than 20 mm across 7.
4. Calyx lobes triangular; lobes of corolla limb not curved in at base 5.
4. Calyx lobes filiform or subulate; lobes or corolla limb curved in at base 6.
5. Corolla with an obvious enlarged throat below the limb **3. tabacum**
5. Corolla without an obvious throat below the limb, but with a gradual swelling in the middle of the tube which then narrows again below the limb **4. sylvestris**
6. Corolla tube (including throat) 50–100 mm; limb 35–60 mm wide **5. alata**
6. Corolla tube (including throat) 30–33 mm; limb 25–40 mm wide **6. forgetiana**
7. Corolla 3.0–3.5 mm in diameter at the top of the calyx **7. suaveolens**
7. Corolla 1.5–2.5 mm in diameter at the top of the calyx 8.
8. Tube of corolla (including the slight throat) 12–17 mm **8. goodspeedii**
8. Tube of corolla (including the slight throat) 30–50 mm **9. ingulba**

Subgenus **1. Rustica** (Don) Goodsp.

Section *Rustica* Don

Short-lived, monoecious *shrubs*. *Leaves* with petioles. *Corolla* with limb green or yellow, entire or shallowly lobed. *Stamens* equally inserted low on corolla.

1. N. glauca Graham Tree Tobacco

Siphaulax glabra Raf.; *Nicotidendron glauca* (Graham) Griseb.

Loosely branched, monoecious *shrub* up to 6 m. *Stem* when young greenish, bluish or purplish, becoming reddish-brown, glabrous, the young plant rapidly elongating and later usually suckering. *Leaves* alternate; lamina 5–25 × 2–12 cm, glaucous, thickish, rubbery, ovate, elliptical or lanceolate more or less obtuse at apex, entire but sinuate, cuneate, truncate or subcordate at base, glabrous; petiole about half as long as lamina. *Inflorescence* an ovoid panicle, its lower branches more spreading than the upper; flowers bisexual; pedicels 3–10 mm, becoming up to 12 mm and markedly thickened. *Calyx* 10–15 mm, cylindrical, glabrous or minutely hairy, membranes lacking or nearly so; lobes 5, much shorter than tube, angular and sharp at apex. *Corolla* green in bud, later yellowish or greenish, the tube 5–8 × about 3 mm, the throat 3–6 times as long and cylindrical to clavate, the limb 7–8 mm wide, nearly circular to nearly pentagonal, entire. *Stamens* 5, subequal, extending almost to the mouth; filaments glabrous, short-kneed immediately above the insertion on the base of the corolla throat. *Style* 1; stigma just exceeding the anthers. *Capsule* 7–15 mm, broadly elliptical; seeds longer than broad, laterally compressed, brown, reticulate. *Flowers* 6–9. $2n = 24$.

Introduced. A wool casual. Probably originally native of Argentina, but now introduced worldwide and naturalised in many countries.

2. N. rustica L. Wild Tobacco

Annual monoecious *herb* with fibrous roots. *Stems* 1.0–1.5 m, usually solitary, occasionally several, pale green, thick, viscid-hairy, with slender, ascending branches, leafy. *Leaves* alternate; lamina 10–15(–30) × 6–10 cm, medium green on upper surface, paler beneath, ovate, elliptical, elliptic-lanceolate or subrotund, obtuse to acute at apex, entire and often wavy, narrowed and often asymmetrical at base, minutely hairy or viscid-hairy; petiole much shorter than the lamina, not winged. *Inflorescence* a narrow and compact to broad and loose panicle; flowers bisexual; pedicels 3–4 mm, becoming 5–7 mm. *Calyx* 8–15 mm, cylindrical, hairy, with narrow membranes, the 5 lobes more or less broadly triangular and acute at apex. *Corolla* greenish-yellow, the tube about 3 × 2 mm, the throat about 9–14 × 6–8 mm, broadly obconical and with a slight contraction at the mouth, the limb 9–16 mm wide with 5 short, obtuse, entire or apiculate lobes. *Stamens* 5, white-hairy for about 2 mm above their insertion on the base of the corolla throat, 4 extending nearly to the corolla mouth and sigmoid at base, the 5th shorter. *Style* 1; stigma a little above the stamens. *Capsule* 7–16 mm, elliptic-ovoid to subglobose, included in the calyx or nearly so; seeds 0.7–1.1 mm, brown, elliptic, reticulate. *Flowers* 6–9. $2n = 48$.

Very variable in all its parts through much cultivation and selection, its wild state being unknown.

Introduced. Once cultivated for tobacco, now grown in gardens from which it escapes onto tips and also occurs as a wool casual. Few scattered localities in England. Probably of South American origin, but grown in many parts of Central and North America in the pre-Columbian period. The first tobacco grown and exported to Europe by the American colonists for smoking, snuffing and chewing, and is now replaced by *N. tabacum*.

Subgenus **2. Nicotiana**

Subgenus *Tabacum* (Don) Goodsp.; Section *Tabacum* Don

Annual to short-lived *perennial* monoecious *herbs*. *Leaves* with broadly winged petioles. *Corolla* with limb white, pink or red, broadly lobed. *Stamens* more or less equally inserted.

3. N. tabacum L. Tobacco

N. fruticosa L.; *N. alba* Mill.; *N. angustifolia* Mill.

Annual to short-lived *perennial* monoecious *herb*. *Stems* up to 3 m, pale green, erect, thick, viscid-hairy, with few branches, leafy. *Leaves* alternate; lamina 10–50 × 5–30 cm, medium green on upper surface, paler beneath, ovate, elliptical or lanceolate, acute or shortly acuminate at apex, entire, sinuate or shallowly crenate, narrowed to a broadly winged petiole, shortly viscid-hairy. *Inflorescence* a panicle with a distinct rhachis and several, usually compound branches; flowers bisexual; pedicels 5–10(–15) mm, lengthening to 10–20(–25) mm. *Calyx* 12–20(–25) mm, cylindrical to cylindrical-campanulate, viscid-hairy, with 5 triangular, acuminate lobes shorter than or equalling the tube. *Corolla* hairy on outer surface, the tube (7–)10–15 × 2.5–3.0 mm, the throat (23–)25–40 × 3–5 mm, with its lower part half-cylindrical and pale greenish-cream and its upper half similar in colour or pink to red, usually abruptly expanded into a deep cup 7–12 mm wide, the limb 20–30 mm wide, 5-lobed or pentagonal, the lobes not curved in at the base, and white, pink or red. *Stamens* 5, inserted on base of the corolla throat, pale green, erect, orientated to upper side of flower or evenly spaced; anthers deep green, 4 longest near to the mouth of the corolla or a little exserted, the 5th shorter. *Style* 1, pale green, a little larger than the anthers; stigma green. *Capsule* 15–20 mm, narrowly ellipsoid, ovoid or globose, obtuse or acute, exserted or included in calyx; seeds about 0.5 mm, brown, globose or broadly ellipsoid, the ridges fluted. *Flowers* 6–9. $2n = 48$.

Introduced. Now rarely cultivated and found as a relic or on tips; formerly much commoner. Scattered records in England. Origin uncertain, but probably South America. In the pre-Columbian era its use would appear to have been restricted to Central and South America and the West Indies, being replaced by *N. rustica* in North America. Both species were brought to the Old World early in the sixteenth century, but *N. tabacum* is now the species mainly used for smoking-tobacco, but *N. rustica* is still locally grown as a source of snuff and nicotine. *N. tabacum* is extremely variable and different cultivars are grown in different parts of the world.

Subgenus **3. Petunioides** (Don) Goodsp.

Section *Petunioides* Don

Annual or *perennial herbs. Leaves* petioled. *Corolla* with limb whitish or greenish, 5-lobed. *Stamens* equally or unequally inserted.

4. N. sylvestris Speg. Argentine Tobacco

Perennial monoecious *herb. Stems* 1–2 m, single or several, pale green, erect, stout, viscid-hairy, with erect branches, leafy. *Leaves* alternate, the basal in a rosette; lamina 20–50×8–26 cm, medium green on upper surface, paler beneath, elliptical or elliptical-ovate, slowly decreasing in size upwards, obtuse to acuminate at apex, entire or slightly sinuate, auricled and sometimes decurrent at base to a winged petiole, viscid-hairy. *Inflorescence* a congested, shortly columnar panicle with a thick rhachis; pedicels 5–15 mm, extending to 15–25(–30) mm; flowers bisexual, pendulous and mildly fragrant. *Calyx* 10–18 mm, oblong or subglobose, the 5 lobes broadly triangular, more or less equal and shorter than the tube. *Corolla* excluding limb 65–85 mm, white, hairy, the tube 20–25×about 2 mm, cylindrical from a slightly broadened base, the throat cylindrical basally for 8–10 mm, then gradually symmetrically or ventricosely dilated to 6 mm wide and tapered again to 5 mm wide, the limb 30–35 mm wide, divided into 5 broadly triangular or ovate and acuminate lobes not curved in at the base. *Stamens* 5, unequally or subequally inserted on the base of the corolla throat, hairy for a short distance above insertions, erect; 4 anthers barely included in the corolla tube, the 5th slightly lower. *Style* 1, the stigma slightly exceeding the anthers. *Capsule* 15–18 mm, included or exserted from the calyx; seeds 0.5 mm, pale brown, elliptical, reticulate. *Flowers* 6–9. $2n=24$.

Introduced. A persistent relic of cultivation in the Channel Islands. Native of Argentina.

5. N. alata Link & Otto Sweet Tobacco

N. persica Lindl.

Perennial monoecious *herb. Stems* 100–150 cm, pale green, erect, viscid-hairy, with ascending basal branches, leafy. *Leaves* alternate; lamina 20–25×7–11 cm, medium green on upper surface, paler beneath, elliptical, ovate, elliptic-ovate or lanceolate, obtuse to acute at apex, entire or sinuate, narrowed below to a broadly winged, amplexicaul, auriculate, decurrent petiole, viscid-hairy. *Inflorescence* of short, few-flowered panicles; pedicels 5–20 mm, extending to 12–21 mm; bracts lanceolate, acuminate and decurrent; flowers bisexual, very fragrant in evening. *Calyx* 15–25 mm, campanulate, ribbed, sometimes shortly and stiffly hairy, more or less glandular-hairy, the membranes narrow, the 5 lobes subulate and about as long as tube. *Corolla* excluding limb and including throat 50–100 mm, pale greenish, the tube 3–4 mm wide, the throat one-third to half as long as the tube, including short often ventricose dilation, the limb 35–60 mm wide, chalky white within and deeply divided into 5 broadly ovate lobes curved in at the base. *Stamens* 5, inserted at the base of the corolla throat, glabrous, 4 extending to or near the mouth, sometimes kneed, the 5th 4–5 mm shorter, not kneed; anthers purple. *Style* 1, the stigma slightly exceeding the anthers. *Capsule* 12–17 mm, ovoid, included in the calyx; seeds about 0.7 mm, greyish-brown, subglobose-ellipsoid, reticulate, the ridges slightly wavy. *Flowers* 6–9. $2n=18$.

Introduced. Much grown for ornament and a frequent casual on tips and rough ground. Scattered records in much of Great Britain. Native of South America.

× **forgetiana** = **N.×sanderae** W. Watson

This hybrid consists of variable annuals with spathulate, undulate leaves, with the corolla tube greenish-yellow at base and the limb red or occasionally white to rose-carmine or purple. The flowers are heavily fragrant in the evening.

Occurs on tips and waste land as a garden casual. Widespread in England and Wales and Isle of Man. Garden origin.

6. N. forgetiana Hemsl. Red Tobacco

Annual sometimes short-lived *perennial*, monoecious *herb* with fibrous roots. *Stems* 50–100 cm, pale green, erect, slender, hairy, with long branches, leafy. *Leaves* alternate; lamina 15–30×7–9 cm, medium green on upper surface, paler beneath, oblanceolate, obovate-elliptical, lanceolate or oblong-lanceolate, obtuse at apex, sinuate, narrowed to a short broadly winged petiole, more or less auriculate and decurrent, shortly hairy. *Inflorescence* a loosely branched panicle, viscid-hairy; pedicels 6–10 mm, lengthening to 7–12 mm. *Calyx* 12–15 mm, cylindrical to subcampanulate, 5-veined or faintly ribbed, viscid-hairy, with 5 filiform lobes from a triangular base of unequal length, with the longest exceeding the tube. *Corolla* pale green, with an intense purplish-red flush, the tube subclavate, including the throat 20–33×3–4 mm, the abrupt ventricose dilation below the slightly contracted mouth 5–8 mm wide, the limb 25–40 mm wide, purplish-red, deeply divided into 5, ovate, acuminate lobes which are curved in at the base. *Stamens* 5, equal or nearly so, extending to mouth of corolla; filaments purplish, inserted about 6 mm above the base of the corolla, sometimes sparsely hairy and kneed below; anthers purple or purplish-brown. *Style* 1, the stigma slightly exceeding the anthers. *Capsule* 9–12 mm, slightly exserted from the calyx, ovoid to oblong-ovoid; seeds about 0.7 mm, brown, oblong-ellipsoid to subglobose, reticulate. *Flowers* 6–9. $2n=18$.

Introduced. Grown as an ornament in gardens and found as a casual on tips and rough ground. Few records in England. Native of Brazil. Named after Forget, a collector for Messrs Sander and Sons.

7. N. suaveolens Lehm. Australian Tobacco

N. undulata Vent.; *N. australasiae* R. Br.

Annual monoecious *herb* with fibrous roots. *Stems* several, 50–150 cm, pale green, slender, glabrous or nearly so, with long branches, leafy. *Leaves* at first in a rosette, then alternate; lamina 10–25(–35)×3–9 cm, medium yellowish-green on upper surface, paler beneath, oblanceolate, elliptic-ovate or lanceolate, more or less acute at apex, entire or sinuate, narrowed at base to a short-winged petiole, glabrous or nearly so and somewhat fleshy. *Inflorescence* a loose, little-branched panicle; pedicels 3–15 mm, extending to 8–27 mm; flowers bisexual, heavily fragrant in the evening. *Calyx* (7–)10–16 mm, tube cylindrical, slightly 5-ribbed, minutely hairy, membranes narrow, with 5 subequal, rather narrow, pointed lobes about as long as the tube. *Corolla* cream, marked by green veins and purple tinging, the tube 5–10×3.0–3.5 mm, the throat longer than the tube, a little broader and apically flared, the limb 8–15 mm wide, the 5 lobes short, broadly obtuse and nearly entire or obcordate. *Stamens* 5, filaments very pale green, 4 equally inserted close to the mouth, anthers pale green, 2 included, 2 partly exserted, remaining stamen usually inserted in lower half of corolla, its anther 4–6 mm below the others. *Stigma* 1, a little shorter than anthers, green. *Capsule* about 10 mm, exserted or included, narrowly ovoid-oblong; seeds 0.8–1.0(–1.2) mm, brown, more or less reniform. *Flowers* 6–9. $2n=32$.

Introduced. A wool casual. Native of Australia.

8. N. goodspeedii H.-M. Wheeler

Small-flowered Tobacco

N. suaveolens var. *parviflora* Benth.

Annual monoecious *herb* with fibrous roots. *Stems* numerous, 30–100 cm, pale green, erect, slender, glabrous, leafy. *Leaves* alternate; lamina 8–30×4–10 cm, medium green on upper surface, paler beneath, somewhat fleshy, spathulate to narrowly or broadly lanceolate or narrowly elliptical, acute at apex, entire but sinuate, narrowed below to a narrowly winged petiole, glabrous. *Inflorescence* a long, loose panicle, major branches long and ascending; pedicels 3–9 mm, lengthening to 20 mm; flowers fragrant. *Calyx* 5–7 mm, tubular, minutely hairy, membranes long and rather conspicuous, lobes 5, more or less equal, subulate, seldom as long as the tube. *Corolla* greenish, often purplish tinged; tube 12–17×1.5–2.5 mm, shortly hairy, apically slightly ventricose; limb 7–9 mm wide, with 5, obcordate lobes. *Stamens* 5, included; 4 filaments very short, unequal and at or close to 1 level near the mouth of the corolla, the 5th 1–2 mm lower, 5–6 mm and inserted in lower half of corolla. *Style* 1; stigma just exceeding calyx, narrowly elliptic-oblong; seeds 0.6–0.7 mm, dark brown, compressed trigonate-reniform, wrinkled. *Flowers* $2n=40$.

Introduced. Wool casual. Native of Australia.

9. N. ingulba J. M. Black Aborigine Tobacco

N. rosulata (S. Moore) Domin subsp. *ingulba* (J. M. Black) P. Horton

Annual monoecious *herb* with fibrous roots. *Stems* usually several, 30–60 cm, erect, slender, nearly glabrous, little branched, not very leafy. *Leaves* mostly basal, alternate; lamina 8–10 × 1.5–2.0 cm, medium green on upper surface, paler beneath, lanceolate or narrowly elliptical or oblong-elliptical, acute at apex, sinuate, narrowed at base to a narrowly winged petiole, nearly glabrous. *Inflorescence* a sparsely branched panicle; flowers bisexual; peduncles 3–10 mm, extending to 15 mm, shortly hairy. *Calyx* 12–17 mm, subcylindrical, ribbed, puberulous, membranes conspicuous, with 5 linear-lanceolate lobes at least half as long as the tube. *Corolla* tube 30–50×1.5–2.5 mm, green, apically slightly broadened, the limb about 10–20 mm wide, white inside, the 5 lobes very short, broadly obtuse, retuse or emarginate, hairy outside. *Stamens* 5, included in corolla tube, 4 with filaments inserted near mouth of the corolla, the 5th 5–15 mm and inserted near the middle of the corolla. *Style* 1, stigma slightly above anthers. *Capsule* 11–12 mm, shorter than or equalling calyx, narrowly ellipsoid; seeds about 0.9 mm, brown, compressed-reniform, reticulate. *Flowers* 6–9.

Introduced. A wool casual. Native of Australia. The leaves, stem and roots are dried and ground up on a stone by the Aborigine Australians then formed into a ball and chewed. *Angulba* is the Aranda word for tobacco.

15. Petunia Juss.

Annual monoecious *herbs*. *Stems* glandular-hairy. *Leaves* alternate, simple, entire, shortly petiolate or sessile, without stipules. *Inflorescence* of solitary, axillary, bisexual flowers. *Calyx* divided half to two-thirds of the way to the base into 5 linear lobes. *Corolla* 60–80 mm in diameter, white, blue, violet, pink, rose and orange, trumpet-shaped, with 5 broad lobes and softly glandular-hairy. *Stamens* 5, unequal in length. *Style* 1. *Fruit* a capsule.

Contains 3 species from south Brazil to north-east Argentina and Bolivia.

1. P.×hybrida (Hook.) Vilm. Petunia

P. axillaris (Lam.) Britton, Sterns & Poggenb. × **integrifolia** (Hook.) Schinz & Thell.

Annual monoecious *herbs* with fibrous roots. *Stems* up to 60 cm, pale green, procumbent to erect, glandular-hairy, branched, leafy. *Leaves* alternate; lamina 1.5–5.0×0.8–4.0 cm, pale yellowish-green on upper surface, paler beneath, lanceolate to ovate or ovate-elliptical, obtuse at apex, entire, narrowed at base, with unequal glandular hairs on both surfaces; petiole short or absent. *Inflorescence* of solitary, axillary, bisexual flowers. *Calyx* 11–15 mm, pale green, sometimes suffused purple, divided from half to two-thirds of the way to the base into 5 lobes, the lobes linear and rounded at apex, glandular-hairy. *Corolla* 60–80 mm in diameter, white, blue, violet, pink, rose, red and orange and in most cases the veins in the throat of a deeper colour; tube 25–40 mm, narrow at base, gradually widening towards the apex; limb spreading and curved over at the rim, with 5 broad lobes with sinuate margins and softly glandular-hairy. *Stamens* 5; filaments unequal in length, those and the anthers often the same colour as the flower, the anthers darker. Style 1, pale green; stigma equalling anthers. *Capsule* 10–15 mm, elliptic, smooth; seeds numerous, small. *Flowers* 6–9. $2n=14, 28$.

Introduced. Much grown in gardens and frequent on waste land tips where it is sometimes self-sown. Frequent in southern Great Britain. Of garden origin.

124. CONVOLVULACEAE Juss. nom. conserv.

Annual to *perennial* monoecious *herbs*. *Stems* procumbent or twining. *Leaves* alternate, simple, entire and petiolate, without stipules. *Inflorescence* of 1–few bisexual flowers in leaf axils, actinomorphic and hypogynous, usually with 2 bracteoles near their base. *Sepals* 5. *Corolla* funnel to trumpet shaped, the 5 petals sometimes fused only to about halfway. *Stamens* 5, borne on the corolla tube. *Style* usually 1, 2 in *Dichondra*; stigmas 1 or 2 per style, globose to linear or 2- to 3-lobed. *Fruit* a capsule, usually without proper dehiscence mechanism.

Contains 55 genera and some 1,650 species, mainly tropical.

1. Corolla less than 5 mm, divided to about halfway; capsule deeply bilobed **1. Dichondra**
1. Corolla more than 10 mm, normally not or scarcely lobed; capsule more or less not lobed 2.
2. Bracteoles ovate, often pouched, partly or wholly obscuring the sepals **3. Calystegia**
2. Bracteoles linear, not obscuring sepals 3.
3. Sepals obtuse to retuse, rarely acuminate; stigmas linear **2. Convolvulus**
3. Sepals acute to acuminate; stigmas more or less globose **4. Ipomoea**

1. Dichondra J. R. & G. Forst.

Perennial monoecious *herb* with stolons forming extensive mats or low cushions. *Leaves* alternate, subrotund to reniform. *Flowers* bisexual, usually solitary and axillary. *Calyx* 5-lobed. *Corolla* 5-lobed, greenish-white. *Stamens* 5. *Styles* 2; stigma 1 per style, capitate. *Ovary* deeply bilobed. *Fruit* a bilobed capsule, usually 1-seeded; seeds smooth and glabrous.

Contains 4–5 species in warm-temperate and tropical areas.

1. D. micrantha Urb. Kidneyweed
D. repens auct.

Perennial monoecious *herb* with stolons forming extensive mats or low cushions. *Stems* up to 5 cm high, and 50 cm long, purplish, freely branching, appressed-hairy. *Leaves* alternate; internodes 10–20 mm; lamina (0.5–)1.0–3.0 cm, dull medium green on upper surface, paler beneath, subrotund to reniform, rounded or slightly emarginate at apex, entire, broadly cordate at base, glabrous or nearly so on upper surface, with appressed hairs beneath; veins not impressed above, prominent beneath; petioles up to 6 cm, erect or nearly so. *Flowers* bisexual, solitary and axillary; peduncles 5–15 mm; bracts very small and inconspicuous. *Calyx* about 2 mm, 5-lobed, densely hairy, the lobes alternately linear-oblong and ovate-oblong, obtuse to subacute at apex. *Corolla* 1.5–2.5 mm, 4–5 mm in diameter, greenish-white, 5-lobed to about halfway, the lobes narrowly oblong to lanceolate. *Stamens* 5, included; anthers violet or violet at the margin. *Styles* 2, filiform. *Capsule* 2.0–2.5 × 4–5 mm, membranous, usually deeply 2-lobed, slightly to strongly, stiffly hairy; seeds about 1.5 mm, yellow to dark brown. *Flowers* 6–8.

Introduced. Was naturalised on fixed sand-dunes near Hale, Cornwall from 1955 to 1979. Native of east Asia.

2. Convolvulus L.

Annual to *perennial* monoecious *herbs* with rhizomes. *Stems* trailing or climbing. *Leaves* alternate, triangular or ovate-oblong to linear, hastate to sagittate at base. *Inflorescence* a 1- to few-flowered axillary or terminal corymb; with 2 linear bracteoles some way below each flower; flowers bisexual. *Calyx* 5-lobed, divided nearly to base. *Corolla* funnel-shaped, scarcely divided, much longer than calyx. *Stamens* 5, inserted at bottom of corolla tube, included; anthers dilated near the base. *Style* 1, filiform; stigmas 2, linear. *Ovary* unlobed. *Fruit* a capsule, 1- to 2-celled, unlobed, usually with 4 seeds.

Contains about 250 species, chiefly in temperate regions.

Brown, E. O. (1946). Notes on some variations in Field Bindweed (*Convolvulus arvensis* L.). *Iowa State Coll. Jour. Sci.* **20**: 269–276.

Druce, G. C. (1914). *Convolvulus arvensis L. Rep. Bot. Exch. Club Brit. Isles*: **3**: 330.

Grime, J. P. et al. (1988). *Comparative plant ecology*. London.

Hultén, E. & Fries, M. (1986). *Atlas of north European vascular plants north of the Tropic of Cancer.* 3 vols. Königstein.

Proctor, M., Yeo, P. & Lack, A (1996). *The natural history of pollination*. London.

Pugsley, H. W. (1916). *Convolvulus arvensis. Jour. Bot. (London)* **54**: 88.

Weaver, S. E. & Riley, W. R. (1982). The biology of Canadian weeds. 53: *Convolvulus arvensis* L. *Canad. Jour. Pl. Sci.* **62**: 461–472.

1. Corolla deep blue inside with a white eye in centre and white bands on the sutures outside **3. sabatius**
1. Corolla white, pink or white variously marked with pink or a deeper colour 2.
2. Sepals acute at apex **2. erubescens**
2. Sepals obtuse at apex 3.
3. Corolla deeply divided into 5 or 6 lobes **1(11). arvensis** forma **stonestreetii**
3. Corolla not deeply divided but sometimes with a tetragonal or pentagonal outline 4.
4. Inside of corolla pure white except for the yellow throat and sometimes a ring of marks above the yellow 5.
4. Inside of corolla tinted pink or clear pink stripes and sometimes also dark marks above the yellow throat 6.
5. Inside of corolla pure white except for the yellow throat **1(1). arvensis** forma **arvensis**
5. Inside of corolla pure white with a yellow throat and a ring of dark marks just above the yellow throat **1(2). arvensis** forma **notata**
6. Inside of corolla tinted pink all over most of the surface or pink except for a white band above the yellow throat 7.

6. Inside of corolla with pink and white stripes 10.
7. Inside of corolla tinted pink all over the surface except for the yellow throat 8.
7. Inside of corolla deep pink round the top half with a white band above the yellow throat into which point 5 lobes of the pink 9.
8. Inside of corolla without a band of marks above the yellow throat **1(3). arvensis** forma **pallidiroseus**
8. Inside of corolla with a band of marks just above the yellow throat **1(4). arvensis** forma **pallidinotatus**
9. Inside of corolla without a band of marks just above the yellow throat **1(9). arvensis** forma **perroseus**
9. Inside of corolla with a band of marks just above the yellow throat **1(10). arvensis** forma **quinquevulnerus**
10. Inside of corolla with 5 pink stripes and 10 white ones 11.
10. Inside of corolla with 10 pink stripes and 10 white ones 12.
11. Inside of corolla without a band of marks just above the yellow throat **1(5). arvensis** forma **pentarrhabdotus**
11. Inside of corolla with a band of marks just above the yellow throat **1(6). arvensis** forma **pentastictus**
12. Inside of corolla without a band of marks just above the yellow throat **1(7). arvensis** forma **decarrhabdotus**
12. Inside of corolla with a band of marks just above the yellow throat **1(8). arvensis** forma **decemvulnerus**

1. C. arvensis L. Field Bindweed

Perennial monoecious *herb* with deep and extensively spreading, pale brown, fleshy roots. *Stems* up to 1 m, sometimes more when climbing, numerous, pale green, sometimes suffused brownish-purple, trailing along the ground, twining about other plants or climbing hedgerows, with 4 slightly winged angles, twisted, smooth, glabrous or thinly hairy, branched mainly from the base, leafy. *Leaves* alternate, very variable; lamina 1.5–5.0×1.0–3.0 cm, dull greyish-green on upper surface, scarcely paler beneath, mostly oblong-hastate or sagittate, sometimes linear, pointed to obtuse or rounded at apex, the margins slightly deflexed, the basal lobes rather short, more or less acute to obtuse and usually diverging at right angles but sometimes small or even obsolete, truncate or cuneate at base, glabrous or slightly to quite thickly hairy; petiole up to 25 mm, rather flattened, channelled above, glabrous or hairy. *Inflorescence* usually of single axillary, bisexual flowers, often the 2 and sometimes 3 flowers; peduncles up to 5 cm, acutely quadrangular, deflexed at apex in fruit; pedicels 10–20 mm; bracts and bracteoles 2.5–3.0 mm, subulate. *Sepals* 5, 3.5–4.0×2.0–2.5 mm, oblong or oblong-obovate, obtuse at apex, the 2 outer with narrow, scarious margin, the 3 inner membranous at the apex and emarginate. *Corolla* 15–40 mm in diameter, faintly scented, broadly and flatly funnel-shaped, often somewhat tetragonal or pentagonal in outline, rarely deeply lobed, the margins often crenulate, throat always yellow, outside of corolla with 5 pink lobes of various intensity with white lobes between which are sometimes touched pink, inside of corolla white, flushed pale pink, with 5 pink lobes and 5 white, with 10 pink lobes and 10 white, or all 4 of these can have a zigzag or dentate ring of deep crimson marks about the yellow throat and lastly the whole inside of the corolla is deep pink except for a white ring about the yellow throat. *Stamens* 4–5; filaments 8–14 mm, white, sometimes tinted pink, papillose towards the base; anthers white to purple. *Style* 1, white, about 10 mm; stigma white, bilobed. *Capsule* about 6 mm in diameter, pale brown, broadly ovoid or subglobose; seeds about 3.5 mm, dark brown, angular, scabridulous. *Flowers* 6–8. $2n=48$.

(1) Forma **arvensis**
Outside of *corolla* with 5 yellowish-green, tinted purple stripes, the rest white; inside of corolla white above the yellow throat. *Filaments* and *anthers* white.

(2) Forma **notatus** P. D. Sell
Outside of *corolla* with 5 yellowish-green stripes, the rest white; inside of corolla white with a band of zigzag and dentate deep crimson marks just above the yellow throat. *Filaments* and *anthers* white.

(3) Forma **pallidiroseus** P. D. Sell
Outside of *corolla* with 5 purplish stripes and the rest tinted pink; inside of corolla tinted pink above the yellow throat. *Filaments* white; *anthers* purple.

(4) Forma **pallidinotatus** P. D. Sell
Outside of *corolla* with 5 purplish stripes and rest tinted pink; inside of corolla tinted pink with a ring of zigzag and dentate purplish marks just above the yellow throat. *Filaments* white; *anthers* purple.

(5) Forma **pentarrhabdotus** P. D. Sell
Outside of *corolla* with 5 pink stripes with rest of surface tinted pink; inside of corolla with 5 pink stripes and 5 white stripes above the yellow throat. *Filament* pink; *anthers* purple.

(6) Forma **pentastictus** P. D. Sell
Outside of *corolla* with 5 purplish stripes with the rest tinted pink; inside of corolla with 5 pink stripes and 5 white ones with a band of zigzag and dentate purplish marks above the yellow throat. *Filaments* tinted pink; *anthers* purple.

(7) Forma **decarrhabdotus** P. D. Sell
Outside of *corolla* with 5 purplish stripes with the rest pink; inside of corolla with 10 pink stripes with white in between. *Filaments* pink; *anthers* purple.

(8) Forma **decemvulnerus** P. D. Sell
Outside of *corolla* with 5 purplish stripes and the rest pink; inside of corolla with 10 pink stripes with white between, with a band of zigzag and dentate marks above the yellow throat. *Filaments* pink; *anthers* purple.

(9) Forma **perroseus** P. D. Sell
Outside of *corolla* with 5 purple stripes and the rest pink; inside the corolla wholly deep pink except for a white star above the yellow throat. *Filaments* pink; *anthers* purple.

(10) Forma **quinquevulnerus** P. D. Sell
Outside of corolla with 5 purple stripes the rest pink; inside of corolla wholly deep pink except for a white star above the yellow throat and with a band of zigzag and dentate marks above the yellow throat. *Filaments* pink; *anthers* purple.

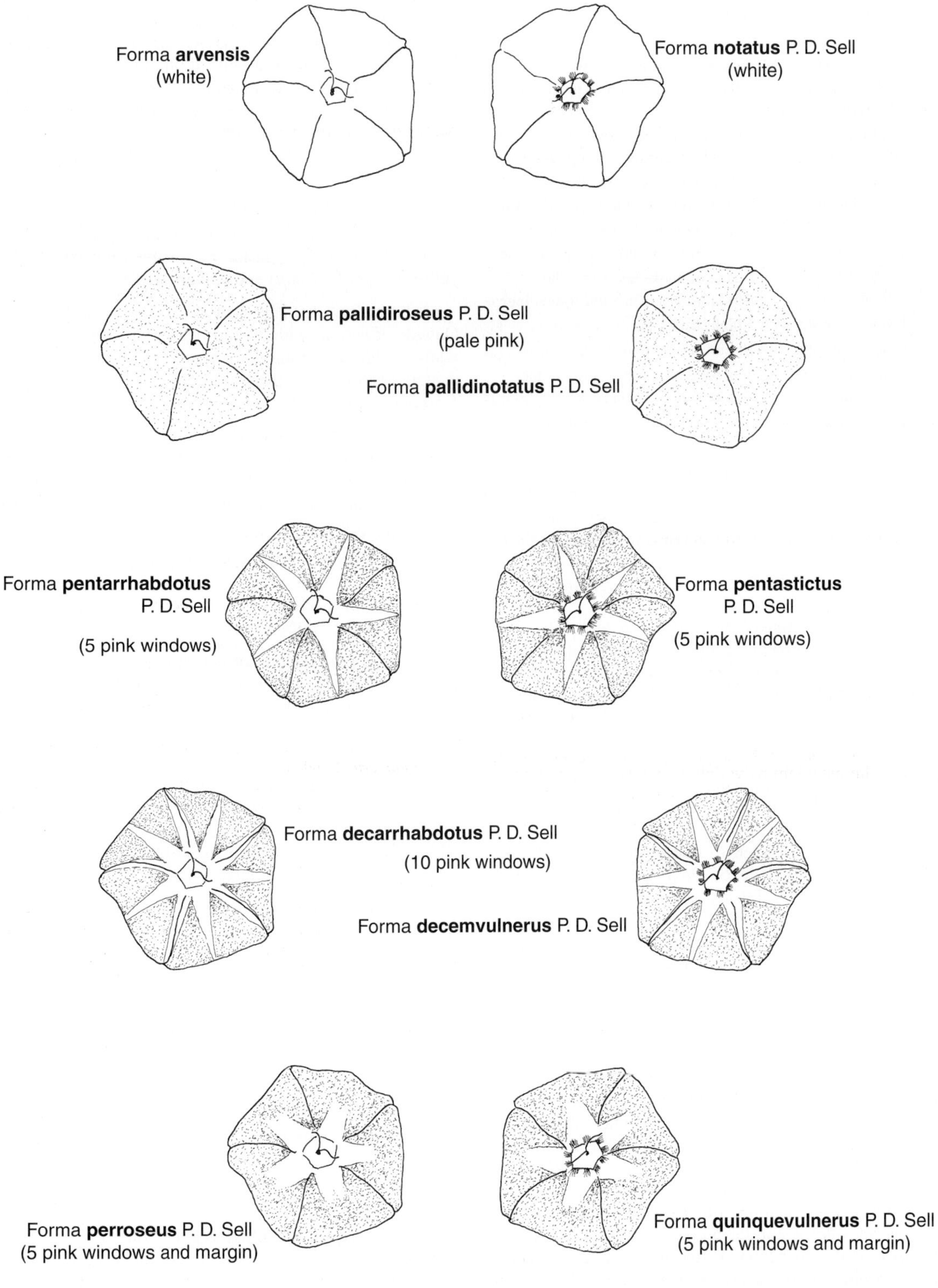
Forma **arvensis**
(white)
Forma **notatus** P. D. Sell
(white)
Forma **pallidiroseus** P. D. Sell
(pale pink)
Forma **pallidinotatus** P. D. Sell
Forma **pentarrhabdotus**
P. D. Sell
(5 pink windows)
Forma **pentastictus**
P. D. Sell
(5 pink windows)
Forma **decarrhabdotus** P. D. Sell
(10 pink windows)
Forma **decemvulnerus** P. D. Sell
Forma **perroseus** P. D. Sell
(5 pink windows and margin)
Forma **quinquevulnerus** P. D. Sell
(5 pink windows and margin)

Flowers of **Convolvulus arvensis** L.

(11) Forma **stonestreetii** (Druce) Druce
C. arvensis var. *stonestreetii* Druce

Corolla divided deeply into 5 or 6 lobes.

Native. Waste or cultivated ground, waysides, railway banks, open scrub, rough and short grassland, including disturbed chalk grassland, and dunes and shingle by the coast. Throughout Great Britain and Ireland, but absent from mountainous areas. The species occurs throughout the temperate regions of both hemispheres. Eurosiberian southern-temperate element. Formerly abundant in arable land where it often got tangled with the sails of the binders at harvest, but now almost completely destroyed by herbicides. Flower colour appears to be constant when grown from seed and one can possibly tell the extent of a single plant by following the line of identical flowers. The longest (possibly one plant) seen was 90 m, along a Cambridgeshire roadside. If this is true the size of leaf and flowers is not constant. Where a plant appears to grow from long grass to a lawn, the size of leaf and flower alters but not flower colour. Proctor, Yeo & Lack (1996) say the degree of variation in corolla colour and other characters suggest that much cross-pollination must take place. P. D. S. has spent a long time watching insects where more than one colour flower occurs and has not seen the insects go from one colour flower to another, but to the same colour even if it is a different species of plant e.g. white-flowered *Convolvulus arvensis* to *Calystegia sepium* or pink-windowed *Convolvulus* to *Epilobium hirsutum*. All flower colour forms are widespread in East Anglia, but there are rarely more than one or two together and all have been seen elsewhere, scattered about Great Britain.

2. C. erubescens Sims — Australian Bindweed

Perennial monoecious *herb* with a thick rootstock. *Stems* long and trailing or rarely climbing, pale green, hairy. *Leaves* alternate; lamina variable, mostly 1–6 cm, medium green on upper surface, paler beneath, narrowly ovate to linear, obtuse to acute at apex, sometimes lobed, the lobes entire, slightly crenate to deeply incised or lobed again, sagittate-cordate at base, glabrous or hairy. *Flowers* bisexual, axillary or in umbel-like cymes; pedicels up to 35 mm; bracts 1.0–1.5 mm. *Sepals* 5, 4–6 mm, linear, acute at apex, hairy. *Corolla* 10–20 mm in diameter, pink or white. *Capsules* 2-locular, 4-valved. *Flowers* 6–8.

Introduced. A wool casual. Native of Australia.

3. C. sabatius Viv. — Great Blue Convolvulus
Including *C. mauritanicus* Boiss.

Perennial monoecious *herb* forming mats. *Stems* numerous, pale green, long and trailing, slender, nearly terete, hairy. *Leaves* alternate; lamina 1.5–4.0 × 1.0–2.6 cm, medium green on upper surface, paler beneath, broadly ovate, elliptical or oblong-elliptical, rounded or obtuse at apex, entire, truncate at base, hairy, especially on the margins; petiole very short. *Flowers* bisexual, axillary, usually solitary; peduncles longer than subtending leaves; pedicels slender, hairy; bracts 6–10 mm, narrowly linear, hairy, positioned about 5 mm below calyx. *Sepals* 5, 5–9 mm, the 3 outer lanceolate, the 2 inner narrower and more or less acuminate at apex, hairy. *Corolla* 17–20 × 23–30 mm, deep blue inside, with a white eye in centre, and white bands on the sutures outside, broadly funnel-like. *Stamens* 5, unequal; filaments glandular-hairy towards the base. *Style* 1; stigma filiform. *Capsule* 2-locular, 4-valved. *Flowers* 5–6.

Introduced. Bird-seed casual and a garden escape on a few walls on Tresco and St Mary's in the Isles of Scilly. Native of the Mediterranean region.

3. Calystegia R. Br.

Perennial monoecious *herbs* with rhizomes. *Stems* climbing or trailing. *Leaves* alternate, triangular and sagittate or reniform. *Inflorescence* usually a single bisexual flower in a leaf axil, with 2 ovate, often more or less pouched bracteoles partly or wholly concealing the calyx. *Sepals* 5, free. *Corolla* scarcely divided, much longer than the calyx. *Stamens* 5, adnate to lower part of corolla tube. *Style* 1; stigmas 2. *Fruit* a capsule.

Contains about 25 species and several subspecies in temperate and subtropical regions. They vary considerably over their ranges and merge geographically one into another. Division into species and subspecies is of necessity somewhat arbitrary, with intermediates in intermediate areas and presumed hybrids when introduced into the range of a different taxon.

Brummitt, R. K. (1967) *Calystegia sepium* (L.) R. Br. subsp. *roseata* Brummitt. *Watsonia* **6**: 298.
Brummitt, R. K. (1973). *Calystegia*: Some British chromosome counts. *Watsonia* **9**: 369–370.
Brummitt, R. K. (1980). Further new names in the genus *Calystegia* (Convolvulaceae). *Kew Bull.* **35**: 327–334.
Brummitt, R. K. (1996). Two subspecies of *Calystegia silvatica* (Kit.) Griseb. Convolvulaceae in the Mediterranean region. *Lagascalis* **18**: 338–340.
Brummitt, R. K. (1998) *Calystegia.* In Rich, T. C. G. & Jermy, A. C. *Plant crib*. London.
Brummitt, R. K. & Heywood, V. H. (1960). Pink-flowered Calystegiae of the *Calystegia sepium* complex in the British Isles. *Proc. B. S. B. I.* **3**: 384–385.
Brummitt, R. K. & Chater, A. O. (2000). *Calystegia* (Convolvulaceae) hybrids in west Wales. *Watsonia* **23**: 161–165.
Grime, J. P. et al. (1988). *Comparative plant ecology*. London.
Hultén, E. & Fries, M. (1986). *Atlas of north European vascular plants north of the Tropic of Cancer.* 3 vols. Königstein.
Stace, C. A. (1961). Some studies in *Calystegia*: compatibility and hybridisation in *C. sepium* and *C. silvatica*. *Watsonia* **5**: 88–105.
Stace, C. A. (1973). Chromosome number in the British species of *Calystegia* and *Convolvulus*. *Watsonia* **9**: 363–367.
Stace, C. A. (1973). *Calystegia*: Inheritance of the schizoflorous character. *Watsonia* **9**: 370–371.

1. Stems not or weakly climbing; at least some leaves reniform — **1. soldanella**
1. Stems usually strongly climbing; leaves ovate and sagittate — 2.
2. Bracteoles 10–18 mm wide when flattened, not or little overlapping at margin, not or little obscuring sepals in lateral vein — 3.
2. Bracteoles 18–45 mm wide when flattened, strongly overlapping at margin, completely or nearly obscuring sepals in lateral view — 7.

3. Base of leaf with a broad, rounded sinus **2(c). sepium** subsp. **spectabilis**
3. Base of leaf with a narrow, acute sinus 4.
4. Corolla pink or with 5 pink windows 5.
4. Corolla white 6.
5. Plant glabrous **2(a,2). sepium** subsp. **sepium** forma **colorata**
5. Plant with sparsely short hairy stems, petioles and pedicels **2(b). sepium** subsp. **roseata**
6. Corolla divided into deep lobes **2(a,3). sepium** subsp. **sepium** forma **schizoflora**
6. Corolla not divided into deep lobes **2(a,1). sepium** subsp. **sepium** forma **sepium**
7. Corolla divided into deep lobes **4(b,i). silvatica** subsp. **disjuncta** forma **quinquepartita**
7. Corolla not divided into deep lobes 8.
8. Corolla pink or with pink windows 9.
8. Corolla white 11.
9. Base of leaves with an oblong or squarish sinus **3. pulchra**
9. Base of leaves with a wide rounded sinus 10.
10. Bracteoles not overlapping **× scanica**
10. Bracteoles overlapping **× howittiana**
11. Bracteoles not overlapping **× scanica**
11. Bracteoles overlapping 12.
12. Bracteoles broadly ovate, the distance from base to apex being about equal to the distance from midrib to margin with the apex broadly rounded and the base very saccate **4(a). silvatica** subsp. **silvatica**
12. Bracteoles ovate-lanceolate, the distance from base to apex being about twice the distance from midrib to margin, obtusely pointed and less saccate at base **4(b). silvatica** subsp. **disjuncta**

1. C. soldanella (L.) R. Br. Sea Bindweed
Convolvulus soldanella L.

Perennial monoecious *herb* with white, slender, fleshy, cylindrical, little-branched, far-creeping rhizome. *Stems* 10–50(–100) cm, pale green, suffused purplish, simple or little branched, prostrate or trailing, glabrous, with several, slightly winged, angled, leafy, containing milky latex. *Leaves* alternate, lamina 1.5–5.0 × 2–5 cm, shiny medium yellowish-green on upper surface with impressed veins, slightly paler beneath with prominent veins, smooth, subrotund or reniform, up to twice as wide as long, rounded with a minute point at apex, more or less bluntly angular, entire, cordate at base, glabrous, but with scattered, sessile glands on both surfaces; petiole usually longer than lamina, glabrous. *Inflorescence* of solitary, axillary, bisexual flowers; peduncle longer than leaves, sharply quadrangular, glabrous; bracteoles 10–15 × 10–15 mm, subrotund to ovate, rounded at tip, shorter than calyx, slightly overlapping. *Sepals* 5, very equal in width, ovate-elliptical, obtuse-apiculate, the 2 outer larger. *Corolla* 32–52 mm, pink or pale purple, with 5 whitish or yellowish folds, yellow in the throat, funnel-shaped. *Stamens* 5; filaments whitish, dilated and glandular-hairy in the lower part; anthers pale. *Style* 1, longer than stamens; stigmas 2, oblong. *Capsule* 15–20 mm, brownish, broadly ovoid or subglobose, obtusely 3- or 4-angled, mucronate; seeds 2–4, dark brown, bluntly triquetrous. *Flowers* 6–8. Pollinated mainly by bumble-bees. Sometimes self-pollinated. $2n=22$.

Native. Sand dunes and shingle by the sea. Coasts of Great Britain and Ireland north to central Scotland. Atlantic and Mediterranean coasts of Europe from Denmark southwards; North Africa; Asia; North and South America; Australia and New Zealand. Mediterranean-Atlantic element.

2. C. sepium (L.) R. Br. Hedge Bindweed
Convolvulus sepium L.

Perennial monoecious *herb* with far-creeping rhizomes. *Stems* up to 3 m, pale green or suffused brownish-purple, trailing to strongly climbing, twining, glabrous or sparsely hairy near the shoot tip. *Leaves* alternate; lamina 4–11 × 4–7 cm, rather pale shiny yellowish-green on upper surface, paler beneath, ovate, obtuse and apiculate or acuminate at apex, entire to wavy at margin, more or less cordate-sagittate at base, glabrous or hairy around the sinus; veins slightly impressed above, prominent beneath; petioles up to 9 cm, pale green, glabrous to shortly hairy. *Flowers* bisexual, solitary and axillary; peduncles up to 15 cm, not winged. *Bracteoles* 15–20 × 10–18 mm, ovate-lanceolate or lanceolate, more or less acute at apex, not or scarcely overlapping, not closely investing the calyx or obscuring it from view, not inflated, sometimes keeled. *Sepals* 5, 15–20 mm, lanceolate, acute. *Corolla* 30–70 mm, white or with pink windows, funnel-shaped. *Stamens* 5, 15–25 mm; filaments white; anthers cream. *Styles* 1, whitish; stigmas 2, cream. *Capsule* 7–12 mm. *Flowers* 7–9. $2n=22$.

(a) Subsp. **sepium**
Plant glabrous. Base of *leaf* with an acute sinus. *Corolla* 30–55 mm, white, rarely with pink windows. *Stamens* 15–25 mm. $2n=22$.

(1) Forma **sepium**
Corolla white and undivided.

(2) Forma **colorata** (Lange) Dörfl.
C. sepium var. *colorata* Lange

Corolla with 5 pink windows and undivided.

(3) Forma **schizoflora** (Druce) Stace
Corolla white divided into several deep lobes.

(b) Subsp. **roseata** Brummitt
Base of *leaf* with an acute sinus. *Stems*, petioles and pedicels sparsely short hairy. *Corolla* 40–55 mm, with 5 pink windows. *Stamens* 17–25 mm. $2n=22$.

(c) Subsp. **spectabilis** Brummitt
Plants glabrous. Base of *leaf* with a broad, rounded sinus. *Corolla* 50–70 mm, white. *Stamens* 20–30 mm.

Native. Hedges, scrub, woodland edges, fens, railway banks, waste ground, coastal sands and dunes and riversides. Throughout Great Britain and Ireland. Widely distributed in temperate regions. Circumpolar Temperate element. Subsp. *sepium* forma *sepium* is the common plant. There are scattered records of both forma *colorata* and forma *schizoflora*. Subsp. *roseata* typically occurs in brackish habitats at the edge of salt-marshes, in reed-beds and in grassy waste places in coastal regions. It also seems to occur on the sites of former fens and moors in East Anglia. It occurs round many of our coastal regions but seems to be more common in the west than in the east. It also occurs on the Atlantic coasts of western continental Europe, temperate South America, Easter Island, New Zealand and Australia. Subsp. *spectabilis* has been collected only once, at Arthog in Merionethshire, where it was naturalised but now seems to be extinct. It is native of northern Eurasia and widely naturalised in Scandinavia.

× **silvatica** = **C.** × **lucana** (Ten.) G. Don
Convolvulus lucanus Ten.

This hybrid shows a complete transition of characters between the two parents.

It is widespread where the parents grow together and sometimes occurs in the absence of either parent. It is also widespread in continental Europe.

3. C. pulchra Brummitt & Heywood Hairy Bindweed
C. sepium subsp. *pulchra* (Brummitt & Heywood) Tutin nom. inval.; *C. dahurica* auct.

Perennial monoecious *herb* with far-creeping rhizomes. *Stems* up to 3(–5) m, trailing to strongly climbing, pale green or suffused brownish-purple, twining, sparsely hairy at least when young, leafy. *Leaves* alternate; lamina 5–14 × 3–5 cm, matt medium yellowish-green on upper surface, paler beneath, ovate to oblong-ovate, obtuse to acute at apex, with rounded lobes at the cordate base, the sinus oblong or squarrose with almost parallel sides, glabrous; veins slightly impressed above, prominent beneath; petiole up to 6 cm, pale green, channelled above, rounded beneath, sparsely minute hairy. *Flowers* bisexual, solitary and axillary, odourless, 50–75 mm in diameter; pedicels up to 14 cm, pale yellowish-green, minutely hairy at least when young, often with a narrow, wavy wing. *Bracteoles* 25–27 × 15–25 mm, pale yellowish-green, broadly ovate, rounded or emarginate, overlapping and closely investing and almost concealing the calyx, strongly inflated. *Sepals* 5, 15–18 mm, pale yellowish-green, triangular-ovate, subacute. *Corolla* 50–75 mm, with 5 pink windows separated by white folds, funnel-shaped. *Stamens* 5; filaments 25–35 mm, white; anthers cream. *Style* 1, longer than anthers, white; stigmas 2, cream. *Capsule* 7–12 mm, seeds 4–7 mm, dark brown. *Flowers* 7–9. $2n = 22$.

Introduced. Naturalised in hedges and on waste ground, usually close to habitation. Cultivated in Great Britain since 1823, the earliest collections being from London in 1867, though perhaps as a garden plant, and Edinburgh in 1884. It was not recognised in the wild until 1956. Origin unknown, but could be wild or a native of north-east Asia.

× ?**silvatica** = **C.** × **howittiorum** Brummitt
Leaves glabrous with wide rounded basal sinus. *Bracteoles* broad, overlapping and obtuse. *Corolla* large, 53–64 mm, and pink. *Stamens* 26–30 mm. *Pollen* partially fertile.

Introduced. Common in Nottinghamshire and Lincolnshire and recorded from Yorkshire and Hampshire. Named after Mr R. D. L. Howitt and Mrs B. M. Howitt.

× ?**sepium** = **C.** × **scanica** Brummitt
Stems and young *leaves* hairy, with a wide, rounded sinus. *Bracteoles* 19–29 × 14–24 mm, not overlapping. *Corolla* (46–)52–64 mm, white to pale pink. *Stamens* 23–28 mm. *Pollen* partially fertile.

Introduced. Recorded from a number of localities in southern England and Wales. Originally described from the province of Skåne in south-western Sweden.

4. C. silvatica (Kit.) Griseb. Large Bindweed
Convolvulus silvaticus Kit.; *C. sepium* subsp. *silvatica* (Kit.) Batt.

Perennial monoecious *herb* with far-creeping rhizomes. *Stems* up to 3 m, pale green or suffused brownish-purple, trailing to strongly climbing, twining, glabrous or sparsely hairy. *Leaves* alternate; lamina 4–15 × 3–9 cm, rather pale shiny yellowish-green, paler beneath, ovate, obtuse to acuminate at apex, entire or wavy at margin, more or less cordate at base, with the sinus wide and rounded, glabrous; veins slightly impressed above, prominent beneath; petioles up to 15 cm, pale green, channelled above, rounded beneath, glabrous. *Flowers* solitary and axillary; peduncles up to 30 cm, usually narrowly winged. *Bracteoles* 20–40 × 14–48 mm, ovate, obtuse to acute, strongly overlapping and strongly pouched at base. *Sepals* 5, 15–18 mm, ovate-lanceolate. *Corolla* 52–80 mm long × 55–80 mm in diameter, white, the lobes very shallow and inconspicuous, funnel-shaped. *Stamens* 5; filaments up to 30 mm, whitish; anthers pale. *Style* 1, glabrous. *Capsule* 10–15 mm, subglobose; seeds 4–5 mm in diameter, more or less triangular-ovoid, with rounded outer surface and slightly concave inner faces, black, smooth. *Flowers* 7–9. $2n = 22$.

(a) Subsp. **silvatica**
Bracteoles broadly ovate, the distance from base to apex being about equal to the distance from midrib to margin, with the apex broadly rounded and the base very saccate.

(b) Subsp. **disjuncta** Brummitt
Bracteoles ovate-lanceolate, the distance from base to apex being about twice the distance from midrib to margin, obtusely pointed and less saccate at base.

(i) Var. **quinquepartita** N. Terracc.
Corolla divided into several lobes.

Introduced. Subsp. *silvatica* is widespread in Great Britain and Ireland, particularly on the Celtic fringe of Cornwall, Wales, Ireland and parts of Scotland where it may be more common than subsp. *disjuncta*. It is native of the east Mediterranean region. Subsp. *disjuncta* is the more common plant in most of the country and is native of the west Mediterranean region. The two subspecies seem to have hybridised with us so that intermediates occur, or they could be natural intermediates that reproduce themselves, which they will certainly do vegetatively. The var. *quinquepartita* is recorded a few times and might be referable to either subspecies. The species has been cultivated since 1815 and was first collected from the wild in 1863. It occurs in gardens, hedgerows, roadsides, fences and waste grounds and also moves into other areas where you would not expect it. In eastern England it is possibly moved around vegetatively on farm vehicles.

4. Ipomoea L.

Annual monoecious *herbs* or tuber-bearing perennials with usually strongly climbing stems. *Leaves* alternate, ovate, entire to deeply 3-lobed, cordate at base. *Inflorescence* a single flower or few-flowered panicle in the axils of leaves; flowers bisexual; bracteoles 2, linear, some way below the flowers. *Sepals* 5, often unequal. *Corolla* 5-lobed, funnel shaped or campanulate. *Stamens* 5. *Style* 1; stigma 1, 2- to 3-lobed. *Fruit* a capsule.

Contains about 500 species throughout tropical and subtropical areas.

Easy, G. M. S. (1983). Morning Glories. *B.S.B.I. News* **35**: 28–29.

1. Peduncles shorter than petioles, with few patent or forwardly directed hairs or glabrous; corolla usually white, rarely purple **3. lacunosa**
1. Peduncle longer than petioles, with reflexed hairs; corolla usually blue or drying pinkish-purple, rarely white 2.
2. Leaves entire; sepals acute at apex; corolla more than 5 cm **1. purpurea**
2. Most leaves deeply 3-lobed; sepals abruptly long acuminate; corolla less than 5 cm **2. hederacea**

1. I. purpurea (L.) Roth Common Morning-glory
Convolvulus purpureus L.; *Pharbitis purpurea* (Roth) Voigt

Annual monoecious *herb* with fibrous roots. *Stems* 5–8 cm, pale green, robust, twining, thinly strigillose, branched, leafy. *Leaves* alternate; lamina 5–16×5–11 cm, medium green on upper surface, paler beneath, ovate, shortly cuspidate at apex, entire, cordate at base; petiole 2–5 cm, slightly channelled, with retrorse hairs. *Inflorescence* axillary, flowers bisexual, usually in pairs or in few-flowered cymes; peduncle 2.5–4.0 cm; pedicels about 7 mm; bracteoles about 5 mm, linear. *Sepals* 5, 12–15×4–5 mm, narrowly ovate, with an acute apical appendage, hairy. *Corolla* 40–50 mm wide at the mouth, white, pink or purple. *Stamens* 5; filaments 18–23 mm, hairy below; anthers oblong. *Styles* 1, about 2.5 mm, glabrous; stigma capitate. *Capsule* about 8 mm in diameter, enclosed within the persistent calyx; seeds about 5×3 mm, brown, ovoid, glabrous. *Flowers* 7–8. $2n=30$.

Introduced. Tips and waste places where it is casual from soya bean waste, bird-seed and other sources. Sporadic in southern England. Native of North and South America from Virginia to Argentina; common as a weed throughout the tropics and subtopics.

2. I. hederacea Jacq. Ivy-leaved Morning-glory
Pharbitis hederacea Choisy

Annual monoecious climbing *herb*. *Stem* up to 4 m, pale green, twining, branched, leafy. *Leaves* alternate; lamina up to 12 cm, medium green on upper surface, paler beneath, ovate, pointed at apex, simple or 3-lobed, cordate at base. *Inflorescence* dense, 1–6-flowered; flowers bisexual; peduncle shorter than petiole, with few, patent or forwardly directed hairs. *Sepals* 5, 15–30 mm, abruptly long-acuminate at apex, densely long-hairy at base. *Corolla* 20–40 mm, usually blue when fresh, funnel-shaped. *Stamens* 5, not protruding. *Style* 1; stigma 3-lobed. *Fruit* a capsule. *Flowers* 7–9. $2n=30$.

Introduced. Tips and waste land. Casual from soya beans, bird-seed and other sources. Sporadic in southern England. Native of North America.

3. I. lacunosa L. White Morning-glory

Annual monoecious *herb*. *Stems* up to 2 m, pale green, twining, with few patent or forwardly directed hairs. *Leaves* alternate; lamina 5–10 cm, medium green on upper surface, paler beneath, broadly ovate, acute or acuminate at apex, entire to rather shallowly 3-lobed, the lobes acute, cordate at base; petioles slender. *Inflorescence* of 1–4 bisexual flowers on axillary peduncles shorter than the petioles. *Sepals* 6, 6–13 mm, oblong or lanceolate, shortly acuminate at apex, hairy. *Corolla* 15–25 mm, usually white, rarely purple, funnel-shaped. *Stamens* 5. *Style* 1; stigma 2-lobed. *Capsule* globose. *Flowers* 7–8. $2n=30$.

Introduced. A constant casual from soya bean waste. Very local and sporadic in southern England. Native of south-east North America.

125. CUSCUTACEAE Dumort. nom. conserv.

Annual to *perennial* monoecious *herbs* which are rootless parasites without visible chlorophyll. *Stems* twining and adherent to host plants by haustoria. *Leaves* alternate, reduced to minute scales, sessile; without stipules. *Inflorescence* a dense, sessile, globose head; flowers very small, bisexual, actinomorphic and hypogynous. *Calyx* 4- to 5-lobed. *Corolla* 4- to 5-lobed. Stamens 4–5, borne on the corolla tube with a small scale just below each. *Styles* 2; stigmas linear or capitate. *Fruit* a capsule, dehiscing transversely.

Consists of a single genus.

1. Cuscuta L.

As family.

Contains about 150 species in temperate and tropical regions, especially the warmer regions of the New World.

Feinbrun, N. (1970). A taxonomic review of European *Cuscuta*. *Israel Jour. Bot.* **19**: 16–29.
Hultén, E. & Fries, M. (1986). *Atlas of north European vascular plants north of the Tropic of Cancer.* 3 vols. Königstein.
Stewart, A., Pearman, D. A. & Preston, C. D. (1994). *Scarce plants in Britain*. Peterborough. [*C. europaea*.]
Verdcourt, B. (1948). Biological flora of the British Isles. No. 24. *Cuscuta europaea* L. *Jour. Ecol.* **36**: 358–365.
Yuncker, T. G. (1932). The genus *Cuscuta*. *Mem. Torrey Bot. Club* **18**: 13–331.

1. Stems yellow; stigmas capitate, the style ending in a distinct knob **1. campestris**
1. Stems reddish; stigmas linear, the style scarcely thickened distally 2.
2. Inflorescence 5–10 mm in diameter; styles (including stigmas) as long as to longer than ovary **4. epithymum**

2. Inflorescence 10–15 mm in diameter; style (including stigmas) shorter than ovary 3.
3. Stems nearly simple; calyx lobes acute at apex **2. epilinum**
3. Stems much branched; calyx lobes obtuse at apex **3. europaea**

1. C. campestris Yunck. Yellow Dodder

Parasitic, monoecious *herb* without or with only a trace of chlorophyll. *Stems* filiform, twining, attached to the host plant by haustoria, yellow, slender, much-branched. *Leaves* reduced to minute scales. *Inflorescence* of a rather irregular cluster of bisexual flowers 6–10 mm in diameter; pedicels short; bracts about 1.5 mm, ovate, acute at apex. *Calyx* about 1.5 mm, glabrous and not very fleshy, with 5 overlapping, subrotund lobes. *Corolla* slightly exceeding calyx, cup-shaped, with 5, triangular, acute, erect or reflexed lobes; throat scales extending to the base of the stamen, obtuse, long-fimbriate. *Stamens* 5, inserted in the throat of the corolla tube neat the apex; filaments white, tapering upwards; anthers yellow, exserted. *Styles* 2, free; stigma capitate. *Capsule* about 4 mm in diameter, depressed-globose, circumscissile towards the base; seeds about 1.5 mm, dark brown, broadly oblong or subglobose, minutely and obscurely rugulose. *Flowers* 7–9. $2n=56$.

Introduced. Occurs on a range of cultivated plants especially *Daucus*. Scattered records in England and Wales. Probably native of North America, but now found in almost every part of the world as a weed of cultivation.

2. C. epilinum Weihe Flax Dodder

Parasitic, monoecious *herb* without or with only a trace of chlorophyll. *Stems* reddish, attached to the host plant by haustoria, twisting, unbranched, glabrous. *Leaves* reduced to minute scales. *Inflorescence* of ebracteate heads of bisexual flowers about 10–15 mm in diameter. *Calyx* 3.5–4.0 mm, about as long as corolla tube, divided about halfway to the base into 5 lobes, the lobes broadly ovate and acute at apex. *Corolla* 3.0–3.5 mm, yellowish, divided about halfway to the base into 5 lobes, the lobes ovate and incurved at apex; throat scales 5, shorter than corolla tube, truncate, usually bifid, shortly fimbriate, sometimes reduced to short wings near the base of the corolla. *Stamens* 5, included; anthers orange. *Styles* 2; stigmas linear. *Capsule* 2.8–3.0 mm, ovoid, 2-celled, cells 2-seeded; seeds 1–2 mm, ellipsoid. *Flowers* 8–9. $2n=42$.

Introduced. Parasitic on *Linum usitatissimum*. Formerly occurring sporadically in Great Britain and Ireland as a casual in flax fields, but has been rare since 1900 and is now probably extinct. Native of Europe.

3. C. europaea L. Great Dodder

Parasitic, monoecious rootless *herb* without or with only a trace of chlorophyll. *Stems* about 1.0 mm in diameter, reddish, slender, attached to the host plant by haustoria, twisting in a counter-clockwise direction, branched, glabrous. *Leaves* reduced to minute scales. *Inflorescence* of bracteate heads of bisexual flowers about 10–15 mm in diameter; pedicels very short and fleshy. *Calyx* 3.5–4.0 mm, rose-pink, tubular, divided one-third of the way to the base into 4 or 5 lobes, the lobes broadly ovate and obtuse at apex, glabrous. *Corolla* 3.0–3.5 mm, pinkish-white, divided about halfway to the base into 4 or 5 lobes, the lobes ovate, rounded at apex, spreading; throat scales 5, very small, attached to the corolla tube but not closing it, deeply bilobed, each lobe entire or deeply laciniate. *Stamens* 4 or 5, included; anthers pink. *Styles* 2, shorter than ovary; stigmas linear. *Capsule* 2.8–3.0 mm, ovoid, 2-celled, cells 2-seeded; seeds 1.5–1.7 mm, ellipsoid. *Flowers* 8–9. $2n=14$.

Native. On a range of hosts, but primarily on *Urtica dioica* and often near water. Scattered records in lowland England north to Northamptonshire; formerly more common. Most of Europe but scattered; North Africa; introduced in North America.

4. C. epithymum (L.) L. Dodder

C. europaea var. *epithymum* L.; *C. trifolii* Bab.

Parasitic, monoecious *herb* without or with only a trace of chlorophyll. *Stems* about 1.0 mm in diameter, reddish, very slender, attached to the host plant by haustoria, twisting in a counter-clockwise direction, glabrous. *Leaves* reduced to minute scales. *Inflorescence* of sessile flowers in dense heads 5–10 mm in diameter. *Calyx* 1.5–2.0 mm, white, flushed pale crimson, campanulate, divided for three-quarters of its length into 5 lobes, the lobes broadly triangular-ovate, mucronate at apex, glabrous. *Corolla* 3.5–4.0 mm, flushed rose-pink or pale crimson, 5-lobed, the lobes broadly ovate and mucronate at apex; throat scales 5, narrowly obovate, incurved at the rounded apex closing the corolla tube, the spaces between their bases narrow and acute. *Stamens* 5, exserted, but shorter than corolla; anthers pink. *Styles* 2, longer than ovary, scarcely thickened distally; stigmas linear. *Capsule* 2.0–2.5 mm in diameter, 2-celled, cells 2-seeded; seeds 1.0–1.2 mm, ellipsoid, angled. *Flowers* 7–9. $2n=14$.

Native. On a wide range of hosts, most commonly on *Ulex* species and *Calluna* on heathland. Frequent in southern Great Britain and the Channel Islands, scattered elsewhere in Great Britain and Ireland north to central Scotland, but mostly casual. Most of Europe, North Africa; temperate Asia; introduced in North America and South Africa.

126. MENYANTHACEAE Dumort. nom. conserv.

Bog or aquatic *perennial* monoecious *herbs* with creeping rhizomes. *Leaves* alternate, simple or ternate, or with entire leaflets, with flat, sheathing bases to the petioles; without stipules. *Inflorescence* an axillary cluster or elongated raceme; flowers showy, actinomorphic, bisexual, hypogynous. *Calyx* divided nearly to base into 5 lobes. *Corolla* divided nearly to base into 5 lobes. *Stamens* 5, borne on the corolla tube, sometimes alternating with short fringed scales. *Style* 1; stigma 2-lobed. *Ovary* 1-celled, with many ovules on 2 parietal placentas. *Fruit* a capsule.

Contains about 5 genera and 33 species; cosmopolitan.

1. Leaves ternate; corolla white or pink **1. Menyanthes**
1. Leaves simple; corolla yellow **2. Nymphoides**

1. Menyanthes L.

Aquatic or bog *perennial* monoecious *herbs* with creeping rhizomes. *Leaves* all alternate, ternate, all held above water. *Inflorescence* an erect raceme of monomorphic, bisexual flowers. *Calyx* deeply 5-lobed. *Corolla* white to pink, 5-lobed, the lobes with many long fringes on their inner edge. *Stamens* 5. *Style* 1. *Fruit* a capsule dehiscing by 2 valves.

Contains 1 species.

Hewett, D. G. (1964). Biological flora of the British Isles. No. 97. *Menyanthes trifoliata* L. *Jour. Ecol.* **52**: 723–735.
Hultén, E. & Fries, M. (1986). *Atlas of north European vascular plants north of the Tropic of Cancer.* 3 vols. Königstein.
Nic Lughadha, M. & Parnell, J. A. N. (1989). Heterostyly and gene-flow in *Menyanthes trifoliata* L. (Menyanthaceae). *Bot. Jour. Linn. Soc.* **100**: 337–354.
Preston, C. D. & Croft, J. M. (1997). *Aquatic plants in Britain and Ireland.* Colchester.

1. M. trifoliata L. Bogbean
M. palustris Gray nom. illegit.

Aquatic or semi-aquatic *perennial* monoecious *herb* with a creeping rhizome. *Stems* up to 1.5 m, pale green, procumbent or floating, glabrous. *Leaves* all alternate and held above water level, ternate; leaflets 3.5–7.0 cm, bright light green on upper surface with a pale yellowish-green midrib, paler, somewhat glaucous green beneath, obovate or elliptical, rounded at apex, entire, rounded at base and sessile, glabrous; petiole 7–20 cm, pale green, with long sheathing base, glabrous. *Flowering* stem 12–30 cm, pale green, scapose. *Inflorescence* a terminal raceme of 10–20 bisexual flowers; pedicels 5–10 mm; bracts ovate, acute at apex, entire. *Calyx* 4–5 mm, deeply divided into 5 lobes, the lobes ovate, rounded at apex and somewhat recurved, glabrous. *Corolla* 15–20 mm in diameter, white on the inside, pink on the outside, deeply divided into 5 lobes, lanceolate, acute at apex, fimbriate, with glandular hairs, caducous. *Stamens* 5, inserted on the corolla tube; filaments white, anthers reddish-purple. *Style* 1; stigma 2-lobed, yellowish-green. *Capsule* 8–10 mm, ovoid to subglobose. *Flowers* 5–7. Pollinated by various insects. Heterostylous. $2n=54$.

Native. In shallow water, bogs, dune slacks and fens up to 100 m in Scotland, and pure stands of this species are often found where the emergent zone gives way to open water. Throughout Great Britain and Ireland, but local in many parts of England where it has decreased in the south and east. Most of Europe, but rare in the Mediterranean region; north and central Asia; north Morocco; Greenland, North America. Circumpolar Boreo-temperate element.

2. Nymphoides Ség.

Limnanthemum S. G. Gmel.; *Limnanthes* Stokes, non R. Br. nom. conserv.

Aquatic herbs with creeping rhizomes. *Leaves* alternate on vegetative stems, opposite on flowering stems, simple, cordate, floating on water. *Inflorescence* of small axillary groups of flowers; flowers dimorphic on long pedicels. *Calyx* deeply 5-lobed. *Corolla* yellow, 5-lobed with fringes on their margins. *Stamens* 5. *Style* 1. *Capsule* dehiscing irregularly.

Contains about 20 species, mainly in the tropics and subtropics.

Cook, C. D. K. (1990). Seeds dispersal of *Nymphoides peltata* (S. G. Gmel.) O. Kuntze (Menyanthaceae). *Aquat. Bot.* **37**: 325–340.
Hultén, E. & Fries, M. (1986). *Atlas of north European vascular plants north of the Tropic of Cancer.* 3 vols. Königstein.
Ornduff, R. (1966). The origin of dioecism from heterostyly in *Nymphoides* (Menyanthaceae). *Evolution* **20**: 309–314.
Preston, C. D. & Croft, J. M. (1997). *Aquatic plants in Britain and Ireland.* Colchester.
Smits, A. J. M., Van Avesaath, P. H. & Velde, G. van der (1990). Germination requirements and seed banks of some nyphacid macrophytes: *Nymphaea alba* L., *Nuphar lutea* (L.) Sm. and *Nymphoides peltata* (Gmel.) O. Kuntze. *Freshw. Biol.* **24**: 315–326.
Stewart, A., Pearman, D. A. & Preston, C. D. (1994). *Scarce plants in Britain.* Peterborough.
Velde, G. van der & Heijden, L. A. van der (1981). The floral biology and seed production of *Nymphoides peltata* (Gmel.) O. Kuntze (Menyanthaceae). *Aquat. Bot.* **10**: 261–293.

1. N. peltata Kuntze Fringed Water-lily
Limnanthemum peltatum S. G. Gmel. nom. illegit.; *Menyanthes nymphoides* L.; *Limnanthes nymphoides* (L.) Stokes; *Limnanthemum nymphoides* (L.) Hoffmanns. & Link; *Limnanthes peltata* Gray; *Nymphoides orbiculata* Druce nom. illegit.; *Nymphoides flava* Druce nom. illegit.; *Nymphoides nymphoides* (L.) Druce nom. illegit.

Aquatic *perennial herb* with creeping rhizome. *Stems* up to 1.5 m, pale green, floating, glabrous. *Leaves* alternate; lamina 3–12 ×2–10 cm, shining medium green, often blotched with blackish-purple on upper surface, dull, purplish-green or bronze-green beneath and closely dotted with glands, subrotund, rounded at apex, cordate at base, glabrous; petioles long, glabrous. *Inflorescence* of 2- to 5-flowerd axillary fascicles at the end of a long, floating flowering stem with opposite leaves; pedicels 3–7 cm, glabrous. *Calyx* 5-lobed, the lobes about 10 mm, oblong-lanceolate, acute at apex, glabrous. *Corolla* 30–40 mm in diameter, bright golden yellow, 5-lobed, the lobes obovate, emarginate at apex and the margins fringed, curved outwards. *Stamens* 5; filaments and anthers bright golden yellow. *Capsule* 23–26 mm, ellipsoid, dehiscing irregularly. *Flowers* 7–8. Pollinated by various insects. Heterostylous. Individual plants with pin or thrum flowers, each flower lasting for only 1 day but plants flower over a long period. $2n=54$.

Possibly native. Lowland, calcareous and eutrophic water over inorganic substrates at the edge of lakes, slowly flowing rivers and fenland lodes, often in dense masses in water up to 2 m deep; also grown as an ornamental plant in ponds, natural and ornamental lakes, flooded gravel pits, clay pits and quarries and rivers, canals and ditches. It is thought to be native in the Thames Valley where it was recorded as frequent in 1570 and in East Anglia where John Ray found it plentiful in the Fens in 1660. It is now naturalised in scattered places elsewhere north to central Scotland. As a native plant widespread in Europe from England and the Baltic southwards and extending eastwards through Asia to Japan; naturlised in North America. Eurasian Temperate element.

127. POLEMONIACEAE Juss. nom. conserv.

Perennial monoecious *herbs. Leaves* opposite and simple or alternate and pinnate, petiolate or more or less sessile, without stipules. *Inflorescence* a terminal, corymbose raceme of showy, actinomorphic, mostly bisexual, hypogynous flowers. *Calyx* 5-lobed. *Corolla* 5-lobed, white, purple or blue. *Stamens* 5, borne on the corolla tube. *Styles* 1; stigmas 3; ovary 3-celled, with several ovules per cell on axil placentas. *Fruit* a capsule opening by 3 valves.

Contains about 15 genera with 300 species, chiefly in North America.

1. Leaves simple and entire	**2. Phlox**
1. Leaves pinnate or pinnatisect	2.
2. Leaves glabrous	**1. Polemonium**
2. Leaves glandular-viscid	**3. Navarretia**

1. Polemonium L.

Perennial monoecious *herbs. Leaves* pinnate, with 13–31 entire leaflets, petiolate. *Inflorescence* a terminal corymbose raceme. *Calyx* 5-lobed. *Corolla* 5-lobed, the lobes much longer than tube. *Stamens* 5, more or less of equal length, well exserted, base of filaments hairy. *Style* 1. *Fruit* a capsule.

Hultén, E. & Fries, M. (1986). *Atlas of north European vascular plants north of the Tropic of Cancer*. 3 vols. Königstein.
Pigott, C. D. (1958). Biological flora of the British Isles. No. 65. *Polemonium caeruleum* L. *Jour. Ecol.* **46**: 507–525.
Wigginton, M. J. (Edit.) (1999). *British red data books*. Vol. 1. *Vascular plants*. Ed. 3. Peterborough.

1. P. caeruleum L. Jacob's-ladder

Perennial monoecious *herb* with short, creeping rhizome. *Stems* 30–90 cm, pale green, erect, rigid, hollow, angled, glandular-hairy in upper part, simple, leafy. *Leaves* alternate; lamina 10–40 cm, oblong-elliptical in outline, imparipinnate; leaflets 13–31, with lamina 20–40 mm, bright medium green on upper surface, paler beneath, ovate-lanceolate or oblong, obtuse to acute at apex, entire, rounded at base and sessile, glabrous; petiole long, winged, upper leaves more or less sessile. *Inflorescence* a terminal corymb, more or less bractless; flowers 20–30 mm in diameter, bisexual, drooping. *Calyx* 7–8 mm, 5-lobed, the lobes ovate, acute at apex. *Corolla* 20–30 mm in diameter, bright deep blue or mauvish-blue, the throat pale greenish-yellow with purple veins, rotate, with a short tube and 5 lobes, the lobes subrotund or ovate, apiculate, spreading. *Stamens* 5; filaments bright deep blue; anthers orange. *Style* 1, bright deep blue; stigmas 3, slender; nectar secreted from a fleshy ring round the base of the ovary. *Capsule* 4.5–5.0 mm, erect, included in the calyx-tube; seeds 3.0–3.5 mm, angular, rugose, shortly winged. *Flowers* 6–7. Visited by hoverflies and bumble-bees. Protandrous. The population at Malham Cove is gynodioecious. $2n=18$.

Native. Limestone grassland, scree, rock-ledges, wood borders. Locally frequent in the Peak District, Yorkshire Dales and one place locally in Northumberland as a native plant; sporadic garden escapes occur elsewhere in Great Britain. North and central Europe; Caucasus; Siberia; North America. Eurosiberian Boreal-montane element. Our native plant is glandular-hairy above, has almost glabrous leaves, a sparingly glandular-hairy, short-toothed calyx and obtuse petals with orange pollen. Garden escapes have often pale purplish-blue flowers, narrower leaflets and yellow pollen. The native Malham Cove plant often has white flowers and a smaller corolla.

2. Phlox L.

Perennial monoecious *herb. Leaves* opposite, simple, entire, more or less sessile. *Inflorescence* a terminal corymbose raceme; flowers bisexual. *Calyx* 5-lobed. *Corolla* 5-lobed, the lobes shorter than tube. *Stamens* 5, with anthers of different heights, the longest at the mouth of the corolla tube. *Style* 1. *Fruit* a capsule.

Contains more than 60 species in Alaska and west Canada to north Mexico.

Wherry, E. T. (1955). *The genus* Phlox. Philadelphia.

1. P. paniculata L. Phlox

Perennial monoecious *herb* with a woody rootstock forming large clumps. *Stems* up to 1 m, pale green, erect, simple, glabrous. *Leaves* opposite; lamina 3.5–9.0 × 1.0–3.5 cm, medium green on upper surface, paler beneath, lanceolate, ovate-lanceolate or elliptic-lanceolate, more or less acute at apex, entire, rounded to subcordate at base, glabrous except for ciliate margins. *Inflorescence* a large, pyramidal, terminal panicle; branches and pedicels shortly eglandular-hairy; flowers bisexual, fragrant; bracts ovate to lanceolate. *Calyx* 7–9 mm, with simple eglandular hairs, 5-lobed, with transparent membrane between the lobes, the lobes linear-subulate, not reflexed. *Corolla* usually white, pink, rose or purplish, the tube 15–20 mm, cylindrical, with eglandular, curly hairs, the limb 20–30 mm in diameter, glabrous, 5-lobed, the lobes subrotund to obovate. *Stamens* 5, inserted at different levels. *Style* 1; stigmas 3. *Capsule* ovoid. *Flowers* 7–9. $2n=14$.

Introduced. Much grown in gardens and naturalised on rough and waste ground. Scattered records in England. Native of North America.

3. Navarretia Ruiz & Pav.

Annual monoecious *herbs* with viscid glandular hairs. *Leaves* alternate or the lower opposite, pinnatifid, spiny. *Inflorescence* a dense, bracteate head. *Calyx* 5-lobed. *Corolla* 5-lobed. *Stamens* 5. *Styles* 1; stigmas 3. *Fruit* a capsule, opening by 3 valves.

Contains about 30 species mostly in western North America; one species in Chile and Argentina. Named after Ferdinand Navarrete a Spanish physician.

Curtis, S. (1830). *Bot. Mag.* no. 2977 as *Gilia pungens*.

1. N. squarrosa (Esch.) Hook. & Arn. Skunkweed
Holtzia squarrosa Esch.; *Gilia squarrosa* (Esch.) Hook. & Arn.; *Gilia pungens* Douglas ex Hook.; *Aegochloa pungens* (Douglas ex Hook.) Benth.; *N. pungens* (Douglas ex Hook.) Hook.

Annual monoecious *herb* with fibrous roots. *Stems* up to 40 cm, pale green to purplish, simple or branched, viscid glandular-hairy. *Leaves* alternate or the lower opposite, up to 6 cm, pinnatifid or bipinnatifid, strongly spiny, glandular-hairy. *Inflorescence* a dense, bracteate head; bracts pinnatifid; flowers bisexual. *Calyx* 8–14 mm, 5-lobed, the lobes slightly unequal, entire or with a few, sharp teeth. *Corolla* 9–12 mm, pale to deep blue, 5-lobed. *Stamens* 5; filaments more or less equal, 1–4 mm. *Style* 1; stigmas 3. *Capsule* 3-locular; seeds (6–)8–9 per locule. *Flowers* 6–9.

Introduced. Casual in gardens, probably from bird-seed, formerly a wool and grain alien. Recorded from Berkshire and Lanarkshire. Native of western North America.

128. HYDROPHYLLACEAE R. Br. nom. conserv.

Annual monoecious *herbs. Leaves* alternate or opposite, pinnately lobed to more or less pinnate or dentate, petiolate, without stipules. *Inflorescence* a terminal, clustered, spiralled cyme or flowers solitary in the axils; flowers actinomorphic, bisexual, hypogynous. *Calyx* divided nearly to the base into 5 lobes. *Corolla* 5-lobed, the tube longer than the lobes, bluish. *Stamens* 5, borne on the corolla tube, well exserted. *Style* 1, divided into 2 distally; stigmas minute. *Fruit* a 1- to 2-valved capsule.

Contains about 20 genera and 250 species, widely distributed, but most common in the western United States.

1. Inflorescence a cyme **1. Phacelia**
1. Inflorescence of solitary axillary flowers **2. Nemophila**

1. Phacelia Juss.

Annual monoecious *herbs. Leaves* alternate, occasionally the lowest ones opposite, lobed or compound. *Inflorescence* an open or congested cyme; flowers bisexual. *Calyx* 5-lobed almost to base. *Corolla* campanulate, saucer- or funnel-shaped, blue or purple, 5-lobed. *Stamens* 5, usually projects from the corolla. *Ovary* 1-celled, but apparently 2-celled by intrusion of the placentas. *Style* 1, bifid. *Fruit* a capsule.

Contains about 100 species, mostly from western North America, a few in eastern North America and in south America.

Briggs, M. (1997). *Phacelia tanacetifolia. B.S.B.I. News*. **74**: 48.
Francis, S. (2005). *British field crops*. Bury St Edmunds.
Leach, S. J. (1997). *Phacelia tanacetifolia* as a 'green manure'. *B.S.B.I. News* **75**: 40–41.

1. P. ciliata Benth. Ciliate Phacelia
P. acanthominthoides Elmer.

Annual monoecious *herb* with a taproot and fibrous side roots. *Stem* up to 50 cm, pale green, erect or ascending, with numerous glandular and simple eglandular hairs. *Leaves* alternate; lamina 3–10 cm, medium green, ovate or oblong in outline, pinnate or pinnatifid into dentate or incise-dentate lobes, hairy; petiole usually rather short. *Inflorescence* of a few, crowded cymes; flowers bisexual, subsessile. *Calyx* 4–16 mm, pale green, with 4, oblong to ovate, stiffly ciliate lobes, with a conspicuous venation. *Corolla* 8–10 mm, blue with a paler centre, broadly campanulate and divided into 5 lobes. *Stamens* 5; filaments 9–13 mm, longer than corolla. *Styles* 1, bifid at apex, 6–8 mm, hairy below. *Capsule* 4–5 mm, stiffly hairy; seeds 4, 2.5–3.0 mm, dark brown, deeply pitted. *Flowers* 5–6. $2n=22$.

Introduced. A grain casual. Native of western North America.

2. P. tanacetifolia Benth. Phacelia

Annual monoecious *herb* with a tap-root and fibrous side-roots. *Stems* up to 70(–100) cm, pale green, erect or ascending, with numerous, short to long, pale, stiff simple eglandular hairs. *Leaves* alternate; lamina 2–20 × 1–6 cm, bright medium green on upper surface, greyish beneath, ovate to ovate-oblong in outline, pinnate, the pinnae pinnatisect with obtuse lobes, with short, curly, bulbous-based simple eglandular hairs on both surfaces, the short petioles and rhachis pale green and hairy. *Inflorescence* scorpioid; flowers bisexual, 10–14 mm in diameter, strongly sweet-smelling. *Calyx* 9–10 mm, pale green, divided nearly to the base into 5 lobes, the lobes unequal, linear, with numerous, short to long, pale, stiff simple eglandular hairs. *Corolla* 6–12 mm, mauve, rotate, divided one-third to halfway to the base into 5 lobes, the lobes ovate and rounded at apex. *Stamens* 5; filaments 15–17 mm, much longer than the corolla, mauve; anthers purple. *Style* 2, a little longer than the stamens, connate for about 1 mm, whitish. *Capsule* 3–4 mm, ovoid, shortly hairy at apex. *Flowers* 5–6. $2n=22$.

Introduced. Grown in gardens for ornament and on a small scale in fields for bees; also a contaminant of crops and grass seed; casual on tips, waste ground, among crops and in newly sown grass. In grass strips sown round fields by farmers where it may be a contaminant or in wild flower seed sown by the same farmers. Scattered records in England and Wales. Native of California.

2. Nemophila Nutt.

Annual monoecious *herbs. Leaves* opposite, pinnately lobed or dentate. *Inflorescence* of solitary flowers in the axils of leaves; flowers bisexual. *Calyx* deeply 5-lobed, with small appendages in the sinuses between the lobes. *Corolla* campanulate to saucer-shaped, exceeding the calyx, nearly white to bright blue, 5-lobed, the lobes sometimes dark-tipped. *Stamens* 5, not exceeding the corolla. *Ovary* 1-celled, not bristly-hairy. *Styles* 1, bifid in the upper half. *Fruit* a capsule.

Contains about 17 species from western North America of which the following two are commonly grown in gardens.

Constance, L. (1941). The genus *Nemophila* Nutt. *Univ. Calif. Publ. Bot.* **19**: 341–398.

1. Leaves lanceolate, dentate; corolla lobes with a dark blotch at apex **1. maculata**
1. Leaves pinnately lobed; corolla not blotched at apex **2. menziesii**

1. N. maculata Benth. ex Lindl. Fivespot

Annual monoecious *herb* with a tap-root and fibrous side-roots. *Stems* up to 30 cm, pale green, slightly hairy, much-branched, leafy. *Leaves* opposite, yellowish green, paler beneath; lamina 20–35 × 14–16 mm, oblanceolate or obovate, obtuse at apex, dentate, the teeth large, narrow at base, hairy on both surfaces and the margins, sessile or shortly petiolate. *Inflorescence* of solitary, bisexual flowers in leaf axils; flowers 25–40 mm in diameter; peduncles 30–60 mm, pale green, hairy. *Calyx* 8–10 mm, medium green, deeply 5-lobed, the lobes lanceolate, more or less acute, with small appendages in the sinuses between the lobes. *Corolla* 15–20 mm, pale lilac blue with deep purple blotches at the end of each lobe, deeply 5-lobed, obovate, broadly rounded at apex, narrowed at base. *Stamens* 5, not exceeding the corolla; filaments white, tinted pink; anthers dark brown. *Styles* 1, bifid, white; stigmas black. *Capsule* spherical or ovoid. *Flowers* 5–6. $2n = 18$.

Introduced. Commonly grown in gardens from which it escapes. Native of California.

2. N. menziesii Hook. & Arn. Baby-blue-eyes
N. insignis Douglas ex Benth.

Annual monoecious *herb* with a tap-root and fibrous side-roots. *Stems* up to 40 cm, pale yellowish-green, glabrous, much-branched, leafy. *Leaves* opposite, yellowish-green, paler beneath; lamina 15–30 × 6–20 mm, pinnately lobed, the lobes oblong and obtuse at apex, shortly hairy on both surfaces and the margins; petiole short, pale green. *Inflorescence* of solitary flowers in leaf axils; flowers 30–35 mm in diameter, bisexual; peduncles 20–50 mm, dirty brown, appressed-hairy. *Calyx* 7–10 mm, medium green, deeply 5-lobed, the lobes lanceolate, obtuse or subacute at apex, hairy, with small appendages in the sinuses between the lobes. *Corolla* 13–15 mm, medium blue, with a white throat, 5-lobed, the lobes obovate, broadly rounded at apex, narrowed at base. *Stamens* 5, not exceeding the corolla; filaments white; anthers dark brown. *Styles* 1, bifid, white; stigmas black. *Capsule* spherical to ovoid. *Flowers* 5–6. $2n = 18$.

Introduced. Commonly grown in gardens from which it escapes, also a grain casual. Native of Oregon and California.

Order 3. LAMIALES Bromhead

Herbs, less frequently *shrubs* and *trees*. *Leaves* usually opposite, without stipules. *Flowers* bisexual or rarely unisexual, usually hypogynous, actinomorphic or zygomorphic. *Calyx* 4- to 5-lobed, sometimes 2-lipped. *Corolla* 4- to 5-lobed, often 2- lipped. *Stamens* as many as or fewer than the corolla lobes. *Style* 1. *Fruits* various.

Consists of 4 families and about 7,800 species.

129. BORAGINACEAE Juss. nom. conserv.

Annual to *perennial* monoecious *herbs* often roughly hairy and the hairs with tuberculate bases. *Leaves* alternate, simple, entire or more or less so, sessile or petiolate; without stipules. *Inflorescence* often a scorpioid cyme; flowers actinomorphic or weakly zygomorphic, bisexual, hypogynous. *Calyx* 5-lobed, sometimes deeply so. *Corolla* 5-lobed, mostly blue to pink, rarely white or yellow, throat often closed by scales or hairs. *Stamens* 5, borne on the corolla tube. *Style* 1, usually simple, sometimes with 2 stigmas. *Ovary* 4-celled, deeply 4-lobed, with 1 ovule per cell. *Fruit* a cluster of 4, 1-seeded nutlets; seeds usually without endosperm, embryo straight or curved.

Contains about 100 genera with 2,000 species, cosmopolitan but especially abundant in the Mediterranean region and east Asia.

1. Style bifid at apex (stigma 2-lobed) 2.
1. Style simple (stigma single) 3.
2. Inflorescence of strongly coiled cymes; flowers creamy-white **1. Heliotropium**
2. Inflorescence of cymes forming a narrow panicle; flowers usually pink to blue or purple, rarely white **4. Echium**
3. All anthers completely exserted 4.
3. All anthers completely included or only tips exserted 5.
4. Annual herbs with a stout tap-root; calyx divided nearly to base; filaments glabrous; anthers longer than filaments **11. Borago**
4. Perennial with a rhizome; calyx divided about halfway to base; filaments hairy; anthers shorter than filaments **13. Trachystemon**
5. Calyx lobes with appendages in the sinuses of the lobes, strongly accrescent in fruit to form an envelope round the nutlets **17. Asperugo**
5. Calyx lobes 5, entire, not or only slightly enlarging in fruit 6.
6. Nutlets with hooked or barbed bristles 7.
6. Nutlets smooth to warty, ridged or hairy 8.
7. Flowers and fruits all or mostly with a bract; nutlets less than 4.5 mm **19. Lappula**
7. Flowers and fruits all or mostly without a bract, nutlets more than 4.5 mm **21. Cynoglossum**
8. Plant glabrous, often very glaucous **14. Mertensia**
8. Plant bristly to hairy, not or scarcely glaucous 9.
9. At least lower leaves opposite **16. Plagiobothrys**
9. All leaves alternate (rarely uppermost pair opposite in *Myosotis*) 10.
10. Open flowers pendulous; stigma exserted **6. Symphytum**
10. Open flowers erect; stigma included or at throat of corolla tube 11.
11. Ripe nutlets smooth (sometimes hairy or with a keel round the edge) 12.
11. Ripe nutlets tuberculate to strongly warty and/or with variously branched ridges 18.
12. Basal and all or most stem leaves petiolate **20. Omphalodes**
12. Basal and all or most stem leaves sessile 13.
13. Corolla more than 10 mm 14.

13. Corolla less than 10 mm 16.
14. Calyx divided nearly to the base **3. Buglossoides**
14. Calyx divided to about halfway 15.
15. Throat of corolla hairy without scales or folds **5. Pulmonaria**
15. Throat of corolla with scales or folds **8. Nonea**
16. Calyx hairs hooked; corolla tube usually shorter than lobes **18. Myosotis**
16. Calyx hairs straight; corolla tube longer than lobes 17.
17. Throat scales of corolla trapeziform **2. Lithospermum**
17. Throat of corolla without scales, but with longitudinal folds of hairs **3. Buglossoides**
18. Basal leaves strongly cordate at base **7. Brunnera**
18. Basal leaves gradually to abruptly cuneate at base 19.
19. Leaves ovate to obovate, at least most basal ones more than 5 cm wide 20.
19. Leaves lanceolate to oblanceolate or linear-oblong, less than 5 cm wide 21.
20. Corolla lobes rounded; scales closing throat of corolla tube; nutlets stalked **11. Pentaglottis**
20. Corolla lobes pointed; scales not closing throat of corolla tube; nutlets sessile **12. Borago**
21. Nutlets tuberculate to strongly warty, not ridged apart from marginal keel, without collar-like base 22.
21. Nutlets tuberculate and with strongly branching ridges, with distinct collar-like base at point of attachment 23.
22. Corolla white to bluish-purple; nutlets minutely tuberculate **3. Buglossoides**
22. Corolla yellow to orange; nutlets coarsely warty or muricate **15. Amsinckia**
23. Corolla tube longer than lobes **9. Anchusa**
23. Corolla tube shorter than lobes **10. Cynoglottis**

1. Heliotropium L.

Annual monoecious *herbs. Leaves* alternate, entire, petiolate, without stipules. *Inflorescence* of strongly coiled terminal and lateral cymes. *Calyx* 5-lobed. *Corolla* 5-lobed, creamy-white. *Stamens* 5, inserted just above the base of the corolla tube. *Style* 1, terminal; stigma shortly bifid. *Fruit* of 4 nutlets.

Contains about 250 species from tropical and temperate areas.

1. H. europaeum L. European Turnsole

Annual monoecious *herb* with fibrous roots. *Stems* 15–40 cm, pale green, erect or spreading, terete, with dense, subappressed hairs, divaricately branched, leafy. *Leaves* alternate; lamina 2–8 × 0.8–3.0 cm, medium green on upper surface, paler beneath, ovate, oblong or obovate, rounded or obtuse at apex, entire, narrowed abruptly at base, with tuberculate-based hairs on upper surface and more densely so below; petiole 0.5–3.0 cm; veins obscure above, prominent below. *Inflorescence* of strongly coiled terminal and lateral cymes, at first dense, elongating to 10 cm or more after anthesis and the lowermost flowers becoming rather remote; flowers bisexual, sessile, in 2 ranks, scentless. *Calyx* 2.0–2.5 mm, divided almost to base into 5 narrowly lanceolate lobes, the lobes at first erect, but soon spreading and becoming stellate-patent after the nutlets are shed, with dense, soft, spreading, shaggy hairs. *Corolla* creamy-white; tube 2.5–3.0 mm, greenish, hairy; lobes 5, spreading, about 1.0 × 0.8 mm, broadly ovate, translucent and distinctly veined, the sinuses plicate and often minutely toothed. *Stamens* 5, inserted about 0.5 mm above the base of the corolla tube; anthers broadly lanceolate. *Style* 1, very short, with a basal glandular-papillose zone; stigma 2-lobed. *Nutlets* 4, about 1.5 × 1.0 mm, brown, ovoid, glabrous or slightly hairy, coarsely rugose-tuberculate. *Flowers* 7–9.

Introduced. A wool and oil-seed casual. Native of Europe; North Africa; south-west Asia; introduced in Australia.

2. Lithospermum L.

Perennial monoecious *herbs* with short rhizomes. *Leaves* alternate, lateral veins conspicuous. *Inflorescence* of terminal and axillary cymes. *Calyx* 5-lobed. *Corolla* yellowish, greenish or nearly white, 5-lobed; throat scales 5, trapeziform. *Stamens* 5, attached to the middle of the corolla tube; filaments glandular-hairy at base. *Style* 1; stigmas simple. *Fruit* of 4, pale pearly grey, shining, ovoid nutlets.

Johnston, I. M. (1952). Studies in the Boraginaceae XXIII: a survey of the genus *Lithospermum*. *Jour. Arn. Arb.* **33**; 299–366.
Hultén, E. & Fries, M. (1986). Atlas of north European vascular plants north of the Tropic of *Cancer*. 3 vols. Königstein.

1. L. officinale L. Common Gromwell

Perennial monoecious *herb* with a short rhizome. *Stems* 30–80(–100) cm, pale green, erect, closely strigose with appressed simple eglandular hairs, much-branched above, leafy. *Leaves* alternate; lamina 2–8 × 1–2 cm, dull bluish-green on upper surface, paler beneath, lanceolate to ovate, acute at apex, entire, narrowed or rounded at the sessile base, with short to medium simple eglandular hairs with bulbous bases on both surfaces; lateral veins conspicuous. *Inflorescence* of terminal and axillary cymes, elongating after flowering; flowers bisexual. *Calyx* 4.5–5.0 mm, divided nearly to base into 5 lobes, the lobes subrotund, rounded or retuse at apex; throat scales 5, trapeziform, inturned, velvety hairy becoming glandular-hairy below, glands extending on to the corolla tube. *Stamens* 5, attached to middle of corolla tube and filaments yellow, glandular-hairy at base; anthers yellow. *Style* 1, 1–2 mm; stigmas 2, simple. *Nutlets* 2.7–3.8 mm, pale pearly grey, shiny, ovoid. *Flowers* 6–7. $2n = 28$.

Native. Grassy and bushy places, hedgerows and wood borders mostly on basic soils. Locally frequent in England, very local in Wales and Ireland and a rare casual elsewhere. Most of Europe, eastwards to the Caucasus and Lake Baikal.

3. Buglossoides Moench

Aegonychon Gray; *Margarospermum* Opiz

Annual or *perennial* monoecious, hairy *herbs. Leaves* alternate, entire, sessile or with an indistinct petiole. *Inflorescence* a terminal or axillary, bracteate cyme, open flowers erect. *Calyx* 5-lobed nearly to base. *Corolla* blue or white, 5-lobed, tube longer than lobes with distinct longitudinal folds of hairs in the throat. *Stamens* 5 included; filaments glabrous. *Style* 1; stigma simple. *Fruits* of 4, ellipsoid or ovoid, smooth nutlets.

Contains 7 species in Europe and Asia.

Johnston, I. M. (1954). Studies in the Boraginaceae XXVI: further revelations of the genera of the Lithospermae. *Jour. Arn. Arb.* **35:** 1–81.

Hultén, E. & Fries, M. (1986). *Atlas of north European vascular plants north of the Tropic of Cancer.* 3 vols. Königstein.

Svensson, R. & Wigren, M. (1986). Sminkrotens historia och biologi i Sverige. *Svensk Bot. Tidskrift* **80**: 107–131.

White, J. W. (1884). Life history of Lithospermum purpureocaeruleum Linn. *Jour. Bot.* (London) **22**: 74–76.

1. Non-flowering stems creeping, flowering stems erect; flowers at first reddish-purple, then deep blue **1. purpureocaeruleum**
1. All stems erect; flowers white, occasionally pink or bluish **2. arvense**

1. B. purpureocaerulea (L.) I. M. Johnst. Purple Gromwell

Lithospermum purpureocaetuleum L.

Perennial monoecious *herb* with creeping, non-flowering shoots. *Flowering stems* 20–60 cm, pale yellowish-green, erect, rough with numerous, pale, medium simple eglandular hairs, branched, leafy. *Leaves* alternate; lamina 3–8 × 0.5–1.5 cm, dull, medium yellowish-green on upper surface, paler beneath, lanceolate or oblong-lanceolate, gradually narrowed to an acute apex, entire, cuneate or attenuate at base, sessile, glabrous on upper surface, with numerous, short or medium, pale simple hairs beneath, midrib impressed above, prominent beneath. *Inflorescence* of groups of flowers at the ends of stems or branches in a bracteate cyme, flowers 12–18 mm in diameter, bisexual, erect, scentless. *Calyx* 8–10 mm, divided almost to base into 5 lobes, the lobes dull green, linear, with short and medium, pale, stiff hairs down each margin. *Corolla* 15–17 mm, divided one-third of the way to the base into 5 lobes, the tube red and appressed-hairy, the lobes reddish-purple, then deep blue, triangular, obtuse at apex and appressed-hairy, with longitudinal folds of hairs in the throat. *Stamens* 5, anthers brownish, sessile, included. *Styles* 1, pale greenish, about equalling the anthers; stigma simple pale greenish. *Nutlets* 4, 3.0–3.5 mm, white, ovoid, obtuse at apex, smooth, somewhat shining. *Flowers* 5–6. Visited by bees. $2n=16$.

Native. Scrub and wood margins on chalk and limestone. Very local in south-west England, south and north Wales; formerly in Kent; rare casual elsewhere. South and central Europe; Caucasus; south-west Asia. European Southern-temperate element.

2. B. arvensis (L.) I. M. Johnst. Field Gromwell

Lithospermum arvense L.

Annual monoecious *herb* with fibrous roots. *Stems* (2–) 6–30(–50) cm, pale green, often suffused brownish-purple, erect or decumbent, thinly or densely appressed-strigose, simple or branched from near the base, leafy. *Leaves* alternate; lamina of basal leaves 1.5–7.0 × 0.5–1.0 cm, dull medium green on upper surface, paler beneath, oblong or lanceolate, obtuse or rounded at apex, entire, tapered at base to an indistinct petiole, the cauline similar but up to 10 × 2 cm, becoming smaller, narrower and more acute upwards; all thinly appressed-strigose on both surfaces. *Inflorescence* simple and terminal or several arising from the axils of the uppermost leaves in a bracteate cyme; flowers bisexual; pedicels lengthening and sometimes becoming asymmetrically thickened in fruit; bracts leafy, oblong-lanceolate, subacute at apex, with appressed, whitish bristles. *Calyx* 4–15 mm, campanulate, divided into 5 lobes, nearly to the base, the lobes erect at first becoming patent or spreading in fruit, linear-subulate, obtuse to acute, with dense, pale bristles. *Corolla* 5–9 mm, milky white, occasionally pink, rarely bluish, infundibuliform or almost hypocrateriform; tube narrowly cylindrical with slightly expanded base, longer than lobes, lobes 5, 1.5–2.5 × 1–2 mm, oblong, rounded or subtruncate at apex, suberect, with longitudinal folds of hairs in the throat. *Stamens* 5, inserted about 1.0 mm above the base of the corolla tube; filaments pale, very short; anthers narrowly oblong, apex with a small appendage. *Style* 1, about 1.0 mm; stigma simple, capitate. *Nutlets* 4, 2–3 × 1.5–2.0 mm, whitish or pale brown, with a hard pericarp, minutely or coarsely rugulose-tuberculate. *Flowers* 5–7. $2n=28$.

Native. Arable fields, rough ground and open grassy places. Locally frequent in England and formerly much more common before herbicides, very scattered localities and much more casual in Wales, Scotland, Ireland and the Channel Isles.

4. Echium L.

Usually *biennial* monoecious *herbs*. *Leaves* alternate, entire or toothed, petiolate. *Inflorescence* of terminal and lateral cymes, forming a narrow panicle; flowers bisexual. *Calyx* divided nearly to base into 5 lobes. *Corolla* funnel-shaped with a straight tube and open throat, 5-lobed, the lobes unequal. *Stamens* 4 or 5, unequal, at least some exserted. *Style* 1; stigma bifid. *Fruits* of 4 nutlets attached to the flat, receptacle by their flat nearly triangular bases.

Contains about 40 species in Europe, the Mediterranean region and Macaronesia.

Butterfield, L. (1999). Boscregan – last refuge of the Purple Viper's-bugloss? *British Wildlife* **10**: 166–171.

Hultén, E. & Fries, M. (1986). *Atlas of north European vascular plants north of the Tropic of Cancer.* 3 vols. Königstein.

Wigginton, M. J. (Edit.) (1999) *British red data books.* Vol. 1. *Vascular plants.* Ed. 3. Peterborough.

1. Shrubs with unbranched woody stems up to 75 × 3–5 cm and with a terminal panicle up to 3.5 m; leaves up to 50 cm, crowded below panicle **4. pininana**
1. More or less herbs, less than 1 m (including panicle); leaves scattered up the stem, the panicle not sharply delimited from rest of stem 2.
2. Corolla hairy on veins and margins only; usually 2 stamens exserted **2. plantagineum**

2. Corolla uniformly hairy on outside; usually 3–5 stamens exserted 3.
3. Stems stiffly erect with narrower, rather dense inflorescence, the bracts scarcely exceeding the cymes; corolla blue when fully open **1. vulgare**
3. Stems ascending with rather loose inflorescence with conspicuous leafy bracts; corolla pinkish-violet when fully open **3. rosulatum**

1. E. vulgare L. Viper's Bugloss

Biennial monoecious *herb* with a tapering root. *Stems* 20–90 cm, pale green, erect, rarely diffuse, covered with stiff, white bristles interspersed with longer bristles which have a bulbous base, simple or branching from the base, leafy. *Leaves* alternate; lamina 6–15 × 1.6–3.0 cm, deep greyish-green on upper surface, paler beneath, ovate, oblong or lanceolate, obtuse at apex, entire or obscurely toothed, narrowed at base, with rather stiff hairs on both surfaces, the cauline usually narrower than the basal; petioles short, hairy. *Inflorescence* a paniculate, unilateral cyme, the cymes short and dense and elongating after flowering; flowers bisexual. *Calyx* 5-lobed, the lobes 5–7 mm, shorter than the corolla tube, linear-lanceolate, acute at apex. *Corolla* 10–19 mm, at first pinkish, afterwards bright deep blue, rarely white, rose-coloured, purple or violet, 5-lobed, the lobes unequal, hairy all over. *Stamens* 4 or 5, long-exserted; filaments reddish; anthers yellow. *Style* 1, long, white, hairy; stigma small and bifid. *Nutlets* 1.8–3.0 mm, angular with irregular ridges, rugose. *Flowers* 6–9. Flowers protandrous and visited by a great variety of insects. $2n = 32$.

Native. Open grassy places, cliffs, dunes, shingle and rough ground, usually on calcareous soils. Locally frequent to common in Great Britain, especially in south and east England, rare in central and north Scotland and introduced, mainly along the east coast in Ireland. Most of Europe and temperate Asia. Eurosiberian Temperate element.

2. E. plantagineum L. Purple Viper's Bugloss
E. maritimum Willd.; *E. lycopsis* auct.

Annual or *biennial* monoecious herb with a tap-root and fibrous side-roots. *Stems* 20–75 cm, pale yellowish-green, more or less robust, terete, branched, often to the base, with numerous, short to long, fairly stiff, colourless, spreading, minutely forked eglandular hairs, leafy. *Leaves* all cauline; lamina 2–14 × 0.5–3.0 cm, medium yellowish-green on upper surface with main veins impressed, paler beneath with prominent midrib and fairly prominent side veins, ovate, lanceolate, lanceolate-oblong to oblong or linear, obtuse to acute at apex, entire, the lower gradually narrowed to a short petiole, the upper rounded or cordate at base and sessile; all with colourless, short to long, slightly stiff hairs on both surfaces and the margins but not rough to the touch, the hairs mostly minutely forked. *Inflorescence* a paniculate, unilateral cyme; flowers bisexual; pedicels with similar hairs to the stem. *Calyx* 7–10 mm in flower, up to 15 mm in fruit, divided nearly to the base into 5 lobes; lobes linear-lanceolate, more or less obtuse at apex, with numerous, short to long, colourless, minutely forked eglandular hairs. *Corolla* 18–30 mm, nearly as broad as long, pink to purple, infundibuliform, hairy on the veins and margins only. *Stamens* 5; filaments pale reddish; anthers purple, 2 long exserted, the others included or only slightly exserted. *Style* and stigma reddish, bifid. *Nutlets* 4, ovoid-trigonous, warty, attached to the flat receptacle by their flat, triangular bases. *Flowers* 5–8. $2n = 16$.

Possibly native. Cliffs and disturbed grassy or sandy ground near the sea. Frequent in Jersey, where it has been known since 1690, very locally in the Isles of Scilly and Cornwall, rare escape or casual elsewhere in Great Britain. South and west Europe; North Africa, Macaronesia; Caucasus.

3. E. rosulatum Lange Lax Viper's Bugloss
E. humile auct.

Perennial monoecious *herb*. *Stems* 1–several, 30–75 cm, pale green, ascending or erect, stiffly hairy, branched above, leafy. *Leaves* alternate; lamina 2–8 × 0.5–2.5 cm, medium green on upper surface, paler beneath, lanceolate to ovate, obtuse at apex, entire, narrowed to a sessile base, with more or less short, appressed to patent simple eglandular hairs. *Inflorescence* of bracteate, very laxly paniculate cymes, the bracts prominent and leaf-like; flowers bisexual. *Calyx* 6–9 mm at anthesis, 10–14 mm in fruit, divided almost to the base into 5 lobes, the lobes linear-lanceolate, obtuse at apex. *Corolla* 11–25 mm, pinkish-violet, usually with a narrow tube, 5-lobed, uniformly hairy on outside. *Stamens* 3–4, exserted, unequal, inserted below the middle of the corolla. *Style* exserted; stigma bifid. *Nutlets* 4, ovoid-trigonous, irregularly and densely ridged, erect, rugose, with a flat base. *Flowers* 5–8. $2n = 32$.

Introduced. Naturalised on waste ground at Barry Docks, Glamorganshire since 1927, but may now be extinct. Native of Portugal and north-west Spain.

4. E. pininana Webb & Berthel. Giant Viper's Bugloss

Giant, monocarpic, stiffly hairy *herb*. *Stem* up to 5 m, elongating before flowering and forming a trunk 1–2 m below flowers which become leafless and woody. *Leaves* at first in a basal rosette; lamina up to 80 × 10 cm, dark green on upper surface, paler beneath, more or less oblong, very long-acuminate at apex, entire, attenuate at base; midrib very prominent beneath; petiole short. *Inflorescence* a large panicle up to 3.5 m high, with many, densely arranged lateral branches each bearing pedunculate, bracteate cymes. *Calyx* 4–6 mm, 5-lobed nearly to base, irregular, with 4 lanceolate or ovate-lanceolate lobes, and 1 larger ovate lobe. *Corolla* 12–15 mm, pink in bud, opening to pale blue, funnel-shaped, 5-lobed, the upper lobes slightly larger. *Stamens* 5, long-exserted, filiform, glabrous, usually pink in upper part. *Style* 1, exserted; stigma bifid. *Nutlets* 2–3 mm, prominently lobed and tuberculate towards the base. *Flowers* 5–6; plants usually existing for several years before flowering, seeding and dying. $2n = 15$.

Introduced. Self-sown garden escapes in rough ground. Channel Islands, Isle of Man, Caernarvonshire, Cornwall,

Isles of Scilly and Devonshire. Native of La Palma in Canary Islands.

5. Pulmonaria L.

Perennial monoecious, bristly *herbs. Leaves* alternate, lanceolate to ovate, entire, narrowed at base. *Inflorescence* a dense, terminal cluster of cymes with open flowers erect; flowers bisexual. *Calyx* 5-lobed, divided to about halfway. *Corolla* more than 10 mm, blue, red or purple, tube slightly longer than limb, with 5 conspicuous hair-tufts forming a ring in the throat, without throat scales. *Stamens* 5, equal, included. *Style* 1; stigma simple, included. *Fruit* of 4 nutlets with a collar-like base, smooth.

About 20 species in Europe and west Asia.

Birkinshaw, C. R. & Sanford, M. N. (1966). *Pulmonaria obscura* Dumort. (Boraginaceae) in Suffolk. *Watsonia* **21**: 169–178.
Hultén, E. & Fries, M. (1986). *Atlas of north European vascular plants north of the Tropic of Cancer.* 3 vols. Königstein.
Leslie, A. C. (1980). A new alien *Pulmonaria. B.S.B.I. News* **25**: 16–17.
Stewart, A., Pearman, D. A. & Preston, C. D. (1994). *Scarce plants in Britain*. Peterborough. [*P. longifolia*.]
Wigginton, M. J. (Edit.) (1999) *British red data books.* Vol. 1. *Vascular plants.* Ed. 3. Peterborough.

1. Basal leaves developing at flowering, cordate to broadly cuneate at base and abruptly contracted into the petiole 2.
1. Basal leaves developing at flowering to gradually narrowed at base and tapered into the petiole 4.
2. Basal leaves abruptly cuneate to rounded at base; corolla bright red when open; inside of corolla tube hairy below the hair-tufts **3. rubra**
2. Basal leaves cordate at base; corolla reddish to bluish-violet when open; inside of corolla tube glabrous below the hair-tufts 3.
3. Basal leaves with large white spots **1. officinalis**
3. Basal leaves unspotted or with faint green spots **2. obscura**
4. Leaves spotted or blotched **5. longifolia**
4. Leaves unspotted or without blotches 5.
5. Stalked glandular hairs frequent in inflorescence **4. Cv. Mawson's Blue**.
5. Stalked glandular hairs few or none in inflorescence **6. angustifolia**

1. P. officinalis L. Lungwort

Perennial monoecious *herb* with a creeping rhizome ending in non-flowering shoots. *Stems* 10–30 cm, pale yellowish-green, erect, with numerous to dense hairs, a mixture of short to medium, spreading, pale, slender, very unequal glandular hairs and thicker, down-curved, pale, thicker-based simple eglandular hairs, branched above, leafy. *Leaves* alternate, dull medium green with paler spots and blotches on upper surface, much paler beneath with a very prominent midrib; basal 10–20 × 4–10 cm, elliptical to broadly ovate, acute to cuspidate at apex, entire but undulate, narrowed or cordate at base, with a winged petiole; cauline much smaller, lanceolate to ovate, acute at apex, entire, semiamplexicaul at base; all with numerous to dense, short to medium, pale, stiff, bulbous-based hairs on both surface and the margins making the leaves very rough to the touch. *Inflorescence* a short cyme, scarcely elongating after flowering; peduncles pale yellowish-green, clothed like the stem; flowers 15–20 mm in diameter, bisexual, erect. *Calyx* 13–14 mm, clothed with a mixture of short to long, very unequal, slender glandular hairs and simple eglandular hairs with thickened bases; lobes 5, divided quarter of the way to the base, triangular, acute at apex. *Corolla* with tube 10–12 mm, whitish; lobes 5, erect or slightly spreading, pinkish-red to violet-blue in the same inflorescence, broadly rounded, with a ring of 5 tufts of hairs round the inside of the tube and glabrous below it. *Stamens* 5, included in the corolla tube; filaments white; anthers black. *Style* 1, whitish, included; stigma simple, cream with red spots. *Nutlets* 4, about 4 × 3 mm, ovoid, acute at apex, smooth, contracted at base above a distinct collar-like ring. *Flowers* 3–5. $2n = 16$.

Introduced. Much grown in gardens and naturalised on banks, in scrub, woods and on rough ground. Scattered records throughout Great Britain and a few in Ireland. Europe, from the Netherlands and south Sweden to north Italy and Bulgaria.

2. P. obscura Dumort. Suffolk Lungwort

P. officinalis subsp. *obscura* (Dumort.) Murb.

Perennial monoecious *herb* with a creeping rhizome ending in non-flowering shoots. *Stems* 10–30 cm, pale yellowish-green, erect, with a mixture of few, short to medium, spreading, pale, slender glandular hairs and thicker simple eglandular hairs, branched above, leafy. *Leaves* alternate; dark green without spots or blotches or with faint green spots and blotches, paler beneath; basal with lamina 8–16 × 3–8 cm, elliptical to broadly ovate, more or less acute at apex, entire, narrowed or cordate at base, with a winged petiole up to 25 cm; cauline much smaller, lanceolate to ovate, acute at apex, entire, semiamplexicaul at base; all with scattered, unequal stiff hairs, occasional glandular hairs and dense, small, stiff hairs. *Inflorescence* a short cyme, scarcely elongating after flowering; flowers bisexual, erect; peduncles pale green, clothed like the stem. *Calyx* 13–14 mm, with numerous simple eglandular hairs and few glandular hairs, divided quarter of the way to base into 5 lobes, the lobes triangular and acute at apex. *Corolla* with tube 10–12 mm, whitish; lobes 5, reddish to bluish-violet, subrotund, without throat scales but with 5 tufts of hairs forming a ring in the throat and glabrous below the ring. *Stamens* 5, inserted in the corolla tube, included. *Style* 1; stigma simple, included. *Nutlets* 4, about 4 × 3mm, ovoid, acute at apex, smooth, contracted at base above a distinct collar-like ring. *Flowers* 3–5. $2n = 14$.

Native. Woods and woodland rides and clearings. Known only from three sites in Burgate Wood and Stubbing's within 4 square kilometres in Suffolk. Widespread in Europe. European Temperate element.

3. P. rubra Schott Red Lungwort

Perennial monoecious *herb* with a creeping rhizome ending in non-flowering shoots. *Stems* 10–20 cm, pale yellowish-green, erect, with numerous simple eglandular hairs and some glandular hairs, branched above, leafy. *Leaves* alternate, dark green, unspotted or rarely somewhat spotted, paler beneath; basal 8–15×4–7 cm, lanceolate to ovate, acute at apex, entire, rounded to very broadly cuneate at base, the petioles up to 13 cm; cauline smaller, decurrent at base up to 5 mm; all rough with long, stiff eglandular hairs, short eglandular hairs and some glandular hairs. *Inflorescence* a short, ebracteate cyme; flowers bisexual, erect; pedicels with long glandular hairs and short stiff hairs. *Calyx* clothed like pedicels, 5-lobed about quarter of way to base, the lobes lanceolate and acute at apex. *Corolla* 10–12 mm, remaining red; 5-lobed, the lobes rounded, without throat scales, but with 5 tufts of hairs making a ring in the throat and hairy below the ring. *Stamens* 5, inserted in the corolla tube, included. *Style* 1; stigma simple, included. *Nutlets* 4, up to 4.5×3.0 mm, ovoid, smooth, contracted at the base above a distinct collar-like ring. *Flowers* 3–5. $2n=14$.

Introduced. Grown in gardens and naturalised in grassy places, hedges and scrub. Scattered records in England and central and south Scotland. Native of south-east Europe.

4. P. Cv. **Mawson's Blue** Mawson's Blue Lungwort

Perennial monoecious *herb* with creeping rhizome. *Stems* up to 30 cm, pale green, ascending to erect, simple, leafy. *Leaves* alternate, medium green and unspotted; basal elliptic to narrowly elliptic, pointed at apex, entire, narrowed at base; cauline narrow and semiamplexicaul; all softly hairy on both surfaces. *Inflorescence* a terminal cluster of cymes, flowers always thrum with frequent glandular hairs. *Calyx* 5-lobed, the lobes lanceolate, obtuse. *Corolla* blue, tinged red with age, funnel-shaped, 5-lobed, the lobes broadly rounded, glabrous on inside of tube below hair-tufts. *Stamens* 5, inserted in throat of corolla. *Style* 1; stigma simple. *Flowers* 4–5.

Introduced. Grown in gardens, escaping and naturalised in shady places. Scattered records in south and central England and south and central Scotland. Of uncertain garden origin.

5. P. longifolia (Bastard) Boreau Long-leaved Lungwort
P. angustifolia auct.

Perennial monoecious *herb* with thick, fleshy roots. *Stems* 20–40 cm, pale green, erect, with dense, spreading, pale simple eglandular hairs, simple, leafy. *Leaves* alternate, medium green, usually spotted white on upper surface, paler beneath; basal with lamina 12–22(–50 in autumn)×2–3(–5) cm, lanceolate or elliptical, acute at apex, entire but wavy, long attenuate at base; cauline lanceolate, oblong-lanceolate or ovate-lanceolate, long acute at apex, entire, rounded at base and semiamplexicaul; all with short to medium simple eglandular hairs and few glandular hairs on both surfaces and the margins. *Inflorescence* a rather dense, terminal cluster of cymes, flowers bisexual, erect. *Calyx* 10–12 mm, divided one-third of the way to the base into 5 lobes, the lobes triangular and more or less acute at apex, with dense, long eglandular hairs, with a few, long glandular hairs. *Corolla* more than 10 mm, deep rose changing to brilliant blue-violet, fading to purple, funnel-shaped, 5-lobed, the lobes rounded and overlapping, with 5 tufts of hairs forming a ring in the throat alternating with the stamens and glabrous below the ring. *Stamens* 5, equal, included; anthers dark. *Style* 1; stigma simple, with a few hairs. *Nutlet* about 4×3 mm, ovoid, strongly compressed, smooth, shining, with a raised collar round the base, the receptacle flat. *Flowers* 4–5. $2n=14$.

Native. Woods, pasture and scrub in ancient countryside. Extremely local in Dorsetshire, Isle of Wight and Hampshire; perhaps a rare escape elsewhere. Western Europe northwards to England and south to Spain and Portugal. Oceanic Temperate element.

6. P. angustifolia L. Narrow-leaved Lungwort
P. azurea Besser

Perennial monoecious *herbs* with creeping rhizome. *Stems* 20–40 cm, pale green, erect, simple, leafy. *Leaves* alternate, dark green and unspotted, paler beneath; basal 20–40×4–8 cm, lanceolate, acute at apex, entire, very gradually narrowed at base; cauline narrowly lanceolate, acute at apex, entire, narrowed to a sessile base; all with equal, stiff eglandular hairs. *Inflorescence* with a rather open terminal cluster of cymes; flowers bisexual, erect; pedicels with stiff eglandular hairs and scattered glandular hairs. *Calyx* in fruit very short and slender, hairy like pedicels, 5-lobed, about halfway to base, the lobes triangular and more or less acute at apex. *Corolla* 10–12 mm, bright blue, 5-lobed, the lobes rounded, without throat scales, but with 5 tufts of hairs making a ring in the throat, area below ring glabrous. *Stamens* 5, inserted in the throat of the corolla, included. *Style* 1; stigma simple, included. *Nutlets* 4, about 4.5×3.5 mm, ovoid, smooth, contracted at the base above a collar-like ring. *Flowers* 4–5. $2n=14$.

Introduced. Garden plant much confused with *P. longifolia* which probably occurs as an escape. Native of eastern Europe, west central European and the south-west Alps.

6. Symphytum L.

Perennial monoecious *herbs,* with bristles and often hooked (uncinate) hairs. *Leaves* alternate, the lower petiolate, upper usually sessile, shortly petiolate or decurrent. *Inflorescence* an ebracteate, terminal, scorpioid cyme, the flowers nodding. *Calyx* campanulate or tubular, 5-lobed, accrescent. *Corolla* funnel-shaped or subcylindrical, shortly and broadly 5-lobed; throat scales 5, linear or subulate, ciliate, connivent, usually included. *Stamens* 5, included. *Style* 1; stigma simple, exserted. *Fruit* of 4, ovoid, smooth or tuberculate nutlets, with a collar-like rim at the base.

The amount of the division of the calyx is a very difficult character to use as the calyx is accrescent. Comfreys have been much used as healing poultices for sprains, bruises and abrasions and with more apparent success than almost any other herbal medicine.

Contains about 25 species in Europe, extending to the Caucasus.

Basler, A. (1972). Cytotaxonomische Untersuchungen an der Boraginaceen-Gattung *Symphytum* L. *Bot. Jahrb. Syst.* **92**: 508–553.
Bucknall, C. (1912). Some hybrids of the genus *Symphytum*. *Jour. Bot. (London)* **50**: 332–337.
Bucknall, C. (1913). A revision of the Genus *Symphytum* Tourn. *Jour. Linn. Soc. London (Bot.)* **41**: 491–556.
Faegri, K. (1931). Über die in Scandinavien gefundenen *Symphytum* – Arten. *Bergens Mus. Aarb. Naturvidensk.* **4**: 1–47.
Gadella, T. W. J. (1972). Cytological and hybridisation studies in the genus *Symphytum Symp. Biol. Hung. (Budapest)* **12**: 189–199.
Gadella, T. W. J. & Kliphuis, E. (1967). Cytotaxonomic studies in the genus *Symphytum*. I. *Symphytum officinale* in the Netherlands. *Proc. Kon. Ned. Akad. Wet. Amsterdam (C)* **70**: 378–391.
Gadella, T. W. J. & Kliphuis, E. (1969). Cytotaxonomic studies in the genus *Symphytum*. II. Crossing experiments between *S. officinale* L. and *S. asperum* Lepech. *Acta Bot. Neerl.* **18**: 544–549.
Gadella, T. W. J. & Kliphuis, E. (1978). Cytotaxonomic studies in the genus *Symphytum*. VIII. Chromosome numbers and classification of ten European species. *Proc. Kon. Ned. Akad. Wet Amsterdam (C)* **81**: 162–172.
Harmata, K. (1977). Pollen morphology and taxonomy n the genus *Symphytum* L. and *Procopiana* Gusuleac. *Zestyty Nankowe Univ. Jagiellonskiego* 462. *Prace Bot.* **5**: 7–29.
Harmata, K. (1981). A supplement to the pollen morphology and taxonomy of the genus *Symphytum* L. and *Procopiana* Gusuleac. *Zestyty Nankowe Univ. Jagiellonskiego* 466. *Prace Bot.* **8**: 7–16.
Hultén, E. & Fries, M. (1986). *Atlas of north European vascular plants north of the Tropic of Cancer.* 3 vols. Königstein.
Ingram, J. (1961). Studies in the cultivated Boraginaceae. 5. *Symphytum*. *Baileya* **9**: 92–99.
Kurtto, A. (1982). Taxonomy of the *Symphytum asperum* aggregate (Boraginaceae) especially in Turkey. *Acta Bot. Fenn.* **19**: 177–192.
Leslie, A. C. (1982). A new alien *Symphytum*. *B.S.B.I. News* **30**: 16–17.
Leaney, B. (2007). Probable hybrid between *Symphytum × uplandicum* and *S. orientale* in Norfolk. *B.S.B.I. News* **105**: 6–9.
Pawlowski, B. (1961). Observations ad genus *Symphytum* L. pertinantes. *Fragm. Flor. Geobot.* **7**: 327–356.
Perring, F. H. (1994). *Symphytum* – Comfrey. In Perry, A. R. & Ellis, R. G. (Edits.) *The common ground of wild and cultivated plants*, pp. 65–70. B.S.B.I. Conference Report No. 22. Cardiff.
Sandbrick, J. M., Van Brederode, J. & Gadella, T. W. J. (1990). Phylogenetic relationships in the genus *Symphytum* L. (Boraginaceae) *Proc. Kon. Ned. Acad. Wet. Amsterdam (C)* **93**: 295–334.
Wade, A. E. (1958). The history of *Symphytum asperum* Lepech. and *S. × uplandicum* in Britain. *Watsonia* **4**: 117–118.
Wickens, G. E. (1969). A revision of *Symphytum* in Turkey and adjacent areas. *Notes Roy. Bot. Garden. Edinb.* **29**: 157–180.

1. Plant with procumbent to decumbent, leafy stolons 2.
1. Plant without stolons 3.
2. Larger flowering stems branched; corolla predominantly blue or pink when open **6. × hidcotense**
2. Flowering stem unbranched; corolla pale yellow when open, often flushed reddish outside **7. grandiflorum**
3. Stem leaves strongly decurrent, forming wings on the stem and extending down more than one internode; nutlets more or less smooth and shining 4.
3. Stem leaves not to moderately decurrent, the wings rarely extending to more than one internode; nutlets minutely tuberculate 5.
4. Flowers purplish or pink **1(i). officinale** var. **officinale**
4. Flowers white or creamy-yellow, sometimes tinged purplish **1(ii). officinale** var. **ochroleucum**
5. Corolla pink, purple or blue 6.
5. Corolla pale yellow to white 10.
6. When fully out flowers clear blue or clear blue with white at base 7.
6. When fully out flowers pale mauve or pinky-red 9.
7. Lower leaves long attenuate at base **11. caucasicum**
7. Lower leaves rounded or cordate at base 8.
8. With numerous, long, rigid, bulbous-based hairs on upper stem and inflorescence, intermixed with few short uncinate ones **4. asperum**
8. With few, long, rigid bulbous-based hairs on upper stem and inflorescence and numerous uncinate ones **3. savvalense**
9. Fully open flowers whitish at bottom and pale mauve or pinkish red above **10. × perringianum**
9. Fully open flowers pinkish-blue, violet or purplish **2. × uplandicum**
10. Throat scales of corolla exserted for more than 1 mm **12. bulbosum**
10. Throat scales of corolla included 10.
11. Calyx divided less than halfway to base; corolla pure white **9. orientale**
11. Calyx divided more than halfway to base; corolla pale to deeper yellow 11.
12. Rhizome with swollen tubers present; stems not or little branched; middle and upper cauline leaves sessile, the upper shortly decurrent **5. tuberosum**
12. Rhizome absent; stems well-branched; cauline leaves petiolate, the uppermost ones sessile but not decurrent **8. tauricum**

1. S. officinale L. Common Comfrey
S. molle Janka

Perennial monoecious *herb* with a thick, fleshy, fusiform, branched root. *Stem* 10–120 cm, medium green, erect, with long, conical, deflexed, white, stiff simple eglandular hairs and much smaller, uncinate ones, branched above, leafy. *Leaves* alternate, medium green on upper surface, paler and bluish beneath; lower 15–25 × 1–8 cm, ovate-lanceolate or lanceolate, acute at apex, entire but wavy, narrowed at base; upper smaller and narrower, oblong-lanceolate, acute at apex, narrowed at base and broadly decurrent on the stem; all with short, stiff simple eglandular hairs, rarely tuberculate, on both surfaces and the margins making the leaf very rough, with numerous, smaller, uncinate ones. *Inflorescence* an ebracteate, terminal, scorpioid cyme with nodding, bisexual flowers; pedicels with numerous, stiff, long eglandular hairs and much smaller uncinate ones. *Calyx* 7–8 mm, green, flushed purple, with numerous, long, stiff eglandular hairs and much smaller uncinate ones, deeply divided into 5 lobes, the lobes lanceolate and acute at apex. *Corolla* 12–18 mm, creamy-white tinged greenish-yellow or purplish or pink, shortly 5-lobed, the lobes triangular and obtuse at apex; throat scales scarcely longer than the stamens, triangular-subulate. *Stamens* 5, included. *Style* 1; stigma simple, exserted. *Nutlets* 4, 4.5–5.0 mm, shining black, ovoid, smooth. *Flowers* 5–6. $2n=24, 44, 48$.

(i) Var. **officinale**
S. patens Sibth.; *S. officinale* var. *purpureum* Pers.

Flowers purplish or pink.

(ii) Var. **ochroleucum** DC.
Flowers white or creamy-yellow, sometimes tinged purplish.

Native. By streams and rivers, in fens and marshy places, also roadsides and rough ground. Locally frequent in Great Britain and Ireland, but often over-recorded for *S.* × *uplandicum*. Var. *ochroleucum* is the common plant of East Anglia. Var. *officinale* occurs in the south and west, but the total distribution of the two varieties is not known. The species occurs in most of Europe and temperate Asia and is widely naturalised elsewhere. European Temperate element.

× **tuberosum**
Plants found by R. Cropper and sent to Perring by I. P. Green, now in (**CGE**) from Nunney Combe in Somersetshire could be this hybrid.

2. S. × **uplandicum** Nyman — Russian Comfrey
S. asperum × **officinale**
S. officinale var. *patens* Syme; *S. peregrinum* auct.

Perennial monoecious *herb* with a thick, branched, fusiform stock. *Stems* 100–200 cm, pale green, erect, with numerous, stiff, retrorse, tuberculate-based hairs and numerous, shorter uncinate ones, branched, leafy. *Leaves* alternate, medium to dark green on upper surface, paler beneath; basal with lamina 20–28 × 7–13 cm, elliptic-lanceolate, ovate or oblong, acute or acuminate at apex, entire, cordate, rounded or slightly attenuate at base, long petiolate; upper 5–15 × 1.5–5.0 cm, ovate, acuminate at apex, entire, sessile and shortly decurrent or amplexicaul; all slightly rigid-hairy, sometimes the hairs tuberculate-based on both surfaces and with numerous, shorter uncinate ones. *Inflorescence* a many-flowered terminal, scorpioid cyme with nodding, bisexual flowers; pedicels with rigid, patent hairs and numerous, much smaller uncinate ones. *Calyx* 5–7 mm, up to 11 mm in fruit, stiffly hairy with some tuberculate-based and numerous, minute uncinate ones, divided one-third of the way to the base into 5 lobes, the lobes triangular-lanceolate, and acute at apex. *Corolla* 12–18 mm, pinkish-blue, violet or purplish, 5-lobed, minutely hairy; throat scales triangular-subulate and obtuse at apex. *Stamens* 5, included. *Style* 1; stigma simple, exserted. *Nutlets* 4, 4.0–5.0 × 2.5–3.0 mm, ovoid, curved, tuberculate. *Flowers* 6–7. $2n=36, 40$.

The following colour forms sometimes refer to whole populations. Their distributions are not known.

(1) Forma **uplandicum**
Corolla rose at length turning bluish.

(2) Forma **lilacinum** (Bucknall) P. D. Sell
S. × *lilacinum* Bucknall

Corolla purplish, with a greenish-yellow tip when in bud, then turning pale purplish-rose.

(3) Forma **densiflorum** (Bucknall) P. D. Sell
S. × *densiflorum* Bucknall

Corolla large and reddish-violet.

(4) Forma **discolor** (Bucknall) P. D. Sell
S. × *discolor* Bucknall

Corolla whitish or more or less tinted with pale rose and blue.

(5) Forma **caeruleum** (Petitmengin ex Thell.) P. D. Sell
S. × *caeruleum* Petitmengin ex Thell.

Corolla rose, tipped with green when in bud, but bright blue or rose and blue later, the tip of the lobes sometimes yellowish.

Introduced. Probably first introduced as a fodder plant, now widely naturalised on roadsides, rough and damp ground and wood borders. Frequent over much of Great Britain and Ireland. Presumably originated from Russia, but may occasionally arise anew where *S. asperum* is naturalised near to *S. officinale*. It is fertile and backcrosses with *S. officinale*. Known from the Caucasus and widely naturalised in temperate Europe. The distribution of the colour forms is not known. No attempt has been made to key them out but the information may be helpful to those who find different coloured colonies.

3. S. savvalense Kurtto — Norfolk Comfrey

Perennial herb with an oblique or vertical rootstock. *Stem* 50–150 cm, pale to medium green, angled, with scattered, long, bulbous-based, curved bristles and numerous, shorter uncinate hairs. *Leaves* dark green on upper surface, paler beneath; basal 14–18 × 9–12 cm, broadly lanceolate to ovate, acute or slightly cuspidate at apex, shallowly crenate, cordate or truncate to rounded at base, with long, curved bristles and small uncinate hairs on upper surface, with numerous, long, bulbous-based, curved bristles and shorter uncinate hairs especially on the veins beneath, with petioles up to 30 cm, covered with long, bulbous-based, curved bristles and short uncinate hairs; cauline lanceolate or ovate, acute or acuminate at apex, entire, more or less rounded at base, the short petioles winged but not decurrent, clothed like basal. *Inflorescence* an ebracteate, terminal, scorpioid cyme, with nodding bisexual flowers; pedicels with few, bulbous-based bristles and numerous, shorter, uncinate hairs. *Flower buds* pinkish-red. *Calyx* 3–4 mm, 5-lobed to just over halfway, the lobes lanceolate, obtuse at apex, with few, curved bristles and numerous, uncinate hairs. *Corolla* 15–18 mm, red in bud becoming sky blue, shortly 5-lobed; throat scales 5, lanceolate. *Stamens* 5, included; filaments purple-tinted; anthers pale. *Style* 1, exserted. *Nutlets* 4, about 4.0 × 2.0 mm, curved. *Flowers* 6–7.

Introduced. Roadside opposite Intwood church, near East Carlton and Berg Apton and Sustead all in Norfolk. A specimen collected by C. C. Babington in the valley between Oakford and The Rocks at Bath in 1833, in **CGE**, was named by Babington *S. asperrimum*. It appears to be this plant. In his discussion of *S. asperum*, C. Bucknall (1913) points out that he has two different plants under that name one of which appears to be this plant. From descriptions and illustrations it is probable that both *S. asperum* and *S. asperrimum* could be this plant. Typification is necessary before anything can be done about it. The type specimen of *S. armeniacum* Bucknall is in **CGE** and is not this plant.

4. S. asperum Lepech. Rough Comfrey
S. echinatum Ledeb.; *S. asperrimum* Bieb.

Perennial monoecious *herb* with a thick, branched root. *Stems* 90–180 cm, pale to medium green, erect, angular, with numerous long strong, uncinate, subretrorse, tuberculate-based hairs and few shorter uncinate ones, well-branched, leafy. *Leaves* medium to dark, dull green on upper surface, paler beneath; lamina of basal 15–19 × 7–12 cm, of cauline 10–20 × 4–10 cm, ovate or elliptical, acuminate at apex, entire, the lower with a rounded base and petiolate, the upper cuneate, not decurrent and shortly petiolate; all with tuberculate-based, stiff or rigid hairs especially on the midrib beneath and numerous small uncinate ones. *Inflorescence* an ebracteate, terminal, scorpioid cyme with nodding, bisexual flowers; pedicels with dense, bulbous-based bristles and shorter, rigid uncinate hairs. *Calyx* 3–5 mm, divided two-thirds of the way to the base into 5 lobes, the lobes linear-oblong, obtuse at apex, with rather long, tuberculate-based, stiff hairs and dense, finer, short, uncinate ones. *Corolla* 9–14 mm, red in bud, turning bright, clear sky blue, 4–5 times longer than the calyx, shortly 5-lobed; throat scales 5, lanceolate. *Stamens* 5, included; filaments purple-tinted, subequal; anthers pale yellow. *Style* 1; stigma simple, exserted. *Nutlets* 4, about 4.0 × 2.5 mm, curved, constricted above the collar-like base, tuberculate. *Flowers* 6–7. $2n = 32$.

Introduced. Naturalised in rough and waste ground, formerly occasional, now very rare. Scattered records over Great Britain, and Co. Sligo in Ireland. Probably much over-recorded for *S.* × *uplandicum* and *S. savvalense* Kurtto needs also to be considered. Possibly in wild flower seed. Native of south-west Asia.

× **caucasicum**
This hybrid is recorded for Cambridgeshire and Edinburghshire.

5. S. tuberosum L. Tuberous Comfrey
S. foliosum Rehmann

Perennial monoecious *herb* with a fibrous root and a creeping rhizome, with alternate, thick, tuberous and thin portions. *Stems* 20–50 cm, pale green, erect, with numerous, reflexed bristles and smaller, uncinate hairs, simple or with 1–2 short branches at the top, leafy. *Leaves* alternate, shining bluish-green on upper surface, paler beneath; lower with lamina small, ovate or spathulate, acute at apex, entire, narrowed at base and petiolate; medium 10–14 × 2–3 cm, ovate-lanceolate or elliptical, acute at apex, entire but wavy, narrowed at base, shortly petiolate; upper smaller and sessile and shortly decurrent; all with numerous, stiff hairs on both surfaces. *Inflorescence* an ebracteate, terminal, scorpioid cyme, with nodding, bisexual flowers. *Calyx* 5–8 mm, divided nearly to base into 5 lobes, the lobes lanceolate and acute at apex, with numerous, stiff hairs and dense, small uncinate ones. *Corolla* 13–19 mm, very pale creamy-yellow, shortly 5-lobed, the lobes rounded; throat scales broadly triangular-subulate, acuminate at apex, somewhat exceeding the stamens. *Stamens* 5, included; anthers deep yellow. *Style* 1, creamy-white; stigma simple, exserted. *Nutlets* 4, dull, ovoid, minutely tuberculate, with a collar-like rim at base. *Flowers* 6–7. $2n = 96$.

Native. Damp woods, ditches and river-banks. Frequent in lowland Scotland; scattered records of introduced plants in England, Wales and Ireland. West, central and south Europe; north-west Anatolia. Our plant is subsp. **tuberosum**. European Temperate element.

× **uplandicum**
asperum × officinale × tuberosum

This hybrid is intermediate in all its characters, with yellow corolla tinged with blue or purple and tuberous rhizomes, and is partly sterile.

Recorded from Gloucestershire, Forfarshire and Morayshire. Some plants may be *S. officinale* × *tuberosum*.

6. S. × hidcotense P. D. Sell Hidcote Comfrey
S. asperum × grandiflorum × officinale
S. grandiflorum* × *uplandicum

Perennial monoecious *herb* with decumbent, vigorous, non-flowering stolons from branched rhizomes with tuberous swellings. *Stems* up to 100 cm, pale green, ascending to erect, with numerous, stiff, bulbous-based hairs and shorter, slender, uncinate ones, branched, leafy. *Leaves* alternate; lamina 5–19 × 2–10 cm, dull medium green on upper surface, paler beneath, ovate-oblong, obtuse to acute at apex, entire, truncate at base, mostly with petioles but not or scarcely decurrent, with rough uncinate hairs and slightly bullate. *Inflorescence* a terminal, scorpioid cyme with nodding, bisexual flowers. *Calyx* 5–7 mm, lobed from three-fifths to two-thirds of the way to the base into 5 lobes, the lobes linear-lanceolate, obtuse at apex, with long, rigid hairs and short uncinate ones. *Corolla* 18–22 mm, red in bud, changing to blue or pink at the base and white at the apex, 5-lobed, with throat scales not or just exceeding the stamens with papillae most dense near the apex. *Stamens* 5, included; filaments and anthers white. *Style* 1; stigma simple, exserted. Some fertile *nutlets* form. *Flowers* 5–7.

The variant with a blue base to the blue and white corolla has been called cv. Jubilee. It is probably a hybrid between *S.* × *uplandicum* and *S. grandiflorum* (female), and cv. Hidcote Blue, cv. Lilacinum and cv. Wisley Blue are probably of a similar origin. The form with pink-based corollas is probably *S. grandiflorum* × *S.* × *uplandicum* (female). The holotype is cv. Hidcote Blue.

Introduced. Grown in gardens and naturalised in hedges and woodland. Scattered records in Great Britain and more frequent in the south. Of garden origin.

7. S. grandiflorum DC. Creeping Comfrey
S. cordatum M. Bieb.; *S. ibericum* Steven

Perennial monoecious *herb* with a rather thick root and procumbent to decumbent stolons arising from rhizomes and bearing leaves. *Stems* 20–40 cm, pale green, erect with rather long, reflexed, rigid hairs and numerous, shorter, uncinate hairs, simple, unbranched, leafy. *Leaves* alternate, medium green on upper surface, paler beneath; lamina of basal 6–10 × 3–6 cm, ovate, acuminate at apex, entire, cordate, with a long, narrowly winged petiole; cauline 2–4 × 1.5–2.5 cm, ovate or ovate-lanceolate, contracted or narrowed to the petiole or sessile and shortly decurrent; all with sparse to numerous, slender, stiff, tuberculate-based hairs. *Inflorescence* a few- to many-flowered, terminal, scorpioid cyme with nodding bisexual flowers, erect and elongating in fruit; pedicels curved and with long, slender and shorter, uncinate hairs. *Calyx* 6–7 mm, divided more than two-thirds of the way to the base into 5 lobes, the lobes linear and obtuse at apex, with some long slender hairs and short uncinate ones. *Corolla* 12–20 mm, pale yellow, often flushed reddish outside, shortly 5-lobed, 3 times longer than the calyx, with dense minute hairs; throat scales 5, linear and obtuse at apex. *Stamens* 5, included; filaments white, subequal; anthers pale yellow, obtuse at base. *Style* 1; stigma simple, exserted. *Nutlets* 4, about 3 × 2.4 mm, fuscus, broadly ovate, curved, constricted above the collar-like rim at base, tuberculate. *Flowers* 6–7. $2n = 24$, 60.

Introduced. Common in gardens and well naturalised in woods and hedges. Scattered records throughout much of Great Britain, rarely north to central Scotland. Native of the Caucasus.

8. S. tauricum Willd. Crimean Comfrey
S. bullatum Hornem.

Perennial monoecious *herb* with a fusiform root. *Stem* 20–60 cm, pale green, rather stout, with short, uncinate hairs and few tuberculate-based, stiff hairs, much branched, leafy. *Leaves* alternate; lamina 5.5–7.5 × 2.0–4.5 cm, ovate or ovate-oblong, acute at apex, entire but slightly undulate, cordate or rounded at base, not decurrent, the lower petiolate, the median slightly petiolate, the upper acuminate at apex and sessile but not decurrent, all bullate and with dense, stiff, uncinate hairs on both surfaces. *Inflorescence* a terminal, scorpioid cyme with nodding, bisexual flowers; pedicels with some, long, rigid and numerous, shorter uncinate hairs. *Calyx* 4–7 mm, divided three-fifths to five-sixths of the way to the base into 5 lobes, the lobes lanceolate, acute or rather obtuse at apex, with dense, long and short uncinate hairs. *Corolla* (8–)9–12(–15) mm, pale yellow, 5-lobed, the lobes not recurved; throat scales broadly linear, obtuse at apex, their marginal papillae up to twice as long as wide and scattered towards the base. *Stamens* 5 included; filaments subequal; anthers 2–3 mm, with apiculate base. *Style* 1; stigma simple. *Nutlets* 4, about 2.5 × 1.5 mm, ovoid, curved, tuberculate. *Flowers* 6–7. $2n = 14$, 18, 80.

Introduced. Naturalised on a hedgebank in one place in Cambridgeshire since 1973 and recorded for Suffolk, Worcestershire and Warwickshire. Native of south-east Europe.

9. S. orientale L. White Comfrey
S. jacquinianum Tausch; *S. tauricum* auct.

Perennial monoecious *herb* with a fusiform root. *Stems* up to 70 cm, pale green, with numerous, long and short, uncinate hairs, much-branched, leafy. *Leaves* alternate, dull medium green on upper surface, paler beneath; lamina of radical up to 14 × 9 cm, ovate or oblong, subacute at apex, entire, truncate, cordate or rounded at base, petiolate, the cauline up to 12 × 8 cm and sessile; all densely and rather softly, long eglandular-hairy with numerous smaller uncinate hairs on both surfaces, often slightly tomentose beneath. *Inflorescence* a terminal, scorpioid cyme with nodding, bisexual flowers; pedicels with dense, long and short, uncinate hairs, curved. *Calyx* 7–9 mm, divided about one-quarter to two-fifths of the way to the base into 5 lobes, the lobes ovate-oblong and obtuse at apex with dense, long and short uncinate hairs. *Corolla* (13–)14–18(–19) mm, white, 5-lobed, the lobes not recurved; throat scales included, lingulate, the marginal papillae up to 2.5 times as long as wide. *Stamens* 5, included; filaments about three-fifths as long as anthers; anthers 2.5–3.5 (–4.0) mm, (2.5–)3–5(–5.5) times as long as wide. *Style* 1; stigma simple, exserted. *Nutlets* 4, about 3.0 × 2–3 mm, blackish-brown, ovoid, curved, constricted above the collar-like rim at base, tuberculate. *Flowers* 6–7. $2n = 32$.

Introduced. Naturalised in hedgerows and other shady places, often self-sown. Frequent in east and southern England and north to central Scotland, very scattered elsewhere. Native of western Russia and north-west Turkey and Caucasus.

× **savvalense**
Some plants growing with *S. savvalense* in Norfolk, with a lot of white in the corolla, may be this hybrid.

10. S. perringianum P. H. Oswald & P. D. Sell
Perring's Comfrey

S. orientale × uplandicum
= *S. asperum × officinale × orientale*

Perennial monoecious *herb*. *Stems* up to 1 m, pale green, angled, the angles sharp, with pale, rather stiff, long eglandular hairs and many, short, uncinate eglandular hairs, branched, the branches ascending or slightly spreading, leafy. *Leaves* dull, dark green on upper surface, paler beneath; basal ovate, cordate at base, soon withering; cauline alternate, 7–15 × 4–9 cm, ovate or oblong-ovate, narrowed at apex but obtuse, entire or very shallowly crenate, undulate, the upper cuneate at the base, the lower more or less cordate; all with long simple eglandular hairs and short uncinate hairs on both surfaces; all veins bullate above and prominent beneath and forming a strong network; upper leaves sessile or nearly so, the lower with petioles up to 5(–10) cm, pale green, channelled above, rounded beneath, all with long and short, uncinate eglandular hairs, not decurrent. *Inflorescences* of dense, terminal cymes at the ends of stem and branches; flowers bisexual; pedicels 2–7 mm in flower, with long and short, eglandular hairs. *Calyx* 7–8 mm, green, 5-lobed, the lobes unequal lanceolate, obtuse at apex, with long and short, uncinate hairs. *Corolla* 9–11 mm, whitish at base, pale mauve or pinkish-red above, 5-lobed. *Stamens* 5; filaments white; anthers pale yellow, included. *Pollen* apparently infertile. *Style* 1; stigma simple, exserted. *Flowers* 4–5 (–12).

Growing with *S. orientale* in the region of Barton Road, Cambridge, where it was discovered by Philip Oswald. Named after Franklyn Hugh Perring (1927–2003), who grew this plant in his garden and supplied the material grown in the Botanic Garden, Cambridge, from which the holotype was taken.

11. S. caucasicum M. Bieb. Caucasian Comfrey
S. racemosum Roem. & Schult.

Perennial monoecious *herb* with a branched, fusiform stock. *Stems* 30–60 cm, pale green, erect, with soft, uncinate hairs, branched, leafy. *Leaves* alternate, medium green on upper surface, paler beneath; basal up to 20×8 cm, ovate, oblong or lanceolate, subacute at apex, entire, long attenuate to a winged petiole; cauline 8–15×4–6 cm, narrower and shorter and slightly decurrent; all softly uncinate-hairy. *Inflorescence* a terminal, scorpioid cyme with nodding bisexual flowers; pedicels stiffly uncinate-hairy. *Calyx* 6–9 mm, stiffly uncinate-hairy, divided one-quarter to one-half of the way to the base into 5 lobes, the lobes broadly linear or triangular and obtuse at apex. *Corolla* 12–14 mm, blue, 5-lobed; throat scales linear, obtuse at apex. *Stamens* 5, included. *Style* 1; stigma simple, exserted. *Nutlets* 4, ovoid, minutely tuberculate. *Flowers* 6–7. $2n=24$, 36, 48.

Introduced. Naturalised in hedgerows and other shady places. Very scattered records in Great Britain. Native of Caucasus.

12. S. bulbosum K. F. Schimp. Bulbous Comfrey
S. filipendulum Bisch.; *S. clusii* C. C. Gmel.;
S. punctatum Gaudin; *S. macrolepis* Gay

Perennial monoecious *herb* with a slender, creeping rhizome producing subglobose tubers. *Stems* 15–50 cm, pale green, erect, with very small uncinate hairs and scattered stiff hairs up to 1.5(–2.0) mm, simple or little branched, leafy. *Leaves* alternate, medium green on upper surface, paler beneath; lamina of lower ovate to elliptic-lanceolate, subacute at apex, gradually attenuate or abruptly contracted into a long petiole; remaining cauline 7–17×3–9 cm, the uppermost sessile and slightly decurrent; all with dense, very small, uncinate hairs and scattered stiff hairs on both surfaces. *Inflorescence* a terminal scorpioid cyme with nodding, bisexual flowers; pedicels with dense hairs. *Calyx* about 5 mm, uncinate-hairy, divided one-third to six-sevenths of the way to the base into 5 lobes, the lobes lanceolate, subacute at apex and long ciliate. *Corolla* (7–)8–11(–12) mm, pale yellow, 5-lobed, the erect lobes one-sixth to one-third as long as the tube; throat scales lanceolate-subulate, rarely triangular-lanceolate, acute at apex and exserted for 1–4(–5) mm, the marginal papillae dense and about as long as wide. *Stamens* 5, included, with filament one-fifth to one-half as long as anther; anthers 2.5–4.0 mm, minutely apiculate. *Style* 1; stigma simple, exserted. *Nutlets* 4, about 3×3 mm, broadly ovate, constricted above the collar-like rim at base, reticulate-rugose and minutely tuberculate. *Flowers* 6–7. $2n=72$, 84, 120.

Introduced. Naturalised in woods and by streams. Very scattered records in central and south Great Britain. Native of south-central and south-east Europe.

7. Brunnera Steven

Tufted, hairy *perennial* monoecious *herbs*. *Leaves* alternate, basal and lower ovate and strongly cordate at base, petiolate. *Inflorescence* of dense, bractless cymes in terminal sub-corymbose panicles; open flowers erect. *Calyx* divided nearly to base into 5 lobes. *Corolla* blue, with 5 equal lobes. *Stamens* 5, included. *Style* 1; stigma simple, included. *Fruit* of 4, rugose, ridged nutlets with a collar-like base.

Contains 3 species from western Europe to eastern Siberia.

1. B. macrophylla (Adams) I. M. Johnst.
Great Forget-me-not
Myosotis macrophylla Adams; *Anchusa myosotidiflora* Lehm.; *B. myosotidiflora* (Lehm.) Steven

Perennial monoecious *herb* with densely tufted stems. *Stems* 20–50 cm, pale yellowish-green, erect, angled, with short and medium, rather stiff, spreading, pale simple eglandular hairs, leafy, branched only in inflorescence. *Leaves* alternate, dull medium yellowish-green on upper surface, paler beneath; basal 5–20×5–16 cm, numerous, broadly ovate, acute at apex, entire but undulate, strongly cordate at base, the petioles up to 30 cm, pale yellowish-green, angled and with short to medium, rather stiff, pale simple eglandular hairs; cauline leaves 9–12, 2.5–4.0×1–3 cm, ovate, acute at apex, entire but undulate, the lower cordate and petiolate, the upper cuneate and sessile; veins 4–7 pairs, impressed above, prominent beneath; all covered on both surfaces and the margins with short and medium, pale, rather stiff simple eglandular hairs. *Inflorescence* of ebracteate cymes in terminal panicles; flowers bisexual, erect; pedicels 2–5(–8) mm, pale green and slender, with very short, pale, appressed simple eglandular hairs. *Calyx* about 2 mm, medium green, 5-lobed almost to the base, the lobes linear-lanceolate, acute at apex. *Corolla* 6–7 mm in diameter, 3–4 mm long, bright blue, 5-lobed, the lobes broadly obovate, rounded at apex. *Stamens* 5, equal, included, inserted at the middle of the tube; filaments white; anthers blue. *Style* 1, green, included; stigma simple, capitate. *Nutlets* 2.5–4.0 mm, oblong-obovoid, slightly asymmetrical, erect, rugose and somewhat longitudinally ribbed, the collar-like basal ring ribbed. *Flowers* 4–5. $2n=12$.

Introduced. Much grown in gardens and a very persistent throw-out, sometimes self-sown in woods and on rough ground and tips. Scattered records in Great Britain and a solitary record from the north of Ireland. Native of the Caucasus and North Asia.

8. Nonea Medik.

Annual or *perennial* monoecious, hairy *herbs*. *Leaves* alternate, without stipules. *Flowers* bisexual. *Calyx* divided to halfway into 5 lobes, slightly enlarged in fruit. *Corolla* yellow, white, pink or bluish, with a long, slender tube or funnel-shaped, the throat with 5, small, semicircular, hairy scales cut into lobes and sometimes a ring of hairs. *Stamens* 5, included in the corolla and attached to the tube. *Style* 1, simple. *Fruit* of 4, glabrous or hairy nutlets.

Contains about 35 species in the Mediterranean area and south-west and central Asia.

1. Corolla 15–18 mm, pinkish-purple **1. rosea**
1. Corolla 7–12 mm, yellow **2. lutea**

1. N. rosea (M. Bieb.) Link Pink Nonea
Anchusa rosea M. Bieb.

Annual monoecious *herb. Stems* 20–30 cm, pale green, erect or ascending, with stiff eglandular hairs. *Leaves* alternate; lamina up to 11.0×0.5–1.0 cm, medium green on upper surface, paler beneath, linear-lanceolate. *Inflorescence* a terminal, bracteate cyme; flowers bisexual. *Calyx* 4–6 mm at anthesis, up to 12 mm in fruit, 5-lobed, the lobes subulate, acuminate at apex. *Corolla* 15–18 mm, pinkish-purple, 5-lobed; scales semilunar. *Stamens* 5. *Style* 1; stigma simple. *Nutlets* 4, 3–4 mm, grey, smooth or rugose, shortly hairy. *Flowers* 4–5.

Introduced. A casual or persistent garden escape, formerly a grain alien. Native of the Caucasus.

2. N. lutea (Desr.) DC. Yellow Nonea
Lycopsis lutea Desr.

Annual monoecious *herb. Stems* 10–60 cm, pale green, erect or ascending, branched, with stiff eglandular and glandular hairs. *Leaves* alternate; lamina 2–7×0.5–2.5 cm, medium green on upper surface, paler beneath, oblong-lanceolate to linear-lanceolate, obtuse to acute at apex, entire or dentate, the cauline semiamplexicaul, hairy. *Inflorescence* a terminal, bracteate cyme. *Flowers* bisexual. *Calyx* 6–10 mm at anthesis, becoming 10–20 mm in fruit, 5-lobed, the lobes narrowly triangular. *Corolla* 7–12 mm, yellow, 5-lobed, the lobes subrotund-ovate; scales shortly hairy. *Stamens* 5; anthers 1.5–2.0 mm. *Style* 1; stigma simple. *Nutlets* 4, 3.5–6.0×2.0 mm, brown with white flecks, oblong-ellipsoid, longitudinally ribbed, shortly hairy; basal ring smooth. *Flowers* 4–6.

Introduced. A casual garden escape and a persistent garden weed; formerly established in Caernarvonshire and recorded from Berkshire, Suffolk and Worcestershire. Native of east Europe and south-west Asia.

9. Anchusa L.

Annual to *perennial* monoecious *herbs*. *Leaves* alternate, sessile or with a short petiole. *Inflorescence* of terminal, branched cymes forming a panicle. *Flowers* bisexual. *Calyx* deeply divided into 5 lobes. *Corolla* blue to purple or yellow, 5-lobed, the lobes sometimes slightly unequal, with tube longer than lobes; throat with scales or hairs. *Stamens* 5, included. *Style* 1; stigma simple, included. *Fruit* of 4, ridged and tuberculate nutlets with a collar-like base.

Contains about 50 species in Europe and Asia.

Hultén, E. & Fries, M. (1986). *Atlas of north European vascular plants north of the Tropic of Cancer.* 3 vols. Königstein.

1. Corolla with straight tube and 5, slightly unequal lobes **4. arvensis**
1. Corolla with straight tube and 5 equal lobes 2.
2. Calyx lobes obtuse to rounded; corolla yellow **1. ochroleuca**
2. Calyx lobes acute; corolla blue to purple 3.
3. Calyx divided about halfway to nearly to base; nutlets less than 5 mm **2. officinalis**
3. Calyx divided nearly to base; nutlets more than 5 mm **3. azurea**

1. A. ochroleuca M. Bieb. Yellow Alkanet

Perennial, rarely *biennial*, monoecious *herb. Stems* 30–80 cm, pale green, erect, with dense, short, soft, appressed hairs and some longer stout hairs. *Leaves* alternate; lamina 4–8(–20)×0.3–1.0(–2.5) cm, lanceolate or oblong-lanceolate, gradually narrowed to an acute apex, entire, semiamplexicaul at base, with dense, short, soft, appressed hairs. *Inflorescence* a terminal, bracteate cyme; flowers bisexual; pedicels very short; bracts mostly shorter than calyx. *Calyx* 4–6(–8.5) mm, up to 15 mm in fruit, divided to one-third to half of the way to the base into 5 lobes, the lobes linear-oblong and obtuse to rounded at apex. *Corolla* pale yellow; tube 5–10 mm, straight; 5-lobed, the lobes equal, oblong or rounded; with 5, oblong, obtuse, densely and shortly hairy scales in the throat. *Stamens* 5, inserted in upper half of tube and not or scarcely overlapping the scales of the throat, included. *Style* 1, included; stigma simple, capitate. *Nutlets* 4, about 2.5×3.5 mm, obliquely ovoid, ridged and tuberculate with a thickened collar-like ring at the base. *Flowers* 6–9. $2n=24$.

Introduced. Naturalised on rough ground on Phillack Towans in Cornwall since at least 1922, probably from animal feed; a rare escape in Kent, Norfolk, Gloucestershire, Lancashire and Edinburghshire. Native of east Europe.

× **officinalis** = **A.** × **baumgartenii** (Nyman) Gusul. Recorded for Cornwall.

2. A. officinalis L. Alkanet

Perennial, rarely *biennial*, monoecious *herb. Stems* 20–80(–170) cm, pale green, erect, with uniform simple eglandular hairs, simple, leafy. *Leaves* alternate; lamina 5–12×1–2 cm, medium bluish-green on upper surface, paler beneath, lanceolate or oblong-lanceolate, acute at apex, entire, the lower petiolate, the upper sessile, uniformly hairy on both surfaces. *Inflorescence* of dense cymes; flowers bisexual; pedicels very short, up to 5 mm in fruit; bracts equalling or shorter than calyx. *Calyx* 5–7 mm, up to 10 mm in fruit, divided to halfway to the base into 5 lobes, the lobes lanceolate and acute at apex, stiffly hairy. *Corolla* violet or reddish, rarely white or yellow; tube 5–7 mm, equalling or up to 1.5 times as long as the calyx, straight, 5-lobed, the lobes more or less equal and rounded; throat scales 5, ovate. *Stamens* 5, inserted at top of tube and overlapping scales, included. *Style* 1, included; stigma simple, capitate. *Nutlets* 4, about 2×4 mm, obliquely ovate, ridged and tuberculate with thickened collar-like base. *Flowers* 6–9. $2n=16$.

Introduced. Grown in gardens and escaping onto waste and rough ground, but not usually permanent. Scattered records over lowland Great Britain and north-west Ireland. Much of Europe, but absent from the extreme north, much of the west and parts of the Mediterranean region.

3. A. azurea Mill. Garden Anchusa
A. paniculata Aiton

Perennial monoecious *herb. Stems* 20–100(–150) cm, pale green, erect, subterete or obscurely angled, sparingly to densely covered with long, spreading, white, bristly hairs. *Leaves* alternate; basal forming a loose tuft, the lamina 8–15(–30) × 1.5–6.0(–10.0) cm, dull medium green on upper surface, paler beneath, narrowly obovate-elliptical, acute at apex, entire, attenuate at base to flattened, narrowly winged petiole 2–8 cm; cauline 6–20 × 0.8–5.0 cm, linear-lanceolate or narrowly oblong-lanceolate, acute or acuminate at apex, often obscurely erose-dentate, sessile; all thinly to densely strigose hairy on both surfaces. *Inflorescence* a lax terminal panicle consisting of numerous, branched cymes; flowers bisexual; pedicels up to 10 mm at anthesis, elongating to up to 30 or more in fruit, hairy, remaining erect after anthesis; bracts leafy, the lowermost ovate, acuminate at apex, the upper much smaller. *Calyx* divided almost to base into 5 lobes, the lobes 10–12 × 0.8–1.0 mm at anthesis, accrescent to 15–20 × 3 mm in fruit, linear and acute at apex. *Corolla* bright blue, tube straight, usually a little longer than the calyx and flower 10–20 mm in diameter, hypocrateriform, divided into 5 equal lobes, the lobes 5–8 × 5–8 mm, broadly oblong to subrotund and glabrous, scales of the throat 5, about 2 × 1.3 mm, oblong, obtuse and densely papillose-hairy. *Stamens* 5, inserted between and a little above the scales of the throat, included; filaments 1.0–1.5 mm, glabrous; anthers white, narrowly oblong. *Style* 1, 10–11 mm, glabrous; stigma simple, capitate. *Nutlets* 4, 5–7 × 2–3 mm, greyish-brown, oblong, erect, ribbed and minutely verruculose, with a thickened, collar-like base. *Flowers* 6–9. $2n=32$.

Introduced. The most common *Anchusa* grown in gardens and an infrequent escape on to waste ground and tips where it is rarely naturalised; also a bird-seed alien. Scattered records over Great Britain; naturalised in Cornwall from about 1922 to 1972. Native of southern Europe and the Mediterranean region eastwards to Pakistan and central Asia; Atlantic Islands.

4. A. arvensis (L.) M. Bieb. Bugloss
Lycopsis arvensis L.

Annual or *biennial* monoecious *herbs. Stems* 15–20 cm, pale green, erect, with dense, spreading, stiff, bulbous-based simple eglandular hairs, shortly branched above, leafy. *Leaves* alternate; lamina 3–15 × 0.5–2.0 cm, pale green on upper surface, even paler beneath, obovate-lanceolate to linear-oblong, more or less obtuse at apex, strongly undulate, distinctly and irregularly toothed, the lower narrowed into a petiole, the upper sessile and semiamplexicaul; all with numerous, short to medium, pale, bulbous-based hairs on both surfaces and the margins. *Inflorescence* of simple or forked cymes, at first subcapitate, but elongating somewhat after flowering; flowers bisexual, subsessile; bracts leaf-like. *Calyx* divided nearly to base into 5 lobes, the lobes linear-lanceolate, acute at apex, enlarging in fruit. *Corolla* 4–6 mm in diameter, bright blue, the tube abruptly curved about the middle, 5-lobed, the lobes broadly rounded; throat scales 5, white. *Stamens* 5, included. *Style* 1; stigma simple. *Nutlets* 4, 3–4 mm, ovoid, reticulate, deeply concave, with a collar-like base. *Flowers* 6–9. $2n=48$.

Introduced. Weed of arable land and waste places on light acid or calcareous soils. Locally common throughout lowland Great Britain, Channel Islands, eastern and northern Ireland. Throughout the greater part of Europe and in Asia. Our plant is subsp. **arvensis**.

10. Cynoglottis (Guşul.) Vural & Kit Tan

Perennial monoecious *herbs. Leaves* alternate, entire. *Inflorescence* of many cymes forming a panicle; flowers erect, bisexual. *Calyx* divided almost to base into 5 lobes. *Corolla* blue or bluish-violet, 5-lobed, the tube much shorter than the lobes; throat scales 5, with short, cylindrical papillae. *Stamens* 5, included. *Style* 1, very short; stigma simple, included. *Fruit* of 4, ovoid, tuberculate nutlets with strong ridges and a distinct collar-like base.

Contains 2 species from east Europe and west Asia.

Vural, M. & Kit Tan (1983). New taxa and records from Turkey. *Notes Roy. Bot. Garden. Edinb.* **41:** 65–76.

1. C. barrelieri (All.) Vural & Kit Tan False Alkanet
Anchusa barrelieri (All.) Vitman; *Buglossum barrelieri* All.

Perennial monoecious *herb. Stems* (20–)50–80 cm, pale green, erect, slender to stout, patent or appressed, often with tubercular bases, simple, leafy. *Leaves* alternate; lamina 3–7(–12) × (0.5–)1.0–1.5(–3.0) cm, medium green on upper surface, paler beneath, oblong-spathulate to oblong-lanceolate or almost linear, acute at apex, subentire, narrowed below, with stiff hairs on both surfaces. *Inflorescence* of many cymes in a panicle; flowers bisexual; pedicels 2–5(–7) mm; bracts mostly shorter than calyx. *Calyx* 2–3 mm, up to 6 mm in fruit, divided almost to base into 5 lobes, the lobes lingulate or spathulate, very obtuse at apex. *Corolla* blue or bluish-violet; tube 1.0–1.5 mm, much shorter than limb; limb 7–10 mm in diameter, 5-lobed, the lobes rounded and equal, throat scales 5; with short, cylindrical papillae. *Stamens* 5, included. *Style* 1, very short; stigma simple. *Nutlets* 4, 2–4 mm, ovoid, erect, tuberculate, with strong ridges and a distinct collar-like base. *Flowers* 6–9. $2n=16, 18, 24$.

Introduced. Grown in gardens and an infrequent escape. Few scattered records in Great Britain. Native of eastern Europe.

11. Pentaglottis Tausch

Perennial monoecious *herbs* with deep, thick roots, stiffly hairy. *Leaves* alternate, shallowly undulate. *Inflorescence* of bracteate, scorpioid, axillary cymes, the flowers bisexual, subsessile and erect. *Calyx* divided almost to base into 5 lobes. *Corolla* bright blue, 5-lobed, the lobes acute, the throat closed by 5, white scales. *Stamens* 5, included. *Style* 1; stigma simple, included. *Fruits* of 4, ridged, tuberculate; nutlets concave at base and attached by a short stalk.

Contains 1 species in Europe.

1. P. sempervirens (L.) Tausch ex L. H. Bailey Green Alkanet

Anchusa sempervirens L.; *Caryolopha sempervirens* (L.) Fisch. & Trautv.

Perennial monoecious *herb* with deep, thick roots. *Stem* 30–100 cm, pale green, sometimes tinted brownish-purple, erect, with stiff, pale, reflexed simple eglandular hairs, branched, leafy. *Leaves* alternate, the lower sometimes very large, becoming slowly smaller upwards; lamina 4–40 × 4–20 cm, medium yellowish-green on upper surface with impressed veins, paler beneath with prominent veins, ovate or elliptical, more or less acute at apex, with a shallowly undulate margin, rounded or cordate at base, with short, appressed simple eglandular hairs on both surfaces; petioles up to 15 cm, pale green, channelled above, with reflexed hairs. *Inflorescence* of bracteate, scorpioid, axillary cymes; peduncles long, hairy; flowers 8–9 mm in diameter, bisexual, subsessile and erect. *Calyx* 8–10 mm, pale green, divided almost to the base into 5 lobes, the lobes lanceolate, more or less acute at apex, with stiff, ascending hairs. *Corolla* bright blue with white scales in throat, with a short tube and 5 spreading lobes, the lobes obovate, rounded at apex. *Stamens* 5, included; filaments white; anthers dark brown. *Style* 1, included, colourless; stigma simple, colourless. *Nutlets* 4, 2.0–2.5 mm, ridged, tuberculate, concave at base, attached by a short stalk. *Flowers* 5–7. $2n = 22$.

Introduced before 1700. Naturalised in hedges, wood borders and rough ground. Frequent over much of Great Britain and Ireland. Native of south-west Europe from central Portugal to south-west France.

12. Borago L.

Annual or *perennial* monoecious *herbs. Leaves* alternate, the basal long-petiolate. *Inflorescence* of dense terminal and lateral cymes; flowers bisexual. *Calyx* divided nearly to base into 5 lobes. *Corolla* blue, with tube shorter than limb; throat scales 5, notched. *Stamens* 5, inserted in throat of corolla, exserted; filaments flattened, glabrous, anthers longer than filaments. *Style* 1; stigma simple, included. *Fruits* of 4, ridged nutlets with a collar-like base.

Contains 3 species in the Mediterranean region, one extending into central Europe.

1. Calyx 8–20 mm; nutlets 7–10 mm **1. officinalis**
1. Calyx 4–8 mm; nutlets 3–4 mm **2. pygmaea**

1. B. officinalis L. Borage

Annual monoecious *herb* with a short, thick tap-root. *Stems* 10–60 cm, pale green, rather fleshy, longitudinally striate, branched, leafy, with a thin clothing of white, prickly bristles. *Leaves* alternate; lamina 4–15 × 2.5–10.0 cm, dull medium green on upper surface, paler beneath with rather prominent venation, ovate, obtuse to subacute at apex, entire, rounded or subcordate at base, thinly clothed with white, prickly bristles on both surfaces; petioles up to 9 cm, channelled. *Inflorescence* a much-branched, terminal panicle; pedicels 15–30 mm at anthesis, lengthening to 6 cm or more in fruit and becoming pendulous, bristly; bracts leafy, narrowly ovate or lanceolate, acute at apex; flowers bisexual, nodding. *Calyx* 8–20 × 2.0–2.5 mm, divided almost to base into 5 lobes, the lobes erect, lanceolate, acute at apex, bristly. *Corolla* 20–30 mm in diameter, blue, pink or white, stellate-rotate with patent lobes; tube about 2 mm, shallowly concave; lobes 5, 10–13 × 8–10 mm, broadly obovate, acute at apex; scales in throat about 2.5 × 2 mm, deltoid or truncate-deltoid, emarginate at apex, glabrous. *Stamens* 5, inserted in the throat of the corolla, exserted; filaments about 2 mm, pale, flattened, with a deltoid-subulate apical appendage, glabrous; anthers blackish, connivent into a cone, with a short apical cusp, longer than filaments. *Style* 1, 6–7 mm; stigma simple and capitate. *Nutlets* 4, enveloped by the persistent calyx, 7–16 × about 3 mm, oblong-ovoid, attached to a short, peg-like extension of the disc, with a concave base surrounded by a well-marked, ribbed annulus. *Flowers* 6–8. $2n = 16$.

Introduced. Still grown as a herb in gardens and as an occasional crop in fields, and is persistent on tips, in rough ground and waysides. Scattered records over much of Great Britain and mostly coastal in Ireland. Southern Europe and the Mediterranean region; frequently cultivated and naturalised in many temperate regions of the world. Still considered to be valuable in the treatment of fevers and pulmonary complaints.

2. B. pygmaea (DC.) Chater & Greuter Slender Borage

B. laxiflora (DC.) Fritsch, non Poir.; *Campanula pygmaea* DC.

Perennial monoecious *herb. Stems* 15–60 cm, pale green, decumbent, slender, with stiff hairs, branched, leafy. *Leaves* alternate; lamina of lower 5–20 cm, medium green on upper surface, paler beneath, oblong to obovate, entire and petiolate; cauline smaller, sessile and amplexicaul; all with stiff hairs on both surfaces. *Inflorescence* of lax, branched, bracteate cymes; flowers bisexual; pedicels 10–40 mm, filiform, deflexed after anthesis. *Calyx* 4–6 mm at anthesis, up to 8 mm in fruit; 5-lobed, the lobes lanceolate, acute at apex, not connivent in fruit. *Corolla* clear blue, campanulate, the tube short, the lobes 5, 5–8 mm, ovate, acute at apex. *Stamens* 5, exserted, inserted near the base of the corolla; filaments with a long, narrow appendages at apex; anthers connivent and mucronate. *Style* included; stigma simple, capitate. *Nutlets* 4, 3–4 mm, obovoid, concave and with a thickened collar-like ring at the base. *Flowers* 6–8. $2n = 32$.

Introduced. Naturalised on heathy ground in Jethon in the Channel Isles since 1932; less permanent on rough ground and by paths in very scattered localities in Wales, England and Kirkcudbrightshire. Native of Corsica and Sardinia.

13. Trachystemon D. Don

Perennial monoecious *herbs* with rhizomes. *Leaves* alternate, ovate, cordate, entire, the basal long-petiolate. *Inflorescence* of dense cymes in a terminal panicle; flowers bisexual. *Calyx* 5-lobed halfway to base. *Corolla* blue, 5-lobed, the lobes equal and longer than rest of corolla, throat scales in 2 series of 5, white, notched. *Stamens* 5, exserted; filaments hairy; anthers shorter than filaments. *Style* 1, included between closely appressed stamens; stigma simple. *Fruit* of 4 nutlets, with a collar-like base.

Contains 2 species in the eastern Mediterranean region.

1. T. orientalis (L.) G. Don — Abraham–Isaac–Jacob
Borago orientalis L.

Perennial monoecious *herb* with rhizomes. *Stems* 20–60 cm, pale green, sparsely hairy, erect, branched, leafy. *Leaves* alternate, dark green on upper surface, paler beneath; lamina of basal 15–50 cm, ovate, subacute to acuminate at apex, entire, cordate at base, long-petiolate; cauline ovate to lanceolate, acute at apex, entire, sessile; all rather sparsely hairy on both surfaces. *Inflorescence* of dense, bracteate cymes in a lax panicle, the cymes with 5–15 bisexual flowers; pedicels long. *Calyx* 3–6 mm at anthesis, 6–9 mm in fruit, divided halfway to base into 5 lobes, the lobes ovate and obtuse at apex. *Corolla* bluish-violet; tube 5–7 mm, infundibuliform and exceeding the calyx; 5-lobed, the lobes revolute; throat scales in 2 series of 5, the lower at about the middle, the upper at the apex of the tube and very short, notched and villous. *Stamens* 5, long exserted, inserted between the 2 series of scales; filaments hairy; anthers shorter than filaments. *Style* 1; stigma simple, included. *Nutlets* 4, about 2×4 mm, ovoid, carinate, rugose-reticulate, more or less flat and with a slightly ribbed, collar-like ring at the base. *Flowers* 4–5. $2n = 30$.

Introduced. Naturalised on shady banks and dampish woods. Scattered records in Great Britain north to central Scotland and two records in east Ireland. Native of the Caucasus and north Turkey.

14. Mertensia Roth

Pneumaria Hill; *Steenhammera* Reichb.

Perennial monoecious *herbs*, glabrous and often glaucous. *Leaves* alternate, elliptic, ovate or obovate, lower narrowed to the petiole, the upper with short, winged petioles or sessile, glabrous, but sometimes papillose on upper surface. *Inflorescence* of bracteate, terminal cymes; flowers bisexual. *Calyx* divided nearly to base into 5 lobes. *Corolla* pink, then blue and pink, 5-lobed, without or with inconspicuous throat scales, or conspicuous scales, but usually with 5 folds. *Stamens* 5, equal, included or slightly exserted. *Style* 1, filiform; stigma simple, included. *Fruit* of 4, ovoid, flattened, smooth nutlets, succulent then papery on outside. Named after Franz Karl Mertens (1764–1831).

Contains about 50 species in north temperate regions.

Hultén, E. & Fries, M. (1986). *Atlas of north European vascular plants north of the Tropic of Cancer.* 3 vols. Königstein.

Scott, G. A. M. (1963). Biological flora of the British Isles. No. 89. *Mertensia maritima* (L.) S. F. Gray. *Jour. Ecol.* **51**: 743–754.

Stewart, A., Pearman, D. A. & Preston, C. D. (1994). *Scarce plants in Britain.* Peterborough.

Welch, D. & Innes, M. (1999). Southward recolonisation by *Mertensia maritima* (L.) Gray on the coast of north-eastern Scotland. *Watsonia* **22**: 424–426.

Williams, L. D. (1917). A monograph of the genus *Mertensia* in North America. *Ann. Missouri Bot. Gard.* **24**: 17–159.

1. Leaves minutely hairy (papillose) on the margins, often papillose on upper surface — **3. ciliata**
1. Leaves glabrous or sparsely papillose on the upper surface — 2.
2. Corolla tube about 5 mm; nutlets smooth — **1. maritima**
2. Corolla tube 10–20 mm; nutlets rugose — **2. virginica**

1. M. maritima (L.) Gray — Oysterplant
Pulmonaria maritima L.; *Pneumaria maritima* (L.) Hill; *Lithospermum maritimum* (L.) Lehm.; *Steenhammera maritima* (L.) Reichb.

Perennial monoecious *herb* with a large tap-root. *Stems* up to 60 cm, purple-pruinose, decumbent, rather fleshy, glabrous, branched, leafy. *Leaves* alternate; lamina 0.5–6.0×1–2 cm, glaucous, rather fleshy, elliptical, ovate or obovate, obtuse or apiculate at apex, entire, the lower attenuate into the petiole, the upper with short, winged petioles or sessile; all glabrous, but often papillose on the upper surface. *Inflorescence* of bracteate, terminal cymes; flowers bisexual; pedicels 2–10 mm, elongating and becoming somewhat curved in fruit; bracts leaf-like but smaller. *Calyx* 5–6 mm, divided nearly to base in 5 lobes, the lobes ovate, shortly acuminate, papillose. *Corolla* 4–6 mm, pink, then blue and pink, 5-lobed, the lobes broadly rounded, without throat scales but usually with 5 hairy folds. *Stamens* 5, inserted towards the top of the corolla tube, slightly exserted; anthers yellow. *Style* 1, filiform; stigma simple. *Nutlets* 4, 5–6 mm, ovoid, flattened, acute at apex, fleshy, the outer coat becoming inflated and papery. *Flowers* 6–8. Usually self-pollinated. $2n = 24$.

Native. On bare shingle or shingly sand by the sea. Local on the coasts of northern Ireland and northern Great Britain south to Denbighshire and formerly south to Norfolk, Cardiganshire and Co. Kerry, but sporadic. Atlantic coast of Europe from Jutland northwards, Iceland, Greenland, North America. Boreo-arctic Montane element. Called Oysterplant because if eaten it tastes like oysters. Our plant is subsp. **maritima**. This occurs throughout the range of the species except the east coast of Asia where if is replaced by subsp. **asiatica** Takede.

2. M. virginica (L.) Pers. — Virginia Bluebells
Pulmonaria virginica L.; *Pneumaria virginica* (L.) Hill; *M. pulmonarioides* Roth; *Steenhammera virginica* (L.) Kostel.

Perennial monoecious *herb*. *Stems* 20–70 cm, green, fleshy, erect, glabrous, branched, leafy. *Leaves* alternate, green, fleshy; lamina of basal 7–20×3–12 cm, elliptical or ovate, rounded at apex, entire with long petioles; the middle and upper smaller, ovate to oblong, obtuse at apex, entire, short-petioled or sessile; all smooth and glabrous. *Inflorescence* a 1-sided scorpioid cyme; flowers bisexual; pedicels up to 10 mm. *Calyx* 2–10 mm, purplish, 5-lobed, enlarged in fruit, the lobes ovate-oblong, rounded at apex, glabrous. *Corolla* 18–25 mm, pink when young, becoming purplish-blue, rarely pink or white; tube 18–20 mm, slender, densely hairy within at base, throat scales inconspicuous, limb campanulate and 5-lobed. *Stamens* 5, inserted below the throat; filaments 4–8 mm; anthers 12–17 mm. *Style* 1, 5–10 mm; stigma small, simple. *Nutlets* 4, 2.5–3.0 mm, ovoid, rugose. *Flowers* 5–7. $2n=24$.

Introduced. A rare garden escape. Native of North America.

3. M. ciliata (James) G. Don Hairy Bluebells
Pulmonaria ciliata James

Perennial monoecious *herb. Stems* 10–120 cm, pale green, erect or ascending, striate, glabrous, branched above, leafy. *Leaves* alternate, medium green on upper surface, paler beneath; lamina of basal 4–15 × 3–10 cm, oblong to ovate or lanceolate, more or less acute at apex, entire, rounded at base; cauline lanceolate to ovate, the upper sessile; all minutely hairy (papillose) on the margins and often papillose on the upper surface. *Inflorescence* a 1-sided, scorpioid cyme; flowers bisexual; pedicels 1–10 mm, papillose or rarely with a few bristle-like hairs. *Calyx* up to 3 mm, 5-lobed, the lobes oblong, rounded at apex and minutely hairy on the margins. *Corolla* rose in bud, becoming bright blue; tube 6–8 mm, glabrous or with crisped hairs within, 5-lobed, the lobes 4–10 mm, erect, scales conspicuous, papillose or hairy. *Stamens* 5; filaments white; anthers 1.0–2.5mm, yellow. *Style* 1; stigma simple. *Nutlets* 4, strongly 3-angled, flattened, fleshy. *Flowers* 5–7. $2n=24$.

Introduced. Garden escape on a roadside at Pett, near Rye in Sussex. Native of eastern North America.

15. Amsinckia Lehm.

Annual, bristly, monoecious *herbs* with fibrous roots. *Leaves* alternate, linear to oblong or lanceolate, sessile. *Inflorescence* of terminal, bracteate or bractless cymes, open flowers erect, bisexual. *Calyx* divided nearly to base into 5 lobes. *Corolla* yellow to orange, with 5 equal lobes, tube longer than limb, without throat scales. *Stamens* 5, equal in length, included. *Style* 1, included; stigma simple, capitate. *Fruit* of 4, keeled and warty nutlets, without a collar-like base.

Contains about 50 species in temperate America.

Beckett, G., Bull, A. & Stevenson, R. (1999). *A flora of Norfolk*. Norwich.
Clements, E. J. (1999). Misconceptions about *Amsinckia lycopsoides* Lehm. *B.S.B.I. News* **80**: 44–45.
Ray, P. M. & Chisaki, H. F. (1957). Studies on *Amsinckia*. I. A synopsis of the genus with a study of heterostyly in it. II. Relationships among the primitive species. III. Aneuploid diversification in the *muricatae*. *Amer. Jour. Bot*. **44**: 529–554.

1. A. lycopsoides (Lehm.) Lehm. Scarce Fiddleneck
Lithospermum lycopsoides Lehm.

Annual monoecious *herb* with fibrous roots. *Stems* 20–50 cm, pale green, erect or ascending, with stout, patent, stiff simple eglandular hairs, often branched, leafy. *Leaves* alternate; lamina 3–5 × 0.5–1.5 cm, medium green on upper surface, paler beneath, linear to oblanceolate, acute at apex, entire, narrowed at base, with erecto-patent, stiff eglandular hairs on both surfaces, the lower shortly petiolate, the upper sessile. *Inflorescence* an ebracteate, terminal cyme or with one bract at base; flowers bisexual, erect, subsessile. *Calyx* 3–5 mm in flower, becoming 6–11(–15) mm in fruit, divided almost to the base into 5 lobes, the lobes ovate. *Corolla* 5–8 mm, deep yellow, 5-lobed, with scales in the throat, which is hairy. *Stamens* 5, inserted in the upper part of the tube, included. *Style* included; stigma simple, capitate. *Nutlets* 4, 2–3 mm, ovoid-trigonous, slightly flattened, strongly muricate, warty, keeled, without a collar-like base. *Flowers* 5–7. $2n=30$.

Introduced. Naturalised on rough ground in the Farne Islands since 1922, and becoming increasingly frequent over much of England, especially in the east on sandy soils. Native of western North America.

2. A. micrantha Suksd. Hairy Fiddleneck
A. menziesii auct.; *A. intermedia* auct.; *A. calycina* auct.

Annual monoecious *herb* with fibrous roots. *Stems* up to 1 m, pale green, erect, often branched, with short to long, curly simple eglandular hairs. *Leaves* alternate; basal and lower cauline with lamina up to 20 × 30 cm, medium green on upper surface, paler beneath, linear-lanceolate to linear-oblong, more or less obtuse and mucronate at apex, entire, attenuate at base; upper leaves similar but smaller; all with stiff, erecto-patent, bulbous-based hairs on both surfaces. *Inflorescence* terminal with basal bracts; pedicels very short; flowers bisexual. *Calyx* (1–)2–4(–4.5) mm, to 5–6(–9) mm at fruiting, 5-lobed nearly to base, the lobes linear-lanceolate, obtuse at apex, stiffly hairy. *Corolla* 4–8 mm, yellow, tube glabrous within, without scales; limb 2–3 mm in diameter. *Stamens* 5, included, borne on upper half of tube. *Style* 1; stigma simple, capitate. *Nutlets* 4, about 2.5 mm, with longitudinal, muricate ridges and transversely rugose furrow. *Flowers* 7–8. $2n=32$.

Introduced. Casual in rough ground and sometimes an abundant weed in fields where it is often persistent. Frequent in eastern England and east Scotland north to Ross and Cromarty, especially common in East Anglia; casual in central and western Great Britain. Native of western North America.

16. Plagiobothrys Fisch. & C. A. Mey.

Annual monoecious *herbs* with fibrous roots and bristly hairy. *Leaves* alternate above, opposite below, linear-oblong, sessile. *Inflorescence* of terminal, semi-bractless cymes; flowers bisexual. *Calyx* divided more than halfway to base into 5 lobes. *Corolla* white, with 5 equal lobes, the tube shorter than limb. *Stamens* 5, equal in length, included. *Style* 1; stigma simple, included. *Fruit* of 4, keeled, ridged, minutely tuberculate nutlets, without a collar-like base.

Contains about 100 species in North and South America.

1. P. scouleri (Hook. & Arn.) I. M. Johnst.
White Forget-me-not

Myosotis scouleri Hook. & Arn.; *Allocarya scouleri* (Hook. & Arn.) Green; *Erythrichium scouleri* (Hook. & Arn.) A. DC.; *Krynitzkia scouleri* (Hook. & Arn.) A. Gray

Annual monoecious *herb* with a tap-root and fibrous side-roots. *Stems* several, up to 20 cm, prostrate to ascending to suberect, more or less appressed-hairy, branched, leafy. *Leaves* mostly cauline, the lower opposite, the upper alternate; lamina up to 6.5 × 0.5 cm, linear, acute at apex, sessile, appressed-hairy. *Inflorescence* a terminal, semi-bractless, spiralled cyme; flowers bisexual. *Calyx* 2–4 mm, divided more than halfway to base into 5 lobes, the lobes lanceolate. *Corolla* 2–4 mm in diameter, white, divided to about halfway to base in 5 rounded lobes. *Stamens* 5, included; filaments short. *Style* 1; stigma simple, included. *Nutlets* 4, 1.2–2.0 mm, ovoid, more or less rugose, with or without minute bristles, keeled, ridged and without a collar-like base. *Flowers* 7–9.

Introduced. Casual, sometimes persistent with grass-seed. Aberdeenshire, Caithness, Shetland Islands, Wiltshire, Hampshire, Berkshire, Buckinghamshire and Staffordshire. Native of western North America.

17. Asperugo L.

Annual monoecious *herbs* with fibrous roots. *Leaves* alternate below, opposite above, lanceolate, oblanceolate or obovate, narrowed to base. *Inflorescence* a lax, dichotomously branched cyme, the flowers bisexual and usually solitary in bracts. *Calyx* 5-lobed, with appendages in the sinuses, strongly accrescent in fruit to form an envelope round the nutlets. *Corolla* violet-blue, 5-lobed, with 5 small scales in the throat. *Stamens* 5, equal in length, included. *Style* 1, included; stigma simple, capitate. *Fruit* of 4 nutlets; the nutlets ovate-pyriform, strongly compressed laterally, minutely and bluntly tuberculate and without a collar-like base.

Contains 1 species in Europe and Asia.

Hultén, E. & Fries, M. (1986). *Atlas of north European vascular plants north of the Tropic of Cancer.* 3 vols. Königstein.

1. A. procumbens L. Madwort

Annual monoecious *herb* with fibrous roots. *Stems* 10–50(–70) cm, pale green, decumbent or sprawling, angled, thinly and retrorsely stiff-hairy, often branched at the base and sparingly so above, leafy. *Leaves* alternate below, opposite above; lamina 4–15 × 1–4 cm, medium green on upper surface, paler beneath, lanceolate, oblanceolate or obovate, acute to obtuse at apex, entire, narrowed at base, thinly and stiffly hairy on both surfaces; the lower petiolate, the upper sessile or subpetiolate. *Inflorescence* a lax, dichotomously branched cyme, the bisexual flowers usually solitary in bracts, occasionally 2 or more; sessile or shortly pedicellate; bracts leafy, 10–40 × 5–20 mm. *Calyx* at first cupular with 5 lobes, the lobes about 1.5 × 0.5 mm, oblong and stiffly hairy, strongly accrescent and deflexed in fruit and strongly compressed laterally to form a envelope round the nutlets, reticulate-veined, with lobes 4–5 mm, erect and acute, long, alternating with bifid, recurved appendages in the sinuses. *Corolla* violet-blue; tube about 1 mm, pale; lobes 5, erect or spreading, subrotund, about 1 mm in diameter; scales in throat about 0.2 × 0.4 mm, broadly oblong, truncate, papillose or shortly hairy. *Stamens* 5, attached about 0.5 mm above the base of the corolla tube; filaments equal but very short; anthers yellow, glabrous. *Style* 1, about 0.3 mm; stigma simple, capitate. *Nutlets* 4, about 3 × 2 mm, pale brown, ovate-pyriform, strongly compressed laterally, minutely and bluntly tuberculate. *Flowers* 6–8.

Introduced. Arable fields, waste land and rough ground and was first recorded in the wild in 1660. Scattered records over Great Britain especially Scotland, formerly more common. Widely distributed in temperate Europe, Asia and the Mediterranean region.

18. Myosotis L.

Annual or *perennial* monoecious *herbs. Leaves* alternate, entire. *Inflorescence* a terminal, scorpioid cyme, sometimes bracteate at the base; flowers bisexual. *Calyx* 5-lobed, sometimes divided nearly to base. *Corolla* rotate, usually pink in bud and ultimately blue, rarely yellow, 5-lobed, the lobes emarginate, retuse or entire, flat or concave and contorted in bud, the throat closed by 5 short, notched scales. *Stamens* 5, included. *Style* 1, short; stigma simple, capitate. *Fruit* of 4, small, shining nutlets, often with a distinct rim, the attachment area usually small and rarely peg-like.

The flowers contain nectar and may be cross-pollinated by insects, though self-pollination is possible, probably more frequent and known to result in good seed production in several species. *M. arvensis*, *ramosissima* and *sylvatica* run into one another and are difficult to distinguish on precise characters but can usually be recognised in the field when all taxa involved have become familiar.

About 50 species in the temperate regions of both hemispheres.

Elkington, T. T. (1964). Biological flora of the British Isles. No. 96. *Myosotis alpestris* F. W. Schmidt. *Jour. Ecol.* **52**: 709–722.
Grau, J. (1965). Cytotaxonomische Bearbeitung der Gattung *Myosotis* L. 1. Atlantische Sippen um *Myosotis secunda* A. Murr. *Mitt. Bot. München* **5**: 675–688.
Grime J. P. et al. (1988). *Comparative plant ecology.* London. [*M. scorpioides, arvensis*, and *ramosissima.*]
Halliday, G. (1997). *A flora of Cumbria.* Lancaster. [*M. alpestris.*]
Hooker, J. D. (1898). *Myosotis dissitiflora* var. *dyerae. Bot. Mag.* tab. 7589.
Hultén, E. & Fries, M. (1986). *Atlas of north European vascular plants north of the Tropic of Cancer.* 3 vols. Königstein.
Le Sueur, F. (1984). *Flora of Jersey.* Jersey.
Preston, C. D. & Croft, J. M. (1977). *Aquatic plants in Britain and Ireland.* Colchester. [*M. scorpioides.*]
Rich, T. C. G. & Jermy, A. C. (1998). *Plant crib.* London.
Salmon, C. E. (1926). A new *Myosotis* from Britain. *Jour. Bot. (London)* **64**: 289–295.
Schuster, R. (1967). Taxonomische Untersuchungen über die Series *Palustres* M. Pop. der Gattund *Myosotis* L. *Feddes Repert.* **74**: 39–98.

Stewart, A., Pearman, D. A. & Preston, C. D. (1994). *Scarce plants in Britain*. Peterborough. [*M. stolonifera*.]
Vestergren, T. (1930). Über den Verwandtschaffskreis der *Myosotis versicolor* (Pers.) J. E. Sm. *Svensk Bot. Tidskr.* **24**: 449–467.
Welch, D. (1967). Notes on *Myosotis scorpioides* agg. *Watsonia* **6**: 276–279.
Wigginton, M. J. (Edit.) (1999). *British red data books*. Vol. 1. *Vascular plants*. Ed. 3. Peterborough.
Wilmott, A. J. (1923). *Myosotis sicula* Gussone in Jersey. *Jour. Bot. (London)* **61**: 212–215.

1. Calyx with hairs all more or less straight and closely appressed 2.
1. Calyx with some hairs patent and distally hooked or at least strongly curved 8.
2. Annual or biennial without sterile shoots; calyx divided less than halfway to base at flowering; nutlets less than 1 mm, shining olive brown **5. sicula**
2. Calyx often divided more than halfway to base at flowering; nutlets more than 1.2 mm, mid-brown to black, shining or not, or if less than 1.2 mm then plants with axillary stolons 3.
3. Calyx with broad lobes forming an equilateral triangle; style longer than calyx tube and often exceeding calyx lobes at flowering 4.
3. Calyx with narrow lobes forming an isosceles triangle with the base shorter than the sides; style shorter than calyx tube at flowering 5.
4. Lower part of stem glabrous or with spreading hairs, upper part of stem with appressed hairs; flowers 5–8 mm in diameter **1(i). scorpioides** var. **scorpioides**
4. Stem with appressed hairs throughout; flowers 4–6 mm in diameter **1(ii). scorpioides** var. **strigulosa**
5. Lower part of stem with more or less patent hairs 6.
5. Stem with only appressed or more or less ascending hairs 7.
6. Pedicels eventually 2.5–5.0 times as long as fruiting calyx; nutlets less than 2 mm **2. secunda**
6. Pedicels less than twice as long as fruiting calyx; nutlets usually more than 2 mm **6. alpestris**
7. Stolons produced from lower nodes; leaves rarely more than 3 times as long as wide **3. stolonifera**
7. Stolons absent; larger leaves more than 4 times as long as wide **4. laxa** subsp. **caespitosa**
8. Perennial, with sterile basal shoots at fruiting 9.
8. Annual, without sterile shoots at fruiting 14.
9. Fruiting calyx narrowed and more or less acute at base; nutlets obtuse to rounded at apex; only on hills over 685 m **6. alpestris**
9. Fruiting calyx rounded to broadly obtuse at base; nutlets more or less acute at apex; lowland or upland 10.
10. Calyx lobes erecto-patent, exposing the ripe nutlets; corolla 5–11 mm in diameter, with a more or less flat rim 11.
10. Calyx lobes erect, more or less appressed and concealing ripe nutlets; corolla less than 5 mm in diameter 12.
11. Flowers up to 8 mm in diameter, often much less **7(i). sylvatica** var. **sylvatica**
11. Flowers 8–11 mm in diameter, often pink or brighter blue **7(ii). sylvatica** var. **culta**
12. Corolla 3–5 mm; nutlets up to 2.5 mm **9(b). arvensis** subsp. **umbrata**
12. Corolla up to 3 mm; nutlets not more than 2.0 mm 13.
13. Plant often not exceeding 15 cm, central stem longer than branches; racemes usually not more than 10 cm **9(a,i). arvensis** subsp. **arvensis** var. **arvensis**
13. Plant often up to 30 cm, with numerous, long branches from base as long as central stem; racemes up to 15 cm **9(a,i). arvensis** subsp. **arvensis** var. **dumetorum**
14. Corolla 10–14 mm in diameter **8. dynerae**
14. Corolla less than 11 mm in diameter 15.
15. Pedicels 1.2–2.0 times as long as fruiting calyx 16.
15. Pedicels shorter than to about as long as fruiting calyx 18.
16. Corolla 3–5 mm; nutlets up to 2.5 mm **9(b). arvensis** subsp. **umbrata**
16. Corolla up to 3 mm; nutlets not more than 2.0 mm 17.
17. Plant often not exceeding 15 cm, central stem longer than branches; racemes usually not more than 10 cm **9(a,i). arvensis** subsp. **arvensis** var. **arvensis**
17. Plant often up to 30 cm, with numerous, long branches from base as long as central stem; racemes up to 15 cm **9(a,i). arvensis** subsp. **arvensis** var. **dumetorum**
18. Calyx lobes narrowly to broadly triangular; corolla blue from start, with tube shorter than calyx 19.
18. Calyx lobes oblong-lanceolate; corolla cream to yellow at first, with tube eventually longer than calyx 20.
19. Calyx up to 4 mm in fruit, the lobes narrowly triangular; corolla scarcely exceeding the calyx **10(a). ramosissima** subsp. **ramosissima**
19. Calyx up to 2.5 mm in fruit, the lobes broadly triangular; corolla distinctly exceeding the calyx **10(b). ramosissima** subsp. **globularis**
20. Corolla up to 4 mm, cream or yellow at first **11(a). discolor** subsp. **discolor**
20. Corolla not more than 2 mm, cream at first **11(b). discolor** subsp. **dubia**

1. M. scorpioides L. Water Forget-me-not
M. palustris (L.) Hill

Perennial monoecious *herb* with short rhizomes and/or stolons. *Stems* 15–70 cm, pale green, ascending to erect, angular, more or less hairy. *Leaves* alternate; lamina 3–10×0.5–2.0 m, soft, pale green on upper surface, even paler beneath, oblong to oblong-lanceolate, usually obtuse at apex, the upper sometimes apiculate, entire, attenuate at base, but scarcely petiolate, with short, appressed hairs on both surfaces to nearly glabrous. *Inflorescence* a terminal, scorpioid, ebracteate cyme; flowers bisexual; pedicels up to 10 mm, spreading or reflexed, with minute hairs. *Calyx* 5.5–6.5 mm, campanulate, divided up to one-third of the way to the base into 5 lobes, the lobes ovate-lanceolate, subacute at apex, forming an equilateral triangle, with short, straight, appressed hairs. *Corolla* (3–)4–8 mm in diameter, bright sky blue, rarely white, with an orange-yellow rim at the throat, with 5 short, notched throat scales, 5-lobed, the lobes very broadly elliptical and emarginate. *Stamens* 5, included; anthers yellow. *Style* 1, equalling calyx tube to longer than calyx; stigma simple. *Nutlets* 4, up to 1.8 × 1.2 mm, narrowly ovoid, obtuse, slightly bordered, not keeled, black and shining. *Flowers* 5–9. Submerged patches may not flower, but emergent and terrestrial plants flower freely. Self-incompatibility differs from plant to plant. Vegetative reproduction is probably more important than that from seed. $2n = 64$.

(i) Var. **scorpioides**
Lower part of stem glabrous or with spreading hairs; upper part of stem with appressed hairs. *Flowers* 5–8 mm in diameter.

(ii) Var. **strigulosa** (Rchb.) Schinz & Keller.
M. strigulosa Rchb.; *M. coronaria* var. *strigulosa* (Rchb.) Dumort.

Stem with appressed hairs throughout. *Flowers* 4–6 mm in diameter.

Native. Most frequent at the edge of still and flowing water, but occurs as an aquatic forming submerged patches or floating rafts at the edges of lakes, rivers, streams, canals and ditches. Common throughout most of lowland Great Britain and Ireland up to 600 m, but rare in the Channel Islands. Widely distributed in temperate Eurasia and introduced in North and South America and New Zealand. Eurosiberian Temperate element. Var. *scorpioides* is the common plant. Var. *strigulosa* is said to be more of an upland plant

2. M. secunda Al. Murray Creeping Forget-me-not
M. repens auct.

Annual to *biennial* monoecious *herb* perennating by means of stolons, and with a short, scarcely creeping rhizome. *Stems* 20–60 cm, pale green, arising from base and decumbent or prostrate, the non-flowering ones rooting at the nodes, with spreading hairs below and upwards-pointing hairs above, sparsely branched, leafy. *Leaves* alternate or the lower opposite; lamina 0.5–4.0 × 0.3–1.5 cm, dull medium green on upper surface, paler beneath, ovate-spathulate to oblong-lanceolate, obtuse at apex, entire, narrowed at base, sparsely hairy on both surfaces and ciliate. *Inflorescence* a terminal, scorpioid cyme, bracteate below; flowers bisexual; pedicels 2.5–5.0 times as long as calyx in fruit and reflexed. *Calyx* 3.0–4.5 mm, with straight, appressed hairs, campanulate in fruit, divided just over halfway to base into 5 lobes, the lobes lanceolate, acute at apex and forming a isosceles triangle with the base shorter than the sides. *Corolla* 4–8 mm in diameter, pale blue with a yellow rim at the throat, 5-lobed, the lobes subrotund, emarginate at apex; throat scales 5, short and notched. *Stamens* 5, included; anthers yellow. *Style* 1, shorter than calyx tube at flowering; stigma simple, capitate. *Nutlets* 4, about 1.8–1.2 mm, black, ovoid, acute, with rim, rhombic attachment area and spongy appendages. *Flowers* 5–8. 2n = 48.

Native. Wet, often acidic places by streams and in pools and bogs up to 800 m. Common in many parts of Great Britain and Ireland, especially upland areas in the north and west, rare or absent in most of central and east England. West Europe; Azores; Madeira. Oceanic Temperate element.

3. M. stolonifera (DC.) Gay ex Leresche & Levier
Pale Forget-me-not
M. caespitosa var. *stolonifera* DC.; *M. brevifolia* C. E. Salmon; *M. secunda* subsp. *stolonifera* (DC.) Lainz

Perennial monoecious *herb* producing roots and stolons from the nodes of the lower stem, the stolons bearing small leaves. *Stems* 12–20(–30) cm, pale bluish-green, erect, with hairs slightly spreading-ascending towards the base, appressed above, sparingly branched, leafy. *Leaves* alternate; lamina 0.5–2.0 × 0.2–1.0 cm, dark bluish-green on upper surface, paler beneath, elliptical or oblanceolate, rounded at apex, entire, the lower narrowed at base, shortly hairy on both surfaces. *Inflorescence* in the upper half of stem and usually ebracteate; flowers bisexual; fruiting pedicels spreading or recurved, 1–2 times as long as the calyx. *Calyx* 2.5–3.0 mm, narrowly campanulate, with straight, appressed hairs, divided over halfway to base into 5 lobes, the lobes oblong and rounded at apex, forming an isosceles triangle. *Corolla* 4–6 mm in diameter, pale blue with a yellow rim at the throat, 5-lobed, the lobes subrotund and retuse at apex, throat closed by 5 short, notched scales. *Stamens* 5, included; anthers yellow. *Style* 1, slightly shorter than the calyx tube; stigma simple. *Nutlets* 4, about 1.2 × 0.7 mm, black, ovoid, obtuse at apex. *Flowers* 6–8. $2n = 24$.

Native. Wet flushes, base-poor springs and seepage areas and streamsides on hills. Local in northern England and southern Scotland. Spain and Portugal. Oceanic Boreo-montane element. It was first collected in 1892, but not recognised as a new British species until described by C. E. Salmon in 1926.

4. M. laxa Lehm. Tufted Forget-me-not

Annual or *biennial* monoecious *herb* with fibrous roots. *Stems* 20–40 cm, pale green, terete, striate, with straight, appressed hairs, simple or branched from the base, leafy. *Leaves* alternate; lamina 1–8 × 0.3–1.5 cm, soft, pale green on upper surface, paler beneath, oblong, lanceolate or oblanceolate, obtuse at apex, entire, narrowed at base, the lower petiolate, with short, appressed hairs on both surfaces. *Inflorescence* a terminal cyme, usually bracteate at base; flowers bisexual; pedicels up to 6 mm, spreading, with ascending-subappressed hairs. *Calyx* 3–5 mm, campanulate in fruit, with straight, appressed hairs, 5-lobed, the lobes broadly triangular, acute at apex and forming an isosceles triangle with the base shorter than the sides. *Corolla* 4–5 mm in diameter, bright sky blue with a yellow rim at the throat, rarely white, 5-lobed, the lobes subrotund, retuse at apex, the throat closed by 5 notched scales. *Stamens* 5, included; anthers yellow. *Style* 1, short; stigma simple, capitate. *Nutlets* 4, up to 1.5 × 1.0 mm, dark brown, broadly ovoid, truncate at base, obtuse, with a spongy attachment area, shining. *Flowers* 5–8. $2n = 88$.

Native. By or in edges of ponds and rivers and in damp fields. Fairly common throughout most of Great Britain and Ireland. Our plant is subsp. **caespitosa** (Schultz) Hyl. ex Nordh. (*M. caespitosa* Schultz). It occurs in most of

Europe; Asia, north Siberia and east to the Himalaya; North Africa. Subsp. *laxa* is in North America. Circumpolar Boreo-temperate element.

× **scorpioides** = **M.** × **suzae** Domin

This hybrid is intermediate in characters between the parents, with the corolla 5–7 mm in diameter and the style as long as the calyx tube.

Native. Recorded in widely scattered localities in Great Britain growing with the parents. Also recorded for central Europe.

5. M. sicula Guss. Jersey Forget-me-not

Annual or *biennial* monoecious *herb* with fibrous roots and without sterile shoots. *Stems* solitary or several, 5–10(–30) cm, pale green, erect or decumbent, more or less forward-pointing, appressed hairs, simple or with divaricate, flexuous branches, leafy. *Leaves* alternate, medium green on upper surface, paler beneath, the lower 2–6 × 0.4–0.8 cm, oblong-spathulate, obtuse at apex, with scattered, appressed hairs on upper surface and more or less glabrous beneath, the upper 1–2 × 0.3–0.4 cm, linear-oblong, obtuse at apex, entire and with appressed-forward-pointing hairs on both surfaces. *Inflorescence* elongate, flexuous, rarely short, the lower branches usually divaricate; bracts sometimes present; flowers bisexual; lower pedicels 1–3 times as long as calyx, spreading or reflexed, thickened above. *Calyx* 4–6 mm, oblong-campanulate, usually with a few, straight, appressed hairs at base, divided less than halfway to base into 5 lobes, the lobes oblong and obtuse at apex. *Corolla* up to 3 mm in diameter, blue, 5-lobed, the lobes concave and entire, the throat closed by 5 short, notched scales. *Stamens* 5, included; anthers yellow. *Style* 1, equalling the calyx tube; stigma simple, capitate. *Nutlets* 4, about 1 mm, narrowly ovoid, obtuse at apex, slightly bordered, not keeled, shining olive brown. *Flowers* 4–6. $2n=36$, 46.

Native. Damp grassland and by a pond in two localities in Jersey where it was first discovered in 1922; now only known from the very small population at Noirmont. South and west Europe. Mediterranean-Atlantic element.

6. M. alpestris F. W. Schmidt Alpine Forget-me-not
M. sylvatica subsp. *alpestris* (F. W. Schmidt) Gams

Perennial monoecious *herb* with a rhizome and sterile basal shoots at flowering. *Stems* 5–20(–35) cm, pale green, stiffly erect, with spreading hairs below, hardly branched, leafy. *Leaves* alternate; lamina 1–3 × 0.2–1.0 cm, dull medium green on upper surface, paler beneath, oblong or oblong-lanceolate, more or less acute at apex, entire, the lower long-petiolate, the upper sessile, all with more or less spreading hairs on both surfaces. *Inflorescence* a rather short, terminal, scorpioid, ebracteate cyme; fruiting pedicels not longer than the calyx, ascending; flowers bisexual. *Calyx* 4–7 mm, campanulate, rather silvery, hairs more or less spreading, with or without a few, short, stiff, hooked bristles on the tube, divided halfway to three-quarters towards the base into 5 lobes, the lobes narrow lanceolate, acute at apex, and forming an isosceles triangle, erect or slightly spreading. *Corolla* 4–9 mm in diameter, bright sky blue, with a yellow rim at the throat, 5-lobed, the lobes obovate and rounded at apex, the throat closed by 4 short, notched scales. *Stamens* 5, included; anthers yellow. *Style* 1, short; stigma simple, capitate. *Nutlets* 4, up to 2.5 mm, roundish-ovoid, with rim only in the distal half, black. *Flowers* 7–9. $2n=24$, 48, 72.

Native. Mountain slopes and ledges 685–1180 m. Very local in about 16 sites in Yorkshire, Westmorland and Perthshire; probably introduced in Forfarshire. Mountains of Europe from Scotland and the Carpathians to northern Spain and Bulgaria; ? Caucasus. Circumpolar Arctic-montane element.

7. M. sylvatica Hoffm. Wood Forget-me-not

Biennial or *perennial* monoecious *herb* with sterile shoots at base at fruiting. *Stems* up to 50 cm, pale green, erect or decumbent, rather thinly crispate-hairy with spreading hairs in the lower part and appressed hairs above, usually branched, leafy. *Leaves* alternate; basal usually withered at anthesis, the lamina medium green on upper surface, pale beneath, 1–4 × 0.2–2.0 cm, obovate-spathulate, rounded at apex, entire, attenuate at base and obscurely petiolate; cauline decreasing in size upwards, oblong, obtuse or subacute at apex, entire, sessile or petioled below; all with pale, spreading hairs. *Inflorescence* terminal, flowers bisexual, without bracts, the rhachis densely appressed-hairy; pedicels 1–3 mm, lengthening to 6 mm or more in fruit. *Calyx* 1.5–2.0 mm, narrowly campanulate and broadly obtuse at base, divided to halfway or more into 5 lobes, the lobes subulate or narrowly oblong, acute or subacute at apex, and erecto-patent, thinly to densely grey-hairy with appressed straight hairs and spreading or deflexed, curved or hooked hairs. *Corolla* bright blue with a pale eye; tube about as long as calyx; limb 4–10 mm in diameter, divided into 5 rounded lobes, the lobes 1.0–1.5 mm in diameter and flat. *Stamens* 5, inserted about two-thirds way down the corolla tube; filaments very short; anthers yellow, oblong, with a recurved ligulate appendage, the base shortly sagittate. *Style* 1, about 1 mm; stigma simple, capitate. *Nutlets* 4, 1.2–2.0 × 1.0–1.5 mm, shining dark brown, ovoid, more or less acute at apex, strongly compressed, with a distinct marginal rim and a small, impressed attachment scar. *Flowers* 5–6(–9). $2n=18$.

(i) Var. **sylvatica**
Flowers up to 8 mm in diameter, often much less.

(ii) Var. **culta** Voss
Flowers 8–11 mm in diameter, often pink or brighter blue.

Native. Woods, scree and rock-ledges. Locally common in central and north Great Britain; very local in southern Great Britain; frequent garden escape elsewhere and in Northern Ireland. Widespread in Europe and western Asia. Our plant is subsp. **sylvatica**, which occurs throughout the range of the species except the south. Eurasian Temperate element. The native plant is var. *sylvatica*, but it frequently occurs as an introduction. Var. *culta* is a common garden weed and escape.

8. M. dynerae (E. J. Lowe ex Hook. fil.) P. D. Sell
Large-flowered Forget-me-not
M. dissitiflora var. *dynerae* E. J. Lowe ex Hook. fil.

Annual monoecious *herb* with fibrous roots and without sterile shoots at flowering. *Stem* 10–70 cm, pale yellowish-green, fairly robust, angled, with numerous short to long, rarely stiff, colourless simple eglandular hairs throughout, with long branches almost to base and sometimes taller than main stem and suberect or arching, leafy. *Leaves* alternate; lamina 2–12 × 0.5–2.0 cm, alternate, dull medium green with an impressed midrib on upper surface, much paler with a prominent midrib beneath, the lower oblong or oblong-spathulate, the upper lanceolate or oblong-lanceolate, rounded-obtuse at apex, entire but undulate, the lower narrowed below, the upper rounded at base; all with numerous, short to medium, soft, pale simple eglandular hairs on both surfaces and the margins. *Inflorescence* of terminal, scorpioid cymes, ebracteate, lax, much elongated after flowering; flowers bisexual; fruiting pedicels 8–10 mm, slender, spreading, about twice length of calyx, ascending-appressed-hairy. *Calyx* 4–5 mm, with triangular teeth, with numerous, short, straight and hooked hairs. *Corolla* 10–14 mm in diameter, a distinctive pale, clear blue; lobes flat and rounded; throat closed by 5 short, notched scales. *Stamens* 5; anthers with the connective terminating in a blunt process. *Style* 1; stigma simple. *Nutlets* 4, 2.0–2.5 mm, black, ovate, acute at apex, dorsally convex, obscurely keeled, provided at the base with a short, stout, white pedicel. *Flowers* 4–6.

Introduced. Widespread and common as a garden weed, in waste areas, on old railway lines and roadsides in Cambridgeshire and probably elsewhere. It is exactly the plant in *Bot. Mag. tab.* 7589. The species *M. dissitiflora* has the same peg-like base to the nutlet, but appears to be a very different plant in size and shape of leaves and it seems safer to raise the variety, which is named after Mrs Thistleton-Dyer, to the rank of species. It is thought to have originated from Switzerland, but other species of that group with the peg-like base to the nutlet belong to eastern Europe and western Asia. It is difficult to take out the nutlet from the calyx without breaking off the peg-like structure.

9. M. arvensis (L.) Hill — Field Forget-me-not
M. scorpioides var. *arvensis* L.

Biennial monoecious *herb* with fibrous roots. *Stems* 15–60 cm, pale green, erect, the lower part with spreading hairs, the upper part with straight appressed hairs, branched, leafy. *Leaves* alternate; lamina 1–8 × 0.3–2.0 cm, medium greyish-green on upper surface, paler beneath, the lower oblanceolate, or oblong, more or less obtuse at apex, entire, long-drawn-out at base, but hardly petiolate, the upper narrower and sessile, all with more or less spreading hairs on both surfaces. *Inflorescence* an ebracteate, terminal, scorpioid cyme; flowers bisexual; pedicels up to 10 mm in fruit, forward-pointing. *Calyx* up to 7 mm in fruit, closed, with many, spreading, hooked hairs, deciduous, divided over half-way towards the base into 5 lobes, the lobes triangular-ovate, acute at apex. *Corolla* about 3–5 mm in diameter, pale sky blue with an orange-yellow rim at the throat, 5-lobed, the lobes subrotund, retuse at apex and concave. *Stamens* 5, included; anthers yellow. *Style* 1, short; stigma simple, capitate. *Nutlets* 4, up to 2.5 × 1.2 mm, greenish-black to black, ovoid, with rim attachment area small. *Flowers* 4–9.

(a) Subsp. **arvensis**
Calyx not more than 5 mm in fruit, hooked hairs not more than 0.4 mm. *Corolla* up to 3 mm. *Nutlets* not more than 2.0 mm. $2n = 52$.

(i) Var. **arvensis**
Plant often only with 1 stem and not more than 15 cm, if longer and branched usually with central branch longer than rest. *Racemes* usually not more than 10 cm.

(ii) Var. **dumetorum** (Crépin) P. D. Sell
M. intermedia var. *dumetorum* Crépin

Plant often up to 30 cm with numerous long branches from base equalling central stem. *Racemes* up to 15 cm with flowers regularly spaced.

(b) Subsp. **umbrata** (Mert. & W. D. J. Koch) O. Schwarz
M. arvensis var. *umbrata* Mert. & W. D. J. Koch;
M. arvensis var. *sylvestris* Schltdl.; *M. arvensis* var. *umbrosa* Bab.

Calyx up to 7 mm in fruit, hooked hairs up to 0.7 mm. *Corolla* 3–5 mm. *Nutlets* up to 2.5 mm. $2n = 66$.

Native. Weed of cultivated and waste places and other open, well-drained ground. Common throughout most of Great Britain and Ireland. Europe; North Africa; temperate Asia; naturalised in North America. The species is a Eurosiberian Boreo-temperate element. Subsp. *arvensis* occurs throughout the range of the species. Var. *arvensis* is the common weed of arable land and waste places which may be introduced. Var. *dumetorum* occurs in more natural often sandy places and may be the native plant. The distribution of subsp. *umbrata* in Great Britain and Ireland is not known, but it is apparently widespread and is said to be confined to western Europe.

10. M. ramosissima Rochel — Early Forget-me-not
M. hispida Schltdl.; *M. collina* auct.; *M. discolor* auct.

Annual monoecious *herb* with fibrous roots. *Stems* 2–25(–30) cm, pale green, erect or spreading, thinly to rather densely clothed with white, straight hairs, subappressed or irregularly spreading towards the base, closely appressed above, simple or more or less branched from near the base, leafy. *Leaves* alternate; basal forming a loose or dense tuft, the lamina 0.5–3.5 × 0.3–1.3 cm, medium green on upper surface, paler beneath, narrowly obovate, rounded or obtuse at apex, tapering at base to an indistinct petiole or subsessile; cauline similar, but becoming smaller and more acute at apex upwards; all with thin to dense, stiff, subappressed hairs on upper surface, more thinly hairy beneath. *Inflorescence* at first dense, becoming lax and elongating to 7–14 cm or more after anthesis; rhachis with appressed, rigid hairs; flowers bisexual, distinctly pedicellate, the pedicels appressed-hairy, elongating to 2–3 mm, becoming patent or recurved in fruit. *Calyx* up to 4 mm in fruit, campanulate, thinly to densely clothed with short, appressed hairs and longer, spreading or reflexed, hooked hairs, divided slightly more than halfway towards the base into 5 lobes, the lobes erect or slightly spreading but not connivent in fruit. *Corolla* blue, rarely white, tube about 1.5 mm, 5-lobed, the lobes about 0.4 mm in diameter, rounded and spreading, throat swellings papillose. *Stamens* 5, inserted about two-thirds of the way down the corolla tube; filaments very short; anthers yellow, oblong, with a conspicuous, rounded apical appendage. *Pollen* 10–13 × 7–9 µm. *Style* 1, 0.6 mm; stigma simple, capitate. *Nutlets* 4, about 1.3 × 1.0 mm, dark shining brown, ovoid, strongly compressed, margin without or with an indistinct rim at apex. *Flowers* 4–6. $2n=48$.

(a) Subsp. **ramosissima**

Flowers only in upper part of the stem. *Calyx* up to 4 mm in fruit, the lobes narrowly triangular. *Corolla* scarcely exceeding the calyx. *Nutlets* smooth.

(b) Subsp. **globularis** (Samp.) Grau
M. globularis Samp.; *M. collina* var. *mittenii* Baker

Flowers present almost to base of stem. *Calyx* up to 2.5 mm, the lobes broadly triangular. *Corolla* distinctly exceeding the calyx. *Nutlets* with a indistinct rim at apex.

Native. Dry open places on sandy or limestone soils, sometimes transported with sand from these places to other localities. Locally common over much of lowland Great Britain and the perimeter of Ireland, especially England and the Channel Islands. Most of Europe; south-west Asia; North Africa. Subsp. *globularis* occurs in sandy places near the sea from Portugal to England but its total distribution in Great Britain is not known. It ought to be in Ireland. Subsp. *ramosissima* occurs throughout the range of the species. The species is a European Southern-temperate element.

11. M. discolor Pers. Changing Forget-me-not
M. collina Ehrh. ex Hoffm.; *M. versicolor* (Pers.) Sm.

Annual monoecious *herb* with fibrous roots. *Stems* up to 30 cm, pale green, erect, slender, with spreading hairs below and appressed hairs above, simple, leafy. *Leaves* alternate. sometimes with 1 pair opposite, lamina 0.7–4.0 × 0.3–0.8 cm, dull greyish-green on upper surface, paler beneath, oblong, oblong-lanceolate or oblong-elliptical, the lower obtuse, the upper acute at apex, entire, narrowed below but sessile, with numerous, short hairs on both surfaces. *Inflorescence* a terminal, scorpioid, lax, ebracteate cyme, not much longer than the leafy part of stem; flowers bisexual; fruiting pedicels shorter than calyx, ascending. *Calyx* up to 4.5 mm, campanulate, closed or nearly so in fruit, the tube covered with soft, deflexed, hooked hairs, divided to below halfway to the base into 5 lobes, the lobes oblong-lanceolate, acute at apex, ultimately nearly erect. *Corolla* up to 4 mm, pale creamy-yellow or cream at first, usually becoming pink or blue with a yellow rim at the throat, tube at length somewhat longer than calyx, 5-lobed, the lobes subrotund or broadly elliptical, the throat closed by 5, notched scales. *Stamens* 5, included; anthers yellow. *Pollen* 17.5–20 × 17.5–20 µm. *Style* 1, equalling or exceeding calyx; stigma simple, capitate. *Nutlets* 4, about 1.2 × 0.8 mm, dark brown or almost black, ovoid, obtuse at apex, with a wide rim. *Flowers* 5–9.

(a) Subsp. **discolor**

At least the uppermost pair of *cauline leaves* opposite. *Corolla* up to 4 mm, yellow at first. $2n=72$, ?64.

(b) Subsp. **dubia** (Arrond.) Blaise
M. dubia Arrond.

No *cauline leaves* opposite. *Corolla* not more than 2 mm, whitish or cream at first. $2n=24$.

Native. Dry open places on sandy or limestone soils, also damp places, marshes and dune slacks. Common locally throughout Great Britain and Ireland. Most of Europe from Iceland southwards and east to Latvia and central Yugoslavia; Azores. European Temperate element. Subsp. *discolor* occurs throughout the range of the species, but subsp. *dubia* is confined to western Europe. Arthur Chater (pers. comm.) says that in Cardiganshire the subspecies do not grow mixed but that he does not understand their ecology. He also says pollen size is really the only reliable difference between *M. ramosissima* and *M. discolor*.

19. Lappula Gilib.

Annual or *biennial* monoecious *herbs. Leaves* alternate. *Inflorescence* a bracteate terminal cyme; flowers bisexual. *Calyx* divided almost to base into 5 lobes. *Corolla* pale blue, 5-lobed, with 5, short, throat scales. *Stamens* 5, included. *Style* 1, included; stigma simple, capitate. *Fruits* of 4, ovoid-trigonous and flattened nutlets with 1–3 rows of cylindrical, conical or flattened spines on their sides, sometimes also on the back, attached to receptacle for usually only part of their length on the margin.

Contains about 40 species in temperate Eurasia and North America whose centre of diversity is in central Asia and west Siberia.

Hultén, E. & Fries, M. (1986). *Atlas of north European vascular plants north of the Tropic of Cancer.* 3 vols. Königstein.

1. Pedicels of flower up to 4 mm; nutlets with margins with 2–3 rows of spines **1. squarrosa**
1. Pedicel of flowers up to 1.5 mm; nutlets with margins with 1 row of spines **2. patula**

1. L. squarrosa (Retz.) Dumort. Bur Forget-me-not
Myosotis squarrosa Retz.; *Echinospermum squarrosum* (Retz.) Rchb.

Annual or *biennial* monoecious *herb* with fibrous roots. *Stems* 10–70 cm, pale green, erect, with sparse to dense, more or less patent as well as appressed hairs, often branched, leafy. *Leaves* alternate; lamina 2–7 × 0.3–1.0 cm, medium green on upper surface, paler beneath, oblong to linear-lanceolate, obtuse at apex, entire, sessile, or the lower petiolate, with more or less patent and appressed hairs on both surfaces. *Inflorescence* a bracteate, terminal cyme; flowers bisexual; pedicels up to 4 mm, erect. *Calyx* 1.5–3.0 mm in flower, 4–5 mm in fruit, divided almost to the base into 5 lobes, the lobes lanceolate. *Corolla* about 4 mm, pale blue, with a short tube and an almost rotate limb, with 5, short, throat scales. *Stamens* 5, included. *Style* 1, included; stigma simple, capitate. *Nutlets* 4, 2.5–4.0 mm, ovoid-trigonous, flattened, upper surface minutely tuberculate, the margin with 2–3 rows of spines which are usually dilated and flattened at the base. *Flowers* 6–8. $2n=48$.

Introduced. Casual from wool, bird-seed, grass-seed and grain in waste places, rough ground and tips. Scattered records throughout Great Britain and eastern Ireland. Widespread in Europe, but often introduced. Our plant is subsp. **squarrosa** (*L. echinata* Fritsch; *L. myosotis* Moench; *Echinospermum lappula* (L.) Lehm.; *Rochelia lappula* (L.) Roem. & Schult.; *Lappula lappula* (L.) Karsten nom. illegit.; *Myosotis lappula* L.).

2. L. patula (Lehm.) Menyh. Grey Bur Forget-me-not
L. marginata auct.; *L. semiglabra* (Ledeb.) Gürke; *L. stricta* (Ledeb.) Gürke; *L. tenuis* (Ledeb.) Gürke; *Echinospermum patulum* Lehm.; *Myosotis marginata* auct.

Annual monoecious *herb* with fibrous roots. *Stems* 5–70 cm, erect, grey, with dense, erecto-patent hairs, simple or branched at base, leafy. *Leaves* alternate; lamina 2–10 × 0.2–0.8 cm, greyish-green on upper surface, paler beneath, oblong to linear, obtuse at apex, entire, grey with dense hairs, the lower petiolate the upper sessile. *Inflorescence* a bracteate terminal cyme; flowers bisexual; pedicels up to 1.5 mm, erect. *Calyx* 2–3 mm in flower, 3–4 mm in fruit, divided almost to base into 5 lobes, the lobes lanceolate. *Corolla* 3–4 mm, blue, rarely white, 5-lobed, with 5 short throat scales. *Stamens* 5, included. *Style* 1, included; stigma simple, capitate. *Nutlets* 4, 2–4 mm, flattened-trigonous, the upper surface smooth or tuberculate, the margin with 1 row of spines which are dilated, flattened and united at the base. *Flowers* 6–8.

Introduced. A wool casual. Native of east Europe and south-east Spain.

20. Omphalodes Mill.

Perennial monoecious, hairy *herbs*. *Leaves* alternate, basal and many cauline petiolate. *Inflorescence* a short, terminal cyme, open flowers bisexual, erect. *Calyx* divided into 5 lobes almost to base. *Corolla* blue, 5-lobed. *Stamens* 5, inserted about the middle of the corolla tube, included. *Style* 1, included; stigma simple, capitate. *Fruit* of 4, depressed-globose, smooth, hairy nutlets, adorned with a ciliate wing incurved to form an umbilicus.

Contains about 28 species in Europe, Asia, Algeria and Mexico.

1. O. verna Moench Blue-eyed Mary

Perennial monoecious *herb* with long, creeping stolons which occasionally roots from mats. *Stems* 5–20(–40) cm, pale green, slender, erect or ascending, flexuous, simple, leafy. *Leaves* alternate, bright green on upper surface, paler beneath; basal with lamina 5–20(–35) × 2–6 cm, ovate, acute at apex, entire, rounded or cordate at base, long petiolate; cauline smaller, ovate to elliptical, acute, entire, rounded at base, petiolate or sessile; all sparsely hairy. *Inflorescence* a terminal, short, lax cyme; flowers bisexual; bracteate at base, open flowers erect; pedicels 8–12(–30 in fruit) mm, slender, recurved, glabrous. *Calyx* 3.5–4.0 mm, divided almost to base into 5 lobes, the lobes elliptical, appressed-hairy. *Corolla* 8–10 mm, blue with yellow folds, 5-lobed, the lobes obovate, rounded at apex. *Stamens* 5, inserted at about the middle of the corolla tube, included; anthers yellow. *Style* 1, simple, included; stigma simple, capitate. *Nutlets* 4, about 2 mm, depressed-globose, smooth, hairy, with a ciliate wing incurved to form an umbilicus, attached to the receptacle for almost their entire length. *Flowers* 3–5. $2n=42, 48$.

Introduced. Grown in gardens and a persistent relic or throw-out mostly in woods. Scattered records over Great Britain especially in the south and in Wales and Isle of Man, much rarer than formerly. Native of south-east Europe.

21. Cynoglossum L.

Biennial monoecious *herbs*. *Leaves* alternate. *Inflorescence* of ebracteate cymes. *Flowers* bisexual. *Calyx* divided almost to the base into 5 lobes. *Corolla* dull reddish-purple, 5-lobed, with 5 scales closing the throat. *Stamens* 5, included. *Style* 1; stigma simple included. *Fruit* of 4, nutlets more than 4.5 mm, and covered with barbed bristles.

Contains about 60 species in temperate and subtropical regions, especially Asia.

Hultén, E. & Fries, M. (1986). *Atlas of north European vascular plants north of the Tropic of Cancer.* 3 vols. Königstein.

1.	Leaves glabrous on upper surface	**2. germanicum**
1.	Leaves hairy on upper surface	2.
2.	Calyx 3.5–4.5 mm; corolla 5–6 mm	**1. officinale**
2.	Calyx 1.5–3.0 mm; corolla 3–5 mm	3.
3.	Calyx with acute lobes and bulbous-based hairs	**3. amabile**
3.	Calyx with obtuse lobes and stiff appressed hairs	**4. australe**

1. C. officinale L. Hound's-tongue

Biennial monoecious *herb* smelling of mice. *Stems* (20–)30–60(–90) cm, pale green, erect, grey-hairy, branched, leafy. *Leaves* alternate; basal with lamina up to 30 cm, lanceolate to ovate, usually acute at apex, entire, petiolate, the median up to 15 × (1.0–)1.6–2.5(–3.7) cm, oblong to lanceolate, more or less acute at apex, entire, narrowed at base, the upper lanceolate, usually acute at apex, entire, sessile and amplexicaul; all with rather silky, appressed hairs on both surfaces, the upper surface sometimes rough with papillae. *Inflorescence* usually of ebracteate cymes; flowers bisexual; pedicels about 10 mm, stout, recurved in fruit. *Calyx* 3.5–4.5 mm, 5-lobed almost to base, the lobes ovate, obtuse at apex and stiffly hairy. *Corolla* 5–6 mm, dull red-purple, 5-lobed, the lobes obovate and retuse, with 5 scales closing the throat. *Stamens* 5, included, inserted in upper part of corolla tube; anthers yellow. *Style* 1, included; stigma simple, small. *Nutlets* 4, 5–8 mm in diameter, ovoid, flattened, with a distinct border, external face with dense, uniform, barbed bristles. *Flowers* 6–8. $2n=24$.

Native. Rather open ground mostly on shingle, sand or limestone and waste ground. Locally frequent in south, central and east of Great Britain, the Channel Islands and eastern Ireland; rare casual elsewhere. Europe, except the extreme north and south; Asia; North America. Eurosiberian Temperate element.

2. C. germanicum Jacq. Green Hound's-tongue
C. montanum auct.

Biennial monoecious *herb* with fibrous roots. *Stems* (20–)30–50(–60) cm, pale green, slender, erect, densely hairy, branching above, leafy. *Leaves* alternate; basal with lamina up to 30 cm, shining deep green, paler beneath, lanceolate to ovate, usually acute at apex, entire, petiolate; cauline lanceolate or ovate, acute at apex, entire, narrowed below to a sessile, semiamplexicaul base; all more or less glabrous and shining above, with sparse, stiff, short, spreading hairs beneath. *Inflorescence* of ebracteate cymes; flowers bisexual; pedicels about 5 mm, hairy. *Calyx* 4.0–6.5 mm, divided almost to base into 5 lobes, the lobes oblong and obtuse at apex. *Corolla* 5–6 mm, reddish-violet, 5-lobed, the lobes rounded, with 5 scales closing the throat. *Stamens* 5, inserted in the middle of the tube, included; anthers yellow. *Style* 1; stigma simple, included. *Nutlets* 4, 6–8 mm in diameter, ovoid, without a distinct border; external face convex, with dense, barbed bristles. *Flowers* 5–7. $2n=24$.

Native. Woods and hedgerows. Rare and decreasing in Gloucestershire, Oxfordshire, Buckinghamshire and Surrey; formerly widespread but very local in south and central England; very rare casual elsewhere. West and central Europe. European Temperate element.

3. C. amabile Stapf & J. R. Drumm.
Chinese Hound's-tongue

Biennial to *perennial* monoecious *herb. Stems* up to 60 cm, pale green, erect, with dense, appressed, bulbous-based hairs, much-branched, leafy. *Leaves* alternate; lamina of lower 7–20 × 1.5–4.0 cm, dull greyish-green on upper surface, paler beneath, oblong or lanceolate or elliptical, obtuse at apex, entire, narrowed at base and petiolate; upper similar but smaller and sessile; all with numerous, bulbous-based, stiff hairs on upper surface and dense beneath. *Inflorescence* a terminal, scorpioid cyme, flowers bisexual, numerous and in upper leaf axils. *Calyx* 2.5–3.0 mm, divided nearly to the base into 5, ovate, acute lobes, with dense, bulbous-based hairs. *Corolla* up to 5 mm, white, pink or intense blue, 5-lobed, with 5 scales closing the throat. *Stamens* 5. *Style* 1; stigma small, simple, capitate. *Nutlets* 4. *Flowers* 6–8.

Introduced. A casual garden escape. Native of Tibet and China.

4. C. australe R. Br. Australian Hound's-tongue

Perennial monoecious *herb. Stems* up to 1 m, pale green, erect, with numerous, short, stiff, bulbous-based, branched hairs, with long, spreading branches, leafy. *Leaves* alternate; lamina of lower 4–10 × 0.8–2.0 cm, medium green on upper surface, paler beneath, narrowly ovate or rarely a few oblong, acute at apex, entire but wavy, narrowed at base, with petioles up to 8 cm; upper leaves gradually reducing in size and more or less sessile; all with short, stiff, bulbous-based hairs, appressed on both surfaces and the margins. *Inflorescence* of terminal and lateral cymes; pedicels up to 7 mm in fruit and recurved; flowers bisexual, slightly scented. *Calyx* about 1.5 mm at anthesis, about 2 mm in fruit, deeply divided into 5 lobes, the lobes oblong, obtuse at apex, with appressed, stiff hairs. *Corolla* about 3 mm, pale blue, rarely white or pinkish, 5-lobed, with 5 scales closing the throat. *Stamens* 5; anthers enclosed in a tube. *Style* 1; stigma small, simple, capitate. *Nutlets* 4, ovoid, flattened, with prickles on the centre of outer face which has an upcurved prickly wing. *Flowers* 7–10.

Introduced. A wool casual. Native of Australia.

130. VERBENACEAE A. St-Hil. nom. conserv.

Perennial, rarely *annual,* monoecious *herbs. Stems* square in section. *Leaves* opposite, simple, serrate to deeply pinnately lobed, sessile to petiolate, without stipules. *Inflorescence* terminal or elongated to corymbose spikes, zygomorphic, bisexual, hypogynous. *Calyx* 5-lobed. *Corolla* with 5, slightly unequal lobes, 2 lipped, the upper lip 2-lobed, the lower 3-lobed, lilac to purple or blue. *Stamens* 4, included, 2 borne at each of 2 levels in the corolla tube. *Style* 1, terminal; stigma capitate. *Ovary* 4-celled, scarcely lobed, with 1 ovule per cell. *Fruit* a cluster of 4, 1-seeded nutlets.

Contains about 75 genera with 3,000 species, chiefly topical and subtropical. These include trees, shrubs and woody climbers. Their fruit is usually a drupe with 1, 2 or 4 stones and sometimes a capsule.

1. Perennial herbs; corolla up to 10 mm	**1. Verbena**
1. Shrub or tree; corolla about 15 mm	**2. Clerodendron**

1. Verbena L.

Perennial or rarely *annual herbs. Stems* square in section. *Leaves* opposite, serrate to pinnately lobed. *Inflorescence* terminal. *Sepals* 5. *Corolla* 5-lobed. *Stamens* 4. *Style* 1. *Fruit* a cluster of 4, 1-seeded nutlets.

About 250 species, chiefly in America, many weeds of arable land and waste places in North America.

1. Plants rooting at the nodes	**7. × hybrida**
1. Plants not rooting at the nodes	2.
2. At least some of the leaves pinnatisect or pinnatifid	3.
2. Leaves serrate or dentate	5.
3. Petioles not winged	**3. supina**
3. Petioles winged	4.
4. Inflorescence without or sparse glandular hairs	**2. menthifolia**
4. Inflorescence densely glandular-hairy	5.
5. Plant erect or ascending	**1(i). officinalis** var. **officinalis**
5. Plant prostrate	**1(ii). officinalis** var. **prostrata**
6. Calyx without glandular hairs	**6. rigida**
6. Calyx with glandular hairs	6.

7. Leaves more or less connate at base **4. bonariensis**
7. Leaves not connate at base **5. litoralis**

1. V. officinalis L. Vervain

Perennial monoecious *herb* with a tap-root. *Stems* 30–100 cm, pale green, erect, decumbent or prostrate, 4-angled, channelled, nearly glabrous or sparsely scabridulous, branches lax and spreading, leafy. *Leaves* opposite; lamina of lower 2–7 × 0.8–4.5 cm, dull medium green on upper surface, paler beneath, ovate in outline, pinnatisect or pinnatifid, the lobes broad, acute at apex, bluntly serrate with prominent nerves, appressed stiffly hairy on upper surface, scabridulous beneath, tapered below to an indistinct winged petiole 1–2 cm; lamina of upper cauline often narrower than those of lower, lanceolate or oblanceolate, coarsely but bluntly serrate, without or with very few lobes. *Inflorescence* lax, of numerous, elongate spikes 5–30 cm long; flowers bisexual; rhachis angular, glandular-hairy; bracts about 2 × 1 mm, narrowly ovate, acuminate at apex, glandular hairy. *Calyx* about 2 mm, lobes 5, minute, deltoid, glandular-hairy. *Corolla* pale lilac-purple; tube about 3 mm; limb 5-lobed, the abaxial lobe 1 mm in diameter, the lateral and adaxial a little smaller, glabrous. *Stamens* 4, inserted at 2 levels in the corolla tube; filaments very short; anthers ovate. *Style* 1, short and stout; stigma minute, unequally 2-lobed. *Nutlets* 4, about 2.0 × 0.6 mm, bright brown, narrowly oblong, glabrous, rugulose-reticulate dorsally, scabridulous-papillose ventrally. *Flowers* 7–9. Homogamous; visited by bees, hoverflies and butterflies, also automatically self-pollinated. $2n = 14$.

(i) Var. **officinalis**
Stems erect.

(ii) Var. **prostrata** Gren. & Godr.
Stems decumbent or prostrate.

Native. Bare ground and open rough grassy places and by tracks, on well-drained, often calcareous soils. Locally common in southern Great Britain, scattered records elsewhere north to northern England and in central Ireland. Widely distributed in temperate and tropical regions. Var. *officinalis* is our common plant. Var. *prostrata* occurs at the edge of a salt-marsh on the south side of Arnside Knott, Westmorland. It was described from a sandy area on the outskirts of Bayonne on the west coast of France and probably occurs elsewhere.

2. V. menthifolia Benth. Mint-leaved Vervain

Annual or short-lived *perennial* monoecious *herb. Stems* 30–60 cm, pale green, erect, decumbent or ascending, with appressed simple eglandular hairs. *Leaves* opposite; lamina 2–6 cm, medium green on upper surface, paler beneath, obovate to oblong in outline, acute at apex, irregularly pinnatifid or serrate, narrowed at base, with whitish, appressed hairs; petioles up to 10 mm, winged. *Inflorescence* a spike 5–20 cm, long and slender; flowers bisexual; bracteoles up to 2 mm, ovate or ovate-lanceolate. *Calyx* 2.5–3.0 mm, 5-lobed, with sparse glandular hairs. *Corolla* purple, the tube 3.5–4.0 mm, the limb 5–6 mm wide. *Stamens* 4. *Style* 1. *Nutlets* 4, 2.0–2.5 mm, striate, reticulate above. *Flowers* 7–9.

Introduced. A wool casual. Native of western North America.

3. V. supina L. Trailing Vervain

Annual or *perennial* monoecious *herb*, sometimes with an elongate rhizome and rooting at the nodes. *Stems* 5–30(–40) cm, pale green, straggling, slender or robust, glabrous to more or less densely appressed-strigose, 4-angled, branched or unbranched, leafy. *Leaves* opposite; lamina 1–2 × 0.8–1.5 cm, medium green on upper surface, paler beneath, broadly rhomboid, pinnatisect into numerous, narrow, blunt or subacute lobes, tapering at base, subglabrous to densely strigose, with prominent veins beneath; petiole short and channelled. *Inflorescence* a simple or sparingly branched, rather crowded spike 1.5–7.0 cm at anthesis elongating to 10 cm in fruit; flowers bisexual; rhachis densely glandular-hairy; bracts about 1.5 × 0.4 mm, narrowly oblong, prominently keeled, thinly hairy. *Calyx* about 2.5 mm, with 5 minute lobes, the lobes purplish and deltoid, appressed-hairy. *Corolla* pale blue; tube about 2.5 mm, thinly hairy, 5-lobed, the abaxial lobe about 1.7 mm, subrotund-oblong, hairy basally around the throat. *Stamens* 4, inserted about halfway up the tube; filaments very short; anthers oblong. *Style* 1, about 0.9 mm; stigma subobovate, obscurely 2-lobed. *Nutlets* 4, about 2.5 × 1.2 mm, rich brown, narrowly oblong, obscurely reticulate-veined dorsally, paler and sparingly papillose ventrally. *Flowers* 7–9. $2n = 14$.

Introduced. Wool and bird-seed casual. Native of south Europe and the Mediterranean region eastwards to Iraq and Arabia.

4. V. bonariensis L. Argentinian Vervain

Annual to *perennial* monoecious *herb. Stems* several, up to 2 m, dull, dark greyish-green, stiffly erect, quadrangular, rough with numerous, short, colourless glandular hairs, branched above, leafy below. *Leaves* opposite; lamina 4–13 × 0.5–2.0 cm, dull dark green on upper surface with impressed veins, paler beneath with prominent veins, linear, linear-oblong or oblong-lanceolate, obtuse to acute at apex, serrate or crenate-serrate, narrowed at base and each pair more or less connate, with numerous eglandular hairs, some minutely forked, and numerous glandular hairs. *Inflorescence* of corymbose heads; peduncles short, reddish-purplish, with dense glandular hairs; flowers 4.0–4.5 mm in diameter, bisexual, scentless; bracts about 3.0 mm, linear, obtuse, with long-ciliate margins. *Calyx* 4.0–4.5 mm, green, with 5 linear lobes and a red web in between the teeth, with dense eglandular and glandular hairs. *Corolla* 8–9 mm, various shades of blue, violet, purple and lavender; tube 6–7 mm, less than twice as long as calyx, hairy in the upper half, 5-lobed, the lobes rectangular, truncate or retuse at apex. *Stamens* 4, included; filaments white; anthers yellow. *Style* 1, terminal; stigma 1, capitate. *Nutlets* 4. *Flowers* 8–9. $2n = 28$.

Introduced. Wool alien and garden escape on tips. Scattered records in England and southern Scotland. Native of South America.

5. V. litoralis Kunth Brazilian Vervain
V. brasiliensis Vell.; *V. bonariensis* auct.

Short-lived, *perennial* monoecious *herb. Stem* up to 1 m, pale green, square, slightly scabrid. *Leaves* opposite; lamina 8–10×1.5–2.5 cm, medium green on upper surface, paler beneath, lanceolate, oblong or rhomboid, acute at apex, usually coarsely and deeply serrate, attenuate at base, with appressed hairs slightly swollen above and below; veins not impressed above; petioles short, sometimes without. *Inflorescence* loosely paniculate, spikes up to 50 mm at flowering, up to 15 cm at fruiting, slender, the flowers bisexual, at first dense soon becoming distant; bracts more or less equalling calyx at flowering, lanceolate, acuminate at apex, keeled and ciliate. *Calyx* 2–3 mm, glandular-hairy, 5-lobed, the lobes green and acute at apex. *Corolla* 3.0–3.5 mm, in diameter, the tube longer than the calyx and rather sparingly hairy, bluish or mauve, 5-lobed and more or less 2-lipped, the lobes obtuse at apex. *Stamens* 4, included, borne well above the middle. *Style* 1, included. *Nutlets* 4, about 1.5 mm, brown, oblong, faintly dorsally ribbed, finely white-papillate on flattened ventral surface. *Flowers* 6–8. $2n=28$.

Introduced. A casual wool alien. Native of South America.

6. V. rigida Spreng. Slender Vervain
V. venosa Gillies & Hook.

Perennial monoecious *herb. Stem* up to 60 cm, pale green, erect or ascending, stiff, with stiff simple eglandular hairs, little-branched, leafy. *Leaves* opposite; lamina 4–8 × 1–2 cm, medium green on upper surface, paler beneath, oblong, acute at apex, irregularly dentate, rigid, sessile, with stiff simple eglandular hairs. *Inflorescence* a lax, terminal corymb, with long branches each ending in a dense spike 2–5 cm in fruit; flowers bisexual; bracts longer than calyx, lanceolate-subulate, scabrid and ciliate. *Calyx* 1–3 mm, tubular, 5-ribbed, unequally 5-lobed. *Corolla* 2–3 times as long as calyx, reddish-purple, weakly 2-lipped. *Stamens* 4, included. *Style* 1, terminal; stigma capitate. *Nutlets* 4. *Flowers* 7–9. $2n=42$.

Introduced. Garden escape on tips. Occasional records in England and southern Scotland. Native of South America.

7. V.×hybrida Voss Garden Vervain

Perennial monoecious *herb. Stems* up to 1 m, procumbent, compact, creeping and mat-like, or lax and trailing, branching freely, rooting at nodes, tetragonal and densely hairy. *Leaves* opposite; lamina up to 8×6 cm, dark green on upper surface, paler beneath, lanceolate to ovate, acute at apex, irregularly incised and toothed, the lobes acute and often doubly toothed, cuneate or truncate at base. *Inflorescence* terminal, large and showy, from a flat corymb to an elongated spike; flowers bisexual, dense and fragrant; bracteoles to 6 mm, shorter than calyx, lanceolate, softly hairy. *Calyx* up to 5.5 mm, variously lobed. *Corolla* from white to shades of purple, blue, scarlet, pink and yellow, usually with a white or yellow mouth. *Flowers* 7–10. $2n=10$.

Garden plant of unknown origin but thought to be a multiple hybrid of *V. incisa* Hook., ***V. peruviana*** (L.) Britton, ***V. phlogiflora*** Cham. and ***V. teurcroides*** Gillies & Hook. Commonly grown in gardens as a bedding plant and sometimes found growing on tips and in waste places.

2. Clerodendron L.

Monoecious *trees, shrubs* and *woody vines. Leaves* opposite, simple, entire or indistinctly dentate, smelling unpleasant if crushed. *Inflorescence* a terminal cyme or panicle; flowers bisexual. *Calyx* tubular, 5-lobed. *Corolla* strongly bilaterally symmetrical, 5-lobed. *Stamens* 4, protruding. *Fruit* a fleshy-coloured drupe with 4 hard stones containing the seeds, surrounded by the persistent calyx.

Contains 400–450 species from the tropics and subtropics.

Bean, W. J. (1970). *Trees and shrubs hardy in the British Isles.* Ed. 8. London.
Rushforth, K. (1999). *Trees of Britain and Europe.* London.

1. C. trichotomum Thunb. Chance Tree
Siphonanthus serotinum Carrière; *Siphonanthus trichotomus* (Thunb.) Nakai

Deciduous monoecious *shrub* or *tree* up to 6 m, with an open crown. *Trunk* up to 25 cm in diameter, extending into crown. *Bark* brown, fissured with orange. *Branches* spreading or ascending; twigs brown, shining, glabrous; young shoots green or greyish-green, with brownish hairs. *Buds* in opposite pairs, about 2 mm, conical, pointed at apex; scales dark brown, ovate, obtuse at apex. *Leaves* opposite; lamina 5–22×4–12 cm, dull medium yellowish-green on upper surface, paler beneath, foetid when crushed, ovate to oblong, acute at apex, entire or with faint, rounded teeth, broadly cuneate at base, glabrous on upper surface, pectinately hairy on the veins beneath; veins 3–4 pairs, impressed above, prominent beneath; petiole 2–6 cm, grooved above, rounded beneath, with numerous, spreading hairs. *Inflorescence* terminal in cymes on current season's growths; pedicels up to 25 mm, green to purple, with curly hairs; flowers 25–30 mm in diameter, bisexual, strong-smelling. *Calyx* 13–15 mm, purple, 5-lobed nearly to base, forming ridges where they meet, the lobes ovate, acute at apex, with short, curly hairs. *Corolla* about 15 mm, with a slender, brownish, tubular base and 5, cream, linear or spathulate lobes, rounded at apex. *Stamens* 4; filaments pale brownish-cream; anthers pale brown. *Style* whitish; stigma green. *Drupe* 6–10 mm, bright blue, ripening to black, subglobose, surrounded by the persistent calyx lobes. *Flowers* 7–9. $2n=46$, 92.

Introduced. Planted in gardens, parks and amenity areas. Formerly occurred as an escape on a railway bank near Dorking Town station in Surrey and by a footpath at Winsley, Bradford-on-Avon in Wiltshire. Native of eastern Asia.

131. LAMIACEAE Lindl. nom. conserv.

Annual to *perennial* monoecious *herbs* or dwarf *shrubs*, often with glandular hairs and aromatic or foetid. *Stems* often quadrangular in section. *Leaves* opposite, simple, entire to serrate or lobed, without stipules. *Inflorescence* a contracted cyme in the axils of leaf-like to much reduced bracts, the opposite pairs of cymes often forming a false whorl of flowers, in this account called a whorl, but technically a verticillaster, sometimes the whole forming a terminal, spike-like inflorescence; flowers bisexual or male-sterile, hypogynous and zygomorphic; bracteoles usually small, sometimes absent. *Calyx* usually 5-lobed, often 2-lipped, with the upper lip 3-lobed and the lower lip 2-lobed, rarely with the lips entire. *Corolla* with a more or less well-developed tube, straight or curved, usually 5-lobed, but with the 2 upper lobes nearly always joined to form a single lip, its origin made clear by the venation; the lower 3 lobes often forming a distinct lower lip, the middle lobe usually larger than the 2 lateral; rarely the 5 lobes form a single lower lip. *Stamens* 4, 2 long and 2 short, sometimes reduced to 2, alternating with the corolla lobes and borne on the corolla tube; anthers introrse. *Style* 1, branched into 2 above, usually gynobasic. *Ovary* 4-celled, each cell with 1 ovule, deeply 4-lobed. *Fruit* of 4, 1-seeded nutlets. *Nectar* is secreted at the base of the ovary.

Contains about 180 genera and 3,500 species in tropical and temperate regions.

Morton, J. K. (1973). A cytological study of the British Labiatae (excluding *Mentha*). *Watsonia* **9**: 239–246.

1. Fertile stamens 2 — 2.
1. Fertile stamens 4 — 4.
2. Evergreen shrub; leaves linear with entire margins — **29. Rosmarinus**
2. Usually not an evergreen shrub, if so leaves not as above — 3.
3. Calyx with 5 more or less equal lobes, rarely the upper lobe larger and the remaining 4 equal — **26. Lycopus**
3. Calyx distinctly 2-lipped or with the 3 lobes or teeth of the upper lip different from the 2 lobes or teeth of the lower — **30. Salvia**
4. Upper corolla lip absent or minute and represented by 2 small teeth; style arising from the top of ovary — 5.
4. Corolla 2-lipped with upper lip present or corolla of more equal lobes; style arising from base of ovary — 6.
5. Corolla without an upper lip, 5-lobed; tube without a ring of hairs inside — **12. Teucrium**
5. Lower lip of corolla 3-lobed; upper lip of 2 short teeth; tube with a ring of hairs inside — **13. Ajuga**
6. Calyx 2-lipped, enlarging in fruit, upper lip bearing a scale; upper corolla lip hooded — **14. Scutellaria**
6. Calyx not bearing a scale on the upper lip, or if so, the upper corolla lip not hooded; calyx enlarging or not in fruit — 7.
7. Calyx teeth 5 or 10, spiny, often hooked — **10. Marrubium**
7. Calyx teeth or lobes 4 or 5, if 10 not spiny or hooked, or calyx 2-lipped — 8.
8. Corolla not or weakly 2-lipped — **27. Mentha**
8. Corolla distinctly 2-lipped — 9.
9. Upper lip of corolla distinctly hooded or strongly concave, often longer than lower lip — 10.
9. Upper lip of corolla flat or corolla flat or convex, sometimes slightly concave, usually equal to or shorter than lower lip — 23.
10. Shrubs — **8. Phlomis**
10. Herbs, sometimes woody at base — 11.
11. Annuals; calyx with spiny teeth — **7. Galeopsis**
11. Annuals or perennials; calyx usually without spiny teeth; if teeth spiny, plant not an annual — 12.
12. Low-growing herbs rooting at the nodes — 13.
12. Plants erect, not rooting at the nodes — 15.
13. Flowers in dense terminal cylindrical spikes — **16. Prunella**
13. Flowers usually axillary; if inflorescence terminal, flowers not in a dense spike — 14.
14. Corolla yellow — **5. Lamiastrum**
14. Corolla pink, red or white — **6. Lamium**
15. Plant strongly aromatic, sometimes unpleasantly so — 16.
15. Plant not or slightly aromatic — 20.
16. Corolla tubular at base, inflated beyond calyx — **17. Dracocephalum**
16. Corolla not as above — 17.
17. Calyx tubular to conical, 10-veined, more or less expanded in upper half, often at right angles with 5–10 teeth — **3. Ballota**
17. Calyx without above combination of characters — 18.
18. Calyx net-veined — **9. Melittis**
18. Calyx with 5, 10 or 15 veins — 19.
19. Flowers not pedicelled; flowers and bracteoles with broad hardened bases; nutlets compressed-trigonous with flattened lateral and apical winged margins, the apical wing usually irregularly lobed — **1. Betonica**
19. Flowers with pedicels; flowers and bracteoles with narrow unhardened bases; nutlets usually obovoid, without flattened lateral and apical winged margins, apices irregularly lobed — **2. Stachys**
20. Corolla tubular at base, inflated beyond calyx — **17. Dracocephalum**
20. Corolla not inflated beyond calyx — 21.
21. Leaves often palmately lobed; calyx teeth spine-tipped, lower 2 deflexed — **4. Leonurus**
21. Leaves not palmately lobed; calyx teeth not as above — 22.
22. Style branches unequal — **8. Phlomis**
22. Style branches more or less equal — **6. Lamium**
23. Plant a shrub — 24.
23. Plant a herb sometimes woody at base — 29.
24. Clusters of flowers aggregated into short, dense, terminal or axillary spikes which are often arranged in panicles or false corymbs — **24. Origanum**
24. Flowers not aggregated into small spikes as above — 25.
25. Creeping shrub often rooting at the nodes — 26.
25. Upright shrub or subshrub — 27.
26. Bracteoles minute; bracts leaf-like, enlarged and coloured — **25. Thymus**
26. Bracteoles 3 mm or more; bracts not leaf-like, enlarged or coloured — **19. Satureja**
27. Upper middle lobe of calyx more or less equalling other lobes, broadly triangular or modified into an appendage — **28. Lavandula**
27. Calyx not modified as above — 28.
28. Bracts leaf-like, enlarged and coloured — **25. Thymus**

28. Bracts not leaf-like, enlarged or coloured **19. Satureja**
29. Calyx distinctly 2-lipped or lower teeth longer than upper 30.
29. Calyx with 5 more or less equal lobes 35.
30. Calyx swollen at one side, the tube constricted **22. Acinos**
30. Calyx not swollen or tube constricted 31.
31. Calyx-tube curved **20. Clinopodium**
31. Calyx-tube straight 32.
32. Flowers yellow becoming pinkish or white; plant strongly lemon-scented **18. Melissa**
32. Flowers white, pink, blue or purple; plant not lemon-scented 33.
33. Annual herb **19. Satureja**
33. Perennial herb 34.
34. Bracts usually similar in shape and size to leaves; stems soft and fleshy **15. Glechoma**
34. Bracts usually smaller and not similar in shape to leaves; stems not fleshy **20. Calamintha**
35. Leaves not linear or linear-lanceolate, or if so then margins toothed and petioles present **14. Nepeta**
35. Leaves linear to linear-lanceolate, margins entire; petioles absent **23. Hyssopus**

1. Betonica L.

Perennial monoecious *herbs. Leaves* in a well-marked basal rosette with few opposite cauline leaves. *Inflorescence* compact above, with distant whorls below; flowers bisexual. *Calyx* 5-lobed, the lobes aristate. *Corolla* reddish-purple, rarely white; tube long, usually without a ring of hairs, but with scattered hairs within. *Stamens* 4, 2 long and 2 short, the outer not diverging from the corolla after flowering; anther-cells nearly parallel. *Style* 1. *Fruit* of 4, trigonous nutlets.

Davy, F. H. (1909). *Flora of Cornwall*. Penryn.
Grime, J. P. et al. (1988). *Comparative plant ecology*. London.
Hultén, E. & Fries, M. (1986). *Atlas of north European vascular plants north of the Tropic of Cancer.* 3 vols. Königstein.

1. Stems 4–15 cm; leaves up to 3 cm **1(i). officinalis** var. **nana**
1. Stems up to 80 cm; leaves up to 9 cm **1(ii). officinalis** var. **officinalis**

1. B. officinalis L. Betony
Stachys officinalis (L.) Travis; *S. betonica* Benth.

Perennial monoecious *herb* with a short, woody rhizome and a well-marked basal rosette. *Stem* 5–80 cm, pale yellowish-green, erect, quadrangular, with numerous, reflexed simple eglandular hairs, simple or somewhat branched below, sparingly leafy. *Leaves* opposite, the cauline distant; lamina 1–9×0.5–2.5 cm, dark green on upper surface, paler beneath, oblong or ovate-oblong, obtuse at apex, coarsely crenate, more or less cuneate at base, sparingly simple eglandular hairy on both surfaces and the margins and glandular-punctate; petioles up to 7 cm, pale green, with simple eglandular hairs. *Inflorescence* compact above, the whorls of numerous, bisexual flowers very distant below; bracts ovate or lanceolate, entire or crenate-serrate; bracteoles lanceolate, aristate at apex and about equalling the calyx. *Calyx* 7–9 mm, green, campanulate, 5-lobed, the lobes subulate-aristate with a red rim, with glandular and simple eglandular hairs. *Corolla* 12–18 mm, reddish-purple; sometimes white; tube without or with scattered simple eglandular hairs within; 2-lipped, the upper lip slightly hooded, the lower lip 3-lobed, the 2 outer lobes smaller than the central lobe, all lobes with a wavy margin. *Stamens* 4, 2 long and 2 short; filaments white; anthers orange. *Style* 1, white. *Nutlets* 4, 2.5–3.0 mm, trigonous. *Flowers* 6–9. Pollinated mainly by bees; protandrous or homogamous. $2n=16$.

(i) Var. **nana** (Druce ex Ham. Davy) P. D. Sell
Stachys officinalis var. *nana* Druce ex Ham. Davy; *Stachys betonica* var. *alba* Lynch; *Stachys officinalis* var. *nana* Druce; *Stachys officinalis* forma *alba* (Lynch) Druce

Stems 4–15 cm. *Leaves* up to 3 cm. Whole plant with more numerous eglandular hairs than var. *officinalis*.

(ii) Var. **officinalis**
Stems up to 80 cm. *Leaves* up to 9 cm. Plant rather sparingly simple eglandular hairy.

Native. Hedgebanks, grassland, heaths, woodland rides and coastal cliffs, avoiding heavy soils. Common in England and Wales and local in Jersey, very local in Scotland and Ireland. Most of Europe northwards to south Sweden and north-west Russia; Caucasus; Algeria. European Temperate element. The common plant is var. *officinalis*. Var. *nana* occurs on a few cliff-tops by the sea in western Great Britain. Reddish-purple and white-flowered plants occur in both varieties.

2. Stachys L.

Annual to *perennial* monoecious *herbs. Leaves* opposite, crenate to serrate. *Inflorescence* a terminal spike of whorls, lax to dense; flowers bisexual. *Calyx* with 5 more or less equal lobes. *Corolla* yellow, pink to purple or white, 2-lipped, the upper lip entire and concave, the lower lip 3-lobed. *Stamens* 4, 2 long and 2 short. *Style* 1. *Fruit* of 4, obovoid, nutlets rounded at apex.

Contains about 300 species, cosmopolitan except Australia and New Zealand.

Dunn, A. J. (1987). Observations on *Stachys germanica* at a new site in Oxfordshire. *Watsonia* **16**: 430–431.
Dunn, A. J. (1991). Further observations on *Stachys germanica* L. *Watsonia* **18**: 359–367.
Dunn, A. J. (1997). Biological flora of the British Isles. No. 196. *Stachys germanica* L. *Jour. Ecol*. **85**: 531–539.
Grime, J. P. et al. (1988). *Comparative plant ecology*. London.
Lousley, J. E. (1934). *Stachy alpina* L.×*S germanica* L. (× *S. divenea* Legué). *Rep. Bot. Soc. Exch. Club Brit. Isles* **10**: 539.
Hultén, E. & Fries, M. (1986). *Atlas of north European vascular plants north of the Tropic of Cancer.* 3 vols. Königstein.
Kay, Q. O. N. & John, R (1995). The conservation of scarce and declining plant species in lowland Wales: population genetics, demographic ecology and recommendation for future conservation in 32 species of lowland grassland and related habitats. *Science Report* No. 110 Countryside Council for Wales. [*S. alpina.*]

Marren, P. R. (1988). The past and present distribution of *Stachys germanica* L. in Britain. *Watsonia* **17**: 59–68.

Wigginton, M. J. (Edit.) (1999). *British red data books*. Vol. 1. *Vascular plants*. Ed. 3. Peterborough. [*S. alpina* and *germanica*.]

Wilcock, C. C. & Jones, B. M. G. (1974). The identification and origin of *Stachys* × *ambigua* Sm. *Watsonia* **10**: 139–147.

1. Corolla white to yellow 2.
1. Corolla purplish, rarely white 3.
2. Plant biennial to perennial; flowers usually more than 6 in a whorl; corolla 15–20 **7. recta**
2. Plant annual; flowers less than 6 in a whorl; corolla 10–16 mm **8. annua**
3. Annual; corolla less than 10 mm **9. arvensis**
3. Perennial; corolla more than 10 mm 4.
4. Bracteoles absent or minute 5.
4. Bracteoles mixed in with flowers, at least as long as calyx 7.
5. Middle and upper stem leaves sessile **6. palustris**
5. All leaves up to inflorescence petiolate 6.
6. Petioles of middle and upper stem leaves one-tenth to one-fifth the total leaf plus petiole length; few or no fruits ripening **5. × ambigua**
6. Petioles of middle and upper stem leaves one-quarter to two-thirds the total leaf plus petiole length; all or most fruits ripening 7.
7. Stem, leaves and inflorescence with few to numerous simple eglandular and glandular hairs; cauline leaves 5–12 × 3–8 cm **4(i). sylvatica** var. **sylvatica**
7. Stem, leaves and inflorescence with dense simple eglandular and glandular hairs; cauline leaves 2–7 × 0.7–5.0 cm **4(ii). sylvatica** var. **subsericea**
8. Upper part of stem and calyx with glandular hairs as well as eglandular hairs **3. alpina**
8. Stem and calyx with only eglandular hairs 9.
9. Leaves cuneate at base; stem and both leaf surfaces densely white-tomentose **1. byzantina**
9. At least the lower leaves cordate at base; green upper surface showing through hairs **2. germanica**

Section **1. Eriostomum** (Hoffmanns. & Link) Dumort.

Perennial herbs without a persistent rosette of leaves. *Whorls* usually many-flowered; bracteoles as long as or longer than the calyx tube.

1. S. byzantina K. Koch — Lamb's-ear

S. lanata Jacq., non Crantz

Perennial monoecious *herb* with thick rhizomes near the surface and many sterile rosettes, forming large patches. *Stems* 30–75 cm, brownish-purple, ascending to erect, quadrangular, densely grey-woolly-hairy, unbranched, leafy. *Leaves* medium yellowish-green, paler beneath; basal 3–10 × 1.2–4.0 cm, narrowly elliptical to elliptical, rounded at apex, entire, attenuate at base, densely woolly on both surfaces, the petioles up to 8 cm, woolly; cauline in pairs, opposite, widely spaced, similar to basal but smaller with shorter petioles and steadily decreasing in size upwards. *Inflorescence* of distant whorls in the upper part of stem; bracts oblong, turned downwards; flowers bisexual. *Calyx* 8–12 mm, with 5, triangular, acute lobes, densely woolly. *Corolla* 14–25 mm, pinkish-purple, 2-lipped, the upper lip elliptical and hooded, the lower lip 3-lobed, with the middle lobe longest. *Stamens* 4, 2 long and 2 short; filaments white, flushed pink; anthers mauve. *Style* 1; stigmas 2. *Nutlets* 4, 1-seeded, obovoid, rounded at the apex. *Flowers* 6–7. Visited by bees. $2n = 30$.

Introduced in 1782. Much grown in gardens and occurring as a throw-out on tips and in waste places. Widely recorded in Great Britain especially in the south and a few records in east Ireland. Native of south-west Asia.

2. S. germanica L. — Downy Woundwort

Perennial or *biennial* monoecious *herb* without rhizomes. *Stems* 30–80(–100) cm, pale green, erect, quadrangular, densely covered with white, silky eglandular hairs, branched, leafy. *Leaves* opposite; lamina 5–12 × 2–5 cm, medium green on upper surface, paler beneath, elliptical to oblong-elliptical, rounded to subacute at apex, crenate, truncate at base, densely covered on both surfaces with white, silky eglandular hairs; petioles up to 8 cm, pale green, slightly channelled above, rounded beneath, silky eglandular hairy. *Inflorescence* a dense, terminal spike, interrupted below, the whorls many-flowered; flowers bisexual; bracts lanceolate, passing into the leaves; bracteoles linear, nearly as long as the calyx. *Calyx* 9–11 mm, with dense, silky eglandular hairs, 2-lipped, the upper lip 2-lobed and concave, the lower lip 3-lobed with all the lobes triangular-mucronate and subequal. *Corolla* 12–20 mm, pinkish-purple, with simple eglandular hairs outside, 2-lipped, the upper lip wavy and hooded, the lower lip 3-lobed, the 2 outer lobes much smaller than the central lobe. *Stamens* 4, 2 long and 2 short; filaments white; anthers purple. *Style* 1, white. *Nutlets* 4, 2.0–2.5 mm, obovoid, rounded at apex. *Flowers* 7–8. Pollinated by bumble-bees; protandrous; small female flowers occur. $2n = 30$.

Native. At present only by green lanes and in game-crop fields cultivated from ancient grassland; formerly in woodland margins, old hedgerows, road verges, ditches and stony fields. Now known only from four sites in Oxfordshire; formerly also known in Hampshire, Northamptonshire and Lincolnshire. Widespread throughout central, western and south Europe, the Mediterranean region, North Africa and eastwards to central-southern Russia. European Temperate element.

3. S. alpina L. — Limestone Woundwort

Perennial monoecious *herb*. *Stems* 30–100 cm, green, quadrangular, with numerous, soft simple eglandular hairs throughout and glandular hairs above, unbranched, leafy. *Leaves* opposite; lamina 5–18 × 3–9 cm, medium bright greyish-green on upper surface, paler beneath, upper flushed purplish, ovate or oblong-ovate, acute at apex, crenate-serrate, cordate at base, with sparse eglandular hairs on the upper surface, more numerous beneath; petioles 3–10 cm, with simple eglandular hairs. *Inflorescence* a lax terminal spike, the whorls many-flowered and distant;

flowers bisexual; bracts lanceolate or ovate and subsessile, the lower crenate-serrate and leaf-like, the uppermost smaller and entire; bracteoles nearly as long as the calyx, entire. *Calyx* 6–12 mm, reddish-purple, tubular, 5-lobed, the lobes unequal, the upper 2 about half as long as the tube, rounded-cuspidate, with dense glandular and eglandular hairs. *Corolla* 15–22 mm, the tube cream, with a ring of hairs within, the lobes spotted, flushed and marked round the edges with dull violet or purple, rarely tinged with yellow, 2-lipped, the upper lip concave and 2-lobed, the lower lip 3-lobed, the lobes rounded and retuse, hairy. *Stamens* 4, 2 long and 2 short; often diverging laterally from corolla after flowering. *Style* 1. *Nutlets* 4, 3.0–3.5 mm, obovoid, rounded at apex. *Flowers* 6–8. $2n=30$.

Native. Open woods. Known only from single sites in Denbighshire and Gloucestershire. West, central and south Europe. European Temperate element.

× **germanica** = **S. × digenes** Legué
This hybrid has occurred in gardens from parents of British origin (see Lousley, 1934).

Section **2. Stachys**

Perennial herbs without a persistent rosette of leaves. *Whorls* usually not more than 12-flowered; bracteoles usually absent or minute.

4. S. sylvatica L. Hedge Woundwort

Perennial monoecious *herb* with a long, creeping rhizome, not producing tubers, foetid when bruised. *Stems* 30–100 cm, pale green often blackish purple, erect, solid, quadrangular, with glandular and bulbous-based, eglandular hairs, often branched, leafy. *Leaves* opposite; lamina 2–12 × 0.7–8.0 cm, soft medium green on upper surface, paler beneath, ovate, acute or acuminate at apex, coarsely crenate-serrate, cordate at base, sparsely to densely hairy on both surfaces with simple eglandular and glandular hairs and glandular-punctate beneath; petioles 1.5–7.0 cm, with glandular and eglandular hairs. *Inflorescence* an interrupted terminal spike formed of up to 6(–8)-flowered whorls; flower bisexual; bracts of lower whorls ovate-lanceolate, acute at apex, crenate-serrate and shortly petiolate, the upper lanceolate, acute and entire; bracteoles minute, linear, scarcely reaching the base of the calyx. *Calyx* 6–8 mm, flushed purple, tubular, 5-lobed, the lobes equal, as long as or slightly longer than the tube, lanceolate and with a long, sharp apex, with numerous to dense eglandular and glandular hairs. *Corolla* 13–18 mm, reddish-purple or magenta with white lines on the lobes of the lower lip, rarely white or pale pink, tube with a ring of hairs inside, 2-lipped, the upper lip slightly hooded, the lower lip 3-lobed, the lateral lobes elliptic, the central lobe broadly ovate, hairy outside. *Stamens* 4, 2 long and 2 short, included. *Style* 1, branches subequal. *Nutlets* 4, 2.0–2.5 mm, obovoid, rounded at apex. *Flowers* 7–8. Pollinated by bees; protandrous. $2n=66$.

(i) Var. **sylvatica**
Stem, leaves and *inflorescence* with few to numerous simple eglandular and glandular hairs. *Cauline leaves* 5–12 × 3–8 cm.

(ii) Var. **subsericea** Grogn.
Stems, leaves and *inflorescence* with dense simple eglandular and glandular hairs. *Cauline leaves* 2–7 × 0.7–5.0 cm.

Native. Woods, hedgerows, streamsides and rough ground. Common throughout most of Great Britain and Ireland. Most of Europe, but rare in the Mediterranean; Caucasus, Kashmir and the Altai. Eurosiberian Temperate element. The common plant is var. *sylvatica*. Var. *subsericea* is recorded from Cornwall, Devonshire, Glamorganshire, Pembrokeshire, the Channel Islands, Norfolk and Cambridgeshire. It also occurs in France.

5. S. × ambigua Sm. Hybrid Woundwort
S. palustris × sylvatica

Perennial monoecious *herb* with long, creeping rhizome, strongly foetid when bruised. *Stems* up to 80 cm, pale green, often flushed reddish-purple, erect, quadrangular, not usually branched, with more or less numerous, bulbous-based, reflexed, stiff simple eglandular hairs and glandular hairs. *Leaves* opposite; lamina 2–11 × 0.5–3.5 cm, medium green on upper surface, oblong, oblong-lanceolate or lanceolate, more or less acute at apex, serrate-crenate, cordate at base, with numerous, appressed simple eglandular hairs especially on the veins on both surfaces; petioles very short with eglandular and glandular hairs. *Inflorescence* a terminal spike, the whorls 4- to 10-flowered, the lowermost often distant; flowers bisexual; bracteoles absent or minute. *Calyx* 6–8 mm, tubular, 5-lobed, the lobes equal, as long as or slightly shorter than the tube, lanceolate and long acute at apex, long hairy, some of the hairs usually glandular. *Corolla* 12–14 mm, the tube cream, the lobes bright red, 2-lipped, the upper lip slightly hooded, the lower lip 3-lobed, the lobes broad and emarginate, hairy outside. *Stamens* 4, 2 long and 2 short; anthers purple. *Style* 1, branches subequal. Normally sterile. *Flowers* 7–8. $2n=83$.

Native. Beside streams, on river-banks and roadsides, on the edge of woodland, on waste ground and as a weed of cultivated ground. Scattered throughout Great Britain and Ireland especially in the west. Widespread in Europe.

6. S. palustris L. Marsh Woundwort

Perennial monoecious *herb* with long, creeping rhizome producing small tubers at the apex in the autumn, almost odourless. *Stems* 30–120 cm, pale green, often flushed brownish-purple, with sparse to dense eglandular and glandular hairs, erect, quadrangular, shortly branched above, leafy. *Leaves* opposite; lamina 3–12 × 0.7–3.5 cm, medium green on upper surface, oblong or oblong-lanceolate, acute at apex, crenate, cordate at base, with appressed simple eglandular hairs on both surfaces, the lower shortly petiolate, the upper sessile. *Inflorescence* a terminal spike, the whorls 4- to 10-flowered, the upper crowded, the lower distant; bracteoles absent or minute. *Calyx* 6–8 mm, tubular, 5-lobed, the lobes equal, as long as or slightly shorter than tube, lanceolate and long acute at apex, eglandular. *Corolla* (11–)12–15 mm, the tube cream, the lobes rose-purple, the lower and lateral ones variegated with cream, 2-lipped, the upper lip slightly hooded, the lower lip 3-lobed, the lobes broad and emarginate at apex, hairy outside. *Stamens* 4, 2 long and 2 short; anthers very dark purple. *Style* 1, branches subequal. *Nutlets* 4, 2.0–2.2 mm, ellipsoid, rounded at apex. *Flowers* 7–9. Pollinated mainly by bees; protandrous. $2n = 102$.

Native. Damp places by rivers, streams and pounds, marsh and fen and in rough places. Common over most of Great Britain and Ireland. Most of Europe, but rare in the Mediterranean region; temperate Asia to Japan; North America. Circumpolar Boreo-temperate element.

7. S. recta L. Perennial Yellow Woundwort

Perennial monoecious *aromatic herb* without rhizomes. *Stems* 15–70 cm, pale green, erect or ascending, quadrangular, subglabrous or with few to numerous eglandular and glandular hairs, shortly branched, leafy. *Leaves* opposite; lamina 1–8 × 0.5–2.0 cm, medium green on upper surface, paler beneath, the lower oblong to ovate, the upper oblong or oblanceolate, acute at apex, entire to crenate-serrate, rounded or cuneate at base, glabrous or with few to numerous eglandular and glandular hairs. *Inflorescence* a terminal spike, the whorls with 6–16 bisexual flowers, crowded or the lower distant; bracteoles minute. *Calyx* 5–7 mm, 5-lobed, the lobes shorter than the tube and more or less equal, with numerous eglandular and glandular hairs. *Corolla* 15–20 mm, pale yellow with purplish streak, 2-lipped, the upper lip slightly hooded, the lower lip 3-lobed, the 2 lateral lobes much smaller than the central lobe. *Stamens* 4, 2 long and 2 short, slightly exserted. *Style* 1. *Nutlets* 4, obovoid, rounded at apex. *Flowers* 7–9. $2n = 32, 34, 48$.

Introduced. Waste ground. Naturalised at Barry Docks, Glamorganshire since 1923 and recorded for Durham. Native of much of Europe.

Section **3. Olisia** Dumort.

Annual herbs, rarely short-lived *perennials*. *Whorls* not more than 6-flowered; bracteoles usually absent or minute.

8. S. annua (L.) L. Annual Yellow Woundwort
Betonica annua L.

Annual monoecious *herb* with fibrous roots, rarely a short-lived *perennial*. *Stems* 10–40 cm, pale green, erect, quadrangular, with eglandular and sometimes glandular-hairs, shortly branched, leafy. *Leaves* opposite; lamina 1–6 × 0.5–2.0 (–3.0) mm, medium yellowish-green on upper surface, paler beneath, lanceolate or oblong, acute or obtuse at apex, crenate or crenate-dentate, rounded or cuneate at base, glabrous or with more or less numerous, eglandular hairs, with punctate glands beneath, the lower petiolate, the upper subsessile. *Inflorescence* a terminal spike, the whorls 2–6-flowered, the upper crowded, the lower distant; flowers bisexual; bracts lanceolate, acute at apex; bracteoles linear. *Calyx* 5–8 mm, 5-lobed, the lobes equal, lanceolate, sharply acute at apex, with numerous eglandular and glandular hairs. *Corolla* 10–16 mm, white or pale yellow, sometimes with red spots, tube exceeding the calyx, 2-lipped, the upper lip entire, the lower lip 3-lobed, the lobes broad and rounded. *Stamens* 2 long and 2 short, included. *Style* 1, branchcs subequal. *Nutlets* 4, obovoid, rounded at apex. *Flowers* 6–10. Visited by bumble-bees. $2n = 34$.

Introduced. Waste places. A rather rare casual in central and south Great Britain; formerly a common weed of arable land through much of the area. Most of Europe except the north, but doubtfully native in much of the centre and west; west Siberia; Orient.

9. S. arvensis (L.) L. Field Woundwort
Glechoma arvensis L.

Annual monoecious *herb* with fibrous roots. *Stems* 10–40 cm, pale green, ascending or erect, slender, with bulbous-based eglandular hairs, usually branched at the base, leafy. *Leaves* opposite; lamina 1–4 × 0.8–3.0 cm, dull medium green on upper surface, paler beneath, ovate, sometimes broadly so, obtuse at apex, crenate-serrate, truncate or cordate at base, with eglandular hairs on both surfaces; petiolate, except the uppermost, the lower with them up to 3 cm, hairy. *Inflorescence* a lax, terminal spike, the whorls with 4–6 bisexual flowers, the upper crowded, the lower distant; bracts resembling the leaves but smaller and sessile above; bracteoles very small and linear. *Calyx* 5–7(–8) mm, often purplish, 5-lobed, the lobes about as long as the tube, lanceolate and sharply pointed at apex, with dense eglandular hairs. *Corolla* 6–8 mm, white, pale pink or purple, 2-lipped, upper lip entire, lower lip 3-lobed, the central lobe much larger than the lateral ones and emarginate. *Stamens* 2 long and 2 short, included. *Style* 1, shortly branched. *Nutlets* 4, 1.5–1.7 mm, ovoid, rounded at apex. *Flowers* 4–11. Insect visits few; self-pollinated. $2n = 10$.

Introduced. Arable land and gardens on non-calcareous soils up to 380 m. Scattered throughout much of Great Britain and Ireland. South, west and central Europe; North Africa; Palestine; naturalised in North America. Suboceanic Southern-temperate element.

3. Ballota L.

Perennial monoecious *herbs* with shortly creeping, rhizomatous stock. *Stems* quadrangular. *Leaves* opposite, foetid when bruised, crenate or crenate-serrate. *Inflorescence* of many-flowered whorls; flowers bisexual. *Calyx* tubular-infundibuliform, with 5, more or less equal, acuminate lobes. *Corolla* purple, pink or rarely white, 2-lipped, upper lip hooded, lower lip 3-lobed. *Stamens* 4, didynamous, shorter than upper lip of corolla. *Style* 1; stigmatic lobes subequal. *Fruit* of 4, oblong-obovate nutlets.

Contains about 35 species, mainly Mediterranean, and one in South Africa.

Davis, P. H. (1952). Additamenta ad floram Anatoliae III. *Notes Roy. Bot. Garden Edinb.* **21**: 61–65.
Druce, G. C. (1918). *Ballota nigra* var. *mollissima* Druce. *Rep. Bot. Soc. Exch. Club Brit. Isles* **5**: 49.
Hultén, E. & Fries, M. (1986). *Atlas of north European vascular plants north of the Tropic of Cancer.* 3 vols. Königstein.

1. Calyx lobes longer than broad **1(a). nigra** subsp. **nigra**
1. Calyx lobes broader than long 2.

2. Stem leaves and calyx with dense, long soft hairs **1(b,iii). nigra** subsp. **foetida** var. **mollissima**
2. Stem leaves and calyx with numerous short hairs 3.
3. Leaves 2–5 × 1–5 cm, serrate-crenate with small, rather pointed teeth **1(b,i). nigra** subsp. **foetida** var. **foetida**
3. Leaves 3–6 × 3–5 cm, crenate, with large, round teeth **1(b,ii). nigra** subsp. **foetida** var. **membranacea**

1. B. nigra L. Black Horehound

Perennial monoecious *herb* with a shortly creeping, rhizomatous rootstock. *Stems* up to 100 cm, pale green, often suffused purplish, quadrangular, erect or decumbent, with stiff, retrorse, eglandular hairs and minute glandular hairs, simple or branched, leafy. *Leaves* opposite, foetid when bruised; lamina 2.5–8.0 × 1.5–7.0 cm, dull medium to dark green on upper surface, paler beneath, ovate, obtuse to acute at apex, bluntly to sharply, regularly or irregularly crenate-serrate or crenate, sometimes obscurely lobed, truncate, broadly cuneate or shallowly cordate at base, thinly to densely eglandular hairy on both surfaces and with minute glandular hairs; petiole 10–30 mm, channelled above, with eglandular and minute glandular hairs. *Inflorescence* branched or unbranched, pyramidal or spike-like, usually rather lax, with up to 12 remote, lax or crowded whorls; flowers bisexual; bracts leafy, closely resembling the leaves, but becoming smaller upwards; bracteoles 5–9 mm, subulate to filiform, hairy; flowers more or less sessile in dense, shortly pedunculate cluster of 5–8 or more. *Calyx* tubular-infundibuliform, with 10 veins, with eglandular hairs and scattered, minute glands; tube 5–11 mm, with 5 more or less equal, erect or spreading, shortly spiny-acuminate or aristate-acuminate lobes. *Corolla* 12–18 mm, purple, pink or rarely white; tube 8–11 mm, equalling or exceeding calyx, with an oblique bearded zone internally, 2-lipped, the upper lip about 5 × 3 mm, hooded, with eglandular and minute glandular hairs externally, the lower lip 3-lobed, the lateral lobes oblong and obtuse, the median lobe subrotund and emarginate. *Stamens* 4, didynamous, inserted 1–2 mm down the corolla tube; filaments 4–6 mm, thinly hairy; anthers about 1.5 mm wide. *Style* 1, 11–12 mm, glabrous; stigmatic lobes subequal. *Nutlets* 4, 2.0–2.3 × 1.0–1.5 mm, shining dark brown, oblong-obovate, rounded at apex, smooth or minutely granulose. *Flowers* 6–10. Pollinated mainly by bees; protandrous. $2n = 22$.

(a) Subsp. **nigra**
B. ruderalis Sw.; *B. vulgaris* Hoffmanns. & Link; *B. nigra* subsp. *ruderalis* (Sw.) Briq.; *B. foetida* Lam. nom. illegit.

Calyx lobes longer than broad, gradually tapered at apex.

(b) Subsp. **foetida** (Vis.) Hayek
B. alba L.; *B. borealis* Schweiger; *B. nigra* subsp. *meridionalis* auct.

Calyx lobes broader than long, abruptly acuminate at apex.

(i) Var. **foetida** Vis.
Stem with numerous, short hairs. *Leaves* 2–5 × 1–5 cm, teeth more pointed than in the other 2 varieties with numerous, short hairs on both surfaces. *Calyx* with numerous hairs on the veins.

(ii) Var. **membranacea** Druce
Stem with numerous, short hairs. *Leaves* 3–6 × 3–5 cm, teeth mostly large and round, with numerous, short hairs on both surfaces. *Calyx* with numerous hairs on the veins.

(iii) Var. **mollissima** Druce
Stem densely and softly hairy. *Leaves* 3–7 × 1.5–8.0 cm, teeth large and round, densely and softly long-hairy on both surfaces. *Calyx* densely long-hairy.

Native. Hedgerows, waysides, waste land, gardens and fields. Common in most of England, Wales and the Channel Islands, very local in Scotland and Ireland and mostly introduced. Europe, eastwards to Iran, Mediterranean region; Atlantic Islands. Subsp. *nigra* is a plant of northern Europe which occurs in Great Britain only as a casual, mainly near ports. Our common and probably native plant is subsp. *foetida* which is widespread in western and central Europe, becoming rare in the Mediterranean region. The distribution of the three varieties is unknown. They seem to be able to reproduce themselves from seed even when growing in the same area. We once saw all three scattered over a Cambridgeshire field of potatoes in which there seemed to be no intermediates. Normally they occur in uniform colonies and var. *foetida* is the most frequent. Other subspecies and varieties occur in the Mediterranean area and eastern Europe.

4. Leonurus L.

Perennial monoecious, foetid *herbs* with rhizomes. *Leaves* opposite, at least some palmately or ternately lobed. *Inflorescence* of distant whorls in leaf axils; flowers bisexual. *Calyx* with 5 more or less equal lobes, the lobes triangular patent and spine-tipped. *Corolla* pinkish-purple, 2-lipped, the upper lip hooded, the lower lip 3-lobed. *Stamens* 4, shorter than upper lip of corolla. *Style* 1. *Fruit* of 4, trigonous nutlets, truncate at apex.

Contains about 14 species in Europe and temperate Asia.

Hultén, E. & Fries, M. (1986). *Atlas of north European vascular plants north of the Tropic of Cancer.* 3. vols. Königstein.
Mabey, R. (1996). *Flora Britannica*. London.

1. L. cardiaca L. Motherwort

Perennial monoecious, foetid *herb* with fibrous roots and a rhizome. *Stems* up to 120 cm, pale green often suffused brownish-purple, erect, quadrangular, with short, reflexed hairs, mostly unbranched, leafy. *Leaves* forming a basal rosette, the cauline opposite, medium yellowish-green on upper surface, paler beneath; basal 3.0–3.5 × 3.0–3.5 cm, subrotund, rounded at apex, incise-dentate to palmately lobed, cordate at base; cauline regularly spaced, 2–7 × 1.5–5.0 cm, narrowly to broadly ovate, obtuse to acute at apex, obtuse-dentate to deeply palmately or ternately lobed, rounded at apex; all densely and softly short hairy on both surfaces, especially beneath; petioles 2–4 cm, pale green, densely hairy. *Inflorescence* of distant whorls in leaf axils; flowers bisexual. *Calyx* 2–3 mm, with 5 more or less equal lobes, the lobes triangular and spine-tipped. *Corolla* 8–12 mm, pinkish-purple, 2-lipped, the upper lip hooded and hairy on back, the lower lip 3-lobed, the middle lobe very narrow, the outer lobes incurved. *Stamens* 4, shorter than upper lip of corolla; filaments white; anthers pink. *Style* 1; stigmas usually 2. *Nutlets* 4, 1.5–2.0 mm, trigonous. *Flowers* 7–9. Pollinated by bees. $2n=18$.

Introduced. Formerly grown medicinally and widely naturalised in waste places and waysides. Now thinly scattered records over much of Great Britain and rare in Ireland. Most of Europe, but absent from the extreme north, the islands and much of the Mediterranean region. It is said to ease childbirth, hence its vernacular name.

5. **Lamiastrum** Heist. ex Fabr.

Galeobdolon Adans.

Perennial monoecious *herbs* with long, leafy stolons. *Leaves* opposite. *Inflorescence* of dense, axillary whorls of bisexual flowers; bracts similar to leaves. *Calyx* with 5 nearly equal lobes. *Corolla* yellow, 2-lipped, the upper lip laterally compressed and helmet-shaped, the lower lip 3-lobed, the middle slightly larger than the lateral, the tube longer than the calyx, straight and dilated above, with a ring of hairs within. *Stamens* 4, shorter than the corolla; anthers glabrous. *Style* 1, curved; stigma bifid. *Fruit* of 4, obovoid-quadrangular nutlets.

A single species.

Grime, J. P. et al. (1988). *Comparative plant ecology*. London.
Hultén, E. & Fries, M. (1986). *Atlas of north European vascular plants north of the Tropic of Cancer.* 3 vols. Königstein.
Packham, J. R. (1978). Biological flora of the British Isles. No. 155. *Lamiastrum galeobdolon* (L.) Ehrend. & Polatschek. *Jour. Ecol.* **71**: 975–997.
Rich, T. C. G. & Jermy, A. C. (1998). *Plant crib*. London.
Rutherford, A. & Stirling, A. McG. (1987). Variegated Archangels. *B.S.B.I. News* **46**: 9–11.
Smejkal, M. (1975). *Galeobdolon argentatum* sp. nov. ein neuer Vertreter de Kollektivort *Galeobdolon luteum* (Lamiaceae). *Preslia* **47**: 241–248.
Wegmuller, S. (1971). A cytotaxonomic study of *Lamiastrum galeobdolon* (L.) Ehrend. & Polatschek in Britain. *Watsonia* **8**: 277–288.

1. L. galeobdolon (L.) Ehrend & Polatschek
Yellow Archangel

Perennial monoecious *herb* with long leafy stolons sometimes not produced until after flowering. *Stems* 20–60 cm, pale green, quadrangular, erect, slightly and stiffly eglandular hairy and with minute glandular hairs. *Leaves* opposite; lamina 4–7 × 1.5–4.0 cm, pale to medium green, sometimes with large, conspicuous white blotches on upper surface, paler beneath, ovate, acute or acuminate at apex, irregularly crenate-serrate or serrate, truncate or rounded at base, with short eglandular and minute glandular hairs on both surfaces; petiolate. *Inflorescence* of dense, axillary whorls of bisexual flowers; bracts similar to leaves. *Calyx* 10–14 mm, tubular-campanulate, with 5 nearly equal lobes, the lobes triangular-ovate and with a cuspidate apex, with short eglandular and minute glandular hairs. *Corolla* 18–20 mm, yellow, the lower lip with reddish-brown markings, 2-lipped, the upper lip laterally compressed and helmet-shaped, the lower lip 3-lobed, the middle lobe slightly longer than the lateral, the tube longer than the calyx, straight and dilated above, with a ring of hairs within. *Stamens* 4, shorter than the corolla; filaments hairy; anthers glabrous. *Style* 1, gently curved; stigmas bifid. *Nutlets* 4, 3.8–4.0 mm, obovoid-quadrangular. *Flowers* 4–6. Pollinated by bees. Homogamous.

(a) Subsp. **galeobdolon**
Galeopsis galeobdolon L.; *Lamium galeobdolon* (L.) L.; *Galeobdolon luteum* Huds.

Flowering stems with hairs more or less confined to the 4 angles, less robust than subsp. *montanum*. *Leaves* rarely with white blotches. *Bracts* 1–2(–2.2) times as long as wide, obtusely serrate. *Flowers* less than 8(–9) per node at less than 4–5 nodes. *Corolla* with upper lip less than 8.0(–8.5) mm wide. *Fruiting calyx* less than 12.0(–12.5) mm. $2n=18$.

(b) Subsp. **montanum** (Pers.) Ehrend & Polatschek
Lamium galeobdolon subsp. *montanum* (Pers.) Hayek; *Galeobdolon luteum* subsp. *montanum* (Pers.) Dvoráková; *Pollichia montanum* Pers.

Flowering stem with hairs on faces as well as angles, more robust than subsp. *galeobdolon*. *Leaves* ovate, sharply serrate, marked with flecks or stippling between the veins. *Bracts* (1.5–)1.7–3.6 times as long as wide, acutely serrate. *Flowers* more than (7–)10 per node at more than (3–)4 nodes. *Corolla* with upper lip less than 8.0(–8.5) mm wide. *Fruiting calyx* less than 12.0(–12.5) mm. $2n=36$.

(c) Subsp. **argentatum** (Smejkal) Stace
Galeobdolon argentatum Smejkal; *Lamium galeobdolon* forma *argentatum* (Smejkal) Mennema; *Lamium galeobdolon* subsp. *argentatum* (Smejkal) Duvign.

Flowering stem with hairs on faces as well as angles. *Leaves* broadly ovate, crenate, with large conspicuous white blotches on all leaves at all seasons. *Bracts* 1.5–3.6 times as long as wide, crenate. *Flowers* more than 10 per node at 4 nodes. *Corolla* with upper lip more than (7.5–)8 mm wide. *Fruiting calyx* more than (11.5–)12.0 mm. $2n=36$.

Native. Woods, wood borders, hedgerows, shrubberies and waysides. Common in most of England and Wales, very local in south-west Scotland, eastern Ireland and the Channel Islands, introduced in Scotland and north-west Ireland. Most of Europe, but rare in the Mediterranean region and the north; Iran. European Temperate element. Subsp. *montanum* is the common native plant. Subsp. *galeobdolon* occurs in a very small area in north Lincolnshire and one locality in Kircudbrightshire. Subsp. *argentatum* is introduced, being much grown in gardens and naturalised in shrubberies and waysides. There are scattered records over Great Britain and Ireland. Its origin is uncertain, but is perhaps only a cultivar of subsp. *montanum*.

6. **Lamium** L.

Annual or *perennial* monoecious *herbs*. *Leaves* opposite, serrate to incise-serrate, rarely more or less entire. *Inflorescence* of whorls in the upper part of the stem. *Flowers* bisexual. *Calyx* of 5 more or less equal, narrowly triangular, acuminate lobes. *Corolla* white, pink, purple or mauve, with hooded upper lip and more or less

1-lobed lower lip with much reduced lateral lobes which are pointed or rounded. *Stamens* 4, shorter than upper lip of corolla. *Style* 1; stigmas 2. *Fruit* of 4 trigonous nutlets truncate at apex.

Contains about 40 species in Europe, temperate Asia and North Africa; introduced elsewhere.

Bernström, P. (1955). Cytogenetic studies on relationships between annual species of *Lamium*. *Heredity* **41**: 1–121.
Druce, G. C. (1924). *Lamium molucellifolium* × *purpureum*, nova hybr. *Rep. Bot. Soc. Exch. Club Brit. Isles* **7**: 53.
Druce, G. C. (1929). Notes on the second edition of the British Plant List. *Rep. Bot. Soc. Exch. Club Brit. Isles* **8**: 867–877.
Grime, J. P. et al. (1988). *Comparative plant ecology*. London. [*L. album* and *purpureum*.]
Hultén, E. & Fries, M. (1986). *Atlas of north European vascular plants north of the Tropic of Cancer*. 3 vols. Königstein.
Mennema, J. (1989). *A taxonomic revision of* Lamium (*Lamiaceae*). Leiden.

1. Upper leaves amplexicaul; bracts absent **1. amplexicaule**
1. Upper leaves usually sessile, but not amplexicaul; bracts present 2.
2. Corolla tube abruptly or gradually dilated 3.
2. Corolla tube straight 4.
3. Corolla white; pollen pale yellow **2. album** subsp. **album**
3. Corolla purple; pollen orange **3. maculatum** subsp. **maculatum**
4. Calyx tube shorter than its teeth; corolla 15–25 mm **4. confertum**
4. Calyx tube as long as or longer than the teeth; corolla 5–17(–20) mm 5.
5. Leaves regularly and mostly faintly crenate **5(iii). purpureum** var. **purpureum**
5. Leaves irregularly and usually deeply incised-crenate, especially the upper ones 6.
6. Upper leaves rhomboid; whorls of flowers remote, especially the lower ones **5(ii). purpureum** var. **molucellifolium**
6. Upper leaves obovate or ovate-cordate; whorls of flowers crowded 7.
7. Stem usually glabrous; corolla (10–)15(–20) mm, tube conspicuously exserted from the calyx **5(i). purpureum** var. **incisum**
7. Stem often hairy; corolla 7–15 mm, tube not exserted from the calyx **6. hybridum**

Section **1. Amplexicaule** Mennema

Upper leaves amplexicaul. *Bracts* absent. *Corolla* tube straight.

1. L. amplexicaule L. Henbit Dead-nettle

L. mesogaeum Heldr. ex Boiss.; *Lamiopsis amplexicaulis* (L.) Opiz; *L. rumelicum* Velen.; *Pollichia amplexicaulis* (L.) Willd.; *Galeobdolon amplexicaule* (L.) Moench; *Lamiella amplexicaulis* (L.) Fourr.

Annual monoecious, foetid *herb* with fibrous roots. *Stems* (5–)10–30(–40) cm, pale green, decumbent to erect, quadrangular, glabrous below to appressed-eglandular hairy above and with minute glandular hairs, branched especially at the base, leafy. *Leaves* opposite; lamina (0.5–)1–2(–3) × 1–3 cm, dull, slightly bluish-green on upper surface, paler beneath, reniform or rhomboid, rounded at apex, coarsely and irregularly crenate, sometimes deeply divided, rounded or cordate at base, sparsely to moderately appressed-eglandular hairs on both sides and with minute glandular hairs; the lower leaves with petioles about as long as the lamina or longer and hairy, the upper leaves sessile and amplexicaul. *Inflorescence* of distant whorls of (6–)8–16(–20) bisexual flowers; bracts not present, the adjacent foliage to the whorls of flowers leaf-like. *Calyx* 5–7 mm, longer after anthesis, with 5 subequal lobes, the lobes narrowly triangular and narrowly and sharply pointed at apex, moderately to densely eglandular-hairy, some of the hairs glandular. *Corolla* (10.2–)10.5–20.0(–25.0) mm, pale rose to deep pink with darker spots, tube abruptly dilated, exserted from the calyx, glabrous at base, hairy above, no internal ring of hairs, 2-lipped, the upper lip arched, rounded at apex, entire to faintly undulate and hairy, the lower lip about as long as the upper or longer, glabrous, the lateral lobes rounded, the medium lobe bipartite with an undulate margin. *Stamens* 4; filaments without glandular hairs; anthers up to 1 mm. *Style* 1, 2–3 mm, shorter than the corolla, glabrous. *Nutlets* 4, 2–3(–3.2) mm. *Flowers* 4–8. Insect visitors rare, mainly self-pollinated. $2n = 18$.

Introduced. Weed on open, cultivated and waste ground. Over most of Great Britain and Ireland, common in eastern Great Britain and the Channel Islands, scattered localities in the west of Great Britain and Ireland. All Europe except the extreme north, where it is probably introduced, extending to the Azores, Canaries, Madeira; North Africa; extending through southern Asia to Japan. Eurosiberian Southern-temperate distribution, but widely naturalised so it is now Circumpolar Southern temperate. Our plant is subsp. **amplexicaule** which occurs throughout much of the range of the species.

Section **2. Lamiotypus** Dumort.

Upper leaves sessile or not, or amplexicaul. *Bracts* present. *Corolla* tube sigmoid, abruptly dilated or ventrally saccate.

2. L. album L. White Dead-nettle

L. vulgatum var. *album* (L.) Benth.; *L. capitatum* Sm.

Perennial monoecious, foetid *herb* with creeping, woody rhizomes and numerous, whitish, fibrous roots, often forming large patches. *Stems* (10–)20–60(–95) cm, pale yellowish-green, sometimes suffused brownish-purple, erect, quadrangular, striate, with medium, pale, spreading or deflexed simple eglandular hairs particularly on the angles and minute glandular hairs, simple, leafy. *Leaves* opposite and widely spaced; lamina 3–10 × 1–6 cm, dull, medium yellowish-green on upper surface, slightly paler beneath, soft, narrowly to broadly ovate, narrowed to apex, the lower obtuse, the upper acute, doubly serrate or crenate-serrate, more or less cordate at base, with numerous to dense, short to medium, pale simple eglandular hairs and minute glandular hairs on both surfaces; petioles 10–80 mm, usually pale yellowish-green, sometimes suffused brownish-purple, with short to long, few to numerous, pale, spreading or deflexed hairs and minute glandular hairs. *Inflorescence* of distant whorls in the upper part of the stem; flowers bisexual; bracts large and similar to upper cauline leaves, but always acute at apex and sessile or shortly petiolate. *Calyx* 9–20 mm, pale green and brownish-purple towards base, tubular-campanulate, the 5 lobes lanceolate, with a very fine drawn-out apex as long as or longer than the tube, with short to medium, pale, spreading simple eglandular hairs and minute glandular ones. *Corolla* 20–25(–39) mm, white, cylindrical at the base for a short distance, then suddenly enlarged and gibbous, the upper part curved, the lateral lobes of the lower lip with 2–3 small teeth, the inner surface near the base with an oblique ring of hairs, the upper lip long-ciliate. *Stamens* 4; filaments white, with short glandular hairs; anthers purplish-black, hairy. *Style* 1, 3–4 mm, simple below, branched into 2 above, white. *Nutlets* 4, 3.0–3.2 mm, brown, trigonous, truncate at apex. *Flowers* 5–12. Pollinated by long-tongued Hymenoptera, mainly bumble-bees; homogamous. $2n=18$.

Introduced. Hedgebanks, waysides and arable and waste land, more frequently growing in grassy places than the other *Lamium* species. Common in most of lowland Great Britain except north and west Scotland, rare in the Channel Islands, local in Ireland and mainly in the east. Most of Europe, but rare in the south and absent from many islands; Himalaya and Japan; only in the mountains in the southern part of its range; introduced in Iceland, North America and New Zealand. Our plant is subsp. **album** which occurs throughout much of the range of the species.

3. L. maculatum (L.) L. Spotted Dead-nettle

L. album var. *maculatum* L.; *L. laevigatum* L.; *L. maculatum* var. *laevigatum* (L.) Mutel; *L. melissifolium* Mill.; *L. foliosum* Crantz; *L. maculatum* (Crantz) Beck; *L. maculatum* subsp. *foliosum* (Crantz) Á. & D. Löve; *L. rubrum* Jenk.; *L. rugosum* Aiton; *L. maculatum* var. *rugosum* (Aiton) Opiz; *L. hirsutum* Lam.; *L. maculatum* var. *hirsutum* (Lam.) Mutel; *L. columnae* Ten.; *L. affine* Guss. ex Ten.; *L. tillii* Ten.; *L. grenieri* Mutel; *L. niveum* Schrad.; *L. gundelsheimeri* K. Koch; *L. truncatum* Boiss.; *L. cupreum* Schott, Nyman & Kotschy; *L. maculatum* var. *cupreum* (Schott, Nyman & Kotschy) Briq.; *L. maculatum* subsp. *cupreum* (Schott, Nyman & Kotschy) Hadač ex Sychowa; *L. elegantissima* Schur; *L. tomentosum* Benth.; *L. urticifolium* Schur

Perennial monoecious, foetid herb with numerous rhizomes forming patches. *Stems* numerous, 15–50(–80) cm, pale yellowish-green, often brownish-purple in lower half, quadrangular in section, rather weak, with numerous, medium to long, pale, spreading or down-turned simple eglandular hairs and minute glandular hairs, not branched, leafy. *Leaves* opposite, all cauline; lamina 1–8 × 1–7 cm, dull medium green with a wide pale green area either side of the midrib on the upper surface making it blotched and often splashed or flecked reddish-purple later, the network of veins impressed, pale yellowish-green beneath with the network of veins prominent, ovate or triangular-ovate, acute or acuminate at apex, but the terminal tooth often blunt, coarsely and irregularly crenate-dentate, cordate at base, with numerous, short to medium, soft, pale simple eglandular hairs on both surfaces and the margins and minute glandular hairs; petioles 20–60 mm, pale yellowish-green, square in section, with numerous, short to medium, pale simple eglandular hairs. *Inflorescence* of 2–5 whorls near the top of the stem and close together; flowers bisexual; bracts mostly large and like the upper cauline leaves, mostly shortly petiolate. *Calyx* 8–15 mm, pale green, tubular-campanulate, the lobes lanceolate, with a fine drawn-out apex shorter than to longer than the tube, with short to medium, pale, spreading simple eglandular hairs and minute glandular hairs. *Corolla* 20–35 mm, usually pinkish-purple, rarely white or brownish-purple, the tube 10–18 mm, curved, equalling or longer than the calyx and with a transverse ring of hairs within, the upper lip 7–14 mm, ciliate and entire, the lateral lobes subulate, the lower lip 4–6 mm and obcordate, its lateral lobes with a single tooth. *Stamens* 4; filaments white; anthers purple, hairy. *Style* branched above, white. *Nutlets* 4, 2.0–2.5 mm, brown, trigonous, truncate at apex. *Flowers* 3–10. Pollinated by bumble-bees; homogamous. $2n=18$.

Very variable, as can be seen from the long list of synonyms, but difficult to recognise anything ecological or geographical. Variation is also found in gardens.

Introduced. Commonly cultivated in gardens and naturalised on waste ground and tips. Scattered records through much of Great Britain and in the Channel Islands. Widespread in Continental Europe but absent from most of the islands; Turkey, Syria and Lebanon; introduced in North America and New Zealand.

Section **3. Lamium**

Section *Lamiopsis* Dumort.; Section *Pollichia* (Willd.) Briq. nom. illegit.

Upper leaves sessile, not amplexicaul. *Bracts* present. *Corolla* tube straight.

4. L. confertum Fr. Northern Dead-nettle

L. intermedium Fr. nom. illegit.; *L. purpureum* var. *intermedium* Wahlenb.; *L. amplexicaule* var. *intermedium* Mutel; *L. hybridum* subsp. *intermedium* Gams; *L. incisum* subsp. *intermedium* Gams ex Garcke; *L. purpureoamplexicaule* Meyer nom. illegit.; *L. molucellifolium* Fr., non. Schum.; *L. amplexicaule* var. *molucellifolium* Fiori

Annual monoecious, foetid herb with fibrous roots. *Stems* (15–)20–40(–70) cm, pale green, erect or decumbent, more or less robust, quadrangular, with few simple eglandular and minute glandular hairs, simple or branched, leafy. *Leaves* opposite; lamina (1.0–)2.0–3.5(–6.5) cm, medium yellowish-green on upper surface, paler beneath, reniform or ovate, sometimes rhomboid, rounded at apex, deeply and coarsely crenate or double crenate, rarely irregularly incised, cordate at base, eglandular-hairy on both surfaces and with minute glandular hairs; petiole (5–)10–25(–40) mm, with few simple eglandular and minute glandular hairs, the upper leaves sessile. *Inflorescence* of 3–4(–5) crowded whorls above and more remote ones below; bracts (2–)3–4(–6) mm, like leaves, with stiff eglandular and short glandular hairs. *Calyx* (7.5–)10(–12) mm, much longer after anthesis, many veined, the main veins often purple, sparsely appressed-hairy, with stiff eglandular and minute glandular hairs; lobes 5, always longer than the tube, narrowly triangular. *Corolla* (15–)20(–25) mm, reddish-purple; tube exserted from the calyx, abruptly dilated, without an internal ring of hairs, 2-lipped, the upper lip arched, rounded at apex, the margin entire or faintly undulate, hairy, the lower lip equalling the upper or longer, the lateral lobes rounded, the median lobe bipartite with an undulate margin. *Stamens* 4; filaments without glandular hairs; anthers dark. *Style* 1, shorter than the corolla, glabrous. *Nutlets* 4, (2.5–) 2.2 -3.0 mm. *Flowers* 5–9. $2n=36$.

Introduced. Cultivated and waste ground. Locally frequent near the coast in north, west and east Scotland and Isle of Man; rare in north-east England and few records further south, scattered records in Ireland. Northern Europe to about 67° N in Fennoscandia and southwards to 52° 30′ in Germany.

5. L. purpureum L. Red Dead-nettle

L. foetidum Garsault nom. illegit.; *L. nudum* Moench nom. illegit.

Annual, slightly foetid, monoecious *herb* with fibrous roots. *Stems* (5–)10–30(–55) cm, pale green, sometimes flushed red, erect or decumbent, quadrangular, with more or less numerous eglandular and minute glandular hairs, simple or branched, leafy. *Leaves* opposite; lamina 0.5–3.5(–6.0) × 0.5–3.0 (–4.0) cm, pale to medium green on upper surface, paler beneath, ovate, rhomboid, reniform or obovate, rounded at apex, faintly crenate to deeply and irregularly incised, cordate at base, with eglandular hairs on both surfaces and minute glandular ones; petiole of lower usually longer than the lamina, the upper sessile. *Inflorescence* of crowded or remote whorls of (4–)12–18 (–20) bisexual flowers; bracts hardly distinguished from the leaves. *Calyx* 6–7 mm, pale green, often flushed purplish, the 5 lobes subequal, narrowly triangular, acute and narrow at apex, nearly glabrous at base, densely eglandular hairy at apex and with minute, glandular hairs. *Corolla* (5–)7–17(–20) mm, bright reddish-violet to pale pinkish mauve, the lower lip spotted with rose-purple, the tube abruptly dilated, not always exserted from the calyx, usually with a perpendicular ring of hairs dorsally inside, 1–2 mm from the base; 2-lipped, the upper lip arched, rounded at apex, entire to faintly undulate and hairy, the lower lip about as long as the upper lip and glabrous, with the lateral lobes rounded and the median lobe bipartite with an undulate margin. *Stamens* 4; filaments without glandular hairs; anthers red. *Style* 1, about as long as the corolla, glabrous. *Nutlets* 4, 2–3(–3.2) mm. *Flowers* 3–10. Pollinated by bees or rarely other insects, or selfed. Homogamous. $2n=18$.

(i) Var. **incisum** (Willd.) Pers.

L. incisum Willd.; *L. hybridum* subsp. *incisum* (Willd.) Janch.; *L. dissectum* With.; *L. hybridum* var. *dissectum* (With.) Mutel; *L. westphalicum* Weihe

Stem (10–)20–30(–55) cm. *Leaves* with lamina ovate or obovate, deeply and irregularly incised, cordate at base, hairy on both surfaces. *Inflorescence* of crowded whorls. *Calyx* (5–)7(–10) mm, lobes as long as the tube. *Corolla* (10–)15(–20) mm, the tube exserted from the calyx, internal ring of hairs present or absent.

(ii) Var. **molucellifolium** Schum.

L. molucellifolium (Schum.) Fr.

Stem 10–20(–35) cm. *Leaves* with lamina reniform or rhomboid, coarsely crenate with the upper deeply and irregularly incised, hairy on both sides. *Inflorescence* of remote whorls, especially the lower. *Calyx* (5–)7(–10) mm, lobes mostly shorter than the tube. *Corolla* (10–)12–17(–20) mm, the tube exserted from the calyx, internal ring of hairs usually present.

(iii) Var. **purpureum**

L. purpureum var. *integrum* Gray; *L. purpureum* var. *albiflorum* Dumort.; *L. purpureum* var. *lumbii* Druce; *L.* × *boreale* Druce

Stem (5–)15–30(–40) cm. *Leaves* with lamina reniform, ovate or obovate, crenate, hairy. *Inflorescence* of crowded whorls. *Calyx* 5–7 mm, lobes usually as long as the tube, internal ring of hairs always present.

Introduced weed of cultivated and waste ground. More or less common throughout Great Britain and Ireland. Most of Europe; western Asia; North Africa; introduced in New Zealand and North and South America. Var. *purpureum* occurs throughout the range of the species as does var. *incisum*. Var. *molucellifolium* occurs in southern England and central and south Scandinavia, Netherlands, Belgium, France, north and west Germany, Switzerland and Baltic Russia. Some biotypes of var. *purpureum* are apparently obligatory winter annual, other biotypes are summer or probably facultative winter annuals. Var. *molucellifolium* is homogeneous. Var. *purpureum* and var. *incisum* both rather variable.

6. L. hybridum Vill. Cut-leaved Dead-nettle

L. incisum auct.

Annual monoecious *herb* with fibrous roots, often growing in large patches. *Stems* 10–45 cm, pale yellowish-green, rarely tinted brownish-purple, erect or ascending, quadrangular, striate, nearly glabrous to rather sparingly hairy, with spreading or deflexed, medium, pale simple eglandular hairs, particularly on the angle and minute glandular hairs, simple to much branched at base, sparsely leafy. *Leaves* opposite and widely spaced; lamina 1–5 × 1–5 cm, dull medium green on upper surface, paler beneath, broadly ovate or reniform, obtuse at apex, deeply incise-crenate-dentate, the teeth or lobes rounded-obtuse, broadly cuneate to truncate at base, the upper more or less decurrent down the petiole, with eglandular hairs on both sides and minute glandular hairs, with prominent veins on the lower surface; petioles pale yellowish-green, rounded below, flat above, sparsely hairy. *Inflorescence* of dense whorls in the upper part of the stem; flowers bisexual; bracts large and similar to upper cauline leaves, shortly petiolate. *Calyx* 5–7(–10) mm, yellowish-green with brownish-purple ribs, divided less than halfway into 5, narrowly triangular, acute lobes with a narrowly campanulate base, with short, pale simple eglandular hairs and minute glandular hairs. *Corolla* 7–15 mm, pink, the tube not exceeding the calyx, cylindrical below, enlarged above into a wide throat, without or with a faint ring of hairs towards the base, 2-lipped, the upper lip 3–4 mm, laterally compressed forming a hood, the lower lip 3-lobed, the lateral lobes small. *Stamens* 4; filaments pale; anthers dark. *Style* simple below, branched into 2 above, white. *Nutlets* 4, 1.0–2.5 mm, trigonous, truncate at apex. *Flowers* 3–10. Visited by bees. $2n=36$.

Introduced. Cultivated and waste places. Scattered over most of lowland Great Britain and Ireland. Most of Europe except the south-east; Morocco and Algeria; Canary Islands.

7. Galeopsis L.

Annual monoecious *herbs. Leaves* opposite. *Inflorescence* of terminal and axillary whorls; bracts like the cauline leaves but smaller; flowers bisexual. *Calyx* tubular or campanulate, with 5 subequal, spiny-pointed lobes. *Corolla* with tube longer than the calyx, straight, dilated at the apex and with a ring of hairs within; 2-lipped, the upper lip laterally compressed and helmet-shaped, the lower lip 3-lobed, with 2 conical projections at the base. *Stamens* 4 shorter than the upper lip and the corolla. *Style* 1; stigma bifid. *Fruit* of 4 trigonous nutlets, rounded at apex.

Contains about 10 species in Europe and temperate Asia.

Briquet, J. (1893). *Monographie du genre* Galeopsis. Brussels.

Grime, J. P. et al. (1988). *Comparative plant ecology*. London.

Hultén, E. & Fries, M. (1986). *Atlas of north European vascular plants north of the Tropic of Cancer*. 3 vols. Königstein.

Müntzing, A. (1930). Outlines to a genetic monograph of the genus *Galeopsis* with special reference to the nature of inheritance or partial sterility. *Hereditas* **13**: 185–341.

O'Donovan, J. T. & Sharma, M. P. (1987). The biology of Canadian weeds. 78. *Galeopsis tetrahit* L. *Canad. Jour. Pl. Sci.* **67**: 787–796.

Rich, T. C. G. & Jermy, A. C. (1998). *Plant crib*. London.

Stewart, A., Pearman, D. A. & Preston, C. D. (1994). *Scarce plants in Britain*. Peterborough.

Townsend, C. C. (1962). Some notes on *Galeopsis ladanum* L. and *G. angustifolium* Ehrh. ex Hoffm. *Watsonia* **5**: 143–149.

1. Stems with soft hairs, not swollen at the nodes — 2.
1. Stems with stiff hairs, swollen at the nodes — 5.
2. Corolla (20–)25–30(–35) mm, predominantly pale yellow — **1. segetum**
2. Corolla 15–25 mm, magenta-purple or rarely white — 3.
3. Hairs of calyx translucent, microscopically smooth to finely dotted — **2. ladanum**
3. Hairs of calyx opaque, microscopically densely papillose — 4.
4. Calyx with appressed eglandular hairs, glandular hairs sparse of absent — **3(i). angustifolium** var. **angustifolium**
4. Calyx densely villous, with spreading, long, white eglandular hairs, usually mixed with numerous glandular hairs — **3(ii). angustifolium** var. **calcarea**
5. Corolla (22–)27–35 mm, yellow, with a purple patch on lower lip — **4. speciosa**
5. Corolla 13–20(–25) mm, purplish-red — 6.
6. Lower lip of corolla with markings falling well short of the tip of the middle lobe which is entire — **5. tetrahit**
6. Lower lip of corolla with markings almost reaching the tip of the middle lobe and sometimes almost covering it, the middle lobe clearly emarginate — **6. bifida**

Subgenus **1. Ladanum**

Stem softly hairy to nearly glabrous, not swollen at the nodes.

1. G. segetum Neck. — Downy Hemp-nettle

G. dubia Leers; *G. ochroleuca* Lam.

Annual monoecious *herb* with fibrous roots. *Stems* 10–50 cm, pale green, erect, quadrangular, softly eglandular hairy with some minute glandular hairs, branched, leafy. *Leaves* opposite; lamina 1.5–8.0 × 1.0–3.5 cm, pale to medium green on upper surface, paler beneath, ovate-oblong or ovate-lanceolate, more or less acute at apex, crenate-serrate or serrate, cuneate or rounded at base, silky eglandular hairy, especially beneath; with some uncinate glandular hairs; petioles short and hairy. *Inflorescence* of terminal and axillary whorls; bracts like cauline leaves, but smaller; flowers bisexual. *Calyx* 7.5–8.0 mm, elongating to 12–13 mm in fruit, tubular, with 5 subequal lobes, the lobes triangular, with a spiny tip with silky eglandular and glandular hairs. *Corolla* (20–)25–30(–35) mm, pale yellow; tube longer than calyx, with a ring of hairs within; 2-lipped, the upper lip laterally compressed and helmet-shaped, the lower lip 3-lobed, the middle lobe largest, the margins of the lobes fringed. *Stamens* 4, shorter than the upper lip of the corolla; filaments pale; anthers yellow. *Style* 1, curved near the apex; stigma shortly 2-lobed. *Nutlets* 4, 2.8–3.0 mm, trigonous. Flowers 7–10. Insect pollinated, but not self-sterile. $2n=16$.

Native. Arable land. Caernarvonshire, where once constant, now sporadic; casual in arable and waste land in other scattered localities in England and Wales. Denmark to France and north and east Spain, eastwards to north-east Italy.

2. G. ladanum L. — Broad-leaved Hemp-nettle

Annual monoecious *herb* with fibrous roots. *Stems* 10–50 (–80) cm, pale green, erect, quadrangular, eglandular hairy to nearly glabrous and with minute glandular hairs, branched, leafy. *Leaves* opposite; lamina 1.5–8.0 × 1–3 cm, pale to medium green on upper surface, paler beneath, ovate, ovate-oblong or ovate-lanceolate, acute at apex, with 3–7 prominent teeth on each side, cuneate at base, usually with numerous appressed eglandular hairs and minute glands on both surfaces; petioles short. *Inflorescence* of terminal and axillary whorls; flowers bisexual; bracts like the cauline leaves but smaller. *Calyx* 8.0–8.5 mm, elongating in fruit, green, tubular, with 5 subequal lobes, the lobes triangular, with a spine-like apex, with translucent, microscopically smooth to finely dotted, transparent eglandular hairs and minute glands. *Corolla* 15–25 mm, magenta-purple; tube scarcely longer than the calyx, with a ring of hairs within; 2-lipped, the upper lip laterally compressed and helmet-shaped, the lower lip 3-lobed, the lobes fringed. *Stamens* 4, shorter than upper lip of corolla; filaments pale; anthers yellow. *Style* 1, curved near the apex; stigma shortly 2-lobed. Nutlets 4, 2.3–2.5 mm, trigonous, rounded at apex. *Flowers* 7–10. Insect- or self-pollinated. $2n=16$.

Introduced. Rare casual of waste or cultivated ground. Has been much confused with *G. angustifolia*. Most of Europe except the islands and the extreme south; Russia in Asia.

3. G. angustifolia Ehrh. ex Hoffm. Red Hemp-nettle
G. ladanum auct.

Annual monoecious *herb* with fibrous roots. *Stems* 10–50 (–80) cm, pale green, erect, quadrangular, nearly glabrous to softly eglandular hairy or canescent and with minute glands, branched, leafy. *Leaves* opposite; lamina 1.5–8.0 × 0.5–1.2 cm, pale to medium green on upper surface, pale green beneath, linear-lanceolate, oblong-lanceolate or broadly lanceolate, more or less acute at apex, with 1–4 small serrations on each side, attenuate at base, usually with numerous appressed eglandular hairs on both surfaces and minute glands; petioles short. *Inflorescence* of terminal and axillary whorls; flowers bisexual; bracts like the cauline leaves, but smaller. *Calyx* 8.0–8.5 mm, elongating in fruit, tubular, with 5 subequal lobes, the lobes triangular with a spine-like apex, sparingly shortly hairy to long-villous, the hairs microscopically densely papillose and not opaque, mostly eglandular, but few to numerous can be glandular. *Corolla* 15–25 mm, magenta purple, marked with bright yellow on the lower lip, and the lower part of the tube cream; tube usually much longer than the calyx, with a ring of hairs within; 2-lipped, the upper lip laterally compressed and helmet-shaped, the lower lip 3-lobed, the lobes fringed. *Stamens* 4, shorter than upper lip of corolla; filaments pale; anthers yellow. *Style* 1, curved near the apex; stigma shortly 2-lobed. *Nutlets* 4, 2.3–2.5 mm, trigonous, rounded at apex. *Flowers* 7–10. Insect- or self-pollinated. $2n = 16$.

(i) Var. **angustifolia**
G. angustifolia var. *campestris* Timb.-Lagr.;
G. angustifolia var. *inermis* (Posp.) Fiori; *G. angustfolia* subvar. *longidentata* Henrard; *G. angustifolia* var. *kerneri* Briq.; *G. angustifolia* subvar. *microda* Henrard;
G. angustifolia var. *odontata* Briq.; *G. angustifolia* var. *oreophila* Briq.; *G. angustifolia* var. *spinosa* Benth.

Plant short to tall, branched above or below, nearly glabrous to shortly hairy. *Calyx* with appressed eglandular hairs, glandular hairs sparse or absent.

(ii) Var. **calcarea** (Schönh.) C. E. Salmon
G. calcarea Schönh.

Plant short, branched almost from the base, canescent. *Calyx* densely villous, with spreading, long, white eglandular hairs, usually mixed with numerous glandular hairs.

Native. Arable land or waste places and other open ground, mostly on calcareous soils, or on maritime sand or shingle. Scattered through much of England, Wales and eastern Ireland. West, central and south Europe westwards to Poland and Bulgaria. European Temperate element. Var. *calcarea* is probably native on sand and shingle by the sea and also occurs on chalk inland. Var. *angustifolia* could be derived from it or introduced.

Subgenus **2. Galeopsis**

Stem with stiff hairs, swollen at the nodes.

4. G. speciosa Mill. Large-flowered Hemp-nettle
G. versicolor Curtis

Annual monoecious *herb* with fibrous roots. *Stems* 10–100 cm, pale green, erect to ascending, robust, quadrangular, swollen at the nodes, uniformly covered with stiff simple eglandular hairs, yellow-tipped glandular hairs mainly on the upper half of the internodes, branched, leafy. *Leaves* opposite; lamina 2.5–10.0 × 1.5–3.0 cm, rather pale green on upper surface, paler beneath, ovate to ovate-lanceolate, acute or acuminate at apex, crenate-serrate, rounded at base with dense eglandular hairs on both surfaces; petioles rather short with eglandular hairs. *Inflorescence* of dense axillary and terminal whorls; flowers bisexual; bracts like the cauline leaves but smaller. *Calyx* 12–17 mm, enlarging to 20–22 mm in fruit, tubular, with 5 rather unequal lobes, the lobes linear-lanceolate with spine-like apex, with dense eglandular hairs and minute glands. *Corolla* (22–)27–35 mm, pale yellow, with deeper yellow protuberances and a purple patch on the lower lip or yellow with the lower lip variegated white; tube longer than calyx, straight, dilated above, with a ring of hairs within; 2-lipped, the upper lip laterally compressed and helmet-shaped, the lower lip entire or slightly emarginate, with 2 conical projections at the base. *Stamens* 4, shorter than the upper lip of the corolla; filaments glabrous. *Style* 1, slightly curved above; stigma shortly 2-lobed. *Nutlets* 4, 3.0–3.5 mm, trigonous, rounded at the apex. *Flowers* 7–9. Insect-pollinated, but not self-sterile. $2n = 16$.

Introduced. A weed of arable land, often on peaty soil with root crops, and in waste places. Locally common in central and north Great Britain, the Fenland and north Ireland, rare with scattered records in southern Great Britain and central and southern Ireland. Europe, except many of the islands and much of the south; Siberia (Yenisei region).

5. G. tetrahit L. Common Hemp-nettle

Annual monoecious *herb* with fibrous roots. *Stems* 20–100 cm, pale green, erect, robust, quadrangular, swollen at the nodes, with stiff simple eglandular hairs especially below the nodes and red-tipped glandular hairs in a group below the nodes, branched, leafy. *Leaves* opposite, lamina 2.5–10.0 × 1.5–3.0 cm, rather pale bluish-green on upper surface, paler beneath, ovate to ovate-lanceolate, acute or acuminate at apex, crenate-serrate, cuneate or rounded at base, with more or less numerous, stiff, appressed simple eglandular hairs on both surfaces; petiole short to medium, with eglandular hairs. *Inflorescence* of dense terminal and axillary whorls; flowers bisexual; bracts like cauline leaves but smaller. *Calyx* 12–13 mm, elongating in fruit, tubular with 5 rather unequal lobes, the lobes linear-lanceolate, with spine-like apex, with dense stiff eglandular hairs and minute glands; veins prominent. *Corolla* 13–16 mm, purplish-red, pinkish or rarely white, hairy, the lower lip more or less flushed and marked with purple, the darker markings falling well short of the tip of the terminal lobe; tube usually scarcely longer than the calyx, straight, dilated above, with a ring of hairs within; 2-lipped, the upper lip laterally compressed and helmet-shaped, the lower lip 3-lobed with the middle lobe flat, entire and often nearly as broad as long, with 2 conical projections at the base. *Stamens* 4, shorter than the upper lip of the corolla; filaments glabrous, white; anthers brownish. *Style* 1, slightly curved above; stigma shortly 2-lobed. *Nutlets* 4, 3.8–4.0 mm, dark brown, ellipsoid, rounded at apex. *Flowers* 7–9. Usually self-pollinated. $2n=32$.

Native. Weed of arable land and rough ground, woodland clearings and damp places. Common over most of Great Britain and Ireland, rare in the Channel Islands. Most of Europe, but rare in the south-east; naturalised in North America. European Boreo-temperate element.

6. G. bifida Boenn. Bifid Hemp-nettle
G. tetrahit var. *bifida* (Boenn.) Lej. & Courtois

Annual monoecious *herb* with fibrous roots. *Stems* 20–100 cm, pale green, erect, robust, quadrangular, swollen at the nodes, with stiff simple eglandular hairs especially below the nodes and red-tipped glandular hairs in a group below the nodes, branched, leafy. *Leaves* opposite; lamina 2.5–10.0 × 1.5–3.0 cm, rather pale bluish-green on upper surface, paler beneath, ovate to ovate-lanceolate, acute or acuminate at apex, crenate-serrate, cuneate or rounded at base, more or less eglandular hairy on both surfaces; petiole short to medium, eglandular hairy. *Inflorescence* of dense terminal and axillary whorls; flowers bisexual; bracts like cauline leaves but smaller. *Calyx* 12–13 mm, elongating in fruit, tubular, with 5 rather unequal lobes, the lobes linear-lanceolate, with spine-like apex, densely eglandular hairy. *Corolla* 15–20(–25) mm, purplish-red, pinkish, yellowish or white, the lower lip more or less flushed or extensively marked with purple to near the margin and sometimes the whole terminal lobe dark; tube usually scarcely longer than the calyx, straight, dilated above, with a ring of hairs within; 2-lipped, the upper lip laterally compressed and helmet-shaped, the lower lip 3-lobed, with the middle lobe clearly emarginate and convex with revolute sides and 2 conical projections at the base. *Stamens* 4, shorter than the upper lip of the corolla; filaments glabrous, pale; anthers yellowish. *Style* 1, slightly curved above; stigma shortly 2-lobed. *Nutlets* 4, 3.8–4.0 mm, ellipsoid, rounded at apex. *Flowers* 7–9. Usually self-pollinated. $2n=32$.

Native. Weed of arable land and rough ground, woodland clearings and damp hollows. Distribution not known but probably throughout Great Britain and Ireland. Most of Europe, but absent from many of the islands and the south-west; northern Asia. Eurasian Boreo-temperate element.

× **tetrahit** = **G.** × **ludwigii** Hausskn.
This hybrid is intermediate in corolla shape between the parents, has 20–70 per cent pollen fertility and a low seed set.

It has been recorded from Wales, and central England and Kirkcudbrightshire, but is probably frequent.

8. Phlomis L.

Perennial monoecious *herbs* or small *shrubs*. *Leaves* opposite, entire, white-hairy. *Whorls* few- to many-flowered, crowded or distant; flowers bisexual. *Calyx* tubular, with 5 more or less equal, subulate teeth. *Corolla* yellow or purple, 2-lipped, the upper lip strongly hooded, the lower lip patent and 3-lobed. *Stamens* 2 long and 2 short, shorter than upper lip of corolla; anther cells divergent. *Style* 1, branches unequal. *Fruit* of 4 trigonous nutlets, glabrous or hairy.

Contains about 250 species from the Mediterranean region to China and Korea.

Vierhapper, F. (1915). Beiträge zur Kenntnis der Flora Kretas. *Österr. Bot. Zeitschr.* **65**: 205–236.

1. Plant with glandular hairs mixed with stellate hairs; corolla purple **1. samia**
1. Plant without glandular hairs but dense stellate hairs; corolla yellow 2.
2. Leaves 8–25 × 4–15 cm; petioles up to 20 cm; calyx 15–25 mm; corolla 30–35 mm **2. russelliana**
2. Leaves 4–10 × 1.5–6.0 cm; petioles 1–3 cm; calyx 10–20 mm; corolla 23–35 mm **3. fruticosa**

1. P. samia L. Greek Sage

Perennial monoecious *herb* without root tubers. *Stems* up to 100 cm, pale green, erect, quadrangular, with glandular hairs, unbranched, leafy. *Leaves* opposite; lamina 8–18 cm, subcoriaceous, medium to dark dull green on upper surface, paler beneath, ovate-lanceolate, pointed at apex, crenate or serrate, cordate or sagittate at base, stellately tomentose on upper surface, with whitish stellate hairs and glandular hairs beneath; petiole up to 18 cm. *Inflorescence* a terminal spike; whorls with (6–)12–20 bisexual flowers; bracts leaf-like, ovate or lanceolate and acuminate at apex; bracteoles 20–26 mm, subulate, with stellate and glandular hairs. *Calyx* 18–25 mm, tubular, 5-lobed, the lobes 6–12 mm and subulate, with stellate and glandular hairs. *Corolla* (26–)30–35 mm, purple, 2-lipped, the upper lip hooded and emarginate, the lower lip patent and 3-lobed. *Stamens* 2 long and 2 short, included. *Style* 1, branches unequal. *Nutlets* 4, trigonous, glabrous. *Flowers* 5–7.

Introduced. Grown in gardens and formerly a persistent garden escape. Native of south-east Europe and Turkey.

2. P. russeliana (Sims) Benth. Turkish Sage
P. lunariifolia Sm. var. *russeliana* Sims; *P. superba* K. Koch; *P. samia* auct.

Perennial monoecious *herb*. *Stems* up to 100 cm, pale green, quadrangular, erect, with dense stellate hairs and minute glands, leafy. *Leaves* opposite; lamina 8–25 × 4–15 cm, medium to dark, dull, green on upper surface, much paler beneath, ovate or oblong-ovate, gradually narrowed to an acute apex, crenate, rounded or cordate to shortly cuneate at base, with numerous simple eglandular and some stellate hairs on upper surface and dense stellate hairs beneath with minute glands; veins impressed on upper surface, prominent beneath; petiole up to 20 cm on lower leaves, upper usually sessile. *Inflorescence* of dense whorls at each pair of leaves; flowers bisexual; bracts linear, stellately hairy. *Calyx* 15–25 mm, pale green, divided about one-third of the way to the base into 5 lobes, the lobes more or less equal, subulate and with minute glands and stellate hairs. *Corolla* 30–35 mm, pale yellow with a dark yellow lower lip; 2-lipped, the upper lip hooded, the lower lip 3-lobed, the side lobes small and ovate, the centre lobe much larger and broadly ovate. *Stamens* 4; filaments very pale green; anthers yellow. *Style* 1, very pale green; stigma pale orange. *Nutlets* 4, 5.2–5.5 mm, dark brown, trigonous, rough, glabrous. *Flowers* 5–7.

Introduced. Banks of roads and railways and rough ground where it is persistent and more or less naturalised. Recorded in a few places in England and Scotland. Native of Turkey.

3. P. fruticosa L. Jerusalem Sage

Evergreen, monoecious *shrub* up to 1.5 m. *Stems* and branches with pale brown, lightly fissured bark; young shoots quadrangular, tomentose with greyish stellate hairs. *Leaves* opposite; lamina 4–10 × 1.5–6.0 cm, greyish-green on upper surface, paler beneath, ovate-oblong, rounded at apex, minutely crenulate, rounded, truncate or shallowly cordate at base, densely stellate-tomentose on both surfaces, rugulose-bullate with finely reticulate venation impressed on upper surface and prominent beneath; petiole 1–3 cm, stout, channelled above, tomentose. *Inflorescence* consisting of 1 or 2, dense, many-flowered heads or whorls 30–35 mm in diameter; flowers bisexual; bracts leafy, similar to the leaves but generally subsessile, narrower and more acuminate; bracteoles 10–18 × 4–7 mm, appressed, oblong-elliptical or oblong-lanceolate, densely stellate-hairy with a fringe of longer hairs. *Calyx* 10–20 mm, tubular, with 5 slender, spreading lobes, densely stellate with scattered longer eglandular hairs, hairy along the nerves and at the mouth. *Corolla* yellow; tube 23–35 mm, dilated at apex, thinly eglandular hairy, with an oblique bearded zone internally; adaxial lip about 18 mm, foliate, strongly compressed laterally, densely eglandular hairy; abaxial lip about 25 mm, hairy, the lateral lobes small, oblong and with caudate appendages, the median lobe abruptly expanded and shallowly emarginate. *Stamens* 4, didynamous, inserted near the mouth of the corolla tube; filaments arcuate, thinly hairy. *Styles* about 40 mm; stigmatic lobes very unequal. *Nutlets* 6–7 mm, dark brown, narrowly oblong, sharply triquetrous, smooth and a densely hairy truncate apex. *Flowers* 5–6. $2n = 20$.

Introduced. Grown in gardens and more or less naturalised on sea-cliffs, banks and rough ground. Recorded in a few scattered localities in England, Isle of Man and Ireland, a rare throw-out elsewhere in Great Britain. Native of the Mediterranean region of Sardinia to Cyprus.

9. Melittis L.

Strong-smelling, *perennial* monoecious *herbs. Leaves* opposite, crenate or crenate-serrate. *Inflorescence* of 2- to 6-flowered, distant axillary whorls; flowers bisexual; bracts not differentiated from the leaves. *Calyx* 2-lipped, the upper lip with 2–3 short lobes, the lower lip with 2 deeper lobes. *Corolla* 2-lipped, the upper lip flat or slightly hooded, the lower lip 3-lobed. *Stamens* 4, shorter than the upper lip of the corolla. *Style* 1; stigma 2-lobed. *Fruit* of 4, ovoid nutlets.

Contains only 1 species.

Brewis, A. et al. (1996). *The flora of Hampshire*. Colchester.
Stewart, A. Pearman, D. A. & Preston, C. D. (1994). *Scarce plants in Britain*. Peterborough.

1. M. melissophyllum L. Bastard Balm

Strong-smelling, *perennial* monoecious *herb* with fibrous roots. *Stems* 20–70 cm, pale green, erect, quadrangular, with dense hairs and numerous, smaller glands, usually simple, leafy. *Leaves* opposite; lamina 5–8 × 1.4–4.0 cm, bright green on upper surface, paler beneath, ovate, obtuse to acute at apex, crenate or slightly crenate-serrate, rounded to subcordate at base, densely eglandular hairy on both surfaces, with prominent venation; petioles rather short, very eglandular hairy. *Inflorescence* of 2- to 6-flowered axillary whorls; flowers bisexual; bracts not differentiated from the leaves. *Calyx* 15–17 mm, 2-lipped, the upper lip with 2–3, short, irregular lobes, the lower lip with 2 deeper, rounded lobes, with eglandular hairs and minute glands. *Corolla* 25–40 mm, pink, creamy-white or white, with a broad band and spots of light reddish-purple on the lower lip, the broad tube much exceeding the calyx, glabrous within, 2-lipped, the upper lip flat or slightly hooded, the lower lip 3-lobed, the margins of the lobes wavy. *Stamens* 4, shorter than upper lip of corolla; filaments hairy; anthers with cells diverging. *Style* 1; stigma 2-lobed. *Nutlets* 4, 4.0–4.5 mm, ovoid, hairy. *Flowers* 5–7. Pollinated by bumble-bees and hawkmoths. Protandrous. $2n = 30$.

Native. Woods and hedgerows. Very local in south-west Wales, south-west and southern England east to Sussex; rare escape from gardens elsewhere in Wales and central England. West, central and south Europe. European Temperate element.

10. Marrubium L.

Strongly scented, *perennial* monoecious, woolly *herbs* with short, stout rhizomes. *Leaves* opposite. *Inflorescence* of many-flowered, distant, axillary whorls; flowers bisexual bracts not differentiated from the leaves; bracteoles linear. *Calyx* with 10 equal lobes, each hooked at the end and patent in fruit. *Corolla* white; tube shorter than calyx, glabrous or with a poorly developed ring of hairs within; 2-lipped, the upper lip 2-lobed and nearly flat, the lower lip 3-lobed. *Stamens* 4, all included in the corolla tube. *Style* 1; stigma shortly bifid. *Fruit* of 4, ovoid, smooth, nutlets.

Contains about 40 species in Europe, temperate Asia and North Africa.

Hultén, E. & Fries, M. (1986). *Atlas of north European vascular plants north of the Tropic of Cancer*. 3 vols. Königstein.
Stewart, A., Pearman, D. A. & Preston, C. D. (1994). *Scarce plants in Britain*. Peterborough.

1. M. vulgare L. White Horehound

Strongly-scented, *perennial* monoecious *herb* with a short, stout rhizome. *Stems* 30–60 cm, erect or ascending, quadrangular, white-tomentose with eglandular hairs and minute glands, branched, leafy. *Leaves* opposite; lamina 1.5–4.0 × 1.5–2.5 cm, medium green when not too densely hairy, subrotund, ovate or subrotund-ovate, rounded at apex, crenate, cuneate to cordate at base, rugose, usually densely white-tomentose, but often becoming less so and green on upper surface; petioles long on lower leaves, short on upper leaves. *Inflorescence* of many-flowered, distant axillary whorls broader than high; flowers bisexual; bracts not differentiated from the leaves; bracteoles linear. *Calyx* 5.0–6.5 mm, tubular, 10-veined, with 10, equal, yellowish, lobes, each lobe hooked at the end and becoming patent in fruit, woolly with eglandular hairs and with minute glands. *Corolla* 12–15 mm, white, tube shorter than calyx, glabrous or with a poorly developed ring of hairs within, 2-lipped, the upper lip 2-lobed and nearly flat, the lower lip 3-lobed. *Stamens* 4, all included in the corolla tube; filaments pale; anthers with cells diverging. *Style* 1; stigma shortly bifid. *Nutlets* 4, 2.0–2.5 mm, ovoid, smooth. *Flowers* 6–11. Pollinated mainly by bees, or selfed. Homogamous or weakly protandrous; small female flowers occur. $2n=34$.

Native. Short grassland, open or rough ground and waste places. Scattered records in Great Britain and Ireland north to Morayshire but much rarer than formerly. Probably only native near the sea in south and west Great Britain from Sussex to Denbighshire and in the Breckland of East Anglia. Europe from southern Sweden and central Russia southwards; central and west Asia; North Africa; Macaronesia. Eurosiberian Southern-temperate element. It can still be found as an ingredient in herbal cough medicines, a use that goes back for at least 2,000 years.

11. Scutellaria L.

Perennial monoecious *herbs* with rhizomes. *Leaves* opposite. *Inflorescence* of pairs of bisexual flowers in a long raceme. *Calyx* with campanulate tube and 2-lipped, both lips entire, the upper with a dorsal scale-like outgrowth, lips closed in fruit. *Corolla* blue or purple, tube long and more or less curved upwards from the base. *Stamens* 4, subglobose, smooth or tuberculate. *Fruit* of 4 nutlets.

Contains about 300 species; cosmopolitan except South Africa.

Hultén, E. & Fries, M. (1986). *Atlas of north European vascular plants north of the Tropic of Cancer*. 3 vols. Königstein.
Pigott, C. D. (1951). *Scutellaria hastifolia* in Britain. *Watsonia* **2**: 18–21.
Rich, T. C. G. & Jermy, A. C. (1998). *Plant crib*. London.

1. Leaves 5–15 cm, with petioles more than 1 cm; flowers in axils of bracts, markedly smaller than foliage leaves **1. altissima**
1. Leaves very rarely more than 5 cm, with petioles less than 1 cm 2.
2. Corolla 6–10 mm, pale pinkish-purple, with nearly straight tube **4. minor**
2. Corolla 10–20 mm, blue, with strongly bent tube 3.
3. Leaves crenate along entire length, cordate to rounded at base; calyx without or with only eglandular hairs **2. galericulata**
3. Leaves entire or at least in distal half, hastate at base; calyx with glandular hairs **3. hastifolia**

1. S. altissima L. Somerset Skullcap

Perennial monoecious *herb* with rhizomes. *Stems* up to 100 cm, pale green, erect, quadrangular, with sparse eglandular hairs and glands above, simple or branched, leafy. *Leaves* opposite; lamina 5–15 × 2–5 cm, medium green on upper surface, paler beneath, ovate, pointed at apex, serrate, truncate or cordate at base, glabrous or with eglandular hairs on the veins beneath, long petiolate. *Inflorescence* in pairs of bisexual flowers forming an oblong raceme 10–12 mm apart; bracts 6–10 mm, shorter than the flowers, ovate to ovate-lanceolate, acute at apex and entire. *Calyx* 4.5–5.0 mm, campanulate, 2-lipped, both lips entire, the upper with a dorsal scale-like outgrowth, without or with a few, long, white eglandular hairs. *Corolla* 12–16(–18) mm, blue with a white lower lip; tube subglobose, hairy outside; 2-lipped, the upper lip hooded, the lower lip obscurely 3-lobed. *Stamens* 4, shorter than to as long as upper lip of corolla. *Style* 1. *Nutlets* 4, subglobose. *Flowers* 6–9. $2n=34$.

Introduced. Naturalised in hedgerows and wood borders since 1929 in Somersetshire and since 1972 in Surrey. Native of south-east Europe and south-west Asia.

2. S. galericulata L. Skullcap

Perennial monoecious *herb* with a slender, creeping rhizome. *Stems* 15–50 cm, pale green, erect to decumbent, quadrangular, glabrous or with scattered eglandular hairs and minute glands, simple or branched, leafy. *Leaves* opposite; lamina 2–5 × 1–2 cm, bright medium green on upper surface, paler beneath, ovate, lanceolate, ovate-lanceolate or oblong-lanceolate, obtuse to more or less acute at apex, shallowly crenate, with 8–23 long, rounded teeth, cordate at base, glabrous or with eglandular hairs and minute glands; shortly petiolate. *Inflorescence* of pairs of axillary bisexual flowers in a poorly marked raceme; bracts similar to the leaves, but gradually decreasing in size upwards; pedicels very short. *Calyx* 4.0–4.5 mm, campanulate, 2-lipped, the lips entire, the upper bearing a small scale on the back, closed after flowering, glabrous or with simple eglandular hairs and minute glands. *Corolla* 10–20 mm, violet-blue, the lower lip variegated with white and spotted with purple, the tube nearly white, straight or curved below, dilated above and naked within, 2-lipped, the upper lip 2-lobed and hooded, the lower lip 3-lobed, the central much larger than the lateral and itself with 3 lobules. *Stamens* 4, shorter than to around as long as upper lip; anthers of outer pair with 2 nearly parallel cells, of the inner pair with one cell. *Style* 1, curved; stigma shortly 2-lobed. *Nutlets* 4, about 1.8 mm, subglobose, tuberculate. *Flowers* 6–9. Visited by various insects. Homogamous. Small female flowers rare. $2n=30, 32$.

Native. Fen, wet meadows and by ponds and rivers. Locally common throughout most of Great Britain and Ireland, but has declined through drainage. Almost throughout Europe, but absent from some islands; north and west Asia; Algeria; North America. Eurosiberian Boreo-temperate element. Very variable in branching and pubescence, and although varietal names have been used we have found difficulty in defining them.

× **minor** = **S.** × **hybrida** Strail
S. × *nicholsonii* Taube

Intermediate in corolla (7–12 mm. and leaf characters (6–13 teeth). Parents are highly sterile, but vegetatively vigorous.

Local in southern England and southern Ireland.

3. S. hastifolia L. Norfolk Skullcap

Perennial monoecious *herb* with a slender, creeping rhizome. *Stems* (5–)15–40(–50) cm, pale green, more or less erect, subglabrous or with sparse eglandular hairs, simple or branched, leafy. *Leaves* opposite; lamina (8–)15–25(–40) × (4–)5–12(–20) mm, medium green on upper surface, paler beneath, lanceolate to ovate, obtuse at apex, entire at least in the distal half, sometimes with 1–3 small crenations proximally, more or less hastate, rarely subcordate at base, more or less glabrous; petioles up to one-quarter as long as lamina. *Inflorescence* of pairs of bisexual flowers forming oblong racemes, the flowers remote or close together; bracts like the leaves but subsessile and shorter than the flowers, the upper truncate at base. *Calyx* 3–4 mm, campanulate, 2-lipped, both lips entire, the upper with a dorsal scale-like outgrowth, with minute glandular hairs. *Corolla* (10–)15–20(–23) mm, violet-blue; tube curved throughout, 2-lipped, the upper lip hooded, the lower lip obscurely 3-lobed. *Stamens* 4. *Style* 1. *Nutlets* 4, subglobose. *Flowers* 6–9. $2n = 32$.

Introduced. Naturalised in woodland in one place in Norfolk from 1940 to 1980, but now gone. Native of much of Europe, but absent from the south-west, the extreme north and the islands; Turkey.

4. S. minor Huds. Lesser Skullcap

Perennial monoecious *herb* with a slender, creeping rhizome. *Stems* 10–15(–30) cm, pale green, erect to decumbent, quadrangular, more or less glabrous, simple or branched, leafy. *Leaves* opposite; lamina 1–3 × 0.3–1.0 cm, pale green on upper surface, paler beneath, lanceolate to ovate, obtuse to acute at apex, entire or with 1–4 shallow teeth towards the base, cuneate or rounded at base, glabrous or with a few eglandular hairs, shortly petiolate. *Inflorescence* of pairs of axillary bisexual flowers in a poorly marked raceme; bracts oblong or lanceolate-oblong, acute at apex, entire; pedicels very short. *Calyx* 2.8–3.0 mm, blackish-purple on upper lip, campanulate, 2-lipped, the lips entire, the upper forming a small scale on the back, closed after flowering, glabrous or with simple eglandular hairs. *Corolla* 6–10 mm, pale pinkish-purple on the upper lip and lateral lobes of the lower lip, lower lip almost white spotted with magenta-violet, 2-lipped, the tube nearly straight, the upper lip small and 2-lobed, the lower lip with 2 small lateral lobes and a large central lobe which contains 4 broad lobules. *Stamens* 4, shorter than to around as long as upper lip; anthers of outer pair with 2 nearly parallel cells, of the inner pair with one cell. *Style* 1, curved above; stigma minutely 2-lobed. *Nutlets* 4, 1.0–1.2 mm, subglobose, tuberculate. *Flowers* 7–10. $2n$ = (28) c. 32.

Native. Wet heaths, bogs, marshes and open woodland on acid soils. Locally frequent in south and west Great Britain and southern Ireland; rare and very scattered records elsewhere in England and Jersey. Western Europe, eastwards to eastern Germany and western Italy; south-west Sweden; Azores. Southern-temperate element.

12. Teucrium L.

Annual or *perennial* monoecious, aromatic or foetid *herbs* or very low *shrubs*. *Leaves* opposite, serrate or crenate-serrate to pinnatisect. *Inflorescence* of terminal racemes or axillary whorls; flowers bisexual. *Calyx* 5-lobed, not 2-lipped. *Corolla* with a 5-lobed lower lip and no upper lip, the middle lobe of the lower lip much larger than the lateral lobes. *Stamens* 4, shorter than the lower lip. *Style* 1; stigma bifid. *Fruit* of 4, obovoid, smooth or reticulate nutlets.

Contains about 300 species, cosmopolitan, but particularly Mediterranean.

T. fruticans L. has been recorded as a garden escape and **T. resupinatum** Desf. as a bird-seed casual.

Beecroft, R. C., Cadbury, C. J. & Mountford, J. O. (2007). Water Germander *Teucrium scordium* L. in Cambridgeshire: back from the brink of extinction. *Watsonia* **26**: 303–316.
Grime, J. P. et al. (1988). *Comparative plant ecology*. London.
Hall, L. B. (1928). *Teucrium scorodonia* L.: a new variety. *Jour. Bot. (London)* **66**: 299–300.
Hultén, E. & Fries, M. (1986). *Atlas of north European vascular plants north of the Tropic of Cancer*. 3 vols. Königstein.
Hutchinson, T. C. (1968). Biological flora of the British Isles. No. 115. *Teucrium scorodonia* L. *Jour. Ecol.* **56**: 901–911.
Wigginton, M. J. (1999) (Edit.) *British red data books*. Vol. 1. *Vascular plants*. Peterborough.

1. Annual; leaves pinnatisect nearly to midrib **4. botrys**
1. Perennial; leaves serrate or crenate-serrate 2.
2. Upper calyx lobe much wider than other 4; corolla greenish-cream **1. scorodonia**
2. Upper calyx lobe scarcely different from the other 4; corolla pinkish-purple or reddish-mauve 3.
3. Very dwarf evergreen shrub; calyx tube not saccate 2. **chamaedrys**
3. Perennial herb; calyx tube saccate at base on lower side 4.
4. Plant fairly hairy; internodes up to 7 cm; leaves up to 7 cm, widely spaced **3(i). scordium** var. **scordium**
4. Plant densely hairy; internodes 0.5–3.0 cm; leaves up to 3 cm, dense **3(ii). scordium** var. **dunense**

Section **1. Scorodonia** (Hill) Benth.
Scorodonia Hill

Inflorescence of pairs of flowers in terminal racemes; bracts very different from the leaves. *Calyx* campanulate, the upper lobes rounded-ovate, much broader than the other 4; tube gibbous.

1. T. scorodonia L. Wood Sage
Scorodonia heteromalla Moench; *Scorodonia solitaria* Stokes nom. illegit.; *Scorodonia sylvestris* Link; *T. salviaefolium* Salisb. nom. illegit.

Perennial monoecious *herb* with a creeping rhizome. *Stems* 15–50 cm, pale green, erect, quadrangular, with eglandular hairs and minute glands, branched, leafy. *Leaves* opposite; lamina 3–7 × 1.5–4.0 cm, bright slightly yellowish-green on upper surface, paler beneath, ovate, obtuse to subacute at apex, crenate or crenate-serrate, cordate at base, rugose and eglandular hairy on both surface and with minute glands; petioles short to medium and hairy. *Inflorescence* of pairs of bisexual flowers in terminal racemes; bracts very different from the leaves, ovate-lanceolate, entire, shorter than the flowers; pedicels short, with eglandular hairs and minute glands. *Calyx* 4.5–5.0 mm, campanulate, the tube gibbous, 5-lobed, the upper lobe subrotund-ovate and much broader than the other 4 which are triangular ovate and acute, with eglandular hairs and minute glands. *Corolla* 10–12 mm, pale greenish-cream, the tube about 8 mm and hairy, 1-lipped, the lip with 5 lobes and deflexed, the central lobe large and subrotund, the lateral lobes small and short, without a ring of hairs within. *Stamens* 4, shorter than the lower lip; filaments hairy below; anthers brown. *Style* 1, straight, glabrous; stigma shortly bifid. *Nutlets* 4, 1.6–1.8 mm, obovoid, smooth. *Flowers* 7–9. Pollinated by bees. Protandrous. $2n = 32, 34$.

Shows variation in shape and dissection of leaves. Var. *dentatum* Bab. and var. *acrotomum* L. B. Hall have been described but only single specimens of each have been seen.

Native. Woods, hedgerows, hilly areas and fixed shingle and dunes, on acidic or alkaline, usually well-drained soils. Common in suitable places throughout most of Great Britain and Ireland, but rare in central Ireland and central-east England. South, west and central Europe, northwards to southern Norway and eastwards to western Poland and north-west Yugoslavia. Suboceanic Southern-temperate element.

Section 2. **Chamaedrys** (Mill.) Schreb.
Chamaedrys Mill.

Inflorescence of 2- to 6-flowered whorls forming terminal spikes. *Calyx* tubular-campanulate, the lobes subequal, the tube not saccate.

2. T. chamaedrys L. Wall Germander
Chamaedrys officinalis Moench

Very dwarf, monoecious, evergreen *shrub*, without a creeping rhizome, forming low tufts. *Stems* many, 5–50 cm, pale green, ascending, with eglandular hairs and minute glands, simple, leafy. *Leaves* opposite; lamina 1–3 × 0.5–1.5 cm, dark green on upper surface, paler beneath, ovate, oblong or oblong-obovate, obtuse at apex, incise-serrate or crenate-dentate, attenuate at base, with more or less numerous eglandular hairs and minute glands; petiole short. *Inflorescence* a terminal spike, formed of 2- to 6-flowered whorls; flowers bisexual; bracts similar to leaves but smaller, the upper shorter than the flowers and subsessile. *Calyx* 5–8 mm; tube curved; 5-lobed, the lobes equal, triangular and acute at apex, with eglandular and minute glandular hairs. *Corolla* 9–16 mm, pinkish-purple, 5-lobed, the upper 4 lobed, acute, the middle lobe broad, obovate-cuneate and crenulate. *Stamens* 4; shorter than the lower lip. *Style* 1, bifid. *Nutlets* 4, obovoid, *Flowers* 7–9. Pollinated by bees, self-pollination possible; protandrous. $2n = 32, 58, 60, 62, 64, 96$.

Native. Chalk grassland at one site in Sussex discovered in 1945. Also grown in gardens and is naturalised on old walls and dry banks; with scattered records in Great Britain and Ireland, north to central Scotland. West, central and south Europe; south-west Asia; North Africa.

Section **3. Scordium** (Mill.) Benth.
Scordium Mill.

Inflorescence of 2- to 6-flowered axillary whorls. *Calyx* tubular; lobes subequal; tube saccate at base on the lower side.

3. T. scordium L. Water Germander
Scordium palustre Fourr.; *Scordium officinale* Gueldenst. ex Ledeb.; *Chamaedrys palustris* Gray nom. illegit.

Perennial monoecious *herb* with creeping rhizome or leafy stolons. *Stems* 10–60 cm, pale green, ascending to decumbent, quadrangular, with more or less numerous eglandular hairs and minute glands, branched, leafy. *Leaves* opposite; lamina 1–7 × 0.8–2.0 cm, deep shining bluish-green on upper surface, greyish-green to whitish beneath, ovate to oblong, obtuse or subacute at apex, coarsely serrate or crenate-serrate, rounded and sessile at base, with more or less dense eglandular hairs and minute glands on both surfaces. *Inflorescence* of 2- to 6-flowered, distant axillary, secund whorls; flowers bisexual; bracts scarcely different from the leaves; pedicels short, with numerous eglandular hairs and minute glands. *Calyx* 5–6 mm, tubular, the tube saccate at the base on the lower side, with 5 subequal lobes, the lobes triangular and acute, with numerous eglandular hairs and minute glands. *Corolla* 7–12 mm, delicate pinkish-mauve, with bright reddish-purple markings, 1-lipped with 5 lobes, the middle lobe large and obovate, the lateral lobes much smaller, the tube without a ring of hairs within. *Stamens* 4; filaments glabrous; anthers pale orange-brown. *Style* 1, slightly curved; stigma shortly bifid. *Nutlets* 4, 1.4–1.6 mm, obovoid, reticulate. *Flowers* 7–10. Pollinated by bees, but self-pollination possible. Protandrous. $2n = 32$.

(i) Var. scordium
Plant fairly hairy. *Internodes* up to 7 cm. *Leaves* up to 7 cm, widely spaced.

(ii) Var. dunense Druce
Plant densely hairy. *Internodes* 0.5–3.0 cm. *Leaves* up to 3 cm, dense.

Native. Fens, slacks and river-banks on calcareous soils. Var. *scordium* used to occur in a number of sites in eastern England in tall vegetation. Its last known site is at Upware in Cambridgeshire in which it is now scarce. It also occurs in central and northern Europe. Var. *dunense* occurs at Braunton Burrows in Devonshire and a number of sites in the west of Ireland in more open situations. It appears to be the same taxon that has been called *T. scordium* subsp. *scordioides* in Portugal and Spain. It has also been called *T. lanuginosum* Hoffmans. or *T. scordium* var. *lanuginosum* (Hoffmans.) Brot. which may be its proper name. The Channel Islands plant may also be this but no specimens have been seen. The plant called subsp. *scordioides* in south-central and south-east Europe is quite a different taxon. When the site at Kingfisher Bridge near Upware was turned into a wetland nature reserve, plants were transferred from the Upware locality and the species became common. Unfortunately the plant that is common there now is var. *dunense* which may have been brought in by migrating birds.

4. T. botrys L. Cut-leaved Germander

Scorodonia botrys (L.) Ser.; *Chamaedrys laciniata* Gray nom. illegit.

Annual monoecious *herb* with fibrous roots. *Stems* 10–30 cm, pale green, erect, softly eglandular hairy and with minute glands, branched below, leafy. *Leaves* opposite; lamina 1.0–2.5 × 1.0–2.5 cm, bright medium green on upper surface, paler beneath, ovate in outline, pinnatifid, the segments oblong, obtuse at apex and often with obtuse lobes, with eglandular hairs and minute glands; petioles short and eglandular hairy. *Inflorescence* of 2- to 6-flowered, distant, axillary, secund whorls along stem and branches; flowers bisexual; bracts small, but longer than the flowers, 1-pinnatifid; pedicels 3–4 mm, with eglandular hairs and minute glands. *Calyx* 9–10 mm, flushed with crimson-purple, tubular, saccate at base on the lower side, reticulate-veined, with 5 subequal lobes, the lobes triangular and acute at apex, with eglandular hairs and minute glands. *Corolla* 15–20 mm, bright pale reddish-mauve, variegated white and dotted with red, 1-lipped with 5 lobes, the middle lobe large and wider than long, the lateral lobes small, the tube without a ring of hairs within. *Stamens* 4; filaments glabrous; anthers brownish. *Style* 1, curved; stigma 2-lobed. *Nutlets* 4, 1.6–1.8 mm, obovoid, reticulate. *Flowers* 7–9. Pollinated by bees; self-pollination possible. $2n = 32$.

Introduced. Bare chalk and chalky fallow fields on downs. Rare in Wiltshire, Hampshire, Kent, Surrey, Gloucestershire, Shropshire and Edinburghshire; formerly elsewhere in southern England. It has more recently benefited from disturbance and in some sites there are thousands of plants. West, central and south Europe, northwards to southern Poland and eastwards to Roumania: Algeria.

13. Ajuga L.

Perennial or *annual* monoecious *herbs*, sometimes with stolons. *Stems* to 30 cm. *Leaves* opposite. *Inflorescence* terminal, formed of 2- to many-flowered, lax whorls; bracts sometimes differentiated from leaves. *Calyx* campanulate, with 5 more or less equal lobes. *Corolla* 2-lipped, the upper very short, the lower 3-lobed, the central lobe much larger than the lateral; tube more or less exserted, with a ring of hairs within. *Stamens* 4 included in the corolla tube. *Style* 1, included in the corolla tube; stigma shortly 2-lobed. *Fruit* of 4 obovoid or ellipsoid, reticulate nutlets.

Contains about 40 species in temperate regions of the Old World.

Gulliver, R. L. (1997). Pyramidal Bugle (*Ajuga pyramidalis*). on Colonsay (v.c. 102). *Glasgow Naturalist* **23**: 55.
Hultén, E. & Fries, M. (1986). *Atlas of north European vascular plants north of the Tropic of Cancer*. 3 vols. Königstein.
Rich, T. C. G., Kay, G. M. & Sydes, C. (1999). Distribution and ecology of Pyramidal Bugle *(Ajuga pyramidalis)*, Lamiaceae in the British Isles. *Bot. Jour. Scotland* **51**: 181–193.
Stewart, A., Pearman, D. A. & Preston, C. D. (1994). *Scarce plants in Britain*. Peterborough.
Wigginton, M. J. (Edit.) (1999) *British red data books*. Vol. I. *Vascular plants*. Peterborough.

1. Leaves divided into 3 linear lobes; corolla yellow **4. chamaepitys**
1. Leaves not divided, entire or toothed; corolla blue, pink, white or purple 2.
2. Plant with stolons; stems hairy on 2 sides only **1. reptans**
2. Plant without stolons; stems hairy all round 3.
3. Bracts all longer than flowers; upper lip of corolla entire; filaments glabrous **2. pyramidalis**
3. Bracts in upper part shorter than flower; upper lip of corolla 2-lobed; filaments hairy **3. genevensis**

Section **1. Ajuga**

Inflorescence a terminal spike of lax whorls each of 6–many flowers. *Corolla* blue, pink, white or purple.

1. A. reptans L. Bugle

Perennial monoecious *herb* with a short rhizome and long, leafy, rooting stolons. *Stems* 10–30 cm, pale green, simple, erect, with more or less numerous eglandular hairs and minute glandular hairs, the hairs often only on two opposite sides, quadrangular, leafy. *Leaves* opposite, bright green on upper surface, paler beneath, more or less glabrous; basal numerous, forming a rosette, the lamina 4–7 × 2–4 cm, obovate, elliptical or oblong, obtuse at apex, entire to obscurely crenate, cuneate or attenuate at base, with a petiole up to 6 cm; cauline few, similar but smaller and with a shorter petiole. *Inflorescence* terminal, formed of a series of fairly distant whorls; flowers bisexual; bracts green, blue-tinged or rarely copper-coloured, ovate, entire, the upper shorter than their flowers. *Calyx* 4–6 mm, enlarging in fruit, tinged crimson-purple, tubular-campanulate, divided nearly halfway into 5 subequal lobes, the lobes triangular-ovate and subobtuse at apex, with long eglandular hairs and minute glandular ones. *Corolla* 10–18 mm, blue, rarely pink or white, 2-lipped, upper lip very short, lower lip 3-lobed, the central lobe largest and emarginate, marked with nectar guides, hairy; tube more or less exserted with a ring of hairs within. *Stamens* 4, shorter than lower lip; filaments blue or pink; anthers blue or pink. *Style* 1, slightly curved; stigma shortly 2-lobed. *Nutlets* 4, 2.0–2.5 mm, ellipsoid, reticulate. *Flowers* 5–7. Pollinated by bees. Usually homogamous and self-pollination is possible. $2n = 32$.

Native. Woods, shady places and damp grassland on neutral or acidic soils. Common throughout most of Great Britain and Ireland. Most of Europe northwards to about 61° N; Caucasus; Algeria, Tunisia; south-west Asia. European Temperate element.

2. A. pyramidalis L. Pyramidal Bugle

Perennial monoecious *herb* with a short rhizome and without stolons. *Stems* 10–30 cm, simple, erect, quadrangular, eglandular hairy all round and with minute glands, leafy. *Leaves* opposite, a bright deep green on upper surface, paler beneath, usually eglandular hairy on both surfaces with some minute glands; basal fairly numerous, persisting at flowering, the lamina 3–8 × 2–4 cm, ovate, obovate or broadly elliptical, rounded at apex, obscurely crenate, attenuate at base into a short petiole; cauline much smaller than basal, ovate or obovate-oblong, obtuse at apex, mostly entire, rounded at base to a short petiole or sessile. *Inflorescence* terminal, formed of a series of fairly distant whorls; flowers bisexual; bracts flushed with purple, all much longer than flowers, ovate or elliptical, obtuse at apex, entire or slightly crenate, hairy, rarely lobed. *Calyx* 5–6 mm, elongating in fruit, campanulate, divided over half way into 5 lobes, the lobes lanceolate and acute at apex, densely eglandular hairy and with minute glands. *Corolla* 9–13 mm, pale bluish-purple to reddish-violet, curved, 2-lipped, the upper lip very short, the lower lip 3-lobed, the central lobe larger than the lateral and emarginate, hairy, with a ring of hairs within. *Stamens* 4, shorter than lower lip; filaments blue or pink, glabrous; anthers blue or pink. *Style* 1, straight most of length and slightly curved at apex; stigma shortly 2-lobed. *Nutlets* 4, 2.0–2.5 mm, ellipsoid, reticulate. *Flowers* 5–7. Pollinated by bees. Protandrous or homogamous with self-pollination possible. $2n = 32$.

Native. Rock crevices and shallow peat in open heathland in hilly areas. Very local in north, north-west and southern Scotland; Westmorland; west-central Ireland and Co. Antrim. Europe southwards to north Portugal, north Italy and Bulgaria. European Boreal-montane element.

× **reptans** = **A.** × **pseudopyramidalis** Schur
A. × *hoppeana* A. Braun & Vatke

Intermediate in most characters between the parents, sterile and forming stolons late in the season.

Sutherland, Ross and Cromarty, Orkney Islands and Co. Clare.

3. A. genevensis L. Blue Bugle

Perennial monoecious *herb* with rhizomes. *Stems* 10–40 cm, pale green, suffused purplish, erect, quadrangular, nearly glabrous to lanate-villous all round, with eglandular hairs mixed with minute glands, simple, leafy. *Leaves* opposite, the basal usually dead at anthesis; lamina of lower 3–12 × 0.8–5.0 cm, medium green on upper surface, paler beneath, obovate to oblong-obovate, rounded at apex, crenate-dentate or shallowly lobed, narrowed or rounded at base, more or less eglandular hairy. *Inflorescence* a terminal spike of distant or somewhat crowded whorls, with 6 to many flowers in a whorl; bracts tinged with blue and violet, obovate, usually lobed and the upper shorter than the flowers. *Calyx* 4–6 (–7) mm, 5-lobed, the lobes about as long as the tube, lanceolate and acute at apex with eglandular hairs mixed with minute glands. *Corolla* 12–20 mm, bright blue, rarely pink or white; tube exceeding calyx, 2-lipped, the upper lip 2-lobed, the lower lip 3-lobed. *Stamens* 4, exserted; filaments pale, hairy; anthers blue or pink. *Style* 1, slightly curved; stigma shortly 2-lobed. *Nutlets* 4, ellipsoid. *Flowers* 5–7. $2n = 32$.

Introduced. Formerly naturalised in chalk grassland in Berkshire and Cornwall. Europe except the south-west.

Section **2. Chamaepitys** (Hill) Benth.

Inflorescence of 2-flowered axillary whorls. *Corolla* yellow.

4. A. chamaepitys (L.) Schreb. Ground Pine
Teucrium chamaepitys L.

Annual monoecious *herb* with tap-root and fibrous side-roots, smelling of pine when crushed. *Stems* 5–20 cm, pale green, sometimes suffused purple, erect to decumbent, with eglandular hairs and with minute glands, with ascending branches from the base, leafy. *Leaves* opposite, bright medium green, sometimes greyish, eglandular hairy; basal withering early, entire or toothed; cauline 3–6 × 1–2 cm, ovate in outline, divided nearly to midrib into 3 linear, obtuse lobes, attenuate at base into a fairly long petiole. *Inflorescence* terminal, formed of a series of 2-flowered whorls, many much shorter than the leaves flowers bisexual; bracts not differentiated from the leaves. *Calyx* 4–6 mm, elongating in fruit, campanulate, divided to about halfway into 5 lobes, the lobes triangular-lanceolate and acute at apex, very eglandular hairy and with minute glands. *Corolla* 10–11 mm, bright yellow, the lower lip spotted with red, 2-lipped, upper lip very short, the lower lip 3-lobed, the central lobe much larger than the lateral and emarginate, hairy, tube included, with a ring of hairs within. *Stamens* 4, shorter than lower lip; filaments pale, hairy in upper parts; anthers yellow. *Style* 1, slightly curved; stigma shortly 2-lobed. *Nutlets* 4, 2.5–3.0 mm, obovoid, reticulate. *Flowers* 5–9. Visited by bees. $2n = 28$.

Native. Bare chalk and chalky arable fields on downland. Local in south and south-east England north to Suffolk, formerly to Northamptonshire, decreasing. Most of Europe except the north; west Asia; North Africa. European Southern-temperate element.

14. Nepeta L.

Perennial monoecious *herbs* with a strong mint-like smell. *Leaves* opposite. *Inflorescence* a terminal spike of lax or dense, many-flowered whorls; flowers bisexual. *Calyx* tubular, 2-lipped, the upper lip 3-lobed, the lower lip 2-lobed. *Corolla* white or violet-blue, 2-lipped, the upper lip 2-lobed, the lower lip 3-lobed. *Stamens* 4, didynamous. *Style* 1. *Fruit* of 4 nutlets.

Cats love to gambol and roll in plants of this genus, hence its vernacular name.

Contains about 250 species in Europe, Asia, North Africa and mountains of tropical Africa.

Hultén, E. & Fries, M. (1986). *Atlas of north European vascular plants north of the Tropic of Cancer*. 3 vols. Königstein.
Stearn, W. T. (1950). *Nepeta mussinii* and *N.×faassenii*. *Jour. Roy. Hort. Soc.* **75**: 403–406.
De-Wolf, G. P., Jr (1955). Notes on cultivated labiates. 6. *Nepeta*. *Baileya* **3**: 98–107.

1. Bracteoles more or less half the length of calyx **1. cataria**
1. Bracteoles very short and inconspicuous 2.
2. Leaves ovate to oblong-ovate, the base cordate **2. racemosa**
2. Leaves narrowly oblong-ovate or lanceolate, the base cuneate **3.×faassenii**

1. N. cataria L. Cat-mint

Perennial monoecious *herb* with several, large, thick, fleshy roots and a strong mint-like scent. *Stems* 40–100 cm, pale green, erect, hollow, quadrangular, with dense, white eglandular hairs and minute glands, branched, leafy. *Leaves* opposite; lamina 3–7 cm, dull medium green on upper surface, paler beneath, ovate, pointed at apex, coarsely serrate, cordate at base, with soft white eglandular hairs and minute glands beneath; petiole short. *Inflorescence* of numerous, many-flowered whorls forming a spike-like raceme, the upper whorls crowded, the lower more widely spaced; flowers bisexual; bracteoles more or less half length of calyx. *Calyx* 5.0–6.5 mm, tubular, with about 15 ribs, 2-lipped, the upper lip 3-lobed, lanceolate-subulate, with eglandular hairs and minute glands. *Corolla* 7–12 mm, white with small, purple spots; tube curved and dilated at the middle; 2-lipped, the upper lip with 2 rounded, nearly erect lobes, the lower lip with the middle lobe broad and rounded, its turned-up edges deeply and irregularly notched, the 2 smaller or lateral lobes spreading or reflexed, hairy within the throat. *Stamens* 4, the outer pair shorter than the inner; filaments pale; anthers rose. *Style* scarcely as long as the stamens; stigma short. *Nutlets* 1.5–1.7 mm, dark brown, rounded-oblong or elliptical, slightly compressed, obtusely triquetrous. *Flowers* 6–11. Pollinated mainly by bees, or selfed. Homogamous or weakly protandrous; small female flowers occur. $2n=34$, 36.

Introduced. Open grassland, waysides and rough ground on calcareous soils. In scattered localities in England and Wales and Isle of Man, once more common, rarely naturalised or casual in Ireland. South, east and perhaps west Europe; west and central Asia to Kashmir; naturalised in North America and South Africa.

2. N. racemosa Lam. Heart-leaved Cat-mint
N. mussinii Spreng. ex Henckel

Perennial monoecious *herb* with spreading rhizomes forming patches, strongly smelling of mint. *Stems* numerous, 30–80 cm, mostly decumbent, pale yellowish-green, sometimes suffused brownish-purple, with dense, very short and short, colourless simple eglandular hairs, shortly branched, leafy. *Leaves* 1.5–2.5 × 0.7–2.3 cm, dull medium green with the veins much impressed on upper surface, paler beneath with prominent veins, ovate, rounded at apex, undulate-crenate, cordate at base to a short petiole, all with dense, very short, grey simple eglandular hairs on both surfaces and the margins. *Inflorescence* of dense, rather distant whorls; flowers bisexual; bracts leaf-like; bracteoles linear, acute at apex, minutely hairy. *Calyx* pale green with a purple upper part, 5–8 mm, urceolate, with 5 triangular, acute teeth, densely eglandular hairy, in the lower part the hairs white, in the upper part the hairs purple. *Corolla* 12–17 mm, violet-blue, white and blue-spotted in the throat, much exserted, curved, hairy, 2-lipped, the tube slender, the upper lip patent, flat and 2-fid, the lower lip 3-lobed. *Stamens* 4, didynamous, parallel; filaments pale purplish; anthers dark. *Style* pale reddish-purple. *Nutlets* 4, obovoid. *Flowers* 5–7.

Introduced. Much grown in gardens and many plants occur as a throw-out on tips and in waste ground. Native of the Caucasus and Caspian region.

3. N.×faassenii Bergmans ex Stearn Garden Cat-mint
N. nepetella L.×**racemosa**

Perennial monoecious, aromatic, rhizomatous *herb*. *Stems* up to 50 cm, slightly quadrangular, greyish-green, much branched from the base, leafy, short tomentose. *Leaves* opposite; lamina 1–3 cm, medium green on upper surface, paler beneath, oblong-ovate or lanceolate, acute at apex, coarsely serrate, truncate or cuneate at base, with silver-grey eglandular hairs on both surfaces; veins impressed above and prominent beneath; petiole up to 1 cm. *Inflorescence* loose and elongated; flowers bisexual; bracteoles short, linear-lanceolate. *Calyx* 5–6 mm, tubular, 5-lobed, the lobes 1.0–1.5 mm, narrowed at apex. *Corolla* up to 12 mm, violet-blue with darker spots, the tube longer than calyx, very slender and slightly bent. *Stamens* 4. *Style* 1. Usually sterile.

Introduced. A garden throw-out or escape on roadsides, rubbish-tips and waste ground. It was first raised in gardens, where it is popular, in 1784 but was not recorded as an escape until 1928. It has now been recorded throughout Great Britain.

15. Glechoma L.

Perennial monoecious *herbs* with erect flowering stems and long trailing sterile ones which root at the nodes. *Leaves* opposite, crenate. *Inflorescence* of 2- to 4-flowered, secund, axillary whorls; flowers bisexual. *Calyx* somewhat 2-lipped, but with 5 subequal lobes. *Corolla* pale to deep mauve, 2-lipped, the upper lip flat or slightly hooded, the lower lip 3-lobed. *Stamens* 4, shorter than the upper lip of the corolla. *Style* 1, curved near the top. *Fruit* of 4, oblong, smooth nutlets.

Contains about 10 species in Europe and temperate Asia.

Armitage, E. (1913). *Nepeta glechoma* var. *parviflora* Benth. *Jour. Bot. (London)* **51**: 253–254.
Grime, J. P. et al. (1988). *Comparative plant ecology*. London.
Hultén, E. & Fries, M. (1986). *Atlas of north European vascular plants north of the Tropic of Cancer.* 3 vols. Königstein.
Hutchings, M. J. & Price, E. A. C. (1999). Biological flora of the British Isles. No. 205. *Glechoma hederacea* L. (*Nepeta glechoma* Benth., *N. hederacea* (L.) Trev.). *Jour. Ecol.* **87**: 347–364.
Turrill, W. B. (1920). *Glechoma hederacea* L. and its sub-divisions. *Rep. Bot. Soc. Exch. Club Brit. Isles* **5**: 694–701.

1. Leaves up to 4.0×4.5 cm, with sparse hairs; corolla 15–20 mm **1(iii). hederacea** var. **grandiflora**
1. Leaves up to 3.0×3.5 cm, with numerous to dense hairs 2.
2. Leaves up to 1.2 (–1.5)×1.2 (–1.5) cm, often densely hairy; corolla 8–10 (-15) mm **1(i). hederacea** var. **minor**
2. Leaves up to 3.0×3.5 cm, with few to numerous hairs; corolla 12–15 mm **1(ii). hederacea** var. **hederacea**

1. G. hederacea L. Ground Ivy

Calamintha hederacea (L.) Scop.; *Chamaeclema hederacea* (L.) Moench; *G. borealis* Salisb. nom. illegit.; *Nepeta hederacea* (L.) Trevis.; *Nepeta glechoma* Benth.

Perennial monoecious *herb* with fibrous roots. *Stems* 10–30 cm, pale green, sometimes tinted reddish or purplish, the flowering stems erect, the sterile decumbent, spreading or creeping and rooting at the nodes, with few to dense, rigid, reflexed simple eglandular hairs, leafy. *Leaves* opposite; lamina 1–4×1–4.5 cm, medium to dark yellowish or greyish-green on upper surface, paler beneath, subrotund, ovate or reniform, rounded at apex, crenate, cordate at base, with few to dense eglandular hairs on both surfaces and dotted with sessile glands beneath; petioles up to 5 cm, eglandular hairy. *Inflorescence* of 2- to 4-flowered, secund, axillary whorls; flowers bisexual or unisexual, the bracts not differing from the leaves. *Calyx* 5–10 mm, tubular, somewhat 2-lipped, with 5 subequal lobes, short to long eglandular hairy. *Corolla* 8–20 mm in bisexual flowers usually larger than female ones, but at least some bisexual-flowered plants are small, pale to deep mauve, rarely pink or white, spotted in the tube and on the lower lip with crimson-purple, 2-lipped, upper lip flat to slightly hooded, lower lip 3-lobed. *Stamens* 4, shorter than upper lip of the corolla; filaments pale; anthers pale pinkish mauve. *Style* 1, curved near the top; stigma bifid. *Nutlets* 4, about 2 mm, oblong, smooth. *Flowers* 3–5. Pollinated mainly by bees, protandrous; small female flowers occur commonly. $2n=36$.

(i) Var. **minor** Gilib.

G. micrantha Boenn. ex Rchb.; *G. hederacea* var. *parviflora* Sond.; *G. hederacea* var. *praecox* Schur; *G. hederacea* var. *hirsuta* auct.

Stems often purple, with more or less dense eglandular hairs. *Leaves* up to 1.2(–1.5)×1.2 (–1.5) cm, greyish-green, densely eglandular hairy. *Corolla* 8–10(–15) mm.

(ii) Var. **hederacea**

Stems pale green, sometimes tinted purplish, with few to fairly numerous eglandular hairs. *Leaves* up to 3.0×3.5 cm, dark green, more or less shortly eglandular hairy. *Corolla* 12–15 mm.

(iii) Var. **grandiflora** H. Mart.

G. hederacea var. *major* Gaudin; *G. magna* Mérat; *G. hederacea* var. *majus* Peterm.

Stems pale green, with sparse eglandular hairs. *Leaves* up to 4.0×4.5 cm, dark green, with sparse eglandular hairs. *Corolla* 15–20 mm.

Native. Woods, hedgerows, banks and waste ground, often on heavy soils. Common throughout Great Britain and Ireland except in north Scotland. Almost throughout Europe; west and north Asia to Japan, and naturalised in North America. Circumpolar Boreo-temperate element. The distribution and ecology of the varieties is unknown, though var. *minor* may be the plant of dryer habitats.

16. Prunella L.

Perennial monoecious *herbs. Leaves* opposite, entire to divided almost to midrib. *Inflorescence* a dense, terminal, cylindrical spike; flowers bisexual. *Calyx* 2-lipped, the upper lip more or less truncate, with 3 very short lobes, the lower lip with long lobes. *Corolla* yellow, blue, pink or white, 2-lipped, the upper lip strongly hooded and more or less entire, the lower lip 3-lobed. *Stamens* 4, shorter than upper lip of corolla. *Style* 1. *Fruit* of 4, obovoid or oblong nutlets.

Contains 7 species; *P. vulgaris* almost cosmopolitan, the rest of Europe and Mediterranean.

Clement, E. J. (1985). Selfheals (*Prunella* spp.) in Britain. *B.S.B.I. News* **41**: 20.
Druce, G. C. (1918). *Prunella vulgaris var. dunensis* Druce. *Rep. Bot. Soc. Exch. Club Brit. Isles* **5**: 48
Grime, J. P. et al. (1988). *Comparative plant ecology*. London.
Hultén, E. & Fries, M. (1986). *Atlas of north European vascular plants north of the Tropic of Cancer.* 3 vols. Königstein.
Morton, J. K. (1973). A cytological study of the British Labiatae (excluding *Mentha*). *Watsonia* **9**: 239–246.

1. Inflorescence not subtended by leaves; corolla usually more than 18 mm **1. grandiflora**
1. Inflorescence usually subtended by leaves; corolla not more than 17 mm 2.
2. At least the upper leaves pinnatifid or lobed; lobes of the lower lip of the calyx linear-lanceolate; corolla yellowish-white, rarely pink or purplish **3. laciniata**
2. Leaves entire or crenulate; lobes of lower lip of calyx lanceolate; corolla violet rarely, pink or white 3.
3. Corolla white **2(3). vulgaris** forma **leucantha**
3. Corolla violet or pink 4.
4. Corolla violet **2(1). vulgaris** forma **vulgaris**
4. Corolla pink **2(2). vulgaris** forma **rubriflora**

1. P. grandiflora (L.) Scholler Large Selfheal

Perennial monoecious *herb. Stems* up to 60 cm, pale green, erect, with sparse eglandular hairs, simple, leafy. *Leaves* opposite; lamina 5–9×3.5–4.0 cm, medium green on upper surface, paler beneath, ovate to ovate-lanceolate, obtuse at apex, entire or crenulate, truncate or cuneate at base, slightly eglandular hairy; petiole up to 4(–9) cm, pale green. *Inflorescence* a dense, terminal, cylindrical spike not subtended by leaves; flowers bisexual; bracts 15–20×15–20 mm. *Calyx* 12–15 mm, tubular-campanulate, 2-lipped, the upper lip 3-lobed, the lobes subequal, the lower lip with 2 larger lobes, the lobes 3–4 mm, broadly lanceolate and ciliate. *Corolla* (18–)25–30 mm, the lips deep violet, the tube whitish, tube exceeding the calyx and straight, obconical with a ring of hairs inside, 2-lipped, the upper lip distinctly hooded, the lower lip denticulate. *Stamens* 2 long and 2 short; filaments with a subulate appendage below the apex; anther cells divergent. *Nutlets* 4, oblong. *Flowers* 6–9. $2n=28$.

Introduced. Grown in gardens and occurs as a short-term relic or escape in a few places. Recorded from Hampshire, Surrey, Hertfordshire, Berkshire and Morayshire. Native of Europe except most of the north and the islands.

2. P. vulgaris L. Selfheal

Perennial monoecious *herb* with a short rhizome. *Stems* 5–30(–50) cm, pale green, often suffused purplish, decumbent, ascending or erect, quadrangular, at first with eglandular hairs, but often soon more or less glabrous and often purplish, simple or branched, leafy. *Leaves* opposite; lamina 1.5–7.0 × 0.7–3.5 cm, pale to medium green on upper surface, paler beneath, oblong-ovate, obtuse at apex, entire, subentire or obscurely or irregularly toothed or lobed, broadly cuneate at base, glabrous or thinly eglandular hairy along the veins which are usually distinct; petiole 1.5–3.0 cm, slender. *Inflorescence* 2–4 cm, cylindrical, dense, subsessile or shortly pedunculate; flowers bisexual or sometimes female; bracts 5–10 × 8–15 mm, papery, subrotund, cuspidate at apex, conspicuously reticulate-veined, ciliate with long white eglandular hairs; flowers subsessile. *Calyx* 6–9 mm, bronze-purple, tubular, 3–4 mm wide at apex, 5-veined with prominent anastomosing venation, eglandular hairy and with minute glands; 2-lipped, the upper lip almost truncate with 3 small lobes, the lower lip divided into 2 narrow, deltoid, acuminate lobes. *Corolla* violet, rarely pink or white; tube 11 mm, shortly exserted, with an internal hairy zone about 4 mm from base; 2-lipped, the upper lip oblong and shallowly lobed, the lower lip 3-lobed, the lateral lobes subrotund-oblong, the middle lobe subrotund, deflexed and with the margins denticulate. *Stamens* 4, 2 inserted at the mouth of the corolla tube, 2 inserted about 5 mm down the tube; filaments glabrous; anthers small. *Style* 1, 13–14 mm, glabrous; stigma 2-lobed, the lobes subequal. Nutlets 4, about 2 × 1.3 mm, yellowish-brown, obovoid, obscurely trigonous, minutely granulose, with 2 fine parallel stripes running longitudinally down the 2 faces and sides, with a minute conical elaiosome; the persistent calyx is closed at the mouth in dry weather and opens in wet weather to release them. *Flowers* 6–9. Pollinated mainly by bees; protandrous or homogamous; small female flowers occur. $2n=32$.

(1) Forma **vulgaris**
Corolla violet.

(2) Forma **rubriflora** Beckh.
Corolla pink.

(3) Forma **leucantha** (Schur) Hegi
P. vulgaris var. *leucantha* Schur

Corolla white.

Native. Grassland, lawns, wood clearings, and rough ground. Common throughout Great Britain and Ireland. Widespread in Europe; temperate Asia; North Africa; North America; probably not indigenous elsewhere. Circumpolar Wide-temperate element.

3. P. laciniata (L.) L. Cut-leaved Selfheal
P. vulgaris var. *laciniata* L.

Perennial monoecious *herb. Stems* up to 30 cm, pale green, erect, quadrangular, densely eglandular hairy, simple, leafy. *Leaves* opposite; lamina 4–7 × 2–3 cm, medium green on upper surface, paler beneath, ovate in outline, obtuse to acute at apex, at least the upper deeply pinnatifid or lobed, eglandular hairy; petiole absent or up to 4 cm. *Inflorescence* a dense, terminal, cylindrical spike subtended by leaves; flowers bisexual or female; bracts 10 × 15 mm. *Calyx* 8–10 mm, tubular-campanulate, 2-lipped, the upper lip 3-lobed and more or less truncate, the lobes almost obsolete, the lower lip 2-lobed, the lobes 2.0–2.5 mm, linear-lanceolate and ciliate. *Corolla* 15–17 mm, yellowish-white, rarely rose-pink or purplish; tube exceeding calyx, straight, obconical with a ring of hairs inside, 2-lipped, upper lip distinctly hooded, lower lip denticulate. *Stamens* 5; filaments with a subulate appendage below the apex, those of the long stamens curved. *Style* 1. *Nutlets* 4, oblong, smooth. *Flowers* 6–8. Pollinated by bees; protandrous; small female flowers occur. $2n=32$.

Formerly thought to be native, but now considered introduced. Calcareous grassland and roadsides. Scattered localities in south and central England north to Lincolnshire; recorded in Ireland in 1999. South, west and central Europe; North Africa; west Asia.

× **vulgaris** = **P. × intermedia** Link
P. × hybrida Knaf

Variably intermediate between the parents in leaf division and flower colour. Said to be sterile, but intermediates of all degrees occur suggesting backcrossing.

Occurs in south and central England wherever *P. laciniata* grows with *P. vulgaris*.

17. Dracocephalum L.

Annual or *biennial* monoecious *herbs. Leaves* opposite. *Inflorescence* of whorls of flowers crowded into a terminal head; flowers bisexual; bracts spine-tipped. *Calyx* tubular; 13- to 15-veined, 5-lobed. *Corolla* pale blue to violet, 2-lipped, the upper lip slightly arched and notched, the lower lip 3-lobed, with the middle lobe largest. *Stamens* 4; anthers sessile. *Style* 1. *Fruit* of 4 nutlets.

Contains about 40 species in Europe, Asia, North Africa and North America.

Clement, E. J. (1977). Aliens and adventives. *B.S.B.I. News* **17**: 14–19.

1. D. parviflorum Nutt. American Dragon-head

Annual or *biennial* monoecious *herb. Stems* 10–80 cm, solitary or clustered, erect, simple or branched. *Leaves* opposite; lamina 0.5–5.0 × about 1.0 cm, medium green on upper surface, paler beneath, lanceolate or ovate-lanceolate, acute at apex, sharply serrate, cuneate at base, shortly hairy; petiole short, hairy. *Inflorescence* of whorls of bisexual flowers crowded into a terminal head; bracts lanceolate, spine-tipped and serrate, very hairy. *Calyx* tubular, 13–15-veined, 2-lipped, upper lip 3-lobed, lower lip 2-lobed. *Corolla* 12–20 mm only slightly exserted from the calyx, pale blue to violet, the upper lip slightly arched and notched, the lower lip 3-lobed, with the middle lobe larger. *Stamens* 4; anthers sessile. *Style* 1. *Nutlets* 4.

Introduced. Grain, bird-seed and wool casual, which sometimes persists for a few years. Native of North America.

18. Melissa L.

Perennial monoecious *herbs* with rhizomes and stolons, smelling of lemon. *Leaves* opposite. *Inflorescence* of axillary whorls; flowers bisexual; bracts leaf-like; bracteoles small. *Calyx* with tube campanulate, 2-lipped, the upper lip flattened and 3-lobed, the lower lip 2-lobed. *Corolla* with tube curved upwards and dilated above the middle, 2-lipped. *Stamens* 4, shorter than the corolla. *Style* 1, branches subequal. *Fruit* of 4 obovoid, smooth nutlets.

Contains 3 species in Europe to central Asia and Iran; North Africa.

Mabey, R. (1996). *Flora Britannica*. London.
Vaughan, J. G. & Geissler, C. (1997). *The new Oxford book of food plants*. Oxford.

1. M. officinalis L. Balm

Perennial monoecious *herb* spreading rapidly by rhizomes and stolons, and smelling of lemon. *Stem* 20–150 cm, pale yellowish-green, quadrangular, shallowly channelled on each face, glabrous below, erect or sprawling, with numerous, long, erect-patent branches, above with numerous, very short glandular hairs and few, medium, pale simple eglandular hairs. *Leaves* opposite; lamina yellowish-green on upper surface or golden, paler beneath, sometimes variegated, soft, the veins impressed above and prominent beneath; leaves of 2 kinds, the larger only on the main stem, 6–12 × 4–7 cm, ovate, more or less acute at apex, crenate or crenate-serrate, and truncate or cordate at base, the smaller, particularly on the inflorescence, 1.5–5.0 × 1.0–3.5 cm, ovate or ovate-lanceolate, more or less acute at apex, crenate-serrate and cuneate at base, all with more or less numerous, short, soft eglandular and small glandular hairs on both surfaces, the rather long petioles with few hairs. *Inflorescence* of axillary whorls, 4- to 12-flowered; flowers bisexual; bracts leaflike; bracteoles 2–3 mm, linear to ovate, entire. *Calyx* 7–9 mm, medium green, tube campanulate, 2-lipped, 13-veined, the upper lip flattened and 3-lobed, the lower lip with 2, long, triangular-lanceolate lobes, with short to medium, patent simple eglandular hairs and numerous, minute glandular hairs. *Corolla* 8–15 mm, pale yellow, becoming white or pinkish, the tube curved and dilated above the middle, 2-lipped, the upper lip erect or deflexed, sometimes slightly hooded, and emarginate, the lower lip 3-lobed. *Stamens* 4, didynamous, included; filaments white; anthers white. *Style* 1, white; branches subequal. *Nutlets* 4, 1.5–2.0 mm, obovoid, smooth. *Flowers* 6–8. $2n=32$.

Introduced. Grown in gardens as a herb or for decoration, especially the gold-leaved form. The young shoots are used fresh for flavouring fruit salads and iced drinks. The dried leaves retain their fragrance and are used for pot-pourri. Frequently thrown out from gardens and becoming naturalised and self-sown. Many records in Great Britain north to central Scotland, and in southern Ireland. Our plant is subsp. **officinalis** which is native of southern Europe, but it is widely introduced elsewhere.

19. Satureja L.

Low, aromatic, monoecious evergreen *shrubs*. *Leaves* opposite, entire. *Inflorescence* of 1–3 flowers in contracted cymes in the leaf-axils all along the branches; flowers bisexual. *Calyx* with 5 subequal lobes. *Corolla* white, tinted pink, 2-lipped, the upper lip emarginate and more or less flat, the lower lip 3-lobed. *Stamens* 4, shorter than the upper lip. *Style* 1; stigmas 2, more or less equal. *Fruit* of 4, ovoid, smooth nutlets.

Contains about 30 species in temperate and warm regions.

Vaughan, J. G. & Geissler, C. (1997). *The new Oxford book of food plants*. Oxford.

1. S. montana L. Winter Savory

Low, sweet-scented, monoecious, evergreen *shrub*. *Stems* numerous, up to 50 cm, pale brown, erect, quadrangular, with dense, short, curled eglandular hairs and minute glands, much branched above, the branches pale green and ascending, leafy. *Leaves* opposite; lamina 10–30 × 1.5–4.0 mm, medium green on upper surface, paler beneath, linear, obtuse to acute at apex, entire, margins curved upwards, narrowed at base, with numerous, curled, short eglandular hairs and glandular-punctate on both surfaces, 1-veined, sessile. *Inflorescence* of 1–3 bisexual flowers in contracted cymes in the leaf axils all along the branches; peduncles up to 3 mm, pedicels up to 1.0 mm, both minutely curled eglandular hairy. *Calyx* 5–6 mm, with 5, subequal lobes, the lobes subulate, acute at apex, with numerous, small, curly eglandular hairs and minute glands. *Corolla* 6–12 mm, white, tinted pink, with purple spots, 2-lipped, the upper lip oblanceolate, emarginate at apex, the lower lip 3-lobed, the lobes obovate and rounded. *Stamens* 2 long and 2 short; filaments white; anthers white. *Style* 1, white; stigma branched. *Nutlets* 4, ovoid, smooth. *Flowers* 7–9. $2n=30$.

Introduced. Grown in gardens and more or less naturalised on old walls. Somersetshire and Hampshire, less persistent elsewhere in southern England. Native of southern Europe. Winter Savory was well known to the Romans who mixed it with vinegar and used it as a sauce during feasts. It is used today as a flavouring for poultry, meats and soups. The essential oil may be used in processed foods.

20. Clinopodium L.

Perennial monoecious, aromatic *herbs* with short, creeping rhizome. *Leaves* opposite. *Inflorescence* a terminal spike of distant or crowded, many-flowered whorls; flowers bisexual. *Calyx* tube curved, 5-lobed. *Corolla* bright pink; tube narrowly infundibuliform, 2-lipped, the upper lip slightly concave and distantly emarginate, the lower lip 3-lobed. *Stamens* 4, inserted in the corolla tube. *Style* 1; stigmatic lobes very unequal. *Nutlets* 4, bluntly trigonous.

Contains about 10 species in north temperate zone.

Grime, J. P. et al. (1988). *Comparative plant ecology*. London.
Hultén, E. & Fries, M. (1986). *Atlas of north European vascular plants north of the Tropic of Cancer.* 3 vols. Königstein.

1. C. vulgare L. Wild Basil

Satureja vulgaris (L.) Fritsch; *Melissa clinopodium* Benth.; *Calamintha clinopodium* Spenn.; *Calamintha vulgaris* (L.) Karst.

Perennial monoecious, aromatic *herb* with a short, creeping rhizome. *Stems* 20–80 cm, pale green, sharply quadrangular, erect, with numerous, retrorse, white eglandular hairs, simple or sparingly branched, leafy. *Leaves* opposite; lamina 1.0–4.5 × 1.0–2.5 cm, medium green on upper surface, paler beneath, ovate, obtuse or subacute at apex, sometimes apiculate, subentire or obscurely crenate-serrate, broadly cuneate or rounded at base, shortly eglandular hairy on both surfaces with some minute glands; petiole 2–7 mm, channelled above, eglandular hairy. *Inflorescence* of 2–3, remote, compact, many-flowered whorls; flowers bisexual or sometimes female; bracts closely resembling the leaves, but usually smaller; pedicels 1.0–1.5 mm, terete, densely eglandular and minutely glandular hairy; bracteoles up to 10 mm, filiform, with spreading bristles. *Calyx* with tube 6–8 mm, prominently 13-veined, slightly curved but not noticeably gibbous at base, obscurely 2-lipped, the upper lobes 2.0–2.5 mm, subulate, the lower 3.5–5.5 mm, with a mixture of minute glandular and long, spreading eglandular hairs. *Corolla* bright pink; tube 11–16 mm, narrowly infundibuliform, shortly hairy, 2-lipped, the upper lip 4–5 mm, slightly concave and distinctly emarginate, the lower lip 3-lobed, the median lobe subrotund, emarginate, and concave, the lateral lobes smaller and convex; palate bearded with thick, white hairs. *Stamens* 4, inserted about 2.5 mm down the corolla tube; filaments flattened, glabrous, the upper 2.5–3.0 mm, included, the lower 4–5 mm, shortly exserted; anthers purplish. *Style* 1, 11–15 mm; stigmatic lobes very unequal. *Nutlets* 4, about 1.0 × 0.8 mm, brown, broadly oblong, bluntly trigonous, granulose. *Flowers* 7–9. Pollinated by bees and Lepidoptera; protandrous, small female flowers occur. $2n = 20$.

Native. Hedgerows, wood-borders and grassland, coastal cliffs and sand dunes, typically on light soils. Frequent in Great Britain north to central Scotland; very few, scattered records in Ireland; Alderney. Widespread in Europe, eastwards to central Asia and India; North America; introduced elsewhere. Circumpolar Temperate element. Our plant is subsp. **vulgare.**

21. Calamintha Mill.

Perennial monoecious, more or less foetid *herbs*. *Leaves* opposite. *Inflorescence* of rather distant whorls; flowers bisexual. *Calyx* tubular, with 13 veins, the tube straight, hairy within. *Corolla* with a straight tube, naked within; 2-lipped. *Stamens* 4, shorter than the corolla, curved and convergent. *Style* 1, branches unequal, the upper subulate, the lower longer and broadened. *Fruit* of 4, ovoid and smooth nutlets.

About 30 species in north temperate regions.

Bromfield, W. A. (1843). Notice of a new British *Calamintha*, discovered in the Isle of Wight. *Phytologist* **1**: 768–770.
Bromfield, W. A. (1845). Observations on a description of *Calamintha sylvatica*, a new British plant. *Phytologist* **2**: 49–52.
Easy, G. M. S. (1993). Calamints in Cambridgeshire. *Nature in Cambridgeshire* **35**: 63–65.
Easy, G. M. S. (1998). In Rich, T. C. G. & Jermy, A. C. *Plant crib.* London.
Wigginton, M. J. (Edit.) (1999). *British red data books.* Vol. 1. *Vascular plants.* Ed. 3. Peterborough.
Winship, H. R. (1994). *The conservation of Wood Calamint,* Clinopodium menthifolium *(Host) Stace on the Isle of Wight.* Peterborough.

In the Botanic Garden at Cambridge *C. menthifolium, ascendens* and *nepeta* grow mixed with apparent fertile intermediates between them. *Clinopodium vulgare* and *Acinos arvensis* also grow with them and remain distinct. This confirms the work by Easy (1993, 1998) who found intermediates in Cambridgeshire, and adds *C. menthifolium* to the complex.

1. Calyx 10–16 mm; corolla 25–40 mm **1. grandiflora**
1. Calyx 4–10 mm; corolla 10–22 mm 2.
2. Lower lobes of calyx 1–2 mm, with both eglandular and glandular hairs all 0.1 mm or very few longer; hairs in throat of calyx protruding beyond tube **4. nepeta** subsp. **glandulosa**
2. Lower lobes of calyx 2–4 mm, with many eglandular hairs more than 0.2 mm; hairs in throat of calyx usually entirely included 3.
3. Leaves often more than 4 cm, with 6–10 teeth of each side; lower calyx lobes 3–4 mm; corolla 12–22 mm **2. menthifolium**
3. Leaves rarely more than 4 cm, with 3–8 teeth on each side; lower calyx lobes 2–3(–3.5) mm; corolla 10–16 mm **3. ascendens**

1. C. grandiflora (L.) Moench Greater Calamint

Clinopodium grandiflorum (L.) Stace; *Satureja grandiflora* (L.) Scheele

Perennial monoecious, foetid *herb*. *Stems* 20–60 cm, pale green, erect, sparingly eglandular hairy and with dense minute glands, quadrangular, branched, leafy. *Leaves* opposite; lamina 3–8 × 2–5 cm, medium green on upper surface, paler beneath, ovate or ovate-oblong, acute at apex, coarsely dentate or serrate, with 6 or more teeth on each side, sparsely eglandular hairy and with minute glands. *Inflorescence* of opposite axillary cymes, each cluster of 1 to 5 flowers; flowers bisexual. *Calyx* (10–)12–16 mm, with dense minute glands, 2-lipped, the upper lip with 3 lobes 2–3 mm, the lower lip with 2 lobes 3.5–5.0 mm, the lobes long-ciliate. *Corolla* 25–40 mm, pink, 2-lipped, the upper lip entire, the lower lip 3-lobed with the middle lobe the longest, the tube straight. *Stamens* 2 long and 2 short, included. *Style* 1, branches unequal. *Nutlets* 4, ovoid. *Flowers* 7–9. $2n=22$.

Introduced. Grown in gardens and persists rarely as a throw-out. Native of southern and south-central Europe from Spain eastwards; south-west Asia.

2. C. menthifolium Host Wood Calamint
C. sylvatica Bromf.; *C. officinialis* subsp. *sylvatica* (Bromf.) Hook.; *Satureja calamintha* subsp. *sylvatica* (Bromf.) Briq.; *Satureja sylvatica* (Bromf.) K. Maly; *Satureja menthifolia* (Host) Fritsch; *Clinopodium menthifolium* (Host) Stace

Perennial monoecious *herb* with a short, creeping rhizome, strongly scented. *Stem* 30–60 cm, pale green, often suffused brownish-purple, quadrangular, with numerous, long, pale, unequal spreading simple eglandular hairs and minute glands. *Leaves* opposite; lamina 1.5–7.0 × 1.0–4.5 cm, medium yellowish-green on upper surface, paler beneath, sometimes dark red tinted, ovate, more or less obtuse, crenate or crenate-serrate, with 6–10 teeth on either side, rounded or broadly cuneate at base, shortly eglandular hairy on both surfaces and the margins and glandular-punctate beneath; petiole up to 2 cm, eglandular hairy. *Inflorescence* of opposite axillary bisexual flower clusters with a common peduncle up to 10 mm; both peduncles and pedicels with minute glandular hairs and short, stiff simple eglandular hairs. *Calyx* 7–10 mm, with minute glandular hairs and longer eglandular hairs on the veins, 2-lipped, upper lip with 3 lobes, the lobes triangular-ovate and acute at apex, the lower lip with 2 subulate lobes 3–4 mm, the margins of all lobes with hairs up to 1 mm. *Corolla* 12–22 mm, pink or lilac with darker spots, 2-lipped, the upper lip shortly 2-lobed, the lower lip 3-lobed, hairs of throat included. *Stamens* 2 short and 2 long; filaments white; anthers lilac. *Styles* 1, pale lilac. *Nutlets* 4, 1.0–1.2 mm, ovoid, tuberculate. *Flowers* 7–9. Visited by bumble-bees. $2n=24$.

Native. Woodland edges and scrubby thickets on chalk. Known only from a dry chalk valley on the Isle of Wight where it formerly occupied a few hectares but is now known only in a few square metres. From France eastwards to the south-west Ukraine. European Temperate element.

3. C. ascendens Jord. Common Calamint
C. sylvatica subsp. *ascendens* (Jord.) P. W. Ball; *Satureja ascendens* (Jord.) K. Malý; *Clinopodium ascendens* (Jord.) Samp.; *Satureja ascendens* var. *briggsii* (Syme) Druce; *C. menthifolia* var. *briggsii* Syme

Perennial monoecious *herb* with slender roots and horizontal rhizomes, strongly mint-scented. *Stems* several, 30–60 cm, pale yellowish-green, sometimes suffused brownish-purple, ascending or reclining, quadrangular, with numerous to dense, spreading, pale simple eglandular hairs and dense, minute glandular hairs, more or less branched, leafy. *Leaves* opposite; lamina 1.5–4.0 × 1.0–3.5 cm, rather pale green on upper surface, even paler beneath, ovate, rounded at apex, shallowly crenate-serrate, cuneate to rounded at base, with numerous simple eglandular hairs on the margins and veins beneath and scattered ones elsewhere, and with minute glandular hairs and covered with punctate glands beneath; petiole up to 1.5 cm, densely eglandular hairy. *Inflorescence* of distant whorls; flowers bisexual; pedicels and peduncles short and eglandular hairy. *Calyx* 5–8 mm, green, cylindrical, ribbed, 2-lipped, the lower lip with 2 subulate lobes 2–3 mm, the upper lip with 3 triangular-ovate, acute lobes, with numerous stiff eglandular hairs and dense minute glandular hairs. *Corolla* 10–16 mm, white with purple spots, hairy outside and in the throat, 2-lipped, the upper lip broadly oblong and emarginate at apex, the lower lip 3-lobed, the outer lobes rounded, the centre lobe broader than long with a wavy margin. *Stamens* 4, 2 long and 2 short; filaments white; anthers deep pink. *Style* 1, pale pink, glabrous. *Nutlets* 4, minute, pale brown, ovoid. *Flowers* 7–9. $2n=48$.

Native. Dry banks and rough grassland, usually calcareous. Local in Great Britain and Ireland north to Donegal, Isle of Man and Durham. West, south and south-central Europe; Caucasus; North Africa. European Temperate element.

× nepeta
See notes under genus.

4. C. nepeta (L.) Savi Lesser Calamint
Melissa nepeta L.; *Satureja nepeta* (L.) Fritsch; *Clinopodium calamintha* (L.) Stace

Perennial monoecious *herb* with a long, creeping rhizome, smelling strongly of mint. *Stems* 30–60 cm, pale green, erect, grey with dense, long, unequal spreading eglandular hairs and dense, minute glands, much branched, leafy. *Leaves* opposite; lamina 1–3 × 0.8–2.5 cm, medium yellowish-green, paler beneath, ovate, rounded at apex, shallowly crenate or crenate-serrate, broadly cuneate or truncate at base, soft eglandular hairs and minute glands on both surfaces and glandular-punctate beneath; petioles up to 1.5 cm, densely eglandular hairy. *Inflorescence* 5–11(–15)-flowered, rather distant whorls; flowers bisexual; peduncles up to 5 mm, shortly and densely glandular-hairy; pedicels short, densely glandular-hairy. *Calyx* 4–6 mm, with dense, minute glandular hairs and fewer longer simple eglandular hairs, 2-lipped, the upper lip with 3 straight or somewhat spreading, triangular-lanceolate, acute lobes, the lower lip with 2 subulate teeth 1.5–2.0 mm, hairs on margin of lobe less than 0.2 mm, hairs in throat protruding beyond tube of calyx. *Corolla* 10–15 mm, lilac, with few spots on lower lip, 2-lipped, the upper lip with 2, oblong, rounded lobes, the lower lip with 3 ovate, rounded lobes. *Stamens* 2 long and 2 short; filaments white; anthers lilac. *Styles* 1, lilac; stigma bifid. *Nutlets* 4, ovoid, smooth. *Flowers* 7–8.

Native. Dry banks and rough grassland, usually calcareous. Local in central-east England, extending west to Pembrokeshire; Channel Islands; formerly in Yorkshire; decreasing. South, west and south-central Europe from France to south Russia; North Africa; north Syria and north Iran. Submediterranean Subatlantic element. Our plant is subsp. **glandulosa** (Req.) P. W. Ball (*Thymus glandulosus*

Req.; *C. glandulosa* (Req.) Benth.; *C. officinalis* Moench; *Satureja calamintha* subsp. *glandulosa* (Req.) Gams; *Melissa calamintha* L.; *Satureja calmintha* (L.) Steele; *Clinopodium calamintha* (L.) Stace.

22. Acinos Mill.

Annual monoecious *herbs*. *Leaves* opposite. *Inflorescence* of axillary whorls of 3–8, bisexual flowers forming a lax terminal spike; bracts not differentiated from leaves; bracteoles minute. *Calyx* tubular, 13-veined, the tube curved and contracted in the middle, hairy within, gibbous at base, 2-lipped, the upper lip 3-lobed, the lower lip 2-lobed, the lobes longer than those of the upper. *Stamens* 2 long and 2 short, curved and converging. *Style* 1, branches unequal. *Fruit of* 4, ovoid, smooth nutlets.

Contains about 10 species in Europe and the Mediterranean region to central Asia and Iran.

Hultén, E. & Fries, M. (1986). *Atlas of north European vascular plants north of the Tropic of Cancer.* 3 vols. Königstein.

1. A. arvensis (Lam.) Dandy Basil Thyme
Clinopodium acinos (L.) Kuntze; *Satureja acinos* (L.) Scheele

Annual monoecious *herb*, or rarely a short-lived perennial, with fibrous roots. *Stems* 10–40 cm, pale green, ascending, quadrangular, eglandular hairy with minute glands, often branched from the base, leafy. *Leaves* opposite; lamina 0.5–1.5 × 0.4–0.7 cm, medium green on upper surface, paler beneath, lanceolate, ovate or elliptical, obtuse to acute at apex, entire or obscurely crenate, cuneate at base, glabrous or with few eglandular hairs and minute glands, conspicuously veined beneath. *Inflorescence* of axillary whorls of 3–8 bisexual flowers forming a lax, terminal spike; bracts not differing from the leaves and usually not exceeded by the flowers; bracteoles minute. *Calyx* 5–7 mm, eglandular hairy, with dense minute glands, the tube contracted in the middle, 2-lipped, the upper lip 3-lobed, the lobes 0.7–1.5 mm, the lower lip 2-lobed, the lobes 1.5–2.5 mm, all lobes subulate. *Corolla* 7–10(–12) mm, usually violet, with white marks on the lower lip, the tube straight and naked within, 2-lipped, the upper lip entire, the lower lip 3-lobed, the middle lobe being the largest. *Stamens* 2 long and 2 short; shorter than the corolla, curved and converging. *Style* 1, branches unequal. *Nutlets* 4, ovoid, smooth. *Flowers* 5–9. Pollinated by bees. $2n = 18$.

Native. Bare or rocky ground and arable fields on dry, usually calcareous soils. Rather local in Great Britain north to central Scotland, sparse in the north and decreasing, very local and introduced in central and south-east Ireland. Most of Europe except the extreme north and parts of the south; Turkey; Caucasus. European Temperate element.

23. Hyssopus L.

Low, evergreen, slightly aromatic, monoecious *shrubs*. *Leaves* opposite, entire. *Inflorescence* with congested whorls above and distant ones below in the axils of reduced leaves; flowers bisexual. *Calyx* with 5 equal lobes. *Corolla* blue, rarely white, 2-lipped, the upper lip shortly 2-lobed and more or less flat, the lower lip 3-lobed. *Stamens* 2 long and 2 short, longer than corolla. *Style* 1; stigma bifid. *Fruit* of 4, trigonous-ellipsoid nutlets.

Contains about 15 species in southern Europe and North Africa to central Asia.

1. H. officinalis L. Hyssop

Low, evergreen, slightly aromatic, monoecious *shrub*. *Stems* up to 60 cm, pale green, erect to ascending, quadrangular, minutely curly eglandular hairy, branched, leafy. *Leaves* opposite; lamina 10–25 × 4–7 mm, dark green on upper surface, paler beneath, narrowly elliptical-oblong to more or less linear, obtuse to acuminate at apex, entire, narrowed at base, with short, curly eglandular hairs and densely glandular-punctate on both surfaces, sessile, 1-veined. *Inflorescence* with congested whorls above and distant ones below in the axils of reduced leaves; flowers bisexual; peduncles and pedicels very short and minutely eglandular hairy. *Calyx* 8–10 mm, green, suffused brownish-red, with 5 equal lobes, the lobes triangular-lanceolate and aristate, with short, crisped eglandular hairs and scattered, shining, sessile glands. *Corolla* 7–12 mm, deep bluish-purple, rarely white, 2-lipped, the upper lip shortly 2-lobed, the lower lip 3-lobed, all lobes with wavy margins. *Stamens* 2 long and 2 short, all longer than the corolla; filaments mauve; anthers dark red. *Style* 1, mauve; stigma bifid. *Nutlets* 4, about 2 mm, chestnut-brown, trigonous-ellipsoid. *Flowers* 7–8. Pollinated by bees; protandrous. $2n = 12$.

Introduced. Grown in gardens and naturalised on old walls. Very scattered records through Great Britain. Native of south Europe.

24. Origanum L.

Perennial monoecious, aromatic *herbs* or *shrubs*. *Leaves* opposite. *Inflorescence* of whorls grouped into short spicules arranged in a corymb or panicle; flowers bisexual; bracts exceeding the calyx, imbricate, distinct from the leaves. *Calyx* campanulate, with 5 nearly equal lobes. *Corolla* 2-lipped, the upper lip 2-lobed, the lower lip 3-lobed. *Stamens* 4, straight, longer than corolla. *Stigma* 1. *Fruit* of 4, ovoid, smooth nutlets.

In Europe when buying herbs called Oregano or Marjoram you might get *O. majorana*, *O. onites* L., *O. syriacum* L., *O. vulgare*, hybrids between *O. majorana* and *O. vulgare* or a mixture of these species. In the American market not only *Origanum* species are found, but other genera and sometimes other families (Ietswaart, 1980).

Contains about 38 species in the north temperate Old World, mainly in the Mediterranean region.

Hultén, E. & Fries, M. (1986). *Atlas of north European vascular plants north of the Tropic of Cancer.* 3 vols. Königstein.
Ietswaart J. H. (1980). *A taxonomic revision of the genus* Origanum (*Labiatae*). Leiden Botanical Series **4**.

1. Calyx split almost to the base into 2 lips **1. majorana**
1. Calyx with 5 subequal teeth 2.

2. Leaves and calyx usually conspicuously glandular-punctate; bracts 1.5–6.0 × 1–3 mm **2(b). vulgare** subsp. **hirtum**
2. Leaves and calyx usually not conspicuously glandular-punctate; bracts 2–11 × 1–7 mm 3.
3. Bracts usually partly purple; flowers pink **2(a). vulgare** subsp. **vulgare**
3. Bracts usually green; flowers usually white 4.
4. Bracts 3.5–11 × 2–7 mm, glabrous or becoming so, yellowish-green; inflorescence compact **2(c). vulgare** subsp. **virens**
4. Bracts 2–8 × 1–4 mm, often densely hairy, usually green; inflorescence not compact **2(d). vulgare** subsp. **viride**

1. O. majorana L. Pot Marjoram
Marjorana tenuifolia Gray, *Amaracus vulgaris* (Mill.) Hill; *O. odorum* Salisb.; *Majorana hortensis* Moench; *Majorana vulgaris* Mill.; *Majorana ovalifolia* Stokes

Bushy, aromatic, monoecious *shrub* 50–100 cm or often grown as an annual; old branches woody, subterete and with flaking greyish-brown bark; young stems sharply tetragonal and greyish-hairy with eglandular crispate hairs, leafy. *Leaves* opposite; lamina 2.0–2.2 × 2–15 mm, medium green on upper surface, paler beneath, oblong, obovate, subrotund or subspathulate, rounded or subacute at apex, entire, broadly to narrowly cuneate at base, thinly to densely grey-hairy with short, appressed, eglandular crispate hairs, rarely nearly glabrous; veins very obscure on both surfaces; petiole up to 8 mm, slender. *Inflorescence* narrow, consisting of 5–7 pairs of clustered partial inflorescences extending far down the flowering stem, the branches of lateral inflorescences divaricate, subequal and usually less than 20 mm; flowers bisexual in compact, subglobose or very shortly oblong, 4-ranked spikes; bracts 3.0–3.5 × 2.0–2.5 mm, broadly obovate or spathulate, gland-dotted and minutely tomentose. *Calyx* about 2 mm, sheath-like, split almost to the base into 2 lips, conspicuously gland-dotted and minutely tomentose. *Corolla* white; tube about 2 mm, infundibuliform; lobes 5, about 1 mm, deltoid; hairy and gland-dotted. *Stamens* 4, inserted in a thinly lanuginous zone about 0.5 mm below the base of the corolla lobes; filaments 4–5 mm, glabrous, exserted; anthers pale, reniform. *Style* 3–4 mm, glabrous, exserted; stigma shortly and acutely 2-lobed. *Nutlets* about 1 mm in diameter, suborbicular, rich brown, distinctly compressed, minutely granulose. *Flowers* 7–9.

Introduced. Grown as a herb and occasionally escaping on to tips and waste places. Probably native of Cyprus and southern Turkey but cultivated and naturalised in many parts of the world. It has been cultivated for at least two centuries in the gardens of Europe as a medicinal herb.

2. O. vulgare L. Wild Marjoram

Perennial monoecious, aromatic *herb* with a rhizome. *Stems* 30–100 cm, pale green, often flushed purple, erect, quadrangular, with curved eglandular hairs, branched above and often with very short, axillary, sterile branches and often somewhat woody, leafy. *Leaves* opposite; lamina 1.5–4.5 × 1–2 cm, pale green on upper surface and even paler beneath, ovate, obtuse at apex, entire or obscurely crenate-serrate, rounded or subcordate at base, glabrous or with scattered appressed hairs on both surfaces, and glandular punctate, sometimes densely, shortly petiolate. *Inflorescence* of whorls grouped into short spicules arranged in a corymb or panicle; flowers bisexual or female; bracts exceeding the calyx, purple, ovate, imbricate, distinct from the leaves. *Calyx* 2.8–3.0 mm, reddish-purple, campanulate, 13-veined, with 5 short, nearly equal lobes, with eglandular hairs and yellow-glandular-punctate and hairy within. *Corolla* 4–8 mm, pinkish-mauve to rose-purple, tube longer than calyx, hairy, 2-lipped, upper lip 2-lobed, lower lip 3-lobed, all lobes rounded at apex. *Stamens* 4, straight, diverging, longer than the corolla; filaments 4.0–5.5 mm, anthers reddish-purple, turning brown. *Style* 1; stigma with 2 short branches. *Nutlets* 4, about 1 mm, ovoid, smooth. *Flowers* 7–9. Pollinated by various insects. Protandrous. Small female flowers occur. $2n = 30$.

(a) Subsp. **vulgare**
O. latifolium Mill.; *O. orientale* Mill.; *O. anglicum* Mill.; *O. floridum* Salisb.

Stems erect or ascending and rooting at the bases, branches up to 10 pairs. *Leaves* 0.3–5.0 × 0.2–3.3 cm, ovate, obtuse to acute at apex, eglandular hairy or becoming glabrous, sessile glands not conspicuous; petioles up to 15 mm. *Spikes* ovoid to cylindrical; bracts 6 pairs, 2–7 × 1–4 mm, more or less vividly purple, membranous, glabrous or becoming so. *Calyx* 2–4 mm. *Corolla* 4–10 mm, pink or purple. *Stamens* with filaments 4–5 mm. $2n = 30, 32$.

(b) Subsp. **hirtum** (Link) Ietsw.
O. hirtum Link; *O. megastachyum* Link; *O. hirtum* var. *prismaticum* Vogel

Stems more or less erect, usually long hairy; branches up to 10 pairs. *Leaves* 0.2–3.3 × 0.1–3.0 cm, ovate or elliptical, obtuse to acute at apex, densely eglandular hairy and with conspicuous sessile glands; petioles up to 12 mm. *Spikes* ovoid or cylindrical; bracts 5 pairs, 1.5–5.0 × 1.0–3.0 mm, green, sometimes slightly purple, green and hairy. *Calyx* 2.0–3.5 mm. *Corolla* 3.0–7.5 mm, white, or slightly pink. *Stamens* with filaments 4–5 mm. $2n = 30$.

(c) Subsp. **virens** (Hoffmanns. & Link) Ietsw.
O. virens Hoffmanns. & Link; *O. virescens* Poir.

Stems erect; branches up to 10 pairs. *Leaves* 0.3–3.5 × 0.1–2.2 cm, ovate or elliptical, obtuse or acute at apex, eglandular hairy becoming glabrous, sessile glands not conspicuous; petioles up to 15 mm. *Spikes* cylindrical or ovoid; bracts 7 pairs, 3.5–11 × 2–7 mm, membranous, usually glabrous, usually yellowish-green. *Calyx* 3–5 mm. *Corolla* 7–11 mm, white, rarely pink. *Stamens* with filaments 4.5–5.5 mm. $2n = 30$.

(d) Subsp. **viride** (Boiss.) Hayek
Stems erect; branches up to 20 pairs. *Leaves* 0.2–5.5 × 0.2–3.0 cm, ovate or subrotund, obtuse, eglandular hairy, sessile glands not conspicuous; petioles up to 15 mm. *Spikes* ovoid or cylindrical; bracts 5 pairs, 2–8 × 1–4 mm, membranous, hairy to glabrous. *Calyx* 2.5–5.0 mm. *Corolla* 5–9 mm, white. *Stamens* with filaments 3.5–5.0 mm.

Native. Dry grassland, hedgebanks and scrub, usually on calcareous soils. Locally common in Great Britain and Ireland, north to central Scotland. Most of Europe, North-west Africa and the Atlantic Islands. Subsp. *viride* occurs from Corsica to east China. All these subspecies are grown in gardens and occasionally escape. Subsp. *vulgare* is the native plant.

25. Thymus L.

Small, aromatic, monoecious *shrubs*. *Leaves* opposite, small. *Inflorescence* a terminal spike or head of few-flowered whorls; bracts more or less differentiated; bracteoles minute; flowers bisexual. *Calyx* 2-lipped, the upper lip with 3 more or less equal teeth, the lower 2-lobed. *Corolla* 2-lipped. *Stamens* 4, straight, diverging, longer than the corolla. *Style* 1, 2-branched. *Fruit* of 4 ovoid, smooth nutlets.

Contains about 300 species in temperate Eurasia.

Grime, J. P. et al. (1988). *Comparative plant ecology*. London.

Hultén, E. & Fries, M. (1986). *Atlas of north European vascular plants north of the Tropic of Cancer.* 3 vols. Königstein.

Marhold, K. (1998). Typification of the name *Thymus serpyllum* L. (Lamiaceae). *Bot. Jour. Linn. Soc.* **128**: 271–276.

Pigott, C. D. (1955). Biological flora of the British Isles. No. 50. *Thymus* L. (pp. 365–368); *Thymus drucei* Ronniger emend. Jalas (pp. 369–379); *Thymus serpyllum* Linn. emend. Mill. subsp. *serpyllum* (pp. 379–382); *Thymus pulegioides* Linn. (pp. 383–387). *Jour. Ecol.* **43**: 365–387.

Ronniger, K. (1924). Contributions to the knowledge of the genus *Thymus:* the British species and forms. *Rep. Bot. Soc. Exch. Club Brit. Isles* **7**: 226–239.

Ronniger, K. & Druce, G. C. (1924). *Thymus zetlandicus* Ronn. & Druce. *Rep. Bot. Soc. Exch. Club Brit. Isles* **7**: 649.

Trist, P. J. O. (1979). *An ecological flora of Breckland.* East Ardley. [*T. serpyllum.*]

Wigginton, M. J. (Edit.) (1999). *British red data books.* Vol. 1. *Vascular plants.* Ed. 3. Peterborough. [*T. serpyllum.*]

1. Plant without procumbent stems rooting at nodes; leaf margin revolute so that leaves are linear to narrowly oblong-elliptic in outline **1. vulgaris**
1. Plant usually with procumbent stems rooting at the nodes; leaf margins not or scarcely revolute 2.
2. Lower internodes of flowering stems with eglandular hairs all or nearly on the angles **2. pulegioides**
2. Lower internodes of flowering stems with eglandular hairs mainly on 2 or 4 faces 3.
3. Lower internodes of flowering stems with eglandular hairs on all faces more or less evenly distributed **3. serpyllum**
3. Lower internodes of flowering stems with eglandular hairs mainly on 2 opposite faces, the other 2 faces without them or nearly so 4.
4. Inflorescence up to 1.5 mm in diameter; leaves 1.5–3.0 mm wide 5.
4. Inflorescence 1.5–2.0 cm in diameter; leaves 3.0–3.5 mm wide 6.
5. Leaves eglandular hairy on upper surface **4(i). polytrichus** var. **britannicus**
5. Leaves without eglandular hairs or nearly so **4(ii). polytrichus** var. **neglectus**
6. Leaves very eglandular hairy on upper surface **4(iii). polytrichus** var. **zetlandicus**
6. Leaves without eglandular hairs or nearly so on upper surface **4(iv). polytrichus** var. **drucei**

1. T. vulgaris L. Garden Thyme

T. aestivus Reut. ex Willk.; *T. ilerdensis* Medrano ex Costa; *T. valentinus* Rouy; *T. webbianus* Rouy

Small, monoecious *shrub* with a stout root. *Stems* 10–40(–50) cm, pale brown, woody, glabrous or very shortly eglandular hairy, with erect to semi-patent, woody branches, leafy; young shoots pale yellowish-green, often tinted brownish-purple towards base, with dense, very short, white eglandular hairs. *Leaves* opposite, 3–8 × 0.5–2.5 mm, dull medium green on upper surface, paler and greyer beneath, linear, elliptical or lanceolate, obtuse at apex, entire, covered with sessile glands on both surfaces and giving off the characteristic smell of thyme, the margins revolute and the midrib prominent beneath, shortly petiolate, scarcely exceeding the axillary leaf clusters. *Inflorescence* capituliform or interrupted, many-flowered; flowers bisexual; *bracts* greyish-green, similar to the leaves but wider and with almost flat margins. *Calyx* 3–4 mm, pale yellowish-green, the tube campanulate, the 3 upper teeth as long as wide, the lower teeth long-linear, with short, spreading, pale simple eglandular hairs, the upper teeth not ciliate, the lower ciliate. *Corolla* 3.5–4.0 mm in diameter, white to pale pink, distinctly longer than the calyx, tube straight, the upper lip entire, the lower lip 3-lobed, the lobes more or less equal and emarginate. *Stamens* 4; filaments white; anthers whitish. *Style* exceeding the corolla, white; stigma branches white. *Nutlets* 4, ovoid, smooth. *Flowers* 5–8. $2n=30$.

Introduced. A garden herb used fresh or dried in stuffing for rich meats, fish and casseroles. The flower is mild, a little sweet and slightly spicy. Commercially, thyme is used in soaps, in antiseptic preparations and in the liqueur Benedictine. It is also used for ground cover, edging or in crevices in rock gardens. It becomes naturalised on old walls or stony banks. Scattered records in Great Britain from the Channel Islands to southern Scotland. Native of the west Mediterranean region extending to south-east Italy.

2. T. pulegioides L. Large Thyme

T. chamaedrys Fr.; *T. ovatus* Mill.; *T. glaber* Mill.

Tufted, strongly aromatic, monoecious dwarf *shrub. Stems* up to 25(–44) cm, woody at base, greenish, often tinged red, with short, creeping, sometimes prostrate branches and ascending flowering branches; flowering stems below the inflorescence sharply 4-sided, with long eglandular hairs only on the angles, 2 opposite faces narrow and very shortly eglandular hairy, the other 2 wider and glabrous. *Leaves* opposite; lamina 6–10(–18) × 3–6(–10) mm, medium green on upper surface, paler beneath, ovate to elliptical, obtuse at apex, entire, cuneate at base, ciliate at base, otherwise with only punctate glands on both surfaces, marginal vein absent; lateral veins not prominent beneath when dry; petiole short. *Inflorescence* elongated and usually interrupted below when normally developed, but capitate in grazed plants; flowers bisexual; bracts similar to leaves. *Calyx* 3–4 mm, campanulate; upper lobes usually longer than wide, at least one-sixth as long as calyx, usually eglandular ciliate and glandular-punctate. *Corolla* about 6 mm, pinkish-purple, 2-lipped, the upper lip with 3 more or less equal teeth, the lower 2-lobed. *Stamens* 4, straight, diverging, longer than the corolla. *Style* 1, 2-branched. *Nutlets* 4, ovoid, smooth. *Flowers* 7–8. $2n=28, 30$.

Native. Short, fine turf on rather bare places on well-drained chalky or sandy soils. Locally frequent in south and central England with scattered records north to Yorkshire. Europe except parts of the north east and many of the islands. European Temperate element.

3. T. serpyllum L. Breckland Thyme

Strongly aromatic, monoecious dwarf *shrub*. *Stems* with long, slender, creeping non-flowering branches, woody at base and rooting at nodes, sometimes with a terminal inflorescence; flowering stems rarely more than 10 cm, densely eglandular hairy all round. *Leaves* 5–10 × (1–) 2–3(–4) mm, all more or less equal, medium green on upper surface, paler beneath, linear to elliptical, obtuse at apex, entire, subsessile, eglandular ciliate at base, glandular-punctate, the lateral veins disappearing towards margin. *Inflorescence* usually capitate; flowers bisexual. *Calyx* 3–4 mm, campanulate; upper lobes about as long as wide, usually ciliate. *Corolla* 6–7 mm, purple, 2-lipped, the upper lip with 3 more or less equal teeth, the lower lip 2-lobed. *Stamens* 4. *Style* 1, 2-branched. *Nutlets* 4, ovoid, smooth. *Flowers* 7–8. $2n=24$.

Native. Sandy heaths of Breckland. Over about 30 km in Norfolk and Suffolk, also formerly in Cambridgeshire. Europe northwards from north-east France, north Austria and north Ukraine.

4. T. polytrichus A. Kern. ex Borbás Wild Thyme
T. praecox auct.; *T. serpyllum* auct.

Mat-like, aromatic, monoecious dwarf *shrub*. *Stems* to 7 cm, with long, creeping branches, the flower-stems in rows on the branches; flowering stems below inflorescence 4-angled, with 2 opposite sides densely eglandular hairy, with hairs of varying length, the other 2 sides less eglandular hairy or glabrous, very rarely all equally eglandular hairy. *Leaves* opposite; lamina 4–8(–11) × 1.5–4.0 mm, usually borne horizontally, rarely upright, subrotund to elliptical or oblanceolate, obtuse at apex, entire, glabrous or eglandular hairy as well as eglandular ciliate and glandular-punctate, flat, lateral veins prominent beneath when dry, curved along leaf margin and anastomosing at apex. *Inflorescence* capitate, rarely somewhat elongated; flowers bisexual; bracts similar to the leaves. *Calyx* 3.5–4.0 mm, the tube campanulate, about equalling upper lip; upper lobes longer than wide with eglandular hairs and glandular-punctate. *Corolla* rose-purple, 2-lipped, the upper lip with 3 more or less equal teeth, the lower 2 lobed. *Stamens* 4. *Style* 1, 2-branched. *Nutlets* 4, ovoid, smooth. *Flowers* 5–8.

(i) Var. **britannicus** (Ronniger) P. D. Sell
T. britannicus Ronniger

Leaves 1.5–3.0 mm wide, with eglandular hairs on upper surface. *Inflorescence* subglobose or somewhat elongated, up to 1.5 cm in diameter.

(ii) Var. **neglectus** (Ronniger) P. D. Sell
T. neglectus Ronniger

Leaves 1.5–2.5 mm wide, without or with very few eglandular hairs. *Inflorescence* globose, up to 1.5 cm in diameter.

(iii) Var. **zetlandicus** (Ronniger & Druce) P. D. Sell
T. zetlandicus Ronniger & Druce

Leaves about 3 mm wide, with numerous eglandular hairs on upper surface. *Inflorescence* 1.5–1.8 cm in diameter.

(iv) Var. **drucei** (Ronniger) P. D. Sell
T. drucei Ronniger

Leaves 3.0–3.5 mm, without eglandular hairs or with a few. *Inflorescence* 1.5–2.0 cm in diameter.

Native. Short grassland on heaths, downland, sea-cliffs and sand-dunes and around rock outcrops and hummocks in calcareous mires; upland grassland, montane cliffs, rocks and ledges. Throughout Great Britain and coastal Ireland on calcareous or base-rich substrates. South, west and central Europe. European Boreo-temperate element. Var. *britannicus* is the most widespread plant. Var. *neglectus* seems to occur mainly in coastal areas. Var. *zetlandicus* occurs in Shetland. Var. *drucei* seems to occur in mountainous northern regions. These four varieties are all thought to be part of subsp. **britannicus** (Ronniger) Kerguélen (*T. praecox* subsp. *britannicus* (Ronniger) Holub). It is considered that *T. serpyllum* var. *ligusticus* Briq. (*T. pycnotrichus* subsp. *ligusticus* (Briq.) Stace) is a variety of *T. pycnotrichus* subsp. *pycnotrichus*, and is a European mountain plant.

26. Lycopus L.

Perennial monoecious, odourless *herbs* with creeping rhizomes. *Leaves* opposite. *Inflorescence* of many-flowered, distant whorls; flowers bisexual; bracts not differentiated from the leaves; bracteoles small. *Calyx* campanulate, with 5 equal lobes, with 13 veins. *Corolla* small, tube shorter than calyx, with 2 lips. *Stamens* 2, longer than corolla. *Style* 1. *Fruit* of 4 tetrahedral, nutlets truncate at apex.

Contains about 14 species in temperate regions of the northern hemisphere.

Hultén, E. & Fries, M. (1986). *Atlas of north European vascular plants north of the Tropic of Cancer.* 3 vols. Königstein.

1. L. europaeus L. Gypsywort

Perennial monoecious, odourless *herb* with a creeping rhizome. *Stem* 30–100 cm, erect, pale green, sometimes tinged purple, quadrangular, glabrous or slightly eglandular hairy, usually with ascending branches, leafy. *Leaves* opposite; lamina 3–10 × 1–4 cm, pale to medium green on upper surface, paler beneath, lanceolate to ovate or elliptical, acute at apex, dentate to pinnately lobed, the teeth triangular, grading to narrow, acute lobes, cuneate to rounded at base, slightly eglandular hairy, shortly petiolate. *Inflorescence* of many-flowered, distant, axillary whorls; flowers bisexual or female; bracts not different from the leaves; bracteoles small. *Calyx* 3.5–4.0 mm, pale green, campanulate, divided over halfway into 5 subequal lobes, the lobes lanceolate and with drawn-out spiny points, with 13 veins, shortly eglandular hairy. *Corolla* 3–5 mm, white, with magenta-purple spots on the lobes of the lower lip, tube shorter than calyx, 2-lipped, the upper lip with 2 rounded lobes, the lower lip with 3 subequal lobes. *Stamens* 2, diverging, longer than the corolla; filaments pale; anthers pink, turning brown, cells parallel. *Style* 1, stigma with 2 short branches. *Nutlets* 4, 1.8–2.0 mm, tetrahedral, truncate at apex, hooded. *Flowers* 6–9. Pollinated by various insects. Protandrous. Small female flowers occur. $2n=22$.

Native. Fens, wet fields, by lakes and waterways, the top of beaches and dune slacks. Common over most of England and Wales, localities much more scattered in Channel Islands, Scotland and Ireland. Europe northwards to 64° 30′ N in Fennoscandia; North Africa; north and central Asia; naturalised in North America. Eurosiberian Temperate element.

27. Mentha L.

Perennial, rarely *annual, herbs* with creeping rhizomes and or stolons and scented foliage. *Leaves* opposite. *Inflorescence* composed of dense, many-flowered whorls sometimes forming a long spike or a terminal head; flowers bisexual or female on the same or different plants. *Calyx* actinomorphic or weakly 2-lipped, tubular or campanulate, 10- to 13-veined, with 5(4) subequal or rarely unequal teeth. *Corolla* weakly 2-lipped, with 4 subequal lobes, the upper lobe wider and usually emarginate; tube shorter than the calyx. *Stamens* 4, about equal, divergent or ascending under the upper lip of the corolla, exserted except in *M. pulegium*, some hybrids and female flowers. *Style*-branches subequal. *Fruit* of 4, smooth, reticulate or tuberculate nutlets.

Contains about 25 species in north and south temperate regions of the Old World.

The mints have been arranged into major taxa according to their evolution. This has put together fundamentally different-looking plants under the same main taxon. We have tried to give these lesser taxa varietal names so that they can be more easily keyed out. In any case we do not know how often these plants actually hybridise and many populations are probably clones which have spread vegetatively.

Briquet, J. (1891). *Les Labiées des Alpes Maritimes.* Geneva and Basle.
Fraser, J. (1927). *Menthae Britannicae. Rep. Bot. Soc. Exch. Club Brit. Isles* **8**: 213–247.
Fraser, J. (1934). *Mentha × niliaca* Jacq. var. webberi Frazer, var. nov. *Rep. Bot. Soc. Exch. Club Brit. Isles* **10**: 479–480.
Graham, R. A. (1948). Mint notes. 1. *Mentha rubra. Watsonia* **1**: 88–90.
Graham, R. A. (1950). Mint notes. 2. *Mentha gracilis* Sole and its relationship to *Mentha cardiaca* Baker. *Watsonia* **1**: 276–278.
Graham, R. A. (1951). Mint notes. 4. *Mentha piperata* L. and the British peppermints. *Watsonia* **2**: 30–35.
Graham, R. A. (1951). Mint notes. 5. *Mentha aquatica*, and the British water mints. *Watsonia* **3**: 109–121.
Graham, R. A. (1958). Mint notes. 7. *Mentha x maximiliana* F. Schultz in Britain. *Watsonia* **4**: 72–76.
Graham, R. A. (1958). Mint notes. 8. A new mint from Scotland. *Watsonia* **4**: 119–121.
Grime, J. P. et al. (1988). *Comparative plant ecology*. London.
Harley, R. M. (1967). The spicate mints. *Proc. B.S.B.I.* **6**: 369–372.
Hultén, E. & Fries, M. (1986). *Atlas of north European vascular plants north of the Tropic of Cancer.* 3 vols. Königstein.
Kay, Q. O. N. & John, R. (1994). *Population genetics and demographic ecology of some scarce and declining vascular plants of Welsh lowland grassland and related habitats.* Science Report no. 93. Bangor. [*M. pulegium.*]
Morton, J. K. (1956). The chromosome numbers of the British Menthae. *Watsonia* **3**: 244–252.
Murray, J. M. (1958). *Evolution in the genus Mentha. Proc. 10th Int. Congr. Genetics* **2**: 201.
Olsson, U. (1967). Chemotaxonomic analysis of the cytotypes in the *Mentha × verticillata* complex (Labiatae). *Bot. Not.* **120**: 255–267.
Pugsley, H. W. (1935). A new British mint. *Jour. Bot. (London)* **73**: 75–78.
Sole, W. (1798). *Menthae Britannicae*. Bath.
Smith, J. E. (1800). Observations on the British species of *Mentha. Trans. Linn. Soc.* **5**: 171–217.
Stewart, A., Pearman, D. A. & Preston, C. D. (1994). *Scarce plants in Britain.* Peterborough. [*M. pulegium.*]
Still, A. L. (1938). *Mentha verticillata* L. var. *trichodes* Briq. *Rep. Bot. Soc. Exch. Club Brit. Isles* **11**: 663.
Wigginton, M. J. (Edit.) (1999). *British red data books.* Vol. 1. *Vascular plants.* Ed. 3. Peterborough. [*M. pulegium.*]

1. Stems filiform, procumbent, rooting, mat-forming; flowers less than 6 per node **14. requienii**
1. Stems sometimes procumbent but not rooting along length and mat-forming; flowers usually more than 6 per node 2.
2. Whorls usually all axillary, the axis terminated by leaves, or by a reduced whorl; bracts like the leaves but reduced 3.
2. Upper whorls contracted into a terminal long or rounded head; upper bracts much reduced, unlike leaves 10.
3. Calyx with hairs in throat, lower 2 lobes narrower and slightly longer than 3 upper 4.
3. Calyx without hairs in throat; calyx lobes more or less equal 5.
4. Plant prostrate **13(i). pulegium** var. **pulegium**
4. Plant erect **13(ii). pulegium** var. **erecta**
5. Calyx 1.5–2.5 mm, including triangular teeth which are less than 0.5 mm, hairy all over; usually fertile **1. arvensis**
5. Calyx 2–4 mm, including narrowly triangular to subulate lobes which are 0.5–1.5 mm, glabrous or hairy; usually sterile 6.
6. Calyx 2.5–4.0 mm, including lobes which are more than 1 mm, the tube about twice as long as wide 7.
6. Calyx 2.0–3.5 mm, including teeth which are usually less than 1 mm, the tube less than twice as long as wide 8.
7. Plant hairy; calyx mostly less than 3.5 mm; stamens usually included **4. × verticillata**
7. Plant nearly without hairs; calyx mostly more than 3.5 mm; stamens usually exserted **5. × smithiana**
8. Leaves variegated **2(i). × gracilis** var. **variegata**

8. Leaves all green 9.
9. Leaves pale green, lanceolate or ovate-lanceolate, hairy **2(ii). × gracilis** var. **gracilis**
9. Leaves deep green, elliptic-oblong, glabrous or sparingly hairy **2(iii). × gracilis** var. **cardiaca**
10. Leaves distinctly petiolate; flower-head 12–25 mm in diameter 11.
10. Leaves sessile or more or less so; flower-head 5–15 mm in diameter 16.
11. Leaves and calyx tube without or with few simple eglandular hairs 12.
11. Leaves and calyx with numerous simple eglandular hairs 14.
12. Leaves crisped at margin **6(iv). × piperita** var. **crispa**
12. Leaves not crisped at margin 13.
13. Inflorescence an oblong spike; smelling of peppermint **6(i). × piperita** var. **piperata**
13. Inflorescence globose; smelling of lemon **6(iii). × piperita** var. **citrata**
14. Leaves lanceolate to narrowly ovate **6(ii). × piperita** var. **hirsuta**
14. Leaves ovate to narrowly ovate or subrotund 15.
15. Leaves ovate to ovate-lanceolate; inflorescence a head **3. aquatica**
15. Leaves ovate to subrotund; inflorescence a spike **7. × suavis**
16. Leaves subrotund to ovate, strongly rugose, hairy, obtuse to more or less rounded at apex, with teeth bent under and thus appearing as crenations from above; corolla whitish; fresh plant with a sickly scent **12. suaveolens**
16. Leaves lanceolate to ovate-oblong, rugose or not, with or without eglandular hairs, acute to subobtuse at apex, with teeth not bent over and thus appearing acute from above; corolla pinkish; fresh plant with a sweet scent 17.
17. Plant hairy to densely hairy; corolla pinkish; leaves often with acuminate teeth that curve outwards and become patent, especially near leaf-base, never broadly ovate to subrotund 18.
17. Plant with few to dense hairs; corolla white to pinkish; leaves with acute, usually forwardly directed teeth unless leaves broadly ovate to subrotund 20.
18. Leaves lanceolate-elliptical, broadest near the middle, narrowed into rounded to subcurvate base; plant sterile **11. × villosonervata**
18. Leaves oblong-lanceolate, ovate or subrotund, with a broad rounded base; plant normally fertile 19.
19. Leaves subrotund or ovate, lanate beneath **10(i). × rotundifolia** var. **rotundifolia**
19. Leaves oblong-ovate, with dense eglandular hairs beneath but not lanate **10(ii). × rotundifolia** var. **webberi**
20. Leaves without or with very few eglandular hairs beneath 21.
20. Leaves with dense eglandular hairs beneath 22.
21. Leaves lanceolate or ovate-lanceolate, acute or acuminate at apex **8(i). spicata** var. **spicata**
21. Leaves broadly oblong-ovate, rounded-acute at apex **9(i). × villosa** var. **cordifolia**
22. Leaves subrotund or broadly ovate **9(iv). × villosa** var. **alopecuroides**
22. Leaves elliptical or oblong-ovate to lanceolate 23.
23. Leaves elliptical, ovate or ovate-oblong, with large teeth, with very short, dense hairs beneath **9(ii). × villosa** var. **villosa**
23. Leaves broadly oblong-ovate or lanceolate 24.
24. Leaves lanceolate **8(ii). spicata** var. **scotica**
24. Leaves oblong-ovate **9(iii). × villosa** var. **nicholsoniana**

Section **1. Mentha**

Bracts variable. *Whorls* usually many-flowered. *Calyx* tubular or campanulate, with 5 more or less equal lobes, the throat glabrous within. *Corolla* tube straight.

1. M. arvensis L. Corn Mint

Vary variable *perennial* or rarely *annual*, monoecious *herb* with a sickly scent and a rhizome. *Stems* 10–60 cm, pale green, 4-angled, ascending or erect, quadrangular, more or less eglandular hairy, simple or branched, leafy. *Leaves* opposite; lamina (1.5–)2.0–5.0(–7.0) × 1–3 cm, medium green on upper surface, paler beneath, elliptical, elliptic-lanceolate to broadly ovate to subrotund, obtuse to acute at apex, shallowly crenate or serrate, cuneate or rounded at base, more or less eglandular hairy on both surfaces and glandular-punctate, petiolate. *Inflorescence* of distant, axillary, compact whorls forming a terminal spike; flowers bisexual; bracts like the leaves, gradually decreasing in size upwards, but always much longer than the flowers; pedicels glabrous or with eglandular hairs. *Calyx* 1.5–2.5 mm, broadly campanulate, eglandular hairy all over and glandular-punctate; 5-lobed, the lobes less than 0.5 mm, equal, about as long as wide, broadly triangular, often deltate, obtuse to acuminate at apex. *Corolla* 5.0–5.5 mm, lilac, white or rarely pink; tube straight, hairy outside; 5-lobed, the lobes ovate and obtuse at apex. *Stamens* 4, more or less equal, exserted; anther purplish-rose. *Style* 1, branches equal. *Nutlets* 4, about 1 mm, pale brown, ellipsoid. *Flowers* 5–10. Usually fertile. $2n = 72$.

Very variable in size and shape of leaf and general eglandular hairiness. Many varieties have been named but there seems to be numerous intermediates between all of them. All plants in a locality may, however, be uniform which is perhaps due to rhizomatous spread.

Native. Arable fields, woodland rides, marshy pastures and waste places. Throughout Great Britain and Ireland, but has substantially declined on arable land since 1950. Most of Europe, north Asia and the Himalaya. Circumpolar Boreo-temperate element widely naturalised outside its native range.

2. M. × gracilis Sole Bushy Mint
M. arvensis × spicata
M. gentilis auct.

Variable *perennial* aromatic *herb* with creeping rhizomes. *Stems* 30–90 cm, pale green, sometimes red, 4-angled, erect, usually glabrous, sometimes eglandular hairy, leafy. *Leaves* opposite; lamina (1.5)3.0–7.0(–9.0) × (0.8–)1.5–4.0(–4.5) cm, medium green on upper surface, paler beneath, sometimes variegated, ovate-lanceolate, lanceolate or elliptic-oblong, acute at apex, serrate, cuneate at base, in cultivars sometimes sparsely eglandular-hairy but otherwise more densely so on both surfaces and glandular punctate. *Inflorescence* of remote whorls; flowers bisexual; bracts like the leaves or decreasing distinctly in size upwards, the uppermost sometimes shorter than the flowers; pedicels with punctate glands, but without simple eglandular hairs. *Calyx* 2.0–3.5 mm, campanulate, with punctate glands, usually without simple eglandular hairs, the tube less than twice as long as wide; lobes 5, 0.5–1.0 mm, more or less subulate or rarely triangular, acuminate or with a setaceous point, often conspicuously ciliate. *Corolla* about 3 mm, lilac. *Stamens* 4, usually included, usually sterile. *Style* 1. *Nutlets* not usually produced. *Flowers* 8–10. $2n=54$, 60, 84, 96, 108, 120.

There are three fairly easily recognisable variants but it is not known how they might fit in with chromosome number. Although often sterile some seed-setting occurs with some back-crossing and segregation. Clones by rhizomatous spread also occur.

(i) Var. **variegata** (Sole) Sm.
M. variegata Sole

Leaves variegated, ovate or ovate-lanceolate, sparingly eglandular-hairy.

(ii) Var. **gracilis**
M. sativa subsp. *gracilis* (Sole) Hook. fil.;
M. × *dalmatica* auct.

Leaves pale green, lanceolate or ovate-lanceolate, eglandular-hairy.

(iii) Var. **cardiaca** Gray
M. cardiaca (Gray) Baker

Leaves deep green, elliptic-oblong, glabrous or sparingly eglandular-hairy. Often smelling of *M. spicata.*

M. arvensis is native, *M. spicata* introduced. Damp places and waste ground. Recorded throughout much of Great Britain and scarce in Ireland. Occurs over most of Europe. It is thought that var. *gracilis* occurs spontaneously, but var. *cardiaca* is an escape from gardens. Only old records have been seen of var. *variegata*, which presumably came from gardens.

3. M. aquatica L. Water Mint
M. hirsuta Huds.; *M. glomerata* Stokes nom. illegit.; *M. palustris* Mill. nom. illegit.; *M. aquatica* subsp. *hirsuta* (Huds.) Hook. fil.

Perennial monoecious, aromatic *herb. Stem* 15–90 cm, pale green often purplish in exposure, 4-angled, more or less erect, simple or branched, without eglandular hairs to tomentose and glandular punctate. *Leaves* opposite; lamina (1.5–)3.0–9.0 × (1.0–)1.5–4.0 cm, medium green on upper surface, paler beneath, ovate to ovate-lanceolate, obtuse to acute at apex, serrate or crenate-serrate, cuneate to subcordate at base, more or less eglandular hairy on both surfaces or occasionally nearly glabrous but with punctate glands, strongly but not pungently scented when crushed; petiole short. *Inflorescence* a terminal head about 20 mm across, composed of 2–3 congested whorls and usually 1–3 distant axillary whorls; flowers bisexual; pedicels eglandular hairy; bracts below leaf-like, those above lanceolate or linear-lanceolate, hidden by flowers. *Calyx* (2.5–)3–4 mm, tubular, the veins distinct, hairy; lobes 5, subulate or narrowly triangular, with eglandular hairs and punctate glands. *Corolla* lilac. *Stamens* 4, normally exserted. *Style* 1. *Nutlets* 4, pale brown. *Flowers* 7–10. $2n=96$.

Very variable in habit, leaf shape and indumentum. Many infraspecific taxa have been recognised, at least some of which retain their character in cultivation. It also spreads clonally by extensive rhizomes. We have followed the view of R. A. Graham (1954), that variation is so great and intermediates so numerous that none is worth recognising.

Native. Typically associated with permanently wet habitats adjacent to open water, often partially submerged, it occurs by ditches, ponds and rives, marshes, wet pasture, dune slacks and fens, and in wet woods. Throughout Great Britain and Ireland. Europe except the extreme north; south-west Asia; North and South Africa; Madeira. European Temperate element.

4. M. × **verticillata** L. Whorled Mint
M. aquatica × **arvensis**
M. sativa L.; *M. stricta* Stokes nom. illegit.

Very variable *perennial herb. Stem* 30–90 cm, pale green, 4-angled, erect, robust, with numerous eglandular hairs. *Leaves* opposite; lamina 2.0–6.5(–8.0) × (1.0–)1.5–4.0 cm, rather dark green on upper surface, paler beneath, usually ovate to elliptical, oblong or lanceolate, acute or obtuse at apex, serrate, cuneate to subcordate at base, with simple eglandular hairs and punctate glands on both surfaces, with a sickly scent when crushed; petiole short. *Inflorescence* of distant whorls below or crowded in the upper part, the whorls often stalked; bracts like the leaves often decreasing in size, the upper sometimes shorter than the flowers; pedicels eglandular hairy. *Calyx* 3.0–3.5 mm, tubular, the tube twice as long as wide, eglandular hairy and punctate glandular; lobes 5, more than 1 mm, usually about twice as long as wide, more or less equal and triangular, acuminate at apex, sometimes with a subulate point. *Corolla* lilac, hairy. *Stamens* 4, usually included, very rarely exserted. *Style* 1. Usually sterile. *Flowers* 8–10. The occasionally highly fertile plants are probably the result of backcrossing with the parents. Sometimes this occurs with only 1 or neither parent. $2n=42$, 84, 120, 132.

Very variable in leaf size, shape, dentation and hairiness.

Native. Damp places in arable fields, wet grassland, tracksides, marshes, woodland rides, river-banks, shores of lakes and disturbed ground. Throughout Great Britain and Ireland except for the west and north of Scotland and central Ireland. Frequent in mainland Europe though rare in the Mediterranean region.

5. M. × **smithiana** R. A. Graham Tall Mint
M. aquatica × **arvensis** × **spicata**
M. rubra Sm., non Mill.

Perennial monoecious *herb*, with creeping rhizomes. *Stems* 50–150 cm, 4-angled, pale green, usually red-tinged, erect to scrambling, simple or branched above, glabrous or more rarely sparsely eglandular hairy and with punctate glands, leafy. *Leaves* opposite; lamina 3.0–7.0(–9.0)×2.0–4.0 cm, rather dark green on upper surface, paler beneath, often purple-veined, sweetly pungent when pressed, ovate, obtuse to acute at apex, serrate, rounded to broadly cuneate at base, usually without eglandular hairs or nearly so, but sometimes thinly eglandular hairy on both surfaces, glandular punctate; petiole short. *Inflorescence* of remote whorls, sometimes crowded at apex of stem; flowers bisexual; bracts decreasing upwards, usually longer than the flowers, the uppermost subrotund, cuspidate at apex, usually serrate; pedicels without simple hairs. *Calyx* 3.5–4.0 mm, tubular, sparsely eglandular hairy above, glandular punctate; lobes (0.7–)1.0–1.5 mm, weakly ciliate. *Corolla* about 5 mm, lilac. *Stamens* 4, usually exserted. *Style* 1. Usually sterile, but sometimes producing a few variable seeds which give rise to segregating progeny. *Flowers* 8–10. $2n=120$.

M. aquatica and *M. arvensis* are native while *M. spicata* is introduced. Damp places and waste ground. Widespread in England and Wales, scattered records in Scotland and southern Ireland. Sometimes arises spontaneously where *M. spicata* and *M.×verticillata* grow together, but more typi cally a garden throw-out. Occurs widely in central Europe.

6. M.×piperita L. Peppermint
M. aquatica×spicata
M.×dumetorum auct.; *M. nigricans* Mill.

Perennial monoecious *herb. Stems* 30–80 cm, pale green, flushed reddish to purple, 4-angled, erect, usually branched, with minute glands, but occasionally also with eglandular hairs to grey tomentose. *Leaves* opposite; lamina 4.0–8.0 (–9.0)×1.5–4.0 cm, medium green on upper surface, paler beneath, ovate-lanceolate or lanceolate, rarely ovate, acute at apex, usually serrate, cuneate to subcordate at base, with or without eglandular hairs, with numerous minute glands, with a pungent smell and taste of peppermint or lemon; petioles long. *Inflorescence* a terminal, oblong spike (2.0)3.5–8.0 cm, usually interrupted at the base, or a globose head; flowers bisexual; pedicels with or without simple eglandular hairs, but with minute sessile glands. *Calyx* 3–4 mm, tubular, the tube with minute glands and sometimes eglandular hairs, often strongly tinted with red; lobes 5, subulate, long-ciliate. *Corolla* lilac-pink. *Stamens* 4, included. *Style* 1. *Sterile*. *Flowers* 7–10. $2n=66, 72, 84, 120$.

(i) Var. **piperita**
Leaves not crisped at margin. *Inflorescence* a terminal oblong spike. *Pedicels* and calyx with few simple eglandular hairs. *Smelling* of peppermint.

(ii) Var. **hirsuta** Fraser
Leaves not crisped at margin. *Inflorescence* a terminal oblong spike. *Pedicels* and calyx with numerous eglandular hairs. *Smelling* slightly of peppermint.

(iii) Var. **citrata** (Ehrh.) Boivin
M. citrata Ehrh.

Leaves not crisped at margin. *Inflorescence* globose. *Pedicel* and calyx without or with occasional simple eglandular hairs. *Smelling* of lemon.

(iv) Var. **crispa** (L.) W. D. J. Koch
M. crispa L.

Leaves crisped at margin. *Inflorescence* a terminal oblong spike. *Pedicels* and calyx more or less glabrous or the calyx shortly eglandular hairy.

Introduced. Peppermint came into general use in the medicine of western Europe only about the middle of the eighteenth century. In the nineteenth century it was extensively cultivated in some parts of England. Peppermint is a source of essential oil, which is obtained by distillation from the fresh flowering plants. Noted as a mild antiseptic and a remedy for flatulence. It is found in damp ground and waste places. The var. *hirsuta* is thought to be produced as a spontaneous hybrid while the other variants are garden escapes or throw-outs. This hybrid occurs throughout Great Britain and is much less frequent in Ireland. It is widespread in Europe both as a spontaneous hybrid and as a garden escape.

7. M.×suavis Guss. Sweet Mint
M. aquatica×suaveolens
M.×maximilianea F. W. Schultz

Perennial monoecious *herb. Stems* up to 60 cm, pale green, sometimes tinted reddish, 4-angled, erect, sometimes branched, with dense, long simple eglandular hairs and minute glands. *Leaves* opposite; lamina 2.5–5.0×1.5–3.5 cm, medium green on upper surface, paler beneath, ovate to subrotund, obtuse to acute at apex, very sharply serrate, rounded or subcordate at base, with numerous, long, white simple eglandular hairs, some branched hairs and minute glands; veins very prominent beneath; petioles up to 10 mm, with long white simple eglandular hairs. *Inflorescence* a short terminal spike up to 5 cm; flowers bisexual; bracts 5–9 mm, ovate, acute at apex, crenate-serrate, rounded at base, with numerous simple eglandular hairs and minute glands; pedicels reddish, with numerous, long, white simple eglandular hairs and minute glands. *Calyx* 2.5–3.0 mm, reddish, with white, simple eglandular hairs and minute glands; 4-lobed, the lobes linear, acute at apex. *Corolla* about 5 mm, reddish-pink. *Stamens* 4. *Style* 1. *Nutlets* 4. *Flowers* 8–9. $2n=72, 84$.

Native. This rare plant is recorded from only a few localities in Cornwall, Devonshire and Jersey where it grows with the parents. It occurs in scattered localities in Continental Europe.

8. M. spicata L. Spear Mint

Perennial monoecious *herb* with a rhizome. *Stems* 30–100 cm, pale green, erect, usually branched, without simple eglandular hairs, but with minute glands. *Leaves* opposite; lamina (3.0–)5.0–9.0×(0.7–)1.5–3.0 cm, medium green on upper surface, paler beneath, smooth or rugose, lanceolate or ovate-lanceolate, acute or acuminate at apex, sharply serrate with regular teeth, without or with dense eglandular hairs, the hairs beneath both simple and branched, sunken glands numerous beneath, subsessile or the petiole not more than 3 mm, usually with a strong, sweet scent, less frequently pungent or musty. *Inflorescence* a terminal, cylindrical spike 30–60×5–10(–15) mm, the whorls often becoming more or less separate, with spikes often clustered at the top of the main axis; flowers bisexual; bracts longer than the flowers, linear-setaceous, the lowest pair sometimes leaf-like; pedicels with or without simple eglandular hairs. *Calyx* 1–3 mm, campanulate, glabrous or eglandular hairy; lobes 5, subequal. *Corolla* lilac, pink or white, glabrous. *Stamens* 4, exserted; fertile anthers 0.3–0.5 mm. *Style* 1. *Nutlets* 4, 0.7–0.9 mm, reticulate in hairy plants, smooth in glabrous plants. *Flowers* 8–9. $2n = 36, 48$.

(i) Var. **spicata**
Leaves glabrous or nearly so.Smelling of spearmint.

(ii) Var. **scotica** (R. A. Graham) P. D. Sell
M. × *niliaca* var. *sapida* auct.; *M. scotica* R. A. Graham

Leaves densely short eglandular hairy beneath.Smell unpleasant.

Introduced. Naturalised in a variety of damp or wet habitats and on rough and waste ground often close to habitation. Common throughout much of Great Britain, less common in Ireland. Naturalised throughout much of Europe. Thought to be of garden origin. The common form is var. *spicata*. Var. *scotica* as its name implies is known only from Scotland.

9. M. × **villosa** Huds. Apple Mint
M. spicata × **suaveolens**
M. × *cordifolia* auct.; *M.* × *niliaca* auct.

Perennial monoecious *herb* with a rhizome. *Stems* up to 60 cm, pale green, sometimes tinted red, erect, quadrangular, glabrous or nearly so to densely eglandular hairy. *Leaves* 2–9 × 1–6 cm, medium to dark green on upper surface, paler beneath, elliptical, ovate, oblong-ovate or subrotund, acute to rounded at apex, serrate or serrate-dentate the teeth small to large, rounded at base, without, with few or with dense eglandular hairs on both surfaces and numerous punctate glands, sessile. *Inflorescence* a dense terminal spike; flowers bisexual; bracts 5–6 mm, lanceolate, acute at apex, with few eglandular hairs. *Calyx* 2–3 mm, with few to dense eglandular hairs. *Corolla* about 4 mm, pink. *Stamens* 4. *Style* 1. *Nutlets* 4.

(i) Var. **cordifolia** (Opiz ex Fresen.) P. D. Sell
M. cordifolia Opiz ex Fresen.

Leaves broadly oblong-ovate, rounded-acute at apex, sharply serrate, subcordate at base, with none or very few simple eglandular hairs and numerous punctate glands.

(ii) Var. **villosa**
Leaves elliptical, ovate or ovate-oblong, acute at apex, serrate with rather large teeth, rounded at base, with fairly numerous eglandular hairs on upper surface and dense short eglandular hairs beneath, with punctate glands on both surfaces.

(iii) Var. **nicholsoniana** (Strail) R. Harley
M. nicholsoniana Strail

Leaves fairly broadly oblong-ovate, acute at apex, sharply serrate, rounded at base, with some simple eglandular hairs on upper surface and more or less dense, long ones beneath, with numerous punctate glands on both surfaces.

(iv) Var. **alopecuroides** (Hull) Briq.
M. alopecuroides Hull

Leaves subrotund or broadly ovate, rounded at apex, serrate-dentate with large teeth, rounded at base, with numerous simple eglandular and glandular hairs on upper surface and dense, short eglandular hairs beneath, with punctate glands on both surfaces.

M. spicata is introduced. *M. suaveolens* is native. The hybrid in its various varieties is grown in gardens and is naturalised in rough and grassy places. It is recorded throughout Great Britain but less so in Ireland. It is widespread in Europe.

10. M. × **rotundifolia** (L.) Huds. False Apple Mint
M. longifolia × **suaveolens**
M. × *niliaca* Juss. ex Jacq.

Perennial monoecious *herb* with rhizomes. *Stems* up to 90 cm, pale green, often suffused reddish, stout, quadrangular, erect, with dense, long, wavy simple eglandular hairs and long-stalked glandular hairs. *Leaves* opposite; lamina 1–9 × 1–5 cm, medium green on upper surface, paler beneath, subrotund, ovate, oblong-ovate or oblong-lanceolate, more or less acute to rounded at apex, crenate-serrate to crenate, rounded below, with numerous simple eglandular and stalked glandular hairs on upper surface, both type of hairs more densely so beneath, leaves sessile or nearly so. *Inflorescence* a dense spike at the end of stems and branches; flowers bisexual; bracts 3–4 mm, linear-lanceolate to ovate, acuminate at apex, with dense stalked glandular and simple eglandular hairs. *Calyx* 1.5–2.0 mm, lobes narrow and long acute, with dense glandular and simple eglandular hairs. *Corolla* 3.5–4.0 mm, pale pink or lilac, sometimes fading white. *Stamens* 4, included. *Style* 1. *Nutlets* 4. *Flowers* 7–9. Fertile. $2n = 24$.

(i) Var. **rotundifolia**
Leaves subrotund or ovate, rounded at apex, crenate-serrate, with numerous simple eglandular and glandular hairs above and lanate beneath including numerous, stalked glands.

(ii) Var. **webberi** (J. Fraser) R. Harley
M. × *niliaca* var. *webberi* Fraser

Leaves broadly oblong-ovate, rounded-acute at apex, sharply serrate, rounded at base, with some simple eglandular hairs on upper surface and dense ones beneath, numerous stalked glands on both surfaces.

M. suaveolens is native and *M. longifolia* introduced. Grown in gardens and a throw-out on rubblish-tips, waste ground and roadsides. Scattered throughout Great Britain and Ireland. Frequent in both Europe and south-west Asia with the parents.

11. M. × **villosonervata** Opiz Sharp-toothed Mint
M. longifolia × **spicata**
M. longifolia var. *horridula* auct.

Perennial monoecious *herb* with rhizomes, slightly sweet-scented. *Stems* up to 60 cm, pale green, often suffused red, erect, quadrangular, with few, minute hairs. *Leaves* opposite; lamina 2–8 × 0.5–3.5 cm, medium green on upper surface, paler beneath, linear-lanceolate, lanceolate or elliptic-lanceolate, acute at apex, sharply serrate, narrowed below, sessile, without or with very few simple eglandular hairs, with numerous punctate glands on both surfaces. *Inflorescence* a terminal cylindrical spike, some of the lower whorls sometimes becoming separate; flowers bisexual; bracts 4–5 × 1.0–1.5 mm, linear to linear-lanceolate, long-acute at apex, with numerous glandular and simple eglandular hairs. *Calyx* about 2 mm, with dense simple eglandular and glandular hairs. *Corolla* about 3 mm, white or pink. *Stamens* 4, included or exserted. *Style* 1. *Nutlets* 4. *Flowers* 8–9. $2n = 36, 48$.

Introduced. Often established in rough grassland, waste ground and on rubbish-tips as a garden escape and throw-out or as a relic of cultivation. Scattered records throughout Great Britain and a few records in Ireland. Probably widespread in Europe.

12. M. suaveolens Ehrh. Round-leaved Mint
M. rotundifolia auct.

Perennial monoecious *herb. Stems* 60–100 cm, pale green, erect, usually branched above the middle, sparsely to densely clothed with white, simple eglandular hairs. *Leaves* opposite; lamina (1.5–)3.0–4.5 × (1.0–)2.0–4.0m, medium green on upper surface, paler beneath, strongly rugose, ovate-oblong to subrotund, obtuse or minutely cuspidate at apex, crenate-serrate, the teeth sometimes cuspidate, rounded or subcordate at base, eglandular hairy on upper surface, usually grey- or white-tomentose beneath, the hairs on the lower surface branched, sessile or very shortly petiolate. *Inflorescence* of numerous whorls, usually congested forming a terminal spike 4–9 × 0.5–1.0 cm, often interrupted below and usually branched; pedicels eglandular hairy. *Calyx* 1–2 mm, campanulate, hairy; lobes 5, subequal. *Corolla* whitish or pink, hairy outside. *Stamens* 4, usually exserted; fertile anthers 0.2–0.3 mm. *Style* 1. *Nutlets* 4, 0.5–0.7 mm. *Flowers* 8–9. $2n = 24$.

Native. In damp places in coastal regions, elsewhere naturalised on roadsides and in waste places inland. Native in the southwest peninsular, Isle of Wight and coastal Wales. Widespread elsewhere as an introduction in Great Britain and Ireland. South and west Europe, North Africa and the Azores. Submediterranean Subatlantic element.

Section **2. Pulegium** (Mill.) DC.
Pulegium Mill.

Bracts similar to the leaves. *Calyx* tubular, weakly 2-lipped with unequal teeth, throat hairy. *Corolla* tube gibbous.

13. M. pulegium L. Pennyroyal
Pulegium vulgare Mill.

Perennial monoecious *herb* with a pungent odour and epigeal rhizomes. *Stems* 10–40 cm, pale green, often tinged or suffused with red, prostrate or erect, relatively stout, quadrangular, nearly glabrous to densely villous, branched, leafy. *Leaves* opposite; lamina 8–30 × 4–12 mm, medium green on upper surface, paler beneath, narrowly elliptical, rarely subrotund, obtuse at apex, entire to obscurely crenate-serrate, cuneate to attenuate at base, minutely eglandular hairy and glandular punctate; petiole short and eglandular hairy. *Inflorescence* of many-flowered whorls with bracts similar to the leaves but usually smaller, the whorls distant; pedicels eglandular hairy. *Calyx* (2.0–)2.5–3.0 mm, tubular, weakly 2-lipped, the upper lip 3-lobed, the lobes linear-lanceolate, the lower 2-lobed, the lobes longer and narrower than the upper, the lobes ciliate and hairy in the throat. *Corolla* (4.0–)4.5–6.0 mm, mauve-lilac, tubular, gibbous on the lower side, 4-lobed, the lobes lanceolate, rounded at apex, hairy outside, glabrous within. *Stamens* 4, more or less equal in length, usually exserted, rarely included; filaments pale; anthers rose fading to cream. *Style* 1; stigma bifid. *Nutlets* 4, about 0.7 mm, pale brown, ovoid, obtuse at apex. *Flowers* 8–10. $2n = 20$.

(i) Var. **pulegium**
Plant prostrate.

(ii) Var. **erecta** (Mill.) Syme
Pulegium erectum Mill.

Plant erect.

Native. Damp grassy or heathy places and by ponds, very local and often near the sea. Southern and central England, south Wales, Isle of Man, Jersey and Ireland, formerly much more widespread north to central Scotland; now sometimes naturalised and increasing as a grass-seed contaminant. Widespread in Europe, except in north Mediterranean area, and east to Iran. European Southern-temperate element.

Section **3. Audibertia** (Benth.) Briq.

Bracts like the leaves. *Whorls* 2- to 6-flowered. *Calyx* turbinate-campanulate, weakly 2-lipped; throat hairy within. *Corolla* tube straight.

14. M. requienii Benth. Corsican Mint
Menthella requienii (Benth.) Perard

Perennial monoecious *herb* with a pleasant, pungent scent. *Stems* 3–12 cm, pale green, sometimes tinted brownish-purple, filiform, diffuse, usually procumbent, rooting at the nodes and mat-forming, quadrangular, very shortly hairy, branched, leafy. *Leaves* opposite; lamina 2–7 × 2–7 mm, pale to medium yellowish-green on upper surface, paler beneath, ovate-subrotund, rounded at apex, entire or sinuate, rounded or truncate at base, sparsely eglandular hairy; petiole up to 5 mm, slender, eglandular hairy. *Inflorescence* of 2- to 6-flowered whorls; flowers bisexual, 1.6 mm in diameter; bracts like the leaves. *Calyx* 1.0–1.5 (–2.5) mm, turbinate-campanulate, weakly 2-lipped, the upper 3-lobed, the lower 2-lobed, the lobes triangular and subacute, hairy in the throat. *Corolla* about 2 mm, scarcely exserted, rose, not gibbous, tube straight, 4-lobed, the lobes 0.6 × 0.6 mm, triangular, ovate-lanceolate, subacute to rounded at apex. *Stamens* 4; filaments pink; anthers pink to purple. *Style* 1, pale green, stigma greenish. *Nutlets* 4, pale brown, ovoid, smooth. *Flowers* 6–8. $2n = 18$.

Introduced. Grown in gardens and escaping on to damp paths and rocky places. Scattered records in southern Great Britain and Ireland. Native of Corsica, Sardinia and the Italian island of Montecristo.

28. Lavandula L.

Evergreen, monoecious *shrubs* with a characteristic scent. *Leaves* opposite, entire. *Inflorescence* a spike of lax to crowded whorls above a long leafless part of the upper stem; flowers bisexual; bracts rhombic-obovate. *Calyx* with 5 subequal lobes, the uppermost with an obcordate appendage at apex. *Corolla* purple, rarely white, 2-lipped, the upper lip shallowly 2-lobed, the lower lip with 3 short lobes. *Stamens* 4, included within the corolla tube. *Style* 1. *Fruit* of 4, oblong, smooth nutlets.

Contains about 30 species, mainly in the Mediterranean region and eastwards through Asia to India; Atlantic islands; north temperate and tropical Africa.

Chaytor, D. A. (1938). A taxonomic study of the genus *Lavandula*. *Jour. Linn. Soc. London (Bot.)* **51**: 153–204.
Upson, T. & Andrews, S. (2004). *The Genus* Lavandula. Kew.

1. Leaves 0.3–0.5 cm, linear to narrowly ovate; bracts ovate; bracteoles absent or minute **1. angustifolia**
1. Leaves 0.2–1.5 cm, narrowly elliptic to ovate; bracts narrowly ovate; bracteoles obvious **2. × intermedia**

1. L. angustifolia Mill. English Lavender
L. spica L. nom. rejic.; *L. minor* Garsault; *L. officinalis* Chaix ex Vill.; *L. vulgaris* Lam.; *L. fragrans* Salisb.; *L. vera* DC.; *L. fragrans* Jord. ex Billot; *L. delphinensis* Jord. ex Billot

Evergreen, monoecious *shrub* with a characteristic scent. *Stems* 40–80 cm, pale brown, rough and scaling, terete; young shoots greyish-green to pale brown, erect, quadrangular, with stellate hairs. *Leaves* opposite; lamina 3–4 × 0.3–0.5 cm, greyish-green on upper surface, paler beneath, linear to narrowly ovate, sometimes revolute, obtuse at apex, with dense stellate hairs and sessile glands. *Inflorescence* a spike of crowded whorls; bracts about half as long as calyx; bracteoles about 1.0 mm, linear and scarious. *Calyx* 4.5–7.0 mm, the upper tooth with an obcordate appendage at the apex. *Corolla* 10–12 mm, purple, rarely pink or white, the upper lobes typically twice the size of the lower. *Stamens* 4; filaments white; anthers orange. *Style* 1. *Nutlets* 4, shining brown. *Flowers* 6–7.

Introduced. This is the lavender which is found as a garden escape growing from seed. presumably the plant found in scattered localities throughout Great Britain, but all records need checking including those in the *New Atlas of the British and Irish Flora* (2002). Native of the Mediterranean region. This species is the most valued of the lavenders both as a garden plant and for its high quality oil and it has a large number of cultivars.

2. L. × intermedia Loisel. Garden Lavender
L. angustifolia Mill. × **latifolia** Mill.
L. × burnatii Briq.; *L. × spica-latifolia* Albert; *L. hortensis* Hy; *L. × feraudii* Hy

Evergreen, monoecious *shrub* with a characteristic scent. *Stems* up to 1.0(–1.7) m, pale brown, rough and scaling, terete; with numerous, long, spreading branches; young shoots greyish-green to pale brown, erect, quadrangular, with numerous, minute stellate hairs. *Leaves* opposite, 2–6 × 0.2–1.5 cm, greyish-green on upper surface, slightly paler beneath, narrowly elliptic to ovate, obtuse at apex, entire, slightly narrowed at the sessile base, densely, minutely stellate-hairy when young, less so when mature, with numerous, shining, sessile glands; midrib impressed on upper surface, prominent beneath. *Inflorescence* a spike of lax to crowded whorls above a long, leafless part of the upper stem; bracts 3–8 mm, greyish-green, rhombic-obovate, acuminate at apex; bracteoles obvious. *Calyx* 5–7 mm, pale yellowish-green, flushed with purple, about 13–veined, with 5, small, subequal, obtuse lobes, the uppermost with an obcordate appendage at apex. *Corolla* 7–9 mm, pale to dark purple, rarely white, hairy, 2-lipped, the upper lip short, flat and shallowly 2-lobed, the lower lip short and 3-lobed. *Stamens* 4, included in the corolla tube; filaments white; anthers orange. *Style* 1, whitish; stigma pale brown. Usually sterile. *Flowers* 6–8. Much visited by bees. $2n = 48, 50$.

Many cultivated plants, of which there are numerous cultivars, may be this or a fertile hybrid.

Introduced. Much grown in gardens and a persistent throw-out which may take root. All records need to be checked.

29. Rosmarinus L.

Evergreen, monoecious *shrubs*. *Leaves* opposite, linear, entire with a revolute margin. *Inflorescence* of few-flowered, short, axillary racemes; flowers bisexual. *Calyx* campanulate, 2-lipped, the upper lip entire or minutely 3-lobed, the lower lip 2-lobed. *Corolla* 2-lipped, exserted, pale to deep mauvish-blue; upper lip strongly concave and 2-lobed; lower lip 3-lobed with an obovate middle lobe. *Stamens* 2, exserted, parallel, the filaments with a small, lateral recurved tooth near the base; anthers 1-locular. *Style* 1, long, incurved, unequally lobed. *Fruit* of 4, obovoid-oblong nutlets.

Contains 3 species in the Mediterranean region.

Bean, W. J. (1980). Trees and shrubs hardy in the British Isles. Ed. 8. **4**. London.

1. R. officinalis L. Rosemary

Evergreen, monoecious *shrub* up to 2 m with a very irregular outline. *Bark* pale brown and rough. *Branches* numerous, long and of various lengths, spreading and then erect, pale brown; young shoots pale yellowish-green, shortly woolly. *Leaves* opposite, crowded, strong-smelling, especially if crushed, 15–40 × 1.0–3.5 mm, dull medium yellowish-green on upper surface and glandular-punctate, green on the revolute sides beneath with a white channel down the centre, linear, obtuse at apex, entire, sessile, sparingly and minutely eglandular hairy. *Inflorescence* of bisexual flowers in 2s all the way down the long terminal shoots; bracts ovate, acute at apex; bracteoles minute. *Calyx* 7–8 mm, pale green or purplish, campanulate, 2-lipped, the upper lip minutely 3-lobed, the lower lip with 2 deltoid lobes, covered with whitish sessile glands. *Corolla* 10–15 mm, pale lilac, to deep mauvish-blue, rarely pink or whitish, tube infundibuliform, gibbous, upper lip 2-lobed, the lower lip 3-lobed, with narrow side lobes and an obovate, emarginate central lobe with darker markings. *Stamens* 2, longer than the corolla; filaments pale lilac, curved forwards, with a small lateral tooth near the base; anthers pale with deeper colour round the edge. *Style* 1, pointing upwards and incurved, long-exserted, lilac; stigma with unequal lobes. *Nutlets* 4, about 2.5 mm, dark brown, obovoid-oblong, minutely granulose, with an oblique, white basal attachment area and a conspicuous circular scar. *Flowers* 4–5. Visited by bees. $2n=24$.

Introduced since at least 1375. Much grown in gardens and self-sown on walls and in rough ground. Scattered records in Great Britain and the Channel Islands. Native of the north Mediterranean region to Lebanon.

30. Salvia L.

Annual to *perennial* monoecious *herbs*, rarely small *shrubs*. *Leaves* opposite, simple, crenate to serrate, rarely pinnatisect. *Inflorescence* of axillary whorls of bisexual flowers forming a more or less interrupted terminal spike. *Calyx* tubular or campanulate, 2-lipped, the upper lip entire or with 3 short lobes, the lower lip with 2 longer lobes. *Corolla* blue to purple or pinkish, or yellow, rarely red or white; 2-lipped, the upper lip shortly 2-lobed and strongly hooded, the lower lip 3-lobed. *Stamens* 2, shorter than upper lip of corolla, the connective much elongated. *Style* 1. *Fruit* of 4, ovoid-trigonous, smooth nutlets.

Contains about 700 species in tropical and temperate zones.

Druce, G. C. (1906). *Salvia marquandii* sp. n. *Jour. Bot. (London)* **44**: 405–407, *tab*. 483.

Druce, G. C. (1909). Salvia horminoides Pourret. Jour. Bot. (London) **47**: 87–88.

Hultén, E. & Fries, M. (1986). Atlas of north European vascular plants north of the Tropic of *Cancer.* 3 vols. Königstein.

McClintock, D. (1975). Marquand's Clary. *The wild flowers of Guernsey*. London.

Pugsley, H. W. (1908). The forms of *Salvia verbenaca* L. *Jour. Bot. (London)* **46:** 97–106; 141–151.

Pugsley, H. W. (1909). *Salvia horminoides. Jour. Bot.* (London) **47**: 89–91.

Pugsley, H. W. (1927). The nomenclature of the group *Salvia verbenaca* L. *Jour. Bot. (London)* **65**: 185–195.

Reales, A., et al. (2004). Numerical taxonomy study of *Salvia* sect. *Salvia*. (Labiatae). *Bot. Jour. Linn. Soc.* **145**: 353–371.

Wigginton, M. J. (Edit.) (1999). *British red data books*. Vol. 1. *Vascular plants*. Ed. 3. Peterborough.

1. Corolla more than 18 mm — 2.
1. Corolla less than 18 mm — 6.
2. Corolla yellow with reddish markings; leaves often hastate at base — **4. glutinosa**
2. Corolla blue to violet, pink or white; leaves never hastate — 3.
3. Bracts green, sometimes tinged violet-blue — 4.
3. Bracts scarlet, pink or white — 5.
4. Stems without glandular hairs — **2. officinalis**
4. Stems with glandular hairs — **5. pratensis**
5. Bracts scarlet — **1. splendens**
5. Bracts pink or white — **3. sclarea**
6. Flowers (8–)15–30 in each whorl — **10. verticillata**
6. Flowers 1–6(–8) in each whorl — 7.
7. Inflorescence with a conspicuous tuft of green or coloured flowerless bracts at apex — **9. viridis**
7. Inflorescence without a terminal tuft of conspicuous flowerless bracts at apex — 8.
8. Plant without glandular hairs — 9.
8. Plant with glandular hairs — 10.
9. Leaves rounded or cordate at base — **6. nemorosa**
9. Leaves cuneate at base — **8. reflexa**
10. Leaves crenate or serrate; corolla 15–30 mm — **5. pratensis**
10. Leaves crenate-serrate to irregularly pinnatipartite, with serrate-crenate lobes; corolla 6–17 mm — 11.
11. Leaves incise-crenate-dentate but not pinnately-lobed; corolla blue or purple usually without spots at the base of the lower lip; with rather few glandular hairs on stem — **7(a). verbenaca** subsp. **verbenaca**
11. At least some leaves pinnatipartite; corolla blue, often with 2 white spots at base of lower lip; with numerous to dense glandular hairs on upper stem — 12.
12. Leaves 2–14 × 1–8 cm; corolla 6–10 mm, cleistogamous — **7(b,i). verbenaca** subsp. **horminoides** var. **horminoides**
12. Leaves 2–5 (–8) × 1.0–3.0 cm; corolla 10–15 mm, chasmogamous — **7(b,ii). verbenaca** subsp. **horminoides** var. **anglica**

1. S. splendens Ker Gawl. — Scarlet Sage

Annual to *perennial* monoecious *herb*. *Stems* up to 50 cm, woody at base, erect, glabrous to eglandular hairy, branched, leafy. *Leaves* opposite; lamina 4–8 × 2–5 cm, medium green on upper surface, paler beneath, ovate, acute at apex, slightly serrate, rounded or cordate at base, glabrous to minutely eglandular hairy, glandular-punctate on lower surface, petiolate. *Inflorescence* of whorls of 2–6 bisexual flowers forming a terminal spike; pedicels up to 4 mm, eglandular hairy; bracts up to 12 × 5 mm, scarlet, ovate to lanceolate, acute at apex, deciduous. *Calyx* up to 22 mm, scarlet, tubular to campanulate, 8-veined, membranous, glabrous or hairy, 2-lipped, the upper lip entire, the lower lip 3-lobed, the lobes ovate, with an acute apex. *Corolla* scarlet, salmon-pink, lilac, lavender, purple, dark red or white; tube 30–40 mm, glabrous within; 2-lipped, the upper lip up to 13 mm and 2-lobed, the lower lip shorter and 3-lobed, minutely hairy. *Stamens* 2. *Style* 1. *Nutlets* 4, ovoid, smooth. *Flowers* 7–10.

Introduced. Frequently used as an annual bedding plant, but actually a low shrub, which occurs as a throw-out on tips and waste places. Native of Brazil.

2. S. officinalis L. — Sage

Small, monoecious *shrub* but often treated as an annual or biennial. *Stems* up to 60 cm, erect, patent-tomentose, with numerous branches, leafy. *Leaves* opposite, simple; lamina 2–5 × 0.5–1.5 cm, oblong, rounded at apex, crenate, more or less narrowed at base, greenish and rugose above, white-hairy beneath, petiolate. *Inflorescence* of axillary whorls of 5–10 bisexual flowers; bracts green. *Calyx* 10–14 mm, not or scarcely accrescent, eglandular hairy and glandular-punctate, 2-lipped, the upper lip with 3 lobes, the lower lip deeply 2-lobed. *Corolla* up to 35 mm, violet-blue, pink or white, 2-lipped, the upper lip straight and hooded, the lower lip 3-lobed, the middle lobe the largest, tube with a ring of hairs inside. *Stamens* 2, the connective shorter than or equalling the filament, arms more or less equal, one with a more or less sterile cell. *Style* 1, usually branched. *Nutlets* 4. *Flowers* 6–10.

This is probably an aggregate species for which Reales et al. (2004) should be consulted. Introduced. It is grown commercially in parts of the Midlands and southern England and in gardens over a much wider area. It is a persistent escape on to waste ground and tips. Native of parts of north and central Spain and southern France and widely naturalised in parts of south and south-central Europe. The non-flowering form is propagated by cuttings or by division and renewed every three or four years. The flowering form is raised from seed and treated as an annual or biennial. Sage is sold, fresh and dried in bunches, but the bulk of the crop is sold in packets alone or as part of mixed herbs; it contains a pungent oil. Sage and onion is a tradition British stuffing for meats and game dishes. It is added to sausages as a flavouring and preservative and makes a classic tisane.

3. S. sclarea L. Clary

Biennial or *perennial* monoecious *herb* with a strong smell. *Stems* several, up to 150 cm, pale green, brownish-purple towards the base, very robust, quadrangular, erect, with numerous medium to long, pale simple eglandular hairs in the upper part and mixed with short glandular hairs, branched only in the inflorescence, leafy. *Leaves* opposite; the lamina 7–22 × 5–15 cm, dull yellowish-green on upper surface, paler beneath, ovate, narrowed to an obtuse or subacute apex, crenate, the teeth wide but shallow, cordate at base, shortly eglandular hairy on both surfaces, especially the veins beneath, the upper surface rough and rugose, the lower surface with very prominent pale veins; petioles up to 15 cm, pale green, grooved above, rounded beneath, with long simple eglandular hairs. *Inflorescence* 20–45 cm, broadly oblong or ellipsoid, its branches ascending; flowers bisexual, 6–7 mm in diameter, smelling slightly; bracts 15–50 mm, pale pinkish-purple with deeper colouring round the margin, broadly ovate, with a cuspidate but double-pointed tip, the margin wavy, the base cordate, covered with simple eglandular and glandular hairs; branches with dense glandular and eglandular hairs. *Calyx* 12–15 mm, pale green, tinted on ribs and margins brownish-red, with short, triangular, cuspidate lobes, covered with numerous, short glandular hairs and longer, curved, stiff eglandular hairs. *Corolla* 25–27 mm, the tube pale yellow and very short, 2-lipped, the upper lip pale yellow and forming a folded hood, the lower lip pale lilac, folded and curved upwards, the whole shortly glandular-hairy. *Stamens* 4, 2 long and 2 short, the 2 longest in the folded lower lip; filaments white, tinged lilac; anthers yellow. *Style* 1, tinted lilac with a blue stigma, protruding between the stamens in the lower lip. *Nutlets* 4, 1-seeded, ovoid-trigonous, smooth. *Flowers* 7. Much visited by bumble-bees. $2n = 22$.

Introduced. Commonly grown in gardens and rather frequently a relic or throw-out on tips and in rough ground. Scattered records in southern Great Britain. Native of south Europe.

4. S. glutinosa L. Sticky Clary

Perennial monoecious *herb*. *Stems* up to 1 m, pale green, often suffused brownish-purple, erect, quadrangular, with numerous eglandular hairs below and with dense viscid glandular hairs above, branched, leafy. *Leaves* opposite; lamina 4–13 × 2–8 cm, medium yellowish-green on upper surface with impressed veins, paler beneath with prominent veins, ovate, acute or acuminate at apex, crenate or serrate-crenate, cordate or hastate at base, shortly eglandular hairy on both surfaces and the margins; petioles up to 8 cm, pale green, often suffused brownish-purple, densely reflexed-hairy. *Inflorescence* branched, of widely spaced whorls of 2–6 bisexual flowers; bracts 6–10 mm, green, ovate, acute at apex, entire, densely glandular-hairy. *Calyx* tubular, ridged, densely glandular-hairy; 2- lipped, the upper lip with 1 triangular-acute lobe, the lower lip with 2 triangular-acute lobes. *Corolla* 30–40 mm, yellow with reddish markings, glandular-hairy; tube narrow; 2-lipped, the upper lip narrowly hooded, the lower lip 3-lobed, the outer lobes entire, the central lobe frilled. *Stamens* 2, shorter than upper lip of corolla; filaments white; anthers purple. *Style* 1, white; stigma bifid. *Nutlets* 4, 4–5 mm, ellipsoid, smooth. *Flowers* 8–9. $2n = 16$.

Introduced. Grown in gardens and naturalised in woods, hedges and on road and river-banks. Few scattered records in England and Perthshire in Scotland. Native of southern Europe.

5. S. pratensis L. Meadow Clary

Perennial aromatic *herb*. *Stems* 30–100 cm, pale green, erect, quadrangular, with numerous simple eglandular hairs below and glandular hairs above, branched, leafy. *Leaves* opposite; lamina of basal 7–15 cm, bright dark green on upper surface with a pale midrib, paler beneath, ovate or ovate-oblong, obtuse at apex, crenate or serrate, cordate at base, petiolate; cauline few, smaller and sessile; all rugose and nearly glabrous to more or less hairy. *Inflorescence* of axillary whorls of 4–6 flowers forming a more or less interrupted terminal spike; bracts green, ovate, acuminate at apex, entire, shorter than the calyx; bracteoles small; flowers bisexual or female, usually on different plants, never cleistogamous; pedicels about 2 mm, glandular-hairy. *Calyx* 7–11 mm, flushed crimson, glandular-hairy, and with short eglandular hairs, but without long eglandular hairs, campanulate, 2-lipped, the upper lip 3-lobed, the lower lip 2-lobed. *Corolla* of bisexual flowers 15–20 mm, violet-blue, the tube exserted, 2-lipped, the upper lip laterally compressed, forming a hood and falcate, the lower lip 3-lobed, glandular-hairy outside, without a ring of hairs inside; female flowers much smaller, sometimes only

10 mm. *Stamens* 2, connective longer than filament, arms equal. *Style* 1, shortly branched. *Nutlets* 4, about 2.5 mm, ellipsoid. *Flowers* 6–7. Pollinated by long-tongued bees. Protandrous or homogamous. $2n=18$.

Native. Calcareous grassland, scrub and borders of woods. Very local in southern England from Kent to Gloucestershire and Monmouthshire; naturalised elsewhere in the centre and south of Great Britain. Most of Europe, northwards to north Germany and north-central Russia; Morocco. European Temperate element.

6. S. nemorosa L. Balkan Clary
S. sylvestris auct.

Perennial monoecious *herb. Stems* 30–60 cm, pale green, quadrangular, with spreading or appressed simple eglandular hairs, branched, leafy. *Leaves* opposite; basal with lamina 3.5 × 10 cm, medium green on upper surface, oblong, long acute at apex, crenate, rounded or cordate at base, petiolate; cauline smaller and more or less sessile; all more or less hairy. *Inflorescence* of axillary whorls of 2–6 bisexual flowers forming a more or less dense terminal spike; bracts as long as or longer than calyx, violet, ovate, rounded at base, imbricate in bud; pedicels about 2 mm. *Calyx* 6–7 mm, campanulate, 2-lipped, upper lip concave in fruit, 3-lobed, the lower lip 2-lobed, with very fine, closely appressed hairs. *Corolla* 8–12 mm, violet-blue, rarely pink or white, 2-lipped, the upper lip straight, the lower lip 3-lobed, tube without a ring of hairs inside. *Stamens* 2, staminal connective longer than filament, the arms unequal. *Style* 1, shortly branched. *Nutlets* 4, ellipsoid. *Flowers* 6–7. $2n=12$.

Introduced. Formerly long naturalised on dunes at Phillack Towans in Cornwall; now only a rare casual. Native of south-east Europe and south-west Asia.

× **pratensis** = **S.** × **sylvestris** L.
Recorded for Norfolk.

7. S. verbenaca L. Wild Clary

Perennial monoecious herb with a stout stock and a characteristic aroma. *Stems* 10–80 cm, pale green, sometimes purplish below, erect, quadrangular, with numerous simple eglandular hairs throughout and glandular hairs intermixed in the upper part, branched above, leafy. *Leaves* mostly basal, the cauline opposite; lamina 1.5–14.0 × 1–8 cm, medium green on upper surface, paler beneath, oblong to broadly ovate or oblong-elliptical, rounded at apex, varying from crenate-serrate to irregularly pinnatipartite, with serrate-crenate lobes, truncate or cordate at base, glabrous or with a few eglandular hairs on the veins beneath; petioles up to 9 cm, eglandular hairy, the uppermost cauline leaves sessile. *Inflorescence* of lax whorls, with 6–10 bisexual flowers in each whorl; pedicels with glandular hairs; bracts shorter than the calyx. *Calyx* 7–10 mm, often purple-tinted, with a mixture of long, white simple eglandular and short glandular hairs, 2-lipped, the upper lip 2-lobed and broad, the lower lip 3-lobed. *Corolla* 6–17 mm, blue, lilac or violet, sometimes with 2 white spots at the base of the lower lip, 2-lipped, the upper lip compressed and falcate, the lower lip 3-lobed with the middle lobe the largest and spreading and deflexed, with few to numerous or no glandular hairs. *Stamens* 2; filaments pale; anthers dark. *Style* 1; stigma branched, slightly exserted. *Nutlets* 4, ovoid-trigonous, smooth. *Flowers* 5–8. $2n=64$.

(a) Subsp. **verbenaca**
S. marquandii Druce; *S. pallidiflora* St Amans; *Gallitrichum pallidiflorum* (St Amans) Jord. & Fourr.; *S. verbenaca* var. *oblongifolia* Benth.

Stems 25–40 cm, slender, with numerous eglandular hairs and rather few glandular hairs. *Leaves* 1.5–7.0 × 1.0–2.5 cm, oblong, incise-crenate-dentate, but not pinnately lobed. *Calyx* tinted purple above, with eglandular and glandular hairs. *Corolla* 12–17 mm, blue or purple, usually without spots at the base of the lower lip. Usually chasmogamous.

(b) Subsp. **horminoides** (Pourr.) Nyman
S. horminoides Pourr.

Stems 10–80 cm, slender to robust, with numerous eglandular hairs and few to numerous glandular hairs. *Leaves* 2–14 × 1–8 cm, ovate-oblong to ovate, sinuate-dentate to irregularly pinnatipartite. *Calyx* suffused with dull bluish-purple, with numerous to dense glandular hairs. *Corolla* 6–15 mm, blue, often with 2 white spots at base of lower lip, cleistogamous or chasmogamous.

(i) Var. **horminoides** (Pourr.) Briq.
Stems 10–80 cm, with numerous eglandular hairs and fairly numerous glandular hairs. *Leaves* 2–14 × 1–8 cm, oblong to ovate, sinuate-dentate to incise-dentate. *Calyx* suffused with dull bluish-purple, with numerous glandular hairs. *Corolla* 6–10 mm, usually cleistogamous.

(ii) Var. **anglica** (Jord. & Fourr.) P. D. Sell
Gallitrichum anglicum Jord. & Fourr.

Stems 10–40 cm, with dense eglandular and glandular hairs. *Leaves* 2–5(–8) × 1.0–3.0 cm, ovate-oblong, deeply incise-dentate or pinnately lobed. *Calyx* suffused with dull bluish-purple, with dense glandular hairs. *Corolla* 10–15 mm, usually chasmogamous.

Native. Dry grassy and rather bare rough ground, roadsides and dunes. Rather frequent in south and east Great Britain north to central Scotland, north Wales, south and east Ireland and the Channel Isles. Subsp. *verbenaca* is known only from dunes at Vazon Bay, in Guernsey. Also in central France. Our common plant is subsp. *horminoides* var. *horminoides* which also occurs in south Europe north to central and west France and in Algeria. Var. *anglica* occurs in central areas of Jersey, Devonshire, Cornwall and Dorsetshire. The species is a Mediterranean-Atlantic element.

8. S. reflexa Hornem. Mintweed

Annual monoecious *herb. Stems* 30–40 cm, pale green, erect, quadrangular, glabrous or slightly hairy with eglandular hairs, simple or branched, leafy. *Leaves* opposite; lamina 3–8 × 0.4–1.2 cm, medium green on upper surface, paler beneath, linear to oblong-lanceolate, obtuse at apex, entire to crenate, cuneate at base, hairy. *Inflorescence* in axillary whorls of 2–4 bisexual flowers, forming a slender interrupted terminal spike. *Calyx* 6–8 mm, deeply 2-lipped, the upper lip entire, the lower lip 2-lobed. *Corolla* 7–11, pale blue, the tube not exceeding the calyx, 2-lipped, the lower lip twice as long as the upper. *Stamens* 2. *Style* 1. *Nutlets* 4, ovoid-trigonous. *Flowers* 8–10. $2n=20$.

Introduced. Casual from bird-seed, grain, grass seed and wool in waste and rough ground. Scattered records in England and Morayshire in Scotland. Native of North America.

9. S. viridis L. Annual Clary

S. horminum L.; *S. viridis* subsp. *horminum* (L.) Holmboe

Annual, scarcely aromatic, monoecious *herb* with fibrous roots. *Stems* 10–80 cm, pale green, sometimes purplish at base, erect, quadrangular, with many, white, tangled, eglandular hairs below, similar ones above with some sessile glands and glandular hairs, branched from the base, leafy. *Leaves* opposite; lamina 2–10 × 0.7–3.5(–5.0) cm, dull, pale or medium green on upper surface with impressed veins, paler beneath with prominent veins, elliptical to broadly oblong, rounded at apex, regularly crenulate to serrulate, rounded or cordate at base, with simple eglandular hairs on upper surface, usually in clusters, with numerous eglandular hairs below, mostly on the veins and some sessile glands; petiole 1.0–5.5 cm. *Inflorescence* of distant to conferted whorls of 4–6 bisexual flowers; bracts 18–30 mm, the lower green, the upper deep violet, pink, purple, rose or yellow, broadly obovate or ovate and acuminate or rounded at apex, often forming a sterile top to the flowering shoots; pedicels 2–3(–4) mm, erect-spreading in flower, strongly deflexed in fruit, flattened. *Calyx* 8–13 mm, elongating to about 12 mm in fruit, tubular, 13-ribbed, with sparse to fairly dense eglandular hairs, glandular hairs and sessile glands, 2-lipped, the upper lip truncate with 2 long, lateral lobes, the lower lip with 2 acuminate lobes, more or less equalling the upper lip. *Corolla* (12–)14–16 mm, with a lilac to violet hood and a paler lip sometimes all white, the tube straight and included within the calyx, slightly wider at throat; upper lip falcate, entire and hairy, the lower lip shorter than upper, with a cuculate, reflexed median lobe and distinct oblong lateral lobes. *Stamens* 2; filaments short, white; anthers yellow. *Style* 1, glabrous, pink; stigma with unequal lobes. *Nutlets* 4, 2.5–3.0 mm, dark brown, rounded- trigonous. *Flowers* 6–9. $2n=16$.

Introduced. Grown in gardens and is casual or rarely naturalised on tips and waste ground. Scattered records in Great Britain. Native of south Europe and Mediterranean region east to Iran.

10. S. verticillata L. Whorled Clary

Perennial monoecious, foetid *herb*. *Stems* 30–80 cm, pale green, erect, quadrangular, with numerous simple eglandular hairs, often simple, leafy. *Leaves* opposite; lamina 5–15 cm, triangular-ovate, obtuse to acute at apex, irregularly crenate-dentate to pinnatifid with 1–2 pairs of lateral segments, truncate to cordate at base, petiolate, with eglandular hairs. *Inflorescence* of axillary whorls forming a more or less interrupted terminal spike; bracts small, brown or purple-tinged and reflexed; pedicels 3–6 mm; whorls of (8–)15–30 bisexual or female flowers. *Calyx* 6–7 mm, with eglandular hairs and sessile glands, tubular-campanulate, 12-veined, 2-lipped, the upper lip with 3 conspicuous lobes, the middle lobe slightly broader and shorter than the lateral, the lower lip with 2 lobes. *Corolla* 8–15 mm, lilac-blue, tube slightly exserted, with a ring of hairs within, 2-lipped, the upper lip weakly convex and narrowed at base, the lower lip 3-lobed. *Stamens* 2, the filaments short, the connective much elongated. *Style* 1; stigma with unequal lobes. *Nutlets* 4, ovoid-trigonous, smooth. *Flowers* 6–8. Pollinated by bees; protandrous; small female flowers occur. $2n=16$.

Introduced. Casual or sometimes naturalised in rough ground and by roads and railways. Scattered records throughout Great Britain. Native of south, east and east-central Europe and south-west Asia.

Order 4. CALLITRICHALES Lindl.

Submerged or emergent *aquatic* monoecious or dioecious *herbs*. *Leaves* opposite, alternate or whorled; without stipules. *Flowers* unisexual or bisexual. *Perianth* a rim round the top of the ovary or absent. *Stamen* 1. *Style* 1. *Fruit* a 1-seeded nutlet or a schizocarp separating into 4 mericarps.

Consists of 3 families and over 50 species.

132. HIPPURIDACEAE Lindl. nom. conserv.

Perennial aquatic or mud-dwelling *herbs*. *Stems* with conspicuous air cavities. *Leaves* in whorls of 6–12, simple, linear, sessile, without stipules. *Inflorescence* a single flower in a leaf axil; often male, female and bisexual flowers on the same plant with female flowers more apical; flowers epigynous. *Perianth* reduced to a minute rim at apex of ovary. *Stamen* 1, borne on top of ovary. *Style* 1, filiform. *Ovary* 1-celled with 1 apical ovule. *Fruit* a 1-seeded nutlet.

Consists of a single genus.

1. Hippuris L.

As family.

Treated by some authors as monotypic, while others recognise 2 or more closely related species.

The genus has given its name to the 'Hippuris syndrome' which describes plants that are erect, unbranched and with simple, elongate leaves borne in symmetrical whorls which occur in several unrelated families (Cook, 1978).

Bodkin, P. C., Spence, D. H. N. & Weeks D. C. (1980). Photoreversible control of heterophylly in *Hippuris vulgaris* L. *New Phytol.* **84**: 533–542.

Cook, C. D. K. (1978). The *Hippuris* sydrome. In Street, H. E. (Edit.) *Essays on plant taxonomy,* pp. 163–176. London.

Hultén, E. & Fries, M. (1986). *Atlas of north European vascular plants north of the Tropic of Cancer.* 3 vols. Königstein.

Preston, C. D. & Croft, J. M. (1997). *Aquatic plants in Britain and Ireland*. Colchester.

1. H. vulgaris L. Mare's-tail

Perennial aquatic *herb* with a stout creeping rhizome. *Stems* when wholly or partly submerged 25–75(–150) cm, in terrestrial forms 7–20 cm, pale green, glabrous, densely leafy, and with conspicuous air cavities. *Leaves* 6–12 in a whorl, 1.0–7.5 × 0.1–0.3 cm, pale green, linear, with a hard, acute tip, entire, sessile, glabrous; on submerged shoots the leaves are longer, thinner, more flaccid and more translucent and the internodes are longer and less rigid, so that the name mare's-tail appears very appropriate, especially to the luxuriant submerged shoots of flowing water. *Inflorescence* of a single flower per leaf axil, often male, female and bisexual on the same plant, the female ones more apical. *Perianth* reduced to a minute rim at the top of the ovary. *Stamen* 1, borne on the top of the ovary; anthers reddish. *Style* 1, filiform. *Nutlet* greenish, ovoid, smooth. *Flowers* 6–7. $2n = 32$.

Native. Edges of ponds, lakes, swamps and upland flushes or sometimes abundant and submerged in clear calcareous water, up to 900 m but rare above 400 m. Throughout Great Britain and Ireland. Europe; west and north Asia; North Africa; southernmost South America. Circumpolar Boreo-temperate element.

133. CALLITRICHACEAE Link nom. conserv.

Perennial or facultative *annual* monoecious (rarely dioecious) *herbs*, aquatic or mud-dwelling. *Leaves* bright or dark green, opposite, simple, lingulate or spathulate, emarginate or notched at apex, often forming a terminal rosette, petiolate or sessile; without stipules. *Flowers* very small and unisexual, 1, 2 (usually one of each sex) or rarely 3 or 4 per leaf axil; each flower with 2 bracts, or bracts absent; perianth absent. *Stamen* 1. *Styles* or *stigmas* 2. *Ovary* 1- to 4-celled, each cell with 1 apical ovule. *Fruit* a schizocarp with (when fully developed) 4, 1-seeded mericarps, in 2 pairs each more closely united, each mericarp often with a wing running down the distal edge.

Contains a single genus, almost cosmopolitan.

1. Callitriche L.

by R. V. Lansdown

As family.

Contains about 75 species worldwide.

Scales may occur in three main locations, on the stem particularly below the nodes, in the leaf axils and on the leaves themselves. Peltate scales are composed of a disc of radiating cells on a stalk and occur on stem and leaves. The scales in the leaf axils are composed of a fan of digitate cells or a series of cells in digitate rows on a 2-celled stalk. In most species there is a single flower in each leaf axil, with either 1 male and 1 female, 2 males or 2 females in a pair of axils, or there may be a male and 1 or more females in a single axil. In some species the stamens are much reduced and often translucent and can be very difficult to see. The styles are produced in pairs from each schizocarp and are smooth and topped by a strongly papillose stigma. Pollen grains are either unornamented, ornamented with only a few granules or they have a very characteristic ornamentation. The fruit is a schizocarp formed of 4 mericarps. In some species the mericarp has a ridge running down the outer edge, here termed a *wing*. When the fruit is viewed from the side, the outline formed by each pair of the mericarps is more or less circular or in some taxa narrowed towards the base, and sometimes they are slightly wider than long or longer than wide. The colour of fully ripe fruits is useful. Microscopic characters of the fruit are helpful. The nature of the internal structure of the cells of the outer edge of the mericarp or wing is important. In most taxa structures are present that are an extension of the tissue forming the testa cell rings. They are referred to as *fibrils*, as in some cases they resemble the fibrils present in cells of *Sphagnum* species. The arrangement of these cells is different in different species.

Adams, K. (1988). *Callitriche truncata* Guss. (Short-leaved Water-starwort) new to Essex. *Bull. Essex Field Club* **38**: 39–41.

Barry, R. & Wade, P. D. (1986). Biological flora of the British Isles. No. 162. *Callitriche truncata* Guss. *Jour. Ecol.* **74**: 289–294.

Booth, E. M. (1975). *Callitriche truncata* Guss. in Ireland. *Irish Nat. Jour.* **18**: 228.

Briggs, M (1978). The identification of *Callitriche* spp. *B.S.B.I. News* **20**: 23.

Clarke, B. (1859). Notes on the structure and affinities of Callitrichaceae. *Trans. Linn. Soc. London* **22**.

Cooper, R. L., Osborn, J. M. & Philbrick, C. T. (2000). Comparative pollen morphology and ultrastructure of the Callitrichaceae. *Amer. Jour. Botany* **87**: 161–175.

David, R. W. (1958). An introduction to the British species of *Callitriche*. *Proc. B. S. B. I.* **3**: 28–32.

Demars, B. O. L. & Gornall, R. J. (2003). Identification of British species of *Callitriche* by means of isozymes. *Watsonia* **24**: 389–399.

Duvigneaud, J. & Schotsman, H. D. (1977). Le genre *Callitriche* en Belgique et dans les régions avoisinantes: nouvelles contributions et clé de détermination. *Natura Mosana* **30** (1): 1–22.

Fassett, N. C. (1951). *Callitriche* in the New World. *Rhodora* **53**: 137–155; 161–182; 185–194; 209–222.

Godfrey, B. E. S. (1982). Microfungi as epiphytes on the leaves of *Callitriche intermedia* Hoffm. *Bull. B.M.S.* **16** (1): 57–58.

Godwin, Sir H. (1975). *The history of the British flora: a factual basis for phytogeography*. 2 Ed. Cambridge.

Haldimann, G. (1982). Contribution à l'étude cytologique de *Callitriche hamulata* Kutz. (Angiospermae). *Bull. Soc. Neuchâteloise Sci. Nat.* **105**: 60–63.

Haldimann, G. (1989). Agamospermie chez la forme submergée de *Callitriche palustris* L. et contribution à l'étude cytologique de cette espèce. *Bull. Soc. Neuchâteloise Sci. Nat.* **112**: 29–32.

Hanbury, F. J. (1889). *Callitriche truncata* Guss. in Gloucestershire. *Jour. Bot.* (*London*) **27**: 95.

Hegelmaier, F. (1864). *Monographie der Gattung* Callitriche. Stuttgart.

Hegelmaier, F. (1867). Zur Systematik von *Callitriche*. *Verh. Bot. Vereins Prov. Brandenburg* **9**: 1–41.

Hegelmaier, F. (1868). Beitrag zur Kenntnis der Wassersterns. *Verh. Bot. Vereins Prov. Brandenburg* **10**: 100–121.

Jäggi, M. & Cook, C.D.K. (1998). Reproductive biology of *Callitriche cophocarpa* Sendt (Callitrichaceae). *Candollea* **53**: 101–115.

Jones, H. (1952). Variation in leaf form in *Callitriche intermedia*. *Nature* **170**: 848–849.

Jones, H. (1955). Heterophylly in some species of *Callitriche*, with especial reference to *Callitriche intermedia*. *Ann. Bot.* **19**: 225–245.
Jones, H. (1955). Further studies in heterophylly in *Callitriche intermedia*: leaf development and experimental induction of ovate leaves. *Ann. Bot.* **19**: 369–388.
Jones, H. (1955). Notes on the identification of some British species of *Callitriche*. *Watsonia* **3**: 186–192.
Lansdown, R. V. (1999). A terrestrial form of *Callitriche truncata* Guss. subsp. *occidentalis* (Rouy) Braun-Blanquet (Callitrichaceae). *Watsonia* **22**: 283–286.
Lansdown, R. V. (2006). Notes on the water-starworts (*Callitriche*) recorded in Europe. *Watsonia* **26**: 105–120.
Lansdown, R. V. (2008). *Water-starworts* (Callitriche) *of Europe*. London.
Lansdown, R. V. & Jarvis, C. E. (2004). Linnaean names in *Callitriche* L. (Callitrichaceae) and their typification. *Taxon* **53**: 169–172.
Lewis-Jones, L. J. & Kay, Q. O. N. (1977). The cytotaxonomy and distribution of water-starworts (*Callitriche* spp.) in West Glamorgan. *Nat. Wales* **15**: 180–183.
Martinsson, K. (1991). Natural hybridization within the genus *Callitriche* (Callitrichaceae) in Sweden. *Nordic Jour. Bot.* **11**: 143–151.
Martinsson, K. (1991). Geographical variation in fruit morphology in Swedish *Callitriche hermaphroditica* (Callitrichaceae). *Nordic Jour. Bot.* **11**: 497–512.
Martinsson, K. (1993). The pollen of Swedish *Callitriche* (Callitrichaceae): trends towards submergence. *Grana* **32**: 198–209.
Martinsson, K. (1996). Growth forms and reproductive characters in six species of *Callitriche* (Callitrichaceae). *Acta Univ. Upsaliensis* **31** (3): 123–131.
Mason, R. (1959). *Callitriche* in New Zealand and Australia. *Austral. Jour. Bot.* **7**: 295–327.
Orchard, A. E. (1980). *Callitriche* (Callitrichaceae) in South Australia. *Jour. Adelaide Bot. Garden* **2**: 191–194.
Osborn, J. M. & Philbrick, C. T. (1994). Comparative pollen structure and pollination biology in the Callitrichaceae. *Acta Bot. Gallica* **141**: 257–266.
Pearsall, W, H. (1934). The British species of *Callitriche*. *Rep. B. S. B. I.* **10**: 861–871.
Philbrick, C. T. (1984). Pollen tube growth within vegetative tissues of *Callitriche* (Callitrichaceae). *Am. Jour. Bot.* **71**: 882–886.
Philbrick, C. T. (1993). Underwater cross-pollination in *Callitriche hermaphroditica* (Callitrichaceae): evidence from random amplified polymorphic DNA markers. *Am. Jour. Bot.* **80**: 391–394.
Philbrick, C. T. (1994). Chromosome counts for *Callitriche* (Callitrichaceae) in North America. *Rhodora* **96**: 383–386.
Philbrick, C. T. & Anderson, G. J. (1992). Pollination biology of the Callitrichaceae. *Syst. Botany USA* **17**: 282–292.
Philbrick, C. T. & Jansen, R. K. (1988). Phylogenetic relationships in *Callitriche* (Callitrichaceae) using chloroplast DNA restriction site mapping. *Am. Jour. Bot.* **75**: 199–200.
Philbrick, C. T., Aakjar, R. A. Jr. & Stuckey, R. L. (1998). Invasion and spread of *Callitriche stagnalis* (Callitrichaceae) in North America. *Rhodora* **100**: 25–38.
Pijnacker, L. P. & Schotsman, H. D. (1988). Nuclear DNA amounts in European *Callitriche* species (Callitrichaceae). *Acta Bot. Neerl.* **37**: 129–135.
Preston, C. D., Pearman, D.A. & Dines, T.D. (2002). *New atlas of the British and Irish flora.* Oxford.
Savidge, J. P. (1958). Distribution of *Callitriche* in North-West Europe. *Proc. B. S. B. I.* **3**: 103.
Savidge, J. P. (1960). The experimental taxonomy of European *Callitriche*. *Proc. Linn. Soc. London* **171**: 128–130.
Savidge, J. P. (1966). Recognition of *Callitriche* spp. in Britain. Papers presented at the local flora writers' conference, Bristol.
Schotsman, H. D. (1967). *Les Callitriches: espèces de France et taxa nouveaux d'Europe*. Paris.
Schotsman, H. D. (1972). Note sur la répartition des Callitriches en Sologne et dans les régions limitrophes. *Bull. Cent. Etud. Rech. Sci. Biarritz* **9**: 19–52.
Schotsman, H. D. (1977). *Callitriche* de la région mediterranéenne: nouvelles observations. *Bull. Cent. Etud. Rech. Sci., Biarritz* **11**: 241–312.
Schotsman, H. D. (1982). Biologie florale des *Callitriche*: étude sur quelques espèces d'Espagne méridionale. *Adansonia* **3–4**: 111–160.
Schotsman, H. D. (1992). Notas taxonomicas y corologicas sobre la flora de Andalucia Occidental. *An. Jard. Bot. Madrid* **49**: 288: 141–257.
Trimen, H. (1870). *Callitriche truncata*, Guss., as a British plant. *Jour. Bot.* (*London*) **8**: 154–157.
Wade, P. M. & Barry, R. (1982). The distribution and autecology of *Callitriche truncata* Guss. *Watsonia* **14**: 234.
Wade, P. M., Vanhecke, L. & Barry, R. (1986). The importance of habitat creation, weed management and other habitat disturbance to the conservation of the rare aquatic plant *Callitriche truncata* Guss. subsp. *occidentalis* (Rouy) Schotsm. *Proc. EWRS/AAB 7th Symposium on Aquatic Weeds*, 15–19 September 1986, Loughborough, Leicestershire, UK: 389–394.
Williams, F. N. (1912). *Prodromus florae Britannicae*. Part 9. *Comprising the 14 families included in the four orders of Rhamnales, Gruinales, Hippocastanales and Tricoccales*. Brentford.

1. Leaf bases not joined; pollen lacking ornamentation, translucent or appearing whitish; no peltate scales on stem and leaves; mericarps divergent 2.
1. Leaf bases joined by a ridge of tissue extending across the node; pollen ornamented or not, yellow or translucent; peltate scales on stem and leaves; mericarps subparallel 4.
2. Mericarps unwinged **2. truncata** subsp. **occidentalis**
2. Mericarps winged 3.
3. Dried schizocarps less than 1.7 mm wide or long **1(a). hermaphroditica** subsp. **hermaphroditica**
3. Dried schizocarps more than 1.8 mm wide or long **1(b). hermaphroditica** subsp. **macrocarpa**
4. Pollen grains weakly ornamented or lacking ornamentation, translucent; styles strongly reflexed, appressed to side of schizocarp, with remains appearing to emerge from below apex of ripe schizocarp 5.
4. Pollen grains strongly ornamented, appearing yellow; styles erect to recurved, never appressed to side of fruit, appearing apical 6.
5. Schizocarps subsessile to long-pedunculate (usually more than 2 mm); leaf apices shallowly to deeply notched, with notch narrow and often irregular **7(i). brutia** var. **brutia**
5. Schizocarps sessile to shortly pedunculate (less than 2 mm); leaf apices shallowly to deeply notched, with notch broad, irregular or regular **7(ii). brutia** var. **hamulata**
6. Most pollen grains more than twice as long as wide, straight to gently curved, strongly ornamented; radiating fibrils present throughout cells immediately surrounding testa; mericarps unwinged **4. obtusangula**
6. Pollen grains less than twice as long as wide, ornamented or not; fibrils only in wing cells or absent; mericarps winged at least in part 7.

7. Ripe schizocarps dark brown to black when fresh; mericarps winged mainly or only at apex; many axils containing both male and female flowers **6. palustris**
7. Ripe schizocarps pale brown or greyish when fresh; mericarps winged throughout and wing only slightly wider at apex; most axils containing a solitary flower 8.
8. Pollen grains ellipsoid to elongate-ellipsoid, appearing subrotund to elliptical in outline; ripe schizocarps greyish, broadly winged **3. stagnalis**
8. Most pollen grains appearing bluntly triangular in outline; ripe schizocarps pale brown, narrowly winged **5. platycarpa**

1. C. hermaphroditica L. Autumnal Water-starwort
C. palustris var. *bifida* L.; *C. autumnalis* L.; *C. angustifolia* Gilib. nom. illegit.; *C. bifida* (L.) Morong; *C. aquatica* subsp. *autumnalis* (L.) Bonnier; *C. hermaphroditica* var. *bicarpellaris* (Fenley) Mason

Monoecious *water-plant*, possibly long-lived; terrestrial form not known. *Stems* whitish, contrasting strongly with the colour of the leaves, little to much branched; internodes gradually decreasing in length upwards; stem scales absent; axillary scales fan-shaped, composed of 2 basal cells and a number of digitate rows of 5–6 cells. *Leaves* opposite, none forming a rosette, spathulate ones absent; lingulate ones 5.5–15.3×0.6–1.8(–2.2) mm, translucent, slightly to markedly narrower at apex than base, 1-veined; scales absent. *Flowers* unisexual, solitary, a male or female in 1 or both of a pair of axils; bracts absent. *Stamen* 1; filament when undehisced up to 0.2 mm, when dehisced 0.5–0.9 mm, erect, continuing to grow after dehiscence; anther 0.1–0.6×0.2–0.7 mm, reniform, translucent; pollen grains 20–30×20–30 µm, more or less globose, colourless, 86(50–100) per cent developed, smooth, not ornamented, intectate; pollination submerged. *Styles* 2, up to 5 mm, initially erect, later recurved and eventually appearing to emerge below the apex of the ripe fruit. *Schizocarp* 1.2–2.4×1.2–3.0 mm, more or less as wide as long, sessile, when ripe dark brown; mericarps 1–4, divergent, so that, viewed from above, 4 mericarps appear as a cross; testa cells in rings; mericarps with wing throughout, 0.1–0.5 mm wide on side, 0.2–0.8 mm wide at top; fibrils only in wing cells. *Flowers* 5–9. $2n=6$.

(a) Subsp. **hermaphroditica**
Schizocarp 1.2–1.7×1.2–1.7 mm when fresh; mericarps 1–4, with wing throughout, 0.1–0.3 mm wide on side, 0.2–0.4 mm at top; fibrils complex, a series of fairly fine linear stems ending in a complex series of radiating loops, occasionally absent.

(b) Subsp. **macrocarpa** (Hegelm.) Lansdown
C. autumnalis forma *macrocarpa* Hegelm.

Schizocarp 1.5–2.4×1.6–2.8(–3.0) mm when fresh; mericarps 3–4, with wing throughout, 0.2–0.5 mm wide on side, 0.2–0.8 mm at top; fibrils simple, a thick linear trunk leading from the testa ending in a series of radiating lines, occasionally absent.

Native. Lakes, gravel-pits, canals and slow-flowing reaches of rivers. It is most typical of lowland, mesotrophic lakes, often occurring in species-rich communities characterised by diverse assemblages of *Potamogeton* species, *Chara aspera*, *Elodea nuttallii*, *Littorella uniflora*, *Luronium natans*, *Myriophyllum alterniflorum* and *Nitella flexilis*. In Great Britain north of a line from the Wash to the Bristol Channel and in Ireland in the north and west. The species is circumboreal and in Europe is predominantly northern, extending south to the Czech Republic. Circumpolar Boreal montane element. The distribution of the subspecies is poorly known, but subsp. *macrocarpa* is more abundant in the northern part of the range.

2. C. truncata Guss. Short-leaved Water-starwort
C. cruciata Lebel nom. nud.; *C. graminea* Link nom. nud.; *C. autumnalis* subsp. *truncata* (Guss.) Arcang.; *C. aquatica* subsp. *truncata* (Guss.) Bonnier; *C. palustris* var. *truncata* (Guss.) Fiori; *C. hermaphroditica* subsp. *truncata* (Guss.) Jahand. & Maire

Monoecious *water-plant*, possibly long-lived; annual terrestrial form known, probably sterile. *Stems* whitish, contrasting strongly with the colour of the leaves, little to much branched; internodes more or less equal in length until abruptly shortening towards the shoot apex; stem scales absent; axillary scales fan-shaped, composed of 2 basal cells and a number of digitate rows of 5–6 cells. *Leaves* opposite, none forming a rosette; spathulate ones absent; lingulate ones (1.8–)2.4–11.0×0.7–1.8 mm, green, parallel-sided to slightly elliptical, truncate or shallowly emarginate at apex, 1-veined; leaves of terrestrial form 1.4–2.5×0.8–1.4 mm, truncate to bluntly emarginate, more or less rigid; scales absent. *Flowers* unisexual, solitary, most frequently 2 female flowers in a pair of axils, less often a single female or occasionally 2 males, a male and a female, or 2 females; bracts absent. *Stamen* 1; filament when undehisced up to 0.9 (–1.0) mm when dehisced, 1.1–1.7 mm, erect, continuing to grow after dehiscence; anther 0.5–1.0×0.7–1.1 mm, reniform, translucent, sometimes appearing white; those of terrrestrial forms usually abort; pollen grains 20–30×20–30 µm, more or less globose, colourless, 85(60–100) per cent developed, smooth, appearing to be slightly ornamented in terrestrial plants (but this requires confirmation); pollination submerged. *Styles* 0.4–6.1 mm. *Schizocarps* frequent in water form, rare in terrestrial form, 1.0–1.4×1.2–1.9 mm, distinctly wider than long, when young dark greyish-brown, when ripe dark brown, on peduncles 0.2–0.9 mm; mericarps 1–4, divergent, so that, viewed from above, 4 mericarps appear as a cross; wing absent; fibrils absent. *Flowers* 5–9. $2n=6$.

Native. Coastal lagoons, mesotrophic to eutrophic lakes, canals, ditches, streams and rivers. Scattered throughout southern England, Wales and Ireland, in five main areas: Lincolnshire and through the Midlands, Dorsetshire, Somersetshire and Devonshire, Kent, Essex and Suffolk, Anglesey, and Co. Wexford. Formerly also in Guernsey. Our plant is subsp. **occidentalis** (Rouy) Schotsman (*C. truncata* proles *occidentalis* Rouy; *C. truncata* race *occidentalis* (Rouy) Braun-Blanq.), which occurs from Great Britain and Ireland south to the Iberian Peninsula and east to Turkey. Mediterranean-Atlantic element. The species extends into southern Russia (i.e. subsp. *fimbriata*).

3. C. stagnalis Scop. Common Water-starwort
C. palustris subsp. *stagnalis* (Scop.) Schinz & Thell.

Monoecious *water-plant*, possibly long-lived; terrestrial form known, usually abundantly fertile. *Stems* white to pale green, not contrasting strongly with the colour of the leaves, much branched; internodes decreasing in length gradually along shoots below rosettes; stem scales consisting of a circular disc of 5–12(–16) cells; axillary scales consisting of a fan of 4–7(–8) cells. *Leaves* opposite; lingulate ones very rare and usually absent; spathulate ones with lamina 2.7–21.4(–21.7) × 1.4–8.3 mm, variable, from broadly parallel-sided, through obovate-spathulate to subrotund, with petiole 0.7–6.5(–7.5) mm; those of rosette 6–12(–13) mm, usually very broad with a short petiole; subsidiary venation characterised by a pair of secondary veins; most leaves with at least part of a tertiary vein, which can be represented by a partially formed vein and additional veins arising from the secondary veins, either as loops (joined at both ends) or as short slightly curved veins pointing more or less directly to the leaf edge; broader ones may have large numbers of short, anastomosing veins leading from secondary veins towards the leaf margin; leaves of terrestrial form 2.6–4.4 × 1.6–3.3 mm, narrowly elliptical, bluntly emarginate, more or less rigid, with petiole 0.8–2.1 mm, scales formed of a circular disc of 6–13(–14) cells. *Flowers* unisexual, usually solitary, generally a male opposed by a female in a pair of axils, rarely a single female opposed by a male and female together; bracts subtending male flowers 0.7–2.6(–2.9) mm, those subtending female flowers 0.6–1.1 mm, translucent, appearing whitish, falcate and persistent. *Stamen* 1; filament when undehisced up to 10.3 mm, when dehisced 8.5–14.8(–16.2) mm, erect, becoming recurved and continuing to grow after dehiscence; anther 0.3–0.8 × 0.3–0.9 mm, reniform, appearing yellow; pollen grains 20–40 × 20–30 µm, with 1 leptoma, globose to slightly ellipsoid, yellow, 79(40–100) per cent developed, strongly ornamented with a reticulate pattern of walls with supratactal spinulae at their intersections; pollination aerial and through internal geitonogamy. *Styles* 2, up to 6 mm, erect, becoming recurved in fruit. *Schizocarp* 1.1–1.8 × 1.1–2.1 mm, more or less circular in profile to slightly wider than long, sessile or very slightly pediculate, when ripe greyish; mericarps 1–4, parallel, with testa cells in rings; wing throughout, 0.1–0.3 mm wide on side and 0.2–0.5 mm at top; fibrils in a complex spiral, only in wing cells. *Flowers* 5–9. $2n = 10$.

Native. Ephemeral pools on woodland rides, heathland, wet corners of pasture, seepages, flushes, lake and river margins, and the margins of ditches, occasionally perennial in rivers. Widespread and locally abundant throughout Great Britain and Ireland. It is found throughout Europe, from the Azores to Iceland and east to western Russia, and in North Africa; it has been introduced in North America, Australia, New Caledonia and New Zealand. Small terrestrial forms have been called var. *serpyllifolia* (Kütz.) Lönnr. but are simply the end of more or less continuous variation. It is a member of the European Temperate element, widely naturalised elsewhere.

4. C. obtusangula Le Gall Blunt-fruited Water-starwort
C. aquatica subsp. *obtusangula* (Le Gall) Jahand. & Maire; *C. palustris* subsp. *obtusangula* (Le Gall) Bonnier

Monoecious *water-plant*, possibly long-lived; annual terrestrial form known, rarely fertile. *Stems* white to pale green, not contrasting strongly with the colour of the leaves, the 2-year and older stems often blackish; internodes gradually decreasing in length up the stem; scales a circular or elliptical disc of (5–)6–10(–13) cells; axillary scales a fan of 4–9 cells. *Leaves* opposite; lingulate ones 9.3–32.2 × 0.2–2.2 mm, with a single central vein and a shallow triangular notch at the apex; spathulate ones with lamina 14.9 × 29.2(–29.7) × 1.8–7.0 mm, rhombic; venation characterised by a pair of secondary veins arising around the point where the lamina widens and rejoining the central vein abruptly just below the leaf apex, with petiole (2.7–)3.3–11.1 mm, colourless; rosette ones 13–24, with ridged veins, the centre often forced clear of the water; most expanded leaves have a few additional short veins which are either free or looped to join the secondary veins; leaves of terrestrial form 4.7–10.2 × 1–3 mm, with ridges along the veins, generally very dense, almost pointed and in a single plane, appressed to substrate; scales with (7–)8 cells in the disc, grading into the stem scales on the petiole. *Flowers* unisexual, usually solitary, generally a male and a female, less often 1 or 2 males or females in a pair of axils; bracts subtending male flowers (1.1–)1.7–3.6 mm, those subtending female flowers 1.4–3.2 mm, translucent, whitish, persistent. *Stamen* 1; filament of undehisced stamen up to 6 mm, of dehisced stamen 4.0–7.6(–12.3) mm, erect before dehiscence, becoming strongly recurved after dehiscence; anther 0.6–1.0 × 0.6–1.3 mm, reniform, bright yellow; pollen grains 30–60(–70) × 20–30(–40) µm, with 1 or 2 leptomata, elongate-ellipsoid and curved, although some grains in each anther are globose to ovoid-elliptical, yellow, 79(34–100) per cent developed; exine strongly ornamented (nature of ornamentation not known); pollination aerial. *Styles* 2, up to 8.3(–8.6) mm, initially erect, eventually becoming strongly recurved, persistent. *Schizocarp* 1.1–1.8 × 1.1–1.7 mm, ellipsoid, appearing longer than wide when ripe, sessile, pale brown when ripe; mericarps 1–4, parallel, with testa cells in rings; wing completely absent; fibrils linear, radiating out from the testa in all cells immediately surrounding the testa. *Flowers* 5–9. $2n = 10$.

Native. Chalk rivers, calcareous fens and ditches in coastal marshes, generally growing over gravels, clay or peat but tolerating deep beds of fine silt; it can tolerate brackish water. It tends to form large dense beds in still or slow-flowing water, but it also forms dense, submerged, sterile beds in fast-flowing reaches of rivers. It occurs through most of England, Wales and Ireland and north to the Great Glen in Scotland. In continental Europe, it occurs from the Netherlands to Austria and Yugoslavia to the east, and westwards along the Mediterranean coast, including Corsica and Sardinia, to Spain and Portugal; elsewhere it occurs in North Africa, including Morocco and Tunisia. Suboceanic southern-temperate element. Schotsman (1967) identified a number of genotypes showing clear geographical separation but could establish no relationship between these variants and morphological differences.

5. C. platycarpa Kütz. Various-leaved Water-starwort
C. platycarpa var. *latifolia* Kütz.; *C. platycarpa* var. *undulata* Kütz.; *C. platycarpa* var. *sterilis* Kütz.; *C. platycarpa* var. *rigidula* Kütz.; *C. platycarpa* var. *elongata* Kütz.; *C. aquatica* subsp. *platycarpa* (Kütz.) Bonnier & Layens; *C. font-queri* P. Allorge

Monoecious *water-plant*, possibly long-lived; annual terrestrial form known, rarely fertile. *Stems* white to pale green, not contrasting strongly with the colour of the leaves, robust and often much-branched, 2-year and older stems often blackish; internodes gradually decreasing in length along shoots to just below rosettes; scales a disc of 6–11(–13) cells, initially regular, then irregular and asymmetrical, with occasional distorted cells; axillary scales a fan of 4–9 cells, generally long and digitate, some cells occasionally distorted and angular. *Leaves* opposite; lingulate ones 10.4–28.5×0.1–2.0 mm, more or less parallel-sided or tapering from base, with a single vein and a shallow notch at apex; lamina of spathulate ones 11.1–15.9×2.0–3.8 mm, elongate-spathulate to elliptical, with petiole (0.3–)1.1–5.5 mm; rosette ones 7–18; subsidiary venation characterised by a pair of secondary veins departing on either side of the primary vein at the apex of the petiole and rejoining it just before the apex of the leaf, often with 1 or 2 short processes arising from the secondary veins; occasionally one of these will rejoin the secondary vein; leaves of terrestrial form 1.5–3.8×0.5–1.8(–2.3) mm, narrowly elliptical, bluntly emarginate, more or less rigid; scales a disc of 8–9 cells. *Flowers* unisexual, solitary, generally a single female, 2 females, a male and a female or 2 males in a pair of axils; bracts subtending male flowers 0.3–2.6 mm, those subtending female flowers 0.3–1.9(–2.2) mm, translucent, appearing whitish, falcate, persistent. *Stamen* 1; filament of undehisced stamen up to 3.6(–3.8) mm, of dehisced stamen 1.5–5.1(–8.3) mm, initially erect, then continuing to grow and recurving after anthesis; anther 0.5–0.8×0.6–1.0 mm, reniform, appearing yellow; pollen grains (18–)21–40(–43)×15–32(–35) µm, with 1 to 3 leptomata, usually appearing bluntly triangular in outline, yellow, 72(10–100) per cent developed, strongly ornamented with a reticulate pattern of walls with supratectal clavae at their intersections; pollination aerial. *Styles* up to 6.5(–7.8) mm erect and slightly curved *Schizocarp* 1.3–1.7×1.4–1.8 mm, more or less as wide as long, sessile or very occasionally with a peduncle up to 0.8 mm, when ripe pale brown; mericarps 1–4, parallel, with testa cells in rings; wing throughout, 0.05–0.14 mm wide on side, 0.07–0.3 mm at top; fibrils complex and dendroid, only in wing cells. *Flowers* 5–9. $2n=20$.

Native. Lowland ditches, canals, streams, ponds and river backwaters with a pH of 6–9. Throughout Great Britain and Ireland. Throughout France, extending north to southern Sweden and east to the Czech Republic; it has also been reported from a few sites in the Iberian Peninsula. European Temperate element.

6. C. palustris L. Narrow-fruited Water-starwort
C. palustris var. *minima* L.; *C. palustris* var. *natans* L.; *C. verna* var. *genuina* Kütz.; *C. androgyna* L.; *C. verna* L.; *C. fontana* Scop. nom. illegit.; *C. aquatica* Huds.; *C. latifolia* Gilib. nom. illegit.; *C. verna* var. *fontana* Kütz.; *C. verna* var. *latifolia* Kütz.; *C. caespitosa* Schultz nom. illegit.; *C. aquatica* var. *caespitosa* Willd.; *C. pallens* Gray nom. illegit.; *C. euverna* Syme; *C. verna* var. *genuina* Kütz.

Monoecious *water-plant*, possibly long-lived; terrestrial form known, probably perennial, usually abundantly fertile. *Stems* white to pale green, not contrasting strongly with the colour of the leaves; internodes gradually decreasing in length along the shoots to just below the rosettes; scales irregular in outline, with 13–16 cells; axillary scales a fan of 4–7(–9) cells. *Leaves* opposite, green; lingulate ones 5.6–9.7×0.2–1.2 mm, linear to narrowly expanded, emarginate, with a single vein; spathulate ones 3.6–9.9(–10.3)×1.2–4.3(–4.5) mm, elliptical to more or less subrotund; rosette ones 7–15; subsidiary venation characterised by a pair of secondary veins and most leaves having at least part of a tertiary vein; they generally also have additional veins arising from the secondary veins, either as loops or as short, slightly curved veins pointing more or less directly to the leaf edge; leaves of terrestrial form 1.6–7.5×0.2–1.4 mm, narrowly elliptical, bluntly emarginate, more or less rigid; scales similar to stem scales, but forming a disc of 7–8 cells. *Flowers* unisexual, usually represented by a single female, a female in each of a pair of axils or a male and female opposed by a single female flower; bracts 0.5–1.0 mm, translucent, whitish, falcate, persistent, occasionally more than 2 in an axil. *Stamen* 1; filament up to 1.1 mm, erect; anther 0.2–0.4×0.2–0.4 mm, elongate-reniform, yellow or whitish-yellow, often aborted; pollen grains 6–20×5–20 µm, probably inaperturate, more or less globose; exine strongly ornamented with a reticulate pattern of walls, lacking supratectal elements; pollination by internal geitonogamy, probably also aerial. *Styles* 2, up to 1.1 mm, erect or slightly spreading, often poorly developed. *Schizocarp* 0.9–1.4×0.8–1.1 mm, longer than wide, widest above the middle, sessile, when ripe black; mericarps (2–)3–4, parallel, reticulate, with the reticulations appearing more or less clearly in vertical rows; testa cells in rings; wing mainly towards apex, 0.05–0.16 mm wide; fibrils very simple, a small number of short, tapering processes, only in wing cells, restricted to widest part of wing. *Flowers* 5–9. 2n=20.

Native. Wet ruts on woodland rides, turloughs, lakes and river margins. It has been recorded from a number of sites in southern Scotland and one in western Ireland. *C. palustris* is the most widespread *Callitriche* species, occurring throughout Europe except for the Mediterranean basin and parts of France, from Iceland and Finland south to the Pyrenees and Apennines and east throughout Russia, with isolated populations in the mountains of Corsica; it occurs throughout most of northern North America and from western Russia east to Kamchatka, Japan and most of China and has been reported from Australia. Boreal-temperate element. A very variable species within which a number of varieties have been recognised.

7. C. brutia Petagna Intermediate Water-starwort

Monoecious *water-plant*, possibly long-lived; terrestrial form known. *Stems* white to pale green, not contrasting strongly with the colour of the leaves; internodes gradually decreasing in length along shoots to just below the rosettes; scales a circular, elliptical or irregular disc of 7–18(–20) cells; axillary scales a fan of 3–8(–10) cells. *Leaves* opposite, green; lingulate ones 4–16×0.1–1.7 mm, narrowly linear, more or less parallel-sided, with apex expanded around a broad or narrow notch and with a single vein; spathulate ones with lamina 1.8–17.8×0.8–3.6 mm, narrowly elliptical to oblong, and petiole 0.1–10.2 mm; those of rosette 7–14(–15), often with ridges along the veins, with petiole as long as lamina; subsidiary venation characterised by 2 secondary veins which anastomose with the primary vein near the leaf apex; there may be short branches arising from 1 or both of the secondary veins and in some populations the secondary veins are sinuous; leaves of terrestrial form 2.4–5.5×0.3–0.8 mm, linear-elliptical to elliptical and 1(–3)-veined; scales a circular disc of 8–12(–16) cells. *Flowers* unisexual, generally represented by 1 male and 1 female in a pair of axils, or a single female; bracts subtending male flowers 0.2–1.8 mm, those subtending female flowers 1.0–1.5 mm, translucent, whitish, falcate, caducous. *Stamen* 1; filament of undehisced stamen up to 0.1 mm, of dehisced stamen 0.2–1.2 mm, erect; anther 0.1–0.5×0.1–0.6 mm, reniform; pollen grains 30–40(–50)×20–40 µm, more or less globose, colourless, translucent, 98(90–100) per cent developed, inaperturate, intectate, not or poorly ornamented, with exine granular or absent; pollination submerged or aerial. *Schizocarp* (0.7–)1.0–1.5(–1.8)×(0.7–)1.0–1.6 mm, more or less as wide as long, globose or subglobose, when ripe blackish; mericarps 1–4, parallel, with testa cells in rings; wing throughout, 0.03–0.2 mm wide on side, 0.04–0.3 mm at top; fibrils complex and dendroid, only in wing cells. *Flowers* 5–9.

(i) Var. **brutia**
C. pedunculata DC.; *C. hamulata* subsp. *pedunculata* (DC.) Syme; *C. capillaris* Parl.; *C. aquatica* subsp. *pedunculata* (DC.) Bonnier; *C. palustris* subsp. *pedunculata* (DC.) Jahand. & Maire

Lingulate leaves may have the apex expanded around a long fine notch, which is irregular and more or less as wide as the lamina. *Pollen grains* lacking ornamentation. *Schizocarp* subsessile or with a peduncle up to 12 mm. $2n=28$.

(ii) Var. **hamulata** (Kütz. ex W. D. J. Koch) Lansdown
C. hamulata Kütz. ex W. D. J. Koch; *C. intermedia* Hoffm.; *C. autumnalis* var. *goldbachii* Kütz.; *C. autumnalis* var. *brutia* Kütz.; *C. autumnalis* var. *callophylla* Kütz.; *C. autumnalis* var. *heterophylla* Kütz.; *C. autumnalis* var. *platyphylla* Kütz.; *C. aquatica* subsp. *hamulata* (W. D. J. Koch) Bonnier & Layens

Lingulate leaves may have the apex expanded around a long fine notch, irregular or regular, more or less as wide as the lamina or much wider. *Pollen* ornamentation finely granular. *Schizocarp* sessile or with a peduncle up to 2 mm. $2n=38$.

Native. Ephemeral pools, lakes and canals to fast-flowing upland rivers. Var. *hamulata* appears to be more frequent in permanent, swift-flowing water, while var. *brutia* typically occurs in ephemeral water-bodies and at the margins of lakes from sea level to a least 950 m on Ben Lawers. *C. brutia* occurs in most of Great Britain and Ireland, though it is more frequent in the west and north. It occurs throughout Europe from Iceland and the Faeroes, through Scandinavia to Poland and the Czech Republic, and throughout the Mediterranean basin east to Greece. It has been recorded from Greenland, Morocco, Iran and the Caucasus. It occurs as an introduction in western North America, Australia and New Zealand. European Wide-temperate element.

Order 5. PLANTAGINALES Lindl.

Monoecious *herbs*. *Leaves* simple, often sheathing at base. *Flowers* actinomorphic, hypogynous, bisexual. *Calyx* 4-lobed. *Corolla* 3- to 4-lobed, scarious. *Stamens* usually 4. *Style* 1. *Ovary* 1- to 4-celled; ovules 1–several in each cell; placentation axile or basal. *Fruit* an achene, capsule or nut; seeds with endosperm.

Consists of the single family Plantaginaceae.

134. PLANTAGINACEAE Juss. nom. conserv.

Annual to *perennial herbs*. *Leaves* usually all basal, sometimes opposite on stems, simple, entire to deeply dissected, sessile or with an ill-defined petiole; stipules absent. *Inflorescence* of solitary flowers or flowers in dense terminal spikes on unbranched basal or axillary stalks; flowers actinomorphic, unisexual or bisexual, hypogynous. *Sepals* (2–)4, papery, persistent in fruit, nearly free to fused at base. *Corolla* 2- to 4-lobed, greenish or brownish. *Stamens* 4, on long filaments. *Style* 1, terminal; stigma 1, linear. *Fruit* a 2- or 4-celled transversely dehiscing capsule or a nut.

Consists of 3 genera and some 250 species, cosmopolitan.

1. Plant terrestrial, without far-creeping stolons; flowers bisexual **1. Plantago**
1. Plant aquatic, with far-creeping stolons; flowers male or female **2. Littorella**

1. Plantago L.

Terrestrial, monoecious or rarely dioecious *herbs* without far creeping stolons. *Leaves* usually all basal, rarely cauline. *Inflorescence* in the form of heads or cylindrical spikes; flowers mostly bisexual. *Calyx* 4-lobed. *Corolla* 3- to 4-lobed, scarious. *Stamens* 4. *Style* 1. *Fruit* a transversely dehiscing capsule.

Contains about 250 species in temperate regions of both hemispheres and a few in the tropics.

Akeroyd, J. R. & Doogue, D. (1988). *Plantago major* L. subsp. *intermedia* (DC.) Arcangeli (Plantaginaceae) in Ireland. *Irish Nat. Jour.* **22**: 441–443.
Cardew, R. M. & Baker, E. G. (1912). Notes on *Plantago serraria* Linn. *Rep. Bot. Soc. Exch. Club Brit. Isles* **3**: 28–29.

Cavers, P. B. et al. (1980). The biology of Canadian weeds. 47. *Plantago lanceolata* L. *Canad. Jour. Pl. Sci.* **60**: 1269–1282.
Dodds, J. G. (1953). Biological flora of the British Isles. No. 38. *Plantago coronopus* L. *Jour. Ecol.* **41**: 467–478.
Druce, G. C. (1913). *Plantago media* var. *lanceolatiformis* Druce. *Rep. Bot. Soc. Exch. Club Brit. Isles* **3**: 173.
Druce, G. C. (1915). *Plantago coronopus* var. *sabrinae* Baker & Cardew. *Rep. Bot. Soc. Exch. Club. Brit. Isles* **4**: 73.
Grime, J. P. et al. (1988). *Comparative plant ecology*. London.
Hawthorn, W. R. (1974). The biology of Canadian weeds. 4. *Plantago major* and *P. rugellii*. *Canad. Jour. Pl. Sci.* **54**: 383–396.
Hultén, E. & Fries, M. (1986). *Atlas of north European vascular plants north of the Tropic of Cancer.* 3 vols. Königstein.
Piliger, R. (1937). Plantaginaceae. In Engler, A. (Edit.) *Das Pflanzenreich* **IV.269**.
Preston C. D. & Whitehouse, H. L. K. (1986). The habitat of *Lythrum hyssopifolia* L. in Cambridgeshire, its only surviving English locality. *Biol. Conservation* **35**: 41–62.
Sagar, G. R. & Harper, J. L. (1964). Biological flora of the British Isles. No. 95. *Plantago major* L. (pp. 189–205), *Plantago media* L. (pp. 205–210) and *Plantago lanceolata* L. (pp. 211–218). *Jour. Ecol.* **52**: 189–221.

1. Plants dioecious **4. virginica**
1. Plants monoecious 2.
2. Leaves opposite 3.
2. Leaves all basal or alternate 4.
3. Plant not or minutely glandular-hairy; lowest bracts very different in shape from upper, with lateral veins at the base **11. arenaria**
3. Plant usually strongly glandular-hairy above; bracts all similar in shape, without lateral veins **12. afra**
4. Corolla tube hairy 5.
4. Corolla tube glabrous 7.
5. Leaves regularly and distinctly toothed to pinnatifid **2. coronopus**
5. Leaves entire or with a few, small irregular teeth 6.
6. Stock unbranched, with a solitary rosette; capsule 3-locular **2. coronopus**
6. Stock branched, with several rosettes; capsule 2-locular **3. maritima**
7. Anterior sepals connected for more than half their length 8.
7. Anterior sepals free for more than half their length 13.
8. Bracts and sepals densely long-hairy (villous) **8. lagopus**
8. Bracts and sepals glabrous or very shortly hairy 9.
9. Leaves 2–7 × 0.2–0.8 cm 10.
9. Leaves 6–20 × 0.3–5.0 cm 11.
10. Plant glabrous or slightly hairy **7(i). lanceolata** var. **pusilla**
10. Plant covered with short grey hairs **7(ii). lanceolata** var. **angustifolia**
11. Leaves 0.3–3.0 cm wide, narrow to broadly lanceolate **7(iv). lanceolata** var. **lanceolata**
11. Leaves 2.5–5.0 cm wide, lanceolate, elliptical or ovate-elliptical 12.
12. Leaves more or less elliptical to ovate-elliptical; inflorescence ovoid **7(iii). lanceolata** var. **latifolia**
12. Leaves lanceolate, long and attenuate at base; inflorescence narrowly conical **7(v). lanceolata** var. **bakeri**
13. Anterior sepals almost entirely scarious, with midrib extending to not more than halfway **10. loeflingii**
13. Anterior sepals with midrib extending to apex 14.
14. Annual with 2 seeds **9. ovata**
14. Perennial 15.
15. Leaves linear, linear-lanceolate or lanceolate, usually less than 1.5 cm wide 16.
15. Leaves lanceolate, elliptical or subrotund, usually more than 1.5 cm wide 17.
16. Bracts and sepals long white-hairy (villous) **5. varia**
16. Bracts and sepals glabrous to shortly hairy **9. ovata**
17. Seeds 2–5(–7), plano-convex or cymbiform 18.
17. Seeds (4–)6–34, ellipsoid or ellipsoid-trigonous 19.
18. Leaves (2–)3–8 × 1.5–6.0 cm; petioles up to 3 cm **6(i). media** var. **media**
18. Leaves 6–15 × 3.5–8.0 cm; petioles up to 15 cm **6(ii). media** var. **urvilleana**
19. Leaves mostly with 5–9 veins; capsules with 4–15 seeds **1(a). major** subsp. **major**
19. Leaves mostly with 3–5 veins; capsules with (9–)14–25 seeds 20.
20. Leaves 0.3–6.0 cm wide; petioles up to 2 cm; inflorescence up to 2 cm 21.
20. Leaves 1.5–10 cm wide; petioles up to 20 cm; inflorescence up to 27 cm 22.
21. Leaves 0.5–2.5 × 0.3–1.5 cm; inflorescence up to 2cm **1(b, i). major** subsp. **intermedia** var. **minima**
21. Leaves 2–9 × 1.5–6.0 cm; inflorescence up to 4 cm **1(b, ii). major** subsp. **intermedia** var. **salina**
22. Leaves 2–9 × 1.5–6.0 cm, entire or with small teeth, glabrous or slightly hairy; inflorescence up to 8 cm **1(b, iii). major** subsp. **intermedia** var. **scopulorum**
22. Leaves 3–15 × 2–10 cm, markedly wavy and toothed, hairy; inflorescence up to 27 cm **1(b,iv). major** subsp. **intermedia** var. **sinuata**

Subgenus **1. Plantago**

Leaves in basal rosettes or alternate.

1. P. major L. Greater Plantain

Perennial monoecious *herb* with a tap-root and fibrous side-roots, and one rosette. *Leaves* in a basal rosette; lamina (1.5–)5–15(–40) cm, dull medium green on upper surface, paler beneath, broadly ovate or broadly elliptical, rounded-apiculate to subacute at apex, entire or shallowly and irregularly toothed, rounded to a shortly cuneate or subcordate base, glabrous to shortly hairy on both surfaces, with 3–9, parallel, longitudinal veins; petiole up to 15(–40) cm, flat, glabrous or hairy. *Flowering stem* equalling or exceeding the leaves, not furrowed, glabrous or hairy; inflorescence (1–)10–15(–50) cm, a cylindrical spike; flowers bisexual; bracts 1–2 mm, brownish with a green keel, ovate, glabrous. *Sepals* 4, 1.5–2.5 mm, green, broadly ovate, rounded at apex, rounded at base. *Corolla* yellowish-white; tube about 2 mm, glabrous; lobes 4, about 1 mm, ovate, subobtuse at apex, glabrous. *Stamens* 4, exserted 2–3 mm; anthers at first lilac, later dirty yellow. *Style* 1, terminal; stigma 1, linear. *Capsule* 2–4 mm, ellipsoid, 2-celled; seeds 4–25(–36), 0.6–2.1 mm, reniform. *Flowers* 5–9. Wind-pollinated.

(a) Subsp. **major**
Leaves mostly with 5–9 veins, usually obtuse at apex, rounded to subcordate at base and subentire. *Capsules* mostly with 4–15 seeds; seeds (1.0–)1.2–1.8(–2.1) mm. $2n=12$.

(b) Subsp. **intermedia** (Gilib.) Lange
P. intermedia Gilib.

Leaves mostly with 3–5 veins, usually subacute at apex, broadly cuneate at base, more or less undulate-dentate near the base. *Capsules* with (9–)14–25(–36) seeds; seeds (0.6–)0.8–1.2(–1.5) mm. $2n=12$.

(i) Var. **minima** Wimm. & Grab.
P. major var. *agrestis* Fr.

Dwarf annual mostly flat to the ground. *Leaves* 0.5–2.5 × 0.3–1.5 cm, with a wavy margin, more or less hairy; petioles up to 1 cm, very hairy. *Flowering stem* very hairy. *Inflorescence* spike up to 2 cm.

(ii) Var. **salina** Wirtg.
P. winteri Wirtg.

Probably mostly annual, more or less erect. *Leaves* 2–9 × 1.5–6.0 cm, fleshy, with a wavy margin and often some teeth, hairy; petioles up to 2 cm, very hairy. *Flowering* stem very hairy. *Inflorescence spike* up to 4 cm, often curved upwards.

(iii) Var. **scopulorum** Fr. & Broberg
P. major var. *pubescens* Lange; *P. major* var. *intermedia* (Gilib.) Lange; *P. pauciflora* var. *scopulorum* (Fr.) Domin

Probably annual in agricultural habitats, perhaps perennial elsewhere. *Leaves* 2–9 × 1.5–6.0 cm, entire or with small teeth, glabrous or sparingly hairy; petioles up to 18 cm. *Flowering* stem glabrous or sparingly hairy. *Inflorescence* spike up to 8 cm, erect or curved upwards.

(iv) Var. **sinuata** (Lam.) Decne
P. sinuata Lam.; *P. pauciflora* var. *sinuata* (Lam.) Domin; *P. dregeana* Decne; *P. gigas* Lév.

Mostly large and erect, annual or perennial according to habitat. Leaves 3–15 × 2–10 cm, margins markedly wavy and toothed, hairy; petioles up to 20 cm, more or less hairy. *Inflorescence* up to 27 cm, usually erect, sometimes curved upwards.

Native. Open habitats such as trampled ground, waste places, roadsides, gardens, grassland, edges of streams and rivers, by ponds and reservoirs, winter flooded areas on a wide range of soils, avoiding only acidic areas. Throughout Great Britain and Ireland. Europe; North Africa; north and central Asia. Eurasian Wide-temperate element. Subsp. *major* occurs throughout the range of the species mainly in disturbed habitats and it may be introduced. Var. *minima* seems to be a plant of dry places where it sometimes occurs in thousands. Var. *salina* is a plant of damp coastal sands and by streams and pools. Var. *scopulorum* is characteristic of wet hollows in fields, commons and the sides of old fens. Var. *sinuata* is the plant of waste places and cultivated ground. It is possible that these variants of subsp. *intermedia* are the native plants which are connected to subsp. *major* by var. *sinuata*. All seem to occur in Continental Europe. J. F. Parker (pers. comm.) says they can be self-pollinated which may help to keep the various populations distinct.

2. P. coronopus L. Buck's-horn Plantain

Annual to *perennial* monoecious *herb* with a persistent tap-root. *Leaves* all basal; lamina 2–20 × 0.1–0.5 (–1.0) cm, very variable, linear to lanceolate, dentate to 1- to 2-pinnatifid, rarely entire, the lobes entire or dentate and more or less distant, gradually narrowed into the winged petiole, glabrous or hairy. *Flowering stems* numerous, 2–25 cm, decumbent or ascending, shorter than or exceeding the leaves, glabrous to hairy, not furrowed. *Inflorescence* 1–6 cm; flowers bisexual; bracts ovate, subacute to long acuminate, shorter than or equalling calyx. *Sepals* 4, about 2 mm, the posterior with a weakly ciliate wing. *Corolla* brownish, 4-lobed, the lobes ovate and acute to acuminate at apex, the tube hairy. *Stamens* 4; filaments pale; anthers pale yellow. *Style* 1, filiform, hairy. *Capsule* 3-locular, 1.5–2.5 mm, broadly ellipsoid or subglobose, usually 4-seeded; seeds 0.8–1.0 mm, pinkish-brown, ovoid-ellipsoid, convex on both sides, the margin hyaline. *Flowers* 5–7. Protogynous and wind-pollinated. $2n=10$.

Varies from small plants with narrow, hardly lobed leaves to broad-leaved, long-toothed plants with long spikes, and from glabrous to very hairy. There are intermediates between the extremes and they may all occur in one locality. There seems to be greater variation in coastal localities than there is on inland heaths. Var. *sabrinae* Cardew & Baker is a very striking plant with the broadest leaves and long sharp teeth but it is connected to the remainder by fertile intermediates.

Native. Dry, open, often heavily trampled habitats on acidic to basic, stony or sandy soils and rock crevices, from open grassland on heaths, sand dunes and shingle to sea-cliffs, sea-walls, waste ground and by paths. Coastal areas throughout Great Britain and Ireland and inland in middle and southern England with scattered localities elsewhere. It has always been known inland, but now occurs also on salt-treated roadsides. Coasts of central and south Europe from southern Sweden southwards; North Africa, west Asia and Macaronesia; introduced in North America, Australia and New Zealand. Eurosiberian Southern Temperate element.

3. P. maritima L. Sea Plantain

Perennial monoecious *herb* with a stout, branched, woody stock and usually several rosettes, the apex of the rootstock often long hairy. *Leaves* alternate; lamina 3–20(–30) × 0.2–1.0(–1.5) cm, medium green, rather fleshy, linear, obtuse to acute at apex, entire or remotely toothed, narrowed at base, but petiole indistinct or wanting, glabrous, faintly 3- to 5(–7)-veined. *Flowering stems* few or numerous, equalling or exceeding leaves, not furrowed, erect, terete or almost so, generally clothed with numerous, appressed hairs; inflorescence 2–6(–12) cm, a narrow, lax or crowded spike; flowers bisexual; bracts 2–3 mm, ovate, obtuse to subacute, strongly keeled with a thickened nerve, membranous at margin, glabrous or sparsely ciliate. *Sepals* 4, 2.0–2.5 mm, with a wide membranous margin, ovate, the adaxial with a narrow or broad, ciliate dorsal wing. *Corolla* brownish; tube about 2 mm, more or less hairy; lobes 4, about 1.5 mm, ovate, acute at apex, and spreading or reflexed. *Stamens* 4; filaments up to 3 mm, glabrous; anthers pale yellow. *Style* 1, very short; stigma filiform. *Capsule* 3–4 mm, ovoid, 2-celled; seeds usually 2, 1.5–2.5 mm, oblong-elliptical, dull brown, smooth. *Flowers* 6–8. Wind-pollinated. $2n = 12$.

Native. Salt-marshes, rock crevices and short turf near the sea and wet, rocky places on mountains. Common round the coasts of Great Britain and Ireland and inland on mountains in Scotland, northern England and north and west Ireland, rare in inland salt-marshes or by salt-treated roads. Most of Europe but rare in the south; North Africa; Greenland; North America and southern South America. Wide Boreal element.

4. P. virginica L. Virginian Plantain

Annual or *biennial* dioecious *herb*. Leaves in a basal rosette; lamina 1–15 cm, spreading or ascending, medium green on upper surface, paler beneath, thin, spathulate to obovate or elliptic, obtuse to rather acute at apex, entire or repand-denticulate, narrowed to a short petiole or sessile, 3- to 5-veined, nearly glabrous to hairy. *Flowering stem* 50–100 cm, much longer than the leaves, erect or ascending, glabrous to rather coarsely hairy. *Inflorescence* a dense spike, sometimes interrupted below; bracts linear-lanceolate to lanceolate. *Sepals* 4, 2.0–2.5 mm, longer than the bracts, with a scarious margin, oblong or ovate, obtuse at apex. *Corolla* variable, those of staminate flowers with 4 spreading lobes, those of pistillate flowers with lobes usually unequally erect after fertilisation. *Stamens* 4. *Style* 1. *Capsule* 1.2–2.0 mm, ovoid or ellipsoid-ovoid; seeds 2–4, about 1.3 mm, golden yellow.

Introduced. A wool casual. Native of North America.

5. P. varia R. Br. Variable Plantain

Perennial monoecious *herb* with a short, vertical stock. *Leaves* in a rosette; lamina 9–15(–20) × 0.8–1.5 cm, medium green on upper surface, paler beneath, thin, narrowly lanceolate or lanceolate, obtusish at apex, undulate or shallowly dentate, narrowed at base to the petiole, with numerous, long, white, rigid hairs. *Flowering stem* 7–15(–24) cm, erect or arcuate, with copious hair. *Inflorescence* a sparse spike; flowers bisexual; bracts 2.5–2.7 mm, ovate, obtuse at apex, gibbose at base, with dense white hair. *Sepals* 4, about 3mm, elliptical or ovate-elliptical, sparse to very hairy. *Corolla* 4-lobed, the lobes rounded-ovate, shortly acute-mucronate. *Stamens* 4. *Style* 1. *Capsule* ovoid.

Introduced. A wool casual. Native of Australia.

6. P. media L. Hoary Plantain

P. concinna Salisb. nom. illegit.; *P. incana* Stokes nom. illegit.; *Arnoglossum incanum* Gray nom. illegit.

Perennial monoecious *herb* with a stout rootstock and 1 to few rosettes. *Leaves* alternate; lamina (2–)5–15 × (1.5–) 2.5–8.0 cm, dull medium green on upper surface, paler beneath, not blackening on drying, broadly ovate to broadly elliptic, abruptly acute or acuminate at apex, entire or remotely crenate or dentate, narrowed at base, more or less densely crispate-hairy; veins (5–)7–9; petiole usually less than half as long as lamina but sometimes up to 15 cm. *Flowering stems* greatly exceeding the leaves, pale green, striate, with appressed or ascending simple eglandular hairs, not furrowed. *Inflorescence* a dense spike, (10–)20–60(–100) mm, up to 150 mm in fruit; flowers bisexual; bracts 2–3 mm, ovate or ovate-lanceolate, acute at apex, with membranous margins, glabrous or shortly hairy. *Sepals* 4, about 2 mm, subequal, almost free, green or purplish with scarious margins, glabrous. *Corolla* brownish; tube about 2 mm; lobes 1.5–2.0 mm, ovate-lanceolate, subacute at apex. *Stamens* 4, exserted 8–13; filaments purple; anthers lilac or white. *Style* 1. *Capsule* 2–4 mm; seeds 2–4(–6), about 2 mm, oblong-elliptical, plano-convex. *Flowers* 7–8. $2n = 12, 24$.

(i) Var. media

Leaves (2–)3–8 × 1.5–6.0 cm, spreading, broadly ovate to broadly elliptical; petiole up to 3 cm.

(ii) Var. urvilleana Rapin

P. media var. *longifolia* G. F. Mey.; *P. urvillei* Opiz; *P. media* var. *oblongata* Opiz; *P. monnieri* Girard; *P. media* var. *monnieri* (Girard) Rouy; *P. oblongifolia* Schur; *P. media* var. *lanceolatiformis* Druce

Leaves 6–15 × 3.5–8.0 cm, erect, oblong-elliptical or broadly elliptical to oblong or oblong-lanceolate; petioles up to 15 cm.

Native. Characteristic of chalk and limestone soils but also on heavy clay. Its main habitats are downland grassland and tracks, calcareous pasture and mown grassland, particularly churchyards and cemeteries; less frequent in hay meadows and on dunes. Common throughout England and getting into Wales and southern Scotland. Introduced elsewhere in Scotland and in Ireland. Most of Europe and temperate Asia. Eurasian Temperate element. Var. *media* is the common variant. Var. *urvilleana* is recorded from a few localities in southern England and may be introduced.

7. P. lanceolata L. Ribwort Plantain

Perennial monoecious *herb* with a thick, tough rootstock, often densely long hairy at the apex and usually divided into several rosettes. *Leaves* in a rosette; lamina 2–30 × 0.5–5.0 cm, dull medium, often rather greyish-green on upper surface, paler beneath, linear, lanceolate or broadly to narrowly elliptical, acute or acuminate at apex, entire or remotely serrulate, long attenuate at base, glabrous to more or less densely hairy; veins 3–5(–7), prominent; petiole up to 12 cm, flattened or canaliculate. *Flowering stem* 10–60(–100) cm, pale green, strongly ribbed or angled, appressed-hairy. *Inflorescence* a dense ovoid, oblong, cylindrical or conical spike 0.7–5.0 cm, lengthening up to 12 cm in fruit; flowers bisexual; bracts 3–5 × 1.5–3.0 mm, obovate, acuminate or with a long attenuate cusp, membranous and brownish, glabrous. *Sepals* 4; 2.5–3.0 × 2.0–2.5 mm, yellowish or fuscous, membranous, the anterior connate almost to apex, the posterior deeply concave or boat-shaped, distinctly veined, the vein sometimes narrowly winged, glabrous or hairy. *Corolla* with tube 1.5–3.0 mm, the lobes 1.5–2.0 mm, ovate, acuminate at apex, spreading or reflexed. *Stamens* 5; filaments up to 5 mm, slender, glabrous; anthers yellowish. *Style* about 2 mm; stigma filiform. *Capsule* 2.0–2.5 mm, subglobose to ovoid; seeds 2, 2.0–2.5 mm, oblong, bright brown, minutely rugulose. *Flowers* 4–8. Wind-pollinated, $2n = 12$.

(i) Var. **pusilla** Baumg.
P. lanceolata subvar. *spaerostachya* (Mert. & W. D. J. Koch) Pilg.; *P. lanceolata* subsp. *spaerostachya* (Mert. & W. D. J. Koch) Hayek; *P. spaerostachya* (Mert. & W. D. J. Koch) A. Kerner; *P. lanceolata* var. *spaerostachya* Mert. & W. D. J. Koch; *P. lanceolata* var. *minima* Gaudin; *P. lanceolata* var. *pumila* Neil.; *P. lanceolata* var. *nana* Hardy; *P. lanceolata* var. *minor* Schltdl.

Leaves 2–7(–10) × 0.2–0.7 cm, narrow, glabrous or slightly hairy, shortly petiolate. *Inflorescence* globose.

(ii) Var. **angustifolia** Poir. ex Lam.
P. dubia L.; *P. lanceolata* var. *dubia* (L.) Wahlenb.; *P. lanceolata* var. *villosa* Mey.; *P. lanceolata* var. *capitellata* Sond. ex W. D. J. Koch; *P. lanceolata* var. *eriophylla* Decne

Leaves 3–5 × 0.4–0.8 cm, linear-lanceolate or narrowly lanceolate, shortly but obviously petioled, covered with short grey hairs and the top of the rootstocks lanuginose. *Inflorescence* ovoid.

(iii) Var. **latifolia** Wimm. & Grab.
P. lanceolata subvar. *latifolia* (Wimm. & Grab.) Pilg.; *P. lanceolata* var. *elliptica* Cop.; *P. lanceolata* var. *elliptica* Druce, non Cop.; *P. lanceolata* var. *platyphylla* Druce

Leaves 6–19 × 2.5–5.0 cm, more or less elliptical to ovate-elliptical, shortly petiolate, nearly glabrous. *Inflorescence* ovoid.

(iv) Var. **lanceolata**
Leaves 10–20(–30) × 0.3–3.0 cm, narrow to broadly lanceolate, attenuate to a rather short petiole, glabrous or thinly hairy. *Inflorescence* subglobose, ovoid or shortly oblong.

(v) Var. **bakeri** C. E. Salmon
P. lanceolata var. *mediterranea* Pilg.; *P. lanceolata* var. *timbali* auct.

Plant up to 100 cm. *Leaves* 10–30 × 2.5–5.0 cm, lanceolate, long and attenuate at base to a long petiole, usually glabrous, often serrulate. *Inflorescence* elongate, narrowly conical.

Native. Grassy and open places. Abundant throughout Great Britain and Ireland. Europe except the extreme north; North Africa; north and central Asia; introduced in most other temperate countries. Southern Temperate element now Circumpolar Southern-temperate. Var. *pusilla* is a plant of dry grassland. Var. *angustifolia* is a plant of dunes and shingle and is probably all round our coasts. Var. *latifolia* is the plant of fens and damp meadows, also lawns. Var. *lanceolata* is rather variable and widespread. Var. *bakeri* seems to have been introduced in hay crops and is native of the Mediterranean region eastwards to Pakistan and introduced in western Europe, Australia and North America.

8. P. lagopus L. Hare's-foot Plantain

Perennial monoecious *herb* with several rosettes or sometimes annual with a single rosette. *Leaves* all basal; lamina 4–15 × 0.6–4.0 cm, medium green on upper surface, paler beneath, linear-lanceolate to lanceolate, acute or obtuse at apex, entire or minutely denticulate, gradually narrowed to a winged petiole, 3- to 5-veined, shortly hairy to villous. *Flowering stem* 3–35 cm, pale green, with appressed hairs; inflorescence a dense spike 0.8–3.0 cm, bracts ovate-lanceolate, acute, densely long hairy; flowers bisexual. *Sepals* 4, 2.5–3.0 mm, obovate, the anterior connate for more than half their length, densely long hairy. *Corolla* 4-lobed, the lobes sharply acuminate. *Stamens* 4; anthers yellowish. *Style* 1. *Capsule* about 2.5 mm; seeds 2, about 1.8 mm. *Flowers* 7–8. $2n = 12$.

Introduced. A bird-seed, grain and esparto casual. Native of the Mediterranean region and south-west Asia.

9. P. ovata Forsskål Blond Plantain

Perennial or *annual* monoecious *herb* with one or few rosettes. *Flowering stems* only slightly exceeding leaves, terete and shortly villous. *Leaves* with lamina 2.5–12.0 × 0.1–0.8 cm, medium green on upper surface, paler beneath, linear to linear-lanceolate, pointed at apex, entire or remotely denticulate, narrowed at base, sparsely to densely villous-lanate. *Inflorescence* a spike 5–35 mm, with dense bisexual flowers; bracts about 3 mm, subrotund to ovate, sometimes shortly hairy. *Sepals* 4, about 2.5 mm, subequal, almost free, keeled to the apex, with wide scarious margins, at least the anterior usually shortly hairy. *Corolla* tube 1.5–2.0 mm, glabrous; lobes about 2.5 mm, ovate-subrotund, subobtuse or very shortly acuminate. *Stamens* 4, inserted on corolla, exserted up to 1.0 mm. *Style* 1, stigma 1. *Capsule* about 3 mm; seeds 2, 2.0–2.5 mm, cymbiform. *Flowers* 7–8. $2n = 8$.

Introduced. An esparto casual; possibly also from seed imported for sale in Asian-owned grocery shops (Clement and Forster 1995). Native of the Mediterranean region.

10. P. loeflingii L. Loefling's Plantain

Annual monoecious *herb* with one or several rosettes. *Flowering stems* shorter than leaves, terete, appressed-hairy. *Leaves* with lamina 2–7(–10) × 0.1–0.7 cm, medium green on upper surface, paler beneath, linear to linear-lanceolate, pointed at apex, entire or remotely dentate, narrowed at base, with sparse to dense, patent rather stiff hairs, 3-veined. *Inflorescence* a spike 5–25 mm; flowers bisexual, bracts 2.5–3.0 mm, wider than long, glabrous. *Sepals* 4, 1.5–2.0 mm, equal, almost free, scarious, subrotund, mid-vein only in lower one-quarter, glabrous. *Corolla* tube about 1.5 mm, glabrous; lobes 4, about 1.0 mm, ovate-lanceolate, subacute at apex. *Stamens* 4, inserted on corolla tube. *Style* 1; stigma 1. *Capsule* about 3 mm; seeds 2, 2.0–2.5 mm, very narrowly cymbiform. *Flowers* 7–9.

Introduced. An esparto casual. Native of the west Mediterranean region to south-west Asia.

Subgenus **2. Psyllium** (Mill.) Harms.
Psyllium Mill.

Leaves opposite on branched stems.

11. P. arenaria Waldst. & Kit. Branched Plantain
P. psyllium L. nom. rejic.; *P. scabra* Moench nom. illegit.; *P. indica* L. nom. illegit.

Annual monoecious *herb. Stems* up to 50(–80) cm, pale green, erect, usually with straight, ascending branches, with patent or ascending simple eglandular hairs and more or less minutely glandular-hairy above. *Leaves* opposite; lamina 3–8 × 0.1–0.3(–0.4) cm, medium green on upper surface, paler beneath, linear or linear-lanceolate, not fleshy, acute at apex, entire. *Inflorescence* a spike 5–15 mm; lower 2 bracts 6–10 mm, with green midrib and wide scarious margins, ovate-subrotund with linear-subulate apex, straight and suberect, with divergent lateral veins at base; upper bracts 3.5–4.5 mm, ovate-subrotund or wider than long; flowers bisexual. *Sepals* 4, unequal, the anterior 3.5–4.0 mm, obovate-spathulate, the posterior 3.0–3.5 mm, ovate-lanceolate. *Corolla* tube 3.5–4.0 mm; lobes 4, about 2 mm, ovate-lanceolate, acute at apex. *Stamens* 4, inserted on corolla tube. *Style* 1; stigma 1. *Capsule* about 2.5 mm; seeds about 2.5 mm, cymbiform, oblong-elliptical in outline. *Flowers* 7–8. $2n = 12$.

Introduced. Open and rough ground on sandy soil, casual or sometimes naturalised. Scattered records throughout Great Britain, especially East Anglia and less common than formerly, now often confused with *P. afra*. Native. of south Europe.

12. P. afra L. Glandular Plantain
P. psyllium L. (1762), non (1753); *P. indica* auct.

Annual monoecious *herb* with fibrous roots. *Stems* 2–20(–30) cm, pale green, often tinged purple, erect or occasionally decumbent, subterete or obscurely angled, usually rather densely glandular-hairy, often unbranched, but sometimes branched from the base, leafy. *Leaves* opposite; lamina 1.0–5.0 × 0.1–0.3 cm, medium green on upper surface, paler beneath, linear or narrowly lanceolate, tapering to apex, entire or remotely serrulate, attenuate at base, thinly glandular-hairy and sometimes clothed with long white hairs near the base, sessile. *Inflorescence* crowded near the apex of the stems, subglobose or shortly ovoid; flowers bisexual; peduncles 1–4 cm, slender, usually viscid-glandular-hairy; bracts all more or less similar, about 5 mm, with a concave, ovate base, tapering to an acuminate-caudate apex, with a distinct median nerve. *Sepals* 4, about 4 mm, subequal, membranous, ovate, acuminate at apex, spreading or strongly reflexed. *Stamens* 4; filaments slender, glabrous; anthers yellowish. *Style* 1, very short; stigma filiform. *Capsule* about 2 mm, ovoid; seeds 2, about 2.2 mm, shining brown, narrowly oblong, convex dorsally, concave ventrally, almost smooth. *Flowers* 7–8. $2n = 12$.

Introduced. Bird-seed and grain alien of tips and waste places. Very scattered records through Great Britain. Mediterranean region eastwards to Pakistan; Atlantic Islands.

2. Littorella P. J. Bergius

Perennial, monoecious, stoloniferous, aquatic *herbs. Leaves* in rosettes, half-cylindrical and linear-subulate. *Flowering stem* usually shorter than the leaves, rarely equalling them, with a solitary male flower at the top and 1–few female at the base and bracts in between. *Sepals* 4, green with a scarious margin. *Petals* 4, white and scarious. *Stamens* 4, up to 20 mm. *Style* 1; stigma terminal, linear. *Fruit* a nut, ovoid.

Contains 3 species, 1 in Europe including the Mediterranean region and the Azores, 1 in North America and 1 in temperate South America.

Arts, G. H. P. & Heijden, R. A. J. M. van der (1990). Germination ecology of *Littorella uniflora* (L.) Aschers. *Aquat. Bot.* **37**: 139–151.
Dietrich, H. (1971). Blütenmorphologische und palynologische Untersuchungen an *Littorella. Feddes Repert.* **82**: 155–165.
Preston, C. D. & Croft, J. M. (1997). *Aquatic plants in Britain and Ireland.* Colchester.
Robe, W. E. & Griffiths, H. (1992). Seasonal variation in the ecophysiology of *Littorella uniflora* (L.) Ascherson in acidic and autrophic habitats. *New Phytologist* **120**: 289–304.

1. L. uniflora (L.) Asch. Shoreweed
Plantago uniflora L.

Perennial, monoecious, scapigerous, glabrous, aquatic *herb* with abundant stolons forming an extensive tuft in shallow water. *Stolons* slender, far-creeping, rooting and producing rosettes of leaves at the nodes. *Leaves* 2–10(–25) cm, dark green, half-cylindrical and linear-subulate, or sometimes flattened and broader, subacute at apex, entire, sheathing at base, glabrous, not septate. *Flowering stem* usually shorter than the leaves, rarely equalling them, slender, with a bract below the middle; bract lanceolate, acute at apex, entire, sheathing; male flower solitary at apex of stem, female flowers 1–few, near the base. *Sepals* 4, 5–6 mm, green, with scarious margins, triangular-ovate, acute at apex. *Petals* 4, 9–11 mm, white and scarious, oblong, obtuse at apex. *Stamens* 4; filaments up to 20 mm, pale; anthers pale yellow. Style 1, about 10 mm, curved; stigma terminal, linear. *Nut* 1, 2.8–3.0 mm, ovoid. *Flowers* 6–8. Wind-pollinated. $2n = 24$.

Native. Shallow, oligotrophic or mesotrophic waters of lakes, rivers, streams, ponds and dune slacks on a wide range of substrates including stones, gravel, sand, peat, marsh and soft mud to a depth of 4 m. In suitable places through much of Great Britain and Ireland, but very local and decreasing in the lowlands. Widespread in west, central and north Europe, becoming rare in the south but extending to Spain, Portugal and Sardinia; recorded from North Africa. Suboceanic Temperate element.

Order **6. SCROPHULARIALES** Lindl.

Herbs, shrubs and *trees*. *Leaves* opposite, alternate and sometimes whorled; usually without stipules. *Corolla* often 2-lipped. *Stamens* only 4 or 2 functional; most families with a nectary disc. *Carpels* 2, rarely 3 or 4. *Ovary* with axile, parietal or free-central placentation. *Fruit* often a capsule, sometimes a drupe or berry.

Contains 12 families and more than 11,000 species.

135. BUDDLEJACEAE Wilh. nom. conserv.

Deciduous or semi-evergreen monoecious *shrubs*. *Leaves* alternate or opposite, simple, entire to serrate, more or less petiolate, exstipulate. *Flowers* in more or less dense panicles, small but showy in dense masses, fragrant, actinomorphic, bisexual, hypogynous. *Calyx* with 4 lobes. *Corolla* 4-lobed, of various colours. *Stamens* 4, borne on the corolla tube. *Style* 1; stigmas 2, linear. *Ovary* 2-celled, each cell with many ovules on axile placentas. *Fruit* a 2-valved capsule; seeds many, narrowly winged.

Contains about 10 genera and 150 species, mainly tropical, extending to temperate America, China, Japan and South Africa.

1. Buddleja L.

As family.

Contains about 100 species found mainly in eastern Asia, but also occurring in South America and eastern Asia. Named after Adam Buddle (died 1715).

1. Leaves alternate; flowers borne in small clusters along previous year's wood **1. alternifolia**
1. Leaves opposite; flowers borne in panicles on current year's growth 2.
2. Flowers in dense, globose, stalked clusters about 16–20 mm in diameter arranged in large open panicles **4. globosa**
2. Flowers in long, dense, narrowly pyramidal panicles, often interrupted mostly below and often the segments more or less globose 3.
3. Flowers lilac to purple or white **2. davidii**
3. Flowers dull yellow to greyish- or purplish-yellow **3.× weyeriana**

1. B. alterniflora Maxim. Alternate-leaved Butterfly bush

Monoecious *shrub* up to 3 m. *Stems* with old bark peeling in thin flakes; branches slender; young shoots with minute stellate hairs. *Leaves* alternate; lamina 4–6 × 0.6–1.5 cm, green on upper surface, silvery grey beneath, lanceolate or oblong, acute to subacute at apex, entire, narrowed below and subsessile, with few, scattered stellate hairs on upper surface and minute papillae and stellate hairs beneath; veins 4–5 pairs, prominent beneath. *Inflorescence* of many, globose, spirally arranged clusters borne on very short, leafy or leafless spurs on long arching or pendulous branches; flowers bisexual; pedicels up to 12 mm, silvery grey. *Calyx* 2.5–3.0 mm, 5-lobed, the lobes ovate-oblong or elliptical, subacute at apex, with a fine silvery covering of minute stellate hairs. *Corolla* salvar-shaped; tube 6–8 mm, reddish-purple, orange in the eye, yellow towards the base inside, glabrous at base, weakly stellate-haired upwards; limb 5–7 mm, rose-purple, 4-lobed, the lobes roundish. *Stamens* 4, inserted at middle of tube; filaments adnate and decurrent as ridges to the level of the stigma. *Style* 1. *Capsule* about 6 mm, green, cylindrical; seeds about 1 mm, brown, ellipsoid to oblong, variously winged. *Flowers* 7–9. $2n = 38$.

Introduced. Grown in gardens, sometimes self-sown and persistent in woodlands, hedges and on banks. Very scattered records in England north to Yorkshire. Native of China.

2. B. davidii Franch. Butterfly bush
B. variabilis Hemsl.

Evergreen, monoecious, semi-evergreen or deciduous *shrub* or small *tree* up to 6 m, with a rounded crown. *Main stems* often several. *Bark* pale brown, fibrous. *Branches* long spreading and arching below, erect above; twigs round, brown; young shoots fast-growing, long, quadrangular, densely grey-woolly, flattened at the nodes. *Buds* opposite, hidden in new foliage, green. *Leaves* opposite, 7–30 × 2–8 cm, dull dark green on upper surface, pale greenish-cream beneath, linear, lanceolate or ovate-lanceolate, gradually tapered to a more or less acute apex, minutely crenate-dentate, the teeth mammiform, rounded at base, glabrous on upper surface, felted beneath; veins 8–14 pairs of primary and alternate subsidiary ones, impressed on upper surface, prominent beneath; petiole up to 7 mm, woolly. *Inflorescence* 15–70 cm, a slender, long, dense pyramidal panicle on the current season's growth; flowers dense, bisexual, 7–10 mm in diameter, strongly scented; rhachis, peduncles and pedicels woolly. *Calyx* 3–4 mm, pale green, divided into 5, shallow, triangular obtuse teeth above. *Corolla* 10–11 mm, usually lilac and orange in throat, sometimes purple or white; tube taking up most of corolla; lobes 5, 4–5 mm, very broadly ovate, rounded with a crinkly margin. *Stamens* 4, borne on and enclosed in the corolla tube; filaments orange; anthers orange. *Style* 1; stigmas 2. *Capsule* 5–6 mm, brown, erect, opening by 2 sutures. *Flowers* 7–9. Visitors include many insects, especially butterflies during the day and moths at night. $2n = 76$.

Introduced. Grown in gardens, estates, parks and on roadsides, naturalised on waste ground, walls, banks, scrub and streets. Common in Great Britain and Ireland but decreasing northwards. Native of China. Named after Abbé Armand David (1826–1900).

3. B. × weyeriana Weyer Weyer's Butterfly bush
B. davidii × globosa

Deciduous *shrub* up to 6 m, with a round crown. *Leaves* opposite; lamina lanceolate, acute at apex. *Inflorescence* a terminal panicle of globose heads. *Corolla* usually dull yellow, but varying through greyish to purple tinged with yellow.

Introduced. Grown in gardens and often named *B. globosa*, escaping and becoming naturalised. Scattered records in southern England. *Flowers* 7–9. Of garden origin. Named after W. van der Weyer (fl. 1920).

4. B. globosa Hope Orange-ball Tree

Deciduous or partly evergreen *tree* up to 5 m with a broad, fairly dense crown. *Trunk* fairly short and much branched, the stiff, erect stems angular and covered with tawny felt, the secondary branches arching. *Bark* pale brown, fissured and scaling. *Twigs* pale brown, thick; young shoots pale brown, angled and tomentose. *Buds* opposite and hidden in new foliage. *Leaves* opposite, 9–30 × 2–8 cm, dull, dark green on upper surface, greenish-white tomentose beneath, lanceolate or narrowly ovate, long-acute at apex with a fine tip, minutely crenate, attenuate at base and sessile; veins impressed on upper surface, prominent beneath. *Inflorescence* 16–20 mm in diameter in diameter, arranged in a terminal, composed panicle of globose heads on the current year's wood; flowers 4–5 mm in diameter, bisexual fragrant. *Calyx* 2.0–2.5 mm, pale green, divided halfway to the base into 4 lobes, the lobes triangular and acute at apex, grey-woolly. *Corolla* 4.0–4.5 mm, orange, divided to nearly halfway to base into 4 lobes, the lobes broadly rounded. *Stamens* 4, sessile; anthers cream. *Style* 1, green; stigma capitate, 2-lobed. *Capsule* brown; seeds numerous. *Flowers* 5–6. $2n = 38$.

Introduced. Grown in gardens, estates and parks and rarely naturalised on roadsides and rough ground. Scattered records in Great Britain north to Dunbartonshire. Native of Chile and Peru.

136. OLEACEAE Mirb. ex DC. nom. conserv.

Trees, *shrubs* or woody *climbers*, monoecious or dioecious. *Leaves* opposite or rarely alternate, simple, with 3 leaflets or pinnate, exstipulate. *Inflorescence* terminal or axillary; flowers small and dull to large and shiny, variously coloured, actinomorphic, bisexual or unisexual, hypogynous. *Calyx* absent, or a 4-lobed tube. *Corolla* absent or a 4-lobed tube. *Stamens* 2, borne on the corolla tube when that is present. *Ovary* 2-celled, each cell with 2 ovules on axile placentas. *Style* 1; stigma 1 and 2-lobed or 2. *Fruit* a 2-lobed capsule, a samara or a drupe or berry.

Contains about 28 genera and 900 species, almost worldwide, but with its greatest representation in east and south-east Asia; many species are widely grown as ornamentals.

1. Fruit a winged samara **5. Fraxinus**
1. Fruit a capsule, drupe or berry 2.
2. Fruit a capsule 3.
2. Fruit a drupe or berry 4.
3. Flowers not yellow; corolla lobes edge to edge in bud **4. Syringa**
3. Flowers yellow; corolla lobes overlapping in bud **6. Forsythia**
4. Leaves pinnate or ternate **7. Jasminum**
4. Leaves simple 5.
5. Flowers yellow **7. Jasminum**
5. Flowers white or greenish-white 6.
6. Flowers terminal; corolla lobes not overlapping in bud **3. Ligustrum**
6. Flowers mostly axillary; corolla lobes overlapping in bud 7.
7. Corolla more than 3 mm, white **1. Osmanthus**
7. Corolla about 2 mm, greenish-white **2. Phillyrea**

1. Osmanthus Lour.

Evergreen, monoecious *shrubs*. *Leaves* opposite, simple, toothed, exstipulate. *Flowers* in an axillary cluster, bisexual, strongly scented. *Calyx* irregularly 4-lobed. *Corolla* tube campanulate, with 4 lobes, the lobes overlapping in bud, white. *Stamens* 2, filaments often short. *Style* 1; stigma bilobed. *Ovary* usually flask or bottle-shaped. *Fruit* a drupe, ellipsoid, dark blue or purple when ripe; endocarp hard.

Contains about 35 species in Asia, especially China, eastern North America and New Caledonia.

Bean, W. J. (1976). *Trees ad shrubs hardy in the British Isles*. Ed. 8. **3**. London.

Green, P. S. (1958). A monographic revision of *Osmanthus* in Asia and America. *Notes Roy. Bot. Garden Edinb.* **22**: 439–542.

1. Leaf teeth 0.2–0.5 mm **1. × burkwoodii**
1. Leaf teeth 3–10 and spine-like at least on some leaves **2. heterophyllus**

1. O. × burkwoodii (Burkwood & Skipw.) P. S. Green Burkwoods's Sweet Olive

O. decorus (Boiss. & Balansa) Kasapligil × **delaveyi** Franch.

× *Osmarea burkwoodii* Burkwood & Skipw.

Evergreen, monoecious *shrub* up to 4 m. *Branches* dense, spreading; twigs brown, knobbly and wrinkled; young shoots pale brown, minutely puberulous. *Leaves* opposite, leathery, 1.5–5.0 × 1–2 cm, dark glossy green on upper surface, pale green beneath, lanceolate, elliptical or narrowly elliptical, acute to acuminate at apex, with small, sharp teeth 0.2–0.5 mm, cuneate at base, glabrous; veins 10–12 lateral, midrib prominent in basal half below; petiole 6–8 mm, pale green, glabrous. *Flowers* bisexual, borne in an axillary cluster, strongly scented. *Calyx* 3.0–3.5 mm, pale yellowish-green with a white margin, divided halfway into 4 lobes, the lobes obtuse at apex. *Corolla* white, the tube 3–5 mm and campanulate, divided halfway into 4 lobes, the lobes 4–5 mm, oblong, obtuse at apex and spreading. *Stamens* 2; filaments short, white; anthers pale yellow or cream. *Style* 1, pale green; stigma brown. *Drupe* unknown. *Flowers* 3–4.

Introduced. A garden hybrid raised by Messrs Burkwood and Skipwith of Kingston-upon-Thames and now widely planted in gardens, parks and round amenity areas and reported as persistent outside cultivation at Abbotsbury in Dorset.

2. O. heterophyllus (G. Don) P. S. Green Prickly Sweet Olive

Ilex heterophyllus G. Don; O. *aquifolium* Sieber; O. *illicifolius* (Hassk.) Carrière; *Olea ilicifolia* (Hassk.) Hassk.

Evergreen, monoecious *shrub* with a dense, rounded, bushy habit up to 8 m. *Branches* spreading; twigs grey and very rough; young shoots pale brown. *Leaves* opposite, of 2 kinds; upper with lamina 3–7 × 1–3 cm, dark shining green with a faint pale midrib on upper surface, paler beneath, ovate to elliptical, acute or bluntly acuminate, entire, cuneate at base and without hairs but gland-dotted on both surfaces, the petioles up to 10 mm and minutely scabrid; lower with lamina 3–5 × 2–4 cm, dark shining green on upper surface and paler beneath, elliptical, sharply acute at apex, with a few spine-like teeth 3–10 mm, and cuneate at base, the petioles very shortly hairy when young; midrib beneath prominent in lower third. *Flowers* bisexual, in an axillary cluster, 5–7 mm in diameter, strongly scented. *Calyx* green, irregularly 4-lobed, the lobes broadly ovate and pointed. *Corolla* white, the tube 1.0–1.5 mm and campanulate, the 4 lobes 2.5–3.5 mm, obovate and rounded at apex. *Stamens* 2; filaments white; anthers pale yellow or cream. *Style* 1, pale green; stigma deeper green, 2-lobed. *Drupes* about 15 mm, slightly obliquely ellipsoid. *Flowers* 9–10.

Introduced. Widely grown in gardens, and on roadsides, in estates and round amenity areas. Native of southern Japan and Taiwan.

2. Phillyrea L.

Evergreen, monoecious *trees* or *shrubs*. *Twigs* terete. *Leaves* opposite, simple. *Flowers* bisexual, in short, axillary racemes. *Calyx* small, more or less 4-lobed. *Corolla* greenish-white, subrotate, 4-lobed, with the lobes longer than wide and overlapping in bud. *Stamens* 2, epipetalous; filaments short; anthers large and exserted. *Styles* 1; stigma 2-lobed. *Fruit* a dry, bluish-black drupe with a crustaceous endocarp; seeds 1–2.

Contains 4 species in southern Europe and western Asia.

Bean, W. J. (1976). *Trees ad shrubs hardy in the British Isles*. Ed. 8. **3**. London.

Sébastion, C. (1956). *Phillyrea L. Trav. Inst. Sci. Chérif. Ser. Bot.* **6**: 1–100.

1. Leaves all similar, with 4–6 pairs of distant, nearly obsolete veins; calyx with short, rounded lobes; fruit apiculate **1. angustifolia**
1. Leaves dimorphic, with 7–11 pairs of close veins, usually distinct; calyx with triangular lobes; fruit not apiculate **2. latifolia**

1. P. angustifolia L. Narrow-leaved Phillyrea

Evergreen, monoecious *shrub* up to 2.5 m. *Stems* with grey, smooth bark. *Branches* subfastigiate; twigs pale yellowish-green; young shoots pale brown. *Buds* narrowly ovoid, acute at apex; scales brown, ovate, acute, shortly hairy. *Leaves* opposite, all similar (2–)3–8(–10) × 0.3–1.5 cm, dull medium green on upper surface, paler beneath, linear to lanceolate, acute at apex, entire or rarely remotely serrulate, cuneate at base, glabrous on upper surface, covered beneath with sessile, punctulate glands; midrib prominent beneath, the lateral veins 4–6 pairs, distant, most visible on upper surface and making a small angle to the midrib, long, straight, not or slightly forked distally; petiole 2–4 mm, glabrous. *Inflorescence* a short, axillary raceme; flowers bisexual. *Calyx* about 0.5 mm, pale yellowish-green, thick, lobed for quarter of its length, the 4 lobes rounded at apex. *Corolla* about 1.5 mm, greenish-white, subrotate, 4-lobed, the lobes longer than wide. *Stamens* 2; filaments short, pale yellowish-green; anthers large, pale yellowish-green, exserted. *Style* 1, pale green, persistent; stigma pale green, rounded with 2 obtuse lobes. *Drupe* 6–8 × 5–7 mm, black, ovoid when young, becoming subglobose and apiculate; seeds 1–2. *Flowers* 3–4.

Introduced. Planted in large gardens, parks, by roads and in amenity areas. Native of evergreen scrub in the west and central Mediterranean region to Portugal.

2. P. latifolia L. Broad-leaved Phillyrea

P. media L.

Evergreen, monoecious *shrub* or small *tree* up to 15 m with a rounded crown. *Trunk* well into crown with smooth, grey bark. *Branches* fastigiate when young; twigs grey; young shoots pale or greyish-brown and densely hairy. *Buds* about 2 mm, narrowly ovoid, acute at apex; scales brown, ovate, acute at apex, densely hairy. *Leaves* opposite, dimorphic, dull dark green on upper surface, paler beneath; juvenile 2–7 × 1–4 cm, ovate-cordate to ovate-lanceolate, rarely lanceolate, acute at apex, more or less dentate or serrate; adult 1–6 × 0.4–2.0 cm, lanceolate to elliptical, entire or finely serrulate; midrib prominent beneath, lateral veins 7–11 pairs and distant; petioles 2–3 mm, shortly hairy; all leaves glabrous. *Inflorescence* a short, axillary raceme; flowers bisexual. *Calyx* about 0.5 mm, pale yellowish-green, thin, lobed to three-quarters of its length, the 4 lobes triangular. *Corolla* about 2 mm, yellowish-green, cup-shaped, 4-lobed, the lobes triangular and longer than wide. *Stamens* 2; filaments short, pale greenish-yellow; anthers large, pale yellowish-green turning brown, exserted. *Style* 1, pale green, caducous; stigma pale green, elongate, with acute lobes. *Drupe* 7–10 × 5–8 mm, bluish-black, subglobose, not apiculate, seeds 1–2. *Flowers* 3–4. $2n = 46$.

Introduced. Gardens, estates, parks and amenity areas, particularly along the south coast of England and scattered elsewhere. Native of evergreen woods in the Mediterranean region to Portugal.

3. Ligustrum L.

Evergreen to deciduous, monoecious *shrubs* or small *trees*. *Leaves* opposite, simple, entire, exstipulate. *Inflorescence* a terminal cymose panicle on side-shoots; flowers bisexual, strongly scented, appearing after the leaves. *Calyx* campanulate, with 4 shallow lobes. *Corolla* funnel-shaped, with 4 lobes, the lobes not overlapping in bud, creamy white. *Stamens* 2, attached near the top of the corolla tube. *Style* 1; stigma more or less 2-lobed. *Ovary* with 2 ovules per cell. *Fruit* a drupe, with oily flesh, 2-celled, the cells 1- to 2-seeded.

Contains about 40 species, mainly from eastern and south-eastern Asia, but reaching north-eastern Australia and with 1 species in Europe.

Bean, W. J. (1973). *Trees ad shrubs hardy in the British Isles*. Ed. 8. **2**. London.

Mansfeld, R. (1924). Vorarbeiten zu einer Monographie der Gattung *Ligustrum*. *Bot. Jahrb.* **132**: 19–75.

1\. Young shoots hairy; leaves 2–6 × 1–2 cm, lanceolate or narrowly elliptical — 2.
1\. Young shoots glabrous; leaves 2–15 × 1–6 cm, ovate or broadly elliptical — 4.
2\. Stems more or less prostrate — **1(ii). vulgare** var. **coombei**
2\. Stems erect or spreading — 3.
3\. Berry black when ripe — **1(i,1). vulgare** var. **vulgare** forma **vulgare**
3\. Berry yellow when ripe — **1(i,2). vulgare** var. **vulgare** forma **xanthocarpum**
4\. Evergreen tree up to 20 m.; leaves 3–15 × 2–6 cm; berry 8–12 × 4–6 mm. — **3. lucidum**
4\. More or less evergreen to deciduous shrub up to 3(–5) m; leaves 2–10 × 1–5 cm; berry 6–10 mm — 5.
5\. Leaves entirely green — **2(1). ovalifolium** forma **ovalifolium**
5\. Leaves marked with cream or yellow or entirely yellow — 6.
6\. Leaves marked with yellow or entirely yellow — **2(2). ovalifolium** forma **aureum**
6\. Leaves dull green in centre with broad cream margins or patches — **2(3). ovalifolium** forma **argenteum**

1. L. vulgare L. Wild Privet

Semi-deciduous, monoecious *shrub* with a sprawling habit if in open woodland or scrub, in hedgerows can be more compact; and if grown as a hedge and regularly trimmed can be dense and compact, on the coast it can be more or less prostrate. *Stems* up to 3(–5) m, sometimes straight, sometimes flexuous or angled, when young sparingly to much branched. *Bark* pale greyish-brown, rough. *Wood* white and hard. *Branches* spreading at all angles, often whip-like; twigs pale greyish-brown, rough, with numerous lenticels, glabrous; young shoot pale green, sometimes tinted brownish-purple, very shortly hairy. *Buds* about 2 mm, ovoid, obtuse at apex; scales reddish-brown, ovate, acute at apex, glabrous. *Leaves* 2–6 × 1–2 cm, dark green on upper surface, paler beneath, lanceolate or narrowly elliptical, acute at apex, entire, cuneate at base, glabrous; midrib pale above, prominent beneath, lateral veins 4–5 pairs, but obscure; petioles up to 3 mm, pale green, glabrous. *Inflorescence* an erect, terminal panicle; flowers 4–6 mm in diameter, bisexual, strongly scented; rhachis, peduncles and pedicels pale green or suffused brownish-purple, with minute hairs. *Calyx* 2–3 mm, pale green, shallowly 4-lobed, the lobes triangular and pointed at apex, minutely hairy, caducous. *Corolla* 5–7 mm, cream, funnel-shaped, divided one-third of the way to the base into 4 lobes, the lobes triangular and pointed. *Stamens* 2; filaments white; anthers pale yellow. *Style* 1; pale green; stigma 2-lobed. *Drupe* 6–9 mm in diameter, green turning shining black, rarely yellow, globose, with oily flesh. *Flowers* 6–7. Pollinated by various insects. $2n=46$.

(i) Var. **vulgare**
Plant upright and spreading.

(1) Forma **vulgare**
Fruit black when ripe.

(2) Forma **xanthocarpum** (Loudon) P. D. Sell
L. vulgare var. *xanthocarpum* Loudon

Fruit yellow when ripe.

(ii) Var. **coombei** P. D. Sell
L. vulgare forma *prostratum* Ostenf. nom. nud.

Plant more or less prostrate. *Leaves* rather broader than in inland plants.

Native. Hedges, scrub and open woodland, especially on base-rich soils. Throughout most of Great Britain and Ireland except in parts of north Scotland. Central and west Europe and North Africa. Var. *vulgare* forma *vulgare* is the common variant. Forma *xanthocarpum* is rare. Var. *coombei* occurs on the coast of the Lizard in Cornwall and in the Isles of Scilly. It retains its prostate habit after 20 years cultivation in the Cambridge Botanic Garden and may well be found elsewhere on southern and western coasts. Often however, when *L. vulgare* grows in coastal localities it retains its upright habit and looks very dense. Privet is poisonous, but was included in some boundary hedges at enclosure.

2. L. ovalifolium Hassk. Garden Privet

More or less evergreen or deciduous, monoecious *shrub* forming a dense hedge or sometimes occurring as a sprawling shrub. *Stems* up to 3(–5) m, erect or ascending usually flexuous or angled, much branched. *Bark* pale greyish-brown, rough. *Wood* white and hard. *Branches* spreading at all angles; twigs pale greyish-brown, rough; young shoots rather shiny brownish-purple, glabrous, with decurrent ridges from the leaf-base. *Buds* opposite, 2–3 mm, ovoid, pointed at apex; scales green and purplish-brown, ovate, pointed. *Leaves* 2–10 × 1–5 cm, glossy dark green on upper surface or in garden forms yellow, yellow and green of dull green with cream margins or patches, only slightly paler beneath, ovate or broadly elliptical, rounded-mucronulate or more or less acute to acuminate at apex, entire, cuneate or rounded at base, glabrous; midrib prominent beneath, the lateral 4–5 pairs, rather obscure, curved upwards; petiole up to 10 mm, pale green, grooved, glabrous. *Inflorescence* an erect terminal panicle; flowers 4–6 mm in diameter, bisexual, strongly and unpleasantly scented; rhachis peduncles and pedicels pale green and glabrous. *Calyx* 2–3 mm, pale green, tubular, shallowly 4-lobed, the lobes triangular and pointed at apex, glabrous, caducous. *Corolla* 6–8 mm, creamy-white funnel-shaped, divided one-third of the way to the base into 4 lobes, the lobes triangular and pointed. *Stamens* 2, attached to upper part of corolla; filaments short, white; anthers pale yellow. *Style* 1, pale green; stigma 2-lobed. *Drupe* 6–10 mm, green turning shining black, globose, with oily flesh. *Flowers* 6–7. Pollinated by various insects. $2n=46$.

(1) Forma **ovalifolium**
Leaves dark green.

(2) Forma **aureum** (Carrière) Rehder Golden Privet
L. ovalifolium var. *aureum* Carrière

Leaves yellow or mostly yellow with some green.

(3) Forma **argenteum** (Bean) Rehder Cv. Argenteum
L. ovalifolium var. *argenteum* Bean

Leaves dull green in centre with broad cream margins and or patches.

Introduced. Abundantly planted as hedges and persistent relics in hedges, waste places, parks, estates, amenity areas and waste places. All three forms occur in hedgerows, *L. ovalifolium* being preferable to *L. vulgare* because it is more evergreen. Scattered throughout much of Great Britain and Ireland, but rarely self-sown. Native of Japan.

3. L. lucidum W. T. Aiton Chinese Privet
L. japonicum auct.

Evergreen, monoecious *tree* up to 20 m, with a rounded, domed crown. *Trunk* often short, up to 50 cm in diameter, often with several large main branches radiating from it. *Bark* pale greyish-brown, smooth at first, becoming, fissured and slightly rough. *Branches* spreading or ascending and arching, drooping at the tip; twigs pale brown, much divaricate, with prominent oval lenticels; young shoots green, flattened, straight, glabrous. *Buds* more or less opposite, 10–15 mm, conical, pointed; scales green, ovate, pointed, glabrous, on a short stalk. *Leaves* decussate and dense, 3–15 × 2–6 cm, coriaceous, dark green on upper surface, paler and glaucous beneath, sometimes blotched yellow, elliptical or ovate, with a short, blunt point at apex, entire, cuneate at base, glabrous on upper surface, with white dots and brown glands below; veins of 4–6 pairs, midrib prominent beneath; petiole to 14 mm, green or purple. *Inflorescence* a large spreading or erect, terminal, pyramidal panicle; 12–16 × 12–16 cm, flowers 5.5–6.0 mm in diameter, bisexual, intensely fragrant; rhachis, peduncles and pedicels pale green and glabrous. *Calyx* 2.5–3.0 mm, pale green, shallowly 4-lobed, the lobes triangular, pointed at apex and glabrous, caducous. *Corolla* 7–9 mm, cream, funnel-shaped, divided one-third of the way to the base into 4 lobes, the lobes lanceolate and becoming reflexed. *Stamens* 2; filaments about 3 mm, white; anthers pale yellow. *Style* 1, pale green or white; stigma 2-lobed. *Drupe* 8–12 × 4–6 mm, green turning bluish-black or purplish with a bloom, oblong; seed one, 6–7 × 4–5 mm, brown, ellipsoid. *Flowers* 8–9. Visited by bees.

Introduced. Planted in parks, estates and by roads, and recorded on waste ground in Kent. Native of China.

4. Syringa L.

Deciduous, monoecious *shrubs* or small *trees*, suckering freely. *Leaves* opposite, simple or rarely lobed, entire, exstipulate. *Inflorescence* a large, dense, pyramidal panicle; flowers bisexual, sweet-smelling, appearing after the leaves. *Calyx* 4-lobed. *Corolla* pale to dark lilac, blue or purple and pink or white, 4-lobed, the lobes edge to edge in bud. *Stamens* 2, adnate to and enclosed in the corolla. *Style* 1. *Ovary* superior, 2-celled. *Fruit* a coriaceous, loculicidal, 2-celled capsule, each cell containing 2 winged seeds.

Contains about 30 species ranging from Europe to the Orient.

Bean, W. J. (1980). *Trees and shrubs hardy in the British Isles.* Ed.. 8. **4**: 544–551.
Fiala, J. L. (1988). *Lilacs, the Genus* Syringa. London.
McKelvey, S. D. (1928). *The lilac: a monograph.* New York.

1. Leaves 2.0–6.0 × 0.8–1.5 cm, lanceolate, entire or rarely lobed or pinnatifid, cuneate or rounded at base **2. × persica**
1. Leaves 4–12 × 2.5–8.0 cm, ovate, sometimes broadly so, entire, truncate or slightly cordate at base 2.
2. Flowers pale to dark lilac, blue or purple **1(3). vulgaris** forma **vulgaris**
2. Flowers white or pink 3.
3. Flowers white **1(1). vulgaris** forma **alba**
3. Flowers pink to reddish **1(2). vulgaris** forma **rubra**

1. S. vulgaris L. Lilac

Deciduous, monoecious *shrub* or small *tree* up to 7 m, with rounded crown, suckering freely. *Stems* 1–numerous, straight and erect. *Bark* pale brown, rather rough and fibrous. *Branches* ascending or erect; twigs ascending, erect or divaricate, pale brown, covered with lenticels, rough, glabrous; young shoots yellowish-green, covered with lenticels, with slight ridges, glabrous. *Buds* opposite, 7–10 × 5–7 mm, ovoid or ovoid-conical, pointed at apex; scales green with brown margins, broadly ovate, rounded at apex, covered with minute glands. *Leaves* opposite and decussate; lamina 4–12 × 2.5–8.0 cm, dull dark green with a pale midrib towards the base on the upper surface, much paler beneath, ovate, sometimes broadly so, acute or acuminate at apex, entire, truncate to subcordate at base, glabrous; veins 6–8 pairs, slightly impressed on upper surface, fairly prominent beneath; petioles up to 3 cm, pale green, glabrous. *Inflorescence* 10–20 cm, a dense, pyramidal panicle; branches and pedicels with minute glandular hairs; flowers 15–20 mm in diameter, bisexual, strongly sweet-smelling, often double. *Calyx* 1.5–1.7 mm, yellowish-green, divided one-third of the way to the base into 4 lobes, the lobes triangular and obtuse at apex. *Corolla* pale to dark lilac, blue, purple, pink or white, the tube 8–12 mm, the 4 lobes 7–8 mm, obovate and rounded at apex. *Stamens* 2, adnate to and enclosed in the corolla; filaments pale; anthers yellow. *Style* 1, green, within the corolla tube; stigma 2-lobed. *Capsule* 8–12 mm, ovoid, pointed at apex, smooth; seeds 2 in each cell, winged. *Flowers* 4–5. Pollinated mainly by bees. $2n = 44, 46, 48$.

Much hybridisation and selection has been carried out by gardeners resulting in very variable flower colour and smell and often double flowers. The following infraspecific classification will not be detailed enough for gardeners who have named many cultivars.

(1) Forma **alba** (Aiton) Morariu
S. vulgaris var. *alba* Aiton

Flowers white.

(2) Forma **rubra** (Lodd. ex Loudon) Morariu
S. vulgaris var. *rubra* Lodd. ex Loudon

Flowers pink to reddish.

(3) Forma **vulgaris**
Flowers pale to dark lilac, blue or purple.

Introduced. Much grown in gardens, parks, estates and forestry areas and also found as a garden relic, in hedgerows, on railway and roadside banks and in copses where its suckers sometimes form large, dense patches. Scattered throughout most of Great Britain and Ireland in all the forms described above. Native from north-central Roumania to central Albania and north-east Greece. All forms are common.

2. S. × persica L. Persian Lilac
?S. afghanica × laciniata Mill.
S. persica var. *pinnatifida* Jacques

Deciduous, monoecious *shrub* up to 2 m, of dense, rounded habit. *Branches* erect or arching, glabrous; twigs ascending, erect or divaricate; young shoots slender, glabrous. *Buds* opposite, ovoid, pointed at apex; scales green, broadly ovate, pointed at apex. *Leaves* opposite; lamina 2–6 × 0.8–1.5 cm, dark green on upper surface, paler beneath, lanceolate, long acuminate at apex, entire or some leaves lobed or pinnatifid, cuneate or rounded on upper surface, fairly prominent beneath; petioles 5–12 mm, pale green, glabrous. *Inflorescence* 5–10 cm, a broad, loose panicle; branches and pedicels glabrous; flowers 12–15 mm in diameter, bisexual; sweet-smelling. *Calyx* about 2mm, divided about one-third of the way to base into 4 lobes, the lobes short and pointed. *Corolla* pale lilac or white, the tube 7–9 mm, the lobes 4–5 mm, ovate and subacute at apex. *Stamens* 2, adnate to and enclosed in the corolla; filaments pale; anthers yellow. *Style* 1, green, within the corolla tube; stigma 2-lobed. *Capsule* 8–10 mm, cylindrical, 4-angled, obtuse at apex, smooth; seeds winged. *Flowers* 4–5. Pollinated by bees.

Introduced. Occurs in gardens, parks and estates and may be missed in the great variation in *S. vulgaris*. Cultivated from time immemorial in Persia and India and reached Europe by the early seventeenth century. A hybrid of ancient origin, possibly as above.

5. Fraxinus L.

Deciduous, monoecious or dioecious *trees* or *shrubs*. *Leaves* opposite, usually imparipinnate, rarely simple. *Inflorescence* a terminal or lateral panicle from buds of the previous year usually appearing before the leaves sometimes with them; flowers small, unisexual or bisexual. *Calyx* if present small and campanulate, 4-lobed. *Corolla* often absent, if present with 4 lobes. *Stamens* 2. *Style* 1. *Ovary* 2-celled. *Fruit* a 1-seeded samara with an apical wing.

Contains about 65 species found throughout northern temperate regions, most species in China and North America, a few reaching Central America and south-east Asia.

Bean, W. J. (1973). *Trees and shrubs hardy in the British Isles*. Ed. 8. **2**. London.

de Jong, P. C. (1989). Het geslacht *Fraxinus*. *Dendroflora* **26**: 31–39.

Lingelsheim, A. (1920). *Fraxinus*. In Engler, A. (Edit.) *Das Pflanzenreich* **72**: 9–65.

Lombarts, P. (1989). *Fraxinus* (Keuringsrapport). *Dendroflora* **26**: 7–30.

Miller, G. N. (1955). The genus *Fraxinus*, the Ashes, in North America, north of Mexico.*Cornell Univ. Agricult. Experiment Stn Memoirs* **335**: 1–64.

Oliver, J. (1998). Variability in the Common Ash (*Fraxinus excelsior*). *B.S.B.I. News* **78**: 28–29.

Rushforth, K. (1999). *Trees of Britain and Europe*. London.

Sargent, C. S. (1894). *The silva of North America*. **6**. Boston.

Scannell, M. J. P. (2007). *Is Fraxinus angustifolia* naturalised in Britain? *B.S.B.I. News* **106**: 38–40.

Scheller, H. (1977). Kritische Studien über die kultivierten *Fraxinus*-arten. *Mitt. Deutsch. Dendrol. Gesell.* **69**: 49–162.

1. Leaves simple or with occasionally 1–2 leaflets **5(7). excelsior** forma **diversifolia**
1. Leaves pinnate, with 5–15 leaflets 2.
2. Buds dull black 3.
2. Buds brown or greyish-brown 8.
3. Branches pendulous or hanging **5(6). excelsior** forma **pendula**
3. Branches ascending or spreading 4.
4. Leaves and twigs yellow for much of the year **5(3). excelsior** forma **aurea**
4. Leaves often falling green or turning slightly yellow before falling, and twigs grey 5.
5. Leaflets 11–15 **5(4). excelsior** forma **multifoliolata**
5. Leaflets 7–11 6.
6. Leaflets 4–10 × 2.0–4.5 cm, broadly elliptical or ovate **5(5). excelsior** forma **latifolia**
6. Leaflets 3–10 × 0.8–3.0 cm, linear-lanceolate, lanceolate or narrowly ovate 7.
7. Leaflets 0.8–2.0 cm, wide, linear-lanceolate **5(1). excelsior** forma **angustifolia**
7. Leaflets 2.0–3.0 cm wide, lanceolate or narrowly ovate **5(2). excelsior** forma **excelsior**
8. Inflorescence terminal; corolla present **1. ornus**
8. Inflorescence lateral; corolla absent 9.
9. Calyx absent 10.
9. Calyx present, irregularly toothed 12.
10. Leaflets completely glabrous **6(a). angustifolia** subsp. **angustifolia**
10. Leaflets hairy along the proximal part of the midrib on the lower surface 1.
11. Leaves remaining green **6(b,1). angustifolia** subsp. **angustifolia** forma **oxycarpa**
11. Leaves turning deep purple in autumn **6(b,1). angustifolia** subsp. **angustifolia** forma **purpurescens**
12. Lateral leaflets with petiolules absent or up to 2 mm **4. latifolia**

12. Lateral leaflets with petiolules up to 12 mm 13.
13. Terminal buds obtuse, lateral buds triangular; wing of samara extending one-third of the way along the seed body; seeds obovate, rounded at both ends, one-third the length of the samara **2. americana**
13. Terminal buds acute, lateral buds reniform; wing of samara extending halfway or more along the seed; seeds linear, narrowed at both ends, one-third or more the length of the samara **3. pennsylvanica**

Subgenus **1. Ornus** (Pers.) Vassileva

Ornus Pers.

Inflorescences terminal or from buds of the current year's growth.

1. F. ornus L. Manna Ash

Ornus europaea Pers.; *F. paniculata* Mill.; *F. florifera* Scop.; *F. montana* Salisb.; *Ornus ornus* (L.) Karst. nom. illegit.

Deciduous, monoecious *tree* with a hemispherical or rather flattened crown, sometimes with a more domed crown. *Trunk* up to 1 m in circumference, fairly straight and extending well into the crown. *Bark* dark grey or greyish-brown and very smooth, usually fissured at base in older trees. *Branches* spreading or arching and rather sinuous; twigs yellowish-grey to greyish-brown, finely speckled whitish, glabrous or with few hairs, flattened at the nodes and in the 2nd year with raised, round, orange lenticels. *Buds* opposite, 9–17 × 3–10 mm, ovoid to mitre-shaped, obtuse to acute at apex; 2 outer scales very dark greyish-brown, ovate, acute, shortly hairy, 2 inner scales pale brown, ovate or lanceolate, obtuse at apex, with dense, brown or grey short eglandular hairs. *Leaves* opposite, 20–30 × 20–30 cm, medium green above, paler beneath, turning yellow to brownish-red in autumn, broadly ovate in outline, imparipinnate with 5–9 leaflets on a slender rhachis which is widely grooved above and shortly brown-hairy at the joints; leaflets on juvenile trees 1.5–3.0 × 1.0–2.0 cm, lanceolate to ovate, cuspidate at apex, crenate-serrate, subsessile, cuneate at base, midrib prominent beneath and glabrous; leaflets on a mature tree 3–8 × 1.8–4.5 cm, lanceolate to oblong-ovate, cuspidate at apex, irregularly serrulate, cuneate at base, midrib prominent beneath, distinctly petiolulate and glabrous; petiole up to 40 mm, pale green, grooved, glabrous or with some brown hairs; leaf scars semilunar or hemispherical. *Inflorescence* a dense, conical terminal panicle up to 15 × 20 cm; pedicels 3–5 mm, pale green. *Calyx* campanulate, deeply 4-lobed, the lobes lanceolate and acute at apex, glabrous. *Petals* 4, 8–10 × 1.0–1.2 mm, white to cream, linear. *Stamens* 2; filaments pale green or whitish; anthers pale green turning brownish-yellow. *Style* 1, about 1.5 mm; stigma yellow, bifid. *Samara* 15–25 × 4–6 mm, pale yellowish-green turning reddish-brown just before leaf-fall, with a seed in the lower half and on a slender pedicel. *Flowers* 4–5. $2n = 46$.

Introduced. Planted in gardens, parks, estates and along roads; occasionally solitary trees found in hedgerows or copses. Native of the Mediterranean region and south-central Europe. A sugary gum or 'manna' exudes from the branches and is collected and used medicinally, for which purpose the tree is cultivated in southern Italy and Sicily. It is sometimes grafted onto *F. excelsior* and if it dies the stock will sometimes grow.

Subgenus **2. Fraxinus**

Inflorescences lateral from bud's of the previous year's growth.

2. F. americana L. White Ash

Calycomelia americana (L.) Kostel.; *F. alba* Marshall; *F. acuminata* Lam.; *F. caroliniana* Wagenhein; *F. epiptera* Michx; *F. juglandifolia* Lam.; *F. canadensis* Gaertn.

Deciduous, usually dioecious *tree* 15–35 mm, with ovoid, domed, open crown. *Trunk* up to 1 m in diameter, extending well into the crown. *Bark* brown, turning grey with age, finely fissured at first, developing deep furrows and forking ridges. *Wood* with a pale brown sapwood and a darker heartwood, heavy, hard, strong and coarse-grained. *Branches* terete, ascending or spreading; twigs grey or brown, glabrous; young shoots shiny olive brown greenish or purplish, straight and slender, smooth with a few, circular lenticels, flattened between the buds, usually glabrous. *Buds* opposite, 5–6 × 3–4 mm, terminal broadly ovoid, obtuse at apex, lateral triangular; scales brown to nearly black, ovate, shortly hairy. *Leaves* 20–36 × 7–20 cm, broadly obovate in outline, imparipinnate; leaflets 5–9, 4–21 × 1.0–9.0 cm, dull to shiny medium green on upper surface, paler whitish-green beneath, turning yellow or purple in autumn, elliptical, ovate, obovate or lanceolate-oblong, acute or acuminate at apex, with small, rather distant, rounded teeth, cuneate or rounded at the often unequal base, with distinct petiolules up to 15 mm, the terminal one even longer, the lower surface often with whitish hairs, the veins 12–15 pairs, slightly impressed on the upper surface and prominent beneath; rhachis pale yellowish-green, terete, faintly channelled, glabrous; petiole up to 15 mm, pale yellowish-green, faintly channelled, glabrous. *Inflorescence* a small lateral cluster of flowers appearing before or with the leaves. *Calyx* about 1.5 mm, campanulate, 4-lobed or laciniate. *Corolla* absent. *Stamens* 2; filaments short and stout; anthers blackish at first becoming reddish-purple. *Style* 1, greenish, long and slender; stigma purple, 2-lobed. *Samara* with an obovate seed rounded at both ends; wing 25–50 × 8–12 mm, green, ripening brown, narrow linear-elliptical, oblong or subspathulate, rounded at apex, extending one-third of the way along the seed. *Flowers* 4–5.

Introduced. Parkland, roadsides, amenity areas and estates. It probably occurs more widely than it has been recorded. Native of eastern North America from Ontario and Newfoundland south to Florida, west to Texas and north to Minnesota. Our plant is subsp. **americana** which is variable in leaf shape and indumentum.

3. F. pennsylvanica Marshall Red Ash

F. pubescens Lam.; *F. americana* subsp. *pennsylvanica* (Marshall) Wesm.; *Calycomelia pennsylvanica* (Marshall) Nieuwl.; *F. subpubescens* Pers.

Deciduous, dioecious *tree* 15–30 m, with an ovoid crown when young, later becoming tall and domed. *Trunk* up to 80 cm in diameter, extending well into the crown. *Bark* greyish-brown with shallow, reddish fissures, in old trees becoming scaly. *Wood* pale brown, hard, heavy and coarse-grained. *Branches* terete, erect or spreading; twigs grey or pale brown with a bluish bloom, with pale lenticels; young shoots olive brown, slender and terete, glabrous to softly hairy. *Buds* in pairs, opposite, 3–8 × 3–4 mm, terminal ovoid and acute at apex, lateral reniform; scales golden brown, ovate, acute, covered with rufus hairs. *Leaves* opposite; lamina 20–35 × 10–18 cm, obovate in outline, imparipinnate; leaflets 5–9, 1.5–16 × 0.6–7.6 cm, dull medium green on upper surface, pale greyish-green beneath, turning yellow in autumn, lanceolate to ovate-lanceolate or elliptical, narrowed to a slender, pointed apex, with fine, rounded, minute teeth or entire, rounded at base, the petiolules up to 12 mm, glabrous or slightly hairy on upper surface and hairy on the veins beneath, the veins 9–12 pairs and prominent beneath; rhachis pale green, channelled and hairy; petiole up to 8 cm, pale green, channelled, hairy. *Inflorescence* a compact, lateral panicle, opening before the leaves or with them. *Calyx* about 1 mm, campanulate, 4-lobed or laciniate. *Corolla* absent. *Stamens* 2; filaments green, short, slender; anthers pale green, tinged purple. *Style* 1, green; stigma green, 2-lobed. *Samara* with 1 linear seed, narrowed at both ends, one-third or more the length of the samara; wing 30–60 × 4–10 mm, green, turning brown, narrowly oblong-elliptical rounded at apex, extending halfway or more along the seed. *Flowers* 4–5. Wind-pollinated.

Introduced. Parkland, roadsides and amenity areas. Probably occurs more widespread than has been recorded. Native of North America from Nova Scotia to Alberta and south to Texas and Florida. Our plant is subsp. **pennsylvanica.**

4. F. latifolia Benth. Oregon Ash

F. oregona Nutt.; *F. pennsylvanica* subsp. *oregona* (Nutt.) G. N. Mill.

Deciduous, usually dioecious *tree* 15–25 m with an ovoid crown when young becoming narrow with dense foliage when old. *Trunk* up to 80 cm in diameter, extending well into the crown. *Bark* brownish-green and finely fissured in young trees, in old trees grey or brown with deep forking fissures and scaly ridges. *Wood* brown with pale sapwood, light, brittle and coarse-grained. *Branches* erect or spreading; twigs grey, thick, slightly flattened; young shoots olive green to chocolate brown, with whitish stellate hairs, shiny and stout. *Buds* 3–6 mm, in pairs, opposite, broadly ovoid-conical, pointed at apex; scales of 2 or 3 pairs, brown, pale hairy. *Leaves* opposite; lamina 12–35 × 6–18 cm, broadly ovate in outline, pinnate; leaflets 5–7(–9), 5–13 × 2–7 cm, medium rather shiny yellowish-green on upper surface, paler green beneath, turning yellow in autumn, elliptical, ovate or oblong, shortly narrowed to a broad, pointed tip, mainly entire but with some, small, rounded teeth especially on the terminal leaflet, rounded to cuneate at base and sessile or with petiolule to 1 mm, glabrous or shortly hairy on upper surface, with dense stellate and simple hairs beneath; veins slightly impressed above and prominent beneath; rhachis channelled and densely hairy; petiole up to 8 cm, widely channelled and widened at base, densely hairy. *Inflorescence* of small, lateral clusters of flowers on a long rhachis, male and female usually on different trees. *Calyx* campanulate and irregularly toothed. *Corolla* absent. *Stamens* 2; filaments short, pale, yellowish-green; anthers pale yellowish-green. *Style* 1, pale green; stigma greenish-yellow, 2-lobed. *Samara* with nutlet green, ripening brown, slightly flattened, the wing 30–50 × 7–10 mm, with rim almost surrounding the entire seed and rounded or notched at apex. *Flowers* 4–5.

Introduced. Planted in large gardens, parks, roadsides and amenity areas. Native of the western United States from Washington south to central California. This may be best as a subspecies of *F. pennsylvanica*. The two species formerly overlapped in distribution and hybrids occurred. *F. latifolia* probably no longer occurs in southern California but has left behind intermediates with *F. pennsylvanica*.

5. F. excelsior L. Common Ash

F. excelsa Salisb. nom. illegit.

Deciduous, monoecious or functionally dioecious *tree* 15–40 m, with a tall, rounded, open crown. *Trunk* up to 1 m in diameter, sometimes stretching through much of the crown, but sometimes short and often pollarded, if coppiced may form several stems from 1 base, often subcylindrical. *Bark* grey or pale greyish-brown, smooth when young, becoming fissured and forming a network of ribs and hollows in old trunks. *Wood* uniform pale brown, sometimes with a darker heart, hard, dense and strong. *Branches* ascending or spreading, sometimes pendulous or hanging; twigs grey, smooth, thick and flattened at the nodes; young shoots greenish-grey or greenish-brown, rarely yellow, stout, smooth, glabrous. *Buds* 5–7 × 4–8 mm, ovoid, blunt; scales dull black, broadly ovate, obtuse to acute, glabrous or minutely hairy. *Leaves* opposite and decussate, appearing about the end of April and one of the first trees to shed them in October; lamina, 20–35 cm, ovate or oblong-ovate in outline, imparipinnate or rarely simple; leaflets 7–15, 3–11 × 1.5–4.5 cm, dull, medium green on upper surface, paler beneath, rarely yellow, sometimes turning yellow in autumn, but often falling green, narrowly linear-lanceolate to broadly ovate or elliptical, rounded-apiculate to acute or acuminate, serrate with small teeth, rarely biserrate, cuneate or rounded at base, the lateral sessile or nearly so, the terminal petiolulate, glabrous except for the lower part of the midrib beneath; rhachis channelled, glabrous or with reddish hairs at base of the leaflets; petiole up to 7 cm, pale green, channelled, glabrous. *Inflorescence* a lateral axillary panicle, appearing before the leaves or with them, it may be male, female or bisexual and the same tree may contain all three and may be different in different years. *Perianth* absent. *Stamens* 2; filaments pale green, short and thick;

anthers dark purple. *Style* 1, pale green; stigma large and fleshy. *Samara* with 1 seed; wing (20–)25–50×5–9 mm, pale green at first, later turning brown, more or less elliptical and rounded at apex; they hang on the tree much of the winter. *Flowers* 4–5. Wind-pollinated. Many bisexual flowers are infertile. *Fruits* 10–11. $2n=46$.

(1) Forma **angustifolia** (Schelle) P. D. Sell
F. excelsior var. *angustifolia* Schelle

Branches erect or spreading; twigs grey. *Leaves* green; leaflets 7–11, 3–6×0.8–2.0 cm, linear-lanceolate, acute or acuminate at apex.

(2) Forma **excelsior**

Branches erect or spreading; twigs grey. *Leaves* green; leaflets 7–11, 5–10×2–3 cm, lanceolate or narrowly ovate, acute or acuminate at apex.

(3) Forma **aurea** (Pers.) Schelle Cv. Aurea
F. excelsior var. *aurea* Pers.; *F. aurea* (Pers.) Willd.

Branches erect or spreading; twigs yellow. *Leaves* yellow; leaflets 7–11, 5–10×2–3 cm, lanceolate or narrowly ovate, acute or acuminate at apex.

(4) Forma **multifoliolata** P. D. Sell
Cv. Westhof's Glorie

Branches erect or spreading; twigs grey. *Leaves* green; leaflets 11–15, 6–10×3–5 cm, narrowly ovate or elliptical, rounded to acute at apex.

(5) Forma **ovata** (Boenn.) Soó
F. excelsior var. *ovata* Boenn.; *F. excelsior* var. *rotundifolia* Elwes

Branches erect or spreading; twigs grey. *Leaves* green; leaflets 9–13, 4–10×2.0–4.5 cm, broadly elliptical or ovate, acute or slightly acuminate at apex.

(6) Forma **pendula** (Aiton) Schelle Weeping Ash
F. excelsior var. *pendula* Aiton; *F. pendula* (Aiton) Hoffmanns.

Branches pendulous and often forming an umbrella-like arbour; twigs grey. *Leaves* green; leaflets 9–11, 2–10×0.6–2.5 cm, narrowly acute at apex.

(7) Forma **diversifolia** (Aiton) Lingelsh.
F. excelsior var. *diversifolia* Aiton; F. *monophylla* Desf.; *F. excelsior* var. *monophylla* (Desf.) Kuntze; *F. heterophylla* Vahl; *F. simplicifolia* Willd.; *F. excelsior* var. *integrifolia* Pott; *F. excelsior* var. *simplicifolia* (Willd.) Pers.

Branches erect or spreading; twigs grey. *Leaves* usually simple with occasional leaves with 1–2 small leaflets, 6–16 × 2–8 cm, elliptical, obtuse, acute or acuminate at apex.

Native. Forming woods on calcareous soils in the wetter areas of Great Britain in the west and north and in Ireland, and in mixed woodland with *Acer campestre, Corylus avellana* and *Quercus*; less frequent on acid soils; common in scrub, copse and hedgerow and widely planted, sometimes as one of the forms described above. Throughout Great Britain and Ireland except the Orkney and Shetland Islands. Most of Europe; North Africa; western Asia. European Temperate element. Our plant is subsp. **excelsior**, which occurs throughout the range of the species except south-east Europe and parts of western Asia. Up to the Second World War it was much coppiced for straight, stout poles and some stools are probably 1,000 years old. It is now regarded as a second-class timber tree, but is an attractive wood for furniture and is used for tool handles, billiard cues, tennis rackets, hockey sticks and oars, but has no durability for outside use. The name Ash comes from the Anglo-Saxon *aesc* and is a common element is place-names. In Scandinavian mythology it was called *Yggdrasil*, the tree of life, and until fairly recently its burning was considered to drive out evil spirits, possibly a relic from the time of the Danish invasions. Ash-bud moth, *Prays curtisellus* (Donovan), and bacterial canker, *Nectria galligena*, renders great areas of ash woodland unattractive scrub, the trees being much branched, bushy and stunted.

6. F. angustifolia Wahl Narrow-leaved Ash

Deciduous, dioecious *tree* 15–30 m, with a dense, irregular dome. *Trunk* up to 110 cm in diameter, extending well into the tree. *Bark* pale grey, dark grey or blackish-grey, becoming reticulate fissured, sometimes deeply fissured and knobbly when old. *Wood* uniform pale brown, hard, dense and strong. *Branches* spreading or ascending; twigs dark grey, glabrous; young shoots olive green, olive brown or greyish-green, turning brown, glabrous. *Buds* opposite or in threes, 4–6×4–6 mm, ovoid, rounded to pointed at apex; scales dark brown, ovate, acute at apex, slightly hairy. *Leaves* opposite; lamina 20–30×8–10 cm, more or less broadly oblong, obovate-oblong or elliptical-oblong in outline, imparipinnate; leaflets of mature trees 5–13, deep dull to slightly shiny green or bluish-green on upper surface, paler and sometimes whitish beneath, usually staying green but sometimes turning purple in autumn, narrowly to broadly lanceolate to lanceolate-oblong, tapering to a slender, acute apex, with forward-pointing, sometimes hooked teeth, cuneate to rounded at base, the lateral sessile or nearly so, the terminal with petiolule up to 15 mm, glabrous or hairy on midrib beneath, the veins 8–11 pairs, fairly prominent beneath; leaflets of juvenile tree 0.8–3.0×0.5–1.7 cm, ovate to obovate, obtuse to acute at apex; rhachis pale yellowish-green, channelled, glabrous; petiole up to 10 cm, pale green, channelled, glabrous. *Inflorescence* a lateral panicle appearing before the leaves. *Calyx* and corolla absent. *Stamens* 2; filaments pale green; anthers green. *Style* 1, green, short and thick; stigma large and fleshy. *Samara* with 1 seed; wing 20–50×6–10 mm, green, ripening brown, lanceolate or oblong, rounded at apex, glabrous. *Flowers* 4–5. Wind-pollinated.

(a) Subsp. **angustifolia**
F. excelsior subsp. *angustifolia* (Vahl) Westmael; *F. oxycarpa* var. *angustifolia* (Vahl) Lingelsh. nom. illegit.; *F. tamariscifolia* Vahl

Leaflets completely glabrous beneath.

(b) Subsp. **oxycarpa** (Bieb. ex Willd.) Franco & Rocha Alfonso
F. oxycarpa Bieb. ex Willd.; *F. pojarkoviana* V. N. Vassil.; *F. syriaca* auct.

Leaflets hairy along the proximal part of the midrib beneath.

(1) Forma **oxycarpa** (Bieb. ex Willd.) P. D. Sell
Leaves green or turning slightly yellow.

(2) Forma **purpurescens** P. D. Sell Cv. Raywood
Leaves turning dark purple in autumn.

Introduced. All three variants are planted along roads and in estates, squares and amenity areas. Native of the Mediterranean region and east-central Europe and south-west Asia. Subsp. *angustifolia* occurs in the western part of its range and subsp. *oxycarpa* from north-east Spain eastwards. Forma *purpurescens* originated in cultivation in Australia. Scannell (2007) raises the question as to whether this species could have been introduced accidentally with weapon handles at an early date.

6. Forsythia Vahl

Deciduous, monoecious *shrubs*. *Leaves* opposite, simple or with 2 small leaflets at the base, entire or serrate. *Inflorescence* axillary; flowers bisexual. *Calyx* 4-lobed. *Corolla* 4-lobed, the lobes overlapping in bud, usually appearing before the leaves. *Stamens* 2. *Style* 1, variable in length, crosses between long- and short-styled plants giving better seed production. *Ovary* with several ovules. *Fruit* a capsule.

Contains about 6 species, 1 native in Europe, the remainder in eastern Asia. Widely cultivated. Named after William Forsyth (1737–1804).

Bean, W. J. (1973). *Trees and shrubs hardy in the British Isles*. Ed. 8. **2**. London.
DeWolf, G. P. & Hebb, R. S. (1971). The story of *Forsythia*. *Arnoldia* **31**: 41–68.
Lingelsheim, A. (1920). *Forsythia*. In Engler, A. *Pflanzenreich* **IV. 243**: 109–113.

Forsythia species make great splashes of yellow in spring throughout the gardens, parks, amenity areas, various waste places, roadsides and railway banks. Most people including botanists just call them *Forsythia*. They are difficult to key out and you really need both flowers and mature leaves, which appear at different times of the year, to make a certain identification.

1. Leaves usually entire, with veins on lower surface and petioles hairy; corolla 15–20 mm **1. giraldiana**
1. Leaves usually serrate, glabrous; flowers 15–35 mm, at least some flowers on a shrub over 20 mm 2.
2. Branches with lamellate pith throughout both nodes and internodes; leaves not usually 3-foliolate; calyx almost half as long as the corolla tube; corolla rather pale yellow with a greenish tinge **2. viridissima**
2. Branches with solid masses of pith at the nodes, the internodes hollow or with lamellate pith; leaves often 3-foliolate; calyx shorter than to about as long as the corolla tube 3.
3. Branches usually with lamellate pith in the internodes; calyx shorter than the corolla tube; corolla 15–35 mm **3. × intermedia**
3. Branches hollow in the internodes without pith of any kind; calyx about as long as the corolla tube; corolla 18–25 4.
4. Stems with long, slender, pendulous branches **4(i). suspensa** var. **suspensa**
4. Stems with erect or spreading, rigid, robust branches **4(ii). suspensa** var. **fortunei**

1. F. giraldiana Lingelsh. Giraldi's Forsythia

F. giraldii Pamp.

Fairly open, monoecious, deciduous *shrub* up to 4 m. *Stems* and *branches* erect and arching, with thin, papery lamellae of pith in the internodes and at the nodes; twigs pale grey and glabrous; young shoots reddish-brown or green and glabrous. *Buds* 5–7 mm, narrowly ovoid, acute at apex; scales medium brown, lanceolate, acute at apex, glabrous. *Leaves* dense, opposite; lamina 2.0–4.5 × 1.4–2.0 cm, dull medium green on upper surface, paler beneath, ovate to ovate-lanceolate, acute at apex, entire or with a few shallow teeth, cuneate or rounded at base, the veins beneath sometimes shortly hairy; veins 4–6 pairs, impressed above, prominent beneath; petiole up to 8 mm, shortly hairy, green. *Flowers* usually solitary, bisexual, paired at each node, scentless. *Calyx* 5–6 mm, divided two-thirds of the way to the base into 4 lobes, the lobes medium yellowish-green in the centre, getting paler towards the margin and with a narrow red margin, ovate, aciculate and keeled on the back, glabrous. *Corolla* 15–20 mm, rather pale yellow and slightly orange in the throat, divided nearly to the base into 5 lobes, the lobes lanceolate, obtuse or acute at apex, becoming twisted. *Stamens* 2; filaments 1.0–1.5 mm, very pale green; anthers clear yellow. *Style* 1, much exceeding the anthers, pale green; stigma slightly green. *Capsule* 15–18 × 6–8 mm, elliptical, with a curved beak at apex, glabrous. *Flowers* 3–4.

Introduced. Discovered in China in 1897 by Giuseppe Giraldi (1848–1901) after whom it is named. A persistent garden escape on Dartford Heath and Biggin Hill in Kent and formerly at Hyndland in Lanarkshire. It may well be among the *Forsythias* planted on roadsides, in parks and around amenity gardens. Native of China.

2. F. viridissima Lindl. Greenish-flowered Forsythia

Dense, monoecious, deciduous *shrub* up to 3 m. *Stems* and *branches* mainly arching, with thin, papery lamellae of pith in the internodes and at the nodes; twigs pale greyish-brown, glabrous; young shoots reddish-brown, glabrous. *Buds* 4–5 mm, ovate, acute at apex; scales dark brown, ovate, acute at apex, glabrous. *Leaves* opposite; lamina 3–6 × 1.0–2.5 cm, dark green on upper surface, paler beneath, lanceolate to elliptic-oblong or oblong, acute at apex, minutely serrate above the middle; cuneate at base, glabrous; veins 3–5 pairs, impressed above, prominent beneath; petiole up to 8 mm, green, glabrous. *Flowers* 1–3 in a cluster, bisexual, 2–6 at a node. *Calyx* 3–4 mm, divided two-thirds of the way to the base into 4 lobes, the lobes ovate and rounded-obtuse at apex, 2 of them green and 2 dark purple, glabrous. *Corolla* 15–25 mm, rather pale yellow with a tinge of green and slightly orange in the throat, divided nearly to the base into 4 lobes, the lobes oblong or oblong-lanceolate, rounded at apex and curved back. *Stamens* 2; filaments 1.0–1.5 mm, yellow; anthers yellow. *Style* 1, exceeding the anthers, yellow; stigma green. *Capsule* 13–15×8–10 mm, broadly elliptical, acuminate at apex, glabrous. *Flowers* 3–5.

Introduced. Discovered by Robert Fortune in China in 1844 and 1845 and sent to the Royal Horticultural Society. For 20 years or more it was the only *Forsythia* cultivated in Great Britain. Its present distribution is not known and it is included here because it is one of the parents of the widely planted hybrid *F.* × *intermedia*.

3. F. × intermedia Zabel Garden Forsythia
F. suspensa × viridissima

Dense to rather open, monoecious, deciduous *shrub* up to 3 m. *Stems* and *branches* erect and spreading or arching and hanging, with usually thin, papery lamellae of pith in the internodes and solid masses of pith at the nodes; twigs pale brown and glabrous; young shoots reddish-brown and glabrous. *Buds* 5–7×2–3 mm, cylindrical, acute at apex; scales brown, sometimes with a green base, ovate, acute at apex, glabrous. *Leaves* opposite; lamina 2.0–3.5×1.2–2.0 cm, medium green on upper surface, slightly paler beneath, ovate to obovate or elliptical, shortly acute at apex, entire or serrate in the upper half, often 3-partite below on the shoots, cuneate or rounded at base, glabrous; veins 4–5 pairs, prominent beneath; petiole up to 14 mm, green, glabrous. *Flowers* 2–5 in each cluster, bisexual, 4–10 at each node. *Calyx* 6–7 mm, shorter than corolla tube, divided two-thirds of the way to the base into 4 lobes, the lobes medium green, lanceolate-oblong, long-acute at apex, glabrous. *Corolla* 15–35 mm, pale to deep golden yellow, sometimes with orange in throat, divided two-thirds of the way to the base into 5 lobes, the lobes ovate-oblong, rounded at apex. *Stamens* 2; filaments 2–3 mm, pale green; anthers yellow. *Style* 1, pale green; stigma pale green. *Capsule* 13–15×7–8 mm, narrowly elliptical, acuminate at apex, glabrous. *Flowers* 3–5.

Introduced. In the summer of 1878 Hermann Zabel found a seedling of *Forsythia* in the Botanic Garden of Göttingen which was apparently the result of a cross between *F. suspensa* var. *fortunei* and *F. viridissima*. This hybrid has been the source of many garden variants. Beginning in 1899 a number of selections were made at the Späth Nurseries in Berlin. Cv. Vitellina was more upright and had larger and more profuse deep yellow flowers. Cv. Densiflora had spreading pendulous branches and crowded pale yellow flowers. Cv. Spectabilis was especially noted for its larger vivid yellow flowers produced in a cluster, and is one of the most popular of all *Forsythias*. The production of variants then switched to America. Cv. Primulina, produced at the Arnold Arboretum, had the same habit as Cv. Spectabilis but the flowers were pale yellow. Cv. Spring Glory is similar to Cv. Primulina but with larger, more densely arranged flowers. In the 1940s, Karl Sax, Director of the Arnold Arboretum and certain of his students became interested in breeding *Forsythias*. Cv. Arnold Giant was produced by treating a seedling of *F.* × *intermedia* with colchicine and doubling its chromosomes. It was crossed back to *F.* × *intermedia* Cv. Spectabilis which produced variable seedlings among which was the large, deep yellow-flowered Cv. Karl Sax. *F.* × *intermedia* in its many forms occurs as a relic or throw-out on rough ground or on road or river banks. It is the most commonly grown garden taxon and is also found in parks, shrubberies and amenity areas.

4. F. suspensa (Thunb. ex Murray) Vahl Golden Bell
Syringa suspensa Thunb. ex Murray

Dense, deciduous *shrub* up to 3 m with a rambling habit. *Stems* and *branches* sometimes stout and erect, sometimes up to 4 m and slender and arching, hollow in the internodes without pith of any kind, and with solid masses of pith only at the nodes; twigs divaricate at every angle and intertwined, 4-angled, pale greyish-brown and glabrous; young shoots reddish-brown, slender and glabrous. *Buds* 4–5×1–2 mm, cylindrical; scales green, tipped brown, ovate, acute at apex, glabrous. *Leaves* 2–8×1.5–5.0 cm, dull medium to dark green on upper surface, paler beneath, elliptical or obovate, rounded to acute at apex, often 3-foliolate or 3-parted, serrated in upper two-thirds, cuneate at base, glabrous; veins 4–5 pairs, faint above, prominent beneath, petiole up to 8 mm, green, glabrous. *Flowers* 1–3(–6) in each cluster, bisexual, 2–6(–12) at each node, scentless. *Calyx* 6–10 mm, as long as corolla tube, divided nearly to base into 4 lobes, the lobes pale yellowish-green to medium green with red tips, ovate-lanceolate to oblong-lanceolate, rounded-obtuse to acute at apex, glabrous. *Corolla* 18–25 mm, pale, sometimes slightly greenish-yellow, with pale orange in the throat, divided nearly to the base into 4–5 lobes, the lobes triangular-lanceolate to triangular-ovate or oblong-lanceolate, rounded-obtuse or slightly emarginate at apex, spreading or slightly twisted. *Stamens* 2; filaments 3–4 mm, white to yellow, longer than style; anthers yellow. *Style* 1, white to greenish; stigma green. *Capsule* 13–20×7–8 mm, narrowly elliptical or ovoid, acuminate at apex, glabrous, but covered with brown lenticels. *Flowers* 3–5.

(i) Var. **suspensa**
F. suspensa var. *sieboldii* Zabel; *F. sieboldii* (Zabel) Dippel

Stems with long, slender, pendulous branches.

(ii) Var. **fortunei** (Lindl.) Rehder
F. fortunei Lindl.

Stems with erect, spreading, rigid, robust branches.

Introduced. This plant first became known to Europe when Carl Pehr Thunberg visited Japan in 1775–1776, and was originally thought to be a *Syringa*. Var. *suspensa* was in English nurseries in 1857 and var. *fortunei* in 1864. Both varieties are now widely planted. It has been recorded as a consistent garden escape on Stone Marshes and Orpington in Kent and Fornham in Suffolk. Both varieties may be under-recorded, both on roadside and railway banks and in parks and amenity gardens. Native of China and perhaps Japan. It is one of the parents of the common hybrid *F.* × *intermedia*.

7. Jasminum L.

Deciduous, monoecious, scrambling *shrubs*, rooting along the stems. *Leaves* simple, ternate or pinnate, opposite.

Flowers solitary or in terminal panicles, bisexual. *Calyx* 5-lobed. *Corolla* white, red or yellow, with a narrow tube and 4–6 lobed. *Stamens* 2, included in the corolla tube. *Style* 1, 2-lobed. *Ovary* 2-celled, each cell with 2 ovules. *Fruit* a 1-seeded, fleshy berry.

Contains about 300 species from the Old World, some are warm temperate species, but most tropical.

Bean, W. J. (1973). *Trees and shrubs hardy in the British Isles*. Ed. 8. **2**. London.
Green, P. S. (1965). Studies in the genus *Jasminum* III: The species in cultivation in North America. *Baileya* **13**: 137–172.
Green, P. S. (1988). *Jasminum. The Plantsman* **10**: 148–159.
Rushforth, K. (1999). *Trees of Britain and Europe*. London.

1. Leaves pinnate; corolla white, often tinged pinkish-purple	**1. officinale**
1. Leaves simple or ternate; corolla yellow	2.
2. Leaves simple; corolla deep rose, produced with the leaves	**2. beesianum**
2. Leaves ternate; corolla yellow, produced before the leaves	**3. nudiflorum**

1. J. officinale L. Summer Jasmine

Deciduous, monoecious *shrub* up to 3(–13) m. *Stems* climbing or scrambling anticlockwise, dark green, turning brown with age, channelled on 2 sides, glabrous, branched. *Leaves* opposite; lamina 4.5–11.0 × 3.0–8.5 cm, ovate in outline, pinnate; leaflets 5–9, terminal larger than lateral, 2–7 × 0.9–2.0 cm, dull medium green on upper surface, paler beneath, lanceolate or ovate-lanceolate, acute or acuminate at apex, entire, rounded and asymmetrical at base, shortly hairy on upper surface and margin, more sparsely so beneath, the midrib prominent and hairy beneath, the upper sessile, the lower shortly petiolulate; petioles up to 3 cm, channelled, with minute appressed hairs, the rhachis also channelled and minutely hairy. *Inflorescence* a terminal almost corymbose cyme of 1–5 bisexual flowers; pedicels up to 15 mm, with minute appressed hairs; flowers fragrant. *Calyx* 9–10 mm; tube about 4 mm; limb divided into 5–6 lobes, the lobes filamentous and acute; minutely hairy. *Corolla* pale green, becoming flushed pink; tube 1.5–1.8 mm, narrowly cylindrical and slightly wider above; lobes 5–6, 9–15 × 5–8 mm, ovate and mucronate. *Stamens* 2; filaments white; anthers green. *Style* 1, white; stigma green, 2-lobed. *Berry* 7–10 × 5–9 mm, globose or ellipsoid, black, 2-lobed. *Flowers* 6–11. $2n = 26$.

Introduced. Much grown in gardens, parks and estates and naturalised on walls and in marginal ground. Southern England and Ireland. Native from the Caucasus to China.

2. J. beesianum Forrest & Diels Red Jasmine

Deciduous, monoecious *shrub* up to 3 m. *Stems* climbing or scrambling, wiry, dull medium green, channelled on 2 sides and rounded on 2 sides, twisted, branched, glabrous, becoming woody and pale brown when old. *Leaves* opposite; lamina 5–32 × 2–15 mm, dull medium to dark green on upper surface, paler beneath, ovate or lanceolate, cuspidate at apex, entire, rounded or subcordate at base, with minute hairs on the upper surface and beneath on the prominent midrib; petiole very short, hairy. *Inflorescence* of 1–3 flowers in terminal leaf axils; pedicels 5–15 mm, glabrous; flowers bisexual, scented. *Calyx* 6–7 mm, pale green, tinted rose, glabrous, divided just over halfway to base into 6 lobes, the lobes subulate and acute. *Corolla* 12–15 mm, deep rose with a ring of deeper colour in the throat; tube 9–10 mm, narrow, with longitudinal paler lines, hairy inside, divided into 6 lobes, the lobes rounded. *Stamens* 2; filaments pale green; anthers pale yellow. *Style* 1, whitish; stigma pale green. *Berry* 5–12 × 5–9 mm, globose or ellipsoid. *Flowers* 6–7. Visited by small bumblebees. $2n = 26$.

Introduced. Not commonly grown, but naturalised on walls and marginal ground. Scattered records in southern England north to Cheshire. Native of China. The name commemorates the seed and nursery firm of Bees, founded by A. K. Bulley of Neston, Cheshire and based on a nickname for Bulley and his sister, the 'busy Bs'.

3. J. nudiflorum Lindl. Winter Jasmine

Deciduous, monoecious, scrambling *shrub*. *Stems* up to 5 m, numerous and intertwined, arching, divaricately branched. *Bark* pale brown, rough. *Young shoots* dark green, slender, 4-angled, the angles ridged, glabrous. *Buds* 3–4 × 2.0–2.5 mm, lanceolate to ovate, acute at apex; scales green with red tips and sometimes margins, lanceolate or ovate, acute or acuminate at apex, glabrous. *Leaves* opposite; lamina broadly ovate in outline, ternate; leaflets 1–3, dark glossy green on upper surface, paler beneath, ovate to oblong-ovate, acute at apex, entire, glabrous. *Flowers* bisexual, produced between autumn and spring while leaves are absent, axillary and solitary, scentless; peduncles short, covered with numerous, imbricate bracts, the bracts green with red margins, lanceolate to ovate, acute or acuminate at apex. *Calyx* 7–9 mm, divided into 6 lobes about twice as long as the tube, yellowish-green, the lobes linear-lanceolate, acute at apex, with a prominent midrib and minutely hairy on the margin. *Corolla* rather pale yellow; tube 13–15 mm, slender; lobes 5–7, spreading, unequal, 10–13 mm, obovate or elliptical, rounded-mucronulate at apex. *Stamens* 2, inserted on and enclosed in the corolla tube; filaments white; anthers yellow. *Style* 1, pale yellow; stigma green, above the anthers. *Berry* 5–6 × 3–4 mm, ovoid to ellipsoid, black. *Flowers* 11–2. $2n = 26, 52$.

Introduced. Very commonly grown in gardens and naturalised on walls and waste places in southern England. Native of Asia from the Caucasus to China.

137. SCROPHULARIACEAE Juss. nom. conserv.

Annual to *perennial* monoecious *herbs* or very rarely small *shrubs* or *trees*. *Stems* usually round in section. *Leaves* alternate, opposite, whorled or rarely all basal, without stipules. *Inflorescence* a terminal or axillary raceme, sometimes in racemose or cymose panicles; flowers sometimes single, zygomorphic, sometimes almost actinomorphic, bisexual and hypogynous. *Calyx* (2–)4- to

5-lobed, never well differentiated into upper and lower lips. *Corolla* strongly 4- to 5-lobed and sometimes nearly actinomorphic to 2-lipped. *Stamens* usually 4, sometimes with a staminode, sometimes 2 or 5 or rarely 3, borne on the corolla tube. *Style* 1, terminal; stigmas 1–2, short. *Ovary* 2-celled, each cell with numerous ovules on axile placentas. *Fruit* a 2-celled capsule.

Consists of about 190 genera and 4,000 species, cosmopolitan but most abundant in temperate regions and in tropical mountains.

1. Stamens 2 — 2.
1. Stamens 4 or 5 — 5.
2. Corolla 2-lipped, the lower lip large and inflated, yellow to reddish-brown; stamens included — **7. Calceolaria**
2. Corolla 4-lobed, without inflated portion, white to blue or pink, stamens exserted — 3.
3. Stems woody; leaves evergreen, entire — **20. Hebe**
3. Herbaceous plants; stems not woody or woody only at base and then with serrate leaves — 4.
4. Leaves opposite — **18. Veronica**
4. Leaves in 3s — **19. Veronicastum**
5. Deciduous tree; flowers purplish blue to violet — **4. Paulownia**
5. Herbs; stems not woody, then flowers red — 6.
6. Stems woody; flowers 25–40 mm, red pendulous in terminal panicles — **3. Phygelius**
6. Stems herbaceous; flowers often less than 20 mm, if more than 20 mm and red then in racemes — 7.
7. Corolla with conspicuous basal spur or pouch on lowerside — 8.
7. Corolla not spurred or pouched at base — 15.
8. Leaves palmately veined and lobed — 9.
8. Leaves with single midrib, often also with lateral pinnate veins, entire to serrate — 10.
9. Plant glandular-hairy; corolla more than 30 mm, yellow with purple veins, pouched at base — **11. Asarina**
9. Plant glabrous or minutely hairy but not glandular-hairy; corolla less than 30 mm, mauve to purple, often with a yellow centre, spurred at base — **12. Cymbalaria**
10. Corolla tube with broad, rounded pouch at base — 11.
10. Corolla tube with narrow, often pointed, spur — 13.
11. Calyx-lobes ovate — **8. Antirrhinum**
11. Calyx-lobes linear — 12.
12. Corolla with a rounded or elongate swelling or bulge at the base on one side — **10. Misopates**
12. Corolla with a distinct conical or parallel-sided spur at base — **15. Nemesia**
13. Leaves ovate to obovate, rounded to cordate at base; capsule opening by detachment of 2 oblique lids leaving large pores — **13. Kickxia**
13. Leaves linear to lanceolate or oblanceolate, narrowed to base, rarely ovate to obovate and rounded at base and then capsule without detachable lids — 14.
14. Mouths of corolla completely closed by boss-like swelling on lower lip — **14. Linaria**
14. Mouth of corolla incompletely closed by a small swelling — **9. Chaenorhinum**
15. Fertile stamens 5, at least in most flowers — **1. Verbascum**
15. Fertile stamens 4, sometimes with a sterile staminode representing a 5th — 16.
16. Leaves all in a basal rosette — **6. Limosella**
16. At least some leaves borne on stems — 17.
17. Stems procumbent, rooting at nodes; leaves reniform, long-stalked; flowers solitary in leaf axils — **21. Sibthorpia**
17. Stems not procumbent nor rooting at nodes; leaves not reniform; flowers in terminal inflorescences — 18.
18. Calyx 5-lobed — 19.
18. Calyx 4-lobed, rarely 2- to 5-lobed with toothed lobes and inflated tube — 22.
19. Leaves opposite — 20.
19. Leaves alternate — 21.
20. Calyx tube shorter than lobes; corolla purplish-brown to dull yellowish-green with tube scarcely longer than wide — **2. Scrophularia**
20. Calyx tube longer than lobes; corolla bright yellow, often with red spots or blotches, with tube much longer than wide — **5. Mimulus**
21. Corolla distinctly zygomorphic, with tube more than twice as long as calyx, lobes not strongly patent — **16. Digitalis**
21. Corolla scarcely zygomorphic, with tube less than twice as long as calyx, with strongly patent lobes — **17. Erinus**
22. Calyx irregularly 2- to 5-lobed, the lobes toothed; leaves divided almost to base, the lobes toothed — **28. Pedicularis**
22. Calyx regularly 4-lobed, with entire lobes; leaves entire to simply toothed up to around halfway to base — 23.
23. Calyx tube inflated, especially at fruiting; seeds discoid, with marginal wing — **27. Rhinanthus**
23. Calyx tube not inflated; seeds not discoid, without marginal wing — 24.
24. Lower lip of corolla with 3 distinctly emarginate lobes — **23. Euphrasia**
24. Lower lip of corolla with entire lobes, or sometimes middle lobe slightly emarginate — 25.
25. Mouth of corolla partially closed by boss-like swellings on lower lip; capsules with 1–4 seeds — **22. Melampyrum**
25. Mouth of corolla open; lower lip without swellings; capsules with more than 4 seeds — 26.
26. Corolla yellow — 27.
26. Corolla pink to dark purple, rarely white — 28.
27. Leaves entire or more or less so; seeds more than 1 mm, ridged and grooved — **24. Odontites**
27. Leaves serrate; seeds about 0.5 mm, more or less smooth — **26. Parentucellia**
28. Annual; corolla pink to reddish-purple, rarely white, less than 12 mm — **24. Odontites**
28. Perennial; corolla dark purple, more than 12 mm — **25. Bartsia**

1. **Verbascum** L.

Celsia L.

Usually *biennial*, less often annual or perennial, monoecious *herbs*. *Stems* tall, erect, terete and ridged. *Leaves* alternate, cauline numerous. *Inflorescence* a terminal raceme or panicle; flowers bisexual. *Calyx* deeply 5-lobed, nearly regular. *Corolla* usually yellow, rotate, with 5 nearly equal lobes and a very short tube. *Stamens* 5; filaments all or only the upper 3 hairy, the 2 lower longer than the 3 upper; 2 types of anther occur, the upper are always reniform and transversely medifixed, the lower may be similar, or may be elongate, longitudinally inserted and decurrent on the filament. *Style* 1. *Fruit* a septicidal capsule; seeds numerous.

Contains about 360 species, mainly in the Mediterranean region, but extending to north Europe and north and central Asia.

Where more than 1 species is found growing together hybrids will almost certainly be found which are more or less sterile.

Arts-Damler, T. (1960). Cytogenetical studies on six *Verbascum* species and their hybrids. *Genetica* **31**: 241–328.
Bagnall, J. E. (1895). Notes on the flora of Warwickshire. *Midl. Nat.* **1892–3**: 22 (1892–3).
Druce, G. C. (1918). *Verbascum nigrum L.×olympicum* Boiss. hybr. nov. *Rep. Bot. Soc. Exch. Club Brit. Isles* **5**: 39.
Druce, G. C. & Robinson, F. (1919). *Verbascum nigrum×pulverulentum Rep. Bot. Soc. Exch. Club Brit. Isles* **5**: 511–512.
Dunn, S. F. (1894). Yellow-flowered *Verbascum lychnitis* L. *Jour. Bot. (London)* **32**: 23.
Easy, G. M. S. (1998). In Rich, T. C. G. & Jermy, A. C., *Plant crib.* London.
Franchet, M. A. (1868). Essai sur les espèces du généra *Verbascum* et plus particulièrement sur leurs hybrides. *Mém. Soc. Acad. Maine–Loire* **22**: 65–204.
Ferguson, I. K. (1978). *Verbascum speciosum* Schrader×*thapsus* L. new to Britain. *Watsonia* **12**: 160–162.
Gross, K. L., Werner, P. A. & Hawthorn, W. R. (1980). The biology of Canadian weeds. 28. *Verbascum thapsus* L. and *V. blattaria* L. *Canad. Jour. Pl. Sci.* **58**: 401–413.
Hultén, E. & Fries, M. (1986). *Atlas of north European vascular plants north of the Tropic of Cancer.* 3 vols. Königstein.
Lousley, J. E. (1934). Two interesting hybrids in the British flora. *Jour. Bot. (London)* **72**: 171–175.
Murbeck, S. (1933). Monographie der Gattung *Verbascum. Acta Univ. Lund* nov. ser. **29**(2): 1–630.
Pugsley, H. W. (1934). New variety of *Verbascum nigrum* L. *Jour. Bot. (London)* **72**: 278–279.
Stewart, A., Pearman, D. A. & Preston, C. D. (1994). *Scarce plants in Britain.* Peterborough.

1. Each bract with a single flower in its axil 2.
1. At least the lower bracts each with a cluster of several flowers in its axil 8.
2. Flowers subtended by bracteoles as well as by a bract **2. virgatum**
2. Bracteoles absent 3.
3. Anthers all reniform 4.
3. Anthers of lower stamens decurrent or obliquely inserted 5.
4. Pedicels longer than the subtending bract **4. phoeniceum**
4. Pedicels shorter than subtending bract **5. pyramidatum**
5. Pedicels at least 12 mm **1. blattaria**
5. Pedicels 2–10 mm 6.
6. Calyx 8–15 mm, with serrate lobes; capsule 9–15 **3. creticum**
6. Calyx 3–9 mm, with entire lobes; capsule 5–9 7.
7. Plant glabrous below, more or less glandular-hairy above; pedicel usually longer than calyx **1. blattaria**
7. Plant usually glandular-hairy throughout; pedicel shorter than calyx **2. virgatum**
8. Anthers of lower stamens decurrent or obliquely inserted 9.
8. Anthers all reniform and transversely inserted 14.
9. Axis of inflorescence conspicuously glandular **2. virgatum**
9. Axis of inflorescence more or less eglandular 10.
10. Upper cauline leaves distinctly decurrent 11.
10. Upper cauline leaves not or scarcely decurrent 12.
11. Stigma spathulate, decurrent on style **8. densiflorum**
11. Stigma capitate **9. thapsus**
12. Filaments of 2 lower stamens entirely glabrous **7. phlomoides**
12. Filaments of 2 lower stamens hairy at least towards their bases 13.
13. Pedicels absent or up to 5 mm. anthers of the 2 lower stamens more than 1.5 mm **6. bombyciferum**
13. Pedicels up to 16 mm. anthers of the 2 lower stamens 1.0–1.5 mm **16. olympicum**
14. Basal leaves distinctly lobed 15.
14. Basal leaves not lobed 16.
15. Leaves with a short tomentum on upper surface **11. sinuatum**
15. Leaves minutely stellately hairy and with a network of veins on upper surface **14(a). chaixii** subsp. **chaixii**
16. Basal and lower cauline leaves truncate, cordate or broadly cuneate at base 17.
16. Basal and lower cauline leaves tapered gradually to the petiole 19.
17. Basal leaves truncate or very shortly cuneate at base; pedicels about as long as calyx **14(b). chaixii** subsp. **austriacum**
17. Basal leaves cordate at base; pedicels 2–3 times as long as calyx 18.
18. Leaves with sparse stellate hairs on upper surface and more numerous beneath **15(i). nigrum** var. **nigrum**
18. Leaves with dense stellate hairs on both surfaces **15(ii). nigrum** var. **tomentosum**
19. Leaves densely and persistently whitish- or greyish-tomentose on both surfaces 20.
19. Mature leaves sparsely tomentose to glabrous, green at least above 21.
20. Pedicels longer than calyx **10. speciosum**
20. Pedicels shorter than calyx **12. pulverulentum**
21. Flowers white **13(ii). lychnitis** var. **album**
21. Flowers yellow 22.
22. Indumentum not persisting; pedicels shorter than calyx **12. pulverulentum**
22. Indumentum persistent; pedicels longer than calyx **13(i). lychnitis** var. **lychnitis**

1. V. blattaria L. Moth Mullein

V. rhinanthifolium Davidov; *V. carduifolium* Murb. ex Hayek

Biennial, rarely *annual*, monoecious *herb* with a stout root. *Stems* 30–120(–150) cm, pale green, often flushed brownish-purple, erect, subterete, obscurely striate, glabrous below, with sparse to numerous glandular hairs above, usually unbranched, leafy. *Leaves* forming a loose rosette below, the cauline alternate; lamina 3–15(–25) × 1–5(–7) cm, dull medium to rather dark green on upper surface with a brownish-purple midrib, paler beneath with prominent veins, lanceolate-oblong to ovate-oblong, narrowed to a rounded point at apex, sinuate-crenate or sinuately lobed, the lower with an attenuate base, the upper sessile and amplexicaul, with dense, soft, forked hairs on the surfaces and margins. *Inflorescence* a lax, slender, unbranched raceme, often 30–50 cm with a glandular-hairy rhachis; flowers bisexual; pedicels up to 14 mm usually longer than calyx, patent, glandular-hairy; bracts up to 10 × 8 mm each with a single flower in its axil, ovate, obscurely dentate, glandular-hairy; bracteoles absent. *Calyx* 5–8 mm, with 5 narrowly oblong or lanceolate lobes, subacute or obtuse at apex and glandular-hairy. *Corolla* yellow, rarely white, 20–30 mm in diameter, the lobes about 10–12 mm in diameter, subrotund and subequal. *Stamens* 5, the 3 upper 5–7 mm, the filaments clothed densely with long purple hairs, the 2 lower 7–8 mm, glabrous towards the apex; anthers of the upper stamens reniform and medifixed, those of the lower stamens shortly decurrent and densely glandular. *Style* 1, about 6–10 mm, filiform, glandular-hairy towards the base; stigma capitate. *Capsule* 5–8 mm in diameter, subglobose, thinly glandular-hairy; seeds 0.8 mm, oblong, dull brown, minutely reticulate with transverse rugosities. *Flowers* 6–10. Pollinated by various insects or selfed, homogamous. $2n = 30, 32$.

Introduced by 1596. Waste and rough ground, field borders, old quarries and gravel pits. Scattered records throughout much of Great Britain; formerly in Ireland and the Channel Islands. Widespread in Europe and eastwards to Afghanistan and central Asia; North Africa; introduced elsewhere. Eurosiberian Southern-temperate element.

× **nigrum** =**V.** × **intermedium** Rupr. ex Bercht. & Pfund

V. blattaria is introduced, *V. nigrum* is native. This hybrid is intermediate between its parents in many characters, the leaves and stems are more or less glandular-hairy, the inflorescence is usually simple with the flowers solitary or in groups of 2–4 and the pedicels are longer than the calyx. The anthers of the lower 2 stamens are decurrent on their filaments.

Recorded only from Wiltshire and Lincolnshire. In continental Europe it is recorded from France and Roumania.

2. V. virgatum Stokes Twiggy Mullein

Biennial, rarely *annual*, monoecious *herb* with a stout root. *Stems* 30–100 cm, pale green, often flushed brownish-purple, erect, subterete, obscurely striate, usually glandular-hairy throughout, unbranched, leafy. *Leaves* forming a loose rosette below, the cauline alternate; lamina of basal (10–)15–33 × 4.5–9.0 cm, dull medium to rather dark green on upper surface, sometimes brownish-purple on midrib, paler beneath with prominent veins, oblong to lanceolate, narrowed at apex, sinuate-crenate or sinuately lobed, the lower attenuate at base, the upper sessile and amplexicaul, all glabrous or nearly so with a few, scattered glandular hairs especially on the margin. *Inflorescence* a lax, slender, unbranched raceme, often 30–50 cm, with a glandular-hairy rhachis; pedicels 2–5(–7) mm, much shorter than calyx, patent, glandular-hairy; bracts 8–20 mm, serrate-dentate, each with 1–5 bisexual flowers in their axil, glandular-hairy; bracteoles usually present. *Calyx* 5–8 mm, with 5 narrowly oblong or lanceolate lobes, subacute or obtuse at apex and glandular-hairy. *Corolla* yellow, 30–40 mm in diameter, the lobes subrotund and subequal. *Stamens* 5, the 3 upper stamens with filaments clothed with long purple hairs, the 2 lower glabrous towards the apex; anthers of the upper stamens reniform and medifixed, those of the lower obliquely inserted or shortly decurrent. *Style* 1, filiform, glandular-hairy towards the base; stigma capitate. *Capsule* subglobose, thinly glandular-hairy; seeds oblong, dull brown, minutely reticulate. *Flowers* 6–10. $2n =$ c. 64, (32, 64, 66).

Introduced and first recorded in the wild in 1787. Fields, waste places and dry banks. Locally frequent in Devonshire and Cornwall, where it has been considered to be native, and a fairly frequent casual in waste places elsewhere in Great Britain, north-west and southern Ireland and the Channel Islands. Western Europe, northwards to south-west England, widely naturalised elsewhere. Suboceanic Southern-temperate element.

3. V. creticum (L.) Cav. Crete Mullein

Celsia cretica L.; *Celsia sinuata* Cav.

Biennial monoecious *herb* with a stout root. *Stems* 40–150 cm, pale green, subterete, erect, with short glandular hairs and longer simple eglandular hairs, simple, leafy. *Leaves* forming a loose rosette below, the cauline alternate; lamina of basal 15–25 × 4–7 cm, rather dark green on upper surface, paler beneath, oblong, narrowed at apex, crenate-dentate to somewhat pinnatifid, sessile and subcordate at base; clothed with glandular and simple eglandular hairs. *Inflorescence* a simple raceme; pedicels 3–6 mm. bracts 12–25 mm, ovate, acuminate at apex, serrate, each with a single, bisexual flower; bracteoles absent. *Calyx* 8–15 mm, with 5 lanceolate, serrate lobes. *Corolla* 40–50 mm in diameter, yellow; lobes 5, rounded at apex. *Stamens* 4, without a staminode, unequal, the lower with decurrent anthers 5–6 mm. *Style* 1, 20–35 mm. *Capsule* 9–15 mm, ellipsoid-globose. *Flowers* 6–9.

Introduced. Grain casual and garden escape. Native of the west Mediterranean region.

4. V. phoeniceum L. Purple Mullein

Celsia rechingeri Murb.

Perennial herb with a strong tap-root. *Stem* up to 150 cm, pale yellowish-green, sometimes suffused purplish towards the base, robust, solid, strongly ridged, shortly crisped-hairy near the base, with numerous, short, colourless glandular hairs and some short, anchor-shaped hairs intermixed above, leafy. *Leaves* rather dark, dull green on upper surface with an impressed network of veins, pale yellowish-green beneath with a very prominent midrib and network of veins; basal numerous, with lamina 4–22 × 2.5–12 cm, broadly ovate to ovate-oblong, obtuse to acute at apex, entire to weakly crenate and undulate, rounded or cuneate at base to a short petiole which is flat on the front and rounded on the back; cauline few, alternate, similar to basal but much smaller, acute at apex, subcordate at base, sessile or shortly petiolate; all glabrous or nearly so on the upper surface, very shortly hairy on the veins beneath or nearly glabrous. *Inflorescence* a long, narrow raceme with numerous lax, bisexual flowers; pedicels up to 30 mm, yellowish-green, sometimes flushed brownish-purple, with numerous, short, colourless glandular hairs in the upper half, with small, green, ovate, acute, dentate, glandular-ciliate bracts at the base bearing a single flower; bracteoles absent. *Calyx* 4–8 mm, deeply 5-lobed, glandular-hairy; lobes lanceolate to obovate-elliptical, acute at apex. *Corolla* 20–38 mm in diameter, violet to purple, rarely yellow, divided two-thirds of the way to the base; lobes 5, broadly rounded. *Stamens* 4 or 5; filaments 3–4 mm, with dense, long, purple glandular hairs; anthers all reniform, purple. *Style* 1, 5–8 mm. *Capsule* 6–8 mm, ovoid. *Flowers* 5–6. $2n = 32$, 36.

Introduced. Grown in gardens and casual on tips and waste ground, and as a bird-seed alien. Scattered records over much of Great Britain. Native of south-east Europe and western Asia.

5. V. pyramidatum M. Bieb. Caucasian Mullein

Perennial monoecious *herb* with a strong tap-root. *Stems* 60–250 cm, medium green, often suffused brownish-purple, erect, robust, angled, with sparse stellate hairs below, more numerous above, branched only in the inflorescence, very leafy. *Leaves* alternate, medium to dark green on upper surface with a pale sometimes purple midrib, paler beneath; basal with lamina 32–45 × 5–12 cm, oblong or oblong-elliptical, more or less obtuse at apex, incise-dentate to weakly pinnatifid, with large, irregular teeth, long petiolate; lower cauline similar to basal and petiolate, the upper gradually becoming smaller in size, ovate, asymmetrical and twisted, acute at apex, the median irregularly incise-dentate, the upper dentate, cordate and sessile; all with numerous, short glandular hairs on upper surface and stellate hairs beneath. *Inflorescence* occupying one-third of plant, with 1 flower per bract, without bracteoles; much branched, the branches arcuate-ascending, the inflorescence very narrow; pedicels very short 3–5(–8) mm, with dense, short glandular hairs; bracts longer than pedicels, 6–10 mm, lanceolate, acuminate at apex; flowers bisexual, yellow, 20–30 mm in diameter, faintly honey-smelling. *Calyx* 3–5 mm, pale green, divided about one-third of way to base into 5 lobes, the lobes broadly triangular, acute at apex, with numerous, very short glandular hairs. *Corolla* 20–30 mm in diameter, yellow, the 3 upper lobes 10–12 × 10–12 mm, the 2 lower lobes 9–10 × 9–10 mm, obovate, rounded at apex. *Stamens* usually 5, sometimes 4; filaments all with purple glandular hairs; anthers all reniform, orange. *Style* 1, green; stigma orange, capitate. *Capsule* 4–5 mm, elliptic-ovoid, woolly; seeds numerous, about 1 mm, oblong. *Flowers* 6–7. Visited by bees. $2n = 32$.

Introduced. Waste and rough ground. Naturalised in a few places in southern and central England, otherwise a rare casual. Native of the Caucasus.

× **speciosum**

Very difficult to distinguish from *V. pulverulentum* × *pyramidatum* but its narrower leaves and rather different hairiness are helpful.

× **thapsus**

This hybrid resembles *V. pyramidatum* but has decurrent upper leaves.

V. pyramidatum is introduced, *V. thapsus* is native. The hybrid has been recorded for Cambridgeshire at Fordham and Burwell.

6. V. bombyciferum Boiss. Broussa Mullein

Biennial monoecious *herb. Stems* 50–150 cm, pale green, erect, robust, terete, with a very dense indumentum of long, soft simple and branched eglandular hairs, simple or rarely with a few branches above, leafy. *Leaves* alternate; basal with lamina up to 40 × 20 cm, medium green on upper surface, paler beneath, ovate or obovate, rounded at apex, crenate, with a petiole 1–5 cm; cauline similar, but smaller, the upper sessile and entire, not decurrent with dense long, soft simple eglandular hairs on both surfaces. *Inflorescence* a simple, very dense terminal spike; flowers bisexual, with 3–7 in a cluster, immersed in long, white, eglandular indumentum; bracts ovate to triangular-lanceolate, with several flowers in an axil; bracteoles 2; pedicels absent or up to 5 mm. *Calyx* 6–10 mm, divided nearly to base into 5 lobes, the lobes lanceolate, acute at apex. *Corolla* 30–40 mm in diameter, yellow, hairy outside, divided into 5 nearly equal lobes. *Stamens* 5; filaments with dense, whitish-yellow hair, the 2 lower glabrous above the middle; anthers 1.5–4.0 mm, the upper reniform, the 2 lower obliquely inserted or decurrent. *Style* 1; stigma capitate, spathulate. *Capsule* 6–8 × 4–6 mm, broadly ovoid, densely white-tomentose. *Flowers* 6–9.

Introduced. Waste and rough ground. Casual or naturalised in a few places scattered over England, Isle of Man and Guernsey. Native of north-west Anatolia.

× **nigrum**

This hybrid is more like *V. nigrum* but with thick woolly flower heads.

V. bombyciferum is introduced, *V. nigrum* is native. The hybrid is recorded with both parents at Trumpington, Cambridgeshire.

× **phlomoides**

Introduced. Recorded with both parents at Trumpington, Cambridgeshire.

× **pulverulentum**

This hybrid is similar to *V. pulverulentum* but the filaments are less hairy, the lower more or less decurrent.

× **pyramidatum**

This hybrid often grows to a great size. The purple hairs on the filaments and glandular hairs on the calyx show some *V. pyramidatum* characters while the general wooliness and leaf shape are similar to *V. bombyciferum.*

× **thapsus**

This hybrid similar to *V. bombyciferum* but shows some of the decurrentness of the leaves of *V. thapsus*.

7. V. phlomoides L. Orange Mullein

Biennial monoecious *herb. Stem* 30–200 cm, pale green, erect, stout, yellowish- or grey-woolly, simple or with a few, long branches in the upper part, very leafy. *Leaves* yellowish-green with pale midrib on upper surface, slightly paler beneath; lamina 7–45 × 5–15 cm, the lower broadly oblong, acute at apex, crenate with shallow teeth and undulate, long attenuate at base, gradually decreasing in size upwards, the median smaller versions of the lower, the upper broadly ovate, acuminate to caudate at apex, cordate at base and sometimes slightly decurrent, softly yellowish- or greyish-woolly on both surfaces; veins prominent beneath. *Inflorescence* a dense, terminal, spike-like raceme; flowers 30–45 mm in diameter, bisexual, scentless; pedicels 3–15 mm. bracts 9–15 mm, lanceolate, with several flowers in each axil. *Calyx* 5–15 mm, divided two-thirds of the way to the base into 5 unequal lobes, the lobes triangular and acute at apex, with dense, yellowish glandular hairs and longer simple eglandular hairs. *Corolla* 20–55 mm, golden yellow, divided nearly to base into 5 lobes, the lobes broadly obovate, with a rounded apex, overlapping, tomentose. *Stamens* 5; 2 glabrous; anthers orange, the upper reniform, the lower decurrent. *Style* 1, 6–15 mm, green; stigma orange, spathulate. *Capsule* 5–12 mm, elliptical, ovoid, woolly; seeds numerous, about 1 mm, oblong. *Flowers* 6–7. Visited by bees. $2n=32$, 34.

Introduced by 1739. Waste and rough ground, roadsides and rubbish tips. Frequent casual and sometimes naturalised. Central and south Great Britain and the Channel Islands, rare in northern Great Britain. Native of most of Europe except the north. European Temperate element.

× **pulverulentum** = **V. × murbeckii Borbás**

V. phlomoides is introduced, *V. pulverulentum* is native. The hybrid has been recorded at Chesterton in Cambridgeshire with the parents.

× **thapsus** = **V. × kerneri** Fritsch

This hybrid is intermediate between the parents in most characters, and is very robust. The calyx hairs are not stellate and the pollen is mostly infertile and the ripe seeds are rarely set. *V. phlomoides* is introduced, *V. thapsus* is native. The hybrid is recorded from a few places in southern England. It is widespread through central Europe.

8. V. densiflorum Bertol. Dense-flowered Mullein
V. thapsiforme Schrad.

Biennial monoecious *herb. Stems* 50–150 cm, pale yellowish-green, erect, yellowish-woolly with branched hairs, usually unbranched, very leafy. *Leaves* yellowish-green with a pale midrib on upper surface, slightly paler beneath with prominent veins; basal 15–25 × 4–8 cm, elliptical or elliptical-oblong, more or less acute at apex, entire to crenate, sessile or with a short, winged petiole; cauline gradually decreasing in size upwards, all decurrent to a winged stem, the lower similar to basal, the upper 2–11 × 2–5 cm, ovate-oblong, more or less acute at apex, entire narrowed at base; all yellowish-woolly with branched hairs. *Inflorescence* occupies about one-third of stem; pedicels 2–10 mm, woolly; 2–3 bisexual flowers in each bract axil; bracts 15–40 mm, ovate, usually long acuminate and decurrent at base. *Calyx* 5–12 mm, divided two-thirds of the way to the base into 5 unequal lobes, the lobes triangular and acute at apex, woolly. *Corolla* 20–25 mm, golden yellow, divided nearly to base into 5 lobes, the lobes broadly obovate, with a rounded apex. *Stamens* 5; filaments yellow, 3 with dense, long yellow hairs and 2 glabrous; anthers of the 2 lower decurrent. *Style* 1; stigma spathulate. *Capsule* ovoid. *Flowers* 6–7. $2n=32$, 34, 36.

Introduced. Waste and rough ground. Frequent casual and garden escape and sometimes naturalised throughout England and south-east Ireland and the Channel Islands. Most of Europe to central Russia. European Temperate element. Perhaps over-recorded for *V. phlomoides* and its hybrids.

× **nigrum** = **V. × ambiguum** Lej.

V. densiflorum is introduced, *V. nigrum* native. Recorded by W. O. Focke for England, but nothing further is known about the record.

9. V. thapsus L. Great Mullein

Biennial or rarely *annual* monoecious *herb. Stems* 30–200 cm, pale green, erect, more or less densely greyish- or whitish-tomentose, with dense long hairs, rarely shortly branched, leafy. *Leaves* alternate, medium green on upper surface, paler beneath; basal with lamina 15–× – 14 cm, obovate-lanceolate to oblong, acute or subacute at apex, entire or finely crenate, attenuate at base, the petiole narrowly winged; cauline smaller but similar to basal, the base decurrent nearly to the next leaf below; all more or less densely greyish or whitish tomentose with long hairs. *Inflorescence* a dense, terminal, spike-like raceme, rarely with axillary racemes from the upper leaves; rhachis with eglandular hairs; flowers bisexual; pedicels 0–2 mm. bracts 12–18 × 3–10 mm, triangular-lanceolate, long acuminate at apex, usually longer than the flowers, with several flowers in their axils. *Calyx* (5–)8–12 mm, deeply divided into 5 lobes, the lobes lanceolate or ovate and acuminate at apex, with stellate hairs. *Corolla* 12–35 mm in diameter, yellow, divided nearly to base into 5 nearly equal lobes, the lobes ovate. *Stamens* 5; the 3 upper with filaments clothed with whitish or yellowish hairs, the 2 lower glabrous or sparingly hairy; upper anthers reniform, the 2 lower obliquely inserted and decurrent. *Stigma* 1, capitate. *Capsule* 7–10 mm, rather longer than calyx, elliptical-ovoid. *Flowers* 6–8. Pollinated by various insects or selfed. $2n=32$, 36.

Native. Waste and rough ground, banks and grassy places, road and rail sides, mostly on sandy or chalky soils.

Common in central and south Great Britain and Channel Islands, locally frequent elsewhere and by far our commonest species. Most of Europe except the extreme north and much of the Balkan peninsula; Asia south to Caucasus and Himalaya and east to western China; naturalised in North America. Eurosiberian Temperate element. Our plant is subsp. **thapsus**.

× **virgatum** = **V.** × **lemaitrei** Boreau

This hybrid is intermediate between its parents. The leaves are crenate and less hairy than *V. thapsus* and only weakly decurrent. The flowers are large and solitary or in fascicles, the filament hairs of the lower 2 stamens are purple, while those of the upper 3 are a mixture of white and purple. The anthers of the lower stamens are obliquely inserted, and those of the upper ones reniform and medifixed. The capsules are abortive.

Native. Recorded from Devonshire, Worcestershire, Warwickshire and Edinburghshire. Known elsewhere only from France.

10. V. speciosum Schrad. Hungarian Mullein

Biennial monoecious *herb. Stems* 50–200 cm, pale green, erect, angled, persistently soft tomentose with stellate hairs, branched, leafy. *Leaves* alternate; lamina medium green on upper surface, paler beneath; basal 12–40 × 3–14 cm, oblong-oblanceolate to obovate, more or less acute at apex, entire, gradually tapered at base to shortly winged petioles; cauline similar, gradually getting smaller, the uppermost almost ovate and acuminate to caudate at apex, all with dense, persistent stellate hairs on both surfaces. *Inflorescence* freely branched, often ellipsoid in outline; pedicels (3–)5–12 mm, with dense stellate or branched hairs; bracts (5–)8–15(–20) mm, the lower ovate-lanceolate, the upper lanceolate; flowers bisexual, 18–30 mm in diameter, several per bract. *Calyx* 3–6 mm, divided nearly to base into 5 lobes, the lobes linear-lanceolate, with dense stellate hairs. *Corolla* yellow, rotate, with 5, broad obovate, rounded lobes, stellate-hairy outside. *Stamens* 5; filaments all clothed with white hairs; anthers all reniform and transversely inserted. *Capsule* 3–7 × 3–4 mm, ovoid-oblong. *Flowers* 7–8. $2n=28$.

Introduced. Grown in gardens and escaping into open areas. Scattered records throughout England and Cardiganshire. Native of south-east and east-central Europe; Transcaucasia; north Iran.

× **thapsus** = **V.** × **duernsteinense** Teyber

Native. Recorded for Norfolk, Cambridgeshire and Nottinghamshire.

11. V. sinuatum L. Mediterranean Mullein

Biennial monoecious *herb. Stems* 50–150 cm, pale green, erect, terete or obscurely angled, with a short tomentum of greyish or yellowish stellate hairs, commonly branched from the base, leafy. *Leaves* alternate, medium green on upper surface, paler beneath; basal with lamina 10–35 × 2.5–15 cm, oblong or oblong-obovate, obtuse or subacute at apex, undulate-sinuate, crenate to pinnatilobed, narrowed at base to a short petiole; cauline rapidly diminishing upwards, subentire, irregularly toothed or lobed, sessile, amplexicaul and decurrent; all with a short tomentum of greyish or yellowish hairs on both surfaces. *Inflorescence* a lax panicle with numerous, spreading branches; bracts deltoid, acute at apex, cordate at base, the lowermost 2.5–3.0 cm, becoming smaller upwards and the uppermost minute, each with several flowers; pedicels 2–5 mm, very short glandular-hairy; bracteoles minute, ovate; flowers bisexual, in clusters of 2–7. *Calyx* 2.0–4.0 mm, deeply divided into 5 lobes, the lobes ovate-lanceolate or lanceolate, densely stellate-tomentose. *Corolla* yellow, 15–30 mm in diameter, divided into 5 lobes, the lobes about 10 × 8–9 mm, obovate or subrotund, rounded at apex. *Stamens* 5; filaments about 8 mm, densely clothed to near the apex with purple, woolly hairs; anthers all reniform. *Style* 1, about 6 mm, glabrous, slightly swollen near the apex; stigma subcapitate. *Capsule* 2.5–4.0 × 2.0–3.5 mm, dark brown, broadly ovoid; seeds about 0.4 mm, dull brown, irregularly oblong, minutely reticulate, conspicuously rugose. *Flowers* 7–9. $2n=30$.

Introduced. A wool casual. Native of the Mediterranean region eastwards to Iran.

12. V. pulverulentum Vill. Hoary Mullein

V. acutifolium Halácsy; *V. floccosum* Waldst. & Kit.

Monocarpic *perennial herb. Stems* 50–200 cm, pale yellowish-green, erect, terete, thickly white or greyish woolly hairy which is easily rubbed off, and becoming glabrous, copiously branched in the upper half, very leafy. *Leaves* greyish-green with a pale midrib on upper surface, pale beneath with pale prominent veins; basal with lamina 20–50 × 6–25 cm, elliptical or elliptical-oblong, more or less acute at apex, entire or crenate, undulate and sinuate, broadly narrowed to a short, winged petiole; lower cauline like basal but slowly getting smaller upwards, the upper cauline with lamina 2–20 × 2–12 cm, ovate or ovate-oblong, with a characteristic caudate tip up to 15 mm, undulate, cordate at base and sessile; all with forked and stellate arachnoid hairs on upper surface giving a mealy appearance, grey-woolly beneath becoming less hairy. *Inflorescence* occupying about half the length of the plant, much branched, the branches ascending giving the total inflorescence an ellipsoid shape; pedicels shorter than calyx; bracts lanceolate, acute at apex; flowers bisexual, very numerous, 20–25 mm in diameter, usually 4–10 at each bract. *Calyx* 5–7 mm, divided halfway to base into 5 lobes, the lobes linear, obtuse at apex and woolly. *Corolla* yellow, with 3 upper lobes and 2 similar lower lobes, the lobes narrowly obovate. *Stamens* 5; filaments pale orange, all with dirty white glandular hairs the 2 lower only at base; anthers deep orange, equal, all reniform and transversely inserted. *Style* 1, pale yellowish-green; stigma deep orange, spathulate. *Capsule* 3–6 × 3–4 mm, ovoid, slightly longer than calyx. *Flowers* 7–8. $2n=32$.

Native. Local on rather bare ground on chalky or sandy soils on heathland, waste places and roadsides. East Anglia, mostly Norfolk, a rare casual or naturalised elsewhere in Great Britain. West, south and south-central Europe. Submediterranean Subatlantic element.

× **pyramidatum**

This hybrid is very similar to *V. pulverulentum* but has a purplish base to the whitish hairs of the stamens, some glandular hairs on the calyx, more toothed leaves and larger flowers.

V. pulverulentum is native, *V. pyramidatum* is introduced.

× **thapsus** = **V.** × **godronii** Boreau
V. × *lamottei* Franch.

This hybrid is usually a large plant intermediate between its parents in most characters, but the indumentum is persistent and not conspicuously floccose, the cauline leaves are sessile and scarcely decurrent and the fascicles of flowers are crowded above but more interrupted below. The hairs of the calyx and filaments resemble those of *V. pulverulentum* and the plant is usually completely sterile.

Native. It has occurred in a few places in southern England and is widespread in central Europe. Named after Dominique Alexandre Godron (1807–1880).

× **virgatum**

Native. Recorded at Chesterton and Milton in Cambridgeshire with both parents.

13. V. lychnitis L. White Mullein

Biennial monoecious *herb. Stems* 50–150 cm, pale green, erect, angled, with short, powdery stellate hairs, branched above, leafy. *Leaves* alternate, dark yellowish-green on upper surface, paler beneath; basal with lamina 15–30 × 6–15 cm, ovate to oblanceolate-oblong, obtuse at apex, coarsely crenate, long tapered at base, long-petiolate; cauline smaller, ovate, acuminate at apex, the upper often narrower and sessile, all green on upper surface, with dense, white, powdery stellate hairs beneath. *Inflorescence* a narrow panicle with 2–7 bisexual flowers in the axils of each bract, usually freely branched; the branches ascending; bracts and bracteoles 8–15 mm, linear-lanceolate or linear, shorter than the flowers; pedicels 6–11 mm, from as long as to twice as long as calyx. *Calyx* (1.5–)2.5–4.0 mm, divided nearly to base into 5-lobes, the lobes linear-lanceolate, acute at apex and very stellate woolly. *Corolla* 12–22 mm in diameter, yellow or white, rotate, with 5 nearly equal lobes. *Stamens* 5; filaments green, all clothed with yellowish or whitish hairs; anthers equal, orange, all reniform and transversely inserted. *Style* 1, pale green, terminal and capitate. *Capsule* 4–5 mm, ovoid-elliptical or pyramidal, longer than calyx. *Flowers* 7–8. Pollinated by various insects or selfed; homogamous. $2n$ = c. 64.

(i) Var. **lychnitis**
Flowers yellow.

(ii) Var. **album** (Mill.) Druce
V. album Mill.

Flowers white.

Native. Fields, waste places and dry banks. Locally frequent in Devonshire and Cornwall; a fairly frequent casual elsewhere in waste places in Great Britain, formerly in Ireland and the Channel Isles. Most colonies are var. *album*. Europe southwards from central Russia, but rare in the Mediterranean region; Caucasus; western Siberia; Morocco; naturalised in North America. European Temperate element.

× **nigrum** = **V.** × **incanum** Gaudin
V. schiedeanum W. D. J. Koch

This hybrid is sterile and intermediate between its parents in leaf characters, the petioles varying in length, the stems simple or branched and the filament hairs usually purple.

Native. Occurs sporadically in southern and central England when the parents grow together. Widespread in Europe.

× **pulverulentum** = **V.** × **regelianum** Wirtg.

This hybrid is intermediate between its parents, is very robust, has low pollen fertility and is apparently sterile, its leaves are petiolate and hairy, but not often stellate hairy.

Native. It has been recorded in Devonshire as a casual. It is widespread in Continental Europe.

× **thapsus** = **V.** × **thapsi** L.
V. × *foliosum* Franch.; *V.* × *spurium* W. D. J. Koch

This hybrid is intermediate to a varying degree between its parents, some having simple inflorescences other branched ones. It is usually very robust, the hairs of the calyx and filaments resemble those of *V. lychnitis*, the pollen is mostly infertile and the capsules have very few seeds.

Native. Occurs with the parents in central and southern Great Britain. It is widespread in Continental Europe.

14. V. chaixii Vill. Nettle-leaved Mullein

Perennial monoecious *herb. Stems* 50–200 cm, pale yellowish-green, sometimes tinted brownish- or reddish-purple, angled with ridges, robust, erect, solid, with numerous to dense stellate hairs, shortly branched above, leafy. *Leaves* medium yellowish-green with pale midrib and a network of veins on upper surface, paler beneath with very prominent pale veins; basal with lamina 10–40 × 4–14 cm, elliptic-oblong or ovate-oblong, more or less acute at apex, undulate and sinuate-crenate, sometimes slightly lobed towards the base, truncate or shortly narrowed to a short, pale petiole; lower cauline similar to basal, remaining cauline with lamina ovate or ovate-oblong, gradually decreasing in size upwards, acute or with a tail at apex, undulate-crenate, truncate or cordate to an amplexicaul base; all minutely stellately hairy on upper surface, green but minutely stellately hairy or grey tomentose beneath. *Inflorescence* occupying half the stem, shortly branched but very narrow, the branches erect-ascending; pedicels 3–6 mm, with very short hairs; bracts 2–5 mm, linear to linear-lanceolate. *Flowers* 15–25(–30) mm in diameter, bisexual, 2–4 per bract. *Calyx* 3–5 mm. lobes linear to lanceolate. *Corolla* 12–22 mm in diameter, yellow; lobes unequal, obovate. *Stamens* 5; filaments yellow, hairs all violet, but lower sometimes glabrous above; anthers all reniform, orange. *Style* 1, green; stigma greenish. *Capsule* 3–6 mm, ellipsoid. *Flowers* 6–8. Visited by various insects. $2n$ = 30, 32.

(a) Subsp. **chaixii**
Basal leaves usually slightly lobed towards the base, greyish-tomentose beneath.

(b) Subsp. **austriacum** (Schott ex Roem. & Schult.) Hayek
V. austriacum Schott ex Roem. & Schult.

Basal leaves not lobed, green beneath but minutely stellately hairy.

Introduced. Grown in gardens and a casual or naturalised escape on waste or rough ground. Scattered records in Great Britain, but may be overlooked as *V. nigrum*. Both subspecies occur. Native of south and central Europe and western Asia.

15. V. nigrum L. Dark Mullein

Perennial monoecious *herb. Stems* 50–200 cm, pale green in the inflorescence area, deep brownish-purple below and sometimes into the inflorescence area, erect, stout, angled, rather sparingly stellately and simply hairy, branched in area of inflorescence, very leafy. *Leaves* alternate, medium yellowish-green on upper surface with the midrib sometimes brownish-purple, paler green beneath; basal with lamina 15–50×7–25, ovate or oblong-ovate, more or less acute at apex, crenate-dentate, sometimes incise, undulate and sinuate, more or less cordate at base; the lower cauline similar to basal, the upper cauline quickly decreasing in size, the lamina 4–15×2–8 cm, ovate, often asymmetrical and twisted, acute to caudate at apex, dentate, the teeth rounded, cordate-amplexicaul; all with sparse stellate hairs on upper surface and more numerous beneath, sometimes more dense and woolly on both surfaces. *Inflorescence* occupying one-third to half the plant, much branched, the branches almost erect, making the inflorescence narrowly ellipsoid; pedicels 5–12(–15) mm, dense stellate hairs, usually 1–2 flowers in each bract; bracts 4–7(–15) mm, linear, flowers bisexual, numerous, 22–25 mm in diameter, faintly honey-smelling. *Calyx* 2.5–5.0 mm, divided halfway to the base into 5 lobes, the lobes lanceolate and acute at apex, shortly stellately hairy. *Corolla* 12–22 mm in diameter, lobes 9–10×8–9 mm, broadly ovate and rounded at apex, 2 lower lobes 7–9×6–8 mm, narrowly obovate, rounded at apex. *Stamens* 5; filaments purple, all with dense purple glandular hairs; anthers orange, all reniform and transversely inserted. *Style* 1, pale green; stigma orange. *Capsule* 4–5 mm, ovoid-ellipsoid, woolly; seeds numerous, about 1 mm, elliptical. *Flowers* 6–7. Visited by bees. $2n=30$.

(i) Var. **nigrum**
V. nigrum var. *bracteosum* Pugsley

Leaves with sparse stellate hairs on upper surface, and more numerous beneath.

(ii) Var. **tomentosum** Bab.
Leaves with dense stellate hairs on both surfaces.

Native. Waste and rough ground, open places in grassland, usually on calcareous soil, and on railway and roadside banks. Locally common in central and southern Great Britain and the Channel Islands, a scarce casual elsewhere. Europe from Scandinavia and northern Russia to north Spain, north Italy and Macedonia, east to the Caucasus and Siberia to the Yenisei region. Our plant is subsp. **nigrum.** The common plant is var. *nigrum*. The distribution of var. *tomentosum* in not known.

× **olympicum** = **V.** × **oxoniense** Druce nom. nud.
This hybrid has the leaves linear-lanceolate or lanceolate and crenate.

V. nigrum is native, *V. olympicum* introduced. Recorded from a garden at Oxford.

× **phoeniceum** = **V.** × **ustulatum** Celak.
V. nigrum is native, *V. phoeniceum* introduced. The hybrid is recorded for Cambridgeshire at Babraham and Stapleford.

× **pulverulentum** = **V.** × **mixtum** Ramond ex DC.
V. × *wirtgenii* Franch.; *V.* × *schottianum* Schrad.

This hybrid has its leaves intermediate in hairiness between the parents, is usually branched and has purple hairs on the filaments.

Native. It occurs rarely with the parents in East Anglia and has occasionally been found elsewhere as a casual. It also occurs in central Europe.

× **pyramidatum**
Very similar to *V. pyramidatum* but with 2–5 flowers to each bract.

V. nigrum is native, *V. pyramidatum* introduced. Recorded from Suffolk and Cambridgeshire.

× **thapsus** = **V.** × **semialbum** Chaub.
V. × *collinum* Schrad., non. Salisb.

This hybrid is intermediate between its parents in most characters, but is variable, the leaves being less hairy than *V. thapsus*, crenate and usually petiolate as in *V. nigrum*. The inflorescence is simple or weakly branched, the filament hairs of the upper stamens are purple, of the lower white and the anthers are all reniform and medifixed. The hybrids are highly sterile with shrivelled pollen and ripe seeds are rarely formed.

Native. Frequent with its parents in much of Great Britain and by far our commonest hybrid. It is widespread in Continental Europe.

16. V. olympicum Boiss. Turkey Mullein

Biennial monoecious *herb. Stems* 100–150 cm, pale green, robust, angular, erect, with dense, appressed, white stellate hairs, often becoming glabrous, without glandular hairs, with numerous branches, leafy. *Leaves* alternate, medium green on upper surface, paler beneath, with white, adpressed stellate hairs sometimes becoming glabrous; basal 15–70×5–13 cm, narrowly to broadly lanceolate, acute or acuminate at apex, entire, sessile or with a short petiole; cauline similar but smaller, the upper long-acuminate at apex. *Inflorescence* with numerous thick branches, forming a broadly ovate-pyramidal panicle, with more or less congested clusters of 5–11 bisexual flowers; pedicels up to 16 mm. bracts linear to lanceolate, long-acuminate or caudate; bracteoles linear. *Calyx* 3–5 mm.

lobes linear to linear-lanceolate, acute at apex. *Corolla* 20–30 mm in diameter, without pellucid glands, stellate-tomentose outside. *Stamens* 5; filaments all with whitish-yellow wool, 2 anterior glabrous above; upper anthers reniform, 2 lower 1.0–1.5 mm, obliquely inserted. *Capsule* 4–7 × 3–5 mm, oblong-ovate, tomentose, becoming glabrous. *Flowers* 6–9.

Introduced. Garden escape. Oxfordshire, Cambridgeshire and Wiltshire. Endemic to north-west Turkey.

2. Scrophularia L.

Perennial, or less often *biennial*, monoecious *herbs* often with rhizomes. *Stems* erect, usually 4-angled and square in section. *Leaves* opposite, simple or rarely with lobes near the base, exstipulate. *Inflorescence* a panicle made up of cymes in the axils of bracts; flowers bisexual. *Calyx* 5-lobed, often with a scarious margin to the lobes. *Corolla* dull purplish, brownish or greenish-yellow, obscurely 2-lipped, the upper lip with 2 lobes, the lower lip with 3 lobes, or lobes more or less equal, the tube more or less globose. *Fertile stamens* 4, the 5th usually represented by a large staminode. *Style* 1, terminal; stigmas capitate. *Fruit* a 2-celled capsule.

Contains about 300 species in temperate Eurasia and 12 in North and tropical America.

1. Corolla lobes more or less equal and not forming 2 lips; staminode absent **5. vernalis**
1. Corolla lobes unequal, forming 2 weak lips; staminode present and longer than anthers 2.
2. Stems and leaves shortly hairy **4. scorodonia**
2. Stems and leaves more or less glabrous 3.
3. Stem angles not or scarcely winged; calyx lobes with a narrow, often obscure scarious margin **1. nodosa**
3. Stem angles distinctly winged; calyx lobes with a conspicuous scarious border 4.
4. Leaves cordate or broadly cuneate at base; staminode more or less suborbicular **2. auriculata**
4. Leaves cuneate to rounded, never cordate at base; staminode bifid or with 2 divergent lobes **3. umbrosa**

1. S. nodosa L. Common Figwort

Perennial monoecious *herb* with a short, swollen, nodular rhizome. *Stems* 40–80 cm, pale yellowish-green, often suffused dark purple, sharply quadrangular but not or scarcely winged, erect, glabrous, leafy. *Leaves* opposite; lamina 6–13 × 2–5 cm, dark slightly bluish-green on upper surface, paler beneath, more or less ovate, acute at apex, coarsely and unequally serrate, more or less truncate or cordate at base, but usually slightly and often unequally decurrent on the petiole, glabrous; petiole up to 30 mm, pale green, not winged, glabrous. *Inflorescence* a panicle made up of cymes in the axils of bracts; pedicels 2–3 times as long as the flowers; bracts with lowest pair similar to the leaves, 1 or 2 more pairs often leaf-like, the upper small, linear and alternate, rarely all leaf-like; rhachis and pedicels minutely glandular-hairy; flowers bisexual, 6–7 mm in diameter. *Calyx* 3.0–3.5 mm, 5-lobed, the lobes ovate, rounded-obtuse at apex, with a very narrow, often scarcely visible, scarious margin. *Corolla* 7–10 mm. tube greenish, nearly globose; 2-lipped, the upper purplish-brown and 2-lobed, the lower greenish and 3-lobed, rarely both lips greenish. *Fertile stamens* 4, bent downwards; filaments 5–6 mm, minutely glandular-hairy; anthers yellow; staminode 1, scale-like, brownish-purple, obovate, broader than long, retuse at apex. *Style* 1, 3.0–3.5 mm. stigma capitate. *Capsule* 6–8 mm, brown, ovoid, acuminate at apex; seeds 0.5–1.0 mm, pale brown, oblong, rugose. *Flowers* 6–9. Pollinated by wasps, less frequently by bees. $2n = 36$.

Native. Damp and wet woods, hedgerows and other open and shady places. Common throughout most of Great Britain and Ireland. Most of Europe southwards from 69° 48′ N in Norway; temperate Asia eastwards to the Yenisei region. Eurosiberian Temperate element.

2. S. auriculata L. Water Figwort

S. aquatica auct.

Perennial monoecious *herb* with a short, swollen rhizome which is not nodular. *Stems* 50–100 cm, pale yellowish-green, quadrangular and winged on each angle, erect, glabrous, leafy. *Leaves* opposite; lamina 6–12 × 2–5 cm, dark green on upper surface, paler beneath, ovate to elliptical, obtuse at apex, crenate, often with 1 or 2 basal lobes, cordate to broadly cuneate at base, glabrous or very shortly and sparsely hairy; petiole pale green, winged, glabrous. *Inflorescence* a panicle, made up of cymes in the axils of bracts; pedicels about as long as the flowers; bracts with the lower somewhat leaf-like but differing markedly from the leaves, the upper oblong; rhachis and pedicels somewhat glandular-hairy; flowers 5.5–6.0 mm in diameter. *Calyx* 5-lobed, the lobes ovate, rounded at apex, with a broad, conspicuous scarious margin, bisexual. *Corolla* 5–9 mm, tube greenish, globose, 2-lipped, the upper brownish-purple, and 2-lobed, the lower greenish and 3-lobed. *Fertile stamens* 4, bent downwards; filaments 4–5 mm, minutely glandular-hairy; anthers yellow; staminode scale-like, more or less suborbicular, or rather broader than long, entire. *Style* 1, 3.5–4.0 mm. stigma capitate. *Capsule* 4–6 mm, subglobose, apiculate at apex; seeds 0.8–1.0 mm, pale brown, oblong, rugose. *Flowers* 6–9. Pollinated by wasps, less frequently by other insects. $2n = 80$.

Native. Edge of ponds, streams, ditches and in wet woods and meadows. Common throughout England, Wales and the Channel Islands, frequent in Ireland, rare in Scotland and Isle of Man. Western Europe northward to the Netherlands; Italy, Sicily and Crete; Morocco and Tunisia; Azores. Suboceanic Southern-temperate element.

3. S. umbrosa Dumort. Green Figwort

S. ehrhartii Stevens; *S. alata* Gilib.

Perennial monoecious *herb* with a short, swollen rhizome which is not nodular. *Stems* 40–100 cm, pale yellowish-green, often suffused brownish-purple, quadrangular, the angles broadly winged, erect, glabrous. *Leaves* opposite; lamina 6–12 × 1.5–4.0 cm, dark green on upper surface, paler beneath, ovate to elliptical, obtuse to acute at apex, serrate, broadly cuneate to rounded, never cordate at base, glabrous; petiole up to 15 mm, pale green, winged, glabrous. *Inflorescence* a panicle made up of lax cymes in the axils of bracts; pedicels about as long as the flowers; bracts large and leaf-like; rhachis and pedicels glabrous; flowers 5–6 mm in diameter, bisexual. *Calyx* 4–5 mm, 5-lobed, the lobes broadly ovate, rounded at apex, with a toothed, scarious margin. *Corolla* 5–9 mm. tube greenish below, purplish above, globose, 2-lipped, the upper brownish-purple and 2-lobed, the lower greenish and 3-lobed. *Fertile stamens* 4, bent downwards; filaments 3.5–4.0 mm, minutely glandular-hairy, anthers yellow; staminode brownish-purple, with 2 diverging lobes so that it is much broader than long. *Style* 1, 2.0–2.5 mm. stigma capitate. *Capsule* 4–6 mm, subglobose, apiculate at apex; seeds 0.5–0.8 mm, pale brown, oblong, rugose. *Flowers* 7–9. $2n = 52$.

Native. Damp shady places. Scattered records in Great Britain and Ireland especially the Welsh borders, East Anglia, northern England, southern Scotland and east Ireland. Europe from Denmark and Latvia southwards, but absent from much of the west; Asia east to Tibet and south to Palestine. Eurosiberian Temperate element.

4. S. scorodonia L. Balm-leaved Figwort

Perennial monoecious *herb* with a short rhizome. *Stems* 60–100 cm, pale green, quadrangular but not winged on the angles, erect, densely clothed with short, greyish simple eglandular hairs, leafy. *Leaves* opposite; lamina 4–10 × 1.5–5.0 cm, pale to rather dark green on upper surface, paler beneath, ovate, obtuse to acute at apex, doubly dentate with mucronate teeth, cordate at base, rugose and covered with dense, very short, greyish simple eglandular hairs on both surfaces; petiole up to 50 mm, pale green, not winged, hairy. *Inflorescence* a panicle of few, lax-flowered cymes in the axils of bracts; pedicels 2 to 3 times as long as the flowers; bracts large and leaf-like; rhachis and pedicels with dense, greyish simple eglandular hairs mixed with glandular hairs; flowers 5–7 mm in diameter, bisexual. *Calyx* 4.5–5.5 mm, 5-lobed, the lobes ovate, rounded at apex, with a broad scarious margin. *Corolla* 8–11 mm, tube purple, globose, 2-lipped, the upper dull purple and 2-lobed, the lower dull purple and 3-lobed. *Fertile stamens* 4, bent downwards; filaments 3.5–4.5, minutely glandular-hairy, staminode suborbicular, entire. *Style* 1, about 3.0 mm. *Capsule* 6–8 mm, subglobose or ovoid, acute-apiculate; seeds 0.5–0.8 mm, pale brown, oblong, rugose. *Flowers* 6–8. $2n = 58$.

Introduced, but known in Jersey since 1689 and in Cornwall since 1712. Grassy field borders and hedgerows, waste places, quarries and old walls, very local and mostly near the coast. Devonshire and Cornwall, south Wales and a few localities in southern England, the commonest species of the genus in the Channel Islands. West France, Spain, Portugal, Madeira, Azores and north-west Morocco. Oceanic Southern-temperate element.

5. S. vernalis L. Yellow Figwort

Biennial or *perennial* monoecious, foetid *herb* with a short rhizome. *Stems* up to 50(–80) cm, pale yellowish-green, obscurely quadrangular, erect, with dense, soft glandular hairs, leafy. *Leaves* opposite, the lamina 4–15 × 2–11 cm, medium yellowish-green on upper surface, paler beneath, thin, broadly ovate, acute at apex, deeply dentate, cordate at base, softly hairy; petiole up to 12 cm, pale green, not winged, hairy. *Inflorescence* of cymes in the axils of the upper leaves, the cymes many-flowered and compact on long peduncles and with leaf-like bracts; pedicels shorter than calyx; flowers 5–7 mm in diameter, bisexual. *Calyx* 5-lobed, the lobes oblong-lanceolate, subacute at apex, with a scarious margin. *Corolla* 6–8 mm, greenish-yellow, not 2-lipped, the lobes very small and nearly equal. *Stamens* 4, bent downwards; filaments 4–7 mm, pale yellowish-green; anthers yellow; staminode absent. *Style* 1, greenish; stigma capitate. *Capsule* 8–10 mm, ovoid-conical; seeds, pale brown, ovoid, rugose. *Flowers* 4–6. Visited by bees. $2n = 28, 40$.

Introduced. Waste places and rough ground, hedges and woodland clearings. Scattered records throughout much of Great Britain. Native of the mountains of central and south Europe from south-central Germany and central Russia southwards to the Pyrenees, Sicily and central Yugoslavia; naturalised in the plains.

3. Phygelius E. Mey. ex Benth.

Small, evergreen, monoecious *shrubs*. *Stems* 4-angled, little branched. *Leaves* opposite, simple, crenulate-serrulate. *Inflorescence* a large, terminal, secund, erect panicle, bracteate and bracteolate. *Calyx* with 4 equal lobes. *Corolla* deep red outside, yellow within, with 5 nearly equal lobes. *Stamens* 4, exserted. *Style* 1. *Fruit* an ovoid, septicidal, more or less oblique capsule; seeds numerous small, reticulate.

Contains 2 species in South Africa.

1. P. capensis E. Mey. ex Benth. Cape Figwort

Evergreen, monoecious *shrub* up to 2 m. *Stems* 4-angled, brown, very square on vigorous vegetative shoots, little branched, glabrous. *Leaves* opposite, occasionally the upper alternate; lamina 2.5–6.5 × 1.5–3.0 cm, medium to dark green on upper surface, paler beneath, ovate or ovate-lanceolate, obtuse or subacute at apex, evenly crenulate-serrulate, broadly cuneate to subcordate at base, glabrous; petioles up to 3 cm, glabrous. *Inflorescence* a large, secund terminal, erect panicle; peduncles 10–25 mm, dark reddish, shortly glandular-hairy; pedicels shorter than peduncles, but similar; bracts up to 10 mm. bracteoles smaller; flowers bisexual. *Calyx* 4–6 mm, with 5 equal lobes, the lobes lanceolate to ovate, purplish, glabrous. *Corolla* 25–40 mm, deep red outside, yellow inside, the tube narrowly trumpet-shaped, curved or nearly straight, with 5 nearly equal lobes. *Stamens* 4; exserted; filaments red. *Style* 1, terminal; stigma capitate. *Capsule* 10–13 mm, shining black, ovoid, more or less oblique, furrowed; seeds 1.0–1.2 mm, oblong to ovoid, sometimes angled, reticulate. *Flowers* 7–10.

Introduced. Naturalised by rivers in Co. Wicklow since at least 1970, and more or less naturalised by garden boundaries in Kent, Dunbartonshire and Co. Galway. Native of South Africa.

4. **Paulownia** Siebold & Zucc.

Deciduous, monoecious *tree* with a broad, open crown. *Leaves* opposite, large, ovate-cordate. *Inflorescence* a large, terminal panicle; flowers bisexual. *Calyx* 5-lobed. *Corolla* trumpet-shaped, mauve, with 5 nearly equal lobes, just organised into 2 lips with 2 lobes above and 3 below, the tube much longer than the lobes. *Stamens* 4, included. *Style* 1, about equalling the anthers. *Fruit* a capsule.

Contains about 17 species in eastern Asia. Named after Princess Anna Paulowna (1795–1865), daughter of Czar Paul I of Russia.

Bean, W. J. (1976). *Trees and shrubs hardy in the British Isles.* Ed. 8. **3**. London.

1. P. tomentosa (Thunb.) Steud. Foxglove Tree.
Bignonia tomentosa Thunb.; *Incarvillea tomentosa* (Thunb.) Spreng.; *P. imperialis* Siebold & Zucc.; *P. tomentosa* var. *lanata* (Dode) C. D. Schneid.; *P. japonica* Réveil; *P. recurva* Rehder

Deciduous, monoecious *tree* up to 26 m, with open, broad, rather domed, rounded crown. *Trunk* up to 150 cm in circumference, soon forked, with large branches. *Bark* often purplish-grey with fine striations and broken orange blisters in young trees, grey and smooth when old or with shallow, longitudinal channels. *Main branches* usually spreading, few and fragile, subsidiary branches often upturned; twigs greyish-brown, thick and rigid, with numerous lenticels, felted; young shoots paler brown and more densely felted. *Buds* minute, purplish, above the large leaf-scars, terminal ones lacking. *Leaves* opposite; lamina 15–30 × 8–27 cm, dark, rather shiny green on upper surface, paler and duller beneath, turning yellowish in autumn, ovate, shortly acute or obtuse-pointed at apex, entire but sinuate and undulate on the margins or very shallowly palmately lobed, cordate at base with rounded lobes, minutely hairy on upper surface with a mixture of short simple eglandular and glandular hairs, also below, but more numerous especially on the veins; main veins almost palmately arranged, impressed above, prominent beneath; petioles 4–40 cm, pale yellowish-green, with numerous, unequal, slender glandular hairs. *Inflorescence* a large, terminal, pyramidal panicle up to 30 cm; flowers bisexual, forming in autumn, over wintering and opening before the leaves in May. *Calyx* 15–20 mm, divided two-thirds of the way to the base into 5 lobes, the lobes 8–10 × 5–7 mm, lanceolate, subacute at apex, yellowish-brown felted. *Corolla* 45–60 mm, trumpet-shaped, strong-smelling; tube 40–45 × 18–20 mm. limb with 2 lobes above and 3 below, mauve on outer side, paler on lower lobes inside and cream in the throat. *Stamens* 4; filaments unequal, 2 are about 22 mm and 2 are about 15 mm, white; anthers creamy-beige. *Style* 1, white, tinted mauve, about equalling anthers. *Capsule* 25–40 × 15–30 mm, brown, woody, ovoid, acute at apex. seeds 2, 25–35 × 9–11 mm, pale brown. *Flowers* 4–5. $2n = 40$.

Introduced in 1838. Planted in parks and by roads for ornament and naturalised in rough ground in Middlesex since at least 1990 and is recorded from Somersetshire, Hampshire, Kent and Yorkshire. Native of China.

5. **Mimulus** L.

Perennial monoecious *herbs* with leafy stolons. *Stems* branched. *Leaves* opposite, dentate. *Inflorescence* of bisexual flowers in the axils of leaves or bracts. *Calyx* tubular, 5-angled and shortly 5-lobed. *Corolla* with a long tube and markedly 2-lipped, the upper lip 2-lobed, the lower longer and 3-lobed, the lobes all flat, not spurred or pouched. *Stamens* 4; anthers 2-celled. *Style* 1; stigma with 2 flat lobes. *Fruit* a loculicidal capsule, included in the calyx; seeds small and numerous.

Contains about 100 species, mostly in temperate America, with a few in the Old World in eastern Asia, Australia, New Zealand and South Africa.

Grant, A. L. (1924). A monograph of the genus *Mimulus*. *Ann. Missouri Bot. Gard.* **11**: 99–388.
Hultén, E. & Fries, M. (1986). *Atlas of north European vascular plants north of the Tropic of Cancer.* 3 vols. Königstein.
Parker, P. F. (1975). *Mimulus* in Great Britain – a cytotaxonomic note. *New Phytologist*. **70**: 155–160.
Rich, T. C. G. & Jermy, A. C. (1998). *Plant crib.* London.
Roberts, R. H. (1964). *Mimulus* hybrids in Britain. *Watsonia* **6**: 70–75.
Roberts, R. H. (1968). The hybrids of *Mimulus cupreus*. *Watsonia* **6**: 371–376.
Silverside, A. J. (1990). A new hybrid binomial in *Mimulus* L. *Watsonia* **18**: 210–212.
Silverside, A. J. (1998). *Mimulus* section *Simiolus*. In Rich, T. C. G. & Jermy, A. C. *Plant Crib,* pp. 258–260. London.
Silverside, A. J. (2000). In Cullen, J. et al. (Edits.) *The European Garden Flora*. **6**: 275–280. Cambridge.

1. Calyx lobes more or less equal in length **1. moschatus**
1. Upper tooth of calyx distinctly longer than lower and lateral lobes 2.
2. Throat of corolla more or less closed by palate; corolla yellow, throat red- spotted but lobes unmarked 3.
2. Throat of corolla open; corolla cream or yellow to red or brown; lobes unmarked or variously spotted or blotched 4.
3. Capsule produced with copious good seed **2. guttatus**
3. Capsule undeveloped and ripe seeds not produced **3. × robertsii**
4. Capsule remaining undeveloped and very few or no ripe seed produced, minute white bristles present on at least the angles and base of the calyx, glandular hairs sometimes also produced 5.
4. Plant partly to fully fertile, though capsule may not develop in unpollinated flowers; calyx sometimes with minute glandular hairs or microscopic bristles, but otherwise without hairs 6.
5. Corolla yellow, though lobes often spotted or blotched with red or brown **3. × robertsii**
5. Corolla orange to bright red-brown, lobes unmarked **4. × burnetii**
6. Corolla lobes with multiple often confluent spots and blotches, rarely unmarked; lower pedicels of well-grown plants rarely exceeding 6 cm **6. × maculosus**
6. Central lobe of lower lip, of all 5 lobes with single red or brown blotches or, rarely blotches absent; lower pedicels of

well-grown plants usually exceeding 6 cm as the capsule matures 7.
7. Corolla unblotched or marked **5(i). luteus** var. **luteus**
7. Corolla blotched or marked 8.
8. Corolla with all lobes blotched with cherry or wine-red **5(ii). luteus** var. **youngianus**
8. Corolla with central lobe of lower lip and commonly all 5 lobes marked with a deep red or red-brown blotch **5(ii). luteus** var. **vivularis**

1. M. moschatus Douglas ex Lindl. Musk

Perennial monoecious *herb* with rhizomes and mat-forming. *Stems* 5–40 cm, pale green, decumbent to ascending, viscid, glandular-hairy, branched, leafy, rooting at the lower nodes. *Leaves* opposite; lamina 1–5 × 0.5–3.0 cm, medium green on upper surface, paler beneath, ovate or elliptical, with small, distant, shallow teeth or entire, narrowed at base, viscid glandular-hairy, shortly petiolate. *Inflorescence* a few-flowered, leafy raceme or bisexual flowers solitary in the axils of upper leaves; pedicels shorter than the leaves. *Calyx* 8–10 mm, campanulate, 5-lobed, the lobes 2–4 mm, somewhat unequal, lanceolate, glandular-hairy. *Corolla* 15–25 mm, pale yellow, equally 5-lobed; throat open, narrow, usually marked with red. *Stamens* 4, included. *Stigma* 2-lobed. *Capsule* 2-valved, oblong; seeds small, tuberculate. *Flowers* 6–9.

Introduced in 1826. Casual or naturalised in damp, often shaded places, including the muddy edges of ditches, wooded swamps, damp woodland rides, by ponds and in damp pastures. Scattered records throughout Great Britain and a very few records in Ireland. Native of western North America.

2. M. guttatus DC. Monkeyflower
M. langsdorffii Donn ex Greene; *M. luteus* auct.

Clumped or creeping *perennial* monoecious *herb* with leafy stolons produced in autumn. *Stems* up to 75 cm, pale yellowish-green, erect to ascending, hollow, terete to quadrangular in section, glabrous below, with numerous to dense, fine, unequal, pale glandular hairs above, sparingly branched or unbranched, leafy. *Leaves* opposite; lamina 2.5–5.5(–10.0) × 1.5–4.0(–7.0) cm, medium yellowish-green with paler veins on upper surface, paler beneath with prominent veins, broadly ovate to subrotund, obtuse to acute at apex, irregularly and sharply dentate, the teeth deltoid or becoming more irregular towards the base, rounded or cuneate at base, glabrous on both surfaces, the lower with petioles up to 6 cm, the upper sessile and subconnate. *Inflorescence* an elongate, bracteate raceme of bisexual flowers; pedicels pale green, 12–25 mm, short and with dense glandular hairs; uppermost bracts sessile, subrotund, cuspidate at apex and entire or finely serrate. *Calyx* 15–20 mm, densely glandular-hairy, campanulate, 5-toothed, the upper tooth larger than the other 4. *Corolla* 25–45 mm, bright yellow, often with red spots in the throat; 2-lipped, the upper lip divided halfway into 2 equal lobes, the lower lip with a large central lobe and 2 small side lobes, mouth of tube nearly closed by 2 large, hairy bosses. *Stamens* 4, didynamous, included; filaments white; anthers yellow. *Style* 1, white; stigma pale yellow. *Capsule* 7–9 mm, oblong, obtuse at apex, glabrous; seeds 0.4–0.5 mm, brown, oblong, finely striate. *Flowers* 7–9. Pollinated by bees. $2n = 28$.

Introduced. By lowland rivers and lakes and in damp meadows and marshy ground. Of scattered occurrence and locally common throughout Great Britain and the perimeter of Ireland, especially in the lowlands. Native of western North America from Alaska to Mexico.

3. M. × robertsii Silverside Hybrid Monkeyflower
M. guttatus × luteus
M. luteus auct.; *M. hybridus* auct.

Perennial monoecious *herb* with rhizomes or stolons, the stolons short, thick and brittle. *Stems* 10–60 cm, pale green, ascending to erect, rooting at the nodes, glabrous or sometimes sparsely downy above. *Leaves* opposite; lamina 1–10 × 1–7 cm, medium green on upper surface, paler beneath, ovate to broadly diamond-shaped, rounded to a short point at apex, dentate-serrate, often with acute, more or less forward-directed teeth alternating with shallow, indistinct teeth, with 5–7 veins from the base, the lower leaves with petioles, the upper sessile. *Inflorescence* a few- to many-flowered, leafy raceme of bisexual flowers; bracts small, broadly ovate, acute at apex, dentate; lower pedicels up to 3 times as long as their calyces at maturity; pedicels and calyx bases with sparse, minute, bristly hairs less than 1 mm, even when the rest of the inflorescence is hairless. *Calyx* 15–20 mm, campanulate, occasionally becoming inflated after corollas fall. *Corolla* 25–45 mm, bright yellow, 5-lobed, palate red-spotted, lobes commonly unmarked, or else blotched with red on the lower central lobe, or on all 5 lobes; lower lip held angled downwards, throat not closed by the ridges of the palate. *Stamens* 4, included. *Style* 1. *Capsule* undeveloped. *Flowers* 7–9. Sterile. $2n = 44, 45, 46, 54$.

Introduced. Along streamsides, particularly upland ones. Widespread in northern and western Great Britain and rare in the east and in Ireland, Garden origin. Named after Richard Henry Roberts (1910–2003).

4. M. × burnetii S. Arn. Coppery Monkeyflower
M. cupreus × guttatus
M. cupreus auct.

Perennial monoecious *herb* with rhizomes, sometimes becoming mat-forming. *Stems* up to 40 cm, sprawling to ascending, sometimes erect up to 60 cm. *Leaves* opposite; lamina 0.7–4 × 0.3–2.3 cm, diamond shape to ovate or elliptical, acute at apex, dentate, petiolate. *Inflorescence* a few-flowered, leafy raceme of bisexual flowers; pedicels 1.5–2.0 times the length of the calyx; pedicels and calyx bases with sparse, minute, bristly hairs less than 0.1 mm, even when the rest of the inflorescence is hairless. *Calyx* 15–20 mm, campanulate, not recorded as becoming inflated after corolla-fall. *Corolla* 35–50 mm, usually bright brownish-orange, rarely yellow or apricot, 5-lobed, the lobes unmarked; throat red-spotted, not closed by the ridge of the palate. *Stamens* 4, included. *Style* 1. *Capsule* undeveloped. *Flowers* 7–9. Sterile. $2n = 45$.

Introduced. Scattered records in west and north Great Britain. Of garden origin.

× **luteus**

This triple hybrid differs from *M.* × *burnetii* in its corolla lobes having dark blotches and from *M.* × *maculosus* in being glandular-hairy above.

Introduced. A garden escape of which there are scattered records throughout Great Britain. Of garden origin.

5. M. luteus L. Blood-drop-emlets
M. variegatus Lodd. nom. nud.; *M. rivularis* auct.; *M. guttatus* auct.; *M. punctatus* auct.

Perennial monoecious *herb* with a creeping rootstock forming mats. *Stems* up to 40 cm, pale green, prostrate and rooting at the nodes or ascending to erect, terete to obscurely 4-angled, brittle, glabrous or developing a covering of minute glandular hairs. *Leaves* opposite; lamina 1.5–4.0 × 1–3 cm, diamond-shaped or ovate, acute at apex, coarsely and irregularly dentate, the teeth often oblong and twisted, rarely more even, usually with 5–7 veins from the base, petiolate. *Inflorescence* a loose, few-flowered leafy raceme; pedicels 2.5–5.0 times the length of the calyx; flowers bisexual. *Calyx* 14–20 mm, campanulate, glabrous, rarely with short bristles, glossy, spotted; 5-lobed, the lobes unequal, broadly triangular, the upper lobe about 4 mm, the lower and lateral lobes 1–2 mm, becoming inflated in fruit with lower lobes curved upwards over lateral lobes. *Corolla* 35–45 mm, pale to mid-yellow, strongly 2-lipped, the lobes unmarked, or at least the central lobe of the lower lip or more commonly all 5 lobes each marked with a round, deep red to red-brown blotch; throat golden-yellow, dotted with red, not closed by the ridges of the palate. *Stamens* 4, included. *Style* 1. *Capsule* oblong; seeds finely furrowed. *Flowers* 7–9. Fertile. $2n=60, 62, 64$.

(i) Var. **luteus**
Leaves evenly toothed. *Corolla* unblotched. *Calyx* with very short bristles

(ii) Var. **youngianus** Hook. fil.
M. smithii Paxton

Leaves broadly and shallowly toothed. *Corolla* with all lobes blotched with cherry or wine-red.

(iii) Var. **rivularis** Lindl.
Leaves coarsely and irregularly dentate. *Corolla* with central lobe of lower lip or commonly all 5 lobes marked with a deep red or red-brown blotch.

Introduced. Rather uncommon in northern Great Britain south to Durham. Our most frequent plant is var. *rivularis*. Var. *luteus* appears only to occur in gardens. Var. *youngianus* is intermediate between the other two varieties and is naturalised occasionally.

6. M. × **maculosus** T. Moore Scottish Monkeyflower
M. cupreus × **luteus**

Perennial, monoecious *herb* with a creeping rootstock. *Stems* up to 40 cm, pale green, prostrate to ascending or erect, terete to obscurely 4-angled. *Leaves* opposite; lamina diamond-shaped or ovate, coarsely and irregularly dentate. *Inflorescence* a loose, few-flowered, leafy raceme; pedicels 1.5–3.0 times the length of the calyx; pedicels and calyx hairless or with minute glandular hairs; flowers bisexual. *Calyx* campanulate. *Corolla* 35–50 mm, yellow, rarely unmarked but usually red-blotched on the central lobe of the lower lip and with all lobes heavily marked with confluent, bright orange-brown to red spots; upper lobes often with a ring-shaped, unspotted area; throat red-spotted. *Stamens* 4. *Style* 1. *Flowers* 7–9. Partially fertile.

Introduced. Scattered records throughout Great Britain. Of garden origin.

6. Limosella L.

Annual monoecious *herbs*, usually with thin, leafless stolons which produce fresh rosettes at the nodes, occasionally perennating. *Leaves* all basal, narrow and entire. *Inflorescence* of inconspicuous, solitary bisexual flowers from the leaf-rosettes. *Calyx* with 5 small lobes. *Corolla* rotate, with a short tube and 4–5 lobes. *Stamens* 4; anthers 1-celled. *Style* 1. *Fruit* a septicidal capsule, the septum incomplete.

Contains about 15 species, almost cosmopolitan.

Hultén, E. & Fries, M. (1986). *Atlas of north European vascular plants north of the Tropic of Cancer.* 3 vols. Königstein.
Leach, S. J., Stewart, N. F. & Ballantyne, G. H. (1984). *Limosella aquatica* L. in Fife, a declining species making a come-back. *Watsonia* **15**: 118–119.
Stewart, A., Pearman, D. A. & Preston, C. D. (1994). *Scarce plants in Britain*. Peterborough. [*L. aquatica*.].
Wigginton, M. J. (Edit.) (1999). *British red data books*. Vol. 1. *Vascular plants*. Peterborough. [*L. australis*.]

1. Leaves with lamina of upper narrowly elliptical; calyx longer than corolla tube; corolla 2.5–3.0 mm, white to pale mauve, scentless **1. aquatica**
1. Leaves all with lamina linear-subulate; calyx shorter than corolla tube; corolla 3.5–4.0 mm, white with an orange tube, scentless **2. australis**

1. L. aquatica L. Mudwort

Annual monoecious *herb* with thin, leafless stolons, which are at first upright and then become horizontal and produce fresh rosettes at the nodes. *Leaves* dark green 5–20 × 2–6(–10) mm, elliptical, obtuse at apex, entire, narrowed at base and with a petiole several times as long as the lamina, the lamina of lower lanceolate-spathulate to subulate, all glabrous. *Inflorescence* of solitary bisexual flowers on pedicels from the leaf-rosettes; flowers 2.0–5.0 mm in diameter. *Calyx* about 2.0 mm campanulate, longer than the corolla tube, green, with 5 lobes, the lobes ovate, more or less acute at apex, glabrous. *Corolla* 2.5–3.0 mm, white to pale mauve, scentless, rotate, with the tube 1.5 mm and 4–5 lobes, the lobes triangular, acute at apex and with a few, long hairs. *Stamens* 4; filaments short; anthers 1-celled. *Style* 1, short; stigma medium-sized. *Capsule* about 2.5 mm, globose-ovoid to ellipsoid. *Flowers* 6–10. $2n=40$.

Native. Wet, sandy mud at the edges of pools, rivers, lakes, reservoirs, ditches, tracks and roadsides where water has stood. Scattered records in Great Britain and Ireland north to Aberdeenshire. Most of Europe except the Mediterranean region, but often rare and local; Egypt; Arctic and temperate Asia to India and Japan; Greenland; North America from Labrador to the North West territory and south in the mountains to Colorado and California. Circumpolar Boreo-temperate element.

× **australis**

This hybrid has a mixture of subulate and elliptical leaves, and is often more vigorous and perennial with intermediate flower characters. $2n = 30$.

It occurs with both parents at Morfa Pools, Glamorgan and is often abundant. Endemic.

2. L. australis R. Br. Welsh Mudwort

L. subulata E. Ives; *L. aquatica* var. *tenuifolia* auct.

Annual monoecious *herb*, with stolons or rhizomes above and below the soil, which occasionally perennate. *Leaves* pale green, in rosettes; the lamina 10–25 × 1.0–2.5 mm, all linear-subulate, acute at apex, attenuate into the petiole, glabrous. *Inflorescence* of solitary, bisexual flowers on pedicels from the leaf-rosettes; flowers 2–4 mm in diameter. *Calyx* about 2 mm, campanulate, shorter than corolla tube, green, with 5 lobes, the lobes triangular, acute at apex. *Corolla* 3.5–4.0 mm, white, with an orange tube, scented, the tube about 3 mm and contracted below the insertion of the stamens, with 4–5 lobes, the lobes ovate-ligulate at apex and with numerous short hairs. *Stamens* 4; filaments short; anthers pale. *Style* 1, long; stigma small. *Capsule* about 2.5 mm, nearly globose. *Flowers* 6–10. $2n = 20$.

Native. Wet sandy mud at the edges of pools where water has stood. Very local in Glamorganshire, Merionethshire and Caernarvonshire. Eastern North America from Labrador to Maryland, and widely in the southern hemisphere and north Pacific; but not known elsewhere in Europe. Oceanic Boreo-temperate element.

7. Calceolaria L.

Annual monoecious *herbs*. *Leaves* opposite, deeply dissected. *Inflorescence* a many-flowered cyme; flowers bisexual. *Calyx* lobes 4. *Corolla* strongly zygomorphic, 2-lipped, the upper lip arched or hooded, the lower lip inflated, much larger than the upper with infolded lower lobes, without a pouch. *Stamens* 2. *Style* 1. *Fruit* a dry capsule opening by 2 valves.

Contains about 300 species occurring from Mexico to South America.

C. integrifolia L. and **C. biflora** Lam. have both been recorded as garden escapes.

Molau, U. (1988). *Flora Neotropica* **47**: 1–246.
Kranzlin, F. (1907). In Engler, A. (Edit.), *Das Planzenreich* **28**: 21–122.

1. C. chelidonioides Kunth Slipperwort

Annual monoecious *herb*. *Stems* 10–200 cm, pale green, somewhat succulent, erect or ascending, well-branched. *Leaves* opposite; lamina 5.5–25 × 4–19 cm, deeply pinnately dissected to the midrib, serrate on the margin; petiole 1.3–12.0 cm, narrowly winged. *Inflorescence* a few-flowered cyme; flowers bisexual; pedicels 7–23 cm. *Calyx* 4-lobed, green, the lobes ovate, dentate on margin, sometimes tinged purple. *Corolla* 10–14 mm, 2-lipped, the upper lip 2–4 × 2–4 mm, the lower lip 6–17 × 4–15 mm, pale to bright yellow, inflated. *Stamens* 2; connective of anther without an outer tooth. *Style* 1. *Capsule* 5–9 mm, pale green to purplish, opening by 2 valves. *Flowers* 6–8. $2n = 60$.

Introduced. Casual on tips, waste places and cultivated ground. Scattered records in Great Britain north to southern Scotland, especially in East Anglia. Native of Central and South America.

8. Antirrhinum L.

Short-lived, monoecious *perennial herbs* often killed in winter. *Stems* usually much branched. *Leaves* entire, the lower opposite, the upper alternate. *Inflorescence* a terminal raceme; flowers bisexual. *Calyx* equally 5-lobed, shorter than the corolla tube. *Corolla* tubular with a broad, rounded pouch at the base, strongly 2-lipped, the upper lip 2-lobed, the lower lip 3-lobed with a projecting palate closing the mouth of the tube. *Stamens* 4, completely enclosed in the corolla. *Style* 1. *Fruit* a capsule with 2 unequal cells, opening by 3 apical pores.

Contains over 40 species, mainly in the Mediterranean region, a few extending further north and in western North America. Widely cultivated for decoration.

1. A. majus L. Snapdragon

Tufted *perennial* monoecious *herb* with a tap-root, spreading by basal shoots but not long-living, and often behaving as annuals. *Stems* 30–80 cm, pale, shiny yellowish-green, fairly robust and fleshy, often woody at base, glabrous below and with numerous, pale glandular hairs above, branched at base and forming a bushy habit, very leafy. *Leaves* opposite below, alternate above; lamina 2–6 × 0.7–2.5 cm, rather dark, shiny green on upper surface, paler and dull beneath, lanceolate or linear-lanceolate, acute at apex, entire, glabrous and cuneate to attenuate at base, the midrib impressed above and prominent beneath, the petioles up to 10 mm, pale green and glabrous. *Inflorescence* a terminal raceme, the bisexual flowers in the axils of bracts; bracts 1.0–2.5 × 0.5–1.0 cm, ovate or lanceolate, more or less acute at apex, entire, narrowed at base, sessile, with unequal, pale glandular hairs; peduncle up to 8 mm, pale green, with dense, pale glandular hairs. *Calyx* 8–10 mm, equally 5-lobed, the lobes ovate and obtuse at apex, covered with dense, pale eglandular hairs. *Corolla* up to 45 mm, pale or dark, yellow, through orange to pink, red and purple, rarely white or a combination of colours, tubular below, saccate at base, strongly 2-lipped, the upper 2-lobed, the lower up to 40 mm wide, 3-lobed, and with a projecting palate closing the mouth of the corolla, glandular-hairy. *Stamens* 4, completely enclosed in the corolla; filaments white; anthers yellow. *Style* 1, white tinted pink; stigma whitish. *Capsule* 10–12 mm, ovoid, with 2 unequal cells, opening by 3 apical pores. *Flowers* 7–10. Pollinated by bumble-bees. Self-fertile or self-sterile, homogamous. $2n = 16$.

The garden varieties can be divided into 4 groups. Tall-growing plants with semi-double flowers and ruffled petals. Intermediate plants with numerous branches and dense flower spikes often with double flowers. Low-growing plants with spreading branches and large flowers. Dwarf plants have a profusion of blossom. It is not possible to fit the garden plants into wild infra-specific taxa due to hybridisation and introgression.

Introduced. Garden escapes on rough ground, tips, walls, rocks and buildings. Frequent throughout Great Britain and Ireland, but often killed by winter conditions. Native of south-west Europe extending to Sicily.

9. Chaenorhinum (DC. ex Duby) Rchb.

Linaria section *Chaenorhinum* DC. ex Duby

Annual or tufted *perennial* monoecious *herbs* with basal new shoots. *Stems* simple or branched. *Leaves* opposite below, mostly alternate above, entire, with a single midrib. *Inflorescence* of solitary flowers in the axils of upper leaves forming a terminal raceme. *Calyx* slightly unequally 5-lobed, more than half as long to as long as the corolla tube. *Corolla* tubular with a pouch at the base, 2-lipped, the upper lip 2-lobed, the lower lip 3-lobed, with a narrow, conical spur at base and a palate which does not completely close the opening to the throat. *Stamens* 4. *Style* 1. *Fruit* a capsule opening by large, irregular apical pores; seeds longitudinally ridged.

Contains about 20 species, all except 1 confined to the Mediterranean region.

Grime, J. P. et al. (1988). *Comparative plant ecology*. London.
Hultén, E. & Fries, M. (1986). *Atlas of north European vascular plants north of the Tropic of Cancer.* 3 vols. Königstein.

1. Annual; leaves linear to oblong-lanceolate; corolla 6–9 mm, purple **1. minus**
1. Perennial; leaves ovate or subrotund; corolla 8–15(–20) mm, bluish-lilac **2. origanifolium**

1. C. minus (L.) Lange Small Toadflax
Antirrhinum minus L.; *C. viscidum* (Moench) Simonk.

Annual monoecious *herb* with fibrous roots. *Stems* 8–40 cm, pale green, sometimes tinted brownish-purple, erect, usually shortly glandular-hairy, rarely glabrous, with ascending branches, leafy. *Leaves* mostly alternate; lamina 0.5–3.5 × 0.1–0.5 cm, medium green on upper surface, paler beneath, often suffused purple, linear to oblong-lanceolate, obtuse at apex, entire, narrowed at base, with a single midrib, more or less glandular-hairy; petiole short, glandular-hairy. *Inflorescence* of solitary flowers in the axils of the upper leaves; pedicels 3–20 mm in fruit, erect or ascending; flowers bisexual. *Calyx* 2–5 mm, medium green, divided almost to the base into 5 slightly unequal lobes, the lobes more or less linear, obtuse at apex, glandular-hairy. *Corolla* including spur 6–9 mm, purple outside, paler within, tube cylindrical, with a pouch at the base, with a short, narrow, obtuse spur 1.5–2.5 mm, the limb 2-lipped, the upper lip 2-lobed, the lower lip 3-lobed, with a pale yellow palate, but the mouth fairly open. *Stamens* 4, included; filaments pale; anthers yellow. *Style* 1, terminal, green; stigma capitate. *Capsule* 3–6 mm, ovoid, shorter than the calyx, opening by large, irregular apical pores; seeds 0.5–1.0 mm, pale brown, ovoid, longitudinally ridged. *Flowers* 5–10. Probably self-pollinated. $2n = 14$.

Introduced. Weed of cultivated and waste places, old walls, quarries and particularly common along railways. Frequent over most of Great Britain and Ireland except north Scotland. Europe except the Mediterranean islands, Iceland, Faeroes and Russia; western Asia east to the Punjab. Our plant is subsp. **minus** which occurs throughout the range of the species.

2. C. origanifolium (L.) Kostel. Malling Toadflax
Antirrhinum origanifolium L.; *Linaria origanifolia* (L.) DC.

Perennial monoecious *herb*. *Stems* up to 35(–50) cm, pale green, numerous, ascending to erect, glabrous below, sometimes hairy above, simple or branched, leafy. *Leaves* opposite below, alternate above; lamina up to 2.2(–2.8) × 1.0(–1.3) mm, medium green on upper surface, subrotund to ovate, rarely lanceolate, acute to obtuse at apex, entire, glandular-hairy. *Inflorescence* a terminal, lax, bracteate raceme; flowers bisexual; bracts linear to lanceolate; pedicels usually erecto-patent in fruit, glandular-hairy. *Calyx* 5–8 mm, deeply and somewhat unequally 5-lobed, the lobes linear-spathulate to oblanceolate, glandular-hairy. *Corolla* 8–15(–20) mm, bluish-lilac with violet lines and pale yellow palate, 2-lipped, the upper lip 2-lobed, the lower lip 3-lobed; spur 2–5(–6) mm, cylindrical or somewhat inflated. *Stamens* 4, didynamous, included. *Style* 1, green, terminal. *Capsule* (2.5–)3.0–5.0 mm, subglobose; seeds 0.5–0.9 mm, numerous, the ribs low or prominent, smooth or denticulate. *Flowers* 5–10. $2n = 14$.

Introduced. Well naturalised on old walls at West Malling in Kent since 1880, rare and impermanent in a few similar places in Devonshire, Somersetshire, Sussex, Surrey, Caernarvonshire and Westmorland. Native of south-west Europe. Our plant is subsp. **origanifolium**.

10. Misopates Raf.

Annual monoecious *herbs*. *Stems* simple or branched. *Leaves* opposite, entire, with a single midrib. *Inflorescence* of subsessile bisexual flowers in the axils of the upper leaves, forming a leafy, terminal raceme. *Calyx* unequally 5-lobed, all lobes longer than the corolla tube. *Corolla* tubular, with a broad, rounded pouch at the base, strongly 2-lipped, the upper lip 2-lobed, the lower lip 3-lobed with a projecting palate closing the mouth of the tube. *Stamens* 4. *Style* 1. *Fruit* a capsule with 2 unequal cells, opening by 3 apical pores; seeds somewhat flattened, with 1 face smooth, keeled and produced into a narrow wing, the other finely tuberculate and with a wide, raised, sinuate border.

Contains 2 species in the Mediterranean region to Cape Verde Islands to Ethiopia and north-west India.

Hultén, E. & Fries, M. (1986). *Atlas of north European vascular plants north of the Tropic of Cancer.* 3 vols. Königstein.
Wilson, P. J. (1991). Britain's arable weeds. *British Wildlife* **3**: 149–161.

1. Plant glandular-hairy; corolla 10–17 mm, usually bright pink, rarely white **1. orontium**
1. Plant glabrous; corolla 18–22 mm, pale pink to white **2. calycinum**

1. M. orontium (L.) Raf. Weasel's-snout
Antirrhinum orontium L.

Annual monoecious *herb* with a tap-root and fibrous side-roots. *Stems* 20–50 cm, greyish-green, erect, with dense, short glandular hairs above, simple or branched sometimes from base, leafy. *Leaves* all cauline, opposite below, alternate above; lamina 30–70 × 1.5–5.0 mm, dull medium green on upper surface, paler beneath, linear or narrowly elliptical, obtuse at apex, entire, attenuate at base, glabrous or ciliate with simple eglandular hairs, sessile. *Inflorescence* of subsessile, bisexual flowers in the axils of the upper leaves forming a leafy, terminal raceme. *Calyx* 18–22 mm, greyish-green, divided nearly to the base into 5 linear lobes, all longer than the corolla tube, shortly glandular-hairy. *Corolla* 10–17 mm, pinkish or rarely white; upper lip 8–10 mm wide, 2-lobed, marked with dark red streaks which converge at the opening of the flower, which is closed by the palate; lower lip 3-lobed, coloured similar to upper but the streaks are all well marked and at its summit is a pale yellow nectar-guide, from either side of which a white zone runs along the lower edge of the closed entrance; the red streaks of both lips continue into the corolla tube which is 6–8 mm and produces at its base a sort of spur, the upper lip interlocks by means of a keel-like projection with a corresponding depression in the lower lip, thus closing the flower more firmly. *Stamens* 4, 2 long projecting beyond the stigma, and 2 short. *Style* 1, bent at the end. *Capsule* 9–10 mm, pale green, ovoid, with dense, short glandular hairs and longer, pale simple eglandular hairs; seeds about 1.5 mm, numerous, pale green. *Flowers* 7–10. Pollinated by bees or selfed; homogamous. $2n = 14$, 16.

Introduced. Weed of cultivated or waste ground. Locally frequent in southern England and Wales, few records and decreasing in central and north Wales and central and north England, casual in Scotland and Ireland. South, west and central Europe, extending east to the Himalaya and south to the Canary Islands and Abyssinia.

2. M. calycinum Rothm. Pale Weasel's-snout

Annual monoecious *herb* with a tap-root and fibrous side-roots. *Stems* 30–80 cm, pale green, erect, usually glabrous, simple or branched, leafy. *Leaves* all cauline, opposite below, alternate above; lamina 3–7 × 0.1–0.5 cm, dull medium green on upper surface, paler beneath, linear to oblong-elliptical, subacute at apex, entire, attenuate at base, glabrous, sessile. *Inflorescence* of subsessile, bisexual flowers in the axils of the upper leaves, forming a dense, leafy, terminal raceme in flower, elongating in fruit. *Calyx* 15–20 mm, greyish-green, divided nearly to the base into 5 lobes, the lobes linear, glabrous. *Corolla* 18–22(–27) mm, exceeding the calyx, pale pink to white; 2-lipped, the upper lip 2-lobed, the lower lip 3-lobed. *Stamens* 4, 2 long and 2 short. *Style* 1. *Capsule* 6–8 mm, ovoid, sometimes glabrous. *Flowers* 7–10.

Introduced. Bird-seed alien on tips and waste ground. A few scattered records in Great Britain. Native of the west Mediterranean region.

11. Asarina Mill.

Perennial monoecious *herbs. Stems* procumbent, woody at base. *Leaves* alternate, crenate-dentate to shallowly lobed. *Inflorescence* of solitary bisexual flowers in leaf axils. *Calyx* divided nearly to base into 5 lobes. *Corolla* with whitish tube, veined purple, 2-lipped, the lips pale yellow, the upper lip 2-lobed, the lower lip 3-lobed, with a deep yellow palate in the throat. *Stamens* 4, 2 long and 2 short. *Style* 1. *Fruit* a capsule, opening by 2 apical pores.

One species in Europe.

Briggs, M. (1986). *Asarina procumbens* Miller. *B.S.B.I. News* **43**: 20–21.

1. A. procumbens Mill. Trailing Snapdragon
A. lobelii Lange; *Antirrhinum asarina* L.

Perennial monoecious *herb. Stems* 10–60 cm, pale green, woody at base, procumbent, glandular-hairy, leafy. *Leaves* alternate; lamina up to 5 × 6 cm, medium green on upper surface, paler beneath, ovate to reniform, rounded at apex, crenate-dentate, sometimes slightly palmatifid, cordate at base, glandular-hairy; petiole about as long as lamina, glandular-hairy; palmately veined. *Inflorescence* of solitary bisexual flowers in leaf axils; pedicels 12–20 mm, glandular-hairy. *Calyx* 10–13 mm, divided almost to base into 5 lobes, the lobes lanceolate, acute at apex, slightly unequal, glandular-hairy. *Corolla* 30–35 mm, tube whitish, slightly veined with purple; 2-lipped, the upper lip 2-lobed, the lower lip 3-lobed, the lips pale yellow, the palate in the throat deep yellow, glabrous except for the palate. *Stamens* 4, 2 long and 2 short. *Style* 1. *Capsule* 8–10 mm, shorter than the calyx, subglobose, glabrous. *Flowers* 6–8. $2n = 18$.

Introduced. Naturalised on dry banks, walls and cliffs. Scattered records throughout Great Britain. Native of the mountains of the south of France and north-east Spain.

12. Cymbalaria Hill

Perennial monoecious *herbs* with trailing or drooping stems. *Leaves* alternate, entire or shallowly lobed, palmately veined and petiolate. *Inflorescence* of solitary, bisexual, axillary flowers. *Calyx* 5-lobed. *Corolla* with a cylindrical tube and a 2-lipped limb, the upper lip 2-lobed, the lower lip 3-lobed with a palate closing the mouth of the tube. *Stamens* 4, included. *Style* 1. *Fruit* a capsule opening by 2 lateral pores, each pore with 3 valves.

Contains about 15 species in the Mediterranean region and western Europe.

Chevalier, A. (1856). Les espèces élémentaires françaises du genre *Cymbalaria*. *Bull. Soc. Bot. Fr.* **83**: 638–653.

1. Stems long and trailing; corolla 9–15 mm (including spur); spur 1.5–3.0 — 2.
1. Stem usually not trailing; corolla 15–25 mm (including spur); spur 4–9 mm — 3.
2. Plant glabrous or slightly hairy on calyx and young parts — **1(a). muralis** subsp. **muralis**
2. Plant hairy on more or less all its parts — **1(b). muralis** subsp. **visianii**
3. Stem, leaves, petioles and calyx with dense, short hairs — **2. pallida**
3. Plant glabrous or nearly so — **3. hepaticifolia**

1. C. muralis P. Gaertn., B. Mey. & Scherb. — Ivy-leaved Toadflax
Linaria cymbalaria (L.) Mill.; *C. cymbalaria* (L.) Druce nom. illegit.; *Antirrhinum cymbalaria* L.

Perennial monoecious *herb* with fibrous roots. *Stems* 10–80 cm, trailing or drooping, pale green often suffused purple, glabrous or hairy and rooting. *Leaves* nearly all alternate; lamina 1.0–5.5 × 1.5–6.5 cm, pale yellowish-green on upper surface, often similar beneath, but sometimes suffused purple, thick, ovate, reniform or subrotund, obtuse at apex, (3–)5(–9)-lobed, more or less cordate at base, palmately veined, glabrous or hairy; petiole 1.0–9.5 cm, pale green, glabrous or hairy. *Inflorescence* of solitary bisexual flowers in the axils of leaves; pedicels in flower bending towards the light, when in fruit bending the other way, glabrous or hairy. *Calyx* 2.0–2.5 mm, pale green or purplish, 5-lobed, the lobes linear-lanceolate, pointed at apex, glabrous or hairy. *Corolla* including spur 9–15 mm, lilac to violet, rarely white, tube cylindrical with a curved spur 1.5–3.0 mm at base, the limb 2-lipped, the upper lip 2-lobed with darker lines and a yellow spot, the lower lip 3-lobed with a yellowish or white palate closing the mouth of the tube. *Stamens* 4, included; filaments pale; anthers cream. *Style* 1, terminal; stigma 2-lobed. Capsule 3.5–4.0 mm, globose, opening by 2 lateral pores, each pore with 3 valves; seeds fairly numerous, about 1 mm, ovoid, with thick, flexuous ridges. *Flowers* 5–9. Pollinated by bees; self-compatible. $2n = 14$.

(a) Subsp. **muralis**
Plant glabrous or sparsely hairy on the calyx and young parts. $2n = 14$.

(b) Subsp. **visianii** (Kümmerle ex Jáv.) D. A. Webb
C. muralis forma *visianii* Kümmerle ex Jáv.; *Linaria cymbalaria* var. *pilosa* Vis.; *C. muralis* var. *pilosa* (Vis.) Degen

Plant hairy in more or less all its parts.

Introduced. Naturalised on walls, pavements, rocky or stony banks and waste places. Subsp. *muralis* was first recorded in 1640 and now occurs throughout most of Great Britain and Ireland. Subsp. *visianii* has been naturalised on waste ground at Wisley in Surrey since 1970. The species is native in the southern Alps, west Yugoslavia, central and southern Italy and Sicily; widely naturalised elsewhere. Subsp. *muralis* occurs throughout the range of the species except parts of Sicily. Subsp. *visianii* is native to central and south Italy and west Yugoslavia.

2. C. pallida (Ten.) Wettst. — Italian Toadflax
Antirrhinum pallidum Ten.

Perennial monoecious *herb* with fibrous roots. *Stems* up to 20 cm, pale yellowish-green, usually more or less hairy, leafy. *Leaves* mostly opposite; lamina up to 2.5 × 3.0 cm, medium green on upper surface, paler beneath, subrotund to semirotund-deltate, rounded to subacute at apex, entire to 5-lobed, the lobes deltate and rounded to subacute, the central usually much the largest, usually more or less hairy; petioles long and as upper internodes are short the leaves overtop the stem apex. *Inflorescence* of single bisexual flowers in the axils of leaves. *Calyx* 3–4 mm, densely hairy, deeply 5-lobed, the lobes somewhat unequal in length. *Corolla* 15–25(–30) mm, pale lilac-blue, tube cylindrical, with a spur 6–9 mm, 2-lipped, the upper lip 2-lobed, the lower lip 3-lobed. *Stamens* 4, included. *Style* 1, terminal; stigma 2-lobed. *Capsule* slightly exceeding calyx, globose, glabrous; seeds about 1 mm, ovoid, with acute, longitudinal ridges. *Flowers* 5–9. $2n = 14$.

Introduced. Naturalised on walls, shingle and stony places. Scattered records in Great Britain, mainly in northern England and southern Scotland. Native of the mountains of central Italy.

3. C. hepaticifolia (Poir.) Wettst. — Corsican Toadflax
Antirrhinum hepaticifolium Poir.; *Linaria hepaticifolia* (Poir.) Steud.

Perennial monoecious *herb* with fibrous roots. *Stems* up to 20 cm, pale yellowish-green, glabrous, leafy. *Leaves* opposite below, usually alternate above; lamina up to 2.5 × 3.0 cm, medium green on upper surface, paler beneath, subrotund to deltate, rounded to subacute at apex, entire to 5-lobed, the lobes deltate and rounded to subacute, glabrous; petioles long. Inflorescence of single, bisexual flower in the axils of leaves. *Calyx* 4–5 mm, glabrous or slightly hairy, deeply 5-lobed, the lobes somewhat unequal in length. *Corolla* 15–18 mm, pale lilac-blue, tube cylindrical with a spur 4–5 mm, 2-lippled, the upper lip 2-lobed, the lower lip 3-lobed. *Stamens* 4, included. *Style* 1, terminal; stigma 2-lobed. *Capsule* slightly shorter than calyx, globose; seeds about 1 mm, ovoid, with longitudinal ridges. *Flowers* 5–9. $2n = 98$.

Introduced. Marginally naturalised in and near gardens and nurseries, perhaps overlooked for *C. pallida*. Few scattered records in England and Scotland. Native of Corsica.

13. Kickxia Dumort.

Annual monoecious *herbs. Stems* decumbent or procumbent. *Leaves* opposite or alternate, entire to remotely dentate, with pinnate venation. *Flowers* solitary, bisexual and axillary. *Calyx* with 5 equal lobes about as long as the corolla tube. *Corolla* tube with a narrowly conical spur at base, the mouth completely closed by a boss-like swelling (palate) on the lower lip; limb 2-lipped, the upper lip 2-lobed, the lower lip 3-lobed. *Stamens* 4. *Style* 1. *Fruit* a subglobose or depressed globose capsule, opening by 2 large, oblique lids; seeds alveolate.

Contains about 25 species in Europe, Africa and western Asia.

Michalet and Ascherson (Knuth, P. (1909) *Handbook of flower pollination,* vol. 3, p. 177) describe thin, short, twisted branches which arise is the axils of the lower leaves, which bury themselves in the earth, and bear subterranean cleistogamous flowers with a reduced corolla. These are usually present in both our species.

Hultén, E. & Fries, M. (1986). *Atlas of north European vascular plants north of the Tropic of Cancer.* 3 vols. Königstein.

1. Leaves ovate, sagittate or hastate; pedicels usually hairy only just under the flower 2.
1. Leaves broadly ovate, ovate-lanceolate or subrotund; pedicels hairy throughout 3.
2. Usually rather sparsely hairy, secondary branches usually none; leaves mostly hastate, more or less acute **1(a). elatine** subsp. **elatine** var. **elatine**
2. Plant densely hairy, with few to numerous secondary branches; leaves more rounded **1(b). elatine** subsp. **crinita** var. **prestandreae**
3. Stems usually simple; calyx lobes ovate-oblong, slightly cordate; capsule 5 mm wide **2 (a). spuria** subsp. **spuria**
3. Stems with slender, flexuous, small-leaved, lateral flowering branches; calyx lobes ovate, distinctly cordate; capsule 3–4 mm wide **2 (b). spuria** subsp. **integrifolia**

1. K. elatine (L.) Dumort. Sharp-leaved Fluellen
Antirrhinum elatine L.; *Linaria elatine* (L.) Mill.

Annual monoecious *herb* with fibrous roots. *Stems* 20–50 cm, pale green, decumbent or procumbent, slender, with short, slender glandular hairs and long, pale simple eglandular hairs, sometimes rather sparse, sometimes dense, branched from base, leafy. *Leaves* steadily getting smaller towards end of branches; lamina 0.5–2.0 × 0.3–1.5 cm, medium yellowish-green on upper surface, paler beneath, ovate, sagittate or hastate, acute at apex, entire, truncate or cordate at base, with numerous sparse to dense, very short to long, pale simple eglandular hairs and a few, short glandular hairs; petioles not more than half as long as the lamina, pale green, with short and long hairs. *Flowers* bisexual, solitary in the leaf axils; pedicels hairy only just under the flower. *Calyx* 4–5 mm, divided almost to the base into 5 subequal lobes, the lobes medium green, lanceolate, acute at apex, clothed with short glandular and simple eglandular hairs. *Corolla* 7–12 mm, the tube cylindrical with a straight spur at the base, the limb 2-lipped, the upper lip pale purple and 2-lobed, the lower lip yellow and 3-lobed with a yellow palate at the base closing the tube. *Stamens* 4, didynamous, included; filaments white; anthers cream. *Style* 1; stigma cream. *Capsule* 4.0–4.5 mm, subglobose, thin-walled; seeds 0.6–0.7 mm, ellipsoid, alveolate. *Flowers* 7–10. Probably self-incompatible. $2n = 36$.

(a) Subsp. **elatine**
Main branches decumbent, weak, usually rather sparsely hairy, secondary branches very occasional or none. *Leaves* mostly hastate, acute to mucronate. *Pedicels* glabrous except just below the flower, 3–6 times as long as the calyx in flower, up to 3 cm in fruit. *Corolla* 7–10 mm. $2n = 36$.

(b) Subsp. **crinita** (Mabille) Greuter
K. elatine subsp. *sieberi* (Rchb.) Hayek; *Linaria sieberi* Rchb.; *Linaria crinita* Mabille

Main branches ascending, relatively stout, densely hairy, with few to numerous short, patent, flowering secondary branches. *Leaves* more or less obtuse, the lower ovate to indistinctly hastate, the middle and upper usually hastate. In our plant which is var. **prestandreae** (Guss.) P. D. Sell the pedicels are glabrous or nearly so, almost as long as in subsp. *elatine*.

Introduced. Arable fields and waste ground, usually on calcareous soils. Formerly common in Great Britain north to North Wales and Lincolnshire, but much reduced by the use of herbicides; rather local in south and west Ireland. West and central Europe; Macaronesia; naturalised in North America. Our common plant is probably subsp. *crinita* var. *prestandreae* which seems to be midway between subsp. *crinita* and subsp. *elatine*. Only a few plants have been seen which seem to be subsp. *elatine*. Var. *prestandreae* occurs also in Spain, Portugal and France. It is named after Antonio Prestandrea (1817–1854).

2. K. spuria (L.) Dumort. Round-leaved Fluellen
Antirrhinum spuria L.; *Linaria spuria* (L.) Mill.

Annual monoecious *herb* with fibrous roots. *Stems* 20–50 cm, pale green, decumbent or procumbent, slender to fairly robust, with numerous, short glandular hairs and long, pale simple eglandular hairs, sometimes simple, sometimes with long, lateral flowering branches, leafy. *Leaves* with lamina 2.0–7.5 × 2–6 cm, medium yellowish-green on upper surface, paler beneath, broadly ovate, ovate-lanceolate or subrotund, rounded or shortly acute at apex, entire or the lowest remotely denticulate, truncate, rounded or cordate at base, with short to long, dense, pale, soft simple eglandular hairs and some short glandular hairs; petioles very short, pale green and long-hairy. *Flowers* bisexual, solitary in the leaf axils; pedicels hairy throughout. *Calyx* divided almost to base into 5 lobes, the lobes 3–5 × 1.5–3.0 mm in flower, accrescent and 4.5–8.0 × 2–4 mm in fruit, ovate to ovate-oblong, acute at apex, more or less cordate at base in fruit. *Corolla* 10–15 mm, tube cylindrical with a curved spur at the base, the limb 2-lipped, the upper lip deep purple and 2-lobed, the lower lip yellow and 3-lobed. *Stamens* 4, didynamous, included; filaments pale greenish; anthers cream. *Style* whitish; stigma cream. *Capsule* 3–5 mm wide, depressed-globose, sometimes slightly emarginate; seeds 0.7–1.2 mm, oblong-ellipsoid, alveolate. *Flowers* 7–10. $2n = 18$.

(a) Subsp. **spuria**
Plant sparsely to moderately hairy. *Stems* usually simple. *Calyx lobes* 4–5 × 2–3 mm in flower, 5–8 × 2–4 mm in fruit, ovate-oblong, slightly cordate. *Capsule* 5 mm wide; seeds 1.0–1.2 mm.

(b) Subsp. **integrifolia** (Brot.) R. Fernandes
Antirrhinum spurium var. *integrifolium* Brot.

Plant densely hairy throughout. *Stem* with numerous slender, flexuous, small-leaved lateral flowering branches. *Calyx lobes* 3–4 × 1.5–2.5 mm in flower, 4.5–6.0 × 2–3 mm in fruit, ovate, distinctly cordate. *Capsule* 3–4 mm wide; seeds 0.7–1.0 mm.

Introduced. Arable fields and waste places; formerly abundant all over fields of cereal crops, but much reduced by use of herbicides and now usually only found round the margins of fields or in root crops. South-east England north to Lincolnshire and a few records in south Wales, casual elsewhere. South, west and central Europe; North Africa; Macaronesia. Subsp. *spuria* occurs throughout most of the range of the species, but is rare in the Mediterranean region. Subsp. *integrifolia* is native of south Europe. Both subspecies occur, sometimes in the same field and can usually be recognised at a glance. Peloric flowers sometimes occur.

14. Linaria L.

Annual to *perennial* monoecious *herbs*. *Leaves* simple, entire, sessile, usually verticillate below and alternate above. *Inflorescence* a terminal, bracteate raceme or spike; flowers bisexual. *Calyx* deeply, often unequally 5-lobed. Corolla with a cylindrical tube, produced at the base into a conical or cylindrical spur, the limb 2-lipped, the upper lip 2-lobed, the lower lip 3-lobed, with at its base a more or less prominent, usually hairy palate, which usually closes the mouth of the tube. *Stamens* 4, didynamous, included. *Style* 1, terminal. *Fruit* a more or less globose capsule with equal loculi dehiscing by several meridional fissures in the apical half; seeds numerous.

Contains about 150 species, mainly Mediterranean and a few in central Europe, temperate Asia and temperate America.

Grime, J. P. et al. (1988). *Comparative plant ecology*. London.
Hultén, E. & Fries, M. (1986). *Atlas of north European vascular plants north of the Tropic of Cancer.* 3 vols. Königstein.
Saber, M. A. et al. (1995). The biology of Canadian weeds. 105. *Linaria vulgaris* Mill. *Canad. Jour. Pl. Sci.* **75**: 525–537.
Stace. C. A. (1982). Segregation in the natural hybrid *Linaria purpurea* (L.) Mill. × *L. repens* (L.) Mill. *Watsonia* **14**: 53–57.
Valdés, B. (1970). Revisión de las especies europeas de *Linaria* con semillas aladas. *Publ. Univ. Sevilla, Series Ciencias* **7**. Seville.

1. Spur longer than rest of corolla **8. maroccana**
1. Spur much shorter than to nearly as long as rest of corolla 2.
2. Whole plant glandular-hairy; corolla, including spur 4–7 mm **6. arenaria**
2. Plant glabrous below, glabrous to glandular-hairy above; corolla, including spur, usually more than 8 mm 3.
3. Corolla predominantly yellow, sometimes very pale or with a purplish tinge or with darker veins 4.
3. Corolla predominantly mauve, violet, purple or pink, sometimes very pale but then with dark veins, sometimes with yellow or orange palate 7.
4. Corolla with darker veins **4. repens × 1. vulgaris**
4. Corolla without darker veins 5.
5. Annual, with stems decumbent and a conspicuous area below the inflorescence bare of leaves **5. supina**
5. Perennial with stems normally erect and leaves more or less up to inflorescence 6.
6. Leaves linear to narrowly elliptical-oblanceolate, cuneate at base; stem often glandular-hairy above; seeds disc-like, with a broad wing round the circumference **1. vulgaris**
6. At least some leaves lanceolate to ovate and subcordate at base; stem always glabrous; seeds angular, scarcely winged **2. dalmatica**
7. Annual; capsule shorter than calyx; seeds disc-like, with broad wing **7. pelisseriana**
7. Perennial; capsule longer than calyx; seeds angular, not winged 8.
8. Corolla without orange, but sometimes with white, patch on palate; spur more than half as long as rest of corolla and usually curved and acute at tip **3. purpurea**
8. Corolla with an orange patch on palate; spur less than half as long as rest of corolla, straight, with a subacute to rounded tip **4. repens**

1. L. vulgaris Mill. Common Toadflax
Antirrhinum linaria L.

Perennial monoecious *herb* with a creeping rhizome. *Stems* numerous, 30–80 cm, glaucous, erect or ascending, rigid, glabrous, or with glandular hairs above, often branched in the upper part, leafy. *Leaves* opposite below, alternate above; lamina 2–8 × 0.1–0.5 cm, greyish-green on upper surface, paler beneath, linear to very narrowly elliptical or linear-oblanceolate, obtuse at apex, entire, cuneate to a sessile base, glabrous, midrib prominent beneath. *Inflorescence* a long, dense raceme, terminal on stems and branches; bracts linear, obtuse at apex, shorter than the pedicels; pedicels and rhachis pale green, glabrous or shortly glandular-hairy; flowers 5–30, bisexual, 10–12 mm in diameter. *Calyx* 3–6 mm, medium green, divided nearly to base; lobes ovate or lanceolate, acute at apex, glabrous. *Corolla* 15–35 mm, yellow, sometimes very pale, tube cylindrical, with a more or straight spur about 7–17 mm, the limb 2-lipped, the upper lip 2-lobed, the lower lip 3-lobed, with an orange palate which closes the tube. *Stamens* 4, included; filaments pale; anthers yellow. *Style* 1, terminal, greenish; stigma greenish. *Capsule* 5–11 mm, ovoid; seeds 2–3 mm, discoid, tuberculate, with a broad, marginal wing. *Flowers* 7–10. Pollinated by large bees; self-incompatible. $2n = 12$.

Native. Open grassland, banks, rough and waste ground, hedge banks, road verges and railway banks. Common over much of Great Britain and Ireland, but absent from parts of Ireland and central and north Scotland. Most of Europe except the extreme north and much of the Mediterranean region; west Asia east to the Altai; naturalised in North America. Eurasian Boreo-temperate element.

2. L. dalmatica (L.) Mill. Balkan Toadflax
Antirrhinum dalmaticum L.; *L. genistifolia* (L.) Mill. subsp. *dalmatica* (L.) Maire & Petitm.

Perennial monoecious *herb*. *Stems* (20–)30–100 cm, erect, branched, especially above, leafy up to inflorescence, glabrous. *Leaves* alternate; lamina 20–60 × up to 40 mm, medium green on upper surface, paler beneath, lanceolate to ovate, acute at apex, entire, more or less amplexicaul, glabrous. *Inflorescence* a raceme; flowers bisexual; pedicels 1–13 mm, more or less equalling bracts. *Calyx* 2–12 mm, 5-lobed, the lobes linear-lanceolate to triangular-ovate, more or less acute at apex, usually subequal. *Corolla* 20–55 mm, yellow; spur 4–25 mm. *Stamens* 4, included. *Style* 1. *Capsule* 3–7 mm. seeds about 1.2 mm, compressed-tetrahedral, rugulose, black, with a narrow often whitish flange on the angles. *Flowers* 6–8. $2n = 12$.

Introduced. Naturalised in waste places, waysides and by railways. Scattered records in Great Britain. Native of south-east Europe.

3. L. purpurea (L.) Mill. Purple Toadflax
Antirrhinum purpureum L.

Perennial monoecious *herb* with stout roots. *Stems* 20–90 cm, medium rather glaucous green, erect, rigid, glabrous, often much branched, especially above, the lower spreading at right angles, the upper ascending, very leafy. *Leaves* opposite below, alternate above; lamina 1.5–6.0 × 0.1–0.4(–0.8) cm, slightly bluish or greyish-green on upper surface, paler beneath, linear or narrowly linear-lanceolate, acute at apex, entire, narrowed to a sessile base, glabrous, midrib prominent beneath. *Inflorescence* a dense raceme, terminal on stems or branches; flowers bisexual; bracts linear, obtuse at apex, longer than pedicels; pedicels and rhachis pale green and glabrous. *Calyx* 2.5–3.5 mm, medium green, divided nearly to base; lobes linear, obtuse at apex. *Corolla* 9–12 mm, purplish-violet, rarely pinkish or whitish, tube cylindrical, with a curved, acute spur about 5 mm, the limb 2-lipped, the upper lip 2-lobed, the lower lip 3-lobed, with a purple, shortly hairy palate which closes the tube. *Stamens* 4, included; filaments pale purple; anthers yellow. *Style* 1, terminal, pale purplish; stigma greenish. *Capsule* about 3 mm, more or less globose, longer than calyx; seeds 1.0–1.3 mm, blackish, trigonous, rugose-ruminate, not winged. *Flowers* 6–8. Pollinated by bees and self-sterile. $2n = 12$.

Introduced. Much grown in gardens and naturalised on rough ground, walls, banks and other waste places. Frequent to sparse throughout much of Great Britain and Ireland. Native of Italy and Sicily.

× **repens** = **L.** × **dominii** Druce

This hybrid is intermediate in corolla characters and is fertile, sometimes forming hybrid swarms.

Occurs in scattered localities in Great Britain with one or both parents north to Cumberland.

4. L. repens (L.) Mill. Pale Toadflax
Antirrhinum repens L.

Perennial monoecious *herb* with a creeping, underground rhizome. *Stems* numerous, 30–80 cm, pale glaucous, erect, glabrous, usually branched above, leafy. *Leaves* opposite below, alternate above; lamina 1–4 × 0.1–0.2 cm, glaucous on upper surface, paler beneath, linear, acute at apex, entire, narrowed at base, glabrous. *Inflorescence* of rather long terminal racemes which are dense in flower and lax in fruit; bracts linear, acute at apex; glabrous; flowers bisexual. *Calyx* 2–3 mm, medium green, divided nearly to the base; lobes 5, linear-lanceolate, acute at apex, glabrous. *Corolla* 8–15 mm including spur, white to pale mauve or lilac with violet veins, tube cylindrical, with a straight spur 1–4 mm, the limb 2-lipped, the upper lip 2-lobed, the lower lip 3-lobed, the palate with an orange spot which closes the tube. *Stamens* 4, included; filaments pale; anthers yellow. *Style* 1, terminal, greenish; stigma greenish. *Capsule* 3–4 mm, longer than calyx, subglobose; seeds 1.2–1.7 mm, dark grey, angular, wrinkled, not winged. *Flowers* 6–9. Pollinated by bees, self-incompatible. $2n = 12$.

Introduced. Stony places, rough ground, banks and walls and along railways. Scattered over much of Great Britain, Channel Islands and east Ireland, but absent from many places and frequent only in parts of the south and west of Great Britain. Native from north Spain and north-west Italy to north-west Germany; naturalised elsewhere in north-west and central Europe.

× **supina** = **L.** × **cornubiensis** Druce

This hybrid is said to be intermediate between the parents in corolla characters and is sterile. It was collected at Par in Cornwall with both parents in 1925 and 1930 and is endemic.

× **vulgaris** = **L.** × **sepium** G. J. Allman

This hybrid most often has a pale yellow corolla with violet veins and the corolla is intermediate in size and shape between those of the parents. It often, however, forms hybrid swarms with a range of plants between the two parents. $2n = 12$.

Frequent within the range of *L. repens* in Great Britain and Ireland north to Ayrshire.

5. L. supina (L.) Chaz. Prostrate Toadflax
Antirrhinum supinum L.

Annual monoecious *herb* with a tap-root and fibrous side-roots. *Stems* 5–30 cm, pale glaucous, glabrous, sometimes branching at the base, the branches decumbent, the ends ascending, leafy below but a bare area below the inflorescence. *Leaves* opposite below, alternate above; lamina 10–30 × 0.5–0.1 mm, glaucous on upper surface, paler beneath, linear to linear-oblanceolate, acute at apex, entire, narrowed to a sessile base, glabrous. *Inflorescence* a short, dense raceme, terminal on stems and branches; flowers bisexual; bracts linear, acute at apex, longer than pedicels; pedicels and rhachis shortly glandular-hairy. *Calyx* 3.5–5.0 mm, medium green, divided nearly to base; lobes 5, linear, obtuse at apex. *Corolla* 15–25 mm (including spur), yellow, sometimes purple-tinged, tube cylindrical, with an almost straight spur 7–11 mm, the limb 2-lipped, the upper lip 2-lobed, the lower lip 3-lobed, with an orange palate which closes the tube. *Stamens* 4, included; filaments pale; anthers yellow. *Style* 1, terminal, greenish; stigma greenish. *Capsule* 3–7 mm, subglobose, not much longer than calyx; seeds 1.7–2.8 mm, grey or blackish, discoid, smooth, with a broad marginal wing. *Flowers* 6–9. Self-incompatible. $2n=12$.

Introduced. Cultivated since 1728 and first recorded in the wild at Par in Cornwall in 1848, and at Hayle. There are no records from Hayle after 1909 and it was last seen at Par in 1987. It is now a rare casual in a few scattered localities. France, Spain, Portugal, north-west Italy and Morocco.

6. L. arenaria DC. Sand Toadflax

Bushy, *annual* monoecious *herb*, sometimes perennating by root-buds. *Stems* 5–15 cm, pale green, erect to ascending, densely glandular-hairy, branched from the base, leafy. *Leaves* verticillate below, alternate above; lamina 0.5–1.4 × 0.1 × 0.2 cm, medium green on upper surface, paler beneath, oblanceolate-elliptical, acute at apex, entire, narrowed at base, glandular-hairy, sessile. *Inflorescence* of solitary bisexual flowers in bracts forming a lax raceme; pedicels 1.0–1.5 mm, much shorter than the bracts, glandular-hairy; bracts similar to the leaves but smaller. *Calyx* 4–5 mm, deeply divided into 5 lobes, the lobes subequal and oblanceolate, glandular-hairy. *Corolla* 4–8 mm, yellowish, the tube cylindrical, with a spur 2–3 mm which is sometimes violet, 2-lipped, the upper lip the limb 2-lobed, the lower lip 3-lobed with an orange palate. *Stamens* 4, included. *Style* 1, terminal. *Capsule* 3–5 mm, ovoid; seeds 1.0–1.5 mm, black, reniform-orbicular, flat, the disc smooth, the wing rather narrow. *Flowers* 6–9.

Introduced. Planted at Braunton Burrows in north Devonshire about 1893 and now well naturalised on semi-fixed dunes. Also recorded for Dorsetshire and Yorkshire. Native of maritime sands in west and north-west France.

7. L. pelisseriana (L.) Mill. Jersey Toadflax
Antirrhinum pelisserianum L.

Annual monoecious *herb* with fibrous roots. *Stems* 8–30(–35) cm, erect, slender, obscurely ridged, glabrous, usually unbranched, leaves few. *Leaves* medium green on upper surface, paler beneath; basal with lamina 10–20 × 3–6 mm, narrowly obovate or oblanceolate, acute at apex, tapering at base to a short, indistinct petiole; cauline remote, alternate, the lamina 10–30 × up to 2 mm, linear to almost filiform, acute, entire, sessile; all glabrous. *Inflorescence* a short, crowded raceme, elongating to 10 cm in fruit; flowers bisexual; bracts 2–3 mm, linear, glabrous or with sessile glands; pedicels 2–3 mm, thinly glandular-hairy. *Calyx* with 5 subequal lobes, the lobes 3.0–3.5 mm, linear-lanceolate, acuminate at apex, with a narrow, membranous margin, keeled and thinly glandular-hairy. *Corolla* violet-purple with a paler palate; tube about 10 mm, thinly glandular-hairy, with a slender, acuminate, down-curved spur 5–8 mm, the limb 2-lipped, the upper lip about 5 mm, narrowly oblong and deeply 2-lobed, the lower lip about 3 mm, recurved, 3-lobed, with a prominent, sulcate palate. *Stamens* 4; filaments of longest pair about 3.5 mm, those of the shortest pair about 2 mm. anthers yellow. *Style* 1, about 2.5 mm. stigma slightly swollen. *Capsule* about 4–5 mm in diameter, shorter than calyx, subglobose, strongly emarginate, somewhat compressed, dehiscing at the apex into irregular teeth; seeds about 0.8 mm in diameter, brown, disc-like, papillose, with a conspicuous fimbriate-laciniate marginal wing. *Flowers* 5–7. $2n=24$.

Native. Rough ground, rocky places and hedgebanks. Very rare and sporadic in Jersey where it was last seen in 1955; a rare casual elsewhere in Great Britain. Widespread in the Mediterranean region and south-west Europe and eastwards to the Caucasus.

8. L. maroccana Hook. fil. Annual Toadflax

Annual monoecious *herb* with a tap-root. *Stems* up to 40 cm, pale green, erect, much-branched, glabrous. *Leaves* opposite; lamina 1.5–4.0 cm, medium green on upper surface, paler beneath, narrowly linear, more or less obtuse at apex, entire, narrowed at base, glabrous. *Inflorescence* a rather lax, elongated raceme of many bisexual flowers; pedicels longer than calyx, glandular-hairy; bracts 3–5(–10) mm, linear to lanceolate. *Calyx* 4–6 mm, divided into 5 lobes almost to the base, the lobes linear-lanceolate or lanceolate, acute or acuminate at apex, usually glandular-hairy, sometimes glabrous. *Corolla* 25–30 mm including spur, usually crimson to purplish, sometimes pale yellow to orange, pink or bicoloured, except for the yellow or orange, hairy pouch; the limb 2-lipped, upper lip 10–15 mm. spur about 15 mm, slender, straight and acute at apex. *Stamens* 4. *Style* 1. *Capsule* 3–5 mm in diameter, globose; seeds 1.7–1.9 mm, black, curved, wingless, strongly ribbed. *Flowers* 6–8. $2n=12$.

Introduced. Grown in gardens and a frequent casual on tips and in waste places. Scattered records throughout Great Britain. Islands. Native of Morocco.

15. Nemesia Vent.

Annual monoecious *herbs*. *Leaves* opposite. *Inflorescence* a terminal raceme; flowers bisexual; bract entire or dentate. *Calyx* 5-lobed, the lobes scarcely overlapping. *Corolla* membranous; tube short, with a pouch; limb 2-lipped, upper lip 4-lobed, lower lobe emarginate, with a palate almost enclosing the throat. *Stamens* 4. *Style* 1. *Ovary* with 2 equal cells. *Fruit* a capsule, laterally compressed; seeds flattened, conspicuously winged.

Contains about 65 species mainly from South Africa, with a few from tropical Africa.

N. melissifolia Benth. and **N. versicolor** E. Mey. ex Benth. have also been recorded as rare casuals.

Hooker, J. D. (1893). *Nemesia strumosa*. *Bot. Mag*. tab. 7272.

1. N. strumosa Benth. Cape-jewels

Annual monoecious *herb*. *Stems* 15–50 cm, pale green, erect, glabrous below, with glandular hairs above, leafy and branched at base. *Leaves* opposite; lamina 1.2–7.5 cm, the lower oblanceolate-spathulate, obtuse at apex and entire or dentate, the cauline lanceolate to linear, pointed at apex and dentate, all glabrous. *Inflorescence* a terminal raceme 2.5–10.0 cm, crowded at first and becoming elongate, glandular-hairy; bracts entire or dentate, glandular-hairy. *Calyx* glandular-hairy, 5-lobed, the lobes linear, obtuse at apex, spreading in flower. *Corolla* yellow or purple to white, sometimes veined with purple on the outside, 5-lobed, the lower lobe emarginate at apex, the pouch short and broad. *Stamens* 4. *Style* 1. *Capsule* laterally compressed; seeds flattened, conspicuously winged. *Flowers* 6–8.

Introduced. A casual garden escape in a few localities. Native of South Africa.

16. Digitalis L.

Biennial, rarely *perennial,* monoecious *herbs. Leaves* alternate, with pinnate venation. *Inflorescence* a terminal, unilateral, bracteate raceme; flowers bisexual; the bracts entire. *Calyx* deeply 5-lobed. *Corolla* large and showy, weakly 2-lipped, with a long, campanulate tube, constricted near the base, shortly 5-lobed, the 2 upper lobes forming an emarginate upper lip, the lowest usually longer and more prominent. *Stamens* 4, included in the corolla tube. *Style* 1. *Fruit* a septicidal capsule.

Contains about 20 species in Europe and the Mediterranean region to central Asia.

D. ferruginea L., Rusty Foxglove, **D. grandiflora** Mill., Yellow Foxglove, and **D. lanata** Ehrh., Grecian Foxglove, have all been recorded as garden escapes but have not persisted.

Grime, J. P. et al. (1988). *Comparative plant ecology*. London.
Herner, K. (1960). Zur nomenklatur und taxonomie von *Digitalis* L. *Bot. Jahrb*. **79**: 218–254.
Hultén, E. & Fries, M. (1986). *Atlas of north European vascular plants north of the Tropic of Cancer.* 3 vols. Königstein.

1. Leaves rugose on upper surface; corolla 40–55 mm, pale to deep pinkish-purple, rarely white **1. purpurea**
1. Leaves smooth on upper surface; corolla 9–25 mm, pale yellow **2. lutea**

Section **1. Digitalis**

Leaves more or less rugose. *Corolla* purple or white with a campanulate tube.

1. D. purpurea L. Foxglove

Biennial, rarely *perennial,* monoecious *herb*, sometimes flowering in its 1st year. *Stems* 5–200 cm, pale green, often suffused brownish-purple, erect, bluntly angular, with dense, grey simple eglandular hairs, very leafy, usually simple. *Leaves* alternate, the lowest forming a rosette; lamina 15–30 × 1–11 cm, gradually, decreasing in size upwards, medium green and rugose on upper surface, paler and greyer beneath, ovate to lanceolate, obtuse at apex, crenate, attenuate at base into a winged petiole, with dense, short, soft, pale simple eglandular hairs and smooth on upper surface, grey-tomentose beneath. *Inflorescence* a terminal, unilateral, bracteate raceme, with 20–80 flowers; bracts decreasing in size upwards, lanceolate, entire and sessile; flowers 15–20 mm in diameter, bisexual, nodding; pedicels longer than calyx, tomentose. *Calyx* 11–13 mm, deeply 5-lobed, campanulate, the lobes, unequal, ovate to lanceolate, acute at apex, with dense glandular hairs. *Corolla* 40–55 mm, pale to deep pinkish-purple, with deeper purple spots on a white ground inside, the lower part of the tube, rarely white and spotted or unspotted, with a long campanulate tube, constricted near the base, shortly 5-lobed, the 2 upper lobes forming an emarginate upper lip, the lowest usually longer and more prominent, shortly ciliate and with a few, long hairs within. *Stamens* 4, included in the corolla tube; filaments white; anthers bright yellow. *Style* 1, purplish, filiform; stigma terminal. *Capsule* 11–13 mm, longer than calyx, ovoid; seeds numerous, almost 1 mm, reddish-brown, ellipsoid. *Flowers* 6–9. Pollinated by bumble-bees, protandrous. $2n = 56$.

Native. Open places, especially woodland clearings, heaths and mountainsides and waste places on acid soils up to 885 m. Throughout Great Britain and Ireland where it also occurs as a garden escape. West, south-west and west-central Europe; Morocco. Suboceanic southern-temperate element. Extensively cultivated elsewhere for ornament and as a medicinal plant. The drug digitalin has long been used for heart conditions.

Section **2. Tubiflorae** Benth.

Leaves smooth. *Corolla* yellow or white with a cylindrical tube.

2. D. lutea L. Straw Foxglove

Perennial monoecious *herb. Stems* 60–100 cm, pale green, erect, glabrous or slightly hairy, simple, leafy. *Leaves* alternate; lamina 2–12 × 1–3 cm, gradually decreasing in size upwards, medium to dark green on upper surface, paler beneath, oblong-lanceolate, long-acute at apex, subentire to serrate, narrowed at base, more or less hairy. *Inflorescence* a terminal, unilateral, bracteate raceme with many more or less dense bisexual flowers; bracts decreasing in size upwards, lanceolate to linear-lanceolate, acute at apex. *Calyx* 4.5–5.5 mm, deeply 5-lobed, the lobes linear-lanceolate, acute at apex, glandular-ciliate. *Corolla* 15–25 mm, pale yellow to whitish, the tube cylindrical, shortly 5-lobed, the 2 upper lobes forming an emarginate upper lip, the lower lip has 3 lobes, the central lobe 5–6 arms and larger than the 2 lateral which are ovate and recurved. *Stamens* 4, included in the corolla tube; filaments white; anthers yellow. *Style* 1, filiform; stigma terminal. *Capsule* 8–10 mm, ovoid. *Flowers* 6–9. $2n = 16, 56, 96, 112$.

Introduced. Naturalised on waste ground, roadsides and walls. Scattered records in Great Britain. West and west–central Europe.

× **purpurea** = **D.** × **fucata** Ehrh.
D. × *purpurascens* Roth

This hybrid occurs in gardens where the two species grow together, but it is not known to escape.

17. Erinus L.

Perennial monoecious *herbs. Stems* numerous. *Leaves* alternate, with pinnate venation. *Inflorescence* a terminal bracteate raceme; flowers bisexual; bracts entire. *Calyx* deeply 5-lobed. *Corolla* 5-lobed, the lobes more or less equal, emarginate and not 2-lipped. *Stamens* 4, didynamous, included. *Style* 1; stigma capitate. *Fruit* a loculicidal capsule; seeds numerous.

Contains 1 species from south and south-central Europe; Algeria and Morocco.

Curtis, W. (1794). *Erinus alpinus*. *Bot. Mag*. **9**: tab. 310.

1. E. alpinus L. Fairy Foxglove

Tufted *perennial* monoecious *herb*. *Stems* numerous, 5–20(–30) cm, medium green, sometimes suffused brownish-purple, ascending to suberect, with numerous glandular hairs, leafy, simple. *Leaves* alternate; lamina 0.5–2.0 × 0.2–0.5 cm, medium green on upper surface, paler beneath, spathulate to oblanceolate, obtuse at apex, crenate or dentate, attenuate at base, glabrous or with simple eglandular hairs, the venation pinnate; petiole 2–5 mm glabrous. *Inflorescence* a terminal, bracteate raceme; bracts oblanceolate, obtuse at apex, entire; flower 6–9 mm in diameter, bisexual. *Calyx* 4–5 mm, deeply 5-lobed, the lobes linear-oblong, obtuse at apex, with glandular hairs. *Corolla* 6–9 mm, purple, rarely white; tube 3–7 mm; 5-lobed, the lobes a little shorter than the tube, more or less equal, patent, broad, deeply emarginate at apex. *Stamens* 4; filaments pale; anthers yellow. Style greenish; stigma terminal. *Capsule* 3.5–4.5 mm, shorter than the calyx, ovoid; seeds smooth. *Flowers* 5–10. Probably mainly pollinated by Lepidoptera, self-pollination possible. $2n = 14$.

Introduced. Naturalised on walls and in stony places. Scattered records throughout Great Britain and Ireland. Native of the mountains of south and south-central Europe from north and east Spain to central Italy and western Austria; Algeria and Morocco.

18. Veronica L.

Cardia Dulac; *Cochlidiospermum* Opiz

Annual to *perennial* monoecious *herbs* sometimes woody at the base. *Leaves* opposite, at least below, with pinnate or sometimes more or less palmate venation. *Inflorescence* an axillary or terminal raceme or flowers solitary in the leaf axils; flowers bisexual. *Calyx* divided almost to the base into 4 or 5 lobes. *Corolla* with a short tube and 4 lobes, the upper lobe slightly the largest, the lower the smallest. *Stamens* 2. *Style* 1. *Fruit* a capsule, more or less laterally compressed.

Contains about 300 species in temperate regions of both hemispheres, but most in the north.

Britton, C. E. (1928). *Veronica anagallis* L. and *V. aquatica* Bernh. *Rep. Bot. Soc. Exch. Club Soc. Brit. Isles* **8**: 548–550.
Burnett, J. (1997). Notes on *Veronica anagallis–aquatica* aggr. *B.S.B.I. News* **75**: 15–17.
Drabble, E. & Little, J. E. (1931). The British *Veronicas* of the *agrestis* group. *Jour. Bot. (London)* **69**: 180–185; 201–205.
Druce, G. C. (1918). *Veronica anagallis–aquatica* L. *Rep. Bot. Soc. Exch. Club Brit. Isles* **3**: 26–27.
Grime, J. P. et al. (1988). *Comparative plant ecology*. London. [*V. arvensis, beccabunga, chamaedrys, montana* and *persica*.]
Hultén, E. & Fries, M. (1986). *Atlas of north European vascular plants north of the Tropic of Cancer.* 3 vols. Königstein.
Lacaita, C. C. (1917). *Veronica buxbaumii*. *Jour. Bot. (London)* **55**: 271–276 and (1918) **56**: 55.
Lehmann, E. (1909). Einige sur kenntnis der Gattung *Veronica*. 1. Unterarten von *V. Tournefortii* und *V. polita* Fr. *Öst. Bot. Zeitschr*. **59**: 249–261.
Les, D. H. & Stuckey, R. L. (1985). The introduction and spread of *Veronica beccabunga* (Scrophulariaceae) in eastern North America. *Rhodora* **87**: 503–515.
Marchant, N. G. (1970). *Experimental taxonomy of* Veronica *section* Beccabungae *Griseb*. Ph.D. thesis, University of Cambridge.
Öztürk, A. & Fischer, M. A. (1982). Karosystematics of *Veronica* sect. *Beccabunga* (Scrophulariaceae) with special reference to the taxa in Turkey. *Pl. Syst. Evol*. **140**: 307–319.
Preston, C. D. & Croft, J. M. (1997). Aquatic plants in Britain and Ireland. Colchester. [*V. beccabunga, anagallis–aquatica, catenata* and × *lackschewitzii*.]
Stewart, A., Pearman, D. A. & Preston, C. D. (1994). *Scarce plants in Britain*. [*V. alpina* and *spicata* subsp. *hybrida*.]
Williams, I. A. (1929). A British *Veronica* hybrid. *Jour. Bot. (London)* **67**: 23–24.

1. Flowers solitary in a leaf axil 2.
1. Flowers in racemes 11.
2. Calyx lobes apparently 2, each bilobed at apex **22. crista-galli**
2. Calyx lobes 4, each rounded to acute at apex 3.
3. Perennial herb; stems rooting at the nodes along its length; pedicels more than twice as long as leaves plus petioles **23. filiformis**
3. Annual herb; stems not rooting at the nodes or doing so near the base; pedicels less than twice as long as leaves plus petioles 4.
4. Leaves with 3–7 shallow lobes or teeth; calyx lobes cordate at base 5.
4. Leaves crenate-serrate, most with more than 7 teeth; calyx lobes cuneate to rounded at base 6.
5. Leaves with apical lobe usually wider than long; fruiting pedicels usually 2–4 times as long as calyx; corolla 6–9 mm in diameter, pale blue with a white centre; anthers 0.7–1.2 mm, blue; style 0.7–1.0 mm **24(a). hederifolia** subsp. **hederifolia**
5. Leaves with apical lobe usually longer than wide; fruiting pedicels usually 3.5–7.0 times as long as calyx; corolla 4–6 mm in diameter, whitish to pale lilac; anthers 0.4–0.8 mm, whitish to pale lilac-blue; style 0.3–0.5 mm **24(b). hederifolia** subsp. **lucorum**
6. Lobes of capsule with apices diverging at about 90° 6.
6. Lobes of capsule with apices more or less parallel or diverging at a very narrow angle 9.
7. Plant slender; leaves 6–15 × 5–10 mm, crenate, the teeth single and obtuse **21(i). persica** var. **kochiana**
7. Plant robust; leaves 10–30 × 5–25 mm, crenate or dentate, the teeth either double and obtuse, or dentate and more or less acute 8.
8. Leaves 10–20 × 5–15 mm, crenate, the teeth more or less double and obtuse **21(ii). persica** var. **corrensiana**
8. Leaves 10–30 × 7–25 mm, dentate, the teeth more or less acute **21(iii). persica** var. **persica**
9. Capsule with only patent glandular hairs **19. agrestis**

9. Capsule with short arched simple eglandular hairs and variable numbers of patent glandular hairs 10.
10. Flowers 5–8 mm in diameter **20(i). polita** var. **polita**
10. Flowers 8–12 mm in diameter **20(ii). polita** var. **grandiflora**
11. Flowers in terminal racemes 12.
11. Flowers in axillary racemes 28.
12. Annual herbs with 1 root system, easily uprooted 13.
12. Perennial herbs with non-flowering shoots and/or stems rooted more than just at the base 21.
13. Plant glabrous **18. peregrina**
13. Plant hairy, sometimes minutely so, at least on capsules and inflorescence axis 14.
14. Bracts much longer than fruiting pedicels 15.
14. Bracts shorter than to more or less as long as fruiting pedicels 19.
15. At least some upper leaves lobed more than halfway to midrib **17. verna**
15. All leaves entire to crenate-serrate, toothed less than halfway to midrib 16.
16. Leaves oblanceolate to narrowly oblong, all hairs glandular **18. peregrina**
16. Leaves ovate, many hairs at least below, non-glandular 17.
17. Plant spreading or decumbent; leaves 10–25 mm, more or less acutely toothed **16(iii). arvensis** var. **polyanthos**
17. Plant erect; leaves 3.5–15 mm, more or less bluntly toothed 18.
18. Plant 2–8 cm, densely glandular-hairy; leaves 3.5–6.0 mm, entire or bluntly toothed; inflorescence short and dense **16(i). arvensis** var. **nana**
18. Plant 10–40 cm, less glandular-hairy; leaves 5–15 mm, bluntly toothed; inflorescence up to 30 cm and very open **16(ii). arvensis** var. **arvensis**
19. Lower bracts and upper leaves lobed much more than halfway to base **15. triphyllos**
19. Bracts and leaves toothed less than halfway to midrib 15.
20. Pedicels more than twice as long as calyx; capsule divided to about halfway into 4 lobes; seeds flat **13. acinifolia**
20. Pedicels less than twice as long as calyx; capsule divided to less than a quarter of the way; seeds cup-shaped **14. praecox**
21. Racemes dense, long and many-flowered; corolla-tube usually more than 2 mm, longer than wide 17.
21. Racemes lax, short and/or few flowered; corolla-tube usually less than 2 mm, wider than long 19.
22. Leaves usually widest in basal one-third, serrate to biserrate with acute to subacuminate teeth, glabrous to sparsely and often minutely hairy on both surfaces **25. longifolia**
22. Leaves usually widest in middle one-third, crenate to serrate with usually obtuse teeth, hairy on both surfaces 23.
23. Plant 8–30 cm; leaves 1.5–3.0 × 0.8–1.2 cm, sparingly crenate, gradually narrowed into a narrow petiole **26(a). spicata** subsp. **spicata**
23. Plant 15–60(–80) cm; leaves 2–4 × 1–2 cm, more deeply crenate or crenate- serrate, abruptly narrowed into a broad petiole **26(b). spicata** subsp. **hybrida**
24. Corolla pink; style about twice as long as capsule **2. repens**
24. Corolla white to blue; style shorter than capsule 25.
25. Capsule wider than long 26.
25. Capsule longer than wide 27.
26. Stems shortly prostrate with flowering branches ascending; racemes with 20–40 flowers; pedicels about as long as calyx; corolla 5–8 mm in diameter, white or pale blue with slaty-violet lines **1(a). serpyllifolia** subsp. **serpyllifolia**
26. Stems decumbent, rooting for most of their length; racemes with 8–15 flowers; pedicels much longer than calyx; corolla 7–10 mm in diameter, bright blue **1(b). serpyllifolia** subsp. **humifusa**
27. Stems herbaceous; racemes with glandular hairs; corolla less than 10 mm in diameter; style less than 2 mm **3. alpina**
27. Stems woody at base; racemes with eglandular hairs; corolla more than 10 mm in diameter; style more than 2 mm **4. fruticans**
28. Stems and leaves glabrous, except stems sometimes glandular-hairy in inflorescence 29.
28. Stems and leaves hairy 32.
29. Racemes 1 per node; capsule dehiscing into 2 valves **9(i). scutellata** var. **scutellata**
29. Racemes mostly 2 per node; capsule dehiscing into 4 valves 30.
30. Flowering stems procumbent or decumbent, rarely ascending; leaves all shortly petiolate **10. beccabunga**
30. Flowering stems erect, rarely ascending; upper leaves sessile 31.
31. Pedicels erecto-patent in fruit; corolla pale blue; capsule 2.5–4.0 mm, more or less globose **11. anagallis–aquatica**
31. Pedicels patent in fruit; corolla pinkish; capsule 2–3 mm, wider than long **12. catenata**
32. Stems hairy along 2 opposite lines only; capsule shorter than calyx **7. chamaedrys**
32. Stems hairy all round; capsule longer than calyx 33.
33. Petioles more than 6 mm. capsules more than 6 mm wide **8. montana**
33. Petioles less than 6 mm. capsule less than 6 mm wide 34.
34. Leaves lanceolate-oblong to lanceolate; pedicels more than 6 mm in fruit, longer than bracts **9(ii). scutellata** var. **villosa**
34. Leaves lanceolate-oblong to ovate or elliptical; pedicels less than 6 mm in fruit, up to as long as bracts 35.
35. Leaves sessile; calyx lobes usually 5, 1 much shorter than the other 4 **5. austriaca**
35. At least the lower leaves petiolate; calyx lobes 4 **6. officinalis**

Section 1. **Veronicastrum** W. D. J. Koch
Section *Berula* Dumort.

Perennial monoecious *herbs. Inflorescence* a terminal raceme, the lower bracts often leaf-like. *Corolla* tube wider than long. *Capsule* dehiscing into 2 valves, usually only apically; seeds flat.

1. V. serpyllifolia L. Thyme-leaved Speedwell
Veronicastrum serpyllifolium (L.) Fourr.; *Cardia multiflora* Dulac nom. illegit.

Perennial herb with fibrous roots more than just at the base. *Stems* 10–30 cm, pale yellowish-green, prostrate or decumbent and rooting at the nodes, or the flowering ones ascending, slender, with minute, pale, curved simple eglandular hairs. *Leaves* opposite; lamina 5–20 × 4–15 mm, medium green on upper surface, paler beneath, broadly ovate, subrotund or oblong-ovate, rounded at apex, entire or shallowly crenate, rounded at base, glabrous or nearly so; midrib and 1 pair of lateral veins which curve upwards, impressed above and fairly prominent beneath; subsessile or with a very short pale petiole. *Inflorescence* a lax, terminal raceme with up to 40 flowers; peduncles and rhachis pale yellowish-green with very short simple eglandular or glandular hairs; upper bracts narrowly oblong, the lower longer and broader and grading into the leaves, longer than the pedicels; flowers 5–10 mm in diameter, bisexual. *Calyx* 4.0–4.5 mm, divided nearly to base into 4 lobes, the lobes white, ovate and rounded at apex, with very short hairs or nearly glabrous. Corolla 3–4 mm, divided nearly to the base into 4 lobes, the lobes pale blue or bright blue, with darker lines, obovate and rounded at apex, the tube wider than long. *Stamens* 2; filaments white; anthers slatey violet. *Style* 1, white, shorter than capsule. *Capsule* about as long as calyx, pale green, obcordate and wider than long, very shortly glandular-hairy, dehiscing into 2 valves, usually apically; seeds 0.7–0.9 mm, flat. *Flowers* 3–10. Visited by flies.

(a) Subsp. **serpyllifolia**
Stems shortly prostrate with flowering branches ascending. *Racemes* with 20–40 flowers, nearly glabrous or with eglandular hairs; pedicels about as long as calyx. *Corolla* 5–8 mm in diameter, white or pale blue with slatey-violet lines. $2n = 14$.

(b) Subsp. **humifusa** (Dicks.) Syme
V. tenella All.; *V. humifusa* Dicks.; *V. borealis* (Laest.) Hook. fil.; *V. serpyllifolia* var. *borealis* Laest.

Stems decumbent, rooting for most of their length. *Racemes* with 8–15 flowers, glandular-hairy; pedicels much longer than calyx. *Corolla* 7–10 mm in diameter, bright blue. $2n = 14$.

Native. Subsp. *serpyllifolia* occurs in waste and cultivated ground, paths, lawns, open grassland, woodland rides and on mountains and is common throughout Great Britain and Ireland. It is found almost throughout Europe; the mountains of North Africa; Madeira and Azores; temperate Asia; North and South America. Circumpolar Boreo-temperate element. Subsp. *humifusa* occurs in damp places in mountains on rock ledges, flushes and wet gravel. It is local in Monmouthshire, Breconshire, Caernarvonshire, Yorkshire, Northumberland, Cumberland and through much of Scotland. The Scottish plants are more extreme than those of northern England and Wales and intermediates occur with subsp. *serpyllifolia* in other mountainous areas. Similar plants to subsp. *humifusa* occur on mountains through much of Europe.

2. V. repens Clarion ex DC. Corsican Speedwell
V. reptans auct.

Perennial monoecious *herb* rooting in more than just at the base. *Stems* 4–10 cm, pale green, slender, glabrous below, shortly glandular-hairy in upper part, creeping and rooting with short, lateral flowering branches. *Leaves* opposite; lamina 4–8 × 4–5 mm, medium green on upper surface, paler beneath, ovate or subrotund to elliptical, rounded at apex, entire or obscurely crenulate, rounded at base; petiole short, pale green, glabrous. *Inflorescence* a lax, terminal raceme with 3–6 flowers; pedicels 4–7 mm, longer than bracts or calyx; bracts 3–4 mm, linear; flowers about 10 mm in diameter, bisexual. *Calyx* up to 3 mm, green, divided into 4 lobes, the lobes subequal, elliptical and obtuse at apex. *Corolla* divided nearly to the base into 4 lobes, the lobes pink, ovate, obtuse or rounded at apex, the tube wider than long. *Stamens* 2; filaments pale; anthers yellow. *Style* 1, about 4 mm, nearly twice as long as capsule. *Capsule* 2.0–2.5 mm, obovate, emarginate at apex, dehiscing into 2 valves, usually apically; seeds about 0.7 mm, flat. *Flowers* 4–5 . $2n = 14$.

Introduced. Naturalised weed in lawns in a very few places in Hampshire, northern England and Scotland and less common than formerly. Native of the mountains of Corsica and southern Spain.

3. V. alpina L. Alpine Speedwell
Cardia alpina (L.) Dulac; *Vermicastrum alpinum* (L.) Fourr.

Perennial monoecious *herb* with roots other than just at the base. *Stems* 5–15 cm, pale green, shortly creeping at base and the flowering ones ascending, glabrous below, with short glandular hairs in upper part, leafy, with very few branches. *Leaves* opposite; lamina 1.0–2.5 × 0.4–1.2 cm, medium green on upper surface, paler beneath, ovate or ovate-elliptical, entire to serrulate, cuneate at base, sessile or almost so, more or less glabrous. *Inflorescence* a terminal raceme, with 4–12, dense flowers; rhachis and pedicels with glandular hairs; bracts longer than the pedicels, not passing into the leaves; flowers 5–10 mm in diameter, bisexual. *Calyx* 3–4 mm, green, divided nearly to base into 4 lobes, the lobes elliptical and subacute at apex, hairy at least on the margins. *Corolla* 6.5–7.0 mm, divided nearly to base into 4 lobes, the lobes dull blue fading to paler blue, with a white eye, obovate and rounded at apex, the tube wider than long. *Stamens* 2; filaments white; anthers yellow. *Style* 1, about 1 mm. stigma capitate. *Capsule* 6–7 × 4–5 mm, longer than wide, obovate, slightly emarginate at apex, glabrous, dehiscing into 2 valves, usually apically; seeds 0.7–0.9 mm, brown, ellipsoid, flat. *Flowers* 7–8. Visited by a few flies, but probably self-pollinated. $2n = 18$.

Native. Damp alpine rocks above 500 m where the snow lies late. Central Scotland from Perthshire and Argyllshire to Aberdeenshire and Inverness-shire. Arctic Europe and North America; high mountains of Europe southwards to south Spain, south Italy and south Bulgaria; Siberia; Manchuria; Korea; mountains of North America southwards to New England and Colorado. Eurosiberian Arctic-montane element.

4. V. fruticans Jacq. Rock Speedwell
V. saxatilis Jacq.

Perennial monoecious *herb* with a woody base. *Stems* 5–20 cm, pale green, erect, glabrous below, with short, curly hairs above, leafy, with numerous, ascending branches. *Leaves* opposite; lamina 5–20 × 3–10 mm, dark green on upper surface, paler beneath, coriaceous, obovate, obovate-elliptical, elliptical or oblong, rounded at apex, entire or slightly crenate, cuneate at the subsessile base, glabrous on the surface and often ciliate on the margins. *Inflorescence* a lax, terminal raceme with up to 10 flowers; rhachis and pedicels with minute, curly eglandular hairs; pedicels 8–10 mm, slender, bracts narrow, not passing into the leaves, shorter than the pedicels; flowers 10–15 mm in diameter, bisexual. *Calyx* 4–5 mm, divided nearly to the base into 4 lobes, the lobes narrowly oblong or oblong-elliptical and obtuse at apex, with minute, curly hairs. *Corolla* 5–7 mm, divided nearly to the base into 4 lobes, the lobes deep bright blue with darker blue veins and reddish towards the middle, ovate and broadly rounded at apex, the tube wider than long. *Stamens* 2; filaments pale; anthers pale violet. *Style* 1, more than 2 mm, shorter than capsule, pale violet; stigma violet. *Capsule* 6–8 × 4–5 mm, longer than wide, brown, ovate or elliptical, entire at apex, minutely hairy, dehiscing into 2 valves, usually apically; seeds 0.5–0.6 mm, brown, elliptical, flat, smooth. *Flowers* 7–8. Visited by various insects, but apparently often selfed; homogamous. $2n=16$.

Native. Alpine rocks above 500 m. Perthshire, Forfarshire, Aberdeenshire, Argyllshire and Inverness-shire, very local. Arctic Europe from Iceland to Russia; alpine Europe south to the Pyrenees, Corsica, the Apennines and Bosnia; Greenland. European Arctic-montane element.

Section **2. Veronica**
Coerulinia Fourr.; *Limnaspidium* Fourr.

Perennial herbs. Inflorescences opposite or alternate, axillary racemes. *Corolla* tube wider than long. *Capsule* dehiscing into 2 valves, usually only apically; seeds flat.

5. V. austriaca L. Large Speedwell

Perennial monoecious *herb* with non-flowering shoots. *Stem* 30–100 cm, pale green, erect to ascending, more or less hairy all round, branched above, leafy. *Leaves* opposite below, alternate above; lamina 2–7 × 0.6–4.5 cm, medium green on upper surface, paler beneath, lanceolate to oblong in outline, acute at apex, crenate to incise-serrate, truncate or subcordate at base, sessile, hairy. *Inflorescence* a long, dense raceme in the axils of branches; flowers bisexual; pedicels 2–5 mm, usually equalling or shorter than bracts. *Calyx* (3–)4–6(–8) mm, divided almost to the base into (4)5 lobes, 1 much shorter than the other 4, the lobes narrow, glabrous or hairy. *Corolla* (8–)10–13(–17) mm in diameter, bright blue, divided nearly to the base into 4 lobes, the lobes broadly ovate, rounded at apex. *Stamens* 2; filaments pale; anthers yellow. *Style* 1. *Capsule* 4–6 × 3–5 mm, longer than calyx, suborbicular, to broadly elliptical, cordate at base, usually longer than wide, glabrous or hairy. *Flowers* 5–8. $2n=64$, 68.

Introduced. Grown in gardens and naturalised in open and rough ground and dunes. Scattered records in Great Britain north to Forfarshire. Native of most of Europe. Our plant is subsp. **teucrium** (L.) D. A. Webb (*V. teucrium* L.) which occurs throughout much of the range of the species.

6. V. officinalis L. Heath Speedwell
Cardia officinalis (L.) Dulac; *V. hirsuta* Hopkirk;
V. officinalis var. *hirsuta* (Hopkirk) F. N. Williams;
V. setigera D. Don nom. illegit.

Perennial monoecious *herb* often creeping and forming large mats. *Stems* 10–40 cm, green, creeping and rooting below, ascending above, hairy all round with short to medium simple eglandular hairs, leafy, branched. *Leaves* opposite; lamina 5–30(–60) × 3–30 mm, dull medium to rather deep green on upper surface, paler beneath, obovate-elliptical, ovate or elliptical, obtuse to subacute at apex, crenate-serrate, cuneate at base and subsessile or with a short petiole, short or medium, rather appressed simple eglandular hairs on both surfaces and the margins. *Inflorescence* an axillary raceme from the axil of 1, or more rarely both of a pair of leaves, the raceme dense, pyramidal, long-stalked and with 15–25 flowers; peduncle and rhachis with dense, short glandular hairs; pedicels 1–2 mm, glandular-hairy; bracts linear, about twice as long as the pedicels; flowers 5–9 mm in diameter, bisexual. *Calyx* 2.0–2.5 mm, divided almost to the base in 4 lobes, the lobes ovate or lanceolate, obtuse at apex, with dense, short glandular hairs. *Corolla* 5–6 mm, divided, nearly to the base into 4 lobes, the lobes lilac or pale violet, rarely white, with deeper violet veins, tube wider than long. *Stamens* 2; filaments lilac; anthers lilac. *Style* 1, 2.5–4.0 mm, lilac. *Capsule* 3–5 × 4–5 mm, brown, obtriangular-obovate, obovate or obcordate, dehiscing into 2 valves apically, glandular-hairy; seeds 0.8–1.2 mm, elliptical to subrotund, flat, smooth. *Flowers* 5–8. Pollinated by various Diptera and Hymenoptera; homogamous, protandrous or protogynous. $2n=36$.

A very variable species in size and hairiness. The plants with the largest leaves and the most hair seem to be plants of open woodland. Plants on the coast, in the mountains and on acid heaths tend to be small.

Native. Banks, open woods, grassland and heathland on well-drained soils and by streams in mountains up to 900 m. Common throughout Great Britain and Ireland. Almost throughout Europe; Asia Minor and Caucasus to north-west Iran; Azores; eastern North America where it is probably introduced. European Boreo-temperate element.

7. V. chamaedrys L. Germander Speedwell
Cardia ciliata Dulac nom. illegit.; *Veronicella chamaedrys* (L.) Fourr.

Perennial monoecious *herb* with long, slender, brown, wiry rhizomes. *Stems* 20–40 cm, pale yellowish-green, often suffused brownish-purple, prostrate and rooting at the nodes below, ascending or erect above, often flexuous, with 2 sides glabrous and 2 sides with short, curly hairs, leafy, branched. *Leaves* opposite and decussate; lamina

10–35 × 8–30 mm, dull medium yellowish-green on upper surface, slightly paler beneath, ovate or triangular-ovate, more or less obtuse at apex, incise-crenate, the teeth rounded, truncate to more or less cordate at base, with short to medium, pale simple eglandular hairs on both surfaces and especially the margins and veins beneath; sessile or with a petiole up to 5 mm. *Inflorescence* an axillary raceme from 1 or more rarely both of a pair of leaves, the raceme lax and 10–20-flowered; peduncle and rhachis with numerous, short and medium, pale simple eglandular hairs intermixed with very short glandular hairs; pedicels 4–8 mm, with short and medium glandular hairs; bracts lanceolate, about as long as or shorter than the pedicels; flowers 8–15 mm in diameter, bisexual. *Calyx* 4–6 mm, divided almost to the base into 4 lobes, the lobes medium green, lanceolate, more or less obtuse at apex, with numerous, short glandular hairs. *Corolla* 5–7 mm, divided almost to base into 4 unequal lobes, the lobes pale to medium bright blue, the veins darker and with a toothed cream area at the base inside, obovate and rounded at apex, the tube wider than long. *Stamens* 2, exserted; filaments blue; anthers bluish. *Style* 3.5–4.5 mm, blue; stigma cream. *Capsule* 3.5–4.0 mm, shorter than calyx, obtriangular-obovate or obcordate, wider than long, shortly hairy and ciliate, dehiscing into 2 valves apically; seeds 1.1–1.7 × 0.8–1.5 mm, elliptical, flat, subrugulose. *Flowers* 3–7. Pollinated by various Diptera and Hymenoptera, homogamous. $2n = 32$.

Native. Woods, hedgerows and grassland in damp places. Common throughout Great Britain and Ireland. Throughout Europe except for some islands and much of the Arctic; north and west Asia; naturalised in North America. Eurosiberian Boreo-temperate element.

8. V. montana L. Wood Speedwell
Cardia montana (L.) Dulac; *Coerulinia montana* (L.) Fourr.

Perennial monoecious *herb* with roots other than at the base of the main stem. *Stems* 20–40 cm, pale green, sometimes suffused brownish-purple, creeping and rooting at the nodes, the ends ascending, densely glandular-hairy all round, leafy, branched. *Leaves* opposite and decussate; lamina 10–40 × 10–35 mm, dull, pale to medium green on upper surface, paler beneath, ovate or subrotund-ovate, more or less obtuse to subacute at apex, coarsely crenate-serrate, broadly cuneate to rounded or truncate at base, with short to medium, pale, appressed simple eglandular hairs on both surfaces; petioles up to 20 mm, pale green, hairy. *Inflorescence* a raceme from the axile of 1 or both of a pair of leaves, lax, long-stalked and with 2 to 8 flowers; peduncles and rhachis with numerous, short and medium glandular hairs; pedicels 4–12 mm, with numerous glandular hairs; bracts small, linear, much shorter than the pedicels; flowers 8–10 mm in diameter, bisexual. *Calyx* 3–4 mm, divided almost to the base into 4 lobes, the lobes green, obovate to subrhombic, obtuse at apex, with numerous glandular hairs. *Corolla* 5–6 mm, divided almost to the base into 4 lobes, the lobes lilac-blue, obovate or ovate and rounded at apex, the tube wider than long. *Stamens* 2; filaments pale blue; anthers blue. *Style* 1, 3.5–5.0 mm. stigma subcapitate. *Capsule* 5–6 × 7–8 mm, subrotund to reniform, nearly flat, retuse or emarginate at apex, ciliate with short glandular hairs, dehiscing into 2 valves apically; seeds 2.0–2.3 × 1.5–2.0 mm, yellow, subrotund, flat, smooth. *Flowers* 4–7. Pollinated by various Diptera and Hymenoptera. $2n = 18$.

Native. Rather damp woods. Of scattered occurrence and locally frequent throughout Great Britain and Ireland except northern Scotland. West, central and south Europe northwards to northern Denmark and eastwards to Latvia and the western Ukraine; Caucasus; mountains of Algeria and Tunisia. European Temperate element.

9. V. scutellata L. Marsh Speedwell
Limnaspidium scutellatum (L.) Fourr.; *V. angustifolia* Gray, non Bernh.; *Cardia scutellata* (L.) Dulac

Perennial monoecious *herb* with roots other than at the main stem. *Stems* 10–60 cm, green, creeping and rooting at the nodes, then ascending or erect, glabrous or hairy, leafy, simple or slightly branched or with numerous branches from base. *Leaves* opposite and decussate; lamina 20–55 × 2–10 mm, pale yellowish-green, often tinged with purple on upper surface, paler beneath, linear-lanceolate to linear-oblong, acute at apex, entire or remotely denticulate, sessile and semi-amplexicaul or very shortly petiolate, the midrib impressed and conspicuous on upper surface, glabrous or hairy. *Inflorescence* a raceme from the axil of 1 of a pair of leaves, very lax, slender and with up to 10 flowers; peduncle and rhachis glabrous or with simple eglandular and glandular hairs; pedicels 7–10 mm, slender, spreading at right angles in fruit, glabrous or with simple eglandular and glandular hairs; bracts linear, acute at apex; flowers 5–6 mm in diameter, bisexual. *Calyx* 2.0–2.5 mm, divided almost to the base into 4 lobes, the lobes ovate, subobtuse at apex, glabrous or with glandular and eglandular hairs. *Corolla* 4–5 mm, divided almost to the base into 4 lobes, the lobes white or pale blue with purple lines, sometimes pinkish, ovate and rounded at apex, the tube wider than long. *Stamens* 2; filaments whitish or purplish; anthers pinkish or pale. *Style* 1, *Capsule* 3.0–3.5 × 3.5–4.0 mm, reniform, flat, deeply emarginate, dehiscing into 2 valves apically; seeds 0.8–1.0 mm, pale brown, broadly elliptical, flat, smooth. *Flowers* 6–8. $2n = 18$.

(i) Var. **scutellata**
V. scutellata var. *procumbens* P. B. O'Kelly

Stems, inflorescence and leaves glabrous.

(ii) Var. **villosa** Schumach.
V. scutellata var. *pubescens* Hook. fil.

Stems, inflorescence and leaves hairy, usually a mixture of glandular and simple eglandular hairs.

Native. Bogs, marshes, wet meadows, by ponds and lakes, on bare ground or among tall vegetation. Scattered throughout Great Britain and Ireland and locally frequent. Most of Europe but rare in the Mediterranean region; northern Asia east to Kamchatka. European Boreo-temperate element. The common plant is var. *scutellata*. Var. *villosa* is rarer and often in drier habitats.

Section **3. Beccabunga** (Hill) Dumort.
Beccabunga Hill

Usually *perennial*, sometimes *annual herbs*. *Inflorescences* opposite, axillary racemes. *Corolla tube* wider than long. *Capsule* dehiscing into 4 valves; seeds flat on 1 side, convex on the other.

Whole populations of non-flowering plants of the *V. anagallis–aquatica* complex cannot be identified unless brought into cultivation and made to flower.

10. V. beccabunga L. Brooklime
Cardia beccabunga (L.) Dulac; *V. fontinalis* Salisb. nom. illegit.; *V. beccabunga* var. *longibracteata* Schur; *Beccabunga vulgaris* Fourr.; *V. beccabunga* var. *repens* Bosch; *V. beccabunga* var. *minor* Rost

Perennial monoecious *herb* creeping and shallowly rooting other than at base of main stem. *Stems* 20–60, pale yellowish-green, sometimes tinted brownish-purple, fleshy, creeping then ascending, glabrous, leafy, small fragments readily rooting. *Leaves* opposite; lamina 15–60 × 10–35 mm, bright medium bluish-green on upper surface, paler beneath, rather thick and fleshy, elliptical, ovate, oblong or ovate-oblong, obtuse to rounded at apex, shallowly crenate-serrate and rounded at base, glabrous; petiole short, pale yellowish-green, glabrous. *Inflorescence* an erect, axillary raceme from both of a pair of leaves, rather lax, with 10–30 flowers; peduncle and rhachis pale green and glabrous; pedicels 4–7 mm, slender, glabrous, more or less patent in fruit; bracts more or less equalling the pedicels, linear-lanceolate; flowers 5–7 mm in diameter, bisexual. *Calyx* 3.0–3.5 mm, divided almost to the base into 4 lobes, the lobes pale green, narrowly ovate, acute at apex, glabrous. *Corolla* 3–4 mm, divided almost to the base into 4 lobes, the lobes bright blue with a white base and dark veins, rarely white, obovate and rounded at apex, the tube wider than long. *Stamens* 2; filaments 2.0–2.5 mm, pale blue; anthers blue. *Style* 1, 1.3–3.5 mm. stigma subcapitate, entire. *Capsule* (2.5–)3.0–4.0(–5.5) × 3.0–4.0 (–4.5) mm, pale brown, subglobose, retuse at apex, dehiscing into 4 valves, shorter than the calyx; seeds 0.5–0.6 mm, pale brown, ellipsoid, flat on 1 side, convex on the other, smooth. *Flowers* 5–9. Pollinated by various Diptera and Hymenoptera; protogynous, often selfed. $2n = 18$.

Native. Streams, ditches, wet hollows in meadows, marshes, damp woodland rides, pondsides and riverbanks particularly where there is open wet marsh and shallow water, up to 850 m. Throughout Great Britain and Ireland, but rare in north-western Scotland. Most of Europe southwards to 65° N; North Africa, temperate Asia extending to Japan and the Himalaya. Eurosiberian Temperate element. Introduced in North America where its spread is probably curtailed by many suitable habitats being already occupied by the closely allied *V. americana* (Raf.) Schwein ex Benth. Our plant is subsp. **beccabunga** which is widespread in Europe. *V. beccabunga* was formerly eaten as a salad plant in northern Europe. The unusual epithet *beccabunga* is derived from the Low German *beckbunga*, *beck* meaning stream and the origin of *bunga* being disputed.

11. V. anagallis-aquatica L. Blue Water Speedwell
Beccabunga anagallis Fourr.; *V. aquatica* Gray nom. illegit.

Perennial, sometimes *annual*, monoecious *herb*, rooting other than at the main stem and found as submerged stands in shallow water and as an emergent or terrestrial species in swamp and disturbed areas by the waters edge. *Stems* 20–60 cm, pale green, sometimes tinted brownish, shortly creeping, rooting and then ascending, glabrous, leafy and branched. *Leaves* opposite; lamina 5–12 × 0.5–2.5(–4.0) cm, medium to dark green on upper surface, paler beneath, the lower ovate, subentire and often petiolate, the upper ovate-lanceolate or lanceolate, acute at apex, remotely serrulate and more or less amplexicaul, all glabrous. *Inflorescence* an axillary raceme, present from both of a pair of leaves, rather lax, ascending and with 10–50 flowers; peduncle and rhachis glabrous or with very short glandular hairs; pedicels 5–7 mm, glabrous or with glandular hairs, ascending to erect after flowering, linear to lanceolate and acute at apex; flowers 5–10 mm in diameter, bisexual. *Calyx* 2.0–2.5 mm, divided almost to the base into 4 lobes, the lobes ovate-lanceolate, acute at apex, glabrous or shortly glandular-hairy. *Corolla* 4–5 mm, divided almost to the base into 4 lobes, the lobes pale blue or lilac with violet veins, broad, rounded at apex, the tube wider than long. *Stamens* 2; filaments whitish; anthers pale blue. *Style* 1, 1.5–2.0 mm, stigma subcapitate, entire. *Capsule* (2.5–)3.0–3.5(–4.0) mm, pale brown, more or less globose, inflated, slightly longer than wide, not or slightly emarginate, dehiscing into 4 valves; seeds 0.5–0.7 mm, pale brown, elliptic to globular, flat on 1 side, convex on the other, smooth. *Flowers* 6–8. Visited by Diptera although easily self-pollinated. $2n = 36$.

Native. By ponds, rivers, ditches and streams, lakes and pits and in marshes and wet meadows up to 380 m. Common throughout most of Great Britain and Ireland, except the Scottish Highlands. Most of Europe except the extreme north; Asia to Japan and the Himalaya; north-east Africa; introduced in central and southern Africa, North and South America and New Zealand. Eurasian Southern-temperate element. Our plant is subsp. **anagallis–aquatica** and is widespread in Europe. In cultivation plants made to flower under water do not open, but nevertheless set seed. The fruits of *V. anagallis–aquatica*, *V. catenata* and their hybrid are often galled by the weevil *Gymnetron villosulum* Gyllenhal.

× catenata = **V. × lackschewitzii** Keller

This hybrid is robust with clumps of up to 70 flowering stems 1.2 m high with longer racemes than either parent and usually sterile, but occasionally partially fertile, otherwise is intermediate in characters between the parents.

It is frequent in Great Britain and Ireland throughout the range of the parents but is most common in England. It is also widespread in mainland Europe and is recorded from North America where *V. catenata* is native, but *V. anagallis–aquatica* a naturalised alien. Many populations of this hybrid consist of highly sterile plants and may represent a single clone reproducing vegetatively by rooting readily at the nodes, sometimes high up the stem. Some wild populations, however, show a range of fertility and F_2 and later generations can be more fertile. Named after Paul Lackschewitz (b. 1865).

12. V. catenata Pennell Pink Water Speedwell
V. aquatica Bernh., non Gray

Perennial, sometimes *annual,* monoecious *herb,* rooting at other than the main stem. *Stems* 20–30 cm, hollow, pale green, usually purplish-tinged, shortly creeping, rooting and then ascending, glabrous, or rarely hairy, leafy and branched, the lower branches often growing into flowering stems. *Leaves* opposite; lamina 1–15 × 0.3–3.0 cm, medium to dark green on upper surface, paler beneath, linear to oblong or linear-lanceolate to lanceolate, more or less acute at apex, subentire to remotely serrulate, sessile and semi-amplexicaul at base, glabrous. *Inflorescence* an axillary raceme, present in both axils of a pair of leaves, lax, spreading and with 10–50 flowers; peduncle and rhachis glabrous or with glandular hairs; pedicels 4–6 mm, spreading at right angles after flowering, bracts lanceolate, subacute, longer than pedicels at flowering; flowers 5–10 mm in diameter, bisexual. *Calyx* 3.0–4.5 mm, divided almost to the base into 4 lobes, the lobes elliptical or oblong, widely spreading after flowering. *Corolla* 3–5 mm, divided almost to the base into 4 lobes, the lobes pink with darker lines, ovate and rounded at apex, the tube wider than long. *Stamens* 2; filaments whitish; anthers bluish. *Style* 1, 1.0–1.2 mm. stigma subcapitate. *Capsule* 2.0–3.0 mm, pale brown, more or less globose, emarginate, dehiscing into 4 valves, usually wider than long; seeds 0.6–0.7 mm, pale brown, ellipsoid, flat on one side, convex on the other, smooth. *Flowers* 6–8. $2n=36$.

Native. Mostly in open muddy places with little or no flowing water, but can be in streams up to 1 m deep. Locally frequent in England, Wales and Ireland, rare in Scotland and the Channel Islands. Most of Europe from about 58° southwards; temperate North America; introduced in central Africa and Australia. Circumpolar Temperate element.

Section **4. Pochila** Dumort.
Omphalospora Bartl.; *Agerella* Fourr.; *Diplophyllum* Lehm.; *Odicardis* Raf.

Annual monoecious *herbs* with fibrous roots, except for *V. filiformis* with creeping fibrous roots. *Inflorescence* a terminal raceme or flowers solitary in leaf axils. *Corolla* tube wider than long. *Capsule* dehiscing into 2 or 4 valves; seeds fairly flat, flat to convex on 1 side, flat to concave on the other.

13. V. acinifolia L. French Speedwell
Agerella acinifolia (L.) Fourr.; *Cardia orbicularis* Dulac nom. illegit.

Annual monoecious *herb* with fibrous roots. *Stems* 5–15 cm, pale green, erect or ascending, more or less glandular-hairy throughout or only above, leafy, branched from the base. *Leaves* opposite; lamina 4–12 mm, medium green on upper surface, paler beneath, ovate, obtuse at apex, obscurely crenate to almost entire, rounded at base, sparsely hairy or nearly glabrous, shortly petiolate. *Inflorescence* a terminal raceme, lax, with numerous flowers; peduncles and rhachis with numerous, unequal glandular hairs; pedicels 3–7(–10) mm, slender, spreading; upper bracts elliptical-oblong and entire, the lower passing into the leaves, as long as or shorter than the pedicels; flowers 2–3 mm in diameter, bisexual. *Calyx* 2–3 mm, divided almost to the base into 4 lobes, the lobes ovate-oblong, obtuse at apex, glandular-hairy. *Corolla* 2–3 mm, divided almost to the base into 4 lobes, the lobes blue, obovate, rounded at apex, the tube wider than long. *Stamens* 2; filaments slender; anthers blue. *Style* 1, about 1 mm. stigma subcapitate. *Capsule* 2–3 mm, wider than long, slightly longer than calyx, 2-lobed, with a deep sinus, with patent glandular hairs; seeds 20–40, 0.7–1.2 × 0.5–0.8 mm, pale brown, broadly elliptical, flat. *Flowers* 4–6. $2n=14, 16$.

Introduced. Casual or persistent weed of gardens, nurseries and public flower beds. Recorded from Somersetshire, Dorsetshire, Sussex, Surrey and Edinburghshire. Native of south, south-central and western Europe northwards to north-central France; Turkey.

14. V. praecox All. Breckland Speedwell
Omphalospora praecox (All.) Fourr.; *Cardia praecox* (All.) Dulac; *Cochlidiospermum praecox* (All.) Opiz

Annual monoecious *herb* with fibrous roots. *Stems* 5–20 cm, pale green, erect, shortly glandular-hairy, leafy, with erect or ascending branches from the base. *Leaves* opposite; lamina 4–12 × 4–10 mm, metallic medium green on upper surface, pale and often purple-tinted beneath, ovate, obtuse at apex, deeply serrate, rounded at base, glandular-hairy; petiole short. *Inflorescence* a terminal raceme, lax, with numerous flowers; peduncle and rhachis glandular-hairy; pedicels 5–8 mm, slender, glandular-hairy; upper bracts elliptic-oblong, entire, the lower passing into the leaves, slightly shorter than the pedicels; flowers 2.5–4.0 mm in diameter, bisexual. *Calyx* 2.5–4.5 mm, divided almost to the base into 4 lobes, the lobes oblong, obtuse at apex, with glandular hairs. *Corolla* 2.5–3.0 mm, divided almost to the base into 4 lobes, the lobes deep blue, obovate, rounded at apex, the tube wider than long. *Stamens* 2; filaments pale; anthers blue. *Style* 1, about 2 mm. stigma subcapitate. *Capsule* 3–4 mm, longer than wide, obovate, emarginate, rather longer than calyx, with numerous glandular hairs; seeds 0.7–1.0 mm, cup-shaped, broadly elliptical. *Flowers* 3–6. $2n=18$.

Introduced. Sandy arable fields. Very local in the Breckland of Suffolk and Norfolk where it was first recorded in 1933. South, central and west Europe, extending northwards locally to south-east Sweden; Turkey, Caucasus; North Africa. A European Southern-temperate species absent as a native from much of western Europe.

15. V. triphyllos L. Fingered Speedwell
Cochlidiospermum digitatum Opiz nom. illegit.; *Omphalospora triphylla* (L.) Fourr.

Annual monoecious *herb* with fibrous roots. *Stems* 5–20 cm, green, sometimes suffused with brownish-purple, erect or decumbent, terete, glandular-hairy, leafy, with spreading or decumbent branches from near the base. *Leaves* opposite; lamina 8–12×5–8 mm, medium green, sometimes suffused brownish-purple, paler beneath, broadly ovate in outline, digitately deeply 3- to 7-lobed, the lobes spathulate to oblong and obtuse at apex, and rounded at base, glabrous or slightly hairy, the lower shortly petiolate, the upper divided almost to the base and sessile. *Inflorescence* a terminal raceme, with 8–20(–30) flowers; peduncle and rhachis with short eglandular hairs and longer glandular hairs; pedicels 5–15 mm, slender, with numerous glandular hairs; bracts shorter than the pedicels, the uppermost entire, passing gradually but quickly into the lower leaf-like ones; flowers 3–4 mm in diameter, bisexual. *Calyx* 5–6 mm, divided almost to the base into 4 lobes, the lobes shorter than pedicels, spathulate and obtuse at apex, with sparse glandular hairs. *Corolla* 2.0–2.5 mm, bright blue, divided almost to the base into 4 lobes, the lobes shorter than the calyx, deep blue, obovate and obtuse at apex, the tube wider than long. *Stamens* 2; filaments slender; anthers violet. *Style* 1, about 1 mm. stigma subcapitate. *Capsule* 5–6 mm, about as wide as long, subrotund, compressed, deeply lobed, with patent glandular hairs; seeds about 1.5 mm, cup-shaped, dark brown, verruculose. *Flowers* 4–6. Visited by small bees, but often selfed. $2n=14$.

Introduced. Sandy arable fields or open patches in waste places. Very local in Norfolk and Suffolk; formerly in scattered localities from Surrey to Yorkshire, decreasing with conservation management ensuring its survival in remaining sites. Europe from southern Sweden and Latvia southwards, but rather rare in the Mediterranean region; Turkey; Caucasus; North Africa.

16. V. arvensis L. Wall Speedwell
Agerella arvensis (L.) Fourr.; *Cardia arvensis* (L.) Dulac

Variable, *annual,* monoecious *herb* with fibrous roots. *Stems* 2–40 cm, green, sometimes suffused brownish-purple, erect to decumbent, shortly hairy, usually more or less glandular-hairy, leafy, simple or branched at base with ascending branches. *Leaves* opposite; lamina 3.5–25×3–15 mm, dull medium green on upper surface, paler beneath, ovate or triangular-ovate, obtuse or subacute at apex, serrate to coarsely crenate-serrate, rounded at base, the lowest petiolate, the upper sessile, with few or no hairs on upper surface and more numerous ones on the veins beneath. *Inflorescence* a terminal raceme, short and dense to long and lax and occupying the greater part of the stem, with 10–30 bisexual flowers; peduncle and rhachis with numerous hairs; pedicels up to 2 mm, hairy; upper bracts 4–7 mm, lanceolate, entire, ciliate, longer than the flowers, the lower passing into the leaves; flowers 2–3 mm in diameter. *Calyx* 3–5 mm, divided almost to the base into 4 unequal lobes, the lobes lanceolate, acute at apex, hairy. *Corolla* 1.0–1.2 mm, divided almost to the base into 4 lobes, the lobes shorter than the calyx, blue, ovate or subrotund, subacute at apex, the tube wider than long. *Stamens* 2; filaments about 0.4 mm. anthers pale. *Style* 1, up to 1 mm. stigma subcapitate. *Capsule* 2.5–3.0 mm, about as long as wide, shorter than calyx, obcordate, glandular-hairy; seeds about 0.8 mm, brown, oblong, flat, rugulose. *Flowers* 3–10. Visited by small bees, probably often selfed. $2n=16$.

(i) Var. **nana** Poir.
V. arvensis var. *glandulosa* Legr.; *V. arvensis* var. *eximia* Towns.; *V. arvensis* var. *perpusilla* Bromf.

Plant 2–8 cm, erect, densely glandular-hairy, without branches or with an occasional branch, rarely with several branches from base. *Leaves* 3.5–6.0 mm, numerous and rather dense, entire or bluntly toothed. *Inflorescence* short and dense.

(ii) Var. **arvensis**
Plant 10–40 cm, erect, glandular-hairy, with long branches from the base. *Leaves* 5–15 mm, fairly widely spaced, bluntly toothed. *Inflorescence* very long (–30 cm) and very open.

(iii) Var. **polyanthos** (Thuill.) Mathieu
V. polyanthos Thuill.

Plant spreading or decumbent, more or less glandular-hairy, branched from base. *Leaves* 10–25 mm, fairly widely spaced, more or less acutely toothed. *Inflorescence* short.

Native. Walls, banks, open acid or calcareous grassland, and cultivated land, usually on dry soils. Common throughout Great Britain and Ireland. Almost throughout Europe; central and west Asia; North Africa; Macaronesia; introduced and naturalised in North America. European southern-temperate element, now Circumpolar southern-temperate. Var. *nana* probably occurs on sandy areas all round our coast and on sandy heaths inland particularly in Breckland. It is probably our native plant. Var. *arvensis* and var. *polyanthos* are particularly weeds of cultivated and waste places and may be introduced. All are recorded for continental Europe.

17. V. verna L. Spring Speedwell
Agerella verna (L.) Fourr.; *Cardia verna* (L.) Dulac

Annual monoecious *herb* with fibrous roots. *Stems* 5–15 cm, pale green, often suffused purplish, erect, with simple eglandular hairs below and glandular hairs above, simple or branched from low down, leafy. *Leaves* opposite; lamina 4–12×5–7 mm, medium green on upper surface, paler beneath, lanceolate or ovate, subacute at apex, coarsely crenate, the lower shortly petiolate, the upper pinnatifid with narrow lobes and sessile glandular hairs. *Inflorescence* a dense terminal raceme; bracts exceeding pedicels, the upper lanceolate, the lower pinnatisect; flowers bisexual. *Calyx* 3.5–4.0 mm, glandular-hairy, divided nearly to base into 4 lobes, the lobes lanceolate and obtuse at apex. *Corolla* about 3 mm in diameter, blue, with a white apex, divided almost to the base into 4 lobes, the upper lobe largest, broadly elliptical and rounded at apex, the lateral lobes are broadly ovate and rounded at apex, the lower lobe is oblanceolate and obtuse at apex. *Stamens* 2; filaments slender; anthers bluish. *Style* 1; stigma capitate. *Capsule* about 3×4 mm, reniform, flat, glandular-hairy; seeds about 1.0×0.7 mm, broadly elliptical, flat. *Flowers* 5–6. $2n=16$.

Native. Open places in poor grassland on dry sandy soils. Very local in Suffolk; formerly in Norfolk. Much of

Europe, but absent from most of the north and parts of the west and of the Mediterranean region; western Asia east to the Altai; Morocco. Eurosiberian Temperate element.

18. V. peregrina L. American Speedwell

Annual monoecious *herb* with fibrous roots. *Stems* 5–25 cm, green or flushed brownish-purple, erect, glabrous, leafy, simple or with spreading branches. *Leaves* opposite; lamina 10–30×2–10 mm, medium green on upper surface, paler beneath, oblanceolate to narrowly oblong, rounded at apex, entire to distantly crenate, narrowed at base, glabrous; petiole short. *Inflorescence* a terminal raceme, long and lax; peduncle and rhachis glabrous, rarely glandular-hairy; pedicels 1–2 mm, glabrous; bracts lanceolate, entire, much longer than the flowers; flowers 2–3 mm, in diameter, bisexual. *Calyx* 4.0–4.5 mm, divided almost to the base into 4 lobes, lanceolate, obtuse at apex. *Corolla* 2–3 mm, divided almost to the base into 4 lobes, the lobes shorter than calyx, whitish or bluish, rounded at apex; the tube wider than long. *Stamens* 2; filaments pale; anthers bluish. *Style* 1, very short. *Capsule* about 3–4 mm, about as wide as long, scarcely emarginate, glabrous; seeds 0.7×0.4 mm, narrowly elliptical, flat. *Flowers* 4–7. Self-pollinated. $2n=52$.

Introduced in 1680. Casual or persistent weed in gardens, nurseries and flower beds in parks and in damp waste places. Scattered records in Great Britain and Ireland. Native of North and South America; widely naturalised in west and central Europe.

19. V. agrestis L. Green Field Speedwell

Pocilla agrestis (L.) Fourr.; *V. agrestis* var. *micrantha* Drabble; *Cardia agrestis* (L.) Dulac; *Cochlidiospermum agreste* (L.) Opiz

Annual monoecious *herb* with fibrous roots. *Stems* 5–30 cm, pale green, erect or decumbent, or ascending, with short eglandular hairs, leafy, branched at base. *Leaves* opposite; lamina 5–15×10–14 mm, dull pale or medium green on upper surface, paler beneath, ovate, obtuse at apex, serrate or irregularly crenate-serrate, truncate at base, with numerous eglandular hairs; petiole short. *Inflorescence* of a solitary flower in a leaf axil; pedicels 5–15 mm, densely hairy; flowers 3–8 mm in diameter, bisexual. *Calyx* 3–7 mm, divided almost to the base into 4 lobes, the lobes oblong or ovate-oblong, more or less obtuse at apex, faintly veined, with numerous hairs. *Corolla* 3–6 mm, divided almost to the base into 4 lobes, the lobes whitish to pale blue or lilac, the upper broadly elliptic, rounded at apex, the lateral ovate and obtuse at apex, the lower narrower, the tube wider than long. *Stamens* 2; filaments pale; anthers bluish *Style* 1, very short. *Capsule* 3–4×4–6 mm, reniform in outline, 2-lobed, the lobes obscurely keeled, with long glandular hairs, often with shorter simple eglandular hairs mixed but without short crispate hairs; seeds 1.5×1.2 mm, broadly elliptical, concave 1 face. *Flowers* 1–12. Visited by Diptera and Hymenoptera; homogamous, often selfed. $2n=28$.

Introduced. Cultivated and waste ground preferring soils that are well-drained and acidic. Frequent throughout most of Great Britain and Ireland. Most of Europe except the extreme north and parts of the south-east; mountains of North Africa; Turkey. The var. *micrantha* seems to differ only in its very small flowers.

20. V. polita Fr. Grey Field Speedwell

Pocilla polita (Fr.) Fourr.

Annual monoecious *herb* with fibrous roots. *Stems* 4–20 cm, pale green, erect or decumbent to ascending, with crispate hairs, leafy, branched at base. *Leaves* opposite; lamina 5–15×4–9 mm, dull medium green on upper surface, paler beneath, broadly ovate, obtuse or subacute at apex, coarsely and regularly crenate-serrate, more or less truncate at base, the lower wider than long, the upper longer than wide, with few or no hairs on upper surface and more numerous on the veins below; petioles short. *Inflorescence* of solitary flowers in a leaf axil; pedicels 5–15 mm, as long as or shorter than the bracts, decurved in fruit, crispate-hairy; flowers 4–8 mm in diameter, bisexual. *Calyx* 3.5–4.0 mm, divided almost to the base into 4 lobes, the lobes ovate, more or less acute, conspicuously 5-veined, accresent and with hairs along the margins and veins. *Corolla* 5–12 mm in diameter, divided almost to the base into 4 lobes, the lobes usually uniformly bright blue, rarely the lower lobe paler, ovate and rounded at apex, the tube wider than long. *Stamens* 2; filaments about 1.5 mm. anthers violet. *Style* 1, about 1.5 mm. stigma capitate. *Capsule* about 3.5–4.5 mm, wider than long, slightly compressed, 2-lobed, the lobes erect and not keeled, clothed with short, crispate eglandular hairs and some longer glandular ones; seeds about 1.3 mm, pale brown, cup-shaped, rugose with parallel ridges on convex surface, smooth or concave surface. *Flowers* 1–12. Usually self-pollinated. $2n=14$.

(i) Var. **polita**

Flowers 5–8 mm in diameter.

(ii) Var. **grandiflora** Bab.

Flowers 8–12 mm in diameter.

Introduced. Cultivated and waste places. Frequent in most of Great Britain, but rare in central and north Scotland and absent from much of west and central Ireland. Most of Europe except the Arctic; temperate Asia; North Africa. Eurosiberian Southern-temperate species. Var. *polita* is the common plant. Var. *grandiflora* is recorded from a few scattered localities.

21. V. persica Poir. Common Field Speedwell

V. buxbaumii Ten., non F. W. Schmidt; *V. tournefortii* C.C. Gmel. nom. illegit.; *Cochlidiospermum buxbaumii* Opiz; *Pocilla persica* (Poir.) Fourr.

Annual monoecious *herb* with fibrous roots. *Stems* 10–50 cm, pale green, sometimes tinted reddish, decumbent to erect, with numerous, crispate hairs, leafy, branched at the base. *Leaves* opposite; lamina 5–30×4–15 mm, medium green on upper surface, paler beneath, triangular-ovate, obtuse to subacute at apex, coarsely crenate-serrate, truncate at base, glabrous or nearly so on upper surface, shortly hairy on the veins beneath and minutely ciliate; petioles up to 5 mm, hairy. *Inflorescence* of solitary flowers in leaf axils; pedicels up to 40 mm, longer than the leaves but not twice as long, decurved in fruit, with crispate hairs;

flowers 8–12 mm in diameter, bisexual. *Calyx* 5–6 mm, divided almost to the base into 4 lobes, the lobes ovate, obtuse at apex, accrescent and strongly divaricate in fruit, ciliate. *Corolla* 8–14 mm, in diameter, divided almost to the base into 4 lobes, the lobes uniformly bright blue, rarely the lower lobe paler, ovate, rounded at apex, the tube wider than long. *Stamens* 2; filaments about 1.5 mm. anthers violet. *Style* 1, about 1.5 mm. stigma capitate. *Capsule* about 3.5 × 4.5 mm, 2-lobed, the lobes sharply keeled and divergent, so that the capsule is nearly twice as wide as long, ciliate with patent glandular and short eglandular hairs; seeds about 1.3 mm, pale brown, cup-shaped, rugose on convex surface, smooth on concave surface. *Flowers* 1–12. Visited by various insects, often selfed. $2n=28$.

(i) Var. **kochiana** Godr.
Plant slender. *Leaves* 6–15 × 5–10 mm, crenate, the teeth obtuse.

(ii) Var. **corrensiana** (E. B. J. Lehm.) Hegi
V. tournefortii subsp. *corrensiana* E. B. J. Lehm.

Plant robust. *Leaves* 10–20 × 5–15 mm, crenate, the teeth more or less double and obtuse. *Corolla* usually with pale lower lip.

(iii) Var. **persica**
V. tournefortii subsp. *aschersoniana* E. B. J. Lehm.;
V. agrestis var. *aschersoniana* (E. B. J. Lehm.) Hegi

Plant robust. *Leaves* 10–30 × 7–25 mm, dentate, the teeth more or less acute. *Corolla* uniform blue.

Introduced. Cultivated and waste land. First recorded in 1825, now common throughout most of Great Britain and Ireland. Native of south-west Asia. All three varieties are common here and on the continent of Europe. Drabble and Little (1931) say they have grown all three varieties side by side for several years in soils of various nature and humidity and they have retained their distinctive features.

22. V. crista-galli Steven Crested Field Speedwell
Diplophyllum crista-galli (Steven) Otto ex Walp.;
Odicardis crista-galli (Steven) Raf.

Annual monoecious *herb* with fibrous roots. *Stems* up to 50 cm, pale green, ascending or decumbent, with slender branches and patent, simple eglandular hairs. *Leaves* opposite below, alternate above; lamina 10–18 mm, pale green on upper surface and more so beneath, ovate or subrotund, obtuse to subacute at apex, crenate, rounded to cordate at base, ciliate, more or less petiolate. *Inflorescence* of 1–3 bisexual flowers in the axils of leaves; pedicels long and erect. *Calyx* segments fused in pairs forming conspicuous bilobed structures completely concealing capsule. *Corolla* small, half as long as the calyx, pale blue. *Stamens* shorter than corolla. *Style* short. *Capsule* shorter than calyx, emarginate, lobes not diverging, shortly ciliate; seeds ovate-orbicular, minutely rugose. *Flowers* 6–8. $2n=18$.

Introduced. Occasional casual in cultivated and rough ground and waste places. Scattered records in England and Wales, naturalised in Somerset and formerly Sussex; well naturalised in south-west Ireland. Native of the Caucasus.

23. V. filiformis Sm. Slender Speedwell

Perennial monoecious *herb* with numerous, creeping, fibrous roots, forming large patches. *Stems* up to 20 cm, creeping and rooting, pale green, usually suffused brownish-purple, slender, with dense, short, pale simple eglandular hairs, leafy. *Leaves* opposite; lamina 5–10 × 5–15 mm, dull medium yellowish-green on upper surface, paler beneath, ovate, reniform or subrotund, broadly rounded at apex, crenate, more or less cordate at base, with short, soft simple eglandular hairs on both surfaces and the margins, the veins impressed on upper surface and prominent beneath; petioles up to 5 mm, pale green, with short, spreading simple eglandular hairs. *Inflorescence* a solitary flower in a leaf axil; pedicels up to 40 mm, filiform, several times as long as the leaves, minutely glandular-hairy; flowers 5–15 mm in diameter, bisexual. *Calyx* 3–4 mm, divided nearly to base into 4 lobes, the lobes green, lanceolate, obtuse at apex, with medium simple eglandular hairs. *Corolla* 5–6 mm, divided nearly to base in 4 unequal lobes, the lobes pale lilac-blue and white at base with darker veins, broadly ovate, rounded at apex, the tube wider than long. *Stamens* 2; filaments white; anthers blue. *Style* 1, green. *Capsules* 3.5–5.0 × 5.5–6.5 mm, 2-lobed, with subparallel lobes and a rather wide sinus, with marginal patent glandular hairs, rarely produced in Great Britain and Ireland; seed about 1.3 × 1.2 mm, pale brown, rugose. *Flowers* 4–6. $2n=14$.

Introduced. First recorded in Essex in 1835, then from 1927 onwards. Now well naturalised in lawns, playing fields, streamsides, cemeteries, grassy paths, banks and roadsides. Throughout most of Great Britain and Ireland. It is often transported from one locality to another on grass mowers and is almost impossible to eradicate from lawns. Native of Turkey and Caucasus.

24. V. hederifolia L. Ivy-leaved Speedwell
Cardia quadriloba Dulac nom. illegit.; *Cochlidiospermum hederifolium* (L.) Opiz; *Pocilla hederifolia* (L.) Fourr.

Annual monoecious *herb* with fibrous roots. *Stem* 10–60 cm, pale green, decumbent or prostrate, with fairly numerous simple eglandular hairs, leafy, usually much-branched at the base. *Leaves* opposite; lamina 5–20 × 5–25 mm, yellowish-green on upper surface, slightly paler beneath, ovate to more or less reniform, obtuse at apex, with 3–7 large teeth or small lobes on each side near the base, more or less truncate at base, ciliate and with a few, scattered, short hairs beneath and 3-veined; petioles 5–10(–15) mm, flattened, and hairy. *Inflorescence* of solitary flowers in leaf axils; pedicels 5–15 mm, usually rather shorter than leaves, with few simple eglandular hairs; flowers 4–9 mm in diameter, bisexual. *Calyx* 5–7 mm, divided almost to the base into 4 lobes, the lobes ovate-deltoid, rounded at apex, more or less enlarging after flowering, glabrous or hairy round the margin. *Corolla* 2.5–3.0 mm, divided almost to the base into 4 lobes, the lobes pale lilac or blue, ovate, obtuse at apex, shorter than the calyx, the tube wider than long. *Stamens* 2; filaments whitish; anthers blue, pale lilac or whitish. *Style* 1, 0.3–1.0 mm, whitish; stigma capitate. *Capsule* 2.5–3.0 × 4–5 mm, pale brown, subglobose, scarcely compressed and scarcely emarginate, glabrous; seeds 2.0–2.5 mm, pale brown, cup-shaped, almost smooth. *Flowers*(3–)4–5(–8). Visited by various insects, often selfed.

(a) Subsp. **hederifolia**

Leaves rather thick, apical lobe usually wider than long, 3- to 5-toothed or lobed. *Fruiting pedicels* usually 2–4 times as long as calyx. *Calyx* enlarging strongly after flowering, with marginal hairs of lobes mostly more than 0.9 mm. *Corolla* 6–9 mm in diameter, pale blue with a white centre. *Anthers* 0.7–1.2 mm, blue. *Style* 0.7–1.0 mm. $2n=54$.

(b) Subsp. **lucorum** (Klett & Richt.) Hartl

V. sublobata M. A. Fisch.

Leaves thin, apical lobe usually longer than wide, shallowly 5- to 7-toothed or lobed. *Calyx* enlarging slightly after flowering, with marginal hairs of lobes mostly less than 0.9 mm. *Corolla* 4–6 mm in diameter, whitish to pale lilac. *Anthers* 0.4–0.8 mm, whitish to pale lilac-blue. *Style* 0.3–0.5 mm. $2n=36$.

Possibly native. Cultivated and waste ground, open woods, hedgerows, walls and banks and gardens. Throughout Great Britain and Ireland except some parts of Scotland and Ireland. Europe, temperate Asia to Japan; North Africa; Madeira; naturalised in North America. European Temperate element. Subsp. *hederifolia* is widespread, but more local than subsp. *lucorum* in our area. It is the most common subspecies of sandy heaths and the coast and could be native. It occurs throughout the European range of the species. Subsp. *lucorum* is found in woodland, shady places and gardens. It is widespread and common in Great Britain and Ireland and occurs in north, north-west and central Europe, extending into the northern part of the Balkan Peninsula. Single plants are frequently difficult to name as regards subspecies, but populations can usually be managed.

Section **5. Pseudolysimachium** W. D. J. Koch

Hedystachys Fourr.

Perennial monoecious *herbs* with shortly creeping, woody rhizome. *Inflorescence* a terminal, spike-like raceme with many flowers. *Corolla* tube longer than wide. *Capsule* dehiscing into 2 valves, usually only apically; seeds flat on 1 side, convex on the other.

25. V. longifolia L. Garden Speedwell

Pseudolysimachion longifolium (L.) Opiz

Perennial monoecious *herb. Stems* 40–120 cm, pale green, robust, erect, glabrous or shortly hairy, usually simple below, branched above, leafy. *Leaves* opposite; lamina 30–120 × 5–20 mm, dark green on upper surface, paler beneath, linear-lanceolate to lanceolate, acuminate at apex, acutely biserrate, truncate to cuneate at base, glabrous or sparsely hairy; petiole usually less than 10 mm. *Inflorescence* a very dense, terminal raceme with one or more small, lateral branches; flowers bisexual; pedicels 1–2 mm, shorter than bracts; bracts linear-filiform. *Calyx* 2–3 mm, divided almost to base into 4 lobes, the lobes unequal, triangular-ovate, acute at apex. *Corolla* 6–8 mm in diameter, lilac or pale blue, divided nearly to base into 4 lobes, the upper large and rounded, the lateral ovate and rounded and the lower narrower. *Stamens* 2; filaments pale; anthers bluish. *Style* 1, 4–10 mm. stigma small. *Capsule* about 3 × 3 mm, broadly ovoid to globose, emarginate at apex, glabrous; seeds about 0.7 mm. *Flowers* 7–9. $2n=68$.

Introduced. Much grown in gardens and naturalised on waste and rough ground, banks and roadsides. Scattered records throughout Great Britain mainly in the centre and south. North, east and central Europe.

× **spicata**

A single record from Sussex.

26. V. spicata L. Spiked Speedwell

Cardia spicata (L.) Dulac; *Hedystachys spicata* (L.) Fourr.; *Pseudolysimachion spicatum* (L.) Opiz

Perennial monoecious *herb* with shortly creeping, somewhat woody rhizome. *Stems* 8–60(–80) cm, pale green, erect, with short simple eglandular hairs, leafy, branched. *Leaves* opposite; lamina 15–40 × 8–20 mm, slightly greyish, medium green on upper surface, paler beneath, ovate, lanceolate or linear, obtuse to acute at apex, more or less crenate or crenate-serrate, with obtuse teeth, the uppermost more or less entire, attenuate at base, with more or less numerous simple eglandular hairs; petiole up to 20 mm, pale, hairy. *Inflorescence* a terminal, spike-like raceme with many flowers; peduncle and rhachis with numerous glandular hairs; pedicels 0.5–1.0 mm. bracts more or less lanceolate; flowers 4–8 mm in diameter, bisexual. *Calyx* 3.0–3.5 mm, divided almost to base into 4 lobes, the lobes more or less lanceolate, obtuse at apex, with numerous glandular hairs. *Corolla* 5–6 mm, divided almost to the base into 4 lobes, the lobes violet-blue, lanceolate, obtuse at apex, the tube longer than wide. *Stamens* 2; filaments pale; anthers bluish. *Style* 1, 4–10 mm. stigma small, terminal. *Capsule* 2–4 × 2–4 mm, pale brown, more or less globose, retuse or emarginate, about as long as calyx, dehiscing into 2 valves, usually only apically; seeds about 1.0 mm, ellipsoid, flat on 1 side, convex on the other, glabrous. *Flowers* 7–9. Pollinated by various insects. $2n=34$, 68.

(a) Subsp. **spicata**

Plant 8–30 cm. *Lowest leaves* 15–30 × 8–12 mm, usually widest near the middle, sparingly crenate, mostly near the middle, gradually narrowed into a narrow petiole.

(b) Subsp. **hybrida** (L.) Gaudin

V. hybrida L.

Plant 15–60(–80) cm. *Lowest leaves* 20–40 × 10–20 mm, usually widest below the middle, more deeply crenate or crenate-serrate, abruptly narrowed into a broad petiole.

Native. Subsp. *spicata* is a rare plant in dry grassland on basic soils in four sites in Cambridgeshire, Norfolk and Suffolk. Subsp. *hybrida* occurs on limestone rocks in a few places in Wales and western England. The species occurs in Europe from southern Scandinavia to central Spain, Italy and Greece; west and central Asia to China and Japan. Eurosiberian Temperate element. Both subspecies occur in Europe, but there are several additional subspecies. It is often difficult to place individual plants but there is not usually any trouble in placing a population.

19. Veronicastrum Moench

Perennial monoecious *herbs*. *Leaves* in whorls of 6–8. *Inflorescence* of long, numerous-flowered, erect racemes at the ends of stems and branches; flowers bisexual. *Calyx* 5-lobed, the 3 upper lobes longer than the 2 lower. *Corolla* tubular, whitish below, lilac above, with 5 unequal lobes. *Stamens* 4. *Style* 1. *Fruit* an ovoid capsule opening by 4 apical slits.

Contains 2 species native to North America and east Asia.

1. V. virginicum (L.) Farw. Culver's-root
Veronica virginica L.; *Leptandra virginica* (L.) Nutt.

Perennial monoecious *herb*. *Stems* up to 1 m, pale green, suffused brownish-purple, erect, ridged, with numerous simple eglandular hairs below, becoming glabrous above, branched only in inflorescence, very leafy. *Leaves* in whorls of 6–8; lamina 5–9 × 1.5–2.5 cm, dull medium to dark green on upper surface, much paler and bluish-green beneath, lanceolate, sharply acute to acuminate at apex, serrate, cuneate at base, glabrous or nearly so with impressed veins on upper surface, shortly hairy with prominent veins beneath; petiole very short, pale green, hairy. *Inflorescence* of long, numerous-flowered, erect racemes at the ends of the stem and branches; flowers bisexual; peduncles brownish-purple, glabrous; rhachis brownish-purple, glabrous. *Calyx* 5-lobed, the 3 upper lobes longer than the 2 lower, lanceolate. *Corolla* tubular, whitish below, lilac above, with 5 unequal lobes, the lobes ovate and subacute at apex. *Stamens* 4; filaments white; anthers yellow, exserted. *Style* 1. *Capsule* 2–5 mm, ovoid, acute at apex, scarcely flattened, opening by 4 short, apical slits. *Flowers* 6–8.

Introduced. A garden escape sometimes persistent for a few years. Native of eastern North America.

20. Hebe Comm. ex Juss.

Evergreen, monoecious *shrubs*. *Leaves* opposite, with pinnate venation, but usually with the lateral veins obscured, more or less entire. *Inflorescence* of opposite, axillary racemes; flowers bisexual. *Calyx* 4-lobed. *Corolla* white, blue or purple, 4-lobed. *Stamens* 2. *Style* 1. *Fruit* a septicidal capsule, compressed parallel to the septum.

Contains about 100 species confined to New Zealand and adjacent islands.

The leaf-bud is composed of successively smaller developing leaves without any modified bud-scales. The outermost 2 leaves enclose all the inner ones and their margins meet along 2 sides, but towards their base a gap is sometimes left on either side, due to the presence or absence of a distinct petiole, which is a terminal leaf sinus.

Allan, H. H., Simpson, G. & Thompson, J. S. (1926). A wild hybrid *Hebe* community in New Zealand. *Genetica* **5**: 375–388.
Allan, H. H. (1961). *Flora of New Zealand*. **1**. Reprinted. 1982. Wellington.
Hutchins, G. (1997). *Hebes here and there*. Reading.

This is another genus where the taxa are mainly geographically or ecologically replacing and when the ecology is upset, or two taxa are brought together in cultivation, hybrid swarms occur. It is fascinating to read what Allan, Simpson and Thomson (1926) wrote all that time ago and realise how the same thought has occurred to us about many groups when writing this flora.

1. Leaf-buds with distinct sinuses; leaves more or less petiolate 2.
1. Leaf-buds without sinuses; leaves more or less sessile 4.
2. Leaves 1.6–3.5 cm wide, broadly elliptical **3. × franciscana**
2. Leaves 0.5–1.8 cm wide, oblong, narrowly elliptical or lanceolate-oblong 3.
3. Leaves 2–7 cm **1. salicifolia**
3. Leaves 2–3 cm **4. brachysiphon**
4. Leaves 1.6–2.5 cm wide **2. × lewisii**
4. Leaves 1.0–1.8 cm wide 5.
5. Leaves 4.5–8.0 cm, medium green; stem glabrous to minutely hairy, green **5. dieffenbachii**
5. Leaves 3–6 cm, greyish-green; stem glabrous turning purple **6. barkeri**

1. H. salicifolia (G. Forst.) Pennell Koromiko
Veronica salicifolia G. Forst.

Evergreen, monoecious *shrub* about 2 m high, strong growing with a more or less bushy habit. *Stems* pale green, glabrous. *Bud* about 50 mm, green, narrowly oblong-elliptical, pointed at apex, with a distinct elliptical sinus. *Leaves* opposite; lamina 7–12 × 1.5–1.8 cm, pale green on upper surface, even paler beneath, narrowly oblong-elliptical or lanceolate-oblong, acute at apex, minutely denticulate, narrowed at base, minutely hairy on both surfaces and the margins, midrib pale and prominent beneath; petiole very short. *Inflorescence* a raceme about 12 cm, often drooping; flowers about 8 mm in diameter, bisexual, not dense; pedicels about 4 mm, pale green, minutely hairy; peduncle 2–3 cm, pale green, minutely hairy; bracts 2–10 mm, green, lanceolate. *Calyx* about 3 mm, divided nearly to base into 4 lobes, the lobes triangular-ovate, acute at apex, the margins ciliate. *Corolla* about 10 mm, pale lilac-blue fading to white; tube 3.5 × 2.5 mm, white, hairy inside; 4-lobed, the upper lobe ovate and slightly hooded, the lateral lobes broadly ovate, the lower lobe cup-like. *Stamens* 2; filaments pale lilac; anthers purple. *Style* 1, about 10 mm, pale purple, with a few minute hairs. *Capsule* about 4 × 3 mm, often drooping, obscurely hairy at the apex. *Flowers* 6–12. $2n = 40$.

Introduced. Naturalised by the sea in Devonshire, Cornwall and the Isle of Man and scattered records throughout Great Britain. Native of South and Stewart Islands, New Zealand and the islands and coasts of southern Chile.

2. H. × lewisii (J. B. Armstr.) A. Wall Lewis's Hebe
H. elliptica × salicifolia
Veronica lewisii J. B. Armstr.; *Veronica amabilis* Cheeseman; *Veronica salicifolia* var. *gracilis* Kirk; *Hebe amabilis* var. *blanda* Cheeseman; *H.* × *blanda* (Cheeseman) Pennell

Evergreen, monoecious, erect *shrub* up to 4 m. *Stems* up to 20 mm in circumference. *Branches* slender to stout, terete, glabrous or with short, greyish-white hairs. *Leaves* opposite, spreading; lamina 5–10×1.6–2.5 cm, pale green on upper surface, more so beneath, oblong-lanceolate, elliptic-lanceolate, oblong or elliptic-oblong, more or less acute at apex, entire, flat, smooth, glabrous or sometimes shortly hairy on the midrib beneath, margins hairy, shortly petiolate. *Inflorescence* a lax, axillary raceme 5–15 cm; pedicels up to 1 mm, slender; flowers about 8 mm in diameter, bisexual. *Calyx* 4-lobed, the lobes ovate, ovate-lanceolate or oblong, acute at apex, ciliate. *Corolla* white or pale blue; tube short and broad; 4-lobed, the lobes about 6×4 mm, oblong, spreading, the upper and lateral similar, the lower smaller. *Stamens* 2; filaments pale; anthers purple. *Style* 1, purplish. *Capsule* about 6 mm, ovate, acute at apex, compressed. *Flowers* 6–12.

Introduced. Naturalised in Isle of Man, Somersetshire, Essex, Suffolk, Cornwall and Dunbartonshire. Native of South and Stewart Islands, New Zealand.

3. H.×franciscana (Eastw.) Souster Hedge Veronica
H. elliptica×speciosa
H. lewisii auct.; *H. elliptica* auct.; *H. speciosa* auct.

Evergreen, monoecious *shrub* to 1 m with a bushy habit. *Twigs* pale green to slightly purplish, minutely hairy in 2 broad lines. *Bud* about 2×1 cm, yellowish with a pale green to purplish centre, the sinus large, elliptical or oblong. *Leaves* opposite; lamina 3–7×1.6–3.5 cm, thick, dark green in centre with pale cream margins or glossy dark green, paler beneath, broadly elliptical, rounded at apex, entire, rounded at base, the margins more or less hairy and sometimes purple with a few minute hairs; petiole up to 5 mm, sometimes purple. *Inflorescence* a raceme 3–8×2.0–3.5 cm; flowers 10–14 mm in diameter, bisexual, more or less dense; pedicels 2–5 mm, hairy; peduncle 1–2 cm, green, hairy all round; bracts 1–3, obtuse at apex, ciliate. *Calyx* 3–4 mm, green, 4-lobed, the lobes subacute at apex, ciliate. *Corolla* violet, violet-purple or violet-blue, sometimes fading to pale pinky-lilac or white; tube about 2–3 mm, sometimes with a few hairs inside; lobes 4, spreading, obtuse at apex. *Stamens* 2; filaments 7–9 mm, violet or violet-purple; anthers purple or violet-purple. *Style* 1, 8–12 mm, purple, glabrous. *Capsule* 5–8×4–6 mm, glabrous.

Introduced. Well naturalised by the sea in Devonshire, Cornwall, south and north Wales, south and west Ireland and the Channel Islands and widely recorded elsewhere in Great Britain. The most grown and naturalised taxon and often used for seaside hedging in south-west England.

4. H. brachysiphon Summerh. Hooker's Hebe

Evergreen, monoecious *shrub* to 1 m with a bushy habit. *Bud* about 15×4 mm, spathulate, obtuse at apex, with a large, oblong sinus. *Leaves* opposite; lamina 2–3×0.5–0.7 cm, shiny medium green on upper surface, paler and duller beneath, oblong or narrowly elliptical, subacute at apex, slightly recurved, entire, attenuate at base, glabrous or with minute hairs on the margin, midrib pale; petiole short. *Inflorescence* a raceme 5–7 cm, occasionally branched; flowers about 9 mm in diameter, bisexual, not dense; pedicels 1–3 mm, green, hairy; peduncle up to 15 mm, minutely hairy; bracts about 2×3 mm, green. *Calyx* green, with 4 lobes 2–3 mm, with pale ciliate margins. *Corolla* white; tube 4–5 mm, cylindrical, glabrous, 4-lobed, the lobes spreading, upper lobe obovate, rounded at apex, the 2 lateral ovate, rounded at apex and downturned, the lower oblong, rounded at apex. *Stamens* 2; filaments about 6 mm, white; anthers pale purple. *Style* 1, about 10 mm, white to pale purple, glabrous. *Capsule* about 4×3 mm, obtuse, glabrous; seeds 2 mm in diameter. *Flowers* 6–8. $2n=120$.

Introduced. Naturalised in the Isles of Scilly and Dorsetshire, and recorded from Cornwall, Somersetshire, Essex and Isle of Man. Native of New Zealand.

5. H. dieffenbachii (Benth.) Cockayne & Allan Dieffenbach's Hebe
Veronica dieffenbachii Benth.

Evergreen, monoecious *shrub* up to 150 cm. *Stems* green, glabrous to minutely hairy, with stout, wide-spreading, grey branches. *Bud* 40–50 mm, greyish-green, narrowly elliptical-oblong, obtuse at apex, diamond-shaped in cross-section and without a sinus. *Leaves* opposite; lamina 4.5–8.0×1.2–1.8 cm, thick to almost fleshy, medium green, narrowly elliptical-oblong, obtuse at apex, entire, narrowed below and amplexicaul, the bases of opposite pairs almost meeting round the twig, subglabrous to minutely hairy on both surfaces, the pale margins distinctly ciliate and midrib prominent beneath. *Inflorescence* a raceme 6–9 cm; flowers about 7 mm in diameter, bisexual, rather dense; pedicels about 2 mm, green; peduncle pale green, hairy; bracts less than 1 mm. *Calyx* 2–3 mm, green, divided nearly to the base into 4 lobes, the lobes acute at apex, ciliate. *Corolla* white to purplish-lilac; tube 3×2 mm, hairy inside, 4-lobed, the lobes obtuse, spreading. *Stamens* 2; filaments 5 mm, white; anthers dark purple. *Style* 1, about 4 mm, white, hairy. *Capsule* about 4×3 mm, acute at apex, hairy; seeds about 1 mm in diameter, flat. *Flowers* 6–7. $2n=40$.

Introduced. Naturalised in Cornwall, Dorsetshire and Worcestershire. Native of Chatham Island, New Zealand.

6. H. barkeri (Cockayne) A. Wall Barker's Hebe
Veronica barkeri Cockayne

Evergreen, monoecious *shrub* up to 100 cm. *Stems* purple, glabrous, with erect branches. *Bud* 20–40×5–8 mm, greyish-green, narrowly elliptical-oblong, diamond-shaped in cross-section, without a sinus. *Leaves* opposite; lamina 3–6×1.0–1.6 cm, pale greyish-green, thick, narrowly elliptical, subacute at apex, entire, broad at base, glabrous but for a few hairs on the midrib on the upper surface, midrib not very prominent beneath. *Inflorescence* a raceme 6–9 cm; flowers 5–7 mm in diameter, bisexual, dense; pedicels 2–3 mm, purplish, hairy; peduncles green to purplish; bracts about as long as the pedicels, narrow, acute at apex, ciliate. *Calyx* about 3 mm, purple at the base, divided halfway into 4 lobes, the lobes triangular, acute at apex, the margins purplish. *Corolla* purplish-blue with a white throat, the whole flower fading to white; tube 3×2 mm, white, hairy inside; lobes 4, the upper broadly ovate, the lateral ovate, the lower much smaller and lanceolate. *Stamens* 2; filaments about 4 mm, white; anthers grey. *Style* 5–6 mm, pale violet to white, very hairy. *Capsule* about 5×4 mm, acute at apex; seeds about 1 mm in diameter. *Flowers* 6–8. $2n=40$.

Introduced. Naturalised in Cornwall, Somersetshire, Doresetshire and Yorkshire. Native of Chatham Island, New Zealand.

21. Sibthorpia L.

Stoloniferous, *perennial,* monoecious *herbs. Leaves* clustered at nodes below, alternate above, with palmate venation, exstipulate. *Inflorescence* of solitary, axillary bisexual flowers. *Calyx* divided nearly to base into 4–5 lobes. *Corolla* 5-lobed, small and inconspicuous, rotate, more or less actinomorphic. *Stamens* (3–)4(–5). *Style* 1. *Fruit* a loculicidal capsule.

Contains about 5 species in south and west Europe, Azores, Madeira and the mountains of tropical Africa, Central America and the Andes of South America.

Stewart, A., Pearman, D. A. & Preston, C. D. (1994). *Scarce plants in Britain.* Peterborough.

1. S. europaea L. Cornish Moneywort

Stoloniferous, *perennial,* monoecious *herb. Stems* 10–40 cm, pale green, procumbent and rooting at the nodes, filiform, with spreading eglandular hairs, leafy. *Leaves* with lamina medium yellowish-green on upper surface, paler beneath, the basal 0.5–2.5×0.5–2.0 cm, in clusters at the nodes, reniform to subrotund, rounded at apex, divided less than halfway to base in 5–7 lobes, the lobes crenate, cordate at base, with palmate venation, the upper similar but smaller and alternate, all with short, pale simple eglandular hairs on both surfaces and margins; petioles up to 30 mm, pale green, with spreading hairs. *Inflorescence* of solitary, axillary, bisexual flowers 1.0–2.5 mm in diameter. *Calyx* divided nearly to base in 4–5 lobes, the lobes broadly ovate, acute at apex, with simple eglandular hairs on margin. *Corolla* 1.5–2.5 mm, rotate, divided more than halfway into 5 lobes, the upper lobes whitish or cream, subrotund and rounded at apex, the 3 lowest lobes pinkish, broadly ovate, rounded at apex. *Stamens* (3–)4(–5); filaments pale; anthers yellowish. *Style* 1, pale; stigma capitate. *Capsule* about 1.0 mm, 2-celled, loculicidal; seeds about 0.1 mm, ellipsoid. *Flowers* 7–10. $2n=18$.

Native. Damp shady places up to 520 m. Locally frequent in south-west England, south Wales, Co. Kerry, Sussex and the Channel Islands; naturalised in the Outer Hebrides and rarely on damp lawns elsewhere. West France, west Spain, Portugal and the mountains of Greece and Crete. Oceanic Temperate element.

22. Melampyrum L.

Annual, hemiparasitic, monoecious *herbs. Leaves* opposite, entire or nearly so. *Inflorescence* a terminal, bracteate raceme or spike, the bracts not inflated and often toothed; flowers bisexual. *Calyx* tubular, not inflated, divided to more than halfway to the base into 4 entire lobes. *Corolla* 2-lipped, the upper lip laterally compressed, the lower lip shorter and 3-lobed with a boss-like palate nearly closing the mouth. *Stamens* 4, included under the upper lip of the corolla. *Style* terminal. Nectary at base of ovary. *Fruit* a compressed capsule with 1–4, ovoid seeds with an oil body.

Contains about 35 species in Europe, temperate Asia and eastern North America.

Horrill, A. D. (1972). Biological flora of the British Isles. No. 125. *Melampyrum cristatum* L. *Jour. Ecol.* **60**: 235–244.
Hultén, E. & Fries, M. (1986). Atlas of north European vascular plants north of the Tropic of Cancer.3 vols. Königstein.
Rich, T., FitzGerald, R. & Sydes, C. (1998). Distribution and ecology of Small Cow-wheat (*Melampyrum sylvaticum* L. Scrophulariaceae) in the British Isles. *Bot. Jour. Scotland* **50**: 29–46.
Smith, A. J. E. (1963). Variation in *Melampyrum pratense* L. *Watsonia* **5**: 336–367.
Stewart, A., Pearman, D. A. & Preston, C. D. (1994). *Scarce plants in Britain.* Peterborough.
Wigginton, M. J. (Edit.) (1999). *British red data books.* Vol. 1. *Vascular plants.* Peterborough.

1. Floral bracts densely overlapping, concealing inflorescence axis at least in the upper part, pink or purple at least near the base, usually with more than 3 teeth on either side at base 2.
1. Floral bracts not or scarcely overlapping, with inflorescence axis well exposed, green, usually with less than 3 teeth on either side at base 3.
2. Bracts cordate at base, strongly recurved, folded inwards along the midrib **1. cristatum**
2. Bracts rounded to cuneate at base, not recurved, not folded inwards along midrib **2. arvense**
3. Lower lip of corolla strongly reflexed; lower 2 calyx lobes patent, not upswept; capsule with 2 seeds **4. sylvaticum**
3. Lower lip of corolla not reflexed, its underside forming a straight line with the lower edge of the tube; lower 2 calyx lobes appressed to corolla and upswept; capsule with 4 seeds 4.
4. Uppermost leaves below bracts (3–)4–7(–10)× (0.4–)0.8–2.0(–2.7) cm **3(b). pratense** subsp. **commutatum**
4. Uppermost leaves below bracts (1–)2–8(–11)× (0.1–)0.2–1.0(–2.0) cm 5.
5. Corolla whitish or pale yellow **3(a,i). pratense** subsp. **pratense** var. **pratense**
5. Corolla deep yellow **3(a,ii). pratense** subsp. **pratense** var. **hians**

1. M. cristatum L. Crested Cow-wheat

Annual, hemiparasitic, monoecious *herb. Stems* 20–50 cm, medium green, often suffused brownish-purple, erect, with few hairs, leafy, simple or with a few, spreading branches. *Leaves* opposite; lamina 5–10×0.3–1.5 cm, bright medium green on upper surface, paler beneath, linear-lanceolate to lanceolate, long acute at apex, entire or nearly so, narrowed at base, sessile, with few hairs or glabrous. *Inflorescence* a dense, 4-sided, terminal spike, concealing its axis; bracts 13×15 mm, bright rosy-purple at base, green above, ovate, folded inwards along midrib, strongly recurved at apex, with fine, pectinate teeth below, cordate at base; flowers bisexual. *Calyx* 5–8 mm, tube with 2 lines of hairs; lobes 4, shorter than tube, unequal, sharply acute at apex. *Corolla* 12–16 mm, 2-lipped, the upper lip pale yellow and laterally compressed, the lower lip shorter, pale yellow tinged with purple, 3-lobed, with a large, deeper yellow palate nearly closing the throat, the tube longer than the calyx. *Stamens* 4, included under the upper lip of the corolla; filaments yellow; anthers purplish. *Capsule* 9–10 mm, suborbicular, compressed, dehiscing along 1 margin; seeds usually 4, 4.5–5.0 mm, pale brown, ovoid. *Flowers* 6–9. Pollinated by bumble-bees. $2n=18$.

Native. Wood borders, their clearings and rides and scrub. Very local in East Anglia and adjacent central England, formerly further north and west. Much of Europe except parts of the north and south; western Asia east to the Yenisei region. Eurosiberian Temperate element.

2. M. arvense L. Field Cow-wheat

Annual, hemiparasitic, monoecious *herb* with fibrous roots. *Stems* 20–60 cm, pale green, often suffused brownish-purple, erect, with few hairs or scabrid, leafy, with erect-spreading branches. *Leaves* opposite; lamina 3–8 × 0.2–1.5 cm, medium green on upper surface, paler beneath, lanceolate, long-acute at apex, the upper usually with a few, long, narrow teeth at the base, rounded at base and sessile, with few or no hairs or scabrid. *Inflorescence* a lax, cylindrical or conical, flat, terminal spike; bracts 20–35 × 10–25 mm, pink at first, lanceolate, pinnatifid, with long (up to 8 mm), slender teeth and without leaf-like points; flowers 10–12 mm in diameter, bisexual. *Calyx* 15–17 mm, tube hairy; divided halfway to base into 4 lobes, the lobes nearly equal, with slender teeth longer than the tube. *Corolla* 20–25 mm, 2-lipped, the upper lip purplish-pink, laterally compressed, the lower lip purplish-pink, sometimes marked yellow and 3-lobed, yellow in the throat and tube as long as the calyx. *Stamens* 4, included under the upper lip of the corolla; filaments yellow; anthers purplish with green backs. *Capsule* 7.0–7.5 mm, ellipsoid, compressed; seeds normally 2, 4.5–5.0 mm, pale brown, ovoid. *Flowers* 6–9. Pollinated by bumble-bees, but can be selfed when not visited. $2n=18$.

Possibly native. Cornfields and grassy field margins. Very local in Isle of Wight, Essex and Bedfordshire; formerly more widespread but still decreasing. Europe from Finland southwards to Spain, Italy and Turkey; Caucasus; western Siberia.

3. M. pratense L. Common Cow-wheat

Annual, hemiparasitic, monoecious *herb. Stems* 8–60 cm, medium green, erect, glabrous or with few hairs, leafy, with suberect or spreading branches. *Leaves* opposite; lamina 1.5–10.0 cm, medium green on upper surface, paler beneath, linear to ovate, acute at apex, entire, rounded at base, glabrous or with few hairs, sessile or shortly petiolate. *Inflorescence* of a single flower in each axil, opposite, green, leaf-like bracts, both bisexual flowers turned to the same side of the stem to form a lax, secund spike; upper bracts pectinate or dentate, except in small plants, the lower often leaf-like. *Calyx* 7–8 mm, divided more than halfway to the base into 4 lobes, the lobes linear-setaceous and appressed. *Corolla* 10–18 mm, 2-lipped, the upper lip whitish to deep yellow, sometimes tinged with red or purple and laterally compressed, the lower lip whitish to deep yellow, with 3 lobes, with a large yellow palate nearly closing the throat, the tube about twice as long as calyx, the mouth more or less closed, the lower lip straight. *Stamens* 4, included under the upper lip of the corolla; filaments pale; anthers whitish. *Style* 1, white; stigma subcapitate. *Capsule* 10–12 mm, ovoid, compressed, dehiscing along one margin; seeds normally 4, 3.0–3.5 mm, pale brown, ovoid. *Flowers* 5–10. Pollinated by bumble-bees, but can be selfed when not visited. $2n=18$.

(a) Subsp. **pratense**

Uppermost leaves below the bracts (1–)2–8(–11) × (0.1–)0.2–1.0(–2.0) cm, mostly 7–15 times as long as wide.

(i) Var. **pratense**

Corolla whitish to pale yellow.

(ii) Var. **hians** Druce

Corolla deep yellow.

(b) Subsp. **commutatum** (Tausch ex A. Kern.) C. E. Britton

M. commutatum Tausch ex A. Kern.

Uppermost leaves below bracts (3–)4–7(–10) × (0.4–)0.8–2.0(–2.7) cm.

Native. Woods, scrub and heathland. Scattered throughout much of Great Britain and Ireland, but local in east-central England and south-central Ireland. Most of Europe and western Asia, east to the Yenisei region. Eurosiberian Boreo-temperate element. Subsp. *pratense* var. *pratense* is on acid soils over much of Great Britain and Ireland. Var. *hians* is said to predominate on the palaeogenic areas of western Great Britain and western Ireland. Subsp. *commutatum* occurs on calcareous soils in southern England and south-east Wales north to Herefordshire.

4. M. sylvaticum L. Small Cow-wheat

Annual, hemiparasitic, monoecious *herb*. *Stems* 8–60 cm, medium green, erect, glabrous or with sparse hairs, leafy, with suberect or spreading branches. *Leaves* opposite; lamina 1.5–10.0 × 0.4–1.5 cm, bright medium green on upper surface, paler beneath, linear-lanceolate, long-acute at apex, entire, narrowed at base, glabrous or with sparse hairs; sessile or shortly petiolate. *Inflorescence* of a single bisexual flower in the axil of each, distant, opposite, green, leaf-like bract, both flowers turned to the same side of the stem to form a lax, secund spike; upper bracts entire except in some large plants which may have a few teeth on the uppermost bract. *Calyx* 5.0–5.5 mm, divided more than halfway to the base into 4 lobes, the lobes longer than the tube, spreading and not appressed to the corolla. *Corolla* 6–12 mm, usually deep, often brownish-yellow, 2-lipped, the upper lip laterally compressed, the lower lip with 3 lobes, deflexed, the tube as long as or shorter than the calyx. *Stamens* 4, included under the upper lip of the corolla; filaments yellow; anthers 1.5–2.2 mm, orange.

Capsule 7.0–7.5 mm, ovoid, compressed, dehiscing along 2 margins; seeds 2, 4.0–4.5 mm, pale brown, ovoid. *Flowers* 6–8. Pollinated by bumble-bees, but can be selfed when not visited. $2n=18$.

Native. Upland woods and moorland. Local in Scotland north to Ross and Cromarty, Yorkshire, Co. Londonderry, Co. Antrim and formerly Durham. North Europe, extending southwards in the mountains to the Pyrenees, central Italy and south Bulgaria. European Boreal-montane element.

23. Euphrasia L.

Annual, hemiparasitic, monoecious *herbs* with fibrous roots. *Leaves* opposite or the upper irregularly alternate, toothed. *Inflorescence* a terminal spike formed of sessile, bisexual flowers in the axils of floral leaves or bracts, the floral leaves differing from the cauline leaves. *Calyx* campanulate, more or less 4-lobed. *Corolla* white or various shades of blue or purple with a yellow blotch and deeper purple lines, 2-lipped, the upper lip slightly concave with 2 porrect or reflexed lobes, the lower lip flat and 3-lobed, the lobes emarginate. *Stamens* 4. *Style* 1. *Fruit* a loculicidal capsule; seeds small, numerous, oblong or fusiform, furrowed.

Contains about 200 species in the temperate northern hemisphere, temperate South America, Australia and New Zealand.

It is difficult to decide what is a species in this genus. It is probable that all recognisable taxa in the northern hemisphere, all of which belong to the Section *Euphrasia*, can hybridise. It is possible to divided them into diploids and tetraploids, but even then they cross sometimes. The taxa are usually separated geographically or ecologically and sometimes seasonally and are really all subspecies or varieties of one or two species, and are equivalent of taxa in *Rhinanthus* and *Odontites*. However, to do this would make a mess of the nomenclature and would prevent us having any easy way of naming the numerous hybrids. It is therefore proposed to treat all the taxa that occur in distinct populations as species, and treat the hybrids in the normal way. Admittedly some taxa have more characters in common than others, but that is true of species in many genera like the replacing species of *Larix* round the northern hemisphere. We are going to have to realise that so-called 'good species' are often a figment of the imagination especially when a problem is looked at on a world scale. Once the mind has realised this the problem of *Euphrasia* becomes fascinating. Sometimes you see for example vast areas of *E. micrantha* on a *Calluna* heath and no other species present, and on another occasion little areas of pure *E. foulaensis, rotundifolia, marshallii* and *ostenfeldii* on the north coast of Sutherland forming a mosaic with scattered hybrids in between. The only taxa which have not been recognised as species in this account are the varieties of *E. nemorosa*, which we do not know enough about. They have, however, been left in the account to draw attention to them for further study.

Twenty-three of the species have been seen in the field and there is a vast collection of specimens of the genus in CGE.

We agree with Yeo that the best way to work out hybrids is in the field where you can see which species are present. Unfortunately no descriptions of the hybrids seem ever to have been made of the living plants and many subtleties of colour and facies are lost in the herbarium.

It is likely that self-pollination is frequent in the genus especially when conditions give a poor pollinator service, but that cross-pollination does take place especially in the larger-flowered taxa. In some cases it appears that taxa of apparent hybrid origin are reproducing themselves.

Bucknall, C. (1917). British *Euphrasiae*. *Jour. Bot. (London)* **55** (append.): 1–28.
Callen, E. O. (1940). Studies in the genus *Euphrasia* L. I. *Jour. Bot. (London)* **78**: 213–218.
Callen, E. O. (1940). Studies in the genus *Euphrasia* L. II. *Jour. Bot. (London)* **79**: 11–13.
Colgan, N. (1897). *Euphrasia salisburgensis* Funk in Ireland. *Jour. Bot. (London)* **35**: 196–199.
Druce, G. C. & Lumb, D. (1924). *Euphrasia atroviolacea* nov. sp. & *Euphrasia variabilis* nov. sp. *Rep. Bot. Soc. Exch. Club Brit. Isles* **7**: 49–51.
French, C. N. et al. (1999). *Flora of Cornwall*. Camborne. [*E. vigursii*.]
French, G. C., Ennos, R. A., Silverside, A. J. & Hollingsworth, P. M. (2005). The relationship between flower size, inbreeding coefficient and inferred selfing rate in British *Euphrasia* species. *Heredity* **94**: 44–51.
Granados, L. & Lane, S. D. (2007). A fine-scale study of selected environmental and floristic parameters in three populations of *Euphrasia vigursii* (Davey) a rare annual endemic to Devon and Cornwall. *Watsonia* **26**: 347–358.
Grime, J. P. et al. (1988). *Comparative plant ecology*. London.
Hultén, E. & Fries, M. (1986). *Atlas of north European vascular plants north of the Tropic of Cancer.* 3 vols. Königstein.
Joergensen, E. (1919). Die *Euphrasia* – Arten Norwegens. *Bergens Museums Aarbok* 1916–1917, *Naturvidenskabelig Raekke* **5**: 1–337.
Pugsley, H. W. (1919). Notes on British *Euphrasias* I. *Jour. Bot. (London)* **57**: 169–175.
Pugsley, H. W. (1922). Notes on British *Euphrasias* II. *Jour. Bot. (London)* **60**: 1–5.
Pugsley, H. W. (1930). A revision of the British Euphrasiae. *Jour. Linn. Soc. London (Bot.)* **48**: 467–542.
Pugsley, H. W. (1933). Notes on British *Euphrasias* III. *Jour. Bot. (London)* **71**: 83–89.
Pugsley, H. W. (1936). Notes on British *Euphrasias* IV. *Jour. Bot. (London)* **74**: 71–75.
Pugsley, H. W. (1936). Enumeration of the species of *Euphrasia* L. sect. *Semicalcaratae* Benth. *Jour. Bot. (London)* **74**: 273–288.
Pugsley, H. W. (1940). Notes on British *Euphrasias* V. *Jour. Bot. (London)* **78**: 11–13.
Pugsley, H. W. (1940). Notes on British *Euphrasias* VI. *Jour. Bot. (London)* **78**: 89–90.
Pugsley, H. W. (1945). The Eyebrights of Rhum. *The Naturalist (Leeds)* **1945**: 4–44.
Scott, W. & Palmer, R. (1987). *The flowering plants and ferns of the Shetland Islands*. Lerwick.
Sell, P. D. & Yeo, P. F. (1970). A revision of the North American species of *Euphrasia* L. (Scrophularoaceae). *Bot. Jour. Linn. Soc.* **63**: 189–234. [Many typifications are made here.]
Silverside, A. J. (1991). The identity of *Euphrasia officinalis* L. and its nomenclatural implications. *Watsonia* **18**: 343–350.
Silverside, A. J. (1998). *Euphrasia.* In Rich, T. C. G. & Jermy, A. C. (Edits.) *Plant crib*, pp. 269–272. London.

Stewart, A., Pearman, D. A. & Preston, C. D. (1994). *Scarce plants in Britain*. Peterborough.

Townsend, F. (1896). *Euphrasia salisburgensis* Funck, native in Ireland. *Jour. Bot. (London)* **34**: 441–444.

Townsend, F. (1897). Monograph of the British species of *Euphrasia*. *Jour. Bot. (London)* **35**: 321–336; 395–406; 417–426; 465–477.

Wettstein, R. von (1896). *Monographie der Gattung* Euphrasia. Leipzig.

Wigginton, M. J. (Edit.) (1999). *British red data books*. Vol. 1. *Vascular plants*. Peterborough.

Yeo, P. F. (1954). The cytology of British species of *Euphrasia*. *Watsonia* **3**: 101–108.

Yeo, P. F. (1956). Hybridisation between diploid and tetraploid species of *Euphrasia*. *Watsonia* **3**: 253–269.

Yeo, P. F. (1961). Germination, seedlings and the formation of haustoria in *Euphrasia*. *Watsonia* **5**: 11–22.

Yeo, P. F. (1964). The growth of *Euphrasia* in cultivation. *Watsonia* **6**: 1–24.

Yeo, P. F. (1966). The breeding relationships of some European *Euphrasiae*. *Watsonia* **6**: 216–245.

Yeo, P. F. (1968). The evolutionary significance of the speciation of *Euphrasia* in Europe. *Evolution* **22**: 736–747.

Yeo, P. F. (1970). New chromosome counts in *Euphrasia*. *Candollea* **25**: 21–24.

Yeo, P. F. (1970). *Euphrasia brevipila* and *E. borealis* in the British Isles. *Watsonia* **8**: 41–44.

Yeo, P. F. (1971). Revisional notes on *Euphrasia*. *Bot. Jour. Linn. Soc.* **64**: 355–361.

Yeo, P. F. (1972). *Euphrasia* L. In Tutin, T. G. et al. (Edits.), *Flora Europaea*, vol. 3. pp. 257–266. Cambridge.

Yeo, P. F. (1973). The Azorean species of *Euphrasia*. *Bol. Mus. Municipal Funchal* **17**: 74–83.

Yeo, P. F. (1975). The Yorkshire records of Euphrasia salisburgensis. The Naturalist (Leeds) **934**: 83–87.

Yeo, P. F. (1976). Artificial hybrids between some European diploid species of *Euphrasia*. *Watsonia* **11**: 131–135.

Yeo, P. F. (1978). A taxonomic revision of *Euphrasia* in Europe. *Bot. Jour. Linn. Soc.* **77**: 223–334.

1. Middle and upper leaves bearing glandular hairs with a stalk 10–12 times as long as the gland 2.
1. Middle and upper leaves without glandular hairs, or glandular hairs with a stalk not more than 6 times as long as the gland 10.
2. Capsule more than twice as long as wide 3.
2. Capsule not more than twice as long as wide 4.
3. Internodes short, the upper floral leaves often imbricate; floral leaves with dense simple eglandular hairs as well as glandular ones **6. reayensis**
3. Internodes up to 4 times as long as leaves; floral leaves with few, short bristles as well as the glandular hairs **7. notata**
4. Corolla not more than 7 mm 5.
4. Corolla more than 7 mm 6.
5. Lowest flower at node 5–8; branches usually 1–5 pairs **3. anglica**
5. Lowest flower at node 2–6(–7); branches usually 0–2 pairs **4. rivularis**
6. Lowest flower at node 2–5(–6) 7.
6. Lowest flower at node 5 or above 8.
7. Lower floral leaves 5–12(–20) mm. corolla (6.5–)8–13 mm **1. montana**
7. Lower floral leaves not more than 7 mm. corolla not more than 9 mm **4. rivularis**
8. Leaves dull greyish-green with more or less dull violet on the margins and veins beneath, or dull violet to blackish-purple all over the upper surface; corolla usually lilac to deep reddish-purple **6. vigursii**
8. Leaves light or dark green, usually with much purplish tinting; corolla with usually at least the lower lip white 9.
9. Stems with erect or divergent branches; lower floral internodes mostly 1.5–3.0 times as long as leaves; corolla (6.5)8–12 mm **2. rostkoviana**
9. Stems with flexuous or arcuate branches; lower floral internodes mostly less than 1.5 times as long as the leaves; corolla (5.0–)6.5–8.5(–10.0) mm **3. anglica**
10. At least some leaf teeth distant; capsule glabrous, or with a few, small cilia **27. hibernica**
10. All leaf teeth usually contiguous; capsule ciliate with long, fairly numerous hairs 11.
11. Corolla more than 7.5 mm 12.
11. Corolla less than 7.5 mm 19.
12. Basal pairs of teeth of lower floral leaves directed towards their apex 13.
12. Basal pairs of teeth of lower floral leaves patent 15.
13. Stems and branches rigidly erect; leaves with subacute to aristate teeth **14. stricta**
13. Stems and branches flexuous; leaves with obtuse to acute teeth 14.
14. Corolla lilac, white or rarely yellowish **12(1). confusa** forma **confusa**
14. Corolla deep reddish-purple **12(2). confusa** forma **atroviolacea**
15. Lowest flower at node 8 or lower, capsule usually elliptical to obovate **9. arctica**
15. Lowest flower at node 9 or higher; capsule oblong to elliptical-oblong 16.
16. Stems and branches usually flexuous; leaves near base or branches usually very small 17.
16. Stems and branches usually not flexuous; leaves near base of branches not much smaller than the others 18.
17. Corolla lilac, white or rarely yellowish **12(1). confusa** forma **confusa**
17. Corolla deep reddish purple **12(2). confusa** forma **atroviolacea**
18. Stems and branches usually rather stout; teeth of floral leaves acute to acuminate; capsule only slightly shorter than calyx or equalling it **11(iv). nemorosa** var. **nemorosa**
18. Stems and branches relatively slender; teeth of floral leaves mostly aristate; capsule much shorter than calyx **13. pseudokerneri**
19. Calyx tube often papery and whitish with veins and margins often blackish **22. campbelliae**
19. Calyx tube green 20.
20. Lowest flower at node 6 or higher 21.
20. Lowest flower at node 5(–6) or lower 50.
21. Basal pairs of teeth of lower floral leaves directed towards the apex of leaf 22.

21. Basal pair of teeth of lower floral leaves patent 23.
22. Teeth of lower floral leaves acute to aristate; corolla usually at least 7 mm **8. borealis**
22. Teeth of lower floral leaves acute to acuminate and fairly short; corolla not more than 6.5 mm **24. micrantha**
23. Corolla at least 6.5 mm 24.
23. Corolla not more than 6.5 mm 26.
24. Lower floral leaves larger than upper cauline; leaves usually with glandular hairs; lowest flower at node 8 or lower **8. borealis**
24. Lower floral leaves smaller than upper cauline, leaves usually without glandular hairs; lowest flower at node 9 or higher 25.
25. Leaves with aristate teeth; capsule often exceeding, equalling or shorter than calyx **11(iv). nemorosa** var. **nemorosa**
25. Leaves with less pointed teeth; capsule often exceeding calyx **11(iii). nemorosa** var. **collina**
26. Leaves glabrous or sparsely hairy 27.
26. Leaves densely hairy 31.
27. Stems and branches very slender, deep brownish-purple; leaves strongly tinted with purple; corolla usually lilac to purple **24. micrantha**
27. Stems and branches either stout or slender and slightly purple; leaves moderately or weakly tinged with purple; corolla usually white 28.
28. Leaves usually pale green on upper surface and purplish beneath; lowest flower at node 7 or lower; capsule usually longer than calyx **25. scottica**
28. Leaves not darker beneath than on upper surface; lowest flower at node (6–)9 or higher; capsule usually shorter than calyx 29.
29. Stems 5–10 cm, greenish; corolla 4–5 mm, lower lip scarcely exceeding upper **11(i). nemorosa** var. **sabulicola**
29. Stems (10–)15–20(–40) cm, tinted purplish; corolla 5–6 mm, lower lip obviously longer than upper 30.
30. Stems and branches moderately robust; leaves dark green; corolla white or with a bluish upper lip **11(iv). nemorosa** var. **nemorosa**
30. Stems of more slender habit, with long flexuous branches; corolla usually purple-tinted **11(v). nemorosa** var. **transiens**
31. Stems up to 40 cm; lower floral leaves often longer than wide; lowest flower at node 9 or higher **11(iv). nemorosa** var. **nemorosa**
31. Stems not more than 15 cm; lower floral leaves about as long as wide; lowest flower at node 8 or lower 32.
32. Leaves hairy mainly towards the apex, the cauline obovate to narrowly ovate or elliptical **22. campbelliae**
32. Leaves more or less uniformly hairy, usually subrotund, ovate or ovate-oblong 37.
33. Branches not more than 3 pairs; teeth of lower floral leaves mostly wider than long **21. rotundifolia**
33. Branches up to 5 pairs; teeth of lower floral leaves mostly as long as wide 34.
34. Corolla 4.5–6.0 mm. capsule not more than twice as long as wide **18. ostenfeldii**
34. Corolla 5.5–7.0 mm. capsule usually more than twice as long as wide **20. marshallii**
35. Basal pairs of teeth of lower floral leaves directed towards the apex 36.
35. Basal pairs of teeth of lower floral leaves patent 41.
36. Teeth of lower floral leaves much longer than wide 37.
36. Teeth of lower floral leaves not much longer than wide 38.
37. Flowers lilac, white or rarely yellowish **12(1). confusa** forma **confusa**
37. Flowers deep reddish-purple **12(2). confusa** forma **atroviolacea**
38. Leaves hairy mainly towards the apex, the cauline obovate to narrowly ovate or elliptical **22. campbelliae**
38. Leaves more or less uniformly hairy, usually subrotund, ovate or ovate-oblong 39.
39. Branches not more than 3 pairs; teeth of lower floral leaves mostly wider than long **21. rotundifolia**
39. Branches up to 5 pairs; teeth of lower floral leaves mostly as long as wide 40.
40. Corolla 4.5–6.0 mm. capsule not more than twice as long as wide **18. ostenfeldii**
40. Corolla 5.5–7.0 mm. capsule usually more than twice as long as wide **20. marshallii**
41. Lowest flower at node 10 or higher 42.
41. Lowest flower at node 9 or lower 44.
42. Stems erect, stout, with stout ascending branches; lower floral leaves mostly opposite **11(iv). nemorosa** var. **nemorosa**
42. Stems and branches slender and flexuous; lower floral leaves mostly alternate 43.
43. Flowers lilac, white or rarely yellowish **12(1). confusa** forma **confusa**
43. Flowers deep reddish-purple **12(2). confusa** forma **atroviolacea**
44. Leaves with few eglandular hairs 45.
44. Leaves with numerous eglandular hairs 47.
45. Lower lip of corolla obviously longer than upper **11(ii). nemorosa** var. **calcarea**
45. Lower lip of corolla scarcely longer than upper 46.
46. Capsule 4.5–5.5(–6.0) mm, usually shorter than calyx **10. tetraquetra**
46. Capsule (4.5–)5.5–7.0 mm, as long as or longer than calyx **26. heslop-harrisonii**
47. Leaves hairy mainly towards the apex, the cauline obovate to narrowly ovate or elliptical **22. campbelliae**
47. Leaves more or less uniformly hairy, usually subrotund, ovate or ovate-oblong 48.
48. Branches not more than 3 pairs; teeth of lower floral leaves mostly wider than long **21. rotundifolia**
48. Branches up to 5 pairs; teeth of lower floral leaves mostly as long as wide 49.
49. Corolla 4.5–6.0 mm. capsule not more than twice as long as wide **18. ostenfeldii**

49. Corolla 5.5–7.0 mm. capsule usually more than twice as long as wide **20. marshallii**
50. Cauline internodes mostly at least 2.5 times as long as leaves 51.
50. Cauline internodes mostly less than 2.5 times as long as leaves 59.
51. Capsule broadly elliptical to obovate-elliptical 52.
51. Capsule oblong to narrowly elliptical 54.
52. Teeth of lower floral leaves mostly subacute and not longer than wide; corolla 4.5–7.0 mm, the lowest flower at node 3–4(–5) **15. frigida**
52. Teeth of lower floral leaves usually acute or acuminate and longer than wide; corolla at least 6.5 mm. the lowest flower at node 4 or higher 53.
53. Upper leaves with short glandular hairs **8 borealis**
53. Upper leaves glabrous or scabridulus **9. arctica**
54. Upper cauline leaves elliptic-ovate to narrowly obovate **25. scottica**
54. Upper cauline leaves subrotund to broadly ovate or broadly obovate 55.
55. Lower floral leaves often much larger than the upper cauline; lowest flower at node 4 or lower **15. frigida**
55. Lower floral leaves often larger than the upper cauline; lowest flower at node 4 or higher 56.
56. Leaves hairy mainly towards the apex, the cauline obovate to narrowly ovate or elliptical **22. campbelliae**
56. Leaves more or les uniformly hairy, usually subrotund, ovate or ovate-oblong 57.
57. Branches not more than 3 pairs; teeth of lower floral leaves mostly wider than long **21. rotundifolia**
57. Branches up to 5 pairs; teeth of lower floral leaves mostly as long as wide 58.
58. Corolla 4.5–6.0 mm. capsule not more than twice as long as wide **18. ostenfeldii**
58. Corolla 5.5–7.0 mm. capsule usually more than twice as long as wide **20. marshallii**
59. Corolla at least 6 mm 60.
59. Corolla not more than 6 mm 62.
60. Teeth of lower floral leaves usually very acute, all directed towards apex **14. stricta**
60. Teeth of lower floral leaves subacute to acute, the basal pairs patent 61.
61. Flowers lilac, white or rarely yellowish **12(1). confusa** forma **confusa**
61. Flowers deep reddish-purple **12(2). confusa** forma **atroviolacea**
62. Leaves with numerous hairs, all eglandular 63.
62. Leaves with few eglandular hairs, short glandular hairs sometimes present 68.
63. Lower floral leaves much larger than the upper cauline **15. frigida**
63. Lower floral leaves scarcely larger than upper cauline 64.
64. Leaves hairy mainly towards the apex, the cauline obovate to narrowly ovate or elliptical **22. campbelliae**
64. Leaves more or less uniformly hairy, usually subrotund, ovate or ovate-oblong 65.
65. Branches not more than 3 pairs; teeth of lower floral leaves mostly wider than long 66.
65. Branches up to 5 pairs; teeth of lower floral leaves mostly as long as wide 67.
66. Corolla 3–4 mm. capsule 3.5–4.0 mm and nearly as broad **19. eurycarpa**
66. Corolla 5–6 mm. capsule 5.5–6.0 mm and about twice as long as wide **21. rotundifolia**
67. Corolla 4.5–6.0 mm. capsule not more than twice as long as wide **18. ostenfeldii**
67. Corolla 5.5–7.0 mm. capsule usually more than twice as long as wide **20. marshallii**
68. Capsule elliptical to obovate, emarginate **17. cambrica**
68. Capsule oblong to elliptic-oblong, usually truncate 69.
69. Distal teeth of lower floral leaves not incurved; capsule usually shorter than calyx **10. tetraquetra**
69. Distal teeth of lower floral leaves more or less incurved; capsule as long as or longer than calyx 70.
70. Upper cauline leaves only obscurely petiolate, the teeth without sinuate margins; capsule 4.5–5.5(–7.0) mm, about twice as long as wide, straight **16. foulaensis**
70. Upper cauline leaves more or less distinctly petiolate, the teeth with sinuate margins; capsule (4.5–)5.5–7.0 mm, 2–3 times as long as wide **26. heslop-harrisonii**

Section **1. Euphrasia**

All our species belong to this section.

Subsection **1. Euphrasia**
Subsection *Ciliatae* Jörg.

Floral leaves generally more than half as long as wide. *Capsule* ciliate, with long, fine hairs.

Series **1. Euphrasia**
Series *Hirtellae* Pugsley; Series *Grandiflorae* Wettst.

Flowering beginning at nodes 2–14. *Leaves* more or less densely clothed with long glandular hairs having the stalks about 12 times as long as the glandular heads, usually mixed with eglandular hairs. *Corolla* 3–12 mm. *Capsule* 4.0–5.5(–6.5) mm.

1. E. montana Jord. Mountain Eyebright
E. rostkoviana subsp. *montana* (Jord.) Wettst.;
E. officinalis subsp. *montana* (Jord.) Berher; *E. officinalis* subsp. *monticola* Silverside

Annual, hemiparasitic, monoecious *herb* with fibrous roots. *Stems* 5–35 cm, green or slightly reddish, erect, rather slender, with crisped, whitish hairs; branches 0–3(–4) pairs, slender, from near the middle; cauline internodes 2–6(–10) times as long as the leaves. *Leaves* with lamina 6–12(–20) mm, bright green, rather thin, the lower cauline oblong or elliptical, rounded-obtuse at apex, with obtuse teeth and a cuneate base, the upper cauline ovate, deltate or subrotund, rounded-obtuse at apex, with obtuse teeth and a rounded base, the floral deltate or subrotund, obtuse or rarely more or less acute at apex, with obtuse, acute or rarely acuminate teeth and rounded at base, the upper floral sometimes narrower and more acutely toothed;

all with more or less dense multicellular glandular hairs, some at least with the stalk about 10 times as long as the gland, on both surfaces. *Lowest flower* at node 2–6. *Calyx* with long, triangular-subulate teeth, slightly accrescent. *Corolla* (6.5–)8–13 mm, the tube elongating during flowering, white, often with a lilac upper lip; and with purple lines and a yellow blotch on lower lip, upper lip with broad, retuse or denticulate lobes; lower lip much longer, with broad, spreading, emarginate lobes, the median the longest. *Stamens* 4. *Stigma* capitate. *Capsule* 4.5–6.0 mm, often longer than the calyx, elliptical, with broad retuse or emarginate apex, with fine hairs on the margin and sparingly on the surface. *Flowers* 6–7. $2n=22$.

Native. This diploid species occurs in upland pastures including meadows cut for hay, where it may have developed its early flowering so it could set seed before being turned into hay. Northern England, southern Scotland and Wales but declining where grassland is improved. Throughout much of Europe, but mainly in the mountains and absent from parts of the south and east and many of the islands. European Boreo-temperate element.

× **rivularis**

This hybrid, growing with its parents, has been recorded for Cumberland.

2. E. rostkoviana Hayne Hill Eyebright

E. rostkoviana var. *obscura* Pugsley

Annual, hemiparasitic, monoecious *herb* with fibrous roots. *Stems* 10–40(–50) cm, green or slightly reddish, erect, with crisped, whitish hairs; branches 5(–12) pairs, slender, flexuous, erect or divergent from near the base; cauline internodes mostly not more than 3 times as long as the leaves, the lower shorter than the upper. *Leaves* with lamina 6–15 mm, bright green, rather thin and often conspicuously veined, the lower cauline oblong or oblong-ovate, rounded-obtuse at apex, with obtuse teeth and a cuneate base, the upper cauline broader, deltate or oblong-ovate and more acutely toothed, the floral smaller, 5–12(–20) mm, ovate, cuspidate to subacute at apex and with acute or aristate teeth; all with dense, multicellular glandular hairs, some at least with the stalk about 10 times as long as the gland, on both surfaces. *Lowest flower* at node (3–)6–10(–14). *Calyx* with long, finely acuminate teeth, scarcely accrescent. *Corolla* (6.5–)8–12 mm, the tube elongating during flowering, white, the upper lip often lilac-tinted and with purple lines and a yellow blotch on lower lip; upper lip with broad, emarginate lobes; lower lip much longer than upper with broad, spreading, emarginate lobes, the median much the longest. *Stamens* 4; filaments yellow, violet above; anthers greyish brown to black. *Style* white; stigma brownish, capitate. *Capsule* 4.0–5.5 mm, not longer than the calyx, elliptical or oblong-elliptical, emarginate or retuse at apex, with fine hairs on the margin and sparingly on the surface. *Flowers* 7–8. $2n=22$.

Native. This diploid species occurs in most grassland which is not cut early for hay and is lightly grazed. By rivers in usually hilly, but sometimes lowland areas. Locally frequent in Ireland, Wales, northern England and central and south Scotland. Throughout a large part of Europe, but absent from parts of the south and east and many of the islands; extending into western Asia. European Boreo-temperate element. Our plant is subsp. *rostkoviana*, named after Friedrich Wilhelm Gottlieb Rostkovius (1770–1848).

3. E. anglica Pugsley English Eyebright

E. officinalis subsp. *anglica* (Pugsley) Silverside;
E. hirtella auct.; *E. anglica* var. *gracilescens* Pugsley

Annual, hemiparasitic, monoecious *herb* with fibrous roots. *Stems* 5–16(–30) cm, green or purplish, usually ascending from a decumbent base, rather robust but flexuous, with crisped, whitish hairs; branches (0–)1–4(–6) pairs, long, arcuate or flexuous, the lowest less than 2(–3) cm above the cotyledons; cauline internodes up to 2.5(–3) times, the lower floral up to 1.5(–2.5) times as long as the leaves. *Leaves* with lamina 2–12 mm, greyish-green, sometimes brownish or purplish on upper surface, rather thick, both lower and upper cauline oblong to elliptical or ovate, rounded-obtuse at apex and with obtuse teeth and a rounded, cuneate or subcordate base, the lamina broadly ovate or ovate-triangular to subrotund, obtuse or acute at apex, and with 4–7 pairs of obtuse to acute teeth, the teeth usually not much longer than wide and the basal pair patent; all with more or less dense, multicellular glandular hairs, some at least with the stalk about 10 times as long as the gland, on both surfaces and some short, stiff hairs above. *Lowest flower* at node 5–8. *Calyx* with triangular-subulate or aristate teeth, scarcely accrescent. *Corolla* (5.0–)6.5–8.5(–10.0) mm, whitish or lilac-tinted with lilac upper lip and purple lines and a yellow blotch on the lower lip; upper lip with broad emarginate lobes; lower lip porrect with broad, emarginate lobes with the median longest. *Stamens* 4. *Stigma* capitate. *Capsule* 4.0–5.5 mm, up to twice as long as wide, elliptical or oblong, truncate or retuse at apex, with fine hairs on the margin. *Flowers* 5–9. $2n=22$.

Native. This diploid species occurs in short turf on damp, acidic substrates in old pastures, heaths, moorland, and disused quarries. Widespread but local in southern England, scattered north to south-west Scotland and in the Isle of Man and south-east and central Ireland. It mostly replaces *E. rostkoviana* in southern Great Britain. Endemic, or possibly in north-west France, but this needs confirmation.

× **borealis**

This hybrid resembles *E. rostkoviana* but is a larger plant with very large corollas and the leaves are relatively small with fine teeth. Plants approaching *E. anglica* also occur.

Native. On the Somersetshire peat moors near Street. Endemic.

× **confusa**

This hybrid resembles *E. anglica* but has more flexuous and sometimes more numerous branches and smaller, narrower leaves with all the teeth forwardly directed.

Native. Occurs occasionally where the parents meet, particularly in the Mendips in Somersetshire and on Exmoor. It has also been recorded when *E. confusa* is absent as in Windsor Great Park. Endemic.

× **micrantha**

A population of fertile diploids with narrow, small leaves and whitish corollas had a single individual nearer to *E. micrantha* which was a highly sterile triploid, $2n=33$ with 11 bivalents and 11 univalents at meiosis.

Native. Near Withypool in Somersetshire. Plants near St David's Head in Pembrokeshire are similar. The species known in this account as *E. vigursii* probably also has this origin. Endemic.

× **nemorosa** = **E.**×**glanduligera** Wettst.

This fertile hybrid has the tall habit of *E. nemorosa* and differs from *E. anglica* in its slightly less broad leaves with longer, finer teeth. It differs from *E. rostkoviana* in its more numerous and more widely divergent branches and in the relatively small upper leaves. It has $2n=22$, with 11 bivalents or 10 bivalents and 2 univalents at meiosis, and is fertile.

Native. It has been found growing in rather dry, heathy places, often in longish grass. Recorded from Cornwall, Devonshire, Sussex and Surrey. Endemic.

× **rostkoviana**

This hybrid is intermediate in corolla size between the two parents.

Native. It occurs where the two parents grow adjacent. Endemic.

× **vigursii**

This hybrid is intermediate in characters between the parents where they grow mixed.

Native. Known only from Cornwall. Endemic.

4. E. rivularis Pugsley Rivulet Eyebright

E. rivularis forma *compacta* Pugsley

Annual, hemiparasitic, monoecious *herb* with fibrous roots. *Stems* 5–10(–15) cm, more or less purplish, erect, very slender, with crisped, whitish hairs; branches 0–2 pairs, short and slender; cauline internodes shorter than or up to 4 times as long as leaves. *Leaves* with lamina 2–6(–7) mm, dark green or purple-tinted, the lower cauline elliptical, rounded at apex, with 1–4 pairs of obtuse teeth and more or less cuneate at base, the upper cauline broader and rounded below, the floral broadly ovate, not more than 7 mm, rounded to subacute at apex, 3–5(–6) pairs of acute teeth and rounded at base; all often sparingly with multicellular glandular hairs, some at least with the stalk about 10 times as long as the gland, on both surfaces. *Lowest flower* at node (2–)3–5(–6). *Calyx* with rather short, triangular-subacute teeth, not accrescent. *Corolla* 6.5–9.0 mm, white-tinted with lilac and with purplish upper lip and purple lines and yellow blotch on lower lip; upper lip with retuse lobes; lower lip longer than upper, with broad, spreading, retuse lobes, the median the longest. *Stamens* 4. *Stigma* capitate. *Capsule* 3.5–5.0 mm, usually much less than twice as long as wide, elliptical, retuse at apex, with fine hairs on the margin and sparingly on the surface. *Flowers* 5–7. $2n=22$.

It has been suggested that this is a stabilised hybrid segregate of *E. micrantha*×*rostkoviana*, but if so it has regained the diploid chromosome number.

Native. Damp, grassy slopes on mountains particularly where small rivulets trickle down. Very local in north-west Wales and the Lake district. Endemic.

× **roskoviana**

Recorded only from Cumberland where it occurs in several localities with the parents. Endemic.

5. E. vigursii Davey South-western Eyebright

Annual, hemiparasitic, monoecious *herb* with fibrous roots. *Stems* 6–18(–25) cm, dark purplish, erect, usually flexuous, with crisped, whitish hairs; branches 0–5(–7) pairs, flexuous, the lowest (2–)3–7 cm above the cotyledons; cauline internodes 1–3 times as long as leaves, and lower floral internodes 1–2(–3) times as long as leaves. *Leaves* with lamina 5–12 mm, dull greyish-green on upper surface, paler beneath and with more or less dull violet on the margin and veins beneath, or dull violet to blackish-purple all over the upper surface, more or less black-spotted, the lower cauline broadly elliptical or oblong-elliptical and broadly rounded at apex, the upper cauline ovate, oblong-ovate or subrotund and obtuse at apex, crenate or serrate, with 1–6 pairs of very obtuse to acute teeth, rounded or truncate at base and more or less contracted into a very short petiole; lower floral 5–10 mm, ovate, subacute to acute at apex, crenate-serrate, with 4–6 pairs of subacute to acute teeth, the teeth usually much longer than wide, the basal pair patent and rounded or subcordate at base; all slightly to stiffly hairy, the upper also sparsely to densely glandular-hairy, the longest glands with stalks 6–10 times as long as its gland, some plants with all leaves scabrid. *Lowest flower* at node 7–10(–12). *Calyx* sometimes spotted or blotched with black, the teeth more or less narrowly deltate and acute or acuminate at apex. *Corolla* (6.0–)7.0–8.5 mm. upper lip lilac to reddish-purple with emarginate lobes; lower lip usually lilac to deep reddish purple or rarely white, with a yellow blotch and purple lines, and broad emarginate lobes, the median longest. *Stamens* 4; anther cells pointed. *Stigma* capitate. *Capsule* (3.0–)4.0–5.5 mm, shorter than to about as long as the calyx, elliptical or obovate-oblong, not more than twice as long as wide, truncate to emarginate at apex, with dense, medium simple eglandular hairs on the margin. *Flowers* 6–9. $2n=22$.

Perhaps originating from hybridisation between *E. anglica* and *E. micrantha.*

Native. This diploid species occurs on *Agrostis curtisii – Ulex gallii* heaths in Cornwall and Devonshire. Endemic. Named after Chambré Corker Vigurs (1867–1940).

6. E. reayensis (Pugsley) P. D. Sell Reay Eyebright

E. brevipila var. *reayensis* Pugsley

Annual, hemiparasitic, monoecious *herb* with fibrous roots. *Stems* 8–30 cm, green or suffused reddish, erect, with crisp whitish hairs; branches 0–3(–4) pairs, slender, flexuous, erect or divergent from the base; cauline internodes often not longer than the leaves but up to twice as long, the upper sometimes shorter than the lower. *Leaves* with lamina 7–11 mm, bright green, rather thick, the lower cauline oblong or oblong-ovate, roundish-obtuse at apex, with obtuse to acute teeth and a cuneate base, upper cauline ovate, more acutely toothed, the floral broad, more acutely toothed; all with dense, long-stalked glands mixed with simple eglandular hairs on both surfaces. *Lowest flower* at node 5–10. *Calyx* with long, finely acuminate teeth. *Corolla* white, upper lip 9–11 mm, lilac, lower lip 9.5–11.5 mm. *Stamens* 4. *Style* 1; stigma capitate. *Capsule* 4.0–5.5 mm, oblong-elliptical, emarginate, glandular-hairy. *Flowers* 7–8. $2n=44$.

Native. Grassy places and sand dunes. Along the north coast of Sutherland and Caithness. Endemic.

× **scottica**

This hybrid has the long glands and round leaves of *E. reayensis* and the smaller leaves and flowers of *E. scottica*.

Native. Growing by Loch Mor at Bettyhill, Sutherland with both parents. Endemic.

7. E. notata Towns. Large Eyebright
E. brevipila var. *notata* (F. Towns.) Pugsley

Annual, hemiparasitic, monoecious *herb* with fibrous roots. *Stems* 15–40 cm, green or suffused reddish, erect, with crisped, whitish simple eglandular hairs; branches up to 6 pairs, long, slender, ascending; cauline internodes mostly not more than 3 times as long as the leaves. *Leaves* with lamina up to 12 mm, bright green, conspicuously veined, the lower cauline broadly ovate, deeply and sharply toothed, cuneate at base, the upper cauline similar, the floral smaller, but similar and sharply toothed; all with dense, multicelled glandular hairs on both surfaces. *Lowest flower* at about node 8. *Calyx* with long, finely acuminate teeth, scarcely accrescent, with dense, long-stalked glandular hairs. *Corolla* 8–9 mm, white, with the upper lip tinted with lilac or blue. *Stamens* 4. *Style* 1. *Capsule* elliptic-oblong.

Native. A conspicuous feature of grassy slopes in Perthshire. Endemic.

Series **2. Boreales** P. D. Sell & Yeo

Flowering beginning at node 3–10. *Plants* robust with long internodes and large leaves. *Leaves* commonly with short glandular hairs and eglandular hairs, sometimes without glandular hairs or subglabrous, upper cauline and lower floral mainly ovate to deltoid, with obtuse to aristate teeth, the basal teeth of the lower floral leaves patent. *Corolla* 6.5–11(–13) mm. *Capsule* 4.5–7.5 mm.

8. E. borealis (F. Towns.) Wettst. Northern Eyebright
E. rostkoviana forma *borealis* F. Towns.; *E. brevipila* auct.; *E. arctica* subsp. *borealis* (Towns.) Yeo.

Annual, hemiparisitc, monoecious *herb* with fibrous roots. *Stems* up to 30 cm, green or purplish, often flexuous, with crisped, whitish hairs; branches 0–5(–6) pairs, straight and often short; cauline internodes up to 4(–5) times as long as leaves, lower floral internodes usually 1–4 times as long as leaves. *Leaves* with lamina 3–14 mm, bright to dark green, the lower often tinged with brown or purple on upper surface; cauline ovate or oblong, more or less acute at apex, crenate to serrate, with 1–5(–6) pairs of obtuse to acute teeth, cuneate to truncate at base; lower floral larger than upper cauline, ovate, to deltate, more or less acute at apex, serrate, with (3–)4–6(–7) pairs of subacute to aristate teeth, the teeth usually much longer than wide, the basal pair usually patent but sometimes incurved or directed towards the apex, rounded, truncate or subcordate at base; all scabridulous or with short, stiff hairs, the upper usually also with short glandular hairs, 2–6 times as long as the gland. *Lowest flower* at node 4–8(–10). *Calyx* with tube occasionally blotched with black, the veins and margins frequently blackish; deltate, acute to aristate at apex. *Corolla* 6–9(–10) mm. upper lip lilac, broad, with retuse or denticulate lobes; lower lip much longer than upper, lilac or white, with violet lines and yellow blotch, with broad, emarginate lobes, the median the longest. *Stamens* 4; anther cells pointed; stigma capitate. *Capsule* (4.0–)4.5–6.5(–7.0) mm, shorter than to about as long as calyx, elliptical to oblong, truncate to emarginate at apex, sparsely hairy. *Flowers* 6–8. $2n=44$.

Native. This tetraploid taxon occurs in meadows, pastures, roadsides and disturbed ground. Rather local from Cornwall to the Isle of Wight and Buckinghamshire northwards to Yorkshire, more widespread and rather common further north and in north Wales; throughout Ireland but more frequent in the centre. North France and Norway. Oceanic Boreal-temperate element.

× **confusa** = **E.** × **variabilis** Druce & Lumb

This hybrid differs from *E. borealis* in its shorter lower internodes, more profuse branching, usually narrower leaves and smaller capsules, and sometimes smaller corollas, and from *E. confusa* in its coarser growth, larger leaves and emarginate capsules. It often forms hybrid swarms.

Native. A fertile hybrid on pastures, dunes, roadsides and grasslands. Wales, northern England, Scotland and east Ireland.

× **foulaensis**

Native. This hybrid is recorded for Sutherland and Caithness.

× **heslop-harrisonii**

Native. This hybrid is recorded only from Inverness-shire. Endemic.

× **hibernica**

This hybrid has the habit and leaf teeth of *E. hibernica* and the larger flowers of *E. borealis* and sometimes its glandular hairs.

Native. It probably occurs throughout the range of *E. hibernica*. Endemic.

× **marshallii**

This hybrid has the habit of *E. marshallii* and the simple eglandular hairs of that species mixed with the glandular hairs of *E. borealis*.

Native. Recorded from North Ebudes and Sutherland. Endemic.

× **micrantha** = **E.** × **difformis** F. Towns.

This hybrid has the finer branching and smaller leaves of *E. micrantha* and the glandular hairs of *E. borealis*.

Native. Probably widespread in Wales, northern England and Ireland.

× **nemorosa**

This hybrid mostly looks like *E. nemorosa* with the glandular hairs of *E. borealis*.

Native. Probably widespread in Great Britain and Ireland.

× **ostenfeldii**

There seems to be no precise record of this hybrid but it almost certainly occurs.

× **rostkoviana**

This hybrid is like *E. rostkoviana* but its leaf shape and large flowers are like *E. borealis*.

Native. Recorded from Breconshire and Cumberland.

× **rotundifolia**

Was recorded by E. F. Warburg but is improbable.

× **scottica** = **E.** × **venusta** F. Towns.

This hybrid differs from *E. borealis* by its small and few leaf teeth and from *E. scottica* by its numerous branches and glandular hairs.

Native. Probably widespread in Wales, northern England, Scotland and Ireland. Endemic.

× **tetraquetra** = **E.** × **pratiuscula** F. Towns.

This hybrid looks like *E. tetraquetra* in size and habit with the glands of *E. borealis*.

Native. Recorded from Haddingtonshire, Kintyre, Isle of Man and Co. Cork.

9. E. arctica Lange ex Rostr. Island Eyebright
E. borealis auct.; *E. borealis* var. *speciosa* Pugsley

Annual, hemiparasitic, monoecious *herb* with fibrous roots. *Stems* up to 30 cm, green or purplish, often flexuous, with crisped, whitish hairs; branches 0–2(–5) pairs, long and often flexuous; cauline internodes up to 4(–5) times as long as leaves, lower floral internodes usually 1–4 times as long as leaves. *Leaves* with lamina 4–11(–16) mm, bright or dark green, the lower often tinged with brown or purple on upper surface; cauline subrotund to broadly ovate or oblong, obtuse at apex, crenate or crenate-serrate, with 1–4 pairs of obtuse to subacute teeth, the proximal margins of the distal teeth usually with a distinct obtuse angle, rounded or broadly cuneate at base, sessile or shortly petiolate; lower floral subrotund to broadly ovate, obtuse to acute at apex, crenate to serrate, with 3–5(–6) pairs of obtuse to acuminate teeth, the teeth usually not much longer than wide, the basal pair patent, rounded, truncate or subcordate at base, usually with a very short petiole; all glabrous or scabridulous. *Lowest flower* at node (3–)4–6(–7). *Calyx* with tube occasionally blotched with black, the veins and margins frequently blackish; teeth elliptical and obtuse to deltate and acute or acuminate. *Corolla* 7–11(–13) mm. upper lip lilac; lower lip relatively large, white to purple with violet lines and a yellow blotch. *Stamens* 4; anther cells pointed. *Stigma* capitate. *Capsule* (5.5–)6.0–7.5(–8.0) mm, about as long as calyx, elliptical to obovate, emarginate or retuse at apex, sparsely hairy. *Flowers* 5–7.

Native. Damp meadows and marshes. Orkney and Shetland Islands. Faeroes. Oceanic Boreal-montane element.

× **borealis**

E. borealis is absent from Orkney and Shetland and *E. arctica* absent from mainland Scotland, but there are some plants on mainland Scotland intermediate between the two species.

× **confusa**

E. borealis var. *zetlandica* Pugsley; ? *E. confusa* var. *grandiflora* Pugsley

This hybrid differs from *E. arctica* in its shorter lower internodes, more profuse branching, generally narrower leaves with fewer more acute teeth, smaller and less elliptic capsules and sometimes smaller corollas, and from *E. confusa* in its coarser growth, larger leaves, large usually emarginate capsules and sometimes larger corollas.

Native. Pastures, roadsides and stabilised dunes. Orkney and Shetland Islands where hybrid swarms are probably not infrequent. Endemic.

× **micrantha**

E. micrantha var. *johnstonii* Pugsley

This hybrid has the smaller flowers and leaves of *E. micrantha*, but shows the leaf shape and capsule of *E. arctica*.

Native. It occurs with the parents in Orkney and Shetland.

× **nemorosa**

Native. This hybrid is recorded for Shetland.

Series **3. Nemorosae** Pugsley

Flowering usually begins at node 4–14. *Leaves* subglabrous or hairy, occasionally with shorter glandular hairs, upper cauline and lower floral mainly oblong-ovate to broadly ovate or deltoid, with subacute to acute or aristate teeth, the basal more or less patent. *Corolla* 5–9(–11) mm. *Capsule* usually 3.5–6.5 mm.

10. E. tetraquetra (Bréb.) Arrond. Coastal Eyebright
E. occidentalis Wettst.; *E. officinalis* subvar. *tetraquetra* Bréb.; *E. americana* var. *canadensis* (F. Towns.) B. L. Rob.; *E. canadensis* F. Towns.

Annual, hemiparasitic, monoecious *herb* with fibrous roots. *Stems* 6–15(–20) cm, usually purplish, erect, with crisped, whitish hairs; branches 0–5(–8) pairs, frequently rather short, erect or ascending and sometimes branched again; cauline internodes up to 1.5(–3.0) times as long as the leaves, lower floral internodes up to 1.5(–2.0) times as long as the leaves, frequently all the internodes shorter than the leaves the plant then often showing its tetraquetrous habit. *Leaves* with lamina 2–10(–14) mm, pale to dark, glossy green, sometimes tinged with dull violet or purple, rugose; cauline ovate to obovate or subrotund, broadly obtuse at apex, crenate or crenate-serrate, with 1–3(–4) pairs of obtuse to subacute teeth, the basal pair usually distant from the leaf-base, cuneate or rounded at base with a short petiole; lower floral ovate, deltate to occasionally oblong-ovate or trullate, obtuse to acute at apex, crenate-serrate to serrate, with (3–)4–6(–9) pairs of subacute to acuminate or aristate teeth, the teeth usually slightly longer than wide, the basal pairs patent or slightly retrorse, truncate or occasionally broadly cuneate, rounded or subcordate at base; all the lower with short more or less stiff simple hairs, the upper also often having glandular hairs with their stalks up to 4 times as long as the gland. *Lowest flower* at node (3–)5–7(–9). *Calyx* with teeth deltate, acute or aristate at apex. *Corolla* (4–)5–7(–8) mm. upper lip white or lilac, with short, retuse lobes; lower lip white or sometimes lilac, with violet lines and a yellow blotch, longer than the upper with rather broad, emarginate lobes, the median the longest. *Stamens* 4; filaments white to yellowish with blackish-violet streaks; anthers red, cells pointed. *Style* white; stigma pale yellowish, capitate. *Capsule* 4.5–5.5(–6.0) mm, usually shorter than the calyx, at least twice as long as wide, oblong or elliptic-oblong, retuse or truncate at apex, shortly hairy and ciliate. *Flowers* 5–8. $2n=44$.

Native. This tetraploid species occurs on short turf on sea-cliffs, maritime dunes and inland limestone pastures. Coasts of Great Britain and Ireland except most of eastern England and northern Scotland, inland in parts of south-west England. Coasts of France and perhaps Germany; introduced in Canada. Oceanic Temperate element.

× **vigursii**

Plants more dwarf, with more rounded leaves and denser inflorescences than are usual in *E. vigursii* appear to be introgressed with *E. tetraquetra*.

Native. Known only from south of Perranporth in Cornwall. Endemic.

11. E. nemorosa (Pers.) Wallr. Common Eyebright
E. officinalis var. *nemorosa* Pers.

Annual, hemiparasitic, monoecious *herb* with fibrous roots. *Stems* 10–40 cm, green or flushed purplish, erect, with crisped, whitish simple eglandular hairs; branches numerous, long, slender, erect or ascending, some of which are often branched again; cauline internodes often shorter than the leaves, the upper and floral leaves generally shorter than the much longer internodes. *Leaves* with lamina up to 12 mm, dark green or sometimes purplish, with more or less prominent veins; lower cauline oblong or elliptical, obtuse at apex, with 1–3, ascending, acute teeth on each side; upper cauline large, ovate, subacute at apex, with 3–6 acute teeth on each side; floral shorter than the upper cauline, ovate, acute at apex, with 4 to 6 shortly aristate teeth on each side; leaves of branches commonly smaller and narrower; all typically glabrous but frequently with some minute marginal bristles, sometimes with scattered longer hairs. *Lowest flower* usually about node 10. *Calyx* clothed like the leaves with acuminate teeth, scarcely accrescent in fruit. *Corolla* 5–6 mm, white or with bluish upper lip, with a yellow spot on the lower lip and dark lines, often not strongly marked, with small, notched, porrect or reflexed lobes; lower lip longer, deflexed, with three emarginate rather narrow lobes, the median exceeding the lateral. *Stamens* 4. *Style* 1. *Capsule* 5–6 mm, oblong, rounded-truncate or retuse, generally slightly hairy, subequalling the calyx teeth.

(i) Var. **sabulicola** Pugsley
E. nemorosa var. *imbricata* Callen

Stems 5–10 cm, greenish, with several long, erect branches. *Lowest flower* node 6–7. *Corolla* about 4.5 mm, lower lip scarcely exceeding upper.

(ii) Var. **calcarea** Pugsley
Stems 5–20 cm, branching near base of stem, with internodes mostly shorter than leaves. *Calyx* somewhat inflated or accrescent in fruit. *Corolla* 5–6 mm.

(iii) Var. **collina** Pugsley
Stems 5–30 cm, with spreading branches, upper internodes longer than leaves. *Calyx* somewhat inflated in fruit. *Corolla* 6–7 mm.

(iv) Var. **nemorosa**
E. nemorosa var. *obtusata* Pugsley

Stems 10–40 cm, with numerous erect or ascending branches, some frequently again branched. *Corolla* 5–6 mm.

(v) Var. **transiens** Pugsley
Stems up to 40 cm, with long, flexuous branches and relatively small leaves. *Corolla* about 6 mm, violet or purplish.

Native. This tetraploid species is found in short grasslands, on heaths, downs and dunes, in open scrub, woodland rides and upland moorlands; it is absent from improved agricultural land. Throughout much of Great Britain and Ireland. North and central Europe, extending southwards to north-east Spain. European Temperate element. Very variable and careful work might find that the varieties fit in with the ecology. Var. *sabulicola* is probably the plant of sands. Var. *calcarea* is the plant of chalk grassland. Var. *nemorosa* is scattered throughout our area in open woods and on banks, but generally not on calcareous soils. Var. *collina* occurs in the hilly districts of Wales, the west and north of England and southern Scotland. Var. *transiens* seems to occur in rather damp places in North Wales, East Anglia and Middlesex and Surrey.

× **ostenfeldii**

This hybrid probably occurs but we cannot find a definite record.

× **pseudokerneri**

This hybrid has the large flowers and sharp leaf teeth of *E. pseudokerneri*, but the habit of *E. nemorosa*.

Native. It is frequent where the two grow together in chalk grassland in southern England, and is recorded for Flintshire. Endemic.

× **reayensis**

This hybrid has the long glands and round leaves of *E. reayensis* and the smaller leaves and flowers of *E. nemorosa*.

Native. Growing at Bettyhill, Sutherland with both parents. Endemic.

× **rostkoviana** = **E.** × **glanduligera** Wettst.

This hybrid has the appearance of *E. nemorosa* but with the glandular hairs of *E. rostkoviana.*

Native. Recorded for Monmouthshire.

× **scottica**

This hybrid has the small flowers of *E. scottica* with branching and larger leaves of *E. nemorosa.*

Native. Scattered records in Scotland and northern England and a solitary record from Wales.

× **stricta** = **E.** × **haussknechtii** Wettst.

This hybrid has the branching habit of *E. nemorosa* and the sharper leaf teeth of *E. stricta.*

Native. It is recorded with both parents on the margin of Housfield Wood, Medmenham, Buckinghamshire.

× **tetraquetra**

This hybrid often has the taller habit of *E. nemorosa* but with the leaves being closer together and with the glands of *E. tetraquetra.*

Native. Recorded from along the southern and western coasts of Great Britain and in the Channel Islands.

12. E. confusa Pugsley Branched Eyebright
E. confusa var. *maciana* Callen

Annual, hemiparasitic, monoecious *herb* with fibrous roots. *Stems* 5–20(–45) cm, greenish, usually ascending or erect, sometimes decumbent, flexuous, with crisped, whitish hairs; branches (0–)2–8(–10) pairs, usually long, slender and flexuous and usually bearing numerous secondary branches; cauline internodes up to 2.5 times and the lower floral 2(–3) times as long as the leaves. *Leaves* with lamina 2–10 mm, deep or greyish-green, often flushed brown or dark violet, especially towards the margins and on the upper surface, the veins prominent beneath; the cauline ovate, oblong or lanceolate-ovate, broadly obtuse to subacute at apex, crenate to serrate, with 1–5 pairs of obtuse to acute teeth, cuneate or rounded at base and usually shortly petiolate; the lower floral often alternate and with a flower only in every other axil, ovate to oblong-ovate, obtuse to acute at apex, crenate-serrate to serrate, with 2–6 pairs of subacute to obtuse teeth, the teeth as long as or much longer than wide, often incurved and the basal pair patent or directed towards the apex, and distant from the leaf-base, usually rounded or cuneate at base or occasionally truncate and usually very shortly petiolate; all glabrous, or the lower scabridulous, or all scabridulous or finely and stiffly hairy, occasionally with a few long bristles, sometimes also with more or less sparse, very short glandular hairs. *Lowest flower* at node (2–)5–12(–14). *Calyx* with narrowly deltate, finely acuminate teeth, scarcely accrescent, the tube and the veins and margins sometimes blackish. *Corolla* 5–9 mm, upper lip lilac or sometimes white, the lower white or lilac with purple lines and a yellow blotch and longer than upper, sometimes both lips more or less deep reddish-purple or very rarely yellowish. *Stamens* 4; anther cells pointed. *Stigma* capitate. *Capsule* 3.5–5.5(–6.5) mm, usually about as long as calyx and 2–3 times as long as wide, oblong or elliptical-oblong, truncate, retuse or emarginate, ciliate and slightly hairy on the surface; seeds small, numerous, oblong, furrowed. *Flowers* 7–9. $2n=44$.

(1) Forma **confusa**
Flowers lilac, white or rarely yellowish.

(2) Forma **atroviolacea** (Druce & Lumb) P. D. Sell
E. atroviolacea Druce & Lumb
Flowers deep reddish-purple.

Native. This tetraploid species occurs on acidic or calcareous grassy cliffs, rough grassland, moorland, heaths and dunes. Throughout much of Great Britain and Ireland but local in the Channel Islands, Ireland and south and east England. Faeroes, and probably introduced in west Norway. Forma *confusa* is the common plant. Forma *atroviolacea* was described from Orkney.

× **foulaensis**

Native. This hybrid is recorded from the North Ebudes and Shetland.

× **frigida**

This hybrid has the capsules of *E. frigida* and the flowers of *E. micrantha.*

Native. It is recorded from the North Ebudes and there are a few other questionable records from Scotland and Ireland.

× **heslop-harrisonii**

Native. This hybrid is recorded from North Ebudes. Endemic.

× **micrantha**

This hybrid has the appearance of an elongated *E. confusa* or a much-branched *E. micrantha.* It often has the larger flowers of *E. confusa.*

Native. It is widely recorded throughout Great Britain and Ireland.

× **nemorosa**

This hybrid usually has the tall habit of *E. nemorosa* but bears flexuous, small-leaved branches and finer leaf teeth suggestive of *E. confusa.* It sometimes resembles *E. confusa* but has a stouter stem and less flexuous branches.

Native. A widespread fertile hybrid occurring in pastures, on roadsides and in disturbed ground. Sometimes occurs in large uniform colonies in absence of either parents and may form a complete continuum between the parents. Mainly western Great Britain and eastern Ireland. Endemic.

× **ostenfeldii**

This hybrid is a small unbranched plant with the hairiness of *E. ostenfeldii* and the flowers of *E. confusa.*

It is recorded from Carmarthenshire, Caernarvonshire and North Ebudes.

× **pseudokerneri**

Native. This hybrid is recorded only for Devonshire. Endemic.

× rostkoviana

This hybrid resembles *E. anglica* × *confusa* but has very large corollas.

Native. Found in Hobcarton Ghyll in Cumberland.

× scottica

This hybrid differs from *E. confusa* in its relatively tall, narrow habit, its internodes being usually longer in relation to the leaves and its short, little branched branches, and from *E. scottica* in its more flexuous stem and branches, its long, finer leaf teeth and sometimes its shorter, broader leaves and larger corollas.

Native. By damp stream edges in hill country. Throughout the range of *E. scottica* in Wales, northern England and Scotland.

× tetraquetra

This hybrid is a variable, low-growing plant differing from *E. tetraquetra* in its more numerous branches with very small leaves which have longer, finer teeth.

Native. Cliff-tops and stabilised dunes. Coast of western Great Britain from south-west England to south-west Scotland and inland in limestone pasture in Somerset. It forms hybrid swarms. Endemic.

13. E. pseudokerneri Pugsley Large-flowered Eyebright
E. kerneri auct.

Annual, hemiparasitic, monoecious *herb* with fibrous roots. *Stems* (5–)10–20(–30) cm, usually purplish, erect, sometimes flexuous, with crisped, whitish hairs; branches (0–)3–8(–10) pairs, spreading or ascending, often branched again; cauline internodes shorter than to 2.5(–3.0) times as long as leaves, lower floral internodes shorter than to 1.5(–2.0) times as long as leaves. *Leaves* with lamina 2–10(–15) mm, green, usually tinged brown, deep reddish- or blackish-purple on upper surface, especially the cauline; lower cauline oblong-ovate, obtuse at apex, with obtuse to subacute teeth and rounded or cuneate at the base, the upper cauline ovate, subacute to acute at apex, serrate, with up to 5(–7) pairs of acute to acuminate teeth and rounded or truncate at base; lower floral deltate, ovate or oblong-ovate, acute or acuminate at apex, serrate, with 3–6(–7) pairs of acute to aristate teeth, the teeth about as long as or much longer than wide, the basal pair patent and usually aristate and truncate, rounded or broadly cuneate at base; the lower usually scabridulous or with stiff hairs, the upper glabrous. *Lowest flower* at node (5–)10–16(–18). *Calyx* with the veins and margins usually blackish, the teeth narrowly deltate to sublinear, acuminate or aristate at apex. *Corolla* (6–)7–9(–11) mm. upper lip white or bluish, long, about equalling tube, with reflexed, emarginate lobes; lower lip white or bluish, with purple lines and a yellow blotch, longer than upper, with broad, deeply emarginate lobes, the median longest. *Stamens* 4; anther cells pointed. *Stigma* capitate. *Capsule* 3.5–5.0(–6.0) mm, much shorter than calyx, about 2.5 times as long as wide, oblong or elliptical-oblong, truncate or retuse at apex, ciliate and slightly hairy on surface. *Flowers* 7–9. $2n = 44$.

Native. This tetraploid species occurs in calcareous grassland, especially on chalk and in calcareous fens. Southern England from Devonshire and Kent to Lincolnshire and Yorkshire. Endemic.

× tetraquetra

Native. This hybrid has been recorded from Devonshire, Dorsetshire and Sussex.

Series **4. Majoriflorae** Jörg.
Series *Brevipilae* Pugsley; Series *Pectinatae* Pugsley

Flowering usually beginning at node 4–14. *Leaves* glabrous or hairy, occasionally with short glandular hairs, upper cauline and lower floral lanceolate or ovate-oblong to rhomboidal, with subacute to acute or aristate teeth, the basal antrorse. *Corolla* 5–10(–11) mm. *Capsule* 4–7 mm.

14. E. stricta D. Wolff ex J. F. Lehm.
Sharp-toothed Eyebright
E. brevipila Burnat & Gremli; *E. condensata* Jord.; *E. rigidula* Jord.

Annual, hemiparasitic, monoecious *herb* with fibrous roots. *Stems* 10–35 cm, usually strongly tinged with purple, erect, with crisped, whitish hairs; branches (0–)2–6(–13) pairs, usually rigid and erect; cauline internodes up to 4 times as long as the leaves, lower floral internodes 2(–3) times as long as the leaves. *Leaves* with lamina 2–14(–16) mm, glossy green, nearly always tinted purple; cauline narrowly ovate to ovate-lanceolate, obtuse to acute at apex, serrate, with 1–6 pairs of subacute to aristate teeth, rounded to cuneate at base; lower floral ovate to trullate, acute at apex, serrate or pectinate, with 4–6 pairs of subacute to aristate teeth, the basal pairs markedly directed toward the apex, rounded to cuneate at base; all glabrous or sometimes with short simple or glandular hairs. *Lowest flower* at node (3–)7–14(–18). *Calyx* with teeth deltate, aristate at apex. *Corolla* (6.0–)7.5–10.0 (–11.0) mm. upper lip lilac or white, with retuse lobes; lower lip white or lilac, with violet lines and a yellow blotch, longer than the upper with emarginate lobes, the median longest. *Stamens* 4; anther cells pointed. *Stigma* capitate. *Capsule* 4.0–5.5(–7.0) mm, 2.5–3.5 times as long as wide, truncate or rarely emarginate at apex, shortly hairy and ciliate. *Flowers* 6–8. $2n = 44$.

Introduced. This tetraploid species occurs in meadows and pastures. Guernsey, where it may be native, formerly in Buckinghamshire and possibly central Scotland. Native of most of Europe, but absent from the extreme north and much of the west and south.

× tetraquetra

This hybrid is recorded from the Channel Islands.

Series **5. Parviflorae** Wettst.
Series *Minoriflorae* Jörg.; Series *Latifoliae* Pugsley

Flowering beginning at the node 2–8, or occasionally higher. *Leaves* subglabrous or hairy, occasionally with short glandular hairs, upper cauline and lower floral mainly subrotund or elliptical, with obtuse to acute teeth. *Corolla* 4.5–8.0 mm. *Capsule* 4–8 mm.

15. E. frigida Pugsley Alpine Eyebright
E. latifolia auct.

Annual, hemiparasitic, monoecious *herb* with fibrous roots. *Stems* 5–20(–30)cm, green or purplish, erect, flexuous, with crisped whitish hairs; branches 0–2 pairs, erect, occasionally branched again; cauline internodes about 1–5(–10) times as long as leaves, lower floral internodes 1–4(–5) times as long as leaves, usually decreasing upwards gradually. *Leaves* with lamina 3–12(–18)mm, pale green, the lower sometimes tinged with brown or purple; cauline subrotund to elliptical, ovate, oblong or obovate, truncate, rounded or subacute at apex, shallowly to deeply crenate or crenate-serrate, with 1–4(–5) pairs of obtuse to subacute, or rarely acute teeth, more or less broadly cuneate at base and shortly petiolate; lower floral leaves frequently alternate, elliptical to ovate or deltate to subrotund, truncate to subacute at apex, crenate or crenate-serrate, with 2–5(–8) pairs of obtuse to acute teeth, the teeth usually not longer than wide, having the proximal margin at least strongly curved and the basal pairs more or less patent or sometimes directed at apex, cuneate to truncate or subcordate at base; all glabrous to hairy with sparse to dense, straight, slender bristles and occasionally glandular hairs having a stalk up to about 5 times as long as the gland. *Lowest flower* at node 2–4(–5). *Calyx* tube occasionally blotched black and the veins sometimes black; teeth deltate, acute or acuminate at apex. *Corolla* 4.5–7.0mm, upper lip white or lilac and rather narrow, lower lip white with purple lines and a yellow blotch, longer than upper, with emarginate lobes, the median rather longer and narrower than the lateral. *Stamens* 4; anther cells pointed. *Stigma* capitate. *Capsule* (4–)5–7mm, about as long as or longer than calyx, less than to slightly more than twice as long as wide, obovate, emarginate to retuse or truncate at apex, more or less hairy on surface and margin. *Flowers* 7–8. $2n=44$.

Native. This tetraploid species occurs in damp alpine rock-ledges, grassland and mountain tops, usually above 400 m and ascending to 1040 m. Lake district, rare in southern Scotland, frequent in central and north Scotland, rare in western Ireland. Northern Europe and North America. Eurosiberian Arctic-montane element and in North America.

× **micrantha**

Native. Recorded only for the North Ebudes.

× **scottica**

This hybrid has the capsules of *E. frigida* and the flowers of *E. scottica*. Its habit can be like either parent.

Native. Scattered records in Scotland and northern England, and possibly Co. Cork.

16. E. foulaensis F. Towns. ex Wettst. Seaside Eyebright

Annual, hemiparasitic, monoecious *herb* with fibrous roots. *Stems* 2–6(–9)cm, usually purplish, erect, with crisped whitish hairs; branches up to 3(–4) pairs, short, ascending, occasionally branched again; cauline internodes up to 1–2(–4) times as long as the leaves, lower floral internodes usually slightly shorter than leaves, the upper much shorter, sometimes all the internodes shorter than the leaves. *Leaves* with lamina 2–10mm, deep green, often tinted dull violet, easily blackening on drying, with thickened margins; cauline subrotund, broadly obovate or oblong, obtuse at apex, crenate, with 1–4 pairs of shallow, obtuse teeth, broadly cuneate or rounded at base, contacted into a very short petiole; lower floral subrotund to broadly ovate, crenate to serrate, with 2–5(–6) pairs of obtuse to acute or sometimes acuminate teeth, the teeth usually not much longer than wide with both margins usually convex, the basal pair patent, and the distal usually strongly incurved, broadly cuneate or rounded at base and contracted into a very short petiole; all glabrous or sparsely hairy. *Lowest flower* at node (2–)4–6. *Calyx* with tube sometimes blotched or striated with black, the veins and margins sometimes blackish; teeth deltate, acute at apex. *Corolla* 4–6mm, white or violet; upper lip with emarginate, subequal lobes; lower lip with violet lines and a yellow blotch, rather longer than upper, with emarginate, subequal lobes. *Stamens* 4; anther cells pointed. *Stigma* capitate. *Capsule* 4.5–5.5(–7.0)mm, sometimes blotched with black, usually as long as or longer than calyx, about twice as long as wide, oblong or elliptic-oblong, retuse or emarginate at apex, hairy on margins and sometimes on surfaces. *Flowers* 7–8. $2n=44$.

Native. This tetraploid species occurs on open turf of the tops of sea-cliffs, in coastal pastures and edges of salt-marshes. From Morayshire to Shetland and Outer Hebrides. Faeroes. Oceanic Boreal-montane element.

× **marshallii**

This hybrid has the flowers of *E. foulaensis* and the hairiness and leaf shape of *E. marshallii.*

Native. It is recorded from Sutherland, Caithness and the Outer Hebrides. Endemic.

× **micrantha**

This hybrid usually has the habit of *E. foulaensis* and the flowers and general coloration of *E. micrantha.*

Native. It is recorded from Sutherland, Outer Hebrides and Shetland.

× **nemorosa**

Native. This hybrid has been recorded from Sutherland, Caithness and Outer Hebrides.

× **ostenfeldii**

Native. This hybrid probably occurs in Shetland.

× **rotundifolia**

This hybrid has the flowers of *E. foulaensis* and the hairiness and leaf shape of *E. rotundifolia.*

Native. This hybrid has been recorded only from Caithness. Endemic.

× **scottica**

Native. This hybrid has been recorded from Shetland.

17. E. cambrica Pugsley Welsh Eyebright
E. cambrica forma *elatior* Pugsley

Annual, hemiparasitic, monoecious *herb* with fibrous roots. *Stems* 1–8(–15)cm, greenish, flexuous, slender, with crisped, whitish hairs; branches 0–2 pairs, slender, flexuous; cauline internodes much shorter than or up to 1.5(–2.0) times as long as leaves, lower floral internodes shorter than or up to 1.0(–1.5) times as long as leaves. *Leaves* with lamina 2–9 mm, pale green, sometimes purplish; cauline broadly ovate or oblong to obovate, obtuse at apex, crenate, with 1–3 pairs of obtuse teeth, rounded at base and shortly petiolate; lower floral broadly ovate, subrotund to deltate, obtuse at apex, crenate or crenate-serrate, with 2–4 pairs of obtuse to subacute teeth, the teeth about as long as wide and the basal pair patent, truncate or rounded at base; all with rather sparse medium simple eglandular hairs. *Lowest flower* at node 2–4. *Calyx* with teeth deltate, acute or acuminate at apex. *Corolla* 4.0–5.5 mm, tube longer than lips; upper lip whitish, sometimes violet tinted, with subentire lobes; lower lip whitish with violet lines and a yellow blotch, scarcely longer than upper, with narrow retuse or emarginate lobes, the median longest. *Stamens* 4; anther cells pointed. *Stigma* capitate. *Capsule* (4–)5–7 mm, longer than the calyx, up to twice as long as wide, elliptical or obovate, emarginate at apex, shortly hairy and ciliate. *Flowers* 6–7. $2n=44$.

Native. This tetraploid species occurs in mountain grassland and on rocky ledges. Caernarvonshire and Merionethshire. Endemic.

× **ostenfeldii**

Native. Recorded only from Caernarvonshire.

18. E. ostenfeldii (Pugsley) Yeo Ostenfeld's Eyebright
E. curta var. *ostenfeldii* Pugsley; *E. curta* auct.

Annual, hemiparasitic, monoecious *herb* with fibrous roots. *Stems* 2–12(–15) cm, erect or flexuous at base, with crisped, whitish hairs; branches 0–4(–6) pairs, erect or ascending, sometimes branched again; cauline internodes shorter than or up to 3(–5) times as long as leaves, lower floral internodes shorter than or up to 1.5(–2.5) times as long as leaves, the upper often very short. *Leaves* with lamina 2–10(–14) mm, more or less purplish on upper or both surfaces, the margins recurved; cauline subrotund or ovate to oblong-ovate or obovate, obtuse at apex, crenate to crenate-serrate, with 1–4(–5) pairs of obtuse to subacute teeth, rounded or cuneate at base; lower floral subrotund, broadly ovate or oblong-ovate, obtuse or subacute at apex, crenate-serrate to serrate, with 3–5 pairs of subacute to acute teeth, the teeth usually about as long as wide, the basal pair more or less patent; all more or less densely clothed with rather long simple eglandular hairs. *Lowest flower* at node (3–)4–7(–9). *Calyx* with tube often tinged with purple, sometimes with black spots, sometimes blackish on veins; teeth deltate, acute or acuminate at apex. *Corolla* (3.5–)4.5–6.0 mm. upper lip white or lilac, with retuse lobes; lower lip white or lilac, with violet lines and a yellow blotch, with subequal, emarginate lobes, longer than upper and more or less hairy beneath. *Stamens* 4; anther cells pointed. *Stigma* capitate. *Capsule* 4.0–5.5 (–6.0) mm, usually about as long as or longer than calyx, up to twice as long as wide, elliptical-oblong or oblong, retuse or emarginate at apex, hairy on surface and margin. *Flowers* 7–9. $2n=44$.

Native. This tetraploid species occurs in grassland, stony and sandy places and rock ledges, often near the sea. North and west Great Britain from central and north Wales to Shetland. Faeroes and Iceland. Named after Carl Emil Hansen Ostenfeld (1873–1931).

× **scottica**

This hybrid probably occurs in Shetland.

19. E. eurycarpa Pugsley Broad-fruited Eyebright

Annual, hemiparasitic, monoecious *herb* with fibrous roots. *Stems* 2–6 cm, purplish, suberect, slender, more or less flexuous, with crisped, whitish hairs; branches absent; internodes equalling or exceeding the leaves, except the uppermost. *Leaves* with lamina 2.5–5 mm, dark or dull green, thick; cauline elliptical to subrotund, obtuse at apex, with obtuse teeth; floral broader with the teeth sometimes less obtuse; all with numerous whitish bristles on both surfaces and the margins. *Lowest flower* at node 3–5. *Calyx* with broad teeth. *Corolla* 3–4 mm. upper lip bluish, rounded with obscure lobes; lower lip white with violet lines and a yellow blotch, scarcely longer than upper, with subequal, retuse lobes. *Stamens* 4; anther cells all pointed. *Stigma* capitate. *Capsule* very broad, 3.5–4.0 mm and nearly as wide, rounded-obovate, deeply emarginate, slightly exceeding calyx. *Flowers* 8.

This little-known plant may be a hybrid derivative of *E. ostenfeldii*, but is kept as a distinct species until more is known about it.

Native. Known only from the Isle of Rhum. Endemic.

20. E. marshallii Pugsley Marshall's Eyebright
E. marshallii var. *pygmaea* Pugsley

Annual, hemiparasitic, monoecious *herb* with fibrous roots. *Stems* 5–12(–15) cm, purplish, erect, robust, with crisped, whitish hairs; branches (0–)1–5 pairs, rather long and erect, the branches sometimes branched again; cauline internodes shorter than or up to 2.5(–3.0) times as long as leaves, the lower floral internodes shorter than or up to 1.5 times as long as leaves, the upper floral internodes very short. *Leaves* with lamina 2–11(–14) mm, dull greyish-green and rather thick; the cauline ovate or ovate-oblong to elliptical, obtuse to subacute at apex, crenate to serrate, with 1–5 pairs of obtuse to acute teeth and rounded or cuneate at base; lower floral broadly ovate or rhombic, obtuse to subacute at apex, serrate, with 4–5 pairs of acute teeth, the teeth mostly as wide as long, the basal pair patent or directed towards the apex, and truncate to cuneate at base; all densely hairy, with wavy, strong, whitish bristles. *Lowest flower* at node (3–)4–7(–9). *Calyx* with deltate, acute to aristate teeth, accrescent in fruit. *Corolla* 5.5–7.0 mm, hairy, white or lilac, the upper lip with short, denticulate lobes, the lower lip with purple lines and a yellow blotch, much exceeding the upper, with emarginate lobes, the median the longest and narrowest. *Stamens* 4; anther cells pointed. *Stigma* capitate. *Capsule* 4.5–5.5(–6.5) mm, usually slightly shorter than to about as long as calyx, rarely longer, about twice or more as long as wide, oblong or elliptic-oblong, retuse to truncate at apex, ciliate and hairy on surface; seeds small, numerous, oblong, furrowed. *Flowers* 7–8. $2n=44$.

Native. This tetraploid species is very local on grassy sea-cliffs and coastal rocks. Ross and Cromarty, Sutherland, Caithness, Outer Hebrides, Orkney and Shetland Islands. Endemic. Named after Edward Shearburn Marshall (1858–1919).

× **micrantha**

Native. Recorded only for Sutherland. Endemic.

× **nemorosa**

This hybrid has the hairiness and often the habit of *E. marshallii* but its leaf shape and toothing is that of *E. nemorosa*. Sometimes the habit is that of *E. nemorosa* and the hairiness that of *E. marshallii*.

Native. It is recorded from North Ebudes, Sutherland and Outer Hebrides. Endemic.

× **rotundifolia**

This probably occurs in Sutherland, but is almost impossible to distinguish.

× **scottica**

Native. Recorded only for Sutherland. Endemic.

21. E. rotundifolia Pugsley Round-leaved Eyebright

Annual, hemiparasitic, monoecious *herb* with fibrous roots. *Stems* 4–10 cm, purplish, erect, strict, with crisped, whitish hairs; branches 0–3 pairs, short and erect; cauline internodes 1.0–2.0(2.5) times as long as leaves, the floral internodes shorter than or up to 2.5 times as long as leaves. *Leaves* with lamina 2–12 mm, dull greyish-green; the cauline subrotund or broadly ovate to oblong-ovate, obtuse at apex, crenate, with 1–4(–5) pairs of obtuse teeth, rounded or truncate to broadly cuneate at base and more or less sessile; the lower floral subrotund to broadly ovate, obtuse at apex, crenate or crenate-serrate, with 3–5 pairs of obtuse to subacute teeth, the teeth mostly wider than long the basal pair patent, and rounded at base; all densely hairy especially beneath with stout-based, white bristles of varying length. *Lowest flower* at node 6–8(–9). Calyx with rather broadly deltate, subacute to acute teeth, accrescent in fruit. *Corolla* 5–6 mm, the upper lip white or purplish, hairy and with emarginate lobes, the lower lip exceeding the upper, with emarginate lobes, the median the longest and dilated at apex, white or lilac and with purple lines and a yellow blotch. *Stamens* 4; anther cells pointed. *Stigma* capitate. *Capsule* 5.5–6.0 mm, slightly longer than calyx, elliptic-oblong, about twice as long as wide, retuse at apex; seeds small, numerous, oblong, furrowed. *Flowers* 7–8.

Native. Turf above sea-cliffs or stabilised dunes. Sutherland, Caithness and Orkney Islands. Endemic.

22. E. campbelliae Pugsley Campbell's Eyebright

Annual, hemiparasitic, monoecious *herb* with fibrous roots. *Stems* 5–10 cm, green or purplish, erect, with crisped, whitish hairs; branches 0–2 pairs, short and erect; cauline internodes 1.0–2.5(–4.0) times as long as leaves, lower floral internodes about as long as leaves. *Leaves* with lamina 2–8 mm, green, more or less tinged with brownish-purple near the edge on the upper surface, sometimes purple beneath and the veins sometimes blackish; cauline obovate to rather narrowly ovate or elliptical, obtuse at apex, crenate or crenate-serrate, with 1–3 pairs of obtuse to subacute teeth and cuneate at base; lower floral ovate, obtuse to subacute at apex, crenate to serrate with 2–5(–6) pairs of obtuse to acute teeth, the teeth as long as or longer than wide, the basal pairs directed towards the apex or patent, cuneate at base, tapered into a short petiole; all with sparse to moderately dense, long bristles, especially distally. *Lowest flower* at node 5–7. *Calyx* tube often papery and whitish, the veins and margins often blackish; teeth deltate, acute or acuminate at apex. *Corolla* 5.5–7.0 mm. upper lip usually purple, with narrow, retuse lobes; lower lip white or tinged lilac with purple lines and a yellow blotch, longer than the upper, with narrow, apically dilated, emarginate lobes, the median the longest. *Stamens* 4; anther cells pointed. *Stigma* capitate. *Capsule* 4.5–5.5(–6.5) mm, usually slightly shorter than to about as long as calyx, more than twice as long as wide, oblong or elliptic-oblong, retuse or emarginate at apex, ciliate and sparingly hairy on surface. *Flowers* 7.

Perhaps originating from hybridisation between *E. marshallii* and *E. micrantha* or *E. scottica*.

Native. This tetraploid taxon occurs on damp grassy moors near the sea. Isle of Lewis in the Outer Hebrides. Endemic. Named after May Sherwood Campbell (1903–1982).

× **confusa**

This hybrid has the hairiness and habit of *E. campbelliae* and the leaf toothing of *E. confusa*.

Native. It is recorded only from the Outer Hebrides. Endemic.

× **marshallii**

Native. Recorded for Outer Hebrides. Endemic.

× **micrantha**

Native. Recorded for Outer Hebrides. Endemic.

× **nemorosa**

Native. Recorded for Outer Hebrides. Endemic.

23. E. rhumica Pugsley Rhum Eyebright

E. rhumica var. *fionchrensis* Pugsley

Annual, hemiparasitic, monoecious *herb* with fibrous roots. *Stems* 7–10 cm, usually purplish, suberect, very slender, with crisped, whitish hairs; branches very slender, suberect, near the middle; internodes all longer than the leaves except the uppermost and the lowest very long. *Leaves* with lamina 4–7 mm, dull green; cauline oblong to oblong-obovate, very obtuse at apex, with narrow, but obtuse teeth; floral sometimes broader, with less obtuse or subacute teeth; with many, fine, whitish bristles on both sides and on the margins. *Lowest flower* at node 4 or 5. *Calyx* with short, acute teeth, not accrescent. *Corolla* 4–5 mm. upper lip white, tinged blue, obscurely lobed; lower lip white with purple lines and a yellow blotch, scarcely longer than upper, with 3 narrow, retuse lobes, the median equalling or longer than the lateral. *Stamens* 4; anther cells pointed. *Stigma* capitate. *Capsule* 4.0–5.5 mm, equalling or exceeding the calyx teeth, obovate-oblong, retuse or emarginate at apex, ciliate and sparingly hairy on surface. *Flowers* 8.

This may be a hybrid segregate of *E. micrantha*. The var. *fionchrensis* differs only in its slightly larger leaves, flowers and capsules.

Native. Endemic to the Isle of Rhum.

24. E. micrantha Rchb. Purple-stemmed Eyebright

E. gracilis (Fr.) Drejer; *E. officinalis* var. *gracilis* Fr.; *E. micrantha* forma *simplex* Pugsley

Annual, hemiparasitic, monoecious *herb* with fibrous roots. *Stems* (2–)5–25(–30) cm, usually strongly suffused brownish-purple, erect, slender, strict, with short, white, crisped hairs, with (0–)2–7(–10) pairs of slender, erect branches, the cauline internodes usually 2–4 times and the lower floral 1.5–2.5(–3.0) times as long as the leaves. *Leaves* with lamina dark green, not darker beneath, often suffused purplish; the cauline 2–8(–11) mm, narrowly ovate to obovate, at apex, crenate to serrate with 1–6 pairs of usually subacute to acute teeth; the lower floral 3.5–7.0(–8.0) mm, ovate to rhombic, serrate, with 3–6 pairs of acute to acuminate teeth, the basal pair usually directed toward apex, truncate or cuneate at base; all glabrous or with minute bristles on the margins and veins beneath. *Lowest flower* at node (4–)6–14(–16). *Calyx* with short, finely acuminate teeth, scarcely accrescent in fruit. *Corolla* 4.5–6.5 mm, usually lilac to purple, rarely white, the upper lip with very small, entire or retuse lobes, the lower lip longer than the upper with 3 narrow, emarginate segments, the middle the longest. *Stamens* 4; anther cells pointed. *Stigma* capitate with violet lines and a yellow blotch. *Capsule* 3–5(–6) mm, usually shorter than the calyx, more than twice as long as wide, oblong to elliptic-oblong, rounded to slightly emarginate at apex. *Flowers* 7–9. $2n=44$.

Native. This tetraploid species occurs on acid heaths, where it is normally associated with *Calluna vulgaris*, sometimes in small bogs. Throughout most of Great Britain and Ireland, but absent from much of central and eastern England. North and central Europe, extending to north Spain and northern Italy. European Temperate element.

× **nemorosa**

This hybrid has the characteristic branching and smaller leaves of *E. micrantha* and the larger flowers of *E. nemorosa*.

Native. It is widely recorded in Great Britain.

× **ostenfeldii**

This hybrid has the condensed hairy leaves of *E. ostenfeldii* and the smaller leaves and branching of *E. micrantha*.

Native. It is recorded from as far apart as Helvellyn in Westmorland and Shetland Islands.

× **rostkoviana**

This hybrid has the long glands of *E. rostkoviana* and the small leaves and flowers of *E. micrantha*.

Native. It is recorded only for acid grassland near Dundrennan in Kirkcudbrightshire.

× **scottica** = **E.** × **electa** Towns.

This hybrid has the larger, green leaves of *E. scottica* and flower colour of *E. micrantha*.

Native. It is widely distributed in Scotland and is recorded for Wales.

× **tetraquetra**

This hybrid has the congested habit and bristles of *E. tetraquetra* and the small purplish flowers of *E. micrantha*.

Native. It is recorded from Cornwall, Anglesey and Co. Cork.

25. E. scottica Wettst. Scottish Eyebright

E. paludosa F. Towns., non R. Br.; *E. scottica* forma *estriata* Pugsley

Annual, hemiparasitic, monoecious *herb* with fibrous roots. *Stems* 5–25 cm, pale green, sometimes tinged with red, erect, slender, with short, crisped whitish hairs, with 0–4 pairs of long, arcuate-erect branches, the cauline internodes 2–5 times and the lower floral 1.5–3.0(–5.0) times as long as the leaves. *Leaves* with lamina pale green, often purplish beneath, the cauline 5–9 mm, narrowly ovate to obovate, subacute at apex and with up to 5(–6) pairs of shallow, obtuse or subacute teeth; the lower floral 4–10(–17) mm, often alternate, oblong or rhombic, acute at apex, with 3–5(–6) pairs of subacute to acute teeth and more or less cuneate at base; all glabrous or with few to numerous short simple eglandular hairs, rarely with some minute glandular hairs. *Lowest flower* at node (2–)3–6(–8). *Calyx* with short, finely acuminate teeth, scarcely accrescent in fruit. *Corolla* (3.5–)4.5–6.5 mm, usually white, rarely violet, the upper lip with very small, entire or retuse lobes, the lower lip with 3 subequal lobes, scarcely exceeding the upper lip. *Stamens* 4; anther cells pointed. *Stigma* capitate. *Capsule* 4.0–5.5(–9.0) mm, as long as or longer than the calyx, more than twice as long as wide, oblong to elliptic-oblong, more or less emarginate at apex. *Flower* 7–8. $2n=44$.

Native. This tetraploid species occurs on wet moorland and flushes, ascending to 915 mm. Wales, north-west England, Scotland and Ireland. Northern Europe. European Boreal-montane element.

26. E. heslop-harrisonii Pugsley Heslop-Harrison's Eyebright

Annual, hemiparasitic, monoecious *herb* with fibrous roots. *Stems* 6–15 cm, green, sometimes tinted purplish, erect, usually flexuous at the base, with crisped, whitish hairs; branches 0–4(–5) pairs, erect or spreading, sometimes branched again; cauline internodes shorter than or 1–2(–3) times as long as leaves, lower floral internodes shorter than or up to 2.0(–2.5) times as long as leaves, sometimes all internodes shorter than leaves. *Leaves* with lamina 2.0–11.5 mm, pale green, occasionally tinged with brown or purple; the cauline ovate or elliptical to elliptical-oblong or narrowly obovate, crenate or crenate-serrate, obtuse at apex, with 1–4 pairs of obtuse to subacute teeth, the teeth with the proximal margin usually sinuous, the basal pair usually distant from the leaf-base, rounded or cuneate at base; lower floral broadly ovate to oblong-ovate, acute at apex, crenate to serrate, with 2–5 pairs of obtuse to acute or acuminate teeth, the teeth about as long as wide, the basal pairs patent or directed towards the apex, rounded, truncate or broadly cuneate at base, sometimes contracted into a short or very short petiole; upper floral with teeth not aristate; all scabridulous or sparsely stiffly hairy, or the upper often glabrous. *Lowest flower* at node 4–7(–8). *Calyx* with veins occasionally purplish, the teeth deltate and acute at apex, accrescent in fruit. *Corolla* 4.5–6.0(–6.5) mm, white or occasionally lilac; upper lip rounded, with short, subtruncate lobes; lower lip with purple lines and a yellow blotch, scarcely longer than upper, with equal, subtruncate or retuse lobes. *Stamens* 4; anthers cells pointed. *Stigma* capitate. *Capsule* (4.5–)5.5–7.0 mm, as long as or longer than calyx, 2–3 times as long as wide, oblong or elliptical, often slightly curved, truncate, retuse or occasionally emarginate. *Flowers* 8.

Perhaps derived from an *E. scottica* hybrid.

Native. This tetraploid species occurs in coastal grassland and salt-marshes. Isle of Rhum in the Outer Hebrides, Ross and Cromarty, Sutherland, Orkney and Shetland Islands. Endemic. Named after John William Heslop-Harrison (1881–1967).

Subsection **2. Angustifoliae** (Wettst.) Jörg.
Section *Angustifoliae* Wettst.

Floral leaves generally less than half as wide as long. *Capsule* glabrous or with weak marginal hairs.

27. E. hibernica (Pugsley) P. D. Sell Irish Eyebright
E. salisburgensis var. *hibernica* Pugsley

Annual, hemiparasitic, monoecious *herb* with fibrous roots. *Stems* 4–12 cm, usually purplish, erect, flexuous; branches (0–)1–7 pairs, slender, erect or spreading and often branched again; cauline and floral internodes shorter than or up to twice as long as leaves, sometimes all internodes shorter than leaves, with crisped, whitish hairs. *Leaves* with lamina 2–8(–10) × up to 3 mm, pale green and glossy, often strongly tinged brown or purple; the cauline mostly oblanceolate to oblong or ovate-lanceolate, obtuse to acuminate at apex, with 1–3 distant pairs of obtuse, deltate or narrowly falcate, acuminate teeth, the basal distant from the leaf-base; lower floral lanceolate to ovate-lanceolate or oblong-lanceolate, obtuse to acuminate at apex, with 2–4 pairs of deltate, subacute to sublinear, aristate teeth, the apical pair distant from the others, the basal pair distant from the leaf-base; all glabrous or scabridulous. *Lowest flower* at node 5–13. *Calyx* with veins and margins sometimes blackish; teeth deltate, acute to acuminate at apex. *Corolla* 4.5–6.5 mm, white, or upper lip sometimes lilac; upper lip with small, entire lobes; lower lip with retuse lobes, the median longest. *Stamens* 4; anther cells pointed. *Stigma* capitate. *Capsule* 3.5–6.0 mm, much shorter than to longer than calyx, 2–4 times as long as wide, oblong or elliptical-oblong to elliptical-obovate or cuneate-obovate, truncate, retuse or emarginate at apex, glabrous or a few marginal bristles. *Flowers* 7–8. $2n = 44$.

Native. This tetraploid species occurs among limestone rocks and on maritime dunes and mountain cliffs. Widespread in western Ireland from Limerick to Donegal. Endemic. Specimens from Yorkshire in 1885–1886 are probably mislabelled.

× **micrantha**

This hybrid has the habit and small leaves and purple flowers of *E. micrantha* but the sharp leaf teeth of *E. hibernica*.

It is scattered through Ireland wherever *E. hibernica* is found. Endemic.

× **scottica**

This hybrid almost certainly occurs in Ireland but is difficult to distinguish from the hybrid with *E. micrantha*.

24. Odontites Ludw.

Annual monoecious *herbs* with fibrous roots. *Leaves* opposite, entire to dentate, sessile. *Inflorescence* a terminal raceme on the stem and upper branches. *Calyx* with 4 entire lobes, not inflated. *Corolla* yellow or pinkish-purple, rarely white, 2-lipped, the upper lip entire or emarginate, the lower 3-lobed, with entire to retuse lobes. *Stamens* 4. *Style* 1. *Fruit* a capsule; seeds few, furrowed longitudinally, without an oil body, hilum basal.

Contains about 30 species in Europe, western Asia and North Africa.

Pedersen, A. (1963). *Odontites verna* subsp. *pumila* ememd., en ny underart fra svenske og danske strandenge. *Bot. Tidskr.* **58**: 290–296.
Sell, P. D. (1967). *Odontites verna* (Bellardi) Dumort. *Watsonia* **6**: 301–303.
Snogerup, B. (1982). *Odontites litoralis* Fr. subsp. *litoralis* in the British Isles. *Watsonia* **14**: 35–39.

1. Corolla yellow — **1. jaubertianus** subsp. **chrysanthus**
1. Corolla purplish or pink — 2.
2. Bracts shorter than or equalling flowers — **2(c). verna** subsp. **serotinus**
2. Bracts longer than flowers — 3.
3. Often unbranched, but sometimes with 1–2 short branches; internodes short giving the plant a condensed habit — **2(a). vernus** subsp. **pumilus**
3. Plant with long, straight branches coming off at an angle of 45°; internodes longer than subsp. *pumila* giving the plant a more open habit — **2(b). vernus** subsp. **vernus**

1. O. jaubertianus (Boreau) D. Dietr. ex Walp. French Bartsia
Euphrasia jaubertiana Boreau; *O. luteus* auct.

Annual monoecious *herb* with fibrous roots. *Stems* 20–50 cm, pale green, sometimes suffused brownish-purple, erect, appressed or crispate-hairy, sparingly branched, the branches ascending to erect and flexuous, usually short, leafy. *Leaves* opposite; lamina 0.5–1.8 × 0.1–0.3 cm, medium green on upper surface, paler beneath, sometimes flushed brownish-purple, linear to linear-lanceolate, obtuse to acute at apex, entire, narrowed at base and sessile, with few hairs. *Inflorescence* a terminal raceme on the stem and upper branches; bracts 6–12 mm, linear-lanceolate and with 1–2 teeth on each side; flowers bisexual, sessile. *Calyx* 4–5 mm, tubular-campanulate, divided into 4 lobes equalling the tube, the lobes triangular to deltate, subacute at apex. *Corolla* 7–9 mm, bright yellow, tinged with red, shortly hairy, 2-lipped, the upper concave, the lower 3-lobed and more or less porrect. *Stamens* 4; filaments pale; anthers yellow. *Style* 1, long and filiform. *Capsule* 3.5–4.0 mm, elliptical, shortly hairy. *Flowers* 6–8. $2n=40$.

Introduced. Naturalised on rough gravelly ground, near Aldermaston in Berkshire since 1965. Native of west, central and south France. Our plant is subsp. **chrysanthus** (Boreau) P. Fourn. which is native to north-central France.

2. O. vernus (Bellardi) Dumort. Red Bartsia

Annual monoecious *herb* with fibrous roots. *Stems* 10–50 cm, pale green, often suffused brownish-purple, erect or ascending, somewhat 4-angled, shortly hairy, leafy, simple or much branched, leafy. *Leaves* opposite; lamina 1.2–4.0 × 0.3–0.8 cm, medium green on upper surface, paler beneath, sometimes suffused with brownish-purple, lanceolate, ovate-lanceolate or linear-lanceolate, obtuse to acute at apex, remotely dentate, rather narrowed at base and sessile, with simple hairs on both surfaces. *Inflorescence* of terminal racemes on the stem and upper branches; bracts like the leaves but smaller; flower 6–7 mm in diameter, bisexual. *Calyx* 5–8 mm, often flushed reddish, tubular-campanulate, divided into 4 lobes, the lobes triangular and obtuse at apex, with numerous simple eglandular hairs. *Corolla* 8–10 mm, purplish-pink, hairy, the tube 5–6 mm, 2-lipped, the upper lip concave, entire or emarginate, the lower 3-lobed with entire lobes. *Stamens* 4; filaments pale; anthers yellow, coherent, slightly exserted. *Style* 1, hairy, long, filiform; stigma capitate. *Capsule* 6–8, oblong, shortly hairy; seeds small, rather few, oblong or fusiform, furrowed, hilum basal. *Flowers* 6–8. Pollinated by bees.

(a) Subsp. **pumilus** (Nordst.) A. Pedersen
O. serotinus forma *pumilus* Nordst.; *O. litoralis* auct.; *Bartsia odontites* var. *litoralis* auct.; *O. vernus* subsp. *litoralis* auct.; *O. vulgaris* subsp. *pumilus* (Nordst.) Soó

Plant up to 17(–21) cm, often unbranched, but sometimes with 1 or 2 pairs of short branches. *Internodes* short, giving the plant a condensed habit. *Leaves* lanceolate to ovate-lanceolate and distinctly toothed. *Bracts* longer than flowers. *Flowers* apparently darker red than other subspecies. *Flowers* 7–8(–9).

(b) Subsp. **vernus**
Euphrasia verna Bellardi; *O. vernus* subsp. *longifolius* Corb.; *Bartsia odontites* var. *verna* (Bellardi) Druce; *Bartsia odontites* var. *longifolia* (Corb.) Druce; *Bartsia odontites* var. *rotundata* auct.

Plant 10–30 cm, with branches coming off at an angle of less the 45° and more or less straight. *Internodes* longer than in subsp. *pumilus* giving the plant a more open habit. *Leaves* lanceolate, distinctly toothed. *Bracts* longer than the flowers. *Flowers* 6–7. $2n=40$.

(c) Subsp. **serotinus** (Syme) Corb.
Euphrasia odontites L.; *Euphrasia serotina* Lam. nom. illegit.; *O. rubra* Gilib. non rite publ.; *Euphrasia rubra* Baumg.; *O. vulgaris* Moench; *O. serotinus* Dumort. nom. illegit.; *Bartsia odontites* var. *serotina* Syme; *Bartsia odontites* (L.) Huds.; *Euphrasia divergens* Jord.; *Bartsia odontites* var. *divergens* (Jord.) Druce

Plant 20–50 cm, with branches spreading at a wide angle, sometimes nearly at right angles, their tips often upcurved. *Internodes* long, giving the plant an open habit. *Leaves* linear-lanceolate, somewhat narrowed at base, obscurely toothed. *Bracts* shorter than or equalling the flowers. *Flowers* July and August. $2n=20$.

Native. Grassy and waste places, arable land, old tracks, waysides and gravelly and rocky shores and salt-marshes. Throughout Great Britain and Ireland. Most of Europe except some of the islands; north and west Asia. Subsp. *serotinus* is common in southern England and rarer further north and in the hills. It is common in the lowlands of Continental Europe but rare in the north and sometimes the only subspecies in the south. Subsp. *vernus* is common in the north of Great Britain and rarer further south. This also applies to Europe as a whole where it is also the plant of the mountains. In some localities there seems to be a complete mixture of these two plants as regards morphology, despite having different chromosome numbers. Subsp. *pumilus* is a plant of gravelly and rocky seashores and salt-marshes. Coasts of north and west Scotland, Netherlands, Denmark and south Sweden. It seems to be a dwarf coastal ecotype of subsp. *vernus* with which it integrates in the Hebrides. Subsp. **litoralis** (Fr.) Nyman (*O. litoralis* Fr.) is the dwarf plant of the Baltic coast which is quite distinct in being unbranched, and having long internodes and much smaller leaves. It has not been seen in our area.

25. Bartsia L.

Perennial monoecious *herbs* with a short rhizome. *Leaves* opposite, crenate-serrate, sessile. *Inflorescence* a terminal, spike-like raceme formed by flowers in the axils of bracts similar to leaves. *Calyx* divided halfway into 4 lobes. *Corolla* dull purple, 2-lipped, the upper lip entire or emarginate and forming a hood, the lower lip longer than the upper, with 3 entire lobes. *Stamens* 4, included in the upper lip. *Style* 1. *Fruit* a capsule; seeds few, with strong longitudinal ribs or wings and a lateral hilum.

Contains about 30 species in Europe, North Africa and tropical mountains.

Hultén, E. & Fries, M. (1986). Atlas of north European vascular plants north of the Tropic of Cancer.3 vols. Königstein.
Wigginton, M. J. (Edit.) (1999). *British red data books.* Vol 1. *Vascular plants.* Peterborough.

1. B. alpina L. Alpine Bartsia

Perennial monoecious *herb* with a short rhizome. *Stems* 10–25 cm, green, sometimes flushed with brownish-purple, erect, with numerous glandular hairs, leafy, simple.

Leaves opposite; lamina 1.0–2.5 × 1.0–1.5 cm, dark green on upper surface, paler beneath, often suffused brownish-purple, ovate, obtuse to subacute at apex, crenate-serrate, narrowed at base and sessile, glandular hairy. *Inflorescence* a terminal, spike-like raceme formed by flowers in the axils of purplish, glandular-hairy bracts similar to the leaves; flowers 11–13 mm in diameter, bisexual. *Calyx* 6–8 mm, dull purple, divided to halfway into 4 lobes, the lobes ovate-lanceolate, obtuse at apex, with glandular hairs. *Corolla* 15–20 mm, dull purple, 2-lipped, the upper lip entire or emarginate, and forming a hood, the lower lip longer than the upper, with 3 entire lobes. *Stamens* 4, included in the upper lip; filaments yellow; anthers yellow, hairy. *Style* 1, terminal, glandular-hairy; stigma dull purple. *Capsule* 8–10 × 1.5–5.0 mm, oblong, apiculate at apex, glandular-hairy; seeds few, 1.0–1.5 mm, with strong longitudinal ribs or wings and a lateral hilum. *Flowers* 6–8. Pollinated by bumble-bees. $2n=24$.

Native. Grassy places and rock-ledges on basic, often damp soils in mountains. Very local in Yorkshire, Durham, Westmorland, Perthshire and Argyllshire. Northern Europe southwards to the mountains of the Pyrenees, southern Alps and south-west Bulgaria; Greenland; eastern North America southwards to Labrador. European Arctic-montane element; also in North America.

26. **Parentucellia** Viv.

Annual monoecious *herbs* with fibrous roots. *Leaves* opposite, serrate, sessile. *Inflorescence* a terminal, spike-like raceme formed by flowers in the axils of bracts similar to the leaves. *Calyx* tubular, divided halfway into 4 lobes, not inflated. *Corolla* yellow or rarely white; 2-lipped, the upper lip entire or emarginate, forming a hood, the lower lip with 3 entire lobes. *Stamens* 4, included in the upper lip; anthers aristate at base. *Fruit* a capsule; seeds numerous, minute, smooth, without an oil body, with hilum basal.

Contains 4 species in western Europe, the Mediterranean region and central and western Asia.

Stewart, A., Pearman, D. A. & Preston, C. D. (1994). *Scarce plants in Britain*. Peterborough.

1. P. viscosa (L.) Caruel Yellow Bartsia
Bartsia viscosa L.; *Eufragia viscosa* (L.) Benth.; *Euphrasia viscosa* Siebert, non L.

Annual monoecious *herb* with fibrous roots. *Stems* 10–50 cm, pale green, often suffused with brownish-purple, erect, rigid, with numerous, viscid glandular hairs, leafy, usually unbranched. *Leaves* opposite or alternate; lamina 1.5–4.0 × 0.5–1.5 cm, medium green on upper surface, paler beneath, oblong to lanceolate, more or less acute to obtuse at apex, coarsely serrate, rounded at base, sessile, with numerous glandular hairs. *Inflorescence* a terminal spike-like raceme formed by flowers in the axils of bracts similar to the leaves; flowers 11–13 mm in diameter, bisexual. *Calyx* tube 8–12 mm, cylindrical, divided halfway into 4 lobes, the lobes 5–8 mm, linear-lanceolate and subacute at apex, with glandular hairs. *Corolla* 16–24 mm, yellow or rarely white, 2-lipped, the upper lip entire or emarginate and forming a hood, the lower lip longer than the upper, with 3 entire lobes. *Stamens* 4, included in the upper lip; filaments free near apex, anthers yellow, aristate at base, hairy. *Style* 1, 16 mm, terminal, shortly hairy; stigma strongly compressed. *Capsule* 7–9 mm, slightly longer than the calyx tube, lanceolate; seeds many, about 0.4 mm, pale brown, oblong-ovoid, smooth, without an oil body, the hilum basal. *Flowers* 6–10. $2n=48$.
Native. Damp grassy places mostly near the coast. Locally frequent in the south and west of Great Britain from Kent to Cornwall and north to Dunbartonshire; north-west, south-west and west Ireland and the Channel Islands, frequent elsewhere as an introduction. South and west Europe, northwards to Great Britain; North Africa; Azores and Canary Islands. Mediterranean-Atlantic element.

27. **Rhinanthus** L.

Alectorolophus Zinn nom. illegit.

Annual, hemiparasitic, monoecious *herbs*. *Leaves* opposite, simple, crenate-dentate. *Inflorescence* a terminal, bracteate spike. *Calyx* flattened and inflated in flower, accrescent, with 4 entire lobes. *Corolla* yellow or brown, 2-lipped; upper lip laterally compressed with 2 teeth at end; lower lip 3-lobed. *Stamens* 4, included in the upper lip. *Style* 1. *Fruit* a compressed capsule; seeds few, large, usually winged. *Nectar* secreted at the base of the ovary.

Contains about 50 species in the north temperate regions.

Intercalary leaves are the leaves on the main stem between the topmost branches and the lowest flower of the terminal spike.

Chabert, A. (1899). Étude sur le Genre *Rhinanthus* L. *Bull. Herb. Boiss*. **7**: 497–517.
Druce, G. C. (1903). Notes on *Rhinanthus*. *Jour. Bot. (London)* **41**: 359–361.
Grime, J. P. et al. (1988). *Comparative plant ecology*. London.
Hambler, D. J. (1953). Prochromosomes and supernumery chromosomes in *Rhinanthus minor* Ehrh. *Nature* **172**: 629–630.
Hambler, D. J. (1958). Some taxonomic investigations on the Genus *Rhinanthus*. *Watsonia* **4**: 101–116.
Hultén, E. & Fries, M. (1986). *Atlas of north European vascular plants north of the Tropic of Cancer*. 3 vols. Königstein.
Marshall, E. S. (1903). On the British forms of *Rhinanthus*. *Jour. Bot. (London)* **41**: 219–300.
Sell, P. D. (1967). *Rhinanthus* in Taxonomic and nomenclatural notes on the British flora. *Watsonia* **6**: 298–301.
Soó, R. de (1929). Die mittel-und südosteuropäsischen Arten der Gattung *Rhinanthus*. *Feddes Repert*. **26**: 179–219.
Sterneck, J. von (1900). Monographie der Gattung *Alectorolophus*. *Abh. Zool.-Bot. Gesell. Wien* **1**(ii): 1–150.
Wilmott, A. J. (1940). Some British species of *Rhinanthus*. *Jour. Bot. (London)* **78**: 201–213.
Wilmott, A. J. (1942). Some remarks on British *Rhinanthus*. *Rep. Bot. Soc. Exch. Club Brit. Isles* **12**: 361–379.
Wilmott, A. J. (1948). Another British *Rhinanthus* with pubescent calyx. *Watsonia* **1**: 84–85.

1. Teeth of upper lip of corolla about 2 mm, twice as long as wide; corolla tube curved upwards **1. angustifolius** subsp. **grandiflorus**
1. Teeth of upper lip of corolla less than 1 mm and not longer than wide; corolla tube straight 2.

2. Calyx hairy all over 3.
2. Calyx hairy only on the margins 4.
3. Leaves of main stem 1.5–3.0(–4.0) mm wide, narrowly linear-lanceolate; intercalary leaves 0–3 pairs; lowest flower at node 7–10 **2(e). minor** subsp. **lintonii**
3. Leaves of main stem 3–7 mm, oblong or oblong-linear; intercalary leaves 0: lowest flower at node 5–7(–8) **2(f). minor** subsp. **borealis**
4. Leaves of main stem mostly linear; intercalary leaves (2–)3–6 pairs; lowest flower at node 14–19 **2(c). minor** subsp. **calcareus**
4. Leaves of main stem oblong to linear-lanceolate; intercalary leaves 0–2(–4) pairs; lowest flower at node 6–13(–15) 5.
5. Leaves of main stem parallel-sided for most of their length; intercalary leaves 0(–1) pairs; lowest flower at node 6–9 **2(a). minor** subsp. **minor**
5. Leaves of main stem mostly tapering from near base; intercalary leaves (0–)1–2(–4) pairs; lowest flower at node (7–)8–13(–15) 6.
6. Stem usually with several pairs of long flowering branches from basal and middle parts; leaves of main stem 15–45 mm. corolla usually yellow **2(b). minor** subsp. **stenopyllus**
6. Stem with 0–3 pairs of long flowering branches from near the base; leaves of main stem 10–25 mm. corolla dull brown or treacle brown **2(d). minor** subsp. **monticola**

1. R. angustifolius C. C. Gmel. Greater Yellow-rattle
R. serotinus (Schönh.) Oborny; *R. major* auct.; *Alectorolophus major* auct.; *Alectorolophus serotinus* Schönh.

Annual, monoecious, hemiparasitic *herb* with fibrous roots. *Stems* 20–60 cm, medium green, sometimes tinted reddish, black-spotted, more or less robust, glabrous to slightly hairy, simple or more frequently with short or long branches from about the middle. *Leaves* opposite; lamina 25–70 × 8–15 mm, medium green on upper surface, paler beneath, lanceolate or linear-lanceolate, obtuse at apex, often strongly crenate-serrate with prominent teeth but sometimes with the teeth less prominent, scabrous, the branch leaves usually smaller than the stem leaves; intercalary leaves 0–1(–2) pairs. *Inflorescence* a terminal bracteate spike of 8–18 bisexual flowers; bracts 20 × 30 mm, yellowish-green, ovate or rhombic-ovate, usually acuminate at apex, at least the basal teeth deep and sharp. *Calyx* 17–20 mm, pale brown with prominent veins; subrotund, flattened and inflated in flower, accrescent, with 4 entire lobes, glabrous except for the hairy margin. *Corolla* 15–20 mm, yellow with violet teeth; tube curved upwards, mouth closed; upper lip laterally compressed with 2 teeth about 2 mm, conical and twice as long as broad; lower lip 3-lobed. *Stamens* 4, included in the upper lip of the corolla; filaments pale greenish-yellow; anthers green. *Style* 1, greenish, bending over the anthers. *Capsule* 10–12 mm, brownish, ovate, compressed; seeds 3–5(–6) mm, winged or unwinged. *Flowers* 6–9. Pollinated by bumble-bees, self-pollination not possible. $2n = 22$.

Introduced. Arable and grassy fields, rough ground, sandy open places on heathland and near the sea. Extremely local in Surrey, Lincolnshire and Forfarshire and occasionally casual elsewhere; formerly widely scattered over Great Britain but over-recorded. Most of Europe, but absent from the Mediterranean region, the south-west and most islands and extending into Siberia. Our plant is subsp. **grandiflorus** (Wallr.) D. A. Webb (*Alectorolophus grandiflorus* Wallr.; *R. serotinus* subsp. *apterus* (Fr.) Hyl.; *R. major* var. *apterus* Fr.; *R. apterus* (Fr.) Ostenf.; *R. major* subsp. *apterus* (Fr.) Schinz & Thell.; *R. major* subsp. *vernalis* Zinger; *R. serotinus* subsp. *vernalis* (Zinger) Hyl.; *R. borbasii* auct.; *R. polycladus* auct.). It occurs throughout the range of the species.

2. R. minor L. Yellow-rattle

Annual, monoecious, hemiparasitic *herb* with fibrous roots. *Stems* 5–50 cm, medium green, often tinged or suffused brownish-purple, slender to fairly robust, ridged, glabrous to slightly hairy, especially below the nodes, simple or branched, leafy. *Leaves* opposite; lamina 5–50 × 1.5–7.0 mm, medium yellowish-green on upper surface, paler beneath, oblong to linear or linear-lanceolate, obtuse at apex, crenate-dentate with more or less sharp teeth, sessile, scabrid. *Inflorescence* a terminal, bracteate spike; flower bisexual; bracts 10–25 × 10–15 mm, lanceolate to triangular-ovate, acute at apex, with triangular, acute or shortly aristate teeth. *Calyx* 10–15 mm, pale brown with prominent veins, subrotund, flattened and inflated in flower, accrescent, with 4 entire lobes, glabrous or hairy. *Corolla* 12–15 mm, yellow or brown with violet, rarely white, teeth; tube straight, mouth somewhat open; upper lip laterally compressed with 2 short, rounded teeth at the end about 1 mm and not longer than wide; lower lip 3-lobed. *Stamens* 4, included in the upper lip of the corolla; filaments pale greenish-yellow; anthers green. *Style* 1, greenish, bending over the anthers. *Capsule* 12–17 mm, brownish, ovate, compressed; seeds few, large, winged. *Flowers* 5–8. Pollinated by bumble-bees, selfed if not visited. $2n = 22$.

Variation is a combination of ecotypes, geographical races and seasonal variants.

(a) Subsp. **minor**
R. drummond-hayi subsp. *salmonii* Soó

Usually with only very short flowerless branches, but sometimes with longer, suberect flowering branches from the middle and upper part of the stem. *Internodes* (except the lowest) more or less equal. *Leaves* of main stem with lamina (10–)20–40(–50) × (3–)5–7 mm, usually oblong and parallel-sided for the greater part of their length. *Intercalary leaves* 9(–11) pairs. *Calyx* hairy only on the margins. *Lowest flowers* usually from the node 6–9. *Corolla* yellow with a violet (or rarely white) tooth. *Flowers* 5–7.

(b) Subsp. **stenophyllus** (Schur) O. Schwarz; *R. minus* var. *stenophyllus* Schur; *R. crista-galli* subsp. *stenophyllus* (Schur) Soó; *R. stenophyllus* (Schur) Druce

Usually with long arcuate-ascending flowering branches and shorter flowerless branches from the lower and middle part of the stem. *Lower internodes* usually short, equalling or shorter than the leaves, the upper much longer. *Leaves* of main stems with lamina 15–45 × 2–5(–7) mm, narrowly lanceolate or linear-lanceolate, more or less tapering from near the base. *Intercalary leaves* (0–)1–2(–4) pairs. *Calyx* hairy only on the margins. *Lowest flowers* usually from node (8–)10–13(–15)·. *Corolla* yellow with a violet tooth, sometimes becoming brown. *Flowers* 7–9.

(c) Subsp. **calcareus** (Wilmott) E. F. Warb.
R. calcareus Wilmott

Usually with long arcuate-ascending flowering branches from about the middle of the stems. *Lower internodes* short, upper very long. *Leaves* of main stems with lamina 10–25 × 1.5–3.0 mm, linear, more or less spreading. *Intercalary leaves* usually (2–)3–6 pairs. *Calyx* hairy only on the margins. *Lowest flowers* usually from node 14–19. *Corolla* yellow, tooth violet. *Flowers* 7–9.

(d) Subsp. **monticola** (Sterneck) O. Schwarz
R. minor var. *monticola* Lamotte; *Alectorolophus monticola* Sterneck; *R. monticola* (Sterneck) Druce; *R. crista-galli* subsp. *monticola* (Sterneck) Soó; *R. spadiceus* Wilmott; *R. spadiceus* subsp. *orcadensis* Wilmott

Usually with short or moderate flowerless branches from near the base, sometimes with 1–3 longer branches in addition. *Lower internodes* usually very short, upper much longer. *Leaves* of main stems with lamina 10–20(–25) × 2–4 mm, linear-lanceolate, more or less tapering from near the base, tending to be more erect than in subsp. *stenophyllus*. *Intercalary leaves* usually 1–2(–3) pairs. *Calyx* hairy only on the margin. *Lowest flowers* usually from node (7–)8–11(–12). *Corolla* dull yellow, becoming treacle brown, or treacle brown from the first, with a violet tooth. *Flowers* 7–8.

(e) Subsp. **lintonii** (Wilmott) P. D. Sell
R. gardineri Druce; *R. lintonii* Wilmott; *R. lochabrensis* Wilmott; *R. vachelliae* Wilmott; *R. drummond-hayi* auct.

Simple or with 1–2 pairs of *branches* from the lower or middle part of the stem. *Internodes* more or less equal in length, or the lower shorter. *Leaves* of main stem with lamina (8–)10–20(–30) × 1.5–3.0(–4.0) mm, very narrow linear-lanceolate, more or less tapering from the base. *Intercalary leaves* 0–3 pairs. *Calyx* hairy all over. *Flowering* from node 7–10. node. *Corolla* yellow or orange-yellow. *Flowers* 7–8.

(f) Subsp. **borealis** (Sterneck) P. D. Sell
R. crista-galli var. *drummond-hayi* F. B. White; *Alectorolophus borealis* Sterneck; *Alectorolophus drummond-hayi* (F. B. White) Sterneck; *R. drummond-hayi* (F. B. White) Druce; *R. borealis* (Sterneck) Druce; *R. borealis* var. *calvescens* Wilmott

Usually unbranched, but occasionally with short, flowerless (very rarely flowering) branches, and then *intercalary leaves* 0. *Internodes* more or less equal. *Leaves* with lamina 10–30 × 3–7 mm, oblong or oblong-linear, more or less parallel-sided. *Calyx* hairy all over. *Lowest flowers* from node 5–7(–8). *Corolla* bright yellow with violet or white teeth. *Flowers* 7–8.

Native. Common in grassland throughout Great Britain and Ireland. Most of Europe, but rare in the Mediterranean region; Caucasus; western Siberia; Alaska; south Greenland; Newfoundland. European-temperate element. Subsp. *minor* occurs in grassy places on dry basic soils. It is common in England and rarer in Scotland where it often grades into subsp. *stenophyllus*. Throughout Europe extending to Caucasus, western Siberia, Greenland and Newfoundland. European Boreao-temperate element. Subsp. *stenophyllus* occurs in damp grassland and fens. It is common in Scotland, north and west England, Wales and probably Ireland, and is mainly in fens in south-east England. It occurs throughout much of Europe particularly in upland regions. European Boreo-montane element. Subsp. *calcareus* is local on chalk and limestone downs from Dorsetshire and Sussex to Gloucestershire and Northamptonshire. Endemic. Plants from Ireland which resemble it are best placed in subsp. *stenophyllus*. It seems to be endemic. Subsp. *monticola* occurs in grassy places in mountainous regions from Yorkshire to the Shetland Islands and is rare in Ireland. It is also in the Alps. European Boreo-montane element. Subsp. *borealis* is local in grassy places on mountains up to 1000 m in Scotland, Caernarvonshire in Wales and Co. Kerry in Ireland. It is also in Alaska, Greenland, Iceland and possibly Scandinavia. European Boreo-arctic Montane element. Subsp. *lintonii* occurs in grassy places on Scottish mountains. It appears to have originated from hybridisation between subspecies *borealis*, *monticola* and *stenophyllus* and now often occupies a zone, where it is uniform, between *borealis* higher up and *monticola* and *stenophyllus* lower down. It seems to be endemic.

28. Pedicularis L.

Annual to *perennial* monoecious *herbs*. *Leaves* alternate, pinnatisect. *Inflorescence* a terminal, leafy-bracted spike or raceme. *Calyx* tubular or cylindrical in flower, becoming inflated in fruit, sometimes 2-lipped, sometimes with leaf-like lobes. *Corolla* 2-lipped the upper lip laterally compressed, entire or with teeth near the end, the lower lip 3-lobed. *Stamens* 4, included in the upper lip. *Style* 1; stigma capitate. *Fruit* an ovoid, compressed capsule, with a few, large seeds at the lower end. *Nectar* secreted from a swelling at the base of the ovary.

Contains about 500 species in north temperate regions, especially the mountains of central and east Asia and the Andes.

Hultén, E. & Fries, M. (1986). *Atlas of north European vascular plants north of the Tropic of Cancer*. 3 vols. Königstein.

Rich, T. C. G. (1994). *Pedicularis sylvatica* L. subsp. *hibernica* Webb D. A. (Scrophulariaceae) new to Wales. *Watsonia* **20**: 70–71.

Webb, D. A. (1956). A new subspecies of *Pedicularis sylvatica* L. *Watsonia* **3**: 239–241.

1. Calyx clearly 2-lipped; upper lip of corolla with a tooth on each side **1. palustris**
1. Calyx entire or indistinctly 2-lipped; upper lip of corolla with entire margin 2.
2. Stems glabrous or with 2 lines of hairs; calyx and pedicels glabrous **2(a). sylvatica** subsp. **sylvatica**
2. Stems hairy; calyx and pedicels hairy **2(b). sylvatica** subsp. **hibernica**

1. P. palustris L. Marsh Lousewort

Biennial, sometimes *annual*, monoecious, rather fleshy *herb*. *Stem* single, 8–60 cm, pale green usually suffused with brownish-purple, glabrous or sparsely hairy, branching from near the base to the middle, the lower branch longer than the upper giving it a pyramidal appearance, leafy. *Leaves* alternate; lamina 2–4 × 0.6–0.8 cm, bright medium green on upper surface, paler beneath, triangular-lanceolate to oblong in outline, pinnatisect, the lobes oblong and crenate to pinnatifid, glabrous or nearly so; petiole pale green. *Inflorescence* a lax terminal raceme often interrupted below; flowers bisexual; pedicels short; bracts similar to leaves, but smaller. *Calyx* 10–15 mm, brownish, tubular in flower, soon becoming inflated, strongly 2-lipped, without distinct lobes, hairy. *Corolla* 20–25 mm, reddish-pink, rarely pale yellow or white, the tube 10–11 mm, horizontal, 2-lipped, the upper lip straight, laterally compressed and with a tooth on each side some way below the apex, the lower lip 3-lobed. *Stamens* 4, included in the upper lip; filaments white; anthers yellow. *Style* 1, white; stigma capitate. *Capsule* 15–17 mm, ovoid, curved, longer than the calyx; seeds few in the lower part, 2.0–2.5 mm, oblong, reticulate. *Flowers* 5–9. Pollinated by bumble-bees. $2n = 16$.

Native. Wet heaths and bogs. Throughout Great Britain and Ireland, common in the north and west but rare in central and east England. Europe southwards to the Pyrenees, north Italy, southern Bulgaria and south Ural; Caucasus. European Boreo-temperate element. Our plant is subsp. **palustris** which occurs throughout the range of the species except parts of the north.

2. P. sylvatica L. Lousewort

Biennial or *perennial*, monoecious, rather fleshy *herb* with a thick tap-root. *Stems* numerous, 6–25 cm, the central one erect, the remainder decumbent or prostrate, pale yellowish-green, sometimes tinted brownish-purple, glabrous or hairy, simple or branched, leafy. *Leaves* alternate; lamina 2.5–3.5 × 0.6–0.8 cm, medium to dark green on upper surface, paler beneath, linear to oblong in outline, 2-pinnatisect, the lobes ovate and incise-dentate and glabrous or slightly hairy; petiole pale green. *Inflorescence* a short, lax, terminal raceme; pedicels glabrous or hairy; bracts similar to leaves but smaller. *Calyx* 14–15 mm, pale green with brownish veins, cylindrical in flower and soon becoming inflated, 5-angled with 4 small, leaf-like 2- to 3-toothed lobes, the 5th upper tooth small and linear, arising at a lower level, glabrous or hairy. *Corolla* 20–25 mm, pink or red, rarely white, the tube up to twice as long as the calyx, 2-lipped, the upper lip obtuse, slightly curved and with entire, laterally compressed margins, the lower lip with 3 subrotund lobes. *Stamens* 4, included in the upper lip; filaments white; anthers yellow. *Style* 1, purplish; stigma yellowish, capitate. *Capsule* 12–15 mm, ovoid, shortly acuminate at apex, shorter than calyx; seeds few in the lower part, about 1.8 mm, oblong, reticulate. *Flowers* 4–7. Pollinated by bumble-bees. $2n = 16$.

(a) Subsp. **sylvatica**
Stems glabrous or with 2 lines of hairs, the central one up to 15 cm. *Calyx* and pedicels glabrous.

(b) Subsp. **hibernica** D. A. Webb
Stems long hairy, the central one not more than 10 cm. *Calyx* and pedicels long hairy.

Native. Damp heaths, bogs and marshes up to 915 m. Fairly common throughout Great Britain and Ireland. West and central Europe extending northwards to central Sweden and eastwards to Lithuania and western Russia. European Temperate element. Subsp. *sylvatica* occurs throughout the range of the species except in that of subsp. *hibernica* and Portugal. Subsp. *hibernica* occurs in western Ireland from Co. Cork to Co. Donegal and also in Co. Waterford and Co. Wicklow; and in west Wales, the Lake District and west Scotland from Kintyre and Lewis; it is also in north-west Europe north to Norway. Oceanic Boreal-montane element. Intermediates occur between the two subspecies especially in east Ireland.

138. OROBANCHACEAE Vent. nom. conserv.

Annual to *perennial* monoecious *herbs* devoid of chlorophyll, brown to whitish, reddish or bluish. *Stems* mostly erect and fleshy. *Leaves* more or less scale-like, alternate, simple, entire, sessile, exstipulate. *Inflorescence* a terminal raceme or spike. *Flowers* zygomorphic, bisexual, hypogynous, in the axils of bracts. *Calyx* with 4, more or less equal lobes fused to about halfway, or with 2 lateral lips each, not to more or less deeply bifid. *Corolla* tubular, with upper and lower lips, the upper lip more or less entire to 2-lobed, the lower lip 3-lobed. *Stamens* 4, borne on the corolla tube. *Style* 1, terminal; stigmas 2, more or less capitate. *Ovary* 1-celled, with numerous ovules borne on 4, inward-thrusted parietal placentas. *Fruit* a 2-celled loculicidal capsule, with numerous small seeds.

Contains about 14 genera and about 180 species mainly in temperate Eurasia, with a few tropical and American species.

The family is separated from the Scrophulariaceae by the aerial parts having no chlorophyll and a 1-celled ovary. Either the whole family or just *Lathraea* are often placed in the Scrophulariaceae.

Almost all the species are rooted in the soil, but few have extensive rooting systems. There is either a congested mass of short, thick roots or a large, single or complex swollen organ. These underground structures are connected via swollen, clamp-like haustoria to the root of the host, from which nearly all its nourishment is obtained.

1. Plant rhizomatous; flower pedicellate; calyx with 4 equal lobes **1. Lathraea**
1. Plant not rhizomatous; flower sessile, except rarely some near base of inflorescence; calyx with 2–4(–5) teeth arranged in 2 lateral lips **2. Orobanche**

1. Lathraea L.

Perennial, monoecious root-parasitic *herbs* devoid of chlorophyll, with branched rhizomes covered with wide, whitish, fleshy, imbricate scales and bearing rootlets which are swollen where they are attached to the roots of the host plant. *Stem* fleshy. *Leaves* alternate, scale-like, exstipulate. *Flowers* borne singly in the axils of scales, bisexual, with pedicels. *Calyx* campanulate, equally 4-lobed. *Corolla* 2-lipped, the lips held nearly parallel to one another. *Stamens* 4, borne on the corolla tube, slightly exserted. *Styles* 1, terminal, exserted, curved downwards near its lip. *Nectary* crescent-shaped at base of ovary. *Fruit* 2-celled, capsule opening elastically at the apex by 2 valves.

Contains 7 species in temperate Europe and Asia.

Atkinson, M. D. (1996). The distribution and naturalisation of *Lathraea clandestina* L. (Orobanchaceae) in the British Isles. *Watsonia* **21**: 119–128.

Hultén, E. & Fries, M. (1986). *Atlas of north European vascular plants north of the Tropic of Cancer*. 3 vols. Königstein.

1. Aerial shoot 8–30 cm; flowers 14–20 mm, cream to dark purple, the pedicel 3–6(–10) mm **1. squamaria**
1. Aerial shoot absent; flowers 40–50 mm, dark purple, the pedicel 10–30 mm **2. clandestina**

1. L. squamaria L. Toothwort

Perennial, monoecious, root-parasitic *herb* without chlorophyll, with branched, creeping rhizome covered with wide, whitish, fleshy, imbricate scales and bearing rootlets which are swollen where they are attached to the host plant. *Stem* 8–30 cm, white, purplish or pale pink, stout, erect, simple, slightly long-glandular hairy above and with a few, whitish, ovate scales below. *Scale-leaves* alternate, 10–12 × 10–15 mm, cream, broadly ovate or subrotund, rounded at apex, entire, cordate-amplexicaul, glandular-hairy. *Inflorescence* a rather dense, 1-sided, spike-like raceme occupying much of the height of the plant, at first drooping, but straightening later. *Flowers* borne singly in the axils of scales, 8–10 mm in diameter, bisexual; pedicels 3–6(–10) mm. *Calyx* 10–16 mm, cream, tinted purplish, campanulate, 4-lobed, the lobes lanceolate and obtuse at apex, sparsely glandular-hairy. *Corolla* 14–20 mm, cream, tinged with dark purple, tubular, slightly longer than the calyx, 2-lipped, the lips subequal, the upper strongly concave, the lower flat, smaller and 3-lobed. *Stamens* 4; filaments white, tinted purple; anthers cream. *Style* 1, much exceeding corolla, white, tinted purplish, exserted and curved downwards near the tip; stigma pale yellow. *Capsule* 8–10 mm, ovoid, acuminate at apex; seeds numerous, 1–2 mm, subglobose, reticulate. *Flowers* 3–5. Visited by bumble-bees. $2n = 36$.

Native. Parasitic on the roots of various trees and shrubs, most frequently on species of *Alnus*, *Corylus* and *Fagus*. Woods and hedgerows, usually on moist rich soils. Locally frequent in Great Britain north to central Scotland and in Ireland. Much of Europe, but absent from parts of the north and south; western Asia east to the Himalaya. European Temperate element and east in Asia.

2. L. clandestina L. Purple Toothwort

Perennial, monoecious, root-parasitic *herb* without chlorophyll, with branched, creeping rhizome covered with wide, fleshy imbricate scales and bearing rootlets which are where they are attached to the host plant, the plant without an aerial shoot, the flowers arising singly in the axils of the fleshy rhizome scales at or just beneath the soil surface. *Scale-leaves* 8–10 × 11–12 mm, whitish, tinted purple, reniform or very broadly rounded, amplexicaul, glabrous. *Inflorescence* a short, corymbose raceme with 4–8 flowers. *Flowers* 8–10 mm in diameter, bisexual; pedicels 10–30 mm, whitish or tinted red. *Calyx* 18–20 mm, campanulate, the tube tinted purple, the teeth 6–7 mm, blackish-purple, with white margin and mid-vein, triangular and acute at apex. *Corolla* 40–50 mm, twice the length of the calyx or more, bright purple, tubular, 2-lipped, the upper lip hooded, the lower lip much shorter than the upper with its margins deflexed. *Stamens* 4; filaments 20–25 mm, suffused purple; anthers cream. *Style* 1, exceeding corolla, curved downwards near the tip, deep purple; stigma purple. *Capsule* 8–10 mm, ovoid; seeds 4–5, up to 5 mm, grey, reticulate-rugose. *Flowers* 3–5. Visited by bumble-bees. $2n = 42$.

Introduced. Naturalised in damp places where it is parasitic on the roots of species of *Salix*, *Populus* and *Alnus*, and rarely other species. Scattered records in north Wales, England, southern Scotland, east Ireland and Guernsey. Native of western Europe from Belgium and north Spain, extending eastwards to central and south Italy. Suboceanic Temperate species.

2. Orobanche L.

Annual to *perennial* monoecious *herbs* devoid of chlorophyll and completely parasitic on other species; without rhizomes. *Stems* erect, fleshy. *Leaves* alternate, scale-like, simple, entire, exstipulate, sessile. *Inflorescence* a terminal spike. *Flowers* mostly sessile, in the axils of bracts or bracteoles, zygomorphic, bisexual, hypogynous. *Calyx* with 2–4(–5) teeth arranged in 2 lateral lips, the lips usually open to the base on upper or both sides. *Corolla* tubular, with upper and lower lips, the 2 lips held apart, the lower turned down, the upper lip more or less entire to 2-lobed, the lower lip 3-lobed. *Stamens* 4, borne on the corolla tube. *Style* 1, terminal; stigmas 2, more or less capitate. *Ovary* 1-celled, with numerous ovules borne on 4 or 2-lobed, inwards-thrusted parietal placentas. *Fruit* a 2-celled capsule.

Contains about 100 species, chiefly in north temperate and subtropical regions.

Ballantyne, G. H. (1992). *Orobanche alba* Steph. ex Willd. in Fife (v.c. 85). *Watsonia* **19**: 39–40.

Beck-Mannagetta, G. (1930). Orobanchaceae. In Engler, A. (Edit.), *Das Pflanzenreich* **IV**. **261**. Leipzig.

Beck von Mannagetta, G. R. (1890). Monographie der Gattung *Orobanche*. *Biblioth. Bot.* **19**.

Foley, M. J. Y. (1993). *Orobanche reticulata* Wallr. populations in Yorkshire. *Watsonia* **19**: 247–257.

Hultén, E. & Fries, M. (1986). *Atlas of north European vascular plants north of the Tropic of Cancer*. 3 vols. Königstein.

Jones, M. (1987). *Orobanche hederae* Duby in the British Isles. In Weber, H. C. & Fortstreuter, W. (Edits.) *Parasitic flowering plants*, pp. 457–471. Marburg. (Proceedings of the 4th International Symposium on Parasitic Flowering Plants.)

Jones, M. (1991). Studies in the pollination of *Orobanche* species in the British Isles. *Progress in* Orobanche *research,* pp. 6–17. Tübingen.

Kreutz, C. A. J. (1995). *Orobanche*: Die Sommerwurzarten Europas. Maastricht.

Newlands, C. & Smith, H. (1998). Management and conservation status of sites with *Orobanche reticulata* Wallr. populations. *The Naturalist (Leeds)* **123**: 70–75.
Pugsley, H. W. (1926). The British *Orobanche* list. *Jour. Bot. (London)* **64**: 16–19.
Pugsley, H. W. (1940). Notes on *Orobanche* L. *Jour. Bot. (London)* **78**: 105–116.
Rumsey, F. J. & Headley, A. (1998). Are British *Orobanche* species in decline? *The Naturalist (Leeds)* **213**: 76–85.
Rumsey, F. J. & Jury, S. L. (1991). An account of *Orobanche* L. in Britain and Ireland. *Watsonia* **18**: 257–295.
Stewart, A., Pearman, D. A. & Preston, C. D. (1994). *Scarce plants in Britain*. Peterborough. [*O. alba, hederae, minor* var. *maritima* and *rapum-genistae*.]
Wigginton, M. J. (Edit.) (1999). *British red data books*. Vol. 1. *Vascular plants*. Peterborough.

1. Each flower subtended by 2 bracteoles more or less adnate to the calyx and more or less similar to the 4-toothed calyx, so that 1 bracteole and 2 calyx teeth show on each side of the flower — 2.
1. Bracteoles absent, each flower with a 2- to 4-toothed calyx — 3.
2. Corolla 10–17 mm — **1. ramosa**
2. Corolla 18–25(–30) mm — **2. purpurea**
3. Stigma yellow, whitish or orange at anthesis — 4.
3. Stigma purple, reddish, brownish or violet at anthesis — 9.
4. Lobes of lower lip of corolla plicate — 5.
4. Lobes or lower lip of corolla not plicate — 6.
5. Corolla 10–30 mm. filaments densely hairy below, sparsely hairy above, rarely glabrous — **3. crenata**
5. Corolla 10–22 mm. filaments hairy below, glabrous above — **11. hederae**
6. Stamens inserted 3 mm or less from the base of the corolla — 7.
6. Stamens inserted more than 4 mm from the base of the corolla — 8.
7. Corolla 10–19 mm — **8(i). minor** var. **flava**
7. Corolla 20–25 mm — **14. rapum-genistae**
8. Lobes of lower lip of corolla more or less equal; style densely hairy — **12. elatior**
8. Middle lobe of lower lip of corolla usually larger than side lobes; style sparsely hairy or glabrous — **13. flava**
9. Margin of lower lobe of corolla glandular-hairy — 10.
9. Margin of lower lobe of corolla not glandular-hairy — 11.
10. Corolla 10–30 mm. middle lobe of lower lip of corolla much longer than outer — **4. alba**
10. Corolla 15–36 mm. lobes of lower lip of corolla more or less equal — **10. caryophyllacea**
11. Corolla with sparse, dark violet glandular hairs, evenly curved so that its mouth is almost at right angles to its base — **5. reticulata**
11. Corolla without dark violet glandular hairs, but sometime pale ones, straight to slightly curved — 12.
12. Lower lip of corolla with large yellow bosses; stigma lobes partly fused — **9. maritima**
12. Lower lip of corolla without large yellow bosses; stigma lobes separate — 13.
13. Stamens inserted more than 3 mm above base of corolla tube — 14.
13. Stamens inserted less than 3 mm above base of corolla tube — 16.
14. Middle lobe of lower lip of corolla usually longer than side lobes — **3. crenata**
14. Lobes of lower lip of corolla more or less equal — 15.
15. Corolla bent downwards near the insertion of the stamens — **6. amethystea**
15. Corolla not bent downwards near the insertion of the stamens — **7. picridis**
16. Corolla 5–8 mm wide, erecto-patent, not appressed to stem — **8(ii). minor** var. **minor**
16. Corolla 3.5–5.0 mm wide, suberect and more or less appressed to the stem — **8(iii). minor** var. **compositarum**

Section **1. Trionychon** Wallr.

Stems simple or branched. *Flowers* with 2 bracteoles adnate to the calyx. *Calyx* with a campanulate tube and 4 subequal teeth and sometimes with a 5th smaller tooth. *Corolla* usually white, cream, blue or violet.

1. O. ramosa L. Hemp Broomrape

Phelipaea ramosa (L.) C. A. Mey.; *Kopsia ramosa* (L.) Dumort.

Perennial monoecious *herb* devoid of chlorophyll and completely parasitic on the roots of *Cannabis sativa, Nicotiana tabacum, Solanum* species and various other plants. *Stem* 8–45 cm, bright yellow, yellowish, brownish-violet, bluish or pale violet, usually slender, erect and usually branched, strongly glandular-hairy. *Scale-leaves* sparse and lax, dark brown, lanceolate, erect or spreading, shortly glandular-hairy or glabrous. *Inflorescence* very lax with numerous bisexual flowers, but sometimes cylindrical and relatively dense with few flowers; bract about one-third as long as the corolla, yellowish-brown or brown, not deflexed, lanceolate, with numerous, white glandular hairs; bracteoles 2, violet to pale violet or yellowish, about one-third the length of the corolla, broadly lanceolate, finely acute at apex, with white glandular hairs. *Calyx* about one-third the length of the corolla, violet, pale violet or yellow, tubular-campanulate, 4-dentate, densely glandular-hairy. *Corolla* 10–17 mm, bluish, violet or bluish-violet, rarely yellow or entirely white with dark violet veins, tubular, slightly constricted just above the ovary, then gradually opening out, the dorsal line almost evenly curved from the base onwards, slightly deflexed near the upper lip and sometimes bent upwards at the tip, conspicuously hairy; upper lip bilobate, with porrect lobes, hairy; lower lip deflexed, with 3 elliptical lobes of almost equal size, with white folds, densely glandular-hairy. *Stamens* 4, 2 long and 2 short, included in the corolla tube, inserted 2–4 mm above the base of the tube; filament almost glabrous or sparsely glandular-hairy at the base; anthers glabrous or with occasional hairs. *Style* 1, down-curved near the tip, sparsely glandular-hairy; stigma white or bright bluish, rarely yellowish, of several globose lobes. *Capsule* 6–7(–10) mm,

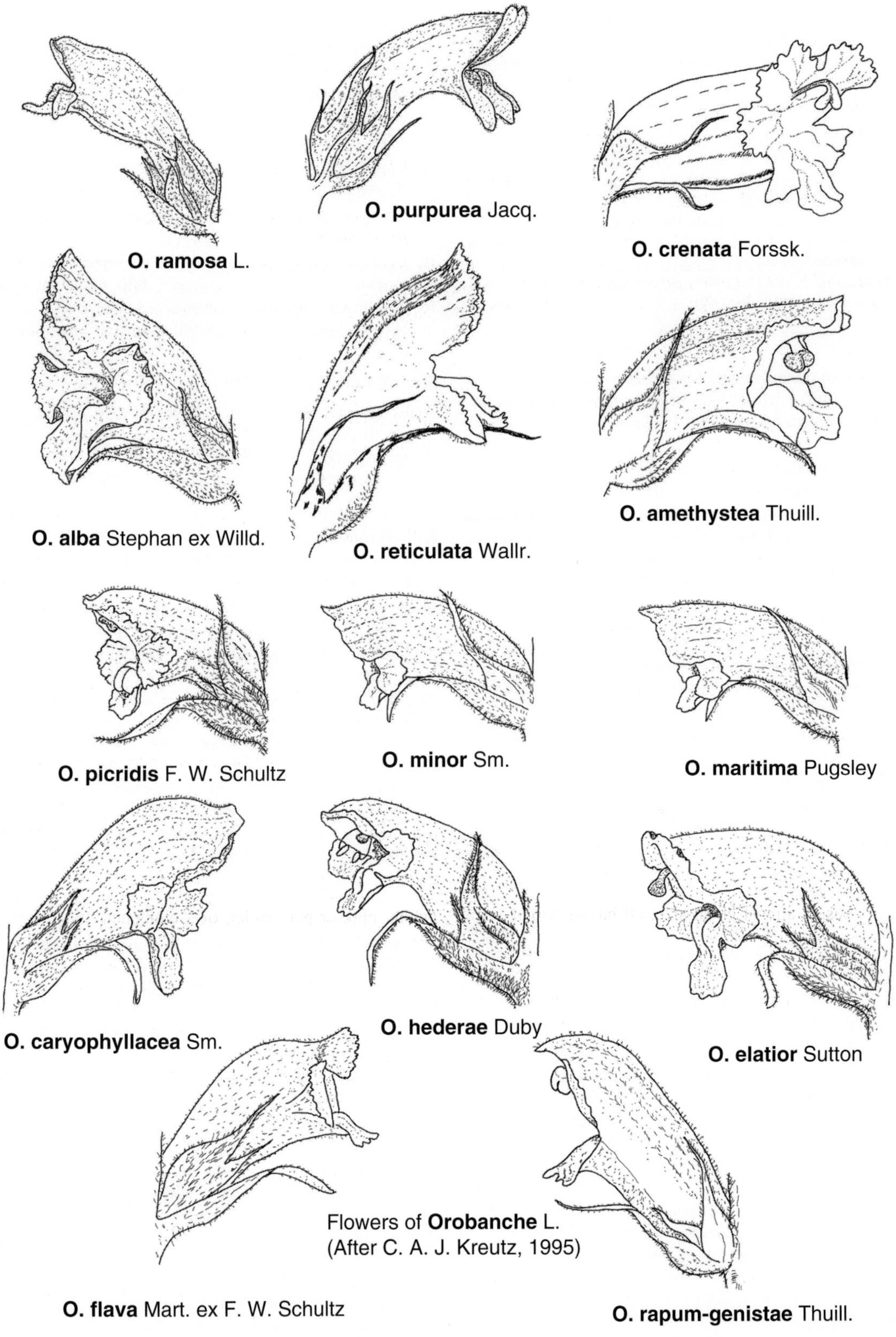

Flowers of **Orobanche** L.
(After C. A. J. Kreutz, 1995)

ellipsoid or globose, with persistent style; seeds numerous, very small, ellipsoid. *Flowers* 6–9. $2n = 24$.

Perhaps formerly native, but now extinct. In the Channel Islands and possibly also in southern England from Devonshire to Kent where it once grew in a few, scattered localities. Perhaps native, but almost certainly introduced with its host, *Cannabis*, in the former hemp-growing areas of Norfolk and Suffolk. Europe except the north, large areas of Asia, northern Africa, South Africa and North America.

2. O. purpurea Jacq. Yarrow Broomrape
O. caerulea Vill.; *Philipaea purpurea* (Jacq.)Asch.; *Philipaea caerulea* (Vill.) C. A. Mey.; *O. arenaria* auct.

Perennial monoecious *herb* devoid of chlorophyll and completely parasitic on the roots of *Achillea millefolium* and perhaps on other Asteraceae. *Stem* 15–65 cm, pale violet, violet or bluish-grey tinged with violet, usually slender, rarely stout, erect, with scattered, short glandular hairs. *Scale-leaves* sparse below, lax above, erect, bluish-grey or pale violet, elliptic-lanceolate, densely glandular-hairy. *Inflorescence* a cylindrical spike, dense at first, becoming lax, frequently elongated, with bisexual flowers down to lower part of stem; bract a third as long as the corolla, bluish grey or violet, lanceolate, acute at apex, not deflexed, very glandular-hairy; bracteoles about a one-third as long as the corolla, yellowish-brown, bluish-grey or violet, lanceolate, acute at apex, with white glandular hairs. *Calyx* tubular-campanulate and 5-toothed, with 1 segment very small, the others about one-third the length of the corolla, yellowish-brown or bluish-grey, densely glandular hairy. *Corolla* 18–25(–30) mm, bluish, pale violet or bluish-violet, yellowish or yellowish-white at the base and with dark violet veins, tubular and slightly inflated in the lower third, constricted just above and gradually opening out, the dorsal line almost evenly curved from the base to the apex, slightly deflexed near the upper lip and sometimes bent upwards at the tip, with short, pale glandular hairs; upper lip bilobate, with violet veins, shortly hairy, with erect lobes; lower lip deflexed, with 3 crenate, acute, hairy lobes of almost equal size, all having violet veins. *Stamens* 4, 2 long and 2 short, inserted 6–8 mm above the base of the corolla tube; filaments are almost glabrous or sparsely glandular-hairy at the base and usually sparsely glandular-hairy up to the anthers; anthers yellow, glabrous or sparsely short-hairy only near the top. *Style* 1, strongly glandular hairy; stigma white or violet, 2-lobed. *Capsule* 8–9 mm, ovoid-oblong, hairy in upper part, with persistent style; seeds numerous, very small, ellipsoid, reticulate and light. *Flowers* 6–7. Apparently self-pollinated. $2n=24$.

Native. Very local from Pembrokeshire and Dorsetshire to Lincolnshire and in the Channel Islands where the Alderney plant has been wrongly called *O. arenaria* Borkh. Central and southern Europe, northwards to England, Holland, Denmark, southern Sweden, the Baltic states and central Russia, south to the Mediterranean region, Morocco and the Canary Islands, east to eastern India and in North America. It is rare in most parts of Europe. European Temperate element.

Section **2. Orobanche**

Stem simple. *Flowers* without bracteoles. *Calyx* split above and below almost completely to base, and thus divided into 2 lateral segments, which may be entire or equally or unequally bifid. *Corolla* usually tinged with yellow, brown or red.

3. O. crenata Forssk. Bean Broomrape
O. speciosa DC.; *O. pruinosa* Laper.; *O. segetum* Spruner; *O. klugei* Schmitz & Regel; *O. picta* Wilms.; *O. pelargonii* Caldesi

Perennial monoecious *herb* devoid of chlorophyll and completely parasitic on the roots of species of Fabaceae. *Stem* 15–50(–100) cm, reddish-brown, yellowish-white, golden yellow, or reddish, usually stout to very stout, erect, densely glandular-hairy to almost glabrous. *Scale-leaves* few up to the inflorescence, erect or spreading, almost glabrous, elliptic-oblong to lanceolate. *Inflorescence* of numerous bisexual flowers smelling of carnations in a dense cylindrical or lax elongated spike; bract about as long as the corolla, dark brown, lanceolate, deflexed at the tip, glandular-hairy; bracteoles absent. *Calyx* 10–30 mm, about two-thirds as long or as long as the corolla, of 2 bifid or unequally bidentate or totally separated parts, yellowish-white with a violet margin or violet all over, and almost glabrous. *Corolla* 10–30 mm, white or yellowish-white, tinged with violet or pink especially near the upper lip, with pale to dark violet veins, tubular-campanulate, inflated above the insertion of the stamens, the dorsal line curved forward at the base, almost straight or slightly curved in the middle and often erect near the tip; upper lip rounded-emarginate, with very broad, porrect, spreading or deflexed lobes, the tips of the lobes glabrous; lower lip consisting of 3, deeply crenate, rounded and plicate lobes, the middle lobe usually being larger than the side lobes. *Stamens* 4, 2 long and 2 short, included in the corolla tube, inserted 2–5 mm above the base of the tube; filaments densely hairy at the base and sparsely hairy above, rarely glabrous; anthers often hairy at the line of fusion. *Style* 1, glandular-hairy; stigma white, orange, yellowish-white, pink or pale violet, of 2 elongated lobes. *Capsule* 10–12 mm, oblong, with persistent style; seeds numerous, reticulate and very small. *Flowers* 5–7. $2n=38$.

Introduced. Naturalised and casual in an area of south Essex about Cranham and Upminster since 1950 and recorded for Somersetshire and Gloucestershire. Native of southern Europe from Portugal and Spain to Greece and Turkey to the Caucasus; also in North Africa and the Canary Islands.

4. O. alba Stephan ex Willd. Thyme Broomrape
O. epithymum DC.

Annual to *perennial* monoecious *herb* devoid of chlorophyll and completely parasitic on the roots of *Thymus* and perhaps other Lamiaceae. *Stems* 7–35(–60) cm, brownish tinged with red, rarely yellow, slender, erect, sometimes slightly bent, densely glandular-hairy. *Scale-leaves* dense below and sparse above, elliptical-oblong, erect to spreading, the lower glabrous, the upper glandular-hairy. *Inflorescence* usually with few bisexual flowers, in a lax and cylindrical spike; bract about as long as the corolla, deflexed in the middle, with a dry brown or black, deflexed tip, sparsely glandular-hairy; bracteoles absent. *Calyx* about one-third to half as long as the corolla, yellowish to bright red, of 2 nearly always entire segments which are elliptical-oblong, conspicuously veined and glandular-hairy. *Corolla* 10–30 mm, yellowish to bright red on the outside, rarely white with conspicuous violet veins and brighter inside, fragrant, campanulate, slightly inflated above the insertion of the stamens, the dorsal line about evenly curved from the base onwards, almost straight in the middle and bent forward again at the upper lip, with red or purple glandular hairs; upper lip keeled, usually entire or emarginate, rarely bilobate with very broad elliptical lobes and densely hairy; lower lip with the central lobe clearly longer, almost twice the length of the outer, emarginate and glandular-hairy ciliate. *Stamens* 4, 2 long and 2 short, included in the corolla tube, inserted near the base of the tube or up to 3 mm from it; filaments hairy at base, glabrous in the middle and glandular-hairy up to the anthers; anthers brown and are usually glabrous at the line of fusion. *Style* 1, with many dark glandular hairs, especially in the upper half; stigma dark red or purple, rarely yellow or orange, of 2 globose lobes. *Capsule* 10–12 mm, oblong-obovoid, with persistent style; seeds numerous, obovoid, reticulate and very small. *Flowers* 5–8. $2n=38$.

Native. Local plant of rocky slopes on basalt, limestone and serpentine in coastal areas in western Scotland, Fifeshire and the Lizard, Cornwall. It is also a rare plant on inland limestone in Yorkshire; also recorded for north and west Ireland. Our plant is mostly, if not always, var. **rubra** (Sm.) (*O. rubra* Sm.). Northern, central and southern Europe from the Atlantic to the Caucasus, eastwards to the Himalaya with isolated localities in Gotland and Öland.

5. O. reticulata Wallr. Thistle Broomrape
O. platystigma Rchb.; *O. scabiosae* W. D. J. Koch; *O. cardui* Saut.

Perennial monoecious *herb* devoid of chlorophyll and completely parasitic on the roots of *Carduus* and *Cirsium* species. *Stems* 10–70 cm, pale to dark yellow to yellowish brown or brownish, usually stout, rarely slender, erect, densely glandular-hairy. *Scale-leaves* dense below, more sparse above, the lower triangular to elliptic and glabrous, the upper lanceolate, glandular-hairy and erect to spreading. *Inflorescence* at first a dense cylindrical spike, later becoming elongated and lax, especially in the lower part, the bisexual flowers covering most of the stem; bract about as long as the corolla, reddish-brown, narrow and deflexed, sparsely glandular-hairy; bracteoles absent. *Calyx* about half as long as the corolla, reddish-brown to violet, the segments elliptical, usually entire, rarely unequally bidentate and very faintly veined. *Corolla* 14–25 mm, whitish at the base, yellowish-white to pale yellow in the middle and yellowish-brown, reddish-brown or violet-purple near the upper lip, tubular or campanulate, slightly inflated above the insertion of the stamens, the dorsal line evenly curved over its entire length or slightly inflected at one-third of its length, almost straight to slightly concave in the middle and bent forwards near the upper lip, with sparse, dark violet glandular hairs; upper lip helmet-shaped, flexed at a right angle and consisting of 2 very broad, rounded, erect to spreading lobes; lower lip deflexed, with the middle lobe only a little longer than the side lobes, the lobes elliptical, usually plicate, crenate and with violet veins. *Stamens* 4, 2 long and 2 short, included in the corolla tube, inserted 3–4 mm above the base of the tube with a small nectar-gland at that point; filaments glabrous or sparsely hairy at the base, almost glabrous in the middle and sparsely glandular-hairy or glabrous just below the anthers; anthers brown, sparsely hairy at the line of fusion. *Style* sparsely glandular-hairy or almost glabrous; stigma of 2 brown, reddish-violet, brownish-violet or purple globose lobes. *Capsule* 10–12 mm, oblong-obovoid, slightly hairy at top, with persistent style; seeds numerous, ovoid, reticulate and very small. *Flowers* 6–8(–11). $2n=38$.

Native. Rough grassland, road verges and especially on river margins and flood plains. Very local in the Magnesian limestone districts of Yorkshire. The species occurs through much of Europe eastwards through the Caucasus to the Himalaya and possibly in North Africa. Our plant is Subsp. **pallidiflora** (Wimm. & Grab.) Hayek (*O. pallidiflora* Wimm. & Grab.; *O. cirsii* Fr.) which is described above and occurs in the lower-altitude regions of central Europe, north to southern Sweden and south to the central Mediterranean region. European Temperate element.

6. O. amethystea Thuill. Amethyst Broomrape
O. amethystina Rchb.; *O. eryngii* Duby

Perennial monoecious *herb* devoid of chlorophyll and completely parasitic on the roots of species of *Eryngium, Daucus* and *Ballota* and Asteraceae. *Stems* 15–50 cm, reddish-brown or violet, rarely yellowish or yellowish-brown, slender to stout, erect, glandular-hairy. *Scale-leaves* numerous below, usually fewer above, brownish, the lower triangular-ovate and glabrous, the upper lanceolate, spreading and glandular-hairy. *Inflorescence* with numerous bisexual flowers in a dense cylindrical spike, becoming more lax below; bract longer than the corolla, deflexed from the middle, bright to dark brown, lanceolate, acute at apex and sparsely glandular-hairy; bracteoles absent. *Calyx* about three-quarters the length of the corolla, usually reddish or violet, with darker veins, the segments entire or unequally bidentate with long acuminate upper part which is almost filiform at the tip, densely glandular-hairy. *Corolla* 12–20 mm, yellow to golden yellow or yellowish-white, tinged with violet and with dark violet veins, tubular and bent downwards above the insertion of the stamens, the dorsal line curved from the base upwards, almost straight in the middle, bent downwards near the upper lip and slightly raised at the tip, sparsely glandular-hairy; upper lip of 2 halves with spreading lobes, becoming deflexed; lower lip with 3 crenate, elliptical lobes with violet veins and deflexed. *Stamens* 4, 2 long and 2 short, included in the corolla tube, inserted 3.0–4.5 mm above the base of the tube; filaments with short hairs and sparsely glandular-hairy or glabrous below the anthers; anthers sparsely hairy at the line of fusion. *Style* 1, sparsely glandular-hairy; stigma violet, brown, black or reddish-brown, of 2 globose lobes. *Capsule* 7–10 mm, oblong, with persistent styles; seeds numerous and very small. *Flowers* 5–7. $2n=38$.

Introduced. On Apiaceae and Asteraceae in the Cambridge Botanic Garden where it was presumably introduced with imported plants. Native of central and southern Europe from Portugal to the Balkans, north to Germany; North Africa. Plants from Kent called this species are apparently *O. minor*. Populations of *O. minor*, especially if introduced, should be looked at carefully with this species in mind.

7. O. picridis F. W. Schultz — Ox-tongue Broomrape
O. artemisiae-campestris auct.

Perennial monoecious *herb* devoid of chlorophyll and completely parasitic on the roots of species of Asteraceae, most frequently *Picris hieracioides* and *Crepis* species. *Stem* 10–70 cm, yellow, yellowish-white to pink, violet or reddish brown or pale to dark yellow to yellowish-brown, usually slender, sometimes stout, erect, densely glandular-hairy. *Scale-leaves* lax below and very sparse above up to the inflorescence, the lower broadly lanceolate and almost glabrous, the upper narrowly lanceolate, erect to spreading and glandular-hairy. *Inflorescence* with numerous flowers in a dense cylindrical spike, rarely lax and with a few bisexual flowers; bract about as long as the corolla or extending beyond it, pale yellow at base with a brown to black tip, lanceolate, deflexed or recurved from the middle; bracteoles absent. *Calyx* about two-thirds the length of to as long as the corolla, the segments elliptical at base and narrowly lanceolate to filiform above, unequally bidentate, same colour as the corolla but slightly darker at the tip, glandular-hairy. *Corolla* 15–20 mm, yellowish-white to pale pink with pink or violet veins, tubular and slightly inflated above the insertion of the stamens, the dorsal line evenly curved over its entire length, almost straight or slightly curved in the middle and flexed forward near the upper lip, often with a raised tip, with numerous, pale glandular hairs; upper lip entire, flatly emarginate or bilobate, usually with porrect, flexed or slightly raised lobes, the tips of which are densely glandular-hairy; lower lip slightly deflexed, with almost equal, evenly crenate, plicate lobes. *Stamens* 4, 2 long and 2 short, included in the corolla tube, inserted 3–5 mm above the base of the tube; filaments sparsely or densely glandular-hairy at the base, densely glandular-hairy up the middle and sparsely glandular-hairy or glabrous above; anthers hairy at the line of fusion. *Style* 1, sparsely glandular-hairy; stigma dark red, purple or pale pink, with 2 lobes close together. *Capsule* 8–10 mm, oblong, with persistent style; seeds numerous, ellipsoid, reticulate and very small. *Flowers* 6–7. $2n=38$.

Native. Very local in Isle of Wight, Sussex and Kent. Records for elsewhere are said to be errors. Mediterranean region from Portugal to Turkey and Transcaucasia, central and eastern Europe north to Denmark and southern Sweden. European Southern-temperate element.

8. O. minor Sm. — Common Broomrape
O. barbata Poir.; *O. nudiflora* Wallr.; *O. apiculata* Wallr.

Perennial monoecious *herb* devoid of chlorophyll and completely parasitic on a wide range of dicotyledonous species but particularly on members of the Fabaceae. *Stems* 10–70 cm, reddish-brown or violet-purple or yellow tinged with red or purple, rarely yellow, usually slender, sometimes stout, erect or slightly curved, more or less glandular-hairy. *Scale-leaves* fairly numerous below, few above, brownish, the lower elliptical or ovate, the upper lanceolate or oblong-lanceolate, erect or spreading, glandular-hairy. *Inflorescence* usually with numerous bisexual flowers in a dense, cylindrical spike at first which elongates and becomes lax, usually with the flowers covering three-quarters of the stem; bract about as long as the corolla, pale to dark brown, narrowly lanceolate, usually slightly deflexed and densely glandular-hairy; bracteoles absent. *Calyx* about half as long as the corolla, sometimes longer, brownish-red with darker veins, the segments entire or unequally bidentate, usually deeply bifid, elliptical at base and almost filiform at tip, glandular-hairy. *Corolla* 10–19 mm long, 3.5–8.0 mm wide, yellowish, yellowish-white or violet to reddish-violet with darker veins, tubular, gradually widening and slightly inflated near the throat, the dorsal line usually evenly curved over its entire length or slightly bent in the lower half, almost straight in the middle and upper part and slightly raised near the upper lip, with numerous, pale glandular hairs; upper lip entire or bilobate with emarginate, porrect lobes; lower lip with 3 rounded, plicate, unevenly crenate, deflexed lobes of almost equal size, without prominent bosses. *Stamens* 4, 2 long and 2 short, included in the corolla tube, inserted 2–3 mm above the base of the tube; filaments sparsely hairy; anthers brown, often hairy at the line of fusion. *Style* 1, sparsely hairy or glabrous; stigma usually more or less purple, sometimes yellow, of 2 separate, hemispherical lobes. *Capsule* 8–10 mm, oblong-obovoid, with persistent style; seeds numerous obovoid, reticulate and very small. *Flowers* 5–8. $2n=38$.

(i) Var. **flava** Regel
Stems yellow. *Spike* dense and short. *Corolla* 5–8 mm wide, yellow, erect-spreading. *Stigma* yellow. On *Hypochaeris* and relatives.

(ii) Var. **minor**
Stems with various amounts of purple colouring. *Spike* variable in density and length. *Corolla* 5–8 mm wide, with some purple coloration, erect-spreading. *Stigma* more or less purple. On a very wide range of dicotyledons, including those occupied by other species.

(iii) Var. **compositarum** Pugsley
Stems with various amounts of purple colouring. *Spike* variable in density and length. *Corolla* 3.5–5.0 mm wide, with some purple coloration, suberect and more or less appressed to the stem. *Stigma* more or less purple. Mostly on Asteraceae.

?Native. Scattered over England, Ireland, Wales and the Channel Islands, but absent from many areas. It is possible *O. minor* was originally a plant of the Mediterranean region and that it has spread with imported seeds of *Trifolium, Medicago* and *Onobrychis*, so that it has spread through central to northern Europe. It is also found in large areas of Africa, North America and New Zealand. Var. *minor* occurs throughout the range of the species. Var. *compositarum* is of scattered occurrence but is most frequent in East Anglia. Var. *flava* occurred at Newport Docks in Monmouthshire and formerly in the Channel Islands.

9. O. maritima Pugsley Coastal Broomrape
O. minor var. *maritima* (Pugsley) Rumsey & Jury;
O. amethystea auct.

Annual or weakly *perennial* monoecious *herb* devoid of chlorophyll and completely parasitic on *Daucus carota* subsp. *gummifer* and very rarely on *Plantago coronopus* and *Ononis repens*. *Stem* 10–50 cm, purple throughout, slender to rather stout, densely glandular-hairy. *Scale-leaves* rather few below and nearly absent above, brownish-purple, elliptic-oblong to lanceolate, erect or spreading, glandular-hairy. *Inflorescence* of numerous bisexual flowers in a dense, cylindrical spike, elongating and becoming lax later; bract about as long as the corolla, purple, lanceolate-subulate from a broad base, glandular-hairy; bracteoles absent. *Calyx* about half as long as the corolla, the segments entire or sometimes unequally bidentate and usually deeply bifid, elliptical at base and almost filiform at tip, brownish-red, glandular-hairy. *Corolla* 12–17 mm long, 5–8 mm wide, dingy pale yellow with purple veins, cylindrical, strongly curved near the base and then more or less straight, sparsely glandular-hairy; upper lip bilobate with spreading lobes; lower lip unequally 3-lobed, with large, yellow bosses, the reniform middle lobe the largest. *Stamens* 4, 2 long and 2 short, included in the corolla tube, inserted 2–3 mm above the base of the corolla tube; filaments hairy below, glabrous above; anthers often hairy on the line of fusion. *Style* 1, sparsely hairy or glabrous; stigma purple, of 2 partially united lobes. *Capsule* 8–10 mm, oblong-obovoid, with persistent style; seeds numerous, obovoid, reticulate and very small. *Flowers* 6–7. Probably mainly self-pollinated.

Native. Steeply sloping coastal grassland on the brink of south-facing sea cliffs and fixed dunes. Southern England from Cornwall to Kent, and the Channel Islands, formerly south Wales. Western France.

10. O. caryophyllacea Sm. Bedstraw Broomrape
O. vulgaris Poir.; *O. galii* Duby

Perennial monoecious *herb* devoid of chlorophyll and completely parasitic on the roots of *Galium mollugo,* with a strong scent of carnations. *Stems* 20–60 cm, pale yellow to violet, usually slender but sometimes stout, erect, densely glandular-hairy. *Scale-leaves* dense below, sparse above, erect, oblong-elliptical to lanceolate, glandular-hairy. *Inflorescence* usually short, with relatively few bisexual flowers in a lax spike; bract about as long as the corolla, reddish-brown, lanceolate, deflexed in the middle, with a shrivelled, reddish-brown, recurved tip; with long, pale glandular hairs; bracteoles absent. *Calyx* about half as long as the corolla, of elliptical, veined parts, brownish-violet like the corolla and densely glandular-hairy. *Corolla* 15–36 mm, brownish-violet, sometimes pale or yellowish and rarely pink or reddish, distinctly widening towards the mouth, the dorsal line evenly curved from the base, less curved or straight in the middle and bent forward near the upper lip, glandular-hairy; upper lip very broad, keeled, slightly emarginate, its lobes erect at first and porrect later; lower lip slightly deflexed and with 3 crenate, plicate lobes of equal size, the margin glandular-hairy. *Stamens* 4, 2 long and 2 short, included in the corolla tube, inserted at the base of the tube or up to 3 mm above it; filaments swollen at the base, very hairy up to the middle and sparsely glandular-hairy above; anthers brown, glabrous or with wart-like hairs at the line of fusion. *Style* 1, sparsely hairy at the base and very glandular-hairy from the middle upwards; stigma brownish-violet to dark purple, of 2 lobes. *Capsule* 9–12 mm, oblong, with persistent style; seeds numerous, ellipsoid, reticulate and very small. *Flowers* 5–8. $2n=38$.

Native. Very local in Kent with a few unconfirmed records elsewhere. Central and eastern Europe to Turkey, the Caucasus and Iran, in the mountains of southern Europe and North Africa. European Temperate element.

11. O. hederae Duby Ivy Broomrape
O. medicaginia Rchb.; *O. laurina* Rchb. fil.

Annual or *perennial* monoecious *herb* devoid of chlorophyll and completely parasitic on the roots of *Hedera helix* and perhaps occasionally on other species. *Stems* 10–55 cm, frequently deep purplish-brown, sometimes yellowish, yellowish-brown, sometimes reddish-brown below and yellowish above, seldom yellow over its entire length, slender to stout, with dense, short and medium glandular hairs. *Scale-leaves* sparse or dense up to the inflorescence; the lower elliptical to lanceolate and glabrous; the upper rather broad lanceolate, erect and glandular-hairy. *Inflorescence* a spike, at first cylindrical and dense above, many-flowered, later lax below with the lower flowers set far apart and very low on the stem; flowers bisexual; bract as long as or longer than the corolla, reddish-brown to brown or black, lanceolate, with brown to black apex, in the upper parts with deflexed or recurved apex, densely glandular-hairy; bracteoles absent. *Calyx* tubular, about two-thirds as long as the corolla, dark red or dark brown, at the base usually paler than the corolla, the segments entire, bifid or unequally bidentate, elliptical at the base, and narrowly lanceolate to filiform above, densely glandular-hairy. *Corolla* 10–22 mm, pale yellow or yellowish, rarely yellowish-brown, its back tinged with violet or red, especially near the upper lip, the whole with reddish-violet veins especially near the upper lip, narrowly tubular, inflated above the insertion of the stamens, contracted near the lower lip and widening into a funnel near the margin, the dorsal line is evenly curved from the base, often almost straight in the middle and sometimes raised near the upper lip, the whole with bright glandular hairs; upper lip entire or emarginate, usually with porrect or spreading lobes with bright glandular hairs near the tips; lower lip usually not deflexed, with 3 dentate, rounded and plicate lobes.

Stamens 4, 2 long and 2 short, inserted 3–4 mm above the base of the corolla tube; filaments are hairy below and glabrous above; anthers brown, almost glabrous and protruding from the corolla. *Style* 1, with sparse, short glandular hairs or glabrous; stigma golden yellow, dark yellow, pale yellow, orange or reddish-yellow turning dark brown, of 2 elongated lobes. *Capsule* 10–12 mm, broadly obovoid, with persistent style; seeds numerous, oblong, reticulate, very small and light. *Flowers* 5–10. Probably self-pollinated. $2n=38$.

Native. Essentially a maritime plant of coastal cliffs, undercliff woodlands and hedgebanks, but extending inland in sheltered sites such as gorges. We also grow it as a curiosity and it persists in botanic gardens, as at Cambridge, parklands and cemeteries, often at wall bases and under hedges. Local in south and western Great Britain from Isle of Wight to Wigtownshire, much of Ireland and the Channel Islands. Western and central Europe north to southern Scotland, Holland, Germany and Austria, east through the Mediterranean region to Turkey, the Caucasus and Iran, and in north-west Africa and the Canary Islands. Submediterranean Subatlantic element.

12. O. elatior Sutton — Knapweed Broomrape
O. fragrans W. D. J. Koch; *O. stigmatodes* Wimm.; *O. kochii* F. W. Schultz; *O. major* auct.

Perennial monoecious *herb* devoid of chlorophyll and completely parasitic on the roots of *Centaurea scabiosa* and occasionally other Asteraceae and *Thalictrum* species. *Stems* 20–70 cm, brown to dark reddish-brown, yellowish or pink, stout, erect. usually densely glandular-hairy. *Scale-leaves* even and dense up to the inflorescence, erect and glandular-hairy, triangular-lanceolate below, the upper lanceolate. *Inflorescence* of numerous bisexual flowers in a dense, cylindrical spike; bract about as long as the corolla, reddish-brown to dark brown or black, lanceolate, with a shrivelled, dark brown, recurved tip, sparsely glandular-hairy to glabrous, deflexed from the middle; bracteoles absent. *Calyx* one-third to half the length of the corolla, of 2, unequally bidentate parts, touching or fused at the base, pale yellowish-brown and strongly glandular-hairy. *Corolla* 15–18 mm, yellowish-brown or reddish-brownish-yellow later, rarely entirely bright yellow, with dark veins, gradually widening above the insertion of the stamens, the dorsal line is evenly curved from the base onwards, sometimes more clearly bent forwards from nearly the middle, with glandular hairs; upper lip usually entire or slightly emarginate, with 2 rounded, erect lobes; lower lip consisting of 3 crenate, deflexed lobes of almost equal size. *Stamens* 4, 2 long and 2 short, included in the corolla tube, inserted 4–6 mm above the base of the tube, with a golden yellow nectar spot at the base; filaments densely hairy at the base and more sparsely hairy up to the anthers; anthers pale brown, often glabrous. *Style* 1, down-curved near the tip, densely glandular-hairy especially above; stigma yellow, of 2 lobes. *Capsule* 8–10 mm, oblong-obovoid, with persistent style; seeds numerous, ellipsoid, reticulate and very small. *Flowers* 6–8. Nectar-secreting, but scentless. $2n=38$.

Native. On chalk and limestone. South and east England north to Yorkshire and in Glamorganshire. Europe northwards to southern Sweden and the Baltic states, southward to north-east Spain, central Italy and southern Greece, east to the Himalayas and northern India. Eurosiberian Temperate element.

13. O. flava Mart. ex F. W. Schultz — Butterbur Broomrape
O. tussilaginis Mutel; *O. froehlichii* Rchb. fil.

Perennial monoecious *herb* devoid of chlorophyll and completely parasitic on species of Asteraceae, particularly *Petasites* species. *Stems* 15–65 cm, orange-yellow, yellowish-white or brownish, usually stout, rarely slender, with dense glandular hairs. *Scale-leaves* numerous below, more sparse above, the lower triangular or oblong-elliptical to lanceolate, the upper lanceolate, all glandular-hairy. *Inflorescence* with numerous bisexual flowers in a dense cylindrical spike which becomes elongated and lax later; bract about as long as or slightly longer than the corolla, reddish at first, becoming brown to black, lanceolate, acute at apex, often slightly deflexed in the middle, densely glandular-hairy; bracteoles absent. *Calyx* tubular, shorter than the corolla, reddish-brown, the segments entire or unequally bidentate, long-acute, glandular-hairy. *Corolla* 15–25 mm, yellow to yellowish-white or reddish-brown, tubular, inflated above the insertion of the stamens, the dorsal line evenly curved from the base onwards, often more clearly bent forward in the middle and sometimes raised a little near the upper lip, with pale glandular hairs; upper lip bilobate, deep emarginate, the lobes porrect at first and deflexed later; lower lip with the middle lobe usually larger than the side lobes, unevenly crenate and sparsely glandular-hairy at the margin. *Stamens* 4, 2 long and 2 short, inserted 4–6 mm above the base of the corolla tube; filaments densely hairy from the base to halfway and more sparsely hairy up to the anthers; anthers hairy at the line of fusion. *Style* 1, sparsely glandular-hairy or glabrous; stigma orange or wax-coloured, consisting of 2 rounded, globose lobes. *Capsule* 10–14 mm, cylindrical, with persistent style; seeds numerous, ovoid, very small and light. *Flowers* 6–8. $2n=38$.

Introduced. Established on *Petasites* and related genera in the Oxford Botanic Garden where it may have been introduced with imported plants. Native of central and eastern Europe from the Jura to Hungary and Roumania, north to Poland.

14. O. rapum-genistae Thuill. — Greater Broomrape
O. major auct.; *O. rapum* auct.

Perennial monoecious *herb* devoid of chlorophyll and completely parasitic on the roots of Fabaceae species, particularly *Cytisus scoparius* and *Ulex* species and occasionally *Genista tinctoria*. *Stems* 30–85 cm, yellow, yellowish-brown, orange-yellow, brownish or reddish-brown, stout, conspicuously swollen below, erect, densely glandular-hairy. *Scale-leaves* dense and imbricate below, commonly numerous above, erect to spreading, glandular-hairy, the lower triangular to broadly lanceolate, the upper lanceolate. *Inflorescence* a cylindrical spike, flowers bisexual, dense in the upper part, more lax and elongated in the middle and lower part with the flowers down to the lower parts of the stem; bract much longer than the corolla, dark brown with a yellowish base, lanceolate, acute at apex, deflexed in the upper part, sparsely glandular-hairy; bracteoles absent; flowers with an unpleasant smell. *Calyx* shorter than to half as long as the corolla and of the same colour, consisting of 2 unfused, unequally bidentate, rarely entire halves, the teeth with drawn out, acute tips, densely glandular-hairy. *Corolla* 20–25 mm, pale yellowish, reddish-brown to violet-brown or dark brown, with dark red veins, widely tubular to campanulate, inflated above the insertion of the stamens, the dorsal line evenly curved over its entire length and helmet-shaped near the top, with numerous glandular hairs; upper lip helmet-shaped, entire, with porrect to slightly raised lobes; lower lip deflexed, with almost equal, rounded, crenate lobes, the middle lobe is somewhat larger. *Stamens* 4, 2 long and 2 short, included in the corolla tube, inserted up to 2 mm above the base of the tube; filaments glabrous at the base, densely glandular-hairy just below the anthers; anthers almost glabrous. *Style* 1, down-curved near the tip, glabrous below, glandular-hairy above; stigma wax-coloured or golden yellow, of 2 distant globose lobes. *Capsule* 10–11(–14) mm, oblong, hairy above, with persistent style; seeds numerous, broadly ellipsoid, reticulate, very small and light. *Flowers* 5–7. Visited by bees, the corolla often perforated at the base by *Bombus terrestris* L. $2n=38$.

(**1**) Forma **rapum-genistae**
Stems brownish or reddish-brown. *Corolla* reddish-brown to violet-brown or dark brown.

(**2**) Forma **flavescens** Durand
O. rapum-genistae forma *hypoxantha* Beck

Stems and corolla pure yellow.

Native. Hedgebanks and scrubby areas on rough hillsides where its hosts occur, and its vigour is often increased after burning of the host plants. Formerly widespread in Great Britain and Ireland, when it was the most commonly encountered broomrape, but is has declined considerably during the present century and is now most often found in coastal areas in the south and west north to southern Scotland, and in the Channel Islands and south and south-east Ireland. At the present time it is almost certainly under-recorded, but even so new sites are being discovered faster than old ones are lost. Even in localities where a host is abundant, however, this broomrape is very local and populations small. Western Europe north to southern Scotland, Holland and northern Germany, eastwards to western Germany, the western part of Switzerland and Italy, southwards to the Mediterranean and Corsica and Sardinia, and in north-west Africa. Suboceanic Southern-temperate element. Forma *rapum-genistae* is the most common form.

139. GESNERIACEAE Dumort. nom. conserv.

Perennial monoecious *herbs. Stems* more or less absent. *Leaves* in a basal rosette, simple, dentate, petiolate, exstipulate. *Flowers* 1–few, on long peduncles arising from a rosette, nearly actinomorphic, bisexual, hypogynous. *Sepals* 5, fused into a tube proximally. *Petals* 5, fused into a tube proximally. *Stamens* 5, borne on the corolla tube, exserted. *Style* 1; stigma 1, capitate. *Ovary* 1-celled, with many ovules on 2 parietal placentas. *Fruit* a capsule, dehiscing into 2 valves.

Contains 133 genera and about 2,500 species mostly pantropical, but some temperate.

1. Ramonda Rich.

As family.

Contains 3 species, 1 in the Pyrenees and 2 in the Balkans. Named after L. F. E. von Ramond de Carbonnières (1753–1827).

1. R. myconi (L.) Rchb. Pyrenean Violet
Verbascum myconi L.; *Chaixia myconi* (L.) Lapeyr.; *Myconia borraginae* Lapeyr.; *R. pyrenaica* Rich.

Perennial monoecious *herb* with a fibrous rhizome. *Leaves* in a dense rosette, 2–6 × 2–5 cm, dark green, broadly ovate or elliptical, rounded-obtuse at apex, dentate or crenate-dentate with more or less regular, rounded teeth, rounded or cuneate at base, densely clothed with long, white, bulbous-based simple eglandular hairs on the upper surface and dense, long, brown simple eglandular hairs beneath; petiole winged and it and the stock with very dense, long, brown simple eglandular hairs. *Flowers* 1–few, bisexual, 25–40 mm in diameter; peduncles up to 11 cm, pale green or suffused brownish-purple, with numerous, short glandular hairs and some longer eglandular ones; pedicels when present short, with short glandular hairs and some longer eglandular ones. *Sepals* 5, 4–6 mm, pale green, oblong, rounded at apex, with numerous glandular and eglandular hairs. *Petals* 5, 12–20 mm, violet with a yellow base, obovate, broadly subacute at apex, narrowed at base, ciliate with short glandular hairs, throat yellow-villous. *Stamens* 5; filaments white; anthers yellow. *Style* 1, pale; stigma capitate. *Capsule* 12–14 mm, ovate-oblong, with numerous, short glandular hairs; seeds small, pale brown, oblong, papillose. *Flowers* 7–8. $2n=48$.

Introduced. Planted on a rock-face in Cwm Glas, Caernarvonshire in 1921 and still there, though often not flowering and also persistent in an old estate in Yorkshire. Native of the Pyrenees.

140. ACANTHACEAE Juss. nom. conserv.

Perennial monoecious *herbs. Stems* erect. *Leaves* in a basal rosette and a few opposite or alternate on stems, pinnate or deeply pinnately divided; exstipulate. *Flowers* large, in robust terminal spikes, zygomorphic, bisexual, hypogynous, each in the axil of a spiny-toothed bract and 2 bracteoles. *Calyx* 4-lobed, with large upper and lower lobes and 2 smaller, narrower lateral ones, fused only at base, persistent in fruit. *Corolla* with a short tube and 3-lobed lower lip, the upper lip absent. *Stamens* 4, borne on the corolla tube, the filaments free, but the 1-celled anthers fused in pairs. *Style* 1; stigmas 2, slightly unequal, linear. *Ovary* 2-celled, each cell with many ovules on axile placentas. *Fruit* a 2-celled capsule with persistent style.

Contains about 250 genera and 2,500 species in tropical and warm temperate regions.

1. Acanthus L.

Like family.

Contains about 50 species in southern Europe and tropical and subtropical Asia and Africa.

Rix, M. (1980). The genus *Acanthus*. 1. An introduction to the hardy species. *The Plantsman* **2**: 132–140.

1. Basal leaves not spinous-dentate	**1. mollis**
1. Basal leaves spinous-dentate	**2. spinosus**

1. A. mollis L. Bear's-breeches

Perennial monoecious *herb. Stems* several, up to 1 m. pale green, erect, glabrous, unbranched, leaves few and small. *Leaves* yellowish-green with a pale midrib on upper surface, slightly paler beneath and veins prominent beneath; basal numerous, with lamina 20–60 × 5–25 cm, oblong-elliptical in outline, obtuse at apex, pinnately divided into 4–7 pairs of lobes and a terminal one, the lobes oblong or ovate-oblong, acute at apex, irregularly lobulate, the lobes acute at the apex and crinkled, the petioles up to 35 cm, very pale green, smooth and glabrous; cauline 1–4, miniture versions of the basal; all glabrous. *Inflorescence* a large terminal spike up to 60 cm; flowers bisexual; bracts 35–40 × 15–25 mm, pale green at base and dull purplish-brown above with a red margin and dark veins, ovate, spinous-acute at apex, with long, spiny teeth on margin and minutely hairy; bracteoles 23–25 mm, dull purple, linear, sharply acute at apex and minutely hairy. *Calyx* 40–60 mm, of 2 large lobes, the upper and lower, and 2 small, the upper pale green at base, dull brownish-purple above with darker veins, obovate, the margins incurved to form a hood and takes the place of the missing upper lip of the corolla and toothed at apex, the lower pale green below with dark veins, dull brownish-purple above, spathulate, sharply toothed at apex and with a shallow channel on outer surface, the 2 lateral lobes are narrower. *Corolla* 45–50 × 40–45 mm, cream with red veins towards the base, lower lip 3-lobed, the lobes broadly rounded, the upper lip absent. *Stamens* 4; filaments cream, curved upwards and then outwards; anthers orange-brown with dense white hairs. *Style* 1, pale yellow, curved to just above anthers, forked at tip. *Capsule* ovoid, glabrous; seeds 2–4. *Flowers* 6–8. Protandrous. Visited by bumble-bees and hoverflies. $2n = 56$.

Introduced. Grown in gardens and long naturalised in waste places, roadsides, railway banks and scrub. Scattered records in southern England, south Wales and the Channel Islands with a few records farther north and a solitary record in south-east Ireland. Native of the west and central Mediterranean region.

2. A. spinosus L. Spiny Bear's-breeches

Perennial monoecious *herb. Stems* several, up to 1 m. pale green, often suffused with brownish-purple, erect, slightly angled, unbranched, leafy below inflorescence. *Leaves* dull, dark green with a pale midrib on upper surface, paler beneath with prominent veins; basal numerous, 30–60 × 16–20 cm, ovate-oblong in outline, acute at apex, pinnately divided to the rhachis into 6–8 lobes, the lobes ovate, acute at apex and irregularly spinous-dentate or spinous-lobate, the petioles very pale green and smooth; cauline several, like basal but smaller and slowly getting smaller upwards; all glabrous. *Inflorescence* a large terminal spike up to 75 cm; flowers bisexual; bracts 40–45 × 20–25 mm, green below, dull purplish-brown near the apex with prominent dull purple veins, broadly ovate, spinous-acute at apex, with long spines on the margin and shortly hairy; bracteoles 23–25 mm, flushed dull purple, linear, spinous-acute at apex and shortly hairy. *Calyx* of 2 large lobes, the upper and lower, and 2 small, the upper 40–50 mm, green with dull purplish-brown round the apex, obovate, sharply toothed at apex, and hooded to take the place of the upper lip of the corolla, the lower 38–42 mm, green below, dull purplish-brown round the apex with darker veins, spathulate, toothed at apex and shortly hairy, the 2 lateral lobes are narrower. *Corolla* 30–35 × 35–40 mm, cream tinted rose, lower lip 3-lobed, the lobes broad and rounded, the upper lip absent. *Stamens* 4; filaments cream, curved upwards and then outwards; anthers orange with white hairs. *Style* 1, pale yellow, forked at tip. *Capsule* ovoid; seeds 2–4. *Flowers* 6–8. Protandrous. Visited by bumble-bees and hoverflies. $2n = 112$.

Introduced. Grown in gardens and naturalised in a few waste and grassy places. Scattered records through much of Great Britain. Native of the central Mediterranean regions.

(**Sesamum orientale** L. of the Pedaliaceae has been recorded as a bird-seed and spice casual.)

140A. BIGNONIACEAE Juss.

Large, deciduous, monoecious *trees. Trunks* stout. *Leaves* in whorls of 3, large, simple and entire. *Inflorescence* a large, terminal panicle of dense bisexual flowers. *Calyx* 5-lobed. *Corolla* campanulate, 2-lipped. *Stamens* 5, 2 fertile and 3 infertile. *Style 1*; stigma 2-lobed. *Ovary* superior. *Fruit* a capsule.

Contains about 250 genera, with 2,500 species, worldwide, but centred in the tropics.

Eccremocarpus scaber (D. Don) Ruiz & Pav. has been recorded as a garden escape and relic and **Campsis radicans** (L.) Seem ex Bureau has occurred as self-sown seedlings at Marble Arch, Middlesex.

1. Catalpa Wolf

As family.

Contains about 10 species in North America and eastern Asia.

1. C. bignonioides Walter Indian Bean Tree

Deciduous, monoecious *tree*, 10–18 m. with a domed spreading crown, wider than high. *Trunk* up to 1.1 m in diameter, usually short, stout and often bent, sometimes taller and often leaning when old. *Bark* pink and pale brown, smooth in young trees, but soon becoming grey and scaly and some fissured into flat ridges. *Branches* wide-spreading and crooked; twigs brown, thick; young shoots green with raised orange lenticels, becoming greyish-brown, glabrous. *Buds* about 1 mm, in pairs or whorls of 3, ovoid, acute at apex; scales purplish-brown, ovate, pointed, glabrous. *Leaves* in whorls of 3 with an unpleasant odour when crushed; lamina 12–25 × 10–22 cm, purple on young trees when emerging, becoming dull green on upper surface, pale green beneath, not changing colour in autumn, but a cultivar with bright yellow leaves, not appearing until late in June, thin, ovate or broadly ovate, with an abrupt, slender point at apex, entire, broadly cuneate or truncate at base, glabrous on upper surface, densely hairy beneath; veins 3 at base, with 6 pairs of secondary ones, extra-floral nectaries in their axils; petiole 5–15 cm, shiny green, tinted purplish above, terete, glabrous. *Inflorescence* a large terminal panicle 20–30 × 20–30 cm; flowers numerous 40–50 mm in diameter, bisexual; pedicels up to 20 mm, dark brownish-red, with scattered sessile glands. *Calyx* 8–11 mm, dark brownish-red, 2-lipped, covered with minute spots. *Corolla* 40–50 mm, white, with purple spotted lines and reddish and orange marks where the lower lip joins and leads into the opening to the throat; tube campanulate, 2-lipped, the upper lip with 2 fringed lobes, the lower lip with 3 fringed, triangular-ovate lobes. *Stamens* 2 fertile and 3 infertile; filaments white; anthers yellow to orange. *Style* 1 white; stigma 2-lobed. *Capsule* 15–40 cm, purplish-brown, linear and slender, acute at apex, hanging from the tips of branches throughout the winter. *Flowers* 7–8. $2n = 40$.

Introduced in 1726. Planted in parks, gardens, city squares and sometimes along streets. Frequent in southern England, uncommon northwards to southern Scotland and in southern Ireland. Self-sown seedlings have been reported in the London area. Native of eastern North America.

141. LENTIBULARIACEAE Rich. nom. conserv.

Insectivorous *perennial* monoecious *herbs*, rootless and aquatic, or rooted, stemless and rosette-forming. *Leaves* rosette-forming, simple, entire and the upper surface covered with viscid glands, or alternate and divided into filiform to linear segments some of which are modified into insect traps. *Flowers* solitary on long pedicels or in racemes on stalks emerging from water, yellow, bluish or white, zygomorphic, bisexual, hypogynous. *Calyx* obscurely 2-lipped, fused at base. *Corolla* 2-lipped, the upper lip 2-lobed, the lower lip 3-lobed, spurred at base. *Stamens* 2, borne at the base of the corolla. *Style* absent or very short; stigma variously expanded. *Ovary* 1-celled, with many ovules on free-central placentas. *Fruit* a capsule.

Contains 4 genera and about 180 species, worldwide.

1. Plant rooted with a basal rosette of simple, entire leaves covered with viscid glands as insect traps; corolla white to bluish-violet **1. Pinguicula**
1. Plant rootless and aquatic; leaves divided into linear to filiform segments bearing bladder-like animal-catching traps; corolla yellow **2. Utricularia**

1. Pinguicula L.

Perennial monoecious *herbs* with roots. *Leaves* simple, entire, in a basal rosette. *Flowers* solitary on long, erect pedicels, bisexual. *Calyx* slightly 2-lipped, the upper lip 2-lobed, the lower lip 3-lobed. *Corolla* white to violet, 2-lipped, the upper lip 2-lobed, the lower lip 3-lobed, with open mouth, spurred at base. *Stamens* 2, borne at the base of the corolla. *Style* absent or very short; stigma expanded. *Ovary* 1-celled, with many ovules on free-central placenta. *Fruit* a capsule.

Contains about 46 species in the northern hemisphere and South America.

The species of this genus have the upper surfaces of their leaves thickly covered with 2 sets of glandular hairs, differing in the size of their glands and the length of their stalks. The larger glands have a circular outline as viewed from above, and are divided into 16 cells containing a light green homogeneous fluid, which when secreted is extremely viscid. It has been calculated that an ordinary rosette of leaves contains about half a million of these glands. Any small insect alighting on these leaves is thus caught. The edge of the leaf curves over thus preventing the insect from being blown away and by pushing it slightly towards the centre of the leaf where there are more glands to digest.

Caspar, S. J. (1966). Monographie der Gattung *Pinguicula* L. *Biblioth. Bot.* **31** (**127/128**).

Druce, G. C. (1922). *Pinguicula grandiflora* × *vulgaris* mihi. = × *P. scullyi* mihi. *Rep. Bot. Sci. Exch. Club Brit. Isles* **6**: 301.

Ernst, A. (1961). Revision der Gattung *Pinguicula. Bot. Jahrb.* **80**: 145–194.

Hultén, E. & Fries, M. (1986). *Atlas of north European vascular plants north of the Tropic of Cancer*. 3 vols. Königstein.

Praeger, R. L. (1930). Notes on Kerry plants. *Jour. Bot. (London)* **68**: 249–250.

1. Corolla white, with 1–2 yellow spots on the lower lip **2. alpina**
1. Corolla pale lilac to deep bluish-mauve or deep mauve-violet, very rarely whitish, although whitish or yellowish in the throat, never with yellow spots on lower lip 2.
2. Corolla 7–11 mm including spur, which is 2–4 mm, pale pinkish-mauve or pale lilac **1. lusitanica**
2. Corolla 14–35 mm, including tapering spur which is 3–14 mm, deep bluish-mauve or mauve-violet 3.
3. Corolla 14–22(–30) mm including spur which is 3–7(–10) mm; lobes of lower lip separated laterally **3. vulgaris**
3. Corolla 25–35 mm including spur which is 10–12(–14) mm; lobes of lower lip overlapping laterally **4. grandiflora**

1. P. lusitanica L. Pale Butterwort

Perennial monoecious *herb* with fibrous roots, overwintering as a rosette. *Leaves* 5–12; lamina 1.0–2.4(–2.9) × 0.3–0.8 cm, pale translucent olive or greyish-green, sometimes tinged purplish or with reddish veins, ovate or ovate-oblong, rounded-obtuse at apex, entire, strongly inturned on the margin, narrowed at base, soft, fleshy and with dense, minute, viscid glands on the upper surface. *Flowers* solitary, bisexual, on long pedicels; pedicels 1–8, 6–15(–25) cm, pale green, sometimes tinged brownish, very slender, glandular-hairy. *Calyx* 2–6 mm, slightly 2-lipped; upper lip deeply 2-lobed, the lobes elliptical and rounded at apex; lower lip 3-lobed, the lobes similar. *Corolla* 7–11 mm, pale pinkish-mauve to pale lilac, with reddish lines on throat and spur and often yellow in the throat, tube cylindrical, 2-lipped, the upper lip 2-lobed with the lobes elliptical and rounded at apex, the lower lip 3-lobed with the lobes elliptic and emarginate; spur 2–4 mm, slender, subcylindrical, down-curved at base. *Stamens* 2; filaments about 1.5 mm, white, thick and curved; anthers pale yellow. *Style* absent; stigma subrotund. *Capsule* subglobose; seeds 0.5–0.6 mm, brown, oblong, reticulate. *Flowers* 6–10. Self-pollinated. $2n=12$.

Native. Bogs and wet heaths. Frequent in Ireland, west and north Scotland, Isle of Man, south-west Wales and south-west and south central England. West France, west Spain, Portugal; north-west Morocco. Oceanic Temperate element.

2. P. alpina L. Alpine Butterwort

Perennial monoecious *herb* overwintering as a bud, developing long, relatively thick, brownish roots. *Leaves* 5–8 in a rosette; lamina (15–)25–45(–60) × 8–14 mm, pale yellowish-green, elliptic-oblong to lanceolate-oblong, rounded-obtuse at apex, entire, strongly inturned on the margin, narrowed at base, soft, fleshy and with dense, minute, viscid glands on the upper surface. *Flowers* bisexual, solitary on pedicels; pedicels 1–8, 5–11(–13) cm, pale green, slender, sparsely glandular-hairy. *Calyx* 3–4 mm, pale brownish, 2-lipped, the upper lip with 2 triangular lobes, the lower lip lobed for half its length into 3 obovate lobes. *Corolla* 8–10 mm, white; 2-lipped; tube short and thick; upper lip with 2 subrotund lobes; lower lip with 3 lobes, the central lobe wide and emarginate, the lateral lobes much smaller and subrotund; throat with 1–2 deep yellow spots; spur 2–3(–5) mm, yellowish, curved. *Stamens* 2; filaments white; anthers yellow. *Style* absent; stigma pale, expanded. *Capsule* ovoid-oblong. *Flowers* 5–8. $2n=32$.

Native. Boggy places. Formerly recorded in Ross and Cromarty from 1831 to about 1900. Northern Europe from Iceland to Finland and the Baltic States, Pyrenees and mountains and uplands of central Europe; Himalaya. Eurosiberian Arctic-montane element.

3. P. vulgaris L. Common Butterwort
P. borealis Salisb. nom. illegit.

Perennial monoecious *herb* overwintering as a rootless bud. *Leaves* 5–11 in a rosette; lamina 2.0–4.5(–9.0) × 1.4–2.0(–2.6) cm, shining yellowish-green, paler on midrib and beneath, oblong to obovate-oblong, rounded-obtuse at apex, entire, somewhat inturned along the margin, narrowed at base, soft, fleshy and with dense, minute, viscid glands on the upper surface. *Flowers* bisexual, solitary on long pedicels; pedicels 1–6, 7.5–18(–27) cm, pale green, sometimes tinged brownish, glandular-hairy. *Calyx* 3–4 mm, slightly 2-lipped; upper lip 2-lobed, the lobes ovate and subacute at apex; lower lip 3-lobed, the lobes ovate and subacute at apex. *Corolla* 14–22(–30) mm, deep mauve-violet, usually whitish in the throat, 2-lipped, the lips unequal, the upper lip 2-lobed and the lobes oblong, the lower lip 3-lobed, the lobes oblong, much longer than wide and separated laterally; spur 3–7(–10) mm, cylindrical-subulate, straight. *Stamens* 2; filaments about 2.0 mm, short and thick; anthers 2-lobed. *Style* absent; stigma broad and fan-like. *Capsule* ovoid; seeds numerous, about 1.0 mm, brown, oblong, reticulate. *Flowers* 5–7. Pollinated by small bees. $2n=64$.

Native. Bogs, wet heathland and limestone flushes. Locally common over much of Great Britain and Ireland especially in the north and west; absent from most of central and southern England. North-west and central Europe extending eastwards to the western Ukraine; northern Asia; north Morocco; North America south to New York and British Columbia. Circumpolar Boreal-montane element.

4. P. grandiflora Lam. Large-flowered Butterwort

Perennial monoecious *herb* overwintering as a bud, but leaves often persisting or precociously developing. *Leaves* 5–8 in a rosette; lamina 3.0–4.5(–6.5) cm, pale yellowish-green, oblong to obovate-oblong, rounded-obtuse at apex, entire, narrowed at base, soft, fleshy and with dense, minute, viscid glands on the upper surface. *Flowers* bisexual, solitary on long pedicels; pedicels 1–5, 6–15(–23) cm, pale green, sometimes flushed mauve, glandular-hairy. *Calyx* 5–6 mm, 2-lipped; upper lip 2-lobed, divided nearly to base, the lobes narrowly elliptical and rounded at apex; lower lip 3-lobed, the lobes elliptical. *Corolla* 25–35 mm, deep bluish-mauve, with darker veins and a whitish throat, 2-lipped; upper lip 2-lobed the lobes broadly elliptical and rounded at apex; lower lip 3-lobed, the median lobe largest, obovate and rounded at apex, the lateral small, the lobes overlapping laterally; spur 10–12(–14) mm, straight and rather stout, sometimes slightly bifid. *Stamens* 2; filaments about 3.0 mm, white; anthers pale yellow. *Style* absent; stigma broad and fan-like. *Capsule* ovoid; seed numerous; stigma broad and fan-like. *Capsule* ovoid; seeds numerous, 0.7–0.8 mm, brown, oblong, reticulate. *Flowers* 5–8. $2n=32$.

Native. Bogs and damp moorland. Locally common in south-west Ireland from Co. Cork to Co. Clare; planted and persistent in scattered localities in England and Merioneth. Mountains of south-west Europe from the Cordillera Cantabrica to the Swiss Jura. Oceanic Temperate element. Our plant is subsp. **grandiflora** which occurs throughout the range of the species.

× vulgaris = **P. × scullyi** Druce

This hybrid is intermediate between the parents, particularly in flower colour and size of flower parts, with considerable variation suggesting backcrossing. It is recorded occasionally in the area of *P. grandiflora* in Ireland. It also occurs in the Pyrenees.

2. Utricularia L.

Perennial monoecious *herbs* without roots, free-floating or with the lower stems in substratum. *Leaves* divided into linear or filiform segments, some or all bearing tiny animal catching bladders (traps). *Flowers* bisexual, in erect racemes emerging from water and in some species occurring rarely. *Calyx* of 2 obscurely lobed lips. *Corolla* yellow, with mouth more or less closed by swollen upfolding of the lower lip. *Stamens* 2, borne at the base of the corolla. *Style* 1, very short; stigma expanded. *Fruit* a capsule.

Contains about 215 species occurring in almost every country in the world although absent in general from arid regions and oceanic islands.

Caspar, S. J. & Manitz, H. (1975). Beitrage zur Taxonomie und Chorologie der mitteleuropäischen *Utricularia*. 2. Androsporo genese, Chromosomenzahlen und Pollenmorphologie. *Feddes Repert.* **86**: 211–232.

Clarke, W. G. & Gurney, R. (1921). Notes on the genus *Utricularia* and its distribution in Norfolk. *Trans. Norfolk Norwich Nat. Soc.* **11**: 128–161.

Druce, G. C. (1911). The British Utriculariae. *Irish Nat.* **38**: 117–123.

Fineran, B. A. (1985). Glandular trichomes in *Utricularia*: a review of their structure and function. I*srael Jour. Bot.* **34**: 295–330.

Friday, L. E. (1988). *Utricularia vulgaris*, an aquatic carnivore at Wicken Fen. *Nature Cambridgeshire* **30**: 50–54.

Friday, L. E. (1989). Rapid turnover of traps in *Utricularia vulgaris* L. *Oecologia* **80**: 272–277.

Friday, L. E. (1991). The size and shape of traps of *Utricularia vulgaris* L. *Funct. Ecol.* **5**: 602–607.

Friday, L. E. (1992). Measuring investment in carnivory: seasonal and individual variation in trap number and biomass in *Utricularia vulgaris* L. *New Phytol.* **121**: 439–445.

Glück, H. (1913). Contributions to our knowledge of the species of *Utricularia* of Great Britain with special regard to the morphology and distribution *Utricularia ochroleuca. Ann. Bot. (London)* **17**: 607–620.

Gurney, R. (1922). *Utricularia* in Norfolk 1921: the effects of drought and temperature. *Trans. Norfolk Norwich Nat. Soc.* **11**: 260–266.

Hall, P. M. (1939). The British species of *Utricularia. Bot. Soc. Exch. Club Brit. Isles* **12**: 100–117.

Preston, C. D. & Croft, J. M. (1997). *Aquatic plants in Britain and Ireland.* Colchester.

Taylor, P. (1989). The genus *Utricularia*: a taxonomic monograph. *Kew Bull.* (Addit. Series) **14**.

Thor, G. (1978). *Utricularia*: Sverige, främst de förbisedda arterna *U. australis* R. Br. Och *U. ochroleuca* R. Hartm. Stokholm.

Thor, G. (1979). *Utricularia* i Sverige, speciellt de förbisedda arterna *U. australis* och *U. ochroleuca. Svensk Bot. Tidskr.* **73**: 381–395.

Thor, G. (1987). Sumpblräddra, *Utricularia stygia*, en ny svensk art. *Svensk Bot. Tidskr.* **81**: 273–280.

Thor, G. (1988). The genus *Utricularia* in the Nordic countries, with special emphasis on *U. stygia* and *U. ochroleuca. Nordic Jour. Bot.* **8**: 381–395.

The presence of bristle-bearing teeth on the leaf segment margin, the shape of the 4-armed hairs (quadrifids) and the distribution of the glands on the inside of the spur of the corolla are of diagnostic importance.

The trap consists of a globose or ovoid body variously positioned in relation to the stalk. On them occur a variety of appendages. Entrance is through a door which the prey enters and then is unable to get out. Inside the traps the walls are invested with glands of a special kind. Their function is discussed by Fineran (1985). Thor (1978, 1979, 1987, 1988) demonstrates that the precise shape of these glands is a useful taxonomic character. The glands are usually of 2 different kinds within a trap. Those just inside the door are usually 2-armed and more widely spaced. Those on the rest of the inner surface of the trap are 4-armed and more widely spaced. It is the length, shape and relative disposition on the arms of the quadrifids that provide the taxonomic character.

The authors have no detailed knowledge of *Utricularia*, though they have seen living plants in many parts of Great Britain and Ireland, and have based this account on the superb monograph by Taylor (1989) and the papers by Thor (1979, 1987, 1988). An understanding of this genus cannot be made by collecting poor material. Either specimens preserved in alcohol need to be made or detailed notes taken in the field.

1. Leaf segments with small apical bristles but without lateral teeth with bristles; spur of corolla very short, saccate or shortly and obtusely conical — 2.
1. Leaf segments with lateral teeth with bristles as well as apical bristles; spur of corolla more or less cylindrical, at least half as long as the lower lip — 3.
2. Leaves with up to 5 ultimate segments; flowers up to 14; lower lip of corolla 8–10 mm wide — **6. bremii**
2. Leaves with up to 22 ultimate segments; flowers up to 6; lower lip of corolla about 6 mm wide — **7. minor**
3. Stolons ('stems') of 1 kind, all bearing green leaves and traps and free-floating; ultimate leaf segments filiform — 4.
3. Stolons of 2 kinds, free-floating ones bearing green leaves and no or few traps, and ones anchored in substrate and bearing very reduced, non-green leaves and many traps; ultimate leaf segments linear — 5.
4. Lower lip of corolla with lateral margins deflexed, the palate slightly shortly hairy; spur distally cylindrical or conical from a broadly conical base — **1. vulgaris**
4. Lower lip of corolla more or less flat, with the lateral margins not deflexed, the palate glabrous; spur not normally conical — **2. australis**
5. Green leaves totally without traps; apex of ultimate leaf segments obtuse; spur of corolla 8–10 mm, about as long as lower lip — **3. intermedia**
5. Green leaves usually with some traps; apex of leaf segments subulate; spur of corolla 3–5 mm, about half as long as the lower lip — 6.
6. Margin of leaf segments with 2–7 teeth with bristles; lower lip of corolla 9–12 × 12–15 mm, with flat or slightly upturned margins — **4. stygia**
6. Margin of leaf segments with 0–5 teeth with bristles; lower lip of corolla about 8 × 9 mm, with flat margins at first, later with reflexed margins — **5. ochroleuca**

1. U. vulgaris L. Greater Bladderwort
Lentibularia major Gilib. nom. illegit.; *Lentibularia vulgaris* (L.) Moench; *U. officinalis* Thornton nom. illegit.; *U. major* Cariot & St Leg. nom. illegit.; *U. biseriata* H. Lindb.

Perennial monoecious, free-floating aquatic *herb*; rhizoids usually present, few, a few cm × about 0.5 mm, filiform, bearing numerous, short, dichotomously divided branches with narrowly ovoid, papillose, apically setulose ultimate segments; stolons up to 1 m, 0.5–1.5 mm thick, filiform, branched, terete, glabrous with internodes 5–20 mm. *Leaves* very numerous, 1.5–6.0 cm, divided into 2 unequal primary segments, each more or less pinnately divided, the secondary segments dichotomously divided into further segments, the ultimate segments filiform, somewhat flattened, apically and laterally with minute bristles and without marginal teeth. *Traps* of 2 kinds, usually moderately numerous, arising laterally from the secondary to penultimate segments and also at the base of the primary segments; lateral traps 1.5–5.0 mm, ovoid, the mouth lateral, stalked, with 2 dorsal setiform, simple or usually branched appendages and usually with further simple lateral setae; local traps ovoid, stalked, the mouth basal, naked or with 2 very short, setiform, simple appendages; the internal glands of both kinds of trap 2- to 4-armed, the arms subulate with the apex rounded, up to 70 µm long and up to 10 times as long as wide, the quadrifids with the longer pair parallel or divergent with an included angle of up to 45°, the shorter pair divergent with an included angle of 90–120°. *Inflorescence* erect, emergent, 10–25 cm; peduncle filiform, terete, glabrous; scales 2–4(–5), always present, mostly in the upper half of the peduncle, similar to the bracts; bracts 3–5 mm, basifixed, broadly ovate, acute to obtuse at apex, entire, more or less cordate or shortly auriculate at base, many nerved; bracteoles absent; flowers 6–12, bisexual, the raceme axis initially short, elongating with age; pedicels 6–12 mm, filiform, terete, erect at anthesis, strongly decurved in fruit. *Calyx* 2-lipped, the lips slightly unequal, 3–5 mm, ovate and glandular-hairy, the upper lip with the apex acute or subacute, the lower lip shorter with the apex obtuse and emarginate. *Corolla* 14–19 mm, yellow, with reddish-brown streaks on the swollen basal part of the lower lip; upper lip about 11 × 10 mm, very broadly ovate and retuse at apex; lower lip with limb very broadly ovate, the base with a very prominent swelling, the lateral margins strongly deflexed, retuse at apex, about 12–15 × 14 mm when flattened; palate covered inside the distal half with short hairs and stipitate glands; spur 7–8 mm shorter than the lower lip, with a broad conical base and a narrow cylindrical or narrowly conical, acute apex, the distal two-thirds, when viewed from the side, with the ventral surface typically straight, sometimes slightly concave or convex, with internal glands on the dorsal surface only. *Stamens* 2; filaments about 2 mm, curved; anthers with thecae subdistinct. *Style* distinct; stigma with upper lip short and truncate and lower lip approximately subrotund and ciliate. *Ovary* globose and densely glandular-hairy. *Capsule* about 5 mm in diameter, globose; seeds prismatic, 4–6 angled, granulose. *Flowers* 7–8. Pollinated by bees. $2n = 36$, 44.

Native. Base-rich waters of shallow, sheltered bays in limestone lakes, pools and ditches in calcareous fens, ditches in grazing marshes, fenland lodes and flooded clay and marl-pits, and as a colonist in flooded gravel-pits. In southern Great Britain and Ireland some flowering plants can usually be found, and are prolific in some years. In northern England and Scotland it flowers much less regularly. This makes knowledge of its distribution, past and present, difficult to assess. It probably occurs in scattered localities throughout Great Britain and Ireland, but is becoming scarcer in the south due to loss of suitable localities. It is widespread in Europe, but rare in both the Mediterranean and towards its northern limit in Scandinavia; North Africa; temperate Asia east to Siberia and Tibet. Eurosiberian Temperate element. Detailed studies of its population at Wicken Fen, Cambridgeshire show that the traps account for about half the biomass of a *Utricularia* plant. An individual plant possesses a mixture of large traps on the midline of each leaflobe and smaller traps on the finer segments. After 6 days their ability to catch prey declines rapidly. About 15,000 traps are produced by a single plant in a season and they catch 230,000 crustaceans, midge larvae and oligochaete worms (Friday, 1988, 1989, 1991, 1992).

2. U. australis R. Br. Bladderwort
U. negleca Lehm.; *U. sacciformis* Benj.; *U. spectabilis* Madauss ex Scribner; *U. protrusa* Hook. fil.; *U. vulgaris* var. *mutata* Döll; *U. vulgaris* var. *neglecta* (Lehm.) Coss. & Germ.; *U. pollichii* F. W. Schultz; *U. mutata* (Döll) Leiner; *U. galloprovincialis* J. Gay ex Webb nom. nud.; *U. dubia* Rosell. ex Ces., Pass. & Gibelli, non Benj.; *U. vulgaris* forma *tenuis* Saelán; *U. jankae* Velen.; *U. incerta* Kamieñski; *U. japonica* Makino; *U. mairii* Cheeseman; *U. tenuicaulis* Miki; *U. siakujiiensis* Nakajima; *U. vulgaris* var. *japonica* (Makino) T. Yamanaka; *U. vulgaris* var. *formosana* Kuo; *U. vulgaris* var. *tenuicaulis* (Miki) Kuo; *U. vulgaris* forma *tenuicaulis* (Miki) Komiya; *U. australis* forma *tenuicaulis* (Miki) Komiya & Shibata

Perennial monoecious, free-floating aquatic *herb*; rhizoids usually present, few, a few cm × about 0.5 mm, filiform, bearing numerous, short, dichotomously divided branches, with narrowly ovoid, papillose, apically setulose ultimate segments; stolons up to 50 cm or longer, 0.3–1.0 mm thick, filiform, branched, terete, glabrous, with internodes 5–20 mm. *Leaves* very numerous, 1.5–4.0 cm, divided to the base into 2 more or less equal primary segments, each more or less pinnately divided, the secondary segments dichotomously divided into further segments, the ultimate segments filiform, somewhat flattened, apically and laterally with minute bristles, the lateral bristles each arising from the apex of a short, more or less acute tooth. *Traps* of 2 kinds, usually moderately numerous, arising laterally from the secondary to penultimate segments and also at the base of the primary segments; lateral traps 0.5–2.5 mm, ovoid, the mouth lateral, stalked, with 2 dorsal, setiform, simple or branched appendages and usually with further simple lateral setae; basal traps ovoid, stalked, the mouth basal, naked or with 2, very short, setiform simple appendages; the internal glands of both kinds of trap 2- and 4-armed, the arms subulate with a subacute apex, up to 70 µm long and up to 12 times as long as wide, the quadrifids with the longer pair subparallel to divergent with an included angle up to 45°, the shorter pair typically divergent with an included angle of about 180°, but the angle varying from 30° to 200°. *Inflorescence* weakly erect, emergent, 10–30(–100) cm; peduncle filiform, terete, glabrous; at first straight, becoming flexuous; scales 1(2–3), always present in the upper half of the peduncle, similar to the bracts; bracts 3–5 mm, basifixed, more or less subrotund, rounded or obscurely tridentate at apex, auriculate at base, many nerved; bracteoles absent; flowers 4–10, bisexual, the raceme axis initially short, elongating with age; pedicels 15–30 mm, filiform, terete, erect at anthesis, spreading later, recurved in fruit. *Calyx* 2-lipped, the lips slightly unequal, 3–4 mm, ovate, the upper lip with rounded apex, the lower lip with emarginate apex. *Corolla* 12–20 mm, yellow, with the basal swollen part of the lower lip much darker and with reddish-brown lines and spots; upper lip 11 × 10 mm, very broadly ovate, with retuse apex; lower lip 12–15 × 14 mm, with a reniform or transversely elliptic limb, with a prominent swelling at the base, the distal part more or less flat, up to 18 mm wide, rounded or retuse at apex; palate glabrous; spur 7–8 mm, broadly conical, obtuse at apex, slightly curved, considerably shorter than the lower lip, covered inside with the whole of the distal half with regularly distributed subsessile glands. *Stamens* 2; filaments about 2 mm, curved; anther thecae distinct. *Style* distinct; stigma with lower lip subrotund and ciliate, the upper lip very short or obsolete. Ovary globose, densely covered with sessile glands. *Capsule* seen only in China and Japan, about 4 mm in diameter, globose; seeds about 0.5 mm, prismatic 4–6 angled, narrowly winged on all the angles, the tecta cells elongate with the anticlinal boundaries raised, smooth. *Flowers* 7–9. $2n = 44$.

Native. Usually in acidic, still or slow-flowing water in lakes, reservoirs, ponds, canals, ditches and swamps over inorganic or peaty substrates, but it has been recorded for calcareous sites in ponds and sand dunes. It is scattered in localities throughout Great Britain and Ireland, but is commonest in western and northern Great Britain. It is widespread in Europe north to central Scandinavia and is found in temperate and tropical regions throughout the Old World and in Australia and New Zealand. Eurasian Boreo-temperate element. There is no evidence that any populations in Great Britain and Ireland ever set seed and reproduction seems to be entirely by turions. Throughout its extensive world range it has been known to fruit only in China and Japan.

3. U. intermedia Hayne Intermediate Bladderwort

U. alpina Georgi, non Jacq.; *U. media* Schum.; *U. millefolium* Nutt. ex Tuck. nom. in syn.; *U. grafiana* W. D. J. Koch; *Lentibularia intermedia* (Hayne) Nieuwl. & Lunell

Perennial, monoecious, aquatic *herb*; rhizoides usually present, few, a few cm long, 0.2–0.7 mm thick, filiform, bearing numerous, short, dichotomously divided branches, the ultimate segments minutely papillose, shortly cylindrical, obtuse at apex and sometimes apically setulose; stolons up to 30 cm, 0.4–0.6 mm thick, sparsely branched and markedly of 2 kinds, some green growing on the surface of the substrate or floating, while others without chlorophyll and more or less lurid in the substrate. *Leaves* very numerous, very variable; those on stolons above the substrate, complanate, imbricate, approximately subrotund in outline, 1–20 mm, palmato-dichotomously divided into up to 15 segments, the ultimate segments 0.1–0.7 mm wide, flattened, narrowly linear, obtuse at apex, typically entire and bearing throughout its length up to 20 short bristles or sometimes (of leaves produced at the beginning and end of the growing season) sparsely denticulate, the teeth acute and each with an apical, solitary bristle or fascicle of up to 4 bristles; the leaves on stolons beneath the substrate fewer and more or less reduced to a single elongate primary segment with few, very reduced, short ultimate segments near the base and at the apex. *Traps* rather few, lateral on the segments, absent or usually so on those above substrate, 1–3 normally present on those below substrate, 1.5–4.0 mm, ovoid, the mouth lateral, stalked, with 2 long, much-branched, setiform, dorsal appendages and a few, lateral, simple setae, the internal glands 2- and 4-armed, the arms narrowly cylindrical, subacute at apex and up to 14 times as long as wide, the quadrifids with the 2 pairs typically both parallel, or sometimes slightly divergent. *Inflorescence* erect, emergent, 10–20 cm; peduncle filiform, terete, glabrous and straight; scales 2 or 3, more or less equally spaced on the peduncle, similar to the bracts; bracts about 3 mm, broadly ovate or ovate-deltoid, acute at apex, conspicuously auriculate at base, several nerved and basifixed; bracteoles absent; flowers 2 or 3, bisexual, the raceme axis short; pedicels 5–15 mm, erect, filiform and terete. *Calyx* 3–4 mm, 2-lipped, the lips subequal, ovate, the upper lip acute at apex, the lower lip shorter and obtuse, shortly bifid or truncate at apex. *Corolla* 10–16 mm, yellow; upper lip 7–8 × about 7 mm, broadly ovate, rounded at apex; lower lip 8–10 × 12–13 mm, with limb transversely elliptic, rounded at apex, and the base with a prominent rounded swelling; palate glabrous; spur 8–10 mm, subulate, acute at apex, a little shorter than the lower lip, the internal surface dorsally and to a lesser extent ventrally glandular. *Stamens* 2; filaments about 2 mm, curved; anthers thecae more or less confluent. *Style* relatively long; stigma with lower lip subrotund and ciliate, the upper lip much smaller, deltoid and bifid or acute at apex. *Ovary* globose, glandular-hairy. *Capsule* 2.5–3.0 mm in diameter, globose; seed not seen, but probably similar to *U. minor*. *Flowers* 7–9. $2n = 44$.

Native. Still shallow water in acidic peaty bogs and marshes. Very scattered records in Ireland, Scotland, north and south-central England, East Anglia and Caernarvonshire, but much over-recorded for *U. ochroleuca*. Reproduction is by turions which usually begin to develop on the green stolons in later July or August. Circumboreal. Circumpolar Boreal-montane element.

4. U. stygia G. Thor Nordic Bladderwort

Small, *perennial,* monoecious, aquatic *herb*; rhizoids rare, 0–1, filiform, branched, ultimate segments cylindrical, acute at apex, setulose with 1 bristle; stolons of 2 kinds, some green growing on the surface of the substrate or suspended or floating, others without chlorophyll are more or less buried in substrate. *Leaves* numerous, the ultimate segments (0.1–)0.2–0.3(–0.4) mm wide, subulate, midrib sometimes indistinct, acute at apex, margin with (2–)3–6(–7) teeth, the teeth terminating in 1–2 bristles. *Traps* on both green and colourless stolons, 1–3 mm, ovoid, the mouth lateral, the internal glands 2- and 4-armed, the quadrifids with the longer arms with an included angle of (16°–) 25.6°–56.0°(–90°) and the shorter arms with an included angle (30°–)52.0°–96.8°(–140°). *Inflorescence* erect, emergent, 5–15 cm, purplish at least at base; peduncle filiform, terete, glabrous; pedicels 3–6 mm, straight. becoming recurved after flowering. *Calyx* 2-lipped, the lips subequal. *Corolla* 10–15 mm, yellow, with a reddish tinge; upper lip about 8×6 mm. lower lip (9–)10–11×(12–)13–15 mm. spur 4–5 mm, directed downwards from the lower lip at an acute angle and with internal glands on both the abaxial and adaxial side. *Capsule* not known. *Flowers* 6–8.

Native. Shallow water in peaty bogs and marshes. West Scotland from Wigtownshire to Sutherland, probably under-recorded. Scandinavia, Germany, Canada and United States.

5. U. ochroleuca R. W. Hartm. Pale Bladderwort
U. occidentalis A. Gray; *U. brevicornis* Čelak.; *U. litoralis* Melander; *U. intermedia* forma *ochroleuca* (R. W. Hartm.) Komiya

Small, *perennial,* monoecious, aquatic *herb*; rhizoids usually present, few, 0.2–0.7 mm thick, filiform, bearing numerous, short, dichotomously divided branches, the ultimate segments shortly cylindrical, obtuse at apex, papillose and rarely setulose; stolons up to 15 cm, filiform, terete, glabrous, sparsely branched, markedly of 2 kinds, some green growing on the surface of the substrate or suspended or floating, others without chlorophyll are more or less buried in substrate. *Leaves* very numerous, very variable; those on stolons above the substrate 2–15 mm, complanate, imbricate, approximately subrotund in outline, palmato-dichotomously divided into about up to 20 segments, the ultimate segments 0.1–0.5 mm wide, flattened, narrowly linear, acute at apex, sparsely denticulate, the teeth up to 9 in number but usually fewer and each terminating in a bristle or sometimes a fascicle of 2 or more bristles, the leaves on stolons beneath the substrate fewer and more or less reduced to a single elongate segment, with a few, very much reduced, short segments at the base and near the apex. *Traps* rather few, lateral on the segments, usually 1 or sometimes more present on those above substrate, 1–3 normally present on those below substrate, 2.0–3.5 mm, ovoid, the mouth lateral with 2 long, much-branched, setiform, dorsal appendages and a few lateral simple setae, stalked, the internal glands 2- to 4-armed, the arms narrowly cylindrical-subulate, obtuse or subacute at apex, up to 50 μm long and up to 10 times as long as wide, the quadrifids with both pairs with an included angle of 45o–70o. *Inflorescence* erect, emergent, 8–15 cm; peduncle filiform, erect, terete, glabrous; scales 3–4, more or less equally spaced, similar to the bracts; bracts about 2.5 mm, broadly ovate, obtuse to acute at apex, auriculate at base and basifixed, 1–several-nerved; bracteoles absent; flowers 2–5(–7), the axis short or slightly elongate; pedicels 5–8 mm, erect, spreading post anthesis, filiform, terete. *Calyx* 2-lipped, the lips slightly unequal, 3–4 mm, ovate, the upper lip acute at apex, the lower lip shorter and broader and shortly bifid at apex. *Corolla* 10–15 mm, pale yellow; upper lip 7×5 mm, broadly ovate, rounded at apex; lower lip transversely elliptical, the base with a prominent, rounded swelling, rounded at apex; palate glabrous; spur about 3 mm, half as long as the lower lip, conical, sometimes with the distal part shortly cylindrical, more or less acute at apex, the internal surface dorsally and to a lesser extent ventrally glandular-hairy. *Stamens* 2; filaments about 2 mm, curved; anther thecae more or less distinct. *Style* short; stigma with lower lip circular and ciliate, the upper lip much smaller and semicircular. *Capsule* and seeds not known. *Flowers* 7–9. Probably mostly, if not always reproduced vegetatively. $2n=44$.

Native. Shallow water in peaty bogs and marshes. Locally frequent in Scotland, scattered records in Ireland and north and south-central England. This species is probably commoner than *U. intermedia* or *U. stygia* in Scotland. Circumboreal.

6. U. bremii Heer ex Kölliker Brem's Bladderwort
U. pulchella C. B. Lehm.; *U. minor* var. *bremii* (Heer ex Kölliker) Franch.; *U. minor* subsp. *bremii* (Heer ex Kölliker) K. & F. Bertsch.

Small, *perennial*, monoecious, aquatic *herb*; rhizoides absent; stolons up to 25 cm, 0.3–0.5 mm thick, filiform, terete, glabrous, sparsely branched, of 2 kinds, some green, on the surface of the substrate or suspended or floating, others without chlorophyll and more or less buried in the substrate. *Leaves* very numerous, very variable; those of the stolons above the substrate subrotund to ovate in outline, 0.5–2.0 cm, palmato-dichotomously or pinnato-dichotomously divided into rather numerous segments, the ultimate segments up to 50, 0.1–0.5 mm wide, flattened, filiform to linear, the margins entire or distally sparsely denticulate, the teeth acute and with bristles at apex; those on stolons beneath the substrate more or less reduced to 1 or 2 capillary, primary segments, bearing a few, reduced, very short secondary segments at the base and apex. *Traps* lateral on the segments, usually present, but few on those above substrate, more numerous and larger on those below substrate, 1–2 mm, ovoid, the mouth lateral with 2 long, much-branched, setiform, dorsal appendages and a few, lateral , simple setae, stalked, the internal glands 2- and 4-armed, the arms narrowly cylindrical with apex subacute, up to 100 μm long and up to 15 times as long as wide, the quadrifids with the longer pair parallel or slightly divergent and the shorter pair very widely divergent to slightly reflexed, with an included angle of 180°–200°. *Inflorescence* erect, emergent, 5–50 cm; peduncle filiform, terete, glabrous, straight; scales 2–4, more or less equally spaced on the peduncle, similar to the bracts; bracts 1.5–2.0 mm, broadly ovate, obtuse at apex, conspicuously auriculate at base and basifixed, several nerved; bracteoles absent; flowers 2–14, bisexual, the raceme axis more or less elongate; pedicels erect at anthesis, late spreading and distally decurved, 0.4–1.5 cm, filiform, terete. *Calyx* 2-lipped, lips 2–3 mm, subequal, ovate; upper lip obtuse at apex; lower lip slightly smaller, with shortly bifid apex. *Corolla* 8–10 mm, yellow, about as long as wide; upper lip broadly ovate and retuse at apex; lower lip 7–8 × 12–14 mm, with limb transversely elliptical or subrotund, rounded at apex; palate elongate with a raised, distally glandular, emarginate rim; spur 3–4 mm, shortly conical, obtuse, in lateral view about as wide as long, the internal surface ventrally densely glandular. *Stamens* 2; filaments about 1.5 mm, curved; anther thecae subdistinct. *Style* relatively long; stigma with lower lip semicircular, ciliate, reflexed at apex, the upper lip much smaller, deltoid, with apex bifid. *Capsule* and seeds not known. *Flowers* 7–9.

This species is questionably in Great Britain and Ireland, but as it may occur should be looked out for. It is native of western and central Europe. The pollen is usually malformed and may not be functional. As capsule and seeds are not known it would seem to survive vegetatively. Named after Jacob Bremi.

7. U. minor L. Lesser Bladderwort

Lentibularia minor (L.) Raf.; *Xananthes minor* (L.) Raf.; *U. rogersiana* Lace; *U. minor* var. *multispinosa* Miki; *U. multispinosa* (Miki) Miki; *U. nepalensis* Kitamura

Perennial, monoecious, aquatic *herb*; rhizoids absent; stolons up to 30 cm, 0.1–0.5 mm thick, sparsely branched, of 2 kinds, some green on the surface of the substrate or suspended or floating, others without chlorophyll and more or less buried in the substrate. *Leaves* very numerous; lamina very variable, those on the stolons above the substrate approximately subrotund in outline, 2–15 mm, palmato-dichotomously divided into 7–22 segments, the ultimate segments 0.1–1.0 mm wide, flattened filiform to linear, acute at apex, with or without a microscopic bristle, and the margins entire or sometimes very sparsely denticulate but the teeth not or only microscopically with bristles, the leaves on the stolons beneath the substrate more or less reduced to 1 or 2, elongate primary segments with a few very much reduced, very short further segments at the base and apex. *Traps* lateral on the segments, rather few or absent on those above substrate, more numerous and longer on those below substrate, 0.8–2.5 mm, ovoid, the mouth lateral with 2 long, much-branched, setiform, dorsal appendages and a few, lateral, simple setae, stalked, the internal glands 2- and 4-armed, the arms up to 100 μm long and up to 14 times as long as wide, subulate, tapering to a rounded apex, the quadrifids with the longer pair subparallel or divergent with an included angle of up to about 25°, the shorter pair reflexed, with an included angle of 280°–300°. *Inflorescence* erect, emergent, 2.5–25.0 cm; peduncle filiform, terete, glabrous, straight; scales 2–4, more or less equally spaced, similar to the bracts; bracts 1.5–2.0 mm, basifixed, broadly ovate or ovate-deltate, acute or obtuse at apex, 1–several-nerved; bracteoles absent; flowers 2–6, bisexual, the raceme axis initially short, elongating with age; pedicels erect at anthesis, later spreading and distally recurved in fruit, 4–8 mm, filiform, terete. *Calyx* 2-lipped, 2–3 mm, the lips subequal and broadly ovate, the upper lip with a subacute apex and cucullate, the lower lip somewhat smaller and the apex narrowly truncate. *Corolla* 6–8 mm, lemon yellow, usually longer than wide, upper lip about 4 × 3 mm, ovate or ovate-oblong, retuse at apex; lower lip about 7 × 6 mm with limb broadly obovate, rounded or retuse at apex, the lateral margins curved downwards; palate elongate, with a raised marginal rim, distally narrowed and glandular; spur 1–2 mm, saccate or obtusely broadly conical, in lateral view wider than long, the internal surface densely glandular. *Stamens* 2; filaments about 1.5 mm, curved; anther thecae subdistinct. *Style* relatively long; stigma with lower lip broadly ovate, ciliate and with reflexed apex, the upper lip much smaller, deltoid and with apex acute or 2- to 3-fid. *Ovary* broadly ellipsoid. *Capsule* 2–3 mm in diameter, globose; seeds around 1 mm wide and about two-thirds as long, lenticular-prismatic, scarcely winged on the angles. *Flowers* 6–9. $2n = 36$.

Native. Peat-stained, acidic and oligotrophic water in bog pools, old peat cuttings, edges of lakes, ditches and wet flushes, but it can grow in nutrient-poor but base-rich water in fens and ditches. Scattered over Great Britain and Ireland, but there has been a marked decline in the eastern part of its range because of the destruction of its habitat. It is, however, almost certainly under recorded. It has a circumboreal distribution extending south to the Himalaya, Japan and California and is also found in high altitudes in Papua New Guinea. In Europe it occurs from Iceland and northern Scandinavia southwards but is rare in the Mediterranean region. Circumpolar Boreo-temperate element. It flowers occasionally throughout its range in Great Britain and Ireland and in some years it does exceptionally well. Reproduction is by turions which are formed at the end of the green stolons from late July onwards.

NEW TAXA AND COMBINATIONS

The new taxa and combinations are to be attributed to the authors whose names follow the names of the taxa and who have supplied the supporting information.

Oxytropis halleri Bunge ex W. D. J. Koch, *Syn. Fl. Germ. Helv*. ed. 2, 200 (1842).
var. **grata** P. D. Sell var. nov.
Holotype: Cliffs, Melvich, W. Sutherland, v.c. 108, 10 July 1897, E. S. Marshall (**CGE**).
Planta ad 30 cm, pilis numerosis vel densis ad 3 mm vestita. *Foliola* 18–31, 8–15 × 4–7 mm, ovata vel lanceolata vel elliptica, ad apicem acuta. *Calyx* 9–12 mm, lobis 2.5–3.0 mm.

Oxytropis campestris (L.) DC., *Astragalogia* 74 (1802).
var. **kintyrica** P. D. Sell var. nov.
Holotype: Rock ledges by the sea, between Rudha Duin Bhain and Uamh Ropa, southwest of Cambeltown, Kintyre, v.c. 101, 16/593146, 22 June 1968, P. D. Sell & A. G. Kenneth, Sell no. 68/211 (**CGE**).
Planta 8–15 cm, pilis numerosis vel densis ad 3.5 mm vestita. *Foliola* 23–33, 7–13 × 2.5–3.5 mm, lanceolata. *Calyx* 7–8 mm, lobis 1.5–2.0 mm triangularibus vel linearibus.
var. **perthensis** P. D. Sell var. nov.
Holotype: Cliffs above Loch Loch, below Beinn a 'Ghlo, Atholl Forest, Perthshire, Aug. 1951, C. D. Pigott (**CGE**).
Planta 10–15 cm, pilis numerosis ad 1.5 mm vestita. *Foliola* 21–25, 10–20 × 1.5–3.0 mm, linearia vel oblonga. *Calyx* 7–10 mm, lobis 2–3 mm linearibus.
var. **scotica** (Jalas) P. D. Sell stat. nov.
O. campestris subsp. *scotica* Jalas, *Ann. Bot. Fenn.'Vanamo'* **24**: 59 (1950).
Holotype: Glen Phee, Clova, Forfarshire, v.c. 90, 17 July 1910, R. & M. Corstorphine (**CGE**).

Onobrychis viciifolia Scop. *Fl. Carniol*. ed. 2, **2**: 76 (1772).
subsp. **collina** (Jord.) P. D. Sell stat. nov.
O. collina Jord., *Cat. Hort. Grenoble* (1851); *Pugillus Pl. Nov*. 63 (1852).
subsp. **decumbens** (Jord.) P. D. Sell stat. nov.
O. decumbens Jord., *Pugillus Pl. Nov*. 64 (1852).

Vicia cracca L. *Sp. Pl*. 735 (1753).
var. **pulchra** (Druce) P. D. Sell stat. nov.
V. cracca forma *pulchra* Druce in *Rep. Bot. Soc. Exch. Club Brit. Isles* **6**: 484 (1922).

Vicia sativa L., *Sp. Pl*. 736 (1753).
subsp. **uncinata** (Rouy) P. D. Sell stat. nov.
V. angustifolia var. *uncinata* Rouy, *Fl. Fr*. **5**: 213 (1899).
var. **bobartii** (E. Forst.) P. D. Sell stat. nov.
v. bobartii E. Forst., *Trans. Linn. Soc*. **16**: 442 (1830).

Lathyrus japonicus Willd., *Sp. Pl*. **3**: 1092 (1802).
subsp. maritimus (L.) P. W. Ball, *Feddes Repert*. **79**: 45 (1968).
Pisum maritimum L., *Sp. Pl*. 727 (1753).
var. **acutifolius** (Bab.) P. D. Sell comb. nov.
Lathyrus maritimus var. *acutifolius* Bab., *Man. Brit. Bot*. ed. 1, 82 (1843).
Lectotype designated here: Barra Firth, Unst, Shetland, v.c. 112, 1837, G. McNab in Herb. C. C. Babington (**CGE**).

Lathyrus linifolius (Reichard) Bässler, *Feddes Repert*. **82**: 434 (1971).
Orobus linifolius Reichard, *Hanauisches Mag*. **4, 5**: 26 (1782).
var. **varifolius** (Matrin-Donos) P. D. Sell comb. nov.
L. macrorhizus var. *varifolius* Matrin-Donos, *Fl. Tarn*. 186 (1864).

Medicago lupulina L., *Sp. Pl*. 779 (1753).
var. **eriocarpa** (Rouy) P. D. Sell stat. nov.
M. lupulina subvar. *eriocarpa* Rouy in Rouy & Fouc., *Flore France* **5**: 9 (1899).

Trifolium campestre Schreber in Sturm, *Deutschl. Fl. Abt*. 1, Band 4, Heft 16 (1804).
var. **majus** (W. D. J. Koch) P. D. Sell comb. nov.
T. procumbens var. *majus* W. D. J. Koch, *Syn. Fl. Germ. Helv*. ed. 2, 194 (1843).

Trifolium dubium Sibth., *Fl. Oxon*. 231 (1794).
var. **microphyllum** (Ser.) P. D. Sell comb. nov.
T. minus var. *microphyllum* Ser. in DC., *Prodr*. **2**: 206 (1825).

Lupinus arboreus Sims, *Bot. Mag*. **18**: t. 682 (1803).
forma **albus** P. D. Sell forma nova
Holotype: By car park, Walberswick, E. Suffolk, v.c. 25, 62/501748, 28 June 1990, P. D. Sell with J. G. Murrell, Sell no. 90/190 (**CGE**).
Flores albi.
forma **aureus** P. D. Sell forma nova
Holotype: Above the beach, Sizewell, E. Suffolk, v.c. 25, 62/476629, 22 June 1965, P. D. Sell no. 65/377 (**CGE**).
Flores aurei.

Elaeagnus pungens Thunb., *Fl. Jap*. 68 (1784).
forma **variegata** (Bean) P. D. Sell stat. nov.
E. pungens var. *variegata* Bean, *Trees Shrubs Brit. Isles* **1**: 508 (1914).

Elaeagnus × ebbingei Doorenbos, *Jaarb. Nederl. Dendr. Ver*. **17**: 109 (1952).
forma **flavistriata** P. D. Sell forma nova
Holotype: Winter Garden, Botanic Garden, Cambridge, v.c. 29, 52/456572, 26 Feb. 2007, P. D. Sell no. 07/2 (**CGE**).
Folia in pagina superiore flavistriata.

Oenothera biennis L., *Sp. Pl.* 346 (1753).
forma **leptomeres** (Bartlett) P. D. Sell stat. nov.
O. biennis var. *leptomeres* Bartlett, *Amer. Jour. Bot.* **1**: 242 (1914).

Swida alba (L.) Holub
var. **flaviramea** (Späth ex Koehne) P. D. Sell comb. nov.
Cornus alba var. *flaviramea* Späth ex Koehne, *Mitt. Deutsch. Dendrol. Ges.* **12**: 39 (1903).
var. **sibirica** (Lodd. ex Loudon) P. D. Sell comb. nov.
Cornus sibirica Lodd. ex Loudon, *Cat.* (1836) nom. nud.
Cornus alba var. *sibirica* Lodd. ex Loudon, *Encycl. Trees Shrubs* 503 (1842).

Frangula alnus Mill., *Gard. Dict.* ed. 8, no. 1 (1768).
var. **prostrata** P. D. Sell var. nov.
Holotype: Limestone south of Lough Mask, Co. Mayo and Galway, 11 July 1895, E. S. Marshall (**CGE**).
Frutex prostratus. *Folia* 1.5–4.0 × 1.3–2.5 cm.

Polygala serpyllifolia J. A. C. Hosé, *Ann. Bot. (Usteri)* **21**: 39 (1797).
var. **vincoides** (Chodat ex Ham.-Davy) P. D. Sell comb. nov.
P. *serpyllacea* var. *vincoides* Chodat ex Ham.-Davy, *Jour. Bot. (London)* **44**: 35 (1906).

Geranium × johnsonii P. D. Sell nothospecies nova
G. himalayense Klotzsch × **pratense** L.
Holotype: Cult. Botanic Garden, Cambridge, v.c. 29, no. 19710284B, 52/454572, 27 May 2005, P. D. Sell no. 05/69 (**CGE**).
Herba perennis rhizomis subterraneis modice extendens. *Caules* ad 70 cm, erecti, tenues, supra ramosi foliosi, pallide flavivirides, pilis simplicibus eglandulosis numerosis brevibus reflexis appressis pallidis vestiti. *Folia* alterna, sursum gradatim minora; lamina 2–7 × 5–20 cm, in pagina superiore obscure mediocriter viridis, in pagina inferiore pallidior, subrotunda vel reniformis, ad apicem acuta, per tres quartas partes vel septem octavas partes basin versus in septem lobos divisa, ad basin cordata, in pagina superiore inque marginibus inque nervis paginae inferioris pilis simplicibus eglandulosis brevibus pallidis vestita; lobi obovati vel oblanceolati, in lobulos divisi; lobuli magis minusve acuti, dentibus parvis magis minusve acutis praediti; nervi supra impressi, subter prominentes; petioli ad 12 cm, pilis simplicibus eglandulosis brevibus reflexis appressis pallidis vestiti. *Inflorescentia* laxa vel floribus bisexualibus plerumque geminatis; pedunculi 3–10 cm, erecti, pilis glanduliferis numerosis brevibus mediocribusque pallidis, cum pilis simplicibus eglandulosis pallidis immixtis, vestiti; pedicelli sepalis circa duplo longiores, patentes, pilis glanduliferis densis brevibus brevissimisque pallidis ad apicem rubris vestiti; bracteae ut folia superiora; bracteolae linearilanceolatae, ad apicem acutae, pilis simplicibus eglandulosis vestitae. *Sepala* 5, 7–11 × 5–6 mm, viridia labe purpurascenti ad basin notata, lanceolata vel anguste ovata, mucrone 1–2 mm, nervata, pilis glanduliferis densis brevissimis vel brevibus pallidis ad apicem rubris vestita. *Petala* 5, 15–25 × 10–17 mm, saturate caerulea, basin versus pallidiora magisque rosea, venis translucidis paene incoloratis notata, obovata, ad apicem rotundata, ungue brevi praedita. *Stamina* 10, sepalis aliquantum longiora; filamenta magis minusve saturate rosea, infra pilis subtiliter fimbriata; antherae atroviolaceae vel caesionigrae vel atrofuscae, non aperientes. *Styli* longiores quam 4 mm; stigmata circa 3 mm. *Mericarpium* ad basin rotundatum, rostro ad 28 mm et parte stylari circa 9 mm; semina non effecta. *Floret* 5–8.
In honore A. T. Johnson ex Tyn-y-Groes in Cambria Septentrionali nominatur.

Hedera helix L. *Sp. Pl.* 2002 (1753).
forma **lobatomajor** (G. H. M. Lawr.) P. D. Sell stat. nov.
H. helix var. *lobatomajor* G. H. M. Lawr., *Gentes Herb.* **6(3)**: 143 (1942).
forma **hibernica** (Kirschner) P. D. Sell stat. nov.
H. hibernica Kirschner in Petzold & Kirschner, *Arb. Muscav.* 419 (1864).
forma **aureovariegata** (Weston) P. D. Sell stat. nov.
H. helix var. *aureovariegata* Weston, *Univ. Bot. Nurs.* **1**: 123 (1770).
forma **irica** P. D. Sell forma nova
Holotype: In the *Symphoricarpos* thicket, near Station Road entrance, Botanic Garden, Cambridge, v.c. 29, 52/456573, 21 March 2000, P. D. Sell no. 00/21 (**CGE**).
Folia juvenilia obscure atroviridia, crassa, coriacea, quinqueloba, latissima.

Hedera colchica (K. Koch) K. Koch, *Wochenschr. Gärtn. Pflanzenk.* **2**: 74 (1859).
forma **dentatovariegata** (Schulze) P. D. Sell stat. nov.
H. colchica var. *dentatovariegata* Schulze, *Gentes Herb.* **6(3)**: 124 (1942).
forma **flavovariegata** P. D. Sell forma nova
Holotype: Botanic Garden, Cambridge, v.c. 29, 52/455573, 26 January 2000, P. D. Sell no. 00/49 (**CGE**).
Folia lamprosmaragdina, centro nervisque colore flavo maculatis et variegatis, integra.

Chaerophyllum temulum L., *Sp. Pl.* 258 (1753).
var. **canescens** (Baenitz ex Thell.) P. D. Sell stat. nov.
C. temulum forma *canescens* Baenitz ex Thell. in Hegi, *Fl. Mitt. Eur.* **5(2)**: 997 (1926).

Pimpinella saxifraga L., *Sp. Pl.* 263 (1753).
var. **nana** (Druce) P. D. Sell stat. nov.
P. saxifraga var. *poteriifolia* forma *nana* Druce, *Rep. Bot. Soc. Exch. Club Brit. Isles* **5**: 108 (1918).

Aethusa cynapium L., *Sp. Pl.* 256 (1753).
subsp. **gigantea** (Lej.) P. D. Sell stat. nov.
A. cynapium var. *gigantea* Lej., *Fl. Spa* **1**: 141 (1811).

Apium graveolens L., *Sp. Pl.* 264 (1753).
subsp. **rapaceum** (Mill.) P. D. Sell stat. nov.
A. rapaceum Mill., *Gard. Dict.* ed. 8, no. 6 (1768).

Heracleum sphondylium L., *Sp. Pl.* 249 (1753).
forma **stenophyllum** (Gaudin) P. D. Sell stat. nov.
H. sphondylium var. *stenophyllum* Gaudin, *Fl. Helv.* **2**: 318 (1828).
forma **latifolium** (Mert. & W. D. J. Koch) P. D. Sell stat. nov.
H. sphondylium var. *latifolium* Mert. & W. D. J. Koch, *Synop.* ed. 2, 373 (1846).

Daucus carota L., *Sp. Pl.* 242 (1753).
var. **acaulis** (Bréb.) P. D. Sell comb. nov.
D. gummifer var. *acaulis* Bréb., *Flore Normandie* ed. 3, 124 (1859).

Gentianella amarella (L.) Börner, *Fl. Deutsch. Volk.* 542 (1912).
Gentiana amarella L., *Sp. Pl.* 230 (1753).
var. **praecox** (F. Towns.) P. D. Sell comb. nov.
Gentiana amarella var. *praecox* F. Towns., *Fl. Hampshire* 216 (1883).
var. **cornubiensis** (N. M. Pritch.) P. D. Sell stat. nov.
G. anglica subsp. *cornubiensis* N. M. Pritch., *Watsonia* **4**: 184 (1959).
var. **uliginosa** (Willd.) P. D. Sell stat. nov.
Gentiana uliginosa Willd., *Sp. Pl.* **1(2)**: 1347 (1798).
var. **calycina** (Druce) P. D. Sell comb. nov.
Gentiana amarella var. *calycina* Druce, *Rep. Bot. Soc. Exch. Club Brit. Isles* **3**: 329 (1914).
var. **druceana** (N. M. Pritch.) P. D. Sell stat. nov.
Gentianella amarella subsp. *druceana* N. M. Pritch., *Watsonia* **4**: 236 (1960).

Vinca major L., *Sp. Pl.* 209 (1753).
forma **variegata** (Loudon) P. D. Sell stat. nov.
V. major var. *variegata* Loudon, *Encycl. Trees and Shrubs* 657 (1842).

Convolvulus arvensis L., *Sp. Pl.* 153 (1753).
forma **notatus** P. D. Sell forma nova
Holotype: South side of Fen Road, Bassingbourn, Cambridgeshire, v.c. 29, 52/322449, 25 July 1990, P. D. Sell 90/301 (**CGE**).
Corollae pars exterior vittis 5 luteoviridibus notata, cetero alba; pars interior alba, supra faucem luteam macularum fractiflexarum dentatarumque saturate carmesinarum annulo notata. *Filamenta* et *antherae* albae.
forma **pallidiroseus** P. D. Sell forma nova
Holotype: Old gateway to Newcut, Fen Road, Bassingbourn, Cambs, v.c. 29, 52/323449, 24 July 1990, P. D. Sell no. 90/299 (**CGE**).
Corollae pars exterior vittis 5 purpurascentibus notata, cetero roseotincta; pars interior supra faucem luteam roseotincta. *Filamenta* alba; *antherae* purpureae.
forma **pallidinotatus** P. D. Sell forma nova
Holotype: Grassy roadside near Boy Bridge, Fen Road, Bassingbourn, Cambridgeshire, v.c. 29, 52/318452, 15 July 2006, P. D. Sell no. 06/232 (**CGE**).
Corollae pars exterior vittis 5 purpurascentibus notata, cetero roseotincta; pars interior roseotincta, supra faucem luteam macularum fractiflexarum dentatarumque purpurascentium annulo notata. *Filamenta* alba; *antherae* purpureae.
forma **pentarrhabdotus** P. D. Sell forma nova
Holotype: West side of road, bottom of Rancell Hill, between Litlington and Abington Pigotts, Cambridgeshire, v.c. 29, 52/309434, 25 July 1990, P. D. Sell no. 90/300 (**CGE**).
Corollae pars exterior vittis 5 roseis notata, cetero roseotincta; pars interior supra faucem luteam vittis 5 roseis et 5 albis notata. *Filamenta* rosea; *antherae* purpureae.
forma **pentastictus** P. D. Sell forma nova
Holotype: Track between Poplar Field and Bull's Fen, Bassingbourn, Cambridgeshire, v.c. 29, 52/319456, 11 August 1990, P. D. Sell no. 90/345 (**CGE**).
Corollae pars exterior vittis 5 purpurascentibus notata, cetero roseotincta; pars interior vittis 5 roseis et 5 albis et macularum fractiflexarum dentatarumque purpurascentium annulo supra faucem luteam notata. *Filamenta* roseotincta; *antherae* purpureae.
forma **decarrhabdotus** P. D. Sell forma nova
Holotype: Gateway to field by Mill House, Bassingbourn, Cambridgeshire, v.c. 29, 52/322450, 24 July 1990, P D. Sell no. 90/298 (**CGE**).
Corollae pars exterior vittis 5 purpurascentibus notata, cetero rosea; pars interior vittis 10 roseis notata, colore albo interposito. *Filamenta* rosea; *antherae* purpureae.
forma **decemvulnerus** P. D. Sell forma nova
Holotype: Side of Spring Lane, Bassingbourn, Cambridgeshire, v.c. 29, 52/338483, 8 July 2006, P. D. Sell no. 06/2003 (**CGE**).
Corollae pars exterior vittis 5 purpurascentibus notata, cetero rosea; pars interior vittis 10 roseis, colore albo interposito, et macularum fractiflexarum dentatarumque annulo supra faucem luteam notata. *Filamenta* rosea; *antherae* purpureae.
forma **perroseus** P. D. Sell forma nova
Holotype: Bank of Fillance, back of 15 Fen Road, Bassingbourn, Cambridgeshire, v.c. 29, 52/327448, 5 August 1990, P. D. Sell no. 90/324 (**CGE**).
Corollae pars exterior vittis 5 purpureis notata, cetero rosea; pars interior omnino perrosea, excepta stella alba supra faucem luteam. *Filamenta* rosea; *antherae* purpureae.
forma **quinquevulnerus** P. D. Sell forma nova
Holotype: North margin of Pit Field, Bassingbourn, Cambridgeshire, v.c. 29, 52/318445, 12 August 2006, P. D. Sell no. 06/286 (**CGE**).
Corollae pars exterior vittis 5 purpureis notata, cetero rosea; pars interior omnino perrosea, exceptis stella alba supra faucem luteam et macularum fractiflexarum dentatarumque annulo proxime supra faucem luteam. *Filamenta* rosea; *antherae* purpureae.

Symphytum × uplandicum Nyman, *Syll. Fl. Eur.* 80 (1854).
forma **lilacinum** (Buckn.) P. D. Sell comb. & stat. nov.
S. × *lilacinum* Buckn., *Jour. Bot. (London)* **50**: 334 (1912).
forma **densiflorum** (Buckn.) P. D. Sell comb. & stat. nov.
S. × *densiflorum* Buckn., *Jour. Bot. (London)* **50**: 334 (1912).
forma **discolor** (Buckn.) P. D. Sell comb. & stat. nov.
S. × *discolor* Buckn., *Jour. Bot. (London)* **50**: 333 (1912).
forma **caeruleum** (Petitm. ex Thell.) P. D. Sell comb. & stat. nov.

S. × *caeruleum* Petitm. ex Thell., *Vierteljahrsschr. Nat. Ges. Zurich* **3**: 459 (1907).

Symphytum × **hidcotense** P. D. Sell nothospecies nova
Holotype: Cult. Botanic Garden, Cambridge, v.c. 29, no. 20040212A, 52/456571, 22 May 2006, P. D. Sell no. 06/70 (**CGE**).

Herba perennis monoecius stolonibus decumbentibus, fortibus, non florentibus, ex rhizomis ramosis tumores tuberosos ferentibus, praedita. *Caules* ad 1 m, pallide virides, ascendentes vel erecti, pilis rigidis ad basin bulbosis numerosis pilisque brevioribus tenuibus uncinatis vestiti, ramosi, frondosi. *Folia* alterna; lamina 5–19 × 2–10 cm, in pagina superiore obscure mediocriter viridis, in pagina inferiore pallidior, ovatoblonga, ad apicem obtusa vel acuta, integra, ad basin truncata, plerumque petiolo praedita sed non vel vix decurrens, pilis asperis uncinatis vestita, leviter bullata. *Inflorescentia* cyma terminalis scorpioidea, floribus nutantibus bisexualibus praedita. *Calyx* 5–7 mm, per tres quintas partes vel duas tertias partes basin versus quinquelobus, lobis linearilanceolatis, ad apicem obtusis, pilis longis rigidis brevibusque uncinatis vestitus. *Corolla* 18–22 mm, in alabastro rubra, postea ad basin caerulescens vel erubescens et ad apicem albescens, quinqueloba, faucis squamis stamina non vel vix excedentibus, papillis prope apicem densissimis. *Stamina* 5, inclusa; filamenta et antherae albae. *Stylus* 1; stigma simplex, exsertum. Nuculae nonnullae fertiles efficiuntur. *Floret* 5–7.

Probabiliter *S. grandiflorum* × *S.* × *uplandicum* (= *S. asperum* × *S. officinale*); propter Hortum Hidcote nominatur.

Symphytum × **perringianum** P. H. Oswald & P. D. Sell nothospecies nova
Holotype: Origin: Barton Road, Cambridge, v.c. 29, 52/442575, 16 April 2002, F. H. Perring. Cult. Botanic Garden, Cambridge, v.c. 29, no. 20040215A, 52/456571, 19 May 2006, P. D. Sell no. 06/57 (**CGE**).
Herba perennis monoecius. *Caules* ad 1 m, pallide virides, argute angulati, pilis eglandulosis longis surrigidis pallidis multisque brevibus uncinatis vestiti, ramosi, ramis ascendentibus vel aliquantum patentibus, frondosi. *Folia* in pagina superiore obscure atroviridia, in pagina inferiore pallidiora; basalia ovata, ad basin cordata, mox marcescentia; caulina alterna, 7–15 × 4–9 cm, ovata vel oblongovata, ad apicem decrescentia sed obtusa, integra vel levissime crenata, undulata; caulina superiora ad basin cuneata, inferiora magis minusve cordata; omnia in paginis ambabus pilis simplicibus eglandulosis longis uncinatisque brevibus vestita; venae omnes supra bullatae, subtus prominentes, reticulum firmum facientes; folia superiora sessilia vel paene sessilia, inferiora petiolis ad 5(–10) cm praedita, supra canaliculatis, subtus rotundatis, pilis eglandulosis longis brevibusque uncinatis vestitis, non decurrentibus. *Inflorescentiae* cymarum confertarum terminalium in caulis ramorumque extremitatibus; flores bisexuales; pedicelli sub anthesi 2–7 mm, pilis eglandulosis longis brevibusque vestiti. *Calyx* 7–8 mm, viridis, quinquelobus, lobis inaequalibus lanceolatis, ad apicem obtusis, pilis longis brevibusque uncinatis vestitis. *Corolla* 9–11 mm, ad basin exalbida, supra malvicolor vel roseirubra, quinqueloba. *Stamina* 5; filamenta alba; antherae pallide luteae, inclusae. *Pollen*, ut videtur, infertile. *Stylus* 1; stigma simplex, exsertum. *Floret* 4–5(12).

Cum *S. orientali*, quod parens alter cum *S.* × *uplandico* (= *S. aspero* × *S. officinali*) esse videtur, a Philippo Oswald inventum est; in memoriam amicabilem Franklyn Hugh Perring (1927–2003) propter generis *Symphyti* studium suum nominatur.

Myostatic dynarae (E. J. Lowe) P. D. Sell stat. nov.
M. dissitiflora var. *dynarae* E. J. Lowe *Bot. Mag.* ser. 3. 54: t. 7589 (1898).

Myosotis arvensis (L.) Hill, *Veg. Syst.* **7**: 55 (1764).
var. **dumetorum** (Crépin) P. D. Sell comb. nov.
M. intermedia var. *dumetorum* Crépin, *Not. Pl. Belg.* **2:** 49 (1862).

Mentha spicata L., *Sp. Pl.* 576 (1753).
var. **scotica** (R. A. Graham) P. D. Sell stat. nov.
M. scotica R. A. Graham, *Watsonia* **4**: 119 (1958).

Mentha × villosa Huds., *Fl. Angl.* ed. 2, 250 (1778).
var. **cordifolia** (Opiz ex Fresen.) P. D. Sell stat. nov.
M. cordifolia Opiz ex Fresen., *Syll. Raticb.* **2**: 232 (1828).

Salvia verbenaca L., *Sp. Pl.* 25 (1753).
var. **anglica** (Jord. & Fourr.) P. D. Sell stat. nov.
Gallitrichum anglicum Jord. & Fourr., *Icones ad Floram Europae* **2**: 19, pl. 263 (1869).

Kickxia elatine (L.) Dumort., *Fl. Belg.* 35 (1827).
Antirrhinum elatine L., *Sp. Pl.* 612 (1753).
subsp. crinita (Mabille) Greuter, *Boissiera* **13**: 108 (1967).
Linaria crinita Mabille, *Rech. Pl. Corsica* fasc. **1**: 30 (1867).
var. **prestandreae** (Tineo ex Guss.) P. D. Sell stat. et comb. nov.
Linaria prestandreae Tineo ex Guss., *Fl. Sic. Syn.* **2**: 842 (1844).

Ligustrum vulgare L., *Sp. Pl.* 7 (1753).
var. vulgare
forma **xanthocarpum** (Loudon) P. D. Sell stat. nov.
L. vulgare var. *xanthocarpum* Loudon, *Arb. Frut. Brit.* **2**: 1199 (1838).
var. **coombei** P. D. Sell var. nov.
Holotype: Origin: Vellan Head, The Lizard, Cornwall, v.c. 1, 1979, D. E. Coombe. Cult. The Mound, Botanic Garden, Cambridge, 52/455572, 7 June 1999, P. D. Sell no. 99/161 (**CGE**).
Planta magis minusve prostrata. *Folia* quam in plantis non maritimis nonnihil latiora.

In piam memoriam David Edwin Coombe (1927–1999), qui hanc plantam in Horto Botanico Cantabrigiensi diutius quam per triginta annos colebat, nominatur.

Fraxinus excelsior L., *Sp. Pl.* 1057 (1753).
forma **angustifolia** (Schelle) P. D. Sell stat. nov.
F. excelsior var. *angustifolia* Schelle, *Handb. Laubholzhen* (1903).
forma **multifoliolata** P. D. Sell forma nova

Holotype: Hedge by the side of the Histon Road, Impington, Cambridgeshire, v.c. 29, 52/443615, 30 June 1999, P. D. Sell no. 99/233 (**CGE**).
Rami erecti vel expansi; ramunculi cinerei. *Folia* viridia; foliola 11–15, 6–10 × 3–5 cm, anguste ovata vel elliptica, ad apicem rotundata vel acuta.

Fraxinus angustifolia Vahl, *Enum. Pl.* **1**: 52 (1804).
subsp. oxycarpa (Bieb. ex Willd.) Franco & Rocha Alfonso, *Bot. Jour. Linn. Soc.* **64**: 377 (1971).
forma **oxycarpa** (M. Bieb. ex Willd.) P. D. Sell comb. nov.
F. oxycarpa M. Bieb. ex Willd., *Sp. Pl.* **4**: 1100 (1806).
Folia in autumno viridia perdurantia vel nonnihil lutescentia.
forma **purpurascens** P. D. Sell forma nova
Holotype: Planted by Perne Road, Cambridge, v.c. 29, 52/473566, 18 October 2002, P. D. Sell & J. G. Murrell, Sell no. 02/536 (**CGE**).
Folia in autumno atropurpurascentia.

Nicandra physaloides (L.) Gaertn. *Fruct. Sem. Pl.* **2**: 237 (1791).
var. *violacea* (Andre ex Lemoine) P.D. Sell stat. nov.
N. violacea Andre ex Lemoine, *Rev. Hort.* **1906**: 208 (1906).

Thymus polytrichus A. Kerner ex Borbas, *Term. Közl.* **24**: 105 (1890).
var. **britannicus** (Ronn.) P.D. Sell stat. nov.
T. britannicus Ronn., *Rep. Bot. Exch. Cl. British Isles* **7**: 237 (1924).
var. **drucei** (Ronn.) P.D. Sell stat. nov.
T. drucei Ronn., *Rep. Bot. Exch. Cl. British Isles* **7**: 235 (1924).
var. **neglectus** (Ronn.) P.D. Sell stat. nov.
T. neglectus Ronn., *Rep. Bot. Exch. Cl. British Isles* **7**: 235 (1924).
var. **zetlandicus** (Ronn. & Druce) P.D. Sell stat. nov.
T. zetlandicus Ronn. & Druce, *Rep. Bot. Exch. Cl. British Isles* **7**: 648 (1924).

Euphrasia reayensis (Pugsley) P. D. Sell stat. nov.
E. brevipila var. *reayensis* Pugsley, *Jour. Linn. Soc. Lond. (Bot.)* **48**: 519 (1930).

ABBREVIATIONS

Authors of taxa are consisted with the abbreviations in: Brummitt, R. & Powell, C. E. (1992). *Authors of plant names*. Kew.

Abbreviations of journals in the references follow: Lawrence, H. M. et al. (1968). *Botanico–Periodicum–Huntianum.* Pittsburgh.

Metric measurements follow standard abbreviations:

µm	=	micrometre
mm	=	millimetre
cm	=	centimetre
m	=	metre
km	=	kilometer

Infraspecific taxa are abbreviated:

subsp.	=	subspecies
var.	=	variety
cv.	=	cultivar

GLOSSARY

abaxial Of a lateral organ, the side away from the axis, normally the lower side.
accrescent Becoming larger after flowering, usually applied to the calyx.
achene A dry, indehiscent, one-seeded fruit, more or less hard, with a papery to leathery wall.
achene pits See **receptacular pits**.
acicle A slender prickle with a scarcely widened base.
acladium The first-opening capitulum in the inflorescence in the Asteraceae.
actinomorphic Radially symmetrical, having more than one plane of symmetry.
aculeate Having prickles or sharp points.
acumen The tip of an acuminate point.
acuminate Curved inwards on both sides to a point. Often wrongly used for gradually narrowed to a point (see **acute**).
acute With a point. Gradually narrowed to a point is *long-acute*, but is often called *acuminate*.
adaxial Of a lateral organ, the side towards the axis, normally the upper side.
adherent Joined or fused.
adnate Joined to another organ of a different kind.
adpressed Closely applied to the surface (same as appressed).
aerial Above ground or above water.
alien Not native. Believed on good evidence to have been introduced by Man and now more or less naturalised.
allopolyploid A polyploid derived by hybridisation between two different species with doubling (or quadrupling, etc.) of the chromosome number.
allotetraploid A doubling of a diploid.
alternate Lateral organs on an axis, one per node, successive ones on opposite sides. Commonly used also to include spiral arrangements.
alveoles See **receptacular pits**.
amphimixis Reproducing by seed resulting from normal sexual fusion. Adjective *amphimictic*.
amphistaminal The arrangement of stamens on a honey disc when the disc is entire and stamens inserted in the centre, or lobed and the stamens in the centre of the lobes, or more or less enclosed between the lobes.
amplexicaul Clasping the stem.
anastomosing Joining up to form loops, usually referring to veins.
anatropous (of an ovule) Bent over against the stalk.
androecium The male parts of the flower, the stamens.
andromonoecious Having male and bisexual flowers on the same plant.
anemophilous Wind-pollinated.
angustiseptate A fruit with the septum across the narrowest diameter.
annual Completing its life in under 12 months, but often not within one calendar year.
annular Ring-shaped.
annulus Special thick-walled cells forming part of the opening mechanism of a fern sporangium, often forming a ring.
anther The pollen-bearing part of a stamen, usually terminal on a filament.
anthesis At the time of flowering.
anticlinal Line of division of cells at right angles to surface of apex of a growing point.
aphyllopodous Without basal leaves.
apiculate With an apiculus.
apiculus A small, broad point at the apex.
apogamous See **apomictic**.
apomictic Reproducing by seed not formed by sexual fusion without pollen stimulation (apogamous), or by pollen stimulation (pseudogamous), or sterile and vegetatively over wide areas.
appendage A small extra protrusion or extension such as on a petal, sepal or seed.
appressed Pressed close to another organ, but not united with it.
arachnoid Appearing as if covered with cobwebs.
archegonium The structure containing the female sexual cell in many land plants.
arcuate Curved so as to form a quarter of a circle or more.
aril The succulent covering around a seed, outside the testa, but not the pericarp.
aristate Extended into a long bristle.
ascending Sloping or curving upwards.
asperous Rough to the touch.
attenuate Gradually tapering.
auricles Small ear-like projections at the base of a leaf, especially in grasses.
autopolyploid A polyploid derived from one diploid species, by multiplication of its chromosome sets.
autotrophic Neither parasitic nor saprophytic.
awn A stiff, bristle-like projection from the tip or back of the glumes and/or lemma in grasses, or from a fruit, usually the indurated style, e.g. *Erodium*, or less frequently the tip of a leaf.
axil The angle between the main and lateral axis.
axile Of a placenta formed by the central axis of an ovary, that is connected by a septa to the wall.
axillary Arising in the axil of a leaf or bract.
barbellate With stiff, hooked hair-like bristles.
base-rich Soils containing a relatively large amount of free basic ions, e.g. calcium, magnesium, etc.
basic number See **chromosomes**.
basifixed (of anthers). Joined by the base to the filament and not capable of independent movement.

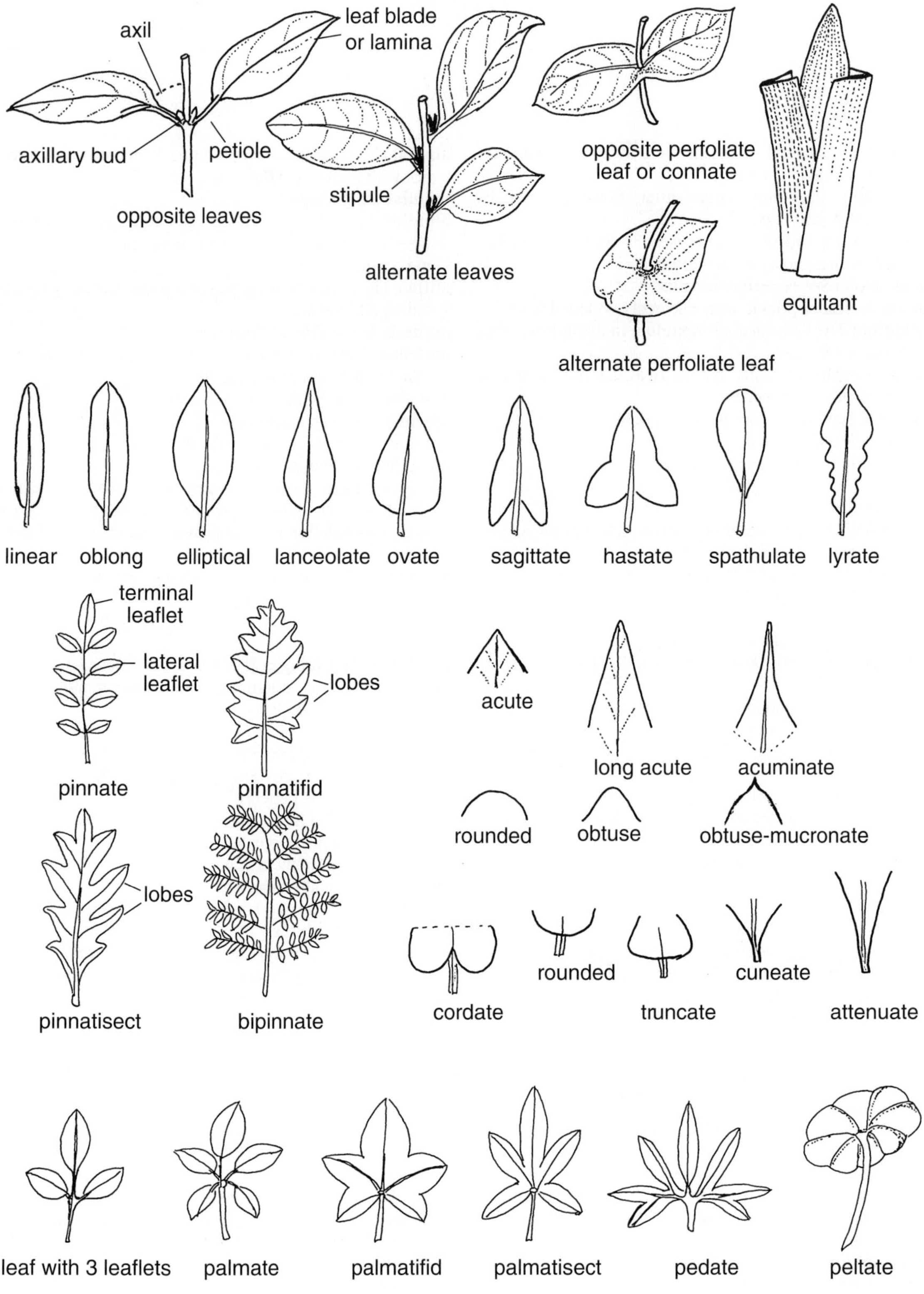
axil
leaf blade
or lamina
axillary bud
petiole
opposite leaves
stipule
alternate leaves
opposite perfoliate
leaf or connate
equitant
alternate perfoliate leaf
linear
oblong
elliptical
lanceolate
ovate
sagittate
hastate
spathulate
lyrate
terminal
leaflet
lateral
leaflet
pinnate
lobes
pinnatifid
acute
long acute
acuminate
rounded
obtuse
obtuse-mucronate
lobes
pinnatisect
bipinnate
cordate
rounded
truncate
cuneate
attenuate
leaf with 3 leaflets
palmate
palmatifid
palmatisect
pedate
peltate

Leaves

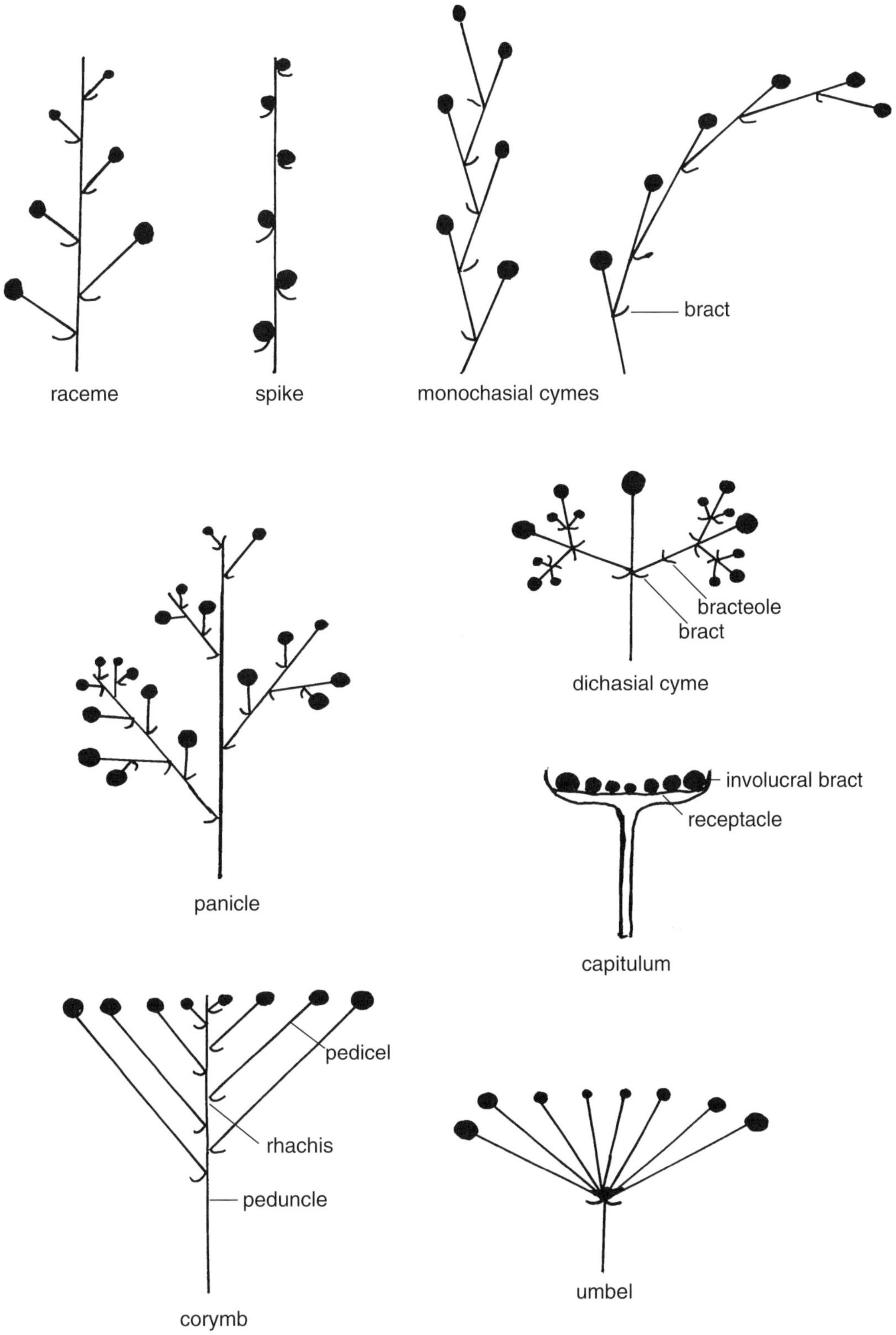
raceme
spike
monochasial cymes
bract
bracteole
bract
dichasial cyme
panicle
involucral bract
receptacle
capitulum
pedicel
rhachis
peduncle
corymb
umbel

Inflorescences

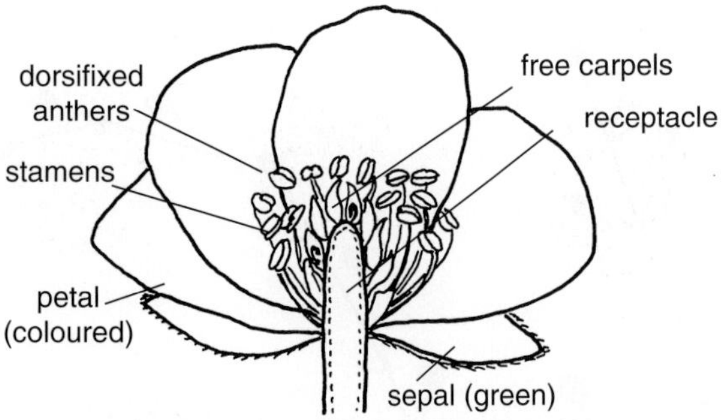
dorsifixed
anthers
stamens
petal
(coloured)
free carpels
receptacle
sepal (green)

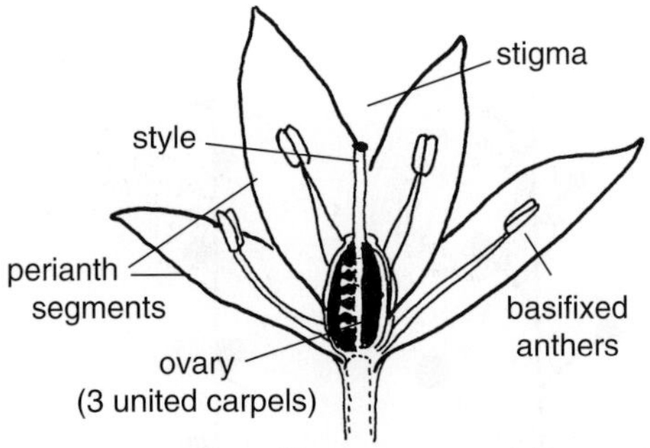
stigma
style
perianth
segments
ovary
(3 united carpels)
basifixed
anthers

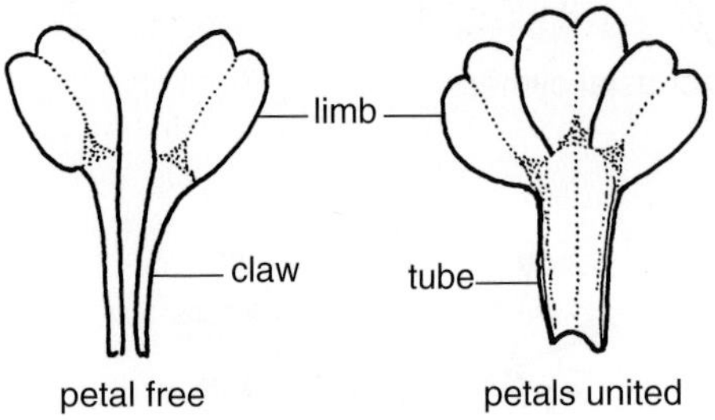
limb
claw
tube
petal free
petals united

Corollas

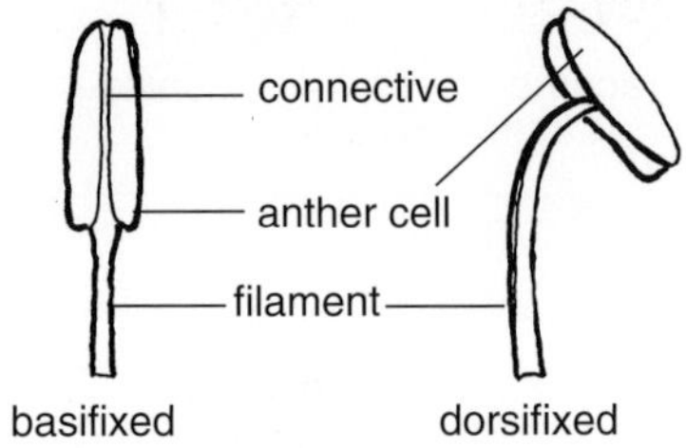
connective
anther cell
filament
basifixed
dorsifixed

Stamens

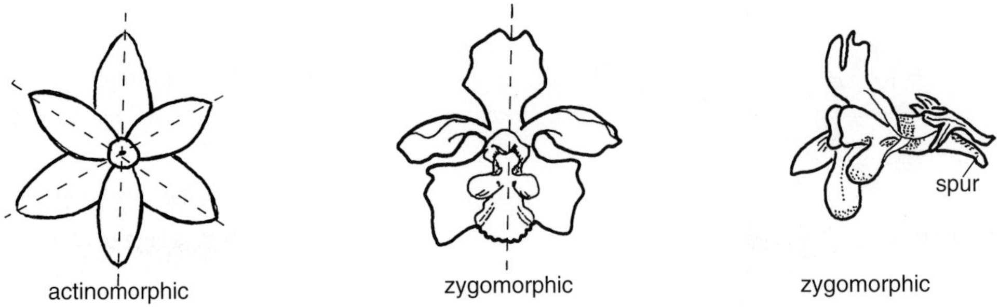
actinomorphic
zygomorphic
spur
zygomorphic

superior ovaries

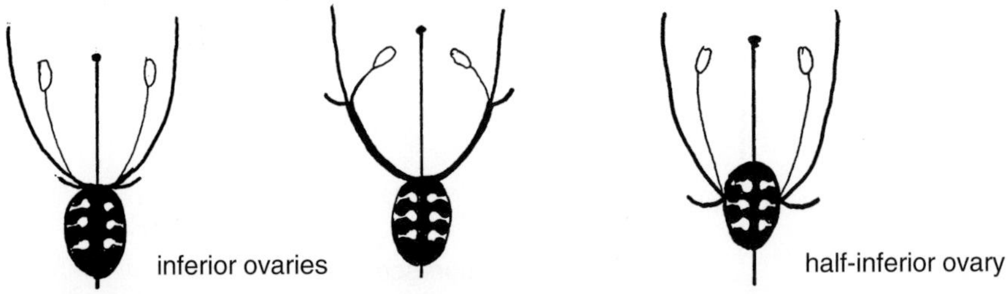
inferior ovaries
half-inferior ovary

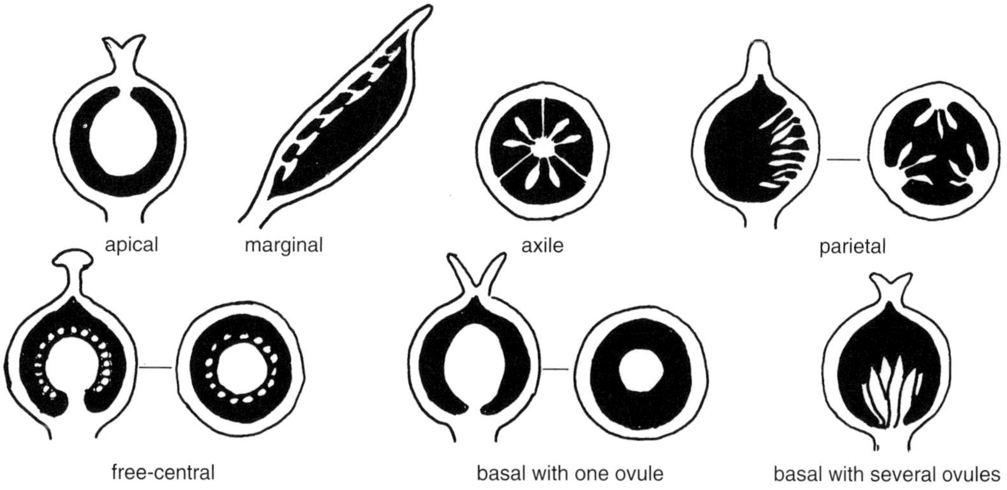
apical
marginal
axile
parietal
free-central
basal with one ovule
basal with several ovules

Ovules and placentation

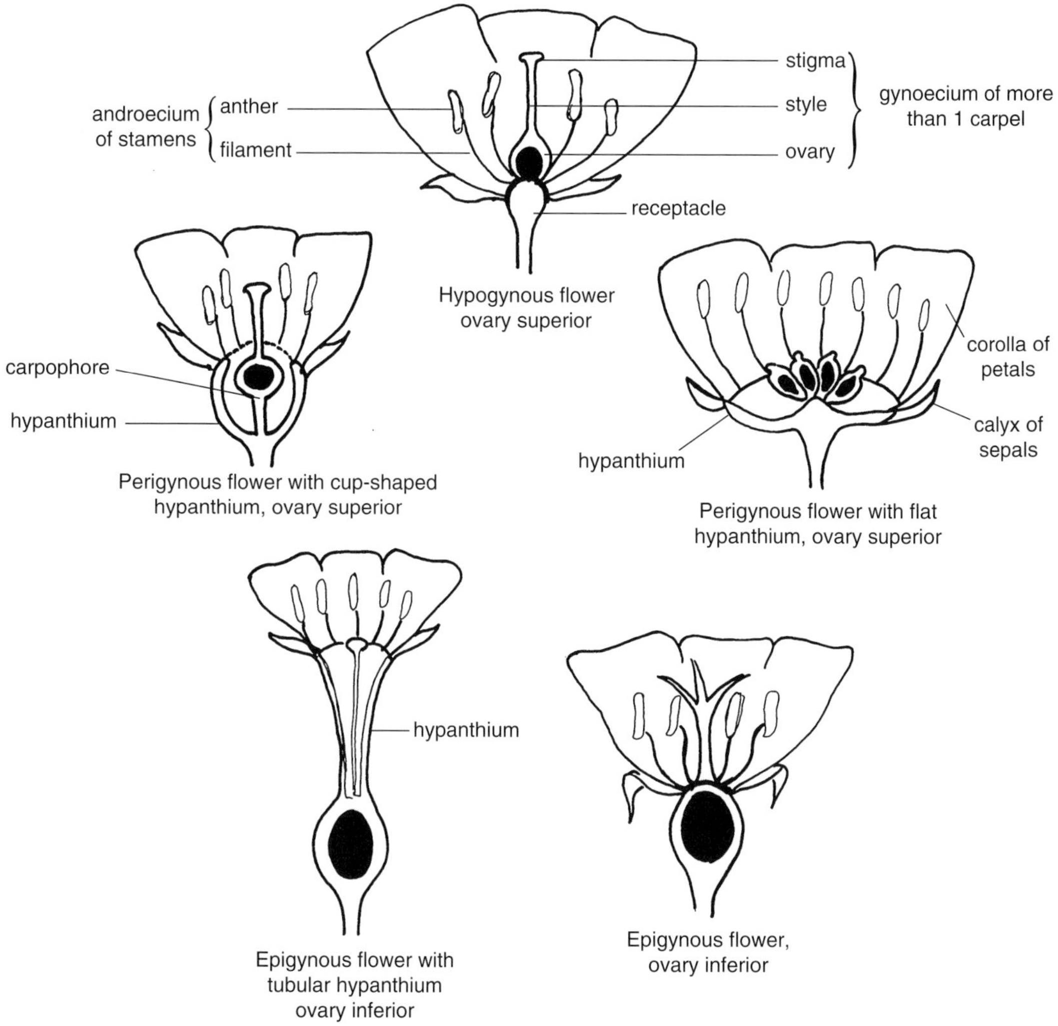
androecium of stamens
anther
filament
stigma
style
ovary
gynoecium of more than 1 carpel
receptacle
Hypogynous flower ovary superior
carpophore
hypanthium
Perigynous flower with cup-shaped hypanthium, ovary superior
corolla of petals
calyx of sepals
hypanthium
Perigynous flower with flat hypanthium, ovary superior
hypanthium
Epigynous flower with tubular hypanthium ovary inferior
Epigynous flower, ovary inferior

beak A narrow, usually apical projection, sometimes at the top of an achene.
berry A fleshy fruit, usually several-seeded, without a stony layer surrounding the seeds.
biennial Completing its life cycle in more than one year, but less than two years; not flowering in the first year.
bifid Divided into two, usually deeply at the apex.
bifurcate Dividing into two branches.
biotype A genetically fixed variant of a taxon particularly to some condition.
bird-seed alien An alien introduced as a contaminant of bird-seed.
bisexual Of a flower, bearing both sexes.
bivalents Paired chromosomes.
blade The main part of a flat organ, e.g. petal, leaf.
bloom A delicate, waxy, easily removed covering to fruit or leaves.
bog A community on wet, very acid peat.
bract Modified, often scale-like leaf subtending a flower, or less often a branch.
bracteate With a bract or bracts.
bracteole A supplementary or secondary bract, or a bract once removed.
bud-scales Scales enclosing a bud before it expands.
bulb A swollen, underground organ consisting of a short stem bearing a number of swollen, fleshy leaf-bases or scales with or without a tunic, the whole enclosing the next year's bud.
bulbil A small bulb or tuber arising in the axil of a leaf or in an inflorescence on the aerial part of the plant.
bullate With the surface raised into blister-like swellings.
caducous Falling off at an early stage.
caespitose Tufted.
calcicole More frequently found upon or confined to soils containing free calcium carbonate.
calcifuge Not normally found on soils containing free calcium carbonate.
callus A horny region at the base of the lemma in grasses.
calyx The sepals as a whole, the outer whorl of the perianth if different from the inner, including the calyx tube and calyx lobes.
campanulate Bell-shaped, widest at the mouth.
campylotropous When the ovule is bent so that the stalk appears to be attached to the side midway between the micropyle and chalata.
canescent Becoming grey.
capillary Hair-like.
capitate Head-like, such as in a tight inflorescence, a knob-like stigma or style, or a stalked gland.
capitulescence Sometimes used for the inflorescence formed by capitula in Asteraceae.
capitulum An aggregate head of sessile flowers in the Dipsacaceae and Asteraceae.
capsule A dry, many-seeded dehiscent fruit formed from more than one carpel.
carpel One of the units of which the gynoecium is composed, and is the basic reproductive unit of the Magnoliophyta; one to many per flower, if more than one then separate or fused.
carpophore A stalk-like sterile part of a flower between the receptacle and the carpels as in the Apiaceae and Caryophyllaceae.
cartilaginous Hard, not green, and easily cut with knife.
caryopsis A fruit in the Poaceae with the ovary wall and seed coat united.
casual An alien plant not naturalised.
catkin A dense spike of reduced flowers on a long axis, often wind-pollinated.
cauline Pertaining to the stem.
cell (of an ovary) The chambers into which the ovary may be divided (often each one corresponding to a carpel).
chartaceous Of papery texture.
chasmogamous Of flowers which open normally.
chlorophyll The green colouring matter of leaves, etc.
chromosomes Small deeply staining bodies, found in all nuclei, which determine most or all of the inheritable characters of organisms. Two similar sets of these are normally present in all vegetative cells, the number (diploid number $2n$) usually being constant for a single species. The sexual reproductive cells normally contain half this number, or a multiple of a common basic number.
cilia Small whip-like structures by means of which some sexual reproductive cells swim.
ciliate With hairs projecting from the margin.
circinately Rolled on the axis, so that the apex is in the centre.
circumscissile Dehiscing transversely, the top of the capsule coming off like a lid.
cladode A green leaf-like lateral shoot in the Liliaceae which, as the leaves are reduced to scales, functionally act as leaves.
clava A club-shaped structure.
clavate Club-shaped, slender and thickened towards the apex.
claw The narrow part of a flat organ of which the broader part is the blade or lamina.
cleistogamous Of flowers which never open and are self-pollinated in the bud stage. Opposite of **chasmogamous**.
clone An area of plant from vegetative spread, or from seeds from an apomictic plant.
column A stout stalk formed by fusion of various floral parts in the Orchidaceae, Geraniaceae, etc.
columnar Column-like.
commissure The faces by which the two carpels are joined together in the Apiaceae.
complanate To flatten, make level.
compound Of an inflorescence with the axis branched; of a leaf made up of several distinct leaflets.
compressed Flattened.
concolorous Of approximately the same colour throughout.
cone A compact body composed of an axis with lateral organs bearing spores or seeds, like the Lycopodiophyta and Pinophyta.
cone-scales The lateral organs of a cone.
conferted Crowded; closely assembled or packed.
connate Organs of the same kind growing together and becoming joined, though distinct in origin.

connective Part of an anther connecting its two halves.
connivent Of two or more organs with their bases wide apart, but their apices approaching one another.
contiguous Touching at the edge with no gap between.
contorted With each lobe of the perianth overlapping the next with the same edge and appearing twisted.
convergent Of more than two organs, with their apices closer together than their bases.
convolute Rolled together, coiled.
cordate A flat object with incurved base on both sides.
coriaceous Of a leathery texture.
corm A short, usually erect and tunicated, swollen underground stem of one year's duration, that of the next year arising at the top of the old one and close to it.
corolla The inner whorls of the perianth, if different from the outer; all the petals, corolla tube and corolla lobes.
corona A long or short tube within the perianth segments in the Liliaceae.
corymb A raceme with the pedicels becoming short towards the top, so that all the flowers are approximately at the same level.
corymbose Corymb-like, but strictly not a corymb but a flat-topped cyme.
cotyledon The first leaf or leaves of a plant already present in the seed and usually differing in shape from the other leaves. Cotyledons may remain within the testa or may be raised above the ground and become green during germination.
crenate With round teeth on the margins of a flat organ.
crenulate The diminutive of **crenate**.
crisped Curled.
cuculate Hooded.
culm The stem of a grass.
cuneate Wedge-shaped, that is flat and narrowed at the base of an organ.
cuneiform Wedge-shaped with the thin end at the base of a solid object.
cupule A fruit which is a nut surrounded by a husk formed of fused scales in the Fagaceae and Corylaceae.
cuspidate Abruptly narrowed to a point.
cyathium The peculiar inflorescence of *Euphorbia*, where a cup-shaped involucre with stamens and stalked gynoecium has each stamen and the gynoecium being a separate flower.
cymbiform Boat-shaped.
cyme An inflorescence to which each flower terminates the growth of a branch, the more distal flowers being produced by longer branches lateral to it.
cymose In the form of a cyme.
cypsela Name sometimes given to achenes in Asteraceae.
decaploid See **polyploid**.
deciduous Not persistent, the leaves falling in autumn or petals falling after anthesis.
decumbent Lying on the ground but turning up at the ends.
decurrent Having the base prolonged down the axis as in leaves where the blade continues down the petiole or stem as a wing.
decurved Curved downwards.
decussate Opposite, with successive pairs at right angles to each other.
deflexed Bent sharply downwards.
dehiscent Opening naturally to shed its seeds or pores.
deltate Shaped like the Greek letter Δ.
deltoid Like a delta, more or less triangular in shape.
dentate With patent teeth at the margin of a flat organ.
denticulate With small teeth.
depressed-globose Like globose but wider than long.
diadelphous With one stamen free, and the rest connate (Fabaceae).
dichasium A cyme in which the branches are opposite and more or less equal.
didymous Formed of two similar parts attached to each other by a small portion of their surface.
didynamous With 4 stamens, 2 long and 2 short.
digitate See **palmate**.
dimorphic Occurring in two forms.
dioecious Having the sexes on different plants.
diploid Having two matching sets of chromosomes.
disc Anything disc-shaped, as the fleshy sometimes nectar-secreting portion of the receptacle, surrounding or surmounting the ovary.
disc flower The central, eligulate flowers on the receptacle in the Asteraceae.
dissected Deeply divided up into segments.
distal At the end away from the point of attachment.
distichous Arranged in two diametrically opposed rows.
divaricate Divided into widely divergent branches.
divergent With organs having their apices further apart than their bases.
dominant The chief constituents of a particular plant community, e.g. oaks in an oak wood or heather on a moor.
dorsifixed (of anthers) Attached by their back.
dorsiventral With a distinct upper side and lower side.
drupe A more or less fleshy fruit with one or more seeds each surrounded by a stony layer.
dry Not succulent.
e Without, e.g. eglandular, ebracteate.
ebracteate Without bracts.
ectomycorrhizal An association of roots with a fungus which may form a layer outside the root.
effuse Spreading widely.
eglandular Without glands of any kind.
elaiosome An oily appendage to seeds offering food bodies to ants.
ellipsoid A solid, elliptical in side view; broadly and narrowly can be used as in elliptical.
elliptical A flat shape widest in the middle and 1.2–3.0 times as long as wide; if less, broadly so, if more, narrowly so. By some authors this shape is called lanceolate.
emarginate Shallowly notched at the apex.
embryo-sac See **gametophyte**.
endemic Confined to one particular area, i.e. in this book to the British Isles.
endomycorrhizal An association of roots with a fungus which may form a layer inside the root.
endosperm In the Magnoliophyta the nutritive tissue for the embryo in the developing seed; it might or might not remain as the food store in the mature seed.
ensiform A flat, narrow shape broadest in the middle, both ends with straight sides but gradually narrowed.

entire The margin of a flat shape not toothed or lobed.
entomophilous Insect-pollinated.
epicalyx Organs on the outside of a flower, calyx-like but outside and additional to the calyx.
epicormic (of new shoots) Borne direct from the trunk of a tree.
epigeal Above ground; in epigeal germination, the cotyledons are raised above the ground.
epigynous (of a flower) With an inferior ovary.
epipetalous Inserted upon the corolla.
equitant (of distichous leaves) Folded longitudinally and overlapping in their parts.
erect Upright.
erecto-patent Between erect and patent.
erose Appearing as if gnawed.
escape A plant growing outside a garden, but having spread vegetatively or by seed from one.
estrophiolate The opposite of strophiolate.
evergreen Retaining its leaves throughout the year.
exceeding Longer than.
exserted Protruding from.
exstipulate Without stipules
extrastaminal The arrangement of stamens on the outer margin of the honey disc.
extravaginal (of branches) Breaking through the sheaths of the subtending leaves.
extrorse (of anthers) Opening towards the outside of the flower.
falcate Sickle-shaped.
false fruit An apparent fruit actually formed from tissue in addition to the real fruit.
fasciculate In tight bundles.
fastigiate A plant with upright branches forming a narrow outline.
fen A community on alkaline, neutral or slightly acid wet peat.
fertile Producing seed capable of germination or (of anthers) containing viable pollen.
fibrous roots A root system in which there is no main axis.
filament The stalk of an anther, the two together forming the stamen.
filiform Thread-like.
fimbriate With the margins divided into a fringe.
fistular Hollow and cylindrical; tube-like.
flabelliform Fan-shaped.
flexuous (of a stem or hair) Wavy.
floccose A short, whitish indumentum often of stellate hairs.
flore pleno A double flower with many more petals than normal, usually due to conversion of the stamens to petals; in the Asteraceae, a double capitulum with all or many disc flowers converted to ray flowers.
floret A small flower.
flush Wet ground, often on hillsides, where water flows but not in a definite channel.
foliaceous Leaf-like, of an organ not usually a leaf.
follicle A dry, usually many-seeded fruit dehiscent along its side, formed from one carpel.
foveolate Having regular small depressions.
free Separate, not fused to another organ or to one another except at the point of origin.
free-central (of a placenta) Formed by the central axis of any ovary which is not connected by septa to the wall.
fruit The ripe seeds and structure surrounding them, whether fleshy or dry; strictly the ovary and seeds, but often used to include other associated parts such as the fleshy receptacle, as in the rose and strawberry.
fugacious Withering or falling off very rapidly.
funicle The stalk of the ovary.
fusiform Spindle-shaped.
gametophyte The haploid generations of a plant that bears the true sex organs; in the Pteridophyta the prothallus, in the Pinophyta and Magnoliophyta the pollen grains (male) and embryo sac (female).
gamopetalous Having the petals joined into a tube, at least at the base.
geitonogamy Fertilisation of a flower by another from the same plant.
geniculate Bent abruptly to make a knee.
gibbous With a rounded swelling on one side as on the base of the calyx in *Acinos*.
glabrescent Becoming glabrous; sometimes wrongly used for sparsely hairy.
glabrous Hairless, including all types of glands and spines.
gland A small globose or oblong vesicle often containing oil, resin or other liquid, sunk in on the surface of or protruding from any part of a plant. When furnished with stalks they are called *glandular hairs* (see also **hairs**).
glandular Furnished with glands.
glaucous Bluish-white or bluish-green.
globose Spherical.
glochidiate Furnished with barbed hairs.
glumaceous Resembling a glume.
glume A flat bract-like organ(s) subtending a flower in Cyperaceae and Poaceae.
grain alien Alien introduced as a contaminant of grain.
granulose With a fine, sand-like surface texture.
gymnodioecious Having female and bisexual flowers on the same plant.
gynobasic A style arising from the base of a carpel.
gynoecium The group of female parts of a flower made up of one or more ovaries, with their styles and stigmas.
hairs Generally, hairs can be described in terms of *pubescent, tomentose, lanate, hispid, strigose,* etc. Usually it is indicated if the hairs are *eglandular, glandular, stellate*, *tufted* or *plumose*. If precise comparisons are made of indumentum it is often diagnostic of a taxon, but very difficult to put into few words. In some critical genera, where the indumentum is a very important character, a more detailed account of the hairs is given. There are two main types of hair, *branched* and *simple*. Branched hairs are either more or less *stellate*, or *plumose* or *subplumose*, that is pinnately branched with the branches projecting longer than the diameter of the hair. *Simple* hairs with a capitate tip are *glandular*, or when more or less the same throughout the hairs are *eglandular*. Eglandular hairs usually include those hairs with minute side-projections not longer than the diameter of the hairs, although these are sometimes called

denticulate hairs. Two rarer types of hair are *medifixed* hairs in the form of a T and *glochidiate* hairs in the form of an anchor. Sessile glands should also be thought of as hairs and some glands are so minutely stalked it is difficult to see. Spines are really rigid hairs and there is every intermediate between a stiff hair, acicle and pricklet all of which can be gland-tipped, to a large spine. The abundance of hairs is indicated by: *few* or *sparse* when the hairs in question form only a small proportion of the total indumentum or are scattered; *numerous* when the hairs are abundant but separate enough to be individually distinct; *dense* when they form a continuous indumentum. The length of the hairs is referred to as *very short* up to 0.3 mm, *short* 0.3–0.7 mm, *medium* 0.7–1.5 mm, and *long* 1.5–4.0 mm. If more than 4 mm the length is given. *Colour*, *rigidity*, *angle* and *waviness* can be added. In those cases when an attempt has been made to be very precise in a particular group of plants it is so stated after the generic description.

half-epigynous Of a flower with a semi-inferior ovary.

haploid Having only one set of chromosomes as in gametophytic tissue.

hastate A leaf with two spreading basal lobes and a much longer erect central lobe.

haustorium An outgrowth of stem, root or hyphae of certain parasitic plants, which serves to draw food from the host plant.

heath A lowland community dominated by heath or ling, usually on sandy soils with a shallow layer of peat.

hemiparasite See **parasite**.

heptoploid See **polyploid**.

herb A plant dying down to ground level each year.

herbaceous Not woody, dying down each year; leaf-like as opposed to woody, horny, scarious or spongy.

hermaphrodite Bisexual.

heterochlamydeous Having the perianth segments in two distinct series which differ from one another.

heterophyllous Having leaves of more than two distinct forms.

heterosporous Having spores of two sorts, megaspores female and microspores male, as in all Pinophyta and Magnoliophyta and a few Pteridophyta.

heterostylous Having two forms, not sexes, of flowers on different plants, the two sorts with different styles and pollen.

heterozygous Bearing two dissimilar alternative genetical factors.

hexaploid Plants having six sets of chromosomes; see *polyploid*.

hilum The scar on a seed where it left its point of attachment.

hip the fruit of *Rosa* for which there appears to be no other name; it is a **hypanthium**.

hirsute Clothed with long stiff hairs.

hispid Coarsely and stiffly hairy.

hispidulous With small stiff hairs.

homochlamydeous or **homolochlamydeous** Having all the perianth segments similar.

homogamous Having the anthers and stigmas maturing simultaneously.

homosporous Having spores of approximately the same size, as in most Pteridophyta.

hyaline Thin and translucent.

hybrid A plant originating by the fertilisation of one species by another.

hybrid swarm A series of plants originating by hybridisation between two or more species and subsequently recrossing with the parents and between themselves, so that a continuous series of forms arise.

hypanthium The extension of the receptacle above the base of the ovary in perigynous and epigynous flowers.

hypocrateriform Saucer-shaped.

hypogeal Below ground; in hypogeal germination, the cotyledons remain below ground.

hypogynous Of a flower with a superior ovary, the calyx, corolla and stamens inserted at the base of the ovary.

imbricate Overlapping at the edges.

imparipinnate Pinnate with an unpaired terminal leaflet.

impressed Sunk below the surface.

Inaperturate Lacking apertures.

included Not exserted.

incurved Curved inwards.

indehiscent Not dehiscent.

indumentum The hairy coverings as a whole.

indurated Hardened and toughened.

indusium A small flap or pocket of tissue covering a group of sporangia in many Pteridophyta.

inferior An ovary which is borne below the point of origin of the petals, sepals and stamens and is fused with the receptacle (hypanthium) surrounding it.

inflated Of an organ which is dilated, leaving a gap between it and its contents.

inflexed Bent inwards.

inflorescence A group of flowers with their branching system and associated bracts and bracteoles.

infundibulum In the form of a funnel.

insertion The position and form of the point of attachment of an organ.

intercalary leaf The leaves on the main stem between the topmost branches and the lowest flowers of the terminal spike in *Rhinanthus*.

internode The stem between adjacent nodes.

interpetiolar Between the petioles.

interrupted Not continuous.

intrapetiolar Between the petiole and the stem.

intravaginal Within the sheath.

introduced A plant which owes its existence in our country to importation by Man, deliberate or not.

introgression The acquiring of characteristics by one species from another by hybridisation followed by backcrossing.

introrse (of anthers) Opening towards the middle of the flower.

involucel An involucre at the base of a flower in the Dipsacaceae, formed by the united bracteoles.

involucral Forming an involucre.

involucre Bracts forming a more or less calyx-like structure round or just below the base of a usually condensed inflorescence, i.e. *Anthyllis*, Asteraceae.

involute With the margins rolled upwards.

isodiametric Of any shape or organ more or less the same distance across in any plane.
isomerous Having the same number of parts in two or more different floral whorls.
jaculator The indurated funicle in Acanthaceae which acts as an ejector of its seeds.
keel A longitudinal ridge on an organ, like the keel of a boat; the lower petal or petals when shaped like the keel of a boat as in *Fumaria* and Fabaceae.
labellum The central inner perianth segment which appears to be the lower one, but actually the uppermost due to the flower twisting through 180° and usually different from all the other perianth segments.
lacerate Deeply and irregularly cut and appearing as if torn.
laciniae Segments of anything incised.
laciniate Irregularly and deeply toothed.
lamina The blade of a leaf
laminar In the form of a flat leaf.
lammas growth Extra, usually abnormal growth put on in summer by some trees.
lanate Woolly; covered with hairs giving a woolly appearance.
lanceolate Very narrowly ovate.
lanuginose With woolly indumentum.
latex Milky juice.
latiseptate Fruit with the septum across the widest diameter.
lax Loose or diffuse, not dense.
leaflet A division of a compound leaf.
leaf-opposed A lateral organ borne on the stem on the opposite side from a leaf, not in a leaf axil as usual.
leaf rosette A radiating cluster of leaves often at the base of the stem at soil level.
leaf teeth Pointing outwards (*dentate*), pointing towards the apex (*serrate*), breast-like with a small nipple-like apex (*mammiform*), similar to last but narrower and curved upwards (*aquiline-mammiform*), deeply and jaggedly cut (*laciniate*).
legume A usually dry, mostly many-seeded fruit dehiscent along two sides, formed from one carpel.
lemma A bract in the Poaceae borne on the abaxial side of the flower.
lenticel Ventilating pore.
lenticular Convex on both faces and more or less circular in outline.
leptoma An aperture in a pollen grain.
ligulate Strap-shaped.
ligule A minute membranous flap at the base of the leaves of *Isoetes* and *Selaginella*; the strap-shaped part of a ligulate flower in the Asteraceae; the short projection in the axil of a leaf in the Cyperaceae and Poaceae.
limb The flattened, expanded part of a calyx or corolla the base of which is tubular.
linear Long and narrow with more or less parallel margins.
lip Part of the distal region of a calyx or corolla sharply differentiated from the rest due to fusion or close association of its parts.
lobe Divided substantially, but not into separate leaflets.
loculicidal Splitting down the middle of each cell of the ovary.
lodicule Two minute scales at the base of the ovary in the Poaceae.
lomentaceous Resembling or having a lomentum, that is a legume or pod constricted between seeds.
long-shoot Stem of potentially unlimited growth, especially in trees or shrubs.
lower side The under surface of a flat organ.
lunate Crescent-moon-shaped.
lyrate More or less lyre-shaped.
macrostylous Having long styles.
mammiform Breast-shaped.
marsh A community on wet or periodically wet, but not peaty soil.
meadow A grassy field cut for hay.
mealy With a floury texture.
medifixed Joined in the middle.
megasporangium In a heterosporous plant, the sporangia bearing megaspores.
megaspore In a heterosporous plant the female spores which give rise to female gametophytes.
meiosis Special form of cell division (in sporangia, pollen sacs or ovules) in which the chromosome number is halved, producing haploid spores.
membranous Like a membrane in consistency.
mericarp A one-seeded portion split off from a syncarpus ovary at maturity.
-merous Denoting the number of parts in a structure, e.g. 5-merous, having the parts in fives.
mesostylous Having a style of intermediate length.
mesotrophic Providing moderate nourishment.
microsporangium In a heterosporus plant, the sporangia bearing microspores.
microspore The male spores in a heterosporus plant which give rise to the male gametophytes.
microstylous Having short styles.
midrib The central main vein.
monadelphous (of stamens) United into a single bundle by the fusion of the filaments.
monocarpic Living for one year, flowering, fruiting and then dying.
monochasium A cyme with one lateral branch at each node.
monochlamydeous Having only one series of perianth segments.
monoecious Having male and female organs on the same plant.
monomorphic Occurring in one form; not dimorphic or trimorphic.
monopodial Of a stem in which growth is continued from year to year by the same apical growing point.
moor Upland communities, often dominated by heather, on dry or damp but not wet peat.
mucro The tip of a mucronate object.
mucronate Having a very short, bristle-like tip.
mucronulate The diminutive of **mucronate**.
mull soil A fertile woodland soil with no raw humus layer.
muricate Rough with short firm projections.

muticous Without an awn or mucro.
mycorrhiza An association of roots with a fungus which may form a layer outside the roots (endomycorrhizal) or within the outer tissues (ectomycorrhizal).
naked Devoid of hair or scales, or not enclosed.
napiform Turnip-shaped.
native A plant growing in an area where it was not put by the hand of Man.
naturalised An alien plant which has become self-perpetuating in the British Isles, or a native plant which is transferred to a new locality by Man and is self-perpetuating.
nec Nor, nor of.
nectariferous Nectar-bearing.
nectar pit A nectariferous pit.
nectary Any nectariferous organ, usually a small knob or a modified petal or stamen.
nerve See **vein**.
nodding Bent over and pendulous at the tip.
node The position of a stem where leaves, flowers or lateral stems arise.
nodule A small, more or less globose swelling.
non Not, not of.
nonaploid See **polyploid**.
nothomorph One of more than two variants of a particular hybrid.
nucellus The tissue between the embryo and the integument in an ovule.
nut A dry, indehiscent, one-seeded fruit with a hard, woody wall.
nutlet A small nut or a woody-walled mericarp.
ob- The other way up from normal, usually flattened or widened at the distal rather than the proximal end.
obdiplostemonous Having the stamens in two whorls, the outer opposite the petals, the inner opposite the sepals.
oblong A flat shape with the middle part parallel-sided, 1.2–3.0 times as long as wide; if less, broadly so, if more, narrowly so.
obpyriform Shaped like a pear upside down; depending on which end of the pear the stalk is found, the fruit could be erect or pendulous.
obtuse Blunt.
ochreae Sheathing stipules in the Polygonaceae.
octoploid See **polyploid**.
oligotrophic Providing little nourishment.
opposite Of two organs, arising laterally at one node on opposite sides of the stem.
orbicular Like an orb.
orthotropous (ovule). Straight and with the axis of the ovule in the same line as that of the funicle.
oval Elliptical.
ovary The basal part of the gynoecium containing the ovule.
ovate A flat shape widest nearer the base 1.2–3.0 times as long as wide; if less, broadly so, if more, narrowly so.
ovoid A solid shape, ovate in side view.
ovule An organ inside the ovary of the Magnoliophyta, naked in the Pinophyta, that contains the embryo sac, which in turn contains the egg developed into seed after fertilisation.
palate A projecting part of the lip closing the mouth of the corolla in *Antirrhinum*.
palea A bract on the adaxial side of a flower in the Poaceae.
palmate Consisting of more than three leaflets arising from the same point.
panduriform Lanceolate in outline with a dip in each side below the middle and with an obtuse apex.
panicle A compound or much-branched inflorescence, either racemose or cymose.
papilla A small nipple-like projection.
pappus The hairs, scales or scarious margin at the top of a fruit (achene or cypsela) in the Asteraceae.
papyraceous Of papery texture.
parasite Plant which gets all or part of its nourishment by attachment (often under the ground) to other plants.
parietal (of a placenta) Formed by a central axis of an ovary that is connected by septa to the wall.
paripinnate Pinnate without an unpaired terminal leaflet.
partial septum A septum which is incomplete.
pasture A grassy field grazed during summer.
patent Projecting at more or less right angles, spreading.
pectinate Lobed with the lobes resembling and arranged like the teeth of a comb.
pedate With five leaflets of a leaf arising from the same point, the leaflets obovate.
pedicel Flower stalk.
peduncle Stalk of a group of flowers.
peloric A flower which is normally irregular becomes regular.
peltate (of a flat organ) With its stalk inserted in the lower surface.
pelviform In the form of a pelvis, a bony cavity.
pendant Pendulous.
pentaploid See **polyploid**.
perennial Living for more than two years and usually flowering each year.
perianth The floral leaves including petals and sepals.
perianth segments, **lobes** or **tube** Petals and sepals, usually used when they are not or little differentiated.
pericarp The wall of a fruit, originally the ovary wall.
perigynous A flower with a superior ovary, but with the calyx, corolla and stamens inserted above the base of the ovary on an extension of the receptacle (hypanthium) that is not fused with the ovary.
perigynous zone The annular region between the gynoecium and the other floral parts in perigynous or epigynous flowers.
perisperm The nutritive tissue derived from the nucellus in some seeds.
perispore A membrane surrounding a spore.
persistent Remaining attached longer than normal.
petal One of the segments of the inner whorl(s) of the perianth.
petaloid Brightly coloured and resembling petals.
petiole The stalk of a leaf.
petiolate Having a petiole.
phyllary One of the involucral bracts surrounding the capitulum in the Asteraceae, which in this account are called **involucral bracts**.

phyllode A green, flattened petiole resembling a leaf.
pilose Hairy.
pinna The primary division of a more than two-pinnate leaf.
pinnate A compound leaf, with more than three leaflets arising in opposite pairs along the rhachis; two-pinnate with the pinnae pinnate again.
pinnatifid Pinnately cut, but not into separate portions, the lobes connected by the lamina as well as the midrib or stalk.
pinnatilobed Divided into lobes arranged like a feather.
pinnatisect Like pinnatifid but with some of the lower divisions reaching very nearly or quite to the midrib.
pinnule The ultimate division of a more than two-pinnate leaf, usually applied only in ferns.
pistillate Bearing female reproductive organs.
placenta The part of the ovary to which the ovules are attached.
placentation The position of the placentae in the ovary. The chief types of placentation are: *apical*, at the apex of the ovary; *axile*, in the axils formed by the meeting of the septa in the middle of the ovary; *basal*, at the base of the ovary; *free-central*, on a column or projection arising from the base in the middle of the ovary, not connected with the wall by septa; *parietal*, on the wall of the ovary or on an intrusion from it; *superficial*, when the ovules are scattered uniformly all over the inner surface of the wall of the ovary.
plano-convex Having one surface plane and the other convex.
plastic Varying in form according to environmental conditions, not according to genetic characteristics.
plicate Folded.
pollen The microspores of Pinophyta and Magnoliophyta.
pollen sac The microsporangium of a species of Pinophyta or Magnoliophyta; one of the chambers in an anther in which the pollen is formed.
pollinia Regularly shaped masses of pollen formed by a large number of pollen grains cohering as in Orchidaceae.
polygamodioecious Having male, female and bisexual flowers.
polygamous Having male, female and bisexual flowers on the same or different plants.
polyploid Having more than two sets of chromosomes e.g. 3, triploid; 4, tetraploid; 5, pentaploid; 6, hexaploid; 7, heptaploid; 8, octoploid; 9, nonaploid; 10, decaploid.
pome A fruit in which the seeds are surrounded by tough but not woody or stony layers, derived from the inner part of the fruit wall, and the whole fused with the deeply cup-shaped, fleshy receptacle, e.g. apple.
porrect Directed outwards and forwards.
premorse Ending abruptly and appearing as if bitten off at the lower end.
pricklet A small spiny outgrowth without a broadened base.
prickly Spiny outgrowth with a broadened base.
procumbent Trailing along or loosely lying on the ground.
proliferating Inflorescences bearing plantlets instead of flowers and fruits.
pro parte Partly; in part.
prostrate Lying closely along the surface of the ground.
protandrous Stamens maturing before the ovary.
prothallus The small gametophyte generation of a plant bearing the true sex organs, mostly applied to the free-living gametophytes of Pteridophyta.
Protogynous Having the ovary maturing before the stamens.
proximal At the end near the point of attachment.
pruinose With a bloom.
pseudogamous See **apomictic**.
pteridophytes Ferns and fern allies, i.e. Lycopodiophyta and Pteridophyta.
puberulous With very short hairs.
pubescent With short, soft hairs; but sometimes used as a word for generally hairy.
punctate Marked with dots or transparent spots.
punctiform A small more or less circular dot.
pungent Sharply and stiffly pointed so as to prick.
pyrenes Fruit stones.
pyriform Pear-shaped.
raceme An unbranched racemose inflorescence in which the flowers are borne on pedicels.
racemose Having an inflorescence, usually conical in outline, whose growing points commonly continue to add to the inflorescence and in which there is usually no terminal flower; a consequence of this mode of growth is that the youngest and smallest branches or flowers are normally nearest the apex.
radical (of leaves) Arising from the base of the stem or a rhizome.
radiate A central region of tubular flowers and an outer region of ligulate flowers in the Asteraceae.
rank A vertical file of lateral organs; two-ranked, etc., with two ranks of lateral organs.
raphe The united portions of the funicle and outer integument in an ovule.
ray Anything that radiates outwards, e.g. branches of an umbel; stigma ridges in *Papaver* or *Nuphar*.
ray flowers The outer ligulate flowers in the capitulum in the Asteraceae.
receptacle The flat, concave or convex part of the stem or peduncle from which the parts of the flower arise; often used to include the perigynous zone; especially in Asteraceae.
receptacular pits The pits in the receptacle of the capitulum in the Asteraceae in which the flowers are seated.
receptacular scales Scales or hairs on the receptacle of the capitulum in the Asteraceae adjacent to each flower.
recurved Curved down or back.
reflexed Bent down or back.
regma A seed vessel whose valves open by elastic movement.
regular Actinomorphic.
reniform Kidney-shaped.
repand With undulated margin.
replum The adjacent wall tissue to the placentae.
resilient Springing sharply back when bent out of position.
resiniferous Producing resin.

reticulate Marked with a network, usually of veins.
retrorse Turned or directed backwards.
retuse Notched at the apex.
revolute Rolled downwards.
rhacilla The short, slender axis of the flower in the Poaceae.
rhachis The axis of an inflorescence or pinnate leaf.
rhizomatous Bearing a rhizome.
rhizome An underground stem lasting more than one growing season.
rhombic Having the shape of a diamond in a pack of playing cards.
rigid Stiff.
rostellum A beak-like process formed by the sterile stigma in the flower of the Orchidaceae.
rounded Without a point or angle.
rugose With a wrinkled surface.
rugulose Finely rugose.
ruminate Looking as though chewed.
runcinate Pinnately lobed with the lobes directed backwards towards the base of the leaf.
saccate Pouched.
sagittate The base of a leaf which is cut by straight lines upwards on either side from the margin to the petiole to leave an inverted V.
salt-marsh The series of communities growing on intertidal mud or sandy mud in sheltered places on coasts and in estuaries.
samara A dry indehiscent fruit, part of the wall of which forms a flattened wing.
saprophyte A plant deriving its nourishment from decaying organisms.
scaberulous Slightly rough with a covering of small hairs.
scabrid Rough to the touch.
scabridulous The diminutive of scabrid.
scale-leaf A leaf reduced to a small scale.
scape A flowering stem of a plant in which all the leaves are basal with none on the stem.
scapigerous Bearing scapes, i.e. stems without leaves.
scapose Having a stalk without leaves.
scarious Of thin papery texture and not green.
schizocarp A fruit which breaks into one-seeded portions or mericarps.
sclerenchyma Woody tissue in a partly or mostly non-woody organ.
scorpioid A monochasial cyme that is coiled up like a scorpion's tail when young.
scrambler A plant sprawling over other plants, fences, etc.
scrub A community dominated by shrubs.
secund All directed towards one side.
seed A fertilised ovule.
self-compatible Self-fertile; able to self-fertilise.
self-incompatible Self-sterile; not able to self-fertilise.
semi-inferior Of an ovary of which the lower part is inferior but the upper part is free and projects above the sepals etc.
semilunar Half-moon-shaped.
sensu lato In the broad sense.
sensu stricto In the narrow sense.
sepal One of the segments of the outer whorls of the perianth.
sepaloid Resembling sepals.
septicidal Dehiscing along the septa of the ovary.
septum A wall of membrane dividing the ovary into cells.
sericeous Having silky, appressed, straight hairs.
serrate Toothed with the teeth pointing towards the apex.
serration With serrate teeth.
sessile Not stalked.
setaceous Shaped like a bristle, but not necessarily rigid.
sheath Long stem, sheathing and often cylindrical round the lower part of a leaf in the Poaceae.
short-shoot A short stem of strictly limited growth usually lateral on a long-shoot, especially on trees and shrubs.
shrub A woody plant branching abundantly from the base and not reaching a very large size.
silicula A dehiscent, two-valved, two-celled capsule less than three times as long as wide in the Brassicaceae.
siliqua A dehiscent, two-valved, two-celled capsule more than three times as long as wide.
simple Not compound.
sinuate Having a wavy outline.
sinus The space of indentation between a lobe or teeth; the space at the base of a leaf both sides of the petiole.
solitary Borne singly.
sorus A group of sporangia in the Pteridophyta.
spadix Sterile axis on which the flowers of an inflorescence in the Araceae are packed which often extends distinctly as a succulent appendix.
spathe An ensheathing bract in the Lemnaceae, Araceae and Hydrocharitaceae.
spathulate Paddle- or spoon-shaped.
spermatophyte A seed plant belonging to the Pinophyta or Magnoliophyta.
spermatozoid A male reproductive cell capable of moving by means of cilia.
spike A racemose inflorescence in which the flowers (or spikelets in Poaceae) have no stalks.
spikelet One to many flowers in a discrete group in *Limonium*, Cyperaceae and Poaceae.
spine A sharp, stiff, straight, woody outgrowth, usually not greatly widened at the
base (see **hairs**).
spinose Spine-like.
spiny With spines.
spiral Lateral organs on the axis, one per node, successive ones not at 180° to each other.
sporangiophore A structure, not leaf-like, bearing sporangia.
sporangium A structure containing spores.
spore The haploid product of meiotic division produced on the sporophyte and developing into the gametophyte (see **meiosis**).
sporophyll A leaf-like structure or one regarded as homologous with a leaf, bearing sporangia.
spreading Growing out divergently, not straight or erect.
spur A protrusion or tubular or pouch-like outgrowth of any part of the flower.
squarrose Rough with projecting scales.

stamen The basic male reproductive unit of the Magnoliophyta, one to many per flower, sometimes fused.
staminode A sterile stamen, sometimes modified to perform some other function.
standard The large, often erect adaxial petal of the zygomorphic flowers of the Fabaceae.
stellate Star-shaped with radiating arms.
stem-leaves Leaves borne on the stem as opposed to basally.
sterile Not producing seed capable of germination, or anthers not viable of pollen.
stigma The receptive surface of the gynoecium to which the pollen grains adhere.
stipel A structure similar to a stipule but at the base of the leaflets of a compound leaf.
stipitate Having a short stalk or stalk-like base.
stipule A scale-like or leaf-like appendage usually at the base of a petiole, sometimes adnate to it.
stipulate Having stipules.
stolon An aerial or procumbent stem, usually not swollen.
stoloniferous Having stolons.
stoma (Pl. **stomata)** A pore in the epidermis which can be closed by changes in shape of the surrounding cells.
stomium The part of the sporangium wall in the ferns, which ruptures during dehiscence.
striate Marked with long narrow depressions or ridges.
strict Growing up at a small angle to the vertical.
strigose With stiff, appressed hairs.
strophiolate Having excrescences round the hilum.
strophiole A small, hard appendage outside the testa of a seed.
style The part of the gynoecium connecting the ovary with the stigma.
stylopodium The enlarged base of the style in the Apiaceae.
sub- Almost, as in subacute, subglabrous, subglobose, subentire, subequal.
subshrub A perennial with a short woody surface stem producing aerial herbaceous stems.
subtended Of a lateral organ, having another organ in its axil.
subulate Awl-shaped, narrow, pointed and more or less flattened.
succulent Fleshy and juicy or pulpy.
sucker A shoot arising adventitiously from the root of a tree or shrub often at some distance from the main stem.
suffruticose A dwarf shrub or undershrub.
superior (of any ovary) Borne above the calyx, corolla and stamens, or if below or partly below them then not fused laterally to the receptacle.
supratectal Emergences on pollen grains borne on the outer wall of the exine.
suture The line of junction of two carpals.
sympetalous Having a tubular corolla formed by union of petals.
sympodial Of a stem in which the growing point either terminates in an inflorescence or dies each year, growth being continued by a new lateral growing point.
syncarpous (of an ovary) Having the carpels united to one another.
tap-root A main descending root bearing laterals.
taxon Any taxonomic grouping such as family, genus or species.
tendril A spirally coiled, thread-like outgrowth from a stem or leaf, used by the plant to climb and support itself.
tepal One of the segments of the perianth, sometimes used when sepals and petals are not differentiated.
terete Round, not ridged, grooved or angled.
terminal Borne at the end of a stem and limiting its growth.
ternate A compound leaf with three leaflets, which may be similarly divided again; two-ternate, etc.
testa The outer coat of a seed.
tetrad A group of four spores cohering in a tetrahedral shape or as a flat plate and originating from a single spare mother cell.
tetraploid See **polyploid**.
tetraquetrous Square in section.
thallus The plant body when not differentiated into a stem, leaf, etc.
theca A spore or pollen case; a sporangium.
thorn A woody, sharp-pointed structure formed from a modified branch.
throat The opening where the tube joins the limb of the corolla or calyx.
thrum Short-styled with long stamens.
thyrsoid Resembling a thyrsus in shape.
thyrsus A mixed inflorescence with the main axis racemose and later axes cymose, with the cluster almost double cone-shaped.
tiller Leafy shoot.
tomentose Having a dense covering of short cottony hairs.
tooth A shallow division of a leaf, calyx or corolla or the apex of a capsule.
transverse Lying crossways.
trapeziform In the form of a flap.
tree A woody plant usually more than 5 m with a single trunk.
triangular A flat shape with three sides, widest at the base and 1.2–3.0 times as long as wide with the two sides gradually narrowing to a point.
trifid Split into three, but not to the base.
trifoliolate A term used in the Fabaceae for ternate.
trigonous Triangular in section, with obtuse to rounded angles.
trimorphic Occurring in three forms.
tripartite Divided into three parts.
triploid See **polyploid**.
triquetrous A solid body triangular in section and acutely angled.
trullate Having a flat shape, widest nearer the base and more or less angled (not rounded) there, 1.2–3.0 times as long as wide; if less, broadly so, if more, narrowly so.
truncate Of the base or apex of a flat organ, straight or flat.

tube The fused part of a corolla or calyx, or a hollow, cylindrical, empty prolongation of an anther.
tuber Swollen root or subterrancous stem.
tuberous Tubercle-like.
tubercle A small more or less spherical or elliptical swelling.
tuberculate With a surface texture covered in minute tubercles.
tubular In the form of a hollow cylinder.
tufted Of elongated organs or stems which are clustered together.
tunic A dry, usually brown and more or less papery covering round a bulb or corm.
turbinate Top-shaped.
turion A detachable winter bud, by means of which many water plants perennate.
twig Ultimate branch of a woody stem.
unarmed Devoid of thorns, spines or pricklets.
umbel An inflorescence in which all the pedicels arise from one point.
umbilicus A depression.
umbo A conical or rounded protuberance.
uncinate Hooked
undulate Wavy at the edge in the plane at right angles to the surface.
unifacial With only one surface, not with a lowerside and underside.
unijugate Having one pair of leaflets.
unilocular Having a single cavity.
unisexual (of a flower) Bearing only one sex.
univalent An unpaired chromosome.
upperside The upper surface of a flat organ.
urceolate More or less globular to cylindrical but strongly contracted at the mouth.
vallecula A depression or groove.
valve A deep division or lobe or a lobe of a capsule apex.
valvate Of perianth segments with their edges in contact with, but not overlapping in bud.
vein A strand of vascular tissue consisting of more than one vascular bundle.
velutinous Velvety, covered with very fine, short, dense upright hairs.
ventricosely Swelling in the middle.
vermiculate Marked with numerous sinuate, fine lines or bands of colour or by irregular depressed lines.
verrucose Covered in small wart-like outgrowths.
versatile Moving freely.
vespertine Blossoming or active in the evening; crepuscular; flying before sunrise or in twilight.
villous Shaggy.
viscid Sticky.
viscidium Two viscid bodies to which the pollinia are attached in the Orchidaceae.
vitta(e) The resin canals on the fruits of Apiaceae.
viviparus With flowers proliferating vegetatively and not forming seed.
waste place Uncultivated more or less open habitat much influenced by Man.
whorl More than two organs of the same kind arising at the same level.
wing Extension of an organ.
woody Hard and wood-like.
wool alien An alien introduced as a contaminate of raw wool imports.
woolly Clothed with shaggy hairs.
zygomorphic Having only one plane of symmetry.
zygote A cell formed by the union of two gametes or reproduction cells; the fertilised ovum.

Index

Accepted latin names and the page numbers on which the account of them occurs are in **bold** type. Synonym latin names are in *italic*. Vernacular names are in roman. When a vernacular name is the same as the latin genus, the latin genus and species are placed first.